FLORA OF ETHIOPIA

VOLUME 3

MAP OF ETHIOPIA SHOWING THE REGIONS USED TO DESCRIBE THE DISTRIBUTION OF PLANTS IN THE FLORA

EW - Eritrea West above 1000 m contour
TU - Tigray region above 1000 m contour
WU - Welo region above 1000 m contour
SU - Shewa region above 1000 m contour
GD - Gondar region
WG - Welega region
KF - Kefa region
AR - Arsi region
BA - Bale region

EE - Eritrea East below 1000 m contour
AF - Afar region below 1000 m contour to Eritrean border in the east and Harege border in the south

GJ - Gojam region
IL - Ilubabor region
GG - Gamo Gofa region
SD - Sidamo region
HA - Harege region

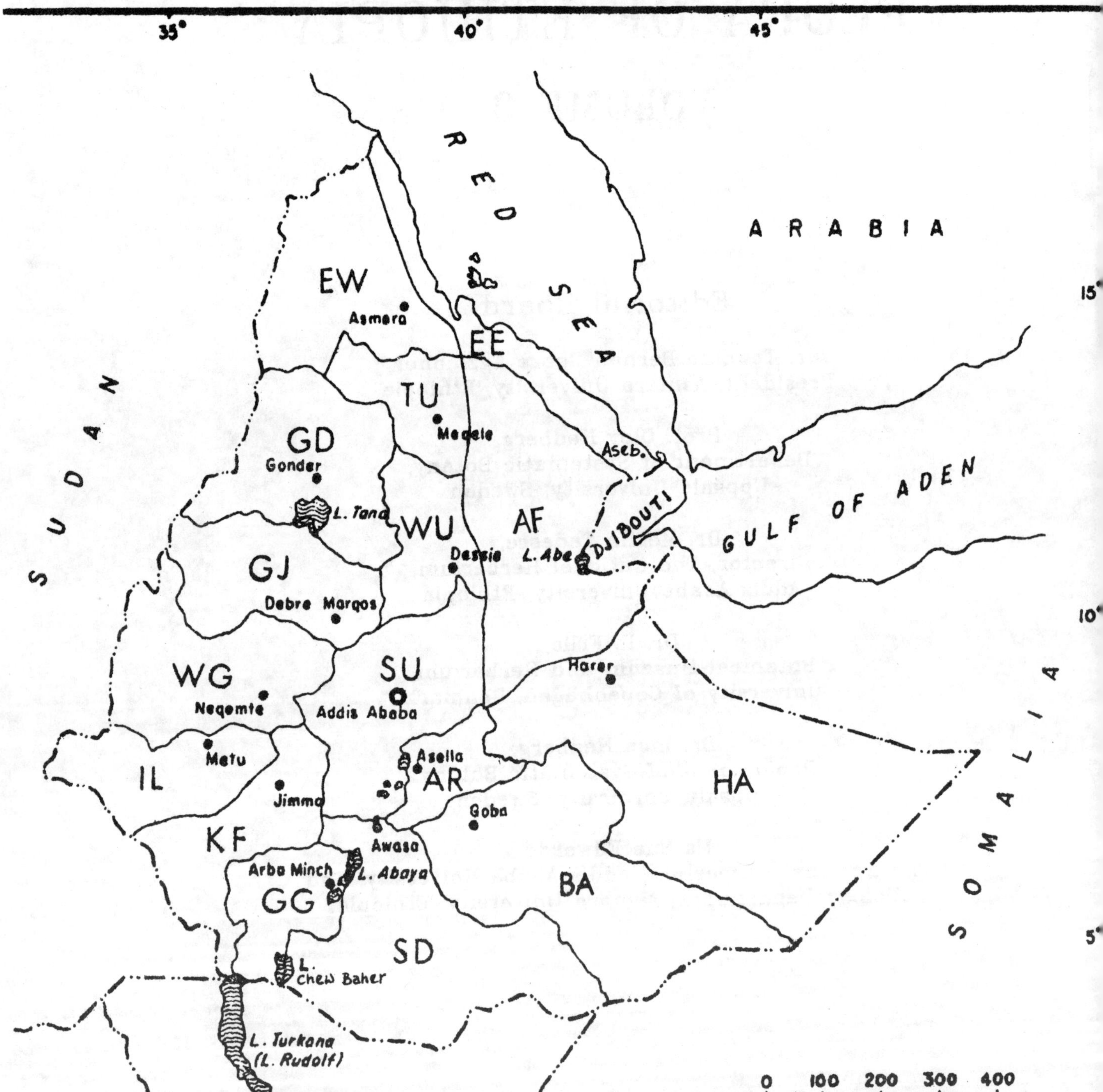

FLORA OF ETHIOPIA

VOLUME 3

PITTOSPORACEAE
TO
ARALIACEAE

Editors
Inga Hedberg & Sue Edwards

Addis Ababa and Asmara, Ethiopia
Uppsala, Sweden
1989

Published by:
The National Herbarium, Biology Department,
Science Faculty, Addis Ababa University, Ethiopia,
and
The Department of Systematic Botany,
Uppsala University, Sweden.

July, 1989

Prepared by the Ethiopian Flora Project, funded by Addis Ababa and Asmara Universities, and the Swedish Agency for Research Cooperation with Developing Countries (SAREC) through the Ethiopian Science and Technology Commission. Project Leader: Dr. Tewolde Berhan Gebre Egziabher; Associate Project Leader, Dr. Mesfin Tadesse; European Coordinator, Professor Olov Hedberg.

Final pages prepared in Asmara University and the National Herbarium, Addis Ababa University using HP 150A Personal Computers, LaserJet+ and LaserJet series II printers. Text formatted with MicroSoft Word, version 1.1 for the HP 150, and printed in TmsRmn 8, 10 and 12 point. List of vernacular names prepared using an IBM XT and LaserJet series II printer with MLS Amharic/English Word Processor, Version 3.0, developed by Gamma Productions and Dashen Engineering.

Printed by the Educational Materials Production and Distribution Agency (EMPDA), Textbook Production Department, Ministry of Education, Peoples' Democratic Republic of Ethiopia, Addis Ababa, Ethiopia.

ISBN 91-971285-0-3

Distributed by:

The National Herbarium,
Addis Ababa University,
P.O. Box 3434,
ADDIS ABABA,
ETHIOPIA

Department of Systematic Botany,
Uppsala University,
Box 541,
S-751 21 UPPSALA,
SWEDEN

Cover illustrations:
Front - *Hypagophytum abyssinicum*, known only from northern Ethiopia;
Back - *Cyphostemma pannosum*, known only from southern Ethiopia;
Spine - details from *Commiphora corrugata*;
all drawn by Eleanor Catherine.

FLORA OF ETHIOPIA, VOL. 3

CONTENTS

List of contributors vii
Preface ix
Foreword xiii
List of abbreviations xv
Notes on plant identification and collection
by Sue Edwards & O. Ryding xvii
Key to Families xxvii

Flora

86. Pittosporaceae by I. Friis 1
Hydrangeaceae 4
87. Escalloniaceae 4
88. Crassulaceae by M.G. Gilbert 5
89. Saxifragaceae by I. Hedberg 27
90. Vahliaceae by D. Bridson 28
91. Rosaceae by O. Hedberg 31
92. Neuradaceae by M.G. Gilbert 45
93. Chrysobalanaceae 46
94. Dichapetalaceae by F.J. Breteler 47
95. Fabaceae by M. Thulin 49
- Caesalpinioideae by R.M. Polhill & M. Thulin 49
- Mimosoideae by Asfaw Hunde & M. Thulin 71
- Papilionoideae by M. Thulin 97
96. Hamamelidaceae by B. Verdcourt 252
97. Myrothamnaceae 254
98. Buxaceae by B. Verdcourt 255
Platanaceae by M.G. Gilbert 257
99. Salicaceae by R.D. Meikle 258
100. Myricaceae by R.M Polhill 261
101. Casuarinaceae by C.M. Wilmot-Dear & M.G. Gilbert 262
Fagaceae by I. Friis 265
102. Ulmaceae by R.M. Polhill 266
103. Barbeyaceae by B. Verdcourt 270
104. Moraceae by I. Friis 271
105. Urticaceae by I. Friis 302
106. Cannabaceae by B. Verdcourt 327
107. Aquifoliaceae by B. Verdcourt 329
108. Celastraceae by N. Robson and Sebsebe Demissew 331
109. Icacinaceae by K. Vollesen 348
110. Salvadoraceae by B. Verdcourt 353
111. Olacaceae by K. Vollesen 356
112. Opiliaceae by P. Hiepko 358
113. Octoknemaceae 360

114.	Loranthaceae by M.G. Gilbert	361
115.	Viscaceae by M.G. Gilbert	375
116.	Santalaceae by A.G. Miller	379
117.	Balanophoraceae by B. Hansen	384
118.	Rhamnaceae by K. Vollesen	385
119.	Vitaceae by K. Vollesen	399
120.	Rutaceae by M.G. Gilbert	419
121.	Balanitaceae by M. Sands	433
122.	Simaroubaceae by B. Stannard	437
123.	Burseraceae by K. Vollesen	442
124.	Meliaceae by B.T. Styles & F. White	479
125.	Sapindaceae by K. Vollesen	490
126.	Melianthaceae by B. Verdcourt	511
127.	Anacardiaceae by M.G. Gilbert	513
128.	Connaraceae	533
	Juglandaceae by Sue Edwards	533
129.	Cornaceae	534
130.	Alangiaceae by B. Verdcourt	535
131.	Araliaceae by P. Bamps	537
	A Glossary of Botanical Terms	543
	Index to Scientific Names	567
	Vernacular Names for Volume 3 edited by Tewolde Berhan Gebre Egziabher and the staff of the National Herbarium	599
	- Introduction	599
	- Arranged alphabetically	602
	- Ethiopic names	621
	- English names	657

LIST OF CONTRIBUTORS

Dr. Asfaw Hunde, Uppsala Universitet - Institutionen för Systematisk Botanik, P.O. Box 541, S-751 21 Uppsala, Sweden

Dr. P. Bamps, Jardin botanique national de Belique, Département des Spermatophytes et des Ptéridophytes, Domaine de Bouchout, B-1860 Meise, Belgium.

Dr. Ir. F.J. Breteler, Department of Plant Taxonomy, Wageningen Agricultural University, P.O.B. 8010 / 6700 ED Wageningen, The Netherlands

Dr. Diane Bridson, The Herbarium, Royal Botanic Gardens, Kew, Richmond, Surrey TW9 3AB, UK.

Ms. Sue Edwards, The National Herbarium, Biology Department, Science Faculty, Addis Ababa University, P.O. Box 3434, Addis Ababa; also Biology Department, Asmara University, P.O. Box 1220, Asmara, Ethiopia.

Dr. Ib Friis, Botanical Museum and Herbarium, University of Copenhagen, Gothersgade 130, DK-1123 Copenhagen K, Denmark.

Mr. Mike Gilbert, The Ethiopian Flora Project, c/o The Herbarium, Royal Botanic Gardens, Kew, Richmond, Surrey TW9 3AB, UK.

Dr. B. Hansen, Botanical Museum and Herbarium, University of Copenhagen, Gothersgade 130, DK-1123 Copenhagen K, Denmark.

Dr. Inga Hedberg, Uppsala Universitet - Institutionen för Systematisk Botanik, P.O. Box 541, S-751 21 Uppsala, Sweden

Prof. Olov Hedberg, Uppsala Universitet - Institutionen för Systematisk Botanik, P.O. Box 541, S-751 21 Uppsala, Sweden

Prof. P. Hiepko, Botanischer Garten und Botanischer Museum, Königin-Luise-Str. 6-8, D-2000 Hamburg 36, Federal Republic of Germany.

Mr. R.D. Meikle, Ranscombe Lodge, Wootton Courtenay, Minehead, Somerset TA24 8RA, England.

Dr. Tony Miller, Royal Botanic Garden, Edinburgh EH3 5LR, Scotland.

Dr. Roger Polhill, The Herbarium, Royal Botanic Gardens, Kew, Richmond, Surrey TW9 3AB, UK.

Dr. Norman Robson, Botany Department, British Museum (Natural History), Cromwell Road, Kensington, London SW7 5BD, UK.

Dr. Olof Ryding, Biology Department, Asmara University, P.O. Box 1220, Asmara, Ethiopia.

Mr. Martin Sands, The Herbarium, Royal Botanic Gardens, Kew, Richmond, Surrey TW9 3AB, UK.

Dr. Sebsebe Demissew, The National Herbarium, Biology Department, Science Faculty, Addis Ababa University, P.O. Box 3434, Addis Ababa, Ethiopia.

Mr. Brian Stannard, The Herbarium, Royal Botanic Gardens, Kew, Richmond, Surrey TW9 3AB, UK.

Dr. B.T. Styles, Department of Plant Sciences, University of Oxford, Southparks Road, Oxford OX1 3RA, UK.

Dr. Mats Thulin, Uppsala Universitet - Institutionen för Systematisk Botanik, P.O. Box 541, S-751 21 Uppsala, Sweden

Dr. Bernard Verdcourt, The Herbarium, Royal Botanic Gardens, Kew, Richmond, Surrey TW9 3AB, UK.

Dr. Kaj Vollesen, The Herbarium, Royal Botanic Gardens, Kew, Richmond, Surrey TW9 3AB, UK.

Dr. Frank White, Department of Plant Sciences, University of Oxford, Southparks Road, Oxford OX1 3RA, UK.

Dr. Melanie Wilmot-Dear, The Herbarium, Royal Botanic Gardens, Kew, Richmond, Surrey TW9 3AB, UK.

PREFACE

The Ethiopian Flora Project was launched nine years ago in April 1980. A short summary of what has happened in these nine years, therefore, makes a fitting start for the Flora.

The need for a modern Flora of Ethiopia has long been felt both inside and outside the country. In 1967, one of us (O.H.) was asked by an Ethiopian government representative to assist in the preparation of such a Flora. The task of producing such a work seemed rather forbidding. However, the size was already indicated in *Enumeratio Plantarum Aethiopiae* by G. Cufodontis. After preliminary discussions involving botanists in Genoa, Wageningen and Uppsala, the idea to write a Flora of Ethiopia was introduced at the seventh plenary meeting of A.E.T.F.A.T. (Association pour l'étude taxonomique de la flore d'Afrique tropicale) that was held in Munich in 1970. At the end of this meeting a minisymposium was devoted to the project and the following two committees were set up.

Advisory Committee	**Working Committee**
R.W.G. Pichi-Sermolli (Genoa, Italy)	O. Hedberg (Uppsala, Sweden, chairman)
G. Cufodontis (Vienna, Austria)	I. Friis (Copenhagen, Denmark)
H.C.D. de Wit (Wageningen, The Netherlands)	M.G. Gilbert (Addis Ababa, Ethiopia)
J.B. Gillett (Nairobi, Kenya)	R.M. Polhill (Kew, England)
W.H. Lewis (Missouri, U.S.A.)	G.E. Wickens (Kew, England)

The working committee came up with a draft proposal for a Flora of Ethiopia. This was then circulated and it reached the Addis Ababa University (AAU) and the Institute of Agricultural Research (IAR).

An Ethiopian *ad-hoc* working committee was then established to modify the proposal in order to safeguard Ethiopia's interests and to solicit its acceptance by the Ethiopian Government so that funds could be raised for it from abroad. This *ad-hoc* committee consisted of people from the IAR, the Ethiopian Plant Genetic Resources Centre (PGRC/E), the State Forest Development Agency, the then College of Agriculture at Alemaya (now the Alemaya University of Agriculture) and the Faculty of Science, AAU. The chairman was Dr. Taddesa Ebba of the Ethiopian Plant Genetic Resources Centre.

After several meetings, the *ad-hoc* committee worked out the detailed modifications it felt were needed to the AETFAT proposal. A final project proposal was then made. A decision was also made that the project be housed in the Faculty of Science, AAU. The Dean of Science, who was also a member of the *ad-hoc* committee, was asked to push the project through to acceptance by the Ethiopian Government and to have it launched. One of us was then the Dean of Science, and it has been his pleasure to have remained leader of the project.

The project proposal was finally pushed through the necessary channels and it is now being funded by the Ethiopian Government and the Swedish Agency for Research Cooperation with Developing Countries (SAREC). The funds from SAREC were solicited by, and given to, the Ethiopian Science and Technology Commission, which has kindly decided to include to completion the Ethiopian Flora Project among the projects it funds. The Ethiopian Flora Project was, therefore, born in 1980 after a long gestation period.

The objectives of the Ethiopian Flora Project are:

- to write up an Ethiopian Flora within the shortest possible time;
- to build up a National Herbarium and a related library to be used as reference centres for taxonomists, pharmacognosists, agriculturists, foresters, wildlife specialists, etc; and
- to promote scientific activities in taxonomic botany, economic botany, forestry, plant ecology, etc.

The completed Flora is expected to include between six and seven thousand species of plant. Such a task is expected to take between ten and fifteen years to complete.

The production of a written Flora for a flora of the size and complexity found in Ethiopia requires a prolonged effort and the involvement of many botanists. Many have indeed been involved and we are grateful to all of them. The names of the contributors to this volume are given before this preface and with the accounts they have written.

When the Flora Project started in April, 1980, all the herbarium collections and equipment were very inadequately housed in a small space on the upper floor of a building designed originally as student living quarters. In 1981, the New Science Building complex was completed and, with the reassignment of building space to various departments, the laboratories and offices originally occupied by the Biology Department were allotted to the National Herbarium and Natural History Museum with their associated teaching and technical staff. The National Herbarium acquired 516 square metres of space which now houses the herbarium cabinets, preparation room, store, and office-laboratories. The herbarium has grown much since.

Although some parts of Ethiopia are botanically better known than others, there are no areas which have been over-collected. It has, therefore, been part of the Flora Project's mission to provide equipment and transport to make plant collecting an integral part of the National Herbarium activities. In 1979, an inventory recorded around 14,000 incorporated collections in the National Herbarium. Now in 1989, it contains around 50,000, most of these having been made since the Project started.

A major contribution to the development of the National Herbarium through the Flora Project has been the establishment and growth of the Herbarium Library. At the start of the Project, the Project Leader with the European Coordinator visited Kew, England, where they were able to obtain for the National Herbarium much needed Flora works written in the first half of the twentieth century. The Herbarium Library has been growing since then. In order to help all contributors to the Flora of Ethiopia, in 1982 Achille Richard's *Tentamen Florae Abyssinicae* was reprinted in Uppsala by the Ethiopian Flora Project. The Herbarium Library has developed an exchange relationship for publicatons with a number of institutions in Ethiopia.

The National Herbarium has developed active working relations with other herbaria in Ethiopia, in particular those at the Alemaya University of Agriculture, the International Livestock Centre for Africa (ILCA), IAR and the Forestry and Wildlife Training Institute at Wondo Genet.

There is little use of improved facilities and products, even publications, if there are no local staff members trained to maintain and use the facilities and to continue the research work. Training has thus been seen as a vital component of the Project with the trainees being sent to Swedish and Danish Universities and, as far as possible, contributing some portion towards the written Flora and carrying out the field work needed for their training in Ethiopia. The Project has thus, so far, trained four, and is training three more, young Ethiopians to the Ph.D. level. One technician has also been trained in the art of botanical illustration.

Even training, though a necessary prerequisite, is not a sufficient condition for active scientific work. The problems that hinder young scientists from an active scientific life have been summarized in the First Symposium of the Ethiopian Flora Project. Active support from the international scientific community, which the Ethiopian Flora Project has been fortunate to enjoy so far, will go a long way to solve these problems. We would, therefore, like to take this opportunity to appeal to the international body of botanists to continue and intensify its support for the development of the scientific careers of Ethiopian botanists in Ethiopia.

The Flora Project supports research in local medicinal plants. A limited number of copies of the compiled information covering over three hundred species with scientific names and reported local medicinal uses has been prepared in a mimeographed form. Through the National Herbarium the project also gives plant identification services. The number of organizations and individuals requesting identification services has increased steadily. Procedures have now been established for recording all such requests and ensuring, as far as possible, that material received for identification is adequate to enter the Herbarium. This is to help acquire material from areas hitherto neglected by professional collectors as well as material from all over the country of plants of economic interest, which have been little collected so far.

An international symposium of the Ethiopian Flora took place in Uppsala, in May, 1984, and the proceedings have now been published. A second international symposium took place in Addis Ababa in October, 1986, and the proceedings are being published. These symposia have summarized some of the achievements of the Ethiopian Flora Project.

The first, really exciting achievement is, however, the present volume, *Volume 3* of the projected 8-volume *Flora of Ethiopia*. We are all aware that this present volume has many shortcomings which arise from the poor information base of Ethiopian botany. But, since young Ethiopian botanists have been trained and are still being trained to improve this base, and since we have no reason to deny ourselves the priviledge of expecting the world fraternity of botanists to continue its present active support for broadening this base and for training and helping the further development of Ethiopian botanists, we expect this volume to bring about its own demise by enhancing the aquisition of more and precise botanical information and make a second edition inevitable in a few years' time. The fact that the whole volume has been stored on computer discs will make the production of future editions very easy.

The activities of the Ethiopian Flora Project have had four centres (Addis Ababa and Asmara in Ethiopia; Uppsala, Sweden, and Kew, England in Europe) as well as a large diffuse region of cooperation (Africa, Europe and North America). The official seat of the Project since its inception in 1980, including its herbarium, other facilities and funds in Ethiopia, has been in The Department of Biology, Faculty of Science, Addis Ababa University. Since 1983, the office of the Project Leader and the co-editor based in Ethiopia, has been in Asmara University, the day to day activities of the Addis Ababa office being managed by the Associate Project Leader. The European and North American activities have been coordinated from Uppsala University by the European Coordinator. The European based co-editor has also worked from Uppsala University. Because of the importance of the Kew Herbarium for plants of sub-Saharan Africa, the Ethiopian Flora Project has had an office there with one (formerly two) full time and one part time botanists. Virtually all the major herbaria of Europe and some in North America as well as some universities in Europe and America have been involved in making their plant specimens and scientists available for writing up the Flora of Ethiopia. It will become immediately obvious that coordinating the activities of such a widespread and diffuse network could not have been easy. The frustrations arising from the combination of weaknesses in the postal and telecommunications systems of the various countries involved have, inevitably, made this inherently difficult task even more difficult. The fact that the task is at all being accomplished, as testified by this volume, is a witness to the tremendous goodwill and patience of the various individuals and institutions involved in the exercise, as well as to the performance of the telecommunications and postal services of the various countries which would inevitably be affected by the worst component in each country of these services but which have, nevertheless, functioned effectively enough.

Given this difficult setting, it gives us great pleasure to thank all the individuals and institutions involved in this truly international and inter-institutional exercise. This proudly shows that, given the socio-political framework to make it possible, all peoples, be they from developed or developing countries, will cooperate in the service of science and humanity. This notwithstanding, specific thanks are called for.

The Swedish Agency for Research Cooperation with Developing Countries (SAREC) through the Ethiopian Science and Technology Commission (ESTC), and the Government of the People's Democratic Republic of Ethiopia through Addis Ababa University and Asmara University have funded the Ethiopian Flora Project. To the SAREC side our special thanks go to Governor Camilla Odhnoff, Chairman, SAREC Board of Directors, for her very valuable encouragement at the beginning of the project and for her continued support; Professor Knutsson who was Director General of SAREC when the project started and Professor Bo Bengtson the present Director General of SAREC for their favourable disposition towards the project; Dr. P.A. Ströberg who was the first officer in SAREC who understood what we were trying to achieve with the project and worked hard to have it supported by SAREC; Dr. M. Bhagavan who has taken over from Dr. Ströberg and continued the same level of understanding and support; and the many other officers of SAREC who have not only supported us both financially and technically, but also encouraged us when, in the 9 long years of effort, we have sometimes felt overwhelmed. On the ESTC side, our thanks for financial, administrative and moral support go to Comrade Haile Leul Tebike, the then Commissioner who stood solidly behind us when we tottered in the first few years of our efforts, and to Comrade Abebe Muluneh, the Present Commissioner, who has continued giving us very dependable support with plenty of encouragement, and the many council chairmen and members as well as administrators who have made possible the transmission of this support into specific actions. The project would have been still-born without the support and assistance of our colleagues in Addis Ababa University, ranging from the administrative staff at all levels through all the academic staff members of the Department of Biology; the erstwhile Department Head, Dr. Amha Belay; the present Head, Dr. Beyene Petros; the present Dean of the Faculty of Science, Dean Besrat Dilnessahu; the erstwhile Research and Development Officer, Professor Shibru Tedla; the present Officer, Dr. Aregay Waktola; the President when the Project started, Dr. Duri Mohammed; his Vice President for Business and Development, Dr. Asmelash Beyene; the present Vice President for Business and Development, Dr. Haile Gabriel Dagne; the Academic Vice President, Dr. Makonnen Dilgassa; the Academic Vice President in the early days of the project and the President during its maturity now, Dr. Abiy Kifle. It has been the pleasure of the Project Leader, as President of Asmara University, to mobilize the support of the administrative and secretarial staff as well as the academic staff and the facilities of Asmara University for the activities of the project. We would like to thank all concerned very warmly. Conversely, we would like to point out with satisfaction that the exercise of producing the camera-ready copies of the Flora of Ethiopia has helped develop the capacity to produce publications of high quality at Asmara University.

The project would equally have been still-born if support in terms of man-power and facilities had not come foreward from Uppsala University, the birth place of the taxonomic sciences. The Department of Systematic Botany, under Prof. O. Hedberg who has been the European Coordinator assisted by Dr. Inga Hedberg as co-editor, has been responsible for most of the original contacts with contributors, soliciting loans of herbarium material, receiving and forwarding manuscripts, buying equipment and literature, and - not least - paying all extra-Ethiopian bills. The Department has also included Ethiopian students in its post-graduate training programme.

The contribution of the Herbarium of the Royal Botanic Gardens, Kew, its leadership and its scientists to the development of taxonomic botany in sub-Saharan Africa and elsewhere is very substantial. When compared to all the undertakings in the field by Kew, the support it has given to the Ethiopian Flora Project as an institutional base and source of information as well as directly in scientific work may seem to the Kew management quite small, but it has been a major and indispensable component for the Project. Our thanks go to the Director of Kew at the beginning of the project, Prof. J.P.M. Brennan, to the present Director of the Herbarium, Mr. G. LL. Lucas, and to the many active scientists who have gone out of their way to help us and contributed to the writing of the Flora.

Many institutions and individuals have helped in making information and material available for the scientists involved in the project. Our thanks go to all of them. The following should, however, be singled out for contributions which go far beyond those involved in their own specific taxonomic specialization. Prof. Dr. Christian Puff of the University of Vienna in Austria has, single handed, been instrumental in making large numbers of reprints of articles available in the National Herbarium Library in Addis Ababa University and, to a lesser extent, in Asmara University. More than fifty percent of the reprints requested with Dr. Puff's assistance are reaching Addis Ababa, often accompanied by papers on topics similar to those requested together with notes of good wishes for the Project. This has enabled all the botany staff and the post-graduate students in the Biology Department to keep abreast with current developments and reduced the feeling of isolation common to academics attempting to develop a career in a developing country. The indefatiguable Dr. Ib Friis has helped in compiling information on botanical collectors and their collections from Ethiopia and in various editorial chores. These two have also contributed greatly to the growth of the National Herbarium in Addis Ababa in particular, and to Ethiopian collections in general, by raising funds and joining in with the National Herbarium staff in extensive plant collection expeditions in Ethiopia.

Many European herbaria have supported our effort by making their collections available to our scientists as well as to scientists from other institutions working on Ethiopian plants. We cannot possibly thank all of them; but we feel we should specifically thank those which have large collections of Ethiopian plants and who, therefore, have had to go out of their way to accomodate our demands. These include the East African Herbarium in Nairobi, the Kew Herbarium, the Florence Herbarium, the Uppsala Herbarium, the Paris Herbarium, the Wageningen Herbarium and the Bruxelles Herbarium.

Because of the inevitable clumsiness, and hence delays, introduced by the international nature of the project, those individuals directly administered by the project have been subjected to varying degrees of discomfort and annoyance. We are proud to say, however, that they have borne it all with patience. In this context, we would like to express our appreciation of Dr. Kaj Vollesen and Mr. Mike Gilbert, who have been working for the project based at Kew. Dr. Vollesen left the Project at the end of 1986 but continues to assist it as a member of the Kew staff. Both taxonomists have contributed greatly to the production of the present volume. The technicians in the National Herbarium in Addis Ababa have also contributed much to the growth and care of the collections in the National Herbarium.

We also wish to thank the editors and publishers of taxonomic works who have so willingly allowed us to take illustrations from their printed works. We particularly wish to thank the editors of the *Flora of Tropical East Africa* and the Bentham Moxon Trustees for permission to use many illustrations from *Flora of Tropical East Africa*, and *Flora Zambesiaca* and one from the *Flora of Cyprus*, including the preparation of high quality printer's copies at Kew. We are also grateful to Oklahoma Agricultural Experiment Station for permission to use and rearrange plates from *Families of Flowering Plants in Ethiopia* by Dr. W.C. Burger; to the Editorial Committee of the Wageningen Agricultural University Papers to reproduce illustrations from *Meded. Landbouwhogeschool Wageningen* 79-3; to Academic Press for permission to reproduce an illustration from the *Botanical Journal of the Linnean Society* 58; to the Association de Botanique Tropical of the Muséum National d'Histoire Naturelle, Paris, for permission to reproduce plates from *Flore du Cameroun*; and to the Jardin Botanique National de Belgique for permission to use any original illustration we require from *Flore de Congo Belge*. Without these generously given permissions it would have been impossible to produce this highly illustrated volume. We are also most grateful to the several botanical illustrators who have produced such high quality illustrations of the several taxa peculiar to Ethiopia. Each source and/or illustrator is acknowledged in the captions to the illustrations.

Compiling vernacular names has been a difficult and thankless job. Trying to identify the vernacular names recorded in the myriads of ways of spelling resulting from the combinations of the numerous Ethiopian languages and the various European transliteration systems followed by the various European collectors, compounded by the nature of the Ethiopian consonants which are radically different from those of the Europeans recording the names, has made the job difficult. The feeling that valuable time is being taken up in recording vernacular names in the field when new plant specimens could be being collected has made the new recording of names thankless. Our thanks to the National Herbarium staff in Addis Ababa, specifically Dr. Mesfin Tadesse, Dr. Sebsebe Demissew, Dr. Zerihun Woldu, Ato Zemede Asfaw and Ato Ensermu Kelbessa for doing this job are, therefore, very warm.

But perhaps the most thankless job in producing the Flora of Ethiopia fell to the two editors, Dr. Inga Hedberg in Uppsala and Ms Sue Edwards in Asmara. It is not only the irkesome endless toing and froing of correspondence about seemingly trivial points that make this job thankless, but also the exacting demands on accuracy of entering texts into the computer and the laying out of texts and illustrations. Our thanks to these two tireless workers are heartfelt.

The editors prepared camera-ready copies for the printers. The printers have, however, continually interacted with the editor based in Ethiopia to help produce a satisfying finish. We thank the printers very much for the care and trouble they have taken.

Dr. Getachew Aweke, of the National Herbarium, also assisted in compiling vernacular names.

Permission to use the printers of the Educational Materials Production and Distribution Agency was given by the Ministry of Education of the People's Democratic Republic of Ethiopia. Our special thanks go to the Minister, Comrade Yayehyirad Kitaw, who gave not only the permission, but also encouragement. Our special thanks go also to the previous Minister, now the Procurator-General, Comrade Bilillign Mandefro, with whom the possibility of using the printers was first discussed, and who still keeps encouraging us in our endeavour.

Encouraged by all the support we have been fortunate to enjoy so far, we dedicate this volume to both future Ethiopian botanists, to whose making this volume will contribute and who, we hope, will contribute to making progressively more accurate and complete editions of it, and to the international solidarity among scientists and statesmen which will have to keep growing if the example set by the Ethiopian Flora Project is to be widely applied and our presently divided world is to be one: a habitat suitable for man and all life.

O.H. & T.B.G.E., April, 1989

*****

FOREWORD

The preface to this volume gives the background to the writing of a modern Flora of Ethiopia. Here, then, is the first volume to appear in print. We hope it will be widely used and found useful by students, teachers, agriculturists, foresters, planners and others whose work brings them into contact with the natural and introduced flora of Ethiopia. We also hope it will stimulate more people, whatever their background, to take an interest in studying and caring for the plants around them. Finding the correct identity for a plant is the beginning of any such study.

Plant identification begins by finding the correct family to which a specimen belongs. To help in this task, the front of this book contains notes on plant identification and collection where the different parts of a plant are described with the terms used for these parts. The notes then explain the steps needed to identify a plant and are followed by a key to families. The notes also outline the procedure needed to make good quality herbarium specimens. However, there are many more terms used in describing plants than could be covered by a set of notes. We have attempted to provide adequate descriptions for all botanical terms in a glossary at the end of this book. Some terms are highly peculiar being needed to describe a situation found in only a few groups of plant. These terms are also explained in footnotes the first time they appear in a family.

The accounts for the dicotyledon families follow the sequence proposed by J. Hutchinson in the first edition (1926) of his work *The Families of Flowering Plants: Volume 1, Dicotyledons arranged according to a new system based on their probable phylogeny*, with a small modification to bring together families considered to be closely related to the Rosaceae - the Rosales of Cronquist's classification of 1968. However, the delimitation of the families has been left to the discretion of the authors.

This volume starts with Pittosporaceae and finishes with Araliaceae. Altogether 51 families are included with detailed information on over 1100 species. Of these families three cover plants which have been introduced into Ethiopia for their ornamental and/or economic importance. Another five families cover plants which have, so far, not been recorded within the country. It is expected that representatives of at least some of these families could be found as plant exploration in Ethiopia expands. This volume contains one such example - *Alangium chinense* of Alangiaceae was not known from Ethiopia until 1984 when the Harenna forest on the southern slopes of the Bale massive was studied systematically. The other families in volume 3 range from Fabaceae, the second largest family in the Flora with 620 species, to several families represented by single species.

The information given for each group (family, genus or species) is as follows.

At the start is the correct name for the group printed in bold type. This is followed, in italics, by the authority for that name. This is the person or persons who first published the name for the plant group according to a set of internationally recognised rules. The authority is followed by the year of publication given in brackets.

The correct name is followed by other names - synonyms - which have been used for the group. The synonyms are printed in italics followed by the authority and year of publication. For synonyms of species and infraspecific groups, information on specimens collected from Ethiopia used when the name was published are also given. The information includes the region and place of collection, the collector's name and number and the herbarium or herbaria where the specimen or specimens can be found.

In some cases, studies have shown that previously widely used scientific names are incorrect. For example, *Dodonaea viscosa* has been found to be the name of a plant confined to the sea shores of east Africa and not yet recorded from Ethiopia. The correct name for the species of *Dodonaea* widely distributed in Ethiopia is *D. angustifolia*.

The synonyms may be followed by references to important literature. This is always the case following the name of the family. The references for a family always include Cufodontis *Enumeratio Plantarum Aethiopicae* (abbreviated to *Enum.*) and the most recent relevant Floras and/or revisions. Papers concerning genera or species are referred to under the group concerned.

The first full paragraph of text is the description of the group. There is a description for each family, genus and species. However, infraspecific groups (subspecies, variety or form) are not described. Where there are two or more infraspecific groups they are separated by a key which also gives the distinguishing characters. If there is only one infraspecific group, the name is followed by a statement on the distinguishing characters. Any reference to an illustration is given at the end of the description.

For species and infraspecific groups, the description is accompanied by a paragraph on habitat and distribution. The distribution within Ethiopia is given according to the 14 Regions of the country with the lowlands of Eritrea (Eritrea East - EE) and the Afar (AF) also considered distinct. The map at the front of this book shows these regions and the abbreviations used for them in the text. Maps of the 14 Administrative Regions and major towns, and the main physical features are given at the back of the book. The distribution of the group within Ethiopia is followed by its regional and world wide distribution. The last part of this paragraph cites up to three, occasionally four, herbarium specimens which have been seen by the author of the account. As far as possible, at least one of these specimens can be found in the National Herbarium in Addis Abeba. The specimens are recognised by the name(s) and collection number of the person(s) who made the collection.

For families and genera, the statement on distribution usually includes the number of genera or species found throughout the world and in Ethiopia.

Where appropriate, additional paragraphs give information on uses and comments on the taxonomy. The information on uses has come from a number of sources not always available to the author of the account. These include the data base at the Royal Botanic Gardens, Kew, and the personal notes of Dr. Tewolde and one of use (S.E.).

Vernacular names are given separately at the end of the book.

It would not have been possible to have completed this volume of the Flora without the support of many institutions and individuals. In addition to those acknowledged in the preface we would like to thank the following. Ms Eva Persson and W/zt Miletitsega Mesfin who typed in and corrected most of the text. W/zt Miletitsega also compiled the index to scientific names assisted by W/zt Tsega Isaac. Mr. J. W. Durkin, former Computer Manager of the International Livestock Centre for Africa (ILCA), who was always ready with help from the selection of appropriate computer equipment to finding the right software and making ILCA facilities available so that we could make the best use of our computer equipment. One of us has also enjoyed support and encouragement from her colleagues and friends in both Addis Ababa and Asmara Universities. Those in Addis Ababa University have already been mentioned in the preface. In Asmara University she has had support and encouragement from Dr. Endashaw Bekele former Dean of the Faculty of Natural Sciences, W/zo Bisrat Ghebru and Ato Kiflemariam Melake of the Biology Department. Other colleagues who have helped in proof-reading and many other ways to bring out this volume are Ato Akalekrstos Taddese, Dr. Olof Ryding, Ato Tilahun Abebe, Ato Sileshi Nemomissa, Ato Endalkatchew Wolde Meskel, Ato Yoseph Gebre Meskel, Ato Solomon Ambaye, W/zt Sara Tewolde and several students from the Biology Department.

In the development of the production of this Flora we wish to thank Mr. Mike Gilbert who, when director of the herbarium in Addis Ababa, brought the need for a modern Flora of Ethiopia to the attention of taxonomists through the forum of AETFAT and has been able to actively support its appearance as one of the major contributors to this volume. Mike has also been of major assistance in getting the most out of our computer equipment. Dr. Kaj Vollesen, as the first taxonomist to work full time on the Flora patiently helped us define the format for presenting the different categories of information. Last, but not least, we wish to thank our husbands, Prof. Olov Hedberg and Dr. Tewolde Berhan Gebre Egziabher, who have given us constant support and encouragement, and our families who have understood when work on the Flora has had to take precedence over other activities.

I.H. & S.E., April, 1989

ABBREVIATIONS
(excluding authorities for names)

All units of measurement are given with SDI abbreviations.
Herbarium abbreviations are according to the latest edition of *Index Herbariorum*.

A - Herbarium, Arnold Arboretum of Harvard University, Cambridge, Massachusetts - U.S.A.
AF - Afar region, Ethiopia
Afr. - Africa
alt. - altitude
AR - Arsi region, Ethiopia
B - Botanischer Garten und Botanisches Museum, Berlin, West Germany
BA - Bale region, Ethiopia
BC - Institut Botánic de Barcelona, Barcelona, Spain
BERN - Systematisch-Geobotanisches Institut der Universität Bern, Bern, Switzerland
BG - Herbariet, Botanisk Institutt, Universitetet i Bergen, Bergen, Norway
BM - Herbarium, British Museum (Natural History), London, U.K.
Bot. - botany
BR - Herbarium, Jardin Botanique National de Belgique, Meise, Belgium
Bull. - Bulletin
C - (before a place name) central
C - Botanical Museum and Herbarium, Copenhagen, Denmark
c - 'circa' meaning about or approximately
consv. - 'conservandum' latin for conserved
Cuf. - Cufodontis
dbh - diameter at breast height, used to measure the size of tree trunks
destr. - destroyed
diam. - diameter
E - east
E.A. - East Africa (Kenya, Tanzania, Uganda)
e.g. - 'exempli gratia', latin meaning for example
EA (also sometimes mistakenly as EAH) - East African Herbarium, Nairobi, Kenya
ed. - edition
EE - Eritrea east, below 1000 m contour to east, Ethiopia
Enum. - *Enumeratio Plantarum Aethiopiae*
et al. - 'et alii' latin meaning and others
etc. - 'et cetera' latin meaning and the rest
ETH - The National Herbarium, Addis Ababa University, Ethiopia
EW - Eritrea west, above 1000 m contour to west, Ethiopia
f. - 'fil' latin for son
FI - Herbarium Universitatis Florentinae, Museo Botanico, Firenze, Italy
FI (Webb) - Herbarium Universitatis Florentinae (Webb collection), Museo Botanico, Firenze, Italy
fig. - figure
Fl. Trop. E. Afr. - *Flora of Tropical East Africa*
Fl. Trop. East Afr. - *Flora of Tropical East Africa*
Fl. Zam. - *Flora Zambesiaca*
FR - Forschuninstitut und Naturmuseum Senckenberg, Frankfurt, West Germany
FT - Erbario Tropicale di Firenze, Firenze, Italy
G - Herbarium, Conservatoire et Jardin botaniques, Geneve, Switzerland
GD - Gondar region, Ethiopia
GE - Erbario dell'Instituto Botanico "Hanbury" e Orto Botanic dell'Università di Genova, Genova, Italy
GG - Gamu Gofa region, Ethiopia
GJ - Gojam region, Ethiopia
GRA - Herbarium, Albany Museum, Grahamstown, South Africa
HA - Harerge region, Ethiopia
holo. - holotype
HUJ - Herbarium, Department of Botany, The Hebrew University, Jerusalem, Israel
I.A.R. - Institute of Agricultural Research, Ethiopia
IECAMA (also given as IECA) - Imperial Ethiopian College of Agriculture and Mechanical Arts, now Alemaya University of Agriculture, Ethiopia
IL - Ilubabor region, Ethiopia
indign. - indigenous
ined. - 'ineditus' latin indicating that the item is being prepared for publication
iso. - isotype
isolect. - isolectotype
isoneo. - isoneotype
isosyn. - isosyntype
K - The Herbarium, Royal Botanic Gardens, Kew, U.K.
Kew Bull. - *Kew Bulletin*
KF - Kefa region, Ethiopia
KR - Landessammlungen für Naturkunde, Karlsruhe, West Germany
L - Rijksherbarium, Leiden, Netherlands
LE - Herbarium of the Department of Higher Plants, V.L. Komarov Botanical Institute of the Academy of Sciences of the U.S.S.R., Leningrad, U.S.S.R.
lecto. - lectotype
leg. - 'legit' latin for collector
LISU - Museum, Laboratório e Jardim Botanico, Lisboa, Portugal
loc. cit. - 'loco citato' latin for in the place cited
M - Museum, Botanische Staatssammlung, München, West Germany
MO - Herbarium, Missouri Botanical Garden, Saint Louis, Missouri, U.S.A.
MPU - Institut de Botanique, Montpellier, France
Mt. - mountain
N - north
NE - north east
neo. - neotype
NEU - Institut de Botanique de l'Université, Neuchâtal, Switzerland

nom. - 'nomen' latin for name
nom. cons. or nom. conserv. - 'nomen conservandum' Latin for conserved name
nom. illegit. - 'nomen illegitimum', latin for illegitimate name
nom. nud. - 'nomen nudum' latin for naked name, a name published without a description of the taxon
non sens. str. - 'non sensu stricto' latin meaning not in the strict sense, not in the same sense as the original author of the group
nr. - near
NW - north west
op. cit. - 'opere citato' latin meaning in the work already quoted
P - Muséum National d'Histoire Naturelle, Laboratoire de Phanérogamie, Paris, France
p.p. - 'pro parte' latin for in part; used to show that only a part of the group as circumscribed by the original author is being used by the later author
PAL - Erbario Siculo e Erbario Generale, Palermo, Italy
PDRY - People's Democratic Republic of Yemen
PI - Herbarium Horti Pisani, Pisa, Italy
PRC - Herbarium, Universitatis Carolinae Facultatis Scientiae Naturalis, Institutum Botanicum, Praha, Czechoslovakia
PRE - National Herbarium, Botanical Research Institute, Pretoria, South Africa
q.v. - 'quod vide' latin meaning which see
R. - river
RO - Erbario dell'Instituto Botanico dell'Università di Roma, Roma, Italy
S - south
S - Herbarium, Naturhistoriska Riksmuseet, Stockholm, Sweden
s.n. - 'sine numero', latin meaning without a number
SD - Sidamo region, Ethiopia
SE - south east
Sect. - section
sens. str. - 'sensu stricto' latin meaning in the strict or narrow sense as circumscribed by the original author
sp. - species (singular)
spp. - species (plural)
SU - Shewa upland, above 1000 m contour to west, Ethiopia
subsp. - subspecies (singular)
subspp. - subspecies (plural)
syn. - synonym
TO - Herbarium, Museum Botanicum Horti Taurinensis, c/o Istituto ed Orto Botanico dell' Università, Torino, Italy
TOM - Herbarium, Istituto Missioni Consolata, Torino, Italy
Trop. - tropical
TU - Tigray region, Ethiopia
TUB - Herbarium, Institut für Biologie I, Tübingen, West Germany
UPS - The Herbarium, University of Uppsala, Uppsala, Sweden
var. - variety
vol. - volume of a book
W - Naturhistorisches Museum, Wien, Austria
W - west
WAG - Herbarium Vadense, Laboratory for Plant Taxonomy and Plant Geography, Wageningen, Netherlands
WG - Welega region, Ethiopia
WRSL - Herbarium, Department of Botany, Meseum of Natural History, Wrocklaw, Poland
WU - Welo region, Ethiopia
YAR - Yemen Arab Republic
Z - Institut für Systematische Botanik der Universität Zürich, Zürich, Switzerland
± - more or less

NOTES ON PLANT IDENTIFICATION AND COLLECTION

by Sue Edwards & Olof Ryding*

1. PLANT IDENTIFICATION

Identification is carried out simply by determining the similarities and/or differences between two elements, i.e., by asking the question, is the unknown material the same as or different from a group of known material. The determination of similarities and differences involves description which means understanding the nature of the unknown material and that presented by identified specimens, illustrations and written descriptions. Establishing that two elements (the unknown and the known) are the same or similar also involves naming and classification. By deciding that the unknown material belongs to the same taxon as already determined materials, the information stored in the classification system becomes available and can be applied to the previously unknown material. The key to the information in the classification system is the name chosen as a result of the identification process. For example, by identifying a specimen as a member of the family Asteraceae (Compositae) it is possible to read descriptions and look at other material of this family and so learn more about the specimen collected. It is also possible to select the appropriate literature plus (when available) illustrations and reliably named herbarium specimens to determine the genus and species of the previously unknown material.

Both processes - identification and classification - involve comparison and judgement which require that the terms used for describing similarities and differences between taxonomic groups are used in a precise and readily understood fashion. Naming requires that the name selected and applied is in accordance with the internationally recognised procedures of the International Code of Botanical Nomenclature.

Identification is one of the basic objectives and activities of systematics. According to Blackwelder (1967) "identification enables us to retrieve the appropriate facts from the (classification) system to be associated with the specimen at hand" and is ".. better described as the recovery side of taxonomy".

In practice the following procedures can be used for identification.

Expert recognition and determination: using the experience of a professional taxonomist for a given group and/or floristic area. This produces the most reliable and accurate identification because the expert will have been working on these plants and probably have prepared publications on them. The disadvantage of this method is that there are few experts available in Ethiopia and their time is limited for such activities, while getting material to experts outside the country can be costly as well as time consuming.

Comparison: where the unknown material is compared with either named specimens and/or illustrations, and/or descriptions. This can be a reliable method if the named specimens being compared have been reliably determined and adequate descriptions are available to be used alongside the materials. However, apart from being a time-consuming method, identification through comparison can result in someone unfamiliar with the terms used and/or the peculiarities of the groups being compared making quite erroneous judgements. In order to get reliable results a great deal of time and care must be used in using this method. However, when comparison is combined with the following method, it adds greatly to the reliability of the identification produced.

Use of Keys and similar devices. This is by far the most widely used method. It has the advantage that keys and similar devices have usually been prepared by professional taxonomists who are experts for the group/groups concerned. Thus keys make this expertise available to the identifier in a published form. Once the mechanics of using keys have been learnt, this method for identification can be used readily even in the field, and is not so time consuming as either of the other methods described.

2. BASIC FEATURES AND TERMS USED IN IDENTIFYING FLOWERING PLANTS

The classification of the angiosperms is mostly based upon comparative studies of the structure of their flowers and fruits. However, features of their ecology and physiology, as well as vegetative characters can also be extremely useful in identification.

Habitat

Certain habitats are associated with distinctive adaptations and some families are known for occurring in restricted habitats and not in others. For example, plants associated with aquatic habitats have often become greatly modified and there are several families which are only found in such habitats - Lemnaceae and Lentibulariaceae in fresh water, and Cymodoceaceae in the sea.

Lifeforms and growth forms

This includes the length of time a plant lives. **Ephemerals** complete their life cycles in a very short period (a matter of weeks rather than months) and are characteristic of habitats such as disturbed areas (weeds in fields) and dry areas (deserts and semi-deserts). **Annuals** only live for one growing season and usually do not develop woody stems or roots, or storage organs. Under tropical conditions the distinction between ephemerals and annuals is often arbitrary. **Biennials** grow for two growing seasons and develop special storage organs to carry them over unfavourable periods between two seasons. Typically a biennial reproduces only in its second season. Under tropical conditions where two or more growing seasons can occur within 12 months, a biennial may complete its life cycle within one calendar year. **Perennials** are plants that continue to grow for two or more growing seasons and usually develop woody parts and/or specialised perennating organs such as bulbs and rhizomes. Generally perennials do not start sexual reproduction until several growing seasons have passed and the plant has reached a certain minimum size. Most perennials produce flowers and fruits at regular intervals throughout their lives and are termed **polycarpic**. However, some such as enset, *Ensete ventricosa*, and sisal, *Agave sisalana*, only flower once and then die; such plants are termed **monocarpic** or **hapaxanthic**.

* Biology Department, Asmara University, P.O.Box 1220, Asmera.

Most flowering plants are **autotrophic**, manufacturing their own organic compounds using sunlight energy trapped by chlorophyll to combine water and carbondioxide in the process called photosynthesis. Sometimes the green colour of the cholophyll is masked by other colours such as red anthocyanin pigments or heavy wax deposits on the epidermis of plants adapted for semi-arid and arid conditions. The normal green colour can, however, be seen by peeling off the outer, coloured layer.

Not all flowering plants are completely autotrophic. Various degrees of parasitism are found and are typical of some families. The Loranthaceae are partial parasites which grow on other woody plants but still produce green leaves and/or stems. Many of the Scrophulariaceae are root parasites taking nutrients from other plants but also producing some vegetative shoots with green leaves. This family includes the devastating witchweeds, *Striga* spp. which can completely destroy crops such as sorghum. The Orobanchaceae are completely parasitic having lost the ability to manufacture chlorophyll.

Habit. Although there are no clear-cut distinctions between each of the following habit forms, it is convenient to use the terms as they describe the overall impression given by a plant. **Trees** are plants which usually have a single main stem, the trunk arising from the ground and growing for some distance before it branches: trees are always woody and perennial. **Shrubs** and **bushes** are also woody plants but here a few to several branches arise from or just above ground level. **Herbs** are generally considered to be plants which do not develop woody stems. Most herbs are small plants, but under tropical conditions where growth can continue throughout the year, some herbs can be large. The mountains of East Africa and Ethiopia carry some striking giant herbs, species of *Lobelia* and *Senecio*, which can be up to 4 m tall. At the other extreme, semi-arid areas are often covered with very short plants, the size of herbs but woody and much branched like shrubs. These are called subshrubs or **suffrutescent** plants. **Climbers** can be woody, when they are called **lianas**, or herbaceous, and are plants that cannot support themselves in an upright position but which grow over other plants and/or nonliving structures. Many climbers have organs which assist in the plant catching and holding onto its support. These include spines and prickles, tendrils and twining stems. Some climbers do not produce specialised organs but simply grow up and fall over other plants.

Morphology

Root. A root is an organ which does not bear other structures and which arises from within an axis next to the vascular tissue in contrast with a stem which arises from the cortex and which bears appendages - the leaves. Roots do not contribute many characters for identification, partly because they are often ignored being under the ground. However, peculiar and striking adaptations like the aerial roots of mangroves, *Avicennia marina* and *Rhizophora* spp., can be useful in identification.

Stem. Stems are typically divided into sections called **internodes** which lie between two adjacent attachment points of leaves, the **nodes**. Stems can also be modified forming perennating organs such as tubers, rhizomes, corms and bulbs.

Leaves. These organs contribute many useful characters for identification at the family, genus and species levels. Leaf arrangement is a very convenient character for use in identification keys and is often a constant feature in a family. The commonest arrangement, generally termed **alternate**, is one in which only one leaf is found at each node (Fig. 1.1-4). However, 'alternate' covers both spirally arranged leaves and those which are arranged in two opposite rows, as in some monocotyledons belonging to the families Amaryllidaceae, Liliaceae and Poaceae. When two leaves are found opposite each other at a node they are called **opposite** (Fig. 1.5). Often the stem bears two sets of opposite leaves at right angles to each other resulting in a stem which has four angles and is square in cross-section. This type of arrangement is typical of the families Lamiaceae (Labiatae), Verbenaceae, Rubiaceae and Acanthaceae. If more than two leaves occur at a node the arrangement is described as **whorled**. This arrangement must not be confused with the situation where the leaves are in irregular clusters, such an arrangement is called **fasciculate**. When leaves are whorled they occur all round the stem while fasciculate leaves are usually clustered on one side of the stem.

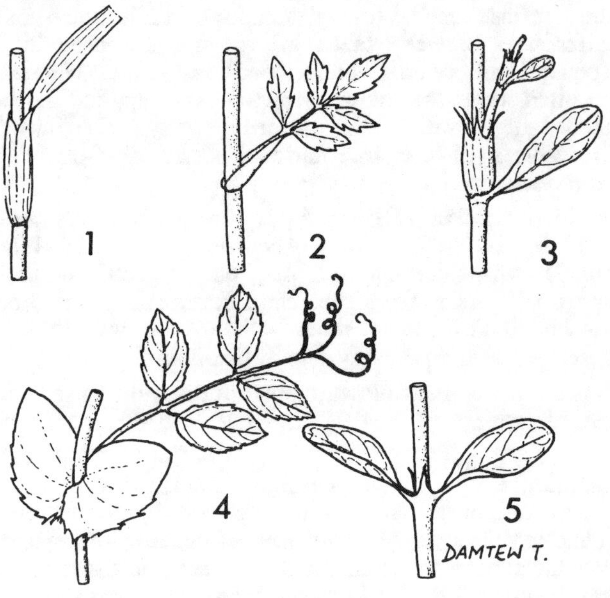

Figure 1. Stems each showing one node with leaves having different types of bases: 1 - a sheath that tightly wraps the stem, with a ligule between sheath and lamina (Poaceae); 2 - an open sheath (Apiaceae); 3 - an ochrea surrounding the base of a lateral shoot (Polygonaceae); 4 - two stipules at the base of a paripinnate leaf with a 3-branched tendril at the apex (Fabaceae); 5 - interpetiolar stipules placed between two opposite leaves, one is shown and the other is hidden by the stem (Rubiaceae).

There are two basic types of leaf - simple and compound. A **simple leaf** has only one leaf stalk (Fig. 2.1), **petiole**, in the axil of which can be found a bud, branch, flower or inflorescence. The lamina of a simple leaf is never completely divided into parts each with its own stalk, as is the case with a **compound leaf**. The parts of such a leaf are called **leaflets**; each leaflet may have its own stalk, the **petiolule**. If the leaflets are arranged along a continuation of the petiole, care must be taken to differentiate between such compound leaves which are large with large leaflets and young branches with soft stems and simple leaves. The best way of telling the difference between these two situations is to look at the apex - if it is a compound leaf growth will have stopped at the end of the leaf, but if it is a branch one will find a bud or evidence of continued growth. Leaflets in angiosperms are never associated with buds, or branches, or flowers or inflorescences.

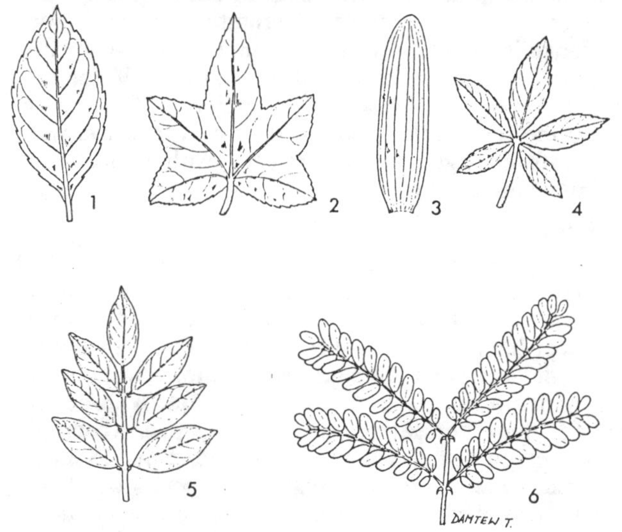

Figure 2. Different leaf types - Simple leaves: 1 - entire leaf with pinnate venation (*Maytenus*); 2 - petiolate, palmately lobed leaf with palmate venation (*Malva*); 3 - sessile leaf with parallel venation (Orchidaceae). Compound leaves: 4 - palmately compound leaf with five leaflets (*Gynandropsis*); 5 - imparipinnate leaf with small stipels at the base of each leaflet (Fabaceae-Papilionoidae); 6 - bipinnate leaf with four pinnae, divided into pinnules, and with prickles on the rachides (*Acacia*).

In **simple leaves**, the venation pattern can be important in identification. Most dicotyledons have a **pinnate venation** pattern in which there is a midrib and a series of secondary veins to either side of this (Fig. 2.1). In **palmate** or **digitate venation** three or more main veins arise from one point at the base of the lamina (Fig. 2.2). **Parallel venation** is typical of many families of monocotyledons. Here there are several veins arising in parallel from the base of the leaf (Fig. 2.3), and there may or may not be a midvein.

The venation patterns found in simple leaves are paralleled by the arrangement of leaflets in compound leaves. When compound leaves are **pinnate**, the leaflets are arranged on either side of the rachis. The end of the rachis may end in a pair of leaflets, in which case the leaf is described as **even-pinnate** or **paripinnate**, or in a single leaflet, when it is described as **odd-pinnate** or **imparipinnate** (Fig. 2.5). In some groups of plants, such as *Acacia* and *Albizia* of the Fabaceae (Leguminosae), each leaflet is replaced by a rachis bearing small leaflets, i.e. the leaf is twice compound and is termed **bipinnate** (Fig. 2.6). In pinnate leaves, each leaflet is termed a **pinna,** while in bipinnate leaves, the leaflets are called **pinnules**. A special and distinctive case of imparipinnate is the **trifoliolate** leaf which has three leaflets. This situation must not be confused with a group of three leaves, **trifoliate**, at a node. When the leaflets are arranged as in the fingers of a hand, the arrangement is called **digitate** or **palmate** (Fig. 2.4) and twice compound leaves with this arrangement are termed **bidigitate** or **bipalmate.**

The base of the leaf stalk can be associated with a pair of small flattened appendages called **stipules** (Fig. 1.4). Stipules can be small or large, attached firmly to the leaf stalk as in many Rosaceae, or falling off as or soon after the leaf unfolds as in *Ficus* (Moraceae). If stipules are associated with opposite leaves they may join-up across the stem forming an **interpetiolar stipule** (Fig. 1.5). This is a distinctive feature of the Rubiaceae where such interpetiolar stipules may develop to the same size as the leaves, as in *Galium*, or be reduced to a line or linear scar. Stipules might be fused to surround the entire stem above the node, and such a structure is called **ochrea** or **stipular sheath** (Fig. 1.3). The base of a petiole or sessile leaf can surround or tightly wrap a part of the stem, and such a base is called **sheath** (Fig. 1.1 & 1.2), while the free flat part of the leaf is called blade or lamina. In compound leaves, each leaflet may bear a pair of small **stipels** (Fig. 2.5).

Stems, leaves, leaflets and stipules can all be modified to form specialised structures. Tendrils are slender coiling organs. Sharp leaves are called spines, sharp shoots are called thorns and sharp outgrowths at the sides of stems or petioles are called prickles.

In some plants, notably those adapted for extreme ecological conditions, leaves may be greatly reduced or lacking. The functions of the leaf can then be carried out by leaf-like structures called **cladodes** or **cladophylls** if they are modified stems, or **phyllodes** if they are modified leaf stalks. Cladodes are found in many Cactaceae such as *Opuntia* while phyllodes are typical of Australian species of *Acacia* introduced into Ethiopia, e.g. *A. melanoxylon*. Other extreme adaptations of leaves include the large petal-like bract, the **spathe**, typical of the Araceae. The leaves associated with inflorescences are usually reduced and/or modified and are refered to as **bracts**. Bracts may be colourful and petal-like as in *Bougainvillea* and Poinsettia, *Euphorbia pulcherima*, or highly reduced as **phyllaries** enclosing and surrounding the capitulum of the Asteraceae (Compositae) and the **glumes** enclosing the florets of Cyperaceae and Poaceae.

Both **texture** and the covering, **indumentum**, of leaves and other parts can assist in identification, particularly at the genus and species level. But these characters can also be useful at family level. Thick, leathery-textured leaves are typical of the families Balanitaceae and Salvadoraceae: **star-shaped (stellate)** hairs are a conspicuous feature of plants belonging to the Malvaceae, Tiliaceae and some genera of Solanaceae. **Smell** is another very useful character in identification. Some families are easily recognised by their smell, for example the Lamiaceae (Labiatae) and Apiaceae (Umbelliferae) contain many plants used as spices while the Burseraceae is the family which provides incense and myrrh. The juice or sap produced by a plant can be important for identification: plants of *Euphorbia* have milky sap while in the Cactaceae the sap is clear.

Reproductive characters

These characters are all associated with the flower. They can be divided into characters associated with the arrangement of flowers, the **inflorescence**, the sexuality of a flower, and the parts of the flower. The parts of the flower can be further divided into two groups; the accessory or sterile appendages and the essential or fertile ones.

Inflorescence. A flower may be **solitary** if it is produced by itself from the axil of a leaf or at the end of a branch. However, flowers are usually grouped together into inflorescences which can be either terminal and/or axillary. The terms for different types of inflorescence are explained in the 'Glossary of Terms'. The stalk to an individual flower is termed a **pedicel** while that of an inflorescence is a **peduncle**. Leaves in an inflorescence are called **bracts**. They are usually smaller than the ordinary vegetative leaves but might also be large and coloured (see above). Characters from inflorescence types are more important in identifying genera and species than families, but there are some families with distinctive inflorescence types. These are the **capitulum**, where many small flowers are densely grouped together to resemble a single flower (typical of the Asteraceae (Compositae) and Dipsacaceae); the **umbel** where the small flowers are all borne at the same level (typical for Araliaceae); several umbels can, in their turn, be arranged into a larger umbel, called **compound umbel** (typical of the Apiaceae (Umbelliferae)); the **spadix** with its thickened axis bearing densely crowded and reduced flowers inside the spathe (typical of the Araceae); the **spikelet** of a few to several very small and reduced, wind-pollinated flowers (typical of the Cyperaceae and Poaceae); and the **fig** (**syconium**) of unisexual flowers embedded in a swollen receptacle (typical of *Ficus*).

Most families have flowers which contain both stamens and pistil and are, therefore, **bisexual** or **hermaphrodite**, but other families have **unisexual** flowers. Unisexual flowers are either **staminate** when they have functional stamens, or **pistillate** when they do not have functional stamens. Staminate flowers may have a non-functional pistil called a pistillode while pistillate flowers may have non-functional stamens called **staminodes**. One way to distinguish between unisexual flowers is to see how they mature. Flowers which are only male in function soon wither and drop off after the pollen is shed and nothing resembling a developing fruit can be found, while in flowers which are functionally female some evidence of a ripening ovary will be found after the perianth parts have withered.

A plant may have three kinds of flowers, male and female as well as bisexual. Neutral flowers can also be found in some plants; these are particularly frequent in ornamental plants which have been selected by man for their showy flowers. **Monoecious** plants have unisexual flowers of both kinds on the same plant. This situation is typical of some families including the Euphorbiaceae and Urticaceae. In **dioecious** plants an individual plant is either male or female. This situation can cause special problems in identification as the male and female plants may appear somewhat different from each other or one may find plants of only one sex which restricts the number of characters available for the identification exercise. Dioecious plants are characteristic of the Cucurbitaceae and Salicaceae and are common in families of woody plants more typical of temperate than tropical regions.

Many botanists believe that the flower has been derived from a modified branch which has been greatly compressed and its leaves modified into the parts of the flower. Thus plants which have an elongated central axis, the **receptacle** or **torus** bearing numerous, spirally arranged parts are considered primitive. Such flowers are found in the Magnoliaceae, Annonaceae, Nymphaeaceae and Ranunculaceae and are found in Volume 2 of the Flora of Ethiopia. In the majority of flowering plants the receptacle is flattened or even concave with the parts arranged in whorls, each with a fixed and definite number of components.

Perianth

The accessory or sterile parts of a flower are called the **perianth**. If all the parts of the perianth look alike they are called **tepals**. This situation is found in some monocotyledonous families like the Liliaceae, Amaryllidaceae and Orchidaceae. In most dicotyledons the perianth is made up of two distinctly different whorls - an outer **calyx** composed of **sepals** which are usually green or dull-coloured and protect the flower in bud, and an inner **corolla** composed of **petals** which are usually colourful and arranged to attract pollinators. The individual sepals and petals may be free from each other, then the flower is discribed as **polypetalous** (Fig. 3.1-3) or **polysepalous**, or united forming a sepal and/or petal tube, when the terms are **gamopetalous/sympetalous** or **gamosepalous/synsepalous**. Similarly all parts of the perianth may be fused to form a perianth tube (Fig. 3.5).

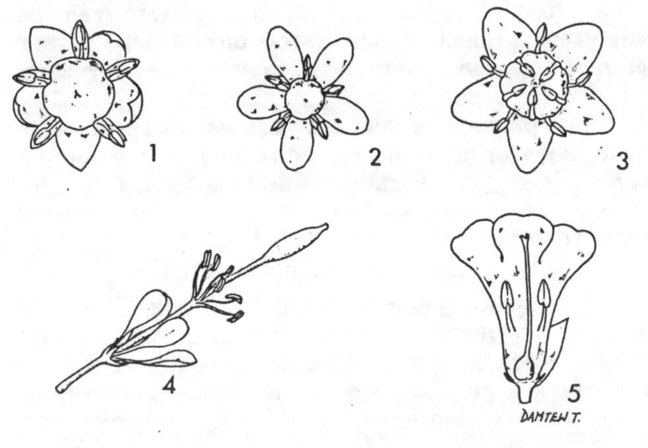

Figure 3. Flower types - regular/actinomorphic flowers with a disc in the center, showing different positions of the stamens: 1 - antipetalous (*Rhamnus*, Rhamnaceae); 2 - antisepalous (Celastraceae); 3 - diplostemonous (*Schinus*). 4 - irregular/zygomorphic flower with a long receptacle, with a pedicel to the right, small sepals and four petals pointing upwards, with six stamens and an androgynophore between the petals and the stamens, a pistil at the apex and a gynophore between the stamens and the pistil (*Gynandropsis*). 5 - a sympetalous flower in longitudinal section showing epipetalous stamens (*Thunbergia*, Acanthaceae).

When the fused, tubular part of the flower appears to be formed from receptacle tissue carrying sepals, petals and stamens, it can be referred to as a

hypanthium as found in *Punica*, the pomegranate ('roman' in Amharic). The receptacle can form a pad of tissue around the base of the ovary or from the base of the stamens. This is called a **disc** (Fig. 3.1-3) and may or may not be responsible for the production of nectar. The receptacle can also form additional structures within the flower - a **gynophore** (Fig. 3.4) is a stalk-like structure found under the ovary while an **androphore** is a column bearing only stamens (in male flowers) and an **androgynophore** (Fig. 3.4) is a column bearing both stamens and pistil. All these structures are found within the perianth and are always associated with a superior ovary.

Other important characters of the perianth useful in identification of families are the position of the parts in the bud (**aestivation**), and their symmetry when the flower has opened. **Contorted** aestivation is having several parts in a whorl or close spiral with one margin overlapping and the other underlapping the adjacent structures (Fig. 4.3), while in **convolute** one perianth part is rolled inside another and usually twisted apically. **Imbricate** is a general term for all arrangements where the parts overlap each other (Fig. 4.1-2) while in **valvate**, the margins of the adjacent parts touch only at the edges (Fig. 4.4). A few flowers have no plane of symmetry (**amorphic**) or all the parts are spirally arranged (**haplomorphic**), but most can be divided along one or more lines into two equal and opposite halves. When a flower is only bilaterally symmetrical it is termed **zygomorphic** or **irregular** (Fig. 3.4) while **actinomorphic** or **regular** flowers are radially symmetrical (Fig. 3.1-3).

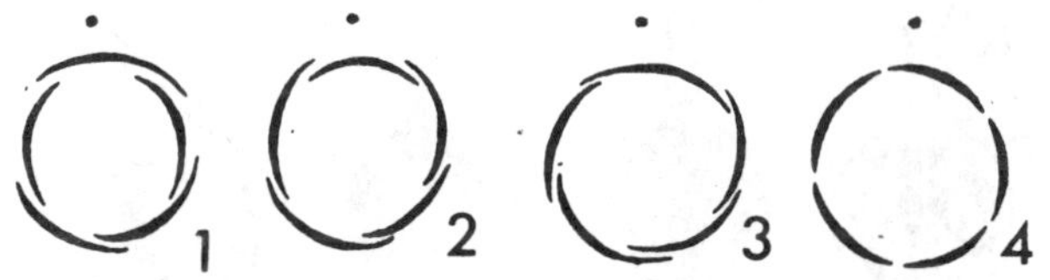

Figure 4. Buds in transverse section showing different aestivation types. The lines are tepals (petals or sepals) and the dot shows the position of the main axis. 1-2 - imbricate. 3 - contorted. 4 - valvate. From Davis & Cullen, 1979.

Androecium

The **androecium** (male house) is made up of the **stamens** which produce the male spores (**pollen grains**) which contain the male gametes. Typically a stamen consists of a stalk - the **filament** - bearing an **anther** made up of 4 **pollen sacs** arranged in 2 **thecae**. The tissue between the thecae is called the **connective**. The filament may be attached to the base of the anther (**basifixed**) or the back of the anther. If the anther stays rigid it is termed (**dorsifixed**), but if it moves readily on the filament it is called **versatile**. Most thecae open by longitudinal slits but in some families they open by transverse slits, or pores (Ericaceae and some Solanaceae and Viscaceae) or valves (Lauraceae). Stamens may have their filaments free or united. If united this can be in groups as in *Hypericum* or to form a tube around the pistil as in the *Pisum* and *Vicia*. In the Asteraceae and *Lobelia* the filaments are free but the anthers are united (**syngenesious**). In **epipetalous** flowers (Fig. 3.5), the stamens are borne on the corolla, and they may be sessile. The position of the stamens in relation to the perianth parts is important in identification: they may be **antisepalous** (opposite the sepals, Fig. 3.2) or **antipetalous** (opposite the petals, Fig. 3.1). When there are two whorls of stamens they may be **diplostemonous** with the outer whorl opposite the sepals and the inner opposite the petals (Fig. 3.3), or **obdiplostemonous** in which the reverse arrangement is found. The stamens can be all the same length with the anthers borne at the same level or the lengths can vary. In the Lamiaceae (Labiatae) there are usually two long and two short stamens (**didynamous**) while in the Brassicaceae (Cruciferae) there are 4 long and 2 short stamens (**tetradynamous**).

Gynoecium

The **gynoecium** contains one or more pistils each made up of a basal **ovary** containing one or more **ovules** where the female gametes are produced and fertilization takes place. Pollen is caught on the **stigma** which is usually held by the style above the ovary in an appropriate position to receive pollen. The ovules develop into **seeds** while the ovary forms a **true fruit** (in a false fruit other plant parts are also included). The position of the ovary in relation to the other parts of the flower is important in both identification and classification of families. The relationship between ovary position and the other floral parts is given in Figure 5 and Table 1.

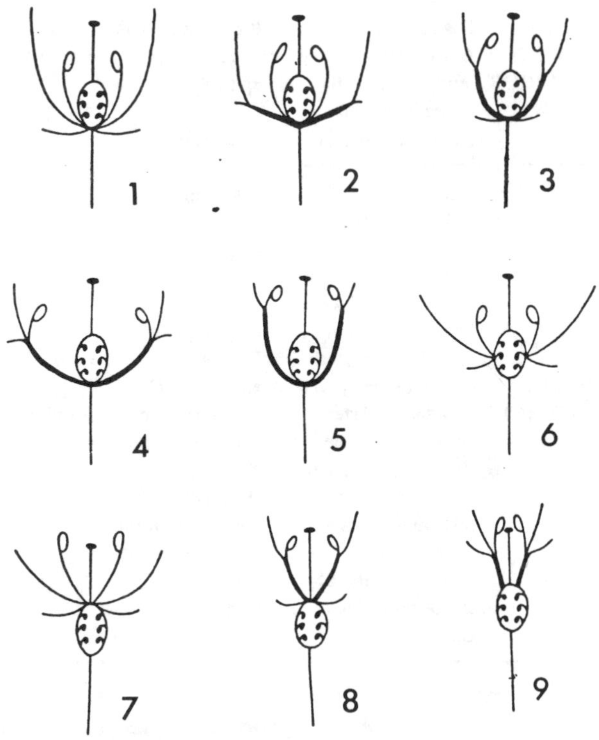

Figure 5. Diagram illustrating the use of the terms superior/hypogyny, partly inferior/perigyny and fully inferior/epigyny. Perigynous and epigynous zones are indicated by the use of heavy lines. For further information see table 1. From Davis & Cullen, 1979.

Table 1. Relationships of floral parts

Ovary (G) position	Insertion of perianth (P) or calyx (K) and corolla (C), and androecium (A)	Description using hypogynous, perigynous, and epigynous	Other descriptive terms used.
	PA or KCA inserted independently on the torus (*Ranunculus*)	PA or KCA hypogynous (Fig. 5.1)	Flower hypogynous
	K & C apparently adnate at the base. A inserted independently on the torus (*Tropaeolum*)	K & C perigynous, borne on a perigynous zone; A hypogynous. (Fig. 5.2)	Various
SUPERIOR	C & A apparently adnate at the base, K inserted independently on torus (*Jasminum, Carissa*)	K hypogynous, C & A perigynous, borne on a perigynous zone	Flower hypogynous, A epipetalous (Fig. 5.3)
	K, C & A inserted on a ring or collar of tissue which is inserted on the torus (*Punica, Rubus*)	K, C & A perigynous, borne on a perigynous zone	Flower perigynous (Fig. 5.4)
	P & A apparently adnate, C absent. (*Viscum*)	P & A perigynous (Fig. 5.5)	Various
PARTLY INFERIOR	P & A or K, C & A inserted independently, apparently on walls of ovary (*Lobelia, Ziziphus*)	P & A or K, C & A partly epigynous (Fig. 5.6)	Various
	P & A or K, C & A inserted independently apparently on top of the ovary (Apiaceae, *Ferula, Daucus*)	P & A or K, C & A epigynous (Fig. 5.7)	Flower epigynous
FULLY INFERIOR	K, C & A inserted on top of ovary, C & A adnate (Asteraceae, Rubiaceae)	K, C & A epigynous, C & A borne on an epigynous zone	Flower epigynous, A epipetalous
	(9) K, C & A inserted on a ring or collar of tissue itself inserted on top of the ovary (*Gladiolus, Crinum*)	K, C (P) & A (Fig. 5.8) epigynous, borne on an epigynous zone (Fig. 5.9)	Flower epigynous

From Davis and Cullen, 1979.

The basic unit of the gynoecium is the **carpel**. In **apocarpous** pistils, the carpels are free from each other and the gynoecium may consist of a single carpel, as in the Fabaceae (Leguminosae), or several carpels as in the Ranunculaceae and Magnoliaceae. In most families, two or more carpels are joined to form a **syncarpous** pistil (Fig 6.1-4). An indication of the number of carpels involved in forming a pistil can be found from the number of stigmas and/or styles borne on top of the ovary and the number of locules and/or placentas found in the ovary. Generally the number of stigmas is a more reliable indication of carpel number than the number of locules in the ovary as many ovaries of several carpels have a single locule. But many plants with more than one carpel in the gynoecium can also have united single stigmas and unilocular ovaries. Then the only way to determine carpel number is from the number of placentas and venation of the ovary wall.

A widely used character of the ovary in the identification and description of families, is the **placentation** of the ovules. The placenta is the part of the ovary to which the ovules are attached. In order to determine the placentation type it is necessary to cut both a cross-section and a longitudinal section of the ovary. It is usually easier to see this character in older flowers after fertilization has taken place and fruit development has started. However, the number and arrangement of the seeds in a mature fruit may not necessarily be the same as the number and arrangement of the ovules in an ovary, but generally there is a close correlation between the two. Following are the terms used to describe the different types of placentation and diagrams to show these in Figure 6.

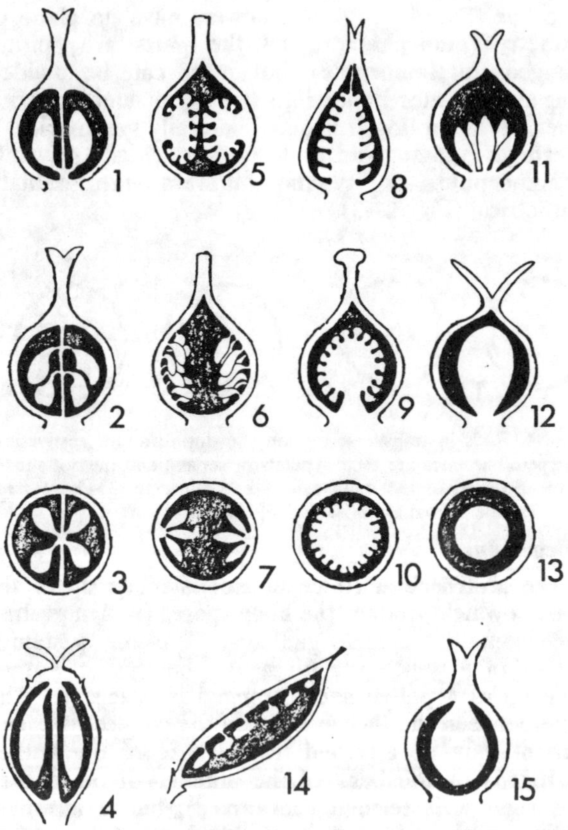

Figure 6. Pistils in longitudinal section (L.S.) and ovaries in transverse section (T.S.) to show different placentation types. Pistil structures and ovules are white and the cavity inside is black. 1-4 - axile type with two locules (carpels): 1-2,4 - pistil in L.S; 3 - ovary in T.S. 5-7 - parietal type originating from two carpels: 5 & 6 - pistil in L.S; 7 - ovary in T.S. 8-10 - free-central type: 8 & 9 - pistil in L.S; 10 - ovary in T.S. 11-13 - basal type: 11 & 12 - pistil in L.S; 13 - ovary in T.S. 14 - simple pistil with marginal type in L.S. 15 - apical type in L.S. Selected and rearranged from Davis & Cullen, 1979.

Marginal (Fig. 6.14): where the pistil is made of a single carpel or free carpels and the ovules are borne along the adaxial margin (*Pisum, Vicia, Delphinium, Nigella*).

Axile (Fig. 6.1-4): where the ovary is always syncarpous and divided by septa into two or more locules and the ovules are borne on the septa, but the position on the septa can vary. The ovules can be found along the central axis (*Aloe*)(Fig. 6.2), on the swollen septa (*Buddleia*) or the placenta may grow out into the locule - become intrusive (*Lobelia, Lantana*). In some families the ovules are reduced to 1 or 2 in each locule: these can be hanging from the top of the septum, as in the Apiaceae (Umbelliferae, Fig. 6.4) and Araliaceae, or ascending from the bottom of the septum (Fig. 6.1), as in the Vitaceae, *Cissus*. The ovules can be side by side, as in the Burseraceae (*Commiphora*), or overlapping each other, as in the Acanthaceae (*Adhatoda, Acanthus*). Sometimes axile ovaries are further divided by false septa which grow out from the ovary wall as the fruit develops resulting in twice as many locules as carpels, as in the Lamiaceae (Labiatae) (*Leonotis, Ocimum*) and Boraginaceae (*Heliotropium, Cynoglossum*).

Parietal (Fig. 6.5-7): where the ovary is always syncarpous with a single locule, the ovules being found on the walls of the ovary or outgrowths from these. In most plants with parietal placentation, the ovules are restricted to definite zones on the walls (*Gynandropsis, Cleome, Moringa*) (Fig. 6.6) or on outgrowths from the walls (Fig. 6.9)- intrusive placentas - (*Carica papaya*). Intrusive placentas can sometimes almost meet in the centre of the ovary making it difficult to differentiate between axile and parietal placentation (*Cucumis, Cucurbita pepo*), while in the Brassicaceae (Cruciferae) the ovary may appear to have two locules due to the growth of a false septum or replum across the locule. The parietal nature of the placenta can be seen from the attachment of the funicle of the ovule to the join between the replum and the ovary wall (*Brassica*).

Free-central (Fig. 6.8-10): where the ovules are borne on a central globose or cylindrical column that grows from the base of the ovary (*Portulaca, Talinum, Maesa*). Often there is a thin thread of tissue connecting the central column to the top of the ovary.

Basal (Fig. 6.11-13): where the ovules arise from the base of a unilocular ovary from a basal placental cushion (all Asteraceae, e.g. *Vernonia, Senecio, Guizotia*). Basal placentation can be found in both syncarpous ovaries (*Polygonum, Cyperus*) and one-carpel ovaries (*Clematis, Ranunculus*).

Apical (Fig. 6.15): where the ovules hang from the top of the ovary (*Pyrenacantha, Osyris, Pterocephalus*).

Ovule type is considered an important character in distinguishing and describing families, but it is not essential for identification when dealing with a restricted number of families as in a regional flora such as that for Ethiopia. Fruit type is rarely useful for family identification, but it is important at lower taxonomic ranks, particularly genera and species. In endospermic seeds, the embryo is generally small and the bulk of the food supply in the seed is retained in the special nourishing tissue developed after fertilization the **endosperm**. In non-endospermic seeds all the food reserve has been deposited in the **cotyledons** (seed leaves) Both types are important as food sources for man - the Fabaceae (Leguminosae) are the source of non-endospermic pulses, peas, beans, etc., while the Poaceae supply the cereals where the starchy endosperm forms the bulk of the seed.

3. COLLECTING AND MAKING HERBARIUM SPECIMENS

Making a full and correct identification of a plant can take quite some time. It is, therefore, very useful to have some permanent record of the work done and the best way to do this is to gradually build up a collection of identified herbarium specimens. Such a collection can also be used to check the results of your own identification efforts with an expert should the time and opportunity to do so arise. In Ethiopia, the National Herbarium in Addis Ababa will always welcome good quality herbarium specimens to add to its collections. In return the staff of the herbarium will provide the sender with the identifications of the plants received. But it must be rembered that only well collected material including adequate collecting notes, can be received and added to the National Herbarium collections. The aim of any collector should be to increase knowledge and not just aquire a set of inadequate scraps of material and information. If the following procedures are followed, anyone and everyone can make good herbarium collections.

How and what to collect

The aim is QUALITY not quantity. First get to know the area where you are going to make your collections and then you will be able to choose good quality specimens which are representative of the plant populations in the area. Common plants are just as important to collect as rare ones. In fact professional collectors often ignore common plants and they can be poorly represented in herbaria. Material should be collected in flower and/or fruit. Sterile material is only valuable if collected for a particular purpose such as an ecological survey and is often difficult or impossible to identify completely.

If possible whole plants, including underground parts, should be collected. If plants are very small, several can be collected to show the range of size in the population. Enough material should be collected to fill two sheets (or more) of newspaper 28 x 40 cm. Large herbs can be folded or cut into portions to represent the basal, middle and upper parts of the plant. For woody plants and climbers representative branches should be cut off with a sharp knife. Older books recommend putting material into a metal container called a vasculum. Nowadays it is easier to put material into polythene bags unless one is carrying a portable plant press. At the same time as the plants are collected they should be given a reference number and notes on the plant recorded in the field notebook - see below. Each polythene bag can contain just one species or a small number of species from the same habitat. As soon as the material has been put into the bag it should be tied tightly to prevent moisture loss so the plants do not wilt. Full polythene bags should also be placed in a shady place - in full sun the plants soon wilt and fragile parts such as petals can be shed.

As soon as possible after collecting, the plant specimens should be put into a plant press. A plant press need not be expensive or elaborate. Basic parts are.

- 2 firm metal or wooden lattice frames or rigid pieces of board 28 x 40 cm.

- sheets of folded newspaper to hold the specimens, called **flimsies,** 28 x 40 cm.

- some absorbent paper, newspaper or other, to absorb moisture from the specimens, called **blotters,** 28 x 40 cm

- if possible, sheets of corrugated cardboard with the corrugations running across to help in removing moist air from the plant press, called **ventilators**, and

- two strong straps made from inner tyre tubing, leather or canvas, to tie up and hold the press together.

The collected materials are arranged in the flimsies, each flimsie containing only one species (kind). The reference number is attached to the specimen on a label written in pencil (biro and ink run and fade easily). The number can also be written on the outside of the flimsie in the bottom right-hand corner to aid in sorting the specimens later. Care should be taken to arrange the specimens carefully so that parts do not overlap too much and delicate features like flowers are not squashed and crumpled. Very delicate flowers can be detached and placed in soft tissue or toilet paper. Once arranged in the flimsie, the plant specimen need not be handled again until it comes for identification and finally for mounting and incorporation into the herbarium. Bulky plant material, bulbs, rhizomes, juicy stems, should be sliced open before being pressed.

Once the material has been placed in flimsies, these are put into the press using the following sequence - ventilator (if possible), blotter, flimsie(s), blotter, ventilator. The number of flimsies that can be put between two blotters will depend on the wetness of the specimens - very juicy material needs plenty of blotters to draw out the moisture while drier material, like grasses and sedges, can be packed closer together. Finally the pile of flimsies, blotters and ventilators are placed between the frames or boards and tied tightly using the straps. It is often helpful to get a friend to sit or stand on the press while it is being tied up to make sure the material is squashed as flat as possible. The press should then be put into a warm dry, and preferably windy place. After 24 hours the press should be opened and the following checked:

- very wet blotters and ventilators should be removed and replaced with dry ones;

- each flimsie should be opened and the specimen checked for insect pests, (caterpillars can survive for remarkably long periods inside a plant press and do a lot of damage, particularly to flowers and fruits even making a specimen worthless);

- the specimen will now be limp and it will be easier to arrange leaves and flowers to aid identification later and to give a pleasing appearance to the specimen.

When the checking is complete, the plants are again tied up in the press. Checking for wet blotters should be done again after 48 hours and then again at the end of a week by which time quite a number of the specimens should be dry enough to take out of the press in their flimsies. A dry specimen feels cool but not cold when touched and should stay more or less rigid when it is picked up. If any parts still feel soft or fall over limply when the specimen is held up, then it is not dry. Flimsies containing dried specimens should be tied firmly into bundles and labeled on the outside with their reference numbers so they are easy to find when identification can be started.

Field notes

A vital part of good collecting is the making of adequate field notes. The most beautifully collected and preserved specimen is scientifically useless unless it is accompanied by adequate notes. The following information has to be put on the label which accompanies any permanent botanical specimen, and the only way to get this information is to make the notes for the label at the time of collection.

A. General information for all specimens collected at one time and place:

Date (preferably according to the Gregorian calendar);

Locality, which is the name of the place, if it can be found on a map, or the distance and direction from a place that can be found on a map, plus the name of the Awraja and Administrative Region. This information has to be good enough to find the collecting site again if more material is required or if another botanist wants to look for the same material;

Altitude (if possible), as this is an important feature in Ethiopia;

(**Latitude and Longitude** are also very useful in identifying locations, but they are added later using a map or gazeteer);

Habitat, which should include both a comment on the topography, (steep slope, gently sloping grassland) and the vegetation type (forest, *Acacia* woodland, thornscrub, etc.)

B. Information to be recorded for each specimen:

Reference number. Each set of specimens of the same kind (species) and collected at the same time from the same locality MUST be given a reference number. This number is written on a piece of paper or label or tag in pencil and put with the specimens in the collecting bag and then in the flimsie and also with the field notes for that specimen in the field note book. It is only by using a reference number system that field notes can be reliably correlated with the collected material. This number plus the name of the collector(s) can be used to refer to that specimen and will distinguish that specimen from any other herbarium specimen anywhere in the world.

Notes on the specimen. These should record anything of importance and interest and which will help in the identification of the material later, and which are likely to be spoilt or disappear during preservation of the specimen. These include:

- plant habit (herb, tree, shrub, climber) and size if only a portion of the plant has been collected;

- colour of parts, texture, smell (crush some of the leaves), sap abundance and colour (if any);

- pollinators (if seen) and microhabitat if it appears important.

Local name and language. A local name should only be recorded if you are sure the person giving the name is familiar with the local plants. It is preferable to write the name in Ge'ez script and the language must also be recorded. The uncritical use of local names in Ethiopia where there are over 80 languages can create more confusion than clarity.

All the information recorded in a field note book is transferred to a herbarium label similar to the one given below.

FLORA OF ETHIOPIA
ADMINISTRATIVE REGION

Locality:

Alt: m. Lat: N, Long: E

SCIENTIFIC
NAME:

Family
Local name: Lang:
Habitat:

Notes:

Collector(s): No:
Date:

One such label must be completed for each flimsie of specimens. When these are identified and mounted the label is placed on the bottom right-hand corner of the sheet of mounting paper.

Any material collected according to the instructions given above can be sent to the Director, The National Herbarium, Addis Ababa University, P.O. Box 3434, Addis Ababa, for identification.

4. EXAMINING A PLANT FOR IDENTIFICATION

The equipment needed to carry out most identifications are a good hand-lens (x 10) and a sharp razor blade. However for examining more detailed parts and very small flowers, a pair of dissecting needles and a dissecting microscope are also useful. Basically the same techniques are used for both fresh and dried herbarium material. Dried herbarium material can be dissected if it is first boiled in water for a minute or two, or soaked in a weak soap solution so the parts become softened.

A flower is usually examined from the outside inwards. For fresh material the parts can be counted before the flower is dissected or sectioned. For wetted herbarium material, it is best to dissect the flower carefully in a drop of water on a slide (for small flowers) or glass plate (for large ones). The parts are carefully arranged and then transferred to a piece of white card to make a permanent record. The easiest way to transfer the dissected parts is to cover the white card with a paper glue (**not** UHU type) and press this down over the dissected parts. If the slide or glass plate is lifted off carefully the flower parts should remain on the card. They can be slightly rearranged while the glue is wet. Once the card with its dissection is dry it can be kept in an envelope with the herbarium specimen. During the dissection of both fresh and herbarium specimens, a record should be made of the character states of the parts found. The following notes are to assist in determining the character states of features that can cause problems in using keys.

Vegetative characters: If a stem has lost its leaves, the arrangement of these can be determined from the leaf-scars and/or branching pattern. When leaves are crowded or clustered together, branching pattern can indicate whether the basic arrangement is alternate or the opposite.

Stipules can be both very small and deciduous. It is best to examine new growth or young stems for the presence of stipules looking at the join between the leaf stalk and the stem. In older leaves, the presence of deciduous stipules can be found by looking for scars by the leaf stalk. In some families the stipules may be between a pair of opposite leaves or form a joined structure across the stem - interpetiolar stipules.

Other important features to check are: the presence or absence of transluscent spots (often containing aromatic oils) in the leaves, best seen by holding a leaf up to the light; milky or watery sap (only possible in fresh material); and a peculiar type of indumentum such as stellate hairs.

Floral characters: The best stage at which to examine a flower is at anthesis, i.e., when it has opened and the stamens have matured and are shedding pollen. However, for features of the gynoecium a slightly older flower in which the fruit has started to develop is often very useful.

To determine the sex of a flower, observe whether the stamen is producing pollen grains and the ovary contains ovules. If stamens are lacking or very difficult to observe and the flower is functionally female, the plant can be keyed out to family in Group 8. Plants which appear to have flowers which are functionally only male and lack any pistil-like structure can be keyed out in Group 9.

The symmetry of a flower, by convention, does not include the gynoecium and usually refers to the perianth. Regular, actinomorphic flowers, usually also have regularly arranged stamens, but this is not always the case. For example the Brassicaceae (Cruciferae) have only two planes of symmetry and 6 stamens but are generally considered to have regular/actinomorphic flowers. If the stamens differ in number from the petals/perianth parts and are also assymmetrically arranged in the flower, the flower is generally considered to be zygomorphic with bilateral symmetry. This situation is often found in the families keyed out in Group 15.

Perianth: A few of the families found in Ethiopia have naked flowers, i.e., they lack a perianth. These families are keyed out in Group 6. Most flowers do have perianth parts. These should be examined carefully to find out if there are one or more whorls and whether the parts in each whorl are the same or different. Difficulties arise when the calyx is very small or even reduced to a rim around the base of the corolla as in many Apiaceae (Umbelliferae), Rubiaceae and Asteraceae (Compositae). In other families the perianth may be shed as a cap when the flower opens, as in *Eucalyptus* and *Syzygium* of the Myrtaceae, or the calyx may come off as a cap leaving only a ring of petals as in *Argemone mexicana* of the Papaveraceae. The best way to find out if petals are fused or free is to gently try and pull them off. If they are all united, the whole corolla will come off as one piece. If the petals are free or only a few are loosely fused, they will come off individually.

Androecium: In a few families the stamens are equal in number to and opposite the petals. These families are keyed out in Group 20. Stamen number and arrangement is also important in selecting the Group key for families keyed out in Group 17 - stamens greater or fewer in number than the petals; Group 18 - stamens epipetalous and opposite the perianth parts in flowers with only one perianth whorl; and Group 19 - stamens alternate with the petals and opposite the sepals.

Gynoecium: The position of the ovary is an important feature used in family identification. This has been fully explained in Table 1 in Basic features and terms. Similarly the different types of placentation found and how to observe these are given in the same section. Free-central placentation is characteristic of the following families: Caryophyllaceae, Portulacaceae, Myrsinaceae, Primulaceae, and Lentibulariaceae, and occurs occassionally in others. Ovule attachment can be seen from transverse and/or longitudinal sections, but with small ovaries with very few ovules it is sometimes easier to see the ovule position by slitting open the side of the ovary. The number of ovules can be found by cutting off the top (or bottom) or the ovary and gently squeezing out the ovules. Suggestions for finding out the number of carpels are given in the section on the gynoecium under Basic features and terms.

5. HOW TO USE THE KEYS

An identification key is a special type of taxonomic literature. Each statement consists of short phrases which attempt to describe the characters to be observed in as concise a way as possible. The type of key used here is called a dichotomous bracketed type because it is made up of pairs of contrasting statements, each pair called a couplet, which appear one under the other. The key to families provides information for the identification of 237 families of plants. In order to simplify the identification process, the key has been divided into 21 parts; a Primary key to 20 artificial groups followed by 20 keys, each to a restricted number of families. Within each part of the key, each couplet has been numbered in sequence. A similar system is used in the keys found in the accounts for the different families. There is first a key to the genera of that family (or subfamily) and then keys to species within each genus.

To identify the family to which a plant belongs, one starts with the **Primary key to the 20 artificial groups**, and compares the plant specimen with **both** statements at couplet 1. If the plant was taken from an aquatic habitat it can then be keyed out in Group 1. However, if the specimen does not fit this requirement the second half of the couplet leads to number 2. This couplet is read in full and a choice is made - either to the key in Group 2 or the couplet at 3 in the Primary key. This process is continued until one finds a statement, or set of statements, which matches the specimen in one or more of its characteristics which leads to one of the Group keys. Each group key is prefaced by a short statement on the characteristics of the plants which can be keyed out in that group. You should compare this statement with your specimen as a check that you have found the correct group for identifying the family. If this is so, you proceed in a similar manner to that used for identifying the artificial group, until you find the name of a family. Usually the name for a family is preceeded by a fairly long statement of characteristics. This is to help you check that the specimen you have actually belongs to the family. Additional checking should be done by comparing the specimen with a description of the family and illustrations.

When the correct family has been found, the full identification can be found first by using the key to genera, and then the key to the species for the identified genus. The keys in the family accounts use the same format as that for the key to families. Final checking is done by comparing the specimen with both the generic and species description, and an illustration if one is available. When the identification is complete the name of the family, and the specific name (genus plus specific epithet) are put onto the herbarium label.

It sometimes happens that the specimen does not agree with all the characters given in one half of a couplet. When that happens, it is usually the first stated characteristic that is the most useful in choosing which of the pair of statements to use as the lead to the next couplet. The only exception to this is when the first part of the couplet contains two or more statements and the second states "combination of characteristics not as above". If this is the case the second lead is taken.

The key to families presented here was prepared by Dr. William C. Burger in 1965, and is still the best available for Ethiopia as it contains information for identifying both indigenous and introduced families of flowering plants and gymnosperms. However, a considerable amount of collecting has been done in Ethiopia since 1965 and the number of plants now known to occur has correspondingly increased. Fortunately this was anticipated by Dr. Burger as he included families which had not been recorded for Ethiopia when he prepared his book, but are now known to occur. But there are still many parts of the country where little or no collecting has been done. It is thus very possible to find plants that do not key out using the key in its present form. Anyone who uses this key and finds such problems should contact the editor of the Ethiopian Flora Project so that the key can be corrected. The address is:

The Editor, Ethiopian Flora Project,
The National Herbarium, P.O. Box 3434,
Addis Ababa, Ethiopia.

KEY TO FAMILIES

(Reproduced with some modifications with permission of Oklahoma Agricultural Experiment Station from Experiment Station Bulletin No. 45, *Families of Flowering Plants in Ethiopia*, by W.C. Burger, 1967)

Families found in this volume are given in bold.

PRIMARY KEY TO 20 ARTIFICIAL GROUPS

(Each artificial group contains keys which will lead to the correct family name).

1. Plants floating on the surface of water or submerged beneath the surface of the water. These plants may be found lying flat on the mud during periods of low water level. **Group 1**
- Plants growing upright out of the water (above the water level) or growing on soil, on rocks, or on other plants; neither floating nor submerged. 2
2. Plants without green pigment (chlorophyll), parasites or saprophytes; the leaves absent or reduced to scale-like or bract-like structures. (Scratch the surface of the stem to make sure there is no green pigment.) **Group 2**
- Plants possessing green pigment (chlorophyll), at least during the growing season. (For leafless plants, green pigment is usually present and can be seen on the slender stems by scratching away the outer surface.) 3
3. Plants with leaves more than one metre long or trees and shrubs with grass-like or palm-like leaves (the leaf or leaf-segments with parallel veins or the veins all arising at one point and spreading out like a fan). **Group 3**
- Plants with leaves less than one metre long, if trees or shrubs then leaves not grass-like nor palm-like. 4
4. Grass-like herbs. Plants with alternate or basal leaves, the leaf-blades narrow and with parallel veins; the flowers either small and enclosed in dry bracts, or large and colourful with the perianth parts in 3's or 6's. **Group 4**
- Plants not grass-like. The leaves simple, deeply divided, or compound, usually with pinnate or palmate venation and usually with a leaf-stalk (petiole). Herbs, shrubs, trees, or climbers, with various kinds of leaf arrangement. 5
5. Flowers densely crowded into compact inflorescences. The individual flowers closely pressed against each other or against their subtending bracts and often difficult to separate from each other. The inflorescence itself may look like a single flower. **Group 5**
- Flowers not densely crowded into compact inflorescences. The individual flowers easy to separate from each other if in a dense inflorescence. 6
6. Flowers or reproductive structures naked; not subtended by 3 or more bracts, without a perianth of 3 or more parts or a perianth tube. Sepals or petals never present, but the inflorescence or flowers may be subtended by 1 or 2 bracts. **Group 6**
- Flowers or reproductive structures subtended or surrounded by three or more bracts or perianth parts (sepals and/or petals) or within a cup-like or tubular structure. If only fruits present, check the base and top for remains of perianths parts. 7
7. Flowers with one or more spurs or with a sac-like pouch at the base of the flower, usually colourful. (A spur is a tube-like structure that projects out from the base of the flower, usually near the flower stalk, and which usually contains nectar). **Group 7**
- Flowers without spurs or without a sac-like pouch at the base of the flower. 8
8. Flowers apparently only female; stamens absent, or lacking anthers and modified into staminodes, or stamens united with each other and/or with other parts of the flower and difficult to identify. Check carefully below the corolla to see if there is an inferior ovary present which may be apparently part of the flower-stalk. **Group 8**
- Flowers or reproductive structures with free and distinct anthers or pollen sacs, apparently either bisexual or only functionally male. The anthers usually on a slender stalk (the filament) or sometimes sessile on the perianth or on the united filaments. 9
9. Whole plant or some flowers on the plant functionally only male; such flowers never develop fruits, after flowering they wither and drop off. Flowers or reproductive structures with distinct anthers or pollen sacs; with or without a central pistil or group of pistils or a pistil-like structure. **Group 9**
- Flowers or reproductive structures with a central pistil or group of pistils or with a pistil-like structure; flowers usually bisexual. If the ovary is inferior only the style may be visible in the centre of the flower. 10
10. Each flower with two or more pistils or if only one pistil is present then the ovary with two or more styles arising from separate areas on the ovary. Two or more styles present and arising from separate ovaries (when more than one pistil is present) or from separate areas on the single ovary (when only one pistil is present). **Group 10**
- Each flower with only one pistil; the pistil with only one style or if two or more styles are present then the styles all arising from one point on the top of the ovary. Pistil never more than one and the styles all arising together from one point on the ovary when two or more styles are present. 11
11. Flower with the pistil (or ovary) borne on a stalk. (Be sure to examine the flower carefully as some slender pistils borne on a stalk may

appear to be the stigma borne on the style - the stalk appears between the receptacle and the ovary). The fruit often on a stalk with a joint or articulation between the base of the fruit and the point of attachment on the stem or inflorescence. **Group 11**

- Flower with the pistil (or ovary) borne flat on the receptacle at the same level as stamens and perianth or borne on a disc or raised receptacle or beneath the level of perianth and stamen attachment. The pistil never on a distinct stalk and the fruit without a stalk that has a joint or articulation between the base of the fruit and the point of attachment to stem or inflorescence. 12

12. Flower with an inferior or half-inferior ovary; at least some part of the ovary below the level of sepal, petal, or perianth-tube attachment. The top of the fruit often with scars where the perianth has fallen off or with the persisting dry perianth. **Group 12**

- Flower with a superior ovary; the ovary borne above the level of sepal, petal, or perianth-tube attachment. The sepals, petals, or stamens may be above the ovary if a perianth-tube is present (the ovary is superior if it is free from the base of the tube as *Buddleia* 'anfer'). The fruit with scars or dry perianth parts usually at the base. 13

13. Flowers with 13 or more than 13 stamens, or with 13 or more than 13 anthers when the filaments are united or absent. **Group 13**

- Flowers with 12 or fewer than 12 stamens, or with 12 or fewer than 12 anthers when the filaments are united or absent. 14

14. Flowers with stamens with united filaments; the filaments united with each other to form a short or long tube. Do not confuse a staminal tube with united filaments with a perianth-tube that bears sessile anthers. The staminal tube is found within the usually colourful perianth parts. **Group 14**

- Flowers with stamens with free filaments; the filaments never forming a tube; if present, always free from each other or united only at the base. The filaments may be absent and then the anthers are sessile on the perianth parts or on a perianth-tube. 15

15. Flowers with the petals united to form an irregular or bilaterally symmetrical (zygomorphic) corolla-tube. An irregular corolla is one that can be divided into two equal halves which are mirror images of each other only on one plane, that is, by only one cut; or it cannot so be divided. Sometimes the petals have become united to form two petal-like 'lips' with a short tube at the base; these flowers are included in this group. The corolla-tube in this group always comes off as a single piece and never as separate petals. **Group 15**

- Flowers with the petals forming a regular or radially symmetrical (actinomorphic) corolla. A regular corolla is one that can be divided into two equal halves along two or more planes, that is, by two or more cuts. Petals may be present or absent, but they never form two 'lips'. 16

16. Flowers with the petals united to form a regular or radially symmetrical (actinomorphic) corolla-tube. The corolla-tube in this group comes off as a single unit and never as separate petals. Sepals that are united to form a colourful corolla-like tube are included in this group. **Group 16**

- Flowers with the petals free from each other or united only at the base or the petals absent. A corolla-tube is absent and the petals usually come off separately and not together as a single unit. The sepals free or united but not forming a colourful tube. 17

17. Flowers with the stamens greater in number than the petals or fewer in number than the petals. If the petals are absent then the stamens are greater or fewer in number than the sepals. The stamens never the same in number as either the petals or the sepals. **Group 17**

- Flowers with the stamens equal in number to the number of petals. If the petals are absent then the stamens equal in number to the number of sepals. 18

18. Flowers with only one perianth whorl (either sepals or petals, but not both) with the stamens opposite to the perianth parts. The stamens attached near the centre or the base of the perianth parts. **Group 18**

- Flowers with two or more perianth whorls or if only one whorl is present then the stamens alternate with the perianth parts; that is with the stamens attached between the perianth parts. 19

19. Flowers with the stamens alternate with the petals; the stamens opposite the sepals if the petals are absent or have fallen off. **Group 19**

- Flowers with the stamens opposite the petals; the stamens alternate with the sepals if the petals are absent or have fallen off. **Group 20**

GROUP 1[1]

Plants growing on or under water, freely floating or if rooted in the soil then not growing more than a few (5) centimetres above the water level, lying flat upon the mud during dry periods and not self-supporting. Herbaceous plants without woody stems.

1. Plants of ocean water or of salty water close to the ocean shore. 2

- Plants of fresh or of only slightly salty water. 5

2. Flowers borne on the lower part of the leaf-base and at first enclosed between the edges of the sheathing leaf-base; sepals and petals

1. The group keys have not been modified to match the Dahlgren and Clifford classification system for mono-cotyledons.

absent. The plants and their flowers below the surface of the water. *ZOSTERACEAE*

- Flowers borne on stalks or in the axils of leaves (between leaf and stem) and not as above. 3

3. Flowers enclosed in 2 opposite bracts; each ovary with 2 to 12 styles. Often flowering below the surface of the water; the pollen thread-like. *HYDROCHARITACEAE*
- Flowers not enclosed in 2 opposite bracts; each ovary with usually only one style. 4

4. Flowers in the axils of leaves, bisexual or unisexual; perianth of 3 scales or absent; stamens 1, 2, or 3, with thread-like pollen; pistil 1 to 9. *ZANNICHELLIACEAE*
- Flowers in spikes, bisexual; perianth and bracts absent; stamens 2 with powdery pollen; pistils 4 or more. *POTAMOGETONACEAE*

5. Plants without a stem or stem-like portion, flat and circular or oval in outline, each little plant less than 1 cm long (sometimes the round plants attached to each other and forming a group longer than 1 cm), floating flat upon the water surface and not attached to rocks or soil. *LEMNACEAE*
- Plants with a stem or stem-like portion or if the stem is absent, then attached to rocks; if less than 1 cm long then with small overlapping leaves. 6

6. Plants cabbage-like and freely floating, the rosette of leaves above the water level, leaves up to 15 cm long; the roots not attached to the soil. Male flowers reduced to several anthers on a stalk above the single ovary with many ovules (reduced female flower) and both enclosed in a single tubular bract. *(Pistia) ARACEAE*
- Plants not cabbage-like and freely floating; the leaves usually flat on the surface or beneath the surface. 7

7. Leaves floating and round in outline but with 4 fan-shaped parts attached to the central leaf-stalk, with dark-coloured nut-like bodies near the base of the leaf-stalk under the water. This is a fern. (*Marsilea*) MARSILEACEAE
- Leaves not floating, round in outline, and without 4 fan- or wedge-shaped parts. Plants either non-flowering or looking like non-flowering plants. 8

8. Plants usually only a few centimetres long with 2 or 3 rows of small (less than 1 cm long) overlapping leaves or leafless and attached to rocks. 9
- Plants usually more than a few centimeters long or if less than a few centimetres long then the leaves not small and in 2 or 3 overlapping rows. 10

9. Plants attached to rocks in fast moving water or in the spray of moving water, superficially looking like a liverwort or moss. Leaves alternate or absent, flowers very small with usually 1 or 2 stamens and an ovary with 2 or 3 styles. *PODOSTEMONACEAE*
- Plant floating on the surface of calm water. Leaves opposite. These are the water ferns *Azolla* with each leaf divided into an above water and below water portion, and *Salvinia* with leaves that float on the water surface. FILICOPHYTA

10. Underwater leaves much divided and very narrow, with small (2 to 5 cm) round bladders that catch very small aquatic organisms. Flowers usually on a stalk above the water, with 2 sepals, a 2-lipped corolla-tube, and 2 stamens. *LENTIBULARIACEAE*
- Underwater leaves without small bladders. 11

11. Styles 2 or more coming from separate ovaries (if more than 1 pistil is present) or from separate areas on the ovary (when only 1 pistil is present). If styles are absent then the stigmas 2 or more on separate ovaries or from separate areas on the solitary ovary. 12
- Style 1 or if 2 or more then coming from one point on the top of the ovary (ovary or pistil never more than 1). Ovary, style and stigma may be absent in male flowers and these are included here. 21

12. Leaves large floating, over 20 cm broad, with the petiole attached at or near the centre of the blade. Flowers usually over 2 cm in diameter. 13
- Leaves usually less than 20 cm broad with the petiole at or near the edge of the blade. Flower usually less than 2 cm in diameter. 14

13. Petals 3, pistils 3 or more, separate from each other and from the receptacle of the flower. **CABOMBACEAE*
- Petals many, pistil 1 and made up of many parts, united below with the receptacle and separate only near the top. *NYMPHAEACEAE*

14. Pistils usually 4 or the single pistil with 4 styles and 4 locules. Leaves borne along the stem. 15
- Pistils 3, 6, or more. Leaves usually all arising from near the base of the plant. 18

15. Flowers in spikes, bisexual; sepals 4, narrow at the base and bract-like; stamens 2 or 4; leaves broad and floating. *POTAMOGETONACEAE*
- Flowers in the axils of leaves; leaves narrow. 16

16. Leaves in whorls, much divided with thread-like leaflets; pistil 1 with ovary 4-lobed. *(Myriophyllum) HALORRHAGACEAE*
- Leaves in opposite pairs or alternate, simple, narrow; pistils 4 or more. 17

17. Plant growing erect out of the water, often in dense masses; sepals 4, petals present. *(Crassula)* 88 **CRASSULACEAE**
- Plant growing submerged; sepals 3, petals absent. *(Zannichellia) ZANNICHELLIACEAE*

18. Flowers in 2-branched spike-like inflorescences, sepals united and petal-like or spathe-like, petals absent; leaves with a few parallel

* Families not recorded for Ethiopia, but which could be found as plant collection and exploration increases.

primary veins and many transverse secondary veins. *APONOGETONACEAE*
- Flowers in umbels or racemes or solitary in the leaf-axils and in whorls around the stem. 19

19. Flowers usually solitary and on a long stalk, with many pistils and stamens in each flower, sepals present petals 5 or more; leaves lobed or much divided. *RANUNCULACEAE*
- Flowers usually in umbels or whorls along the stem, with 3 sepals, the petals 3 or absent; leaves entire. 20

20. Stamens 3 or 6 or more, ovules basal within the ovary, fruit an achene; leaves often sagitate at the base (with 2 backward pointing lobes). *ALISMATACEAE*
- Stamens usually 9, ovules scattered over the walls within the ovary, fruit a follicle. **BUTOMACEAE*

21. Leaves alternate and bipinnate; flowers in heads, each flower with 5 or 10 stamens or the asexual flowers with 10 petal-like staminodes. (*Neptunia*) *MIMOSOIDEAE* 95 **FABACEAE**
- Leaves not alternate and bipinnate and the flowers not in heads. 22

22. Flowers lacking a pistil or pistil-like structure, occasional bisexual flowers absent, all flowers unisexual. 23
- Flowers with a pistil or pistil-like structure or at least some bisexual flowers present. 29

23. Stamen 1 in each flower. 24
- Stamens 2 or more in each flower. 27

24. Flowers in dense spikes, each spike on a stalk; leaves with many opposite divisions. **HYDROSTACHYACEAE*
- Flowers in the axils of opposite or whorled leaves, each flower sessile or on a short stalk. 25

25. Filament absent, the anther sessile and with 1 to 4 thecae (pollen chambers), the anther enclosed in a tubular bract. (*NAIACACEAE*) *NAJACACEAE*
- Filament present, anther on a long and slender stalk (the filament). 26

26. Anthers with 1 theca (chamber) each, the anther opening by a slit. *CALLITRICHACEAE*
- Anthers with 2 thecae (chambers) each, the anther opening by 2 slits. *ZANNICHELLIACEAE*

27. Stamens 10 to 20 in each flower; leaves in whorls along the stem, with many slender divisions that usually have serrulate edges. *CERATOPHYLLACEAE*
- Stamens 3 to 9 (rarely 10) in each flower; leaves in whorls or from the base of the plant. 28

28. Stamens 3 or 6 in each flower (if more than 6 stamens present the plant should have a few bisexual flowers), flowers in a whorl around the stem or sometimes in umbels; all leaves from the base of the plant. *ALISMATACEAE*
- Stamens 8 in each flower; leaves and flowers in whorls along the stem with usually 4 leaves and 4 flowers at each node, the leaves with many slender divisions with entire edges. (*Myriophyllum*) *HALORRHAGACEAE*

29. Ovary-like structure minute (<3mm) with a wall with parallel spiral grooves. Leaf-like structures borne in whorls along the stem (rough to the touch in *Chara* and smooth in *Nitella*). The algae CHARACEAE
- Ovary not minute nor with a wall of parallel spiral grooves, the ovary producing seeds, each seed containing an embryo. 30

30. Flower without a perianth; the reproductive structures without subtending bracts or with only 1 or 2 subtending bracts, and not enclosed by a tubular bract. 31
- Flower with a perianth or the reproductive structures subtended by 3 or more bracts or within a tubular bract. 33

31. Flowers in densely flowered spikes; ovary with 2 styles, 1-locular with 2 parietal placentas and many ovules. **HYDROSTACHYACEAE*
- Flowers in the axils of leaves or branches; ovary with 2 or 4 styles and 1 to 4 ovules. 32

32. Pistil without subtending bracts, ovary with 2 or 4 styles and 1 locule with 1 basal ovule; the fruit not breaking up. (*NAIADACEAE*) *NAJADACEAE*
- Pistil subtended by 2 bracts, ovary with 2 styles and 4 locules, each locule with 1 pendulous ovule; the fruit breaking up into 4 parts. *CALLITRICHACEAE*

33. Flowers enclosed within 2 opposite bracts or within a tubular bract, ovary 1-locular and with 3 to 6 styles, the style often 2-branched. *HYDROCHARITACEAE*
- Flowers not enclosed within 2 opposite bracts nor enclosed within a tubular bract. The reproductive parts may be enclosed with a tubular perianth but the flower itself (with its perianth) is not enclosed by a bract. 34

34. Leaves opposite or whorled along the stem. 35
- Leaves alternate along the stem or all from the base. 39

35. Flowers with a 2-lipped corolla-tube or with an inferior ovary; the leaves usually opposite and not in whorls. 36
- Flowers lacking a 2-lipped corolla-tube and the ovary not inferior; the leaves often in whorls, usually very narrow or with narrow divisions. 37

36. Petals untied to form a 2-lipped corolla-tube; stamens 2 or 4; the ovary superior. *SCROPHULARIACEAE*
- Petals free from each other and not forming a tube; stamens usually 4, 5, 8, or 10; the ovary inferior. *ONAGRACEAE*

37. Leaves with linear segments at the end and with a folded structure near the base that traps minute animals; ovary 1-locular; sepals 5, petals 5. (*Aldrovanda*) *DROSERACEAE*
- Leaves without a folded or lens-like structure near the base. 38

38. Leaves simple (not divided) and narrow; sepals 2 to 5, petals 2 to 5; ovary 3- to 5-locular. *ELATINACEAE*
- Leaves deeply divided into narrow blades with serrulate margins; sepals 8 to 12, petals absent; ovary 1-locular. *CERATOPHYLLACEAE*

39. Floating leaves round with a deep notch on one side, the petiole attached at the centre of the blade; sepals 5, petals 5, and stamens 5, the petals with hairs. (*Nymphoides*) *MENYANTHACEAE*
- Floating leaves with the petiole attached at the edge. 40

40. Floating leaves with a leaf-stalk (petiole) that has a thicker portion at its midpoint made up of spongy tissue. 41
- Floating leaves with a leaf-stalk that does not have a swollen portion near its midpoint (between leaf base and blade). 42

41. Leaves more or less triangular in outline; sepals 4, petals 4; ovary 2-locular and half-inferior. **TRAPACEAE*
- Leaves rounded in outline (sometimes up above the surface of the water); sepals and petals alike in form and colour, 3 and 3, together 6; ovary 1 or 3 locular and superior. (*Eichhornia*) *PONTEDERIACEAE*

42. Ovary inferior, the ovary below the level of sepal and petal attachment; fruit often with a dry perianth or the scars where the perianth has fallen off at the top. 43
- Ovary superior, the ovary above the level of sepal and petal attachment; fruit often with a dry perianth or the scars where the perianth has fallen off at the bottom (near the attachment of the fruit-stalk). Perianth parts usually 3 or 6. 44

43. Styles 2, ovary with 2 locules and 1-ovule in each locule; flowers in umbels of heads. (*UMBELLIFERAE*) *APIACEAE*
- Style 1 or none, ovary with usually 4 locules and 1 to many ovules in each locule; flowers solitary in the leaf axils or in recemes or panicles. *ONAGRACEAE*

44. Leaves linear (very narrow) and without a definite leaf-stalk (petiole); the ovary with 3 to 6 locules and the fruit breaking up into 3 to 6 parts. (*Triglochin*) *JUNCAGINACEAE*
- Leaves not linear, narrowly elliptical to rounded or arrow-shaped, a distinct leaf-stalk usually present. 45

45. Style 1 on the ovary, ovary with 1 or 3 locules and usually many ovules; leaves often arrow-shaped with palmate or curved almost parallel venation. *PONTEDERIACEAE*
- Styles (or stigmas) 2 or 3 on the ovary; ovary with only 1 locule and 1 basal ovule; leaves usually elliptic in outline and with pinnate venation. *POLYGONACEAE*

GROUP 2

Plants lacking green pigment (chlorophyll); the leaves absent or reduced to scale-like or bract-like structures. Mostly herbaceous plants growing on the ground and parasitic on the underground parts of the host plant or saprophytic.

1. Plant climbing and twining about the host, attached to the host plant by root-like structures along the length of the slender stem, stems yellowish to purple or red. 2
- Plant not climbing or twining, attached to the host at usually only one point. 3

2. Stamens 12 in each flower, anthers opening by flaps; petals separate and arising from a cup-shaped receptacle, fruit with 1 locule and 1 seed. (*Cassytha*) *LAURACEAE*
- Stamens 4 or 5 in each flower, anthers opening by slits; petals united and the stamens borne on the short corolla-tube; fruit with 2 locules and 1 to 4 seeds. (*Cuscuta*) *CONVOLVULACEAE*

3. Plants growing upon the above ground stems or aerial parts of their hosts. 4
- Plants growing on the ground (attached to the roots of their hosts if parasitic). 6

4. Flowers borne on the stem of the parasite and not directly on the stem of the host (that is the parasite with a stem of its own). Leaves usually present; flower very small with 8 or 4 sepals; ovary inferior and with 1 locule and 1 ovule both of which may be difficult to see. **115 VISCACEAE**
- Flowers or reproductive structures growing directly from the stem of the host, the parasite without a stem or with only a short stem and without leaves. 5

5. Flowers with sepals or subtended by bract-like structures; male flowers with 8 to many anthers, the filaments free or united to form a column; female flowers with an inferior ovary and many ovules in a single locule. **RAFFLESIACEAE*
- Flowers absent, the underside of the reproductive structure with many pores or slits from which minute dust-like spores are shed. These are fungi. Kingdom FUNGI

6. Flowers few to very many, rarely solitary and then borne on a very slender (<1 cm) stalk; the flower always borne on a stalk above ground. 7
- Flowers or reproductive structure solitary, borne on a thick (>1 cm) stalk or a broad base. 12

7. Flowers very numerous (more than 50) and densely crowded together, the flowers usually unisexual. **117 BALANOPHORACEAE**
- Flowers fewer than 50 and usually loosely arranged in spikes or racemes, the flowers usually bisexual. 8

8. Flowers unisexual; ovary inferior with 1 locule and 8 to 14 parietal placentas; anthers 8 to 20 and borne on a single column; perianth of 1

whorl forming a short tube but the flower subtended by two perianth bracts. (*Cytinus*) *RAFFLESIACEAE
- Flower bisexual and not as above. 9

9. Ovary inferior and 1- or 3-locular; stamens 3 or 1 or completely united to the style and not easily seen. 10
- Ovary superior and 1- or 2-locular; stamens 4 or 2 and borne on the irregular corolla-tube. 11

10. Flowers mostly regular; anthers 3 and borne on 3 very short filaments which are themselves borne on the corolla-tube. *BURMANNIACEAE
- Flowers irregular; anthers or pollinia 1, 2, or 4 and borne on the end of a thick column in the centre of the flower or the anther not visible. ORCHIDACEAE

11. Ovary 1-locular with 4 parietal placentas. OROBANCHACEAE
- Ovary 2-locular with axile placentas. SCROPHULARIACEAE

12. Reproductive parts subtended and at first enclosed in a single large (sometimes petal-like) bract, the reproductive parts borne on a thick central column. These plants are neither parasites not saprophytes but produce green leaves after flowering. ARACEAE
- Reproductive parts not subtended and at first enclosed in a single large bract. 13

13. Plant represented by a single flower in which only the 3 or 4 perianth parts rise above the soil; anthers borne on the perianth tube, very large and with many transverse ridges; ovary inferior with a single locule with pendulous placentas at the top. HYDNORACEAE
- Plant with the flower or reproductive structure coming to the surface or above soil level, not remaining partly sunken into the ground. 14

14. Flowers unisexual or rarely bisexual, with 3 to 10 sepals or the sepals sometimes absent in the female flowers; anthers 8 to many, on separate filaments or on a single column. *RAFFLESIACEAE
- Flowers absent; plants variously shaped but often umbrella-shaped and the broad upper part with many small pores or slits beneath that produce minute dust-like spores. These are fungi. Kingdom FUNGI

GROUP 3

Plants with leaves over one metre long or trees and shrubs with palm-like leaves or grass-like leaves (the leaves or the leaf-segments with parallel veins or the veins arising all at one point and spreading out like a fan). Herbs, shrubs, trees, or climbers.

1. Leaves simple and nearly all the secondary veins parallel with the primary vein or midrib of the leaf (in mature leaves). 2
- Leaves simple, cut into segments, or compound, the secondary veins arising at an angle from the primary vein or midrib of the leaf, or if arising together from the base of the blade then the leaf fan-shaped or round in outline. 13

2. Plants with a trunk or with a thick stem; the leaves borne in a spiral at the top or along the branches. 3
- Plants without a trunk or thick stem; the leaves coming from the base of the plant. 6

3. Plants growing near the sea coast or along stream and river edges near the sea coast (ocean); male flowers crowded together, each with a single anther. *PANDANACEAE
- Plants not growing near the sea coast; male flowers not as above. 4

4. Leaves with sheathing bases; flowers very reduced, enclosed in opposite dry bracts and without sepals or petals, ovary with 2 hairy styles and stigmas; the leaves clasping the stem at their base with an open sheath. (GRAMINEAE) POACEAE
- Leaves without sheaths; flowers with sepals or petals. 5

5. Flowers solitary and on long stalks (pedicels), large (>2 cm); ovary inferior. VELLOZIACEAE
- Flowers numerous in usually branched inflorescences, usually small (<2cm) the inflorescences often large; ovary superior. AGAVACEAE

6. Leaves long and narrow (up to 3 m long and less than 10 cm wide) in two opposing ranks and with air-chambers within; growing along stream edges and lake shores; inflorescence a brown cylinder of thousands of very slender tightly packed flowers. TYPHACEAE
- Leaves and inflorescences not as above. 7

7. Plants with white or brightly coloured flowers, the stamens and ovary not enclosed in dry bracts. 8
- Plants lacking coloured flowers, the stamens and ovary enclosed in small (<2 cm) dry bracts; the leaf-base enclosing the stem. 12

8. Ovary superior; the fruit with the dry perianth or the scars where the perianth has fallen off at the base near the attachment of the fruit stalk. 9
- Ovary inferior; with the dry perianth or the scars where the perianth has fallen off at the top of the fruit - opposite the fruit stalk. 10

9. Leaves tough and fibrous, very difficult to cut across the strong fibres; flowers usually in many flowered, often branched, inflorescences. AGAVACEAE
- Leaves slender or thick and succulent but lacking strong fibres, usually easy to cut across the long axis of the leaf; flowers mostly in spikes or racemes. 'LILIACEAE'[1]

10. Leaves thick and fibrous, very difficult to cut across the strong fibres (to cut across the long axis of the leaf); inflorescence to 4 m tall; plants planted for fiber or ornament. (*Agave*) AGAVACEAE

---

1 = 'Liliaceae' in the broad sense, not as circumscribed by Dahlgren and Clifford.

- Leaves not thick and without strong fibres. 11

11. Stamens 6; perianth parts usually all alike, 6 in number; inflorescence an umbel; first enclosed in 1 (or more) dry bract. *AMARYLLIDACEAE*
- Stamens 3; perianth parts usually of 2 different kinds, 3+3; flower solitary or in a cyme, first enclosed in 2 dry bracts. *IRIDACEAE*

12. Style branches and stigmas 2, usually feathery; the filament attached to a notch at the base of the anther; the stamens and pistil enclosed in two opposite bracts; the leaves in 2 vertical ranks or rows (when seen from above), the leaf base enclosing the stem by a usually split sheath. *(GRAMINEAE) POACEAE*
- Style branches and stigmas 2 or 3, not feathery; the filament attached to the base of the anther; the stamens and pistil enclosed in a single bract or subtended by a whorl of 3 or 6 bract-like structures; the leaves in 3 vertical ranks or rows (when seen from above), the leaf base usually enclosing the stem by a closed sheath. *CYPERACEAE*

13. Trunk made up of the large (+20 cm) and spongy leaf bases; leaves curled up length-wise in the bud, simple and entire at first but becoming torn into strips with age; flowers in unbranched inflorescences over 1 m long; plants mostly cultivated. *MUSACEAE*
- Trunk absent or a trunk present and the leaf-bases not broad and spongy. 14

14. Leaves palm-like and often with a spiny stalk, the leaf divided into pinnate segments (feather-like) or with the parts divided only near the outer portion (fan-like or V-shaped), the leaf-segments usually sharp tipped and with parallel venation. 15
- Leaves not palm-like, the leaves not divided or if divided then the segments with pinnate venation. 16

15. The midrib of the leaf more or less flattened in cross-section, the leaf curled up in the bud from apex to base like a fern leaf, the leaflets very stiff and with veins that branch into 2. Reproductive structures in cones or on a leaf-like part, the young cones not enclosed in a woody boat-shaped structure or in a stiff bract. **CYCADACEAE*
- The midrib of the leaf triangular in cross-section or a midrib absent and the leaf fan or V-shaped; the leaf folded as a V or O in bud, the leaflets or blade stiff but flexible with mostly parallel veins arising from near the base. Flowers usually on branched inflorescences, the young infloresence enclosed in a woody boat-shaped structure or within a stiff bract. *(PALMAE) ARECACEAE*

16. Leaves divided 2 or 3 times, the leaf-stalk usually with scale-like hairs, the leaves curled up with the tip innermost when young. Flowers never present and the plant never producing seeds, reproductive structures usually brownish areas on the underside of the leaves which produce minute powdery spores. These are ferns. Phylum FILICOPHYTA
- Leaves undivided or 2 or 3 times divided but never curled up with the tip innermost when young. Flowers and/or fruits and seeds usually present. 17

17. Plants growing in wet places, the leaf-stalk attached near the centre of the blade, the leaves round in outline or if otherwise then flowers and fruits absent. 18
- Plants growing in dry or moderately wet places, if growing in wet places (swamps and lake or pond edges) then the leaf-stalk attached on the edge of the blade and not near the centre. 20

18. Leaves usually arrow-shaped, with a deep notch where the leaf-stalk is attached to the blade, the plants rarely with flowers (the flowers enclosed in a tubular bract when present) the plants grown for the starch filled rhizomes. *(Colocasia) ARACEAE*
- Leaves round in outline; the plants often with flowers or inflorescences; the plants grown for ornament at the edges of lakes and ponds. 19

19. Leaves completely round and without a notch at one side and with an entire edge; the flowers large (>4 cm) and borne singly on long narrow stalks. *(Nelumbo) NYMPHAEACEAE*
- Leaves with a notch on one side or with a round and lobed edge; the flowers small (<1 cm) and inconspicuous, borne close together on a densely flowered spike or panicle. *(Gunnera) HALORRHAGACEAE*

20. The inflorescence resembling a flower, with a large, often tubular bract that encloses (at least in the early stages) a thick spike-like axis on which pistils and/or stamens are borne, the bract may resemble a single large petal. The leaves simple or divided; sap sometimes white. *ARACEAE*
- The inflorescence or flower not as above. 21

21. Flowers with the anthers united to each other and forming a tube around the style; ovary inferior. 22
- Flowers with the anthers free from each other (the anthers sometimes borne on a petal-like structure and solitary); ovary superior or inferior. 23

22. Filaments free below the anthers, ovary with 1 locule and 1 ovule; flowers in dense heads. *(COMPOSITAE) ASTERACEAE*
- Filaments united below the anthers, ovary with 2 locules and many ovules; flowers in spikes and racemes. *(Lobelia) LOBELIACEAE*

23. Flowers with only one anther or pollen sac which is often borne on a petal like structure, the flowers usually large (>2 cm); ovary inferior. Herbs or bamboo-like shrubs. 24
- Flowers with 2 or more anthers or if each flower has only one anther then the flowers very small (<1 cm); ovary superior or inferior. Trees. 26

24. Ovary with 1 or 3 locules and only 1 ovule in each locule; leaf stalk with a thickened area just below the blade. *MARANTACEAE*
- Ovary with 1, 2 or 3 locules and many ovules in each locule; leaf stalk without a thickened portion just below the blade. 25
25. Anther at the edge of a petal-like structure with the connective not apparent; plant parts not spicy-aromatic when crushed. *CANNACEAE*
- Anther on a slender grooved filament and with an enlarged, often 2-pronged connective; plant parts spicy aromatic when crushed. *ZINGIBERACEAE*
26. Sap white or watery-white; the leaves palmately divided or palmately compound. 27
- Sap not white; leaves pinnate or bipinnate. 28
27. Flowers large (>2 cm) and not tightly congested; each ovary with many ovules; plants often cultivated for their edible fruits. *CARICACEAE*
- Flowers small (<1 cm) and densely congested in spike-like inflorescences; each ovary with 1 ovule in the individual female flowers, whole inflorescence often forming a fig. *MORACEAE*
28. Leaves with stellate hairs; the ovary inferior and 2-locular. *ARALIACEAE*
- Leaves without stellate hairs; the ovary superior. 29
29. Leaves alternate, stamens 5 to many, petals free. *(LEGUMINOSAE)* 95 **FABACEAE**
- Leaves opposite or in threes; stamens 4, petals united. *BIGNONIACEAE*

GROUP 4

Grass-like herbs. Plants with alternate or basal leaves, the leaf-blades narrow and with parallel veins; the flowers either small and enclosed in dry bracts, or large and colourful with the perianth parts in 3's or 6's.

1. Petals and sepals absent, if a perianth is present then composed of 6 small dry brown or yellowish parts all alike; leaves not ending in coiled tendrils. 2
- Petals and/or sepals present, greenish or brightly coloured, not dry and brown except when of 2 kinds (both sepals and petals present) or on old flowers; leaves sometimes with tendrils at the tips. 6
2. Reproductive structures (spore-filled cavities) present in the base of the very narrow grass-like leaves just below a small tongue-like ligule. *(Isoetes) ISOETACEAE*
- Reproductive structures not present in the tissue of the leaf-base beneath a tongue-like ligule; plants producing flowers and fruits. 3
3. Flowers and/or fruits enclosed or subtended by 6 sepal-like or bract-like structures, these often in 2 whorls of 3 each. 4
- Flowers and/or fruits enclosed or subtended by 1 or 2 dry bracts, if 3 or 6 perianth-like structures are present they are smaller than the bract or bracts. 5
4. Anthers borne on slender filaments; each flower subtended by a small bract; the fruit (a capsule) breaking open at maturity. *JUNCACEAE*
- Anthers without slender filaments and almost sessile; each flower not subtended by bracts; the fruit breaking into 3 to 6 separate 1-seeded parts. *JUNCAGINACEAE*
5. Leaves in 2 vertical rows when viewed from above; leaf-base enclosing the stem in a sheath which is usuallly split; stems usually round and hollow between the nodes; style branches and stigmas 2, usually feathery; filaments attached to a notch in the base of the anther; stamens and pistil enclosed within 2 opposite bracts. *(GRAMINEAE) POACEAE*
- Stems leafless or with scale leaves, or with leaves in 3 vertical ranks when viewed from above; leaf-base enclosing the stem with a closed sheath; leafy stems usually 3-angled and solid or pithy between the nodes; style branches and stigmas 2 or 3, not feathery; filament attached to the base of the anther without a notch; stamens and pistil enclosed in a single bracts or subtended by 3 or 6 small bract-like structures. *CYPERACEAE*
6. Flowers in heads or spikes and densely crowded togethr, the flowers usually difficult to separate from each other; the petals partly united to form a corolla-tube. (Note that an inflorescence of many small flowers with large petals on the outside of the head may look like a single large flower.) 7
- Flowers not densely crowded into heads or spikes or the petals not united to form a tube if the flowers are closely crowded together. 10
7. Flower heads or short spikes surrounded or subtended by an involucre of empty bracts and often with a whorl of petal-like structures around the outer edge. (The head of flowers and whorl of petal-like structures may look like an individual flower.) 8
- Flower heads or spikes not as above. Each bract subtending or enclosing a flower, flower bud or fruit. 9
8. Stamens 5 or apparently only 1, the anthers united to form a slender tube around the style, filaments free and borne on the corolla tube; ovary inferior and 1-locular; fruit not opening when ripe. *(COMPOSITAE) ASTERACEAE*
- Stamens 2, 3, 4 or 6, the anthers free from each other, filaments free or united, often borne on a stalk; ovary superior and 2- or 3-locular; fruit opening when ripe. *ERIOCAULACEAE*
9. Corolla tube 3-lobed; stamens 3 in each flower and free from each other but borne on the corolla-tube; ovary superior and 1- or 3-locular, fruit opening by a slit. *XYRIDACEAE*

- Corolla 4-lobed; stamens 4 in each flower and free from each other but borne on the corolla tube; ovary superior and usually 2-(3- or 4-) locular; fruit opening by a circular lid. *PLANTAGINACEAE*

10. Ovary inferior or half-inferior, at least some part of the ovary beneath sepal, petal or perianth tube attachment; the top of the fruit often with scars where the perianth has fallen off or with the persisting dry perianth. (Note that the ovary may appear to be the flower stalk.) 11
- Ovary superior, the ovary borne above the level of sepal, petal or perianth tube attachment (if a tube is present); the base of the fruit or the fruit-stalk with scars where the perianth has fallen off or with the remains of the dry perianth. 18

11. Perianth parts all alike with little or no difference between the inner and outer whorls, or the perianth parts of a single whorl, usually 6 in number; stamens usually 6; ovary usually 3-locular. 12
- Perianth parts of at least 2 different kinds, the inner and outer whorls differing in form or colour or both, the perianth usually with 3 sepals and 3 petals; stamens 1 or more; ovary with 1 or more locules. 15

12. Anthers opening by pores or by short pore-like slits at the top; ovary completely inferior or tightly enclosed within the perianth tube, with 3 locules and 2 ovules in each locule. *(Cynastrum)* **CYNASTRACEAE*
- Anthers opening by long slits along the side and not near the top; ovary completely inferior and usually with many ovules. 13

13. Flowers in umbels or solitary, subtended and at first enclosed in a large thin bract; ovary usually smooth and hairless on the outside. *AMARYLLIDACEAE*
- Flowers in racemes or corymbs, or if the flower is solitary then without a large bract enclosing the bud; the outer surface of the ovary usually covered with short or long hairs. 14

14. Leaves and inflorescence with thin white hairs; flowers in racemes and corymbs or sometimes solitary. *HYPOXIDACEAE*
- Leaves and pedicels without thin white hairs; flower solitary on a long leafless stalk. *VELLOZIACEAE*

15. Sepals 5, petals usually united to form a 5-lobed and bell-shaped corolla tube. (If the plant in question does not fit the family given here, try the key in Group 12.) *CAMPANULACEAE*
- Sepals 3, petals usually 3; corolla not bell-shaped. 16

16. Stamens 1 or 2 or apparently absent, the pollen sacs (when present) borne near the end of a thick column in the centre of the flower; one petal usually larger than the others and often with a backward projecting tube (spur) that contains nectar; ovary slender and looking like the flower stalk, usually twisted. *ORCHIDACEAE*
- Stamens 3 or 6, flower not as above; ovary easily identified, not looking like the flower stalk. 17

17. Leaves usually over 10 cm long; stamens with the filaments free or partly united to each other; flowers usually irregular. *IRIDACEAE*
- Leaves usually less than 10 cm long, often very small and scale-like; stamens with very short free filaments or the filaments absent and the anthers sessile; flowers usually regular. **BURMANNIACEAE*

18. Ovary with only 1 locule and basal ovule; fruit with only 1 seed and usually 3-winged; leaf-base with a thin membranous collar-like stipule around the stem. *POLYGONACEAE*
- Ovary with 2 or more locules or if 1-locular, then with more than 1 ovule; fruit with more than 1 seed and not winged; leaf base without a collar-like stipule around the stem. 19

19. Perianth of 2 whorls that differ in form and/or colour, sepals and petals both present, stamens 2 to 6 or 10. 20
- Perianth parts all alike in form and colour; stamens usually 6. 22

20. Petals 4 or 5 and free from each other; ovary with 4 or 5 locules and with 2 ovules in each locule. *LINACEAE*
- Petals 3, or apparently 2; ovary with usually 2, 3 or 6 locules. 21

21. Flowers small and not brightly coloured, without subtending bracts; the fruit breaking up when mature into 3 to 6 separate follicles. *JUNCAGINACEAE*
- Flowers with coloured petals (blue, yellow, white, brown), but the petals quickly wilting, the flower and inflorescence subtended by bracts (these can be boat-shaped); fruit a capsule not breaking into separate follicles. *COMMELINACEAE*

22. Ovary with 1 ovule in each locule; the perianth parts small and often dry. *FLAGELLARIACEAE*
- Ovary with 1 to many ovules in each locule; perianths parts usually colourful 23

23. Anthers opening by pores or pore-like slits near the top. **TECOPHILAEACEAE*
- Anthers opening by slits. *'LILIACEAE'*[1]

1 'Liliaceae' in the broad sense, not as circumscribed by Dahlgren and Clifford.

GROUP 5

Plants with the flowers densely crowded into compact inflorescences. The individual flowers closely pressed against each other and/or against their subtending bracts and difficult to separate from each other. Sometimes these inflorescences can look like large single flowers.

1. Inflorescence subtended and at first enclosed in one large bract, the bract splitting open usually along one side only. 2
- Inflorescence not subtended by a single large bract which encloses the young inflorescence. 5
2. Leaves or the segments of the leaves with parallel veins. 3
- Leaves or the segments of the leaves with palmately or pinnately arranged veins. 4
3. Subtending bract woody or very stiff; leaves large and stiff with sharp-tipped segments, hooks or spines often present along the stem and/or petiole. Trees, shrubs or climbers. *(PALMAE) ARECACEAE*
- Subtending bract thin and paper-like; leaves and stems soft and flexible, without spines or hooks. Herbs. *AMARYLLIDACEAE*
4. Herbs or subshrubs, stems without hard wood; inflorescence a single thick axis, often looking like a single flower; leaves always smooth and hairless. *ARACEAE*
- Trees or shrubs with woody stems; inflorescence usually compound, often with a swollen axis in which the flowers are embedded and forming a fig (*Ficus*); leaf surface usually hairy and sometimes rough. 104 **MORACEAE**
5. Anthers united to form a slender tube around the style; ovary inferior. 6
- Anthers free from each other, not united to form a tube round the style, or the anthers missing; ovary position various. 7
6. Filaments free from each other and borne on the corolla tube; ovary with a single locule, fruit with a single seed; flowers usually in heads surrounded by an involucre of sterile bracts. *(COMPOSITAE) ASTERACEAE*
- Filaments united with each other and arising from the base of the corolla tube; ovary with 2-5 locules, fruit with many seeds; flower in large racemes. *(Lobelia) LOBELIACEAE*
7. Inflorescence subtended by an involucre of sterile bracts; sterile bracts do not subtend or enclose solitary flower-buds, flowers or fruits. The involucre forming a conspicuous, and often petal-like, base around the inflorescence. 8
- Inflorescence not subtended by an involucre of sterile bracts; each bract holding a flower bud, flower or fruit. 11
8. Flower with 2 to 5 stigmas or with 2 to 5 styles above the ovary; herbs, sometimes with woody stems. 9
- Flower with only 1 style and stigma; trees, shrubs or small woody herbs. 10
9. Petals free; ovary inferior and 2-locular, with 2 short styles or with a short 2-lobed stigma. *(UMBELLIFERAE) APIACEAE*
- Petals and sepals united into a perianth tube; ovary superior with 1 locule, style 1 with 5 branches or 5 styles. *PLUMBAGINACEAE*
10. Perianth tube split into 2 parts, 1 with 3 small lobes and the other with only 1 lobe; stamens 4. *PROTEACEAE*
- Perianth tube not split into 2 parts, usually with 5 lobes at the top; stamens usually 10. *THYMELAEACEAE*
11. Leaves palmately lobed and often folded like a fan, stipules joined to the petiole. 91 **ROSACEAE**
- Leaves not palmately lobed and folded like a fan, stipules not joined to the petiole or absent. 12
12. Flowers very small, with anthers that open with flap-like valves; leaves alternate and simple; leaves and stems aromatic when crushed; fruit with wings. *(Gyrocarpus) HERNANDIACEAE*
- Anthers not opening by flap-like valves, or anthers absent. 13
13. Inflorescence unisexual, flowers of only one sex, either anthers or pollen sacs absent or ovules (ovaries containing ovules) absent. (Be sure to check unopened flowers for anthers as these can fall off quickly after bisexual flowers open. Male inflorescences quickly wither and fall off after the pollen is shed. Female inflorescences persist with one or more ovaries developing into fruits.) 14
- Inflorescences bisexual, flowers either bisexual or both male and female flowers present, both functional anthers and functional ovaries present. 51
14. Inflorescence male, anthers or pollen sacs present, ovaries and ovules absent, inflorescence not producing fruits or seeds. 15
- Inflorescence female, anthers or pollen sacs absent, ovules and/or fruits and seeds present. 35
15. Flowers with 20 or more stamens; trees and shrubs with bipinnate leaves, (or with thick, leathery phyllodes formed from the leaf stalk on introduced trees). *MIMOSOIDEAE* 95 **FABACEAE**
- Flowers with less than 20 stamens; leaves not bipinnate or thick leathery phyllodes. 16
16. Herbs or slender plants, stems not woody. 17
- Trees or shrubs with woody stems. 21
17. Plants with leaves reduced to whorls of united, pointed scales around the joints of the stem. Cone-like structures borne on the ends of the stems, made up of small stalks with caps, each of which carries 4 or more small sacs that contain pollen-like spores. The spores with 4 arm-like projections. *(Equisetum) EQUISETACEAE*
- Plants not as above, if stems and leaves appear as above, the plant is a seedling of a tree. 18

18. Plants with white sap; flowers sunk into a flat, thick inflorescence with slender finger-like processes around the edge. *(Dorstenia) MORACEAE*
- Plants without white sap; flowers not sunk into the stem of the inflorescence. 19

19. Flowers with only 1 perianth whorl of usually 4 or 5 parts, the stamens opposite the perianth parts and bent inwards in bud, a non-functional pistil-like structure present in the centre of the flower. 105 **URTICACEAE**
- Flowers with or without a perianth whorl; stamens not 4 or 5 arranged as above. 20

20. Plants growing in wet places or along the edge of water; leaves often large with toothed edge; inflorescence arising separately from the leaves; flowers usually interspersed with bracts, perianth present; inflorescence usually branched. *(Gunnera) HALORRHAGACEAE*
- Plants growing in moist forests and often found on other plants or rocks, usually in shade; leaves small, somewhat thick, with an entire edge; inflorescence from a leafy stem; flower often naked, with or without bracts, on a thick, unbranched spike-like inflorescence, or with an umbel of spikes. *PIPERACEAE*

21. Leaves reduced to whorls of membranous scales on thin, ridged, needle-like stems; male flowers in dense terminal spikes only a little thicker than the normal stems. 101 **CASUARINACEAE**
- Leaves and stems not as above; male inflorescences not terminal spikes looking like the end of a vegetative stem. 22

22. Leaves needle-shaped, solitary or in bunches of 2 or more together with a membranous sheath around the base, round or triangular in cross-section; trees planted for wood or ornament, not native. *(Pinus) PINACEAE*
- Leaves scale-like, or normal and flat, not in bunches of 2 or more together and without a membranous sheath at the base; trees or shrubs, native or introduced. 23

23. Leaves scale-like with a broad clasping base and overlapping each other, at least on the young shoots. 24
- Leaves flat and well developed, not scale-like with a broad base, internodes clearly visible between leaves. 26

24. Leaves on mature, exposed shoots in opposite pairs, very small (<2 mm), on young trees and branches in deep shade, leaves larger with a long slender point. *CUPRESSACEAE*
- Leaves on mature branches spirally arranged. 25

25. Leaves >4 mm wide, usually with a hooked tip; branches usually borne in whorls at regular intervals along the main stem. *(Araucaria) ARAUCARIACEAE*
- Leaves <4 mm wide, tip not hooked; branches borne alternately, not in whorls, along the main stem. *TAXODIACEAE*

26. Leaves with a single vein, flat and somewhat stiff, evergreen, persisting for up to 4 growing seasons. *(Podocarpus) PODOCARPACEAE*
- Leaves with pinnate or palmate venation; deciduous, shed after each growing season. 27

27. Leaves opposite along the stem and folded like a fan in bud or when dry; stamens usually 5 (4-8) with united filaments. 97 ***MYROTHAMNACEAE**
- Leaves alternate along the stem and not folded in bud; anthers 1 to many, filaments usually free. 28

28. Plants with white or coloured sap. 29
- Plants with clear sap, or without conspicuous sap. 30

29. Flowers without a disc or series of glands at the base of the anthers; sap without a strong unpleasant smell. 104 **MORACEAE**
- Flowers with a disc or series of glands at the base of the anthers; sap with a strong unpleasant smell. *EUPHORBIACEAE*

30. Each group of stamens subtended by a single bract and without a perianth whorl of 3 or more parts. 31
- Each group of stamens surrounded or subtended by 2 or more bracts, or by a whorl of 3 or more perianth parts. 32

31. Axis of spike slender (<4 mm wide); leaves with stipules and without bright dots when viewed against the light. 99 **SALICACEAE**
- Axis of spike thick (>4 mm wide); leaves without stipules but with bright dots when viewed against the light. *(Myrica)* 100 **MYRICACEAE**

32. Bark flaking in round scales; flowers hanging in globose heads. *(Platanus)* **PLATANACEAE**
- Bark smooth or rough, not flaking in round scales; flowers in spikes or catkins, not in globose heads. 33.

33. Leaves with bright dots when viewed against the light and pleasantly aromatic when crushed; filaments united to form a column on which the anthers are borne. **MYRISTICACEAE*
- Leaves without bright dots but often with light coloured patches with crystaline patches in the leaf surface, not pleasantly aromatic when crushed; filaments free. 34

34. Perianth of 4-5 parts; stamens 4-5; leaves often with a rough surface; hairs stiff, sometimes stinging. 105 **URTICACEAE**
- Perianth with 6 lobes; stamens 6-12; leaves usually with a smooth surface; hairs soft, never stinging. *(Quercus)* **FAGACEAE**

35. Seeds produced in woody cones made up of woody bracts arranged close together on a central axis. 36
- Seeds produced in other types of structures which are dry or fleshy but not composed of woody bracts. 40

36. Leaves reduced to whorls of membranous scales on thin, ridged, needle-like stems. 101 CASUARINACEAE
- Leaves and stems not as above. 37

37. Leaves needle-shaped, solitary or in bunches of 2 or more together with a membranous sheath around the base, round or triangular in cross-section; trees planted for wood or ornament, not native. *(Pinus)* PINACEAE
- Leaves scale-like, not in bunches of 2 or more together and without a membranous sheath at the base; trees or shrubs, native or introduced. 38

38. Leaves on mature, exposed shoots in opposite pairs, very small (<2 mm), on young trees and branches in deep shade, leaves larger with a long slender point. CUPRESSACEAE
- Leaves on mature branches spirally arranged. 39

39. Leaves >4 mm wide, usually with a hooked tip; branches usually borne in whorls at regular intervals along the main stem. *(Araucaria)* ARAUCARIACEAE
- Leaves <4 mm wide, tip not hooked; branches borne alternately, not in whorls, along the main stem. TAXODIACEAE

40. Leaves on mature branches very small scales (<2 mm), in overlapping, opposite pairs; seeds in fleshy cones looking like green or black berries and smelling strongly of resin. *(Juniperus)* CUPRESSACEAE
- Leaves on mature branches well developed; seeds in dry or fleshy fruits, not in berry-like cones. 41

41. Leaves with a single vein, flat and stiff, persisting for upto 4 growing seasons; seed in a drupe-like cone with a fleshy base below the seed. *(Podocarpus)* PODOCARPACEAE
- Leaves with pinnate or palmate venation, not persisting for upto 4 growing seasons; seeds in fruits, not in drupe-like cones. 42

42. Plants with white sap; fruits made up of many flowers grown together and usually fleshy, in *Ficus* fruits enclosed in the swollen, fleshy receptacle forming a fig. 104 MORACEAE
- Plants without white sap; fruits not as above, easily separated or solitary. 43

43. Trees or shrubs with opposite leaves that are folded fan-wise in bud and when dry; fruit a capsule with 3 to 6 locules. *(Myrothamnus)* 97 MYROTHAMNACEAE
- Leaves mostly alternate and not folded fan-wise in bud or when dry; fruits achenes or drupes or capsules with 1 locule. 44

44. Leaves with bright dots seen when held against the light and aromatic when crushed. 45
- Leaves without bright dots when held against the light, not aromatic when crushed. 46

45. Fruits small papillate drupes 3-4 mm across; aril absent. *(Myrica)* 100 MYRICACEAE
- Fruits drupes >4 mm across, often with a delicate coloured aril. *MYRISTICACEAE

46. Small herbs with small, entire, somewhat stiff and fleshy leaves; growing in shade on other plants or rocks. PIPERACEAE
- Large herbs, shrubs or trees; leaf margins usually serrate; not growing on other plants or rocks. 47

47. Bark flaking in round scales; fruits in hanging globose heads which readily break up. *(Platanus)* PLATANACEAE
- Bark smooth or rough, not flaking in round scales; fruits not as above. 48

48. Leaves usually with 1 main and 2 lateral veins from the base and often with white, crystoline patches on the upper surface; stinging hairs usually present on either stems, leaves or inflorescences. 105 URTICACEAE
- Leaves with either pinnate venation or 5 or more veins from the base, without white crystaline patches on the upper surface; stinging hairs absent. 49

49. Herb; leaves large, appearing independently from the flowers and fruits; fruit a small nut or drupe breaking into separate parts. *(Gunnera)* HALORRHAGACEAE
- Trees or shrubs; leaves on the same stem as the flowers and fruits; fruit a capsule or large nut. 50

50. Fruit a capsule, seeds with silky hairs; leaves not lobed, ovate or lanceolate. 99 SALICACEAE
- Fruit a large 1-seeded nut in a cup-shaped structure; leaves usually lobed, ± a serrate margin. FAGACEAE

51. Flowers with an inferior ovary or half-inferior, at least part of the ovary borne below the level of sepal and petal attachment, or the ovary united with the perianth tube if one is present. The top of the fruit usually with scars where the perianth has fallen off or with a persisting dry perianth at the top, or the fruit a globose structure containing many flowers (like a fig). 52
- Flowers with a superior ovary, the ovary borne above the level of sepal and petal attachment, or the ovary free from the perianth tube if one is present. The fruit usually with scars where the perianth has fallen off or with the persisting perianth at the base near the fruit stalk. 61

52. Leaves alternate along the stem (sometimes opposite the inflorescence); if the leaves are in clusters look to see if all the branches are alternate along the stem. 53
- Leaves opposite or whorled along the stem; if the leaves are in clusters look for the presence of at least a few opposite branches. 58

53. Plants without woody stems; flowers with 1 or 2 very short styles above the ovary, petals or petal lobes 5, stamens 5, ovary with 2 locules. 54
- Plants with woody stems; flowers not as above. 55

54. Petals united to form a tube, ovary half-inferior and with many ovules in each locule; fruit opening by a lid at the top. *(Sphenoclea)* *SPHENOCLEACEAE*
- Petals free, ovary inferior and with only 1 ovule in each locule; fruit splitting into 2 parts. *(UMBELLIFERAE) APIACEAE*

55. Leaves large (>15 cm long) and palmately compound, inflorescences usually more than 15 cm long. 131 **ARALIACEAE**
- Leaves and inflorescences smaller (<15 cm long). 56

56. Inflorescences as short axillary racemes when young with overlapping bracts making them look like small cones; disc of 4-5 fleshy lobes between stamens; fruit a small drupe. *(Opilia)* 112 **OPILIACEAE**
- Inflorescences when young not looking like small cones; disc between the stamens and ovary or absent; fruit dry. 57

57. Each flower with a single style, ovary very slender and 1-locular, looking like the flower stalk; stamens usually 10; leaves without star-shaped hairs. *COMBRETACEAE*
- Each flower with 2 styles, ovary with 2 locules, easily distinguished from the flower stalk; stamens 4 to 8; leaves with a dense covering of star-shaped hairs. *(Trichocladus)* 96 **HAMAMELIDACEAE**

58. Each ovary with only 1 locule. 59
- Each ovary with 2 or more locules. 60

59. Stamens 10 in each flower; fruit with wings; trees or shrubs found below 2000 m. *COMBRETACEAE*
- Stamens 2 to 4 in each flower; fruit often with hairs on the top; herbs or subshrubs found above 2000 m. *DIPSACACEAE*

60. Ovary 3-locular, 2 empty and 1 with 1 ovule; fruit 1-seeded, dry; leaves without stipules across the stems, or basal. *VALERIANACEAE*
- Ovary with 2 or more locules, all with ovules; fruit with 2 or more seeds, usually fleshy; leaves with stipules joined across the stems. *RUBIACEAE*

61. Inflorescence with many stiff dry bracts; perianth of only 1 whorl, usually dry and papery or stiff. *AMARANTHACEAE*
- Inflorescence without stiff dry bracts; if dry bracts present, perianth of 2 whorls, both sepals and petals or corolla tube present. 62

62. Leaves usually alternate and pinnately compound; fruit a legume with 1 locule and seeds arranged along one side; stamens 5 to many. *(LEGUMINOSAE)* 95 **FABACEAE**
- Leaves always simple; fruit not a legume; stamens 4 or fewer. 63

63. Leaves alternate. 64
- Leaves opposite, at least on the main stem. 66

64. Flowers with 2 stamens; small trees or shrubs of upland forest. *(*Dekindtia) OLEACEAE*
- Flowers with 4 stamens; herbs or shrubs of dry lowland areas. 65

65. Fruit a capsule opening by a circular lid; sepals and petals 4, flowers ± regular. *PLANTAGINACEAE*
- Fruit a nutlet enclosed in the dry calyx; calyx and corolla 5-lobed, flowers irregular. **GLOBULARIACEAE*

66. Trees of mangove formations on the sea coast, with vertical woody structure coming up from the roots; leaves with salt crystals formed on the surface, thick and somewhat leathery. *(Avicennia) AVICENNIACEAE*
- Plants not found in mangrove formations, without vertical woody structures from the roots; leaves without salt crystals on the surface, not leathery. 67

67. Leaves spicy aromatic when crushed; fruit with 1 seed in each part. 68
- Leaves not spicy aromatic when crushed; fruit a capsule with 2 or more seeds in a locule. 69

68. Ovary with 4 lobes at the top, style arising from a depression between the lobes; fruit dry and splitting into 4 nutlets. *(LABIATAE) LAMIACEAE*
- Ovary round at the top or with 2 to 8 lobes, style arising from the top of the ovary; fruit usually fleshy. *VERBENACEAE*

69. Ovary with 2 to 6 ovules in each locule; fruit usually broad above and narrowed at the base with the seeds borne on stiff stalks. *ACANTHACEAE*
- Ovary with usually more than 6 ovules in each locule; fruit usually broadest at the middle and the seeds small without conspicuous stiff stalks. *SCROPHULARIACEAE*

GROUP 6

Flowers or reproductive structures (stamens and pistils) naked; not subtended by 3 or more bracts, without a perianth of 3 or more parts or a perianth tube. Sometimes inflorescences or flowers subtended by 1 or 2 bracts.

1. Inflorescence a thick fleshy spike or spikes, the inflorescence axis a thick cylinder with small flowers embedded in it. 2
- Inflorescence not a thick fleshy spike, the inflorescence axis either thin and the flowers not embedded in it, or a thick flat or hollow receptacle (fig) with flowers embedded in it. 5

2. Fruits berries; inflorescence usually with upper portion with only male flowers and lower portion with only female flowers; leaf-bases clasping the stem. *ARACEAE*
- Fruits drupes or achenes; flowers bisexual or male and female flowers mixed on the spike; leaves scale-like or leaf bases not clasping the stem. 3

3. Plants of hot, dry areas, often with salty soils; leaves scale-like or succulent and almost round in cross-section. *CHENOPODIACEAE*
- Plants of moist to wet habitats, never found in salty soils; leaves flat, not succulent, scale-like or round in cross-section. 4

4. Leaves small, entire, somewhat fleshy and stiff, on the same stems as the inflorescence; plants growing on shrubs or trees or rocks in shade. *PIPERACEAE*
- Leaves large, with toothed margins, not fleshy and stiff, growing from an underground rhizome and not on the same stem as the inflorescence; plants of wet habitats. *(Gunnera) HALORRHAGACEAE*

5. Plants with white milk-like or sticky sap (at least during the growing season). 6
- Plants with clear sap or without abundant sap during the growing season. 7

6. Fruit fleshy, made up of the whole inflorescence with either fleshy perianth parts or a swollen flat, cup-shaped or round receptacle; ovary with 1 locule and 1 style; anthers usually opening by longitudinal slits. 104 **MORACEAE**
- Fruit a capsule, berry or drupe; ovary with usually 3 (2-4) locules and 1 or 2 ovules in each locule; anthers often opening by transverse slits. *EUPHORBIACEAE*

7. Leaves needle-shaped, solitary or in bunches of 2 or more together with a membranous sheath around the base, round or triangular in cross-section; trees planted for wood or ornament, not native. *(Pinus) PINACEAE*
- Leaves scale-like, or normal and flat, not in bunches of 2 or more together and without a membranous sheath at the base; trees or shrubs, native or introduced. 8

8. Plants with cone-like structures borne on the ends of the stems, made up of small stalks with caps, each of which carries 4 or more small sacs that contain pollen-like spores. The spores with 4 arm-like projections. *(Equisetum) EQUISETACEAE*
- Plants with cones but not as above, or with flowers; anthers producing pollen. 9

9. Leaves reduced membranous scales either in whorls or opposite pairs on thin flexible green stems, or leaves absent. 10
- Leaves fleshy scales or larger and flatter; stems not thin and flexible, or if flexible, not thin and green. 11

10. Branches with longitudinal grooves; male flowers borne in whorls in spikes at the end of stems; female flowers borne in woody cone-like groups. *(Casuarina)* 101 **CASUARINACEAE**
- Branches without longitudinal grooves; male and female 'flowers' in fleshy cones along the stem. *(Ephedra) EPHEDRACEAE*

11. Leaves scale-like with a broad clasping base and overlapping each other, at least on the young shoots. 12
- Leaves flat and well developed, not scale-like with a broad base, internodes clearly visible between leaves. 14

12. Leaves opposite; female cone with opposite scales. *CUPRESSACEAE*
- Leaves spirally arranged; female cones with spirally arranged scales. 13

13. Leaves >4 mm wide, usually with a hooked tip; branches usually borne in whorls at regular intervals along the main stem. *(Araucaria) ARAUCARIACEAE*
- Leaves <4 mm wide, tip not hooked; branches borne alternately, not in whorls, along the main stem. *TAXODIACEAE*

14. Leaves with a single vein, flat and somewhat stiff, evergreen, persisting for up to 4 growing seasons. *(Podocarpus) PODOCARPACEAE*
- Leaves with pinnate or palmate venation; deciduous, shed after each growing season. 15

15. Leaves opposite or in whorls along the stem; if the leaves are in close bunches look for at least a few opposite branches. 16
- Leaves alternate or spiral along the stem; if the leaves are in close bunches look at the arrangement of the branches, there should not be any opposite branching. 18

16. Shrubs with resinous sap; the leaves triangular in shape, folded like a fan in bud or when dry; stamens 4 to 8; styles 3 to 6. *(Myrothamnus)* 97 ***MYROTHAMNACEAE**
- Herbs without resinous sap; leaves not triangular and not folded fanwise in bud; flowers with 1 or 2 styles. 17

17. Ovary with 4 lobes at the top and 2 styles; male flower with a single stamen; flowers solitary in the leaf axils and usually subtended by 1 or 2 bracts. *CALLITRICHACEAE*
- Ovary not lobed at the top, with only 1 style; male flowers usually with 4 or 5 stamens; flowers borne in cymes or clusters in the leaf axils with each flower usually subtended by 2 slender bracts. 105 **URTICACEAE**

18. Ovary with 2 to 8 locules; male flowers (if present) not arranged in spikes or catkins. 19
- Ovary with only 1 locule; flowers usually unisexual, male flowers often in spikes or catkins. 22

19. Fruit a 3-locular capsule with perianth scars at the base; ovary superior; stamens with united filaments or anthers sessile and filaments absent. *EUPHORBIACEAE*
- Fruit with perianth scars as a rim or on top; ovary inferior; stamens free. 20

20. Plants usually climbing with tendrils opposite the leaf attachment on the stem. 119 **VITACEAE**
- Plants without tendrils, not climbing. 21

21. Fruit dry and usually breaking into 2 parts; herbs or soft-stemmed shrubs often spicy aromatic when crushed. *(UMBELLIFERAE) APIACEAE*
- Fruit usually fleshy, not breaking into 2 parts; trees or climbers, not spicy aromatic when crushed. 131 **ARALICACEAE**

22. Ovary with a solitary style and stigma (the stigma sometimes finely divided); herbs or vines or shrubs with soft wood. 23
- Ovary with 2 styles or with 1 style and 2 to 4 stigmas; trees and shrubs with hard wood. 24

23. Vines with the leaves usually round and the petiole attached near the centre of the blade; leaf surface smooth; plants without stinging hairs. *(Cissampelos) MENISPERMACEAE*
- Herbs, shrubs, small trees or vines with leaves of various shapes and the leaf stalk attached to the base of the leaf blade; leaf surface often with pale crystalline patches; plants often with stinging hairs. 105 **URTICACEAE**

24. Ovary with 2 styles; fruit fleshy and covered with small protruberances; leaves with bright dots when viewed against the light. 100 **MYRICACEAE**
- Ovary with 1 style; fruit a capsule and seeds with hairs; leaves without bright dots when veiwed against the light. 99 **SALICACEAE**

GROUP 7

Flowers which have a spur or sac-like pouch, usually containing nectar, and which are colourful. A spur is a tubular structure that projects out from the base of the flower, usually near the flower stalk.

1. Trees or sometimes shrubs or climbers, wood hard. 2
- Herbs or soft-stemmed or succulent shrubs. 4

2. Flower with a white or pink corolla tube over 4 cm long, with 4 anthers on long filaments. *(Sesamothamnus) PEDALIACEAE*
- Flower not as above. 3

3. Leaves alternate; stamens with anthers 5 to many. 93. ***CHRYSOBALANACEAE**
- Leaves opposite; stamen with 1 anther in each flower. **VOCHYSIACEAE*

4. Flowers with the anthers united with each other or to other structures, 1 or 2 or absent. 5
- Flower with the anthers free from each other, filaments free or partly united, but not united to other structures. 6

5. Filaments united to the style to form a thick column in the centre of the flower; ovary inferior, usually twisted, looking like the flower stalk. *ORCHIDACEAE*
- Filaments free from the ovary and style, but partly united with each other; ovary superior, not looking like part of the flower stalk. *BALSAMINACEAE*

6. Leaf blade circular in outline with the leaf stalk attached to the centre; stamens 8; ovary 3-locular. *TROPAEOLACEAE*
- Leaf blade not circular, the leaf stalk attached to the base of the blade; stamens either less than or more than 8; ovary not 3-locular. 7

7. Styles 2 or more arising from separate areas on the top of the ovary or the flower with 2 or more separate pistils. *RANUNCULACEAE*
- Style 1 arising from the top of the ovary; pistil of 2 or more united carpels. 8

8. Stamens 1 or 2 in each flower. 9
- Stamens 3 or more in each flower. 10

9. Plants often with small bladder-like pouches on the leaves; 2 stamens free from the corolla tube; ovary superior. *LENTIBULARIACEAE*
- Plants never with bladder-like pouches on the leaves; 1 or 2 stamens borne on the corolla tube; ovary inferior. *VALERIANACEAE*

10. Stamens arising from near the base of the ovary, not borne on the corolla tube. 11
- Stamens borne on the corolla tube or on a cup-like receptacle, not arising near the base of the ovary. 12

11. Flowers with 2 spurs or sacs at the base of the perianth; stamens united in 2 groups, each with 3 anthers; sepals 2, petals 4, leaves divided into small segments. *FUMARIACEAE*
- Flowers with 1 spur or sac at the base of the perianth; filaments free or partly united but not in 2 groups, anthers 5; sepals 5, petals 5; leaves not deeply divided. *VIOLACEAE*

12. Stamens borne on the corolla tube; flowers irregular. *SCROPHULARIACEAE*
- Stamens borne on the edge of a cup-like or tubular receptacle; petals not forming a corolla tube, free and borne on the receptacle cup or tube; flowers regular or slightly irregular. *LYTHRACEAE*

GROUP 8

Flowers apparently only female; stamens absent, or lacking anthers and modified into staminodes, or stamens united with each other or with other parts of the flower and difficult to identify. Be sure to look below the flower as an inferior ovary may look like part of the flower stalk.

1. Inflorescence a thick spike with female flowers embedded in it and subtended and surrounded by a single large bract; fruits closely packed berries on a thick axis with the remains of the bract or a scar below the fruits. *ARACEAE*
- Inflorescence and fruits not as above. 2

2. Ovary inferior or half-inferior, at least part of the ovary borne below the level of sepal and petal attachment, or the ovary united to the perianth tube, if a perianth tube is present. Fruit usually with the dry perianth or scars from the perianth at the top. 3
- Ovary superior, the ovary borne above the level of sepal and petal attachment, or the ovary free from the perianth tube if one is present. 20

3. Ovary slender, green and usually twisted, looking like part of the flower stalk, ovules minute or undeveloped at flowering time; centre of the flower with a thick column with stigma and pollen in sticky clusters in special pockets

in the column; perianth parts 6 with one much larger than the others. *ORCHIDACEAE*
- Flowers not as above. 4

4. Climbers, stems twining or with tendrils; ovary with 2 or more locules; fruit with 2 or more seeds. 5
- Herbs, shrubs or trees; if climbers, ovary with 1 locule and fruit with 1 seed. 8

5. Tendrils present; flower with a disc above and surrounding the ovary. 6
- Tendrils absent, stems twining; flower without a disc. 7

6. Tendrils arising from the stem opposite the leaves; inflorescence also opposite the leaves; petals free from each other or coming off as a cap. 119 **VITACEAE**
- Tendrils arising from the leaf axils, on the same side of the stem as the leaves; petals united into a long or short corolla tube. *CUCURBITACEAE*

7. Perianth tube long and usually bent above a bottom, swollen part; ovary with 4-6 locules. *ARISTOLOCHIACEAE*
- Perianth parts separate from each other, not forming a tube; ovary with 3 locules. *DIOSCOREACEAE*

8. Anthers present but united to form a slender tube around the style. 9
- Anthers absent or solitary and united with another part of the flower, not forming a tube around the style. 10

9. Filaments free from each other below the anthers and borne on the corolla tube; ovary with 1 locule and 1 ovule; flowers in dense heads. *(COMPOSITAE) ASTERACEAE*
- Filaments united with each other below the anthers and borne at the base of the corolla tube; ovary with 2 locules and many ovules; flowers not in dense heads. *(Lobelia) LOBELIACEAE*

10. Ovary with only 1 ovule or the ovule not visible; locule 1; fruit with only 1 seed. 11
- Ovary with 2 or more ovules, with 1 or more locules; fruit with more than 1 seed. 15

11. Plants with white or coloured sap at least during the growing season; fruit fleshy, an aggregate of many flowers often sunken in a fleshy receptacle. 104 **MORACEAE**
- Plants with clear sap or no sap during the growing season; fruit not an aggregate of many flowers, the ovary not sunken into the tissue of the inflorescence. 12

12. Herbs with the leaves all basal, the leaf blade usually deeply indented at the attachment to the leaf stalk. *(Gunnera) HALORRHAGACEAE*
- Trees, shrubs, or herbs, the leaves not all basal. 13

13. Plants parasitic, growing on trees or shrubs; stems green, breaking easily at the nodes; fruit a 1-seeded berry producing a very sticky substance. 115 **VISCACEAE**
- Plants not parasitic herbs, stems not breaking easily at the nodes; fruit an achene or drupe, not producing a very sticky substance. 14

14. Flowers with 3 or 4 styles or with 1 style and 3 or 4 stigmas; perianth 3- or 4-lobed. *(Osyris)* 116 **SANTALACEAE**
- Flowers with only 1 style and stigma for each ovary; perianth of 3-5 parts, free or united only at the base. 105 **URTICACEAE**

15. Flowers large and asymmetical (without any line along which the flower can be divided into 2 equal halves); anther long and slender and united to a petal-like structure; large herbs without woody stems. 16
- Flowers more or less symmetrical (regular or irregular), anthers absent. 17

16. Ovules many in each locule. *CANNACEAE*
- Ovules 1 in each locule. *MARANTACEAE*

17. Herbs or small shrubs, with succulent stems and with usually swollen nodes; ovary with usually 3 locules and 6 stigmas that are twisted and have a velvet-like surface. *BEGONIACEAE*
- Shrubs or trees, with woody stems; ovary with 1 or 2 locules. 18

18. Leaves imparipinnate; flowers small, in many-flowered hanging panicles. *(Hagenia)* 91 **ROSACEAE**
- Leaves simple; flowers not in many-flowered haning panicles. 19

19. Leaves with star-shaped hairs; flowers without petals; stigmas 3 to 5. 113 ***OCTOKNEMACEAE**
- Leaves without star-shaped hairs; flowers with petals; stigmas 2. **MONTINIACEAE*

20. Styles or stigmas 2 or more and arising from separate areas on the top of the ovary or arising from separate ovaries; each flower with more than 1 pistil. 21
- Style 1 or if 2 or more styles are present then the styles all arising from 1 point on the top of the ovary, the ovaries sometimes 2 and separate but with the styles united above to form a single stigma; each flower with only 1 pistil. 32

21. Herbs without woody stems, not climbers. 22
- Trees, shrubs or climbers, or herbs with some woody stems. 23

22. Leaves compound (2 to 4 times pinnate), the blades of the leaflets lobed towards the end and usually small (>3 cm), whole leaf looking somewhat like a fern-frond. *(Thalictrum) RANUNCULACEAE*
- Leaves simple and entire, the blade of the leaf over 5 cm long, often arrow-shaped. *ALISMATACEAE*

23. Leaves pinnately compound (sometimes simple near the inflorescence. 24
- Leaves simple or palmately compound. 26

24. Climbing plants with opposite leaves, the leaflets usually 5 in number; flowers with usually

4 petal-like sepals and with many stamens and pistils. *(Clematis) RANUNCULACEAE*
- Plants not climbing, with alternate or opposite leaves; flowers with 1 to 5 pistils. 25

25. Leaflets with small bright dots when viewed against the light and usually lemony or spicy aromatic when crushed. 120 **RUTACEAE**
- Leaflets lacking small bright dots when viewed against the light, often with an unpleasant smell when crushed. 122 **SIMAROUBACEAE**

26. Climbers or scrambling herbs and shrubs; perianth parts in whorls of 3; pistils (and ovaries) quite separate. *MENISPERMACEAE*
- Trees or shrubs, if climbing then the pistil (and ovary) solitary. 27

27. Ovary solitary with 1 to several locules and with 2 to several styles. 28
- Ovaries 2 or more, each with 1 locule and 1 style; leaves often aromatic when crushed. 31

28. Fruit 1-seeded with a thin fleshy covering and a hard endocarp; leaves ovate to lanceolate with 3-5 veins from the base. 102 **ULMACEAE**
- Fruit and capsule or berry with 2 or more seeds; leaves with pinnate venation or compound. 29

29. Ovary usually 1-locular with several intruding parietal placentas; styles 4 or more; leaves alternate. *FLACOURTIACEAE*
- Ovary with 2 or 3 locules and 1 or 2 ovules in each locule on axile placentation (rarely 4-locular); styles as many as locules. 30

30. Leaves opposite, simple and entire, evergreen and usually stiff; stems and leaves without white sap. *(Buxus)* 98 **BUXACEAE**
- Leaves alternate or rarely opposite and then usually with white sap. *EUPHORBIACEAE*

31. Leaves opposite (or sometimes appearing alternate through the loss of one of the leaves of a pair, look for opposite branches); pistils borne in a cup-shaped receptacle. **MONIMIACEAE*
- Leaves alternate; pistils borne in a cup-shaped receptacle. *ANNONACEAE*

32. Filaments with anthers which are united with each other, filaments slender free borne on the corolla tube, the corolla tube with 4 or 5 small lobes and with usually 2 prominent lips; each ovary with many ovules. 33
- Filaments without anthers or stamens free or anthers sessile on a tube formed from the united filaments. 35

33. Ovary with 1 locule. *GESNERIACEAE*
- Ovary with 2 or more locules. 34

34. Flowers irregular; fruit a capsule, seeds small and round. *SCROPHULARIACEAE*
- Flowers regular; fruit a berry, seeds flat. *SOLANACEAE*

35. Ovary with only 1 locule. 36
- Ovary with 2 or more locules. 61

36. Leaves alternate along the stem; if the leaves are clustered in bunches, look for the absence of opposite branches (there should be none). 37
- Leaves opposite or in whorls along the stem; if the leaves are clustered in bunches look for the presence of opposite branches (a few should be present). 56

37. Plants climbing or at least the branches climbing if a thick stem is present. 38
- Plants herbs, shrubs, or trees, not climbing or the branches not climbing. 39

38. Tendrils present on the stem and coming from the leaf-axil; ovary on a stalk with several to many ovules on 3 parietal placentas. *(Adenia) PASSIFLORACEAE*
- Tendrils absent; ovary sessile and without a stalk, ovules 1 or 2 and pendulous from the apex of the locule. 109 **ICACINACEAE**

39. Leaves pinnately compound (the leaflets without axillary buds, flowers, or fruits at their base). 40
- Leaves simple, trifoliolate, or palmately compound. 41

40. Ovary with 2 to 4 styles with usually thick stigmas and with only 1 ovule; wood and fruit usually with aromatic resin. 127 **ANACARDIACEAE**
- Ovary with 1 style and 1 stigma and with usually 2 or more ovules borne along one side of the ovary/fruit wall; wood and fruit without resin. *(LEGUMINOSAE)* 95 **FABACEAE**

41. Leaves with small bright dots when viewed against the light (x 10), usually spicy aromatic when crushed. 42
- Leaves without bright dots when viewed against the light, not spicy aromatic when crushed. 44

42. Anthers present and united to a tube formed by the filaments which surround the pistil; leaves lustrous above; trees of wet lowland forests. *CANELLACEAE*
- Anthers absent, the pistil not surrounded by a staminal tube. 43

43. Sepals borne on a cup-like receptacle, petals absent; leaves simple. **MYRISTICACEAE*
- Sepals not borne on a cup-like receptacle, petals absent; leaves simple. 120 **RUTACEAE**

44. Ovary with only 1 ovule; fruit with only 1 seed. 45
- Ovary with 2 or more ovules; fruit often with 2 or more seeds. 54

45. Trees or shrubs with white sap; leaves often with palmate venation (sometimes palmately compound); ovary with 1 or 2 styles. 104 **MORACEAE**
- Trees. shrubs, or herbs, without white sap. 46

46. Ovary with 1 style and 1 stigma. 47
- Ovary with 2 to 6 styles or with 2 to 6 stigmas if the styles are absent. 50

47. Leaves less than 2 cm long with 3 narrow leaflets at the apex and 2 broad stipules at the base; stigma with a feathery margin. *(*Cliffortia)* 91 **ROSACEAE**
- Leaves without 3 narrow leaflets at the apex. 48

48. Perianth parts of 1 type and size and in 1 whorl, usually 4 or 5 in number or united to form a tubular structure (and then often difficult to see); style slender. 105 **URTICACEAE**
- Perianth parts of 2 types and sizes and in whorls of 3, 4 or 5 parts each; style short and thick. 49

49. Petals densely hairy on their inner surface; plants usually with large axillary spines; fruit a drupe with a single large seed. *(Ximenia)* 111 **OLACACEAE**
- Petals not densely hairy; plants without spines; fruit; fruit a berry or drupe with a small seed. *(Myrsine)* MYRSINACEAE

50. Trees or shrubs with woody stems. 51
- Herbs or if shrubs then with weak fibrous or succulent stems. 52

51. Styles 2, ovule pendulous from near the apex of the locule; leaves usually lanceolate but with palmate venation. 102 **ULMACEAE**
- Styles usually 3, ovule usually borne at the base of the locule; leaves with one major vein (midrib). 127 **ANACARDIACEAE**

52. Leaves very deeply lobed or apparently palmately compound; ovary with 2 slender styles; leaves with a sharp odour when crushed. *(Cannabis)* 106 **CANNABACEAE**
- Leaves not deeply lobed or palmately compound. 53

53. Leaves with a thin membranous stipule that encloses the stem at the base of the leaf. POLYGONACEAE
- Leaves without a thin membranous stipule and leaf base not enclosing the stem; leaves sometimes absent. CHENOPODIACEAE

54. Ovules 2 or 3 in the locule of each ovary; sap clear or sometimes white. EUPHORBIACEAE
- Ovules usually more than 3 in the locule of each ovary. 55

55. Plants with white sap, ofte5 planted for their fruit; leaves large, over 30 cm long and palmately divided; petals present. *(Carica)* CARICACEAE
- Plants without white sap, rarely planted for their fruits; leaves less than 30 cm long and not palmately divided; petals often absent. FLACOURTIACEAE

56. Flowers with sepals and petals; trees and shrubs with resinous sap in the bark. 57
- Flowers without petals; trees, shrubs, herbs or climbers usually without resinous sap. 58

57. Staminodes usually present, ovules 1 to many, style 1 or none. *(GUTTIFERAE)* CLUSIACEAE
- Staminodes usually absent, ovule 1 in each locule, styles usually 3. 127 **ANACARDIACEAE**

58. Herbs or climbers or shrubs or small trees with soft wood. 59
- Trees or shrubs with hard wood. 60

59. Ovary with 2 slender styles; climbing cultivated plants. *(*Humulus)* 106 **CANNABACEAE**
- Ovary with 1 style; plants not cultivated. 105 **URTICACEAE**

60. Flower with usually 3 separate sepals; ovary with a long and slender style and stiga, ovule pendulous from the top of the locule. 103 **BARBEYACEAE**
- Flower with the sepals borne on a cup-like receptacle and apparently united at the base; ovary with a broad stigma, ovule borne at the base of the locule. MONIMIACEAE

61. Climbing plants with tendrils or thorns. 62
- Herbs, shrubs, or trees, or if climbing then without tendrils or thorns. 64

62. Tendrils solitary and arising from the stem opposite the leaves; ovary 2-locular. 119 **VITACEAE**
- Tendrils paired (borne in groups of 2 each); ovary usually 3-locular. 63

63. Tendrils borne on the stalk of the inflorescence; leaves usually compound. 125 **SAPINDACEAE**
- Tendrils borne on the stalk of the leaf; thorns often present, leaves simple. *(Smilax)* SMILACACEAE

64. Leaves with bright dots (at least along the edg: of the blade) when viewed against the light (x 10), aromatic when crushed, usually trifoliolate or pinnately compound; flowers usually with petals. 120 **RUTACEAE**
- Leaves without bright dots when viewed against the light. 65

65. Leaves trifoliate or pinnately compound, rarely with only 1 blade per leaf and then the leaves usually clustered on short side branches and aromatic when crushed. 66
- Leaves simple or palmately compund. 69

66. Trees or shrubs with resin in the bark; leaves often spicy aromatic when crushed; flowers with staminodes; fruit usually fleshy. 67
- Trees, shrubs or herbs without resin in the bark; leaves not spicy aromatic but sometimes with a strong or unpleasant smell when crushed; flowers usually without staminodes; fruit dry or lobed if fleshy. 68

67. Ovary with 1 style and 2 ovules in each locule; leaves usually strongly aromatic when crushed, usually trifoliolate or odd-pinnate. 123 **BURSERACEAE**
- Ovary with 1 to 3 styles and 1 ovule in each locule (usually only 1 locule present); leaves usually weakly aromatic when crushed, usually regularly trifoliolate or simple. 127 **ANACARDIACEAE**

68. Ovary with usually 1 style and with 1 ovule in each locule; leaves usually weakly or not at all aromatic when crushed, mostly even-pinnate (sometimes simple or bipinnate). 125 **SAPINDACEAE**

- Ovary with 4 or 5 styles or the styles united and aparently 1, with 1 ovule in each locule; leaves often with an unpleasant odour when crushed, odd-pinnate. 122 **SIMAROUBACEAE**

69. Flowers with the perianth parts united to form a short or long tube or a cup-like structure. 70
- Flowers with the perianth parts free from each other or united only at the base. 75

70. Each flower with 2 ovaries that are separate at the base and united above by the thick stigmas (the ovaries may by difficult to see because of the structures that surround them in the flower, look for a young fruit), leafless succulent plants or herbs, climbers, or shrubs with simple opposite leaves; sap often white. *ASCLEPIADACEAE*
- Each flower with a single ovary, not usually surrounded by complicated structures. 71

71. Petals forming a short thick tube above the sepals; ovary with 2 to many locules and 2 to 8 styles; leaves simple and entire, usually stiff. *EBENACEAE*
- Petals forming a slender tube or a thin walled tube or the petals absent and only sepals or calyx tube present. 72

72. Leaves opposite or in whorls along the stem, or the leaves very small or crowded and the branches opposite. 73
- Leaves alternate along the stem, or the leaves crowded and the branches alternate. 74

73. Stems green, slender and drooping, the leaves very small (<1 cm) or absent, not folded fan-like in the bud; each 'flower' with 2 styles and 2 ovules not enclosed in a carpel but surrounded by swollen bracts. *(Ephedra)* *EPHEDRACEAE*
- Stems not green and slender; leaves over 1 cm long and folded fan-wise in bud or when dry; ovary with 3 to 6 styles and 3 locules. 97 ***MYROTHAMNACEAE**

74. Small shrubs, leaves small (<3 cm) and not lobed or parted; star-shaped hairs absent; sepals forming broad wings in fruit. *CONVOLVULACEAE*
- Shrubs or trees, leaves usually over 5 cm long and lobed or palmately divided, star-shaped hairs present; ovaries becoming separate in fruit. *STERCULIACEAE*

75. Style 1 or absent and the single stigma sessile on top of ovary. 76
- Styles 2 or more on the top of the ovary. 78

76. Leaves opposite; ovary with 1 to many ovules in each locule. *(GUTTIFERAE) CLUSIACEAE*
- Leaves alternate along the stem; ovary with 1 or 2 ovules in each locule. 77

77. Fruit a drupe; leaves often spiny along the margin or with a sharp point, stiff. *(Ilex)* 107 **AQUIFOLIACEAE**
- Fruit usually dry; leaves without a spiny margin and without a sharp point. 125 **SAPINDACEAE**

78. Ovary with 3 (sometimes 2 or 4) locules separated by well developed septa and with 1 or 2 ovules in each locule, the ovules attached to the central axis of the ovary. *EUPHORBIACEAE*
- Ovary with 2 or more locules formed by the intrusion of the placentas, the septa often not clear, with usually 3 to many ovules in each locule, the ovules attached to extensions of the outer wall of the ovary. *FLACOURTIACEAE*

GROUP 9

Whole plant or some flowers on the plant functionally only male. Male flowers never develop fruits; after flowering they wither and drop off. Flowers or reproductive structures with free and distinct anthers or pollen sacs; with or without a central pistil or group of pistils or a pistil-like structure. (Note that flowers with a slender inferior ovary may appear to be without a pistil; be sure to examine the base of the flower and the flower stalk for the presence of an ovary with locule(s) and seeds. If you find an inferior ovary in the flower - go to Group 12).

1. Plants climbing or the branches climbing and coming from a thick trunk, the plants rooted in the ground. 2
- Plants not climbing, but may be epiphytic without roots reaching the ground; trees, shrubs, or herbs. 10

2. Tendrils present on the stem, leaves, or inflorescence. 3
- No tendrils found on the plant. 7

3. Tendrils borne on the stem and arising opposite the leaves, the tendril usually absent at every third node; inflorescence also opposite the leaves. 119 **VITACEAE**
- Tendrils borne on the stem next to the leaf (and not opposite) or on the leaf-stalk or inflorescence stalk. 4

4. Tendrils borne on the leaf stalk or on the stalk of the inflorescence, the tendrils paired (in groups of 2). 5
- Tendrils borne on the stem next to the leaves, the tendrils solitary but sometimes with 2 or more branches. 6

5. Tendrils borne on the stalk of the inflorescence; thorns absent along the stem. 125 **SAPINDACEAE**
- Tendrils borne on the stalk of the leaf; thorns and spines often present along the stem. *SMILACACEAE*

6. Leaves with a gland at the base of the blade (where the petiole is attached to the blade); anthers usually 5 and free from each other. *PASSIFLORACEAE*
- Leaves without a gland at the base of the blade; anthers 3 or 5 and usually united to each other or apparently 1. *CUCURBITACEAE*

7. Stamens 4 or 5 in each flower, flower with only 1 perianth whorl. 8

- Stamens 2, 3 or more or apparently only 1 in each flower, flower with 1 or 2 perianth whorls. 9
8. Leaves opposite along the stem; stamens 5 and opposite the 5 sepals. *(Cannabis)* 106 **CANNABINACEAE**
- Leaves alternate along the stem; stamens 4 or 5 and alternating with the perianth parts. *(Pyrenacantha)* 109 **ICACINACEAE**
9. Anthers solitary on each filament and with 2 pollen sacs opening by vertical slits; leaves opposite and/or compound, usually without hairs. *DIOSCOREACEAE*
- Anthers solitary on each filament but with 4 pollen sacs, or sometimes 2 or more anthers borne on a single stalk, the anthers opening by vertical or transverse slits or pores; leaves never opposite or compound, always alternate (spiral) and simple. *MENISERPMACEAE*
10. Leaves with bright dots when viewed against the light (at least dots along the margin of the blade), usually spicy-aromatic when crushed. 11
- Leaves without bright dots when viewed against the light. 13
11. Sepals not forming a cup and not borne upon a cup-like receptacle, petals usually present; leaves pinnately compound, trifoliolate, or sometimes simple, usually very strongly aromatic (lemon scented). 120 **RUTACEAE**
- Sepals united at the base to form a cup or borne upon a cup-like receptacle, petals absent; leaves simple and entire. 12
12. Stamens united by their filaments and the anthers borne upon a single column; leaves alternate; cultivated trees. **MYRISTICACEAE*
- Stamens free from each other; leaves opposite or sub-opposite. **MONIMIACEAE*
13. Inflorescence (which may appear to be a single flower) subtended and at first enclosed by a single sepal-like or petal-like bract; anthers or male flowers borne on a central axis that is usually thick and fleshy; herbs often with large leaves. *ARACEAE*
- Inflorescence not resembling a single flower and not at first enclosed by a single large bract. 14
14. Flowers with 3 sepals and 3 petals (1 petal usually much larger than the other two), irregular, with a single central column on which the pollen sacs are borne; herbs or epiphytes. *ORCHIDACEAE*
- Flowers not as above and without an inferior ovary. 15
15. Leaves alternate along the stem, simple or compound; if leaves are clustered close together look for the absence of opposite branches. 16
- Leaves opposite or in whorls along the stem, simple or sometimes palmately compound; if the leaves are clustered close together look for the presence of opposite branches. 47
16. Centre of the flower with a slender filament-like structure that bears a small thickened portion at the apex which may be mistaken for a false stamen (staminode), which is actually a small ovary with a long slender stalk from the centre of the flower. *(CAPPARIDACEAE) CAPPARACEAE*
- Centre of the flower without a small ovary on a long slender stalk. 17
17. Stamens with the filaments completely or partly united; flowers often without petals; the leaves simple or palmately compound, often with star-shaped hairs. 18
- Stamens with the filaments free from each other or united only near the base; leaves simple or compound, generally without star-shaped hairs. 19
18. Filaments completely united, with the anthers sessile on the end of a column. *STERCULIACEAE*
- Filaments partly united, with the anthers borne on the ends of the short free filaments. *EUPHORBIACEAE*
19. Flowers with only 1 perianth whorl (petals absent) with the stamens the same number as the perianth parts and borne opposite them. 20
- Flowers with 2 perianth whorls, or if only 1 whorl is present then the stamens alternate with the perianth parts or greater in number than the perianth parts. 26
20. Leaves pinnately compound or palmately compound, or very deeply palmately lobed and the lobes resembling separate leaflets. 21
- Leaves simple, entire or lobed but not palmately lobed. 22
21. Leaves pinnately compound; trees. *(Ceratonia)* 95 **FABACEAE**
- Leaves palmately compound or deeply palmately divided; herbs or slender stemmed shrubs. *(Cannabis)* 106 **CANNABINACEAE**
22. Calyx 3- or 4-lobed, sepals united near the base; leaves without stipules, evergreen, stiff and entire; shrubs with woody stems. 116 **SANTALACEAE**
- Sepals 3 to 6, free or united near the base; leaves with stipules (at least near the new growth) or if the stipules are absent then the plants herbs of weak-stemmed shrubs. 23
23. Trees of shrubs with hard wood; stipules present, at least on the new leaves. 24
- Herbs, or if shrubs or trees, then with weak succulent stems or with soft wood. 25
24. Plants with white sap; all flowers unisexual. 104 **MORACEAE**
- Plants without white sap; bisexual flowers often present. 102 **ULMACEAE**
25. Stipules present on the young leaves; plants usually growing in shaded or moist places. 105 **URTICACEAE**
- Stipules absent; plants often growing in dry areas and on salty soils. *CHENOPODIACEAE*

26. Each flower with 13 or more stamens. 27
- Each flower with 12 or fewer stamens. 35

27. Leaves bipinnate (or sometimes simple, entire, stiff, and with sub-parallel veins in planted trees); flowers with the petals valvate in bud; an ovary usually present but very slender and difficult to see among the many stamens. (*MIMOSOIDEAE*) 95 **FABACEAE**
- Leaves never bipinnate; if planted trees then the leaves not simple, entire, stiff and with sub-parallel veins. 28

28. Herbs or shrubs with weak juicy stems, growing in wet places or planted for ornament. 29
- Trees or shrubs with woody stems. 30

29. Perianth parts in whorls of 3 each; leaves borne at the base of the plant and the plants usually growing in water. *ALISMATACEAE*
- Perianth parts in whorls of 2 each; leaves borne along the stem; plants grown for ornament or from moist evergreen forest. *BEGONIACEAE*

30. Leaves pinnately compound or trifoliolate. 31
- Leaves simple. 32

31. Stamens usually borne on the outside of a disc, wood and bark with resin. 127 **ANACARDIACEAE**
- Stamens borne toward the inside of a disc or the disc on only 1 side, without resin in wood and bark. 125 **SAPINDACEAE**

32. Sepals partly united or the petals partly united; a short calyx tube or corolla tube present. 33
- Sepals free or united only at the base, the petals (if present) free or united only at the base; a short calyx tube or corolla tube absent. 34

33. Leaves and young stems without star-shaped hairs; leaves usually entire. *EBENACEAE*
- Leaves and young stems with star-shaped hairs; leaves usually lobed or serrate. *TILIACE AE*

34. Perianth parts in groups of 3; stamens usually without glands or a disc at their base. *ANNONACEAE*
- Perianth parts 4 or 5 or more, usually in only one whorl (the petals often absent); stamens often with glands or a disc at their base. *FLACOURTIACEAE*

35. Petals united to form a tube. 36
- Petals free from each other or united only near the base and not forming a tube, or the petals absent. 37

36. Leaves large (>20 cm) and deeply lobed; plants with white sap, cultivated for their fruits. *CARICACEAE*
- Leaves rarely large, usually entire or with an undulating margin, never deeply lobed; plants without white sap and not grown for fruit. *EBENACEAE*

37. Leaves compound (trifoliate or pinnate) or if simple then clustered on short side branches and very aromatic when crushed. 38
- Leaves simple and not strongly aromatic when crushed. 42

38. Wood and bark containing resin. (Be sure to compare all the following 4 families as they are very difficult to separate using male flowers alone). 39
- Wood and bark without resin. 40

39. Flowers in unbranched axillary clusters or in racemes; stamens usually twice as many as the petals and of two different lengths; small trees and shrubs often with bent and twisted branches; the bark usually peeling off in thin papery flakes. 123 **BURSERACEAE**
- Flowers in branched inflorescences; stamens as many as the petals or more (sometimes the petals absent); the bark not peeling off in thin paper-like flakes. 127 **ANACARDIACEAE**

40. Leaves imparipinnate with strongly serrate margins to leaflets; flower small, wind pollinated, in large hanging panicles. (*Hagenia*) 91 **ROSACEAE**
- Leaves imparipinnate with entire or undulate margins, or trifoliolate with toothed margins to leaflets; flowers insect or bird pollinated, not hanging in large panicles. 41

41. Flowers mostly in unbranched inflorescences; stamens mostly twice as many as the petals (petals sometimes absent), a disc often present on one side of or outside the stamens; wood usually hard and the bark not peeling in thin sheets. 125 **SAPINDACEAE**
- Flowers mostly in branched inflorescences or in a spike-like inflorescence stamens mostly as many as the petals, filaments sometimes with scale-like appendages at the base, a disc often present on the inside of the stamens; wood usually soft, the bark not peeling in thin flakes. 122 **SIMAROUBACEAE**

42. Leaves clasping the stem at their base with a thin membranous stipule round the stem. *POLYGONACEAE*
- Leaves not clasping the stem at their base with a thin membranous stipule. 43

43. Stamens 9 to 12 in each flower, the filaments united into 3 or 4 groups near the base. *FLACOURTIACEAE*
- Stamens not united near the base to form 3 or 4 groups. 44

44. Stamens 3, dark red in colour and borne on very short filaments; leaves usually less than 3 cm long. (*Myrsine*) *MYRSINACEAE*
- Stamens 4 or more; leaves usually more than 4 cm long. 45

45. Petals with long hairs on the inner surface; disc inconspicuous; branches with large axillary spines. (*Ximenia*) 111 **OLACACEAE**
- Petals without long hairs; disc conspicuous; branches without axillary spines. 46

46. Stamens borne on the outside of the disc; bark and wood with aromatic resin. 127 **ANACARDIACEAE**

- Stamens borne on the inside of the disc or the disc on only 1 side or the disc absent (do not mistake a small pistillode for a disc); bark and wood without aromatic resin. **125 SAPINDACEAE**

47. Stamens united by their filaments to form separate groups or borne on a single stalk; sap often white or coloured. 48
- Stamens free from each other, the filaments not united. 49

48. Stamens forming separate groups; sap often coloured but not white. *GUTTIFERAE*
- Stamens borne on a single stalk or forming a single group; sap sometimes white. *EUPHORBIACEAE*

49. Leaves glabrous and shiny dark green above, below densely covered with white tightly curled hairs; perianth of 3-4 lobes; stamens 6-12 with very short filaments. (*Barbeya*) 103 **BARBEYACEAE**
- Leaves and male flowers not as above. 50

50. Flowers without petals, the stamens equal in number to the sepals and opposite to them. 51
- Flowers with petals or if the petals are absent then the stamens not equal in number and opposite the sepals. 54

51. Herbs or climbers, if shrubs then the wood soft; leaves usually thin, flexible, and with a lobed or serrated edge. 52
- Trees or shrubs with woody stems or growing upon and attached to trees and shrubs; leaves usually stiff and with an entire edge. 53

52. Flowers with the stamens bent inwards in bud; leaves often with conspicuous white patches on the surface. 105 **URTICACEAE**
- Flowers with the stamens straight in bud; leaf surface smooth. (*Cannabis*) 106 **CANNABINACEAE**

53. Shrubs parasitic on other trees and/or shrubs, the leaves differing from the leaves of the host plant, at least in texture; flower without a large central disc. 115 **VISCACEAE**
- Plants not parasitic and not attached to the branches of another tree or shrub, leaves on the plant all alike; male flower with a large central disc or with female flowers nearby. (*Buxus*) 98 **BUXACEAE**

54. Flowers with petals that are united to form a short tube; a pistillode (false pistil) usually present. (*Euclea*) *EBENACEAE*
- Flowers with free petals or the petals absent. 55

55. Petals absent, the sepals borne on a cup-shaped receptacle; stamens with very short filaments; stipules absent. **MONIMIACEAE*
- Petals present or absent, the sepals not borne on a cup-shaped receptacle. 56

56. Leaves simple and entire; stamens as many as the petals and alternate with them; flowers without a pistillode; shrubs or small trees. *MONTINIACEAE*
- Leaves various or sometimes absent; flowers of various types; abundant sap present, either white or clear; herbs or succulent shrubs or trees. *EUPHORBIACEAE*

GROUP 10

Each flower with 2 or more separate pistils or with the pistils partly united and with free styles, or with 2 or more styles arising separately from the top of the ovary when only one pistil is present. Two or more styles present and arising from separate ovaries or from separate areas on the single ovary. Flowers usually bisexual with both anthers and pistils present. (If only pistils present, look for anthers in flower buds; if none are found use key to Group 8).

1. Each flower with a single pistil that has only a single locule (one ovary made of 2 or more carpels but with a single locule). 2
- Each flower with 2 or more pistils or if only 1 pistil is present then the ovary with 2 or more locules. 4

2. Ovules 3 or more in each ovary (the ovary sometimes open at the top), styles or stigmas 2 to 5; leaves simple; herbs or small shrubs. RESEDACEAE
- Ovules 1 in each ovary; trees or shrubs. 3

3. Styles 2 and covered with very fine projections or short hairs; leaves simple and without star-shaped hairs. 102 **ULMACEAE**
- Styles 3 or 4, with smooth stigmas; leaves compound or simple and with star-shaped hairs. 127 **ANACARDIACEAE**

4. Herbs; flowers with perianth parts in groups of 3 or 6 similar parts; leaves arising only from the base of the plant. 5
- Trees, shrubs, or herbs; if herbs the perianth absent, or parts in multiples of 4 or 5, sepals and petals distinct; leaves arranged along the stem with or without basal leaves. 6

5. Ovules numerous in each ovary and scattered over the walls of the locule; pistils usually 6 and stamens usually 9; sap often white. **BUTOMACEAE*
- Ovules 1 or a few in each ovary and borne at the base of the locule; pistils 3 to many, stamens 3 to many; sap clear. *ALISMATACEAE*

6. Pistils borne on a thick axis which is at first enclosed in a single tubular sepal-like or petal-like bract (this is actually an inflorescence of unisexual flowers); herbs or subshrubs. *ARACEAE*
- Pistils not borne on a long thick axis which is at first enclosed in a tubular bract. 7

7. Pistils more than 7 in each flower, separate from each other or united only near the base; stamens usually more than 15 in each flower. 8
- Pistils fewer than 7 in each flower or united and not separate from each other. 14

8. Herbs with succulent leaves and stems; leaves simple with entire or ciliate margins. 9

- Herbs with normal stems and leaves (not succulent); leaves compound or simple with toothed margins; or trees, shrubs or climbers with various types of leaves. 10
9. Flowers without petals; leaves in opposite pairs. *(Gisekia)* MOLLUGINACEAE
- Flowers with petals; leaves alternate or in whorls of 3. 88 **CRASSULACEAE**
10. Leaves opposite; climber with twining stems and leaf stalks. *(Clematis)* RANUNCULACEAE
- Leaves alternate; tree, shrub or herb, stems and leaf-stalks not twining. 11
11. Trees or shrubs without thorns; leaves simple and entire. 12
- Herbs or shrubs, sometimes with thorns; leaves simple or compound, often deeply lobed or with a serrate edge. 13
12. Leaves without stipules, the stem-tip not covered by a stipule or cap-like structure; sepals and petals usually in whorls of 3 each; fruit a berry or a fleshy aggregrate of the pistils. *ANNONACEAE*
- Leaves at the tip of the branch with a cap-like stipule that covers the tip of the stem and young leaves, falling off to leave a circular scar around the young stem; sepals and petals borne in a spiral of many parts; fruit of follicles attached to the main axis. *MAGNOLIACEAE*
13. Flowers with the petals and sepals free from each other and not borne on a cup-like receptacle; herbs or climbers without thorns; the leaves often clasping the stem at their base, without stipules. *RANUNCULACEAE*
- Flowers with the sepals and petals united near the base or borne on a cup-like extension of the receptacle or on the edge of a disc; herbs, shrubs, or climbers, often with thorns; the leaves often with stipules united with the lower part of the leaf stalk. 91 **ROSACEAE**
14. Herbs or climbers with simple leaves (sometimes succulent). 15
- Trees, or shrubs, if herbs or climbers the leaves compound (not succulent) but stems sometimes succulent and with white sap. 18
15. Ovary inferior with 2 styles and 2 locules; plants sometimes with glandular hairs. 16
- Ovary superior, of 3 or more separate pistils; plants without glandular hairs. 17
16. Climbing or trailing herbs, often with sticky hairs; leaves in whorls; style not persistent on top of fruit. *RUBIACEAE*
- Herbs not climbing, with glandular but not sticky hairs; leaves opposite; style persistent on top of fruit. *(Vahlia)* 90 **VAHLIACEAE**
17. Petals absent; each flower with 3 or more pistils; leaves only slightly succulent. *(Gisekia)* MOLLUGINACEAE
- Petals present, 4 or more; pistils as many as the petals; leaves succulent. 88 **CRASSULACEAE**
18. Sepals, petals, and stamens borne on a cup-like receptacle or at the edge of a disc that surrounds the pistil or pistils, (the petals sometimes absent); leaves often with stipules united to the base of the leaf stalk and the leaf stalk clasping the stem at its base. 91 **ROSACEAE**
- Sepals, petals, and stamens not borne on a cup-like receptacle; stipules and leaf bases not as above. 19
19. Trees or shrubs with pinnately compound or trifoliolate leaves. 20
- Trees, shrubs or herbs with simple or palmately compound leaves, or herbs with pinnately compound leaves. 23
20. Pistils more than 1 and covered with short hairs (at least along 1 side), the ovaries separate from each other or united only near the base, ovules usually 2 in each locule; leaflets entire. 128 ***CONNARACEAE**
- Pistil 1 and mostly without a covering of hairs, the ovaries sometimes separate but then united by the styles or united for half their length. 21
21. Leaves with small bright dots (x10) when viewed against the light, fruit of 2-5 follicles with gland pitted walls. *(Zanthoxylum)* 120 **RUTACEAE**
- Leaves lacking bright dots when viewed against the light; fruit a drupe. 22
22. Bark with aromatic resin, the leaves slightly aromatic when crushed; pistil with 1 to 3 locules. 127 **ANACARDIACEAE**
- Bark without aromatic resin, leaves often with an unpleasant smell when crushed; pistil with 4 or 5 locules or with the pistils more than 1 and separate in each flower. 122 **SIMAROUBACEAE**
23. Leaves opposite or borne in whorls along the stem. 24
- Leaves alternate along the stem or borne at the base of the plant from the top of the root or the stems green and the leaves absent. 27
24. Trees or shrubs; flowers with more than 13 stamens; styles usually absent and the stigmas sessile on the top of the ovary. *GUTTIFERAE*
- Herbs, small shrubs, or climbing plants; flowers with 12 or fewer stamens; styles usually present. 25
25. Stipules usually joined between the leaf pairs to form an interpetiolar line or sheath; petals joined at the base; ovary inferior. *RUBIACEAE*
- Stipules not as above; petals free; ovary superior. 26
26. Herbs or small shrubs, often with many-celled gland-like hairs; stamens 3 to 10. *ELATINACEAE*
- Climbing plants or sometimes shrubs, with unicellular hairs which may be branched and/ or attached in the middle; stamens usually 10. *MALPIGHIACEAE*
27. Each pistil with 3 (rarely 2 or 4) locules and styles, each locule with 1 or 2 ovules, the

ovules attached on the inner angle of the locule and pendulous; leaves simple or palmately compound or absent and the stems green; sometimes with milky sap. *EUPHORBIACEAE*

- Each flower with 2 or more pistils, or if the pistil is solitary then other features not as above. 28

28. Herbs or small shrubs; sepals and petals present; leaves deeply lobed or compound. 29

- Trees or shrubs; leaves simple and entire or serrate. 31

29. Stamens usually of 2 different sizes within the flower and with the filaments united at the base; leaves pinnately compound or trifoliolate; wild plants, not cultivated for spices. *OXALICACEAE*

- Stamens usually all the same length; leaves deeply divided but not compound; plants grown for spices or ornament. 30

30. Leaves with small bright dots when viewed against the light; petals with a fringed margins; stamens 10 or fewer. *(Ruta)* 120 **RUTACEAE**

- Leaves without bright dots when viewed against the light; petals bifid at the tip; stamens more than 12 in number. *(Nigella) RANUNCULACEAE*

31. Ovule 1 in each locule and borne at the base of the locule; leaves smooth; climbing plants small shrubs, or trees. *PHYTOLACCACEAE*

- Ovules 2 or more in each locule. 32

32. Plants with star-shaped hairs; leaves simple or sometimes palmately compound; herbs, shrubs, or trees with soft wood; stamens 5 to many and free or united. *STERCULIACEAE*

- Plants without star-shaped hairs; leaves simple; trees or shrubs with hard wood. 33

33. Stamens 15 to many, free from each other or united to form groups; pistils 1 in each flower, ovary with 2 to 5 locules. *THEACEAE*

- Stamens many and free from each other; each flower with 2 to 5 separate pistils, each ovary with 1 locule. *DILLENIACEAE*

GROUP 11

Flower with the pistil (or ovary) borne on a stalk which comes from the centre of the flower above the sepals and petals. Be sure to examine the flower carefully as some small ovaries may look like a fat stigma on a style. Check older flowers in which the fruit has started developing. The fruit often borne on a stalk with a joint or articulation between the base of the fruit and the point of attachment to the stem or inflorescence.

1. Ovary with only 1 locule. 2

- Ovary with 2 or more locules. 15

2. Ovary with 1 style and 1 stigma (or with only 1 stigma if the style is absent). 3

- Ovary with 2 or more styles and stigmas or with a 2-lobed stigma or with 2 or more stigmas if the styles are absent. 11

3. Ovary with ovules borne along 2 or more sides of the locule (on the inner ovary wall), ovules 2 or more within each ovary; seeds borne along 2 or more sides of the inner wall of the fruit. 4

- Ovary with the ovules borne along only 1 side of the locule or in only 1 place within the locule (base or apex), ovules 1 or more within each ovary; seeds borne along only 1 side of the inner wall of the fruit or in only 1 place. 5

4. Leaves tripinnately, bipinnately, or sometimes pinnately compound; stamens 5 alternating with 5 staminodes; ovules along 3 sides of the locule; fruit a long capsule. *MORINGACEAE*

- Leaves simple, trifoliolate, or palmately compound, never pinnate; stamens 3 to many; ovules usually on only 2 sides of the locule; fruit with the seeds surrounded by pulpy tissue in woody plants, or a long dry capsule in the herbaceous species. *CAPPARACEAE*

5. Stamens borne high up on the inside of a long slender perianth tube, the perianth tube not subtended by a whorl of sepals at its base near the flower stalk; petals absent or very small and borne near the top of the perianth tube. 6

- Stamens free from the perianth tube and borne near the base of the ovary or the stamens borne on the base of the perianth tube; sepals and/or petals well developed. 7

6. Perianth tube without petals at the top near the mouth of the tube, the perianth tube splitting open along one side during flowering, stamens 4. *PROTEACEAE*

- Perianth tube with small petals at the top of the tube, the tube not splitting open along 1 side during flowering, stamens 2, 4, 8 or 10 and often inserted at 2 levels within the tube. *THYMELAEACEAE*

7. Ovary with the ovules borne along 1 side of the locule; fruit with the seeds attached along only 1 edge (sometimes only 1 seed in the fruit); stamens free or partly united, borne below the ovary or on a cup-shaped receptacle; flowers often irregular; leaves usually compound. *(LEGUMINOSAE)* 95 **FABACEAE**

- Ovary with the ovules basal or only 1 or 2 borne on the ovary wall; stamens free or united only near the base; leaves simple or compound. 8

8. Flowers with a long slender perianth tube or with a short tube, the perianth not subtended by a whorl of sepals although dry or colourful bracts may be present; ovule solitary within the locule and borne on a slender stalk from the base of the locule; leaves opposite or alternate, simple. 9

- Flowers without a perianth tube or of a perianth tube is present then it is subtended by a whorl of sepals, but not a colourful bract or series of bracts, or somewhat irregular with the stamens united near the base; ovule not borne on basal stalk; leaves alternate, simple or compound. 10

9. Flowers usually subtended by stiff dry bracts, perianth usually dry; stamens united near the base. *AMARANTHACEAE*
- Flower subtended by a large colourful bract or without subtending bracts; perianth a long tube and not stiff and dry; stamens usually free from each other. *NYCTAGINACEAE*

10. Stamens usually many and arising from the base of the ovary; leaves 3 or 4 times divided into pinnules. (*Thalictrum*) *RANUNCULACEAE*
- Stamens 4 to many, arising from the edge of a disc or from a cup-like or tubular receptacle; leaves simple or once pinnate. 91 **ROSACEAE**

11. Climbing plants or with climbing branches, tendrils borne on the stems next to the leaf stalk; ovary with 3 styles and the ovules along 3 sides of the locule; leaves with palmate venation; wild or planted for their edible fruit or ornament. *PASSIFLORACEAE*
- Herbs, shrubs, or trees, or if climbing then without tendrils. 12

12. Flowers each with more than 15 stamens; small trees of shrubs. *FLACOURTIACEAE*
- Flowers with 15 or fewer stamens; herbs or shrubs. 13

13. Flowers usually surrounded by dry bracts, the perianth usually dry and of only 1 whorl; ovule 1 and borne on a basal stalk within the ovary. *AMARANTHACEAE*
- Flowers not surrounded by dry bracts, the perianth of 1 or 2 whorls; ovule not as above. 14

14. Stamens 10 or more in each flower; ovules 3 or more at the base of the locule; leaves alternate or crowded together; herbs or shrubs. *RESEDACEAE*
- Stamens 10 or fewer in each flower; ovule 1 and basal or more than 1 and central or free central; leaves usually opposite, sometimes alternate; herbs. *CARYOPHYLLACEAE*

15. Flowers very irregular, the stamens united to form a tube that is split open on the upper side; ovary with usually 2 (3 to 5) locules and 1 ovule in each locule. *POLYGALACEAE*
- Flowers regular or if irregular then not as above. 16

16. Herbs with usually opposite leaves, the sap not white; stamens as many or twice as many as the petals, filaments often united near the base; ovary with 2 to 5 locules and 1 to many ovules in each locule. *CARYOPHYLLACEAE*
- Trees, shrubs, herbs, or climbers, the leaves mostly alternate or absent and the stems green (if herbs with opposite leaves then the sap often white). 17

17. Ovary with 3 (rarely 2 or 4) locules and 1 or 2 ovules in each locule; plants with or without star-shaped hairs, the sap sometimes white. *EUPHORBIACEAE*
- Ovary with 5 locules and with 3 or more ovules in each locule; plants with star-shaped hairs, the sap never white. *STERCULIACEAE*

GROUP 12

Flowers with an inferior or half inferior ovary; at least a part of the ovary borne below the level of the perianth attachment or a part of the ovary attached to the perianth tube (when a perianth tube is present), or the ovary borne below the perianth or perianth tube. Fruit with a dry perianth or the scars where the perianth has fallen off at or near the top of the fruit and opposite the stalk.

1. Trees and shrubs of mangrove formations, found on the ocean shore and bays along the edge of the sea. 2
- Herbs, or trees and shrubs not forming mangrove formations. 4

2. Small shrubs with alternate leaves; flowers irregular with a slender corolla tube which may be slit on 1 side and have 2 lips; stamens 5. **GOODENIACEAE*
- Trees with opposite leaves; flowers regular and the petals not united to form a tube, stamens 8 to many. 3

3. Ovary half inferior with 10 to 12 locules; stamens more than 20. **SONNERATIACEAE*
- Ovary inferior with 2 to 5 locules; stamens less than 20. *RHIZOPHORACEAE*

4. Each flower with 13 or more than 13 stamens. 5
- Each flower with 12 or fewer than 12 stamens. 15

5. Plants with thick and fleshy leaves or with fleshy green stems if the leaves are absent; the petals numerous (more than 10) or absent in each flower. 6
- Plants without thick fleshy leaves; petals usually 10 or fewer. 7

6. Plants with spines grouped in small clusters; leaves usually absent and the plant with thick green stems which are often flat and oval; ovary surrounded by stem-like tissue and often with spines or sharp hairs, sepals and petals often intergrading. *CACTACEAE*
- Plants without spines; leaves present or absent; ovary not surrounded by stem-like tissue; petals absent or, if present, intergrading with the stamens. *AIZOACEAE*

7. Leaves with bright dots when viewed against the light, stiff and leathery, usually aromatic when crushed, the leaves opposite or alternate in the introduced *Eucalyptus;* the petals usually united and coming off the top of the flower as a cap when it opens. *MYRTACEAE*
- Leaves without bright dots when viewed against the light; other features not as above. 8

8. Leaves opposite or sub-opposite or whorled along the stem (with 2 or more leaves at each node). 9

- Leaves alternate along the stem (only 1 leaf present at each node). 12

9. Flowers large (>3 cm) with the petals crumpled in bud and red or yellow in colour; plants grown for their fruit, ornament and twigs used as tooth sticks. *(Punica)* LYTHRACEAE
- Flowers usually <3 cm, petals not crumpled in bud; plants wild or grown for ornament. 10

10. Plants with minute hooked hairs on the surface of leaves and stems; ovary with the ovules on parietal or apical placentas. *(Kissenia arabica)* LOASACEAE
- Plants without minute hooked hairs; ovules often axile (the flowers sometimes sterile). 11

11. Sterile flowers with large showy sepals; fertile flowers small. **HYDRANGEACEAE**
- Sterile flowers absent; fertile flowers with showy petals. **PHILADELPHACEAE*

12. Stamens many, united near the base and also to the petals. *(Barringtonia)* LECYTHIDACEAE
- Stamens usually free from each other, not united with the petals. 13

13. Leaves with stipules which are sometimes united to the base of the leaf stalk; stamens usually borne on the edge of a cup-like receptacle or on the edge of a disc-like structure; ovaries sometimes separate from each other within the receptacle cup (or hypanthium). 91 **ROSACEAE**
- Leaves without stipules, simple; stamens not borne on a cup-like receptacle but sometimes borne on the edge of a disc-like structure. 14

14. Small shrubs or climbers, the leaves and stems with minute hooked hairs; plants of dry habitats. *LOASACEAE*
- Trees or shrubs, often with spines; growing in moist forests. *(Alangium)* 130 **ALANGIACEAE**

15. Plants climbing or twining with stiff or woody stems; herbs with slender twining stems are not included here. 16
- Plants not climbing or twining (though they may climb up and then fall over other plants) or if climbing then slender weak-stemmed herbs. 24

16. Ovary with only 1 locule. 17
- Ovary with 2 or more locules. 18

17. Tendrils and hook-like structures absent; leaves usually opposite; ovary with style and stigma; plants wild or grown for ornament. *COMBRETACEAE*
- Tendrils and hook-like structures present; leaves alternate; ovary with 3 styles or with 1 style and 3 stigmas. **ANCISTROCLADACEAE*

18. Flowers with the petals united to form a long tube, the tube over 1 cm. 19
- Flowers with the petals free, united only near the base, or the petals united and falling off as a cap when the flower opens. 21

19. Leaves alternate along the stem; sepals not found below the perianth-tube; the perianth-tube usually bent; stamens often united to the thick style; sap clear (not white). *ARISTOLOCHIACEAE*
- Leaves opposite or in whorls along the stem; corolla-tube subtended by a whorl of sepals; stamens free from the style (except sometimes near the base). 20

20. Plants usually with white sap; corolla-tube usually bell-shaped and broad at the mouth; plants mostly wild. *(Canarina)* CAMPANULACEAE
- Plants without white sap; corolla-tube narrow at the mouth. *(Lonicera)* CAPRIFOLIACEAE

21. Flowers with the ovary completely inferior and beneath the level of sepal and petal attachment; stamens alternate with the petals or corolla-lobes, or stamens 6 and opposite the petals. 22
- Flowers with the ovary only half-inferior; ovary covered with a disc-like structure and the petals and sepals attached to the edge of this disc; stamens 4 or 5 and opposite the petals (which may be small and difficult to see). 23

22. Leaves with 3 to 5 main veins running from the base to the apex of the leaf; sepals and petals alike in form and colour, 6 in number, stamens 6 or 3, usually non-functional; ovary often 3-angular or 3 winged, with 3 locules. *DIOSCOREACEAE*
- Leaves entire with a single main vein or palmately lobed and palmately veined; sepals and petals different in form and colour or the sepals reduced to a rim and apparently absent; stamens 5; plants grown for ornament and climbing by means of aerial roots. *(Hedera)* 131 **ARALIACEAE**

23. Tendrils arising from the stem opposite the leaves; leaves compound or simple and with palmate venation. 119 **VITACEAE**
- Tendrils absent or if present arising from the axil between the stem and the leaf or from the end of the stem; leaves simple. 118 **RHAMNACEAE**

24. Trees or shrubs with the leaves opposite or whorled along the stem (with more than 1 leaf at each node), stems woody, not herbs. 25
- Trees or shrubs with the leaves alternate along the stem (with only 1 leaf at each node), or herbs with soft stems and various types of leaf arrangement. 38

25. Ovary with only 1 locule or the locule difficult to see. 26
- Ovary with 2 or more locules. 31

26. Plants semi-parasitic, growing on other trees or shrubs; fruits a 1-seeded berry surrounded by a very sticky substance. 27
- Plants free-living, not semi-parasitic, fruits not as above. 28

27. Flowers with colourful corolla, growing in racemes or umbels; plants usually shrubby, stems brittle but not jointed. 114 **LORANTHACEAE**

- Flowers without colourful corolla, solitary or in axillary cymes; plants herbaceous with brittle jointed stems. **115 VISCACEAE**

28. Stamens as many as the petals or corolla lobes; petals or petal-like perianth parts present. 29
- Stamens twice as many as the petals or twice as many as the sepals if petal-like perianth parts absent. 30

29. Leaves with 3 to 7 palmate primary veins and with many parallel and transverse secondary veins; ovules basal with the locules. *MELASTOMATACEAE*
- Leaves with pinnate venation; ovules pendulous from the apex of the locule, ovary often very narrow and difficult to distinguish from the flower-stalk. *COMBRETACEAE*

30. Stems with interpetiolar stipules, the stipule extending from the base of 1 leaf stalk across the stem to the base of the other leaf stalk or forming a line across the stem; stamens with very short filaments and borne upon the corolla tube. *RUBIACEAE*
- Stamens without interpetiolar stipules or interpetiolar lines; stamens longer. *(Viburnum) CAPRIFOLIACEAE*

31. Stems with interpetiolar stipules or with a line across the stem where the stipules have fallen off; leaves always simple. *RUBIACEAE*
- Stamens without interpetiolar stipules or lines; leaves compound or simple. 32

32. Leaves with 3 to 7 palmately arranged primary veins with many parallel transverse secondary veins; stamens with bent anthers that open near the ends by pores. *MELASTOMATACEAE*
- Leaves not veined as above; anthers usually straight and opening by slits. 33

33. Stamens borne on the corolla tube or perianth tube; ovary completely inferior. 34
- Stamens borne at the base of the corolla tube or perianth tube or at the edge of or upon a disc but not upon the tube itself (a perianth tube often absent); ovary inferior or half-inferior. 35

34. Leaves simple and entire, stiff and somewhat leathery; sepals petal-like and forming a tube, petals small and borne upon the tube; wild plants with 4-angled stems. *OLINIACEAE*
- Leaves simple or compound, sepals present below the corolla tube (petals united to produce a showy corolla); shrubs planted for ornament or wild. *CAPRIFOLIACEAE*

35. Stamens 10 in each flower; plants usually with star-shaped hairs. **PHILADELPHIACEAE*
- Stamens fewer than 10 in each flower; plants without star-shaped hairs. 36

36. Ovary completely inferior and often with a ring-like disc above; sepals reduced to a rim; petals 4; ovule pedulous from the top of the locule. **129 *CORNACEAE**
- Ovary half-inferior, partly covered by the disc; petals 4, 5 or 6; ovules basal and erect within each locule. 37

37. Stamens alternate with the sepals and opposite the usually small petals (or the petals absent). **118 RHAMNACEAE**
- Stamens opposite the sepals and alternate with petals. **108 CELASTRACEAE**

38. Each flower with 1, 2 or 3 stamens, anther sometimes borne on a petal-like structure. 39
- Each flower with 4 or more stamens (if only 3 stamens present then flower without a corolla tube). 45

39. Anthers borne on filaments, stamens 1, 2 or 3; leaves less than 20 cm long. 40
- Anther borne on a petal-like structure or on a grooved filament, stamen 1; leaves usually more than 20 cm long. 43

40. Leaves alternate with stipules united with the leaf-stalks; flowers in clusters either in the leaf axils or opposite the leaves. **91 ROSACEAE**
- Leaves opposite, whorled or in basal rosettes, stipules absent or free; flowers solitary or in cymes or heads. 41

41. Semi-parasitic herb growing on trees or bushes; perianth a whorl of 5 green lobes; seed surrounded by a very sticky substance. **115 VISCACEAE**
- Free-living herbs; perianth with calyx and corolla; seed not as above. 42

42. Each flower with 2 stamens and anthers usually united with each other; ovary with many ovules; fruit a many-seeded capsule. *GESNERIACEAE*
- Each flower with 1, 2 or 3 stamens, the anthers free from each other; ovary with 1 ovule; fruit a single-seeded achene, often winged or hairy. *VALERIANACEAE*

43. Flower bilaterally symmetrical; anther borne on a grooved filament that usually surrounds the style, the connective enlarged at the top of the anther; plants with a spicy odour. *ZINGIBERACEAE*
- Flower without a plane of symmetry, completely irregular; anther usually borne on a petal-like structure. 44

44. Ovules numerous in each locule of the ovary. *CANNACEAE*
- Ovules 1 in each locule. *MARANTACEAE*

45. Trees or woody stemmed shrubs with alternate leaves. 46
- Herbs with various kinds of leaf arrangement. 64

46. Each ovary with only 1 locule or the locule not visible. 47
- Each ovary with 2 or more locules. 56

47. Stamens same in number as the petals and opposite the petals, or same in number as sepals and opposite the sepals if the petals are absent. 48
- Stamens more than the petals in number or alternate with the petals when the same in number. 52

48. Ovule or ovules difficult to see within the locule; perianth of only 1 whorl or if 2 whorls

appear to be present then the lower one forming a rim; leaves usually stiff. 49
- Ovules visible within the locule (use x 10 lens); ovary only half inferior, perianth of 2 whorls. 50

49. Perianth of only 1 whorl; style with 3 to 5 stigmas; plants not growing parasitically upon branches of other plants; small trees or shrubs. 116 **SANTALACEAE**
- Perianth of 1 or 2 whorls (colourful petal-like parts often present), style with a single stigma; plants growing parasitically on branches of trees and shrubs; leaves simple and leathery, different from those of the host. 114 **LORANTHACEAE**

50. Petals united near the base to form a short tube; stamens borne upon the corolla tube; common in evergreen woodland and forest. *(Maesa)* *MYRSINACEAE*
- Petals not united to form a tube; stamens borne at the base of the petals. 51

51. Petals with hairs on their inner surface; style 1 with single stigma. *(Ximenia)* 111 **OLACACEAE**
- Petals without hairs on the inner surface; styles 2 or more or 1 with 2 or more branches; glands present at base of ovary. *FLACOURTIACEAE*

52. Stamens twice as many as the petals or twice as many as the sepals when petals are absent; ovary usually slender and often hairy at the base of the style. *COMBRETACEAE*
- Stamens as many as the petals or as many as the sepals when petals are absent. 53

53. Each ovary with only 1 ovule; each fruit with only 1 seed. 54
- Each ovary with 2 or more ovules; each fruit with 2 or more seeds. 55

54. Anthers opening by pores or flaps, filaments slender and hairless; leaves palmately compound or simple, usually aromatic when crushed. *(Gyrocarpus)* *HERNANDIACEAE*
- Anthers opening by slits, filaments broad and hairy at base; leaves simple and not aromatic when crushed. *(Alangium)* 131 **ALANGIACEAE**

55. Fruit a berry; plants cultivated for ornament or fruit. **GROSSULARIACEAE*
- Fruit a capsule; plants growing wild or planted. 87 **ESCALLONIACEAE***

56. Each locule with many ovules; fruit usually with many seeds. 57
- Each locule with 1 or 2 ovules; fruit with only a few seeds. 58

57. No disc above the ovary; stamens usually twice as many (8 or 10) as the petals; shrubs planted for ornament. **HYDRANGEACEAE**
- Disc present above the ovary; stamens usually as many as the petals. 87 ***ESCALLONIACEAE**

58. Leaves compound or simple and deeply lobed, the base of the leaf stalk clasping the stem; flowers in umbels or other compound inflorescences; stamens bent inwards in bud, sepals very small or reduced to a rim; wood soft. 59
- Leaves simple, the leaf stalk not clasping the stem; flowers not in umbels; sepals usually distinct; wood hard. 60

59. Fruit berry-like; flowers in racemes of umbels or in long spike-like or raceme-like inflorescences. 131 **ARALIACEAE**
- Fruit dry and splitting into 2 parts; flowers in simple or compound umbels. *(UMBELLIFERAE) APIACEAE*

60. Ovary usually only half-inferior, often surrounded and enclosed by a disc; style 1. 61
- Ovary completely inferior, the disc (if present) above the ovary; styles 2 to 5. 63

61. Petals with hairs on the inner surface; ovules pendulous within the locules; stamens opposite the petals. 111 **OLACACEAE**
- Petals without hairs on the inner surface or the petals absent; ovules basal and erect within the locules. 62

62. Stamens alternate with petals and opposite sepals; petals usually larger than sepals. 108 **CELASTRACEAE**
- Stamens opposite petals and alternate with sepals, petals often smaller than sepals and wrapped round filaments, sometimes petals absent. 118 **RHAMNACEAE**

63. Flowers in heads, each flower with 2 styles and usually 6 stamens; leaves densely covered with star-shaped hairs; trees of evergreen forests. *(Trichocladus)* 96 **HAMAMELIDACEAE**
- Flowers in spikes, each flower with 4 or 5 styles and 8 or more stamens; leaves without star-shaped hairs. *(*Anisophyllea) RHIZOPHORACEAE*

64. Each ovary with only 1 locule; fruit with only 1 locule and a single seed or the seeds completely filling the fruit. 65
- Each ovary with 2 or more locules; fruit usually (but not always) with 2 or more locules or splitting up into 2 or more parts. 77

65. Ovary with usually many ovules on 3 parietal placentas; each flower with 3 or 6 stamens and with the perianth parts in whorls of 3; leaves with parallel or sub-parallel curving veins and with the leaf base clasping the stem. 66
- Ovary and leaves not as above. 67

66. Leaves with a leaf stalk, the leaf blade entire to deeply lobed; each flower with 6 stamens on very short filaments and with hood-like tips; plants of wet habitats. *TACCACEAE*
- Leaves without a distinct leaf stalk, the leaf blade narrow and entire; each flower with 3 stamens - sometimes borne on petal-like structures; plants from grassland, woodland and rocky areas. *IRIDACEAE*

67. Each flower with 2 sepals and 3 or more petals; ovary half-inferior with ovules on a free central placenta; leaves thick and juicy

within; usually small or sometimes lacking. *PORTULACACEAE*
- Each flower with 3 or more sepals; petals present or absent. 68

68. Semi-parasitic herb growing on trees or shrubs; stems jointed; fruit with 1 seed surrounded be a very sticky substance. 115 **VISCACEAE**
- Free-living herbs, shrubs or trees; stems not jointed; fruit not as above. 69

69. Flowers without petals, only a single perianth whorl present, without a slender corolla tube; flowers usually small. 70
- Flowers with petals, 2 perianth whorls present in each flower or if only 1 perianth whorl present then a colourful corolla tube present; flowers small to large. 72

70. Leaves palmately lobed with toothed margins; stipules usually conspicuous and united to the leaf stalk. 91 **ROSACEAE**
- Leaves simple, or pinnately lobed, or scale-like; stipules absent or inconspicuous and not united to the leaf stalk. 71

71. Fruit an achene, nut or dry compound structure; ovary with a single ovule borne at the base of the locule; leaves present or absent; plants grown for their roots or leaves or wild plants of salty soils. *CHENOPODIACEAE*
- Fruit a nut or drupe; ovary with 1 to 4 locules, the ovules pendulous from a free-central placenta; leaves small, sometimes scale-like; plants partly parasitic. 116 **SANTALACEAE**

72. Petals united to form a narrow corolla tube, the sepals absent or represented by hairs; ovules pendulous and solitary within the locules. 73
- Petals free from each other or united only near the base to form a short broad tube; sepals present. 74

73. Stamens 2 or 3; fruit often winged or with a tuft of hairs. *VALERIANACEAE*
- Stamens usually 4; fruit without wings or hairs. *DIPSACACEAE*

74. Stamens as many as the petals and borne opposite them, petals united at base; style and stigma 1. *(Samolus) PRIMULACEAE*
- Stamens as many as the petals and alternating with them or greater in number than the petals; usually free from each other; styles usually 2 in each flower. 75

75. Fruit discoid with straight spines; leaves pinnately lobed and densely hairy, hairs simple. *(Neurada)* 92 **NEURADACEAE**
- Fruit a capsule, not spiny; leaves palmately lobed, not densely hairy, some hairs glandular. 76

76. Leaves opposite, simple; ovules numerous on 2 pendulous apical placentas. *(Vahlia)* 90 **VAHLIACEAE**
- Leaves alternate or basal, palmately lobed or dissected; ovules not borne on pendulous apical placentas. 89 **SAXIFRAGACEAE**

77. Flower with perianth in whorls of 3 parts each; stamens 3 or 6; ovary with usually 3 locules; leaves with secondary veins more or less parallel with the midvein. 78
- Flower with perianth in whorls of 4 or 5 or more parts in each; leaves with secondary veins usually arising from an angle with the midvein. 82

78. Flowers with the anthers opening by pores or very short slits near the top; ovary half-inferior with only 1 or 2 ovules in each locule; fruit with lobes at the top. *TECOPHILAEACEAE*
- Flowers with anthers opening by slits. 79

79. Leaves very stiff and usually with strong fibres; plants grown for fiber or ornament. 80
- Leaves thin and flexible; plants wild or planted for ornament. 81

80. Leaves <50 cm long; perianth of 2 different whorls (sepals and petals differing from each other in form and colour); each flower with 6 stamens; plants only found growing as ornamentals in pots. *BROMELIACEAE*
- Leaves >50 cm, often much larger; perianth whorls alike (sepals and petals not differing from each other) each flower with 6 stamens. *AGAVACEAE*

81. Perianth whorls alike (sepals and petals not differing from each other) each flower with 6 stamens. *AMARYLLIDACEAE*
- Perianth of 2 different whorls (sepals and petals differing from each other in form and colour); each flower with 3 stamens. *IRIDACEAE*

82. Flowers with petals united to form a tube; sepals present at the base of the corolla tube; stamens usually borne on corolla tube. 83
- Petals free above the ovary and forming only a short tube or the petals absent, if a tube is present formed from the calyx or hypanthium, then without sepals at the base though 1 or 2 bracts may be present. 85

83. Leaves opposite or whorled along the stem; stamens on very short filaments and these borne on the corolla tube, stamens free from each other; interpetiolar stipules or line present between leaf bases. *RUBIACEAE*
- Leaves alternate along the stem; stamens borne at base of corolla tube, filaments often long, free or united; white sap often present. 84

84. Corolla lobes valvate (meeting edge to edge in bud); ovary with 2 to 10 locules; fruit not opening by a circular lid - but opening in various other ways. *CAMPANULACEAE*
- Corolla lobes imbricate (overlapping in bud); ovary with 2 locules; fruit opening by a circular lid at the top. *(Sphenoclea) SPHENOCLEACEAE*

85. Style and stigma 1 in each flower; petals, sepals and stamens often borne on a short calyx tube or hypanthium. 86
- Style and stigmas usually more than 2. 87

86. Stamens usually bent, anthers opening by pores; leaves opposite, each blade with 3 to 9 primary veins and many transverse secondary veins. *MELASTOMATACEAE*
- Stamens usually straight, anthers opening by slits; leaves opposite, alternate, or whorled, usually with pinnate venation and without transverse secondary veins. *ONAGRACEAE*

87. Leaves usually thick or reduced to scales; petals many or absent; ovary with 2 to many styles. *AIZOACEAE*
- Leaves not thick and succulent (or only slightly) and not reduced to scales, often all basal; petals 5 or absent; ovary with usually 2 styles. 89 **SAXIFRAGACEAE**

GROUP 13

Each flower with 13 or more than 13 stamens, or with 13 or more than 13 anthers when the filaments are united or absent; anthers always free from each other. The ovary superior and sessile.

1. Leaves usually thick and juicy; flower either with only 2 sepals or with 3 sepals and many petals or petals absent; herbs or soft-stemmed little shrubs. 2
- Leaves not thick and juicy; flowers with 3 or more sepals or the sepals united to form a rim, petals fewer than 8 or absent; herbs or woody shrubs or trees. 3

2. Each flower with 2 sepals; ovary 1-locular and with 2 to 8 styles, either 1 basal ovule or many ovules on a free-central or basal placenta. *PORTULACACEAE*
- Flowers with 3 or more sepals; ovary with either 2 or more locules or 1 locule and 1 style, ovules borne on a single parietal placenta. *AIZOACEAE*

3. Leaves opposite or in whorls along the stem, 2 or more leaves at each node, each leaf simple (never compound). 4
- Leaves alternate along the stem, only 1 leaf at each node (but sometimes the leaves closely congested), the leaves simple or compound. 8

4. Trees growing along the ocean shore in mangrove formations; roots producing vertical structures that rise above the ground or water level; ovary with 10 to 20 locules. *(Sonneratia alba)* **SONNERATIACEAE*
- Herbs or shrubs or if trees then not growing in mangrove formations along the ocean shore; ovary with less than 10 locules. 5

5. Stamens borne on a prominent calyx tube or receptacle tube or upon a cup-like receptacle, the petals borne on the edge of the tube or cup, crumpled in bud. *(Punica* and *Lagerstoxemia) LYTHRACEAE*
- Stamens and petals not borne on a cup or tube. 6

6. Stems with interpetiolar stipules (stipules that extend from the base of one leaf-stalk to the base of the opposite leaf stalk); ovary with 1 style and 2 or locules. *(Cassipourea) RHIZOPHORACEAE*
- Stems without interpetiolar stipules between the leaves. 7

7. Ovary with a single style that is bent near the base, usually 1-locular with intruding parietal placentas; leaves and twigs usually with star-shaped hairs. **CISTACEAE*
- Ovary with 2 or more styles or if with only 1 style, then the style not bent near the base, ovary with 2 or more locules. *GUTTIFERAE*

8. Leaves usually bipinnate, sometimes pinnate (simple and stiff only in introduced *Acacia* spp. from Australia); ovary with 1 locule, the ovules attached along 1 side, pod often splitting open along both sides. *(MIMOSOIDEAE)* 95 **FABACEAE**
- Leaves not bipinnate; ovary and fruit not as above. 9

9. Stamens borne on the edge of a cup-like structure or from a thick area (resembling a disc) from which sepals and petals also arise; stipules usually present and often attached to the base of the leaf stalk. 91 **ROSACEAE**
- Stamens not borne on a cup-like structure (but sometimes from the edge of a disc). 10

10. Leaves pinnately compound with 4 or more leaflets. 11
- Leaves palmately compound, or trifoliolate, or simple, sometimes divided into small, thin segments. 13

11. Herbs; fruit with spines. *(Tribulus) ZYGOPHYLLACEAE*
- Trees or shrubs with woody stems; fruit without spines. 12

12. Each ovary with 1 locule and usually 3 styles; bark and wood with resin. *ANACARDIACEAE*
- Each ovary with 2 or 4 locules and usually 1 style; bark and wood without resin. 125 **SAPINDACEAE**

13. Flowers with the filaments united to each other for much of their length and forming a short or long tube that surrounds the style or styles, each anther with only 1 theca and opening by a single slit; plants often with star-shaped hairs on stems and leaves. 14
- Flowers with the filaments free from each other or united to form separate bundles or united near the base but not forming a tube around the style; each anther with 2 thecae; plants without star-shaped hairs. 15

14. Herbs or shrubs or small trees; the pollen grains spiny (x 20). *MALVACEAE*
- Trees with large trunks; the pollen grains smooth (x 20). *BOMBACACEAE*

15. Ovary with only 1 locule but the placentas sometimes intruding deeply into the locule and appearing to fill the locule, the ovules borne on the outer wall of the locule or on placentas projecting inward from the outer wall. 16
- Ovary with 2 or more locules, the ovules borne along the central axis of the ovary or from the apex or base of the locule. 21

16. Stamens with anthers that open by short pore-like slits near the top; ovary with a single style; leaves simple; trees, shrubs, or herbs with a woody root, with coloured sap; often planted for ornament. 17
- Stamens with anthers that open lengthwise by slits; sap not coloured except in some herbs with a woody root. 18

17. Leaves not deeply lobed; fruit splitting into 2 parts, seeds covered with a red powder; trees or shrubs planted for ornament. *(Bixa orellana)* BIXACEAE
- Leaves deeply lobed; fruit splitting into 3 to 5 parts, seeds not covered in a red powder; herb-like or a small shrub or a small tree planted for ornament. *(Cochlospermum)* *COCHLOSPERMACEAE

18. Herbs; leaves often palmately compound or lobed and with spiny margins. 19
- Trees or shrubs with woody stems; leaves without spiny margins but the stems may have spines. 20

19. Sap clear; leaves usually palmately compound; wild plants rarely found in cultivation. *(CAPPARIDACEAE) CAPPARACEAE*
- Sap often coloured; leaves simple or deeply divided or with a spiny margin; plants often grown for their showy flowers or with spiny leaves and found wild. *PAPAVERACEAE*

20. Flower with the perianth parts in whorls of 3, petals and sepals valvate in bud. *(Monodora) ANNONACEAE*
- Flower with the perianth parts in whorls of 4 or 5, petals and sepals usually imbricate in bud. *FLACOURTIACEAE*

21. Ovary deeply lobed above, each lobe developing into a separate shiny black fruit or a whorl of petal-like structures present between the outer sterile stamens and the inner fertile ones; stellate hairs absent. *(Ochna) OCHNACEAE*
- Ovary not deeply lobed and forming separate fruits, or if the ovary is lobed then star-shaped hairs usually present. 22

22. Sepals completely united to form a cup or a lobed rim around the base of the flower, or the sepals enlarging in fruit to form 5 wings on the fruit. 23
- Sepals not completely united and forming a rim or cup and not forming wings in the fruit. 24

23. Sepals united to form a lobed rim, not enlarging in fruit. **SCYTOPETALACEAE*
- Sepals not forming a rim, enlarging to form 5 wings on the fruit. **DIPTEROCARPACEAE*

24. Flowers with the petals united to form a tube; ovary with 1 or 2 ovules in each locule; leaves alternate, stiff and with entire margins. *EBENACEAE*
- Flowers with petals free from each other or united only at the base. 25

25. Stamens in 2 groups on opposite sides of the flower, one fertile and the other sterile; flowers with a conspicuous receptacle tube. **93 CHRYSOBALANACEAE**
- Stamens not as above; flowers without a conspicuous receptacle tube. 26

26. Stamens with the anthers opening by longitudinal slits; common wild plants or sometimes planted. 27
- Stamens with the anthers opening by short pore-like slits at the top or opening by pores; plants sometimes planted for ornament. 28

27. Anthers with 2 chambers and usually opening by 2 slits; seeds not embedded in hairs; herbs, shrubs or small trees with simple, usually palmately veined leaves, common in drier woodland. *TILIACEAE*
- Anthers usually with only 1 chamber and opening by 1 slit; seeds embedded in soft hairs within the fruit; leaves palmately compound. *BOMBACACEAE*

28. Flowers with the petals valvate in bud; plants known only from the island of Socotra (in our area). *(Elaeocarpus)* *ELAEOCARPACEAE
- Flowers with the petals imbricate in bud; plants grown by man. 29

29. Leaves not deeply lobed; the fruit splitting into 2 parts, seeds covered in red powder. *BIXACEAE*
- Leaves deeply lobed; the fruit splitting into 3 to 5 parts, the ovary with 3 to 5 placentas; herb-like or a small shrub, or a small tree. *(Cochlospermum) COCHLOSPERMACEAE*

GROUP 14

Flowers with 12 or fewer stamens with united filaments; the filaments united with each other to form a short or long tube. (Do not confuse a staminal tube of united filaments with a perianth tube). The staminal tube is found within the usually colourful perianth parts.

1. Leaves opposite or in whorls along the stem or all the leaves coming from the base of the plant at the top of the root, with 2 or more leaves at each node (if the leaves are crowded at the ends of the stem look for long shoots on which the leaves are not crowded or for the presence of opposite branches). 2
- Leaves alternate along the stem, with only 1 leaf at each node (if the leaves are crowded at the ends of the stem look for long shoots on which the leaves are not crowded or look for the absence of opposite branches). 13

2. Plants climbing and usually with a woody stem. 3
- Plants not climbing or if climbing for short distances, stem herbaceous. 4

3. Stamens 3 (with 3 anthers), the filaments united for half their length. **108 CELASTRACEAE**
- Stamens 10 or rarely 5, the filaments united only near the base. *MALPIGHIACEAE*

4. Leaves spicy-aromatic when crushed, the stems usually 4-angled in cross-section; corolla tubular near the base and usually 2-lipped; stamens 4, ovary 4-lobed at the top; herbs or shrubs. *(LABIATAE) LAMIACEAE*
- Leaves not spicy-aromatic when crushed and flowers not as above. 5

5. Trees and shrubs of dry decidous woodland, the leaves remaining on the plant during the dry season and not falling off, stiff and with an entire margin; flowers with 4 sepals, 4 petals, and 4 stamens, the ovary with 1 or 2 locules. **110 SALVADORACEAE**
- Herbs, or if shrubs then leaves and flowers not as above. 6

6. Ovary with only 1 locule, fruit with only 1 locule; herbs or shrubs. 7
- Ovary with 2 or more locules, fruit usually with 2 or more locules or breaking up into 2 or more parts; herbs. 10

7. Ovary with 2 to 4 parietal placentas and 2 to many ovules on each placenta; sepals united, petals 4 or 5 and free from each other, stamens 6. **FRANKENIACEAE*
- Ovary with a single basal or free-central placenta and with 1 to many ovules. 8

8. Flowers usually subtended and surrounded by dry bracts, often as large as the flowers; each flower with a perianth of a single petal-like whorl but the perianth usually dry; ovule solitary within the locule and borne on a slender basal stalk. *AMARANTHACEAE*
- Flowers not subtended and surrounded by dry bracts; each flower with usually 2 perianth whorls (sepals and petals both present). 9

9. Stamens as many as the petals and opposite them, the filaments borne on the base of the corolla tube. *PRIMULACEAE*
- Stamens more numerous than the petals or equal in number to the petals and alternate with them, the filaments borne on the corolla tube or free from the petals. *CARYOPHYLLACEAE*

10. Leaves often thick and juicy, the leaf blade usually entire; ovary with 2-3 styles and 2-3 locules; petals absent or, if present, numerous within each flower. *AIZOACEAE*
- Leaves not thick and juicy or if so lobed and with palmate venation; ovary with 4-5 styles and 4-5 locules; petals usually present and the same number as the stamens. 11

11. Fruit with a long beak at the top, the parts of the fruit becoming separate at the bottom and curling upwards; leaves with palmate venation or the leaves palmatelt compound, with a strong pleasant smell of gernaium when crushed. *GERANIACEAE*
- Fruit without a long beak at the top; leaves simple and entire with pinnate venation; without a strong pleasant smell when crushed. 12

12. Flowers with the sepals free from each other, petals present; styles with capitate stigmas; including the crop *Linum usitatissimum*. *LINACEAE*
- Flowers with the sepals free or united, petals present or absent; stigmas not capitate or globose; plants wild or grown for ornament. *CARYOPHYLLACEAE*

13. Trees or shrubs with trifoliolate or pinnately compound leaves. 14
- Trees or shrubs with simple or palmately compound leaves or herbs with simple or compound leaves. 19

14. Flowers with the filaments united for more than half their length, the anthers 10 and often sessile on the staminal tube (the filaments often completely united and resembling a corolla tube); ovary with 2 to 5 locules. **124 MELIACEAE**
- Flowers with either the filaments united for less than half their length (not resembling a corolla tube) and ovary with 2 to 6 locules, or the filaments united for more than half their length and the ovary with only 1 locule, or the stamens of different lengths. 15

15. Ovary with 2 to 6 locules. 16
- Ovary with only 1 locule. 18

16. Inflorescence borne opposite the leaves along the stem, the leaves bipinnately compound and with large (>4 cm long) leaflets; ovary 3- to 6-locular with 1 ovule in each locule; shrubs with soft wood. *(Leea guineensis) *LEEACEAE*
- Inflorescence not borne opposite the leaves, the inflorescence or flower borne in the axils of the leaves or at the ends of branches. 17

17. Fruit ovoid; ovary with 2 to 4 locules; stamens 5 and all about the same length; wild shrubs and small trees. *(Bersama)* **126 MELIANTHACEAE**
- Fruit star-shaped; ovary with 5 locules; stamens 10 with filaments of 2 different lengths; small trees planted for fruit. *(Averrhoa) OXALIDACEAE*

18. Leaves without stipules, leaflets without stipels; flowers regular, filaments of 2 lengths; fruit a follicle with a single seed (opening along only 1 side). **128 *CONNARACEAE**
- Leaves usually with stipules (look at new leaves), the leaflets often with stipels at their base; flowers regular or irregular; fruit usually a legume (opening along 2 sides) and usually with 2 or more seeds. *(LEGUMINOSAE)* **95 FABACEAE**

19. Ovary with only 1 locule. 20
- Ovary with 2 or more locules. 28

20. Anthers 6 and borne on 2 thick filaments which are free below but united above the ovary; flowers irregular with 2 sac-like lobes at the base; herbs with the leaves divided into fine segments. **FUMARIACEAE*
- Anthers not borne on 2 thick filaments; flowers without sac-like lobes at the base; leaves not divided in to fine segments. 21

21. Leaves with small bright dots when viewed against the light, simple and entire; the filaments completely united to form a staminal tube around the ovary, ovules borne on 2 to 6 perietal placentas. **CANELLACEAE*
- Leaves without small bright dots when viewed against the light. 22

22. Ovary with the ovules borne along 3 different sides of the locule wall, the fruit usually splitting into 3 parts; mostly herbs and small shrubs. 23
- Ovary with the ovules borne along only 1 side of the locule wall or at the apex or base of the locule. 24

23. Flowers usually irregular and with only 2 perianth whorls; ovary with a thick and usually bent style; anthers often with appendages. *VIOLACEAE*
- Flowers regular and with what appear to be 3 different perianth whorls; ovary with a straight and slender style. *(*Vausagesia) OCHNACEAE*

24. Ovary borne on the outside edge of a tube-like receptacle, style arising from the side of the ovary; fruit with only 1 seed; trees. 93 **CHRYSOBALANACEAE**
- Ovary borne in the centre of the flower or in the centre within a cup-like or tubular receptacle and not at the edge; style usually terminal on the ovary. 25

25. Herbs; the ovary with only 1 or a few ovules borne on a basal or free-central placenta or on a basal stalk. 26
- Herbs, shrubs, trees, or climbers; ovary with 1 or more ovules borne on the wall of the locule and not at the base. 27

26. Flower with only 1 perianth whorl (the flower subtended by dry bracts that may resemble a second perianth whorl), the stamens opposite the perianth segments. *AMARANTHACEAE*
- Flowers with 2 perianth whorls or sometimes only 1 perianth whorl present, the stamens alternate with the petals. *CARYOPHYLLACEAE*

27. Leaves with stipules (at least on new leaves); flowers usually irregular and the fruit with 2 or more seeds. *(LEGUMINOSAE)* 95 **FABACEAE**
- Leaves without stipules; flowers regular; fruit with only 1 seed and usually opening along only 1 side. 128 **CONNARACEAE**

28. Anthers sessile on the top of the staminal column formed by the united filaments; ovary with 1 style and stigma; shrubs and small trees. *(Turraea)* 124 **MELIACEAE**
- Anthers on the ends of short or long filaments - at least the filaments free from each other at the top of the staminal tube. 29

29. Flowers very irregular, the lower petal the largest and usually with a fringe at the end; anthers 8 opening by pores. *POLYGALACEAE*
- Flowers usually regular and without a fringed lower petal. 30

30. Star-shaped hairs usually present on stems and leaves; leaves mostly with palmate venation, simple; stamens with filaments of several different lengths. 31
- Star-shaped hairs absent; leaves with palmate or pinnate venation, simple or compound; stamens often with filaments of 2 different lengths. 32

31. Each anther 1-thecous and opening by 1 slit, pollen grains large and spiny (use x20 lens), stamens united for most of their length. *MALVACEAE*
- Each anther 2-thecous and usually opening by 2 slits, pollen small and smooth (x20), stamens united only near base. *TILIACEAE*

32. Flowers without petals; ovary with 2 styles and 2 locules, fruit splitting into 2 parts; plants sometimes with viscid glandular hairs. *AIZOACEAE*
- Flowers with petals usually present; ovary and fruit not as above, either splitting into 5 parts or a capsule; plants without viscid glandular hairs, or glandular hairs only on leaves. 33

33. Fruit with a long beak at the top, parts of fruit separating from the base and curling upwards; leaves with palmate venation or compound. *GERANIACEAE*
- Fruit without a long beak at the top; leaves with pinnate or simple venation. 34

34. Petals 2-lobed or 2-fid at apex, at base joined to stamens; inflorescence often joined to leaf-stalk. *(Tapura)* 94 **DICHAPETALACEAE**
- Petals not as above; inflorescence borne separate from the leaf stalk. 35

35. Petals with small appendages on their inner face or the branches with opposite woody tendrils. 36
- Petals without appendages; tendrils absent; herbs. 37

36. Petals with appendages; tendrils absent. **ERYTHROXYLACEAE*
- Petals without appendages; tendrils present. **HUGONIACEAE*

37. Leaves simple and entire; petals free from each other. *LINACEAE*
- Leaves usually compound, of 3 or 4 fan-shaped parts or pinnate; petals free or united to each other; plants wild, often weeds. *OXALIDACEAE*

GROUP 15

Flowers with the petals united to form an irregular or bilaterally symmetrical (zygomorphic) corolla tube, or perianth tube. In some plants the petals have become united to form two petal-like lips with a short tube at the base. The corolla or perianth tube always comes off as a single unit and never as separate pieces.

1. Trees of the mangrove formations along the ocean shore with vertical woody structures (to 0.5 m tall) coming up from the roots. *(Avicennia) AVICENNIACEAE*
- Herbs, climbers, shrubs, or if trees then not growing in mangrove formations nor with vertical woody structures coming from the roots. 2

2. Stamens in 2 groups, one fertile, the other sterile; style from the base of the ovary; leaves with 2 glands at the base of the blade. **93 CHRYSOBALANACEAE**
- Stamens not as above; style or styles from the top of the ovary; leaves without 2 glands at the base of the blade. 3

3. Ovary with only 1 locule; fruit usually with only 1 locule or the seed(s) filling the fruit and a locule not apparent (but sometimes a space may develop round each seed which looks like a locule). 4
- Ovary with 2 or more locules; fruit usually with 2 or more chambers. 10

4. Sepals absent below the perianth tube; ovary with only 1 ovule; stamens 4 in each flower. 5
- Sepals present below the perianth tube (corolla tube); ovary with 1 to many ovules; stamens 2 or more in each flower. 6

5. Perianth tube long and narrow and split open along 1 side; stamens borne high up on the tube; trees or shrubs. *PROTEACEAE*
- Perianth tube short and broad; stamens borne at the base of the tube; herbs. *(*Hilleria) PHYTOLACCACEAE*

6. Ovary with the ovules borne along 1 side of the locule, seeds attached to 1 side of fruit (fruit often opens along 2 sides); corolla tube usually less than half the length of the flower; leaves often pinnate or bipinnate or trifoliolate; herbs, climbers, shrubs, or trees. *(PAPILIONOIDEAE)* 95 **FABACEAE**
- Ovules borne along 2 or more sides of the locule, or ovules borne at the base or apex of the locule. 7

7. Trees with pinnate leaves; each flower with 4 stamens and anthers free from each other; fruit large and cylindrical, woody, hanging down from the tree on long stalks. *(Kigelia) BIGNONIACEAE*
- Herbs, small shrubs or trees with simple leaves, fruit not woody. 8

8. Each flower with 4 fertile stamens, the anthers free from each other; ovary with only 1 ovule. **GLOBULARIACEAE*
- Each flower with only 2 fertile stamens, the anthers free or partly united with each other; ovary with 2 or more ovules. 9

9. Plants not found in very wet habitats, either native *(Streptocarpus)* or cultivated; leaves opposite or all basal; plants without special structures to catch small animals. *GESNERIACEAE*
- Plants of very wet habitats, frequently flooded or next to water; leaves alternate or basal or absent; plants usually with special structures to catch small animals. *LENTIBULARIACEAE*

10. Petals 2-lobed or 2-fid at the apex; stamens and petals joined at the base; inflorescense joined to the leaf-stalk. *(Tapura)* **94 DICHAPETALACEAE**
- Petals and stamens not as above; inflorescense separate from the leaf-stalk. 11

11. Each flower with 5 or more stamens. 12
- Each flower with 4 or fewer stamens. 13

12. Ovary with 2 or 4 locules and 1 or 2 ovules in each locule; stamens 5; herbs with rough hairs planted for ornament *(Echium)* or wild *(Echiochilon)*. *BORAGINACEAE*
- Ovary with 5 or more locules and with many ovules in each locule; stamens 5 or 10; cultivated ornamental shrubs. *ERICACEAE*

13. Trees or shrubs with pinnately compound leaves or climbers with 2- or 3-foliolate leaves, the leaves opposite or in whorls along the stem; flowers showy and usually large (>2 cm). *BIGNONIACEAE*
- Trees or shrubs with simple or palmately compound leaves or climbers with simple or deeply divided leaves. 14

14. Leaves on the lower stems opposite or whorled with 2 or more leaves at a node (sometimes leaves below flowers tend to be alternate), stems often with 4-angles. 15
- Leaves on all parts alternate with only one leaf at a node, stems usually round, not 4-angled. 21

15. Leaves spicy-aromatic when crushed; fruit often lobed but never with horns, a nutlet, 1-2-seeded drupe or berry. 16
- Leaves not spicy-aromatic when crushed; fruit not lobed but it may have horns, a capsule, 2-3-seeded drupe or many seeded berry. 17

16. Ovary with 4 lobes at the top, the lobes usually almost separate to the base and the style coming from between the lobes, the ovary usually with 4 locules and 2 ovules in each locule (rarely with 2 locules and 2 ovules in each locule); fruit usually breaking up into 4 (or fewer) separate parts. *(LABIATAE) LAMIACEAE*
- Ovary without lobes or with 2, 4, or 8 lobes at the top, the style borne at the top of the ovary, ovary with 2 or 4 (rarely 1 or 8) locules and with 1 or 2 ovules in each locule; fruit a single drupe or berry (from each flower) but sometimes deeply lobed. *VERBENACEAE*

17. Mature fruit woody with ridges, spines or horns; glandular hairs usually present, at least on the flower stalk. 18
- Mature fruit not woody, a drupe, berry or capsule without horns or ridges; plant without glandular hairs. 19
18. Plants, including flowers, with mucilage glands which become slimy when wet and rubbed; fruit with spines or ridges. *PEDALIACEAE*
- Plants with sticky hairs; fruit woody with 2 or more long horns. **MARTYNIACEAE*
19. Fruit a capsule that is usually broader near the apex than the base, with the seeds borne on stiff stalks; flowers often subtended by conspicuous bracts; plants sometimes with straight spines. *ACANTHACEAE*
- Fruit an ovoid capsule, or berry; flowers not subtended by conspicuous bracts; plants without straight spines, but sometimes with hooked prickles. 20
20. Seeds usually small and round; sepals overlapping (imbricate) in bud; flowers and leaves often turning dark when dried; plant without a strong unpleasant smell when crushed. *SCROPHULARIACEAE*
- Seeds usually flattened, often quite large; sepals held edge to edge (valvate) or folded in bud; flowers and leaves not turning dark when dried; plant with a strong unpleasant smell when crushed. *SOLANACEAE*
21. Mature fruit woody with ridges, spines or horns; glandular hairs usually present, at least on the flower stalk. 22
- Mature fruit not woody, a drupe, berry or capsule without horns or ridges; plant without glandular hairs. 23
22. Plants, including flowers, with mucilage glands which become slimy when wet and rubbed; fruit with spines or ridges. *PEDALIACEAE*
- Plants with sticky hairs; fruit woody with 2 or more long horns. **MARTYNIACEAE*
23. Small trees, leaves with stipules; fruit a 2- or 3-seeded drupe; flowers greenish. *(Tapura) 94* **DICHAPETALACEAE**
- Shrubs or herbs, leaves without stipules; fruit a capsule or berry; flowers usually colorful. 24
24. Fruit 1-seeded; herb with small linear leaves; corolla split along one side. *(Hebenstretia) SELAGINACEAE*
- Fruit many-seeded; leaves scale-like or well developed, not linear; corolla not split along one side. 25
25. Seeds usually small and round; sepals overlapping (imbricate) in bud; flowers and leaves often turning dark when dried; plant without a strong unpleasant smell when crushed. *SCROPHULARIACEAE*
- Seeds usually flattened, often quite large; sepals held edge to edge (valvate) or folded in bud; flowers and leaves not turning dark when dried; plant with a strong unpleasant smell when crushed. *SOLANACEAE*

GROUP 16

Flowers regular (actinomorphic), with a corolla or perianth tube, the tube long or short, the petals (or sepals) always coming off as a single unit and never as separate parts. The stamens borne either upon the tube or free from the tube. Sepals that are united to form a colourful perianth tube are included in this group.

1. Plants with compound leaves (the blades of the leaflets do not have an axillary bud, flower, or fruit in the axils at their base). 2
- Plants with simple leaves but sometimes the leaf deeply divided (the axil between the stem and the leaf stalk usually has an axillary bud, flower, or fruit). 8
2. Leaves opposite or whorled along the stem, with 2 or more leaves arising from each node (the leaflets may be either opposite or alternate along the leaf rachis). 3
- Leaves alternate along the stem, with only 1 leaf at each node; if the leaves are crowded look for a branch with longer internodes or the absence of opposite branches. (The leaflets may be either alternate or opposite along the leaf rachis). 5
3. Leaves palmately compound (with all the leaflets attached at 1 point); each flower with 4 stamens; trees. *(Vitex) VERBENACEAE*
- Leaves pinnately compound or trifoliolate; each flower with 2 or 4 stamens; trees, shrubs, or climbers. 4
4. Each flower with 2 stamens, the flower regular and often small with a tube less than 6 mm in diameter. *OLEACEAE*
- Each flower with 4 stamens, the flowers slightly irregular (with 1 or 2 corolla lobes larger than the others) and large with a corolla tube more than 6 mm in diameter. *BIGNONIACEAE*
5. Ovary with only 1 locule and with the ovules attached along 1 side; fruit usually a pod splitting open along 2 sides. *(MIMOSOIDEAE* or *Cadia)* 95 **FABACEAE**
- Ovary with 2 to 5 locules. 6
6. Trees or shrubs (both wild and planted for ornament and shade); anthers borne on a staminal tube that may be mistaken for a corolla tube. 124 **MELIACEAE**
- Climbing plants or herbs; anthers free on separate filaments or united into a ring. 7
7. Climbing plants with tendrils, planted for ornament; filaments of stamens free. *(Cobaea) *POLEMONIACEAE*
- Herbs, often grown for food; anthers united. *SOLANACEAE*
8. Sepals (or whorl of 3 or more sepal-like structures) absent below the colourful perianth tube. 9
- Sepals (or a whorl of 3 or more sepal-like structures) present below the colourful tube or upon the upper parts of the tube and alternating with the petals. 21

9. Ovary with only 1 locule and with 1 or a few ovules or the locule not visible; style and stigma 1; fruit with usually only 1 seed. 10
- Ovary with 2 or more locules or if with only 1 locule then with 5 or more ovules; style and stigma 1 or more; fruit usually with more than one seed. 15

10. Ovary without an apparent locule above the level of perianth attachment, the ovary very narrow and present well below the perianth, the locule present in what at first appears to be the stalk of the flower; inner part of the perianth tube usually with hairs; stamens 8 or 10. *COMBRETACEAE*
- Flower with an ovary different from that given above. 11

11. Leaves with the lower surfaces covered with silvery or golden hairs that are star-shaped or broad and flat. **ELEAGNACEAE*
- Leaves not covered with silvery or golden hairs on the lower surface. 12

12. Flowers with the stamens borne high up of at least half-way up on the perianth tube. 13
- Flowers with the stamens borne below the ovary or on the base of the perianth tube. 14

13. Stamens usually twice as many as the perianth lobes and borne on two different levels within the tube; very small petals often present at the top of the tube; ovule pendulous from the apex. *THYMELAEACEAE*
- Stamens 4, usually as many as the perianth lobes and opposite them, borne at the same level within the tube; the perianth tube usually split down 1 side and without small petal-like structures near the top; ovule usually attached to the wall or base of the locule and more than 1. *PROTEACEAE*

14. Perianth usually dry and papery and surrounded by stiff dry bracts. *AMARANTHACEAE*
- Perianth tube usually flexible and not dry, often subtended by a single large colourful bract or with sticky hairs. *NYCTAGINACEAE*

15. Lobes of the perianth tube 5 or fewer, or if 6 then the stamens 12 in number; mostly plants with hard wood. 16
- Lobes of the perianth tube 6; stamens 6 or 3 and opposite the perianth lobes; herbs, if large with spongy wood. 18

16. Each flower with 8 to 12 stamens; petals very either small and borne on the upper rim of the perianth tube, or absent, or the petals large and crumpled in bud; leaves usually opposite or in whorls along the stem. (If a succulent herb, check *AIZOACEAE*). *LYTHRACEAE*
- Each flower with fewer than 8 stamens; leaves usually alternate; trees or shrubs. 17

17. Stamens alternate with the sepal-lobes and opposite and often wrapped in the small petals or the petals absent. 118 **RHAMNACEAE**
- Stamens opposite the sepals or sepal lobes and alternate with the petals. *(Dichapetalum)* 94 **DICHAPETALACEAE**

18. Stamens with the anthers opening by pores or short apical slits; herbs often with orange or red sap. **HAEMODORACEAE*
- Stamens with the anthers opening by longitudinal slits; sap clear or somewhat mucilaginous. 19

19. Plants growing in shallow water; stamens borne upon the perianth tube. **PONTEDERIACEAE*
- Plants not growing in shallow water (except during flooding); stamens usually borne at the base of the perianth tube. 20

20. Leaves with very strong fibres, leaves very difficult to cut across. *AGAVACEAE*
- Leaves without strong fibres, not difficult to cut, but the leaf may be very thick and juicy. *'LILIACEAE'*[1]

21. Stamens twice as many as the lobes of the corolla tube or twice as many as the petals, usually more than 7 stamens. 22
- Stamens as many as the lobes of the corolla tube or as many as the petals, usually fewer than 7 stamens. 24

22. Flowers with a long (>3 cm) tube with the 10 anthers borne on very short filaments near the top of the tube; plants with soft wood and white sap and planted mainly for fruit. *CARICACEAE*
- Flowers with a shorter (<3 cm) tube; stamens borne near the base of the corolla tube; herbs, or shrubs or trees with normal hard wood. 23

23. Anthers opening by pores and with appendages near the base; leaves often scale-like. *ERICACEAE*
- Anthers opening by slits and without appendages at the base; leaves well developed. *EBENACEAE*

24. Stamens opposite the corolla lobes and borne on the lobes or at their base, but not between them; corolla tube usually short (<1 cm), or if longer then usually with sticky glandular hairs. 25
- Stamens alternate with the corolla lobes and borne between the lobes, or stamens fewer in number than the corolla lobes; corolla tube short or long. 28

25. Trees and shrubs with alternate stiff leaves. 26
- Herbs or small shrubs with alternate or opposite leaves; ovary with only 1 locule. 27

26. Ovary with 2 to 10 locules; stamens often borne on petal-like structures; sap often white. *SAPOTACEAE*
- Ovary with only 1 locule; stamens borne near the base of the corolla tube; sap clear. *MYRSINACEAE*

27. Ovary with only 1 ovule, the fruit with 1 seed; leaves alternate or basal; corolla tube often long and narrow with sticky hairs on the outside. *PLUMBAGINACEAE*

---
1 = 'Liliaceae' in the broad sense, not as circumscribed by Dahlgren and Clifford.

- Ovary with 3 to many ovules, the fruit with usually more than 1 seed; leaves alternate, opposite or in whorls; corolla tube usually short. *PRIMULACEAE*

28. Leaves opposite or whorled along the stem, with more than 1 leaf at each node, or the leaves all basal and coming from near the top of the root. 29
- Leaves alternate along the stem or only those near the flowers opposite, usually only 1 leaf at each node. If leaves are crowded look for the absence of opposite branching. 45

29. Trees of mangrove formations along the ocean shore, roots producing vertical woody structures that rise above the ground or water level. *(Avicennia) AVICENNIACEAE*
- Herbs, climbers, or if trees, then not from mangrove formations. 30

30. Corolla tube very short (<2 mm); leaves thick and stiff; sap clear; trees and shrubs of dry areas with evergreen leaves. 110 SALVADORACEAE
- Corolla tube 2 mm or longer; plants not as above. 31

31. Ovary with only one locule containing many ovules. 32
- Ovary with 2 or more locules or each flower with 2 ovaries that are united above by the stigma. 33

32. Each flower with 2 stamens; corolla usually forming a narrow tube. *GESNERIACEAE*
- Each flower with 4 or more stamens, the corolla usually forming a broad bell-shaped tube. *GENTIANACEAE*

33. Plants with white or clear sap or leafless plants with thick juicy stems; ovaries usually 2 in each flower and separate at the base (or rarely only 1 ovary present with 2 locules), ovules many in each locule; seeds often with long, silky hairs or winged. 34
- Plants without white sap or not leafless succulents; each flower with only 1 ovary and the ovary with 2 or more locules; seeds never with long silky hairs or winged. 35

34. Centre of flower occupied by a complicated structure made up of both stamens and pistil; pollen in waxy masses; herbs or climbers with leaves or succulents lacking leaves. *ASCLEPIACACEAE*
- Centre of flower with clearly separate stamens and pistil; pollen not in waxy masses; trees, shrubs or climbers. *APOCYNACEAE*

35. Leaves usually spicy-aromatic when crushed, the stems often 4-angled; ovary with usually 4 ovules in 2 or 4 locules; stamens usually 4 in each flower. 36
- Leaves not spicy-aromatic when crushed; other characters not combined as above. 37

36. Ovary with 4 lobes at the top, the style borne between the lobes; locules 4; fruit breaking into 4 parts. *(LABIATAE) LAMIACEAE*
- Ovary with 2, 4 or 8 lobes without lobes at the top, the style borne on the top of the ovary and not from between the lobes, ovary usually with 2 locules; fruit not breaking into 4 separate parts, usually a berry. *VERBENACEAE*

37. Trees or shrubs or climbers, the wood usually hard. 38
- Herbs or sometimes small shrubs or weak-stemmed climbers. 40

38. Each flower with only 2 stamens; ovary usually with 2 locules; stipules absent. *OLEACEAE*
- Each flower with 4 or more stamens; stipules present or represented by a line, the stipule of line interpetiolar (extending from the base of one leaf stalk to the opposite leaf stalk). 39

39. Each flower with 4 corolla lobes and 4 stamens; ovary 2-locular with many ovules in each locule, the stigma usually simple; fruit a capsule; star-shaped or flat hairs often present. *(Buddleia) LOGANIACEAE*
- Each flower usually with 5 (sometimes 4 or up to 20) corolla lobes and 5 (sometimes 4 or as many as 20) stamens; ovary 2- or 4-locular with 2 to many ovules in each locule, the stigma often with 2 or 4 lobes; fruit a capsule or berry; star-shaped or glandular hairs absent. (other genera) *LOGANIACEAE*

40. Leaves thick and juicy within; flowers with more than 8 petals and usually with some non-functional stamens; herbs planted for ornament. *AIOZOACEAE*
- Leaves not thick and succulent; petals usually fewer than 8. 41

41. Ovary with 3 locules; herbs planted for ornament. **POLEMONIACEAE*
- Ovary with 3 or 4 or more locules; wild or cultivated plants. 42

42. Stamens usually fewer than the corolla lobes in number; corolla tube usually slightly irregular and often very narrow. 43
- Stamens usually the same number as the corolla lobes; corolla tube usually regular and often broad. 44

43. Ovary with 2 to 8 ovules in each locule; fruit a capsule usually broader above than below and with few, large seeds, the seeds borne on stiff stalks; leaves and flowers not turning black or dark blue on drying. *ACANTHACEAE*
- Ovary with 6 to many ovules in each locule; fruit a capsule that is usually broadest in the middle or near the base with many small seeds; seeds not borne on stiff stalks; leaves and flowers often turning black or dark blue on drying. *SCROPHULARIACEAE*

44. All leaves of the plant opposite; fruit a capsule; inflorescence usually terminal; usually small herbs. *GENTIANACEAE*
- At least some of the leaves alternate or sub-opposite; fruit a capsule or berry; inflorescence often arising outside the leaf axil; herbs or woody plants. *SOLANACEAE*

45. Ovary with only 1 locule or the locule absent (or not visible). 46
- Ovary with 2 or more locules. 49

46. Ovary and/or style covered with short hairs. 47
- Ovary and styles not covered with short hairs. 48
47. Ovary with 3 styles and many ovules; petals free and borne on the calyx tube. *TURNERACEAE*
- Ovary with 1 style and 1 or 2 ovules; petals united and forming a tube. *(*Leptaulus)* 109 **ICACINACEAE**
48. Herbs, climbers, or shrubs, sometimes with white sap; ovary with 1 or 2 basal ovules or the ovary without a locule and non-functional. *CONVOLVULACEAE*
- Herbs or cultivated ornamental shrubs, without white sap, ovary with many ovules on 2 or more parietal placentas. **HYDROPHYLLACEAE*
49. Ovary with 3 locules; corolla tube usually very narrow. 50
- Ovary with 2 or 4 or more locules or 2 separate ovaries each with 1 locule present in the same flower. 51
50. Leaf-base clasping the stem, the secondary veins more or less parallel with the primary veins; corolla lobes 2 or 3 or rarely 4. *COMMELINACEAE*
- Leaf base not clasping the stem, the secondary veins usually arising at an angle from the primary veins; corolla lobes 5 or sometimes 4; planted ornamental herbs. **POLEMONIACEAE*
51. Ovary with 1 to 4 ovules in each locule. 52
- Ovary with 5 to many ovules in each locule. 54
52. Each flower with 5 corolla lobes and 4 stamens; shrubs planted for ornament. **MYOPORACEAE*
- Each flower with the same number of stamens as corolla lobes, either 4 or 5; native plants. 53
53. Herbs, shrubs, or more often climbers, often with white sap; corolla lobes very small and the corolla tube not deeply lobed, or if deeply lobed, usually shrubs with small (<2 cm) leaves and the corolla folded in bud; plants without rough hairs. *CONVOLVULACEAE*
- Herbs, shrubs, or trees, without white sap; corolla tube deeply divided, corolla lobes overlapping or crumpled in bud; plants often with rough hairs. *BORANGINACEAE*
54. Leaves often with an unpleasant smell when crushed; corolla lobes held edge to edge or folded in bud; sepals often becoming enlarged in fruit. *SOLANACEAE*
- Leaves and flowers not as above. 55
55. Shrubs or small trees; flowers purple or yellow-orange, with 4 corolla lobes; leaves usually hairy. *(Buddleia) LOGANIACEAE*
- Herbs, if shrubs or trees then not as above. 56
56. Plants usually with white sap; leaves with an entire margin; anthers often cohering together around the stigma; stamens borne high on the corolla tube. *APOCYNACEAE*
- Plants without white sap; leaves usually without an entire margin and with rough hairs; stamens usually borne low on the corolla tube and the anthers free; ovary solitary. 57
57. Each flower with 4 (rarely 2 or 5) stamens; ovary with a single style; flowers and leaves often turning dark blue or black on drying. *SCROPHULARIACEAE*
- Each flower with 5 (rarely 5) stamens; ovary with a 2-parted style or with 2 styles. **HYDROPHYLLACEAE*

GROUP 17

Flowers usually regular, with petals free from each other or united only at the base, or absent; sepals free or united but not forming a colourful tube. Stamens greater or fewer in number than the petals (or the sepals when the petals are absent), never equal in number to the petals or sepals.

1. Stamens fewer in number than the petals or sepals if petals are absent or only 1 perianth whorl present. 2
- Stamens more in number than the petals or sepals if petals are absent or if only 1 perianth whorl is present. 9
2. Ovary with only 1 locule. 3
- Ovary with 2 to 5 locules. 5
3. Herbs or small weak-stemmed shrubs, the stems often succulent; flowers green, without petals. *CHENOPODIACEAE*
- Trees; leaves simple or compound; flowers white or colourful, with sepals and petals. 4
4. Fruit a fleshy, aromatic drupe; each flower with 1 fertile stamen and 2 or 3 small staminodes, these borne together with the ovary on a large disc; trees with simple leaves. *(Mangifera)* 127 **ANACARDIACEAE**
- Fruit a pod, usually dry but sometimes with a fleshy pulp around the seeds; each flower with 2 or more stamens, the stamens and ovary not borne on a large disc; leaves compound. *(LEGUMINOSAE)* 95 **FABACEAE**
5. Herbs, the leaves alternate or opposite. 6
- Trees, shrubs, or woody climbers; leaves opposite. 8
6. Leaf lamina with secondary veins parallel with the midvein; flowers and fruits often borne in boat-shaped bracts. *COMMELINACEAE*
- Leaf lamina with pinnate or palmate venation; flowers and fruits not borne in boat shaped bracts. 7
7. Flowers with 5 sepals and without petals; fruit a more or less spherical capsule. *(Mollugo) MOLLUGINACEAE*
- Flowers with 4 sepals and 4 petals; fruit an elongated capsule with 2 locules and with the seeds borne in the angle between the inner and outer walls of the ovary; called a silique or silicula. *(CRUCIFERAE) BRASSICACEAE*

8. Each flower with 2 stamens, filaments free from each other and borne on a short corolla tube, petals usually 4; trees and shrubs. *OLEACEAE*
- Each flower with 3 stamens, filaments united with each other and borne on a disc that surrounds the ovary; woody climbers. 108 **CELASTRACEAE**

9. Each flower with 6 stamens and 4 (rarely 5) petals; ovary with 2 or 4 axile or parietal placentas. 10
- Each flower with 7 or more stamens. 11

10. Sepals free from each other or united only near the base, petals 4; fruit an elongated capsule with 2 locules and with the seeds borne in the angle between the inner and outer walls of the ovary; called a silique or silicula; leaves usually alternate. *(CRUCIFERAE) BRASSICACEAE*
- Sepals united with each other to form a long tube, petals 4 or 5; fruit a capsule; leaves opposite or in whorls. **FRANKENIACEAE*

11. Ovary with only 1 locule or the ovary very small (<3 mm) and the locule absent; fruit with 1 or only a few seeds. 12
- Ovary with 2 or more locules; fruit often with 2 or more chambers or breaking into 2 or more parts; fruit with many seeds. 31

12. Trees without leaves or the leaves represented by small alternate scales along the slender green twigs; flowers pink or white; ovary with 3 thick stigmas; plants of dry areas growing near water courses. *TAMARICACEAE*
- Herbs, shrubs, climbers, or if trees then with conspicuous leaves. 13

13. Leaves with only 2 leaflets, tough and leathery; plants with very large spines; fruit a large (>1 cm), 1-seeded drupe. *(Balanites)* 121 **BALANITACEAE**
- Leaves simple or compound with 3 or more leaflets; plants without spines or with small spines or prickles; fruit various, if a drupe then <1 cm long. 14

14. Leaves compound and alternate along the stem (check for the absence of an axillary bud or flower at the base of the stalk carrying the leaflet). 15
- Leaves simple, alternate, opposite, or whorled along the stem (look for a small axillary bud or flower or group of flowers at the base of the leaf stalk). 18

15. Ovary usually with 2 or 3 stigmas; flowers usually with a disc that surrounds the ovary or is borne on only 1 side; stipules very small or absent. 16
- Ovary with only 1 style and stigma; ovary not usually surrounded by a disc but sometimes borne within a cup-like receptacle; stipules usually present at the base of young leaves and sometimes attached to the base of the leaf stalk. 17

16. The disc usually on only 1 side of the flower or outside the stamens; bark without resin. 125 **SAPINDACEAE**
- The disc surrounding the ovary and within the stamens; the bark usually with resin. 127 **ANACARDIACEAE**

17. Ovary usually with 2 or more ovules attached along 1 side of the locule and not at the base or the apex; fruit a pod with the seeds attached along 1 side and usually opening along 2 sides; leaves mostly compound. *(LEGUMINOSAE)* 95 **FABACEAE**
- Ovary with only 1 or 2 ovules; fruit usually round and not breaking open along 2 sides; leaves simple or compound and often with prominent stipules attached to the leaf stalk. 91 **ROSACEAE**

18. Leaves alternate along the stem, with 1 leaf per node. 19
- Leaves opposite or whorled along the stem, with 2 or more leaves at each node. (If leaves are clustered look for opposite branches). 28

19. Anthers opening by flaps, the stamens often with brightly coloured gland-like nectaries at their base; trees or shrubs, wild or cultivated. *LAURACEAE*
- Anthers opening by slits or pores but not by flaps. 20

20. Trees, shrubs, or woody climbers. 21
- Herbs, or sometimes small shrubs with soft wood; leaves often succulent. 26

21. Locule not visible above the level of sepal and petal attachment, the locule in what appears to be the flower stalk; fruit with wings. *COMBRETACEAE*
- Locule present above the level of petal and sepal attachment, not found in what appears to be the flower stalk. 22

22. Fruit a drupe or dry achene with 1-2 seeds; receptacle wide, cup-like with perianth parts and stamens round the edge. 91 **ROSACEAE**
- Fruit a berry or capsule with 2 to many seeds; receptacle not wide and cup-like but a disc present at the base of the stamens. 23

23. Ovary with the ovules borne along 2 or more sides of the locule; flower often with a ring-like structure at its base or with glands. *FLACOURTIACEAE*
- Ovary with ovules borne at the base or along only 1 side of the locule; flower without a ring-like structure or glands at the base. 24

24. Petals densely hairy on the inner surface; fruit a drupe with 1 large seed. *(Ximenia)* 111 **OLACACEAE**
- Petals smooth or with short hairs outside; fruit dry and usually breaking open. 25

25. Leaves with stipules (at least on new leaves), leaf-blade often with a deep notch at the tip; flowers usually irregular; ovules 2 to many; fruit a pod usually opening along 2 sides. *(LEGUMINOSAE)* 95 **FABACEAE**
- Leaves without stipules and leaf blade never with a deep notch at the apex; flowers regular;

ovules 1 or 2 in a locule; fruit usually opening along only 1 side. 128 ***CONNARACEAE**

26. Leaves clasping the stem at their base with a thin membrane; ovary very small and with only 1 ovule; perianth parts usually in whorls of 3. *POLYGONACEAE*
- Leaves not clasping the stem at their base with a thin membrane; perianth parts in whorls of 2, 4, or 5 parts. 27

27. Sepals 2, petals present. *PORTULACACEAE*
- Sepals 3 or more, petals usually absent. *AIZOACEAE*

28. Locule not visible above the level of sepal and petal attachment, locule in what appears to be the flower stalk; fruit often with 2 or 4 wings; trees, shrubs, or climbers. *COMBRETACEAE*
- Locule visible above the level of sepal and petal attachment; herbs or succulent or weak-stemmed small shrubs. 29

29. Each flower with 2 sepals and 4 to 6 petals; leaves often thick and succulent. *PORTULACACEAE*
- Each flower with 3 or more sepals and 3 or more petals or the petals absent (only 1 perianth whorl present). 30

30. Each flower with 4 to 6 petals or rarely the petals absent; ovary with ovules borne at the base of the locule or on a central column. *CARYOPHYLLACEAE*
- Each flower without petals; ovary with ovules borne along 1 side of the ovary wall; sepals sometimes brightly coloured like petals but only 1 perianth whorl present. *(Trianthema) AIZOACEAE*

31. Leaves opposite or whorled along the stem. If the leaves are clustered look for opposite branches or for shoots with long internodes. 32
- Leaves alternate along the stem, only 1 leaf at each node. If leaves clustered then look for absence of opposite branches. 40

32. Anthers opening by pores at the top and usually bent near the middle and with appendages at the base; leaves with 3 to 7 palmate primary veins and many prominent transverse secondary veins. *MELASTOMATACEAE*
- Anthers opening by slits and not as above; leaf venation not as above. 33

33. Trees, shrubs, or climbers, the stem woody. 34
- Herbs or small shrubs, the stems without hard wood. 36

34. Stems with interpetiolar stipules between the leaves, the stipule extending from the base of one leaf stalk across the stem to the base of the opposite leaf stalk; petals fringed at the tip and falling off early; ovary with 1 style; trees of forest and river edges. *(Cassipourea) RHIZOPHORACEAE*
- Stems without interpetiolar stipules; flowers not as above; shrubs or climbers. 35

35. Petals crumpled in bud and usually narrowed at the base, without appendages at the base; ovary with 1 style and stigma; shrubs of stream edges in dry areas. *(Lawsonia) LYTHRACEAE*
- Petals not crumpled in bud, narrowed and with appendages at the base; ovary with 3 styles or with 1 style and 3 stigmas; climbers. *MALPIGHIACEAE*

36. Leaves pinnately compound or with 2 leaflets or if the leaves simple, then with palmate venation and serrate margins; fruit with a beak. 37
- Leaves simple and often entire; fruit without a beak at the top formed from the united styles. 38

37. Fruit with a single large beak at the top, parts of the fruit separating from the bottom and curling upwards; style with 3 to 5 separate stigmas at the top. *GERANIACEAE*
- Fruit without a beak but often with spines; style without separate stigmas. *ZYGOPHYLLACEAE*

38. Leaves usually thick and fleshy; flowers either without petals or with more than 8 petals in each flower. *AIZOACEAE*
- Leaves usually thin; flowers with 4 to 10 petals. 39

39. Flowers usually sessile in the axils of leaves; ovary with 3 to 5 locules throughout its length; fruit opening along the sides. *(Bergia) ELATINACEAE*
- Flowers usually on long stalks or borne at the end of the stem; ovary often with 1 locule at the top; fruit usually opening at the top. *CARYOPHYLLACEAE*

40. Herbs or shrubs with weak stems. 41
- Trees or woody shrubs. 46

41. Leaves clasping the stem and enclosing it in a sheath; secondary veins more or less parallel with the midvein; flowers with 4 or 6 stamens and 2 or 3 petals; ovary with 3 locules. *COMMELINACEAE*
- Leaves not clasping nor enclosing the stem as above; flower characters not combined as above. 42

42. Fruit with a long beak at the top, the parts of the fruit breaking up from the base and curling upwards; ovary with a central column that separates the locules. *GERANIACEAE*
- Fruit without a long beak at the top but sometimes with spines. 43

43. Leaves often thick and juicy within, simple and usually with an entire margin; flowers either without petals or with more than 8 petals. *AIZOACEAE*
- Leaves not thick and juicy; flower with both sepals and petals, the petals fewer than 8 in each flower. 44

44. Leaves and stems with star-shaped hairs (a dense covering of simple hairs may also be present), leaves simple and with palmate venation. *TILIACEAE*
- Leaves and stems (and flower parts) without star-shaped hairs though simple hairs may be

present; leaves simple with pinnate venation or compound. 45

45. Ovary with 5 separate styles; stamens of 2 different lengths; leaves pinnate or with 3 or 4 fan-shaped lobes, folding up at night or sometimes when touched. *OXALIDACEAE*
- Ovary with a single style; stamens usually of the same length; leaves without fan-shaped lobes and not folding up when touched. *ZYGOPHYLLACEAE*

46. Leaves pinnately compound or 2- or 3-foliolate. 47
- Leaves simple (each leaf with only 1 blade and with a bud, flower, or inflorescence between the stem and the stalk of the blade). 52

47. Leaves with small bright dots when viewed (x 10) against the light and with a spicy-aromatic (lemon) odour when crushed; a disc usually present at the base of the ovary. 120 **RUTACEAE**
- Leaves without bright dots when viewed against the light; aromatic odour may be present but not lemony. 48

48. Leaves, bark, and fruit with resinous aromatic sap; leaves trifoliolate, pinnate, or rarely simple, often absent at flowering time; small trees or shrubs often with crooked branches, bark usually peeling in thin, paper-like strips. 123 **BURSERACEAE**
- Plants without resinous sap but sometimes with a strong and unpleasant odour in bark and leaves. 49

49. Each leaf with only 2 leaflets (bifoliolate), the stem often with green spines over 3 cm long; ovary with 1 style, surrounded by a lobed disc; trees of dry woodland and scrub. 121 **BALANITACEAE**
- Each leaf with 3 or more leaflets. 50

50. Trees planted for their fruit which are borne on the older stems; ovary with 5 styles; fruit a large berry with many seeds. *(Averrhoa) OXALIDACEAE*
- Trees or shrubs usually found wild, not planted for fruit; ovary with 1 to 4 styles or stigmas. 51

51. Style usually 2 or 4 or sometimes united and apparently only 1; disc usually present all around the ovary; wood usually soft. 122 **SIMAROUBACEAE**
- Style usually 1 or 2 (rarely 3 or 4); disc often present outside of the stamens; wood usually hard. 125 **SAPINDACEAE**

52. Anthers opening near the top by pores; leaves simple. *(Ouratea) OCHNACEAE*
- Anthers not opening by terminal pores; leaves compound or. 53

53. Leaves with small bright dots when viewed (x 10) against the light and with a spicy-aromatic (lemon) odour when crushed; a disc usually present at the base of the ovary. 120 **RUTACEAE**
- Leaves without bright dots when viewed against the light; aromatic odour may be present but not lemony. 54

54. Leaves, bark, and fruit with resinous aromatic sap; leaves trifoliolate, pinnate, or rarely simple, often absent at flowering time; small trees or shrubs often with crooked branches, bark usually peeling in thin, paper-like strips. 123 **BURSERACEAE**
- Plants without resinous sap but sometimes with a strong and unpleasant odour in bark and leaves. 55

55. Petals with hairs upon the inner surface, the sepals very small and almost reduced to a rim; ovary with 1 style and stigma. 111 **OLACACEAE**
- Petals without hairs on the inner surface, sepals prominent and not reduced to a rim. 56

56. Leaves, stems, and flower parts with star-shaped hairs (a dense covering of simple hairs may also be present); ovary with a single styele; stipules small and usually falling off early. *TILIACEAE*
- Leaves and stems without star-shaped hair. 57

57. Shrubs or climbers with opposite woody tendrils; ovary with 3 style; stipules divided and not covering the shoot-tip. *(Hugonia) *HUGONIACEAE*
- Large trees without tendrils; ovary with a single style; stipules forming a cap over the stem tip. **IRVINGIACEAE*

GROUP 18

Flowers with only 1 perianth whorl, either sepals or petals present but not both; stamens same in number as perianth parts and opposite these borne at their base (not borne either between or on the perianth parts but can be united to each other).

1. Ovary with 3 locules or rarely with only 1 locule and the ovules on 3 parietal placentas; petals usually 6, stamens usually 6. 2
- Ovary with 1 or 2 locules and with 1 or 2 ovules in each locule or the locule absent and the ovary non-functional; flower parts in 3's, 4's or 5's. 7

2. Climbers with thorns and often with paired tendrils coming from the leaf stalk, the leaves usually stiff and with palmate venation. *(Smilax) SMILACACEAE*
- Not as above. 3

3. Plants growing in water; inflorescence subtended by a bract-like sheath. *PONTEDERIACEAE*
- Plants not growing in water (except during flooding); inflorescence not subtended by a bract-like sheath. 4

4. Anthers opening by pores or by pore-like slits at their tips. **TECOPHILAEACEAE*
- Anthers opening by longitudinal slits. 5

5. Leaves very tough and often thick, with very strong fibres, very difficult to cut or tear

across, often used as a source of fibre. *AGAVACEAE*
- Leaves not very tough and usually without strong fibres. 6

6. Ovary with 2 or more ovules in each locule; perianth parts usually showy and not dry; fruit a capsule or berry; leaves ending in tendrils only in which has large showy petals. *'LILIACEAE'*[1]
- Ovary with only 1 ovule in each locule; perianth parts small and dry; fruit a drupe; leaves ending in a tendril. *FLAGELLARIACEAE*

7. Climbing plants without tendrils; leaves opposite or alternate, simple with palmate venation or palmately compound. *DIOSCOREACEAE*
- Trees, shrubs, herbs, or if climbing then leaves usually with pinnate venation or pinnately compound. 8

8. Leaves opposite or whorled along the stem, with 2 or more leaves at each node. 9
- Leaves alternate along the stem, with only 1 leaf at each node. 11

9. Leaves with serrate margins and cystoliths on the surface; stamens bent inwards in bud. 105 **URTICACEAE**
- Leaves with entire margins, surface smooth; stamens not bent inwards in bud. 10

10. Stamens usually 5 and often united at the base; leaves herbaceous, not shiny. *AMARANTHACEAE*
- Stamens usually 4 in each flower, filaments free; leaves stiff and shiny. *BUXACEAE*

11. Trees or shrubs with white or pink sap; leaves often somewhat leathery; fruit usually a swollen fleshy structure formed from the whole inflorescence. 104 **MORACEAE**
- Plants without coloured sap; leaves usually herbaceous or fleshy; fruit not formed from the whole inflorescence. 12

12. Ovary very small (<3 mm) and surrounded by 5 gland-like structures, with only 1 stigma; fruit fleshy, a drupe. 112 **OPILIACEAE**
- Ovary not surrounded by 5 gland-like structures; fruit dry. 13

13. Ovary with 2 thick styles, with 1 locule and a single pendulous ovule; trees with hard wood; stipules on new leaves. 102 **ULMACEAE**
- Ovary with 1 or 3 styles, if 2 styles present, ovule basal in locule; herbs or shrubs with soft wood. 14

14. Leaves sheathing the stem at their base with a thin membrane-like stipule that surrounds the stem; flower parts often in multiples of 3. *POLYGONACEAE*
- Leaves not sheathing the stem at their base and without a thin membrane-like structure; flower parts not in multiples of 3's. 15

15. Plants with stipules at least on new growth; leaves mostly thin; plants of wet or shaded places. 105 **URTICACEAE**
- Plants without stipules; leaves often thick and fleshy; plants of drier and/or more open habitats. 16

16. Trees and shrubs; stems and leaves with star-shaped hairs. 113 **OCTOKNEMACEAE**
- Herbs or slender stemmed climbers, without star-shaped hairs. 17

17. Climbers with thin but fleshy leaves, growing from an irregularly shaped tuber; anthers opening by pore-like slits. *BASELLACEAE*
- Herbs or subshrubs, never growing from tubers; leaves fleshy or thin; anthers opening by longitudinal slits. *CHENOPODIACEAE*

GROUP 19

Flowers with two perianth whorls, both sepals and petals present; the stamens equal in number to the petals and alternate with them. The stamens arising from between the petals or opposite the sepals if the petals have fallen off.

1. Tree or shrub with the leaves reduced to minute alternate scales or the leaves apparently absent, the slender branchlets green; plants growing in dry regions near water courses; ovary with 1 locule and 3 short styles. *(Tamarix) TAMARICACEAE*
- Tree with leaves which are well developed (at least during the growing season) or herb, shrub, climber, or succulent. 2

2. Small herb growing in wet places; leaf blades with sticky glandular hairs that catch small insects over the entire surface ; flower with 5 sepals, 5 petals, 5 stamens and a 1-locular ovary with 3 to 5 styles. *(Drosera) DROSERACEAE*
- Plant not as above - without sticky glandular hairs that catch small insects. 3

3. Leaves opposite or whorled along the stem, with 2 or more leaves at each node. If leaves clustered, look for opposite branches. 4
- Leaves alternate along the stem, with only 1 leaf at each node. If leaves clustered, look for the absence of opposite branches. 14

4. Each flower with 2 separate ovaries that are united by the broad stigma; fruit of 2 long follicles; plants with white or clear sap. *ASCLEPIADACEAE*
- Each flower with a single ovary with 1 or more locules; fruit various but never of 2 follicles; plants without milky sap. 5

5. Stems often with swollen nodes and usually breaking easily at the nodes, leaves often joined at the base. 6
- Stems and leaves not as above. 7

6. Ovary with the ovules at the base of the locule or on a central column; petals usually without a scale on the inner surface; leaves never covered in salt crystals. *CARYOPHYLLACEAE*

1 = 'Liliaceae' in the broad sense, not as circumscribed by Dahlgren and Clifford.

- Ovary with the ovules on 2 to 4 parietal placentas; petals with a scale-like structure on the inner surface near the base; leaves often covered in salt crystals. **FRANKENIACEAE*
7. Trees, shrubs or climbers with hard stems. 8
- Herbs or small shrubs with soft stems. 12
8. Flowers with petals that are crumpled in the bud and have a wide blade with a narrow base; fruit a capsule with many small seeds. *(Lawsonia) LYTHRACEAE*
- Petals not crumpled in the bud and not shaped as above; fruit a drupe, berry or capsule with a few winged seeds. 9
9. Leaves leathery, with entire margins. 10
- Leaves herbaceous, with serrate or crenulate margins. 11
10. Leaves with very many parallel veins running from the midrib to the margin; no axillary spines. *(Ozoroa)* 127 **ANACARDIACEAE**
- Leaves with a few, rather indistinct lateral veins; axillary spines may be present. 110 **SALVADORACEAE**
11. Flowers with large discs between the ovary and the stamens; fruit a hard drupe or capsule with winged seeds. 108 **CELASTRACEAE**
- Flowers with small discs above the ovary; fruit a soft drupe or berry. 129 ***CORNACEAE**
12. Fruit with a long beak at the top formed by the united styles; parts of the fruit separating at the bottom and curling upwards; ovary with a solid central portion that separates the locules; leaves strongly scented when crushed; spines never present. *GERANIACEAE*
- Fruit without a long beak and not breaking up as above; leaves not scented when crushed; plant sometimes covered in spines. 13
13. Leaves often compound; ovary usually with only 1 style and without lobes, or only slightly lobed above; plants often found in dry areas. *ZYGOPHYLLACEAE*
- Leaves simple; ovary with 2 to 5 styles and deeply lobed above; plants growing in wet places. **ELATINACEAE*
14. Leaves palmately lobed or palmately compound. 15
- Leaves simple, or pinnately lobed or pinnately compound. 16
15. Herb with strong pleasant smell of geranium when crushed; fruit with an elongated beak and up to 5 mericarps which twist off the beak. *GERANIACEAE*
- Small tree or shrub with an unpleasant smell when crushed; fruit a capsule without an elongated beak. *EUPHORBIACEAE*
16. Leaves pinnately compound. 17
- Leaves simple. 22
17. Plants with aromatic oils and resins in either the bark and/or the leaves and/or fruit wall. 18
- Plants without aromatic oils and resins in bark or leaves or fruits, but white latex may be present. 20
18. Bark peeling in large or small papery flakes; seeds usually partly or completely covered by a fleshy, colourful, pseudaril; flowers without a disc at the base of the ovary. 123 **BURSERACEAE**
- Bark smooth or rough but not peeling in papery flakes; seeds without an aril; flowers with a disc at the base of the ovary or stamens. 19
19. Leaves with small bright dots when viewed against the light giving a smell of lemon when crushed; style 1, often thick; ovary with 2 or more ovules. 120 **RUTACEAE**
- Leaves without small bright dots giving a peppery spicy smell when crushed; styles usually 3, with thick stigmas; ovary with 1 ovule. 127 **ANACARDIACEAE**
20. Ovary with 2 or 3 locules; leaves without stipules. 125 **SAPINDACEAE**
- Ovary with 4 or 5 locules. 21
21. Flowers slightly irregular and usually over 15 mm long; stipules present (at least on new growth). 126 **MELIANTHACEAE**
- Flowers regular and usually less than 10 mm long; stipules absent. 122 **SIMAROUBACEAE**
22. Ovary with only 1 locule or a locule not apparent. 23
- Ovary with 2 or more locules. 30
23. Small shrubs of dry areas, the leaves simple, entire, and usually less than 2 cm long; ovary with 2 long styles and often with thick, twisted stigmas; stamens usually 2 different lengths. *(Ipomoea) CONVOLVULACEAE*
- Trees, herbs, climbers, or if shrubs then not as above. 24
24. Ovary with the ovules borne on 2 or more parietal placentas, the ovules usually more than 2. 25
- Ovary with 1 or 2 ovules borne at the base or apex of the locule or the ovary without ovules and non-functional. 27
25. Ovary with usually 2 parietal placentas and the fruit splitting up into 2 parts, the seeds surrounded by a coloured pulp; trees and shrubs with simple and usually entire leaves. 86 **PITTOSPORACEAE**
- Ovary with 3 or more parietal placentas; seeds not surrounded by a coloured pulp. 26
26. Ovary with 3 parietal placentas; anthers on short filaments and with basal appendages; fruit usually a capsule breaking into 3 parts. *VIOLACEAE*
- Ovary with 4 to 7 parietal placentas; a disc present between the ovary and stamens; fruit a drupe and not breaking up. *BREXIACEAE*
27. Ovary with 1 or 2 ovules pendulous from the apex of the locule, style 1 or none, the stigma with 1 to many lobes; fruit a drupe. 109 **ICACINACEAE**
- Ovary often without ovules and non-functional, or with 1 or 2 ovules from the wall of the locule or basal. 28

28. Leaves with a spiny margin or with an entire margin and with a sharp point at the tip; ovary without a locule and non-functional; each flower with 4 petals and 4 stamens; stipules absent. 107 **AQUIFOLIACEAE**
- Leaves without a spiny margin or sharp tip; ovary functional or non-functional; stipules present or absent. 29

29. Ovary functional with a single basal ovule; stipules absent. 127 **ANACARDIACEAE**
- Ovary usually non-functional; stipules usually present, at least in the new growth. *EUPHORBIACEAE*

30. Herbs or small trees with soft wood. 31
- Trees or small shrubs with hard wood. 32

31. Stems and leaves usually with star-shaped hairs (a dense covering of simple hairs may also be present), leaves usually serrate or lobed; petals strap-shaped, often recurved. *TILIACEAE*
- Stems and leaves without star-shaped hairs, leaves usually entire; petals obovate to lanceolate, not recurved. *LINACEAE*

32. Flowers with a conspicuous or ring-like disc between the ovary and stamens. 33
- Flowers without a disc between the ovary and stamens. 35

33. Flower with the petals twisted in bud; ovary 5-locular with 1 style and 1 or 5 stigmas. **IXONANTHACEAE*
- Flower with the petals straight and not twisted in bud. 34

34. Ovary with 1 or 2 erect ovules borne at the base of each locule, the locules 2 to 5. 108 **CELASTRACEAE**
- Ovary with 3 or more ovules in each locule, the locules 4 to 7 with axile placentas. *BREXIACEAE*

35. Each petal usually with a deep notch at the apex; inflorescence often united with the leaf stalk; leaves often glandular. 94 **DICHAPETALACEAE**
- Petals without a deep notch at the tip; inflorescence separate from the leaf stalk; leaves not glandular. 36

36. Stigmas sessile on top of the ovary; fruit a berry; leaves leathery, margin entire or serrate or spiny. *(Ilex)* 107 **AQUIFOLIACEAE**
- Stigma on top of a style; fruit a capsule; leaves not leathery, margin entire to crenate or undulate. 37

37. Style 1 with 1 stigma or the stigma somewhat lobed; fruit with 2-4 seeds in each locule. 86 **PITTOSPORACEAE**
- Style 1 or 2, stigmas 2 or more; fruit with many seeds in each locule. 87 ***ESCALLONIACEAE**

GROUP 20

Flowers with the stamens the same number as the petals and opposite them; if the petals have fallen off or if the petals are absent, the stamens alternate with the sepals.

1. Leaves sheathing or clasping the stem at their base and alternate along the stem; herbs or soft-stemmed shrubs often growing in wet places. 2
- Leaves not sheathing or clasping the stem at their base, alternate, opposite, or whorled along the stem; herbs, shrubs or trees usually growing in mesic or dry habitats. 3

2. Ovary very small and 1-locular with a single basal ovule; the base of the leaf with a thin membranous stipule that encloses the stem. *POLYGONACEAE*
- Ovary 3-locular with many ovules; leaf base clasping the stem and not thin and membranous. *PONTEDERIACEAE*

3. Plants with star-shaped hairs on the surface of stems and leaves; filaments sometimes with wings or cilia; herbs, shrubs or trees. *STERCULIACEAE*
- Plants without star-shaped hairs. 4

4. Each flower with only 2 sepals, the petals (or an inner perianth whorl present); ovary with only 1 locule. 5
- Each flower with 3 to 8 sepals, the petals present or absent; ovary with 1 to many locules. 6

5. Slender climbing plants, the leaves only slightly succulent; ovary with only 1 ovule and the fruit with only 1 seed. *BASELLACEAE*
- Herbs or subshrubs usually with thick succulent leaves, rarely climbing; ovary with 1 to many ovules, the fruit with 1 to many seeds. *PORTULACACEAE*

6. Herbs or small shrubs with weak stems, never with tendrils; leaves always simple. 7
- Trees, shrubs, or climbers, a strong woody stem usually present, or weak-stemmed climbers usually with tendrils; leaves simple or compound. 8

7. Flowers with 4 or 5 petals and sepals; ovary with 1 locule and a single style with a thick stigma; herbs with leaves that are not thick and succulent. *PRIMULACEAE*
- Flowers without petals; ovary usually with 2 locules and 2 styles; if petal-like structures present, these are numerous and ovary always with 2 or more locules and 2 or more styles; herbs or subshrubs often with thick succulent leaves. *AIZOACEAE*

8. Each flower with 5 gland-like structures surrounding the ovary, the ovary very small

(up to 3 mm in diameter) and with a single locule; sepals very small or reduced to a rim, petals present; shrubs or woody climbers. 112 **OPILIACEAE**

- Each flower not surrounded by 5 gland-like structures and not as above. 9

9. Ovary with only 1 locule or without a locule and the ovary non-functional (not producing seeds); flowers functionally male. 10
- Ovary with 2 or more locules, usually functional; flowers functionally bisexual or female. 14

10. Anthers opening by flaps, each flower with 6 stamens; leaves often with a spiny margin or with a sharp tip; shrubs of high altitudes. *BERBERIDACEAE*
- Anthers opening by slits or pores. 11

11. Stamens usually borne on the petals; leaves stiff and often with bright dots or streaks when viewed against the light. *MYRSINACEAE*
- Stamens borne at the base of the petals or the petals absent. 12

12. Ovary without a locule and non-functional, no fruit formed; leaves various. *EUPHORBIACEAE*
- Ovary with a locule, fruit a drupe; leaves simple and entire. 13

13. Ovary with 1 or 2 ovules; plant with twining stems. 109 **ICACINACEAE**
- Ovary with 3 to 5 ovules; a woody tree or shrub. *(Ximenia)* 111 **OLACACEAE**

14. Climbing plants with the inflorescences and the tendrils borne opposite the leaves along the stem, leaves alternate; ovary usually with 2 locules. 119 **VITACEAE**
- Trees or shrubs or if climbers, then not as above. 15

15. Each flower with 8 sepals, 8 petals, 8 stamens, and an ovary with 8 locules; plants found only on the island of Socotra. *(Dirachma socotrana) DIRACHMACEAE*
- Each flower with fewer than 8 parts in each whorl, or if with 8 parts, then not confined to Socotra. 16

16. Flower with the ovary surrounded by a conspicuous disc or by cup-like or tube-like calyx (or receptacle); petals often very small or absent. 118 **RHAMNACEAE**
- Ovary not surrounded by a conspicuous disc or borne within a cup-like structure. 17

17. Ovary with 5 styles and 5 locules, with 2 or more ovules in each locule; fruit a capsule. *STERCULIACEAE*
- Ovary with 1 style and with 2 to 10 locules, each locule with only 1 ovule; fruit a berry. *SAPOTACEAE*

86. PITTOSPORACEAE

by I. Friis*

Cufodontis, *Enum.*: 173-175 (1954); Friis in *Kew Bull.* **42**: 319-335 (1987).

Evergreen trees, shrubs (or climbers) with resin ducts in bark and fruits, rarely spiny. Leaves alternate, often crowded towards the ends of the branches, simple, entire, petiolate or subsessile, lamina dentate or lobed, ± leathery. Stipules absent. Flowers few to many in terminal and/or axillary panicles or cymes, sometimes in clusters on the old wood, or solitary in the leaf axils, regular or sometimes slightly zygomorphic, hypogynous, bisexual or functionally (rarely morphologically) unisexual. Sepals 5, free and imbricate, at least in bud, or ± united. Petals 5, imbricate in bud, generally free, rarely with claws somewhat touching. Stamens 5, alternate with petals, free or with somewhat united filaments; anthers dithecous, introrse, opening by slits or pores. Ovary superior, sessile or shortly stipitate, with 2-5 carpels and parietal placentas; style simple, usually short, with capitate or somewhat lobed stigma. Fruit a berry or a capsule with generally entire valves. Seeds without aril, but mostly covered by a viscid resin, rarely dry and winged; testa thin and smooth; endosperm copious and firm; embryo small.

In Africa only the genus *Pittosporum*; however, there are 8 other genera in the family, chiefly occurring in S.E. Asia and Australia.

PITTOSPORUM *Soland.* in *Gaertn.* (1788).

Cufodontis in *Öster, Bot. Zeitschr.* **98**: 105-137 (1951), in *Fedde Repert.* **55**: 27-113 (1952), in *Fl. Zambesiaca* **1**: 298-303 (1960), in *Bol. Soc. Brot., Sér.* **2**, **34**: 159-177 (1960) & in *Fl. Trop. East Afr.*, **Pittosporaceae**: 14 pp. (1966); Friis in *Kew Bull.* **42**: 319-335 (1987).

Trees and shrubs, never climbing or spiny. Leaves petiolate or subsessile, lamina entire (outside Africa, sometimes with undulate or sinuate margin), glabrous or hairy. Inflorescence usually many-flowered, sub-racemose or sub-umbellate panicles, terminal and/or in the axils of the uppermost leaves, (outside Africa, the flowers sometimes solitary or clustered on the old wood). Flowers regular, up to 15 mm long, white, greenish white or yellow in Africa, (of various colours elsewhere), sweet scented, very often, perhaps always, functionally unisexual. Male flowers with long filaments, fertile anthers and short and slender sterile pistil: female flowers with short filaments, reduced, sterile anthers, and longer and stouter fertile pistil. Ovary unilocular, 2(-5)-merous (outside Africa); style ± as long as ovary, with globular, truncate or sublobed stigma, splitting in fruit according to number of carpels. Fruit a capsule; valves leathery or woody, yellow or brown, finally suberect, spreading or reflexed. Seeds variously deformed by mutual pressure, 2-4 (to many outside Africa) in two rows on each placenta, ripening orange or red, covered with a sticky slow-drying resin, which remains sticky for many years in herbarium specimens.

An old world tropical and subtropical genus of between 70 and 150 species, some of which produce useful timber.

1. Capsule valves thick, leathery to woody, 8-10 mm across, held upright to slightly spreading, never reflexed; lamina subsessile, with petiole up to 5 mm long. **1. P. abyssinicum**
- Capsule valves thin, occasionally thickened in the central part, leathery, 6-8 mm across, spreading or reflexed; lamina clearly petiolate, with at least some petioles more than 5 mm long. **2. P. viridiflorum**

1. **P. abyssinicum** *Del.* (1843)

- type: GD, Semien, Addeselam, *Ferret & Galinier* s.n. (MPU, holo. not seen).

P. tomentosum Engl. (1909), nom. illeg., non Bonpl. (1813); *P. engleri* Cuf. (1951); *P. abyssinicum* subsp. *engleri* (Cuf.) Cuf. in *Fedde Repert* **55**: 98 (1952) - type: Woramgarubo, *Ellenbeck* 1898/a (B holo. destr.).

P. lanatum Hutch. & Bruce (1941); *P. abyssinicum* subsp. *lanatum* (Hutch. & Bruce) Cuf. in *Fedde Repert.* 55: 96 (1952) - type: HA, Mt. Kondudo, *Gillett* 5228 (K holo.).

P. cardiocarpum Cuf. (1951); *P. abyssinicum* subsp. *cardiocarpum* (Cuf.) Cuf. in *Fedde Repert.* **55**: 96 (1952) - type: SU, Debra Sina, *Giordano* 2347 (FT holo.).

Shrub or tree to about 8 m tall, bole up to 30 cm in diam. near base; bark dark grey, smooth, not peeling noticeably. Leaves very noticeably crowded towards the end of the branches; lamina spathulate, narrowly ovate or lanceolate, not very well defined from the short petiole, into which the lamina narrows, 4-8 x 1.5-3.5 cm, pale green and glabrous on the upper surface, usually paler, glabrous or with a yellowish or golden, cobwebby indumentum on the lower surface; apex rounded, rarely subtruncate, mostly ± acuminate; margin entire, base cuneate, lateral nerves up to 6 on either side of

* Botanical Museum and Herbarium, University of Copenhagen, Gothersgade 130, DK - 1123 Copenhagen K, Denmark.

midnerve, with a usually indistinct net of anastomosing secondary nerves; petiole usually less than 5 mm long. Inflorescences short racemose or sub-umbellate panicles, up to 4 cm long and broad, inflorescence branches glabrous to puberulous. Flowers greenish or yellowish, on 4-6 mm long pedicels. Sepals 2-5 mm long, glabrous or puberulous, ± free. Petals spathulate, 5-8 x 1.5-2.5 mm, erect or spreading. Fertile stamens 4-6 mm long, sterile stamens 3-5 mm long. Ovary and style 4-5 mm long if fertile, 5-7 mm long and slender if sterile. Capsule with 4-8 seeds; valves thick, usually erect or slightly speading, leathery to almost woody, 8-13 mm in diam.

In dry upland evergreen forest and upland evergreen bushland; 2300-3200 m. GD SU BA HA. *Mooney* 7017; *Burger* 1456, 2499.

Although this is one of the most distinctive of the many taxa which are more or less clearly recognizable in the genus in Africa, it may sometimes be difficult to distinguish it from forms here referred to *P. viridiflorum*, and some doubt remains as to the distinctiveness of the two taxa at the species level.

2. **P. viridiflorum** *Sims* (1814).

P. feddeanum Pax (1907); *P. viridiflorum* subsp. *feddeanum* (Pax) Cuf. in *Fedde Repert.* **55**: 66 (1952) - types: SU, Gennet, *Rosen* s.n. (B, syn. destr.), SU, between Mumitscha and Urga, *Rosen* s.n. (B, syn. destr.), HA, Geldid, *Gillett* 5389 (K, neo. selected by Cufodontis).

P. dalziellii Hutch. (1927); *P. viridiflorum* subsp. *dalziellii* (Hutch.) Cuf. in *Fedde Repert.* **55**: 65 (1952).

P. ripicola J. Léon. (1950), as "*P. ripicolum*"; *P. mannii* subsp. *ripicola* (J. Léon.) Cuf. in *Fl. Zambesiaca* **1**: 302 (1960).

P. quartinianum Cuf. (1951); *P. viridiflorum* subsp. *quartinianum* (Cuf.) Cuf. in *Fedde Repert.* **55**: 69 (1952) - type: EE/EW, Mt. Lesa, *Pappi* (216) 4660 (FT, holo., K, W, iso.).

P. ustulatum Cuf. (1960) - type: SD, Mega, *Gillett* 14306 (holo.).

Shrub or tree to about 15 m tall, bole up to 40 cm in diam. near base; bark dark grey, smooth, not peeling noticeably. Young branches sometimes rusty pubescent. Leaves mostly crowded towards the end of the branches; lamina spathulate, obovate, ovate or lanceolate, rarely suborbicular, often not very well defined from the petiole, into which the lamina narrows, 4-15 x 1.5-7 cm, dark to pale green and glabrous on the upper surface, usually paler, glabrous or with a cobwebby indumentum on the lower surface; apex rounded, rarely subtruncate, mostly ± acuminate; margin entire; base cuneate; lateral nerves 6-10 on either side of midnerve, with a ± distinct net of anastomosing secondary nerves, sometimes uniform throughout the lamina, sometimes laxer towards the midnerve; petiole 8-20 mm long. Inflorescences short racemose or sub-umbellate panicles, up to 8 cm long and broad, inflorescence branches glabrous to puberulous, rarely tomentose, sometimes with small inflorescences in the upper leaf axils. Flowers whitish, greenish or yellowish, on 4-10 mm long pedicels. Sepals 2-5 mm long, glabrous or puberulous, ± free or united into a tube (which may split along one side) and then with irregular calyx-lobes or teeth. Petals spathulate, 5-8 x 1.5-2.5 mm, erect or spreading. Fertile stamens 4-6 mm long, sterile stamens 3-5 mm long. Ovary and style 4-5 mm long if fertile, 5-7 mm long and slender if sterile. Capsule with 4-8 seeds; valves usually thin, occasionally somewhat thickened in the middle, leathery, 6-8 mm in diam, spreading or reflexed. Fig. 86.1.

In upland rain forest, riverine forest and upland evergreen bushland; 1400-2700(-3000) m. EW ?TU GD WU GJ WG SU AR HA IL KF GG SD.

This species is very polymorphic, especially outside the Flora area. In the works cited above under the genus, Cufodontis has proposed a number of species, subspecies and varieties to accomodate this variation. Friis in *Kew Bull.* **42**: 326-331 (1987) has found that these taxa are extremely difficult to diagnose unambiguously because of the many intermediate specimens. A more informal classification has therefore been proposed, based on only the most characteristic forms. The following characteristic forms occur in the Flora area.

"*viridiflorum*", with ovate to obovate, glabrous lamina, the secondary nervation of which is uniform, found typically in the drier areas of the country and in parts of the central plateau; *Burger* 3437; *Mooney* 8072; *Gilbert & Thulin* 935.

"*ripicola*", with large, spathulate to ovate, glabrous lamina, the secondary anastomosing nervation of which is laxer towards the midnerve, found typically in the humid southern and southeastern parts of the country; *Burger* 3412; *Mooney* 5329; *Friis* et al. 1950.

"*feddeanum*", with smaller, lanceolate to ovate lamina with uniform secondary nervation and usually a cobbwebby tomentum on the lower surface, found chiefly in the more central parts of the plateau and Arsi; *Burger* 2632, 2686; *W. de Wilde* 6183.

"*ustulatum*", with almost suborbicular lamina, known only from a few collections (perhaps from the same tree) from Mega; *J. de Wilde & Gilbert* 388.

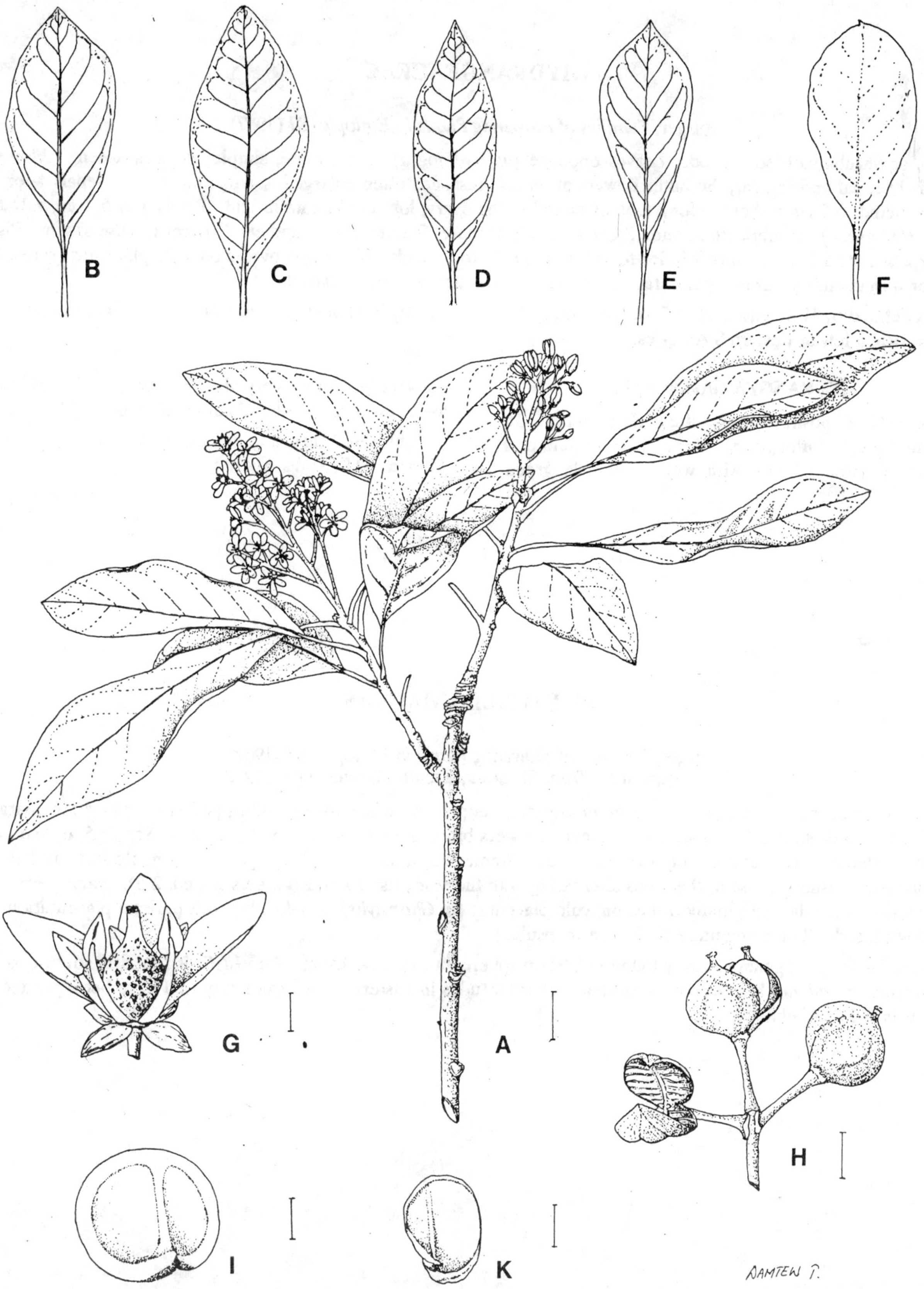

Figure 86.1. *PITTOSPORUM VIRIDIFLORUM*: A - "feddeanum", flowering branch from *W. de Wilde* 10414, scale 1 cm; B, C, D, E, & F - range of leaf shapes, B & F from *W. de Wilde* 10414, C from *Friis* et al. 1140, D from *W. de Wilde* 7673B, E from *W. de Wilde* 5569; G - flower, 2 front petals removed, from *W. de Wilde* 10414, scale 1 mm; H - mature fruits from *Friis* et al. 1950, scale 5 mm; I - valve of capsule from *Friis* et al. 291, scale 1 mm; J - seed from *Friis* et al. 291, scale 1 mm. Drawn by Damtew Teferra.

HYDRANGEACEAE

Burger, *Families of Flowering Plants in Ethiopia*: 87 (1967)

Herbs or shrubs with soft wood. Leaves opposite (in *Hydrangea*) or alternate, simple; stipules absent. Flowers usually bisexual and regular, the outer flowers of an inflorescence often enlarged, sterile and showy. Sepals 4- or 5-lobed, united to form a short or long tube or united to the ovary; lobes imbricate in bud. Petals 4 or 5, free, valvate (as in *Hydrangea*) or imbricate in bud. Stamens 8, 10, or more, filaments free; anthers 2-thecous. Disc absent. Pistil 1 (carpels united, 2 to 6) ovary inferior to half-inferior, 2- to 6-locular with many ovules on axile placentas (rarely 1-locular with intruding parietal placentas); styles and stigmas 2 to 6. Fruit a capsule.

As defined by Hutchinson, *Kew Bull.*: 100 (1927), 7 genera mostly in N hemisphere, from Himalayas to Japan and N America; a few in western S America.

HYDRANGEA *Gronov.* ex *L.*

Shrubs with opposite leaves, often gland-dotted. In cultivated plants, inflorescence showy, blue or pink or a mixture of these colours with white and pale green. Flowers in cymose corymbs, the outer, or in cultivated forms all, sterile with 4-5 enlarged, ± petaloid sepals.

Cultivated shrubs in gardens of hotels and houses, *c* 2400 m. SU (Addis Abeba).

87. ESCALLONIACEAE

Burger, *Families of Flowering Plants in Ethiopia*: 87 (1967);
Verdcourt in *Fl. Trop. E. Africa*, **Escalloniaceae**: 4 pp. (1973).

Shrubs or small trees. Leaves alterate or in clusters, simple and often with gland-tipped teeth along the margin; stipules absent or stipule-like structures present. Flowers bisexual or some unisexual, regular. Sepals 5, united near the base. Petals 5, free and usually valvate in bud. Stamens 5, alternate with the petals, filaments free; anthers 2-thecous: a disc usually present, the lobes alternating with the stamens. Pistil 1 (carpels united, 2-5), ovary inferior to half-inferior, 2-locular with many ovules on axile placentas (in *Choristylis)* or 1-locular with parietal placentation (in *Escallonia*); styles 2 or 1, stigmas 2-5. Fruit a capsule.

Family with 10-12 genera mainly found in S Hemisphere, but rare in Africa. *Choristylis* is found in south-eastern Africa while *Escallonia* has been cultivated at higher altitudes in eastern Africa (and may well have been planted in gardens in Addis Abeba).

88. CRASSULACEAE

by M. G. Gilbert*

Cufodontis, *Enum*. 165-172 (1954); R. Fernandes, **Crassulaceae** in *Fl. Zamb*. 7(1):3-74 (1983); Wickens, **Crassulaceae** in *Fl. Trop. E. Afr.* (1987).

Succulent herbs or subshrubs, occasionally annual or monocarpic. Leaves alternate, opposite or whorled, simple, rarely pinnately divided or compound, usually very fleshy; stipules absent. Inflorescences terminal or axillary, mostly cymose - branches often 1-sided with the flowers in 2 rows (a 'cincinnus'), occasionally racemose (*Umbilicus*) or flowers solitary, axillary. Flowers regular, usually bisexual, 3-12(-*c* 30)-merous. Sepals free or joined into tube, often fleshy. Petals as many as sepals, free or joined into tube. Stamens as many or twice as many as petals, rarely more, free or attached to corolla. Nectaries at bases of carpels, usually scale-like. Gynoecium a whorl of carpels equal in number to petals, free or shortly joined at base, ovules 1-many. Fruit a whorl of follicles, mostly dehiscing adaxially, sometimes breaking transversely in species with few seeds. Seeds minute.

Cosmopolitan with perhaps 35 genera and 1500 species, with major centres of diversity in South Africa, Mexico and eastern Asia, rare in wet tropics. The delimitation of genera is controversial and subject to constant revision. There are 8 genera and 29 species recorded as indigenous in the Flora area and 2 further genera and at least 8 species, the number increasing rapidly, introduced as ornamentals.

The highly succulent nature of most members of this family makes the preservation of good herbarium material difficult. Material is best killed by dipping in petrol, kerosene or boiling water, or by freezing or microwaving, and then dried as rapidly as possible over a heat source. Spirit material and photographs are invaluable. Living plants are easily collected, either as seed or cuttings, but growth is strongly affected by the environment and material of African species grown in Europe is often very atypical. Dimensions given are mostly from dried herbarium material and considerable allowances must be made when using them, especially for living material.

Many members of the family are cultivated as ornamentals, the number increasing all the time. The genera most likely to be met with are *Aeonium* from the Canary Islands, *Bryophyllum* from Madagascar, *Cotyledon* and *Crassula* from South Africa and *Echeveria* and *Sedum*, mainly from Mexico. Some species of *Bryophyllum* (q.v.) have become widely established as escapes elsewhere in Africa and may be expected to do the same in Ethiopia.

Key to Genera

1. Leaf-margin pectinate-ciliate, leaves on non-flowering shoots always in dense rosettes. 2
- Leaf-margins glabrous or with a soft indumentum similar to that of rest of leaf, leaves of non-flowering shoots often spaced along stem. 4

2. Leaves opposite, joined into sheath at base; flowers 5-merous; stamens as many as petals. **1. Crassula**
- Leaves alternate, not sheathing; flowers 6-10-merous; stamens twice as many as petals. 3

3. Prostrate herb with red stolons and leaves in sessile rosettes; flowers 6-merous; petals pink, joined into tube, erect. **4. Rosularia**
- Erect, sprawling or hanging subshrub with leaves in rosettes at tips of stout stems; flowers 8-10-merous; petals yellow, free, spreading. **5. Aeonium**

4. Inflorescence axillary or lateral. 5
- Inflorescence terminal. 6

5. Leaves in dense rosettes; inflorescence conspicuous, much longer than subtending leaves; petals bright red, orange or yellow, erect (cultivated). **10. Echeveria**
- Leaves spaced along stem; inflorescences inconspicuous, often shorter than subtending leaves; petals white or dull red, spreading. **1. Crassula**

6. Leaves alternate, occasionally opposite towards base and then white- or pink-flowered annuals. 7
- Leaves strictly opposite or in whorls of 3. 8

7. Leaves sessile; inflorescence cymose; petals free, spreading. **3. Sedum**
- Leaves petiolate, mostly peltate; inflorescence racemose; petals joined. **6. Umbilicus**

8. Petals ± free, spreading, white, occasionally tinged pink; stamens as many as petals. 9
- Petals joined into obvious corolla tube or if ± free, spreading, bright yellow; stamens twice as many as petals. 10

9. Leaves in whorls of 3; flowers 10-12-merous; follicles spiny-papillose on sides, 2-seeded. **2. Hypagophytum**
- Leaves opposite; flowers 5-merous; follicles smooth, many seeded (cultivated spp.). **1. Crassula**

10. Leaves opposite; petals joined into obvious corolla tube. 11
- Leaves in whorls of 3; petals ± free, spreading. **3. Sedum**

11. Leaf-margin entire; flowers 5-merous, pendent. **7. Cotyledon**
- Leaf-margin toothed or crenate, rarely entire, sometimes leaves lobed to pinnately compound; flowers 4-merous, erect or pendent. 12

12. Calyx small, usually divided almost to base; flowers erect; filaments connate to corolla-tube for over half its length. **8. Kalanchoe**
- Calyx with conspicuous inflated tube; flowers pendent; filaments free or inserted below middle of corolla-tube. **9. Bryophyllum**

---

* Ethiopian Flora Project, c/o Royal Botanic Gardens, Kew, Richmond, Surrey, TW9 3AB, UK.

1. **CRASSULA** *L.* (1753).
Tillaea L. (1753); *Bulliardia* DC. (1801); *Combresia* A. Rich. (1848)

Wickens & Bywater, *Kew Bull.* **34**: 629-637 (1980); Wickens, *Kew Bull.* **36**: 665-671 (1982).

Annual or perennial herbs, rarely shrubs. Leaves opposite, often joined at base into sheath, entire, often with submarginal hydathodes[1]. Inflorescence a terminal cyme or a 1-few-flowered axillary cluster. Flowers (3-) 4-5(-8)-merous. Sepals free or shortly joined. Petals usually spreading and very shortly joined, white or reddish (yellow or bluish outside Ethiopia). Stamens as many as sepals, inserted on corolla tube. Nectary scales usually very small. Carpels 4-many-ovulate, styles short. Follicles with 1-many seeds.

About 200 species, nearly 150 of which are restricted to South Africa, the rest mainly in Africa except for the small, often semiaquatic, species formerly included within the genus *Tillaea* which are almost cosmopolitan.

1. Rootstock not enlarged, stems less than 5 mm thick, plants often ephemeral; flowers axillary. 2
- Rootstock enlarged, fleshy; flowers in conspicuous, terminal cymes. 6
2. Leaves linear-lanceolate to oblong, sessile; petals up to 2 mm long; seeds 1-8 per carpel. 3
- Leaves ovate, attenuate into petiole; petals 5-6 mm long; seeds many per carpel. **1. C. alsinoides**
3. Flowers (4-)5-merous; sepals lengthened to a point; seeds smooth or longitudinally ridged; plants growing in well-drained habitats. 4
- Flower 4-merous; sepals subacute to obtuse; seeds papillate; plant growing in wet habitat, often aquatic. **2. C. granvikii**
4. Plant ± perennial, stem usually with peeling brown bark at base; pedicels mostly shorter than flowers; seeds smooth. **3. C. schimperi**
- Plant ephemeral, stem slender, without peeling bark; pedicels mostly much longer than flowers; seeds longitudinally ridged. **4. C. alata**
5. Leaves in dense basal rosette from fleshy rootsock, margins pectinate; plants usually papillate hairy. 6
- Leaves regularly distributed along erect stems, plant eventually forming erect thick-stemmed shrub to 75 cm high, leaf margins smooth; plants glabrous (cultivated). *C. ovata*
6. Lower leaf-sheaths 1-3 cm long; anthers 0.3-0.5 mm long; plant very sparsely hairy. **5. C. vaginata**
- Lower leaf-sheaths up to 1 cm long; anthers (0.5-) 0.6-0.9 mm long; plant often densely hairy. **6. C. alba**

C. ovata (Mill., 1768) Druce (1917) (more widely known as *C. argentea* Thunb. (1778)), a thick-stemmed shrub to 75 cm high with dark green leaves and small cymes of whitish flowers, is grown in dwarf hedges and as an outdoor pot plant in Asmera and Addis Ababa.

1. Hydathodes are water-secretory structures visible as dark dots at or near the tips or margins of leaves.

1. **C. alsinoides** *(Hook.f.) Engl.* (1892);
Tillaea alsinoides Hook.f. (1864); *C. pellucida* L. subsp. *alsinoides* (Hook.f.) Tölken, *J. S. Afr. Bot.* **41**: 114 (1975) - type: Fernando Po, *Mann* 1450.

Creeping perennial herb, rooting from lower nodes, glabrous except for 2 lines of hair-like papillae along internodes. Leaves ± ovate, 0.6-2(-3) x 0.4-1.2(-1.5) cm, with ill-defined petioles joined at base into short sheath, tip usually acute, apiculate, margin ± entire, often hyaline, often with submarginal dark dots (hydathodes) on underside. Flowers usually solitary in leaf axils, 5(-6)-merous; pedicels 3-17(-21) mm long. Sepals 3.5-5 x 0.7-1 mm, joined for *c* 0.5 mm, tip acute to acuminate. Petals oblong-lanceolate, 5-6 x 1.2-1.8 mm, tip acute, white or pale pink. Seeds many, *c* 0.5 x 0.3 mm, yellow-brown, minutely tuberculate. Fig. 88.1.1&2.

Usually in upland forest with *Juniperus* and/or *Hagenia* , sometimes on boulders or fallen logs, mostly in shade; (1500-)1900-3100 m. SU AR KF SD BA HA; Sudan, Cameroun, Fernando Po, Zaire, south-east to east South Africa; Madagascar; Yemen. *W. de Wilde* 5880; *M. & S. Gilbert* 1795; *Mooney* 6382.

2. **C. granvikii** *Mildbr.* (1922)
- type: Mt. Elgon (?Uganda side), *Granvik*.
Bulliardia abyssinica A. Rich. (1848) [*non Crassula abyssinica* A. Rich. (1848)]; *C. hedbergii* Wickens & Bywater (1980) - type: TU, Arbati Inseca[2], *Quartin-Dillon* (P holo.).
C. rivalis (Peter) Hutch. & Bruce (1941).
C. vaillantii auct. non (Willd.) Roth.
Tillaea aquatica sensu Britten (1871), *non* L. (1753).
C. intricata auct. non Nees.

Glabrous annual herb, often aquatic, forming mats 2-4.5 cm high in drier sites, up to 15 cm long or more in water. Leaves sessile, linear to oblong, rarely elliptic near stem tip, 3-15(-20) x 0.5-3(-5) mm, tip acute to obtuse. Flowers 4-merous, axillary, 1(-4) per axil. Pedicels 1.5-5 mm long. Calyx-lobes 0.9-1.3 x 0.7-1 mm, tip subacute to obtuse. Petals oblong, 1-1.8(-2) x 0.7 mm, tip rounded, white to pink or pale purple; filaments and carpels purple. Follicles 1-7-seeded. Seeds oblong-elliptic, 0.5-1 mm long, brown, minutely papillate. Fig. 88.1.3-6.

Still or slow-flowing water, wet mud; (1650-) 2000-3900 m. EW TU GD GJ SU AR KF SD BA; south and west to Zaire, Tanzania & Malawi; Yemen. *W. de Wilde* 6901; *Gilbert & Thulin* 1292; *Hedberg & Getachew* 5335.

A polymorphic species that should repay more detailed study in the field. The habit varies considerably with environment, plants growing in water are very much larger than those growing out of water. Material from northern Ethiopia was separated from *C. granvikii* as *C. hedbergii* on the basis of having shorter,

2. Arbati Inseca [Arba'ete Insisa], meaning four holy beasts, is the name of a church or village named after it and is on the label of the holotype but the protologue gives Chiré [Shire].

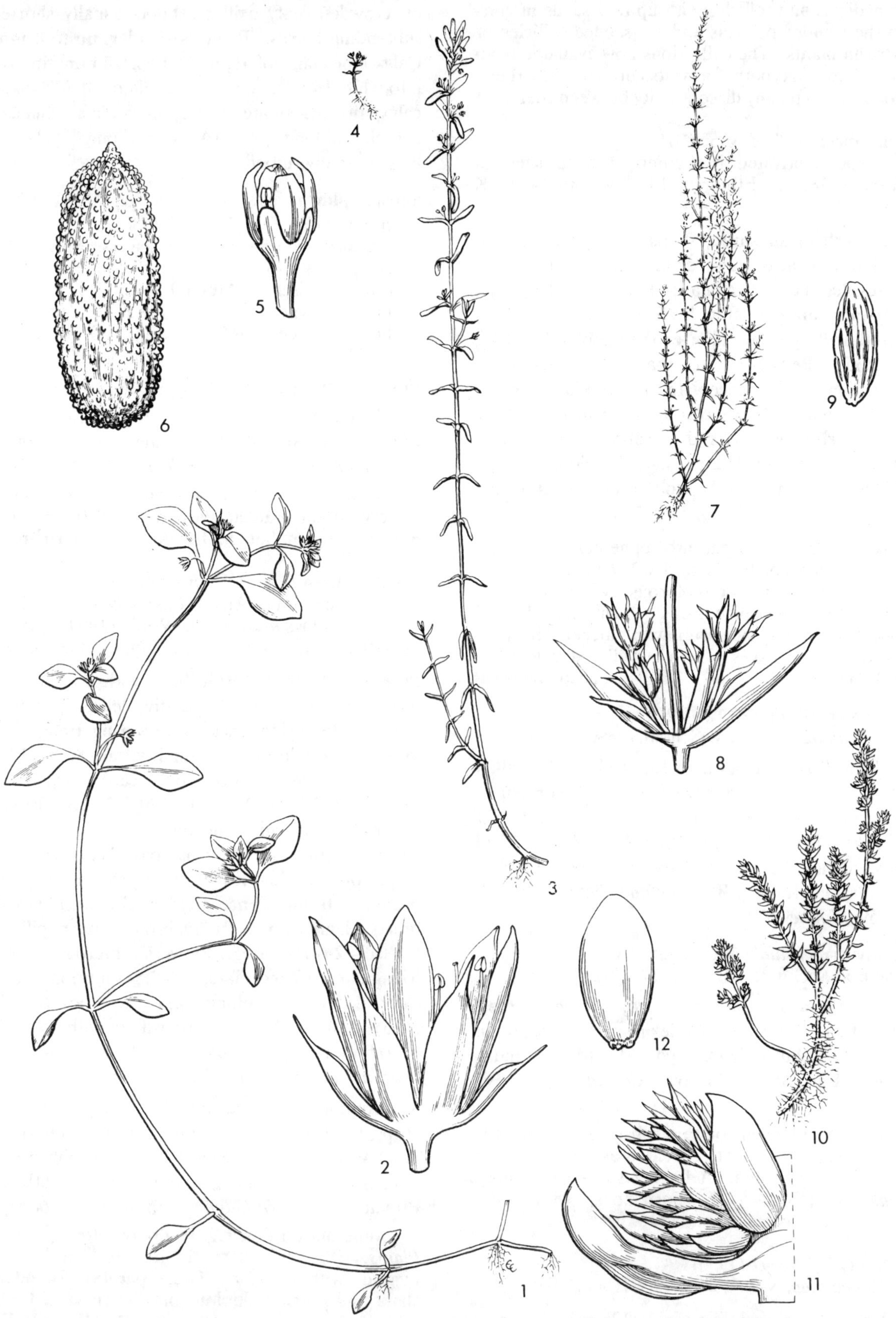

Figure 88.1. *CRASSULA* spp.: *C. ALSINOIDES*: 1 - flowering branch x 2/3; 2 - flower x 9. *C. GRANVIKII*: 3 & 4 habit x 2/3; 5 - flower x 9; 6 - seed x *c* 50. *C. ALATA* subsp. *PHARNACEOIDES*: 7 - habit x 2/3; 8 - inflorescence x 9; 9 - seed x *c* 50. *C. SCHIMPERI* subsp. *SCHIMPERI*: 10 - habit x 2/3; 11 - inflorescence x 9; 12 - seed x *c* 50. 1 & 2 from *Gillett* 14348; 3 & 5 from *W. de Wilde* 6901; 4 from *Hedberg & Aweke* 5380; 6 from S.E.M. of *Gilbert & Thulin* 947; 7 & 8 from *Meyer* 8529; 9 from S.E.M. of *Mann* 1991; 10 & 11 from *Hedberg* 5344; 12 from isotype. Drawn by Eleanor Catherine.

stouter pedicels and follicles with up to 7 seeds in contrast to the slender pedicels and 1-2-seeded follicles of East African plants. The collections now available from Ethiopia show that both forms occur and that there does not seem to be any discontinuity between them.

3. **C. schimperi** *Fisch. & Mey.* (1841)
  - type: cultivated in Leningrad from seed of *Schimper* '183' ex Ethiopia (LE holo. not seen; K iso.).

Glabrous herb or subshrub, stems usually with brown peeling bark towards base, dead leaves ± persistent. Leaves sessile, linear-triangular to lanceolate, 3-8(-12) x 0.7-2(-3) mm, margin often hyaline. Flowers in very condensed axillary cymes shorter than subtending leaf, (4-)5-merous. Pedicel shorter than calyx. Calyx-lobes narrowly triangular to lanceolate, 0.7-1.8 x 0.2 mm, tip acute to attenuate. Petals oblong-ovate to lanceolate, equalling or slightly shorter than calyx, tip acute, dull white to pale pink. Seeds 2 per carpel, oblong-ovoid, *c* 0.5 x 0.3 mm, brown, smooth or obscurely longitudinally ribbed.

1. Herb, sometimes apparently ephemeral; leaves linear to lanceolate, up to 8 x 2 mm, tip acute-setose, thin, not discolorous when dried, hydathodes inconspicuous. subsp. **schimperi**
- Robust sprawling subshrub with thicker obtuse discolorous leaves with well developed hydathodes. subsp. **phyturus**

Subsp. **schimperi**. Fig. 88.1.10-12.
C. pentandra (Edgew.)Schönl. (1890).

Moist, often shaded sites on rocks, earth banks, fallen logs etc., often in *Juniperus procera* forest, sometimes epiphytic; (1500-)2000-4200 m. EW GD GJ WU SU AR GG SD BA HA; Cameroun Mt., Zaire, Chad (Tibesti), Sudan, Kenya, Uganda, Tanzania, Zambia; tropical Arabia, India. *W. de Wilde* 7990; *Gilbert & Thulin* 459; *Mooney* 6392.

Subsp. **phyturus** *(Mildbr.) R. Fernandes* in *Bol. Soc. Brot.* Sér. 2 **52**: 172 (1978)

Habitat in Flora area not recorded; elsewhere most characteristic of basement complex rock outcrops, often seasonally moist but fully exposed. SD (Mega); Sudan (Imatongs), East Africa, Socotra. *Corradi* 7296, 7298, 7302.

A possible collection of this subspecies from EW, *Pappi* 2852, is probably better treated as a more robust form of subsp. *schimperi*. A further subspecies, subsp. *transvaalensis* (Kuntze) R. Fernandes, with long-attenuate leaves, occurs in southern Africa.

4. **C. alata** *(Viv.) Berger* (1930)
Tillaea alata Viv. (1830) - type from Egypt.

Erect slender glabrous ephemeral to 12 cm high, simple or with branches from lower nodes, all parts often reddish. Leaves sessile, lanceolate, 3-5(-7) x 0.5-1(-2) mm, tip formed into pale bristle. Flowers 3-5-merous on crowded leafy axillary shoots usually shorter than subtending leaves. Pedicels slender, up to 4 mm long. Calyx-lobes lanceolate, up to 1.5 x 0.2 mm, tip extending into white bristle. Petals lanceolate, 50-70% as long as calyx, tip acuminate, white or reddish. Seeds 2 per carpel, oblong, *c* 0.4 mm long, yellow-brown, longitudinally ribbed.

Subsp. **pharnaceoides** *(Fisch. & Mey.) Wickens & Bywater Kew Bull.* **36**: 633 (1980);
Crassula pharnaceoides Fisch. & Mey. (1841); *C. campestris* (Eckl. & Zeyh.) Walp. subsp. *pharnaceoides* (Fisch. & Mey.) Tölken, *J. S. Afr. Bot.* **41**: 99 (1975) - type: TU, near Adwa, *Schimper* I:104 (LE holo. not seen; BM K P iso.; B G GRA S iso. not seen).

Flowers 5-merous. Fig. 88.1.7-9.

Shaded bare or moss-covered earth banks; 2050-2550 m. TU SU WG GG; Cameroun, Sudan, Kenya, Uganda, Somalia. *Gilbert & Thulin* 457, 813, 1008.

Subsp. *alata* with 3-4-merous flowers occurs in the eastern Mediterranean east to NE India, south to the Red Sea Hills in Sudan. It could occur in Eritrea.

5. **C. vaginata** *Eckl. & Zeyh.* (1837)
  - type: South Africa, *Ecklon & Zeyher* 1903.
  C. schweinfurthii De Wild. (1900) - type: cultivated plant of Ethiopian origin, location not known.

Perennial or monocarpic herb with swollen fleshy or subwoody rootstock. Vegetative leaves forming stemless rosette, dying back at flowering time, similar to lower cauline leaves but slightly narrower. Flowering stems 10-40(-90) cm high, lower cauline leaves ovate to lanceolate, 2.5-14(-24) x 0.8-1.5(-2.2) cm, tip subacute to obtuse, margin shortly and densely pectinate-papillate, otherwise glabrous, basal sheath 0.8-2(-3) cm long; upper cauline leaves closer, shorter and relatively broader. Inflorescence a terminal hemispherical cyme, bracts leafy, axes subglabrous to papillose-hairy. Pedicels 2-6 mm long. Flowers 5-merous. Calyx-lobes triangular to lanceolate, 1.5-2.5 mm long, obscurely toothed. Petals oblong-elliptic to obovate, (2.5-)3-4 (-4.5) x (0.7-)1-1.5(-2) mm, white; anthers ± square, 0.3-0.5 mm long. Seeds many, 0.5 x 0.3 mm, minutely longitudinally rugulose. Fig. 88.2.B1-B3.

Montane grassland, mainly on slopes or rocky outcrops, or clearings in montane forest; (1500-)1900-3300 m. EW GD GJ WG SU GG SD BA; Nigeria, Cameroun, Sudan (Imatongs), eastern Africa south to Cape Province. *Ash* 587; *Gilbert & Thulin* 428; *Mooney* 6275.

Some authorities, e.g. Hedberg (*Afroalpine Vascular Plants*: 101, 1957) have treated this species as synonymous with *C. alba*. The separation is indeed very slight and possibly dubious but the views of the 2 most recent treatments, Fernandes (1983) and Wickens (1987), have been followed, mainly because of the apparent coexistence of the two taxa in some areas (eg Mega (SD) where *Bally* 9138 is a mixed collection within which the two elements look very distinct).

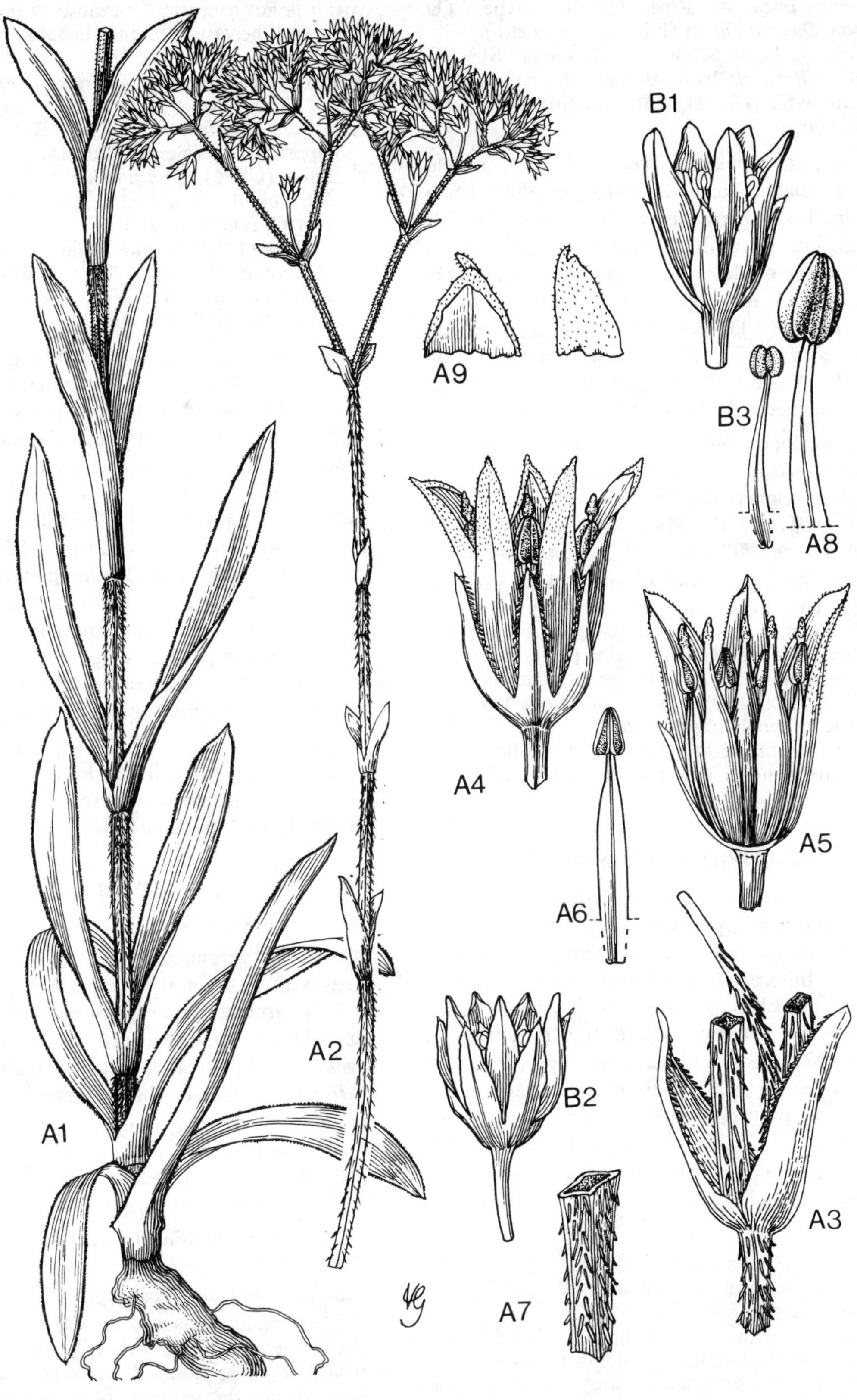

Figure 88.2. *CRASSULA* spp.: A - *C. ALBA*: A1 & A2 - habit x 4/5; A3 - base of inflorescence showing first dichotomy x 5; A4 - flower, A5 flower with nearest petals and sepals removed to show carpels x 7; A6 - stamen after dehiscence x 10; A7 - portion of stem x 5; A8 - stamen x 12; A9 - tip of petal in adaxial and side view. B - *C. VAGINATA*: B1 & B2 - flowers x 7; B3 - stamen x 12. A1 from *Balsinhas* 1343; A2-A9 from *Torre* 6693: B1 & B3 from *Chase* 1285; B2 from *Torre & Paiva* 10792. Drawn by Victoria Goaman. Reproduced by permission of Bentham Moxon Trustees from *Flora Zambesiaca*.

6. **C. alba** *Forssk.* (1775)
- type: Yemen, *Forsskål* in Herb. Forsskål 691.
C. abyssinica A. Rich. (1848) - type: TU, Scholoda, *Quartin-Dillon* (P lecto. + isolecto.).
C. ellenbeckiana Schönl. (1909) - type: SU, Mt. Zuquala, *Ellenbeck* 1665 (B holo. destr.); neotype (Wickens, 1982: 669): SU, Mt. Furi (Mt. Furé), *Ash* 3019 (K neo.).

Very similar to *C. vaginata*. Plant 10-30(-60) cm high, rather more slender, internodes usually papillose-hairy; lower cauline leaves linear-oblong to ovate, 3.5-14(-20) x 0.3-2(-2.5) cm, tip acute to acuminate, glabrous or papillose-hairy below, less often also above. Calyx-lobes 2.5-4(-6.5) x 1-1.3 mm, tip acute, toothed or ciliate-papillate. Petals (3-)3.5-8 x (1.2-)1.5-2(-2.5) mm, tip subacute to obtuse, thickened and mucronate, white or pink to red. Anthers oblong, (0.5-)0.6-0.9 mm long, often dark red. Seeds 8-30 per carpel. Fig. 88.2.A1-A9.

Dry open slopes and rock crevices in evergreen bushland or clearings amongst *Juniperus* or *Erica*, rarely wetter sites at higher altitudes; 1700-3200(-3900) m. EW TU SU GG SD BA HA; East Africa south to Cape, Arabia. *Ash* 1707; *Friis* et al. 1344; *Mooney* 6480.

A polymorphic species with distinctive local variants. Wickens (p. 669, 1982) equated collections with very narrow (3-5 mm wide), densely papillate-hairy leaves with *C. ellenbeckiana*. Rather similar plants are found scattered throughout the area and separation does not seem justified, especially as other collections from Mt. Zuquala, the original locality, are typical of *C. alba*. Plants from Hararge mostly have deep red flowers in contrast to the white or pale pink flowers usually seen in other areas.

2. HYPAGOPHYTUM *Berger* (1930).

Herb with tuber or tuberous rhizome. Stems 1-few, simple, dying back in dry season. Leaves in whorls of 3, obscurely spurred at base, with submarginal hydathodes. Inflorescence terminal, whorled at base with cymose branches. Flowers 10-12-merous. Calyx divided to below half-way. Petals free or nearly so, spreading, white. Stamens as many as petals, free. Carpels as many as petals, distinctly constricted into 2 segments, upper part spiny-papillose, style long. Mature follicle 2-seeded, breaking transversely, upper seed dispersed within upper part of carpel, lower seed released separately.

Monospecific.

A genus of most uncertain affinity. The unusual habit and high petal and sepal number are ± matched in specialised species of both *Crassula* and *Sedum*. The stamens, as many as the petals, suggest a relationship to *Crassula*, as does the occurrence of hydathodes. Conversely, the peculiar fruits are most similar to those of the Mediterranean *Sedum caeruleum* L., as are the seeds, but *S. caeruleum* follicles dehisce in a normal manner and the plant otherwise is very different - a blue flowered annual with alternate leaves.

The suggestion by Ohba (*J. Fac. Sc. Tokyo, Sec. III* **12**: 1431, 1975) that it might be included within *Sempervivum* is not acceptable because of both the distinctive fruits and the vast difference in habit.

H. abyssinicum *(A. Rich.) Berger* (1930);
Sempervivum abyssinicum A. Rich. (1848); *Sedum abyssinicum* (A.Rich.) Raym.-Hamet (1912) - types: TU, Wejerat (Ougerate), *Petit* (P syn.) & Shire (Chiré), *Quartin-Dillon* (P syn. not seen) & Mt. Selloda (Scholloda), nr. Adwa, *Schimper* I:138 (P syn.; BM K isosyn.)
Sedum malladrae Chiov. (1911); *Crassula malladrae* (Chiov.) Chiov. (1919) - types: GD, Debarek, *Chiovenda* 3042 & Lemalmo (Lumalmò), *Chiovenda* (both FT holo.).

Stems up to 26 cm long with 4-6 nodes. Leaves obovate, 10-35 x 8-25 mm, sessile or ± cuneate into petiole, not joined at base, often streaked with red. Inflorescence with basal whorl of small bracts, terminal flower and usually 3 branches up to 5 cm long, branches lax, irregularly 1-sided. Pedicels up to 17 mm long. Calyx-lobes lanceolate, 1.5-2.5 x 0.5-1 mm, tip obtuse. Petals lanceolate, 5.5-6.5(-12) x 1.3-1.5 mm, tip acuminate, white, sometimes faintly pink-tinged. Stamens about as long as petals, anthers *c* 0.3 mm long, dark. Carpels *c* 1 mm broad, laterally compressed, inner margin ciliate, sides densely spiny-papillate, style filiform, as long as stamens. Seeds cylindrical ovoid, dark brown, obscurely longitudinally ridged. Fig. 88.3.

Steep rock faces, usually seasonally wet and moss-covered, exposed or in shade; 2250-3500 m; EW TU GD SU; not known elsewhere. *Ash* 1834; *J. de Wilde & Gilbert* 126; *W. de Wilde* 8634.

3. SEDUM *L.* (1753).

Gilbert, *Bradleya* **3**: 48-52 (1985)

Annual or perennial herbs, occasionally subshrubby. Leaves alternate or in whorls of 3, rarely partly opposite in Flora area, terete or flat, usually sessile, base spurred and glabrous in Flora area. Inflorescence a terminal cyme (rarely lateral outside Flora area). Flowers 4-5 (-9)-merous. Petals shortly joined at base or free. Stamens inserted on base of petals, usually twice as many as petals. Nectaries entire, emarginate or dentate. Carpels as many as petals, base thick, fused laterally. Follicles 1-many seeded.

A large predominantly north-temperate genus with the number of species variously estimated at 300-600 depending on generic concepts.

The genus appears to occupy a core position within the family as a whole with other genera probably best regarded as specialised derivatives from it. As a consequence the genus is often defined negatively - it includes those species left after more specialised genera have been removed. The species found in Africa appear to be a group of relicts, some with very uncertain affinities.

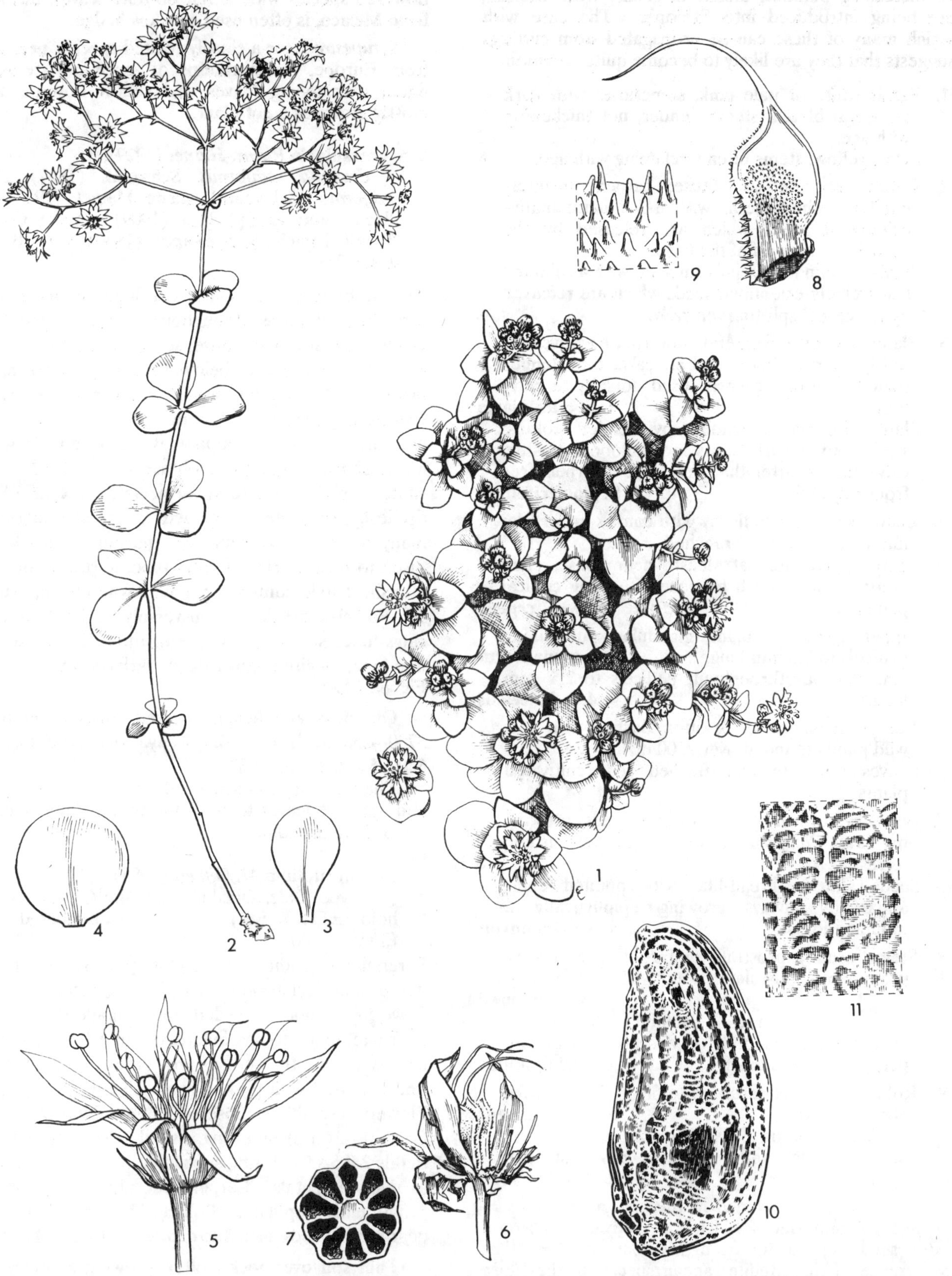

Figure 88.3. *HYPAGOPHYTUM ABYSSINICUM*: 1 - habit x 2/3; 2 - flowering stem x 2/3; 3 & 4 - leaves x *c* 1; 5 - flower, 6 - infructescence (most follicles dispersed) x 5; 7 - top view of receptacle after follicles have been dispersed x 16; 8 - follicle (seed dispersed from bottom) x 10; 9 - detail of ornamentation on side of upper part of follicle x 64; 10 - seed x 73; 11 - detail of surface of seed x *c* 240. 1 from slide of *J. de Wilde & Gilbert* 126; 2, 3, 5 & 7 from *W. de Wilde* 8634; 4, 6, 8 & 9 from *Ash* 1834; 10 and 11 from S.E.M. of *Ash* 1834. Drawn by Eleanor Catherine.

A number of species are grown as ornamentals and an increasing number, mostly originally from Mexico, are being introduced into Ethiopia. The ease with which many of these can be propagated from cuttings suggests that they are likely to become quite common.

1. Petals white or pale pink, sometimes with dark subapical blotch; stems slender, not thickening with age. 2
- Petals yellow; stems often thickening with age. 6
2. Petals subacute to obtuse; carpels minutely papillate when fresh, with 1-3 longitudinally-orientated seeds which are released by the carpel breaking off at the base. 3
- Petals acuminate; carpels smooth, with 4 to many transversely-orientated seeds which are released by the carpel splitting ventrally. 4
3. Plants strictly ephemeral, not rooting adventitiously; sterile shoots absent; calyx tube longer than free lobes; carpels fused dorsally to calyx tube. **1. S. crassularia**
- Plant perennial, spreading by slender stolons; sterile shoots with leaves in subglobose cluster; calyx tube shorter than free lobes; carpels free from calyx tube. **2. S. glomerifolium**
4. Leaves sessile, very fleshy with convex underside; inflorescences little branched, ± flat. 5
- Leaves petiolate at least near base, flat; inflorescences much branched, forming conical panicles. (see note below) *S. cepaea*
5. Upper parts of plant glandular puberulent; pedicels to 0.5 mm long. **3. S. hispanicum**
- Plant glabrous throughout; pedicels to 2.5 mm long. **4. S. baleensis**
6. Leaves in whorls of 3, often flushed with red; wild plants found at over 2500 m. 7
- Leaves alternate, not flushed red; cultivated plants. 9
7. Seeds cylindrical-ovoid with blunt ends; growing in *Erica-Philippia* zone or above, usually terrestial. 8
- Seeds almost thread-like with pointed ends; apparently always growing epiphytically in montane forest. **5. S. epidendron**
8. Stems usually prostrate and rooting at nodes, prominently pustulate; petals 7.5-9 mm long. **6. S. mooneyi**
- Stems more erect, usually not rooting adventitiously, smooth or obscurely pustulate; petals 10-14 mm long. **7. S. churchillianum**
9. Robust succulent shrub up to 50 cm high; leaves obovate, ± flat. *S. dendroideum*
- Low herb forming dense tussocks up to 10 cm high excluding inflorescences; leaves slender, cylindrical. *S.* sp. cf. *S. rupestre*

S. cepaea L. is included in the key on the basis of *Pappi* 9067, material of which was cultivated in Rome from seed supposedly from Mt. Savour, EW. The occurrence of this Mediterranean species in the Flora area needs confirmation as the record could be the result of mislabelling in the botanic garden or of contamination of a planting of other Ethiopian seed by this somewhat weedy species.

S. dendroideum DC., a robust subshrubby yellow-flowered species with ± flat obovate leaves, introduced from Mexico, is often used as a low hedge.

S. rupestre L. or a closely related species, presumably from Europe, forms clumps of ascending stems with linear overlapping leaves and yellow flowers, and is mostly grown as a pot plant.

1. **S. crassularia** *Raym.-Hamet* (1929);

Crassula sediformis Schweinf. (1867); *Sedum sediforme* (Schweinf.) Raym.-Hamet (1912) nom. illegit., non (Jacq.) Pau (1909) - type: GD, Mt Buahit ['Bachit'], *Schimper* (1850) 126 (B holo. destr., P iso.).

Annual herb up to 5(-10) cm high, stems erect or spreading; glabrous throughout. Leaves opposite towards base, alternate above, sessile, obovate, 4-5 x 2-2.5 mm, flat above, convex below, blue- or grey-green, often flushed red. Cymes usually numerous, leafy, few-flowered; pedicels 2-2.5 mm long. Flowers (3-)4-5-merous. Calyx ± campanulate, 2-3 mm long, lobes much shorter than tube, obtuse or rounded. Petals white or pink, ovate to suborbicular, *c* 2 x 1.5-1.8 mm, tip fleshy, subacute to rounded. Stamens sometimes as many as petals, anthers dark purple. Carpels erect, fused to level of rim of calyx-tube, abruptly contracted into short style, minutely papillate when fresh, ovules 1-3. Follicles breaking transversely at level of rim of calyx-tube. Seeds arranged longitudinally, ovoid, *c* 1.2 mm long, longitudinally ridged, dark brown.
Fig. 88.4.E-F.

On moss and lichen covered rocks in moorland; 3700-4350 m. GD AR BA; Kenya. *Hedberg & Getachew* 5408, 5432; *Thulin* 1657.

The tendency to have a reduced number of stamens and opposite leaves towards the base is more typical of *Crassula* than *Sedum*.

2. **S. glomerifolium** *M. Gilbert* (1985)

- type: Bale, Dinshu Flats, *Sue Gilbert* 100 (UPS holo. ETH K iso.) (sometimes distributed as *Sue Edwards* 100).

Perennial spreading by slender, pale stolons to form dense mats; glabrous throughout. Leaves of non-flowering shoots crowded into subglobose clusters, alternate, ovoid to spathulate, *c* 5 mm long, flat above, convex below, tip obtuse, grey-green often flushed or finely spotted with pink. Cymes slender, to 5 cm high. Flowers resembling those of *S. crassularia* but smaller. Calyx 0.7-1(-1.5) mm long, usually free almost to base. Petals 2-2.5 x 0.8-1 mm, subacute. Stamens always twice as many as petals. Carpels free almost to base, more prominently papillate. Follicles breaking transversely near base. Seeds as in *S. crassularia*. Fig. 88.4.A-D

Thin soil over rocks, often growing amongst moss and lichens, in *Erica-Philippia* zone, less often in damp depressions at higher altitudes, sometimes abundant along roadside ditches; 3000-3800 m. BA; not known elsewhere. *J. de Wilde* 6807; *Hedberg* 5700; *Mesfin* 643.

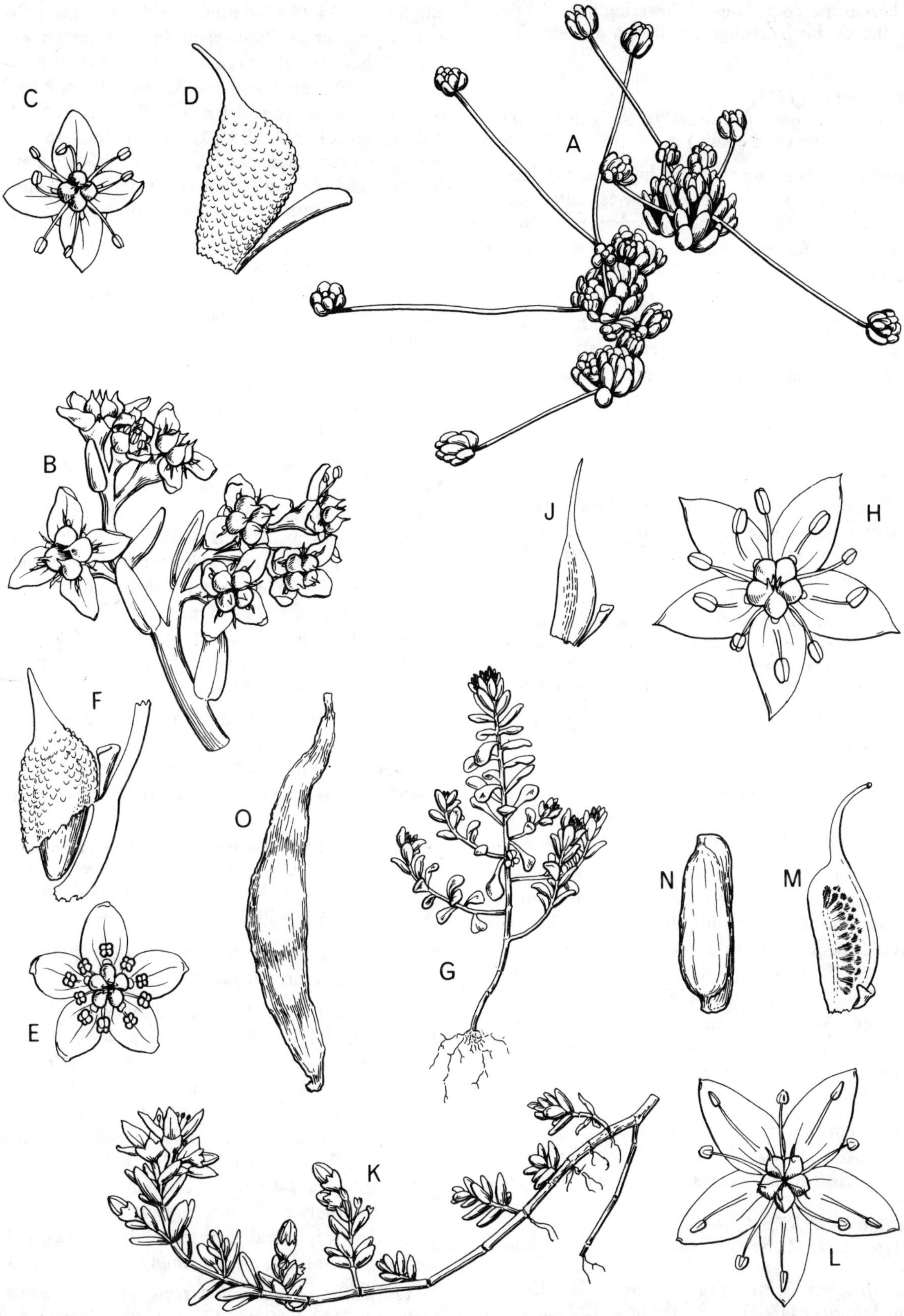

Figure 88.4. *SEDUM* spp.: *S. GLOMERIFOLIUM*: A - habit x 1; B - partial inflorescence x 4; C - flower x 6; D - carpel x 30. *S. CRASSULARIA*: E - flower x 6; F - carpel x 3. *S. BALEENSIS*: G - habit x 1; H - flower x 6; J - carpel x 10. *S. MOONEYI*: K - habit x 1; L - flower x 3; M - carpel x 6; N - seed x 24. *S. EPIDENDRON*: O - seed x 24. A from photo taken along road south from Goba (BA); B - D from *S. Gilbert (Sue Edwards)* 100; E & F from *Friis* et al. 3716; G-J from *Friis* et al. 3812; K - M from *M. Gilbert* s.n.; N from S.E.M. of *Mooney* 7119; O from S.E.M. of *Mooney* 5769 (all except N & O from spirit material). Drawn by Eleanor Catherine.

Most herbarium collections of flowering material do not show the distinctive congested leaves and slender stolons.

3. S. hispanicum *L.* (1755)
- type 'from Spain' (apparently mislabelled as the species is not known in Spain.)

Little-branched annual, up to 5 cm high, usually less, minutely glandular-puberulent. Leaves alternate, linear, semiterete, 4-18 mm long, tip subacute. Cyme unbranched or 2-3-forked, pedicels up to 0.5 mm long. Flowers 5-merous. Sepals triangular, 1-1.5 mm. Petals white with reddish midrib, lanceolate, 4-6 mm long, acuminate. Stamens subequal to petals, anthers dark. Follicles 2-3 mm long, erect, many-seeded, styles (0.5-) 0.7-1 mm long. Seeds ovoid with one end pointed, *c* 0.4 mm long, longitudinally ridged, dark brown.

Winter rainfall area of the eastern escarpment, habitat details not recorded; *c* 1400 m. EW; (species as a whole: Arabia, SE Europe, SW Asia east to India). *Baldrati* 3508; *Pappi* 5889; *Schweinfurth & Riva* 2032.

The above description applies only to Ethiopian collections which are of a distinctive erect, often completely unbranched, ephemeral with 5-merous flowers and styles 0.7 mm long. The species as a whole also includes biennials and perennials, glabrous plants and plants with 5-7(-9)-merous flowers. Several varieties have been described, none of which correspond with the Ethiopian material and a more detailed revision of this polymorphic species would probably lead to the formal recognition of the Ethiopian plants as a distinct taxon.

4. S. baleensis *M. Gilbert* (1985)
- type: Bale, 66 km N of Menna on road to Goba, *Friis* et al. 3812 (K holo. C ETH UPS iso.).

Related to *S. hispanicum*, differing as follows: Leaves at base of stem opposite, spathulate when dried, up to 5.5 x 1.5 mm, flushed deep pink. Inflorescence lax and few flowered, pedicels up to 2.5 mm long. Flowers 5-merous. Calyx 1.2 mm long, lobes triangular, joined for *c* 0.3 mm. Petals white or pale pink, sometimes midrib darker, ovate, 3-3.5 x 2 mm, markedly acuminate. Carpels erect, *c* 1 mm long tapered into slender style *c* 1.2 mm long, transversely 3-8 seeded. Mature seeds not seen. Fig. 88.4.G-J

On moss-covered rocks at forest margin, 'grassy slope'; 2700-3400 m. BA ?HA; not known elsewhere. *Gilbert & Ermias* 8480; *Mooney* 8384; *Thulin* et al. 3714.

5. S. epidendron *A. Rich.* (1848)
- type: GD, Mt Selki, *Schimper* II:1291 (P holo. K iso.).
S. schimperi Britten (1871) - type: GD, Debre Tabor, *Schimper* (1863) 1469 (BM holo. K iso.).

Epiphytic subshrub, to 15 cm high, stems thickening with age, smooth or obscurely papillate when young, often pendent except for inflorescence; glabrous throughout. Leaves in whorls of 3, narrowly oblong-elliptic, flat, 14-35 x 3-6 mm, tip subacute, often flushed red. Cymes erect, flat-topped, up to 4 cm across, with several short few-flowered branches; pedicel up to 5-8 mm long. Flowers 5-merous. Calyx 2-3 mm long, lobes triangular, acute, joined for 0.5-0.8 mm. Petals bright yellow, lanceolate, 4.5-8 x *c* 1.5 mm, acuminate. Stamens slightly shorter than petals. Carpel 4.5-5.5 mm long, ventrally gibbous, with many seeds, style 1-1.2 mm long. Seeds slender fusiform, pointed at both ends, *c* 2.5 mm long, dark brown. Fig. 88.4.O

Epiphytic on trunks and larger branches of trees in *Juniperus* and *Hagenia* forest; 2500-3000 m. GD GJ SU AR; not known elsewhere. *W. de Wilde* 9076; *Friis* et al. 1261; *Thulin* 1633.

6. S. mooneyi *M. Gilbert* (1985)
- type: BA, Angasu, near Goba, *Mooney* 7156 (K holo., ETH iso.).

Closely related to *S. epidendron*, differing as follows: Plants usually prostrate and rooting at nodes, more densely branched, stems very densely papillate when young. Leaves more congested, 7-12 x 3-5 mm, more fleshy with convex lower side, often flushed dark red. Cymes usually smaller, rather few-flowered; pedicels up to 5.5-6.5 long. Calyx 2-3.5 mm long. Petals 7.5-9 x 2.5-3 mm, often flushed red at tip; style 1.5-2 mm long. Seeds oblong-ovoid, subacute at one end only, 0.7-0.8 mm long. Fig. 88.4.K-M

In rock crevices or on moss-covered rocks among *Erica* and *Philippia* in afro-alpine moorland, mostly in exposed situations; 3400-4100 m. AR BA; not known elsewhere. *W. de Wilde* 9096; *Hedberg* 4222; *Mooney* 8319.

Formerly confused with the East African species *S. ruwenzoriense* Bak. and *S. meyeri-johannis* Engl. but these always have some leaves alternate and smaller flowers with free sepals.

7. S. churchillianum *Robyns & Boutique* (1945)
- type: Zaire, *Bequaert* 4511.

Closely related to *S. epidendron* and *S. mooneyi*, differing as follows: Stems up to at least 24 cm long, usually erect, apparently not rooting adventitiously, smooth or obscurely papillate. Leaves up to 25(-30) x 8(-14) mm, tip obtuse. Cymes up to 8 cm across; pedicels 10-17 mm long. Calyx lobes lanceolate-triangular, 4.5-6.5 x 2.5 mm, joined for *c* 1 mm. Petals lemon yellow, 10-13 x 4-4.5 mm. Stamens just shorter than petals. Carpels c 6.5 mm long, styles *c* 2.5 mm long. Seeds narrowly ovoid, pointed at both ends, *c* 1 mm long, reddish-brown, longitudinally ridged. Fig. 88.4.N.

On cliffs or rock outcrops amongst *Erica* and *Philippia*; 3150-4050 m. SU SD BA; Uganda & Zaire (Ruwenzori Mts). *Gilbert & Tewolde* 3226; *Mooney* 7119; *Puff* et al. 820930-2/7.

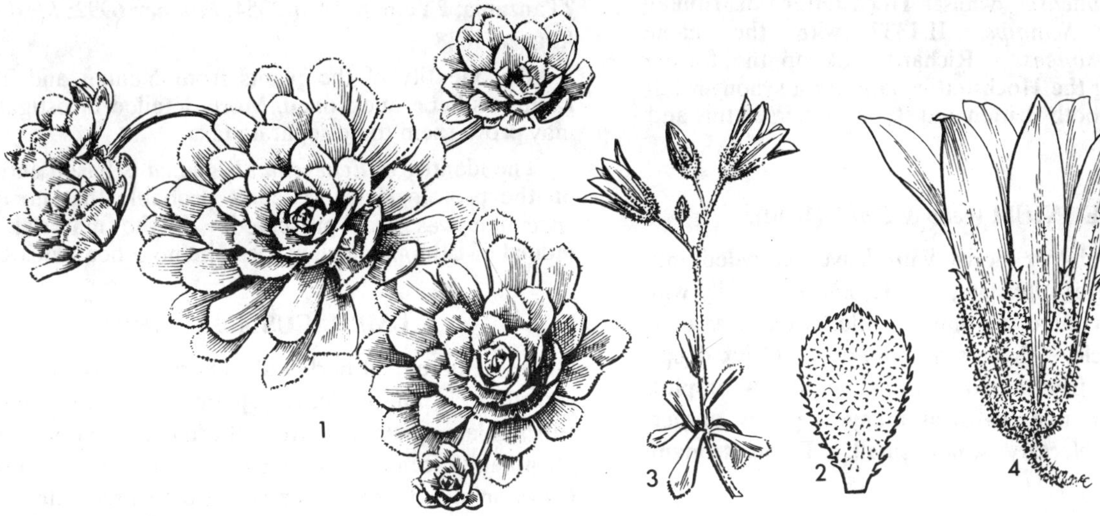

Figure 88.5. *ROSULARIA SEMIENSIS*: 1 - habit x c 2/3; 2 - leaf x 2; 3 - inflorescence x c 2/3; 4 - flower x 2. 1 from photo taken in Jinbar valley, Simen Mts. Nat. Park, 2 from *Scott* 278, 3 from drawing by Urs Eggli, 4 from *Schimper* II:1337. Drawn by Eleanor Catherine.

Imperfectly known species:

S. erlangerianum *Engl.* (1907)
- type: BA, Ladjo (7°10´N 39°40´E), *Ellenbeck* 1888 (B holo. destr.).

Stems to 1 m. Leaves lanceolate, 25-30 x 5-6 mm, obtuse. Cyme flat-topped, few-flowered; pedicels 3-4 mm long. Calyx 4.5 mm long. Petals yellow, ovate-lanceolate, 10 x 4 mm, acute. Filaments 6 mm long. Ovary 4 mm long, style 2 mm long.

Known only from the type lost in Berlin in 1943.

The description is based entirely on the protologue. The overall dimensions approach those of *S. churchillianum* but there is a marked discrepency in pedicel length and important details of stem texture and leaf arrangement are not given. Both *S. mooneyi* and *S. churchillianum* are recorded from near the type locality but nothing fitting the description exactly has been seen. It is possible that there was a mistake in the original description but there is no way of proving this and thus the species must be regarded as imperfectly known.

4. **ROSULARIA** *(DC.) Stapf* (1923).
Afrovivella Berger (1930).

Ohba, *J. Fac. Sci. Univ. Tokyo, Bot.*, **12**: 139-198 (1975); Eggli, *Bradleya* (in press).

Perennial herbs, often stoloniferous, with leaves crowded into rosettes. Leaves alternate, sessile, flat, entire but often ciliate along margin. Inflorescence usually axillary but shoots sometimes producing a large terminal panicle, branches cymose with reduced cauline leaves. Flowers 5-8-merous. Calyx-lobes deeply divided, fleshy. Corolla campanulate, lobes joined for 1/6 to 2/5 of length, suberect, pink, purplish, yellow or white, sometimes patterned. Stamens twice as many as petals, inserted on corolla tube. Nectary scales oblong to square. Carpels many-seeded.

About 30 species from Morocco through the northern Mediterranean to the Himalayas and central Asia. Only one species known from Tropical Africa.

R. semiensis *(A. Rich.) Ohba* (1978)
Umbilicus semiensis A. Rich. (1848) - type: GD, Semien, Mt. Buahit ['Bachit'], *Schimper* II:1337 (P holo.; BM K iso.; BERN G L NEU W iso. not seen).
Cotyledon simensis Britten (1871); *Afrovivella simensis* (Britten) Berger (1930) - type as above (BM holo.).
Sempervivum coccinolepis Steudner *nom. nud.*

Stems short and thick, producing conspicuous red stolons. Leaf-rosettes very crowded, pale green, 3-4 cm diameter, leaves obovate to spathulate, *c* 15-20 x 6-13 mm, tip cuspidate, margin conspicuously white-pectinate. Inflorescence ± pendent, up to 5 cm long, 1-8-flowered, axes glandular-hairy; pedicels 7-10 mm long. Flowers (5-)6(-7)-merous. Calyx-lobes oblong-lanceolate, *c* 8 x 1.4 mm, tip with bristles, hairy. Corolla-lobes joined for *c* 2/5 of length, *c* 13-14 x 3-3.5 mm, tip obtuse-mucronate, pink. Stamens *c* 7 mm long, anthers 0.7 mm long. Follicles slender, 10-12 mm long. Seeds pointed at one end, 1-1.3 mm long, longitudinally ridged, brown. Fig. 88.5.

Shaded, vertical moss-covered rocks. 3600-4200 m. GD (Semien Mts.); not known elsewhere. *Lemma* 684; *Pichi-Sermolli* 2639; *Scott* 285.

A very striking but little-collected endemic sometimes placed within the monotypic genus *Afrovivella* because of the distinctive large campanulate corolla. The quality of the illustration has been limited by the lack of good herbarium material.

There has been confusion as to what the correct epithet for this species is. Gay used the manuscript name

Umbilicus semiensis whilst Hochstetter distributed exsiccata of *Schimper* II:1337 with the name *Sempervivum simense*. Richard took up the former name, treating the Hochstetter name as a synonym but unfortunately both Britten and Berger reversed this and used the latter.

5. AEONIUM *Webb & Berth.* (1840).

Perennial shrubs or herbs with leaves crowded into rosettes. Leaves alternate, sessile, entire, usually with ciliate or pectinate margin. Inflorescence a large terminal panicle. Flowers 6-12-merous. Calyx deeply lobed. Petals free, narrow, spreading, yellow or pink, rarely red. Stamens twice as many as petals, anthers subcircular. Nectary scales small. Follicles many-seeded.

About 39 species, mostly restricted to the Canary Islands and Madeira with only 1 species in Morocco plus the following in tropical Africa and Arabia.

Closely related to the Eurasian genus *Sempervivum* but usually kept separate because of the very markedly different habit of nearly all species - *Sempervivum* always forms mats of sessile, stoloniferous rosettes.

Aeonium arboreum (L.) Webb & Berth., especially the form with dark purplish, almost black, leaves (var. *atropurpurea* (Nich.) Berger), is quite widely grown. It differs from the native species most obviously by the more robust erect habit.

A. leucoblepharum *A. Rich.* (1848);

Sempervivum chrysanthum Britten (1871); *S. chrysanthum* var. *glabrum* Chiov., *Nuov. Giorn. Bot. Ital.* **26**: 54 (1919) - types: TU, Wejerat (Ouogerate), *Quartin-Dillon* (P syn.; ?K isosyn.) & GD, Mt. Aber, *Schimper* II:838 (P syn.; K isosyn.).

Echeveria abyssinica Otto (1873) - type: based on cultivated material said to have been propagated from plants introduced from Ethiopia by a Major Leveson, probably no material preserved.

S. chrysanthum var. *glandulosum* Chiov., *loc. cit.* (1919) - type: EW, Soira, *Dainelli & Marinelli* (FT holo.).

Shrub, always sprawling or hanging, with ascending branches; stems thick with prominent leaf-scars. Leaf-rosettes very tight in dry season, more lax during rains and prior to flowering; leaves sessile, obovate to spathulate, (1.5-)2.5-9(-11) x 1.5-2.5(-5) cm, base cuneate, tip obtuse or cuspidate, glabrous or minutely glandular-puberulous, margin conspicuously ciliate-pectinate. Inflorescence hemispherical to conical, 8-27 x 8-20 cm overall, peduncle leafy, glabrous to glandular-pubescent. Pedicels up to 7 mm long. Flowers 8-10-merous. Calyx-lobes 2-3.5 mm long. Petals lanceolate, 6-9 x 1.5-2 mm, tip acute, bright yellow. Seeds oblong-clavate, *c* 0.8 mm long, striate. Fig. 88.6.

Usually on steep to vertical banks or rock-faces, occasionally epiphytic in *Juniperus* forest; 2000-3350 m. EW GD GJ SU HA; Somalia, Kenya, Uganda, ?Tanzania; ?Yemen. *Ash* 2354; *Mooney* 6992; *Mesfin & Kagnew* 1638.

The identity of the plants from Yemen and Tanzania has been questioned. More detailed investigation may prove them to be separate taxa.

The identity of *Echeveria abyssinica* is based entirely on the protologue as it seems probable that no type specimen was ever preserved. If the material was indeed of Ethiopian origin it can be no other species.

6. UMBILICUS *A. DC.* (1801).

Glabrous perennial herb with tuberous or rhizomatous rootstock. Basal leaves petiolate, often peltate, subcircular; cauline leaves alternate. Inflorescence terminal, racemose, bracteate. Flowers 5-merous. Calyx small. Corolla campanulate to urceolate, lobes short, ± erect. Stamens (5-)10, short, inserted on corolla. Nectary scales narrow oblong. Follicles many-seeded.

About 10 species in the Atlantic Islands, North Africa, Europe and W Asia with 2 species in Tropical Africa.

U. paniculiformis Wickens, from the Red Sea Hills in Sudan, could well occur in northern Eritrea; it is distinguished by the laxer, branched inflorescence and more deeply divided corolla.

U. botryoides *A. Rich.* (1848)

- types: TU, Chiré (Shire), *Quartin-Dillon* (P syn.) & GD, Mt. Silke, *Schimper* II:1300 (P syn.; K isosyn.).

Cotyledon umbilicus auct. non L.

Rootstock globose to cylindrical, up to 4 cm diam., covered by fibrous roots. Basal leaves peltate, petiole up to 13 cm long, leaf-blade up to 9 cm diameter, margin crenate to sinuate. Raceme erect, up to 30 cm high overall, unbranched, rarely with short lateral branches, cauline leaves intergrading with bracts. Flowers horizontal or pendulous; calyx-lobes ovate to oblong-lanceolate, 1.5-3 x 0.7-1(-2) mm. Corolla urceolate, 5.5-8 x 4-5 mm overall, lobes ovate, acuminate, 2-3 mm long, greenish-white, sometimes flushed purple. Stamens: filaments *c* 0.5 mm, anthers 0.5 mm long. Fig. 88.7.5-8.

Well-drained, moist, shady sites on rocks or earth banks, frequently epiphytic in forest with *Juniperus* and/or *Hagenia*, less often in grassland at higher altitudes; (1350-)2400-4150 m. EW TU GD GJ SU AR BA HA; Egypt, Sudan, Cameroun, Zaire, Somalia, Kenya, Uganda, Tanzania. *Ash* 586; *Mooney* 6974, 8329.

Closely related to the European *U. rupestris* (Salisb.) Riddelsd. which has a narrower corolla.

7. COTYLEDON *L.* (1753).

Shrubs, rarely herbs. Leaves opposite, entire, often with white waxy coat. Inflorescence a terminal cyme. Flowers 5-merous, pendulous. Calyx deeply lobed.

Figure 88.6. *AEONIUM LEUCOBLEPHARUM*: 1 - leafy shoot, 2 - flowering shoot x 2/3; 3 - section of flower x 4; 4 - sepal, 5 - petal, 6 - stamen, 7 - nectary, 8 - carpel x 5; 9 - seed x 24. 1, 3 - 7 from *Bally & Melville* 16213 (Somalia); 2 from *Ash* 2354; 8 & 9 from *Gillett* 13894 (Kenya). Drawn by Eleanor Catherine. (Reproduced with permission from *Fl. Trop. E. Af.*, Crassulaceae, 1987.)

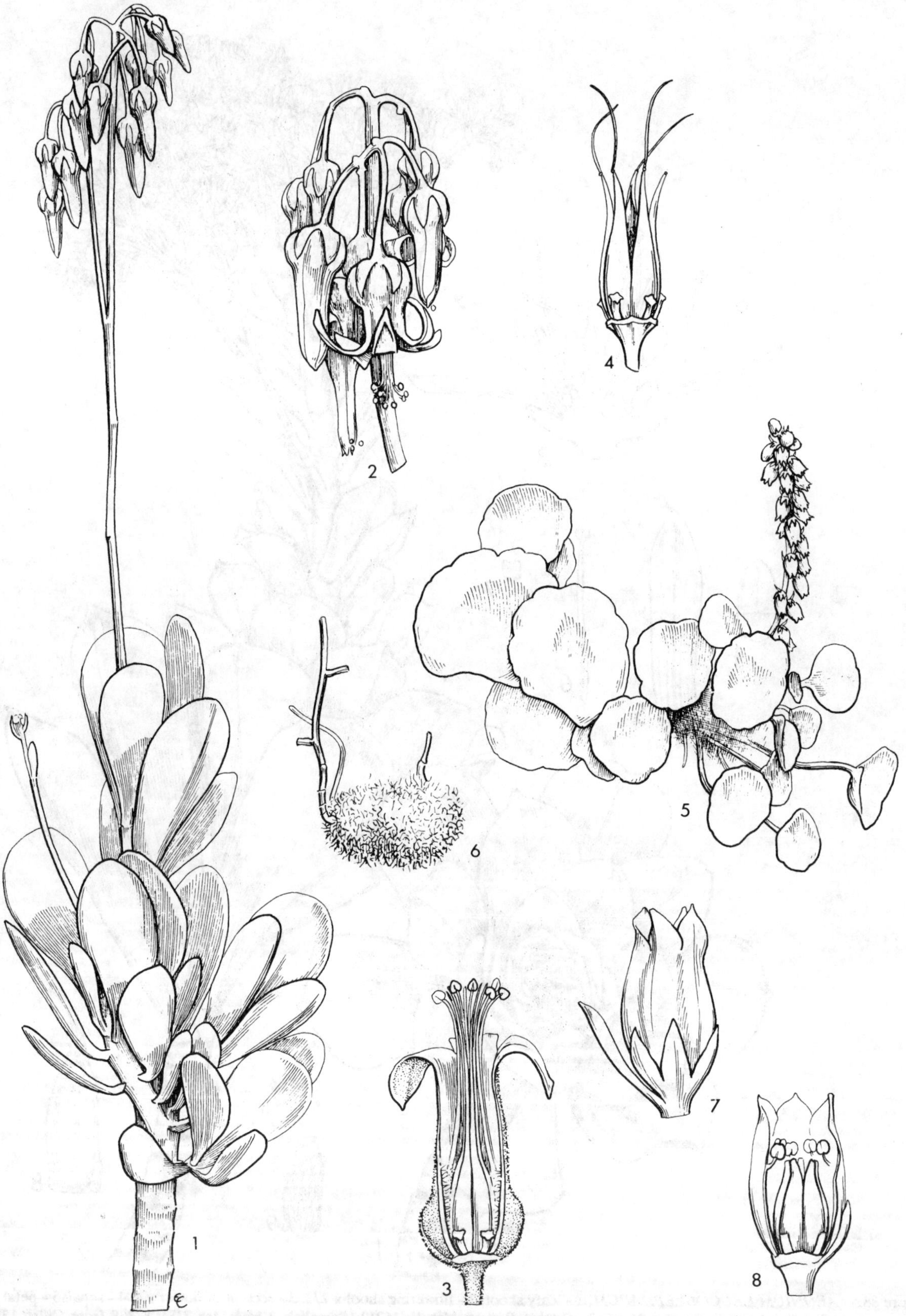

Figure 88.7. *COTYLEDON BARBEYI*: 1 - flowering branch x *c* 1/2; 2 - inflorescence x *c* 1; 3 - flower, front sepals and petal removed, 4 - fruit x 1.5. *UMBILICUS BOTRYOIDES*: 5 - habit, 6 - rootstock x *c* 1; 7 - bract and flower, 8 - flower with front sepals and part of corolla removed x 5. 1 & 2 from photograph taken near Welenchiti (SU); 3 from *Bally* 10234; 4 from *Gillett* 4571; 5 from photograph taken near Dinshu (BA); 6 - 8 from *Mooney* 8324. Drawn by Eleanor Catherine.

Corolla tubular or urceolate, bright red or yellow, lobes twisted in bud, spreading or recurved. Stamens (5-)10, inserted on corolla tube, exserted, anthers oblong. Nectary scale linear to square. Follicles many seeded.

Genus of 20-30 species in South Africa with 1 species extending north to Ethiopia and Arabia.

C. barbeyi *Bak.* (1893)
- lectotype: Yemen, *Schweinfurth* 1493.
C. wickensii Schönl. (1915).
?*C. wickensii* var. *glandulosa* Poellnitz in Feddes Rep. **42**:37 (1937).

Erect soft-wooded shrub up to 2(-4) m high, very variable in stature. Leaves sessile, obovate, 4.5-10 x 2-6 cm, base cuneate, tip obtuse to shortly acuminate, glabrous, usually very glaucous grey-green. Cyme glandular-pubescent, on stout peduncle to 50 cm long, subglobose, irregularly branched; pedicels up to 4 cm long. Flowers at first erect, later pendulous. Calyx-lobes ovate-triangular, 5.5-7 x 3-4 mm, acute, joined for *c* 2 mm, reflexed and eventually basally circumcissile. Corolla-tube 8-13(-15) mm long, swollen at base, lobes oblong-oblanceolate, 13-17 x 3-5 mm, acute, reflexed, bright red, less often yellow. Stamens hairy at base, slightly shorter than lobes, straight, anthers *c* 2 mm long. Seeds oblong-ovoid, *c* 1 mm long, longitudinally striate. Fig. 88.7.1-4.

Deciduous bushland with *Acacia & Commiphora*, on termite mounds and lava flows; 1350-2200(?-2600) m. EW SU SD; Yemen, Sudan, Somalia, East Africa, Zimbabwe, South Africa (Transvaal). *Ash* 2284; *Mooney* 5523; *Schweinfurth & Riva* 2203.

This species is rather uniform in the northern part of its range but in the south Fernandes (1983) regards it as intergrading with *C. wickensii* Schönl., with narrow, almost terete, leaves, and possibly also *C. wickensii* var. *glandulosa* Poellnitz which has very densely hairy leaves.

8. KALANCHOE *Adanson* (1763).

Cufodontis, *Webbia* **19**: 711-744 (1965); Raadts, *Willdenowia* **8**: 101-158 (1977).

Perennial or monocarpic herbs or subshrubs. Leaves opposite, mostly simple, occasionally deeply divided to pinnately compound, often crenate or serrate. Inflorescence terminal, branches cymose, one-sided with flowers in 2 rows ('cincinnal monochasia'). Flowers erect, (3-)4(-5)-merous. Calyx usually much shorter than corolla-tube, lobes very shortly joined at base to almost free. Corolla-tube swollen towards base, narrow above, lobes less than half as long as tube, spreading or reflexed, often mucronate. Stamens twice as many as corolla-lobes, inserted high on corolla-tube. Nectary scales narrow oblong. Carpels articulated at junction with style, many seeded. Seeds small, longitudinally ribbed.

About 60 species, mostly in Africa and Madagascar, extending into tropical Arabia and Asia east to the Philippines, 1 species extending into tropical America.

Recent cytotaxonomic studies (Raadts, unpublished) show the existence of polyploidy which may account for some of the problems in delineating taxa in this genus.

1. Corolla tube 3-12 cm long, lobes pure white. 2
- Corolla tube 0.7-3 cm long, lobes yellow, orange, red or pink, rarely if ever pure white. 4
2. Plant glabrous throughout; calyx up to 20 mm long. 3
- Plant hairy, at least on inflorescence; calyx 25-40 mm long. **1. K. schimperiana**
3. Leaves sessile, glaucous, often marked with dark blotches; corolla-tube (4.5-)6-12 cm long. **2. K. marmorata**
- Leaves petiolate, unmarked; corolla-tube 3-5 cm long. **3. K. quartiniana**
4. Corolla-tube 15-30 mm long; anthers included or exserted. 5
- Corolla-tube 7-14(-15.5) long; anthers included. 7
5. Inflorescence glabrous to densely short-white-pubescent; anthers exserted. 6
- Inflorescence densely brown-tomentose; anthers included. **5. K.** sp. = *Buscalioni* 1278
6. Leaves dark green, often flushed dark red; corolla pale pink to almost white; found at over 2000 m alt., common. **4. K. petitiana**
- Leaves grey-green, often with dark marks near margins on underside; corolla pale yellow, rarely pinkish; found below 2000 m alt., little collected. **6. K. prittwitzii**
7. Plant glabrous throughout or very sparsely glandular-puberulent on calyx only. 8
- Inflorescence and often entire plant densely pubescent. 12
8. Glaucous perennials with swollen roots branching from base to form clumps of stems. 9
- Plants monocarpic or nearly so - dying or producing only occasional weak shoots from base after flowering; stems and leaves not glaucous. 11
9. Leaves lanceolate to ovate. 10
- Leaves linear. **7. K. angustifolia**
10. Inflorescence compact, pedicels rarely more than 5 mm long; corolla-tube (5-)8-12 mm long, limb clear yellow, orange, red or pink. **8. K. glaucescens**
- Inflorescence lax, some pedicels over 10 mm long; corolla-tube *c* 15 mm long, limb yellow-green. **9. K. stenosiphon**
11. Pedicels thickened below flowers. **11. K. densiflora**
- Pedicels uniformly slender. (see note below) *K. crenata*
12. Indumentum mostly glandular, often restricted to inflorescence; leaves mostly glabrous, entire or lobed, sometimes bluntly serrate or crenate but not sharply serrate. 13
- Indumentum non-glandular; leaves densely hairy throughout, sharply serrate or dentate. **10. K. citrina**

13. Leaves sessile, undivided; stem typically 4-angled near base. **12. K. lanceolata**
- Leaves petiolate, at least some prominently lobed to pinnately 3-5-foliolate; stem always rounded. **13. K. laciniata**

The almost pantropical *K. crenata* (Andr.) Haw. is in Sudan just outside the Flora area and can be expected to be found, most likely, in the west.

1. **K. schimperiana** *A. Rich.* (1848).
- type: TU, Mt. Semayata (Mt Semajata), nr. Adwa, *Schimper* I:388 (P holo.; BM K iso.; B G FT M W iso. not seen.)

Perennial subshrub up to 1 m high, stems often decumbent at base; all parts glandular pubescent. Leaves ovate to suborbicular, up to 14 x 13 cm, base abruptly contracted into petiole up to 3 cm long, tip obtuse, margin crenate to almost serrate, pale green. Inflorescence to 20 cm long. Pedicels 5-20 mm long. Calyx lobes lanceolate, 25-40 x 3-6 mm, almost free. Corolla pure white, glandular-pubesent, tube 5-6.5 cm long, lobes ovate, 15-25 x 8-16 mm, acuminate with 2 mm mucro. Upper anthers partly or completely exserted. Follicles 18-22 mm long; styles 32-40 mm long. Fig. 88.8.7.

Open rocky places, often at cliff tops or margins of thickets; 2400-2700 m. EW TU GD SU BA; not known elsewhere. *Ash* 1643; *W. de Wilde* 8640; *Schimper* (1862) 704.

Raadts (p. 148, 1977) placed collections from the Uluguru Mts., Tanzania, including the type of *K. schliebenii* Werderm., within this species. The collections differ from Ethiopian material by the much sparser indumentum - some plants are glabrous, and the smaller sepals, *c* 10 mm long.

2. **K. marmorata** *Bak.* (1892)
- type: ex hort. La Mortola, coll. *Penzig* in EW, Mt Lalamba, nr. Keren (K holo.).

K. grandiflora A. Rich. (1848) *nom. illegit.*, non Wight & Arn., (1834); *K. macrantha* Bak. (1892) - type: ?SU, 'Choa'[1], *Petit* (P holo.).

K. somaliensis Bak. (1892).

K. kelleriana Schinz (1896).

K. grandiflora A. Rich. var. *angustipetala* Engl., *Ann. Ist. Bot. Roma* 9: 251 (1902) - type: HA, Harar, *Robecchi-Briccheti* 74 (B lecto.; FT isolecto., selected by Raadts).

Perennial subshrub up to 1.2 m high, stems decumbent at base; all parts glabrous. Leaves sessile or nearly so, obovate, up to 20(-25) x 13 cm, glaucous grey-green, often with large purple spots on both sides, base cuneate to semiamplexicaule, tip rounded, margin crenate or serrate. Inflorescence to 18 cm long. Pedicels *c* 15(-25) mm long. Calyx-lobes linear-triangular, 10-17 x 3-6 mm, joined for 1-2 mm. Corolla pure white, rarely flushed very pale pink, tube (4.5-)6-12 cm long, lobes lanceolate to ovate, 10-25 x 6-13 mm, acute with *c* 2 mm mucro. Upper anthers partly exserted. Follicles 15-30 mm long; styles 30-80 mm long. Fig. 88.8.1&2.

Open rocky sites, grassland or clearings in evergreen bushland or *Juniperus* forest, often on heavy black soils; (1340-)1900-2500 m. EW TU SU SD BA HA; Sudan, Somalia, south to Rwanda, Zaire & N Tanzania; naturalised in New Zealand. *Ash* 636; *Mooney* 5300, 8181.

Often cultivated both as an ornamental and as a medicinal plant. This has obscured the distribution in E Africa where it is possibly largely introduced.

3. **K. quartiniana** *A. Rich.* (1848)
- type: TU, Mai Guagua(Mai Gouagoua), nr Adwa, *Quartin-Dillon* (P holo. + iso.).

Apparently closely related to *K. petitiana* and probably of similar habit. All parts glabrous. Flowers larger; corolla apparently pure white, tube 3-5 cm long, lobes *c* 10 x 7 mm with 1 mm mucro. Follicles 16-19 mm long, styles 14-18 mm long. Fig. 88.8.3.

No information on habitat; 2150 m. TU GD; not known elsewhere. *Lemma* 665; *Schimper* (1862) 953; *Scott* 290.

A very poorly collected species, well preserved and annotated specimens are needed.

4. **K. petitiana** *A. Rich.* (1848)
- type: WU, Sanka in prov. Yedjou, *Petit* (P holo.).

K. integra sensu Cufodontis (1965), *non* (Forssk.) Asch. & Schweinf.

Perennial, forming dense stands, stems decumbent at base, up to 3 cm thick, *c* 70 cm high. Petiole 2.5-4 cm long; leaf-blade broadly elliptic to ovate, 6-16 x 4-11 cm, base abruptly cuneate to subcordate, tip rounded, margin irregularly crenate to dentate. Inflorescence (0.6-)1.2-1.5(-2) m high overall, ± flat-topped, branches long, flowers terminally congested. Pedicels 5-10 mm long. Calyx-lobes linear-triangular, 7-11 x 1.3-2 mm, joined for *c* 1 mm, basally circumcissile in fruit. Corolla magenta-pink to white with pink flush, tube 14-20 mm long, lobes 6-7 x 3.4-4.5 mm, obtuse with *c* 1 mm mucro. Upper anthers exserted. Follicles 7-12 mm long; styles slightly shorter.

1. Plant glabrous throughout. var. **petitiana**
- Plant shortly glandular pubescent throughout, usually densely so. var. **neumannii**

Var. **petitiana**;

K. quartiniana A. Rich. var. *micrantha* Pampanini, *Bull. Soc. Bot. Ital.* **1909**: 54 (1909) - type: EW, Embaloca, nr. Mai Atal, *Dainelli & Marinelli* (FT holo.).

Fig. 88.8.11.

1. On the only collection in the Richard herbarium, the protologue gives 'Ifat' (GD).

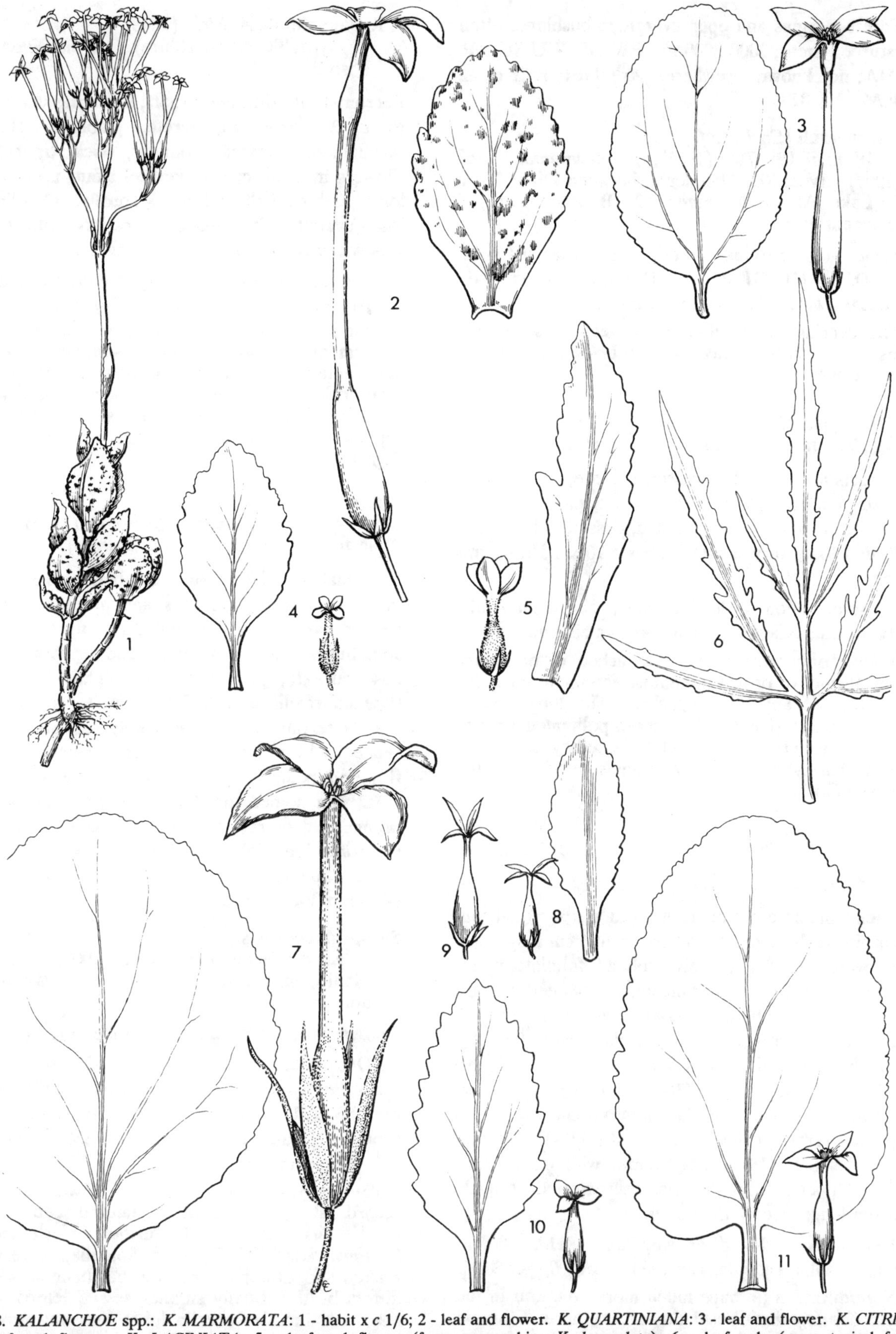

Figure 88.8. *KALANCHOE* spp.: *K. MARMORATA*: 1 - habit x *c* 1/6; 2 - leaf and flower. *K. QUARTINIANA*: 3 - leaf and flower. *K. CITRINA*: 4 - leaf and flower. *K. LACINIATA*: 5 - leaf and flower (form approaching *K. lanceolata*); 6 - leaf only (more typical form). *K. SCHIMPERIANA*: 7 - leaf and flower. *K. GLAUCESCENS* subsp. *GLAUCESCENS*: 8 - leaf and flower. *K. STENOSIPHON*: 9 - flower only. *K. DENSIFLORA* var. *SUBPILOSA*: 10 - leaf and flower. *K. PETITIANA* var. *PETITIANA*: 11 - leaf and flower. All leaves x 4/9, all flowers x 1. 1 from photo taken near Dodola (BA); 2 from *Perdue* 6375; 3 from *Scott* 290; 4 from *Gillett* 14319; 5 from *Friis* et al. 523; 6 from *Greenway* H16/30/14;7 from H885/33, cultivated at Kew; 8 from *Ash* 2263; 9 from *Tewolde Berhan* 1109; 10 from *Ash* 642; 11 from *Mooney* 8291. Drawn by Eleanor Catherine.

Forest margins and open evergreen bushland, often in disturbed areas; 2000-3000 m. EW GD WU SU AR BA HA; not known elsewhere. *Ash* 1761; *Friis* et al. 1194; *Mooney* 8291.

Var. **neumannii** *(Engl.) Cuf.*
Webbia **19**: 716 (1965); *K. neumannii* Engl. (1907) - type: SD, Uatederra, between Djam Djam & Lake Abera, *Neumann* 82 (B holo. destr., P fragment of type).

Montane forest margins, sometimes in rather wet sites; 2400-3000 m. GD GJ SU GG SD BA; not known elsewhere. *Ash* 2417; *Chaffey* 355; *Evans* 475.

The overlap in distribution between these 2 taxa seems small and they may prove to be better regarded as subspecies.

5. **K.** sp. = *Buscalioni* 1278
'*K. cyanonilotica*' Chiov. nom. nud.

Basal parts not seen. Inflorescence dense, *c* 6 cm wide, all parts densely brown-tomentose. Sepals *c* 7 x 1.5 mm. Corolla colour not recorded, possibly white; tube 25-30 mm long, very slender; lobes *c* 10 x 2.5 mm, tips acute.

No information on habitat. SU ('road from Addis Abeba to Blue Nile Gorge'); not known elsewhere.

Known only from a single detached inflorescence with a distinctive brown-tomentose indumentum quite unlike that of any other species. The long slender corolla suggests that this is a moth-pollinated species with white flowers. The epithet proposed by Chiovenda suggests that the collection was made somewhere in the Blue Nile Gorge.

6. **K. prittwitzii** *Engl.* (1907)
- type: Tanzania, *von Prittwitz & Gaffron* 267.
K. lugardii Bullock (1932).

Perennial, stems erect or decumbent at base; glabrous and pruinose throughout. Petiole up to 6 cm long; leaf-blade ovate to obovate, rarely 3-lobed/ -foliolate, up to 20(-25) x 10(-13) cm, base cuneate, tip rounded, margin coarsely crenate to bluntly serrate, sometimes with dark spots near margin on underside. Inflorescence 0.4-1.5(-2.3) m high overall, relatively lax. Pedicels 5-15(-20) mm long. Calyx lobes 6-13(-22) x (1.5-)2-3(-4) mm, joined for 0.5-2 mm. Corolla yellow, cream with yellow centre or, rarely, yellowish-pink, tube (13-)15-28 mm long, lobes 6-15(-18) x 3-6(-7) mm with 0.5-1.5 mm mucro. Upper anthers sometimes ± exserted. Follicles 7-14 mm long; styles 4-13 mm long.

Deciduous bushland or woodland. HA; Sudan, Kenya, Somalia, Uganda, Tanzania, Zaire. *Burger* 3596.

K. prittwitzii is perhaps much more common in the *Acacia-Commiphora* bushland of southern Ethiopia than the single collection seen suggests - most succulents are badly undercollected.

7. **K. angustifolia** *A. Rich.* (1848)
- type: TU, Chiré (Shire), *Quartin-Dillon* (P holo. + iso.).

Perennial with thickened roots; flowering stems slender, to *c* 25 cm high, terete, glabrous, glaucescent throughout. Leaves sublinear, thick, tip subobtuse. Flowers in small cymes. Pedicel stout, up to *c* 10 mm long. Calyx *c* half as long as corolla. Corolla 14 mm long overall, lobes narrow, possibly with brownish-yellow centre. Carpels 10 mm long, style 2 mm long.

TU; known only from the type, collected in 1840.

The type collection consists of 3 plants now all lacking intact leaves and flowers so it is not possible to confirm or augment existing descriptions. The habit and remaining fragments of flowers suggest a close relationship to the next two species, particularly *K. stenosiphon*, but it would seem best to keep this species separate until more collections are available from Tigray.

8. **K. glaucescens** *Britten* (1871)
- type: TU, Musana, *Schimper* (1862) 724A (BM lecto.; K isolecto. chosen by Raadts).

Perennial, stems erect or decumbent at base and forming large clumps; glabrous and pruinose throughout. Leaves narrowly to broadly ovate, up to 10 x 7 cm, sometimes with dark spots beneath or with red margin, base cuneate; petiole distinct slightly constricted at base and readily detached, tip acute to rounded, margin bluntly serrate or crenate, rarely subentire. Inflorescence 0.3-1.2 m high overall, branches erect, rather few flowered, pedicels 2-5(-8) mm long. Calyx lobes (2-)3-6 x 1-2 mm, joined for (0.2-)0.5-1 mm. Corolla yellow, orange, red or red with yellow centre, tube (5-)10-12 (-15) mm long, lobes (3-)5-6.5(-7.5) x (1-)1.8-2.5 mm, acute with mucro 0.2-1 mm long. Follicles (4.5-)6-8 (-10) mm long; styles 1-2 mm long.

Subsp. **glaucescens.**
K. marinellii Pampanini (1909) - type: EW, Mahio on Hadas river, *Dainelli & Marinelli* (FT holo.).

Anthers included within corolla tube. Fig. 88.8.8.

Deciduous woodland/open bushland with *Acacia* and *Commiphora*, sometimes on black soils; 1000-1900 m. EW TU (WU) GG SD ?HA; Sudan, Somalia, south to N Tanzania & Zaire. *Ash* 2263; *Beals* 418; *M. & S. Gilbert* 1545B.

Burger 3241 from near Dire Dawa, the only HA record, differs from all other material seen by the hairy calyx - its status needs further investigation. *K. flammea* Stapf (1897), from Somalia, is very closely related and Ethiopian material has been so named. It differs by the usually smaller sessile leaves, caducous calyx and broader corolla lobes but may prove to be better regarded as conspecific.

Subsp. *arabica* Cuf., from Yemen, differs by the exserted anthers.

9. **K. stenosiphon** *Britten* (1871)
- type: TU, Serriro, *Schimper* (1862) 726 (BM holo.; K iso.).

Closely related to *K. glaucescens*. Leaves lanceolate, up to *c* 10 x 1.5 cm, base cuneate, relatively broad, tip acute, margin subentire, reddish-green. Inflorescence up to 0.6 m high overall, very lax with long-pedunculate ascending (2-)3-4(-5)-flowered lateral branches. Pedicels 5-12 mm. Calyx-lobes *c* 6.5 x 2.8 mm, almost free. Corolla yellow-green, tube conspicuously inflated tapering to very narrow mouth, 14-16 mm long, lobes lanceolate, 7-11 x 1.3-1.7 mm, long acuminate. Follicles 8-10 mm long; styles *c* 2.5 mm long. Fig. 88.8.9.

Limestone slope; (1600-)2050-2300 m. EW TU; not known elsewhere. *Tellini* 1902; *Tewolde Berhan* 1109.

10. **K. citrina** *Schweinfurth* (1896)
- type: Yemen, *Schweinfurth* 1831.
K. citrina var. *ballyi* Wickens, *Kew Bull.* **36**: 674 (1982) - type: SD, Mega Mt., *Bally* 9151 (K holo.; EA iso. not seen).

Perennial with rather swollen roots, stems erect or decumbent at base; densely pubescent throughout, not glandular. Leaves lanceolate to broadly elliptic, up to 12 x 4 cm, often much smaller, base cuneate, sometimes into distinct narrow based petiole, tip acute or obtuse, irregularly dentate to subentire, rarely dissected. Inflorescences 20-90(-150) cm high overall, cymes relatively small, dense, flat-topped; pedicels 1-5 mm. Calyx lobes 4-8 x 1-1.8 mm, joined for up to 1 mm; corolla yellow, tube 10-12 mm long, lobes 5-7 x 2.8-3.5 mm, rounded with mucro 0.3-1 mm long, exterior pilose; anthers included. Follicles 6-8.5 mm long; styles 1-1.5 mm long. Fig. 88.8.4.

Acacia woodland or margins of *Juniperus* forest, usually on rocky slopes; 1000-2000 m. SD HA; Kenya, Somalia, Uganda, Arabia. *Bally* 9151; *Burger* 3379; *Robertson* 1423.

Polymorphic: plants resembling the type from higher altitudes more erect with lanceolate subentire sessile leaves; those from lower altitudes with more decumbent stems, broader, coarsely dentate, more distinctly petiolate leaves and longer pedicels. The latter form has been described as var. *ballyi*.

11. **K. densiflora** *Rolfe* (1919)
- type: Kenya, *Snowden* 556.

Erect annual or short-lived perennial occasionally producing new shoots from base; glabrous except sometimes for pedicels and calyx. Leaves oblong, ovate or suborbicular, rounded, crenate, up to 18 x 10 cm, never glaucous or pruinose, base cuneate; petiole broad-based. Inflorescences (0.3-)0.6-3 m high overall, panicle to 70 cm long, flowers congested. Pedicels 1-4 mm long, thickened below flower. Calyx lobes 2-8 x 1-1.5 mm, fused for 0.1-0.3 mm, basally circumcissile but persisting round fruit. Corolla yellow, orange or red, tube (8-)9-12 mm long, lobes 2-5 x 1.5-4 mm, rounded or retuse with mucro 0.2-1 mm long. Anthers included. Follicles 5-9 mm long; styles 1-3.5 mm long.

1. Plant glabrous throughout. var. **densiflora**
- Pedicel and calyx very sparsely glandular-hairy. var. **subpilosa**

Var. **densiflora**.

Disturbed or open habitats, forest margins, near streams, often along roadsides, rarely epiphytic; (1000-) 1400-2300 m. GD/GJ SU AR IL KF SD HA; Sudan (Imatongs), Kenya, N Uganda, Tanzania, Zaire (Ruwenzori). *Mooney* 5344, 8262; *Robertson* 1327.

Var. **subpilosa** *Cuf.*
Webbia **19**: 715 (1965) - type: GD, Abbena, Ataba Valley, *Steudner* 1247 (K holo.).
K. lentiginosa Cuf. (1958) - type: cultivated, *Wettstein* (WU holo. not seen).
Fig. 88.8.10.

Mostly, but not always, in sheltered sites among trees, less often in open rocky areas. 1300-2550 m. GD SU IL; not known elsewhere. *Ash* 642; *Friis* et al. 1961; *Mooney* 5034.

Probably just a sporadic variant of var. *densiflora*. A group of collections from near Gondar (*Chiovenda* 2504 bis, 2852, etc.) are probably just extreme variants of this variety that differ by the conspicuously denser indumentum on most of the inflorescence.

Var. *minor* Raadts, flowers small, orange-red with narrower corolla lobes, is found in the Kenya Highlands.

12. **K. lanceolata** *(Forssk.) Pers.* (1805)
Cotyledon lanceolata Forssk. (1775); *Verea lanceolata* (Forssk.) Spreng. (1825) - type: Yemen, *Forsskål*.
K. glandulosa A. Rich. (1848); *K. lanceolata* var. *glandulosa* (A. Rich.) Cuf., *Webbia* **19**: 730 (1965) - types: TU, Chiré (Shire), *Quartin-Dillon* (P syn.) & Tchélatchekanne (Dscheladscheranne), *Schimper* II:904 (P syn.; K isosyn.; B FT LE W isosyn. not seen).

Erect annual, stems typically 4-angled towards base, lower parts glabrous, upper parts, especially inflorescence, densely glandular pubescent, rarely almost pilose. Leaves sessile, lanceolate to oblong-ovate, up to 15(-30) x 6(-9) cm, never glaucous or pruinose, base decurrent, tip obtuse or acute, entire, bluntly serrate or crenate. Inflorescence 0.2-1.5(-2) m high overall with few often long branches with many regularly arranged flowers; pedicels 1-3(-10) mm long. Calyx lobes ovate to oblong-elliptic, 2.5-13 x 1.5-4.5 mm, joined for 1-8(-11) mm. Corolla yellow or orange, rarely pinkish, tube 8-15 mm, lobes 2.6-6.5 x (1.1-) 1.5-3.5 mm, rounded with mucro up to 0.7 mm long. Anthers included. Follicles (5-)6-9 mm long; styles 0.5-1(-1.5) mm long. Fig. 88.9.

Open woodland, bushland or grassland on deep or rocky soils including black clay, occasionally epiphytic; 1100-2000 m. EW TU GD ?WU GJ WG SU ?AR GG

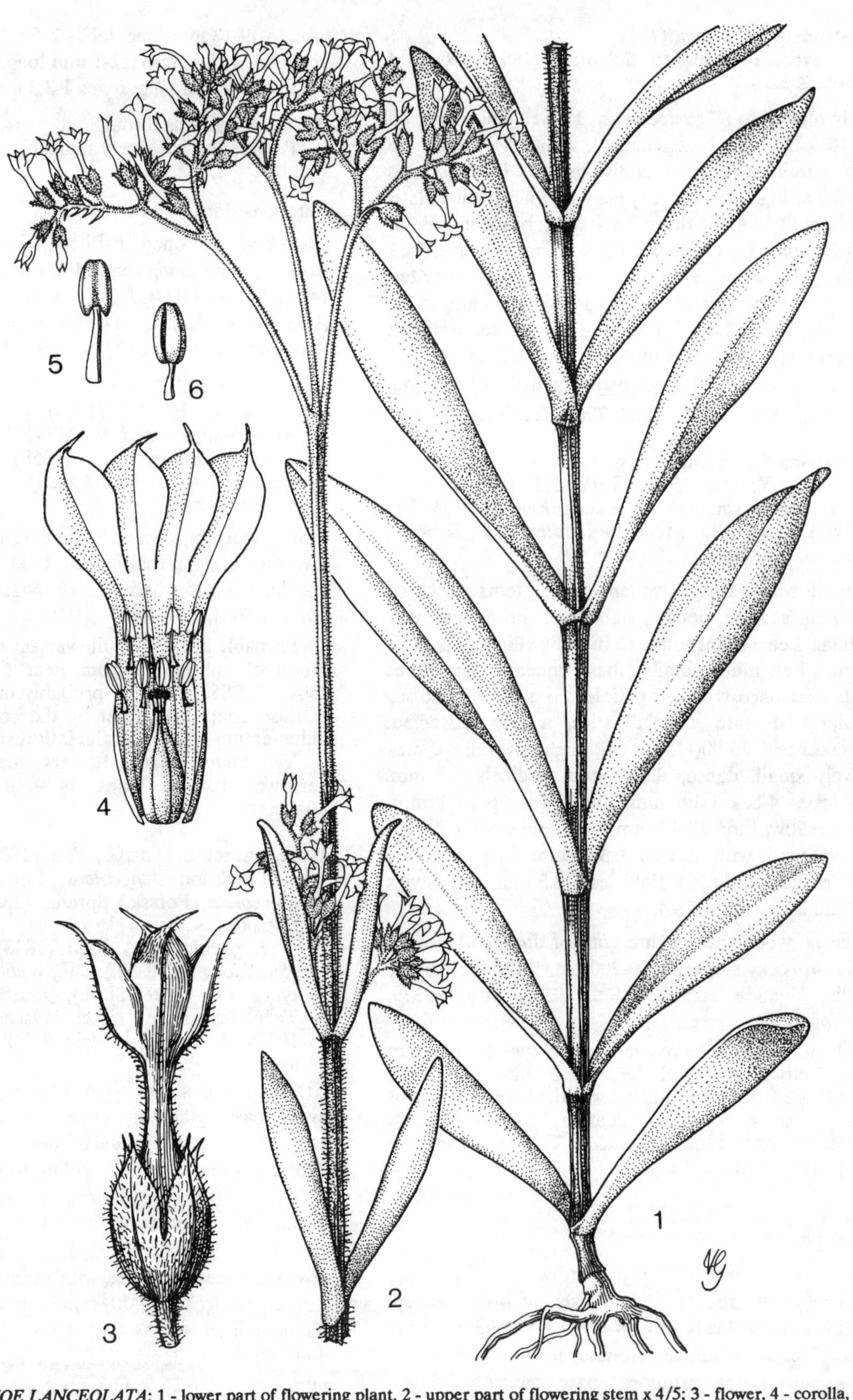

Figure 88.9. *KALANCHOE LANCEOLATA*: 1 - lower part of flowering plant, 2 - upper part of flowering stem x 4/5; 3 - flower, 4 - corolla, opened out, and carpels x 5; 5 & 6 - stamens x 10. 1 from *Teixeira* 982; 2 - 6 from *Polhill & Paulo* 2103. Drawn by Victoria Goaman. (Reproduced by permission of Betham Moxon Trustees from *Flora Zambesiaca*.)

HA; Guinea east to Arabia and India, south to South Africa; Madagascar. *Burger* 1419, 3601; *Westphal* 2913.

Material from the north of Ethiopia has better developed calyx-tubes than elsewhere but the variation seems to be clinal with the occasional long-tubed calyx occurring further to the south, such that the maintenance of var. *glandulosa* cannot be supported.

The separation from the following species is difficult and needs further investigation, particularly within the Flora area. Typical specimens with either 4-angled stems plus sessile, undivided leaves or with petiolate, divided to compound leaves present no problem but most Ethiopian material tentatively placed in *K. lanceolata* has rounded stems and might be better treated as an entire-leaved form of *K. laciniata*. The situation is worsened by the tendency for the leaves of both species to die back during flowering.

Mesfin & Vollesen 4334 from Mega has rounded stems and densely pilose, only obscurely glandular, inflorescences; it is tentatively included in *K. lanceolata* as here defined but does look rather distinctive.

13. **K. laciniata** *(L.) DC.* (1802)
Cotyledon laciniata L. (1753) - type: cultivated, Hort. Sic. Cliff.
?K. brachycalyx A.Rich. (1848); *K. laciniata* (L.) DC. var. *brachycalyx* (A. Rich.) Chiov., *Res. Scient. Miss. Stefanini-Paoli* 1:75 (1916) - type: TU, Mai Guagua (Mai Gouagoua), *Quartin-Dillon* (P holo.)
K. schweinfurthii Penzig (1893) - type: EW, Gheleb, *Penzig* (?GE holo. not found).
K. rohlfsii Engl. (1902) type: EW, Katz[1], *Rohlfs & Stecker* (B lecto.: chosen by Raadts (1977) not seen, P fragment).

Closely related to *K. lanceolata*. Stems consistently rounded. Leaves petiolate, very variable in shape but typically pinnately 3-5-foliolate or at least deeply lobed, leaflets up to 8 x 2.5(-6) cm, coarsely serrate to laciniate. Inflorescence 0.4-1(-3) cm high overall, panicle up to 40 cm long, relatively dense with short erect terminal branches. Pedicels 1-2(-4) mm. Calyx lobes 2.5-5.5 (-6.5) x 1-1.8 mm, joined for 0.4-1 mm. Corolla cream, dull pale yellow or pink, tube 8.5-11(-13) long, lobes 3.5-5(-7) x 2-3(-5.5) mm, rounded. Fig. 88.8.5&6.

Deciduous bushland or woodland or dry *Juniperus* forest, usually in light shade, on sandy or clay soils; (1350-)1600-2200 m. EW GD WG SU KF SD BA; Kenya, Uganda, south to Mozambique and Namibia, Arabia, S India. *Ash* 2283; *Friis* et al. 523; *Mooney* 5485.

The problem of separation from *K. lanceolata* in Ethiopia has been noted above. The plants with compound leaves are most distinctive and the variation in leaf form deserves careful investigation.

The type of *K. brachycalyx* lacks leaves and it is debatable whether it belongs to this circumscription of *K. lanceolata*. It is placed here following R. Fernandes, on the basis of the rather shorter indumentum and rounded stems.

1. The fragment in P is labelled 'Karen' (or possibly Kasen), altitude 2450 m.

9. BRYOPHYLLUM *Salisb.* (1805)

Often included within *Kalanchoe* but differing by the pendent flowers with well developed, often ± inflated, calyx-tubes and the stamens free or inserted below the middle of the corolla tube.

A genus of 30 or more species in Madagascar, one of which has become ± pantropical whilst several other species are widely cultivated, sometimes naturalising.

Increasingly likely to be met with as escapes, especially as several species propagate very readily by means of 'bulbils' - tiny plantlets on leaf margins or within the inflorescence.

1. Leaves obviously pinnately divided or compound. 2
- Leaves terete except for a few teeth at tip, usually bearing small plantlets between teeth. **3. B. tubiflorum**

2. Plant usually propagated from seed and rarely forming large stands; leaves pinnately compound; pedicels smooth; calyx-tube 20-30 mm long. **1. B. pinnatum**
- Plant usually propagated vegetatively from leaf-cuttings and plantlets produced within the inflorescence and eventually forming large stands; leaves pinnatisect; pedicels papillose; calyx-tube 13-16 mm long. **2. B. proliferum**

1. **B. pinnatum** *(Lam.) Oken* (1841)
Cotyledon pinnata Lam. (1786); *Kalanchoe pinnata* (Lam.) Pers. (1805) - type from Madagascar.

Erect annual or short-lived perennial to 2 m high, often unbranched; all parts glabrous. Leaves pinnately 3-5-foliolate; petiole 2-10 cm; leaf-blade 6-20 x 4-12 cm overall, leaflets circular to oblong-spathulate, crenate to crenate-dentate. Inflorescence a very lax panicle. Pedicels 10-25 mm long. Calyx-tube 2-3 cm long, lobes 0.7-1 cm long. Corolla up to 5 cm long, lobes bright red or purplish.

An attractive pantropical adventive originating in Madagascar. Found throughout Africa and grown in public gardens in Addis Ababa. *Sue Edwards* 4207.

2. **B. proliferum** *Hook.* (1859)
K. prolifera (Hook.) Raym.-Ham. (1908) - type from Madagascar.

Robust perennial to 4 m high, often branching from base and eventually forming large stands; all parts glabrous. Leaves pinnatisect; petiole up to 16 cm long, lobes 7-15 x 1.5-5 cm, crenate to subpinnatisect. Inflorescence a large panicle of dense cymes, usually with some flowers replaced by plantlets. Pedicels 8-15 mm long, very slender, densely papillate. Calyx-tube 13-16 mm long, lobes 3-4 mm long. Corolla tube 18-24 mm long, lobes 2.5-3.5 mm long, yellow.

Cultivated as an ornamental, readily propagating from broken leaves or from the plantlets produced by the inflorescences and thus establishing itself in the wild. Sight record (Addis Abeba).

3. **B. tubiflorum** *Harvey* (1862);
Kalanchoe tubiflora (Harvey) Raym.-Ham. (1912) - type from Madagascar.
K. delagoense Eckl. & Zeyh. (18..) *nom. nud.*; *Bryophyllum delagoense* (Eckl. & Zeyh.) Schinz. (19..) *nom. illegit.*

Erect perennial up to 1.2 m high, eventually branching from base; all parts glabrous. Leaves sometimes alternate, terete except for a few teeth near to tip, plantlets produced from notches between these, grey green. Inflorescence 10-15 cm diam. Pedicels 6-20 mm long. Calyx up to 13 mm long overall. Corolla reddish, tube 22-35 mm long, lobes 7-10 mm long.

Cultivated, often for the novelty of the plantlet-bearing leaves which facilitate its propagation. Native of Madagascar now often naturalised in drier regions such as found in Kenya and Mexico. Sight records (Addis Abeba, Alemaya, Asmera).

Tölken & Leistner, (*Flowering Plants of Africa* pl. **1938**, 1986) claim that the correct name for this species is *Bryophyllum delagoense*. Most authors have rejected that name as a *nomen nudum*, correctly in the view of this author.

Note added in press.

In Asmera, there is a fourth species found in both private and public gardens. It has simple leaves bearing plantlets around the toothed margin and a terminal panicle of hanging flowers with dark red to purple flowers. It seems nearest to *Bryophyllum daigremontiana*

10. ECHEVERIA *DC.* (1828).

Walther, '*Echeveria*', California Acad. Science (1972).

Perennial herbs or subshrubs. Leaves alternate, usually crowded into rosettes, usually sessile and entire. Inflorescence normally lateral, ± pedunculate, branches usually 1-few, either a cincinnus or apparently racemose (probably a reduced panicle). Flowers 5-merous. Sepals free or almost so. Petals almost free, erect, touching to form an apparent corolla tube, mostly red, orange and/or yellow, usually keeled with prominent nectar cavity inside at base. Stamens 10; carpels erect with prominent nectaries at base. Ripe follicles spreading; seeds numerous.

A New World genus of *c* 140 species from the USA (Texas) south to Argentina but with all but *c* 25 species restricted to Mexico.

Many species are grown as ornamentals and numerous hybrids and cultivars have been raised in cultivation. Other species, apart from the following, are likely to be met with in private gardens.

E. elegans *Rose*

Stemless herb forming mats of leaf-rosettes; all parts glabrous. Rosettes up to 20 cm wide, leaves numerous, very glaucous grey-green without any red. Inflorescence a dense unbranched cincinnus up to 22 cm long of red and yellow flowers.

Widely used as a bedding plant in many parts of the world, including Addis Abeba, Asmera and many other towns; native of Mexico.

89. SAXIFRAGACEAE

by I. Hedberg*

Cufodontis, *Enum.*:172 (1954); Rampi in *Webbia* **28(2)**:522 (1973).

Herbs, mostly perennials. Leaves usually alternate, simple, exstipulate or with small sheathing stipules. Flowers regular, (rarely somewhat irregular), bisexual. Sepals usually 5. Petals usually 4-5, (sometimes 0) often narrow at the base. Stamens usually twice as many as the sepals, rarely equal in number or fewer. Ovary superior to partly inferior, with 2 carpels united at the base, styles free; ovules numerous, placentation various. Fruit a capsule.

Some 30 genera and 600 species, cosmopolitan but mainly found in cool temperate regions. One genus in Ethiopia. The family was earlier taken in a wider sense, including **Escalloniaceae, Hydrangeaceae** and **Vahliaceae.**

This family contains a number of beautiful ornamental plants such *Astilbe*, 'Spiraea' and *Saxifraga umbrosa* L. 'London pride'. The genus *Saxifraga* is best known for the very large number of species and cultivars which can be grown in rock gardens. However, these are not often found in gardens in Ethiopia.

SAXIFRAGA *L.* (1753)

Herbs, usually perennial. Leaves simple but often dissected, often thick, usually alternate or basal. Flowers 5-merous, usually in cymes or panicles. Petals present. Stamens 10, ovary superior or partly inferior. Capsule opening along the inner edges of the carpels.

Nearly 400 species mostly of the north temperate regions, the Arctic and the Andes, typically found in alpine habitats. One species endemic to high altitudes in Ethiopia.

S. hederifolia *A. Rich.* (1847).
S. abyssinica A. Rich. ex Engler (1872) - type: Gonder, Mt. Buahit (Ras Dejen) in Simien, *Schimper* II:1305 (K UPS isolecto.).

Perennial herb, 1 cm to *c* 10 cm high, sometimes with trailing stems. Leaves 0.5-1 cm in diameter, slightly succulent, reniform-orbiculate, crenate with (3)7-9 evenly rounded or apiculate crenules. Stem, petioles, leaf edges and pedicels with scattered, long or short translucent or brownish glandular hairs. Leaves and sepals sometimes with brownish streaks. Flowers solitary; sepals green, petals yellow. Fig. 89.1

On rocky slopes along streams or on marshy ground, 3000-4600 m, endemic to the high mountains in Northern and Central Ethiopia. TU GD GJ SU AR. *De Wilde* 9169; *Evans* 336; *Mooney* 6448.

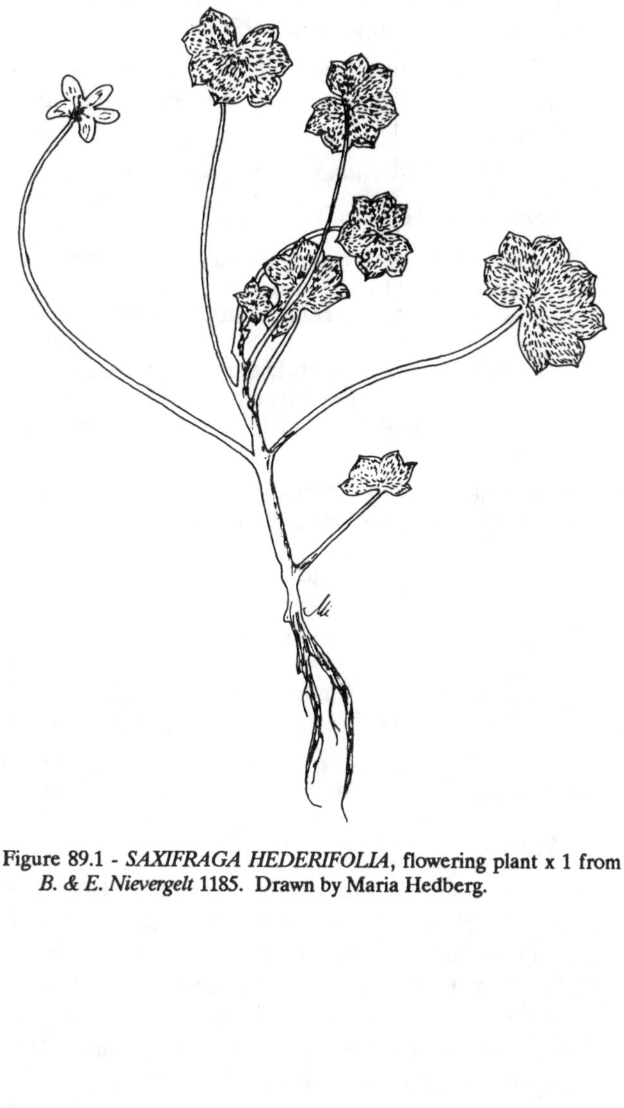

Figure 89.1 - *SAXIFRAGA HEDERIFOLIA*, flowering plant x 1 from *B. & E. Nievergelt* 1185. Drawn by Maria Hedberg.

* Uppsala University - Department of Systematic Botany, P.O. Box 541, S-751 21 Uppsala, Sweden.

90. VAHLIACEAE

by D. Bridson*

Cufodontis, *Enum.*: 172 (1954); Rampi in *Webbia* **28(2)**: 527 (1973); Bridson in *Kew Bull.* **30**: 163-182 (1975); in *Fl.Trop.E.Afr.* (1975); in *Fl. Zamb.* **4**: 48-53 (1978) & *Fl. Egypt, fam.* **80**: 13 pp. (1983).

Annual or perennial herbs or subshrubs, glabrous or pubescent with multicellular, often gland-tipped hairs. Leaves opposite, sessile to subsessile, simple, ovate to linear, entire, exstipulate, midrib apparent, lateral nerves obscure. Inflorescence axillary, erect, sessile to subsessile (or pedunculate outside Flora area), bearing paired sessile to subsessile (or pedicellate outside Flora area) flowers. Flowers hermaphrodite, regular. Calyx-tube adnate to ovary, campanulate to subglobose, often distinctly 5-veined; lobes 5, valvate, persistent. Petals 5, valvate, alternate with calyx-lobes, shorter than or exceeding calyx-lobes, lanceolate to round, entire to crenate or irregularly dentate. Stamens 5, alternate with petals, inserted on margin of epigynous disk; filaments free, subulate with ± widened membraneous area or scale-like appendage at the base; anthers introrse, dorsifixed. Ovary inferior, 1-locular; styles 2 (or ?3), shorter than or exceeding the calyx-lobes, separating shortly above the base (or at base outside Flora area); stigmas capitellate; ovules numerous, attached to 2 thick, pendulous, flattened rounded placentas (see fig. 90.1.6). Capsule, subglobose or obovoid, dehiscing by 2 (?3) valves between style-bases, all floral parts (except anthers) persistent. Seeds numerous, oblong minute.

The family, formerly included in **Saxifragaceae**, was recently revised by Bridson (1975). It consists of only one genus with five species which are distributed on the African continent, and in Madagascar, Iraq, Iran, India and Vietnam.

VAHLIA *Thunb.* (1782), *nom. conserv.*

(Description as for family)

1. Peduncles and pedicels clearly present, the former up to 2.2 cm long; leaves linear to lanceolate; gynoecium with style-bases free to the disk. *V. dichotoma*
- Peduncles and pedicels absent or not exceeding 2 mm; leaves ovate-lanceolate to ovate; gynoecium with style-bases dividing 0.15-0.8 mm above the disk. 2
2. Petals distinctly shorter than calyx-lobes; style-arms never exceeding calyx-lobes, upto 1.3 mm long. **1. V. digyna**
- Petals ± same length as calyx-lobes; style-arms often exceeding calyx-lobes, over 2 mm long. 3
3. Style-arms up to 3.5 mm long; leaves glandular pubescent. **2. V. geminiflora**
- Style-arms over 4 mm long; leaves glabrous to sparsely hairy. **3. V. somalensis**

V. dichotoma (Murray) Kuntze, is not recorded from Ethiopia but could be found in future collections.

1. V. **digyna** *(Rets.) Kuntze* (1891).

Erect, sparsely to much-branched annual herb, 5-33.5 cm tall. Stems covered with patent or crisped, often glandular, hairs. Leaves ovate to ovate-lanceolate, 0.5-2.5(-4.8) x 0.2-0.6(-0.9) cm, apex acute, base tapering to rounded, usually only a few hairs but occasionally pubescent. Calyx-tube 1.1-2.0 mm long, hairs patent or crisped, sparse; lobes ovate, 1.3-2.0 x 0.6-1.2 mm, acute, glabrous or sparsely pubescent outside. Petals yellow, fading to white, always shorter than sepals, round to ovate, 1.0-1.3(-1.6) x 0.55-1.0 mm, usually distinctly apiculate, narrowed to base; only midvein distinct. Stamens: filaments 0.3-1.1 mm long, translucent with vein clearly seen, with a small membranous hairy scale-like appendage at the base; anthers 0.15-0.4 mm long. Styles 0.7-1.3 mm long, glabrous. Capsule 1.4-2.2 mm long, sparsely pubescent. Seed straw-coloured, ovoid-cylindrical, 0.15-0.27 mm long. Fig. 90.1.1-8.

Near rivers, ponds or wells. GG SD HA; west to Mauritania and south to Mozambique and Botswana, north to Egypt, also in Madagascar, India and Pakistan. *Corradi* 5763, 8366; *Gilbert* et al. 8196; *Riva* 900.

2. V. **geminiflora** *(Del.) Bridson* (1975).

Erect, branching annual herb, 10-37 cm tall. Stems and leaves ± densely covered with long, patent, usually gland-tipped, hairs. Leaves lanceolate-ovate to ovate, 1-4 x 0.4-1 cm, apex acute, base rounded or tapering. Calyx-tube 1.5-3 mm long, densely covered with patent, gland-tipped hairs; lobes ovate, 2.4-4.1 x 1.3-2.3 mm, apex acute to shortly acuminate, practically glabrous or with a few short hairs on the outside. Petals yellow, usually drying with a maroon spot at the base, ovate to round, 2.1-3.7 x 1.3-3.4 mm, acute to acuminate, gradually or abruptly narrowed to the base; mid- and lateral veins easily seen. Stamens: filaments 1.2-3.6 mm long, usually hairy, with a conspicuous basal appendage; appendage membranous, ± semicircular to oblong, top projecting forward and free of filament, hairy; anthers

* Royal Botanic Gardens, Kew, Richmond, Surrey, TW9 3AB, UK.

Figure 90.1 *VAHLIA DIGYNA*: 1 - flowering stem from *Conrads* 5906; 2 - flower with 2 calyx-lobes and 2 petals removed; 3 - petal (from inside); 4 - calyx-lobe (from outside); 5 - stamen (front & side view); 6 - calyx-tube and ovary cut open to show ovules; 7 - valve of capsule (style base in fruiting stage); 8 - seed, 2-8 from *Robson* 1309. *V. GEMINIFLORA*: 9 - flowering stem; 10 - petal; 11 - calyx-lobe; 12 - stamen; 13 - styles; all from *Carr* 552. *V. SOMALENSIS* ssp. *GODDINGII*: 14 - flowering stem; 15 - petal; 16 - calyx-lobe; 17 - stamen; 18 - filament appendage from behind; 19 - styles; all from *Hemming* 1481. Drawn by D. Bridson.

1.2-1.6 mm long. Styles 2.2-3.5 mm long, hairy. Capsule subglobose, 1.3-2.5 mm long, hairy. Seeds straw-coloured, ovoid, 0.2-0.3 mm long. Fig. 90.1.9-13.

Near rivers, ponds or wells; 500-1200 m. EW GD GG; Mali, Nigeria, Sudan, Egypt, Iraq & Iran. *Benedictis* 201; *Corradi* 8358; *Pappi* 9050.

3. V. **somalensis** *Chiov.* (1929).

Much-branched herbaceous to woody annual or perennial. Stems ascending or decumbent, 11-50 cm long; glabrous or covered with patent or crisped hairs. Leaves ovate-lanceolate or ovate, 0.4-3.1 x 0.25-1.3 cm, apex acute, base narrow or rounded, glabrous or with a few hairs confined to the margins and midrib below, occasionally a sparse cover of short hairs all over. Calyx-tube 1.6-2.5 mm long, glabrous or densely hairy; lobes lanceolate to ovate, 3.8-7.2 x 1.2-3.0 mm, acute to rostrate, glabrous or with few hairs on margin. Petals yellow, sometimes drying with a maroon spot at base, ovate 4-7 x 2.2-5 mm, short or long apiculate, narrowed to base; mid and lateral veins easily seen. Filaments 3.5-6.3 mm long, hairy, with conspicuous basal appendage; appendage membranous, hairy, ± round, with top projecting forward and free of the filament; anthers 1.3-2.8 mm long. Styles 4.1-7.6 mm long, sparsely hairy. Capsule subglobose, 2.5-3.6 mm long, glabrous or hairy. Seeds orange-brown or grey-brown, ovoid, 0.2-0.3 mm long. Fig. 90.1.14-19.

1. Calyx-tube densely covered with crisped hairs; stems covered with hairs; plants generally small. subsp. **somalensis**
- Calyx-tube sparsely covered with crisped hairs near base; stems glabrous; plants robust. subsp. **goddingii**

subsp. **somalensis**

Seasonal ponds ('ballehs'); 215 m. SD HA; Somalia & Kenya. *Peck* 41; *IECAMA*. I-99; *Gilbert* et al. 8195.

subsp. **goddingii** *(Bruce) Bridson* in Kew Bull. 30: 181 (1975) - type: *Godding* 237(34) (K holo!)

Seasonal ponds ('ballehs'); 610-980 m. HA; Somalia. *Ellis* 157; *Hemming* 1481; *Popov* 1145A.

91. ROSACEAE

by O. Hedberg*

Cufodontis, *Enum*.: 175-183 (1954); Graham, *Fl. Trop. E. Afr.*, **Rosaceae**: 61 pp. (1960); Valentine & Chater (eds.), *Fl. Europaea*, vol. 2: 3-80 (1968); Browicz, *Fl. Turkey* 4: 1-172 (1972); Hedberg, *Nord. J. Bot.* 6:573-579 (1986).

Trees, shrubs and herbs, usually perennial. Leaves alternate, stipulate, simple or compound, often with a toothed margin. Flowers very variable. Receptacle well developed, flat, convex or concave, often hollow. Calyx with 5 or 4 sepals, sometimes with an outer alternating row of lobes forming an epicalyx. Petals 5 or absent. Stamens 1 to many, sometimes biseriate. Carpels 1 to many, free or fused to each other and sometimes to the receptacle. Ovules 1-2. Fruit various, dry or fleshy. Seeds usually exalbuminous (non endospermic).

A cosmopolitan family with about 2000 species, largely concentrated in temperate regions, and containing many herbs, shrubs and trees of economic value, both for food and as ornamentals.

A number of species with fruits of potential commercial importance have been introduced into Ethiopia and these have been included in this account.

Key to Genera

1. Herbs or unarmed low shrubs. 2
- Trees, shrubs or scramblers. 5

2. Flowers with distinct petals, receptacle convex. 3
- Flowers apetalous; receptacle concave. 4

3. Receptacle much enlarged in fruit and fleshy; petals white; leaves ternate (strawberry). **4. Fragaria**
- Receptacle neither enlarged nor fleshy in fruit; petals white, yellow or reddish; leaves pinnate, digitate or ternate. **3. Potentilla**

4. Perennial, stems sometimes woody; stamens 4; epicalyx present. **5. Alchemilla**
- Annual, no woody basal parts; stamen only 1; epicalyx absent. **6. Aphanes**

5. Prickly or thorny shrubs or scramblers. 6
- Trees or shrubs without prickles or thorns. 7

6. Receptacle convex; fruit a ± tight collection of many one-seeded drupelets. **1. Rubus**
- Receptacle concave; fruit of nutlets inside a flask-shaped fleshy receptacle (roses). **2. Rosa**

7. Leaves imparipinnate; flowers small, unisexual, in large, pendulous inflorescence. **13. Hagenia**
- Leaves simple, flowers hermaphrodite, often showy, not in pendulous inflorescence. 8

8. Ovary superior; carpel 1; fruit dry or fleshy with a single stone. **8. Prunus**
- Ovary ± inferior; carpels 2-5; fruit fleshy with horny endocarp enclosing 2 or more seeds. 9

9. Flowers solitary. 10
- Flowers in 2 to many flowered inflorescences. 11

10. Flowers less than 1 cm diam.; fruit small, red or black; leaf margin entire. **7. Cotoneaster**
- Flowers 4-6 cm diam.; fruit large, 3 or more cm long, green, yellow or brown; leaf margin dentate (quince). **10. Cydonia**

11. Leaf margin entire; petals 5 mm or less long; fruit small, red or black. **7. Cotoneaster**
- Leaf margin serrate or dentate; petals more than 5 mm long; fruit usually large, green, yellow or red. 12

12. Leaves up to 25 cm long; inflorescence a terminal panicle; fruit yellow (loquat). **9. Eriobotrya**
- Leaves up to 10 cm long; inflorescence an umbel or corymb; fruit green or red. 13

13. Fruit globose (but not umbilicate) to obconate with numerous grit cells in the flesh; styles free (pears). **12. Pyrus**
- Fruit globose and umbilicate without or with only a few grit cells in the flesh; styles connate at base (apples). **11. Malus**

1. RUBUS *L.* (1753)

Shrubs (in Ethiopia), usually scrambling, but sometimes erect, with ± arching sterile shoots, often rooting if they touch the ground, carrying ± erect flowering branches. Stems armed with prickles and various amounts of hair and stalked glands. Leaves petiolate, stipulate, imparipinnate or trifoliolate. Flowers with 5 calyx lobes and 5 petals, sometimes reduced. Stamens many. Pistil superior: carpels many, each containing 2 ovules. Fruit a tight collection of many one-seeded drupelets crowded on an enlarged receptacle.

An almost cosmopolitan genus of about 250 species most abundant in temperate regions. All fruits are edible and some are very popular.

This genus is renowned for taxonomic difficulties caused, amongst other things, by apomixis in the *R. fruticosus* group, from which more than 2000 agamospecies have been described. Additional difficulties were caused by the fact that species descriptions and revisions from tropical material were made almost exclusively by European taxonomists who had not seen the plants in the field but based their descriptions on fragmentary collections available in

* Uppsala University - Department of Systematic Botany, P.O. Box 541, S-751 21 Uppsala, Sweden.

Figure 91.1.1 - *RUBUS APETALUS*: leaf and fruit. 2 - *R. STEUDNERI*: leaf, part of inflorescence and cross section of flower (bottom centre). 3 - *PRUNUS AFRICANA*: leafy stem with old inflorescence and a few fruits at top, longitudinal section of flower and cross section of ovary (lower left). Centimetre scale for leaves and inflorescences - closed bar; millimetre scale for floral details - open bar. Drawn by W.C. Burger. (Reproduced with permission of Oklahoma Agricultural Experiment Station from Experiment Station Bulletin No. 45, *Families of Flowering Plants in Ethiopia*, figure 21, 1967.)

herbaria together with scant collectors' notes. The present treatment has tentatively reduced a number of insufficiently known taxa to synonymy. Further collecting and field studies may justify the revival of some of these taxa, but the material at present available does not do so. A definitive treatment of the Ethiopian material cannot be achieved without a world-wide revision of the genus.

R. idaeus L. is the cultivated raspberry which has sometimes been grown in private gardens, as have forms of *R. fruticosus* L. which is the cultivated blackberry. Neither of these species has become naturalised, but *R. niveus*, an Indian and Malayan bramble has been introduced and become naturalised in and around Addis Ababa.

1. Sepals with a prolonged, filiform apex. 2
- Sepals without a prolonged, filiform apex but apex sometimes tapering. 5
2. Carpels more than 100, each at most 1 mm long. **1. R. rosifolius**
- Carpels less than 50, each at least 2 mm long. 3
3. Stem and inflorescence with numerous stipitate glands. 4
- Stem and inflorescence without (or with few) stipitate glands. **2. R. aethiopicus**
4. Leaves densely white-tomentose below. **3. R. erlangeri**
- Leaves glabrous or sparsely puberulous below. **4. R. volkensii**
5. Sepals reflexed after anthesis; petals longer than sepals. **5. R. steudneri**
- Sepals clasping or spreading, never reflexed after anthesis; petals shorter than sepals or absent. 6
6. Stems densely and softly white or yellowish pilose; inflorescence paniculate, usually exceeding the basal leaves. **6. R. apetalus**
- Stems glabrescent; inflorescence corymbose, leafy, usually somewhat covered by the leaves. **7. R. niveus**

1. **R. rosifolius** *Sm.* (1791)

Scrambling shrub; branches up to 2.5 m long, covered in white hairs and straight spines. Leaflets 3-5-pairs with a terminal leaflet. Inflorescence few-flowered; petals white turning pink; carpels very numerous; fruit pink, edible.

Secondary growth in forest, bushland and wooded grassland, also escaped from cultivation, 1950 m. KF; a native of India and Sri Lanka, widely cultivated in the tropics and often escaped. *Friis* et al. 2256; *Westphal* 3347.

Often spelt as *R. rosaefolius.*

2. **R. aethiopicus** *R.A. Grah.* (1958)
- type: SD, Dekano on Walayita-Soddu (Sodo) road, *Mooney* 6177 (K holo., EA ETH P iso.).

Erect shrub to 2 m, not climbing. Stems smooth with whitish waxy covering and few prickles. Leaves imparipinnate or trifoliolate, green above and densely white tomentose below, with ovate-elliptic to elliptic leaflets. Inflorescence axillary or terminal, less than 10 cm long (relatively small for the genus). Peduncle, pedicel and calyx densely covered by stalked glands, intermingled with small prickles. Flowers *c* 1.5 cm wide. Sepals 1-1.5 cm, erect, caudate, often purplish. Petals *c* 1 cm long and 0.8 cm wide, cream-coloured. Ripe fruit unknown.

In *Juniperus* forest or in scrub, sometimes on moist ground, *c* 2600-3000 m. SU SD, only known from Ethiopia and then only from 4 collections. *Friis* et al. 1208, *Gillett* 15026, *Mooney* 6177.

3. **R. erlangeri** *Engl.* (1911)
- type: SD, Arbe (Arbegona), *Ellenbeck* 1857 (B destr.; K photo); AR, E slope of Mt. Borulucca, 10 km W of Ticcio (Ticho), *W. de Wilde* et al. 10069 (K neo., C EA WAG isoneo.).

Shrub to 2 m, covered in sticky hairs; stems purplish brown, densely covered by stalked glands intermixed with loose hairs and scattered prickles. Leaves imparipinnate or trifoliolate, leaflets ovate-acuminate up to 7 x 5 cm, sharply serrate, green above and very densely white-tomentose below, with stalked glands along the veins. Inflorescence paniculate, few-flowered, axillary or terminal. Flowers 2-3 cm diameter. Peduncles, pedicels and calyx densely covered by purplish stalked glands, intermingled with small prickles. Sepals ovate, long-caudate, 1.5-2 cm long, appressed to fruits. Petals creamy-white, sometimes slightly pinkish, 1.5 cm long, obtuse. Fruit pale orange.

Open places in *Hypericum* forest, 3600 m. BA SD; only known from Ethiopia and then only from the 2 type collections.

Further material of this taxon is urgently required.

4. **R. volkensii** *Engl.* (1894).
R. chiesae Chiov. (1911) - type: SU, Gubba-Gatterà near M. Uaciaccia (Mt. Wechecha), *Negri* 416 (FT syn. not seen); SU, M. Mangascià (Mt. Menagesha), near Addis Ababa, *Negri*, 495 (FT syn.).

Viscid shrub to 4 m, often smaller with erect stems to 1.5 m. Stems reddish brown, densely covered by a mixture of stalked glands and loose hairs with scattered prickles. Leaves imparipinnate or trifoliolate, leaflets ovate-elliptical to 9 x 6 cm, often long acuminate, sharply serrate or biserrate, green on both sides; rachis glandular, as hairy and prickly as the stem. Inflorescence paniculate, few-flowered, axillary or terminal. Flower to 3 cm in diameter. Peduncles, pedicels and calyx densely covered by stalked glands intermingled with prickles. Sepals with ovate base, caudate, 1.5-2.5 cm long, appressed to the maturing fruits. Petals yellow-white, broadly ovate, 1.5-2.5 cm long, obtuse. Fruit orange, acid.

Montane forest and pastures, often along stream sides, 2500-3800 m. GJ SU AR BA KF; also on a few mountains in Uganda, Kenya and Tanzania. *Ash* 1646; *Gilbert* 1798; *Mooney* 8302.

5. **R. steudneri** *Schweinf.* (1868)
- type: GD, Gonder, *Steudner* 921 (B holo. destr.); SU/WG, 92 km NW of Ambo, on new access road to Fincha, *Ash* 1391 (ETH neo., K MO isoneo.).
R. scioanus Chiov. (1911) - types: SU, Cabanna (Kebena) valley below Entoto, *Negri* 234 (FT syn., K isosyn.) & *Negri* 306 (FT syn. not seen).
R. steudneri var. *aberensis* C.E. Gust. in *Ark. f. Bot.* **26A**: 25 (1934) - type: SD, Abera, *Neumann* 41 (B holo. destr.).
R. steudneri var. *sidamensis* C.E. Gust. in *Ark. f. Bot.* **26A**: 25 (1934) - type: SD, *Ellenbeck* 1748 (B holo. destr.).

Scandent shrub up to 4 m with deeply furrowed stems. Flowering shoots ± thickly covered by stellate hairs or sometimes glabrous, with ± recurved prickles. Leaves simple, trifoliolate or 5-foliolate, usually white-tomentose below (but sometimes glabrous except for the midrib and primary veins); leaflets obovate-elliptic, shallowly serrate-biserrate, rhachis prickly. Inflorescence usually a large ± pyramidal panicle with tomentose branches and pedicels. Sepals ovate-lanceolate 4-7 mm long, acute, tomentose (except for lowest adaxial part), reflexed after anthesis. Petals pink, ovate-oblanceolate, 7-10 x 5-8 mm. Carpels about 25, glabrous or with apical tuft of hairs. Fruit orange to dark red, glabrous. Fig. 91.1.2

Forest fringes and secondary scrub, 2300-3000 m. TU GD WG SU AR HA KF GG SD; also on a few mountains in Congo, Uganda, Kenya and Tanzania. *Ash* 1391; *Friis* et al. 1436; *Mooney* 5081.

6. **R. apetalus** *Poir.* (1804).
R. rigidus Sm. (1819).
R. exsuccus A. Rich. (1847) - type: GD, Simien, Mt. Aber near Dschenausa, *Schimper* II:867 (P holo., BM K syn.).
R. petitianus A. Rich. (1847) - types: TU, Oudjerate, *Quartin Dillon & Petit* s.n. (P syn.); TU, near Adoua, *Quartin Dillon* s.n. (P syn.).
R. quartinianus A. Rich. (1847) - type: TU, Chiré (Shire), *Quartin Dillon* s.n. (P holo.).
R. assaortinus Chiov. (1915) - type: EE, Caribozzo wood in Assaorta, *Pappi* 2763 (FT holo.).
R. assaortinus var. *erythraeus* C.E. Gust. in *Ark. f. Bot.* **26A**: 37 (1934) - type: EW, Mogo near Mt. Sabur in Hamasen, *Fiori* 95 (FT holo.).
R. petalabigens C.E. Gust. (1934) - type: EW, Medrizien in Hamasen, *Pappi* 3543 (FT syn.); EW, Guna Guna in Scimezana, *Pappi* 688 (FT syn.).
R. chiovendae C.E. Gust. (1934) - types: GD, Gondar, *Chiovenda* 1076, 1505, 1317 (all FT syn.).
R. quartinianus var. *pappianus* C.E. Gust. in *Ark. f. Bot.* **26A**: 52 (1934) - type: EW, Hamasen, Mai Amus near Ad Nefas, *Pappi* 3596 (FT syn.); GD, Amhara-Dembia, *Chiovenda* 1266 (FT syn.).
R. quartinianus var. *hararensis* C.E. Gust. in *Ark. f. Bot.* **26A**: 52 (1934) - type: HA, Harar, *Robecchi-Brichetti* 78, 79 (FT syn., not seen).

Scrambling shrub to 2.5 m, stem of flowering shoots glabrous or usually ± densely pubescent with scattered hooked or deflexed spines. Leaves imparipinnate or trifoliolate; leaflets lanceolate to oblong-elliptic, up to 10 x 7 cm, shallowly serrate or biserrate, green and thinly pubescent above, pale and usually ± densely pubescent below, rachis with ± hooked prickles. Inflorescence paniculate, pyramidal, up to 35 x 13 cm, branches and pedicels ± whitish tomentose with scattered stiff hairs and prickles (and occasionally a few stalked glands). Sepals 4-7 mm long, lanceolate, apiculate, usually clasping. Petals absent or very short, occasionally to 5 mm, white or pink. Fruit black when ripe. Fig. 91.1.1.

Forest fringes, secondary scrub, stream-sides, etc., 1500-2500 m. EW TU GD WU GJ SU AR HA WG IL KF GG SD; also in Cameroon, Central African Republic, Uganda, Kenya, Tanzania, and Mascarene islands. *Friis* 215; *Gilbert* 501; *Hedberg* 9062.

A number of specimens of *Rubus* from KF, IL and GJ have been identified as *R. pinnatus* Willd. differing most markedly from *R. apetalus* in the sparse hairs on the lower surface of the leaves. This is a complex taxon in East Africa with three varieties recognised by Graham (1960). The Ethiopian material seems difficult to differentiate from the variable *R. apetalus* and further collections of *Rubus* from forests are needed to determine if *R. pinnatus* does occur in Ethiopia or if these are just almost glabrous specimens of *R. apetalus.*

7. **R. niveus** *Thunb.* (1913).

Erect, bushy shrub; stems glabrescent. Leaves with 2-3 pairs of leaflets, green and thinly hairy above, whitish-grey tomentose beneath; terminal leaflet broadly ovate, base cordate, larger than lateral leaflets. Inflorescence short, ± corymbose, leafy and usually somewhat hidden by leaves. Calyx lobes ± 5 mm long, exceeding the petals, densely pilose. Carpels densely tomentose, ripening dark red.

In larger gardens and probably naturalised among bushes in and around Addis Ababa, *c* 2400 m. SU; India and Malaya, also introduced into East Africa. *Bolus* et al. 11545; *Friis* et al. 3817.

2. ROSA *L.* (1753)

by Ö. Nilsson*

Erect or climbing shrubs (occasionally small trees), deciduous or evergreen. Stems and branches usually with prickles and/or thorns. Leaves imparipinnate, leaflets serrate; stipules usually sheathing with free

* Uppsala University - Botanical Gardens, P.O. Box 541, S-751 21 Uppsala, Sweden.

auricles. Flowers solitary or in corymbs, hermaphrodite. Sepals 5, entire or 3 outer lobed. Petals 5. Stamens, styles and carpels numerous; ovule 1 per carpel, pendulous. Receptacle flask-shaped, becoming coloured and fleshy when ripe. Styles free or connate in a column protruding through the opening of the receptacle.

About 100 species found mainly in temperate but also in subtropical areas of the Northern Hemisphere.

Some species and many cultivars of *Rosa* are grown very successfully for ornament throughout the highlands of Ethiopia.

1. Prickles abundant, of varying size, outer sepals with pinnately lobed extension, styles densely villous. **1. R. x richardii**
- Prickles rather sparse, uniform in size, sepals usually entire, soon deciduous, styles only slightly hairy. **2. R. abyssinica**

1. **R.** x **richardii** *Rehd.* (1922)

R. sancta A. Rich. (1848), non Andr., *R. centifolia* L. var. *sancta* (A. Rich.) Zab. - type: TU, collected around a church, *Quartin-Dillon* s.n. (P not seen)

Erect shrub, to 2 m. Prickles straight or slightly curved, from a narrow base, interspersed with stiff bristles. Leaves deciduous; leaflets 3-5, broadly elliptic or ovate, 2-4 cm long, obtuse, rounded or subcordate at base, simply glandular serrate with blunt teeth, dark green and distinctly rugose above, paler and pubescent beneath; petiole and rachis bristly and glandular-hispid; stipules rather narrow with short auricles. Flowers solitary or few in corymbs. Pedicels 3-6 cm straight, glabrous to densely glandular and bristly. Sepals shortly ovate with a pinnately lobed extension, reflexed. Petals 2-3.5 x 2-2.5 cm, pale rose (pink), fragrant. Disc rather flat, narrow with a wide orifice. Styles densely villous, free in upper part, long exerted, to *c* 1 cm. Receptacles ± globose, glabrous or densely glandular-hispid, not seen when ripe.

Cultivated since ancient times, particularly around churches; occasionally escaped into dry forests, around villages and on way-sides, 1000-2100 m. EW SU and certainly throughout the Christian highlands; Egypt. *Brodie* s.n., *Plowden* s.n.

R. x *richardii* is very probably the hybrid, *R. gallica* x *phoenicia*, but its origin is uncertain. It has been planted in churches and graveyards in Ethiopia for many centuries. The same rose was used to decorate the mummies during 500-200 B.C. in Egypt.

2. **R. abyssinica** *Lindley* (1820);

R. moschata Herrm. var. *abyssinica* (Lindley) Crépin *Bull. Soc. Bot. Belg.* **28**:47 (1899) - type: Abyssinia, *Salt* s.n. (BM holo.).

R. schimperiana Hochst. (1841) - type: TU, Mt. Selleuda (Mt. Scholoda), *Schimper* I:189 (P syn.)

R. schweinfurthi Bouleng. (1933).

Shrub, creeping or often climbing, sometimes forming a small tree 0.5-7 m; prickles usually slightly curved from a wide base. Leaves evergreen, somewhat leathery; leaflets usually 7, shortly petiolate, 0.8-6.5 cm, ovate to narrowly ovate to almost lanceolate, acute, rounded or cuneate at base, usually simply serrate, occasionally glandular biserrate with acute, low, numerous teeth, glabrous or hairy on main veins beneath or sometimes glandular beneath; petiole and rachis glabrous or often with some small prickles and glands; stipules narrow with short glandular auricles. Flowers usually 3-20 in rather dense corymbs occasionally solitary; bracts small, narrow, entire, deciduous. Pedicels 1.5-3.5 cm, straight, glabrous or sparsely hairy at least above and often sparsely glandular hispid. Sepals ovate, rather gradually tapering into a long, narrow tip, rather villous on back, deflexed. Petals 1.2-2.6 x 0.8-2 cm, white to pale yellow, slightly emarginate, fragrant. Disc conical, rather wide with a narrow orifice. Styles usually somewhat hairy sometimes glabrous, connate into a long, protruding column, *c* 0.5 cm. Receptacle globose to broadly ovoid, 0.7-1.8 cm, glabrous or usually ± hairy, at least above, and with sparse glands, red to orange red when ripe.

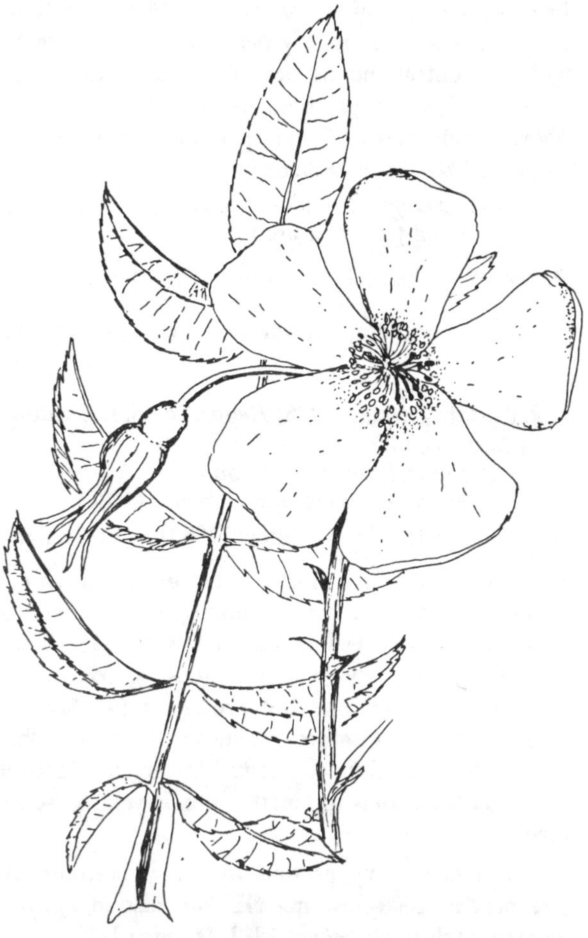

Figure 91.2 - *ROSA ABYSSINICA*: detached leaf and flowering shoot, x 1 from slide of Sue Edwards. Drawn by Sue Edwards.

Common, forming thickets in upland dry evergreen forest and margins and clearings; upland bushland, rocky places, dry grassland and riparian formations; also in different types of man-made habitats, sometimes standing alone as a small tree; 1900-3300 m. EW TU GD WU GJ SU AR HA BA; also in Arabia (Yemen), Somalia and Sudan. *Pichi-Sermolli* 664; *Robertson* 1321; *Gilbert* et al. 1102.

The fruits are edible and are collected and eaten frequently by children. There is a report of the fruits being used against hookworm.

The species is variable in many features such as the size of the leaves and the inflorescence. The small-leaved forms have been described as var. *microphylla* Almagia (FI holo), synonym *R. bottaiana* Bouleng. However, all intermediates between the extremes have been seen and there seems not to be any correspondence with ecological or geographical features.

R. abyssinica R. Br. in Salt (1814) *nom. nud.* was the first name applied to the Salt collection and taken up by Lindley. *R. schimperiana* Hochst. (1841) was used for *Schimper* I:189 collected from TU, below Mt. Scholoda.

3. **POTENTILLA** *L.* (1753)

Perennial herbs or shrubs, rarely annuals or biennials. Leaves pinnate, digitate or ternate. Flowers solitary or in cymes, bisexual, usually parts in 5's. Receptacle flat with a central hemispherical portion carrying the carpels. Epicalyx present, green as the calyx. Petals obovate, orbicular or linear-spathulate. Stamens 10-30, carpels 10-80. Fruits achenes.

About 500 species; almost cosmopolitan, but most species are temperate or arctic.

1. Leaves pinnate; flowering stems erect. **1. P. dentata**
- Leaves digitate; flowering stems procumbent. **2. P. reptans**

1. **P. dentata** *Forssk.* (1775) *Hepper* in ed. *(Kew Bull. Add. Series)*.
P. pennsylvanica auct. non L.
P. hispanica auctt. non Zimmet. (1884).

See Ball, Pawtowski & Walters (1968) in *Fl. Eur.* 2:40.

Perennial herb with thick, often branched rhizomes crowned by leaf rosettes containing flowering stalks up to 40 cm tall. Leaves imparipinnate with 7-15 oblong to lanceolate, coarsely dentate lobes, green or grey; stipules adnate at base but producing two lanceolate lobes at top. Flowers numerous. Sepals oblong-lanceolate, epicalyx segments linear to lanceolate, equalling the sepals in length. Petals yellow, 8-12 mm long.

On moist stony ground in valley bottoms along streamsides; 2600-3900 m. GD BA; also in Kenya and Yemen. *Ash* 1652; *Gilbert* 1817; *Mooney* 7122.

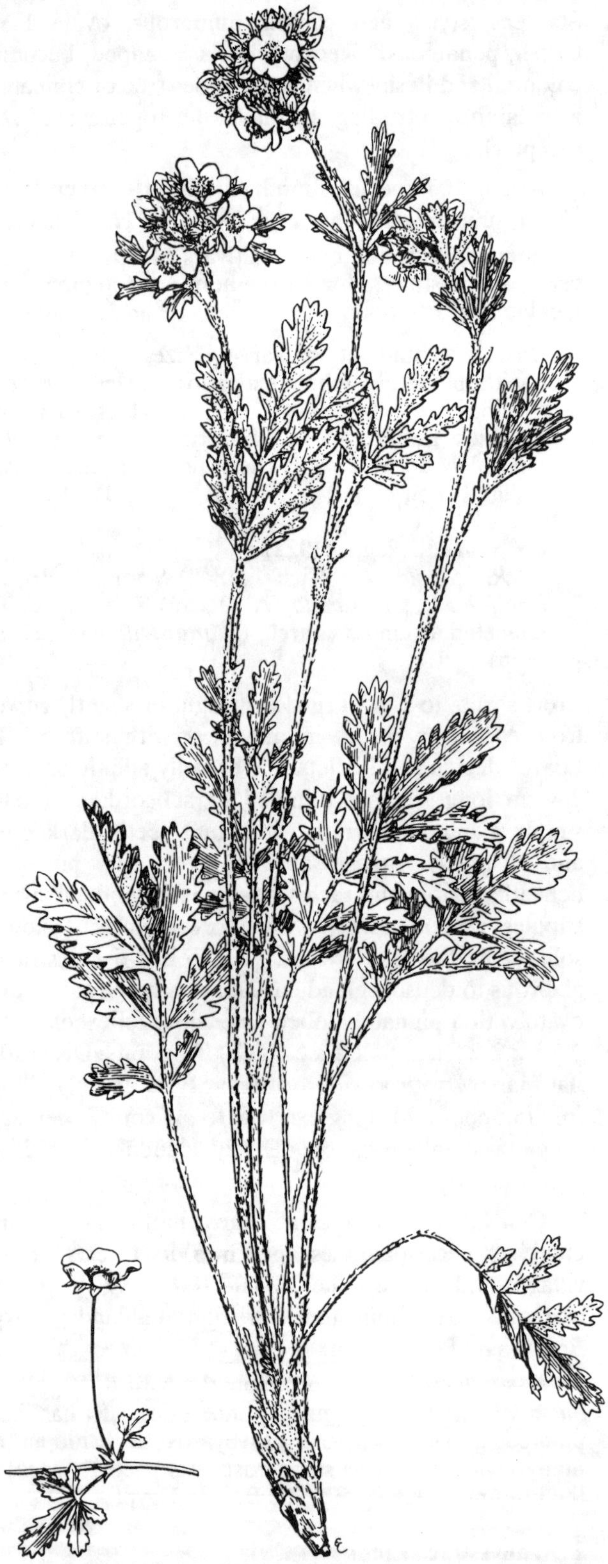

Figure 91.3.1 - *POTENTILLA DENTATA*: whole plant from *Ash* 2913. 2 - *POTENTILLA REPTANS*: node of rhizome with leaves and flower from *Burger* 2646, x 2/3. Drawn by Eleanor Catherine.

2. **P. reptans** *L.* (1753).

P. abyssinica A. Rich. (1847) - type: TU, Chéle-got (Tchéleukote), *Petit* s.n. (P holo. not seen).

Perennial with stout rhizome and persistent leaf rosette giving rise to procumbent stems up to 1 m long, rooting at the nodes and producing leaves and flowers. Leaves digitate with 5-7 leaflets up to 56 x 23 mm, dentate or serrate; stipules basally adnate, apically produced with 2 lanceolate lobes. Sepals and epicalyx segments variable. Petals 8-12 mm, much longer than sepals. Stamens *c* 20. Carpels 60-120.

Moist grassy places along stream-sides and in forest remnants; 2000-2500 m. EW TU GJ SU HA; also in N Africa, the Orient, Europe, E Asia and N America. *Burger* 2646; *Amare Getahun* A-29; *Mooney* 6389.

From the description *P. abyssinica* would seem to be a rather large specimen of *P. reptans* of which more material is now available: *Schweinfurth & Riva*, 1238 from EW, *Mooney* 6389 from Jem Jem forest, *Schimper* I:1840 from Adua.

4. FRAGARIA *L.* (1753)

Tutin in *Fl. Europaea* vol. 2: 47-48.

Perennial stoloniferous herbs. Leaves 3-foliolate, densely clustered in a basal rosette. Flowers very similar to *Potentilla* except that the petals are white and the receptacle becomes swollen and fleshy turning red when ripe.

A temperate and subtropical genus with edible fruits. Many species have been taken into cultivation and a very large number of hybrids and cutivated varieties developed. Both the following species have been introduced into Ethiopia where they can be grown successfully on a commercial scale.

1. Sepals patent or deflexed after flowering; leaves bright green, sparsely hairy on the upper surface. **1. F. vesca**
- Sepals appressed after flowering; leaves blue-green, nearly or quite glabrous. **2. F. x ananassa**

1. **F. vesca** *L.* (1753).

Perennial herb, with long stolons rooting at the ends. Leaves 3-foliolate in a basal rosette; leaflets 1-6 cm, ovate or obovate to rhombic, coarsely serrate, bright green and sparsely hairy on the upper surface. Peduncle 5-30 cm, little longer than the leaves, erect in fruit. Pedicels appressed-pubescent. Flowers *c* 15 mm in diameter, white usually hermaphrodite. Sepals patent or deflexed in fruit. Achenes uniformly scattered over and projecting from the usually red, glabrous receptacle.

Cultivated in gardens in Asmera and probably also Addis Abeba; *c* 2300 m. EW ?SU; Europe, temperate Asia, N America.

2. **F. x ananassa** *Duchesne* (1766).

Perennial herb producing numerous stolons. Leaves 3-foliolate in a basal rosette, rather leathery, blue-green and nearly or not quite glabrous above. Peduncle up to 25 cm, decumbent in fruit. Flowers white, hermaphrodite. Sepals appressed after flowering. Receptacle up to 3 cm wide after flowering, covered all over with achenes on the surface or slightly sunken into the flesh.

Cultivated in gardens and on a commercial scale; 1700-2000(-2400) m. ?EW, SU; worldwide.

This species is a hybrid between *F. virginiana* Duchesne and *F. chiloensis* (L.) Duchesne, which has given rise to numerous cultivars grown throughout the world. In central Ethiopia, commercial production has its greatest potential under irrigation at higher elevations in the Rift Valley, such as around Nazret.

5. ALCHEMILLA *L.*

Herbs or low shrubs. Stems erect or decumbent, often arising from a basal leaf rosette. Leaves radical and cauline, petiolate, divided into a varying number of lobes; petioles ± adnate to foliaceous or membranous stipules, which may differ between radical and cauline leaves. Inflorescence axillary, few- to many-flowered, exceeding or ± hidden within stipules. Flowers small, bisexual, parts in 4's. Receptacle flask-shaped with its mouth closed by a fleshy disc, carrying an inner series of 4 free "calyx" lobes alternating with 4 outer "epicalyx" lobes. Petals absent. Stamens 4, inserted on the rim of the disc, alternate with calyx lobes. Carpels 1-20, inserted at the base of the receptacle; style basal. Achenes included within the receptacle.

About 250 species, mostly in temperate areas and in tropical mountains. Some species have apomictic seed formation, but the type of variation found in Ethiopian material indicates sexual reproduction - as does the occurrence of at least one clear hybrid between widely different species - see Hedberg (1986).

1. Without basal leaf rosettes; stems erect or ascending. 2
- With basal leaf rosettes; stems usually procumbent. 5
2. Stems woody; stipules forming tubular sheaths. **1. A. haumannii**
- Stems herbaceous; stipules not tubular and sheathing. 3
3. Full-grown leaves more than 5 cm wide; stipules membranous. **3. A. fischeri**
- Full-grown leaves at most 4 cm wide; stipules green, not membranous. 4
4. Stipules adnate to the petiole for half their length; inflorescence usually concealed within stipules. **5. A. cryptantha**
- Stipules free from the petiole. **2. A. ellenbeckii**
5. Inflorescence few-flowered; not or barely exceeding the stipules. 6

- Inflorescence many-flowered, much exceeding the stipules. 8
6. All stipules essentially membranous. 7
- Stipules of cauline leaves entirely green and leaf-like. **5. A. cryptantha**
7. Inflorescence slightly exceeding stipules; whole plant with dense silvery indumentum. **8. A. abyssinica**
- Inflorescence concealed within stipules. **6. A. microbetula**

8. All stipules membranous, adnate throughout; leaf segments with 25-35 teeth each. **3. A. fischeri**
- Stipules of cauline leaves green, partly free; leaf segments with at most 23 teeth each. 9
9. Achenes 5-8 in a fruit; pedicels short (0.5-2 mm). 10
- Achenes 1-3 in a fruit; pedicels usually longer (2-15 mm). 11
10. Leaves deeply dissected, with lobes more long than wide; flowers usually emerging beyond the stipules. **4. A. kiwuënsis**
- Leaves not deeply dissected, with lobes more wide than long; flowers mostly remaining concealed in the stipular sheath. **5. A. cryptantha**
11. Leaves deeply dissected, with lobes more long than wide; whole plant with appressed silvery pubescence. **7. A. pedata**
- Leaves not deeply dissected, with lobes more wide than long; whole plant with spreading hairs but not silvery pubescent. **8. A. abyssinica**

1. **A. haumannii** *Rothm.* (1936);
A. geranioides Rolfe subsp. *chillaloensis* Haum. & Balle in *Rev. Zool. Bot. Afr.* **24**: 351 (1934); *A. chillaloensis* (Haum. & Balle) Haum. & Balle (1936), nom. illeg. - type: AR, Mt. Chilalo (western slope facing Lake Zway), *Scott* s.n. (K holo.).

Small shrub with distinctly woody, often richly branched stems, sometimes rooting at some upper nodes when these touch the ground; stem surface densely pubescent and largely covered by stipular sheaths. Leaves reniform, 5-7-9-lobed, usually densely pubescent on both surfaces, but sometimes glabrous above. Stipules adnate to the petiole in their lower part, fused in their upper part to form an ocrea-like structure opposite to the leaf. Inflorescence barely exceeding the leaves, ± compact. Calyx lobes ovate, 1-2 mm long, epicalyx lobes much smaller, 0.5-0.8 mm long. Achene 1 (rarely 2).

Moist moorland in the ericaceous and Afroalpine belts, 3000-4250 m. AR BA GG; endemic to S Ethiopia. *Ash* 2329; *Hedberg* 4255; *Mooney* 7149.

There is some considerable variation in the lobing, folding and pubescence of the leaves, but a taxonomic subdivision seems so far not recommendable. There are indications of hybridization between this species and *A. pedata*, and further field studies are desirable.

The name *A. haumannii* Rothm. has one month's priority over *A. chillaloensis* Haum & Balle.

2. **A. ellenbeckii** *Engl.* (1911)
- type: SD, Abera, *Ellenbeck* 1784 (B destr. BR iso.).
A. gracilis Pax (1907), nom. illeg., non Opiz (1836) nec Buser (1895) - type: *F. Rosen* s.n. (BRSL holo. not seen).

Scrambling herb without leaf rosette, stems terete, purplish, erect-ascending, ± branched and forming thick clumps; internodes 1-2.5 cm long. Leaves reniform-rounded, ± deeply 5-lobed, 1-2.5 cm wide, pale below with darker veins, petiole 0.2-1.5 cm long. Stipules

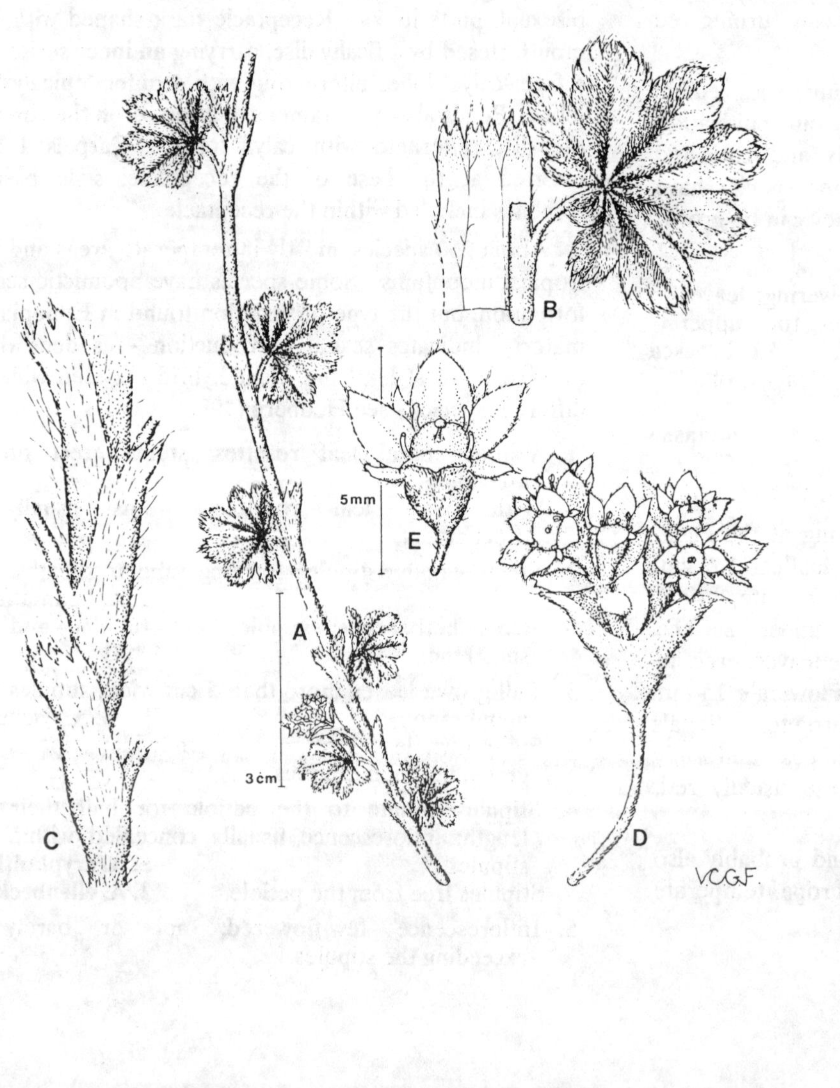

Figure 91.4. *ALCHEMILLA HAUMANNII*: A - portion of stem x 1; B - detail of leaf and stipule enlarged; C - detail of stipules; D - enlargement of inflorescence; E - detail of flower; all from *Smeds* 558. Drawn by Victoria Friis.

bilobed, dentate, leaf-like, free from petiole. Inflorescence simple or richly branched, exceeding the subtending leaf. Calyx lobes 1-1.2 mm long, epicalyx lobes 0.6-1.2 mm. Achenes 1-3.

On moist ground along small streams, especially in the ericaceous belt, 2600-4000 m. AR BA GG; Uganda, Kenya, Tanzania, with another subspecies in Malawi and Zambia. *E.F. Gilbert* 511; *Hedberg* 4142; *Mooney* 7149.

3. **A. fischeri** *Engl.* (1892).

A. erlangeriana Engl. (1911) - type: SD, Jem Jem (Djam-Djam), *Ellenbeck* 1778 (B syn. destr.); SD, Busaftu, *Ellenbeck* 1871 (B syn. destr.).

Herb with trailing, often branched stem up to 1.2 m long and 0.9 cm thick. Each shoot may produce either a ± condensed leaf rosette or carry widely separate leaves and inflorescences. Leaves to 12 cm wide, pubescent on both sides, 9-19-lobed, lobes oblong-obovate, dentate to the base; stipules all membranous, to 7 cm long, truncate at apex, adnate throughout to petiole. Inflorescence to 40 cm long, paniculate and richly branched; pedicels 0.5-2 mm long. Calyx lobes ovate-triangular, 1.5 mm long, epicalyx lobes equalling or shorter, narrower. Achenes 1-3.

On moist ground in swamps and along brooks in montane forest, sometimes in fallow fields, 1600-3200 m. GD WG BA IL KF GG; Kenya, Tanzania. *Ash* 1487; *Friis* et al. 318; *Mooney* 8643.

4. **A. kiwuënsis** *Engl.* (1911).

Weak herb with leaf rosettes producing stolons. Leaves long-petiolate, reniform-orbicular; rosette leaves 5-7 (-9)-lobed; division usually deeper than half the lamina; lobes oblong-elliptic, terminally rounded, dentate all round with 12-22 teeth. Stem leaves smaller, usually 5-lobed, with shorter petiole. Stipules of rosette leaves entirely membranous, those of stolon leaves apically foliaceous. Inflorescences in axils of rosette and stolon leaves, exceeding the leaves, sparsely branched. Pedicels less than 1 mm. Epicalyx lobes longer and wider than calyx lobes. Achenes 5-8.

Moist grassland along stream sides and forest edges, 2300-3300 m. EW SU BA KF SD; Uganda, Kenya, Tanzania, Central African Republic W to Fernando Po. *E.F. Gilbert* 383; *Gillett* 14337; *Hedberg* 9036.

Occasionally inflorescences covered in the stipular sheath occur, as in *A. cryptantha*. The distinction between these two taxa is not sharp.

5. **A. cryptantha** *A. Rich.* (1847)

- type: GD, between Enschedcap and Schoata in Semien (Simen), *Schimper* II:566 (P lecto.; BM K S isolecto.).

Weak herb with short-lived leaf rosette and prostrate, furrowed, green stems, rooting at the (lower) nodes; internodes 2.5-7 cm long; whole plant loosely pubescent from patent hairs. Leaves 1.5-5 cm wide, reniform, 5-lobed, each lobe truncate with 8-15 teeth; stipules of rosette leaves membranous, those of stem leaves foliaceous, each half with 4-10 teeth. Inflorescences usually small and few-flowered, concealed in, or barely protruding from the stipules, but occasionally fairly long, much exceeding the leaves. Calyx lobes less than 1 mm long, epicalyx lobes slightly longer. Achenes (2-)5-8.

On moist shaded ground in montane rain forest and along streams, sometimes as a weed in lawns; 1500-3200 m. EW GD WU GJ WG SU AR HA KF SD; Sudan, Uganda, Kenya, Tanzania, S to Transvaal, also Cameroons and Madagascar. *Gillett* 5281; *Sebsebe Demissew* 18; *de Wilde* 6260.

6. **A. microbetula** *Th. C.E. Fries* (1923).

A. sessiliflora Rothmaler (1935) - type: GD, Mt. Buahit in Semien (Simen), *Schimper* 648 (S holo., P iso.).

A. commutata Rothmaler (1938) - types: GD, Mt. Buahit, *Schimper* 648 bis (P holo. not seen); GD, Mt. Dedschen (Mt. Ras Dejen), *Schimper* s.n. (S syn.).

A. abyssinica var. *schimperi* Engl. in *Engl. Bot. Jahrb.* 46:130 (1911) - type: *Schimper* s.n. (B destr.).

A. abyssinica f. *muscoidea* Haum. & Balle (1936); *A. commutata* f. *muscoides* (Haum. & Balle) Rothmaler (1938) - type: GD, Mt. Buahit in Semien (Simen), *Schimper* 664 (P holo.)

A. abyssinica auct. non Fresen. 1837: Engler 1892:237

Dwarf, prostrate herb with basal rosettes mostly producing herbaceous or suffrutescent stolons, rooting at the nodes. Leaves on petioles 5-15 (-25) mm long or almost sessile, broadly ovate, 5-13(-22) mm wide, 5-lobed with partitions reaching halfway or almost to the base; stipules entire or with 2-3 teeth, entirely membranous or greenish in the upper part only, adnate for most of their length to the petiole. Inflorescence of 1-3 flowers, usually hidden within the stipules. Flowers 2-3 mm long with calyx and epicalyx lobes 0.5-1 mm long, the latter usually slightly narrower. Achenes 1-4 (in Ugandan material up to 20) per flower, 1.3-2 x 0.9-1.2 mm, greyish brown.

On moist ground kept open by solifluction, erosion, rodent grazing, road building, etc., sometimes on thin soil over rocks; mainly in the afroalpine belt; 3200-4400 m. GD SU AR BA; Uganda, Kenya and Tanzania. *Hedberg & Aweke* 5434; *Hedberg* 4148; *S. Gilbert* 44.

A deviating specimen, *Puff* et al. 810927-2 (ETH) has inflorescences to 4 cm long with 7 flowers.

7. **A. pedata** *A. Rich.* (1847)

- type: GD, Enchetkab (Enschedcap) in Semien (Simen), *Schimper* II:1166 (P holo., BM S iso.).

A. pedata var. *argentea* Fiori in *N. Giorn. Bot. Ital.* **24**:481 (1916) - type: EW, Paradiso in Asmera, *Baldrati* s.n. (FI holo. not seen).

Stoloniferous herb with primary basal leaf rosette and sometimes secondary rosettes produced by stolons; side roots thick and furrowed. Stems and leaves (particularly below) with densely appressed silvery silky hairs. Rosette leaves up to 5 cm wide, usually (7-)9-(11-)lobed with membranous stipules adnate throughout; stolon leaves (5-)7-lobed with leaf-like stipules adnate in lower half. Inflorescences in leaf axils of stolons, much exceeding subtending leaves, usually elongate but sometimes condensed, unbranched or sparsely branched, with one flower (and sometimes an inflorescence branch) per bract; pedicels 3-15 mm. Calyx lobes ovate-triangular, 2-2.5 mm long, epicalyx lobes equalling or exceeding, but narrower than the calyx lobes. Achenes 1-3.

Moist ground in grazed montane grassland; 1800-4000 m. EW TU GD WU GJ SU AR BA SD; Kenya and Tanzania. *De Wilde* 7984, *Ensermu Kelbessa* 99, *Hedberg* 9105.

8. **A abyssinica** *Fresen.* (1837)
- type: GD/TU, between Tembien (Temben) and Semien (Simen), *Rüppell* s.n. (FR holo.).

A. gunae Schweinf. (1868) - type: GD, Mt. Guna, *Steudner* 923 (B holo. destr., K iso.).
A. rothii Oliv. (1871) - type: SU, Ankober, *Roth* s.n. (K holo.).
A. steudneri Th. C.E.Fries (1923) - type: WU, Jennija near Magdala (Mekdela), *Steudner* 922 (B holo. destr., UPS iso.).

Kalheber (1980) in *Senckenbergiana biol.* **61**:127-142.

Robust herb with basal leaf rosette producing long stolons with 1-8 cm internodes, often rooting at the nodes and forming secondary rosettes. Leaves reniform-orbicular, with 7-9 terminally truncate or broadly rounded lobes each with 9-13 subacute teeth. Stipules bilobed, apically dentate, those of rosette leaves membranous, brownish, truncate; those of stem leaves leaf-like, apically dentate. Inflorescences mostly much exceeding the leaves, usually lax with pedicels 2-12 mm long but sometimes compact. Calyx lobes ovate-triangular, 1.7-2 mm long, epicalyx lobes shorter and narrower. Achenes 1-2.

In moist montane forests as well as on moist places in somewhat overgrazed moorland and on rocky slopes; 2500-4400 m. TU WU GD GJ SU BA HA SD; Kenya. *Gilbert* 1788; *Hedberg* 5535; *de Wilde* 6814.

Taxa of uncertain status.

It has not been possible to ascertain the identity of *A. commutata* f. *robusta* Rothmaler (1938), based on the type *Rochet d'Hericourt* s.n. collected from Ras (Mount) Buahit in Simen.

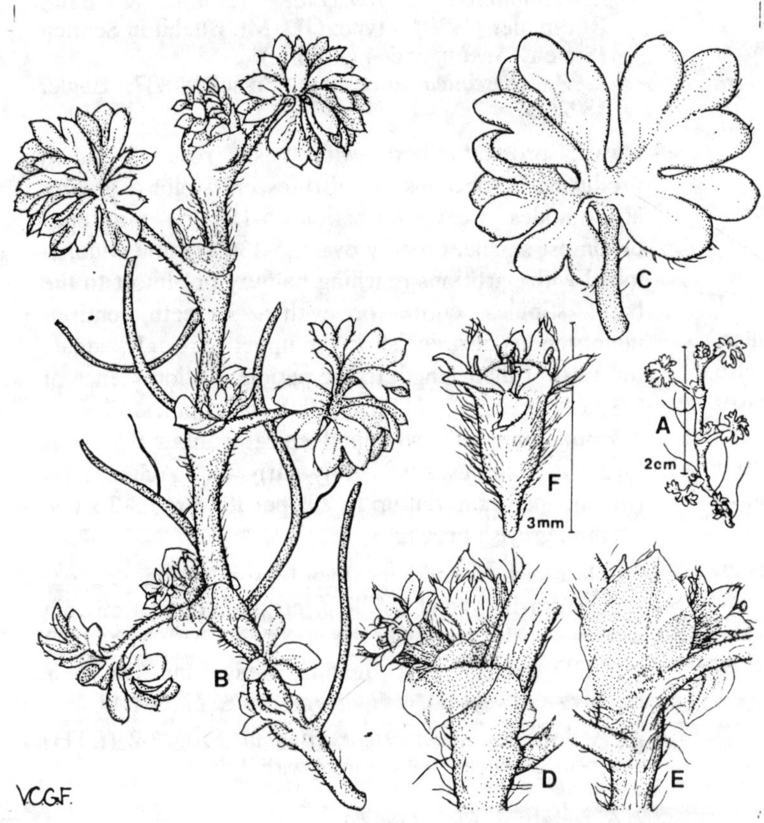

Figure 91.5. *ALCHEMILLA MICROBETULA*: A - portion of stem x 1; B - same much enlarged; C - leaf enlarged; D - enlargement of stipules with inflorescence in leaf axil, front and back view; E - detail of flower. All from *Sue Gilbert (Edwards)* 44. Drawn by Victoria Friis.

6. APHANES *L.* (1753)

Small annual herbs with deeply dissected leaves and connate stipules. Flowers in condensed cymes, opposite to the leaves. Stamen 1, on the inner margin of the disc.

About 20 species in America, the Mediterranean area, Central Asia and Australia, with one species in Ethiopia.

A. bachiti *(Hauman & Balle) Rothm.* (1937),
Alchemilla bachiti Hauman & Balle (1936), *A. arvensis* Scop. f. *alpina* Engl. in *Abh. Preuss. Akd. Wiss.* 1891:237 (1892) - type: GD, Mt Buahit (Bachit) in Semien (Simen), *Schimper* 124 (P not seen, S).

Minute herb, 1-3 cm tall, with internodes of less than 5 mm. Leaves irregularly 4-7-lobed; stipules adnate to the petiole and connate, covering at least the base of the inflorescence. Epicalyx missing completely.

On bare patches in tussock grassland in the upper part of the ericaceous belt, about 3700 m. GD, only known from Ethiopia. *Hedberg & Getachew* 5414.

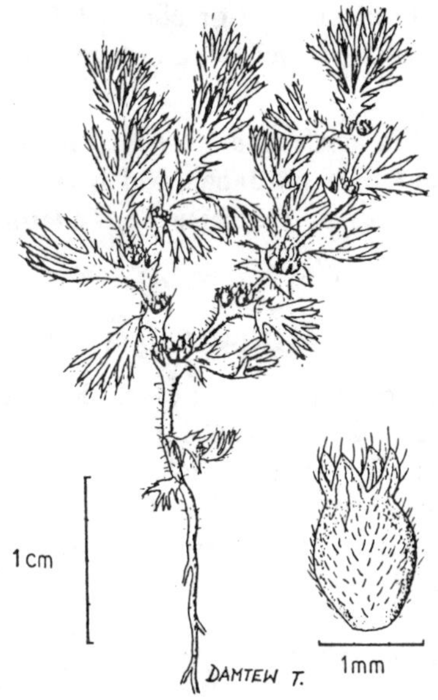

Figure 91.6 - *APHANES BACHITI*, whole plant and a single flower from *Hedberg & Getachew* 5414. Drawn by Damtew Tefera.

7. COTONEASTER *Medik* (1789)

Unarmed shrubs or small trees. Leaves entire, leathery, deciduous (in Ethiopian species); stipules falling quickly. Flowers small, solitary or in cymes or corymbs. Sepals persistent, petals white or pink, stamens *c* 20, carpels 2-5, partly free. Fruit red or black with mealy flesh; pyrenes 2-5.

About 50 species mainly in the N temperate zone.

C. nummularia *Fischer & C. A. Meyer* (1835).

Erect deciduous shrub to 2.5 m, sometimes low and spreading. Leaves broadly elliptic to orbicular, mucronate, glabrescent above and tomentose below; petioles 2-4 mm. Inflorescence compact, 3-7-flowered, dense; petals white, spreading; pedicels, receptacle and sepals white tomentose. Fruit subglobose, *c* 8 mm, red, with 2 pyrenes.

Mountain scrub; 2500-2600 m. EW; also in Crete, Cyprus, Syria, Lebanon, N. Iraq, Caucasia, Turkey and W Iran. *Pappi* 1311, 1576.

Only known in Ethiopia from these two collections.

8. PRUNUS *L.* (1753)

Kalkman in *Blumea* **13**:1-115 (1965)

Trees or shrubs with entire or serrate petiolate leaves and usually bisexual flowers. The receptacle is wide and cup-like carrying at the rim 5-14 segments which are sometimes uniseriate with subequal members, and sometimes biseriate and differentiated into petals and sepals. The number of stamens varies between 10 and 85. There is one carpel with 1-2 ovules which gives rise either to a 1-2 seeded drupe or to a dry achene.

Several species have been taken into cultivation but only the peach, *Prunus persica,* is widely grown in Ethiopia. With the delimitation followed here (including the genera *Amygdalus, Cerasus, Laurocerasus, Padus, Persica* and *Pygeum*) it includes about 430 species and is subcosmopolitan.

Both Richard (1847) and Cufodontis, *loc. cit.*, record *Armeniaca vulgaris* Lam, synonym *Prunus armeniaca* L., the apricot, as cultivated in Eritrea, but no further evidence of its cultivation has been found.

1. Flowers small, petals *c* 2 mm long; fruit 5-8 mm long. **1. P. africana**
- Flowers and fruits larger (fruit trees). 2
2. Ovary and fruit pubescent to tomentose; stones distinctly pitted. 3
- Ovary and fruit glabrous; stones ± smooth. **3. P. x domestica**
3. Fruit dehiscent, pericarp becoming dry (almonds). **4. P. communis**
- Fruit indehiscent, pericarp fleshy (peaches). **2. P. persica**

1. **P. africana** *(Hook. f.) Kalkm.* (1965).
Pygeum africanum Hook. f. (1864).

Evergreen tree to 40 m with dark brown to grey longitudinally fissured or scaly bark. Leaves leathery, elliptic-oblong, to 15 x 5 cm, shallowly crenate-serrate. Inflorescence a 10-30-flowered raceme, 3-8 cm long. Pedicels 2-6 cm long. Receptacle 3-5 mm wide. Sepals 5, triangular, 1-1.5 mm long. Petals 5, *c* 2 mm long, creamy white, pubescent, often revolute. Stamens 25-35, filaments *c* 2 mm long. Style 2 mm long, ovary ovoid. Fruit dry, transversely ellipsoid, *c* 5-8 mm long and 9-11 mm wide, dark reddish brown. Fig. 91.1.3.

In montane or riverine forest; 1700-2500 m. GD GJ WG SU AR BA HA IL KF SD; also in W Tropical Africa (Ivory Coast, Cameroon, Fernando Po, San

Tome), Central Africa ("Congo"), East Africa (Uganda, Kenya, Tanzania), S Tropical Africa (Angola, Zambia, Malawi, Zimbabwe), S. Africa, Madagascar and the Comores. *Burger* 2434; *Friis* et al. 157; *Mooney* 5687.

The stem bark provides a popular medicine against urinary disorders. The wood makes good timber with many uses

2. **P. persica** *(L.) Batsch* (1801).
Amygdalus persica L. (1753).
Persica vulgaris Miller (1768).

Fruit tree up to 6 m; bark grey-brown, splitting; twigs glabrous, reddish. Leaves oblong-lanceolate, 5-15 x 2-4 cm, serrulate. Flowers subsessile, mostly solitary. Petals 10-20 mm long, deep pink. Fruit subglobose, 40-80 mm, velvety.

Widely planted in the highlands in home gardens; *c* 1800-2200 m. Throughout the cultivated highlands in Ethiopia; a native of SW Asia, extensively cultivated for its fruits in the warm temperate parts of Europe, Asia and N America. *Amare Getahun* 44, *Stokes* 4, *Tadesse Ebba* 507.

Although severely affected with peach curl, it produces large quantities of small, rather hard fruits which are eaten fresh and are very popular.

3. **P. x domestica** *L.*

Small deciduous tree up to 12 m, unarmed or with small spines, twigs glabrous or pubescent, especially when young. Leaves up to 10 x 6 cm, elliptic or oblong, pubescent when young, later becoming glabrous or sparsely pilose beneath, hairs often restricted to midrib and veins; petioles up to 20 mm. Flowers white, 15-25 mm diam., in 2(-4)-flowered clusters; pedicels up to 20 mm. Drupes pendent, ovoid or subglobose, up to 8 cm long; greenish, yellow or red to violet and dark blue; flesh greenish, yellowish or reddish, sweet; stone ± rough to smooth.

Cultivated around Addis Abeba and Asmera and with good potential for other highland areas over 1900 m. EW SU; long cultivated in Europe, W. Asia and N. Africa.

4. **P. communis** *(L.) Arc.* (1882).
Amygdalus communis L. (1753).
Prunus amygdalus Batsch. (1801).

Unarmed tree to 8 m; young shoots glabrous. Leaves ovate-lanceolate or elliptic, 2.5-9(-12) x 2-3 cm, dentate-serrate, glandular, entirely glabrous or pubescent beneath when young, leathery, petioles 10-30 mm. Flowers white or pink, up to 4 cm diameter, pedicels up to 5 mm. Drupe obliquely ellipsoid to elongate-ellipsoid, soon dry, compressed, up to 50 x 30 mm, olive-green, velvety, stone pitted and ± grooved, keeled.

Grown in a few gardens in Asmera and at experimental sites in SU; above 1900 m. EW SU; native in SW and C Asia (and possibly N Africa), and cultivated throughout its native range and in Europe both for ornament and for its edible seeds.

9. ERIOBOTRYA *Lindley* (1822)

Evergreen shrubs or trees. Leaves simple. Inflorescence a terminal panicle. Sepals persistent in fruit; petals ovate or suborbicular, clawed, white; stamens 20; carpels 2-5, walls thin in fruit; ovules 2; styles connate only at base. Fruit with 1 or a few large seeds.

A genus with 30 species from the Himalayas to Japan, SE Asia and W Malaysia. One species widely cultivated.

E. japonica *(Thunb.) Lindl.* (1831).

Shrub or small tree up to 6 m; shoots pubescent even at maturity. Leaves lanceolate-oblong to obovate, up to 25 x 7 cm, dentate with lateral veins ending in teeth, glabrous above, reddish-brown tomentose beneath. Panicles up to 17 cm long, broadly pyramidal. Flowers white, 1.5-2 cm diameter, strongly fragrant, sessile; calyx tomentose. Fruit obovate or ellipsoidal, up to 10 cm long in some cultivars, yellow or orange; seeds 1-1.5 cm, ovoid.

Cultivated in gardens; *c* 1800-2400 m. EW SU KF HA; native of China and Japan; widely cultivated for its edible fruits. *Sebsebe Demissew* 61; *Tadessa Ebba* 504; *Ellis* 61.

10. CYDONIA *Miller* (1754)

Large shrub or small tree. Leaves deciduous, entire. Flowers large, solitary; sepals dentate, persistent; petals larger than sepals, white or pink; stamens 15-25; styles 5, free; ovary 5-locular, inferior. Fruit large, obovate to subglobose, with a leathery carpel wall and many seeds.

Monotypic.

C. oblonga *Miller* (1768).

Shrub or tree to 8 m; twigs sparsely tomentose when young, becoming glabrous. Leaves ovate to oblong or occasionally suborbicular, up to 10 x 7 cm, entire, both sides white-felty at first, becoming glabrous above and densely villous beneath later; petiole 1-2 cm. Flowers 4-6 cm diameter, sepals glandular, toothed, reflexed. Fruit obovate to subglobose, (3-)5-12 cm long, yellowish, velvety, flesh fragrant.

Cultivated in some gardens and abandoned or escaped on Entoto, Addis Abeba, also in Asmera; *c* 2400 m. SU EW; native in Caucasia, N Iran and possibly N Iraq; widely cultivated in temperate regions for its edible fruits.

11. **MALUS** *Miller* (1754)

Shrubs or trees, unarmed or sometimes with subspinescent branchlets; twigs slightly tomentose. Leaves simple, sometimes lobed, stipules falling quickly. Flowers in umbels; petals clawed, white, pink or red; stamens 15-50; anthers dehiscing towards the middle, yellow; styles usually 5, united at the base; carpels 3-5, united, walls cartilaginous in fruit. Fruit globose with a depression at the base, green, yellow, red, or with a mixture of these colours, usually without grit cells.

Genus with *c* 35 species in north temperate zone, several of which are cultivated for their fruits and/or for ornament.

M. sylvestris *Miller* (1768).
M. communis Desf. (1798).
M. domestica Borkh. (1803).

Tree to 8(-12) m. Leaves 3-11 x 2.5-5.5 cm, ovate, elliptic to obovate or suborbicular, crenate or serrate, pilose when young, glabrous or almost so when mature, tomentose beneath; petioles tomentose, ¼-½ x blade length. Flowers 3-4 cm diameter; calyx glabrous outside, pilose inside; petals white or pink. Pome subglobose, tomentose to glabrous, sometimes developing a waxy covering, up to 15 cm diameter in some cultivars.

Cultivated in gardens and experimental sites; 1700-2500(-3000) m. EW SU AR HA; native of SW Asia now cultivated throughout the world for its edible fruits. *Plowden* s.n.

Plowden s.n. (K) was identified as *M. pumila* Mill. because of its very small fruits: this is generally conisdered as a synonym for *M. sylvestris*.

12. **PYRUS** *L.* (1753)

Shrubs or trees, often spiny on lower branches or when young. Leaves simple, rarely lobed; stipules falling quickly. Flowers in corymbs; petals white, clawed; stamens 15-30, anthers dehiscing outwards, usually red; carpels 2-5, free. Fruit obovate, flesh containing grit-cells.

Genus with 30 species in E Europe and W Asia.

P. communis *L.* (1753).

Tree up to 20 m, with spines, especially when young; twigs glabrous or slightly pubescent when young. Leaves ovate-elliptic to ovate-orbicular, 3-5(-7) x 1.5-4 cm, usually *c* twice as long as broad, entire or crenate-serrulate, sparsely pubescent when young, becoming glabrous, green and shining above, drying blackish; base rounded or subcordate; petioles up to 5 cm, usually as long as or longer than the blade. Inflorescence 5-9-flowered. Flowers white, 2-3 cm diameter; pedicels 1.5-3 cm. Fruit obovate to subglobose, without a depression at the base, 2-16 x 4-12 cm, flesh with a sweet taste; calyx persistent.

Cultivated in gardens and experimental sites; 1700-2500 m. EW SU AR; native of SW Asia, probably of hybrid origin with several species in its parentage. Now widely grown in all temperate areas.

13. **HAGENIA** *J.F. Gmel.* (1791)
Banksia (Bankesia) Bruce (1790)
non Linnaeus f. (1781) nom. illegit.

Graham (1960) in *Fl.Trop.E.Afr.*

Dioecious (or polygamous) tree (hermaphrodite flowers rare) with imparipinnate leaves. Flowers small, anemophilous, in many-flowered panicles. Receptacle "urn-shaped", apically closed and carrying 2 alternating rows of membranous calyx lobes, which provide a parachute for the fruit. Carpels 2, with one pendant ovule in each. Achenes brown, reticulately rugose, ovoid.

A monotypic genus confined to Africa.

H. abyssinica *(Bruce) J.F. Gmel.* (1791)
Banksia (Bankesia) abyssinica Bruce (1790) - type: illustration in Bruce *Travels,* vol. 5, opposite p. 73 (1790).

Tree up to 20 m tall with brownish bark peeling in strips, sometimes very thick on old stems. Young branches pubescent with stiff hairs 3-4 mm long. Leaves to 40 cm long with 5-8 narrowly oblong leaflets on each side; petiole 5-13 cm long, winged (adnate stipules). Inflorescence a conspicuous drooping panicle up to 60 cm long and 30 cm wide. Male flowers orange to white, female ones reddish.

Montane forest and grassland, 2450-3250 m. EW TU GD WU GJ WG SU AR BA HA KF SD; also in Sudan, Uganda, Kenya, Tanzania, Central African Republic, Zimbabwe and Malawi. *Ash* 1388; *Burger* 1415; *Mooney* 8380.

Hagenia abyssinica may have been one of the dominant trees in the upper part of the montane forest belt in Ethiopia, as it still is in East Africa, but now only scattered trees remain in most areas.

Provides a widely used and strong anthelmintic, 'kosso', from the female flowers.

Figure 91.7 - *HAGENIA ABYSSINICA*, branch with leaves and inflorescence, x ½; detail of inflorescence x 1; detail of flowers x 1.5. Reproduced from Bruce, *Travels to discover the source of the Nile*, ed. 2, vol. VII, No. 22 (1805). Drawn by Luigi Balugani.

92. NEURADACEAE

by M. G. Gilbert*

Zohary, *Flora Palestina* **2**: 24-25, fig. 37 (1972); Meikle, *Flora of Cyprus* **1**: 628-631 (1977).

Prostrate annual herbs; most parts densely pubescent. Leaves alternate, pinnately lobed to compound; stipules minute or obsolete. Flowers axillary, solitary, pedicellate, bisexual, 5-merous; ± epigynous. Receptacle discoid, enlarged and woody in fruit. Sepals valvate, sometimes alternating with epicalyx lobes. Petals free. Stamens free, twice as many as petals, inserted on slightly prolonged receptacle tube. Gynoecium of 3-10 carpels immersed in receptacle, styles subulate, stigma capitate, ovules solitary. Fruit discoid, spiny and/or with winged margin formed from accrescent sepals.

A very homogenous group of 3 genera and *c* 10 species, all restricted to southern African except for the following widespread northern genus.

The South African genera, *Neuradopsis* Bremek. & Obermeyer and *Grielum* L. differ by their much larger flowers, without an epicalyx, and fruits with a marginal wing formed from the accrescent sepals.

This family is often included within the **Rosaceae** as subfamily **Neuradoideae**. It differs only by the very distinctive pollen and the characteristic fruits. Ecologically it is atypical within the **Rosaceae** as it is the only group restricted to the arid subtropics, the family otherwise being largely restricted to moist montane and temperate to cold regions.

NEURADA *L.* (1753).

Sepals small, alternating with similar sized epicalyx-lobes, not accrescent in fruit. Petals small, greenish. Styles ± connivent into central cone, spinescent in fruit. Fruit with straight spines, these not arranged in whorls, and sepals not forming stiff marginal wings.

Monospecific.

N. procumbens *L.* (1753)
- type: 'in Aegypto, Arabia'.

Stems prostrate, rarely more than 14 cm high, most parts densely whitish tomentose. Petiole 5-15 mm long; leaf-blade 8-25(-30) x 5-20 mm, ovate, irregularly bluntly lobed. Pedicels 3-10 mm long; receptacle 5-10 mm diam. at anthesis; epicalyx-lobes and sepals *c* 2.5 x 1-1.5 mm, acute; petals oblong-ovate, 2.5-3 x 1.5-2 mm, tip rounded, greenish with distinct veins; stamens 2 mm long; styles slightly longer, hairy. Fruit 10-18 mm diam., usually discoid, occasionally 5-angled or lobed, densely spiny above, smooth below, often persisting as collar round base of seedlings. Fig. 92.1.

Coastal plain; 60-150 m. EE; Saharo-Sindian region, west to Mali, east to Pakistan, North Africa, Cyprus; introduced into South Africa. *Bally* 6900; *Jannone* 175.

Forms with 5-lobed fruits are sometimes distinguished as var. *stellata* M. & D. Zohary.

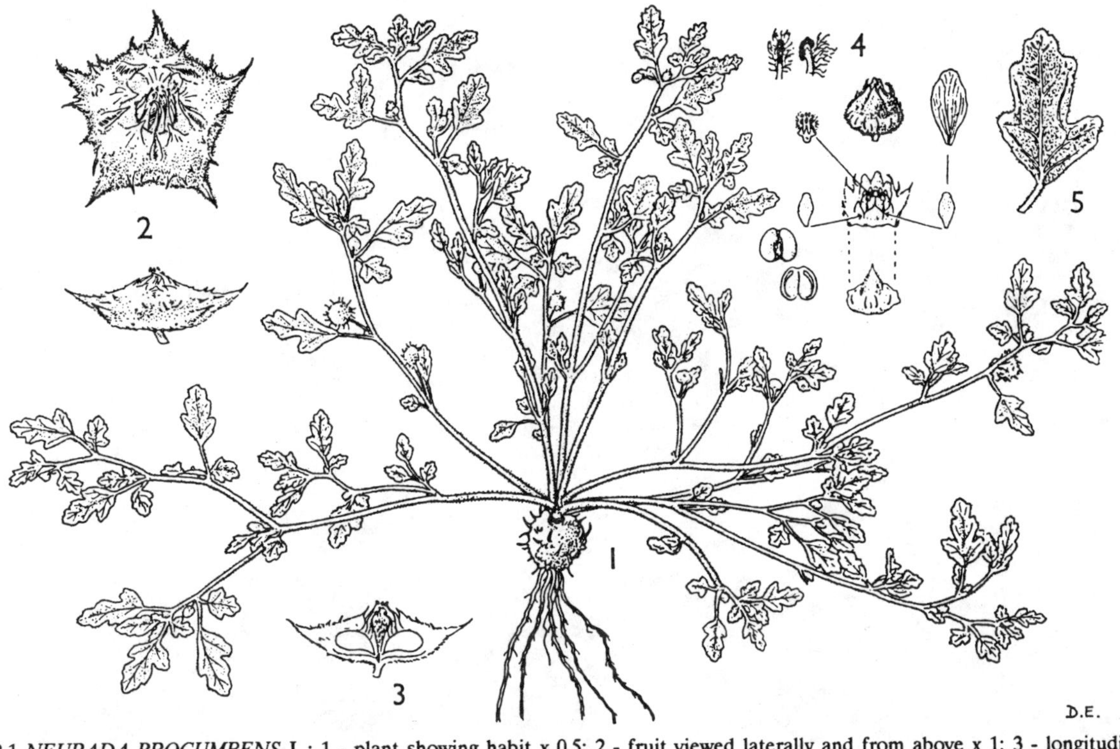

Figure 92.1 *NEURADA PROCUMBENS* L.: 1 - plant showing habit x 0.5; 2 - fruit viewed laterally and from above x 1; 3 - longitudinal section through fruit; 4 - dissection of flower x 4; 5 - leaf x 1. All from *Guest & Rawi* 14244. (Reproduced with permission from *Flora of Cyprus* vol. 1, p. 630 (R.D. Meikle), copyright with the Bentham-Moxon Trustees.)

* Ethiopian Flora Project, c/o Royal Botanic Gardens, Kew, Richmond, Surrey TW9 3AB, UK.

93. CHRYSOBALANACEAE

Graham, *Fl. Trop. E. Afr.* as part of **Rosaceae**: 47-58 (1960); Burger, *Fam. Fl. Pl. in Ethiopia*: 88 (1967).

Trees or shrubs. Leaves alternate in a spiral or sometimes in 2 rows, simple and entire, often with glands at the base of the blade; stipules present. Flowers bisexual and usually irregular or only slightly so, with a conspicuous receptacle-tube (hypanthium or floral tube) at the apex of which are borne the sepals, petals, and stamens. Sepals 5, united near the base. Petals 5, free and often falling off quickly. Fertile stamens 5 to many, often inserted on only one side of the flower opposite a group of sterile stamens, filaments free or united near the base. No disc present. Pistil 1 (carpels united or apparently only 1), ovary superior and borne near the rim of the hypanthium or receptacle-tube (or at the base of the tube in *Chrysobalanus*), 1-locular with 2 erect basal ovules or 2-locular with 1 ovule in each locule (in *Parinari*); style 1 and arising at the side near the base of the ovary and not near the top, stigmas 1 to 3. Fruit usually a drupe.

A tropical and subtropical family of both the Old and New Worlds.

The genera *Chrysobalanus, Parinari, Hirtella* and *Magnistipula* are found in East Africa. *Chrysobalanus* includes the fruit tree Coco plum (*C. icaco*) known for its good quality timber, its pleasant shade and erosion control capability. *Parinari* also contains important timber trees. Plants in this family could be found in the southwestern forests of Ethiopia, the most likely species being *P. curatellifolia* which is found in the Sudan.

This family is often included in the **Rosaceae** as the subfamiliy **Chrysobalanoideae**. It differs in having a basal style, erect basal ovules, and often a deep and symmetric hypanthium.

94. DICHAPETALACEAE

by F. J. Breteler*

Cufodontis, *Enum*.: 412 (1956); Breteler, The African Dichapetalaceae IX: a taxonomical revision. *Ag. Univ. Wag. Papers* **86-3**: 74 pp. (1986).

Climbers, shrubs, or trees. Leaves alternate, usually in two rows, simple, entire, pinnately nerved, stipulate, often glandular. Inflorescence cymose, compact or not, axillary, pedunculate or not, the peduncle adnate to petiole or not. Flowers small, regular or irregular, usually bisexual, 5-merous. Pedicel articulated, rarely not so. Sepals free or shortly united, imbricate, subequal or not. Petals equal in regular flowers, strongly unequal in irregular flowers, usually 2-lobed to 2-fid apically, at base more or less united with the stamens. Stamens 5, all fertile in regular flowers, 2-3 fertile in irregular ones. Disc with up to 5 epipetalous, small staminodes, free or partly united in regular flowers, often less than 5 and forming half a ring in irregular ones. Ovary superior 2-3(4)-celled; ovules 2 in each cell, pendulous. Style usually simple, with as many lobes as cells in the ovary. Fruit a drupe, 1-3(4)-seeded, the exocarp sometimes dehiscent. Seeds without endosperm.

Three genera with about 160 species, circumtropical. In Africa (including Madagascar) 86 species in *Dichapetalum* and 7 in *Tapura*. The third genus, *Stephanopodium*, is confined to tropical America. Most species are dwellers of tropical rainforest.

In Ethiopia the family is, so far, only represented by one species of *Tapura*. As there is some possibility that *Dichapetalum* might be represented in southwest Ethiopia, a key to the two African genera is presented.

Key to Genera

1. Flowers regular, petals equal, fertile stamens 5. *Dichapetalum*
- Flowers irregular, petals unequal, fertile stamens 2-3. **Tapura**

TAPURA *Aubl*. (1775)

Trees or shrubs. Stipules small, deciduous. Inflorescence of usually contracted cymes, axillary, peduncle often adnate to petiole. Flowers bisexual, irregular. Sepals unequal, almost free to shortly united. Petals usually 5, 1-2 distinctly larger, bilobed, the lobes hooded, the other petals smaller, lobed or not. Stamens 2-3 fertile, 3-2 sterile, alternating with petals and basally united with them in a usually oblique tube of principally 10 elements. Ovary 2-3(4)-locular. Style 2-3(-4)-lobed. Fruit a drupe.

Tapura fischeri *Engl*. (1895)
- type from Tanzania

Dichapetalum corradii Chiov. (1952) - types: GG, near Lake Turkana (Rudolph), Murle side of Omo river, *Corradi* 8401, 8402, 8403, 8404 (all FT syn.); Lago di Murle, in the forests on the shores, *Corradi* 8400, 8405, 8406, 8407, 8408, 8409 (all FT syn.)

Breteler, *Meded. Landbouwhogeschool Wag*. **78-10**: 23 (1978); Breteler, *loc. cit*: 56-61 (1986).

Shrub to small tree, up to *c* 3 m, sometimes multi-branched from near ground level. Branches glabrous or nearly so, often minutely lenticellate; branchlets sparsely subappressed-hairy when young, often soon glabrescent. Stipules triangular, usually soon deciduous. Leaves papery; petiole up to 5 mm long, up to 10 mm long when united with peduncle; lamina elliptic to slightly obovate, 4-8 x 2-4 cm, with 4-7 pairs of main lateral nerves, glabrous to glabrescent, but usually with persistent tufts of hairs in the axils of main lateral nerves beneath; glands, when present, beneath only; apex acuminate; base cuneate to rounded. Peduncle partly or entirely adnate to petiole. Flowers small, greenish, pedicel articulated. Fruits subglobose to subellipsoid, shallowly lobed or not, up to 5 mm diameter, glabrous or nearly so, with persistent calyx. Fig. 94.1.

Open woodland, riverine forest; up to c 500 m. GG KF IL; widespread in tropical Africa, from Ivory Coast to Natal. *Carr & Brown* 719; *Carr & Matolo* 906; *Corradi* 8400-8409.

* Department of Plant Taxonomy, P.O. Box 8010, 6700 ED Wageningen, The Netherlands.

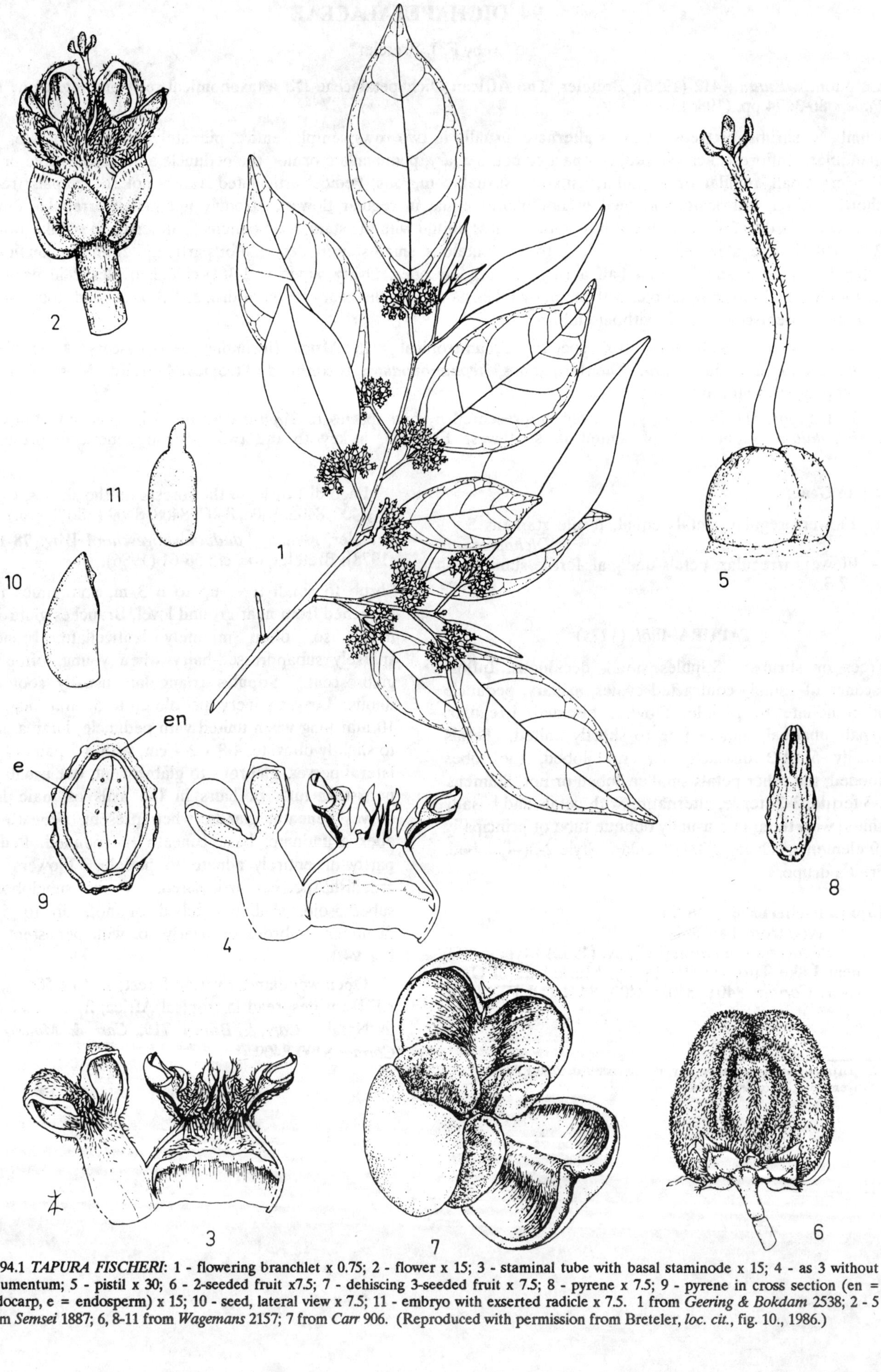

Figure 94.1 *TAPURA FISCHERI*: 1 - flowering branchlet x 0.75; 2 - flower x 15; 3 - staminal tube with basal staminode x 15; 4 - as 3 without indumentum; 5 - pistil x 30; 6 - 2-seeded fruit x7.5; 7 - dehiscing 3-seeded fruit x 7.5; 8 - pyrene x 7.5; 9 - pyrene in cross section (en = endocarp, e = endosperm) x 15; 10 - seed, lateral view x 7.5; 11 - embryo with exserted radicle x 7.5. 1 from *Geering & Bokdam* 2538; 2 - 5 from *Semsei* 1887; 6, 8-11 from *Wagemans* 2157; 7 from *Carr* 906. (Reproduced with permission from Breteler, *loc. cit.*, fig. 10., 1986.)

95. FABACEAE

(LEGUMINOSAE)

by M. Thulin*

Thulin, Leguminosae of Ethiopia, *Opera Bot.* **68**: 14-223 (1983).

Trees, shrubs, lianes or herbs. Leaves alternate, often pinnate or pinnately 3-foliolate, less often bipinnate or digitately 3-foliolate, occasionally 1-foliolate, simple or digitate; stipules usually present. Receptacle usually expanded. Sepals usually united. Petals usually 5, occasionally fewer or 0, free or less often some or all united. Stamens inserted with the petals at the rim of the receptacle, usually 10 but sometimes more or fewer. Ovary almost always a single, superior, (1-)2-several-ovulate carpel, the ovules attached to the adaxial suture. Fruit (pod) usually dehiscent into 2 valves, less often dehiscent only along the upper suture or indehiscent or splitting transversely into 1-seeded segments. Seed without, or with scanty, endosperm.

About 650 genera and 18 000 species, widely distributed over most of the world.

A family of great economic importance containing besides the pulse crops many species used for forage, pasture improvement, charcoal production, timber, gums, medicine and ornament. The ability of most members of the family to form associations with symbiotic bacteria which fix atmospheric nitrogen contributes greatly to improving soil fertility.

Key to Subfamilies

1. Flowers regular; petals valvate in bud, often united at the base; stamens as many as the petals or twice as many or numerous; leaves almost always bipinnate; seeds normally with an areole[1] on each side. **2. Mimosoideae**
- Flowers nearly always irregular; petals imbricate in bud, free or some of them united; stamens usually 10 or fewer; leaves various, rarely bipinnate; seeds usually without an areole. 2
2. Adaxial (upper) petal overlapped by the adjacent lateral petals when these are present; sepals often free; stamens usually free; seed with radicle usually straight. **1. Caesalpinioideae**
- Adaxial petal (standard) outside the adjacent lateral petals (wings); sepals united at the base; stamens usually variously united; seed with radicle usually curved. **3. Papilionoideae**

Subfamily 1. CAESALPINIOIDEAE

by R. M. Polhill** & M. Thulin

Leg. Trop. Afr. (1930); Cuf. *Enum.*: 211-225 (1955); *Adumbr.* **6**: 133-228 (1957) & **22**: 1-99 (1971); *Fl. Trop. E. Afr.* (1967).

Trees, shrubs or lianes, less often herbs, sometimes armed. Leaves usually pinnate, sometimes bipinnate, rarely 1-foliolate or simple. Flowers usually more or less regular. Sepals usually imbricate, free or less often partly united. Petals 5 or fewer, rarely 0, the upper (adaxial) one innermost, the others variously imbricate. Stamens usually 10 or fewer, rarely numerous, free or less often united below; anthers usually dehiscing lengthwise, rarely (*Cassia*, *Senna*, *Chamaecrista*) by terminal pores, without an apical gland. Pods various. Seeds generally without areoles[1] or if present (*Senna*, *Tamarindus*) normally with a continuous margin (closed); radicle generally straight.

About 150 genera and 2500 species, mostly tropical and subtropical.

Records of *Burkea africana* Hook. from Ethiopia in Burtt Davy, *Fl. Pl. & Ferns Transvaal*: 331 (1932) and Cuf. *Enum.*: 211 (1955) seem to be based on *Massey* 68 (K), a sterile specimen of *Albizia malacophylla* (A. Rich.) Walp.

Key to Genera

1. Leaves with 4 or more leaflets. 2
- Leaves simple, bilobed or deeply divided. 14
2. Leaves bipinnate (rhachis occasionally very short and spine-tipped). 3
- Leaves simply paripinnate, sometimes undeveloped at flowering time (if imparipinnate, see *Cadia* in subfam. **Papilionoideae**). 8
3. Stamens 5; seeds with loosely adherent arils. **2. Acrocarpus**
- Stamens 10; seeds without arils. 4
4. Stamens red, long-exserted, 3-12 cm long; pods dehiscent, the valves separating completely, not prickly; shrubs or trees, unarmed or occasionally (*Caesalpinia pulcherrima*) with some small prickles. 5
- Stamens up to 1.5 cm long; pods indehiscent, tardily dehiscent only along upper suture or prickly; shrubs, often scandent, or small trees, armed with many prickles or spines. 6

* Uppsala University - Department of Systematic Botany, P.O. Box 541, S-751 21 Uppsala, Sweden.
** Royal Botanic Gardens, Kew, Richmond, Surrey TW9 3AB, UK.

1. areole = an area on each face of a seed bounded by a fine line or crack in the testa.

5. Sepals valvate, subequal; pods 13-80 cm long, normally 12 to many seeded. **1. Delonix**
- Sepals imbricate, the lower one longer and hooded towards apex; pods up to 12 cm long, *c* 4-10-seeded. **3. Caesalpinia**

6. Petals *c* 3 mm long; fruits samaroid, with 1 seed at base and a broad distal wing. **4. Pterolobium**
- Petals 7-15 mm long; fruits flat or turgid pods without a broad distal wing. 7

7. Plant armed with scattered prickles or sometimes with stipular spines paired at the nodes; sepals imbricate, the lower one usually longer and hooded towards apex; pods not flattened between seeds, without conspicuous oblique or longitudinal venation. **3. Caesalpinia**
- Plants armed with stipular spines paired at the nodes or with single spines produced from the very short leaf rhachides; sepals narrowly imbricate to valvate, subequal; pods flattened or constricted between seeds, papery or thinly leathery, with close longitudinal or oblique striate venation. **5. Parkinsonia**

8. Flowers apetalous. 9
- Flowers with 3-5 conspicuous petals. 10

9. Flowers small; stamens 5; pods oblong, compressed, rather woody, thickened at sutures with pulpy sugary septa between the seeds. **10. Ceratonia**
- Flowers rather large; stamens numerous; pods subglobose, stipitate. **15. Cordyla**

10. Shrub with conspicuous red glands on stems and on the lower surface of the leaves. **6. Cordeauxia**
- Plants without such glands. 11

11. Flowers with 3 well-developed petals, the lower 2 minute; stamens joined half-way, only 3 filaments bearing anthers; pods oblong, somewhat compressed, sometimes irregularly constricted between the seeds, with a dry scurfy outer shell and pulpy inner layer. **11. Tamarindus**
- Flowers with 5 well-developed petals; stamens free, often varying in size but 4-10 fertile; pods diverse. 12

12. Filaments of 3 lower stamens 2-3 cm long, with an S-bend near the base; pod cylindrical, 30-60 cm long, woody. **7. Cassia**
- Filaments all straight or nearly so, much shorter; pods smaller. 13

13. Pedicels without bracteoles; pods indehiscent or ± tardily dehiscent, valves never twisting. **8. Senna**
- Pedicels with 2 bracteoles; pods elastically dehiscent with twisting valves. **9. Chamaecrista**

14. Plants with trailing or climbing mostly herbaceous stems from a large underground tuber; tendrils usually present; calyx with 2 upper sepals partly or completely united, the rest free; pod (1) 2-seeded. **14. Tylosema**
- Plants growing as shrubs or small trees, without tendrils; calyx shortly 4-5-lobed or spathaceous; pod usually more than 2-seeded. 15

15. Calyx turbinate, with 4-5 short broad lobes; flowers normally dioecious; male flowers with 10 fertile stamens; female flowers with a broad sessile stigma; pods indehiscent, rusty tomentellous. **12. Piliostigma**
- Calyx spathaceous (splitting to base down one side only); flowers bisexual; fertile stamens 1-10; style elongate; pods dehiscent. **13. Bauhinia**

1. DELONIX *Raf.* (1837)

Unarmed trees. Leaves bipinnate, without specialised glands (sometimes with glandular hairs at insertions of pinnae); leaflets small, opposite. Flowers usually large, in short axillary corymbose racemes. Sepals 5, valvate. Petals usually 5, similar except sometimes the upper one, which may differ in colour and shape, sometimes 1 or 3. Stamens 10, exserted; filaments hairy below. Ovary subsessile to shortly stipitate, with many ovules. Pods linear-oblong, flattened, dehiscent; valves leathery or woody, septate between the many seeds. Seeds transverse, hard, oblong, sub-cylindrical to flattened and oblong-elliptic.

Two species from E Africa to India and 7 in Madagascar (one widely cultivated).

1. Petals, except upper one, scarlet or rarely orange-red, 5-7 cm long; pods 35-80 cm long; seeds oblong-cylindrical; stipules pinnate, resembling small leaves. **1. D. regia**
- Petals, except upper one, white turning apricot, 1.6-3.8 cm long; pods 13-25 cm long; seeds oblong-elliptic, flattened; stipules subulate, falling quickly. 2

2. Leaflets 1.25-4 mm wide; ovary pubescent or tomentose all over; petals ± rounded in outline or shortly pointed at apex. **2. D. elata**
- Leaflets 4.5-20 mm wide; ovary glabrous or nearly so; bark papery, peeling; petals, especially the upper one, ± distinctly tapering and pointed at apex. **3. D. baccal**

1. D. regia *(Boj. ex Hook.) Raf.* (1837).

An ornamental, spreading tree with numerous small leaflets and racemes of showy scarlet or rarely orange-red flowers and large pods.

Cultivated at several places below 1700 m, e.g. Diredawa (HA) and Ginda (EE/EW); native of Madagascar. *Archer* 10125; *Baldrati* 3095, 4176.

Wood has been used to make gun stocks.

2. D. elata *(L.) Gamble* (1919).

Tree 2.5-15 m high, with rounded-spreading crown; bark rather smooth, buff or grey. Leaves with small deciduous subulate stipules; pinnae 2-12 pairs; leaflets (8-)11-25 pairs, linear-oblong, 4-17 x 1.25-4 mm, appressed puberulous on both surfaces. Racemes *c* 5-20-flowered; bracts small, deciduous before flowers open. Petals 1.6-3.8 cm long; upper one smaller than the others, pale yellow, remainder white and all turning apricot; blades rounded or shortly pointed and lacerate. Filaments (3)6-11 cm long, red. Ovary hairy. Pods 13-

25 x 2-3.7 cm. Seeds oblong-elliptic, flattened, 12-15 x 5-8 mm, mottled. Fig. 95.1.1.

Acacia-Commiphora bushland, often on rocky slopes or by streams; 0-1500(-2200) m. EE EW AF WU HA KF GG SD; Egypt south to eastern Zaire and Tanzania, east to Arabia and India. *Burger* 2335; *Schweinfurth & Riva* 423; *Ash* 2280.

Elsewhere recorded as browse for domestic and wild animals, as producing tannins from the fruit and wood which can be used for making furniture and domestic utensils.

3. **D. baccal** *(Chiov.) Bak. f.* (1930).

Tree 6-18 m high, with umbrella-shaped crown; bark grey with the outer surface peeling off and papery. Leaves with small deciduous subulate stipules; pinnae (1-)2-5 pairs; leaflets (4-)6-15 pairs, elliptic-oblong, 10-35 x 4.5-20 mm, appressed puberulous on both surfaces. Racemes 4-9-flowered; bracts small, deciduous before flowers open. Upper petal 4.2-5.5 cm long, bright yellow; other petals ± 2.8-3.5 cm long, white; all, especially the upper one, ± distinctly tapering and pointed at apex. Filaments 3-4 cm long. Ovary glabrous or nearly so, except for some pubescence on margins and stipe. Pods 15-24 x 2.5-3.5 cm. Seeds oblong-elliptic, flattened.

Deciduous bushland; *c* 600-900 m. HA SD; Somalia, N Kenya. *Hemming* 1504; *Gilbert* et al. 7703.

2. ACROCARPUS *Wight & Arn.* (1839)

Unarmed trees. Leaves bipinnate. Flowers in dense drooping inflorescences. Sepals 5, slightly imbricate. Petals 5, greenish. Stamens 5, exserted. Ovary stipitate, many-ovulate, with style scarcely differentiated. Pods narrowly winged adaxially, dehiscent. Seeds compressed with a small loosely adherent aril.

Two species from India to Thailand and Sumatra.

A. fraxinifolius *Wight & Arn.* (1839).

Large tree; pinnae 3-5 pairs; leaflets 4-9 pairs per pinna, 4-15 x 1.5-7 cm, acute; flowers in long racemes; petals 6-10 mm long, pubescent.

Cultivated at least in SU; native in SE Asia. *Roos* 879.

3. CAESALPINIA *L.* (1753)

Shrubs, often climbing, or trees, often with spines or prickles. Leaves bipinnate, without specialised glands (sometimes with glandular hairs); leaflets usually opposite, often glandular. Flowers in racemes or panicles. Sepals 5, imbricate, the lower one often largest and hooded. Petals 5, subequal except for the generally smaller upper one. Stamens 10; filaments alternately longer and shorter, hairy and often glandular below. Ovary subsessile to shortly stipitate; ovules *c* 2-10. Pods various, usually flattened, not winged, dehiscent or not; valves leathery, woody or pulpy, sometimes spiny. Seeds transverse or nearly so, hard.

A tropical genus of *c* 150 species, most numerous in the New World.

1. Filaments scarlet, long-exserted, 5-12.5 cm long. 2
- Filaments not scarlet, shortly exserted, 0.6-1.5 cm long. 3

2. Plant glabrous except on stamens; petals scarlet, red and yellow or sometimes all yellow. **1. C. pulcherrima**
- Plant conspicuously pubescent and glandular-hairy on the inflorescence and sepals; petals yellow. **2. C. gilliesii**

3. Sepal-margins lacerate. **3. C. spinosa**
- Sepal-margins entire. 4

4. Pods covered with spreading prickles. 5
- Pods not prickly. 6

5. Leaflets acuminate at apex; stipules minute, subulate; sepals 10-14 mm long. **9. C. volkensii**
- Leaflets rounded or emarginate at apex; stipules leaf-like; sepals 6.5-8 mm long. **10. C. bonduc**

6. Anthers densely woolly-pubescent; prickles absent, but short stipular spines present in pairs at nodes. **7. C. erianthera**
- Anthers glabrous or minutely puberulous; prickles scattered over stems. 7

7. Pedicels 1-2.4 cm long; pod shortly pubescent. **8. C. decapetala**
- Pedicels 0.2-0.7 cm long; pod glabrous or puberulous. 8

8. Pinnae 1-2 pairs; leaflets 1-2 pairs. **6. C. oligophylla**
- Pinnae 3-7 pairs; leaflets 4-12 pairs. 9

9. Bracts 3-6 mm long; flowers mauve-pink; leaflets glandular beneath but without prominent venation. **4. C. trothae**
- Bracts 9-10 mm long; flowers white or pinkish white; leaflets prominently veined and eglandular beneath. **5. C. dauensis**

1. **C. pulcherrima** (*L.*) *Sw.* (1791);
Poinciana pulcherrima L. (1753).

Glabrous ornamental shrub, unarmed or with small prickles; pinnae 3-10 pairs; leaflets 5-11(-13) pairs per pinna, oblong-elliptic, 5-18(-28) x 4-9(-15) mm; flowers in long racemes, with scarlet, red and yellow, orange-red or yellow petals 15-25 mm long; filaments scarlet, 5-6.5 cm.

Cultivated at lower altitudes, e.g. in EE (Ginda and Mitsiwa) WU (Dese) HA (Dire Dawa) IL (Gambella) and SD; probably native of tropical America. *Getahun* in *IECAMA* E-22 & RS-198; *Thomerson* 603.

Elsewhere with many uses including production of a red dye from the roots, a good source of pollen and nectar for bees, gums and resins and edible seeds.

Figure 95.1.1 - *DELONIX ELATA*, flowering and fruiting branch, from *Gilbert* 1856. 2 - *CAESALPINIA DECAPETALA*, a: flowering branch, from *Thulin* 1615, b: pod, from *Burger* 992. 3 - *C. OLIGOPHYLLA*, flowering branch, from *Hemming* 134. (All x 0.75) Drawn by Maureen Church. (Reproduced with permission from *Opera Botanica*, 68, fig. 3, 1983.)

2. C. **gilliesii** *(Hook.) Dietr.* (1840).

Ornamental unarmed shrub or small tree; pinnae 8-15 pairs; leaflets 6-10 pairs per pinna, narrowly oblong, 4-10 x 1.25-2.5 mm, glabrous; flowers in pubescent and glandular racemes, with pale yellow broadly spathulate petals 2.5-3 cm long; filaments scarlet, 8-12 cm long.

Cultivated, e.g. around Dire Dawa (HA) and naturalised below Nefasit (EW); native of South America. *Hemming* 1071; *Ebba* 730; *Sue Edwards* 3875.

3. C. **spinosa** *(Molina) Kuntze* (1898);
Poinciana spinosa Molina (1782).
C. tinctoria (Kunth) Taub. (1892).

Tree with short paired and scattered prickles; pinnae 2-4 pairs; leaflets 4-8 pairs per pinna, oblong-ovate, asymmetric at base, 2-4(-6) x 1-2.5 cm, glabrous; flowers in racemes; lower sepal hooded and fimbriate-pectinate, rest slightly lacerate; petals yellow, *c* 1 cm long; stamens 6-8 mm; pods often reddish, oblong, 6-10 x 1.7-2.5 cm, rather spongy and wrinkled.

Cultivated in EW, SU and SD; native of South America. *Giordano* 377; *Senni* 1952; *Jackson* 693.

Grown for ornament, hedging or for the pods which are used for dyeing and tanning.

4. C. **trothae** *Harms* (1899).

Erect or somewhat climbing shrub, 0.3-3(-5) m high, armed with scattered prickles on the branches and often on leaf rhachides. Leaves: pinnae 3-17 pairs; leaflets 4-33 pairs per pinna, narrowly oblong to slightly obovate, (2.5-)4-10(-13) x 1-4(-5) mm, ± pubescent or puberulous on midrib and margins or glabrescent. Flowers mauve-pink in terminal and lateral racemes. Sepals 8-12 mm long. Petals obovate, 9-20 mm long. Anthers minutely puberulous. Pods shortly boat-shaped, 1.5-5 x 1.5-2.2 cm, compressed, glabrous. Seeds obovoid, 10-11 x 7-9 mm.

Subsp. **erlangeri** *(Harms) Brenan* in *Kew Bull.* **17**:201 (1963);
C. erlangeri Harms (1902) - types: SD, Tarro Gumbi, *Ellenbeck* 2071 (B holo. destr.); Ethiopia/Somalia border, Dolo, *Riva* 1104 (FT neo.).

Young stems and axis of raceme puberulous; pinnae 3-6(-7) pairs; leaflets 4-12 pairs.

Dry bushland; *c* 800 m. SD BA; Somalia, N Kenya; not collected in Ethiopia since 1901.

Subsp. *trothae* occurs in Kenya and Tanzania.

5. C. **dauensis** *Thulin* (1980).

Shrub, *c* 2.5 m tall, armed with scattered prickles on branches and leaf rhachides. Leaves: pinnae (3-)4-5 pairs; leaflets 7-11 pairs per pinna, narrowly oblong or elliptic-oblong, (4-)6-12 x (1.5-)2-4.5 mm, puberulous on both surfaces, eglandular, prominently veined. Flowers white or pinkish white in terminal and lateral racemes; bracts greenish white or pinkish white, ovate-elliptic, 9-10 x 7-8 mm. Sepals 9-10 mm long. Petals narrowly obovate, *c* 9-10 mm long. Anthers minutely puberulous. Pods shortly boat-shaped, compressed, puberulous (only young ones seen).

Rocky slope in deciduous bushland; *c* 900 m. SD; NE Kenya. *Gilbert* et al. 7695.

6. C. **oligophylla** *Harms* (1902)
- type: SD, "Arussi Galla", *Ellenbeck* 2038 (B holo. destr.); neotype from Somalia.

Shrub, up to 4 m high, branches armed with scattered prickles. Leaves: pinnae 1-2 pairs; leaflets 1-2 pairs per pinna, obovate, 5-14 x 3-11 mm, sparsely puberulous or glabrous; venation prominent. Flowers reddish purple in axillary and terminal racemes. Sepals 7-10 mm long, entire. Petals obovate to spathulate, 11-12 mm long. Anthers glabrous. Pods shortly boat-shaped, glabrous (only young ones seen). Fig. 95.1.3.

Dry bushland. BA-SD; S Somalia.

In Ethiopia only known from the holotype collected in 1901 near Ganale river.

7. C. **erianthera** *Chiov.* (1929).

Shrub up to 2.5 m high, armed with stipular spines in pairs at the nodes. Leaves: stipules often foliaceous, rounded or reniform; pinnae 2-6 pairs; leaflets (2-)3-5 pairs per pinna, elliptic to sub-circular, 2-9.5(-18) x 1.5-6(-13) mm, glabrous or pubescent. Flowers greenish-yellow in lateral or terminal short racemes. Sepals 3-6 mm long (lower one up to 10 mm long). Petals 4-6.5 mm long, with suborbicular lamina. Anthers densely woolly-pubescent. Pod shortly boat-shaped, *c* 27 x 11 mm, flattened, glabrous.

Var. **pubescens** *Brenan* in *Kew Bull.* **17**: 203 (1963).

Stems and leaves pubescent.

Dry bushland, rocky slopes; *c* 250-1000 m. AF SD; Somalia, N Kenya, Arabia. *Gilbert* 2557A; *Corradi* 4011; *Gilbert* et al. 8154.

8. C. **decapetala** *(Roth) Alston* (1931).

Climbing or straggling shrub 2.5-10 m high, armed with scattered prickles on the branches and leaf rhachides. Leaves: pinnae 4-10 pairs; leaflets 8-12 pairs per pinna, elliptic-oblong to oblong-obovate, (0.8-)1-2(-2.7) x 0.3-0.8(-1.7) cm, pubescent on both surfaces. Flowers yellow or yellowish white, in long terminal racemes; pedicels 1-2.4 cm long. Sepals 9-10 mm long, entire, densely pubescent. Petals rather shortly clawed, with broadly rounded blade except smaller upper one, 1-1.5 cm long. Filaments 1.2-1.5 cm long, densely hairy. Pods oblong-elliptic or oblong, 6-10(-12) x (2-)2.4-2.7 (-3.5) cm excluding conspicuous beak, straight or slightly curved, compressed but the leathery valves convex, ultimately dehiscent along the upper suture, shortly pubescent. Seeds ellipsoid, 9-11 x 6-8 mm, brown to black, sometimes mottled. Fig. 95.1.2.

Wooded grassland and upland evergreen bushland, frequently established on dry hillsides and valley slopes; 1700-2100 m. WU SU AR HA KF; native of tropical and subtropical Asia. *Burger* 992; *Perdue* 6490; *Thulin* 1615.

Widely grown as a hedge plant and often becoming naturalised; seeds used to make necklaces.

9. C. **volkensii** *Harms* (1910).

C. bonduc sensu Cuf. in *Senckenberg. Biol.* **39**: 294 (1958) non (L.) Roxb.

Bush or liane, armed with scattered prickles on the branches and leaf rhachides. Leaves: pinnae 3-6 pairs; leaflets 3-7 pairs per pinna, ovate to ovate-elliptic, (2-) 3.2-8 x (1-)1.6-4.3 cm,acuminate at the apex, glabrous or nearly so, except when very young. Flowers yellow in axillary racemes. Sepals 10-14 mm long. Petals *c* 16 mm long. Pods broadly oblong- to obovate-elliptic, (6-) 7-13 x 3.5-6.5 cm, covered with prickles. Seeds subglobose, hard, 1.7-2.3 cm in diameter, brown.

Forest, forest margins at intermediate altitudes. BA GG; Uganda, Kenya, Tanzania. *Kuls* 537(sterile); *Friis* et al. 3516.

10. **C. bonduc** *(L.) Roxb.* (1832).

Spreading or clambering shrub copiously armed with scattered straight or recurved prickles; stipules with several leafy lobes; pinnae 3-9 pairs; leaflets 6-9 pairs per pinna, ovate-oblong to elliptic-oblong, (0.8-)1.5-5 x (0.5-)0.8-2.5 cm; apex rounded to emarginate; racemes long-pedunculate; sepals entire, tomentellous; petals yellow or greenish yellow, 1-1.4 cm long; pods covered with prickles.

Cultivated as a hedge plant in AF; widespread on tropical sea-shores, salt tolerant. *Brown* 1580/A.

4. **PTEROLOBIUM** *Wight & Arn.* (1834), *nom. conserv.*

Similar to *Caesalpinia* except the small petals all equal or almost so, ovary with 1 ovule near top and fruit samaroid. Fruit samaroid, with a shortly stipitate basal seed-containing portion (with a single seed) and a much prolonged upper suture, which is broadly winged on its lower side, the wing usually becoming wider distally.

About 10 species, one in Africa, the rest in Asia.

P. stellatum *(Forssk.) Brenan* (1954);

Mimosa stellata Forssk. (1775); *Cantuffa stellata* (Forssk.) Chiov. (1915).

Cantuffa exosa J.F. Gmel. (1791); *Pterolobium lacerans* R. Br. (1814), nom. illegit.; *Acacia kantuffa* Poir. (1817); *Mimosa kantuffa* DC. (1825); *Fillaea brucei* Fresen. (1839); *Pterolobium exosum* (J.F. Gmel.) Bak.f. (1930) - type: based on "Kantuffa" in Bruce, *Travels*, **ed. 1,5**:49 (1790).

Quartinia abyssinica A. Rich. (1840); *Pterolobium abyssinicum* (A. Rich.) A. Rich. (1847) - type: TU, *Quartin Dillon* (P holo.).

Climbing or straggling shrub 2-15 m high; branches armed with recurved prickles paired at the nodes and often also scattered on the internodes. Leaves with 5-13 pairs of pinnae, with recurved prickles at their insertion beneath, often with a single straight one on the upper side and others scattered on rhachis; leaflets 7-16 pairs per pinna, narrowly oblong or elliptic-oblong (terminal ones more obovate), 4-12 x 2-5mm, pubescent to almost glabrous. Flowers small, sweetly scented, pale creamy yellow, in dense racemes which are aggregated into extensive terminal panicles. Petals oblanceolate-oblong, *c* 3 mm long. Fruits brick-red to scarlet, ultimately brown, 3-6 cm long; wing 2.3-4.5 x 0.9-1.6 cm. Seeds ovoid-ellipsoid, *c* 11 x 6.5 mm, olive. Fig. 95.2.1.

Common, forming thickets in upland dry evergreen forest margins and clearings, upland bushland, *Acacia* woodland and riparian formations; (750-)1200-2500 m. EW TU GD WU GJ WG SU AR BA HA KF GG SD; Arabia and eastern Africa from Sudan to South Africa (Transvaal). *Burger* 797; *Pappi* 160; *Schimper* I:361.

Leaves formerly used locally in the production of ink.

5. **PARKINSONIA** *L.* (1753)

Shrubs or small trees, not climbing, armed with stipular spines paired at the nodes or with single spines produced from the very short leaf rhachides (in Flora area), eglandular. Leaves bipinnate, with generally opposite leaflets, sometimes with very short rhachis and very reduced leaflets. Flowers in axillary often short racemes, yellow or whitish. Sepals 5, narrowly imbricate to valvate. Petals 5, subequal except upper one, which is usually somewhat modified (often with longer claw). Stamens 10; filaments alternately longer and shorter, pubescent below. Ovary stipitate, with 2-6 ovules. Pods flat or turgid, sometimes constricted between the seeds, tardily dehiscent or indehiscent, with generally papery or thinly leathery brown valves. Seeds usually oblique or longitudinal, hard.

About 14 species, mostly in the drier parts of America, but one native in S and SW Africa and three in E and NE Africa.

1. Rhachides of pinnae slender, terete, with 3-6 pairs of leaflets. **1. P. scioana**
- Rhachides of pinnae flattened, with many opposite to alternate leaflets. **2. P. aculeata**

1. **P. scioana** *(Chiov.) Brenan* (1963);

Peltophoropsis scioana Chiov. (1915); *Peltophorum scioanum* (Chiov.) Bak. f. (1930) - type: *Negri* 1358 (FT holo.).

Caesalpinia gillettii Hutch. & Bruce (1941).

Figure 95.2.1 - *PTEROLOBIUM STELLATUM*, a: flowering branch, from *Mooney* 6078, b: pod, from *Pichi-Sermolli* 642. 2 - *PARKINSONIA SCIOANA*, a: flowering branch, from *Burger* 2731, b: pod, from *Friis* et al. 997. 3 - *CORDEAUXIA EDULIS*, a: flowering branch, from *Hemming* 375, b: pod, from *Field, Bally & Clarke* 79. (All x 0.75) Drawn by Maureen Church. (Reproduced with permission from *Opera Botanica*, 68, fig. 4, 1983.)

Shrub or small tree 0.6-5 m high, branching from near base, with straight or hooked stipular spines paired at the nodes. Leaves with 2-8 pairs of rather short pinnae; leaflets 3-8 pairs per pinna, elliptic or oblong-elliptic, 2-7 x 1.25-3.5 mm, puberulous. Racemes short, but up to 25-flowered. Sepals 4-5 mm long. Petals yellow; upper one rhombic, 7-9 x 4-4.5 mm, scarcely clawed, the others lanceolate, 5-7 x 1.5-2 mm. Pods mostly narrowly elliptic to oblanceolate, 4.5-9.5 x 1-2.5 cm, beaked, attenuate-stipitate, 1-4-seeded; valves thin, stiffly papery, closely and longitudinally or obliquely veined; upper suture slightly flanged. Seeds somewhat compressed, ellipsoid-oblong to ovoid-ellipsoid, 7-9 x 4-6 mm, brown or mottled. Fig. 95.2.2.

Acacia-Commiphora bushland and semi-desert open bushland; *c* 900-1250 m. AF SU HA SD; Somalia, N Kenya. *Burger* 2731; *Ellis* 144; *Thulin, Hunde & Tadesse* 3602.

2. **P. aculeata** *L.* (1753).

Evergreen shrub or small tree up to 6 m high, with slender often pendulous branches. Petiole and leaf rhachis very short, produced into a straight well-developed spine, with 1-3 pairs of long flattened pinnae-rhachides; leaflets many, opposite to alternate, oblong-elliptic to obovate, 2-7 x 1-2.5 mm, glabrous. Racemes lax, up to 20 cm long. Sepals 6-7 mm long. Petals yellow, ovate-rhombic, 8-12 x 6-8 mm; upper one marked red, with more rounded blade and longer claw. Pod constricted into oblong or elliptic segments, pointed or beaked, *c* 3-15 x 0.6-0.8 cm; valves leathery, diagonally striate, eventually separating along upper suture. Seeds oblong-ellipsoid, 7.5-10 x 3.5-4 mm, brownish, sometimes mottled.

Cultivated e.g. in EE EW TU SU HA SD and naturalised at least in EW TU and HA; native of tropical and subtropical America. *Fiori* 123; *Hagos* 136; *Burger* 928.

Widely cultivated for ornament, also planted for shade and wind-breaks, and often becoming an escape. Known to be salt tolerant. Elsewhere with a wide range of uses including production of fibre pulp paper, food for domestic animals both direct as browse and in concentrates, and a source of wood for charcoal, construction and carving.

6. CORDEAUXIA *Hemsl.* (1907)

Unarmed shrub with conspicuous red glands on leaves and stems. Leaves paripinnate, exstipulate. Flowers in few-flowered racemes. Sepals 5, valvate in bud. Petals 5, subequal. Stamens 10, free; filaments hairy below. Ovary 2-ovulate. Pods dehiscent with hard valves, 1-seeded, beaked.

Monotypic.

C. edulis *Hemsl.* (1907).

Much-branched shrub or small tree, up to 2.5 m high. Leaflets 2-12, elliptic-oblong, up to 30(-50) x 15(-25) mm, densely covered with red glands beneath. Sepals oblong, obtuse, c 10 mm long, glandular. Petals yellow, *c* 15 mm long. Pods 4-6 cm long, ovoid with an upcurved beak. Seed 2.5-4.5 cm long, ovoid. Fig. 95.2.3.

Semi-desert bushland and scrub on sandy soil; 400-500 m. HA(eastern Ogaden); S Somalia. *Bally* 9611, 10146, 20134.

The seeds are edible and nourishing and highly appreciated as food. The plant has recently attracted much interest as a potential food crop for arid areas (see Nat. Acad. Sci. (1979), *Tropical Legumes; resources for the future*: Nat. Acad. Sci. Washington DC. 103-107).

7. CASSIA *L.* (1753)

Bentham in *Trans. Linn. Soc.* **27**: 503-592 (1871); Brenan in *Kew Bull.* **13**: 231-252 (1958) & *Fl. Trop. E. Afr.* (1967); *Adumbr.* **22**: 1-99 (1971). For cultivated species see also de Wit in *Webbia* **11**: 197-292 (1955).

Unarmed trees or shrubs. Leaves paripinnate, without glands on petiole or rhachis. Flowers in many-flowered racemes; bracteoles 2, at base of pedicels. Sepals 5, imbricate. Petals 5, usually yellow. Stamens 10; filaments of 3 lower stamens with an S-bend near the base, many times longer than their small anthers which are dorsifixed and dehiscent by slits; filaments of upper stamens ± straight, shorter, their anthers dehiscing by basal pores. Pods cylindric or elongate, usually indehiscent, with septa between the seeds. Seeds ± compressed, without areole.

About 30 species, pantropical.

The genus is treated here in a restricted sense, excluding *Senna* and *Chamaecrista,* following Irwin & Barneby in *Adv. Leg. Syst.*: 97-106 (1981) and in *Mem. New York Bot. Gard.* 35 (1982), and Lock in *Kew Bull.* 43 (1988).

1. Racemes erect; bracts persistent; pods dehiscent. **1. C. arereh**
- Racemes long, pendulous; bracts falling quickly; pods indehiscent. **2. C. fistula**

1. **C. arereh** *Del.* (1826).

Tree 6-10 m high or a shrub. Leaves 25-40 cm long, eglandular, glabrous or nearly so; stipules minute, falling quickly; leaflets (5-)7-10 pairs ovate-elliptic, acuminate, 4.5-8 x 2.5-4.5 cm. Racemes erect, corymbose; bracts lanceolate-caudate, (0.8-)1.5-2.5 cm long, persistent; lower pedicels up to 8-10 cm long. Petals bright yellow, 2-3 x 1-1.5 cm. Filaments of 3 lower stamens relatively long, with an S-bend near base and a slight swelling near middle. Pods cylindrical, 30-60 x 1.5-2.5 cm, transversely septate, splitting longitudinally at maturity at least along one suture. Seeds very many, ovate, rather compressed, *c* 10 x 7 mm, slightly pitted, brown. Fig. 95.3.1.

Deciduous woodland and wooded grassland, often in rocky places or riparian; 1200-2000 m. EW TU GD GJ SU KF; Sudan, N Nigeria. *Mooney* 6677; *Schimper* III:1770; *W. de Wilde* 10116.

Crushed pods or roots may be used as a fish poison; the roots also have purgative and other medicinal properties.

2. C. **fistula** *L.* (1753).

Tree; leaves eglandular; leaflets 3-6(-8) pairs, acute or bluntly pointed, minutely puberulous beneath; racemes lax, pendulous, 1-3 together, 15-60 cm long; bracts falling quickly; petals yellow, *c* 2-3 cm long; pods cylindrical, 20-50 x 1.5-2.5 cm, black, indehiscent, with hard woody walls.

Cultivated as an ornamental at least in EW and IL; native of tropical Asia. *Baldrati* 4334; *Jansen* 4956.

Elsewhere the wood is used for tool and implement handles, as construction timber and for making wheels.

8. SENNA *Mill.* (1754)

For references, see under *Cassia*.

Unarmed trees, shrubs or herbs. Leaves paripinnate, often with globose to clavate or cylindrical glands on petiole and/or rhachis. Flowers in few- to many-flowered racemes, without bracteoles. Sepals 5, imbricate. Petals 5, yellow. Stamens usually 10; filaments straight, shorter than or not over twice as long as the basifixed anthers which are dehiscent by apical pores or short slits. Pods terete or compressed, indehiscent or ± tardily dehiscent, valves not twisting. Seeds ± compressed, often with a closed areole on each face.

About 240 species, pantropical, most numerous in the New World, some extending into temperate regions.

Several species of *Senna* have important medicinal properties and are used in both traditional and modern medicine.

1. Leaves with a conspicuous projecting gland on rhachis between 1 or more pairs of leaflets (particularly the lower ones) or near base of petiole (the only species likely to be encountered flowering before the leaves appear is *S. singueana* which comes in this group); pods terete or somewhat compressed. 2
- Leaves without large specialised glands (minute reddish clustered glandular bodies may occur at leaflet-insertions); pods flat. 10

2. Gland inserted near base of petiole. 3
- Glands inserted on rhachis at insertion of leaflet-pairs. 4

3. Leaflets usually in 4-5 pairs; bracts acute; gland on petiole ± ovoid or globose; peduncles 3-5 mm long. **7. S. occidentalis**
- Leaflets in 4-10 pairs; bracts ± obtuse; gland on petiole clavate or cylindrical; peduncles 8-25 mm long. **8. S. sophera**

4. Leaflets gradually narrowed to an acute or subacuminate apex. 5
- Leaflets abruptly narrowed to an obtuse, rounded or retuse apex. 6

5. Leaflets 4-13 pairs; stipules semi-cordate to reniform; pods linear, somewhat compressed, 12-25 cm long. **1. S. petersiana**
- Leaflets 3-4 pairs; stipules linear, falling quickly; pods subcylindrical, 6-10 cm long. **2. S. septemtrionalis**

6. Leaflets 4-12 pairs; racemes corymbose, 2 to many flowered; pedicels of flowers 1-5 cm long. 7
- Leaflets (2-)3 pairs; flowers either 1-2 on short axillary peduncles or in longer racemes with pedicels only 0.3-0.5 cm long. 9

7. Pedicels 2.2-4(-5) cm long in flower; pods subcylindrical, tough-walled, often rather torulose. **3. S. singueana**
- Pedicels 1-1.8 cm long in flower; pods not as above. 8

8. Leaflets puberulous; pods with papery valves, without expanded sutural margins. **4. S. baccarinii**
- Leaflets woolly beneath; pods with smooth expanded sutural margins. **5. S. multiglandulosa**

9. Flowers in elongate racemes; pods cylindrical, (5-)8-15 x 1-1.5 cm; seeds transversely arranged. **6. S. bicapsularis**
- Flowers 1-2 on very short peduncles; pods linear, (6-)13-23 x 0.4-0.6(-0.7) cm; seeds longitudinally arranged. **9. S. obtusifolia**

10. Stipules broadly ovate-cordate, 8-12 mm wide. **11. S. didymobotrya**
- Stipules linear to ovate-triangular, less than 3 mm wide. 11

11. Racemes mostly aggregated into terminal panicles; pedicels 2-3.5 cm long; pods 15-30 cm long; seeds smooth. **10. S. siamea**
- Racemes mostly axillary; pedicels less than 1 cm long; pods up to 7.5 cm long; seeds reticulate or rugose. 12

12. Pods with a ridge of longitudinal crests along middle of each valve. 13
- Pods not ridged along middle. 14

13. Branches and leaves spreadingly puberulous; pods straight. **15. S. truncata**
- Branches and leaves appressed puberulous; pods curved. **16. S. italica**

14. Leaflets emarginate or retuse at the apex, not mucronate, broadly elliptic to suborbicular, 1.8-6 cm wide. **12. S. ruspolii**
- Leaflets rounded to acute at the apex, often mucronate, shape various, 0.25-1.6(-2.2) cm wide. 15

15. Branches glabrous; leaflets glabrous except for sometimes ciliate margins. **14. S. ellisiae**
- Branches and leaflets pubescent. 16

16. Pods, excluding stipe, 4-5 times as long as wide, straight or nearly so. **13. S. longiracemosa**
- Pods 2-3 times as long as wide, often curved. 17

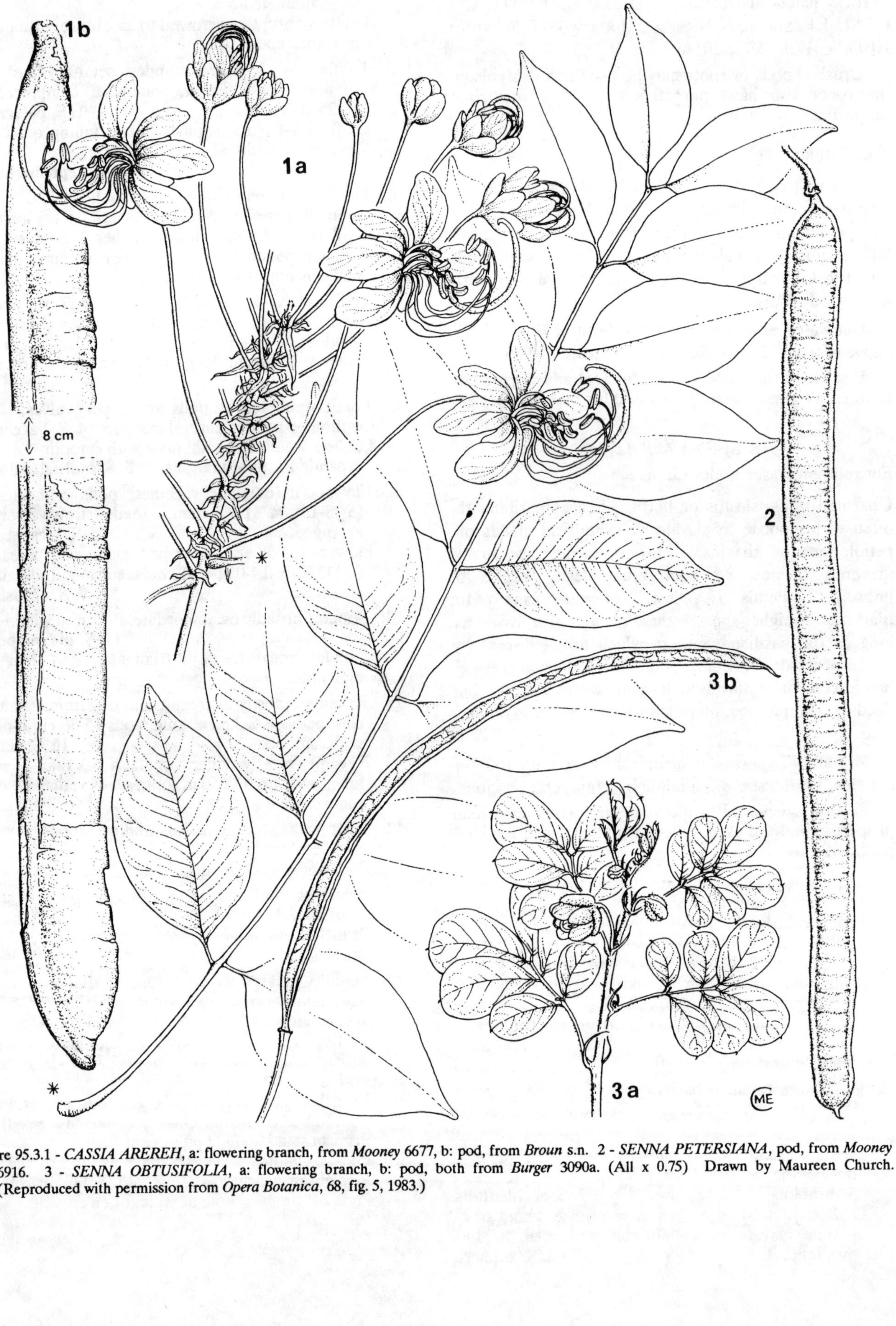

Figure 95.3.1 - *CASSIA AREREH*, a: flowering branch, from *Mooney* 6677, b: pod, from *Broun* s.n. 2 - *SENNA PETERSIANA*, pod, from *Mooney* 6916. 3 - *SENNA OBTUSIFOLIA*, a: flowering branch, b: pod, both from *Burger* 3090a. (All x 0.75) Drawn by Maureen Church. (Reproduced with permission from *Opera Botanica*, 68, fig. 5, 1983.)

17. Hairs densely arranged and spreading on branches, leaf-axes, pedicels, etc,; petals 4-8 mm long; pods 2.5-3(-3.5) x 1.2-1.5 cm. **17. S. holosericea**
- Hairs appressed at least on the pedicels, usually also elsewhere, sometimes sparse; petals 7-17 mm long; pods (3-)4-7 x 1.6-2.6 cm. **18. S. alexandrina**

1. **S. petersiana** *(Bolle) Lock* (1988);
Cassia petersiana Bolle (1861).

Shrub or tree 0.6-12 m high. Leaves 20-40 cm long, with columnar glands on rhachis between leaflets; stipules semi-cordate to reniform often caudate, 1-2.5 cm long; leaflets 4-10(-13) pairs, lanceolate to ovate, somewhat acuminate, (2-)3-10 x (0.7-)1-4 cm, subglabrous to tomentose beneath (usually coarsely pubescent in Ethiopia). Racemes corymbose, aggregated into terminal panicles. Sepals rounded at apex. Petals bright yellow, obovate, 1.5-3.2 cm long, the largest somewhat oblique and toothed. Stamens: 3 large, 4 medium-sized, 3 small. Pods linear, somewhat compressed, 12-25 x 1-1.5 cm, indehiscent; sutures thickened, transversely cracked. Seeds many, ovate to subcircular, somewhat compressed, 5-7 x 4-6 mm, brown, with a well-developed olive areole. Fig. 95.3.2.

Forest edges, evergreen bushland and thicket, deciduous woodland, sometimes riparian; 1200-2000 m. WU GD WG SU IL KF SD; westwards to Cameroun and south to South Africa (Transvaal), also in Madagascar. *Gillett* 14761; *Mooney* 8625; *Ash* 2228.

Elsewhere roots and leaves with a number of medicinal and veterinary uses; fruits and seeds said to be edible and used to make alcoholic beverages.

2. **S. septemtrionalis** (*Viv.*) *Irwin & Barneby* (1982);
Cassia septemtrionalis Viv. (1802).
C. laevigata Willd. (1809).
C. floribunda sensu auctt., non Cav. (1802).

Woody herb, shrub or small tree 1-5 m high, glabrous. Leaves 15-30 cm long, with columnar glands on rhachis between each pair of leaflets except sometimes the uppermost; stipules linear, falling quickly; leaflets 3-4 pairs, lanceolate to ovate, tapering or acuminate to an acute apex. Racemes mostly axillary but aggregated near branchlet-ends. Sepals rounded at apex. Petals bright yellow, obovate to obovate-circular, 1-1.5(-2) cm long. Stamens: 2 with large anthers and long filaments, 1 with medium anther and filament, 4 with rather small anthers and very short filaments, 3 with usually reduced empty anthers and very short filaments. Pods brown, subcylindrical, 6-10 x 1-2.5 cm, tardily dehiscent, longitudinally and transversely septate within. Seeds many, olive or brown, asymmetrically ovate, 5-6 x 3.5-4 mm, without areoles.

Deciduous woodland, evergreen thicket and bushland, in hedges and disturbed places; 1700-2400 m. SU AR HA IL KF SD; probably native of tropical America, but now widespread throughout the tropics; locally naturalized. *Getahun* in IECAMA J-67; *Mooney* 6744; *Thulin* 1597.

3. **S. singueana** *(Del.) Lock* (1988);
Cassia singueana Del. (1826).
?*C. sabak* Del. (1826).
C. goratensis Fresen. (1839) - type: Rüppell (FR, holo., not seen).

Shrub or tree 1-15 m high. Leaves 10-20(-30) cm long, with fusiform or stipitate glands on rhachis between each pair of leaflets except often the uppermost; stipules subulate, falling quickly; leaflets (5-)6-10(-12) pairs, oblong-elliptic to obovate-elliptic, (1.5-)2.5-6 (-7.5) x (0.7-)1.5-3 cm, usually rounded and often emarginate at the apex, glabrous to densely pubescent beneath (usually moderately to rather densely pubescent in Ethiopia). Racemes often produced before the leaves, corymbose, aggregated towards branchlet-ends; bracts small, falling quickly (stalked glands often present also). Sepals rounded at apex. Petals yellow, obovate to suborbicular, 1.5-3(-3.5) cm long. Stamens: 3 large, 4 medium-sized, 3 upper small with reduced anthers. Pods linear, torulose, sub-cylindrical or slightly compressed, 5.5-26 x 0.7-1 cm, tough-walled, indehiscent, transversely septate within. Seeds numerous, dull brown, subcircular, flattened, 5-7 mm across, with a small areole on each face.

Deciduous woodland, bushland and wooded grassland of various types, often riparian; 1500-2000 (-2400)m. EW TU GD WU GJ SU SD; widespread in tropical Africa, also Comoro Is. *Schimper* 908; *Hagos* 88; *Friis* et al. 1104.

The bark is used in tanning, particularly in the removal of hairs. Elsewhere the wood is used in hut frames.

4. **S. baccarinii** *(Chiov.) Lock* (1988);
Cassia baccarinii Chiov. (1915).

Shrub or small tree, 1-4 m high. Leaf rhachis with a fusiform or stipitate gland between the lowest pair or several pairs of leaflets; stipules subulate, falling quickly; leaflets (2-)4-6 pairs, elliptic to obovate-elliptic, 1-3 x 0.5-1.6 cm, puberulous. Racemes corymbose. Sepals rounded at apex. Petals yellow, obovate, 1.3-2 cm long. Stamens: 3 large, 4 medium-sized and 3 small. Pods oblong-linear, slightly curved, flattened, 5-7.5 x 1-1.8 cm, indehiscent, with papery valves. Seeds olive, with a small areole on each face. Fig. 95.4.1.

Dry woodland or bushland; *c* 1150-1500 m. HA SD; Somalia, N Kenya. *Gilbert* 3207; *Burger* 2292.

5. **S. multiglandulosa** *(Jacq.) Irwin & Barneby* (1982);
Cassia multiglandulosa Jacq. (1783).
C. tomentosa L. f. (1782), non *Senna tomentosa* Batka (1854).

Ornamental shrub or small tree, with tomentose branches; leaflets 4-8(-11) pairs, mostly elliptic-oblong, 2-4 x 0.7-1.5 cm, unequal-sided at base, woolly beneath, with glands between at least lower pairs; racemes axillary, corymbose; bracts ovate; sepals rounded at apex; petals golden or orange-yellow, 1.5-2 cm long; pods 10-14 x 0.8-1.2 cm, with smooth expanded sutural margins, coarsely pubescent, indehiscent.

Cultivated e.g. in WG, SU(Addis Abeba) and SD, apparently sometimes naturalised; native of tropical America, occasionally grown for ornament elsewhere in the tropics. *Scott* 26; *W. de Wilde* 7232, 10305.

6. **S. bicapsularis** *(L.) Roxb.* (1832);
Cassia bicapsularis L. (1753).

Spreading shrub, 1-3(-9) m high, glabrous in Ethiopia. Leaves 4-10 cm long, with a clavate or subglobose gland between lowest pair of leaflets; stipules minute; leaflets (2-)3 pairs, obovate-elliptic to subcircular, 1-3(-4) x 0.7-2(-2.5) cm, rounded at apex. Racemes elongate, axillary, 3-15-flowered. Sepals rounded at apex. Petals yellow or orange, obovate, 1-1.3 cm long. Stamens: 3 lower with large anthers, central with short, laterals with longer filaments; 4 with medium-sized anthers and very short filaments; 3 upper with reduced anthers, at least the lateral ones obhastate. Pods cylindrical, (5-)8-15 x 1-1.5 cm, not or tardily dehiscent, with thin transverse septa within. Seeds many, olive-brown, elliptic to subreniform, 5-6 x 3.5-4.5 mm, without clearly marked areoles.

Evergreen bushland and wooded grassland, often riparian, also in hedges and disturbed places; 1000-1900 m. AR HA KF; native of West Indies and South America. *Burger* 1340, 1509, 2243.

Grown for ornament and naturalised in other parts of the tropics. Elsewhere the wood is used to make paper pulp; fruits are said to be edible.

7. **S. occidentalis** *(L.) Link* (1831);
Cassia occidentalis L. (1753).

Erect, sometimes slightly woody herb, 0.2-2 m high. Leaves 10-25 cm long, with a large squat ovoid or globose sessile gland near base of petiole; leaflets (3-)4-5(-6) pairs, lanceolate to ovate-elliptic, (2.5-)-12 x (1.5-)2-4 cm, acute or acuminate, glabrous except for ciliolate margins and inconspicuous scattered glands beneath. Racemes from upper axils, very short, almost umbellate; peduncles 3-5(-8) mm long; bracts acute. Sepals obtuse. Petals yellow, obovate, 0.9-1.5 cm long. Stamens: 3 large, 4 medium-sized, 3 small. Pods linear, usually slightly upcurved, (5-)8-13 x 0.5-1 cm, not or tardily dehiscent, subglabrous, transversely septate within. Seeds many, transversly arranged, grey-brown, ovate-circular, compressed, 4.5-5 x 3.75-4.5 mm, minutely pimpled and with an elliptic areole on each face.

Usually a weed of cultivation, roadsides and waste places, also in wooded grassland and near lakes or streams; (250-)500-1700(-2400) m. EE AF EW TU GD WG SU AR HA KF GG SD; pantropical, possibly originating in America. *Burger* 205; *Mooney* 5111.

Roasted seed used as a coffee substitute; leaf-infusions with pain-killing and other medicinal properties. Elsewhere all parts with a wide range of medicinal uses and some parts used as famine food.

8. **S. sophera** *(L.) Roxb.* (1832);
Cassia sophera L. (1753).

Shrub or shrublet, 0.5-2(-3) m high. Leaves 10-25 cm long, with a sessile clavate or cylindrical, often somewhat pointed, gland near base of petiole; leaflets 4-10 pairs, lanceolate to narrowly ovate, 2-7 x 0.8-2 (-2.5) cm, acute, glabrous except for ciliolate margins and sometimes a few scattered glands beneath. Racemes from upper axils, very short, almost umbellate; peduncles 8-25 mm long; bracts ± obtuse. Sepals obtuse. Petals yellow, obovate, 1-1.5 cm long. Stamens: 3 large, 4 medium-sized, 3 small. Pods linear, usually straight, 5-10.4 x 0.6-1.1 cm, not or tardily dehiscent, subglabrous, septate within. Seeds many, transversely arranged, grey-brown, ovate-circular, compressed, 4-4.5 x 3.5-4.5 mm, almost smooth and with an elliptic areole on each face.

Dry, rocky ground; *c* 600 m. AF; pantropical, occurring especially in Asia, rare in East Africa. *W. de Wilde* 9814.

9. **S. obtusifolia** *(L.) Irwin & Barneby* (1982);
Cassia obtusifolia L. (1753).
C. tora sensu auctt., non L.

Erect herb or shrub 0.6-2.5 m high. Leaves 5-13 cm long, with a cylindrical gland between each of 1-2 pairs of leaflets; stipules linear; leaflets 3 pairs, obovate, (1-) 1.5-6 x (0.5-)1-4 cm, subglabrous to pubescent. Flowers 1-2 on generally very short axillary peduncles; pedicels (0.5-)1.5-3.5(-4.5 in fruit) cm long. Sepals rounded at apex. Petals yellow, obovate, 1-2 cm long. Stamens: 3 lower ones longest with the anthers narrowed into a neck towards their tips, 4 somewhat smaller and 3 very small. Pods linear, straight, or curved, slightly angular in section, (6-)13-23 x 0.4-0.6(-0.7) cm, dehiscent, glabrescent. Seeds many, longituinally arranged, brown, sub-rhombic or ovoid, 4.5-6.5 x 2-4 mm, with a very narrow areole on each face. Fig. 95.3.3.

Weed of cultivation, roadsides and waste places, also in wooded grassland and often near lakes and streams; (0-)500-2000 m. EE EW TU WG SU HA IL GG; pantropical. *Burger* 3090a; *Schimper* 1034.

Elsewhere claimed to have antibiotic properties and used in treating a number of complaints.

10. **S. siamea** *(Lam.) Irwin & Barneby* (1982);
Cassia siamea Lam. (1783).

Ornamental tree; leaves eglandular; leaflets 4-16 pairs, ovate-oblong to elliptic-oblong, 3-7.5 x 1.2-2.5 cm, puberulous beneath; racemes dense, aggregated into erect terminal panicles; pedicels 2-3.5 cm; petals bright yellow, 1.2-2 cm long; pods 15-30 x 1.2-1.6 cm, tardily dehiscent.

Cultivated in EW(Keren), EE(Ginda), HA, IL (Gambela), and no doubt elsewhere; native in SE tropical Asia. *Desert Locust Survey PT* 68; *Ellis* 341; *Jansen* 5036.

11. S. **didymobotrya** *(Fresen.) Irwin & Barneby* (1982);
Cassia didymobotrya Fresen. (1839) - type: Rüppell (FR holo., not seen).

Bushy shrub 1-5(-9) m high. Leaves 10-30(-36) cm long, eglandular (except sometimes for tiny reddish papillae clustered at leaflet-insertions); stipules ovate-cordate, acuminate, 1-2.5 x 0.8-1.2 cm; leaflets 8-18 pairs, oblong-elliptic, 2-6.5 x 0.6-2.5 cm, pubescent. Racemes 11-40 cm long; bracts 0.9-2.7 x 0.5-1.4 cm. Petals bright yellow, elliptic-ovate, 1.8-2.7 cm long. Stamens: 2 with large anthers, 5 medium-sized and 3 small. Pods oblong, flattened, 8-12 x 1.5-2.5 cm, dehiscent, transversely septate within, pubescent. Seeds transversely arranged, compressed, nearly oblong, apiculate near hilum, 8-9 x 4-5 mm, with an oblong areole on each face.

Montane wooded grassland, evergreen thicket and bushland, often riparian or in disturbed places; 1450-2400 m. WU WG SU AR SD; eastern tropical Africa south to Zimbabwe and west to Sudan, Zaire and Angola. *Bally* 9172; *Meyer* 8123.

12. S. **ruspolii** *(Chiov.) Lock* (1988);
Cassia ruspolii Chiov. (1940) - type: HA, *Ruspoli & Riva* 1663 (FT holo.).

Shrublet, up to 1.5 m high. Leaves large, rosetted at apex of short woody stem; stipules small, ovate-triangular, 2-3.5 x 1.5-2.5 mm; leaflets 4-10 pairs, broadly elliptic to suborbicular, 2.5-9 x 1.8-6 cm, emarginate or retuse at the apex, puberulous at least beneath. Racemes, including peduncles, up to at least 50 cm long. Petals yellow, 2-2.5 cm long. Stamens: 2 anthers large, 5 medium, 3 small. Pods oblong, flattened, straight, 7.5-9 x 1.5-2.2 cm, slightly venose. Seeds compressed, ± ovate, 7-9 x 3.5-5mm, reticulate, with an areole on each face.

Dry woodland and bushland; *c* 350-1000 m. HA SD; Somalia, N Kenya. *Gilbert* et al. 7602.

This very striking plant is, in Ethiopia, only known from the type collected in the 'Ueb Karanle valley at Ito Danna' (*c* 42°45'E, 6°20'N) and from near Dolo in SE Sidamo.

13. S. **longiracemosa** *(Vatke) Lock* (1988);
Cassia longiracemosa Vatke (1880).

Shrub or small tree 1-7 m high; branches covered with short spreading, almost scabrous, pubescence. Leaves (3.5-)9-20 cm long, eglandular except sometimes for tiny bodies clustered at leaflet-insertions; stipules subulate above broader base, 5-11 mm long; leaflets 5-13 pairs, elliptic or elliptic-obovate or subcircular, 0.7-2.5(-3.5) x 0.4-1.4(-2.2) cm with generally slightly scabrid pubescence. Racemes 10-37 cm long; bracts ovate-triangular, caudate, falling quickly. Petals bright yellow, 1.4-2.8 cm long. Stamens 9-10: 2 with large anthers, 4-5 medium-sized, 3 small. Pods oblong, flattened, 3.5-7.5 x 1-1.8 cm, not crested, at most rounded over each seed, dehiscent, transversely septate within. Seeds transversely arranged, compressed, oblong-ovate, apiculate near hilum, often emarginate at opposite end, 6-7.5 x 4-4.5 x 1.5-2.5 mm, reticulate, with a small areole on each face.

Semi-desert open bushland, *Acacia-Commiphora* bushland and wooded grassland, often by streams and lakes; *c* 300-1000 m. AF/HA BA SD GG; Somalia, Uganda, Kenya and N Tanzania. *Burger* 2818; *Corradi* 4073; *Ash* 1927.

14. S. **ellisiae** *(Brenan) Lock* (1988);
Cassia ellisiae Brenan (1958) as "ellisae" - type: HA, *Ellis* 327 (K, holo.).

Shrub or small tree, 1-3 m high; branches glabrous. Leaves 4-8 cm long, eglandular; leaflets 5-7(-9) pairs, obovate-elliptic, 3.5-11 x 2.5-6 mm, rounded to subacute at the apex, glabrous or ciliate. Racemes 10-30 cm long. Petals yellow, 20-32 mm long. Stamens: 2 anthers large, 5 medium-sized, 3 small. Pods oblong, flattened, 4.5-7.5 x 1.5-2.5 cm, not crested, glabrous. Seeds transversely arranged, compressed, oblong, 8 x 4 mm, reticulate, with an areole on each face.

Dry bushland or woodland, usually on sandy soil; 400-750 m. HA; Somalia. *Gilbert* 2077; *Hemming* 1462, 1660.

15. S. **truncata** *(Brenan) Lock* (1988);
Cassia truncata Brenan (1958) - type: HA, *Ellis* 325 (K holo.).

Shrublet, up to 1.2 m tall; branches spreadingly puberulous. Leaves 10-25 cm long, eglandular; leaflets in (3-)4-10 pairs, obovate-elliptic to obovate-oblong, 0.8-3.5 x 0.6-2 cm, truncate or retuse at apex, pubescent. Racemes, including peduncles, up to 15-50 cm long. Petals yellow, obovate, 1.7-2.1 cm long. Stamens: 2 anthers large, 5 medium-sized, 3 small. Pods oblong, flattened, straight, 2-7 x 1.5-2.2 cm, transversely venose, with a crest along middle of each valve, pubescent. Seeds tranversely arranged, compressed, oblong-ovate, 5-6 x 3.5-4 mm, reticulate, with a small areole on each face.

Dry bushland and woodland, roadsides, usually on sandy soil; 250-750 m. HA; Somalia. *Mooney* 7643; *Ellis* 153; *Bally* 9613.

Used as a purgative.

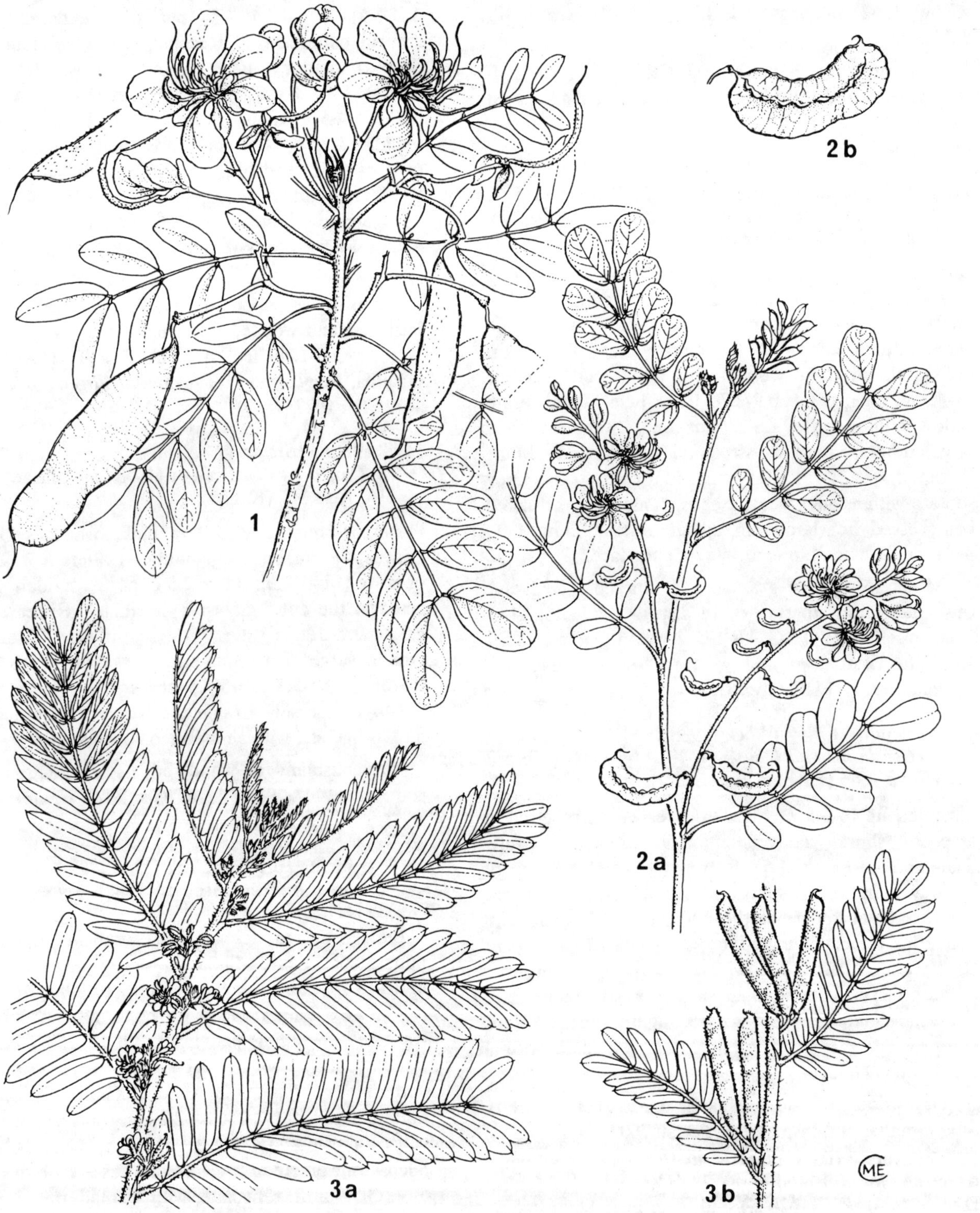

Figure 95.4.1 - *SENNA BACCARINII*, flowering and fruiting branch, from *Gilbert* 3207. 2 - *SENNA ITALICA* subsp. *MICRANTHA*, a: flowering branch, b: pod, both from *Burger* 1129. 3 - *CHAMAECRISTA NIGRICANS*, a: flowering branch, b: pods, both from *Burger* 2877. (All x 0.75) Drawn by Maureen Church. (Reproduced with permission from *Opera Botanica*, 68, fig. 6, 1983.)

16. **S. italica** *Mill.* (1768);
Cassia italica (Mill.) Spreng. (1800).

Erect to prostrate herb or small shrub; branches appressed puberulous in Flora area. Leaves 3-12 cm long, eglandular; stipules triangular to ovate-triangular, 3-9 mm long; leaflets (3-)4-6(-7) pairs, obovate-elliptic to obovate-oblong, 1-4.3 x 0.4-2(-2.7) cm, minutely appressed puberulous. Racemes 2-25 cm long; bracts broadly ovate or elliptic, shortly acuminate, 3-5 mm long, falling quickly. Sepals usually blackish or brownish except for hyaline margins. Petals yellowish-white to bright yellow, 0.8-2 cm long. Stamens arranged as in *S. longiracemosa*. Pods shortly oblong-falcate, flattened, 3-6 x 1.3-2 cm, transversely venose, with a ridge of raised crests along middle of each valve. Seeds similar to *S. longiracemosa*.

One source of senna, used as a purgative.

1. Petals 9-20 mm long; racemes (including peduncle) 6-23 cm long, mostly exceeding subtending leaf. subsp. **italica**
- Petals 8.5-9 mm long; racemes 2-8 cm long, shorter than subtending leaf. subsp. **micrantha**

Subsp. **italica**.

Wooded grassland, *Acacia-Commiphora* bushland and semi-desert, open bushland, often by streams or in disturbed places; 0-1850 m. EE AF EW TU GD; widespread in northern Africa from Cape Verde Is. to Egypt, extending south to Nigeria, Sudan and Somalia, also in Israel and Arabia to S Iran and Pakistan. *Mooney* 8059; *Schimper* 433; *Pappi* 2512.

Subsp. **micrantha** *(Brenan) Lock* (1988);
Cassia italica subsp. *micrantha* Brenan in *Kew Bull.* **13**:241 (1958). Fig. 95.4.2.

In similar habitats to subsp. *italica*. EE AF SU AR HA GG; Senegal and Mali east to Somalia and Socotra, south to South Africa and Namibia, also India and Pakistan. *Burger* 1129, 2123; *Corradi* 4138.

Although most flowering specimens can be referred readily to one subspecies or the other, the ranges of subsp. *italica* and *micrantha* meet in the Flora area and intermediate forms may be encountered. For comments on nomenclature, see Mabberley in *Taxon* **30**:9 (1981).

17. **S. holosericea** *(Fresen.) Greuter* (1986);
Cassia holosericea Fresen. (1839) - type; EE, Mitsiwa (Massawa), *Rüppell* s.n. (FR holo., not seen).

Spreading or prostrate herb or small shrub; branches densely spreading pubescent. Leaves 5-15 cm long, eglandular; stipules subulate above broader base or triangular, (2.5-)4-7 mm long; leaflets (3-)4-8(-9) pairs, mostly oblong-elliptic with upper ones more obovate, 1-2.5 x 0.5-1(-1.5) cm, conspicuously to densely pubescent. Racemes 2-8(-17) cm long; bracts broadly ovate to elliptic, obtuse to shortly acuminate, 5-8(-10) mm long. Petals yellow, 4-8 mm long. Stamens arranged as in *S. alexandrina*. Pods shortly oblong-falcate, flattened, 2.5-3(-3.5) x 1.2-1.5 cm, conspicuously hairy and without a ridge of crests. Seeds similar to *S. longiracemosa*, 4.5-5 x 2.8-3.5 mm.

Semi-desert open bushland; less than 1000 m. EE ?EW HA; near Red Sea from Sudan to Somalia and east through Arabia to Pakistan (Sind), also Socotra and possibly in the Saharo-Sindian region as far west as Mali. *Baldrati* 62; *Fiori* 1065; *Simmons* 130.

Very similar to *S. italica* and *S. alexandrina* var. *obtusata*, differing from both in the consistently spreading rather dense indumentum and smaller pods, from the former in the uncrested pods and the latter in the smaller flowers and generally smaller size of the plants.

18. **S. alexandrina** *Mill.* (1768);
Cassia alexandrina (Mill.) Thell. (1908).
C. senna L. (1753).
C. angustifolia Vahl (1790).

Shrub, rarely procumbent, 0.2-3 m high; young branches with short appressed or spreading hairs. Leaves (3-)5-15 cm long, eglandular; stipules subulate, linear or narrowly triangular, 3-5 mm long; leaflets (3-) 4-8(-13) pairs, lanceolate to elliptic, 1.5-5(-6) x 0.4-1 (-1.5) cm, appressed puberulous or pubescent. Racemes 5-30 cm long; bracts elliptic to obovate, obtuse to shortly acuminate, 5-11 mm long. Sepals greenish or hyaline. Petals yellow or orange-yellow, 0.7-1.7 cm long. Stamens 10: 2 anthers large, 5 medium-sized, 3 small. Pods shortly oblong, slightly curved to almost straight, flattened, (3-)4-7 x 1.6-2.6 cm, dehiscent, transversely septate within. Seeds transversely arranged, compressed, oblong or oblong-ovate, apiculate near hilum, often emarginate at opposite end, 5-7 x 3.5-4 mm, reticulate or rugose, with a small areole on each face.

Leaves and pods used as a purgative, known as senna, produced commercially from var. *alexandrina* principally in India.

1. Leaflets mostly pointed, with a slender mucro 0.5-1 mm long; pedicels glabrous to sparingly pubescent; pods sparingly puberulous or pubescent. var. **alexandrina**
- Leaflets mostly rounded to obtuse at apex, the mucro 0-0.5 mm long; pedicels and pods mostly rather densely pubescent or puberulous. var. **obtusata**

Var. **alexandrina**.

Semi-desert scrub and grassland, particularly drainage areas, flood plains and river banks; 0-800 (?-1400) m. EE AF EW; central Sahara (Silet) to Arabia and India, Sudan along the Red Sea coast to Somalia and southwards to E Kenya, also (probably introduced) in Mozambique. *Bally* 6927; *Burger* 3769; *Mooney* 8567.

Var. **obtusata** *(Brenan) Lock* (1988);
Cassia senna var. *obtusata* Brenan in *Kew Bull.* **13**:244 (1958).

Acacia-Commiphora bushland and semi-desert open bushland; 0-1000 m. EE (Anfile Bay Is.) HA; Somalia, N Kenya. *Burger* 3387; *Buscalioni* 527; *Terracciano* 1816.

9. CHAMAECRISTA *Moench.* (1794)

For references, see under *Cassia*.

Herbs (in Ethiopia) to unarmed shrubs or trees. Leaves paripinnate, often with ± cup-shaped, sessile or stalked glands on petiole and/or rhachis. Flowers in 1 to many flowered racemes; pedicels with 2 bracteoles near or above middle. Sepals 5, imbricate. Petals 5, yellow, often heteromorphic. Stamens usually 10, rarely only 4-5; anthers basifixed, longer than their short filaments, ciliolate along lateral sutures, dehiscent by apical pores or short slits. Pods compressed, elastically dehiscent with twisting valves. Seeds smooth or pitted, without areole.

About 250 species, pantropical, most numerous in the New World, some extending into temperate regions.

1. Plants sticky on account of numerous, glandular-based, spreading bristly hairs in the indumentum; leaflets in 2 pairs. **1. C. absus**
- Plants not sticky, without glandular-based bristly hairs; leaflets in 4 or more pairs. 2
2. Petiolar gland raised on a distinct stalk; perennials developing a woody rootstock. 3
- Petiolar gland sessile; annuals or short-lived perennials. 4
3. Midrib of leaflets excentric but giving off lateral nerves towards both margins; stalk of petiolar gland 0.3-0.7 mm long; pods (3.5-)4.5 mm wide. **2. C. hildebrandtii**
- Midrib of leaflets close to anterior margin, with very inconspicuous or no lateral nerves on this side; stalk of petiolar gland 1-2.25 mm long; pods 3-4 mm wide. **3. C. fallacina**
4. Leaf rhachis with a definite crenate-crested wing running along upper side (rub off leaflets and examine from side with x 10 lens). **8. C. mimosoides**
- Leafrhachis longitudinally channelled, but not or scarcely crenate-crested along upper side. 5
5. Petals 3.5-5 mm long; pedicels 2-7(-10) mm long; apex of leaflets rounded on both sides to nearly median mucro. 6
- Petals 8-19 mm long; pedicels (6-)10-23 mm long; apex of leaflets very asymmetric, with mucro near virtually straight anterior margin. 7
6. Stamens 10; petiolar gland 2-4 mm long; leaflets (10-)12-25(-35) x 2.5-5(-7.5) mm. **4. C. nigricans**
- Stamens 4-5; petiolar gland 0.2-0.7 mm long; leaflets 4-7.5(-10) x 1-2 mm. **7. C. hochstetteri**
7. Petiolar gland elliptic, depressed in middle, (0.75-)1.2-1.8(-2) mm long, yellow; petals 8-15 mm long. **5. C. kirkii**
- Petiolar gland circular or very broadly elliptic, flattened or somewhat concave on upper side, 0.5-1 mm long, dark purplish red; petals 10-19 mm long. **6. C. wittei**

1. **C. absus** *(L.) Irwin & Barneby* (1982);
Cassia absus L. (1753).

Erect or procumbent much-branched annual or short-lived perennial c 0.1-1.2 m high, sticky on account of glandular-based setae in the indumentum. Leaves, apart from glandular hairs, with only a scale-like scarcely glandular structure between each of the 2 pairs of leaflets; leaflets obliquely elliptic or obovate, 1-4.5 x 0.8-3 cm. Racemes 1-13 cm long, with few to numerous small flowers. Sepals obtuse. Petals yellow to red or occasionally white, 5-7 mm long. Stamens 5, subequal. Pods linear-oblong, flat, (2.5-)4-5.5 x 0.6-0.8 cm, elastically dehiscent, pubescent and glandular hairy. Seeds brown to black, glossy, ovate or subrhombic, compressed, 4-5.5 x 3.4-4.5 mm.

Cultivated ground, waste places, river banks and in a variety of wooded grassland and scrub associations; up to 1700 m. EE EW TU GD SU BA HA KF SD; widespread in the Old World tropics. *Pappi* 4738; *Gilbert & Getachew* 3086; *Thulin, Hunde & Tadesse* 3514.

Seeds contain alkaloids which have powerful actions on the nervous and vascular systems, and are used accordingly for a variety of purposes in folk medicine.

2. **C. hildebrandtii** *(Vatke) Lock* (1988);
Cassia hildebrandtii Vatke (1880).
Cassia hildebrandtii var. *crispata* Serrato in *Webbia* **26**:79 (1971) - type: SD, *Mooney* 5441 (K holo.).
Cassia grantii Oliv. var. *pilosula* sensu Cuf. *Enum*.: 217 (1955), non Oliv.

Perennial herb with woody rootstock; stems prostrate or ascending, shortly crisped pubescent to densely spreading pilose. Leaves (1-)1.5-6.5 cm long; gland near top of petiole flattened or shallowly cup-shaped, raised on a 0.3-0.7 mm long stalk; leaflets sessile, in *c* 4-14 pairs, oblong or lanceolate-oblong (uppermost often more obovate), straight or slightly curved, *c* (4-)5-15 (-18) x (1.25-)1.5-5.5(-7) mm, glabrous to densely hairy; midrib generally distinctly excentric but giving off lateral nerves towards both margins. Inflorescences 1-4-flowered; pedicels (0.4-)1-3 cm long. Sepals acute. Petals yellow, 4-12 mm long. Pods linear-oblong, flat, 2.5-5.2 x (0.35-)0.4-0.5 cm, elastically dehiscent. Seeds brown to blackish, ovate-rhombic, up to 4 x 2.5 mm, not areolate.

Upland grassland, bushland and degraded dry evergreen forest; 1000-2350 m. EW AR BA HA SD; N Somalia, Uganda, Kenya, N Tanzania, Zaire. *Gillett* 14307; *Mooney* 8425; *Schweinfurth & Riva* 1892.

This species varies considerably in habit, indumentum and size of leaflets, flowers, etc. Var. *crispata*, based on one specimen from Sidamo near Aghare-Mariam, is notably depauperate with short crisped hairs on the branches and sepals and the leaflets glabrous except for the minutely ciliolate margins. More extreme glabrescent forms occur in Kenya (see Brenan 1967:85) and, over the range as a whole, the features of var. *crispata* occur in various degrees and combinations, such that it does not seem useful to recognise this category. In general, however, within the Flora area, plants with the longest and densest indumentum tend to occur throughout the area, whereas those with shorter and sparser indumentum appear to occur only in central and southern Ethiopia.

3. C. **fallacina** *(Chiov.) Lock* (1988);
Cassia fallacina Chiov. (1916).

Perennial herb developing a narrow woody rootstock; stems prostrate or ascending, shortly crisped pubescent. Leaves *c* 1-5 cm long; gland on upper part of petiole, peltate and flattened, raised on a slender 1-2.25 mm long stalk; leaflets sessile, in *c* 9-27 pairs, oblong, asymmetric, straight or slightly curved, *c* 3-10 x 1-2 mm, ciliolate and sometimes sparsely hairy beneath; midrib close to anterior margin, with no or very inconspicuous lateral nerves on this side. Inflorescences 1-2-flowered; pedicels 4-6 mm long. Sepals acute. Petals yellow, 5-8 mm long. Pods similar to *C. hildebrandtii*, 2-4 x 0.3-0.4 cm.

Woodland and bushland; *c* 1300-2500 m. EW WU SU HA BA SD; S Somalia, Kenya, Tanzania. *Pappi* 7226; *Gilbert & Thulin* 126; *Thulin, Hunde & Tadesse* 3826.

4. C. **nigricans** *(Vahl) Greene* (1899);
Cassia nigricans Vahl (1970).

Erect annual or short-lived perennial; stems with both short crisped and longer spreading hairs. Leaves 3-10 cm long; gland near top of petiole, sessile, cushion-shaped, 2-4 mm long; rhachis channelled but not crenate-crested; leaflets sessile, in c 7-15(-18) pairs, narrowly oblong, straight, (10-)12-25(-35) x 2.5-5(-7.5) mm, ciliate, glabrous to pubescent on the surfaces; midrib median or nearly so. Inflorescences mostly supra-axillary, 3-8-flowered, pedicels 2-4(-5) mm long. Sepals acute. Petals yellow, 3.5-4.5 mm long. Pods erect, similar to *C. hildebrandtii*, 1.7-2.4(-4) x 0.4-0.5 cm. Fig. 95.4.3.

Semi-desert and *Acacia-Commiphora* bushland, usually by rivers, drainage lines and hot springs, or wooded grassland of various types including upland associations; 50-1300 m. EE AF EW TU GD WU IL KF GG SD; Senegal to Central African Republic and south to Angola and N Zambia, Sudan, Uganda, W Kenya, NW Tanzania, Arabia and India. *Burger* 2877; *Schimper* II:1024; *Ash* 3646.

Throughout most of its range the leaflets are distinctly pubescent on the surfaces as well as at the margins, but in the Flora area plants with practically no hairs on the surface or only on the midrib beneath are relatively common. Similar forms have been seen elsewhere only from the Niger Republic.

5. C. **kirkii** *(Oliv.) Standley* (1917);
Cassia kirkii Oliv. (1871).

Erect annual or short-lived perennial, with diverse indumentum. Leaves 2.5-13(-17) cm long; gland near top of petiole, sessile, ± elliptic, cushion-shaped, depressed in middle, (0.75-)1.2-1.8(-2) mm long, yellow (sometimes a second smaller gland below); rhachis channelled but not crenate-crested; leaflets sessile, in (12-)18-43(-51) pairs, linear-oblong, (4-)5.5-17(-20) x (1-)1.25-3(-5) mm, glabrous to pubescent; midrib somewhat excentric but giving off lateral nerves towards both margins. Inflorescences mostly supra-axillary, (1-)2-5-flowered; pedicels (0.6-)1-2.3 cm long. Sepals acute. Petals yellow, 8-15 mm long. Pods similar to *C. hildebrandtii*, (3-)5-9 x 0.5-0.6 cm.

Var. **guineensis** *(Steyaert) Lock* (1988);
Cassia kirkii var. *guineensis* Steyaert in *Bull. Jard. Bot. Brux.* **20**: 256, t. 9 (1950).

Stems shortly pubescent with small incurved hairs and without longer spreading hairs. Leaflets glabrous or nearly so except ciliolate margins.

Upland wooded grassland in rather damp places; *c* 1200-1800 m. WG KF; W Africa, Gabon, Zaire, Sudan, Uganda, Kenya, Tanzania and Zambia. *Ist. Agr. Ital.* 63; *J. De Wilde* 6240.

6. C. **wittei** *(Ghesq.) Lock* (1988);
Cassia wittei Ghesq. (1932).
Cassia mimosoides sensu *Adumbr.* **22**:83 (1971), in part.

Very similar to C. kirkii but petiolar gland circular or very broadly elliptic, flattened or somewhat concave on upper side, 0.5-1 mm long, dark purplish red at least when dry, the petals often larger, 1-1.9 cm long, and in Ethiopia the leaflets more distinctly mucronate and a little less asymmetric.

Upland forest margins, bushland and wooded grassland; 1830-2400 m. EW GD GJ SU IL KF; Cameroun, Zaire, Uganda, Tanzania, Malawi, Mozambique and Zimbabwe. *Mooney* 5971, 8673; *Ash* 79.

7. C. **hochstetteri** *(Ghesq.) Lock* (1988);
Cassia hochstetteri Ghesq. (1932)
- type: TU, *Schimper* I:66 (BR lecto., FT G K LISU P W isolecto.).
Cassia dimidiata sensu auctt., non D. Don.

Erect or prostrate annual; stem crisped pubescent. Leaves 1.5-6 cm long; gland near top of petiole, sessile, circular or broadly elliptic, 0.2-0.7 mm long, purplish to blackish at least when dry; rhachis channelled but not crenate-crested; leaflets sessile, in c 9-31 pairs, linear-oblong to elliptic-oblong, 4-7.5(-10) x 1-2 mm, glabrous except sometimes for ciliolate margin; midrib only slightly excentric. Inflorescences mostly supra-axillary, 1-3-flowered; pedicels 3.5-7(-10) mm long. Sepals acute. Petals yellow, 4-5 mm long. Stamens 4-5. Pods as *C. hildebrandtii*, (1.5-)2.2-4.3 x 0.3-6 mm.

Upland grassland and bushland often by rivers; 1300-2000 m. EW TU GD SU; Tanzania, Madagascar, India, China, Japan. *Pappi* 1773, 7302, 8576.

8. **C. mimosoides** *(L.) Greene* (1899);
Cassia mimosoides L. (1753).

Erect or prostrate annual or short-lived perennial; stem crisped pubescent to densely spreading pilose. Leaves 0.6-10 cm long; gland near top of petiole, sessile, circular, 0.4-1.0 mm in diameter; rhachis serrate- or crenate-crested along upper side; leaflets sessile, in 16-76 pairs, obliquely oblong to elliptic-oblong or linear-oblong, (2-)2.5-9 x 0.5-1.25(-2) mm, glabrous or subglabrous except sometimes for ciliolate margins; midrib somewhat excentric. Inflorescences axillary or supra-axillary, 1-3-flowered; pedicels 0.3-2.5(-3) cm long. Sepals acute. Petals yellow, 4-13 mm long. Pods linear to linear-oblong, (1.5-)3.5-8 x 0.3-0.5 cm, flat, elastically dehiscent. Seeds brown, subrhombic, 2-3 x 1-2 mm, without areoles.

Upland grassland and bushland, cultivated and disturbed ground, often along rivers, ditches, etc.; 1000-2800 m. ?EE EW TU GD GJ SU WG AR HA KF GG SD; widespread in the Old World tropics. *Burger* 3048; *Schimper* III:1443; *Gilbert, Thulin & Getachew* 408.

This is the most common and variable species of *Chamaecrista*. It is easily recognised by the definite raised wing-like projection running along the upper side of the leaf rhachis, whose outline seen from sideways is more or less distinctly crenate or serrate, the sinuses corresponding to the points of insertion of the leaflets. It does not seem possible at present to devise any very satisfactory infraspecific classification despite the numerous differences between extreme forms, but Brenan (1967:100) has devised a system of unnamed groups which, despite interlinking forms, are related in part to geography and habitat. It is desirable that much more material be collected from Ethiopia, with notes on habit, habitat and variation, from which an analysis of the divergence patterns in the Flora area can be established.

10. **CERATONIA** *L.* (1753)

Trees or shrubs. Leaves paripinnate. Flowers small, greenish, in short axillary racemes, polygamous, dioecious or rarely bisexual, apetalous, with a fleshy hypogynous disc. Calyx shortly 5-lobed. Stamens 5. Pods oblong, compressed, thickened at sutures, indehiscent, many-seeded.

Two species; besides *C. siliqua* there is a second, recently described species in N Somalia and Oman.

C. siliqua *L.* (1753).

Evergreen, broad-crowned tree to 10 m; leaflets 2-4(-6) pairs, leathery, elliptic-oblong to obovate, *c* 2-8 x 1-4.5 cm, puberulous beneath; calyx-lobes 1.5-2.5 mm, shortly joined, usually reddish; disk *c* 4 mm across; stamens variably developed, up to 7 mm long; pod *c* 10-30 x 1.5-3.5 cm, with pulpy sugary septa between the hard brown seeds.

Cultivated, e.g. in Asmara and at Godofelasi (near Mendefera, EW) and in Gondar (GD); native of E Mediterranean. *Baldrati* 4172; *O.Ryding & S. Edwards* 1030.

Pods can be used as livestock fodder; cultivated in the Middle East and Mediterranean since ancient times as a source of food for people and animals. The fruit can be processed into a substitute for chocolate called carob.

11. **TAMARINDUS** *L.* (1753)

Tree. Leaves paripinnate. Flowers in racemes; young buds enclosed by 2 valvate bracteoles, falling quickly. Sepals 4, imbricate. Petals: upper 3 well developed, lower 2 minute, setiform. Stamen filaments united c half-way, with 3 upcurved anther-bearing filaments alternating with 5 sterile teeth (1-2 of these rarely elongated into short filaments). Pods oblong, somewhat compressed, sometimes irregularly constricted between seeds, usually curved, with a dry scurfy outer shell and pulpy inner layer, indehiscent. Seeds sub-rhombic to trapeziform, compressed, with a continuous marginal areole on each face.

Monotypic.

T. indica *L.* (1753).
T. somalensis Mattei (1908).
T. erythraea Mattei (1908), as "*erythraeus*" - type: Eritrea, *Senni* 250 (?PAL holo.).
T. indica var. *emarginata* Chiov. in *Fl. Somala* 2:178 (1932).

Evergreen tree 3-24 m high. Leaves 5-12(-16) cm long; leaflets *c* 10-18 pairs, narrowly oblong, (0.8-)1.2-3.2 x 0.3-1.1 cm, usually glabrous or nearly so. Racemes 1-15(-22) cm long. Sepals 0.8-1.2 cm long, reddish outside. Large petals elliptic or obovate-elliptic, 1-1.3 cm long, golden with red veins. Pods (3-)6.5-14 x 2-3 cm, 1-10-seeded. Seeds chestnut brown, 11-17 x 10-12 mm. Fig. 95.5.1.

Grassland, woodland and *Combretum* bushland, frequently riparian; up to 1500 m. EE EW TU GD WU GJ SU HA IL KF GG SD; native of tropical Africa (doubtfully also Asia), now widely cultivated in tropical and subtropical regions for ornament and for the fruits. *Burger* 1427, 1621; *Friis* et al. 1931.

The acid pulpy part of the fruits is used for preserves, in cooking and as a cooling, mildly laxative drink; the seeds are also edible. The very hard durable wood is useful for various constructional and domestic purposes. Now recognised as an important species for conservation and use in agroforestry systems.

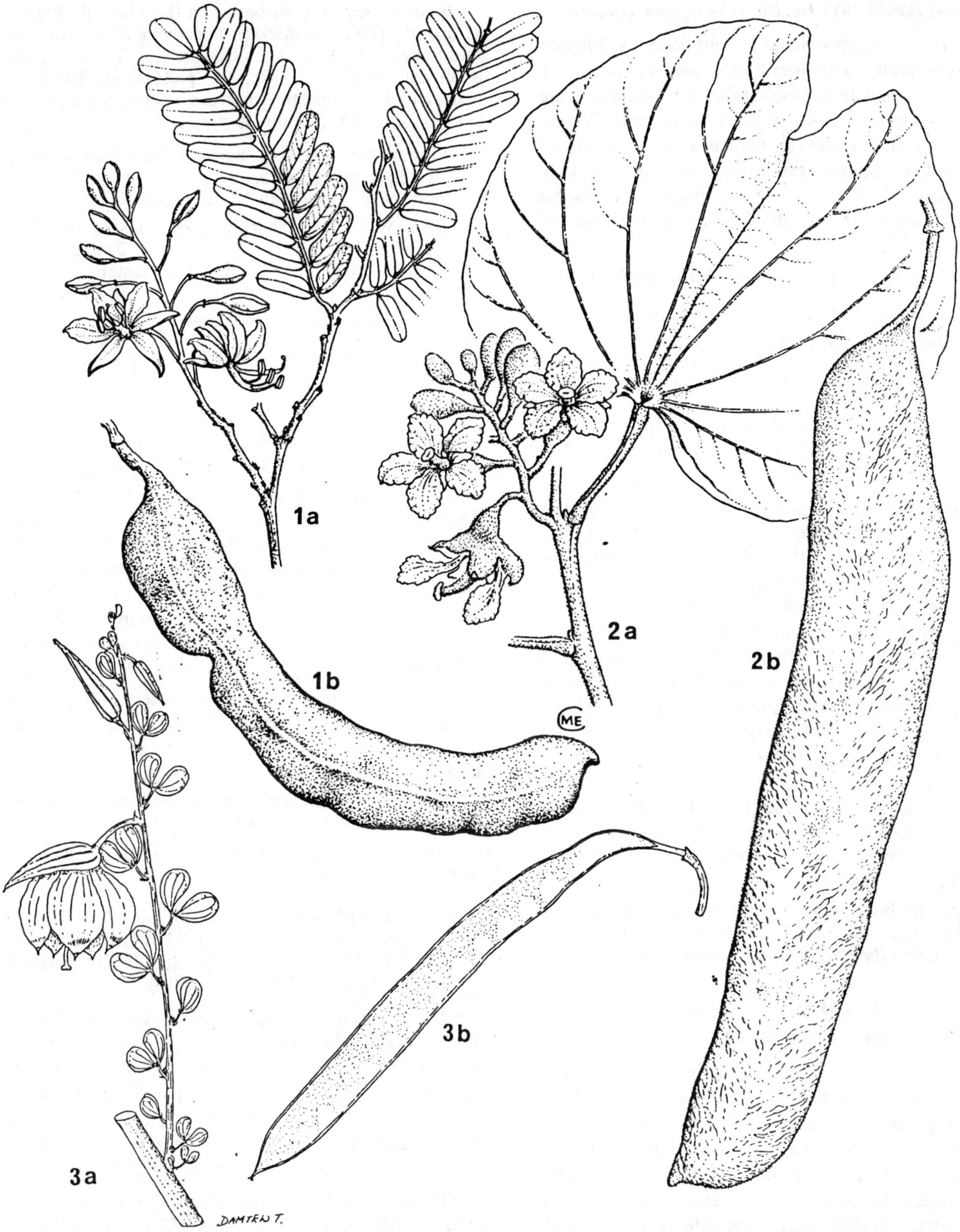

Figure 95.5.1 - *TAMARINDUS INDICA*, a: flowering branch, from *Wickens* 2942, b: pod, from *Burger* 1427. 2 - *PILIOSTIGMA THONNINGII*, a: flowering branch, from *Mooney* 9020, b: pod, from *Friis* et al. 292. 3 - *BAUHINIA ELLENBECKII*, a: flowering branch, b: pod, both from *Gilbert* 3383. 1 and 2 drawn by Maureen Church. 3 drawn by Damtew Tefera. (All x 0.75) (Reproduced with permission from *Opera Botanica*, 68, fig. 7, 1983.)

12. **PILIOSTIGMA** *Hochst.* (1846), *nom. conserv.*

Small trees. Leaves simple, conspicuously bilobed. Flowers medium-sized, usually dioecious, in racemes or panicles. Calyx with turbinate tube and 4-5 short acute lobes. Petals 5. Fertile stamens 10 in male flowers, reduced to staminodes in female flowers; filaments villous below. Stigma sessile, flattened-globose. Pods many-seeded, leathery or woody, indehiscent. Seeds irregularly arranged, embedded in pulp, with a u-shaped line on one side.

Three species in tropical Africa, Asia and Australia.

P. thonningii *(Schumach.) Milne-Redh.* (1947);
Bauhinia thonningii Schumach. (1827).
P. pyrrhocarpum Hochst. (1846) - types: TU, *Schimper* II: 712 & 1658 (both B syn., FT G K P W isosyn.).
Bauhinia abyssinica A. Rich. (1847) - types: TU, *Quartin Dillon* s.n. (P syn.) & *Schimper* II:712 (P syn., FT G K W isosyn.).

Deciduous tree, rarely shrubby, (2-)3-10 m high, with rough bark; branchlets rusty or rarely greyish tomentose. Leaves mostly 5-17 x 6-19 cm, bilobed for 15-30% of way down, densely reticulate and rusty pubescent beneath. Panicles usually alternately leaf-opposed and axillary along branches, male 5-19 cm, female 2-7 cm long. Calyx 1.3-2.3 cm long, rusty tomentellous. Petals white or pinkish, obovate, 1.4-2.6 cm long, wrinkled, hairy outside. Pod stipitate, oblong, *c* 13-26 (-37) x 3-6.3 cm, rusty tomentellous. Seeds dark brown to blackish, *c* 7-9 x 5-7 x 3-4 mm. Fig. 95.5.2.

Deciduous woodland and wooded grassland, often in river valleys; 500-2000 m. EW TU GD GJ WG SU BA IL KF SD; widespread in tropical Africa. *Mooney* 7803, 9020; *Ash* 160.

Root and bark with a variety of medicinal properties, the latter also used for cordage and tanning. Wood suitable for carpentry and construction.

13. **BAUHINIA** *L.* (1753)

de Wit in *Reinwardtia* 3:390 (1956).

Shrubs or small trees. Leaves simple, conspicuously bilobed, or rarely divided to base. Flowers usually large and showy, bisexual, in short racemes or solitary. Calyx spathaceous (sepals more or less cohering after calyx has opened). Petals 5. Fertile stamens 1-10, sometimes accompanied by staminodes; filaments glabrous or hairy. Style elongate; stigma capitate or small. Pods oblong to linear, few to many seeded, usually woody, dehiscent (in Flora area).

Some 200 species, pantropical, often treated in a broader sense to include such genera as *Piliostigma* and *Tylosema*, see most recently Wunderlin, Larsen & Larsen in *Adv. Leg. Syst.*: 107-116 (1981).

B. farek Desv. (1814) based on the plate of "Farek" in Bruce's *Travels to discover the source of the Nile* was said to originate from Mt. Giesh in GJ. However, this is almost certainly a fake, the plate being based on material of a tropical American species of *Bauhinia* cultivated in Paris (Thulin, in prep.).

1. Flowers large, purple, partly white or all white; fertile stamens 3 or 5. 2
- Flowers yellow or white; fertile stamens 10. 3

2. Buds winged or sharply ridged above; fertile stamens 3. **3. B. purpurea**
- Buds not winged or ridged; fertile stamens 5. **4. B. variegata**

3. Leaves divided to half way down or less. **1. B. tomentosa**
- Leaves divided to the base. **2. B. ellenbeckii**

1. **B. tomentosa** *L.* (1753).

Shrub or small tree 1-8 m high. Leaves 1-7(-10) x 1.2-6.5(-11) cm, bilobed for 30% (50%) of way down, glabrous to finely pubescent beneath; lobes rounded. Racemes 1-2(-7)-flowered. Calyx 1.2-2.6 cm long before anthesis, glabrous or pubescent; hypanthium short, funnel-shaped. Petals sulphur yellow, 1-3 of them often maroon blotched at base, subcircular to obovate or elliptic, (2.5-)3-5.5 cm long. Fertile stamens 10, free. Pods thinly woody, 6.5-13 x (1-)1.3-2 cm. Seeds brown or blackish, ovate-elliptic, compressed, 7-9 x 5-6 mm.

Habitat not recorded. GG; eastern Africa from Somalia and Kenya south to South Africa and westwards to Zaire and Angola, also in Asia. *Corradi* 4313-4316.

Also grown for ornament; elsewhere leaves and flowers used medicinally and the wood for rafters.

The record in *Adumbr.* 6:153 (1957) for Djibouti is based on a specimen from Kenya.

2. **B. ellenbeckii** *Harms* (1904)
- types: HA, *Ellenbeck* 1107 & SD, *Ellenbeck* 2048 (B syn. destr.); neotype: fig. 169 in Engl. & Drude, *Veg. Erde*, **9(1/1)**:193 (1910).

Shrub or small tree up to 3 m high. Leaves 4-15(-27) mm long and wide, bilobed to the base, glabrous to finely puberulous beneath. Flowers solitary. Calyx 15-22 mm long, finely pubescent; hypanthium obconical. Petals white or pale yellow, obovate, 2.4- 3.5 cm long. Fertile stamens 10, free. Pods 9-15 x 1-1.4 cm. Seeds brown, compressed, *c* 7 x 5 mm. Fig. 95.5.3.

Dry woodland and scrub, often on limestone; *c* 700-1500 m. HA SD; Somalia. *Burger* 3346; *Thulin, Hunde & Tadesse* 3584; *J. De Wilde* 7271.

3. **B. purpurea** *L.* (1753).

Small tree; leaves large, bilobed for up to one-third way down; buds winged or sharply ridged above; petals purple, up to 6 cm long; fertile stamens 3; pods *c* 20 x 2 cm.

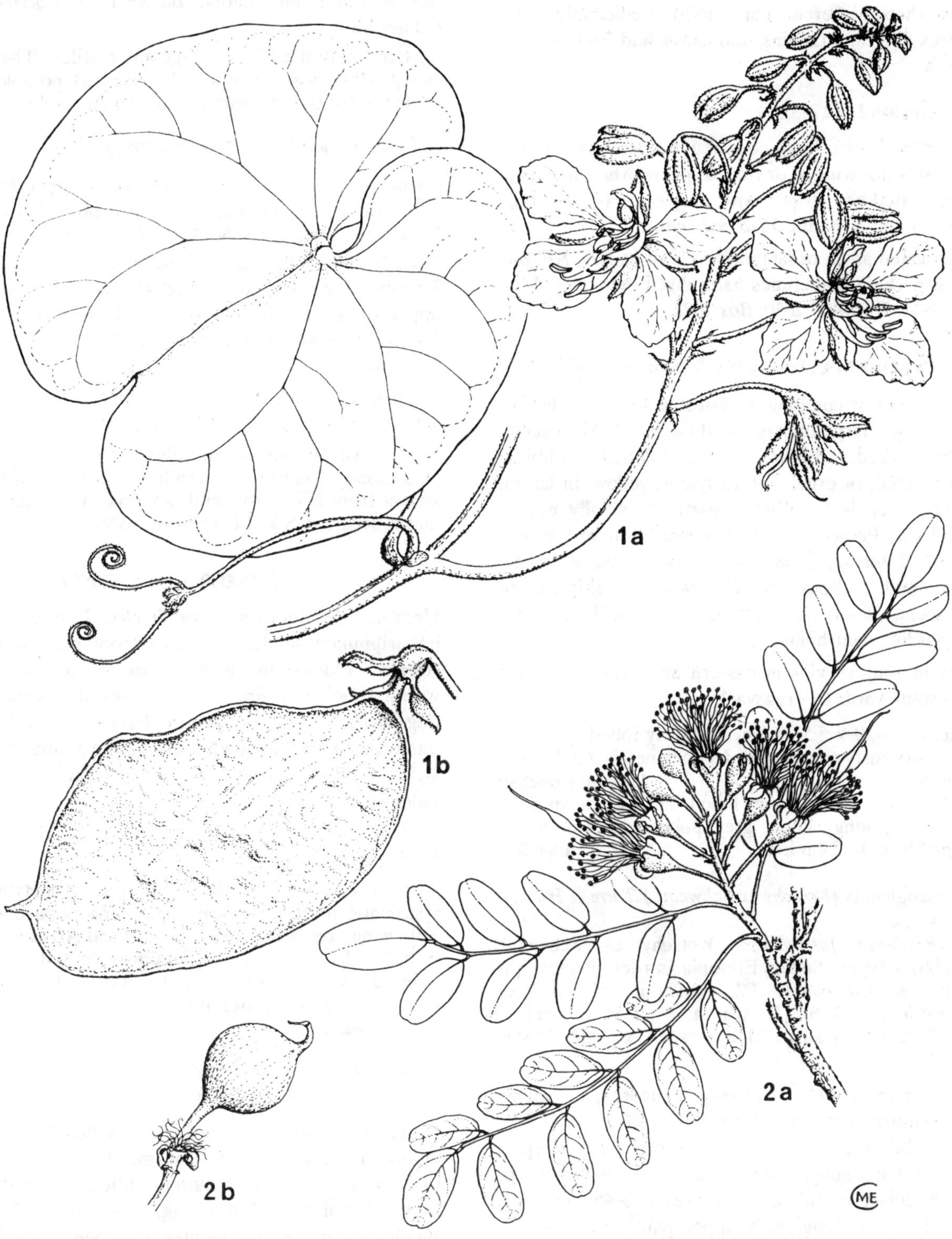

Figure 95.6.1 - *TYLOSEMA FASSOGLENSIS*, a: flowering branch, from *Ash* 911, b: pod, from *Beals* 422. 2 - *CORDYLA SOMALENSIS*, a: flowering branch, from *Hemming* 1458, b: pod, from *Hemming* 1557. (All x 0.75) Drawn by Maureen Church. (Reproduced with permission from *Opera Botanica*, 68, fig. 8, 1983.)

Cultivated as an ornamental and sometimes naturalised as near Gambela (IL); native of Asia. *Ash* 523; *Jansen* 4935, 5010.

Elsewhere different parts used medicinally, bark produces fibre and tannins, and leaves and fruits said to be edible.

4. **B. variegata** *L.* (1753).

Small tree; leaves large, bilobed to one-quarter way down; buds not winged or ridged; petals white or partly yellow, sometimes striped with purple, *c* 4-5 cm long; fertile stamens 5; pods *c* 15-20 x 2 cm.

Cultivated as an ornamental at least in HA, KF and SD and perhaps sometimes naturalised; native of Asia. *Meyer* 8834; *Westphal* 2630; *Bos* 7672.

14. **TYLOSEMA** (*Schweinf.*) *Torre & Hillc.* (1955)

Stems trailing or climbing, herbaceous or woody below, arising from a large underground tuber; tendrils usually present, forked. Leaves simple, bilobed. Flowers medium-sized, bisexual, heterostylous, yellow, in lateral racemes. Sepals 5, upper 2 partly or wholly united, others free. Petals 5; upper one smaller than the rest. Stamens: 2 fertile, (7-)8 staminodial. Style elongate; stigma small. Pods stipitate, woody, dehiscent or indehiscent, 2-seeded. Seeds large, with a short u-shaped line near hilum.

About four species in eastern and central tropical Africa southwards to Transvaal.

1. Leaves 5-20 x 5.5-23 cm, shallowly lobed; young shoots rusty tomentose; 4 larger petals 2.2-4 cm long. **1. T. fassoglensis**
- Leaves up to 4 x 4 cm, lobed half-way down or more; young shoots grey pubescent; 4 larger petals up to 2 cm long. **2. T. argentea**

1. **T. fassoglensis** *(Kotschy ex Schweinf.) Torre & Hillc.* (1955);
Bauhinia fassoglensis Kotschy ex Schweinf. (1868) - types: Sudan/Ethiopia border, *Boriani* (W syn.) & *Cienkowski* 92 (?LE or W syn.) & *Schweinfurth* 2250 (B, syn. destr., BM isosyn.) & 2252 (B syn. destr., BM K isosyn.) & 2253 (B syn. destr.).

Stems to 6 m or more, with rusty tomentum on young parts; tendrils present. Leaves 5-11.5(-20) x 5.5-12 (-23) cm, bilobed for 10-30(-50)% of way down, deeply cordate at base, subglabrous to densely rusty pubescent beneath, lobes rounded. Racemes *c* 5-45 cm long. Sepals 1-1.7 cm long, with upper pair wholly united. Petals yellow, fading to pink; 4 with distinct claws and obovate-subcircular crinkled blades 2.2-4 cm long; upper one much smaller. Pods 7-12 x 4.5-7.3 cm. Seeds brown to blackish, elliptic to ovate-circular, somewhat compressed, 1.7-2.8 x 1.5-2 mm. Fig. 95.6.1.

Deciduous woodland and wooded grassland; 500-1950 m. GD GJ WG SU HA IL KF GG SD; eastern Africa from Sudan to South Africa (Transvaal) and west to Zaire and Angola. *Burger* 1737; *Mooney* 6825; *Gilbert* 1465.

Elsewhere used for cordage and textiles. The shoots can produce sap which can be used as potable water while the fruits and seeds are said to be edible.

2. **T. argentea** *(Chiov.) Brenan* (1963).

Stems to 1.5 m or more, with short appressed grey pubescence on young parts. Tendrils present. Leaves 1-4 x 1-4 cm, bilobed for 50-70% of way down, cordate at base, ± densely grey pubescent beneath, lobes rounded. Racemes *c* 2-7 cm long. Sepals *c* 6-10 mm long, the upper ones free to near base. Petals yellow; 4 with distinct claws and obovate blades, 1.5-2 cm long. Pods *c* 6 x 4 cm.

Acacia-Commiphora bushland; 1000 m. SD; Somalia, N Kenya. *Gilbert & Jones* 125.

A third species, *T. humifusa* (Pic.-Serm. & Roti-Michelozzi) Brenan, in which tendrils are absent, is known from NE Kenya and Somalia. It is highly likely that it occurs in S and E Ethiopia as well.

15. **CORDYLA** *Lour.* (1790)

Unarmed deciduous trees or shrubs. Leaves alternate, imparipinnate; leaflets with numerous pellucid dots or streaks. Flowers in racemes, bisexual or male. Calyx with a subglobose limb entire before dehiscence, splitting into 3-5 lobes on opening. Petals lacking. Stamens numerous. Ovary long-stipitate, tapering into a subulate style. Pods stipitate, ellipsoid to subglobose, indehiscent, 1-6-seeded.

Five or six species in tropical Africa and in Madagascar.

Cordyla belongs to the tribe **Swartziae**, the systematic position of which has long been uncertain. Following common practice in tropical African floras it is here placed in **Caesalpinioideae**, but it should be pointed out that according to recent research it is better placed in **Papilionoideae**, see Cowan in *Adv. Leg. Syst.*: 209-212 (1981).

C. somalensis *Gillett* (1960)
- type: HA, *Ellis* 328 (K holo.).

Small tree or shrub, 1-5 m high. Leaflets 9-13, elliptic-oblong, 11-26 x 6-13 mm. Racemes terminal or axillary, to 6 cm long. Calyx minutely pubescent; tube 5 mm long; 3-lobed, each 9 mm long. Stamens 90-120. Pods subglobose, *c* 3 cm in diameter, 1-seeded. Fig. 95.6.2.

Dry bush/woodland, usually on sandy soil; 500-900 m. HA; S Somalia. *Hemming* 1458, 1557; *Ellis* 126.

The Ethiopian plant is subsp. *somalensis*; subsp. *littoralis* Gillett occurs on coastal dunes in S Somalia.

Subfamily 2. MIMOSOIDEAE

by Asfaw Hunde* & M. Thulin

Leg. Trop. Afr. (1930); Cufodontis *Enum.*: 183-192 (1954) & 193-211 (1955); *Fl. Trop. E. Afr.* (1959).

Trees, shrubs, lianes or rarely herbs, often prickly or spiny. Leaves bipinnate or (in exotic species only) simply pinnate or modified to phyllodes or absent. Flowers regular, small, arranged in spikes, racemes or heads. Sepals valvate (in Ethiopia), usually basally united into a tube, rarely free. Petals (3-)5(-6), valvate in bud, basally united into a tube, rarely free. Stamens 4-10 (as many as or twice as many as the petals) or numerous, free or adnate below to the corolla or the filaments united below into a tube, usually ± exserted; anthers dehiscing lengthwise, sometimes with an apical gland. Pods various. Seeds generally with areoles which normally have a discontinuous margin (open); radicle straight.

About 50 genera and 2800 species, mostly tropical and subtropical, especially numerous in the southern hemisphere.

Key to Genera

1. Leaves reduced to simple entire phyllodes. **22. Acacia** (exotic spp.)
- Leaves bipinnate. 2
2. Plant armed with prickles, thorns or spines. 3
- Plant unarmed (except for some spinescent branches in *Albizia anthelmintica*). 7
3. Spines terminating short branchlets, plants otherwise unarmed; inflorescence two-coloured, upper part yellow, lower white to mauve. **19. Dichrostachys**
- Spines replacing stipules and paired at nodes or plant armed with prickles either scattered or 1-3 together at nodes. 4
4. Pinnae and leaflets in one pair. **24. Pithecellobium**
- Pinnae and/or leaflets in more than one pair. 5
5. Stamens numerous. **22. Acacia**
- Stamens 8 or 10. 6
6. Flowers in spikes; pod not bristly. **18. Prosopis**
- Flowers in heads; pod ± bristly. **20. Mimosa**
7. Aquatic herb; stems creeping. **16. Neptunia**
- Trees or shrubs. 8
8. Stamens 10; petals free or almost so. 9
- Stamens numerous; petals basally united. 10
9. Leaflets obtuse at apex although sometimes mucronate; anthers glabrous; pod at maturity splitting transversely into segments each containing a seed. **17. Entada**
- Leaflets acute at the apex; anthers hairy; pod splitting into 2 valves. **21. Leucaena**
10. Stamens free; anthers with apical gland. **22. Acacia**
- Stamens basally united into a tube; anthers eglandular. **23. Albizia**

16. NEPTUNIA *Lour.* (1790)

Aquatic or terrestrial herbs. Leaves bipinnate. Flowers in solitary axillary globose to ellipsoid heads; upper part of head of bisexual flowers, lower part of male or neuter ones. Calyx 5-lobed. Corolla lobes 5, free or ± united. Stamens 5 or 10, free; anthers with or without an apical gland; staminodes of neuter flowers elongate, without anthers. Pod compressed, dehiscent, not contorted.

About 12 species mainly in tropical and subtropical America, only one in Africa.

N. oleracea *Lour.* (1790).

Aquatic herb; stems creeping, usually floating, swollen and rooting at nodes. Leaves sensitive, with 2-4 pairs of pinnae; rhachis including petiole up to 13 cm long; leaflets 7-22 pairs, oblong, 5-20 x 1.5-4 mm, basal ones smaller, glabrous or almost so. Flowers yellow, in long-pedunculate heads 1.5-2.5 cm long. Calyx 1-3 mm long. Corolla 3-4 mm long. Stamens 10; anthers eglandular; staminodes 17-21 mm long. Pod stipitate, shortly oblong, 1.3-2.7(-3.8) x 1-1.2 cm.

In and by fresh water of pools, lakes and swamps; *c* 500-600 m. GG; widespread in the tropics. *Corradi* 4214, 4215, 4216.

17. ENTADA *Adans.* (1763), *nom. conserv.*

Brenan in *Fl. Trop. E. Afr.* (1959).

Trees, shrubs or lianes, unarmed. Leaves bipinnate; leaflets obtuse, sometimes mucronate. Flowers in spiciform racemes or spikes, bisexual or polygamous. Calyx 5-lobed. Petals 5, free or nearly so. Stamens 10, free; anthers with a deciduous apical gland. Pod straight or curved, flat, with thickened sutures; at maturity the valves splitting transversely into 1-seeded segments with the inner layer (endocarp) persisting round the seed and separating from the outer layer (exocarp); the sutures persisting as a continuous but empty frame.

Some 30 species, mainly in the Old World tropics.

1. Climbing shrub or tree; leaves with 2-4 pairs of pinnae, each with 7-10 pairs of leaflets. **3. E. leptostachya**
- Erect shrubs or trees; leaves usually with pinnae and leaflets more numerous. 2
2. Leaflets 3-13(-15) mm wide, with (8-)10-24 pairs per pinna, glabrous or occasionally ± puberulous, midrib ± central from shortly above leaflet-base. **1. E. africana**
- Leaflets 1-3(-3.5) mm wide, mostly with 22-55 pairs per pinna, usually ± pubescent, midrib excentric. **2. E. abyssinica**

* Uppsala University - Department of Systematic Botany, P.O. Box 541, S-751 21 Uppsala, Sweden.

1. **E. africana** *Guill. & Perr.* (1832).
E. sudanica Schweinf. (1868); *Entadopsis sudanica* (Schweinf.) Gilb. & Bout.(1952) - types from Nigeria, Sudan and Sudan/Ethiopia border: GD, Gallabat, Metema ("Matamma"), *Schweinfurth* 1891 & 1953 (B syn. destr., K isosyn.).

Shrub or small tree up to 10 m high; young branchlets glabrous. Leaves with pinnae 3-9 pairs; leaflets 10-24 pairs, linear-oblong to obovate-oblong, 9-45 x 3-13 mm, rounded at the apex, normally glabrous. Racemes 1-4 together, 6.5-15 cm long, sometimes aggregated on short lateral shoots; axis glabrous or almost so. Flowers yellowish to whitish. Calyx glabrous, 0.75-1.25 mm long. Corolla 1.5-3 mm long. Filaments 4-5 mm long. Pods up to *c* 38 x 5-7.3 cm, straight or nearly so, subcoriaceous. Seeds 12 x 9-10 mm.

Woodland; 450-1100 m. GD(Sudan border) GJ IL; from Sudan to Senegal, also in N Uganda and Zaire. *Thulin & Hunde* 4028, 4040; *Thomerson* 614.

2. **E. abyssinica** *Steud. ex A. Rich.* (1847);
Pusaetha abyssinica (Steud. ex A. Rich.) Kuntze (1891); *Gigalobium abyssinicum* (Steud. ex A. Rich.) Hiern (1896); *Entadopsis abyssinica* (Steud. ex A. Rich.) Gilb. & Bout. (1952) - types: TU, *Schimper* II:520 (P syn., K isosyn.), *Quartin Dillon* s.n. (P syn.).
Entada abyssinica var. *intermedia* Fiori in *L'Agricoltura Colon.* **5**: 170 (1911) - types: EW, *Pappi* 966, 4194, 5387, *Terracciano & Pappi* 2578 (all FT syn.).

Tree up to 10 m high; young branchlets glabrous or pubescent. Leaves: pinnae 4-22 pairs; leaflets 18-50 pairs, linear-oblong, 4-12 x 1-3 mm, rounded at the apex, pubescent on both surfaces. Racemes 1-4 together, 7-16 cm long; axis pubescent. Flowers creamy-white. Calyx glabrous, 0.75-1 mm long. Corolla 1.5-2 mm long. Filaments 3.5-4 mm long. Pods *c* 15-39 x 5-8 cm, straight or nearly so, subcoriaceous. Seeds 10-13 x 8-10 mm. Fig. 95.7.1.

Woodland, wooded grassland and scrub; 1300-2050 m. EW TU GJ WG SU HA IL KF GG SD; widespread in tropical Africa. *Burger* 3485; *Thulin & Hunde* 4006; *W. de Wilde* 10773.

3. **E. leptostachya** *Harms* (1915);
Entadopsis leptostachya (Harms) Cuf. (1969).

Climbing shrub or tree up to 10 m high; young branchlets glabrous or almost so. Leaves: pinnae 2-4 pairs, on long shoots with modified ± hooked pinnae; leaflets 7-10 pairs, narrowly oblong, 9-25(-38) x 3-9(-14) mm, rounded and sometimes emarginate at the apex, puberulous or sometimes glabrous. Racemes 1-3 together, 3-8 cm long, often aggregated on short lateral shoots; axis glabrous or almost so. Flowers yellowish. Calyx glabrous, 0.5-0.75 mm long. Corolla *c* 2 mm long. Filaments *c* 2.5-4 mm long. Pods 17-23 x 4.3-8.4 cm, straight or nearly so, subcoriaceous. Seeds *c* 11-14 x 9 mm.

Dry deciduous woodland and bushland; 1150-1450 m. HA BA SD; Somalia, Kenya, Tanzania. *Burger* 3340; *J. De Wilde & Gilbert* 339; *Thulin, Hunde & Tadesse* 3768.

18. PROSOPIS *L.* (1767)

Shrubs or trees, spiny, prickly or unarmed. Leaves bipinnate, rarely reduced or absent. Flowers in spikes, spiciform racemes or heads, bisexual. Calyx 5-lobed. Petals 5, free or united below. Stamens 10; anthers with an often inconspicuous apical gland. Pod straight, curved or spirally coiled, woody or coriaceous, subcylindrical or ± compressed, internally septate between the seeds, indehiscent.

Some 44 species, one native in tropical Africa, three in the Middle East and the remainder American.

P. juliflora *(Sw.) DC.* (1825).

Tree or shrub, armed with stipular spines 0.5-2.5(-5) cm long. Leaves with 1-2(-4) pairs of pinnae; leaflets 6-29 pairs, 6-23 x 1.5-5.5 mm, glabrous. Flowers yellowish in spiciform racemes, 5-10(-15) cm long. Pod pale brown, linear, straight or slightly curved, 8-29 x 0.8-1.7 cm, compressed, with a sugary-pulpy mesocarp.

Introduced and naturalised, salt-tolerant, growing near waterholes and along wadis at low altitude in EE, AF; native of America. *Bally* 6926; *Greathead* 86.

Now widely cultivated in the tropics for shade, timber and forage; pods can be ground into flour and used to make a flat bread. This species is also reported to have medicinal uses and edible lipids can be extracted from the seeds.

19. DICHROSTACHYS *(DC.) Wight & Arn.* (1834), *nom. conserv.*

Shrubs or small trees, usually armed with spines terminating short lateral spreading twigs, prickles absent. Leaves bipinnate; rhachis glandular at insertion of pinnae. Flowers in axillary spikes; upper part of spike of bisexual flowers, lower part broader of differently coloured neuter flowers. Calyx shortly 5-lobed. Corolla-lobes 5, ± united below. Stamens 10, free; anthers with a deciduous apical gland; staminodes of neuter flowers elongate, without anthers. Pod contorted or straight, indehiscent or dehiscent.

A small genus in the tropics of the Old World.

1. Pinnae 2-3 pairs with 5-8 pairs of leaflets per pinna; pod dehiscing from apex downwards, the valves recurving. **1. D. kirkii**
- Pinnae and leaflets usually more numerous; pod indehiscent, ± contorted. **2. D. cinerea**

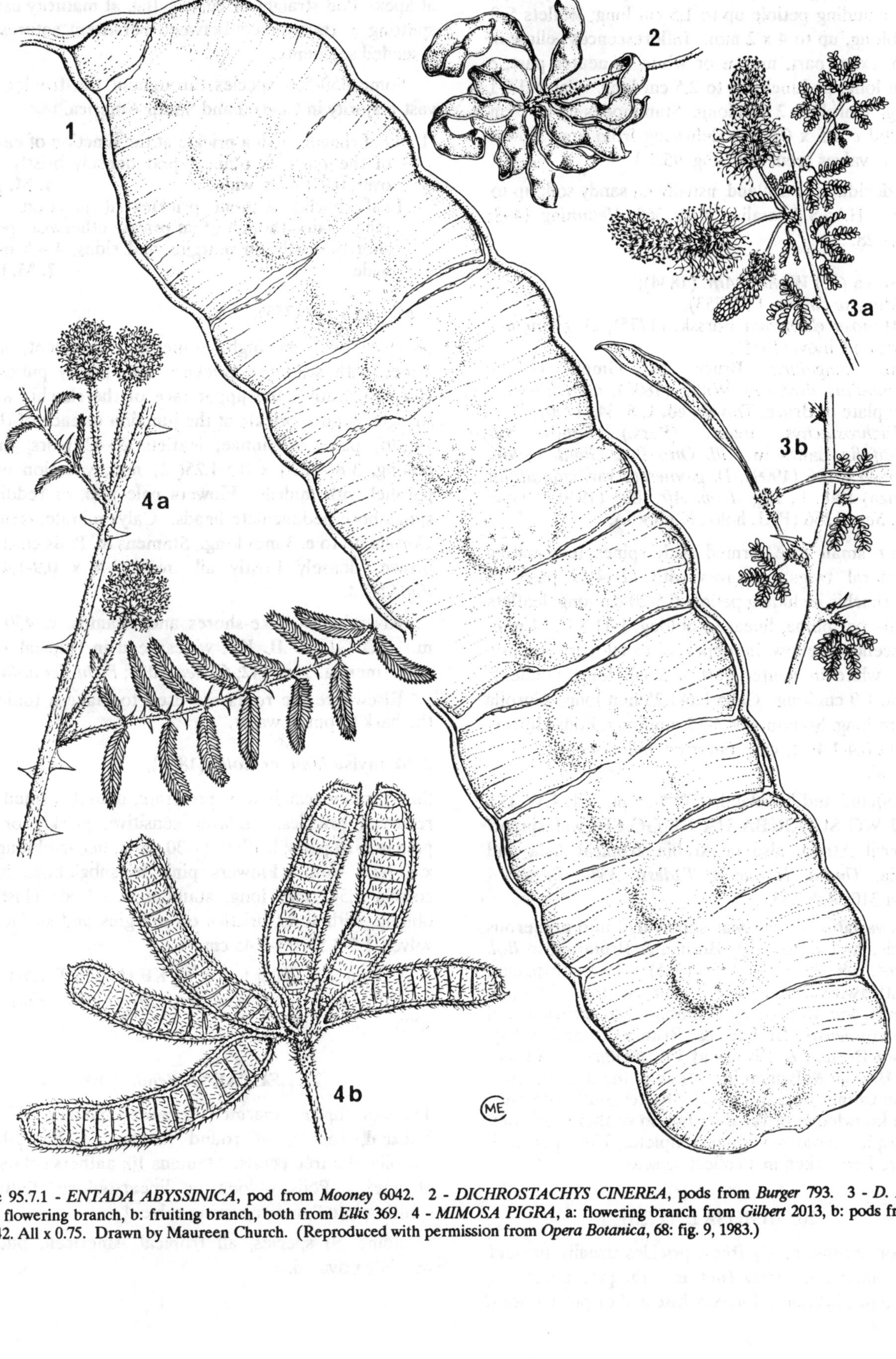

Figure 95.7.1 - *ENTADA ABYSSINICA*, pod from *Mooney* 6042. 2 - *DICHROSTACHYS CINEREA*, pods from *Burger* 793. 3 - *D. KIRKII*, a: flowering branch, b: fruiting branch, both from *Ellis* 369. 4 - *MIMOSA PIGRA*, a: flowering branch from *Gilbert* 2013, b: pods from *Ash* 142. All x 0.75. Drawn by Maureen Church. (Reproduced with permission from *Opera Botanica*, 68: fig. 9, 1983.)

1. **D. kirkii** *Benth.* (1875).

Shrub or small tree, armed with spines terminating short lateral twigs. Leaves with 2-3 pairs of pinnae; rhachis including petiole up to 1.5 cm long; leaflets 5-8 pairs, oblong, up to 4 x 2 mm. Inflorescences yellow in upper bisexual part, mauve or white in neuter part, *c* 1.5-2 cm long; peduncle up to 2.5 cm long. Calyx 0.5-1 mm long. Corolla *c* 2 mm long. Staminodes up to 1 cm long. Pod *c* 4-5 x 0.8 cm, dehiscing from apex downwards, the valves recurving. Fig. 95.7.3.

Dry deciduous bushland, usually on sandy soil; up to *c* 700 m. HA; S Somalia. *Ellis* 369; *Hemming* 1498; *Simmons* 28.

2. **D. cinerea** *(L.) Wight & Arn.* (1834);
Mimosa cinerea L. (1753).
Mimosa glomerata Forssk. (1775); *D. glomerata* (Forssk.) Chiov. (1915).
M. sanguinea Bruce ex Gmel. (1791); *Desmanthus divergens* Willd. (1806), *nom. illegit.* - type: plate in Bruce, *Travels*, **ed. 1, 5**: 34 (1790).
Dichrostachys nutans (Pers.) Benth. var. *grandifolia* Lanza in *Boll. Orto Bot. Giard. Colon. Palermo* **8**:106 (1909); *D. glomerata* var. *grandifolia* (Lanza) Bak. f., *Leg. Trop. Afr.*: 808 (1930) - type: EW, *Senni* 256 (PAL holo., K photo.).

Shrub or small tree, armed with spines terminating short lateral twigs. Leaves with (2-)4-19 pairs of pinnae; rhachis including petiole 0.5-20 cm long; leaflets 9-41 pairs per pinna, linear to oblong, 1-11 x 0.3-4 mm. Inflorescences yellow in upper bisexual part, mauve, pink or white in neuter part, 2-5 cm long, pendent; peduncle 1-9 cm long. Calyx 0.6-1.25 mm long. Corolla 1.5-3 mm long. Staminodes 4-17 mm long. Pods clustered, 2-10 x 0.4-1.4(-2) cm, contorted, indehiscent. Fig. 95.7.2.

Woodland and bushland; 450-2000 m. EE EW TU WU GJ WG SU AR BA HA KF GG SD; widespread in tropical Africa, also in Arabia, tropical Asia and Australia. *Thulin, Hunde & Tadesse* 3387; *Burger & Getahun* 310; *Ash* 255.

This variable species was subdivided into numerous subspecies and varieties by Brenan & Brummitt in *Bol. Soc. Brot., sér.* **2, 39**: 77-115 (1965). Two subspecies, each with two varieties, were recorded from Ethiopia: subsp. *africana* Brenan & Brummitt var. *africana* and var. *karamojensis* Brenan & Brummitt, and subsp. *argillicola* Brenan & Brummitt var. *argillicola* and var. *hirtipes* Brenan & Brummitt. Much of the material now available cannot be named with any certainty, however, and the knowledge of the distribution of these varieties in Ethiopia remains very incomplete. The species is therefore here taken in a collective sense.

20. MIMOSA *L.* (1753)

Herbs or shrubs, rarely trees; prickles usually present. Leaves bipinnate, rarely (not in Ethiopia) absent or modified to phyllodes. Flowers bisexual or polygamous, sessile, in heads or (not in Ethiopia) spikes. Calyx very small. Corolla (3-)4(-6)-lobed. Stamens as many as or twice as many as the corolla-lobes; anthers eglandular at apex. Pod straight or curved, flat, at maturity usually splitting ± transversely between persistent sutures into 1-seeded segments.

Some 450-500 species throughout the tropics, the vast majority in Central and South America.

1. Leaf rhachis with a prickle at the junction of each of the pairs of pinnae; pod densely bristly all over, 0.9-1.4 cm wide. **1. M. pigra**
- Leaf rhachis without prickles at junctions of pinna-pairs, though often prickly otherwise; pod shortly bristly on margins and sides, 4-4.5 mm wide. **2. M. invisa**

1. **M. pigra** *L.* (1755).

Shrub, 0.6-4.5 m high, sometimes scandent; stems armed with scattered prickles, also bristly pubescent. Leaves sensitive, the upper face of the rhachis with a straight slender prickle at the junction of each of the 6-14(-16) pairs of pinnae; leaflets 20-42 pairs, linear-oblong, 3-8(-12.5) x 0.5-1.25(-2) mm, venation nearly parallel with midrib. Flowers pale pink or reddish in subglobose pedunculate heads. Calyx minute, laciniate. Corolla up to c. 3 mm long. Stamens 8. Pods clustered, brown, densely bristly all over, 3-8 x 0.9-1.4 cm. Fig. 95.7.4.

River banks, lake-shores and swamps; c. 450-2000 m. GD GJ SU IL KF; widespread in tropical Africa and America. *Ash* 142; *Gilbert* 2013; *Pichi-Sermolli* 582.

Elsewhere the roots are used to make a tonic and the bark to purify water.

2. **M. invisa** *Mart. ex Colla* (1834).

Shrub, often scandent or prostrate, densely armed with recurved prickles. Leaves sensitive, prickly or not; pinnae 3-10 pairs; leaflets 11-30 pairs, linear-oblong, 2-6 x 0.7-1.5 mm. Flowers pink in subglobose heads; corolla 1.5-2 mm long; stamens 8. Pods clustered, oblong, with short bristles on margins and surfaces of valves, 1.5-3.5 x 0.4-0.45 cm.

Grown as a living fence; IL KF (Misan Teferi), 1400 m; native of tropical America. *Meyer* 8980; *Friis* et al. 4025.

21. LEUCAENA *Benth.* (1842)

Trees or shrubs, unarmed. Leaves bipinnate. Flowers bisexual, sessile, in round heads. Calyx 5-lobed. Corolla of 5 free petals. Stamens 10; anthers eglandular at apex. Pods oblong or linear-oblong, flattened, splitting into 2 non-recurving valves.

Some 50 species, all tropical American, but one widely cultivated.

L. leucocephala *(Lam.) De Wit* (1961).
L. glauca sensu auctt. mult.

Shrub or small tree; young branchlets densely grey-puberulous. Leaves with pinnae (2-)3-7 pairs; leaflets (5-)7-17(-21) pairs, obliquely oblong-lanceolate, 7-18 x 1.5-5 mm, acute at apex, puberulous on margins and sometimes also on midrib beneath. Calyx 2-3.5 mm long. Corolla 4-5.25 mm long, pale green. Filaments 6.5-7.5 mm long, white; anthers hairy. Pod 8-18 x (1.4-) 1.8-2.1 cm.

Cultivated at least in EE EW WU HA KF and GG and sometimes naturalised; widespread in the tropics, native in the New World. *Meyer* 8835; *Milchersich* 2007; *IECAMA* J-15.

Introduced as shade tree, wood crop or browse plant. Now being widely planted in soil and water conservation programmes at lower altitudes. As browse it is suitable in moderate amounts only for cattle and goats, while poisonous to other domestic animals. This species is recorded as having a wide range of other uses.

22. **ACACIA** *Mill.* (1754)

Brenan in *Fl. Trop. E. Afr.* (1959); Ross in *Mem. Bot. Surv. S. Afr.* **44** (1979).

Trees or shrubs, sometimes climbing; the native species almost always armed with prickles or stipular spines, the introduced ones usually unarmed. Leaves bipinnate or, in introduced species, modified to phyllodes; gland usually present on upper side of petiole, glands often also present at insertion of pinnae. Flowers in spikes, spike-like racemes or heads, hermaphrodite or polygamous, sometimes enclosed in bud by a ring of bracts, the involucel, the remnants of which remain on the peduncle (see Fig. 95.8/4). Calyx usually 4-5-lobed, or subtruncate. Corolla 4-5(-7)-lobed. Stamens many; filaments free, or (in *A. albida*) connate at the base; anthers glandular at apex, or (in *A. albida* and all introduced species) eglandular. Pods very variable, dehiscent or sometimes indehiscent, flat to ± thickened, straight, curved, spiral or contorted.

Some 1100 species, mostly tropical and subtropical, the majority in Australia, some 130 in Africa.

Acacia albida is sometimes placed in the monotypic genus *Faidherbia* A. Chev. Pedley, in *Bot. J. Linn. Soc. Lond.* 92: 219-254 (1986), suggested the subdivision of the remaining species of *Acacia* into three genera - the species belonging to subgenus *Aculeiferum* Vassal (species 1-19 of this account) being placed in *Senegalia* Raf.

A record of *A. somalensis* Vatke from Eritrea in Cuf. *Enum.*: 203 (1955) has not been verified and is omitted here.

The collections from HA previously referred to *A. cheilanthifolia* seem better placed in *A. senegal*. Still, it is not unlikely that the species occurs in E Ethiopia. It is therefore included in the key, although at present it is only known from Somalia.

This is an important genus for its many uses to man and his domestic animals. Most species produce abundant nectar and pollen which are good for honey bees. Both domestic and wild animals browse on the leaves, shoots and pods. Roots and bark are fibrous and can be used as rope substitutes. Many species produce gums and resins which may be used medicinally and are eaten in times of food shortage. The wood of many *Acacia* spp. is hard and durable and is used in making local ploughs, building houses and making the best grade of charcoal while the smaller branches provide firewood. Branches of spiny species are often cut to make temporary fencing round fields of growing crops and stacks of harvested crops. Living plants are important in stabilizing soil and improving soil fertility by both nitrogen fixation and their attraction to animals as browse and shade. Uses for individual species are given only where they differ from these general uses.

1. Leaves simple, modified to phyllodes by dilation of the petiole and rhachis (introduced species). 2
- Leaves bipinnate. 7

2. Phyllodes each with 3-7 longitudinal nerves. 3
- Phyllodes each with one main longitudinal nerve. 4

3. Flowers bright yellow; petals free; phyllodes usually ± straight, sometimes slightly falcate, with anastomosing longitudinal venation between the main longitudinal nerves. **55. A. cyclops**
- Flowers pale yellowish white; petals united to above the middle; phyllodes usually ± falcate, sometimes ± straight with conspicuous reticulate venation between the main longitudinal nerves. **56. A. melanoxylon**

4. Phyllodes less than 4 cm long. 5
- Phyllodes 5-22 cm long. 6

5. Young branchlets and phyllodes densely grey-pubescent; phyllodes mostly 10-20 mm wide; pods 15-20 mm wide. **53. A. podalyriifolia**
- Young branchlets and phyllodes glabrous; phyllodes mostly 5-11 mm wide; pods 5-7 mm wide. **57. A. cultriformis**

6. Flower-heads more than 6.5 mm in diameter; peduncles 6-22 mm long. **54. A. saligna**
- Flower-heads less than 6 mm in diameter; peduncles 3-6 mm long. **58. A. retinodes**

7. Inflorescences spicate. 8
- Inflorescences capitate or somewhat oblong-clavate. 30

8. Stipules spinescent, straight, rarely somewhat curved. 9
- Stipules not spinescent; prickles (sometimes absent or obsolete) borne below the stipules, short, usually ± hooked. 12

9. Leaf rhachis with a conspicuous gland between each pair of pinnae; anthers eglandular; pods orange-yellow, falcate or coiled, thick, indehiscent. **19. A. albida**
- Leaf rhachis without conspicuous glands; anthers with apical gland falling off; pods not as above. 10

10. Stipular spines not inflated; inflorescence axis with numerous sessile or subsessile reddish glands. **22. A. lahai**

- Stipular spines; at least some of them, inflated or fusiform below; inflorescence axis with few or no glands. 11
11. Inflated spines deeply constricted at the base; corolla 3-5 times as long as the calyx; pods straight; tree, usually with a well-defined trunk. **21. A. bussei**
- Inflated spines not or only slightly constricted at the base; corolla 2-2.5 times as long as the calyx; pods subreniform or shortly falcate; shrub, branching from near ground. **20. A. horrida**
12. Leaflets 18-35 x 11-31 mm; plant apparently unarmed, with smooth grey powdery bark. **17. A. pseudonigrescens**
- Leaflets usually much smaller; plant usually conspicuously armed; bark various. 13
13. Prickles in pairs near nodes, sometimes absent or obsolete. 14
- Prickles in threes or single near nodes, almost always present. 20
14. Leaflets in 1-2 pairs per pinna; pedicels (0.5-)0.75-1.5 mm long. **11. A. mellifera**
- Leaflets in 3 or more pairs per pinna; pedicels up to 0.5 mm long, mostly subsessile. 15
15. Pinnae of all leaves 2-3 pairs; leaflets 3-5 pairs per pinna. **12. A. laeta**
- Pinnae of all or most leaves 4 or more pairs; leaflets mostly 5 or more pairs per pinna. 16
16. Calyx 1-1.4 mm long, red or purplish; often flowering before the leaves are folded out. 17
- Calyx 1.5-3 mm long, normally not red or purplish; flowers produced with the leaves. 18
17. Leaflets mostly 8-10 x 2.5-3.5 mm, with lateral venation clearly visible beneath. **13. A. venosa**
- Leaflets smaller, with lateral venation almost invisible. **14. A. persiciflora**
18. Pinnae of well-developed leaves up to 8(-10) pairs. **18. A. goetzei**
- Pinnae of well-developed leaves 12-20 or more pairs. 19
19. Leaflets 1.25-2.5(-3.5) mm wide, conspicuously pale and greyish beneath. **16. A. hecatophylla**
- Leaflets 0.4-0.9(-1.25) mm wide, not conspicuously pale beneath. **15. A. polyacantha**
20. All or most leaflets less than 3 mm wide. 21
- All or most leaflets more than 3 mm wide. 25
21. All leaves with 1 or 2 pinna pairs. **10. A. oliveri**
- All or most leaves with 3-8 pinna pairs. 22
22. Leaf rhachillae each with a recurved prickle on the lower surface at or just below the apex. **5. A. hamulosa**
- Leaf rhachillae without a recurved prickle at the apex, but sometimes with a prickle terminating the leaf rhachis, or with prickles along its lower face. 23
23. Leaflets 3-5 pairs per pinna; stem whip-like. **A. cheilanthifolia**
- Leaflets usually 7-25 pairs per pinna; stem not whip-like. 24
24. Pinnae mostly less than 5 mm apart; leaflets 0.5-1.75 wide, usually ± pubescent; pods (1.2-)1.5-3.4 cm wide, mostly ± pubescent, ± yellowish-brown. **4. A. senegal**
- Pinnae mostly more than 7 mm apart; leaflets 1.25-3.75 mm wide, mostly glabrous; pods (0.9-)1.1-1.5(-1.8) cm wide, glabrous, dark brown. **6. A. asak**
25. Pinnae one pair. 26
- Pinnae 2-6 pairs. 27
26. Calyx 1-1.5 mm long; filaments up to 5 mm long; leaflets 4-9 pairs per pinna, usually less than 1.7 mm wide. **10. A. oliveri**
- Calyx 3-5 mm long; filaments *c* 13-18 mm long; leaflets 2-4 pairs per pinna, 3-13 mm wide. **7. A. ogadensis**
27. Leaflets 1 or 2 pairs per pinna, 8-19 x 5-13 mm; pods (2.6-)3.2-4.8 cm wide. **9. A. zizyphispina**
- Leaflets 3-20 pairs per pinna, mostly smaller than above; pods up to 2.1 cm wide. 28
28. Pinnae 2 pairs. **10. A. oliveri**
- Pinnae 3-6 pairs. 29
29. Leaflets 3-9 x 1.25-3.75 mm, mostly glabrous; petiole with a small slightly raised gland less than 1 mm in diameter. **6. A. asak**
- Leaflets 9-23 x 4-7(-9) mm, puberulous on both sides; petiole with a large flattened ± discoid gland, 1.5-3 mm in diameter, at the base. **8. A. condyloclada**
30. Plant unarmed; anthers eglandular. 31
- Plant armed with prickles or stipular spines; anthers with apical gland falling off quickly (except in *A. farnesiana*). 33
31. Leaflets 6-12 mm long. **52. A. decurrens**
- Leaflets 1.5-5.5 mm long. 32
32. Leaf rhachis with a gland at the junction of each pinna pair and usually also with additional glands between the pinna pairs; pods ± moniliform. **50. A. mearnsii**
- Leaf rhachis with a gland at the junction of all or most pinna pairs, but not between them; pods not or only slightly moniliform. **51. A. dealbata**
33. Prickles scattered along the internodes of the stem; stipular spines absent. 34
- Prickles absent; stipular spines in pairs at or near nodes. 36
34. Petiole 0.4-1.5 cm long; prickles arising from longitudinal bands which are paler than the intervening bark. **1. A. brevispica**
- Petiole 1.5-7 cm long; prickles mostly arising from longitudinal bands which are darker than the intervening bark. 35
35. Young branchlets ± densely pubescent and with ± numerous dark reddish glands; pods dehiscent, with margins 1-1.5 mm thick. **2. A. montigena**
- Young branchlets sparsely puberulous to glabrous, normally eglandular; pods indehiscent, thick and hard, with margins 2-4 mm thick. **3. A. pentagona**
36. Heads of flowers arranged in mostly leafless panicles; leaves 10-20 cm wide. **28. A. amythethophylla**

- Heads of flowers on peduncles which are axillary and clustered or borne singly, sometimes racemosely aggregated, not paniculate; leaves smaller. 37
37. Some pairs of stipular spines fused at base into an inflated ± rounded or bilobed, ultimately woody structure ("ant-galls"*). 38
- Spines not inflated. 40
38. Flowers white or cream; pods 4-7 cm long. **35. A. drepanolobium**
- Flowers bright yellow; pods mostly 7-20 cm long. 39
39. Calyx 0.8-1.5 mm long. **24. A. zanzibarica**
- Calyx 2-2.5 mm long. **25. A. seyal var. fistula**
40. Flowers bright or golden yellow. 41
- Flowers white, cream, pale yellow or pink. 46
41. Involucel at apex of peduncle, not subtending additional flowers; pods subterete, indehiscent. **29. A. farnesiana**
- Involucel below or about the middle of the peduncle, often subtending a few additional flowers below the main inflorescence; pods flattened, dehiscent. 42
42. Inflorescence somewhat oblong-clavate, longer than wide. **23. A. dolichocephala**
- Inflorescence spherical. 43
43. Pods straight or slightly curved, indehiscent, 1-2.2 cm wide; spines, some at least, ± deflexed. **30. A. nilotica**
- Pods distinctly curved, dehiscent, 0.3-0.9 cm wide; spines suberect to spreading. 44
44. Pinnae 1-2(-3) pairs; leaflets 8-10(-12) pairs per pinna; spines usually longer than the leaves. **27. A. ehrenbergiana**
- Pinnae mostly 3-11 pairs; leaflets mostly 11-29 pairs per pinna; spines usually shorter than the leaves. 45
45. Bark on trunk powdery; epidermis of twigs conspicuously flaking off to expose a greyish or ± reddish powdery under-surface; calyx 2-2.5 mm long; pods glabrous. **25. A. seyal var. seyal**
- Bark not powdery; epidermis of twigs not conspicuously flaking; calyx 1.5-2 mm long; pods usually ± puberulous. **26. A. hockii**
46. Spines all or many short and hooked or recurved, often with some longer straight ones intermixed. 47
- Spines all straight or nearly so, varying in length. 49
47. Pods contorted or spirally twisted. **31. A. tortilis**
- Pods straight. 48
48. Normally a tree with a well-marked trunk; very often with long straight and short curved spines intermixed; pinnae 1-9(-17) pairs, some leaves usually with 4 or more pairs. **33. A. etbaica**
- Shrub branching from the base; spines all short and hooked; pinnae 1-3 pairs. **34. A. reficiens**
49. Involucel at apex of, or above middle of, peduncle; pods thick and almost woody, very slowly dehiscent, glabrous or almost so. **40. A. sieberiana**
- Involucel at base of or below middle of peduncle; pods various. 50

* ant galls = swellings which certain species of ants colonize

50. Young branchlets villous with rather long spreading hairs, 0.75-3 mm long. 51
- Young branchlets with shorter hairs or glabrous. 53
51. Well-developed leaves usually each with more than 10 pairs of pinnae; corolla-lobes glabrous; pods glabrous or nearly so. **38. A. pilispina**
- Well-developed leaves usually with less than 10 pairs of pinnae; corolla-lobes hairy outside; pods villous. 52
52. Pods dehiscent, coriaceous, attenuate for *c* 2-3 cm at base and apex, 6-12.5 x 0.6-1 cm; calyx and bracteoles glabrous or almost so. **45. A. paolii**
- Pods indehiscent, thick and woody, less attenuate than above, 2-4 x 1-1.5 cm (in Ethiopia); calyx ± pubescent outside; bracteoles ciliate or pubescent. **46. A. stuhlmannii**
53. Pinnae 15-36 pairs on well-developed leaves; young twigs conspicuously longitudinally ridged. **39. A. abyssinica**
- Pinnae less than 15 pairs; twigs mostly not conspicuously ridged. 54
54. Leaflets 6.5-15 x 4-9 mm; pinnae 2 pairs, the lower much shorter than the upper. **49. A. walwalensis**
- Leaflets up to 7(-9) x 2(-3.2) mm; pinnae usually more than 2 pairs. 55
55. Corolla-lobes conspicuously hairy outside. 56
- Corolla-lobes glabrous or almost so outside. 58
56. Pods flat, straight or almost so, ± winged; branchlets greyish-white or yellowish-green. **44. A. oerfota**
- Pods thick, falcate to almost straight, unwinged; branchlets yellowish to purplish brown. 57
57. Pods densely pubescent; leaflets with lateral venation invisible; peduncles up to 1.5 cm long, eglandular. **47. A. edgeworthii**
- Pods minutely puberulous; leaflets with somewhat conspicuous lateral venation beneath; peduncles up to 0.6 cm long, glandular. **48. A. bricchettiana**
58. Pods distinctly curved. 59
- Pods straight or almost so. 60
59. Young branchlets glabrous; pods glabrous. **37. A. robusta**
- Young branchlets pubescent; pods usually puberulous to tomentellous. **36. A. gerrardii**
60. Mature pods 6-12 mm wide; plants found below 2000 m altitude. 61
- Mature pods more than 15 mm wide; plants usually found above 2000 m altitude. 62
61. Leaflets 5-8 mm long with both midrib and lateral veins prominent on both faces. **32. A. prasinata**
- Leaflets up to 4 mm long with only the midrib visible. **33. A. etbaica subsp. etbaica**
62. Branchlets pale, yellowish- or greenish-brown, ± pubescent or puberulous; young pods ± pubescent or puberulous and with sessile reddish glands. **42. A. origena**
- Branchlets dark, purplish- or blackish-brown, mostly glabrous or almost so; young pods glabrous, eglandular. 63

63. Bark of older branchlets exfoliating in papery yellowish flakes without a powdery under-surface; peduncles 2.5-4.6 cm long. **41. A. negrii**
- Bark of older branchlets flaking off to expose a yellowish powdery inner layer; peduncles 1.2-2 cm long. **43. A. bavazzanoi**

1. **A. brevispica** *Harms* (1923).
A. pennata sensu Cuf. *Enum.*:198 (1955) non (L.) Willd.

Shrub or slender tree, 1-7 m high, often scandent; twigs pubescent and with numerous reddish glands. Prickles scattered, ± recurved, arising from longitudinal bands which are usually paler than the intervening lenticellate bands. Leaves: petiole 0.4-1.5 cm long; pinnae up to 18 pairs; leaflets up to 50 or more pairs, 3-5 x 0.5-1 mm, ± ciliolate, midrib nearer one margin at the base. Flowers white to yellowish white, in heads which are arranged racemosely or in long panicles. Calyx 2-2.5 mm long. Corolla up to 3.5 mm long. Pods straight, dehiscent, 6-15 x 1.5-3.3 cm, glabrous or puberulous, with many minute reddish glands. Seeds elliptic, compressed, 8-12 x 6-8 mm; areole 6-8 x 3-5 mm. Figs. 96.8.3, 96.10.2.

Bushland, scrub; 900-2000 m. WU SU BA HA KF GG SD; widespread in tropical Africa; subsp. *dregeana* (Benth.) Brenan in southern Africa. *Ash* 1134; *Fukui* 959; *Thulin, Hunde & Tadesse* 3815.

Subsp. *dregeana* (Benth.) Brenan occurs in southern Africa.

Elsewhere the twigs are used as brushes.

2. **A. montigena** *Brenan & Exell* (1968);
A monticola Brenan & Exell (1957), *nom. illegit.*, non J.M. Black (1937).

Scandent shrub up to 30 m high; young branchlets dark brown, ± densely pubescent with yellowish hairs and many scattered red-purple glands. Prickles scattered, deflexed, arising from longitudinal bands usually darker than the intervening ones. Leaves: petiole 1.5-3.5 cm long; pinnae 7-19 pairs; leaflets 35-50 pairs, 2.5-6 x 0.5-1.25 mm, midrib nearer one margin at the base. Flowers cream or white, in heads which are usually arranged in pyramidal panicles. Calyx *c* 2.4 mm long. Corolla *c* 3 mm long. Pods straight, oblong, dehiscent, 8-18 x 2.7-4.5 cm, dark brown, with margins 1-1.5 mm thick. Seeds elliptic, compressed, 9-12 x 6-7 mm; areole *c* 5-7 x 2.5-3.5 mm.

Upland rain-forest and riverine forest; 1150-2100 m. IL KF ?SD; Uganda, Kenya, Tanzania, Zaire, Zambia and Malawi. *Jansen* 2162; *Mooney* 9149; *Friis* et al. 1710.

3. **A. pentagona** *(Schumach.) Hook. f.* (1849).

Liane; young branchlets red brown to deep purplish, sparsely puberulous to glabrous. Prickles scattered, deflexed, arising from longitudinal bands usually darker than the intervening ones. Leaves: petiole 2-7 cm long; pinnae 8-15 pairs; leaflets 24-58 pairs, 3-7 x 0.7-1.8 mm, midrib nearer one margin at the base. Flowers white, in heads which are arranged in panicles. Calyx *c* 2.4 mm long. Corolla *c* 3.2 mm long. Pods thick, hard, indehiscent, 7.5-16 x 1.8-3.5 cm, with markedly thickened margins, dark brown. Seeds ellipsoid, 10-13 x 6-8 mm; areole 7-10 x 4.5-6 mm.

Lowland rain-forest and riverine forest; *c* 500-600 m. IL; widespread in the rain-forest regions of tropical Africa. *Chaffey* 960; *Jansen* 5099.

4. **A. senegal** *(L.) Willd.* (1806).

Shrub or tree up to 10(-15) m high; branchlets subglabrous to densely pubescent. Prickles in threes, the central one hooked downwards and the two laterals ± curved upwards, or solitary. Leaves: pinnae 3-8 pairs, mostly less than 0.5 mm apart; leaflets 7-25 pairs, *c* 1-5 x 0.5-1.75 mm, pubescent. Flowers white or cream, in up to 12 cm long spikes. Calyx 2-3 mm long. Corolla 2.75-4 mm long. Pods straight, dehiscent, 4-19 x (1.2-) 1.5-3.4 cm, yellowish brown to brown, veined, densely to sparsely appressed pubescent. Seeds subcircular, 8-12 mm in diameter; areole 2.5-6 x 2.5-5 mm.

This species is the source of GUM ARABIC of international trade. It also has many uses as given for the genus and is an important tree for agroforestry systems in arid and semi-arid areas of Sudan and western Ethiopia.

1. Shrub, branching from near the base. var. **kerensis**
- Tree, with a distinct main stem. 2
2. Bark not papery and peeling; young branchlets sparingly to densely pubescent inflorescence axis mostly pubescent. var. **senegal**
- Bark yellow, papery and peeling; young branchlets glabrous or almost so; inflorescence axis glabrous or almost so. var. **leiorhachis**

Var. **senegal**.
?*A. cufodontii* Chiov. (1939) - type; SD, *Cufodontis* 114 (FT holo.). Fig. 95.9.8.

Wooded grassland, deciduous bushland, dry scrub; *c* 600-1700 m. AF EW WU SU AR BA GG SD; widespread in tropical Africa. *Burger* 675; *W. de Wilde* 7346; *Thulin, Hunde & Tadesse* 3623.

Var. **kerensis** *Schweinf.* in *Bull. Herb. Boiss.* **4**, app. **2**: 216 (1896)
- types: EW, *Schweinfurth* 745 (B syn. destr., FT K isosyn.), 741 (B syn. destr.), 998 (B syn. destr.).

Wooded grassland, deciduous bushland, dry scrub; *c* 700-1300 m. EW HA; Somalia, Uganda, Kenya, Tanzania. *Burger* 2135, 2324; *Simmons* S15.

Var. **leiorhachis** *Brenan* in *Kew Bull.* **8**:98 (1953).
A. circummarginata Chiov. (1915).

Dry bushland; *c* 1000-1700 m. SU HA; eastern Africa southwards to Transvaal. *W. de Wilde* 10027; *Hemming* 1512.

Figure 95.8.1 - *ACACIA LAHAI*, flowering branch from *Chaffey* 250. 2 - *A. MELLIFERA*, flowering branch from *Hagos* 138. 3 - *A. BREVISPICA*, flowering branch from *Ash* 1134. 4 - *A. NEGRII*, flowering branch from *W. de Wilde* 6052. All x 1.125. (Reproduced with permission from *Opera Botanica*, 68: fig. 10, 1983.)

5. **A. hamulosa** *Benth.* (1842).
A. paradoxa Chiov. (1929), *nom. illegit.*, non DC. (1813).

Shrub up to 3 m high; young branchlets grey or greyish brown, glabrous or almost so. Prickles in threes, the central one hooked downwards and the two laterals curved upwards. Leaves: pinnae 3-4 pairs, each pinna with a recurved small prickle on the lower face of the rhachis at or just below the apex; leaflets 5-8 pairs, 2-5 x 0.75-2 mm (in Ethiopia). Flowers in up to 4 cm long spikes. Calyx 1.5-2.5 mm long. Corolla 2.5-3.75 mm long. Pods ± ovate with thin and papery valves, dehiscent, 4-8.5 x 2.6-5 cm, pale brown or yellowish, glabrous.

Deciduous bushland; *c* 750 m. HA; Somalia, Kenya and in Arabia. *Simmons* S22.

6. **A. asak** *(Forssk.) Willd.* (1806).
A. glaucophylla Steud. ex A. Rich. (1847) - types: TU, *Schimper* II:725 (P, syn., S, isosyn.) & III:1710 (P, syn., K, isosyn.) & *Quartin Dillon* s.n. (P, syn.).
A. triacantha Hochst. ex A. Rich. (1847) - type: EE, *Schimper* III:1746 (P, holo.).

Shrub or tree up to 10 m high; bark on young stems yellow, peeling, on older stems dark grey and fissured. Prickles single or in threes, the central one curved downwards. Leaves with 3-6, widely spaced, pairs of pinnae; leaflets 7-20 pairs per pinna, 3-9 x 1.25-3.75 mm, glaucous. Flowers white to pale yellowish in up to 11 cm long spikes. Calyx *c* 2 mm long. Corolla 2.5-3 mm long. Pods straight or almost so, dehiscent, 3.8-12 x (0.9-)1.1-1.5(-1.8) cm, dark brown, transversely veined, sparingly puberulous at base. Fig. 95.9.4.

By water courses, on rocky ground and in deciduous bushland; *c* 400-1900 m. EE EW TU WU HA; Sudan, Somalia, Arabia. *Burger* 2975; *Mooney* 8079; *W. de Wilde* 9725.

7. **A. ogadensis** *Chiov.* (1915);
Albizia ogadensis (Chiov.) Bak. f. ex Chiov. (1932) - types from Somalia and Ethiopia: HA, Ogaden, Robecchi-Bricchetti 594 (FT syn.).

Small tree up to 5 m high, with whip-like stem; branchlets grey to purplish. Prickles in threes, the central one hooked downwards and the two laterals spreading or sometimes also slightly hooked downwards. Leaves with 1 pair of pinnae; leaflets 2-4 pairs, 5-17 x 3-13 mm, veined. Flowers in up to 5.5 cm long spikes. Calyx 3-5 mm long. Corolla 4-6 mm long. Filaments up to 18 mm long, lilac. Pods straight, dehiscent, 6-10 x 1.3-2.2 cm, brown, densely puberulous.

Bushland on limestone; *c* 500-1250 m. HA SD; Somalia, NE Kenya. *Friis* et al. 3316.

8. **A. condyloclada** *Chiov.* (1915)
- type: HA, between Sassaban and Carbaden, *Riva & Ruspoli* 1079 (FT holo.).

Tree (2-)3-11 m high with smooth yellowish peeling bark; branchlets dark-brown, puberulous, internodes 5-12 cm long, often enlarged towards the apex. Prickles solitary or in threes, slightly hooked or straight. Leaves with a large flattened gland, 1.5-3 mm in diameter, at the base of the petiole; pinnae 3-5 pairs; leaflets 6-9 (-12) pairs, 9-23 x 4-7(-9) mm, ± puberulous on both sides. Flowers sessile in up to 13 cm long spikes. Calyx *c* 2.5 mm long, with a dark stripe running up each lobe. Corolla *c* 3.5 mm long, glabrous. Pods straight, dehiscent, 6-13 x 1.8-2.2 cm, brownish, transversely veined, densely puberulous.

Dry bushland on limestone outcrops; up to *c* 1200 m. HA BA SD; Somalia, NE Kenya. *Burger* 3396, 3397; *Gilbert* 3996.

9. **A. zizyphispina** *Chiov.* (1929).
A. impervia Gilliland (1951) - type: HA, Ogaden, *Glover & Gilliland* 386 (K holo.).

Shrub up to 3 m high; young branchlets dark reddish brown, glabrous or almost so. Prickles in threes, the central one strongly hooked downwards, the two laterals strongly curved upwards. Leaves: pinnae 2-3 pairs; leaflets 1-2 pairs, obliquely obovate, 8-19 x 5-13 mm, minutely appressed puberulous on both surfaces. Flowers in up to 5 cm long spikes. Calyx up to 2.5 mm long, glabrous. Corolla up to 4 mm long. Pods 4.5-9.3 x (2.6-)3.2-4.8 cm, dehiscent, yellowish brown to purplish, apically rounded, valves thin, transversely veined, glabrous. Fig. 95.9.11.

Dry bushland; *c* 800 m. HA; Somalia. *Hemming* 1463, 1514; *Popov* 1124.

10. **A. oliveri** *Vatke* (1880)
- type: EE, Danakil, *Hildebrandt* 729 c (BM drawing).
A. asak (Forssk.) Willd. var. *unispinosa* Fiori in *L'Agricoltura Colon.* **5**:163 (1911); *A. unispinosa* (Fiori) Chiov. (1929) - type: EE, *Fiori* 135 b (FT lecto.).

Shrub or small tree up to 4 m high with rounded crown; young branchlets reddish brown, sparsely pubescent. Prickles solitary or in threes, recurved. Leaves: pinnae 1-2 pairs; leaflets 4-9 pairs, 3-5(-9) x 1.25-1.75(-3.75) mm, glaucous, glabrous. Flowers in spikes. Calyx 1-1.5 mm long. Corolla 1.75-2.25 mm long. Filaments up to 5 mm long. Pods straight, dehiscent, 3.2-6 x 1-1.6 cm, olive-brown, inconspicuously veined, ± sparsely pubescent on margins and towards the stipe. Fig. 95.9.5.

Bushland, on rocky ground or wadi bottoms; up to 520 m. EE; Somalia. *Gilbert* 2548; *Gilliland* 4063; *Schweinfurth & Riva* 538.

Material from N Somalia which is partly intermediate between *A. senegal* and *A. oliveri* may be only an extreme variant of *A. oliveri* .

11. **A. mellifera** *(Vahl) Benth.* (1842).

Shrub or tree up to 8 m high; young branchlets brownish, pubescent or glabrous. Prickles in pairs, hooked, up to 5 mm long. Leaves: pinnae 2-3 pairs; leaflets 1-2 pairs, ± obovate, up to 20 x 15 m, veined, glabrous (in Ethiopia). Flowers cream to white in spiciform racemes up to 3.5 cm long; pedicels 0.5-1.5 mm long. Calyx up to 1 mm long. Corolla 2.5-3.5 mm long. Pods oblong, straight, dehiscent, 2.5-8 x 1.5-2.5 cm, pale brown to straw-coloured, transversely veined. Seeds subcircular, up to 10 x 8 mm; areole *c* 2.5 x 2.5 mm. Figs. 96.8.2, & 96.9.10.

Deciduous bushland, dry scrub, often forming thickets and an indicator of overgrazing; 400-2500 m. EE EW WU SU HA BA KF GG SD; from Egypt to Tanzania and in Angola and Namibia, also in Arabia; subsp. *detinens* (Burch.) Brenan also in southern Africa. *Ash* 1584; *Bally* 7043; *Burger* 3598.

Elsewhere the roots have been used to make baskets and the bark sacking. The wood has been used as spear shafts, fuelwood and in hut frames.

12. **A. laeta** *R. Br. ex Benth.* (1842)
- type: Ethiopia, without precise locality, *Salt* s.n. (BM holo.).

Shrub or tree up to 6 m high; young branchlets brownish, glabrous. Prickles in pairs, hooked, up to 5.5 mm long. Leaves: pinnae 2-3 pairs; leaflets 3-5 pairs, ± obovate-elliptic, 4-15 x 2-7 mm, veined. Flowers yellowish-white in 3.5-5 cm long spiciform racemes. Calyx 1.25-2 mm long, glabrous. Corolla 2.7-4 mm long. Pods straight, dehiscent, 3.5-8 x 1.7-2.8(-4) cm, pale brown, veined, glabrous or slightly puberulous basally. Seeds subcircular, 8-10 mm in diameter; areole 2 x 2 mm. Fig. 95.9.9.

Deciduous bushland; up to 1800 m. EW TU GD; Somalia, Sudan and westwards to Niger and Mali, Kenya, Tanzania, also in Arabia and the Dead Sea region. *Beccari* 81; *Baldrati* 385; *Schimper* 2170.

The leafy branches can be used as green manure and the larger ones as fencing or in hut frames.

13. **A. venosa** *Hochst. ex Benth.* (1846)
- type: TU, *Schimper* II:524 (K holo.).
A. sanguinea Hochst. ex A. Rich. (1847) - types: TU, *Schimper* III:1567 (P syn., K isosyn.), *Quartin Dillon* s.n. (P syn.).

Small tree; young branchlets glabrous. Prickles in pairs, sometimes absent, recurved, up to 5 mm long. Leaves: pinnae 3-6 pairs; leaflets (3-)6-14 pairs, 6-12 x 2-4 mm, glabrous, with lateral venation somewhat prominent beneath. Flowers in up to 4.5 cm long spikes, produced before or with the new leaves. Calyx c 1 mm long, red or purplish. Corolla *c* 3 mm long, red or purplish. Pods straight or almost so, dehiscent, *c* 13.5 x 2.5 cm, straw-coloured to brownish, transversely veined, glabrous. Seeds up to *c*7 x 8 mm; areole *c* 1 x 2.5 mm. Fig. 95.9.3.

Woodland, evergreen bushland; 1890-2400 m. EW TU GD; not known elsewhere. *Chiovenda* 2620; *Pichi-Sermolli* 596; *Aweke & Gilbert* 975.

14. **A. persiciflora** Pax (1907)
- type: SU, *Rosen* s.n. (WRSL holo.).
A. eggelingii Bak.f. (1935).

Tree 4.5-9(-15) m high; young branchlets pubescent to puberulous. Prickles in pairs, often absent from branchlets, recurved, up to 3 mm long. Leaves: pinnae 4-8(-12) pairs; leaflets 11-17 pairs, 3-5.5(-10) x 0.75-1.5(-2.5) mm. Flowers with white filaments, in 1.5-3 cm long spikes, usually accompanied by very young leaves which are clothed in a yellow indumentum. Calyx 1-1.4 mm long, red or purplish, glabrous. Corolla 2.5-3.5 mm long, red or purplish. Pods straight or slightly curved, dehiscent, 6-15 x 1.6-2.5 cm, brown, transversely veined, with minute dark glands, sparsely puberulous to subglabrous. Seeds subcircular-lenticular, 7-8 mm in diameter, areole *c* 2 x 2 mm. Fig. 95.9.12.

Woodland and wooded grassland; *c* 1700-2100 m. GD GJ SU HA KF SD; Sudan, Uganda, Kenya, Zaire. *Ash* 197; *Burger* 3724; *Gillett* 14862.

15. **A. polyacantha** *Willd.* (1806).

Tree up to 20 m high; young branchlets pubescent or puberulous. Prickles in pairs, 4-15 mm long. Leaves: pinnae 13-60 pairs; petiole with a conspicuous gland, 1.5-4 mm long; leaflets 25-68 pairs, 2-6 x 0.4-0.9(-1.25) mm, usually only the midrib visible beneath. Flowers yellowish-white, sessile, in up to 13 cm long spikes. Calyx 1.5-2.25 mm long, puberulous or pubescent. Corolla 2-3 mm long. Pods straight, dehiscent, 6.5-18 x 1-2.1 cm, brown, veined, glabrous or nearly so. Seeds subcircular, 7-9 x 6-8 mm; areole 3-4 x 2.5-3.5 mm.

A species with many uses as for the genus and as a source of a salt substitute, and gums used in confectionary and adhesives.

Subsp. **campylacantha** *(Hochst. ex A. Rich.) Brenan* in *Kew Bull.* **11**: 195 (1956);
A. campylacantha Hochst. ex A. Rich. (1847); *A. caffra* (Thunb.) Willd. var. *campylacantha* (Hochst. ex A. Rich.) Aubrév., *Fl. Forest. Soudano-Guin.*: 272(1950) - types: TU, *Schimper* II:639 (P syn., K isosyn.), II: 893 (P syn., K isosyn.).
A. erythrantha Steud. ex A. Rich. (1847) - type: GD, *Schimper* II:1120 (P syn., K isosyn.).

Bark whitish to yellowish or grey. Prickles ± hooked. Fig. 95.9.14.

Wooded grassland, deciduous woodland and bushland, riverine and ground water forest; 500-1600 m. EW TU GD GJ SU IL KF GG SD; widepread in tropical Africa. *Thulin & Hunde* 4023; *Chaffey* 929; *Gillett* 14768.

Subsp. *polyacantha* grows in India.

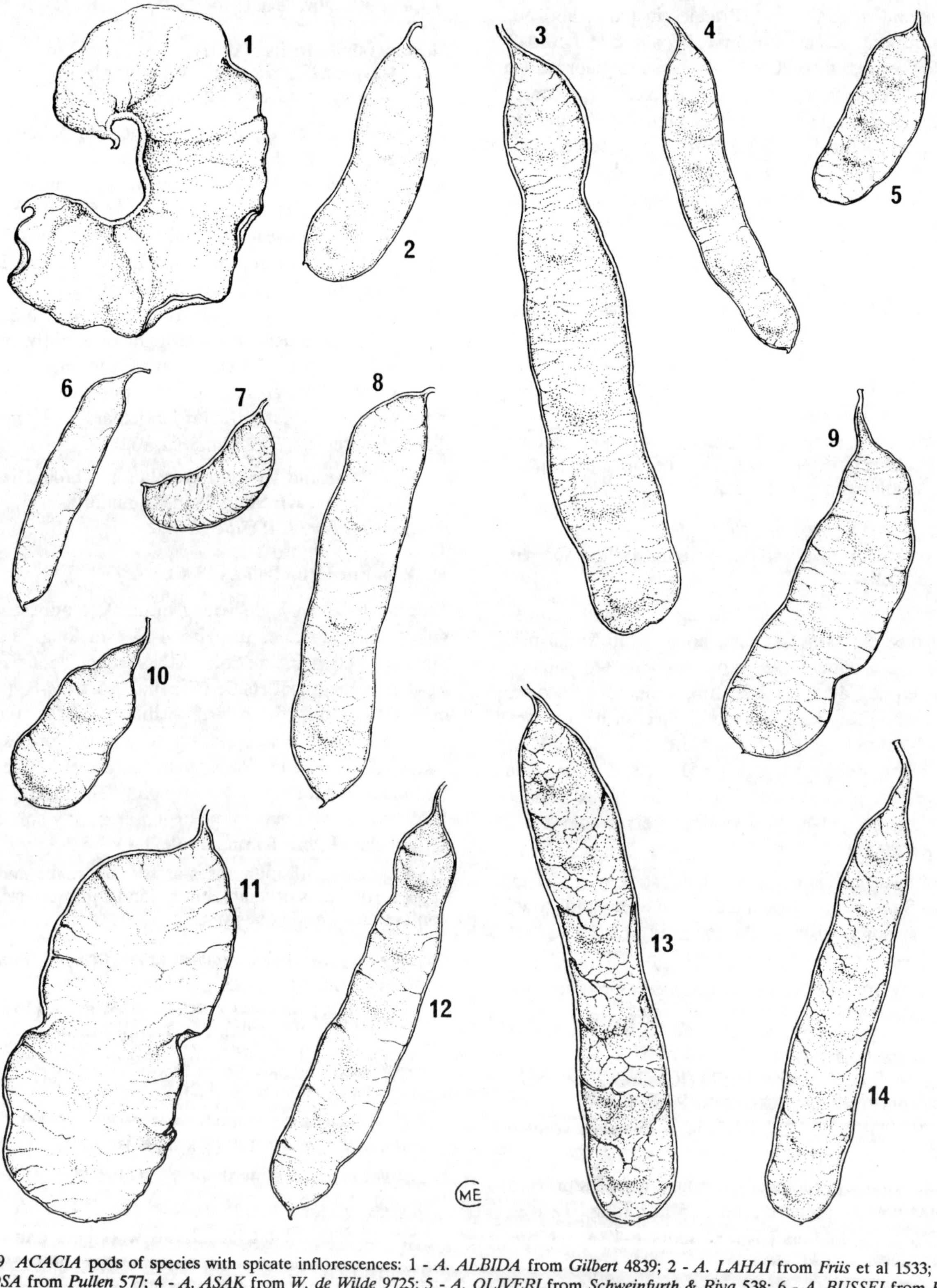

Figure 95.9 *ACACIA* pods of species with spicate inflorescences: 1 - *A. ALBIDA* from *Gilbert* 4839; 2 - *A. LAHAI* from *Friis* et al 1533; 3 - *A. VENOSA* from *Pullen* 577; 4 - *A. ASAK* from *W. de Wilde* 9725; 5 - *A. OLIVERI* from *Schweinfurth & Riva* 538; 6 - *A. BUSSEI* from *Burger* 2134; 7 - *A. HORRIDA* subsp. *BENADIRENSIS* from *Hemming* 1484; 8 - *A. SENEGAL* var. *SENEGAL* from *Beals* 274; 9 - *A. LAETA* from *Schimper* 2170; 10 - *A. MELLIFERA* from *Ash* 1584; 11 - *A. ZIZYPHISPINA* from *Popov* 1124; 12 - *A. PERCIFLORA* from *Gillett* 14862; 13 - *A. HECATOPHYLLA* from *Schweinfurth* 254; 14 - *A. POLYCANTHA* subsp. *CAMPYLACANTHA* from *Gillett* 14768. All x 0.75. (Reproduced with permission from *Opera Botanica*, 68: fig. 11, 1983.)

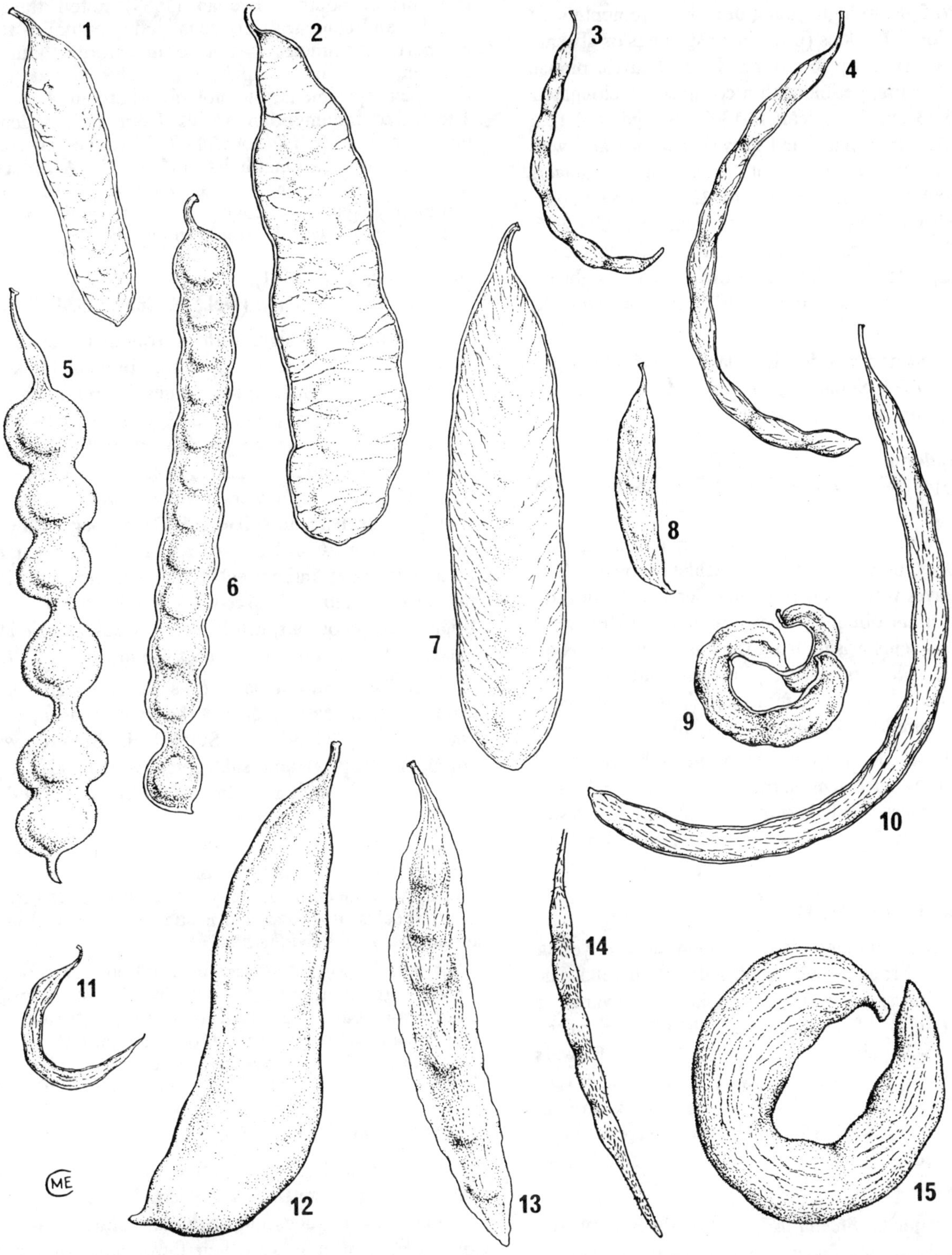

Figure 95.10 *ACACIA* pods of species with capitate inflorescences. 1 - *A. DOLICHOCEPHALA* from *Thulin* 1599; 2 - *A. BREVISPICA* from *Hemming* 83; 3 - *A. EHRENBERGIANA* from *Gilbert* 2554; 4 - *A. SEYAL* from *Pichi-Sermolli* 2232; 5 - *A. NILOTICA* subsp. *INDICA* from *Burger* 1602; 6 - *A. NILOTICA* subsp. *LEIOCARPA* from *Thulin* 1598; 7 - *A. ABYSSINICA* from *Thulin* 1685; 8 - *A. ETBAICA* subsp. *UNCINATA* from *Robertson* 1198; 9 - *A. TORTILIS* subsp. *SPIROCARPA* from *Mooney* 8254; 10 - *A. ROBUSTA* subsp. *USAMBARENSIS* from *Monod* 13427; 11 - *A. DREPANOLOBIUM* from *Burger* 2198; 12 - *A. SIEBERIANA* var. *WOODII* from *Gillett* 14772; 13 - *A. OERFOTA* from *Ash* 649; 14 - *A. PAOLII* from *Hemming* 1511; 15 - *A. EDGEWORTHII* from *Hemming* 1518. All x 0.75. (Reproduced with permission from *Opera Botanica*, 68: fig. 12, 1983.)

16. **A. hecatophylla** *Steud. ex A. Rich.* (1847)
- types: TU, *Schimper* II:628 (P syn.) & II:884 (P syn., K isosyn.).

Tree up to 5(-8) m high; young branchlets tomentellous to puberulous. Prickles (often inconspicuous or absent) in pairs, recurved, up to *c* 6 mm long. Leaves: pinnae up to 11-20 pairs; petiole with a conspicuous elongated gland, 3.5-13 mm long; leaflets 13-40(-50) pairs, 4-12 x 1.25-2.5(-3.5) mm, paler and greyish beneath and with the lateral nerves somewhat prominent. Flowers yellowish-white, sessile, in up to 12 cm long spikes. Calyx 2-2.7 mm long, densely pubescent. Corolla 2.75-3.5 mm long. Pods straight or almost so, dehiscent, coriaceous, 6-15 x 1.7-2.7 cm, brown, veined, glabrous or nearly so. Seeds subcircular, 10-12 mm in diameter; areole *c* 3 x 4 mm. Fig. 95.9.13.

Woodland and wooded grassland; *c* 1450-1700 m. EW TU GD GJ; Sudan, Uganda and Zaire. *Schweinfurth* 1937; *Thulin & Hunde* 4010.

17. **A. pseudonigrescens** *Brenan & Ross* (1974)
- type: BA, Ogaden, *Gilbert* 2129 (K holo., EA iso.).

Slender tree up to 5 m high, apparently unarmed; bark smooth, grey, powdery; young branchlets brown puberulous. Leaves with 1 pair of pinnae; leaflets 2 pairs, 18-35 x 11-31 mm, obliquely obovate-elliptic, with several conspicuous veins arising from the base, puberulous on both sides. Flowers sessile, yellowish-white in 2.5-3.2 cm long spikes. Calyx *c* 2 mm long, densely pubescent. Corolla *c* 3 mm long, pubescent. Pod unknown.

Limestone or gypsum hillocks with cover of predominantly succulent shrubs; *c* 300 m. BA; not known elsewhere. Only known from the type which was collected 8 km W of Mustahil on western track to Kelafo.

18. **A. goetzei** *Harms* (1900).

Tree up to 20 m high with rounded crown; young branchlets glabrous to pubescent, hairs often yellowish. Prickles in pairs, 1-7 mm long, hooked downwards or almost straight. Leaves: pinnae (3-)4-8(-10) pairs; leaflets (2-)5-23 pairs, 3-25 x (0.75-)1-12.5 mm. Flowers yellowish-white in (2-)3-12 cm long spikes. Calyx 1.5-2.75 mm long. Corolla 2-3.75 mm long. Pods straight or almost so, dehiscent, (5-)8-18 x (1.8-)2-3.5 cm, transversely veined, glabrous or nearly so. Seeds subcircular, compressed, 8-11 mm in diameter; areole *c* 5-6 x 4 mm.

Subsp. **microphylla** *Brenan* in *Kew Bull.* **11**: 204 (1956).

Leaflets nearly all less than 3 mm wide, oblong or linear-oblong, with (8-)10-17(-23) pairs per pinna.

Woodland; 1300-1880 m. GG SD; Kenya and southwards to Zimbabwe and Mozambique, also in Zaire and Angola. *Gillett* 14764; *J. de Wilde & Gilbert* 325; *Thulin, Hunde & Tadesse* 3384.

There is a fairly large geographic discontinuity between the specimens from S Ethiopia and N Kenya and the main range of subsp. *microphylla* in Tanzania and further south. Brenan (1959) noted the few prickles and comparatively dense indumentum of these northern specimens. An even more significant difference is the ± pubescent calyx found in all Ethiopian specimens, but not elsewhere in *A. goetzei*. The calyx indumentum varies from rather densely pubescent in *J. de Wilde & Gilbert* 325 to the occurrence of just a few occasional hairs in *Kuls* 332. A collection from NE Elgon in Kenya, *Tweedie* 1800, has a completely glabrous calyx, however, and forms a connecting link with the southern populations.

19. **A. albida** *Del.* (1813);
Faidherbia albida (Del.) A. Chev. (1934).

Tree up to 30 m high, with ± rounded crown; bark rough, dark brown or greyish; branchlets straw-coloured or whitish. Stipular spines up to *c* 2 cm long, straight. Leaves with pinnae 3-10 pairs; rhachis with a conspicuous gland at the junction of each pinna pair, but petiole eglandular; leaflets 6-23 pairs, 3.5-6(-12) x 0.7-2.25(-4) mm. Flowers yellowish-white, sessile or shortly pedicellate, in spikes 3.5-17 cm long. Calyx up to 2 mm long. Corolla up to 3.5 mm long. Filaments basally connate; anthers eglandular even in bud. Pods falcate or coiled, indehiscent, 6-25 x 2-3.5 cm, cream, bright orange or purplish-brown. Seeds elliptic-lenticular, 9-11 x 6-8 mm; areole 7-9 x 4-6 mm. Fig. 95.9.1.

Woodland and grassland, on seasonally waterlogged ground, often leafless during the rainy season; up to 2600 m. EW TU WU GD SU AR HA SD GG; widespread in tropical and subtropical Africa, also in W Asia. *Pappi* 389; *Mooney* 5106; *Thulin, Hunde & Tadesse* 3474.

A. albida displays a number of unusual characters, such as eglandular petiole, basally connate filaments, and eglandular anthers. It is not closely related to any other species of *Acacia* and is sometimes placed in the monotypic genus *Faidherbia.*

A very important species in traditional agroforestry systems. In addition to the uses given for the genus, *A. albida* pods have been used as a soap substitute, bark extracts are used to treat toothache and the boiled seeds are edible. Previously in Ethiopia, planks for making large secure doors were made from *A. albida*.

20. **A. horrida** *(L.) Willd.* (1806).

Shrub, 1-3.6 m high or occasionally higher, obconical and branching from the base; young branchlets brownish, glabrous or pubescent. Stipular spines up to 9 cm long, some normally with their lower part swollen and not or only slightly constricted at base. Leaves: pinnae 2-8 pairs; leaflets 5-14 pairs, 2-6 x 0.75-1.8 mm. Flowers whitish, sessile, in 1-4.5 cm long spikes. Calyx 0.3-1 mm long. Corolla 2-2.5 mm long. Pods ± boat-shaped to falcate, dehiscent, 3-6 x 1.5-2.5 cm, purplish brown, glabrous or slightly puberulous. Seeds 4-5 x 4-4.5 mm, flattened; areole *c* 2 x 1.5 mm.

Subsp. **benadirensis** *(Chiov.) Hillc. & Brenan* in *Kew Bull.* **13**: 40 (1958).

Inflorescence-axis 1-2.5(-4.5) cm long. Calyx 0.8-1 mm long. Corolla 2 mm long. Pods usually glabrous or almost so. Fig. 95.9.7.

Deciduous bushland and dry scrub; *c* 500-1700 m. HA BA KF SD GG; Sudan, Somalia, Uganda and Kenya. *Ash* 3626; *Burger* 3006; *Thulin, Hunde & Tadesse* 3833.

A form from BA and SD and also known from NE Kenya (e.g. *Gilbert* et al. 7564) differs in having ± densely hairy pods, leaves, branches and inflorescence-rhachides.

Subsp. *horrida* grows in India.

21. **A. bussei** *Harms ex Sjöstedt* (1908).

Tree up to 4(-10) m high with ± flattened crown; young branchlets grey-brown to purplish, glabrous or pubescent. Stipular spines up to 5(-9) cm long, some with their lower part strongly swollen but much constricted basally. Leaves: pinnae 2-8 pairs; leaflets 10-18 pairs, 1.5-5 x 0.5-1.6 mm, ± pubescent. Flowers yellowish-white, sessile, in 1.5-6 cm long spikes. Calyx *c* 0.7 mm long. Corolla 2.5-3.5 mm long, glabrous. Pods narrowly oblong, straight, dehiscent, up to 8 x 1.5 cm, puberulous. Seeds flattened, *c* 5 x 4 mm; areole *c* 2 x 1.5 mm. Fig. 95.9.6.

Deciduous bushland, dry scrub; 330-1800 m. HA BA SD; also in Somalia, Kenya and Tanzania. *J. de Wilde* 4155; *Bally* 9212; *Thulin, Hunde & Tadesse* 3832.

The bark produces tannins used in treating hides and skins.

22. **A. lahai** *Steud. & Hochst. ex Benth.* (1842)
- type: TU, *Schimper* I:119 (K holo.).

Shrub or tree, 0.5-15 m high; young branchlets brownish purple, pubescent. Stipular spines up to 7 cm long, slender and straight. Leaves: pinnae 6-15 pairs; leaflets 10-28 pairs, 1.5-4.5 x 0.3-0.75 mm. Flowers cream in up to 9 cm long spikes; axis ± pubescent and with many reddish, sessile or subsessile glands; involucel inserted at or near the base of the peduncle. Calyx 0.5-1.25 mm long. Corolla 2-3 mm long. Pods straight or ± falcate, dehiscent, 4-7 x 1.5-3 cm, brown, glabrous, or puberulous on the stipe. Seeds 6-7 x 5-5.5 mm, compressed; areole 1.5-2.5 x 1.5 mm. Fig. 95.8.1, & 96.9.2.

Upland woodland and scrub; 1700-2600 m. EW TU GD WU GJ WG SU HA KF; Uganda, Kenya and Tanzania. *Mooney* 8111; *W. de Wilde* 6731; *Thulin & Hunde* 3906.

A. lahai is often thicket-forming in Ethiopia and is then flowering and fruiting as a low shrub, often only 0.5-1 m high.

23. **A. dolichocephala** *Harms* (1897)
- type: SD, *Riva* 599 (FT holo.).

Tree, 3-13 m high; branchlets glabrous, often longitudinally ridged. Stipular spines up to 5(-16) mm long, straight. Leaves: pinnae 4-21 pairs; leaflets 12-35 pairs, 2.5-7.5 x 0.6-1.5 mm, glabrous or almost so, dark green above, paler beneath. Flowers bright yellow in somewhat oblong-clavate heads; peduncles up to 9 cm long, puberulous; involucel up to 5 mm long, inserted at varying height on the peduncle, usually subtending a few additional flowers. Calyx *c* 1 mm long. Corolla 3-4 mm long, glabrous. Pods straight or slightly falcate, dehiscent, 5.5-11 x 1.3-2.5 cm, purplish brown, reticulately veined, glabrous. Seeds ± elliptic, compressed, up to *c* 6 x 5 mm; areole 2.5 x 1-1.25 mm. Fig.96.10.1.

Woodland and grassland, often at lake margins and riversides; 1100-2130 m. GJ SU AR GG SD; S Sudan, E Uganda, W Kenya and NE Tanzania. *Ash* 2695; *Gilbert & Tewolde* 194; *Thulin, Hunde & Tadesse* 3326.

24. **A. zanzibarica** *(S. Moore) Taub.* (1895).

Tree or shrub, 3-9 m high; bark yellowish green or whitish, turning brownish and powdery with age; young branchlets glabrous or nearly so, brownish, older ones with minutely flaking or powdery yellowish to brown bark. Stipular spines mostly straight, up to 7.5 cm long, some basally inflated and fused into ± deeply bilobed "ant-galls". Leaves glabrous or nearly so; pinnae 1-4 (-6) pairs; leaflets 3-10 pairs, 2-13(-20) x 0.5-6(-10) mm. Flowers bright yellow in heads which are typically arranged along short lateral branches; peduncles up to 2.5 cm long with the involucel at or near the base. Calyx 0.8-1.5 mm long. Corolla 2.25-2.5(-4) mm long. Pods linear, falcate to almost straight, 5.5-12 x 0.4-0.7 cm, finely longitudinally veined, blackish brown, glabrous. Seeds compressed, 6-7 x 3-4 mm.

Var. **microphylla** *Brenan* in *Kew Bull.* **12**: 75 (1957).

Leaflets 0.5-1.5 mm wide, with normally only the midrib raised beneath.

Dry scrub; 300-1000 m. HA SD; Somalia, Kenya. *Bally* 9309, 10129.

Var. *zanzibarica*, with wider leaflets, occurs in Somalia, Kenya and Tanzania.

25. **A. seyal** *Del.* (1813).

Tree up to 9 m high with usually flattened crown; bark on trunk powdery, white to greenish yellow or orange red; young branchlets almost glabrous, epidermis reddish, conspicuously flaking off to expose a greyish or reddish powdery under-surface. Stipular spines up to 8 cm long, some sometimes basally inflated and fused into bilobed "ant-galls". Leaves: pinnae 3-7 pairs; leaflets 11-20 pairs, 3-8 x 0.7-1.5 mm. Flowers bright yellow, in heads; involucel in lower half of peduncle, often subtending a few additional flowers. Calyx 2-2.5 mm long. Corolla 3.5-4 mm long. Pods ± falcate and constricted between the seeds, dehiscent, 7-20 x 0.5-0.9 cm, finely longitudinally veined, glabrous. Seeds elliptic, compressed, 7-9 x 4.5-5 mm; areole 5-6 x 2.5-3.5 mm.

The bark and exudate have medicinal uses and a red dye can also be extracted from the bark. Other uses as given for *Acacia* as a whole.

1. Spines all slender, inflated spines absent; bark colour variable. var. **seyal**
- Some pairs of spines fused and inflated at base; bark mostly white or greenish yellow. var. **fistula**

Var. **seyal**.
A. stenocarpa Hochst. ex A. Rich. (1847) - types: TU, *Schimper* III:1948 (P syn.), I:218 (P syn., K isosyn.), *Petit* s.n. (P syn.), *Quartin Dillon* s.n. (P syn.).

Woodland, wooded grassland; 1200-2100 m. EW TU WU GD GJ SU AR HA IL KF SD; widespread in northern tropical Africa, extending to Egypt. *Burger* 1364; *Meyer* 7468; *Thulin, Hunde & Tadesse* 3854.

Var. **fistula** *(Schweinf.) Oliv.* in *Fl. Trop. Afr.* **2**: 351 (1871);
A. seyal forma *fistula* (Schweinf.) Cuf., Cuf. *Enum*.: 202 (1955) and *Zweiter Nachtrag*: 23 (1969), *nom. invalid*.

Woodland, wooded grassland; 500-2000 m. EW SU HA IL SD; Sudan, Somalia and southwards to Malawi and Mozambique. *Baldrati* 904; *Jansen* 5081; *W. de Wilde* 6082.

26. **A. hockii** *De Wild.* (1913).

Shrub or tree up to 6(-12) m high; bark on stem red-brown to greenish, peeling off in papery layers; branchlets ± densely puberulous with numerous reddish sessile glands (the bark not peeling to expose a powdery layer as in *A.seyal*). Stipular spines short, straight, mostly up to *c* 2 cm long. Leaves: pinnae 2-11 pairs; leaflets 9-29 pairs, 2-6.5 x 0.5-1(-1.25) mm, ciliolate. Flowers bright yellow, in heads; involucel 1/3-2/3 way up the peduncle, often subtending a few additional flowers. Calyx (1-) 1.5-2 mm long. Corolla 2.5-3.5 mm long. Pods ± falcate and constricted between the seeds, dehiscent, 5-14 x 0.3-0.6 cm, finely longitudinally veined, ± puberulous. Seeds elliptic, compressed, 5-7 x 3-4 mm; areole *c* 4 x 2 mm.

Deciduous woodland and wooded grassland; *c* 700 m. GG; widespread in tropical Africa. *Fukui* 1239; *Turton* 81.

The bark is reported to be used as a famine food.

27. **A. ehrenbergiana** *Hayne* (1827).
A. flava (Forssk.) Schweinf. (1896), *nom. illegit*., non Spreng. ex DC. (1825).

Shrub or small tree 2-4(-7) m high; young branchlets reddish. Stipular spines slender, straight, up to 5 cm long, usually longer than the leaves. Leaves: pinnae 1-2(-3) pairs; leaflets 8-10(-12) pairs, up to 5 x 1 mm. Flowers bright yellow, in heads; peduncles 1.5-3 cm long, glabrous or sparingly pubescent, with the involucel at the middle or on the lower half. Calyx up to *c* 1.5 mm long. Corolla 2-3 mm long, glabrous. Pods linear, falcate, ± constricted between the seeds, dehiscent, 6-15 x 0.3-0.5 cm, reddish-brown, finely longitudinally veined, glabrous. Seeds 5-6.5 x 2.75-3.25 mm; areole *c* 4 x 2 mm. Fig. 95.10.3.

On sandy and rocky soils in dry areas, frequently along seasonal water courses; up to *c* 450 m. EE AF; Sudan, Egypt and westwards to Mali and Mauritania, also in Arabia. *Bally* 7086; *Smith & Gilliland* 4066; *Gilbert* 2554.

A useful species for the hot arid environment in which it is found.

28. **A. amythethophylla** *Steud. ex A. Rich.* (1847)
- type: TU, *Quartin Dillon & Petit* s.n. (P lecto.).
A. macrothyrsa Harms (1900).

Hunde in *Bot. Notiser* **132**: 393 (1979).

Tree up to 12 m high, with rounded crown; bark greyish, rough and fissured. Stipular spines mostly up to 0.8(-1.6) cm long. Leaves large, 10-20 cm wide; pinnae 7-35 pairs; leaflets 33-70 pairs, mostly 6-11 x 1-4 mm. Flowers orange-yellow, in heads which are arranged in ample, loose, leafless panicles; peduncles up to 3.6 cm long, ± puberulous, with the involucel at about the middle. Calyx *c* 1 mm long. Corolla 3-5 mm long. Pods straight, dehiscent, 9.5-20.5 x 1.7-3 cm, reddish brown or blackish, glossy, glabrous. Seeds elliptic, compressed, 8-10 x 7-9 mm; areole 4-5 x 3-3.5 mm.

Woodland; 1300-1450 m. EW TU GD; widepread in tropical Africa. *Pappi* 2386; *Gilbert & Getachew* 2968; *J. de Wilde* 7135.

Elsewhere the leaves are used as a cooked vegetable.

29. **A. farnesiana** *(L.) Willd.* (1806).

Shrub 1.5-4 m high; young branchlets glabrous or nearly so; stipular spines up to 1.8(-3) cm long, straight, slender. Leaves with pinnae 2-7 pairs; leaflets 10-21 pairs, 2-7 x 0.75-1.75 mm, with the midrib and lateral veins somewhat raised beneath. Flowers bright yellow; involucel at apex of peduncle. Pods straight or curved, indehiscent, subterete, 4-7.5 x 0.9-1.5 cm, dark brown to blackish, glabrous; seeds elliptic, slightly compressed, *c* 8 x 5.5 mm, areole *c* 6.5 x 4 mm.

Occasionally cultivated at least in AF; probably a native of tropical America. *Shantz* 48.

A plant with many uses, it is grown elsewhere for its flowers from which cassie perfume is made.

30. **A. nilotica** *(L.) Willd. ex Del.* (1813).

Tree, 2.5-14 m high; branchlets glabrous to tomentose. Stipular spines up to 8 cm long, straight or almost so, often deflexed. Leaves: pinnae 2-14 pairs, leaflets (7-) 12-27(-36) pairs; 1.5-7 x 0.5-1.5 mm. Flowers bright yellow, in heads; peduncles up to 4.5 cm long; involucel from near the base to about halfway up the peduncle, often subtending a few additional flowers. Calyx 1-2 mm long. Corolla 2.5-3.5 mm long. Pods straight or slightly curved, indehiscent, 8-17(-22) x 1-2.2 cm, with entire margin or ± deeply constricted between the seeds, glabrous to tomentose. Seeds 6.5-9 x 5-8 mm; areole 5-7 x 4-7 mm.

1. Pods necklace-like, narrowly and regularly constricted between the seeds. 2
- Pods not necklace-like, their margins straight or crenate, or if narrowly constricted, then only irregularly so here and there. 3
2. Pods glabrous or almost so. subsp. **nilotica**
- Pods white tomentellous. subsp. **indica**
3. Pods glabrous or subglabrous even when young; branchlets glabrous or nearly so. subsp. **leiocarpa**
- Pods pubescent at least when young; branchlets ± densely pubescent. 4
4. Pods persistently pubescent all over, 1.3-2.2 cm wide, with straight or slightly crenate margins. subsp. **subalata**
- Pods glabrescent and later ± shining on the raised part over each seed, 1-1.7 cm wide, with shallowly crenate margins. subsp. **kraussiana**

Subsp. **nilotica.**

Woodland; *c* 600 m. EW; Sudan, Egypt and westwards to Senegal. *Baldrati* 4327; *Pappi* 8837.

Subsp. **indica** *(Benth.) Brenan* in *Kew Bull.* **12**:84 (1957). Fig. 95.10.5.

Cultivated and sometimes naturalized in EE, AF, SU and HA; native of India. *Ash* 1161; *Baldrati* 624; *Burger* 3274.

Burger 3274 was erroneously referred to subsp. *tomentosa* (Benth.) Brenan by Ross (1979).

Elsewhere with many uses as for the genus plus the production of a soothing extract from the bark also sometimes used to treat diabetes; dyes can be extracted from the seeds.

Subsp. **subalata** *(Vatke) Brenan* in *Kew Bull.* **12**: 85 (1957).

Woodland and scrub; *c* 900-1000 m. GG; Uganda, Kenya, Tanzania. *Corradi* 3942, 3943, 3970.

Subsp. **kraussiana** *(Benth.) Brenan* in *Kew Bull.* **12**: 84 (1957).

Woodland and scrub; *c* 700-1700 m. KF GG SD; Tanzania to Natal and in Angola. *Mooney* 5576; *Strecker* 56; *Turton* 74.

Turton 74 was referred to subsp. *subalata* (Vatke) Brenan by Ross (1979) but the pods are glabrous over the seeds and it seems better placed in subsp. *kraussiana*.

Subsp. **leiocarpa** *Brenan* in *Kew Bull.* **12**: 84 (1957). Fig. 95.10.6.

Woodland and scrub; 1100-1700 m. SU AR HA; Somalia, Kenya, Tanzania. *Ash* 1136; *Burger* 1064; *Thulin* 1598.

Specimens of *A. nilotica* without pods and not referable to subspecies have been collected in SU, HA and SD at altitudes up to 1900 m.

31. **A. tortilis** *(Forssk.) Hayne* (1827).

Tree 4-21 m, mostly with flat crown; young branchlets glabrous to densely pubescent. Stipular spines in pairs, some short, hooked and up to 5 mm long, mixed with other straight slender ones up to 10 cm long. Leaves: pinnae 2-10 pairs; leaflets 6-22 pairs, 0.5-4(-6) x 0.5-1 mm. Flowers whitish, in heads; peduncles 0.4-2.4 cm long, with the involucel in the lower half. Calyx 1-2 mm long. Corolla 1.5-2.5 mm long. Pods contorted or spirally twisted, indehiscent or very slowly dehiscent, 0.6-1 (-1.3) cm wide, longitudinally veined, glabrous to tomentellous or villous. Seeds 4-7 x 3-6 mm, compressed; areole 3-6 x 2-4 mm.

1. Pods 3-5 mm wide, shortly pubescent, eglandular. subsp. **tortilis**
- Pods 6-9(-13) mm wide, appressed puberulous to tomentellous and with numerous dark red glands. subsp. **spirocarpa**

Subsp. **tortilis.**

Dry bushland; *c* 600 m. EE; Sudan, Somalia, Arabia and Israel. *Gilbert* 2561.

The Ethiopian record needs confirmation. The only specimen seen, collected between Edd and Assab, is without fruits.

Subsp. **spirocarpa** *(Hochst. ex A. Rich.) Brenan* in *Kew Bull.* **12**: 88 (1957);

A. spirocarpa Hochst. ex A. Rich. (1847) - types: TU, *Schimper* II:502, 612, 658 (all P syn., K isosyn.), *Quartin Dillon* s.n. (P syn.).

A. pappii Gand. (1913) - type: EE, *Pappi* 7388 (K iso.).

Fig. 95.10.9.

Woodland, wooded grassland, dry scrub; *c* 600-1900 m. AF EW TU WU SU AR HA BA; Sudan, Somalia and southwards to southern tropical Africa. *Burger* 3471; *Friis* et al. 717; *Simmons* S1.

An important shade and browse tree whose seeds are said to be edible and used as a famine food. The wood has many uses.

32. **A. prasinata** *Hunde* (1982)
- type: SU, Awash Nat. Park, *Thulin, Hunde & Tadesse* 3865 (UPS holo., ETH K iso.).

Tree up to 6 m high with rounded crown; young branchlets brownish, glabrous. Stipular spines usually up to 8 mm long, occasionally up to 4 cm. Leaves glabrous throughout; pinnae 2-6 pairs; leaflets 8-30 pairs, 5-8 x 1.3-3.2 mm, with the midrib and lateral veins prominent on both faces. Flowers yellowish-white, in heads; peduncles 1-3.8 cm long, glabrous; involucel *c* ¼ way up the peduncle. Calyx *c* 1.5 mm long, glabrous. Corolla *c* 3 mm long. Pods straight, dehiscent, 3.3-8.4 x 0.6-1.1 cm, ± purplish brown and glossy, with oblique veins which converge to become longitudinal in the middle of the valves, glabrous. Seeds ± quadrangular or subcircular, compressed, 5-6 x 4-5.5 mm; areole *c* 2.6 x 2 mm.

Dry woodland and semi-desert bushland; (500-) 900-1300 m. AF SU; not known elsewhere. *Bally* 13004; *Ash* 192; *Burger* 2913.

33. **A. etbaica** *Schweinf.* (1867-68).

Normally a tree 2.5-12 m high; branchlets grey-brown to purplish-brown, glabrous to puberulous. Stipular spines all short and hooked or straight and up to 7 mm long, or with long straight spines up to 6 cm long intermixed, or sometimes all spines long and straight. Leaves: pinnae 1-9(-17) pairs; leaflets 4-35 pairs, 0.5-4 x 0.25-1.25 mm, glabrous or puberulous. Flowers white or cream, in heads; peduncles up to 2.5(-3.5) cm long, with the involucel below or above the middle. Calyx up to 1 mm long. Corolla up to *c* 3 mm long. Pods straight, linear-oblong, dehiscent, 2-12 x 0.7-1.5 (-2.2) cm, brown to purplish, with oblique veins which converge to become longitudinal in the middle of the valves, glabrous or puberulous especially near the base. Seeds subcircular, up to 8 mm in diameter; areole up to 4.5-5 mm.

Provides the pillars and beams to hold the heavy earthen roofs of houses in northern Ethiopia.

1. Spines all straight, mostly 0.6-2.8 cm long; rhachides of leaves and pinnae mostly glabrous. subsp. **etbaica**
- Spines all short and ± hooked or straight, up to 0.7 cm long, or sometimes with long straight spines up to 6 cm long intermixed; rhachides of leaves and pinnae pubescent. 2
2. Pods 7-11(-12) mm wide; rhachides of leaves and pinnae puberulous. subsp. **uncinata**
- Pods 11-15(-22) mm wide; rhachides of leaves and pinnae with short but distinct ± spreading hairs. subsp. **platycarpa**

Subsp. **etbaica.**

Bushland; 600-2000 m. EE EW; Sudan, Somalia. *Pichi-Sermolli* 601; *Gilbert* 2563.

Subsp. **uncinata** *Brenan* in *Kew Bull.* **12**: 91 (1957). Fig. 95.10.8.

Wooded grassland; *c* 1200-2000 m. EW WU AR HA SD; Somalia, Uganda, Kenya. *Friis* et al. 729; *Burger* 3732; *Gillett* 14769.

Thulin, Hunde & Tadesse 3590 from SD has leaves with up to 17 pairs of pinnae, but otherwise agrees completely with *A. etbaica* subsp. *uncinata*.

Subsp. **platycarpa** *Brenan* in *Kew Bull.* **12**: 93 (1957).

Bushland; *c* 1550 m. SD; Kenya and Tanzania. *Thulin, Hunde & Tadesse* 3477; *Mesfin & Tewolde* 2658.

34. **A. reficiens** *Wawra* (1859).

Shrub 1-6 m high, obconical, branching from base; young branchlets grey to brown, shortly puberulous. Stipular spines (in Ethiopia) all short, hooked, up to 6 mm long. Leaves: pinnae 1-3 pairs; leaflets 5-11 pairs, 2-4.5 x 0.5-1.25 mm. Flowers whitish in heads; involucel basal or in lower third of peduncle. Pods straight, dehiscent, 3-6.5(-8.8) x 0.6-1.2 cm, brown or deep purplish, finely longitudinally veined, ± pulverulent to glabrous or nearly so. Seeds mostly elliptic, compressed, 5-7.5 x 4-5 mm; areole 4-4.5 x 2.5-3 mm.

Subsp. **misera** *(Vatke) Brenan* in *Kew Bull.* **12**: 90 (1957).
A. stefaninii Chiov. (1915).

Peduncles 0.4-1(-1.3) cm long. Pods 0.6-0.95 cm wide.

Dry scrub; probably *c* 500 m. HA KF SD; Sudan, Somalia, N Uganda, Kenya. *Hemming* 1526.

Subsp. *reficiens* occurs in Angola and Namibia.

35. **A. drepanolobium** *Harms ex Sjöstedt* (1908).

Shrub or small tree, 1-5(-7.5) m high; young branchlets shortly pubescent to glabrous. Stipular spines 1.5-4.5 (-7.5) cm long, some basally fused and enlarged into ± globose, purplish-brown *ant-galls*. Leaves: pinnae 3-13 pairs; leaflets 11-22 pairs, 1.5-5.5 x 0.7-1.25 mm, glabrous or minutely ciliolate. Flowers white or cream, in heads; peduncles glabrous or pubescent, with the involucel at or near the base. Calyx 0.75-1.5(-2.5) mm long. Corolla 3-4 mm long. Pods falcate, mostly attenuate at both ends, dehiscent, 4-7 x 0.5-1 cm, brownish, finely longitudinally veined, glabrous or ± puberulous. Seeds ± elliptic, compressed, 10-12 x 4.5-5.5 mm; areole up to 6 x 3.5 mm. Fig. 95.10.11.

Grassland, often gregarious over large areas liable to flooding; *c* 1000-1700 m. HA GG SD; Sudan, Somalia, Uganda, Kenya, Tanzania, E Zaire. *Burger* 2072; *Carr* 297; *Gillett* 13661.

Reported to be tolerant of alkaline soils.

36. **A. gerrardii** *Benth.* (1875).

Usually a tree up to 8(-15) m high, with flat, umbrella-shaped or irregular crown; young branchlets ± densely grey-pubescent, epidermis usually flaking to expose a rusty-red inner layer. Stipular spines straight or nearly so, up to *c* 1 cm long, occasionally much longer. Leaves: pinnae mostly 5-10 pairs; leaflets 12-23 pairs, 3-7.5 x 1-2 mm, ± ciliate. Flowers white or cream, in heads; peduncles *c* 2-4.5 cm long with the involucel at or near the base. Calyx *c* 1-2 mm long. Corolla *c* 2.4-4 mm long. Pods falcate, linear or linear-oblong, dehiscent, *c* 7-16 x 0.6-1.1 cm, usually puberulous to tomentellous. Seeds olive-brown, ± irregularly 4-angled, compressed, 9-12 x 7 mm; areole up to 7 x 4.5 mm.

Woodland and wooded grassland; *c* 1500-2000 m. SU; Sudan to Natal, westwards to Nigeria. *Ash* 997; *Giordano* 708; *Jansen* et al. 4582.

The Ethiopian material is subsp. *gerrardii* var. *gerrardii*: there is another subspecies in the Middle East.

37. **A. robusta** *Burch*. (1824).

Tree 5-25 m high; young branchlets grey to grey-brown, glabrous (in Ethiopia), epidermis not flaking. Stipular spinesstraight or nearly so, usually short, up to *c* 8 mm long, occasionally up to 6(-11) cm. Leaves: pinnae (2-) 3-8(-10) pairs; leaflets 9-27 pairs, mostly *c* 3.5-7.5 x 1-4 mm, usually glabrous. Flowers white, in heads; peduncles 1.2-5.4 cm; involucel near the base to 1/3 way up the peduncle. Calyx 1.5-3.2 mm long. Corolla 3-4 mm long. Pods straight to falcate, dehiscent, *c* 7-19 x 0.7-3 cm, brown, longitudinally veined, glabrous. Seeds compressed, 8-15 x 5-9 mm; areole 6.5-9 x 3.5-6.5 mm.

Subsp. **usambarensis** *(Taub.) Brenan* in *Fl. Zamb*. 3(1): 104 (1970);
A. clavigera E. Mey. subsp. *usambarensis* (Taub.) Brenan in *Kew Bull*. **12**: 369 (1958).

Pods ± falcate, 0.7-1.2 cm wide. Leaflets 1-2 mm wide; leaf rhachis glabrous. Peduncles shortly pubescent. Fig. 95.10.10.

Woodland, usually riparian; *c* 500-1450 m. SU AR HA BA SD; Kenya, Tanzania and Mozambique. *Ash* 2115; *Burger* 3091; *Thulin, Hunde & Tadesse* 3858.

Subsp. *robusta* and subsp. *clavigera* (E. Mey.) Brenan are both widespread in southern Africa.

38. **A. pilispina** *Pic.-Serm.* (1951)
- type: GD, *Pichi-Sermolli* 2696 (FT holo.).

Shrub or tree, 1-15 m high; young branchlets brownish, densely clothed with long grey to yellowish spreading hairs mostly 0.5-2 mm long, the epidermis on older branchlets exfoliating in thin flakes exposing a powdery inner layer. Stipular spines up to 0.7-5 cm long, straight or almost so, hairy except towards the tip. Leaves: pinnae (5-)8-16(-26) pairs; leaflets 8-28 pairs, 1.5-4.5 x 0.5-1 mm. Flowers cream, in heads; peduncles 1.5-3.5 cm long, hairy; involucel hairy, basal or in lower third of peduncle. Calyx *c* 2.5 mm long. Corolla *c* 3 mm long. Pods straight or slightly curved, dehiscent, 4-12.5 x 1.4-2.9 cm, grey to brown, finely and ± longitudinally veined, glabrous or nearly so. Seeds subcircular, compressed, 6-7 x 5-7 mm; areole up to *c* 5 x 4 mm.

Woodland, wooded grassland, upland bushland; 1650-3100 m. TU GD WU GJ SU HA; Tanzania, Zaire, Zambia, Malawi and Mozambique. *Meyer* 7444; *Seegeler* 2983; *Thulin & Hunde* 4105.

Gillett 5187 (K) is a deviating specimen from Sarerta Mt (NE of Harer in HA), which may represent an undescribed taxon close to *A. pilispina* and *A. negrii*. The indumentum is reminiscent of *A. pilispina* but the leaves have fewer pinna pairs (4-8) and larger (up to *c* 6 x 1 mm) and more discolorous leaflets than this species. From *A. negrii* it differs markedly by the dense indumentum.

39. **A. abyssinica** *Hochst. ex Benth*. (1846)
- type: TU, *Schimper* III:1813 (K holo.).
A. xiphocarpa Hochst. ex Benth. (1846) - type: TU, *Schimper* II:522 (K holo.).

Tree up to 20 m high, flat-crown up to 30 m wide; bark dark, fissured; branchlets conspicuously longitudinally ridged, ± pubescent to shortly villous with grey or yellowish hairs. Stipular spines variable in length, up to 4 cm long, straight. Leaves: pinnae (9-)15-38 pairs; leaflets up to 40 pairs, up to 4 x 1 mm, ciliolate. Flowers white,tinged red, in heads; peduncles up to 3 cm long, puberulous, with the involucel at or near the base. Calyx *c* 2 mm long, reddish. Corolla *c* 3 mm long, reddish. Pods straight or slightly curved, dehiscent, 5-12 x 1.2-2(-2.8) cm, purplish brown or grey, longitudinally veined, ± glandular and sometimes puberulous. Seeds elliptic, compressed, 7-10 x 4-6 mm; areole up to 7 x 3.5 mm. Fig. 95.10.7.

Woodland, wooded grassland, forest margins and along sides of streams and rivers; 1500-2800 m. EW TU GD WU GJ WG SU AR BA IL KF SD. *Friis* et al. 89, 1941; *Thulin, Hunde & Tadesse* 3665.

Subsp. *calophylla* Brenan in Sudan, Uganda, Kenya, Tanzania, Zaire, Malawi, Zimbabwe and Mozambique.

40. **A. sieberiana** *DC*. (1825).

Tree 3-18 m; young branchlets glabrous to tomentose. Stipular spines up to 9(-12.5) cm long, straight. Leaves: pinnae 8-35 pairs; leaflets 13-45(-52) pairs, 2-6.5 x 0.6-1.5 mm. Flowers white or very pale yellow, in heads; peduncles 1.5-5 cm long; involucel apical or in upper half of peduncle. Calyx *c* 2 mm long. Corolla *c* 3.5 mm long. Pods straight or almost so, thick and almost woody in texture, very slowly dehiscing, 8-21 x 1.5-3.5 cm, ± smooth and glossy, glabrous or pubescent. Seeds 9-12 x 7-8 mm; areole 7-9.5 x 5-6 mm.

A species with many uses including the use of the wood in making mills and presses.

1. Young branchlets glabrous or nearly so; branches of crown usually ascending. var. **sieberiana**
- Young branchlets ± densely hairy with often golden hairs; branches of crown usually widely spreading. var. **woodii**

Var. **sieberiana**.

Deciduous woodland and riverine forest; 500-2200 m. EW WU SU WG IL KF SD; from Sudan to Senegal, southwards to Malawi and Mozambique. *Meyer* 7510, *Friis* et al. 419; *Thulin, Hunde & Tadesse* 3312.

Var. **woodii** *(Burtt-Davy) Keay & Brenan* in *Kew Bull.* **5**: 364 (1951).
Inga nefasia Hochst. ex A. Rich. (1847); *A. nefasia* (Hochst. ex A. Rich.) Schweinf. (1896) - types: TU, *Schimper* II:940 (P syn., K S isosyn.), *Quartin Dillon* s.n. (P syn.).
A. sieberiana var. *vermoesenii* (De Wild.) Keay & Brenan in *Kew Bull.* **5**: 364 (1951).
Fig. 95.10.12.

Woodland and wooded grassland; *c* 1700-2100 m. TU GD WU SU ?AR KF SD; widespread in eastern and southern tropical Africa. *Ash* 2845; *Gillett* 14772; *Gilbert & Tewolde* 2441.

41. **A. negrii** *Pic.-Serm.* (1951)
- types: SU, *Negri* 335, *Senni* 383 & 1589 (all FT syn.).

Shrub or tree up to 6(-10) m high with ± umbrella-shaped crown; young branchlets purplish brown to black densely lenticellate, glabrous or occasionally puberulous, the epidermis on older branchlets exfoliating in thin yellowish flakes. Stipular spines 0.5-8.5 cm or more long, straight, when well developed with a longitudinal furrow above. Leaves: pinnae (3-)6-10 (-14) pairs; leaflets 13-31 pairs, mostly 4-7.5 x 0.9-2 mm, paler beneath. Flowers cream, in heads 12-18 mm in diameter; peduncles 2.5-4.8 cm long, densely to sparsely pubescent; involucel densely pubescent below, inserted on the lower half of peduncle. Calyx 1.75-2.5 mm long. Corolla 3-3.5 mm long. Pods straight or slightly curved, oblong, dehiscent, 5-14 x 1.7-2.8 cm, brown, glabrous, eglandular, longitudinally veined. Seeds compressed, 8-10 x 5-6 mm; areole *c* 6.5 x 4 mm. Fig. 95.8.4.

Upland wooded grassland; 2000-3100 m. GD WU GJ SU HA; not known elsewhere. *Ash* 1639; *Burger* 2638; *Thulin & Hunde* 3884.

42. **A. origena** *Hunde* (1982)
- type: EW, Ad Rassi, *Pappi* 4946 (FT holo., K iso.).

Tree up to 8 m high; young branchlets pale yellowish- to greenish-brown, ± puberulous, the epidermis on older branchlets flaking. Stipular spines 0.5-4.7 cm long, straight, not or scarcely furrowed above, ± puberulous. Leaves: pinnae 2-10 pairs; leaflets 11-30 pairs, 3-7 x 0.75-1.5(-1.8) mm, paler beneath. Flowers cream, in heads 9-12 mm in diameter; peduncles 1.2-4.3 cm long, ± pubescent; involucel near base of peduncle. Calyx *c* 2 mm long. Corolla *c* 3 mm long. Pods straight, oblong, dehiscent, *c* 6-8.5(-12) x 1.4-2 cm, brown, longitudinally veined, mostly ± puberulous or pubescent and with sessile reddish glands. Seeds *c* 7 x 5 mm, compressed; areole *c* 5 x 2.5 mm.

Upland wooded grassland; 1700-2600 m. EW WU; Yemen. *Pappi* 4871; *Schweinfurth* 253; *Schweinfurth & Riva* 2216.

43. **A. bavazzanoi** *Pic.-Serm.* (1951)
- type: GD, *Pichi-Sermolli* 2253 (FT, holo.).

Tree up to 6 m high; stem and branches yellowish, powdery; branchlets reddish brown, glabrous, exfoliating to reveal a powdery yellowish inner layer. Stipular spines up to 1(-3) cm long. Leaves: pinnae 8-14 pairs; leaflets 16-26 pairs, *c* 3.5-5 x 1 mm. Flowers in heads; peduncles 1.2-2 cm long, sparsely pubescent, with the involucel in the lower third. Calyx *c* 1.5 mm long. Corolla up to 3.5 mm long. Pods straight or slightly curved, dehiscent, 10-15 x 1.5-2 cm, purplish brown, longitudinally veined, glabrous. Seeds elliptic-oblong, compressed, *c* 7 x 5 mm; areole *c* 4 x 3 mm.

Woodland; *c* 1680-2400 m. EW TU GD GJ SU; not known elsewhere. *Fiori* 153; *Pichi-Sermolli* 2247, 2249, 2252.

44. **A. oerfota** *(Forssk.) Schweinf.* (1896);
Mimosa oerfota Forssk.(1775).
A. nubica Benth. (1842).
A. pterygocarpa Hochst. ex Benth. (1846); *A. nubica* var. *erythraea* Schweinf. in *Linnaea* **35**: 338 (1867-68) - type: without precise locality, *Schimper* II:1036 (K, holo.).
A. gorinii Chiov. (1932).

Shrub 1-5 m high, branching from the base; young branchlets greyish-white to yellowish green, glabrous to spreading pubescent. Stipular spines straight or almost so, 0.4-1.7(-2.7) cm long. Leaves: pinnae (2-)3-7(-11) pairs; leaflets 5-16 pairs, 2.5-6(-9) x (0.5-)0.75-2.5 mm. Flowers whitish or greenish, in heads; peduncles 0.5-1.5 cm long, pubescent; involucel below or sometimes about middle of peduncle. Calyx 1.5-2.5 mm long, pubescent. Corolla up to 3.5 mm long, with the lobes conspicuously pubescent outside. *Pods* straight or sometimes slightly curved, 4-15 x 0.9-2.2 cm, with a narrow wing-like margin 1-3.5 mm wide, straw-coloured, shortly pubescent. Seeds globose or ellipsoid, 4.5-6.5 x 3.5-6 mm, shallowly and closely wrinkled; areole 4-5 x 3-3.5 mm. Fig. 95.10.13

Deciduous bushland and semi-desert scrub; 100-1600 m. EE AF EW TU WU SU BA HA SD; Egypt, Sudan, Somalia, Uganda, Kenya, Tanzania and in Arabia. *Mooney* 9666; *Ash* 649; *Simmons* S50.

For comments on nomenclature, see Wood in *Kew Bull.* 37 (1983).

An extract of the bark is used to treat rheumatism and the wood is used in making hut frames.

45. **A. paolii** *Chiov.* (1915).

Shrub up to 2.5 m high, branching from the base, or sometimes a tree up to 5 m high; young branchlets villous with spreading hairs. Stipular spines straight, slender, up to 5 cm long. Leaves: pinnae (1-)4-9 pairs; leaflets 7-15 pairs, 2-7 x 0.8-1.75 mm. Flowers white, in heads; peduncles 0.7-2 cm long; involucel basal or in lower quarter of peduncle; bracteoles glabrous or almost so. Calyx *c* 2 mm long, glabrous or almost so. Corolla *c* 2.5 mm long, with the lobes conspicuously pubescent outside. Pods somewhat falcate, coriaceous, attenuate basally and apically, dehiscent, 6-12.5 x 0.6-1 cm, densely clothed with whitish spreading hairs up to 4 mm long. Seeds ellipsoid, 6-8 x 4.5-5 mm, minutely punctate. Fig. 95.10.14.

Dry bushland at low altitude. HA KF; Sudan, Somalia, Kenya. *Hemming* 1511; *Kaji* s.n.

Hemming 1511 from HA is subsp. *paolii*. *Kaji* s.n. from Omo National Park, with hairs on pods only *c* 1.5 mm long, may represent subsp. *paucijuga* Brenan, previously only known from Kenya.

Elsewhere the gum is used in making masticatories.

46. **A. stuhlmannii** *Taub*. (1895).

Shrub or small tree up to 6 m high; young shoots villous with spreading golden hairs up to 1.5-3 mm long. Stipular spines 0.7-4.5(-6.5) cm long, straight. Leaves: pinnae 4-12 pairs; leaflets 7-25 pairs, *c* 2-5.5 x 1-1.5 mm. Flowers white, in heads; peduncles 0.4-3 cm, densely hairy; involucel basal or in lower half of peduncle; bracteoles ciliate or pubescent. Calyx 1.7-2.5 mm long, ± pubescent outside. Corolla 2.5-3.5 mm long, with the lobes conspicuously pubescent outside. Pods straight or somewhat curved, thick and hard, attenuate basally, indehiscent, up to 9 x 2.5 cm (in Ethiopia 2-4 x 1-1.5 cm), densely clothed with long spreading hairs. Seeds ± ellipsoid, 4.5-9 mm in diameter, minutely punctate; areole 4-7 x 2.5-4.5 mm.

Dry bushland at low altitude. HA; Somalia, Kenya, Tanzania, Zimbabwe, Botswana and Transvaal. *Hemming* 1513.

47. **A. edgeworthii** *T. Anders*. (1860).
A. socotrana Balf. f. (1882).
A. sultani Chiov. (1929).
A. pseudosocotrana Chiov. (1929).
A. humifusa Chiov. (1929).
A. erythraea Chiov. (1929).

Low flat-topped, spreading shrub, 0.3-2 m high; young branchlets puberulous or sometimes tomentose. Stipular spines straight or almost so, *c* 1-3.5 cm long on mature shoots. Leaves: pinnae 4-10 pairs; leaflets 6-15(-20) pairs, 0.75-3.5(-5) x 0.5-1.5(-2) mm, ciliolate, lateral veins invisible. Flowers white, in heads; peduncles up to c 1.5 cm long, tomentose or densely pubescent, eglandular, with the involucel basal or in the lower half. Calyx 1-2 mm long. Corolla 2.5-3 mm long. Pods falcate to almost straight, thick and woody, ultimately dehiscent, 7-15(-20) x 1.3-2.5 cm, brownish, longitudinally veined, velvety pubescent or tomentellous. Seeds blackish, ellipsoid to subglobose, 9-13 mm in diameter; areole up to 10 x 4.5mm. Fig. 95.10.15.

Acacia-Commiphora bushland; 360-700 m. HA BA SD; Somalia, NE Kenya, Socotra and SW Arabia. *Bally* 12982; *Hemming* 1518; *Gilbert* 4003.

Hemming 1473 from 16 km S of El Dere is a deviating specimen with leaflets up to 5 x 2 mm and pods up to 20 cm long.

Elsewhere the leaves are used as a cooled vegetable. This species has been used in sand stabilization.

48. **A. bricchettiana** *Chiov*. (1915)
- type: HA, Ogaden, *Robecchi-Bricchetti* 533 (FT holo.).
A. gloveri Gilliland (1951) - type: HA, Ogaden, between Warder and Welwel (Wardere and Walwal), *Glover & Gilliland* 388 (K holo., BM iso.).

Shrub up to 3 m high; branchlets reddish- to dark purplish-brown, ± puberulous and densely lenticellate. Stipular spines up to 4.5 cm long, straight or almost so. Leaves: pinnae 6-13 pairs; leaflets 10-17 pairs, paler beneath, 3-6 x 1-1.75 mm, lateral venation somewhat conspicuous on lower surface, ciliate and sometimes with scattered hairs beneath. Flowers in heads; peduncles up to 6 mm long, densely pubescent, glandular; involucel basal. Calyx up to 2 mm long. Corolla *c* 3 mm long. Pods falcate, dehiscent, 10-17 x 2-2.4 cm, pale brown, thinly woody, conspicuously longitudinally veined, minutely puberulous. Seeds black, 12 x 8 mm (immature).

Dry scrub at low altitude. HA; ?Somalia.

A. bricchettiana is a badly understood species only known from the two type collections cited.

49. **A. walwalensis** *Gilliland* (1951)
- type: HA, Ogaden, *Glover & Gilliland* 392 (K holo.).

Shrub up to 3 m high; young branchlets conspicuously lenticellate, glabrous or almost so. Stipular spines straight or slightly curved, 1.5-2.2 cm long. Leaves; pinnae 2 pairs, the lower pair much shorter than the upper; leaflets 1-2 pairs on lower pinnae, 3-5 pairs on upper, 6.5-15 x 4-9 mm, obliquely obovate or elliptic, veined, glabrous. Flowers white, in heads; peduncles 2-4 cm long, finely pubescent; involucel at or somewhat below the middle of peduncle. Calyx 3-3.5 mm long, pubescent. Corolla 4.5-5.5 mm long, pubescent. Pods straight, turgid, ± cylindrical and constricted between the seeds, dehiscent, 9.5-16 x 1-1.4 cm, dark brown, longitudinally veined, puberulous. Seeds blackish, ellipsoid to subglobose, 10-12 x 6.5-7.5 mm, punctate; areole 8-10 x 4-5 mm.

Dry bushland at low altitude. HA; Somalia. *Ashall* CA/1/58.

50. **A. mearnsii** *De Wild*. (1925).

Unarmed tree; young branchlets densely short pubescent. Leaves with 12-21 pairs of pinnae; leaf rhachis with numerous raised glands all along its upper side; leaflets 16-70 pairs, 1.5-4 x 0.5-0.75 mm. Flowers pale yellow in heads which are arranged in panicles Pods jointed, almost moniliform, dehiscent, usually 3-10 x 0.5-0.8 mm, ± grey-puberulous.

Cultivated in EW WU SU WG HA KF and probably elsewhere; native of Australia. *Jansen* 2333; *Seegeler* 2958; *W. de Wilde* 10443.

The most important source of commercial tannins used in treating hides and skins, this species has many other uses.

51. **A. dealbata** *Link* (1822).

Unarmed shrub or tree; young branchlets usually densely short pubescent, rarely subglabrous. Leaves with 10-26 pairs of pinnae; leaf rhachis with a gland at the junction of all or most pairs of pinnae, but not between them; leaflets 17-50, 2-5.5 x 0.4-0.7 mm, ± pubescent. Flowers bright yellow, in heads. Pods straight or slightly curved, not or only slightly moniliform, dehiscent, 3-8 x 0.7-1.3 cm.

Cultivated at least in EW and SU; native in Australia. *Baldrati* 4139, 4329; *Albers* 61007a.

The buds and flowers are used in cosmetics and for the extraction of essential oils.

52. **A. decurrens** *Willd.* (1806).

Unarmed tree; young branchlets prominently angular, sometimes with wing-like ridges, glabrous or the very young shoots slightly tomentose. Leaves with 8-15 pairs of pinnae; leaflets 15-35 pairs, 6-12 x 0.3-0.75 mm. Flowers bright yellow, in heads. Pods not or only slightly moniliform, dehiscing along one margin, 3.5-10 x 0.4-0.7 cm, brown or dark brown.

Cultivated in EW, WG and SU; native in Australia. *Senni* 12; *Milchersich* 110.

53. **A. podalyriifolia** *A. Cunn. ex G. Don* (1832).

Unarmed shrub or small tree; young branchlets densely grey-pubescent. Leaves phyllodic, simple, ovate to elliptic, mostly 1.5-4 x 1-2 cm, with one main longitudinal nerve and finely penninerved, grey-pubescent. Flowers bright yellow in heads. Pods flattened, dehiscent, 4-8.5 x 1.5-2 cm, glabrous.

Cultivated at least in SU (Addis Ababa); native in Australia. *Beshew* s.n.; *Sebsebe Demissew* 540.

54. **A. saligna** *(Labill.) Wendl.* (1820).
A. cyanophylla Lindl. (1839).

Unarmed shrub or tree up to 10 m high; young branchlets glabrous. Leaves phyllodic, simple, mostly 8-22 x 0.5-1.4 cm, straight or slightly falcate, with a single main longitudinal nerve and finely penninerved. Flowers bright yellow in heads more than 6.5 mm in diameter on peduncles 6-22 mm long; heads arranged in axillary racemes. Pods straight or slightly falcate, margins slightly constricted between some of the seeds, 5.5-15 x 0.5-0.6 cm.

Introduced to most medium and higher altitudes of the country through reafforestation programmes; native in Australia. *Meyer* 8615; *Chaffey* 514; *Westphal* 1906.

The gums have been used elsewhere as food preservatives and the seeds have been used in making flour and animal feed concentrates.

55. **A. cyclops** *A. Cunn. ex G. Don* (1832).

Unarmed shrub or small tree; young branchlets usually angular and glabrous. Leaves phyllodic, simple, usually ± straight or sometimes slightly falcate, 3-9 x 0.6-1.5 cm, with 3-5 prominent longitudinal nerves with anastomosing almost longitudinal venation between. Flowers bright yellow in heads; petals free. Pods ± falcate, not constricted between the seeds, dehiscent, 5-15 x 0.8-1.3 cm.

Cultivated at least in EW (Asmera); native in Australia. *Baldrati* 4140, 4330.

56. **A. melanoxylon** *R. Br.* (1813).

Unarmed tree; young branchlets angular, glabrous or minutely pubescent. Leaves phyllodic, simple, straight to falcate, mostly 6-12 x 0.6-1.2(-2.5) cm, with 3-7 prominent longitudinal nerves with a conspicuous reticulate venation between. Flowers pale yellowish white in heads; petals united to above the middle. Pods oblong, falcate, coiled or spirally twisted, not constricted between the seeds, dehiscent.

Introduced to most medium and higher altitudes of the country through reafforestation programmes; native in Australia. *Bos* 7673, 9088; *Ash* 12.

Elsewhere used to make fish poison. The wood has many uses.

57. **A. cultriformis** *A. Cunn. ex G. Don* (1832).

Unarmed shrub or small tree; young branchlets angular, glabrous. Leaves phyllodic, simple, ovate-triangular, mostly 5-11 mm wide, glaucous, with a single main longitudinal nerve and finely penninerved, glabrous. Flowers yellow, in small round heads arranged in axillary racemes which are longer than the phyllodes and are often ± aggregated terminally. Pods linear-oblong, dehiscent, 5-9 x 0.5-0.7 cm, glabrous.

Cultivated at least in EW (Asmera); native in Australia. *Baldrati* 4146, 4331.

58. **A. retinodes** *Schlechtend.* (1847).

Unarmed glabrous shrub or small tree. Leaves phyllodic, simple, 4.5-17 x c 1.5 cm, narrowing gradually towards both ends, with a single main longitudinal nerve and finely penninerved Flowers cream or pale yellow, in heads up to c 6 mm in diameter on 3-6 mm long peduncles, heads arranged in short axillary racemes. Pods linear-oblong, dehiscent, 7-12 x 0.5-0.7 cm.

Cultivated in EE and EW; native in Australia. *Baldrati* 4153; *W. de Wilde* 7299.

Elsewhere the seeds are said to be edible.

23. **ALBIZIA** *Durazz*. (1772)

Brenan in *Fl. Trop. E. Afr.* (1959).

Trees, sometimes shrubs, the African species unarmed or almost so. Leaves bipinnate. Flowers mostly bisexual, in heads (in all native species), spikes or spiciform racemes, the central flower often larger and different in form from the others. Calyx and corolla normally 5-lobed. Stamens numerous, the filaments united below into a tube; anthers eglandular at the apex. Pods oblong, straight, flat, usually dehiscent. Seeds usually flattened.

Some 100-150 species throughout the tropics.

Many of the species in this genus are important sources of nectar and pollen for bees and provide shade for coffee plants and for people and animals in small towns at medium altitudes.

1. Inflorescences spicate. **1. A. lophantha**
- Inflorescences capitate. 2

2. Staminal tube exserted more than 1 cm beyond the corolla. 3
- Staminal tube not or scarcely exserted beyond the corolla. 4

3. Leaflets in 9-16 pairs, mostly *c* 10-20(-25) x 4-8(-13) mm; stipules and bracts up to *c* 6-7 x 2-2.5 mm. **10. A. gummifera**
- Leaflets in 3-6 pairs, much larger; stipules and bracts much larger. **11. A. grandibracteata**

4. Leaflets 0.5-3(-4) mm wide. 5
- Leaflets 4-33 mm wide. 8

5. Midrib of leaflets running along their upper margin to the base. **2. A. chinensis**
- Midrib of leaflets not marginal. 6

6. Leaflets *c* 2-6 x 0.5-1.5 mm, symmetrical at the apex; pedicels up to 1.5 mm long. **7. A. amara**
- Leaflets 5.5-12 x 1.25-4 mm, asymmetrical at the apex; pedicels 1.5-6 mm long. 7

7. Leaflets 1.25-4 mm wide; pedicels 1.5-1.75 mm long; indumentum on outside of calyx and corolla grey. **8. A. isenbergiana**
- Leaflets mostly 4 mm or more wide; pedicels 2-6 mm long; indumentum on outside of calyx and corolla brown. **9. A. schimperiana**

8. Rhachides of leaves and pinnae mostly projecting at ends as short rigid deflexed hooks or claws; leaflets usually about 3 pairs. **6. A. anthelmintica**
- Rhachides of leaves and pinnae without projections as above; leaflets usually more than 3 pairs. 9

9. Filaments 0.5-1.3 cm long. **9. A. schimperiana**
- Filaments 1.5-4 cm long. 10

10. Filaments red above and white below. **3. A. coriaria**
- Filaments white to green or yellowish. 11

11. Pedicels up to 0.75 mm long; leaflets glaucescent beneath, ± pubescent to subtomentose; filaments white. **4. A. malacophylla**
- Pedicels 1.5-4.5(-7.5) mm long; leaflets usually not glaucous beneath, usually subglabrous; filaments green or greenish-yellow. **5. A. lebbeck**

1. **A. lophantha** *(Willd.) Benth.* (1844).

Tree. Leaves with pinnae 6-14 pairs; leaflets 20-40 pairs, linear-oblong from a very unequal base, brown-pubescent when young. Flowers in axillary spikes; corolla 7-9 mm long, green, brown-pubescent; filaments yellowish, 1-1.3 cm long. Pod *c* 10 x 0.8-2 cm, swollen over the seeds.

Planted as an ornamental at least in EW SU HA and BA; native of Australia. *Baldrati* 4116; *Mooney* 7104.

2. **A. chinensis** *(Osbeck) Merr.* (1916).

Tree. Leaves with pinnae 5-20 pairs; leaflets 12-45 pairs, *c* 6-13 x 1.5-4 mm, very unequal at base, obliquely acute at apex, hairy when young, the midrib running along the upper margin from the base. Flowers pubescent, sessile or subsessile, in heads; corolla 6-7 mm long; filaments 3-3.5 cm long, joined into a short tube at the base. Pod flat, 10-17.5 x 1.7-3.5 cm.

Planted at least in HA; native of tropical Asia. *Bos* 7518.

3. **A. coriaria** *Welw. ex Oliv.* (1871).

Tree; bark rough, flaking off; young branchlets puberulous or shortly pubescent. Leaves: pinnae (2-)3-6 pairs; leaflets (4-)6-11(-12) pairs, oblong to ovate-oblong, 13-33 x 5-14(-17) mm, rounded at apex, subglabrous or thinly puberulous beneath. Flowers subsessile or on pedicels 0.5-2 mm long. Calyx 3.5-6.5 mm long, puberulous outside. Corolla 8-13.5 mm long, whitish puberulous outside. Staminal tube not or scarcely exserted beyond corolla; filaments 1.7-4 cm long, red above, white below. Pod oblong, (10-)14-21 x (2.3-)3.2-3.7(-4.8) cm, glabrous or almost so, brown or purplish brown. Seeds *c* 9-12 x 8-9 mm.

Forest and wooded grassland; *c* 500 m. IL; Uganda, W Kenya, N Tanzania, Sudan and westwards to the Ivory Coast and southwards to Angola. *Chaffey* 918.

The roots and bark have medicinal and veterinary uses.

4. **A. malacophylla** *(A. Rich.) Walp.* (1852);

Inga malacophylla A. Rich. (1847) - type: TU, *Schimper* 521 (P lecto., K UPS isolecto.).

Inga quartiniana A. Rich. (1847); *Albizia quartiniana* (A. Rich.) Walp. (1852) - type: TU/EW, Mereb Valley, *Quartin Dillon* s.n. (P holo., K iso.).

A. elliptica Fourn. (1860) - type: *Schimper* 1087 (P holo.).

A. pallida Fourn. (1860) - type: TU, *Schimper* III:1578 (P holo., K iso.).

Tree; bark rough; young branchlets densely grey- to pale brown-pubescent. Leaves: pinnae 2-8 pairs; leaflets 4-15 pairs, oblong to oblong-elliptic, 10-40 x 4-30 mm, rounded andmucronate or slightly emarginate at apex, glaucescent beneath, ± pubescent to subtomentose. Flowers sessile or on pedicels up to 0.75 mm long. Calyx 3-5 mm long, ± densely and shortly grey-pubescent outside. Corolla 5-7 mm long, pubescent. Staminal tube not or scarcely exserted beyond corolla; filaments white, c 2-2.5 cm long. Pod oblong, *c* 10-21 x 2-4 cm, subglabrous, brown. Seeds *c* 9 x 7 mm.

1. Leaflets in 8-14 pairs, oblong, mostly less than 15 x 7 mm, densely spreading pubescent above, subtomentose beneath. var. **malacophylla**
- Leaflets in 3-9(-12) pairs, oblong-elliptic, mostly more than 15 x 10 mm, ± appressed pubescent above, sparsely to densely pubescent beneath. var. **ugandensis**

Var. **malacophylla**

Wooded grassland; *c* 1400-1900 m. EW TU GD; not known elsewhere. *Chiovenda* 2677; *Schweinfurth* 260; *Pappi* 965.

Var. **ugandensis** *Bak. f.* in *Leg. Trop. Afr.*: 860 (1930).

Wooded grassland, riverine forest; 550-2200 m. GD GJ WG IL; Sudan, Uganda and westwards to West Africa. *W. de Wilde* 8944, 10760; *Friis* et al. 1930.

5. **A. lebbeck** *(L.) Benth.* (1844).

Tree; bark grey, rough; young branchlets puberlous. Leaves: pinnae (1-)2-4(-5) pairs; leaflets 3-11 pairs, oblong or elliptic-oblong, 15-48(-65) x (6-)8-24(-33) mm, rounded at apex, mostly glabrous above and sub-glabrous or puberulous beneath. Flowers on pedicels 1.5-4.5(-7.5) mm long. Calyx (2.5-)3.5-5 mm long, ± puberulous. Corolla 5.5-9 mm long, glabrous except for puberulence on outside of lobes. Staminal tube not or scarcely exserted beyond corolla; filaments 1.5-3 cm long, pale green or greenish-yellow in upper part, white below. Pod oblong, (12-)15-33 x (2.4-)2.9-5.5(-6) cm wide, glabrous except near base, coriaceous, veined, pale straw-coloured. Seeds 7-11.5 x 7-9 mm.

Planted in streets and gardens at least in EW (Ginda) and HA (Dire Dawa), and becoming naturalised; probably native in tropical Asia. *Baldrati* 4118; *W. de Wilde* 11042.

6. **A. anthelmintica** *(A. Rich.) Brogn.* (1860);
Besenna anthelmintica A. Rich. (1847) - type: TU, *Quartin Dillon & Petit* s.n. (P holo., K iso.).

Shrub or small tree, deciduous; bark smooth, grey to brown; young branchlets mostly glabrous. Rhachides of leaves and pinnae in all or most leaves projecting at ends as a short rigid deflexed hook or claw; pinnae 1-2(-4)pairs; leaflets 1-4(-5) pairs,obliquely obovate to suborbicular, (7-)10-36(-42) x (4-)6-31 mm, mucronate at apex, glabrous to sparsely short pubescent. Flowers usually on leafless twigs, pale greenish, on pedicels 0.5-5.5 mm long. Calyx 3-5 mm long, glabrous or sparsely pubescent, irregularly denticulate. Corolla 6-12 mm long, glabrous or puberulous at lobe-margins. Staminal tube not or scarcely exserted beyond corolla; filamentş *c* 1.5-2 cm long, white. Pod oblong, 7-18 x 1.5-2.9 cm, glabrous or occasionally puberulous, straw-coloured. Seeds 9-13 mm in diameter.

Bushland, woodland and dry scrub; *c* 500-1350 m. EE EW TU HA KF GG; Sudan and southwards to Transvaal and Natal. *Bally* 6668; *J. De Wilde* 7289; *Gilbert* 1530.

A vermifuge and taenicide is extracted from the bark which also contains tannins. The timber is useful in construction and making furniture.

7. **A. amara** *(Roxb.) Boiv.* (1834).

Tree or shrub, deciduous; bark rough; young branchlets with rather short dense spreading grey to golden pubescence. Leaves: pinnae 4-46 pairs; leaflets 12-48 pairs, oblong to linear, 2-5 (-7.5) x 0.5-1.5(-2.5) mm, apex symmetrical and ± obtuse, ± appressed-pubescent on one or both sides or on margins only. Flowers white or flushed pink; pedicels up to 1.5 mm long. Calyx 1-2 mm long, puberulous or pubescent outside. Corolla 3.5-7 mm long, puberulous or pubescent outside. Staminal tube not or scarcely exserted beyond corolla; filaments *c* 0.5-1.2 cm long. Pod oblong, 10-28 x 2-5 cm, puberulous, brown. Seeds 8-13 x 7-8 mm.

Subsp. **sericocephala** *(Benth.) Brenan* in *Kew Bull*. **10**: 190 (1955)
- types from Sudan and Ethiopia: TU, *Schimper* 818 (K syn.) & 883 (K syn., UPS isosyn.).

Pinnae (7-)14-46 pairs; leaflets (12-)21-48 pairs, 2-4 (-4.5) x 0.5-1(-1.25) mm, usually green beneath.

Wooded grassland, scrub; 1400-1700 m. EW TU GD SU SD; Sudan and southwards to Transvaal. *Ash* 2383; *Ellis* 28; *Thulin, Hunde & Tadesse* 3634.

Subsp. *amara* occurs mainly in India and Ceylon.

Saponins have been extracted from the roots and leaves and tannins and gums from the bark. Poles are taken to make digging sticks and frames for huts.

8. **A. isenbergiana** *(A. Rich.) Fourn.* (1860);
Inga isenbergiana A. Rich. (1847) - type: TU, *Schimper* I:275 (P holo., K iso.).

Tree; young branchlets densely and rather shortly pubescent. Leaves: pinnae (3-)6-18 pairs; leaflets (12-)16-33 pairs, slightly falcate to oblong, 5-12(-14.5) x 1.25-4 mm, with apex usually slightly asymmetric, ± appressed-pubescent on both surfaces or glabrescent above, lower surface paler. Flowers white, on pedicels 1.5-1.75 mm long. Calyx 1.5-2 mm long, grey, densely short-pubescent outside. Corolla 4-5 mm long, grey, densely short-pubescent outside. Staminal tube not or scarcely exserted beyond corolla; filaments *c* 1 cm long.

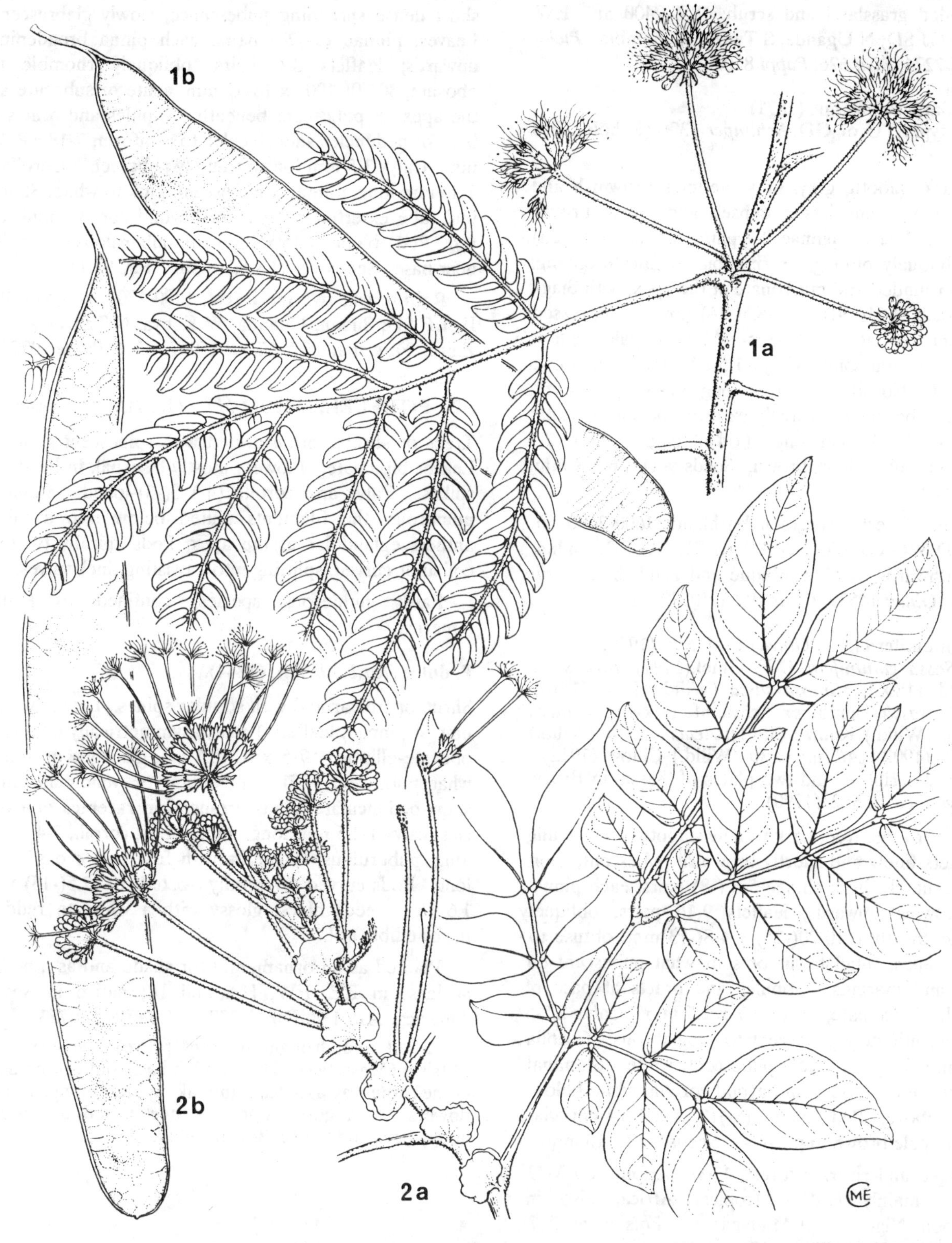

Figure 95.11.1 - *ALBIZIA SCHIMPERIANA* 1a: flowering branch from *Meyer* 9083; 1b: pod from *Gillett* 14618. 2 - *A. GRANDIBRACTEATA* 2a: flowering branch from *Gilbert & Tewolde* 2439; 2b: pod from *Friis* et al. 1736. All x 0.75. Drawn by Maureen Church. (Reproduced with permission from *Opera Botanica*, 68: fig. 13, 1983.)

Pod oblong, 10-18 x 2-3 cm, puberulous. Seeds *c* 12 x 7 mm.

Wooded grassland and scrub; 1850-2400 m. EW TU GD GJ SD; N Uganda, S Tanzania, Zambia. *Pichi-Sermolli* 2274; *Riva* 626; *Pappi* 810.

9. **A. schimperiana** *Oliv*. (1871)
- type: TU or GD, *Schimper* 1396 (K holo., BM iso.).

Tree; bark smooth, grey, or sometimes brownish and rough; young branchlets ± densely and shortly brown-pubescent. Leaves: pinnae 2-7 pairs; leaflets 6-21(-23) pairs, obliquely oblong, or rhombic to falcate-oblong, acute to rounded and mucronate at the apex, with diagonal midrib, 7-21(-30) x 3.5-8.5(-16) mm, ± appressed-pubescent beneath. Flowers white or pale yellow; pedicels 2-6 mm long. Calyx 1.5-2.5 mm long, brown pubescent. Corolla 4-7.5 mm long, brown pubescent. Staminal tube not or scarcely exserted beyond corolla; filaments *c* 0.7-1.2 cm long. Pod oblong, 18-34 x (2-) 2.8-5.9 cm, puberulous, brown. Seeds 9-11 x 6.5-8 mm. Fig. 95.11.1.

Upland forest, evergreen bushland; 1600-2600 m. ?TU GD WU GJ WG SU AR KF IL GG SD; Sudan, and southwards to Mozambique and Zimbabwe. *Friis* et al. 79; *Gillett* 14539; *W. de Wilde* 10379.

10. **A. gummifera** *(J. F. Gmel.) C. A. Sm*. (1930); *Sassa gummifera* J. F. Gmel.(1791); *Inga sassa* Willd. (1806); *Mimosa sassa* (Willd.) Poir. (1810); *Zygia sassa* Benth. ex Schweinf. (1868); *Feuilleea sassa* (Willd.) Kuntze (1891); *Acacia sassa* (Willd.) Baill. (1902); *Albizia sassa* (Willd.) Chiov. (1912) - type: description and two plates of 'Sassa' in Bruce, *Travels, ed. 1*, **5**: 27(1790).

Medium or large tree; bark usually smooth, grey; young branchlets finely and shortly brownish-pubescent, soon glabrescent. Leaves: pinnae 5-6 (-8) pairs, each pinna ± narrowing upwards; leaflets 9-16 pairs, obliquely rhombic, mostly c 10-20(-25) x 4-8(-13)mm, obtuse to acute at apex, subglabrous or somewhat pubescent on midrib and margins. Stipules and bracts at base of peduncles lanceolate, up to c 7 x 2.5 mm. Calyx 2.5-5 mm long, minutely pubescent to subglabrous. Corolla 6.5-12 mm long, minutely pubescent, white. Staminal tube exserted c 1.5-2.8 cm beyond corolla, white below, crimson above. Pod oblong, (8-)10-21 x 2-3.4 cm, glabrescent, pale brown to purplish. Seeds 9-12 x 10 mm.

Upland and riverine forest; 1700-2400 m. GJ WG IL KF; mainly eastern tropical Africa, also in Cameroon, Nigeria and Madagascar. *Friis* et al. 277; *Mooney* 6866; *W. de Wilde* 10711.

11. **A. grandibracteata** *Taub*. (1895).

Tree, deciduous; bark smooth; young branchlets with short dense spreading pubescence, slowly glabrescent. Leaves: pinnae (1-)2-3 pairs, each pinna broadening upwards; leaflets 3-6 pairs, obliquely rhombic to obovate, 30-70(-100) x 16-32 mm, acute or subacute at the apex, ± pubescent beneath. Stipules and bracts at base of peduncles suborbicular to reniform, 7-18 x 8-20 mm. Calyx 3-5 mm long, minutely pubescent. Corolla *c* 7-10 mm long, minutely pubescent, pink to white. Staminal tube exserted c 1.2-2 cm beyond corolla, pink or red. Pod oblong, 7-15 x 1.5-3 cm, puberulous especially near base. Seeds *c* 8-11 x 6.5-8 mm. Fig. 95.11.2.

Rain-forest, riverine forest; 1200-1700 m. WG SU IL KF SD; Sudan, Uganda, W Kenya, N Tanzania and Zaire. *Friis* et al. 592; *Gillett* 15067; *W. de Wilde* 10770.

24. **PITHECELLOBIUM** *Mart.* (1837), *nom. conserv.*

Trees or shrubs, often armed with spinescent stipules. Leaves bipinnate. Flowers mostly bisexual, in heads or spikes. Calyx and corolla (4-) 5(-6)-lobed. Stamens numerous, the filaments united below into a tube; anthers eglandular at the apex. Pods spirally twisted, coiled or curved, splitting into 2 twisting thick valves.

Fairly numerous species, confined to South America.

P. dulce *(Roxb.) Benth*. (1844).

Shrub or tree armed with stipular spines. Leaves with 1 pair of pinnae; leaflets 1 pair, asymmetrically elliptic to obovate-elliptic, 0.7-5 x 0.3-2.3 cm, glabrous or somewhat pubescent. Flowers cream or yellow in small short-pedunculate heads arranged in racemes or panicles; calyx 1-1.5 mm long, puberulous; corolla 3-4.5 mm long, puberulous; filaments with free part *c* 6.5-7 mm long. Pods curved or spirally twisted, 10-12.5(-15) x 1-1.6 cm; seeds black, glossy with a white to reddish fleshy edible aril.

Planted as an ornamental for shade and as a hedge at least in EE, EW, HA and IL; native in South America. *Ash* 138, *Burger* 3275; *W. de Wilde* 9753.

A salt and drought tolerant provenance grows in Mitsiwa (Massawa). The wood is hard and can be used in the same way as *Acacia* in making local ploughs. The flowers are a good source of nectar and pollen for honey bees and the seeds are edible.

Subfamily 3. PAPILIONOIDEAE
(*FABOIDEAE*)

by M. Thulin

Leg. Trop. Afr. (1930); Cufodontis *Enum.*: 225-334 (1955); *Fl. Trop. E. Afr.* (1971).

Trees, shrubs, lianes or most often herbs, mostly unarmed. Leaves usually pinnate or pinnately 3-foliolate, less often digitately 3-foliolate, rarely 1-foliolate, simple or digitate. Flowers zygomorphic. Sepals basally united into a tube; lobes imbricate or valvate. Petals 5, imbricate, the adaxial petal (standard or vexillum) outermost, the 2 lateral (wings or alae) parallel with each other, the 2 lower and innermost cohering at the margin, forming the keel or carina. Stamens 10 or rarely fewer, variously united, rarely free; anthers usually dehiscing lengthwise. Pods various. Seeds without areoles; aril sometimes well developed; radicle generally curved.

About 440 genera and 12 000 species, widely distributed in most parts of the world, but with the greatest diversity in tropical and subtropical regions.

A record of *Ptycholobium plicatum* (Oliv.) Harms from Ethiopia by Brummitt in *Kew Bull.* **35**:462 (1980) is apparently a mistake.

Key to Genera

1. Leaves consisting of a tendril only, but with large leaf-like stipules. **108. Lathyrus**
- Leaves not as above. 2
2. Leaves all simple or 1-foliolate. 3
- Leaves, anyway some of them, compound. 14
3. Tree; stamens free. **29. Baphia**
- Herbs or small shrubs; stamens united. 4
4. Calyx spathaceous, deeply split on the upper side, the teeth very short; flowers fairly large, yellow. **83. Spartium**
- Calyx not spathaceous. 5
5. Pods curved or contorted, segmented; leaves with 3-5 parallel veins. **90. Scorpiurus**
- Pods, if segmented, not curved or contorted; leaves not as above. 6
6. Anthers alternately long and short; keel beaked; pod inflated. **79. Crotalaria**
- Anthers uniform; keel and pod not as above. 7
7. Anthers apiculate; hairs, anyway some of them, biramous; corolla usually reddish and falling soon. **37. Indigofera**
- Anthers not apiculate; hairs simple or lacking; corolla variously coloured. 8
8. Pods constricted between the seeds, mostly breaking up into 1-seeded segments. 9
- Pods neither constricted between the seeds, nor breaking into 1-seeded segments. 11
9. Stipels absent; the withered corolla persistent around the pod. **92. Taverniera**
- Stipels present; corolla not persistent around the pod. 10
10. Calyx herbaceous; pods flattened. **47. Desmodium**
- Calyx scarious; pods subcylindrical or somewhat flattened. **50. Alysicarpus**
11. Undersurfaces of leaves and in many cases other organs, e.g. calyx, with mostly yellow or orange gland dots; pods 2-seeded. 12
- Undersurfaces of leaves, etc., without gland dots; pods more than 2-seeded. 13
12. Funicle inserted at middle of ± rounded hilum; flowers spreading; pod broadened upwards, often 3-4 times as long as wide, pubescent, sometimes with longer hairs interspersed. **73. Rhynchosia**
- Funicle attached at one end of linear hilum; flowers deflexed; pod abruptly contracted to stipe, up to twice as long as wide, covered with long hairs. **75. Eriosema**
13. Leaflets with many closely parallel, almost straight lateral nerves running up to the margins; no glands beneath the anthers. **36. Tephrosia**
- Leaflets with more open venation; vexillary and 4 short stamens with a pair of fused glands below the anthers. **67. Vigna**
14. Leaves digitately 2-11-foliolate, the leaflets all arising from the same point and without any extension of the axis beyond them. 15
- Leaves pinnate, either 3-many-foliolate with leaflets arising from separate points, or sometimes 2-foliolate, with the rhachis prolonged beyond them. 29
15. Leaflets toothed. 16
- Leaflets entire. 17
16. Corolla falling soon; pod 3-4 times as long as calyx, 10-15 seeded. **98. Parochaetus**
- Corolla persistent; pod less than twice as long as calyx, 1-9 seeded. **99. Trifolium**
17. Leaflets 2 or 4; stipules and bracts conspicuous, extended below the point of insertion. **46. Zornia**
- Leaflets 3 or 5 or more; stipules and bracts not so extended. 18
18. Leaflets and calyx (often also corolla and pod) with sessile rounded resinous glands. 19
- Leaflets and calyx without such glands. 20
19. Herbs, mostly with pedunculate inflorescences; pods oblong-elliptic, flattened, without red globules of secretion. **75. Eriosema**
- Subshrub, with fasciculate dense sessile axillary inflorescences; pods oblong-ovoid, somewhat

inflated, 9-12 x 6 mm, densely but irregularly covered with dark red globules of secretion. **76. Flemingia**

20. Anthers uniform, none aborted. 21
- Anthers 4-5 basifixed, alternating with 5-6 shorter, dorsifixed, or 5 only. 24

21. Anthers apiculate; hairs, anyway some of them, biramous. **37. Indigofera**
- Anthers not apiculate; hairs simple or lacking. 22

22. Pods not breaking up into segments; annual herb with sessile clusters of small flowers. **77. Rothia**
- Pod breaking up into 1-seeded segments; other characters not combined as above. 23

23. Annual; pods without spines, curved, not constricted between the seeds. **88. Coronilla**
- Small shrubs or subshrubs; pods spiny, ± straight, constricted between the seeds. **92. Taverniera**

24. Calyx 2-lipped, the lower lip 3-fid. 25
- Calyx not 2-lipped, or if slightly so the lower lobes much longer than the united part. 27

25. Leaflets 5-11; keel beaked. **80. Lupinus**
- Leaflets 3; keel obtuse. 26

26. Pods without glandular tubercles; lip formed by upper calyx-lobes more than twice as long as tube, and as long as lower lip. **81. Argyrolobium**
- Pods with conspicuous glandular tubercles; lip formed by upper calyx-lobes about as long as tube, and shorter than lower lip. **82. Adenocarpus**

27. Stipels present; pods linear, compressed with an upcurved beak; trailing or twining herb. **57. Teramnus**
- Stipels absent; pods various, if linear without an upcurved beak, never twining. 28

28. Keel obtuse; lateral calyx-lobes more united with the upper than the lower lobe, which is often narrower; 4 anthers large, 6 anthers small; pods only slightly inflated. **78. Lotononis**
- Keel beaked; calyx-lobes subequal or the 3 lower slightly more united or with the upper 2 also slightly united to form 2 lips; 5 large anthers alternating with 5 small; pods usually markedly inflated. **79. Crotalaria**

29. Leaves paripinnate (rhachis sometimes extended as a bristle or tendril) or imparipinnate with a much reduced terminal leaflet. 30
- Leaves imparipinnate, terminating in a leaflet about as large or larger than the lateral leaflets. 39

30. Pods developing underground; flowers subsessile but receptacle long and filiform and easily mistaken for a pedicel; anthers alternately long and short; leaves 4-foliolate. **45. Arachis**
- Pods not developing underground; receptacle not developed as above; anthers mostly uniform; leaves 2-many foliolate. 31

31. Calyx 2-lipped; pods with 1-28, 1-seeded segments. 32
- Calyx not distinctly 2-lipped; pods not segmented. 34

32. Inflorescence ± lax, never scorpioid; pods easily visible, well exserted from the calyx. **41. Aeschynomene**
- Inflorescence dense and ± scorpioid; pods hidden, folded like a pleated dress. 33

33. Stipules spurred; leaflets opposite, with only 1 main nerve; inflorescences subumbellate. **42. Smithia**
- Stipules not spurred; leaflets alternate, with several basal nerves; inflorescences distichous, usually dense and strobilate. **43. Kotschya**

34. Stiff erect herbs, softly woody shrubs or small trees or rather woody twiners; tendrils absent. 35
- Weak herbs; leaves usually ending in tendrils. 36

35. Pods oblong or narrowly oblong, compressed, 2-9 x 0.7-1.6 cm, never winged; sometimes twining or trailing; standard without appendages; stamens 9, vexillary one absent. **33. Abrus**
- Pods linear or cylindrical, 7-55 x 0.15-0.9 cm, or if wider then pods broadly 4-winged, always erect; standard with appendages on the inner face; stamens 10, vexillary one free. **39. Sesbania**

36. Pods broadly oblong, very compressed, up to 9-17 x 4-12 mm with 1-2 orbicular flattened seeds; stipules entire; calyx-lobes subequal, at least twice as long as the tube. **107. Lens**
- Pods narrowly oblong or if oblong and *c* 6-10 x 3-3.5 mm then stipules toothed; at least 2 calyx-lobes less than twice as long as the tube. 37

37. Stipules up to 10 cm long, leafy; style-margin winged. **105. Pisum**
- Stipules not so large and leafy; style not winged. 38

38. Style pubescent on the lower side, all round or glabrous; leaflets usually more than 2. **106. Vicia**
- Style pubescent on upper side only; leaflets 2 (in Ethiopia). **108. Lathyrus**

39. Leaves 5 to many foliolate (in *Lotus* the 5-foliolate leaves are sessile and the basal leaflets may be mistaken for stipules - see fig. 96.40.1&2). 40
- Leaves all or mostly 3-foliolate. 65

40. Petals all similar, cuneate or elliptic-obovate. 41
- Petals dissimilar, differentiated into standard, wings and keel-petals. 42

41. Leaflets 20-50; stipules free, small; flowers cream turning wine-red. **25. Cadia**
- Leaflets 9-15; stipules connate, the persistent bases covering ends of twigs; flowers white. **26. Dicraeopetalum**

42. Anthers apiculate; hairs, anyway some of them, biramous; corolla usually reddish. 43
- Anthers not apiculate; hairs simple; corolla variously coloured. 44

43. Vexillary stamen free; standard usually pubescent outside, but if glabrous then pod not ridged. **37. Indigofera**
- Vexillary stamen united with others; standard glabrous; pods stout, longitudinally ridged. **38. Cyamopsis**

44. Pods folded like a pleated dress, constricted between the seeds; flowers crowded in spike-like inflorescences, the pedicels covered with long deciduous hairs and short persistent hooked hairs. **48. Uraria**

- Pods not so folded; pedicels without hooked hairs. 45
45. Standard 2.5-5 cm long, held lowermost, much exceeding other petals; bracteoles ovate, persistent; upper part of style hairy. **53. Clitoria**
- Standard smaller and not much exceeding other petals; bracteoles small or deciduous; style glabrous above except sometimes around stigma. 46
46. Standard pubescent to silky tomentose outside. 47
- Standard glabrous or with small scattered hairs mostly towards apex and margin. 48
47. Tree; lateral nerves of leaflets not very closely parallel; stipels present; pods somewhat woody. **34. Milletia**
- Herbs or shrubs; lateral nerves of leaflets numerous, closely parallel; stipels absent; pods not woody. **36. Tephrosia**
48. Trees or shrubs. 49
- Annual or perennial herbs. 56
49. Stamens free or shortly joined into a ring only at the base. 50
- Stamens united into bundles, a sheath or tube. 51
50. Flowers violet, mostly appearing before the leaves; ovary 1-ovuled; pod elliptic, not winged. **27. Platycelyphium**
- Flowers yellow; ovary with several ovules; pod narrowly oblong, winged along the upper edge. **28. Calpurnia**
51. Pods breaking up transversely into 1-6, 1-seeded segments; branchlets with swollen-based glandular hairs whose bases persist as small tubercles. **40. Ormocarpum**
- Pods not segmented; branchlets without swollen-based glandular hairs. 52
52. Pods bladder-like, papery; style thick, incurved at tip, apically bearded inside. **97. Colutea**
- Pods not bladder-like; style slender, tapered, glabrous or nearly so above. 53
53. Anthers basifixed with apical dehiscence. **32. Dalbergia**
- Anthers dorsifixed with longitudinal dehiscence. 54
54. Corolla yellow; pods with a wing-like expansion around the seed-cavity. **31. Pterocarpus**
- Corolla white, mauve or lilac; pods without such a wing-like expansion. 55
55. Corolla white; leaflets alternate; pods tapering at base, the seeds in the distal half. **35. Craibia**
- Corolla mauve to lilac; leaflets opposite; pods narrowly oblong with seeds distributed more evenly along its length. **30. Lonchocarpus**
56. Leaflets toothed. **104. Cicer**
- Leaflets entire. 57
57. Flowers in dense apparently terminal heads; calyx inflated, enclosing the pod; terminal leaflet of lower leaves much larger than the lateral. **86. Anthyllis**
- Flowers in racemes or umbels, if in heads then clearly axillary; calyx not inflated, nor enclosing the pod; terminal leaflet not or only slightly larger than the others. 58
58. Leaves sessile or subsessile, the basal pair of leaflets sometimes resembling leafy stipules; stipules usually reduced to glands. 59
- Leaves distinctly petiolate and/or stipulate. 61
59. Pods straight, dehiscent, narrowly cylindrical; leaflets 5. **85. Lotus**
- Pods curved, indehiscent or composed of spherical 1-seeded individually dehiscent segments; leaflets 5-11. 60
60. Flowers in heads; pods curved almost in a circle, indehiscent, 2-seeded. **87. Vermifrux**
- Flowers in umbels; pods much less curved, composed of spherical 1-seeded individually dehiscent segments. **89. Antopetitia**
61. Pods breaking up transversely into 1-seeded indehiscent spinulose segments; corolla bright reddish purple. **91. Hedysarum**
- Pods not breaking up into segments; corolla various but not bright reddish purple. 62
62. Pods ± orbicular and flat, toothed or spiny. **93. Onobrychis**
- Pods ± oblong, sometimes toothed. 63
63. Pods dorsiventrally compressed, each valve produced at the side into a flat crest; only 5 stamens fertile. **96. Biserrula**
- Pods laterally compressed or rather inflated; all stamens fertile. 64
64. Pod unilocular with prominent parallel veins running obliquely upwards from each suture; stipules divided into 2-5 narrow lobes; filaments all united. **94. Galega**
- Pod divided lengthwise by a vertical septum, without prominent parallel veins; stipules entire; vexillary filament free. **95. Astragalus**
65. Leaflets toothed. 66
- Leaflets entire or 3-5-lobed. 70
66. Plant glandular-pubescent, at least above; filaments all united. **103. Ononis**
- Plant not glandular-pubescent; upper filament free. 67
67. Petals persistent after flowering, the claws of the wings and keel attached to the staminal tube; pod straight, short, enclosed in the persistent corolla. **99. Trifolium**
- Petals falling soon, not attached to the staminal tube; pod straight or curved, not enclosed in the persistent corolla. 68
68. Pod more than 3 times as long as calyx, often very long, nearly straight, beaked; flowers 1-2, subsessile in leaf-axils. **102. Trigonella**
- Pod less than 3 times as long as calyx or coiled; flowers few-many in axillary racemes. 69
69. Pod curved or spirally coiled; inflorescences short, the flowers few or crowded. **100. Medicago**
- Pod short, straight; inflorescences elongate, many-flowered. **101. Melilotus**
70. Anthers apiculate; hairs, anyway some of them, biramous. 71
- Anthers not apiculate; hairs simple. 72

71. Vexillary stamen free; standard usually pubescent outside, but if glabrous then pod not ridged. **37. Indigofera**
- Vexillary stamen united with others; standard glabrous; pods stout, longitudinally ridged. **38. Cyamopsis**

72. Calyx obliquely truncate, with no trace of lobes or scarcely any. **52. Dumasia**
- Calyx-lobes developed. 73

73. Undersurfaces of leaves, and in many cases other organs, e.g. calyx, with sessile, rounded gland-dots (often difficult to see if indumentum dense). 74
- Undersurfaces of leaves, etc., without gland-dots. 78

74. Ovary 1-ovuled; pod indehiscent; corolla ± blue. **84. Psoralea**
- Ovary (1-)2-7-ovuled; pod dehiscent; corolla ± yellow to orange. 75

75. Calyx scarious, strongly accrescent and up to 45 mm long in fruit, lobes obtuse, the lowermost the largest, boat-shaped. **74. Paracalyx**
- Calyx not as above. 76

76. Ovary 4-7-ovuled; pod obliquely grooved between the seeds. **72. Cajanus**
- Ovary 2-ovuled; pod not grooved between the seeds. 77

77. Funicle inserted at middle of ± rounded hilum; flowers spreading; pod broadened upwards, often 3-4 times as long as wide, pubescent, sometimes with longer hairs interspersed. **73. Rhynchosia**
- Funicle attached at one end of linear hilum; flowers deflexed; pod abruptly contracted to stipe, up to twice as long as wide, covered with long hairs. **75. Eriosema**

78. Style distinctly flattened and spathulate at the apex (Fig. 96.30.5c); standard glabrous. **68. Sphenostylis**
- Style-apex various but not flattened and spathulate; standard glabrous or hairy. 79

79. Pods usually covered with fine irritant bristles; flowers in pendulous false racemes, large, blackish purple and/or white; standard much shorter than other petals; keel-petals horny at the apex. **59. Mucuna**
- Pods not covered by irritant hairs (though often bristly) and without other characters combined. 80

80. Trees or shrubs. 81
- Herbs, subshrubs or herbaceous climbers. 82

81. Trees or shrubs often armed with prickles; corolla large, red or orange (rarely yellow), the standard usually much larger than other petals. **58. Erythrina**
- Unarmed tree; corolla white, the standard as long as other petals. **35. Craibia**

82. Standard hairy outside. 83
- Standard glabrous outside or practically so. 85

83. Corolla *c* 1-2 cm long; inflorescence with long peduncles; style long, with tenuous basal part and curved barbate thickened upper part beaked beyond the lateral stigma; pods linear-cylindric. **67. Vigna**
- Corolla 0.4-1.1 cm long, if much over 1 cm then inflorescences subsessile; style short, not as above; pods linear-oblong to oblong. 84

84. Flowers in subsessile short racemes; pods 1-2-seeded, 14-20 x 6.5-7 mm. **54. Pseudoeriosema**
- Flowers in elongated lax racemes (in Ethiopia); pods 3-5-seeded, 17-35 x 4.5-6.5 mm. **55. Ophrestia**

85. Alternate stamens sterile, lacking anthers; pods linear with an upcurved beak. **57. Teramnus**
- All stamens with anthers; pods not linear with an upcurved beak. 86

86. Pods breaking up transversely into 1-seeded segments. 87
- Pods not breaking up transversely into 1-seeded segments. 89

87. Stipels present. **47. Desmodium**
- Stipels absent. 88

88. Calyx-tube very slender, stalk-like; pods with 1-2 segments, the apical ones narrowed into distinct hooked beaks, not spiny. **44. Stylosanthes**
- Calyx-tube not slender and stalk-like; pods spiny with 1-5 segments, the apical ones without hooked beaks. **92. Taverniera**

89. Pods longitudinally 4-winged; bracteoles large. **63. Psophocarpus**
- Pods not longitudinally 4-winged; bracteoles various or absent. 90

90. Style with a reflexed appendage beyond the stigma (Fig. 96.29.2c); usually a woody climber; keel incurved. **64. Vàtovaea**
- Style sometimes produced beyond the stigma as a beak but never reflexed back as an appendage; habit and keel various. 91

91. Pods large, linear-oblong to oblong, with longitudinal ribs close to the upper margin, *c* 10-35 x 2.5-3.5 cm; seeds *c* 1.5-2 cm long; standard reflexed, *c* 3 cm long. **60. Canavalia**
- Pods, seeds and/or standard smaller. 92

92. Style with a distinct thickening at its junction with the ovary; bracteoles absent; leaflets often 3-lobed. **71. Neorautanenia**
- Style without such a thickening; bracteoles often present; leaflets entire or lobed. 93

93. Vexillary and 4 short stamens with a pair of fused glands below the anthers. **67. Vigna**
- Stamens without such glands. 94

94. Style divided into a thin basal part and a thickened upper part. 95
- Style uniformly thick, uniformly thin or tapering, not divided into two distinct parts. 97

95. Style with thickened part (and keel) curved through more than 360°; stipules not produced beyond point of attachment; pods oblong to linear-lanceolate, *c* 5-20 x 1-3 cm. **66. Phaseolus**
- Style with thickened part curved through no more than 360°, usually up to 180°; stipules sometimes produced beyond point of attachment; pods usually narrower, ± linear or cylindric. 96

96. Thickened part of the style characteristically abruptly curved through 90° just above its junction with the tenous part and narrowed and slightly curved towards the apex, resembling a squarish hook; wings round, large, longer than standard and keel, blackish purple. **65. Macroptilium**
- Thickened part of the style straight or curved but not of the characteristic shape described above; wings not as above. **67. Vigna**

97. Style distinctly thickened, conspicuous. 98
- Style not distinctly thickened, often short and inconspicuous. 99

98. Style conspicuously laterally flattened, straight and blade-like throughout its length, forming an angle of just less than 90° with the ovary and with a line of hairs near top of inner margin; stigma terminal, not penicillate; keel sharply angled at middle; seeds with a long linear hilum with a whitish aril. **70. Lablab**
- Style not conspicuously laterally flattened, more rounded in section, usually curved, nearly always glabrous save for terminal almost invariably penicillate stigma; keel rounded beneath; seeds with short hilum. **69. Dolichos**

99. Erect herbs or subshrubs with somewhat lax to very dense many-flowered panicles of small reddish-purple or white flowers; standard 6-8 mm long; pods narrowly oblong, flat, 7-38 x 3-4(-5) mm. **49. Pseudarthria**
- Climbing or trailing herbs or, if ± erect, other characters not as above. 100

100. Bracteoles absent; corolla mauve or violet, 1-1.6 cm long; pods linear-oblong, 25-35 x 5.5-6 mm, glabrous except for densely ciliate margins. **51. Amphicarpa**
- Bracteoles present; other characters not combined as above. 101

101. Corolla yellowish; standard 0.6-2.4 cm long, inside with 2 long linear appendages. **62. Macrotyloma**
- Corolla blue, mauve, violet or white; standard 0.45-1.2 cm long, devoid of appendages. 102

102. Leaflets elliptic-oblong, obtuse or emarginate; calyx 4-lobed with two upper lobes completely joined; vexillary stamen free. **61. Galactia**
- Leaflets ovate, elliptic or lanceolate, usually acute or acuminate; calyx 5-lobed, the upper lobes not completely joined; vexillary stamen joined with the others or sometimes becoming free with age. **56. Glycine**

25. CADIA *Forssk.* (1775)

Shrubs. Leaves imparipinnate. Flowers solitary in the upper axils or few in short racemes or panicles. Calyx broadly campanulate; lobes broad, subequal. Petals subequal, free, erect. Stamens *c* 10, subequal, free. Pod linear, flat, leathery, dehiscent. Seeds numerous, flattened.

Genus with 6 species, all but 1 in Madagascar.

C. purpurea *(Picc.) Ait.* (1789);
Panciatica purpurea Picc. (1783) - type: a plant cultivated in Italy, from seeds collected in Ethiopia by Bruce.

Unarmed shrub 0.8-5 m tall. Leaves up to 17 cm long; leaflets 20-50, usually opposite or almost so, narrowly oblong, up to 23 x 5 mm, emarginate, glabrous above, puberulous on the midrib and margins beneath; terminal leaflet cuneate-oblong. Flowers on axillary short shoots, hanging. Calyx-lobes triangular, acute, *c* 10 mm long. Petals *c* 30 mm long, cream turning wine-red, falling quickly. Pod straight or slightly curved, 120-130 x 11-13 mm, acuminate, glabrous; valves twisting on dehiscence. Seeds reddish-brown, elliptic, *c* 8 x 5 mm. Fig. 95.12.1.

Rift valley escarpment, bushland, locally codominant; 1300-2700 m. EW TU WU SU(cult.) HA BA; Yemen, Oman, N Somalia, N Kenya. *Gilbert & Thulin* 128; *Burger* 436; *Schweinfurth & Riva* 1063.

26. DICRAEOPETALUM *Harms* (1902)

Tree. Leaves imparipinnate. Flowers in racemes. Calyx-lobes 5, subequal. Petals 5, subequal, free. Stamens 10, all alike, free. Pod flat, papery, indehiscent, 1-2-seeded.

Monotypic.

D. stipulare *Harms* (1902);
Acosmium stipulare (Harms) Yakovlev (1969) - type: HA, Adjabo plateau about 8°N, *Ellenbeck* 1108 (B holo. dest.); neotype from Somalia.

Tree 4-8 m tall; ends of twigs densely covered with persistent stipule-bases. Leaves 2-4 together at the ends of the twigs, 8-10 cm long; leaflets 9-15, subopposite, elliptic-oblong, 20-30 x 9-14 mm, puberulent on both surfaces. Racemes many-flowered, 5-7 cm long. Calyx silky pubescent, 5-6 mm long, divided to just above the middle into lanceolate lobes. Petals 5-6 mm long, white, notched at the apex. Pod narrowly oblong, *c* 3-4 cm long, bluntly pointed, appressed pubescent.

Deciduous bushland. HA; S Somalia, N Kenya. The holotype is the only record from Ethiopia.

27. PLATYCELYPHIUM *Harms* (1905)

Tree. Leaves imparipinnate. Flowers in racemes. Calyx glandular; lobes a little shorter than tube, the upper 2 more united than the others. Corolla violet to deep blue, differentiated into standard, wings and keel. Stamens free. Pod flat, papery, indehiscent, 1-seeded.

Monotypic.

P. voënse *(Engl.) Wild* (1959);
Commiphora voënsis Engl. (1904).
P. cyananthum Harms (1905).

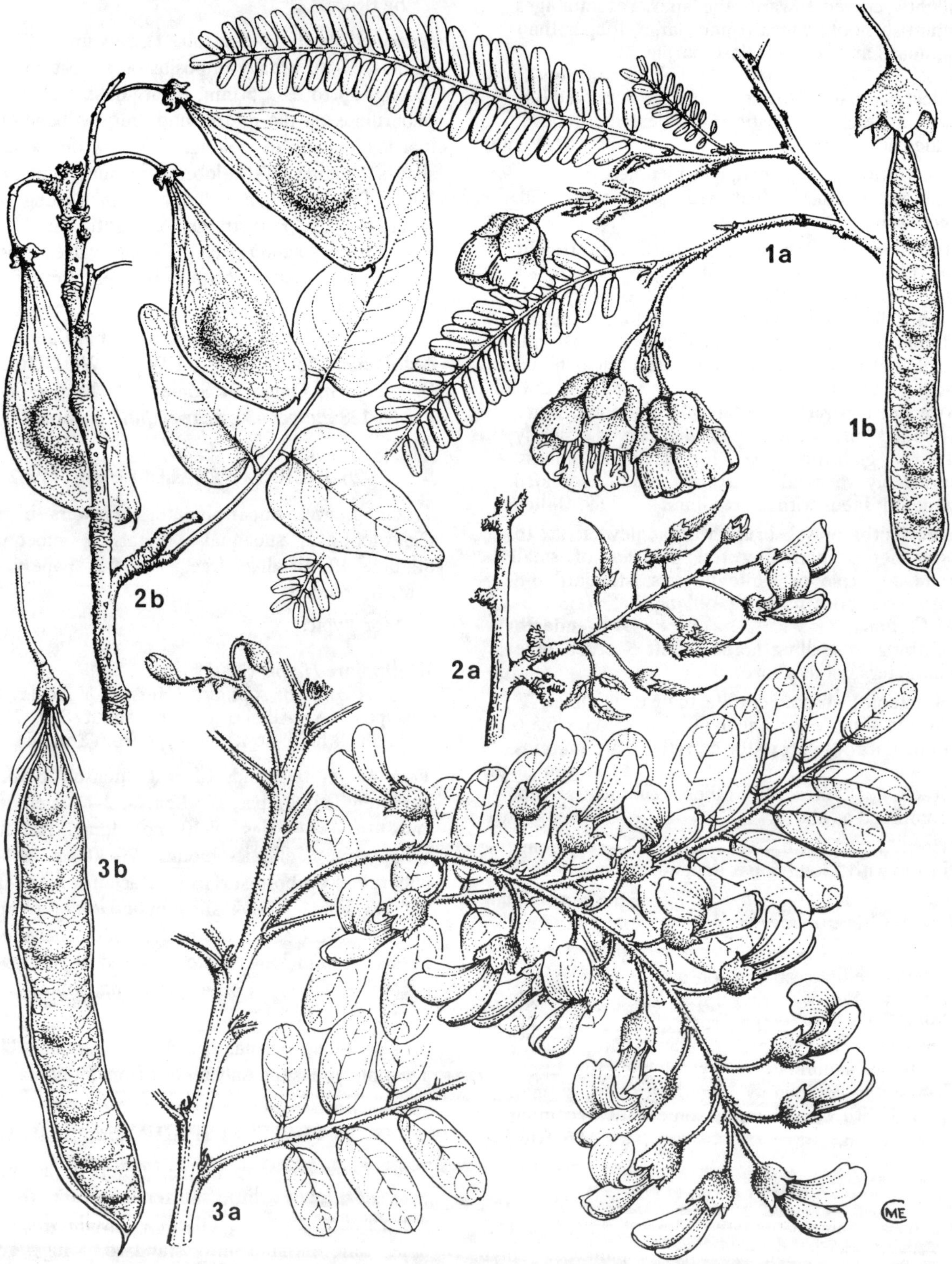

Figure 95.12.1 - *CADIA PURPUREA*, a: flowering branch; b: pod, both from *Collenette* 231. 2 - *PLATYCELYPHIUM VOËNSE*, a: flowering branch, from *Ellis* 104; b: fruiting branch, from *Glover & Gilliland* 1050. 3 - *CALPURNIA AUREA*, a: flowering branch, from *Hillier* 918; b: pod, from *Cheke* 7. All x 0.75. Drawn by Maureen Church. (Reproduced with permission from *Opera Botanica* 68, fig. 14, 1983.)

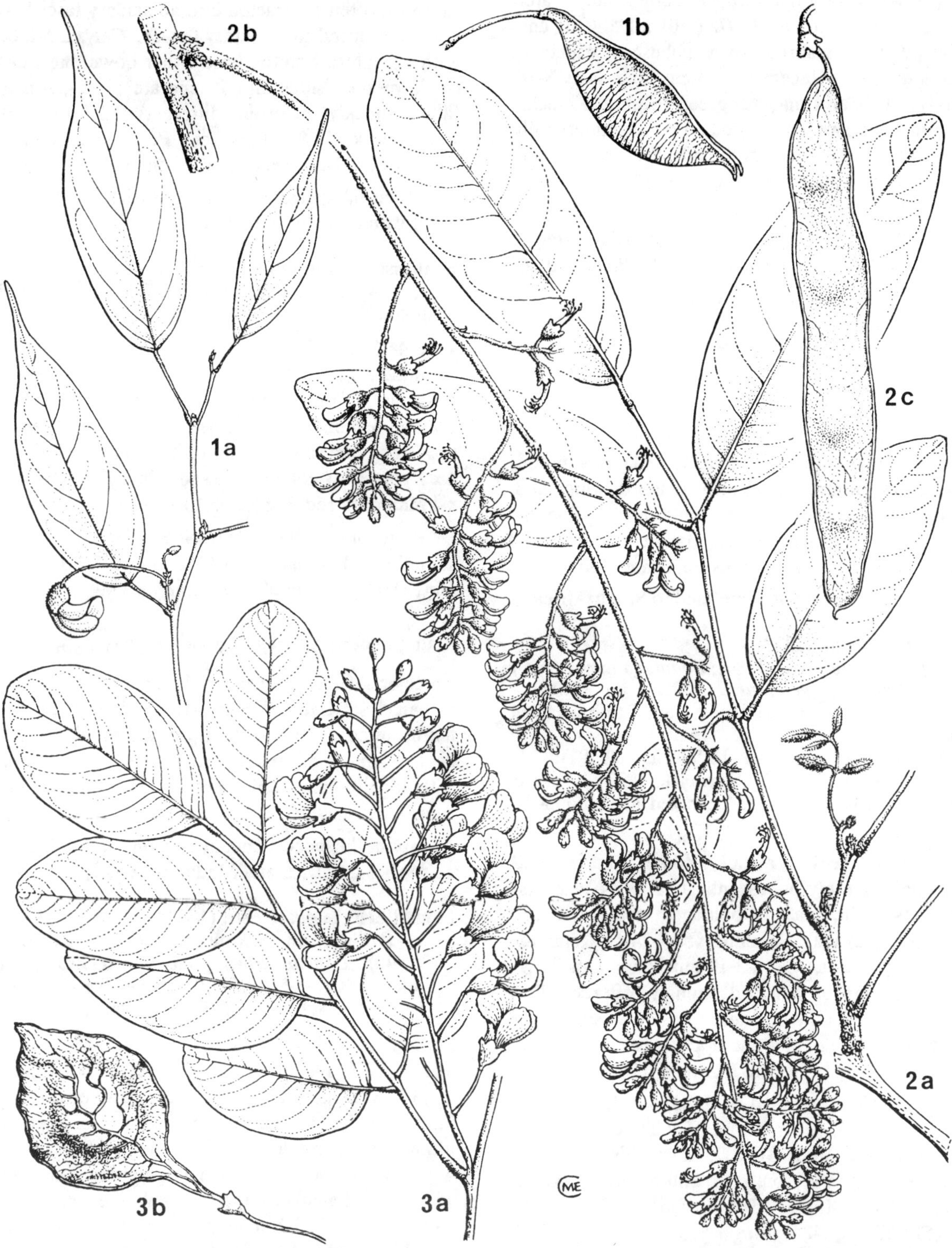

Figure 95.13.1 - *BAPHIA ABYSSINICA*, a: flowering branch, from *Chaffey* 949; b: pod, from *Thomerson* 821. 2 - *LONCHOCARPUS LAXIFLORUS*, a: leafy branch; b: inflorescence; c: pod, all from *Mooney* 6830. 3 - *PTEROCARPUS LUCENS*, a: flowering branch, from *W. de Wilde* 10714; b: pod, from *Wickens* 2940. All x 0.75. Drawn by Maureen Church. (Reproduced with permission from *Opera Botanica* 68, fig. 15, 1983.)

Deciduous tree or shrub 1.5-7.5 m tall; bark peeling when old in papery strips. Leaves up to 10-27 cm long; leaflets (1-)3-7, mostly opposite or subopposite, ovate-lanceolate to ovate-elliptic, 3.5-7.5(-10) x 2.5-4(-6) cm, glandular-punctate, glabrous to ± pubescent especially beneath, glutinous. Racemes 3-13 cm long, laxly 8-16 (-30)-flowered. Calyx purplish-green to almost black, (4-)5-7 mm long, sweetly scented. Pod asymmetrically oblong-elliptic, narrowed to both ends, 5-7.5 x 2-3.7 cm, sparsely to densely pubescent. Figure 95.12.2.

Scattered in *Acacia-Commiphora* deciduous bushland; 600-900 m. HA BA SD; Somalia, Kenya, Tanzania. *Glover & Gilliland* 1050; *Bally* 12986; *Hemming* 1475.

Wood is hard and durable, and is used for making handles for farm implements, etc.

28. **CALPURNIA** *E. Mey.* (1836)

Shrubs or small trees. Leaves imparipinnate. Flowers in racemes. Calyx with the 2 upper teeth largely united. Corolla yellow, differentiated into standard, wings and keel. Stamens free or more commonly shortly joined at the base. Pod narrowly oblong, flat, ± membranous, usually dehiscent. Seeds compressed, ovate-oblong.

About 7 species, all but 1 confined to South Africa.

C. aurea *(Ait.) Benth.* (1837);
Sophora aurea Ait. (1789); *Podalyria aurea* (Ait.) Willd. (1799); *Virgilia aurea* (Ait.) Lam. (1819) - type: grown at Kew, originally from Ethiopia (BM lecto.).
Robinia subdecandra L'Hérit. (1791); *C. subdecandra* (L'Hérit.) Schweick.(1937) - type: grown at Paris from seeds collected in Ethiopia by Bruce (G holo., not seen).

Shrub or small tree 1-10 m tall. Young branches and inflorescences densely pubescent. Leaves (4-)6-24 cm long; leaflets 10-30, ovate or oblong to obovate, *c* 2.5-4(-5) x 1.2-1.6(-2) cm. Racemes (5-)7-24 cm long, 8-30-flowered, pendent. Calyx 3-10 mm long. Corolla 9-20 mm long; limb of standard strongly reflexed. Pod 50-120 x 8-16 mm, straw-coloured or brownish, with a 1-2.5 mm broad wing along the upper edge, 4-8-seeded. Seeds 5-6 mm long, brownish. Figure 95.12.3.

Forest margins, bushland or grassland, favoured by overgrazing; 1650-2550(-3000) m. EW TU GD WU GJ WG SU AR BA HA KF GG SD; westwards to Central African Republic and south to South Africa (Cape Province), also in India. *Schweinfurth & Riva* 1062; *Thulin* 1510; *Gillett* 14244; *Burger* 698.

The seeds may be used as a fish-poison or as a cure for dysentery.

All material from eastern Africa belongs to subsp. *aurea*. Subsp. *indica* Brummitt is confined to S India and subsp. *sylvatica* (Burch.) Brummitt occurs in the Cape Province.

29. **BAPHIA** *Lodd.* (1825)

Shrubs or trees. Leaves unifoliolate, entire. Flowers in racemes, often contracted into an axillary fascicle which may be reduced to a solitary flower. Calyx 2-5-lobed, at anthesis splitting to the base either down one side only or down both sides into 2 separate pieces, sometimes falling quickly. Corolla differentiated into standard, wings and keel. Stamens free. Pod compressed, dehiscing into 2 valves. Seeds few, often only 1 developing.

About 60 species in tropical Africa with 1 in Natal and 1 extending to Madagascar.

B. abyssinica *Brummitt* (1968)
- type: KF, Tepi to Mizan Teferi, *Meyer* 8931 (K holo., WAG iso.).

Tree 4-15 m tall. Leaves with petiole 0.7-1.8 cm long; lamina 2.5-11 x 1.3-5.5 cm, ovate or elliptic-acuminate, glabrous. Flowers in few-flowered axillary fascicles. Calyx 5-6 mm long, appressed brown-pubescent, splitting along one side. Petals white, 9-11 mm long. Pod *c* 4 x 1 cm, straw-coloured. Seed solitary, oblong, *c* 18 x 7 x 4 mm, dark red. Figure 95.13.1.

Rain-forest and riverine forest, woodland, sometimes dominant; 500-1250 m. IL KF GG; Sudan. *Meyer* 8035; *Friis* et al. 1729; *Chaffey* 949.

30. **LONCHOCARPUS** *Kunth* (1824), *nom. conserv.*

Trees or shrubs. Leaves usually imparipinnate. Flowers in panicles. Calyx shortly 5-toothed with the 2 upper teeth united almost to the tips. Wings adhering to the keel. Stamens all united into a tube; anthers dorsifixed with longitudinal dehiscence. Pod flattened, indehiscent, 1-few-seeded.

About 75 species, mostly in tropical America; about 12 species in Africa and Madagascar.

L. laxiflorus *Guill. & Perr.* (1832).
Philenoptera schimperi Hochst. ex A. Rich. (1847); *Lonchocarpus philenoptera* Benth. (1860), nom. illegit. - types: TU, *Schimper* II:897 (P syn., BM isosyn.) & *Schimper* III:1802 (P syn.).

Deciduous tree, 3-12 m tall. Leaves 15-45 cm long; leaflets 5-7, narrowly oblong-elliptic or elliptic, 5.4-15(-18) x 2-5.5 cm, greyish pubescent on both sides first, ± glabrescent at maturity. Panicles erect at first, ultimately pendulous, 18-40(-60) cm long, with numerous flowers. Calyx 3.5-5(-6) mm long, purplish with a short whitish appressed pubescence. Corolla 9-14 mm long, pinkish-mauve to deep lilac. Pod narrowly elliptic-oblong to linear-oblong, slightly sinuate between the seeds, 6-14 x 1.3-2 cm, pale brown, glabrescent. Seeds 1-3, oblong-reniform. Figure 95.13.2.

Wooded grassland; 450-2150 m. EE EW TU GD WG SU BA IL KF; Uganda and westwards to Senegal. *Schimper* III:1778; *Mooney* 6830, 7771; *Scott* 343.

Elsewhere important as browse and as a source of fuelwood, pole timber and charcoal.

31. **PTEROCARPUS** *Jacq.* (1763), *nom. conserv.*

Trees. Leaves imparipinnate. Flowers in racemes or panicles. Calyx shortly 5-toothed with the upper 2 teeth practically united. Corolla yellow to orange; wings free. Stamens usually all united into a tube; anthers dorsifixed with longitudinal dehiscence. Pod compressed, winged, with a thickened or hardened central seed-bearing part, indehiscent, usually 1-seeded. Seeds reniform or oblong-reniform.

Some 20 species in the tropics, most numerous in Africa.

P. lucens *Guill. & Perr.* (1832).
P. abyssinicus Hochst. ex A. Rich. (1847) - types: TU, *Schimper* III:1597 (P syn., BM K S WAG isosyn.) & III:1668 (P syn., BM isosyn.).

Deciduous tree up to 7.5-18 m tall. Leaves (6.5-)10-28 cm long; leaflets 3-11, mostly subopposite, broadly lanceolate to ovate or elliptic, up to 4-10 x 2.4-4.8 cm, glabrous or glabrescent above, appressed pubescent beneath. Racemes 6-20(-30) cm long, few- to many-flowered. Calyx 4-6 mm long, glabrous outside, often coloured with reddish gum-exudate. Corolla 10-14 mm long, yellow, fragrant. Pod suborbicular with a relatively narrow leathery wing running on to a distinct stipe, 4.5-6.5 x 2-3 cm, glabrous. Figure 95.13.3.

Wooded grassland, particularly on rocky hills; 550-1520 m. EW TU GD GJ WG IL; Uganda and westwards to Senegal; subsp. *antunesii* (Taub.) Rojo in S tropical Africa. *Mooney* 6831; *W. de Wilde* 10714; *Thulin & Hunde* 4041.

Elsewhere the leaves are used as a vegetable and animal fodder. The wood has many uses including fuelwood, house construction, telegraph poles, carpentry and furniture.

32. **DALBERGIA** *L. f.* (1781), *nom. conserv.*

Trees, free-standing or scandent shrubs or woody lianes, sometimes with branches modified as thorns or climbing aids. Leaves imparipinnate; leaflets usually alternate. Flowers in panicles or more rarely in racemes. Calyx shortly 5-toothed; lowest lobe usually narrower and the longest, the upper pair usually broader than the others and often more united. Corolla small, usually white and fragrant; wings free. Stamens 9 or 10, usually all united into a tube; anthers basifixed with apical dehiscence. Pod indehiscent, samaroid, 1-few-seeded. Seeds reniform.

Some 100 species throughout the tropics.

1. Leaflets acuminate, 3-5. **5. D. sissoo**
- Leaflets obtuse, usually more than 5. 2

2. No branches thorny, but sometimes coiled as climbing aids; leaflets at least 4 x 2 cm; pod at least 10 x 3 cm. **4. D. lactea**
- At least some branches of previous season's growth remaining short and often modified as thorns; leaflets usually less than 4 x 2 cm; pod less than 7 x 2 cm. 3

3. Inflorescences on cushion-shaped branchlets with tufts of leaves; leaflets 13-17(-25) in number, mostly less than 1 cm long. **3. D. microphylla**
- Inflorescences on at least somewhat elongated shoots; leaflets 3-13(-17) in number, mostly more than 1 cm long. 4

4. Leaves *c* 6-22 cm long, with 9-13(-17) leaflets; panicles laxly branched, up to 4-12 cm long. **1. D. melanoxylon**
- Leaves *c* 2-6 cm long, with (3-)5-7 leaflets; racemes or little-branched panicles up to *c* 3 cm long. **2. D. commiphoroides**

1. **D. melanoxylon** *Guill. & Perr.* (1832).

Shrub or tree, up to 5-12 m tall, typically several-stemmed and much branched. Branchlets clustered at the nodes, some becoming thorns. Leaves *c* 6-22 cm long; leaflets 9-13(-17), mostly oblong-obovate to obovate; 1-5(-5.5) x 0.7-3(-5) cm, shortly pubescent at least beneath, glabrescent. Panicles *c* 3-12 cm long, laxly branched, many-flowered. Calyx 2-3(-4) mm long. Corolla 4-6 mm long, white. Stamens usually 9. Pod ± elliptic-oblong, 3-7 x 0.8-1.4(-2) cm, papery, glabrous, laxly and rather diffusely veined, 1-2-seeded.

Deciduous woodland, wooded grassland or bushland, often in rocky places or valleys of impeded drainage; 600-1900 m. EW TU GD(Sudan border); widespread in tropical Arica south to Transvaal, also in India. *Mooney* 8584; *Schimper* 1737; *Greathead* 106.

The purplish to brownish black heartwood of this species, known as AFRICAN BLACKWOOD or AFRICAN EBONY, is very hard and heavy and of considerable commercial value. Among other things, it is used to make carvings and the best quality clarinets. Recognised elsewhere as a good plant for honey bees.

2. **D. commiphoroides** *Bak. f.* (1929).

Shrub 1.8-5 m tall. Branchlets clustered at the nodes, some becoming thorns. Leaves 2-6 cm long; leaflets (3-)5-7, oblong-obovate to broadly obovate, 0.8-2.4 x 0.7-2.3 cm, pubescent on both surfaces. Racemes or little-branched panicles (1.5-)2-3 cm long. Calyx 2-2.5 mm long. Corolla 5-6 mm long, white. Stamens 9. Pod elliptic-oblong, (2.2-)3-4.5 x 1-1.5 cm, papery, subglabrous, uniformly and prominently veined, 1-2-seeded.

Deciduous bushland; 400-850 m. HA; Somalia and N Kenya. *Ellis* 307; *Simmons* 111; *Bally* 10124.

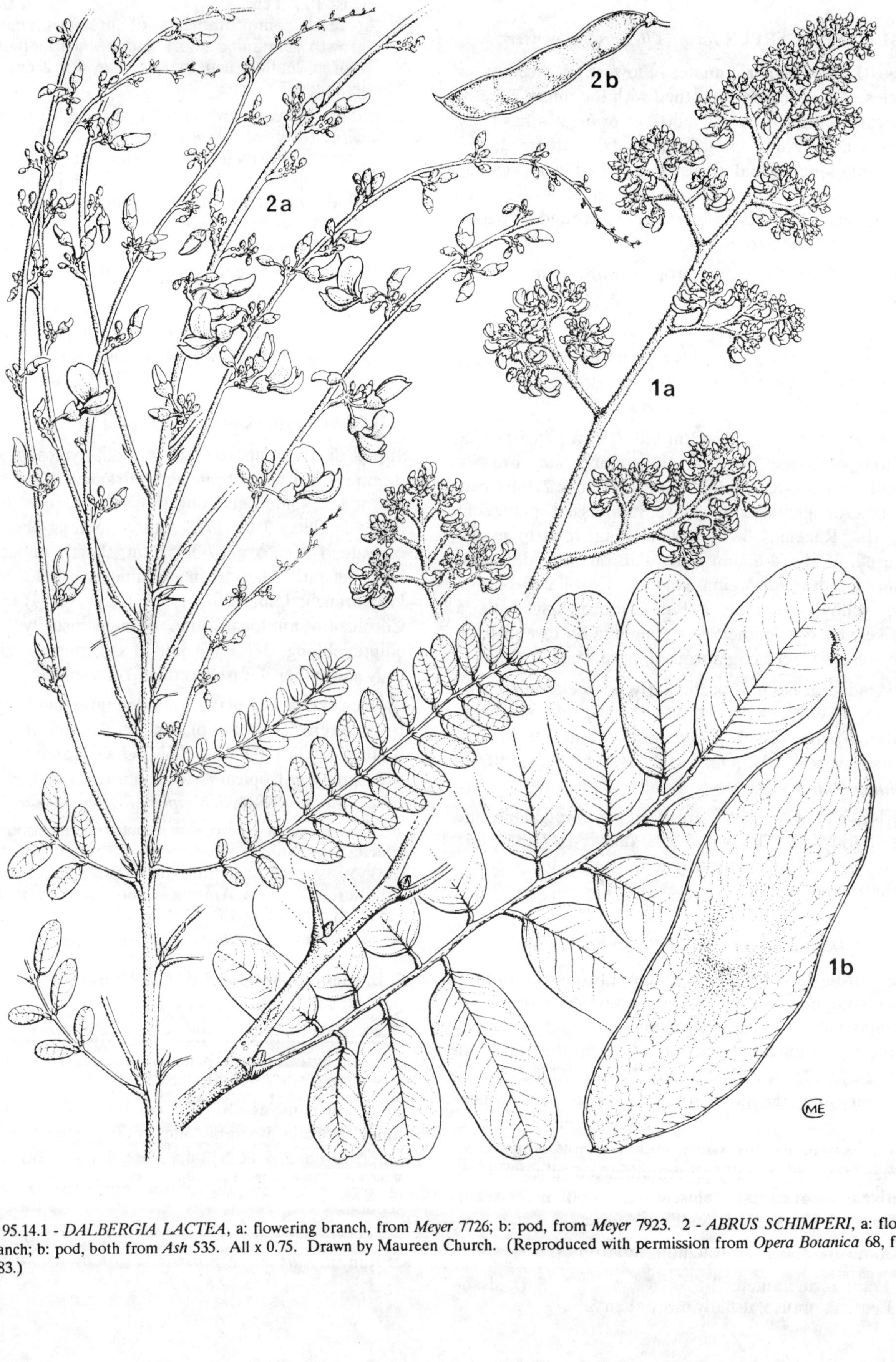

Figure 95.14.1 - *DALBERGIA LACTEA*, a: flowering branch, from *Meyer* 7726; b: pod, from *Meyer* 7923. 2 - *ABRUS SCHIMPERI*, a: flowering branch; b: pod, both from *Ash* 535. All x 0.75. Drawn by Maureen Church. (Reproduced with permission from *Opera Botanica* 68, fig. 16, 1983.)

3. **D. microphylla** *Chiov.* (1915)
- type: SD, *Riva* 404 (FT holo.).

Shrub 2-4 m tall. Branchlets sometimes several per node, mostly relatively short, tapered but scarcely spine-tipped. Leaves 2-3.5 cm long; leaflets 13-17(-25), oblong, up to 4-11 x 1.5-5 mm, pubescent. Inflorescences on cushion-shaped branchlets with tufts of leaves, 1.5-5 cm long. Calyx 4-4.5 mm long. Corolla 7-8 mm long, white. Stamens 9. Pod elliptic-oblong, 2-3 x 0.8-1.3 cm, papery, glabrous, narrowly winged along upper margin, inconspicuously veined, 1-seeded.

Deciduous bushland, rocky slopes; *c* 1200-1500 m. BA SD; Somalia, Kenya, N Tanzania. *Corradi* 3981; *J. De Wilde* 6640, 7323; *Mesfin* et al. 3230.

4. **D. lactea** *Vatke* (1879).
D. sciadendron Chiov. (1937) - type: GD, *Taschdjian* 356 (FT holo.).
D. macrothyrsus sensu Cufodontis, *Enum*. (1955), non Harms.

Small tree or shrub, 3-9(-25) m tall, ± evergreen, usually scandent, sometimes climbing with aid of coiled branchlets or peduncles. Leaves *c* 13-30 cm long; leaflets 13-17(-21), oblong or oblong-elliptic, 4-9 x 2-5.5 cm, glabrous to pubescent. Flowers many, congested on ultimate axes in *c* 5-15 cm long panicles; flowering branches often growing out with leaves ± suppressed and forming compound inflorescences up to 40-50 cm long. Calyx 3.5-5 mm long, usually pubescent with short dark hairs. Corolla 6.5-9(-10) mm long, white flushed mauve to purplish. Stamens (9-)10. Pod oblong, (8-)10-15 x 3-4.5 cm, papery, glabrous, laxly reticulately veined, 1(-2)-seeded. Figure 95.14.1.

Forest margins, bushland and grassland; 1350-2100 m. GD GJ WG SU AR IL KF GG SD BA; widespread in tropical Africa. *Meyer* 7726; *Pichi-Sermolli* 2273; *Burger* 2556; *Friis* et al. 1684.

5. **D. sissoo** *Roxb.* (1832).

Ornamental unarmed tree; leaf rhachis usually conspicuously zigzag with 3-5 leaflets, 3-7 x 2-6.5 cm, acuminate. Flowers yellowish-white in axillary panicles; calyx 3-4 mm long; corolla *c* 8 mm long. Pod coriaceous, 3-8 x 1-1.3 cm, 1-3-seeded.

Cultivated at least in SU(Awash National Park); native in SE Asia. *Gilbert* et al. 7386.

33. ABRUS *Adans.* (1763)

Verdcourt in *Kew Bull.* **24**: 235-253 (1970).

Woody subshrubs or lianes. Leaves paripinnate; leaflets numerous, opposite. Flowers small in fascicles forming axillary or terminal inflorescences. Calyx subtruncate to shortly 5-toothed. Pod linear or oblong, usually compressed.

About 17 species throughout the tropics.

1. Seeds almost globose, black and red; inflorescence thick, usually curved. **1. A. precatorius**
- Seeds much compressed, never black and red; inflorescence more slender and straighter. 2
2. Bracts and bracteoles equalling or exceeding the calyx; corolla very dark red or purple. **4. A. canescens**
- Bracts and bracteoles much shorter than the calyx; corolla cream to pale purplish or pink. 3
3. Shrub; pods woody, tapering at the apex, sparsely to mostly densely covered with tubercles. **2. A. schimperi**
- Twiner; pods thinner, rounded at the apex; pubescent but not tuberculate. **3. A. pulchellus**

1. **A. precatorius** *L.* (1767).

Woody climber 1-4.5 m tall. Leaflets 16-34, deciduous, oblong, obovate-oblong or ovate, 0.6-2.7 x 0.3-1 cm, glabrous above, sparsely appressed pubescent beneath. Inflorescences robust, usually curved, 2-7 cm long with subsessile flowers in dense fascicles; peduncles 1.5-6 cm long; bracts and bracteoles 0.5-1 mm long. Calyx 3 mm long, denticulate. Corolla yellow, white, pink or mauve, 0.9-1.5 cm long. Pod oblong, 2-4(-5) x 1-1.5 cm, with a hooked beak, with dense short reddish-brown appressed pubescence. Seeds (1-)3-7, red and black, almost globose, 5-7 x 4-5 mm, shining.

Subsp. **africanus** *Verdc.* in *Mitt. Bot. München* **7**: 328 (1970).

Pod mostly 2-3.5 cm long, somewhat rough outside.

Grassland, bushland, cultivations, often in rocky places; 500-1300 m. EE EW WG SU-KF(Gibe gorge) IL GG SD; widespread in tropical Africa, and also in the Seychelles, Madagascar, Mauritius, and introduced into Australia and the New World. *Pappi* 4491; *Ash* 900; *Chaffey* 961; *Haugen* 415.

The seeds are used for making necklaces, etc. They contain abrin, a highly toxic blood poison. All parts of the plant are reported to have medicinal properties. Elsewhere the shoots and leaves have been used as a vegetable and to extract a sweetener.

Subsp. *precatorius* occurs in tropical Asia and Australia.

2. **A. schimperi** *Hochst. ex Bak.* (1871)
- type: TU, *Schimper* 1552 (K holo., BM BR FT G L M P S W iso.).

Woody shrub 0.9-3.6 m tall. Leaflets 14-32, oblong-elliptic, glabrous above, appressed puberulous beneath. Inflorescences slender, 13-30 cm long with subsessile flowers in numerous fascicles; peduncle very short; bracts and bracteoles very short. Calyx 3 mm long, truncate or undulate. Corolla cream, yellow or blue and mauve, 1.1-1.4 cm long. Pod oblong to oblanceolate, 5.2-7.2 x 0.7-1.15 cm, rather thick and woody, appressed pubescent and usually markedly tuberculate.

Seeds brown with darker mottling, compressed, rounded-oblong or rhombic, 6-7.5 x 5-6 mm.

Subsp. **schimperi**.

Leaflets 20-32. Indumentum of inflorescence-rhachis and calyx golden. Fig. 95.14.2.

Thickets, bushland, woodland, forest edges, often on rocky edges; *c* 550-1700 m. EW TU GD WG IL KF GG; Sudan, Central African Republic and N Uganda. *Pappi* 7219; *Benedetto* 124; *Ash* 535; *Mesfin & Kagnew* 2373.

Subsp. *africanus* (Vatke) Verdc. with 14-20 leaflets and greyish indumentum on inflorescence-rhachis and calyx is known from Kenya, Tanzania, Zambia and Zimbabwe, and may well occur also in Ethiopia. *Donaldson-Smith* 191 (BM) from Bale, Sheik Husein may represent an additional subspecies of or a new species closely allied to *A. schimperi* (see Verdcourt loc. cit.). More material is required.

3. **A. pulchellus** *Thw.* (1859).

Woody herb or liane. Leaflets 12-34, oblong, glabrous to silky pilose. Flowers subsessile in fascicles on cushion-like reduced branchlets; bracts and bracteoles minute. Calyx 1.5-3 mm long, subtruncate to toothed. Corolla white to purple, 0.8-1.3 cm long. Pods oblong or linear-oblong, compressed, smooth, sparsely to densely appressed pubescent. Seeds greyish or reddish brown, sometimes mottled, compressed, oblong or pyriform, 4-6 x 3-4.5 mm.

Subsp. **tenuiflorus** *(Benth.) Verdc.* in *Kew Bull.* **24**: 250 (1970);
A. tenuiflorus Benth. (1859)

Twiner up to 6 m tall. Leaflets 12-26, 1.4-4.5 x 0.7-1.8 cm; indumentum on stem and leaves usually sparse. Pods 2.2-4.5(-6) x 0.8-1.3 cm.

Gallery forest; *c* 1300 m. WG; Sudan, Uganda, Tanzania, West Africa, Angola, Zaire, Zambia, Mozambique, also in South America. *W. de Wilde* 8924.

A. pulchellus is a pantropical complex divided into 5 subspecies by Verdcourt.

4. **A. canescens** *Bak.* (1871).

Woody climber or subshrub 0.5-3(-6) m tall. Leaflets 20-28, narrowly oblong or elliptic-oblong, (0.6-)1-2(-3) x 0.2-0.7(-1.2) cm, densely pubescent, particularly beneath. Flowers sessile in usually well-separated verticillate fascicles; peduncles 3-8 cm long; bracts and bracteoles 3-6 mm long, equalling or exceeding the calyx. Calyx 2-3 mm long, denticulate. Corolla very dark red or purple, 0.8-1.5 cm long. Pod oblong, (2.4-)4-5.5 x 0.8-1.2 cm, pubescent. Seeds grey-brown to black, compressed-ellipsoid, 4-5.5 x 3-4 mm.

Gallery forest; 1300 m. WG; West Africa, Central African Republic, Uganda, Kenya, Tanzania, Zaire, Burundi and Zambia. *Gilbert & Thulin* 741.

34. **MILLETTIA** *Wight & Arn.* (1834)

Trees, shrubs or lianes. Leaves imparipinnate; leaflets opposite, rarely subalternate. Flowers large in panicles, often contracted to pseudoracemes. Corolla white, pink or violet; standard usually silky outside. Vexillary stamen free at first, but usually becoming adherent in the mature flower. Disc often present between stamens and ovary. Ovary pubescent; ovules 3-11, ± evenly distributed throughout the ovary. Pod dehiscent, usually flat, 2- or more-seeded.

Some 200 species in tropical and subtropical Africa, Asia and Australasia.

M. ferruginea *(Hochst.) Bak.* (1871);
Berrebera ferruginea Hochst. (1846) - types: TU, *Schimper* I:279 (B syn. destr., BM K S isosyn.) & GD, *Schimper* II: 536 (B syn. destr.)

Tree. Leaves up to 40 cm long; leaflets up to 27, oblong-acuminate, up to 5-9 x 1-2.5 cm, ± pubescent on both sides or subglabrous above. Pseudoracemes up to 30 cm long including a peduncle of 5-10 cm. Calyx 6-8 mm long, densely pubescent with golden to blackish brown hairs; lobes *c* 1 mm long. Corolla (17-)20-30 mm long, violet; standard silky outside. Disc tubular 1-1.5 mm long. Pod flat, up to 27 x 3 cm, brown-pubescent first, then glabrescent, 5-10-seeded. Figure 95.15.1.

1. Hairs on lower surface of leaflets copious, spreading, 0.5-0.6 mm long; those on inflorescence and calyx golden brown; pedicels up to 9 mm and standard up to 24 mm long. subsp. **ferruginea**
- Hairs on lower surface of leaflets sparse, appressed, 0.2-0.3 mm long; those on inflorescence and calyx blackish brown; pedicels up to 12 mm and standard up 30 mm long. subsp. **darassana**

Subsp. **ferruginea**.

Upland forest; 1000-2500 m. TU GD GJ SU WG HA IL; not known elsewhere. *Schimper* 526; *Pichi-Sermolli* 502; *W. de Wilde* 6744; *Friis* et al. 56.

Subsp. **darassana** *(Cuf.) Gillett* in *Kew Bull.* **15**: 35 (1961);
M. darassana Cuf. (1958) - type: SD, *Kuls* 188 (FR holo. not seen).

Rain forest, ground water forest, woodland; 1600-2500 m. WG SU HA BA IL KF SD; not known elsewhere. *Gillett* 14620, 14870; *Mooney* 5684.

Typical subsp. *ferruginea* is ± restricted to N Ethiopia (N of 11°30´N), while typical subsp. *darassana* occurs in Sidamo. Most specimens from C and W Ethiopia show some mingling of the characters of the two forms.

This species has also been reported from the Sudan but its occurrence there has not been confirmed.

This is an important shade tree for coffee grown by peasant farmers and Peasants' Associations. The wood is used to make tool handles and other small household utensils. The seeds contain a narcotic and are ground and the flour used to stun fish.

35. **CRAIBIA** *Harms & Dunn* (1911)

Gillett in *Kew Bull.* **14**: 189-197 (1960) & *Fl. Trop. E. Afr.* (1971).

Trees or shrubs with golden brown or black pubescence. Leaves imparipinnate; leaflets usually alternate. Flowers large, in racemes or panicles. Corolla white, or speckled with pink or purple, glabrous or almost so. Vexillary stamen essentially free. Disc absent. Ovary glabrous or with long hairs falling ± quickly; ovules 2-6 in the distal part of the ovary. Pod dehiscent, flat, tapering to the base, markedly asymmetric, 1-2(-3)-seeded.

Ten species, all in tropical Africa.

1. Leaflets 5-7, elliptic-lanceolate; ovary glabrous or nearly so; inflorescence a many-flowered panicle. **1. C. laurentii**
- Leaflets (in Flora area) 3-4, ovate; ovary pubescent below; inflorescence a raceme which is (in Flora area) branched near the base. **2. C. brevicaudata**

1. C. **laurentii** *(De Wild.) De Wild.* (1920).

Tree. Leaflets 5-7, elliptic-lanceolate, acuminate, up to 17 x 5.5 cm. Flowers in many-flowered panicles. Calyx appressed brown pubescent; tube *c* 1.5 mm long. Corolla white, 12-15 mm long. Ovary glabrous or with a few long hairs on the upper suture. Pod up to 12 x 3.5 cm. Seeds 1(-2), up to 20 x 15 x 8 mm.

Dry forest at fairly low altitudes. GG (Omo valley); S Sudan, Uganda, N Kenya, Zaire and E Cameroun. *Fukui* 280.

2. C. **brevicaudata** *(Vatke) Dunn* (1911).

Tree. Leaflets 3-7 (in Flora area 3-4), ovate (in Flora area) to oblong-lanceolate, acuminate, up to 11 x 4 cm (in Flora area up to 6 x 4 cm). Flowers in racemes which may be branched near the base. Calyx pubescent with brown and sometimes also white hairs; tube 3-4 mm long; lobes *c* 2 mm long. Corolla white, 12-19 mm long. Ovary pubescent below, particularly along sutures. Pod 5.5-10.5 x 1.8-2.5 cm (in Flora area). Seeds 1-4, up to 12.5 x 8 x 2.5 mm.

Acacia-Commiphora woodland; *c* 1300 m. BA; E and S tropical Africa. *J. De Wilde* 7325; *Thulin, Hunde & Tadesse* 3735.

The only known Ethiopian locality for this variable species is the limestone gorge at Sof Omar. The material does not quite match other material of the species, but is nearest to subsp. *brevicaudata* which is otherwise known from Kenya and Tanzania.

36. **TEPHROSIA** *Pers.* (1807), *nom. conserv.*

Gillett in *Kew Bull.* **13**: 111-132 (1958) & *Fl. Trop. E. Afr.* (1971).

Annual or perennial herbs, or shrubs. Leaves usually imparipinnate, less often simple or 1-foliolate, rarely palmate; lateral nerves of leaflets usually closely parallel. Flowers in upper leaf-axils or in pseudoracemes. Calyx with the upper 2 teeth often ± united. Corolla usually reddish-purple; standard pubescent outside. Vexillary filament lightly attached to the others or, less often, free. Disc usually present between stamens and ovary. Pod linear or oblong, ± flattened, thinly leathery, dehiscing, often explosively, the valves then becoming twisted. Seeds 1 to many.

Large genus throughout the tropics and in many warm temperate regions.

Several species are recognised as good browse plants for domestic and wild animals. However, some species are also used as fish poison.

1. Leaves 1-foliolate. 2
- Leaves, or most of them, compound. 4
2. Flowers in leaf-axils only; calyx *c* 3 mm long; leaves linear, up to 5(-7) mm wide. **1. T. schweinfurthii**
- Flowers in terminal or leaf-opposed pseudoracemes, often also in upper leaf-axils; calyx *c* 6-8 mm long; leaves wider. 3
3. Lateral calyx-lobes not above 3 mm long, about as long as tube; aril minute. **6. T. hildebrandtii**
- Lateral calyx-lobes over 3 mm long, more than twice as long as tube; aril large. **7. T. holstii**
4. Style glabrous on both sides for most of its length, though often pubescent at the base and penicillate* at the tip. 5
- Style pubescent, at least on one side, for most of its length. 20
5. Flowers in leaf-axils only, never in pseudoracemes. 6
- Flowers in terminal or leaf-opposed pseudoracemes, often also in upper leaf-axils. 9
6. Slender shrub; flowers 1-2 together from clusters of leaves on short shoots. **5. T. obbiadensis**
- Annual or perennial herbs; flowers not from clusters of leaves on short shoots. 7
7. Leaf rhachis hardly half as long as the terminal leaflet; leaflets 3-7, usually 5. **2. T. pentaphylla**
- Leaf rhachis usually longer than the terminal leaflet; leaflets usually 7 or more. 8
8. Style not twisted, ribbon-shaped, 1.5-2.5 mm long. **3. T. subtriflora**
- Style twisted, tapering, 2.5-3.5 mm long. **4. T. uniflora**
9. Ovules 1-4; seeds 1-2. 10
- Ovules 5 or more; seeds 3 or more. 11

* penicillate = with a tuft of hairs shaped into a point like a pencil.

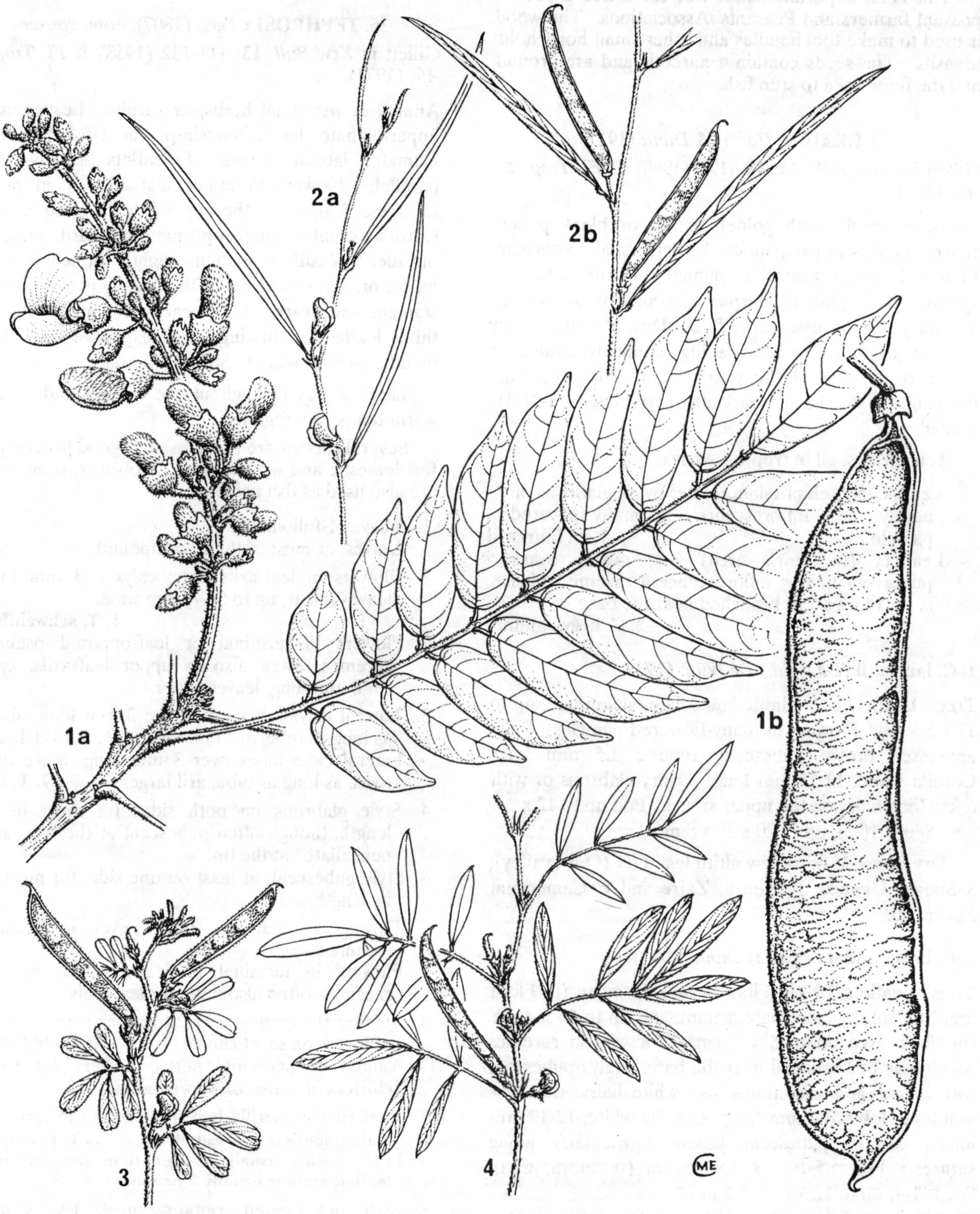

Figure 95.15.1 - *MILLETTIA FERRUGINEA*, a: flowering branch, from *IECAMA* RS-276; b: pod, from *Gillett* 14870. 2 - *TEPHROSIA SCHWEINFURTHII*, a: flowering branch; b: fruiting branch, both from *Gilbert* et al. 3093. 3 - *T. PENTAPHYLLA*, a: branch with flowers and pods, from *Andrews* 88. 4 - *T. SUBTRIFLORA*, a: branch with flowers and pods, from *Andrews* 75. All x 75. Drawn by Maureen Church. (Reproduced with permission from *Opera Botanica* 68, fig. 17, 1983.)

10. Leaflets in (3-)5-6(-8) pairs; hairs on stems up to 2 mm long; pod with the veins of each valve strong near the margins, diffuse and weak at the centre (visible when valve held in front of a strong light). **18. T. polyphylla**
- Leaflets in (2-)4-5(-6) pairs; hairs on stems rarely over 1 mm long; pod with strongly marked veins running diagonally across each valve, not weakening or breaking up at the centre (visible when valve held in front of a strong light). **19. T. nubica**

11. Calyx-tube 3 mm long, about as long as, or a little shorter than the lower tooth of the calyx. **8. T. emeroides**
- Calyx-tube less than 3 mm long, or, if longer, then much shorter than the lower tooth of the calyx. 12

12. Hairs on calyx and pod-sutures, or at least some of them, brown or black. 13
- Hairs on calyx and pod-sutures white or yellowish. 15

13. Aril conspicuous, clasping one corner of the seed; leaflets linear, up to 4(-7) mm wide; pod *c* 3 mm wide. **9. T. linearis**
- Aril minute, inconspicuous; leaflets and pods wider. 14

14. Petioles 3-10 mm long; standard 10-14 mm long; inflorescences dense and short. **10. T. dichroocarpa**
- Petioles practically obsolete; standard 5-7 mm long; inflorescences lax and few-flowered. **11. T. fulvinervis**

15. Pod with a dense felt of hairs, which are often over 1 mm long, obscuring the surface. **17. T. villosa**
- Pod with appressed or spreading hairs, which are rarely over 0.5 mm long and not dense enough to hide its surface. 16

16. Standard *c* 4.5 mm long; leaflets up to 4 mm wide; pod 20-27 mm long. **12. T. gracilipes**
- Standard 6-11 mm long; leaflets usually wider and pod longer. 17

17. Distance between centres of adjacent seeds not more than width of pod; inflorescences rarely more than 5-6-flowered. 18
- Distance between centres of adjacent seeds greater than width of pod; inflorescences usually more than 6-flowered. 19

18. Leaflets 5-7; standard 9-11 mm long. **13. T. lortii**
- Leaflets 7-13, usually 9-11; standard under 8 mm long. **14. T. pumila**

19. Leaflets (9-)11-17(-21). **15. T. purpurea**
- Leaflets 5-9. **16. T. apollinea**

20. Keel pubescent, at least near the lower joined margin; standard 20 mm or more long; pod 10 mm or more wide. **27. T. vogelii**
- Keel glabrous or, if not, other characters not as above. 21

21. Seeds transverse; pod straight or curved slightly downwards near the tip. 22
- Seeds longitudinal; pod straight or curved slightly upward near the tip. 23

22. Standard 12-18 mm long; style *c* 6 mm long; stems rusty villose. **25. T. nana**
- Standard c 10 mm long; style *c* 3 mm long; stems white pubescent. **26. T. humilis**

23. Leaflets 17-31, up to 11 cm long, about 10 times as long as wide; racemes long-pedunculate; bracts ovate-lanceolate, falling soon; lateral calyx-lobes very short; seeds 13-16. **20. T. bracteolata**
- Leaflets fewer or wider than above, or both, not above 6 cm long; other characters various. 24

24. Inflorescence long-pedunculate, few-flowered; bracts ± persistent; leaflets 5-19. 25
- Inflorescence short-pedunculate, rather dense and many-flowered; bracts falling soon; leaflets 9-27. 26

25. Pods 4-5 mm wide, very shortly pubescent, 13-19-seeded; calyx white pubescent; leaflets 5-19, sparsely pubescent above. **21. T. reptans**
- Pods *c* 6 mm wide, densely pubescent, brownish along the sutures, 9-14-seeded; calyx villose; leaflets 7-13, glabrous or almost so above. **22. T. hochstetteri**

26. Inflorescence not markedly interrupted, with numerous nodes; leaflets often 4-6 cm long; style *c* 5 mm long; hairs on calyx and pod not dark brown or black. **23. T. elata**
- Inflorescence markedly interrupted, with fewer nodes; leaflets usually under 3 cm long; style 7-8 mm long; hairs on calyx and pod long, often dark brown or black. **24. T. interrupta**

1. **T. schweinfurthii** *Deflers* (1895).
T. simplicifolia Franch. (1882), *nom. illegit., non* F. Muell. ex Benth. (1864); *T. franchetii* Hutch. & Bruce (1941).

Perennial up to 50 cm tall. Leaves 1-foliolate; petiole (1-)2-5 mm long; leaflet linear to narrowly oblong, up to 90 x 5(-7) mm, glabrous or sparsely strigulose above, usually appressed pubescent beneath. Flowers red, 1-2(-3) together in the leaf-axils. Calyx pubescent, *c* 3-4 mm long with ovate-acuminate lobes about as long as the tube. Standard pubescent, *c* 8 mm long. Style glabrous, 2-2.5 mm long, with a few short hairs at the tip. Pod slightly upcurved, *c* 45 x 4 mm, usually minutely appressed pubescent with 6-7 seeds, *c* 4 x 1.5 mm, wrinkled. Fig. 95.15.2.

Dry bushland, often on limestone; 950-1650 m. SU SD; N Somalia, Djibouti, N Kenya, Yemen. *Gilbert & Thulin* 988; *Gilbert & Getachew* 3093; *Friis* et al. 2756.

T. heterophylla Vatke (1878) may be an older name for this species, but its application remains somewhat uncertain as the type, *Hildebrandt* 835a from Mt Ahl in Somalia, is lost. According to the description it differs from *T. schweinfurthii*, which is also known from Mt Ahl, in having terminal pedunculate racemes instead of flowers in leaf-axils. Records of *T. heterophylla* from Ethiopia by Engler in *Sitzungsber. Königl. Preuss. Akad. Wiss. Berlin* 10:383 (1904) and Cufodontis *Enum.*: 281 (1955) seem to be based on a no longer extant Ellenbeck collection, a drawing of which is however preserved in BM. This agrees completely with *T. schweinfurthii*.

A collection from near Negele in SD (*Friis* et al. 913) differs in having a dense ± rusty-red pubescence with spreading hairs on stems, calyces, pods and undersides of leaves. It is probably just a form of *T. schweinfurthii*.

2. **T. pentaphylla** *(Roxb.) G. Don* (1839).
T. decidua Steud. ex A. Rich. (1847) - types: Eritrea, *Schimper* II:1011 (P syn., BM isosyn.), Petit s.n. (P syn.).
T. similis Chiov. (1932).
T. senticosa sensu Cuf. *Enum.* (1955) in part, non (L.) Pers.

Annual or short-lived perennial, up to 80 cm tall. Leaf rhachis, including a petiole of *c* 2 mm, up to 1 cm long. Leaflets 3-7, usually 5, cuneate-oblanceolate, up to 22(-50) x 7(-13) mm, glabrous above, silvery beneath. Flowers red, 1-3 together in the leaf-axils. Calyx silvery, 5-6 mm long with triangular-acuminate lobes, 3-4 mm long. Standard yellowish silver, *c* 6 mm long. Style glabrous, not twisted, *c* 2 mm long. Pod somewhat upcurved towards the tip, 4-4.5 x 0.4 cm, pubescent, brownish on the sutures. Seeds 7-8, 3-4 x 2-2.5 mm. Fig. 95.15.3.

Grassland, woodland, gravelly hillsides; up to 1800 m. EE AF EW SU HA BA GG; Sudan, S Somalia, N Uganda, Kenya, Tanzania, Mozambique; also in Arabia and S India. *Gilbert, Thulin & Getachew* 246; *Aweke & Gilbert* 1028; *Burger* 2204.

3. **T. subtriflora** *Hochst.* ex *Bak.* (1871)
- type: *Schimper* 2312 (K holo., BM P iso.).
T. encoptosperma Schweinf. (1896) - types: Eritrea, *Schweinfurth & Riva* 319 (B syn. destr., K isosyn.) & 1117 (B syn. destr.).
T. apollinea (Del.) Link var. *subuniflora* A. Terracc. in *Ann. Ist. Bot. Roma* **5**: 109 (1894) - type: Eritrea, *Terracciano* 1738 (FT holo.).
T. graminifolia Chiov. (1915).
T. sulphurea Chiov. (1929).

Annual or short-lived perennial. Leaf rhachis up to 7 cm long, including a petiole of up to 2.5 cm. Leaflets 5-13, narrowly oblong, oblanceolate or broadly linear, up to 40 x 7 mm, glabrous above, appressed silky beneath. Flowers pale pinkish-purple or almost white, 2-6 together in the leaf-axils. Calyx pubescent, 3-6.5 mm long with triangular-acuminate lobes, 2-5 mm long. Standard pubescent, 5-8 mm long. Style glabrous, 1.5-2.5 mm long, up to 0.3 mm wide, not twisted, with a few short hairs at the tip. Pod curved, up to 4.5 x 0.4 cm, pubescent with 5-10 seeds, mottled, *c* 2.7 x 1.8 x 1 mm. Fig. 95.15.4.

Acacia-Commiphora woodland or bushland; up to *c* 1250 m. EE AF EW TU SU HA SD; Sudan, Somalia, Kenya and scattered in some other dry parts of tropical Africa, also in Madagascar, Arabia, Pakistan, India, Burma. *Gilbert & Thulin* 72; *Frahm Leliveld* 62317; *Gilliland* 4064; *Friis* et al. 2876.

4. **T. uniflora** *Pers.* (1807).
T. vicioides A. Rich. (1847), nom illegit., non Schlecht. (1838); *T. quartiniana* Cuf. (1955), nom invalid. - type: TU, *Quartin Dillon* s.n. (P holo.).
T. anthylloides Hochst. ex Webb (1850) - type: *Schimper* III:1614 (K syn., BM S UPS isosyn).
T. uniflora var. *parvifolia* Bak. f., *Leg. Trop. Afr.*: 184 (1926) - type: EW, *Pappi* 358 (K holo.).
T. transjubensis Chiov. (1932).

Perennial, sometimes softly woody, up to 1 m tall. Leaf rhachis *c* 2.5-4.5 cm long, including a petiole of 3-8 mm. Leaflets (3-)5-7(-9), oblanceolate to elliptic or obovate, up to 3 x 1.3 cm, glabrous or pubescent above, silvery beneath. Flowers pink, 2-3 together in the leaf-axils. Calyx pilose, (3-)4-8 mm long with triangular-acuminate lobes, up to 6 mm long. Standard pubescent, (5-)8-10 mm long. Style glabrous, 2.5-3.5 mm long, twisted, penicillate at the tip. Pod upcurved towards the tip, (3-) 4.5-7 x 0.4 cm, pubescent with (6-)9-13 seeds, *c* 2.5 x 2 mm.

Acacia-Commiphora woodland or bushland, dry grassland, roadsides, rocky places; 0-2600 m. EE AF EW TU GD(Sudan border) SU BA HA KF GG SD; widespread in drier parts of tropical Africa south to N Transvaal. *Gilbert & Thulin* 178; *Gilbert & Getachew* 2751; *Bally* 6894.

The African material belongs to subsp. *uniflora*, but several specimens from NE Africa approach subsp. *petrosa* (Blatt. & Hall.) Gillett & Ali which occurs in Arabia, Pakistan and India.

The name *T. quartiniana* (= *T. vicioides*) has been used for ± spreadingly pubescent plants with small obovate or obcordate leaflets, 5-6 mm long standard and up to 35 mm long, 6-9-seeded pods. Characteristic specimens have been seen from Eritrea and Tigray in Ethiopia, as well as from Sudan, Egypt, Mali, Niger and Arabia, but intermediates with *T. uniflora* occur in Ogaden (HA), Somalia and Socotra and the plant would probably best be treated as only a variety of *T. uniflora*. At present, however, no valid and legitimate name seems to exist for it.

5. **T. obbiadensis** *Chiov.* (1929).

Slender shrub. Leaf rhachis up to 1.5 cm long, including a petiole of up to 1 cm. Leaflets 3-7, oblanceolate, up to 17 x 6 mm, ± densely pubescent on both sides or glabrous above. Flowers purple, 1-2 together from clusters of leaves on short shoots. Calyx appressed pubescent, 3.5-5 mm long with short triangular lobes. Standard broadly ovate, pubescent, *c* 12 mm long. Style glabrous, 4-5 mm long, densely penicillate at the tip. Pod slightly upcurved towards the tip, up to 70 x 4.5 mm, pubescent, *c* 8-seeded. Fig. 95.16.1.

Dry bushland and thickets; *c* 500-1000 m. HA; Somalia. *Glover & Gilliland* 394; *Hemming* 1479; *Ashall* CA/8/58.

6. **T. hildebrandtii** *Vatke* (1879).
T. boranensis Chiov. (1939) - type: SD, *Cufodontis* 310 (FT holo.).

Short-lived perennial up to 1 m tall. Leaves 1-foliolate; leaflet narrowly oblong or oblanceolate, 2.5-6 x up to 1.4 cm, glabrous above, silky strigose beneath. Flowers reddish-purple, 2-3 together in the upper leaf-axils and sometimes also in terminal pseudoracemes. Calyx golden silky, 6-7 mm long, with ovate-acuminate lobes about as long as the tube. Standard golden silky, 11-14 mm long. Style glabrous, *c* 5 mm long, straight apart from a sharp bend at the base, with a few short hairs at the tip. Pod slightly upcurved, 5-5.5 x 0.35 cm, silky pubescent, brown at the sutures with *c* 8 seeds, 3-4 x 1.6-2.2 mm; aril minute.

Upland grassland and bushland; 1600-1900 m. SD; Kenya, Tanzania. *Cufodontis* 310; *Thesiger* 2017; *Gilbert & Jefford* 4388; *Sebsebe* 1229.

7. **T. holstii** *Taub.* (1895).
T. rivae Harms (1897) - type: SD, *Riva* 1222 (FT holo.).

Perennial or annual herb up to 80 cm tall. Leaves 1-foliolate; leaflet elliptic-oblong to lanceolate, up to 7 x 2.5 cm, pilose on both sides. Flowers orange or brick red, in short dense terminal pseudoracemes, often also in the uppermost leaf-axil. Calyx hirsute, *c* 8 mm long with lanceolate-acuminate lobes more than twice as long as tube. Standard tomentose outside, *c* 9 mm long. Style glabrous, *c* 6 mm long, curved, penicillate at the tip. Pod slightly upcurved, *c* 5 x 0.5 cm, brown pubescent with 10-11 seeds; aril large. Fig. 95.16.2.

Upland grassland and forest margins, swampy ground; 1500-2000 m. KF SD; Kenya, Tanzania. *Gillett* 14580; *Mooney* 8635; *Stewart* C-40.

8. **T. emeroides** *A. Rich.* (1847)
- type: *Petit* s.n. (P holo., K iso.).
T. ustulata Taub. ex Harms (1897) - type: SD, *Riva* 556 (FT holo.).

Softly woody erect perennial 0.5-2.5 m tall. Leaf rhachis up to 8 cm long, including a petiole of up to 13 mm. Leaflets 5-13(-17), narrowly oblong-lanceolate, up to 3 x 0.7 cm, glabrous above, appressed strigulose beneath. Flowers pink or purple in rather lax terminal pseudoracemes and also in upper leaf-axils. Calyx appressed brown-strigulose, *c* 6-7 mm long with broadly triangular lobes, the lowest 3-4 mm long. Standard appressed golden-pubescent, 10-13 mm long. Style glabrous, *c* 5 mm long and 0.7 mm wide, straight except for a sharp bend at the base, not twisted, not penicillate. Pod slightly upcurved, up to 7 x 0.5 cm, appressed brown-strigulose with 9-11 seeds, *c* 4 x 2.5 mm. Fig. 95.16.3.

Upland wooded grassland and thickets; 1400-2800 m. GD WU SU AR HA BA GG SD; N Somalia, N Uganda, Kenya. *Thulin* 1345; *Gilbert, Thulin & Getachew* 363; *Burger* 3166.

The leaves and young twigs are used to make a local perfume.

9. **T. linearis** *(Willd.) Pers.* (1807).

Erect annual or short-lived perennial herb up to 1.5 m tall. Leaf rhachis up to 5 cm long, including a petiole of 2-7 mm. Leaflets 5-15, usually linear, 2.5-4(-7) x 0.15-0.4(-0.6) cm, usually glabrous above, densely silky beneath. Flowers orange or brick red in terminal and axillary lax pseudoracemes. Calyx brown silky, 2-6.5 mm long with triangular-acuminate 1-5 mm long lobes. Standard silky outside, 5-12 mm long. Style glabrous, 1.6-3 mm long, curved, penicillate at the tip. Pod upcurved near the tip, *c* 5-6 x 0.3 cm, brownish pubescent with 9-12 seeds, *c* 2.5 x 1.5 mm with a large yellowish aril. Fig. 95.16.4.

Grassland, woodland, rocky slopes; 550-2350 m. WG SU IL KF GG SD; widespread in tropical Africa and south to Natal, also in Madagascar. *Gilbert, Thulin & Getachew* 221, 351; *Vàtova* 603; *Jansen* 4980a.

Elsewhere the branches have been used to make brushes.

10. **T. dichroocarpa** *Steud.* ex *A. Rich.* (1847) emend. *Engl.* (1892)
- type: TU, *Schimper* I:102 (P holo., BM K S UPS iso.).
T. brunnea Bak. (1871); type: TU, *Schimper* 292 (K holo., BM iso.).

Shrubby perennial. Leaf rhachis up to *c* 8 cm long, including a petiole of 3-10 mm. Leaflets 7-15, elliptic-oblong, up to 35 x 10 mm, sparsely pilose above, appressed pubescent beneath. Flowers in pedunculate short pseudoracemes. Calyx brownish pubescent, 5-6.5 mm long with lanceolate lobes, the lowest *c* 4 mm long. Standard brownish pubescent outside, 10-14 mm long. Style glabrous, curved, *c* 3 mm long, penicillate at the tip. Pod slightly upcurved towards the tip, *c* 45 x 4-5 mm, densely tomentose, darker along the sutures.

Rocky places, roadsides; 2250-2400 m. TU GD GJ; not known elsewhere. *Gilbert & Getachew* 3065; *Schimper* 797; *Chiovenda* 2241.

11. **T. fulvinervis** *Hochst.* ex *A. Rich.* (1847)
- types: TU, *Schimper* III:1939 (P syn., BM isosyn.), *Quartin Dillon* (P syn.).

Annual up to 1 m tall; stems appressed brownish pubescent. Leaf rhachis up to *c* 7 cm long, petiole practically obsolete. Leaflets 9-17, linear-oblong, up to 70 x 9 mm, glabrous above, appressed silky beneath with brownish midrib. Flowers in axillary, lax, few-flowered, pedunculate pseudoracemes; bracts filiform. Calyx brown-strigose, 3-5 mm long with acuminate lobes, the lowest 2-4 mm long. Standard appressed brownish pubescent outside, 5-7 mm long. Style glabrous, 1-2 mm long, curved, penicillate at the tip. Pods spreading, upcurved, *c* 50-65 x 4.5-5.5 mm, densely appressed pubescent,

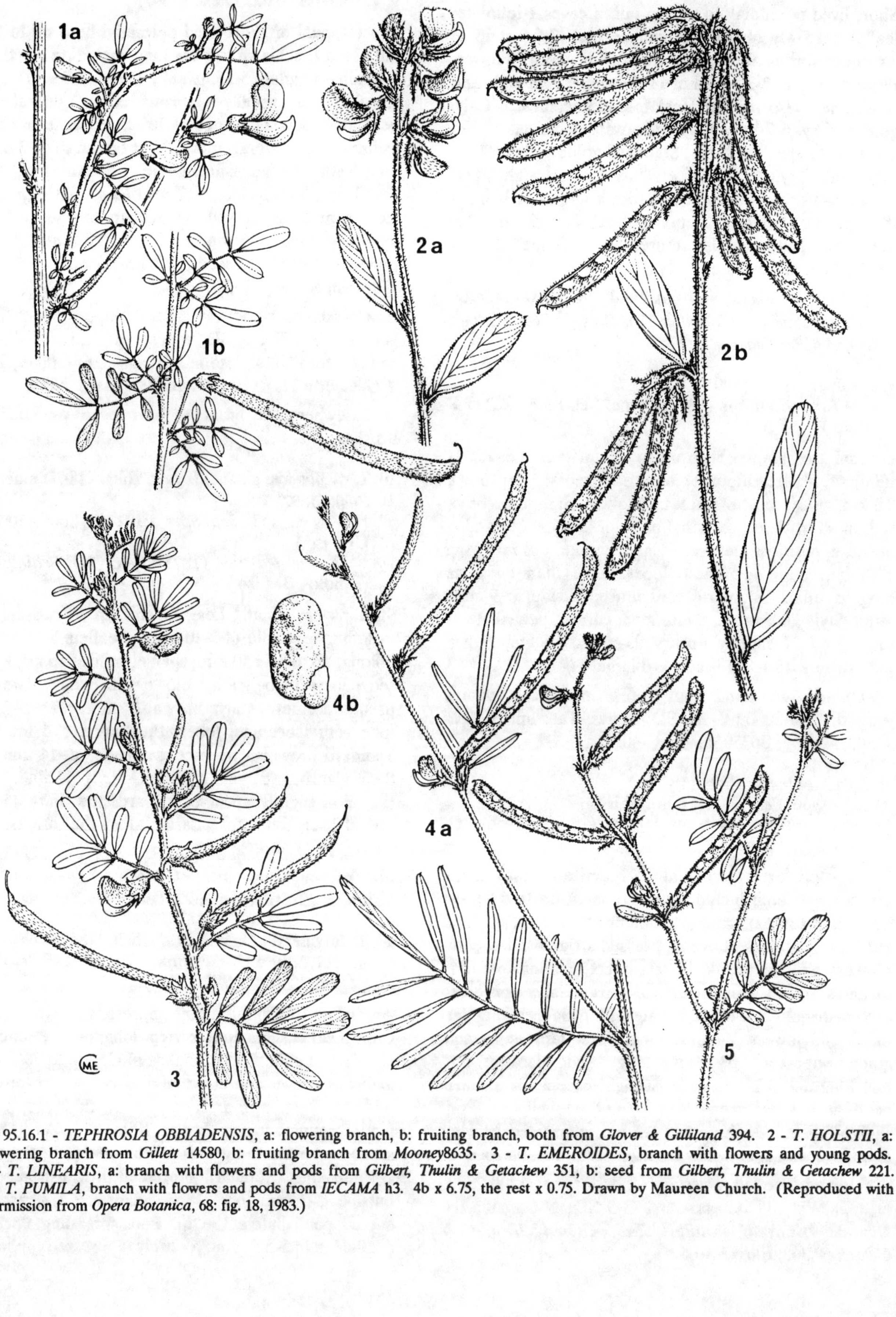

Figure 95.16.1 - *TEPHROSIA OBBIADENSIS*, a: flowering branch, b: fruiting branch, both from *Glover & Gilliland* 394. 2 - *T. HOLSTII*, a: flowering branch from *Gillett* 14580, b: fruiting branch from *Mooney*8635. 3 - *T. EMEROIDES*, branch with flowers and young pods. 4 - *T. LINEARIS*, a: branch with flowers and pods from *Gilbert, Thulin & Getachew* 351, b: seed from *Gilbert, Thulin & Getachew* 221. 5 - *T. PUMILA*, branch with flowers and pods from *IECAMA* h3. 4b x 6.75, the rest x 0.75. Drawn by Maureen Church. (Reproduced with permission from *Opera Botanica*, 68: fig. 18, 1983.)

brownish along the sutures with 8-10 seeds, without aril.

Grassland, woodland; *c* 1100-1700 m. EW TU GD SU(Gibie valley); not known elsewhere. *Ash* 2736; *Eriksson* 115; *Handlos* 95.

12. **T. gracilipes** *Guill. & Perr.* (1832).

Erect slender annual up to 30(-50) cm tall, appressed pubescent. Leaf rhachis up to *c* 1.5-4 cm long, including a petiole of up to 5 mm. Leaflets 3-11(-13), linear to oblanceolate, up to 25(-35) x 4(-6) mm, glabrous or with scattered hairs above, appressed silky beneath. Flowers purplish in lax, few-flowered pseudoracemes. Calyx appressed pubescent, 2.5-4 mm long with subulate lobes, the lowest *c* 2-3 mm long. Standard appressed pubescent outside, *c* 4.5 mm long. Style glabrous, *c* 1.5 mm long, curved, penicillate at the tip. Pod slightly upcurved at the tip, almost straight, 20-35 x 3(-4) mm, shortly and spreadingly pubescent with 5-9 seeds, aril minute.

Grassland and woodland; *c* 900-1600 m. EW TU GD; Sudan, Mauritania, Upper Volta, Senegal. *Schimper* 1999; *Schweinfurth* 1885; *J. De Wilde* 7122.

The number of leaflets as well as the length of the leaf rhachis is more variable in Ethiopia than in West Africa.

13. **T. lortii** *Bak. f.* (1926).

Perennial straggling herb up to 70 cm long. Leaf rhachis up to 4(-6) cm long, including a petiole of up to 1(-2) cm. Leaflets 3-7, cuneate-elliptic or obovate, up to 25 (-45) x 12(-23) mm, pubescent on both sides. Flowers purple in short lax terminal pseudoracemes and in the upper leaf-axils. Calyx pubescent, 4-5 mm long with triangular-acuminate 2-3 mm long lobes. Standard grey pubescent outside, 9-11 mm long. Style glabrous, 3-4 mm long, twisted, penicillate at the tip. Pod upcurved near the tip, 4-5 x 0.4-0.45 cm, shortly and stiffly pubescent with 9-12 seeds, rhombic, *c* 2.5-3 mm long.

Dry bushland; 1350 m. HA; N Somalia, Kenya, N Tanzania. *Gilbert & Thulin* 122.

14. **T. pumila** *(Lam.) Pers.* (1807).

T. vicioides A. Rich. var. *inflexa* Chiov. in *Ann. Ist. Bot. Roma* **8**:422 (1908), pro majore parte; *T. quartiniana* Cuf. var. *inflexa* (Chiov.) Cuf., Cuf. *Enum., Zweiter Nachtrag*: 24 (1969), nom. invalid. - types: Eritrea, *Terracciano & Pappi* 1931 & 2654 (FT syn.) & *Pappi* 7082 (FT syn.) & *Terracciano* 1740 (FT syn. = *T. uniflora*) & TU?, Gapdia, *Schimper* s.n. (FT syn., S isosyn.).

Annual or short-lived perennial with procumbent or straggling branches. Leaf rhachis up to 4 cm long including a petiole of 3-10 mm. Leaflets 7-13, cuneate-oblong or oblanceolate, up to 20 x 8 mm, glabrous or pubescent above, pilose beneath. Flowers white, pale pink or purplish in short pseudoracemes and upper leaf-axils. Calyx pilose, 4-5 mm long with triangular-acuminate lobes, 2.5-3.5 mm long. Standard pubescent, 6-10 mm long. Style glabrous, 2 mm long, twisted, curved, penicillate at the tip. Pod upcurved towards the tip, 3-4 x 0.4 cm, short pubescent with 8-14 seeds, *c* 3 x 2 mm. Fig. 95.16.5.

Short grassland, *Acacia* bushland and as a weed; 550-2000 m. EW AF TU-GD WU SU AR HA IL KF GG SD; widespread in tropical Africa, south to Transvaal, also in Madagascar, Mauritius, S Asia, Central America and West Indies. *Thulin* 1331; *Ash* 1107; *Mooney* 9024.

The Ethiopian material has spreading indumentum and belongs to var. *pumila*.

15. **T. purpurea** *(L.) Pers.* (1807).

Erect to procumbent annual or short-lived perennial, up to 80(-150) cm tall. Leaf rhachis up to 8 cm long, including a petiole of less than 1 cm. Leaflets 9-17(-21), elliptic-oblong to obovate-oblong, *c* 10-20 x 4-8(-10) mm, pubescent on both sides or beneath only. Flowers reddish-purple or pink in lax pseudoracemes, sometimes also in the upper leaf-axils. Calyx pubescent, *c* 3-4.5 mm long, with narrowly triangular lobes, the lowest 2-3 mm long. Standard pubescent, 6-9 mm long. Style glabrous, *c* 2.5 mm long, not twisted, curved. Pod slightly upcurved towards the tip, (25-)40-45(-60) x 3-4.5 mm (in Ethiopia), strigulose with 6-9 seeds, *c* 3 x 1.5-2 mm.

Subsp. **leptostachya** *(DC.) Brummitt* in *Bol. Soc. Brot., sér. 2*, **41**:245 (1968).

1. Hairs on stem, leaf rhachis and pods appressed, sometimes almost absent. var. **leptostachya**
- Hairs more copious, usually spreading. var. **pubescens**

Var. **leptostachya**.

Grassland and cultivations, sandy places; up to 1700 m. EE EW ?TU; throughout tropical Africa south to Transvaal, Yemen. *Pappi* 6259; *Schimper* 2249; *Bally* 6894 A.

The relationship of this variety with *T. desertorum* Scheele (1843) from the area around Jeddah in Arabia requires further investigation.

Var. **pubescens** *Bak.* in *Fl. Trop. Afr.* **2**:125 (1871).

Grassland and cultivations; *c* 1500 m. EW? SU; throughout tropical Africa south to Transvaal. *Schimper* s.n.; *Negri* 899.

Elsewhere this subspecies is used as a source of antihelminthic and brown dye. Cut branches can be used as a green manure.

Subsp. *purpurea* is distributed from Pakistan to Indonesia and three further subspecies are recognized in Africa.

16. **T. apollinea** *(Del.) Link* (1822).

Annual or short-lived perennial, erect, up to 50 cm tall. Leaf rhachis up to 7 cm long, including a petiole of up to 2 cm. Leaflets 5-9, cuneate-oblong or oblanceolate, up to 40 x 20 mm, appressed silvery pubescent, more densely so beneath. Flowers purplish or white in lax pedunculate pseudoracemes and sometimes in upper leaf-axils. Calyx appressed pubescent, *c* 4 mm long with triangular-acuminate lobes, the lowest *c* 2.5 mm long. Standard pubescent, 6-8 mm long. Style glabrous, 2.5-3 mm long, not twisted, curved, penicillate at the tip. Pod upcurved towards the tip, *c* 35-50 x 4 mm, short pubescent to ± glabrescent with 5-8 seeds, *c* 3.5 x 2 mm.

Dry grassland and bushland; up to 150 m. EE; Egypt, Sudan, Somalia, Socotra, Arabia. *Hemming* 1247.

Except for the usually fewer leaflets in *T. apollinea* it seems hardly distinguishable from *T. purpurea* subsp. *dunensis* Brummitt, which occurs mainly along the E coast of tropical Africa. *T. apollinea* will probably ultimately have to be included within *T. purpurea*, but until the group has been more thoroughly revised in NE tropical Africa and Arabia the author prefers to keep it distinct.

A salt tolerant species used elsewhere as a source of blue dye and fish poison.

17. **T. villosa** *(L.) Pers.* (1807).

Annual or short-lived perennial, villous, 0.3-1.3 m tall. Leaf rhachis up to 10 cm long, including a petiole of up to 1 cm. Leaflets 11-15, cuneate-oblong or elliptic, up to 35 x 12 mm, densely appressed villous beneath, more sparsely so above. Flowers purple in terminal pseudoracemes, often also in the upper leaf-axils. Calyx densely villous with long-acuminate lobes, the lowest up to *c* 8 mm long. Standard villous with brownish hairs, up to 11-15 mm long. Style glabrous, up to 3-5 mm long, bent sharply upward at base, twisted, penicillate. Pod ± deflexed, strongly upcurved near the tip, up to 54 x 5-6 mm, densely villous with 6-12 seeds, *c* 4 x 2.5 mm.

Subsp. **ehrenbergiana** (*Schweinf.*) *Brummitt* in *Bol. Soc. Brot., sér. 2*, **41**:225 (1968);
T. ehrenbergiana Schweinf. (1867) - type: Eritrea, *Ehrenberg* (B lecto. destr.), *Pappi* 65 (K neo.).

Stem-hairs usually spreading, brownish. Corolla (11-) 12-15 mm long. Fig. 95.17.2.

Acacia-Commiphora bushland or woodland; 800-1500 m. EE EW SU KF GG SD; from S Egypt to Transvaal and in Madagascar; subsp. *villosa* in S Asia. *Gilbert, Thulin & Getachew* 302; *Ash* 1600, 3629; *Pappi* 4263; *Friis* et al. 2923.

18. **T. polyphylla** *(Chiov.) Gillett* (1958);
T. nubica var. *polyphylla* Chiov. in *Fl. Somala* **2**:162 (1932).

Annual or perennial villous herb up to 1 m tall. Leaf rhachis up to 5(-11) cm long, including a petiole of up to 0.7(-1.7) cm. Leaflets 7-17, usually 11-13, cuneate-oblong or cuneate-obovate, up to 20(-30) x 7(-11) mm, villous, more so beneath. Flowers purple in terminal pseudoracemes, often also in the upper leaf-axils. Calyx densely villous with long-acuminate lobes up to 9 mm long. Standard villous with brownish hairs, 10-13 mm long. Style glabrous, 5-6 mm long, curved, twisted, penicillate. Pod ovate, sessile or almost so, *c* 11 x 6 mm, densely white villous with solitary seeds, *c* 4 x 2 mm.

Acacia-Commiphora bushland; 300-1400 m. HA SD; S Somalia, N Uganda, Kenya, N Tanzania. *Glover & Gilliland* 416; *Hemming* 1545; *Corradi* 3980; *Friis* et al. 2922.

19. **T. nubica** *(Boiss.) Bak.* (1871);
Pogonostigma nubicum Boiss. (1843).
Pogonostigma abyssinicum Jaub. & Spach (1856); *T. nubica* var. *abyssinica* (Jaub. & Spach) Schweinf. in *Bull. Herb. Boiss.* **4**, app. 2:247 (1896) - type: Eritrea, *Schimper* s.n. (P holo.).

Annual or perennial villous herb up to at least 1 m tall. Leaf rhachis up to 8 cm long, including a petiole of up to 2.5 cm. Leaflets 7-11(-13), oblanceolate, up to 45 x 14 mm, villous, more so beneath. Flowers in terminal pseudoracemes, often also in the upper leaf-axils. Calyx densely villous with long-acuminate lobes up to 10 mm long. Standard densely villous, *c* 12 mm long. Style glabrous, 5 mm long, curved, twisted, penicillate. Pod ovate, with a 2 mm long stipe, *c* 11 x 6 mm, densely white villous with solitary seeds, *c* 4 x 2 mm.

Dry bushland, stony places; *c* 300-1500 m. EE AF EW WU SU GG; Sudan, Mauritania, Niger, Djibouti, N Kenya; subsp. *arabica* (Boiss.) Gillett in Arabia. *Pappi* 7170; *Corradi* 3965; *J. De Wilde* 6873; *Boulos* 11814.

20. **T. bracteolata** *Guill. & Perr.* (1832).

Annual herb 0.5-3 m tall. Leaf rhachis up to 25 cm long, including a petiole of *c* 1 cm. Leaflets (15-)17-31, linear-oblong, up to 11 x 1 cm, glabrous above, appressed pilose beneath. Flowers mauve in terminal long-pedunculate pseudoracemes; bracts ovate or ovate-lanceolate, soon falling. Calyx brown-pubescent, *c* 5.5 mm long, lowest lobe *c* 3.5 mm long. Standard brown-pubescent, *c* 16 mm long. Style pubescent, *c* 5 mm long. Pod ascending, slightly upcurved towards the tip, up to 75 x 5 mm , with short stiff pubescence with (9-)13-16 seeds, *c* 2.5 x 2 mm. Fig. 95.17.1.

Grassland or woodland, usually in seasonally wet places; 700-1700 m. EW TU GD SU KF GG SD; Uganda, N Tanzania, westwards to Senegal. *Schimper* III:1678; *Gilbert & Thulin* 963; *W. de Wilde* 7883.

The Ethiopian material is var. *bracteolata*: var. *strigulosa* Brummitt with spreading hairs on stems and

Figure 95.17.1 - *TEPHROSIA BRACTEOLATA*, a: inflorescence, b: leaf, both from *W. de Wilde* 7883. 2 - *T. VILLOSA* subsp. *EHRENBERGIANA*, branch with flowers and pods from *Ash* 1600. 3 - *T. INTERRUPTA*, branch with flowers and pods from *Ash* 76. All x 0.75. Drawn by Maureen Church. (Reproduced with permission from *Opera Botanica*, 68: fig. 19, 1983.)

leaf rhachides is distributed from Tanzania and Malawi west to Angola.

21. **T. reptans** *Bak.* (1871).

Perennial, prostrate or ascending herb, up to 1 m long. Leaf rhachis up to 14 cm long, including a petiole of up to 3 cm. Leaflets 5-19, narrowly elliptic-oblong, up to 5 x 1 cm, pubescent beneath, less so above. Flowers in long-pedunculate pseudoracemes; bracts narrowly triangular with subulate tips, rather persistent. Calyx white-pubescent, *c* 5.5 mm long with triangular-attenuate lobes, the lowest *c* 3.5 mm long. Standard greenish and pubescent outside, purplish within, 14-16 mm long. Style pubescent, *c* 4 mm long. Pod slightly upcurved towards the tip, 60-75 x 4-5 mm, very shortly pubescent with (9-)13-19 seeds, *c* 2.5 x 2.5 mm.

Grassland and woodland, often in rocky places; 1250-1700 m. BA GG SD; eastern tropical Africa south to Zimbabwe and Mozambique, also in Madagascar. *Gilbert, Thulin & Getachew* 299; *Cufodontis* 595; *Friis* et al. 3291.

The Ethiopian material belongs to var. *reptans*: var. *arenicola* Brummitt & Gillett, with appressed pubescence, is widespread in eastern Africa from Tanzania and southwards.

22. **T. hochstetteri** *Chiov.* (1908)
- types: EW, *Tellini* 1294, *Pappi* 1046, 2463, 3011, 6585, 7316 (FT syn.).

Annual or short-lived perennial up to 80 cm tall; stems spreadingly brownish pubescent. Leaf rhachis up to *c* 12 cm long, including a petiole of up to 2.5 cm. Leaflets 7-13, linear-oblong, up to 60 x 12 mm, glabrous or almost so above, appressed silky beneath. Flowers purplish in short, few-flowered, long-pedunculate pseudoracemes; bracts narrowly triangular, rather persistent. Calyx villous, *c* 6 mm long with narrowly triangular lobes, the lowest *c* 4 mm long. Standard appressed pubescent, *c* 14 mm long. Style pubescent, *c* 3 mm long. Pods ascending, upcurved towards the tip, *c* 70 x 6 mm, densely pubescent, brownish along the sutures with 9-14 seeds, *c* 3.5 x 2.5 mm.

Probably in grassland; 1600-1800 m. EW TU; not known elsewhere. *Schimper* 206, 716, 788.

23. **T. elata** *Deflers* (1899).
T. rigida Bak, (1871), nom. illegit., non Spanoghe (1841).

Annual or short-lived perennial up to 2 m tall. Leaf rhachis up to 15(-20) cm long, including a petiole of up to 2.5 cm. Leaflets 9-23, linear-oblanceolate to linear-oblong, 20-70 x 5-17 mm, glabrous above, appressed pilose beneath. Flowers purple in many-flowered short-pedunculate pseudoracemes; bracts lanceolate, soon falling. Calyx pubescent, *c* 6 mm long, with triangular-acuminate lobes, the lowest about as long as the tube. Standard golden-brown outside, 10-16 mm long. Style pubescent, *c* 5 mm long. Pods ascending, slightly upcurved towards the tip, 40-60 x 4-5 mm, densely brown-pubescent with usually 9-11 seeds.

1. Hairs on stems, leaf rhachides and the lower surface of leaflets appressed. var. **elata**
- Hairs on stems, leaf rhachides and the lower surface of leaflets spreading. var. **tomentella**

Var. **elata.**

Grassland, often in rocky or seasonally wet places, or on waste ground; 1100-2100 m. GD WU GJ WG SU KF GG SD; eastern Africa southwards to Zimbabwe, also in Yemen. *Scott* 241; *Mooney* 6254; *Ash* 2169.

Var. **tomentella** *Brummitt* in *Bol. Soc. Brot., sér. 2,* **41**:341 (1968).

In habitats similar to var. *elata*. WG HA KF GG; S Sudan, Uganda, Kenya, Tanzania. *Gillett* 5103; *Meyer* 7901; *Bally* 10029; *W. de Wilde* 8854.

24. **T. interrupta** *Hochst. & Steud.* ex *Engl.* (1892)
- type: TU, *Schimper* I:344 (B holo. destr., BM K S iso.).
Dolichos genistiformis Chiov. (1937) - type: SU, *Taschdjian* 193 (FT holo.).

Shrub or woody herb up to 3(-5) m tall. Leaf rhachis up to 11 cm long, including a petiole of 5-15 mm. Leaflets 9-21(-27), elliptic-oblong, up to 30 x 8 mm, usually glabrous above, appressed pubescent beneath. Flowers purple in grey, brown or blackish tomentose pseudoracemes, with at least the lower nodes very distant from one another; bracts lanceolate to broadly ovate, soon falling. Calyx brown or blackish tomentose with lanceolate or ovate-acuminate lobes, about as long as the tube. Standard brownish tomentose outside, 14-21 mm long. Style pubescent at least on one side, 7-8 mm long. Pod slightly upcurved towards the tip, up to 80(-90) x 5-6 mm, densely grey or brown tomentose with 7-11 seeds, *c* 4.5 x 3 mm. Fig. 95.17.3.

Upland grassland, scrub or forest margins and rocky places; 1350-2800 m. EW TU WU GD WG SU AR HA IL KF SD; Sudan, Uganda, Kenya, Tanzania. *Gillett* 14567; *Thulin* 1555; *Burger* 3323; *Gilbert & Getachew* 2312.

The Ethiopian material belongs to subsp. *interrupta*; the distribution given refers to this taxon only. Two more subspecies found in E and SE tropical Africa.

25. **T. nana** *Schweinf.* (1868)
- type: Sudan/Ethiopia border, *Schweinfurth* 1871 (B syn. destr., BM K isosyn.).

Annual or short-lived perennial, 20-150 cm tall; stems rusty villose. Leaf rhachis up to 13 cm long, including a petiole of 3-50 mm. Leaflets 7-19, oblong-elliptic, up to 55 x 17 mm, softly pubescent beneath, only sparsely so above. Flowers purplish in rather dense pedunculate pseudoracemes, occasionally a few also in the upper leaf-axils. Calyx pubescent, *c* 5.5 mm long, lowest lobe *c*

2.5 mm long. Standard brown silky, 12-18 mm long. Style pubescent, *c* 6 mm long. Pod ascending, curved slightly downwards, 50-60 x 5-6 mm, stiffly pubescent with 12-20 seeds, *c* 3 x 1.5 mm.

Grassland and woodland; 1300 m. GD(Sudan border) GG; widespread in tropical Africa. *Gilbert, Thulin & Getachew* 532.

26. **T. humilis** *Guill. & Perr.* (1832).

Annual or short-lived perennial up to 50 cm tall; stems white pubescent. Leaf rhachis up to 11 cm long, including a petiole of up to 5 cm. Leaflets 5-11, oblong-elliptic, up to 50 x 15 mm, glabrous above, appressed pubescent beneath. Flowers purplish in rather dense short-pedunculate pseudoracemes, usually also in the upper leaf-axils. Calyx pubescent, *c* 4 mm long, lowest lobe *c* 1.5 mm long. Standard brown silky, *c* 10 mm long. Style pubescent, *c* 3 mm long. Pod ascending, curved slightly downwards, *c* 50 x 4-5 mm, appressed pubescent with 8-14 seeds, *c* 3 x 1.5 mm.

Grassland and woodland, stony places; 1300 m. WG KF; Sudan and westwards to Senegal. *Gilbert & Thulin* 646.

27. **T. vogelii** *Hook. f.* (1849).

Shrubby herb 1-4 m tall. Leaf rhachis 10-25 cm long, including a petiole of 1-3 cm. Leaflets 13-29, elliptic-oblong or elliptic-oblanceolate, up to 5 x 2 cm, tomentose above, more densely so beneath. Flowers white or purple in dense rusty tomentose, long-pedunculate pseudoracemes. Calyx brown tomentose, *c* 14 mm long, the lowest lobe *c* 10 mm long. Standard white silky, *c* 23 mm long. Style pubescent, *c* 10 mm long. Pod straight, *c* 11 x 1.3 cm, tomentose 12-16 seeds, *c* 7 x 4.5 mm.

Forest margins, waste ground; 1800-1950 m. SU HA KF; throughout tropical Africa. *Gilbert & Thulin* 903; *Friis* et al. 352; *IECAMA* J-14.

Not native in Ethiopia, but cultivated as a fish poison and naturalised. Elsewhere insecticides and molluscicides have been extracted from the leaves and essential oils from the roots of this species. It has also been used as a windbreak and hedge in erosion control.

37. INDIGOFERA *L.* (1753)

Gillett in *Kew Bull.,* add. ser. 1 (1958) & *Fl. Trop. E. Afr.* (1971).

Herbs or shrubs with biramous hairs. Leaves usually imparipinnate, sometimes 3-foliolate or 1-foliolate. Flowers usually in axillary racemes. Corolla usually red or pink, falling soon, or rarely the standard persistent; standard pubescent or glabrous, usually indistinctly veined. Vexillary filament free; anthers almost always apiculate. Pod cylindrical, 4-angled or flattened, usually dehiscent, 1-many-seeded.

Some 700 species in tropical and subtropical areas, but absent from the Mediterranean region.

Several species, particularly *I. arrecta*, *I. articulata*, *I. coerulea* and *I. tinctoria* were once of international importance as the source of the blue-black dye indigo, but now they are only used locally. Presently species of *Indigofera* are being studied for use as browse and in erosion control.

I. pulchra Willd. which was reported from Ethiopia by Gillett (1958:30) is omitted here. The record is apparently based on *Armbruster* s.n. (K) from "Addis Abeba", a collection which most probably originated from Sudan.

1. Standard pubescent outside, sometimes at the tip only. 2
- Standard glabrous. 74

2. Pod flattened. 3
- Pod ± round or 4-angled in cross-section. 6

3. Standard ± persistent, more than twice as long as calyx. **45. I. dauensis**
- Standard falling quickly, less than twice as long as calyx. 4

4. Perennial; pod *c* 6-11 x 3 mm, appressed pubescent. **1. I. arabica**
- Annual; pod larger or with spreading pubescence. 5

5. Pod 2-seeded, *c* 6 mm long with spreading pubescence. **2. I. chirensis**
- Pod 5-9-seeded, more than 6 mm long, with spreading pubescence. **3. I. hochstetteri**

6. Leaflets opposite, or leaves simple. 7
- Leaflets alternate. 66

7. Inflorescence a much-branched, lax panicle; pod 4-5 x 2 mm. **4. I. paniculata**
- Inflorescence an axillary raceme; pods various. 8

8. Fruiting pedicels erect or spreading, or a few of the lower ones only reflexed. 9
- Fruiting pedicels reflexed. 40

9. Axis of inflorescence a spine. 10
- Axis of inflorescence not a spine. 11

10. Shrub or small tree, up to 4 m tall; pod covered with stiff glandular hairs, leaflets 5-9, 10-18 mm long. **19. I. rothii**
- Shrublet, up to 0.5 m tall; pod without glandular hairs; leaflets 3(-7), less than 10 mm long. **20. I. spinosa**

11. Peduncle short, less than half as long as pod, or, if pod very short, peduncle less than 4 mm long. 12
- Peduncle more than half as long as pod, or, at least over 4 mm long. 20

12. Leaves simple. 13
- Leaves, at least most of them, compound. 14

13. Pod 1-seeded, ± spherical; leaves linear. **5. I. linifolia**
- Pod 2-seeded, oblong; leaves cordate-ovate. **6. I. cordifolia**

14. Pod with a 5-10 mm long, upcurved beak; shrub more than 1 m tall with pinnately 3-foliolate leaves. **42. I. curvirostrata**
- Pod without long, upcurved beak; herbs or small shrubs; leaves various. 15

15. Leaflets with a translucent blister at each side and a third beneath the tip. **22. I. ogadensis**
- Leaflets without translucent blisters. 16

16. Racemes 10-50-flowered, dense. 17
- Racemes 1-5-flowered. 19
17. Racemes ± elongated; leaflets broadly linear or oblong, up to 40 mm long. **9. I. prieureana**
- Racemes rounded, very dense; leaflets obovate, up to 16 mm long. 18
18. Leaflets 3-5, pod somewhat 4-angled, but not torulose[1]; seeds 2, with 4 pits on each face. **7. I. trigonelloides**
- Leaflets 7-11; pod torulose but not 4-angled; seeds 3-6, not pitted. **8. I. sessiliflora**
19. Stipules up to *c* 6 mm long, with a broad scarious base, attached to the petiole for *c* 2 mm, and with it persistent after the leaflets have fallen. **21. I. ruspolii**
- Stipules not as above. **23. I. suaveolens**
20. Plant without glandular multicellular hairs and dark brown biramous hairs; pods and fruiting pedicels ± erect. 21
- Plant with glandular multicellular hairs (sometimes sparse and inconspicuous) and/or dark brown biramous hairs; pods and fruiting pedicels ± spreading or rarely erect. 30
21. Fruiting pedicels 3-5 mm long; leaflets 11-31, *c* 2-4 times as long as wide. **10. I. dendroides**
- Fruiting pedicels up to 2 mm long; leaflets various. 22
22. Leaflets 5-13, with a translucent blister at each side and a third beneath the tip. **12. I. brevicalyx**
- Leaflets without translucent blisters. 23
23. Racemes with 10-25 flowers. **16. I. eremophila**
- Racemes with 1-7 flowers. 24
24. Most hairs on young stems spreading. 25
- Hairs on young stems appressed. 28
25. Leaflet margins channelled and glutinous. **11. I. tanganyikensis**
- Leaflet margins not as above. 26
26. Indumentum of both long spreading basifixed hairs and appressed medifixed hairs; leaves up to 50 mm long; stipules more than 1-nerved. **24. I. pilosa**
- Indumentum of only medifixed hairs, either appressed or with spreading ends; leaves up to 11 mm long; stipules 1-nerved. 27
27. Annual; peduncle up to 30 mm long, almost glabrous; leaflets 5-11; leaf rhachis distinctly prolonged beyond lateral leaflets, up to 20 mm long including a petiole usually less than 1 mm. **13. I. ambelacensis**
- Perennial; peduncle 10-15 mm long, densely pubescent; leaflets (3-)5-7; leaf rhachis not or only shortly prolonged beyond lateral leaflets, up to 10 mm long including a petiole of (1-)2-4 mm. **18. I. zavattarii**
28. Petiole less than 1 mm long; racemes 1-2-flowered; pod 9-11 mm long; slender annual. **14. I. sparsa**
- Petiole 1-6 mm long; racemes 1-7-flowered; pod 13-25 mm long; annual or perennial. 29
29. Leaflets ± strigose above; plant smelling of coumarin (like freshly cut grass or hay); raceme 4-7-flowered; pod 18-25 x 2 mm long. **15. I. vohemarensis**
- Leaflets glabrous above; plant not smelling of coumarin; raceme 1-3-flowered; pod 13-18 x 1.2 mm. **17. I. congolensis**
30. Biramous hairs all white or, rarely, brownish. 31
- Some biramous hairs dark brown or black. 39
31. Pod often rather torulose, 2 mm or more wide, with the suture ± 0.5 mm wide; leaflets glabrous, or nearly so, above. **33. I. vicioides**
- Pod various but not torulose and with a narrower suture; leaflets pubescent above. 32
32. Pod *c* 4 x 2 mm, 2-seeded; racemes dense, many-flowered, the flowers all turned to one side. **30. I. secundiflora**
- Pod longer and/or narrower, usually more than 2-seeded; racemes various but flowers not all turned to one side. 33
33. Leaflets obovate, up to 6 mm long, both sides silvery; pod *c* 2 mm wide, silvery. **31. I. argentea**
- Leaflets elliptic or oblong, usually up to 6 mm long, pubescent on both sides but not silvery. 34
34. Pod up to 10 x 1.2 mm; petiole usually shorter than basal leaflets. **25. I. mimosoides**
- Pod 10-23 x 1.8 mm; petiole either shorter or longer than basal leaflets. 35
35. Glandular hairs on stems up to 2-4 mm long. **29. I. zenkeri**
- Glandular hairs on stems less than 2 mm long, sometimes very difficult to see. 36
36. Petiole longer than basal leaflets. 37
- Petiole shorter than basal leaflets. 38
37. Glandular hairs numerous, also present on leaflet-margins; stamens 3-3.5(-4) mm long; style 1.2-1.6 mm long. **26. I. colutea**
- Glandular hairs not present on leaflet-margins; stamens 2.3-3 mm long; style 0.7-1 mm long. **28. I. brachynema**
38. Glandular hairs numerous on most parts of the plant; leaflets 9-11; calyx up to half as long as stamens. **27. I. dembianensis**
- Glandular hairs very sparse on stems, leaf and inflorescence rhachides, calyces and pods; leaflets 7-17; calyx more than half as long as stamens. **32. I. subargentea**
39. Erect glandular multicellular hairs always present on pods and usually on stems, inflorescence and leaf rhachis; racemes many-flowered, dense. **34. I. atriceps**
- Erect glandular multicellular hairs practically absent; racemes with *c* 10-20 flowers, fairly lax. **35. I. mooneyi**
40. Leaves dotted with impressed glands beneath. **63. I. microcarpa**
- Leaves without impressed glands beneath. 41
41. Calyx about as long as stamens, with long stiff spreading hairs. 42
- Calyx shorter than stamens, without long stiff spreading hairs. 44

1. beaded - pod swollen over the seeds and constricted in between.

42. Leaflets up to 20 x 10 mm; petiole 2-6 mm long; peduncle usually longer than the raceme; pods up to 8 x 2 mm; hairs on calyx dark brown or black. **60. I. longibarbata**
- Leaflets larger; petiole longer; peduncle usually shorter than the raceme; pods larger; hairs on calyx brownish or whitish. 43

43. Peduncle 2.5-8 cm long; corolla brick-red; leaflets 5-7(-9). **61. I. hirsuta**
- Peduncle up to 2.5 cm long; corolla pale red; leaflets 7-9(-11). **62. I. astragalina**

44. Stamens 10 mm or more long; calyx lobes very short, broadly triangular; leaflets 3-9. 45
- Stamens less than 10 mm long, or if 10 mm long or more then leaflets 11 or more. 46

45. Stem strongly 3-4 angled, the angles winged; leaflets 5-9 with spreading pubescence on the nerves below. **36. I. garckeana**
- Stem not, or hardly angular; leaflets 3-7(-9) with appressed pubescence on the nerves below. **37. I. binderi**

46. Hairs on pod mostly dark brown. 47
- Hairs on pod white. 49

47. Indumentum of standard appressed, brown; leaflets 7-11. **40. I. emarginella**
- Indumentum of standard short, crisped, white; leaflets 11-27. 48

48. Stamens 10-12 mm long; hairs on calyx, stems, leaf and inflorescence rhachides stiff and spreading. **38. I. macrantha**
- Stamens 6-9 mm long; hairs on calyx, stems, leaf and inflorescence rhachides ± appressed. **39. I. roseo-coerulea**

49. Stamens 6-8 mm long, indumentum everywhere white; calyx-lobes much longer than the tube. **41. I. boranica**
- Stamens 2-5 mm long, or if up to 6.5 mm, the indumentum mostly brown and calyx-lobes about as long as the tube. 50

50. Leaflets (3-)5-many. 51
- Leaflets always 3. 58

51. Pod 4-angled; calyx lobes much longer than the tube. 52
- Pod not or hardly 4-angled; calyx lobes shorter than or about as long as the tube. 53

52. Pod 3.5-4 mm wide; leaflets 4-20 x 2-12 mm. **43. I. lupatana**
- Pod *c* 2 mm wide; leaflets usually larger than 20 x 12 mm. **53. I. trita var. scabra**

53. Pod torulose in plane of sutures; seeds up to 4; leaflets up to 9(11). 54
- Pod not torulose or, if somewhat so, seeds usually 5 or more; leaflets usually 9 or more. 56

54. Stems not densely silvery and the surface not hidden by appressed hairs; leaflets less than 5 mm wide. **56. I. bogdanii**
- Stems densely silvery with the surface hidden by appressed hairs; leaflets more than 5 mm wide. 55

55. Pod markedly torulose, the segments globular, dorsal and ventral sutures narrow; leaflets 3-5 (-7). **54. I. articulata**
- Pod slightly torulose, the segments oblong-elliptic, dorsal and ventral sutures very wide; leaflets 5-9(-11). **55. I. coerulea**

56. Pod up to 6 x 3 mm, 1-2(-3)-seeded. **57. I. amorphoides**
- Pod more than 6 mm long, more slender and with more than 3 seeds. 57

57. Indumentum on flowers and inflorescence brownish; pod straight, 4-6-seeded, 12-20 mm long. **58. I. arrecta**
- Indumentum on flowers and inflorescence white; pod sometimes ± curved, 8-12-seeded, up to 35 mm long. **59. I. tinctoria**

58. Pod strongly upcurved, sometimes forming a circle. **52. I. cliffordiana**
- Pod straight, sharply bent at the base or somewhat curved. 59

59. Standard falling quickly; woody herbs usually 1-2 m tall. **53. I. trita**
- Standard persistent; shrublets or perennial herbs, usually smaller. 60

60. Axis of the inflorescence becoming spinescent and glabrous; pod torulose, the swellings in the plane of the sutures coinciding with constrictions in the plane at right angles, and vice versa. **44. I. spiniflora**
- Axis of inflorescence ± pubescent, not spinescent; pod not torulose or if so with the constrictions in both planes coinciding. 61

61. Leaf rhachis, including petiole, more than 15 mm long in well-developed leaves; at least some leaflets more than 15 mm long. 62
- Leaf rhachis, including petiole, less than 15 mm long; leaflets less than 15 mm long. 63

62. Leaflets strigose with appressed rough hairs; racemes 3-5 cm long, many-flowered. **46. I. kelleri**
- Leaflets densely grey tomentose with spreading soft hairs; racemes 1-3 cm long, 12-20-flowered. **47. I. ellenbeckii**

63. Pod not torulose. 64
- Pod ± torulose. 65

64. Herb with a woody base; petiole up to 3 mm long; racemes 40-110 mm long including a peduncle of 8-15 mm; pod densely spreading pubescent. **48. I. sp** = *Pappi* 74
- Shrublet; petiole up to 9 mm long; racemes 15-75 mm long including a peduncle of less than 5 mm; pod practically glabrous. **49. I. sabulosa**

65. Peduncle (2-)5-10 mm long; pod 2.5-3 mm wide, somewhat torulose; stipules 0.5 mm long; leaves digitate. **50. I. sp.** = *Hemming* 1471
- Peduncle less than 1.5 mm long; pod *c* 1.5 mm wide, torulose with oblong segments; stipules 1-2.5 mm long; leaves pinnate to subdigitate. **51. I. sp.** = *Friis* et al. 2976

66. Longer leaflets up to 3 cm long, glabrous above. **69. I. conjugata**
- Leaflets all less than 3 cm long or, if some reach 3 cm pubescent above. 67

67. Racemes very short, 1-3-flowered; leaflets 1-5, up to 7 x 4 mm. **21. I. ruspolii**

- Racemes with more than 3 flowers; leaflets various. 68

68. Calyx much less than half as long as corolla, the lobes little longer than the tube; standard densely covered with stiff appressed hairs. 69
- Calyx at least half as long as corolla, the lobes usually much longer than the tube; standard relatively sparsely pubescent with often ± spreading hairs. 70

69. Leaflets 1-5(-7), oblong, *c* 2-4 times as long as wide, the terminal often much longer than the rest; pod curved and ± torulose; racemes usually less than 11 cm long. **64. I. oblongifolia**
- Leaflets 3-10, elliptic or obovate, the terminal as long as, or only slightly longer than the rest; pod straight or somewhat curved, scarcely torulose; racemes often more than 11 cm long. **65. I. schimperi**

70. Racemes rounded, very dense. 71
- Racemes elongated, ± lax. 72

71. Leaflets 3-5; pod somewhat 4-angled, not torulose; seeds 2, with 4 pits on each face. **7. I. trigonelloides**
- Leaflets 7-11; pod torulose, not 4-angled; seeds 3-6, not pitted. **8. I. sessiliflora**

72. Pod *c* 5 mm long; leaf rhachis including petiole up to 6 mm long; the whole plant densely covered in ± spreading silvery hairs. **66. I. semitrijuga**
- Pod more than 5 mm long; leaf rhachis more than 6 mm long; pubescence not as above. 73

73. Stipules with a wide scarious base, glabrescent at the margin; hairs on stems and leaves closely appressed. **68. I. spicata**
- Stipules narrow, hardly scarious, without a glabrescent margin; hairs spreading or appressed. **67 I. volkensii**

74. Bracts not persistent; fruiting pedicels rarely more than 2 mm long, usually reflexed; leaflets 5-15. 75
- Bracts persistent; fruiting pedicels often more than 2 mm long, ± spreading; leaflets 1-5. 76

75. Raceme pedunculate, longer than the subtending leaf, the peduncle longer than a leaflet. **70. I. costata**
- Raceme ± sessile, shorter than the subtending leaf, the peduncle, if present, shorter than a leaflet. **71. I. parviflora**

76. Leaves all 1-foliolate. 77
- Leaves, or at least the upper ones, compound. 78

77. Leaflets about twice as long as wide; peduncles 1.5-6 mm long, much shorter than the flower-bearing part of the racemes; fruiting pedicels 1.5-3 mm long. **73. I. cana**
- Leaflets more than three times as long as wide; peduncles 20-150 mm long; fruiting pedicels 4-5 mm long. **72. I. stipulosa**

78. Upper leaves digitately 3-foliolate. 79
- Upper leaves pinnately 3-foliolate or pinnate. 80

79. Plant silvery, densely appressed pubescent. **74. I. tritoides**
- Plant green, hairs relatively sparse and spreading. **75. I. cufodontii**

80. Perennial; racemes many-flowered; pod strongly curved or spirally coiled, spreading pubescent. **76. I. gyrata**
- Annual; racemes up to 15-flowered; pod not as above. 81

81. Leaflets 3, less than twice as long as wide; inflorescence up to 2 cm long; plant procumbent. **77. I. butayei**
- Leaflets 1-5, more than twice as long as wide; inflorescences often more than 2 cm long; plant ± erect. **78. I. welwitschii**

1. **I. arabica** *Jaub. & Spach* (1856).

Perennial ± woody below, diffusely branched from the base, up to 30 cm high. Leaf rhachis up to 17 mm long including a petiole of up to 10 mm. Leaflets 3-5, opposite, obovate or oblanceolate, up to 15 x 6 mm, downcurved at the apex, appressed silvery strigose, particularly beneath. Racemes dense, almost sessile, shorter than subtending leaf. Calyx up to 4 mm long, deeply divided, silvery. Standard strigulose outside. Stamens *c* 4.5 mm long. Pod flattened, *c* 6-11 x 3 mm, upcurved at the tip, silvery, 4-5-seeded.

Acacia woodland or bushland; up to *c* 1200 m. EE AF EW HA; Somalia, Yemen. *Bally* 6800; *Hummel* 16; *Gilbert & Thulin* 96.

2. **I. chirensis** *Cuf.* ex *Gillett* (1958);

I. ervoides A. Rich. (1847), nom. illegit., non Meisn. (1844) - type: TU, *Quartin Dillon* (P holo., K iso.).

Annual, diffusely branched from the base. Leaf rhachis up to 12 mm long including a petiole of up to 4 mm. Leaflets 3-5, opposite, oblong or oblanceolate, up to 3 cm long, sparsely appressed strigose or glabrescent above, thinly silvery beneath. Racemes dense, almost sessile, shorter than the subtending leaf. Calyx 2.5 mm long, deeply divided. Standard strigulose outside. Stamens 3 mm long. Pod flattened, *c* 6 x 3.5 mm, with long spreading hairs, 2-seeded.

Short grassland; up to *c* 2100 m. EW TU WU SU; Tanzania. *Pappi* 286; *Schimper* 1545; *Albers* 62306.

3. **I. hochstetteri** *Bak.* (1871)

I. arenaria A. Rich. (1847), nom. illegit., non E.Mey. (1836); *I. semhaensis* Vierh. (1907) - type: Eritrea, *Petit* s.n. (P holo.).

I. ornithopodioides Hochst. ex Jaub. & Spach (1856), nom. illegit., non Schumach. & Thonn. (1827) - syntypes from Arabia, Nubia and Ethiopia: TU, *Schimper* 495 (P syn.).

I. anabaptista Steud. ex Bak. (1876).

Spreading annual up to 50 cm tall. Leaf rhachis up to 4 cm long including a petiole of 8-20 mm. Leaflets 3-5 (-7), opposite, oblong-oblanceolate, 10-30 x 3-10 mm, silvery strigose beneath, sparsely strigose to glabrous above. Racemes dense, many-flowered, shortly pedunculate, shorter than or equalling the subtending leaf. Calyx 2-3 mm long, deeply divided, silvery. Stamens 3-4 mm long. Standard strigulose outside. Pod flattened, *c* 15 x 3 mm, deflexed, curved, appearing jointed, strigose, 5-9-seeded. Fig. 95.18.1.

Grassland, sandy ground; 0-1900(-2600) m. EE AF EW TU GD(Sudan border) WU SU HA GG SD; westwards to Mali, N Somalia, Djibouti, N Uganda, Kenya, Tanzania, E Zaire, and Arabia to Pakistan and India. *Gilliland* 4030; *Pappi* 3933; *Gilbert* 2398; *Hemming* 1090.

4. **I. paniculata** *Pers.* (1807).

Annual, up to 2 m tall, strigose. Leaves simple, linear, up to 50 x 8 mm; petiole 1-2 mm long. Inflorescence a much-branched, lax panicle; pedicels 2-4 mm long, jointed. Calyx *c* 1.5 mm long, deeply divided, strigose. Standard strigose outside. Stamens 3-4 mm long. Pod 4-5 x 2 mm, erect or spreading, strigose, 2-seeded.

Grassland, around marshy places; *c* 500 m. IL (Gambela); widespread in tropical Africa. *Seyoum Ayehunie* 16.

The single Ethiopian collection is nearest to subsp. *paniculata*, although the leaves are glabrous above, a character not previously reported in the species.

5. **I. linifolia** (*L. f*). *Retz.* (1786).
Sphaeridiophorum abyssinicum Jaub. & Spach (1856) - type: N Ethiopia, *Schimper* 477 (P holo.).

Much-branched annual up to 25 cm tall. Leaves subsessile, simple, linear, up to 50 x 3 mm, silvery pubescent particularly beneath. Racemes numerous, axillary, dense, sessile, shorter than subtending leaf. Calyx *c* 2 mm long, deeply divided, silvery. Standard pubescent outside. Stamens *c* 2 mm long. Pod *c* 2 mm long, globose, apiculate, silvery, 1-seeded. Fig. 95.18.2.

Dry grassland, bushland; up to 1500 m. EE EW TU GD SU; Sudan, and in S Asia cast to China and in Australia. *Bally* 6895; *Gilbert & Thulin* 180; *Schimper* 2321; *Mooney* 10042.

Elsewhere the seeds have been ground into flour and used as a famine food.

6. **I. cordifolia** *Heyne* ex *Roth* (1821).

Diffusely branched annual up to 35 cm tall. Leaves subsessile, simple, cordate-ovate, up to 20 x 14 mm, spreadingly pubescent on both sides. Racemes many, axillary, dense, sessile, shorter than subtending leaf. Calyx *c* 3 mm long, deeply divided into filiform lobes, densely spreading pilose. Corolla equalling calyx, pubescent outside. Pod *c* 4 x 1.8 mm, oblong, cuspidate, pilose, (1-)2-seeded.

Dry grassland or scrub, in sandy or gravelly places; up to 1000 m. EE EW TU GD(Sudan border) WU; westwards to Mauritania and the Cape Verde Is., Socotra, S Asia. *Bally* 6890; *Schimper* 2273; *Pappi* 3928.

7. **I. trigonelloides** *Jaub. & Spach* (1856)
- type: N Ethiopia, *Schimper* 1523 (P holo.).
I. aeruginis Schweinf. (1867) - type: N Ethiopia, *Schimper* 2272 (B holo. destr., K P iso.).

Erect much-branched herb up to 30 cm tall. Leaf rhachis up to 2 cm long including a petiole of up to 16 mm. Leaflets 3-5, often alternate, obovate, up to 16 x 10 mm, appressed silvery pubescent on both sides. Racemes very dense, rounded, sessile, 12-20-flowered. Calyx *c* 3 mm long, deeply divided into linear lobes, densely silvery. Corolla equalling calyx; pubescent outside. Pod *c* 6 x 2 mm, somewhat tetragonal, not torulose, silky. Seeds 2, with 4 pits on each face.

Sandy slopes; *c* 1000 m. TU-GD (Takeze valley); only known from Ethiopia and Namibia. *Schimper* 157.

8. **I. sessiliflora** *DC.* (1825).

Ascending much-branched herb up to 30 cm tall. Leaf rhachis up to 2.5 cm long including a petiole of up to 10 mm. Leaflets 7-11, often alternate, obovate, up to 7 mm long, silvery pubescent particularly beneath. Racemes very dense, rounded, sessile, *c* 12-flowered. Calyx *c* 2.5 mm long, deeply divided into setaceous lobes, densely silvery. Corolla equalling calyx, pubescent outside. Pod *c* 12 x 1.5-2 mm, not tetragonal, torulose, silky. Seeds 3-6, not pitted.

Coastal plains, also in cultivations; up to 60 m. EE; Sudan and westwards to Senegal and in India. *Bally* 6893; *Popov* 1421.

9. **I. prieureana** *Guill. & Perr.* (1832).
I. stenophylla Guill. & Perr. var. *latifolia* A. Rich., *Tent. Fl. Abyss.* 1:179 (1847) - types: TU, *Quartin Dillon* s.n. (P syn.), *Schimper* III:1711 (P syn., K UPS isosyn.).
I. stenophylla var. *brachypoda* Steud. ex Bak. in *Fl. Trop. Afr.* 2:84 (1871) - type: *Schimper* III:1711 (K holo., P UPS iso.).

Erect annual, up to 80 cm tall, everywhere strigose, the hairs white or, especially on the inflorescence, brown. Leaf rhachis up to 12 cm long including a petiole of up to 1(-2) cm. Leaflets 5-23, broadly linear or oblong, up to 40 x 10 mm. Racemes up to 50-flowered, up to 10 cm long, almost sessile; flowers erect in bud, later reflexed and finally fruits becoming erect again. Calyx *c* 5 mm long, divided almost to the base, brown or white strigose. Corolla equalling calyx, white pubescent. Stamens 5-7 mm long. Pod erect, up to 35 x 4 mm, with prominent ± winged sutures, *c* 10-seeded.

Grassland and woodland; *c* 550-1600 m. EW TU GD(Sudan border) WG IL GG; Sudan and westwards to Senegal. *Gilbert, Thulin & Getachew* 539; *Schweinfurth* 1816; *Pappi* 7666; *Friis* et al. 2546.

I. prieureana is very closely related to and scarcely distinct from *I. stenophylla* Guill & Perr. (1832) which is distributed from Senegal to Zaire, Uganda and Tanzania. The Ethiopian specimens in their few (5-11) leaflets and up to 2 cm long petioles are ± intermediate between the two, but the author follows Gillett (1958) and places them in *I. prieureana*, with which they agree in having the flowers reflexed for a short period after flowering.

10. **I. dendroides** *Jacq.* (1789).

Erect annual up to 1.3 m tall, brownish pubescent especially in the upper part. Leaf rhachis up to 6 cm long including a petiole of *c* 1 mm. Leaflets 11-27, opposite, cuneate-oblong, up to 11 x 5 mm. Racemes axillary, long-pedunculate, 8-30-flowered, longer than subtending leaf. Calyx *c* 2.5 mm long, deeply divided into triangular-subulate lobes. Standard brown strigose outside. Stamens 4-5 mm long. Pod erect, 20-30 x 1.6 mm, rather tetragonal, strigulose, 8-12-seeded.

Grassland or scrub, often in seasonally wet places; 1300-1350 m. WG GG; throughout tropical Africa. *Gilbert, Thulin & Getachew* 222b; *Gilbert & Thulin* 673a; *Vàtova* 1918.

11. **I. tanganyikensis** *Bak. f.* (1932).

Erect or decumbent often shrubby herb, 0.2-1 m tall; young stems with spreading pubescence. Leaf rhachis 10-50 mm long including a petiole of 1-2 mm. Leaflets (3-)7-15, opposite, cuneate-oblong, up to 11 x 3 mm, strigose on both sides or glabrous above, channelled and glutinous at the margins. Racemes 2-3-flowered; peduncle up to 15 mm long, glabrescent. Calyx *c* 1.5 mm long, divided to the middle. Standard pubescent outside. Stamens 4-6 mm long. Pod erect, 10-20 x 1.8 mm, straight, glabrous or strigulose, 7-10-seeded.

Var. **strigulosior** *Gillett* in *Kew Bull.* **24**:469 (1970).

Stipules pilose; leaflets strigulose above; pod strigulose.

Grassland, degraded woodland; 1350-2200 m. HA; Kenya. *Bally* 10078; *Burger* 3047, 2021; *Gilbert & Thulin* 25.

Var. *tanganyikensis* occurs in East Africa, E Zaire, Rwanda and Burundi.

12. **I. brevicalyx** *Bak. f.* (1903)
- type: GD, Dembea, *Schimper* 1397 (K holo.).
I. geminata sensu Cuf. *Enum.* (1955), non Bak.

Decumbent perennial up to 35 cm long from a woody taproot. Leaf rhachis up to 15 mm long including a petiole of *c* 1 mm. Leaflets 5-13, opposite, elliptic-oblong or cuneate-obovate, up to 10 x 3 mm, sparsely appressed strigose on both sides or glabrous above, with a blister at each side and another beneath the tip. Racemes (1-)2-3(-6)-flowered; peduncle up to 2 cm long, glabrous or sparsely strigose. Calyx *c* 1.5 mm long, divided to the middle, each lobe with a blister on it. Standard pubescent outside. Stamens 4-5 mm long. Pod erect, up to 20 x 1.8 mm, straight, strigulose, 8-12-seeded.

Short grassland; 1100-2400 m. EW TU GD WU SU AR BA HA KF SD; N Somalia, Uganda, Kenya, Tanzania, E Zaire, Rwanda, Burundi. *Thulin* 1330; *Mooney* 6089; *Gillett* 14393; *Friis* et al. 3290.

13. **I. ambelacensis** *Schweinf.* (1896)
- type: Eritrea, *Schweinfurth* 293 (B holo. destr.).

Erect or ascending annual up to 80 cm tall; young stems with spreading pubescence. Leaf rhachis up to 20 mm long including a petiole usually less than 1 mm. Leaflets 5-11, opposite, narrowly oblong or oblanceolate, up to 10 x 3 mm, appressed pilose above, strigose beneath. Racemes 2-4-flowered; peduncle up to 30 mm long, almost glabrous. Calyx 2-3 mm long, deeply divided into filiform lobes. Standard pubescent outside. Stamens *c* 4 mm long. Pod erect, 10-25 x 1.7 mm, straight, strigose-puberulent, 8-15-seeded.

Grassland, woodland, rocky places; 1250-1700 m. EW SU HA GG SD; Sudan, Uganda, Kenya, Tanzania, Zaire. *Hildebrandt* 587; *Gilbert & Thulin* 222a, 309; *W. de Wilde* 8333.

14. **I. sparsa** *Bak.* (1871)
- type: N Ethiopia, *Schimper* 204 (K holo., BM iso.).

Erect slender annual up to 25 cm tall; young stems appressed pubescent. Leaf rhachis up to 20 mm long including a petiole less than 1 mm. Leaflets 3-7, opposite, elliptic to cuneate-oblong, up to 12 x 4.5 mm, glaucous and strigose beneath, green and appressed pilose above. Racemes 1-2-flowered; peduncle up to 18 mm long, filiform, glabrescent. Calyx *c* 1 mm long, deeply divided. Standard pubescent outside. Stamens *c* 3 mm long. Pod erect, 9-11 x 1.7 mm, straight, very sparsely strigulose, 6-8-seeded.

Rocky places. TU; not known elsewhere. *Schimper* 2009.

This species has been much misunderstood and often confused with *I. brevicalyx.*

15. **I. vohemarensis** *Baill.* (1883).
I. viscosa Lam. var. *subglabra* A. Rich., *Tent. Fl. Abyss.* **1**:181 (1847) - type: TU, *Quartin Dillon* (P holo., K iso.).
I. suaveolens Jaub. & Spach var. *subquadriflora* Chiov. in *Ann. Ist. Bot. Roma* **8**:414 (1908) - types: numerous syntypes from Eritrea (FT).
I. suaveolens var. *pseudocongolensis* Chiov. in *Miss. Biol. Boran., Racc. Bot.*: 75, fig. 14 (1939) - type: SD, *Cufodontis* 548 (FT holo.).

Erect shrubby annual or perennial up to 1.5 m tall; young stems appressed pubescent. Leaf rhachis up to 20 mm long including a petiole of 1-6 mm. Leaflets (3-)5-7(-9), opposite, elliptic to cuneate-oblong, up to 11 x 5 mm, appressed pilose above, strigose beneath. Racemes 4-7-flowered; peduncle up to 2 cm long, pilose or glabrescent. Calyx *c* 1.5 mm long, divided to the

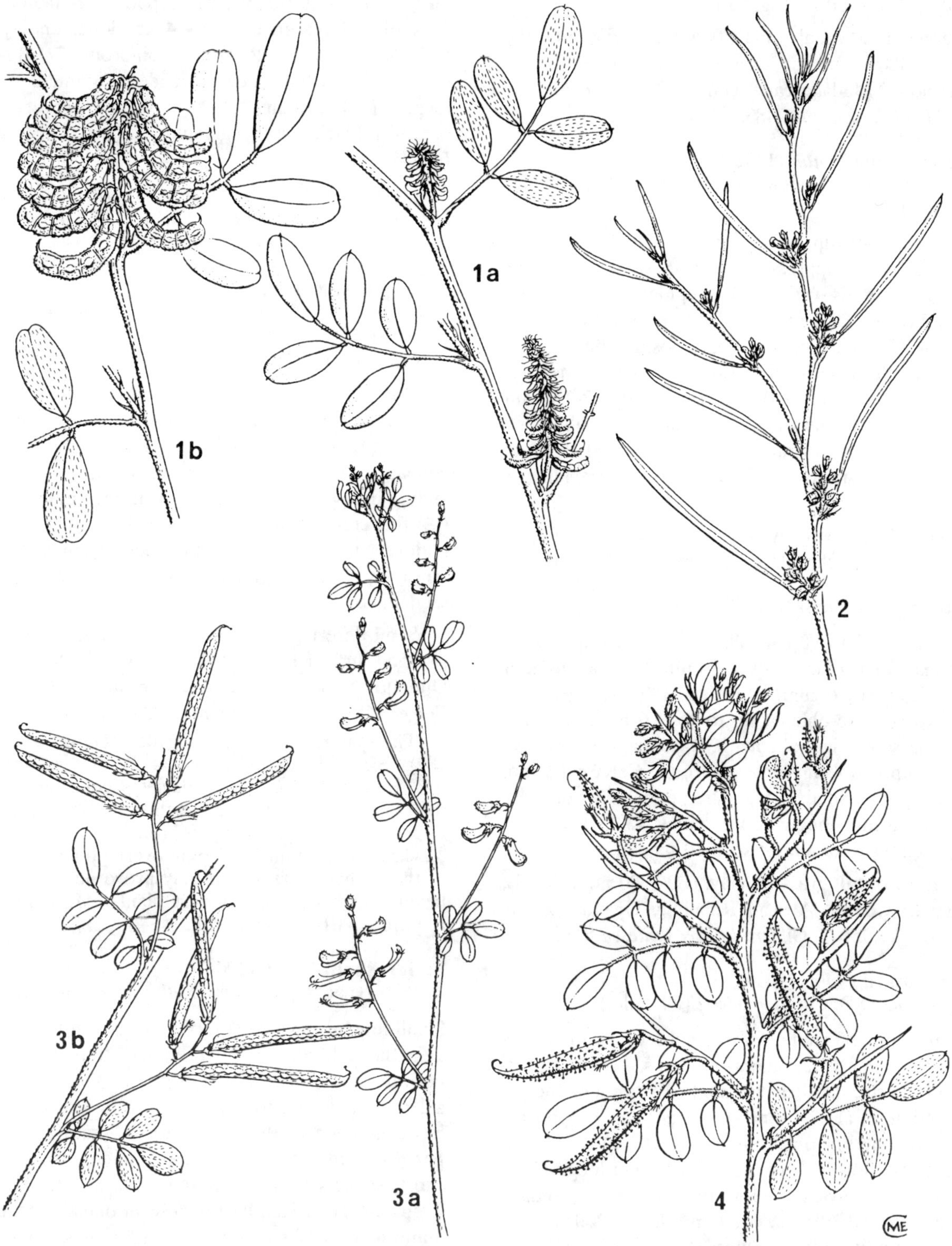

Figure 95.18.1 - *INDIGOFERA HOCHSTETTERI*, a: flowering branch from *Gilbert* 1082, b: fruiting branch from *Pettersson* 16986. 2 - *I. LINIFOLIA*, habit from *Gilbert* 1246. 3 - *I. VOHEMARENSIS*, a: flowering branch from *Gilbert & Tewolde* 2483, b: fruiting branch from *Burger* 874. 4 - *I. ROTHII*, flowering and fruiting branch from *Mooney* 8258. All x 1.75. Drawn by Maureen Church. (Reproduced with permission from *Opera Botanica*, 68, fig. 20, 1983.)

middle. Standard pubescent outside. Stamens 4-5 mm long. Pod ± erect, 18-25 x 2 mm, straight, appressed strigulose, 8-14-seeded. Fig. 95.18.3.

Grassland or woodland; 1350-2000 m. EW TU SU HA BA SD; Uganda, Kenya, Tanzania, E Zaire, Mozambique, Madagascar. *Gilbert & Thulin* 20; *Ash* 2165; *Gilbert & Tewolde* 2483.

16. **I. eremophila** *Thulin* (1982)
- type: HA, Ogaden, *Gilbert* 2090 (K holo., ETH, EA iso.).

Slender shrublet up to 0.5 m tall, silvery strigose throughout with appressed or somewhat spreading hairs; older stems with flaking epidermis, becoming purplish brown. Leaf rhachis 5-14 mm long including a petiole of 3-9 mm. Leaflets 3-5, opposite, obovate, 4-9 x 2-5 mm. Racemes 10-25-flowered; 1.5-8 cm long, including a peduncle of 0.2-1.2 cm. Calyx 2-2.5 mm long, divided to the middle. Standard pubescent outside. Stamens *c* 5 mm long. Young pod ± erect, linear, straight, strigose with the hairs ± spreading at the tips.

Open plains with widely scattered *Acacia* and dominant cover of shrublets; *c* 550 m. HA; Somalia.

17. **I. congolensis** *De Wild. & Th. Dur.* (1901).

Erect annual up to 70 cm tall; young stems appressed pubescent. Leaf rhachis up to 35 mm long including a petiole of up to 6 mm. Leaflets 3-9(-11), opposite, elliptic to cuneate-oblong, up to 15 x 6 mm, glabrous above, sparsely strigose beneath. Racemes 1-3-flowered; peduncle up to 15 mm long, glabrous. Calyx *c* 1.5 mm long, deeply divided. Standard pubescent outside. Stamens *c* 5 mm long. Pod ± erect, 13-18 x 1.2 mm, straight, glabrous or almost so, 6-10-seeded.

Margins of tall grass on clay; 600-1300 m. WG IL; Senegal to Central African Republic, Uganda, Zaire. *Gilbert & Thulin* 673, *Gilbert, Friis & Vollesen* 8387.

18. **I. zavattarii** *Chiov.* (1939)
- type: SD, *Cufodontis* 467 (FT holo.).

Perennial up to 50 cm tall, silvery pubescent with long ± appressed or spreading hairs on stems, leaves, peduncles and calyces. Leaf rhachis up to 10 mm long including a petiole of (1-)2-4 mm. Leaflets (3-)5-7, opposite, cuneate-oblong or oblanceolate, up to 11 x 3 mm. Racemes 1-4-flowered; peduncle 10-15 mm long. Calyx 2-3 mm long, divided below the middle. Standard pubescent outside. Stamens *c* 6 mm long. Pod ± erect, 12-17 x 2 mm, straight, strigulose, 5-8-seeded.

Rocky hillsides, roadsides; 1200-2300 m. ?SU GG SD; N Uganda, N Kenya. *Mooney* 5509; *Thesiger* 1975; *Corradi* 4447.

19. **I. rothii** *Bak.* (1871)
- type: *Roth* s.n. (K holo.).

Shrub or small tree up to 4 m tall, much-branched. Leaflets 5-9, opposite, oblong or obovate, 10-18 mm long, both sides thinly silvery; petiole 2-10 mm long. Axis of inflorescence up to 4 cm long, ending in a glabrescent spine. Racemes numerous, 2-4-flowered. Calyx *c* 4 mm long, deeply divided. Stamens *c* 6 mm long. Pod straight, 15-20 x 4 mm, covered with brownish-black stiff glandular hairs, 5-6-seeded. Fig. 95.18.4.

Upland scrub, forest margins; 2150-2800 m. SU HA; not known elsewhere. *Gillett* 5230; *Mooney* 8258; *Burger* 1367; *Gilbert & Lavranos* 2250.

20. **I. spinosa** *Forssk.* (1775).
I. spinosa var. *microphylla* A. Rich., *Tent. Fl. Abyss*. **1**:185 (1847) - type: Eritrea, *Quartin Dillon & Petit* s.n. (P holo., K iso.).

Shrublet, much-branched, up to 50 cm tall. Leaflets 3 (-7), opposite, elliptic, cuneate-obovate or oblanceolate, up to 9 x 4 mm, strigulose above, densely strigose beneath. Axis of inflorescence 6-25(-40) mm long, ending in a glabrescent spine. Racemes numerous, 2-4 (-8)-flowered. Calyx 1.2-2.5 mm long, divided to the middle. Standard pubescent outside. Stamens (3.2-)5-6 mm long. Pod straight, 15-20 x 1.8 mm, strigulose, 6-9-seeded.

Semi-desert grassland, dry deciduous bushland; up to 1550 m. EE AF EW HA GG SD; N Sudan, Djibouti, N Somalia, Kenya, Tanzania and in Arabia. *Burger* 3456; *Glover & Gilliland* 289; *Gilbert* 1859.

The material seen from Ogaden (*Glover & Gilliland* 289), SE Sidamo (*Corradi* 4412) as well as from NE Kenya (*Gillett* 13282) deviates considerably from the typical form in having 5-8-flowered racemes, gland-tipped calyx-lobes, usually 5-7-foliolate leaves and only 3.2-4 mm long stamens. Some overlapping occurs in all of these characters, but it may prove desirable to give taxonomic recognition to these plants. More material is required, particularly in fruit.

21. **I. ruspolii** *Bak. f.* (1904)
- type: HA, Ogaden, *Keller* s.n. (Z holo.).

Small much-branched shrub, up to 90 cm tall. Leaflets 1-5, the lateral, where present, alternate, cuneate-obovate, up to 7 x 4 mm, silvery pubescent on both sides. Stipules up to 6 mm long, with a broad scarious base, attached to the petiole for *c* 2 mm and, with it, persistent after the leaflets have fallen. Racemes very short, subsessile, 1-3-flowered. Calyx *c* 1.5 mm long, deeply divided. Standard pubescent outside. Stamens *c* 3 mm long. Pod straight, *c* 10 x 1.5 mm, strigulose, *c* 5-6-seeded.

Semi-desert grassland, dry deciduous bushland; 300-800 m. HA BA; Somalia. *Glover & Gilliland* 457; *Ellis* 139; *Simmons* 37.

22. **I. ogadensis** *Gillett* (1958)
- type: HA, Ogaden, *Ellis* 60 (K holo.).

Shrublet, aromatic; young stems appressed silvery strigose. Leaves digitately 3-foliolate or occasionally pinnately 5-foliolate; petiole 1-5 mm long; leaflets obovate, up to 6 x 3 mm, strigose on both sides, with a blister at each side and another beneath the tip. Racemes 3-15-flowered; peduncle 1-4 mm long. Calyx *c* 1.2 mm long, divided to the middle, each lobe with a blister on it. Standard pubescent outside. Stamens *c* 3.5 mm long. Pod up to 20 x 2.5 mm, cylindrical, sparsely strigose, 7-9-seeded.

Dry deciduous bushland and open plains with widely scattered *Acacia* and dominant cover of shrublets; 550-600 m. HA; Somalia. *Gilbert* 2089.

23. **I. suaveolens** *Jaub. & Spach* (1856)
- type: *Schimper* 1221 (P holo.).
I. kerensis Chiov. (1908) - type: Eritrea, *Pappi* 6293 (FT holo.).

Shrublet, much-branched; young stems appressed silvery pubescent. Leaves 5-foliolate or digitately to pinnately 3-foliolate; petiole 2-7 mm long; leaflets opposite, narrowly obovate, up to 12 mm long, sparsely strigose on both sides. Racemes 2-4-flowered; peduncle up to 6 mm long. Calyx 1.5-2 mm long, deeply divided. Standard pubescent outside. Stamens 4-5 mm long. Pod *c* 17-25 x 1.6-2 mm, subtetragonous, thinly strigose, 6-12-seeded.

Woodland and bushland; 700-1200 m. EE EW TU-GD HA-AF; Somalia, Sudan, Niger. *Pappi* 6253; *Schimper* 2288; *Jansen* 7105.

The local name indicates that this plant is used as a source of perfume.

24. **I. pilosa** *Poir.* (1813).

Annual or short-lived perennial woody herb up to 50 cm tall, spreadingly branched; indumentum of both long spreading basifixed hairs and appressed medifixed hairs. Leaves 3-5-foliolate, shortly petiolate; leaflets oblong or elliptic, up to 5 cm long, the lateral smaller and often unequal, both sides densely appressed pubescent. Racemes 1-4-flowered, axillary; peduncle up to 20 mm long. Calyx 2.5 mm long, deeply divided, more than half as long as the stamens. Standard pubescent outside. Pod 12-25 x 2 mm, straight, pubescent, 10-12-seeded.

Probably wooded grassland; *c* 800 m. EW; Sudan and west to Senegal; a variety in Angola. *Pappi* 7176, 7129, 6205.

25. **I. mimosoides** *Bak.* (1871).

Annual or perennial up to 1 m tall; stems reddish, white strigulose, usually also with long slender multicellular hairs. Leaf rhachis up to 40(-60) mm long including a petiole of 1-15 mm. Leaflets 3-11(-15), elliptic, up to 12(-15) x 7 mm, shortly strigulose above, with both long and short biramous hairs beneath. Racemes 6-30-flowered; peduncle 6-25 mm long. Calyx 3-5 mm long, divided nearly to the base. Standard pubescent outside. Stamens 3-3.5(-4.5) mm long. Pod up to 10 x 1.2 mm, spreading, straight, strigulose-pubescent, 3-6-seeded.

Upland grassland, forest margins; 900-2100 m. TU GD GJ WG SD; throughout tropical Africa. *Schimper* 1475, 1540; *Gillett* 14499; *J. De Wilde & Gilbert* 283.

The Ethiopian material belongs to var. *mimosoides*. There are two more varieties in Tanzania and Malawi.

26. **I. colutea** *(Burm. f.) Merr.* (1921);
Galega colutea Burm. f. (1768).
I. viscosa Lam. (1789).

Erect or spreading annual or short-lived perennial herb, 10-90 cm tall, appressed or spreading strigose throughout and in most parts also with erect multicellular glandular hairs. Leaf rhachis up to 7 cm long including a petiole longer than the basal leaflets. Leaflets 9-11, elliptic-oblong, up to 14 x 4 mm, glandular hairs present on margins. Racemes 7-20-flowered; peduncle up to 7-12 mm long. Calyx *c* 2 mm long, deeply divided. Corolla pubescent outside. Stamens 3-3.5(-4) mm long. Style 1.2-1.6 mm long, slightly curved upwards. Pod up to 18-23 x 1.8 mm, spreading, straight, densely covered with erect glandular hairs, 8-14-seeded. Fig. 95.19.1.

1. Glands on stiff erect multicellular hairs. var. **colutea**
- Glands subsessile. var. **somalensis**

Var. **colutea**.

Grassland, bushland, usually in sandy or rocky places; 0-1500 m. EE EW SU HA BA GG SD; throughout tropical Africa south to Transvaal, also in Yemen and through India to Australia. *Gilbert & Thulin* 183; *Gilbert, Thulin & Getachew* 326; *IECAMA* I-46.

Var. **somalensis** (*Bak. f.*) *Cuf.* ex *Gillett* in *Kew Bull., Add. Ser.* 1:66 (1958);
I. viscosa Lam. var. *somalensis* Bak. f. in *Journ. Bot.* 41:242 (1903) - type: Ogaden, *Donaldson Smith* s.n. (BM holo.).

Deciduous woodland or bushland; *c* 600 m. HA; N Somalia. *Glover & Gilliland* 290 (intermediate); *Gilbert* 2087.

27. **I. dembianensis** *(Chiov.) Gillett* (1970);
I. viscosa Lam. var. *dembianensis* Chiov. in *Ann. di Bot.* 9:58 (1911); *I. colutea* (Burm. f.) Merr. var. *dembianensis* (Chiov.) Cuf. ex Gillett in *Kew Bull., Add. Ser.* 1:66 (1958) - type: GD, *Chiovenda* 2391 (FT lecto., K fragment).

Erect or spreading annual up to 25 cm tall, appressed strigose throughout and in most parts also with erect multicellular glandular hairs. Leaf rhachis up to 3 cm long including a petiole shorter than or equalling the basal leaflets. Leaflets 9-11, elliptic-oblong, up to 12 x 4 mm, without glandular hairs on margins. Racemes 5-12-flowered, short; peduncle up to 10 mm long. Calyx *c*

2 mm long, deeply divided. Stamens *c* 4 mm long. Style 2.5-3 mm long, bent abruptly up above the middle. Pod 12-18 x 1.7 mm, erect, straight, glandular hairs sparse, 8-12-seeded.

Poor grassland, hillsides; *c* 1500-2000 m. TU GD SU; not known elsewhere. *Schimper* 1475; *Chiovenda* 1229, 2278; *Piovani* 107.

28. **I. brachynema** *Gillett* (1970).

I. colutea (Burm. f.) Merr. var. *linearis* Bak. f. ex Gillett in *Kew Bull., Add. Ser.* 1:65 (1958).

Erect annual up to 35 cm tall, ± appressed white-pubescent throughout, stem, leaf rhachis and pod also with short erect multicellular glandular hairs. Leaf rhachis up to 4 cm long including a petiole longer than the basal leaflets. Leaflets 7-11, narrowly elliptic, up to 15 x 3 mm, margins not glandular. Racemes 6-20-flowered, about as long as subtending leaf. Calyx *c* 2 mm long, deeply divided. Standard pubescent outside. Stamens 2.3-3 mm long. Style 0.7-1 mm long, bent abruptly up near the base. Pod 10-18 x 1.6 mm, spreading, straight, shortly glandular pubescent, 9-13-seeded.

Dry grassland or bushland. TU; Somalia, Kenya and south to Zimbabwe and Angola. *Schimper* II:1205.

29. **I. zenkeri** *Bak. f.* (1903).

Erect, sticky, bushy herb 0.5-2 m tall, with white ± appressed, pubescence throughout and also, almost everywhere, with erect multicellular glandular hairs up to 2(-4) mm long on the stems, but shorter elsewhere. Leaf rhachis up to 8 cm long including a petiole which is usually longer than the basal leaflets. Leaflets up to 23, opposite, elliptic, up to 12 x 7 mm, margins glandular. Racemes 10-20-flowered, usually longer than the subtending leaf; peduncle 1-1.5 cm long. Calyx *c* 3 mm long, deeply divided. Standard pubescent outside. Stamens *c* 4 mm long. Style 2-3 mm long, bent abruptly up near the middle. Pod *c* 16 x 1.8 mm, spreading, straight, densely covered with erect glandular hairs *c* 1 mm long, 6-11-seeded. Fig. 95.19.2.

Woodland, grassland, on disturbed ground or as weed in cultivations; 1300-2100 m. WG KF; widespread in tropical Africa. *Gilbert & Thulin* 647; *Smeds* 715, 887.

30. **I. secundiflora** *Poir.* (1813).

Erect annual up to 130 cm tall, with somewhat crisped pubescence and often also with red spreading multicellular hairs on stems and leaf and inflorescence rhachides. Leaf rhachis up to 5 cm long including a petiole of 10-15 mm. Leaflets (7-)11-15, opposite, elliptic, up to 15 x 7 mm, pubescent on both sides. Racemes dense, many-flowered, the flowers all turned to one side; peduncle 2-2.5 cm. Calyx *c* 4 mm long, white setose and with dark glandular hairs, deeply divided into blackish lobes with setaceous tips. Standard strigose outside. Stamens *c* 3.5 mm long. Pod ovoid-oblong, *c* 4 x 2 mm, densely white strigulose and with short, black, erect, multicellular hairs, 2-seeded.

Var. **rubripilosa** *De Wild.* in *Bull. Jard. Bot. État* 8:156 (1923).

I. secundiflora var. *gondarensis* Pic.-Serm. in *Missione di Studio al Lago Tana 7, Accad. Naz. dei Lincei*; 64, t. 11 (1951) - type: GD, *Pichi-Sermolli* 2265 (FT holo., K iso.).

Stems as well as leaf and inflorescence rhachides with red spreading multicellular hairs.

Grassland or woodland, often on moist or disturbed ground; 900-2200 m. EW TU GD GJ WG SU KF; widespread mainly in the E parts of tropical Africa. *Schimper* II:761; *Gilbert & Thulin* 648; *Mooney* 6093.

Var. *secundiflora*, which is mainly W African, differs in having stems without red spreading multicellular hairs. It has been recorded from Ethiopia, but the material in question is ± intermediate and the variety is omitted here.

31. **I. argentea** *Burm. f.* (1768).

Shrubby much-branched herb, up to 30 cm or more tall; stems ± ascending, glaucous-silvery and with very short glandular hairs. Leaf rhachis up to 2 cm long including a petiole of up to 1 cm. Leaflets 5-7, opposite, cuneate-obovate, up to 6 mm long, both sides silvery. Racemes 3-12-flowered, lax, exceeding the subtending leaves. Calyx *c* 1.5 mm long, densely pubescent. Standard pubescent outside. Stamens *c* 3 mm long. Pod linear-mucronate, up to 13 x 2 mm, straight, silvery and with very short and sparse yellowish glandular hairs, 2-5-seeded.

Sandy plains; up to 60 m. EE; Somalia, Egypt, Sudan and west to Mauritania, and S Arabia to India. *Pappi* 8227; *Hemming* 1065, 1132.

32. **I. subargentea** *De Wild.* (1923).

Annual, up to 60 cm tall; everywhere appressed strigulose; multicellular hairs sometimes also present on stems, leaf and inflorescence rhachides, calyces and pods. Leaf rhachis up to 5 cm long including a petiole of 1-7 mm. Leaflets 7-17, elliptic-oblong or linear-lanceolate, acute or rounded at the apex, up to 12(-17) x 5 mm, strigulose on both surfaces. Racemes 5-20-flowered, 1-6 cm long, of which about a third is peduncle. Calyx 2.5-4 mm long, more than half as long as stamens, divided almost to the base into subulate-setaceous lobes. Corolla strigulose outside. Stamens 3-4 mm long. Pod erect or spreading, *c* 15-20 x 1.7 mm, strigulose, 7-13-seeded.

Grassland, roadsides; *c* 1500-2000 m. WG HA KF; scattered in tropical Africa southwards to Zambia. *Romagnoli* s.n.; *Benedetto* 242; *Ouren* 21346; *Lundström* 155.

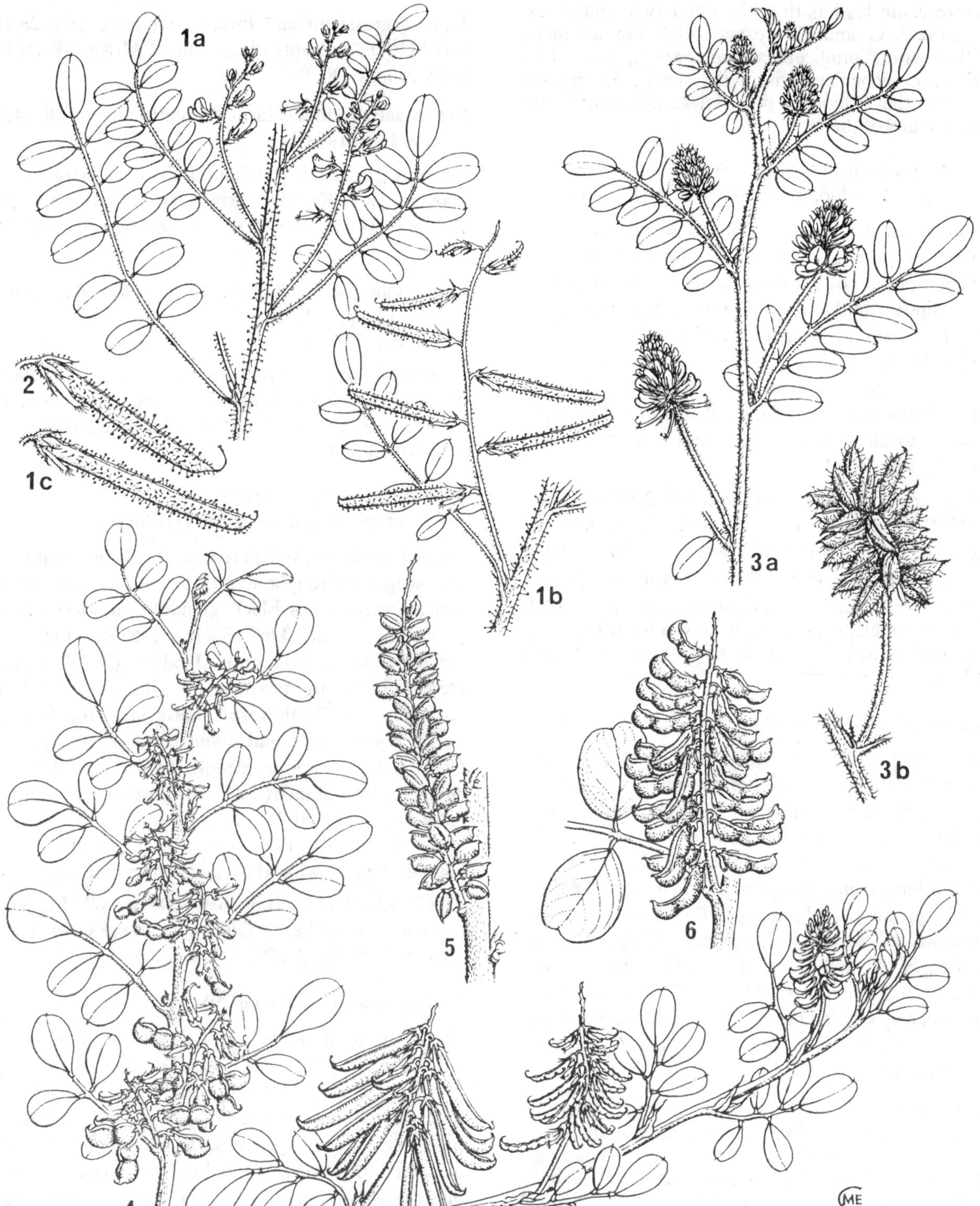

Figure 95.19.1 - *INDIGOFERA COLUTEA*, a: flowering branch, b: fruiting branch, c: pod, all from *Gilbert, Thulin & Getachew* 326. 2 - *I. ZENKERI*, pod, from *Gilbert & Thulin* 647. 3 - *I. ATRICEPS* subsp. *KAESSNERI*, a: flowering branch from *Gilbert, Thulin & Getachew* 510, b: fruiting raceme from *Fukui* 566. 4 - *I. ARTICULATA*, flowering and fruiting branch from *Gilbert & Thulin* 102. 5 - *I. AMORPHOIDES*, fruiting raceme from *Burger* 3004. 6 - *I. COERULEA*, fruiting raceme from *Schweinfurth* 358. 7 - *I. SPICATA*, flowering and fruiting branch from *Friis* et al. 2008. 1c & 2 x 2.25, 5 & 6 x 0.75, the rest x 1.125. Drawn by Maureen Church. (Reproduced with permission from *Opera Botanica*, 68, fig. 21, 1983.)

The Ethiopian material here referred to *I. subargentea* is not quite uniform. *Romagnoli* s.n. from HA has a denser indumentum, proportionally longer calyx-lobes and more acute leaflets than the other two collections cited (from WG and KF respectively). On all three collections some multicellular hairs are present. Two varieties are recognized within the species in tropical Africa, but the Ethiopian material cannot readily be assigned to any of them.

33. **I. vicioides** *Jaub. & Spach* (1856)
- type: N Ethiopia, *Schimper* 1216 (P holo.).

Spreading annual or perennial up to 30 cm tall; stems strigulose to pilose. Leaf rhachis up to 5 cm long including a petiole of 3-7 mm. Leaflets 5-13, opposite, oblong-elliptic, up to 15 x 5 mm. Racemes 5-15-flowered, lax, usually longer than the subtending leaf; peduncle 10-50 mm long. Calyx *c* 2 mm long, lobes subulate. Corolla whitish or yellowish, pubescent outside. Stamens *c* 3 mm long. Pod 8-17 x 2-2.5 mm, somewhat torulose with *c* 0.5 mm broad sutures, pubescent, 4-8-seeded.

Stony grassland and woodland; 900-2000 m. EW TU SU BA HA GG; widespread in the E parts of tropical Africa and south to Transvaal. *Thulin* 1495; *Gilbert & Thulin* 28; *Gilbert, Thulin & Getachew* 242.

Only var. *vicioides* with strigulose stems and leaflets glabrous or subglabrous above is known from Ethiopia. Var. *rogersii* (R.E.Fr.) Gillett is distributed southwards from Kenya.

34. **I. atriceps** *Hook. f.*(1864).

Bushy usually erect herb up to 2 m tall, strigose throughout with biramous hairs, some or many of which are brown or black; erect glandular multicellular hairs always present on the pods and usually on the stems, inflorescence and leaf rhachis. Leaf rhachis up to 15-60 mm long including a petiole of 2-20 mm. Leaflets (3-)5-15, opposite, ± elliptic, up to 25 x 8 mm. Racemes many-flowered, dense, pedunculate. Calyx 2-5 mm long, deeply divided into lobes with subulate tips. Standard dark strigose outside. Style *c* 3 mm long, sharply bent up beyond the middle. Pod 7-11 mm long, straight, with ± dark appressed hairs and erect multicellular hairs, 3-8-seeded.

1. Biramous hairs on stems and leaves ± crisped and spreading at the tips; stamens usually less than 30 % longer than calyx. subsp. **kaessneri**
- Biramous hairs mostly appressed; stamens usually 30 % or more longer than calyx. **2**
2. Stamens 5-6 mm long; pod usually over 8 mm long; multicellular hairs on stems absent, short or long. subsp. **atriceps**
- Stamens *c* 4 mm long; pod rarely over 8 mm long; multicellular hairs always present on stems, often up to 3 mm long. subsp. **setosissima**

Subsp. **atriceps.**
I. alboglandulosa Engl. (1892); *I. atriceps* subsp. *alboglandulosa* (Engl.) Gillett in *Kew Bull.* **10**:580 (1956) - type: GD, *Schimper* 1547 (B holo. destr., BM K iso.).

Upland grassland and forest margins; *c* 1800-2800 m. GD SU KF; widespread in tropical Africa. *W. de Wilde* 8623; *Smeds* 130.

Subsp. **setosissima** (Harms) Gillett in *Kew Bull., Add. Ser.* **1**:79 (1958).

Swampy grassland, bushland and forest margins; 1550-2000 m. WG KF SD; eastern tropical Africa south to Zambia. *W. de Wilde* 7573, 7740; *Mooney* 6080; *Meyer* 7804.

Subsp. **kaessneri** ((Bak. f.) Gillett in *Kew Bull., Add. Ser.* **1**:80 (1958).
Fig. 95.19.3.

Grassland, bushland and forest margins; 1100-2000 m. SU BA KF GG SD; Uganda, Kenya, Tanzania and NE Zambia. *Gillett* 14377; *Gilbert, Thulin & Getachew* 510, 548; *Mooney* 8433.

35. **I. mooneyi** *Thulin* (1982)
- type: KF, *Mooney* 5893 (K holo.).

Annual or short-lived perennial bushy herb up to 1 m tall, strigose throughout with biramous hairs, many of which are brown or black; glandular multicellular hairs practically absent. Leaf rhachis 15-50 mm long including a petiole of 2-12 mm. Leaflets (3-)5-9, opposite, elliptic-oblong to narrowly obovate, 6-18 x 3-9 mm. Racemes *c* 10-20-flowered, fairly lax, the floriferous part *c* 2-4 cm long and about equalling the peduncle. Calyx *c* 2-3 mm long, deeply divided into lobes with subulate tips. Standard dark strigose outside. Stamens 3.2-4 mm long. Style *c* 2.5 mm long, upcurved a little above the middle. Pod 5-8 x 1.6 mm, straight or slightly curved, strigose with white and brown hairs, 2-5-seeded.

Grassland and scrub; 1650-2000 m. WG SU KF SD; not known elsewhere. *Gilbert, Thulin & Getachew* 556; *Stewart* C-25; *Mogk* 270.

36. **I. garckeana** *Vatke* (1879).

Shrub up to 3 m tall; stems 3-4-angled or winged, strigulose or pubescent. Leaf rhachis *c* 10 cm long including a petiole of up to 4 cm. Leaflets 5-9, opposite, oblong-elliptic, up to 5 x 3 cm, strigulose on both sides, spreading pubescent on the nerves beneath. Racemes brownish, spreading pubescent, long and many-flowered; peduncle 2-3 cm. Calyx densely strigulose, *c* 2 mm long, with short, obtuse, broadly triangular lobes. Standard brown strigulose outside. Stamens 12-13 mm long. Pod reflexed, 50-60 x 3.5 mm, usually upturned and pointed at the tip, strigulose, 13-17-seeded.

Woodland or bushland; 550-1950 m. EE EW WU WG BA IL KF GG SD; widespread in tropical Africa.

Pappi 4253; *Gillett* 14491; *Mooney* 8424; *Friis* et al. 3464.

37. **I. binderi** *Kotschy* (1865).

Shrub to 2 m high; stems cylindrical or very slightly 3-angled, ± appressed pubescent. Leaf rhachis up to 9 cm long, including a petiole of up to 2.5 cm. Leaflets 3-7 (-9), opposite, elliptic, up to 3.5 x 2 cm, strigulose on both sides or sometimes glabrous above, ± appressed pubescent on the nerves beneath. Racemes ± appressed pubescent, long and many-flowered; peduncle usually more than 2 cm long. Calyx white or black strigulose, *c* 2 mm long, with short, obtuse, broadly triangular lobes. Standard brown-strigulose outside. Stamens 12-13 mm long. Pod reflexed, 50-60 x 3.5 mm, strigulose, *c* 8-seeded.

Woodland and bushland, rocky hillsides; 1500-1800 m. HA BA GG SD; Sudan, Central African Republic, N Uganda, Kenya. *Gilbert, Thulin & Getachew* 381; *Burger* 3043; *Gilbert* 1470; *Gillett* 13488.

38. **I. macrantha** *Harms* (1897)
- type: SD, *Riva* 1313 (FT holo.).

Erect herb or shrub 1-2 m tall; stems somewhat ridged, like the leaf and inflorescence rhachides with short stiff dark brown hairs. Leaf rhachis up to 17 cm long including a petiole of 2-3 cm. Leaflets 11-27, opposite, oblong-elliptic, up to 23 x 10 mm, white pilose on both sides. Racemes long, many-flowered; peduncles 5-7 cm long. Calyx *c* 4 mm long, dark brown, divided to about the middle into triangular-acuminate lobes. Standard densely crisped white pubescent outside. Stamens 10-12 mm long. Pod reflexed, *c* 40 x 3 mm, finely pubescent, *c* 10-seeded.

Bushland and woodland; 1500-1900 m. BA SD; N Uganda. *J. De Wilde* 6054; *Mooney* 9706.

39. **I. roseo-coerulea** *Bak.* f. (1926).

Shrub, up to 3 m tall; stems ribbed, brownish or whitish strigulose. Leaf rhachis up to 14 cm long including a petiole of 1-3 cm. Leaflets 11-21, opposite, oblong-elliptic, up to 25 x 14 mm, finely white strigulose above, with both brown and white hairs beneath. Racemes long, many-flowered, dark brown strigulose; peduncle up to 6 cm long. Calyx 2-3 mm long, brown strigulose, divided to about the middle into triangular-subulate lobes. Standard densely white puberulent outside. Stamens 6-9 mm long. Pod reflexed, 30-40 x 2.5-3 mm, somewhat upturned at the tip, strigulose, *c* 8-seeded.

Upland bushland; 1800-2700 m. WG SU AR KF SD; Uganda, Tanzania, E Zaire, Malawi. *Cooper* s.n.; *Friis* et al. 1512; *W. de Wilde* 7826.

40. **I. emarginella** *Steud.* ex *A. Rich.* (1847)
- types: TU, *Schimper* II:735 (P syn., K isosyn.), III:1723 (P syn., K UPS isosyn.), III:1893 (P syn.).

I. coluteifolia Jaub. & Spach (1856), nom. illegit. superfl. - type: TU, *Schimper* III:1723 (P holo., K UPS iso.).

Shrubby perennial up to 2 m tall; stems strigulose. Leaf rhachis up to 8 cm long including a petiole of up to 10(-15) mm. Leaflets (5-)7-11, opposite, elliptic-oblong, often emarginate, up to 20-40 x 16-20 mm, strigulose on both sides. Racemes almost sessile, many-flowered, up to 7 cm long, densely brown strigulose. Calyx *c* 1.5 mm long with broadly triangular lobes. Standard densely brown strigulose outside. Stamens 5-6 mm long. Pod reflexed 14-22 x 2 mm, straight, brown strigulose, 4-5-seeded.

Grassland, woodland and forest margins; 900-1800 m. EW TU GD GJ WG KF; throughout tropical Africa. *Gilbert & Thulin* 739; *Aweke & Gilbert* 938; *Schimper* II:735.

41. **I. boranica** *Thulin* (1982).

Slender erect shrublet up to 1-2 m tall; old stems with dark purple bark; young branches white strigose. Leaf rhachis 2-4.5 cm long including a petiole of 0.5-1.3 cm. Leaflets 5-9, opposite, elliptic to obovate, rounded to emarginate at the apex, 4-15 x 2.5-10 mm, strigose on both surfaces. Racemes fairly dense, *c* 10-25-flowered, 3-7.5 cm long including a peduncle of 0.5-2 cm, white strigose. Calyx 3-5 mm long, deeply divided into very narrowly triangular to almost setaceous lobes. Standard white pubescent outside. Stamens 6-8 mm long. Pod reflexed, 20-60 x 3-4 mm, straight, with a forwardly directed beak, somewhat flattened and not markedly tetragonal, strigose, 4-8-seeded.

In *Acacia-Commiphora* bush/woodland on limestone; alt. 780-1450 m. SD; NE Kenya. *Ash* 2434; *Thulin, Hunde & Tadesse* 3597, 3629.

42. **I. curvirostrata** *Thulin* (1988)
- type: SD, 90 km E of Filtu, *Gilbert, Sebsebe & Vollesen* 8126 (K holo., ETH UPS iso.).

Shrub, up to 2 m tall; young stems covered by short, closely appressed hairs. Leaves pinnately 3-foliolate; rhachis up to 35 mm long including a petiole of up to 25 mm; leaflets broadly elliptic to suborbicular, up to 30 x 27 mm, fairly sparsely pubescent with appressed hairs on both sides or subglabrous above. Racemes 8-10-flowered, dense, 10-12 mm long in fruit including a peduncle of 1-4 mm. Calyx 2.5 mm long. Standard (seen in bud) partly pubescent outside. Stamens *c* 6 mm long (seen in fruit). Pods erect or spreading, 40-65 mm long including a 5-10 mm long upcurved beak, *c* 4-5 mm wide, strigulose, *c* 6-10-seeded.

Bushland or woodland on rocky limestone slopes; *c* 900 m. SD; not known elsewhere. *Gilbert, Ensermu & Vollesen* 7693.

43. **I. lupatana** *Bak. f.* (1926).

I. commiphoroides Chiov. (1939) - type: SD, *Cufodontis* 340 (FT holo.).

Shrub up to 2.5 m tall, everywhere sparsely white strigulose. Leaf rhachis up to 45 mm long including a petiole of *c* 10 mm. Leaflets usually 7-9, opposite, oblong-ovate, often emarginate, up to 4-20 x 2-12 mm. Racemes many-flowered, 1-13 cm long, sessile or shortly pedunculate. Calyx 2.5 mm long with triangular-acuminate lobes. Standard white pubescent outside. Stamens *c* 3 mm long. Pod reflexed, 30-40 x 4 mm, pointed, markedly tetragonal, 4-8-seeded.

Bushland, woodland, rocky slopes; 1500-1900 m. BA SD; Kenya and south to Transvaal. *Cufodontis* 340; *Mooney* 7371; *Thulin, Hunde & Tadesse* 3393; *Gillett* 13491.

44. **I. spiniflora** *Boiss.* (1872).

I. spinosa Forssk. var. *spiniflora* (Boiss.) Schweinf. in *Bull. Herb. Boiss.* **4**, app. 2:237 (1896).

I. spinosa var. *spiniflora* f. *densissima* Chiov. in *Ann. Ist. Bot. Roma* **8**:91 (1903) - type: Eritrea, *Terracciano* 1757 (FT holo.).

Much-branched shrublet or woody herb, young stems appressed silvery. Leaves pinnately 3-foliolate; rhachis up to 10 mm long including a 2-8 mm long petiole; leaflets cuneate-obovate to oblanceolate, up to 10(-15) x 4(-6) mm, both sides silvery. Axis of inflorescence up to 5 cm long ending in a green glabrescent weak spine. Racemes up to 20-flowered, sessile. Calyx *c* 2 mm long, divided to below the middle. Standard pubescent outside, persistent. Stamens 3-3.5 mm long. Pod reflexed, torulose, the swellings in the plane of the sutures coinciding with constrictions in the plane at right angles and vice versa, *c* 10 x 2 mm, silvery, almost indehiscent, *c* 4-6-seeded.

Dry grassland or bushland; 0-210 m. EE; Sudan, Somalia and Yemen. *Popov* 1404; *Hemming* 1099; *Terracciano* 443; *Bally* 6962.

45. **I. dauensis** *Gillett* (1958).

Perennial decumbent herb, strigose, with the hairs often spreading at the tips. Leaves pinnately 3-foliolate; rhachis up to 30 mm long including a petiole of up to 24 mm; leaflets elliptic, obtuse at the apex, 10-24 x 4-12 mm. Racemes many-flowered, up to 10-16 cm long including a peduncle of up to 4-5.5 cm. Calyx 1.5-2 mm long; lobes narrowly triangular-acuminate, longer than the tube. Standard narrow, 4-5 mm long, strigose outside, rather persistent. Stamens *c* 4 mm long. Pod reflexed, straight, acute, with a continuous lateral ridge, 10-18 x 2-2.4 mm, strigose, *c* 5-seeded.

Acacia-Commiphora scrub; 350-850 m. SD; NE Kenya. *Friis, Mesfin & Vollesen* 2964; *Gilbert, Ensermu & Vollesen* 7601.

46. **I. kelleri** *Bak. f.* (1904)

- type: Ogaden, *Keller* s.n. (Z holo.).

Shrub; bark becoming blackish; branchlets densely strigose. Leaves pinnately 3-foliolate; rhachis up to 23 mm long including a petiole of up to 17 mm; leaflets elliptic, 15-25 x 6-11 mm, strigose. Racemes slender, many-flowered, 3-5 cm long, subsessile. Calyx 1.5-2 mm long; lobes narrowly lanceolate, longer than tube. Standard narrow, 4.5-5 mm long, strigose outside, persistent. Stamens 3.5-4 mm long. Ovary densely strigose. Pods unknown.

Probably dry bushland. HA; only known from the type from Abdallah at *c* 6°15´N, 43°30´E.

In the original description the ovary of *I. kelleri* was erroneously said to be glabrous. Gillett (1958, 1971), who did not see the type, suggested that it might be conspecific with *I. dauensis*. The shrubby habit, densely white-strigose branchlets and short subsessile racemes of *I. kelleri* are distinctive, however. Gillett (1958) placed *I. dauensis* in subgenus *Amecarpus* Benth. ex Harvey on account of its flattened pods. The affinity of both *I. dauensis* and *I. kelleri* seems however rather to be with *I. spiniflora, I. ellenbeckii* and the following four species in subsection Tinctoriae (Bak.) Gillett within subgenus *Indigofera*. The persistent standard is a characteristic feature of this group of species, of which more material is much needed.

Thulin, Hunde & Tadesse 3781 from Sof Omar in BA is an unmatched collection related here. It differs from *I. kelleri* in having slightly larger flowers with stamens *c* 5 mm long and an indumentum of hairs which are spreading at the tips. A single incomplete pod present in this collection is dorsiventrally flattened and *c* 2.5 mm thick.

47. **I. ellenbeckii** *Bak. f.* (1932)

- type: "Arussi Galla", *Ellenbeck* 1047 (B holo. destr., BM fragment).

Erect shrublet; stems white pubescent. Leaves pinnately 3-foliolate; rhachis up to 20 mm long including a petiole of up to 9 mm; leaflets ovate or elliptic, 10-26 x 7-18 mm, densely grey tomentose with spreading soft hairs. Racemes 12-20-flowered, 1-3 cm long, subsessile. Calyx 4-5 mm long, white pubescent; lobes 3-4 mm long, subulate. Standard 6-7 mm long, pubescent outside, persistent. Stamens *c* 5 mm long. Young pod straight, white pubescent, *c* 12 mm long, many-seeded.

Only known from the type, probably from HA, Gobelli R., *c* 8°33´N, 42°05´E.

48. **I.** sp. = *Pappi* 74.

I. arenaria A. Rich. var. *strigosa* A. Terracc. in *Ann. Ist. Bot. Roma* **5**:109 (1894) - type: EE, Anfile Bay (Baia d'Anfilho), *Terracciano* 1756 (FT holo.).

I. argentea sensu Gillett in *Kew Bull. Add. Ser.* **1**:69 pro parte, quoad specim. *Terracciano*, non Burm. f.

I. sp. *A* sensu Thulin, *Leg. Eth.* (1983).

Perennial decumbent herb up to 25 cm long, everywhere densely covered with appressed or somewhat spreading white hairs. Leaves pinnately to subdigitately 3-foliolate; rhachis up to 4 mm long including a petiole of up to 3 mm; leaflets cuneate-obovate, up to 8 x 6 mm. Racemes many-flowered, 40-110 mm long including a peduncle of *c* 8-15 mm. Calyx *c* 2 mm long, deeply divided into linear-setaceous lobes. Standard pubescent outside, persistent. Stamens *c* 3 mm long. Pod reflexed, straight, dorsiventrally somewhat flattened, *c* 6-13 x 1.6 mm, spreading pubescent, *c* 5-7-seeded.

Coastal arid plains. EE; not known elsewhere. *Pappi* 74.

The possible identity of this with *I. deflersii* Bak. f., described from S Yemen, must await further material. *I. deflersii* is only known from the type, *Deflers* 384 (Z holo.), a young plant without fruits. *I.* sp.= *Pappi* 74 differs mainly in its short petioles and distinctly pedunculate racemes.

48. **I. sabulosa** Thulin (1988);
I. sp. *B* sensu Thulin, *Leg. Eth*. (1983).

Shrublet to 90 cm tall, young stems densely covered with spreading white hairs. Leaves pinnately 3-foliolate; rhachis up to 11 mm long including a petiole of up to 9 mm; leaflets cuneate-obovate, *c* 3-9 x 2-5 mm, rounded to emarginate and mucronate at the apex, with spreading white hairs on both surfaces. Racemes many-flowered, 15-75 mm long including a peduncle of up to 5 mm. Calyx 1-1.6 mm long, divided to or below the middle into narrowly triangular lobes. Standard pubescent outside, somewhat persistent. Stamens *c* 2.3-2.5 mm long. Pod somewhat reflexed, straight, dorsiventrally flattened, 8-20 x 2 mm, glabrous or with a few scattered hairs, 3-6-seeded.

Deciduous bushland in red sandy or silty soil; *c* 500 m. HA (Warder); Somalia. *Barnes* s.n.

50. **I.** sp. = *Hemming* 1471.
I. sp. *C* sensu Thulin, *Leg. Eth*. (1983).

Shrublet *c* 40-60 cm tall, densely covered with appressed or slightly spreading white hairs. Leaves digitately 3-foliolate; petiole up to 5(-7) mm long; leaflets cuneate-obovate, up to 6 x 3.5 mm; stipules narrowly triangular, *c* 0.5 mm long. Racemes many-flowered, 10-70 mm long including a peduncle of (2-)5-10 mm. Calyx 2-2.5 mm long, divided to below the middle into narrowly triangular lobes. Standard pubescent outside, persistent. Filaments 2.4-3.2 mm long. Pod reflexed, straight, dorsiventrally flattened and somewhat torulose, densely strigose, *c* 3-5-seeded.

Deciduous bushland; *c* 500-600 m. HA(El Dere); N Somalia. *Hemming* 1471.

Differs from *I. spiniflora* by the non-spinescent, distinctly pedunculate, not glabrescent inflorescence axis, fatter much less torulose pods, smaller stipules and digitate leaves. Apparently conspecific specimens from N Somalia are *Bally* 7243 and *Gillett* 4532 (both K).

51. **I.** sp. = *Friis* et al. 2976.
I. sp. *D* sensu Thulin, *Leg. Eth*. (1983).

Perennial woody herb up to 15-30 cm tall, everywhere appressed to spreading strigose. Leaves pinnately or subdigitately 3-foliolate; petiole up to 5(-8) mm long; leaflets ± narrowly obovate, up to 10(-12) x 4.5(-6) mm; stipules linear, 1-2.5 mm long. Racemes many-flowered, 10-35 mm long including a peduncle of less than 1.5 mm. Calyx *c* 1.5 mm long, deeply divided to below the middle into narrowly triangular lobes. Standard pubescent outside, persistent. Stamens *c* 3 mm long. Pod reflexed, straight or slightly curved, up to 18 x 1.5 mm, torulose with oblong segments, densely strigose with appressed or spreading hairs, *c* 3-5-seeded.

Deciduous bush/woodland; up to 1200 m. HA SD; Somalia, NE Kenya. *Gilbert & Thulin* 79; *Friis* et al. 2976.

Differs from *I. spiniflora* in the scarcely spinescent inflorescence axis with a more persistent indumentum, and in the pods which are torulose with normal oblong swellings. However, some material, particularly from Somalia, is difficult to place.

52. **I. cliffordiana** *Gillett* (1958).

Shrublet *c* 0.5 m tall; branchlets densely appressed silvery strigose. Leaves pinnately or subdigitately 3(-5)-foliolate; rhachis 2-8 mm long including a petiole of *c* 2-5 mm; leaflets oblong-obovate, apiculate at the apex, 7-24 x 3-12 mm, strigose. Racemes 10-40-flowered, 3-10 cm long, including a short peduncle. Calyx 2-2.5 mm long; lobes narrowly triangular-acuminate, longer than the tube. Standard strigose outside, caducous. Stamens *c* 3.5 mm long. Pod reflexed, strongly upcurved, sometimes forming a circle, up to 20-30 x 2 mm, with flattened gaps between the 5-10 seeds, strigose.

Acacia-Commiphora scrub; *c* 850-1000 m. SD; Kenya. *Friis, Mesfin & Vollesen* 2949; *Mesfin Tadesse & Vollesen* 4280.

Only plants with 3-foliolate leaves are known from Ethiopia.

53. **I. trita** *L. f.* (1781).

Erect or subscandent woody herb, up to 0.5-2 m tall; stems ridged, strigose. Leaf rhachis up to 10 cm long including a petiole of up to 35 mm. Leaflets 3-9(-13), opposite, elliptic, up to 35(-60) x 17(-30) mm, strigose on both sides, more coarsely so beneath. Racemes many-flowered, long; peduncle up to 4 cm. Calyx 1.5-4 mm long; lobes triangular-subulate, much longer than the tube. Standard white strigose outside, caducous. Stamens 3-4 mm long. Pods 20-25(-40) x 2 mm, deflexed, straight or curved, sometimes slightly torulose, white strigose, 6-13-seeded.

1. Pod spreading stiffly outwards from a sharply bent base; stamens *c* 3 mm long; leaflets 3. var. **maffeii**
- Pod hanging alongside inflorescence axis, straight or curved at the tip; stamens *c* 4 mm long. 2
2. Leaflets 3. var. **subulata**
- Leaflets 5-9(-13). var. **scabra**

Var. **subulata** (*Vahl* ex *Poir.*) *Ali* in *Bot. Notiser* 3:558 (1958); *I. subulata* Vahl ex Poir. (1813).
I. subulata var. *microphylla* Chiov. in *Fl. Somala* 1: 138 (1929).
I. pseudosubulata sensu Cuf. *Enum.*: 269 (1955), non Bak. f.

Grassland, woodland, waste ground; 1350-2100 m. EW GJ WG SU KF SD; widespread in tropical and southern Africa. *Gilbert & Getachew* 3060; *Mooney* 8665; *Stewart* B26.

Var. **scabra** (*Roth*) *Ali*, loc. cit. (1958);
I. subulata var. *scabra* (Roth) Meikle in *Kew Bull.* 5:352 (1951).
I. quartiniana A. Rich. (1847) - type: TU, *Quartin Dillon* s.n. (P lecto.).
I. coronilloides Jaub. & Spach (1856), nom. illegit., non Mart.; *I. spachii* Bak. (1871) - type: GD, *Schimper* 858 (P holo.).

Grassland, woodland, stony places; 1200-2000 m. EW TU GD WG SU AR BA HA GG SD; widespread in tropical Africa and in Transvaal, Madagascar, India and Central America. *Gilbert & Thulin* 516, 683; *Gillett* 14373.

Var. **maffeii** (*Chiov.*) *Ali*, loc. cit. (1958);
I. maffeii Chiov. (1936); *I. subulata* var. *maffeii* (Chiov.) Gillett in *Kew Bull., Add. Ser.* 1:101 (1958).
I. spachii Bak. var. *trifoliolata* Schweinf. in *Bull. Herb. Boiss.*, App. 2:240 (1896) - syntypes from Yemen and Eritrea: *Schweinfurth* 64 & 209 (both B syn. destr.), *Schweinfurth & Riva* 648 (B syn. destr., FT K isosyn.).

Deciduous bushland; *c* 150-1000 m. EE EW TU GG; Sudan, Somalia, East Africa, Mozambique and in India. *Pappi* 3944, 4056; *Quartin Dillon* s.n.

Var. *trita* occurs in tropical Asia and Australia.

54. **I. articulata** *Gouan* (1773).

Shrub up to 1 m tall, everywhere densely appressed silvery pubescent. Leaf rhachis up to 10-35 mm long including a petiole of 5-12 mm. Leaflets 3-5(-7), opposite, obovate, up to 25 x 20 mm. Racemes *c* 10-20-flowered, shorter than the subtending leaf, subsessile. Calyx *c* 1.2 mm long; lobes triangular, about as long as the tube. Stamens 3 mm long. Pod 7-12 x 3.5-4 mm, deflexed, straight, or almost so, markedly torulose with globular segments, indehiscent, 2-4-seeded. Fig. 95.19.4.

Dry grassland or bushland; to 1300 m (2700 m in EW). EE AF EW WU SU HA; Sudan, Djibouti, Somalia, Egypt and in Arabia, Yemen and Socotra. *Beccari* 104; *Gilbert & Thulin* 102; *Burger* 3244; *Beals* 438.

Both *I. articulata* and *I. coerulea* are salt tolerant and can be used in land reclamation programmes for both soda salted and marine salted land.

55. **I. coerulea** *Roxb.* (1832).

Shrubby herb up to 1 m tall; stems, leaf and inflorescence rhachides densely appressed silvery pubescent. Leaf rhachis up to 8 cm long including a petiole of up to 2 cm. Leaflets 5-9(-11), obovate or oblong-obovate, up to 32 x 23 mm, strigulose on both sides or glabrous above. Racemes many-flowered, up to 4 cm long, subsessile. Calyx 1.5 mm long; lobes triangular, about as long as the tube. Standard brownish strigulose outside. Stamens 3-3.5 mm long. Pod *c* 15 x 2-3 mm, deflexed, curved, torulose with oblong-elliptic segments, indehiscent, strigulose, 3-4-seeded. Fig. 95.19.6.

1. Leaflets glabrous above. var. **coerulea**
- Leaflets strigulose on both sides. var. **occidentalis**

Var. **coerulea**.

Coastal arid plains. EE; N Sudan, N Somalia, Arabia and eastwards to India and Ceylon. *Terracciano* 12; *Massa* 613.

Var. **occidentalis** *Gillett & Ali* in *Kew Bull., Add. Ser.* 1:102 (1958).

Dry grassland and bushland; up to 1350 m (-2700 m in EW). EE AF EW TU WU SU HA BA GG SD; Sudan and east to Niger, Somalia, Socotra, Uganda, Kenya, Arabia to Pakistan and on Mauritius.

56. **I. bogdanii** *Gillett* (1958).

Much-branched shrublet, up to 0.5 m tall, branching ± zigzag, young stems silvery strigulose. Leaf rhachis 5-15 mm long including a petiole of *c* 2 mm. Leaflets 5-9, cuneate-oblong, up to 5 x 3 mm, strigulose beneath, glabrous above. Racemes 10-12-flowered, sessile, *c* 1 cm long. Calyx *c* 1.5 mm long; lobes shorter than the tube. Standard strigulose outside. Stamens *c* 4 mm long. Pod reflexed, up to 12 x 2.2 mm, straight or slightly curved, strongly torulose with subglobular segments, brown, strigulose, indehiscent, 1-4-seeded.

Rocky and saline places, overgrazed pastures, roadsides; *c* 1600-1800 m. SU; Kenya, Tanzania. *Ash* 456; *Vàtova* 2122.

57. **I. amorphoides** *Jaub. & Spach* (1856)
- type: N Ethiopia, *Schimper* 1474 (P holo.).
I. alta Schweinf. (1867) - type: TU, *Schimper* 2213 (B holo. destr., P iso.).

Shrubby herb up to 1 m tall; stems densely strigulose. Leaf rhachis up to 11 cm long including a petiole of 1-3 cm. Leaflets 7-15, opposite, elliptic-oblong, up to 19 x 9 mm, strigulose beneath, glabrous above. Racemes many-flowered, subsessile, up to 3.5 cm long. Calyx *c* 1.3 mm long with narrowly triangular lobes about as long as the tube. Standard brownish strigulose outside. Stamens *c* 3 mm long. Pod up to 6 x 3 mm, straight, deflexed, not torulose, indehiscent, crisped pubescent, 1-2(-3)-seeded. Fig. 95.19.5.

Bushland or woodland, roadsides, arable land; 400-2500 m. AF EW TU WU SU AR HA BA SD; N Kenya, N Somalia and S Arabia. *Pappi* 3518; *Gilbert & Thulin* 54; *Thulin* 1385.

58. **I. arrecta** *Hochst.* ex *A. Rich.* (1847)
- types: TU, *Schimper* III:1923 (P syn., K isosyn.), *Quartin Dillon* s.n. (P syn.), *Petit* s.n. (P syn.).

Shrub or woody herb 0.4-2(-3) m tall; stems white and brown strigulose. Leaf rhachis up to 6 cm long including a petiole of up to 1.5 cm. Leaflets 7-17, opposite, elliptic-oblong, up to 22 x 10 mm, strigulose beneath, usually glabrous above. Racemes many-flowered, subsessile, up to 5 cm long. Calyx *c* 1.5 mm long with triangular lobes about as long as the tube. Standard brown strigulose outside. Stamens 3-4(-6.5) mm long. Pod 12-20 x 2.5 mm, straight, deflexed, somewhat torulose, strigulose, 4-6-seeded.

Grassland, bushland, forest margins, roadsides, waste ground; (400-)1000-2700 m. EW TU WU GD GJ WG SU AR BA HA IL KF GG SD; throughout tropical Africa, Transvaal, Natal, also S Arabia and Madagascar. *Thulin* 1348; *Pappi* 224; *Gillett* 14528; *Friis* et al. 171.

Some specimens from low-lying arid areas in GG (*Corradi* 4031, 4384, 4385) have the leaflets strigulose also above and a more sparse pubescence of mostly white hairs on stems, flowers and inflorescence. Similar plants also occur in N Kenya and may merit taxonomic recognition (cf. Gillett 1971:307).

This is the most important source of blue-black dye in Ethiopia.

59. **I. tinctoria** *L.* (1753).
I. houer Forssk. (1775).

Shrubby herb 0.5-1.7 m tall; stems strigulose. Leaf rhachis up to 6 cm long including a petiole of up to 1 cm. Leaflets 9-13(-17), opposite, elliptic-oblong, up to 18 x 9 mm, strigulose beneath, glabrous or strigulose above. Racemes many-flowered, subsessile, up to 5 cm long. Calyx *c* 1.2 mm long with triangular lobes about as long as the tube, white strigulose. Standard white strigulose outside. Stamens 4-5 mm long. Pod up to 35 x 2 mm, straight or sometimes ± curved, deflexed, strigulose, 8-12-seeded.

Bush margins, cultivations; 350-400 m. KF GG; widespread in tropical Africa and also in S Arabia, India, Ceylon, SE Asia and introduced in tropical America. *Carr* 474, 724, 845; *Corradi* 4292.

The record for Eritrea in Cuf. *Enum.*: 276 (1955) has not been verified and is omitted here.

60. **I. longibarbata** *Engl.* (1892)
- type: GD, *Schimper* 1389 (B holo. destr., K iso.).

Annual or short-lived perennial up to 1.3 m tall; stems, leaf and inflorescence rhachides, calyces and pods stiffly brown-black pilose. Leaf rhachis up to 5 cm long including a petiole of 2-6 mm. Leaflets 9-13, opposite, oblong-elliptic, up to 20 x 10 mm, white pilose on both sides. Racemes dense, many-flowered, long-pedunculate. Calyx *c* 5 mm long, divided almost to the base into setaceous lobes. Standard white pubescent outside. Stamens *c* 5 mm long. Pod up to 8 x 2 mm, straight, deflexed, *c* 3-seeded.

Upland grassland and woodland, forest margins; 1500-2300 m. GD WG SU BA KF; widespread in tropical Africa and in South Africa. *Gilbert & Thulin* 613; *Mooney* 8705; *Smeds* 718.

61. **I. hirsuta** *L.* (1753).

Shrubby annual or short-lived perennial up to 1.5 m tall; stems, leaf and inflorescence rhachides, calyces and pods stiffly ± brownish pilose. Leaf rhachis up to 9 cm long including a petiole of up to 2 cm. Leaflets 5-7(-90), opposite, elliptic-oblong, up to 40 x 25 mm, pilose on both sides. Racemes dense, many-flowered; peduncle 2.5-8 cm long. Calyx *c* 4 mm long, divided almost to the base into setaceous lobes. Corolla white pubescent outside, brick-red. Stamens *c* 4 mm long. Pod 12-20 x 2 mm, straight, deflexed, 6-9-seeded.

Roadsides, waste or cultivated ground; 500-1200 m. TU WG IL; widespread in tropical Africa, also in S Asia and N Australia and introduced into tropical America. *Gilbert & Thulin* 747; *Germain* 1052; *Ash* 148; *Lundström* 170.

Elsewhere with a number of uses including antihelminthic, source of blue-black dye and as a green manure.

62. **I. astragalina** *DC.* (1825).

Shrubby annual; everywhere whitish or somewhat brownish pilose. Leaf rhachis up to 10 cm long including a petiole of up to 2.5 cm. Leaflets 7-9(-11), opposite, elliptic-oblong, up to 55 x 25 mm. Racemes dense, many-flowered; peduncle less than 2.5 cm long. Calyx *c* 4 mm long, divided almost to the base into setaceous lobes. Corolla white pubescent outside, pale red. Stamens *c* 4 mm long. Pod 12-15 x 3 mm, straight, deflexed, 4-6-seeded.

Waste or cultivated ground; up to 1000 m. EW; west to Senegal and south to Transvaal, also in India, Burma, Pakistan and Ceylon and introduced into the West Indies. *Pappi* 6132, 6472, 7173.

63. **I. microcarpa** *Desv.* (1814).

Prostrate herb, white strigose or pubescent. Leaf rhachis up to 45 mm long including a petiole of up to 10 mm. Leaflets 7-11, opposite, oblanceolate, up to 20 x 7 mm, dotted with impressed glands beneath. Racemes dense, up to 5 cm long, including a peduncle of up to 1 cm. Calyx *c* 2.5 mm long, hirsute, deeply divided into narrow lobes. Stamens *c* 3 mm long. Pod up to 8 x 2.5 mm, torulose, reflexed, indehiscent, pubescent, 2-4-seeded.

Shores of lakes and rivers; *c* 400 m. GG; tropical Africa, tropical America. *Corradi* 4251.

64. **I. oblongifolia** *Forssk.* (1775).

Shrublet up to 1.5(-3) m tall, silvery pubescent. Leaf rhachis up to 3 cm long including a petiole of up to 10 mm. Leaflets 1-5(-7), alternate, oblong, up to 35 x 12 mm, the terminal often much longer than the rest, strigulose on both sides. Racemes many-flowered, usually less than 11 cm long. Calyx 2-3 mm long; lobes triangular, equalling or slightly longer than the tube. Standard silvery outside. Stamens 4-5 mm long. Pod 12-18 x 2 mm, curved and ± torulose, reflexed, white strigulose, 6-8-seeded.

Grassland, bushland, sandy ground; up to 1200 m. EE AF EW HA BA KF GG SD; Sudan and westwards to Senegal, Somalia, Djibouti, Socotra, Angola, and Arabia and eastwards to Java. *Pappi* 6932; *Carr* 753; *Corradi* 4140; *Sue Edwards* 633.

Elsewhere the roots are reported to be toxic and used as a purgative; the stems are used as toothsticks and browse.

65. **I. schimperi** *Jaub. & Spach* (1856)
- type: N Ethiopia, *Schimper* 1218 (P holo.).
I. tettensis Klotzsch (1861).

Shrubby perennial up to 1.5 m tall, silvery pubescent. Leaf rhachis up to 6 cm long including a petiole of *c* 3 mm. Leaflets 3-10, alternate, elliptic or obovate, up to 25 x 20 mm, strigulose on both sides. Racemes many-flowered, up to 20 cm long including a peduncle of 1-2 cm. Calyx 2-3 mm long; lobes triangular, 1-2 times as long as the tube. Standard densely yellowish strigulose outside. Stamens 4-10 mm long. Pod up to 28 x 2 mm, straight or rarely curved, sometimes torulose, reflexed, white strigulose, up to 12-seeded.

Grassland, bushland, woodland, seasonally swampy places; 300-1600 m. AF EW TU GD SU HA BA GG SD; widespread in tropical Africa and in South Africa. *Gilbert & Thulin* 184; *Gilbert, Thulin & Getachew* 245; *Friis* et al. 957.

I. schimperi var. *baukeana* (Vatke) Gillett with stamens 7-10 mm long and pod bent sharply at the base has been reported from Ogaden (Gillett 1971:314). Numerous intermediates, agreeing with var. *baukeana* in one of the characters mentioned, occur in various parts of Ethiopia including Ogaden. No quite typical var. *baukeana* has been seen, however, and the variety seems difficult to maintain in this area.

Elsewhere reported to be salt tolerant and a source of browse.

66. **I. semitrijuga** *Forssk.* (1775).
I. somalensis Vatke (1878).

Procumbent herb, up to 30 cm long, densely covered with ± spreading silvery hairs. Leaf rhachis up to 6 mm long including a short petiole. Leaflets *c* 5, alternate, cuneate-obovate or obcordate, up to 7 x 5 mm, silvery on both sides. Racemes many-flowered, much longer than the subtending leaf. Calyx *c* 2.5 mm long, deeply divided. Standard pubescent outside. Stamens *c* 3 mm long. Pod *c* 5 x 2 mm, straight torulose silvery, 2-seeded.

Sandy or silty, often saline ground; 0-60 m. EE; Somalia, Arabia. *Hemming* 1114, 1133, 1137.

67. **I. volkensii** *Taub.* (1895).
I. insularis Chiov. (1908) - types: EE, *Terracciano* 662, 725, 762 & *Tellini* 424, 537 (all FT syn.).
I. boranensis Chiov. (1939) - type: SD, *Cufodontis* 294 (FT holo.).
I. subhirtella Chiov. (1939) - type: SD, *Cufodontis* 502 (FT holo.).

Perennial, up to 40 cm tall; stems, leaf rhachides and inflorescences ± densely strigose, the hairs often spreading at the tips. Leaf rhachis up to 30 mm long including a petiole of 1-7 mm. Leaflets 2-7, alternate, lanceolate-oblong to broadly elliptic, the terminal largest, up to 45 x 18 mm, strigose on both sides. Racemes up to 7(-15) cm long including the peduncle. Calyx 3.5-6 mm long, divided almost to the base into setaceous lobes. Standard pubescent outside. Stamens 3.5-5 mm long. Pod up to 15 x 2 mm, straight, deflexed, 5-8-seeded.

Bushland or woodland; 0-1800 m. EE SU AR HA BA KF GG SD; Somalia, Sudan, Uganda, Kenya, Tanzania. *Burger* 2779; *Gilbert* 2140; *Gilbert & Thulin* 110.

I. insularis has up to now generally been treated as distinct from *I. volkensii* on account of its fewer and broader leaflets, more appressed indumentum, and slightly shorter stamens. However, the variation seems to be completely gradual in all these features and the author now prefers to unite them.

68. **I. spicata** *Forssk.* (1775).
I. hendecaphylla Jacq. (1788).
I. parvula sensu Hochst. ex A. Rich. (1847), non Del.
I. hendecaphylla var. *angustifolia* A. Rich., *Tent. Fl. Abyss.* **1**:182 (1847), as "*endecaphylla*" - types: TU, *Quartin Dillon* s.n., *Schimper* 366, II:1085 & III:1689 (P syn.).
I. hendecaphylla var. *acutifolia* Chiov. in *Ann. Ist. Bot. Roma* **8**:88 (1903), as "*endecaphylla*" - type: Eritrea, *Terracciano & Pappi* 283 (FT holo.).
I. hendecaphylla var. *parvula* Chiov. in *Ann. Ist. Bot. Roma* 8: 89 (1903), as "*endecaphylla*" - type: Eritrea, *Schweinfurth & Riva* 1752 (FT holo., K iso.).

Prostrate or ascending perennial; stems, leaf and inflorescence rhachides ± sparingly appressed strigose. Stipules with a broad scarious base, glabrescent at least at the margins. Leaf rhachis up to 3 cm long including a petiole of 1-3 mm. Leaflets (3-)5-11, alternate, cuneate-obovate or cuneate-oblong, 3-30 mm long, strigulose on both sides or glabrous above. Racemes dense, many-flowered; peduncle 1-4 cm. Calyx 2-4.5 mm long, divided almost to the base. Standard strigulose outside. Stamens 3-4 mm long. Pod 11-18 x 2 mm, straight, deflexed, strigulose, 5-8-seeded. Fig. 95.19.7.

Grassland or woodland, often on waste or cultivated ground; (380-)1300-2700 m. EW TU GD WU GJ WG SU AR HA BA IL KF GG SD; throughout tropical Africa, Transvaal, Natal, Madagascar, Mascarene Is., also Yemen, India, Ceylon, SE Asia and introduced in America. *Mooney* 5861; *Gilbert & Getachew* 2739; *Gillett* 14380; *Thulin* 1533.

Planting of *I. spicata* has been recommended for the control of erosion and it is also a valuable fodder plant. If eaten in large amounts, however, it is reported to cause abortion in cattle and sheep.

69. **I. conjugata** *Bak.* (1871).

Procumbent, erect or spreading perennial; stems, leaf and inflorescence rhachides appressed strigose. Stipules narrowly triangular-subulate. Leaf rhachis broad, channelled above, up to 4 cm long including a petiole of up to 1 cm. Leaflets 1-5(-8), alternate, narrowly oblong, up to 60 x 12 mm, strigulose beneath. Racemes many-flowered; peduncle 0.5-2 cm long. Calyx *c* 4 mm long, divided almost to the base. Standard pubescent outside. Stamens 3-4 mm long. Pod 12-20 x 2 mm, straight, deflexed, strigulose, 5-8-seeded.

Grassland, wooded grassland, roadsides; *c* 1300 m. KF (Gojeb R. between Jima and Bonga); widespread in tropical Africa. *Jansen* 5815.

The single collection known from Ethiopia belongs to var. *conjugata*.

70. **I. costata** *Guill. & Perr.* (1832).

Annual up to 60 cm tall; stems strigulose. Leaf rhachis up to 7.5 cm long including a petiole of 2-7 mm. Leaflets (3-)5-15, opposite, oblong, up to 35 x 8 mm, appressed strigose beneath, appressed strigulose or glabrous above. Raceme many-flowered, up to 20 cm long and longer than subtending leaf, pedunculate. Calyx 2-3 mm long, divided to the middle into triangular lobes. Corolla glabrous. Stamens *c* 4 mm long. Pod up to 40 x 2.5 mm, upturned towards the tip, deflexed, strigulose, 15-20-seeded.

1. Upper surface of leaflets sparsely appressed strigulose. subsp. **costata**
- Upper surface of leaflets glabrous. subsp. **goniodes**

Subsp. **costata.**

Grassland; *c* 1500 m. EW; westwards to Senegal. *Pappi* 6736, 6737.

Subsp. **goniodes** (*Hochst.* ex *Bak.*) *Gillett* in *Kew Bull., Add. Ser.* 1:125 (1958);

Indigofera goniodes Hochst. ex Bak. (1871) - type: TU, *Schimper* 2242 (K holo., P iso.).

Indigastrum macrostachyum Jaub. & Spach (1857) - type: TU, *Schimper* 1305 (P holo.).

Grassland or woodland; 1100-2100 m. EW TU SU HA AR BA SD; N Uganda, Kenya and Tanzania. *Pappi* 15; *Thulin* 1339; *Gilbert & Thulin* 159.

71. **I. parviflora** *Heyne* ex *Wight & Arn.* (1834).

I. *deflexa* Hochst. ex A. Rich. (1847); *Indigastrum deflexum* (Hochst. ex A. Rich.) Jaub. & Spach (1857) - syntypes from Sudan and Ethiopia: TU, *Quartin Dillon* s.n. (P syn.), *Schimper* III:1467 (P syn., K UPS isosyn.).

I. barcensis Chiov. (1932) - type: Eritrea, *De Benedictis* 200 (FT holo.).

Annual up to 0.5 m tall; stems strigulose. Leaf rhachis up to 5 cm long including a petiole of up to 13 mm. Leaflets 5-13, opposite, oblong, up to 30 x 5 mm, strigose beneath, glabrous above (in Ethiopia). Racemes up to 12-flowered, up to 3 cm long, shorter than subtending leaf, sessile or almost so. Calyx 2-3 mm long, divided to the middle into narrowly triangular lobes. Corolla glabrous. Stamens *c* 3 mm long. Pod up to 35 x 2 mm, upturned towards the tip, deflexed (in Ethiopia), strigulose, 10-20-seeded.

Grassland and woodland; up to 1400 m. EW TU GD SU GG; widespread in eastern and southern Africa, also in India and N Australia (var. *occidentalis* Gillett occurs in western Africa). *Mooney* 8130; *Gilbert & Getachew* 2964.

72. **I. stipulosa** *Chiov.* (1939)
- type: SD, *Cufodontis* 228 (FT holo.).

Decumbent to erect perennial, woody below; stems ± covered by the persistent, linear-lanceolate, 6-11 mm long stipules. Leaves 1-foliolate, linear-lanceolate, up to 70 x 12 mm, ± appressed silvery pubescent on both sides; petiole 4-8 mm long. Raceme up to 12-flowered, lax; peduncle 20-150 mm long. Calyx 3-4 mm long, divided to the middle into narrowly triangular lobes. Corolla glabrous. Stamens *c* 4 mm long. Pod *c* 20-30 x 2 mm, straight, strigulose, *c* 12-seeded.

Bush/woodland, rocky hillsides, on limestone; 1350-1850. HA SD; Somalia. *Bally* 9308; *Friis* et al. 3142; *Gilbert, Sebsebe & Vollesen* 8270.

The single collection seen from HA, *Gilbert* 2028 (ETH K) from 48 km S of Jijiga, is not typical. It has a more closely appressed pubescence, narrower, only 3-7 mm long stipules and shorter, up to 3-5 mm long petioles. More material is required.

73. **I. cana** *Thulin* (1982)
- type: WU, below Bati on Aseb road, *Mooney* 9620 (ETH holo., UPS iso.).

Perennial with prostrate to ascending, rather woody stems up to at least 20 cm long, everywhere densely covered with ± appressed to somewhat spreading silvery pubescence. Stipules 1-4 mm long. Leaves 1-foliolate, narrowly obovate to elliptic, 8-25 x 3-12 mm; petiole 1-6 mm long. Racemes *c* 10-30-flowered, fairly dense; peduncle 1.5-6 mm long. Calyx 1.5-2.5 mm long, divided almost to the middle into narrowly triangular lobes. Corolla glabrous. Stamens 3-3.5 mm long. Pod *c* (7-)15-20 x 1.5 mm, straight or slightly curved, strigose, *c* 3-9-seeded.

Stony or rocky places; 1200-1800 m. WU (near Bati); not known elsewhere. *W. de Wilde* et al. 7348.

74. **I. tritoides** *Bak.* (1895).
I. tenuisiliqua Schweinf. (1896) - types from Arabia and Eritrea: *Schweinfurth* 57, 290 & 320 (B syn. destr.).

Perennial, up to 0.5 m high, erect or decumbent, densely appressed silvery pubescent. Leaves digitately 3-foliolate or some leaves simple; petiole up to 10 mm long; leaflets oblanceolate to broadly obovate, up to 10-35 x 7-25 mm, silvery on both sides. Racemes up to 30-flowered, longer than the leaves; peduncle 5-30 mm long. Calyx *c* 3.5 mm long, divided to the middle into narrowly triangular lobes. Corolla glabrous. Stamens 3-5 mm long. Pod up to 25-45 x 2 mm, straight or upcurved, silvery, more than 14-seeded.

Semi-desert grassland or deciduous bushland, often in rocky places; up to 1200 m. EE SU; Sudan, Djibouti, Somalia, Arabia. *Schweinfurth & Riva* 369a; *Gilbert* 1248; *J. De Wilde* 4673.

75. **I. cufodontii** *Chiov.* (1939)
- types: SD, *Cufodontis* 473 & 613 (FT syn.).

Perennial, up to 0.5 m high, crisped strigose. Leaves digitately 3-foliolate; leaflets oblanceolate or obovate, up to 35 x 15 mm; petiole up to 5 mm long. Racemes lax, very long; peduncle up to 5 cm long. Calyx 3 mm long divided to the middle into subulate lobes. Corolla glabrous. Stamens *c* 4 mm long. Pods up to 45 x 1.5 mm, straight, crisped pubescent, on stiffly spreading pedicels up to 10 mm long, up to 20-seeded.

Woodland or grassland; 1600-2000 m. SD; Kenya, N Tanzania. *Gillett* 14310; *Thesiger* 2020; *Riva* 624.

76. **I. gyrata** *Thulin* (1982)

Perennial, 10-30 cm tall, with spreading branches, covered on all parts with stiff, spreading hairs. Leaves pinnately 3-5-foliolate, the lower ones often 1-foliolate; leaflets elliptic, the terminal 12-40 x 4-10 mm; petiole 10-20 mm. Racemes lax, many-flowered; peduncle up to 3 cm. Calyx 1.8-2.4 mm long, divided to the middle into narrow lobes. Corolla glabrous. Stamens 2.8-3.6 mm long. Pod strongly curved or spirally coiled, flattened, *c* 20-30 x 1.6 mm, spreading pubescent, *c* 20-25-seeded.

Acacia-Commiphora bushland; 325-400 m. SD; NE Kenya. *Gilbert, Ensermu & Vollesen* 7563, 7646.

77. **I. butayei** *De Wild.* (1904).

Procumbent annual up to 30 cm long, sparsely appressed strigose. Leaf rhachis up to 10 mm long including a petiole of up to 6 mm. Leaflets 3, broadly cuneate-obovate, up to 16 x 12 mm. Raceme 4-10-flowered, up to 2 cm long including a peduncle of 3-7 mm. Calyx 1.5-2 mm long, divided to the middle into narrowly triangular lobes. Corolla glabrous. Stamens *c* 2 mm long. Pod *c* 15 x 1 mm, somewhat curved, on pedicels up to 2 mm long, *c* 10-seeded.

Tall grassland; 1300 m. WG; N Nigeria, Central African Republic, N Uganda, Zaire, N Zambia. *Gilbert & Thulin* 665.

78. **I. welwitschii** *Bak.* (1871).

Annual up to 40 cm tall, sparsely and stiffly spreading pubescent. Leaf rhachis up to 15 mm long including a petiole of up to 7 mm. Leaflets 1-5, linear-lanceolate or oblong, up to 35 x 7 mm. Raceme up to 15-flowered, long, lax; peduncle up to 2 cm long. Calyx 1.5-3 mm long, divided to the middle into narrowly triangular lobes. Corolla glabrous. Stamens 2-2.5 mm long. Pod 8-25 x 1.6 mm, usually sparsely strigose, on slender spreading pedicels up to 5 mm long, 4-17-seeded.

Tall grassland; 1300 m. WG; widespread in tropical Africa. *Gilbert & Thulin* 659.

The single collection known from Ethiopia is nearest to var. *welwitschii*, although the leaves are rather too large.

38. CYAMOPSIS *DC.* (1825)

Gillett in *Kew Bull. Add. Ser.* 1:6-8 (1958) & *Fl. Trop. E. Afr.* (1971).

Annual herbs with biramous hairs. Leaves imparipinnate. Racemes axillary. Standard glabrous, markedly veined. Filaments united into a tube or the vexillary one almost free; anthers apiculate. Pod flat, longitudinally ridged, dehiscent, ending in a pronounced beak. Seeds several, flattened, minutely tuberculate.

About three species, two of which are native in tropical Africa.

1. Leaflets 3, toothed, usually more than 1 cm wide. **1. C. tetragonoloba**
- Leaflets 3-7, entire, less than 1 cm wide. **2. C. senegalensis**

1. **C. tetragonoloba** (*L.*) *Taub.* (1894);
Psoralea tetragonoloba L. (1767).

Bushy annual. Leaflets 3, ovate-oblong, usually more than 1 cm wide, toothed. Racemes dense; corolla white and pinkish, *c* 8 mm long. Pod 40-100 mm long, linear, in stiff erect clusters, appressed pubescent. Seeds 5-12, oval, *c* 5 mm long, varying in colour from white to grey or black.

Grown for its edible young pods and seeds and also used as fodder, green manure and for the production of guar gum. EW (Keren); an ancient crop in India, now widely cultivated in the drier tropics, unknown in the wild state; scarcely of any importance in Ethiopia. *Baldrati* 4306.

2. **C. senegalensis** *Guill. & Perr.* (1832).

Herb up to 0.5 m tall. Leaf rhachis up to 7 cm long including a petiole of about half its length. Leaflets 3-7, broadly linear or oblong, the terminal one the longest, up to 50 x 9 mm, glabrous above, appressed pubescent beneath. Racemes 5-25-flowered, up to 12 cm long including a peduncle of up to 2 cm. Calyx 2 mm long, oblique; lobes triangular, the 3 lowest the longest and about as long as the tube. Corolla falling soon, pale mauve, *c* 7.5 mm long. Filaments all strongly united. Pod 40-60 x 5 mm, erect, sparsely appressed pubescent. Seeds 7-9, *c* 3 x 3 mm. Fig. 95.20.1.

Poor grassland. EW; west to Senegal, Tanzania, Namibia, and in Arabia. *Pappi* 6256, 7156, 7482.

39. **SESBANIA** *Scop.* (1777), *nom. conserv.*

Gillett in *Kew Bull.* **17**:91-159 (1963) & *Fl. Trop. E. Afr.* (1971).

Herbs or softly woody shrubs or small trees. Leaves paripinnate. Racemes axillary. Calyx truncate to shortly 5-lobed, sparsely woolly at margin. Corolla glabrous, usually yellow, the standard nearly always streaked and spotted with purple and having on its claw 2 variously shaped appendages. Vexillary stamen free. Pod long, dehiscent, rostrate, 8-50-seeded.

Some 50 species in the tropics and subtropics, more than 30 of which are in Africa. Nearly always growing in, at least seasonally, wet places.

1. Pod with 4 wings, each 4-6 mm wide; peduncle with hooked prickles at the base. **12. S. tetraptera**
- Pod not winged; peduncles usually not with hooked prickles. 2
2. Filament-sheath 35-60 mm long, curved for most of its length; mature leaflets rarely less than 6 mm wide. **1. S. grandiflora**
- Filament-sheath less than 30 mm long, only upcurved near the tip; leaflets usually less than 6 mm wide. 3
3. Appendages of standard with free tips more than 2 mm long. 4
- Appendages of standard without free tips or with free tips up to 2 mm long. 6
4. Filament-sheath 9-13 mm long. **4. S. sesban**
- Filament-sheath 15-24 mm long. 5
5. Leaflets glabrous, rest of plant almost so. **2. S. dummeri**
- Leaflets, stem and inflorescence pubescent. **3. S. goetzei**
6. Appendages of standard with acute free tips 0.5-2 mm long. 7
- Appendages of standard without free tips. 10
7. Pedicel with spreading hairs; calyx-lobes with narrow, ± subulate tips; stem with vertical rows of spots or outgrowths above leaf-axils. **5. S. rostrata**
- Pedicel glabrous; calyx-lobes broad, acuminate; stems without vertical rows of spots or outgrowths. 8
8. Blade of standard broadly cordate, wider than long. **4. S. sesban**
- Blade of standard ± oblong, as long as or longer than wide. 9
9. Inflorescence 3-6-flowered; peduncle less than 2 cm long; pod thicker at the centre than at the edges. **6. S. somalensis**
- Inflorescence usually 5-12-flowered; peduncle more than 2 cm long; pod as thick at the edges as in the middle. **7. S. quadrata**
10. Filament-sheath 16 mm or more long. **9. S. hepperi**
- Filament-sheath less than 16 mm long. 11
11. Filament-sheath over 9 mm long. **8. S. pachycarpa**
- Filament-sheath under 9 mm long. 12
12. Leaflets glabrous. **10. S. leptocarpa**
- Leaflets pubescent. **11. S. sericea**

1. **S. grandiflora** (*L.*) *Poir.* (1806).

Shrub or small tree; leaf rhachis 15-30 cm long including a petiole of *c* 1 cm; leaflets 20-50, linear-oblong, 25-40 x 6-14 mm, sparsely appressed pubescent on both surfaces; racemes 2-4-flowered, rhachis up to 5 cm long; calyx *c* 2 cm long, subtruncate; corolla up to 8 cm long, white, yellowish, greenish or red; filament-sheath 35-60 mm long, curved for most of its length; pod 20-55 x 0.6-0.9 cm with broad sutures, 15-50-seeded.

Cultivated for ornament in EE, HA; probably a native of Indonesia. The red-flowered form has been called var. *coccinea* (L. f.) Bak. ex Chiov. *Baldrati* 2; *Meyer* 8726; *Fiori* 99.

2. **S. dummeri** *Phil. & Hutch.* (1921).

Shrub or small tree up to 5 m tall; stem glabrous or sparsely pilose. Leaf rhachis up to 28(-40) cm long including a short petiole. Leaflets 20-48, linear-oblong, 10-40 x 4-8 mm, glabrous. Racemes 4-15-flowered; rhachis, including peduncle, up to 20 cm long, usually ± aculeate. Calyx *c* 7-9 mm long; lobes broad, *c* 2 mm long. Blade of standard elliptic-oblong, *c* 20 mm long; claw 4-5 mm long; appendages with free tips 4-8 mm long. Filament-sheath 17-23 mm long. Style glabrous, *c* 8 mm long. Pod 20-30 x 0.4 cm, thicker at the centre than at the margin, 40-55-seeded.

Swamps; 1200-1800 m. KF IL; Uganda, Kenya, E Zaire. *Friis* et al. 356, 570 & 1746.

3. **S. goetzei** *Harms* (1901).

Shrub or small tree up to 6 m tall; young stems densely pubescent. Leaf rhachis 1.5-4 cm long including a short petiole. Leaflets 8-18, linear-oblong, 10-18 x 3-5 mm, pubescent on both surfaces. Racemes 1-2-flowered; peduncle 0.5-2 cm long. Calyx 5-6 mm long; lobes broad, acuminate, *c* 1 mm long. Blade of standard suborbicular, emarginate, 18-24 mm long and wide; claw 4 mm long; appendages with free tips 2-4 mm long. Filament-sheath 18-20 mm long. Style glabrous, *c* 6 mm long. Pod up to 26 x 0.4 cm, thicker at the centre than at the margin, 30-45-seeded. Fig. 95.20.2.

Seasonally flooded areas, usually beside ± alkaline lakes; 1150-1650 m. SU AR GG SD; Kenya, Tanzania, NE Zambia (subsp. *multiflora* Gillett in Tanzania also). *Brehme* in *Mooney* 9064; *Ash* 298, 2837; *Lundström* 3.

4. **S. sesban** (*L.*) *Merr.* (1912).

Shrub or small tree, 1-7 m tall; young stems usually ± pubescent. Leaf rhachis 3-12 cm long including a short petiole. Leaflets 20-50, linear-oblong, up to 23 x 5 mm, glabrous or almost so above, ± pubescent beneath at least at the margins. Racemes 4-20-flowered, up to 15 cm long including a peduncle of up to 4 cm. Calyx *c* 5 mm long; lobes broad, acuminate, 0.5-1 mm long. Blade of standard 13-14 x 18-21 mm, cordate at the base; claw 4 mm long; appendages with free tips (1.5-)2-5 mm long. Filament-sheath (9-)12-13(-17) mm long. Style glabrous, *c* 5 mm long. Pod up to 20(-30) x 0.4 cm, thicker at the centre than at the margin, 20-40-seeded.

Var. **nubica** Chiov., *Fl. Somala* 1:145 (1929).

S. leptocarpa DC. var. *confaloniana* Chiov., *Fl. Somala* 2:163 (1932); *S. confaloniana* (Chiov.) Chiov. (1937).

S. punctata sensu Cuf. *Enum*. (1955), in part, non DC.

Plant ± pubescent, at least at leaf-bases. Filament-sheath 9-13 mm long. Pod 2-3.5 mm wide and with septa separating the seeds 4-6 mm apart. Fig. 95.20.3.

By streams or on the borders of freshwater lakes; 300-2000 m. EE AF EW TU GD(Sudan border) WU GJ WG SU AR HA IL KF GG SD; Somalia, west to Lake Chad and south to Natal. *Ash* 164; *Burger* 2417; *Gilbert & Tewolde* 2437.

Var. *sesban*, cultivated in Egypt and Sudan, is probably native in S Asia, var. *zambesiaca* Gillett occurs in S tropical Africa, and subsp. *punctata* (DC.) Gillett from Senegal to Sudan.

Trials are now taking place to introduce this species as a multipurpose tree in agroforestry systems. Branches can be used for either thatching or browse, living trees provide the light shade required by young coffee plants. Elsewhere the seeds are said to have medicinal properties. Some provenances are reported to be salt tolerant.

5. **S. rostrata** *Brem. & Oberm.* (1935).

Softly woody plant 1-3 m tall; stem pilose at first, later glabrous with vertical rows of spots or warty outgrowths above the leaf-axils. Leaflets 30-44, up to 25-30 x 5-6 mm, sparsely pilose on the margins and on the midrib beneath. Racemes 3-15-flowered; rhachis pilose, up to 5 cm long including a peduncle of *c* 1-1.5 cm; pedicels with spreading hairs. Calyx 6-7 mm long; lobes *c* 2 mm long with ± subulate tips. Blade of standard *c* 11-12 x 11 mm; claw *c* 3 mm long; appendages with free tips 0.5-0.6 mm long. Filament-sheath 11-13 mm long. Style glabrous, 5-6 mm long. Pod curved, 15-22 x 0.4-0.5 cm, with a slender beak of 15-35 mm, thicker at the centre than at the sutures, 25-30-seeded.

Temporarily flooded areas; *c* 550 m. IL(Gambela); Sudan to Senegal, Tanzania and S to Botswana, Madagascar. *Mesfin & Kagnew* 2408.

6. **S. somalensis** *Gillett* (1963)

- type: HA, Ogaden, 30 km NE of Welwel (Wal Wal), *Hemming* 1544 (K holo., EA FT iso.).

Semi-woody herb 0.5-1.5 m tall; stem almost glabrous. Leaf rhachis up to 6 cm long, usually aculeate. Leaflets 14-30, linear-oblong, up to 13 x 4.5 mm, glabrous. Racemes 3-6-flowered; rhachis not aculeate, up to 4 cm long including a peduncle of up to 2 cm. Calyx *c* 5 mm long; lobes broad, acuminate, 1 mm long. Blade of standard 12-18 x 12-19 mm; claw 4-5 mm long; appendages with free tips 0.5-1.8 mm long. Filament-sheath 13-17 mm long. Style glabrous, 3-4 mm long. Pod 8-14 x 0.25 cm, thicker at the centre than at margin, 18-25-seeded.

Temporarily flooded areas; 300-600 m. HA; N Somalia, N Kenya. *Peck* 61.

7. **S. quadrata** *Gillett* (1963).

Softly woody herb 1-3 m tall; stem glabrous. Leaf rhachis up to 20 cm long, usually minutely aculeate. Leaflets 12-60, linear-oblong, up to 20 x 4 mm, glabrous or sparsely pilose at the margins. Racemes 5-12-flowered; rhachis 5-14 cm long including a 2-8 cm long peduncle which is often minutely aculeate. Calyx *c* 5 mm long; lobes broad, slightly acuminate, 1 mm long. Blade of standard 15-18 x 12-14 mm; claw 3.5-4.5 mm long; appendages with free tips 1.5-2 mm long. Filament sheath 13-16(-18) mm long. Style glabrous, 4 mm long. Pod 10-18 x 0.25-0.3 cm, as thick at the margin as at the centre, 30-50-seeded. Fig. 95.20.4.

Temporarily flooded areas; 1000-1700 m. SU HA; N Somalia, Uganda, Kenya, Tanzania, Burundi. *Amare Getahun* in *IECAMA* J64, G 14; *Burger* 1324.

Figure 95.20.1 - *CYAMOPSIS SENEGALENSIS*, a: flowering shoot from *Schweinfurth* 677, b: fruiting raceme from *Kotschy* 141. 2 - *SESBANIA GOETZEI*, a: flowering branch from *Jackson* 722, b: pod from *Ash* 298. 3 - *S. SESBAN* var. *NUBICA*, a: flowering branch from *Gillett* 14748, b: pod from *IECAMA* BH-74. 4 - *S. QUADRATA*, pod from *Burger* 1324. 5 - *S. TETRAPTERA*, pod from *Parker* E. 467. 6 - *ORMOCARPUM TRICHOCARPUM*, a: flowering branch, b: pod, both from *Turton* 90. 6b x 1.125, the rest x 0.75. Drawn by Maureen Church. (Reproduced with permission from *Opera Botanica*, 68, fig. 22, 1983.)

8. **S. pachycarpa** *DC.* (1825).

Semi-woody herb 1-2.5 m tall; stem glabrous, sometimes sparsely aculeate. Leaf rhachis up to 40 cm long, aculeate, especially towards the base. Leaflets 60-120, linear-oblong, up to 22(-32) x 5(-7) mm, glabrous. Racemes 3-18-flowered; rhachis not aculeate, up to 5 mm long including a peduncle of up to 1.5 cm. Calyx *c* 5 mm long; lobes broad, slightly acuminate, 1 mm long. Blade of standard 12-13 x 13-16 mm; claw 4 mm long; appendages rounded without free tips. Filament-sheath 11-13(-15) mm long. Style pubescent towards the tip, 5 mm long. Pod 14-30 x 0.3-0.6 cm, thicker at the centre than at the margin, 20-30-seeded.

Temporarily flooded areas; *c* 900-1000 m. EW TU GD(Sudan border); west to Senegal and Cape Verde Is., Uganda, Zaire, Angola (a subspecies in Namibia). *Baldrati* D 90; *Pappi* 6300, 6440; *Schweinfurth* 1869 A.

9. **S. hepperi** *Gillett* (1963).

Slightly woody herb up to 2(-4) m tall; stem glabrous or almost so. Leaf rhachis up to 20 cm long. Leaflets 14-70, linear-oblong, up to 14 x 13 mm, glabrous. Racemes 5-7-flowered; rhachis up to 9 cm long, including a peduncle of up to 2 cm. Calyx *c* 7 mm long; lobes broad, acuminate, 2 mm long. Blade of standard *c* 18 mm long and wide; claw 6 mm long; appendages almost or quite without free tips. Filament-sheath 16-20 mm long. Style pubescent towards the tip, 5-6 mm long. Pod *c* 26-28 cm long, 26-30-seeded.

Probably temporarily flooded areas; *c* 1000 m. GD (Sudan border); Guinea, Mali, Nigeria, Sudan. *Schweinfurth* 1869B.

10. **S. leptocarpa** *DC.* (1825).

Slightly woody herb up to 1(- *c* 2) m tall; stem glabrous or almost so. Leaf rhachis up to 10 cm long. Leaflets 30-60, linear-oblong, up to 14 x 3 mm, glabrous. Racemes 1-6-flowered; peduncle 5-50 mm long. Calyx 2.5-3.5 mm long; lobes triangular *c* 0.8 mm long. Blade of standard *c* 8 mm long and wide; claw *c* 3 mm long; appendages without free tips. Filament-sheath *c* 8 mm long. Style *c* 2 mm long, glabrous. Pod up to 15 x 0.3 cm, torulose, 20-30-seeded.

Temporarily flooded areas; weed in cotton fields; *c* 400-1250 m. AF EW HA GG; Senegal, Nigeria, Sudan, Zambia, Mozambique and Yemen. *Pappi* 6473; *Klingel* 47; *Parker* E. 93; *Handlos* 12.

11. **S. sericea** (*Willd.*) *Link* (1822).

Erect slightly woody herb, 1-3 m tall, pubescent throughout except for the flower and fruit. Leaf rhachis 10-15 cm long, sometimes minutely aculeate near the base. Leaflets 40-50, up to 20 x 4 mm. Racemes 2-6-flowered; rhachis up to 4.5 cm long including a peduncle of up to 2 cm. Calyx 3-4 mm long, glabrous except at the base and the margin; lobes triangular with subulate tips. Blade of standard *c* 6 x 8-10 mm; claw 2 mm long; appendages very narrow, truncate at the tip. Filament-sheath 6-8 mm long. Style glabrous, 2 mm long. Pod up to 16 mm long, thicker at the centre than at the sutures, 15-30-seeded.

Marshy ground; *c* 400 m. GG; widespread in tropical Africa; S Arabia, Ceylon (introduced), West Indies and northern S America (perhaps not native). *Carr* 390; *Corradi* 4128, 4213.

12. **S. tetraptera** *Bak.* (1871).

Erect annual, 1-2 m tall; stems sparsely pilose when young, glabrous later, sometimes aculeate. Leaf rhachis 10-22 cm long. Leaflets 40-60, up to 24 x 6 mm, glabrous. Racemes 2-12-flowered; rhachis up to 11 cm long including a peduncle of up to 3.5 cm with hooked prickles at the base. Calyx *c* 4.5 mm long; lobes acuminate. Blade of standard 7-10 x 9-13 mm; claw 4 mm long; appendages rounded. Filament-sheath 7-9 mm long. Style pubescent, 3-4 mm long. Pod 12-16 cm long, the sutures expanded into 4 wings each 4-6 mm wide, 15-20-seeded. Fig. 95.20.5.

Seasonally flooded areas; weed in cotton fields; *c* 600 m. AF; Sudan, Tanzania, Mozambique, South Africa (Transvaal). *Parker* E. 467.

40. **ORMOCARPUM** *Beauv.* (1806), *nom. conserv.*

Gillett in *Kew Bull.* **20**: 323-355 (1966) & *Fl. Trop. E. Afr.* (1971).

Shrubs or small trees usually with both normal, rather weak, white hairs and stiffer, darker, swollen-based hairs which often exude a yellow viscid fluid when young and whose bases may persist as small tubercles when old. Leaves often fasciculate on short shoots, imparipinnate; leaflets usually ± alternate, glabrous and minutely black-dotted above. Flowers in axillary racemes or solitary. Calyx usually strongly veined, 5-lobed, the lower lobe longest, the 2 upper lobes united for about half their length. Corolla glabrous or the 3 upper petals somewhat pubescent. Stamens divided dorsally into 2 groups of 5 or all united to form a sheath slit above. Pod breaking transversely into 1-6, 1-seeded, indehiscent, elliptic or oblong segments.

Some 20 species, the majority in Africa.

Several species have very fibrous roots and stem which can be used in making alcoholic drinks.

1. Corolla falling off after fertilization; tubercle-based bristles on pod numerous, up to 6(-7) mm long. **1. O. trichocarpum**
- Corolla persistent; tubercle-based bristles on pod less than 2 mm long. 2
2. Ovary and pod straight or almost so; some tubercles on the pod up to 0.8 mm wide at the base. **2. O. trachycarpum**

- Ovary becoming strongly arcuate, fruit strongly curved; tubercles on the pod less than 0.3 mm wide at the base. 3
3. Leaflets 7-11, up to 7 mm long; blade of standard broadly cuneate to truncate but not cordate at the base. **3. O. muricatum**
- Leaflets often 15-17, often more than 7 mm long; blade of standard cordate at the base. **4. O. pubescens**

1. **O. trichocarpum** (*Taub.) Engl.* (1904).
O. mimosoides sensu Cuf. *Enum.* (1955), non S. Moore.

Shrub or small tree 1-5 m tall. Leaf rhachis 1-4 cm long. Leaflets 7-15, elliptic-oblong, up to 7(-11) x 3.5 mm, white pubescent beneath at least on the midrib and margins. Racemes 1-3-flowered; rhachis up to 25 mm. Calyx *c* 8-11 mm long, usually glabrous except at the margin, never with tubercle-based hairs; lower lobe 4-6 mm long, acute. Corolla falling soon, cream or bluish with dark veins; blade of standard 10-14 x 11-17 mm, ± truncate at the base. Ovary straight, 5-8-ovulate; style 6-9 mm long. Pod straight or almost so, up to 5 cm long, densely covered with stiff yellowish tubercle-based bristles up to 6(-7) mm long. Fig. 95.20.6.

Wooded grassland and in dry bushland; 700-2100 m. BA IL GG SD; throughout eastern Africa south to Natal. *Gillett* 14207; *Friis et al.* 1035; *Thulin, Hunde & Tadesse* 3486.

2. **O. trachycarpum** *(Taub.) Harms* (1915).
O. melanodictyotum Chiov. (1939) - type: SD, *Cufodontis* 567 (FT holo.).

Shrub or small tree 1-6 m tall. Leaf rhachis up to 5 cm long. Leaflets 9-15, elliptic-oblong, up to 15 x 6 mm, white pubescent beneath and usually with a few tubercle-based hairs on the margin. Racemes 1-4-flowered; rhachis up to 30 mm long. Calyx up to 9 mm long, sometimes with normal hairs at the margin, often with tubercle-based hairs near the base; lower lobe 4-5 mm long. Corolla persistent, cream or greenish-yellow with dark veins; blade of standard 14-17 x 13-14 mm, broadly cuneate to truncate at the base. Ovary straight, 4-6-ovulate; style 11-13 mm long. Pod straight or almost so, up to 6.5 cm long, pubescent and with many tubercles up to 0.8 mm in diameter.

Deciduous woodland, bushland and grassland; 1200-1900 m. HA SD; N Uganda, Kenya, Tanzania. *Gillett* 5084; *Robecchi-Bricchetti* 258; *Burger* 2095; *Thulin, Hunde & Tadesse* 3443.

3. **O. muricatum** *Chiov.* (1939)
- type: SD, *Cufodontis* 111 (FT holo.).

Shrub 1-3 m tall. Leaf rhachis up to 2.5 cm long. Leaflets 7-11, elliptic-oblong, up to 7 x 3.5 mm, glabrous or with white hairs beneath, sometimes also a few tubercle-based hairs on the margin. Inflorescences 1-2-flowered. Calyx *c* 7 mm long, sometimes with white hairs at the margin and tubercle-based hairs near the base; lower lobe 4 mm long. Corolla persistent, cream or greenish-cream with dark veins; blade of standard 11-14 x 10-16 mm, truncate, or rounded at the base. Ovary soon arcuate, 3-4-ovulate; style 8-12 mm long. Pod strongly curved, pubescent, with or without small tubercles not above 0.3 mm in diameter.

Dry bushland; 400-1300 m. HA SD; Somalia, Kenya. *Burger* 3068; *Riva* 811; *Gilbert* 2096.

4. **O. pubescens** (*Hochst.) Cuf.* ex *Gillett* (1966);
Pictetia pubescens Hochst. (1846) - type: GD, Schoata, *Schimper* II:1298 (B holo. destr., K P iso.).
Acrotaphros bibracteata Steud. ex A. Rich. (1847); *Ormocarpum bibracteatum* (Steud. ex A. Rich.) Bak. (1871) - types: TU, *Quartin Dillon* s.n. (P syn.), *Petit* s.n. (P syn.), GD, *Schimper* II:1298 (P syn., K isosyn.).

Shrub or small tree, up to 3 m tall. Leaf rhachis up to 7 cm long. Leaflets up to 15-17(-21), elliptic-oblong, *c* 6-10 x 4 mm, slightly white pubescent beneath. Inflorescences 1-2-flowered. Calyx *c* 8 mm long, usually pubescent; lower lobe *c* 4 mm long. Corolla persistent, purplish; blade of standard 12-17 x 15-20 mm, cordate. Ovary soon arcuate, 5-6-ovulate; style *c* 10 mm long. Pod curved, pubescent and sparsely and minutely tuberculate.

Woodland or bushland; 1000-2100 m. EE EW TU GD; Senegal, Cameroun, Sudan. *Pappi* 4714; *Bally* 7034; *Gilbert & Getachew* 2843.

The closely related *O. kirkii* S. Moore occurs from S Somalia south to Transvaal.

41. AESCHYNOMENE *L.* (1753)

Verdcourt in *Fl. Trop. E. Afr.* (1971).

Shrubs or herbs, mostly covered with tubercle-based hairs. Leaves paripinnate. Inflorescences axillary or flowers sometimes solitary. Calyx 2-lipped. Corolla mostly yellow and lined with purple. Stamens usually all united to form a sheath. Pod narrow, compressed, straight or curved, breaking transversely into 1-28, 1-seeded, indehiscent, round to rectangular segments.

Some 150 species, half of which are found in Africa.

1. Pod of 1-9 segments joined by narrow necks. 2
- Pod with straight margins and 4-28 segments. 4
2. Pod of 3-9 segments; leaflets ciliate. **8. A. americana**
- Pod of 1-2 segments; leaflets glabrous. 3
3. Segments 5-7 x 4-5 mm. **6. A. abyssinica**
- Segments 9 x 6 mm. **7. A. ruspoliana**
4. Stems spiny; pod spirally coiled; corolla 3-4.5 cm long. **2. A. elaphroxylon**
- Stem not spiny; pod ± straight; corolla usually smaller. 5
5. Pod oblong-elliptic, 3-5 x 1.1-1.5(-1.9) cm, 6-7-jointed. **5. A. pfundii**
- Pod linear or linear-oblong, 2.4-11 x 0.3-1 cm, 4-28-jointed. 6

6. Keel-petals not laciniate along lower margin. **1. A. indica**
- Keel-petals laciniate. 7
7. Standard glabrous; pod with 4-28 segments which are 3-6 mm long; leaves 1-8(-9.5) cm long. **3. A. schimperi**
- Standard hairy outside towards the apex; pod with 6-12 segments which are 8-12 mm long; leaves 4-20 cm long. **4. A. cristata**

1. **A. indica** *L.* (1753).

Subshrubby herb 0.3-2.5 m tall, submerged part of stem becoming thick and spongy. Leaflets 16-50(-70), linear-oblong, 3-13 x 1-3 mm, glabrous. Inflorescences 1-6-flowered. Calyx 4-6 mm long, glabrous, upper lip 2-fid, lower 3-fid. Standard elliptic, 7-10 mm long; keel-petals not laciniate. Pod 24-48 x 4 mm with a stipe 6-9 mm long, 5-13-jointed, slightly constricted between the oblong segments along one margin, covered with sparse short tubercle-based hairs.

In wet places, such as margins of swamps and pools; 700-1200 m. EE EW GD SD GG; widespread in tropical Africa and in Asia, Australia and N America. *Pappi* 2418, 6192; *Gilbert* 3419; *Mesfin & Kagnew* 1755.

Elsewhere the leaves are grazed but the seeds are reported to be toxic.

2. **A. elaphroxylon** (*Guill. & Perr.*) *Taub.* (1894).

Shrub or small tree 2-9(-12) m tall; often ± floating; stems often swollen below, with short and long sticky hairs and short sharp spines. Leaflets 20-40, oblong, 8-26 x 4-10 mm, glabrescent above, short pubescent beneath. Inflorescence 1-4-flowered. Calyx up to 2.5 cm long, densely pubescent, lips entire or 2-3-toothed at the apex. Standard rounded, 3-4.7 cm long and wide, puberulous above on the outside; keel-petals laciniate along their lower margins. Pod spirally coiled, 10-14 x 0.8 cm, 6-17-jointed, densely covered with short and long glandular hairs; segments 6-8 mm long.Fig. 95.21.1.

Lake-sides and pools; 1200-1850 m. SU AR GG SD; widespread in tropical Africa. *Mooney* 7299; *Ash* 379; *Jackson* 748.

The AMBATSCH tree is famous for its very light wood which is used for making canoes, fishing floats, etc.

3. **A. schimperi** *Hochst.* ex *A. Rich.* (1847)
- types: TU, *Quartin Dillon* (P syn.), *Schimper* II:1073 (P syn., BM K S STU UPS WAG isosyn.), *Schimper* 1528 (P syn., BM K isosyn.).

Shrubby herb or small shrub, 0.6-3 m tall; stem usually ± pubescent to hispid. Leaflets (10-)18-68, linear to linear-oblong, 2-12 x 1-2.5 mm, glabrous except for the often ciliolate margins. Inflorescences 1-2(-4)-flowered. Calyx 0.6-1.8 cm long, ± pubescent, lips entire or 2-3-fid. Standard rounded, 0.8-3.3 cm long, glabrous; keel-petals laciniate along their lower margins. Pod linear, straight or somewhat curved, 3-10 x 0.4-0.7 cm with a stipe 0.4-1.2 cm long, (4-)6-28-jointed, glabrous to densely hairy; segments 3-6 mm long. Fig. 95.21.2.

Swampy places; 1350-2200 m. EW TU GD GJ WG SU HA KF GG SD; widespread in tropical Africa. *Gilbert & Thulin* 487; *Gilbert & Getachew* 3024; *Burger* 3230.

4. **A. cristata** *Vatke* (1878).

Shrub or herb 0.9-3(-6) m tall; stems thick and procumbent with erect branches, glabrous to densely hairy. Leaflets 40-86, linear to oblong, 6-22 x 1-4.5 mm, glabrous except for the ciliolate margins. Inflorescences (1-)2-8-flowered. Calyx up to 1.6 cm long, pubescent, upper lip emarginate, lower 3-lobed. Standard round, 1.8-3.3 cm long and wide, pubescent above on the outside; keel-petals laciniate along their lower margins, hairy. Pod linear, straight or slightly curved, 5-11 x 0.7-1 cm with a stipe *c* 2 cm long, 6-12-jointed, glabrous or sparsely hairy; segments 8-12 mm long.

Var. **pubescens** *J. Léon.* in *Bull. Jard. Bot. État* **24**: 67 (1954).

Stems glabrous. Petiole and rhachis sparsely pubescent.

Swampy places; *c* 1250 m. GG SD(Lake Abaya); widespread in tropical Africa. *Eriksson* 492; *Vàtova* 1831; *Aweke & De Wit* 1543.

Var. *cristata*, with hairy stems and petioles, is widespread in tropical Africa.

5. **A. pfundii** *Taub.* (1895).

Shrub 1.2-4.5 m tall, often ± floating; stems often swollen below, densely hispid with tubercle-based glandular hairs. Leaflets 16-31(-40), oblong, 10-35 x 4-12 mm, ciliate on margins and main nerve beneath, ciliate-serrulate towards the apex. Inflorescences 3-7-flowered. Calyx 1.3-1.8 cm long, hairy, upper lip emarginate, lower 3-lobed. Standard rounded, 2-4 cm long, ciliolate at the apex; keel-petals laciniate. Pod oblong-elliptic, straight, 3-5 x 1.1-1.5(-1.9) cm, sessile, 6-7-jointed, hispid with tubercle-based hairs; segments 3-12 mm long.

Swampy places; *c* 1200 m. SD(Lake Abaya); Sudan, Kenya, Tanzania, Malawi, Zambia. *Kuls* 177.

6. **A. abyssinica** (*A. Rich.*) *Vatke* (1879);
Rüppellia abyssinica A. Rich. (1847); *Aeschynomene rüppellii* Bak. (1871), nom. illegit. - type: Tekeze R., *Quartin Dillon* s.n. (P holo.).

Shrub or herb 0.8-4.5 m tall; stems sometimes with viscid tubercle-based hairs when young, later glabrous. Leaflets 10-30(-48), linear to oblanceolate, 2-12 x 0.5-3 mm, glabrous. Inflorescences several-flowered. Calyx 3-7 mm long, glabrous or almost so, upper lip emarginate, lower 3-lobed. Standard violin-shaped, 0.5-1.3 cm long, glabrous; keel-petals not laciniate. Pod of 1-2 segments joined by a very narrow neck, each 5-7 x 4-5 mm, glabrous. Fig. 95.21.3.

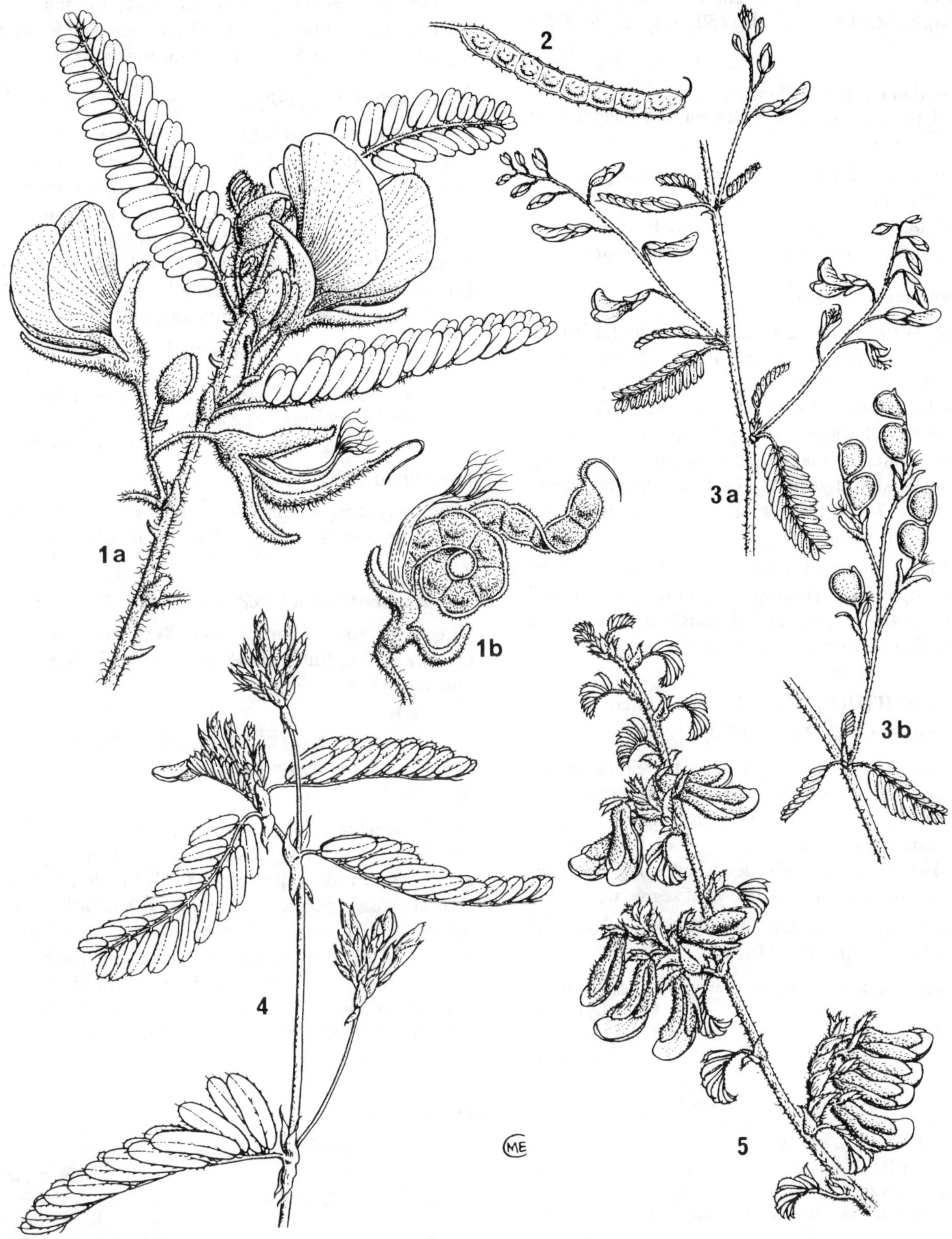

Figure 95.21.1 - *AESCHYNOMENE ELAPHROXYLON*, a: flowering branch from *Jackson* 716, b: pod from *W. de Wilde* 6672. 2 - *A. SCHIMPERI*, pod from *Wickens* 1453. 3 - *A. ABYSSINICA*, a: flowering branch from *Mooney* 5962, b: fruiting raceme from *Meyer* 8855. 4 - *SMITHIA ABYSSINICA*, fertile branch from *Gilbert & Thulin* 905. 5 - *KOTSCHYA RECURVIFOLIA* subsp. *AETIOPICA*, flowering branch from *Friis* et al. 443. 1, 2 & 3a x 0.75, 3b, 4 & 5 x 1.125. Drawn by Maureen Church. (Reproduced with permission from *Opera Botanica*, 68, fig. 23, 1983.)

Grassland, woodland or scrub; 1300-2900(-3300) m. TU GD GJ WG SU IL KF GG SD; N Nigeria, Cameroun, Uganda, Kenya, and south to Zimbabwe and Mozambique. *Gilbert & Thulin* 650, 921; *W. de Wilde* 7596.

7. **A. ruspoliana** *Taub.* ex *Harms* (1897)
- type: SD, between Bidduma and Alghe, *Riva* 1294 (FT holo.).

Similar to *A. abyssinica*, except for the larger pods with segments 9 x 6 mm. Flowers unknown.

Only known from the type which looks distinct but possibly is just a large-fruited variant of *A. abyssinica*.

8. **A. americana** *L.* (1753).

Sprawling herb; stems glabrous or glandular-bristly. Leaflets 20-60, somewhat falcate, 4-15 x 1-2 mm, ciliate. Inflorescences few-flowered. Calyx 3-6 mm long, upper lip 2-toothed, lower 3-toothed. Corolla pale pink; standard with yellowish central spot, 6-8 mm long; keel-petals not laciniate. Pod of 3-9 segments, curved, the upper margin almost straight, the lower deeply indented; segments 3-6 x 2.5-5 mm, ± rugose in centre, glabrous or pubescent.

Roadsides, forest clearings; *c* 1000 m. IL KF; native of tropical and subtropical America, introduced elsewhere in the tropics and obviously naturalised in Ethiopia. *Friis, Gilbert & Vollesen* 4058.

42. **SMITHIA** *Ait.* (1789), *nom. conserv.*

Verdcourt in *Kew Bull.* **24**: 15 (1970).

Herbs or subshrubs. Leaves paripinnate; leaflets with a single main nerve; stipules spurred. Inflorescences axillary, usually dense scorpioid cymes. Calyx 2-lipped. Stamens usually in 2 groups of 5. Pod folded like a pleated skirt, included in the persistent calyx, breaking transversely into 1-seeded, indehiscent segments.

Some 30 species in the Old World tropics, 2 of which are found in tropical Africa.

1. Flowers yellow; bracts spurred, persistent. **1. S. abyssinica**
- Flowers purplish; bracts not spurred, falling soon. **2. S. elliotii**

1. **S. abyssinica** (*A. Rich.) Verdc.* (1970);
S. sensitiva Ait. var. *abyssinica* A. Rich., *Tent. Fl. Abyss.* 1:203 (1847) - types: TU, *Quartin Dillon* s.n. (P syn., BR K isosyn.), *Schimper* III:1838 (P syn., BM K S isosyn.).
S. erubescens sensu Cuf. *Enum.* (1955), non (E. Mey.) Bak.

Annual, 30-70 cm tall; stem glabrous. Leaflets 8-24, linear-oblong, 5-15 x 1.5-3.5 mm, ciliate on margins and midvein beneath. Inflorescences 1-4(-6)-flowered, subumbellate; bracts spurred, persistent. Calyx *c* 10 mm long, striate, glabrous. Corolla yellow with the standard ± 8.5 x 3.5 mm. Pod of *c* 5 rounded segments, each *c* 2.5 x 2.2 mm. Fig. 95.21.4.

Swampy grassland, roadsides, waste ground; 1800-2450 m. TU GD GJ WG SU KF; not known elsewhere. *Gilbert & Thulin* 905; *Mooney* 7593; *Stewart* C33.

2. **S. elliotii** *Bak. f.* (1929).

Decumbent herb, 0.4-1.8 m long; stems bristly to glabrescent. Leaflets 10-28, linear-oblong, 3-15 x 1.2-5 mm, ciliate, bristly on nerves beneath. Inflorescence up to 12-flowered or more, subumbellate; bracts not spurred, falling soon. Calyx up to 10 mm long, striate, densely to very sparsely covered with yellow bristles. Corolla purplish with the standard *c* 10 x 4-8 mm. Pod of 4-7 segments, each *c* 2-3 mm long and wide.

1. Calyx densely covered with long yellow bristly hairs. var. **elliotii**
- Calyx with very sparse short yellow tubercle-based hairs, appearing almost glabrous. var. **sparse-strigosa**

Var. **elliotii**.

Swampy ground, streamsides, *c* 1750 m. WG KF; widespread in tropical Africa. *Friis* et al. 2242; *Giordano* 2443.

Var. **sparse-strigosa** *Verdc.* in *Kew Bull.* **24**: 17 (1970).

Grassland, rocky slopes; 1600-2000 m. WG BA; Uganda, Zaire, Burundi. *W. de Wilde* 6751, 8788, 8808; *Friis* et al. 3590.

43. **KOTSCHYA** *Endl.* (1839)

Verdcourt in *Kew Bull* **24**: 17-54 (1970) & *Fl. Trop. E. Afr.* (1971).

Herbs or shrubs, the stems generally covered with tubercle-based often glandular hairs. Leaves paripinnate; stipules not spurred. Leaflets alternate with several basal nerves, the main nerve submarginal. Inflorescences dense, 1-several-flowered, axillary, ± scorpioid, the flowers reflexed, arranged distichously. Calyx 2-lipped, the lower lip 3-lobed, the upper bifid. Stamens all joined. Pod stipitate, enclosed in the calyx, 1-9-jointed, folded like a pleated skirt; segments rounded, indehiscent.

Some 30 species, all in tropical Africa and Madagascar.

1. Leaflets 20-42, slightly curved at the apex. **1. K. africana**
- Leaflets 8-14, very distinctly curved. **2. K. recurvifolia**

1. **K. africana** *Endl.* (1839)
- type: Sudan/Ethiopia border, Tumat R. near Beni Shangul and Camaril, *Kotschy* 524 (W holo., BM K iso.).

Shrub 1-4.5(-6) m tall. Stems covered with glandular tubercle-based hairs. Leaf rhachis 2-7.6 cm long including the short petiole. Leaflets 20-42, oblong, 2-10(-12.5) x 0.5-3.5(-4) mm, slightly falcate at the apex, acute, glabrous or pubescent; margins denticulate-ciliate. Inflorescences 1-few-flowered. Calyx pubescent and bristly hairy, 1.3-2.2 cm long. Standard orange-yellow, often purplish-veined, obovate or rounded, 9.5-19 mm long, glabrous. Pod 4-9-jointed, pubescent; segments 3-4 x 2-3.2 mm.

Wooded grassland, forest edges, also in cultivations; 1400-2050 m. WG SU KF GG; fairly widespread in eastern tropical Africa, probably also Madagascar. *Mooney* 6907; *Friis et al.* 437; *W. de Wilde* 8879.

Verdcourt (1970) recognized four varieties of this species in tropical Africa, two of which, var. *africana* and var. *bequaertii* (De Wild.) Verdc. were recorded from Ethiopia. They differ mainly in flower size (calyx 1.6-2.2 cm and 1.3-1.8 cm long respectively). Judging from the Ethiopian material seen, the distinction does not seem meaningful in this area and var. *bequaertii* is therefore omitted here.

2. **K. recurvifolia** *(Taub.) F. White* (1962).

Sticky shrub 0.2-3.6(-4.5) m tall. Stems pubescent and with longer tubercle-based hairs. Leaf-rhachis 0.3-2.3 cm long. Leaflets 8-18, narrowly oblong-falcate, distinctly curved, 3-9(-11) x 0.7-2.5 mm; margins ciliate. Inflorescences few- to several-flowered. Calyx pubescent and ciliate, 0.5-1.2 cm long. Standard golden-yellow or whitish, elliptic to rounded or obovate-oblong, 9-16 mm long, glabrous. Pod 1-3-segmented, pilose; segments *c* 3.2 x 2.8 mm.

Subsp. **aetiopica** *Verdc.* in Kew Bull **24**: 30 (1970)
- type: KF, Sodu road above Nada, *Mooney* 6203 (K holo., EA iso.).

Stems ± densely pubescent. Leaves 5-8 mm long; leaflets 8-14, up to 6 x 1.5 mm. Upper calyx-lip divided for about 4 mm into acute diverging lobes. Standard 12-13 x 9-10 mm. Fig. 95.21.5.

Grassy slopes, hillsides, roadsides; 1350-2400 m. BA HA KF SD; not known elsewhere. *Mooney* 7710; *Friis et al.* 1529; *Meyer* 7958.

There are three more subspecies in Kenya, Tanzania, Zambia and Malawi.

44. STYLOSANTHES *Sw.* (1788)

Perennial herbs or subshrubs. Leaves pinnately 3-foliolate; stipules adnate to the petiole for most of their length. Inflorescences dense, with primary and secondary bracts; flowers subsessile with a long filiform receptacle. Calyx 5-lobed, the lowest lobe longest. Corolla yellow or orange. Filaments all joined; anthers 10, alternately long and short. Style long and filiform, breaking off after flowering. Pod oblong, beaked, 1-2-jointed, reticulate.

Some 25 species in the tropics and subtropics.

Elsewhere an important genus of fodder and forage plants being introduced to Ethiopia.

S. fruticosa *(Retz.) Alston* (1931);
Arachis fruticosa Retz. (1788).
S. flavicans Bak. (1871).

Woody herb, 0.1-1 m tall, usually with a thick woody rootstock. Stems pubescent to densely hairy. Leaflets elliptic or lanceolate, 5-33 x 1-9 mm, usually pubescent and ± bristly. Receptacle 5-7 mm long; calyx-lobes 2.5-3 mm long. Standard 5-7 x 4-5 mm. Pod 4-9 mm long with segments 3.5-4 x 2-2.5 mm, usually densely pubescent. Fig. 95.22.1.

Grassland, woodland, scrub, old cultivations; 1200-1900 m. EE EW SU BA HA KF GG SD; widespread in tropical Africa, also in Arabia, Madagascar, India and Ceylon. *Gilbert, Thulin & Getachew* 231; *Cufodontis* 704; *Pappi* 7058.

45. ARACHIS *L.* (1753)

Annuals or perennials. Leaves paripinnate with 4 leaflets; stipules partly adnate to the petiole. Inflorescences axillary, of short dense sessile spikes or flowers solitary, with primary and secondary bracts; flowers subsessile with a long, filiform, pedicel-like receptacle. Calyx 5-lobed. Corolla yellow, often with red nerves. Filaments all joined; anthers 8-10, alternately long and short. Style very long, filiform, soon breaking off. Pod oblong or sausage-shaped, 1-few-seeded, indehiscent, developing below the soil, having been pushed beneath by lengthening, reflexing and stiffening of the gynophore.

Some 20 species in S America, one of which, the groundnut, is widely cultivated in the warmer parts of the world.

A. hypogea *L.* (1753).

Annual, stems *c* 30 cm long. Leaflets obovate or elliptic, 1-7 x 0.7-3.2 cm. Receptacle 0.2-4 cm long, pilose; corolla 0.7-1.3 cm long. Pod 2-6 x 1-1.5 cm; gynophore becoming 1-20 cm long. Fig. 95.22.2.

Grown as an oilcrop mainly in EE, EW, TU and HA, and occasionally found as an escape. *Westphal* 971; *Seegeler* 1749, 1952.

46. ZORNIA *J. F. Gmel.* (1791)

Milne-Redhead in *Fl. Trop. E. Afr.* (1971).

Annuals or perennials. Leaves digitately 2- or 4-foliolate; leaflets ± glandular-punctate; stipules and bracts conspicuous, extended below point of insertion. Flowers in a ± lax spike; bracts in pairs, enclosing the flowers when in bud. Calyx hyaline, ciliate, the 2 upper lobes connate. Filaments all united into a tube; anthers 10, alternately long and short. Pod flattened, jointed; segments almost always with bristles.

Some 80 species in the tropics, 13 of which are native in Africa.

1. Leaflets usually 4, ± obtuse. **4. Z. setosa**
- Leaflets usually 2, acute. 2
2. Annual; flowers much shorter than the bracts. **1. Z. glochidiata**
- Perennial; flowers about as long as or longer than the bracts. 3
3. Leaflets of lowest leaves less than twice as long as broad; bristles on pod glochidiate and retrorsely hispid. **2. Z. apiculata**
- Leaflets of lowest leaves more than twice as long as broad; bristles on pod with short spreading hairs, rarely glabrous. **3. Z. pratensis**

1. **Z. glochidiata** *Reichb.* ex *DC.* (1825).

Annual, up to 40 cm long. Leaflets 2, of upper leaves lanceolate, up to 30 x 8 mm, acute; those of lower leaves smaller and proportionally broader. Flowers much shorter than the bracts. Calyx *c* 3 mm long. Corolla yellow or red; standard *c* 5 mm long. Pod up to 20 mm long, 4(-6)-seeded; segments 3.5 x 3 mm; bristles stiff, 2-3 mm long, glochidiate and retrorsely hispid.

Grassland, rocky places, often also as a weed; 550-2000 m. EE EW TU GD SU BA HA IL SD; widespread in tropical Africa, and in Madagascar. *Gilbert* 2166; *Gilbert & Getachew* 2962; *Ash* 2164.

2. **Z. apiculata** *Milne-Redh.* (1954).

Perennial, up to 45 cm long. Leaflets 2, of upper leaves ovate-lanceolate, up to 25 x 10 mm, apiculate; those of lower leaves smaller, ovate or suborbicular. Flowers about as long as the bracts. Calyx *c* 3.5 mm long. Corolla yellow, sometimes with red veins; standard up to 8 mm long. Pod up to 16 mm long, 4(-6)-seeded; segments 3 x 2.5 mm; bristles stiff, 1.5 mm long, glochidiate and retrorsely hispid.

Grassland, rocky places, scrub; 1500-2100 m. SU AR HA GG SD; Somalia, Kenya, Tanzania. *Gilbert & Thulin* 8; *Mooney* 8226; *Burger* 2022.

3. **Z. pratensis** *Milne-Redh.* (1954).

Perennial, up to 35 cm long. Leaflets 2, of upper leaves linear-elliptic, up to 27 x 5 mm, acute or shortly apiculate; those of lower leaves similar. Flowers longer than the bracts. Calyx *c* 4 mm long. Corolla yellow with red veins; standard up to 10 mm long. Pod up to 17 mm long, 5-6(-7)-seeded; segments 2.5 x 2 mm; bristles with short spreading hairs, rarely glabrous. Fig. 95.22.3.

Grassland, rocky places; 1250-2100 m. TU GJ WG AR IL KF SD; widespread in tropical Africa. *Thulin* 1407; *Schimper* I:305; *W. de Wilde* 10813.

The Ethiopian material belongs to subsp. *pratensis* var. *pratensis*. There is one more subspecies and one more variety in tropical Africa.

4. **Z. setosa** *Bak. f.* (1929).

Prostrate perennial, up to 60 cm long. Leaflets 4, terminal ones obovate or oblanceolate, up to 21 x 11 mm, rounded at the apex. Flowers longer than the bracts. Calyx *c* 3.5 mm long. Corolla yellow, red or mauve; standard up to 9 mm long. Pod up to 20 mm long and 8-seeded; segments 2.5 x 2.5 mm.

Subsp. **obovata** (*Bak. f.*) *J. Léon. & Milne-Redh.* in *Bol. Soc. Brot., sér. 2*, **28**: 99 (1954).

Bristles on the pod *c* 1 mm long or less or absent, glabrous or pubescent.

Grassland, also as a weed; 1150-2200 m. WG AR BA HA KF SD; Uganda, Kenya and south to Zambia. *Thulin* 1383; *Gilbert & Thulin* 558, 841; *Mooney* 5934.

Subsp. *setosa* occurs in southern tropical Africa.

47. DESMODIUM *Desv.* (1813), *nom. conserv.*

Schubert in *Fl. Congo Belge* (1954) & *Fl. Trop. E. Afr.* (1971).

Herbs or shrubs. Leaves pinnately 3-foliolate or 1-foliolate. Inflorescences axillary or terminal, falsely racemose or racemose-paniculate. Calyx 2-lipped with the 2 upper lobes connate (in Ethiopia). Anthers uniform. Pod flattened, (1-)2-many-segmented, glabrous to densely pubescent with straight or hooked hairs.

Some 300 species in tropical and temperate areas, but not in Europe. Several species are good forage or cover crops.

1. Leaves 1-foliolate. 2
- Leaves, at least most of them, 3-foliolate. 3
2. Leaflets thick with ± undulate or sinuate margins, tomentose beneath; inflorescence a dense leafy panicle or raceme. **5. D. velutinum**
- Leaflets more thin with entire margins, pilose beneath; inflorescence composed of lax slender racemes. **6. D. gangeticum**
3. Flowers large, pink to brick red with the standard 8-15 mm long; pod with narrowly oblong, moon-shaped segments up to 8 x 3 mm. **4. D. repandum**
- Flowers smaller, or if standard 8-10 mm long with proportionally much broader segments. 4
4. Pod glabrous (in Ethiopia), dehiscent along the lower suture. **1. D. hirtum**
- Pod pubescent, usually indehiscent, the segments only separating from each other. 5
5. Secondary bracts absent. 6
- Secondary bracts present. 7
6. Calyx subequally 5-lobed, silky pilose. **11. D. barbatum**
- Calyx 2-lipped, if pilose then on lobes only. 8
7. Pod with the upper suture straight or slightly indented between segments; segments ± obliquely oblong; plants creeping or scandent. **2. D. adscendens**

Figure 95.22.1 - *STYLOSANTHES FRUTICOSA*, portion of flowering stem from *Andrews* A3007. 2 - *ARACHIS HYPOGEA*, a: flowering branch, b: fruits, both from *Schweinfurth* 2425. 3 - *ZORNIA PRATENSIS*, portion of flowering and fruiting plant from *Ash* 2002. 4 - *DESMODIUM REPANDUM*, a: flowering stem from *Palmer* 56, b: pod from *Frahm-Leliveld* 62372. 4b x 1.125, the rest x 0.75. Drawn by Maureen Church. (Reproduced with permission from *Opera Botanica*, 68, fig. 24, 1983.)

- Pod with the upper suture slightly indented at the centre of each segment; plants erect. **3. D. ramosissimum**

8. Principal lateral veins of the leaflets few and not prominent; segments of pod ± rhombic in outline, except the terminal 1-2 which are ± oval and flat. **9. D. ospriostreblum**
- Principal lateral veins prominent and parallel, up to 8 pairs on the terminal leaflet; segments of pod equal. 9

9. Pedicels reflexed in fruit; segments of the pod oblong. **7. D. salicifolium**
- Pedicels not reflexed; segments of the pod almost round or rhombic in outline. 10

10. Leaflets obtuse to retuse at the apex; standard *c* 3.5 mm long; pod nearly evenly indented on both sutures in the broad gaps between the segments; segments almost round in outline. **8. D. dichotomum**
- Leaflets acute at the apex; standard *c* 9 mm long; pod with the lower suture much more indented than the upper in the broad gaps between the segments; segments ± rhombic in outline. **10. D. uncinatum**

1. **D. hirtum** *Guill. & Perr.* (1832).

Herb or subshrub, 0.5-2 m long; stems pilose. Leaflets (1-)3, obovate, truncate to retuse at the apex, glabrous to ± pilose above, pilose beneath, the terminal 1.2-2.7 x 1.2-1.7 cm, the lateral somewhat smaller; petiole *c* 1 cm long. Inflorescences composed of lax axillary and terminal racemes; pedicels usually ascending, 6-12.5 mm long, puberulent. Calyx up to 3 mm long, puberulent with pilose teeth. Corolla pale pink to purple; standard *c* 4 mm long, obovate. Pod 2-6-segmented, the upper suture ± straight, the lower slightly indented, dehiscing along the lower suture; segments ± 3 mm long and wide, glabrous to puberulent.

1. Stems spreading-pilose; pods puberulous. var. **hirtum**
- Stems sparsely pilose with ± appressed hairs; pods glabrous. var. **delicatulum**

Var. **hirtum**.

In crevices of basalt pavement rocks; 1300 m. WG; widespread in tropical Africa. *Gilbert & Thulin* 632.

Gilbert & Thulin 632 was previously (Thulin, *Leg. Eth.* (1983)) placed in *D. schweinfurthii*, but the segments of the nearly ripe pods present in some of the individuals are too large for this species. *D. hirtum* is also very closely related to *D. heterophyllum* (Willd.) DC. in Asia. If the two prove to be conspecific *D. heterophyllum* has priority.

Var. **delicatulum** (*A. Rich.) Harms* ex *Bak. f., Leg. Trop. Afr.*: 329 (1929);
D. delicatulum A. Rich. (1847) - type: TU, *Quartin Dillon & Petit* s.n. (P holo.).

Grassland; 600 m. TU IL; Sudan and widespread in tropical Africa, but not known from East Africa. *Gilbert & Friis* 8385.

2. **D. adscendens** (*Sw.) DC.* (1825).

Herb, up to 1 m long; stem creeping, pilose, with ± erect branches. Leaflets 3, elliptic-obovate, usually obtuse and emarginate at the apex, glabrous or pilose above, densely pilose beneath, the terminal 1.7-4.3 x 0.7-3.2 cm, the lateral somewhat smaller. Racemes axillary and terminal; primary bracts with long attenuate tips, falling soon; pedicels puberulent and with longer hairs, 0.4-1.7 cm long. Calyx 2-3 mm long, puberulent with pilose teeth. Corolla pink or purplish; standard 4-5.5 mm long, obovate. Pod 1-5-segmented, the upper suture ± straight, the lower indented; segments oblong, 3.5-5.5 x 2.5-3 mm, indehiscent, pubescent.

Var. **robustum** *Schubert* in *Bull. Jard. Bot. État* **22**: 290 (1952).

Leaflets thick, glabrous above.

Grassland, forest margins; 1300-1800 m. WG KF; widespread in tropical Africa; var. *adscendens* pantropical. *Gilbert & Thulin* 672; *Mooney* 9206; *W. de Wilde* 7771.

3. **D. ramosissimum** *G. Don* (1832).

Perennial herb or small shrub; stem erect, strigose. Leaflets 3, narrowly obovate to elliptic, 1-4.8 x 0.4-1.2 cm, obtuse and apiculate at the apex, appressed pilose at least beneath. Racemes terminal; primary bracts ovate-attenuate; pedicels 4-10 mm long, puberulent and with longer hairs. Calyx (2-)2.5-3 mm long, puberulent with pilose teeth. Corolla pink, reddish or purple; standard 3.5-5 mm long. Pod 1-5-segmented, the upper suture slightly indented above each seed, lower suture indented between the segments; segments 3-5 x 2.5-3.5 mm, indehiscent, puberulent.

Grassland, woodland; 1250-2000 m. WG KF; widespread in tropical Africa, Madagascar, Mascarene Is. *Friis et al.* 2283; *W. de Wilde* 6875, 10755.

4. **D. repandum** (*Vahl) DC.* (1825);
Hedysarum repandum Vahl (1791).
?*Hedysarum abyssinicum* Willd. ex Hoffmgg. (1824); *D. abyssinicum* (Willd. ex Hoffmgg.) DC. (1825) - type: a plant cultivated in Dresden from seeds said to be from Abyssinia, not seen.
D. schimperi Hochst. ex A. Rich. (1847) - type: Simien, *Schimper* II:857 (P holo., BM G K iso.).

Perennial herb or slender shrub up to 1 m tall; stems puberulent and spreading pilose. Leaflets 3, ± rhombic, obtuse to acute at the apex, appressed pilose on both sides, the terminal 1.8-11 x 1.5-8.5 cm, the lateral somewhat smaller. Inflorescences lax, terminal or axillary, racemose or paniculate; bracts falling soon, the secondary ones very small; pedicels 1.2-3.5 cm long, puberulent and with longer hairs. Calyx up to 5.5 mm long, puberulent, with somewhat pilose teeth. Corolla pink to brick red; standard 8-15 mm long, almost orbicular. Pod up to 5-segmented, the upper suture ±

straight, the lower deeply indented; segments narrowly oblong, 5.5-8 x 2-3 mm, indehiscent, puberulent. Fig. 95.22.4.

Forest margins, clearings, stream-banks; (1400-) 1600-2700(-3300) m. EW WU GD GJ WG SU AR HA KF GG SD; widespread in Africa, also in Madagascar, Yemen, India and Malaysia. *Thulin* 1396; *W. de Wilde* 8297; *Burger* 3158.

5. **D. velutinum** (*Willd.*) *DC.* (1825).

Perennial herb or shrub up to 3 m tall; stem not or little branched, densely pubescent. Leaflet single, thick, suborbicular to elliptic or rhombic-ovate, 3-11 x 2-7 cm, ± tomentose on both sides, with ± undulate or sinuate margins. Inflorescence a dense leafy panicle or raceme, with primary and secondary bracts; pedicels pubescent, 1-3 mm long. Calyx *c* 3 mm long, strigose. Corolla white to pink or purple; standard 4.2-6 mm long, suborbicular. Pod 2-6(-7)-segmented, the upper suture slightly indented, the lower more deeply so; segments 2.5-4 mm long and wide, indehiscent, pubescent. Fig. 95.23.1.

Grassland and woodland; 1200 m. WG IL KF GG; widespread in Africa, also in Madagascar, SE Asia and Malaysia. *Gilbert & Thulin* 732; *Meyer* 9019; *Friis* et al. 2494.

6. **D. gangeticum** (*L.*) *DC.* (1825).

Perennial, up to 1.5 m tall; stems often ± woody, puberulent and strigose. Leaflets single, ovate to elliptic-acuminate, 1.2-15.5 x 0.9-6.5 cm, puberulent and ± pilose above, pilose beneath. Inflorescence composed of lax, slender racemes, with primary and secondary bracts; pedicels puberulent, 3-8 mm long. Calyx *c* 3 mm long, puberulent and strigose. Corolla white to purple; standard 3.2-4.5 mm long, obovate. Pod up to 8-segmented, the upper suture slightly indented, the lower deeply so; segments 2-3 mm long and wide, indehiscent, puberulent and with some long hairs on sutures.

Grassland and woodland; 550-1500 m. EW GD (Sudan border) WG IL KF; throughout the Old World tropics. *Pappi* 5439; *Gilbert & Thulin* 713; *W. de Wilde* 8935.

7. **D. salicifolium** (*Poir.*) *DC.* (1825).

Perennial herb or subshrub, up to 3 m tall; stem puberulent and spreading pilose, above. Leaflets 3, lanceolate to ovate, obtuse to acuminate at the apex, sparsely pilose or almost glabrous above, pilose beneath, the terminal 6-17.5(-25) x 2-6.5(-10) cm, the lateral somewhat smaller. Inflorescences racemose-paniculate, with primary and secondary bracts; pedicels puberulent often also with longer hairs, becoming reflexed, 2.5-6 mm long. Calyx 3.5-4 mm long, puberulent and pilose. Corolla yellow, greenish, red or white and blue; standard 4-7 mm long, suborbicular. Pod 2-7-segmented, the upper suture ± straight, the lower slightly indented; segments oblong, 3.5-6.5 x 2-3.5 mm, indehiscent, puberulent. Fig. 95.23.2.

Swampy places, river margins; 1200-2000 m. GD WG BA IL KF GG SD; tropical and South Africa, also in Madagascar and the Mascarene Is. *Gilbert* 1970; *Gillett* 14754; *Mooney* 5402.

8. **D. dichotomum** (*Klein* ex *Willd.*) *DC.* (1825).

Herb or subshrub, up to 1 m tall; stem puberulent and pilose. Leaflets 3, elliptic to ovate or orbicular, obtuse to retuse at the apex, appressed pilose above, velvety with long and short hairs beneath, the terminal 1.8-8 x 2-6 cm, the lateral smaller. Inflorescences racemose or racemose-paniculate, with primary and secondary bracts; pedicels puberulent, 3.5-5 mm long. Calyx *c* 3 mm long, puberulent and pilose. Corolla pink or blue; standard 3.3-3.5 mm long, obovate. Pod 4-6-segmented, indented on both sutures; segments ± quadrate, 2-3 mm long and wide, indehiscent, puberulent and pubescent. Fig. 95.23.3.

Grassland, deciduous woodland, also a weed; 600-1900 m. EW GD WU SU GG; Sudan, Uganda, also in India to China and Malaysia. *Pappi* 369; *Gilbert, Thulin & Getachew* 368; *Parker* E. 466.

9. **D. ospriostreblum** *Steud.* ex *Chiov.* (1908)
- type: Eritrea, Modat, *Schimper* II:1039 (BM G K iso.).
Anarthrosyne abyssinica Hochst. ex A. Rich. (1847); *Meibomia abyssinica* (Hochst. ex A. Rich.) Schindl. (1924); *D. abyssinicum* (Hochst. ex A. Rich.) Hutch. & Dalz. (1928), nom. illegit., non *DC.* - type: TU, *Schimper* III:1434 (P holo., BM G GH K W iso.).

Annual, up to 0.6 m tall; stems pubescent. Leaflets 3, narrowly ovate to rhombic, obtuse to acute at the apex, puberulent and with longer hairs on both sides, the terminal 1.6-7.8 x 1-3.6 cm, the lateral smaller. Inflorescences lax, racemose to racemose-paniculate, with primary and secondary bracts; pedicels puberulent, 4.5-17 mm long. Calyx *c* 2 mm long, puberulent and pilose. Corolla pink, bluish or white and red; standard 3-3.5 mm long, obovate. Pod 3-5(-7)-segmented; segments ± rhombic, except the terminal 1 or 2 which are ± oval and flat, pubescent, sometimes on sutures only, indehiscent, terminal segment 5-9 x 3.5-5.5 mm. Fig. 95.23.4.

Deciduous woodland or bushland; 800-1600 m. EE EW TU GD SU KF; mainly in western and southern tropical Africa. *Gilbert & Thulin* 971; *Pappi* 4088; *J. De Wilde* 7129.

10. **D. uncinatum** (*Jacq.*) *DC.* (1825).

Scrambling herb with ascending flowering stems; stems densely covered with hooked hairs; leaflets 3, usually with a pale silvery marking along the midrib above, ± narrowly ovate, acute, pubescent or pilose on both sides; inflorescence lax, unbranched with primary bracts up to 1 cm long; standard *c* 9 mm long; pod deeply indented along the lower suture; segments ± rhombic, *c* 5-6 x 3-4 mm, pubescent.

Cultivated experimentally as forage or as a cover crop at least in AR (Asela), SD (Soddo) and KF (Jimma); native of tropical America. *Jansen* 5855.

The closely related *D. intortum* (Mill.) Urb., with which *D. uncinatum* often is combined, has also been reported from experimental cultivation in AR, but no material has been seen. It differs mainly in having the leaflets uniformly green above and in the somewhat smaller segments to the pod.

11. **D. barbatum** *(L.) Benth.* (1852).

Perennial. Leaflets 3, suborbicular to obovate, glabrous to appressed pilose above, more densely so beneath, the terminal 1.4-5 x 0.5-2 cm, the lateral ones, when present, smaller. Inflorescences axillary and terminal heads or short racemes; bracts rather long persistent, secondary ones usually lacking; pedicels 3.5-9 mm long, uncinulate-pubescent, becoming reflexed. Calyx subequally 5-lobed, 4-6 mm long, silky pilose. Corolla red, pink or purple; standard 4-5 mm long, obovate. Pod up to 6-segmented, the upper suture ± straight, the lower somewhat indented; segments 2-3.5 x 2-2.5 mm, uncinulate-pubescent, partially separating, then dehiscing along the lower suture.

Var. **prostratum** *Schubert* in *Bull. Jard. Bot. Brux.* **22**: 297 (1952).

Plant procumbent, mat-forming with short internodes. Leaflets 3, ± obovate with obtuse apex. Inflorescences densely capitate.

Woodland; *c* 1400 m. BA; Uganda, Kenya, Tanzania, Zaire, Rwanda, Zambia. *Friis, Gilbert & Vollesen* 3453.

Var. *barbatum* is an American taxon sometimes introduced in Africa. Two further varieties have been recognized in East Africa.

48. **URARIA** *Desv.* (1813)

Verdcourt in *Fl. Trop E. Afr.* (1971).

Perennials. Leaves pinnately (1-)3-9-foliolate. Inflorescences mostly terminal, false racemes or panicles. Calyx 5-lobed with 3 lower lobes equal and longer than the upper pair. Anthers uniform. Pods folded like a concertina, mostly enclosed in the persistent calyx, constricted between the seeds, the segments ovate, inflated, 1-seeded and indehiscent.

Some 20 species in the Old World tropics.

U. picta *(Jacq.) DC.* (1825).

Erect herb or subshrub, 0.2-1.8 m tall; stems scabrid. Lower leaves 1-3-foliolate, upper 5-9-foliolate; leaflets of lower leaves ovate, 2-8 x 2-3.2 cm, of upper leaves lanceolate, 7-25 x 0.2-2.5(-4) cm, shining and often whitish along midrib above, scabrid and strongly reticulate beneath. Inflorescences long, narrow, sessile or short-pedunculate; pedicels becoming inflexed, covered with long deciduous hairs and short hooked hairs; bracts long, deciduous. Corolla pink, bluish or yellowish, glabrous; standard 8-9 mm long. Pods 5-9 mm long; segments 3-6, 2-3 mm wide, shining.

Grassland and wooded grassland; 570 m. IL; widespread in tropical Africa but absent from most of E and NE Africa, also in Asia and N Australia. *Jansen* 4967.

49. **PSEUDARTHRIA** *Wight & Arn.* (1834)

Verdcourt in *Kew Bull.* **24**: 64-67 (1970) & *Fl. Trop. E. Afr.* (1971).

Perennial herbs or subshrubs. Leaves pinnately 3-foliolate. Inflorescences terminal or axillary false racemes or panicles. Calyx 5-lobed with subequal lobes, the upper pair almost entirely joined. Anthers uniform. Pods narrowly oblong, flat, few-to many-seeded, 2-valved, not breaking into segments.

Four to six species in the Old World tropics.

1. Inflorescence ± lax; pods mostly stipitate, 12-38 mm long, 3-12-seeded. **1. P. hookeri**
- Inflorescence very dense; pods sessile, 7-20 mm long, 2-6-seeded. **2. P. confertiflora**

1. **P. hookeri** *Wight & Arn.* (1834).

Herb or subshrub, 0.3-3(-4) m tall; stems pubescent to velvety. Leaflets elliptic or narrowly ovate, 2.5-16.5 x 1.3-10 cm, acute to rounded at the apex, rough above, pubescent to velvety beneath, usually broadly crenulate at the margins; petiole 1-7 cm long. Inflorescence usually much branched and with numerous flowers. Corolla reddish-purple, blue or white; standard 6-8 mm long, rounded. Pod straight, 12-38 x 3-4(-5) mm, mostly puberulous, 3-12-seeded. Fig. 95.23.5.

Grassland, woodland, scrub; 1100-2050 m. GD GJ WG SU BA IL KF GG SD; widespread in tropical Africa. *Gilbert & Thulin* 662; *W. de Wilde* 7814; *Ash* 1432.

The Ethiopian material belongs to var. *hookeri*. Var. *argyrophylla* Verdc. with leaves very thickly silvery hairy beneath, occurs in West Africa, Sudan and Uganda.

2. **P. confertiflora** *(A. Rich.) Bak.* (1871);
Rhynchosia confertiflora A. Rich. (1847); *P. hookeri* var. *confertiflora* (A. Rich.) Schindl. in Feddes Rep., Beih. 2:14 (1914) - type: TU, *Quartin Dillon* s.n. (P holo.).

Herb or subshrub 0.4-1.5 m tall; stems pubescent to velvety. Leaflets lanceolate, elliptic or narrowly

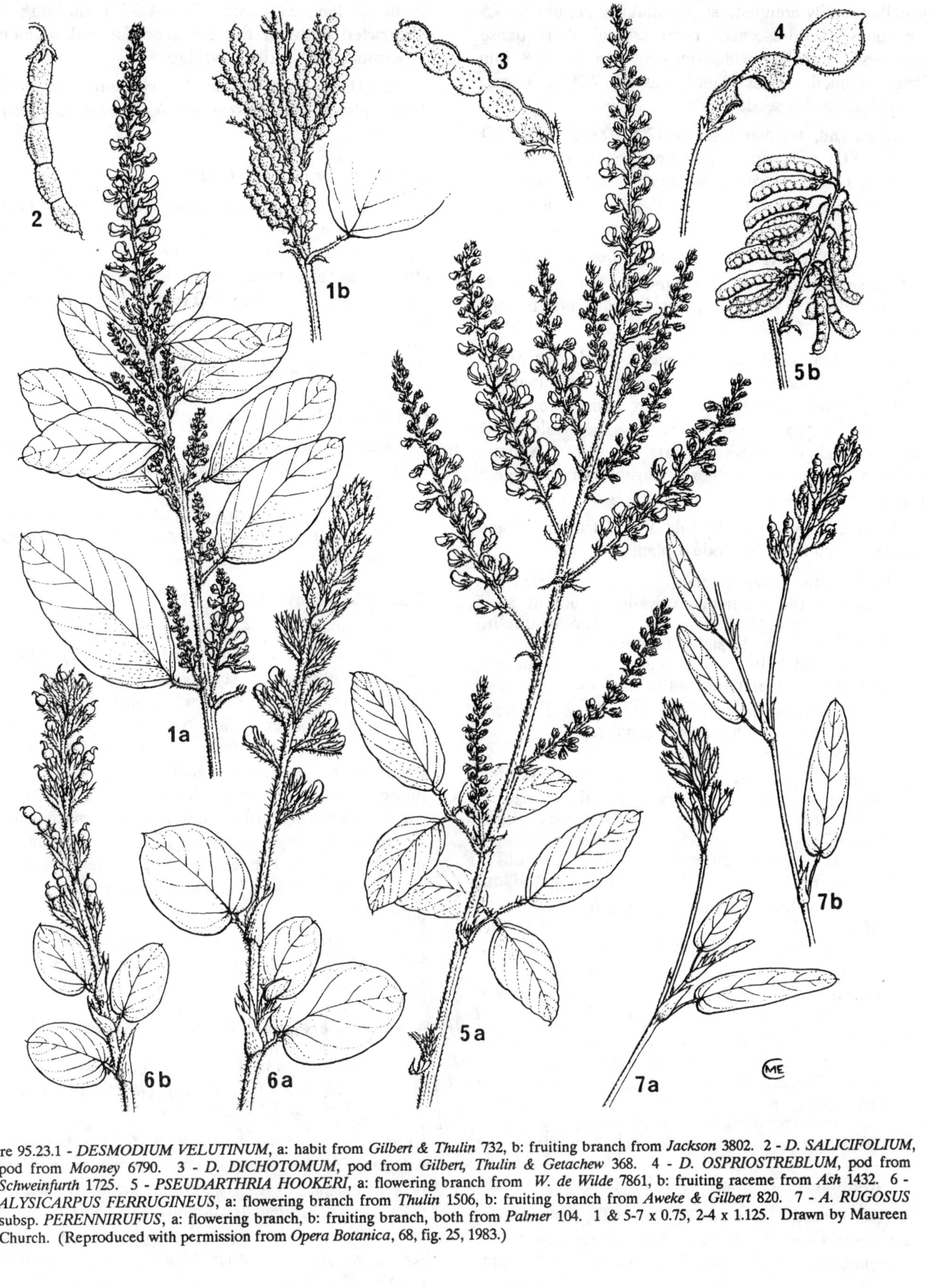

Figure 95.23.1 - *DESMODIUM VELUTINUM*, a: habit from *Gilbert & Thulin* 732, b: fruiting branch from *Jackson* 3802. 2 - *D. SALICIFOLIUM*, pod from *Mooney* 6790. 3 - *D. DICHOTOMUM*, pod from *Gilbert, Thulin & Getachew* 368. 4 - *D. OSPRIOSTREBLUM*, pod from *Schweinfurth* 1725. 5 - *PSEUDARTHRIA HOOKERI*, a: flowering branch from *W. de Wilde* 7861, b: fruiting raceme from *Ash* 1432. 6 - *ALYSICARPUS FERRUGINEUS*, a: flowering branch from *Thulin* 1506, b: fruiting branch from *Aweke & Gilbert* 820. 7 - *A. RUGOSUS* subsp. *PERENNIRUFUS*, a: flowering branch, b: fruiting branch, both from *Palmer* 104. 1 & 5-7 x 0.75, 2-4 x 1.125. Drawn by Maureen Church. (Reproduced with permission from *Opera Botanica*, 68, fig. 25, 1983.)

obovate, 4-14 x 1.5-6.5 cm, acute to rounded at the apex, rough above, pubescent to velvety beneath, usually broadly crenulate at the margins; petiole 0.5-4.5 cm long. Inflorescences composed of short dense racemes. Corolla reddish-purple; standard 6-8 mm long, rounded. Pods slightly curved, 7-20 x 3 mm, mostly pilose, 2-6-seeded.

Grassland, woodland, scrub; 1200-2000 m. TU GD IL KF SD; widespread in tropical Africa. *Gilbert, Thulin & Getachew* 550; *Gillett* 14530; *Schimper* 291.

This is often considered only a variety of *P. hookeri*.

50. **ALYSICARPUS** *Desv.* (1813), *nom. conserv.*

J. Léonard in *Bull. Jard. Bot. État.* **24**: 84 (1954) & *Fl. Congo Belge* (1954); Verdcourt in *Fl. Trop. E. Afr.* (1971).

Annual or perennial herbs. Leaves 1-foliolate or rarely pinnately 3-foliolate. Inflorescences terminal or axillary, falsely racemose or rarely paniculate; bracts scarious. Calyx scarious, the 2 upper calyx-lobes connate and the calyx appears 4-lobed. Anthers uniform. Pods linear-oblong, mostly several-jointed; segments indehiscent.

Some 30 species in the Old World tropics. Some species are cultivated as fodder plants.

1. Pods with straight margins, not constricted between the segments; calyx-lobes not at all overlapping at the base. **1. A. ovalifolius**
- Pods not with straight margins; ± strongly constricted between the segments; calyx-lobes sometimes overlapping at the base. 2
2. Calyx and bracts with scattered very distinct brown or black long spreading hairs. 3
- Calyx and bracts not as above. 4
3. Perennial; pedicels spreadingly brown or black pilose; segments of pod smooth, usually puberulous. **4. A. ferrugineus**
- Annual; pedicels pubescent, hairs hooked; segments of pod with transverse ridges, glabrous or puberulous. **5. A. quartinianus**
4. Calyx equalling or shorter than the first segment of the pod. **2. A. monilifer**
- Calyx longer than the first segment of the pod. 5
5. Segments ± smooth; calyx-lobes overlapping at the base. **3. A. zeyheri**
- Segments with ± strong sculpture. 6
6. Calyx-lobes conspicuously overlapping at the base; inflorescences mostly short; calyx-lobes with white or rusty-red hairs. **6. A. rugosus**
- Calyx-lobes not overlapping at the base; inflorescences elongate and spike-like; calyx-lobes never with rusty-red hairs. **7. A. glumaceus**

1. **A. ovalifolius** *(Schumach.) J. Léon.* (1954).

Erect or spreading annual, 20-60 cm tall; stems pubescent, glabrescent. Leaflet elliptic or oblong to narrowly lanceolate, 1-10 x 0.6-3 cm, finely puberulous beneath. Racemes terminal and leaf-opposed, very lax. Calyx 4.5-6 mm long with narrowly triangular lobes. Corolla about as long as calyx. Pod 1.8-2.5 cm long, not constricted between the 2-8 subcylindrical segments, puberulous, with reticulate ridges.

Grassland; *c* 550-600 m. IL; widespread in tropical Africa, also in Madagascar and Asia. *Ash* 524; *Friis* et al. 2539.

2. **A. monilifer** *(L.) DC.* (1825).

Slender diffuse annual; stem pubescent. Leaflet cordate-ovate, oblong or, in upper leaves, lanceolate, 1.5-6 x 0.5-2 cm, both sides prominently veined and puberulous, also some longer hairs beneath; margins ciliate. Racemes axillary and terminal, very lax. Calyx *c* 4 mm long with linear lobes usually shorter than the first segment of the pod. Corolla about as long as or slightly exceeding calyx. Pod up to 2 cm long, with deep constrictions between the 3-4(-6) oblong segments, slightly wrinkled, pubescent.

Grassland, sandy places or as a weed; up to *c* 1000 m. GD-TU; Sudan, Somalia, also in Pakistan and India. *Schimper* 2251.

A record from Eritrea in Cuf. *Enum.* (1955) is doubtful. It is based on *Pappi* 4069, a very young plant without fruits.

3. **A. zeyheri** *Harv.* (1862).

Erect or spreading perennial herb, 0.1-1.2 m tall, with a woody rootstock; stems glabrous to pubescent. Leaflet linear to elliptic or lanceolate, 1-10 x 0.2-1.8 cm, strongly reticulate on both sides, puberulous and also with some longer hairs beneath; margins ciliate. Racemes usually terminal, lax. Calyx 5-8 mm long, longer than the first segment of the pod, brownish pilose or pubescent; lobes elliptic to lanceolate, slightly overlapping. Standard 6-7 mm long, salmon tinged red, or yellowish; wings purplish. Pod up to 1.5 cm long, much constricted between the 2-6 segments; segments oblong-elliptic, 2.5-3 x 2-2.5 mm, almost smooth, usually puberulous.

Grassland, also in swampy situations; 1500-2200 m. WG SU KF SD; widespread in Africa. *Gilbert & Thulin* 583, 847, 870; *Ash* 1991.

4. **A. ferrugineus** *Hochst. & Steud.* ex *A. Rich.* (1847);
Fabricia ferruginea (Hochst. & Steud. ex A. Rich.) Kuntze (1891) - types: TU, *Schimper* I: 306 (P syn., BM K S UPS isosyn.), *Quartin Dillon* s.n. (P syn.), *Petit* s.n. (P syn.).

Ascending perennial herb up to 50 cm tall, with a woody root-stock; young stems brownish pilose. Leaflet broadly cordate-ovate to suborbicular, up to 4 x 3.5 cm, thick, strongly reticulate on both sides, puberulous and with some longer hairs beneath; margins ciliate. Racemes terminal, rather dense; pedicels spreadingly brown or black pilose. Calyx 7-8 mm long, much longer than the first segment of the pod, spreadingly brown or black pilose; lobes lanceolate-elliptic, overlapping. Standard 7-9 mm long, pale pink; wings purplish. Pod up to *c* 1 cm long, much constricted between the 2-4 segments; segments oblong, *c* 3 x 2.5 mm, smooth, usually puberulous. Fig. 95.23.6.

Grassland, often on clay; 1600-2400 m. EW TU GD GJ WG SU AR; not known elsewhere. *Thulin* 1506; *Gilbert & Thulin* 600; *Gilbert & Getachew* 2800.

5. **A. quartinianus** *A. Rich.* (1847);
A. rugosus (Willd.) DC. var. *quartinianus* (A. Rich.) Bak. in *Fl. Trop. Afr.* **2**: 171 (1871); *Fabricia rugosa* (Willd.) Kuntze var. *quartiniana* (A. Rich.) Taub. in *Pflanzenw. O.-Afr.* C:217 (1895); *A. ferrugineus* var. *quartinianus* (A. Rich.) Schindl. in *Feddes Rep.* **21**: 12 (1925) - type: TU, *Quartin Dillon* (P holo.).

Erect annual up to 60 cm tall; young stems white or brown to black pilose. Leaflet cordate-ovate to suborbicular in lower leaves to linear-lanceolate in upper leaves, up to 8 x 3.5 cm, strongly reticulate on both sides, puberulous and with some longer hairs beneath; margins ciliate. Racemes terminal, rather dense; pedicels pubescent with hooked hairs. Calyx 8-10 mm long, enclosing the pod, spreadingly brown to black pilose; lobes lanceolate-elliptic, overlapping. Standard 7-9 mm long, pale pink; wings purplish. Pod up to *c* 8 mm long, much constricted between the 4-6 segments; segments rounded, *c* 1.4 x 1.6 mm, with ± strong close transverse ridges, glabrous or puberulous.

Grassland, also as a weed in cultivations; 1650-2100 m. EW TU GD GJ WG SU AR KF SD; Cameroun. *Thulin* 1619; *Gilbert* 2226; *Mooney* 6052.

6. **A. rugosus** *(Willd.) DC.* (1825).

Erect to prostrate, annual or perennial herb, 0.15-1 m tall; young stems pubescent to pilose. Leaflet oblong, narrowly ovate or linear, 1.5-5(-11) x 0.2-1.5(-2.3) cm, slightly reticulate on both sides, puberulous and with longer hairs beneath; margins ciliate. Racemes axillary or terminal, rather dense. Calyx 6-8 mm long, ± enclosing the pod, ciliate with white or brownish hairs; lobes lanceolate, overlapping. Standard 6-7 mm long, whitish, pink to purplish. Pod 5-10 mm long, much constricted between the 3-6 segments; segments 1-1.5 x 2-2.5 mm, with strong close transverse ridges, glabrous or pubescent.

1. Plant annual, ± erect; calyx-lobes with white hairs. subsp. **rugosus**
- Plant perennial, ± decumbent; calyx-lobes with brownish hairs. subsp. **perennirufus**

Subsp. **rugosus**.

Grassland, woodland or as a weed in cultivations; 1200-2000 m. EW TU GD WU WG SU HA IL GG SD; widespread in tropical Africa and in Madagascar. *Gilbert, Thulin & Getachew* 536; *Schimper* 2254; *Parker* E. 107; *Friis* et al. 2540.

Subsp. **perennirufus** *J. Léon.* in *Bull. Jard. Bot. État* **24**: 95 (1954). Fig. 95.23.7.

Grassland and woodland; 100-2100 m. EW GD GJ WG SU AR BA HA KF SD; widespread in tropical and southern Africa. *Thulin* 1620; *Friis* et al. 2301; *Gilbert* 1336.

All collections with rusty-red calyx-hairs seen from Eritrea are annuals (e.g. *Pappi* 112, 7336, 8868, all in FT). They can thus be regarded as intermediates with subsp. *rugosus* but as their characters seem constant they could also be treated as an entity of their own. Further studies are required on this point.

7. **A. glumaceus** *(Vahl.) DC.* (1825).
A. hochstetteri A. Rich. (1847) - type: Eritrea, *Schimper* 1049 (P holo., BM iso.).

Erect to decumbent annual, 0.15-1.5 m tall, often somewhat woody at the base; stems pubescent to pilose. Leaflets oblong to linear, 1-12.5 x 0.2-1.6 cm, slightly reticulate on both sides, puberulous and with longer hairs beneath; margins ciliate. Racemes terminal and leaf-opposed, usually elongate and ± lax. Calyx 5-7 mm long, much longer than the first segment of the pod, ciliate with white hairs or glabrous; lobes lanceolate, not or slightly overlapping. Standard 5-6 mm long, pink to red. Pod up to 1.5 cm long, much constricted between the 4-7 segments; segments 1.5-2 x 2-3 mm, with strong close transverse ridges, puberulous.

Grassland and woodland; up to 1600 m. EE TU GD (Sudan border) WU SU HA GG SD; throughout tropical Africa and in Arabia. *Gilbert, Thulin & Getachew* 300, 314, 542; *Gilbert* 1247.

In tropical Africa a distinction is usually made between subsp. *glumaceus* with appressed hairs on the stems and subsp. *hispidocarpus* (Fiori) J. Léon. with spreadingly pilose stems. The latter has been reported from Eritrea, but the difference between the two subspecies is not convincing there. *Gilbert & Jefford* 4500A from SD is the only collection seen of typical subsp. *hispidocarpus*.

51. AMPHICARPA *Nuttall* (1818), *nom. conserv.*

Climbing herbs. Leaves pinnately 3-foliolate. Racemes axillary. Calyx tubular, appearing 4-lobed, the 2 upper lobes being entirely connate. Vexillary stamen free. Pod compressed.

Small genus in N and tropical America, tropical Africa (one species) and Asia.

A. africana *(Hook. f.) Harms* (1921).

Stems with reflexed rusty-red pubescence. Leaflets elliptic or obovate, 2-7 x 1-5 cm, glabrescent above save on nerves, ± densely appressed pubescent beneath; petiole up to 10 cm long; stipules and bracts ± ovate-oblong, closely veined. Racemes lax, many-flowered; bracteoles absent. Corolla violet, 10-16 mm long. Pod linear-oblong, 25-35 x 5.5-6 mm, *c* 4-seeded, glabrous except for densely ciliate margins. Fig. 95.24.4.

Montane forest, forest edges; 2150-2550 m. WG SU AR BA IL KF SD; widespread in tropical Africa. *Thulin* 1584; *Mooney* 8800; *Friis* et al. 1514.

52. DUMASIA *DC.* (1825)

Climbing herbs. Leaves pinnately 3-foliolate. Racemes axillary. Calyx-tube cylindrical, the mouth obliquely truncate. Vexillary stamen free. Style filiform, dilated and hollow near the middle. Pod compressed, constricted between the seeds.

Small genus in the Old World tropics.

D. villosa *DC.* (1825).

Stems usually covered with spreading rusty-red hairs. Leaflets ovate-oblong, 1.5-6 x 1-5 cm, sparsely pilose to almost glabrous above, sparsely to densely appressed pilose beneath; petiole 2-6.5 cm long; stipules linear-lanceolate. Racemes short. Corolla yellow, 16-17 mm long. Pod linear-oblong, 20-40 x 8 mm, velvety pubescent. Fig. 95.24.3.

Evergreen forest, thickets; 1900-2200 m. BA KF; widespread in eastern Africa and Madagascar, extending to India and China. *Mooney* 8476; *Friis* et al. 412; *J. De Wilde* 6216.

53. CLITORIA *L.* (1753)

Climbing or erect herbs or shrubs. Leaves pinnately (1-) 3-9-foliolate. Inflorescences few-flowered racemes, or flowers axillary, solitary or paired; bracteoles large. Calyx 5-lobed. Standard large, much exceeding the other petals. Style bearded inside at the slightly dilated apex. Pod linear-oblong, compressed.

Some 30-40 species in the tropics and subtropics of both hemispheres.

C. ternatea *L.* (1753).

Climbing or trailing perennial herb. Leaflets 5-7, oblong-lanceolate to almost round, 1-6.7 x 0.3-4 cm, glabrescent above, appressed pubescent beneath; petiole 1-3 cm long. Flowers axillary, solitary or paired; bracteoles ovate. Calyx 15-20 mm long. Standard held lowermost, white or greenish-white, often margined with blue or entirely blue, 2.5-5 cm long. Pod 6-12.5 x 0.7-1.2 cm, 8-10-seeded, glabrous or sparsely pubescent.

1. Leaflets elliptic to almost round. var. **ternatea**
- Leaflets oblong-lanceolate. var. **angustifolia**

Var. **ternatea**. Fig. 95.24.2.

Grassland, woodland, old cultivations; 0-1600 m. EE EW TU GD SU HA BA KF GG SD; pantropical. *Gilbert* 1553; *Ash* 816; *Thulin, Hunde & Tadesse* 3784.

Var. **angustifolia** *Hochst.* ex *Bak. f., Leg. Trop. Afr.*:428 (1929)
- type: Tekeze R., *Schimper* III:1617 (BM lecto., K isolecto.).

Grassland, woodland, old cultivations; *c* 1000-1500 m. TU SU; pantropical. *Mooney* 9025; *Westphal* 1497.

54. PSEUDOERIOSEMA *Hauman* (1955)

Herbs or weak subshrubs. Leaves unifoliolate or pinnately 3-foliolate. Inflorescences axillary or terminal, racemose or condensed and paniculate. Calyx 5-lobed, the upper pair of lobes ± joined. Corolla small, scarcely exceeding the calyx. Ovary 2-ovuled. Pod oblong-ovoid, compressed, 1-2-seeded, densely villous.

Small genus restricted to tropical Africa.

P. borianii *(Schweinf.) Hauman* (1955);
Rhynchosia borianii Schweinf. (1868); *Glycine borianii* (Schweinf.) Bak. (1871) - types: Sudan/ Ethiopia border, *Schweinfurth* 1892 (B syn., BM K isosyn.) & *Boriani* 102 (W syn.).

Perennial usually trailing herb; stems usually with whitish deflexed bristly pubescence. Leaflets 3, oblong-elliptic to linear-oblong, the lateral ones shorter, 2.5-10 x 0.6-2.5 cm, sparsely bristly pubescent at least beneath. Racemes axillary, dense and short, 6-20-flowered; peduncle short. Calyx 4-5 mm long, bristly pubescent. Standard pink, red or purplish, 5-6 mm long. Pod 14-20 x 6.5-7 mm, bristly pubescent. Fig. 95.24.1.

Coarse grassland, seasonally flooded plains; *c* 800 m. GD GJ; Sudan, Uganda, Tanzania, E Zaire, Zambia.

Subsp. *longipedunculatum* Verdc. in SE Kenya and NE Tanzania.

The two syntypes, collected before 1870, are still the only collections known from Ethiopia.

55. OPHRESTIA *H. L. Forbes* (1948)

Verdcourt in *Kew Bull.* **24**: 257-262 (1970) & *Fl. Trop. E. Afr.* (1971).

Climbing or suberect herbs. Leaves digitately or pinnately (1-)3-7-foliolate. Flowers in axillary false racemes or almost sessile axillary clusters. Calyx 5-lobed, the upper pair of lobes free or partly joined. Standard hairy outside. Pods oblong, compressed, 2-8-seeded.

Small genus in the Old World tropics.

O. radicosa *(A. Rich.) Verdc.* (1970);
Eriosema radicosum A. Rich. (1847); *Glycine radicosa* (A. Rich.) Bak. f. (1929); *Paraglycine radicosa* (A. Rich.) F. J. Hermann (1962) - type: TU, *Quartin Dillon* s.n. (P holo., BM fragment).

Perennial climber. Leaves pinnately 3-foliolate; leaflets elliptic, 1-7(-8) x 0.4-2.4(-6) cm, pubescent at least beneath. Racemes axillary, lax, up to 10 cm long. Calyx

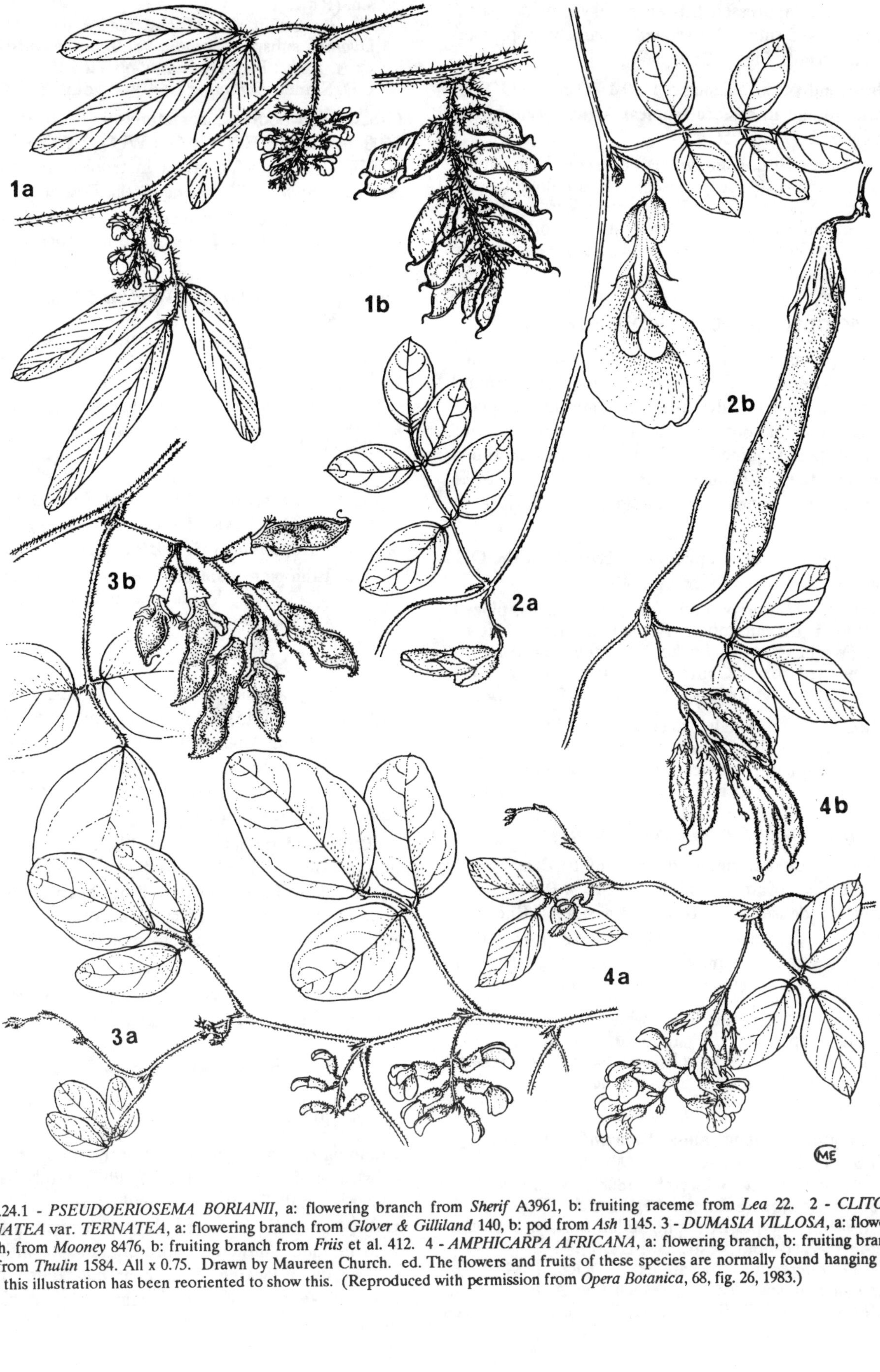

Figure 95.24.1 - *PSEUDOERIOSEMA BORIANII*, a: flowering branch from *Sherif* A3961, b: fruiting raceme from *Lea* 22. 2 - *CLITORIA TERNATEA* var. *TERNATEA*, a: flowering branch from *Glover & Gilliland* 140, b: pod from *Ash* 1145. 3 - *DUMASIA VILLOSA*, a: flowering branch, from *Mooney* 8476, b: fruiting branch from *Friis* et al. 412. 4 - *AMPHICARPA AFRICANA*, a: flowering branch, b: fruiting branch, both from *Thulin* 1584. All x 0.75. Drawn by Maureen Church. ed. The flowers and fruits of these species are normally found hanging and hence this illustration has been reoriented to show this. (Reproduced with permission from *Opera Botanica*, 68, fig. 26, 1983.)

3-5 mm long, the upper pair of lobes joined for ± two-thirds of their length. Standard 4-7 mm long, whitish, tinged purple, appressed brownish silky outside. Pod 17-35 x 4.5-6.5 mm, 3-5-seeded, sparsely appressed black-pubescent. Fig. 95.25.1.

Grassland or woodland; *c* 1600-1800 m. TU WG SU; widespread in eastern tropical Africa. *Lundstrom* 154; *Gilbert & Thulin* 756.

The three Ethiopian collections seen have stems with spreading rusty-red pubescence and belong to var. *radicosa*. Var. *schliebenii* (Harms) Verdc. with ± appressed pubescent stems was also reported from Ethiopia by Verdcourt (1971), but no specimen has been seen.

56. **GLYCINE** *Willd.* (1802), *nom. conserv.*

Twining, procumbent or erect herbs. Leaves usually pinnately 3-foliolate. Flowers small, in axillary racemes or pseudoracemes. Calyx 5-lobed, somewhat 2-lipped; the upper pair of lobes ± joined. Standard glabrous. Vexillary stamen joined to the others or sometimes becoming free with age; anthers uniform. Style glabrous. Pod linear or oblong, subcylindrical or compressed.

Small genus in the tropics and subtropics of the Old World.

Lackey (1977), apparently for good reasons, placed *G. wightii* in the monotypic *Neonotonia* Lackey as *N. wightii* (Wight & Arn.) Lackey. As the new combinations needed for the infraspecific taxa have not yet been published, the species is here retained in *Glycine*.

1. Climbing perennial; racemes more than 20-flowered; pods 15-36 x 2.5-5 mm. **G. wightii**
- Erect annual; racemes less than 20-flowered; pods 25-80 x 8-15 mm. **G. max**

1. **G. wightii** *(Wight & Arn.) Verdc.* (1966).

Climbing perennial; stems glabrescent to densely velvety. Leaflets ovate or elliptic, 1.3-15 x 0.8-12.5 cm, glabrous to densely velvety on both sides. Racemes many-flowered. Standard 4.5-11 mm long. Pod linear-oblong, 15-36 x 2.5-5 mm, glabrous to densely rusty-red pubescent, ± constricted between the seeds.

1. Flowers 4.5-7.5 mm long, white or white with mauve spot on the standard, usually drying orange. subsp. **wightii** var. **longicauda**
- Flowers 8-11 mm long, blue or mauve. subsp. **petitiana**

Subsp. **wightii** var. **longicauda** *(Schweinf.) Verdc.* in *Taxon* **15**: 36 (1966);

G. longicauda Schweinf. (1868); *G. javanica* L. var. *longicauda* (Schweinf.) Bak. in *Fl. Trop. Afr.* **2**: 178 (1871) - type: Ethiopia/Sudan border, *Schweinfurth* 1888 (B holo. destr., BM K iso.).

G. micrantha Hochst. ex A. Rich. (1847); *G. javanica* subsp. *micrantha* (Hochst. ex A. Rich.) F. J. Hermann in *U.S. Dept. Agric. Tech. Bull.* **1268**: 31, fig. 9 (1962) - types: TU, *Quartin Dillon* s.n. (P syn.) & *Schimper* I:8 (P syn., BM K isosyn.) & SU, *Petit* s.n. (P syn.).

G. moniliformis Hochst. ex A. Rich. (1847); *G. javanica* subsp. *micrantha* var. *moniliformis* (Hochst. ex A. Rich.) F. J. Hermann, *op. cit.*: 35 (1962) - type: GD, Simien, *Schimper* II:703 (P holo., BM K iso.).

Grassland, woodland, sometimes in cultivations; 380-2600 m. EE AF EW TU GD WU GJ WG SU AR HA IL KF GG SD; widespread in tropical Africa and in Arabia. *Thulin* 1337; *Mooney* 8606; *Friis* et al. 448.

There is a deviating form near Harer with small (*c* 2.5 x 2 cm), rounded leaflets. Similar forms occur also in East Africa.

Subsp. **petitiana** *(A. Rich.) Verdc.* in *Taxon* **15**: 36 (1966);

Johnia petitiana A. Rich. (1847); *Glycine petitiana* (A. Rich.) Schweinf. (1868) - type TU, *Petit* s.n. (P holo.).

G. petitiana var. *dembianensis* Chiov. in *Ann. di Bot.* **9**: 58 (1911) - types: GD, *Chiovenda* 2382 & 2449 (FT syn.).

Fig. 95.25.2.

Upland forest, scrub and bushland; 2100-2800 m. TU GD WU WG SU AR HA KF; N Kenya, S Tanzania. *Mooney* 6310; *Thulin* 1431; *Friis* et al. 461.

The Ethiopian plant belongs to var. *petitiana*. Var. *mearnsii* (De Wild.) Verdc. occurs in Kenya, Tanzania, Malawi and Zambia.

2. **G. max** *(L.) Merr.* (1917).

Bushy hairy annual 20-200 cm tall; leaflets ovate to lanceolate, 3-10(-15) x 2-6 cm; racemes *c* 5-8-flowered; corolla 5-7 mm long, white or lilac; pod 25-80 x 8-15 mm, slightly curved and usually somewhat compressed, hairy; seeds (1-)2-3(-5), 6-11 mm in diameter, usually ± globose, with small hilum.

Cultivated for its edible seeds, rich in protein and oil, and also as fodder and green manure. Recently introduced to Ethiopia where it has been grown experimentally at a number of medium altitude locations in AR (Asella) and SU, and on some State Farms, particularly in GJ.

The soya bean probably originates from E Asia, where its wild relatives are found.

57. **TERAMNUS** *P. Br.* (1756)

Verdcourt in *Kew Bull.* **24**: 263-284 (1970) & *Fl. Trop. E. Afr.* (1971).

Climbing or trailing perennial herbs, more rarely erect subshrubs. Leaves pinnately or subdigitately 3-foliolate. Flowers small, in axillary pseudo-racemes or fascicles. Calyx 4-5-lobed. Standard glabrous or with a few hairs near middle. Anthers: 5 normal alternate with 5 small and sterile, or lacking. Pod linear, *c* 8-seeded, the style-base thickened, forming a right-angled hook at the apex.

Figure 95.25.1 - *OPHRESTIA RADICOSA*, a: flowering branch from *Gilbert & Thulin* 756, b: fruiting raceme from *Froment* 275. 2 - *GLYCINE WIGHTII* subsp. *PETITIANA*, a: flowering branch from *Mooney* 6310, b: part of fruiting raceme from *Meyer* 7955. 3 - *TERAMNUS LABIALIS* var. *ABYSSINICUS*. flowering and fruiting branch from *Robertson* 1307. 4 - *T. UNCINATUS* subsp. *RINGOETII*, flowering branch from *Gilbert & Thulin* 640. All x 0.75. Drawn by Maureen Church. (Reproduced with permission from *Opera Botanica*, 68, fig. 27, 1983.)

Small pantropical genus with ill-defined species.

1. Stems robust, very angular, densely hairy; leaflets 3.3-13.5 x 1.7-9 cm. **4. T. micans**
- Stems not markedly angular, or if so slender and leaflets narrow. 2
2. Calyx with deep brown to blackish hairs; leaflets narrowly oblong, obtuse; seeds smooth. **3. T. uncinatus** subsp. **ringoetii**
- Calyx with white or brown hairs, but without the other characters combined. 3
3. Pod with spreading hairs. **2. T. repens**
- Pod with appressed hairs. 4
4. Leaflets pubescent above, narrowly oblong, obtuse; seeds smooth. **3. T. uncinatus** subsp. **axilliflorus**
- Leaflets glabrous or pubescent above, if pubescent other characters not as above. **1. T. labialis**

1. **T. labialis** *(L. f.) Spreng.* (1826).

Stems climbing or trailing, appressed to spreading pubescent or glabrescent. Leaflets round to elliptic, ovate or lanceolate, 1-8 x 0.5-4 cm, glabrous to appressed pilose at least beneath. Inflorescences slender and rather few-flowered. Corolla white, pink or purplish, 5 mm long. Pod 25-60 x 2-4 mm, appressed pubescent (in Ethiopia).

1. Seeds usually pale, smooth; leaflets mostly obtuse at the apex, glabrous or almost so above; pod 2-2.5 mm wide. subsp. **labialis** var. **abyssinicus**
- Seeds dark brown, rough; leaflets acute or subacute at the apex, appressed pilose above; pod *c* 3.5 mm wide. subsp. **arabicus**

Subsp. **labialis** var. **abyssinicus** *(Hochst. ex A. Rich.) Verdc.* in *Kew Bull.* **24**: 270 (1970);
Glycine abyssinica Hochst. ex A. Rich. (1847) - type: TU, *Schimper* II:1057 (P lecto., BM K isolecto.).
Fig. 95.25.3.

Grassland, bushland, cultivations; 1600-2100 m. EW TU GD WU SU AR HA KF SD; Uganda, Kenya, Tanzania, Zaire, Zambia, Zimbabwe and Angola. *Mooney* 6221; *Thulin* 1332, 1621.

Var. *labialis* occurs mainly in S Africa and var. *acutus* Verdc. across tropical Africa from Sierra Leone to Kenya.

Subsp. **arabicus** *Verdc.* in *Kew Bull.* **24**: 272 (1970).

Grassland, woodland, scrub; *c* 900-1250 m. EW HA BA GG; widespread in Africa, also in the West Indies and Guayana. *Gilbert & Thulin* 291; *Pappi* 6857, 5440.

2. **T. repens** *(Taub.) Bak. f.* (1928).

Prostrate, often mat-forming herb; stems spreading pilose. Leaflets almost round to oblong, 0.7-4.7 x 0.5-2.4 cm, rounded to subacute at the apex, glabrescent or pilose above, pubescent on the nerves beneath. Inflorescence 2-5-flowered. Corolla white, pink or pale mauve, 6 mm long. Pod 22-40 x 2.5-3 mm, spreading pubescent.

Subsp. **gracilis** *(Chiov.) Verdc.* in *Kew Bull.* **24**: 275 (1970);
T. gracilis Chiov. (1908); types: EE, *Terracciano* 16 & 739 (not 736) & *Pappi* 60 & 4057 (all FT syn.).

Leaflets oblong or obovate; seeds rough.

Bushland near sea-level. EE; Sudan, S Somalia, Socotra, Kenya, Tanzania, Arabia; probably also in India. *Baldrati* 258, 627; *Bettini* s.n.

Subsp. *repens* occurs in tropical Africa from Uganda to Angola and Zimbabwe.

3. **T. uncinatus** (L.) Sw. (1788);
Dolichos uncinatus L. (1763).

Stems climbing, usually with dense dark rusty-red pubescence. Leaflets paler beneath, elliptic to linear-oblong, 1.8-15.2 x 0.5-3.5 cm, rounded and mucronulate at the apex, appressed pubescent at least beneath. Inflorescences usually elongate and dark brown hairy. Corolla greenish, white, mauve or purple, 3.5-5.5 mm long. Pod 25-53 x (2-)2.5-3 mm, densely covered with spreading or appressed dark hairs. Seeds smooth.

1. Inflorescences elongate; calyx and pods with rusty-red or blackish-brown hairs. subsp. **ringoetii**
- Inflorescences mostly rather short, often the flowers in subsessile clusters; calyx and pods usually with a mixture of whitish and rusty-red hairs. subsp. **axilliflorus**

Subsp. **ringoetii** *(De Wild.) Verdc.* in *Kew Bull.* **24**: 279 (1970);
Glycine ringoetii De Wild. (1913).
Fig. 95.25.4.

Grassland and woodland; 1300-1650 m. WG SU KF GG; widespread in tropical Africa. *Gilbert & Thulin* 640; *Mogk* 247; *J. De Wilde* 5334; *Haugen* 336.

Subsp. **axilliflorus** *(Kotschy) Verdc.* in *Kew Bull.* **24**: 278 (1970);
Glycine axilliflora Kotschy (1865).

Grassland and woodland, often on seasonally flooded ground; *c* 500 m. IL; widespread in tropical Africa. *Jansen* 5042.

First record for Ethiopia. Subsp. *uncinatus* occurs in tropical S and C America.

4. **T. micans** *(Bak.) Bak. f.* (1928).

Robust climber; stems angular to terete, densely hairy to glabrescent. Leaflets paler beneath, obovate, rhomboid or elliptic, 3.3-13.5 x 1.7-9 cm, rounded, emarginate or acuminate at the apex, pubescent at least beneath. Inflorescences elongated, densely rusty-red pubescent. Corolla orange or blue, 3.5-6 mm long. Pod 30-46 x 3-4 mm, densely covered with whitish or rusty-red silky hairs.

Var. **cyaneus** *(De Wild.) Hauman* in *Fl. Congo Belge* **6**: 104 (1954).

Stems distinctly angled and with rusty-red pubescence particularly on the angles. Leaflets broadly oblong-elliptic, rather obtuse at the apex, densely hairy beneath.

Grassland, scrub, cultivations; 1400-1600 m. IL GG; Uganda, Kenya, Tanzania, Zaire, Malawi, Zambia. *Friis* et al. 1986; *Fukui* 160.

Var. *micans* is widespread in W and C tropical Africa and var. *fagifolius* Hauman occurs in Cameroun, Sudan, Uganda and Zaire.

58. ERYTHRINA *L.* (1753)

Gillett in *Kew Bull.* **17**: 77-80; Verdcourt in *Fl. Trop. E. Afr.* (1971); Krukoff & Barneby in *Lloydia* **37**: 332-459 (1974).

Trees or shrubs, rarely herbs, often armed with prickles. Leaves pinnately 3-foliolate. Pseudoracemes axillary or terminal, mostly pyramidal and many-flowered, often appearing when the plant is quite leafless. Corolla mostly large, red or orange, standard usually much larger than the other petals. Pods mostly leathery or woody, often much constricted between the seeds, 2-valved or opening only along upper edge or apparently indehiscent. Seeds 1-14, mostly red, orange or yellow with a white or black hilum.

Some 110 species throughout the tropics, but predominantly neotropical.

A genus which has stimulated considerable interest because of chemistry. Several species are very ornamental. Indigenous species can be used as browse.

1. Inflorescence leafy; leaflets elliptic. **1. E. crista-galli**
- Inflorescence not leafy; leaflets (elliptic-) ovate to triangular, rhomboid or oval. 2

2. Calyx 2-lipped. **2. E. lysistemon**
- Calyx spathaceous, not 2-lipped. 3

3. Calyx with 5 distinct linear lobes; pod strongly constricted between the seeds. **4. E. abyssinica**
- Calyx split down one side but otherwise entire or nearly so; pod not, or but slightly constricted between the seeds. 4

4. Indumentum of calyx very short, whitish to rusty-red; valves of pod stiff, woody, up to 30 cm long. **6. E. melanacantha**
- Indumentum of calyx dense, shaggy, brown; valves of pod papery or leathery, shorter. 5

5. Leaflets longer than broad, often more than 10 cm long, acuminate at the tip. **3. E. brucei**
- Leaflets broader than long, less than 10 cm long, truncate at the tip. **5. E. burana**

1. **E. crista-galli** *L.* (1767).

Tree. Leaflets elliptic, minutely reticulately waxy beneath. Inflorescences terminal, leafy or flowers in axils of leaves. Corolla red, *c* 5 cm long. Pods *c* 20 x 1.2-1.5 cm, constricted between the seeds.

Ornamental tree native in S America, but widely cultivated elsewhere. EW. *Baldrati* 4095.

2. **E. lysistemon** *Hutch.* (1933).

Tree or shrub. Leaflets triangular-ovate, glabrescent. Inflorescences dense, rusty-red pubescent. Calyx shortly 2-lipped, 0.8-1.3 cm long, tomentose. Corolla red or orange-red, *c* 4-6 cm long; keel *c* 1-1.5 cm long. Pods stipitate, straight or curved, 5-15 x 1 cm, 1-9-seeded, very markedly constricted between the seeds.

Ornamental tree, native in Southern Africa, but widely cultivated elsewhere. EW HA. *Afeworki G. & Mehret A.* 42; *Jansen* 7152.

3. **E. brucei** *Schweinf.* (1868); emend. *Gillett* in *Kew Bull.* **15**: 428 (1962)
- type: WU, *Steudner* s.n. (B holo. destr.).

Tree up to 15-20 m tall; branches prickly. Leaflets elliptic to ovate or obovate, up to 23 x 16 cm, acuminate at the tip, glabrous or almost so; petiole and nerves sometimes prickly. Calyx split down one side but otherwise entire or nearly so, *c* 3 cm long, densely shaggy brown pubescent. Standard *c* 5 cm long, about twice as long as the keel, orange red (occasionally light-yellow). Pod linear-oblong, up to 15 cm long, dehiscent along both sutures; valves thinly leathery, soon separating from the whitish endocarp and mesocarp which persist as freely pendulous ribbons. Seeds 2-4, red with white hilum. Fig. 95.26.1

Edges and open places of upland forests or woodlands; 1400-2600 m. WU WG GJ SU BA HA IL KF GD GG SD; not known elsewhere. *Friis* et al. 2101; *Mooney* 8873, 8561; *Burger* 1520.

Can be grown easily from cuttings and is used to form live fences. The wood is very soft and hollowed-out branches are used for traditional beehives.

4. **E. abyssinica** *Lam. ex DC.* (1825);

Chirocalyx abyssinicus (DC.) Hochst. (1846) - type: a plant cultivated at the Trianon, Paris, from seeds brought from Ethiopia by Bruce (G holo.).

C. tomentosus Hochst. (1846) - type: TU, *Schimper* II:531 (B holo. destr., K iso.).

E. tomentosa R. Br. ex A. Rich. (1847) - types: *Schimper* II:531 (P syn., K isosyn.) & *d'Hericourt* (P syn.) & *Quartin Dillon* s.n. (P syn.) & *Petit* s.n. (P syn.) & *Bruce*, *Voy. Abyss.*, t. 19 in part (syn.).

Tree or shrub, 1-15 m tall; branchlets prickly, at first tomentose, later glabrescent. Leaflets usually broadly ovate or rhomboid, up to 20 x 21 cm, rounded or emarginate at the apex, ± pubescent; petiole and midrib sometimes prickly. Calyx divided to the base, 5-lobed at the apex, densely tomentose. Standard 2.8-5 cm long, scarlet to pale pink, at least 4 times as long as the keel. Pod woody, 4-16 cm long, strongly constricted between the seeds, tomentose or glabrescent. Seeds 1-10, red with a black hilum. Fig. 95.26.2.

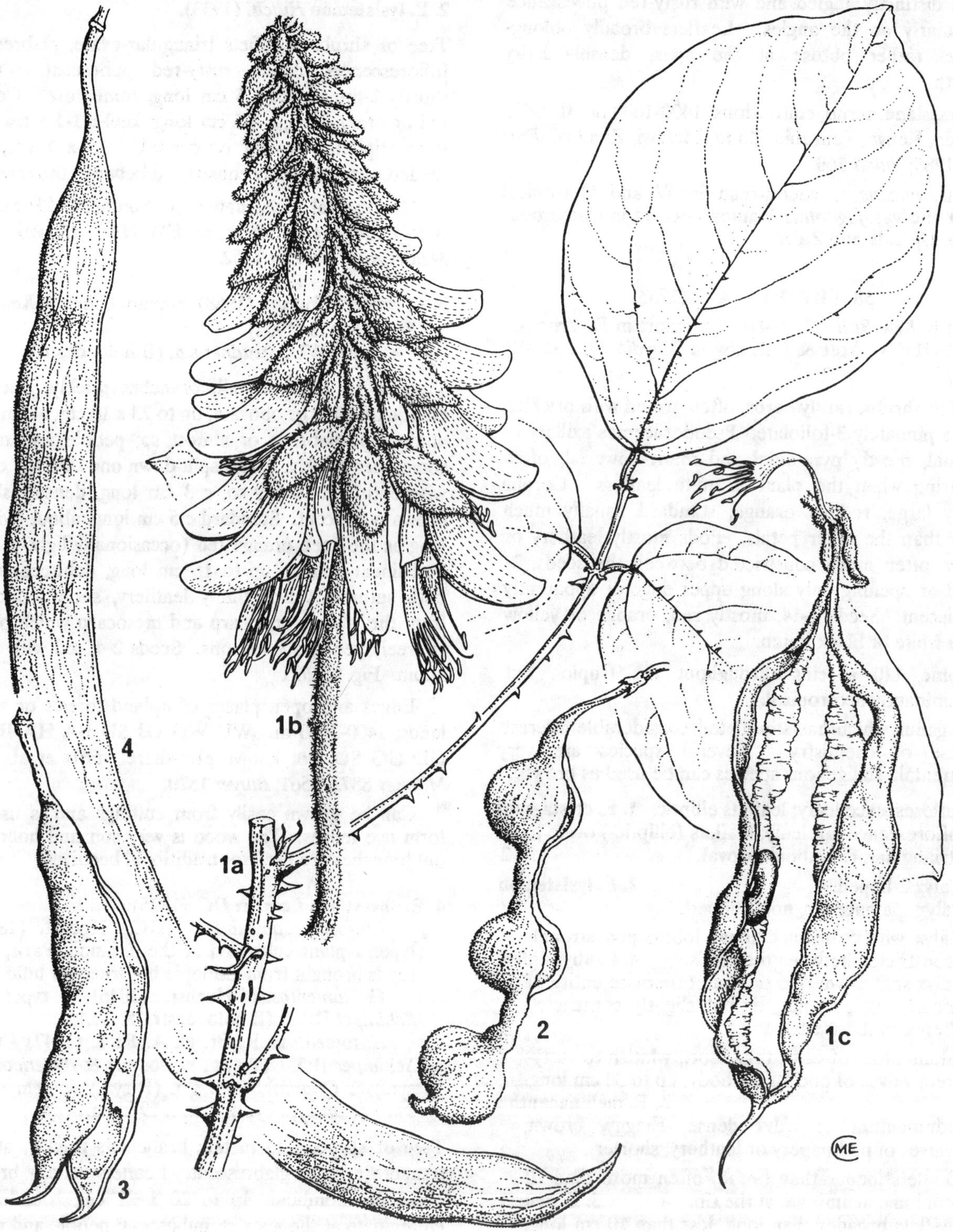

Figure 95.26.1 - *ERYTHRINA BRUCEI*, a: leafy shoot, b: inflorescence, both from *Friis* et al. 2101, c: pod after dehiscence from *Mooney* 8561. 2 - *E. ABYSSINICA*, pod from *Meyer* 8654. 3 - *E. BURANA* pod from *Burger* 1563. 4 - *E. MELANACANTHA* subsp. *SOMALA*, pod from *Hemming* 1491. All x 0.75. Drawn by Maureen Church. (Reproduced with permission from *Opera Botanica*, 68, fig. 28, 1983.)

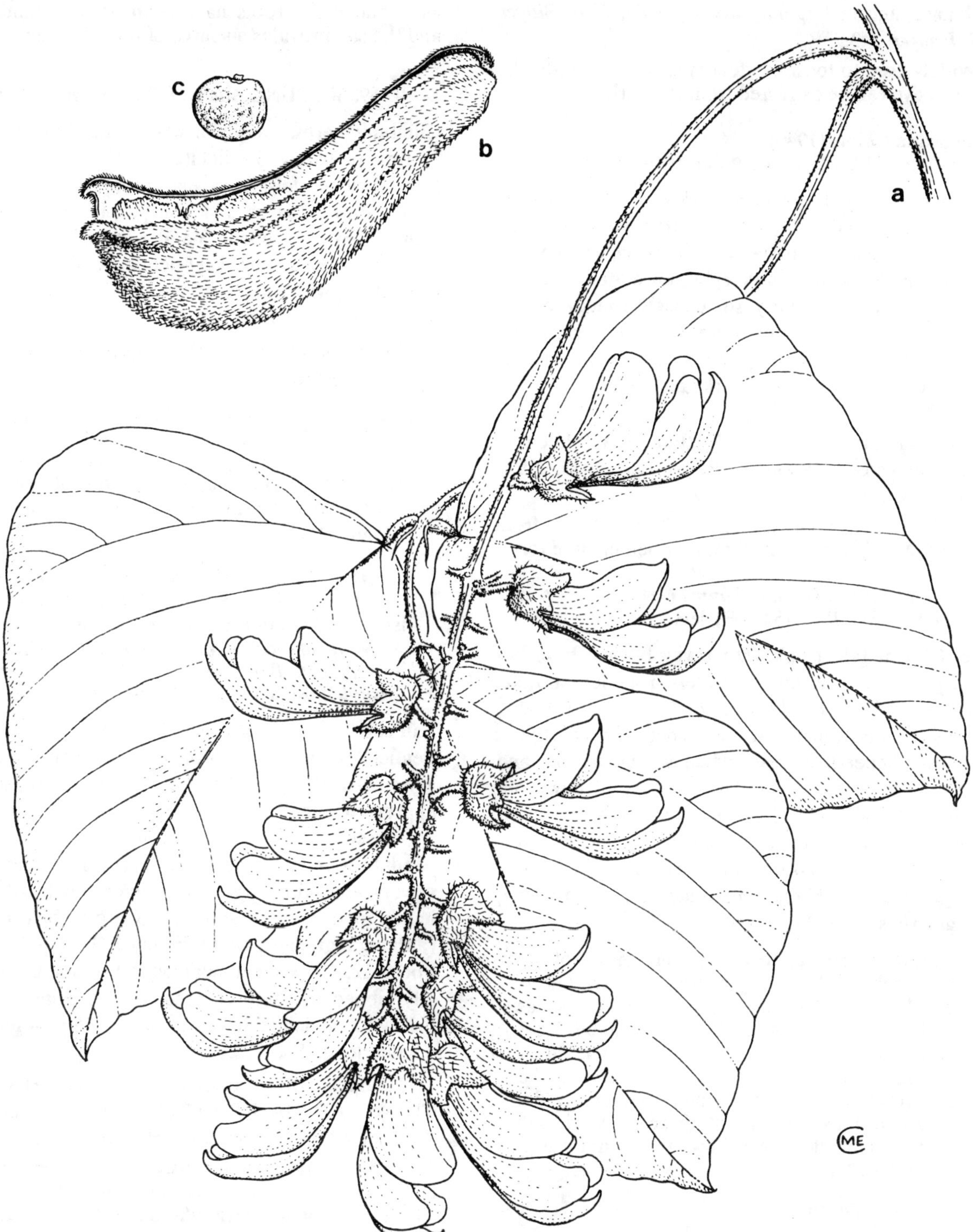

Figure 95.27.1 - *MUCUNA MELANOCARPA*, a: flowering branch (many flowers removed) from *Gilbert & Thulin* 825, b: pod, c: seed, both from *Thulin* 1401. All x 0.75. Drawn by Maureen Church. (Reproduced with permission from *Opera Botanica*, 68, fig. 29, 1983.)

Grassland, woodland, forest edges, rocky places; 1300-2400 m. EW TU WU GD GJ WG SU AR KF IL GG SD; widespread in eastern tropical Africa south to Zimbabwe, and in Angola. *Mooney* 8585, 7731; *Burger* 2567; *Haugen* 543.

Widely used to form live fences around homesteads. A brown dye can be extracted from the bark.

5. **E. burana** *Chiov.* (1940)
- type: HA, *Milizia Forestale* 1619 (FT, holo.).

Tree up to 12 m tall; branches prickly. Leaflets broadly ovate to oblate, up to 10 x 12 cm, truncate at the tip, always with some stellate hairs on the rhachis and at the base of the leaflets. Calyx split down one side but otherwise entire or nearly so, densely shaggy brown pubescent. Standard *c* 4 cm long, red or orange, much longer than the keel. Pod up to *c* 8 cm long, tomentose when young. Seeds red with dark grey hilum. Fig. 95.26.3.

Woodland, bushland; 1350-2100 m. HA (SU cultivated); not known elsewhere. *Burger* 1421, 1435, 2599; *Mesfin* et al. 299.

A sterile specimen, *Gilbert & Ermias* 8468, from BA, S of Dolo Menna at 1650 m, may belong here too.

6. **E. melanacantha** *Taub. ex Harms* (1897);
- type: SD, *Riva* 1547 (FT lecto.).

Tree, 6-20 m tall; trunk with corky bosses; branches bearing curved prickles. Leaflets rounded-obovate, rhomboid or broadly elliptic, 2-9 x 1.5-12 cm, rounded or slightly emarginate at the apex, finely stellate tomentose beneath at least when young; petiole and often midrib of leaflets prickly. Calyx split down one side but otherwise entire or nearly so, with very short whitish to rusty-red stellate pubescence. Standard red, 4.5-7.5 cm long, 3 to 4 times as long as the keel. Pod oblong, curved, 13-30 cm long, tomentose when young, later glabrous, (2-)6-11-seeded.

1. Fruiting pedicels up to 14 mm long; valves of the pod feebly ridged lengthwise, abruptly falcate distally; ventral suture distinctly winged; seeds dull yellowish or reddish brown, the hilum 2.5-3 mm wide. subsp. **melanacantha**
- Fruiting pedicels up to 7 mm long; valves of the pod prominently ridged lengthwise, gently decurved distally; ventral suture scarcely winged; seeds bright sulphur yellow, the hilum rarely more than 2 mm wide. subsp. **somala**

Subsp. **melanacantha.**

Dry wood/bushland; *c* 1200-1300 m. BA GG SD; Somalia, Kenya, Tanzania. *Ash* 2780; *Corradi* 4390; *Thulin, Hunde & Tadesse* 3559.

Subsp. **somala** *(Chiov.) Gillett* in *Kew Bull.* **27**: 289 (1972);
E. melanacantha var. *somala* Chiov. in *Fl. Somala* 1: 53 (1929).
Fig. 95.26.4.

Dry bushland; *c* 600-700 m. HA; Somalia, Socotra. *Hemming* 1491; *Simmons* 194; *Ellis* 323.

The following uses have been noted for the species as a whole: the roots have been used as famine food and the sap provides a source of potable water.

59. MUCUNA *Adans.* (1763), *nom. conserv.*

Climbing herbs, woody lianes or rarely erect shrubs. Leaves pinnately 3-foliolate. Inflorescences axillary, subumbellate, falsely racemose or condensed paniculate. Flowers large, blackish purple, red or greenish-white. Calyx 4-5-lobed. Standard usually much shorter than the other petals; keel-petals usually horny at the apex. Pods with thick valves, often covered with irritant bristles, mostly dehiscent.

Some 100 species in the tropics and subtropics of both hemispheres.

1. Keel 4-5 cm long, blackish purple; calyx with orange bristles; pods 9-12(-15) cm long. **1. M. melanocarpa**
- Keel 3.2-3.7 cm long, pale; calyx silvery pubescent, often without orange bristles; pods less than 9 cm long. **2. M. pruriens**

1. **M. melanocarpa** *Hochst. ex A. Rich.* (1847);
- types: TU, *Schimper* III:1425 (P syn., BM K S isosyn.) & *Quartin Dillon* s.n. (P syn.).
M. melanocarpa var. *somalensis* Taub. ex Harms in *Ann. Ist. Bot. Roma* **7**: 97 (1897) - types: SD, *Ruspoli & Riva* 150 & 1411 (both FT isosyn.).

Tall climber, up to at least 10 m long; stems pubescent. Leaflets rhomboid or broadly ovate, the laterals very asymmetrical, up to 10-18 x 8-14 cm, subglabrous to appressed pilose above, sparsely pubescent to densely appressed silvery pilose beneath. Inflorescences pendulous, many-flowered false racemes. Calyx densely velvety and with appressed orange-red bristly irritant hairs; tube *c* 7-8 mm long. Corolla blackish-purple (although greenish-white in bud); keel 4-5 cm long. Pod 9-12(-15) x 2 cm, somewhat S-shaped, longitudinally ribbed, densely covered with orange-red irritant bristly hairs, dehiscent. Seeds 2-4, *c* 16 x 14 mm; hilum *c* 3 mm long. Fig. 95.27.

Woodland, forest margins, scrub; 1400-2100 m. EW TU WG AR HA KF GG SD; not known elsewhere. *Thulin* 1401; *Gilbert & Thulin* 825; *Mooney* 8709.

Forms with conspicuously hairy leaves have been called var. *somalensis*, but the material now available shows a complete intergradation in this respect.

2. **M. pruriens** *(L.) DC.* var. **utilis** *(Wall. ex Wight) Bak. ex Burck* in *Ann. Jard. Bot. Buitenzorg* **11**: 187 (1893).

Climbing herb; leaflets rhomboid to ovate, rather thin; calyx silvery; standard dark purple to almost white; wings purple; keel whitish, 3.2-3.7 cm long; pods up to 9 x 2 cm, longitudinally ribbed, glabrous or pubescent, ±

irritant bristles, dehiscent; seeds 2-4, up to 19 x 13 mm; hilum 4-5 mm long.

Cultivated at least in WG (Gimbi) and HA. *Westphal* 8626, 8627; *Stewart* 60.

Previous records of the wild variety, var. *pruriens*, which is a lowland plant widespread in the tropics, are based on misidentified material of *M. melanocarpa*. Like *M. melanocarpa* this variety has pods with irritant bristles.

60. **CANAVALIA** *DC.* (1825), *nom. conserv.*

Climbing robust herbs or woody lianes. Leaves pinnately 3-foliolate. Inflorescences axillary, racemose. Flowers mostly large, rather thick in texture. Calyx 5-lobed, 2-lipped, the lower 3 lobes joined to form a trifid or entire lip, upper pair joined to form a bifid or truncate lip. Standard reflexed. Pod linear-oblong, large, usually with longitudinal ribs close to upper margin.

Some 50 species throughout the tropics and subtropics, but mainly American.

1. Plant usually annual, less than 1 m long; seeds white with brown mark near hilum; pod 15-35 x 3-3.5 cm. **1. C. ensiformis**
- Plant perennial, several m long; seeds brown; pod 10-20 x 2.5-3 cm. **2. C. africana**

1. C. **ensiformis** *(L.) DC.* (1825).

Plant usually grown as an annual but capable of becoming a perennial climber, up to at least 1 m long. Leaflets elliptic or ovate-elliptic up to 20 x 11.5 cm, obtuse to shortly acuminate, glabrescent or sparsely pubescent on both surfaces. Calyx-tube 6-7 mm long, upper lip 5 mm long, truncate. Corolla rose to purple, *c* 2.75 cm long. Pod 15-35 x 3-3.5 cm with longitudinal ribs close to the upper margin. Seeds 1.4-2.1 cm long, white with brown mark near hilum, which is 5.5-9 mm long.

Cultivated for its edible seeds and as a cover crop or as a green manure throughout the tropics, native in America; up to *c* 1800 m. GD (Sudan border), SU. *Meyer* 8095; *Schweinfurth* 1902.

2. C. **africana** *Dunn (1922).*
C. virosa sensu auctt., non *(Roxb.) Wight & Arn.* sensu stricto.

Perennial climber or trailer, 3-15 m tall. Leaflets ovate, up to 16.5 x 12 cm, usually acuminate to a short blunt tip, sparsely to densely pubescent on both surfaces. Calyx-tube 6-9 mm long; upper lip 4-5 mm long, rounded or emarginate. Corolla mauve or pale purple, 2.7-3 cm long. Pod 10-20 x 2.5-3 cm, with longitudinal ribs close to the upper margin. Seeds 1.6-2 cm long, brown; hilum 11-15 mm long. Fig. 95.30.1.

Grassland, bushland, also cultivated as a cover and shade on fences and houses; *c* 400-1800 m. EE EW HA IL GG; widespread in tropical Africa, Socotra, S Arabia and India. *Hemming* 1487; *Ebba* 708; *Carr* 747.

For the nomenclature of this species, see Verdcourt in *Kew Bull.* **42**:658 (1987).

61. **GALACTIA** *P. Br.* (1756)

Herbs or subshrubs, prostrate, climbing or erect. Leaves usually pinnately 3-foliolate. Flowers small in axillary few-flowered pseudoracemes. Calyx 4-lobed with the two upper lobes completely joined. Vexillary stamen free; anthers uniform. Style glabrous. Pod linear-oblong, compressed.

Some 50 species, pantropical, mostly New World.

G. **tenuiflora** *(Willd.) Wight & Arn.* (1834).

Perennial slender climber. Leaflets paler beneath, ± oblong, 2-6.5 x 0.8-3 cm, rounded or emarginate at the tip, rounded or subcordate at the base, sparsely appressed pubescent above, glabrescent to densely pubescent beneath. Racemes lax (1-)2-10-flowered. Calyx-tube 2.5-3 mm long; lobes 3.5-6(-7) mm long, the lowest the longest. Standard whitish to mauve and purple, 1-1.2 cm long, glabrescent. Pod 30-60 x 5-7.5 mm, usually 7-8-seeded, finely appressed pubescent.

Var. **villosa** *(Wight & Arn.) Benth.* in *Mart., Fl. Bras.* **15(1)**: 143 (1859).
G. tenuiflora var. *biflora* Schweinf. in *Bull. Herb. Boiss.* **4**, app. 2: 253 (1896); syntypes from Yemen and Eritrea: *Schweinfurth* 1884 (B syn. destr.).

Leaflets pubescent beneath. Fig. 95.30.2.

Grassland or woodland; *c* 1600 m. EW; widespread in Africa, India and China. *Pappi* 5493.

Var. *tenuiflora* in tropical Asia.

62. **MACROTYLOMA** *(Wight & Arn.) Verdc.* (1970), *nom. conserv.*

Verdcourt in *Kew Bull.* **24**: 400-406 (1970), *Fl. Trop. E. Afr.* (1971) & *Hook. Ic. Pl.* **38(4)** (1982).

Annual or perennial herbs, climbing, prostrate or sometimes erect. Leaves pinnately (1-)3-foliolate. Flowers usually axillary and fasciculate. Calyx 4-5-lobed; the upper pair of lobes joined to form an entire or bifid lip. Corolla yellowish, glabrous; standard free. Pod compressed.

Some 24 species native in Africa and Asia.

Dolichos chrysanthus A. Chev. var. *occidentalis* (Harms) Wilczek which is now placed in *Macrotyloma* has been reported from Ethiopia (see Cuf. *Enum.*: 341, 1955). The record has not been verified and is excluded here.

1. Stipules conspicuous, ovate, 5-10 x 3-5 mm, up to 10-veined. **8. M. schweinfurthii**
- Stipules not so large. 2
2. Pod 2-5.5 mm wide. 3
- Pod 6-8 mm wide. 7

3. Pod 3.5-5.5 mm wide. 4
- Pod 2-3.5 mm wide. 6
4. Calyx-lobes all 1.5-3 mm long; plant glabrescent or sparsely pubescent. **3. M. africanum**
- Calyx-lobes, at least the lower one, 3-8 mm long; plant more hairy. 5
5. Pod upcurved towards apex; seeds without elaiosome[1]. **1. M. uniflorum**
- Pod straight; seeds modified into an elaiosome at one end. **5. M. biflorum**
6. Plant completely prostrate; leaflets broadly elliptic, round or obovate; calyx-lobes triangular, 1-2 mm long. **6. M. tenuiflorum**
- Plant climbing; leaflets oblong or lanceolate, calyx-lobes linear-lanceolate; the lower mostly 4-7 mm long. **7. M. stenophyllum**
7. Lower and lateral calyx-lobes quickly attenuate into long filiform points; pod sparsely pilose with an understorey of minute hairs; plant usually annual. **2. M. daltonii**
- Calyx-lobes acuminate, not filiform-pointed; pod pubescent or glabrescent; plant perennial. **4. M. axillare**

1. **M. uniflorum** *(Lam.) Verdc.* (1970);
D. uniflorus Lam. (1786).

Perennial or annual climbing herb up to 0.6 m long; stems pilose. Leaflets ovate-rhomboid, obovate or elliptic, 1-3.5 x 0.7-2.7 cm, rounded or subacute at the tip, glabrescent to pubescent. Flowers in axillary (1-)2-3(-5)-flowered fascicles. Calyx-tube 2 mm long; lobes narrow, 3-8 mm long. Standard 0.6-1.2 cm long. Pod 30-55 x 4-8 mm, upcurved towards the apex, pods pilose with an undercover of very fine hairs, or glabrous.

Var. **stenocarpum** *(Brenan) Verdc.* in *Kew Bull.* **24**: 401 (1970);
Dolichos uniflorus var. *stenocarpus* Brenan in *Mem. N. Y. Bot. Gard.* **8**: 416 (1954).

Pod 4-5.5 mm wide, hairy. Leaves pubescent.

Bushland, thicket, often among rocks; *c* 1300-1700 m. HA; scattered in eastern and southern Africa, also in India. *Bally* 10085; *Westphal* 1860; *Corrà* JF-6.

2. **M. daltonii** *(Webb) Verdc.* (1970);
Dolichos daltonii Webb (1849).

Annual or sometimes perennial climbing or trailing herb up to 0.7 m long; stems softly pilose. Leaflets elliptic, 2-7.5 x 0.7-4.5 cm, acute or acuminate at the tip, appressed pubescent to shortly pilose on both surfaces. Flowers axillary, solitary or 2-3 in fascicles. Calyx-tube 1.5-2 mm long; lobes filiform, 5-10 mm long. Standard 0.9-1.6 cm long. Pod 25-45 x 6-8 mm, straight or slightly curved, laxly pilose with an undercover of minute hairs.

Grassland, woodland, thickets; 750-1700 m. EE EW TU WU SU; widespread in tropical Africa. *Gilbert & Getachew* 2951, 3078; *Gilbert & Thulin* 973.

3. **M. africanum** *(Wilczek) Verdc.* (1970).

Annual or short-lived perennial climbing herb, up to more than 1 m long. Stems glabrous or sparsely pilose. Leaflets ± narrowly ovate, 1-5 x 0.5-2.5 cm, narrowed to an obtuse but mucronulate apex, glabrous or sparsely appressed pubescent. Flowers axillary, solitary or in few-flowered fascicles. Calyx-tube 1.5-2 mm long; lobes 1.5-3 mm long. Standard 0.6-1.3 cm long. Pod linear, straight, 30-48 x 3-4 mm, glabrous or finely pubescent.

Grassland, bushland, cultivations; *c* 1800 m. AR; scattered in tropical Africa. *Thulin* 1329.

4. **M. axillare** *(E. Mey.) Verdc.* (1970);
Dolichos axillaris E. Mey. (1836).

Perennial climber up to 3.5 m long; stems ± pubescent. Leaflets elliptic to ovate, 1.5-9 x 1-5 cm, subacute to acuminate at the apex, pubescent to glabrous on both surfaces. Flowers axillary in 2-4(-10)-flowered fascicles. Calyx-tube *c* 2 mm long; lobes lanceolate, 2.5-5 mm long. Standard 1-2.4 cm long. Pod linear-oblong, 30-80 x 6-8 mm, glabrous to pubescent.

1. Stems with ± dense, spreading hairs; standard 1.2-1.5 cm long, pod with beak 3-7 mm long. **var. axillare**
- Stems with ± sparse, appressed hairs; standard 1.3-2 cm long; pod with beak 5-15 mm long. **var. glabrum**

Var. **axillare.**

Grassland, bushland; 1600-1700 m. EW SD; widespread in tropical Africa, Madagascar, Arabia. *Pappi* 4715; *Cufodontis* 568.

Var. **glabrum** *(E. Mey.) Verdc.* in *Kew Bull.* **24**: 402 (1970);
D. axillaris E. Mey. var. *glaber* E. Mey., *Comm. Pl. Afr. Austr.* **1**: 144 (1836).
Fig. 95.28.3.

Grassland, bushland, forest margins; 1500-2350 m. EW TU WG SU HA KF GG SD; widespread in tropical Africa, extending to Yemen, Madagascar, Mauritius, Ceylon. *Gilbert, Thulin & Getachew* 390; *Friis* et al. 98; *Gillett* 14406.

5. **M. biflorum** *(Schum. & Thonn.) Hepper* (1972).

Perennial prostrate or trailing herb; stems ± densely pubescent. Leaflets elliptic to ovate, 1-4 x 0.5-1.2(-2.1) cm, obtuse and mucronate at the apex, ± densely pubescent on both surfaces. Flowers axillary, solitary or in few-flowered fascicles. Calyx-tube 2 mm long; lobes lanceolate, 3-5.5 mm long. Standard 1.1-1.6 cm long. Pod linear, straight, (22-)35-47 x (3.5-)4-4.5 mm, glabrous or puberulous. Seeds modified into an elaiosome at one end.

1. elaiosome = an outgrowth from a seed containing oil or fat and which is often attractive to ants and aids in the dispersal of the seed.

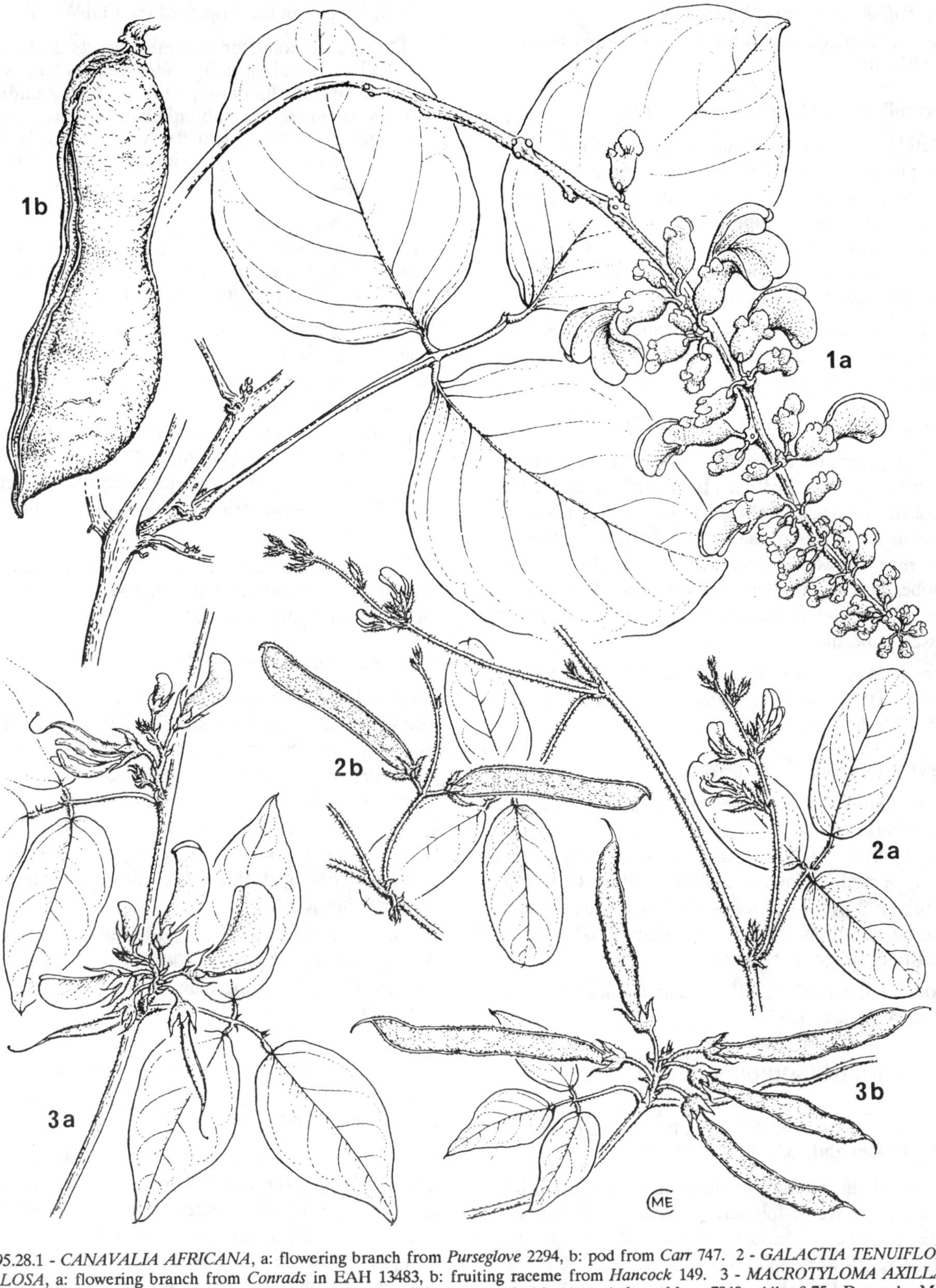

Figure 95.28.1 - *CANAVALIA AFRICANA*, a: flowering branch from *Purseglove* 2294, b: pod from *Carr* 747. 2 - *GALACTIA TENUIFLORA* var. *VILLOSA*, a: flowering branch from *Conrads* in EAH 13483, b: fruiting raceme from *Hancock* 149. 3 - *MACROTYLOMA AXILLARE* var. *GLABRUM*, a: flowering branch from *Gilbert, Thulin & Getachew* 390, b: fruiting branch from *Meyer* 7948. All x 0.75. Drawn by Maureen Church. (Reproduced with permission from *Opera Botanica*, 68, fig. 30, 1983.)

Roadside in red clayey soil; *c* 550 m. IL (Gambela); Sudan and westwards to West Africa, Zaire, Zambia, Angola. *Puff & Ensermu* 820813-7/2.

The single specimen known from Ethiopia belongs to var. *biflorum*.

6. **M. tenuiflorum** *(Micheli) Verdc.* (1970).

Completely prostrate herb up to 0.8 m long; stems densely pubescent. Leaflets broadly elliptic, round or obovate, 1-4.5 x 0.7-3.5 cm, emarginate or rounded at the apex, appressed pubescent on both surfaces. Flowers axillary, solitary or few together. Calyx-tube 1.5-2 mm long; lobes 1-2 mm long. Standard 0.8-1.2 cm long. Pod linear, 40-52 x 2-3 mm, glabrous.

On bare ground in grassland or woodland; *c* 1300 m. WG; Uganda, Togo, Cameroun, Central African Republic, Zaire, Angola. *Gilbert & Thulin* 642.

7. **M. stenophyllum** *(Harms) Verdc.* (1970).

Annual or perennial climber up to 0.9 m long; stems with spreading hairs. Leaflets oblong or lanceolate, 1.8-7 x 0.5-2 cm, subacute to obtuse at the apex, glabrous or puberulous above, appressed pilose beneath. Flowers axillary in 2-4-flowered fascicles. Calyx-tube 2-3 mm long; lobes linear-lanceolate, 4-7 mm long. Standard 1-1.5 cm long. Pod linear, 40-80 x 2-3.5 mm, sparsely appressed puberulous.

Grassland or woodland; *c* 1300 m. WG; Uganda, W Tanzania, Togo, Nigeria, Cameroun, Zaire, Angola. *Gilbert & Thulin* 643.

8. **M. schweinfurthii** *Verdc.* (1982).

Trailing annual; stems densely white pilose. Leaflets oblong-elliptic, up to 5 x 2.8 cm, obtuse at the apex, sparsely appressed pilose beneath; stipules ovate, 5-10 x 3-5 mm, veined. Flowers axillary, solitary or paired. Calyx-tube *c* 3 mm long; lobes 7-13 mm long, acuminate. Standard yellow, *c* 1 cm long. Pods 40-60 x 5.5-6.5 mm, aristate, shaggy pubescent.

Woodland; *c* 1000 m. GD (Sudan border); Nigeria, Togo. *Schweinfurth* 1887.

63. **PSOPHOCARPUS** *Neck. ex DC.* (1825), *nom. conserv.*

Verdcourt in *Fl. Trop. E. Afr.* (1971); Verdcourt & Halliday in *Kew Bull.* **33**: 191-227 (1978).

Herbs or subshrubs, mostly climbing or prostrate, less often erect. Leaves 1-foliolate or pinnately 3-foliolate; stipules spurred. Inflorescences axillary, pseudoracemose or flowers solitary or fasciculate. Calyx 5-lobed; upper pair of lobes forming an entire or bifid lip. Corolla blue or purplish, glabrous. Anthers 5 dorsifixed alternating with 5 basifixed. Style thickened above the ovary, bent, flattened towards the apex, glabrous, bearded longitudinally or with a ring of hairs below the stigma. Pod oblong, ± distinctly 4-winged along the angles, dehiscent.

Nine species in the tropics of the Old World.

1. Flowers in elongated racemes; bracts deciduous; leaflets broad, usually more than 4 cm wide; style laterally hairy to the tip. **1. P. grandiflorus**
- Flowers in short, subumbellate inflorescences; bracts persistent; leaflets narrow, usually less than 3.5 cm wide; style with a ring of hairs below the apex. **2. P. lancifolius**

1. **P. grandiflorus** *Wilczek* (1954).

P. palustris Desv. cultivar. 'Wondo Surprise' Westphal, *Pulses in Ethiopia*: 196, fig. 23 (1974); type: SD, *Westphal* 2666 (WAG holo.).

Perennial climbing herb up to 5 m; stems appressed pubescent, glabrescent. Leaflets 3, ± ovate, 5-15 x 4-12 cm, acute to acuminate, sparsely pubescent along the nerves on both sides but mainly beneath. Flowers in long racemes; bracteoles ovate, 7-10 mm long, puberulous. Standard 2.5-4.5 cm long. Pods square in cross-section, 4-10 x 0.7-1.2 cm, prominently winged, the wings 3-5 mm wide, crinkly, puberulous or pubescent. Fig. 95.29.1.

Upland bushland, forest and grassland, occasionally cultivated; 1600-2300 m. WG KF SD; Uganda, Zaire. *Friis* et al. 2196; *Mooney* 6227; *Gillett* 14655.

2. **P. lancifolius** *Harms* (1911).

Perennial climbing herb up to 5 m; stems appressed yellowish-pubescent to spreading pilose. Leaflets 3, lanceolate or elliptic, 2.2-11 x 0.7-3.5(-6) cm, acute, pubescent to glabrescent on both sides. Inflorescences short, subumbellate; bracteoles ovate-oblong, 8-12 mm long, rusty-red hairy. Standard 1.5-2.5 (-3.2) cm long. Pods rectangular in cross-section, 3.5-7 x 0.5-0.7 cm, prominently winged, the wings 1.5-3 mm wide, crinkly, sparsely pubescent.

Grassland or wooded grassland, particularly in slightly marshy places; *c* 2000 m. SD (near Sodo); eastern Africa S to Zimbabwe and in Nigeria. *J. De Wilde* 5607.

64. **VÀTOVAEA** *Chiov.* (1951)

Gillett in *Kew Bull.* **20**: 104 (1966); Verdcourt in *Fl. Trop. E. Afr.* (1971).

Woody climber. Leaves pinnately 3-foliolate. Calyx 5-lobed, 2-lipped, the lower 3 lobes rounded-triangular, the upper pair of lobes united to form a slightly emarginate lip. Standard with a pair of large stiff erect parallel appendages near the base; keel incurved, scarcely beaked. Vexillary stamen free; anthers uniform. Style long, incurved, usually bearded inside towards the apex and with a reflexed appendage above the stigma (see Fig. 95.29.2c). Pod linear-oblong, dehiscent.

Monotypic.

V. pseudolablab *(Harms) Gillett* (1966);
Vigna pseudolablab Harms (1913); syntypes from Kenya and Ethiopia: Lake Chamo, *Neumann* 100 (B syn. destr.).
Vàtovaea biloba Chiov. (1951); type: SD, *Vàtova* 87 (FT holo.).

Woody climber, although young plants may flower while still quite herbaceous, from a large tuber; stems glabrous to sparsely pubescent. Leaflets ± ovate, up to 6 x 5 cm, sometimes ± 3-lobed, glabrous to sparsely pubescent. Inflorescences long, many-flowered. Calyx glabrous; tube 2.5 mm long, lobes 2.5 mm long. Standard greenish to purplish blue, 1-1.8 cm long. Pods curved, flattened, 45-60 x 5.5-10 mm, widening towards the apex, first silky pubescent, becoming glabrescent. Fig. 95.29.2.

Dry grassland or bushland; 600-1500 m. AF SU HA BA GG SD; Somalia, Sudan, N Uganda, Kenya, Tanzania, and in Oman. *Burger* 2830; *Gilbert & Thulin* 136; *Gilbert* 1645.

65. MACROPTILIUM *(Benth.) Urb.* (1928)

Herbs, erect, prostrate or climbing. Leaves pinnately 3-foliolate. Inflorescences axillary, pseudo-racemose. Calyx 5-lobed. Keel incurved; wings round, large, longer than standard and keel; claws of wings and keel long, partly adnate to the staminal tube. Vexillary stamen free. Style with the upper part thickened and characteristically abruptly curved through 90° just above its junction with the tenuous part, and narrowed and slightly curved towards the apex, resembling a squarish hook, upper portion hairy on the internal face. Pod long, linear, subcylindrical.

Some 20 species in tropical America, two of which are now widely cultivated.

1. Plant creeping; leaflets ± broadly ovate or rhomboid, densely velvety beneath. **1. M. atropurpureum**
- Plant climbing or suberect; leaflets narrowly elliptic to ovate, thinly pubescent beneath. **2. M. lathyroides**

1. **M. atropurpureum** *(DC.) Urb.* (1928);
Phaseolus atropurpureus DC. (1825).

Creeping perennial, ± densely velvety with white or yellowish hairs. Leaflets ± broadly ovate or rhomboid, 1.5-5 cm long and wide, the laterals often lobed on the outer side. Inflorescences fairly lax, long-pedunculate, much exceeding the subtending leaves. Calyx 6-8 mm long, lobes shorter than tube. Corolla 17-20 mm long; wings blackish purple; standard and keel paler. Pod 6-9 x 0.4 cm, nearly straight, densely appressed pubescent, *c* 12-15-seeded.

Recently introduced in Ethiopia and being promoted through regional forage crop trials, locally naturalised; 750-1500 m. AF SU(N of Awash); native of tropical America, introduced elsewhere in the tropics for soil-improvement and fodder. *Seegeler* 2784; *Sue Edwards* 615; *Pülschen* 601.

2. **M. lathyroides** *(L.) Urb.* (1928);
Phaseolus lathyroides L. (1763).

Climbing or suberect annual, sparsely to densely pubescent. Leaflets narrowly elliptic to ovate, 2-8 x 0.7-3.5 cm, not or only very shallowly lobed, glabrous above, thinly appressed pubescent beneath. Inflorescences fairly lax, long-pedunculate, much exceeding the subtending leaves. Calyx 5-6 mm long, lobes shorter than tube. Corolla *c* 20 mm long; wings dark red; standard and keel paler. Pod *c* 6-10 x 0.3 cm, nearly straight, appressed pubescent, *c* 18-30-seeded.

Recently introduced in Ethiopia, locally naturalised along irrigation canals; *c* 300 m. HA-BA (Gode Experimental Farm); native of tropical America, introduced elsewhere in the tropics for soil-improvement and fodder; suited for poorly drained conditions. *Sue Edwards* 692.

66. PHASEOLUS *L.* (1753)

Verdcourt in *Kew Bull.* **24**: 507-526 (1970) & *Fl. Trop. E. Afr.* (1971).

Herbs or subshrubs, erect, prostrate or climbing. Leaves pinnately 3-foliolate, rarely 1-foliolate. Stipules not prolonged below the point of insertion, striate. Inflorescences axillary, pseudo-racemose. Calyx 5-lobed, 2-lipped; lower lip 3-lobed; upper bifid. Keel often narrow, elongated, the apex beaked and forming a spiral of 1-5 complete turns. Vexillary stamen free. Style with the upper part thickened, curved through at least 360°, upper portion glabrous or hairy on the internal face. Pod linear or oblong, compressed or subcylindrical. Seeds ± oblong; hilum short, ± central.

Some 50 species, predominantly American.

An important genus containing a number of widely grown pulses all of which originated in the New World.

P. acutifolius A. Gray has been reported as cultivated in Ethiopia, Harer (see Cuf. *Enum.*: 329, 1955). No specimen has been seen and the record needs confirmation.

1. Pods oblong, curved, 2-4-seeded; flowers mostly whitish; bracteoles 1.5-2 mm long. **2. P. lunatus**
- Pods linear-lanceolate, seeds often 5 or more; flower-colour various; bracteoles 5-6 mm long. 2
2. Flowers white, yellowish, pink or violet-purple, mostly few and 1-1.5 cm long. **1. P. vulgaris**
- Flowers scarlet or white, often numerous, mostly *c* 2 cm long. **3. P. coccineus**

1. **P. vulgaris** *L.* (1753).

Annual climber or suberect herb. Leaflets ovate or ovate-rhombic, 4.5-15 x 2.5-10 cm, acuminate, pubescent. Inflorescences 1-several-flowered; bracteoles

Figure 95.29.1 - *PSOPHOCARPUS GRANDIFLORUS*, a: leafy branch and inflorescence from *Gillett* 14655, b: pod from *Friis* et al. 2196. 2 - *VÀTOVAEA PSEUDOLABLAB*, a: flowering branch from *Gillett* 13848, b: pod from *Milne-Redhead & Taylor* 7136, c: style from *Gillett* 13848. 3 - *PHASEOLUS LUNATUS*, a: flowering branch, b: pod, both from *Meyer* 7813. 2c x 4.5, the rest x 0.75. Drawn by Maureen Church. (Reproduced with permission from *Opera Botanica*, 68, fig. 31, 1983.)

ovate-lanceolate, 5-6 mm long. Calyx 4-5 mm long. Corolla white, yellowish, purple or pink; standard 1-1.5 cm long; keel spirally incurved. Pod linear-lanceolate, *c* 10-20 x 1-1.5 cm, compressed, puberulous or glabrous, *c* 5-10-seeded.

Native in America, now cultivated throughout the world; in Ethiopia commonly grown at intermediate altitudes and perhaps locally naturalised. *Ebba* 583; *Westphal* 525; *Jansen* 2089.

2. **P. lunatus** L. (1753).

Perennial or biennial climber or sometimes subshrub. Leaflets ovate to lanceolate or narrowly rhombic, 3-15 x 1.2-10 cm, acute or acuminate, sparsely pubescent or glabrous. Inflorescences lax, few-many-flowered; bracteoles 1.5-2 mm long. Calyx *c* 3-4 mm long. Corolla whitish; standard *c* 1 cm long; keel spirally incurved for 1.5 turns. Pod oblong, 5-12 x 1.5-3 cm, compressed, glabrous or pubescent, 2-4-seeded. Fig. 95.29.3.

Native in America, now widely cultivated in the tropics; in Ethiopia mainly cultivated in the SW parts and perhaps locally naturalised. *Meyer* 7813; *Westphal* 1971, 2334; *Tewolde Berhan & Gelahun* 1572.

3. **P. coccineus** L. (1753).

Climbing annual or perennial. Leaflets ovate-rhombic, 5-12.5 x 3.5-8.5 cm, sparsely pubescent. Inflorescences many-flowered; bracteoles *c* 6 mm long. Corolla scarlet or less often white; standard *c* 2 cm long; keel spirally incurved. Pods linear-lanceolate, ± curved, *c* 5-20 x 1.5-2 cm, compressed, rough with small oblique ridges, *c* 3-5-seeded.

Native in America, now widely cultivated in temperate countries and in the higher regions of Africa and Asia; in Ethiopia sometimes grown at least in the SW parts. *Ebba* 550; *Westphal* 4067, 5493.

67. **VIGNA** *Savi* (1824), *nom. conserv*.

Verdcourt in *Kew Bull.* **24**: 507-569 (1970) & *Fl. Trop. E. Afr.* (1971); Maréchal et al. in *Boissiera* **28** (1978).

Climbing, prostrate or erect herbs or subshrubs. Leaves pinnately 3-foliolate or 1-foliolate. Stipules truncate, bilobed or spurred at the base, sometimes peltate. Inflorescences axillary or terminal, pseudo-racemose or flowers in dense subumbellate clusters or fasciculate. Calyx 5-lobed, 2-lipped; lower lip 3-lobed; upper lip of 2 lobes completely or partly united. Vexillary stamen free. Style with the upper part thickened, straight or curved, upper portion bearded or hirsute on the internal face, sometimes produced beyond the stigma. Pods linear, cylindrical or flattened.

Some 150 species throughout the tropics.

The genus with many important pulse crops in the tropical and subtropical regions of the Old World.

1. Leaves 1-foliolate. **20. V. monophylla**
- Leaves 3-foliolate. 2
2. Fruit ripening below ground-level, oblong-ovoid; petioles very long, often more than 10 cm. **8. V. subterranea**
- Fruit not ripening below ground-level; petioles usually shorter. 3
3. Keel with an upwardly directed marked ± conical pocket on left-hand side. 4
- Keel without a marked pocket. 6
4. Stipules bilobed at the base; style subcapitate. **16. V. vexillata**
- Stipules peltate; style with distinct slender beak beyond the stigma. 5
5. Plant erect or climbing; leaflets entire or 2-3-lobed; pod 40-90 x 6 mm. **17. V. radiata**
- Plant with trailing branches from a short erect main stem; leaflets deeply divided into 3-5 narrow lobes; pod *c* 30 x 4 mm. **18. V. aconitifolia**
6. Stipules with marked entire spur projecting below the base and narrower than the upper part; peduncles with minute stiff hook-like hairs at the apex. **12. V. unguiculata**
- Stipules bilobed or truncate at the base. 7
7. Style with subulate beak *c* 2 mm long projecting beyond the stigma; leaves often 3-lobed; sometimes flowering before the leaves appear. 8
- Style without such beak. 9
8. Inflorescences usually several-flowered; plant densely hairy; calyx-tube pubescent, lobes 2-10 mm long. **13. V. frutescens**
- Inflorescences 1-4-flowered; plant sparsely pubescent; calyx-tube glabrous, lobes *c* 1.5 mm long. **14. V. debanensis**
9. Vexillary and 4 short stamens with a pair of fused glands below each anther; stems glabrous, ± winged; peduncles winged below. **21. V. triphylla**
- Stamens without such glands. 10
10. Keel with long beak incurved through nearly 360°; stipules truncate. **19. V. macrorhyncha**
- Keel with short beak; stipules cordate or ± bilobed at the base. 11
11. Standard velvety outside. 12
- Standard glabrous outside. 13
12. Calyx-lobes longer and narrower; calyx and fruit indumentum mostly spreading. **5. V. heterophylla**
- Calyx-lobes wider and blunter, much shorter than the calyx-tube; calyx and fruit indumentum mostly appressed. **6. V. ambacensis**
13. Glabrous twiner; leaflets with a very characteristic venation of numerous lateral nerves forming a right-angle with the midrib. **9. V. multinervis**
- Habit various; leaflets without such venation, if tertiary nerves closely parallel then plant hairy. 14
14. Corolla predominantly yellow or greenish. 15
- Corolla predominantly blue or purple. 20
15. Corolla *c* 0.5-1 cm long. 16
- Corolla well exceeding 1 cm. 17
16. Leaflets ovate to linear-lanceolate; pods 2.3-6.5 cm long, rather appressed bristly hairy. **4. V. oblongifolia**

- Leaflets mostly round; pods 0.9-2(-3) cm long, finely pubescent. **7. V. parkeri**

17. Pod densely covered with long spreading rusty-red hairs. **2. V. fischeri**
- Pod glabrescent or pubescent with the hairs appressed or if spreading then short, under 0.5 mm long. 18

18. Small erect or procumbent herbs; keel shaped like the end of a canoe; seeds mostly *c* 2 mm long. **11. V. friesiorum**
- Vigorous climbers; keel shortly beaked; seeds 3-6 mm long. 19

19. Calyx glabrescent or with scattered hairs; inflorescence-rhachis somewhat elongated; pods reflexed. **1. V. luteola**
- Calyx densely and finely appressed pubescent; inflorescence-rhachis under 0.5 cm long; pods ± spreading or erect. **3. V. schimperi**

20. Stems erect or almost so. **11. V. friesiorum**
- Stems twining or prostrate. 21

21. Tertiary venation of leaflets characteristically close and transversely parallel; calyx-lobes mostly much longer than the tube, spreading pilose. **15. V. reticulata**
- Tertiary venation of leaflets more open. 22

22. Leaflets ± round, obtuse at the apex; pod 0.9-2 (-3) cm long. **7. V. parkeri**
- Leaflets ± ovate, acuminate at the apex; pod 3-9.5 cm long. **10. V. membranacea**

1. **V. luteola** *(Jacq.) Benth*. (1859).

Perennial climber; stem pubescent to glabrescent. Leaflets 3, ± ovate, 2.5-10 x 0.4-5 cm, acute or acuminate, sparsely pubescent on both surfaces. Stipules ovate-lanceolate, shortly bilobed at the base. Flowers few on a rhachis 1.5-5 cm long; peduncles 5-40 cm long. Calyx glabrescent or sparsely pubescent; tube 3-4 mm long; lobes 2-4 mm long. Standard yellow, 1.3-2.5 cm long, glabrous. Pod linear, reflexed, 40-80 x 5-6.5 mm, slightly constricted between the 6-9(-12) seeds, sparsely appressed pubescent.

Wet grassland, lake shores; 1150-1300 m. GG SD; Sudan, Uganda, Kenya and throughout tropical Africa, also Egypt, Syria and the tropics of Asia and America. *Gilbert, Thulin & Getachew* 331; *Beals* 412; *Strecker* 55.

2. **V. fischeri** *Harms* (1899).

Twining or trailing perennial; stems densely rusty-red pilose. Leaflets 3, narrowly ovate to lanceolate, 3-10 x 1-4 cm, acuminate, hairy on both surfaces. Stipules ovate to linear-lanceolate, slightly bilobed at the base. Flowers several on a rhachis 2-3 cm long; peduncles 6-27 cm long. Calyx ± hairy; tube 2.5-5 mm long; lowest lobe 5-7 mm long. Standard yellow, 1.5-2.2 cm long, glabrous. Pod linear, 30-55 x 4-5.5 mm, densely covered with spreading rusty-red hairs, *c* 11-seeded.

Upland grassland, old cultivations; *c* 2250 m. KF; Kenya, Tanzania, Cameroun, Burundi, Malawi and Zambia. *Mooney* 8745.

3. **V. schimperi** *Bak*. (1871);
- type: GD, *Schimper* 1539 (K holo., BM EA US iso.).

Twining perennial, up to several meters long; stems densely appressed pubescent when young. Leaflets 3, ± narrowly ovate, (1.2-)3-7 x (0.6-)1.3-4.8 cm, acute, pubescent on both surfaces. Stipules triangular-lanceolate, usually slightly bilobed at the base. Inflorescence subumbellate; rhachis up to 1.3 cm long; peduncle 3-18 cm long. Calyx finely appressed pubescent; tube 3-6 mm long; lowest lobe 2.5-5 mm long. Standard yellow inside, greenish or brownish outside, 1.8-2.5 cm long, glabrous. Pod erect, cylindrical, 40-95 x 3.5-4 mm, appressed rusty-red pubescent, *c* 16-seeded. Fig. 95.30.1.

Grassland, bushland, forest margins; 1850-2800 m. GD GJ SU HA KF; Sudan, Uganda, Kenya, Tanzania, Zaire. *Friis* et al. 466; *Mooney* 8880; *Gillett* 5445.

4. **V. oblongifolia** *A. Rich*. (1847)
- type: TU, *Quartin Dillon* s.n. (P holo., K MPU iso.).

V. lancifolia A. Rich. (1847) - type: TU, *Quartin Dillon* s.n. (P holo., K iso.).
Dolichos dillonii Del. (1847); type: a plant grown at Montpellier from seed sent from Ethiopia by Quartin Dillon (MPU holo., not seen).

Usually annual climbing or suberect herb, ± bristly pubescent. Leaflets 3, ovate to linear-lanceolate, 1.5-12 x 0.2-2.5 (-3.5) cm, appressed pubescent on both surfaces. Stipules ovate-lanceolate, cordate and slightly prolonged below the point of insertion. Inflorescence dense, 2-10-flowered; peduncle 2-35 cm long. Calyx pubescent; tube 1.5-2 mm long; lobes 1-4 mm long, the lowest narrow. Standard yellow or greenish, 6-11 mm long, glabrous. Pod ± reflexed, linear-cylindrical, 23-65 x 3-5 mm, 3-9-seeded, curved or almost straight, rather appressed bristly hairy.

1. Standard 10-11 mm long; pod 40-65 mm long. **var. oblongifolia**
- Standard 6-8 mm long; pod 23-40 mm long. **var. parviflora**

Var. **oblongifolia**.

Damp grassland, bushland; *c* 1600-2600 m. EW TU GD GJ; widespread in tropical Africa, also in Transvaal and Natal. *Pappi* 135, 1417, 7309.

Var. **parviflora** *(Welw. ex Bak.) Verdc.* in *Kew Bull.* 24: 528 (1970);
V. parviflora Welw. ex Bak. (1871).

Grassland, bushland, cultivations; *c* 1500-2000 m. TU GD SU AR HA KF; widespread in tropical Africa. *Thulin* 1327; *IECAMA* RS-210; *Frahm-Leliveld* 62358.

5. **V. heterophylla** *A. Rich.* (1847)
- type: TU, *Quartin Dillon* s.n. (P, holo., K, iso.).

Annual or perennial climbing or sometimes suberect herb; stem with spreading, often rusty-red hairs. Leaflets 3, ovate to linear-lanceolate, 1.7-8.5 x 0.5-3.3 cm, pubescent on both surfaces. Stipules ovate-lanceolate, bilobed at the base. Flowers on a rhachis 0.7-2 cm long; peduncles 1-10 cm long. Calyx spreading pubescent; tube 2-2.5 mm long; lobes narrow, 3-4 mm long. Standard blue or violet, 1-1.3 cm long, appressed velvety outside. Pod linear, 15-50 x 4.5-6 mm, densely covered with spreading dark brown hairs.

Grassland, woodland; 1400-2100 m. EW TU GD WG SU KF GG; N Uganda, Kenya, S Tanzania, Cameroun, Zaire, Zambia. *W. de Wilde* 8919; *Schimper* 243; *Friis* et al. 2306.

6. **V. ambacensis** *Bak.* (1871).

V. abyssinica Taub. (1892) - types: TU, *Schimper* 454 (B syn. destr., BM K isosyn.) & *Schimper* 458 (B syn. destr., BM EA K P isosyn.).

V. micrantha Chiov. (1911), *nom. illegit.*, *non* Harms (1899); *V. chiovendae* Bak. f. - types: GD, *Chiovenda* 2089, 2289 & 2324 (all FT syn.).

Annual or perennial climbing or rarely suberect herb; stem with ± spreading, often rusty-red hairs. Leaflets 3, narrowly ovate to linear-lanceolate, 2.5-15 x 0.3-5 cm, pubescent on both surfaces. Stipules ovate-lanceolate, slightly bilobed at the base. Flowers on a rhachis 1-4 (-8) cm long; peduncle 1.5-15 cm long. Calyx pubescent; tube 2-3 mm long; lobes triangular, 0.6-2 mm long. Standard blue or violet inside (in Ethiopia), yellowish, greenish or brownish outside, 1-2 cm long, appressed velvety outside. Pod linear, 27-60 x 4-7 mm, pubescent.

Grassland or woodland; 1200-2000 m. TU GD KF; widespread in tropical Africa. *Mooney* 8845; *Smeds* 1264 (K); *Schweinfurth* 1901.

Very close to *V. heterophylla* and perhaps not specifically distinct. *V. micrantha* and *V. chiovendae* were treated as synonyms of *V. heterophylla* by Verdcourt (1970, 1971). However, they are decidedly better assigned to *V. ambacensis,* which is also in accordance with Verdcourt's own determinations on the type sheets.

7. **V. parkeri** *Bak.* (1882).

Perennial prostrate or climbing herb, often rooting at the nodes. Leaflets 3, round, ovate or obovate, 1-8.8 x 0.8-5.4 cm, pubescent on both surfaces, ciliate. Stipules lanceolate, prolonged below the point of insertion and bilobed at the base. Flowers 2-5(-10) on a rhachis 0.5-2.5 cm long; peduncle 2-13 cm long. Calyx-tube 1.5-2 mm long, lobes 1-1.5 mm long. Standard blue or yellow, 5-8.5(-12) mm long, glabrous. Pods linear-oblong, (9-) 13-20(-30) x 4.5-5.5 mm, pubescent or glabrescent.

Subsp. **maranguensis** *(Taub.) Verdc.* in *Kew Bull.* **24**: 530 (1970).

V. gracilis sensu Cuf. Enum.: 333 (1955), *non* (Guill. & Perr.) Hook. f.

Leaflets mostly round with rounded apex, up to *c* 4 x 3.5 cm. Flowers mostly blue.

Grassland, woodland, forest margins, cultivations; 1600-2400 m. WU WG SU IL KF GG SD; Uganda, Kenya, Tanzania, Zaire, Rwanda. *Friis* et al. 341, 256; *Mooney* 8799.

Subsp. *parkeri* in Madagascar.

8. **V. subterranea** *(L.) Verdc.* (1980);
Voandzeia subterranea (L.) DC. (1825).

Annual, sparsely pubescent; stems short, creeping, with densely set erect, long-petiolate leaves; leaflets 3, elliptic to oblanceolate, 3-8 x 0.8-4 cm; stipules spurred, striate; racemes 1-3-flowered; peduncle 0.5-2 cm long; fertile flowers small, without petals, functionally male flowers with petals; calyx glabrous, *c* 2 mm long; standard yellow, 4-7 mm long, glabrous; pod 1-2.2 x 1.2-1.8 cm, glabrous, 1(-2)-seeded, ripening below ground-level; seeds ovoid or ellipsoid, 8.5-15 x 6.5-10 mm, slightly keeled, variously coloured; hilum elliptic; aril almost obsolete, thin, white.

Locally grown for its edible seeds, at least in WG (near Didessa R.); native of West Africa, now cultivated throughout the tropics. *IAR* s.n. (ETH).

9. **V. multinervis** *Hutch. & Dalz.* (1928).

Perennial climber; stems slender, glabrous. Leaflets 3, oblong-lanceolate to linear, 2.8-15 x 0.2-1.3 cm, rather thick, glabrous except for midvein and margin beneath; venation strongly raised and reticulate on both surfaces. Stipules ovate-lanceolate, prolonged and cordate at the base. Flowers few, rhachis 1-1.5 cm; peduncle 1-10 cm long. Calyx glabrous except for ciliate margins; tube *c* 2 mm long; lobes triangular, 0.7-2 mm long. Standard yellow inside, brownish outside, 8-10 mm long, glabrous. Pod linear-cylindrical, 30-60 x 3.5-5 mm, glabrous or glabrescent. Fig. 95.30.2.

Grassland; *c* 1300 m. WG; widespread in tropical Africa. *Gilbert & Thulin* 691.

10. **V. membranacea** *A. Rich.* (1847)
- types: TU, *Quartin Dillon* s.n. (P syn., K isosyn.) & *Schimper* I:352 (P syn., BM K isosyn.).

V. mensensis Schweinf. var. *hastata* Chiov. in *Ann. Ist. Bot. Roma* **8**: 438 (1908) - type: EE, *Terracciano & Pappi* 1776 (FT holo.).

Annual or perennial usually climbing herb; stems spreading pilose, later glabrescent. Leaflets 3, ± ovate, 1.2-10 x 1.1-6.5 cm, sometimes 3-lobed, appressed pilose. Stipules lanceolate, bilobed and unequally prolonged at the base. Flowers few to several on a rhachis 0.3-2 cm long; peduncle 1-45 cm long. Calyx glabrous or pubescent; tube 2-3 mm long; lobes 1.5-17 mm long, ciliate. Standard mauve to blue, 1-2.3 cm long, glabrous. Pods erect, linear-cylindrical, 30-95 x 2-3 mm, with short curved tubercle-based hairs, 14-25-seeded.

1. Calyx-lobes longer than the tube, (4-)5-9 mm long; standard *c* 1-1.5 cm long. subsp. **membranacea**
- Calyx-lobes usually shorter than the tube, 1-7 mm long; standard 2-2.3 cm long. subsp. **caesia**

Subsp. **membranacea.**

Grassland, woodland; 100-2400 m. EE EW TU WU GD GJ WG SU AR HA IL KF GG SD; N Sudan, Uganda, Kenya, Tanzania. *Thulin* 1326; *Gilbert & Thulin* 965; *Ash* 173.

Subsp. **caesia** *(Chiov.) Verdc.* in *Kew Bull.* **24**: 536 (1970);
V. caesia Chiov. (1932).

Dry bushland; 1050-1600 m. SU-DN HA SD; Somalia, E Kenya, NE Tanzania. *Gilbert & Thulin* 195; *Burger* 2870; *Westphal* 1477.

11. **V. friesiorum** *Harms* (1927).

Erect or decumbent perennial herb, up to 30 cm tall; stems spreading pubescent. Leaflets 3, oblong or elliptic to linear-lanceolate, 0.8-7 x 0.3-3.2 cm, glabrous to pubescent on both surfaces. Stipules lanceolate, somewhat bilobed at the base. Racemes few-several-flowered, subumbellate; peduncle 3.5-20 cm long. Calyx-tube 2-3.5 mm long; lobes triangular, 2-3.5 mm long. Standard yellowish to mauve or purple, 1-1.8 cm long, glabrous. Pods ± erect, 40-70 x 2-3 mm, pubescent or minutely puberulous, *c* 17-20-seeded.

Grassland or woodland; *c* 1200-1700 m. GG, SD; Uganda, Kenya, Tanzania, Rwanda, Zaire. *Gilbert, Thulin & Getachew* 322; *Gilbert & Jefford* 4611 A.

The two specimens known from Ethiopia are nearest to var. *ulugurensis* (Harms) Verdc., which is one of three rather diffuse varieties distinguished by Verdcourt in *Kew Bull.* **24**: 537 (1970).

12. **V. unguiculata** *(L.) Walp.* (1842)
- neotype: Ethiopia, *Westphal* 8682 (WAG neo.).

Verdcourt in *Kew Bull.* **24**: 542-545 (1970) & *Fl. Trop. E. Afr.* (1971); Westphal, *Pulses in Ethiopia*: 213-232 (1974).

Annual or perennial, erect to climbing herb; stems glabrous or slightly hairy. Leaflets 3, usually ovate or triangular, 1.5-16.5 x 1-12.5 cm, entire or 3-lobed, glabrous or sparsely pubescent on both surfaces. Stipules submedifixed with an entire spur, 2-6 mm long. Inflorescences subumbellate, few-several-flowered. Calyx glabrous. Standard white, greenish, yellow or purplish, 1.2-3.3 cm long, glabrous. Pods 5.5-10 x 0.3-1.1 cm (may be longer in cultivated races), glabrous or scabrous. Seeds ± oblong; hilum one-third to half the longest dimension, eccentric.

1. Calyx-lobes longer than the tube, 0.5-1.4 cm long. subsp. **mensensis**
- Calyx-lobes mostly shorter than the tube, 2.5-5 mm long. 2
2. Pods 5.5-10 x 0.3-0.5 cm, mostly dark and scabrous when mature. subsp. **dekindtiana**
- Pods 7.5-90 x more than 0.5 cm wide. 3
3. Pods 30-90 cm long. subsp. **sesquipedalis**
- Pods shorter. 4
4. Pods *c* 15-20 cm long; plant twining or sometimes erect. subsp. **unguiculata**
- Pods *c* 7.5-13 cm long; plant often suberect, sometimes twining. subsp. **cylindrica**

Subsp. **unguiculata.**

Cultivated mainly in HA at altitudes up to *c* 2000 m for its edible seeds; also the leaves are sometimes used as a green vegetable.

Subsp. **sesquipedalis** *(L.) Verdc.* in *Fl. Turkey* **3**: 266 (1970).

Cultivated in N Ethiopia according to Cuf. *Enum.* (1955).

Subsp. **cylindrica** *(L.) Verdc.* in *Kew Bull.* **24**: 544 (1970).

Cultivated at least in EE, EW and HA. Leaves can be used as a green vegetable. Some landraces appear very drought resistant. *Edwards* et al. 3846.

Subsp. **dekindtiana** *(Harms) Verdc.* in *Kew Bull.* **24**: 544 (1970);
V. dekindtiana Harms (1901).

Grassland, swamps, scrub, also cultivated; 350-1500 m. EE EW TU GD IL; widespread in tropical Africa; cultivated in other parts of the tropics. *Pappi* 7385, 7681; *Schimper* I:50.

Subsp. **mensensis** *(Schweinf.) Verdc.* in *Kew Bull.* **24**: 545 (1970);
V. mensensis Schweinf. (1896) - type: EW, *Schweinfurth* 1820 (B holo. destr.).

Scrub, cultivations; up to *c* 1500 m. EW KF GG; Uganda, Kenya, Tanzania, Zaire, Zambia. *Fukui* 177; *Friis* et al. 2269.

Differing from subsp. *dekindtiana* in its longer calyx-lobes, but there are numerous intermediates in various parts of Africa and the taxon is not recognized by Westphal (1974).

13. **V. frutescens** *A. Rich.* (1847)
- type: SU, *Petit* s.n. (P, holo.).
V. esculenta (De Wild.) De Wild. (1909).
V. glandulosa Chiov. (1915) - type: AR, *Negri* 1103 (FT, holo.).

Perennial, prostrate or climbing, or erect at first when appearing after burns, sometimes flowering before the leaves appear; stems densely pubescent herbaceous or ± woody. Leaflets 3, ± ovate in outline, 1.5-7.5(-9.5) x 1-

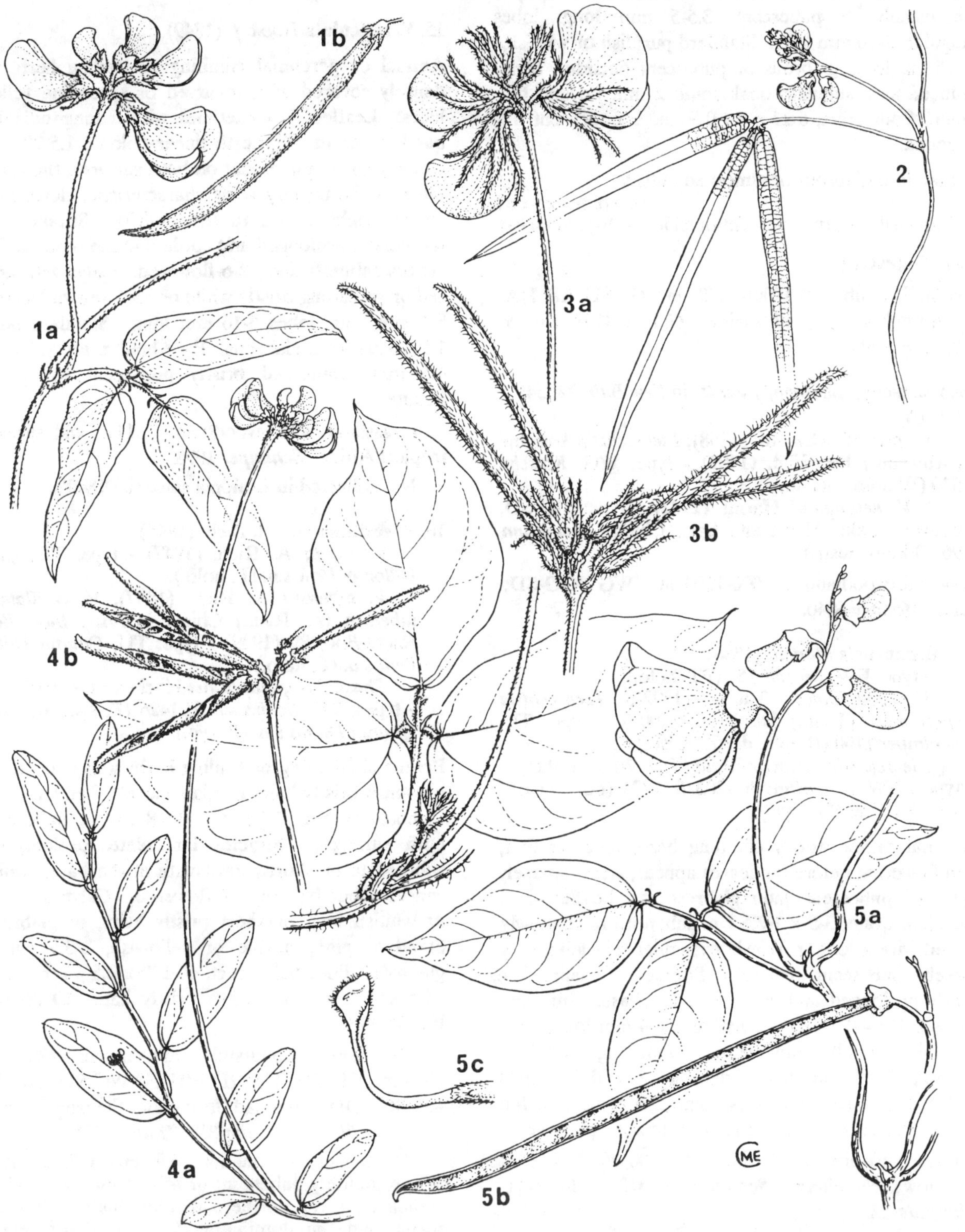

Figure 95.30.1 - *VIGNA SCHIMPERI*, a: flowering raceme and leaf from *Mooney* 6226, b: pod from *Meyer* 7706. 2 - *V. MULTINERVIS*, flowering raceme and leaf from *Gilbert & Thulin* 691. 3 - *V. VEXILLATA*, a: flowering raceme and leaf from *Gilbert & Thulin* 605, b: pods from *Gilbert & Thulin* 853. 4 - *V. MONOPHYLLA*, a: flowering branch from *Mooney* 5851, b: pods from *Ash* 2138. 5 - *SPHENOSTYLIS STENOCARPA*, a: flowering raceme and stem portion from *Gilbert & Getachew* 2980, b: pod from *Schimper* 223, c: style from *Gilbert & Getachew* 2980. 5c x 2.25, the rest x 0.75. Drawn by Maureen Church. (Reproduced with permission from *Opera Botanica*, 68, fig. 32, 1983.)

4.7 cm, entire to 3-lobed, pubescent. Stipules ± lanceolate, slightly prolonged and cordate at the base. Inflorescences subumbellate, several-flowered. Calyx-tube usually ± pubescent, 3.5-5 mm long; lobes triangular, 2-10 mm long. Standard purplish or whitish, 1.5-2.6 cm long, glabrous or pubescent outside. Style produced as a subulate beak some 2 mm beyond the stigma. Pods erect, 6-11 x 0.4-0.5 cm, ± pubescent, 12-16-seeded.

1. Standard glabrous or almost so outside. subsp. **frutescens**
- Standard velvety pubescent outside. subsp. **kotschyi**

Subsp. **frutescens**.

Grassland, scrub; 700-1900(-2400) m. GJ SU AR HA SD; throughout tropical Africa. *Gilbert* 3198; *Burger* 3350; *Bally* 9299.

Subsp. **kotschyi** *(Schweinf.) Verdc.* in *Kew Bull.* 24: 549 (1970);

V kotschyi Schweinf. (1868); *Liebrechtsia kotschyi* (Schweinf.) De Wild. (1902) - type: WG, *Kotschy* 530 (W holo., not seen).

? *V. neumannii* Harms (1902) - type: GG-SD, between Lake Abaya and Lake Chamo, *Neumann* 96 (B holo. destr.).

Grassland, woodland; *c* 1000-1300 m. WG ?GG-SD; Sudan. *Mooney* 6840.

14. V. debanensis *Martelli* (1886)

- type: EW, *Beccari* 152 (FT, holo.).

V. spartioides Taub. (1892); *Liebrechtsia spartioides* (Taub.) De Wild. (1902) - type: TU, *Schimper* 1100 (B holo. destr., K iso.).

Liebrechtsia schweinfurthii De Wild. (1902) - types: EW, *Schweinfurth* 1203 & 1724 (both in BR syn., FT K & P isosyn.).

Perennial rather woody climbing herb, erect at first, often flowering before the leaves appear; stems sparsely appressed pubescent, later glabrescent. Leaflets 3, ± lanceolate-ovate, 3-5 x 1-2.5 cm, subentire to shallowly 3-lobed with a long, narrowly triangular mid-lobe, very sparsely pubescent, ciliate. Stipules ± lanceolate, slightly prolonged and cordate at the base. Inflorescences 1-4(-6)-flowered. Calyx-tube *c* 4 mm long, glabrous; lobes broadly triangular, *c* 1.5 mm long. Standard purplish, 2-2.5 cm long, glabrous outside. Style produced as a subulate beak some 2 mm beyond the stigma. Pods *c* 90 x 4 mm, shortly appressed pubescent.

Grassland and woodland; 1600-2200 m. EW TU; not known elsewhere. *Schweinfurth* 318; *Bally* 7111; *Chiovenda* 82.

Very close to *V. frutescens* and perhaps best regarded as a subspecies of this. Verdcourt in *Kew Bull.* 24: 548 (1970) included *V. spartioides* and *Liebrechtsia schweinfurthii* with a question mark in *V. frutescens* var. *buchneri* (Harms) Verdc., a taxon otherwise not known north of Tanzania, while keeping *V. debanensis* separate. However, all the plants from EW and TU clearly belong to the same taxon and the author prefers to keep it specifically distinct for the present with the above changes in the synonymy.

15. V. reticulata *Hook. f.* (1849).

Annual or perennial climbing or trailing herb, stems densely covered with recurved or spreading yellowish hairs. Leaflets 3, sometimes with a longitudinal pale band, ovate or elliptic to linear-oblong, 1.5-22 x 0.3-4 cm, appressed pubescent on both surfaces, the arrangement of the tertiary veins characteristic, close, parallel and at right angles to the midrib. Stipules ovate-lanceolate, prolonged and cordate at the base. Inflorescences subumbellate, 2-6-flowered. Calyx with appressed or spreading bristly white or dark brown hairs; tube 3-6 mm long; lobes 6-16 mm long. Standard pinkish, 1.5-2.3 cm long, glabrous. Pods held ± erect, 38-70 x 4-4.5 mm, appressed bristly pubescent to spreading hirsute.

Grassland and woodland. TU; widespread in tropical Africa. *Schimper* 1998.

Not collected in Ethiopia since the 1840's.

16. V. vexillata *(L.) A. Rich.* (1845).

V. crinita A. Rich. (1847) - type: TU, *Quartin Dillon & Petit* s.n. (P, holo.).

V. tuberosa A. Rich. (1847); *V. vexillata* var. *tuberosa* (A. Rich.) Chiov. in *Miss. Biol. Boran., Racc. Bot.*: 89 (1939) - type: TU, *Quartin Dillon & Petit* s.n. (P, holo.).

Phaseolus quadriflorus Hochst. ex A. Rich. (1847) - types: TU, *Schimper* III:1426 (P, syn., K, isosyn.), *Quartin Dillon* s.n. (P, syn.).

Perennial climbing or trailing herb; stems covered with brownish bristly hairs or glabrescent. Leaflets 3, ovate to linear-oblong, 2.5-16.5 x 0.4-8 cm, ± pubescent on both surfaces. Stipules lanceolate, prolonged and subcordate or shortly auriculate at the base. Inflorescences subumbellate, 2-6-flowered. Calyx with brown or whitish long or short bristly hairs, or glabrescent. Standard pink, purple or yellowish, 2-3.5 cm long, glabrous. Pods held erect, 40-140 x 2.5-4 mm, covered with short or long brown bristly hairs, 10-18-seeded. Fig. 95.30.3.

Grassland, woodland, cultivations; (800-)1350-2500 m. EE EW TU GD WU GJ WG SU AR HA IL KF GG SD; widespread throughout the tropics. *Friis* et al. 1979; *Gilbert & Thulin* 853; *Thulin* 1451.

All plants seen from WG, AR, HA, KF, IL, GG and SD are densely pubescent or bristly and agree with var. *vexillata.* In the northern regions this variety is more scarce and the dominating form is glabrescent with sometimes more narrow leaflets. This form, which includes the types of *Phaseolus quadriflorus*, comes close to var. *angustifolia* (Schumach. & Thonn.) Bak., but most Ethiopian specimens are rather broad-leaved for this variety. In Shewa the two forms have been found growing together without intergradation (*Gilbert & Thulin* 605 and 609).

17. V. **radiata** *(L.) Wilczek* (1954);
Phaseolus radiatus L. (1753).

Annual erect or climbing herb; stem covered with long spreading hairs. Leaflets 3, elliptic, rhomboid or ovate, glabrous or pilose on both surfaces. Stipules peltate, ± ovate. Calyx-tube glabrous, 3-4 mm long; lobes 1.5-4 mm long, ciliate. Standard greenish-yellow outside, 1.2 cm long, glabrous; keel incurved. Pod 40-90 x 5-6 mm, somewhat constricted between the 8-14 seeds, with brown short spreading bristly pubescence. Seeds greenish, brown or blackish, oblong-cylindrical to globose or cube-like, 2.5-4.2 x 2.5-3.2 x 2.3-2.8 mm; hilum *c* half the length of the seed with aril not developed; testa often with characteristic pattern of rather wavy raised ridges with close parallel riblets between making it minutely rough to the touch.

1. Stems mostly erect; leaflets usually entire; pods 6-9 cm long. var. **radiata**
- Stems twining or prostrate; leaflets often lobed; pods usually smaller. var. **sublobata**

Var. **radiata.**

Ancient Indian crop cultivated for its edible seeds at fairly low altitudes throughout the tropics and subtropics. Apparently recently introduced in Ethiopia. SU HA IL GG. *Jansen* 7025.

Var. **sublobata** *(Roxb.) Verdc.* in *Kew Bull.* **24**: 559 (1970).
? *Phaseolus abyssinicus* Savi (1835) - type: apparently a specimen grown in Pisa from seeds obtained from Ethiopia (?PI holo., not seen).
Vigna opisotricha A. Rich. (1847) - type: TU, *Schimper* III:1718 (P holo., BM K US iso.).

Native in Asia but found as an escape or naturalized elsewhere in the tropics. TU GD. *J. De Wilde* 7125.

18. V. **aconitifolia** *(Jacq.) Maréchal* (1969).

Slender hairy annual with trailing branches from a short erect main stem; leaflets 3, deeply divided into 3-5 narrow lobes; flowers in pedunculate short racemes, yellow, 5-10 mm long; pods *c* 30 x 4 mm, cylindrical, brown with short stiff hairs; seeds rectangular, 5 mm long, with a white, linear hilum.

Cultivated for its edible pods and seeds and as fodder particularly in hot, dry areas; in Ethiopia known from EE and EW; native of Asia. *Pappi* 3939, 5799; *Terracciano* 36.

19. V. **macrorhyncha** *(Harms) Milne-Redh.* (1963);
Phaseolus macrorhynchus Harms (1900).
Phaseolus schimperi Taub. (1892) non *Vigna schimperi* Bak. - type: TU, *Schimper* 738 (B holo. destr.).
V. proboscoidella Chiov. (1911) - type: GD, *Chiovenda* 2704 (FT holo.).

Perennial trailing or twining herb, rarely ± woody; stems glabrous or sparsely pubescent. Leaflets 3, linear-lanceolate to ovate, 1.5-7 x 0.3-2.5 cm, glabrous or sparsely pubescent. Stipules small, not extended below the base. Racemes lax; rhachis 1.5-7 cm long; peduncle 3.5-14 cm long. Calyx glabrous; tube 2-3 mm long; lobes triangular, 2.5 mm long. Standard pale purple inside, dull purple or greenish outside, 0.9-1.3 cm long, glabrous; keel incurved through almost a circle. Pods 48-110 x 2-4 mm, upcurved at the apex, glabrous or rarely pubescent, 8-15-seeded.

Grassland, bushland and woodland; 1300-2050 m. TU GD WG SU HA KF GG SD; Sudan, Somalia and widespread in tropical Africa. *Gilbert, Thulin & Getachew* 422; *Mooney* 6847, 7307.

Three collections from near Negele in SD (*Vàtova* 134 and 163, *Corradi* 8226, all in FT) deviate in having ± woody stems and appressed pubescent pods. The flower structure is identical with *V. macrorhyncha.*

20. V. **monophylla** *Taub.* (1895);
Haydonia monophylla (Taub.) Wilczek (1954).
V. ostinii Chiov. (1911) - type: GD, *Chiovenda* 1704 (FT holo.).

Erect, trailing or somewhat climbing perennial herb; stems glabrous or sparsely pubescent. Leaves unifoliolate; leaflets oblong-lanceolate to ovate, 2-11.5 x 0.5-3 cm; glabrous or with a few hairs on the midrib beneath and along margins. Stipules lanceolate. Racemes with rhachis 5-12 mm long; peduncle 7-35 cm long. Calyx glabrous or with small blackish hairs. Standard usually purplish tinged greenish outside, 1-1.4 cm long, glabrous. Alternate anthers with a pair of fused glands at their base. Pods held ± erect, 38-50 x 3-4 mm, densely covered with short dark hairs. Fig. 95.30.4.

Grassland and woodland, often on waterlogged clay soils; 1800-2400 m. GD, SU, KF; widespread in eastern tropical Africa. *Mooney* 5851, 9267; *Ash* 2138.

21. V. **triphylla** *(Wilczek) Verdc.* (1970).

Prostrate or climbing herb. Stems glabrous or almost so, ± winged. Leaflets 3, linear-oblong to ovate, 2-5.5 x 0.3-2.2 cm, glabrous or almost so. Stipules lanceolate, often reflexed. Racemes with rhachis 0.3-2.5 cm long; peduncle (1.5-)7-9(-16) cm long, winged below. Calyx glabrescent; tube *c* 1.5-2 mm long; lobes triangular, 1-2.5 mm long, ciliolate. Standard yellow or purple, 0.9-1.3 cm long, glabrous. Alternate anthers with a pair of fused glands at their base. Pods held ± erect, *c* 40 x 3 mm, densely covered with short stiff hairs.

Grassland; 2000-2500 m. EW AR; Kenya, Nigeria, Central African Republic, Zaire, Zambia, Angola. *Thulin* 1617; *Wellby* s.n.; *Baldrati* s.n.

Wellby's collection was made "between Harar and Addis Abeba". The species may thus occur also in HA or SU.

68. SPHENOSTYLIS *E. Mey.* (1836)

Herbs or subshrubs, prostrate, climbing or erect. Leaves pinnately 3-foliolate. Stipules not spurred. Pseudoracemes axillary. Calyx 5-lobed, short, 2-lipped, the lower lip of 3 obtuse lobes, the upper pair of lobes joined to form a small rounded or emarginate lip. Corolla glabrous. Style ± twisted, thickened basally, thinning near the middle and then enlarged, flattened and spathulate at the apex, pubescent inside and along the margins; stigma terminal. Pod linear, compressed.

Small genus mainly in tropical and southern Africa. The Ethiopian record of *S. schweinfurthii* Harms in Cuf. *Enum*.: 338 (1955) is based on *Armbruster* s.n. (K) which is said to be from Addis Abeba, but instead most probably originates from Sudan.

S. stenocarpa *(Hochst. ex A. Rich.) Harms* (1899);
Dolichos stenocarpus Hochst. ex A. Rich. (1847) - type: TU, *Schimper* III:1450 (P holo., BM K S US iso.).

Perennial prostrate or climbing herb; stems glabrous or sparsely puberulous. Leaflets ovate to lanceolate, up to 13 x 5.5 cm, acute to acuminate at the apex, glabrous or sparsely pubescent on the midrib beneath. Flowers 4-9 on a rhachis 2-6.5 cm long; peduncles 6-27 cm long. Calyx-tube 3-7 mm long; lobes broad and rounded, 0.5-2 mm long. Corolla creamy to pink or mauve; keel slightly twisted. Pod 10-20 x 0.5-0.8 cm, glabrous. Fig. 95.30.5.

Grassland and woodland; *c* 1000-1500 m. TU GD (Sudan border); widespread in tropical Africa. *Schweinfurth* 1899; *Gilbert & Getachew* 2980; *Schimper* 223.

69. DOLICHOS *L.* (1753), *nom. conserv.*

Verdcourt in *Fl. Trop. E. Afr.* (1971).

Herbs or shrubs, climbing, prostrate or erect. Leaves pinnately 3-foliolate, subdigitate or 1-foliolate. Flowers axillary, fasciculate or in pseudoracemes or umbels. Calyx 4-5-lobed, the upper pair of lobes joined to form an entire to 2-toothed lip. Corolla glabrous; standard with short appendages on the inside; keel not twisted but often beaked. Anthers uniform. Style swollen, thickened, often ± twisted towards the base, usually glabrous and mostly with a ring of hairs around the terminal capitate stigma. Pod compressed. Seeds compressed; hilum short.

Some 60 species mainly confined to Africa but some in India and eastern Asia.

1. Small ascending or suberect herb to 25 cm tall; flowers in umbel-like heads; peduncle 4-20 cm long; stipules 6-10 x 2-5 cm. **1. D. luticola**
- Large herbs or subshrubs; flowers in fascicles or racemes; other characters not combined as above. 2
2. Climbing or scrambling with slender terete herbaceous stems; peduncles 0.5-18 cm long. 3
- Erect herb or subshrubs or if scrambling then stems woody and angular; peduncles up to 1.8 cm long. 4
3. Bracteoles 1-1.5 mm long, 1-3-nerved; corolla 0.9-1.2 cm long. **2. D. trilobus**
- Bracteoles 2.5-7 mm long, 3-7-nerved; corolla 1.5-2.5 cm long. **3. D. sericeus**
4. Leaflets prominently 3-nerved from base to apex; petiole up to 5 mm long; rhachis obsolete. **6. D. schweinfurthii**
- Leaflets with mainly pinnate and spreading venation; petiole and rhachis longer. 5
5. Corolla 1.5-1.8 cm long; bracteoles 3-5-nerved. **4. D. oliveri**
- Corolla 2-2.9 cm long; bracteoles 1-3-nerved. **5. D. kilimandscharicus**

1. **D. luticola** *Verdc.* (1970).

Small suberect or ascending perennial herb; stems sparsely pilose. Leaflets ± elliptic, 1.2-6 x 0.8-4 cm, with scattered hairs on the margins and nerves; stipules leafy, elliptic, up to 10(-16) x 5(-8) mm, 12-14-nerved. Flowers *c* 5-10 in axillary, dense, umbel-like heads; peduncle 4-20 cm long. Calyx-tube 1.5-3 mm long; lobes triangular 1.5-3.2 mm long. Standard greenish with yellow base and lilac tip, 0.7-0.9(-1.2) cm long. Pod linear-oblong, 25-45 x 4.5-5 mm, pubescent.

Grassland, wood/bushland; *c* 1550-1800 m. BA SD; Kenya, N Tanzania. *Thulin, Hunde & Tadesse* 3400, 3440, 3820; *Friis* et al. 3251.

The plants found on limestone have larger flowers than normal, but the difference does not appear to be taxonomically significant.

2. **D. trilobus** *L.* (1753).
D. falcatus Klein ex Willd. (1802).
D. debilis Hochst. ex A. Rich. (1847) - types: TU, *Quartin Dillon* s.n. (P syn.) & *Schimper* III:1440 (P syn., BM K isosyn.) & *Schimper* III:1472 (P syn., BM K US isosyn.).
?*D. formosoides* Harms (1902) - type: HA, *Ellenbeck* 1120 (B holo. destr.).

Perennial climbing herb; stems spreading or appressed pubescent. Leaflets broadly ovate to lanceolate, 1.5-9 x 0.2-9 cm, sometimes 3-lobed, pubescent or glabrescent. Inflorescences 2-3-flowered; pedicels 4-10 mm long; bracteoles 1-1.5(-2) mm long, 1-3-nerved. Calyx-tube 2-3.5 mm long; lobes triangular, 1-3 mm long. Standard pinkish, 0.9-1.2 cm long. Pod oblong, curved, 25-75 x 6-11.5 mm, glabrous or with minute hairs.

Grassland, bushland and forest; *c* 1500 m. TU HA SD; widespread in Africa, also in Arabia and India to Ceylon, China and the Philippines. *Gilbert & Thulin* 43; *Burger* 3188; *Schimper* 450.

The Ethiopian plants have ± ovate leaflets and are var. *trilobus*.

3. **D. sericeus** *E. Mey.* (1836).

Perennial climbing or prostrate herb; stems covered with spreading or appressed hairs. Leaflets ± ovate, 0.9-9.5 x 0.5-8 cm, appressed pilose on both surfaces. Flowers 1-10 in short axillary racemes; pedicels 3-18 mm long; bracteoles 2.5-7 mm long, 3-7-nerved. Calyx glabrous to pubescent; tube 3-4 mm long; lobes 1.5-6 mm long. Standard pink and purplish, 1-2 cm long. Pod oblong, somewhat curved, *c* 45-75 x 8-11 mm, glabrous to densely pubescent.

1. Pod glabrous when mature. subsp. **sericeus**
- Pod pubescent when mature. subsp. **formosus**

Subsp. **sericeus.**

Forest edges and scrub; 1800-2350 m. WG SU KF SD; widespread in tropical Africa extending southwards to Natal. *Mooney* 8693, 8849; *Gillett* 14243.

Subsp. **formosus** *(A. Rich.) Verdc.* in *Kew Bull.* **24**: 419 (1970);
D. formosus A. Rich. (1847) - types: TU, *Quartin Dillon & Petit* s.n. & *Schimper* I: 52 & 226 (all P syn., K isosyn.).

Fig. 95.31.1. Forest edges and scrub; (1350-)1850-2650 m. EW TU GD WU SU AR HA KF SD; Uganda, Kenya. *Thulin* 1428; *W. de Wilde* 8698; *Burger* 3170.

4. **D. oliveri** *Schweinf.* (1868)
- types: WU, *Steudner* 139, 146 & 142 (all B syn. destr., K isosyn.).

Perennial erect to rambling shrub, woody herb or climber, often flowering when quite leafless; stems sparsely pubescent to velvety. Leaflets ovate, 1.2-5.5 x 0.9-4.2 cm, sometimes very obtusely 3-lobed, ± densely pubescent, silvery beneath. Flowers 1-5 in axillary fascicles. Calyx silky pubescent; tube 4-6 mm long; lobes 1.5-5.5 mm long. Standard purplish or greenish, 1.5-1.8 cm long. Pod linear-oblong, falcate, 50-85 x 8-11 mm, densely pubescent.

Grassland, bushland and woodland, often on rocky ground; 1200-2400 m. EW TU GD WU BA HA GG SD; Sudan, Uganda, Kenya, Tanzania, Zimbabwe. *Gilbert, Thulin & Getachew* 244; *Burger* 2451a; *Bally* 9275; *Gilbert* 1464.

5. **D. kilimandscharicus** *Taub.* (1894).
D. malosanus Bak. (1897).
D. oliveri Schweinf. var. *uondensis* Fiori in *Nuovo Giorn. Bot. Ital.* **47**: 31 (1940) - type: SD, *Saccardo* s.n. (FT holo.).

Erect perennial herb or subshrub, up to at least 1 m tall, often flowering when the leaves are undeveloped. Stems simple or sparsely branched, usually densely pubescent. Leaflets ovate to lanceolate, 2.5-18 x 1-7.5 cm, densely silvery pubescent, often with contrasting rusty-red hairs on the nerves. Flowers 1-4 together in fascicles; upper fascicles running together to produce a long terminal inflorescence; upper bracts 3-fid or of 3 separate lobes. Calyx silvery pubescent; tube 4-5 mm long; lobes triangular, 2-6 mm long. Standard purple, 2-2.9 cm long. Pods linear-oblong, 40-90 x 9-14 mm, densely appressed rusty-red pubescent.

Grassland and woodland; 1150-2300 m. GJ WG SU BA SD; widespread in tropical Africa. *Gillett* 14771; *Ash* 1947; *W. de Wilde* 6803.

6. **D. schweinfurthii** *Harms* (1899).

Erect perennial herb up to at least 1 m tall, sometimes flowering when the leaves are undeveloped. Stems simple or sparsely branched, glabrous to pubescent. Leaflets (1-)3, linear-lanceolate to oblong-elliptic, 6-22.5 x 0.7-8 cm, ± pubescent or glabrous, conspicuously 3-nerved from base to apex; petiole up to 5 mm long; rhachis obsolete. Flowers 2-8 together in axillary fascicles. Calyx pubescent or glabrescent; tube 3-4 mm long; lobes narrowly triangular, 1.5-3 mm long. Standard blue, purple or white, 1.7-2.8 cm long. Pods linear-oblong, 6-7 x 1 cm, glabrous except for usually ciliate margins.

Grassland or grassland and bushland with scattered trees; *c* 1800-1900 m. WG, SU; Sudan, N Uganda, Tanzania, W Africa. *Jansen* 6211; *Lemma G.S.* 755.

This is a badly understood species. The Ethiopian material has pubescent leaves which are prominently 3-nerved from the base, but only the midvein quite reaches the apex.

70. **LABLAB** *Adans.* (1763)

Verdcourt in *Kew Bull.* **24**: 409-411 (1970) & *Fl. Trop. E. Afr.* (1971).

Climbing or suberect usually perennial herbs. Leaves pinnately 3-foliolate. Inflorescences axillary, pseudo-racemose. Calyx 2-lipped; upper lip entire or emarginate; lower lip 3-lobed. Corolla glabrous; standard with short appendages on the inside; keel with beak incurved at a right angle. Anthers uniform. Style stiffened, laterally flattened, slightly curved at the apex, hairy on the internal face; stigma terminal. Pod compressed. Seeds ovoid; hilum linear with a whitish aril.

Monotypic genus, widely cultivated throughout the tropics but originating in Africa. Its nomenclature has since long been a matter of dispute. Verdcourt's treatment is followed here.

L. purpureus *(L.) Sweet* (1827);
Dolichos purpureus L. (1763).
D. lablab L. (1753); *L. niger* Medik. (1787).
D. uncinatus Schweinf. (1896), *nom. illegit.*, *non* L.; *D. lablab* var. *uncinatus* (Schweinf.) Chiov. in *Ann. Ist. Bot. Roma* **8**: 440 (1908), *nom. illegit.* - types: EW, *Schweinfurth* 315, 316, 871, 896, 1418 (all B syn. destr.).
L. niger var. *crenatifructus* (Steud.) Cuf. & var. *uncinatus* (A. Rich.) Cuf., Cuf. *Enum*: 339 (1955) & *Zweiter Nachtrag:* **25** (1969), both *nom. invalid.*

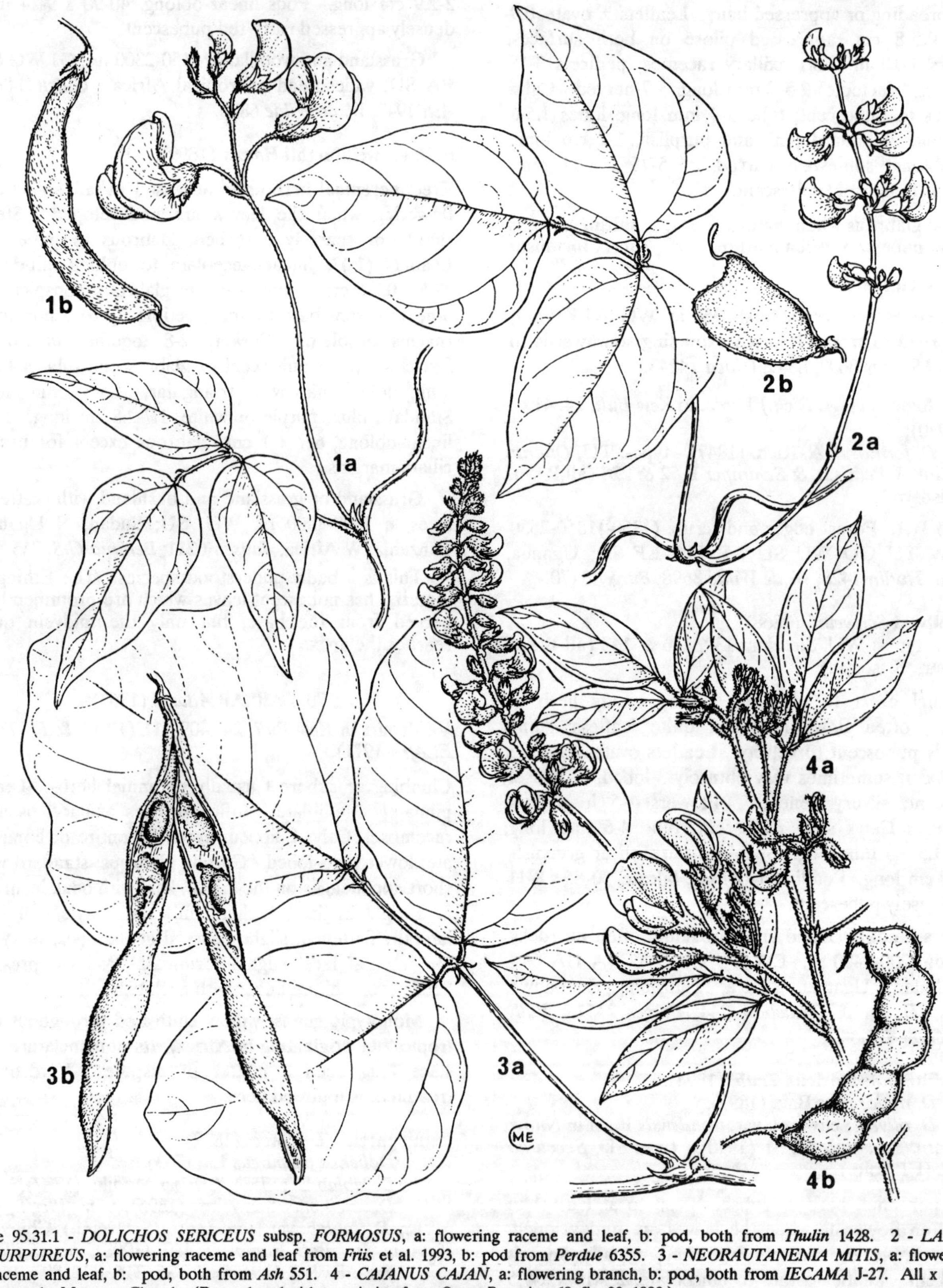

Figure 95.31.1 - *DOLICHOS SERICEUS* subsp. *FORMOSUS*, a: flowering raceme and leaf, b: pod, both from *Thulin* 1428. 2 - *LABLAB PURPUREUS*, a: flowering raceme and leaf from *Friis* et al. 1993, b: pod from *Perdue* 6355. 3 - *NEORAUTANENIA MITIS*, a: flowering raceme and leaf, b: pod, both from *Ash* 551. 4 - *CAJANUS CAJAN*, a: flowering branch, b: pod, both from *IECAMA* J-27. All x 0.75. Drawn by Maureen Church. (Reproduced with permission from *Opera Botanica*, 68, fig. 33, 1983.)

Stems pubescent or glabrous. Leaflets ovate-triangular, 2.5-15 x 1.5-14 cm, acuminate. Rhachis of inflorescence 4-20 cm long; peduncle 2-40 cm long; bracteoles elliptic or round, 4-6(-8) mm long. Calyx glabrous or pubescent; tube 3-4 mm long; lobes 2-4 mm long. Standard crimson, purple or white, 1.2-1.5 cm long. Pods oblong-falcate to linear-oblong, 3.5-14 x 1.2-4 cm, 2-5-seeded, pubescent or glabrescent. Fig. 95.31.2.

Grassland, bushland, gallery forests and also cultivated as a hedging plant and/or for its edible seeds; 400-2350 m. EE EW TU GD WU GJ WG SU HA IL KF GG SD; throughout tropical Africa and south to E Cape Province. *Friis* et al. 1912; *Perdue* 6355; *Meyer* 7832.

The wild, although sometimes cultivated form, subsp. *uncinatus* Verdc., has pods *c* 4 x 1.5 cm and is widespread in Ethiopia. Subsp. *purpureus* with larger pods is widely cultivated in the tropics but plays a minor role in Ethiopia. According to Westphal, *Pulses in Ethiopia* (1974) it is an important crop only in the Konso area in Gamo Gofa, but herbarium material is also present from Eritrea (in FT) where it is a common plant over fences in Asmara and other towns. It is also now being widely planted in soil and water conservation programmes.

71. **NEORAUTANENIA** *Schinz* (1899)

Verdcourt in *Kew Bull.* **24**: 306-307 (1970) & *Fl. Trop. E. Afr.* (1971).

Prostrate trailers, lianes or erect subshrubs. Leaves pinnately 3-foliolate. Inflorescences axillary, pseudo-racemose; bracteoles absent. Calyx 5-lobed, 2-lipped; lower lip 3-fid; upper lip 2-fid. Corolla glabrous, standard without appendages. Anthers uniform. Style with a distinct thickening at its junction with the ovary, bent over at a right angle, glabrous; stigma small, capitate, glabrous. Pod linear-oblong to oblong, dehiscent.

Three species confined to tropical Africa.

N. mitis *(A. Rich.) Verdc.* (1969);

Dolichos mitis A. Rich. (1847) - type: TU, *Quartin Dillon* s.n. (P holo.).

D. pseudopachyrhizus Harms (1899); *N. pseudopachyrhiza* (Harms) Milne-Redh. (1951).

D. pseudopachyrhizus var. *subintegrifolius* Harms in *Engl. Bot. Jahrb.* **26**: 322 (1899) - type: TU, *Schimper* 140 (B lecto. destr., BM K US isolecto.).

D. ellenbeckii Harms (1902); *N. pseudopachyrhiza* var. *ellenbeckii* (Harms) Cuf., Cuf. *Enum., Zweiter Nachtrag*: 25 (1969) - type: AR, *Ellenbeck* 1387 (B holo. destr.).

Pueraria hochstetteri Chiov. (1908) - types: EW, *Pappi* 2998, 3064 bis, 3721, 3728, 5795 (all FT syn.).

Subshrubby herb, climbing, scrambling or erect from a tuberous rootstock; stems glabrous or velvety. Leaflets ovate, oblong or almost round in outline, 5-20 x 4-18 cm, entire to deeply 3-lobed. Flowers numerous on a rhachis 7-34 cm long. Calyx appressed pubescent; tube 3-4 mm long; lobes 3-6 mm long. Standard ± greyish outside, mauve or violet and green inside, 1-1.2 cm long. Pod rather swollen, 7-14 x 1-1.8 cm, densely covered with appressed pale rusty-red hairs.Fig. 95.31.3.

Grassland, bushland, woodland, often in rocky places; *c* 500-1500(-2000) m. EE EW TU GD(Sudan border) GJ AR HA IL; widespread in tropical Africa. *Ash* 551; *J. De Wilde* 4946; *Thulin & Hunde* 4026.

72. **CAJANUS** *DC.* (1813), *nom. conserv.*

van der Maesen in *Agric. Univ. Wageningen Papers* **85-4** (1986).

Shrubs, subshrubs or climbers. Leaves mostly pinnately 3-foliolate; leaflets with small yellow glands. Flowers in terminal or axillary racemes. Calyx 5-lobed; lobes subequal, the upper pair joined. Vexillary stamen free; anthers uniform. Ovary 3-10-ovuled; style thickened above, flattened below the small capitate stigma. Pod linear-oblong, obliquely grooved between the seeds.

Genus now with 32 species, (incl. *Atylosia* Wight & Arn.), the majority in India and Australia, one widely cultivated.

C. cajan *(L.) Millsp.* (1900).

Stems erect, ribbed, densely pubescent. Leaflets elliptic to lanceolate, 2.5-10 x 1.5-3.5 cm, silvery pubescent beneath. Calyx velvety yellowish and glandular; tube *c* 5 mm long; lobes 3-7 mm long. Standard yellow, 1.2-1.7 cm long. Pod straight or upcurved, 4-10 x 0.8-1.5 cm, hairy and glandular. Seeds compressed globose, 6-8.5 x 4.5-7.5 mm, minutely shallowly pitted, with a small elliptic hilum. Fig. 95.31.4.

Cultivated for its edible seeds and an important plant in soil and water conservation work throughout the medium altitude highlands; also found as an escape; 1000-2400 m. EW GD(Sudan border) GJ SU HA KF GG SD; not known in the wild state, cultivated throughout the tropics. *Burger* 2341; *Fukui* 313; *IECAMA* J027; *Taddesa Ebba* 517.

This is a multipurpose species which provides fresh and dried seeds which are edible, browse for domestic animals, and wood for basket work and fuel.

73. **RHYNCHOSIA** *Lour.* (1790), *nom. conserv.*

Verdcourt in *Fl. Trop. E. Afr.* (1971).

Climbing, prostrate or sometimes erect herbs or subshrubs. Leaves pinnately, rarely subdigitately, 3-foliolate, or 1-foliolate; leaflets with yellow or orange gland-dots at least beneath. Inflorescences axillary or terminal, racemose or paniculate, flowers spreading; bracteoles absent. Calyx 5-lobed; lobes unequal, the upper pair ± joined. Corolla usually yellow, lined purplish. Vexillary stamen free; anthers uniform. Ovary (1-)2-ovuled; style glabrous and usually stiffened above; stigma small. Pod almost circular to narrowly oblong, compressed, pubescent. Seeds 1-2 each with rounded to shortly elongate hilum; funicle inserted at middle of hilum.

Some 200 species throughout the tropics and subtropics.

1. Leaves 1-foliolate. 2
- Leaves 3-foliolate. 3

2. Plant not sticky; standard glabrous. **12. R. alluaudii**
- Plant sticky; standard pubescent. **4. R. orthobotrya**

3. Flowers in dense subsessile oblong inflorescences; calyx with very marked dark or orange dots, lobes 10-17 mm long, ± enclosing the corolla; standard glabrous; plant climbing or procumbent, not very sticky. **1. R. densiflora**
- Flowers in laxer pedunculate inflorescences or if dense and subsessile then other characters not combined as above. 4

4. Standard glabrous. 5
- Standard pubescent outside. 10

5. Corolla persistent, drying a brick-red colour; seeds subglobose, blue or black, remaining attached to the placenta; robust climbers usually several m long. 6
- Corolla ± caducous; seeds ± compressed, not blue, not remaining attached to the placenta; habit not as above. 7

6. Calyx *c* 25 mm long in Flora area, longer then corolla. **25. R. hirta**
- Calyx *c* 10 mm long in Flora area, about half as long as corolla. **26. R. congensis**

7. Erect or scrambling woody herb or subshrub; inflorescences mostly short and subsessile. **5. R. nyasica**
- Climbing or prostrate herbs, sometimes with ± woody stems; inflorescences various. 8

8. Leaflets oblong-lanceolate, more than twice as long as wide; inflorescence lax and few-flowered. **11. R. totta**
- Leaflets ovate to rhomboid, much less than twice as long as wide; inflorescences various. 9

9. Stems densely covered with pale rusty-red hairs; standard 9-15 mm long; pedicels *c* 2.5-3 mm long. **7. R. ferruginea**
- Stems usually finely pubescent or glabrescent; standard 13-21 mm long; pedicels 4-19 mm long. **6. R. elegans**

10. Mature leaflets particularly densely gland-dotted beneath, the pale greenish-yellow glands so numerous that they are scarcely separated by their own width and give a greyish-yellow tinge to the foliage; inflorescence mostly shorter than the leaves; standard 4-6.5 mm long; plant erect or scrambling; pod covered with very short and fine white hairs. **17. R. pulverulenta**
- Mature leaflets with gland-dots less closely spac ed and/or without other characters combined. 11

11. Indumentum of stems and inflorescence spreading, sticky and glandular; robust herbs or shrubs, ± erect or climbing. 12
- Indumentum not sticky and glandular; habit various. 15

12. Calyx asymmetrically pouched; leaflets long acuminate at the apex, distinctly paler beneath; stems covered with dense short glandular hairs. **2. R. resinosa**
- Calyx not pouched; leaflets rounded to shortly acuminate at the apex; stems often with longer bristly hairs. 13

13. Stipules small, *c* 4 x 1 mm; stems pubescent, without long yellow bristly hairs. **5. R. nyasica**
- Stipules larger, *c* 10-15 x 2.5-6 mm; stems with long yellow bristly hairs. 14

14. Lowest calyx-lobe elliptic-lanceolate, conspicuously wider than the rest, 6-11 x 3-4 mm; standard 9-11 mm long; plant usually ± erect. **4. R. orthobotrya**
- Lowest calyx-lobe not as above; standard *c* 14-15 mm long; plant climbing. **3. R. stipulosa**

15. Pods characteristically inflated when ripe, clavate, *c* 21-30 x 8-13 mm; leaflets usually slightly lobed. 16
- Pods not inflated; leaflets not lobed. 17

16. Prostrate or climbing herb, ± pubescent; inflorescence lax, unbranched; lower calyx-lobe up to 6 mm long. **13. R. sublobata**
- Shrubby ± erect robust herb, often with climbing branches, densely velvety; inflorescence fairly dense, often branched; lower calyx-lobe 7-10 mm long. **14. R. malacotricha**

17. Erect herbs or shrubs. 18
- Climbing, spreading or prostrate herbs or shrubs. 21

18. Petiole 0.2-0.5 cm long, usually shorter than the rhachis; bracts broadly elliptic to broadly obovate, 4-8 x 3-5 mm; leaflets not silvery pubescent. **22. R. erlangeri**
- Petiole usually longer, longer than the rhachis; bracts smaller; leaflets silvery pubescent at least beneath. 19

19. Standard with wart-like appendages on the inside near the base. **19. R. albissima**
- Standard without wart-like appendages. 20

20. Leaflets up to 3.8 x 2.5 cm, sparsely to densely appressed pubescent above, densely silvery pubescent beneath, venation narrowly reticulate, very prominent beneath. **21. R. erythraeae**
- Leaflets up to 5.5 x 3 cm, densely silvery tomentose above, densely woolly beneath; venation partly hidden by the indumentum beneath. **20. R. splendens**

21. Pod with fine, sparse to dense pubescence of short hairs under 0.5 mm long. 22
- Pod with some longer hairs in addition to fine pubescence and gland dots. 26

22. Leaflets at least twice as long as wide, usually paler green along the midrib above. **10. R. mensensis**
- Leaflets less than twice as long as wide, not paler green along the midrib above. 23

23. Inflorescences equalling or not much exceeding subtending leaves, usually 2-5-flowered; leaves obtuse, truncate or emarginate; standard (5-)8-12 mm long. **23. R. schimperi**

- Inflorescences usually much exceeding subtending leaves, usually more than 5-flowered; leaves and standard various. 24
24. Leaflets densely greyish-white tomentose on both sides; inflorescences often branched; standard *c* 16 mm long. **24. R. ramosa**
- Leaflets not densely tomentose, or if somewhat so other characters not as above. 25
25. Standard 9-17 mm long, often plane yellow; leaflets very finely puberulous, often with silky margins. **15. R. malacophylla**
- Standard 5-8 mm long, often with purplish veins; leaflets finely puberulous or glabrescent to densely pubescent. **18. R. minima**
26. Some leaflets at least with pale area along the midrib above; stems with spreading yellowish hairs. **9. R. usambarensis**
- Leaflets without pale area above; stems without spreading yellowish hairs. 27
27. Standard 13-16 mm long; plant in most parts pubescent with longer usually glandular-based hairs intermixed. **16. R. verdcourtii**
- Standard up to 10 mm long; pubescence without longer glandular-based hairs intermixed. 28
28. Stipules ovate, 3.5-4.5 x 2-2.5 mm; racemes 3-5-flowered. **8. R. oblatifoliolata**
- Stipules narrower, up to 4 x 1 mm; racemes usually more than 5-flowered. **18. R. minima**

1. **R. densiflora** *(Roth) DC.* (1825).

Climbing procumbent or ascending perennial herb; stems pubescent. Leaflets 3, elliptic, ovate or almost round, 1.5-8 x 1-7 cm, ± pubescent, densely covered with small orange or black gland-dots beneath. Inflorescences axillary, dense, subsessile; rhachis 2-16.5 cm long. Calyx pubescent or pilose and glandular dotted; tube *c* 1.5-2 mm long; lobes linear or linear-lanceolate, 5-17 mm long. Corolla persistent, ± enclosed by the calyx; standard yellow with purplish venation, 7-14 mm long, glabrous. Pod 10-14 x 4.5-5 mm, covered with short hairs, sparser long hairs and orange-red glands.

1. Leaflets mostly rounded at the apex; calyx-lobes 10-17 mm long; rhachis of inflorescence 3-16.5 cm long; gland-dots mostly yellow or orange-red. subsp. **chrysadenia**
- Leaflets acute or acuminate; calyx-lobes 5-7.5 mm long; rhachis of inflorescence 2-3 cm long; gland-dots mostly black. subsp. **debilis**

Subsp. **chrysadenia** *(Taub.) Verdc.* in *Kew Bull.* **25**: 73 (1971).
R. glomerulans Fiori (1912) - type: EW, *Fiori* 1140 (FT holo.).

Grassland, bushland, thickets; *c* 1000-1700 m. EW SU AR GG SD; widespread in tropical and southern Africa. *Baldrati* 3501; *Frahm-Leliveld* 62345; *Corradi* 4448.

Frahm-Leliveld 62345 has black gland-dots and approaches subsp. *stuhlmannii* (Harms) Verdc.

Subsp. **debilis** (G. Don) Verdc. in *Kew Bull.* **25**:75 (1971).

Forest edges, stream-sides; *c* 1050 m. KF; Uganda, W. Africa, Congo, Zambia, Malawi, Angola. *Friis, Gilbert & Vollesen* 3926.

Subsp. *densiflora* in India, Ceylon and Burma has ± acute leaflets with yellow gland-dots and subsp. *stuhlmannii* in eastern Africa has obtuse leaflets and black gland-dots.

2. **R. resinosa** *(Hochst. ex A. Rich.) Bak.* (1871);
Fagelia resinosa Hochst. ex A. Rich. (1847) - types: TU, *Quartin Dillon* s.n. (P syn.) & *Schimper* I:317 & III:1547 (both P syn., BM K S isosyn.).

Robust, sticky, mostly woody scrambler or ± erect shrub; stems densely covered with short glandular hairs, later glabrescent. Leaflets 3, distinctly paler beneath, mostly rhomboid or triangular, 1.5-8 cm long and wide, acuminate, pubescent; stipules *c* 7.5 x 1.5 mm. Calyx-tube pouched, 4 mm long; lobes 6-17 mm long, the lowest the longest. Standard yellow, often veined purplish, 11-16 mm long, pubescent. Pod with pale brown reticulation, *c* 15 x 4-4.5 mm, with short and long whitish hairs.

Grassland, bushland, woodland, thickets and waste ground; 1400-2000 m. EW TU GD GJ SU AR BA HA WG KF SD; Sudan to West Africa and south to Angola and Transvaal. *Gilbert* 3196; *Mooney* 8419; *Ash* 1424.

3. **R. stipulosa** *A. Rich.* (1847);
R. viscosa (Roth) DC. var. *stipulosa* (A. Rich.) Verdc. in *Kew Bull.* **25**: 87 (1971) - type: TU, *Quartin Dillon & Petit* s.n. (P holo., K ?iso.).

Sticky robust climber; stems pubescent and also with longer yellow bristly hairs. Leaflets 3, ovate to rhomboid, *c* 7 x 6 cm, acute to acuminate at the apex, pubescent; stipules *c* 10-15 x 4-5 mm. Calyx-tube not pouched, *c* 3.5 mm long; lobes 3.5-7 mm, the lowest the longest. Standard yellow with purplish veins, *c* 14-15 mm long, pubescent outside. Pod elliptic-oblong, *c* 30 mm long, pubescent and also with long yellow bristly hairs.

Woodland and wooded grassland; 1000-1800 m. TU WG GG. *Schimper* 282; *Mesfin Tadesse & Kagnew G.Y.* 2339; *Tewolde Berhan* 1541.

This was considered a variety of *R. viscosa* by Verdcourt, but with its large stipules and flowers it is probably rather conspecific with *R. luteola* (Hiern) K. Schum., a species widespread in tropical Africa, and if so Richard's name has priority.

4. **R. orthobotrya** *Harms* (1917).

Perennial sticky erect herb or undershrub, more rarely a climber, up to 2.5 m long; stems pubescent and also with longer glandular-based hairs. Leaflets (1-)3, oblate or ovate, 2-8 x 1.5-7.5 cm, acute or acuminate at the apex, pubescent and with scattered longer hairs; stipules *c* 10-12 x 2.5-6 mm. Inflorescences fairly dense; rhachis 4.5-18 cm long. Calyx-tube 2-3 mm long; lowest lobe longest and conspicuously wider than the rest, 6-11 x 3-4 mm. Standard yellow, veined purple, 9-11 mm long, pubescent. Pod oblong, 25-30 x 8-9 mm, with long yellow glandular-based hairs and also with short hairs.

Grassland, bushland and woodland; 1300-2400 m. EW GD GJ SU HA KF GG SD; Uganda, Kenya, Tanzania, Sudan and westwards to Senegal. *Ash* 1328, 1428; *Burger* 1220.

Burger 1220 from HA and *Vàtova* 386 from SD deviate in having 1-foliolate leaves.

5. **R. nyasica** *Bak*. (1897).
R. imbricata Bak. (1897).

Sticky erect or scrambling woody herb or subshrub; stems pubescent and glandular. Leaflets 3, oblate or ovate, 1-5 cm long and wide, rounded to shortly acuminate at the apex, pubescent to glabrescent; stipules *c* 4 x 1 mm. Inflorescences mostly short, axillary, often subsessile. Calyx-tube 3 mm long; lobes 2.5-6 mm long, the lowest the longest. Standard yellow veined purplish, 10-13 mm long, glabrous or pubescent. Pod 25-30 x 7-10 mm, finely pubescent and glandular and often with longer hairs as well.

Grassland, woodland, hill-sides; 550-1300 m. GJ WG IL GG; widespread in tropical Africa. *Gilbert, Thulin & Getachew* 531; *Ash* 162; *Mooney* 6837.

6. **R. elegans** *A. Rich*. (1847)
- types: TU, *Petit* s.n. & *Quartin Dillon* s.n. (both P syn.) & *Schimper* I:88 (P syn., BM K S isosyn.) & II:518 (P syn., BM K isosyn.).
Dolichos reticulatus A. Rich. (1847) - type: TU, *Quartin Dillon & Petit* s.n. (P holo.).
R. buramensis Hutch. & E. A. Bruce (1941).

Perennial climbing herb; stems appressed pubescent to glabrescent. Leaflets 3, ovate or rhomboid, 2-8 x 1-6.5 cm, acute to acuminate, pubescent or glabrescent above, pubescent to velvety beneath. Inflorescence lax; rhachis glandular hairy, 3-14 cm long; pedicels 4-19 mm long. Calyx pubescent; tube 3-4 mm long; lobes mostly linear, 5.5-17 mm long. Standard yellow with purplish veining, 13-21 mm long, glabrous. Pod yellowish-green when young with distinctive darker green reticulation, later brown, 20-38 x 6-8 mm, densely covered with viscid glandular hairs and also with shorter hairs and gland-dots.

1. Stipules 3.5-10 x 2.5-4 mm; bracts mostly lanceolate. var. **elegans**
- Stipules larger, *c* 13 x 6 mm; bracts ovate. var. **grandistipulata**

Var. **elegans.**
Bushland, scrub; 1000-2700 m. EW TU WU SU AR HA SD; N Somalia, Uganda, Kenya, Tanzania, Zaire, Yemen. *Mooney* 6309; *Gilbert* 3395; *Schweinfurth* 310.

Var. **grandistipulata** *Verdc*. in *Kew Bull*. **25**: 93 (1971)
- type: KF, Dekano, *Mooney* 6186 (K holo.).

Thicket at swamp edge; 2650 m. KF; N Uganda. In Ethiopia only known from the type.

R. buramensis, described from N Somalia, is less glandular than typical *R. elegans* and has somewhat smaller pods. Similar plants also occur in HA, but I do not think any taxonomic distinction can be maintained.

7. **R. ferruginea** *A. Rich*. (1847)
- type: TU, *Petit* s.n. (P holo., K iso.).
?R. rivae Schweinf. (1896) - type: EW, *Schweinfurth* 1898 (B holo. destr.).

Perennial climbing herb; stems densely covered with pale rusty-red hairs. Leaflets 3, ovate, 0.5-5 x 0.3-4 cm, rounded to acute at the apex, pubescent at least beneath, with mostly rusty-red hairs on the main nerves; stipules 5-8 x 1-3.5 mm. Calyx-tube 2.5-3 mm long; lobes 2-6 mm long, the lowest the longest. Standard dull yellow veined purple, 9-15 mm long, glabrous. Pods 17-23 x 5.5-6.5 mm, puberulous and glandular and also with longer yellow hairs.

Grassland, bushland, forest margins; (1400-)1900-2800 m. EW TU GD SU BA SD; Uganda, Kenya, Tanzania, Zaire. *Mooney* 6305, 8106; *Gillett* 14525; *W. de Wilde* 6688.

R. rivae is based on a single fruiting specimen from Eritrea which is now lost. The description agrees with *R. ferruginea* with which it is most probably conspecific. Schweinfurth (1896) in the same paper also recorded *R. ferruginea* from Eritrea, but the material cited (*Schweinfurth & Riva* 2001) is *R. totta* var. *venulosa.*

8. **R. oblatifoliolata** *Verdc*. (1971)
- type: HA, *IECAMA* RS-248 (K holo.).

Perennial prostrate or climbing herb; stems densely pubescent. Leaflets 3, oblate or ovate-rhomboid, 0.8-2.1 x 0.8-2.4 cm, rounded to truncate at the apex, densely pubescent on both sides; stipules ovate, 3.5-4.5 x 2-2.5 mm. Inflorescences 3-5-flowered; rhachis 1-1.5 cm long. Calyx densely pubescent; tube 2.5 mm long; lobes 2.5-3.5 mm long. Standard 8-9.5 mm long, densely pubescent. Pod *c* 22 x 7 mm with long hairs, short hairs and glands.

Grassland, rocky slopes; 1750-2100 m. HA; Kenya. *IECAMA* H-70; *J. De Wilde* 6405; *Jansen* 6073.

9. **R. usambarensis** *Taub*. (1895).

Perennial climbing herb; stems with long spreading yellowish hairs. Leaflets 3, usually paler green along the midrib above, 1-7.5 x 1-4 cm, pubescent on both sides; stipules 4-11 x 1-3 mm. Calyx pubescent; tube 2-3 mm long; lobes 2-6 mm long, the lowest the longest. Standard yellow veined purplish, 9-12 mm long, pubescent. Pod oblong, 18-24 x 5.5-7 mm, covered with long yellow hairs, short white hairs and glands.

1. Leaflets elliptic or elliptic-ovate, acute. var. **usambarensis**
- Leaflets ovate, ovate-orbicular or rhomboid, obtuse. var. **obtusifoliola**

Var. **usambarensis.**

Bushland, thickets; 1000-2250 m. EW BA HA GG SD; Uganda, Kenya, Tanzania. *Gilbert, Thulin & Getachew* 412; *Mooney* 5432; *Eriksson* 688.

Var. **obtusifoliola** *Verdc.* in *Kew Bull.* **25**: 96 (1971)
- type: SD, Mega, *Gillett* 14251 (K holo., EA FT iso.).

Bushland, thickets; *c* 1900 m. SD; Kenya. In Ethiopia known only from the type.

10. **R. mensensis** *Schweinf.* (1896)
- types: EW, *Schweinfurth* 1354 & 1556 (B syn. destr.).

Perennial climbing herb; stems ± densely pubescent with retrorse whitish hairs. Leaflets 3, usually paler green along the midrib above, ± oblong-lanceolate, up to 5.5 x 2.3 cm, acute, pubescent on both sides; stipules 3-7 x 1-2 mm. Racemes lax, up to 5-15-flowered. Calyx pubescent; tube 2-2.5 mm long; lobes 2-4 mm long, the lowest the longest. Standard yellow veined purplish, *c* 10 mm long, pubescent. Pod oblong, *c* 20-25 x 6 mm, pubescent or tomentose with short hairs.

Grassland, bushland; *c* 1500-2000 m. EW SU GG; Yemen. *Ouren* 20514, 20747; *Gilbert, Thulin & Getachew* 412a.

R. mensensis is a badly understood species the types of which are lost. Some recent material from Ethiopia and Yemen agrees well with the description and is here tentatively associated with the name. It may be just a variant of *R. usambarensis* from which it differs mainly in the narrower leaflets, the whitish retrorse pubescence on the stems and the shortly pubescent pods.

11. **R. totta** *(Thunb.) DC.* (1825).

Twining or ascending perennial; stems slender, glabrescent to densely pubescent. Leaflets 3, oblong-lanceolate, 2-5 x 0.5-1 cm, acute at the apex, glabrescent to silky pubescent; venation prominent and reticulate. Flowers solitary or few in very lax inflorescences. Calyx-lobes lanceolate, 2.5-8 mm long. Standard yellow, flushed reddish, 7-12 mm long, glabrous. Pod *c* 15-25 x 6 mm, with long and short hairs or silky pilose.

1. Stems and leaves sparsely pubescent or glabrescent; flowers usually solitary. var. **totta**
- Stems and leaves densely pubescent to velvety; inflorescences up to 6-flowered. var. **venulosa**

Var. **totta.** Fig. 95.32.2.

Short grassland, bushland; 1350-2800 m. EW SU AR HA GG SD; N Somalia and southwards to South Africa, also in Yemen and Saudi Arabia. *Thulin* 1623; *Gilbert, Thulin & Getachew* 407; *Gillett* 14446.

Var. **venulosa** *(Hiern) Verdc.* in *Kew Bull.* **25**: 99 (1971); *R. venulosa* (Hiern) K. Schum. (1901).

Grassland; *c* 2000 m. EW HA SD; Sudan and southwards to South Africa. *Schweinfurth & Riva* 2001; *J. De Wilde* 7197; *Pappi* 4870.

Var. *totta* and var. *venulosa* are very distinct in NE Africa and it is difficult to believe they are not different species. Both varieties have been collected on Mt Bizen in Eritrea. If var. *venulosa* is considered specifically distinct *R. filicaulis* Welw. ex Bak. seems to have priority.

12. **R. alluaudii** *Sacl.* (1910).

Erect not or sparsely branched subshrub, 0.6-1.8 m tall, from a tuberous rootstock; branches velvety pubescent, later glabrescent. Leaves 1-foliolate; leaflet ovate, 2-6.5 x 1.5-5 cm, rounded to subacute at the apex, cordate, velvety pubescent. Rhachis of inflorescences 2-5 cm long; peduncle 0.4-2 cm long. Calyx silky pilose with lanceolate lobes, 7-13 mm long. Standard yellow with purplish veins, 11-13 mm long, glabrous. Pod *c* 23 x 7 mm, densely covered with long hairs, short hairs and glands.

Grassland, bushland, hill-slopes; 1100-1900 m. BA GG SD; Kenya, N Tanzania. *Gilbert, Thulin & Getachew* 402; *Mooney* 8422; *Gillett* 14729.

13. **R. sublobata** *(Schumach.) Meikle* (1951); *Glycine sublobata* Schumach. (1827). *R. transjubensis* Chiov. (1932).

Perennial prostrate or climbing herb; stem glabrescent to densely pubescent. Leaflets 3, rhomboid, elliptic or ovate, 2-8 x 2-8.5 cm, rounded, often slightly lobed at the base, glabrescent to pubescent. Racemes lax, unbranched. Calyx pubescent; tube 3 mm long; lowest lobe the longest, up to 6 mm long. Standard yellow with purplish veining, 9-22 mm long, pubescent. Pod oblong-falcate, inflated, 21-30 x 8-13 mm, short pubescent and glandular. Seeds 4.3-5(-6.4) x 3.5-3.8(-4.8) x 2-3.2 mm, finely roughened.

Grassland, woodland; 500-1600 m. EE EW SU KF GG; widespread in tropical Africa south to Transvaal, and in Yemen. *Gilbert, Thulin & Getachew* 301; *Gilbert* 1533; *Pappi* 6827.

14. **R. malacotricha** *Harms* (1907)
- types: HA/BA, Luku, *Ellenbeck* 1227 (B holo. destr.), SU, Mt Fantale, *J. De Wilde* 4875 (WAG neo.).

Shrubby ± erect robust herb, often with climbing branches, 1-3 m tall; stems densely velvety with white hairs. Leaflets 3, the terminal oblate to rhomboid, the laterals ± broadly ovate, 2.5-11 x 2-9 cm, often slightly lobed, rounded to acute at the apex, densely velvety on both surfaces; venation reticulate and prominent beneath; petiole 1.5-4 cm long. Racemes fairly dense, often branched, 4-9 cm long, pedunculate. Calyx

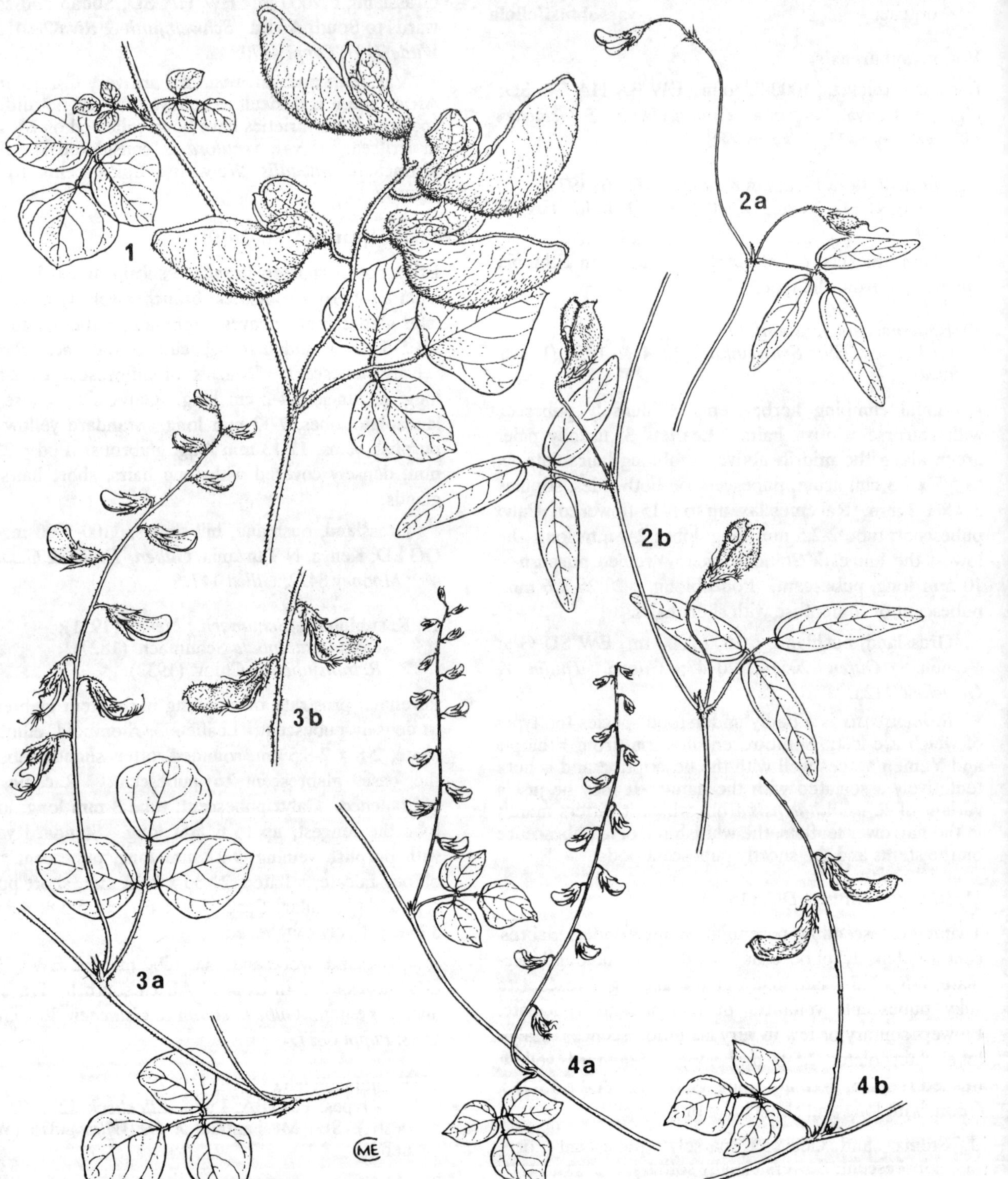

Figure 95.32.1 - *PARACALYX SOMALORUM*, portion of flowering branch from *Burger* 3103. 2 - *RHYNCHOSIA TOTTA* var *TOTTA*, a: flower, young fruit and leaf, b: pods and leaves, both from *Gilbert, Thulin & Getachew* 407. 3 - *R. VERDCOURTII*, a: flowering raceme and leaves, b: pods, both from *Gilbert, Thulin & Getachew* 376. 4 - *R. MINIMA* var. *MINIMA*, a: flowering racemes and leaves, b: fruiting raceme, both from *Gilbert & Thulin* 176. All x 0.75. Drawn by Maureen Church. (Reproduced with permission from *Opera Botanica*, 68, fig. 34, 1983.)

velvety; tube *c* 4 mm long; the lowest lobe the longest, 7-10 mm long. Standard yellow or orange-yellow with purplish veins, 15-17 mm long, densely pubescent. Pod oblong, curved, inflated, *c* 30 x 10 mm, tomentose. Seeds 6-6.4 x 4.4-4.8 x 3.8-4 mm, finely roughened.

Grassland and wood/bushland; 1350-2000 m. SU (Mt Fantale) HA HA-BA; not known elsewhere. *Jansen* 5242; *Bally* 13028; *J. De Wilde* 6474.

15. **R. malacophylla** *(Spreng.) Boj.* (1837).

R. sennaarensis Hochst. ex Schweinf. (1868); syntypes from Sudan, Sudan/Ethiopia border: *Schweinfurth* 1745, 1746, 1917 and EW: *Steudner* 149 (all B syn. destr.).

R. sennaarensis var. *flavissima* Schweinf. in *Verh. Zool.-Bot. Ges. Wien* **18**: 656 (1868) - types from Sudan and N Ethiopia: *Schimper* 192 (B syn. destr., K isosyn.).

R. flavissima Hochst. ex Bak. (1871) - type: N Ethiopia, *Schimper* 2244 (K lecto., S isolecto.).

Perennial trailing or climbing herb from a woody base; stems glabrous or pubescent. Leaflets 3, ovate to rhomboid, 1-5 x 1-5 cm, mostly acuminate at the apex, finely puberulous on both surfaces, often with silky margins; stipules narrow. Flowers on a rhachis 4-16 cm long. Calyx pubescent; tube 1.5-3 mm long; lobes 2-5 mm long. Standard mostly plain yellow but sometimes purplish veined, 9-17 mm long, pubescent and gland-dotted. Pod 15-20 x 4.5-5.5 mm, finely to densely puberulous and gland-dotted.

Bushland, scrub; 150-2350 m. AF EW TU GJ SU HA BA KF GG SD; Sudan, Somalia, N Uganda, Kenya, Tanzania and in Yemen. *Gilbert, Thulin & Getachew* 238, 537; *Ellis* 346.

16. **R. verdcourtii** *Thulin* (1981)
- type: GG, *Gilbert, Thulin & Getachew* 376 (UPS holo., ETH K MO iso.).

Perennial climbing or trailing herb sometimes ± woody below; stems pubescent and with longer usually glandular-based hairs. Leaflets 3, the terminal oblate to rhomboid, the laterals obliquely-ovate, 0.8-3 x 0.8-3.5 mm, mostly rounded at the apex, shortly pubescent and also with longer glandular-based hairs; stipules ± ovate-cuspidate. Flowers on a rhachis 5-18 cm long. Calyx with short and long hairs; tube 2.5-3.2 mm long; lobes 4-7 mm long. Standard yellow with purplish veins, 13-16 mm long, pubescent and often also gland-dotted. Pod *c* 12-18 x 4-5.5 mm, with long glandular-based hairs, short hairs and gland-dots. Fig. 95.32.3.

Grassland, bushland; *c* 1200-1450 m. SU GG; N Uganda, Kenya, N Tanzania. *Mooney* 9029; *Ash* 996.

17. **R. pulverulenta** *Stocks* (1852);

R. memnonia (Del.) DC. forma *pulverulenta* (Stocks) Terracc. in *Ann. Ist. Bot. Roma* **5**: 110 (1893).

R. elachistantha Chiov. (1903) - type: EW, *Terracciano & Pappi* 21 (FT holo.).

Erect or scrambling woody herb; stems densely tomentose with greyish-white hairs. Leaflets 3, rhomboid or ovate-triangular, 0.4-3 x 0.6-3.5 cm, velvety pubescent and densely dotted with pale glands on both surfaces, especially beneath. Inflorescences rather lax, 1-10-flowered. Calyx shortly pubescent and with yellow gland-dots. Standard pale yellow, 4-6.5 mm long, rather densely pubescent. Pod 15-20 x 4.5-6 mm, densely pubescent with very short white hairs and with small pale gland-dots.

Semi-desert grassland and bushland; 0-1000 m. EE EW HA SD; Sudan, N Somalia, Socotra, Kenya, Tanzania, ?SW Africa and Arabia to India. *Simmons* 171; *Terracciano* 728, 560.

18. **R. minima** *(L.) DC.* (1825).

Perennial climbing or prostrate herb; stems glabrous to velvety. Leaflets 3, rhomboid, ovate or suborbicular, 0.9-6 x 0.8-5 cm, glabrescent to velvety. Calyx pubescent; lobes 2-7 mm long. Standard yellow often with purplish veins, 5-10 mm long, pubescent, glandular. Pod oblong-falcate, 6-20(-22) x 3-5(-9) mm, glabrescent to pubescent and often with longer hairs and glands.

1. Calyx longer than or only very slightly shorter than the corolla. var. **macrocalyx**
- Calyx shorter than the corolla. 2
2. Standard 7-10 mm long; leaflets often rounded at the apex, pubescent to velvety on both surfaces. var. **prostrata**
- Standard usually 5-6 mm long; leaflets various. 3
3. Leaflets rhomboid, acute or acuminate at the apex, plant glabrescent. var. **minima**
- Leaflets rhomboid to almost round, mostly blunt at the apex. 4
4. Stems and leaves glabrescent or pubescent; pod *c* 10-12 x 3-3.5 mm. var. **nuda**
- Stems and leaves densely silvery pubescent to velvety; pod usually larger. var. **memnonia**

Var. **minima.** Fig. 95.32.4.

Grassland, bushland, woodland; up to 1100 m. EE EW SU-AF HA-AF HA SD; throughout tropical Africa, Arabia and eastwards to China, also in West Indies and tropical America. *Pappi* 4190; *Gilbert & Thulin* 145, 176.

Var. **nuda** *(DC.) Kuntze, Rev. Gen. Pl.* **1**: 204 (1891).

Grassland, bushland, woodland, 300-1900 m. AF EW TU GD WU SU AR IL KF SD; tropical Africa south to Tanzania, India, Thailand and south to Australia. *Thulin* 1328; *Ash* 270; *Gillett* 14181.

Var. **macrocalyx** *(Chiov.) Verdc.* in *Kew Bull.* **25**: 105 (1971);

R. flavissima Hochst. ex Bak. var. *macrocalyx* Chiov. in *Ann. Ist. Bot. Roma* **8**: 442 (1908); *R.*

sennaarensis Hochst. ex Schweinf. var. *macrocalyx* (Chiov.) Cuf., Cuf. *Enum., Zweiter Nachtrag*: 25 (1969) - type: EW, *Pappi* 1400 (FT holo.).

Grassland, bushland; 600-2600 m. EW WU SD; Sudan, Uganda, Tanzania. *Pappi* 334, 7829; *Jansen* 3861.

Var. **prostrata** *(Harv.) Meikle* in *Kew Bull.* 9: 275 (1954).
R. schoelleri Schweinf. (1896) - type: EW, *Schweinfurth* 307 (B holo. destr., K iso.).

Grassland, bushland, woodland; 1400-2550 m. EW ?TU GD SU AR HA KF GG; widespread in tropical and southern Africa and in Yemen. *Thulin* 1505; *Stewart* 116; *Ash* 2133.

Var. **memnonia** *(Del.) Cooke*, *Fl. Bombay* 1: 389 (1902).

Grassland, bushland; *c* 500-900 m. EE AF EW HA; Sudan, Somalia, Egypt, S Sahara, West Africa, and Arabia. *Ash* 1576; *IECAMA* F-37; *Pappi* 6838.

R. minima in the present sense is a pantropical, taxonomically difficult complex with its centre of variation in Africa. Large-flowered plants (var. *prostrata*) sometimes approach *R. malacophylla*.

19. **R. albissima** *Gand.* (1913).

Erect subshrub or woody herb, 0.3-2 m tall; stems pubescent, the young parts covered with a dense silvery tomentum. Leaflets 3, elliptic to broadly obovate or rhomboid, 0.7-4 x 0.6-3.7 cm, rounded or subacute at the apex, velvety and usually silvery-grey on both surfaces; venation prominent beneath; petiole 0.8-3 cm long, longer than rhachis. Racemes fairly dense, 2-5 cm long, ± pedunculate or subsessile. Calyx grey pubescent; tube 2.5-4 mm long; lobes 2-5 mm long. Standard yellow, veined or flushed purplish, 7-15 mm long, silvery pubescent and gland-dotted, with wart-like appendages on the inside near the base. Pod 12-18 x 5-6.5 mm, silvery tomentose, sometimes with sparse longer yellow hairs intermixed.

Woodland and wooded grassland; *c* 1000 m. GJ; Sudan, N Uganda, Kenya, Tanzania, Mozambique, Zimbabwe and NE South Africa. *Thulin & Hunde* 4038.

20. **R. splendens** *Schweinf.* (1868)
- type: Sudan/Ethiopia border, Metema (Matamma), *Schweinfurth* 1870 (B holo. destr., BM P iso.).

Erect subshrub, *c* 0.6 m tall; stems silvery tomentose. Leaflets 3, obovate-elliptic, up to 5.5 x 3 cm, acute, densely silvery tomentose above, densely woolly beneath; venation partly hidden by the indumentum beneath; petiole 1-3 cm long, much longer than rhachis. Racemes fairly dense, 3-5 cm long, shortly pedunculate. Calyx silvery tomentose; tube *c* 3 mm long; the lowest lobe the longest, 5-6 mm long. Standard *c* 13 mm long, pubescent and gland-dotted, without appendages on the inside. Ovary tomentose. Pod not known.

Probably wooded grassland at *c* 1000 m. GD; only known from the type collection.

21. **R. erythraeae** *Schweinf.* (1896)
- type: EW, *Schweinfurth* 1850 (B syn. destr., K lecto., FT P S isolecto.).

Erect shrub, up to 1 m tall; stems tomentose. Leaflets 3, elliptic to rhomboid or broadly ovate, 1-3.8 x 0.6-2.5 cm, acute or subacute at the apex, sparsely to densely appressed pubescent above, densely silvery pubescent beneath; venation narrowly reticulate, very prominent beneath; petiole 0.3-1.5 cm long, usually much longer than the rhachis. Racemes fairly dense and short; peduncle up to 2.5 cm long; bracts lanceolate to ovate, *c* 2.5-3 x 1-1.5 mm, falling quickly. Calyx appressed silvery; tube *c* 3 mm long; lobes 4-7 mm long, the lowest the longest. Standard yellow with brownish veins, 14-15 mm long, silvery pubescent and gland-dotted, without appendages on the inside. Pod oblong, *c* 13 x 5 mm, silvery pubescent, often ± enclosed by the dry but persistent corolla. Fig. 95.33.1.

Grassland or bushland; 1000-2000 m. EW SU (Mt Fantale); not known elsewhere. *Gilbert* 1395; *Pappi* 3385, 4719.

22. **R. erlangeri** *Harms* (1907)
- types: HA, *Ellenbeck* 537 (B holo. destr.), *Westphal* 2463 (WAG neo., C isoneo.).

Erect shrub, 0.5-1.5 m tall; stems pubescent. Leaflets 3, rhomboid or ovate to suborbicular or oblate, 1-4.5 x 0.8-3 cm, rounded to acute at the apex, puberulous to finely pubescent on both surfaces; venation very prominent beneath, narrowly reticulate; petiole 0.2-0.5 cm long, usually shorter than the rhachis. Racemes fairly dense, 2-3 cm long, subsessile or shortly pedunculate; bracts broadly elliptic to broadly obovate, 4-8 x 3-5 mm, falling quickly. Calyx pubescent; tube 3-3.5 mm long; lobes 5-10 mm long, the lowest the longest. Standard yellow with brownish veins, 14-19 mm long, pubescent and gland-dotted, without appendages on the inside. Young pod ± enclosed by the dry but persistent corolla, spreadingly pilose and densely gland-dotted.

Scrub, on grassy slopes or between rocks; 1840-2600 m. HA; not known elsewhere. *J. De Wilde* 5771.

23. **R. schimperi** *Hochst.* ex *Boiss.* (1872).
R. cliffordii Hutch. & Bruce (1941).

Prostrate or spreading shrub or subshrub; ultimate branches ± herbaceous, shortly grey tomentose. Leaflets 3, obovate to suborbicular or oblate, 0.8-2.5 x 0.6-3 cm, obtuse, truncate or emarginate, densely grey tomentose on both sides, or dark green and more finely pubescent above; venation narrowly reticulate and ± prominent beneath. Racemes 2-5-flowered, unbranched or rarely branched, short, equalling or not much exceeding subtending leaves. Calyx grey pubescent; tube 2-2.5 mm long; lowest lobe up to 4 mm long. Standard yellow,

(5-) 8-12 mm long, pubescent. Pod oblong, 1.5-2.2 x 0.7-0.9 cm, shortly and densely pubescent. Fig. 95.33.2.

Stony ground; 1350-1500 m. HA; N Somalia, also in Arabia, Pakistan and India.. *Hummel* 28.

The NE African plants tend to have larger flowers than normal in *R. schimperi* and previously have been called *R. cliffordii*. With the material now available, particularly from Somalia, the distinction has become less clear and I prefer to regard *R. cliffordii* as conspecific with *R. schimperi*.

24. **R. ramosa** *Verdc.* (1971)
- type: HA, Ogaden, *Peck* 402 (EA, holo., K, iso.).

Perennial probably prostrate herb; stems densely greyish-white tomentose. Leaflets 3, suborbicular to oblate, 0.7-2 x 0.7-2.5 cm, rounded to truncate at the apex, densely greyish-white tomentose, beneath also with numerous pale yellow gland dots. Inflorescences axillary, often branched. Calyx tomentose; tube 3 mm long; lowest lobe *c* 9 mm long. Standard *c* 16 mm long, densely pubescent. Young pods elliptic-oblong, 17 x 8 mm, shortly tomentose.

HA (near Somalia border at *c* 8°22'N, 45°55'E, *c* 900 m); only known from the type.

25. **R. hirta** (Andr.) *Meikle & Verdc.* (1967).

Robust climber or trailer to several m long; young stems densely ferruginous hairy. Leaflets 3, the terminal ovate to rhomboid, 5-18 x 4-14 cm, the laterals obliquely ovate, smaller, all acuminate at the apex, sparsely pubescent above, densely pubescent to velvety beneath; petiole 7-10 cm long. Racemes sometimes branched, 7-35 cm long, pedunculate. Calyx appressed pubescent; tube 5-6 mm long; the lowest lobe usually the longest, 14-23 mm long, the upper pair joined to form a bifid lip. Corolla persistent, mostly shorter than calyx, greenish to orange; standard 18-22 mm long, glabrous. Pod twisted narrowly oblong, 15-30 x 3.5-7 mm, velvety pubescent. Seeds 5-6.5 x 3.5-4.5 mm, subglobose, blue, later black.

Forest edges, coffee plantations; *c* 1250 m. IL; widespread in tropical Africa, also in the Mascarene Is., India and Ceylon. *Friis, Gilbert & Vollesen* 4098.

26. **R. congensis** *Bak.* (1871).

Robust climber or trailer to several m long; young stems finely pubescent. Leaflets 3, the terminal rounded to broadly ovate, 3.5-12.5 x 2.5-10.5 cm, the laterals obliquely ovate, somewhat smaller, all acute to acuminate at the apex, ± sparsely pubescent above, sparsely to densely pubescent beneath; petiole 2-11 cm long. Racemes sometimes branched, 10-36 cm long, shortly pedunculate. Calyx appressed pubescent; tube 3-5 mm long; the lowest lobe 5-15 mm long, the upper pair joined to form a bifid lip. Corolla persistent, longer than calyx, pale yellow; standard 15-20 mm long, glabrous. Pod usually twisted, narrowly oblong, *c* 20-25 x 5.5-7 mm, velvety pubescent. Seeds *c* 5.5 x 3.5-4.5 mm, subglobose, blue, later brownish.

Forest edges and clearings; *c* 1000 m. IL; fairly widespread in tropical Africa. *Friis, Gilbert & Vollesen* 4082.

The single Ethiopian collection has calyces only c 10 mm long and agrees with certain specimens from Uganda, e.g. *Chandler* 1808. It cannot be satisfactorily placed in any of the three subspecies at present recognized within *R. congensis*.

74. PARACALYX *Ali* (1968)

Twining or trailing herbs. Leaves pinnately 3-foliolate; leaflets with yellowish gland-dots beneath. Flowers in axillary racemes; bracts cup-shaped; bracteoles absent. Calyx scarious, strongly accrescent; lobes obtuse, the lowermost the largest, boat-shaped. Vexillary stamen free; anthers uniform. Ovary 2-ovuled; style filiform; stigma small. Pod short, compressed, enclosed in the calyx, 1-2-seeded. Seed with a rounded to shortly elongate hilum.

Six species in the Horn of Africa, Socotra and S Asia.

P. somalorum *(Vierh.) Ali* (1968);
Cylista somalorum Vierh. (1907) - type: Ogaden, *Riva* 852 (B holo. destr., FT iso.).

Stems twining or climbing, tomentose, glabrescent. Leaflets broadly ovate or rhombic, up to 35 x 40 mm, acute, tomentose on both sides. Calyx up to 45 mm long in fruit, tomentose and gland-dotted; lowermost lobe *c* twice as long as the upper. Standard yellow, *c* 15-20 mm long, tomentose. Pod *c* 12 x 7 mm, with glands and dense pubescence of short and long hairs. Fig. 95.32.1.

Dry bushland; *c* 1000-1220 m. HA ?BA; N Somalia. *Burger* 3090, 3103.

75. ERIOSEMA *(DC.) G. Don* (1832), *nom. conserv.*

Verdcourt in *Fl. Trop. E. Afr.* (1971).

Herbs or subshrubs, erect or prostrate, rarely twining. Leaves pinnately or rarely digitately 3-foliolate or 1-foliolate; leaflets often with small glands beneath. Flowers in axillary or terminal racemes, sometimes in panicles. Calyx 5-lobed; lobes equal or unequal, often glandular. Corolla usually yellow; standard with or without appendages inside, glabrous to velvety and usually glandular outside. Vexillary stamen free; anthers uniform. Ovary 2-ovuled; style bent, glabrous above; stigma very small. Pods oblong or oval, compressed, hairy or glabrescent. Seeds 2, with a linear hilum to which the funicle is attached at one end.

Some 130 species throughout the tropics but mainly centered in Africa.

1. Leaves all 1-foliolate. 2
- Leaves 3-foliolate or with some 1-foliolate leaves in addition. 4
2. Leaflets cuneate at the base. **10. E. rhodesicum**
- Leaflets cordate at the base. 3
3. Standard yellow; plant ± robust and erect. **13. E. verdickii**
- Standard black or dark brown; plant ± slender, trailing or ascending. **14. E. cordifolium**
4. Leaf-rhachis longer than or about as long as the petiole or petiole only *c* 2 mm long. 5
- Leaf-rhachis much shorter than the petiole which is always more than 2 mm long. 10
5. Standard glabrous or nearly so; robust woody herb or subshrub with leaves white tomentose beneath. **1. E. psoraleoides**
- Standard sparsely to densely pubescent, if with sparse hairs then other characters not as above. 6
6. Racemes elongated, lax, with some flowers *c* 5 mm apart; indumentum short. **9. E. sparsiflorum**
- Racemes denser, or if rather lax then plant pilose with long hairs. 7
7. Leaves densely appressed silvery pubescent beneath. **8. E. tuberosum**
- Leaves not silvery beneath. 8
8. Leaves 1-3; indumentum on stems, leaves and calyx of spreading hairs, often 3-4 mm long. **12. E. shirense**
- Leaflets 3; indumentum of shorter hairs. 9
9. Racemes many-flowered, elongate; petiole and rhachis together 5-25 mm long. **2. E. nutans**
- Racemes 2-4-flowered, short; petiole and rhachis together less than 4 mm long. **11. E. elliotii**
10. Bracts large, 11-16 x 5-10 mm; calyx 13-20 mm long. **7. E. robustum**
- Bracts and calyx much smaller. 11
11. Claw of keel much shorter than the lamina. 12
- Claw of keel almost as long as or exceeding the lamina. 13
12. Racemes 2-7 cm long; petioles usually more than 1.5 cm long; standard yellow, veined or flushed purplish. **3. E. montanum**
- Racemes *c* 1 cm long; petioles rarely more than 1.5 cm long; standard dark outside. **6. E. longipedunculatum**
13. Petioles mostly about 1 cm long; corolla scarcely exceeding the densely hairy calyx; stems very leafy with rather short internodes. **4. E. jurionianum**
- Petioles mostly much longer; corolla well exceeding the calyx; stems less leafy with longer internodes. **5. E. scioanum**

1. **E. psoraleoides** *(Lam.) G. Don* (1832).

Erect subshrub or woody herb; branches shortly and densely brownish hairy. Leaflets 3, oblanceolate to narrowly elliptic, up to 5-7(-9.5) x 1.5-2.5(-3.5) cm, usually rounded at the apex, pubescent above, white tomentose beneath; venation prominent beneath and PNusually with brownish hairs; petiole 1-5 mm long; rhachis 1-8 mm long. Racemes many-flowered. Calyx appressed pilose, 3.5-7 mm long. Standard yellow, 7-14 mm long, glabrous or rarely with a few hairs outside, glandular. Pod suborbicular to ovate-oblong, 11-18 x 9-11 mm, covered with long rusty-red hairs.

Grassland and wooded grassland; 1300-1800 m. KF GG; westwards to W Africa, southwards to Transvaal and Natal, also in Madagascar. *Friis* et al. 2274; *W. de Wilde* 10202; *Fukui* 392; *Mesfin & Kagnew* 2491.

2. **E. nutans** *Schinz* (1908).

E. richardii Benth. ex Bak. f. & Haydon (1929); *E. buchananii* Bak. f. var. *richardii* (Benth. ex Bak. f. & Haydon) Staner in *Kew Bull.* 1935: 277 (1935) - type: Ethiopia, without precise locality, *Schimper* 708 (K holo.).

Perennial herb; branches erect or decumbent, densely covered with ± spreading white to rusty-red hairs. Leaflets 3, lanceolate or elliptic, up to 8(-9) x 2.5(-3.8) cm, usually acute at the apex, pilose; petiole 2-9(-12) mm long; rhachis 3-17 mm long. Racemes many-flowered, elongate. Calyx pubescent, 4-5 mm long. Standard 5-8 mm long, pubescent. Pod ovate-oblong, 8-10 x 5-8 mm, covered with long hairs, short hairs and glands. Fig. 95.33.3.

Grassland and woodland; 800-1800 m. EW TU WG SD; Sudan and throughout eastern tropical Africa south to Transvaal. *Gilbert & Thulin* 856; *Pappi* 2996; *Steudner* 128; *Friis* et al. 3259.

3. **E. montanum** *Bak. f.* (1895).

Perennial herb, often subshrubby; stems mostly erect, sometimes straggling, rusty-red hairy. Leaflets 3, narrowly ovate or elliptic, 3-11 x 1.7-4.5 cm, usually acute, softly pubescent; petiole 1.2-3.3 cm long, much longer than rhachis; stipules ovate-lanceolate, 0.8-2.5 cm long. Racemes elongated, long-pedunculate. Calyx pubescent, 4.5-5.5 mm long. Standard yellow, veined or flushed purplish, 6.5-9 mm long, sparsely pubescent along the middle outside or glabrescent. Keel with the claw much shorter than the lamina. Pod ovate-oblong, 10-17 x 7-9 mm, covered with long rusty-red hairs and also with very fine pubescence and glands.

Scrub, forest edges; *c* 1650-1950 m. KF; widespread in tropical Africa. *Mooney* 8664; *J. De Wilde* 5313, 6232.

4. **E. jurionianum** *Staner & De Craene* (1934);

Rhynchosia polystachya A. Rich. (1847); *E. polystachyum* (A. Rich.) Bak. (1871), *nom. illegit.*, *non* E. Mey. - type: N Ethiopia, without precise locality, *Petit* (P holo.).

E. humbertii Staner & De Craene (1934).

E. jurionianum var. *fulvum* (Bak. f.) Verdc. in *Kew Bull.* **25**: 128 (1971).

E. glomeratum sensu Cuf. *Enum.*: 327 (1955).

Perennial herb with several stiffly erect stems from a woody rootstock; branches very leafy with short internodes, densely rusty-red hairy. Leaflets 3, elliptic

Figure 95.33.1 - *RHYNCHOSIA ERYTHRAEAE*, flowering stem portion from *Gilbert* 1395. 2 - *R. SCHIMPERI*, flowering and fruiting racemes and leaves from *Gillett* 4864. 3 - *ERIOSEMA NUTANS*, portion of flowering stem from *Gilbert & Thulin* 856. 4 - *E. SCIOANUM*, a: flowering raceme and stem portion, b: fruiting raceme and supporting leaf, both from *Thulin* 1622. 1 & 2 x 1.125, 3 & 4 x 0.75. Drawn by Maureen Church. (Reproduced with permission from *Opera Botanica*, 68, fig. 35, 1983.)

or lanceolate, 2-7 x 1-2.5 cm, acute, sparsely to densely pubescent; petiole 0.7-1.5 cm long, much longer than rhachis. Racemes terminal and in the upper axils, dense, 1.5-5 cm long. Calyx densely hairy, 6-9 mm long. Standard 6.5-9 mm long, glabrescent or sparsely hairy. Keel with thc claw about as long as the lamina. Pod ovate-oblong, 10-14 x 6-7 mm, covered with long, rusty-red hairs, short hairs and glands.

Grassland, bushland, scrub; 1400-2700 m. EW ?TU GD GJ WG SU BA KF IL SD; Uganda, Kenya, N Tanzania, Zaire. *Mooney* 6299; *Friis* et al. 269; *Ash* 339.

Two varieties were recognized in Ethiopia by Verdcourt (1971), the type variety in the north and var. *fulvum* (Bak. f.) Verdc. in the south. The latter was distinguished by its more leafy stems, more bullate leaves and more hairy standard. The variation seems continuous, however, and I prefer to unite them here.

5. **E. scioanum** *Avetta* (1889)
- type: SU, *Antinori* s.n. (FT holo.).

Perennial herb; stems ascending, scrambling or erect, covered with spreading rusty-red hairs. Leaflets 3, elliptic, up to 7.5 x 4 cm, mostly acute, pubescent; petiole 1.1-4.5 cm long, much longer than rhachis. Racemes 2-10 cm long. Calyx pubescent, 4.5-5 mm long. Standard yellow veined purplish, 7-8 mm long, glabrous or almost so. Keel with the claw about as long as the lamina. Pod ovate-oblong, 8-12 x 6-9 mm, covered with long rusty-red hairs and also with very fine pubescence and glands. Fig. 95.33.4.

Grassland, scrub; 1600-2800 m. ?WG SU AR IL KF SD. *Thulin* 1622; *W. de Wilde* 8749; *Mooney* 5925.

The Ethiopian plant is subsp. *scioanum* which is not known elsewhere; subsp. *lejeunei* (Staner & De Craene) Verdc. occurs in Cameroun, East Africa, Rwanda and Burundi.

6. **E. longipedunculatum** *A. Rich.* (1847)
- type: TU, Schimper II: 925 (P holo., BM K P iso.).
E. longipedunculatum var. *hirsutum* Verdc. in *Kew Bull.* **25**: 134 (1971) - type: SD, *Gillett* 14498 (K holo.).

Erect or ascending perennial; stems several, densely pubescent. Leaflets 3, elliptic, 20-55 x 8-17 mm, acute or subacute, appressed pubescent, more densely so beneath; venation prominent beneath; petioles rarely exceeding 1.5 cm, much longer than rhachis. Racemes *c* 1 cm long, pedunculate. Calyx pubescent, 4-6 mm long. Standard dark outside, 8-9 mm long, mostly glabrous. Keel with the claw much shorter than the lamina. Pod oblong-ovate, *c* 12-14 x 8 mm, densely covered with long rusty-red hairs.

Grassland, bushland, rocky places; 1300-2400 m. EW TU WG SU AR GG SD; Yemen. *Gilbert, Thulin & Getachew* 444; *Thulin* 1448; *W. de Wilde* 10753.

Var. *hirsutum* was distinguished by its denser and more shaggy pubescence on stems and leaves and by having coarser stems and peduncles. As the variation in the material now available seems quite continuous in these respects I prefer not to maintain the variety.

7. **E. robustum** *Bak.* (1871)
- type: Ethiopia, without precise locality, *Parkyns* s.n. (K holo.).

Subshrub, 1-3.5 m tall; branches densely silvery or rusty-red pubescent. Leaflets 3, ovate-lanceolate or elliptic, up to 16 x 7.5 cm, usually acuminate at the apex, velvety pubescent on both surfaces but particularly beneath; petioles *c* 2-4 cm long, much longer than rhachis. Upper racemes forming a large panicle; bracts ± ovate, 11-16 x 5-10 mm. Calyx densely pilose, 13-20 mm long, the lower lobe much the longer. Corolla yellow with standard purplish outside, 12-18 mm long, hairy. Pod ovate-oblong, 12-17 x 8-12 mm, covered with long and short hairs and glands.

Grassland and woodland; 520-2300 m. TU GD WG SU IL KF SD GG; Cameroun, Uganda, Kenya, N Tanzania, Zaire, Rwanda, Burundi. *Gilbert & Thulin* 827; *Mooney* 6214; *Gillett* 14595; *Haugen* 335.

8. **E. tuberosum** *A. Rich.* (1847)
- types: TU, *Schimper* II:1202 (P syn., BM K S UPS isosyn.) & 1624 (P syn., BM K isosyn.).

Erect or ascending perennial herb from a tuberous rootstock; stem silvery pilose above. Leaflets (1-)3, oblong-elliptic, up to 7.5(-12) x 2.5 cm, acute, with scattered long hairs or glabrescent above, densely appressed silvery pubescent beneath; petiole obsolete; rhachis 1-6 mm long. Racemes 1-4.5 cm long, ± elongate, dense; peduncle up to 1.5 cm long. Calyx silvery, *c* 5 mm long. Standard *c* 9 mm long, pubescent. Pod ovate, *c* 8 x 7 mm, silvery pilose.

Probably wooded grassland at *c* 1500 m. TU GD; not known elsewhere. *Chiovenda* 645.

E. tuberosum is close to *E. schweinfurthii* Bak. f. (1895) known from Sudan and N Uganda and differs mainly in its predominantly 3-foliolate leaves. Possibly they will turn out to be variants of a single species, but with the limited material now available they are better kept distinct.

9. **E. sparsiflorum** *Bak. f.* (1895).

Erect pyrophytic* herb with many unbranched stems from a woody rootstock; stems spreading pubescent, the foliage mostly developing after the flowers are well developed. Leaflets (1-)3, elliptic-oblong or lanceolate, 1.5-11.5 x 0.45-3 cm, obtuse or subacute, pubescent; petioles 1-3 mm long; rhachis 1-3 mm long. Racemes lax with very deflected flowers. Calyx spreading pubescent, 4-5 mm long. Standard 8-12 mm long, very sparsely or finely pubescent. Pod ovate-oblong, 13-17 x

* pyrophytic = fire loving, found in areas regularly burnt.

8-10 mm, covered with long hairs, short hairs and glands.

Wooded grassland, appearing after fires; *c* 1300 m. WG; Sudan, Kenya, Nigeria. *Smeds* 1257 (K) pro parte (pro parte *E. longipedunculatum* A. Rich.).

The Ethiopian material is var. *sparsiflorum*: var. *sessilifolium* Hauman occurs in Uganda, NE Zaire, Central African Republic and N Nigeria.

10. **E. rhodesicum** *R. E. Fr.* (1914).

Erect herb with usually several stems from a woody rootstock, sometimes flowering when less than 10 cm tall and leafless. Leaves 1-foliolate; leaflet elliptic, 3.5-11 x 0.7-5.5 cm, subacute, cuneate at the base, pubescent above and on the nerves beneath; petiole 2-3 mm long. Racemes 1-3.5 cm long. Calyx pubescent, 4.5-5 mm long. Standard red-veined or flushed red outside, 6.5-10 mm long, pubescent. Pod elliptic-oblong, 10-13 x 9-10 mm, covered with long hairs and glands.

Grassland, appearing after fires; *c* 1800-1900 m. SD; widespread in tropical Africa. *Gillett* 14850; *Gilbert & Jefford* 4348; *Gilbert & Jones* 11.

11. **E. elliotii** *Bak. f.* (1895).

Perennial herb with several decumbent stems from a woody rootstock, eglandular; branches pilose with white spreading hairs. Leaflets 3, ± elliptic, 1-4 x 0.5-1.4 cm, acute to rounded and mucronate at the apex, appressed pilose on both sides or glabrous above; petiole *c* 2 mm long; rhachis short or obsolete. Racemes 1-3 cm long, 2-4-flowered. Calyx pilose, 5-8 mm long. Standard with reddish veins, *c* 9 mm long, pubescent. Pod oblong, 14-20 x 8-9 mm, pilose.

Grassland on red soil; *c* 1800 m. SD; Uganda, Kenya, Tanzania, Nigeria. *Gilbert & Jefford* 3440.

12. **E. shirense** *Bak. f.* (1894).

Perennial erect herb from a tuberous rootstock; young stems with spreading hairs up to 4 mm long. Leaflets 1-3, elliptic to oblong-lanceolate, 2.5-13 x 0.5-2.4 cm, acute, particularly on the margins and the nerves beneath with long spreading hairs; petiole 1-10 mm long. Racemes from upper axils, 1-7 cm long, pedunculate. Calyx 4.5-6 mm long, with long spreading hairs. Standard 7-11 mm long, hairy. Pod oblong-elliptic, 10-13 x 6.5-11 mm, with long rusty-red hairs.

Grassland or wooded grassland; *c* 1800-2000 m. SD; widespread in eastern tropical Africa and in Angola. *Thesiger* 2089; *J. De Wilde* 6076; *Thulin, Hunde & Tadesse* 3378.

Elsewhere the roots are eaten.

13. **E. verdickii** *De Wild.* (1903).

Erect perennial herb, 15-50(-125) cm tall, from a tuberous rootstock; stems ± robust with deflexed hairs. Leaves 1-foliolate; leaflet oblong to ovate-lanceolate, 3-20 x 0.6-4.5 cm, acute, cordate at the base, ± pubescent on both sides. Racemes 1.5-9 cm long; peduncle 1-3.5 cm long. Calyx hairy, 4-11 mm long. Standard yellow, 4-7.5 mm long, pubescent. Pod oblong, 7-10 x 5-7 mm, covered with long hairs, short hairs and scattered glands.

Wooded grassland; *c* 1300 m. WG; Cameroun, Zaire, Tanzania, Malawi, Zambia. *Gilbert & Thulin* 720.

Very close to *E. cordifolium* and mainly differing in its taller, more erect and robust stems, and yellow standard.

14. **E. cordifolium** *Hochst. ex A. Rich.* (1847)
types: TU, *Quartin Dillon* s.n. (P, syn.) & *Schimper* III: 1542 (P, syn., S, isosyn.).

Trailing or ascending perennial herb, 7-15(-30) cm, from a tuberous rootstock; stems slender with deflexed hairs. Leaves 1-foliolate; leaflet ovate to ovate-lanceolate, 1.5-10(-14) x 1-4.5(-7) cm, acute, cordate at the base, ± pubescent on both sides, margins ciliate. Racemes 1-2.5 cm long; peduncle 0.5-1.5 cm long. Calyx hairy, 3-4.5 mm long. Standard black or dark brown, 5-6 mm long, very sparsely pubescent or subglabrous. Pod oblong, 6-9 x 4.5-5.5 mm, covered with long hairs, short hairs and scattered glands. Fig. 95.34.1.

Grassland and woodland, roadsides; 1700-2500 m. TU, GD, WG, SU, KF, SD; Sudan, Nigeria, Uganda, Kenya. *W. de Wilde* 6906; *Ash* 1983; *Gilbert* 1997; *Mooney* 8932.

72. FLEMINGIA *Ait. f.* (1812), *nom. conserv.*

Herbs or subshrubs. Leaves digitately 3-foliolate; leaflets covered with small glands, particularly beneath. Inflorescences dense, fasciculate, sessile. Calyx 5-lobed; lobes longer than the tube. Corolla small. Vexillary stamen free; anthers uniform. Style filiform, thickened above; stigma very small. Pod oblong-ovoid, inflated, 2-seeded.

Some 30-50 species, mainly in Asia, two in tropical Africa.

F. grahamiana *Wight & Arn.* (1834).
Eriosema erythrocarpon Beck (1888) - type: HA, *Paulitschke* s.n. (?W, holo., K, ? iso.).
Moghania rhodocarpa (Bak.) Kuntze (1891).

Herb or subshrub; stems first hairy, later glabrescent. Leaflets 3, elliptic to obovate, 3-15 x 1.5-7.5 cm, obtuse to acuminate, pubescent to velvety. Calyx-tube 1.5 mm long; lobes linear, 3-6 mm long. Corolla greenish, yellowish-white or pink; standard 7-8 mm long. Pod 9-12 x 6 mm, yellowish pubescent and densely but irregularly covered with dark red globules of secretion which stain orange-red on touching. Fig. 95.34.2.

Grassland, woodland, rocky places; *c* 1400-2000 m. EW GD WG BA HA KF SD; widespread in tropical

Figure 95.34.1 - *ERIOSEMA CORDIFOLIUM*, part of complete plant with flowers, fruits, leaves and swollen tap root from *Mooney* 8932. 2 - *FLEMINGIA GRAHAMIANA*, a: flowering branch, b: fruiting raceme and subending leaf, both from *Tweedie* 2430. 3 - *ROTHIA HIRSUTA*, part of flowering and fruiting branch from *Gillett* 13180. 4 - *LOTONONIS LAXA*, part of flowering and fruiting branch from *Thulin* 1508. 1 & 2 x 0.75, 3 & 4 x 1.125. Drawn by Maureen Church. (Reproduced with permission from *Opera Botanica*, 68, fig. 36, 1983.)

Africa south to Natal and Transvaal, also extending to S India. *Chiovenda* 754; *Vàtova* 501; *Baldrati* 4079; *Mesfin & Kagnew* 2384.

Elsewhere a red dye which is used as a cosmetic is extracted from the seed.

77. **ROTHIA** *Pers.* (1807), *nom. conserv.*

Annuals. Leaves digitately 3-foliolate. Racemes terminal or leaf-opposed, 1-5-flowered. Calyx-lobes 5, subequal. Keel rounded at apex, not beaked. Stamens all joined into a tube split on the upper side; anthers 10, uniform. Pod many-seeded.

Genus of two species.

R. hirsuta *(Guill. & Perr.) Bak.* (1871).

Stems prostrate, up to 35 cm long, pilose. Leaflets oblanceolate, pilose, the terminal up to 2.5 cm long. Racemes 2-5-flowered, shortly pedunculate or subsessile. Calyx *c* 5 mm long, pilose. Corolla white or yellowish, often with reddish marks; standard up to 4 mm long; keel 3.5 mm long. Pod oblong, narrowing into an acute apex, ± straight, *c* 12 mm long, slightly inflated, deflexed, densely pilose. Fig. 95.34.3.

Disturbed ground, often on sandy soil; 700-2600 m. EW TU BA; widespread in tropical Africa. *Pappi* 1426, 7041; *Schimper* 2267.

78. **LOTONONIS** *(DC.) Eckl. & Zeyh.* (1836), *nom. conserv.*

Annual or perennial herbs. Leaves digitately 3-foliolate. Inflorescences terminal or leaf-opposed, racemose or subumbellate. Calyx-lobes 5, the upper 4 united higher up than the lower lobe. Keel rounded at the apex, not beaked. Stamens all joined into a tube split on the upper side; 4 long and 6 short anthers. Pod ± many-seeded.

Some 100 species, the vast majority in South Africa.

1. Annual, prostrate herb; stems and leaves pilose. **1. L. platycarpos**
- Perennial, erect or decumbent herb; stems and leaves sericeous with appressed, medifixed hairs. **2. L. laxa**

1. **L. platycarpos** *(Viv.) Pic.-Serm.* (1950);

Lotus platycarpos Viv. (1830); *Amphinomia platycarpa* (Viv.) Cuf. (1969).

Leobordea abyssinica Hochst. ex A. Rich. (1847); *Lotononis abyssinica* (Hochst. ex A. Rich.) Kotschy (1865); *Xerocarpus abyssinicus* (Hochst. ex A. Rich.) Schweinf. (1867); *Lotononis leobordea* Benth. var. *abyssinica* (Hochst. ex A. Rich.) Schweinf. in *Bull. Herb. Boiss., app.* **2**: 223 (1896); *L. platycarpos* var. *abyssinica* (Hochst. ex A. Rich.) Pic.-Serm. in *Webbia* **7**: 332 (1950) - type: TU, *Schimper* III: 1493 (P holo., BM K UPS iso.).

Lotononis leobordea Benth. var. *filicaulis* A. Terracc. in *Ann. Ist. Bot. Roma* **5**: 109 (1894) - type: EW, *Terracciano* 1742 (FT holo.).

Annual; stems prostrate, up to 35 cm long, pilose. Leaflets oblanceolate, pilose, up to 1(-2) cm long. Inflorescence subumbellate, 1-5-flowered, subsessile. Calyx *c* 5 mm long, pilose. Corolla yellow; standard up to 4 mm long; keel 5.5 mm long. Pod oblong, obtuse, straight, *c* 5 x 2 mm, slightly inflated, pilose towards the apex, *c* 10-seeded.

Dry grassland, rocky places, and on cultivated or disturbed ground; 0-2000(-2300) m. EE EW TU WU ?GD SU AR; widespread in Africa, also in Arabia and east to Pakistan. *Ouren* 20703; *Pappi* 1245; *Schimper* 2293; *Bally* 6850.

Most Ethiopian specimens can be referred to var. *abyssinica*, in which the calyx-lobes are all of about the same length. In var. *platycarpos* the lowest tooth is much shorter.

Elsewhere noted as good for grazing.

2. **L. laxa** *Eckl. & Zeyh.* (1836).

Perennial; stems erect or decumbent, up to 20 cm long, silky with appressed, medifixed hairs. Leaflets oblanceolate, somewhat silky, the terminal up to 2.5 cm long. Inflorescence subumbellate, 1-5-flowered; peduncle 1-15 mm long. Calyx *c* 7 mm long, ± silky. Corolla yellow; standard up to 8.5 mm long. Pod oblong, acute at the apex, ± straight, up to 20 x 4 mm, scarcely inflated, silky, *c* 20-seeded. Fig. 95.34.4.

Upland grassland and bushland, often in stony places; 2100-2200 m. AR HA SD; Uganda, Kenya, Tanzania, Malawi, South Africa. *Gillett* 14400; *Thulin* 1508; *Burger* 2025.

79. **CROTALARIA** *L.* (1753)

Polhill in *Kew Bull.* **22**: 169-348 (1968) & *Fl. Trop. E. Afr.* (1971) & *Crot. Afr. Madag.* (1982).

Herbs or shrubs. Leaves simple, 1-foliolate or digitately 3(-7)-foliolate. Inflorescences usually terminal or leaf-opposed racemes. Calyx-lobes (4-)5 or sometimes 3 by union of upper and lateral lobes. Corolla usually yellow; keel rounded or angled, usually with a prominent beak. Stamens all joined; 5 long and 5 short anthers. Pod usually inflated, dehiscent, 1-many-seeded.

Some 600 species, throughout the tropics and subtropics particularly of the southern hemisphere, but most numerous in tropical Africa, from where about 500 are recorded.

Crotalaria lunata Beddome ex Polhill is a native of S India which is grown as an ornamental in gardens in Addis Abeba. It is a shrub with villous simple leaves, large flowers, and glabrous fruits.

In many species the seeds become loose and rattle inside the dry pod when it is shaken. This is recognised in the vernacular name SHASHAHATA (Tign.) which is given to several species of *Crotalaria.*

Several species are noted as having pods and/or seeds which are toxic for domestic animals. However, the foliage of some species is said to be good for browse.

1. Leaves all or mostly 3(-6)-foliolate. 2
- Leaves all simple or 1-foliolate. 85
2. Beak of the keel not twisted or only shortly circumflexed. 3
- Beak of the keel spirally twisted through 180° or more. 62
3. Stipules resembling small leaflets, shortly stalked, with an expanded portion above. 4
- Stipules usually narrow or lacking, sometimes 2-5-partite. 7
4. Standard subcircular; wings broad, exceeding the keel; pod 28-45 mm long; shrubby. 5
- Standard ovate to ovate-elliptic; wings fractionally shorter than the keel; pod 15-30 mm long; herbs. 6
5. Pod glabrous; calyx 6-10 mm long, glabrous to shortly appressed pubescent; keel 10-14 mm long. **9. C. natalitia**
- Pod shaggy tomentose; calyx 11-18 mm long, spreading tomentose; keel (15-)20-24 mm long. **12. C. lachnophora**
6. Pod appressed pubescent; seeds smooth; racemes many-flowered or subdense at least above. **10. C. goreensis**
- Pod glabrous; seeds tuberculate; racemes laxly 2-5(-8)-flowered. **11. C. podocarpa**
7. Calyx becoming basally truncate and deflexed against the pedicel. 8
- Calyx not becoming basally truncate and deflexed. 26
8. Bracts and bracteoles falling before the flowers open. 9
- Bracts persistent. 10
9. Branches pubescent; keel sharply beaked, ± yellow; pod 38-50 mm long, puberulous. **28. C. pallida**
- Branches white tomentellous; keel bluntly beaked, ± purple; pod 16-22 mm long, tomentellous. **29. C. albicaulis**
10. Bracteoles short, not or scarcely exceeding the calyx-tube. 11
- Bracteoles longer than the calyx-tube. 20
11. Keel (14-)17-23 mm long, rounded below the middle, with a projecting beak; pod 40-70 mm long. 12
- Keel 3.5-14 mm long; pod 8-45 mm long. 13
12. Pod (10-)15-20 mm broad; standard usually pale yellow; calyx glabrous. **30. C. ochroleuca**
- Pod 4-7 mm broad; standard usually bright yellow; calyx glabrous to puberulous. **31. C. brevidens**
13. Perennial, with numerous decumbent to ascending stems from a woody rootstock. **39. C. cylindrica**
- Annual or short-lived perennial, with a single erect stem though sometimes much branched from near the base. 14
14. Keel with a very short straight blunt beak. **38. C. onobrychis**
- Keel with a distinctly formed, generally incurved, acute beak. 15
15. Keel 3.5-4 mm long; pod 10-15 mm long. **37. C. impressa**
- Keel 6.5-14 mm long; pod 15-45 mm long (if intermediate see also *C. plowdenii).* 16
16. Leaflets linear-oblanceolate to oblanceolate, 6-15 times as long as petiole, densely covered with short white hairs beneath; petiole 1-12 mm long; keel 10-12 mm long. **36. C. ruspoliana**
- Leaflets not as above or if almost so then keel 7-8 mm long. 17
17. Wings much longer than the keel. 18
- Wings shorter than the keel. 19
18. Stem, inflorescence axis, pedicels and calyx appressed pubescent; petioles 5-15 mm long; keel distinctly beaked. **32. C. polhillii**
- Stem, inflorescence axis, pedicels and calyx practically glabrous; petioles longer; keel with a very short beak. **33. C. sp. = *Benedetto* 122**
19. Calyx glabrous. **34. C. awasensis**
- Calyx pubescent. **35. C. fallax**
20. Stipules present. 21
- Stipules absent. 24
21. Keel 12-17 mm long; pod 33-60 mm long; leaves 3-5-foliolate. **40. C. cleomifolia**
- Keel 6-11 mm long; pod 15-27 mm long; leaves all 3-foliolate. 22
22. Keel shortly rounded, with a practically straight forwardly projecting beak; bracts often little longer than the pedicel. **41. C. vatkeana**
- Keel more strongly rounded, with an incurved beak; bracts usually much longer than the pedicel. 23
23. Pod 22-27 mm long; keel 7.5-11 mm long; standard elliptic, the blade longer than broad. **42. C. intonsa**
- Pod 15-20 mm long; keel 6-8 mm long; standard broadly elliptic to subcircular, the blade not or scarcely longer than broad. **43. C. gillettii**
24. Bracts 1-3 mm long; pod rather narrowly cylindrical, 4-5 mm across; keel 5-7(-8) mm long. **44. C. plowdenii**
- Bracts 3-12 mm long; pod 4-8 mm across. 25
25. Keel 10-12 mm long; pod (15-)18-24 mm long. **45. C. petitiana**
- Keel 6-8 mm long; pod 10-16 mm long. **46. C. comosa**
26. Calyx 0.8-1.2 times as long as the keel. 27
- Calyx not or little more than half as long as the keel. 38
27. Keel rounded about the middle. 28
- Keel angular or subangular. 34
28. Calyx-lobes ovate-cordate, 8-10 mm long. **4. C. quartiniana**
- Calyx-lobes triangular. 29

29. Standard 10-25 mm long; calyx 8-20 mm long. 30
- Standard and calyx 3-6 mm long. 32
30. Keel hairy in a band towards the lower margin; pod cylindrical, furry tomentose. **13. C. lachnocarpoides**
- Keel glabrous; pod oblong-clavate, puberulous to velvety tomentellous. 31
31. Keel 11-14 mm long; pod tomentellous, 24-36-seeded. **16. C. verdcourtii**
- Keel 6-8 mm long; pod shortly, sometimes densely, appressed pubescent, 16-20-seeded. **17. C. deflersii**
32. Intricately branched deciduous densely strigulose shrublet, with weakly spine-tipped branchlets and inflorescence rhachides; pod tadpole-shaped, ovoid or ovoid-ellipsoid with a long stylar beak, 2-4-seeded. **54. C. persica**
- Low herbs, branchlets and inflorescence rhachides not spine-tipped; pods various, 8-24-seeded. 33
33. Perennial; leaflets linear-lanceolate to obovate, 6-35 x 1-9 mm; wings much longer than the keel; pod flattened. **61. C. leptocarpa**
- Annual; leaflets (except lower leaves) linear, 10-25 x 0.3-1 mm; wings little longer than keel; pod inflated. **62. C. boudetii**
34. Bracteoles basally fused to the calyx-tube, the free part directed forwards, 4-12 mm long; pod enveloped by the calyx; racemes short, dense. **15. C. ononioides**
- Bracteoles linear to oblong-lanceolate, (2-)3-10 mm long, curved-ascending from the pedicel; pod much exceeding the calyx; racemes elongate. 35
35. Leaflets elliptic-obovate to subcircular, drying blackish; bracteoles linear, less than 0.5 mm wide; standard elliptic. **8. C. incana**
- Leaflets linear-lanceolate to elliptic, not drying blackish; bracteoles usually lanceolate and more than 1 mm wide; standard obovate to subcircular. 36
36. Flowers blue; pod broadly oblong-clavate, (35-)40-50 mm long. **5. C. polysperma**
- Flowers yellow or whitish; pod ellipsoid-clavate, 15-35 mm long. 37
37. Pod rather densely spreading hirsute; calyx-lobes hirsute with hairs 1-3.5 mm long; racemes 6-20-flowered. **6. C. rufocaulis**
- Pod glabrous or hairy on the upper side; calyx-lobes with long hairs on the margins only; racemes mostly 2-6-flowered. **7. C. barkae**
38. Keel 35-55 mm long; calyx 18-32 mm long, with upper and lateral lobes united on either side. 39
- Keel 4-30 mm long; calyx smaller, the lobes free at least in mature flowers. 40
39. Keel-beak incurved at the tip. **1. C. rosenii**
- Keel-beak short, projecting forwards. **2. C. agatiflora**
40. Keel rounded about the middle, with an upwardly directed truncate beak, 20-30 mm long; pod 45-150 mm long with a stipe 0.5-2 times as long as the seed-bearing part. **3. C. laburnifolia**
- Keel smaller or with a forwardly directed beak; pod much smaller or shortly stipitate. 41
41. Pod and ovary packed inside with long hairs; bracteoles (at base of calyx) triangular to ovate or falling soon; stipules absent. 42
- Pod and ovary glabrous or with a few scattered hairs inside; other characters various. 43
42. Branches ascending to form a bushy herb up to 70 cm tall; keel (8-)10-12 mm long; seeds rugulose. **21. C. saltiana**
- Branches long and procumbent from the base, with the stem up to only some 30 cm tall; keel 6-8(-9) mm long; seeds smooth (?always). **22. C. thomasii**
43. Keel (15-)17-23 mm long; pod 35-50 mm long. **18. C. comanestiana**
- Keel up to 13 mm long; pod smaller (unknown in *C. trifoliolata*). 44
44. Standard extensively pubescent outside; pod 4-7.5 mm long; keel angular, 4.5-7.5 mm long; racemes short, usually dense. **57. C. microcarpa**
- Standard glabrous outside, except sometimes along the midvein. 45
45. Leaves 4-6-foliolate; young pod soon glabrescent except along the sutures. **25. C. pleiophylla**
- Leaves 3-foliolate, uppermost and lowermost occasionally 1-foliolate; pod more persistently hairy. 46
46. Stipules absent. 47
- Stipules present, persistent, sometimes small. 48
47. Pod spreading pilose; bracts deflexed; calyx 5-6 mm long, pubescent with irregularly arranged hairs. **23. C. heterotricha**
- Pod appressed puberulous; bracts falling soon; calyx 4-5 mm long, with short appressed hairs. **24. C. somalensis**
48. Racemes subsessile, dense, with flowering pedicels no more than 2 mm long; branches ± ribbed. 49
- Racemes pedunculate, lax or if subdense when young pedicels longer; branches terete or subterete. 50
49. Perennial; calyx densely pubescent; standard white or almost so, pubescent along midvein outside. **26. C. jijigensis**
- Annual; calyx glabrous or almost so; standard yellow, glabrous. **27. C. pycnostachya**
50. Keel angular; style bent, constricted at the bend; bracteoles on the lower part of the pedicel; stems very slender, numerous, spreading or ascending, with small hairs. 51
- Keel rounded or if subangular style not bent and bracteoles on the calyx or just below; habit various. 52
51. Flowers 1-2 on a short slender axis; pod oblong-obovate to oblong-clavate, 12-15 mm long; calyx puberulous. **58. C. microphylla**
- Flowers 4-12 per raceme; pod subglobose or subglobose-ellipsoid, *c* 9 mm long; calyx subglabrous. **59. C. boranica**
52. Leaflets densely pilose on both surfaces. **19. C. trifoliolata**

- Leaflets glabrous to sparsely hairy above. 53
53. Shrublets, much branched, the younger branches densely white pubescent, the older ones with peeling epidermis; leaflets small or deciduous; pod ovoid to ellipsoid or oblong-obovoid, 2-8-seeded. 54
- Herbs, mostly with long procumbent to weakly ascending branches, without peeling epidermis; pod 8-40-seeded. 55
54. Inflorescence rhachides very slender, not pointed; leaves persistent; pod not beaked, hairs basifixed. **53. C. dumosa**
- Inflorescence rhachides ± rigid, pointed; leaves deciduous; pod prominently beaked; hairs slightly biramous. **54. C. persica**
55. Stipules 1-3-partite, the segments linear-lanceolate to lanceolate. **52. C. sacculata**
- Stipules all simple. 56
56. Flower-buds deflexed almost from inception; pod oblong-clavate, narrowed to the short asymmetrically inserted stipe. **20. C. bogdaniana**
- Flower-buds ascending, then spreading; pod globose to cylindrical, rounded to the short medially inserted stipe. 57
57. Racemes with *c* 4-8 flowers inserted 1-2 cm apart on the filiform rhachis, the peduncle basally less than 1 mm in diameter. **56. C. distantiflora**
- Racemes les lax and less slenderly supported. 58
58. Pod subglobose to oblong-ellipsoid, less than twice as long as broad, often mottled; hispid annual; terminal leaflets mostly (3-)4-12 times as long as the petiole. **55. C. steudneri**
- Pod more than twice as long as broad, not mottled; pubescent to pilose annuals or short-lived perennials; terminal leaflets mostly less than 2.5 times as long as the petiole. 59
59. Pod 20-28 mm long; keel 9-11 mm long, with a narrow incurved beak 4 mm long. 60
- Pod 8-18 mm long; keel 4-8 mm long, with a short straight to slightly incurved beak. 61
60. Pod glabrous to thinly pubescent; racemes lax. **48. C. phillipsiae**
- Pod rather densely pilose with spreading hairs; racemes with flowers ± crowded towards the top. **49. C. ukambensis**
61. Stipules 3-9(-16) mm long; pod 12-18 mm long; keel 6-8 mm long. **50. C. uguenensis**
- Stipules 1-2 mm long; pod 8-11 mm long; keel 4-6(-8) mm long. **51. C. massaiensis**
62. Inflorescence axes spine-tipped. **79. C. spinosa**
- Inflorescence axes not spine-tipped. 63
63. Keel 35-38 mm long, more than twice as long as broad; pod 80-90 mm long. **65. C. exaltata**
- Keel 3-20 mm long, strongly rounded to angled; pod 3-65 mm long. 64
64. Bracts and bracteoles lanceolate to ovate-cordate; calyx as long as the corolla, with triangular-lanceolate to ovate-cordate lobes. 65
- Bracts and bracteoles triangular to linear; calyx shorter than the corolla. 66
65. Pod 36-45 mm long; calyx 10-12 mm long, with ovate-cordate lobes. **4. C. quartiniana**
- Pod 13-25 mm long; calyx 4-5 mm long, with triangular-lanceolate lobes. **63. C. orixensis**
66. Flowers all in axillary clusters; keel rounded, 15-20 mm long. **72. C. axillaris**
- Flowers in racemes, sometimes augmented by axillary clusters but then flowers much smaller. 67
67. Keel 3-5 mm long, deeply rounded, with a small strongly circumflexed beak, shorter than the wings; style apically flattened and slightly bilobed; pod flattened to inflated. 68
- Keel angular or larger, with a neatly twisted beak; style terete, unlobed; pod inflated. 69
68. Perennial, leaflets 6-35 x 1-9 mm; wings much longer than the keel; pod flattened. **61. C. leptocarpa**
- Annual; leaflets mostly 10-25 x 0.3-1 mm; wings little longer than the keel; pod inflated. **62. C. boudetii**
69. Racemes strictly terminal, sometimes augmented by few-many flowers in the axils; stipules absent; pod 3-8 mm long. 70
- Racemes mostly becoming leaf-opposed; stipules usually present; pod larger. 74
70. Flowers in heads surrounded by crowded leaves larger than those below. **80. C. cephalotes**
- Flowers not in heads above an involucre of relatively large leaves. 71
71. Stems several-numerous from a woody root-stock; keel 9-13 mm long; pod 7-8 x 4.5-5 mm. **81. C. stenorhampha**
- Stem single from a taproot; keel and pod smaller. 72
72. Flowers developing in many axils down the branches, the terminal raceme short and few-flowered; pod 7-12-seeded. **84. C. alexandri**
- Flowers all or almost all in terminal racemes; pod 2-seeded. 73
73. Standard brownish tomentellous outside (hairs obscuring the surface); pod 5-6 mm long. **82. C. pseudotenuirama**
- Standard white pubescent outside (hairs not obscuring the surface); pod 4-5 mm long. **82. C. hyssopifolia**
74. Inflorescences of two sorts, the primary racemes supplemented by 1-few flowers on short axes (up to 2 cm, ending in a bristle) from leaf-clusters; pod ellipsoid-obovoid, 8-10 mm long, downy outside, thinly pilose inside. **78. C. oöcarpa**
- Inflorescence uniform, ending in a bud; pod glabrous inside. 75
75. Upper calyx-lobes obliquely oblong-triangular, ± abruptly contracted to a persistent excentric apiculate tip, lanate along the inner margin at maturity. **77. C. senegalensis**
- Upper calyx-lobes triangular. 76
76. Wings much exceeding the rounded keel. 77
- Wings subequalling or shorter than the keel. 79

77. Standard, except midvein, and usually the pod glabrous; stems hollow; petiole broadly grooved. **73. C. recta**
- Standard outside and pod hairy; stems solid; petiole terete. 78
78. Keel 11-14 mm long; pod 24-36-seeded. **16. C. verdcourtii**
- Keel 6-11 mm long; pod 16-20-seeded. **17. C. deflersii**
79. Erect laxly branched annual without leaf-clusters, puberulous; keel sharply angular, 6-13 mm long (usually small in Ethiopia). **75. C. deserticola**
- Shrubs or perennial herbs woody at the base, if herbaceous leaves in clusters on short shoots; keel rounded to subangular, 9-20 mm long. 80
80. Stipules absent; epidermis, except on new shoots, peeling to reveal greyish or purplish bark; racemes 2-12-flowered. 81
- Stipules almost always present at least on new growth; epidermis not readily peeling (sometimes cracked on old stems); flowers often more numerous. 82
81. Pod oblong-clavate, 20-25 mm long; leaflets semisucculent, the nervation obscure; keel 12-15 mm long; pubescence very fine, often rather sparse. **70. C. incrassifolia**
- Pod ovoid, 8-10 mm long; leaflets membranous, the nerves visible; keel 9-12 mm long; pubescence on new shoots white, very dense, obscuring the surface. **71. C. oligosperma**
82. Leaves all or almost all clearly alternate; hairs on leaflets beneath 0.1-0.2 mm long, appressed, uni-directional, scarcely overlapping; pod 25-60 mm long. 83
- Leaves mostly in tufts; hairs on leaflets mostly longer, irregularly or more densely arranged; pod 15-30 mm long. 84
83. Pod *c* 60-80-seeded, 50-60 x 18-25 mm; calyx-tube 5-6 mm long. **66. C. mildbraedii**
- Pod *c* 10-16-seeded, 25-40 x 7-12 mm; calyx-tube 2-4 mm long. **67. C. keniensis**
84. Calyx-lobes subequalling the tube; keel (12-)13-16 mm long, abruptly rounded below the middle; seeds smooth. **68. C. fascicularis**
- Calyx-lobes 1.5-2.5 times as long as the tube; keel 11-14 mm long, subangular; seeds rugulose. **69. C. emarginella**
85. Leaves 1-foliolate, jointed at the top of a well-formed petiole. 86
- Leaves simple, subsessile or with a short petiole and a single pulvinus. 87
86. Flowers in heads; keel rounded. **47. C. anthyllopsis**
- Flowers in racemes; keel angular. **60. C. bongensis**
87. Keel half-round, crested on the upper side behind the short circumflexed beak, 6-8 mm long, shortly exceeding the glabrous to sparsely puberulous calyx. **14. C. glauca**
- Keel not crested, the beak more definitely twisted. 88
88. Calyx as long as or nearly as long as the corolla, deeply divided, or at least slightly 2-lipped. 89
- Calyx much shorter than the corolla, without an upper lip. 91
89. Pod velvety tomentose; calyx brownish tomentellous (introduced). **85. C. juncea**
- Pod glabrous; calyx glabrous to pilose. 90
90. Calyx pilose; flowers few in terminal racemes and upper axils; stipules minute or lacking. **64. C. calycina**
- Calyx glabrous or nearly so; flowers in long racemes; stipules variable, often forming conspicuous wings down the stem (introduced). **86. C. paulina**
91. Keel angled in the lower third, longer than the wings. **76. C. karagwensis**
- Keel rounded about the middle, shorter than the wings. **74. C. retusa**

1. **C. rosenii** *(Pax) Milne-Redh.* ex *Polhill* (1968);
Erythrina rosenii Pax (1907) - type: SU, *Rosen* s.n. (WRSL holo.)
C. bieberi Cuf. (1947 & 1955) - type: KF, *Bieber* s.n. (W holo.).
C. raffillii Milne-Redh. (1950) - type: a plant grown at Kew from seeds collected in Ethiopia by Sheldon (K holo.).

Woody herb or shrub, up to 4 m tall; twigs glabrous. Leaflets 3, elliptic, up to 6(-7) x 3(-4) cm, glabrous above, subglabrous to finely pubescent beneath; petiole up to 7 cm long. Racemes many-flowered. Calyx up to 32 mm long, glabrous or finely pubescent, upper and lateral lobes on each side joined almost to the tips, the calyx thus 3-lobed. Standard ovate, up to 5 cm long, bright yellow, tinged and veined with purple; keel up to 55 mm long, with a long, upturned, untwisted, purplish beak. Pod 7-12 x 1.5-2 cm, gradually narrowed into a distinct stipe, glabrous, many-seeded. Fig. 95.35.1.

Forest margins, secondary scrub, stream-edges; 1350-2800 m. SU AR BA KF SD; not known from elsewhere. *Ash* 1042; *W. de Wilde* 8003, 10450; *Mooney* 6193.

Some collections from Mt Zuquala in SU and Bale Mts have the beak of the keel only slightly curved to almost straight, thus approaching *C. agatiflora*.

2. **C. agatiflora** *Schweinf.* (1892).

Woody herb or shrub, 1-10 m tall; twigs glabrous to tomentose. Leaflets 3, elliptic, up to 3.8-6.7 x 2-3.5 cm, glabrous to pubescent above, glabrous to tomentose beneath; petiole up to 12 cm long. Racemes many-flowered. Calyx up to 30 mm long, glabrous to pubescent, upper and lateral lobes on each side joined almost to the tips, the calyx thus 3-lobed. Standard ovate, yellow; keel 35-45(-55) mm long, with a relatively short, not upturned beak. Pod *c* 7.5-10 x 1.2-2 cm, gradually narrowed into a distinct stipe, glabrous, many-seeded.

1. Bracts ovate-elliptic, (6-)8-20 mm wide; bracteoles up to 6-12 mm long; leaflets elliptic-lanceolate, more than twice as long as broad. subsp. **erlangeri**

- Bracts subulate to narrowly lanceolate, 0.7-6 mm wide; bracteoles less than 5 mm long; leaflets almost always ovate-elliptic, less than twice as long as broad. subsp. **imperialis**

Subsp. **erlangeri** *Bak. f.* in *Journ. Linn. Soc., Bot.* **42**: 316 (1914)
- type: HA, Mt Gara Mullata, Ellenbeck 542 (B holo. destr.).

Forest margins, heath scrub; 2500-3300 m. SU BA HA GG; not known elsewhere. *Gillett* 5308; *Burger* 1466; *Thulin, Hunde & Tadesse* 3693.

Specimens from BU and GG are sometimes atypical, verging towards subsp. *imperialis,* subsp. *engleri* (Harms ex Bak. f.) Polhill or *C. rosenii.*

Subsp. **imperialis** *(Taub.) Polhill* in *Kew Bull.* **22**: 204 (1968).

Forest margins, wooded grassland, evergreen bushland, roadsides; *c* 1800 m. SD; Uganda, Kenya, Tanzania, E Zaire. *Gilbert* 1349.

3. **C. laburnifolia** *L.* (1753).

Bushy herb, up to 0.6-2 m tall; stems glabrous to thinly puberulous. Leaflets 3, narrowly oblong to elliptic or elliptic-obovate, up to 1-10.5 x 0.4-5.5 cm, glabrous or puberulous on the lower or both surfaces; petiole up to (1-)3-7.5(-20) cm long. Racemes lax, few- to many-flowered. Calyx 12-17 mm long, glabrous to sparsely puberulous, 5-lobed in mature flowers. Standard ovate, yellow, often marked red-brown; keel strongly rounded, (19-)22-30 mm long, with a long straight upwardly directed, untwisted beak. Pod 4.5-7.5(-15) x 1-1.3 cm, narrowed abruptly into a long stipe, glabrous, often mottled reddish-brown, 16-42-seeded.

1. Pod 4.5-7.5 x 0.9-1.3 cm, abruptly contracted into the 1.7-3 cm long stipe; ovules *c* 16-32. subsp. **laburnifolia**
- Pod (8-)9.5-15 cm long, narrowed into the 2.5-9.5 cm long stipe; ovules *c* 32-42. subsp. **tenuicarpa**

Subsp. **laburnifolia.** Fig. 95.35.2.

Deciduous woodland, bushland, grassland, and on roadsides and cultivated ground; 1300-1700 m. SU AR BA HA GG SD; widespread in eastern Africa, also from India to NE Australia. *Thulin* 1489; *Mooney* 7979; *Gilbert* 1080.

Subsp. **tenuicarpa** *Polhill* in *Kew Bull.* **22**: 212 (1968).

Deciduous bushland; *c* 1300 m. SD; S Somalia, Kenya, NE Tanzania. *Cufodontis* 713; *Friis* et al. 2817.

Three more subspecies occur in Africa and one of these, subsp. *petiolaris* (Franch.) Polhill may occur also in Ethiopia. It has 6.5-20 cm long petioles and is known from N Somalia and NE Kenya.

4. **C. quartiniana** *A. Rich.* (1847) - type: TU, *Quartin Dillon* s.n. (P holo.).
Phyllocalyx quartinianus A. Rich. (1847); *Crotalaria platycalyx* Steud. ex Bak. (1871) - types: GD, Simien, *Schimper* II:860 (P syn., FT K MO isosyn.); TU, *Quartin Dillon & Petit* s.n. (P syn.).
C. azaïsii Sacl. (1936) - type: Gamogi, *Azaïs* s.n. (P, holo.).

Erect or straggling annual, 0.3-1.5 m tall; stems pilose. Leaflets 3, elliptic to narrowly ovate, up to 3-7 x 1.2-3.5 cm, sparsely pilose on both surfaces; petiole up to 2-8 cm long. Racemes very lax, *c* 3-10-flowered; bracteoles large, cordate-ovate. Calyx ± enveloping the corolla, 10-12 mm long, ciliate on the lobes; upper lobes broadly ovate, cordate. Standard suborbicular, yellow, often with a dark mark at the base inside; keel strongly rounded about the middle, 8-10 mm long, with a slightly incurved shortly twisted beak. Pod 36-45 x 9-12 mm, narrowed into a distinct stipe, glabrous. Fig. 95.35.3.

Forest-margins, upland bushland; 1200-2660 m. EW TU GD GJ SU AR BA HA IL KF GG SD; scattered in tropical Africa, Arabia. *Thulin* 1453; *Gilbert & Thulin* 940; *Friis* et al. 1747.

5. **C. polysperma** *Kotschy* (1868).

Annual, up to 0.3-1 m tall; stems densely pilose. Leaflets 3, narrowly to broadly elliptic, up to 3.5-5.5 x 2-3.6 cm, pilose at least beneath; petiole up to 4-7 cm long. Racemes lax. Calyx accrescent, 14-20(-24) mm long, pubescent; upper lobes lanceolate. Standard suborbicular, blue, pubescent outside; keel bent at right-angles in the lower half, 16-18 mm long, with a distinct straight, untwisted beak. Pod (35-)40-45 x 14-20 mm, shortly stipitate, pilose, *c* 70-100-seeded.

Deciduous woodland and grassland, also on roadsides and cultivated ground; *c* 1000-2000(-2400) m. EW SU HA GG SD; Sudan, Uganda, Kenya, Tanzania, Malawi, Zimbabwe, Mozambique. *Gilbert, Thulin & Getachew* 346; *Ash* 168; *Pappi* 5596; *Mooney* 8048.

6. **C. rufocaulis** *Gilli* (1970).

Spreading, much-branched annual or short-lived perennial, up to 1 m tall; stem hirsute with long brownish hairs. Leaflets 3, elliptic, 10-40 x 7-20 mm, thinly pilose on both surfaces, sometimes long-aristate; petioles longer than leaflets. Racemes lax, 6-20-flowered. Standard suborbicular, greenish-yellow or whitish, densely pilose outside; keel bent at right-angles in the lower half, 13-15 mm long, with a slightly incurved untwisted beak. Pod 25-30 x 11-13 mm, shortly stipitate, rather densely covered with long spreading, reddish hairs, *c* 30-seeded.

Secondary bushland and grassland; *c* 1000-1620 m. SD; N Somalia, N Kenya. *Donaldson Smith* s.n.; *Gilbert* 3398; *Mesfin* et al. 3478.

7. **C. barkae** *Schweinf.* (1896)
- type: Eritrea, *Schweinfurth* 277 (B holo. destr.).

Annual or short-lived perennial, up to 0.5(-1) m tall; stems pilose. Leaflets 3, lanceolate to elliptic, up to 15-

70 x 5-20 mm, sparsely pilose at least beneath; petiole up to 15-55 mm long. Calyx accrescent, (9-)10-20 mm long; upper lobes attenuate-lanceolate to ovate-lanceolate. Standard longer than calyx, yellow to white, sometimes with purplish marks, pubescent medially and apically outside; keel bent at right-angles in the lower half, (9-)10-13(-14) mm long, with a long slightly incurved untwisted beak. Pod (15-)18-26 x 8-16 mm, shortly stipitate, hairy on the upper side or glabrous, up to 30-80-seeded.

Deciduous bushland and grassland, and on roadsides and cultivated ground; 900-2600 m. EW TU GD; widespread in the drier parts of Africa. *Pappi* 2676, 2085; *Schimper* 442.

The Ethiopian material is subsp. *barkae.* Three more subspecies occur in East Africa.

8. C. **incana** *L.* (1753).

Annual or short-lived perennial, up to 1.5 m tall; stems with spreading hairs pilose or shortly pubescent. Leaflets 3, narrowly obovate to suborbicular, up to 25-50 x 17-44 mm, glabrous above, subglabrous to thinly pilose beneath; petiole up to 30-80 mm long. Racemes lax. Calyx deflexed against the pedicel, 7-11 mm long. Standard elliptic, with yellow and purplish veins; keel bent at right-angles in the lower half, 8-11.5 mm long, with a straight untwisted beak. Pod 30-45 x 8-12 mm, sessile, spreading pilose, *c* 40-50-seeded.

1. Stem shortly pubescent; calyx subglabrous. subsp. **incana**
- Stem with long usually yellow-brown spreading hairs; calyx-lobes spreading-pilose. subsp. **purpurascens**

Subsp. **incana.**

Bushland, grassland, and on disturbed ground; 1300-1900 m. EW TU WU SU AR HA KF GG SD; native in tropical America and adventive in other parts of the tropics. *Thulin* 1312; *Ash* 916; *Gilbert & Tewolde* 2484.

Subsp. **purpurascens** *(Lam.) Milne-Redh.* in *Kew Bull.* **15**: 159 (1961);
C. purpurascens Lam. (1786).
C. montana A. Rich. (1847), nom. illegit. - type: TU, *Quartin Dillon* s.n. (P, holo.).

Bushland, grassland, river banks, and on disturbed or cultivated ground; 1200-2700 m. EW TU GD WU WG SU AR BA HA IL KF SD; widespread in tropical Africa, and in Madagascar. *Thulin* 1452; *Gilbert & Thulin* 36, 569; *Pappi* 4096.

9. C. **natalitia** *Meissn.* (1843).

Woody herb or shrub, 1-2.6 m tall; branches densely ± appressed pubescent. Leaflets 3, narrowly oblanceolate to oblanceolate-elliptic, up to 20-45 x 4-15 mm, glabrous above, appressed pubescent beneath; petiole up to 10-30 mm long. Stipules up to 5-15 x 1-4 mm. Racemes few- to many-flowered. Calyx 6-9(-10) mm long. Standard yellow, sometimes tinged red-brown outside; keel abruptly rounded about the middle, 11-14 mm long, with a slightly incurved untwisted beak. Pod 35-40(-45) x 11-13 mm, abruptly contracted into the short stipe, glabrous, often mottled, 25-40-seeded.

1. Calyx and upper part of pedicel glabrous. var. **natalitia**
- Calyx and upper part of pedicel appressed pubescent. var. **rutshuruensis**

Var. **natalitia.**

Woodland, grassland, forest margins, also on roadsides and cultivated ground. HA; widespread in east tropical Africa south to Natal. *IECAMA* F-82.

Var. **rutshuruensis** *De Wild.* in *Bull. Jard. Bot. Etat* **8**: 137 (1923).

Habitat as for var. *natalitia;* 1800-2400 m. EW SU BA HA KF SD; widespread in east tropical Africa, Angola and in Yemen. *Gillett* 14568; *Burger* 2945; *Friis* et al. 1595; *W. de Wilde* 7572.

10. C. **goreensis** *Guill. & Perr.* (1832).
C. macrostipula Steud. ex A. Rich. (1847); *C. goreensis* subsp. *macrostipula* (Steud. ex A. Rich.) Bak. f. in *Journ. Linn. Soc., Bot.* **42**: 413 (1914) - types: TU, *Schimper* II:708 (P syn., BM FT K MO isosyn.); *Quartin Dillon & Petit* s.n. (P syn., K isosyn.).

Annual or short-lived perennial, up to 0.5-2.6 m tall; stem densely pubescent with ± appressed hairs. Leaflets 3, narrowly oblanceolate to obovate, up to 35-85 x 8-25 mm, sparsely appressed pubescent at least beneath; petiole up to 35-80 mm long. Stipules 9-25 x 3-7 mm, oblong-falcate. Racemes densely many-flowered. Calyx 4-5 mm long. Standard yellow, veins darker; keel bent at right-angles in the lower half, 8.5-10(-11.5) mm long, with a distinct, straight, untwisted beak. Pod 15-20 x 7-9 mm, subsessile, thinly pubescent, often mottled, 12-16-seeded; seeds smooth.

Deciduous woodland and grassland, also on roadsides and disturbed or cultivated ground; 500-1300 m. TU WG SU-KF IL; widespread in tropical Africa. *Gilbert & Thulin* 654; *Eriksson* 150; *Schimper* 439.

11. C. **podocarpa** *DC.* (1825).

Erect annual up to 0.7 m tall; stems pilose. Leaflets 3, linear to oblong or narrowly obovate, up to 18-50(-78) x 3-16(-24) mm, thinly hairy at least beneath; petiole 20-45 mm long. Stipules 5-17 x 3-7 mm. Racemes lax, 2-5(-8)-flowered. Calyx 7-9 mm long. Standard yellow, glabrous outside; keel bent at right-angles in the lower half, 12-17 mm long, with a long almost straight untwisted beak. Pod 20-30 x 9-14 mm, shortly stipitate, glabrous; seeds tuberculate.

Deciduous woodland and bushland, also on road-sides and cultivated ground; 800-1800 m. EW TU;

widespread in tropical and southern Africa. *Robertson* 1215; *Pappi* 6233; *Schimper* 85.

12. **C. lachnophora** *Hochst.* ex *A. Rich.* (1847); *C. lachnocarpa* Hochst. ex Bak. (1871), nom. illegit. - types: TU, Schimper II:780 (P syn., BM K L MO UPS isosyn.) & several localities, *Quartin Dillon* s.n. (P syn.).

Bushy herb or shrub, 1-3 m tall; branches brownish tomentose. Leaflets 3, oblanceolate to obovate, up to 30-75 x 8-35 mm, sparsely pubescent or glabrous above, appressed pubescent beneath; petiole 20-35 mm long. Stipules 9-25 x 4-8 mm. Racemes lax. Calyx 11-15 (-18) mm long, tomentose. Standard yellow, glabrous outside; keel abruptly rounded in lower half, (18-)21-24(-26) mm long, with a rather short, blunt, slightly incurved, untwisted beak. Pod 28-45 x 15 mm, subsessile, tomentose, *c* 16-18-seeded. Fig. 95.35.4.

Grassland, woodland, also on roadsides and disturbed or cultivated ground; 900-2200 m. EW TU GD GJ WG SU AR HA SD GG; widespread in tropical Africa. *Gilbert & Thulin* 757; *Mooney* 8164; *Meyer* 7479; *Tewolde & Gelahun* 1571.

13. **C. lachnocarpoides** *Engl.* (1892) - type: GD, near Debra Tabor, *Schimper* s.n. (B holo. destr., BM EA iso.).

Woody herb or shrub, up to 0.5-2(-2.7) m tall; branches tomentose. Leaflets 3, oblanceolate to narrowly obovate, up to 35-70 x 12-25 mm, densely pilose on both surfaces; petiole up to 13 mm long. Stipules 5-8 mm long, filiform. Racemes lax or dense. Calyx 10-15 mm long, tomentose. Standard yellow, usually brownish tinged and ± pubescent outside; keel semiorbicular, 11-15 mm long, hairy on the sides towards the lower margin, with a short untwisted beak. Pod up to 25-30 x 12-13 mm, subsessile, densely long tomentose, (12-)20-26-seeded.

Grassland, bushland, forest margins and clearings; 1600-2900 m. GD GJ WG SU AR BA HA IL KF GG SD; widespread in eastern tropical Africa. *Thulin* 1389; *Gilbert, Thulin & Getachew* 570; *Mooney* 5959; *W. de Wilde* 8395.

14. **C. glauca** *Willd.* (1802).

Annual, up to 1.2 m tall, glabrous. Leaves simple, subsessile, linear to narrowly lanceolate, up to 10-85 x 2-10 mm, glabrous. Racemes lax, 3-12-flowered. Calyx 4-6(-7) mm long, glabrous to sparsely puberulous. Standard yellow, with brownish veins outside; keel semiorbicular, 6-8 mm long, crested behind the small circumflexed blunt beak. Pod cylindrical, *c* 16-32 mm long, glabrous, 12-32-seeded. Fig. 95.36.1.

Grassland, deciduous woodland; 1200-2000 m. TU GD GJ WG SU HA KF GG SD; widespread in tropical Africa. *Gilbert & Thulin* 534; *Ash* 2158; *Burger* 2154.

15. **C. ononoides** *Benth.* (1834).

Spreading annual, up to 0.5(-1) m tall; stem pilose. Leaflets 3, lanceolate to obovate, up to 30-50(-60) x 8-12(-24) mm, pilose at least beneath; petiole 4-9 mm long. Flowers numerous in sessile heads; bracts fused to the pedicel, 3-fid, with linear spreading segments up to 10-13 mm long. Calyx as long as the corolla, 7-10 (-12) mm long, pilose. Standard yellow; keel 7-9 mm long. Pod ellipsoid, 9-12 mm long, sessile, with a few hairs at the apex, 8-14-seeded. Fig. 95.36.2.

Grassland, woodland, also on roadsides and disturbed ground; 1200-1900 m. WG SU BA KF GG SD BA; widespread in tropical Africa and Madagascar. *Gilbert & Thulin* 557, 715; *W. de Wilde* 8811; *Friis* et al. 3477.

16. **C. verdcourtii** *Polhill* (1968).

Bushy herb, up to 0.6(-1.4) m tall; stem densely white pubescent. Leaflets 3, elliptic or obovate, up to 30-66 x 12-36 mm, densely pubescent on both surfaces when young; petiole 40-80 mm long. Racemes long and many-flowered. Calyx 10-14 mm long, densely pubescent; lobes much longer than tube, reflexed. Standard yellow with wine-red centre, pubescent outside; keel strongly rounded, 11-14 mm long, with a rather short incurved beak much shorter than wings. Pod (35-)43-55 x 11-12 mm, tomentellous, 24-36-seeded.

Grassland, woodland at low altitudes. GG; Sudan, Kenya. *Corradi* 3931, 4515-4518, 4566.

17. **C. deflersii** *Schweinf.* (1896).

Bushy perennial, up to 1 m tall; stem densely and shortly white pubescent. Leaflets 3, broadly elliptic or obovate, up to 40-60 x 18-45 mm, usually densely pubescent on both surfaces when young; petiole up to 45-90 mm long. Racemes long and many-flowered. Calyx 8-10 mm long, densely pubescent; lobes much longer than tube, reflexed. Standard yellow with red centre, pubescent outside; keel strongly rounded, 6-8 mm long, with a very short incurved beak much shorter than wings. Pod up to 35-45 x 9-11 mm, puberulous, *c* 16-20-seeded.

Deciduous bushland and grassland; up to 1000 m. AF SD; Djibouti, N Somalia, Kenya, N Tanzania and in Arabia. *Mooney* 9656; *Godding* 205; *Ash* 3327.

18. **C. comanestiana** *Volkens & Schweinf.* (1897) - type: Ogaden (HA-BA), *Ghika-Çomanesti* s.n. (B holo. destr.).

Much-branched perennial, up to 1.3 m tall; branches densely white appressed tomentellous. Leaflets 3(-4), narrowly to broadly elliptic or obovate, up to 38-95 x 27-60 mm, glabrous above, pubescent beneath; petiole *c* 30-70 mm long. Racemes fairly long and many-flowered. Calyx 5-7(-8) mm long, appressed pubescent; lobes usually a little shorter than the tube. Standard yellow with purplish veins, glabrous or with a few hairs

Figure 95.35.1 - *CROTALARIA ROSENII*, a: flowering branch from *Mooney* 7896, b: pod from *W. de Wilde* 6376. 2 - *C. LABURNIFOLIA* subsp. *LABURNIFOLIA*, a: flowering branch from *Thulin* 1489, b: pod from *Ash* 344. 3 - *C. QUARTINIANA*, a: flowering branch from *Gilbert, Jefford & Lundström* 1694, b: pods from *Thulin* 1453. 4 - *C. LACHNOPHORA*, a: flowering branch, b: pod, both from *Gilbert & Thulin* 757. All x 0.75. Drawn by Maureen Church. (Reproduced with permission from *Opera Botanica*, 68, fig. 37, 1983.)

medially outside; keel rounded below the middle, (15-) 17-23 mm long, with a well-developed straight beak. Pod 35-50 x 10-12 mm, appressed pubescent, *c* 20-26-seeded.

Deciduous bushland, stony places, flood plains; 900-1450 m. AF EW SU HA BA GG SD; Somalia, Kenya, Tanzania. *Sebald* 2833; *Burger* 2735; *Gilbert* 1111; *Friis* et al. 3204.

19. **C. trifoliolata** *Bak. f.* (1896)
- type: BA, Walenso, *Donaldson Smith* 213 (BM, holo.).

Fairly robust herb; branches with a rather persistent felted tomentum. Leaflets 3, broadly elliptic-obovate, up to 4 x 3 cm, pilose on both surfaces; petioles about as long as the leaflets. Racemes pedunculate, with ± numerous flowers crowded above, lax below. Calyx *c* 6 mm long, coarsely tomentose; lobes ± twice as long as the tube. Standard glabrous outside; keel abruptly rounded a little below the middle, 10-11 mm long, the beak projecting and only slightly incurved at the tip. Pod unknown.

Probably dry bushland at *c* 1400 m. BA; not known elsewhere.

C. trifoliolata is only known from the type collected on 26 Oct. 1894. According to the maps of Donaldson Smith's expedition in *Geogr. Journ.* **8** (1896) the type locality is at *c* 7°N, 40°50´E in BA, just SE of Ginir.

20. **C. bogdaniana** *Polhill* (1968).

Annual or perennial with the lower branches usually decumbent, up to 0.1-0.5 m long; stems ± appressed pubescent. Leaflets 3, usually elliptic, up to 23-45 x 12-22 mm, glabrous above, appressed pubescent beneath; petiole 7-10 mm long. Stipules linear 2-6 mm long. Racemes shortly pedunculate, ± dense, many-flowered; buds deflexed almost from inception; bracteoles inserted at the base of the calyx, up to 1 mm long. Calyx 2.5-3 mm long, appressed pubescent; lobes ± as long as the tube. Standard pale yellow, sometimes with dark veins, glabrous outside; keel rounded about the middle, (6-)7-9 mm long, with a short untwisted beak. Pod up to 16-20 x 5-6 mm, appressed pubescent, *c* 8-16-seeded.

Grassland and bushland, often on limestone; 1250-1600 m. HA BA SD; Somalia N Uganda, Kenya, N Tanzania. *Burger* 2201; *Rippstein* 412; *Thulin, Hunde & Tadesse* 3570; *Friis* et al. 3337.

21. **C. saltiana** *Andr.* (1812)
- type: plate 648 in Andrews, *Botanical Repository*, of a plant grown from seeds collected by Salt in NE Africa.
C. farcta R. Br. ex Bak. f. (1914) - type: *Salt* s.n. (BM holo.).

Bushy herb, up to 0.7 m tall; stem densely white ± appressed pubescent. Leaflets 3, narrowly elliptic to obovate, up to 45 x 10-25 mm, glabrous to thinly pubescent above, appressed pubescent beneath; petiole *c* 20-40 mm long. Racemes long and many-flowered. Calyx 4.5-6 mm long, pubescent; lobes longer than tube, becoming reflexed. Standard yellow, glabrous or puberulous along the midvein outside; keel rounded about the middle, (8-)10-12 mm long, with an incurved untwisted beak. Pod up to 18-20 x 4.5-5.5 mm, densely and shortly pubescent outside, densely packed with long hairs inside, (2-)8-12-seeded.

Deciduous bushland and semi-desert scrub; up to 1300 m. AF HA BA GG SD; Sudan, Somalia, N Kenya and Arabia. *Burger* 3767; *Gilbert* 2381; *IECAMA* A-67; *Sue Edwards* 704.

Some collections from Ogaden deviate by having fewer (2-8) ovules than normal.

22. **C. thomasii** *Harms* (1899).

Herb up to 0.3 m tall, with numerous procumbent branches from near the base; stem finely white appressed pubescent. Leaflets 3, obovate-elliptic to obovate, up to 25 x 14-18 mm, usually glabrous above, thinly appressed puberulous beneath; petiole up to 15-25 mm long. Calyx 4-4.5 mm long, appressed puberulous; lobes ± as long as the tube. Standard yellow, glabrous or puberulous along the midvein outside; keel rounded about the middle, 6-8(-9) mm long, with a short incurved untwisted beak. Pod up to 14-16 x 4.5-6 mm, densely and finely pubescent outside, densely packed with long hairs inside, *c* 12-14-seeded.

Seasonally waterlogged ground, dry stream beds. GG SD; S Somalia, E Kenya, Tanzania. *Corradi* 4061, 4063.

23. **C. heterotricha** *Polhill* (1982)
- type: Ogaden (HA-BA), Welwel to Sirrouw [Walwal to Sirauw], *Glover & Gilliland* 433 (K holo., BM EA iso.).

Herb; branches pubescent with fine white appressed hairs. Leaflets 3, lanceolate, 45-60 x 8-12 mm, glabrous above, finely appressed pubescent beneath; petiole shorter than the leaflets. Racemes pedunculate, many-flowered. Calyx 5-6 mm long, pubescent; lobes not much longer than the tube. Standard yellow, glabrous except sometimes along midvein outside; keel rounded about the middle, 11-13 mm long, with a narrow attenuate incurved beak. Pod with the narrow 3-4 mm long stipe up to 15 mm long, densely covered outside with long fine spreading hairs, glabrous inside, *c* 8-seeded.

Sandy open places; *c* 500 m. HA; only known from the type.

24. **C. somalensis** *Chiov.* (1916).

Erect or spreading herb, up to 1 m tall; branches shortly appressed pubescent, glabrescent. Leaflets 3, oblanceolate-elliptic to obovate, up to 25-40 x 10-25 mm, usually appressed puberulous on both surfaces; petiole 16-34(-40) mm long. Racemes long, many-flowered, rather lax. Calyx 4-5 mm long, subglabrous to appressed puberulous. Standard yellow, glabrous outside; keel rounded about the middle, 11-12 mm long, with a slightly incurved untwisted beak, ciliate along upper margin. Pod 12-24 x 4.5-6 mm, appressed puberulous, *c* 6-8-seeded.

Subsp. **fusula** *Polhill* in *Kew Bull*. **22**: 256 (1968)
- type: Ogaden (HA-BA), between Welwel and Sirrouw [Walwal and Sirauw], *Glover & Gilliland* 431 (K holo., BM EA iso.).

Pod oblong-fusiform, *c* 22-24 mm long.

Deciduous bushland, sandy soil; 450-600 m. HA; S Somalia, (subsp. *somalensis* occurs in S Somalia and adjacent parts of Kenya). *Bally* 9612; *Simmons* 82.

25. **C. pleiophylla** *Polhill* (1982)
- type: Ogaden (HA), Warder wells, *Glover & Gilliland* 288 (K holo., BM EA iso.).

Herb; branchlets early glabrescent. Leaflets 4-6, oblanceolate to elliptic-obovate, 15-30 x 4-12 mm, sparsely appressed puberulous beneath; petiole ± as long as the leaflets. Racemes 10-40-flowered, rather lax. Calyx 4-4.5 mm long; lobes 1.5 times as long as the tube. Standard yellow, glabrous outside; keel strongly rounded, 7-9 mm long, with a fairly short narrow slightly incurved beak. Pod 17-20 x 3.5-5 mm, glabrescent except along the sutures from an early stage, 14-20-seeded.

On sand; up to 600 m. HA; S Somalia. Known from Ethiopia only by the type.

26. **C. jijigensis** *Thulin* (1982)
- type: HA, near Jijiga, *J. De Wilde* 6407 (WAG holo.).

Perennial, *c* 8-25 cm tall; stems ± ribbed, densely white pubescent with somewhat spreading hairs. Leaflets 3, linear-oblanceolate, up to 15-37 x 3-6 mm, glabrous above, densely pubescent with subappressed or ± spreading hairs beneath; petioles much shorter than the leaflets, up to 10 mm long. Racemes subsessile, dense, many-flowered. Calyx 5.5-6.5 mm long, densely pubescent. Standard white or very pale yellow, sometimes with purplish veins, sparsely pubescent along the midvein outside, at least near the tip; wings white with yellow tips; keel white, subangular, 10-12 mm long, with a long narrow untwisted beak. Pod *c* 12 x 4.5 mm, densely pubescent, *c* 12-seeded.

Rocky or sandy plains with scattered shrub-vegetation on limestone; 1750-1875 m. HA; not known elsewhere. *Jansen* 6059, 6717; *Ash* 1199; *Sandford* in *Mooney* 7960.

27. **C. pycnostachya** *Benth*. (1843).
C. melilotoides Steud. ex A. Rich. (1847) - types: TU, *Quartin Dillon* s.n. (P syn.) & *Schimper* II:1213 (P syn., FT K MO isosyn.).

Much-branched annual, up to 15-130 cm tall; stem ribbed, shortly appressed pubescent. Leaflets 3, oblanceolate, up to 20-50 x 5-16 mm, glabrous above, thinly appressed puberulous beneath; petioles shorter than the leaflets. Racemes subsessile, short, dense, many-flowered. Calyx 1.5-6 mm long, glabrous or almost so. Standard pale to bright yellow, glabrous outside; keel subangular to abruptly rounded, 4-10 mm long, with an untwisted beak. Pod 10-15(-20) x 3-5 mm, appressed puberulous, *c* 8-20-seeded.

1. Calyx 1.5-2 mm long; standard pale yellow, except for a darker basal patch; keel 4-5 mm long, shortly beaked. subsp. **pycnostachya**
- Calyx 3-4(-6) mm long; standard bright yellow; keel 7-10 mm long, with a long narrow beak. subsp. **donaldsonii**

Subsp. **pycnostachya**. Fig. 95.36.3.

Grassland, bushland, stony hillsides and on cultivated or waste ground; (460-)1200-1900(-2600) m. AF EW TU WU SU AR HA BA GG SD; Sudan, Uganda, Kenya, Tanzania, N Somalia and in Arabia. *Pappi* 368; *Thulin* 1289; *Gilbert & Thulin* 267.

Subsp. **donaldsonii** (*Bak. f.*) *Polhill* in *Crot. Afr. Madag*.: **179** (1982);
C. pycnostachya var. *donaldsonii* Bak. f. in *Journ. Linn. Soc., Bot*. **42**: 373 (1914) - type: S Somalia or Ethiopia, Webi Shebeli river[R. Scebeli], *Donaldson Smith* (BM holo.).

Up to 1500 m. BA HA SD; S Somalia, NE Kenya. *Burger* 3346B; *Ellis* 386; *Ruspoli & Riva* 789.

Subsp. *tropeae* (Mattei) Polhill occurs in S Somalia and Kenya.

28. **C. pallida** *Ait*. (1789)
- type: plant cultivated at Kew, from seeds collected by Bruce in Ethiopia (BM holo.).
C. mucronata Desv. (1814).
C. fertilis Del. (1848) - type: *Ferret & Galinier* s.n. (?MPU holo.).

Erect annual or short-lived perennial, up to 1-2(-3) m tall; stem shortly appressed pubescent. Leaflets 3(-4), elliptic to obovate, up to 40-130 x 24-70 mm, glabrous above, appressed puberulous beneath; petiole *c* 20-85 mm long. Racemes up to 15-30 cm long, many-flowered; bracts falling soon. Calyx becoming basally truncate and deflexed against the pedicel, 6-7.5 mm long, appressed pubescent. Standard yellow, often with brownish veins, glabrous outside; keel rounded about the middle, 12-14(-15) mm long, with a narrow untwisted beak. Pod subcylindrical, curved, 38-50 x 6-8 mm, puberulous, *c* 30-40-seeded.

1. Leaflets elliptic, mostly 6-13 cm long, acute to rounded at the apex. var. **pallida**
- Leaflets obovate-elliptic to obovate, *c* 3-7 cm long, rounded or retuse at the apex. var. **obovata**

Figure 95.36.1 - *CROTALARIA GLAUCA*, stem with flowers and pods from *Ash* 2158. 2 - *C. ONONOIDES*, flowering stem from *Gilbert & Thulin* 715. 3 - *C. PYCNOSTACHYA* subsp. *PYCNOSTACHYA*, stem with flowers and pods from *Mooney* 7534. 4 - *C. ALBICAULIS*, flowering stem from *Hemming* 2134. All x 0.75. Drawn by Maureen Church. (Reproduced with permission from *Opera Botanica*, 68, fig. 38, 1983.)

Var. **pallida.**

River banks, roadsides, cultivated ground; 1300-1450 m. TU WG IL; throughout the tropics. *Gilbert & Thulin* 530, 828; *Meyer* 7986; *Schimper* 2003.

Widely cultivated as a cover crop, apparently native in Ethiopia.

Var. **obovata** *(G. Don) Polhill* in *Kew Bull.* **22**: 265 (1968).

Lake shores, river banks, tall grassland; 1200-1350 m. WG SU GG SD; throughout the tropics. *Gilbert, Thulin & Getachew* 530; *Ash* 1372; *Gilbert* 1628A.

29. **C. albicaulis** *Franch.* (1882).

Shrubby perennial, 0.2-1 m tall; branches white tomentellous. Leaflets 3(-4), elliptic-obovate, mostly 25-40 x 12-20 mm, glaucous, ± densely covered with short white appressed hairs beneath; petioles about as long as the leaflets. Racemes rather dense, many-flowered; bracts falling soon. Calyx becoming basally truncate and deflexed against the pedicel, 5-6 mm long, densely white pubescent; lobes a little longer than the tube. Standard purplish; wings usually yellow; keel strongly rounded, 9-11 mm long, purple, with a rather broad blunt incurved beak. Pod 16-22 x 8-10 mm, tomentellous, 6-12-seeded. Fig. 95.36.4.

Dry deciduous bushland, rocky outcrops or along river courses; *c* 600-1600 m. AF SU HA SD; N Somalia. *Gilbert* 2040; *Burger* 453, 613; *W. de Wilde* 9815.

30. **C. ochroleuca** *G. Don* (1832).

Erect annual or short-lived perennial, 0.5-2.7 m tall; stem pubescent with fine appressed hairs. Leaflets 3, linear to lanceolate, 50-130 x 5-30 mm, glabrous above, appressed puberulous beneath; petioles shorter than leaflets. Racemes many-flowered, fairly dense. Calyx becoming basally truncate and deflexed against the pedicel, 6-8 mm long, glabrous. Standard cream or pale yellow, with reddish veins, glabrous outside; keel shortly rounded, (17-)18-21(-23) mm long, with a projecting untwisted beak. Pod 50-70 x 15-20 mm, appressed puberulous, up to 100-seeded.

Damp grassy places in woodland, bushland and grassland, also on roadsides and cultivated ground; 550-1900 m. WG IL KF GG; widespread in tropical Africa. *Ash* 519; *Fukui* 364, 365; *W. de Wilde* 7598.

Difficult to distinguish from *C. brevidens* var. *brevidens* in the absence of fruits, but flowers generally pale, and usually in wetter places.

31. **C. brevidens** *Benth.* (1843).
C. intermedia Kotschy var. *abyssinica* Taub. ex Engl., *Hochgebirgsfl. Trop. Afr.*: 247 (1892) - type: GN, Ifak, *Schimper* 1526 (BM K iso.).

Erect or decumbent annual or short-lived perennial, 0.5-2 m tall; stem short pubescent. Leaflets 3, linear to elliptic, 40-100(-140) x 3-20(-33) mm, glabrous above, appressed puberulous beneath; petiole shorter than the leaflets. Racemes many-flowered, fairly dense. Calyx becoming basally truncate and deflexed against the pedicel, 3-8 mm long, puberulous or glabrous. Standard cream or yellow with brownish veins, glabrous outside; keel shortly rounded, 12-24 mm long, with a long projecting untwisted beak. Pod (35-)40-50 x 4-7 mm, puberulous or pubescent, up to at least 80-seeded.

1. Calyx glabrous, 3.5-5(-7) mm long. var. **brevidens**
- Calyx puberulous, 5-8 mm long. var. **intermedia**

Var. **brevidens.**

Grassland, woodland and bushland; 500-1900 m. GD GJ WG IL KF; Sudan. *Gilbert & Thulin* 854; *Friis* et al. 165; *Siegenthaler* 1484.

See note under *C. ochroleuca.*

Var. **intermedia** *(Kotschy) Polhill* in *Kew Bull.* **22**: 266 (1968).

Grassland, woodland and bushland; 1300-1800 m. SU KF; west to Nigeria and south to Tanzania. *Chojnacki* in *Mooney* 7525; *Smeds* s.n. (H).

32. **C. polhillii** *Thulin* (1982)
- type: SU, S of Thibe, *Gilbert & Thulin* 868 (UPS holo., C EA K MO WAG iso.).

Erect annual up to 1.5 m tall; stem appressed pubescent. Leaflets 3, linear to linear-lanceolate, 20-75 x 1-9 mm, glabrous above, appressed pubescent beneath; petiole 5-15 mm long; stipules minute or obsolete. Racemes many-flowered, fairly dense, axis and pedicels appressed pubescent. Calyx becoming basally truncate and deflexed against the pedicel, 3-4 mm long, appressed pubescent; lobes triangular with subulate tips, about as long as tube. Standard yellow with purplish veins; wings broad, much longer than keel; keel abruptly rounded about the middle, *c* 7 mm long, with an incurved untwisted beak. Pod *c* 25 x 5 mm, appressed pubescent, glabrescent, *c* 25-seeded.

Swampy grassland; 1500-1800 m. SU KF; not known elsewhere. *J. De Wilde* 5328; *Zemede & Birhanu* 555.

33. **C.** sp. = *Benedetto* 122;
C. sp. *A. sensu* Thulin, *Leg. Eth.* (1983)

Apparently an erect annual; stem practically glabrous. Leaflets 3, linear, 70-130 x 4-6 mm, glabrous above, appressed pubescent beneath; petiole 40-50 mm long; stipules absent. Racemes many-flowered, fairly dense, axis and pedicels glabrous. Calyx apparently becoming basally truncate and deflexed against the pedicel, 3.5-4 mm long, glabrous or with a few scattered hairs on the lobes; lobes triangular, about as long as tube. Standard apparently yellow with purplish veins; wings longer than keel; keel rounded about the middle, *c* 8 mm long, with a very short untwisted beak. Pod not known.

Habitat unknown. WG (Dembidolo-Gambela); not known elsewhere.

Differs from *C. polhillii* in the practically glabrous stem, inflorescence-axis, pedicels and calyx, the longer petioles, the very shortly beaked keel and the minute bracteoles which are inserted at the top of the pedicel.

34. **C. awasensis** *Thulin* (1982)
- type: SU, *W. de Wilde* 8340 (WAG holo., ETH K iso.).
C. lanceolata E. Mey. subsp. *contigua* Polhill in *Kew Bull*. **22**: 268 (1968) pro parte, quoad *Vàtova* 2157.

Much-branched annual, up to about 0.6 m tall; stem appressed pubescent. Leaflets 3, linear-lanceolate, 15-65 x 4-12 mm, glabrous above, appressed pubescent beneath; petiole 10-25 mm long; stipules absent. Racemes many-flowered, fairly dense, axis and pedicels appressed pubescent. Calyx becoming basally truncate and deflexed against the pedicel, 3-3.5 mm long, glabrous; lobes triangular, shorter than tube. Standard yellow with purplish veins; wings equalling or slightly shorter than keel; keel rather abruptly rounded about the middle, 8-10 mm long, with a projecting untwisted straight beak. Pod *c* 16 x 4 mm, appressed pubescent.

Wooded grassland; *c* 1700 m. SU SD (Lake Awasa area); not known elsewhere. *Vàtova* 2157; *Ouren* 20930.

35. **C. fallax** *Chiov*. (1939)
- type: SD, Negele, *Cufodontis* 255 (FT holo.).

Erect or ascending herb, up to 0.9 m tall; branches covered with short appressed hairs. Leaflets 3, mostly linear-oblanceolate, 20-60 x 3.5-14 mm, appressed pubescent beneath; petioles shorter than the leaflets. Racemes fairly dense, many-flowered. Calyx becoming basally truncate and deflexed against the pedicel, 3-4 mm long, densely and shortly pubescent. Standard yellow with brownish veins; wings shorter than keel; keel 8-10 mm long, strongly rounded with a narrow untwisted incurved beak. Pod narrowly cylindrical, *c* 15 mm long, appressed pubescent, 20-30-seeded.

Grassland and woodland, *c* 1550 m. SD; not known elsewhere. *J. De Wilde* 6682.

36. **C. ruspoliana** *Chiov*. (1915)
- type: SD, between Rogona and Gobbo-Dubaya, *Ruspoli & Riva* 604 (FT holo.).
C. sidamaënsis Chiov. (1940) - type: SD, *Massa* 36 (FT holo.).

Erect herb; branches densely appressed white pubescent. Leaflets 3, ± linear-oblanceolate, mostly 30-65 x 3-8 mm, densely covered with short hairs beneath; petioles much shorter than leaflets, *c* 1-12 mm long. Racemes many-flowered. Calyx becoming basally truncate and deflexed against the pedicel, 4-5 mm long, densely and shortly appressed pubescent. Standard yellow with brown veins; keel rounded about the middle, 10-12 mm long, with an incurved acute beak. Pod narrowly cylindrical, *c* 22-24 x 3-5 mm, densely and shortly hairy, 12-17-seeded.

Upland grassland. SD; known only from the two collections cited above from E of Lake Chamo.

37. **C. impressa** *Nees* ex *Walp*. (1842)
- type: plant cultivated in Horto Vratislaviensi, from seeds collected in Ethiopia by Schimper (location uncertain).
C. striata A. Br. (1841) nom. illegit.; *C. abyssinica* Dietr. (1847) - type: plant grown in Carlsruhe Botanic Garden from same source of seed.
C. astragalina Hochst. ex A. Rich. (1847) - types: TU, *Quartin Dillon & Petit* s.n. (P syn.) & *Schimper* II:808 (P syn., BM FI-WEBB K isosyn.) & *Schimper* III:1495 (P syn., BM FI-WEBB K MO isosyn.).

Erect herb, 15-100 cm tall; branches ± sparsely appressed puberulous. Leaflets 3, mostly linear to narrowly elliptic-oblong, mostly 10-45 x 3-8 mm, somewhat thinly appressed puberulous beneath, sometimes also above; petioles mostly shorter than the leaflets. Racemes 1-8 cm long with flowers crowded above a relatively long peduncle. Calyx becoming basally truncate and deflexed against the pedicel, 2.5-3 mm long, appressed puberulous. Standard yellow with purplish veins; keel strongly rounded, 3.5-4 mm long, with a sharp incurved beak. Pod 10-15 x 4-5 mm, appressed puberulous, 12-18-seeded.

Woodland and bushland; up to 2200 m. EE EW TU GD SU; Sudan, Djibouti. *Gilbert & Thulin* 980; *Mooney* 8027; *Pappi* 1020.

38. **C. onobrychis** *A. Rich*. (1847);
C. astragalina subsp. *onobrychis* (A. Rich.) Bak. f. in *Journ. Linn. Soc., Bot*. **42**: 375 (1914); *C. impressa* subsp. *onobrychis* (A. Rich.) Cuf., Cuf. *Enum., Zweiter Nachtrag*: 23 (1969) - type: TU, *Quartin Dillon* s.n. (P holo.).

Erect annual, up to 40 cm tall; branches appressed puberulous. Leaflets 3, elliptic to narrowly elliptic-oblong, mostly 20-60 x 4-25 mm, appressed puberulous beneath and sometimes also above; petiole shorter than the leaflets. Racemes 2.5-15 cm long, shortly pedunculate, dense at first, elongating and ± lax. Calyx becoming basally truncate and deflexed against the pedicel, 2-3 mm long, ± puberulous. Standard yellow with purple veins; keel rounded, 4-6 mm long, with a very short blunt beak. Pod 10-14 x 4-5 mm, subglabrous to appressed puberulous, 10-20-seeded.

Grassland and bushland; *c* 1200-1500 m. TU GD (Sudan border) SU KF; Sudan, Zaire, Malawi. *Gilbert & Thulin* 974; *Ouren* 21253.

39. **C. cylindrica** *A. Rich.* (1847)
- type: TU, *Petit* 4 (P holo.).
C. tigrensis Bak. (1871) - type: *Schimper* 259 (K holo., BM iso.).
C. cylindrica var. *pusilla* Chiov. in *Ann. Ist. Bot. Roma* 8: 401 (1908) - type: Eritrea, *Pappi* 5043 (FT, holo.).
C. nigrescens Chiov. (1939) - type: SD, Mega, *Cufodontis* 641 (FT, holo.).

Perennial with numerous ascending stems to *c* 0.4 m long; stems appressed pubescent. Leaflets 3, oblanceolate to obovate-elliptic, mostly 10-25(-40) x 4-10 mm, usually glabrous above, appressed puberulous beneath, drying blackish; petiole 2-10 mm long. Racemes pedunculate with 4-16 flowers generally rather closely arranged towards the top. Calyx becoming truncate at the base and deflexed against the pedicel, 3-4(-4.5) mm long, usually puberulous. Standard yellow with brown veins, glabrous outside; keel rounded, (7-)8-9 mm long, with an incurved (in Ethiopia) untwisted beak. Pod 16-22 x 3-4.5 mm, thinly appressed puberulous to subglabrous, 12-18-seeded.

Grassland, woodland, forest edges; 1400-2450 m. EW TU SU BA GG SD; Kenya. *Mooney* 5768, 8432; *Gillett* 14305; *Friis* et al. 796.

The Ethiopian plant is subsp. *cylindrica;* there is another subspecies in the Kenya highlands. Some specimens from GG and SD (e.g. *Gilbert, Thulin & Getachew* 409 and *Rippstein* 708) are unusually tall and upright plants which appear to be rather short-lived. They are included in *C. fallax* by Polhill (1982).

40. **C. cleomifolia** Bak. (1871).
C. schumanniana Harms (1897) - type: SD, Coromma, *Riva* 1407 (FT iso.).

Shrubby perennial, 1-4 m tall; branches puberulous to yellowish tomentose. Leaflets 3-5, elliptic, 5-10.5(-12) x 2-5.5 cm, hairy at least beneath. Stipules present. Racemes many-flowered, rather dense. Calyx becoming basally truncate and deflexed against the pedicel, 6-7 mm long, puberulous to tomentose. Standard bright yellow usually with brownish veins, glabrous outside; keel rounded about the middle, 12-17 mm long, with a narrow slightly incurved untwisted beak. Pod (33-)38-50(-60) x 7-9 mm, sparsely puberulous to densely pubescent, 30-40-seeded.

Upland bushland and grassland, forest margins, river banks and roadsides; *c* 1600-2300 m. WG IL KF SD GG; widespread in tropical Africa. *Mooney* 7792; *Gillett* 14544, 14571; *Haugen* 480.

41. **C. vatkeana** *Engl.* (1892)
- type: GN, Dschan-Meda, *Schimper* 1221 (B holo. destr., BM K iso.).

Ascending annual, 0.3-1.5 m tall; stem ± thinly appressed pubescent. Leaflets 3, linear-lanceolate to elliptic or narrowly obovate, 20-70 x 4-22 mm, appressed pubescent at least beneath; petiole *c* 10-20 mm long. Stipules present. Racemes long-pedunculate, with many rather crowded flowers towards the top. Calyx becoming basally truncate and deflexed against the pedicel, 4-4.5 mm long, subglabrous to appressed pubescent, with long bracteoles. Standard yellow with purple veins, glabrous outside; keel shortly rounded about the middle, 8-10 mm long, with a straight forwardly projecting untwisted beak. Pod (23-)25-27 x 5-6 mm, pubescent, *c* 30-seeded.

Upland bushland and grassland, forest margins and roadsides. GD; S Sudan, Kenya, N Tanzania. In Ethiopia known only from the type.

42. **C. intonsa** *Polhill* (1982)
- type: KF, *Mooney* 6251 (K holo., BR ETH FT iso.).

Perennial, 0.6-1.5 m tall; branches appressed pubescent. Leaflets 3, mostly elliptic, 50-90 x 20-40 mm, pubescent beneath; petiole 2-4 cm long. Stipules present. Racemes shortly pedunculate, long, densely many-flowered, with long bracts and bracteoles. Calyx becoming basally truncate and deflexed against the pedicel, 5-6 mm long, pubescent. Standard yellow with purplish veins; keel strongly rounded, 7.5-11 mm long, with an incurved attenuate beak. Pod 22-27 x 4-6 mm, spreading pubescent, *c* 24-28-seeded.

Upland grassland, forest edges; 1850-2600 m. GD SU KF SD; not known elsewhere. *Ash* 81; *W. de Wilde* 8296; *Westphal* 2890.

43. **C. gillettii** *Polhill* (1982)
- type: SD, *Gillett* 14509 (K holo.).

Straggling-ascending herb, 0.4-1.5 m tall; branches appressed pubescent. Leaflets 3, mostly elliptic to obovate-elliptic and 20-55(-80) x (5-)8-20 mm, appressed pubescent at least beneath; petiole 1-3 cm long. Stipules present. Racemes long, pedunculate, densely many-flowered, with long bracts and bracteoles. Calyx becoming basally truncate and deflexed against the pedicel, 3.5-4.5 mm long, appressed pubescent. Standard yellow with purplish veins; keel strongly rounded about the middle, 6-8 mm long, with an incurved fairly short beak. Pod 15-20 mm long, densely pubescent with rather long and slightly spreading hairs, 16-28-seeded.

Forest margins, grassland and bushland; 1700-2800 m. WG KF IL SD; not known elsewhere. *Gilbert & Thulin* 811; *Mooney* 8762; *Friis* et al. 2175.

44. **C. plowdenii** *Bak.* (1871)
- type: without precise locality, *Plowden* s.n. (K holo.).
C. brachycephala Harms ex Bak. f. (1914) - type: "Arussi-Galla", *Ellenbeck* 1557 (B holo. destr., BM fragment).

Straggling herb, (0.1-)0.3-1 m tall; branches subappressed pubescent. Leaflets 3, linear-lanceolate to narrowly elliptic-oblong, (10-)20-70 x 4-15(-20) mm, appressed pubescent beneath; petioles shorter than the leaflets. Stipules absent. Racemes shortly pedunculate, short or long, dense. Calyx becoming basally truncate and deflexed against the pedicel, 3-4 mm long, appressed pubescent, with long bracteoles. Standard yellow with purplish veins; keel strongly rounded, 5-7(-8) mm long, with a narrow sharp ± incurved beak. Pod 8-13 x 4-5 mm, puberulous, 10-16-seeded. Fig. 95.37.1.

Upland grassland; 1150-2200 m. GD GJ SU AR HA KF; Yemen. *Thulin* 1377; *Burger* 2031a; *W. de Wilde* 8311; *Ash* 2131.

Closely related to *C. comosa* and critical forms occur particularly in the Lake Tana area.

45. C. **petitiana** *(A. Rich.) Walp.* (1852);
Chrysocalyx petitiana A. Rich. (1847) - type: TU, near R. Mareb, *Quartin Dillon & Petit* s.n. (P holo.).

Erect herb, 0.5-2 m tall; branches short pubescent. Leaflets 3, narrowly lanceolate to broadly elliptic, 30-105(-150) x 8-32 mm, glabrous to sparsely pubescent above, appressed pubescent beneath; petiole shorter than leaflets. Stipules absent. Racemes pedunculate, long, many-flowered, with long bracts and bracteoles. Calyx becoming basally truncate and deflexed against the pedicel, (5-)6-8 mm long, appressed pubescent. Standard yellow usually with brownish veins, glabrous outside; keel abruptly rounded about the middle, 10-12 mm long, with a slightly incurved untwisted beak. Pod (15-)18-24 x 6-8 mm, pubescent, (20-)26-38-seeded.

Grassland, forest edges, swamp margins and on cultivated ground; 1780-2500 m. TU GD GJ; S Sudan, Uganda, Kenya, Tanzania, Zaire. *Hillier* 953.

46. C. **comosa** *Bak.* (1871).
Chrysocalyx quartiniana A. Rich. (1847); *Crotalaria dilloniana* Bak. (1871) - type: TU, *Quartin Dillon* s.n. (P holo.).

Erect herb, 0.6-2 m tall; branches pubescent. Leaflets 3, linear-lanceolate to elliptic, mostly 30-170 x 10-45 mm, pubescent beneath; petioles shorter than the leaflets. Stipules absent. Racemes shortly pedunculate, dense, many-flowered, with long bracts and bracteoles. Calyx becoming basally truncate and deflexed against the pedicel, 3.5-5 mm long, appressed pubescent. Standard yellow with brownish veins; keel strongly rounded about the middle, 6-8 mm long, with a narrow sharp incurved beak. Pod 10-16 x 4-7 mm, shortly pubescent to tomentellous, 10-16-seeded.

Grassland or woodland; (900-)1500-2300 m. EW TU GD GJ WG SU; widespread in tropical Africa. *Gilbert & Getachew* 3104; *Robertson* 1216; *Pappi* 180.

47. C. **anthyllopsis** *Bak.* (1871).

Short erect annual with decumbent lower branches, covered with sparse bristly pubescence. Leaves 1-foliolate; leaflet ovate to oblong-lanceolate, 30-65(-95) x 12-20(-25) mm, sparsely pilose on both surfaces; petiole 4-10(-15) mm long. Racemes dense, up to 4.5 cm long, many-flowered. Calyx becoming basally truncate and deflexed against the pedicel, 5-6 mm long, sparsely hirsute. Standard pale yellow with purplish veins and glabrous outside; keel rounded about the middle, 5-6(-8) mm long, with a short untwisted beak. Pod subglobose, 6-7 mm long, densely hirsute, (1-)3-6-seeded.

Grassland or woodland; *c* 2000 m. GD; widespread in tropical Africa. *Schimper* 1454.

Only once collected in Ethiopia, near Debra Tabor.

48. C. **phillipsiae** *Bak.* (1895).

Procumbent to ascending annual or short-lived perennial, 0.3-2 m tall; branches thinly pilose. Leaflets 3, the lower ones lanceolate to obovate, 24-36 x 16-23 mm, the upper ones usually lanceolate, 50-72 x 14 mm, glabrous above, thinly appressed pilose beneath; petiole 25-48 mm long. Racemes long-pedunculate, many-flowered, lax. Calyx 5-7 mm long, glabrous to sparsely pubescent. Standard cream or yellow with purplish veins, glabrous outside; keel rounded about the middle, 9-11 mm long, with a narrow incurved beak. Pod 20-28 x 7-9 mm, glabrous to thinly puberulous, 30-40-seeded.

Deciduous bushland; up to 1600 m. HA SD; Somalia, Kenya. *Gilbert & Ermias* 8446; *Glover & Gilliland* 420; *Ruspoli & Riva* 1052; *Friis* et al. 2767.

49. C. **ukambensis** *Vatke* (1879).

Annual or short-lived perennial, up to 1 m tall; branches ± densely pilose with spreading hairs. Leaflets 3, the lower ones ovate-lanceolate to obovate, 35-60 x 20-40 mm, the upper ones narrower, glabrous above, thinly pilose beneath; petioles mostly shorter than the leaflets. Racemes long-pedunculate, with many flowers usually rather crowded towards the top. Calyx 4.5-5 mm long, thinly pilose. Standard cream or yellow with purplish veins, glabrous outside; keel rounded about the middle, 9-10 mm long, with a narrow incurved beak. Pod 20-24 x 5-7.5 mm, rather densely pilose with spreading hairs, 26-30-seeded.

Acacia-Commiphora bushland; 1200 m. SD; E Kenya, N Tanzania. *Mesfin Tadesse & Vollesen* 4153.

50. C. **uguenensis** *Taub.* (1895).

Decumbent or spreading annual or short-lived perennial, up to 0.7 m tall; stems pubescent to pilose. Leaflets 3, linear-lanceolate to elliptic, 20-50 x 5-30 mm, usually glabrous above, thinly appressed pilose beneath; petiole 16-42 mm long. Racemes long-pedunculate, many-flowered, fairly dense. Calyx 4-5 mm long, ± thinly appressed pubescent. Standard yellow with purplish veins, glabrous outside; keel rounded about the middle, 6-8 mm long, with a short incurved

Figure 95.37.1 - *CROTALARIA PLOWDENII*, flowering stem, b: fruiting stem, both from *Thulin* 1377. 2 - *C. MICROCARPA*, stems with flowers, fruits and leaves from *Gilbert & Thulin* 846. 3 - *C. LEPTOCARPA*, a: flowering stem from *Gilbert* 3392, b: pod from *Curle* 104. 4 - *C. EXALTATA*, stem with flowers, fruits and leaves from *Gilbert & Tewolde* 3282. 1-3 x 1.125, 4 x 0.75. Drawn by Maureen Church. (Reproduced with permission from *Opera Botanica*, 68, fig. 39, 1983.)

untwisted beak. Pod 12-18 x 4-6 mm, rather densely short pubescent, 24-30-seeded.

Deciduous bushland and grassland; *c* 1000-1400 m. GG SD; Uganda, Kenya, N Tanzania. *Corradi* 3941, 3959, 4037, 4219.

51. C. **massaiensis** *Taub.* (1895).
C. paolii Cuf., Cuf. *Enum.*: 235 (1955), nom. invalid.

Spreading or trailing perennial; branches ± thinly pubescent. Leaflets 3, linear-lanceolate to oblanceolate or lanceolate-elliptic, 7-30 x 2-5(-10) mm, thinly appressed pilose at least beneath; petiole 5-19 mm long. Racemes long and very slenderly pedunculate, *c* 6-18-flowered, lax. Calyx 3-4 mm long, appressed pubescent. Standard cream to yellow with purplish veins, glabrous outside; keel rounded, 4-6(-8) mm long, with a straight untwisted beak. Pod 8-11 x 3.5-4.5 mm, pubescent, 20-24-seeded.

Deciduous woodland or bushland, often in sandy and rocky places; up to 1550 m. SD; S Somalia, Uganda, Kenya. *Gillett* 14468; *Corradi* 3945, 4119.

52. C. **sacculata** *Chiov.* (1939)
- type: SD, Mega, *Cufodontis* 654 (FT holo.).

Small spreading perennial; branches pilose. Leaflets 3, oblanceolate-elliptic to obovate, (6-)10-20 x (3-)5-10 mm, pilose beneath and sparsely so above. Stipules foliaceous, 1-3-partite, the segments linear-lanceolate to lanceolate, 5-11 x 1.5-5 mm. Racemes pedunculate, lax, 8-14-flowered. Calyx (3.5-)4-5 mm long. Standard pale yellow veined and flushed crimson-purple; keel rounded about the middle, 5-7 mm long, with a narrow incurved beak. Pod up to 16 x 6 mm, pubescent, *c* 10-seeded.

Upland grassland; *c* 2100 m. SD; not known elsewhere. *Gillett* 14359; *Corradi* 4424, 4468; *Mesfin* et al. 3370.

53. C. **dumosa** *Franch.* (1882).
C. jamesii Oliv. (1886) - types: Ogaden (HA-BA), *James & Thrupp* (K syn.).
C. parvula Beck (1888), nom. illegit.; *C. paulitschkei* Bak. f. (1914) - type: HA, *Hardegger & Paulitschke* s.n. (W holo.).

Much-branched shrublet, *c* 0.2-0.7 m tall; branches densely appressed white pubescent. Leaflets 3, oblanceolate to obovate, up to 15 x 7 mm, ± thinly appressed pilose on both surfaces; petiole up to 6 mm long. Racemes *c* 2-7 cm long, up to 10(-15)-flowered, lax. Calyx 2-3.5 mm long, pubescent. Standard bright yellow with reddish veins, glabrous or with a few hairs along the midvein outside; keel strongly rounded about the middle, 5.5-8(-9) mm long, with a narrow slightly incurved untwisted beak. Pod up to 7 x 4 mm, often mottled, white pilose, 4-8-seeded.

Deciduous bushland and scrub, in sandy and rocky places; 750-1400 m. HA BA SD; Djibouti, Somalia, N Kenya. *Burger* 3081; *Ellis* 169; *Gilbert & Thulin* 1496.

54. C. **persica** *(Burm.) Merr.* (1921).

Low intricately branched deciduous shrublet; branchlets and inflorescence-rhachides weakly spine-tipped, appressed puberulous with stiff white hairs. Leaflets 3 or the upper leaves reduced, linear to oblong-elliptic, mostly 10-25 x 1-5 mm, shortly appressed pubescent on both surfaces; petiole up to 4 cm long. Racemes numerous, very laxly few-flowered. Calyx 4-5 mm long, appressed puberulous. Standard bright yellow with purplish veins, glabrous to medially puberulous; keel rounded, 5-6 mm long, obtuse to shortly beaked. Pod drawn out into a long stylar beak, 10-15 mm long, densely pubescent, 2-4-seeded.

Semi-desert grassland on sand near the coast. EE; Djibouti, N Somalia, Arabia, Socotra, Iran. *Beccari* s.n.

55. C. **steudneri** *Schweinf.* (1868)
- type: EW, Keren, *Steudner* 114 (B holo. destr., K iso.).
C. steudneri forma *latifolia* Lanza in *Boll. R. Orto Bot. Palermo* **8**: 95 (1909) - types: Eritrea, *Senni* 160 & 161 (PAL syn., not seen).

Ascending and spreading annual, with the lower branches mostly long and decumbent, sometimes up to 0.6(-1) m tall, but usually much less; stem ± thinly covered with rather long hispid spreading hairs. Leaflets 3, linear-lanceolate to oblanceolate, the lower ones sometimes broader, 20-70 x 8-17 mm, glabrous above, sparsely appressed pilose beneath; petiole 6-15 mm long. Racemes long-pedunculate, few- to many-flowered, lax. Calyx 3-4(-4.5) mm long, appressed pubescent or subglabrous. Standard yellow with brownish veins and glabrous outside; keel rounded below the middle, 5-7(-9) mm long, with a short untwisted beak. Pod 10-13(-17) x *c* 8 mm, puberulous, 20-24-seeded.

Bushland, grassland, roadsides; 1800-2600 m. EW; Sudan, Tanzania and widespread in southern tropical Africa. *Pappi* 1378, 1882, 2483.

56. C. **distantiflora** *Bak. f.* (1914).

Short-lived perennial, prostrate to ascending, often forming mats; stems covered with spreading hairs, usually sparsely so. Leaflets 3, elliptic to narrowly obovate, 14-30 x 10-25 mm, subglabrous above, appressed pubescent beneath; petiole 13-23 mm long. Racemes slender, 7-16 cm long, few-flowered, lax. Calyx 3.5-4.5 mm long, thinly pubescent. Standard yellow with purplish veins, glabrous outside; keel rounded about the middle, 6-7 mm long, with a narrow untwisted beak distinctly exceeding the wings. Pod oblong-ellipsoid 13-16 x 4.5-7 mm, pubescent, 14-20-seeded.

Upland grassland or bushland, forest margins, also on roadsides and cultivated ground; 1600-2100 m. SU BA KF SD; Kenya, Tanzania. *W. de Wilde* 10292; *Gillett* 14343; *Thulin, Hunde & Tadesse* 3636; *Friis* et al. 1078.

57. C. **microcarpa** *Hochst.* ex *Benth.* (1843).
C. lotoides A. Rich. (1847), nom. illegit. - types: TU, *Quartin Dillon* s.n. (P syn.) & *Schimper* II:1190 (P syn., BM FI-WEBB FT K isosyn.).
C. lotoides var. *uniflora* Steud. ex A. Rich., *Tent. Fl. Abyss.* **1**: 157 (1847) - type: TU, *Schimper* III:1965 (P holo., BM K iso.).
C. microcarpa forma *subuniflora* Steud. ex Bak. f. in *Journ. Linn. Soc., Bot.* **42**: 403 (1914) - type: TU, *Schimper* III:1965 (BM holo., K P iso.).

Much-branched spreading annual or short-lived perennial herb, 10-70 cm tall; stem pubescent. Leaflets 3, lanceolate to oblanceolate, 12-30 x 4-8 mm, pilose on both surfaces; petiole 2-4 mm long. Flowers few to numerous in pedunculate heads or short usually dense racemes. Calyx 2.5-3.5 mm long, appressed pilose. Standard yellow, usually marked red, pubescent outside; keel bent in right-angles in the lower part, 4.5-7.5 mm long, with a narrow untwisted beak. Pod 4-7.5 x 2-4 mm, pubescent, 8-12-seeded. Fig. 95.37.2.

Deciduous woodland, bushland and grassland, and on cultivated or waste ground; 1575-1770 m. EW TU GD(Sudan border) WG; widespread in the drier parts of tropical Africa. *Pappi* 6066; *Gilbert & Thulin* 846; *Gilbert & Getachew* 2982.

58. C. **microphylla** *Vahl* (1790).

Prostrate or ascending annual; stems many, slender, puberulous with stiff appressed white hairs. Leaflets (1-)3, linear to oblanceolate or obovate, 3-25(-30) x 2-5 mm, puberulous beneath; petioles shorter than the leaflets. Flowers 1-2(-3), leaf-opposed on a short slender axis. Calyx (3-)3.5-5 mm long, thinly puberulous; lobes nearly twice as long as the tube. Standard pale yellow with red veins and sometimes flushed reddish, puberulous medially outside; keel subangular in the lower part, 6-10 mm long, with a rather long straight beak. Pod 12-15 x 6-8 mm, thinly puberulous, 20-30-seeded.

Sandy coastal plains. EE; Somalia, Djibouti, Sudan and westwards to Mauritania, and in Arabia. *Bally* 6953; *Popov* 1390; *Stower, Smith & Gilliland* 4032; *Robertson* 1088.

59. C. **boranica** *Harms* ex *Bak. f.* (1914).

Annual with many slender ascending or prostrate sparsely appressed puberulous stems, 5-25 cm long. Leaflets 3, linear-lanceolate to oblanceolate, 9-30 x 3.5-5 mm, glabrous above, subglabrous to sparsely subappressed puberulous beneath; petiole *c* 3-9 mm long. Racemes pedunculate, lax, up to 5 cm long, 4-12-flowered. Calyx 3.5-5 mm long, subglabrous. Standard whitish or dull yellow with crimson veins, glabrous or puberulous along the midvein outside; keel bent almost at right-angles in the lower part, 6-8 mm long, with a straight untwisted beak. Pod *c* 9 x 7 mm, subglabrous, *c* 22-24-seeded.

Deciduous bushland; up to 900 m. HA BA SD; Somalia, N Kenya. *Glover & Gilliland* 291; *Ellis* 237A; *J. De Wilde* 5935; *Gilbert* et al. 7513.

The Ethiopian material is subsp. *boranica*. There is another subspecies in NE Kenya and S Somalia.

60. C. **bongensis** *Bak. f.* (1914).

Erect annual up to 45 cm tall. Leaves 1-foliolate; leaflets linear to oblong-elliptic, 25-80 x 2-8 mm, pilose at least beneath; petiole 2-4 mm long. Flowers 3-12 in lax racemes. Calyx 2-2.5 mm long, pubescent. Standard yellow with reddish veins, glabrous outside; keel angled in the lower part, 4-5 mm long, with a long narrow untwisted beak. Pod 7-10 x 2-3 mm, pubescent, 12-16-seeded.

Grassland in thin soil overlying rock; *c* 600 m. IL (between Abobo and Gog); widespread but scattered in tropical Africa. *Gilbert & Friis* 8380.

61. C. **leptocarpa** *Balf. f.* (1888).

Perennial herb, with many short slender strigulose much-branched stems. Leaflets 3, linear-lanceolate to obovate, (6-)12-35 x 1-9 mm, usually glabrous above, sparsely to densely appressed puberulous beneath; petiole 4-12 mm long. Racemes very lax, 3-12-flowered. Calyx ± as long as the keel, appressed puberulous; lobes longer than the tube. Standard yellow with reddish veins, medially puberulous outside; wings much longer than the keel; keel strongly rounded, 4-5 mm long, with a short circumflexed beak. Pod flattened, 10-22 x 5-8 mm, strigulose, *c* 10-12-seeded. Fig. 95.37.3.

Deciduous bushland and woodland; 950-1250 m. BA SD; Somalia, Kenya, Arabia and Socotra. *Friis* et al. 1024; *Gilbert & Thulin* 1493; *Gilbert* 3392.

The Ethiopian material is subsp. *leptocarpa*. There are two more subspecies in Somalia.

62. C. **boudetii** *Polhill* (1982)
- type: Ogaden (BA), Kebri Dehar, *Boudet* 8069 (K holo., ALF iso.).

Slender annual, 15-25 cm long, sparsely strigulose. Leaflets 3, linear, 10-25 x 0.3-1 mm, the lower ones shorter and up to 2 mm wide, thinly appressed puberulous beneath; petiole 5-14 mm long. Racemes very lax, 3-6-flowered. Calyx 3 mm long, appressed puberulous; lobes longer than tube. Standard yellow with reddish veins, medially puberulous outside; wings little longer than the keel; keel deeply rounded, 3-4 mm long, with a slightly incurved beak. Pod 8-10 x 4-5 mm, strigulose, *c* 8-seeded.

On limestone; *c* 550 m. HA; only known from the type.

63. C. **orixensis** *Rottl.* ex *Willd.* (1803).
C. macropoda Hochst. ex A. Rich. (1847) - types: TU, *Quartin Dillon* s.n. (P syn.) & Schimper III:1611 (P syn., BM K MO UPS isosyn.).

Annual, (5-)15-50 cm tall, covered with long subappressed to spreading hairs. Leaflets 3, usually elliptic-oblong to obovate, 15-30 x 4-15 mm, pilose on both surfaces; petiole about as long as the leaflets; stipules ovate, up to 2.8 mm wide. Racemes laxly 3-12(-24)-flowered; bracts ovate-cordate. Calyx 2-lipped, 4-5 mm long, pilose; lobes *c* 3 times as long as the tube. Standard yellow, medially pubescent outside; keel subangular, 4-5 mm long, with a straight, twisted beak. Pod 13-20(-25) x 6-10 mm, stipitate, glabrous, 10-12-seeded.

Grassland, in damp places, or ruderal; 1300-1900 m. EW TU GD GJ; India. *Parker* 353; *Pappi* 285, 367; *Mesfin & Kagnew* 1598.

64. C. **calycina** *Schrank* (1817).

Erect little-branched annual, 15-100 cm tall; stem densely hirsute with long ± appressed hairs. Leaves simple, linear to oblong-lanceolate, 65-150 x 6-15 mm, glabrous above, appressed pilose beneath; petiole 1-4 mm long. Flowers few in terminal racemes and in the upper axils. Calyx 2-lipped, longer than the corolla, accrescent in fruit, *c* 22-30 mm long, pilose with long spreading brownish hairs. Standard pale yellow, subglabrous outside; keel subangular, 15-17 mm long, with a straight twisted beak. Pod 20-25 x 8 mm, glabrous, 16-36-seeded.

Wooded grassland, disturbed places; *c* 1400 m. TU SU; widespread in tropical Africa, Asia, N Australia. *Ash* 2238; *Quartin Dillon & Petit* 27.

65. C. **exaltata** *Polhill* (1968)
- type: SD, Mt Delo, *Gillett* 14991 (K holo., BR EA iso.).

Shrub or small tree 3-7.5 m tall; young branches tomentellous. Leaflets 3, oblong-oblanceolate, 40-75 x 12-23 mm, glabrous above, puberulous beneath; petiole 20-36 mm long. Racemes appearing axillary, subsessile, up to 7 cm long, several-flowered. Calyx 16-20 mm long, densely puberulous; lobes longer than the tube. Standard greenish-yellow with crimson veins, medially pubescent outside; keel rounded below the middle, 35-38 mm long, with a long twisted beak. Pod 80-90 x 16-18 mm, appressed puberulous. Fig. 95.37.4.

Margins of upland forest or bamboo thicket; 3000-3400 m. SU BA KF SD; not known elsewhere. *Meyer* 7651; *Eriksson* 367; *Gilbert & Tewolde* 3282.

66. C. **mildbraedii** *Bak. f.* (1911).

Shrub 1-4 m tall; branches puberulous. Leaflets 3, elliptic to oblong-obovate, mostly 30-50 x 20-30 mm, glabrous above, appressed puberulous beneath; petiole *c* 30-55 mm long. Racemes rather lax, many-flowered. Calyx 11-13 mm long, puberulous; tube 5-6 mm long. Standard yellow, glabrous outside; keel rounded about the middle, 17-20 mm long, with a fairly short twisted beak. Pod *c* 50-60 x 18-25 mm, puberulous, 60-80-seeded.

Montane forest margins and bushland; 1700-2400 m. WG SU BA KF GG SD; S Uganda, Zaire, Rwanda. *Mooney* 6005, 9705; *Friis* et al. 750.

67. C. **keniensis** *Bak. f.* (1926).

Bushy or straggling herb or shrub, up to 2(-4) m tall; branches puberulous. Leaflets 3, oblanceolate, elliptic or elliptic-obovate, 16-30 x 5-15 mm, glabrous above, puberulous beneath; petiole 15-25(-30) mm long. Racemes pedunculate, rather dense and many-flowered. Calyx 5-8 mm long, puberulous; tube 2-4 mm long. Standard yellow, glabrous outside except along the midvein; keel rounded below the middle, 11-13(-15) mm long, with a long twisted beak. Pod 25-40 x 7-12 mm, puberulous, 10-16-seeded.

Upland grassland and bushland, forest margins; 1800-2100 m. WG IL KF SD; Uganda, Kenya, Tanzania. *W. de Wilde* 9335; *Mooney* 8794; *Friis* et al. 19.

68. C. **fascicularis** *Polhill* (1968).

Bushy herb or shrub, 1-2 m tall; branches yellowish tomentose. Leaves mostly crowded on very short lateral shoots; leaflets 3, oblanceolate to obovate, 15-37 x 5-20 mm, glabrous above, thinly to rather densely appressed pubescent beneath; petiole *c* 10-42 mm long. Racemes few- to many-flowered, fairly dense. Calyx 5-7 mm long, appressed pubescent; upper lobes ± as long as the tube. Standard yellow, becoming suffused red or brown, sometimes puberulous along the midvein outside; keel abruptly rounded below the middle (12-)13-16 mm long with a long twisted beak. Pod *c* 25-30 x 5-6 mm, puberulous. Seeds smooth.

Upland grassland, bushland and woodland, forest margins, often in rocky places; 1500-1950 m. BA SD; N Uganda, Kenya. *Burger* 1802; *Mooney* 7374; *Thulin, Hunde & Tadesse* 3340; *Tewolde* 172.

69. C. **emarginella** *Vatke* (1878)
- type: EW, Habab, *Hildebrandt* 586 (W holo.).
C. laxa Franch. (1882).
C. nogalensis Chiov. (1929).
C. rathjensiana Schwartz (1939).

Figure 95.38.1 - *CROTALARIA EMARGINELLA*, a: flowering stem from *Burger* 2829, b: fruiting branch from *Gilbert* 3393. 2 - *C. KARAGWENSIS*, stem with flowers, fruits and leaves from *Gilbert, Thulin & Getachew* 549. 3 - *C. SPINOSA*, a: flowering stem, b: fruiting stem, both from *Thulin* 1346. 4 - *C. STENORHAMPHA*, a: flowering stem from *Smeds* 1271, b: fruiting stem from *Mooney* 6812. 1, 2 & 4 x 0.75, 3 x 1.125. Drawn by Maureen Church. (Reproduced with permission from *Opera Botanica*, 68, fig. 40, 1983.)

Spreading subshrub or woody herb, 0.3-1.7 m tall; branches pubescent. Leaves mostly crowded on very short lateral shoots; leaflets 3, obovate, 9-30 x 6-13(-17) mm, pubescent on both surfaces or glabrous above; petiole 10-18(-30) mm long. Racemes fairly lax and many-flowered. Calyx 6-8 mm long, pubescent; lobes mostly *c* twice as long as the tube. Standard yellow, sometimes with reddish veins or becoming suffused red or brown, glabrous to puberulous outside; keel bent at right-angles in the lower half, *c* 11-14 mm long, with a long twisted beak. Pod 15-25 x 5-6 mm, pubescent, 14-20-seeded. Seeds rugulose. Fig. 95.38.1.

Deciduous bushland or woodland, roadsides; 400-1550 m. EE AF EW HA BA SD; Sudan, Djibouti, Somalia, Kenya, Yemen. *Burger* 2829; *Gilbert* 1858; *Thulin, Hunde & Tadesse* 3565.

Friis et al. 1009 from SD (near Filtu) differs in being a distinctly woody shrub with densely appressed white pubescent branches. It also has a reddish veined standard and an unusually short keel (8-9 mm). Similar plants have been found on limestone in NE Kenya. More material is desired.

70. **C. incrassifolia** *Polhill* (1982).

Shrub, 1-1.5 m tall; youngest parts puberulous. Leaflets 3, obovate, 4-15 x 3-9 mm, ± thickened, puberulous beneath; petiole 3-30 mm long. Stipules absent. Racemes laxly 4-12-flowered. Calyx 8-10 mm long, puberulous; lobes twice as long as the tube. Standard obovate, yellow, medially puberulous outside; keel subangular, 12-15 mm long, with a twisted beak. Pod 20-25 x 5 mm, puberulous, 14-20-seeded.

Deciduous bushland; *c* 400 m. HA; Somalia. *Simmons* S 77.

71. **C. oligosperma** *Polhill* (1982)
- type: HA, *Burger* 3277 (K holo., FT iso.).

Small shrub, 0.3-1.2 m tall; branchlets densely white pubescent. Leaflets 3, obovate, (3-)6-15 x (2-)4-8 mm, puberulous at least beneath; petiole (3-)5-15 mm long. Stipules absent. Racemes short, 2-8-flowered. Calyx 4-6 mm long; lobes 1.5-2 times as long as the tube. Standard yellow sometimes with reddish veins, glabrous or medially puberulous outside; keel subangular, 9-12 mm long, with a straight twisted beak. Pod 8-10 x 4-5 mm, densely puberulous, 6-8-seeded.

Deciduous woodland or bushland, often on limestone; *c* 400-1600 m. HA SD; N Somalia, N Kenya. *Ellis* 395; *Friis* et al. 925; *Thulin, Hunde & Tadesse* 3567.

72. **C. axillaris** *Ait*. (1789).

Shrub or woody herb, 1-4 m tall; branches puberulous to densely yellowish pubescent. Leaflets 3, elliptic or lanceolate, 45-100 x 22-55 mm, glabrous above, appressed puberulous to pubescent beneath; petiole *c* 3-10.5 cm long. Inflorescences axillary, very short, 2-6(-12)-flowered. Calyx *c* 10-13(-15) mm long, thinly puberulous to pubescent; upper lobes much longer than the tube. Standard yellow, puberulous along the midvein outside; keel rounded below the middle, *c* 15-18 mm long, with a long twisted beak. Pod *c* 45-60 x 8-15 mm, puberulous or pubescent, 16-20-seeded.

Forest edges, along rivers; 900-1600 m. BA KF GG; widespread in tropical Africa. *Meyer* 9005; *Mooney* 8460.

73. **C. recta** *Steud.* ex *A. Rich.* (1847)
- types: *Schimper* III:1894 (P syn., BM FT K M MO, isosyn.) & *Schimper* II:1083 (P syn.) & *Quartin Dillon* s.n. (P syn.).
C. simplex A. Rich. (1847); *C. recta* subsp. *simplex* (A. Rich.) Bak. f. in *Journ. Linn. Soc., Bot.* **42**: 353 (1914) - type: TU, *Quartin Dillon & Petit* s.n. (P holo.).

Perennial, usually with several erect sparingly branched stems, 1-3 m tall; stems hollow, appressed pubescent. Leaflets 3, linear-lanceolate to elliptic or obovate, 18-140 x 14-60 mm, glabrous above, appressed pubescent beneath; petiole broadly grooved, 25-40(-60) mm long. Racemes usually long and many-flowered, fairly dense. Calyx 9-12 mm long, subglabrous to appressed pubescent. Standard yellow with purplish veins at the base inside, glabrous or with a few hairs at the apex outside; keel rounded about the middle, 12-15 mm long, with a short incurved twisted beak. Pod 45-65 x 15-20 mm, usually glabrous, 26-34-seeded.

Grassland, bushland, forest margins; (1200-)1800-2700 m. TU GD GJ WG SU AR HA KF SD; widespread in tropical Africa. *Thulin* 1435; *Gilbert & Thulin* 849; *Ash* 2068.

74. **C. retusa** *L.* (1753).

Erect annual or short-lived perennial, up to 1.5 m tall; stem shortly pubescent. Leaves simple, oblanceolate to oblong-obovate, 35-110 x 18-40 mm, rounded or emarginate at the apex, glabrous above, finely appressed pubescent beneath; petiole *c* 2-4 mm long. Racemes many-flowered. Calyx 11-14 mm long, glabrous to appressed puberulous. Standard yellow with purplish veins, usually with a few hairs along the midvein outside; keel rounded, 13-15 mm long, with a twisted beak. Pod 40-50 x 13-18 mm, glabrous, 12-20-seeded.

Grassland, woodland, old cultivations and waste places; 1000-1800 m. AF EW HA; pantropical. *Gilbert & Thulin* 17; *Baldrati* 397; *Burger* 1341.

Widely cultivated for its fibers or as an ornamental, also spreading as a weed. Possibly not native in Ethiopia.

75. **C. deserticola** *Bak. f.* (1914).
C. rhopalocarpa Chiov. (1939) - type: SD, *Cufodontis* 565 (FT, holo.).

Annual or short-lived perennial, 0.5-1.5 m tall but often much smaller; branches thinly appressed puberulous.

Leaflets 3, linear to oblanceolate or elliptic, 20-75 x 2-28 mm, glabrous above, thinly appressed puberulous beneath; petiole 12-40 mm long. Racemes lax, usually many-flowered. Calyx 3-7 mm long, glabrous to medially puberulous outside; keel bent at right-angles in the lower half, 6-18 mm long, with a long narrow twisted beak. Pod (15-)20-40 x 4-11 mm, appressed puberulous, 16-30-seeded.

Grassland, woodland and bushland; 1300-1950 m. SU KF GG SD; widespread in the eastern part of tropical Africa. *Gilbert, Thulin & Getachew* 535, 599; *Jansen* 5565.

The Ethiopian material is small-flowered and belongs to var. *deserticola.*

76. C. **karagwensis** *Taub.* (1895).
C. lugardiorum Bullock (1932).

Erect annual, 0.2-1 m tall; stem strigulose. Leaves simple, linear-lanceolate to elliptic, 38-115 x 4-12 mm, glabrous above, strigulose beneath; petiole 1-2 mm long. Racemes lax, *c* (6-)10-24-flowered. Calyx (4.5-)6-8 mm long, strigulose. Standard yellow with purplish veins, puberulous apically outside; keel angular, (7-)8-11 mm long, with a long straight twisted beak. Pod 25-35 x 6 mm, appressed puberulous, 26-34-seeded. Fig. 95.38.2.

Grassland and woodland; 1300-2150 m. WG SU HA KF GG SD; Nigeria, Cameroun, Uganda, Kenya, Tanzania, Zaire, Rwanda, Burundi. *W. de Wilde* 7595; *Gilbert, Thulin & Getachew* 549; *Mooney* 6020.

77. C. **senegalensis** *(Pers.) Bacle* ex *DC.* (1825).
C. carinata Steud. ex A. Rich. (1847); *C. senegalensis* var. *carinata* (Steud. ex A. Rich.) Bak. f. in *Journ. Linn. Soc., Bot.* **42**: 338 (1914) - types: TU, *Schimper* II:519 (P syn., K MO UPS isosyn.) & *Schimper* II:772 (P syn., BM K isosyn.) & *Quartin Dillon* s.n. (P syn.).
C. macilenta Del. (1826); *C. senegalensis* var. *macilenta* (Del.) Chiov. in *Ann. Ist. Bot. Roma* **8**:82 (1903).
C. senegalensis var. *sanguinolenta* Chiov. in *Ann. Ist. Bot. Roma* **8**: 399 (1908) - type: Eritrea, *Tellini* 1179 (FT holo.).

Erect annual or short-lived perennial, 0.3-1 m tall; branches shortly pubescent. Leaflets 3, lanceolate-oblong to elliptic, 20-65 x 13-25 mm, glabrous above, appressed pubescent beneath; petiole 1-5 cm long. Racemes rather many-flowered and lax. Calyx 5-7(-8) mm long, appressed pubescent; upper lobes with an excentric tip, lanate along the inner margin. Standard yellow with reddish veins, pubescent outside; keel angular, (7-)9-13(-15) mm long, lanate along the upper margin and with a twisted beak. Pod 10-17 x 5-7(-8) mm, pubescent, 10-16-seeded.

Woodland and bushland, also on roadsides and cultivated ground; up to 2000 m. EE AF EW TU GD GJ WG SU KF; widespread in tropical Africa. *Pappi* 2576; *Gilbert & Thulin* 966; *Mooney* 8054.

78. C. **oöcarpa** *Bak.* (1897).

Erect or spreading annual or short-lived perennial, up to 1(-2) m tall; branches pilose. Leaves mostly with tufts of smaller leaves in the axils; leaflets 3, oblanceolate to obovate, 12-25 x 4-10 mm, pilose at least beneath; petiole *c* 5-20 mm long. Inflorescences of 2 sorts, terminal several-flowered racemes and groups of 1-few flowers on short axes from leaf-tufts. Calyx 5.5-8 mm long, pilose. Standard yellow with reddish veins and densely pubescent outside; keel angular, 10-12 mm long, with a twisted beak. Pod 8-16 x 6-9 mm, pilose outside, 8-10-seeded.

Subsp. **microcarpa** *Milne-Redh.* in *Kew Bull.* **15**: 163 (1961).

Pod 8-10 mm long, with many of the hairs subappressed.

Upland grassland, roadsides and on cultivated ground; *c* 2000 m. HA; Kenya, N Tanzania. *Robertson* 1304.

Subsp. *oöcarpa* occurs in S Tanzania, Burundi and Malawi.

79. C. **spinosa** *Hochst.* ex *Benth.* (1843)
- type: TU, *Schimper* I:150 (K lecto., BM BR FT MO P UPS isolecto.).
C. minima Bak. f. (1896) - type: *Donaldson Smith* s.n. (BM holo.).

Bushy annual or short-lived perennial, 0.2-0.8 m tall; branches pubescent. Leaves mostly clustered on short shoots; leaflets 3, oblanceolate to obtriangular, 6-15 x 1.5-4 mm, glabrous above, appressed puberulous beneath; petiole *c* 2-8 mm long. Inflorescences from the leaf-clusters, with 1-3 flowers on the spine-tipped axes, sometimes also 1-2 flowers in the axils. Calyx 2.5-3 mm long, pubescent. Standard yellow with reddish veins, pubescent outside; keel angular, 4-6 mm long, with a twisted beak. Pod 7.5-9 x 3.5-5 mm, appressed puberulous, 6-8-seeded. Fig. 95.38.3.

Deciduous bushland and grassland, waste places, favoured by overgrazing; 1350-2600 m. EW TU GD WU GJ SU AR HA KF SD; widespread in tropical Africa, also Yemen and Socotra. *Thulin* 1346; *Mooney* 7533; *Friis* et al. 2288.

80. C. **cephalotes** *Steud.* ex *A. Rich.* (1847)
- types: TU, *Schimper* II:695 (P syn., BM FT K MO UPS isosyn.) & *Quartin Dillon & Petit* 598 (P syn.).

Erect annual, 0.1-0.3(-1) m tall, much branched; stem densely hirsute. Leaflets 3, linear-oblanceolate to oblong-oblanceolate, mostly 15-30 x 4-6 mm, but the upper ones often larger, ± appressed pilose on both surfaces; petiole *c* 6-15 mm long. Inflorescences sessile, head-like, many-flowered, crowded at the top of the shoots and surrounded by the larger spreading leaves. Calyx 3-4 mm long, densely hairy. Standard yellow and red, medially pubescent outside; keel angular, 4-5.5 mm long, with a twisted beak. Pod sessile, subglobose, 5-6 mm long, densely pilose, 2-seeded.

Grassland and woodland; 1400-2100 m. TU GD BA HA IL GG SD; widespread in tropical Africa. *Gilbert, Thulin & Getachew* 410; *Burger* 3265; *Mooney* 8428.

81. C. **stenorhampha** *Harms* (1922).

Perennial, 15-75 cm tall, with several erect strigulose stems from a woody rootstock. Leaflets 3, mostly oblanceolate, 10-25 x 3-10 mm, appressed puberulous beneath; petiole 5-15 mm long. Racemes ± laxly 10-40-flowered. Calyx 4-5.5 mm long, appressed pubescent. Standard yellow, sometimes with red veins, pubescent outside; keel angular, 9-13 mm long, with a narrow twisted beak. Pod 7-8 x 4.5-5 mm, appressed puberulous, 2-4-seeded. Fig. 95.38.4.

Grassland, appearing after burning; 1300-1700 m. WG IL; S Sudan and westwards to Cameroun. *W. de Wilde* 10745; *Mooney* 6812, 6932; *Friis* et al. 2413.

82. C. **pseudotenuirama** *Torre* (1960).

Erect annual, 20-90 cm tall; stem pubescent. Leaflets 3, linear-oblanceolate to oblong-oblanceolate, 12-20 x 2-5 mm, glabrous above, appressed pubescent beneath; petiole 4-9 mm long. Racemes mostly many-flowered, fairly dense. Calyx 2.5-3.5 mm long, brownish tomentose. Standard pale yellow with brown veins, brownish tomentellous outside; keel angular, 4-6 mm long, with a twisted beak. Pod sessile, 5-6 mm long, pubescent, 1-2-seeded.

Swampy grassland; 1400-1650 m. TU GD GJ SU; widespread in tropical Africa. *Gilbert & Thulin* 883; *Ash* 2243; *Chiovenda* 2610; *Sebsebe* 445.

83. C. **hyssopifolia** *Klotzsch* (1861).

Annual or short-lived perennial, 10-70 cm tall; stem appressed pubescent. Leaflets 3, oblanceolate to obovate, 8-25 x 2-8(-12) mm, appressed pubescent beneath; petiole 2-12(-18) mm long. Racemes many-flowered, fairly dense. Calyx 2-3.5 mm long, appressed pubescent. Standard yellow with reddish veins and usually pubescent outside with white hairs; keel angular, 3.5-5 mm long, with a twisted beak. Pod 4-5 mm long, appressed pubescent, 1-2-seeded.

Deciduous woodland, grassland, roadsides; 1400-2000 m. GD WU GJ WG KF IL; widespread in tropical Africa. *Gilbert & Thulin* 829; *Friis* et al. 2313; *Pichi-Sermolli* 2231; *Mesfin & Sebsebe* 1386.

84. C. **alexandri** *Bak. f.* (1914).

Erect annual, 10-40 cm tall; stem densely pilose. Leaflets 3, linear-oblanceolate to oblong-oblanceolate, 8-15 x 1-5 mm, pilose on both surfaces; petiole 4-6 mm long. Racemes sessile, dense and short, with most of the flowers clustered in many of the axils below. Calyx 2.5-5 mm long, pilose. Standard pale yellow with reddish veins and pubescent outside; keel angular, 4.5-6 mm long, with a straight twisted beak. Pod 4-5 x 3-3.5 mm, pilose, 7-12-seeded.

Woodland, bushland, upland grassland; 1800-1900 m. GG, SD; widespread in eastern tropical Africa. *Gilbert, Thulin & Getachew* 398; *Vàtova* 533; *Mooney* 9990.

85. C. **juncea** *L.* (1753).

Erect annual, up to 1.6(-3) m tall; stem appressed pubescent. Leaves simple, oblong-lanceolate, 75-150 x 14-30 mm, thinly appressed pilose above, more conspicuously so beneath. Racemes lax, 6-20-flowered. Calyx 16-20 mm long, brownish tomentellous, with longer hairs interspersed; upper lobes 3-4 times as long as the tube. Standard yellow, with scattered hairs outside; keel subangular, 17-22 mm long, with a long twisted beak. Pod 30-55 x 12-17 mm, tomentose, 6-12-seeded.

Deciduous bushland and grassland, or as a weed; *c* 1000 m. SU; native of India but cultivated throughout the tropics for its fibre or as a cover crop and locally naturalized. *W. de Wilde* 10653; *Gelahun* 486.

86. C. **paulina** *Schrank* (1822).

Erect herb or shrub 1.5-3 m tall; glaucous, nearly glabrous. Leaves simple, elliptic to obovate, 80-160 x 25-65 mm, pubescent beneath; petiole 5-8 mm long. Racemes long, lax, many-flowered. Calyx 15-20 mm long, 2-lipped; standard yellow; keel rounded, 16-20 mm long. Pod 40-60 x 8-10 mm, glabrous, 36-50-seeded.

Introduced as green manure. AR; native in Brazil. *Coady* K6.

Described species of uncertain position

87. C. **hypargyrea** Chiov. (1929) - type: BA, Mt Lagio, *Basile* s.n. (TOM holo., K fragment).

Erect herb. Leaflets 3, lanceolate to linear-lanceolate, glabrous above, densely appressed white-pubescent beneath; petiole shorter than the leaflets; stipules small, linear. Racemes long; bracts linear, deflexed. Pod *c* 22 mm long, deflexed, cylindrical, spreading pubescent, *c* 18-seeded.

Only known from the type from BA. Without flowers the position of this plant remains unclear, but probably it is near to *C. vatkeana, C. plowdenii* and allied species.

80. LUPINUS *L.* (1753)

Abebe Demissie, *The Genus* Lupinus *in Ethiopia,* M.Sc. thesis, Reading University (1983).

Herbs, rarely shrubs. Leaves usually digitate; stipules adnate to base of petiole. Flowers in terminal racemes. Calyx 2-lipped, divided almost to the base. Corolla variously coloured; wings connate at apex; keel beaked. Stamens all joined into a closed tube. Pod dehiscent, compressed, usually constricted between the seeds.

Some 200 species, most numerous in the New World with a second centre around the Mediterranean. In Ethiopia, apart from one native and one cultivated species, at least four species have been introduced for both trial as forage crops or green manure, as well as use in soil and water conservation programmes or as ornamentals.

1. Upper lip of calyx divided into 2 lobes for all or most of its length. 2
- Upper lip of calyx almost entire to bifid. 4
2. Corolla yellow. **2. L. luteus**
- Corolla white, or blue and white, or blue and yellow and white. 3
3. Leaflets oblanceolate-cuneate, pilose above; keel 14-18 mm long; pods 1.5-2 cm across. **1. L. princei**
- Leaflets linear, glabrous above; keel up to 12 mm long; pods *c* 1 cm across. **3. L. angustifolius**
4. Stem glabrous; pedicels 7-10 mm long. **4. L. mutabilis**
- Stem hairy; pedicels 2-6 mm long. 5
5. Stem with long spreading hairs; racemes pedunculate; seeds up to 6 mm long; standard suborbicular. **6. L. mexicanus**
- Stem shortly hairy; racemes mostly sessile; seeds 8-14 mm long; standard elliptic. **5. L. albus**

1. **L. princei** *Harms* (1900).

Erect annual or short-lived perennial 30-120 cm tall; stem softly tomentose. Leaflets 7-11(-13), oblanceolate-cuneate, (20-)40-70 x (4-)7-10(-14) mm, appressed pilose on both surfaces; stipules linear-subulate. Racemes 6-15(-20) cm long with 6-16(-24) irregularly verticillate flowers. Calyx 15-17 mm long; upper lip deeply divided; lower lip slightly longer, 3-fid. Corolla 15-18 mm long, blue with a white and yellow median zone to the standard, rarely all white. Pod broadly oblong, 6-7.5 x 1.5-2 cm, brownish tomentose. Seeds (1-)3-4, 9-10 x 8-9 mm, rough. Fig. 95.39.1.

In upland grassland; *c* 1900 m. SD; also in Kenya and Tanzania. *Gillett* 14199; *Mooney* 9795; *Mesfin* et al. 3352.

The collections seen from Ethiopia are from Mega and have entirely white flowers.

2. **L. luteus** *L.* (1753).

Hairy annual 25-80 cm tall. Leaflets obovate-oblong, 40-60 x 8-12 mm, sparsely villous; stipules dimorphic, those of the lower leaves *c* 8 mm long, subulate, those of the upper leaves 22-30 x 2-4 mm, linear-obovate. Racemes 5-16 cm long with regularly verticillate flowers; upper lip of calyx 6-7 mm long, deeply divided, the lower 10 mm long, 3-dentate; corolla 13-16 mm long, bright yellow. Pod 4-5 x 1 cm, densely villous; seeds 4-6, 6-8 x 4.5-6.5 mm, smooth.

Introduced for cultivation at least in AR at intermediate altitudes; native to the W Mediterranean region, widely cultivated for fodder and green manure.

3. **L. angustifolius** *L.* (1753).

Shortly hairy annual 20-80 cm tall. Leaflets linear to linear-spathulate, 10-50 x 2-5 mm, glabrous above and sparsely villous beneath; stipules linear-subulate. Racemes 10-20 cm long with alternate flowers; upper lip of calyx *c* 4 mm long, deeply divided, the lower 6-7 mm long, 3-dentate to subentire; corolla 11-13 mm long, blue. Pod shortly hirsute; seeds 4-6, ellipsoid, smooth.

Introduced for cultivation as a forage crop at least in AR at intermediate altitudes; native to the Mediterranean region, cultivated elsewhere for forage.

4. **L. mutabilis** *Sweet* (1825).

Subshrubby herb to 150 cm tall; stem glabrous. Leaflets oblanceolate, up to 60 x 14 mm, glabrous above, pubescent and somewhat glaucous beneath. Racemes *c* 6-20 cm long; upper lip of calyx slightly bifid or entire; corolla *c* 18-20 mm long, white, blue and yellow. Pod *c* 80 x 16 mm, appressed pubescent.

Occasionally cultivated in the highlands for its edible seeds or as an ornamental. WU GG SD; native of S America where it is also used as a fish poison. *Kuls* 699; *Hall* 35; *Smeds* 287 (H).

5. **L. albus** *L.* (1753).
L. termis Forssk. (1775).

Shortly hairy, bushy annual to 120 cm tall. Leaflets of lower leaves 25-35 x 14-18 mm, obovate, of the upper leaves 40-50 x 10-15 mm, obovate-cuneate, nearly glabrous above, sparsely villous beneath; stipules setaceous. Racemes 5-10 cm long, flowers alternate; calyx 8-9 mm long, both lips shallowly dentate; corolla 15-16 mm long, white to blue. Pod 60-100 x 11-20 mm, shortly villous, glabrescent; seeds 4-6, 8-14 mm long, smooth.

Cultivated, particularly in GJ; 1800-3000 m. GD GJ SU AR HA; native to the Mediterranean region, widely cultivated. *Meyer* 8664; *Evans & Lythgoe* 3; *Ebba* 736.

The seeds are eaten after soaking for 2-3 days to remove a bitter taste and then boiling. The seeds are also used to make a high quality 'ARAKI', local distilled spirits.

6. **L. mexicanus** *Cerv.* ex *Lag.* (1820).
L. ehrenbergii Schlechtend. (1838).

Short-lived perennial; stem villous with long spreading hairs. Leaflets oblanceolate, 15-70 x 3-15 mm, pilose beneath, usually glabrous above; stipules linear-subulate. Racemes pedunculate with many ± verticillate flowers; calyx *c* 10 mm long, upper lip bifid, the lower entire to minutely 3-fid; corolla 12-15 mm long, mostly blue. Pod villous, 3-6-seeded.

Locally naturalized in grassland, native to Central America; *c* 2000 m. SU. *Ash* 3029.

81. **ARGYROLOBIUM** *Eckl. & Zeyh.* (1836), *nom. conserv.*

Polhill in *Kew Bull.* **22**: 145-168 (1968) & *Fl. Trop. E. Afr.* (1971).

Herbs or small shrubs. Leaves digitately 3-foliolate. Flowers in terminal short racemes or subumbelliform, rarely solitary. Calyx deeply 2-lipped; upper lip deeply 2-fid; lower lip shortly 3-lobed. Corolla yellow; keel obtuse. Stamens all united into a usually closed tube. Pod linear-oblong, compressed, dehiscent.

Some 70 species, mostly in South Africa.

1. Stipules (1.5-)2-5 mm wide; valves of the pod continuous over the seeds, or the seeds apparent only as slight swellings; upper calyx-lobes ± lanceolate, broadest above the base. 2
- Stipules subulate, less than 1 mm wide; valves of the pod transversely constricted between the seeds; upper calyx-lobes subulate, broadest at the base. 3
2. Racemes *c* 6-40-flowered; bushy woody herb or subshrub usually 0.3-2 m tall. **1. A. fischeri**
- Racemes *c* 1-4-flowered; low diffuse herb, prostrate to shortly erect with slender stems. **2. A. rupestre**
3. Shrub; stipules united 1-2 mm with petiole; standard (10-)12-15 mm long. **3. A. schimperianum**
- Herbs, sometimes ± woody; stipules free or rarely slightly united to base of petiole; standard 5-10 mm long. 4
4. Usually annual with a single often much branched stem; flowers solitary or laxly racemose; leaflets usually more than 3 times as long as broad. **6. A. arabicum**
- Perennials, with numerous stems; flowers solitary or subumbelliform; leaflets less than 3 times as long as broad. 5
5. Flowers (1-)3-12(-16), in mostly terminal subumbelliform racemes supported on 10-70 mm long peduncles; leaflets 8-24 x 4-9 mm, subglabrous or with scattered hairs above, sparsely pilose beneath. **4. A. ramosissimum**
- Flowers 1-2(-3), leaf-opposed on a rhachis up to 5(-15) mm long or practically lacking; leaflets *c* 3.5-10 x 3-5 mm, usually densely pilose on both surfaces. **5. A. confertum**

1. **A. fischeri** *Taub.* (1895).
Macrolotus rivae Harms (1897); *A. rivae* (Harms) Cuf. (1969) - type: SD, Coromma, *Riva* 638 (B holo. destr., FT iso.).

Bushy woody herb or subshrub, 0.3-2 m tall, densely hairy. Leaflets mostly elliptic or elliptic-oblong, 20-60 x 9-26 mm, thinly to very densely covered with appressed or ± spreading hairs on both surfaces; petioles mostly 6-20(-40) mm long. Racemes usually subumbelliform, *c* 6-40-flowered. Calyx 10-12 mm long. Corolla 10-13 mm long, bright yellow, becoming orange or reddish flushed with age. Pod (20-)30-50 x 5-7 mm, densely silky pubescent or spreading pilose, *c* 12-18-seeded.

Margins and clearings of upland forest, upland bushland and grassland, stream sides; 1600-2400 m. WG SU BA HA IL KF GG SD; S Sudan, eastern tropical Africa, Angola. *Mooney* 6849; *Gillett* 14556; *Gilbert & Thulin* 391; *Friis* et al. 1542.

2. **A. rupestre** *(E. Mey.) Walp.* (1839).

Perennial with numerous slender, prostrate, straggling and ascending or shortly erect stems, sparsely to densely pilose. Leaflets elliptic-lanceolate to elliptic or obovate-elliptic, 10-34 x 5-14 mm, sparsely pilose to silky tomentose at least beneath; petiole 4-20 mm long. Racemes subumbelliform on 1-6(-9) cm long peduncles, *c* 1-4-flowered. Flowers often cleistogamous. Calyx 5-8 mm long. Corolla 9-13 mm long, bright yellow. Pod 18-30(-45) x 3.5-5 mm, sparsely pilose to densely pubescent, *c* 8-14-seeded.

1. Stems closely and rather densely appressed pubescent, straggling or weakly ascending, mostly 20-50 cm long; petioles *c* 8-16(-20) mm long. subsp. **remotum**
- Stems sparsely pilose with spreading or subappressed hairs, mostly shorter and prostrate; petioles *c* 4-6(-10) mm long. subsp. **aberdaricum**

Subsp. **remotum** *(Hochst. ex A. Rich.) Polhill* in *Kew Bull.* **22**: 159 (1968);
A. remotum Hochst. ex A. Rich. (1847) - type: GD, Simien, Mt Aber, *Schimper* II:1324 (P holo., BM iso.).
A. virgatum Bak. (1871) - type: TU, *Schimper* 570 (K holo., BM iso.).

Upland grassland and moor; 2000-2700 m. EW TU GD GJ SU AR; N Somalia, Yemen. *Pappi* 4722; *Thulin* 1393; *Gilbert & Aweke* 3102.

Subsp. **aberdaricum** *(Harms) Polhill* in *Kew Bull.* **22**: 159 (1968).

Upland grassland and moor; 2300-3400 m. SU AR BA HA SD; in E tropical Africa southwards to Malawi. *Mooney* 6462; *Burger* 1914; *Thulin* 1462.

Subsp. *rupestre* is distributed from E Zimbabwe to the SE Cape Province and subsp. *kilimandscharicum*

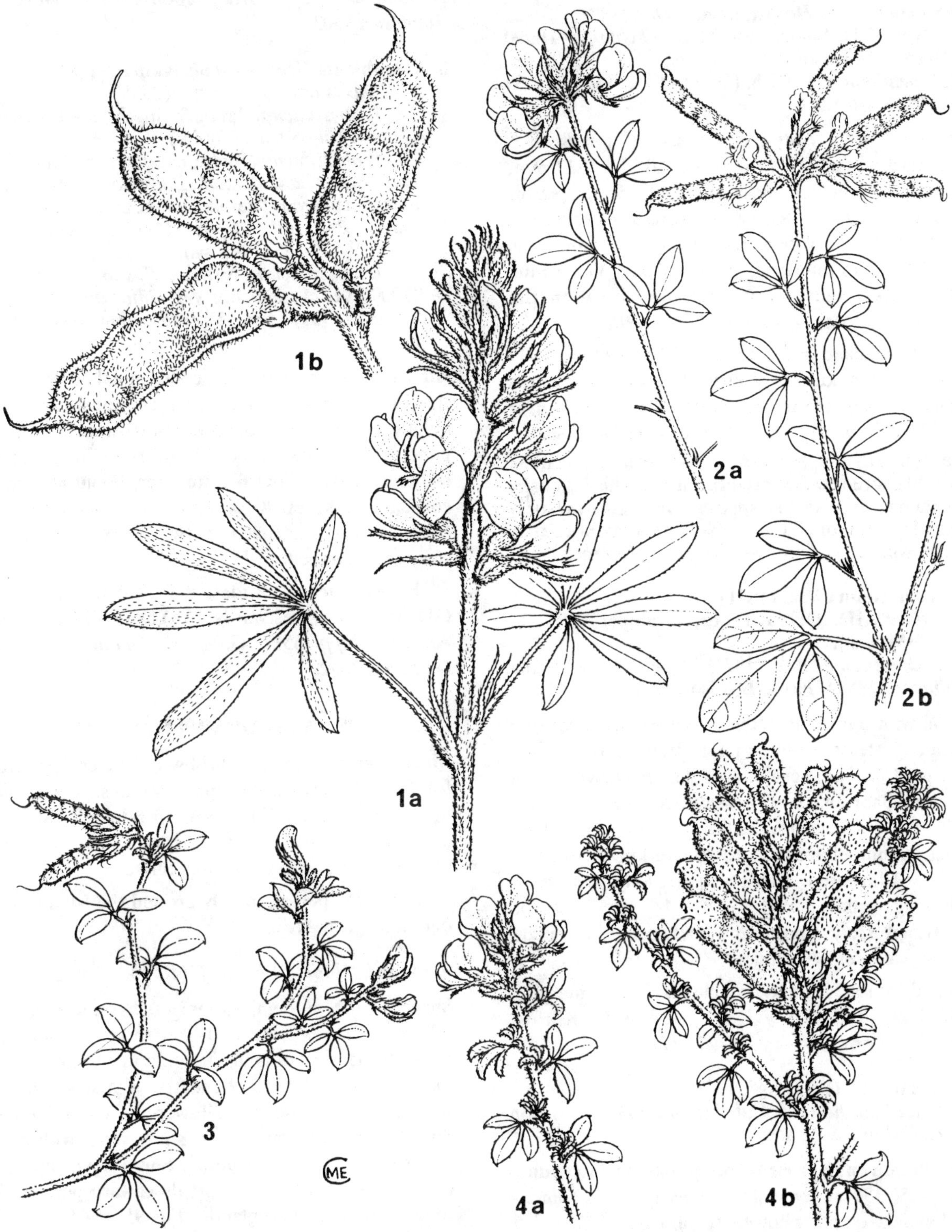

Figure 95.39.1 - *LUPINUS PRINCEI*, a: flowering stem, b: pods, both from *Gillett* 14199. 2 - *ARGYROLOBIUM SCHIMPERIANUM*, a: flowering stem, b: fruiting stem, both from *Ash* 1417. 3 - *A. CONFERTUM*, flowering and fruiting stem from *Robertson* 1303. 4 - *ADENOCARPUS MANNII*, a: flowering stem, b: fruiting stem, both from *Gillett* 14992. 1 & 2 x 0.75, 3 & 4 x 1.125. Drawn by Maureen Church. (Reproduced with permission from *Opera Botanica*, 68, fig. 41, 1983.)

(Taub.) Polhill in E tropical Africa from S Sudan to N Tanzania.

3. **A. schimperianum** *Hochst.* ex *A. Rich.* (1847)
- type: GD, Simien, Mt Buahit, *Schimper* II:990 (P lecto., BM K isolecto.).
A. petitianum A. Rich. (1847) - type: TU, Petit 39 (P holo., K iso.).

Shrub; stem with rusty-red or silvery pubescence. Leaflets narrowly elliptic to obovate, 10-35 x 7-13 mm, appressed pubescent or pilose at least beneath; petiole 3-20 mm long. Racemes subumbelliform, ± pedunculate, 2-10(-20)-flowered. Calyx 8-10 mm long. Corolla (10-)12-15 mm long, yellow. Pod *c* 30 x 5 mm, with the valves transversely constricted between the seeds, densely pilose, *c* 6-seeded. Fig. 95.39.2.

Upland bushland and grassland; 2100-3500 m. TU GD GJ SU; not known elsewhere, previous records from Sudan being erroneous (Polhill 1968). *Schimper* II:1326; *Pichi-Sermolli* 2693; *Gilbert & Tewolde* 3288.

The type of *A. petitianum* has smaller leaflets, shorter and more few-flowered racemes and is less hairy than other material of the species. It is also the only specimen known from TU. Probably it is no more than a form growing in an exceptionally exposed place.

4. **A. ramosissimum** *Bak.* (1871)
- type: GD, Debra Tabor, Schimper 1283 (K, holo., BM, iso.).
A. dorycnoides Bak. (1871) - type: GD, Mt Guna, *Schimper* 1285 (K lecto., BM isolecto.).

Perennial with numerous often rather woody spreading or ascending stems, 10-70 cm tall; stems rather thinly spreading pilose. Leaflets oblanceolate to obovate, 8-24 x 4-9 mm, glabrous or with scattered hairs above, ± sparsely pilose beneath; petiole *c* 2-4 mm long. Racemes subumbelliform on 10-70 mm long peduncles, (1-)3-12(-16)-flowered. Calyx 6-10 mm long. Corolla 7-10 mm long, yellow. Pod 13-28 x 4-5 mm, with the valves transversely constricted between the seeds, spreading pilose, *c* 8-10-seeded.

Upland grassland and bushland, forest margins; 2000-3550 m. GD GJ WG SU AR HA IL KF; W Kenya. *Mooney* 6274; *Evans* 347; *Thulin* 1363.

5. **A. confertum** *Polhill* (1968).
A. uniflorum sensu Cuf. *Enum.*: 243 (1955) non (Decne.) Jaub. & Spach.

Perennial with numerous procumbent or ascending stems, 10-15(-35) cm long; stems densely silvery tomentose. Leaflets elliptic-obovate to obovate, 5-10 x 3-5 mm, usually densely silvery pilose; petiole 1-3(-4) mm long. Flowers 1-2(-3), leaf-opposed on a rhachis up to 5(-15) mm long or practically lacking. Calyx 6-8 mm long. Corolla 7-8 mm long, yellow. Pod 20-24 x 3-3.5 mm, with the valves transversely constricted between the seeds, pilose, 7-9-seeded. Fig. 95.39.3.

Woodland, bushland, rocky grassland or hillsides, waste ground; 1200-2600 m. EW TU GD SU AR HA; N Somalia. *Pappi* 5768; *Mooney* 8108; *Thulin* 1463; *Robertson* 1303.

6. **A. arabicum** *(Decne.) Jaub. & Spach* (1843);
Cytisus arabicus Decne. (1835).
A. abyssinicum Jaub. & Spach (1843) - type: TU, *Quartin Dillon* s.n. (P holo.).
A. abyssinicum forma *caespitosum* Lanza in *Boll. Orto Bot. Giard. Colon. Palermo* **8**: 96 (1909); *A. abyssinicum* var. *caespitosum* (Lanza) Fiori in *Nouvo Giorn. Bot. Ital.* **19**: 451 (1912) - types: Eritrea, Senni 164 & 165 (PAL syn., not seen).
A. abyssinicum forma *diffusum* Lanza, loc. cit. (1909); *A. abyssinicum* var. *diffusum* (Lanza) Fiori, loc. cit. (1912) - type: Eritrea, Senni 166 (PAL holo., not seen).

Annual or rarely short-lived perennial to 50 cm high, spreading and ascending. Leaflets linear-lanceolate to elliptic, 10-25 x 2-8 mm, silvery strigose at least beneath; petioles 10-15 mm long. Flowers yellow, solitary or laxly racemose. Calyx 6-7 mm long, about as long as the corolla. Pod 30-40 x 3.5-4 mm, with the valves transversely constricted between the seeds, silvery strigose, 9-12-seeded.

Rocky places, hillsides; 850-2700 m. EE EW TU GD WU SU; westwards to Tchad (Tibesti) and in Arabia. *Pappi* 3512; *Gilbert & Thulin* 985; *Gilbert & Getachew* 2812.

82. ADENOCARPUS *DC.* (1815)

Much-branched hairy shrubs. Leaves digitately 3-foliolate. Flowers in terminal racemes. Calyx 2-lipped; upper lip deeply 2-fid; lower lip 3-lobed. Stamens all united into a closed tube. Pod linear-oblong, compressed, dehiscent, glandular-tuberculate.

Some 15 species, mainly around the Mediterranean, one in tropical Africa.

A. mannii *(Hook. f.) Hook. f.* (1864).

Shrub to 4.5 m high; stems much branched, densely pilose. Leaves clustered on short shoots, shortly petiolate; leaflets oblanceolate-elliptic, up to 8(-12) mm long, densely pilose beneath, sparsely so above. Inflorescence dense, 6-20-flowered. Calyx with upper lip *c* 6 mm long, lower lip *c* 8 mm long, with narrowly linear lobes. Corolla yellow, *c* 8 mm long. Pod 18-25 x 6-7 mm, straight with a short downward pointing beak, rough with projecting glands. Fig. 95.39.4.

Upland grassland and scrub, forest margins; 2900-3300 m. SU KF SD; widespread in tropical Africa. *Gillett* 14992; *Gilbert & Tewolde* 3291; *Thomerson* 578.

83. SPARTIUM *L.* (1753)

Shrub. Leaves 1-foliolate, falling soon. Flowers in lax, terminal racemes. Calyx spathe-like, split above, irregularly one-lipped, with 5 short lobes. Corolla yellow. Stamens all united into a tube. Pod linear-oblong, dehiscent.

Monotypic.

S. junceum *L.* (1753).

Stems glabrous, to at least 3 m tall. Leaves sparse, oblong-linear, 10-30 x 2-5 mm, appressed pubescent beneath. Corolla 20-25 mm long. Pod flat, 40-80 x 5-7 mm, sericeous, becoming glabrous, 10-18-seeded.

Cultivated for its fibers and as an ornamental, in Ethiopia only as an ornamental at least in most major towns; native in the Mediterranean region. *Smeds* 746 (FT); *Senni* 1428; *IECAMA* B-27.

84. PSORALEA *L.* (1753)

Herbs or shrubs, gland-dotted. Leaves 3-foliolate or rarely imparipinnate or 1-foliolate. Calyx usually campanulate, the lobes equal or somewhat unequal. Corolla glabrous, not persistent after flowering, bluish. Pod, small, indehiscent, 1-seeded.

A heterogeneous group of some 120 species sometimes split into segregate genera, most numerous in South Africa. According to the generic classification proposed by Stirton in *Adv. Leg. Syst.*: 337-343 (1981) *P. plicata* belongs to *Cullen* Medik. and *P. bituminosa* to *Bituminaria* Heist. ex Fabr.

1. Flowers in lax racemes; corolla 4-5 mm long; leaflets undulate-plicate; pod not beaked. **1. P. plicata**
- Flowers in dense heads; corolla 15-20 mm long; leaflets entire; pod with a long beak. **2. P. bituminosa**

1. **P. plicata** *Del.* (1813).

Bushy plant to 1 m high, grey pubescent, weakly spiny. Leaves pinnately 3-foliolate; leaflets oblanceolate or oblong, up to 25 x 10 mm, the terminal the largest; margin repand to undulate or plicate. Racemes stout, axillary, lax, up to 30-flowered. Calyx strongly nerved. Corolla 4-5 mm long, white or white and purple. Pod ellipsoid, about as long as the calyx, not beaked, pubescent. Fig. 95.43.1.

Sandy or silty often saline ground at low altitudes. HA; Somalia, Sudan and west to Senegal, also in India. *Glover & Gilliland* 300.

2. **P. bituminosa** *L.* (1753).

Bushy plant to 1 m high, pubescent, smelling of tar. Leaflets 3, entire; flowers in long-pedunculate heads. Corolla 15-20 mm long, blue-violet. Pod ovoid, compressed, with a falcate beak to 15 mm long.

Introduced in EW and apparently locally naturalized; native in the Mediterranean region. *Pappi* 73, 125.

85. LOTUS *L.* (1753)

Brand in *Engl. Bot. Jahrb.* **25**: 165-232 (1898); Gillett in *Kew Bull.* **13**: 361-381 (1958) & *Fl. Trop. E. Afr.* (1971).

Herbs or softly woody shrublets. Leaves imparipinnate, (3-)5-foliolate, the lowest pair resembling stipules; stipules minute. Flowers in pedunculate heads or umbels, or rarely solitary. Calyx subequally 5-toothed or rarely 2-lipped. Upper filament free, the other 9 united. Pod straight, usually cylindrical, dehiscent, many-seeded.

Old World genus of 50-100 species, most numerous around the Mediterranean.

The Ethiopian record of *L. nubicus* Hochst. ex Bak. in Cufodontis *Enum.*: 259 (1955) is based on *Rosen* s.n. (WRSL) from GJ which is instead *Argyrolobium rupestre* subsp. *remotum*.

1. Corolla yellow with red markings. 2
- Corolla rose, cream or white. 3
2. Plant practically glabrous; rootstock woody. **1. L. schoelleri**
- Plant pilose at least on peduncles and young shoots; rootstock ± herbaceous. **2. L. corniculatus**
3. Ovary and pod pubescent. **9. L. hebecarpus**
- Ovary and pod glabrous. 4
4. Perennials growing above 1200 m altitude. 5
- Annuals mostly growing below 1200 m altitude. 7
5. Calyx-lobes less than 1.3 times as long as the tube. **3. L. discolor**
- Calyx-lobes more than 1.3 times as long as the tube. 6
6. Style 3.3-4 mm long; ovules *c* 17; pod usually less than 3 times as long as calyx; peduncle shorter or longer than subtending leaf. **4. L. quinatus**
- Style 4-5.5 mm long; ovules *c* 30; pod more than 3 times as long as calyx; peduncle longer than subtending leaf. **5. L. lalambensis**
7. Peduncle much shorter than subtending leaf; style *c* 1.5 mm long. **8. L. schimperi**
- Peduncle up to about twice as long as subtending leaf; style 3-5.5 mm long. 8
8. Basal leaflets much wider at the base than the distal leaflets; calyx-lobes about as long as the tube; pod torulose. **7. L. torulosus**
- Basal leaflets of much the same shape as the distal ones; calyx-lobes markedly longer than the tube; pods not or scarcely torulose. 9
9. Leaflets ± oblanceolate; leaf-rhachis usually not prolonged beyond the lateral leaflets. **5. L. lalambensis**
- Leaflets cuneate-obovate; leaf-rhachis usually prolonged 0.5-1.5 mm beyond the lateral leaflets. **6. L. arabicus**

1. **L. schoelleri** *Schweinf.* (1896);
L. corniculatus L. var. *schoelleri* (Schweinf.) Lanza in *Boll. R. Orto Bot. Palermo* **8**: 99 (1909) -

type: Eritrea, *Schweinfurth* 322 (B holo. destr., K iso.).

L. corniculatus var. *eremanthus* Chiov., *Racc. Bot. Miss. Consol. Kenya*: 29 (1935).

L. mearnsii De Wild. (1925), nom. illegit., non Britton (1890).

Perennial herb with ± woody rootstock and erect or straggling stems to 60 cm long, but usually much smaller, whole plant practically glabrous except for a few hairs on the edge of the calyx and the petiolules. Leaves very variable in size; rhachis to 6 mm long; terminal leaflet up to 14 x 7 mm, cuneate-obovate, rounded at the apex; lateral and basal leaflets somewhat smaller, assymmetric. Umbel 1-2(-3)-flowered; peduncle up to 50(-70) mm long. Calyx lobes up to 3 mm long, equalling or somewhat longer than tube. Corolla yellow, *c* 9 mm long. Pod up to 30 x 4 mm, *c* 15-21-seeded. Fig. 95.40.1.

Upland grassland, damp places; 2000-3000 m. EW TU GD WU GJ SU AR HA; Sudan, Kenya. *Thulin* 1575; *Mooney* 6402; *W. de Wilde* 5874.

L. schoelleri is a diploid and therefore, according to current taxonomy of the group, should be treated as specifically distinct from the closely related, tetraploid *L. corniculatus* s. str.

2. **L. corniculatus** *L.* (1753).

Perennial herb with ± herbaceous stoloniferous rootstock and erect or ascending stems to at least 60 cm long, ± pilose at least on the young shoots. Leaf rhachis up to 10 mm long; terminal leaflet up to 20 x 10 mm, cuneate-obovate, rounded at the apex; lateral and basal leaflets somewhat smaller, asymmetric. Umbel 1-3(-4)-flowered; peduncle up to 50(-90) mm long, weakly pilose. Calyx-lobes 3-5 mm long, pilose, longer than tube. Corolla yellow, 9-10 mm long. Pod up to 40 x 3 mm, *c* 20-25-seeded.

Marshy places in grassland; 1900-2400 m. EW TU HA; Europe, temperate Asia. *Burger* 2039; *Aweke & Gilbert* 758.

The material placed here was by Gillett (1958) considered to be partly *L. palustris* Willd. (= *L. lamprocarpus* Boiss.) and partly *L. corniculatus* L. subsp. *major* (Scop.) Gams (= *L. uliginosus* Schkuhr). In the authors opinion there is only one taxon in Ethiopia, but the plants cannot readily be assigned to any of the two European species mentioned. Progeny of *Berhe* 8, from *c* 15 km NW of Harer, has been cultivated in Uppsala and was found to be tetraploid (2n = 24). This number agrees with that of *L. corniculatus* and the Ethiopian plants are probably best referred to this variable species. The description only applies to the Ethiopian material, however.

3. **L. discolor** *E. Mey.* (1836).

L. tigrensis Bak. (1871) - type: GD, 1863, *Schimper* s.n. (K holo., BM ? iso.).

L. goetzei Harms (1901).

Perennial often bushy herb from a woody rootstock, appressed or spreading pubescent. Leaf rhachis 2-12 mm long; leaflets very variable in size and shape, cuneate-ovate to oblanceolate, ± pubescent. Umbels 2-9-flowered; peduncle 0.5-8 cm long. Calyx pubescent; lobes less than 1.3 times as long as the tube. Corolla white with pink markings, *c* 8-15 mm long. Pod *c* 2.5 cm long, *c* 25-ovulate, glabrous.

Upland grassland, scrub or forest edges; 2000-4000 m. GD WU SU AR BA HA KF GG SD; widespread in tropical and southern Africa. *Thulin* 1655; *Mooney* 6975; *Friis* et al. 1404.

L. discolor is here taken in a broad sense. Gillett (1958 and 1971) assigned some Ethiopian material to *L. goetzei* and also recognized several other closely related species in eastern Africa. The author prefers to treat all Ethiopian forms as a single variable species as all the distinguishing characters seem to break down in this area. Here is also included *L.* sp. ? new *'A'* sensu Gillett (1958) from GG. The plants from HA have a characteristic, very dense short greyish indumentum of spreading hairs.

4. **L. quinatus** *(Forssk.) Gillett* (1958).

Prostrate perennial, to 50 cm long, appressed or spreading pubescent. Leaf rhachis up to 4 mm long; leaflets 3-5, ± oblanceolate, up to 25 x 7 mm, the basal pair similar to the others. Umbels 2-6-flowered; peduncle shorter or longer than subtending leaf. Calyx-lobes more than 1.3 times as long as the tube. Corolla white ± flushed pink, 7-8 mm long. Ovules *c* 17; style 3.3-4 mm long. Pod up to 20 x 2 mm, usually less than 3 times as long as calyx, glabrous.

Var. **brachycarpus** *(Hochst. & Steud. ex A. Rich.) Gillett* in *Kew Bull.* **13**: 373 (1958);

L. brachycarpus Hochst. & Steud. ex A. Rich. (1847) - types: TU, *Schimper* I:242 (P syn., BM K isosyn.) & *Quartin Dillon* s.n. (P several syn.).

L. montanus A. Rich. (1847); *L. nubicus* Hochst. ex Bak. var. *montanus* (A. Rich.) Bak. in *Fl. Trop. Afr.* **2**: 62 (1871); *L. brachycarpus* var. *montanus* (A. Rich.) Cuf., Cuf. *Enum.*: 257 (1955) & *Zweiter Nachtrag*: 24 (1969), nom. invalid. - type: TU, *Petit* s.n. (P holo.).

Dorycnium schimperi Jaub. & Spach (1856) - type: *Schimper* 19 (P holo.).

Pubescence of ± spreading hairs. Leaf rhachis 1-4 mm long. Bract usually 1-foliolate. Peduncle often longer than subtending leaf. Fig. 95.40.2.

Grassland, roadsides, old cultivations; (1350-)1900-3000 m. EW TU GD WU SU HA; not known elsewhere. *Gilbert & Getachew* 2849; *Schweinfurth* 226; *Robertson* 1219.

Var. *quinatus* occurs in Yemen and N Somalia, a record from SD (Gillett, loc. cit.), was based on *Riva* 172 which is *L. discolor.*

5. **L. lalambensis** *Schweinf.* (1896)
- type: Eritrea, *Schweinfurth* 861 & 1520 (B syn. destr.).
L. arabicus L. var. *glabrescens* Schweinf. in *Bull. Herb. Boiss.* **4**, app. 2: 234 (1896) - types: Eritrea, *Schweinfurth* 323 (B syn. destr., K isosyn.), 707 & 1498 (both B syn. destr.) & *Schweinfurth & Riva* 1849 (B syn. destr., K isosyn.).
L. hebranicus Hochst. ex Brand (1898) pro parte - types from Arabia, Egypt and Eritrea: *Schweinfurth* 323 (B syn. destr., K isosyn.).
?*L. brachycarpus* var. *major* Brand in *Engl. Bot. Jahrb.* **25**: 222 (1898) - type: N Ethiopia, *Schimper* 251 (B holo. destr.).

Prostrate herb to 50 cm long, annual or short-lived perennial, ± spreading pubescent. Leaf rhachis 2-5 mm long; leaflets 5, ± oblanceolate, up to 25 x 7 mm, the basal pair similar to the others. Umbels 3-7-flowered; peduncle longer than subtending leaf. Calyx-lobes more than 1.3 times as long as the tube. Corolla white ± flushed pink, 7-11 mm long. Ovules *c* 30; style 4-5.5 mm long. Pod up to 23 x 2 mm, more than 3 times as long as calyx, glabrous.

Grassland, cultivations; 1200-2000 m. EW HA; probably also in Djibouti. *Colville* 76; *Schweinfurth & Riva* 1849; *Bally* 11640.

6. **L. arabicus** *L.* (1767).

Annual, spreading or erect herb, to 60 cm tall, pilose throughout. Leaf rhachis 2-8 mm long; leaflets 5, cuneate-obovate, up to 20(-30) x 12(-16) mm, the basal pair similar to the others. Umbel 1-5-flowered; peduncle up to 4 cm long. Calyx-tube *c* 3.5 mm long; lobes linear *c* 7 mm long. Corolla pink, *c* 8-9 mm long. Style *c* 4 mm long. Pod 20-35 x *c* 3.5 mm, subtorulose, *c* 12-18-seeded.

River banks at ± low altitude. EE ?TU; westwards to Senegal, S tropical Africa, and in Arabia. *Quartin Dillon & Petit* s.n.; *Pappi* 8704, 8819.

7. **L. torulosus** *(Chiov.) Fiori* (1912);
L. arabicus L. var. *torulosus* Chiov. in *Ann. Ist. Bot. Roma* **8**: 86 (1903) - types: Eritrea, *Pappi* 39, 64, 1725, 4127, *Scotti* s.n. (all FT syn.).
?*L. brachycarpus* var. *cuscutae* Schweinf. in *Bull. Herb. Boiss.* **4**, App. 2: 233 (1896) - type: Eritrea, *Schweinfurth* 2034 (B holo. destr.).

Procumbent annual to 40 cm long, pilose. Leaf rhachis 2-10 mm long; leaflets ± cuneate-obovate, up to 16 x 9 mm, the basal pair much wider at the base than the others. Umbel 2-5-flowered; peduncle up to 4 cm long. Calyx-tube *c* 3 mm long; lobes 3-4 mm long. Corolla pink, 7-8 mm long. Style 4-5 mm long. Pod up to 30-50 x 2 mm, torulose, glabrous.

Sandy or gravelly places, cultivations; up to 1500 m. EE EW; N Sudan. *Bally* 6834; *Pappi* 2873, 4800; *Sue Edwards* 3746.

L. brachycarpus var. *cuscutae* was placed with a question mark as a synonym of *L. lalambensis* by Gillett (1958). As the pods were described as torulose and 2-3 cm long it is more likely *L. torulosus*.

8. **L. schimperi** *Steud.* ex *Boiss.* (1872).
L. glinoides sensu Cuf. *Enum.*: 258 (1955), non Del.

Small, prostrate annual, sparsely pubescent with spreading hairs. Leaf rhachis 1-5 mm long; leaflets cuneate-obovate, up to 10 x 5 mm. Umbel (1-)2-4(-6)-flowered, sessile or subsessile, shorter than the subtending leaf. Calyx-tube 1.5-2 mm long; lobes 1.5-2 mm long. Corolla pinkish, *c* 4 mm long. Style *c* 1.5 mm long. Pod *c* 7-9 (-11) x 2 mm, glabrous, *c* 6-8-seeded.

Coastal plains. EE; Sudan, Egypt, Socotra, Arabia and E to Pakistan. *Terracciano* 691.

9. **L. hebecarpus** *Gillett* (1958).

Annual or sometimes perennial herb; stems procumbent, to 40 cm long, pilose. Leaf rhachis 1-4 mm long; basal leaflets asymmetrically ovate, apical leaflet oblanceolate to cuneate-obovate, pilose. Umbels 1-3(-5)-flowered; peduncle 3-18 mm long. Calyx-lobes linear-acuminate, *c* 4-5 mm long. Corolla red or white and red, *c* 10 mm long. Pod up to 24 x 2 mm, pilose, *c* 30-ovulate.

On rocky or sandy ground; 0-60 m. EE; Sudan, Djibouti. *Bally* 6887; *Hemming* 1156.

86. ANTHYLLIS *L.* (1753)

Herbs or shrubs. Leaves usually imparipinnate. Flowers usually in dense heads. Pod frequently indehiscent or tardily dehiscent, usually completely included within the persistent calyx.

Some 20 species, mainly in the Mediterranean region.

A. vulneraria *L.* (1753).

Annual or perennial. Lower leaves with a much larger terminal leaflet. Heads many-flowered, subtended by 2 palmatisect bracts. Calyx inflated at anthesis, constricted at the apex, with 5 unequal teeth and the mouth oblique. Corolla variously coloured. Pod 1(-2)-seeded.

Subsp. **abyssinica** *(Sag.) Cullen* in *Notes Roy. Bot. Gard. Edinb.* **35**: 13 (1976);
A. abyssinica (Sag.) Beck. in *Beih. Bot. Centralbl.* **27, 2**: 273 (1910) - type: TU, *Schimper* 616 (WU holo., K iso.).

Plant robust, often 20 cm or more high. Stem indumentum in upper part of plant usually appressed, sericeous. Leaves confined to the lower part of the flowering stem, with conspicuously long-ciliate margins. Corolla purple. Fig. 95.40.3.

Figure 95.40.1 - *LOTUS SCHOELLERI*, flowering and fruiting stem from *Mooney* 5007. 2 - *L. QUINATUS* var. *BRACHYCARPUS*, a: flowering stem, b: fruiting stem, both from *Schweinfurth* 226. 3 - *ANTHYLLIS VULNERIA* subsp. *ABYSSINICA*, flowering stem from *Hagos* 178. 4 - *VERMIFRUX ABYSSINICA*, a: flowering stem from *Mooney* 8030, b: pods from *Pappi* 16. 1, 2 & 4a x 1.125, 3 x 0.75, 4b x 2.25. Drawn by Maureen Church. (Reproduced with permission from *Opera Botanica*, 68, fig. 42, 1983.)

Rocky ground; *c* 2500-3000 m. EW TU WU; not known elsewhere. *Hagos* 178; *Petit* s.n.; *Pappi* 3112.

87. **VERMIFRUX** *Gillett* (1966)

Perennial herb. Leaves imparipinnate, subsessile. Flowers small in axillary heads. Pod falcate, indehiscent, 2-seeded with an imperfect partition between the seeds.

Monotypic.

V. abyssinica *(A. Rich.) Gillett* (1966);
Helmintocarpon abyssinicum A. Rich. (1847) - type: TU, *Schimper* II:1108 (P lecto., BM K isolecto.).

Stems usually numerous, ± procumbent, to 40 cm long, pubescent. Leaflets 5-9, oblanceolate, up to 12 x 4 mm, pubescent. Heads 4-6-flowered, short-pedunculate. Corolla yellowish-red, about twice as long as calyx. Pod *c* 4 mm across, the end nearly or quite brought round to the base, transversely veined. Fig. 95.40.4.

Upland grassland and bushland; 1500-3400 m. EW TU GD WU SU AR HA; Sudan, N Somalia, Yemen. *Gilbert & Thulin* 1011; *Burger* 1179; *Mooney* 8030.

Elsewhere the roots are used to treat gripe.

88. **CORONILLA** *L.* (1753)

Annual or perennial herbs or dwarf shrubs. Leaves imparipinnate, rarely simple or 3-foliolate. Flowers in axillary heads. Calyx-lobes subequal. Upper filament free, the other 9 united. Pod jointed.

Some 20 species mainly in the Mediterranean region.

C. scorpioides *(L.) Koch* (1835).

Annual to 40 cm tall. Leaves simple or 3-foliolate; terminal leaflet elliptic to suborbicular, up to 40 mm long, much larger than the lateral leaflets. Heads 2-5-flowered; corolla yellow, 4-8 mm long. Pod 20-60 mm long, curved, angular, with 2-11 segments.

Weed; *c* 2450 m. EW; widespread in the Mediterranean region and in W Asia. *De Benedictis* 297.

89. **ANTOPETITIA** *A. Rich.* (1840)

Annual herb. Leaves imparipinnate; stipules reduced to glands. Flowers in axillary pedunculate umbels. Calyx-lobes subequal. Upper filament free, the other 9 united. Pod of spherical 1-seeded segments each splitting into 2 valves which separate from the persistent upper suture.

Monotypic.

A. abyssinica *A. Rich.* (1840)
- type: TU, *Quartin Dillon* s.n. (P, holo.).

Ornithopus coriandrinus Hochst. & Steud. ex Field & Gardn. (1844) - type: TU, *Schimper* 113 (OXF holo., BM K iso.).

Leaflets 5-11, alternate or subopposite, linear-lanceolate, 4-20 x 0.4-4 mm, glabrous above. Umbels 2-8-flowered. Corolla yellow, *c* 4 mm long. Pod 2-5-seeded, the segments 2-2.5 mm in diameter. Fig. 95.41.1.

Upland grassland; 1600-2900 m. EW TU GJ WG SU AR HA IL KF GG SD; widespread in tropical Africa. *Thulin* 1499; *Mooney* 6141; *W. de Wilde* 8397.

90. **SCORPIURUS** *L.* (1753)

Annuals. Leaves simple with 3-5 parallel veins; stipules free, linear. Flowers solitary or in axillary heads. Calyx-lobes 5, equal. Corolla yellow or purplish. Upper filament free, the other 9 united. Pod curved or variously contorted, longitudinally ridged, usually with spines or tubercles.

Small genus around the Mediterranean and east to the Caucasus.

S. muricatus *L.* (1753).

Stems to 80 cm long, glabrous or pubescent. Heads (1-) 2-5-flowered. Corolla 5-10(-12) mm long, yellow. Pod smooth (not in Ethiopia) or with tubercles or spines on the outer ridges. Fig. 95.41.2.

Grassland, roadsides, a weed in cultivations particularly on black cotton soil; 1500-3200 m. EW TU GD WU SU AR BA KF; as the genus. *Pappi* 4362; *Thulin* 1616; *Ash* 2133; *W. de Wilde* 6034.

91. **HEDYSARUM** *L.* (1753)

Annual or perennial herbs, rarely shrublets. Leaves imparipinnate. Flowers in axillary racemes. Calyx-lobes subequal. Standard and usually also keel longer than the wings. Upper filament free, the other 9 united. Pod ± flattened, breaking transversely into up to 8 indehiscent 1-seeded segments.

About 100 species in Europe, N Africa, Asia and N America.

H. coronarium *L.* (1753).

Perennial, sparsely pubescent. Leaflets 7-11, elliptic to obovate-orbicular, pubescent beneath. Racemes 10-35-flowered, dense; corolla 12-15 mm long, reddish-purple. Pod with 2-4 segments, spinulose, glabrous.

Grassland; *c* 2000-2500 m. EW; native of the Mediterranean region, cultivated for fodder and perhaps locally naturalized in Eritrea. *Pappi* 154, 367.

92. TAVERNIERA *DC.* (1825)

Thulin in *Symb. Bot. Ups.* **25(1)**:43-95.

Shrubs or shrublets. Leaves 1-foliolate or digitately to pinnately 3-foliolate; leaflets usually rather thick. Flowers in axillary racemes which are sometimes reduced to 1-few-flowered clusters. Standard and keel longer than the wings; the wilted corolla persistent around the pod. Upper filament free, the other 9 united. Pod flattened, breaking transversely into 1-5 indehiscent 1-seeded segments.

Small genus of arid regions from Egypt to India.

1. Leaves digitately 3-foliolate. **4. T. lappacea**
- Leaves 1-foliolate, sometimes with some pinnately 3-foliolate ones present as well. 2
2. Leaflets appressed pubescent on both sides; inflorescences very short with the rhachis obsolete, subsessile or with peduncle up to 2.5 mm long. **3. T. aegyptiaca**
- Leaflets glabrous above; inflorescences ± elongated with a distinct rhachis, peduncle usually more than 2.5 mm long. 3
3. Calyx pubescent outside; leaflets usually about twice as long as broad. **1. T. abyssinica**
- Calyx glabrous outside; leaflets about 1.5 times as long as broad. **2. T. schimperi**

1. **T. abyssinica** *A. Rich.* (1847)
 - type: TU, *Petit* s.n. (P holo.).

Shrub or shrublet to *c* 2 m high; young stems rather densely appressed pubescent. Leaves 1-foliolate, very rarely pinnately 3-foliolate ones present as well; leaflets obovate-oblong, up to 20(-23) x 10(-13) mm, glabrous above, appressed pubescent beneath; petiole 1-10 mm long. Racemes 2-8-flowered; rhachis and peduncle together *c* 3-25 mm long. Calyx 5-7.5 mm long, appressed pubescent outside; lobes equalling or longer than tube. Corolla 12-17 mm long, dark pink to purplish red. Pod with 1-3 segments, stipitate; segments *c* 5-6.5 x 4.5-5 mm, finely pubescent and with spines to *c* 1.5 mm long. Fig. 95.41.3.

Bushland, also on limestone; *c* 1700-2150 m. TU SU; not known elsewhere. *J. De Wilde* 6991; *Negri* 720; *Adome* s.n.; *Samuel Tadesse* s.n.

2. **T. schimperi** *Jaub. & Spach* (1856)
 - type: TU/EW, Tekeze River, Gurrsarfa, *Schimper* 1489 (P holo., BR EA MO S W WU iso.).
 T. floribunda Schweinf. (1867) - type: TU/EW, Tekeze River, Gurrsarfa, *Schimper* 2329 (B holo. destr., K P W iso.).

Shrub or shrublet to *c* 2 m high; young stems sparsely appressed pubescent. Leaves 1-foliolate, sometimes pinnately 3-foliolate ones present as well; leaflets obovate to broadly elliptic, up to 30 x 22 mm, glabrous above, appressed pubescent beneath; petiole up to 18 mm long. Racemes 3-10-flowered; rhachis and peduncle together *c* 8-60 mm long. Calyx 4.5-6 mm long, glabrous outside; lobes equalling or longer than tube. Corolla 11-15 mm long. Pod with 1-3 segments, stipitate; segments *c* 7 x 4.5-5 mm, finely pubescent and with spines up to 1.6 mm long.

Bush/woodland; *c* 1000-1300 m. EW/TU SU(Blue Nile Gorge); not known elsewhere. *Schimper* 201; *Thulin & Hunde* 4096 (sterile).

3. **T. aegyptiaca** *Boiss.* (1849).

Shrublet to *c* 1 m high; young stems densely appressed pubescent. Leaves 1-foliolate; leaflets obovate to obcordate, 4-10 x 4-10 mm, appressed pubescent on both sides or glabrescent above; petiole 0.5-5.5 mm long. Racemes 1-5-flowered, very short with the rhachis obsolete; peduncle up to 2.5 mm long. Calyx 3.5-5 mm long, appressed pubescent outside; lobes shorter to longer than tube. Corolla 9.5-12 mm long, pink or purple. Pod with 2-5 segments, stipitate; segments 3.5-6 x 3.5-5 mm, pubescent and usually spiny.

Sandy, often seasonally flooded, saline ground near sea level. EE; along Red Sea coast northward to the Gulf of Suez. *Hemming* 1186.

The single Ethiopian collection has pods with rudimentary spines.

4. **T. lappacea** *(Forssk.) DC.* (1825).
 T. stefaninii Chiov. (1929).

Small shrub or woody herb to *c* 1 m high, densely woolly pubescent throughout. Leaves digitately 3-foliolate; leaflets obovate to obcordate, 3-12 x 3-12 mm; petiole 0.5-5 mm long. Racemes 1-5(-8)-flowered, very short with the rhachis usually obsolete; peduncle 0.5-6 mm long. Calyx 3-4 mm long; lobes longer than tube. Corolla 5-6.5 mm, yellowish, pink or purple. Pod with (1-)2(-3) segments, practically sessile; segments 6-10 x 6-8 mm, woolly pubescent and with hairy spines 1-4 mm long.

Coastal plains and semideserts on sandy or gravelly, often seasonally flooded, saline ground. EE; along the coasts of the Red Sea and the Gulf of Aden, also Oman and Pakistan. *Hemming* 1187.

93. ONOBRYCHIS *Mill.* (1754)

Annual or perennial herbs, or shrublets. Leaves imparipinnate. Flowers in axillary racemes. Calyx-lobes subequal. Standard and keel much longer than the wings. Pod indehiscent, ± orbicular and compressed, with a distinct, usually toothed margin, 1-3-seeded.

Some 100 species, mainly in the Mediterranean region and W Asia.

1. Stem sparsely pubescent; leaflets 12-28; pod toothed at the margin, otherwise smooth. **1. O. viciifolia**
- Stem densely white-pubescent; leaflets 6-12; pod spiny at the margin and on both faces. **2. O. richardii**

Figure 95.41.1 - *ANTOPETITIA ABYSSINICA*, a: flowering stem from *W. de Wilde* 8397, b: pods from *Berhe* 26. 2 - *SCORPIURUS MURICATUS*, a: flowering stem from *Hansson* 395, b: pods from *Meyer* 7575. 3 - *TAVERNIERA ABYSSINICA*, flowering and fruiting stem from *Harlan* s.n. 4 - *GALEGA SOMALENSIS*, a: flowering stem, b: fruiting raceme, both from *Gillett* 14884. 1, 2b & 3 x 1.125, 2a & 3 x 0.75. Drawn by Maureen Church. (Reproduced with permission from *Opera Botanica*, 68, fig. 43, 1983.)

1. **O. viciifolia** *Scop*. (1772).
O. vulgaris Hill (1756), nom. invalid.

Subglabrous to pubescent perennial, 10-80 cm tall. Leaflets 12-28, ovate to oblong or linear, 10-35 x 4-7 mm. Corolla pink with purple veins, 8-14 mm long, glabrous. Pod 5-8 mm long, pubescent; sides toothed; margin usually with 6-8 teeth, to 1 mm long.

Introduced; 2200-2300 m. EW; mainly Europe and Asia, widely grown for fodder and naturalized. *Pappi* 157, 4400; *Schweinfurth & Riva* 952.

2. **O. richardii** Bak. (1871)
- type: TU, *Petit* s.n. (P holo., W iso.).

Densely pilose herb, to 40 cm high. Leaflets (6-)8-12, lanceolate or oblong, acute, *c* 20 x 6 mm, glabrous above, silky pilose beneath. Racemes long-pedunculate. Corolla *c* 10 mm long; standard hairy. Pod *c* 11 mm long, suborbicular, densely lanate, echinate on the edge and on both faces.

Habitat not known. TU; only known from the type collected in Shelicot (Tchéleucote) at app. 13°21´N, 39°34´E in 1844.

94. GALEGA *L.* (1753)

Gillett in *Kew Bull.* **17**: 81-88 (1963) & *Fl. Trop. E. Afr.* (1971).

Erect herbs. Leaves imparipinnate; stipules divided into 2 or more acute lobes. Flowers in axillary racemes. Calyx-lobes 5, subequal. Corolla purplish-blue or white. Stamens united into a closed tube. Pod unilocular, dehiscent, the sutures not impressed, with prominent parallel oblique veins.

Some three species in the area from Iran to the Adriatic and three more in E Africa.

G. somalensis *(Taub. ex Harms) Gillett* (1963);
Astragalus somalensis Taub. ex Harms (1897) - type: SD, Burgi, *Riva* 1398 (FT holo.).

Perennial, sparsely pilose. Leaflets 9-17, opposite or alternate, up to 22 x 6 mm; stipules 3-5-partite. Racemes lax, many-flowered. Corolla 6-8 mm long, purplish blue. Pod *c* 17 x 4-5 mm, flat, curved downwards and tapering to a short beak, 1-3-seeded. Fig. 95.41.4.

Upland grassland, near streams and in shady places; *c* 1600 m. SD; not known elsewhere. *Gillett* 14884.

95. ASTRAGALUS *L.* (1753)

Herbs or small shrubs. Leaves imparipinnate or paripinnate; stipules entire. Flowers in axillary racemes or clusters. Calyx-lobes 5, subequal. Upper filament free, the other 9 united. Pod ± divided lengthwise into 2 loculi by a vertical membrane, usually dehiscent, 1-many-seeded.

Some 2000 species, mainly north temperate, most numerous in W and C Asia.

A. boeticus L., a Mediterranean species, was said to be cultivated as a substitute for coffee in SW Ethiopia by Fiori in *L'Agricoltura Colon*. **33**, 7(1939). No material has been seen.

1. More or less robust perennial; corolla 8-13 mm long. **1. A. atropilosulus**
- Small annuals or short-lived perennials; corolla up to 6 mm long. 2
2. Plant appressed silvery pubescent; racemes many-flowered; pod 7-9 mm long, not falcate. **2. A. fatmensis**
- Plant villous with spreading hairs; racemes 2-6-flowered; pod 12-25 mm long, ± falcate. **3. A. eremophilus**

1. **A. atropilosulus** *(Hochst.) Bunge* (1868);
Diplotheca atropilosula Hochst. (1846) - type: GD, Simien, *Schimper* II:1235 (B holo. destr., BM K iso.).

Perennial; stem glabrous or nearly so. Leaves imparipinnate up to 25 cm long; leaflets 11-51, opposite or subopposite, narrowly elliptic or lanceolate, up to 30 x 14 mm, usually glabrous above, sparsely pubescent beneath. Racemes many-flowered. Corolla purplish, white or yellow, 8-13 mm long. Pod lanceolate, up to 40 x 7 mm, glabrous or dark strigulose.

1. Ovary densely covered with appressed dark hairs. 2
- Ovary glabrous. 3
2. Standard and wings longer than keel; corolla usually yellow. var. **atropilosulus**
- Standard and wings not longer than keel; corolla purplish. var. **mooneyi**
3. Corolla mainly yellow; standard and wings usually longer than keel. 4
- Corolla purplish; standard and wings not, or hardly, longer than keel. 6
4. Leaflets usually less than twice as long as wide. var. **coerulescens**
- Leaflets usually more than twice as long as wide. 5
5. Stem very stout; whole plant almost glabrous; both upper and lower stipules much larger than leaflets; racemes not, or but little longer than subtending leaf; filament-sheath 8-9 mm long. var. **abyssinicus**
- Stem rather slender; plant ± white-pubescent; upper stipules smaller than the lower, not much larger than the leaflets; raceme much longer than subtending leaf; filament-sheath *c* 6 mm long. var. **venosus**
6. Stipe of pod under 3 mm long; pod, including stipe, less than 25 mm long; racemes usually shorter than subtending leaf; style 1.5-2.5(-3) mm long. var. **bequaertii**
- Stipe of pod over 3 mm long; pod, including stipe, almost always over 25 mm long; racemes usually longer than subtending leaf; style 3-4 mm long. var. **burkeanus**

Subsp. **atropilosulus.**

Var. **atropilosulus.**

Upland grassland; *c* 2800-3550 m. TU GD SU; not known elsewhere. *Schimper* 372; *Sebald* 1301; *J. De Wilde & Gilbert* 83; *Gilbert & Tewolde* 3249; *Ash* 1304.

Var. **mooneyi** *Gillett* in *Kew Bull.* **17**: 416 (1964) - type: BA, *Mooney* 8368 (K holo., ETH FT iso.). Fig. 95.42.1.

Upland grassland, scrub, stream-sides; 2300-3700 m. GJ SU AR BA KF SD; not known elsewhere. *Mooney* 6244; *Thulin* 1677; *Gillett* 14990; *W. de Wilde* 8584.

Subsp. **abyssinicus** *(Hochst.) Gillett* in *Kew Bull.* **17**: 417 (1964);
Diplotheca abyssinica Hochst. (1846) - types: *Schimper* I:252 (B syn. destr.) & II:736 (B syn. destr., BM isosyn.).

Var. **abyssinicus.**

Upland grassland; 1500-2500 m. EW TU GD GJ WG; not known elsewhere. *Pappi* 6601; *Schimper* 711; *Pichi-Sermolli* 529; *Mesfin & Kagnew* 1868, 1626; *Mesfin* 2822.

Var. **venosus** *(Hochst.) Gillett* in *Kew Bull.* **17**: 417 (1964);
Diplotheca venosa Hochst. (1846) - types: TU, *Schimper* I:190 (B syn. destr., BM K isosyn.) & II:1067 (B syn. destr., BM isosyn.).
D. tigrensis Hochst. (1846) - type: TU, Schimper I:241 (B holo. destr., BM K iso.).

Upland grassland; 1800-2700 m. EW TU; not known elsewhere. *Schweinfurth & Riva* 1358; *Pappi* 287; *Schimper* 455; *Robertson* 1158; *Mooney* 9603, 9498.

Var. **coerulescens** *(Chiov.) Gillett* in *Kew Bull.* **17**: 418 (1964);
A. coerulescens Chiov. (1908) - types: Eritrea, *Pappi* 2497 (FT syn.), 3238 (FT syn.),4241 (FT syn., BM K isosyn.), 4574 (FT syn., BM K isosyn.) & 4648 (FT syn., BM K isosyn.).

Upland grassland; 1500-2500 m. EW; not known elsewhere.

Subsp. **bequaertii** *(De Wild.) Gillett* in *Kew Bull.* **17**: 418 (1964).

Var. **bequaertii.**

Grassland, rocky places; 1800-2200 m. SU AR GG SD; Uganda, Kenya, Tanzania, E Zaire, Burundi. *Gillett* 14489; *Thulin* 1573; *Gilbert & Thulin* 494.

Subsp. **burkeanus** *(Harv.) Gillett* in *Kew Bull.* **17**: 420 (1964).

Var. **burkeanus.**

Upland grassland; 2000-2400 m. SU HA; E Africa from N Somalia to Transvaal, and in Yemen. *Mooney* 8243; *IECAMA* C-66; *Gilbert & Thulin* 939.

2. **A. fatmensis** *Hochst. ex Chiov.* (1903);
A. arabicus Ehrenb. ex Bunge (1868), nom. illegit., non Kotschy (1866); *A. vogelii* (Webb.) Bornm. subsp. *fatimensis* Maire in *Mém. Soc. Hist. Nat. Afr. Nord* 3:126 (1933).
A. arabicus Ehrenb. ex Bunge var. *congestus* Schweinf. in *Bull. Herb. Boiss.* **4**, app. 2: 248 (1896) as *"congesta"*; *A. vogelii* (Webb) Bornm. var. *congestus* (Schweinf.) Cuf., Cuf. *Enum., Zweiter Nachtrag*: 24 (1969) - type: Eritrea, *Schweinfurth* 712 (B holo. destr.).

Annual or short-lived perennial herb, densely appressed silvery hairy. Leaves imparipinnate; leaflets *c* 15-19, oblong-oblanceolate, 5-10 mm long. Racemes dense, head-like, shorter than the leaves. Calyx 4-5 mm long; lobes setaceous, longer than the tube. Corolla whitish pink or blue, *c* 4 mm long. Pod inflated, *c* 7-9 x 3-4 mm, pilose.

Woodland, bushland, rocky places; up to 2200 m. EE EW TU SU; N Africa, Arabia, Iran, Pakistan, India. *Thulin* 1494; *Ash* 2107; *Schweinfurth* 295.

A. fatmensis is very difficult to distinguish from *A. vogelii*, particularly in Arabia, and may be best treated as only a variety or subspecies of this (see synonymy). All material seen from Ethiopia is typical *A. fatmensis.*

3. **A. eremophilus** *Boiss.* (1843).
A. schimperi sensu Cuf. *Enum.*: 289 (1955), non Boiss.

Villous decumbent annual. Leaves imparipinnate; leaflets *c* 7-13, oblanceolate to obovate, 4-15 mm long. Racemes rather dense, 2-6-flowered, equalling or longer than the subtending leaves in fruit. Calyx 3-5 mm long. Corolla pale yellow, white or pinkish, 5-6 mm long. Pod *c* 12-25 x 2.5-3 mm, ± falcate, pilose.

Sandy or stony ground near sea level. EE; Sudan, Egypt, Arabia and east to Pakistan. *Pappi* 4470; *Terracciano* 689.

96. **BISERRULA** *L.* (1753)

Annuals. Leaves imparipinnate; stipules small, free. Flowers in axillary racemes. Calyx-lobes 5, subequal. Upper filament free, reduced to a staminode; only 5 stamens fertile. Pod oblong, indehiscent, dorsiventrally compressed, each valve produced at the side into a flat crest.

One species in the Mediterranean region and in E and NE tropical Africa.

B. pelecinus *L.* (1753).

Annual ± pubescent herb, to 40 cm tall. Leaflets 11-31, usually opposite, linear-oblong to obovate-orbicular, emarginate, up to 10 x 5 mm. Corolla blue, pale yellow with blue tip, or reddish, 4-6 mm long. Pod glabrous or appressed pubescent, brown.

1. Crest at the sides of the pod toothed; racemes 2-11-flowered; each half of pod with 5-10 seeds. subsp. **pelecinus**

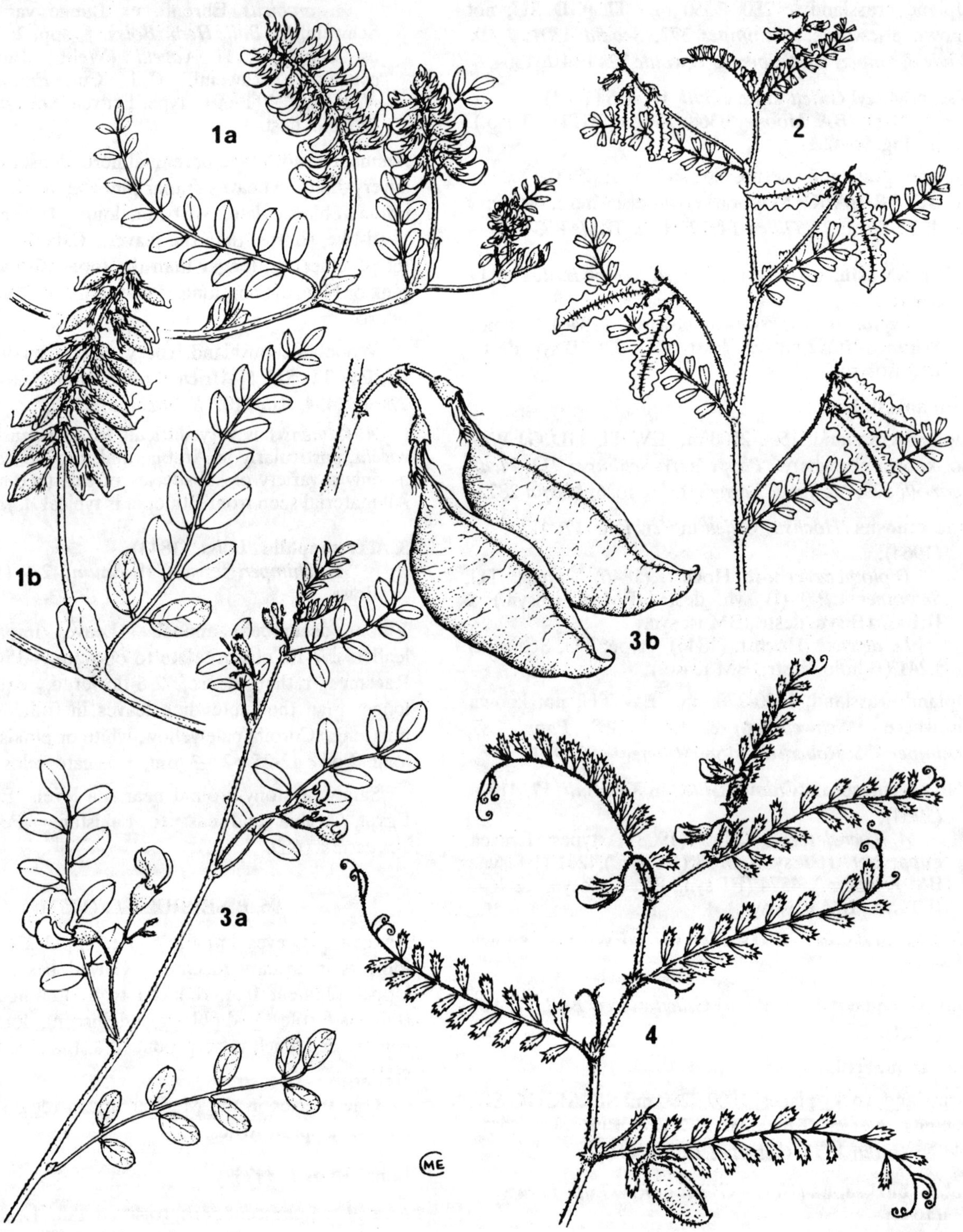

Figure 95.42.1 - *ASTRAGALUS ATROPILOSULUS* var *MOONEYI*, a: flowering branch, b: fruiting node, both from *Thulin* 1677. 2 - *BISERRULA PELICINUS* subsp. *PELICINUS*, flowering and fruiting branch from *Schweinfurth* 298. 3 - *COLUTEA ABYSSINICA*, a: flowering branch from *W. de Wilde* 6440, b: pods from *Ash* 1275. 4 - *CICER CUNEATUM*, flowering and fruiting branch from *Prior* s.n. All x 0.75. Drawn by Maureen Church. (Reproduced with permission from *Opera Botanica*, 68, fig. 44, 1983.)

- Crest at the sides of the pod straight or very slightly undulate; racemes 1-3(-4)-flowered; each half of pod with 5-6(-7) seeds. subsp. **leiocarpa**

Subsp. **pelecinus.** Fig. 95.42.2.

Grassland; 1800-2500 m. EW TU; Djibouti, Mediterranean region, Canary Is. *Pappi* 5737; *Schweinfurth* 298; *Schimper* III:1762; *Mooney* 8070.

Subsp. **leiocarpa** *(A. Rich.) Gillett* in *Kew Bull.* **17**: 504 (1964);

B. leiocarpa A. Rich. (1847), pro majore parte; *B. pelecinus* var. *leiocarpa* (A. Rich.) Chiov. in *Ann. Ist. Bot. Roma* **8**: 96 (1903) - type: TU, *Quartin Dillon* s.n. (P syn.).

B. pelecinus L. var. *subintegra* Bak. f., *Leg. Trop. Afr.*: 271 (1929) - types from Sudan, Tanzania and Ethiopia: TU, *Schimper* 77 (BM syn.).

Grassland, scrub; 1900-2400 m. EW TU WU SU SD; Sudan, N Tanzania. *Pappi* 580; *Gillett* 14369; *J. De Wilde* 7065; *Debru Getinet* 51.

97. COLUTEA *L.* (1753)

Shrubs. Leaves imparipinnate; stipules small. Flowers few together in axillary racemes. Calyx-lobes 5, subequal, short. Corolla yellow or reddish. Upper filament free, the other 9 united. Style thick, incurved at tip; stigma large, on the inner edge of the style, surrounded by hairs. Pod inflated, indehiscent or dehiscent near the tip, many-seeded.

Some 26 species ranging from Morocco to W China.

C. abyssinica *Kunth & Bouché* (1847)

- type: TU, *Schimper* I:240 (B holo. destr., BM FT G K P W iso.).

C. halepica Lam. var. *sericea* A. Rich., *Tent. Fl. Abyss.* **1**: 192 (1847); *C. istria* Mill. var. *sericea* (A. Rich.) Cuf., in Cuf. *Enum., Zweiter Nachtrag*: 24 (1969) - type: TU, Petit s.n. (P holo.).

C. istria var. *macrophysa* Chiov. in *Ann. Ist. Bot. Roma* **9**: 58 (1911); *C. abyssinica* var. *macrophysa* (Chiov.) Browicz in *Monogr. Bot.* **14**: 68 (1963) - type: GD, Chiovenda 1805 (FT holo.).

Shrub 1-4 m tall; young stems with appressed white hairs. Leaflets 9-15, elliptic, up to 18 x 10 mm, mucronate, glabrous or nearly so above, with white appressed hairs beneath. Racemes 2-3-flowered. Corolla usually dark reddish-brown, yellowish-green at the base, the keel almost black at the tip, 13-15 mm long. Pod often purplish or brownish, papery, 3-8 cm long including the 0.5-2.2 cm long stipe, 1.5-2.8 cm wide. Fig. 95.42.3.

Upland grassland, rocky slopes; 1600-4000 m. EW TU GD WU SU AR BA HA SD; N Somalia, Uganda, Kenya, Tanzania, Rwanda. *W. de Wilde* 8218, 6440; *Gilbert* 2231; *Thulin, Hunde & Tadesse* 3831.

98. PAROCHAETUS *D. Don* (1825)

Prostrate herb, rooting at the nodes. Leaves digitately 3-foliolate; stipules free from the petiole. Flowers in axillary, few-flowered long-pedunculate umbels. Calyx 5-lobed, the 2 upper lobes fused almost to the tip. Corolla falling quickly. Upper filament free, the other 9 united. Pod linear, 10-15-seeded.

Monotypic.

P. communis *D. Don* (1825).

P. communis var. *grossecrenatus* Cuf. in *Senckenberg. Biol.* **39**: 295, t. 35/1 (1958) - type: GG, Chencha ("Cencia"), *Kuls* 701 (FR holo., not seen).

Stems glabrous or sparsely pilose. Leaflets cuneate-obovate, up to 35 x 35 mm, emarginate, entire to crenate or coarsely toothed in outer part; petiole 3-25 cm long. Calyx 6-7 mm long. Corolla bright blue, up to 17 mm long. Pod up to 25 x *c* 4 mm, ending in a short acute beak, glabrous. Fig. 95.43.2.

Upland forest, bamboo forest, especially in damp places; 1800-3500 m. GJ, SU, AR, BA, KF, GG, SD; widespread in E tropical Africa, also in S Asia. *Thulin* 1468; *Mooney* 6002, 8193; *Ash* 1384.

99. TRIFOLIUM *L.* (1753)

Fiori in *Nuovo Giorn. Bot. Ital.* **55**: 335-346 (1949); Gillett in *Kew Bull.* **7**: 367-404 (1952) & *Fl. Trop. E. Afr.* (1971); Fröman, Illustrated Guide Pasture Leg. Eth., *Rural Dev. Stud.* **3** (1975).

Annual or perennial herbs. Leaves usually digitately 3-foliolate; leaflets usually toothed; stipules adnate to the petiole. Flowers in heads or short spikes, rarely flowers solitary. Calyx-lobes 5, equal or unequal. Corolla persistent. Upper filament free, the other 9 united. Pod surrounded by the persistent calyx and corolla, 1-10-seeded.

A large, mainly temperate genus containing many important fodder and bee plants. Especially well represented in the Mediterranean region; in tropical Africa with a centre of distribution in Ethiopia.

1. Leaves pinnately 3-foliolate; calyx 5-nerved; corolla yellow. **27. T. campestre**
- Leaves digitately 3-foliolate; calyx usually more than 5-nerved; corolla purple, white or pink. 2
2. Bracts absent; pod 1(-2)-seeded. 3
- Bracts present though often minute; pod usually 2-several-seeded. 5
3. Fertile flowers 2-5(-7); inner flowers sterile, numerous, developing after anthesis; fruiting heads appressed to or buried in the soil by the long, deflexed peduncle. **32. T. subterraneum**
- Fertile flowers numerous; sterile flowers absent; fruiting heads not reflexed and buried in the soil. 4

4. Annual; corolla *c* 4 mm long, whitish or pink, much shorter than the calyx. **31. T. arvense**
- Perennial; corolla 12-15 mm long, usually reddish-purple, longer than the calyx. **30. T. pratense**
5. Calyx 2-lipped with the upper lip greatly inflated and net-veined in fruit. 6
- Calyx ± regularly 5-lobed, not or but little inflated in fruit. 7
6. Perennial, ± hairy; flowers not twisted through 180°. **28. T. fragiferum**
- Annual, glabrous; flowers twisted through 180°. **29. T. resupinatum**
7. Bracts several-nerved, more than half as long as calyx, forming an involucre; blade of wing not auriculate. 8
- Bracts 1- or few-nerved, usually much shorter than the calyx and not forming an involucre; blade of wing auriculate. 9
8. Calyx-lobes 2-3 mm long; corolla 7-8 mm long. **25. T. steudneri**
- Calyx-lobes 4-7 mm long; corolla 11-12 mm long. **26. T. quartinianum**
9. Petiole of all leaves united with stipules throughout its length. 10
- Petiole free for much of its length, at least in the lower leaves. 14
10. Pedicels conspicuously reflexed in fruit; corolla white. **5. T. somalense**
- Pedicels not reflexed in fruit; corolla usually purple. 11
11. Calyx-nerves 17-30; leaflets usually more than 5 times as long as wide. **1. T. simense**
- Calyx-nerves *c* 11; leaflets usually less than 5 times as long as wide. 12
12. Perennial from a ± thick rootstock, with the stems creeping below and rooting at the nodes, often forming large mats; leaflets up to 9(-13) mm long; inflorescences ± globose. **4. T. cryptopodium**
- Annuals or short-lived perennials without a thick rootstock; the stems ± ascending; leaflets 6-30 mm long; inflorescences usually ± oblong. 13
13. Leaflets usually acute at the apex; inflorescences *c* 15-35 mm long, often more than twice as long as broad. **2. T. polystachyum**
- Leaflets usually rounded to emarginate at the apex; inflorescences 9-18 mm long, less than twice as long as broad. **3. T. usambarense**
14. Calyx 15-30-nerved. 15
- Calyx *c* 10-11-nerved. 19
15. Corolla 15-18 mm long, red; calyx-lobes *c* 10 mm long. **24. T. schimperi**
- Corolla 7-14 mm long, purplish; calyx-lobes 2-4.5 mm long. 16
16. Inflorescences subsessile, 1-2-flowered, leaflets cuneate-obovate; calyx-lobes 2-2.5(-3.5) mm long, triangular. **21. T. elgonense**
- Inflorescences ± pedunculate, more than 1-2-flowered, or if so then leaflets ± oblanceolate and calyx-lobes 3-4 mm long, narrowly triangular with subulate tips. 17
17. Leaflets ± ovate, 30-45 x 18-28 mm; corolla 10-14 mm long, twice as long as calyx or more. **22. T. decorum**
- Leaflets ± oblanceolate, up to 25 x 7(-10) mm; corolla 7-10(-12) mm long, less than twice as long as calyx. 18
18. Inflorescences usually numerous, 1-6(-8)-flowered; calyx 20-30-nerved, with pilose lobes as long as or slightly longer than the tube; seeds smooth. **20. T. multinerve**
- Inflorescences few, 5-12-flowered; calyx *c* 15-nerved with glabrous or very sparsely pilose lobes usually about twice as long as the tube; seeds finely warty. **23. T. chilaloense**
19. Pedicels conspicuously reflexed after flowering (also 11. *T.* = *Burger* 3824 may belong here). 20
- Pedicels not reflexed after flowering. 21
20. Stems pilose. **6. T. semipilosum**
- Stems glabrous or almost so. **7. T. repens**
21. Standard abruptly narrowed above the middle into an oblong, truncate, downcurved tip. 22
- Standard not as above. 23
22. Perennial; standard 14-20 mm long. **16. T. calocephalum**
- Annual; standard *c* 9 mm long. **17. T. tembense**
23. Inflorescences 1-6-flowered. 24
- Inflorescences 10-many-flowered. 25
24. Pedicels *c* 1 mm long; corolla not much exceeding calyx. **18. T. spananthum**
- Pedicels *c* 2-4 mm long; corolla more than twice as long as calyx. **19. T. acaule**
25. Perennial; stems usually creeping and rooting at the nodes. **8. T. burchellianum**
- Annuals; stem erect or decumbent, rarely rooting at the nodes. 26
26. Plant, including calyces and both sides of leaflets densely pilose; calyx 10-nerved; flowers subsessile. 27
- Plant usually practically glabrous, calyces and leaflets not densely pilose; calyx usually 11-nerved; flowers usually with at least 1 mm long pedicels. 28
27. Leaflets ovate to elliptic, mostly less than twice as long as wide; inflorescences 13-20 mm across; bracts white, up to 1 mm long. **9. T. bilineatum**
- Leaflets ± oblanceolate, more than twice as long as wide; inflorescences *c* 12 mm across; bracts frequently elongated and green, 5-7 mm long, but sometimes all white and less than 1 mm. **10. T. pichisermollii**
28. Commisural calyx-nerves usually forking at a narrow angle low down the calyx-tube; corolla 3-4 mm long. **15. T. baccarinii**
- Commisural calyx-nerves forking at a wide angle at the base of the sinus; corolla 5-8 mm long. 29
29. Calyx-tube becoming somewhat inflated and constricted at the throat. **13. T. mattirolianum**
- Calyx-tube not inflated or narrowed at the throat. 30

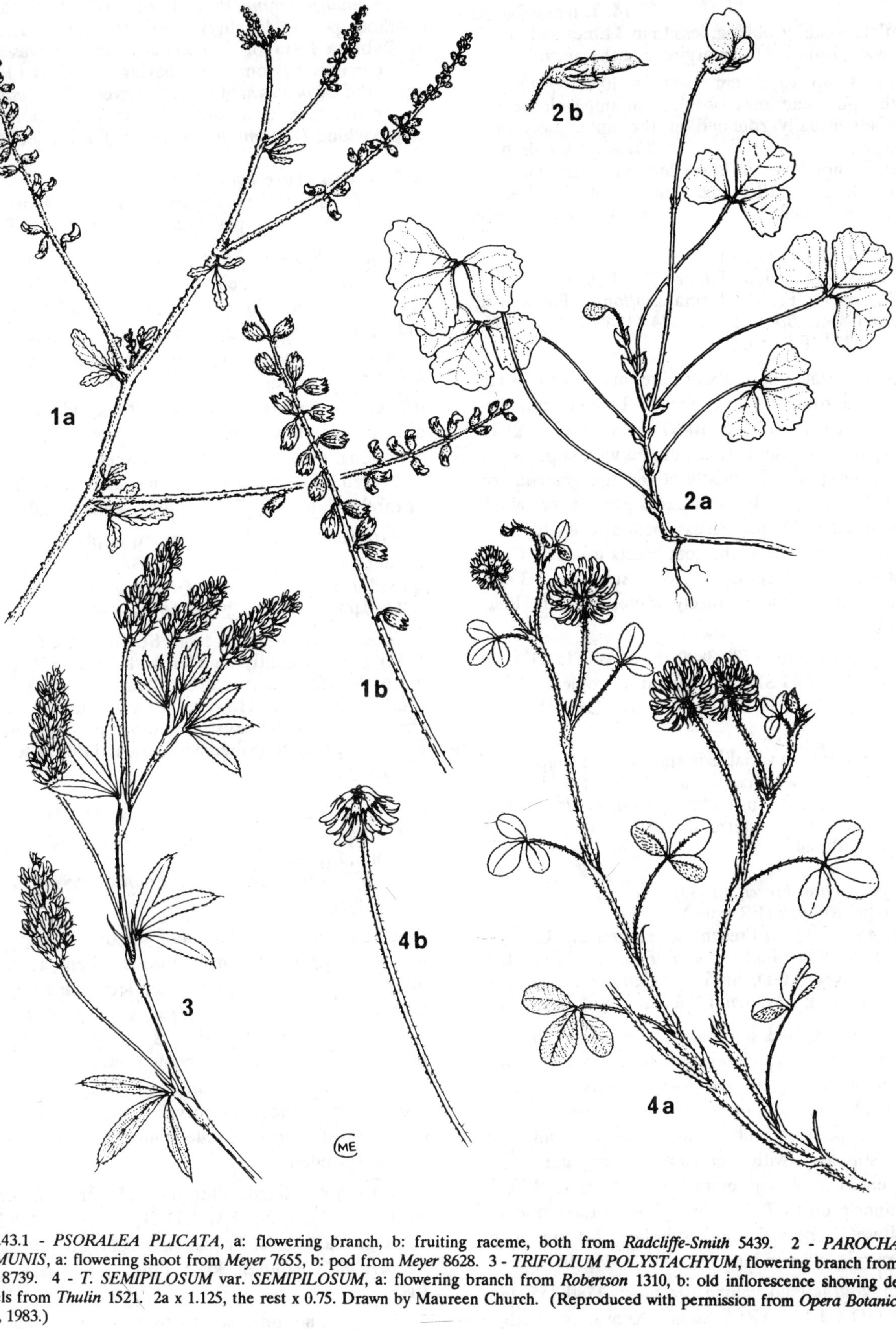

Figure 95.43.1 - *PSORALEA PLICATA*, a: flowering branch, b: fruiting raceme, both from *Radcliffe-Smith* 5439. 2 - *PAROCHAETUS COMMUNIS*, a: flowering shoot from *Meyer* 7655, b: pod from *Meyer* 8628. 3 - *TRIFOLIUM POLYSTACHYUM*, flowering branch from *W. de Wilde* 8739. 4 - *T. SEMIPILOSUM* var. *SEMIPILOSUM*, a: flowering branch from *Robertson* 1310, b: old inflorescence showing deflexed pedicels from *Thulin* 1521. 2a x 1.125, the rest x 0.75. Drawn by Maureen Church. (Reproduced with permission from *Opera Botanica*, 68, fig. 45, 1983.)

30. Leaflets lanceolate to oblanceolate, 3 or more times as long as wide, usually acute at the apex; stipules somewhat laciniate at the margin. **14. T. lanceolatum**
- Leaflets ovate or oblong, less than 3 times as long as wide, rounded to emarginate at the apex. 31
31. Petioles up to 5 cm long in lower leaves, although sometimes obsolete in upper leaves; leaflets usually rounded at the apex; flowers purple. **12. T. rueppellianum**
- Petioles not (even in lower leaves) more than 7 mm long; leaflets ± emarginate; flowers apparently pink. **11. T.** = *Burger* 3824.

1. **T. simense** *Fresen*. (1839)
- type: GD, Simien, *Ruppell* (FR holo.).
T. simense Fresen. forma *albiflorum* Piovano in *Nuovo Giorn. Bot. Ital.* **56**: 474 (1949) - type: SU, *Piovano* 234 (FT holo.).

Erect or sometimes decumbent perennial, up to 60 cm tall; stem glabrous or pilose above. Leaflets cuneate-oblong to linear-lanceolate, 10-60 x 1.5-5.5 mm, glabrous or sparsely pilose; petioles united with stipules for all their length; stipules bristly at the tip, glabrous or pilose at the margin. Inflorescences globose or ovoid, many-flowered, *c* 15 mm across; peduncle up to 7 cm long, pilose, especially at the top; bracts minute. Calyx pilose at margins, 17-30-nerved; lobes subulate, 2-3 mm long. Corolla red-violet, rarely white, 6-8 mm long. Seeds 1-2.

Upland grassland; 1780-3800 m. EW TU GD GJ SU AR BA HA GG SD; widespread in tropical Africa. *Thulin* 1358; *Gilbert & Thulin* 463; *Ash* 2052; *Mooney* 5781.

T. simense is very variable in the size and shape of its leaflets. A form with comparatively short and broad leaflets was called *T.* sp. *E.* by Fröman (1975), but as there are numerous intermediates taxonomic recognition is not advisable.

2. **T. polystachyum** *Fresen*. (1839)
- type: *Ruppell* (FR holo.).
T. polystachyum Fresen. var. *contractum* Lanza in *Boll. Orto Bot. Giard. Colon. Palermo* **8**: 98, t. 5 (1909) - types: GD, Simien, *Schimper* 2422 (PAL syn., not seen); EW, *Senni* 191 (PAL syn., not seen).

Annual or short-lived perennial; stems ascending, to more than 1 m long, glabrous below, sparsely pilose above, often rooting at the lower nodes. Leaflets elliptic or oblanceolate, up to 30 x 9 mm, usually acute at the apex; petioles united with stipules for all their length; stipules with narrowly triangular tips. Inflorescences ± oblong, many-flowered, up to 35 x 15 mm; peduncle up to 7 cm long, pilose; bracts minute. Calyx pilose, 11-nerved, lobes subulate, 3-4 mm long. Corolla purple, *c* 8 mm long. Seeds 1-2. Fig. 95.43.3.

Swampy grassland, along streams; 1780-3550 m. EW TU GD GJ SU AR; S Sudan, Kenya, and with a variety in S tropical Africa. *Schimper* II:1071; *Thulin* 1556; *Gilbert & Thulin* 919; *Gilbert & Getachew* 2794.

T. mauginianum Fiori (1949) of which the type (without precise locality) is lost (Gillett 1952: 373) is probably a form of *T. polystachyum*. It was said to differ mainly by sometimes having 4 leaflets. Furthermore the calyx was said to be 5-nerved. *T. polystachyum* var. *contractum* is a form with shorter inflorescences approaching *T. usambarense*, but with acute leaflets.

3. **T. usambarense** *Taub*. (1895).
T. pseudocryptopodium Chiov. ex Fiori (1949), nom. invalid. - types: *Vàtova* 833 & 1627 (FT syn.).

Annual or short-lived perennial; stems ascending, up to 1 m long, glabrous below, sparsely pilose above, often rooting at the lower nodes. Leaflets cuneate-oblanceolate, 6-22 x 3-7 mm, usually rounded at the apex, glabrous or nearly so; petioles united with stipules for all their length; stipules with the free part triangular-acuminate. Inflorescences globose or ± oblong, many-flowered, 9-18 x 7-13 mm; peduncle up to 6 cm long, pilose above; bracts minute. Calyx ± pilose, 11-nerved; lobes subulate, *c* 3 mm long. Corolla purple or rarely white, 4-5(-7) mm long. Seeds usually 2.

Grassland, usually in swampy places and along streams; 1600-2700 m. SU AR BA KF GG SD; widespread in tropical Africa. *Gilbert & Thulin* 1007; *W. de Wilde* 9468; *Friis* et al. 2164; *Mooney* 8857.

The specimens seen from SU and AR (e.g. *Thulin* 1352) are constantly smaller with smaller leaves and more densely pilose calyces than the material from southern Ethiopia. However, they are matched in these characters by specimens from other parts of the range. Fröman (1975) described and depicted this as form *B* of *T. cryptopodium*.

4. **T. cryptopodium** *Steud. ex A. Rich*. (1847)
type: GD, Simien, Mt Buahit, *Schimper* II:556 (P holo., BM K iso.).
T. stolzii sensu Cuf. *Enum*.: 255 (1955), non Harms.

Perennial from a ± thick and woody root-stock, often forming large mats; stems glabrous, creeping below and often rooting at the nodes. Leaflets cuneate-obovate, 5-13 x 2-9 mm, glabrous; petioles united with stipules for all their length; stipules with lanceolate tips. Inflorescences ± globose, many-flowered, up to 18 mm across; peduncle up to 8 cm long, pilose; bracts minute. Calyx sparsely pilose, 11-nerved; lobes subulate, 4-5 mm long. Corolla purple to pale mauve, 8-11 mm long. Pod usually 1-seeded.

Upland grassland, often on rocky or moist ground; 2000-3950(-4300) m. EW GD GJ SU AR BA KF GG SD; Uganda, Kenya, Tanzania. *Hedberg & Aweke* 5352; *Hedberg* 5570; *Burger* 3655; *Friis* et al. 1607; *Gilbert* 1085.

Some of the more stout Ethiopian specimens much approach the Tanzanian *T. wentzelianum* Harms. *T.* sp.

C sensu Fröman (1975) from moist habitats in S Arsi may be an aberrant form of *T. cryptopodium*, with blue flowers, twisted through 180°. No material has been seen, however.

5. **T. somalense** *Taub*. (1897)
- type: SD, Bidduma-Alghe, *Riva* 1301 (FT holo.).

Perennial; stems erect or ascending from a cluster of tuberous roots. Leaflets oblong-elliptic, 10-35 x 4-10 mm, with 25-40 nerves on each side, glabrous except for a few hairs on midrib; petioles united with stipules for all their length; stipules with short lanceolate tips. Inflorescences ± globose, many-flowered, *c* 20 mm across; flowers reflexed in fruit; bracts minute. Calyx pilose at margins, 11(-14)-nerved; lobes subulate, *c* 3 mm long. Corolla white, *c* 9 mm long.

Upland grassland up to 2000 m. SD; not known elsewhere. *Gillett* 14323; *Vàtova* 430.

6. **T. semipilosum** *Fresen*. (1839)
type: GD, Simien, *Ruppell* s.n. (FR holo.).

Perennial herb with a taproot and prostrate pilose stems, often rooting at the nodes. Leaflets orbicular, obovate or oblong-elliptic, up to 15 x 11 mm, glabrous or pilose above, pilose beneath at least on the midrib; petiole pilose, usually much longer than a leaflet. Inflorescences ± globose, few-many-flowered; peduncle often longer than the leaves, pilose; pedicels reflexed in fruit. Corolla white, pale pink or white with purplish centre, 8-12 mm long. Seeds 2-6.

1. Leaflets rounded to slightly emarginate at the apex, glabrous above, pilose beneath on the outer halves of the 2 lower leaflets and the midribs, teeth short and rounded; stipules narrowly falcate-triangular; inflorescences (7-)10-25-flowered; ovary and pod pilose at least near the tip. var. **semipilosum**
- Leaflets ± emarginate at the apex, pilose to subglabrous above, sparsely to densely pilose over the whole of the lower surface, teeth larger and sharper; stipules usually broader, ± lanceolate; inflorescences 3-15-flowered; ovary and pod pilose or papillose to subglabrous. 2
2. Leaflets shallowly emarginate at the apex, sparsely pilose; inflorescences 10-15-flowered; flowers purplish, pale purple or whitish at the apex. var. **intermedium**
- Leaflets deeply emarginate at the apex, usually densely pilose on both sides; inflorescences 3-12-flowered; flowers white to pink. var. **brunellii**

Var. **semipilosum.**

T. semipilosum var. *microphyllum* Chiov. in *Ann. Ist. Bot. Roma* **8**: 406 (1908) - type: EW, *Pappi* 1132 (FT holo., K iso.).

T. semipilosum var. *kilimanjaricum* Bak. f., *Leg. Trop. Afr.*: 81 (1926).

T. semipilosum var. *sennii* Chiov. ex Fiori in *Nuovo Giorn. Bot. Ital.* **55**: 342, Fig. 10 (1949), nom. invalid. - type: *Senni* 1235 (FT).

Fig. 95.43.4.

Upland grassland and bushland, forest margins; 1800-3150 m. EW TU GD WU SU AR BA HA SD; N Uganda, Kenya, Tanzania and Yemen. *Mooney* 8290; *Thulin* 1521; *Ash* 2095.

Var. **intermedium** *Thulin* in *Nord. J. Bot.* **2**:52 (1982)
- type: SU, *W. de Wilde* 7400 (WAG holo., C ETH K UPS iso.).

Upland grassland in rocky places; 2800-3800 m. SU AR; not known elsewhere. *Hagos* 218; *W. de Wilde* 8084; *Senni* 895.

Var. **brunellii** *Thulin* in *Nord. J. Bot.* **2**:51 (1982)
- type: SD, *Vàtova* 717 (FT holo.).
T. brunellii Chiov. ex. Fiori (1949), nom. nud.

Upland grassland, forest margins, often in rather damp situations; 1800-3700 m. AR BA SD; not known elsewhere. *Mooney* 5157, 8300; *Gillett* 14781.

W. de Wilde 9124 and 10085 from AR represent a highland form from 3700-4000 m altitude which cannot readily be placed in any of the varieties here recognized. It has conspicuously long-pedicelled flowers in practically sessile inflorescences.

7. **T. repens** *L*. (1753).

Glabrous or glabrescent perennial, usually with long creeping stems rooting at the nodes. Leaflets obovate or elliptic, 10-25(-40) mm long; petioles 2-20 cm long; stipules contracted into a subulate tip. Inflorescences ± globose, many-flowered, long-pedunculate; pedicels reflexed in fruit; calyx-lobes narrowly lanceolate, the 2 upper longer than the rest; corolla white or pale pink, 7-10 mm long. Pod linear, 3-4-seeded.

Introduced for experimental cultivation in AR (Fröman 1975); widespread in Eurasia and extensively cultivated for fodder.

8. **T. burchellianum** *Ser*. (1825).

Perennial herb with a taproot; stems glabrous or nearly so, usually creeping and rooting at the nodes. Leaflets cuneate-obovate to oblong, up to 25(-50) x 19 mm, glabrous or nearly so, with a well-developed free petiole; stipules with the free part triangular, short. Inflorescences ± globose, many-flowered, up to 30 mm across; peduncle long, pilose towards the top. Calyx ± pilose at the margin, *c* 11-nerved; lobes triangular-subulate, 4-7 mm long. Corolla purple, 8-13 mm long. Seeds 1-2.

1. Leaflets less than 1.5 times as long as wide, less than 25 mm long, usually emarginate; stems creeping, less than 2.5 mm thick. var. **johnstonii**
- Leaflets more than 1.5 times as long as wide, more than 25 mm long; stems ascending, more than 2.5 mm thick. var. **oblongum**

Subsp. **johnstonii** *(Oliv.) Cuf. ex Gillett* in *Kew Bull.* **25**: 178 (1971).

Var. **johnstonii**.

T. basileianum Chiov. (1929) - type: BA, source of Webbi Shebelli, *Duke of Abruzzi* 112 (TO holo. destr.).

Fig. 95.44.1.

Upland grassland and moor, forest glades and margins; 2000-3900 m. SU AR BA KF GG SD; Uganda, Kenya and Tanzania. *Thulin* 1360; *Gilbert, Thulin & Getachew* 462, 575; *Gilbert* 1822; *Mooney* 5368.

Var. **oblongum** *Gillett* in *Kew Bull*. 7:382 (1952).

Upland grassland and moor, forest margins; 3200-4000 m. AR BA KF GG SD; Kenya. *W. de Wilde* 9042; *Thulin* 1658; *Mooney* 7237.

Subsp. *burchellianum* occurs in South Africa and Angola.

9. **T. bilineatum** *Fresen*. (1839)
type: *Rueppell* s.n. (FR holo.).

Annual herb, pilose above, to 40 cm high. Leaflets ovate to elliptic, up to 23 x 15 mm, densely pilose on both sides; petioles up to 5 cm long; stipules ovate-acuminate. Inflorescences ± globose, many-flowered, 13-20 mm across; flowers subsessile; bracts up to 1 mm long, white. Calyx densely white or brownish pilose, 10-nerved; lobes triangular-subulate, 2.5-3.5 mm long. Corolla purple, *c* 8 mm long. Pod 1-2-seeded. Fig. 95.44.2.

Upland grassland and in cultivations; 1800-2500 m. EW TU GD GJ WG SU; not known elsewhere. *Schimper* 1264; *Gilbert & Thulin* 603; *Gilbert & Getachew* 3025.

10. **T. pichisermollii** *Gillett* (1952)
- type: SU, *Negri* 85 bis (FT holo.).

Annual pilose herb, up to 35 cm tall. Leaflets ± oblanceolate, up to 19 x 9 mm, densely pilose on both sides; petioles up to 3 cm long; stipules ovate-acuminate. Inflorescences ± globose; many-flowered, *c* 12 mm across; flowers subsessile; bracts frequently elongated and green, 5-7 mm long, but sometimes all white and less than 1 mm. Calyx densely white or brownish pilose, 10-nerved; lobes triangular-pointed, 2-2.5 mm long. Corolla purple, *c* 7 mm long. Pod 1-2-seeded.

Upland grassland; 2100-2700 m. GJ SU; not known elsewhere. *Bally* 3070; *Gilbert & Thulin* 1000; *Ash* 2087.

The usually strikingly large bracts are not always developed and such specimens may be difficult to distinguish from the closely related *T. bilineatum*, of which it is perhaps no more than a variety.

11. **T.** = *Burger* 3824

Slender ascending herb, apparently annual; stems glabrous, rooting at lower nodes. Leaflets broadly cuneate-obovate, up to 9 x 8 mm, truncate to emarginate at the apex, densely nerved, sparsely pilose along midvein beneath; petiole up to 7 mm long; stipules ovate-elliptic, entire. Inflorescences *c* 15-flowered, *c* 17 mm across; pedicels up to 2 mm long; peduncle long, pilose; bracts minute. Calyx glabrous except on margins of lobes, 10-nerved; lobes subulate, 2.5-3 mm long. Corolla apparently pink, *c* 8 mm long. Ovules *c* 4.

Mountain slope; 2500 m. HA (Gara Mullata Mt); not known elsewhere.

Probably related to *T. rueppellianum* from which it differs in the short-petiolate leaves with ± emarginate, densely nerved leaflets and in the apparently pink corolla with less well-developed auricle at base of wings. The flower-colour may indicate a relationship also with *T. semipilosum*.

12. **T. rueppellianum** *Fresen*. (1839)
- type: GD, Simien, *Rueppell* s.n. (FR holo.).

T. subrotundum Steud. & Hochst. ex A. Rich. (1847), pro majore parte; *T. rueppellianum* var. *subrotundum* (Steud. & Hochst. ex A. Rich.) Fiori in *Nuovo Giorn. Bot. Ital.* **55**: 339 (1949) - types: TU, *Schimper* I:303 (P syn., K isosyn.) & *Quartin Dillon* s.n. (P, syn.).

Annual up to 50 cm tall; stems glabrous or pilose, erect or decumbent. Leaflets oblong or obovate, up to 24 x 18 mm, rounded to emarginate at the apex, glabrous or sparsely pilose; petiole up to 7 cm long, in upper leaves short or wholly united with stipules; stipules with acuminate tips, entire. Inflorescences ± globose, usually many-flowered, *c* 17 mm across; pedicels up to 2 mm long; peduncle long, pilose at least near the top; bracts minute. Calyx glabrous or pilose, *c* 11-nerved; lobes abruptly narrowed near the base, subulate for most of their length, 2-5 mm long. Corolla purple or rarely white, 5-8 mm long. Seeds usually 2-3. Fig. 95.44.3.

Upland grassland and moor; 1700-3650 m. EW TU GD WU WG SU AR BA HA KF GG SD; widespread in tropical Africa. *Thulin* 1581; *Gilbert & Thulin* 774; *Gillett* 14698; *Friis* et al. 2039; *Burger* 707.

A collection of dwarfed specimens from 3700 m altitude in AR (*Hedberg* 4237) with only 6 nerves on the calyces and 4-7 flowers in the heads, may be a distinct taxon close to *T. rueppellianum*.

Another deviating form has been collected at the Koka Dam in Shewa at *c* 1500 m altitude (*Ash* 272, 758, *Gilbert* 1175). It has *c* 10-flowered inflorescences, calyx-lobes about as long as the corolla and *c* 1 mm wide at the base, peduncles comparatively short in relation to petioles, and unusually large pods.

Some specimens from HA, with somewhat laciniate stipules (e.g. *J. De Wilde* 4147) are, with some hesitation, referred too here.

13. **T. mattirolianum** *Chiov*. (1911)
types: GD, *Chiovenda* 1422 & 1706 (FT syn.).

Annual up to 30 cm tall, erect or decumbent, glabrous. Leaflets obovate, up to 16 x 14 mm, usually truncate or emarginate at the apex; petiole up to 5 cm long, in upper leaves much shorter; stipules with acuminate tips,

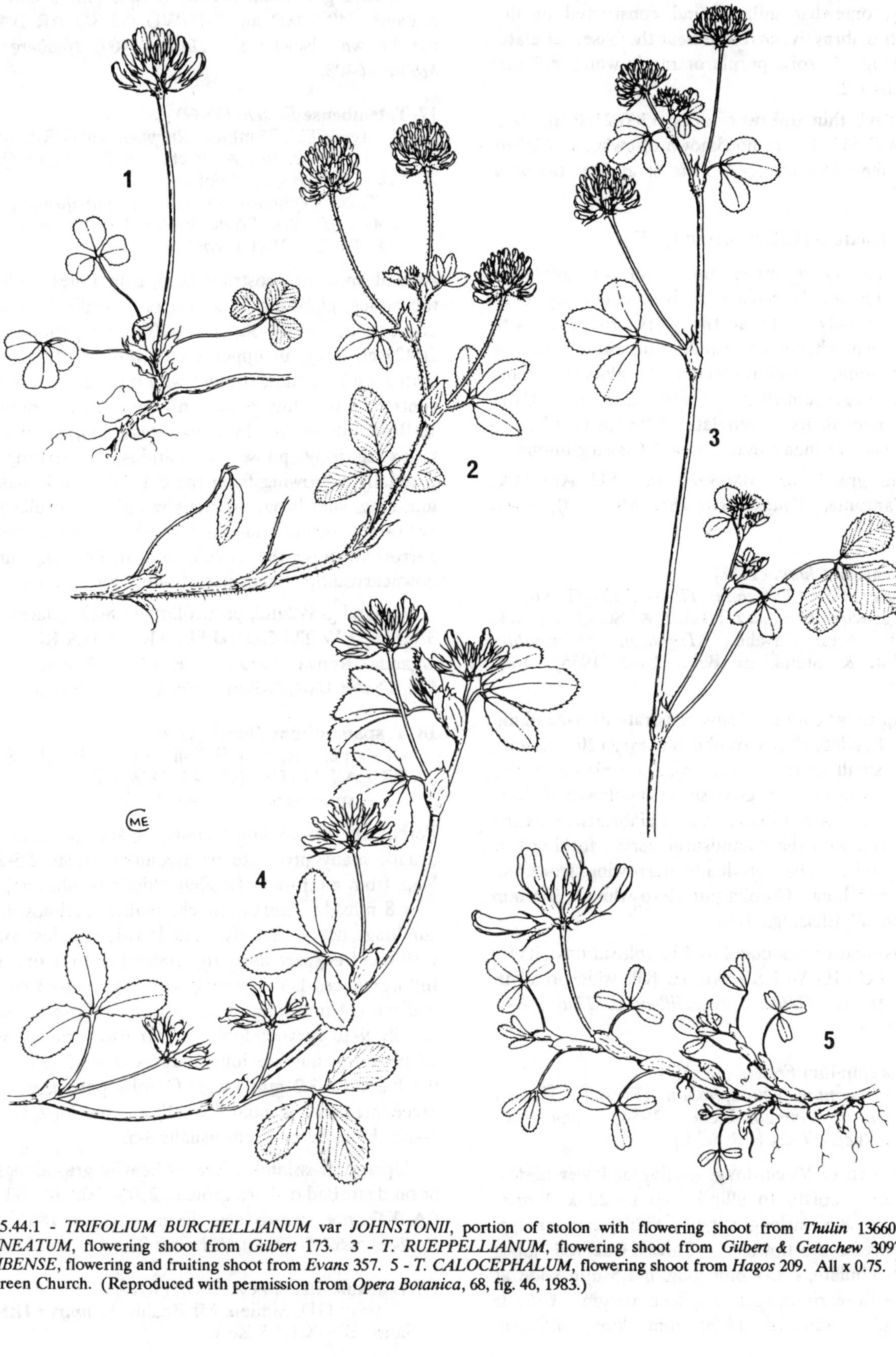

Figure 95.44.1 - *TRIFOLIUM BURCHELLIANUM* var *JOHNSTONII*, portion of stolon with flowering shoot from *Thulin* 13660. 2 - *T. BILINEATUM*, flowering shoot from *Gilbert* 173. 3 - *T. RUEPPELLIANUM*, flowering shoot from *Gilbert & Getachew* 3097. 4 - *T. TEMBENSE*, flowering and fruiting shoot from *Evans* 357. 5 - *T. CALOCEPHALUM*, flowering shoot from *Hagos* 209. All x 0.75. Drawn by Maureen Church. (Reproduced with permission from *Opera Botanica*, 68, fig. 46, 1983.)

usually somewhat laciniate. Inflorescences ± globose, many-flowered, *c* 17 mm across; pedicels *c* 0.5 mm long; peduncle long, glabrous. Calyx glabrous, 11-nerved, becoming somewhat inflated and constricted at the throat; lobes abruptly narrowed near the base, subulate, *c* 2 mm long. Corolla purple or rarely white, *c* 7 mm long. Seeds 1-2.

Grassland, thin soil over rocks; 1300-2100 m. TU GD GJ WG SU IL KF; not known elsewhere. *Gilbert & Thulin* 699; *Mooney* 5864; *Friis* et al. 20; *Aweke & Gilbert* 971.

14. **T. lanceolatum** *(Gillett) Gillett* (1970).

More or less erect annual up to 40 cm tall; stem glabrous. Leaflets lanceolate to oblanceolate, up to 27 x 7 mm, usually acute at the apex, glabrous, with prominent whitish nerves; stipules somewhat laciniate at the margin. Inflorescences ± globose, many-flowered, long-pedunculate; bracts minute. Calyx pilose, 11-nerved; lobes subulate. Corolla purple, *c* 5 mm long, shorter than calyx. Seeds 2. Cleistogamous.

Upland grassland; 2100-3000 m. SU AR HA; Kenya, Tanzania. *Thulin* 1646; *IECAMA* C-90; *Borini* 27.

15. **T. baccarinii** *Chiov*. (1911)
- types: GD, *Chiovenda* 1774 & 2155 (FT syn.).
Trigonella marginata Hochst & Steud. ex Bak. (1871), nom. invalid.; *Trifolium marginatum* (Hochst. & Steud. ex Bak.) Cuf. (1955), nom. invalid.

Annual up to 30 cm tall; stems prostrate or ascending, glabrous. Leaflets elliptic or obovate, up to 20 x 12 mm, with very small teeth; stipules with acuminate, entire tips. Inflorescences ± globose, many-flowered, long-pedunculate; bracts minute. Calyx glabrous or nearly so, 11-nerved with the commisural nerves forking low down the tube; lobes gradually narrowing from the base, 2-3 mm long. Corolla purple to whitish, 3-4 mm long. Seeds 2. Cleistogamous.

In grassland or woodland and in cultivations; 1000-2500 m. TU GD WG SU AR IL KF; widespread in tropical Africa. *Thulin* 1586; *Gilbert & Thulin* 840; *Mooney* 5860.

16. **T. calocephalum** *Fresen*. (1839)
Loxospermum calocephalum (Fresen.) Hochst. (1846) *T. abyssinicum* Heller (1984) - type: GD, Simien, *Rueppell* s.n. (FR holo.).

Perennial herb to 50 cm long, rooting at lower nodes. Leaflets oblanceolate to elliptic, up to 22 x 9 mm; petioles up to 8.5 cm long; stipules with triangular tips. Inflorescences *c* 4-8-flowered. Calyx 11-nerved; lobes triangular-subulate, *c* 6-7 mm long, *c* 1.5 mm wide at the base, with scarious, sparsely pilose margins. Corolla dark purple; standard 14-20 mm long, abruptly narrowed above the middle into an oblong, truncate, downcurved tip. Seeds 7-10. Fig. 95.44.5.

Upland grassland, usually in wet places and along streams; 2400-3600 m. GD WU GJ SU AR BA SD; not known elsewhere. *Thulin* 1469; *Hedberg* 5716; *Mooney* 6403.

17. **T. tembense** *Fresen*. (1839)
- type: TU, Tembien, *Rueppell* s.n. (FR holo.).
T. umbellatum A. Rich. (1847) - type: *Quartin Dillon & Petit* s.n. (P holo.).
T. calocephalum Fresen. var. *parviflorum* Chiov. in *Ann. Ist. Bot. Roma* **8**: 405 (1908) - types: *Pappi* 720, 864 & 2821 (FT syn.).

Annual erect or prostrate herb, sometimes rooting at the nodes, glabrous or nearly so. Leaflets elliptic or obovate, up to 20(-30) x 10(-20) mm; petioles up to 5(-12) cm long, in upper leaves very short or wholly united with stipules; stipules with the free part abruptly contracted to a fine point. Inflorescences ± globose, 3-16-flowered, up to 23 mm across, long-pedunculate. Calyx sparsely pilose at margins, 11-nerved; lobes gradually narrowing from the *c* 1.5 mm wide base, *c* 5 mm long, with broad scarious margins. Corolla purple or rarely white; standard *c* 9 mm long, abruptly narrowed above the middle into an oblong, truncate, downcurved tip. Seeds usually 4-6. Fig. 95.44.4.

Upland grassland, particularly in moist places; 2050-3200 m. EW TU GD GJ SU AR BA HA KF GG SD; Uganda, Kenya, Tanzania, E Zaire, Rwanda. *Pappi* 720; *Thulin* 1568; *Gilbert & Thulin* 574; *Mooney* 5778.

18. **T. spananthum** *Thulin* (1976)
- type: AR, near Bekoji, *Thulin* 1355 (UPS holo., BR EA ETH FT HUJ K MO WAG iso.).
T. sp. *A.* sensu Fröman (1975).

Perennial, often mat-forming glabrous herb, with usually many prostrate or ascending stems 2.5-20 cm long, from a taproot. Leaflets elliptic or obovate, up to 19 x 8 mm, but usually much smaller; petioles up to 4 cm long, never entirely united with stipules; stipules with the free part abruptly contracted to a fine point. Inflorescences 1-6-flowered; peduncle up to 3 cm long; pedicels *c* 1 mm long. Calyx sparsely pilose at margins, usually 9-10-nerved; lobes ± narrowly triangular with subulate tips and scarious margins, *c* 1-1.2 mm wide at the base, 1.8-3.2 mm long. Corolla purple, not much exceeding calyx; standard usually *c* 7 mm long, broadest above the middle. Seeds usually 3-5.

Upland grassland, often in heavily grazed pastures or on disturbed or bare ground; 2300-3900 m. WU AR BA KF; not known elsewhere. *Thulin* 1552, 1354; *Hedberg* 5662, 5678; *Sue Edwards* 73.

19. **T. acaule** *Steud. ex A. Rich*. (1847)
type: GD, Simien, Mt Buahit, *Schimper* II:587 (P holo., BM K UPS iso.).

T. petitianum A. Rich. (1847) - types: TU, *Petit* s.n. (P syn.) & GD, Simien, *Schimper* II:1389 (P syn., BM K UPS isosyn.).

T. petitianum var. *cuneatum* Bak. f., *Leg. Trop. Afr.*: 78 (1926) - type: EW, *Pappi* 1084 (K holo.).

T. acaule var. *emarginatum* Chiov. ex Fiori in *Nuovo Giorn. Bot. Ital.* **55**: 343, Fig. 17 (1949), nom. invalid. - type: Vàtova 2460 (FT).

Small perennial rhizomatous herb with taproot, glabrous or ± pilose. Leaflets obcordate to oblanceolate, *c* 2-12 x 1-5 mm; petioles 15-50 mm long. Inflorescences 1-5-flowered; peduncle 1-5 cm long. Calyx 11-nerved; lobes narrowly triangular, 1-2 mm long. Corolla mauve or blue-violet, *c* 8-12 mm long. Seeds 2-4.

In short grass and rock crevices; 2500-4150 m. EW TU GD WU GJ SU AR BA KF; Mt Elgon on Uganda/Kenya border. *Thulin* 1679; *Hedberg & Aweke* 5356; *Gilbert* 1916.

T. petitianum has up to now been considered an endemic species of N Ethiopia. It has been distinguished from the more widespread *T. acaule* mainly on account of its small emarginate leaflets, long petioles and peduncles and by its stoloniferous habit. Gillett (1952) noted that *T. acaule* and *T. petitianum* are connected by a series of intermediate forms and suggested that there may be only one variable species. The rather rich material now available shows a wide variation in N and C Ethiopia, but no clear distinctions can be made as the characters vary ± independently of one another. The two specific epithets were published at the same time but as *T. acaule* is the more well-known name *T. petitianum* is here treated as a synonym.

Some collections from BA, e.g. *Smeds* 658 (FT) and *Hedberg* 5663, somewhat approach *T. semipilosum.*

20. **T. multinerve** *A. Rich.* (1847)

types: TU, *Schimper* I:300 (P syn., BM K isosyn.) & *Quartin Dillon* s.n. (P syn.).

Loxospermum multinerve Hochst. (1846) - type: *Schimper* I:300 (B holo. destr., BM K P iso.).

T. multinerve var. *debilior* A. Rich., *Tent. Fl. Abyss.* **1**: 175 (1847) - type: TU, *Quartin Dillon & Petit* s.n. (P holo.).

Annual erect or procumbent, subglabrous herb, up to 30 cm tall. Leaflets ± oblanceolate, up to 25 x 6(-10) mm, usually acute, glabrous; petioles 1-6 cm long; stipules up to 1 cm long. Inflorescences 1-6(-8)-flowered, ± pedunculate. Calyx *c* 20-30-nerved; lobes narrowly triangular with subulate tips, 3-4 mm long, as long as or slightly longer than the tube, pilose at the margins. Corolla purple, 7-10(-12) mm long. Seeds 5-9, smooth. Fig. 95.45.1.

Short upland grassland and moorland, especially in moist places; 1850-3550(-4150) m. EW TU GD WU GJ WG SU AR BA GG SD; Sudan, Uganda, Kenya and E Zaire. *Pappi* 2259; *Thulin* 1450; *Gilbert & Thulin* 824; *W. de Wilde* 6347; *Gilbert & Getachew* 2875.

21. **T. elgonense** *Gillett* (1952).

Small procumbent annual, stems to 15(-30) cm long, glabrous. Leaflets cuneate-obovate, up to 7(-13) x 5(-8) mm, glabrous; petioles up to 1.5(-3) cm long. Inflorescences 1-2-flowered, peduncle absent or very short. Calyx with pilose margins, *c* 25-nerved; lobes triangular, 2-2.5(-3.5) mm long, shorter than the tube. Corolla purple, *c* 9 mm long. Seeds *c* 6, smooth.

Wet open places in upland forest, moor or grassland; 2500-3720 m. GD SU SD; Mt Elgon on Uganda/Kenya border. *Gillett* 15023; *Hedberg & Aweke* 5342, 5370.

22. **T. decorum** *Chiov.* (1911)

types: GD, Dembia, *Chiovenda* 1110, 1197, 1199 & 1391 (FT syn.).

T. sp. *D.* sensu Fröman (1975).

Annual or short-lived ascending perennial, glabrous, up to 50 cm tall. Leaflets ovate to suborbicular, 30-45 x 18-28 mm; petioles up to 12 cm long; stipules ovate-acuminate. Inflorescences 6-15-flowered, long-pedunculate. Calyx *c* 25-nerved; lobes triangular, 2-3 mm long, sparsely pilose at the margins. Corolla purple, 10-14 mm long, twice as long as calyx or more. Seeds 3-5. Fig. 95.45.2.

Upland grassland, especially in moist places; 1800-3000 m. TU GD GJ SU AR; not known elsewhere. *Gilbert & Thulin* 918; *Schimper* 1281; *Evans & Lythgoe* 33.

23. **T. chilaloense** *Thulin* (1976)

type: AR, Kersa, *Thulin* 1642 (UPS holo., BR EA ETH FT HUJ K MO WAG iso.).

T. sp. *B* sensu Fröman (1975) excl. Fig.

Annual erect herb up to 12 cm tall; stem glabrous or sparsely pilose. Leaflets ± oblanceolate, up to 25 x 7 mm, acute, ± weakly pilose near the base; petioles up to 4 cm long; stipules 1-2 cm long. Inflorescences few, 5-12-flowered. Calyx glabrous or almost so, 15(-18)-nerved; lobes narrowly triangular with subulate tips, 3.5-4.5 mm long, usually about twice as long as the tube. Corolla purplish-red, *c* 7 mm long. Seeds usually 3-4, finely warty.

Grassland patches in montane forest; 2900 m. AR; only known from the type.

24. **T. schimperi** *A. Rich.* (1847)

types: TU, *Quartin Dillon* s.n. (P syn.), *Schimper* I:86 (P syn., BM K UPS WAG isosyn.) & III: 1988 (P syn., BM K isosyn.).

Loxospermum schimperi Hochst. (1846) - type: TU, *Schimper* I:86 (B holo. destr., BM K P UPS WAG iso.).

Annual erect glabrous herb up to 35 cm tall. Leaflets linear, up to 45 x 3 mm; petioles up to 3.5 cm long; stipules with subulate tips. Inflorescences 1-5-flowered, short-pedunculate. Calyx *c* 25-nerved; lobes narrowly triangular, *c* 10 mm long. Corolla red, 15-18 mm long. Seeds *c* 8-15. Fig. 95.45.3.

Upland grassland, especially in damp places; 1700-3150 m. TU GD GJ WG WU SU AR KF; not known elsewhere. *Ash* 2069; *Thulin* 1587; *Gilbert & Thulin* 820; *Mooney* 7507.

25. **T. steudneri** *Schweinf.* (1868)
type: GD, Simien, *Steudner* 159 (B holo. destr.).
T. bellianum Chiov. (1908) - types: EW, *Pappi* 104 (FT syn., BM isosyn.) & 2315 (FT syn.).

Annual glabrous herb up to 60 cm tall. Leaflets ± narrowly elliptic, up to 40 x 8 mm; petioles up to 5 cm long; stipules with narrowly triangular tips. Inflorescences ± globose, *c* 10-20-flowered, up to 13(-20) mm across, long-pedunculate; bracts several-nerved forming an involucre, outer bracts abruptly and sharply pointed, up to 7 mm long, inner bracts oblong, 3 mm long. Calyx *c* 40-nerved, glabrous or occasionally pilose; lobes triangular, 2-3 mm long. Corolla purplish, 7-8 mm long, sometimes with filiform tips. Seeds *c* 4. Fig. 95.45.4.

Upland grassland and bushland, especially in damp places; 1100-2800 m. EW TU GD WU GJ SU AR KF; Uganda, Kenya. *Gilbert & Thulin* 936; *Mooney* 4780, 5846; *Ash* 2117.

A form with densely pilose calyces is known from GJ (*Kahurananga & Abdulkadir Ahmed* 4482A).

26. **T. quartinianum** *A. Rich.* (1847)
type: TU, *Quartin Dillon* s.n. (P holo.).

Annual subglabrous herb up to 60 cm tall. Leaflets narrowly oblong, up to 52 x 16 mm; petioles up to 4 cm long; stipules with ± narrowly triangular tips. Inflorescences ± globose, long-pedunculate; bracts several-nerved forming an involucre, outer bracts ovate, lobed or toothed, up to 7 mm long. Calyx *c* 40-nerved; lobes 4-7 mm long, ovate and overlapping at the base, drawn out into subulate points. Corolla purplish red, rarely white, 11-12 mm long. Seeds usually 4-5. Fig. 95.45.5.

Upland grassland, especially in damp places; 1500-2600 m. EW TU GD GJ SU KF; Uganda (Mt Elgon). *Pappi* 1369; *Schimper* 642; *Gilbert & Thulin* 900; *Aweke & Gilbert* 980; *Mooney* 5849, 6092.

27. **T. campestre** *Schreb.* (1804).

Annual; stems up to 30(-50) cm tall, hairy, erect or ascending. Leaflets obovate, 8-12 mm long, the terminal one petiolulate. Inflorescences up to 15 mm long, dense, 20-30-flowered. Calyx 5-nerved. Corolla yellow, (3-)4-5(-6) mm long; limb of standard sulcate. Pod 1-seeded.

Grassland in open sunny spots, rock crevices; 1000-3100 m. EE EW TU GD SU; widespread in Europe, N Africa and SW Asia. *W. de Wilde* 9641; *Hedberg & Aweke* 5494; *Mooney* 7876; *Aweke & Gilbert* 968.

28. **T. fragiferum** *L.* (1753).

More or less hairy perennial; stems procumbent, often rooting at the nodes, 10-30(-40) cm long. Leaflets (3-) 8-20 mm long, ovate, elliptic or obcordate; stipules lanceolate-subulate. Inflorescences ± globose, 10-14 mm across, long-pedunculate; bracts 3-4 mm long, forming an involucre. Calyx 2-lipped with the upper lip greatly inflated and net-veined in fruit. Corolla 6-7 mm long, pale pink. Seeds 1-2.

Along streams; 1650-1900 m. EW TU; widespread in Europe, W Asia, N Africa and in Arabia. *Schimper* 718, 720; *O. Ryding & Ermias K.* 1440.

29. **T. resupinatum** *L.* (1753).

Glabrous annual, 10-30(-60) cm tall, procumbent to erect. Leaflets cuneate-obovate, 7-20 mm long. Inflorescences globose; bracts minute; flowers twisted through 180°. Calyx 2-lipped with the upper lip greatly inflated and net-veined in fruit, crowned by the two divergent upper calyx-lobes. Corolla *c* 6 mm long, usually pink. Seeds 1-2.

Habitat in Ethiopia not known, only once collected and probably introduced. EW; widespread in Europe, N Africa and SW Asia. *Pappi* 4874.

30. **T. pratense** *L.* (1753).

More or less hairy perennial, up to 1 m tall; leaflets ± obovate, up to 5 cm long, hairy below, often glabrescent above; stipules triangular above, abruptly contracted into a setaceous, usually ciliate point; inflorescences globose or ovoid, 2-4 cm long, usually sessile and involucrate; calyx 10-nerved; lobes triangular with filiform apex; corolla reddish-purple or pink, 12-15 mm long; seeds 1-2.

Introduced for experimental cultivation in EW and AR (Fiori 1949 & Fröman 1975); widespread in Europe and W Asia and extensively cultivated for fodder. *Pappi* 366.

31. **T. arvense** *L.* (1753).

Annual; stems 4-40 cm, ± erect, pubescent. Leaflets linear-oblong, 5-20 mm long; upper leaves sessile; lower stipules lanceolate-subulate, the upper subulate from an ovate base. Inflorescences ovoid or oblong, numerous, up to 20 mm long, many-flowered; flowers sessile, without bracts. Calyx densely hairy; lobes (1-)2-3 times as long as the tube, setaceous. Corolla whitish or pink, c 4 mm long, much shorter than the calyx. Seeds 1-2.

Upland grassland and moor; 1300-3000 m. EW TU GD SU; widespread in Eurasia and N Africa. *Pappi* 4698; *Schimper* I:322; *Mooney* 7858.

32. **T. subterraneum** *L.* (1753).

Annual; stems up to 20(-30) cm, procumbent; leaflets broadly obcordate; stipules semi-ovate; fertile flowers 2-

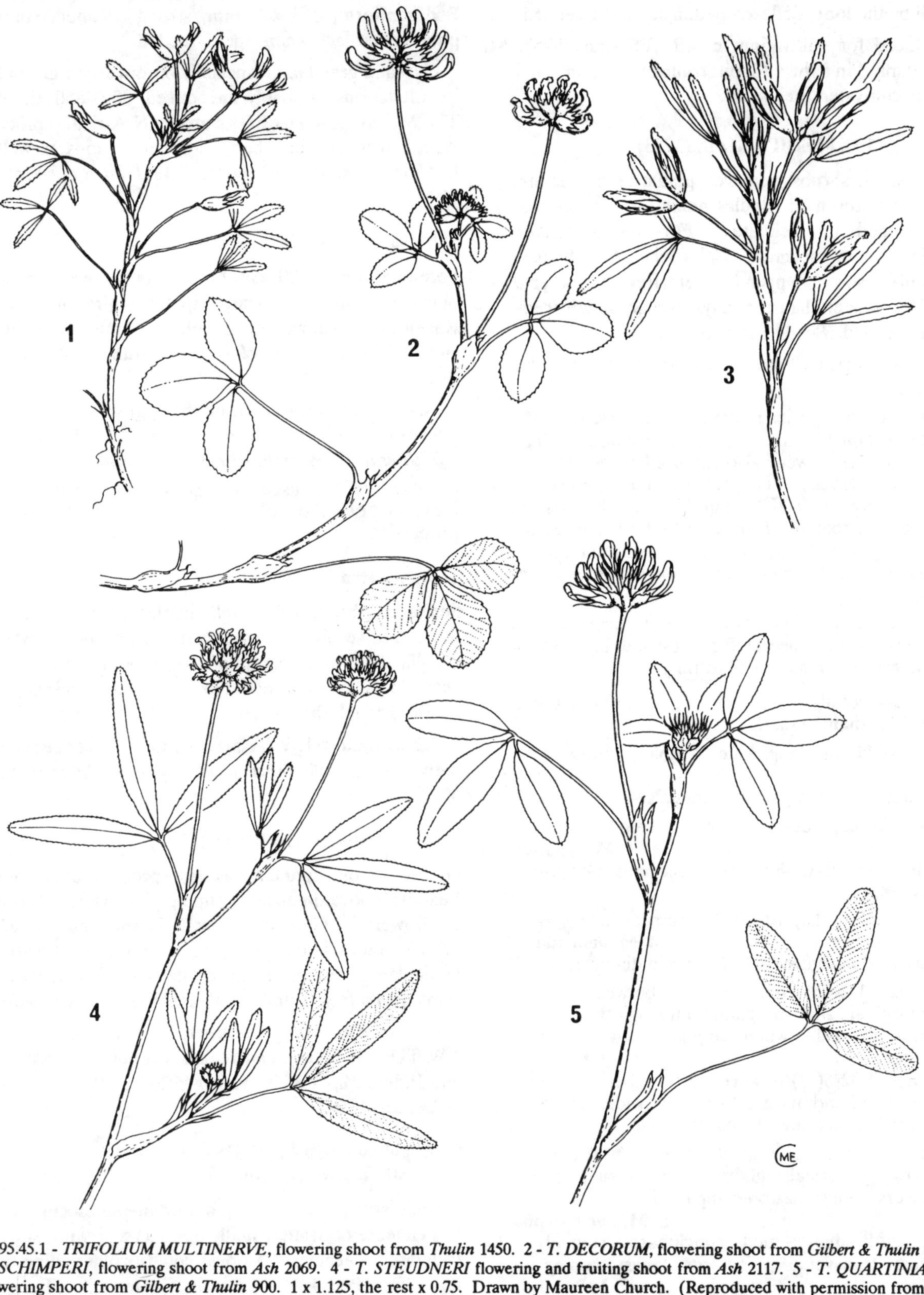

Figure 95.45.1 - *TRIFOLIUM MULTINERVE*, flowering shoot from *Thulin* 1450. 2 - *T. DECORUM*, flowering shoot from *Gilbert & Thulin* 918. 3 - *T. SCHIMPERI*, flowering shoot from *Ash* 2069. 4 - *T. STEUDNERI* flowering and fruiting shoot from *Ash* 2117. 5 - *T. QUARTINIANUM*, flowering shoot from *Gilbert & Thulin* 900. 1 x 1.125, the rest x 0.75. Drawn by Maureen Church. (Reproduced with permission from *Opera Botanica*, 68, fig. 47, 1983.)

5(-7); corolla 8-14 mm long, whitish; sterile flowers numerous, becoming strongly deflexed over the fruiting calyces; fruiting heads globose, appressed to or buried in the soil by the long, deflexed peduncle; pod 1-seeded.

Introduced for cultivation in AR (Fröman 1975); native mainly in the Mediterranean region and extensively cultivated for fodder.

100. MEDICAGO *L.* (1753)

Herbs or small shrubs. Leaves pinnately 3-foliolate; leaflets usually toothed; stipules adnate to the petiole, usually toothed or laciniate. Flowers in axillary pedunculate short racemes. Calyx-lobes 5, subequal. Corolla falling soon. Upper filament free, the other 9 united. Pod longer than the calyx, usually indehiscent and spirally coiled. Seeds 1-several.

Some 50 species, most numerous around the Mediterranean.

Apart from the species treated here *M. praecox* DC. and *M. truncatula* Gaertn. (syn. *M. tentaculata* Willd., *M. tribuloides* Desr.) were also reported from Ethiopia in Cuf. *Enum*. (1955). The record of *M. praecox* is based on *Penzig* s.n. (GE) from Eritrea which is *M. polymorpha* and that of *M. truncatula* derives from an unverified report of *M. tentaculata* by Schweinfurth & Ascherson (1867). *M. truncatula* is now being tested in forage trials.

Medicago includes a number of species important as forage, as sources of nectar for bees and in rotation with cereal crops to restore soil fertility.

1. Shrub, 1-4 m tall. **3. M. arborea**
- Herbs, less than 1 m tall. 2
2. Corolla 7-11 mm long, blue to violet; perennial. **2. M. sativa**
- Corolla 2-5 mm long, yellow; annuals. 3
3. Pod reniform; racemes *c* 10-50-flowered. **1. M. lupulina**
- Pod in a spiral of 3-6 turns; racemes 1-5(-8)-flowered. 4
4. Pod not spiny, flat, 10-17(-20) mm in diameter. **4. M. orbicularis**
- Pod spiny at margins, 3-8(-10) mm in diameter. 5
5. Racemes 1-2-flowered; groove between the submarginal and marginal veins of the pod narrow, not visible when the pod is viewed from the edge. **6. M. laciniata**
- Racemes (1-)2-6(-8)-flowered; groove between the submarginal and marginal vein very wide, visible when the pod is viewed from the edge. 6
6. Plant somewhat pubescent; pod flattened or shortly cylindrical, glabrous or nearly so, transverse veins anastomosing freely. **5. M. polymorpha**
- Plant softly hairy; pod subglobose, sparsely villous, transverse veins not anastomosing. **7. M. minima**

1. **M. lupulina** *L.* (1753).

Annual or short-lived perennial, prostrate or ascending, up to 60 cm long, ± pubescent. Leaflets cuneate-obovate, up to 14 x 12 mm, pubescent. Racemes dense, usually many-flowered. Corolla yellow, 2-3 mm long. Pod reniform, *c* 3 x 2 mm, strongly veined, sparsely pubescent, black when ripe, 1-seeded.

Upland grassland, especially on disturbed ground or in cultivations, probably introduced; 900-2850 m. EW TU WU SU AR HA SD; Europe, N Africa, temperate Asia, introduced in other temperate regions as well as E Africa. *Thulin* 1380; *Pappi* 4343; *Gilbert & Getachew* 3099.

2. **M. sativa** *L.* (1753).

Perennial up to 80 cm tall, ± pubescent. Leaflets obovate to almost linear; stipules entire or dentate. Racemes 5-40-flowered. Corolla 7-11 mm long, blue to violet. Pod in a spiral of 1-3 turns with a hole through the centre.

Cultivated for fodder in the highlands and under irrigation also in more arid areas down to 350 m. *Pappi* 2327; *Berhe* 17; *Westphal* 3998; *Sue Edwards* 687.

The seed is used in making feed concentrates. Extracts of the plant have shown bactericidal properties.

3. **M. arborea** *L.* (1753).

Sericeous shrub, 1-4 m tall; leaflets obovate; stipules lanceolate, entire; racemes very short, 4-10-flowered; corolla 10-15 mm long; pod 10-15 mm in diameter, in a spiral of 1-1.5 turns, not spiny, reticulately veined, with a hole through the centre.

Cultivated in EW (Asmara) and probably elsewhere; native in the Mediterranean region. *Baldrati* 4001, 4037.

4. **M. orbicularis** *(L.) Bartal.* (1776).

Glabrous or sparsely hairy, procumbent annual. Leaflets obovate-cuneate; stipules laciniate. Racemes 1-5-flowered. Corolla yellow, 2-5 mm long. Pod 10-17(-20) mm in diameter, in a spiral of 4-6 turns, lenticular, glabrous or almost so, not spiny; transverse veins with a few, usually weak anastomosing branches.

Grassland, scrub, disturbed ground; 1650-2750 m. EW TU; Mediterranean region, extending to SW Asia and India. *Pappi* 3458; *Parker* E.265; *J. De Wilde* 6989; *Whiteman* 1060.

5. **M. polymorpha** *L.* (1753).
M. hispida Gaertn. (1791).

Weakly spreading annual, somewhat pubescent. Leaflets cuneate-obovate, usually up to 11 x 10 mm; stipules laciniate. Racemes 2-5-flowered. Corolla yellow, 3-4.5 mm long. Pod 4-8(-10) mm in diameter in a lax spiral of *c* 3 turns, glabrous or almost so; transverse veins strong, anastomosing freely, at least near the submarginal vein; submarginal vein conspicuous, separated from the marginal vein by a deep groove, the margin consequently with 3 keels separated by 2 grooves; spines usually hooked. Fig. 95.46.1.

In pastures and as a weed in cultivations, probably introduced; 1400-3000 m. EW TU GD WU GJ SU AR HA KF; E Africa, Europe, N Africa, Asia, and widely introduced elsewhere. *Thulin* 1379; *Pappi* 4585; *Mooney* 5777.

6. **M. laciniata** *(L.) Mill.* (1768).

Spreading annual, subglabrous or sparsely pilose. Leaflets cuneate-obovate or cuneate-oblong, up to 12 x 5 mm, strongly toothed; stipules laciniate. Racemes 1-2-flowered. Corolla yellow, 4-5 mm long. Pod up to 5 mm in diameter, in a globose coil of *c* 5 turns, glabrous; transverse veins sigmoid, sparingly branched; submarginal vein broad, separated from the flat or convex marginal vein by a narrow groove; spines up to 3 mm long, hooked.

Grassland and woodland; 1500-2200 m. HA BA SD; Kenya, Tanzania, N Somalia, Socotra, South Africa, and N Africa to SW Asia; widely introduced in other areas. *Gilbert & Thulin* 52; *Burger* 2026 A; *Thulin, Hunde & Tadesse* 3836.

M. laciniata subsp. *schimperiana* P. Fourn. (syn. *M. aschersoniana* Urb.) with the stipules divided less than halfway to the middle, and the pods 2.5-4 mm in diameter with 7-10(-12) transverse veins per turn of the spiral was believed to occur in Eritrea in Cuf. *Enum*.: 245 (1955). No specimen has been seen.

Elsewhere known as a good plant for bees and used for the extraction of condiments and infusions.

7. **M. minima** *(L.) Bartal.* (1776).

Softly hairy annual. Leaflets obovate; stipules ± dentate. Racemes 1-6(-8)-flowered. Corolla yellow, 4-4.5 mm long. Pod 3-5 mm in diameter in a spiral of 3-5 turns, subglobose, sparsely villous; transverse veins slender, curved, not anastomosing; submarginal vein wide, separated from the narrow, flat or convex marginal vein by a wide groove; spines hooked, short or long.

Pastures, roadsides, usually in sandy places; 1650-3050 m. EW TU GD AR; N Somalia, Socotra, N Africa, Europe, S Asia. *Thulin* 1540; *Pappi* 1936; *Schimper* III:1912.

101. **MELILOTUS** *Mill.* (1754)

Annual to short-lived perennial fragrant herbs. Leaves pinnately 3-foliolate; leaflets usually toothed. Flowers in elongated axillary racemes. Calyx-lobes 5, subequal. Corolla falling soon. Upper filament free, the other 9 united. Pod small, globose to obovoid, straight, indehiscent, 1-2(-4)-seeded.

Some 25 species mainly in the Mediterranean region and central Asia. It is doubtful if any of the species are native to Ethiopia.

Elsewhere several species are important plants for honey bees as well as for browse, hay and grazing. In Ethiopia, as the vernacular name JEMAMBERI (Tign.). indicates, twigs are picked and used to perfume clothes and oinment.

A record of *M. neapolitana* Ten. in Cuf. *Enum*. (1955) has not been confirmed and is omitted here.

1. Corolla white. **1. M. alba**
- Corolla yellow. 2
2. Corolla 2-3 mm long; pod 1.5-3 mm long. **3. M. indica**
- Corolla 3-5 mm long; pod 3-8 mm long. 3
3. Racemes 3-10-flowered, shorter than the subtending leaf; stipules denticulate at the base. **5. M. messanensis**
- Racemes more than 10-flowered, usually at least as long as the subtending leaf; stipules entire. 4
4. Racemes 15-30-flowered; pod with transverse or sigmoid veins, brownish-yellow when ripe. **4. M. elegans**
- Racemes at least 30-flowered; pod with a weak irregular network of veins, black when ripe. **2. M. suaveolens**

1. **M. alba** *Desr.* (1796).

Erect annual to at least 1 m tall. Leaflets cuneate-oblong, serrate; rhachis of larger leaves, excluding the terminal petiolule, often prolonged more than 4 mm beyond the lateral leaflets; stipules entire. Racemes lax and slender, many-flowered. Corolla white, 4-5 mm long. Pod *c* 4 mm long with a weak irregular network of veins, glabrous, almost black when ripe.

Cultivated or disturbed ground; *c* 2000 m. SU HA; Europe, W and C Asia, introduced in E Africa and elsewhere. *IECAMA* A-21; *Westphal* 3996; *Bos* 7543; *Stewart* 13.

Elsewhere the shoots are used in making paper pulp.

2. **M. suaveolens** *Ledeb.* (1824).
M. officinalis sensu auctt., non (L.) Pallas.
M. altissima sensu auctt., non Thuill.

Erect annual often more than 1 m tall. Leaflets narrow, serrate; stipules entire. Racemes rather lax, 30-40-flowered. Corolla yellow, 3-4 mm long. Pod 3.5-5 mm long, with a weak irregular network of veins, glabrous, black when ripe.

Grassland, on disturbed ground and as a weed in cultivations; 1500-2400 m. EW WG SU AR BA HA IL KF SD; Tanzania, N Asia and India. *Thulin* 1436; *Haberland* 2034; *Burger* 755; *Sereke W.-H.* 24.

3. **M. indica** *(L.) All.* (1785).

Erect or ascending annual, 15-50 cm tall. Leaflets lanceolate-oblong, serrate; stipules subentire. Racemes dense, (10-) many-flowered. Corolla pale yellow, 2-3 mm long; wings and keel equal, shorter than standard. Pod 1.5-3 mm long, subglobose, strongly reticulate-veined, glabrous.

Weed of cultivation, disturbed ground; 1600-2500 m. EW TU; Mediterranean region, C Asia, India, introduced in Sudan, Uganda, Kenya and elsewhere in warm-temperate regions. *Pappi* 133, 2331; *Quartin Dillon & Petit* 141.

Elsewhere used to make poultices and emollients.

4. **M. elegans** *Salzm. ex Ser.* (1825).
M. abyssinica Bak. (1871) - types: *Parkyns* s.n. & *Schimper* 439 (both K syn.).

Erect annual to more than 1 m tall. Leaflets obovate-orbicular to oblong, serrate; stipules entire. Racemes lax, 15-30-flowered. Corolla yellow, 4-5 mm long; standard and wings equal, shorter than keel. Pod 3-4 mm long, obovoid, compressed, with transverse or sigmoid veins, brownish-yellow when ripe. Fig. 95.46.2.

Cultivated or disturbed ground; *c* 1700-2800 m. EW, TU, GD, WG, SU, AR, HA, KF; Mediterranean region. *Siegenthaler* 55; *IECAMA* C-33; *Chiovenda* 2387; *Kokwaro* 496.

5. **M. messanensis** *(L.) All.* (1785).
M. sicula (Turra) B. D. Jackson (1894).

Erect or ascending annual, 20-40 cm tall, leaflets obovate-orbicular or lanceolate-cuneate, serrate; stipules entire but denticulate at the base. Racemes 3-10-flowered, shorter than their subtending leaves. Corolla yellow, 4-5 mm long; standard and keel subequal, longer than wings. Pod 5-8 mm long, oblique-ovoid, acute, concentric-striate, yellowish-brown when ripe.

Probably on disturbed ground; 2200 m. EW; Mediterranean region. *Terracciano & Pappi* 393.

102. TRIGONELLA *L.* (1753)

Annuals. Leaves pinnately 3-foliolate; leaflets usually toothed. Flowers solitary or in axillary short racemes. Corolla falling soon. Pod usually linear or oblong, straight or curved, indehiscent or dehiscing along one suture.

Some 70 species, mainly in the Mediterranean region.

T. foenum-graecum *L.* (1753).

Erect aromatic annual to 50 cm tall, sparsely pubescent. Leaflets obovate, 20-50 x 8-17 mm, denticulate. Flowers 1-2 in leaf-axils. Corolla pale yellow or white, 12-18 mm long. Pod (80-)120-150(-190) x 2-4 mm, with a long beak, (7-)10-15(-18)-seeded. Fig. 95.46.3.

Cultivated at altitudes between *c* 1600 and 2300 m all over Ethiopia for various uses but mainly for its edible seeds; native in S Europe and Asia. *Stewart* C-30; *Mooney* 8791; *Schimper* II:959.

Other uses include its use in making an unfermented drink, as a spice or condiment in baking, and in treating skin and stomach disorders. Some bactericidal properties have been found in extracts of FENUGREEK.

103. ONONIS *L.* (1753)

Annuals or perennials, usually glandular-hairy. Leaves usually 3-foliolate; leaflets usually toothed; stipules adnate to the petiole. Flowers in panicles, spikes or racemes. Filaments all united. Pod dehiscent.

Some 70 species mainly in the Mediterranean region.

O. reclinata *L.* (1753).

Erect or procumbent annual, 2-15 cm tall, villous and glandular hairy. Leaflets 3, oblanceolate to obovate-orbicular, 5-10 x 2-6 mm. Corolla pink or purple, 5-10 mm long, about equalling the calyx. Pod 8-14 mm long, pubescent. Seeds up to 20, tuberculate. Fig. 95.46.4.

Grassland; 1600-2500 m. EW TU ?GD; in the Mediterranean region and eastwards to Iran. *Pappi* 2196; *Robertson* 1249; *Schimper* III:1543.

Some Ethiopian material has been placed in var. *minor* Moris, with the corolla shorter than calyx, but this form scarcely merits taxonomic recognition.

104. CICER *L.* (1753)

van der Maesen in *Meded. Landbouwhogeschool Wageningen* 72-10 (1972) & in Saxena & Singh (eds.), *The Chickpea* (1987).

Herbs or subshrubs with glandular hairs. Leaves usually imparipinnate, rarely paripinnate and terminated by a tendril; leaflets toothed; stipules toothed or divided. Inflorescences axillary, few-flowered, or flowers solitary. Calyx gibbous at base, 2-lipped or with subequal lobes. Corolla white or violet. Upper filament free, the other 9 united. Pod ovate or oblong, dehiscent, 1-4-seeded.

Some 40 species from Morocco to Tibet. One species, *C. arietinum* L., the CHICK PEA, is widely cultivated.

1.	Tendrils absent.	**1. C. arietinum**
-	Tendrils present in upper leaves.	**2. C. cuneatum**

1. **C. arietinum** *L.* (1753).

Annual, 0.2-1 m tall, glandular pubescent. Leaves imparipinnate with (7-)11-15(-17) leaflets, 7-19 x 3-10 mm; stipules 2-5-fid. Flowers solitary. Corolla white to purplish, 1(-2.2) cm long. Pod 2-3.5 x 1-1.7 cm.

Cultivated throughout the highlands for its edible seeds. *Gillett* 14605; *Schimper* I:47; *J. De Wilde* 6008; *Mesfin* 860.

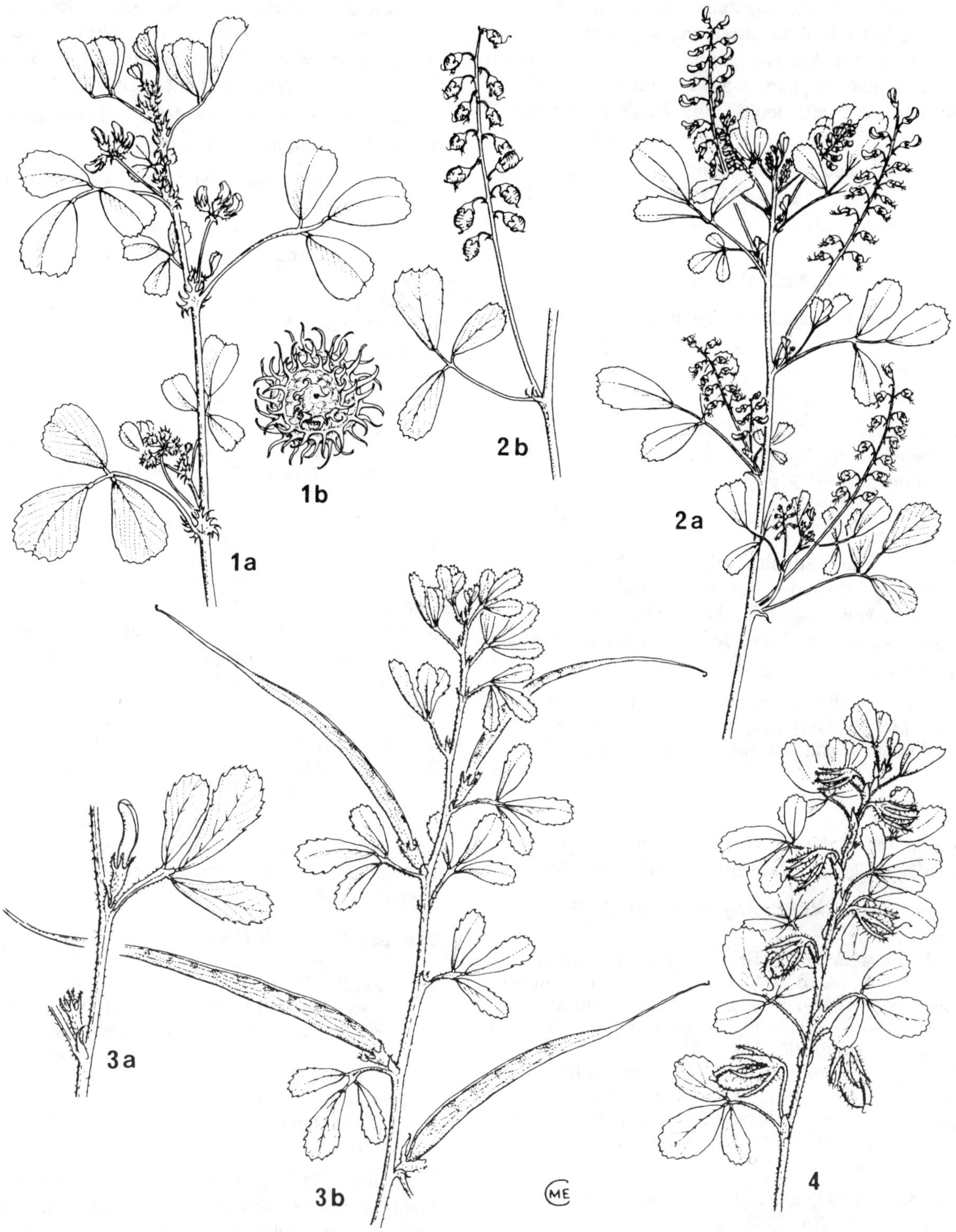

Figure 95.46.1 - *MEDICAGO POLYMORPHA*, a flowering branch from *Mooney* 9093, b: pod from *Meyer* 7576. 2 - *MELILOTUS ELEGANS*, a: flowering branch, b: fruiting raceme, both from *Siegenthaler* 55. 3 - *TRIGONELLA FOENUM-GRAECUM*, a: part of flowering branch, b: fruiting branch, both from *Stewart* C-30. 4 - *ONONIS RECLINATA*, flowering and fruiting branch from *Pappi* 2196. 1a, 2b, 3a and 4 x 1.125, 1b x 2.25, 2a and 3b x 0.75. Drawn by Maureen Church. (Reproduced with permission from *Opera Botanica*, 68, fig. 48, 1983.)

2. **C. cuneatum** *Hochst. ex A. Rich*. (1847)
- type: TU, *Schimper* II:810 (P lecto., BM BR K L M P S W isolecto.)

Annual, 40-60 cm tall, glandular pubescent. Lower leaves with 8-10 leaflets, often imparipinnate, upper leaves with up to 22 leaflets, ending in a tendril; leaflets 5-10 x 2-5 mm; stipules 2-5-fid. Flowers solitary. Corolla pink, *c* 7 mm long. Pod 15-23 x 7-10 mm. Fig. 95.42.4.

Grassland, weed in cultivations; 1000-2200 m. EW TU SU; SE Egypt. *Pappi* 107; *Prior* s.n.; *Seegeler* 157; *Tewolde Berhan* 1101.

105. **PISUM** *L*. (1753)

Glabrous annuals climbing by means of tendrils. Leaves paripinnate with 2-6 leaflets; stipules large. Wings adnate to keel. Style dilated at apex, longitudinally grooved with recurved margins.

A few species in the Mediterranean region and W Asia, one of which, *P. sativum* L., the GARDEN PEA, is widely cultivated.

P. sativum *L*. (1753).

Herb to 2 m long. Stipules larger than the leaflets, up to 10 cm long. Racemes 1-2(-3)-flowered. Corolla 15-35 mm long, white to purple. Pod 3.5-9.5 x 1-1.8 cm, 2-9-seeded. Seeds ± globose, 5-8 mm in diameter; hilum small, elliptic.

1. Leaflets usually in 2 pairs; corolla more than 20 mm long, white or purple. var. **sativum**
- Leaflets in 1 pair; corolla *c* 15 mm long, purple. var. **abyssinicum**

Var. **sativum.**

Grown in the highlands of all regions. *Pappi* 4937; *Chiovenda* 2869; *Sue Edwards* et al. 108; *Burger* 746.

Var. **abyssinicum** *(A. Br.) Alef*. in *Bonplandia* **9**: 126 (1861);

P. abyssinicum A. Br. (1841); *P. arvense* L. var. *abyssinicum* (A. Br.) Alef. ex Engl. in *Abh. Konigl. Akad. Wiss*. Berlin 1891: 265 (1892); *P. sativum* var. *elatius* (Stev.) Beck forma *abyssinicum* (A. Br.) Gams in Hegi, *Ill. Flora Mittel-Europa* **4**: 1613 (1924); *P. sativum* subsp. *abyssinicum* (A. Br.) Gov., *Bull. Appl. Bot*. **24**: 423 (1929-1930) - type: a plant grown from seeds collected by Schimper near Adua (?KR holo. destr.); TU, *Schimper* 1866 (P neo.).

Fig. 95.47.1.

Grown mainly in TU and WU but known also from SU, AR, BA and HA. *Bos* 8340.

106. **VICIA** *L*. (1753)

Annuals or perennials, often climbing by means of tendrils. Leaves paripinnate, usually with a tendril, very rarely imparipinnate. Flowers in axillary racemes or sometimes solitary in leaf-axils. Calyx 2-lipped or calyx-lobes subequal. Style pubescent all round or on the lower side, or glabrous. Pod ± oblong, compressed, dehiscent.

Some 120 species in temperate regions of the N hemisphere and S America; a few species in Africa. A record of cultivated *V. narbonensis* L. in Ethiopia (see Cuf. *Enum*.: 307 (1955) has not been confirmed.

All indigenous species have edible seeds often collected and eaten by children.

1. Flowers subsessile or very shortly pedunculate in the leaf-axils. 2
- Flowers in pedunculate racemes. 3
2. At least the upper leaves with a tendril. **4. V. sativa**
- All leaves without a tendril, the rhachis terminated by a short mucro. **5. V. faba**
3. Pod up to 10 mm long; corolla 3-5 mm long. **3. V. hirsuta**
- Pod 20-40 mm long; corolla 7-20 mm long. 4
4. Leaflets 8-24; racemes up to 30-flowered; pod 2-8-seeded. **1. V. villosa**
- Leaflets (2-)3-4(-7); racemes (1-)2-8-flowered; pod 5-11-seeded. **2. V. paucifolia**

1. **V. villosa** *Roth* (1793).

Straggling, climbing or trailing annual, 0.3-2 m long; stems hairy to glabrous. Leaflets 8-24, linear to elliptic, 3-35 x 1-8 mm, mostly pilose. Racemes elongated with up to 30 flowers. Corolla violet, purple or blue, 10-20 mm long. Pod 20-40 x 4-12 mm, glabrous or appressed pubescent, 2-8-seeded.

Cultivated for fodder and also found as an escape; 2100 m. AR SU EW; Europe, W Asia, widely cultivated. *Thulin* 1601.

Several subspecies, not always easy to distinguish, have been recognized within *V. villosa.* The single Ethiopian collection seen belongs nearest to subsp. *eriocarpa* (Hausskn.) P. W. Ball.

2. **V. paucifolia** *Bak*. (1871)
- type: GD, Debra Tabor, *Schimper* 1218 (K holo., BM iso.).

Lathyrus schimperi Engl. (1892) - type: GD, *Schimper* 1218 (B holo. destr., BM K iso.).

L. kilimandscharicus Taub. var. *scioanus* Fiori in *Rel. e Monogr. agr. -col*. **58**: 38 (1940) - type: SU, *Bartolozzi* s.n. (FT syn.).

L. hygrophilus sensu Cuf. *Enum*.: 309 (1955) non Taub.

Straggling, climbing or suberect perennial, 0.3-1 m long, young stems ± densely pilose. Leaflets (2-)3-4(-7), linear to narrowly elliptic, 10-45 x 1-8 mm, sparsely pilose. Racemes (1-)2-8-flowered; peduncle 0.5-6 cm long. Corolla blue or violet, 7-10 mm long. Pod 20-40 x 6-7.5 mm, glabrous, 5-11-seeded. Fig. 95.47.2.

Upland grassland; 1700-2700 m. TU GD GJ WG SU AR KF SD; Kenya, Tanzania, Zaire. *Thulin* 1447; *Gilbert & Thulin* 932; *W. de Wilde* 8846.

The Ethiopian material is subsp. *paucifolia*. There is another subspecies in Tanzania, Malawi and E Zambia.

3. **V. hirsuta** *(L.) S. F. Gray* (1821).

Trailing or climbing annual, 30-70 cm long. Leaflets 6-20, linear or narrowly oblong, 4-5 x 1.5-3 mm. Racemes 2-7-flowered; peduncles 0.5-4 cm long. Corolla white, rose or pale blue, 3-5 mm long. Pod oblong, 6-10 x 3-3.5 mm, pilose, 1-2-seeded.

Upland grassland, scrub, forest margins; 2700-3450 m. EW GD WU GJ SU AR BA; E and N Africa, Europe, Asia and introduced elsewhere. *Thulin* 1516; *Hedberg & Aweke* 5315; *W. de Wilde* 9668.

4. **V. sativa** *L.* (1753).

Straggling or ascending annual up to 80 cm long. Leaflets 6-16, linear to obovate, 3-40 x 2-15 mm; stipules dentate. Flowers solitary or paired, subsessile. Corolla purplish, 7-25 mm long. Pod 23-60 x 4-8.5 mm, glabrous or pubescent, 6-12-seeded.

1. Leaflets up to 40 x 15 mm; standard 20-25 mm long, often very pale; pod 40-60 x 6-8.5 mm. var. **sativa**
- Leaflets 4-25 x 1.5-6.5 mm; standard 10-18 mm long, not pale; pod 23-40 x 4.5-6 mm. var. **angustifolia**

Var. **sativa.**

Cultivated for forage and fodder in the highlands.

Var. **angustifolia** *L., Fl. Suecica*, ed. 2: 255 (1755).
V. segetalis Thuill. (1799).
V. sativa subsp. *nigra* (L.) Ehrh. in *Hannover. Mag.* **1780 (15)**: 229 (1780).
V. cuneata Guss. (1828).
V. abyssinica Alef. (1861); *V. sativa* var. *abyssinica* (Alef.) Bak. in *Fl. Trop. Afr.* **2**: 172 (1871) - type: TU, *Schimper* ?13 (B holo. destr., ?BM ?K iso.).
V. sativa var. *aristulata* Chiov. in *Atti R. Accad. Ital., Mem. Cl. Sci. Fis.* **11**: 29 (1940) - type: SU, *Piovano* 244 (FT holo.).

Upland grassland and scrub; 1500-3300 m. EW TU GD GJ SU AR HA KF GG SD; widespread in E and N Africa, Europe to NW India. *Thulin* 1382; *Burger* 1013; *Friis* et al. 2041; *Lemma G.-S* 462.

5. **V. faba** *L.* (1753).

Erect robust glabrous annual. Leaves without tendril; leaflets 2-6, 40-80(-100) x 10-20(-40) mm. Raceme subsessile, short, 1-6-flowered. Corolla *c* 25 mm long, white to pale mauve with partly black wings. Pod (3-)5-6(-8) x 1-1.5 cm, sparsely pubescent when mature, generally 2-5-seeded. Seeds ovoid-oblong, compressed, 1-2 cm in diameter; hilum narrowly oblong.

Cultivated for its edible seeds in all Ethiopian highlands; originating from the Mediterranean region or SW Asia. *Mulvany* 171; *Haigh* 50; *Pappi* 7345; *Beals* 171.

107. **LENS** *Mill.* (1753), *nom. conserv.*

Annuals. Leaves mostly paripinnate, the rhachis terminating in an awn or tendril. Flowers small, solitary or in axillary few-flowered racemes. Calyx-lobes 5, subequal, at least twice as long as tube. Upper filament free. Style pubescent on the upper side. Pod flat, 1-2-seeded.

A few species in S Europe, the Mediterranean region, W Asia and E Africa.

1. Stipules entire; pod 10-17 x 6-12 mm; seeds 4-6 mm in diameter; plant 20-60 cm tall; leaves 8-16-foliolate. **1. L. culinaris**
- Stipules semi-hastate or dentate; pod up to 9 x 4 mm; seeds *c* 2 mm in diameter; plant mostly under 20 cm tall; leaves 4-6(-8)-foliolate. **2. L. ervoides**

1. **L. culinaris** *Medik.* (1787).

Herb 20-60 cm tall. Leaflets 8-16, oblong or elliptical, up to 20 x 8 mm; stipules entire. Racemes 1-3-flowered, about equalling leaves. Corolla *c* 5 mm long, pale blue or violet. Pod 10-17 x 6-12 mm, glabrous. Seeds circular, compressed, 4-6 mm in diameter.

Cultivated for its edible seeds throughout the Ethiopian highlands. *J. de Wilde* 5825; *W. De Wilde* 10925; *Smeds* 27 (H); *Amare Getahun* in *IECAMA* 51.

2. **L. ervoides** *(Brign.) Grande* (1918)

Sparsely hairy herb, 15-25 cm tall. Leaflets 4-6(-8), narrowly oblong or linear, 4-14(-23) x 1-4 mm; stipules semi-hastate or dentate. Racemes 1-2-flowered, longer than the leaves. Corolla 3(-6) mm long, pale blue. Pod up to 9 x 4 mm, pubescent at least at margins. Seeds 2-2.5 mm in diameter. Fig. 95.47.3.

Upland grassland; *c* 2500-3300 m. ?TU GD SU; Uganda, widespread in the Mediterranean region and in Caucasus. *Mooney* 5830; *Schimper* 753; *J. De Wilde & Gilbert* 190.

108. **LATHYRUS** *L.* (1753)

Annuals or perennials, often climbing by means of tendrils. Leaves usually paripinnate; leaflets usually distinctly parallel-veined. Flowers in axillary racemes, or solitary. Calyx-lobes subequal or the upper pair shorter. Style pubescent on the upper side, rarely glabrous. Pod ± oblong, compressed, dehiscent.

Some 130 species widely distributed in the N hemisphere and S America; a few species in Africa.

L. annuus L., a Mediterranean species, has also been reported from Ethiopia, without precise locality (see Cuf. *Enum.*: 309 (1955). No material has been seen and the species is omitted here.

1. Leaflets absent; leaves consisting of a tendril only, but with large leaf-like stipules. **5. L. aphaca**
- Leaflets present. 2
2. Stem winged. 3
- Stem not winged. 4
3. Plant pubescent; pod not winged. **2. L. odoratus**
- Plant glabrous; pod with upper margin 2-winged. **4. L. sativus**
4. Corolla yellow; racemes several-flowered. **1. L. pratensis**
- Corolla red or white; racemes 1-flowered. **3. L. sphaericus**

1. **L. pratensis** *L.* (1753).

Glabrous or pubescent perennial. Leaflets 2, 10-30(-40) x 2-9 mm, linear-lanceolate or elliptical. Racemes 5-12-flowered. Corolla yellow, 10-16(-18) mm. Pod 20-40 x 4-7 mm, black, 5-12-seeded.

Upland grassland; *c* 3000-3200 m. SU BA; widespread in Eurasia. *Petit* s.n.; *Duncan* in *Ash* 1977; *Sandford* in *Ash* 2903.

The Ethiopian material is pubescent throughout. The species has been collected only a few times in the country but appears to be native.

2. **L. odoratus** *L.* (1753).

Annual, pubescent climber; stems winged. Leaflets 2, ovate-oblong, or elliptic, 20-60 x 7-30 mm. Racemes 1-3-flowered. Corolla varying from purple to white, 20-35 mm long. Pod 50-70 x 10-12 mm, pubescent.

Cultivated for ornament at least in SU (Addis Abeba) and EW (Asmara); native of S Italy and Sicilia. *Buscalioni* 909.

3. **L. sphaericus** *Retz.* (1783).

Glabrous or pubescent annual, up to 30(-50) cm tall. Leaflets 2, linear to linear-lanceolate, 10-65(-100) x 1-7 mm. Flowers solitary in leaf-axils; peduncles aristate. Corolla orange-red, red or white, 5-9(-10) mm long. Pod 30-70 x 3.5-7 mm, glabrous, with prominent longitudinal veins, 8-15-seeded. Fig. 95.47.4.

Upland grassland, scrub, forest edges; 1800-3000 m. EW TU GD GJ SU; E and N Africa, Europe and eastwards to Afghanistan. *Gilbert & Thulin* 931; *Hedberg & Aweke* 5304; *Mooney* 5816.

4. **L. sativus** *L.* (1753).

Glabrous annual; leaflets 2, narrowly elliptic-oblong, (30-)40-50(-65) x 3-5(-11) mm; corolla violet-blue or white, *c* 15 mm long; pod 25-35 x 10-18 mm, with upper margin 2-winged, glabrous, 2-5-seeded.

Cultivated for its edible seeds and for fodder at medium altitudes up to 2200 m mainly in the highlands north and east of the Rift Valley, also in HA; origin not known. *Mooney* 6303; *Meyer* 7507; *Pappi* 206; *Mesfin* 857.

Particularly well suited to grow on residual moisture in heavy black clay soils and often planted after other crops are harvested. In times of rainfall shortage this can be the main crop harvested and if it makes more than 30% of the diet it severely affects the nervous system resulting in paralysis.

5. **L. aphaca** *L.* (1753).

Glabrous annual; stem angled. Mature leaves without leaflets but with a tendril; stipules large, leaf-like, ovate, hastate. Flowers usually solitary. Corolla yellow, 6-18 mm. Pod 20-30 x 4-6 mm, glabrous.

Introduced, apparently occasionally, in EW; native probably in the E Mediterranean region and W Asia. *Baldrati* 5002.

Figure 95.47.1 - *PISUM SATIVUM* var. *ABYSSINICUM*, a: flowering shoot, b: fruiting shoot, both from *Schimper* 1866. 2 - *VICIA PAUCIFOLIA*, flowering and fruiting branch from *Ash* 2070. 3 - *LENS ERVOIDES*, flowering and fruiting branch from *Mooney* 5830. 4 - *LATHYRUS SPHAERICUS*, flowering and fruiting branch from *Evans* 361. All x 0.75. Drawn by Maureen Church. (Reproduced with permission from *Opera Botanica*, 68, fig. 49, 1983.)

96. HAMAMELIDACEAE

by B. Verdcourt*

Cufodontis, *Enum*.: 175 (1954); Verdcourt, *Fl. Trop. E. Afr*., **Hamamelidaceae**: 6 pp. (1971).

Evergreen or deciduous trees or shrubs, indumentum often stellate. Leaves alternate or less often opposite, simple, often pinnatifid or palmately lobed, entire or serrate, usually stipulate. Inflorescence terminal or axillary, racemose, often spicate or capitate. Flowers small to large, regular or rarely irregular, bisexual or unisexual, rarely without a perianth. Calyx tube usually ± adnate to the ovary; lobes (3-)5(-7), imbricate or valvate. Petals 4-5 or more, sometimes absent in female flowers, free, imbricate or valvate, rarely circinate, mostly linear-spathulate or obovate. Stamens 4-5(-25), rarely fewer, as many as and alternating with the petals. Ovary inferior, half-inferior or rarely superior, (1-)2(-3)-locular; ovules 1 or more in each cell, apical or anatropous, pendulous; placentation axile; styles 2, subulate, usually free, the stigmas terminal or lateral. Fruit a woody dehiscent capsule, with leathery exocarp and bony endocarp, often appearing apically 4-valved. Seeds 1-many, sometimes winged, with thin endosperm.

A fairly small family of 19 to 22 genera and about 50 to 100 species, according to delimitation, mostly in subtropical regions in the northern hemisphere; one genus only in continental tropical Africa.

TRICHOCLADUS *Pers.* (1807)

Trees or shrubs, mostly with conspicuous stellate indumentum. Leaves alternate or opposite, simple, usually entire but ± toothed in one species; stipules small. Flowers in dense terminal or axillary round heads, bisexual or sometimes only female. Calyx tube shallow or campanulate; lobes (3-)4-5, broadly triangular, valvate. Petals 4-5 or often absent in female flowers, linear-spathulate, often free to base, valvate or slightly imbricate, at first sometimes rolled up or folded in the bud. Stamens as many as the petals; anthers opening by valves, mostly beaked; staminodes absent. Ovary inferior to almost superior, 2-locular; ovules pendulous, solitary in each cell. Capsule subglobose, (1-)2-locular, 2-valved, but appearing 4-valved at apex after dehiscence. Seed ellipsoid or ovoid-oblong, the hilar area extending from the proximal end of the seed on each side.

A small genus of 5 species confined to Africa and extending from South Africa up through East Africa to Ethiopia; one species in the Flora area.

Considerably more fieldwork needs to be done to elucidate variation in the sex of the flowers. Various works state that the plants can be monoecious or dioecious and that the petal size varies in the differently sexed flowers. The author has found anthers and ovules in all the long-petaled and short-petaled flowers examined and suspects that the rapid lengthening of the petals during anthesis has been misunderstood so far as the tropical African species are concerned. The author has not found any buds which did not have petals.

T. ellipticus *Eckl. & Zeyh.* (1837).

Evergreen shrub or tree, (2-)5-10(-18) m tall; bark greyish-white or creamy brown, smooth or somewhat rough; young branchlets tomentose with yellow or rusty-brown stellate hairs, older ones glabrescent. Leaves alternate; petiole 0.5-1.8 cm long; stipules linear, soon falling; lamina elliptic, obovate-elliptic to oblanceolate, (1.2)5-28 x 0.8-12 cm, entire, glabrous and glossy green above, densely persistently stellate tomentose beneath; apex acute to acuminate; base cuneate to rounded or rarely subcordate[1]. Inflorescence capitate, 0.7-2 cm in diameter; peduncle 0.5-1.5 cm long. Flowers sessile, sweetly scented. Bisexual flowers: true calyx tube 1 mm long, tomentose; limb 2 mm long, its rounded lobes 1 mm long, recurved; petals white, greenish-yellow or yellow, 0.9-1.7 x 1.7-2, revolute, tomentose outside; stamens 5; connective prolonged into an erect beak; styles 1.3 mm long. Female flowers stated to be similar but with petals reduced, 4 mm long, or even lacking (see note after generic description). Capsule subglobose, 6-7(-8) x 5 mm, the apical valves conspicuous, pubescent, 1-2 seeded. Seeds yellowish or greyish, mottled black, 5 x 3-4 mm; hilar area brownish.

subsp. **malosanus** *(Bak.) Verdc.* (1970).

Leaf-blades more often broadly elliptic, mostly larger, at least 7-10 x 4-6 cm. Inflorescences usually larger than in the typical subspecies.

A local dominant forming a distinct understorey and nearly pure stands in *Juniperus* and *Podocarpus* forest, often on moist, shady, very steep slopes, also in riverine forests of *Mimusops, Garcinia, Oncoba, Pappea, Hymenodictyon*, etc., and on banks by rocky streams; 1250-2100(-?3000) m. WG SU KF GG AR BA HA; Zaire, Sudan (Didinga Mts.), Uganda, Kenya, Tanzania, Malawi, Mozambique, Zambia and Angola. *Burger* 2650; *Gilbert & Thulin* 387; *Mooney* 5710.

Typical *T. ellipticus* with usually more lanceolate, smaller, often more acuminate leaves occurs in South Africa.

The wood is tough and hard but bends easily.

* Royal Botanic Gardens, Kew, Richmond, Surrey, TW9 3AB, UK.

1. Leaves markedly subcordate in some Ethiopian populations.

Figure 96.1 *TRICHOCLADUS ELLIPTICUS* subsp. *MALOSANUS*: 1 - branch with flower heads and buds x1; 2 - stellate hair x 12; 3 - flower x 4; 4 - calyx and corolla opened out to show stamens x 4; 5 - petal x 4; 6 - stamen, front view x 6; 7 - stamen, side view x 6; 8 - longitudinal section of gynoecium x 9; 9 - styles x 6; 10 - infructescence x 1; 11 - dehisced capsule x 2; 12 - seed x 2. 1, 3-9 from *C.G. Rogers* 5; 2, 10-12 from *A.S. Thomas* 3617. (Reproduced with permission from *Fl. Trop. E. Afr.*, Hamamelidaceae: fig. 1, 1971.)

97. MYROTHAMNACEAE

Burger, *Families of Flowering Plants in Ethiopia*: 85 (1967); Bywater, *Fl.Trop. E. Afr.*, **Myrothamnaceae**: 5 pp. (1984).

Shrubs, usually less than 2 m tall, with aromatic resinous sap; the plants either male or female but not both. Leaves opposite and simple, clasping the stem at their base and folded like a fan when in bud or dry; stipules small and borne on the leaf base. Flowers unisexual and regular, without a perianth but each flower subtended by a bract and borne in a dense spike. Sepals absent. Petals absent. Male flower with either 4 stamens with filaments almost free, or (3-) 4-6(-8) stamens with filaments united at the base into a central column; anthers basifixed, dehiscing by longitudinal slits; connective prolonged into a beak; no rudimentary ovary. Female flower lacking staminodes, pistil 1, carpels superior, (2)3 or 4 united below, ± free above and tapering to ± recurved styles; stigmas subspathulate, oblong, facing inwards; ovules numerous. Fruit a capsule opening from above into the centre; seeds with endosperm; embryo minute.

Family with a single genus, *Myrothamnus*, with 2 species, one only known from Africa south of the Sahara, the other from Madagascar. *M. flabellifolius* Welw. occurs in northern Kenya, and could be found in southern Ethiopia in future collections.

98. BUXACEAE

by B. Verdcourt*

Cufodontis, *Enum.*: 465 (1958); Verdcourt, *Fl. Trop. E. Afr.*, **Buxaceae:** 3 pp. (1962)

Trees, shrubs, or rarely herbs, mostly glabrous. Leaves evergreen, alternate or opposite, simple, leathery and stiff, rarely toothed; stipules absent. Flowers unisexual or very rarely a few bisexual, in spikes, fascicles or cymes, bracteate. Sepals usually 4, imbricate or absent. Petals absent. Male flowers: stamens 4 or 6 (rarely more), when 4 then opposite the sepals, when 6 two pairs opposite the inner sepals; anthers large, sessile (or borne on fairly long filaments); rudimentary ovary present or absent. Female flowers often larger and fewer than the males, often pedicellate; ovary superior, mostly 3-locular; styles 2-3; ovules 1-2, pendulous, anatropous, with a ventral raphe. Fruit a capsule, loculicidally dehiscent or drupaceous; seeds often black and shining, mostly with rather fleshy endosperm.

About 50 species in 6 genera in the temperate regions of both hemispheres; in the tropics only at higher elevations; very rare in continental America; absent from Polynesia, Australasia and E. South America.

Hutchinson in his most recent work merged *Notobuxus* Oliv. with *Buxus* but this purely African genus of eight species seems worth maintaining. *Notobuxus* does not occur in Ethiopia.

BUXUS *L.* (1753)

Much-branched glabrous shrubs, sometimes trees. Leaves olive-green, opposite, entire, stiff and tough. Racemes axillary, small, dense, sessile or shortly pedunculate; bracts often numerous, like the sepals but smaller. Flowers monoecious, one to each bract, sessile or shortly pedicellate, the terminal often female, the remainder males. Male flowers: sepals 4, biseriate; stamens 4-6, opposite the sepals; filaments exserted; anthers oblong, at length recurved, opening by slits length-wise; rudimentary ovary truncate or 3-lobed, or absent. Female flowers: sepals 4-6, outer smaller; ovary 3-locular; styles short, thick, arising separately; ovules 2, pendulous from the outer edge of the loculi. Capsule ovoid, 3-horned with the persistent styles, dehiscent, valves (after opening) 2-horned. Seeds oblong, 3-sided, testa black, shining; endosperm rather fleshy; cotyledons oblong.

A genus with around 30 species, only 1 in Ethiopia.

B. hildebrandtii *Baill.* (1874);
Buxanthus hildebrandtii (Baill.) van Tieghem (1897).
Buxus calophylla Pax. (1907) - type: HA, Diredawa, *Rosen* s.n. (10/1/1905) (B holo. destr.)
Garcinia buxifolia Chiov. (1940) - type: BA, between Mt. Elo and El Mara, near El Kere [El Caré], *Reghini* 28 (FT holo, K photo).
Friis, I. in *Nord. J. Bot.* 4: 747-748 (1984).

Shrub 1.8-3 m tall or 4-6 m[1]; bark with vertical stripes and fissures. Leaves thick, obovate, elliptic or oblanceolate, mostly olive or grey-green, 2-6.5 x 0.5-3 cm, venation rather indistinct; apex rounded or emarginate; base acute. Inflorescence sweet-scented, each cyme made up of a terminal female flower with surrounding male flowers or all male; flowers subsessile or very shortly stalked. Male flowers: anthers 4, filaments 3 mm long, rudimentary ovary present. Capsule ovoid, 1-1.5 x 0.7 cm; seeds ellipsoid, brown and shining, 5-7 mm long.

Acacia-Commiphora-Combretum woodland on rocky limestone slopes particularly near stream beds; 1000-1300 m. AF/HA HA BA; N Somali Republic, Djibouti and Socotra. *Gilbert* 2418, *Burger* 1596, *Gilbert* et al. 7852.

Elsewhere *Buxus* is used for scaffolding poles but probably none in Ethiopia reaches sufficient size for this. The wood is extremely hard and compact having been used in the past on the island of Socotra for blocks in ships's tackle.

* Royal Botanic Gardens, Kew, Richmond, Surrey, TW9 3AB, UK.

1. Sometimes, outside Ethiopia, in old *Buxus* forest reaching 9 m tall with trunks 15 cm diameter.

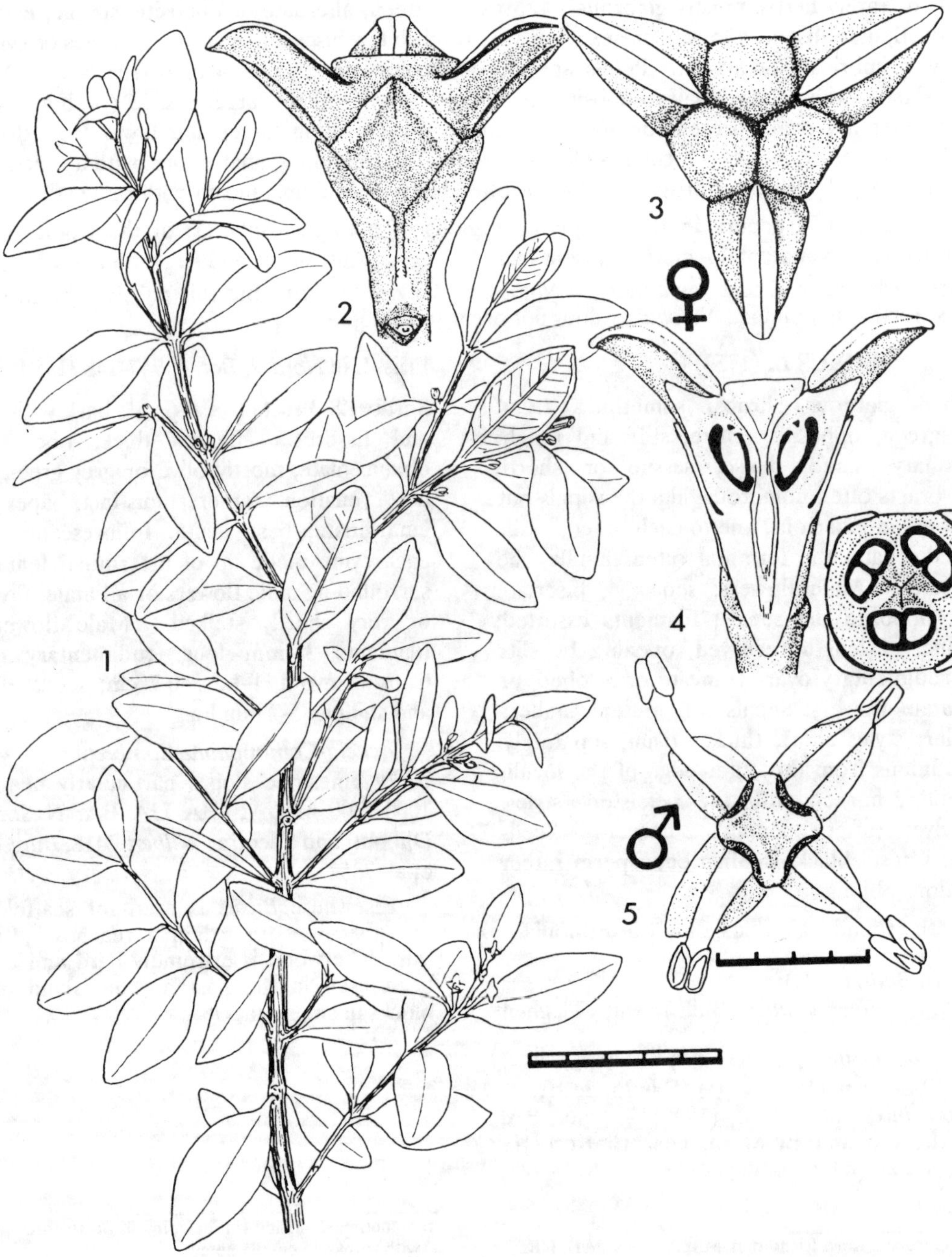

Figure 98.1 *BUXUS HILDEBRANDTII*: 1 - leafy branch; 2 - female flower, side view; 3 - female flower, top view; 4 - longitudinal section of flower and cross section of ovary; 5 - top view of male flower. mm scale as open bar and cm scale as closed bar. (Reproduced with permission of Oklahoma Agricultural Experiment Station from Experiment Station Bulletin No. 45, *Families of Flowering Plants in Ethiopia*, figure 32, 1967.) Drawn by W.C. Burger.

PLATANACEAE

by M.G.Gilbert*

Trees with bark flaking in scales, leaves alternate, palmately lobed, petiole base swollen and enclosing bud, stipulate. Monoecious, flowers in unisexual pendulous globose heads. Tepals in 2 whorls of 3-8, male flowers with 3-8 stamens, filaments very short; female flowers with 3-8 free carpels, styles long. Fruit an obconical nut surrounded by long hairs at base.

Monogeneric, with about 6 species in Europe, Asia and N America.

Platanus orientalis *L.*

Leaves lobed to below middle, cuneate at base, quickly glabrescent. Female heads 3-6 per inflorescence.

Cultivated tree; 2450m. SU. *Gilbert & Jackson* s.n.

* Ethiopian Flora Project, c/o Royal Botanic Gardens, Kew, Richmond, Surrey, TW9 3AE, UK.

99. SALICACEAE

by R.D. Meikle*

Cufodontis, *Enum.*: 4 (1953); Wilmot-Dear, *Fl. Trop. E. Afr.*, **Salicaceae**: 8 pp. (1985).

Dioecious, deciduous or rarely evergreen trees, shrubs and subshrubs. Leaves simple, stipulate, generally alternate, entire or variously toothed or lobed. Flowers in spicate or racemose, erect, patent or pendulous catkins, each subtended by a membranous catkin-scale. Perianth much reduced, cupuliform, cyathiform or consisting of 1 or more nectary scales. Stamens 2-many; filaments free or occasionally connate; anthers 2-thecous, dehiscing by longitudinal slits. Ovary 1-locular, 2-4-valved; placentation basal or parietal; ovules often numerous. Fruit a capsule. Seeds small, enveloped in a tuft of silky hairs.

Threc (or pcrhaps four) genera and more than 500 species, mostly in temperate regions of the northern hemisphere. Only one genus, *Salix*, is represented by indigenous species in Ethiopia.

Exotic *Populus* (poplar) species and hybrids are in cultivation in Addis Ababa and elsewhere. *Populus* differs most obviously from *Salix* in its scaly buds, long petiolate leaves, often with broadly ovate-deltoid laminas; lacinate catkin-scales, and cupular or cyathiform perianths. Some species of *Populus* are quick growing trees which can tolerate heat and can be grown under irrigation to provide shelter belts and pole wood. The timber of many species is light and easily worked being suitable for tool handles and light furniture.

SALIX *L.* (1753)

Trees and shrubs with sympodial branching; bud covered by a single scale. Leaves shortly petiolate, generally longer than wide, often narrowly oblong, lanceolate or linear. Catkin-scales entire. Perianth generally reduced to 1-2 small nectary-scales.

A genus with around 400 species, one indigenous to Ethiopia. The second species is probably introduced and there are undoubtedly other species which have been introduced as ornamentals in private gardens.

1. Catkins terminal on short lateral, leafy shoots; catkin-scales pallid; male flowers with 6-8 stamens. **1. S. subserrata**
- Catkins subsessile with a few, much-reduced basal bracts; catkin-scales dark-tipped; male flowers with 2 stamens. **2. S. pedicellata**

1. **S. subserrata** *Willd.* (1806) Fig. 99.1
 - type: Egypt, near Cairo, Bulak, *Herb. Willd.* 18137.

Salix 'safsaf baelledi' Forsskål (1775); *S. safsaf* Trautv., Salicetum in *Mém. près. à l'Acad. de St. Petersb*: 6-7 & tab.II (1836), not validly published.

S. axillaris A. Rich. (1851) - type: TU, Wejerat [Ouoderate], *Petit* s.n. (P holo.).

S. cyathipoda A. Rich. (1851); *S. subserrata* var. *cyathipoda* (A. Rich.) Cuf. in *Bull. Jard. Bot. Nat. Belg.* **39**, suppl.: **xx** (1969) - type: TU, at Assai river, *Quartin Dillon* s.n. (P holo.).

S. octandra Anderss. ex A. Rich (1851) - type: TU/GD, banks of Tekeze [Tacazze], *Schimper* 700 (P holo., BM iso.).

Bush or small tree 2-10 m high with dark grey, fissured bark. Twigs often thinly villose at first, becoming glabrous or subglabrous and often reddish with age. Leaves variable, oblong-elliptic, narrowly ovate or lanceolate, up to 16 cm long, 3.5 cm wide, olive green and rather shining above, generally glaucous below, soon glabrous on both surfaces; apex acute or acuminate, rarely obtuse; margins of fully developed leaves usually glandular-serrulate; petiole generally less tahn 1 cm long; stipules small, falling quickly, serrulate, glandular on the adaxile surface, often absent. Catkins narrowly cylindrical, spreading, 2-7 x 0.4-0.7 cm, terminal on short, lateral leafy shoots. Catkin-scales falling quickly, broadly ovate, about 1.5-2 x 1-1.3 mm, pallid, woolly. Male flowers with 6-8 stamens; filaments 2-4 mm long, woolly towards base; anthers yellow; nectary-scales usually 2, shortly oblong, truncate, emarginate or crenate at apex. Female flowers with the nectary-scales generally united and cyathiform or horseshoe-shaped; ovary subsessile or shortly pedicellate, ovoid 2-3.5 x 2 mm, glabrous; style very short or almost absent; stigmas 2, short, blunt, deeply bifid.

Beside rivers and streams; 1250-2850 m. EW TU GD WU GJ SU AR HA BA WG IL KF GG; widespread in tropical Africa north to Egypt. *Burger* 599; *Ash* 1353; *Friis* et al. 162.

2. **S. pedicellata** *Desf.* (1799)
 - type: Tunisia, Sbiba

Shrub 2-4 m high. Young shoots subglabrous to densely grey-tomentose, becoming glabrous with age. Leaves oblong, shortly acute or obtuse, 5-9 x 1.5-3 cm, dull green above, ashy-tomentose or glabrous and glaucous below, margins remotely glandular-serrulate; petiole short; stipules generally minute, falling quickly, often missing. Catkins erect, subsessile or very shortly pedunculate, narrowly cylindrical, 3-5 x 0.5 cm, generally with a few small, silky, leaf-like basal bracts;

* Ranscombe Lodge, Wootton Courtenay, Minehead, Somerset TA24 8RA, England.

Figure 99.1 *SALIX SUBSERRATA*: 1 - Branch with immature leaves x 2/3; 2 - male flowering shoot x 2/3; 3 - male flower x 6; 4 - male disc x 16; 5 - male bract x 8; 6 - female flowering shoot x 2/3; 7 - female flower x 6; 8 - female disc x 16; 9 - female bract x 8; 10 - fruiting branch x 2/3; 11 - fruit x 4; 12 - seed x 8. 1 - from *Eggeling* 3543; 2-5 - from *Eggeling* 6081; 6-9 from *Paulo* 277; 10-12 from *Kimera* 21. Drawn by Eleanor Huxley. (Reproduced with permission from *Fl. Trop. E. Afr.*, Salicaceae, fig. 1, 1985.)

rhachis densely woolly. Catkin-scales persistent, oblong-ovate, about 1.8 x 1 mm, obtuse or subacute, upper half reddish-brown, lower half pallid, densely woolly. Male flower with 2 stamens; filaments up to 6 mm long, glabrous; anthers yellow or reddish; nectary-scale 1, oblong. Female flower with a single, oblong nectary-scale; ovary shortly pedicellate (the pedicel lengthening in fruit), glabrous, flask-shaped, tapering to apex, about 3 x 1.5 mm; style short but distinct; stigmas 2, short, blunt, bifid. Fruit distinctly pedicellate.

Wet hillsides and along streams and irrigation ditches; 1000-1900 m. HA; widespread in the Mediterranean region, with an allied species or subspecies in the Canary Islands. *Burger* 2590, 2737; *W. de Wilde* 9851.

Not previously recorded from tropical Africa, and possibly a recent introduction. The Ethiopian specimens differ from typical *S. pedicellata* in having oblong (not oblong-ovate) glabrous or glabrescent leaves, but such is the polymorphism of this species that it would be unwise to add further to the already large list of named infraspecific taxa. Only the female plant has been collected in Ethiopia and the specimens are evidently sterile.

******

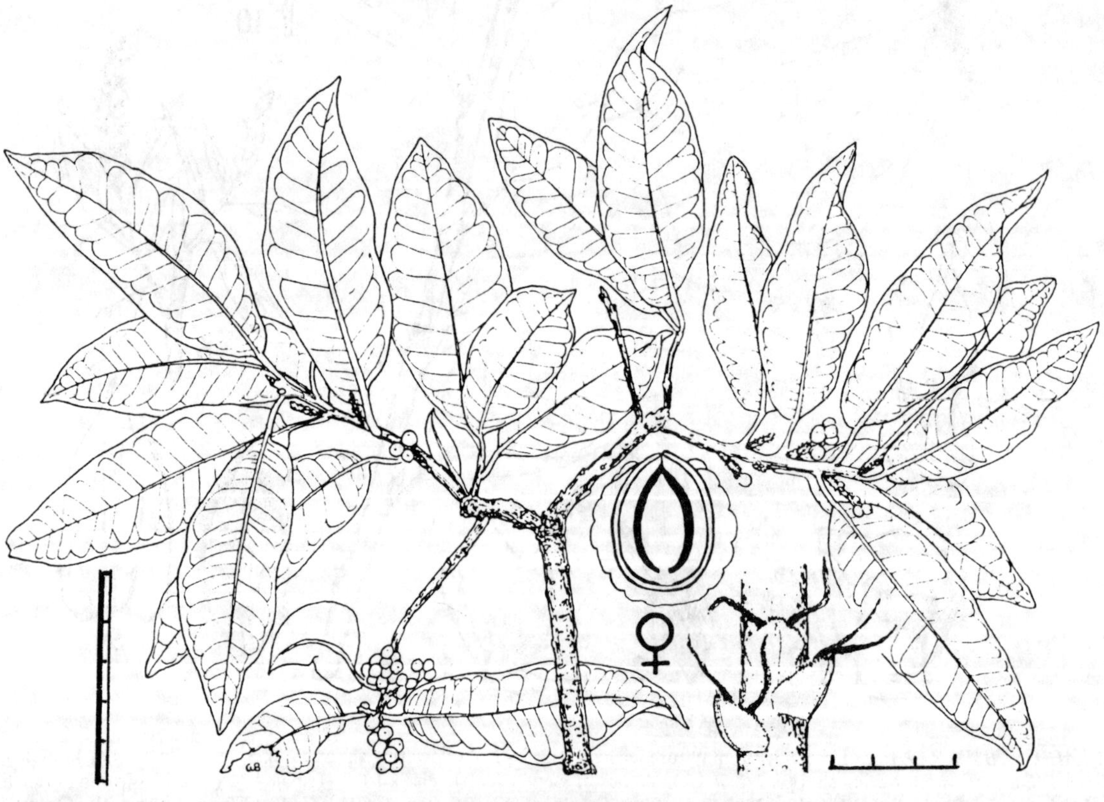

Figure 100.1 - *MYRICA SALICIFOLIA*: branches with flowering and fruiting inflorescences, longitudinal section of a fruit, and part of a female inflorescence with 3 female flowers within their bracts. (Reproduced with permission of Oklahoma Agricultural Experiment Station from Experiment Station Bulletin No. 45, *Families of Flowering Plants in Ethiopia*, figure 4, 1967.) Drawn by W.C. Burger.

100. MYRICACEAE

by R. M. Polhill*

A. Chev., *Monographie des Myricacées* (1901); *Fl. Congo Belge* **1**: 33-38 (1948); Cufodontis, *Enum.*: 5 (1953).

Trees or shrubs, monoecious or dioecious. Leaves alternate, usually without stipules, entire to irregularly toothed or lobed, rarely pinnatifid, usually dotted with resinous aromatic peltate glands. Flowers in dense axillary spikes, mostly unisexual, without a perianth, but subtended by a bract and smaller bracteoles. Male flowers: stamens *c* 2-10(-20); filaments short, free or partly joined; anthers erect, 2-thecous, opening lengthwise. Female flowers: ovary sessile, 1-locular; style short, with 2 slender branches; ovule 1, erect from the base. Fruit an achene or a small papillate drupe; endocarp hard. Seeds with membranous testa and little or no endosperm.

Probably 4 genera, with some 40 species, widely distributed, but mostly in temperate regions and tropical highlands.

MYRICA *L.* (1753)

Leaves without stipules, entire to coarsely lobed. Spikes borne on undifferentiated persistent shoots. Bracteoles 2-4, remaining small, deciduous. Fruit densely covered with bead-like wax-covered papillae.

About 35 species distributed as for the family.

M. salicifolia *A. Rich.* (1851)
- types: SU, Choa (Shewa), *Petit* s.n.; TU, Haramat, *Schimper* II:1093; GD, Enchetkab [Endschetcap], *Schimper* II:1135; GD, Jenausa [Dschenausa], *Schimper* II:1327 (all P syn., K isosyn.)

Tree or shrub 3-10(-15) m tall, with a rounded crown; bark grey; twigs glandular, subglabrous to densely hairy. Petiole 5-15 mm; lamina oblong-lanceolate to oblong-elliptic, 4-14 x 1.5-6 cm, glabrous to sparsely hairy, especially on the midvein, covered with golden-yellow glands on both surfaces at first and rather persistently so beneath, aromatic (spicy smell) when crushed; apex slightly acuminate; base cuneate to rounded; margin entire to coarsely and remotely toothed; primary lateral nerves 8-20 pairs, fine, spreading. Male catkins soon lax, the flowers often irregularly grouped in clusters, 5-35 mm long; bracts triangular-ovate, concave, 1-2 mm long, usually hairy on the back; stamens mostly 5-10. Female catkins becoming lax, 5-20(-50 in fruit) mm long; bracts broadly ovate-triangular, 1.5-2 mm long, ciliate to hairy overall; bracteoles usually 4, 0.4-1 mm long; style 0.8-3 mm long. Fruits ellipsoid-globose to subglobose, 3-4 mm across. Fig. 100.1.

Evergreen montane forest, commonly along rivers and descending beside them to lakes and marshes at lower altitudes, persisting in relic forest patches and in bushland and ericaceous scrub; (1750)2000-3300 m. EW TU GD WU SU AR HA SD; Saudi Arabia, Yemen, Uganda, Rwanda and E Zaire - in a broad sense extending through E Africa to the mountains around Lake Malawi. *Burger* 2637; *W. de Wilde* 9078; *Ash* 1545.

The wood has many uses.

Quite frequently the catkins are found distorted by galls.

The description is based principally on Ethiopian plants; further south the variation is much more complex and *M. kilimandscharica* Engl. (*M. usambarensis* Engl.), *M. meyeri-johannis* Engl. and *M. mildbraedii* Engl. are probably best regarded as subspecies of *M. salicifolia*.

* Royal Botanic Gardens, Kew, Richmond, Surrey TW9 3 AB, UK.

101. CASUARINACEAE

by C. M. Wilmot-Dear* & M. G. Gilbert**

Cufodontis, *Enum.*: 2-3 (1953); Wilmot-Dear, *Fl. Trop. E. Afr.*, **Casuarinaceae**: 8 pp. (1985).

Trees (and shrubs); stems of 2 kinds: normal woody branches from which arise whorls of slender little-branched green deciduous stems with longitudinally grooved or ribbed internodes, both types with leaves reduced to whorls of triangular scales at nodes. Monoecious or dioecious. Inflorescences with closely spaced whorls of bracts, each subtending 2 bracteoles and a sessile flower. Male inflorescences at tips of deciduous stems, spicate; flowers with 1-2 deciduous perianth segments and 1 stamen, anther exserted at anthesis. Female inflorescence a globose or ovoid head on short side branch; flower without perianth; ovary 1-locular with short terminal style and 2 long slender stigmas exserted at anthesis. Infructescence cone-like due to accrescent woody bracts and bracteoles, latter forming a pair of valves enclosing fruit. Fruit a seed-like samara, compressed with upper part produced into translucent wing, up to 8 mm long.

Family with 3(-4) genera and about 70 species native to Australia, SE Asia and Polynesia, occurring mainly in dry, infertile and saline areas.

Easily identified by the *Equisetum*-like stems, quite unlike those of any other trees. The nodular roots fix atmospheric nitrogen which, combined with the rapid growth of some species, means that they are widely grown for timber and fuel wood in many parts of the tropics.

CASUARINA *L.* (1759).
Allocasuarina S. Johnston (1982).

Deciduous stems with narrow grooves. Fruiting bracts relatively inconspicuous, not elaborately thickened.

Some 50 species distributed as for family, several widely cultivated. All the species found in Ethiopia are introduced and are relatively fast growing trees.

1. Leaf tips on deciduous stems free for 0.3-0.7 mm; infructescence valves with longitudinal ridges only; seed-like fruit pale fawnish-brown, glossy but not shiny. **2**
- Leaf tips on deciduous stems free for 1-1.3 mm; infructescence valves with prominent dorsal projection in addition to longitudinal ridges; samara dark brown, shiny. **1. C. verticillata**

2. Scale leaves with pale tip and distinct transverse brown band below; internodes with inconspicuous rounded ribs. **3**
- Scale leaves uniformly pale; internodes with prominent, sharp ribs. **2. C. equisetifolia**

3. Deciduous stems 0.4-0.6 mm thick with whorls of 7-9 leaves and a corresponding number of ribs; infructescence valves with several irregular longitudinal wrinkles. **3. C. cunninghamiana**
- Deciduous stems 0.7-0.8 mm thick with whorls of 10-13 leaves and a corresponding number of ribs; infructescence valve with a single longitudinal ridge. **4. C.** sp. = *Milchersich* 108

1. **C. verticillata** *Lam.* (1788).
Casuarina stricta Ait. (1789). *C. quadrivalvis* Labill. (1806).

Deciduous stems with (7-)9-10(-12) inconspicuous ribs; scale leaves free for 1-1.3 mm, uniformly yellow brown or slightly darker towards tip. Male spikes 2-3 mm thick, bracts slightly reflexed. Infructescence 2-3.3 x 2 cm; valves with prominent dorsal projection and some irregular longitudinal wrinkles. Samara very dark brown, shiny.

Cultivated tree; *c* 2400m. SU; native of Australia. *Senni* 1181.

Johnson (1982) has proposed that this and some 40 other species of *Casuarina* with similarly dorsally thickened infructescence valves and shiny fruits be placed in a separate genus, *Allocasuarina* (as *A. verticillata* (Lam.) S.A.M. Johnson (1982).

2. **C. equisetifolia** *L.* (1759).
Casuarina litorea L. ex Fosberg & Sachet (1975).

Deciduous stems with (6-)7-8 prominent sharp ribs; scale-leaves free for (0.4-)0.5-0.7 mm, uniformly pale green. Male spikes 1.2-2 mm thick; bracts adpressed. Infructescence 0.8-1.7(-2.5) x 1.1-1.6 cm, valves with 2 longitudinal dorsal ridges. Samara pale straw coloured. Fig. 101.1.9.

Cultivated tree; 1000-1400m. EW SU; often presumed to be a native of Australia but a common littoral tree as far west as the East African coast where it behaves as if indigenous. *Baldrati* 2129; *Gilbert* et al. 7396.

See Friis (*Taxon* **29**: 499, 1980) for a discussion of the nomenclature of this species.

3. **C. cunninghamiana** *Miq.* (1848).

Deciduous stems 0.4-0.6 mm thick with 7-9 inconspicuous ribs; scale leaves free for 0.4-0.6 mm, with distinct transverse brown band below pale tip. Male spike 1-1.5 mm thick; bracts adpressed. Infructescence 0.8-1.2 x 0.7-1 cm; valves with several irregularly longitudinal fine wrinkles. Samara pale brown. Fig. 101.1.1-8.

* Royal Botanic Gardens, Kew, Richmond, Surrey TW9 3AB, UK.
** Ethiopian Flora Project, Royal Botanic Gardens, Kew, Richmond, Surrey TW9 3AB, UK.

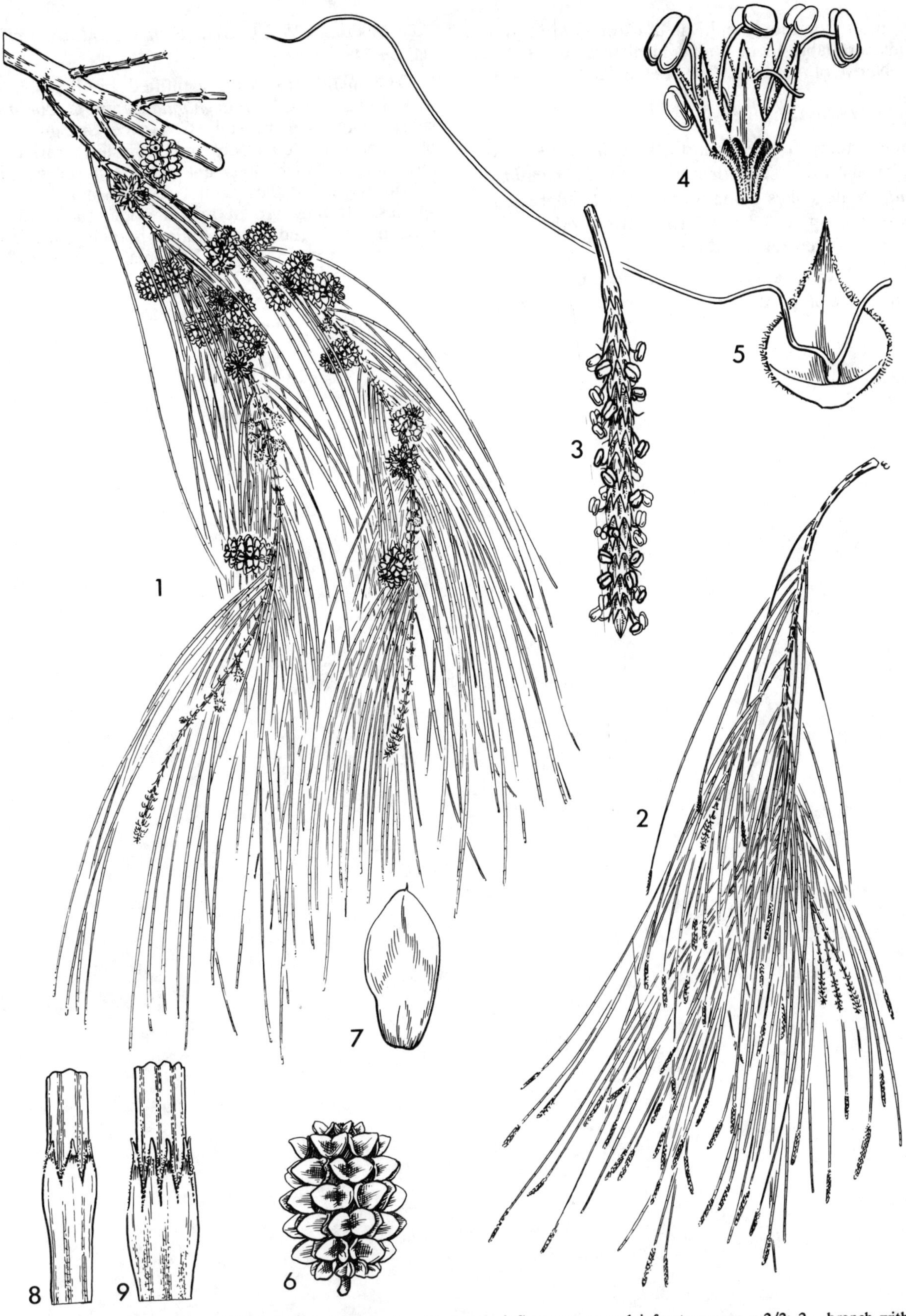

Figure 101.1 *CASUARINA CUNNINGHAMIANA*: 1 - branch with female inflorescences and infructescences x 2/3; 2 - branch with male inflorescences x 2/3; 3 - male inflorescence x 10; 4 - whorl of male flowers; 5 - bracteole with female flower, both x 20; 6 - infructescence x 1.5; 7 - fruit x 4; 8 - whorl of scale leaves x 10. 1, 3-8 from *Gilbert* et al. 7356. 2 from *Parry* 211.*C. EQUISETIFOLIA*: 9 - whorl of scale leaves x 10 from *Gilbert* et al. 7396. Drawn by Eleanor Catherine.

The most commonly cultivated *Casuarina* in the highlands, probably to be found in all regions; 1500-2500m. Native of Australia. *Gilbert* et al. 7356.

4. C. sp. = *Milchersich* 108.

Deciduous stems 0.7-0.8 mm thick with 10-13 very inconspicuous ribs; scale leaves as in *C. cunninghamiana*. Male spikes c 2 mm thick; bracts adpressed. Infructescence up to 1.8 x 1.2 cm; valves with single well-defined off-central dorsal ridge.

Cultivated tree, apparently more common in N Ethiopia where it partly replaces *C. cunninghamiana*; 2000-2400m. EW TU SU. *Fiori* s.n.; *Milchersich* 108; *Wilson* 648.

This plant has been identified by Johnson as a hybrid between *C. cunninghamiana* and *C. obesa* Miq. subsp. *obesa* (or possibly *C. glauca* Sprague). The plants have fruited freely and show little variation in morphology which does not suggest a recent simple hybrid origin but they do not seem to match any wild species. If they are of hybrid origin then this most probably occurred in an Italian botanic garden from whence seed was taken to Ethiopia prior to 1913, the earliest collection seen.

FAGACEAE

by I. Friis*

Trees. Leaves alternate, simple, petiolate, stipulate, margin entire, or serrate. Inflorescence monoecious. Male flowers in erect or pendulous spikes or heads; perianth 4-7-lobed, stamens 4-20, often twice as many as the perianth lobes, rudimentary ovary absent. Female flowers pedunculate or sessile, single or in groups of 2-3, surrounded by a solid common involucrum (cupula) with scales or prickles; hypogynous, with a 4-6-lobed, very inconspicuous perianth; styles 3-6, simple. Fruit a one-seeded nut, single or 2-3 together in the accrescent cupula.

Eight genera, very widespread, but with no native representatives in tropical S America or in tropical and Southern Africa. Several genera are widely cultivated, but only the genus *Quercus* has been recorded with certainty in cultivation from the Flora area. However, the genus *Castanea*, with the species *C. sativa* L., the Sweet Chestnut, which has serrate leaves, bisexual spikes and 2-3 female flowers in each cupula, may possibly be grown for its edible fruits.

QUERCUS *L.* (1753)

Deciduous or evergreen trees. Leaves entire, dentate or deeply lobed. Male flowers in slender, pendent, ± interrupted spikes; perianth green, 6-lobed, with 6-12 stamens. Female inflorescences with one or more cupulas, each with a single female flower. Fruit an oblong nut, partly covered by the accrescent cupula.

About 450 species, mostly northern temperate; only 2 species recorded as planted in the Flora area, but other species may occur in cultivation.

1. Leaves deciduous, not leathery, with sinuate margin. **1. Q. robur**
- Leaves evergreen, leathery, with dentate or entire margin. **2. Q. ilex**

* Botanical Museum and Herbarium, University of Copenhagen, Gothersgade 130, DK 1123 Copenhagen K, Denmark.

1. Q. robur *L.* (1753).
Q. pedunculata Ehrh. (1790).

Deciduous tree, 10-40 m tall. Leaves obovate, sinuately lobed with 5-7 pairs of lobes, petiole up to 5 mm long. Peduncle of female inflorescence up to 5 cm long, with 1-3 cupulas. Nut ovoid, to *c* 2 cm long.

Cultivated; 1800-2400 m. EW SU(sight record). *Ryding* 1211.

2. Q. ilex *L.* (1753).

Evergreen tree, 10-20 m tall, sometimes a shrub. Young twigs and buds grey tomentose. Leaves 3-5 cm long, thick, leathery, ovate with cuneate or rounded base, entire dentate or spinose serrate, dark green above, grey tomentose below; petiole 0.5-1.5 cm long. Peduncle of female infloresence very short, with 1-4 cupulas. Nut ovoid or ellipsoid, 2-3(-4) cm long.

Cultivated; *c* 2300 m. EW. *Ryding* 1204.

102. ULMACEAE

by R.M.Polhill*

Cufodontis, *Enum.*: 5-6 (1953); Polhill *Fl. Trop. E. Afr.*, **Ulmaceae**: 14 pp. (1966); Letouzey, *Fl. Cameroun* **8**:1-65 (1968).

Trees or shrubs. Leaves alternate, distichous, stipulate, often unequal-sided, entire or toothed (always toothed on seedlings and saplings), 3(-5)-veined from the base in Ethiopia, commonly scabrid with epidermal incrustations. Flowers small, regular, apetalous, polygamous, solitary or in axillary cymes or clusters sometimes aggregated into panicles. Sepals (3-)4-6(-8), free or shortly joined, imbricate or valvate. Stamens as many as the sepals and opposite them or rarely a few more, inserted at the base of the calyx; anthers 2-thecous, opening lengthwise. Ovary 1-2-locular; styles 2, divergent; ovule solitary, pendent. Fruit a drupe or samara.

About 14 genera and 120 species, mostly tropical and north temperate, with about 15 species in Africa south of the Sahara and Madagascar.

Key to Genera

1. Leaf-lamina entire to coarsely toothed in the upper two-thirds; calyx-lobes of male flowers imbricate; styles 2-5 mm long. **1. Celtis**
- Leaf-lamina finely serrate almost to base; calyx-lobes of male flowers valvate; styles 0.5-1(-2) mm long. **2. Trema**

1. CELTIS *L.* (1753)

Trees or shrubs. Leaves entire or coarsely-toothed in upper two-thirds, often scabrid. Flowers male, female or hermaphrodite,the pistilloid ones always fewer and often terminal in the inflorescence, longer pedicellate. Sepals (4-)5, free or almost so, imbricate. Stamens as many as the sepals; anthers ovate. Ovary sessile, 1-locular; styles 2-5 mm long, simple or 1-2-branched. Drupes small, thinly fleshy; endocarp hard, sometimes ribbed.

About 50 species in the tropics and north temperate regions, with about 8 species in tropical Africa.

1. Stipules (at youngest nodes) peltate; leaves with basal lateral veins extending into acumen (looping with upper veins), glabrous, punctate and smooth above. **5. C. philippensis**
- Stipules basifixed; leaves with basal lateral veins extending 2/3-3/4 up the lamina, and/or hairy to scabrid. 2
2. Main lateral veins of 1-2(-3) pairs above basal veins, with tufts of hairs in the basal angles beneath (often not obvious in *C. africana*); fruit smooth. 3
- Main lateral veins of (2-)3-5(-6) pairs above basal veins, without axillary tufts of hairs; fruit, when dried, or endocarp angled or ribbed. 4
3. Leaves serrate; styles simple; fruits 5-8 mm long, pubescent. **1. C. africana**
- Leaves entire on mature shoots; styles branched; fruits 8-12 mm long, puberulous to scabrid, with a ring of longer hairs at the base. **2. C. toka**
4. Leaves scabrid above; styles simple; fruits 1-few in the axils, conic-ovoid, 4-angled, glabrous. **3.C. gomphophylla**
- Leaves smooth above; styles branched; fruits in small axillary panicles, subglobose or ovoid, 4-6-ribbed, usually pubescent. **4. C. zenkeri**

1. C. africana *Burm. f.* (1768).
C. kraussiana Bernh. (1845).
C. vesciculosa Planch. (1848) - type: Gonder, Semien, *Schimper* II:1162 (K holo.);
C. kraussiana var. *flavescens* Almagia, *Ann. R. Ist. Bot. Roma* **8**: 119 (1903).
C. australis sensu auctt. non L.

Deciduous tree 5-35 m tall; bark smooth, grey; slash cream to yellow with brown speckling, turning brownish; twigs tawny pubescent to tomentose. Leaves: stipules linear, 4-6 mm long; petiole 1-5 mm long; lamina ovate to ovate-lanceolate, 5-10 x 2-6 cm, acuminate, upper 2/3 below acumen serrate, base a little unequal-sided, pubescent, ultimately subglabrous except on the veins beneath, generally a tuft of hairs in the angles of lateral veins, scabrid above; basal lateral veins extending 3/4-way up lamina, 1-2 pairs of prominent lateral veins above. Cymes axillary, of 3-35 clustered male flowers, pedicels 1-4 mm long; uppermost cymules of 1-several bisexual flowers, pedicels elongating after fertilisation. Sepals 4-5, 1.5-3 mm long. Ovary densely hairy; styles simple, 2.5-4 mm long. Fruits on 1-2.5 cm long pedicels, ellipsoid-globose, 5-8 mm long, yellow or orange, pubescent. Fig. 102.2.1.

Dry and moist evergreen forest, riverine forest, sometimes persisting as a relic in disturbed places and on dry rocky hills; 1300-2300(?-2500) m. EW TU WU GD GJ WG SU AR HA BA IL KF GG SD; east to Yemen, west to Ghana and south to Angola and Cape Province of South Africa. *Gilbert* 1763; *Friis* et al. 156; *Mesfin Tadesse*2543.

Leaves and fruits form an important part of the diet of Black and White Colubus monkeys (ed.). Browsed when present in rangelands. The wood can be used for tool handles, but has little value as timber as it rots and splits easily.

* Royal Botanic Gardens, Kew, Richmond, Surrey, TW9 3AB, UK.

2. C. **toka** *(Forssk.) Hepper & Wood*(1983).
Ficus toka Forssk. (1775).
Celtis integrifolia Lam. (1797).

Deciduous tree to 25 m tall, bole generally short, often strongly buttressed, crown large spreading; bark grey-brown, smooth or scaly; slash cream with streaks and blotches of red-brown, turning brownish; twigs pubescent. Leaves: stipules linear, 3-5 mm long; petiole 3-6 mm long; lamina ovate, 4-9 x 2.5-5 cm, shortly acuminate, base rounded to cordate, unequal-sided, margin usually entire (toothed on seedlings and sucker-shoots), soon subglabrous except on the veins beneath and tufts in the vein-angles, scabrid; basal lateral veins extending 3/4-way up lamina, with 1-2(-3) lateral veins above. Panicles of cymules 1-3(-4 in fruit) cm long, axillary, with many clustered male flowers, upper ones with several terminal female or bisexual flowers on longer pedicels. Sepals 5, 2-3 mm long. Ovary shortly pubescent, with a ring of longer hairs at the base; styles 2-3-branched, 3-5 mm long. Fruits subglobose, 8-12 x 7-10 mm, yellow-orange turning brown, puberulous to scabrid. Fig. 102.1.3.

Riverine forest; 375-1000 m. EW WU WG SU IL KF GG; Sudano-Sahelian zone from Senegal to Yemen. *Carr* 899; *Chaffey* 624; *Sebsebe D.* 273.

Hollowed out trunks used for boats in the lower Omo valley, *Carr* 899.

3. C. **gomphophylla** *Bak.* (1887).
C. durandii Engl. (1900).
C. ugandensis Rendle (1906).

Deciduous tree 5-30 m tall, monoecious or dioecious; bark smooth, grey; slash pale brown with darker granular stripes and flecks; twigs pubescent. Leaves: stipules linear-oblong, 4-5 mm long, falling soon; petiole 5-10 mm long; lamina ovate to oblong-elliptic or oblanceolate, 5-15(-20) x 2.5-8 cm, long-acuminate, base slightly unequal-sided, entire or with a few coarse teeth, glabrescent, becoming scabrid above; basal lateral veins extending a little into upper half of lamina, 3-6 pairs of lateral veins above. Male flowered cymules often precocious, crowded, pedicels 2-6 mm long; female and bisexual flowers axillary or at nodes below, few, short pedicels elongating after fertilisation. Sepals 4-5, 1.5-2.5 mm long. Ovary glabrous or puberulous; styles simple, 2-3 mm long. Fruits on pedicels 5-10 mm long, conic-ovoid, 4-angular when dry, 4-6(-8) x 3-4(-6) mm, yellow, glabrous. Fig. 102.1.1.

Moist evergreen forest; 1200 m. IL; west to Ivory Coast, south to Angola and the Cape Province of South Africa, also Comoro Is. and Madagascar. *Chaffey* 1276; *Friis* et al. 3864.

4. C. **zenkeri** *Engl.* (1900).

Deciduous tree 10-30(-50) m tall, buttressed; bark smooth, grey-brown, finely flaking to give yellowish patches; slash cream with red concentric rings, rapidly turning brown; twigs pubescent to tomentose. Leaves: stipules lanceolate, (5-)7-10 mm long, falling soon; petiole 5-8 mm long; lamina chartaceous, ovate to oblong-elliptic or slightly obovate, 6-15 x 4-8 cm, shortly acuminate, base slightly unequal-sided, entire or more rarely toothed on the upper half, glabrescent and smooth above, puberulous beneath at least along the veins; basal lateral veins extending 2/3-3/4 way up lamina, with (2-)3-4(-5) pairs of lateral veins above. Inflorescences essentially as*C. toka*. Sepals 4-6, 1.5-2.5 mm long. Ovary densely tawny pubescent; styles 1-2-branched, 2.5-4 mm long. Fruits on 3-5 mm long pedicels, subglobose or ovoid, 7-8 x 5-6.5 mm, red, 4-6-ribbed, pubescent to subglabrous. Fig. 102.1.4.

Wet evergreen and semi-deciduous forest; 500-1800 m. IL; west to Guinée, south to Zaire and Angola. *Chaffey* 284, 945; *Friis* et al. 3865;*Thomerson* 737.

Misidentified as *C. integrifolia*, Chaffey(1979), in *S. W. Eth. For. Inv.*: 167, 178, indicates that this species is an important component of the forests of Illubabor.

5. C. **philippensis** *Blanco* (1837).
C. wightii Planch. (1848).
C. mauritiana Planch. (1848).
C. prantlii Engl. (1900).
C. brownii Rendle (1915).

Evergreen tree or shrub, 3-20 m tall, often with short sharp buttresses; bark smooth, grey, hard; slash cream or pale brown, rapidly turning brown; twigs puberulous to nearly glabrous. Leaves: stipules lanceolate, 3-7 mm long, peltate (at youngest nodes); petiole 5-16 mm long; lamina chartaceous to thinly coriaceous, usually elliptic to elliptic-oblong in Africa, (5-)8-18 x 3.5-8 cm, acuminate, base broadly cuneate to rounded or sub-cordate, only a little unequal-sided, glabrous, smooth with small dot-like incrustations above (transluscent when fresh); basal lateral veins extending practically to apex, often looped with a few subprominent upper veins. Panicles of cymules 5-30(-40 in fruit) mm long, with many clusters of male flowers and a few female or bisexual flowers at the top; upper ones with many bisexual flowers. Sepals 4-5, 1.5-2.5 mm long. Ovary glabrous or nearly so, with a basal ring of hairs; styles shortly bifurcate, 1-3.5 mm long. Fruits on 3-7 mm long pedicels, ovoid, shortly beaked, 8-12 x 7-10 mm in Africa, red, glabrous. Fig. 102.1.2.

Wet evergreen forest, sometimes common in the understorey, persisting in secondary associations; 900-1320 m. IL; west to Senegal, south to Angola and Mozambique, also Madagascar, Mascarene Islands and tropical Asia to Australia. *Friis* et al. 3887; *Meyer* 8007, 8929.

For a discussion of the variation see Letouzey, *Fl. Cameroun* **8**: 27 (1968), and the references he cites.

2. **TREMA** *Lour.* (1790)

Trees or shrubs. Leaves finely serrate. Flowers male, female or bisexual in axillary cymes. Sepals (4-)5, shortly joined, with male buds valvate and female buds ± imbricate. Ovary sessile, 1-locular; styles short, simple. Drupes small, thinly fleshed; endocarp hard.

About 15 species in the tropics and subtropics, with only one in Africa.

T. orientalis *(L.) Bl.* (1852).
Celtis orientalis L. (1753).
C. guineensis Schumach. & Thonn. (1827); *T. guineensis* (Schumach. & Thonn.) Ficalho (1884).
Sponia orientalis var.*asperata* Solms in Schweinf. (1867) - type: GD, Dehli-Dikeno between Simien and Lasta,1854, *Schimper* 195 (B holo. destr., K iso).
S. hochstetteri Planch. (1873); *T. guineensis* var. *hochstetteri* (Planch.) Engl. (1895), nom. illegit.; *T. guineensis* var. *asperata* (Solms) Cuf. *Enum.*, 6 (1953, nom. non rite publ.), *Enum.*, XX (1969).

Tree or shrub to 10-15 m tall; bark smooth, grey; slash pale, turning brownish. Leaves: petiole 7-12(-15) mm long; lamina oblong-lanceolate to attenuate-ovate, (2-) 7-14(-22) x 2-7.5 cm, acuminate, base rounded to cordate, evenly and closely serrate, glabrous to puberulous above, pubescent to tomentose beneath, becoming scabrid; lateral veins 3-8 pairs, basal ones ascending 1/2-3/4-way up lamina. Cymes usually congested, up to 1(-3 in fruit) cm long, many-flowered, mostly male with a few female (bisexual) flowers at top. Calyx 1-2 mm long. Ovary pubescent; styles inrolled or divaricate, 1-2 mmlong. Fruits ovoid to globose, 3-5 mm long when dried, black, glabrescent. Fig. 102.2.2.

Pioneer of forest edges and clearings, riverine forest, rarely dry evergreen bushland; (500-)900-1800 m. EE/EW EW GD GJ WG SU AR BA IL KF; throughout tropical Africa south of the Sahara; Madagascar, Mascarene Islands and tropical Asia. *Friis* et al. 1112, 1970; *Mooney* 9551.

The bark is reported to have insecticidal properties, and, in East Africa, it is rubbed on fishing line to preserve it. *Trema* is very fast growing and may have potential as a firewood crop, but there is no record of it being so used.

A polymorphic species tending to have smaller and more hairy leaves in exposed places.

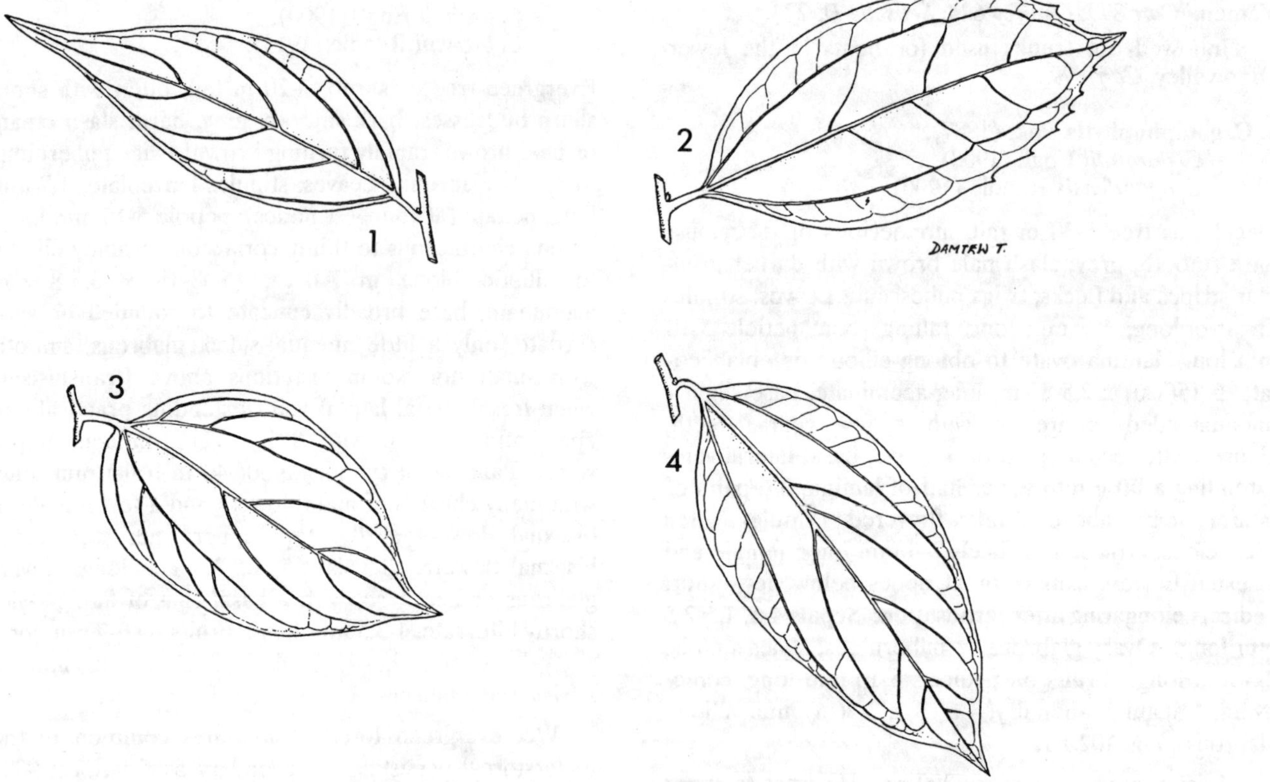

Figure 102.1 Leaves of *CELTIS* spp.: 1 - *C. GOMPHOPHYLLA*, from *Friis* et al. 3864; 2 - *C. PHILLIPPENSIS*, from *Friis* et al. 3887; 3 - *C. TOKA*, from *Sebsebe D.* 273; 4 - *C. ZENKERI*, from *Friis* et al. 3865. All x 2/3. Drawn by Damtew Teferra.

Figure 102.2.1 - *CELTIS AFRICANA*: leafy branch with top view of male flowerand side view of bisexual flower. 2 - *TREMA ORIENTALIS*: leafy branch and longitudinal section and side view of female flower, and cross section of ovary; E - leafy branch. Scale: cm as closed bar, mm as open bars. (Reproduced with permission of Oklahoma Agricultural Experiment Station from Experiment Station Bulletin No. 45, *Families of Flowering Plants in Ethiopia*, figure 5, 1967.) Drawn by W.C. Burger.

103. BARBEYACEAE

by B. Verdcourt*

Rendle, *Fl. Trop. Afr.* **6(2)**: 14-15 (1916); Cufodontis, *Enum.*: 6-7 (1953);
Hutchinson, *Gen. Fl. Pl.* **2**: 195 (1967) & *Fam. Fl. Pl.* ed. 3: 251, fig. 72. (1973).

Trees or shrubs; leaves opposite, simple, pinnately nerved, white tomentose beneath with curled hairs; stipules absent. Flowers dioecious, regular, in fascicles of short cymes. Male flowers: perianth lobes 3-4, valvate; petals absent; stamens 6-12 with very short free filaments and erect anthers with 2 locules opening lengthwise with apiculate connectives. Female flowers: calyx of 3-4 segments which are very accrescent in fruit, becoming membranous and veined; staminodes absent; ovary 1 (rarely 2 or 3) superior, subsessile, 1-locular with solitary pendent ovule; style linear with stigmatic papillae covering the apical 3/4. Fruit dry and indehiscent, compressed ellipsoid, acute with thin, veined pericarp. Seed without endosperm; embryo straight, cotyledons flat, fleshy, oily.

A single species only found in Arabia and NE Africa.

Most authorities consider this family as being closely related to the **Ulmaceae.** Burger (1967), in *Fam. Fl. Pl. Ethiopia*, believes the ovary is a simple carpel and not reduced thus indicating that *Barbeya* is not related to the **Ulmaceae** and its allies **Urticaceae** and **Moraceae**. Burger supplied material for more thorough investigation to W.C. Dickson & E.M. Sweitzer who in 1970 (*Amer. J. Bot.* **57**:468) discussed the morphology and relationships of *Barbeya*. Although Dickson & Sweitzer accepted the evidence which supports *Barbeya*'s retention in a separate family, they did not find any character "to negate the general urticalean affinities of the genus".

Barbeya oleoides *Schweinf.* in *Malpighia* **5**: 332, figs. 24, 25 (1891)

- types: EW, Addi-Beddel near Azegn [Azzaga] and M. Sabbar near Geleb, *Schweinfurth* 621, 622, 1189, 1190 (B syn. destr., K isosyn. except 621).

B. oleoides subsp. *arussorum* Chiov. in *N. Giorn. Bot. Ital.* **36**: 370 (1929) - type: SD, Laga-Hida (Aro-Ida), *Basile* 326 (TO holo. not seen).

Shrub or small tree 3-10.5(-12) m tall, much-branched and rather resembling an olive; branches ± drooping or arched; bark of young trees pale and smooth, on older trees grey, ± rough; slash deep blood-red, sticky; young shoots with dense white pubescence. Leaves olive-like, oblong-elliptic to elliptic-lanceolate, 1.2-6 x 0.35-2.3 cm, acute at the apex, rounded to cuneate at the base, markedly discolorous, soon glabrous and ± shiny above, densely white velvety tomentose beneath with curled hairs. Male cymes subsessile, *c* 6 mm long, densely rusty-brown pubescent; pedicels 2-5 mm long; perianth 3 mm long, divided into ovate or elliptic 3-nerved lobes. Female cymes sessile; pedicels 7-8 mm long, perianth 7-8 mm long; lobes elliptic-oblong to narrowly elliptic, 3-nerved and reticulately veined. Fruiting calyx-lobes lime green, 1.0-2.3 cm long. Fruit 6-10 mm long with obvious suture; pedicels up to 2.3 cm long. Fig. 103.1.

Acacia and *Commiphora* bushland, evergreen scrub with *Euclea schimperi,* grassland with scattered *Terminalia, Acacia*, etc., open *Juniperus* forest with *Pistacia*, on granite, sandstone and limestone slopes, often on red clay, sometimes locally dominant in badly eroded bushland; 1260-2200 m. EW HA BA SD; Somali Republic (N) and Arabia (Yemen and Saudi Arabia). *Burger* 1926, 2824, *Gilbert* et al. 2258.

Newbould and Bally have independently reported being told in the Somali Republic that the plant is poisonous to goats and, presumably, other stock.

Figure 103.1 - *BARBEYA OLEOIDES*: leafy stem with flowers, cm scale as closed bar; female flower, side view and longitudinal section through female flower; side view of male flower, mm scale shown as open bar. (Reproduced with permission of Oklahoma Agricultural Experiment Station from Experiment Station Bulletin No. 45, *Families of Flowering Plants in Ethiopia*, figure 4(1), 1967.) Drawn by W.C. Burger.

* Royal Botanic Gardens, Kew, Richmond, Surrey TW9 3AB, UK.

104. MORACEAE

by I. Friis*

Rendle & Hutchinson in *Fl. Trop. Afr.* **6(2)**: 17-240 (1916-1917); Cufodontis (1953), *Enum.*: 7-17 (1953); Berg et al. in *Fl. Gabon* **26**: 1-276 (1984) & in *Fl. Cameroun* **28**: 1-297 (1986).

Monoecious, rarely dioecious, trees, shrubs, or perennial herbs, often with milky latex. Leaves petiolate, alternate, usually penninerved, with paired lateral or fused intrapetiolar stipules mostly falling quickly. Inflorescence axillary spikes or heads, sometimes disc-shaped with flowers on one side, or hollow receptacles ("figs") with flowers on the inside. Flowers unisexual, regular, usually 4-merous, tepals free or ± united, sometimes reduced. Male flowers with 1-4 stamens opposite the tepals, inflexed or straight in bud, and a ± rudimentary ovary. Female flower with superior or inferior, unilocular ovary; one ovule pendulous from the apex of the ovary; style one, filiform, stigmatic branches long, 2(or 1 by abortion). Fruit a small achene or drupe, enclosed in the persisting accrescent perianth or immersed or enclosed in the fleshy receptacle.

About 53 genera and 1400 species. Most species are found in the tropics, but the family is also represented in subtropical and warm temperate regions.

Key to Genera

1. Stamens inflexed in bud, suddenly becoming reflexed in the open flower; leaves with small, lateral stipules, not leaving a circular scar round the stem. 2
- Stamens erect in bud; leaves with larger, intrapetiolar, free or ± fused stipules leaving a circular scar round the stem. 4
2. Flowers in spicate or capitate catkin-like inflorescences; tall forest trees (or trees in cultivation or escaped). 3
- Flowers on a flattened receptacle; herbs (in Flora area) with fleshy stems up to 2(?2.5) m tall. **3. Dorstenia**
3. Monoecious or dioecious; male inflorescence up to 3 cm long; female inflorescence up to 1.5 cm long; infructescence with laterally fused fruiting perianths, tepals fleshy, greenish, white, orange-red or dark purple; ovary with two functioning style branches; leaves with serrate or markedly crenate margin. **1. Morus**
- Dioecious; male inflorescence up to 15 cm long; female inflorescence 2-3 cm long; infructescence loose in structure, fruiting perianth not laterally fused, only slightly fleshy, greenish; ovary with one functional and one reduced style-branch; leaves with entire or slightly crenate margin. **2. Milicia**
4. Inflorescence a fig, i.e. an ovoid or globular, hollow receptacle with flowers on the inside; the opening ("ostiole") of the receptacle narrow, partly closed by scale-like bracts. **4. Ficus**
- Inflorescence not a fig, i.e. not hollow and without an ostiole. 5
5. Receptacle bisexual; disc-shaped when young, later ovoid, with many male flowers and a single female flower in the middle, all exposed on the upper side of the disc or on the tip of the fruiting receptacle, which encloses the single fruit; leaves smooth. **5. Trilepisium**
- Receptacle unisexual; male flowers many together on flattened, ± hollow receptacles with basal bracts; female flowers solitary on club-shaped receptacles; juvenile leaves ± scabrid. **6. Antiaris**

1. MORUS *L.* (1753).

Monoecious or dioecious trees or large shrubs with milky latex. Leaves alternate, petiolate, 3-nerved from the base, serrate; stipules small, lateral. Inflorescences unisexual, pedunculate spikes, solitary or few together in the leaf-axils. Male inflorescences catkin-like, the female ones short, capitate. Male flowers 4-merous, stamens inflexed in bud. Female flowers 4-merous, tepals completely enclosing the ovary; style short, with 2 equally developed stigmatic branches. Fruit a drupe, enclosed in and partly fused with the persistent fleshy perianth.

A genus with *c* 10 species in the warm temperate and subtropical regions of the northern hemisphere; one species native in tropical Africa, another species widely cultivated and occasionally naturalized.

1. Styles and perianth segments of female flowers with long, soft hairs, ripe fruits dark purple to black; leaves scabrid above. *M. nigra*
- Styles and perianth segments of female flowers glabrous or shortly pubescent, ripe fruits white, red or greenish, rarely purplish black; leaves ± smooth above. 2
2. Leaves broadly ovate or cordate; main lateral nerves (the pair branching off the midrib at the leaf base) not more prominent than other lateral nerves, reaching 2/3 towards the apex before anastomosing; female inflorescence many-flowered. **1. M. alba**
- Leaves ovate to elliptic; main lateral nerves very prominent, reaching 3/4 or more towards the apex before anastomosing; female inflorescence 3-8 flowered. **2. M. mesozygia**

* Botanical Museum and Herbarium, University of Copenhagen, Gothersgade 130, DK-1123 Copenhagen K, Denmark.

M. nigra L., Black Mulberry, is indicated by Cufodontis as cultivated in Eritrea, but no herbarium material has been found. The species is widely cultivated for its fruits in the Middle East and the warmer parts of Europe.

1. **M. alba** *L.* (1753).
M. indica L. (1753); *M. alba* L. var. *indica* (L.) Bur. in DC., *Prodr.* **17**: 243 (1873).

Small monoecious tree to *c* 5 m tall. Leaves very polymorphic, even on the same branch, broadly lanceolate, ovate, or ± trilobed, or sometimes lobed on one side only, 5-15 x 3-8 cm, base truncate or cordate, often unequal, margin serrate, apex acuminate; lamina 3-nerved from base, with smaller lateral nerves from the midnerve, glabrous to hispid on both sides, pubescent on the nerves beneath. Stipules lanceolate, falling quickly. Petiole 3-5 cm long. Inflorescences pedunculate, single; male inflorescence 10-12 mm long; female inflorescence 7-10 mm long, with many flowers. Male flowers 1.5-2 mm long, tepals erect, ciliate. Female flowers 1.5-2 mm long, tepals erect, ± pubescent, style *c* 1 mm long, stigmatic branches *c* 2 mm long. Syncarp juicy, white when ripe, but often dark red on one side, rarely black, *c* 2 cm long, edible, but with very little taste.

Cultivated as a hedge and shade plant around houses and also found left in abandoned cultivations; probably naturalized in hedgerows and secondary bushland and thicket; 1500-2300 m. EW SU SD IL; probably to be found in most towns above 1500 m. A native of warm temperate Asia, now widely cultivated in tropical Africa for its edible fruits and sometimes naturalized. *Fiori* 15; *Friis* et al. 1854; *Westphal* 2628.

This species is the food plant of the silkworm, *Bombyx mori*. Experiments on the rearing of silkworms in Nazareth have shown good results, and could be expanded to parts of the country, such as the Jima area, where *M. alba* is abundant.

2. **M. mesozygia** *Stapf* (1909).
Celtis lactea Sim (1909); *Morus lactea* (Sim) Mildbr. (1922); *M. mesozygia* Stapf var. *lactea* (Sim) A. Chev. in *Rev. Bot. Appl.* **29**: 72 (1949).

Large, apparently dioecious shrub or tree to *c* 30 m tall, with milky latex. Leaves ovate, elliptic to oblong, 2.5-10 x 2-7 cm, base rounded to subcordate, symmetrical, margin serrate, apex acuminate; the two main lateral nerves reach 3/4 towards the apex before anastomosing, 2 or 3 pairs of lateral nerves branch off from the midnerve just below the apex, major nerves connected by parallel tertiary nerves; lamina sparsely pubescent above, with a few hairs in the axils of the main nerves beneath. Petiole 5-22 mm long, glabrescent. Stipules linear-lanceolate, falling quickly, 4-10 mm long. Male inflorescence single, 1.5-3 mm long, tepals pubescent, rudimentary ovary small. Female inflorescences 1-3 together, 3-8 flowered, up to 0.6 cm in diam., on 12-20 mm long peduncles. Female flowers 2-3 mm long, tepals pubescent or ciliate, stigmatic branches sessile, 2-3 mm long. Syncarp subglobose, 1-1.5 cm in diam., greenish, moderately fleshy. Fig. 104.1

Humid forest with *Aningeria altissima* and *Celtis* spp.; 500-1500 m. KF IL WG; distributed along the northern fringe of the lowland forest zone of central Africa, west to Senegal, also Uganda, Kenya, Tanzania, N Angola, Zambia, Malawi, Mozambique, N Natal. *Chaffey* 386; *Friis* et al. 3988, 3989.

2. MILICIA *Sim* (1909).
Chlorophora Gaud.
quoad species ex Africa, excl. typo.

Dioecious trees. Leaves alternate, petiolate, entire, pinnately nerved. Stipules lateral, free, falling quickly. Inflorescences single in the leaf axils, unisexual, catkin-like, pedunculate, male inflorescences long and slender, female inflorescences shorter and thicker. Male flowers 4-merous, tepals broad, stamens inflexed in bud, rudimentary ovary small. Female flowers with 4 partly united, unequal segments, surrounded by long hairs at the base; ovary enclosed, style with one functional and one rudimentary stigmatic branch (the latter may be very small or absent). Female perianth becoming somewhat fleshy during ripening of the fruit, pressed closely together, but not laterally fused. Achene enclosed, compressed.

Two species, the other restricted to W Africa.

M. excelsa *(Welw.) C.C. Berg* (1982);
Morus excelsa Welw. (1869); *Chlorophora excelsa* (Welw.) Bent. & Hook. f.(1880).
Milicia africana Sim (1909).

Tree to 30 m or more, with spreading crown, buttresses, and milky latex; young branches pubescent to glabrescent. Leaves ovate, elliptic or oblong, 5-15 x 3.5-8 cm, dark green, membranous, base obtuse, truncate or subcordate, often very unequal, margin entire, apex acute to shortly acuminate; lamina puberulous or slightly hispid above, slightly pubescent below, with *c* 10-18 pairs of lateral nerves, all reaching the leaf margin and connected with parallel tertiary nerves. Petiole 1-4 cm long, glabrescent. Stipules prominent, up to 4 cm long. Male inflorescences 5-15 cm long, on 0.5-1.5 cm long peduncles. Male flowers *c* 1.5 mm long, tepals ovate. Female inflorescence 2-3 cm long, on 0.5-1 cm long peduncles, thicker than the male inflorescence. Female flower *c* 3 mm long, tepals ovoid, almost free, style not discernable, functional stigmatic branch 5-8 mm long, rudimentary branch 0-1 mm long. Syncarps 3-5 cm long, moderately fleshy, loose in structure. Achene *c* 3 mm long. Fig. 104.2.

Lowland forest with *Aningeria altissima*, mostly in clearings, sometimes left standing in cultivations or at villages; 500-600 m. KF IL; west to Sierra Leone,

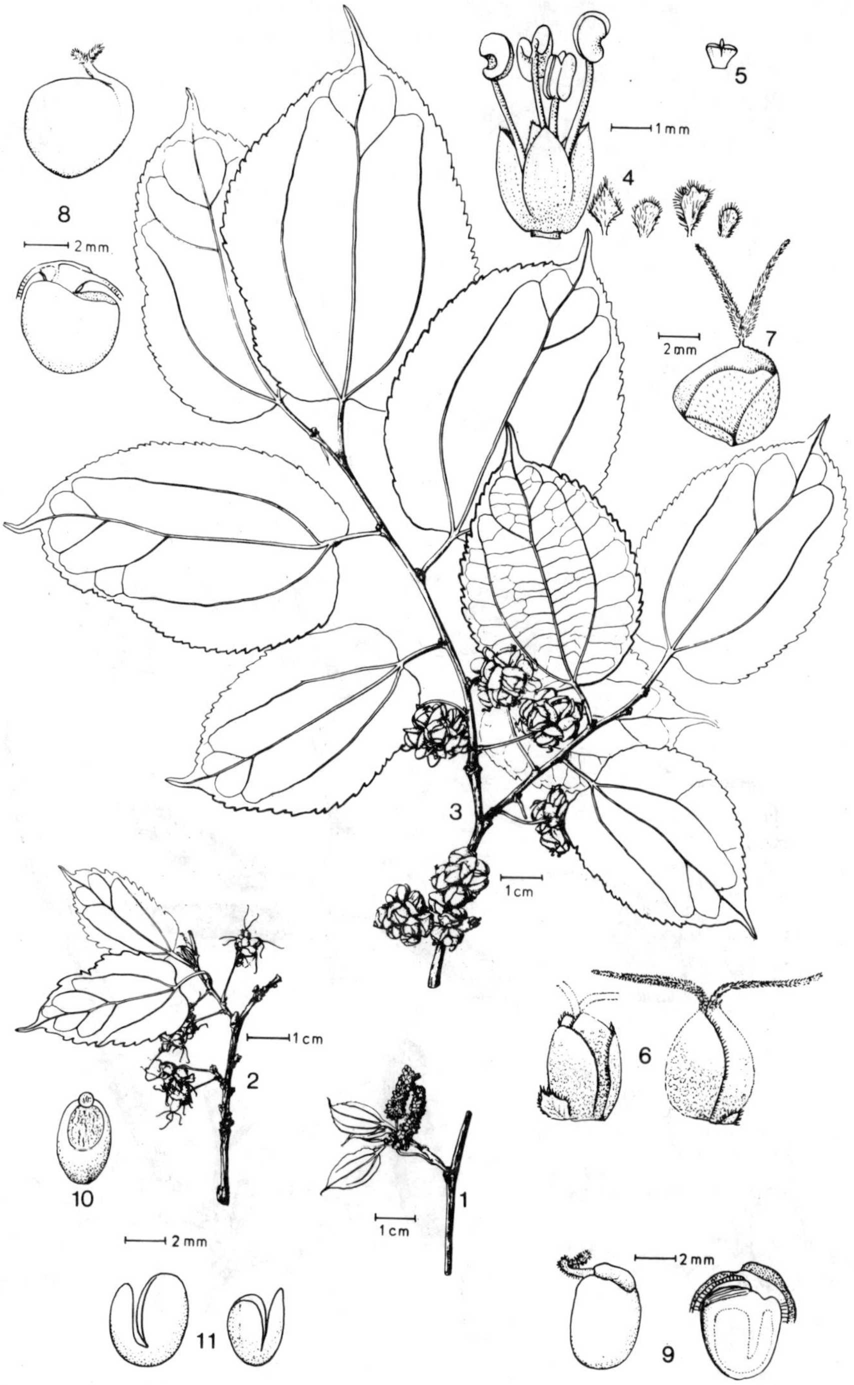

Figure 104.1 *MORUS MESOZYGIA*: 1 - leafy branch with male inflorescences; 2 - leafy branch with female inflorescences; 3 - leafy branch with infructescences; 4 - male flower and tepals from; 5 - pistillode; 6 - female flower and tepals; 7 - female flower in fruit; 8 & 9 - fruits; 10 - seed; 11 - embryo. 1 from *Simâo* 14; 2 from *Fanshawe* 9319; 3, 8 & 11 from *Simâo* 233; 4 & 5 from *Andrada* 14447; 6 from *Gomes E Sousa* 1862; 7 from *Espirito Santo* 1961. (Reproduced with permission from *Bull. Jard. Bot. Nat. Belg.* 47, 1977: *Flore du Cameroun*, 28: pl. 1., 1986.)

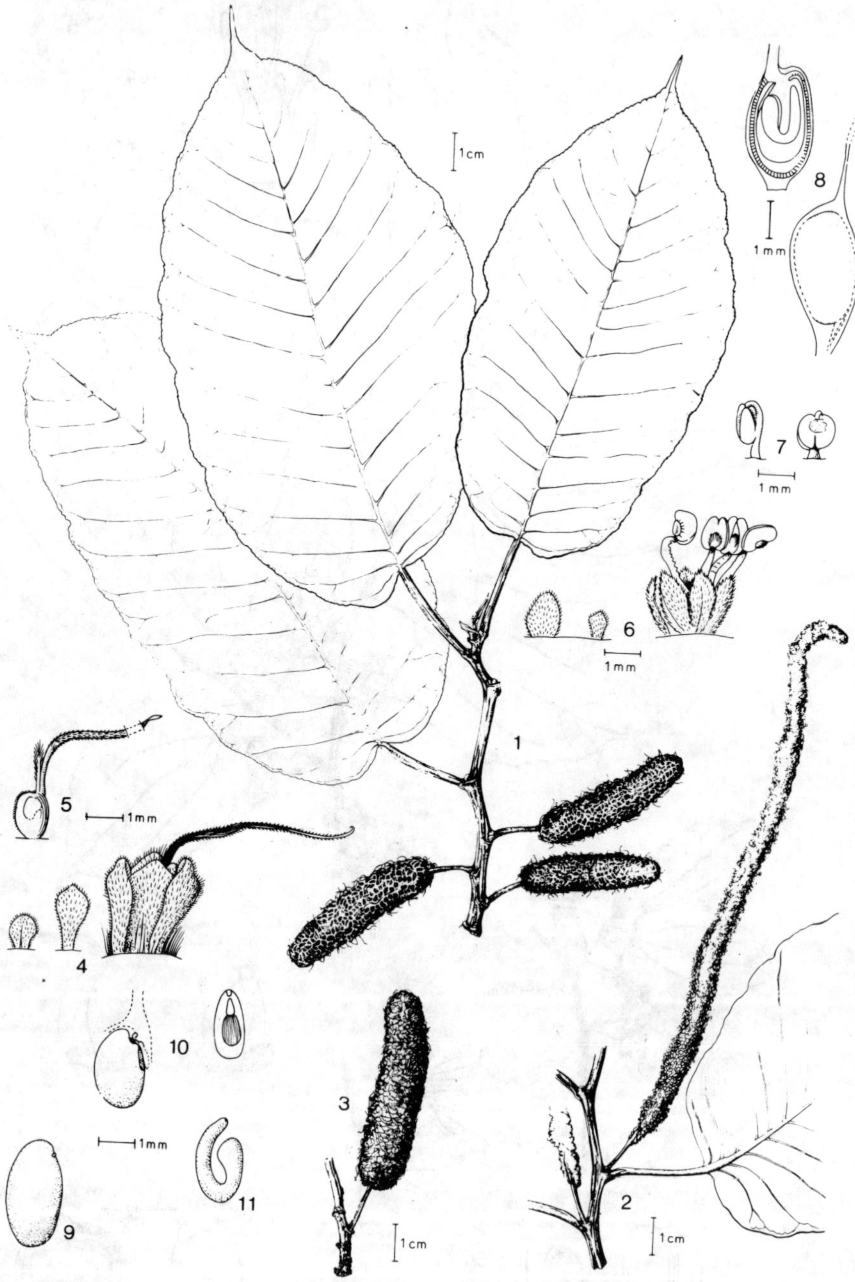

Figure 104.2 *MILICIA EXCELSA*: 1 - leafy branch with female inflorescence; 2 - male inflorescence; 3 - infructescence; 4 - female flower with tepals; 5 - pistil; 6 - male flower with tepals; 7 - stamen; 8 - fruit; 9 - endocarp; 10 - seed; 11 - embryo. 1 from *Simâo* 152, 2 from *Barbosa* 2611; 3 from *Torre & Paiva* 9372; 4 & 5 from *Léonard* 1049; 6 & 7 from *Barbosa* 2278; 8 & 9 from *Léonard* 1049; 10 & 11 from *Simâo* 654. (Reproduced with permission from *Bull. Jard. Bot. Nat. Belg.* 47, 1977: *Flore du Cameroun*, 28: pl. 2., 1986.)

Uganda, coastal areas of Kenya and Tanzania, Malawi, Mozambique. *Thomerson* 731; *Chaffey* 947, 950.

A valuable timber tree.

3. **DORSTENIA** *L.* (1753).
Kosaria Forssk. (1775).

Monoecious perennial herbs, often with fleshy or succulent stems, or stemless, with rhizome or subterranean tuber, usually with milky latex. Leaves alternate, in stemless species arising ± directly from the subterranean tuber, petiolate, entire or with a crenulate, crenate, dentate or irregularly toothed margin. Stipules lateral, free, falling quickly or persisting. Inflorescences single or up to 5 together in the leaf axils, pedunculate, peltate, circular or elongated, with flowers on the upper side, a thinner flowerless margin and mostly with marginal appendages. Flowers densely crowded on or deeply embedded in the upper surface of the receptacles. Male flowers mostly more numerous than the female ones, only partially embedded, with 2-3 lobed perianth and 1-3 stamens, which are inflexed in bud; a rudimentary ovary present or absent. Female flowers deeply embedded, perianth without clearly defined lobes and mostly with a ± constricted opening through which the style protrudes; ovary completely enclosed, with an obliquely placed style, stigmatic branches 1 or 2. Fruit subglobose to ovoid, with a fleshy exocarp and a hard, smooth or rugose endocarp, which at maturity is ejected by the elastic exocarp.

About 175 species, widespread in tropical America and Africa; also on Madagascar, Socotra, in Arabia, and one species in India.

1. Stems obvious, arising from creeping or shallowly rhizomatous base which is never enlarged; stigma with 2 branches; plants of loose soil on floor of humid forest. 2
- Stems absent or arising from enlarged or tuberous base; stigma unbranched; plants of drier situations, often in rock crevices. 3
2. Plant to 2(?2.5) m high; receptacles distinctly elongated, 15-20 x 3-5 mm, appendages over 2 mm and up to 70 mm long. **1. D. soerensenii**
- Plant to 0.5 m high; receptacle circular, up to 10 mm in diam., appendages 1-2 mm long. **2. D. brownii**
3. Plants with obvious aerial stems arising from enlarged or tuberous base. 4
- Plants with subterranean tubers, leaves and inflorescences apparently arising direct from ground. 5
4. Stems slender, fleshy, herbaceous, with leaves spaced along length; receptacle 3-4-(or more) angled, appendage 1 at each angle. **3. D. cuspidata**
- Stems thick, succulent, with dense apical cluster of leaves in growing season and prominent leaf-scars lower down; receptacle circular, appendages 7-10, of different lengths. **4. D. foetida**
5. Tuber depressed discoid, often over 5 cm in diam., often producing a short herbaceous stem which hardly reaches above ground, leaves ovate to elliptic, pinnately veined; receptacle circular. **5. D. ellenbeckiana**
- Tuber higher than wide, up to 5 cm in diam., sometimes with a distinct swelling on top, but never producing a herbaceous stem; leaves as broad or broader than long, sometimes palmately divided, radiately or pedately veined; receptacle elliptic to linear. 6
6. Leaves not peltate (rarely slightly sub-peltate when young), cordate or reniform in outline, entire or variously lobed or divided; receptacles green or tinged purplish brown. **6. D. barnimiana**
- Leaves peltate, ± circular in outline; receptacles tinged white. **7. D. tropaeolifolia.**

1. **D. soerensenii** *Friis* (1974)
- type: KF, Bonga, *Friis* et al. 2113 (C holo., ETH K iso.).

Erect herb to *c* 2 (?-2.5) m from creeping rhizome, with milky latex; younger stems puberulous. Leaves alternate, petiolate, elliptic to obovate, 10-17 x 4-6.5 cm, base cuneate to rounded or subtruncate, margin irregularly serrate, sometimes irregularly toothed towards apex, apex acuminate, often with a 1.5-2.5 cm long tip; lamina with scattered stiff hairs above, minutely puberulous on the nerves below. Petiole 3-7 cm long, minutely puberulous. Stipules 1-1.5 mm long, linear, falling quickly. Receptacles 1-5 together, on peduncles 7-10 mm long; boat-shaped, 1-2.5 x 0.3-0.5 cm (excl. appendages), with 10-15 thin, purple-brown terminal and lateral apendages, up to 7 cm long. Male flowers numerous, with 3 stamens and rudimentary ovary. Female flowers 3-8, along the central axis of the receptacle, stigmatic branches 2. Fruit with thin exocarp and a white, hard, verrucose endocarp, *c* 3 mm in diam. Fig. 104.3.

In upland rain forest, often at small streams; 1950-2200 m. KF IL; not known elsewhere. *Gilbert & Rankin* 4203; *W. de Wilde* 9392, 10243.

2. **D. brownii** *Rendle* (1915).

Ascending perennial herb from creeping rhizome, stems up to 30 cm; young branches with a dense tomentum of dark hairs. Leaves petiolate, ovate to elliptic, 4-8 x 2-3.5 cm, base cuneate to rounded, margin crenate to sinuous, irregularly toothed towards apex, apex obtuse or acuminate; lamina with scattered, short, stiff hairs above, soft hairs below, especially on the nerves. Petiole 1-2.5 cm long, pubescent. Stipules linear, *c* 5 mm long. Receptacles 1-3 together, on peduncles 1-2 cm long; circular in outline, 4-10 mm in diam., marginal appendages 8-15, linear to lanceolate, purplish brown, up to 2 mm long. Male flowers numerous, tepals (?2-) 3, stamens 2-3, ovary rudimentary. Female flowers 4-10, scattered over the upper surface of the receptacles,

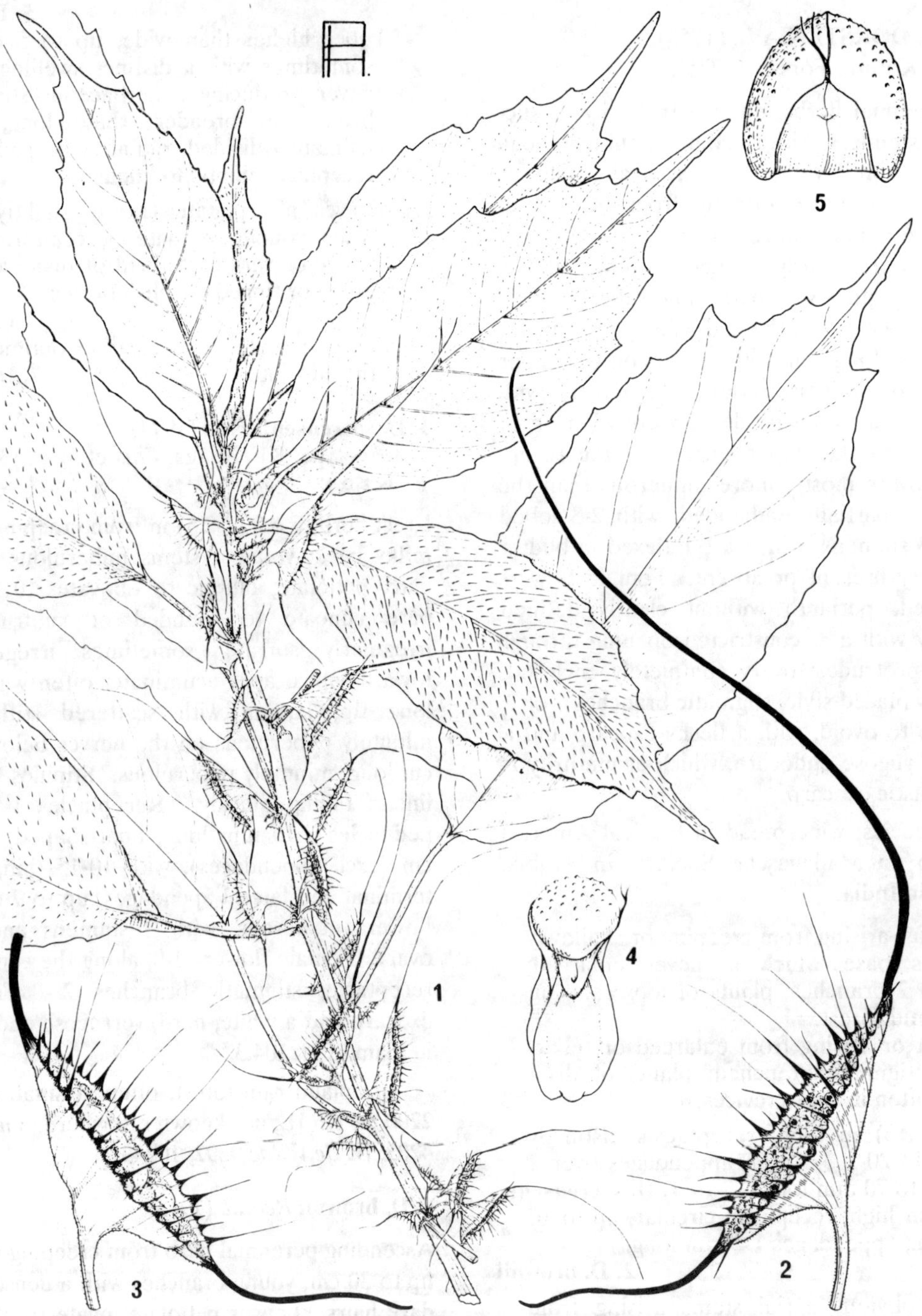

Figure 104.3 *DORSTENIA SOERENSENII*: 1 - leafy branch with inflorescences x 2/3; 2 - flowering receptacle x 2; 3 - fruiting receptacle x 2; 4 - fruit before ejection of endocarp x 4; 5 - endocarp and apical part of exocarp x 8. All from holotype, *Friis* et al. 2113. Drawn by Pat Halliday. (Reprinted from "*Dorstenia soerensenii* (Moraceae) from SW. Ethiopia and related species" by I. Friis from *Norw. J. Bot* 21: 107 by permission of Norwegian University Press (Universitetsforlaget AS), Oslo.)

stigmatic branches 2. Fruit with white, verrucose endocarp, *c* 1 mm in diam. Fig. 104.4.1.

In riverine and lowland forest with *Aningeria altissima* and *Celtis* spp.; 1000-1900 m. KF IL WG; Sudan (Imatong Mts.), E Zaire, Uganda, Kenya, Tanzania. *W. de Wilde* 7208; *Friis* et al. 4042; *Gilbert & Friis* 8404.

3. **D. cuspidata** *A. Rich.* (1851)
- type: TU, Djeladjeranne [Tchélatchékanné] towards Tekeze [Taccazé] River, *Schimper* III:1727 (P holo., K iso.).
D. walleri Hemsl. (1893).

Erect perennial from tuberous rootstock, stems single or a few together, up to 20 cm tall (in flora area); young stems puberulous. Leaves gathered near the top of the stem, petiolate, elliptic, 4-8 x 1.5-4.5 cm, base cuneate, margin crenate, apex acuminate; lamina with very scattered short hairs above and short soft hairs on the nerves below. Petiole 0.5-1.5 cm, pubescent. Stipules linear, up to *c* 5 mm long. Receptacles 1-2 together, on peduncles up to 1.5 cm long; triangular, quadrangular or polygonal, very rarely ± boat-shaped, up to 1.5 cm in diam. (excl. appendages), appendages 1 on each corner, linear, slightly swollen at tip. Male flowers numerous, number of tepals undefined, stamens (1?-)3, rudimentary ovary present (?). Female flowers numerous, scattered over the upper surface of the receptacle; stigma unbranched. Fruit with whitish to pale brown, verrucose endocarp, *c* 1 mm in diam. Fig. 104.4.2.

Between rocks in woodland; altitude probably between 1000 and 1500 m. TU; Upper Volta, Nigeria, Cameroun, Central African Rep., Sudan, Tanzania, Malawi, Mozambique, Zimbabwe, Madagascar. *Gilbert & Aweke* 2914.

4. **D. foetida** *(Forssk.) Schweinf.* (1896);
Kosaria foetida Forssk. (1775).
Friis in *Nordic J. Bot.* 3: 536 (1983).

Perennial herb, base swollen, stems one to several together from a common base, up to 1-1.5 cm thick and up to 10 cm tall, succulent, with ± pronounced markings of leaf- and stipule-scars. Leaves alternate, crowded near the top of the stem, lanceolate to ovate, obovate or elliptic, 1.8-18 x 1.0-2.5 cm, base cuneate, truncate or rounded, margin entire, crenulate, crisp or denticulate, apex rounded, obtuse or acuminate; lamina ± scabrid above, sparsely puberulous below. Petiole 1-3.6 cm long, puberulous. Stipules subulate from a broad base, 1-2 mm long, mostly withering before falling off. Receptacles 1-3 together in the leaf axils, on peduncles 1-6 cm long; circular in outline, 0.6-1.5 cm in diam. (excl. appendages), appendages 7-10, linear, up to 1.5 cm long. Male flowers numerous, tepals 2(-3?), stamens 2-3, rudimentary ovary absent(?). Female flowers numerous, scattered on the upper surface of the receptacle, stigma unbranched. Fruit with elongated, minutely tuberculate, pale brown endocarp, 1-1.4 mm long.

1. Leaves lanceolate, margin entire or crisped; petiole much shorter than lamina. subsp. **lancifolia**.
- Leaves ovate, obovate (sometimes broadly so), elliptic or suborbicular, margin entire, crenulate, crenate, crisped, denticulate or toothed; petiole almost as long as or longer than lamina. 2
2. Stems branched from a tuberous basal part; leaves ovate, obovate or elliptic, margin crenulate, crenate crisped, denticulate or toothed; petiole as long as or slightly shorter than lamina. subsp. **foetida** var. **foetida**.
- Stems unbranched from an only slightly swollen base; leaves broadly ovate, obovate or suborbicular, margin entire; petiole as long as or longer than lamina. subsp. **foetida** var. **obovata**.

Subsp. **foetida** var. **foetida**
D. arabica Hemsl. (1897).
D. crispa Engl.(1898) - type: SD, Gerima at Daua Parma, *Riva* 441 (B holo. dest., BM photo of holo. FT iso.).
D. phillipsiae Hook. f. (1899).

Rocky outcrops in deciduous bushland; 700-2100 m. ?EE/EW HA ?BA SD; N & S Yemen, Somalia. *Burger* 2804, 2844; *Simmons* 161.

Subsp. **foetida** var. **obovata** *(A. Rich.) Schweinf. & Engl.* in Engl., *Monogr.Afr. Pflanzenfam.* 1: 27 (1898);
D. obovata A. Rich. (1851); *Kosaria obovata* (A. Rich.) Schweinf. ex Solms in Schweinf. (1867) - type: TU, Tekeze [Taccazé] Valley, towards Djeladjeranne [Tchélatchékanné], *Schimper* III:1675 (P holo., BM K iso.).
Fig. 104.4.4.

Rocky outcrops in deciduous bushland; 1000-1400 m. TU; Sudan (Red Sea Hills), N. Yemen. *Gilbert & Aweke* 892.

Subsp. **lancifolia** *(Rendle) Friis* in *Nordic. J. Bot.* 3: 538 (1983);
D. crispa Engl. var. *lancifolia* Rendle in *Journ. Bot. (London)* 53: 302 (1915).
D. crispa Engl. var. *pachypoda* Chiov. in *Result. Sci. Miss. Stefanini-Paoli*: 165 (1916) - type: Somalia, Bur Egherta, Paoli 779 (FT, holo.).
Fig. 104.4.3.

Deciduous bushland on limestone, sand, and black cotton soil; 1000-1250 m. SD BA; E Kenya, SW Somalia, NE Tanzania. *Friis* et al. 1016, 3640; *Gilbert & Thulin* 1494.

All infraspecific taxa in the *D. foetida* complex occur in Ethiopia. The distribution of the species as a whole is N & S Yemen, Ethiopia, Somalia, E Kenya and NE Tanzania.

5. **D. ellenbeckiana** *Engl.* (1902);
- type: AR/HA, Burkar, *Ellenbeck* 2018 (B holo. dest.).
Friis in *Nordic J. Bot.* 3: 536 (1983).

Perennial herb with depressed, ovoid underground tuber, a short annual stem sometimes present, but not reaching above ground level. Leaves petiolate, ovate to elliptic, 4-18.5 x 2.5-10.5 cm, base truncate or subcordate, margin crenulate or finely toothed, apex acute or acuminate; slightly scaberulous above, with softer hairs on the nerves below. Petiole up to 8.5 cm long, puberulous. Stipules not observed. Receptacles on 2.5-13 cm long peduncles, orbicular in outline, up to 7 mm long and 1.5 mm thick. Male flowers numerous, number of stamens not known. Female flowers fewer, stigma unbranched. Fruit ovoid to subglobular, 1-1.5 mm in diam., with a white, slightly verrucose endocarp. Fig. 104.4.5.

Deciduous bushland on limestone; 900-1250 m. AR/HA SD; N Kenya. *Friis* et al. 1015; *Ash* 2402.

6. **D. barnimiana** *Schweinf.* (1862).
Kosaria barnimiana (Schweinf.) Schweinf. (1863).
Dorstenia ophioglossioides Bureau (1873); *D. barnimiana* Schweinf. var. *ophioglossioides* (Bureau) Engl., *Monogr. Afr. Pflanzenfam.* 1: 25 (1898) - type: GD, Semien, Debre Eski, *Schimper* II:402 (P holo., K iso.).
D.telekii Schweinf. in von Hoehnl (1892); *D. barnimiana* Schweinf. var. *telekii* (Schweinf.) Engl., *Monogr. Afr. Pflanzenfam.* 1: 25 (1898).
D. palmata Engl. (1894).
D. palmata Engl. var. *integrifolia* Chiov. in *Miss. Biol. Paese dei Borana*, 4, Bot.: 36 (1939) - type: SD, Neghelle, *Cufodontis* 230 & Arero, *Cufodontis* 320 (both FT syn.).
D. triternata Chiov. (1940) - type: BA, El Mara, *Reghini* 10 (FT holo.).
Friis in *Nordic J. Bot.* 3: 534 (1983).

Perennial herb with slightly depressed or subglobose tuber, 2-5 cm in diam., sometimes with small swelling above but aerial stems absent; leaves and receptacles arise from the tuber. Leaves petiolate, reniform, cordate, or palmate, often varying in shape during the season, 2-10 x 3.5-11 cm, base cordate or rounded, margin entire or irregular denticulate, apex bluntly acute to rounded; lamina puberulous to glabrescent on both sides, nerves pedately branched. Petiole 4-15 cm long, glabrescent. Stipules not observed. Receptacles 1-2 together, on peduncles 4-12 cm long; narrowly oblong, 1-3 x 0.3-0.8 cm (excl. appendages), attached to the peduncle near the base; appendages 2 (basal and apical), a few (one apical and 3-7 basal ones) or many (evenly distributed along the margin or usually with longer apical and basal ones), 4-40 x 0.5-1 mm, often purplish brown, rarely greenish. Male flowers numerous, tepals 2, stamens 2, rudimentary ovary not observed. Female flowers fewer, scattered, stigma unbranched. Endocarp ovoid, whitish, verrucose, *c* 1 mm long. Fig. 104.4.6-9.

Deciduous woodland or bushland, upland grassland, evergreen bushland, clearings in evergreen forest; 500-2450 m. GD SU HA SD BA IL GG; Sudan, Somalia, NE Zaire, Uganda, Kenya, Tanzania. *Friis* et al. 860, 861; *Mooney* 7034.

7. **D. tropaeolifolia** *(Schweinf.) Bureau* (1873);
Kosaria tropaeolifolia Schweinf. (1868); *D. barnimiana* var. *tropaeolifolia* (Schweinf.) Rendle in *Fl. Trop. Afr.* 6,2: 71 (1916) - type: Sudan/GD, Gallabat, *Schweinfurth* 564 (B holo., BM K iso.).
D. peltata Engl.(1911).
Friis in *Nordic J. Bot.* 3: 536 (1983).

Perennial herb with subglobose tuber, 2-5 cm in diam., aerial stems absent; leaves and receptacles arise from the tuber. Leaves peltate, circular in outline, 2-5 cm in diam., margin entire or irregular denticulate or crenate; puberulous to glabrescent on both sides, nerves radiating, irregularly forked. Petiole 4-7 cm long, glabrescent. Stipules not observed. Receptacles 1-2 together, on peduncles 4-8 cm long; narrowly elliptic to elongate, 1-2.5 x 0.3-0.8 cm (excl. appendages), attached to the peduncle near the base; appendages few (one apical and 3-7 basal ones) or many (evenly distributed along the margin or usually with longer apical and basal ones), 4-20 x 0.5-1 mm, pale green, whitish or pale brownish. Male flowers numerous, tepals 2, stamens 2; rudimentary ovary not observed. Female flowers fewer, scattered, stigma unbranched. Endocarp ovoid, whitish, verrucose, *c* 1 mm long. Fig. 104.4.10.

Deciduous woodland; 600-1300 m. Sudan/GD GJ; Cameroun, Central African Republic, Sudan, Kenya, Uganda. *Thulin & Hunde* 4076.

4. FICUS *L.* (1753)

Corner, *Ficus* in Asia and Australasia, in *Gardens' Bulletin, Singapore* **17**: 368-485 (1959) & **21**: 1-185 (1965); Aweke, Revision of the genus *Ficus* L. (Moraceae) in Ethiopia, in *Mededelingen Landbouwhogeschool Wageningen* **79-3**: 1-115 (1979); Berg et al. in *Fl. Gabon* **26**: 113-249 (1984).

Monoecious trees or shrubs, mostly epiphytic at first, with watery, colourless, or milky latex. Leaves alternate, sometimes opposite or subopposite, petiolate, pinnately or palmately nerved. Stipules free or ± fused, encircling the stem, covering the terminal bud, falling quickly, leaving a circular scar round the stem. Inflorescence a fig: a fleshy, globular or cone-shaped, hollow receptacle with male, female and gall flowers covering the inside and a narrow opening ("ostiole"), partly closed by bracts. Figs sessile or pedunculate, usually with 2-3 basal bracts, situated where the receptacle narrows into the peduncle, sometimes also with bracts spirally arranged on the outside (sect. *Sycidium*). Figs and peduncles sometimes covered when young by a flat or conical structure ("calyptra") consisting of fused bracts attached at the base of the peduncle (some species of sect. *Galoglychia*). Figs either with male and gall flowers or with female flowers only (subgen. *Ficus*)

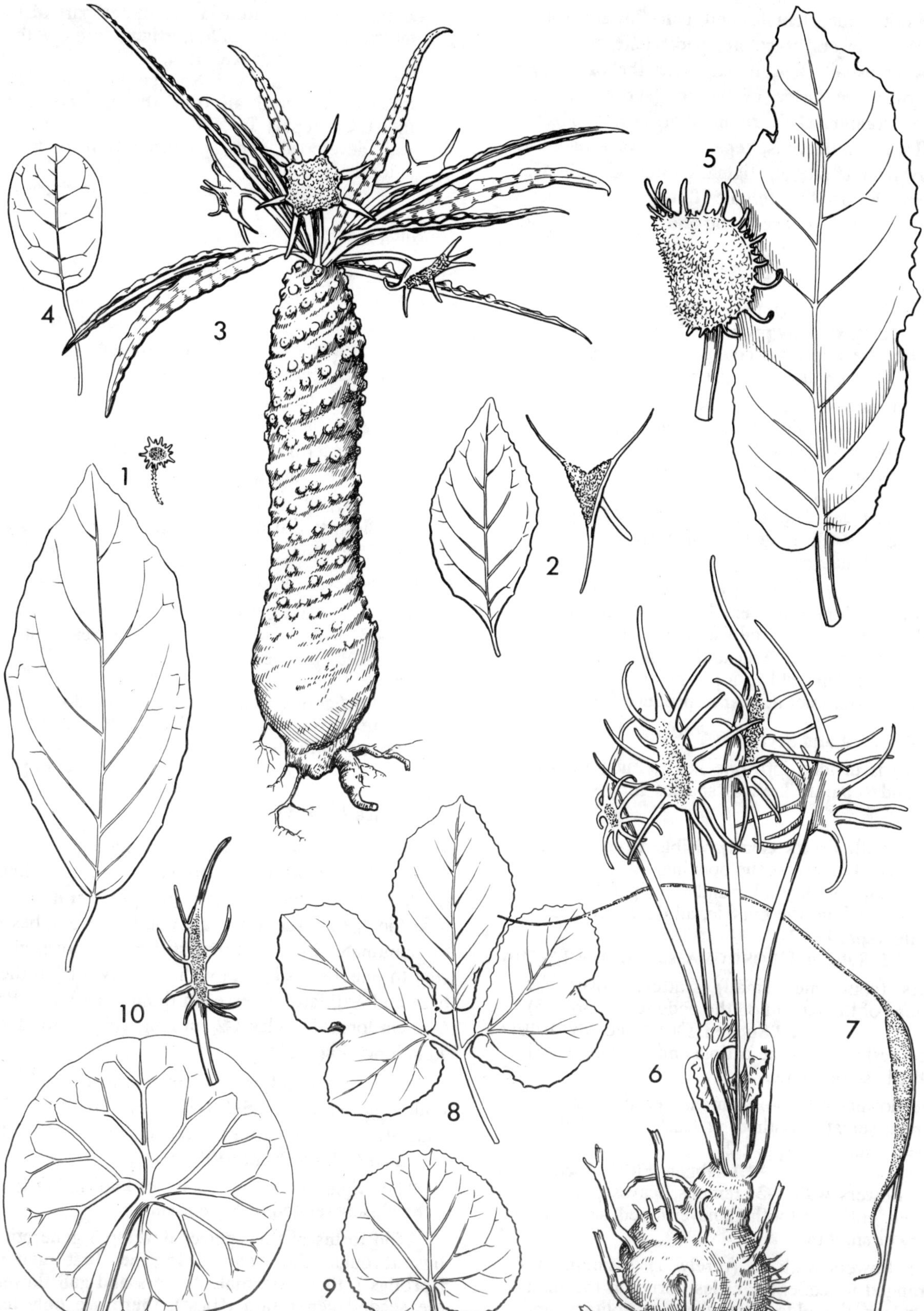

Figure 104.4 *DORSTENIA* spp.: 1 - *D. BROWNII*, leaf and inflorescence, from *Gilbert & Friis* 8408; 2 - *D. CUSPIDATA*, leaf and inflorescence, from *Schimper* III:1727 & photo of *Gilbert & Getachew* 2914; 3 - *D. FOETIDA* subsp. *LANCIFOLIA*, whole plant with leaves and inflorescences, from a painting by P.R.O. Bally for *Fl Pl. of Afr.*, t.1372; 4 - *D. FOETIDA* subsp. *FOETIDA* var. *OBOVATA*, leaf, from *Schimper* III:1675; 5 - *D. ELLENBECKIANA*, leaf and inflorescence, from *Gilbert* et al. 7470 and photo; 6 - *D. BARNIMIANA*, whole plant with tuber bearing fully formed inflorescences and very young leaves, from a lithograph by Schweinfurth; 7 - inflorescence with only 2 appendages, from *Burger* 2708; 8 - palmately lobed leaf; 9 - simple reniform leaf, from *Ash* 2405; 10 - *D. TROPAEOLIFOLIA*, inflorescence and leaf seen from below, from *Cook* 23 & 32. All x 2/3. Drawn by Eleanor Catherine.

or with both male, female and gall flowers (other subgenera). Flowers bracteate, pedicellate or sessile. Male flowers usually few, mostly near the ostiole, or scattered over the inside of the fig (subgen. *Ficus*), tepals 2-4, stamens 1-6; rudimentary ovary absent. Female flowers numerous, tepals 2-6, ovary ovoid, ± asymmetrical, style lateral, stigmatic branches 1-2. Gall flowers mostly as the female flowers, but usually pedicellate and with reduced style and stigma. Fruit an achene.

About 800 species, pantropical.

The division into subgenera and sections used here is based on the taxonomy proposed by Corner (loc. cit.) as applied by Berg in his accounts of the **Moraceae** for the Floras of Gabon and Cameroun (1984, 1986).

The current knowledge of the complicated pollination of figs by wasps that deposit their eggs in the gall flowers has been surveyed by Wiebes in *Gardens' Bull. Singapore* **29**: 207-232 (1977). An account of the fig wasps found in Ethiopian *Ficus* spp. has been published by Wiebes in *Meded. Landbouwhoogeschool Wageningen* **79-3**: 99-102 (1979); 9 genera of wasps have been recorded as pollinators.

Ficus elastica Roxb. is an ornamental species with large ovate to elliptic leathery leaves which are shiny above; the stipules are reddish and very large. It originates from tropical Asia and is now cultivated in parks, public gardens and as a pot plant in most of the large towns. It does not appear to produce fruits.

Key to Subgenera and Sections

The key is divided into one key to subgenera and sections, and separate keys to species for each of the groups.

1. Ostiole with bracts clearly visible from outside and spreading across the opening. 2
- Ostiole slit- or pore-like and 2-4 lipped, bracts not visible from the outside, all descending into the fig. (spp. 13-23) **4.2. Subgen. UROSTIGMA, sect. GALOGLYCHIA**

2. Bracts (sometimes small) scattered over the surface of the fig and on the peduncle. (spp. 3-5) **1.2 Subgen. FICUS, sect. SYCIDIUM**
- Bracts restricted to the ostiole and a single whorl at the base of the fig. 3

3. Male flowers with 3-6 stamens; leaves palmately veined, serrate, toothed, at least some of them lobed, scabrous. (spp. 1-2) **1.1 subgen. FICUS, sect. FICUS**
- Male flowers with 1-2 stamens; leaves pinnately veined, entire or toothed, never lobed, scabrous, hairy or smooth. 4

4. Male flowers with 2 (rarely 1) stamen; figs arranged in leafless panicles borne on the main stem, or the older branches, or solitary and axillary; leaves entire, sinuate to dentate. (spp. 6-9) **2. Subgen. SYCOMORUS**
- Male flowers with 1 stamen; figs always axillary on young branches; leaves always entire. **5**

5. Lenticels concentrated in the upper part of the internodes; tepals united, stigma bifid with 2 equal lobes; leaves scabrid. (sp. 10) **3. Subgen. PHARMACOSYCEA**
- Lenticels scattered all over the internodes or indistinct; tepals free, stigma 1, rarely with 1 well developed and 1 reduced branch; leaves smooth. (spp. 11-12) **4.1 Subgen. UROSTIGMA, sect. UROSTIGMA**

1.1 Subgen. FICUS, sect. FICUS

Shrubs or small trees; dioecious; sap milky. Leaves spirally arranged, usually at least some ± deeply lobed, ± scabrid; margin ± dentate. Stipules semi-amplexicaule, free. Figs often solitary in leaf axils or just below the current leaves, pedunculate, with 3-4 whorled bracts on the peduncle, lateral bracts absent; ostiole round, protruding or flat, with several ostiolar bracts visible; interfloral bracts present. Male flowers near ostiole or spreading over inside of fig; tepals 3-5, ciliolate; stamens (3-)5; rudimentary ovary ± developed. Female and gall flowers distinct; tepals (3-)5, ciliolate; style in female flowers long, with 2 long stigmas. Ripe fig with reddish-pink, brown or blackish, soft wall.

1. Mature figs 3-5 cm in diam.; all leaves somewhat lobed, leaf apex and apex of lobes acute to rounded, not acuminate; leaves up to 20 x 18 cm. **1. F. carica**
- Mature figs less than 3 x 2 cm; usually at least some leaves not lobed; leaf apex and apex of lobes acuminate; leaves up to 7.5 x 9.5 cm. **2. F. palmata**

1. **F. carica** *L.* (1753).

Shrub or tree to *c* 4 m tall, markedly deciduous; bark brown. Leaves 3-5 lobed, broadly ovate in outline, up to 20 x 18 cm, base cordate, margin serrate, at least near the apex of the lobes, apex and apex of lobes acute to acuminate; with scattered stiff hairs above, glabrescent to puberulous below, with 3-5 nerves from the leaf base and *c* 10 lateral nerves in the middle lobe. Petiole 6-10 cm long. Stipules 1-2 cm long. Fig obovoid, up to 7 cm long. Fig. 104.5.

A cultivated and occasionally escaped species. EW (cult. and escaped) SU (cult.) HA (cult.). Widely cultivated and sometimes escaped. *Aweke* 429; *Aweke & Gilbert* 697; *Westphal* 630.

The origin of this species is not known, but it seems to be closely related to *F. palmata*.

Wild forms of this species develop 3 generations of figs through the year. In the first generation ("profigs"), there are male flowers and gall flowers; in the second generation ("figs"), there are only female flowers; and in the third generation ("mamme"), there are only gall flowers. The "figs" only ripen if they are pollinated by a fig wasp (*Blastophaga psenes*). The wasps develop in the gall flowers of the first and third generations; attempts at putting wasp eggs in the female flowers of the second generation are not successful because of the long styles, but the flowers do get pollinated. Cultivated races of *F. carica* with entirely female flowers are widely grown; they need artificial pollination or pollination by wasps from a few male individuals of the wild species which are grown in the orchard.

Figure 104.5 *FICUS CARICA*, leafy branch x 1/2 from *Aweke & Gilbert* 697. (Redrawn with permission from *Meded. Landbouwhogeschool Wageningen* 79-3, Fig. 3, 1979.)

2. **F. palmata** *Forssk.* (1775).
F. morifolia Forssk. (1775).
F. forsskalii Vahl. (1805).
F. pseudocarica Miq. (1848) - type: TU, Adua, *Schimper* I:157 (L holo. BM K MPU P iso.).
F. pseudocarica Miq. var. *tomentosa* A.Rich., *Tent. Fl. Abyss*. **2**: 271 (1851) - type: TU, Wogerate, *Petit* s.n. (P syn.) & Tchelikote, *Petit* s.n. (P syn.).
F. petitiana A. Rich.(1851); *F. palmata* Forssk. var. *petitiana* (A. Rich.) Fiori, *Boschi e piante legn. Eritrea*: 371 (1910) - type: SU, *Petit* s.n. (P holo.).

Shrub or small tree to *c* 5 m tall with erect branches; young branches with purplish brown bark. Leaves ovate in outline, variously lobed, 3-7.5 x 2.5-9.5 cm, base truncate or cordate, margin serrate, apex and apex of lobes acuminate; lamina mostly scabrid above, scabrid to densely pubescent below, with 3-5 nerves from the base and 5-7 lateral nerves on each side of midrib. Petiole up to 4.5 cm, pubescent. Stipules puberulous to pubescent, falling quickly. Figs axillary, 1-2 together, on peduncles up to 1.5 cm long; basal bracts, *c* 1 mm long, ovate, puberulous. Fig ± stipitate, globose to obovoid, up to 2.0 x 1.5 cm, pubescent, pinkish purple when mature. Male flowers numerous, pedicellate or sessile. Female flowers subsessile, numerous. Fig. 104.6.

Hedgerows, secondary scrub, forest edges, common in riverine forest and scrub, one record of epiphytic growth; 1000-2400 m. ?EE EW TU GD GJ WU SU HA AR KF; N Somalia, Saudi Arabia, Yemen, Egypt, Sudan, Afghanistan, Pakistan, NW India. *Aweke & Gilbert* 825; *Friis* et al. 163; *Mooney* 7805.

The figs are, according to the collectors, considered a delicacy in many parts of Ethiopia.

1.2. Subgen. FICUS, sect. SYCIDIUM Miq. (1848).
Ficus subgen. *Sycidium* (Miq.) Mildbr. & Burret (1911).

Shrubs or trees; dioecious; sap watery. Leaves ± distichous, subopposite or subverticillate, especially when juvenile ± lobed or almost divided, mostly scabrid. Stipules lateral to semi-amplexicaul, free. Figs often solitary in the leaf-axils or on short spurs on older wood, pedunculate, with 2-4 ± whorled brachts on the peduncle and several lateral bracts on the fig; ostiole round, with several ostiolar bracts visible; interfloral bracts present. Male flowers near the ostiole or scattered over inside of fig; tepals 3-6, ciliolate; stamens 1-3; rudimentary ovary ± developed. Female and gall flowers distinct; tepals 4-6, ciliolate, style in female flowers long, stigma 1. Ripe fig usually with reddish, orange or yellow, soft wall.

1. Leaves opposite or subopposite, apex often tridentate; a shrub to 5 m. **3. F. capreaefolia**
- Leaves always alternate, apex not tridentate; shrub, climbers or trees to 20 m. **2**
2. Lamina with 3-5 pairs of lateral nerves, ± symmetrical, shrubs or small trees. **4. F. exasperata**
- Lamina with (5-)7-12 pairs of lateral nerves, often very symmetrical, shrubs or climbers. **5. F. asperifolia**

3. **F. capreaefolia** *Del.* (1843);
F. antitheophylla Miq. (1847), nom. illeg. superfl. - type: GD/TU, near the Tacazze River, *Galinier* s.n. (MPU holo.).
F. paoli Pampanini (1915).

Shrub or small tree up to 7 m tall, with a trunk up to 0.8 m in diam. Bark grey or greyish brown. Young branches puberulous to hirsute, with brownish bark. Leaves opposite or subopposite, oblong to lanceolate, 4-16.5 x 1-6 cm, papery, base cuneate to slightly rounded, margin entire or subentire, apex acute, acuminate or tridentate; lamina mostly scabrous on both sides, with 6-10 pairs of lateral veins. Petiole up to 1.3 cm long, puberulous. Stipules free, lateral, linear to narrowly lanceolate, up to 1 cm long, long persistent. Figs solitary in the leaf axils, scabrous, on peduncles up to 2 cm long, slender, puberulous; basal bracts, ovate, up to 2 mm long. Fig obovoid, elipsoid or subglobose, up to 3.5 x 1 cm, ± stipitate; ostiole prominent, with clearly visible, puberulous scales. Male flowers numerous, scattered on the inside of the fig, pedicellate, tepals 4, stamen 1. Female flower, tepals 4; ovary stipitate, style long, stigma short, slightly bilobed. Fig. 104.7.

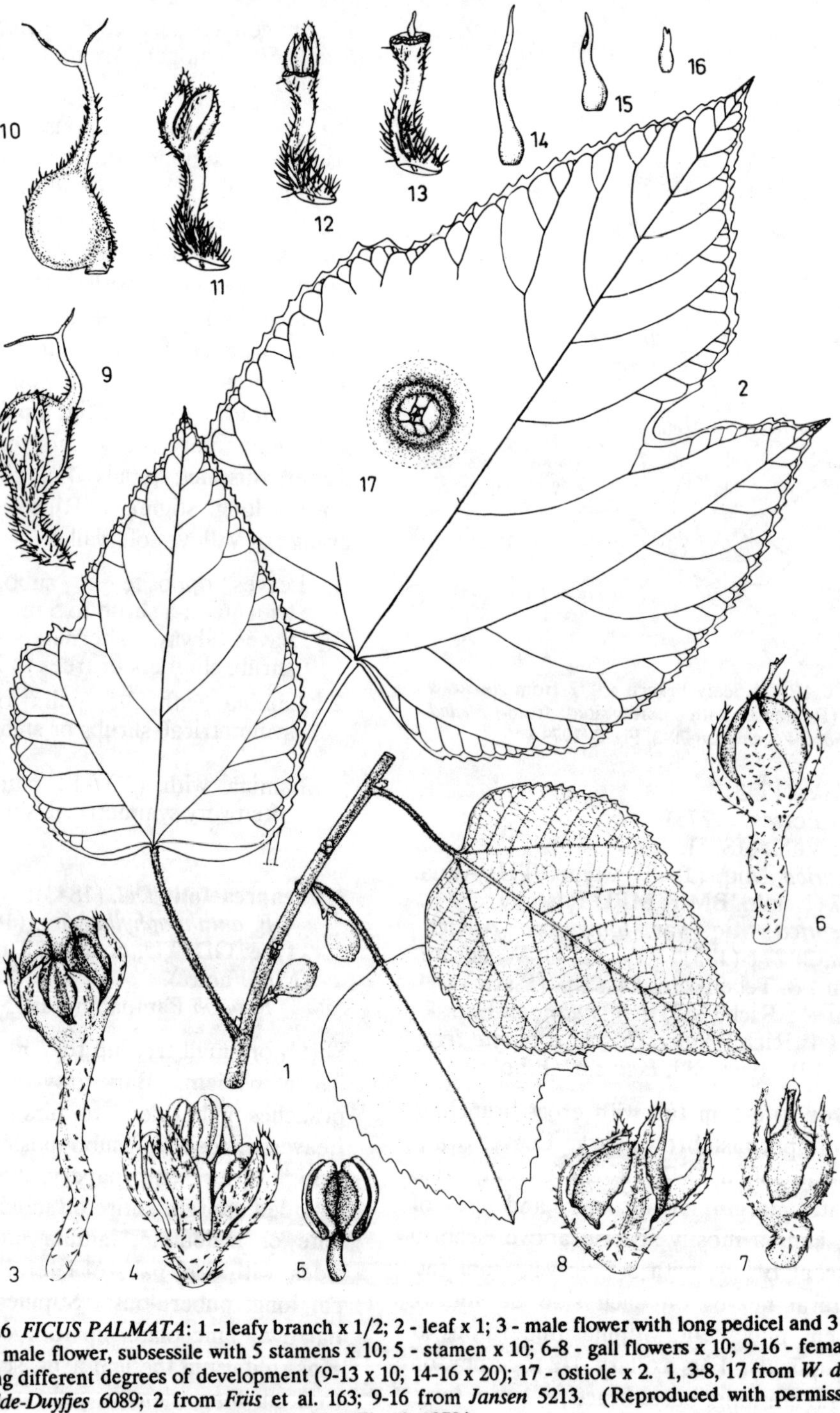

Figure 104.6 *FICUS PALMATA*: 1 - leafy branch x 1/2; 2 - leaf x 1; 3 - male flower with long pedicel and 3 stamens x 10; 4 - male flower, subsessile with 5 stamens x 10; 5 - stamen x 10; 6-8 - gall flowers x 10; 9-16 - female flowers showing different degrees of development (9-13 x 10; 14-16 x 20); 17 - ostiole x 2. 1, 3-8, 17 from *W. de Wilde & de Wilde-Duyfjes* 6089; 2 from *Friis* et al. 163; 9-16 from *Jansen* 5213. (Reproduced with permission from *Meded. Landbouwhogeschool Wageningen* 79-3, Fig 12, 1979.)

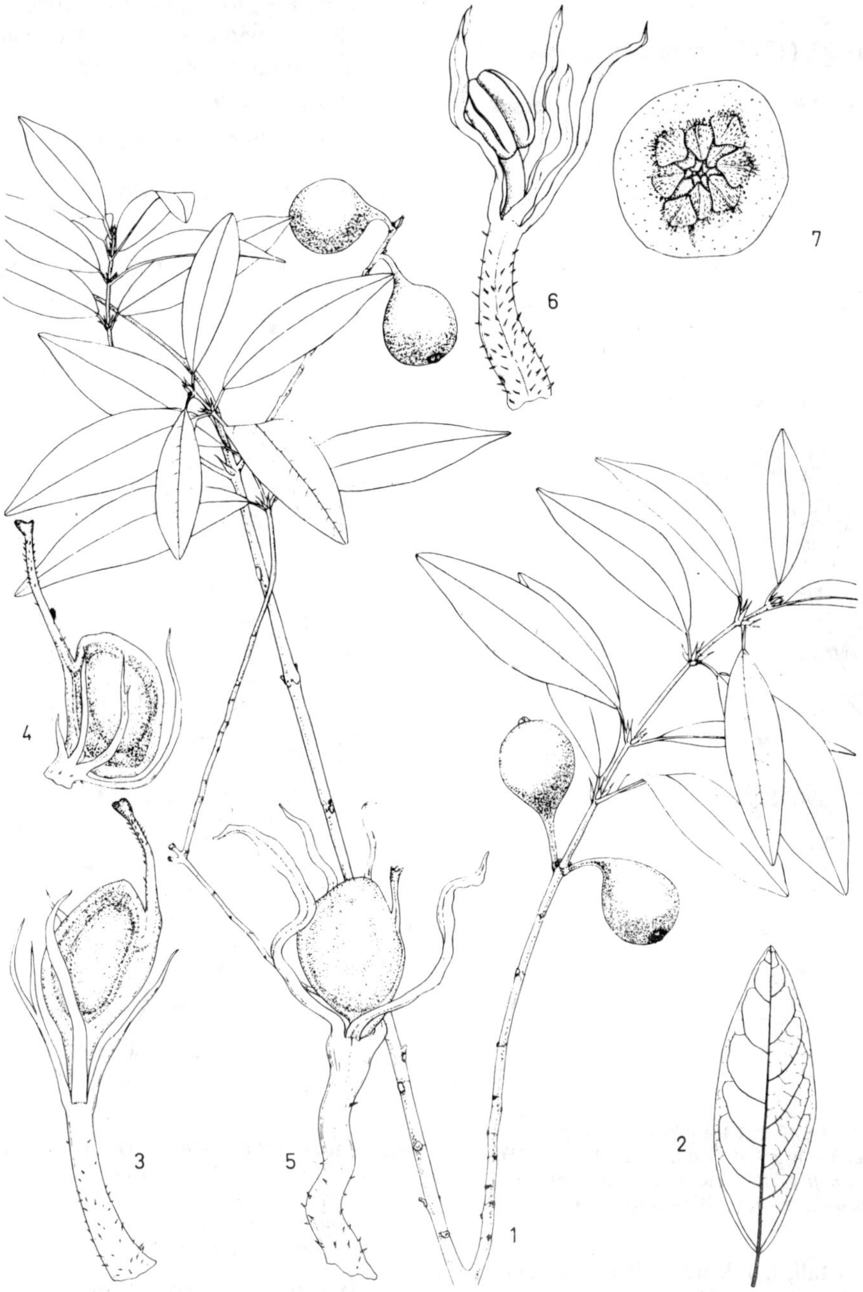

Figure 104.7 *FICUS CAPREAEFOLIA*: 1 - leafy branch with figs x 1/2; 2 - leaf, upper surface x 1/2; 3-4 - female flowers x 10; 5 - gall flower x 10; 6 - male flower x 10; 7 - ostiole x 2. 1 from *Aweke* 696 & *Jansen* 5760; 2, 5-6 from *Aweke* 696; 3-4, 7 from *W. de Wilde & de Wilde-Duyfjes* 7858. (Reproduced with permission from *Meded. Landbouwhogeschool Wageningen* 79-3, Fig. 2, 1979.)

A prominent component of riparian scrub in drier areas, often associated with *Salix subserrata*; 600-2600 m. EE EW TU GD AF SU HA KF GG SD; S Somalia, widespread in tropical Africa, west to Senegal, south to S Afr. Rep. *Mooney* 9095; *De Wilde* 7857; *Burger* 1972.

4. **F. exasperata** *Vahl* (1805).

F. serrata Forssk. (1775), nom. illeg, non L. (1759).

F. scabra Willd. (1801), nom. illeg., non Forssk. (1786).

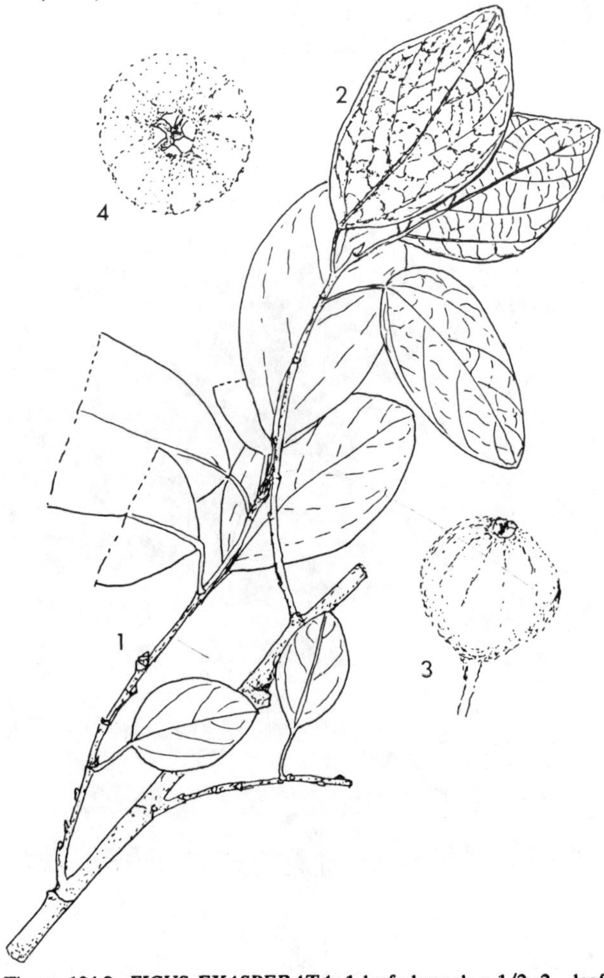

Figure 104.8 *FICUS EXASPERATA*: 1 leafy branch x 1/2; 2 - leaf showing lower surface; 3 - fig x 1; 4 - ostiole x 2. 1-2 from *Friis* et al. 567; 3-4 from *Chaffey* 269. (Redrawn with permission from *Meded. Landbouwhogeschool Wageningen* 79-3, Fig. 5, 1979.)

Shrub or tree to 20 m tall, trunk up to 0.5 m in diam. Bark brownish, slash pale brown. Young branches with scabrous, pale brown bark. Leaves alternate or distichous, ovate to elliptic or obovate, 4-14 x 3-7 cm, papery, base cuneate, rounded, or subcordate, margin entire, irregularly crenate or repand-dentate, apex obtuse, acute or shortly acuminate; juvenile foliage often deeply 3-5 lobed; older foliage very rough or scabrous on both sides, bristly on the nerves below, with 2-5 pairs of mostly opposite lateral nerves, tertiary nerves forming a prominent net below. Stipules free, lanceolate, pointed, *c* 0.5 cm long. Petiole 0.5-2 cm long. Figs 1-2 in the leaf axils, on peduncles 4-7 mm long, scabrous; basal bracts, tringular, spirally arranged on the surface of the fig. Figs ± stipitate, globose to obovoid, up to 1 x 0.8 cm, scabrous; ostiole ± prominent, with many visible bracts. Fig. 104.8.

Secondary scrub and forest edges; 900-2000 m. ?TU GJ KF IL; widespread in tropical Africa, west to Senegal, south to Zambia; Yemen. *Friis* et al. 567; *Chaffey* 1256, 1279.

Leaves used as sand paper.

5. **F. asperifolia** *Miq.* (1848)

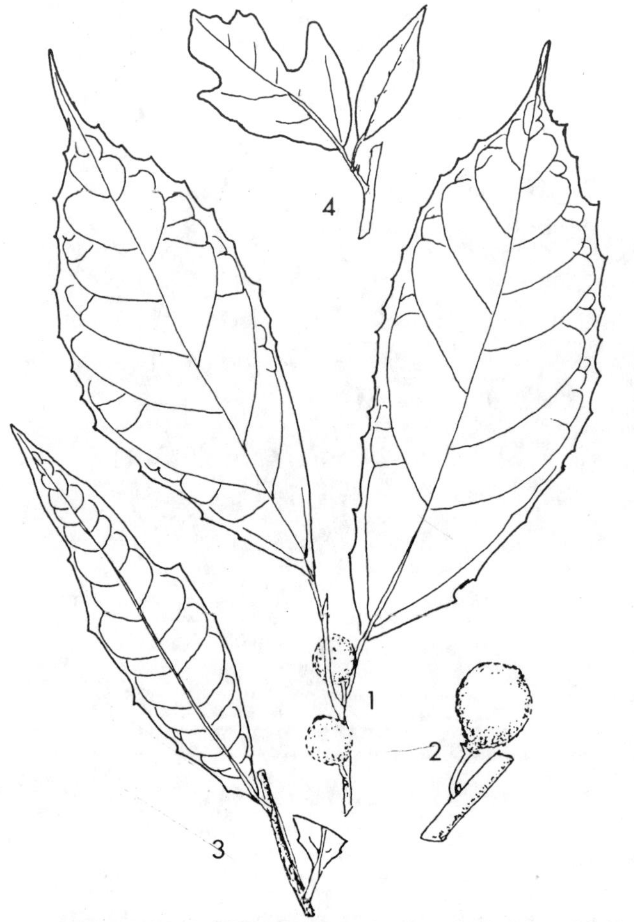

Figure 104.9 *FICUS ASPERIFOLIA*: 1 - branch with leaves and figs x 1; 2 - fig x 3/2; 3-4 - leaves showing variation in shape x 1. 1 from *Letouzey* 7610; 2 from *Raynal J. & A.* 9996; 3 from *Jacques-Félix* 2411; 4 from *Letouzey* 3420. (Redrawn with permission from *Flore du Cameroun*, 28: pl. 40., 1986.)

Shrub, sometimes scrambling to 4 m, or climber to 10 m or more. Stems strigose or hispidulous to almost glabrous. Leaves ± distichous, membranous to somewhat leathery, lanceolate, elliptic or oblong, usually asymmetrical, 3-23 x 1.5-12 cm, base cuneate, rarely rounded, margin with ± coarse teeth or ± unilaterally lobed, apex long acuminate or acute; lamina above scabrous or strigose, below scabrous, basal pair of lateral nerves usually reaching far towards the leaf apex, and anastomosing with the other pairs of lateral nerves to form a submarginal nerve. Petiole 0.5-2 cm long, with same indumentum as the lamina. Stipules free, up to 0.6 cm long, falling quickly. Figs 1-3 in the leaf axils, on peduncles 0.3-1 cm long, scabrous; with 2-4 small bracts scattered on the peduncle, and with 2-4 similar bracts scattered over the surface of the fig. Fig globose, 0.5-3 cm in diam.; ostiole prominent, with clearly visible scales spreading across. Fig. 104.9.

In lowland rain forest, especially along streams; 1000-1200 m. KF; widespread in tropical Africa, west to Senegal, south to Angola and Zambia. *Friis* et al. 3980.

2. Subgen. SYCOMORUS *(Gasparrini) Mildbr. & Burret* (1911).
Sycomorus Gasparrini (1845); *Ficus* sect. *Sycomorus* (Gasparrini) Miq. (1867).

Trees, rarely shrubs; monoecious; sap milky. Leaves spirally arranged, rarely ± distichous, margin mostly dentate, crenate or ± sinuate; stipules fully amplexicaul, free. Figs axillary or, more often, on ± branched, leafless branchlets on the older wood, often from the truk or the large branches, pedunculate; basal bracts 3, whorled; lateral bracts absent; ostiole rounded, with several bracts visible, only the lower ones descending into the fig; interfloral bracts present. Male flowers near the ostiole, initially surrounded and covered by 2(-3) bracts; perianth deeply 3-lobed; stamens (1-)2-3. Female flowers and gall flowers rather distinct; tepals 2-4(-6), irregularly shaped, free or united; style in female flowers long, stigmatic branch 1. Ripe figs with orange to yellow, fleshy walls.

1. Leaves elliptic to suborbicular, usually broadest at or just below the middle, with rounded or acute, rarely shortly acuminate apex; figs axillary or in leafless panicles on the main stem or older branches. 2
- Leaves ovate, broadest near the base, with acute or acuminate, sometimes drawn out apex; figs always in leafless panicles on the main stem or older branches. **8. F. sur**
2. Leaves and stipules with long brown hairs; tree with tall trunk in lowland rain forest. **7. F. mucuso**
- Leaves and stipules with different indumentum; trees with short trunk in woodland, upland rain forest or riverine forest. 3
3. Leaves strigose, entire or rarely repand-sinuate; figs axillary or in leafless panicles on the main stem or older branches. **6. F. sycomorus**
- Leaves glabrous or puberulous, coarsely sinuate-dentate, not strigose; figs always axillary. **9. F. vallis-choudae**

6. **F. sycomorus** *L.* (1753).

A large tree up to *c* 30 m tall, buttressed, with a trunk up to 3.5 m in diam.; crown spreading. Bark greyish-brown, grey or yellowish; slash pale brown, yellowish or pinkish. Young branches with brown bark, puberulous to pubescent or lanate, often ± scaly in herbarium material. Leaves almost distichous, ± leathery, broadly ovate, elliptic or sub-orbicular, 3-15 x 2-10.5 cm, base rounded to narrowly cordate, margin subentire to crenulate, apex rounded, acute or short acuminate; lamina dull, slightly scabrous or smooth above, scabrous or smooth, often pubescent below, midrib with 4-8 pairs of lateral nerves. Stipules covering buds as scarious, lanate to hirsute, grey or brown scales, falling quickly, leaving a circular fringe of whitish or brownish hairs. Figs 1-2 together, either on clusters of leafless branchlets on the trunk or older branches or in the axil of normal leaves, on peduncles up to 1.5 cm long; basal bracts, ovate to triangular, 3-4 mm long, puberulous. Fig obovoid or subglobose, often ± stipitate, 1.8-2.5(-5) x 1-2 cm, puberulous to tomentose; ostiole prominent, with 3-5 clearly visible, tringular bracts. Male flowers with 1-3 stamens.

Two subspecies in Ethiopia, impossible to separate on sterile material; figs of both subspecies edible:

1. Figs 1-2 together in the leaf axils on young branches, often ripening after the leaves have fallen. subsp. **gnaphalocarpa**.
- Figs on special leafless, clustered, pendent branches on the trunk or older branches. subsp. **sycomorus**.

Subsp. **sycomorus**
F. chanas Forssk. (1775).
Sycomorus antiquorum Gasparrini (1845).
Sycomorus rigida Miq. (1848).
Fig. 104.10.

River and lake margins, woodland, wooded grassland, evergreen bushland, forest edges and clearings, "coffee forest"; 500-2000 m. EE EW TU WU GD GJ SU HA IL KF GG SD; widespread in E Africa, north to Egypt, south to S Africa Rep., Yemen. *Friis* et al. 1929; *Burger* 2444.

Subsp. **gnaphalocarpa** *(Miq.) C.C. Berg* in *Adansonia* **20,3**: 271 (1980);
Sycomorus gnaphalocarpa Miq. (1848); *Ficus gnaphalocarpa* (Miq.) A. Rich. (1851) - type: TU, Djeladjeranne, *Schimper* II:847 (L holo. K P iso.).
Sycomorus trachyphylla Miq. (1848); *Ficus trachyphylla* (Miq.) Miq. (1848).

In the same habitats as the typical subspecies; 500-1900 m. TU GD GJ SU HA IL KF SD. Widespread in tropical Africa, west to Senegal, south to S African Rep. *Friis* et al; 128; *Ash* 961; *Westphal* 2588.

Note to species as a whole: The two subspecies proposed by Berg overlap in Ethiopia, where both seem to occur together in the same habitats. Much herbarium material, however, is not referable to subspecies due to lack of figs, and more field work is therefore necessary in order to establish the distribution and ecology of the two taxa in Ethiopia.

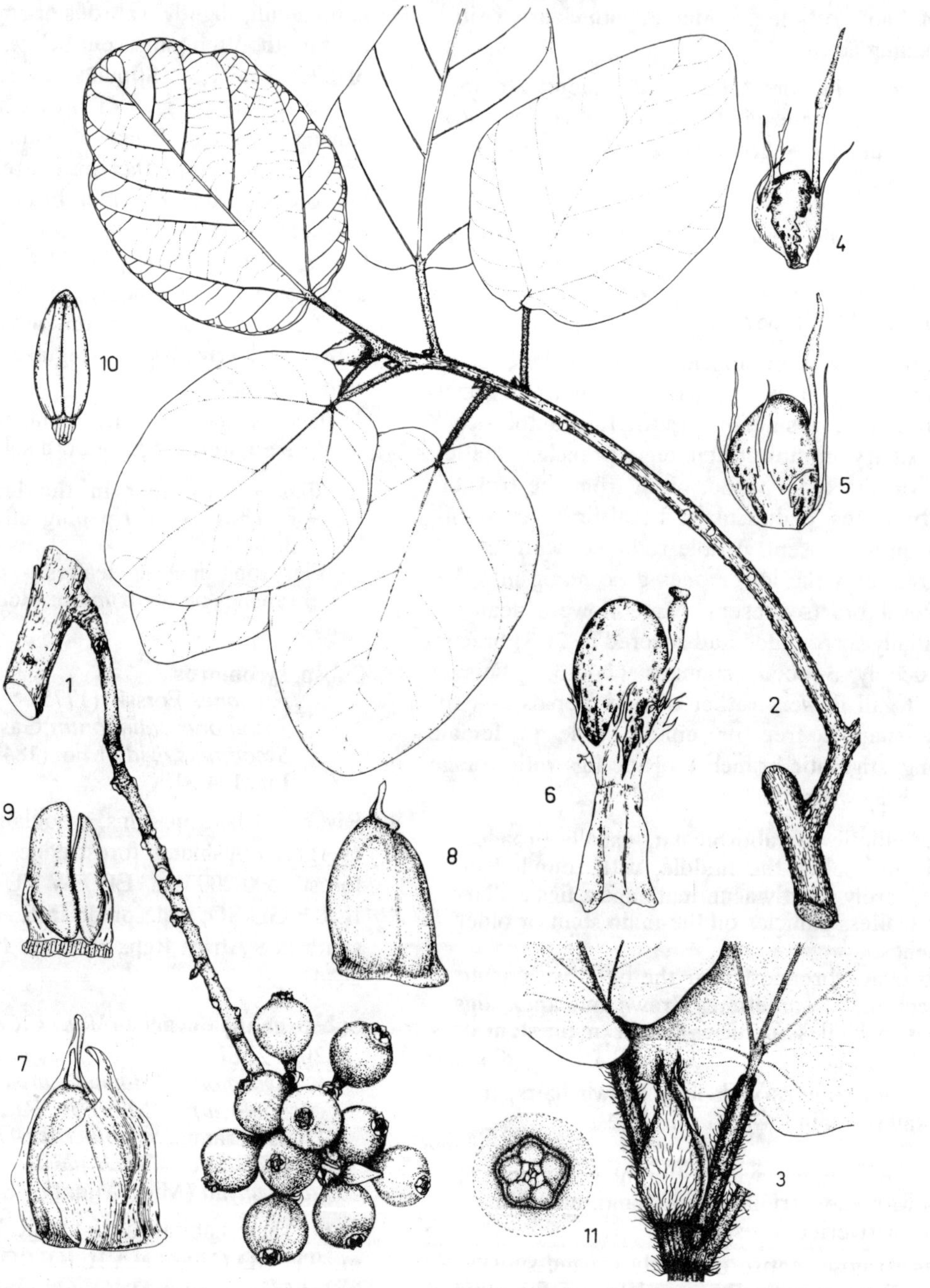

Figure 104.10 *FICUS SYCOMORUS* subsp. *SYCOMORUS*: 1 - portion of inflorescence x 1/2; 2 - leafy branch x 1/2; 3 - stipules x 1; 4-5 - female flowers x 10; 6 - gall flower x 10; 7 - male flower x 10; 8 - male flower with 2 tepals removed to show stamens x 10; 10 - stamen, front view x 10; 11 - ostiole x 1. 1-3, 11 from *W. de Wilde & de Wilde-Duyfjes* 6218; 4-10 from *J. De Wilde* 7372. (Reproduced with permission from *Meded. Landbouwhogeschool Wageningen* 79-3, Fig. 18a, 1979.)

Figure 104.11 *FICUS MUCUSO*: 1 - leafy branch x 1; 2 - detail of undersurface of leaf; 3 - branch with figs x 1. 1-2 from *Polhill* 751; 3 from *Hallé N.* 1790. (Redrawn with permission from *Flore du Cameroun*, 28: pl. 43., 1986.)

7. **F. mucuso** *Ficalho* (1884).
F. sidifolia Hiern (1900).

Tree to 35 m tall, with a tall, clear trunk devoid of side branches, except in the crown; mostly with large buttresses. Bark yellowish brown, smooth. Stems glabrescent; younger stems with long brownish or whitish hairs at the nodes. Leaves suborbicular to elliptic or obovate, 6-15 x 4-12 cm, somewhat leathery, base cordate to deeply cordate with the lobes ± overlapping, margin entire to crenulate, apex rounded, obtuse or minutely acuminate, scabrous; lamina with scattered long hairs on the nerves above, puberulous to hirsute, with long brown hairs on the nerves below, with 3-6 pairs of lateral nerves. Petiole 2-8 cm long, puberulous with long brown hairs. Stipules 1-2 cm long, puberulous with ± appressed long brown hairs. Figs single or paired at the nodes of special, leafless branchlets on the main branches or upper part of the trunk; peduncle 1-2 cm long, puberulous, basal bracts broadly ovate, up to 4 mm long, free or united at the base, puberulous; figs obovate to obovoid, up to 5 cm in diam. (3 cm when dry), densely puberulous or pubescent, orange to red at maturity. Male flowers with 2 stamens. Fig. 104.11.

In lowland rain forest and coffee plantations; 900-1300 m. KF IL; widespread in central Africa, west to Guinea Bissau, south to Angola and Tanzania. *Chaffey* 1290; *Friis* et al. 3860, 4020.

8. **F. sur** *Forssk.* (1775);
Sycomorus sur (Forssk.) Miq. (1848).
F. capensis Thunb. (1786); *Sycomorus capensis* (Thunb.) Miq. (1848).
F. panifica Del. (1843); *Sycomorus panifica* (Del.) Miq. (1848) - type: TU, Gotho, *Galinier* s.n. (MPU, holo.).
Sycomorus riparia Miq.(1848); *Ficus riparia* (Miq.) A. Rich. (1851) - type: TU, Mandel, *Schimper* III:1585 (L, holo., K, P, iso.).
F. mallotocarpa Warb. (1894); *F. capensis* Thunb. var. *mallotocarpa* (Warb.) Mildbr. & Burret in *Engl. Bot. Jahrb*. **46**: 198 (1911).

Tree to 30 m tall, with trunk up to 1 m or more in diam. and large buttresses. Bark brownish to grey; slash greenish to pink. Young branches densely pubescent to glabrescent, with brown bark. Leaves alternate, almost distichous, ovate to elliptic or oblong to lanceolate, 4-20 x 2-12 cm, base rounded, truncate or subcordate, margin entire or serrate to repand-dentate (on shaded

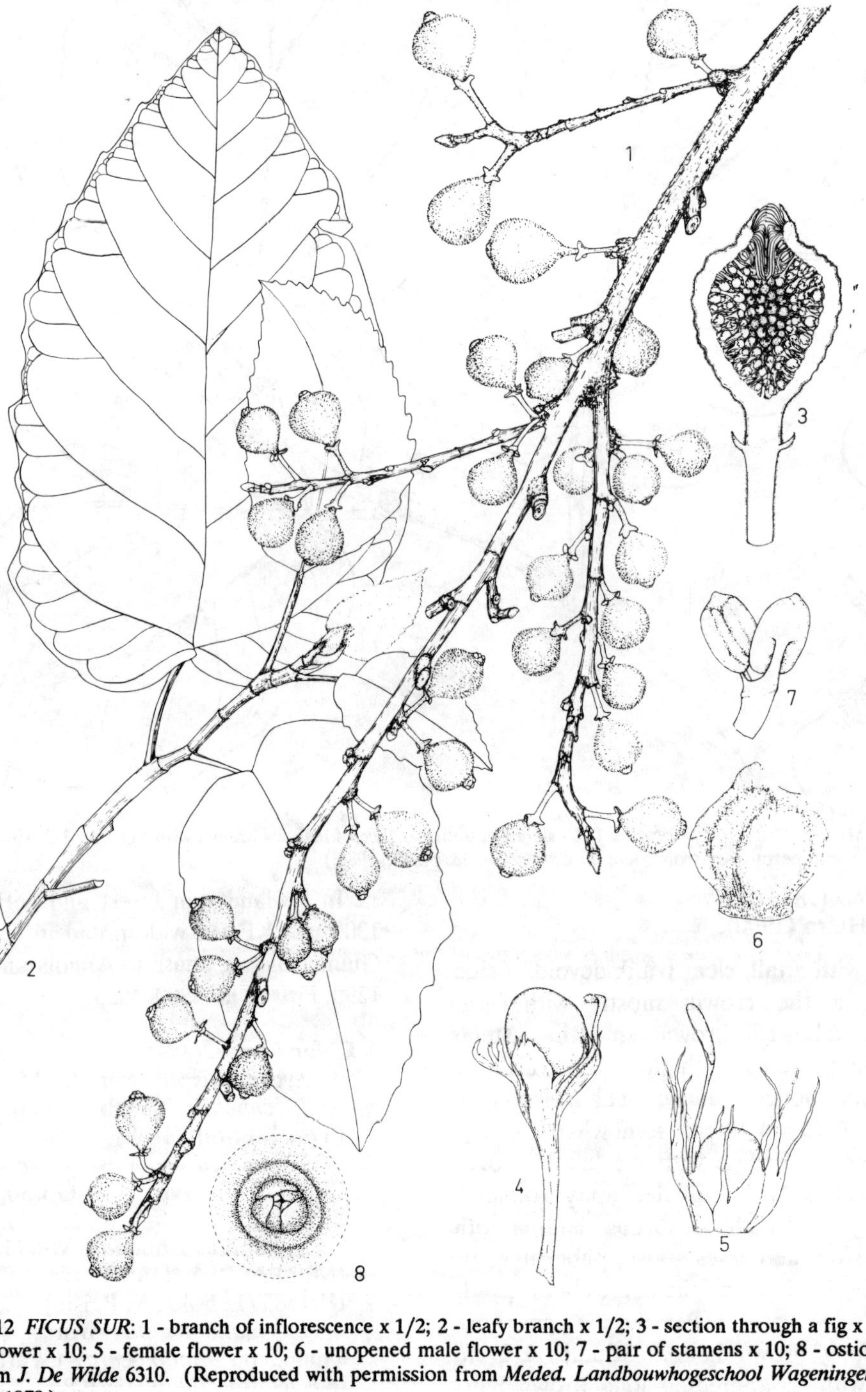

Figure 104.12 *FICUS SUR*: 1 - branch of inflorescence x 1/2; 2 - leafy branch x 1/2; 3 - section through a fig x 3/2; 4 - gall flower x 10; 5 - female flower x 10; 6 - unopened male flower x 10; 7 - pair of stamens x 10; 8 - ostiole x 2. All from *J. De Wilde* 6310. (Reproduced with permission from *Meded. Landbouwhogeschool Wageningen* 79-3, Fig. 10, 1979.)

branches or suckers), apex acuminate or acute; lamina dull, almost glabrous above, sparsely puberulous below, with 5-7 pairs of lateral nerves. Petiole up to 7 cm long, puberulous to pubescent. Stipules falling quickly, villous, up to 1 cm long. Figs on clustered, ± branched panicles up to 70 cm long, hanging from the trunk or older branches, or rarely 1-2 together in the leaf axils, on peduncles 0.5-1 cm long; basal bracts ovoid, pubescent. Figs ± stipitate, up to 2 x 1.5 cm, densely puberulous to lanate, orange-grey when ripe, in dried material sometimes with peeling periderm ("scaly figs"); ostiole prominent, with clearly visible bracts. Male flowers with 1-2 stamens. Fig. 104.12.

Upland rain forest and riverine forest, mountain grassland or secondary scrub, mainly at streams; 1400-2500 m. EW TU WU GD GJ WG SU HA AR IL KF SD. Widespread in tropical Africa, west to Senegal and Cape Verde Islands., south to Cape Prov. of S Afr. Rep., Yemen. *Aweke & Gilbert* 954; *Friis* et al. 246; *Burger* 1490.

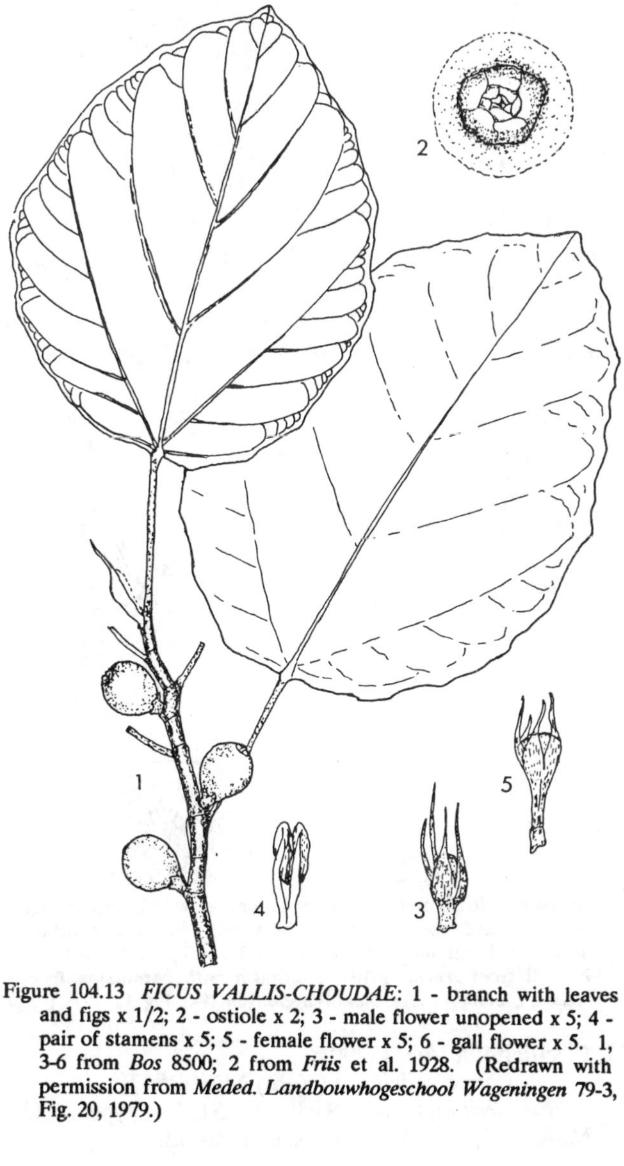

Figure 104.13 *FICUS VALLIS-CHOUDAE*: 1 - branch with leaves and figs x 1/2; 2 - ostiole x 2; 3 - male flower unopened x 5; 4 - pair of stamens x 5; 5 - female flower x 5; 6 - gall flower x 5. 1, 3-6 from *Bos* 8500; 2 from *Friis* et al. 1928. (Redrawn with permission from *Meded. Landbouwhogeschool Wageningen* 79-3, Fig. 20, 1979.)

9. **F. vallis-choudae** *Del.* (1843)
- type: GD, Semien, Chouda valley, *Galinier* s.n. (MPU holo.).
Sycomorus schimperianus Miq. (1848), nom. illeg., includes holotype of *F. vallis-choudae*.
Ficus schweinfurthii Miq. (1867) - type: Sudan/GD, Gallabat, *Schweinfurth* 547 (L holo., BM K P iso.).

A large tree to 25 m, with trunk up to 0.3 (?) m in diam. Bark grey, with crimson slash. Young branches hollow, glabrous or pubescent, with light brown bark. Leaves distichous to spirally arranged, broadly ovate to almost orbicular, 15-25 x 13-18 cm, base broadly cuneate, rounded or cordate base, margin entire, repand, or repand-dentate, apex obtuse, acute or very short acuminate; lamina dull, glabrous and smooth on both surfaces, or rarely puberulous or pubescent on the nerves beneath, with 3-5 pairs of lateral nerves. Petiole 2-13 cm long, minutely puberulous. Stipules minutely puberulous, up to 2 cm long. Figs single in the leaf axils, peduncles thick, up to 0.5 cm long; basal bracts with fused, persistent base. Fig short obovoid to globose, up to 4 x 3 cm, whitish or greyish puberulous; ostiole prominent, with clearly visible bracts. Male flowers with 2 stamens. Fig. 104.13.

Closely associated with riverine forest or scrub; two doubtful records from "coffee forest" or upland rain forest of the SW; 700-1900 m. EW GD SU WG IL KF SD; widespread in tropical Africa, west to Guinean Rep., south to Mozambique. *Aweke* 426; *Friis* et al. 568, 1928.

Aweke, op.cit., p. 87 (1979) records this species from HA. The author has been unable to confirm this record.

3. Subgen. PHARMACOSYCEA (Miq.) Miq. (1867).
Pharmacosycea Miq. (1847).

Trees, monoecious; sap watery, scanty; lenticels concentrated in the uppermost part of the internodes. Leaves spirally arranged, entire (or lobed when juvenile), scabrid; stipules fully amplexicaul, free. Figs solitary or in pairs in the leaf axils, pedunculate; basal bracts 3; lateral bracts absent or rarely present; ostiole circular with 3 bracts visible, only the lower bracts descrnding into the fig; interfloral bracts (?). Male flowers near the ostiole; tepals 2-3, ± united, glabrous; stamen 1; rudimentary ovary ± developed. Female flowers and gall flowers rather distinct; tepals 2-3, united, glabrous, style 1, stigma-branches 2, equal or subequal, linear. Ripe fig with firm wall, yellow to orange.

10. F. **dicranostyla** *Mildbr.* (1911).

Tree to 10 m with spreading crown, trunk up to 1 m or more in diam. Bark greyish brown, rough and fissured; slash pale brown with paler streaks. Leaves papery, drying pale green, ovate, oblong or elliptic-oblong, 6-12.5 x 3-6.5 cm, base rounded, ± asymmetric, margin entire, apex shorly acuminate; lamina glabrous or slightly scabrous, with 6-9 pairs of lateral nerves, which are prominent on both sides, midrib with weakly developed gland at base below. Petiole 1.5-3.5 cm long, pubescent to densely pilose. Stipules *c* 7 mm long, falling quickly. Receptacles axillary, solitary, on peduncle *c* 1 cm long, puberulous; basal bracts ovate to triangular, puberulous. Fig subglobose, *c* 2 cm in diam., puberulous, not stipitate; ostiole ± prominent. Fig. 104.14.

Deciduous woodland; 500-600 m. IL; widespread in W tropical Africa to Senegal. *Chaffey* 625; *Jansen & Aweke* 5087; *Friis* et al. 2501.

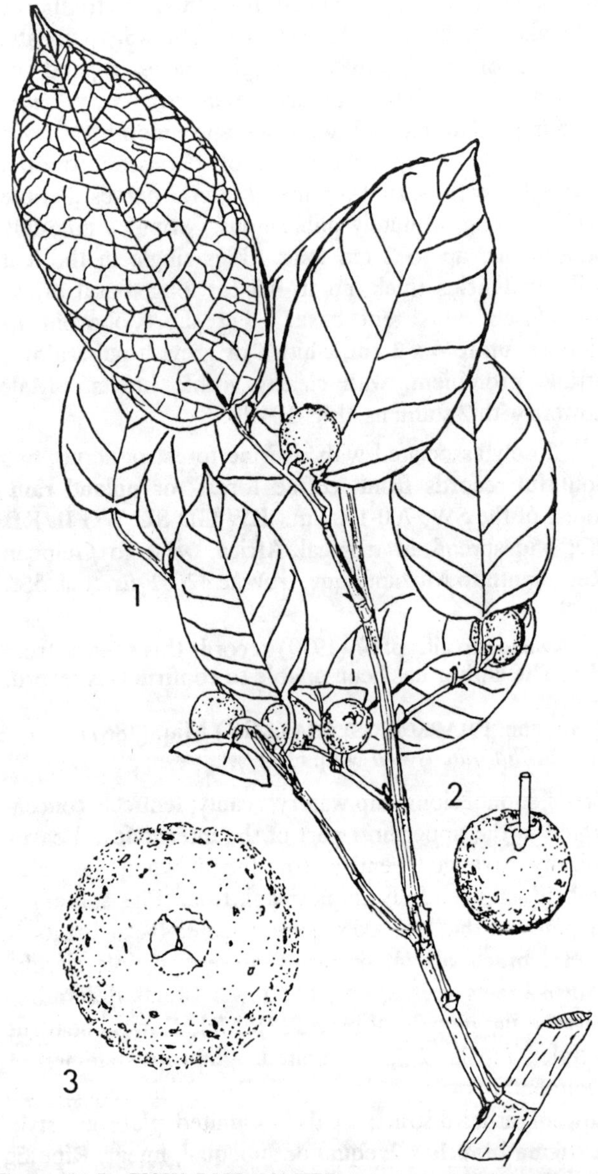

Figure 104.14 *FICUS DICRANOSTYLA*: 1 - branch with leaves and figs x 1/2; 2 - fig showing basal bracts x 1; 3 - ostiole x 2. All from *Jansen & Aweke* 5087. (Redrawn with permission from *Meded. Landbouwhogeschool Wageningen* 79-3, Fig. 4, 1979.)

4.1. Subgen. UROSTIGMA (Gasparrini) Miq. (1867) sect. UROSTIGMA
Urostigma Gasparrini (1844).

Trees or shrubs, often growing over rocks; monoecious; sap milky. Leaves spirally arranged, margin almost entire; stipules fully amplexicaul, free. Figs in pairs, or occasionally more together, in the leaf axils or just below the current leaves, pedunculate to subsessile; basal bracts 3; lateral bracts absent; ostiole circular, with 3 bracts visible; only the lower bracts descend into the fig; interfloral bracts absent ot few. Male flowers usually only near ostiole; tepals 3-4, free; stamen 1, rudimentary ovary sometimes ± devellopped. Female and gall flowers distinct; tepals 3-4(-7), free; stigma 1, filiform; ripe fig with soft wall, red pink or purplish.

1. Leaves lanceolate, glabrous, less than 3 cm wide; figs sessile or subsessile. **11. F. salicifolia**
- Leaves ovate to elliptic, scabrous, pillose or glabrous, 2-7 cm wide; figs sessile or pedunculate. **12. F. ingens**

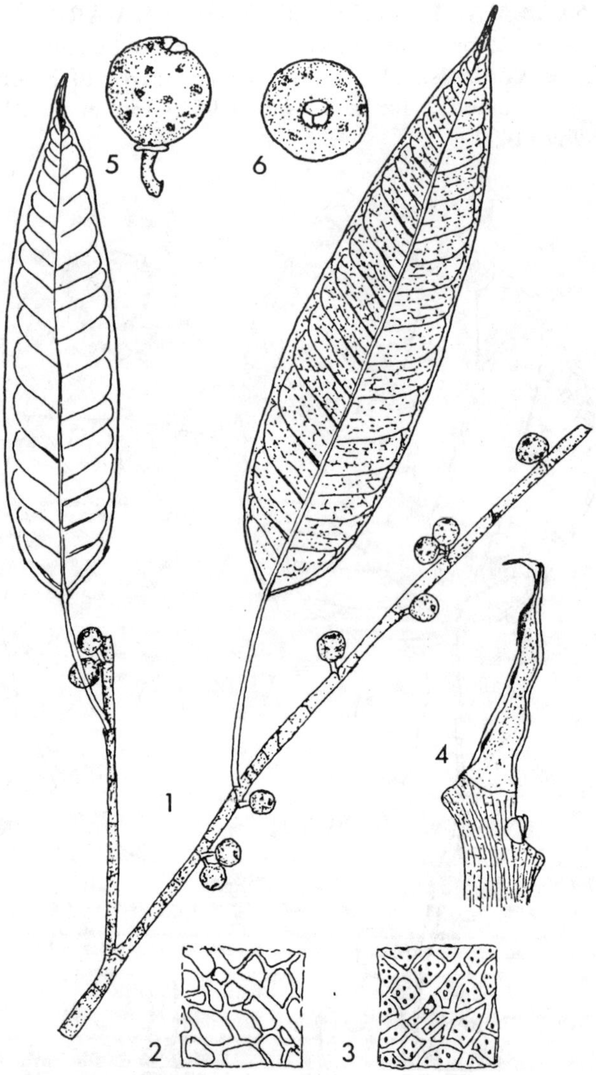

Figure 104.15 *FICUS SALICIFOLIA*: 1 - branch with leaves and figs x 1/2; 2 - leaf, upper surface detail x 5; 3 - leaf, lower surface detail x 5; 4 - stipules enclosing bud x 3; 5 - fig x 2; 6 - ostiole x 2. All from *Aweke* 1016. (Redrawn with permission from *Meded. Landbouwhogeschool Wageningen* 79-3, Fig. 16, 1979.)

11. **F. salicifolia** *Vahl* (1790);
Urostigma salicifolia (Vahl) Miq. (1847).
F. neriifolia A. Rich. (1851) - type: TU, Maigoigoi, *Quartin Dillon* s.n. (P holo.).

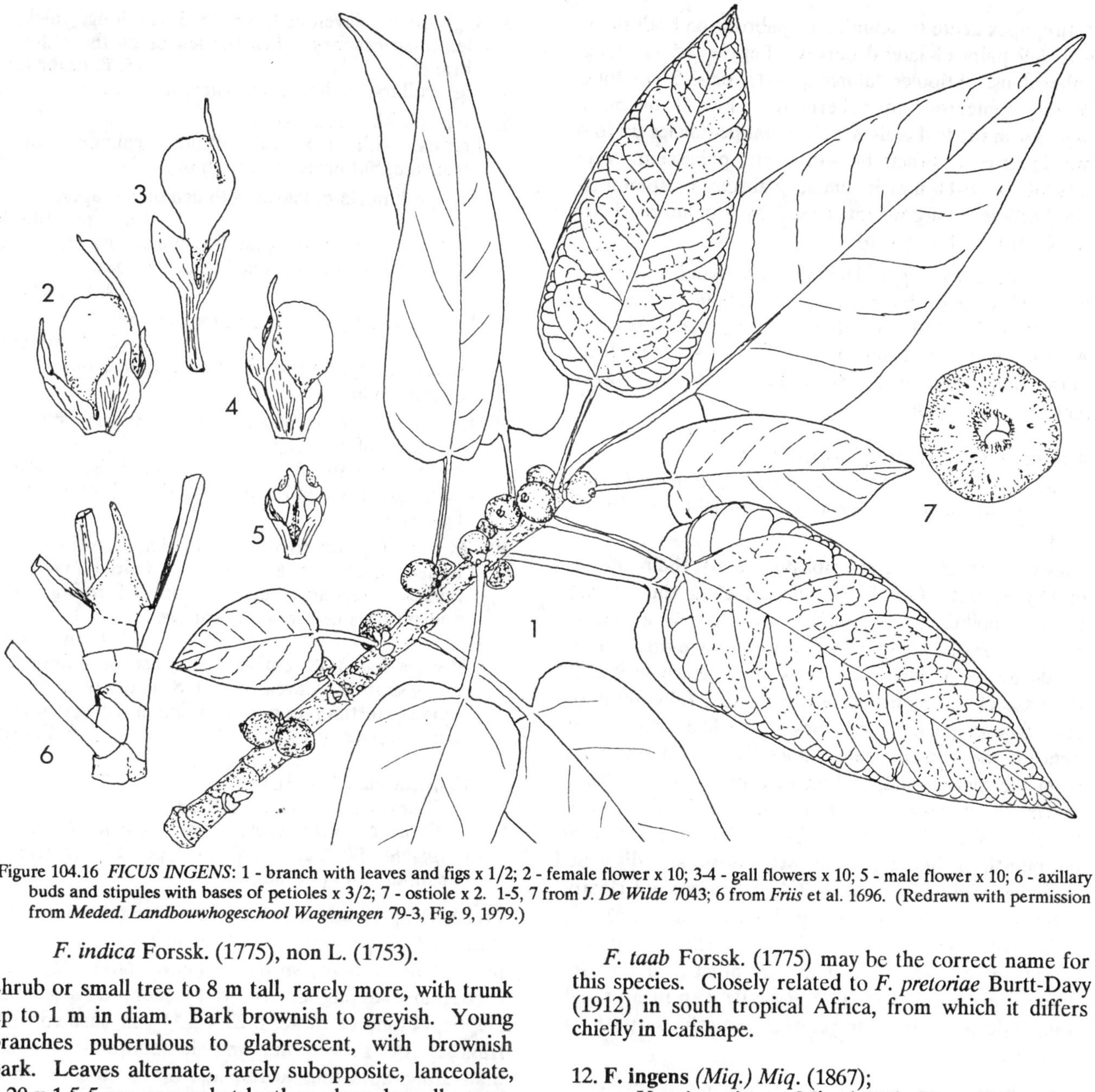

Figure 104.16 *FICUS INGENS*: 1 - branch with leaves and figs x 1/2; 2 - female flower x 10; 3-4 - gall flowers x 10; 5 - male flower x 10; 6 - axillary buds and stipules with bases of petioles x 3/2; 7 - ostiole x 2. 1-5, 7 from *J. De Wilde* 7043; 6 from *Friis* et al. 1696. (Redrawn with permission from *Meded. Landbouwhogeschool Wageningen* 79-3, Fig. 9, 1979.)

F. indica Forssk. (1775), non L. (1753).

Shrub or small tree to 8 m tall, rarely more, with trunk up to 1 m in diam. Bark brownish to greyish. Young branches puberulous to glabrescent, with brownish bark. Leaves alternate, rarely subopposite, lanceolate, 5-20 x 1.5-5 cm, somewhat leathery, base broadly cuneate, rounded or rarely slightly cordate, margin entire, apex long acuminate; lamina often shiny above, glabrous on both sides, with 8-15 pairs of lateral nerves. Stipules broadly lanceolate, *c* 1.5 cm long. Petiole 2-7 cm long, slender, glabrous. Figs 1-2 in the leaf axils, on peduncles 2-4 mm long, puberulous; basal bracts ovate, puberulous. Figs globose, 4-5 mm in diam., not stipitate; ostiole ± prominent, with 3-4 ovate bracts visible from the outside. Fig. 104.15.

On rocks, rocky slopes, and lava flows; on granite, lava or limestone, often on rocks in riverine scrub or along wadis; 900-2300 m. EE EW TU GD GJ SU HA; widespread in tropical Africa, west to Ivory Coast, south to S Afr. Rep.; also widespread on the Arabian peninsula. *Aweke* 1016; *De Wilde* 6506; *Friis* et al. 2606.

F. taab Forssk. (1775) may be the correct name for this species. Closely related to *F. pretoriae* Burtt-Davy (1912) in south tropical Africa, from which it differs chiefly in leafshape.

12. **F. ingens** *(Miq.) Miq.* (1867);

Urostigma ingens Miq. (1847); *Ficus schimperiana* A. Rich. (1850), nom. illeg. superfl. - type: TU, Djeladjeranne, *Schimper* III:1771 (L holo. K iso.).

Urostigma xanthophylla Miq. (1847); *F. xanthophylla* (Miq.) Martelli (1886) - type: Ethiopia, without loc., *Schimper* II:943 (L holo., K iso.).

F. ingentoides Hutch. (1915) - type: EW, Acrour, *Schweinfurth & Riva* 1687 (K lecto., BR isolecto.).

F. lutea auct., *non* Vahl *sensu* C. C. Berg in *Kew Bull.* **36**: 597 (1981)), Aweke, op.cit.: 37-41 (1979), and numerous other authors on the Ethiopian flora.

Shrub or tree to *c* 15 m tall, with large buttresses, trunk up to 0.4 m or more in diam. Bark grey, smooth, slash pale brown to reddish. Young branches pale brown, glabrous to tomentose. Leaves alternate, leathery, usually drying pale green, ovate to oblong-lanceolate, 6-16 x 2-7 cm, base cordate, truncate or rounded, margin entire, apex acute to acuminate; glabrous on both sides, with 6-9 pairs of lateral nerves. Petiole 2-5 cm long, puberulous. Stipules falling quickly, 0.5-3.5 cm long, densely tomentose to puberulous. Figs 1 or more together in the leaf axils, sessile or on peduncles up to 4 mm long; basal bracts broadly ovate, pubescent. Figs subglobose, 6-11 mm in diam., glabrous or tomentellous; ostiole ± prominent, closed by 3 or more clearly visible bracts. Fig. 104.16.

Rocks, rocky slopes, deciduous woodland, evergreen bushland, edge of dry montane forest; 600-2050 m. EE EW TU GD GJ SU HA IL SD; widespread in tropical Africa, west to Gambia, south to S Afr. Rep. (Transvaal); Yemen, Saudi Arabia. *Ash* 316, *Aweke* 559; *Friis* et al. 1696.

4. Subgen. UROSTIGMA *(Gasparrini) Miq.*,
2. sect. GALOGLYCHIA *(Gasparrini) Endlicher* (1847).
Ficus subgen. *Bibracteatae* Mildbr. & Burret (1911).

Trees or shrubs, hemi-epiphytes (start their life as epiphytes, but often finish by strangling their hoast tree) or epilithic; sap milky. Leaves spirally arranged, rarely ± distichous or subopposite. Leaf margin entire or almost so; a glandular spot is often clearly visible at the base of the midrib beneath; stipules fully amplexicaul, free or partly fused. Figs in the leaf axils or just beneath the current leaves, sometimes also on spurs from the older wood, pedunculate or subsessile; basal bracts 2(-3); lateral bracts absent; ostiole slit-shaped (rarely + shaped, all ostiolar bracts descending into the fig; interfloral bracts present. Male flowers ± dispersed over the inside of fig; tepals 2-4 or number not clearly defined, free or united, glabrous; stamen 1(-2); rudimentary ovary not develloped. Female and gall flowers ± different; tepals 2-4 or number not clearly defined, glabrous; style 1(-2). Ripe fig with firm or soft wall, reddish, orange or yellow, greenish, purplish or brownish.

1. Calyptra large, long persistent, leaving an obvious circular scar round the base of the pedunculate figs. **13. F. ovata**
- Calyptra insignificant, soon lost. 2
2. Leaves with a large area of glandular tissue (yellow in fresh material, dark brown or black in dried specimens) on the lower side of the midrib near the base of the blade. 3
- Leaves without such glandular tissue. 5
3. Figs sessile or subsessile. **20. F. vasta**
- Figs pedunculate. 4
4. Leaves cordate, petiole glabrous or almost so; figs glabrous, mottled; peduncles glabrous, basal bracts often reduced to an oblique rim. **19. F. abutilifolia**
- Leaves elliptic or elongate, petiole softly puberulous; figs densely puberulous to glabrous, not mottled; peduncle densely pubescent, basal bracts mostly 2. **18. F. platyphylla**
5. Figs on ± developed, up to 3 cm long, thick, leafless branches below the leaves on the older branches. **16. F. umbellata**
- Figs axillary, in the axils of current leaves. 6
6. Leaves with cordate base. 7
- Leaves with cuneate, obtuse, rounded or truncate, but never cordate base. 8
7. Figs pedunculate; leaves with acuminate apex. **17. F. populifolia**
- Figs sessile or subsessile (peduncle less than 3 mm); leaves with rounded or acute apex. **14. F. glumosa**
8. Leaves oblong, *c* 3 times as long as wide. **23. F. ruspolii**
- Leaves ovate, obovate or elliptic, up to 2 times as long as wide. 9
9. Leaves pubescent, at least on the midrib below, surface often slightly rough; figs shortly pedunculate to subsessile. **22. F. hochstetteri**
- Leaves glabrous and quite smooth on both sides; figs sessile. 10
10. Mature figs up to 0.5 cm in diam., ostiole 2-lipped, linear; leaves with 12-25 pairs of lateral nerves, tertiary nerves not forming a conspicuous net on lower surface. **21. F. thonningii**
- Mature figs 1-1.5 cm in diam., ostiole 4-lipped, cross-shaped; leaves with 6-8 pairs of lateral nerves, tertiary nerves forming a conspicuous net on lower surface. **15. F. lutea**

13. **F. ovata** *Vahl* (1805).
F. brachypoda Hutch. (1915).
F. brachypoda Hutch. var. *scoana* Chiov. in *Malpighia* **34**: 527 (1937) - type: SU, Kachissy, *Taschdjian* 145 (FT holo.).

Tree with spreading crown, up to *c* 15 m tall. Bark grey, slash red to pale brown. Young branches minutely puberulous, with brown bark. Leaves alternate, thinly herbaceous, drying brown to blackish, membranous, elliptic, oblong or ovate, 6-25 x 3-11 cm, base cuneate, truncate or subcordate, margin entire or slightly undulating, apex long acuminate, tip up to 1 cm long; lamina glabrous above, pubescent beneath, at least on the nerves, leaf base 3-nerved, midnerve with 7-14 pairs of lateral nerves. Figs in pairs in the leaf axils, sessile or on peduncles up to 3 mm long, sparsely puberulous, protected by a calyptra when young; basal bracts several, semi-orbicular, pubescent, basally united. Calyptra conical, up to *c* 1 cm long, persisting for some time, leaving circular scar on stem round the base of the peduncles when shed. Figs subglobose, 2-3 cm in diam., sparsely puberulous to glabrous; ostiole small, prominent, closing bracts hardly visible from the outside. Male flowers with 2 stamens. Fig. 104.17.

Riverine forest, upland rain forest, evergreen bushland, sometimes left in cultivations following forest clearing; 1300-2000 m. EW GD GJ WG KF SD GG; widespread in tropical Africa, west to Senegal, south to

Zambia and Angola. *Friis* et al. 531; *Aweke & De Wit* 1531; *Pichi-Sermolli* 2390.

Thomerson 726 from IL, cited as this species by Aweke in op. cit., p. 48 (1979) is *F. lutea* sensu str.

Figure 104.17 *FICUS OVATA*: branch with a leaf and axillary figs x 1 from *Letouzey* 7681. (Redrawn with permission from *Flore du Cameroun*, 28: pl. 83., 1986.)

14. **F. glumosa** *Del.* (1826)

Urostigma glumosa (Del.) Miq. (1847).

Urostigma fazoklense Miq. (1847); *U. fazoklense* f. *minor* Miq. (1847); *Ficus fazoklensis* (Miq.) Miq. (1867).

F. glumosa Del. var. *lanuginosa* Martelli in *Fl. Bogos*: 76 (1886) - type: EW, Bogos, *Beccari* 17 (FT holo.).

F. glumosa Del. var. *intermedia* Martelli in *Fl. Bogos*: 76 (1886) - type: EW, Bogos, *Beccari* 19 (FT holo.).

F. glumosa Del. var. *glaberrima* Martelli in *Fl. Bogos*: 76 (1886) - type: EW, Bogos, *Beccari* 40 (FT holo.).

Shrub or tree to *c* 10 m tall, trunk up to 0.8 m in diam., often with thick, spreading branches from the base. Bark greyish to pale brown, often peeling in big flakes and leaving scars, slash pale brown to reddish. Young branches puberulous, pilose or hirsute. Leaves alternate, papery, broadly ovate to elliptic or oblong, 3-10 x 2-5 cm, base cordate, margin entire, apex rounded, mucronate, acute or acuminate; lamina smooth and glabrous above, densely pubescent to glabrescent beneath, midrib with 5-7 pairs of lateral nerves. Petiole up to 4 cm long, glabrous to pubescent. Stipules pubescent, up to 3 cm long. Figs 1-2 in the leaf axils, sessile or on peduncles up to 3 mm long; basal bracts united below, with upper part falling quickly, left as a flat cup with repand edge. Calyptra present, shed very early. Figs globose, up to 1 cm in diam., white or yellow spotted, hirsute to glabrescent; ostiole ± prominent, 2-lipped, with no visible bracts. Male flowers with 1 stamen. Fig. 104.18.

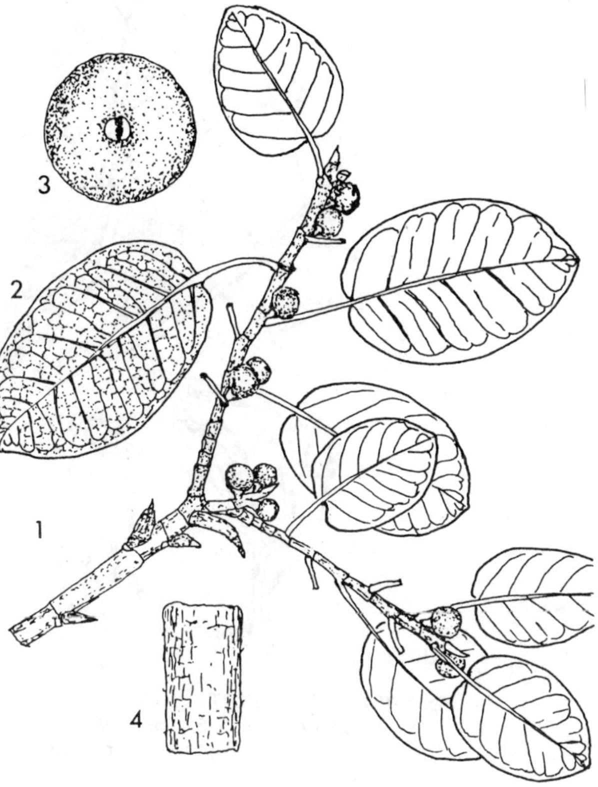

Figure 104.18 *FICUS GLUMOSA*: 1 - leafy branch with figs x 1/2; 2 - lower surface of leaf x 1/2; 3 - ostiole x 3; 4 - portion of bark x 1. 1, 3, 4 from *J. De Wilde* 5458; 2 from *Aweke & Gilbert* 684. (Redrawn with permission from *Meded. Landbouwhogeschool Wageningen* 79-3, Fig. 6, 1979.)

On rocks, boulders and rocky slopes in deciduous woodland or bushland; 500-1650 m. EE EW TU GD GJ HA IL SD; widespread in tropical Africa, west to Senegal, south to Tanzania. *Burger* 3356; *Friis* et al. 1921; *Gillett* 12840.

Further south *F. glumosa* is replaced by the closely related *F. sonderi* Miq. (1867), from Tanzania to S Afr. Rep.

Figure 104.19 *FICUS LUTEA*: 1 - leafy branch with figs x 1/2; 2 - male flower, closed x 15; 3 - male flower, open x 15; 4 - gall flower x 15; 5 - female flower x 15; 6 - ostiole x 5; 7 - stipule covering axillary bud x 1/2. All from *Jansen* 5763. (Reproduced with permission from *Meded. Landbouwhogeschool Wageningen* 79-3, Fig. 22, 1979.)

15. **F. lutea** *Vahl* (1805).
F. neumannii Kunth & Bouché (1847).
Urostigma vogelii Miq. (1849); *Ficus vogelii* (Miq.) Miq. (1867).
F. senegalensis Miq. (1867).
F. quibeba Ficalho (1884).
F. baronii Baker (1883).
F. nautarum Baker (1877).
F. vestito-bracteata Warb. (1894).

Tree to *c* 20 m with spreading crown, trunk up to 2 m in diam. Bark grey, uneven, wrinkled lengthwise, slash pink. Leaves alternate, papery or somewhat leathery, drying brown, elliptic, oblong or obovate, 11-18 x 6-9 cm, base cuneate to narrowly rounded, margin entire, apex broadly acuminate; lamina glabrous on both sides, midnerve with 6-8 pairs of lateral nerves, tertiary nerves forming a fine, clearly visible net below. Figs 1 or more in the leaf axils, sessile or subsessile; basal bracts large and prominent, membranous, triangular, pesistent. Calyptra present, consisting of two partly fused, hirsute bracts, shed early. Figs globose to somewhat depressed globose, *c* 1.5 cm in diam., glabrous; ostiole not prominent, 4-lipped, without visible bracts. Male flowers with 1 stamen. Fig. 104.19.

Riverine forest, upland rain forest or evergreen bushland; 900-1500 m. KF IL GG ?SD; widespread in tropical Africa, west to Senegal, south to S Afr. Rep.

(Natal), Madagascar, Mascarenes. *Jansen* 5763; *Chaffey* 1266; *Thomerson* 726.

The name *F. lutea* Vahl., often used for *F. ingens* (Miq.) Miq., has, according to a neotypification by C. C. Berg in *Kew Bull.* **36**: 397 (1981), to be used for the species hitherto known as *F. vogelii* (Miq.) Miq.

16. **F. umbellata** *Vahl* (1805).

Tree to 20 m, mostly starting as an epiphyte and strangler. Branches glabrous. Leaves broadly ovate to broadly elliptic, 10-26 x 8-18 cm, leathery, base subcordate or truncate, less frequently cordate, margin entire, apex shortly acuminate; lamina glabrous on both sides, with 8-10 pairs of lateral nerves, anastomosing near the margin. Petiole 3-12 cm long, 2-4 mm thick, glabrous. Stipules 2-5 mm long. Figs 2-3 together, developing below the leaves on dwarf shoots, or on thick, ± curved side branches, up to 3 cm long, on the older branches ("spurs"); peduncle 2-8 mm long, robust; basal bracts very broadly ovate, up to 3 mm long, finely puberulous. Figs globose to obovoid, sessile or slightly stipitate, 2-3 cm in diam., glabrescent to puberulous, wrinkled when ripe and dry; ostiole protruding, bilobed. Male flowers with 1 stamen. Fig. 104.20.

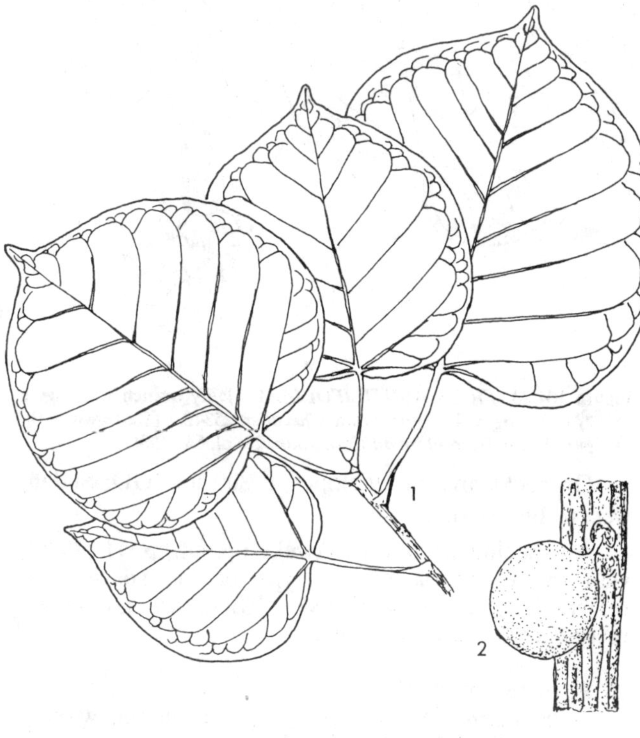

Figure 104.20 *FICUS UMBELLATA*: 1 - leafy branch x 2/3; 2 cauliflorus fig x 1. Both from *Berg* 159. (Redrawn with permission from *Flore du Cameroun*, 28: pl. 80., 1986.)

In lowland rain forest, coffee plantations and left in open land when the forest has been cleared; 1100-1300 m. KF IL; widespread in central Africa, west to Senegal, south to Angola. *Chaffey* 1198; *Friis* et al. 3973, 4128.

17. **F. populifolia** Vahl (1790);
Urostigma populifolium (Vahl) Miq. (1847)
F. intermedia Del. (1826).
Urostigma catalpaefolium Miq. (1847).
F. religiosa Forssk. (1775), non L. (1753).

Shrub or tree to *c* 15 m tall, trunk up to 0.9 m in diam. Bark grey, smooth. Young branches brownish, glabrous. Leaves alternate, petiolate, papery, drying pale green, cordate, 4-15 x 3-15.5 cm, base broadly cordate, margin entire or slightly repand, apex ± long acuminate; lamina glabrous on both sides, leaf base 5-nerved, the lowermost of pair of which give off numerous side nerves in the broad cordate leaf base, midnerve with 5-10 pairs of lateral nerves. Petiole 3-13 cm long, thin (hence the characteristically trembling leaves on this species), glabrous. Stipules falling quickly, glabrous, 0.5-1.5 cm long. Figs 1-2 in the leaf axils, on peduncles 0.5-2 cm long, puberulous to glabrescent; basal bracts broadly ovate, puberulous, basally united, apical part falling quickly, basal part persisting. Figs subglobose, 5-12 mm in diam., puberulous or glabrescent; ostiole not protruding, 3-lipped. Male flowers with 1 stamen. Fig. 104.21.

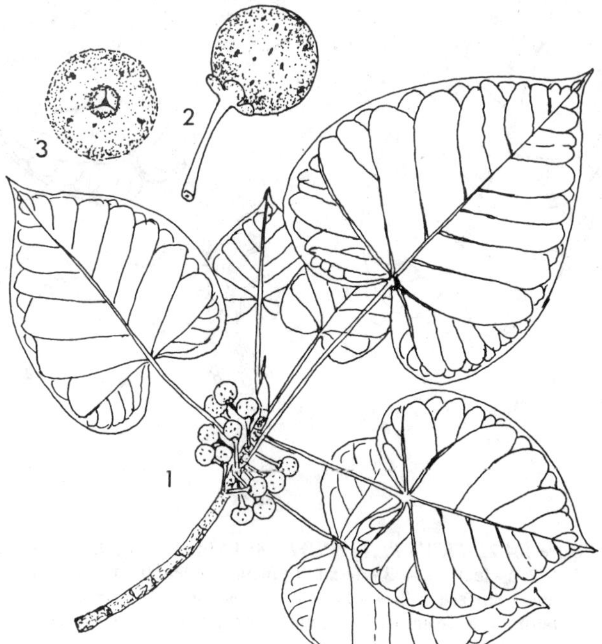

Figure 104.21 *FICUS POPULIFOLIA*: 1 leafy branch with figs x 1/2; 2 - fig x 2; 3 - ostiole x 5. All from *Burger* 3810. (Redrawn with permission from *Meded. Landbouwhogeschool Wageningen* 79-3, Fig. 14, 1979.)

On rocks, boulders, and rocky slopes in deciduous bushland and woodland; 600-1600 m. EE EW TU GD GJ SD; widespread in central tropical Africa, west to Togo, south to Tanzania, Yemen, Saudi Arabia. *Gillett* 12578; *Burger* 3810; *Friis* et al. 2579.

18. **F. platyphylla** *Del.* (1826).
Urostigma kotschyanum Miq. (1847); *Ficus kotschyana* (Miq.) Miq. (1867).

Tree to *c* 20 m tall, with spreading crown, trunk up to 1 m in diam. Bark pale brown, scaly, fissured. Young branches thick, with brown bark, puberulous to glabrescent. Leaves alternate, leathery, drying dark green, ovate-elliptic to rarely obovate, 5-25 x 3-17 cm, base deeply cordate, often with ± overlapping basal lobes, margin entire, apex rounded or obtuse; glabrescent above, pubescent to glabrescent below, venation as in *F. populifolia*, midrib with 9-11 pairs of lateral nerves and glandular tissue below at the leaf base. Petiole up to 12 cm long, pubescent. Stipules up to 2.5 cm long. Figs 1 or more in the leaf axils, on peduncles 2-5 cm long, slender, pubescent; basal bracts ovate, united at the base, apical part falling quickly. Calyptra flattened, conical, very early shed. Figs globose, *c* 1 cm in diam., pubescent; ostiole not prominent, 2-lipped, without visible bracts. Male flowers with 1 stamen. Fig. 104.22.

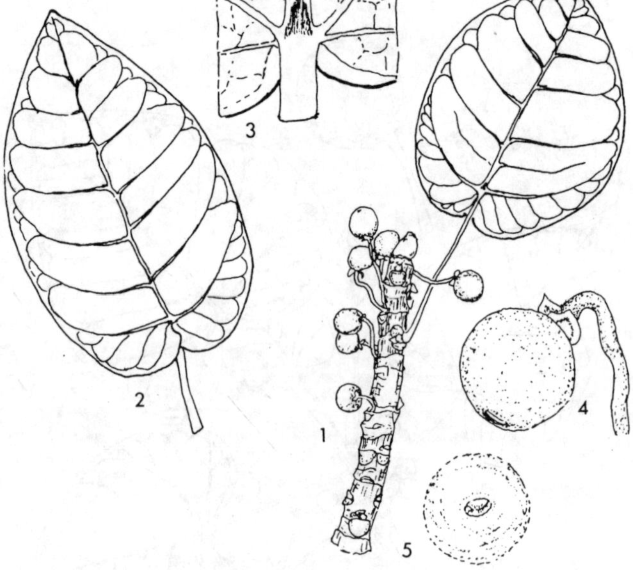

Figure 104.22 *FICUS PLATYPHYLLA*: 1 - branch with leaf and figs x 1/2; 2 - leaf x 1/4; 3 - detail of junction of leaf blade and petiole, lower surface x 2; 4 - fig x 2; 5 - ostiole x 2. (Redrawn with permission from *Meded. Landbouwhogeschool Wageningen* 79-3, Fig. 13, 1979.)

Deciduous woodland; 600 m. GG IL; throughout West Africa to Senegal; N Somalia. *Tewolde* 1548; *Friis* et al. 2511.

Only two Ethiopian records.

19. **F. abutilifolia** *(Miq.) Miq.* (1867);
Urostigma abutilifolium Miq. (1847) - type: Sudan/GD, Fazogli, *Kotschy* 462 (K iso.).
Ficus discifera Warb. (1905).

Tree up to 10 m tall. Young branches glabrous. Leaves alternate, drying dark green or brownish, broadly ovate to cordate, up to 20 x 19 cm, base deeply cordate, mostly with overlapping basal lobes, margin entire, apex acute to short acuminate; lamina glabrous above, pubescent to glabrous below, venation as in *F. popilifolia*, midrib with 5-9 pairs of lateral nerves, which often branch near the margin, glandular tissue on the midrib below at the leaf base. Petiole 5-10cm long, glabrescent to puberulous. Stipules up to 1 cm long, glabrescent. Figs 2 or more in the leaf axils, on 0.5-1 cm long, puberulous peduncles; basal bracts fused at the base, of uncertain number, often forming a ± disc shaped structure. Calyptra small, early shed. Fig ± stipitate, globose to obovoid, 1-1.5 cm in diam., glabrescent to glabrous; ostiole prominent, 2-lipped, without visible bracts. Male flowers with 1 stamen. Fig. 104.23.

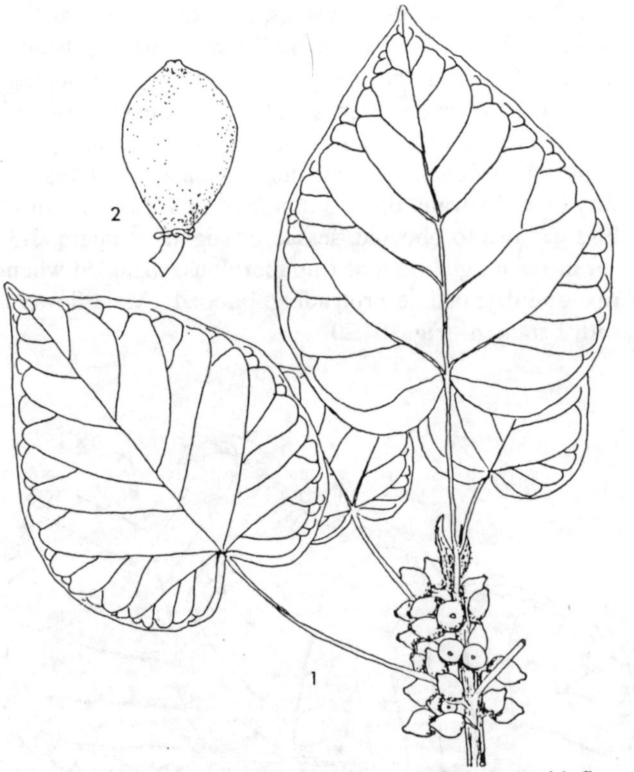

Figure 104.23 *FICUS ABUTILIFOLIA*: 1 - leafy branch with figs x 2/3; 2 - fig x 4. Both from *Chevalier* 25342. (Redrawn with permission from *Flore du Cameroun*, 28: pl. 54., 1986.)

On rocks and rocky slopes; *c* 800 m. ?GD; Sudan, west to Ivory Coast.

All specimens cited by Aweke op. cit., p. 11 (1979), with the exception of the type of the species are referrable to *F. vasta* Forssk. There are no certain records from within Ethiopia.

20. **F. vasta** *Forssk.* (1775).
F. dahro Del. (1843) - type: Ethiopia, without loc., *Galinier* 196 (MPU, holo.).
F. socotrana Balf. f. (1883).
F. rivae Warb. (1905) - type: SD, Giacorso, *Ruspoli & Riva* 230 (?B, holo. destr., not traced in FT).
F.hararensis Warb. (1905) - type: HA, Harar, *Robecchi Brichetti* 156 (FT syn.) & 162 (FT syn.).
F. callabatensis Warb. (1905) - type: Sudan/GD, Gallabat, *Schweinfurth* 556 & 557 (both B, syn. destr., K isosyn.).

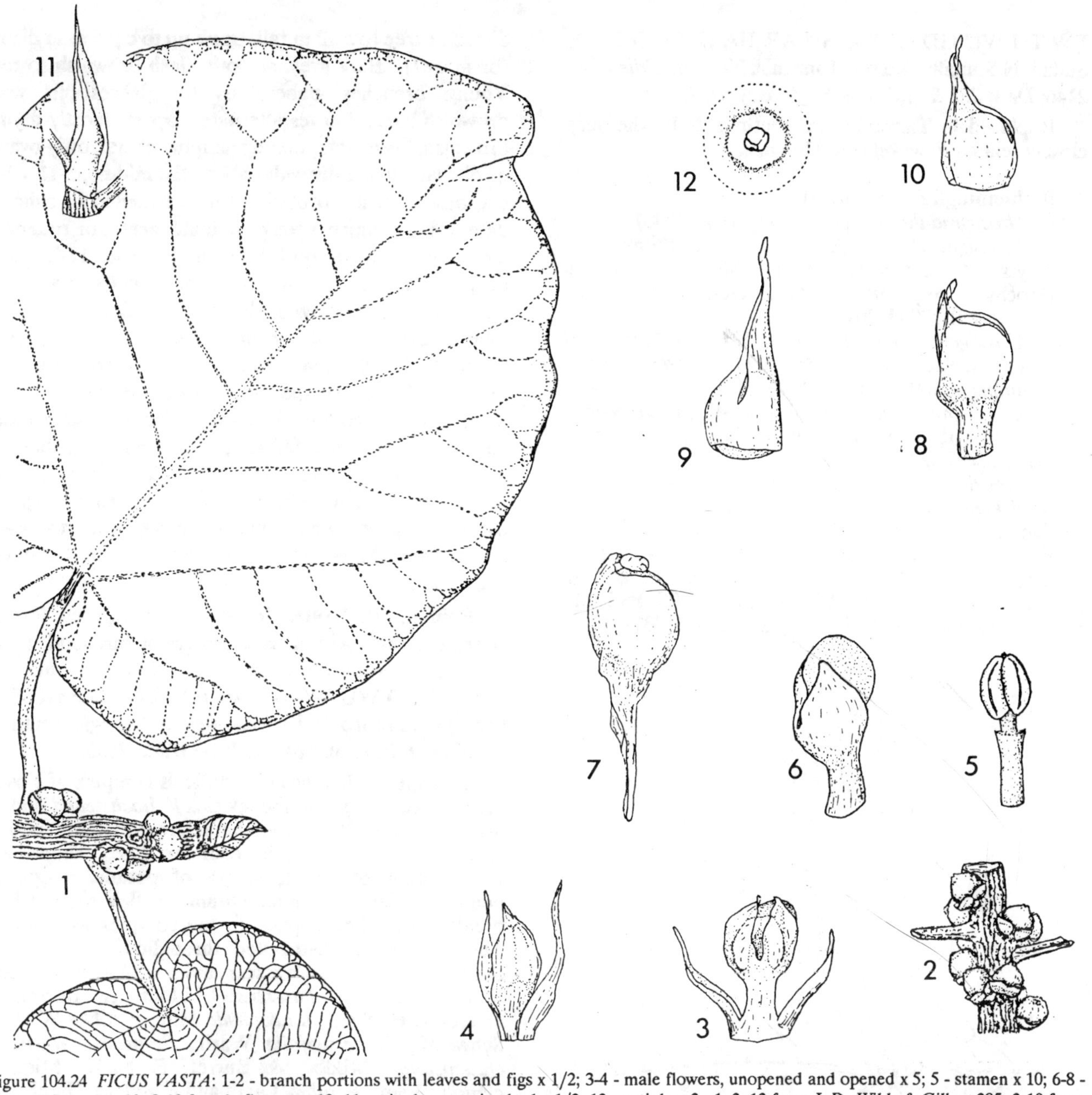

Figure 104.24 *FICUS VASTA*: 1-2 - branch portions with leaves and figs x 1/2; 3-4 - male flowers, unopened and opened x 5; 5 - stamen x 10; 6-8 - gall flowers x 10; 9-10 female flowers x 10; 11 - stipules covering bud x 1/2; 12 - ostiole x 2. 1, 2, 12 from *J. De Wilde & Gilbert* 395; 3-10 from *Jansen* 5246; 11 from *Jansen* 5474. (Reproduced with permission from *Meded. Landbouwhogeschool Wageningen* 79-3, Fig. 21, 1979.)

F. vasta Forssk. var. *velutina* Fiori in *Boll. R. Soc. Tosc. di Ort.* **35**: 166 (1910) - type: EE, Ghinda, *Fiori* 36 & 37 (?FT, syn. not found).

F. vasta Forssk. var. *glabrescens* Hutch. in *Fl. Trop. Afr.* **6(2)**: 195 (1917) - type: EE, Ghinda, *Schweinfurth & Riva* 129 (K lecto.).

Tree up to 20 tall or more, with spreading crown, trunk irregularly shaped, ± buttressed, up to 1.5 m in diam. Bark smooth grey. Young branches with brownish bark, hirsute. Leaves alternate, papery, elliptic, broadly ovate or suborbicular, 10-25 x 10-22 cm, base cordate, margin entire, apex broadly rounded, sometimes apiculate or short acuminate; lamina usually glabrescent above, glabrescent, puberulous, hirtellous, or velutinous below, venation as in *F. populifolia*, midrib with 6-8 pairs of lateral nerves and glandular tissue below at the leaf-base. Petiole 2-8 cm long, hirtellous. Stipules tomentose, up to 2.5 cm long. Figs 1 to few in the leaf axils, sessile or on peduncles up to 5 mm long; basal bracts membranous, pubescent, united below, falling quickly in upper part, lower part persisting, forming hairy rim. Calyptra small, pubescent or velutinous, shed early. Figs globose or depressed globose, 1-1.5(-2) cm in diam., tomentose to densely lanate, sometimes warty; ostiole prominent, 2-lipped, often finally gaping so that some of the bracts descending into the fig are visible from the outside. Male flowers with 1 stamen. Fig. 104.24.

On rocks, in forest clearings and riverine forest, at forest margins and lake shores, occasionalley open woodland, in secondary bushland or farmland, often as single big tree in village market places; 1000-2400 m.

EW TU WU GD GJ WG SU AR HA IL KF GG SD; Sudan, N Somalia, Kenya, Tanzania, Yemen. *Friis* et al. 2146; *De Wilde & Gilbert* 395; *Mooney* 5757.

Replaced in Tanzania and southwards by the very closely related *F. wakefieldii* Hutch.

21. **F. thonningii** *Blume* (1836).

Urostigma thonningii (Blume) Miq. (1848).

F. saligna Hochst. (1842), in exsicc. Schimper, Pl. Abyss.; *F. acrocarpa* (Miq.) Miq. var. *saligna* (Hochst.) Miq. (1867) - type: Gondar, Ferrakob, *Schimper* II:709 (K iso.).

Urostigma schimperi Miq. (1847); *F. schimperi* (Miq.) A. Rich. (1851) - type: TU, Haramat, Geraz, *Schimper* II:1096 (L holo., K iso.).

Urostigma acrocarpum Miq. (1847); *F. acrocarpa* (Miq.) Miq. (1867) - type: TU, Mai Dogale, *Schimper* II:627 (L holo., K iso.).

Urostigma dekdekena Miq. (1847); *Ficus dekdekena* (Miq.) A. Rich. (1851) - type: TU, Mt. Scholoda, *Schimper* I:220 (L, holo., K, iso.).

F. hochstetteri Miq. var. *glabrior* Miq., *Ann. Mus. Bot. Lugd. Batav*. 3: 289 (1867) - type: TU, at Mai Gogale, *Schimper* II:627 (L holo, BM K P UPS iso.).

F. microcarpa Vahl (1805), non L. f. (1781).

Figure 104.25 *FICUS THONNINGII*: 1 - leafy branch with figs x 1/2; 2 ostiole x 3. Both from *W. de Wilde & de Wilde-Duyfjes* 7635. (Redrawn with permission from *Meded. Landbouwhogeschool Wageningen* 79-3, Fig. 19, 1979.)

Shrub or tree to *c* 12 m tall, trunk up to *c* 0.5 m in diam. (or more?). Bark grey, smooth, slash brownish to red. Young branches puberulous to glabrescent, with brownish bark. Leaves alternate, papery, mostly drying brownish, lanceolate, oblong, elliptic, or narrowly ovate, or obovate, generally wider above the middle, 5-11 x 2-4 cm, base cuneate to obtuse, rarely rounded or subcordate, margin entire, apex acuminate, acute, or rounded; lamina glabrous on both sides, midrib with 12-25 pairs of lateral nerves and an often poorly marked zone of glandular tissue below at the base. Petiole 1-3(-5) cm long, glabrous or glabrescent. Stipules lanceolate, up to 0.8 cm long, pubescent, at least on the midrib. Figs 1-2 in the leaf axils, sessile or on peduncles up to 8 mm long; basal bracts ovoid, fused in the persisting basal part, apical part free, falling quickly. Fig subglobose to ± ovoid, 6-9(-14) mm in diam., glabrous to slightly puberulous, often with brownish to reddish spots; ostiole ± prominent, 2-lipped, small, with no bracts visible from the outside. Male flowers with 1 stamen. Fig. 104.25.

Rain forest, "coffee forest", dry evergreen forest, riverine forest, lake shores, evergreen bushland, or as single trees in farmland or villages; 1300-2200 m. EE EW TU GD WU WG SU AR BA HA IL KF GG SD; widespread in tropical Africa, west to Senegal, south to Zambia. *Friis* et al. 524, 2117; *De Wilde* 7635.

This species belongs to a difficult complex of closely related taxa, including the species *F. hochstetteri* and *F. ruspolii* in this Flora. The characters previously used to separate species in this complex (size, shape and indumentum of leaves, length of petiole, length of peduncle, size and indumentum of fig) seem all to break down when applied to the complex as a whole. The complex is distributed throughout tropical and South Africa, on Madagascar and the Mascarenes, and also includes: *F. basarensis* Mildbr. & Burret, *F. natalensis* Hochst., *F. iteophylla* Miq., *F. rokka* Warb. & Schweinf., *F. persicifolia* Warb., *F. petersi* Warb., *F. craterostoma* Mildbr. & Burret, *F. burkei* Miq., *F. pyrifolia* Lam., *F. rubra* Lam., and *F. reflexa* Thunb., the oldest name in the complex. The group, which includes taxa which have been widely cultivated and selected for the production of bark cloth, needs a total revision before the taxonomy can be settled. The treatment presented here is clearly artificial, as most of the poorly defined and very variable forms have been placed under *F. thonningii*, and only two peripheral forms have been maintained as species. The author has assumed that the type of *F. hochstetteri* var. *glabrior* Miq. is as stated above; if so, then the variety belongs as a synonym here, not under *F. hochstetteri*, as stated by Aweke, op. cit., p. 34 (1979).

The widely cultivated street tree of the towns of Eritrea is referable to this complex and keys out as *F. thonningii*. It is vegetatively propagated as in W and E Africa. Cufodontis (*Enum*.: 14, 1953) lists *F. punctata* Lam. as cultivated in Eritrea on the authority of Chevalier; this is probably a misidentification of the species dealt with here.

22. **F. hochstetteri** *(Miq.) A. Rich.* (1851)
Urostigma hochstetteri Miq. (1847); *F. schimperi* (Miq.) A. Rich. var. *hochstetteri* (Miq.) Mildbr. & Burret in *Eng. Bot. Jahrb.* **46**: 253 (1911) - type: TU (?), Shahageni, *Schimper* I:373 (L holo., K iso.).
F. pubicosta Warb. (1904).

As for *F. thonningii*, except for the pubescent leaves, petioles and figs. Peduncles up to 3.5 mm long, figs up to 8 mm in diam.

Riverine forest, single trees in farmland or villages; 1500-2200 m. EW TU SD; Kenya, Tanzania, Zaire, ?elsewhere - the total distribution needs further study. *Schweinfurth* 2072; *Aweke & Gilbert* 826.

It seems likely that this taxon only merits infraspecific rank under *F. thonningii*, but no combination is available, and it is undesirable to add new names to this group before a complete revision is undertaken.

23. **F. ruspolii** *Warb.* (1905)
- type: SD, Ciaffa, Boran Uata, *Ruspoli & Riva* 526 (FT holo., K iso.).

A shrubby tree to 3 m tall or more. Leaves alternate, papery, drying brown, oblong-lanceolate to oblong-elliptic, 6-10 x 1.5-3 cm, base rounded to subcordate, margin entire, apex narrowing, bluntly obtuse or slightly acuminate; lamina glabrous or slightly pubescent towards the base of the midrib above, ± hairy and rufous hirsute on the midrib below, midrib with 8-10 pairs of lateral nerves. Petiole 1-2 cm long, rufous pubescent. Stipules lanceolate, up to 4 mm long, pubescent on midnerve, falling quickly. Figs 1-2 in the leaf axils, on peduncles 2-4 mm long, puberulous; basal bracts united at base, puberulous on outside, falling quickly above. Figs seen immature, globose, up to 4 mm in diam., puberulous to glabrescent; ostiole not prominent, without bracts visible from the outside. Male flowers as in *F. thonningii*.

Deciduous woodland; *c* 1800 m. SD KF; not known elsewhere. *De Wilde* 6824

Close to *F. thonningii*. Only known from the type and one recent collection from the Kaffa side of the Omo River gorge. More material desirable.

5. TRILEPISIUM Thouars (1806).

Bosqueia Thou. ex Baill. (1863).

Berg in Bull. *Jard. Bot. Brux.* **47**: 297-306 (1977).

Monoecious trees with milky latex. Leaves alternate, petiolate, entire. Stipules intrapetiolar, fused, leaving a circular scar round the stem. Inflorescences bisexual, solitary (rarely paired) in the leaf axils, when young enclosed in a conical cover of modified stipules, pedunculate, discoid to turbinate, with an involucrum formed by 6-8 bracts, with one female flower in the centre of the disc and numerous male flowers at the perifery. In bud the margin of the receptacle is bent inwards and fused with the upper surface of the disc round the female flower, thus forming a protective ring covering the male flowers. At anthesis this ring is circumcessile, leaving one torn edge round the perifery of the disc and another, almost tubular, round the central female flower. Male flowers naked, each consisting of 1 to several, ± fused stamens. Female flower sunk in the middle of the disc shaped receptacle, with a tubular perianth; ovary ± fused with the perianth, with one long protruding style, stigmatic branches 2. Infrutescence fleshy, ovoid, crowned by the withered male flowers.

Monotypic.

T. madagascariense *DC.* (1825).
Bosqueia thouarsiana Baill.(1863).
Bosqueia phoberos Baill. (1867).

Tree to 25 m tall, trunk buttressed, up to 0.6 m in diam. Latex milky white. Young brances glabrous. Leaves papery, usually drying brown, ovate, obovate, or lanceolate, 3-10 x 1.5-5 cm, base cuneate to truncate, margin entire, apex usually long acuminate; mostly glabrous on both sides, lateral nerves anastomosing near the leaf margin. Petiole 3-10 mm long. Stipules 2-12mm long. Inflorescence *c* 5 mm in diam. Stamens (male flowers) numerous, up to 10 mm long. Single female flower with 1.5-3 mm long perianth, style up to 9 mm long, stigmatic branches up to 8 mm long. Fruiting receptacle ovoid, often slightly curved, 0.8-1.5 cm long. Fig. 104.26.

Along SW fringe of the Ethiopian upland rain forest; 1050-1600 m. KF WG; widespread in tropical Africa, west to Guinean Rep., south to Angola and S Africa (Natal); Madagascar, Seychelles. *Meyer* 7859; *Friis* et al. 4140; *Thomerson* 717.

6. ANTIARIS *Lesch.* (1810), nom. cons.

Monoecious or dioecious trees with white, milky latex. Leaves alternate or often distichous. Stipules free, semi-amplexicaul. Inflorescences on short, lateral branches, which each carry 1-2 female inflorescences and/or up to 8 male inflorescences. Male inflorescence discoid, pedunculate, with 1-3 rows of bracts at the edge. Male flowers numerous on the upper surface of the disc, tepals 2-7, free, stamens 2-4, straight in bud; rudimentary ovary absent. Female inflorescence subsessile or pedunculate, claviform, with a ring of bracts surrounding the single female flower. Female flower with 4-lobed perianth, partly embedded and fused with the receptacle; ovary erect, style 1, terminal, stigmatic branches 2. Infrutescence fleshy, orange when mature, the accrescent receptacle completely fused with the fruit wall.

Monotypic, with one widespread and very variable species.

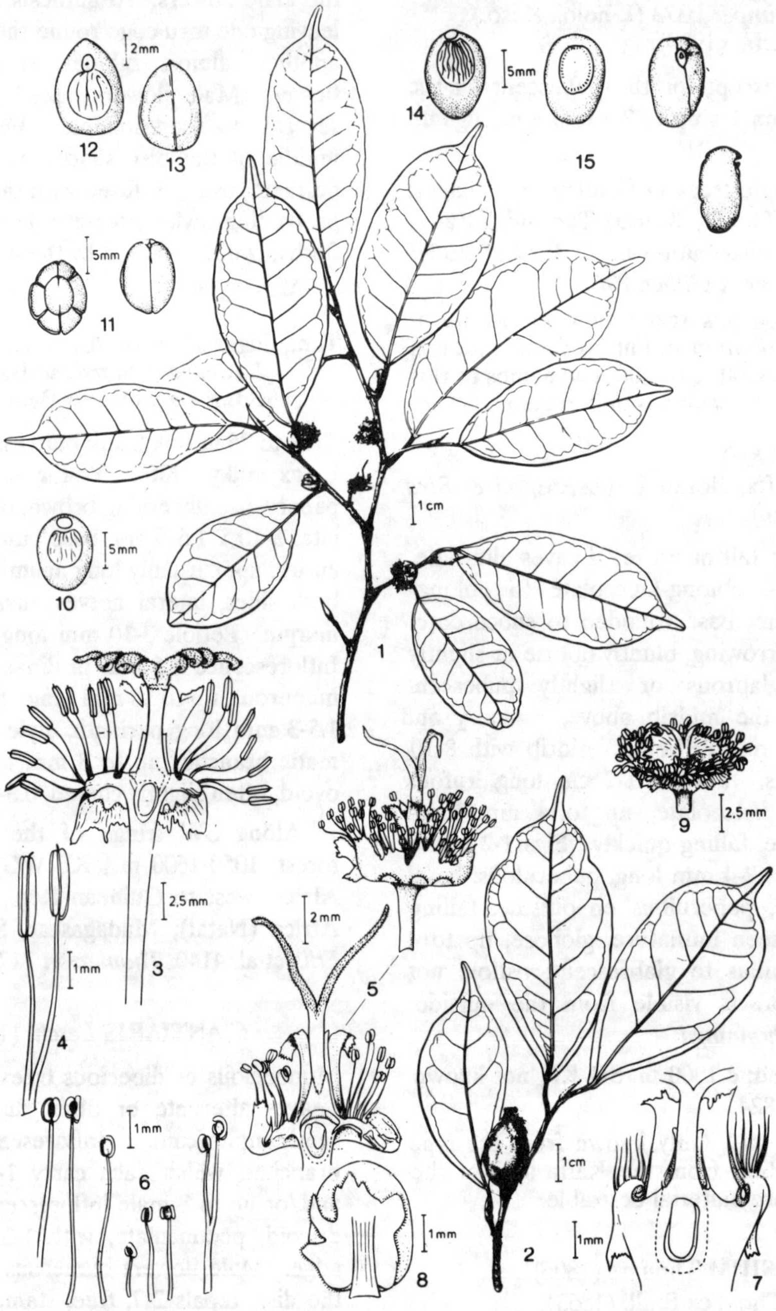

Figure 104.26 *TRILEPISIUM MADAGASCARIENSE*: 1 - leafy branch with inflorescences; 2 - leafy branch with fruit; 3 - inflorescence cut in half with mature stamens; 4 - ripe stamens; 5 - inflorescence complete and cut in half with mature pistil; 6 - old stamens; 7 - inflorescence cut in half to show torn tubular ring of receptacle round central female flower; 8 - outer part of torn receptacle seen from below; 9 - old inflorescence with persistent stamens; 10, 12, 14 - seeds; 11, 13, 15 - embryos. 1, 2, 10, 11 from *Breteler* 2731; 3-4 from *Gossweiler* 4394; 5-6 from *Simão* 506; 7-8 from *Gossweiler* 6514; 9 from *Mendonça* 141; 12-13 from *Capuron* 948; 14-15 from *Devred* 2454. (Reproduced with permission from *Bull. Jard. Bot. Nat. Belg.* 47, 1977: *Flore du Cameroun*, 28: pl. 35., 1986.)

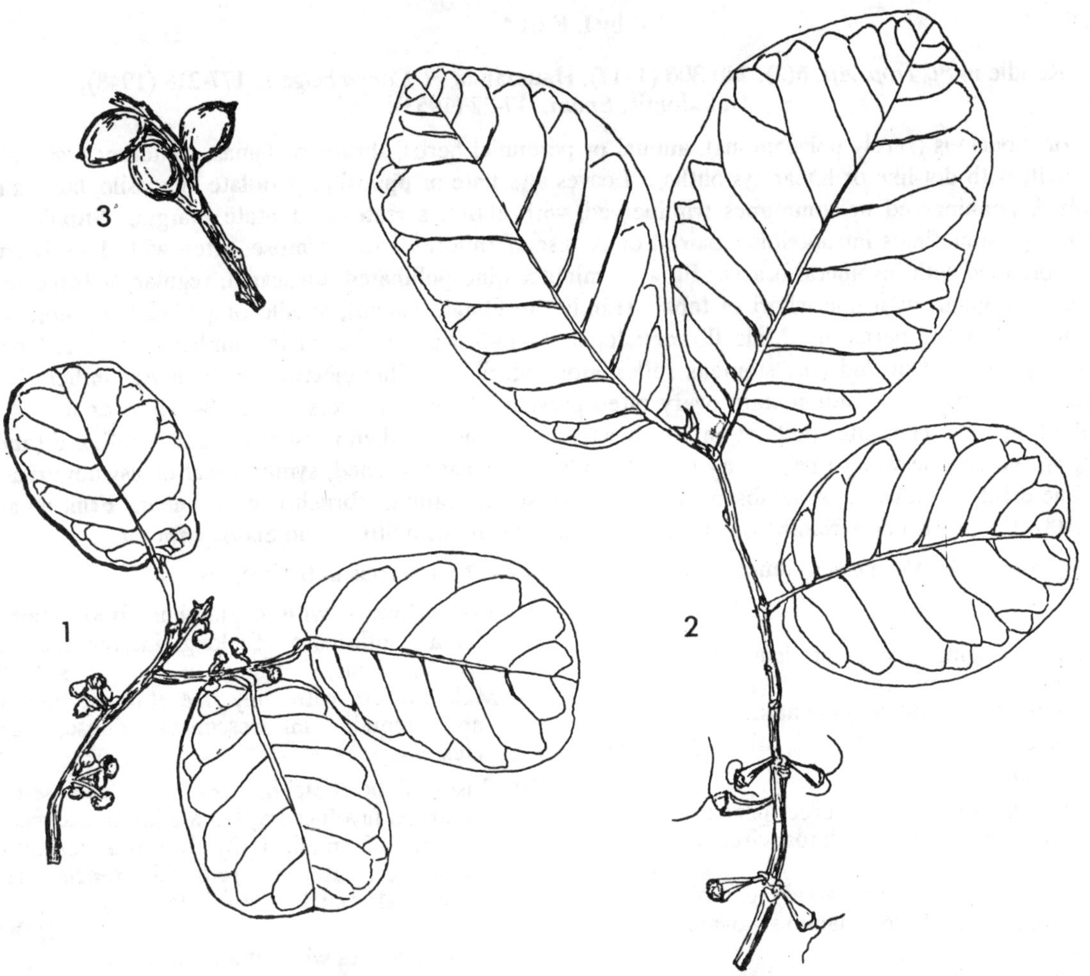

Figure 104.27 *ANTIARIS TOXICARIA* subsp. *WELWITSCHII* var. *WELWITSCHII*: 1 - leafy branch with male inflorescences; 2 - leafy branch with female flowers; 3 - branch with fruits. 1 from *Flamigni* 10490; 2 from *Toka* 26; 3 from *Devred* 762. (Redrawn with permission from *Bull. Jard. Bot. Nat. Belg.* 47, 1977: *Flore du Cameroun*, 28: pl. 36., 1986.)

A. toxicaria *Lesch.* (1810)

subsp. **welwitschii** *(Engl.) C. C. Berg* in Bull Jard. Bot. Nat. Belg. 48: 466 (1978)

Tree to 30 m tall, trunk up to 0.8 m in diam., ± buttressed. Bark dark greyish-brown. Latex milky white or yellowish. Young branches puberulous to hirsute. Leaves ovate, elliptic, elongate or obovate, 6-15 x 2-10 cm, base unequal, subcordate, truncate or rarely broadly cuneate, margin entire or crenulate, apex shortly acuminate or obtuse; immature leaves hispid or scabrous above, pubescent beneath. Petiole 3-9 mm long, puberulous. Stipules 3-10 mm long, falling quickly. Male inflorescence on 0.5-1 cm long peduncles, 0.5-0.8 cm in diam. Male flowers numerous, 1-1.5 mm long. Female inflorescences 0.3-0.5 cm long, *c* 0.3 cm in diam. Infrutescence elipsiod, 1-1.5 cm long, on 0.5-1 cm long peduncle.

var. **welwitschii**

Crown leaves glabrous or glabrescent above, glabrescent to puberulent beneath. Fig. 104.27.

Lowland forest; 500-(?)1900 m. IL; widespread in West tropical Africa, west to Senegal. *Chaffey* 659, 939; *Friis* et al. 4012.

Only this subspecies and variety recorded with certainty in Ethiopia, as only the specimen *Friis* et al. 4012 has the mature crown leaves needed for positive identification of the variety.

A. toxicaria subsp. *welwitschii* var. *africana* A. Chev. in *Veget. ut. Afr. trop. Fr.* 5: 259 (1909) has crown leaves which are scabrous above and hirtellous to subtomentose below. It occurs in drier vegetation than does var. *welwitschii* and is recorded from S Sudan. It is likely to occur also in humid woodlands of SW Ethiopia.

The subspecies as a whole is found throughout large parts of tropical Africa, south to Angola and Zambia. Subsp. *toxicaria* occurs in tropical Asia, Indonesia, the Philippines and Tonga; other subspecies occur on Madagascar and in Australia. There are two old incomplete records of a tree from the Yemen which has been referred to this species, but in absence of more complete material its identity is uncertain.

105. URTICACEAE

by I. Friis.*

Rendle in *Fl. Trop. Afr.* **6(2)**: 240-306 (1917); Hauman in *Fl. Congo belge* **1**: 177-218 (1948); Cufodontis, *Enum.*: 17-22 (1953).

Monoecious or dioecious (rarely polygamous), annual or perennial herbs, shrubs or lianas, sometimes with stinging hairs, and usually with dot-like or linear cystoliths[1]. Leaves alternate or opposite, petiolate or sessile, lamina simple to deeply lobed, penninerved or sometimes triplinerved, with entire, serrate or dentate margin. Stipules usually present, lateral, or sometimes intrapetiolar, paired or ± fused. Inflorescences cymose, often with densely clustered flowers and subtended with involucral bracts. Flowers minute, wind-pollinated, unisexual, regular or (especially the female flowers) irregular, with one whorl of tepals (rarely the female naked), sessile or pedicellate, often with an articulation just below the perianth. Male flowers: tepals (1-)3-5, stamens equal in number and placed opposite tepals or solitary, inflexed in bud (the stamens reflex momentarily, forcibly ejecting the pollen); anthers 2-celled, opening with longitudinal slits; rudimentary ovary often present. Female flowers: tepals 3-5, free or united, often very unequal, often accrescent after pollination, rarely naked; staminodes when present scale like, inflexed (and then actively ejecting the achene when ripe); ovary erect usually somewhat flattened, symmetrical or asymmetrical, one-celled, with one erect, basal ovule; style absent or very short; stigma capitate, brush-like, or linear. Fruit an achene, often enclosed by the persistent perianth. Seed with thin testa, mostly with little or no endosperm.

About 50 genera and 1000 species, almost cosmopolitan, but most numerous in the tropics.

Key to Genera.

1. Plants with stinging hairs, at least on the inflorescences and petioles. 2
- Plants completely without stinging hairs. 6

2. Leaves opposite. **1. Urtica**
- Leaves alternate. 3

3. Stipules free, lateral; shrubby, erect perennial or small shrub; lamina ovoid to almost circular. **3. Obetia**
- Stipules fused, intrapetiolar; woody climber or erect, short-lived herb; lamina ovate to lanceolate. 4

4. Woody climber with axillary roots from the climbing stems, stinging hairs mostly restricted to petioles and inflorescences; female perianth with fused tepals, accrescent, becoming fleshy and red in fruit. **2. Urera**
- Erect, perennial herbs, not climbing and without axillary roots from the stems; usually with stinging hairs on all aerial parts of the plant; female perianth membranous in fruit. 5

5. Female perianth with 3 more or less united tepals (sometimes also with one minute free tepal), forming a one-sided cover on the ovary or achene; stinging hairs often more than 5 mm long. **4. Girardinia**
- Female perianth with 4 free, very unequal tepals; stinging hairs shorter than 5 mm. **5. Laportea**

6. Leaves opposite. 7
- Leaves alternate. 11

7. Inflorescence a pedunculate disk- or bell-shaped receptacle. **7. Lecanthus**
- Inflorescence different (if disk- or bell-shaped then sessile). 8

8. Stipules fused, intrapetiolar. **6. Pilea**
- Stipules free, lateral. 9

9. Male flowers with 4 stamens; both male and female inflorescences long, interrupted, spike-like, pendent. **9. Boehmeria**
- Male flowers with only one stamen; both male and female inflorescences dense, axillary clusters. 10

10. Bisexual or male inflorescences enclosed by a common involucrum; female inflorescences with clusters of small, 1(-2) flowered perianth-like involucres, often on reduced branches (some individuals may be entirely female). **14. Droguetia**
- Inflorescences without an involucrum. 11

11. Leaves opposite; male flowers in a small, pedunculate cluster, female flowers sessile, with tubular perianth. **15. Australina**
- Leaves alternate; male and female flowers usually in the same sessile inflorscence, female flowers without perianth. **16. Didymodoxa**

12. Leaves distichous, (alternate), lamina sessile or subsessile, clearly not symmetrical around the midnerve. **8. Elatostema**
- Leaves spirally arranged, petiolate, symmetrical or almost symmetrical around the midnerve. 13

13. Stipules absent. **12. Parietaria**
- Stipules present, at least on younger stems. 14

14. Flowers enclosed by a campanulate involucrum with (2-)4-6 lobes; flowers surrounded by long, woolly hairs. **13. Forsskaolea**
- Flowers not enclosed in an involucrum, without long, woolly hairs. 15

15. Inflorescence a pedunculate, lax, cymose panicle or an interrupted spike. 16
- Inflorescence a sessile or subsessile, densely capitulate cluster. 17

* Botanical Museum and Herbarium, University of Copenhagen, Gothersgade 130, DK-1123 Copenhagen K, Denmark.

1. cystolith = cell(s) containing one or more crystals of calcium carbonate, often forming a light coloured mark when held up against the light, often seen as a minute pale protuberance on leaf surfaces in dried specimens.

16. Woody climber; female perianth with ± fused tepals which increase in size and become orange or red in fruit. **2. Urera**
- Erect herbaceous plant; female perianth with free tepals, membranous in fruit. **5. Laportea**

17. Stipules fused, intrapetiolar; female perianth increasing in size and becoming red and fleshy in fruit; stigma penicillate[1]. **11. Debregeasia**
- Stipules free, lateral; female perianth not increasing considerably and not becoming red and fleshy in fruit; stigma filiform. 18

18. Male flowers with one stamen; minute annual herb. **16. Didymodoxa**
- Male flowers with 4-5 stamens; annual or perennial herbs or shrubs. **10. Pouzolzia**

1. URTICA *L.* (1753).

Monoecious or dioecious, annual or perennial herbs with stinging hairs and dot-like cystoliths. Leaves opposite, petiolate, lamina simple, base rounded or cordate, margin serrate. Stipules lateral, free. Inflorescence a bisexual or unisexual lax cymose panicle, often in pairs in the upper leaf axils. Male flowers 4-merous, tepals free, in two unequal pairs, stamens inflexed, ovary rudimentary. Female flowers 4-merous, tepals free, in two slightly unequal pairs; staminodes absent; ovary ovoid, stigma sessile, penicillate. Achene enclosed in or released from the persisting perianth.

About 50 species, mostly in the northern temperate regions, a few in the tropics and in the southern temperate regions.

1. Annual, monoecious herb; lamina ovate or elliptic-ovate, with cuneate base. **1. U. urens**
- Perennial, dioecious herb; lamina ovate or cordate, with ± cordate base. 2

2. Stipules broadly ovate-triangular. **2. U. simensis**
- Stipules linear. **3. U. dioica**

1. U. urens *L.* (1753).

Erect, pale green, monoecious, sparsely branched annual to 60 cm tall, glabrous except for the stinging hairs. Leaves: lamina thin, ovate to elliptic ovate, 1-4 x 0.7-2 cm, with 3-5 nerves from the base, base rounded, margin serrate, apex rounded to acute or acuminate: petiole 0.5-2 cm long. Stipules linear, to *c* 4 mm long. Inflorescences up to 1.5(-2) cm long, bisexual. Male flowers *c* 1 mm long. Female flowers *c* 1 mm long, the inner tepals each with a stinging hair. Achene *c* 1.6 mm long.

A weed of cultivations, in villages and along roadsides; 1500-2800 m. EE/EW GD BA/SD; almost cosmopolitan. *Schimper* 843; *Schweinfurth & Riva* 1262; *Terracciano & Pappi* 81.

1. penicillate = with a tuft of hairs shaped like a paint brush.

2. U. simensis *Steudel* (1850)

- type: GD, Debill in Semien, *Schimper* III:1283 (P herb. Steudel holo., K iso.).

Erect, pale green, dioecious, almost unbranched perennial herb to *c* 1 m tall from a creeping rhizome, finely pubescent to glabrescent, with numerous, up to 2.5 mm long stinging hairs on lamina, petioles and inflorescences. Leaves: lamina ovate, 5-12 x 3-8 cm, base rounded to subcordate, margin coarsely serrate, apex broadly acute or acuminate; petiole to *c* 4 cm long. Stipules broadly ovate to triangular, to *c* 1.8 cm long, ± persistent. Inflorescences axillary, 4 at each node, to *c* 5 cm long. Male and female flowers *c* 1.5 mm long. Achene *c* 2 mm long. Fig. 105.1.

Figure 105.1. *URTICA SIMENSIS*: 1 - branch showing leaves and inflorescences x 2/3; 2 - stinging hair x 10; 3 - bract x 2/3; 4 - female flower x 5; 5 - male flower with dehisced stamens x 5. All from *Mooney* 9096. Drawn by Eleanor Catherine.

Upland grassland, most common in disturbed places, often in large quantities near houses; 1500-3400 m. TU GD GJ SU AR BA SD; apparently endemic in Ethiopia, with a closely related, somewhat taller and more robust species with more finely serrate leaves and larger stipules, *U. massaica* Peter, in Kenya, Uganda, Tanzania and E Zaire. *Mooney* 7137, 9096; *Burger* 596.

Edible, eaten as a pot herb in some areas.

The indication from TU is based on a sight record; a collection from this region should be made for confirmation.

3. **U. dioica** *L.* (1753).

Erect, dark green, dioecious, unbranched perennial herb to *c* 1 m tall from a creeeping rhizome, finely pubescent to glabrescent, with numerous stinging hairs on petioles, lamina and inflorescences. Leaves: lamina ovate, 4-10 x 2.5-8 cm, base truncate or subcordate, margin serrate, apex acuminate; petiole to *c* 6 cm long. Stipules linear, to *c* 2 cm long. Inflorescences to *c* 10 cm long. Male and female flowers *c* 1 mm long. Achene *c* 1.2 mm long.

Habitat unknown, probably an introduced garden weed; altitude unknown. EE/EW/TU; widespread in the northern hemisphere.

This record is based entirely on an entry in Cufodontis' *Enum.*: 17 (1953), where this species is stated to occur in "Eritrea", a geographical unit used in a more wide sense by Cufodontis than in the present work. It has not been possible to trace the place of publication of the original record, and no specimen has been seen in the herbaria consulted for the account. A confusion with *U. simensis* is possible, but that species has never been recorded from EE or EW.

2. **URERA** *Gaud.* (1830).

Friis in *Nordic J. Bot.* **5**: 547-553 (1986); Friis & Immelman in *Nordic J. Bot.* **7**: 126 (1986).

Dioecious woody climbers, often fixed to the substratum by adventitous, axillary roots from the stems; stinging hairs on the herbaceous parts or at least a few on the petioles and inflorescence. Cystoliths dot-like. Leaves alternate, petiolate, lamina simple, with finely serrate or crenulate margin. Stipules fused for 1/2-2/3 of the length, intrapetiolar. Inflorescences lax axillary cymes. Male flowers 4-5-merous, with rudimentary ovary. Female flowers with 4 somewhat unequal tepals which are fused for 3/4 of their length or more. Ovary ovoid, with an almost sessile, penicillate stigma. Fruit an achene, enclosed by the persisting perianth which increases in size and becomes fleshy and red in fruit.

About 40 species, widespread in tropical Africa, on Madagascar, in tropical America and on the Pacific Islands.

1. Lamina with crenulate margin, with 3-4(-5) pairs of lateral nerves above the basal pair; female perianth covering about 3/4 of the ovary, with 2 minute outer lobes and 2 larger, almost circular inner ones. **1. U. hypselodendron**
- Lamina with entire margin, never crenulate or dentate, with 2(-3) pairs of lateral nerves above basal pair; female perianth ± tubular, almost entirely enclosing the ovary and the young fruit. **2. U. trinervis**

1. **U. hypselodendron** *(A. Rich.) Wedd.* (1852).

Urtica hypselodendron A. Rich. (1850) - types: GD, Semien, on Mt. Endschetcap towards Shoata, *Schimper* III:1136 (P syn., K isosyn.), TU, Chire, *Quartin Dillon* s.n (P syn.); *Urera schimperi* Wedd. (1856), nom. illeg., based on same type as above.

Woody climber to 20 m or more, with stinging hairs on the younger parts. Leaves: lamina ovate to obovate, 10-18 x 7-13 cm, usually glabrous on both sides, base rounded to subcordate, margin finely serrate or crenulate, apex short acuminate; petiole up to 4-10 cm long, mostly with a few stinging hairs. Stipules broadly triangular to lanceolate, up to 3 mm long. Inflorescence a lax cyme, usually with numerous stinging hairs, 4-18 cm long. Male flowers on *c* 2 mm long pedicels, *c* 1.5 mm in diam., with stinging hairs. Female flowers subsessile, surrounded by stinging hairs, *c* 1 mm long. Achene up to 1.5 mm long, enclosed in the persisting, fleshy perianth. Fig. 105.2.A-G.

A common climber in clearings of upland rain forest and upland dry evergreen forest, along forest edges, in riverine or ravine forest, or sometimes on isolated trees left in farmland; 1700-2800 m. TU GD GJ WG SU HA BA AR KF IL SD; Kenya, Uganda, Tanzania, Malawi, E. Zaire. *Meyer* 8776; *Burger* 579; *Mooney* 6035.

The name has been taken from a *nomen nudum* proposed by C. F. Hochstetter.

2. **U. trinervis** *(Hochst.) Friis & Immelman* (1987).

Elatostema trinerve Hochst. (1846).
Urera cameroonensis Wedd. (1869).

Dioecious liana climbing to 30 m or more. Stems glabrous or rarely pubescent, mostly with numerous adventitious roots, usually without stinging hairs. Leaves: lamina leathery, green (drying pale green), elliptic, ovate or obovate, 6-12 x 3.5-8 cm, triplinerved, with the basal pair of nerves reaching more than 2/3 towards the apex, always with one other pair of nerves (rarely more), glabrous, rarely with scattered stiff hairs above and rarely with stinging hairs; base cuneate, rounded, rarely subcordate; margin entire: apex long acuminate; petiole 2.5-5 cm. Stipules brown, 6-10 mm long. Male inflorescences up to 8 cm in diam., with glabrous branches, rarely with stinging hairs. Male flowers clustered, on *c* 1 mm long pedicels, articulated below the perianth, tetramerous, up to 1 mm in diam., with obconical pistillode. Female inflorescence up to 5 cm in diam., main axis glabrous or puberulous, rarely with a few stinging hairs. Female flowers sessile, clustered, surrounded by stinging hairs. Female perianth up to 1 mm long, tubular, 4-toothed, constricted at apex. Ovary enclosed, with only the penicillate stigma protruding. Achene up to 2 mm long, ± completely enclosed in the accrescent fleshy perianth. Fig. 105.2.H-K.

A climber in lowland rain forest, especially at forest edges and in clearings; 900-1300 m. KF IL; throughout

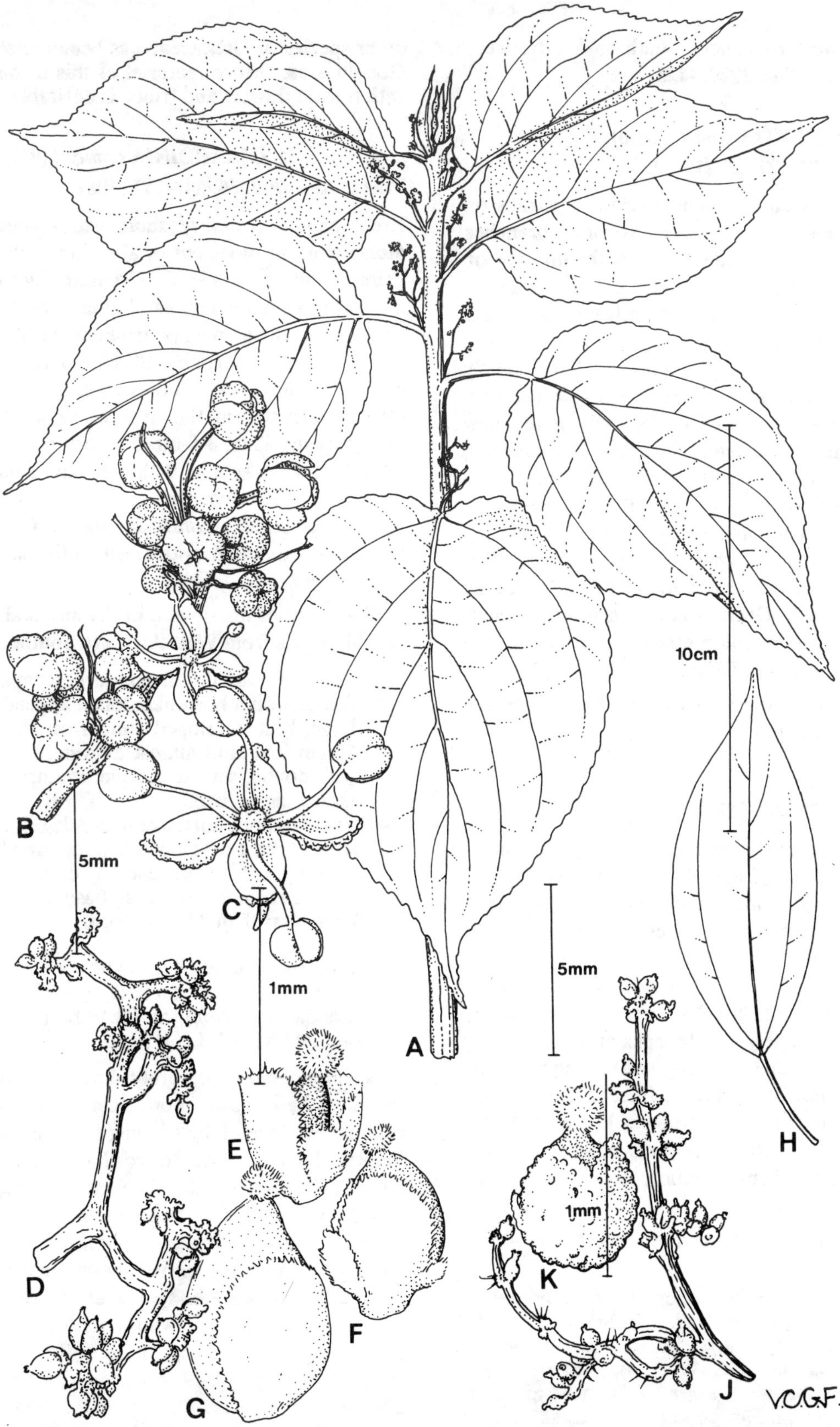

Figure 105.2. *URERA HYPSELODENDRON*: A - leafy branch with inflorescences from *Chiovenda* 2593; B - part of male inflorescence; C - male flower, both from *Wild* 3570; D - part of female inflorescence; E & F - young and older female flowers; G - fruit with persisting perianth, all from *Mooney* 8720. *U. TRINERVIS*: H - leaf from *Hansen* 120; J - part of female inflorescence; K - young fruit with persisting fleshy perianth, both from *Faulkner* 4199. Drawn by Victoria Friis. (Modified and reproduced by permission, Editorial Board, *Flora Zambesiaca*.)

central Africa, west to Ghana, south to S Afr. Rep. (Natal). *Friis* et al. 3845, 3920, 4125.

3. OBETIA *Gaud.* (1830).

Friis in *Kew Bull.* **38**: 221-228 (1983).

Small to medium sized dioecious, deciduous shrubs or small trees with soft, juicy wood; younger branches herbaceous, with stinging hairs. Cystoliths dot-like or slightly elongated. Leaves alternate, petiolate, lamina simple or lobed, mostly with a velvety tomentum beneath, base rounded or cordate. Stipules free, lateral, mostly long persistent. Inflorescences axillary, mostly developed after leaves have fallen and before the new leaves appear, but sometimes in the axils of the new leaves. Flowers pedicellate, clustered in branched, bracteate panicles, the male inflorescences much smaller than the female ones. Male flowers regular, 5-merous, with rudimentary ovary. Female flowers irregular, 4-merous, outer pair of tepals smaller than inner pair, tepals increasing in size and becoming thinly membranous in fruit. Ovary ovoid, with sessile penicillate stigma. Achene compressed, enclosed in the persisting membranous perianth.

Eight species in E and S Africa, on Madagascar, the Mascarenes and Aldabra.

O. sp. = *Burger* 3368.
Friis in *Kew Bull.* **38**: 226 (1983).

A branched shrubby perennial or small shrub, presumably dioecious. Bark of older branches brown. Younger branches, leaves and inflorescences with numerous stinging hairs. Leaves: lamina ovoid to almost circular, 2-6 x 1.5-5 cm, above with stinging hairs and a few stiff hairs, beneath pubescent when young, later glabrescent, base broadly cuneate, truncate or rounded, margin coarsely serrate, apex acute or broadly acuminate; petiole 0.5-3 cm long, with stinging hairs. Stipules up to 3 mm long, lateral, free. Male inflorescences scattered in the axils of younger leaves, up to 1.5 cm long, bracteate; male flowers pedicellate, 5-merous, *c* 1 mm in diam. Female inflorescences and fruits unknown.

In rocky ravine in deciduous bushland on limestone. HA. *Burger* 3368.

Only known from this collection made on the upper slope of the Gobelli River Valley, below Mt. Daletti (8°32'N). Apparently a rather isolated species which seems to be closest to *O. tenax* (N. E. Brown) Friis from Transvaal and Zimbabwe. *Burger* 3368 may be identical with the "tall perennial *Fleurya lanceolata* Engl." which Engler in *Sitzungsber. könichl. Preuss. Akad. Wissensch.* 10: 389 (1904) stated had been collected in deciduous bushland near the Gobelli River in Ennia Galla country by Ellenbeck. The Ellenbeck specimen is now lost, but the leaf shape and the distribution of the male inflorescences along the stems of the Burger specimen shows resemblance with true *Fleurya lanceolata*. No other species of Urticaceae has been collected from the Gobelli area. More material of this taxon, particularly with female flowers and fruits, is desirable.

4. GIRARDINIA *Gaud.* (1830).

Friis in *Kew Bull.* **36**: 143-157 (1981).

Erect annual (or very short lived perennial) herbs, monoecious or dioecious by abortion, with long stinging hairs on all aerial parts. Cystoliths dot-like. Leaves alternate, petiolate, lamina elliptic to ovate, mostly very variously divided, always triplinerved from the base, margin coarsely serrate. Stipules intra-petiolar, fused almost to the apex. Inflorescences unisexual, dense, elongated cymes in the axils of upper leaves. Male flowers pedicellate, 4-5-merous with rudimentary ovary. Female flowers sessile, with 3 almost completely fused tepals, ± enclosing the ovary, and sometimes one minute free tepal. Staminodes absent. Ovary ± reflexed, laterally compressed, with sessile, filiform stigma. Fruit a rugose achene.

Two species distributed in the mountains of the Old World tropics from West Africa to Southern China.

1. Tepals of male flower and largest tepal of female flower with a horn-like dorsal appendage; leaves large, best developed lamina usually more than 15 cm long and almost as wide, bullate, always with double serrate margin; mature stems more than 2 cm in diam. **1. G. bullosa**
- Tepals without horn-like appendages; leaves smaller, best developed lamina often less than 15 cm long, not bullate, usually serrate (if double serrate then only so at the base of lamina); mature stems less than 2 cm in diam. **2. G. diversifolia**

1. G. bullosa *(Steudel) Wedd.* (1854);
Urtica bullosa Steudel (1850). - type: GD, Tschenausa, *Schimper* III:1409 (P herb. Steudel holo., A BM G L LE P iso.).

Erect monoecious or rarely dioecious annual to 3 m tall, stem base mostly more than 2 cm in diam., pubescent, covered with long stinging hairs on all aerial parts. Leaves: lamina ovate to cordate, not conspicuously lobed, (10-)15-40 x (10-)15-35 cm, bullate, glabrescent above apart from the stinging hairs, densely pubescent beneath, with a longer tomentum and numerous stinging hairs on the nerves, base truncate to cordate, margin double serrate, apex acute to short acuminate; petiole 5-20 cm long, with numerous stinging hairs. Stipules up to 2 cm long. Male inflorescences narrow, up to 20 cm long, ± branched panicles of densely contracted cymes. Male flowers on *c* 3 mm long pedicels, 4-5-merous, each tepal with dorsal horn-like appendage. Female inflorescences appear less branched, forming very dense cylinders of small, repeatedly branched dichasia, 5-18 cm long and *c* 3 cm in diam. Female flowers up to 1.5 mm long, with 3 fused sepals,

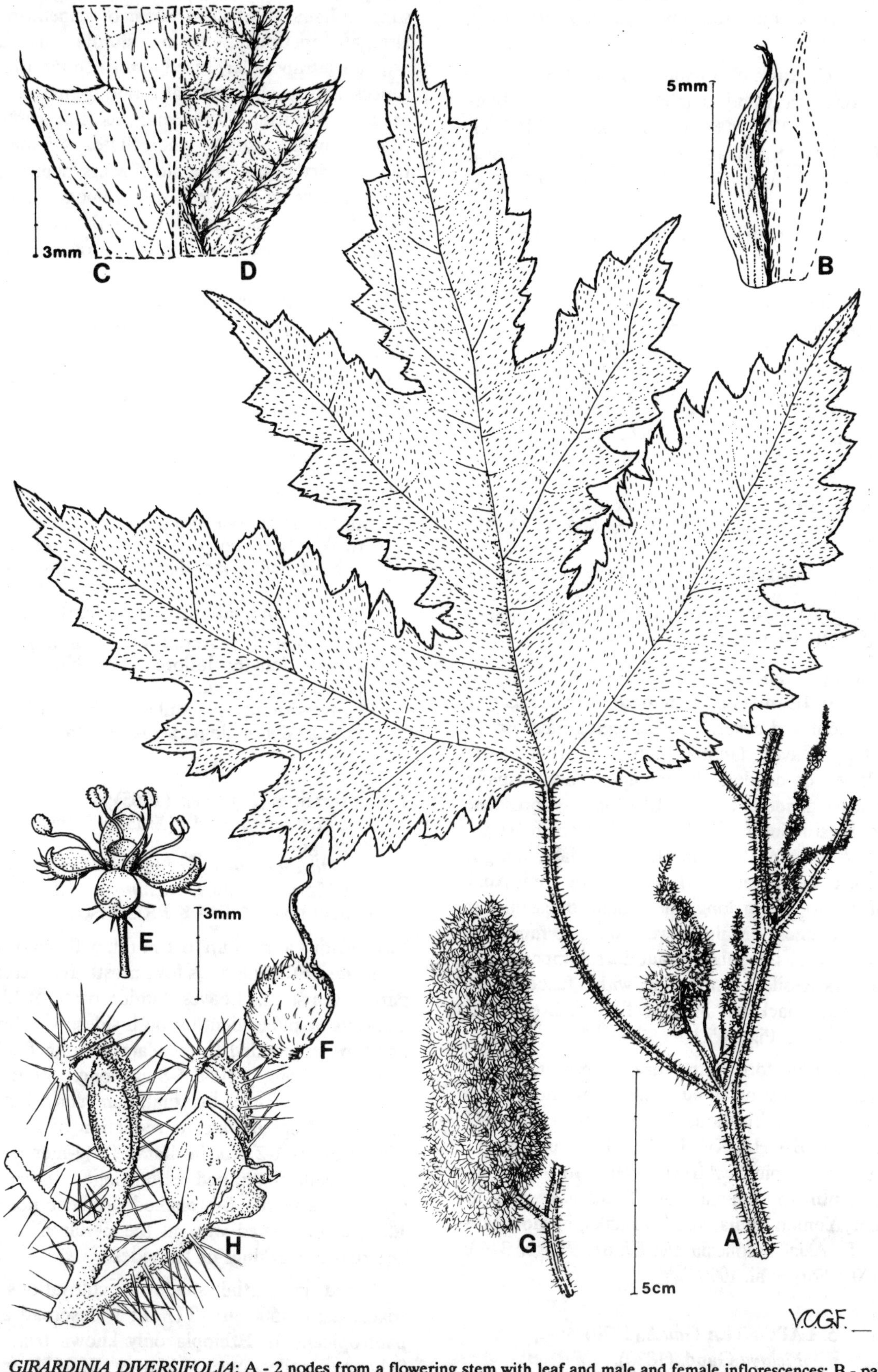

Figure 105.3. *GIRARDINIA DIVERSIFOLIA*: A - 2 nodes from a flowering stem with leaf and male and female inflorescences; B - pair of partly fused stipules; C - detail of leaf from above; D - detail of leaf from below, all from *Biegel* 2889; E - male flower, from *Letouzey* 6922 (redrawn from *Fl. Cameroun* 8: pl. 17,4); F - female flower with perianth on one side, from *Chiovenda* 1563 (redrawn from Friis l.c.); G - ripe infructescence, from *Biegel* 2889; H - detail of female inflorescence with two flowers, from *Jacques-Felix* 2690 (redrawn from *Fl. Cameroun* 8: pl. 17,6). Drawn by Victoria Friis. (Reproduced by permission, Editorial Board, *Flora Zambesiaca*.)

largest sepal with dorsal appendage, minute free tepal present. Achenes compressed, somewhat assymmetrical, rugose.

Edges and clearings of upland rain forest, in upland grassland along roads and near houses, often in abandoned cultivations; 1750-3000 m. GD GJ SU AR WG KF SD BA HA; S Sudan (Imatong Mts.), Kenya, Tanzania, Burundi. *Meyer* 7634; *Burger* 1969; *Friis* et al. 1094.

Fibre for making string can be extracted from the stem.

2. G. **diversifolia** *(Link) Friis* (1981);
Urtica diversifolia Link (1822).
Girardinia heterophylla Decn. (1844).
Urtica adoensis Steudel (1850); *Girardinia adoensis* (Steudel) Wedd. (1854); *G. condensata* (Steudel) Wedd. var. *adoensis* (Steudel) Engl. ex De Wildem. in *Ann. Mus. Congo Belge, Bot.*, ser. **IV, 1**: 173 (1903); *G. heterophylla* Decn. subsp. *adoensis* (Steudel) Cuf. in *Bull. Jard. Bot. Nat. Belg.* **39**, suppl.: XX (1969) - type: TU, near Adua, *Schimper* II:101 (P herb. Steudel holo., A BM G K LE LISU P S iso.).
Urtica condensata Steudel (1850); *Girardinia condensata* (Steudel) Wedd. (1854) - type: TU, Mount Selleuda, *Schimper* III:1888 (P herb. Steudel holo., BM BR FT G K L LE P S iso.).

Erect, sparsely branched monoecious or dioecious annual herb to 1.5 m tall, stem base to *c* 1.5 cm in diam., pubescent and covered with stinging hairs on all aerial parts. Leaves: lamina ovate to cordate, very variously lobed or divided, the younger leaves usually being the least divided, 10-20 x 10-18 cm, glabrescent except for the stinging hairs above, pubescent beneath, not bullate, base cuneate, truncate or cordate, margin serrate, apex acuminate. Inflorescence unisexual, cylindrical, up to 10 cm long, the female ones thicker than the male ones. Male flowers on *c* 1 mm long pedicels, 4-5-merous, tepals without dorsal appendages. Female flowers sessile, *c* 2 mm long, with 3 fused tepals, free tepal usually lacking. Achene flattened, ovate to subcordate, rugose. Fig. 105.3.

In upland rain forest and dry evergreen forest, especially in clearings or along roads, or in moderately shaded areas near cultivations; 1200-2100 m. TU GD GJ WG SU AR BA HA KF IL SD; widespread in the mountains of tropical Africa, west to Guinean Republic, south to Angola and Transvaal, also in Madagascar, Yemen, India, and Sri Lanka, in mountain areas of S. E. Asia, Indonesia and China. *Burger* 3163; *Mooney* 6004; *Friis* et al. 1990.

5. LAPORTEA *Gaud.* (1830).

Fleurya Gaud. (1830).

Chew in *Garden's Bull. Singapore* **25**: 111-178 (1969).

Branched monoecious or dioecious annual or perennial herbs (shrubs outside flora area) with numerous stinging hairs. Cystoliths linear or punctiform. Leaves alternate, petiolate, lamina simple, margin serrate. Stipules intrapetiolar, fused almost to the tip. Inflorescences unisexual, paniculate, but sometimes with very reduced branches, appearing as interrupted spikes. Male flowers usually pedicellate, 4-5-merous, with rudimentary ovary. Female flowers pedicellate, sometimes with winged pedicel, with 4 unequal tepals: lateral pair of tepals usually much larger than the median pair, tepals of the median pair equal or unequal in size; staminodes absent. Ovary asymmetrical, ovoid, compressed; stigma sessile, filiform, or deeply trifid with filiform branches. Achene compressed, often with a characteristic sculpturing on the sides, stipitate and shed without the persistent perianth (sect. *Laportea*), or sessile and shed with the persisting perianth (sect. *Fleurya*).

About 50 species, pantropical, with an extension into the temperate regions of N America and E Asia.

1. Stigma trifid (one branch longer than the other two); inflorescence appears like an interrupted, unbranched spike due to reduction of lateral branches. **1. L. interrupta**
- Stigma simple; inflorescences clearly branched. 2
2. Plants densely covered with glandular hairs of varying length; inflorescence bisexual, female pedicels unwinged. **2. L. aestuans**
- Plant without glandular hairs; inflorescence unisexual, female pedicels winged. **3. L. alatipes**

1. L. **interrupta** *(L.) Chew* (1965);
Urtica interrupta L. (1753); *Fleurya interrupta* (L.) R. Wright (1853).
Urtica lomatocarpa Steudel (1850) - type: TU, in Mai-Mezano valley, *Schimper* III:1471 (P herb Steudel holo., A BM K PRC S iso.).

Monoecious annual up to 1 m tall with sparsely branching stems. Stinging hairs few, mostly restricted to upper part of the stem. Leaves: lamina ovate, 8-12 x 5-7 cm, glabrous or puberulent on both sides, with few stinging hairs; base rounded to truncate; margin finely serrate; apex acuminate; petiole 5-8 cm long. Stipules 3-5 mm long. Inflorescences bisexual, paniculate, but appear as interrupted spikes due to reduction of side branches. Male flowers 1-2 mm in diam., (3-)4-merous. Female flowers pedicellate, pedicel unwinged. Ovary compressed, stigma sessile, 3-branched. Achene not stipitate, *c* 1.2 mm long, shed without the perianth, with circular and rugose markings on the sides.

Weed in partial shade in cultivations or along roadsides; *c* 1500 m. TU IL; widespread and almost pantropical. In Ethiopia only known from two early Schimper collections: *Schimper* III:1471 & *Schimper* I:88, and a recent one, *Gilbert & Friis* 8374.

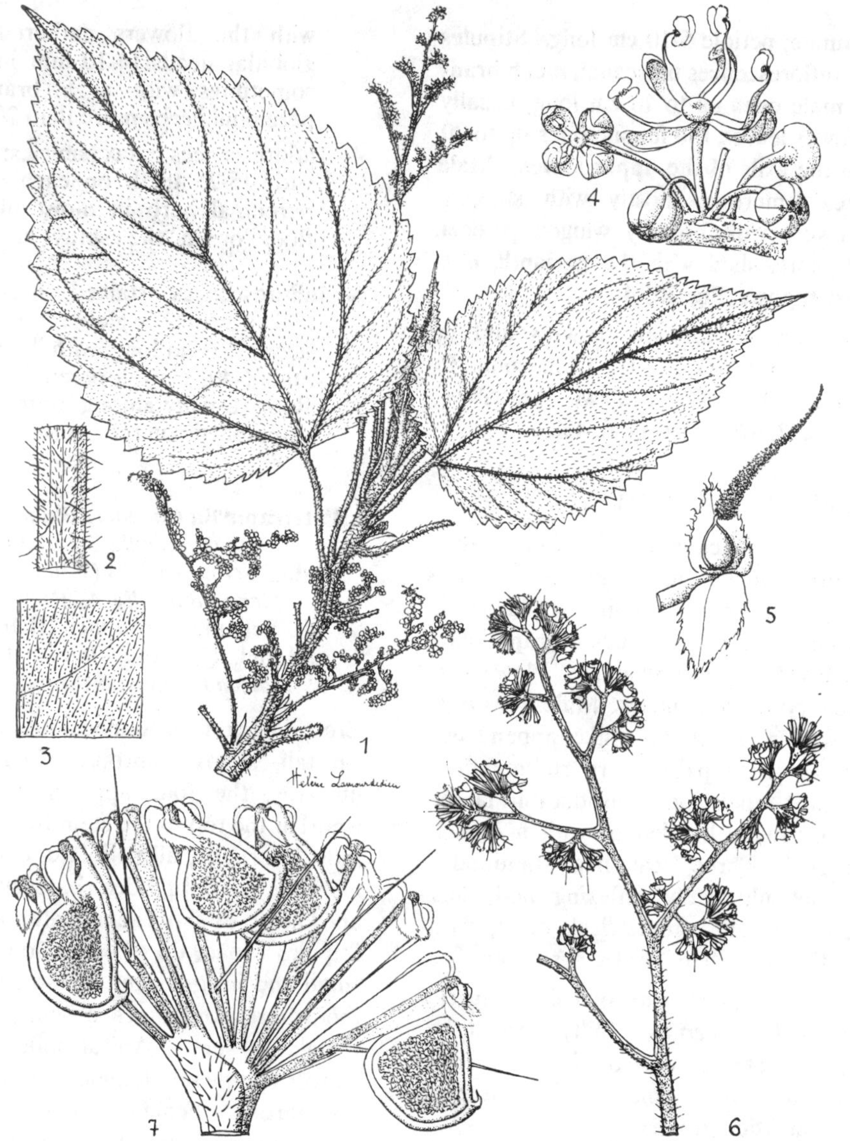

Figure 105.4. *LAPORTEA ALATIPES*: 1 - flowering branch x 2/3; 2 - detail of stem x 2; 3 - detail of upper surface of leaf x 2; 4 - detail of male inflorescence x 8, all from *Maitland* 1368; 5 - female flower x 16; 6 & 7 - details of infructescence, all from *MC* 2252. (Reproduced with permission from *Fl. Cameroun* 8, pl. 18.)

2. **L. aestuans** *(L.) Chew* (1965);
Urtica aestuans L. (1753); *Fleurya aestuans* (L.) Miq. (1853).
Urtica schimperiana Steudel (1850) - type: TU, Modat, *Schimper* III:1739 (P herb. Steudel holo., K iso.).

Erect annual monoecious herb to *c* 1 m tall, covered with stinging hairs and long glandular hairs. Leaves: lamina ovate to broadly ovate, 10-15 x 8-12 cm, usually with stinging hairs on both sides but without glandular hairs; base rounded to truncate, margin regularly serrate, apex acuminate; petiole 5-15 cm long. Stipules up to 1 cm long. Inflorescences bisexual, mainly in the axils of the uppermost leaves, to *c* 20 cm long, densely paniculate, covered with glandular and stinging hairs. Male flowers pedicellate, 4-5-merous, 1-2 mm in diam., usually with glandular hairs on the outside. Female flowers with unwinged pedicel. Achenes 1-2 mm long, shortly stipitate, shed without the perianth, with circular and rugose markings on the side.

Weed in partial shade along roads and in cultivations; sea level to 1300 m. EE EW TU GG; a pantropical weed. *Ehrenberg* 43; *Pappi* 4078; *Schweinfurth & Riva* 475.

Steudel's name for this species was taken from a *nomen nudum* proposed by C. F. Hochstetter, independently taken up with a *Quartin-Dillon* syntype from Shire (TU) by A. Richard (late 1850).

3. **L. alatipes** *Hook. f.* (1864).

Erect annual or shortlived perennial, monoecious herb to 1.5 m, somewhat woody at the base, with numerous stinging hairs on all aerial parts, but without glandular hairs. Leaves: lamina ovate, 8-15 x 4-8 cm, with stinging hairs and a few stiff hairs above, pubescent below; base cuneate, rounded or subcordate, margin coarsely serrate, apex acuminate; petiole 5-10 cm long. Stipules up to 1.5 cm long. Inflorescences unisexual, much branched panicles, the male ones up to 10 cm long, usually in the axils of the lower leaves, the female ones up to 20 cm long, usually in the axils of the upper leaves. Male flowers pedicellate, 4-merous, mostly with stinging hairs. Female flowers with broadly winged pedicel. Achene slightly stipitate, shed with the perianth, *c* 1 mm long, irregularly rugose. Fig. 105.4.

Upland rain forests, *Arundinaria* thickets and in ravines; 1700-2700 m. BA SD KF IL; Cameroun, E Zaire, Uganda, Kenya, Tanzania, Malawi, Zimbabwe, Transvaal. *Mooney* 8487, 8765; *Friis* et al. 4523.

6. **PILEA** *Lindl.* (1821).

Annual or perennial, monoecious or dioecious herbs, with ± juicy stems; stinging hairs absent. Leaves opposite, petiolate or subsessile, lamina triplinerved, usually with serrate margin. Stipules intrapetiolar, almost completely fused. Cystoliths linear. Flowers in unisexual or bisexual cymes or panicles. Male flowers 3-5-merous, tepals with a dorsal horn-like appendage; stamens as many as the tepals; ovary rudimentary. Female flowers with 3 tepals, one considerably larger than the other two, usually hooded or with a dorsal horn-like appendage. Three scale-like staminodes present, inflexed, but ultimately reflexing and thus ejecting the achene. Ovary symmetrical, erect, with sessile, penicillate stigma. Achene ovate, ± smooth.

About 200 species, almost pantropical in distribution. One species, *P. cadieri* A. Guill., with fleshy partly variegated obovate leaves originating from tropical Asia, is cultivated as an ornamental in gardens and as a house plant in Addis Ababa.

1. Female inflorescences restricted to the top of the plant, very dense cymes subtended by the four uppermost leaves in an apparent whorl. **1. P. tetraphylla**
- Female inflorescences not restricted to the leaf axils of the four uppermost leaves, distributed along the stems; all leaves with clearly discernable internodes. 2
2. Inflorescences sessile in apparent whorls at the nodes. **3. P. rivularis**
- Inflorescences clearly pedunculate, not forming apparent whorls. 3
3. Inflorescences always much shorter than the petiole of the subtending leaf, often 2-6 together in the leaf-axil; each inflorescence with flowers clustered in 1-8 small secund cymules up to 3 mm in diam. on the main axis or on small side branches; an annual herb usually less than 20 cm tall. **2. P. angolensis**
- Inflorescences often (and male inflorescences always) as long as or longer than the petiole of the subtending leaf, only 1-2(-4) together in the leaf-axil; inflorescences either panicles, stalked heads 4-10 mm in diam., or interrupted spikes with the flowers gathered together in 2-3 globular heads (4-10 mm in diam.) along axis, sometimes with a ± branched axis; usually perennial herbs more than 20 cm tall. 4
4. Inflorescences ± lax panicles; leaves 8-12 cm long, with 15-25 teeth on each side; 30-60 cm tall perennial herb in moist places on the forest floor or in the spray zone from waterfalls. **4. P. bambuseti**
- Inflorescences stalked heads or interrupted spikes with 2-3 globular heads along the axis; leaves 2-6 cm long, with 7-15 teeth on each side; 10-30(-40) cm tall perennial herbs on the forest floor (usually among moss at the base of trees) or epiphytic on the trunks of trees or tree ferns. **5. P. johnstonii**

1. **P. tetraphylla** *(Steudel) Blume* (1856);

Urtica tetraphylla Steudel (early 1850) - type: TU, Adua, *Schimper* I:74 (P herb. Steudel holo., K iso.).

Pilea quadrifolia A. Rich. (late 1850) - type: TU, Mt. Selleuda, *Schimper* III:1680 (P syn., K isosyn.) and ibid., *Quartin Dillon* s.n. (P syn.).

Adicea tetraphylla (Steudel) O. Kuntze (1891).

Erect dioecious or monoecious annual herb to about 20 cm tall, stems glabrous. Leaves increasing in size upwards, the four uppermost forming an apparent whorl at the top of the stem by shortening of the uppermost internode, all petiolate, ovate, 0.7-2 x 0.4-1.5 cm, glabrous on both sides, base cuneate, margin serrate, apex acute or short acuminate; petiole up to 1 cm long. Stipules broad, up to 4 mm long. Male inflorescences small and insignificant, in the axils of the leaves below the whorl, sometimes absent, sometimes with a few female flowers. Apical inflorescence completely or almost completely female. Male flowers almost sessile, 3-4-merous. Female flowers almost sessile, with one large and 2(-3?) smaller tepals and the same number of staminodes. Stigma sessile, penicillate. Achene *c* 1.5 mm long. Fig. 105.5.A-E.

In humid, shaded areas in upland rain forest and "coffee" forest, under boulders and in rock crevices, or as a weed in gardens; 1350-2400 m. EW TU SU KF IL; in mountain areas of tropical Africa, west to Cameroun and Fernando Po, south to Angola and Malawi, also in Madagascar. *Mooney* 6001, 7399; *Friis* et al. 66.

2. **P. angolensis** *(Hiern) Rendle* (1917).

Adicea tetraphylla (Steudel) O. Kuntze var. *angolensis* Hiern, *Cat. Welw. Afr. Pl.* 1: 990 (1900).

Erect monoecious annual herb with minute reddish glands, up to 15 cm tall, but usually much smaller. Leaves usually largest at or just below the top of the stem, lamina ovate, 2-3 x 1-2 cm, glabrous on both sides; base truncate to rounded, margin serrate with 4-8 teeth (rarely, on very small leaves, entire), apex acuminate, often with a slightly drawn out tip; petioles up to 4 cm long. Stipules small, early deciduous. Inflorescences distributed in the leaf axils along the stem, uni- or

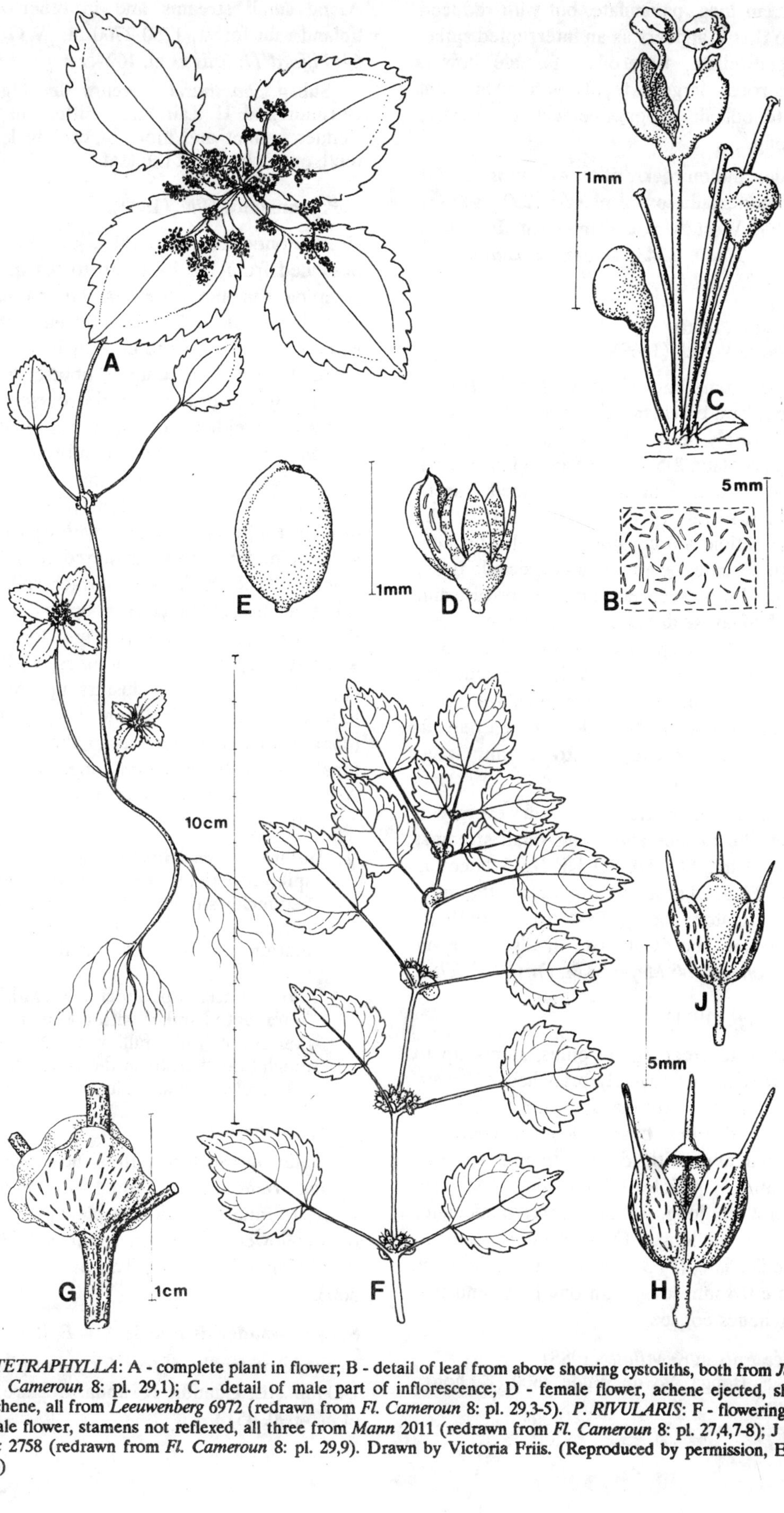

Figure 105.5. *PILEA TETRAPHYLLA*: A - complete plant in flower; B - detail of leaf from above showing cystoliths, both from *Jacques-Felix* 2640 (redrawn from *Fl. Cameroun* 8: pl. 29,1); C - detail of male part of inflorescence; D - female flower, achene ejected, showing reflexed staminodes; E - achene, all from *Leeuwenberg* 6972 (redrawn from *Fl. Cameroun* 8: pl. 29,3-5). *P. RIVULARIS*: F - flowering stem; G - detail of stipules; H - male flower, stamens not reflexed, all three from *Mann* 2011 (redrawn from *Fl. Cameroun* 8: pl. 27,4,7-8); J - female flower, from *Jacques-Felix* 2758 (redrawn from *Fl. Cameroun* 8: pl. 29,9). Drawn by Victoria Friis. (Reproduced by permission, Editorial Board, *Flora Zambesiaca*.)

bisexual, to *c* 1 cm long, paniculate, but with reduced side branches so that it appears as an interrupted spike. Male flowers pedicellate, 4-merous. Female flowers subsessile, 3-merous, largest tepal with prominent appendage on the outside. Stigma penicillate. Achene ovate, *c* 1 mm long.

In shade under and on rocks, often with mosses and other moisture- and shade-loving plants; 1250-1800 m. WG SU AR GG; West Africa, Cameroun, E. Zaire, Uganda, Angola. *Thulin* 1343; *Gilbert & Thulin* 373; *Gilbert & Tewolde* 2527.

3. **P. rivularis** *Wedd.* (1856).
P. ceratomera Wedd. (1869).

Perennial (perhaps sometimes short-lived ?) herb with creeping rhizomes and erect, mostly unbranched stems up to 0.5 m tall, rarely more. Leaves: lamina broadly ovate to ovate-lanceolate, 2-5 x 1.5-3.5 cm, glabrous on both sides; base obtuse, truncate or rounded, rarely subcordate, margin serrate or dentate with 5-15 teeth, apex acuminate; petiole 4-6 cm long. Stipules broadly ovate, often with cordate, or almost amplexicaule base, up to 0.5 cm long. Inflorescences dense uni- or bisexual clusters in the leaf axils, distributed along the stems. Male flowers on pedicels up to 2 mm long, 3-merous, tepals with a very prominent horn-like appendage. Female flowers on pedicels up to 2 mm long, 3-merous, tepals with appendages as in the male. Stigma penicillate. Achene ovate, smooth, up to 2 mm long. Fig. 105.5.F-J.

Upland rain forest and wetter areas in dry evergreen forest, usually in the ravines and along small streams; 1350-2600(-3200 ?) m. SU BA SD KF IL; S Nigeria, Cameroun, São Tomé, Fernando Po, Zaire, Rwanda, Burundi, Sudan (Imatong Mts.), East Africa south to Transvaal, also on the Comoro Ilands. and in Madagascar. *Mooney* 8474; *Meyer* 8876; *Friis* et al. 1786.

4. **P. bambuseti** *Engl.* (1911).

Perennial herbs from creeping rhizomes, stems up to 0.9 m tall, unbranched or sparsely branched. Leaves: lamina oblong-lanceolate to almost ovate, 8-12 x 3-5 cm, glabrous on both sides; base cuneate to rounded, margin serrate to dentate with 15-28 teeth on each side, apex long acuminate; petiole 2-3 cm long. Stipules triangular, up to 2 mm long. Inflorescences paniculate, unisexual, up to 4 cm long. Flowers poorly known. Male flowers pedicellate, (?) 3-merous, *c* 1 mm in diam. Female flowers *c* 0.8 mm long, with one large and two small tepals. Achenes not seen.

subsp. **aethiopica** *Friis* in *Kew Bull.* (1988)
- type: KF, Bonga, W. de Wilde 7800 (K, holo., BR WAG iso.).

Along small streams and in other moist places in upland rain forest; 1750-2700 m. WG KF IL. *Palmer* 64; *Piffard* 77; *Friis* et al. 1624.

Subsp. *bambuseti* occurs in Uganda, Rwanda, Burundi and E Zaire and differs in having a more slender habitat and more lanceolate leaves with 10-15 teeth on each side of the leaf.

5. **P. johnstonii** *Oliv.* (1887).

Erect monoecious, perennial, juicy herb, little to much branched from the base, up to 0.4 m tall, with short, creeping rhizome. Leaves: lamina broadly ovate to ovate-lanceolate 1-6 x 0.8-4.5 cm, with scattered stiff hairs above, glabrous to densely hairy, especially on the nerves below; base cuneate, obtuse, truncate or rounded, rarely subcordate, margin coarsely dentate, with 7-15 teeth on either side, apex acuminate; petiole 1.5-4 cm long. Stipules triangular to broadly ovate, up to 5 x 5 mm, deciduous or persisting. Inflorescence axillary, stalked heads or interrupted spikes, sometimes with a branched inflorescence axis, with up to 4 inflorescences at each node; flowers gathered into heads or in 2-6 glomerules along the axis of each spike, or on the few side branches. Male clusters up to 8 mm in diam.; male flowers on pedicels up to 2 mm long, 4-merous, with apiculate tepals (or sometimes with a ± horn-like appendage). Female clusters up to 6 mm in diam.; female flowers on 1-2 mm long pedicels, 3-merous, tepals with appendages as in the male flower, middle tepal lanceolate, *c* 3 mm long, lateral tepals shorter. Achene compressed, lanceolate, smooth, shiny brown, 1-2 mm long.

1. Inflorescences mostly branched or interrupted spikes, with more than one head on each peduncle; stipules more than 3 mm wide, long persisting; plants usually little branched; leaf margin with 7-15 teeth on each side. subsp. **johnstonii**
- Inflorescences a pedunculate head, very rarely an "interrupted spike" with two heads; stipules less than 3 mm wide, falling soon; plants sometimes much branched from the base; leaf margins with 4-10 teeth on each side. subsp. **rwandensis**

Subsp. **johnstonii**

In shade on the forest floor near the upper limits of upland rain forest, often with mosses and other moisture- and shade-loving plants, sometimes epiphytic on the lower part of tree trunks; *c* 3200 m. BA; E Zaire, Uganda, Kenya, Tanzania, Malawi. *Friis* et al. 3603.

Subsp. **rwandensis** *Friis* in *Kew Bull.* (1988)

In shade on trunks of tree ferns in upland rain forest; *c* 2400 m. IL; Uganda, Rwanda, E Zaire, Tanzania. *Friis* et al. 4090, 4512.

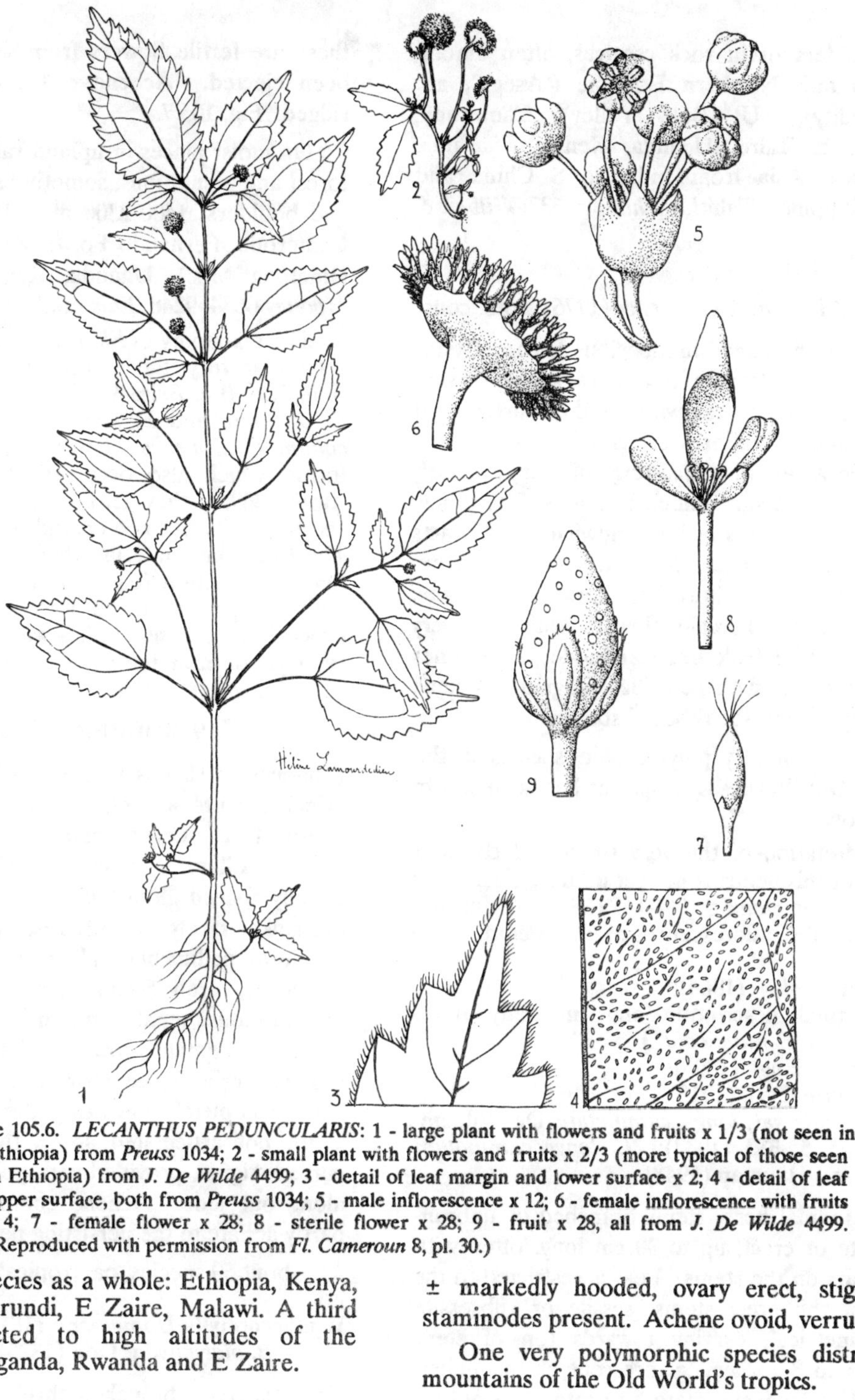

Figure 105.6. *LECANTHUS PEDUNCULARIS*: 1 - large plant with flowers and fruits x 1/3 (not seen in Ethiopia) from *Preuss* 1034; 2 - small plant with flowers and fruits x 2/3 (more typical of those seen in Ethiopia) from *J. De Wilde* 4499; 3 - detail of leaf margin and lower surface x 2; 4 - detail of leaf upper surface, both from *Preuss* 1034; 5 - male inflorescence x 12; 6 - female inflorescence with fruits x 4; 7 - female flower x 28; 8 - sterile flower x 28; 9 - fruit x 28, all from *J. De Wilde* 4499. (Reproduced with permission from *Fl. Cameroun* 8, pl. 30.)

Distribution of species as a whole: Ethiopia, Kenya, Uganda, Rwanda, Burundi, E Zaire, Malawi. A third subspecies is restricted to high altitudes of the mountains between Uganda, Rwanda and E Zaire.

7. LECANTHUS Wedd. (1854).

Annual or short-lived perennial, monoecious or apparently dioecious herbs. Cystoliths elongated. Leaves petiolate, opposite; the two of a pair sometimes markedly unequal in size. Stipules intrapetiolar, fused. Inflorescences single in the leaf axils, unisexual, rarely bisexual, pedunculate receptacles surrounded by fused bracts. Male inflorescences usually small, ± bell or cup-shaped, usually with less than 10 male flowers. Male flowers 4-5-merous, with rudimentary ovary. Female inflorescences larger, with many flowers on a flat, disc-shaped receptacle. Female flowers 4-(3-)merous, tepals ± markedly hooded, ovary erect, stigma penicillate, staminodes present. Achene ovoid, verrucose.

One very polymorphic species distributed in the mountains of the Old World's tropics.

L. peduncularis *(Royle) Wedd.* (1869).

Annual (in Flora area) herb to 10 cm tall, with juicy stems, branched from the lower nodes, with short, stiff hairs. Specimens from Flora area with 2-3 pairs of leaves only. Leaves: lamina ovate, 1-1.5 x 0.5-1 cm, almost glabrous above, puberulous below; base cuneate, margin serrate, apex acuminate; petioles to *c* 1 cm long. Inflorescences on 0.5-1 cm long peduncles; male bell-shaped ones up to 3 mm in diam., discoid female ones up to 6 mm in diam. Male flowers pedicellate, *c* 0.5 mm in diam. Female flowers pedicellate or sessile, up to 0.8 mm long. Achene up to 1 mm long. Fig. 105.6.

Under boulders or in rock crevices, often among moss; *c* 1650 m. Northern Ethiopia ("Asega", an untraced locality), SU (Bole Valley); Cameroun, Fernando Po, E Zaire, Uganda, Kenya, Tanzania, Malawi, tropical Asia from India to S China and Indonesia, Philippines, Tahiti. *Schimper* 637; *Gilbert & Tewolde* 2508.

8. ELATOSTEMA *J. R. & G. Forster* (1776), nom. cons.

Erect annual and perennial monoecious herbs. Cystoliths linear. Leaves alternate, distichous, sessile or very short petiolate, asymmetrical, with fused or partly fused intrapetiolar stipules. Inflorescences unisexual, axillary, sessile (in flora area), consisting of very densely clustered flowers and surrounded by almost free bracts in the male inflorescences or by somewhat fused bracts in the female inflorescences. Male flowers 4-5-merous, tepals generally with a horn-like dorsal appendage, ovary rudimentary. Female flowers with 3-5 very reduced tepals and 3 scale-like staminodes, sometimes sterile. Ovary ovoid, with penicillate stigma. Achene ovate, small, ejected by the reflexed staminodes.

About 200, often very polymorphic, species in the tropics of the Old World. One species has been shown to be apogamous.

The interpretation of the very small and compact inflorescence of this genus is not yet entirely clear. See Letouzey in *Fl. Cameroun* **8**: 144-145 (1968) for further discussion about the inflorescences of *Elatostema*.

E. monticolum *Hook. f.* (1864),
[in original description and in many floras erroneously as "*E. monticola*".]
E. orientale Engl. (1895).
E. kiwuense Engl. (1915).
?E. orientale var. *longeacuminatum* De Wildem., *Pl. Bequaert.* **5**: 383 (1932); *?E. longeacuminatum* (De Wildem.) Hauman (1948).

Annual or perennial herb, stems branched or unbranched, prostrate or erect, up to 30 cm long, often with short crisp hairs on the stems. Leaves restricted to the upper part of the erect stems, sessile or subsessile; lamina asymmetrical, curving towards top of stem, lanceolate, up to 8 x 2 cm, glabrescent above, puberulous below; base cuneate, margin serrate, apex obtuse, acute or long acuminate. Stipules fused to about the middle, up to 4 mm long. Male inflorescences very infrequent, probably mainly in the axils of the lower leaves, up to 3 mm in diam. Male flowers on *c* 2 mm long pedicels, 4-merous, tepals with dorsal horn-like appendages. Female inflorescences frequent, in the axils of the upper leaves, bell- or disc-shaped, up to 0.5 cm in diam, surrounded by bracts. Female flowers almost sessile, but probably becoming pedicellate during ripening of the achene, tepals 3, very reduced, 3 staminodes, ovary erect, stigma penicillate. Sterile pedicellate flowers apparently present, but possibly these are fertile flowers from which the achenes have been ejected. Achene *c* 1 mm long, longitudinally ridged. Fig. 105.7.

On moist places in upland rain forest or in "coffee" forest and plantations, sometimes in shade among rocks and boulders; 1400-2300 m. WG KF IL; S. Nigeria, Cameroun, Fernando Po, E. Zaire, Rwanda, S. Sudan (Imatong Mts.), Uganda, Kenya, Tanzania, Malawi. *Mooney* 6684, 9248; *Friis* et al. 1820.

The taxonomy of *Elatostema* in tropical Africa is not yet satisfactory; the author has followed Lambinon in *Bull. Soc. Bot. Belge* **91**: 209-211 (1959) in considering *E. orientale* and *E. kiwuense* synonyms of *E. monticolum*. *E. kiwuense* seems to represent the small annual forms, which also occur in Ethiopia (e.g. *Gilbert & Aweke* 2899); they may deserve recognition at infraspecific level, but not enough material is yet available. The Ethiopian material which has been identified as *E. longeacuminatum* (e.g. *Friis* et al. 220) belongs, in the author's opinion, to the present species and it is to be expected that closer studies will show that the name is a synonym of or an infraspecific taxon of *E. monticolum*.

9. BOEHMERIA *Jacq.* (1760)

Monoecious shrubs with opposite or alternate leaves. Stinging hairs absent. Cystoliths dot-like. Stipules lateral, free. Leaves petiolate, triplinerved. Inflorescences axillary, usually spike-like with unisexual and equally spaced glomerules of flowers along the axis of the spike, rarely the inflorescences consist of a single sessile or pedunculate glomerule. Male flowers usually 4-merous, rarely 5-merous or 3-merous, with ± fused, membranous tepals; rudimentary ovary present or absent. Female flowers with ± completely fused, tubular perianth, constricted and 2-4 toothed at the apex, completely enclosing the ovary. Style filiform, short, not articulated above the ovary, with a long, curved stigma, provided with pollen-perceptive papillae along one side. Achene completely enclosed in and partly adnate to the persisting perianth.

About 50 species, pantropical.

B. macrophylla *Hornemann* (1815)
B. platyphylla Don (1825).

Erect, sparsely branched shrub to 3 m; younger stems pubescent, older ones glabrous. Leaves opposite: lamina ovate (sometimes very broadly so) to lanceolate, 15-30 x 7-15 cm, with an indumentum of unequally long, rather stiff hairs above and almost equally long, softer hairs below; base cordate, rounded or cuneate, margin serrate, apex acuminate. Inflorescences pendent, axillary, interrupted spikes, with small, dense glomerules of flowers equally spaced along the axis, each glomerule covered by bracts when young, later without bracts. Male clusters 3-5 mm in diam., female clusters 5-8 mm in diam. Male flowers sessile, up to 2 mm long, tetramerous, almost glabrous. Female flowers sessile,

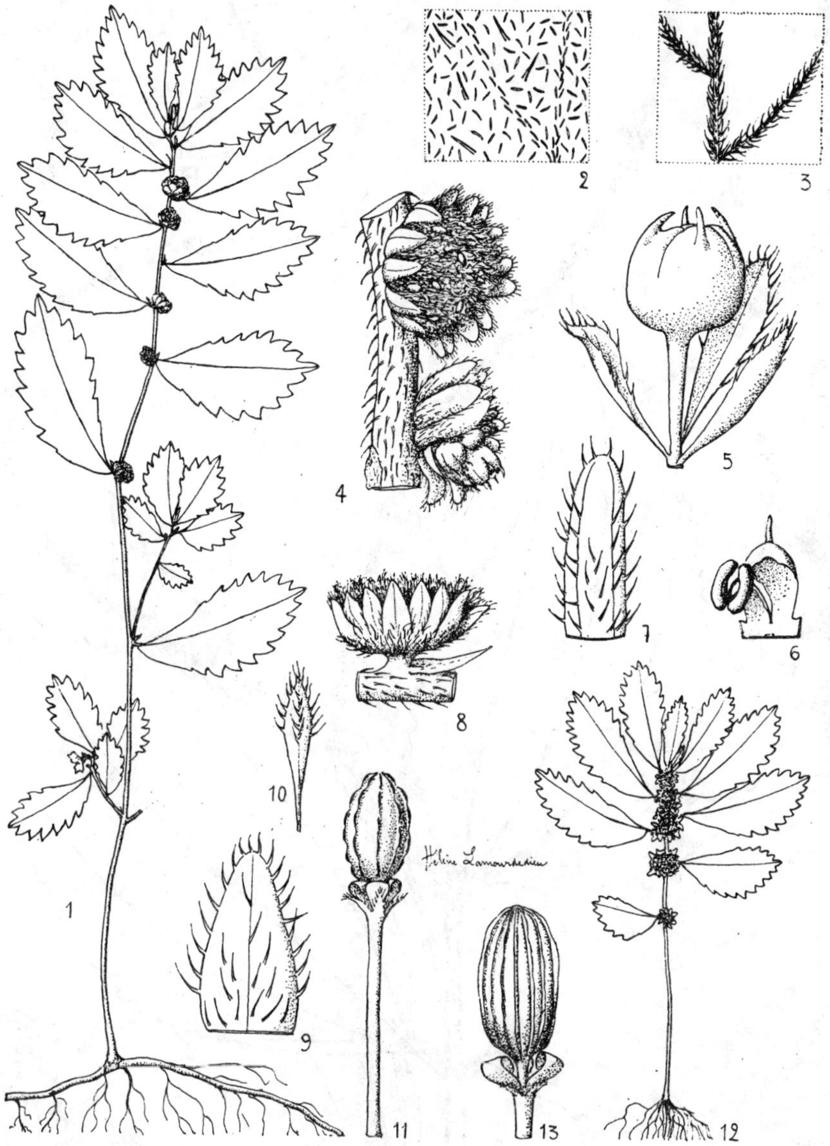

Figure 105.7. *ELATOSTEMA MONTICOLUM*: 1 - flowering branch of form with stolon x 2/3; 2 - detail of upper leaf surface x 4; 3 - detail of lower leaf surface x 4; 4 - male and female inflorescence x 4; 5 - male flower x 16; 6 - sepal and stamen from a male bud x 8; 7 - bract of male inflorescence x 8; 8 - female inflorescence x 4; 9 & 10 - bract and bracteole of female inflorescence x 8; 11 - fruit x 20; 12 - flowering stem of form without stolon; 13 - fruit x 20. 1-3 from *Mann* 2014; 4-8 from *Letouzey* 8899; 9-11 from *Tiku* FHI 22240; 12-13 from *Letouzey* 5890. (Reproduced with permission from *Fl. Cameroun* 8, pl. 23.)

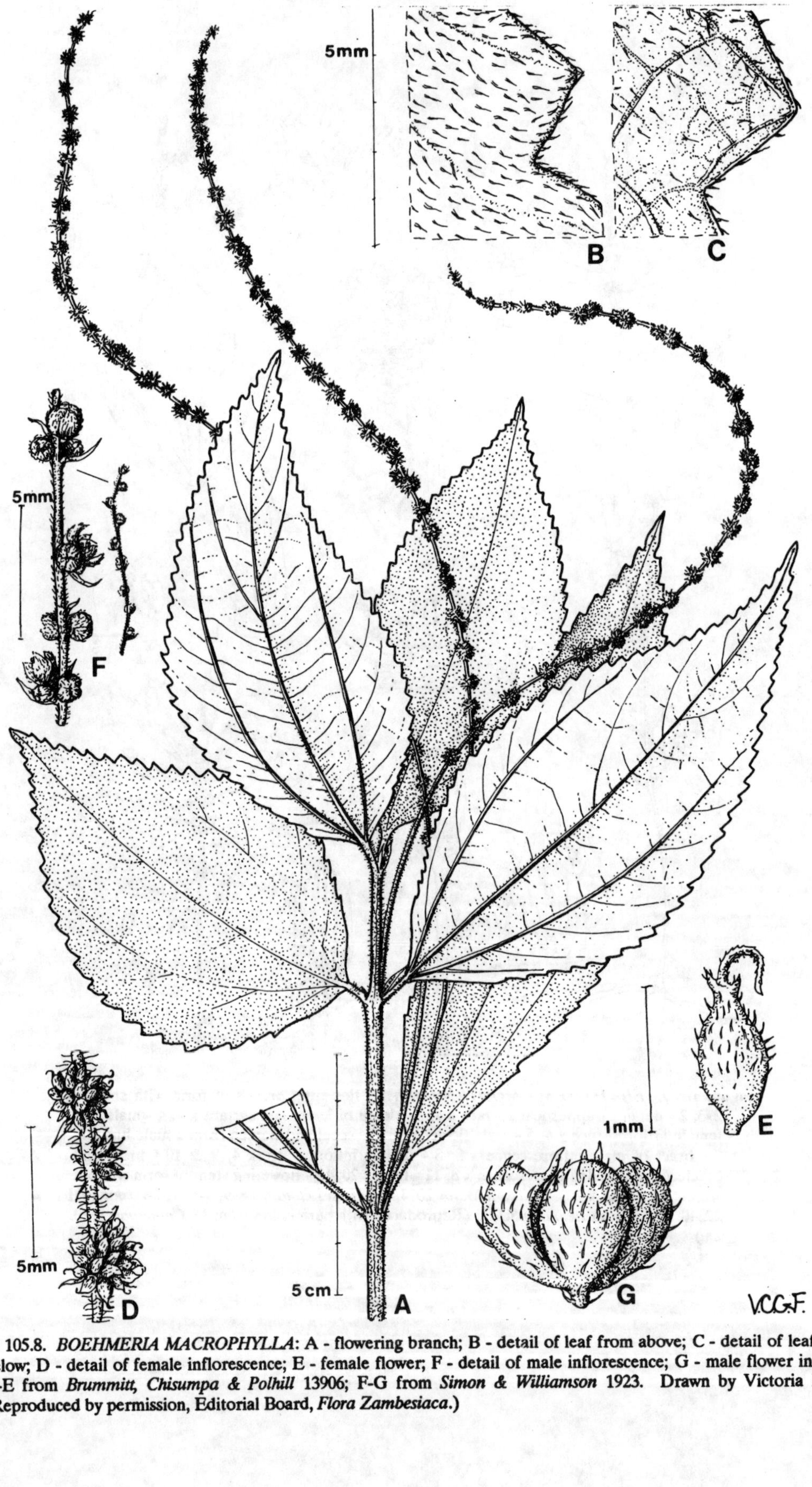

Figure 105.8. *BOEHMERIA MACROPHYLLA*: A - flowering branch; B - detail of leaf from above; C - detail of leaf from below; D - detail of female inflorescence; E - female flower; F - detail of male inflorescence; G - male flower in bud. A-E from *Brummitt, Chisumpa & Polhill* 13906; F-G from *Simon & Williamson* 1923. Drawn by Victoria Friis. (Reproduced by permission, Editorial Board, *Flora Zambesiaca*.)

up to 3 mm long, perianth tubular, bidentate and constricted at apex, pubescent in the upper part. Achene to *c* 3 mm long, enclosed in the persisting perianth. Fig. 105.8.

In lowland rain forest especially along streams; *c* 1000 m. KF IL; widespread lowland rain forest species, west to Guinean Republic, south to Angola, Zimbabwe and Mozambique. *Friis* et al. 4062.

The nomenclature and taxonomy of this species, which in most floras is referred to as *B. platyphylla*, has been discussed by Friis & Marais in *Kew Bull.* **37**:164 (1982).

10. **POUZOLZIA** *Gaud.* (1830).

Friis & Jellis, *Kew Bull.* **39**: 587-601 (1985).

Erect monoecious annual or perennial herbs, or shrubs. Cystoliths dot-like. Leaves alternate, petiolate, lamina entire or dentate; stipules free, lateral. Inflorescences compact, axillary, bisexual glomerules, usually provided with linear to ovate bracts with hooked hairs. Male flowers pedicellate, 4-5-merous, ovary rudimentary or absent. Female flowers sessile, with an indefinite number of completely fused tepals which enclose the erect ovary; staminodes absent, stigma sessile, articulated just above the ovary, filiform with unilateral papillae, deciduous. Achene ovoid, completely enclosed in and often partly fused with the persisting membranous perianth.

About 50 species, mainly in the tropics of the Old World.

1. Leaf margin serrate. **3. P. parasitica**
- Leaf margin entire. 2

2. Low herbs, sometimes slightly woody at the base; ultimate branchlets about 0.5 mm in diam.; male flowers with hooked hairs on the tepals (seen best on buds). **1. P. guineensis**
- Shrubs up to 3 m tall; ultimate branchlets more than 1.5 mm in diam; male flowers without hooked hairs. **2. P. mixta.**

1. **P. guineensis** *Benth.* (1849).
Parietaria abyssinica A. Rich. (1850); *Pouzolzia abyssinica* (A. Rich.) Blume (1856); *Pouzolzia guineensis* Benth. var. *abyssinica* (A. Rich.) Rendle in *Fl. Trop. Afr.* **6(2)**: 288 (1917) - type: TU, Djeladjerane, valley of Mai-Mezano, *Schimper* III:1433 (P holo., K iso.).

Annual or short-lived perennial herb up to 1 m tall, branched from lower parts. Ultimate part of younger branches very thin, covered with a cobwebby (arachnoid) tomentum. Leaves of same size almost to tip of branchlets: lamina ovate to lanceolate, up to 12 x 5 cm, densely white cobwebby (arachnoid) tomentose beneath, less so or glabrescent above; base rounded, margin entire, apex acuminate; petiole up to 6 cm long. Stipules up to 7 mm long. Inflorescences dense, globular, sometimes with broadly ovate bracts and lanceolate bracteoles. Male flowers often few in a cluster, on pedicels up to 1 mm long, globose, covered with hooked hairs. Female flowers sessile, perianth broadly cordate, with *c* 5 longitudinal ridges; stigma filiform, curled up in a ± globular head. Achene *c* 1.5 mm long, pale brown, shiny.

Among rocks in deciduous woodland or in riverine woodland and forest, also at the lower limit of lowland rain forest; 600-1300 m. TU WG IL; throughout western Africa to Senegal, south to Angola and Tanzania. *Gilbert & Thulin* 684; *Friis* et al. 3886.

Two different forms are known from Ethiopia: one ("*P. abyssinica*") has leaves of almost the same size along the stem and a marked, white, cobwebby tomentum on the lower surface of the leaves; it occurs in woodland. The other ("*P. guineensis*" sensu stricto) has leaves which decrease gradually in size towards the apex of the stems and a less striking white tomentum; it occurs along the margins of forests and in clearings of lowland rain forest. Transitional forms exist between the two forms which makes formal taxonomic recognition difficult.

2. **P. mixta** *Solms-Laub.* in *Schweinf.* (1867)
- type: Sudan/GD, Jebel Fazogl, *Cienkowsky* s.n. (B holo. destr., LE iso.).
P. hypoleuca Wedd. (1869).
P. fruticosa Engl. (1902) - type: HA, Erer Valley, *Ellenbeck* s.n. (B holo. destr.).

Shrub to 3 m tall, branchlets with white cobwebby (arachnoid) indumentum, more than 1.5 mm in diam. Older branches with brown bark. Leaves: lamina ovate, 2-6 x 1-3 cm, with scattered hairs above and dense white cobwebby (arachnoid) tomentum beneath; base rounded, margin entire, apex acuminate; petiole 0.8-1.5 cm long. Stipules lanceolate, up to 5 mm long. Inflorescences bisexual axillary clusters, male flowers numerous, female flowers few. Male flowers on pedicels *c* 1 mm long, 5-(4-)merous. Female flowers sessile, perianth ovoid; stigma filiform. Achene *c* 2 mm long, brownish, smooth. Fig. 105.9.A-F.

On volcanic, limestone or granitic rocks or rocky slopes, presumably in deciduous woodland; (?500-) 1300-1500 m. Sudan/GD SU HA; Yemen, Tanzania, Malawi, Zambia, Zimbabwe, Angola, Transvaal. *Mooney* 9117; *Burger* 2874; *West* 5558.

P. fruticosa is placed here as a synonym, though the type has been lost; the description indicates no character in which it differs from *P. mixta*, and recent material from the type locality is certainly typical *P. mixta*. There is a peculiar disjunction in the distribution of this species which seems to be absent from Kenya and Uganda, and then recurs in Tanzania and in southern tropical Africa.

3. **P. parasitica** *(Forssk.) Schweinf.* (1896);
Urtica parasitica Forssk. (1775).

Perennial plant from a woody rhizome, stems to *c* 1 m tall, slightly woody at the base, covered by an indumen-

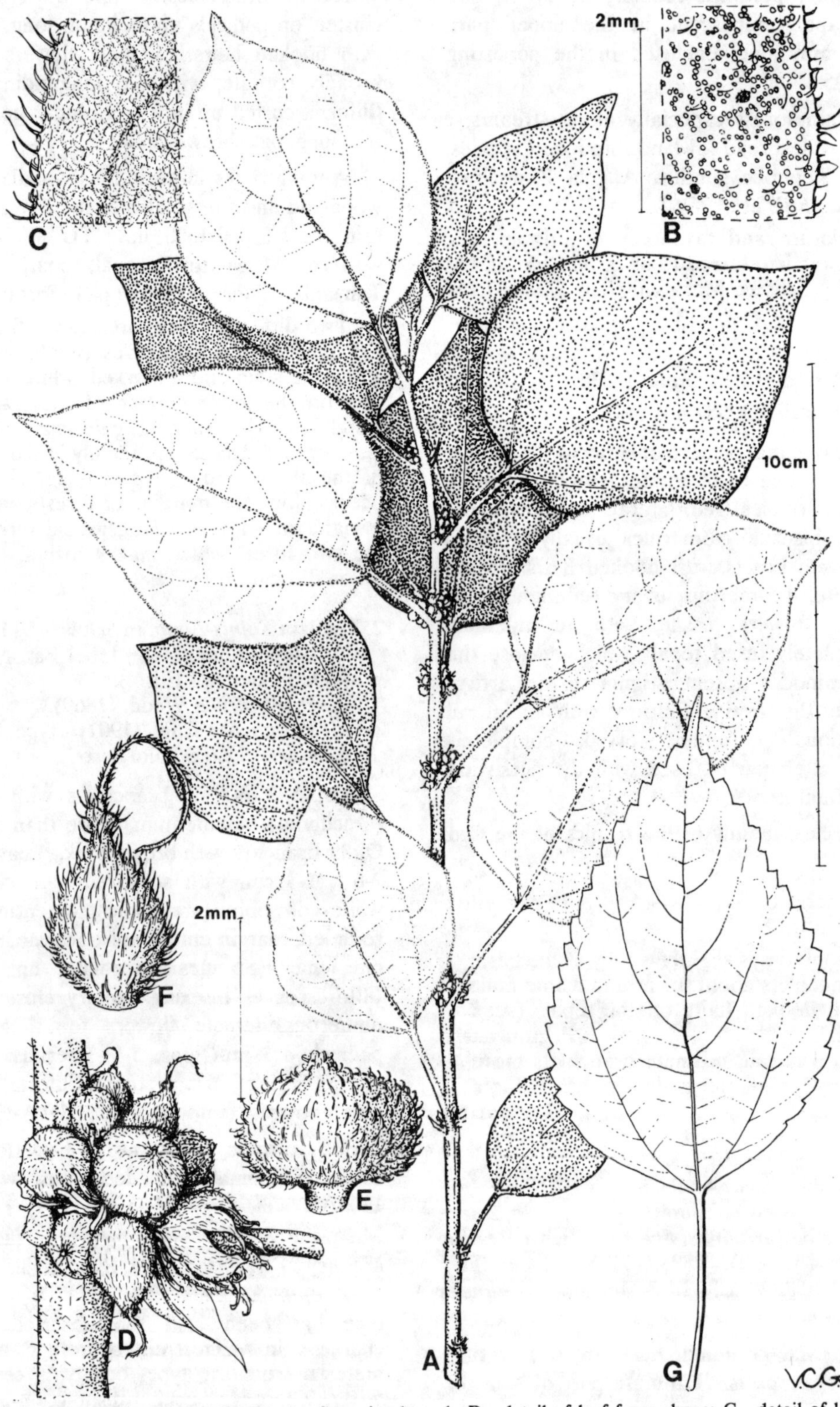

Figure 105.9. *POUZOLZIA MIXTA*: A - flowering branch; B - detail of leaf from above; C - detail of leaf from below; D - inflorescence; E - male flower in bud; F - female flower. All from *Norrgrann* 73. *P. PARASITICA*: G - leaf, from *Salubeni & Masiye* 2391. Drawn by Victoria Friis. (Reproduced by permission, Editorial Board, *Flora Zambesiaca*.)

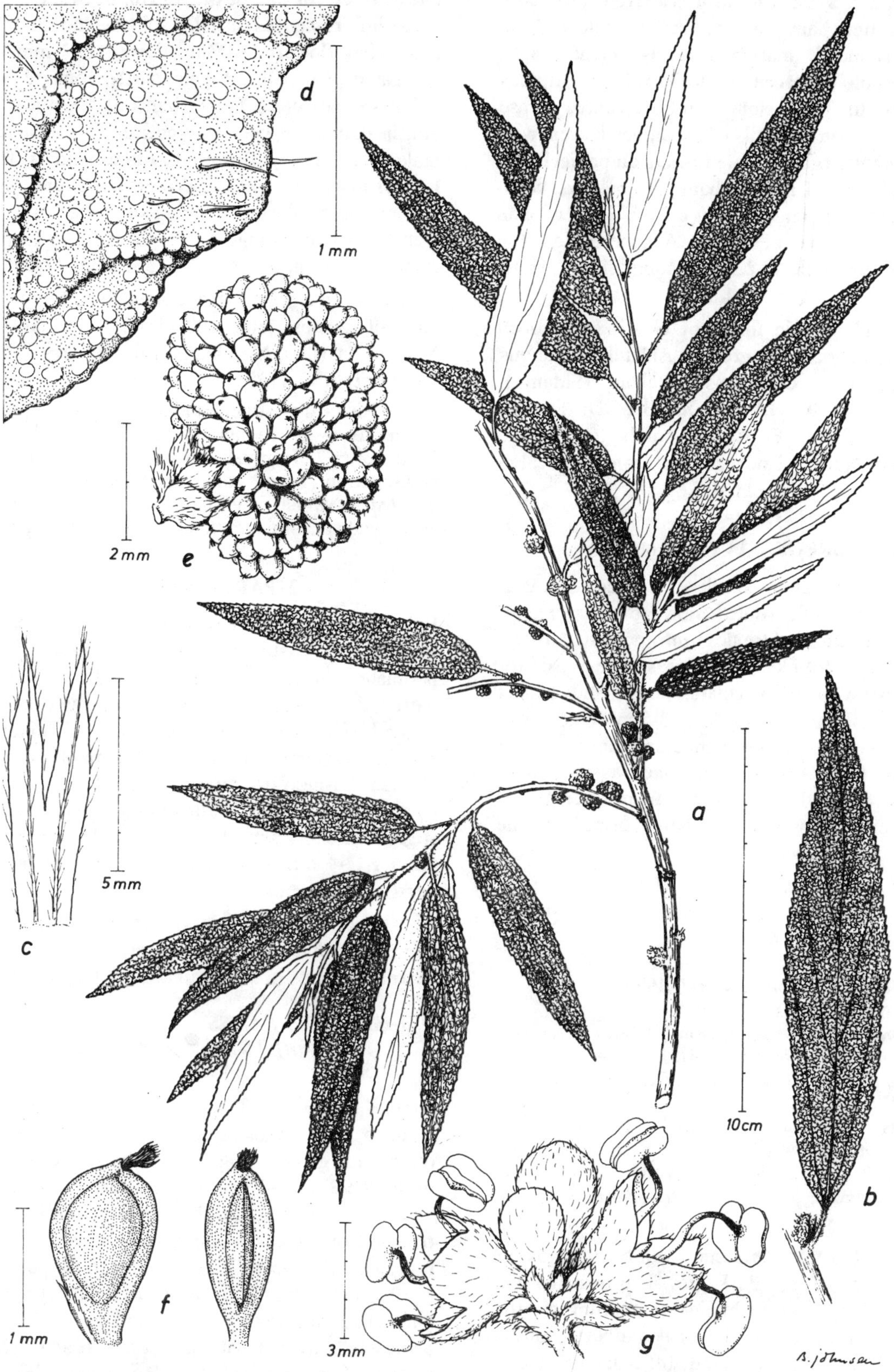

Figure 105.10. *DEBREGEASIA SAENEB*: a - flowering branch; b - leaf; c - stipule; d - detail of upper surface of leaf showing dot-shaped cystoliths; e - female infloresence in early fruiting stage; f - young fruit, side and end views; g - part of male inflorescence. All drawn from *Schimper* 158 by Bent Johnsen.

tum of short, erect hairs. Bark greyish brown. Leaves: lamina ovate, 4-8 x 3-5 cm, with scattered stiff hairs above and softer hairs on the nerves below, base cuneate or rounded, margin coarsely serrate, apex acuminate; petiole pubescent, up to 7 cm long. Stipules lanceolate, up to 7 mm long. Inflorescences dense clusters of flowers in the axils of the upper leaves, with narrowly triangular bracts. Male flowers on pedicel *c* 1.5 mm long, globose, *c* 1 mm in diam., 3-4-merous, vestigial ovary present. Female flowers *c* 2 mm long, with an equally long erect, filiform stigma. Achene enclosed in the persisting perianth, *c* 2.5 mm long, shiny, white. Fig. 105.9.G.

Edges of upland rain forest or clearings, "coffee" forest and plantations, evergreen bushland, deciduous woodland, rarely in deciduous bushland (Sidamo); 1500-2000 m. TU SU HA WG KF BA SD; Yemen, Cameroun, E Zaire, E Africa, south to Angola, Transvaal and Natal; also recorded from tropical S America. *Burger* 3552; *Friis* et al. 57, 1871.

11. **DEBREGEASIA** *Gaud.* (1844).

Monoecious or dioecious shrubs. Leaves alternate, petiolate, lamina with white cobwebby (arachnoid) tomentum beneath. Stipules intrapetiolar, fused. Cystoliths fine, dot-like. Inflorescences axillary, unisexual, sessile or pedunculate, dense and almost globose clusters. Male flowers 3-5-merous, ovary rudimentary. Female flowers with an indefinite number of completely fused tepals; staminodes absent; ovary erect, enclosed in the perianth; stigma penicillate, sessile. Achene enclosed in and partly fused with the persistent, accrescent perianth.

About 15 species, mainly in the Himalayas and the mountains of SE Asia; one species extends to the Yemen and NE tropical Africa.

D. saeneb *(Forssk.) Hepper & Wood* (1983);
Rhus saeneb Forssk. (1775).
Boehmeria salicifolia D. Don (1825); *Debregeasia salicifolia* (D. Don) Rendle (1916).
Urtica bicolor Roxb. (1832); *Debregeasia bicolor* (Roxb.) Wedd. (1869).
Procris hypoleuca Steudel (early 1850); *Boehmeria hypoleuca* (Steudel) A. Rich. (late 1850); *Missiessya hypoleuca* (Steudel) Wedd. (1854) - type: GD, Dschenausa, *Schimper* II:862 (P herb. Steudel holo., K iso.).

Erect shrub to 4 m tall, sometimes with overhanging or ± pendulous branches. Bark greyish brown, covered with white cobwebby (arachnoid) tomentum on the branchlets. Leaves: lamina somehat leathery, lanceolate, 7-15 x 1-3 cm, with scattered stiff hairs above and softer hairs on the nerves below, base truncate to rounded, margin finely serrate or crenate, apex long acuminate, finely tomentose to glabrescent above, with white cobwebby (arachnoid) tomentum beneath; petiole pubescent or arachnoid, 0.5-1.2 cm long. Stipules linear, fused for 2/3 of the length, brown, *c* 5 mm long. Male inflorescences usually solitary, up to 0.5 cm in diam., few flowered. Male flowers *c* 2 mm in diam., elongated, 4-merous, tepals partly connate. Female inflorescences often more than one together, up to 4 mm in diam., but generally noticeably smaller than the male ones, many flowered, often with a few male flowers mixed in. Female flowers 0.5-1 mm long; tepals become fleshy during ripening of the achene. Mature fruit berry-like, orange-reddish, *c* 0.8 mm long, achene crowned by remains of withered stigma. Fig. 105.10.

On rocky slopes in dry upland forest and in humid parts of ravines; 1500-2700 m. EW TU GD SU HA; Yemen, Pakistan, N India, Nepal. *Mooney* 7889; *Burger* 3438; *Friis* et al. 1135.

A. Richard's name, *Boehmeria hypoleuca,* has been taken from a *nomen nudum* proposed by C. F. Hochstetter. It must, however, be taken as a new combination based on Steudel's name of *Procris hypoleuca,* which antedates it. Richard added an additional specimen, *Petit* s.n., to his protologue.

12. **PARIETARIA** *L.* (1753).

Polygamous annual or perennial herbs, often profusely branched. Cystoliths dot-like. Leaves alternate, petiolate, entire. Stipules absent. Inflorescences paired, dense, sessile or very short pedunculate clusters in the leaf axils, sometimes with very prominent bracts. Flowers bisexual or unisexual. Hermaphrodite or male flowers 3-4-merous, tepals almost free. Ovary erect, free from the tepals, rudimentary in the male flowers, stigma sessile, penicillate. Female flowers with ± tubular, 3-4 lobed perianth. Achene enclosed in the persisting perianth.

About 10 species widely distributed in the tropics, subtropics and warm temperate regions; one polymorphic species in the mountains of tropical Africa.

P. debilis *G. Forst.* (1786).
?P. alsinefolia auct. non Gaudich: A. Richard (1850)

Annual or perennial herb, erect, sometimes subscandent, to 1 m or prostrate, finely pubescent on all parts, often with glands or glandular hairs. Leaves: lamina ovate, 0.3-3 x 0.2-2 cm, with scattered fine hairs above and longer soft hairs on the nerves below; base cuneate, truncate or subcordate, margin entire, apex rounded, mucronate, acute or acuminate; petiole pubescent, up to 2.5 cm long. Inflorescences sometimes loosely paniculate but mostly contracted in axillary clusters, profusely provided with lanceolate bracts. Male and bisexual flowers mixed together, subsessile, 4-merous, up to 1.5 mm in diam. Female flowers with 4 tepals which are fused for *c* half their length, 1-1.5 mm long, provided with ± hooked hairs. Achene pale brownish, smooth, compressed, *c* 1 mm long. Fig. 105.11.

Figure 105.11. *PARIETARIA DEBILIS*: 1 - branch with flowers and fruits x 2/3; 2 - a second type of branch x 1; 3 - detail of upper surface of leaf x 6; 4 - detail of inflorescence x 4, all from *MC* 251; 5 - female flower x 14; 6 - fruit with perianth segment removed x 14, both from *Brenan* 9509; 7 - branch from a small form x 1; 8 - detail with male and female flowers x 12; 9 - fruit x 10, all from *MC* 47. (Reproduced with permission from *Fl. Cameroun* 8, pl. 36.)

On the forest floor of high altitude dry evergreen forest and in evergreen, ericaceous bushland, in shade among boulders, or on shaded stream banks in open upland areas; 1350-3750 m. EW TU GD GJ SU AR BA SD; on high mountains throughout Africa, also on mountains in other tropical regions and widespread in warmer temperate regions. *Mooney* 7400; *Hedberg* 4161; *Friis* et al. 1384.

13. FORSSKAOLEA *L.* (1764).

Monoecious annual or perennial herbs, sometimes woody at the base, often covered with hispid hairs. Cystoliths dot-like. Leaves alternate, petiolate, variously serrate, lamina often white tomentose beneath and ± scabrous above. Stipules lateral, free. Inflorescences bisexual, sessile in the leaf axils, enclosed in a bell-shaped involucrum of bracts and almost covered by a dense, long, woolly indumentum; bracts spathulate, lanceolate, ovate or obovate, free or fused at the base. Male flowers pedicellate, mostly near edge of bell-shaped inflorescence, perianth irregular three-lobed, stamen one, inflexed; rudimentary ovary absent. Female flowers few at centre of inflorescence, perianth and staminodes absent; ovary erect, stigma sessile, filiform. Staminodes absent. Achene ovate, ± compressed, completely covered by woolly indumentum.

Five or six species, occurring in the drier parts of Africa, on the Macaronesian Islands, in Spain and extending to Pakistan and NW India.

1. Leaves densely white tomentose beneath, hispid above; involucral bracts erect, apex acute or acuminate, pointed, usually less than 1.5 mm wide. **1. F. tenacissima.**

- Leaves hairy on the nerved below, sometimes with a sparse white tomentum, slightly scabrid above; involucral bracts spreading, apex rounded, blunt or slightly mucronate, 2.5-6 mm wide. **2. F. viridis.**

1. **F. tenacissima** *L.* (1764).
   ?F. tenacissima L. var. *erythraea* Terracc. in *Ann. Ist. Bot. Roma* **8**: 119 (1894) - type: EE, Haressan in Anfila Bay, *Terracciano* 3bis, 23/3/1892, (?FT holo., not traced.)

Annual or probably mostly perennial herb with branching annual stems from a woody rootstock, up to 40 cm tall. Branches hispid with spiny or hooked hairs and a softer tomentum in between. Leaves: lamina obovate or rhomboid, 1-3 x 0.8-2 cm, hispid to glabrescent above, white tomentose below, cystoliths usually not visible; base cuneate, margin entire basally, serrate or crenate in the upper half, apex rounded or acute; petiole not clearly defined, *c* 0.3-1 cm long. Stipules lanceolate, to *c* 2 mm long. Inflorescences two or more together in the leaf axils, involucral bracts (4-)5-6(-7), lanceolate, up to 0.5 x 0.15 cm, densely covered by long woolly hairs. Male flowers up to 10, club-shaped in bud. Female flowers up to 5. Achene ovoid, brownish. Fig. 105.12.8-10.

Sand dunes or sandy soil near the sea, lava flows further inland; 50-300 m. EE AF; Djibouti, Sudan, Egypt, N Africa, S Spain, Arabian Peninsula, S Iran, Afghanistan, Pakistan, NW India. *Schweinfurth & Riva* 195; *Bally* 7059; *Ash* 2318.

F. tenacissima var. *erythraea* seems, from the description, to fall within the variation of *F. tenacissima* as accepted here, and probably does not merit recognition as a distinct infraspecific taxon.

2. **F. viridis** *Webb.* (1849)
   - type: plant grown in the Paris botanical garden from seeds collected at the Red Sea coast of Ethiopia by *Ehrenberg* (FI- WEBB lectotype); Friis in *Taxon* **31**: 727-729 (1982).
   ?F. viridis var. *ehrenbergii* Solms-Laub., in Schweinf., *Beitr. Fl. Aeth.*: 187 (1867) - type: EE, *Ehrenberg* s.n. (? B holo., not traced, probably destroyed).

Branched annual herb, sometimes slightly woody at the base, up to 0.5 m tall. Branches slender, puberulous to glabrescent. Leaves alternate, petiolate: lamina ovate to lanceolate, 1-6 x 0.5-4 cm, sparsely hispidulous above, sparsely white tomentose beneath, with longer hairs on the nerves; cystoliths usually visible above; base cuneate, margin serrate or crenulate, at least in the upper half, apex long acuminate; petiole 0.5-3.5 cm long. Stipules ovate, acuminate, *c* 3.5 x 2 mm. Inflorescences sessile, usually two in each leaf axil. Involucral bracts (3-)4(-6), 0.4-0.6 x 0.3-0.4 cm, base cuneate, apex rounded, slightly mucronate or short acuminate. Flowers completely embedded in woolly hairs. Male flowers usually many, female flowers few. Achene elipsoid, pale brownish. Fig. 105.12.1-7.

Rocky places near the sea, or in rock crevices at higher altitudes, on lava flows; 1-1100 m. EE EW TU ?GD SU; Sudan, Arabian Peninsula, Socotra, Kenya, Tanzania, Angola, SW Africa (Namibia). *Schweinfurth & Riva* 253; *Bally* 7042; *Gilbert & Thulin* 196.

The record from GD is based on a Schimper collection, but this may have a misplaced label. It would seem from the description of *F. viridis* var. *ehrenbergii* that this taxon does not merit recognition even on infraspecific level.

14. DROGUETIA *Gaud.* (1830).

Slender annual or perennial herbs, monoecious or apparently dioecious by abortion. Cystoliths dot-like. Leaves alternate or opposite, petiolate, serrate. Stipules lateral, free. Inflorescences sessile clusters in the leaf axils; either bisexual and surrounded by a common bell-sgaped involucrum of fused bracts or a cluster of very small involucres each with 1(-2) female flowers (the involucres then appear as perianths). Bisexual involucres usually with many male flowers and one to a few female flowers (occasionally all male); bracts and flowers with long woolly hairs. Male flowers pedicellate, club-shaped, perianth 3-lobed, one lobe apiculate and provided with prominent setae, stamen one, inflexed in bud; rudimentary ovary absent. Female flowers naked, pedicellate, consisting of an ovary covered with a woolly tomentum, stigma sessile, filiform; staminodes absent. Female involucres much smaller than the bisexual ones. Achene enclosed in the involucrum, brown, shiny, often crowned by the persistent stigma.

About five species distributed in S Africa, the mountains of tropical Africa, Yemen, India, and Java; also on Madagascar and the Mascarenes. The species in Ethiopia represents the type with the most extreme reduction of the inflorescence.

D. iners *(Forssk.) Schweinf.* (1896);
   Urtica iners Forsskal 1775.
   Urtica pauciflora Steudel (early 1850); *Boehmeria pauciflora* (Steudel) Blume (1850); *Pouzolzia pauciflora* (Steudel) A. Rich. (late 1850); *Didymogyne abyssinica* Wedd. (1854), nom. illeg.; *Droguetia pauciflora* (Steudel) Wedd. (1869) - type: GD, Semien, Mt. Silke, *Schimper* II:682 (P herb. Steudel holo., K iso.).
   Droguetia diffusa Wedd. (1854) - type: from Ethiopia, not specified.
   Boehmeria parvifolia Wedd. (1854).

Perennial herb with creeping or scrambling stems, rooting at the nodes, and erect branched stems to *c* 1.5 m tall. Younger stems pubescent or slightly hispid, later glabrescent. Leaves opposite: lamina ovoid, 3-7 x 2-4 cm, slightly hispid above and with soft hairs on the

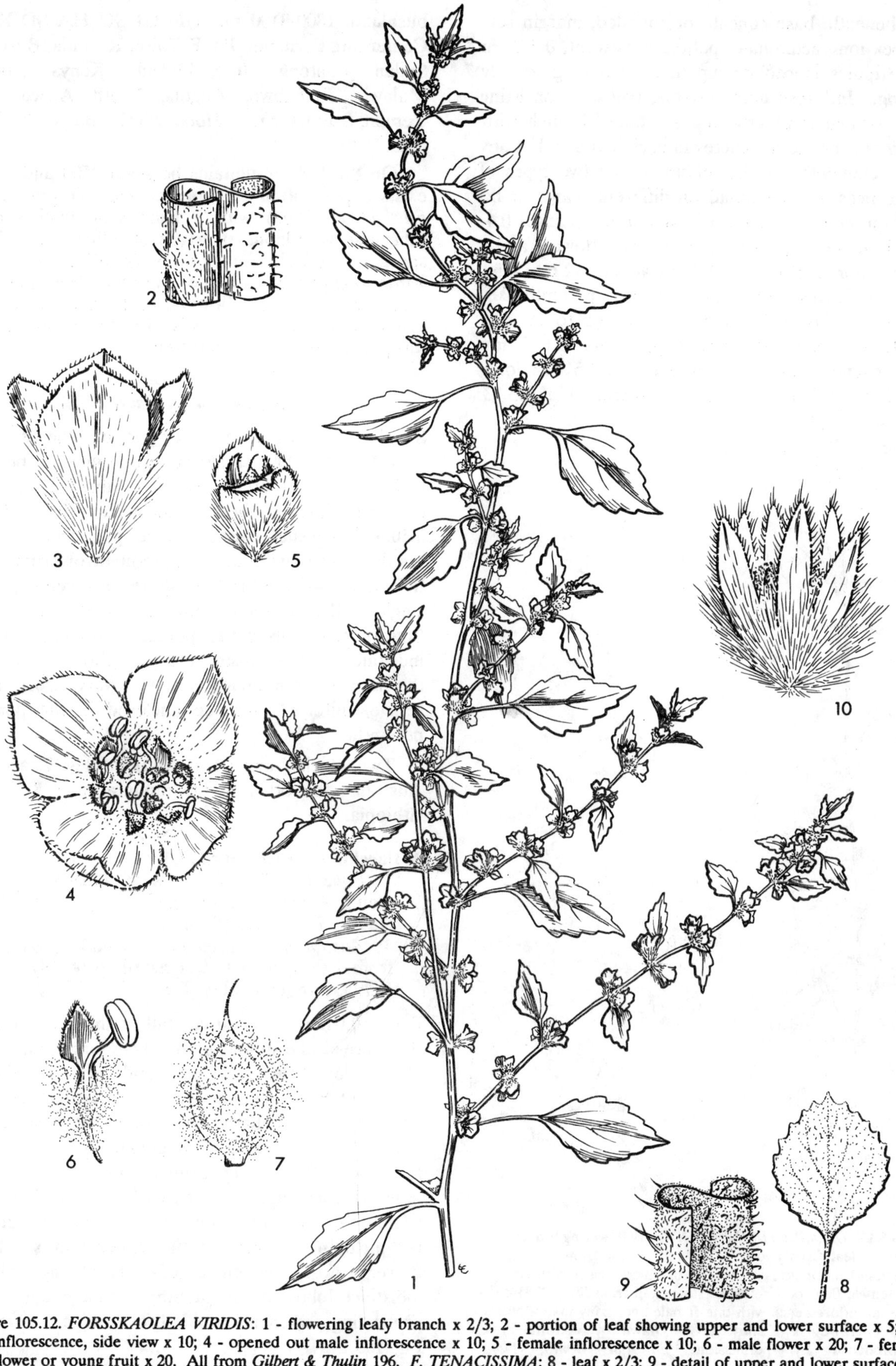

Figure 105.12. *FORSSKAOLEA VIRIDIS*: 1 - flowering leafy branch x 2/3; 2 - portion of leaf showing upper and lower surface x 5; 3 - inflorescence, side view x 10; 4 - opened out male inflorescence x 10; 5 - female inflorescence x 10; 6 - male flower x 20; 7 - female flower or young fruit x 20. All from *Gilbert & Thulin* 196. *F. TENACISSIMA*: 8 - leaf x 2/3; 9 - detail of upper and lower surface of leaf; 10 - side view of inflorescence x 10. All from *Ash* 2318. Drawn by Eleanor Catherine.

nerves beneath; base cuneate or rounded, margin serrate, apex long acuminate; petiole pubescent, 0.4-3 cm long. Stipules lanceolate, up to 0.5 cm long, mostly persisting. Inflorescences variable, typically consisting of two bisexual involucres and a reduced branch with few to many female involucres in each leaf axil, but any of these elements may be missing. The two types of inflorescences may be found on different plants or be present on the same plant. Bisexual inflorescences 0.4-0.6 cm in diam., mostly with many male flowers and a few naked female flowers. Male flowers to *c* 2 mm long. Female involucres often many together, perianth-like, each with usually 1(-2) female flowers, up to 1.5 mm long. Ovary erect, 1-1.5 mm long, stigma to *c* 3 mm long. Achene compressed, brownish, *c* 1.5 mm long, enclosed in the slightly enlarged involucrum. Fig. 105.13

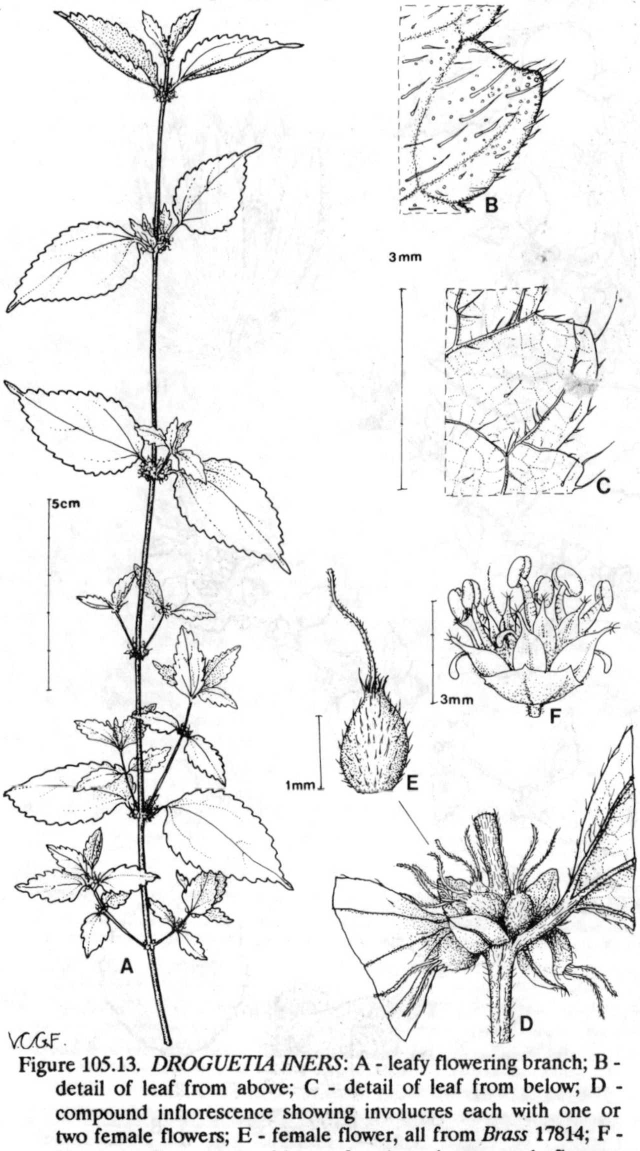

Figure 105.13. *DROGUETIA INERS*: A - leafy flowering branch; B - detail of leaf from above; C - detail of leaf from below; D - compound inflorescence showing involucres each with one or two female flowers; E - female flower, all from *Brass* 17814; F - bisexual inflorescence with one female and many male flowers, from *Greenway & Kanuri* 13882. Drawn by Victoria Friis. (Reproduced by permission, Editorial Board, *Flora Zambesiaca*.)

On forest floor in upland rain forest, in "coffee" forest and plantations or in ravines in upland evergreen bushland; 1800-3100 m. GD GJ SU HA SD KF IL; Cameroun, Fernando Po, E Zaire, Rwanda, Burundi, S Sudan (Imatong Mts.), Uganda, Kenya, Tanzania, Malawi, Zimbabwe, Angola, South African Rep., Yemen, S India, Java. *Mooney* 5728; *Burger* 1045; *Friis* et al. 2019.

On the Bale mountains between 2500 and 3100 m exists a population of *Droguetia iners* (represented by *Friis* et al. 5608) with a fixed modification of the otherwise very labile structure of inflorescence in this species. The Bale plants have male inflorescences on 4-7 mm long peduncles in the upper leaf axils, and female inflorescences of 1-2 flowers, each surrounded by an involucrum, in the lower axils. The taxonomic status of this population is as yet uncertain.

15. **AUSTRALINA** *Gaud.* (1830).

Slender monoecious annual or perennial herbs. Cystoliths dot-like. Leaves alternate or opposite, petiolate, with serrate or toothed margins. Stipules lateral, free or slightly fused in the lowermost part. Inflorescences axillary unisexual clusters, male pedunculate, female sessile, always without a common involucrum. Male flowers club-shaped in bud, shortly pedicellate, stamen single, inflexed, rudimentary ovary absent. Female flowers few, subsessile; perianth tubular, with *c* 4 indistinct teeth enclosing ovary, slightly contracted at the mouth, staminodes absent; ovary erect, stigma sessile, filiform. Achene enclosed in the persistent perianth.

Two species, one (*A. pucilla)* in New Zealand and Australia, the other in the mountains of Kenya and Ethiopia.

A. flaccida *(A. Rich.) Wedd.* (1869)

Pouzolzia flaccida A. Rich. (1850) - type: TU, Adua, *Quartin Dillon* s.n. (P syn.) & Memsah, *Quartin Dillon* s.n. (P syn.).

Australina schimperiana Wedd. (1854); type: from Ethiopia, not designated, probably *Schimper* II:795, without locality (P K).

Perennial herb to *c* 20 cm tall, with erect stems and thin, thread-like stolons up to 40 cm long from the axils of the lower leaves, stems and stolons pubescent. Leaves opposite: lamina ovate to lanceolate, 2-4 x 0.7-1.5 cm, sparsely hairy on both sides; base cuneate, margin serrate, apex long acuminate; petiole absent or rarely up to 3 mm long. Stipules lanceolate, base of the opposite pairs fused, upper part free, up to 3 mm long. Male flowers *c* 1.5 mm long, in pairs on a 2-4 mm long peduncle in the axils of the upper leaves. Female flowers mostly in pairs, 1.5-2 mm long, with 3-4 indistinct lobes at the mouth; stigma *c* 2 mm long, protruding. Achene ovoid, brownish, *c* 2 mm long. Fig. 105.14.H-N.

Among rocks in deciduous woodland, bushland and open dry upland forest; 1950-3150 m. TU GD SU;

Kenya. *Gilbert & Aweke* 2784, 2886; *de Wilde & Gilbert* 219.

16. DIDYMODOXA *Wedd.* (1869)

Friis & Wilmot-Dear in *Nordic J. Bot.* (in press).

Slender monoecious annuals. Cystoliths dot-like. Leaves alternate, petiolate, with serrate or toothed margin. Stipules lateral, free or partly fused to the petiole. Inflorescences axillary, bisexual, or rarely unisexual by abortion, sessile in dense clusters or on reduced branches; always without a common involucrum, but often with ciliate or bristle-bearing bracts. Male flowers shortly pedicillate, boat-shaped in bud with bristly apex; stamen single, inflexed, rudimentary ovary absent. Female flowers without perianth, consisting of a naked ovary only with a +/- developed unilateral wing; stigma sessile, filiform. Achene often with longitudinal lines.

Two species in south Africa and the mountains of eastern tropical Africa.

D. caffra *(Thunb.) Friis & Wilmot-Dear* (1986).
Urtica caffra Thunb. (1796).
D. acuminata Wedd. (1854).
Pouzolzia erythraeae Schweinf. (1896); type: Eritrea, Lawa Valley, *Schweinfurth* 1658 (B holo. destr.)

Annual herb to *c* 30 cm tall, branched from the base, puberulous. Leaves alternate: lamina ovate, 2-4 x 1-2.5 cm, sparsely pilose above, and on the nerves beneath; base rounded, margin serrate, apex long acuminate; petiole 0.8-2 cm long, puberulous. Stipules lanceolate, up to 2 mm long. Inflorescences usually bisexual, consisting of a dense, ebracteate cluster of male flowers in the axil of one stipule and a bracteate cluster of female flowers in the axil of the other stipule; sometimes also a few flowers on a short axillary branch with much reduced leaves, or only female flowers in a cluster in the leaf axil. Bracts narrowly lanceolate, often crowned by bristles. Male perianth *c* 1 mm long, unequally 2-lobed, larger lobe with 2-3 apical bristles. Female flower *c* 2 mm long, stigma *c* 1 mm long, sessile, filiform. Achene ovoid, compressed, with one side characteristically flattened, the other rounded, brownish, *c* 2 mm long. Fig. 105.14 A-G.

In shade among rocks; *c* 1000 m. EE/EW BA; Kenya, Tanzania, South African Republic. *Pappi* 4033; *Friis* et al 3532.

The very disjunct distribution of this species is almost certainly due to under-collecting. More material is desirable.

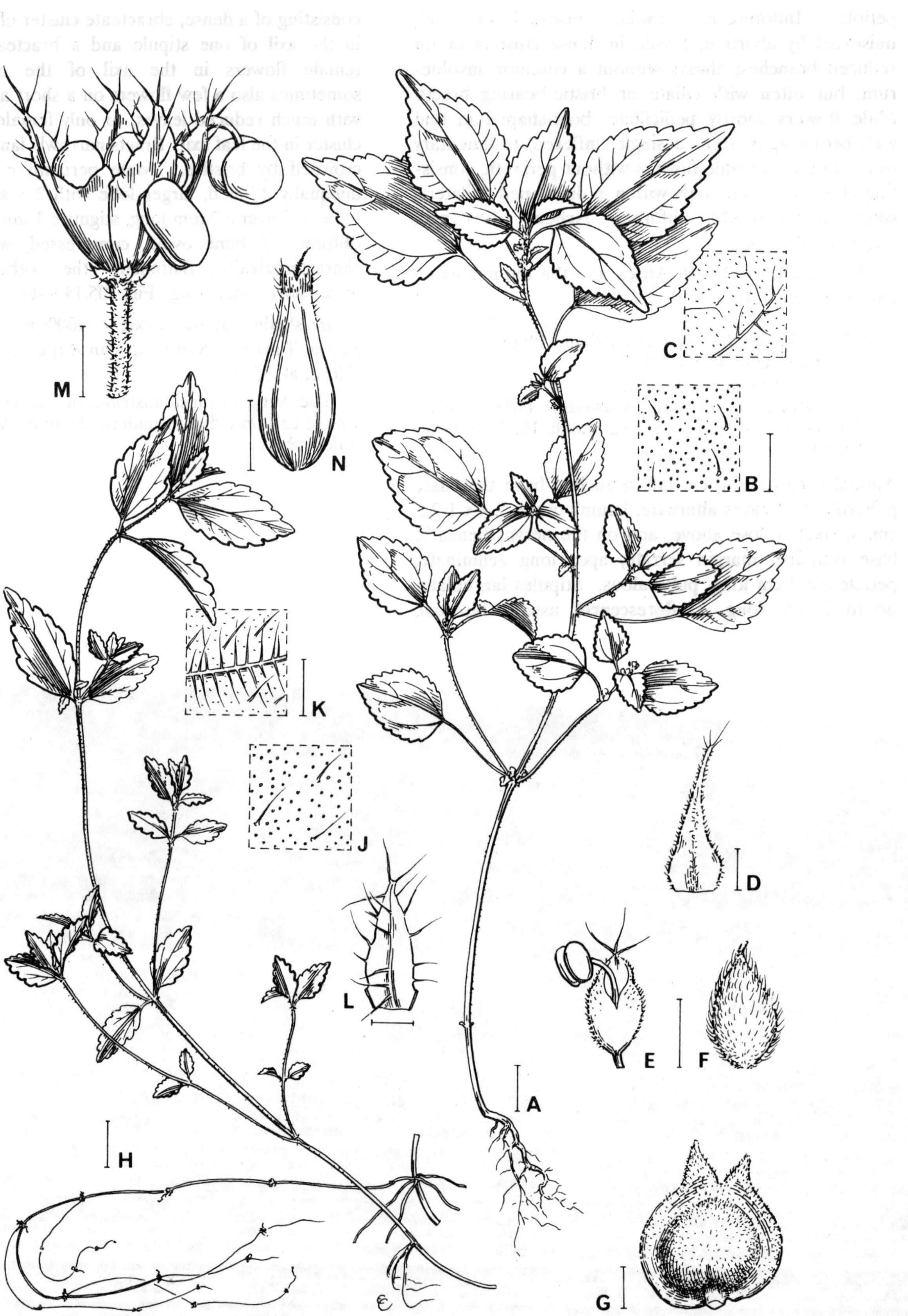

Figure 105.14. *DIDYMODOXA CAFFRA*: A - whole plant; B - leaf, upper surface of lamina; C - leaf, lower surface of lamina; D - stipule; E - male flower; F - female flower, wth remains of withered stigma, all from *Gachathi* 678; G, "double fruit", with two achenes fused along their rounded, unwinged side from *Merxmüller & Giess* 30244. *AUSTRALINA FLACCIDA*: H - erect stem and portion of stoloniferous base; J - leaf, upper surface of lamina; K - leaf, lower surface of lamina; L - stipule, at the base partly fused with its opposite number, all from *Tweedie* 2677; M - male inflorescence from *Gilbert & Getachew* 2784; N - female flower from *Schimper* 795. Bar lengths: A, H = 1 cm. B-G, J-N = 1 mm. Drawn by Eleanor Catherine. (Reproduced from *Nord. J. Bot.*)

106. CANNABACEAE[1]

by B. Verdcourt*

Cufodontis, *Enum.* (in **Moraceae**): 17 (1953); Miller in *J. Arn. Arb.* **51**: 185-203 (1970); Verdcourt, *Fl. Trop. E. Afr.*, **Cannabaceae**: 3 pp. (1975).

Annual or perennial erect or climbing herbs with clear sap. Leaves alternate or opposite, simple undivided or palmately lobed or divided into separate leaflets; petioles well developed; stipules present, free or fused. Flowers mostly dioecious, axillary, wind-pollinated. Male flowers paniculate; perianth 5-partite with imbricate segments; stamens 5, opposite the tepals. Female flowers ± sessile, crowded or strobilate, tightly covered or loosely subtended by small or large conspicuous persistent bracteoles; bracts also present; perianth membranous, entire, covering the ovary; ovary superior, sessile, 1-locular with 1 pendulous anatropous ovule; style terminal, short, with 2 long filiform stigmas. Fruit an achene covered by the persistent perianth; endosperm sparse, fleshy and oily.

A family of only 2 genera occurring naturally in the temperate parts of the northern hemisphere, formerly included in either **Urticaceae** or **Moraceae**. The number of species is in dispute; *Cannabis* is usually considered to be monotypic but a second species, *C. ruderalis* Janisch, has been described; *Humulus* is usually considered to contain 2 species although 3-4 have been claimed.

Although small this family is of considerable economic importance, *Humulus* ('hops' in English) being used to flavour beer and *Cannabis* to produce the fibre hemp and the drug variously known as 'bhang', 'dagga', 'hashish', 'pot', 'marijuana' and 'Indian hemp'.

CANNABIS *L.* (1753)

Erect annual aromatic herb. Leaves alternate, or opposite at the base of the stems, palmately divided into narrow leaflets (rarely palmately lobed[2]); stipules linear, acute. Male flowers numerous, shortly pedicellate, in lax drooping terminal and axillary panicles; tepals free, oblong, spreading or reflexed; anthers at length pendulous. Female flowers fewer, in short erect spikes, leafy below, each flower sessile in the axil of an enveloping bracteole and also with a stipule-like bract; perianth inconspicuous, very thin, closely enveloping the ovoid ovary. Fruit ovoid, compressed with seed closely conforming to the thin crustaceous (hard) pericarp. Embryo strongly curved; cotyledons fleshy.

C. sativa *L.* (1753).
C. indica Lam. (1785).

Joyce & Curry (eds.) *The Botany and Chemistry of Cannabis* (1970); Small & Cronquist, *Taxon* **25**: 405-435 (1976).

A rather rank-smelling herb 0.9-4.5 m tall; stems angular, covered with rather short stiffish hairs. Leaves 3-7(-11)-foliolate; petioles 3-7.5 cm long, pubescent; leaflets sessile, narrowly lanceolate, 2.5-15 x 0.35-2 cm, covered with very short bristly hairs and small yellow glands; apex tapering-acuminate; base narrowly cuneate; margin coarsely toothed. Male inflorescence up to 18 cm long; flowers whitish to yellowish-green; sepals oblong-elliptic, 2.8-4 x 1-1.6 mm, minutely hairy. Female inflorescences more compact, short and few-flowered, not projecting beyond the leaves; bracteole green, acuminate, covering the ovary and a forming a basally swollen tubular sheath 1.8-8.5 mm long, up to 6 mm wide when opened out, covered with slender hairs and short-stalked or stalkless circular, resin-secreting glands. Achene ("seed") ellipsoid or subglobose, slightly compressed, keeled, shiny, 2.5-5 x 2-3.5 mm, greyish to brownish and usually covered with a pale "map-like" pattern. Fig. 106.1.

A frequent weed of cultivation, sometimes illegally cultivated; 0-2350 m. Most likely in all regions; a central Asian plant now widely cultivated and naturalized throughout the world. *Burger* 2408; *IECAMA* E20; *Ryding* 1238.

Used in traditional medicine for treating epilepsy and other forms of mental illness. Elsewhere grown for its fibre which is called jute or hemp.

IECAMA E20 from Harar has been named *C. sativa* L. subsp. *indica* (Lam.) Small & Cronquist var. *indica* by E. Small. The varieties are discussed by Small & Cronquist (1976).

* Royal Botanic Gardens, Kew, Richmond, Surrey, TW9 3AE, UK.

1. Almost invariably spelt as Cannabinaceae in older works, also as Cannabiaceae and Cannabidaceae (used in Willis, *Dict. Fl. Pl.*, ed. 8:197, 1973).
2. According to Burger (1967).

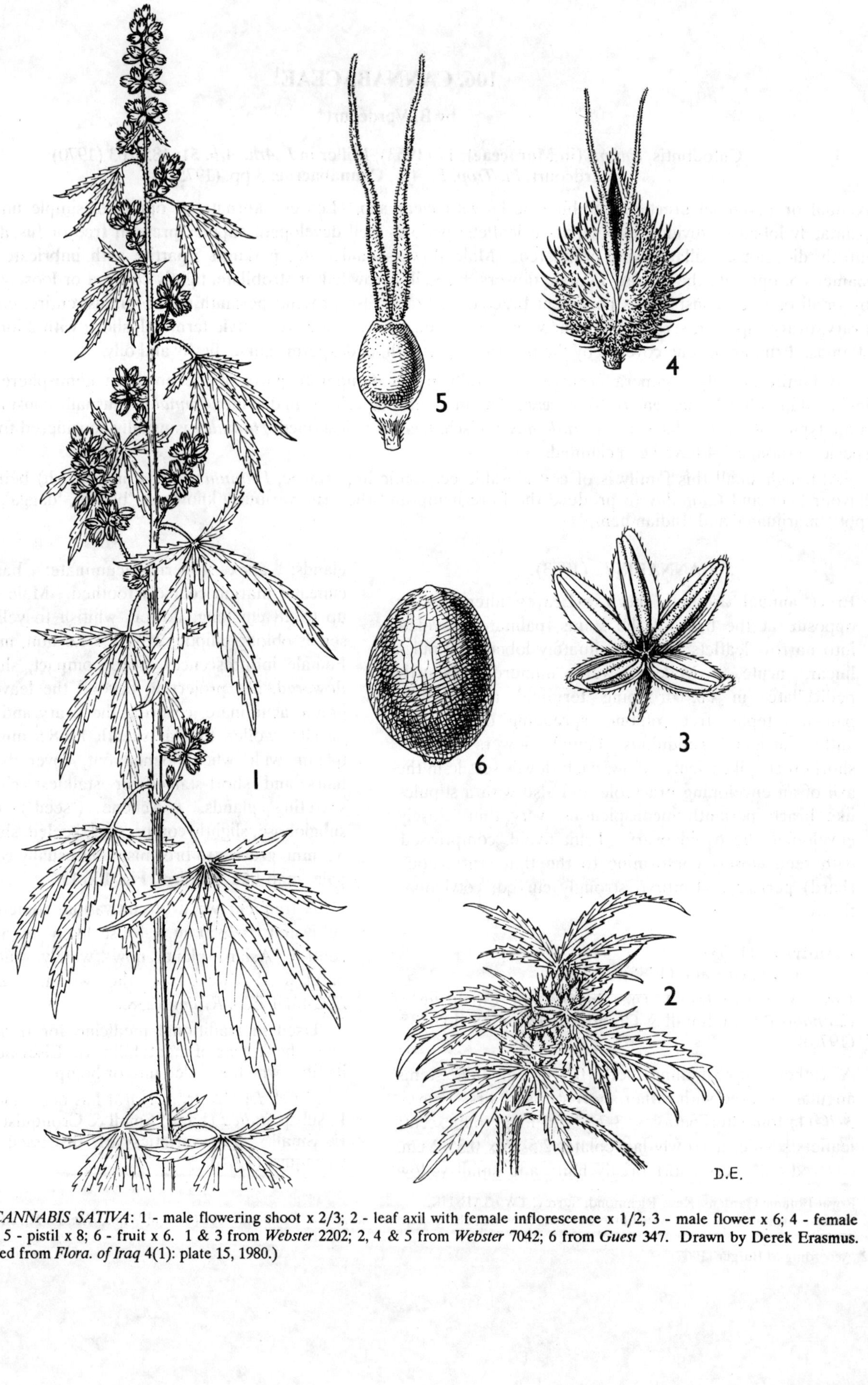

Figure 106.1 *CANNABIS SATIVA*: 1 - male flowering shoot x 2/3; 2 - leaf axil with female inflorescence x 1/2; 3 - male flower x 6; 4 - female flower x 4; 5 - pistil x 8; 6 - fruit x 6. 1 & 3 from *Webster* 2202; 2, 4 & 5 from *Webster* 7042; 6 from *Guest* 347. Drawn by Derek Erasmus. (Reproduced from *Flora. of Iraq* 4(1): plate 15, 1980.)

107. AQUIFOLIACEAE

by B. Verdcourt*

Cufodontis, *Enum.*: 476 (1958); Verdcourt, *Fl. Trop. E. Afr.*, **Aquifoliaceae**: 4 pp. (1968).

Trees or shrubs, mostly evergreen, usually practically glabrous. Leaves alternate or rarely subopposite, simple; stipules minute or absent. Flowers regular, bisexual or sexes separate, in cymes or fascicles, rarely solitary. Calyx-lobes 4-6(-9), imbricate. Corolla rotate; petals 4-5(-9), free or joined at the base, imbricate or valvate; disc absent. Male flowers: stamens 4-5(-9), free or inserted at the base of the corolla-tube, alternating with the petals; ovary rudimentary, 1-locular. Female flowers: ovary superior, (2-)3(-22)-locular; ovules 1-2 in each locule, pendulous; style terminal or absent; stigma capitate or discoid; staminodes present. Fruit a drupe with (1-)2-6(-many) 1(-2)-seeded pyrenes. Seeds with copious endosperm.

A family with 2-3 genera[1] closely allied to the **Celastraceae** and occurring in both the tropical and temperate parts of both hemispheres.

ILEX *L.* (1753)

Trees or shrubs. Leaves mostly leathery, margin entire to spiny; stipules small, persistent or soon falling. Inflorescences solitary or fasciculate, usually composed of 1-few-flowered cymes. Calyx-lobes 4-6(-9), glabrous or ciliolate. Petals 4-6(-9), imbricate, ovate, elliptic or rounded, usually united at the base. Male flowers with 4-6(-9) stamens. Female flowers with 4-6(-9) staminodes; ovary (2) 4-6(-22)-locular; stigma sessile, discoid, ± lobed. Drupes with persistent stigma.

A large cosmopolitan genus sometimes estimated at 400 species, poorly represented in Africa and Australia.

This genus includes the European 'holly' whose dark green spiny leaves and bright red berries are used as decorations at Christmas.

I. mitis *(L.) Radlk.* (1886).

I. mitis var. *erythraea* Fiori, in *Boschi e pianti legnose dell'Eritrea*: 224 (1912) - type: EW, Hamasen, near Vochi, in Mergat Feres Valley, *Fiori* (FT holo. not seen).

Evergreen shrub, or more usually a tree, (1.8-)4-24 (-40) m tall, up to 0.3(-1) m diam. at base of trunk; bark grey, fairly smooth; young branchlets often puberulous. Leaves alternate; lamina somewhat thick and stiff, ± elliptic, rarely almost round, 2.5-14 x 1.3-4.6(-5.3) cm, glabrous; midrib deeply impressed above; apex acute, apiculate or rounded; base obtuse to cuneate; margin entire or sometimes with spiny teeth; petiole 5-10(-20) mm long. Cymes axillary, usually fascicled, rarely solitary, 1-6(-7)-flowered; peduncles 1-15 mm long, puberulous; pedicels 3-8(-14) mm long. Flowers fragrant. Sepals 4-6(-8), rounded or subacute, ± 1.5 mm long and wide, joined at the base. Petals white, 4-6(-8), rounded to elliptic, ± 3.5 x 1.5 mm, joined at the base. Male flowers: stamens 1.5-2 mm long; rudimentary ovary conical. Female flowers: staminodes shorter; ovary globose, 4-6(-8)-locular; stigma discoid, 4-6(-8)-lobed. Drupe yellow-green, turning brownish-violet or crimson, globose, 4-7 mm diam., topped by persistent stigma, containing 4-6(-8) pyrenes.

var. **mitis**.

Leaves elliptic to elliptic-lanceolate. Fig. 107.1.

Frequent and sometimes abundant in moist montane or high rainfall forests, also in hillside thicket or scrub especially along stream valleys; particularly associated with some of the following *Aningeria, Arundinaria, Croton macrostachys, Erica, Hagenia, Juniperus, Maesa, Myrsine, Nuxia, Olea, Olinia, Podocarpus, Polyscias, Prunus africanus, Schefflera* in any one area; 1575-3000 m. WG SU IL KF AR BA HA; Sierra Leone (Loma Mts.), Nigeria, Cameroon and Fernando Po to East Africa and south to Cape Province of South Africa, also Madagascar. *Friis* et al. 1482; *Mooney* 8801, 8475.

An extremely variable and widely distributed tree easily recognised in the field by the rather thick dark green leaves with the clearly impressed midrib above. Has been used as timber but good lengths are not easy to obtain since the bole is usually short.

* Ethiopian Flora Project, Royal Botanic Gardens, Kew, Richmond, Surrey, TW9 3AB, UK.

1. One is sometimes placed in the Araliaceae.

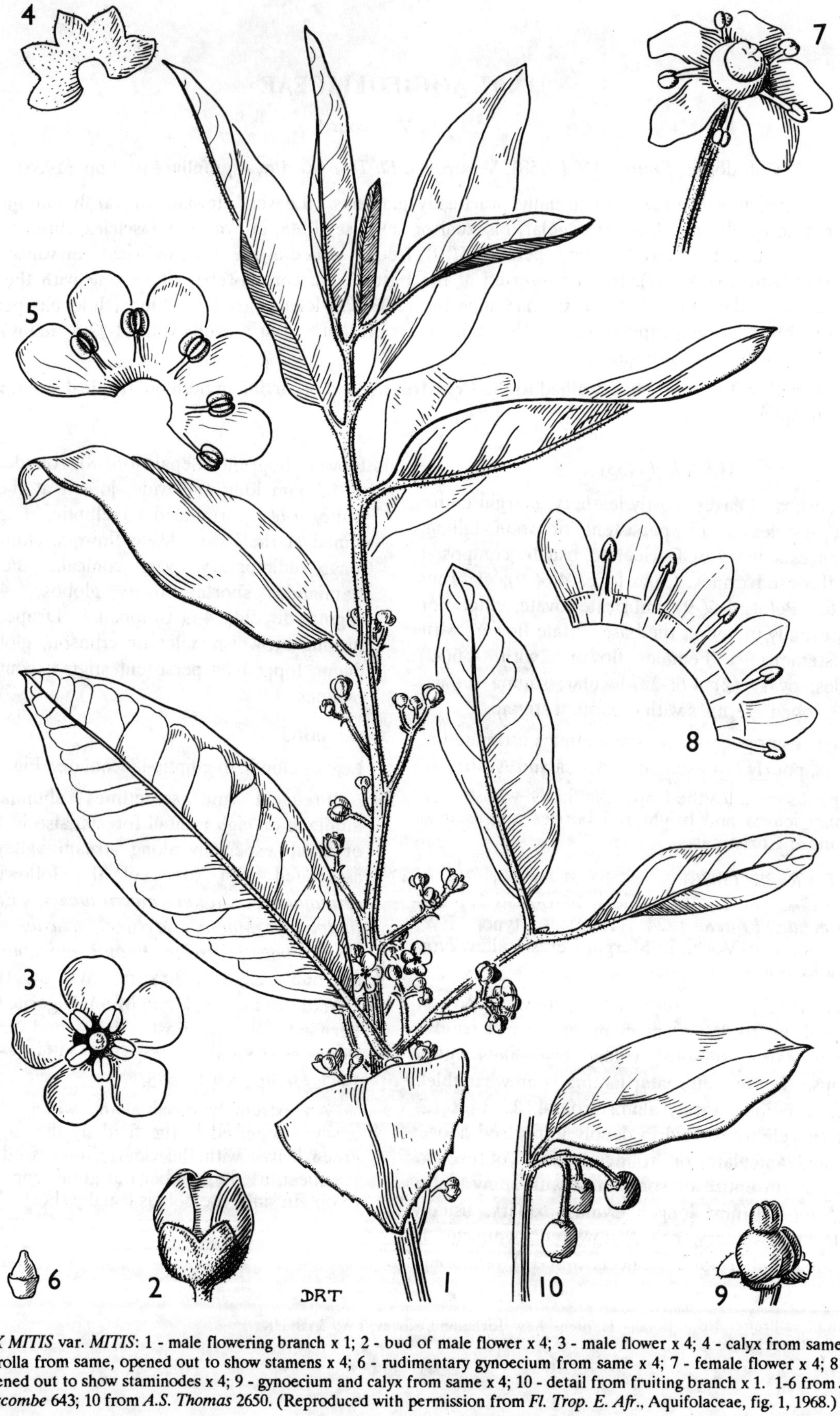

Figure 107.1 *ILEX MITIS* var. *MITIS*: 1 - male flowering branch x 1; 2 - bud of male flower x 4; 3 - male flower x 4; 4 - calyx from same, opened out x 4; 5 - corolla from same, opened out to show stamens x 4; 6 - rudimentary gynoecium from same x 4; 7 - female flower x 4; 8 - corolla from same opened out to show staminodes x 4; 9 - gynoecium and calyx from same x 4; 10 - detail from fruiting branch x 1. 1-6 from *Jessel* 61; 7-9 from *Battiscombe* 643; 10 from *A.S. Thomas* 2650. (Reproduced with permission from *Fl. Trop. E. Afr.*, Aquifolaceae, fig. 1, 1968.)

108. CELASTRACEAE
(incl. *HIPPOCRATEACEAE*)

by N. K. B. Robson*

Cufodontis, *Enum.*: 477-486 (1958); Robson in *Fl. Zamb.*: 355-418 (1966).

Trees, shrubs, shrublets or woody climbers. Leaves alternate or spiral to subopposite or opposite, sometimes fasciculate on short shoots, usually with small stipules, simple, penninerved, entire or crenate to spinose. Flowers regular, bisexual or polygamous to unisexual. Sepals (3)4-5(6), imbricate or rarely valvate in bud, free or united at base, persistent. Petals the same number as sepals, free or rarely united at base, imbricate or rarely valvate in bud. Stamens (2)3-5(6), opposite the sepals, free or with filaments partly united, inserted outside, on or inside disc; pollen simple or in tetrads or polyads. Disc producing nectar, annular and entire or lobed (rarely discontinuous or absent). Ovary superior, free or partly or wholly immersed in disc, sessile or on short androgynophore[1], syncarpous, usually completely 2-5-locular or 1-locular by abortion, with 1 to numerous erect (rarely pendulous) ovules usually in 2 rows in each locule; styles as many as locules, free or ± united or absent; stigmas free or somewhat united. Fruit capsular and dehiscing loculicidally or indehiscent and baccate or drupaceous, sometimes winged. Seeds sometimes arillate or winged, with or without endosperm.

About 60-70 genera and 850-900 species, widely distributed.

The genera with a smaller number of stamens than sepals within the disc have frequently been segregated as the **Hippocrateaceae**; but this segregation is unjustified as the family so formed appears to be polyphyletic[2]: *Hippocratea* and its relatives (with dehiscent fruits) being related to a different part of **Celastraceae** proper from *Salacia* and its relatives (with indehiscent fruits).

Key to Genera

1. Leaves all spirally inserted; fruit a capsule with arillate seeds or a drupe. 2
- Leaves all opposite or sometimes (*Elaeodendron*) subopposite to alternate towards the base of the shoot; fruit a capsule with winged seeds or a drupe. 3
2. Fruit a capsule; anthers (in the Flora area) persistent. **1. Maytenus**
- Fruit a drupe; anthers eventually deciduous. **2. Mystroxylon**
3. Fruit a capsule; seeds winged. 4
- Fruit a drupe; seeds without appendages. 5
4. Capsule elongated, narrow; seed wing without funicular thickening; stamens 5, outside the disc. **3. Catha**
- Capsule expanded laterally, forming 3 dehiscent mericarps; seed wing (in the Flora area) with marginal and funicular thickening; stamens 3, inside the disc. **4. Hippocratea**
5. Stamens 4-5, united with base of the disc; flowers (in the Flora area) unisexual, dioecious; inflorescence (in the Flora area) a thyrse, i.e. a panicle of cymes. **5. Elaeodendron**
- Stamens 3, inside the disc; flowers bisexual; inflorescence (in the Flora area) fasciculate. **6. Salacia**

1. MAYTENUS *Molina* (1782)

Celastrus sect. *Gymnosporia* Wight & Arnott (1834); *Gymnosporia* (Wight & Arnott) Hook. f. (1862)

by Sebsebe Demissew**

Sebsebe Demissew, The genus *Maytenus* (Celastraceae) in NE Tropical Africa and Tropical Arabia. *Symbolae Botanicae Upsalienses* **XXV(2)**: 101 pp. (1985).

Shrubs or trees, glabrous to pubescent, mostly dioecious. Branches round, lined to angled, usually with spines. Leaves alternate, or fasciculate on short shoots, entire to serrate. Stipules free, falling quickly or absent. Inflorescences pedunculate or sessile, cymose, solitary or fasciculate in axils of leaves or on branches; pedicels articulated; bracts persistent. Flowers mostly functionally unisexual with staminodes in female and pistillodes in male flowers, or bisexual. Sepals 5(6), imbricate in bud, entire to fimbriate, sometimes ciliolate. Petals 5(6), imbricate in bud, entire to toothed. Disc intrastaminal, convex to concave, often ± 5-10 lobed. Stamens 5(6) with filaments attached to the base of the disc; anthers introrse, with 2 thecae dehiscing longitudinally. Ovary superior, 1/4-3/4 immersed in the disc; 2-3(-4) locular with 2 erect, collateral ovules in each locule; style short or elongated with 2-3(-4) branched stigma (which may appear capitate on pistillode). Fruit a loculicidal capsule, obconic to globose or pyriform. Seeds 1-4(-8), glossy, red brown to orange; aril white, yellow or purple (mostly white when young), basal or partially to completely covering the seed.

A genus of about 200 species in the tropics and subtropics.

A number of species in *Maytenus* are used in traditional medicine to treat various disorders including

* Botany Department, British Museum (Natural History), Cromwell Road, London SW7 5BD, UK.
** The National Herbarium, Science Faculty, Addis Ababa University, Box 3434, Addis Ababa, Ethiopia.
1. androgynophore = a stalk made up of both stamen and pistil material.
2. polyphyletic = a group which has evolved from two or more different sources.

tumours and a tumour inhibitor, maytansine, has been extracted - see Sebsebe (1975) for more detail.

1. Inflorescence fasciculate, unbranched, rarely with peduncle up to 2 mm; branches unarmed. **13. M. undata**
- Inflorescence pedunculate, branching; branches usually armed with spines. 2
2. Aril basal; ovary 3(4) locular; leaves usually glabrous, sometimes hairy on the lower surface and/or veins above. 3
- Aril investing 1/3 to the whole of the seed; ovary 2-locular, or if 3-locular, then leaves hairy on both surfaces. 11
3. Capsule 8-16 mm long; leaf margin singly or doubly serrate, or if crenate-serrulate, then inflorescence longer than 5.5 cm and pubescent. 4
- Capsule 4-8 mm long (up to 10 mm in *M. cortii* in which leaves are orbicular); leaf margin crenate or crenate-serrulate; inflorescence 5.5 cm or shorter or, if longer, then glabrous. 6
4. Branches glabrous to puberulous; capsule glabrous; inflorescence 2.5-10 cm long. 5
- Branches densely pubescent; capsule pubescent; inflorescence 1-5 cm long. **2. M. serrata**
5. Leaf blade 3-16.5 x 1.6-7.5 cm, margin usually singly or doubly serrate; flowers unisexual, *c* 2-4.5 mm in diameter. **1. M. gracilipes**
- Leaf blade 2-5 x 1.5-4 cm, margin crenate; flowers bisexual, *c* 7 mm in diameter. **7. M. harenensis**
6. Young shoots and inflorescence parts glabrous; aril white to pale yellow. 7
- Young shoots and/or inflorescence parts pubescent; aril white to pale yellow or purple. 9
7. Leaves acuminate at the apex; inflorescence (2-) 2.5-6.2 cm long. **4. M. addat**
- Leaves obtuse to acute at the apex; inflorescence 0.5-2(-2.5) cm long. 8
7. Leaves ± elliptic; petiole 4-10 mm long; capsule 5-6 mm long; inflorescences usually fasciculate; aril ± circular at the base of the seed; petals *c* 2-3.75 mm long. **5. M. obscura**
- Leaves ± orbicular; petiole 1-1.5 mm long; capsule 6-10 mm long; inflorescence solitary; aril protruding on one side of the seed only; petals *c* 1.5 mm long. **6. M. cortii**
8. Branches glabrous or practically so; inflorescence 1.5-5.5 cm long, solitary; leaves with reticulation ± equally prominent on both surfaces. **3. M. parviflora**
- Branches pubescent at least on young parts; inflorescences less than 2.5 cm long, solitary or fasciculate; leaves with reticulation more prominent on lower surfaces. 10
10. Seed aril smooth, yellow; flowers *c* 2.5 mm in diam.; leaves always glabrous. **8. M. buchananii**
- Seed aril rugose, purple; flowers *c* 2.5-7 mm in diam.; leaves sometimes hairy on lower side and/or on veins above. **9. M. arbutifolia**
11. Plant pubescent throughout; ovary 3-locular; flowers 8.25-10.25 mm in diam.; disc red, convex. **10. M. putterlickioides**
- Plant glabrous; ovary 2-locular (very rarely 2- and 3-locular on the same specimen); flowers up to 7 mm in diam.; disc white to yellow, ± concave. 12
12. Leaves glaucous or mottled green with regularly crenulate margin; petiole glaucous or wine red; young branches terete, glaucous to wine red; aril white. **11. M. senegalensis**
- Leaves pale green to green with irregularly serrulate to subentire margin; petiole green; young branches angular, green; aril yellow. **12. M. heterophylla**

1. M. gracilipes ***(Welw. ex Oliv.) Exell*** (1952);
Celastrus gracilipes Welw. ex Oliv. (1868).

Shrub or shrubby tree 0.5-6 m high with spines up to 4(6) cm long; branches green to orange brown, with lenticels, glabrous or puberulous. Leaves: petiole 2-9 mm long; lamina 3-16.5 x (1-)1.6-7.5 cm, elliptic to oblong, ovate or oblanceolate, leathery to membranous, glabrous, apex acute or subacuminate to obtuse or rarely retuse, margin singly or doubly serrate to serrulate-crenate. Cymes (2.5-)3-10 cm long; flowers 3-60 or more in each cyme, 2-4(-4.5) mm in diam., functionally unisexual. Sepals 5, 0.5-1.3 mm x 0.7-1.3 mm triangularly ovate, ciliolate or glabrous to puberulous. Petals 5, 1.2-2.5 x 1-1.5 mm, white to cream, elliptic to oblong. Male flowers with 5 stamens longer than the pistillode. Female flowers with 5 staminodes shorter than ovary; ovary 3-4 locular. Disc convex, 5-lobed. Capsule pendulous, obovoid, pink to red, streaked white or mottled green when young, 10-16 mm long, glabrous in Ethiopia. Seeds 3-4(-8), orange, brown, glossy, with a cup-shaped white to yellowish aril at the base.

Subsp. **arguta** ***(Loes.) Sebsebe*** (1985);
Gymnosporia gracilipes (Welw. ex Oliv.) Loes. var. *arguta* Loes. in Engl. *Bot. Jahrb.* 17: 541 (1893); *Maytenus ovata* (Wall. ex Wight & Arn.) Loes. var. *arguta* (Loes.) Blakelock (1956); *Maytenus arguta* (Loes.) Robson (1965).

Celastrus serratus var. *steudneri* Engl. *Über die Hochgebirgsflora des tropischen Africa*: 291 (1882); *Gymnosporia serrata* var. *steudneri* (Engl.) Loes. in Engl. *Bot. Jahrb.* 17: 546 (1893) - type: GD, Gafat near Debre Tabor, *Steudner* 606 (B holo. destr., K iso.).

Gymnosporia ellenbeckii Loes. (1908); *Maytenus ellenbeckii* (Loes.) Cuf. (1962) - type: SD, Gerbicho near Aleta Wendo, *Ellenbeck* 1745 (B holo. destr.).

Gymnosporia castellii Pic.-Serm. 1951); *Maytenus castellii* (Pic.-Serm.) Cuf. (1958) - type: GJ, Zege peninsula, *Pichi-Sermolli* 2316 (FT holo.).

Leaf-margins usually singly or doubly serrate; capsule glabrous.

Undergrowth in evergreen forest, forest margins, grassland with scattered trees, along rivers; *c* 1250-2500(-2800) m. GD GJ WG SU BA IL KF GG SD; Sudan, Uganda, Tanzania, Rwanda, Burundi, and

Zaire. *Sebsebe D.* et al. 1142, 1319; *Mesfin T.* et al. 1301.

Subsp. *gracilipes* distinguished by its pubescent capsule, and usually crenate to crenate-serrulate leaf margin, occurs in Tanzania, Zaire, Angola and Cameroun.

2. **M. serrata** *(A. Rich.) Wilczek* (1960);
Celastrus serratus A. Rich. (1847); *Gymnosporia serrata* (A. Rich.) Loes. (1893); *G. serrata* var. *typica* Fiori in *Boschi e piantae legn. del. l'Erit.*: 226 (1912) - types: TU, *Schimper* III:1908 (P syn., K S isosyn.), Chire (Shire) *Quarton Dillon* s.n. (P syn.).
Celastrus schimperi A. Rich. (1847); *Gymnosporia serrata* var. *schimperi* Fiori in *Boschi e piantae legn. del. l'Erit.*: 227 (1912) - type: TU, Adwa, *Schimper* I:118 (P holo., BM K S iso.).
Maytenus ovata var. *ovata* forma *pubescens* (Schweinf.) Blakelock (1956 *pro parte* including synonyms above and *Gymnosporia serrata* var. *schimperi*).

Shrub or small tree 0.8-4 m high, with spines up to 2.7 cm long; branches grey to brownish-purple, pubescent, with brown lenticels. Leaves: petiole 2-7 mm long; lamina 3.2-10.5 x 1.9-5.5 cm, elliptic to ovate or obovate, leathery, glabrous, apex acute to rounded or retuse, margin usually singly serrate with acute teeth. Cymes 1.1-5 cm long, pubescent; peduncle 3-15(-20) mm long; pedicels 1-5 mm long; flowers solitary or 2-50 in each cyme, (3.5-)4-7.5 mm in diameter; functionally unisexual. Sepals 5(6), 1 x 1 mm, ovate to triangular, pubescent and with ciliolate margins. Petals 5(6), 3.2 x 1-2 mm, white to yellowish-green, oblong to elliptic, glabrous to puberulous. Male flowers with 5(6) stamens, longer than the pistillode. Female flower with 5 staminodes shorter than the ovary; ovary 3-locular. Disc flat, rounded to ± 5-lobed. Capsule pink to red (greenish purple when young), obovoid, 8-10(-12) mm long, pubescent. Seeds 3-5, brown, with a yellowish (white when young) aril at the base.

Rocky or grassy slopes or in flatter areas in open fields as bushes often in clumps of 2-4 together; *c* 1650-2250 m. EW TU GD GJ; not known elsewhere. *Sebsebe D.* 350, 1221; *Ash* 1411.

3. **M. parviflora** (*Vahl*) *Sebsebe* (1984);
Celastrus parviflorus Vahl (1790).

Shrub up to 2 m high with spreading branches, and spines up to 2.7 cm long; branches brown to greyish with numerous pale lenticels, glabrous or practically so. Leaves: petiole 2-10 mm long; lamina 1.2-12.5 x 0.6-6.8 cm, elliptic to ovate, leathery, glabrous, apex acute to obtuse or rarely retuse, margin crenate or crenate-serrulate. Cymes (1-)1.5-5.5 cm long, pubescent; peduncle 1-20 mm long; pedicels 1-5(-8) mm long, flowers 3-50 in each cyme, 3.2-5.8 mm in diameter, functionally unisexual. Sepals 5, 1 x 1 mm, ovate to triangular, slightly pubescent with ciliolate margins. Petals 5, 2-2.5 x 1-1.5 mm, oblong. Male flowers with 5 stamens longer than the pistillode. Female flowers with 5 staminodes shorter than the ovary; ovary 3-locular. Disc convex, 5-lobed. Capsule pink to red, obconic, trigonous, 6-8 mm long, glabrous. Seeds 3-4, brown, glossy, with a white aril at the base.

subsp. **eritreana** (*Sebsebe*) *Sebsebe* (1985);
Maytenus eritreana Sebsebe (1984) - type: EW, Imbatkala, *Sebsebe Demissew* 1442 (ETH holo., K UPS iso.).

Bushland often on steep slopes; *c* (450-)1400-2000 m. EE EW; not known elsewhere. *Mooney* 9549; *Terraciano & Pappi* 1648; *Pappi* 5496.

Subsp. *parviflora* distinguished by its puberulous and pubescent branches, rounded leaf base, longer peduncles up to 65 mm long and apparently bisexual flowers is restricted to the Yemen Arab Republic.

4. **M. addat** *(Loes.) Sebsebe* (1984);
Gymnosporia addat Loes. (1908) - type: AR, Evano (Ewano), *Ellenbeck* 1864 (B holo. destr.); AR, 17 km SE of Shashemane near Kofele, *Sebsebe Demissew* 1262 (K neo., ETH UPS iso.).

Shrub or large tree 2-30 m high, usually unarmed or with spines up to 1.2 cm long; branches dark brown with pale brown lenticels, glabrous. Leaves with petiole 5-12 mm long; lamina 3-13.6 x 2-4.8(-5.8) cm, elliptic to ovate-oblong, leathery, glabrous, acuminate at the apex, margin crenate-serrulate. Cymes (2-)2.6-6.2 cm long, glabrous; peduncle 1-10 mm long; pedicels 1.5-3.5 mm long; flowers 15-50 in each cyme, 3-6.5 mm in diam., functionally unisexual. Sepals 5, 0.7-1.3 x 1-1.8 mm, oblong or elliptic to suborbicular. Petals 5(-6), white, 2-2.5 x 1.2-1.8 mm, oblong or elliptic to suborbicular, with margin slightly serrate. Male flower with 5 stamens longer than the pistillode. Female flowers with 5 staminodes shorter than the ovary; ovary 3(4)-locular. Disc flat to concave, 5-lobed. Capsule pink to red (greenish when young), trigonous to ± round, 6-8 mm long, glabrous. Seeds brown, 3(4), glossy, with a white aril at the base. Fig. 108.1.7-8.

Forest, forest margins, mountain slopes and along streams; *c* 2200-3000 m. SU AR SD GG; not known elsewhere. *Sebsebe D.* et al. 927, 1248; *J. De Wilde* 5724.

5. **M. obscura** *(A. Rich.) Cuf.* (1962);
Celastrus obscurus A. Rich. (1847); *Gymnosporia obscura* (A. Rich.) Loes. (1893); *G. serrata* (A. Rich.) Loes. var. *obscura* (A. Rich.) Fiori in *Boschi e piantae legn. del. l'Erit.*: 227 (1912) - types: TU, *Schimper* I:129 (P syn., BM K S isosyn.), Arbate Ensesa (Araten Seca), near Axum, *Quarton-Dillon* 18 (P lecto.).
Maytenus ovata (Wall. ex Wight & Arn.) Loes var. *ovata* forma *ovata* sensu Blakelock (1956) *pro parte* including synonyms *Celastrus obscurus* A. Rich. and *Gymnosporia obscura* (A. Rich.) Fiori (1912).

Shrub or tree 2-10 m high with spines up to 4 cm long; branches black or dark grey to dark brown with

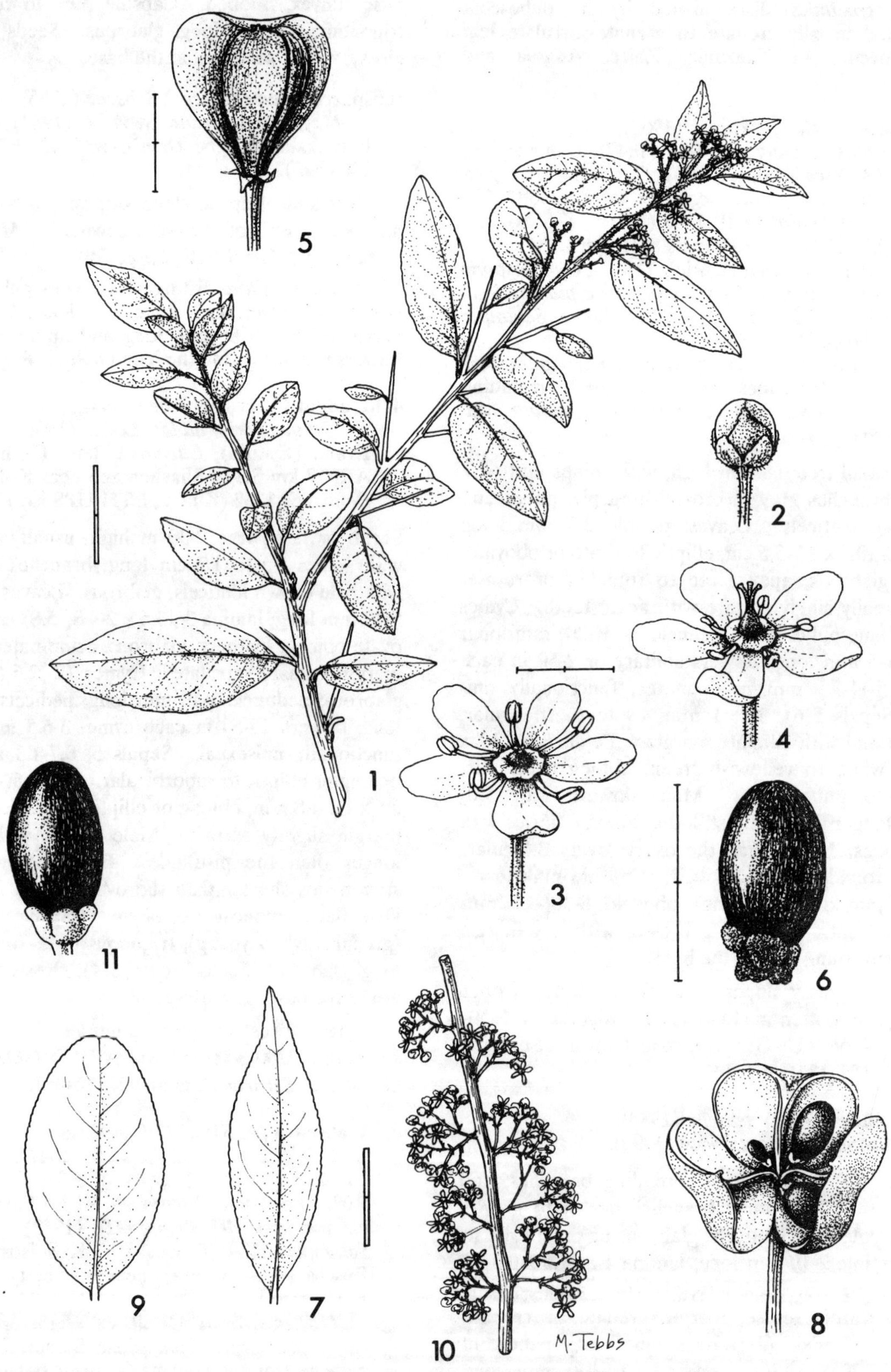

Figure 108.1 *MAYTENUS ARBUTIFOLIA*: 1 - flowering branch; 2 - bud; 3 - male flower; 4 - female flower; 5 - capsule; 6 - seed. *M. ADDAT*: 7 - leaf; 8 - capsule. *M. OBSCURA*: 9 - leaf; 10 - inflorescence; 11 - seed. Scale: closed bar represents cm; single line mm. 1-6 from *Friis* et al. 1888; 7 & 8 from *Friis* et al. 1388; 9-11 *Cheesman* 7440. Drawn by M. Tebbs.

numerous pale lenticels, glabrous. Leaves: petiole 4-10 mm; lamina 4-8.7 x 1.7-4(5.8) cm, elliptic or oblong to ovate or oblanceolate, leathery, glabrous, apex obtuse to rounded or sometimes subacute to retuse, margin crenate. Cymes 1-2(-2.5) cm long, glabrous; peduncle 1-7 mm long; pedicels 2-5 mm long; flowers 2-50 in each cyme, 3.2-6.3 mm in diam., bisexual or functionally unisexual. Sepals 5, 0.7-1.5 x 0.7-1.5 mm, triangular with subentire to shortly ciliolate margin. Petals 5, 2-3.8 x 1-2.3 mm, oblong. Male flowers with 5 stamens longer than the pistillode. Female flowers with 5 staminodes shorter than the ovary. In bisexual flowers the stamens and ovary are ± equal in length. Ovary 3-4 locular. Disc convex, 5-lobed. Capsule red to red-brown when mature, greenish when young, obconic, trigonous, 4-6 mm long, glabrous. Seeds 3-4, reddish-brown, glossy with a fleshy white (turning pale yellow at maturity) aril at the base. Fig. 108.1.9-11.

Open woodland, grassland, forest margins, sometimes along streams; *c* (1700-)2100-3100 m. TU GD GJ WG SU AR BA IL KF SD; Uganda, Kenya, Tanzania and Burundi. *Mooney* 5042; *Sebsebe D.* et al. 1330; *Chaffey* 354.

6. **M. cortii** *(Pich.-Serm.) Cuf.* (1958);
Gymnosporia cortii Pich.- Serm. (1951) - type: GD, Simen, Beleghes, W. of Ambaras, *Pichi-Sermolli* 2599 (FT holo.).

Shrub with spines up to 2 cm long; branches light to dark grey with white lenticels, glabrous. Leaves: petiole 1-1.5 mm long; lamina 0.5-1.5 x 0.4-1.1 cm, broadly elliptic to ovate or ± orbicular, leathery, glabrous, apex obtuse, margin crenate. Cymes 0.5-1 cm long, glabrous; peduncle 3-6 mm long; pedicels 1-1.5 mm long; flowers up to 3 in each cyme, *c* 3 mm in diam., bisexual. Sepals 5, *c* 1.5 x 1.3 mm, ovate-triangular, margin toothed to fimbriate. Petals 5, *c* 1.5 mm, elliptic. Stamens 5, ± equal to the ovary in length. Ovary 3-locular. Disc convex, 5-lobed. Capsule light brown with green spots, 6-8(-10) mm long, trigonous, obovoid, glabrous. Seeds 3(4), dark brown, glossy; aril dirty white to yellow, basal and on one side covering up to half the length of the seed.

On mountainous slopes; *c* 3500 m. GD; not known elsewhere. *Pichi-Sermolli* 2598, 2600.

7. **M. harenensis** *Sebsebe* (in ed.).

Shrub or tree to 6 m high with spines to 8 mm long; branches dark to reddish brown, glabrous to sometimes appressed puberulous in the very young shoots. Leaves: petiole 4-6 mm long; lamina 2-5 x 1.5-4 cm, broadly elliptic to obovate, leathery, glabrous, apex rounded, acute to abruptly acuminate. Cymes 2.5-5cm long, glabrous; peduncle 3-12 mm long; pedicels 5-10 mm long; flowers up to 25 in each cyme, *c* 7 mm in diameter, bisexual. Sepals 5, *c* 1 x 1 mm, triangular with ciliolate margins. Petals 5, *c* 3 x 1.5 mm, obovate with entire margin. Stamens 5, ± equal to the pistil. Ovary 3-locular. Disc concave ± rounded. Capsule pink to red (5-)8-17 mm long, obconic, trigonous, glabrous. Seeds 6-8, red. glossy with a white fleshy aril at the base.

In forest and forest margins; 1600-3050 m. Known only from BA (Harena forest). *Mesfin T.* 5140; *Sebsebe D.* 2064; *Gilbert & Sebsebe D.* 8154.

8. **M. buchananii** *(Loes.) Wilczek* (1960).

Shrub up to 2.5 m high with spines up to 2.8 cm long; branches brown to grey with pale lenticels, pubescent. Leaves: petiole 2-3 mm long; lamina 1-3 x 0.8-2.3 cm, ovate or elliptic to suborbicular, leathery, glabrous, apex obtuse to rounded or sometimes acute. Cymes 0.5-1.8 cm long, pubescent; peduncle absent or up to 3 mm long; pedicels 3-4 mm long; flowers 3-15 in each cyme, *c* 2.5 mm in diameter, bisexual or functionally unisexual. Sepals 5, 0.5-1.3 x 1 mm, triangular with ciliolate margin. Petals 5, 1.5-2 x 1.3 mm, oblong. Stamens 5, exceeding or equal to or shorter than the ovary. Ovary 3-locular. Disc convex, ± unlobed. Capsule pink to red, 5-8 mm long, obconic, trigonous, glabrous. Seeds 3, pinkish when young, turning brown, glossy, with a white (turning yellow) aril at the base.

In forest and forest margins; *c* 2000 m. SD; Uganda, Kenya, Tanzania, Sudan, Rwanda, Burundi, Central African Republic, Zambia, Angola, Nigeria, Ivory Coast. *Cufodontis* 455; *Mesfin T.* et al. 2736; *Sebsebe D.* 1219.

Gymnosporia filamentosa Loes. in Engl. *Bot. Jaharb.* **17**: 542 (1893) might be the earliest name for this species, but the specimens which could confirm this were destroyed. The shortly petiolate leaves, densely fasciculate short inflorescences, long filaments exceeding 3 mm, and the small disc mentioned in the description all more or less agree with *M. buchananii*. Loesener (1893) also proposed two varieties within *G. filamentosa* from E Africa, namely var. *major* and var. *minor*, the types of both of which were destroyed in Berlin. A third variety, *G. filamentosa* var. *brevistaminea* proposed later (Loesener, 1908) from Rwanda is *M. arbutifolia*.

9. **M. arbutifolia** *(A. Rich.) Wilczek* (1960).

Shrub or tree, 0.3-6(-12) m high with spines up to 4-7 cm long; branches grey, brown or sometimes black with white lenticels, pubescent. Leaves: petiole 1-6 mm long; lamina 1.2-6(-9) x 0.6-4.2(-5) cm, elliptic or oblong to oblanceolate, ovate or orbicular, hard or leathery, glabrous or puberulous below and on veins above, apex acute to rounded or emarginate, margin crenate to serrulate or sometimes serrate. Cymes 1-2.5 cm long, pubescent; peduncle 2-16 mm long; pedicel 1-5 mm long; flowers solitary or 2-30 in each cyme, 2.5-7 mm in diam., functionally unisexual. Sepals 5, 0.7-1 mm long, ovate, with ciliolate margin. Petals 5, 1.5-3(-3.7) mm long, oblong to elliptical with subentire to toothed margin. Male flowers with 5 stamens longer than the pistillode. Female flowers with 5 staminodes shorter than the ovary; ovary 3-locular. Disc convex, 5-lobed. Capsule greenish-purple when young, rose-red when mature, obconic, trigonous 4-8 mm long, glabrous. Seeds 1-4, orange-brown, glossy, with a rugose, dark purple (whitish to pink when young) aril at the base. Fig. 108.1.1-6.

1. Leaves glabrous on both sides; widespread in NE Africa, tropical Arabia and elsewhere in Africa. var. **arbutifolia**
- Leaves puberulous or pubescent below and/or on veins above; restricted to SD and adjacent areas of Ethiopia. var. **sidamoensis**

var. **arbutifolia**;

Celastrus arbutifolius A. Rich. (1847); *Gymnosporia arbutifolia* (A. Rich.) Loes. (1893) - type: GD, *Schimper* II:832 (P holo., BR K S syn.).

C. arbutifolius A. Rich. var. *major* A. Rich. (1847) - type: GD, Simen, Ouodgerate, *Quartin-Dillon & Petit* s.n. (P lecto.).

Celastrus atkaio A. Rich. (1847); *Gymnosporia atkaio* (A. Rich.) Loes. (1893) - type: SU, Shewa "Choa", *Quartin-Dillon & Petit* s.n. (P holo.).

Gymnosporia engleriana Loes. (1893); *Maytenus engleriana* (Loes.) Cuf. (1962) - type: GD, Djenda, *Steudner* 603 (K lecto.).

Maytenus engleriana Loes. var. *macrantha* Loes. (1908); *Maytenus engleriana* (Loes.) Cuf. var. *macrantha* (Loes.) Cuf. (1962) in *Senck. Biol.* **43**: 289 (1962) - types: SU, near Addis Abeba, *Ellenbeck* 1577 (B syn. destr.), BA, Ginea-Denek, *Ellenbeck* 1954 (B syn. destr.).

Forest, forest margins, degraded forest, river banks, grassland and scattered thickets and along roads; *c* 1200-3000 m. EE EW TU GD WU GJ WG SU HA AR BA IL KF; Kenya, Uganda, Tanzania, Rwanda, Burundi, Zaire, Somalia, Yemen, Saudi Arabia. *Sebsebe Demissew* 514; *Thulin* 1434; *Hunde & Jacobsen* 94.

var. **sidamoensis** *Sebsebe* (1985) - type: SD, near Dilla, 22 km from Aleta Wendo, *Sebsebe D. & Tewolde B.G.E.* 950 (ETH holo., K UPS iso.).

Gymnosporia filamentosa Loes. (1893) sensu Cufodontis (1958) including *Ruspoli & Riva* 190.

Maytenus prob. *somalensis* sensu Cufodontis (1960, 1962) non *M. somalensis* (Loes.) Cuf. sensu stricto.

Forest and forest margins; 1600-2150 m. SD and adjacent areas of SU, not known elsewhere. *Westphal & Westphal* 1694; *Sebsebe Demissew* 602; *Kuls* 885.

10. **M. putterlickioides** *(Loes.) Exell & Mendonça* (1953).

Shrub or small tree 1-6 m high, often a many-stemmed bush, sometimes straggling, with spines up to 6.7(-8.3) cm long; branches grey to brown with pale lenticels, pubescent. Leaves: petiole 1-3 mm long; lamina 0.7-5.2 x 0.6-3 cm, elliptic to oblanceolate or obovate, leathery, densely pubescent on both sides, apex obtuse to rounded or retuse, margin subentire to crenate with minute glands. Cymes 1.5-4.5(-5.2) cm long, pubescent; peduncle 2-15(-30) mm long; pedicels 2-7 mm long; flowers 10-25 in each cyme, *c* 8.2-10.3 mm in diam., bisexual. Sepals 5, 0.7-1.3 x 1-1.8 mm, ovate to suborbicular, rounded, with toothed margin. Petals 5, 4-4.8 x 2.5-3 mm, obovate to oblong. Stamens 5, ± equal with ovary. Ovary 3-locular. Disc convex, ± 5-lobed. Capsule pink, orange or red, obconic, trigonous, 4-6 mm long, pubescent. Seeds (2)3, reddish-brown, completely enclosed by a thin pale white to reddish aril.

Dry deciduous woodland, scrub, termite hills and riverine areas; 1000-1200 m. SD; Kenya, Tanzania, Angola, Zimbabwe, Zambia, Malawi, Mozambique, Zaire. *Cufodontis* 757; *Sebsebe D.* 1237.

11. **M. senegalensis** *(Lam.) Exell* (1952);

Celastrus senegalensis Lam. (1785).

Celastrus senegalensis Lam. var. *inermis* A. Rich. in *Tent. Fl. Abyss.* 1: 133 (1847); *Gymnosporia senegalensis* (Lam.) Loes. var. *inermis* (A. Rich.) Loes. in Engl. *Bot. Jahrb.* **17**: 541 (1893); *G. senegalensis* (Lam.) Loes. var. *inermis* (A. Rich.) Loes. forma *macrocarpa* Loes. (1893) - type: *Schimper* III:1598 (P lecto., BR K UPS syn.).

Gymnosporia intermedia Chiov. (1911) - type: GD, near Mt. Inceduba, *Chiovenda* 1894 (FT holo.).

Shrublet to tree, up to 8(-13) m high, unarmed or with spines up to 5 cm long; branches wine-red, glaucous to grey with indistinct whitish lenticels, glabrous. Leaves: petiole 2-17 mm long; lamina 3.2-12.5 x 1.2-8 cm, elliptic or oblong to obovate or oblanceolate, leathery to somewhat leathery, glabrous, apex acute to rounded, retuse or truncate, margin crenulate to crenate. Cymes 0.7-3(-5) cm long, glabrous; peduncle 2.6-17(-26) mm long; pedicels 0.5-5 mm long; flowers 3-60 or more in each cyme, 1.25-7 mm in diameter, functionally unisexual. Sepals 5, 0.5-1 x 0.5-1.3 mm, ovate to triangular, with ciliolate margins. Petals 5, 1.2-4 x 0.7-2.5 mm, elliptic to oblong or oblanceolate. Male flowers with 5 stamens longer than the pistillode. Female flowers with 5 staminodes shorter than the ovary, ovary 2(3)-locular. Disc ± concave, 5-10 lobed. Capsule pink to red when mature, green when young, globose, 3-5(-7) mm long, glabrous. Seeds 1-2, dark red to brown, glossy, with a white aril covering 1/3-2/3 of the surface. Fig. 108.2.1-6.

Deciduous woodland, open dry scrub, grassland, dry mountain slopes, river banks, edges of lakes; *c* 380-2440 m. EE EW TU WU GD GJ WG SU HA AR BA IL KF GG SD; in tropical Africa from Somalia to Senegal, and southwards to South Africa, north Africa and the Canary Islands; also in southern Spain, Yemen, Afghanistan, Pakistan, India, Bangladesh and Madagascar. *Mooney* 8417; *Sebsebe D.* 586; *Thulin* 1341.

12. **M. heterophylla** *(Eckl. & Zeyh.) Robson* (1965);

Celastrus heterophyllus Eckl. & Zeyh. (1834).

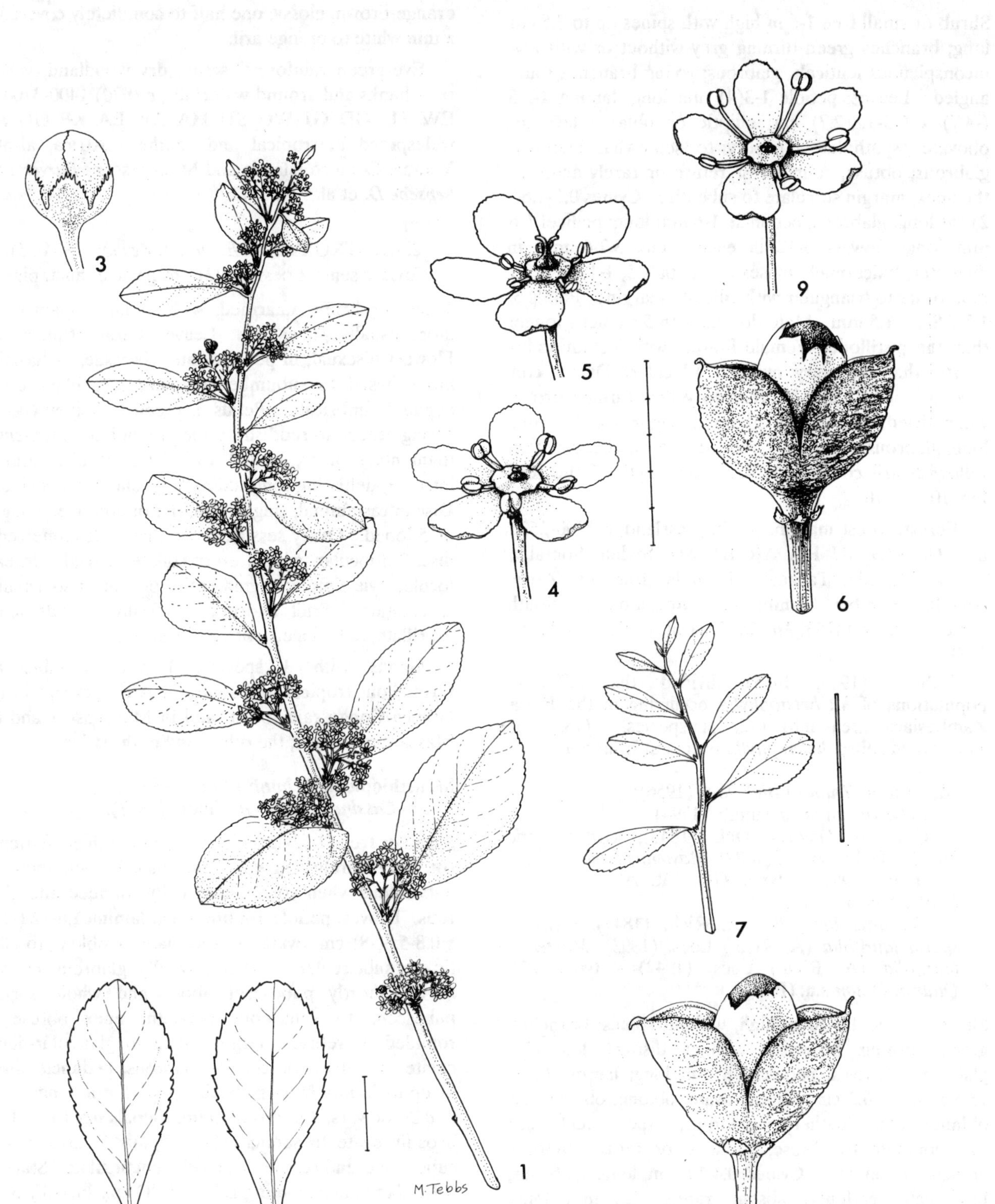

Figure 108.2 *MAYTENUS SENEGALENSIS*: 1 - flowering branch; 2 - leaf; 3 - bud; 4 - male flower; 5 - female flower; 6 - fruit. *M. HETEROPHYLLA*: 7 - young shoot; 8 - leaf; 9 - male flower; 10 - fruit. Scale: closed bar represents cm; single line mm. 1-6 from *Ensor* et al. 7302; 7-10 from *Scott Elliot* 983. Drawn by M. Tebbs.

Shrub or small tree 1-6 m high with spines up to 3.5 cm long; branches green turning grey without or with few inconspicuous lenticels, glabrous; young branches four-angled. Leaves: petiole 1-3(4) mm long; lamina 1-3.5 (-4.7) x 0.3-1.8(2.7) cm, elliptic or oblanceolate to obovate (spathulate), papery to somewhat leathery, glabrous; obtuse, rounded or retuse or rarely acute at the apex, margin serrulate to subentire. Cymes 0.5-1.6(-2) cm long, glabrous; peduncle 1-6 mm long; pedicel 1-6 mm long; flowers 3-12 in each cyme, 3.7-6 mm in diameter, functionally unisexual. Sepals 5, 1-1.3 x 0.5-1 mm, ovate to triangular with ciliolate margins. Petals 5, 1.5-2.8 x 1-1.5 mm. Male flowers with 5 stamens longer than the pistillode. Female flowers with 5 staminodes shorter than the ovary; ovary 2(3)-locular. Disc ± concave, 5-lobed. Capsule green to yellow turning brown when dried, obovoid to globose, trigonous, 3-5 mm long, glabrous. Seeds 2-4, reddish-brown, glossy, with a yellowish aril covering 1/3-2/3 the length of the seed. Fig. 108.2.7-10.

Forest, forest margin, scrub, grassland; *c* 1500-2620 m. GJ WG SU HA AR BA SD; Sudan, Somalia, Kenya, Uganda, Tanzania, Rwanda, Burundi, Zaire, Angola, Zambia, Zimbabwe, Mozambique, South Africa. *Gillett* 14253; *Mesfin T.* et al. 2893; *Sebsebe D.* 1371.

Robson (1963, 1965) divided the different populations of *M. heterophylla* occurring in the Flora Zambesiaca area into four subspecies. Only the widespread subsp. *heterophylla* occurs in Ethiopia.

13. **M. undata** *(Thunb.) Blakelock* (1956);
Celastrus undatus Thunb. (1794).
Celastrus luteolus Del. (1843); *Gymnosporia luteola* (Del.) Szyszy. (1888); *Maytenus luteola* (Del.) Andrews (1952) - type: GD, Mt. Aber, *Galinier* (MPU holo.).
Celastrus laurifolius A. Rich. (1847); *Gymnosporia laurifolia* (A. Rich.) Loes. (1893); *Maytenus laurifolia* (A. Rich.) Loes. (1942) - type: TU, *Quartin-Dillon* s.n. (P holo.).

Shrub or tree 1.5-10 m high, without spines; branches green turning brownish without distinct lenticels, glabrous. Leaves: petiole 3-8 mm long; lamina (2)3-13.5 x (0.8)1-6.2 cm, elliptic, ovate, oblong, obovate or oblanceolate, leathery, glabrous, apex acute or subacuminate to obtuse, rounded or retuse, margin crenate to serrate. Cymes 0.4-1.5 cm long, glabrous, fasciculate; peduncle absent, rarely 1(2) mm long; pedicel 2-9 mm long; flowers 3.2-6.5 mm in diameter, bisexual. Sepals 5, 0.5-1.8 x 1-2.5 mm, ovate, with entire to slightly toothed margins. Petals 5, 2-4 x 1.7-2.8 mm, broadly elliptic to obovate or oblanceolate. Stamens 5, ± as long as the ovary; ovary 3-locular (sometimes 2-locular by abortion). Disc flat to concave, 5-10-lobed. Capsule green to purple-green or yellow, trigonous, obovoid, 5-7 mm long, glabrous. Seeds 2-3, brown to orange-brown, glossy, one half to completely covered by a thin white to orange aril.

Evergreen rainforest, scrub, dry woodland, valleys, river banks and around waterfalls; *c* (920) 1400-3100 m. EW TU GD GJ WG SU HA AR BA KF GG SD; widespread in tropical and southern Africa, also in Yemen, Comoro Islands and Madagascar. *Burger* 2939; *Sebsebe D.* et al. 934, 1271.

2. **MYSTROXYLON** *Ecklon & Zeyher* (1834-35);
Cassine sensu Loes. (1892) pro parte et auct. plur.

Trees or shrubs, unarmed, without latex, glabrous or more usually pubescent. Leaves spiral; stipules free. Flowers bisexual, in pedunculate to sessile, dichasial to monochasial to subumbellate, simple, axillary cymes. Sepals 5, imbricate. Petals 5, green to yellow (usually drying orange to reddish), imbricate in bud. Stamens 5; filaments slender, united with base of disc; anthers introrse, dehiscing longitudinally, eventually deciduous. Disc intrastaminal, single, thin, flat or concave, 5-angled or 5-lobed. Ovary sessile, more than half-immersed in disc, 2-3-locular, with 2 erect collateral ovules in each locule; style simple; stigma entire or almost so, small or ± capitate. Fruit a drupe, ± fleshy. Seeds solitary, exarillate, not winged, with endosperm.

Genus with 3 species; 1 very variable and throughout tropical and southern Africa except in the Guinea-Congo rain forest, and in Madagascar and the Mascarene Islands; the others in South Africa.

M. aethiopicum *(Thunb.) Loes.* (1897);
Cassine aethiopica Thunb. (1818).

Shrub or tree (1-)2-12(-18) m high; branchlets flattened (or very rarely angular) and usually velutinous to puberulous when young, eventually rounded and glabrous. Leaves: petiole 1-8 mm long; lamina 1.6-12 (-15) x 0.8-5.6(-8) cm, ovate or lanceolate or oblong to elliptic or subcircular, leathery, wholly glabrous or with nerves shortly pubescent above and whole surface pubescent to velutinous beneath, apex obtuse to rounded or retuse, margin entire to glandular-denticulate. Cymes pubescent or glabrous; peduncle absent or up to 20 mm long; pedicels up to 15 mm long; (1-2-) 3- *c* 20 flowers, 3-6 mm in diam. Sepals 0.5-1 mm long, broadly ovate to semicircular. Petals 1-2.5 mm long, subcircular and sessile or broadly unguiculate. Stamens with filaments becoming reflexed. Ovary broadly ovoid to subglobose. Drupe bright red to purplish-red, 0.8-2 cm long, globose to ovoid, often apiculate, glabrous or glaucous to puberulous. Fig. 108.3.

Submontane evergreen forest and scrub; *c* 1700 m. SD; from Guinée, Cameroun and Sudan Republic south to the Cape and west to Angola and Namibia, also in Madagascar, Comoro Islands, Aldabra and the Seychelles. *Burger* 1830.

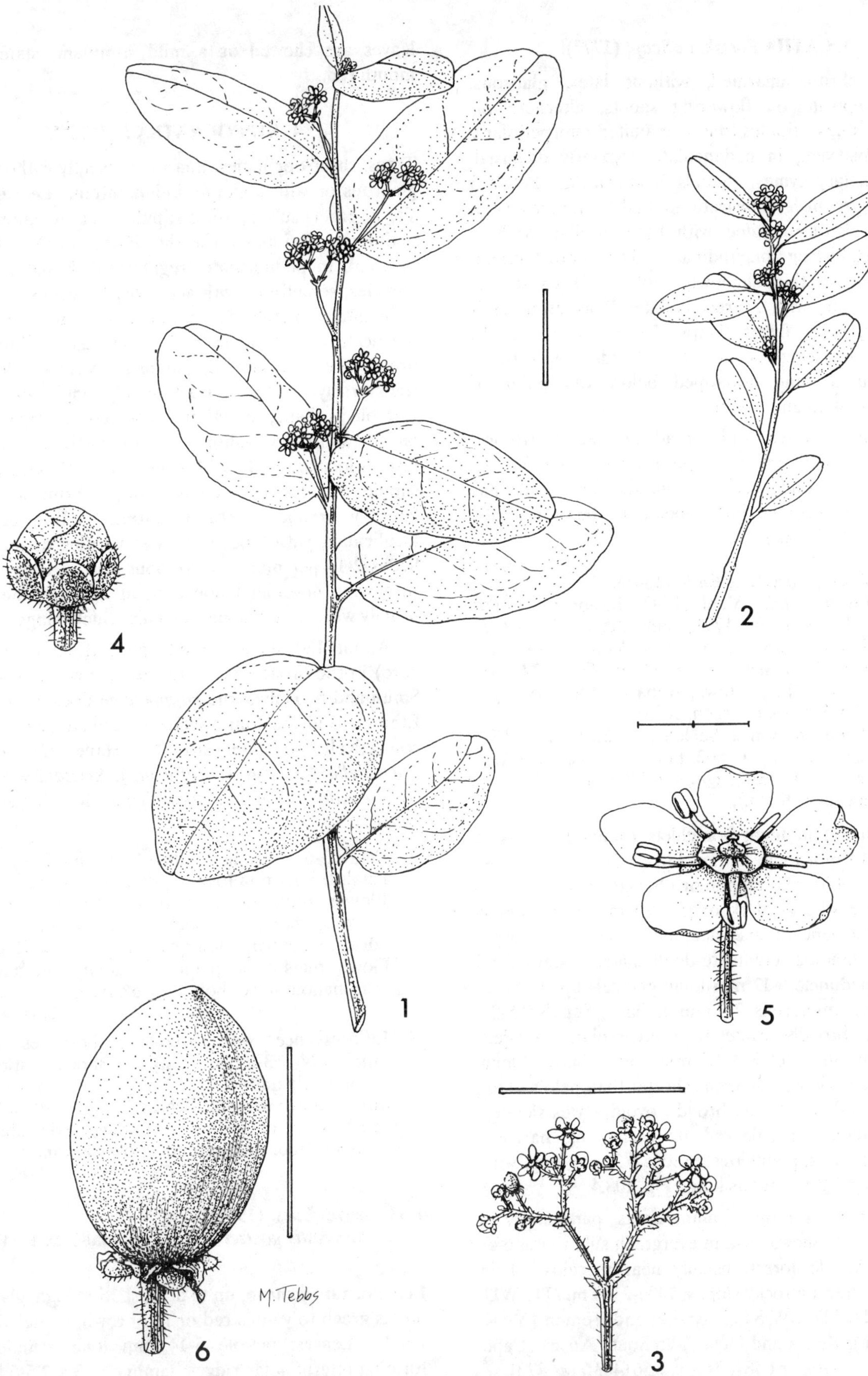

Figure 108.3 *MYSTROXYLON AETHIOPICUM*: 1 - flowering branch; 2 - young shoot; 3 - inflorescence; 4 - bud; 5 - flower (2 anthers fallen); 6 - fruit. Scale: closed bar represents cm; single line mm. 1 from *Gardner* 1119; rest from *Osmaston* 1356. Drawn by M. Tebbs.

3. CATHA *Forssk. ex Scop.* (1777).

Tree or shrub, unarmed, without latex, glabrous. Leaves opposite on flowering shoots, alternate on vegetative ones; stipules free or ± united interpetiolarly. Flowers bisexual, in pedunculate, regularly dichasial, simple, axillary cymes. Sepals 5, imbricate. Petals 5, white or greenish, imbricate in bud. Stamens (4)5; filaments slender, united with base of disc; anthers introrse, dehiscing longitudinally. Disc intrastaminal, single, concave, thin, shallowly 5-lobed. Ovary sessile, free, 3-locular, with 2 erect collateral ovules in each locule; styles 3, free; stigmas 3. Fruit a capsule, elongate, dehiscing loculicidally. Seeds 1-3, without arils, with a wing developed below the point of attachment, with endosperm.

A single species, widespread in eastern tropical Africa from Ethiopia to Cape Province and in SW Arabia. A second species (from the Transvaal) has been added recently, but appears to be generically distinct (*Lydenburgia*).

C. edulis *(Vahl) Forssk. ex Endl.* (1841);
Celastrus edulis Vahl (1790) basionym; *Catha inermis* J.F. Gmelin (1791) nom. illegit; *C. forskalii* A. Rich. (1847) nom. illegit - type: Yemen, *Forsskal.*
Trigonotheca serrata Hochst. in *Flora* **24**: 662 (1841) - type: TU, Adowa, Abba Gerima, *Schimper* II:649 (TUB holo. not seen, K iso.)
Dillonia abyssinica Sacleux, in *Bull. Mus. Hist. Nat. Paris, sér. 2*, **4**: 602 (1932) - type: TU/GD, Tcumze (?= Tekeze), *Quartin-Dillon & Petit* s.n. (P holo. not seen, K iso.).

Tree 2-15(-25) m high; branchlets rounded. Leaves: petiole 3-10 mm long; lamina (3.7-)5.5-11 x (0.8-)1.5-4.5(-6) cm, oblong to elliptic or obovate, ± leathery (or somewhat leathery), dark or greyish green and glossy above, paler beneath, apex acuminate to obtuse, margin markedly glandular-crenulate-denticulate. Cymes 1-2.3 cm long; peduncle 6-12 mm long; pedicels up to 2 mm long; flowers numerous, 2-3 mm in diam. Sepals 0.5-0.7 mm long, broadly ovate to semicircular, rounded, ciliate-fimbriate. Petals 1-1.5 mm long, elliptic-oblong, margin paler, ciliate. Stamens shorter than petals. Disc 1-1.5 mm in diam. Ovary broadly ovoid; styles slender; stigmas small. Capsule red, 0.6-1 cm long, narrowly oblong-trigonous, pendulous. Seeds *c* 3 mm long, with narrow wing 2-2.5 times as long. Fig. 108.4.

An important crop in many areas, particularly in Harerge and S Shewa, also in evergreen submontane or medium altitude forest, usually near margins, or in woodland often on rocky slopes; 1400-2100 m. TU WU WG SU HA SD; SW Saudi Arabia and Yemen (YAR and PDRY), Kenya and Uganda to South Africa (Cape Province). *Schimper* I:781; *Mooney* 5614; *Meyer* 8721.

Grown in the same areas as non-forest coffee and an important source of income for small farmers. The leaves are chewed as a mild stimulant, staining the mouth red.

4. HIPPOCRATEA *L.* (1753).

Trees, shrubs or lianes, unarmed, usually without latex, glabrous or with a simple indumentum. Leaves opposite or rarely subopposite; stipules free or united interpetiolarly, deciduous (in the Flora area). Flowers bisexual, in pedunculate, regularly dichasial cymes or panicles, sometimes with accessory branches. Sepals 5, imbricate. Petals 5, variously coloured or green, imbricate or valvate in bud. Stamens 3; filaments ± broad, free or united; anthers extrorse, dehiscing transversely. Disc extrastaminal, single or double, variable in shape, sometimes dissected. Ovary sessile or on an (andro-)gynophore, free or partly embedded in disc, 3-locular, with 2-14 ovules in each locule; style simple or absent; stigmas small. Fruit a capsule, laterally expanded into 3 flattened lobes, dehiscing loculicidally with lobes usually separating as mericarps. Seeds 2-14 per mericarp, without arils, usually winged, with wing developed above point of attachment and usually with marginal and funicular thickenings.

About 120 species, pantropical if interpreted (as here) in the broad sense. In the narrow sense of A.C. Smith and N. Hallé, true *Hippocratea* does not occur in Ethiopia and the four species described below belong respectively to *Simirestis* N. Hallé (*H. goetzei*), *Loeneriella* A.C. Sm. (*H. africana*), *Reissantia* N. Hallé (*H. parvifolia*) and *Apodostigma* R. Wilczek (*H. pallens*).

1\. Petals 3-6 mm long; seeds 6-14 per mericarp. 2
\- Petals 0.5-1.7 mm long; seeds 2 per mericarp. 3
2\. Flower buds globose; petals oblanceolate to oblong with pale margin; disc elongate, cylindric, united with androgynophore. **1. H. goetzei**
\- Flower buds conic; petals triangular, same colour throughout; disc short, convex, free. **2. H. africana**
3\. Inflorescence with accessory branches; disc shallowly 3-lobed; style short; stigmas completely united. **3. H. parvifolia**
\- Inflorescence simply dichasial; disc interrupted, each lobe enclosing a stamen; style absent; stigmas free, each extending over a stamen. **4. H. pallens**

1\. **H. goetzei** *Loes.* (1901);
Simirestis goetzei (Loes.) N. Hallé ex R. Wilczek (1960).

Liane or rarely shrub, up to at least 15 m high, glabrous. Stems green to vinous red or red-brown, 4-angled or 4-lined. Leaves: petiole 5-14 mm long; stipules not forming interpetiolar ridges; lamina 5-13 x 2.5-6.5(-8.3) cm, ovate to oblong or elliptic, papery to somewhat leathery, dark or greyish green and dull above, paler beneath, apex obtusely acuminate to rounded, margin

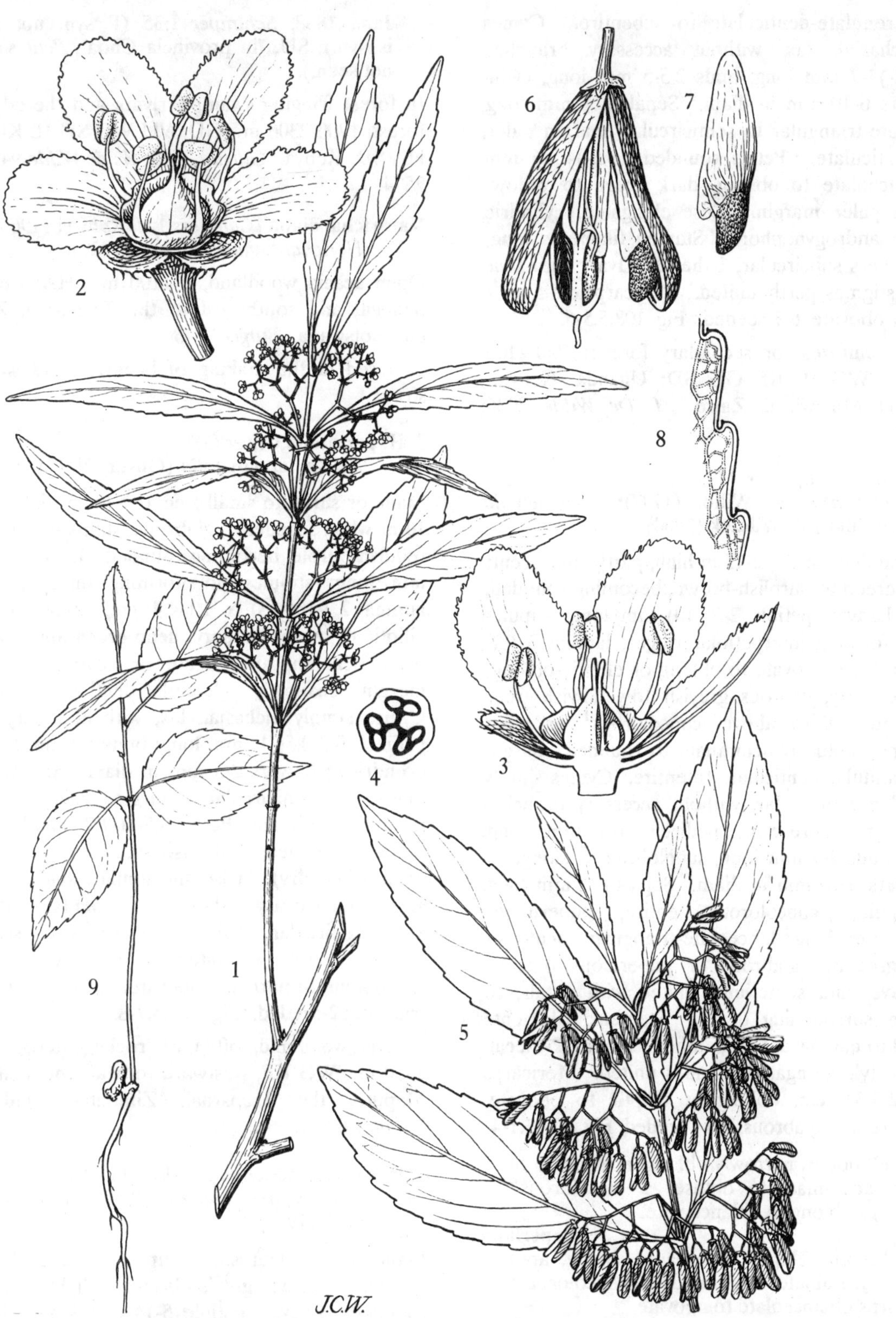

Figure 108.4 *CATHA EDULIS*: 1 - flowering shoot x 2/3; 2 - flower x 12; 3 - vertical section of flower x 12; 4 - transverse section of ovary x 12; 5 - fruiting shoot x 2/3; 6 - fruit x 4; 7 - seed x 4; 8 - detail from leaf margin x 2.5; 9 - seedling x 2/3. 1-4 from *Eyles* 977, 5-8 from *Chase* 37, 9 from *Swynnerton* 102. (Reproduced by permission, Editorial Board, *Flora Zambesiaca*: tab. 80, 1966.)

glandular-crenulate-denticulate to subentire. Cymes simply dichasial, lax, without accessary branches; pedicles (2-)3-7 mm long; buds 2.5-5 mm long, cylindric; flowers 6-10 mm in diam. Sepals *c* 1 mm long, broadly ovate-triangular to semicircular, margin paler, ciliolate-denticulate. Petals rounded, (3.5-)4.5-6 mm long, oblanceolate to oblong, dark green to yellow-green with paler margin. Disc elongated, cylindric, united with androgynophore. Stamen filaments long, narrow; anthers subcircular, 1-thecal. Ovary with style elongated; stigmas partly united. Mericarps *c* 5.5 x 3.6 cm, broadly obovate, 6-7-seeded. Fig. 108.5.5-6.

Upland rainforest or secondary forest; 1750-2100 (-2600) m. WG IL KF GG SD; Uganda, Kenya, Tanzania, N Malawi, E Zaire. *J. De Wilde* 7036; *Mooney* 6834.

2. **H. africana** *(Willd.) Loes.* (1910);
Tonsella africana Willd. (1797); *Loeseneriella africana* (Willd.) R. Wilczek (1960).

Liane or scandent shrub, 2-15 m high, partly pubescent; stems pale green to purplish-brown, becoming rounded, glabrous. Leaves: petiole 2-7(-10) mm long; stipules sometimes forming interpetiolar ridges; lamina 3-10(-20) x 2-5.5(-10) cm, ovate to oblong or elliptic-oblong, papery to leathery, glabrous, greyish- to yellowish-green and sometimes shiny above, concolorous[1] or paler beneath, apex obtusely acuminate to rounded, margin obtusely glandular-denticulate to entire. Cymes simply dichasial, dense or ± lax, without accessory branches, pubescent or glabrescent; pedicels 1-6 mm long, pubescent; buds 2-7 mm long, subglobose to elongate-conic; flowers 5-10 mm in diam. Sepals *c* 1 mm long, ovate-triangular, concolorous, entire, puberulous. Petals 3-6 mm long, lanceolate-triangular, acute or subacute, greenish, glabrous or puberulous outside. Disc concave, and sometimes united with petals, to convex and surrounding ovary. Stamens: filaments long, broad to narrow at base; anthers circular, 2-thecal. Ovary with style elongated; stigmas united. Mericarps 3.5-5.5 x 2.3-3.9 cm, oblong or elliptic to obovate, rounded or retuse, glabrous, 6-14-seeded. Fig. 108.5.1-4.

1. Leaves elliptic to narrowly oblong, apex acute or shortly acuminate to obtuse, ± discolorous[2]; mericarps oblong to oblanceolate. var. **schimperiana**
- Leaves broadly elliptic to broadly oblong, apex shortly acuminate to rounded, concolorous; mericarps oblanceolate to obovate. var. **richardiana**

Var. **schimperiana** *(Steud. & Hochst. ex A. Rich.) Cuf.* in *Enum.*: 484 (1958) & XXVII (1969);
H. schimperiana Steud. & Hochst. ex A. Rich. (1847) - types: TU, Ado, Mt. Scholoda, N. side, 10 Jan., 1838, *Schimper* I:35 (P syn. not seen, BM isosyn.); SU, "in provincia Choa", *Petit* s.n. (P syn. not seen,).

In forest fringing swamps, rivers and the edge of rain forest; 1700-2300 m. GD WU WG SU IL KF GG AR HA SD; Kenya. *Gutetta* 1088; *W. de Wilde* 9478; *Burger* 2574.

Var. **richardiana** *(Cambess.) N. Robson* (1965);
H. richardiana Cambess. (1929).

Open acacia woodland; *c* 1100 m. HA; westward to Senegal and southward to the Transvaal, Zimbabwe and Botswana. *Burger* 1746.

Used in the making of beehives and woven seed stores.

3. **H. parvifolia** *Oliver* (1868);
Reissantia parvifolia (Oliver) N. Hallé (1962).

Liane or shrub to small tree, 0.3-4.5 m high (or up to *c* 10 m when scandent), glabrous; stems greyish-green to reddish-purple, becoming rounded. Leaves: petiole 1-9 mm long; stipules not forming interpetiolar ridge; lamina 2.7-10(-13) x 1.1-4.4 cm, oblong or oblong-elliptic to ovate, leathery, deep green and shiny above, paler beneath, apex acute or acuminate to obtuse, margin shallowly glandular-denticulate to entire. Cymes simply dichasial, lax, with accessory branches; pedicels 0.2-0.5(-1) mm long; buds *c* 1 mm long, ovoid-cylindrical; flowers 2-3 mm in diam. Sepals *c* 0.4 mm long, ± semicircular, margin paler (when dry), irregularly ciliolate. Petals 0.5-1(-1.7) mm long, elliptic-oblong, rounded, yellowish-green to cream (margin paler when dry). Disc subcupular, shallowly 3-lobed, surrounding ovary. Stamens: filaments short, narrow; anthers circular, 1-thecal. Ovary with style short; stigmas completely united. Mericarps 3-5.8 x 1.4-2.1 cm, oblong to narrowly obovate, apiculate or obtuse to truncate, 2-seeded. Fig. 108.5.7-8.

Dry woodland, often on rocky outcrops or sand; 1350 m. WG KF; westward to Mali and southward to Maputo, the Transvaal, Zimbabwe and Angola. *Mooney* 6922.

4. **H. pallens** *Planchon ex Oliver* (1868);
*Apodostigma pallens (*Planchon ex Oliver) R. Wilczek (1960).

Liane or scandent shrub, up to at least 10 m high, glabrous; stems greyish-brown, slightly 4-lined or rounded. Leaves: petiole 5-14 mm long; stipules not forming interpetiolar ridge; lamina 3.5-9.6(-12) x 1.7-3.6(-6.6) cm, oblong to elliptic-ovate or obovate, papery to somewhat leathery, pale dull green, concolorous, apex acuminate, margin entire to glandular-serrulate. Cymes simply dichasial, lax or ± dense, without accessory branches; pedicels 0.2-1(-2.5) mm long; buds *c* 0.6 mm long, globose; flowers 2-3 mm in diam. Sepals *c* 0.5 mm long, ovate to oblong, margin ciliolate to

1. concolorous = both sides of leaf the same colour.
2. discolorous = upper and lower surfaces differently coloured, such as one much paler or darker than the other.

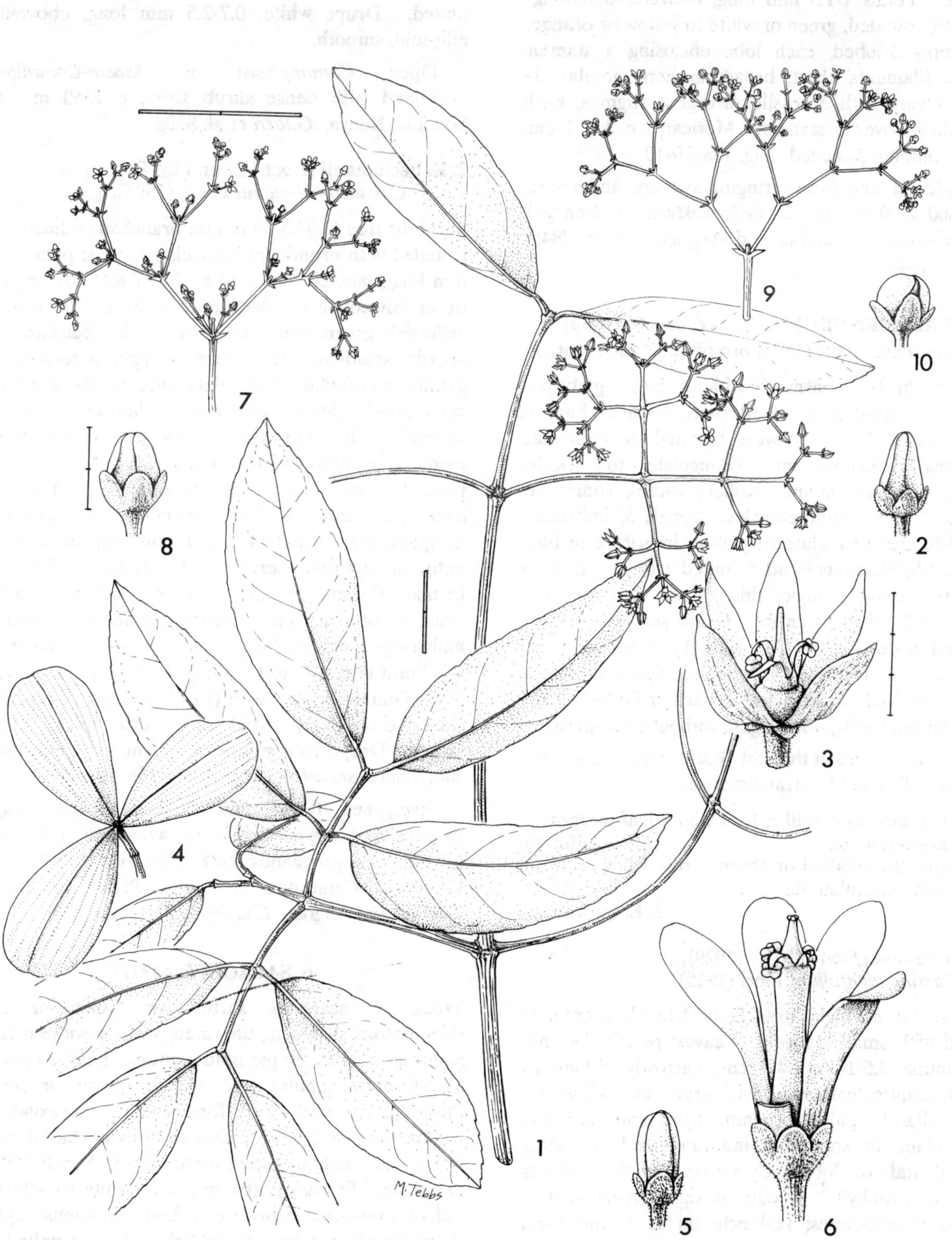

Figure 108.5 *HIPPOCRATEA AFRICANA*: 1 - leafy branch; 2 - bud; 3 - flower; 4 - fruit. *H. GOETZEI*: 5 - bud; 6 - flower (one petal cut off). *H. PARVIFOLIA*: 7 - inflorescence; 8 - bud. *H. PALLENS*: 9 - inflorescence; 10 - bud. Scale: closed bar represents cm; single line mm. 1-4 from *Pichi-Sermolli* 368 and *Dale* 2404; 5 & 6 from *Taylor* 2823; 7 & 8 from *Le Testa* 1472; 9 & 10 from *Vaughan* 2719. Drawn by M. Tebbs.

subentire. Petals 1-1.5 mm long, obovate to oblong-lanceolate, rounded, green or white to yellow or orange. Disc deeply 3-lobed, each lobe enclosing a stamen. Stamens: filaments short, broad; anthers circular, 1-thecal. Ovary with 3 sessile divergent stigmas, each curved down over a stamen. Mericarps *c* 3 x 1 cm, narrowly oblong, 2-seeded. Fig. 108.5.9-10.

Rain-forest and forest fringing swamps and rivers; (400-)1300-1700 m. IL KF GG; westward to Senegal and southward to Zambia and Angola. *Meyer* 7845; *Friis* et al. 1947.

5. ELAEODENDRON *Jacq. f. ex Jacquin* (1782).
Cassine sensu Loes. (1892) pro parte et auct. plur.

Trees or shrubs, unarmed, without latex, glabrous. Leaves all opposite or ± alternate towards base of shoot; stipules free. Flowers bisexual or unisexual, sometimes dioecious, in pedunculate to sessile, regularly dichasial, simple, axillary cymes, sometimes aggregated into pseudo-panicles. Sepals 5, imbricate. Petals 4-5, green to white or yellow, imbricate in bud. Stamens 4-5; filaments slender, united with base of disc. Disc intrastaminal, single, thick, convex to concave, shallowly 4-5 lobed to entire. Ovary sessile, ± deeply immersed in disc, 2-3-locular, with (1)2 erect collateral ovules in each locule; style simple; stigma entire or slightly 2-3-lobed. Fruit a drupe, hard or fleshy. Seeds 1-2(-3), without arils, not winged, without endosperm.

About 15 species in the Old World tropics, S Africa, the West Indies and Central America.

1. Staminodes stamen-like; fruit white; leaf-margins spinose-dentate. **1. E. aquifolium**
- Staminodes petaloid or absent; fruit yellow; leaf-margins crenulate-dentate. **2. E. buchananii**

1. E. aquifolium *(Fiori) Chiov.* (1929);
Cassine aquifolium Fiori (1915).

Shrub to 1.8 m high; branchlets 4-lined, eventually rounded with small lenticels. Leaves: petiole 1-6 mm long; lamina 3.5-14 x 1.4-4 cm, narrowly oblong or narrowly elliptic, leathery, greyish-green and dull above, not or slightly paler beneath, apex spinose-acute, margin plane or somewhat indurated and spreading spinose-dentate or -lobed. Cymes on specialised shoots in axils of prophylls[1] of current shoot, forming 1-3-flowered pseudocymes; peduncle up to 1 mm long; pedicels *c* 2 mm long; flowers 4-5 mm in diam., dioecious. Sepals (4)5, 0.5-1 mm long, oblate-semicircular, eroded-denticulate. Petals (4)5, 1.2-2.5 mm long, ovate-oblong to oblong, not unguiculate or ridged, entire. Male flowers with 5 stamens; disc 2 mm in diam.; pistillode immersed in disc. Female flowers with (4)5 stamen-like staminodes; style thin; stigmas united. Drupe white, 0.7-2.5 mm long, obovoid to ellipsoid, smooth.

Open *Commiphora* or *Acacia-Commiphora* woodland with dense shrub layer; *c* 1350 m. SD; Somalia, Kenya. *Gilbert* et al. 8208.

2. E. buchananii *(Loes.) Loes.* (1897);
Cassine buchananii Loes. (1895).

Shrub or tree (1-)4.5-30 m tall; branchlets 4-lined, soon rounded with prominent lenticels. Leaves: petiole 5-13 mm long; lamina (5-)6.5-14 x (2-)2.7-8.1(-10) cm, elliptic or lanceolate to obovate or oblanceolate, leathery, yellowish green and dull above, paler beneath, apex shortly acuminate to obtuse, margin indurated and glandular-crenulate-denticulate (appressed- or incurved spinulose-denticulate on juvenile shoots). Cymes on specialised shoots in axils of prophylls of current shoots, forming pseudo-panicles; peduncles 1.5-8 mm long; pedicels absent or up to 2 mm long; 1-7(-11) flowers, each 3-5.5 mm in diam., dioecious (or sometimes polygamous). Sepals 4-5, *c* 1 mm long, semicircular, entire or very finely serrulate. Petals 4-5, 2-2.5 mm long in male flowers, or 1.5-2 mm long in female flowers, ovate to oblong, often shortly unguiculate, longitudinally ridged, entire. Male flowers with 4-5 stamens; disc 1.5-2 mm in diam.; pistillode (except style) immersed in disc. Female flowers with (rarely without) 4-5 petaloid staminodes; ovary ovoid-conic; style thick; stigmas united. Drupe pale yellow, 1.3-2 cm long, ellipsoid to subglobose, smooth or almost so. Fig. 108.6.

Evergreen forest and thickets, forest fringing swamps and rivers, deciduous woodland, on termite mounds and grasslands; 1000-2250 m. IL KF; Sierra Leone to Cameroun and from Sudan to Malawi, Zambia and Angola. *Chaffey* 1237; *Meyer* 9075.

6. SALACIA *L.* (1771)

Trees, ± scandent shrubs or lianes or rarely rhizomatous shrublets, unarmed, with or without latex, glabrous or with simple indumentum. Leaves opposite to alternate; stipules free and deciduous or absent. Flowers bisexual or functionally unisexual, in pedunculate or sessile, dichasial to monochasial cymes or fascicles, without accessory branches. Sepals (4)5(6), imbricate. Petals (4)5(6), green or white to yellow or red or brownish, imbricate in bud. Stamens (2)3(4); filaments often ± broadened below, free or united with the ovary or gynophore; anthers versatile, extrorse or apical, dehiscing vertically or transversely. Disc extra-staminal, single, variable in shape. Ovary sessile or on gynophore, usually embedded in disc, (2)3-locular, with 2-8 ovules in each locule; style simple; stigma 3-lobed or entire. Fruit a drupe, 1-3-locular, ± fleshy. Seeds 1- *c* 20, without arils, not winged.

About 150 species, pantropical and subtropical.

1. prophyll = reduced leaf or bracteole; lowermost leaf on side shoots following directly after the subtending leaf.

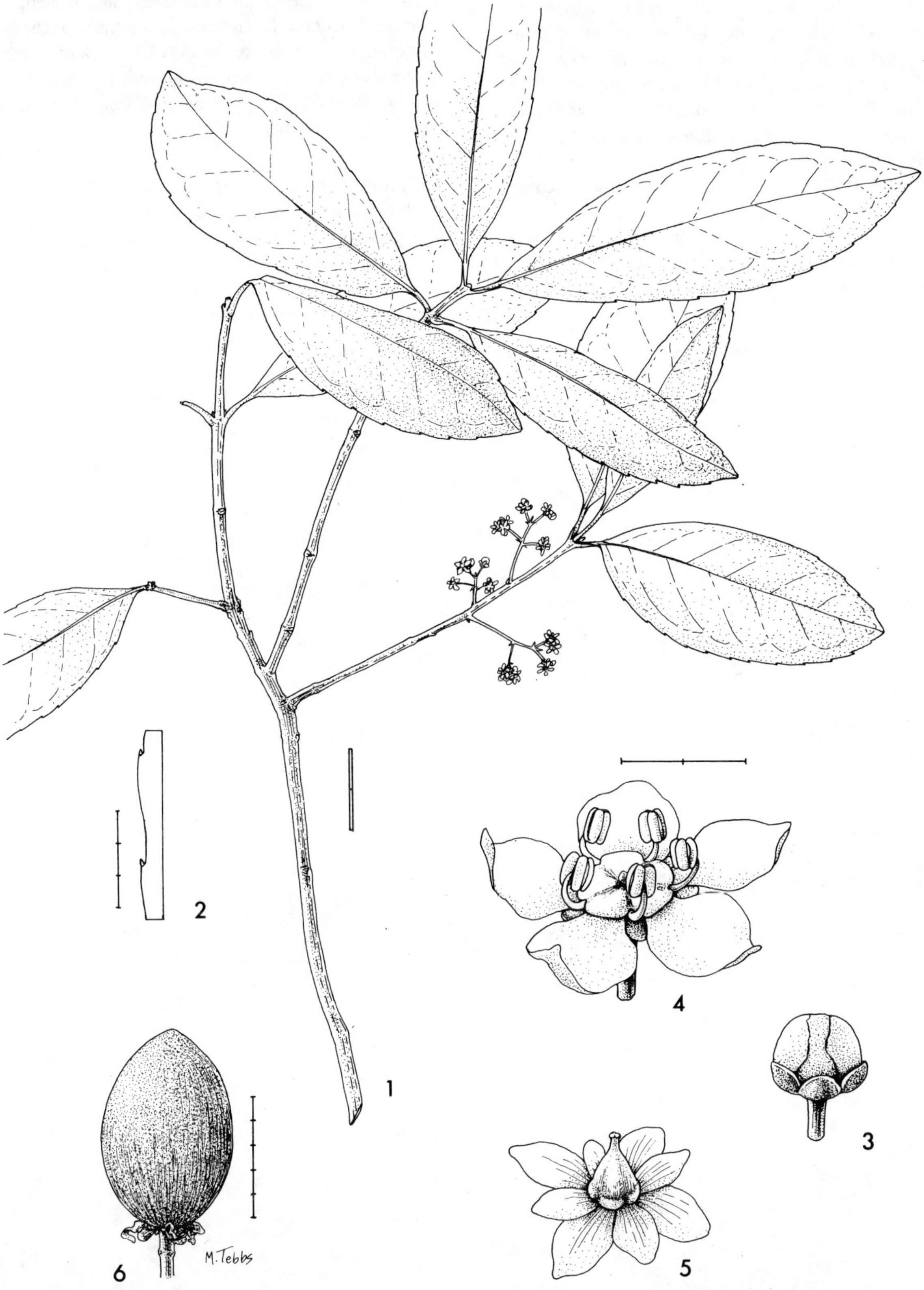

Figure 108.6 *ELAEODENDRON BUCHANANII*: 1 - branch; 2 - section of leaf margin to show glands; 3 - bud; 4 - male flower; 5 - female flower; 6 - fruit. Scale: closed bar represents cm; single line mm. All from *Robson* 555. Drawn by M. Tebbs.

S. congolensis *De Wild & Th. Dur.* (1899).

Liane or shrub, 1 m or more high, glabrous, containing latex; stems glaucous-green and smooth at first, becoming blackish-purple and finely lenticellate-rugulose. Leaves opposite (alternate on climbing shoots): petiole 5-13 mm long; stipules very small, falling quickly; lamina 5-18 x 2-8 cm, oblong or elliptic-oblong to obovate, leathery, deep green and shiny above, paler beneath, apex ± shortly acuminate, margin shallowly rounded-denticulate. Inflorescence of 2-9 (-18) bisexual flowers in sessile or shortly pedunculate fascicles; pedicels 4-8 mm long; buds 2-3 mm long, ovoid-conic to subcylindric; flowers 6-8 mm in diam. Sepals *c* 1 mm long (outer) or *c* 2 mm long and subpetaloid (inner), broadly ovate to subcircular, margin entire. Petals 2-3.5 mm long, ovate to elliptic-oblong, sessile to shortly unguiculate, rounded, orange to yellow. Disc cylindric to conic-cylindric, with margin free and sometimes undulate, surrounding ovary. Stamen filaments broadened at the base; anthers with 2 oblique, sometimes confluent, clefts. Ovary with style *c* 2 mm long, slender; stigmas united, punctiform. Drupe orange to reddish, *c* 2-3 cm in diam., globose, smooth or with wart-like surface swellings. Seeds 2-3. Fig. 108.7.

Forest fringing swamps and rivers; 920-1400 m. WG IL KF; also in Sudan, Central African Republic, Zaire, Congo and Angola. *Mooney* 7774; *W. de Wilde* 10206; *Friis* et al. 1728.

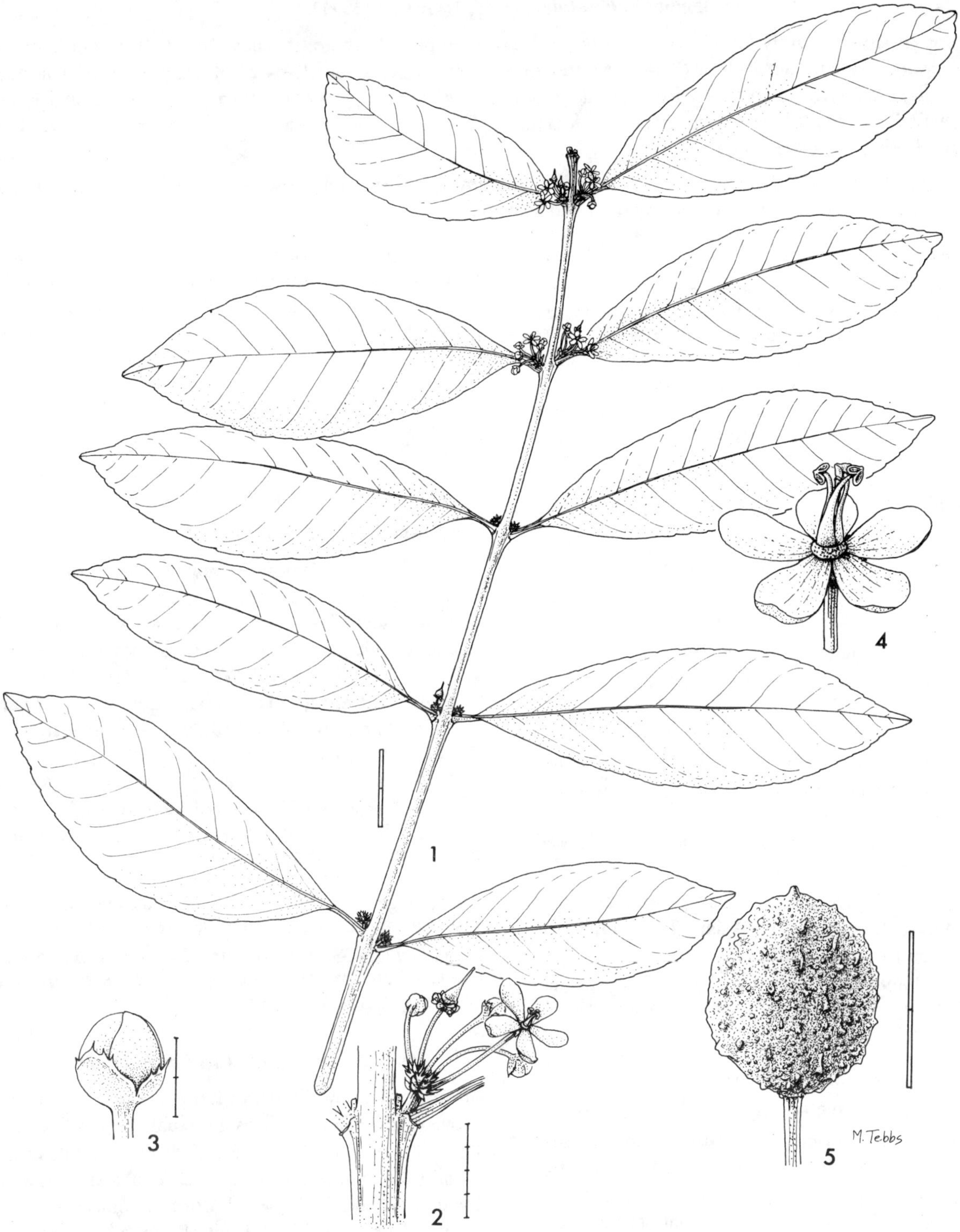

Figure 108.7 *SALACIA CONGOLENSIS*: 1 - flowering branch; 2 - inflorescence; 3 - bud; 4 - flower; 5 - fruit. Scale: closed bar represents cm; single line mm. All from *Sillans* 1564. Drawn by M. Tebbs.

109. ICACINACEAE

by K. Vollesen*

Cufodontis, *Enum.*: 487-488 (1958); Lucas, *Fl. Trop. E. Afr.*, **Icacinaceae**: 18 pp. (1968); Sleumer in *Fl. Males., ser. I*, **7(1)**: 1-87 (1971).

Trees, shrubs or woody (rarely herbaceous) climbers. Leaves simple, alternate, stipules absent. Flowers in spikes, racemes or cymes, regular. Sepals 0 or (3-)4-5(-6), free or united at base. Petals (3-)4-5(-6), free or united. Stamens (3-)4-5(-6), alternate with the petals. Disc present or not. Ovary superior, 1-locular (rarely with additional empty locules); ovules 2, pendulous from apex of locule; style simple or absent; stigma entire, 2-5-lobed or multiradiate. Fruit a 1-seeded drupe.

About 55 genera and 300-400 species: widespread in all tropical regions and a few extending into the subtropical and warm temperate regions. In Ethiopia 3 genera and 5 species.

Key to Genera

1. Tree (rarely a shrub); leaves entire; flowers bisexual, with calyx and corolla. **1. Apodytes**
- Woody or herbaceous climbers or scramblers. 2
2. Flowers bisexual, with calyx and corolla; style filiform; leaves entire; tall woody climber. **2. Raphiostylis**
- Flowers unisexual (rarely bisexual), with single perianth only; stigma sessile; leaves dentate or lobed; herbaceous or woody climbers or scramblers. **3. Pyrenacantha**

1. APODYTES *E. Mey. ex Arn.* (1840)

Sleumer in *Blumea* **17**: 181-265 (1969); Capuron in *Adansonia, ser. 2*, **10**: 507-510 (1970).

Trees or shrubs. Leaves entire, penninerved. Flowers in terminal (rarely axillary) panicles, bisexual. Sepals 5, united at base. Petals 5, free. Stamens 5; anthers sagittate. Disc absent. Ovary 1-locular, asymmetric, with a lateral appendage; style eccentric; stigma slightly oblique. Fruit unilaterally developed, obliquely ellipsoid and compressed; style persistent, finally lateral or almost basal; lateral appendage large and fleshy. Seed ovate to reniform, compressed.

Four species: 1 widespread in eastern and southern Africa, Madagascar and in Asia; 2 endemic in Madagascar and 1 in Australia.

Sleumer (1969) has shown that the 15 or so species described from Asia are all synonymous with *A. dimidiata.*

A. dimidiata *E. Mey. ex Arn.* (1840).

Tree to 25 m; dbh to 0.8 m; bole fluted in larger trees; bark smooth, grey; branchlets glabrous to sparsely pubescent. Leaves dark green and glossy, drying blackish, glabrous; petiole 0.5-3.2 cm; lamina ovate to elliptic, 4-13.5 x 1.5-7.5 cm; apex acuminate to rounded; base cuneate to truncate. Panicles divaricately branched, 3-9 cm long, puberulous; peduncle 0.5-2 cm; pedicels 0.5-2.5 mm. Calyx *c* 1 mm long, with short deltoid lobes, glabrous to sparsely puberulous. Petals linear to oblong, 4-6 x *c* 1 mm, white to cream. Filaments *c* 2 mm; anthers *c* 3 mm. Ovary *c* 1 mm long, glabrous to densely pubes-cent; style 2-3 mm. Fruit 5-11 x 5-9 x 3-4 mm, black with red appendage.

var. **acutifolia** *(Hochst. ex A. Rich.) Boutique* in *Fl. Congo* **9**: 274 (1960);
A. acutifolia Hochst. ex A. Rich. (1847) - type: GD, Mt. Aber nr. Adeselam, Simien, *Schimper* II: 1315 (P holo, BM FT K UPS iso).

Ovary and fruit glabrous. Fig. 109.1.1, 3, 4

Podocarpus-Olea-Syzygium upland rain forest, often remaining as solitary trees in grassland after forest clearing, secondary forest, riverine forest, secondary *Acacia-Combretum* wooded grassland; 1350-2600 m. EW TU GD SU AR GJ WG IL KF GG SD BA HA; S Sudan, Kenya, Tanzania, Zanzibar, Rwanda, Burundi, E and SE Zaire, Malawi, also in S India and Ceylon. *Merga Gutetta* 990; *Mooney* 6924, 6935.

Var. *dimidiata* (fig. 109.1.2, 5, 6) with pubescent ovary and fruit is widely distributed in tropical Africa, west to Nigeria and south to S Africa, also in Madagascar and widespread in tropical Asia. Although the two taxa are not sympatric (var. *dimidiata* has a much larger area and is not in Ethiopia), there is no justification for treating them above varietal level.

Gillett 15053 Amaro Mts., Mt. Delo in Sidamo is intermediate between the two varieties having some ovaries sparsely pubescent.

2. RAPHIOSTYLIS *Benth.* (1849)

Shrubs or woody climbers. Leaves entire, penninerved. Flowers in axillary fascicles, bisexual. Sepals 5, united at base. Petals 5, free. Stamens 5, free; filaments tapering from base, anthers oblong. Disc absent. Ovary 1-locular, style eccentric, filiform, persistent; stigma capitate or discoid. Fruit flattened through the longitudinal axis.

Six species in tropical Africa.

* formerly Ethiopian Flora Project, now Royal Botanic Gardens, Kew, Richmond, Surrey, TW9 3AB, UK.

Figure 109.1 *APODYTES DIMIDIATA* var. *ACUTIFOLIA*: 1 - flowering branch x 2/3; 3 - part of infructescence x 2/3; 4 - fruit showing lateral lobe and persistent style x 2. A. *DIMIDIATA* var. *DIMIDIATA*: 2 - ovary and style x 4; 5 - part of inflorescence x 2; 6 - flower x 4. 1 from *Polhill & Paulo* 1075; 3-4 from *Faulkner* 2498; 2, 5-6 from *Carmichael* 328. Drawn by Derek Erasmus. (Reproduced with permission from *Fl. Trop. E. Afr.*, Icacinaceae: fig. 2, 1968.)

Figure 109.2 *RHAPHIOSTYLIS BENINENSIS*: 1 - flowering branch x 2/3; 2 - detail of same x 2; 3 - flower x 6; 4 - ovary and style x 6; 5 - fruiting branch x 2/3; 6-8 - fruits seen from different angles x 1. 1 from *Chandler* 1953; 2-4 from *E. Brown* 380; 5-8 from *Maitland* 881. Drawn by Derek Erasmus. (Reproduced with permission from *Fl. Trop. E. Afr.*, Icacinaceae: fig. 3, 1968.)

R. beninensis *(Planch.) Benth.* (1849).

Scandent shrub or woody climber to 10 m or more, drying blackish; branches and leaves glabrous. Petiole 3-5 mm; lamina ovate to elliptic 5-15 x 2-8 cm; apex abruptly acuminate to acute; base cuneate. Fascicles few- to many-flowered; pedicels up to 1 cm, with a few scattered hairs. Calyx *c* 1.5 mm long with deltoid ciliate lobes. Petals white, linear to lanceolate, up to 7 x 1 mm. Filaments *c* 4 mm long, subulate; anthers *c* 1.5 mm. Ovary *c* 2 mm long, glabrous or with a few hairs at apex; style filiform, up to 4 mm. Fruit broadly ovoid, obliquely dorsiventrally compressed, up to 1.3 cm long, 2 cm wide and 1 cm thick, red (but drying black), reticulately wrinkled. Fig. 109.2.

Aningeria altissima-Chlorophora-Celtis lowland rain forest, riverine forest; 1000-1250 m. WG IL; widespread in tropical Africa. *Friis* et al. 4079; *Gilbert & Friis* 8359A.

3. PYRENACANTHA *Wight* (1830), *nom. conserv.*
Trematosperma Urb. (1883).

Scandent shrubs or woody (rarely herbaceous) climbers. Leaves penninerved or palmately nerved; major veins ending in conspicuous flask-shaped hydatodes[1] (see fig. 109.3.5). Flowers in axillary spikes or racemes, dioecious (rarely monoecious or bisexual). Sepals 0. Petals (3-)4-5(-6), united at base. Disc absent. Male flowers with (3-)4-5(-6) stamens; pistillode minute. Female flowers with 1-locular ovary; stigma sessile, discoid or multiradiate. Fruit with woody endocarp which is papillose or spiny on the inside.

Between 20 and 30 species: mostly in tropical Africa and Madagascar, but one species widespread in tropical Asia and one in the Philippines.

1. Stems arising from large epiterranean tuber up to 1.5 m in diameter; leaf venation palmate. **1. P. malvifolia**
- Stems without tuber or with subterranean tuber which is much smaller. 2
2. Stems annual, up to 1 m long, from tuber up to 15 cm in diameter; leaf venation palmate; in dry bushland. **2. P. kaurabassana**
- Stems woody, up to 20 m long or more, without tuber; leaf venation pinnate; in rainforest. **3. P. sylvestris**

1. **P. malvifolia** *Engl.* (1893).
P. ruspolii Engl. (1898) - type: BA, Web, *Ruspoli & Riva* 1010 (910) (FT holo, K photo).
Trematosperma cordatum Urb. (1883), non *P. cordata* Thode (1924).

Woody climber or scrambling shrub; stems to 5 m long, arising from the subglobose partly buried tuber-like part of the stem; "tuber" to 1 m tall and 1.5 m in diameter; young branches strigose, older and tuber with greyish white papery flaking bark. Leaves sparsely to densely strigose; petiole 1.5-4.5(-8.5) cm; lamina broadly ovate to reniform, 4-12 x 3-13 cm; apex acute to broadly emarginate; base cordate with broad sinus; margin shallowly lobed. Flowers di- or monoecious (rarely bisexual); petals 4-5, yellowish green, lobes deltoid, *c* 1.5 mm long, pubescent. Male flowers in up to 6 cm long tomentose spikes along an up to 10 cm long axillary branch. Female flowers arranged as males with axillary branch to 10(-30) cm long and spikes with a few flowers at base and several hydatodes apically, or clustered on short branches at base of male spikes (rarely basally on male spikes); ovary densely strigose. Fruit ellipsoid, compressed, 15-19 x 10-13 x 7-8 mm, yellowish, hispid; seed ellipsoid, compressed. Fig. 109.3

Acacia-Commiphora woodland and bushland, on sandy to loamy soil overlying limestone or on rocky limestone slopes; 900-1400 m. SD BA HA; Somalia, E Kenya, NE Tanzania. *Burger* 2488, 3398, 3399.

Plants from Somalia are generally densely hairy. Westwards into Ethiopia and southwards into Kenya and Tanzania the indumentum gets thinner, and especially the Ethiopian material shows a wide range of variation. There seems no justification for recognizing infraspecific taxa on this basis as suggested by Lucas (l.c.).

While the female flowers on male plants always have staminodes, these never occur on purely female plants. Bisexual flowers are occasionally seen basally on male spikes.

2. **P. kaurabassana** *Baill.* (1872).

Climber or scrambler; stems annual, to 1 m long, from a subterranean subglobose tuber up to 15 cm in diameter; all parts hispid. Petiole 3-6(-12) cm; lamina ovate to reniform, 4-10 x 5-15 cm; apex obtuse or mucronate; base cordate to sagittate; margin subentire to deeply 3-5(-7)-lobed. Flowers dioecious, in spikes on erect shoots appearing before the leaves; petals 4, yellowish green, lobes deltoid, *c* 1.5 mm long, pubescent. Male spikes dense, 2-3 cm long, on 3-12 cm long peduncles; female spike 1-1.5 cm, peduncle shorter than in male. Fruit ellipsoid, compressed, 13-20 x 8-13 x 5-8 mm, yellowish to orange.

Anogeissus-Pterocarpus and *Acacia* woodland, on grey sandy soil; 600-1300 m. IL SD; Somalia, Kenya, Tanzania, Mozambique, Malawi, Zimbabwe. *Friis* et al. 2432, *Gilbert & Glyn Jones* 39.

3. **P. sylvestris** *S. Moore* (1920).

Slender woody climber to 20 m or more; stems densely rusty-red pubescent when young, becoming glabrescent. Leaves greyish silky (rusty-red when young); petiole up to 3 cm; lamina ovate to elliptic, up to 13 x 6 cm, undulate to denticulate towards apex; apex acuminate; base truncate to subcordate. Flowers dioecious, in fasciculate or solitary axillary pubescent racemes, subsessile or pedicels up to 1.5 mm. Male flowers with up to 0.5 mm long bracts, petals (3-)4, up to 2 mm long, green, united at base, pubescent. Female flowers with up to 2 mm long bracts, petals up to 1.5 mm long; ovary subglobose. Fruit ovoid to subglobose, up to 1.5 x 1.5 cm, rusty-red tomentose.

Aningeria altissima-Chlorophora-Celtis lowland rain forest; *c* 1100 m. KF; Zaire, W Uganda, S Sudan. *Friis* et al. 4003.

1. hydatode = a water-secreting gland.

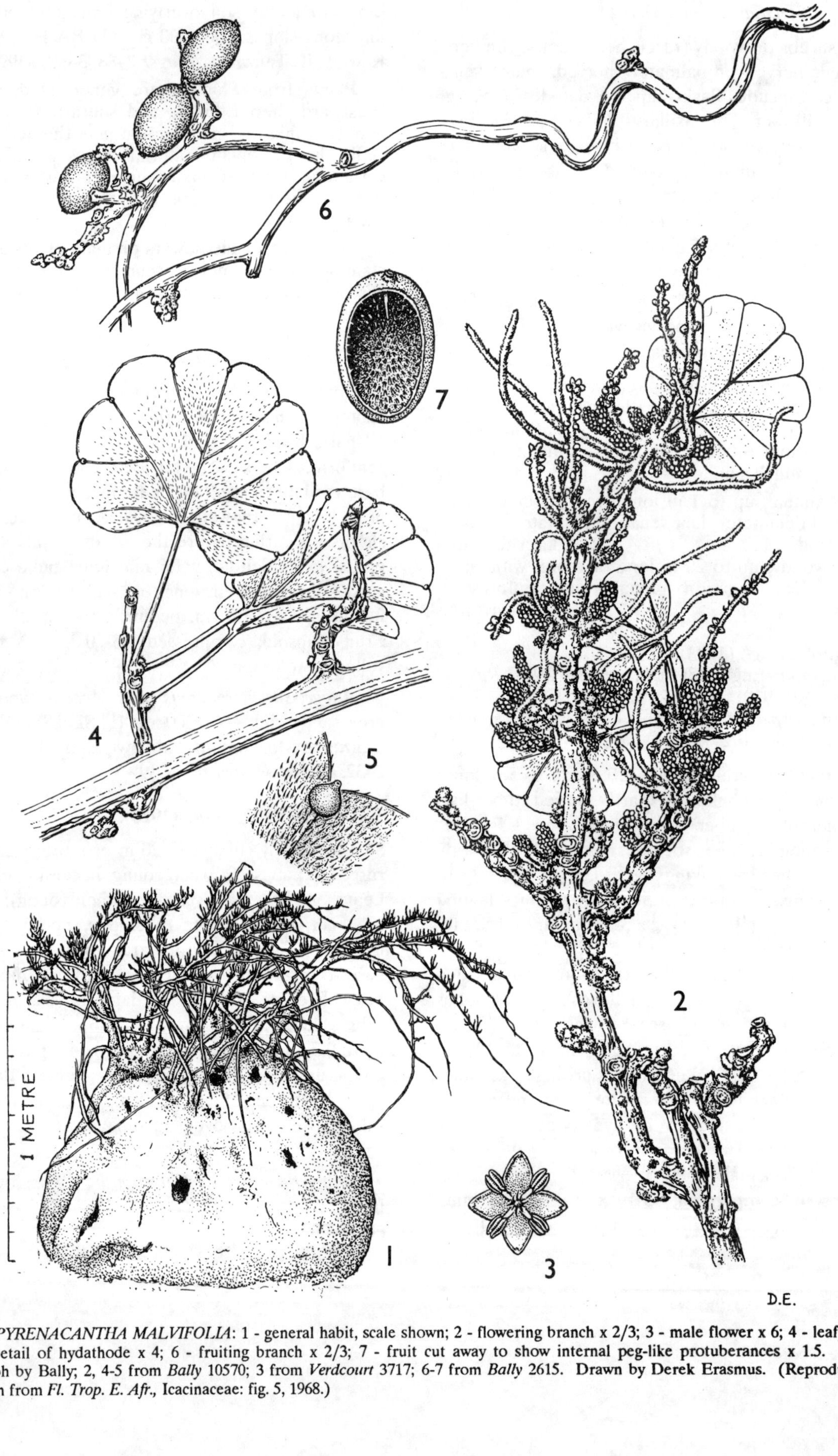

Figure 109.3 *PYRENACANTHA MALVIFOLIA*: 1 - general habit, scale shown; 2 - flowering branch x 2/3; 3 - male flower x 6; 4 - leafy branch x 2/3; 5 - detail of hydathode x 4; 6 - fruiting branch x 2/3; 7 - fruit cut away to show internal peg-like protuberances x 1.5. 1 from a photograph by Bally; 2, 4-5 from *Bally* 10570; 3 from *Verdcourt* 3717; 6-7 from *Bally* 2615. Drawn by Derek Erasmus. (Reproduced with permission from *Fl. Trop. E. Afr.*, Icacinaceae: fig. 5, 1968.)

110. SALVADORACEAE

by B. Verdcourt*

Cufodontis, *Enum.*: 486-487 (1958); Verdcourt, *Fl. Trop. E. Afr.*, **Salvadoraceae**: 10 pp. (1968).

Trees or shrubs, unarmed or with axillary spines. Leaves mostly leathery, opposite, simple, entire; stipules minute or absent. Flowers regular, bisexual or unisexual, in dense axillary or terminal fascicles or panicles. Calyx-lobes 2-4(-5[1]). Petals 4(-5[1]), imbricate[2], free or partially connate. Stamens 4(-5[1]), epipetalous or inserted near base of petals and alternate with them; filaments free or connate at base. Anthers 2-thecous; thecae back to back, opening by longitudinal slits. Disc absent or consisting of separate glands between the filaments. Ovary superior, 1-2-locular; style short; ovules 1-2, erect[2]. Fruit a berry or drupe. Seeds without endosperm; embryo with thick cordate cotyledons.

A small palaeotropical family of 3 genera, 2 of which occur in the Flora area, and are often important constituents of the vegetation in arid areas. The fibrous stems of *Dobera*, and particularly *Salvadora*, are used by many Africans as toothbrushes. Recent research has shown *Salvadora* contains anti-bacterial compounds.

Key to Genera

1. Petals free; stamens hypogynous, united into a tube at the base. **1. Dobera**
- Petals shortly united at the base; stamens epipetalous. **2. Salvadora**

1. DOBERA *Juss.* (1789)

Schizocalyx Hochst. (1844)

More or less glabrous, unarmed, evergreen trees or shrubs. Flowers bisexual, sessile in axillary or terminal panicles. Calyx ovoid, irregularly 3-4(-5)-toothed, sometimes somewhat bilobed. Petals 4 (-5), free, elliptic, oblong or linear-oblong, sometimes ± spathulate, imbricate. Stamens 4(-5), hypogynous; filaments dilated basally, shortly connate into a tube, free and filiform above; anthers ovate. Disc consisting of 4 glands alternating with the stamens, each at the base of a petal and ± attached to it. Ovary 1-locular; style short; stigma obtuse or truncate; ovules 1-2(-3), erect. Fruit a subglobose or ellipsoidal drupe; only 1 of the ovules appears to develop into a seed in 2-ovuled ovaries.

A genus of only 2 species, both occurring in tropical Africa and 1 extending to India. The two are closely related but are certainly distinct: *D. loranthifolia* occurs in Kenya and Somalia and may possibly be found also in Ethiopia.

1. Ovule solitary; calyx often bilabiate, ± (2.2-)2.8-3mm long; petals (3.5-)4.5-5 mm long; filaments (1.2-)2-2.5 mm long; anthers 1.6-2 mm long; fruits ± 2 cm long. **D. glabra**
- Ovules 2; calyx often shortly 4-toothed, ± 2 mm long; petals 3-3.5 mm long; filaments 0.8-1 mm long; anthers 1-1.5 mm long; fruits ± 1-1.5 cm long. *D. loranthifolia*

D. glabra *(Forssk.) Poir.* (1812).

Schizocalyx coriaceus Hochst. (1844); *D. coriacea* (Hochst.) A.DC (1873) - type: EE/EW, Modat valley, *Schimper* 1744 (B holo. destr., K iso.).

A much-branched evergreen shrub or tree, up to 1.8-7.5 m tall, with a spreading crown; bark green to dark grey, fissured into rectangular patches; wood pale butter-yellow when cut. Leaves olive-green, opposite; lamina leathery, elliptic to ovate or obovate, rarely lanceolate or orbicular, 1.5-10 x 0.7-9.2 cm, glabrous, or if glands present, then extremely minute but dense; venation often obscure, though sometimes prominent and reticulate; apex mostly obtuse but often acute, usually mucronulate; petiole 3.5-4 mm long. Flowers white, 3.5-5 mm long (mostly large in the Flora area but smaller in India and Arabia), in axillary and terminal panicles. Calyx minutely papillate. Staminal tube sometimes with small teeth alternating with the upper free parts of the filaments. Ovule solitary. Fruit oblong-ellipsoid or subglobose, (1.8-)1.9-2.6 x 1-1.3 cm. Seeds ovoid, flattened, 1.2-2 x 0.75-1.5 cm. Fig. 110.1.1.

Acacia savanna, *Acacia* open woodland with grassy clearings and *Acacia nubica* scrub, often on rocky hillsides; 400-1250 m. EE SU(Awash) HA BA SD; throughout the dry parts of NE Africa to Sudan and Somalia, N Uganda, N & E Kenya, extending to S Arabia and India (Bombay). *Burger* 1747, 3514; *Mooney* 7696.

The fruits are edible.

D. roxburghii Planch. (in *Ann. Sci. Nat.*, sér. 3(10): 191 (1848)), used by J.G. Baker in *Fl. Trop. Afr.* 4(1): 21 (1902) for this species, is based on some quite different plant which it has not been possible to identify (see *Kew Bull.* **19**: 155 (1964)).

* Royal Botanic Gardens, Kew, Richmond, Surrey, TW9 3AB, UK.

1. According to Baker in *Fl. Trop. Afr.* 4(1): 21 (1902)
2. According to E. Phillips, *The Genera of South African Flowering Plants*, ed. 2: 573 (1951), valvate aestivation can also occur; also the ovule in *Salvadora* is said to be apical.

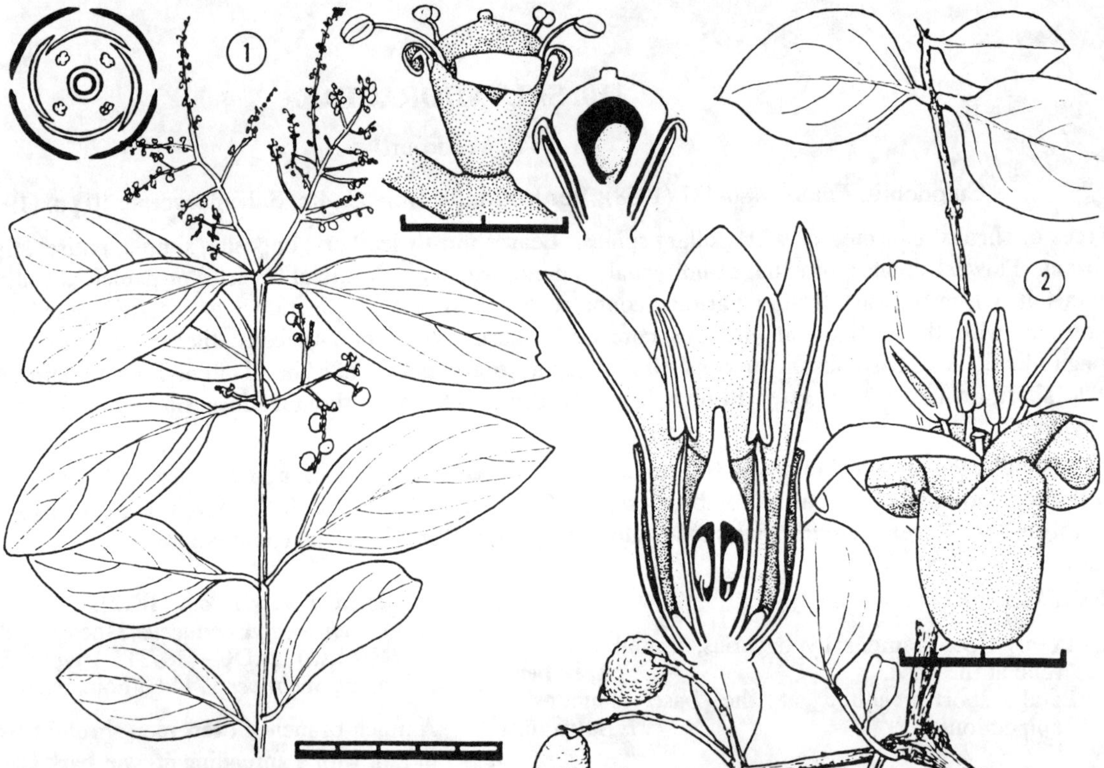

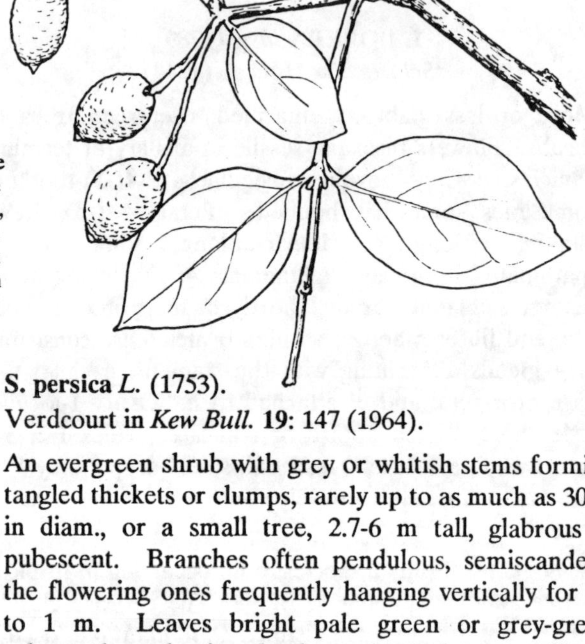

Correction:
1. SALVADORA PERSICA
2. DOBERA GLABRA

Figure 110.1.1. ~~DOBERA GLABRA~~: branch with leaves and fruits, cm scale in closed bar; flower longitudinal section; flower side view, mm scale shown as open bar. 2. ~~SALVADORA PERSICA~~: branch with leaves, flowers and fruits; floral diagram; flower side view; flower longitudinal section, mm scale shown. Drawn by W.C. Burger. (Reproduced with permission of Oklahoma Agricultural Experiment Station from Experiment Station Bulletin No. 45, *Families of Flowering Plants in Ethiopia*, figure 37,1967).

2. SALVADORA *L.* (1753)

Glabrous or pubescent unarmed trees or shrubs, often scrambling. Flowers bisexual[1], small, numerous, sessile or pedicellate in axillary or terminal panicles. Calyx campanulate, 4-toothed. Corolla campanulate; lobes imbricate, elliptic, obtuse, shortly joined at the base. Stamens 4, inserted at the base or middle of the corolla-tube. Disc consisting of 4 glands alternating with the stamens or absent. Ovary 1-locular; style shortly columnar or almost obsolete; stigma subpeltate or broadly truncate; ovary erect, solitary. Fruit a globose drupe.

A small genus of about 4 species ranging from Africa and Arabia to India and China. A single species in tropical Africa.

S. persica *L.* (1753).
Verdcourt in *Kew Bull.* **19**: 147 (1964).

An evergreen shrub with grey or whitish stems forming tangled thickets or clumps, rarely up to as much as 30 m in diam., or a small tree, 2.7-6 m tall, glabrous or pubescent. Branches often pendulous, semiscandent, the flowering ones frequently hanging vertically for up to 1 m. Leaves bright pale green or grey-green, subsucculent; lamina leathery, lanceolate to elliptic, sometimes orbicular, 1.4-10.5 x 1.2-3(-7.5) cm, rounded to subacute or acute and mucronate at apex, cuneate to subcordate at base; petiole 0.3-1.3(-2) cm long. Flowers small, greenish-white, in numerous lateral and terminal panicles, up to 10 cm long, with slender racemose branches. Drupes red or dark purple when ripe, about the size of a pea in life, 5-7 mm diam. when dry.

1. In several works the flowers are described as "subunisexual"!

var. **persica**

Leaf-blades lanceolate to elliptic, not orbicular. Pedicels 0-1.5 mm long. Petals about 3 mm long; plant glabrous or slightly pubescent. Fig. 110.1.2.

Fringing *Acacia-Commiphora* woodland, dense thorn scrub and thickets, especially associated with Capparidaceae (Capparaceae), *Hyphaene* associations near springs and desert flood plains mostly on saline, sandy or loamy soils, also at the coast; 0-1300 m. EE AF SU KF GG SD HA; extending to Sudan and Somali Republics as well as Arabia and East Africa, Mozambique, Malawi, Zambia and South West Africa, Angola and W Africa, also in N Africa through the Middle East to NW India and Ceylon. *Bally* 7025; *Burger* 3422; *Gilbert* 1552.

Fruits can be found in fairly large quantities in local markets in Sudan where they are popular with children, and could well be used in a similar fashion in the lowlands of northern Ethiopia. Also a favourite food of the Dasinich and Donyiro peoples in southwestern Ethiopia. The fruits are much sought after by monkeys and birds. The whole plant grazed and much favoured by camels. It is also reported as being used as a purge.

Several Eritrean specimens are somewhat pubescent and approach var. *pubescens* Brenan (1949) which occurs in East Africa southwest to Angola. Var. *cyclophylla* grows on coral just above the high-tide level in N Kenya and Somalia.

111. OLACACEAE

by K. Vollesen*

Cufodontis, *Enum.*: 25-26 (1953); Lucas in *Fl. Trop. E. Afr.*, **Olacaceae**: 1-16 (1968); Defillips in *Webbia* **30** (*Adumb.* **28**): 177-190 (1976).

Trees or shrubs. Leaves simple, alternate, stipules absent. Flowers in cymes, fascicles or solitary, regular. Sepals 3-5(-6), free or united. Petals 3-5(-6), free or united. Stamens 3-12, free or epipetalous. Disc present, intrastaminal. Ovary superior, semi-inferior (disc enveloping basal part) or inferior, 2-5-locular at base, 1-locular at apex; locules with 1 ovule, pendulous from apex of the central placenta; style simple; stigma 2-5-lobed or capitate. Fruit a 1-seeded drupe or nut.

About 25 genera and 250 species, widespread in all tropics. In Ethiopia 1 genus with 2 species, one of these with 2 varieties.

XIMENIA *L.* (1753)

Shrubs or trees, usually with axillary spines. Leaves entire, fleshy or leathery. Flowers in axillary pedunculate cymes, fascicles or solitary, bi- or unisexual. Sepals 4-5, united at base. Petals 4-5, free, densely bearded inside, recurved. Stamens 8-10, free. Disc inconspicuous. Ovary superior, 3(-4)-locular to near apex; stigma capitate. Fruit a drupe with a single large seed.

Genus with 8 species: 1 pantropical, 1 in eastern and southern Africa and the rest in America.

The fruit is edible and has a refreshing acid taste.

1. Flowers in pedunculate cymes (rarely solitary with bracteolate pedicel), bisexual; style 3-4 mm long. **1. X. americana**
- Flowers in fascicles or solitary with ebracteolate pedicels unisexual; style *c* 1 mm long. **2. X. caffra**

1. **X. americana** *L.* (1753).

X. laurina Del. (1843) - type: TU, Djeladjekanne, *Galinier* 119 (MPU holo. not seen, K photo.).

X. americana var. *inermis* Hochst. in *Flora* **27**: 97 (1844) - type: TU-GD, Taccazze (Tekeze) Valley, *Schimper* II:876 (TUB holo. not seen, K P UPS iso).

Shrub or tree to 7(-10) m; bark dark grey to blackish, reticulately fissured on old trunks; branchlets and leaves glabrous. Petiole 3-10 mm, pubescent; lamina ovate to elliptic, 2-8.5 x 1-5 cm; apex rounded to emarginate; base cuneate to rounded. Flowers bisexual, in few-flowered glabrous to sparsely puberulous racemoid or umbelliform cymes (rarely solitary); peduncle 2-16 mm; pedicels 3-8 mm. Calyx 4(-5)-lobed, lobes triangular, *c* 1 x 1 mm. Petals 4(-5), white to cream, oblong, 7-12 x 1.5-2.5 mm, glabrous outside. Stamens 8, filaments 2-4 mm; anthers 2.5-4 mm. Ovary 2.5-3 mm long; style 3-4 mm. Fruit ellipsoid to subglobose, 1.5-2(-2.5) x 1-2(-2.5) cm, yellow to orange, seed subglobose, 1-1.5 cm in diameter. Fig. 111.1.1-5.

Acacia woodland, *Acacia-Balanites* woodland, *Combretum-Terminalia* wooded grassland; 500-2100(-2450)m. EE EW TU GD WU SU AR GJ WG IL KF GG SD BA HA; pantropical. *Gilbert* 1167, *Mooney* 6697, *W. de Wilde* et al. 7331.

2. **X. caffra** *Sond.* (1850).

Shrub or tree to 3(-7) m; bark pale grey; branchlets glabrous to densely pubescent. Leaves glabrous to puberulous; petiole 3-7 mm; lamina elliptic, 2-7 x 1-4 cm; apex retuse to emarginate; base cuneate to rounded. Flowers functionally unisexual, in few-flowered fascicles or solitary; pedicels 3-13 mm. Calyx 4(-5)-lobed, lobes broadly triangular, *c* 0.5 x 1 mm. Petals 4, white, sometimes tinged red, oblong, 6-9 x 1.5-2 mm, glabrous or puberulous outside. Stamens 8; filaments 2-3.5 mm; anthers 2-4 mm; staminodes in female flowers as functional stamens but without anthers. Ovary *c* 4 mm long; style *c* 1 mm; pistillode in male flowers 2-3 mm long, without style. Fruit ovoid to ellipsoid, 2-3 x 1.5-2 cm, orange to red; seed ellipsoid, *c* 2 x 1 cm.

1. Young leaves hairy above and below; branchlets pubescent. var. **caffra**
- Young leaves glabrous above, glabrous or hairy below; branchlets glabrous or sparsely puberulous (rarely pubescent). var. **natalensis**

var. **caffra.** Fig. 111.1.6 & 7.

Woodland and wooded grassland, alluvial *Acacia* woodland; 1200-1750 m. IL GG SD BA HA; from Uganda and Ethiopia south through eastern Africa to S Africa and Namibia. *Burger* 2537, 2827.

var. **natalensis** *Sond.* in *Linnaea* **23**: 21 (1850). Fig. 111.1.8.

Riverine forest, alluvial *Acacia* woodland and bushland, *Euphorbia* woodland; 400-1300 m. KF GG SD BA; from Uganda and Ethiopia south through eastern Africa to S Africa. *Ash* 1601, *Beals* 520, *Carr* 693.

* formerly Ethiopian Flora Project, Royal Botanic Gardens, Kew, Richmond, Surrey, TW9 3AB, UK.

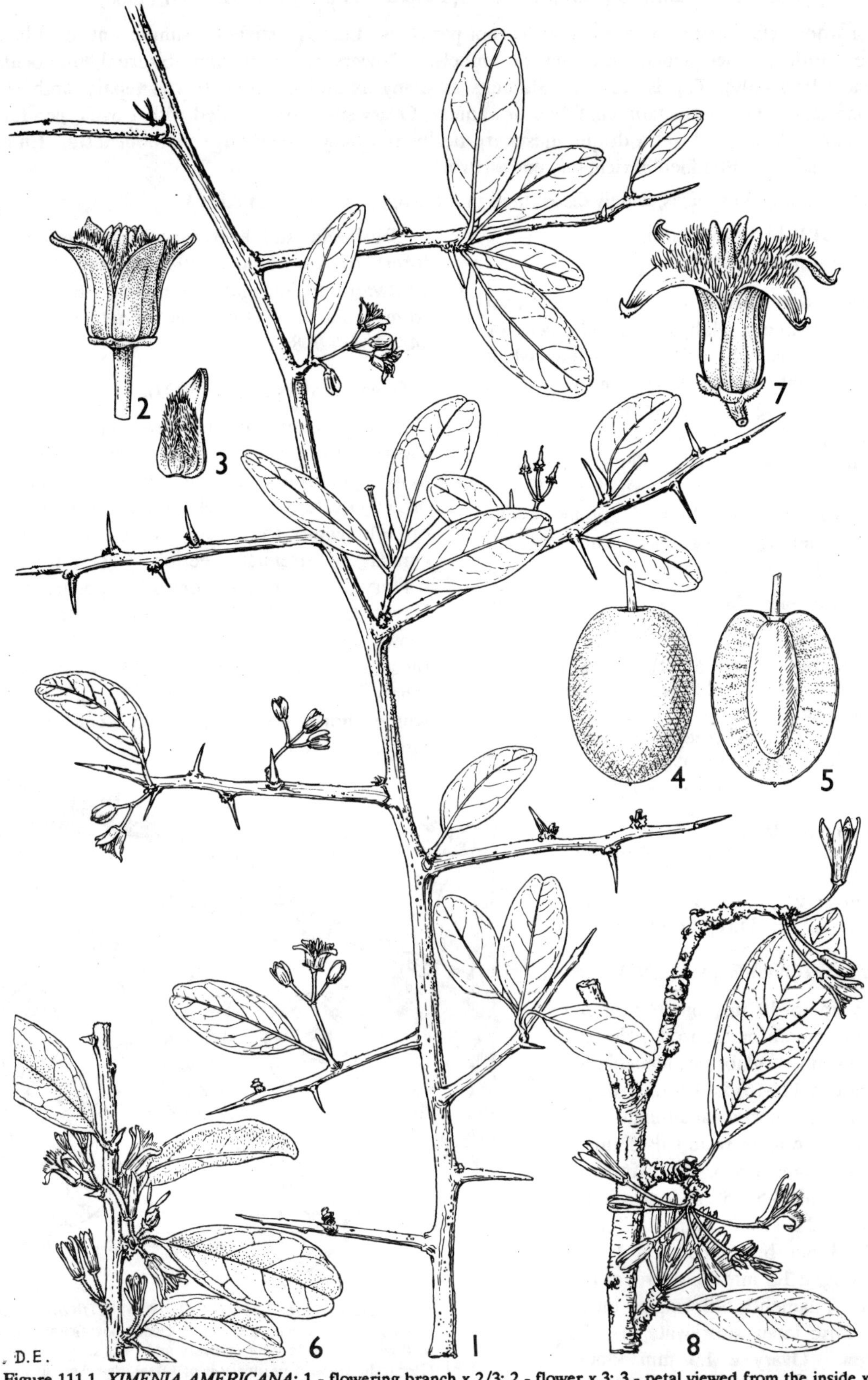

Figure 111.1 *XIMENIA AMERICANA*: 1 - flowering branch x 2/3; 2 - flower x 3; 3 - petal viewed from the inside x 3; 4 - fruit x 2/3; 5 - longitudinal section of fruit showing stone x 2/3. *X. CAFFRA* var. *CAFFRA*: 6 - part of flowering branch x 2/3; 7 - flower x 3. *X. CAFFRA* var. *NATALENSIS*: 8 - part of flowering branch x 2/3. 1-3 from *Semsei* 3406; 4, 5 from *Milne-Redhead & Taylor* 8134; 6,7 from *Tanner* 4433; 8 from *Bullock* 3237. (Reproduced with permission from *Fl. Trop. E. Afr.*, Olacaceae: fig. 2, 1968.)

112. OPILIACEAE

by P. Hiepko*

Cufodontis, *Enum.*: 25 (1953); Lucas, **Opiliaceae** in *Fl. Trop. E. Afr.*: 1-7 (1968).

Trees, shrubs or woody climbers; some are known as root parasites. Leaves distichous, simple, entire and exstipulate; veining pinnate. Inflorescence axillary racemes or umbels. Flowers small, regular, bisexual, monochlamydeous (perianth of one whorl only). Tepals valvate. Stamens as many as and opposite to the tepals; anthers 2-celled, longitudinally dehiscent. Disc intrastaminal, lobed or annular. Ovary superior, 1-celled with 1 ovule pendulous from the apex of a central placenta. Fruit a drupe, mesocarp fleshy and juicy. Seed large, without testa. Embryo small with 3-4 linear cotyledons, embedded in rich, oily endosperm.

Ten genera with about 30 species, widely distributed in the tropics; 4 genera in Africa.

OPILIA *Roxb.*(1802)

Hiepko, in *Willdenowia* **12(2)**: 162-182 (1982).

Woody climbers, erect shrubs or small trees, root parasites. Leaves leathery, glabrous to pubescent or tomentose. Short racemes axillary, conelike when young; rachis pubescent. Bracts broadly ovate, peltate, densely imbricate, falling before anthesis. Flowers solitary or in groups of 2 or 3, sweet-scented. Tepals free, 5(-4), recurved. Stamens exceeding the perianth. Disc with 5(-4) thick, fleshy lobes alternating with the stamens. Ovary cylindrical to ellipsoid, stigma sessile. Drupe ellipsoid, sometimes puberulous. Embryo with 3 cotyledons, nearly as long as the seed.

Two species in tropical Africa; *O. amentacea* Roxb. also from India through Malesia to N Australia.

1. Evergreen climber (rarely erect), leaves mostly longer than 5 cm; rachis of flowering racemes at least 1.5 cm long. **1. O. amentacea**
- Deciduous shrub or small tree, leaves less than 5 cm long; rachis under 1 cm long. **2. O. campestris**

1. O. amentacea *Roxb.* (1802).

O. celtidifolia (Guill. & Perr.) Endl. ex Walp. (1842).

O. celtidifolia var. *sphaerocarpa* Chiov. (1951) - type: SD, El Dire, *Corradi* 8425 (FI lecto.).

O. tomentella (Oliv.) Engler (1895).

Woody climber up to 10 m, sometimes an erect shrub; bark smooth or fissured, pale or dark grey. Leaves mostly glabrous; lanceolate to ovate or elliptic, 5-12(-16) x (1.5-)2-5(-6.5) cm; apex acute or shortly acuminate to obtuse; base attenuate to ± rounded; petiole 3-8(-10) mm. Racemes 1-5 or more in the axil of one leaf, rachis 1.5-3.5 cm long when flowering; bracts 2-3 mm diam., ciliate. Flowers usually 3 per bract; pedicels 1.5-2 mm. Tepals oblong, c 1.5 mm long, cream to yellowish-green. Filaments filiform, as long as the tepals. Disc lobes subclavate, c 0.7 mm long, green. Ovary c 1.3 mm long. Drupe up to 3 cm long, orange-yellow; pedicel 5-7 mm.

Bushland and evergreen forest, often in riverine forest, from lowland to 1935 m. WG IL KF GG SD; westwards to Senegal and south to Mozambique; also from India to N Australia. *Friis* et al. 1922; *Gillett* 14879; *Ash* 968.

2. O. campestris *Engler* (1909).

Deciduous shrub or small tree, up to 5 m high, with short spiny branches; bark grey-brown to blackish. Leaves puberulous to tomentose when young, elliptic to broadly ovate or narrowly obovate, 1-5 x 0.6-3 cm; apex broadly acute to obtuse or retuse (notched); base cuneate to attenuate; petiole 1-2(-4) mm. Racemes solitary or clustered, normally appearing before the leaves, rachis 5-8(-12) mm; bracts 2 mm diam., ciliate. Flowers 1, 2 or 3 per bract; pedicels 1.5-2(-4) mm; like those of *O. amentacea* but disc lobes obovoid and tepals falling less readily. Drupe slightly beaked when young, shortly tomentose, up to 1 cm long; pedicel 5-8 mm long.

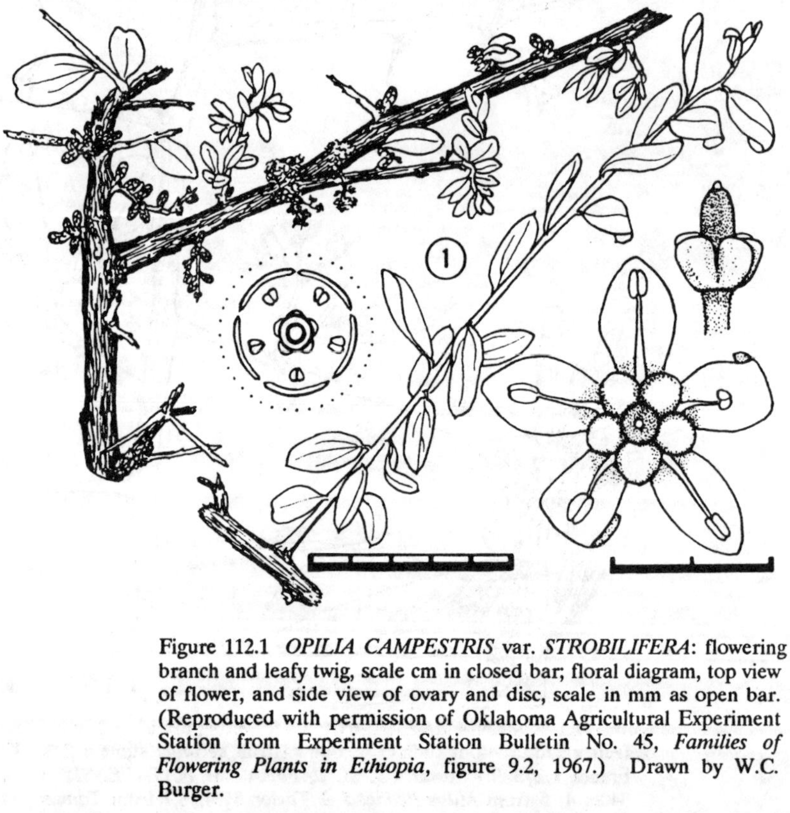

Figure 112.1 *OPILIA CAMPESTRIS* var. *STROBILIFERA*: flowering branch and leafy twig, scale cm in closed bar; floral diagram, top view of flower, and side view of ovary and disc, scale in mm as open bar. (Reproduced with permission of Oklahoma Agricultural Experiment Station from Experiment Station Bulletin No. 45, *Families of Flowering Plants in Ethiopia*, figure 9.2, 1967.) Drawn by W.C. Burger.

* Botanischer Garten und Botanisches Museum, Königin-Luise-Str. 6-8, D-1000 Berlin 33, West Germany.

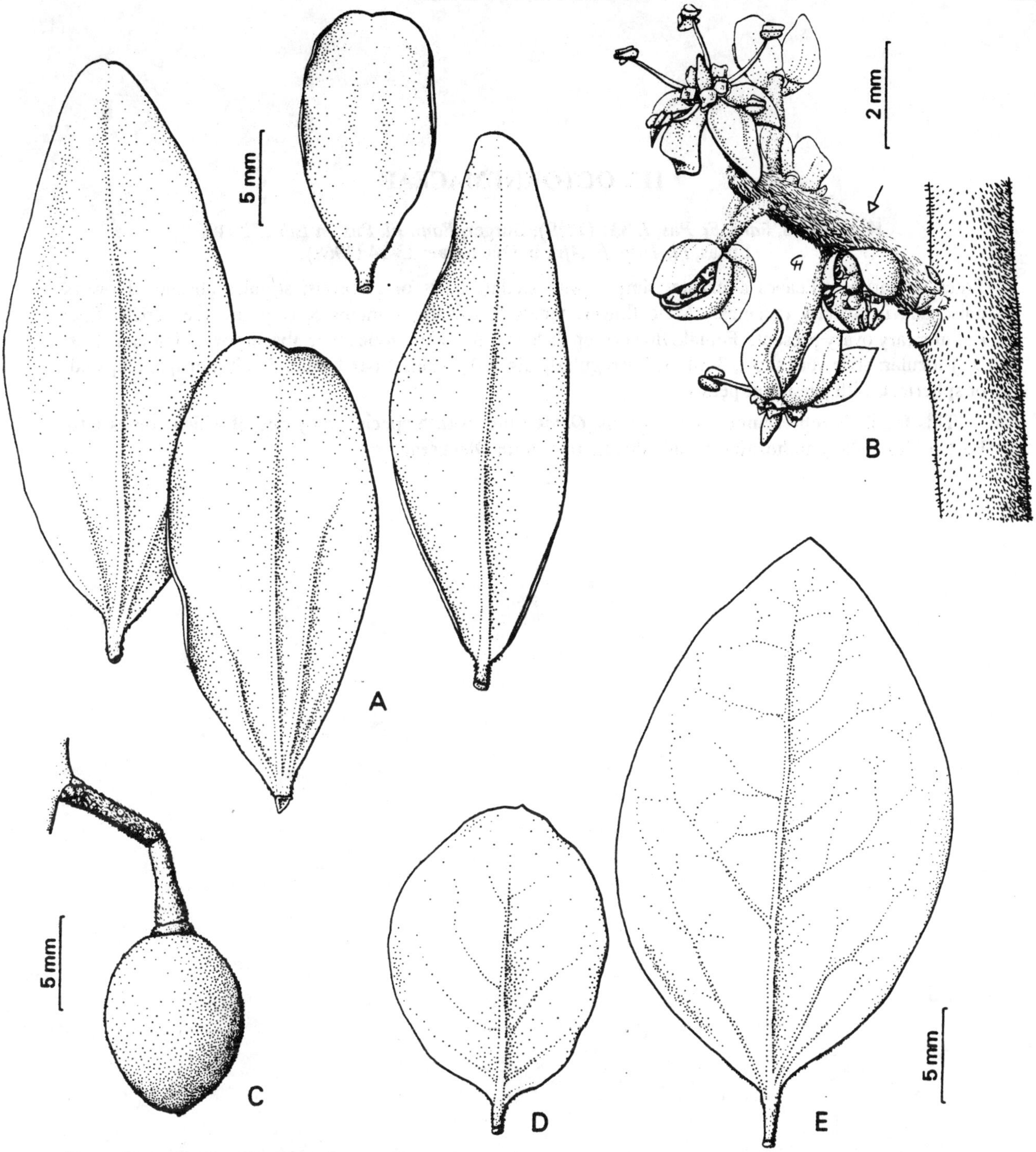

Figure 112.1 *OPILIA CAMPESTRIS* var. *STROBILIFERA*: A - leaves; B - inflorescence, one flower removed (arrow), stamens partly fallen off; *O. CAMPESTRIS* var. *CAMPESTRIS*: C - infructescence with one drupe; D & E - leaves. A from *Hemming* 2279; B *Burger* 2494; C *Gillett* 16502; D *Engler* 1625; E *Gillett* 12657. (Reproduced with permission from *Willdenowia* 12(2): fig. 5 (1982).)

1. Leaves broadly ovate to elliptic, green, (1.5)2-5 x 1-3 cm, apex acute to obtuse; flowers inserted singly or in groups of 2 or 3. var. **campestris**
- Leaves narrowly obovate to elliptic, pale bluish-green, 1-2.5(-3.5) x 0.6-1.5 cm; apex obtuse or retuse; flowers solitary. var. **strobilifera**

Var. **campestris**. Fig. 112.2.C,D & E.

Habitat not recorded, probably deciduous bushland between 200 and 500 m. BA; Kenya, Tanzania, Angola and Namibia. *Riva* 753 (coll. *Ruspoli* 831).

Var. **strobilifera** *(Hutch. & Bruce) Hiepko* in *Willdenowia* **12**: 176 (1982);
O. strobilifera Hutch. & Bruce (1941).
Fig. 112.1. & 2.A&B.

Acacia open woodland, on granite and limestone slopes, above 1000 m. HA; Somali Republic. *Burger* 2494, 2669, 3391.

113. OCTOKNEMACEAE

Hutchinson, *Fam. Fl. Plts. I*: 331 (1959); Burger, *Fam. Fl. Plts. in Eth.*: 72 (1967); Lucas, *Fl. Trop. E. Afr.*, in **Olacaceae**: 13-14 (1968).

Trees or shrubs. Leaves alternate, simple, with stellate hairs or glabrous; stipules absent. Flowers unisexual, in axillary racemes. Male flower: tepals 5, valvate; stamens 5, opposite the sepals, free; rudimentary ovary present. Female flower: tepals 5, valvate; staminodes 5 or absent; ovary inferior, 1 or 3(4)-locular; style very short, 3-5-lobed, irregularly divided; ovules 1 per locule. Fruit a drupe, ellipsoid or spherical, with persistent petals.

This family is represented by one genus, *Octoknema*, with 5 species, only one of which comes into East Africa. Many authorities include this family within **Olacaceae**.

114. LORANTHACEAE

(excluding *Viscaceae*)

by M. G. Gilbert.*

Sprague, **Loranthaceae** p.p. in *Fl. Trop. Afr.* **6(1)**: 255-384 (1910), 385-393 (1911); Danser, *Verhand. Kon. Ned. Akad. Wet. Afd. Natuurk.* sect. **2.29**(6) (1933); Cufodontis, *Enum.* 26-33 (1953); Balle, *Webbia* **11**: 541-585 (1955) & *Fl. Cameroun* **23**: **Loranthaceae** (1982); Gilbert, Polhill & Wiens, *Nordic J. Bot.* **5**: 221-224 (1985).

Perennial shrubs, hemiparasitic on the stems (rarely roots) of woody Dicotyledons (rarely Gymnosperms and exceptionally on woody Monocotyledons). Stems brittle, not articulated. Leaves alternate or opposite, rarely whorled, simple, coriaceous or fleshy, sometimes deciduous, pinnately or palmately veined, stipules absent. Inflorescences axillary or terminal, racemose or umbellate (paniculate outside Africa). Flowers bisexual (dioecious in S. America); pedicels with apical bracts which often partly enclose receptacle (more complex in Asia). Calyx fused, often reduced to a low rim. Corolla usually large and colourful with reds and yellow, usually bird pollinated, petals free or joined into tube, radially or bilaterally symmetrical. Stamens inserted on petals; anthers basifixed (sometimes dorsifixed in Asia). Ovary inferior, 1-locular (multilocular outside Africa), ovules undifferentiated with embryo sacs formed at base of ovary. Style simple, stigma often capitate. Fruit in Africa a 1-seeded berry with a viscin (sticky) layer outside the vascular cylinder; seeds without a testa.

Approximately 65 genera and 900 species, predominantly tropical but also extending into the temperate regions mainly to the south; 10 genera and 28 species so far recorded in Ethiopia.

For a long time African botanists followed Sprague (loc. cit.) in recognising a single all-embracing genus, *Loranthus*, for all the African species. Danser (loc. cit.) gave an account of the Old World Loranthoideae in which he recognised some 10 genera in Africa but this was never taken up. Modern studies initiated by Balle (1955, loc. cit.) support Danser's views, but a continent-wide revision, upon which generic delimitation must be dependent, is only now in progress and until this is completed the generic limits adopted here must be regarded as provisional. Polhill and Wiens are working on a systematic revision of the family in Tropical Africa and the accounts of *Helixanthera*, *Plicosepalus*, *Erianthemum* and, in part, *Oncocalyx* are based on their preliminary manuscripts.

Key to Genera

1. Stamens remaining straight or gently curved after anthesis. 2
- Stamens coiling, reflexing sharply or breaking off at anthesis. 4
2. Inflorescence a 1-sided raceme; corolla straight in bud, up to 3 cm long. **1. Helixanthera**
- Inflorescence an umbel; corolla curved or S-shaped in bud, 2.5-8.5 cm long. 3
3. Stems usually creeping over host by means of adventitious haustoria; corolla 5-merous, inside with a series of paired folds at base. **2. Plicosepalus**
- Stems from single haustorium; corolla 4-merous; lobes without paired folds at base. **3. Emelianthe**
4. Bud tip with conspicuous flattened-globose swelling; stigma conspicuously broad and peltate. *Globimetula*
- Bud tip not swollen or with less conspicuous swelling longer than broad; stigma globose or obovoid, not conspicuous. 5
5. Corolla lobes more or less as long or longer than corolla tube, uniformly thick; style uniformly slender. 6
- Corolla lobes less than one third as long as corolla tube, tips clearly thicker than lower parts; style distinctly swollen below stigma. 10
6. Twigs and/or inflorescence with stellate hairs; corolla pilose, usually densely so; upper part of stamens breaking off at anthesis to leave straight base. **4. Erianthemum**
- Twigs glabrous or with short simple hairs only; corolla glabrous or very minutely papillose-puberulous; filaments conspicuously coiled, not breaking off. 7
7. Calyx tubular, longer than broad; receptacle concealed by cupular bract; corolla yellow, sometimes with green markings or suffused red towards tip, rarely white with red tip. **6. Oncocalyx**
- Calyx reduced to narrow rim, broader than long; receptacle not concealed by bract; corolla not yellow nor green. 8
8. Corolla up to 24 mm long, multi-coloured - tube grey or grey and red outside, very dark purplish and red or possibly orange inside; young fruits with a conspicuously lobed disc. 9
- Corolla 35-50 mm long, uniformly bright orange or red; young fruits with an obscure annular disc. **9. Schimperina**
9. Whole plant hairy; corolla distinctly curved, (4-)5-merous. **7. Oliverella**
- Plant glabrous except for ciliolate calyx and bracteole margins; corolla straight, always 4-merous. **8. Englerina**
10. Corolla with peltate scales, often concealed by verticillate hairs; filaments without apical teeth; anther-locules divided up by 4-6 transverse septa. **5. Phragmanthera**

* Ethiopian Flora Project, Royal Botanic Gardens, Kew, Richmond, Surrey, TW9 3AB, UK.

- Corolla glabrous or with simple hairs only, no scales; filament with a small apical tooth behind anther, anther-locules without transverse septa. **10. Tapinanthus**

An eleventh genus, *Globimetula* Tieghem, is represented in the SE Sudan by the generally widespread *G. braunii* (Engl.) Tieghem which could well turn up in the Flora area. It is easily recognised by the very distinctive capitate tip of the bud, corolla lobes often spirally coiled after anthesis, the emarginate anthers and the very conspicuous large peltate stigma.

1. HELIXANTHERA *Lour.* (1790)

Loranthus sect. *Acrostachys* (Tieghem) Sprague (1910); *Loranthus* sect. *Sycophila* (Tieghem) Sprague (1910).

Small shrubs, sometimes with small branched or glandular hairs. Leaves alternate or opposite, sometimes crowded on short shoots. Flowers in terminal and/or axillary spikes or racemes, 4(-7)-merous, subtended by small bracts. Calyx rim-like to short, truncate or obscurely toothed. Petals separate, narrow, reflexing above, sometimes thickened or appendaged below filament-insertion. Stamens erect, anthers sometimes transversely septate. Style angular, sometimes thickened basally or constricted in upper part; stigma small (to capitate). Berry round or ellipsoid.

A predominantly Asiatic genus of about 50 species; 10 species recognised from Africa.

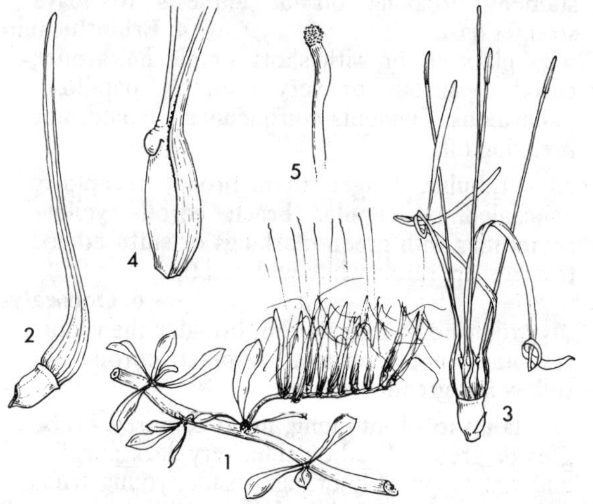

Figure 114.1 *HELIXANTHERA THOMSONII*: 1 - flowering branch x 1; 2 - bud; 3 - flower (1 petal + stamen removed) x 3; 4 - petal base, side view x 12; 5 - stigma x 18. All from *Gilbert & Thulin* 1436. Drawn by Christine Grey-Wilson.

H. thomsonii *(Sprague)Danser* (1933);
Loranthus thomsonii Sprague (1910) - type: N. Somalia, *Thomson* 23.

Branches soon thickened to 4-10 mm. Leaves alternate, mostly crowded on short-shoots; blade fleshy, oblanceolate to spathulate, tip subacute to rounded, 4-25 x 1.5-6 mm. Racemes on short shoots, 1-sided, 1-5 cm long, densely 10-20-flowered. Calyx subentire to obscurely toothed, glandular-ciliolate; corolla in bud linear, slightly swollen below insertion of filaments and around anthers. Petals pale greenish yellow turning yellow, very narrowly linear-spathulate, looped near middle, 1.5-3 cm long, notably thickened at insertion of filaments. Anthers 2-5 mm long, filaments very much longer, pink to red, thickened at base; style as long as petals. Berry obovoid, 7 mm long, bright red. Fig.114.1.

Deciduous woodland dominanted by the host plants, *Commiphora* and *Lannea*; 400-1400 m. SD HA; Somalia, Kenya, Yemen. *Bally* 10125; *Ellis* 109; *J. de Wilde & Gilbert* 370.

2. PLICOSEPALUS *Tieghem* (1894)

Tapinostemma (Benth. & Hook.) Tieghem (1895); *Loranthus* sect. *Plicopetalus* Sprague (1910); *Loranthus* sect. *Tapinostemma* (Benth. & Hook.) Sprague (1910).

Stems a few cm to several m long, often spreading by haustoria-bearing runners. Leaves irregularly opposite to alternate, often clustered on short shoots, sessile to shortly petiolate, leathery to slightly fleshy, palmately veined. Flowers 1-few in pedunculate umbels, axillary or terminal on short shoots, 5-merous; bract saucer-shaped with small limb, sometimes lacerate. Calyx rim-like to cupular. Petals curved in bud, connivent or joined to form basal tube, mostly bright red or yellow, with 2 rows of oblique folds below stamen insertion, these parts connivent, the upper part narrow below the linear tip, reflexing, looped and often twisted. Stamens following curve of style, subequal in length. Style filiform, slightly curved to sinuous, sometimes corrugated near base; stigma capitate. Berry ellipsoid or urceolate, pink, red or yellow, smooth or warty.

Genus with 11 species in Africa, mainly in the drier areas to the east, 2 of which extend into Arabia and the Middle East.

The distinctly zygomorphic buds of this and the next genus, *Emelianthe*, set them apart from most other tropical African genera. The creeping habit has only been seen in this genus within the Flora area though it has been recorded within *Tapinanthus*/*Agelanthus* in Kenya.

1. Corolla slightly S-shaped, dull glaucous-purple or pinkish-red tipped pale green or bluish, tube narrow, slightly flared, about as long as lobes; usually on *Commiphora*, *Boswellia* or *Lannea*, infrequently on *Acacia*. 2
- Corolla strongly curved, bright red and/or yellow; tube absent or very much shorter than lobes and rather broad; often on *Acacia*. 3
2. Corolla pinkish-red tipped pale green or bluish, lobes with obtuse tips; bracts not keeled. **1. P. meridianus**
- Corolla dull glaucous purple, lobes with subulate appendages 2.5 mm long on tips; bracts with prominent winged keels. **2. P. ogadenensis**

3. Sinus between corolla lobes flat; corolla tube poorly defined or absent with internal folds extending virtually to the base. 4
- Sinus between corolla lobes raised into distinct tubercle; corolla tube well defined, interior with prominent whorl of flaps delineating smooth basal chamber. **3. P. acaciae**

4. Stem smooth or obscurely lenticillate; leaves up to 5(-7.2) x 0.6-1.5 cm, obscurely 3-5-veined; petals free to base, red, red and yellow or yellow fading to red. 5
- Stem densely light brown lenticillate; leaves up to 10 x 2.5 cm, prominently 3-5-veined; petals joined into tube at base, uniformly yellow. **4. P. robustus**

5. Leaves linear to linear-oblong, more than 4 times longer than broad, cuneate at base. **5. P. curviflorus**
- Leaves obovate to suborbicular, especially on long shoots, less than 3 times longer than broad, often cordate at base. **6. P. sagittifolius**

1. **P. meridianus** *(Danser) Wiens & Polhill* (1985); *Tapinostemma meridianum* Danser (1931); *Loranthus meridianus* (Danser) Krause (1935) - type: Tanzania, *von Prittwitz* 175.

Spreading by slender runners which produce cushion-like haustoria on which leaves and inflorescences are borne, less often also long shoots with alternate leaves. Leaves mostly in tufts on haustoria, narrowly elliptic to obovate, up to 4 x 2.3 cm, fleshy, obscurely veined. Inflorescence 1-4-flowered; bract limb obscurely keeled. Calyx rim-like; corolla 4.5-6.7 cm long, obscurely S-shaped, pink with greenish or bluish-green lobes, tube slender, 2-3 cm long, lower connivent part of lobes 7-15 mm with *c* 5 pairs of folds on inside; anthers dark purple. Berry orange or pink, obovoid, smooth or obscurely tuberculate. Fig. 114.2.

Acacia-Commiphora-Jatropha rivae bushland, growing on *Commiphora*; *c* 250 m. SD; Somalia, Kenya, Uganda, Tanzania. *Gilbert* et al. 7546; *Ruspoli & Riva* '1860(617)1686'.

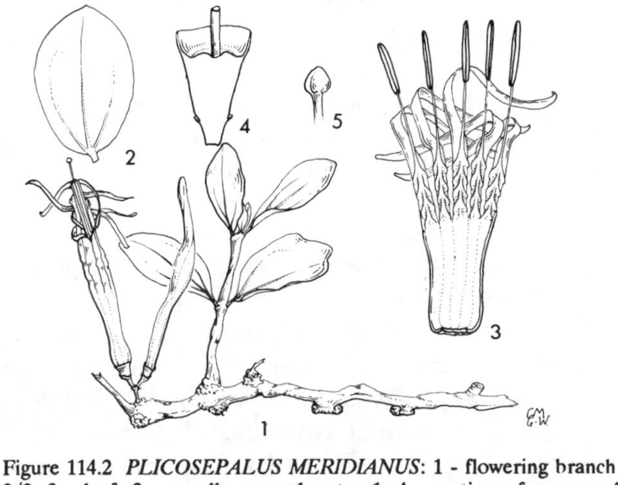

Figure 114.2 *PLICOSEPALUS MERIDIANUS*: 1 - flowering branch x 2/3; 2 - leaf; 3 - corolla opened out x 1; 4 - section of ovary x 4; 5 - stigma x 3. From *Tweedie* 2974. Drawn by Christine Grey-Wilson.

2. **P. ogadenensis** *M. Gilbert* (1985)
- type: HA, between Shillavo & Mustahil (6°10′N 44°50′E), *Gilbert* 2108 (K holo; EA ETH iso).

Similar in habit and corolla form to *P. meridianus*. Leaves very fleshy, glaucous, oblong-elliptic, up to 2.6 x 0.6 cm. Inflorescence 1-flowered, from short-shoot. Bract-limb ovate with prominent winged keel. Calyx shortly tubular with distinct triangular teeth; corolla 4.5-6.5 cm long, dull purple with grey bloom, interior dull purplish-red; tube almost straight, obscurely verrucose, 14-23 mm long, connivent part of lobes 17-23 mm long, gently curved with 6-8 pairs of folds inside; lobes linear, sharply recurved, tips with subulate appendages *c* 2.5 mm long. Anthers bright yellow. Berry glaucous pink, ellipsoidal, 7 x 4 mm, sparsely tuberculate with prominent accrescent calyx *c* 3 mm long.

Medium dense *Acacia-Commiphora-Lannea* bushland, growing on *Commiphora* and *Lannea*; *c* 500 m. HA; known only from the type gathering.

3. **P. acaciae** *(Zucc.) Wiens & Polhill* (1985); *Loranthus acaciae* Zucc. (1840); *Tapinostemma acaciae* (Zucc.) Tieghem (1895) - type: Israel, *Schubert*.
Loranthus gibbosulus A. Rich. (1847) - type: TU, near Adowa, *Schimper* I:256 (P holo. FI FT K iso.).

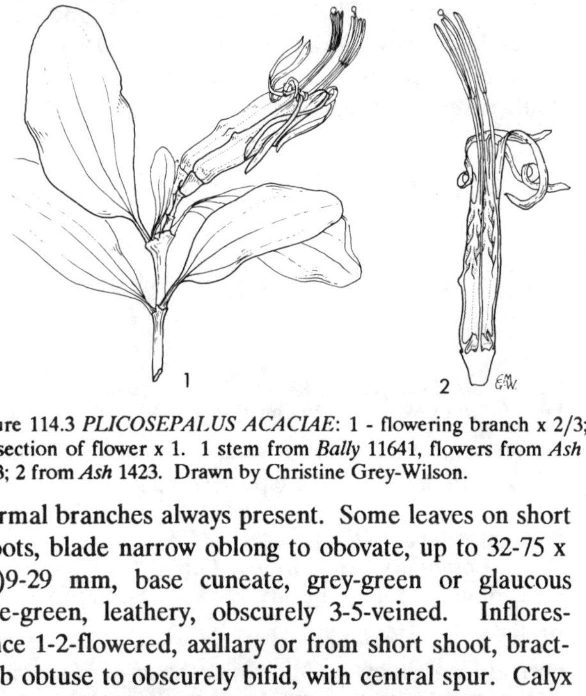

Figure 114.3 *PLICOSEPALUS ACACIAE*: 1 - flowering branch x 2/3; 2 - section of flower x 1. 1 stem from *Bally* 11641, flowers from *Ash* 1423; 2 from *Ash* 1423. Drawn by Christine Grey-Wilson.

Normal branches always present. Some leaves on short shoots, blade narrow oblong to obovate, up to 32-75 x (6-)9-29 mm, base cuneate, grey-green or glaucous blue-green, leathery, obscurely 3-5-veined. Inflorescence 1-2-flowered, axillary or from short shoot, bract-limb obtuse to obscurely bifid, with central spur. Calyx shortly cupular, tending to split; corolla usually bright red, sometimes pink or orange or with some yellow, curved, 25-47 mm long, tube 5-10 mm, interior with whorl of flaps about one third of way up, lower connivent part of lobes 9-12 mm long with 4-5(?-10) paired folds inside and sinus raised into prominent tubercle, sometimes winged along suture, upper part recurved and looped, linear-lanceolate above; anthers red. Berry red, oblong-ellipsoid, 12 x 10 mm, smooth, corolla ± persistent. Fig. 114.3.

Wide range of habitats from very dry *Acacia* bushland to deciduous *Combretum* woodland, hosts include *Combretum*, *Terminalia*, *Acacia*, *Ozoroa*, *Dobera* and *Melia azederach*; 50-2200 m. EE AF EW TU GD SU HA; north to Dead Sea, east to Oman. *Ash* 2515; *Bally* 7088; *Schweinfurth & Riva* 1169.

In spite of the specific epithet, *Acacia* is a minor host in Ethiopia. Collections from Harerge Region (*Bally* 11641, *Burger* 1995, 2285, 3082; *Gilbert & Thulin* 42) have distinctive obovate leaves, darkening when dried, and more robust flowers with the buds having the sutures raised into distinct wings, these may well represent a distinct taxon. *Gilbert* et al. 7995, growing on *Delonix elata* between Ginir and Imi (BA), may be this species but it is sterile and has more robust, less glaucous leaves.

4. **P. robustus** *Wiens & Polhill* (1985);
- type: SD, Keli (5°49´N 37°56´E), *Gillett* 14869 (K holo., EA iso.).

Branches stout, dark, with small close-set pale brown lenticels. Leaves mostly crowded on knob-like short shoots; blade oblong-subfalcate, up to 10 x 2.5 cm, base obliquely cuneate, leathery, 3-5-veined. Umbels 3-4-flowered, on short-shoots; bract limb obtuse-deltoid with prominent dorsal spur. Calyx truncate. Corolla golden yellow, curved, 40-45 mm, tube 5-10 mm long with 5 pairs of folds, 3-4 in a row intergrading with paired folds above, lower part of lobes 10-12 mm with 5 pairs of folds inside, tip recurved, linear-lanceolate. Anthers red. Style corrugated near base. Fig. 114.4.

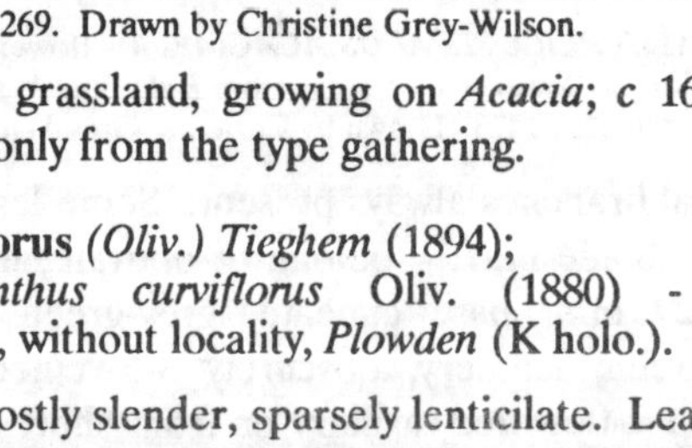

Figure 114.4 *PLICOSEPALUS ROBUSTUS*: section of flower x 1. From *Gillett* 14269. Drawn by Christine Grey-Wilson.

Wooded grassland, growing on *Acacia*; *c* 1600 m. SD; known only from the type gathering.

5. **P. curviflorus** *(Oliv.) Tieghem* (1894);
Loranthus curviflorus Oliv. (1880) - type: Ethiopia, without locality, *Plowden* (K holo.).

Branches mostly slender, sparsely lenticilate. Leaves on normal long shoots and on short shoots similar, blades linear to oblanceolate, rarely obovate, 42-50(-72) x 4-9 (-12) mm, base cuneate, obscurely 3-5-veined, grey- or glaucous blue-green. Umbels 3-7-flowered; bract limb deltoid acuminate, with short (rarely long and slender) dorsal spur. Calyx rim-like; petals free, 25-35 mm long, part below stamen insertion yellow, rarely red (or white or orange), 5-8 mm long, S-shaped, with 4-5 paired folds inside, distal part normally red, rarely yellow, recurved, ± twisted, linear-spathulate; anthers red. Berry red, ellipsoidal or urceolate, 9-12 x 7 mm, smooth. Fig. 114.5.

Acacia-Commiphora bushland/woodland, apparently always growing on *Acacia*, especially *A. mellifera*; 700-1350 m. EE EW GG SD BA HA; north to coastal south Egypt, Somalia, Kenya, Uganda, Tanzania and E. Zaire, also in Middle East and Arabia. *Burger* 1347; *Gilbert* 2101; *Thulin* et al. 3604.

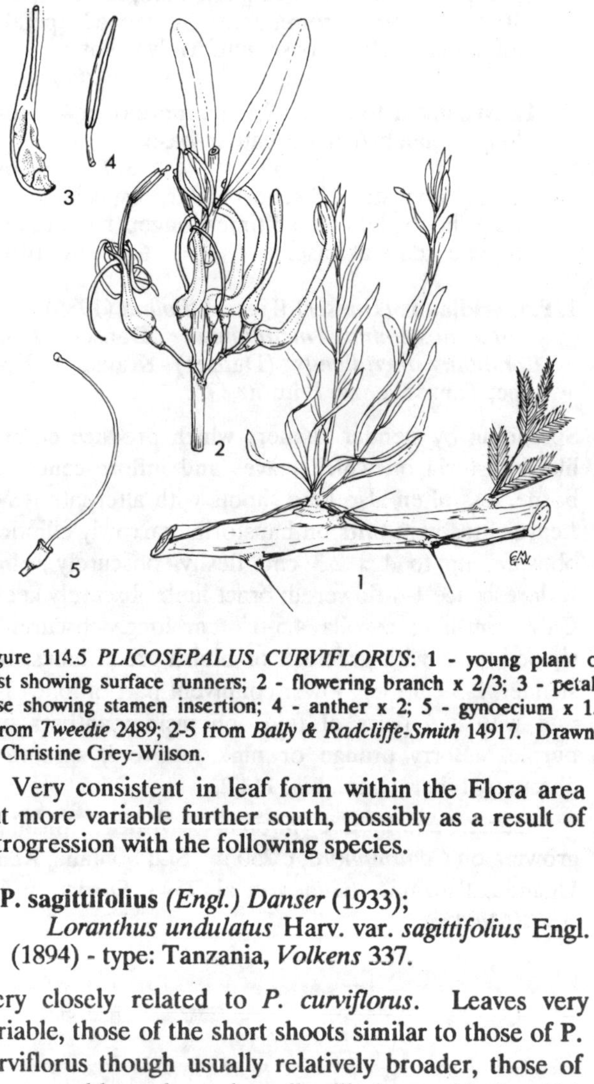

Figure 114.5 *PLICOSEPALUS CURVIFLORUS*: 1 - young plant on host showing surface runners; 2 - flowering branch x 2/3; 3 - petal base showing stamen insertion; 4 - anther x 2; 5 - gynoecium x 1. 1 from *Tweedie* 2489; 2-5 from *Bally & Radcliffe-Smith* 14917. Drawn by Christine Grey-Wilson.

Very consistent in leaf form within the Flora area but more variable further south, possibly as a result of introgression with the following species.

6. **P. sagittifolius** *(Engl.) Danser* (1933);
Loranthus undulatus Harv. var. *sagittifolius* Engl. (1894) - type: Tanzania, *Volkens* 337.

Very closely related to *P. curviflorus*. Leaves very variable, those of the short shoots similar to those of P. curviflorus though usually relatively broader, those of the normal long shoots broadly elliptic to orbicular, (8-) 18-40(-65) x (5-)10-15(-30) mm, base rounded to cordate or sagittate, obscurely 3-7-veined. Petals 30-35 mm, normally ageing from yellow green to yellow at base and through yellow and orange to red or pink at tip (sometimes yellow with redder base).

Acacia-Commiphora bushland/woodland, growing mostly on *Acacia mellifera*, less often on other species of *Acacia* (rarely *Albizia* or *Commiphora*); 600-1500 m. GG SD; East African coast from S Somalia to N Tanzania, inland to Rift Valley from Ethiopia to S

Tanzania. *Carr* 706; *J. de Wilde & Gilbert* 404; *Gilbert & Thulin* 362.

The Ethiopian populations are entirely west of those of *P. curviflorus* within the Flora area, the boundary between them falling somewhere in central Sidamo region. Often they do not have the distinctly cordate or sagittate leaf bases most diagnostic for the species but seem best placed here.

3. EMELIANTHE *Danser* (1933)
Loranthus sect. *Tetrameri* Sprague (1910)

Medium sized shrub with single haustorium, glabrous except for bracts. Leaves alternate to subopposite, pinnately veined. Flowers in small umbels on short shoots, 4-merous. Bract saucer-shaped with limb nearly as long as receptacle. Calyx saucer-shaped; corolla gently S-shaped, tube narrow, slightly swollen at base, expanded up to obliquely inserted lobes, lobes prominently winged along sutures at base, narrowing to linear-spathulate tip, reflexing. Stamens inserted above base of lobes, filaments nearly straight, anthers slender, about as long as filament, connective produced into short appendage which remains adhering to style after anthesis; style filiform, stigma capitate. Fruit not known.

Monotypic.

Balle (loc. cit.: 580) suggested a relationship to the large Asiatic genus *Amyema* but the differences seem too many and large for the relationship to be anything but distant. *Plicosepalus*, particularly *P. meridianus*, shows the most similarities to *Emelianthe* amongst other African loranths.

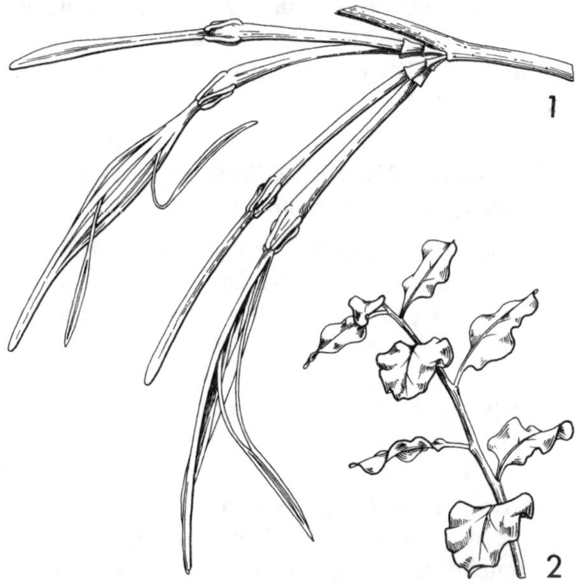

Figure 114.6 *EMELIANTHE PANGANENSIS*: 1 - inflorescence x 1; 2 - leafy shoot x 2/3. 1 from *Gilbert* et al. 5500; 2 from *Milne-Redhead & Taylor* 7114. Drawn by Eleanor Catherine.

E. panganensis *(Engl.) Danser* (1933);
Loranthus panganensis Engl. (1894) - type: Tanzania, *Holst* 3503.

Leaves deciduous, often produced after flowering, rarely collected mature, leaf blade obovate to oblanceolate, up to 3.5 x 2 cm but probably larger when fully mature, strongly undulate at least when young, fleshy. Umbels 2-4-flowered. Bract with acute ovate limb, margin ciliolate. Calyx spreading, margin pale, almost scarious; corolla 6-7(-8.5) cm long, bright ± glaucous pink to red except for grey-green band at top of tube and yellow suture-wings. Filaments and style red. Fig. 114.6.

Acacia-Commiphora bushland, on *Commiphora* and *Lannea*; 1100-1400 m. SD; Somalia, Kenya, Tanzania. *Bally* 9129; *J. de Wilde & Gilbert* 371.

Flowering in the dry season and probably more common in southern Ethiopia than the lack of records suggests. Material with old leaves and mature fruits needed.

4. ERIANTHEMUM *Tieghem* (1895)
Loranthus sect. *Hirsuti* Sprague (1910)

Twigs with stellate and often longer branched hairs. Leaves pinnately veined. Inflorescence a (2-)4-6(-7)-flowered pedunculate head, axillary or terminal on short shoots bearing (1-)2 pairs of leaves, elsewhere sometimes racemose. Bract ovate to suborbicular, concave, often with leafy limb. Flowers 5-merous; receptacle usually long-haired. Calyx cupular to tubular, subentire to shortly toothed. Corolla with some long subsimple hairs, opening explosively, radially symetrical, tube shorter than lobes, constricted above basal swelling, often flushed orange to pink, lobes spreading or reflexed at anthesis. Filaments: lower part straight, slender; upper part thickened, flattened, breaking off with anther at anthesis, anther long. Style filiform, stigma capitate. Berry globose to urceolate, often pilose, with persistent calyx.

16 species from Ethiopia to Cape Province, extending west to Central African Republic, Zaire and Angola.

Usually easily recognised by the characteristically hairy flowers.

1. Leaves well spaced except at base of new shoots, blade (2.5-)3-10 cm long, sharply differentiated from petiole; calyx conspicuously tubular, 2.5-6 mm long. **1. E. dregei**
- Leaves mostly clustered on short shoots, blade 0.5-3 cm long, decurrent into short indistinct petiole; calyx reduced to narrow rim up to 1.5 mm long. 2
2. Corolla uniformly densely hairy; bract up to 10 mm long, sometimes with small leafy limb. **2. E. aethiopicum**
- Corolla sparsely hairy, almost glabrescent; bracts leafy, 10-20 mm long. *E. occultum*

E. occultum (Sprague) Danser, closely related to *E. aethiopicum*, has been twice collected on the Kenyan side of the Ethiopian boundary and should turn up within the Flora area.

1. **E. dregei** *(Eckl. & Zeyh.) Tieghem* (1895);

Loranthus dregei Eckl. & Zeyh. (1837) - type: Cape Province, *Ecklon & Zeyher* 2284.

Loranthus oblongifolius A. Rich. (1848) - type: TU, Gapdiam, *Schimper* II:768 (P holo., B BM FI K iso).

L. dregei Eckl. & Zeyh. forma *subcuneifolius* Engl. in *Bot. Jahrb.* **20**: 104 (1894); *L. dregei* var. *subcuneifolius* (Engl.) Sprague, *loc. cit.*: 312 (1910) - types include *Schimper* II:768 above (B syn.).

L. dregei Eckl. & Zeyh. var. *kerenicus* Sprague, *loc. cit.*: 312 (1910) - type: EW, Abita, Keren, *Beccari* 82 (K holo., FT iso.).

Twigs with white to brownish stellate and shortly branched hairs, glabrescent. Leaves well spaced except at base of new shoots, blade elliptic-oblong to ovate, 3-10 x 1.2-6 cm, base cuneate to cordate, glabrous to subdensely stellate-pubescent. Heads 1-several together in lower axils, 2-6-flowered, peduncle 3-10 (8-18) mm long; bract usually with leafy limb up to 15 mm long, stellate pubescent. Calyx tubular, 2.5-6 mm long, silky villous. Corolla (3-)4-5(-5.5) cm long, mostly pale greenish, usually with long pale hairs masking constriction. Berry 10-15 mm long including calyx, orange to red, pilose. Fig. 114.8.6.

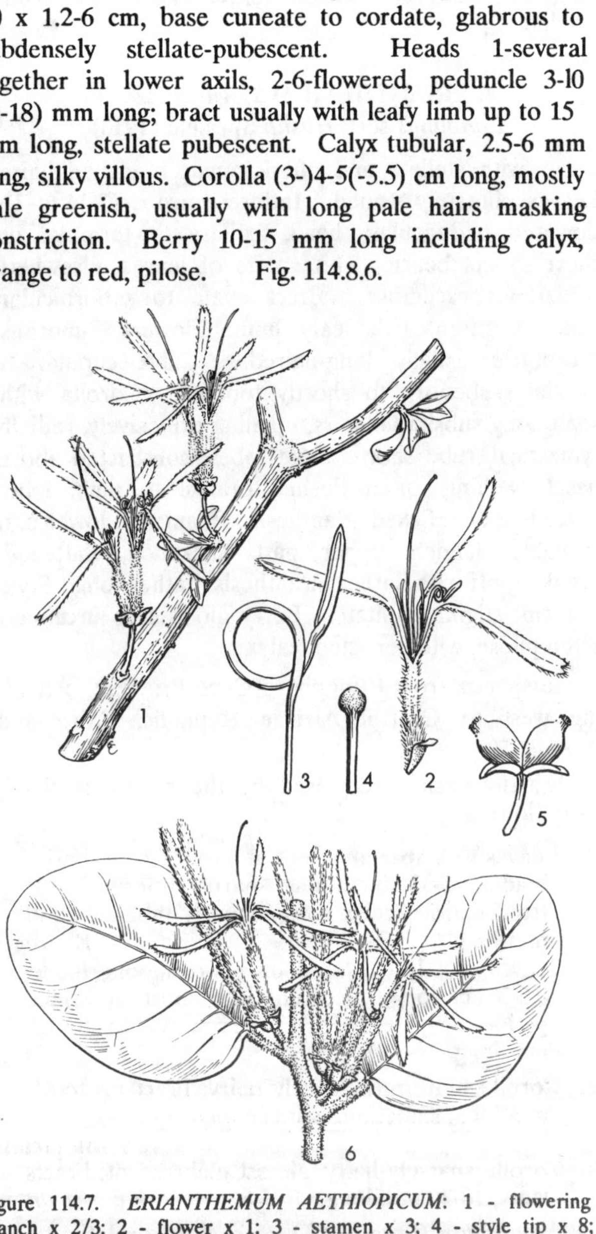

Figure 114.7. *ERIANTHEMUM AETHIOPICUM*: 1 - flowering branch x 2/3; 2 - flower x 1; 3 - stamen x 3; 4 - style tip x 8; 5 - fruit x *c* 2/3. *E. DREGEI*: 6 - part of flowering branch x 2/3. 1-4 from *Ash* 2433; 5 from photo of *Gilbert* et al. 7665; 6 from *J. de Wilde* 6592. Drawn by Eleanor Catherine.

Margins of evergreen forest, often locally abundant, growing on a very wide range of hosts including cultivated trees; 1350-2000 m. EW TU SU AR KF GG SD BA HA; the only widespread species in the genus, extending south to Cape Province and west to Angola. *J. de Wilde* 6592; *Gilbert & Getachew* 2957; *Gillett* 14414.

W. de Wilde 6542, from near Sodere, is placed here but has atypically small leaves and well defined short shoots.

Material from the East African Coast (*Loranthus sodenii* Engl.), distinguished by the shorter peduncles and smaller flowers and bracts, is only subspecifically distinct.

2. **E. aethiopicum** *Wiens & Polhill* (1985);

- type: HA, 15km SW of Werder, *Ellis* 142 (K holo.).

Twigs tomentose with white stellate and short branching hairs, glabrescent. Leaves mostly clustered on short shoots, often few at flowering time; blade elliptic to oblanceolate, 5-25 x 1.5-8 mm, base cuneate into ill-defined petiole, stellate pubescent, ± glabrescent. Heads on short shoots or at older leafless nodes, (2-)4-flowered; peduncle 5-15 mm long; bract 3-6 mm long, sometimes with short leafy limb and up to 10 mm long, stellate-tomentose. Calyx 0.7-1 mm long, tomentose. Corolla 3-4.5 cm long, mostly yellowish, basal swelling villous, rest with dense mixture of stellate, branching and a few long simple hairs. Berry urceolate, 10 mm long including calyx, dark blue-green with red flesh, sparsely hairy. Fig. 114.7.1-5.

Acacia-Commiphora bushland or woodland, growing on *Commiphora* or *Acacia*, rarely on *Ipomoea donaldsonii*; 750-1300 m. SD BA HA; Kenya. *Ash* 1919; *Gilbert* 3386; *Hemming* 1495.

5. PHRAGMANTHERA *Tieghem* (1895).

Loranthus sect. *Rufescentes* Sprague (1910).

Robust shrubs, young shoots conspicuously reddish (or greyish) tomentose with verticillate hairs. Leaves opposite or subopposite. Inflorescence of axillary umbels, often so dense and hairy that bracts and calyx are difficult to see. Corolla erect, swollen at tip and often at base in bud, deeply split unilaterally, lobes much shorter than tube, erect (or reflexed), yellow to red with sessile peltate scales sometimes concealed by dense verticillate hairs. Filaments usually without apical tooth, inrolled through 360° or more at anthesis, anthers multiloculate with transverse septa. Style 'skittle-shaped' (clavately swollen from about half way and then abruptly contracted into short slender tip below globose stigma). Fruit subglobose or obovoid, red or metallic blue.

A very homogenous group of about 30 species extending from Senegal to Tropical Arabia and south to Namibia and Mozambique.

Though hitherto considered closely associated with *Tapinanthus*, current studies suggest that the similarity in flower form is due to convergence and that the genus is nearer to *Erianthemum*. The species are very similar and Balle, mainly by means of annotations of herbarium sheets has suggested the reduction of several species to subspecies of the west African species, *P. rufescens* (DC.) Balle. Studies so far suggest that most of these species are best kept separate.

1. Corolla apparently smooth, with sessile or subsessile scales (use hand lens!) but no obviously verticillate hairs. 2
- Corolla with clearly verticillate hairs that usually hide the sessile scales beneath them. 3
2. Inflorescence pedunculate with up to 5 flowers; bud tip with distinctly winged sutures; corolla indumentum of sparse and hard to see scales. **1. P. macrosolen**
- Inflorescence sessile with many flowers; bud tip smooth; corolla densely and clearly scaly. **2. P. sarertaensis**
3. Leaves linear-lanceolate to broadly elliptic with rounded to cuneate base; corolla with short floccose densely verticillate hairs. 4
- Leaves ovate with cordate to truncate base; corolla with conspicuously spreading laxly verticillate hairs. **5. P. regularis**
4. Leaves on long shoots only, broadly elliptic, up to 3 times as long as broad. 5
- Leaves mostly clustered on short shoots, 4-11 times as long as broad. **6. P. dschallensis**
5. Inflorescences clearly pedunculate with up to 5 flowers. **3. P. erythraea**
- Inflorescences made up of sessile many flowered clusters. **4. P. usuiensis**

1. **P. macrosolen** *(A. Rich.) M. Gilbert* (1985);
Loranthus macrosolen A. Rich. (1848) - types: TU, Schoata, *Schimper* II:529 & II. 1288 (both P syn., BM FI FT K isosyn.).

Leaves linear-lanceolate to lanceolate, up to 7.5-22 x 1.3-4.5 cm, dark to medium green. Inflorescence 3-5-flowered with peduncle up to 7 mm long. Corolla (4.5-) 5-6.5 cm long, slightly enlarged but not bulbous at base, tip acute with narrow but well defined sutural wings, yellow to orange in bud darkening to orange red with red base and tip at anthesis, indumentum of sparse sessile scales only, interior more yellowish except for red filaments. Fruit obovoid pyrriform, soon glabrescent, *c* 17 x 14 mm, red, calyx not persisting. Fig. 114.8.1-3.

Open forest or woodland in higher rainfall areas, nearly always growing on *Acacia* or *Albizia*, rarely spreading onto other hosts in vicinity; (1100-)1350-2500 m. TU GD GJ SU WG IL KF SD; not known elsewhere. *Ash* 1946; *W. de Wilde* 7170; *Gilbert & Jefford* 4683.

2. **P. sarertaensis** *(Hutch. & E.A. Bruce) M. Gilbert* (1985);
Loranthus sarertaensis Hutch. & E.A. Bruce (1941) - type: HA, Sarerta Mountain, *Gillett* 5194 (K holo.).

Closely related to *P. regularis*, differs by the more prominently lenticillate stems, narrower leaves, up to 10.8-17.5 x 3.6-7 cm, with rounded to cuneate base; corolla 4.7-6 cm long, with dense covering of sessile scales and short 2(-3)-verticillate scale-like hairs and more rapidly glabrescent fruits. Fig. 114.8.7.

Montane forests, recorded hosts include *Acacia*, *Buddleja* and *Rapanea*; 2300-2650 m. AR SD HA; not known elsewhere. *J. de Wilde & Gilbert* 310; *W. de Wilde* 6593; *Westphal* 2673.

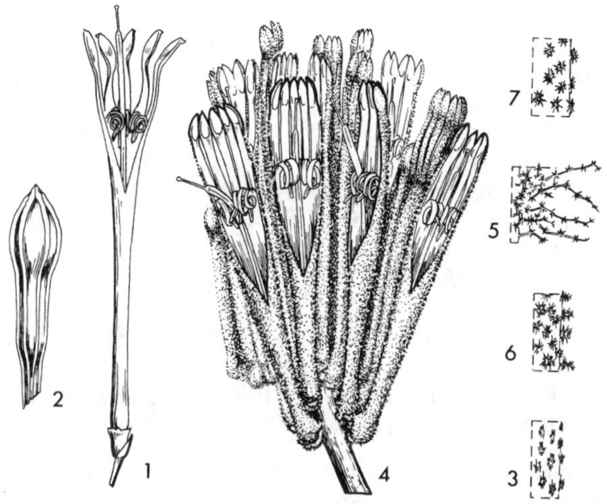

Figure 114.8 *PHRAGMANTHERA MACROSOLEN*: 1 - flower x 1; 2 - bud tip x 2; 3 - surface of corolla x 8. *P. REGULARIS*: 4 - inflorescence x 1; 5 - surface of corolla. *P. ERYTHRAEA*: 6 - surface of corolla. *P. SARERTAENSIS*: 7 - surface of corolla x 8. 1-3 from *J. de Wilde* 5407; 4 & 5 from *M. & S. Gilbert* 1766; 6 from *Gilbert & Lavranos* 2245; 7 from *J. de Wilde & Gilbert* 310. Drawn by Eleanor Catherine.

3. **P. erythraea** *(Sprague) M. Gilbert* (1985);
Loranthus erythraea Sprague (1910) - type: EW, near Saganeiti, *Schweinfurth & Riva* 785 & 1800 (K holo. FT iso.).

Closely related to *P. regularis*, differing by the rather smaller leaves, up to 10 x 5 cm, with rounded bases, inflorescences pedunculate with up to 5 flowers only, more clearly distinguishable bracts and pedicels and the smaller, more slender corolla 4-5 cm long with a rather floccose indumentum of densely verticillate hairs. Fig. 114.8.6.

Drier areas than other members of this genus in Ethiopia but very few details of the exact habitat are available; recorded hosts include *Nuxia*, *Rumex ellenbeckii* and *Steganotaenia*; 1150-2200 m. EW TU SU; not known elsewhere. *Fiori* 477, 481; *Gilbert & Lavranos* 2245.

Balle (1982: 40) indicates that she regards this species as a subspecies of *P. rufescens* but does not seem to have ever validated the combination.

4. **P. usuiensis** *(Oliver) M. Gilbert* (1985);
Loranthus usuiensis Oliver (1873) - type: Tanzania, *Grant*.

Very closely related to *P. regularis*, differing by the relatively narrower leaves, up to 22.5 x 8.5 cm, base cuneate, rather more persistently hairy. Corolla more slender with distinct apical swelling, with indumentum of yellowish orange ± floccose densely verticillate hairs similar to those of *P. erythraea*, fruits 10 x 6-8 mm, blue-green, glabrescent.

Montane forest; record from *Prunus africana*; 1950 m. IL; south to Zaire, Zambia, Malawi & N Mozambique. *Friis* et al. 4147.

Friis et al. 1486 from Mt. Maigudo, KF, probably belongs here but is too immature for positive identification.

5. **P. regularis** *(Sprague) M. Gilbert* (1985);
Loranthus regularis Sprague (1910) - type: TU, near Adowa, *Schimper* II:747 (K holo. BM FI FT P iso.).
Loranthus rufescens sensu A. Rich. (1847) non DC. (1830).
L. rufescens DC. var. *pilosus* Pax in *Bot. Jahrb.* **39**: 615 (1907); *L. regularis* Sprague var. *pilosus* (Pax) Cufod., *Enum.*, 31 (1953) non rite publ. & XX (1969) - type: SU, *Rosen* (B holo destr.).

Robust shrub sometimes dominating canopy of host tree. Leaves ovate, blade up to 10-18 x 6-11 cm, tip acute to acuminate, base cordate to truncate, usually grey green, rather fleshy to leathery, rapidly glabrescent. Inflorescence a group of sessile many-flowered clusters with all parts obscured by very dense reddish hairs. Corolla robust, 4.2-5.5 cm long, base cylindrical, tip smooth, exterior completely concealed by very dense indumentum of laxly verticillate reddish brown or rust coloured, rarely yellowish orange, hairs, interior pale red. Berry ovoid with prominent persistent calyx, densely reddish tomentose, sometimes glabrescent, flesh bright red. Fig. 114.8.4-5.

Montane forest or woodland, usually near margins, growing on *Croton macrostachyus*, *Ficus*, *Rhus* and less frequently on a variety of other hosts; (1100 in SW-) 1500-2800 m. TU GD GJ SU AR WG IL KF; Kenya. *Gilbert* 1766; *W. de Wilde* 10688; *Meyer* 7707.

The authority for this species is usually given as Steudel ex A. Rich. but Richard clearly regarded Steudel's manuscript name as a synonym of the West African *L. rufescens*. The first author to both accept the species and provide a description was Sprague. A closely related plant in the Yemen is often named as this species.

6. **P. dschallensis** *(Engl.) M. Gilbert* (1985);
Loranthus dschallensis Engl. (1895) - type: Tanzania, *Volkens* 1776.

Similar to *P. regularis*. Leaves mostly borne in clusters on short shoots, narrowly oblong lanceolate, (3.5-)5.5-7(-14) x (0.5-)0.7-1.1(-1.5) cm, 4-11 times as long as broad, often conspicuously grey-green. Inflorescences fewer flowered; flowers clearly pedicillate. Corolla 4.2-5 cm long, tip slightly winged, indumentum floccose with densely verticillate hairs.

Typically in open *Acacia* woodland in moderate to low rainfall areas, nearly always growing on *Acacia*; *c* 2000 m. SD; Somalia, Kenya, Uganda, Tanzania. *Ruspoli & Riva* 1037; *Vatova* 2114.

6. ONCOCALYX *Tieghem* (1895).

Odontella Tieghem (1895) non C. A. Agardh. (1832);
Loranthus sect. *Coriaceifolii* Sprague (1910);
Loranthus sect. *Longicalyculati* Sprague (1910).

Small shrubs with simple haustorium, glabrous or with short simple (to slightly branched) hairs. Leaves mostly alternate, palmately 3-5-veined. Flowers in axillary clusters or shortly pedunculate umbels; bract saucer shaped to cupular. Flowers 5-merous. Calyx cupular to tubular. Corolla lobes joined for less than half length, mostly split unilaterally (rarely radially symmetrical), mostly yellow at least in part, tube narrow throughout (or constricted above basal swelling), lobes linear or nearly so, erect or reflexed. Filaments coiling, often explosively, at anthesis, often with small tooth in front of anthers, anthers breaking off during anthesis (or persistent). Style filiform, stigma ovoid-globose to obovoid. Berry usually obovoid, often with subpersistent calyx, red.

Perhaps 16 species in the drier forests and bushlands of eastern and southern Africa and the Yemen.

The delimitation of this genus is only provisional. Wiens and Polhill believe there to be 4 major groups of species, only 1 of which, corresponding to sect. *Longicalyculati* Sprague plus sect. *Coriaceifolii* Sprague, occurs in the Flora area. This is easily recognised by the long tubular calyx, uniformly slender corolla and deciduous anthers and is restricted to the Somali-Masai region of E Africa and Arabia. The other sections extend north to Kenya but seem unlikely to turn up in the Flora area.

1. Corolla lobes strongly reflexed, often coiled; tube without internal appendages. 2
- Corolla lobes erect; tube with ciliate internal appendages. 4
2. Twigs and leaves usually sparsely to densely hairy; margins of bracts & calyx woolly; bract not keeled; peduncle obsolete. 3
- Plant entirely glabrous; bract often keeled; pedunculate inflorescences usually present. **2. O. glabrata**

3. Leaves 25-57 mm long, not glaucous nor clustered on short shoots; stems rounded. **1. O. fischeri**
- Leaves less than 20 mm long, very glaucous and mostly clustered on short shoots; stems angular. *O. ugogensis*

4. Leaves up to at least 25 x 7 mm usually more, glabrous or sparsely puberulent, green; corolla primarily yellow often with green or red. 5
- Leaves up to 11 x 3.3 mm, very densely grey pubescent; corolla white with pink tip. **3. O. ghikae**

5. Older stems not prominently lenticellate; calyx smooth; corolla glabrous outside, often with green markings. 6
- Older stems pale grey with conspicuous dark lenticels; calyx minutely puberulent/papillate outside, lobes flushed red near tip. **4. O. kelleri**

6. Twigs terete, pubescent; corolla 12-23 mm long. **5. O. schimperi**
- Twigs sharply angular, glabrous; corolla *c* 30 mm long. **6. O. angularis**

Cufodontis (loc. cit.) records *O. ugogensis* (Engl.) Wiens & Polhill from the Flora area but this is based on misidentifications of collections by Corradi of *O. kelleri*. The species, more often confused with *O. glabratus*, is found in northern Kenya and could extend into the Flora area.

1. **O. fischeri** *(Engl.) M. Gilbert* (1985);
Loranthus fischeri Engl. (1894) - type: Tanzania, *Fischer* 130.
Loranthus stuhlmannii Engl. (1894).
L. stuhlmannii Engl. var. *somalensis* Sprague, *loc. cit.*: 331 (1910) - type: BA, Webi (River) - Abdallah, *Keller* 218 (Z holo. not seen, K fragment of holo.).

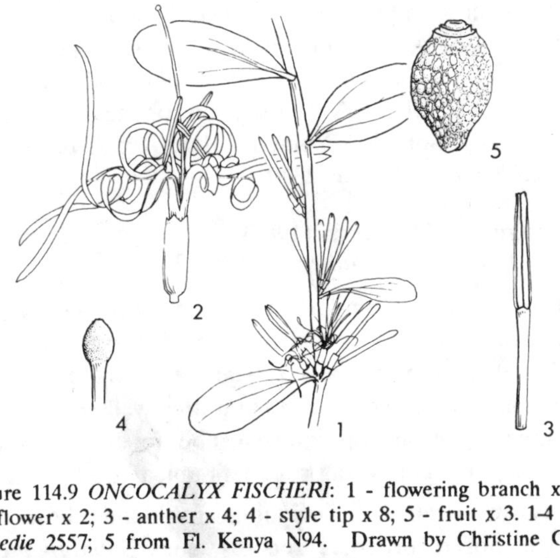

Figure 114.9 *ONCOCALYX FISCHERI*: 1 - flowering branch x 2/3; 2 - flower x 2; 3 - anther x 4; 4 - style tip x 8; 5 - fruit x 3. 1-4 from *Tweedie* 2557; 5 from Fl. Kenya N94. Drawn by Christine Grey-Wilson.

Twigs terete, sparsely puberulent to densely pubescent. Leaves ovate to narrowly elliptic, 25-57 x 7-25(-40) mm, not glaucous, shortly pubescent to glabrous. Umbels sessile, up to 7-flowered, from raised peduncular patch; bracts truncate or very obscurely toothed, glabrous except for woolly margin to almost velutinous. Calyx truncate or nearly so, glabrous except for woolly-ciliolate margin. Corolla 20-24(-27) mm long, yellow, sometimes with reddish tip, becoming orange with age, tube without internal appendages, lobes expanded at base in bud, strongly reflexed, often coiled after anthesis. Filaments without terminal tooth. Berry verrucose, bright pinkish red, up to 7 x 5 mm, calyx usually, but not always, soon lost. Fig. 114.9.

Dry bushland or woodland; hosts in Flora area include *Maytenus senegalensis*, *Flueggea virosa* and *Ximenia*, other hosts recorded elsewhere; 1200-1900 m. SU KF GG SD BA; Sudan, Kenya, Uganda, Tanzania.

The Ethiopian material divides easily into 2 groups, a more southerly one corresponding to *O. fischeri* s.str.: *Burger* 2093; *Friis* et al. 2598; *Gilbert* 1546, and a more northerly one, particularly common on *Maytenus* around Lake Awassa, with narrower leaves and a sparser indumentum corresponding to *L. stuhlmannii* var. *somalensis*: *W. de Wilde* 7071; *Gilbert* 1165; *Thulin* et al. 3314. The two groups do not maintain their distinction in Kenya and Tanzania and it does not seem practicable to recognise them formally.

2. **O. glabratus** *(Engl.) M. Gilbert* (1985);
Loranthus fischeri Engl. var. *glabratus* Engl., *Bot. Jahrb.* **40**: 528 (1908); *L. glabratus* (Engl.) Sprague (1910) - types: BA, Luku, *Ellenbeck* 1229 & (?BA), Abu-el-Kasin, *Ellenbeck* s.n. (both B syn., K fragment of syn.).
Loranthus ugogensis Engl. var. *ganocalyx* Chiov., *Miss. Biol. Borana* 39 (1939) - type: SD, Yavello, *Cufodontis* 602 (FT holo.).

All parts except peduncular patch glabrous; twigs obscurely to sharply angled, later smooth greyish brown with inconspicuous lenticels. Leaves ovate-elliptic, 34-52 x 18-29(-35) mm, dull glaucous green. Umbels often on peduncles to 3 mm long, (2-)3-5-flowered, arising successively from axillary brown-haired peduncular patch up to 7 mm wide; bract with keeled limb. Calyx with ± flared obscurely lobed slightly scarious rim. Corolla 15-23(-32) mm long, yellow becoming orange-brown, very similar to that of *O. fischeri*. Filaments with small apical tooth, sometimes all but absent. Berry smooth, orange-red, up to 10 x 6 mm with persistent calyx *c* 5 mm long.

From *Acacia-Commiphora* bushland up to *Euclea-Rhus* bushland at margins of montane forest, growing on *Catha edulis*, *Dodonea*, *Maytenus*, *Olea europaea* ssp. *africana*, *Rhus* and *Ximenia americana*; 1350-2150 m. SU SD BA HA; Somalia. *Burger* 1689; *J. de Wilde* 4950; *Gilbert* et al. 7406.

Ash 237, from near Debre Zeit [Bishoftu], is described as an erect terrestial shrub, 2 plants seen, 1 m and 1.6m high, perhaps growing on the roots of a nearby plant. The material is otherwise quite typical of this species.

Plants growing on *Catha edulis* are valued by traditional healers.

3. O. **ghikae** *(Volkens & Schweinf.) M. Gilbert* (1985);
Loranthus ghikae Volkens & Schweinf. (1897) - type: HA, Bourka, Kaldesch Mt., *Ghika* (K fragment of holo.).

Twigs rounded, puberulous, soon pale grey without conspicuous lenticels. Leaves mostly on short shoots, oblong-elliptic, up to 12 x 4 mm, fleshy, very densely grey pubescent. Flowers solitary on short shoots; bract obscurely 2-3-toothed, densely grey pubescent. Calyx truncate, dark red, glabrous except for margin. Corolla 27-30 mm long, white or cream with pink tip, tube with ciliate internal appendages, lobes erect. Filaments reddish with obtuse triangular apical tooth. Immature berry smooth, 8 x 4 mm, with persistent accrescent calyx 3.5 mm long.

Along water course in open *Acacia-Commiphora* bushland, growing on *Ziziphus*; 330 m. HA; Somalia. *Gilbert* 2123.

4. O. **kelleri** *(Engl.) M. Gilbert* (1985);
Loranthus kelleri Engl. (1908) - type: BA, Webi [River] - Abdallah, *Keller* 219 (Z holo. not seen, K fragment of holo.).

Twigs rounded or obscurely ridged, glabrous, soon grey with conspicuous small raised dark lenticels. Leaves oblong-lanceolate to broadly elliptic, 2.2-6(-9) x 1-3(-4.5) cm, dull pale green, glabrous or minutely puberulent, prominently 3-5-veined. Umbels sessile, 1-3-flowered, most parts minutely puberulous papillate. Bract-limb obtuse triangular, keeled, reddish. Calyx truncate or obscurely toothed, dull pink. Corolla as in *O. ghikae* and *O. schimperi*, 30-35 mm long, yellow with lobes suffused red. Filaments yellow with obtuse reddish apical tooth. Berry smooth, red or scarlet, up to 12 x 7 mm plus persistent calyx, minutely papillose puberulent.

Acacia-Commiphora bushland, growing mostly on species of *Capparaceae*, especially *Boscia minimifolia*, also recorded from *Acacia*, *Dobera* and *Salvadora*; 400-1250 m. GG SD BA HA; Somalia, Kenya, Tanzania. *Edwards* 641; *Gillett* 4077; *Simmons* 143.

The indumentum of the calyx needs a good hand lens to be distinguished clearly but gives a distinctive texture easily recognised with experience.

5. O. **schimperi** *(A. Rich.) M. Gilbert* (1985);
Loranthus schimperi A. Rich. (1848); *Odontella schimperi* (A. Rich.) Tieghem (1894) - type: TU, Schahageni, *Schimper* I:158 (P holo., BM ETH FI K iso.).
Loranthus schimperi A. Rich. var. *parviflorus* Hutch. & Bruce, in *Kew Bull.* **1941**: 127 (1941) - type: HA, Harar - Kofajalo road, *Gillett* 5271 (K holo.).

Twigs terete, usually densely pubescent, later smooth, grey, lenticels not conspicuous. Leaves elliptic, oblanceolate or obovate, 27-55(-65) x 11-25(-32) mm, rather fleshy, pale dull green, sometimes slightly glaucous, obscurely veined when dry. Umbel 2-4(-6)-flowered, sessile; bract truncate or obscurely 1-3-toothed, with obscure keel, densely puberulent to pubescent. Calyx truncate, greenish, glabrous except for brownish wool on margin; corolla as in *O. ghikae*, 12-25 mm long, pale yellow often with green at base of lobes, becoming deeper, almost orange, interior minutely puberulous; filaments green at base, then yellow, with minute obtuse triangular apical tooth. Berry smooth, orange-red, up to 10 x 7 mm plus persistent calyx. Fig. 114.10.

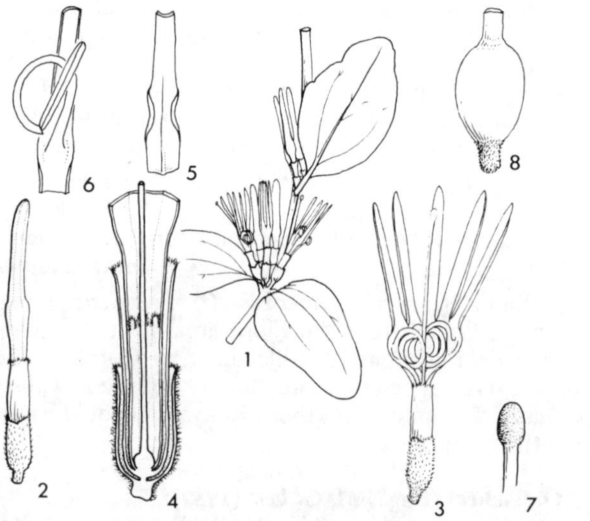

Figure 114.10 *ONCOCALYX SCHIMPERI*: 1 - flowering stem x 2/3; 2 - bud; 3 - flower x 2; 4 - half flower showing ciliate appendage x 4; 5 - outer view of base of corolla lobe x 4; 6 - stamen; 7 - style tip x 8; 8 - fruit x 3. 1-7 from *Ash* 1503; 8 from photo. Drawn by Christine Grey-Wilson.

Dry forest margins, evergreen and deciduous bushland on a wide variety of hosts including *Acacia*, *Balanites*, *Cadia*, *Dobera*, *Euclea*, *Grewia*, *Maytenus*, *Stereospermum*, *Tamarix* & *Ziziphus*; 700-2400 m. AF EW TU WU SU HA (mostly on or near Eastern Escarpment); Somalia, Yemen. *Ash* 726; *J. de Wilde* 7377; *Gilbert & Getachew* 2319.

6. O. **angularis** *M. Gilbert* (1985)
- type: Kenya, *Gilbert* et al. 5172.

Vegetatively most similar to *O. glabratus* but twigs distinctively erect, more sharply angled; leaves smaller, 18-33 x 9-23 mm; umbels sessile. Corolla as in *O. ghikae* but more slender, 23-33 mm long, yellow, usually with green base and tip, internal tube appendages very near base; filaments with prominent apical tooth.

Dry to very dry deciduous bushland, host in Ethiopia not recorded, 2 records from *Euphorbia* cf. *stapfii* in Kenya; 1200 m. SD (near Sidambale Bridge on Genale Doria River); Somalia, Kenya. *J. de Wilde & Gilbert* 341.

The only Ethiopian collection seen, along with *Gillett* 12535 from due south just over the Kenyan border, differs from the typical material from just south of Lake Turkana by the darker older stems, less sharply angled twigs and suborbicular leaves.

7. OLIVERELLA *Tieghem* (1895).

Loranthus sect. *Involutiflori* Sprague (1910).

Small shrubs, all parts with simple hairs. Leaves alternate to opposite, pinnately veined. Inflorescence an axillary, subsessile umbel; bract cupular with narrow limb, often longer than receptacle. Flowers (4-)5-merous. Calyx reduced to narrow rim. Corolla curved in bud, narrowly fusiform with clavately swollen tip, splitting unilaterally at anthesis, lobes longer than poorly delineated tube, not fully separating. Free part of filaments short and thick, without apical tooth, inrolled through 360° at anthesis, anthers oblong, connective produced into short appendage. (4-)5-lobed disc present. Style slender, stigma small, globose.

A small genus of 2-3 species confined to eastern Africa from Ethiopia south to Mozambique.

The genus is usually described as having only 5-merous flowers but occasional 4-merous flowers have been seen. When coupled with similarities in habit, corolla form and, at least for the species in Ethiopia, colouration and disc morphology, this suggests that *Oliverella* could be related to *Englerina woodfordioides*.

O. hildebrandtii *(Engl.) Tieghem* (1895);
Loranthus hildebrandtii Engl. (1894) - types: Kenya, *Hildebrandt* 2852; Tanzania, *Fischer* 534 & Sudan, *Schweinfurth* 2858.

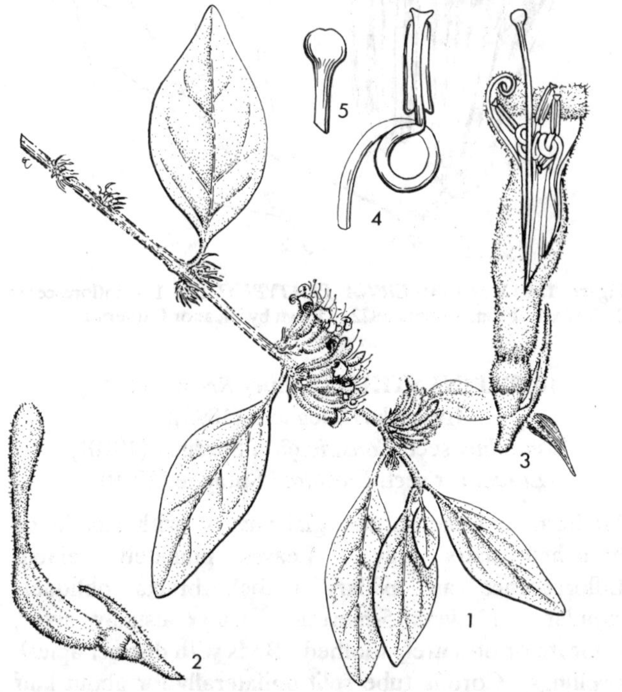

Figure 114.11 *OLIVERELLA HILDEBRANDTII*: 1 - habit x 2/3; 2 - bud, 3 - flower x 3; 4 - stamen, 5 - style tip x 8. 1 from photo of *Gilbert* 5009; 2-5 from *Napier Bax* TNP/G5/20. Drawn by Eleanor Catherine.

All parts densely whitish pilose at least when young; old stems brown. Leaves lanceolate, 3-7 x 1-3 cm, base attenuate into petiole up to 1 cm long, tip acute to obtuse. Umbels up to *c* 20-flowered; peduncle up to 3 mm long; bract-limb subulate, 3.5-4.5 mm long. Corolla 15-18 mm long, exterior red with grey tip, interior reddish with almost black lobes; stamens yellow, anthers *c* 1.5 mm long with short, dark, 2-lobed connective-appendages; style dark red. Fruit not seen. Fig. 114.11.

Ethiopian record from dense *Acacia* bushland with much *Dracaena ellenbeckii*, most Kenyan records from rock outcrops; usually growing on *Grewia*; 1400 m. SD; Kenya, Tanzania. *Mesfin & Vollesen* 4319.

8. ENGLERINA *Tieghem* (1895)

Loranthus sect. *Ischnanthus* Sprague (1910)

Glabrous (or with simple or branched hairs). Leaves opposite or subopposite, pinnately veined. Inflorescence an axillary pedunculate umbel; bract obliquely cupular, shorter than receptacle. Flower 4-merous. Calyx truncate or obscurely lobed. Corolla: bud straight, narrowly fusiform to subcylindrical, tip clavately swollen, splitting unilaterally at anthesis to near base, lobes as long or longer than poorly-delineated tube, linear-spathulate, erect or slightly incurved (or reflexed to revolute). Filaments coiling circinately at anthesis, always with prominent apical tooth, anthers oblong to linear, reflexing through 180°. Disc 4-lobed. Style filiform, stigma small, obovoid to globose. Fruit where known obovoid to pyrriform with persistent disc and calyx.

About 20 species restricted to tropical Africa.

Easily recognised by the combination of consistently 4-merous flowers and coiled filaments.

E. woodfordioides *(Schweinf.) M. Gilbert* (1985);
Loranthus woodfordioides Schweinf. (1892) - type: Kenya, *von Hoehnel* 39.
Loranthus tschertscherensis Pax (1907) - type: HA, Tschertscher, Kuni, *Rosen* (B holo. not seen).
Englerina woodfordioides var. *subintegridisca* Balle in *Senck. Biol.* **39**: 108 (1958) nom. invalid. - based on *Haberland* 2100 from SD.

Stems to 1 m long. Petiole narrowly winged; leaf-blade linear lanceolate to ovate, 4.3-11 x 1.2-3.5 cm, tip acute to acuminate, often flushed red when young. Umbel 6-21-flowered, glabrous except for ciliolate bract and calyx margins, peduncle shorter than pedicels. Calyx spreading, obscurely 4-lobed. Corolla 9-24 mm long, exterior grey with red, rarely pale orange, tip, interior dark, almost black. Filaments with subulate apical tooth up to half as long as anther. Disc lobes shorter (to longer) than calyx. Fruit pyrriform, 5.4-7.5 x 3.2-4 mm, translucent red with orange seed *c* 3.5 mm long. Fig. 114.12.

Margins of wetter montane forests, particularly common in thickets resulting from forest clearance, most common host in Ethiopia *Maesa lanceolata* but also recorded from a wide range of other hosts; 1700-2550(-3000) m. GJ SU AR WG IL KF SD; Kenya, Uganda, Tanzania, Rwanda, Zaire. *J. de Wilde* 5605; *Friis* et al. 467; *Gilbert* 1768.

Material from lower altitudes in the wetter parts of SW Ethiopia (WG, IL, KF) - e.g. *W. de Wilde* 6305, 7543 and *Friis* et al. 248 - has distinctly smaller (9-15 mm) flowers apparently with paler orange lobes, as against the red tipped 17-24 mm flowers at higher altitudes. For the present these seem best regarded as just a local variant as similar though possibly not identical small flowered plants, including the type of *Loranthus bagshawei* S. Moore, occur in W Uganda and scattered through Kenya among normal *E. woodfordioides*.

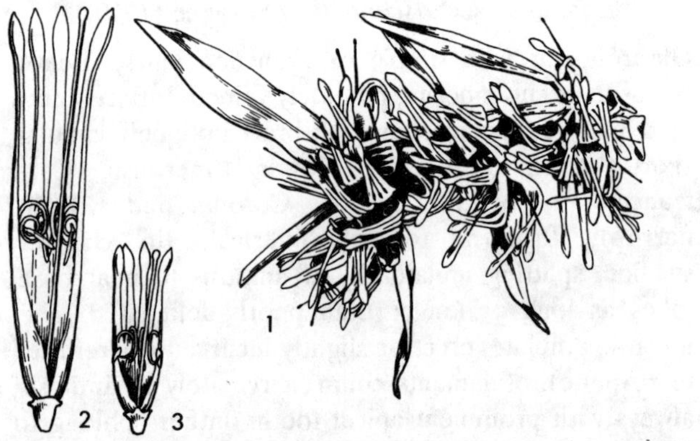

Figure 114.12 *ENGLERINA WOODFORDIOIDES*: 1 - flowering branch x *c* 2/3; 2 - flower; 3 - flower (small-flowered form) x 2. 1 from photo of *M. & S. Gilbert* 1765; 2 from *Friis* et al. 467; 3 from *W. de Wilde* 6305. Drawn by Eleanor Catherine.

Balle (*Senck. Biol.* **39**: 108, 1958) gave the description of 2 new varieties and a new combination within *Englerina woodfordioides* without correctly validating this name - the place of publication of the basionym was not given. The taxa were based on variation in the size of the disc lobes but this appears to be continuous and in the absence of clear-cut geographical separation (see for example Troupin, *Flore plante lign. Rwanda*: 373, 1982) the varieties do not seem worth formal recognition.

9. SCHIMPERINA *Tieghem* (1895)

Loranthus sect. *Longiflori* Sprague (1910)

Moderate sized shrub, all part glabrous. Leaves alternate to subopposite, pinnately veined. Inflorescence a pedunculate axillary umbel; bract semicircular. Flowers 5-merous; calyx reduced to a toothed rim. Corolla without distinct basal or apical swellings, split unilaterally to just below anther insertion, lobes about as long as tube, seperating at base but normally remaining joined along upper half; filaments without apical tooth, inrolled through 360° at anthesis, anthers long, persistent; style filiform, stigma globose. Fruit subglobose-ovoid, with annular terminal disc when young, calyx not accrescent.

Monotypic.

Although Balle (*Fl. Cameroun* **23**: 50, 1982) regarded the following species as conspecific with *Tapinanthus flamignii* (De Wild.) Danser, the difference in corolla facies between *Schimperina* and normal *Tapinanthus*, including *T. flamignii*, seem comparable to those between other generic segregates of *Loranthus*. For the present van Tieghem's genus has been resurrected until the vexed question of the delineation of *Tapinanthus* has been more fully investigated.

S. platyphylla *(A. Rich.) Tieghem* (1895);
Loranthus platyphyllus A. Rich (1848); *Tapinanthus platyphyllus* (A. Rich.) Danser (1933) - type: TU, Djeladjeranne, *Schimper* III:1584 (P holo., BM FI FT K iso.).

Older stems regularly prominently longitudinally fissured. Leaves broadly elliptical, ovate or lanceolate, up to 6.2-12 x 2.5-6 cm, conspicuously blue-green when young. Umbel 6-17-flowered, peduncle up to 1 cm long. Calyx a 5-toothed rim. Corolla 37-47 mm long, brilliant red to orange-red with paler interior; anthers 6.5-8 mm long. Fruit 10 x 9 mm. Fig. 114.13.

Broadleaved deciduous woodland, all records from *Combretum* or *Terminalia*; 900-1800 m. EW TU/GD WG IL; Sudan, Kenya, Uganda. *J. de Wilde* 6368; *W. de Wilde* 10689; *Mooney* 6822.

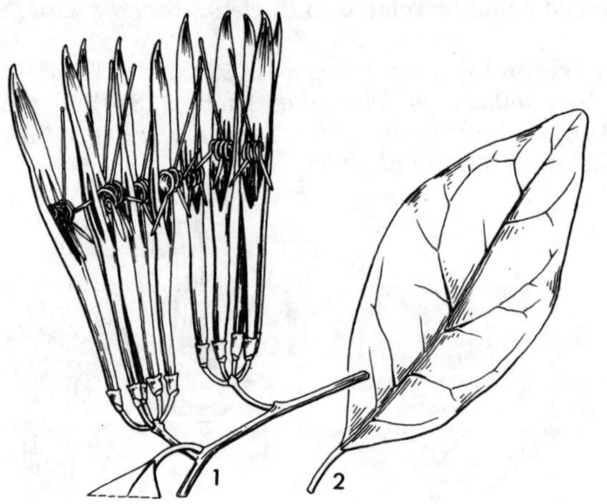

Figure 114.13 *SCHIMPERINA PLATYPHYLLA*: 1 - inflorescences; 2 - leaf x 1. From *Mooney* 6822. Drawn by Eleanor Catherine.

10. TAPINANTHUS *(Blume) Reichb.* (1841)

Agelanthus Tieghem (1895);
Loranthus sect. *Constrictiflori* Sprague (1910);
Loranthus sect. *Erectilobi* Sprague (1910).

Medium to large shrubs, glabrous or with simple or branched hairs only. Leaves pinnately veined. Inflorescence an axillary umbel, bracts obliquely cupular. Flowers 5-merous. Calyx usually short, truncate or obscurely toothed. Buds with distinct apical swellings. Corolla tube split unilaterally for about half length, over twice as long as lobes, base often conspicuously swollen, lobes reflexed (*Tapinanthus* s. str.) or erect (*Agelanthus*), tips thickened, involute below. Filaments short, inflexed through about 200°

after anthesis, usually with apical tooth, anthers short, persistent. Style skittle-shaped (clavately swollen from about half way, then abruptly contracted into short slender tip below globose stigma). Fruit obovoid to subglobose, often with persistent calyx.

The largest and most diverse genus of African loranths, extending into Tropical Arabia. Perhaps some 80-90 species but the group is much in need of revision both at the species level and with regard to the delineation from other genera.

The species fall into 2 main groups differing fundamentally in the mechanism of anthesis. In *Tapinanthus* s. str., including species such as *T. globiferus*, the corolla lobes do not separate before anthesis which is initiated by a visiting bird pecking at or squeezing the bud tip when the corolla lobes reflex rapidly. In species such as *T. heteromorphus* the corolla lobes first separate at the base only and anthesis is initiated when a bird inserts its' beak into one of the small openings between the lobes. Such species may prove to be better separated as the genus *Agelanthus* Tieghem.

1. Corolla, and usually the whole plant, hairy; corolla lobes erect. **1. T. heteromorphus**
- Plant glabrous except for minutely ciliolate bract and calyx margins; corolla lobes reflexed. 2
2. Leaves on long shoots only, margins not coloured. **2. T. globiferus**
- Leaves mostly clustered on short shoots, margins red at least when young. **3. T.** sp. = *Westphal* 635

1. **T. heteromorphus** *(A. Rich.) Danser* (1933);

Loranthus heteromorphus A. Rich. (1848) - type: TU, Shire, ['Chiré'], *Quartin-Dillon* (P holo.).

Loranthus sennii Mattei (1910) - type: EW, Keren [Cheren], *Senni* 626 (PAL holo. not seen; K iso.).

?*L. scassellattii* Chiov. var. *glabrescens* Pichi-Serm., *Miss. Stud. Lago Tana* **7**: 29 (1951) - type: GD, Gorgora Peninsula, *Pichi-Sermolli* 2491 (FT holo. K iso.).

Large shrub, shoots often tomentose when young, soon glabrescent. Leaves subopposite, lanceolate to ovate, 6.2-14 x 2.5-7.2 cm. Umbel 3-10-flowered, peduncle 1.5-2 mm long, all parts usually densely hairy; bract limb as long as receptacle. Calyx reduced to asymmetrical 0.5 mm limb. Corolla 3.5-5 cm long, with conspicuous ovoid basal swelling and cylindrical apical swelling, uniformly greenish or yellowish usually covered with sparse to dense short-branched red to orange hairs, lobes *c* 1/4 the length of the corolla, erect with conspicuously thickened tips and strongly revolute bases seperating before anthesis. Filaments with apical tooth. Fruit not seen. Fig. 114.14.5-6.

Broad-leafed deciduous woodland/wooded grassland growing on *Combretum* and *Terminalia* (less often on various *Caesalpinoideae* in West Africa); 1300-1900 m. EW TU WG IL SD BA; Togo, Nigeria, Cameroun, Sudan. *J. de Wilde* 6363; *Friis* et al. 423; *Gilbert* et al. 7757.

The type of *L. scassellatii* var. *glabrescens* has only immature buds but there does not seem to be any reason for not regarding it as a subglabrous form of *T. heteromorphus*. Mature material from the type locality is needed to confirm this opinion. *L. scassellatii* sens. str. is the correct name for the species widely known in East Africa as *L. sansibarensis* Engl.

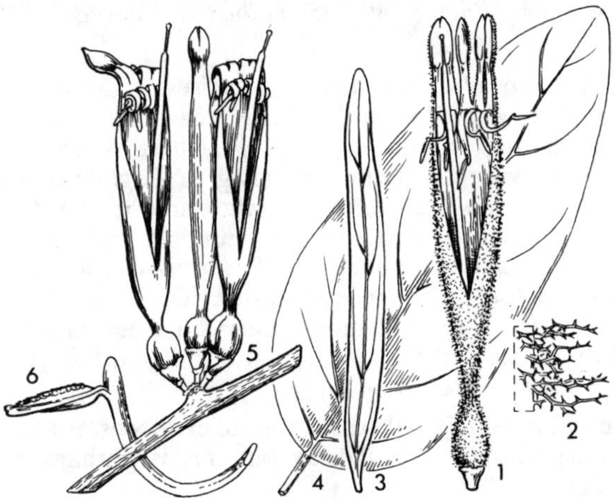

Figure 114.14 *TAPINANTHUS HETEROMORPHUS*: 1 - flower x 1; 2 - corolla indumentum x 8. *T. GLOBIFERUS*: 3 - leaf of dry country form, 4 - leaf of forest form ('*L. verrucosus*') x 2/3; 5 - inflorescence x 1; 6 - stamen x 5. 1 & 2 from *Gilbert* et al. 7757; 3 & 6 from *Gilbert & Tewolde* 2525; 4 from *Ash* 562; 5 from photo of *Gilbert* et al. 7389. Drawn by Eleanor Catherine.

2. **T. globiferus** *(A. Rich.) Tieghem* (1895);

Loranthus globiferus A. Rich. (1847) - type: TU, Mai Gouagoua, near Adowa, *Quartin-Dillon* (P holo.).

Loranthus verrucosus Engl. (1894); *L. globiferus* var. *verrucosus* (Engl.)Sprague, *loc. cit.*:353 (1910).

L. globiferus var. *angustifolius* Schweinf. in *Bull. Herb. Boiss., Append.* **2**: 150 (1896) - types: TU, Gessat Errar, *Schimper* I:414 & GD, Repp Gebriel, *Schimper* (1863) 1491 (both B syn., BM K isosyn.); EW, Keren, *Steudner* 184 (B syn., K isosyn.).

L. lingelsheimii Pax (1907) - type: HA, Hara, *Rosen* (B holo. K fragment of holotype).

L. globiferus var. *salicifolia* Sprague, *loc. cit.*: 352 (1910) - types include *Schimper* (1863) 1491 (K syn.).

L. ostinii Chiov. (1911) - types: GD, Dembia, Cococc Valley near Gonder, *Chiovenda* 1845 & Asoso, *Chiovenda* 2696 (both FT syn.).

Shrub up to 2m across, usually less, glabrous except for minutely ciliolate bracts and calyx. Leaves alternate to subopposite, linear-lanceolate to oblong-ovate or ovate, up to 5-14 x 0.6-4.3 cm, often glaucous when young. Umbel 3-10-flowered, peduncle to 2 mm long; bract up to two thirds as long as receptacle. Calyx truncate, sometimes splitting. Corolla 3-4.8 cm long, base subglobose to ovoid, abruptly contracted into narrow neck, then fusiform with swollen ovoid tip which is usually rounded in bud but is sometimes angular or narrowly winged, lobes spathulate, 6-8 mm long, sharply reflexed, exterior pinkish red, tip pale greenish-grey turning dark at maturity; interior of tube pink, lobes pale with dark purple tip. Fruit subglobose, *c* 9 x 8 mm, bluish green with white spots immature, orange-red or scarlet when mature. Fig. 114.14.3-6.

Margins of montane and submontane forests and thickets on a wide variety of hosts, especially *Rhus*; 540-2350 m. EW TU GD WU GJ SU WG IL KF SD HA; Sahel from Mauretania east to Yemen, south to Cameroun. *Friis* et al. 135; *Gilbert & Tewolde* 2525; *Mooney* 5313, 9098.

The corollas are often parasitised causing malformed growth. Within the Flora area there is a geographical segregation between the narrow to very narrow leaved plants corresponding to *T. globiferus* s. str. and *L. globiferus* var. *salicifolia* found in the drier areas of northern and southern Ethiopia and broad-leaved plants, corresponding to *L. verrucosus*, found in the wetter forests of south-western Ethiopia. However in West Africa the variation seems clinal between wet and dry areas and thus no formal distinction has been made between the two groups.

Balle (1982; 54-60) treats a number of western taxa as subspecies of *T. globiferus* but this is perhaps not justified.

3. T. sp. = *Westphal* 635 (Kew).

Glabrous throughout. Leaves in irregular whorls of 3, mostly borne on congested short shoots, elliptical, up to 3.4 x 1.4 cm, margins distinctively thickened and red. Umbel sessile, up to 5-flowered. Bract and calyx as in *T. globiferus*. All corollas seen probably malformed, *c* 2 cm long with indistinct basal swelling and obscurely verrucose tube, otherwise similar to those of *T. globiferus*. Fruit not seen.

Growing on cultivated *Annona*; *c* 2000 m. HA (near 'Mrs. Diggs' Fruit Farm', between Alemaya and Harer); known only from the single collection.

The only specimen seen has flowers resembling parasitised flowers of *T. globiferus* and it is possible that it should normally produce flowers similar to those of that species. However vegetatively the plant seems quite distinct. The locality was visited in 1983 but only dead remains of a possible *Tapinanthus* could be found.

115. VISCACEAE

by M. G. Gilbert*

Sprague, *Fl. Trop. Afr.* **6(1)**: 393-411 (1911); Cufodontis, *Enum.* 33-35 (1953).

Perennial herbs or shrubs, hemiparasitic on the stems of other woody plants. Stems green, brittle, often with articulated nodes. Leaves opposite simple, evergreen, often reduced to scales, stipules absent. Flowers minute, unisexual, (2-)3-4(-7)-merous with a perianth of a single whorl of valvate lobes, solitary or clustered at nodes or in dichasia (or axillary spikes outside Flora area). Male flowers with stamens opposite tepals, rarely fused into synandria, pistil vestigial or lacking. Female flowers without staminodes, with simple style and capitate or linear stigma; ovary inferior, uniloculate, ovules undifferentiated with embryo sacs originating within short placental column. Fruit a 1-seeded berry with a viscin (sticky) layer inside the vascular cylinder. Seeds without a testa, usually with 2 cotyledons.

About 7 genera and 450 species widely distributed throughout the Tropics and North Temperate zone, most numerous in the Americas: 3 genera and 7 species in Ethiopia.

Traditionally included within the *Loranthaceae* but differing markedly in floral structure and now widely accepted as being of an independent origin e.g. Barlow: *Proc. Linn. Soc. New South Wales* **89**: 268-272 (1964) and Kuijt: *Brittonia* **20**: 163-174 (1968).

Key to Genera

1. Large plants with stems to 60 cm long or more, often with well developed leaves; anthers dehiscing by numerous pores; fruits 4 mm or more long, indehiscent. **1. Viscum**
- Small plants with stems rarely exceeding 15 cm long, never with well developed leaves; anthers dehiscing by single slit or fused into synandrium with single apical pore; fruits up to 3 mm long, explosively dehiscent. 2
2. Scale leaves bifarious (in 2 ranks); anthers fused into synandrium; host in Ethiopia apparently always *Syzygium*. **2. Korthalsella**
- Scale leaves decussate (in 4 ranks); anthers separate; host always *Juniperus*. **3. Arceuthobium**

1. VISCUM *L.* (1753)

Glabrous shrubs with simple haustorium. Stems articulated, internodes rounded or flattened, longitudinally ridged when dry. Leaves decussate, well developed (sect. *Viscum*) or reduced to scales (sect. *Aspidixia*). Inflorescence a dichasium subtended by a more or less fused pair of scale-like bracts, unisexual or with central male flower and lateral female flowers. Flowers (3-)4-merous. Male flowers with anthers inserted on perianth lobes or, rarely, fused into central synandrium, dehiscing by numerous pores. Female flowers with very narrow rim outside perianth and conspicous green globose stigma. Fruit an indehiscent translucent berry.

Genus mainly of the Old World Tropics but with significant extensions into the adjacent Temperate regions, with perhaps as many as 100 species, most numerous in Tropical Africa and Madagascar.

1. Leaves well developed; most infructesctences with more than 1 fruit. 2
- Leaves reduced to scarious-margined scales; most infructescences with solitary fruit. 4
2. Leaves often with obscure veins, margins often minutely crenate; dichasia sessile, usually densely crowded; fruit tuberculate/ warted at least when young, red or orange (rarely yellow) when mature. **1. V. tuberculatum**
- Leaves usually with well defined veins, margins often distinctly undulate, entire; dichasia pedunculate, 1-5 together; fruit smooth, white or very pale green. 3
3. Largest leaves 2.6-4.5(-6.2) cm long; monoecious; dichasia 1-3(-4)-flowered; flowers 2-2.5 mm long. **2. V. triflorum**
- Largest leaves 6.8-11 cm long; dioecious; dichasia (4-)5-flowered; flowers 3.5-5 mm long. **3. V. congolense**
4. Plant pendent from host branch, lateral branches inserted at right angles; all internodes rounded; male flowers with anthers fused into central synandrium. **4. V. bagshawei**
- Plant patent from host branch, lateral branches inserted at acute angles; upper internodes flattened; male flowers (often apparently absent) with anthers separate. **5. V. schimperi**

1. V. **tuberculatum** *A. Rich.* (1848)
- type: Shewa [Choa], *Petit* (P holo.).

V. tuberculatum forma *obovatifolia* Schweinf., *Beitr. Fl. Aeth.*: 82 (1867) - type: TU, Aman-Eski, 5.11.1854, *Schimper* 471 bis (B holo. not seen, K iso.).

V. tuberculatum forma *obovato-oblongifolium* Schweinf., *loc. cit.*: 82 (1867) - type: TU, Bellaka, 7.11.1854, *Schimper* 471 (B holo. not seen, BM iso.).

Stems forming dense globose mass, less often pendent, up to 60 cm long. Internodes rounded. Leaves spath-

* Ethiopian Flora Project, c/o Royal Botanic Gardens, Kew, Richmond, Surrey, TW9 3AB, UK.

ulate, oblanceolate, obovate or round, 1.4-4.2(-5) x 0.9-2.2(-2.5) cm, tip rounded, margin entire or very minutely notched so as to be crenulate, fleshy with very obscure veins visible only when old or dried, dull to yellowish green. Monoecious. Dichasia sessile, in dense axillary clusters, usually unisexual with (1-)3 male flowers or 1-2 female flowers, each axil tending to produce first male and then female dichasia, very rarely dichasia bisexual. Male flowers 2-3 mm long, yellow to orange. Female flowers similar but with conspicuously warted ovary, perianth soon falling. Fruit ovoid with conspicuous apical disc, 5-7 x 4-5 mm, warted to virtually smooth, orange or red, rarely yellow. Fig. 115.1.1.

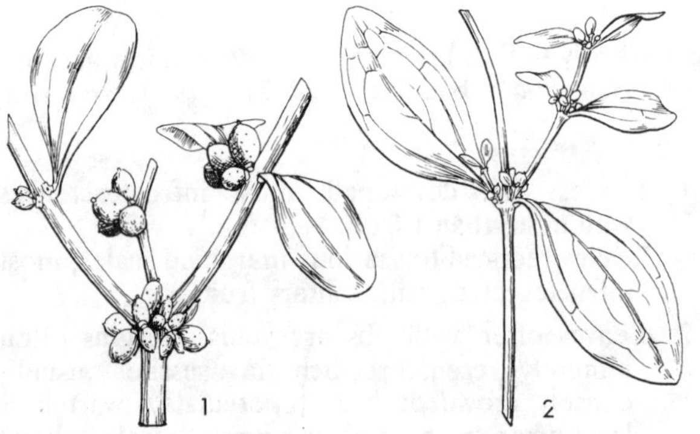

Figure 115.1 *VISCUM TUBERCULATUM*: 1 - part of fruiting plant x 1. *V. TRIFLORUM* subsp. *NERVOSUM*: 2 - part of plant with young fruits x 1. 1 from *Mooney* 8418; 2 from *Jackson* 315a. Drawn by Eleanor Catherine.

Common in relatively dry woodland or evergreen bushland on a variety of hosts, especially *Croton macrostachyus*, *Rhus*, *Maytenus*, *Nuxia congesta*, *Olea europaea* ssp. *africana*, etc.; 1400-2500 m. EW TU GJ WU SU AR WG KF GG SD BA HA; Somalia, Kenya, Uganda, Tanzania, Burundi, Zaire. *J. de Wilde* 6585; *Gilbert* et al. 2301; *W. de Wilde* 9904.

Material from Malawi, Zimbabwe and Angola otherwise typical of this species has consistently bisexual dichasia and is perhaps better regarded as distinct.

2. **V. triflorum** *DC.* (1840)
- type: Mauritius, *Bory* s.n.

Stems usually pendent, rarely erect, to 60 cm long. Internodes rounded or upper slightly flattened or winged. Leaves lanceolate to broadly ovate, 2.6-4.5 (-6.2) x 1.6-3.8 cm, tip rounded or obscurely acuminate, margin entire, often undulate, usually prominently 3-5-veined from base. Monoecious. Inflorescence an axillary cluster of 1-3(-4) pedunculate dichasia. Peduncle 1.5-2.2 mm long at anthesis. Dichasia (1-)3 (-5)-flowered, mostly all female, less often all male or with central male flower and lateral female flowers. Male flowers *c* 2.5 mm long with smooth ovary. Fruit translucent white or pale green with rather persistent perianth and prominent apical disc.

subsp. **nervosum** *(A. Rich.) M. Gilbert*
Nordic J. Bot. **5**: 224 (1985); *V. nervosum* A. Rich. (1848) - type: TU, Schahageni, *Schimper* I:211 (P holo. BM K iso.).
V. nervosum var. *angustifolium* Sprague in Fl. Trop. Afr. **6**(1):398 (1911) - syntypes include Eritrea, near Keren, *Beccari* 62 (K holo. FT iso.).

Drying relatively pale. Fruits distinctly ovoid, to 5 mm long. Fig. 115.1.2.

Montane forest margins through to relatively dry riverine forest; hosts include *Croton macrostachyus*, *Buxus*, *Tamarix*, *Euphorbia abyssinica* and *Coffea*; 1100-2000 m. EW TU GJ WU SU AR KF SD BA HA; Eastern Africa south to Transkei and west to Zaire (Kivu), ?Sao Tome. *W. de Wilde* 9161; *Gilbert* et al. 3514; *Schweinfurth & Riva* 1225.

Subsp. *triflorum*, differing by its tendency to blacken on drying and the more globose fruits, occurs in the Comores, Seychelles and Mascarene Islands.

3. **V. congolense** *De Wild.* (1901)
- syntypes: Zaire, *Gillett* 382 & 573.
V. grandifolium Engl. (1908).

Resembling a very robust *V. triflorum*, stems up to 100 cm long, leaves 6.8-11 x 3.2-4.7 cm. Dioecious. Inflorescence an axillary cluster of up to 5 pedunculate dichasia. Peduncle up to 5.5 mm long at anthesis. Male flowers 3-5 per dichasium, *c* 3.5 mm long, white. Female flowers (2-)3-4(-5) per dichasium, 4-5 mm long with smooth ovary, perianth falling quickly. Fruit ellipsoidal, up to 9 mm long including 1.5-3 mm stipe, translucent greenish white. Fig. 115.2.

Humid forest, usually on large trees; hosts in Ethiopia include *Albizia grandebracteata*, *Croton macrostachyus* and *Mimusops kummel*; 1250-2200 m. WG IL KF; Ivory Coast, Ghana, Nigeria, Cameroun, Gabon, Zaire, Burundi. *Friis* et al. 4023-4; *Mesfin* et al. 235,1336; *Mooney* 8846.

An apparent record from SD, *Brehme* in *Mooney* 9066 from 'Negele', is probably due to mislabelling.

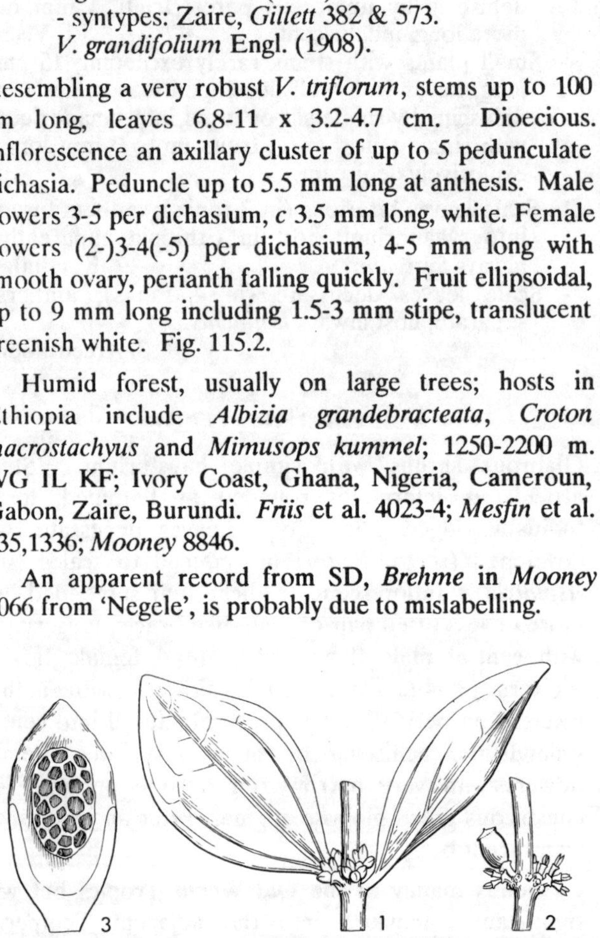

Figure 115.2 *VISCUM CONGOLENSE*: 1 - inflorescence of male plant x 1; 2 - single perianth lobe showing multiporate anther x 12; 3 - female inflorescences with fruit x 1. 1 & 2 from *Friis* et al. 4023; 3 from *Brehme* in *Mooney* 9066. Drawn by Eleanor Catherine.

4. **V. bagshawei** *Rendle* (1905)
type: Uganda, *Bagshaw* 333.

Stems often over 40 cm long, pendent. Internodes rounded, apical mostly over 2 cm long. Leaves reduced to paired scales forming a very shallow scarious-edged collar. Monoecious. Dichasia more or less sessile, sometimes several together. Male flowers solitary, 2-2.5 mm long, anthers fused into a central multiporate synandrium free from perianth. Female flowers 1(-3) per dichasium, 2.5-3 mm long, ovary smooth, perianth soon falling. Fruit globose, *c* 5.5 mm diam., with stipe *c* 1 mm long, red. Fig. 115.3.3.

Habitat in Ethiopia not recorded, in East Africa mostly on *Acacia*, less often hyperparasitic on *Loranthaceae* in dry to moderately high rainfall areas; *c* 1400 m. GG/SD (near Lake Abaya); Kenya, Uganda, Tanzania. *Vatova* 1590.

The type of *V. bagshawei* lacks male flowers but all similar Ugandan material with male flowers shows the very distinctive synandrium described here. The only Ethiopian collection seen also lacks male flowers and so the determination must be regarded as tentative. *V. hildebrandtii* Engl. (1894) from drier areas of SE Kenya and N Tanzania is very similar but has normal anthers.

5. **V. schimperi** *Engl.* (1894)
- type: TU, Harre-Schoa, 1862, *Schimper* 667 (B holo. not seen; BM K iso.).
Aspidixia semiplanum Tieghem (1896); *V. semiplanum* (Tieghem) Engl. (1897) - type: EW, Geleb [Gheleb], *Schweinfurth* 1620 (BM FT iso.).

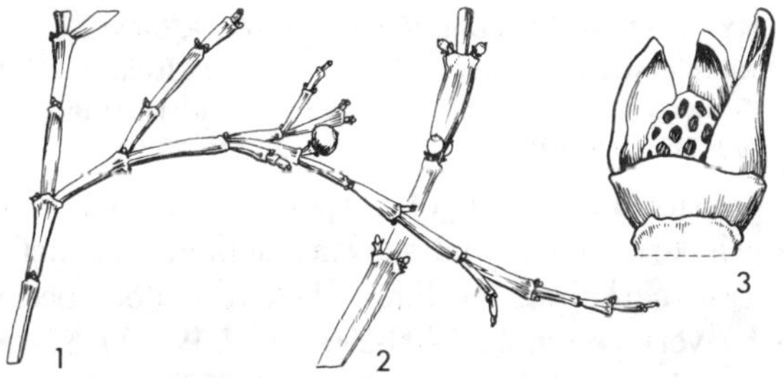

Figure 115.3 *VISCUM SCHIMPERI*: 1 - part of plant with fruit x 1; 2 - detail of stem to show orientation of internodes x 3. *V. BAGSHAWEI*: 3 - male flower with 1 tepal removed to show central synandrium x 12. 1 & 2 from *Gilbert* et al. 7823; 3 from *Lye* 4596. Drawn by Eleanor Catherine.

Stems usually less than 40 cm long, tending to spread patently from branch of host. Lower internodes rounded, upper internodes decussately flattened, up to 2 cm long. Leaves reduced to scales as in *V. bagshawei*. Monoecious but male flowers soon falling and often apparently absent. Male flowers with separate, often unequal, anthers. Female dichasia 1(-3)-flowered, peduncles up to 1.5 mm long; flowers *c* 2.3 mm long, perianth greenish-yellow, soon falling, ovary smooth. Fruit globose, *c* 5 mm diam., with very short but distinct stipe, orange-red. Fig. 115.3.1-2.

Deciduous or evergreen bushland; hosts include *Acacia* spp., *Vepris glomerata* and *Rhus*; 1350-1900 m. EW TU GG SD BA; Somalia, Kenya, Uganda. *Gilbert & Thulin* 107; *Gilbert* et al. 7823; *Schweinfurth & Riva* 671.

2. **KORTHALSELLA** *Tieghem* (1896)

Bifaria van. Tiegh. (1896)

Danser, *Bull. J. Bot. Buitenz. ser.3*, **14**:133, 1937.

Densely branched minute shrubs with single haustoria. Stems articulated, all but the lowermost usually flattened in one plane, rarely rounded. Leaves bifarious (in 2 superimposed ranks), rarely decussate, scale-like. Inflorescence a dense axillary cluster of sessile flowers, male and female intermixed, interspersed with short hairs. Flowers 3-merous. Male flowers globose, anthers fused into globose 6-locular synandrium with locules open by slits into central cavity from which pollen is released by single apical pore. Female flowers clavate, stigma capitate. Fruit clavate with persistent perianth, dehiscing by squirting seed out explosively through apical rupture of skin.

About 20 species from Ethiopia, Kenya, Socotra and Madagascar east to Hawai and Henderson Island (128°W), north to the foothills of the Himalayas and Japan and south to New Zealand.

K. japonica *(Thunb.) Engl.* (1894);
Viscum japonicum Thunb. (1794); *Korthalsella opuntia* (Thunb.) Merrill (1916) - type: Japan, *Thunberg* (UPS-THUNB 23335).
Bifaria abyssinica Tieghem (1896) - types: TU/GD, Tekeze [Takaze] Valley, Addi-Maria near Derakka, *Quartin-Dillon* & TU, Shire [Siré], *Petit* (both P syn).
?*Korthalsella binii* Pichi-Serm., (1951) - type: GD, Gorgora, *Pichi-Sermolli* 2487 (FT holo. not seen).
Viscum opuntia Thunb. (1784) non L. (1753).
Viscum taenioides A. Rich. (1848) non Commers. ex DC. (1830).
Viscum schimperi Sprague (1911) *pro parte, non sensu stricto*.

Stems usually less than 10 cm long, basal internodes rounded, upper usually flattened, often with prominent midrib and up to 3 parallel lateral veins each side, 1-8 mm wide, narrower towards tip, branches single, rarely in groups. All but lowermost axils fertile, flower clusters in opposite axils usually remaining separate. Flowers *c* 0.7 mm across. Fruits *c* 2 mm long. Fig. 115.4.

Parasitic on *Syzygium guineense*; 1800-2200 m. TU GD GJ WG; for species as a whole - Madagascar, Indian Ocean Islands, Ceylon, east to Japan and Australia. *W. de Wilde* 7256; *Pichi-Sermolli* 2488.

A very polymorphic species occurring as far east as Australia and Japan. Danser (loc. cit.) recognised 5 varieties (under *K. opuntia*). Only 4 collections have so far been seen from within the Flora area, which can be ± matched with collections from Madagascar and the Indian Ocean islands corresponding to 2 different varieties within Danser's scheme:

- a narrow stemmed plant with internodes up to 3 mm wide and up to 3 inconspicuous longitudinal veins, corresponding to *K. opuntia* var. *richardii* (Tieghem) Danser and including *B. abyssinica* and *K. binii* (ex desc., type not found in 1983);
- a broad stemmed plant with internodes up to 4.5 mm wide and (3-)5(-7) longitudinal veins, represented by *W. de Wilde* 7256, and matching *K. opuntia* var. *gaudichaudii* (van Tiegh.) Danser.

The combinations for these infraspecific taxa under *K. japonica* have not been made and the complex needs further revision before this is done.

K. binii was separated on the basis of the more slender stems but there is little in the dimensions given in the protologue to justify the seperation from *K. japonica* sensu latu.

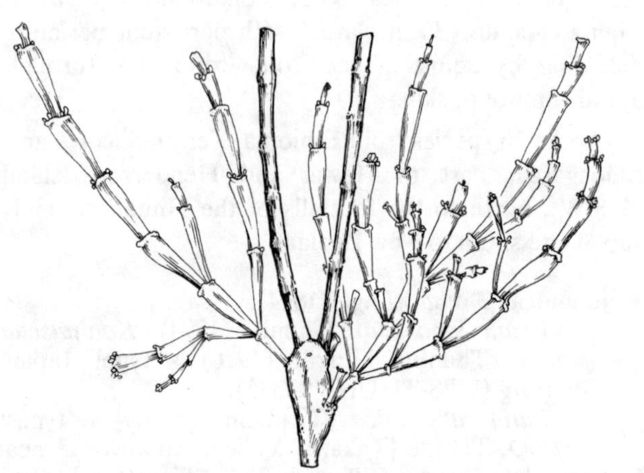

Figure 115.4 *KORTHALSELLA JAPONICA*: showing young and old stems x 1. Drawn from *W. de Wilde* 7256 by Eleanor Catherine.

3. **ARCEUTHOBIUM** *Marsch. Bieb.* (1819) *nom. cons.*
Hawksworth & Wiens, *Kew Bull.* **31**: 71-80 (1976).

Herbs or shrubs parasitic on conifers (*Pinaceae* & *Cupressaceae*), sometimes forming systematic infections and malformed growth. Stems articulated, often with angular internodes. Leaves decussate (in 4 ranks), scale-like. Dioecious. Flowers usually solitary in axil or rarely verticillate. Male flowers (2-)3-4(-7)-merous, perianth lobes spreading, each with circular 1-locular anther dehiscing by transverse slit, central nectary present. Female flower 2-merous, perianth persistent. Fruit ovoid, upper and lower portions differently coloured, dehiscing by breaking off at base and squirting seed out explosively through the resultant opening.

Genus with 28 species, most in the west of Mexico and the U.S.A. but with 5 in the Old World temperate zone, extending from the Azores to China, only the following species in Tropical Africa.

A. juniperi-procerae *Chiov.* (1911)
- types: EW, M. Mamahot & Amba Debra, *Pappi* 1161 & 2554 (both FT syn. not found 1983).
A. oxycedri auct. *non* (DC.) Marsch. Bieb. (1819).

Forming systematic infections along young stems of *Juniperus procera*. Stems more or less herbaceous, up to 13 cm long, branches opposite, not in groups, internodes ± 4-angled, up to 9 mm long; old stems dropping off as infection progresses along host stems which can become swollen and distorted. Male flowers *c* 2 mm diam. Female flowers *c* 1 mm long. Fruit 2.5 x 1.2 mm, lower part *c* 1.5 mm long, smooth, darker, upper part paler, strongly wrinkled when dried. Fig. 115.5.

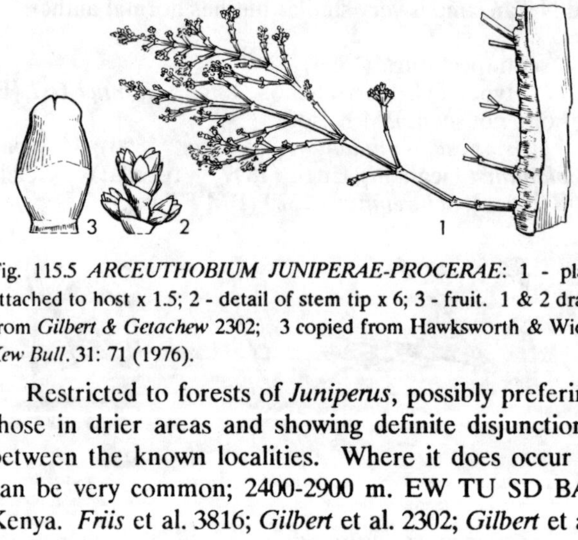

Fig. 115.5 *ARCEUTHOBIUM JUNIPERAE-PROCERAE*: 1 - plant attached to host x 1.5; 2 - detail of stem tip x 6; 3 - fruit. 1 & 2 drawn from *Gilbert & Getachew* 2302; 3 copied from Hawksworth & Wiens, *Kew Bull.* 31: 71 (1976).

Restricted to forests of *Juniperus*, possibly prefering those in drier areas and showing definite disjunctions between the known localities. Where it does occur it can be very common; 2400-2900 m. EW TU SD BA; Kenya. *Friis* et al. 3816; *Gilbert* et al. 2302; *Gilbert* et al. 3514.

Perhaps often overlooked because of the superficial resembleance to the host but usually easily picked out by the yellow-green colouring of plants on exposed branches. Sometimes included in *A. oxycedri* of the Mediterranean and Near East but see Hawksworth & Wiens, *loc. cit.* **31**: 71 (1976).

116. SANTALACEAE

by A.G. Miller*

Baker & Hill, *Fl. Trop. Afr.*: 411-434 (1911); Cufodontis, *Enum.*: 23-25 (1953); Agnew, *Upland Kenya Wild Flowers*: 334-338 (1974).

Herbs, trees or shrubs, often hemi-parasitic. Leaves entire, simple, exstipulate. Flowers, usually small, greenish or white, regular, epigynous, bisexual or plant dioecious, perianth 1-whorled. Stamens opposite to and attached to tube below perianth lobes. Ovary inferior, 1-celled, ovules 1-3, style simple, stigma capitate or lobed. Fruit indehiscent, dry or fleshy; seed 1, with no testa and much endosperm.

About 36 genera and some 400 species, widespread in temperate and tropical regions of the world.

Key to Genera

1. Shrubs or trees, sometimes scandent, more than 0.5 m tall; leaves elliptic to elliptic-oblong; fruit a drupe. 2
- Herbs or undershrubs to 0.5 m; leaves scale-like or ± linear; fruit a nut, rarely fleshy. **1. Thesium**
2. Flowers 3(-4)-merous, plants dioecious. **3. Osyris**
- Flowers (4-)5-merous, bisexual. **2. Osyridocarpos**

1. THESIUM *L.* (1753).

Annual or perennial often woody-based, hemi-parasitic herbs. Leaves alternate, ± linear or scale-like. Inflorescence a spike, raceme or panicle. Flowers bisexual, ± sessile, subtended by 1 bract and 2 bracteoles. Perianth uniseriate, 4-5 lobed. Ovary inferior, style simple, stigma capitate. Fruit a 1-seeded nut, sometimes fleshy, crowned by the persistent perianth.

A genus of some 100 species in the temperate and tropical (mainly montane) areas of the old world with only 2 species in the new world.

1. Plant erect; stems narrowly winged. **2. T. subaphyllum**
- Plant prostrate to erect; stems at most longitudinally ridged. 2
2. Leaves linear to linear-oblong, if scale-like above, at least some below more than 4 mm long. 3
- Leaves scale-like, less than 3 mm long. 6
3. Bracts and bracteoles scale-like, subulate, shorter than fruit. **3. T. hararensis**
- Bracts and bracteoles leaf-like, longer than fruit; if shorter then leaf-like. 4
4. Prostrate, ericoid[1]-like herb, nuts 1.25-1.75 mm long; flowers 4-merous; perianth lobes glabrous within. **1. T. kilimandscharicum**
- Prostrate or erect herb, not ericoid; nuts more than 2 mm long; flowers 5-merous; perianth lobes villous within. 5
5. Prostrate herb with creeping, rhizomatous stems; fruits bright red and fleshy at maturity. **5. T. radicans**
- Erect or ascending herb, stems from ± capitate rootstock; fruit green and dry at maturity. **4. T. schweinfurthii**
6. Persistent perianth *c* half as long as nut. **8. T.** sp. = *Thulin* 1430
- Persistent perianth equalling or longer than nut. 7
7. Stems creeping; persistent perianth ± equalling nut in length. **6. T. matteii**
- Stems erect; persistent perianth *c* 1.5 x as long as nut. **7. T. stuhlmannii**

1. **T. kilimandscharicum** *Engl.* (1891).

Prostrate, spreading, densely leafy perennial ericoid-like herb or under shrub. Stems much branched, 10-20 cm. Leaves linear, flattened, keeled, mucronate, 3-10 mm, overlapping. Inflorescence of 1-5 flowered axillary branches, hidden by leaves; bracts and bracteoles leaf like, longer than fruit. Flowers white; perianth hypercrateriform[2], 4-merous, *c* 0.7 mm long; lobes *c* 0.5 mm, ovate, obtuse, hooded, glabrous within, free almost to base. Nut ovoid, pale orange when dry, 1.25-1.75 x 1.25-1.5 mm, very indistinctly longitudinally-nerved; persistent perianth 1/4-1/3 x as long as nut. Fig. 116.1. 8-10.

Erica arborea scrub; 3100-4000 m. AR; Kenya, Tanzania. *Hedberg* 4228; *Thulin* 1654.

2. **T. subaphyllum** *Engl.* (1895).

Erect, almost leafless annual herb. Stems *c* 50 cm, ± flattened, narrowly winged, branched above. Leaves ± absent, occasionally 1 or 2 flattened, linear leaves below, above with few scale-like, subulate leaves, *c* 1 mm. Inflorescence a lax spike; bracts and bracteoles, scale-like, subulate much shorter than flower. Flowers greenish yellow; perianth, infundibuliform[3], 5-merous, 1.5 mm long; lobes triangular-oblong, obtuse, glabrous within above. Nut ± spherical, pale, yellowish green, *c* 2.5 x 2.5-2.5 mm, conspicuously reticulately-nerved, 10-ribbed, persistent perianth *c* 1/3 x as long as nut. Fig. 116.1.14.

* Royal Botanic Garden, Edinburgh, EH3 5LR, Scotland.

1. ericoid = small, wiry plant with closely packed leaves looking like 'asta', *Erica arborea*.

2. hypercrateriform = salver-shaped, i.e. with a narrow tube opening suddenly into a wide cup-shaped mouth, see fig. 116.1.9.

3. infundibuliform = funnel- or cone-shaped.

Dodonaea-Combretum-Terminalia woodland; *c* 1570 m. GG, SD; Kenya, Tanzania and Malawi. *Gilbert & Thulin* 358; *Riva* 1485 n.v.

3. **T. hararensis** *A.G. Miller* (1983)
- type: HA, S of Harar, *Burger* 2199 (K holo., HARRAR WAG iso.).

Woody-based, glabrous (rarely scabrous) herb. Stems several, 10-30 cm, ascending, much branched above. Leaves linear, ± terete or flattened, mucronate, 5-30 mm, becoming scale-like and smaller above. Inflorescence of many spikes, *c* 2-3 cm long; bracts and bracteoles, scale-like, subulate, shorter than flower and fruit. Flowers, brownish green outside, white inside; perianth broadly campanulate[1], 5-merous, (1.5-)1.75-2 mm long; lobes triangular, *c* 1 mm, hooded, densely villous within. Nut ± spherical, 2.5-3 x 2.5-2.75 mm, conspicuously 10-ribbed and reticulately-nerved; persistent perianth *c* 1/3 x as long as nut.

In open grassland on limestone slopes, probably semi-parasitic on grasses; 1370-2200 m. HA; neighbouring regions of Somalia. *Burger* 2065; *J.J.F.E. de Wilde* 5475, 6376.

4. **T. schweinfurthii** *Engl.* (1895).

Perennial. Stems many from ± vertical, woody rootstock, erect or ascending, 10-30 cm, branched. Leaves linear to linear-oblong, ± flattened, 10-30 x 1-1.5(-2.75) mm, mucronate. Inflorescence of leafy, ill-defined racemes; bracts and bracteoles, leaflike, exceeding fruit. Flowers green or white; perianth campanulate, 5-merous, (1-)1.5-2.5 mm long; lobes triangular, *c* 1 x 0.7 mm, hooded, obtuse, villous within. Nut ovoid, (2-)2.75-3 x (2.25-)2.5-2.75 mm, green and dry at maturity, 10-ribbed and conspicuously reticulately-nerved, persistent perianth *c* 1/4-1/3 x as long as nut. Fig. 116.1.11-13.

Grassland and cleared ground in *Juniperus* and *Combretum* woodland; 1300-2000m. WG SD IL KF; Sudan, Kenya and Malawi. *Gillett* 14401; *Gilbert & Thulin* 615.

A rather variable plant with wide distribution.

5. **T. radicans** *A. Rich.* (1851)
- types: TU, Dagabur church near Adua, *Schimper* I:168 (P syn., K isosyn.); Endchedcab, *Schimper* II:1130 (or 1330) (P syn.); Ouodgerate, *Petit* s.n. (P syn.).
T. wightianum Wall. var. *radicans* DC. (1857).

Prostrate herb with creeping rhizomatous stems. Stems up to 30 cm long, branched. Leaves flattened linear, 4-8(-18) x 0.5-0.1 mm, acute. Inflorescence ill-defined, flowers arranged on 1-3 flowered axillary branch or singly in the axils of leaves; bracts and bracteoles, leaflike, shorter than or rarely longer than fruit. Flowers white, perianth campanulate 5-merous, *c* 1.5 mm long; lobes triangular, obtuse, hooded, *c* 0.7 mm long, internally densely villous. Fruit ovoid, when unripe green, very indistinctly 5-ribbed and reticulate, becoming bright red and fleshy, 3-3.5 mm long x 2.75-3 mm wide, persistent perianth *c* 1/3 x as long as nut. Fig. 116.1.5-7.

Grazed pasture and upland grassland; 1985-3200 m. EW TU GD HA SU KF AR SD BA; Somali Republic, Kenya and N Yemen. *Ash* 367; *Thulin* 1520.

The only Ethiopian *Thesium* with red, fleshy fruits. *T.* sp. *B* in Agnew (1974) is almost certainly referable to this species.

6. **T. matteii** *Chiov.* (1912)
- type: SU, Mt. Oulontai (Alatù) between Lake Zuia and Lake Langano, *Negri* 984 bis (FT not seen).

Creeping plant; stems simple or with few branches 7-17 (-30) cm. Leaves scale-like, ovate, acute, 1-2(-3) x *c* 0.5 mm, ± adpressed to stem. Inflorescence ± dense of 3-7 flowered terminal spikes or singly in axils of leaves; bracts and bracteoles scale-like, equaling or shorter than flower. Flowers greenish yellow or white; perianth campanulate, 5-merous, 2.5(-3.5) mm; lobes narrowly oblong, 1.75(-2.3) x *c* 0.6 mm, hooded, glabrous within. Nut ovoid, 2(-2.5) x 1.5(-2) mm, 5-ribbed, obscurely reticulately-nerved, persistent perianth ± equalling fruit in length. Fig. 116.1.1-4.

Wet pasture in valley and upland grassland; 2000-3050 m. GJ, SU-AR; not found elsewhere. *I. M. Evans* 35; *Thulin & Hunde* 3975.

Unusual amongst the Ethiopian *Thesium* spp. in having the persistent perianth equalling the fruit.

7. **T. stuhlmannii** *Engl.* (1895).

Erect woody-based perennial herb. Stems several, 10-20(-30) cm, ridged, much-branched throughout. Leaves scale-like, narrowly ovate, 1-2 mm, acute, adpressed to stem. Inflorescence of dense, short, terminal and axillary spikes; bracts and bracteoles scale-like, subulate, shorter than flower and fruit. Flowers white; perianth campanulate, 5-merous, 2-2.5 mm long; lobes 1-1.5 mm, linear-ovate, acute, glabrous within. Nut dull green, ovoid, 1.5 x 1.25 mm, obscurely longitudinally veined; persistent perianth *c* 1.5 x as long as nut.

Themeda grassland on basalt hill, on wall of caldera on pumice; 2075-2450 m. SU AR; Kenya, Tanzania, Malawi and N Yemen. *Gilbert* et al. 8287; *Tewolde Berhan* 1312.

8. **T** sp. = *Thulin* 1430.

Herb, ± leafless; stems several, ± vertical, woody rootstock, ascending to 30 cm, branched. Leaves scale-like, 1-3 mm. Inflorescence a lax spike. Flowers unknown. Nut (immature) ovoid, *c* 2 mm, indistinctly-longitudinally veined, persistent perianth *c* 1/4-1/3 x as long as nut.

1. campanulate = bell-shaped with a ± broad funnel and somewhat spreading lobes, see fig. 116.1.12.

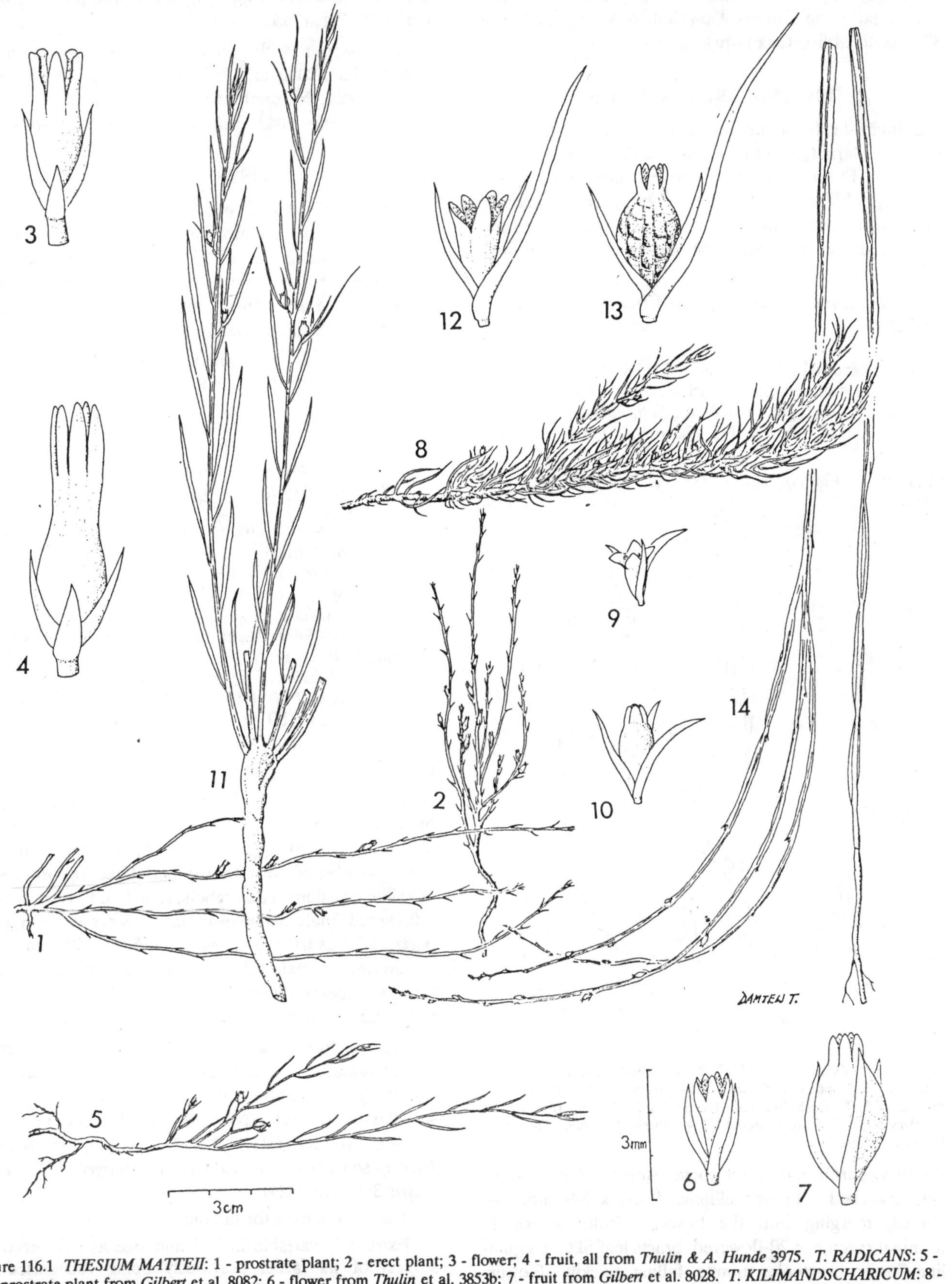

Figure 116.1 *THESIUM MATTEII*: 1 - prostrate plant; 2 - erect plant; 3 - flower; 4 - fruit, all from *Thulin & A. Hunde* 3975. *T. RADICANS*: 5 - prostrate plant from *Gilbert* et al. 8082; 6 - flower from *Thulin* et al. 3853b; 7 - fruit from *Gilbert* et al. 8028. *T. KILIMANDSCHARICUM*: 8 - prostrate plant; 9 - flower; 10 - fruit, all from *Mesfin T.* 709. *T. SCHWEINFURTHII*: 11 - part of plant with stout woody rootstock; 12 - flower; 13 - fruit, all from *Friis* et al. 3106. *T. SUBAPHYLLUM*: 14 - part of plant, stems bent down, from *Gilbert* et al. 358. All drawn to the same scale as shown: 3 cm bar for plants; 3 mm bar for details. Drawn by Damtew Teferra.

Grassy slope along small stream; 2400 m. AR. Only known gathering *Thulin* 1430 (Chilalo Awaraja, *c* 3 km S of Asella at livestock farm).

2. OSYRIDOCARPUS *A. DC* (1857).

Climbing shrub; stems scandent. Leaves elliptic, 3-nerved alternate, entire. Inflorescence a terminal raceme. Flowers bisexual; perianth uniseriate, (4-)5-merous, tubular; subtended by 1 bract and 2 minute bracteoles. Ovary inferior, style simple, stigma capitate. Fruit a drupe, crowned with the persistent perianth tube.

A monotypic genus distributed from Ethiopia to S Africa.

O. schimperanus *(A. Rich.) A. DC.* (1857);

Thesium schimperianum A. Rich. (1851) - types: TU, Mt. Scholoda near Adua, *Schimper* I:404 (P syn. not seen, K isosyn.); TU, *Quartin Dillon* (P syn. not seen).

Stauffer in *Viertelj. Natur. Ges. Zurich* **106**: 400-406 (1961)

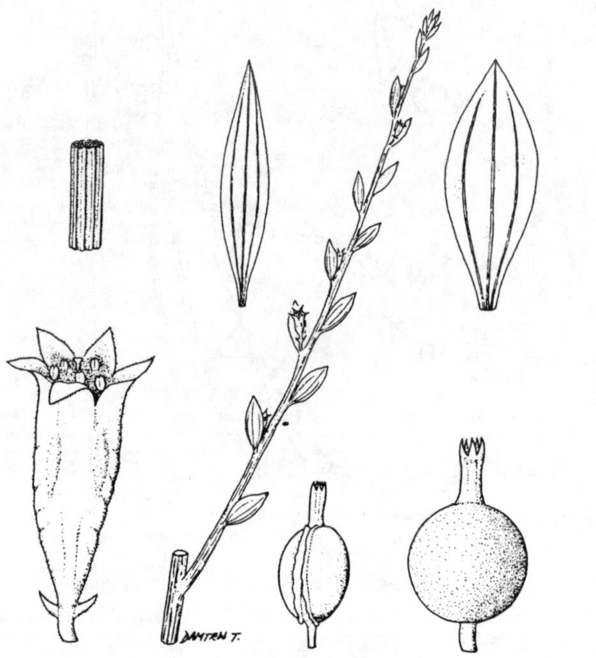

Figure 116.2 *OSYRIDOCARPUS SCHIMPERIANUS*: 1 - flowering branch x 2/3; 2 - portion of stem x 2; 3-4 - leaves x 3; 5 - flower x 6; 6 - dry fruit x 2; 7 - fresh fruit (construction) x 2. 1-2, 4-5 from *Amare Getahun* E-37; 3, 6 from *Mesfin T. & Sebsebe D.* 3709. Drawn by Damtew Tefera.

Climbing shrub, pubescent above; stems, 1-4 m, scandent, branched. Leaves elliptic, 15-50 x 3-9 mm, 3-nerved, merging into the bracts. Inflorescence a terminal raceme, *c* 20-flowered; bracts leaf like, ± equalling flower, bracteoles minute, falling quickly. Flower white or yellowish-green; perianth tubular, 4.5-6.5 mm long; lobes oblong-ovate, *c* 1.5-2.0 mm long. Fruit globose, *c* 6-7 mm long, topped by a *c* 4 mm long persistent perianth. Fig. 116.2.

Dry slopes with open scrub, evergreen scrub, hedges; 1200-2050 m. EW TU GJ SU KF HA SD; South Africa, Mozambique, Malawi, Tanzania, Kenya, Uganda. *Burger* 2043; *Pichi-Sermolli* 1635, *Gillett* 5056.

3. OSYRIS *L. (1753)*

Dioecious trees or shrubs. Leaves alternate, entire. Inflorescence of subumbellate lateral cymes. Flowers with uniseriate perianth, 3(-4)-merous, with a lobed disc; subtended by minute caducous bracts. Female flowers with ovary inferior, style simple, stigma 3-lobed. Male flowers without ovary and style. Fruit a drupe.

Two species distributed throughout Africa, the Mediterranean area and southern Asia.

O. quadripartita *Decn.* (1836).

O. abyssinica A. Rich. (1850) - types: TU, Mt. Scholoda near Adua, *Schimper* I:281 (P syn. not seen, K isosyn.); TU, Tigre, *Quartin-Dillon* s.n. (P syn. not seen).

O. lanceolata Hochst. et Steudel ex DC. (1857).

O. wightiana Wallich ex Wight (1852).

O. arborea Wallich ex A. DC. (1857).

O. rigidissima Engl. (1892).

O. tenuifolia Engl. (1895).

O parvifolia Baker (1910) - type: SU, Domak-Efat, *Roth* (K holo.).

O. abyssinica forma *latifolia* Fiori in *Nuov. Giorn. Bot. Ital.* **47**: 28 (1940) - type: SD, Mega, *Saccardo* s.n. (FT holo. not seen).

Stauffer in *Viertelj. Natur.Ges. Zurich* **106**: 388-399 (1961).

Dioecious tree or shrub, 1-7 m tall, much branched, branches sometimes pendant. Leaves elliptic to elliptic-oblong, (5-)10-70 x (5-)15-42 mm, tip acute, margin slightly thickened, very shortly petiolate. Inflorescence of axillary, pedunculate umbels; female inflorescences 1-3 flowered; male 5-13 flowered. Flowers yellow-green; perianth lobes triangular-ovate, *c* 1.5 x 1-2.25 mm; tube ± absent; female flowers infundibulariform; male flowers hypercrateriform. Fruit flask-shaped, bright red, 7-7.5 x 5-6 mm. Fig. 116.3.

Gallery forest, *Juniperus*, *Podocarpus*, *Combretum* and *Dodonea* woodland, *Erica* scrub, *Acacia nilotica*, *Commiphora* scrub, on rocky slopes, degraded woodland and scrub; 1600-2900 m. EW TU GD WU GJ WG SU KF GG AR HA BA SD; throughout Africa, southern Asia to China. *Burger* 772; *Tekle Hagos* 3; *Gillett* 14241.

The bark is used for tanning.

Extremely variable in leaf and size as evidenced by several of its facets previously being recognised as separate species.

Figure 116.3 *OSYRIS QUADRIPARTITA*: showing the variation in leaf size and shape, bottom branch with fruits; 1 - group of 3 female flowers, top view and section through a female flower; 2 - male inflorescence side view and top view of male flower showing stamens; 3 - top view of female flower, side view of flowers and floral diagram. Scale: cm scale for branches, closed bar: mm scale for floral details, open bar. (Reproduced with permission of Oklahoma Agricultural Experiment Station from Experiment Station Bulletin No. 45, *Families of Flowering Plants in Ethiopia*, figure 10.1-3, 1967.) Drawn by W.C. Burger.

117. BALANOPHORACEAE

by B. Hansen*

Andrews, *The Flowering Plants of the Sudan*, vol. II: 297 (1952);
Burger, *Families of Flowering Plants in Ethiopia*: 73 (1967).)

Herbaceous, fleshy root-parasites, without chlorophyll and roots, often brightly coloured red or blood red. After germination on a suitable host root, a branched, cylindrical tuber develops, from which arise at regular intervals many endogenous inflorescence-bearing stems. Leaves scale-like, overlapping and spirally arranged. Inflorescence like a spadix. Flowers unisexual, greatly reduced. Stamens united into synandria. Ovary poorly differentiated, ovule hard to see, style 1. Fruit a minute, 1-seeded achene.

Family with 18 genera and 43 species widely distributed in the tropics and subtropics of both hemispheres: 4 genera exclusively African, 1 only in Ethiopia.

THONNINGIA *Vahl* (1810)

Plants dioecious. Tuber rhizome-like, cylindrical, branched, swollen at point of contact with host root. Inflorescence-bearing stems breaking through buds on the tuber, leaving a short lobed volva basally. Stem with numerous, spirally arranged triangular-lanceolate, scale leaves. Inflorescences unisexual, unbranched, axis obconical, terminally convex, bearing numerous flowers. Male flowers pedicellate; perianth segments 2-6, linear, spirally arranged. Anthers 3-6, linear-elongate, united but with apices finally free, 2-locular and extrorsely dehiscing. Female flowers linear prismatic, densely crowded, perianth superior, tubular, obscurely 2-3-lobed. Style 1, exerted. Infructescence a swollen hemisphere, bearing minute 1-seeded achenes.

A genus confined to tropical Africa and Madagascar with 1 polymorphic or up to 5 species recognised.

T. sanguinea *Vahl* (1819).
T. ugandensis Hemsl. (1911)

Tuber trailing 10-15 cm below soil surface, 20 cm or more long, 0.6-0.8 (1-1.5 swollen part) cm diam., minutely pilose. Inflorescence-bearing stems appearing above soil surface with their upper parts bright red, 15-20 cm tall. Leaves triangular ovate, lower *c* 8 x 4 mm, upper 20 x 8 cm forming an involucrum around the spadix. Inflorescence 2.8-3 x 3.2-3.9 cm, male spadices depressed-hemispherical, female spadices hemispherical. Flowers apparently with bracts or males subtended by minute reduced bracts. Male flowers up to 1.5 cm long with perianth segments *c* 3 mm long and synandrium 5-6 mm long. Female flowers 3 mm long, ovary with perianth *c* 2mm, style 1-1.5 mm. Infructescence dark red, fleshy, ovoid, with spadix strongly swelling above involucrum. Fig. 117.1.

In shade in rainforest; 1000-1300 m. IL KF; widely distributed all over wet areas of tropical Africa. *Mooney* 9282; *Friis* et al. 4116.

Outside Ethiopia, *T. sanguinea* has been found parasitizing roots of a wide variety of trees, e.g. *Acacia mildbraedii, Baphia nitida* (Fabaceae), *Antiaris africana* (Moraceae), *Bridelia micrantha* (Euphorbiaceae), *Funtumia africana* (Apocynaceae).

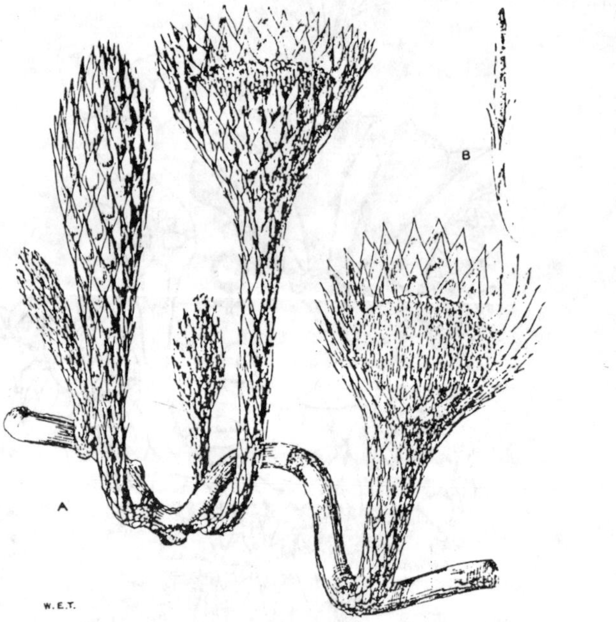

Figure 117.1 *THONNINGIA SANGUINEA*: A - plant with several flower heads; B - male flower. (Reproduced from Andrews, *Fl. Plts. Anglo-Egyp. Sudan*, fig. 107, 1952.)

* Botanical Museum, University of Copenhagen, Gothersgade 130, DK-1123 Copenhagen K, Denmark.

118. RHAMNACEAE

by K. Vollesen*

Cufodontis, *Enum*.: 497-503 (1958); Drummond in *Fl. Zamb*. **2**: 419-439 (1966); Johnston, in *Fl. Trop.E. Afr*., **Rhamnaceae**: 1-40 (1972).

Trees, shrubs (sometimes scandent) or lianas, often spiny. Leaves simple, alternate to opposite; stipules present. Flowers regular, bisexual, solitary or in fascicles, cymes or umbels, these sometimes merging into terminal panicles or thyrses. Sepals, petals and stamens attached at the edge of a flat to hemispherical receptacle. Sepals 4-5, deltoid, falling off quickly. Petals 4-5 or 0, small and clasping the stamens, hooded. Stamens 4-5, alternate with sepals. Disc well-developed, intrastaminal. Ovary superior to inferior (immersed in disc), (1-)2-3(-4)-locular; locules with 1 basal ovule; style entire or 2-4(-10)-lobed. Fruit a capsule, schizocarp or drupe with 1 seed per locule.

About 45 genera and 900 species: widely distributed in all tropical, subtropical and temperate regions. In Ethiopia 8 genera and 15 species.

Key to Genera

1. Coiled tendrils present; scandent shrubs or climbers. — 2
- Coiled tendrils absent. — 3

2. Leaves entire; flowers in umbels; fruit not winged. — **1. Helinus**
- Leaves serrate; flowers in elongated racemoid thyrses; mature fruit winged. — **2. Gouania**

3. Strong woody climber; fruit prolonged apically into a long strap-shaped wing. — **8. Ventilago**
- Erect trees or shrubs; fruit not prolonged into a wing. — 4

4. Leaves mostly opposite or subopposite. — 5
- Leaves all clearly alternate. — 7

5. Unarmed tree or shrub; fruit 15-20 mm long, ellipsoid with a single stone. — **6. Berchemia**
- Spiny shrubs; fruit 5-9 mm long, globose, with 2-3 stones. — 6

6. Spines short, recurved, at the nodes; flowers in axillary fascicles or cymes, pedicellate. — **4. Scutia**
- Spines straight (modified inflorescence axes); flowers in spikes or open panicles, rarely solitary, subsessile. — **3. Sageretia**

7. Leaves penninerved; fruit with 3 stones; plant unarmed or with spine-tipped branches. — **5. Rhamnus**
- Leaves 3-5-nerved from base; fruit with 1 stone; plant unarmed (rarely) or with stipular spines at nodes. — **7. Ziziphus**

1. HELINUS *E. Mey. ex Endl.* (1840), *nom. conserv.*

Woody climbers with coiled axillary tendrils. Leaves alternate, entire, penninerved; stipules linear, caducous. Flowers in (1-)3-15-flowered axillary umbels which sometimes merges into terminal panicles, only 1-2 fertile flowers per umbel. Receptacle broadly campanulate. Sepals 5. Petals 5, oblanceolate. Stamens 5. Disc flat, thickened around and adhering to the 3-locular inferior ovary; style 3-lobed. Fruit a pendulous schizocarp, splitting into 3 cocci[1] which eventually dehisce. Seed 3-angled or plano-convex, black.

Five species: three in Africa, one in Madagascar and one in India.

1. Tendrils and flowers glabrous; fruit smooth and glabrous. — **1. H. integrifolius**
- Tendrils and flowers pubescent to tomentose; fruit glandular-tuberculate and hairy. — **2. H. mystacinus**

1. **H. integrifolius** *(Lam.) O. Ktze*. (1891).
H. scandens (Eckl. & Zeyh.) A. Rich. (1847), *nom. illeg.*

Woody climber to 6 m (usually less in dry places); branchlets glabrous or with scattered hairs, especially near nodes. Leaves subglabrous to sparsely (rarely densely) covered in silky hairs; petiole 2-15 mm; lamina ovate to orbicular, (0.5-) 1-6 x 0.5-5 cm; apex subacute to retuse; base truncate to shallowly cordate. Peduncle 1-4.5 cm; pedicels 3-7 (-10 in fruit) mm; both glabrous. Sepals *c* 2 mm long, glabrous; petals white to yellowish-green, 1.5-2 mm long; filaments 1.5-2 mm; style *c* 2 mm; stigmas *c* 1 mm. Fruit ellipsoid or obovoid to globose, 5-7 mm long, glabrous and smooth.

Acacia and *Acacia-Commiphora* woodland, wooded grassland, and bushland, dry deciduous thickets; 500-1450 m. SU(Awash Valley) GG SD BA HA; from Ethiopia and Somalia through east Africa to S Africa and Namibia, also in Arabia. *Friis* et al. 3171, *Gilbert* 1301, *Glover & Gilliland* 450.

2. **H. mystacinus** *(Ait.) E. Mey. ex Steud*. (1840)
- type: *J. Bruce* s.n., cultivated at Kew from seeds sent by Bruce from Ethiopia (BM-Banks holo.).

H. mystacinus forma *pilosiusculus* Radlk. in *Abh. Naturwiss. Ver. Bremen* **8**: 387 (1883) - type: TU, near Adoa (Adwa), *Schimper* I:155 (M holo., K P iso.).

H. mystacinus forma *tomentosus* Radlk., l.c.: 387 (1883) - types: TU, Mt. Scholoda nr. Adwa, *Schimper* I:363 (M syn., BM FI (Webb) K P isosyn.);

* formerly Ethiopian Flora Project, now Royal Botanic Gardens, Kew, Richmond, Surrey, TW9 3AB, UK.

1. coccus (pl. cocci) = a round, one-seeded part of a fruit.

TU, near Axum, *Schimper* III:1548 (M syn., FI (Webb) K P UPS isosyn.).

Woody climber to 10 m or more; branchlets pubescent to subtomentose. Leaves sparsely pilose to subtomentose; petiole 2-23 mm; lamina ovate or elliptic to orbicular, 1.5-6.5 x 1-5.5 cm; apex acute to retuse; base rounded to cordate. Peduncle (0.5-)1-4 cm; pedicels 3-11(-17 in fruit) mm; both pubescent to tomentose. Sepals *c* 2 mm long, pubescent to tomentose; petals white to yellowish green, 1.5-2 mm long; filaments 1.5-2 mm; style *c* 2 mm; stigmas *c* 1 mm. Fruit ellipsoid to subglobose, 5-9 mm long, with numerous acute glandular tubercles and pilose. Fig. 118.1.

Forest margins and forest clearings, secondary forest and scrub, wetter types of woodland, wooded grassland and bushland, abandoned cultivations; (750-)1200-2200(-2600) m. EE EW TU GD WU SU SU-AR GJ WG IL KF GG SD BA HA; from Ethiopia, Uganda and Somalia through east Africa to Natal. *Burger* 848, *Mooney* 5865, 8450.

2. GOUANIA *Jacq.* (1763)

Scandent shrubs or woody climbers with coiled tendrils. Leaves alternate, entire or dentate, penninerved; stipules linear, caducous. Flowers in elongated axillary racemoid thyrses (sometimes merging into terminal panicles). Receptacle broadly campanulate. Sepals 5. Petals 5, obovate. Stamens 5. Disc flat, thickened around and adhering to the inferior 3-locular ovary; style 3-fid or obscurely lobed. Fruit a schizocarp with 3 longitudinal wings, when mature splitting into 3 woody or coriaceous indehiscent cocci. Seed compressed or plano-convex.

About 25 species found in all tropical regions but mainly in America.

G. longispicata *Engl.* (1895).

Scandent shrub or woody climber to 8(-12) m; branchlets sparsely to densely rusty brown pubescent. Leaves persistently greyish (lamina) to rusty brown (veins) tomentose; petiole 0.5-2 cm, with fine-pointed reddish glands; lamina ovate, 4-11 x 2-7 cm, serrulate or serrate, lowermost teeth glandular; apex acuminate to acute; base truncate to shallowly cordate. Thyrses rusty brown pubescent, (5-)10-35 cm long including a 1-3 cm long peduncle, with 3-10-flowered subsessile cymules; pedicels 2-5 mm. Receptacle pubescent. Sepals 1.5-2 mm long, pubescent; petals white to cream, *c* 1.5 mm long; filaments *c* 1.5 mm; style 3-fid, *c* 0.5 mm. Fruit 6-11 mm long, wings 3-5 mm wide. Seed brown. Fig. 118.2

Lowland and upland forest and riverine forest (mostly at edges and in clearings), secondary forest and scrub, coffee plantations; 1000-2350 m. SU AR GJ WG IL KF SD BA HA; Nigeria and from S Sudan and Ethiopia through east Africa to Zimbabwe. *Friis* et al. 72, *Meyer* 7983, *Mooney* 6037.

3. SAGERETIA *Brongn.* (1827)

Lamellisepalum Engl. (1897)

Johnston in *J. Am. Arb.* **49**: 377-379 (1968).

Erect shrubs or small trees (rarely climbers), usually with straight spines representing modified inflorescence axes. Leaves opposite or subopposite, penninerved; stipules linear, persistent. Flowers in spikes or more or less elaborate panicles composed of spikes, or solitary. Receptacle campanulate. Sepals (4-)5. Petals (4-)5. Stamens (4-)5. Disc thin, annular. Ovary superior, 2-3-locular; style entire or lobed. Fruit a drupe with 2-3 free stones.

About 25 species; mostly Asian with one species reaching NE Africa and two species in America.

S. thea *(Osbeck) M. C. Johnston* (1968).

S. spiciflora (A. Rich.) Chiov. (1939) - type: Ethiopia, no locality, *Quartin-Dillon & Petit* s.n. (P holo.).

S. theezans (L.) Brongn. (1827) var. *schweinfurthii* Chiov. in *Ann. di Bot.* **10**: 445 (1912) - types: EW, Saganeiti, *Schweinfurth & Riva* 1207 (FT syn., K isosyn.); EE-EW, Mensa, *Fiori* 449 (FT syn.); EW, Hamasen, Filfil, *Fiori* 537 (FT syn.).

S. theezans var. *spiciflora* (A. Rich.) Chiov., *l.c.* **10**: 445 (1912).

Lamellisepalum hildebrandtii Engl. (1897) - types: EW, Habab, *Hildebrandt* 648 (B syn. destr.); HA, Ogaden, *Robecchi-Bricchetti* 244, 245, 246, 247 (all FT syn.).

S. theezans var. *hildebrandtii* (Engl.) Chiov., *l.c.* **10**: 445 (1912).

Shrub to 3 m; spines few to numerous; branchlets subglabrous to pubescent or stiffly hairy. Leaves glabrous to stiffly hairy on petiole and midrib; petiole 1-6 mm; lamina ovate to broadly elliptic, 5-38 x 3-32 mm, serrulate to crenulate or subentire; apex acute to retuse, often apiculate; base rounded to truncate. Flowers yellow, subsessile, in pubescent to pilose 2-10 cm long spikes or these sometimes merging into lateral or terminal panicles, rarely solitary in leaf-axils. Receptacle glabrous to sparsely covered with stiff hairs. Sepals 1-1.5 mm long, glabrous; petals 0.5-1 mm long; filaments 0.5-1 mm; ovary 2(-3)-locular; style *c* 1 mm; stigma subcapitate. Fruit subglobose, *c* 6 mm long, with 2(-3) flat stones; stones *c* 6 x 5 mm. Fig. 118.3.

Dry *Combretum-Terminalia* woodland and bushland, dry *Juniperus* forest, montane scrub, mostly on granitic slopes but also on limestone; (1000-)1300-2500 m. EE-EW EW TU WU AR HA; widely distributed from Malesia north to Japan and west through China, India and Iran to Turkey, Arabia, Egypt, Ethiopia and Somalia. *Burger* 980, *Gilbert* 1453, *Gilbert & Thulin* 112.

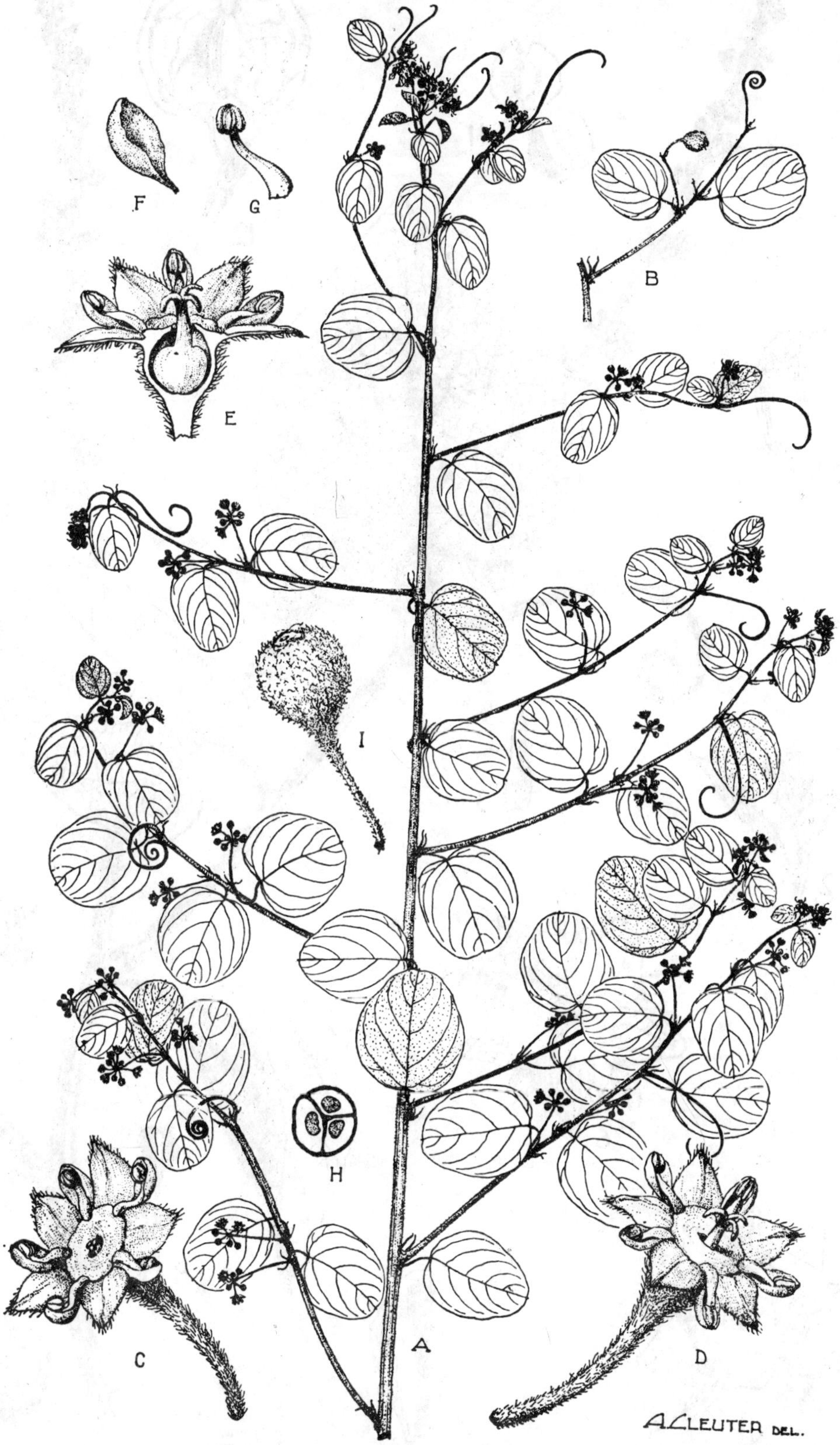

Figure 118.1 *HELINUS MYSTACINUS*: A - branch with flowers; B - branch with a fruit x 1/2; C - flower with short style; D - flower with long style; E - vertical section of flower with long style x 5; F - petal; G - stamen; H - transverse section through ovary x 10; I - fruit x 2. All from *de Witte* 1118 and *Lebrun* 8847. (Reproduced with permission from *Fl. Spermat. Parc. Nat. Albert*, 1: 546, 1948.)

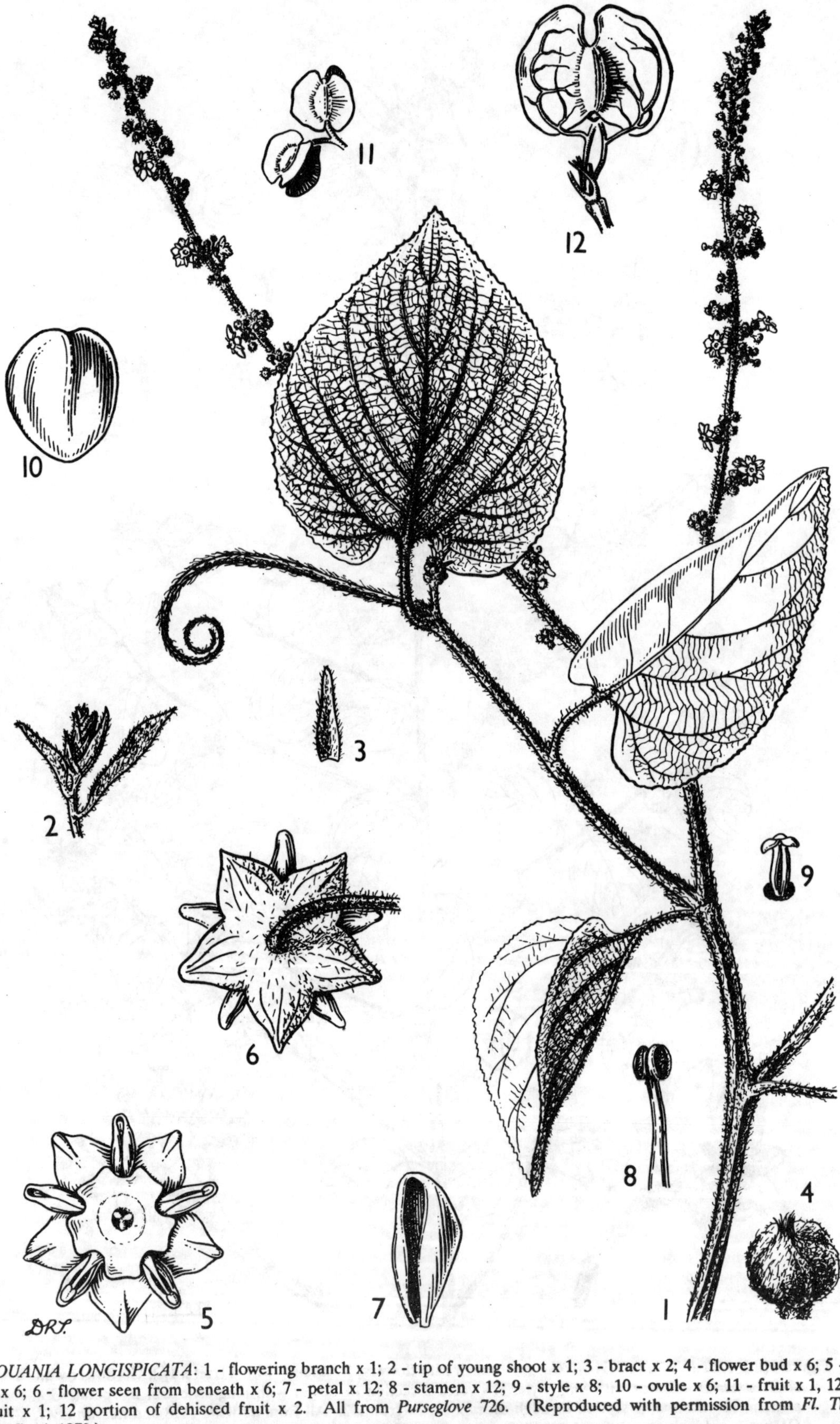

Figure 118.2 *GOUANIA LONGISPICATA*: 1 - flowering branch x 1; 2 - tip of young shoot x 1; 3 - bract x 2; 4 - flower bud x 6; 5 - flower seen from above x 6; 6 - flower seen from beneath x 6; 7 - petal x 12; 8 - stamen x 12; 9 - style x 8; 10 - ovule x 6; 11 - fruit x 1, 12 - portion of dehisced fruit x 1; 12 portion of dehisced fruit x 2. All from *Purseglove* 726. (Reproduced with permission from *Fl. Trop. E. Afr.*, Rhamnaceae: fig. 4, 1972.)

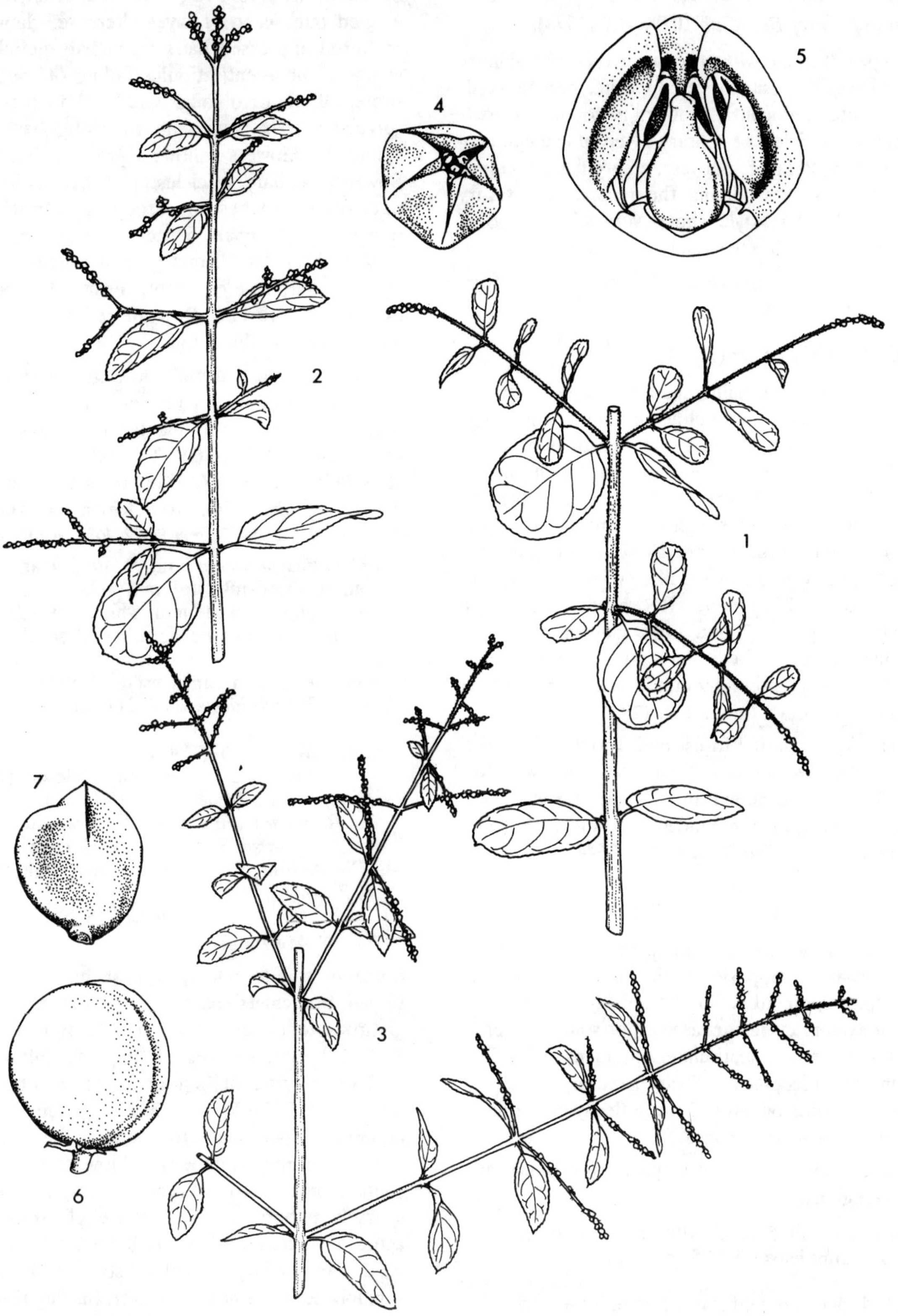

Figure 118.3 *SAGERETIA THEA*: 1, 2 & 3 - flowering branches showing variation in leaves and inflorescences x 1.5; 4 - flower x 15; 5 - flower in section x 30; 6 - fruit x 6; 7 - seed x 6. 1 from *Gilbert* 1453; 2, 4 & 5 from *Gillett* 5266; 3 from *Bally* 11675: 6 & 7 from from *Schweinfurth & Riva* 1207. Drawn by Eleanor Catherine.

4. **SCUTIA** *Commers. ex Brongn.* (1826), *nom. conserv.* Johnston in *J. Torrey Bot. Club* **101**: 64-72 (1974).

Shrubs or small trees, with straight or recurved axillary spines. Leaves opposite or subopposite, penninerved; stipules minute, caducous. Flowers in condensed axillary cymes or fascicles, or solitary. Receptacle campanulate. Sepals 5. Petals 5, obovate, emarginate. Stamens 5. Disc thin, annular, lining the receptacle. Ovary superior, 2-3-locular; style short, very slightly lobed. Fruit a drupe with 2-3 free stones, mesocarp fleshy.

4 species: 3 in S America and 1 from eastern Africa and S Africa to Vietnam.

S. myrtina *(Burm. f.) Kurz* (1875).

Shrub or tree to 5 m, usually scandent; spines recurved; branchlets glabrous to puberulous, usually at right angles to main branches. Leaves glabrous to puberulous on petiole and midrib; petiole 3-10 mm; lamina ovate, elliptic or obovate, 2-7(-9) x 1-4.5 cm, entire to indistinctly crenulate; apex acute to emarginate, apiculate; base cuneate to rounded. Flowers yellowish green, in 2-20-flowered fascicle-like cymes; peduncle to 7 mm; pedicels 1-2(-3) mm. Receptacle glabrous to puberulous. Sepals 1-2 mm long, glabrous; petals *c* 1 mm long, deeply 2-lobed; filaments *c* 1 mm; ovary 2-locular. Fruit subglobose, 7-9 mm in diameter, purplish black; stones flat, *c* 6 x 7 mm. Fig. 118.4

Forest edges, riverine forest and woodland, termite mounds in *Combretum* woodland; 1500-1900 m. SU AR SD BA; from Ethiopia and Uganda through east Africa to S Africa, also in Madagascar and from India to Vietnam. *Jackson* 706, *Thulin* et al. 3644.

5. **RHAMNUS** *L.* (1753).

Shrubs or trees, sometimes with spine-tipped branches. Leaves alternate or opposite, entire or dentate, penninerved; stipules pointed, falling off quickly or persistent. Flowers in axillary cymes or fascicles or solitary. Sepals 4-5. Petals 0 or 4-5, spathulate. Stamens 4-5. Disc thin, lining the receptacle. Ovary superior, (2-)3(-4)-locular; style simple or 3-fid. Fruit a fleshy or leathery drupe with (2-)3(-4) free stones.

About 150 species. In all tropical, subtropical and temperate regions.

1. Sepals 5; petals 5 or usually absent; pedicels 5-15(-20) mm; leaves 3-12.5 cm long. **1. R. prinoides**
- Sepals 4; petals 4; pedicels 1-5 mm; leaves 0.5-5.5 (-6.5) cm long. **2. R. staddo**

1. **R. prinoides** *L'Herit.* (1788).

R. pauciflorus Hochst. ex A. Rich. (1847) - types: *Lefebvre* s.n. (P syn.); TU, Chiré (Shire), *Quartin-Dillon & Petit* s.n., (P syn. not seen); GD, Adeselam, *Schimper* II:1276 (P syn., FI(Webb) K isosyn.).

Shrub or tree to 6 m, unarmed; branchlets sparsely crisped pubescent. Leaves alternate, glabrous but for scattered appressed hairs on midrib; petiole 3-17 mm, sparsely pubescent; stipules falling off quickly; lamina ovate, elliptic or oblong, 3-12.5 x 1.5-4.5 cm, glandular serrulate; apex acuminate to acute; base cuneate to rounded. Flowers yellowish green, solitary or in 2-5-flowered axillary fascicles; pedicels 5-15(-20 in fruit) mm, sparsely pubescent, drooping in fruit. Receptacle puberulous. Sepals 5, acute, *c* 2 mm long; petals usually absent or *c* 1 mm long; filaments *c* 1 mm; ovary 3(-4)-locular, style *c* 1 mm. Fruit subglobose, 5-8 mm in diameter, turning through red to blackish purple; stones obconic. Fig. 118.5.

Upland forest (usually on edges or in clearings and former cultivations), riverine forest, secondary forest and scrub, also widely planted in hedges and gardens; (1175-)1400-3200 m. EW TU GD WU SU AR GJ WG IL KF GG SD BA HA; Cameroun, S Sudan and Ethiopia through east Africa to S Africa and Angola, also in Arabia. *Ash* 2841, *Burger* 3062, *Friis* et al. 2144.

Rhamnus prinoides and *R. staddo* are used to add flavour to the drinks 'tej' and 'tela' - the fermentation starter comes from the malt which is usually germinated barley but can also be germinated sorghum, or finger millet. *R. staddo* is not recommended for 'tela' although it is sometimes used. Both wild and garden plants of *R. prinoides* are used equally for both drinks.

2. **R. staddo** *A. Rich.* (1847)

- type: TU, Asiela in Wajerat (Ouadjerate), *Quartin-Dillon & Petit* s.n., (P holo.).

R. infusionum Del. (1847), *nom. illeg.*

R. deflersii Schweinf. ex Herzog (1903) - type: EW, Acrour, *Schweinfurth & Riva* 1040 (M holo., FT K P iso.).

R. staddo var. *deflersii* (Schweinf. ex Herzog) Chiov. ex Engl. (1921).

Shrub or tree to 5(-6.5) m, branches sometimes spine-tipped; branchlets sparsely pubescent. Leaves alternate, glabrous to pubescent on petiole and midrib; petiole 1-11(-15) mm; stipules often persistent; lamina lanceolate, elliptic, oblong or narrowly obovate (rarely ovate), 0.5-5.5(-6.5) x 0.3-3 cm, glandular serrulate or crenulate; apex acute to rounded; base attenuate to cuneate (rarely rounded). Flowers yellowish green, solitary or in 2-4-flowered fascicles; pedicels 1-5 mm, sparsely puberulous. Receptacle glabrous. Sepals 4, acute, 1.5-2 mm long; petals 4, 0.5-1 mm long; filaments 0.5-1 mm; ovary 3locular; style *c* 1 mm. Fruit subglobose, 4-7 mm in diameter, turning through red to blackish purple; stones ellipsoid to obconic, grooved ventrally.

Upland forest (usually at edges), wooded grassland or scrub-grassland, cultivated areas; 1400-2900 m. EW TU GD SU AR KF GG SD BA HA; Somalia, Uganda, Kenya, Tanzania, Rwanda, Burundi, E Zaire, Zimbabwe. *Burger* 2592, *Friis* et al. 1586, *Mooney* 8448.

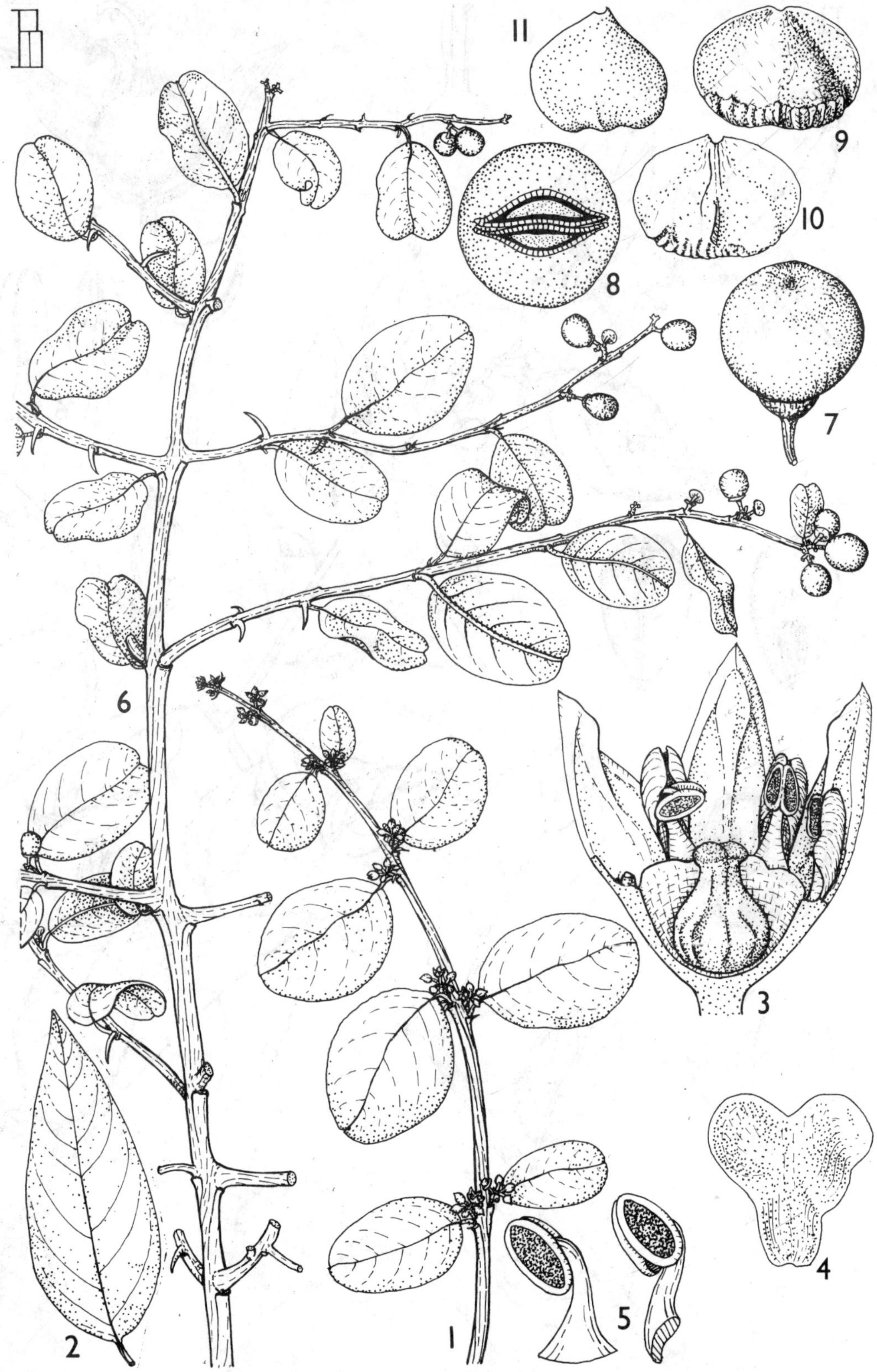

Figure 118.4 *SCUTIA MYRTINA*: 1 - flowering branch x 2/3; 2 - leaf x 2/3; 3 - vertical section of flower x 10; 4 - petal spread out x 20; 5 - stamens x 20; 6 - fruiting branch x 2/3; 7 - fruit x 2; 8 - transverse section of fruit x 3; 9 - stone viewed from outer side x 3; 10 - same viewed from inner side x 3; 11 - seed viewed from outer side x 3. 1, 6 from *Kokwaro* 241; 2-5 from *Glover* et al. 2043. (Reproduced with permission from *Fl. Trop. E. Afr.,* Rhamnaceae: fig. 7, 1972.)

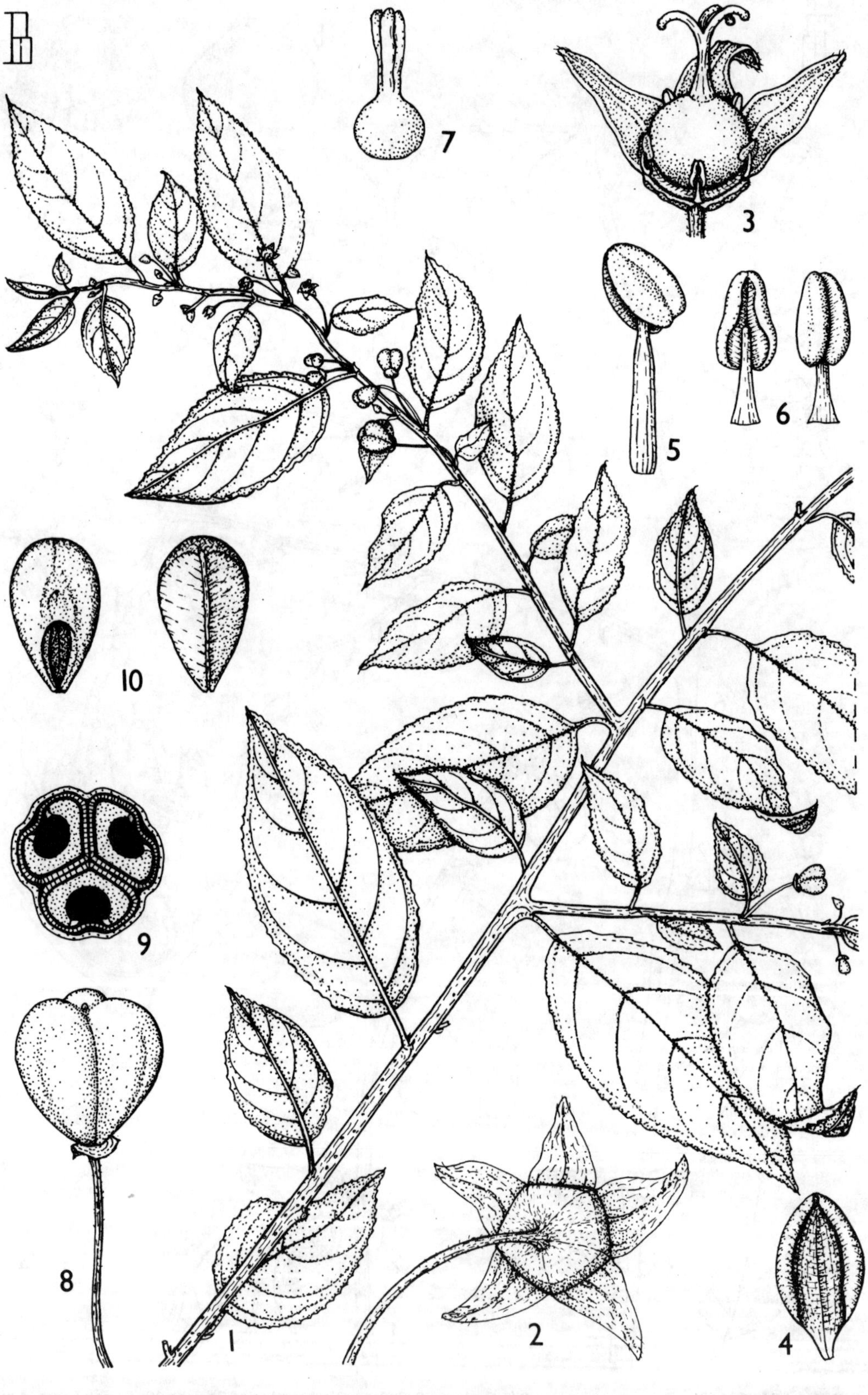

Figure 118.5 *RHAMNUS PRINOIDES*: 1 - branch x 2/3; 2 - flower viewed from beneath x 6; 3 - flower with 2 sepals removed x 10; 4 - petal x 20; 5 - long stamen x 20; 6 - short stamen, back and front views x 20; 7 - ovary x 20; 8 - fruit x 4; 9 - transverse section of fruit x 4; 10 - stone, two views x 4. 1-3 from *Willan* 234; 4 from *Stolz* 451; 5 & 7 from *Glover* et al. 1602; 6, 8-10 from *Trapnell* 60/458. (Reproduced with permission from *Fl. Trop. E. Afr.*, Rhamnaceae: fig. 6, 1972.)

Used in the same way as the preceeding species but less commonly and only for 'tej'.

Gillett 14334 from Mt. Mega, Sidamo differs from all other Ethiopian collections in having pubescent to pilose leaves and densely pubescent branches as well as a pubescent receptacle. Very hairy specimens have also been seen from Kenya and Tanzania, and may occur sporadically all over the species' area. They might eventually prove worthy of varietal rank.

6. **BERCHEMIA** *Neck. ex DC.* (1825), *nom. conserv.*

Trees or shrubs, unarmed. Leaves opposite or subopposite (rarely alternate), entire, penninerved; stipules triangular, intrapetiolar. Flowers (in African species) in axillary cymes or fascicles or solitary. Receptacle flat. Sepals 5. Petals 5, spathulate. Stamens 5. Disc partially enveloping ovary, but free from it. Ovary 2-locular; style 2-fid. Fruit a drupe with fleshy exocarp and woody endocarp, with a single 2-celled and 2-seeded stone.

22 species: 1 in N America, 2 in Africa and the rest in Asia.

B. discolor *(Klotzsch) Hemsl.* (1868);
Scutia discolor Klotzsch (1862);
Phyllogeiton discolor (Klotzsch) Herzog (1903).

Tree to 8 m; bark dark grey to brownish, reticulately fissured; branchlets glabrous to pubescent. Leaves opposite or subopposite (rarely some alternate), glabrous to sparsely puberulous on midrib; petiole 3-9 mm; stipules 2-4 mm long, basal parts fused and persistent; lamina ovate, elliptic or oblong, 2-7.5 x 1-4.5 cm; apex acute to rounded (rarely retuse or emarginate); base cuneate to truncate. Flowers yellowish green, solitary or in 2-5-flowered glabrous fascicles; pedicels 2-8(-10 in fruit) mm. Receptacle glabrous. Sepals acute, 2-3 mm long; petals *c* 2 mm long; filaments *c* 2 mm; style *c* 1 mm. Fruit narrowly ovoid to ellipsoid, 15-20 x 8-10 mm, yellow. Fig. 118.6.

Riverine forest, *Acacia-Commiphora-Balanites* woodland and wooded grassland, *Acacia* woodland and bushland on alluvial soils, termite mounds; (550-) 800-1900 m. EE EW WU SU GG BA HA; through East Africa to S Africa and Namibia, Madagascar, also in Arabia. *Mooney* 8255, *Tadesse Ebba* 739, *W. de Wilde* et al. 7333.

The ripe fruit is edible.

7. **ZIZIPHUS** *Mill.* (1754).

Shrubs or trees, usually with straight or hooked spiny stipules; branches often zigzag. Leaves alternate, crenate to serrate, usually 3-5-nerved from base. Flowers in axillary cymes or fascicles. Receptacle flat to obconical. Sepals 5. Petals 5, spathulate. Disc thickened around - and often completely enveloping - but free from ovary. Ovary 2(-4)-locular; style 2(-4)-fid. Fruit a globose drupe with fleshy exocarp and woody endocarp, with a single 2(-4)-celled and -seeded stone. Seed with a thin shiny testa.

About 85 species widely distributed in all tropical, subtropical and warm temperate regions, mostly in dry areas. A number are cultivated for their edible fruits.

1. All branches without spines. 2
- Some or all branches with spines. 3

2. Leaf-lamina pubescent, each marginal tooth with a tuft of fine hairs; flowers densely pubescent. **1. Z. pubescens**
- Leaf-lamina subglabrous, teeth not with tufts of hairs; flowers glabrous. **2. Z. mucronata**

3. Mature leaves glabrous or subglabrous. 4
- Mature leaves pubescent to tomentose. 5

4. Bark on branchlets reddish to brownish; spines usually few; flowers glabrous; leaves ovate, acute to acuminate. **2. Z. mucronata**
- Bark on branchlets whitish, smooth and shiny; spines numerous; flowers tomentellous; leaves elliptic (rarely ovate), rounded to acute. **3. Z. spina-christi**

5. Mature leaves 0.5-2 cm long, pubescent, symmetric at base; petiole 1-2(-5) mm long; branches arched and recurved. **6. Z. hamur**
- Mature leaves 2-12.5 cm long, tomentose (rarely pubescent); petiole 3-13 mm long; branches not arched. 6

6. Most or all leaves markedly asymmetric at base; pedicels 0.5-2 mm long in flower. **4. Z. abyssinica**
- Most or all leaves symmetric or nearly so; pedicels 2-5 mm long in flower. **5. Z. mauritiana**

1. **Z. pubescens** *Oliv.* (1887).

Tree to 15 m, unarmed; bark grey to dark brown, almost smooth to slightly reticulately fissured; branchlets densely yellowish pubescent to tomentose. Leaves pubescent; petiole 2-10 mm; lamina narrowly ovate to elliptic, 2-12.5 x 1.2-5 cm, serrulate, each tooth with a tuft of fine hairs; apex acute to subacuminate; base cuneate to rounded, slightly (rarely distinctly) asymmetric. Flowers yellowish green, in densely pubescent 2-6-flowered cymes; peduncle 1-3 mm; pedicels 1-3(-5 in fruit) mm. Receptacle and calyx densely pubescent. Sepals *c* 2 mm long; petals 1.5-2 mm long; filaments 1.5-2 mm; style *c* 1 mm. Fruit 6-8 m in diameter, yellow. Fig. 118.7.3.

Lowland dry semi-evergreen forest, ground-water forest, riverine forest; 400-650 m. IL GG; from Sudan and Ethiopia through East Africa to Zimbabwe and Angola. *Carr* 914, *Chaffey* 950.

2. **Z. mucronata** *Willd.* (1809).
Z. adelensis Del. (1846) - type: *Rochet d'Hericourt* 20 (P holo. not seen).
Z. mitis A. Rich. (1847) - type: SU (Choa), *Quartin-Dillon & Petit* s.n., (P holo., K iso.).

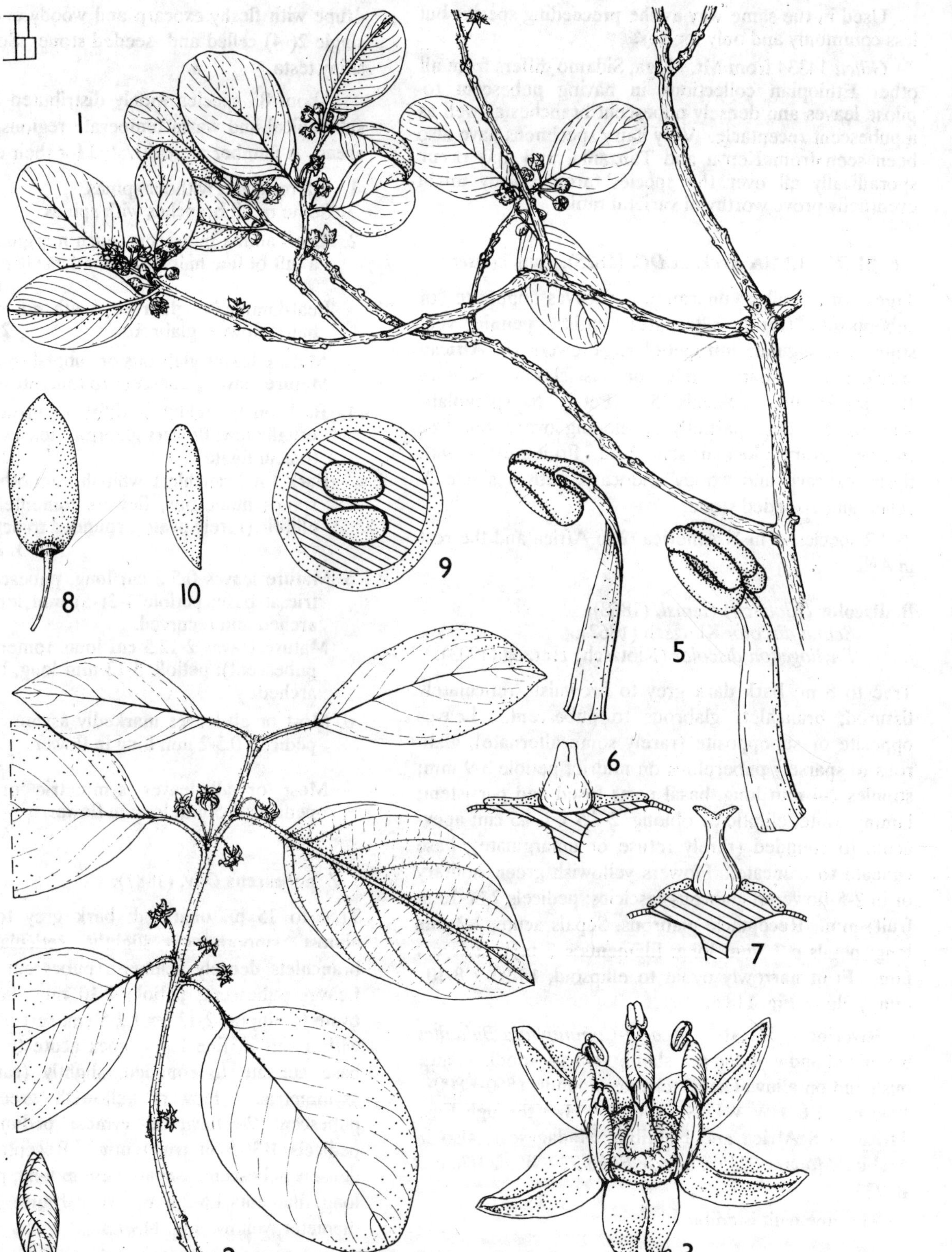

Figure 118.6 ***BERCHEMIA DISCOLOR*: 1 & 2 - flowering branches showing variation x 2/3; 3 - flower x 8; 4 - petal, front view x 10; 5 - stamens x 20; 6 & 7 - gynoecium of short- and long-styled flowers, respectively, with disc and receptacle cut away to show ovary x 10; 8 - fruit x 1; 9 - transverse section of fruit x 3; 10 seed x 2. 1, 3-5 from *Eggeling* 2961; 2 & 7 from *Wigg* 936; 6 from *Gillett* 12566; 8-10 from *Hemming* 325. (Reproduced with permission from *Fl. Trop. E. Afr.*, Rhamnaceae: fig. 10, 1972.)**

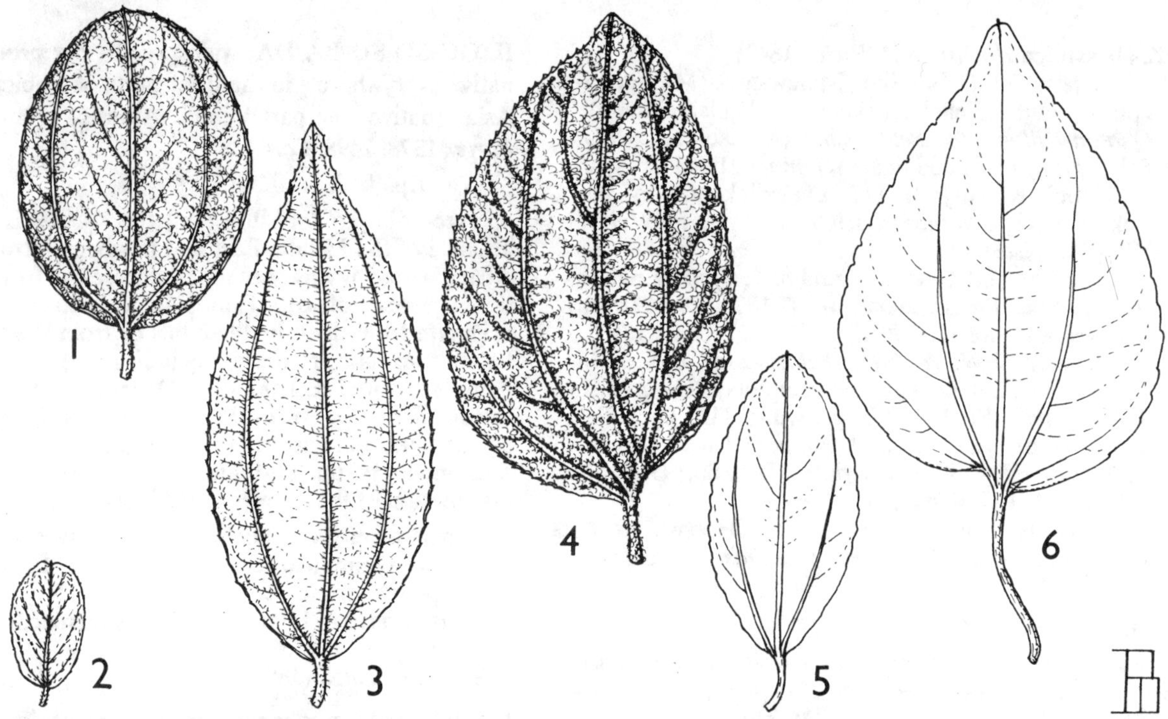

Figure 118.7 Leaves of species of *ZIZIPHUS*, all x 2/3: 1 - *Z. MAURITIANA* from *R.M. Graham* 245; 2 - *Z. HAMUR* from *Gillett* 13314; 3 - *Z. PUBESCENS* from *Dale* K786; 4 - *Z. ABYSSINICA* from *Hazel* 641; 5 - *Z. SPINA-CHRISTI* from *Forest Dept.* 448; 6 - *Z. MUCRONATA* subsp. *MUCRONATA* from *Leippert* 5247. (Reproduced with permission from *Fl. Trop. E. Afr.,* Rhamnaceae: fig. 8, 1972.)

Shrub or tree to 8(-15) m, unarmed or spiny; bark grey, reticulately fissured; branchlets reddish to brownish, subglabrous to pubescent. Leaves glabrous or sparsely puberulous on petiole and midrib; petiole 3-13 mm; lamina ovate, 2-9 x (1-)1.5-6 cm, crenulate to serrulate; apex acute to acuminate, actual tip rounded and apiculate; base cuneate to sub-cordate, almost symmetric to distinctly asymmetric. Flowers yellowish green, in puberulous few- to many-flowered cymes; peduncle 0-4 mm; pedicels 1-3(-5 in fruit) mm. Receptacle and calyx glabrous. Sepals 1.5-2 mm long; petals *c* 1.5 mm long; filaments *c* 1 mm; style *c* 1 mm. Fruit 1-1.5 cm in diameter, reddish. Fig. 118.7.6.

Acacia-Terminalia, *Acacia-Balanites* and *Anogeissus-Boswellia* woodland and bushland, *Acacia* woodland on alluvial soils, dry riverine forest and thicket; 100-1950(-2100) m. EE EW GD GD-GJ WU SU AR IL KF GG SD BA HA; widespread in the drier parts of tropical Africa and S Africa, Madagascar and in Arabia. *Ash* 348, *Burger* 3179, *Desta Hundessa* 49.

The ripe fruit is edible.

The Ethiopian plants belong to subsp. *mucronata*: subsp. *rhodesica* Drummond with denser brownish indumentum occurs from Zambia southwards.

3. **Z. spina-christi** *(L.) Desf.* (1798);

Rhamnus spina-christi L. (1753).

Z. spina-christi var. *microphylla* Hochst. ex A. Rich. in *Tent. Fl. Abyss.* 1: 137 (1847) - type: TU, Djeladjeranne, *Schimper* III:1798 (P holo., BM FI (Webb) FT K iso.).

Z. spina-christi var. *mitissima* Chiov. in *Fl. Somala* 2: 139 (1932).

Shrub or tree to 10 m; bark grey to brown, reticulately fissured; branchlets whitish to pale yellowish, subglabrous to densely pubescent. Mature leaves glabrous to sparsely pubescent on midrib; petiole 3-13 mm; lamina ovate to elliptic, 1.5-7(-8.5) x 0.8-4(-5) cm, finely crenulate to serrulate; apex acute to rounded; base cuneate to truncate, symmetric or nearly so. Flowers yellowish green, in pubescent, usually many-flowered cymes; peduncle 0-5 mm; pedicels 2-5 (-7 in fruit) mm. Receptacle and calyx whitish tomentellous. Sepals 1.5-2 mm long; petals 1.5-2 mm long; filaments *c* 1.5 mm; style *c* 1 mm. Fruit 8-15 mm in diameter, yellow to red. Fig. 118.7.5.

Wooded grassland on limestone slopes, *Acacia*-bushland on alluvial soils, in and along dry riverbeds, edges of cultivations and gardens; 0-1900(-2400) m. EE AF EW TU GD WU SU GG BA HA; widespread in the Near and Middle East and in the dry parts of N Africa, introduced and naturalized in east Africa. *Bally* 7026, *Getachew Aweke & Gilbert* 942, *Tadesse Ebba* 626.

The ripe fruit is edible and when in season, it is found in large quantities in local markets.

It is clear that at least part of the Ethiopian distribution is due to plants which have grown from fruits eaten and the stones discarded by both local inhabitants and travellers. Some of the populations occurring in wadis and riverbeds are probably native, especially those from lower altitudes in EE and AF. Dry riverbeds are also often the habitat of the species in Asia.

4. **Z. abyssinica** *Hochst. ex A. Rich.* (1847)
- types: TU, Debre Sina, Maigwagwa (Mayegoua-goua) and Tchelatchekanne (Djeladjekanne); *Quartin-Dillon & Petit* s.n. (K isosyn.); TU, Selasagwila (Selassaquilla) *Schimper* II:763 (P syn., FI(Webb) K isosyn.); TU, Djeladjekanne by the Tekeze river, *Schimper* III:1694 (P syn., FI(Webb) FT K UPS isosyn.).

Z. jujuba (L.) Gaertn. forma *obliquifolia* Engl. in *Abh. Konigl. Preuss. Akad. Wiss*. 1891: 294 (1892) - types: EW, Keren, Mt. Deban, *O. Beccari* 85 (FT K isosyn.); EW, Habab, *Hildebrandt* 665 (BM isosyn.); TU, Selasagwila (Selassaquilla) *Schimper* II: 763, (P syn., FI(Webb) K isosyn.); TU, Djeladjekanne by the Tekeze river, *Schimper* III:1694 (P syn., FI(Webb) FT K UPS isosyn.); TU, Adua (Adwa), *Schimper* 171 (BM syn.).

Z. jujuba Lam. var. *abyssinica* (Hochst. ex A. Rich.) Fiori in *Boschi e piante Legnose dell'Eritrea:* 234 (1912).

Shrub or tree to 6 m; bark dark grey, reticulately fissured; branchlets densely rusty pubescent to tomentose. Mature leaves (fig. 118.7/4) greyish to rusty tomentose (rarely pubescent); petiole 3-13 mm; lamina narrowly to broadly ovate (rarely elliptic), 3.5-12.5 x 1.5-7 cm, crenulate to serrulate; apex shortly acuminate to acute (rarely rounded); base cuneate to subcordate, usually very asymmetric. Flowers yellowish green, in tomentose, usually many-flowered cymes; peduncle 1-7 mm; pedicels 0.5-2(-4 in fruit) mm. Receptacle and calyx rusty tomentose. Sepals 1.5-2 mm long; petals *c* 1.5 mm long; filaments *c* 1.5 mm; style *c* 1.5 mm. Fruit 1-2 cm in diameter, reddish. Fig. 118.7.5 & 118.8.

Combretum-Terminalia and *Acacia* woodland, wooded grassland and bushland; 450-2000m. EW TU GD SU GJ WG IL KF GG SD BA; widespread in the drier parts of tropical Africa from Senegal to Ethiopia and south to Angola and Zimbabwe. *Ash* 533, *Gilbert & Getachew Aweke* 2663, *Tesfaye Haile* 842.

The ripe fruit is edible, but the flavour is not as pleasant and it is not as widely used as that of *Z. spina-christi.*

5. **Z. mauritiana** *Lam.* (1789).

Shrub or tree to 10(-15) m; bark dark grey, reticulately fissured; branchlets whitish pubescent to woolly. Mature leaves whitish (rarely rusty) tomentose; petiole 3-12 mm; lamina ovate to elliptic (rarely suborbicular), 2-7 x 1-4 cm, minutely serrulate; apex subacute to rounded; base cuneate to subcordate, symmetric or nearly so. Flowers white to yellowish green, in densely whitish pubescent to tomentose, usually many-flowered cymes; peduncle 0-5 mm; pedicels 2-5 (-10 in fruit) mm. Receptacle and calyx whitish (rarely rusty) tomentose. Sepals 1.5-2 mm long; petals 1.5-2 mm long; filaments *c* 1.5 mm; style *c* 1 mm. Fruit 1-2 cm in diameter, red. Fig. 118.7.1.

Riverine thicket and riverbanks, *Acacia* woodland on alluvial soils, *Acacia* wooded grassland; 400-1600 m. IL GG SD SD-BA HA; widespread - but probably not native everywhere - in the drier parts of tropical Africa, Asia (native in parts) and America (introduced). *Burger* 1378, 2498, *Carr* 748.

The ripe fruit is edible.

According to Johnston (l.c.) this species is probably a native of the Middle East and has been introduced to Africa where it now occurs as an escape from cultivation. None of the Ethiopian collections seem to originate from disturbed habitats or from the vicinity of habitations, nor are there any indications of the species being cultivated in Ethiopia. Drummond (l.c.) accepts the species as native in the drier parts of N Africa and as introduced in eastern and southern Africa. This view falls much better into line with the occurrence and distribution of the species in Ethiopia.

Apart from the generally smaller leaves with obtuse apex and symmetric base there is little to distinguish this species from *Z. abyssinica*. The latter may eventually have to be treated as an infraspecific taxon.

6. **Z. hamur** *Engl.* (1898).

Shrub to 3(-6) m, strongly armed and with long straggling arched and recurved branches; branchlets dark brownish; branchlets and mature leaves pubescent. Petiole 1-2(-5) mm; lamina oblong, elliptic or obovate, 5-20 x 3-11 mm, subentire to obscurely crenulate, penninerved; apex rounded to emarginate; base cuneate to rounded, symmetric. Flowers white to yellowish green, in pubescent 2-5-flowered fascicles (rarely in a many-flowered *c* 1.5 cm long racemoid thyrse); pedicels 2-5 mm. Receptacle and calyx pubescent. Sepals 1.5-2 mm long; petals *c* 1.5 mm long; filaments *c* 1.5 mm; style *c* 0.5 mm. Fruit 7-10 mm in diameter, blackish. Fig. 118.7.2.

Acacia-Commiphora-Boswellia bushland on red to grey sandy soil overlying limestone, *Acacia* bushland on alluvial limestone and gypsum soils, riverine forest; 250-400 m. SD BA HA; Somalia, NE Kenya. *Gilbert* et al. 7532, *Simmons* 123, *J. de Wilde* 5979.

The ripe fruit is edible.

8. VENTILAGO *Gaertn.* (1788).

Tall woody climbers, tendrils absent. Leaves alternate, penninerved; stipules minute, subulate, falling quickly. Flowers in dense axillary cymes or these merging into large terminal leafless panicles. Sepals 5. Petals 5, clawed, hooded. Stamens 5. Disc thick, adnate to lower half of ovary and later to fruit. Ovary 2-locular (only 1 locule fertile), half-inferior, the apical half expanding in fruit to a long strap-shaped wing. Fruit a dry samara with a basal sphaerical 1-seeded part and a long terminal wing.

About 30-40 species, mainly in tropical Asia with a few in Africa.

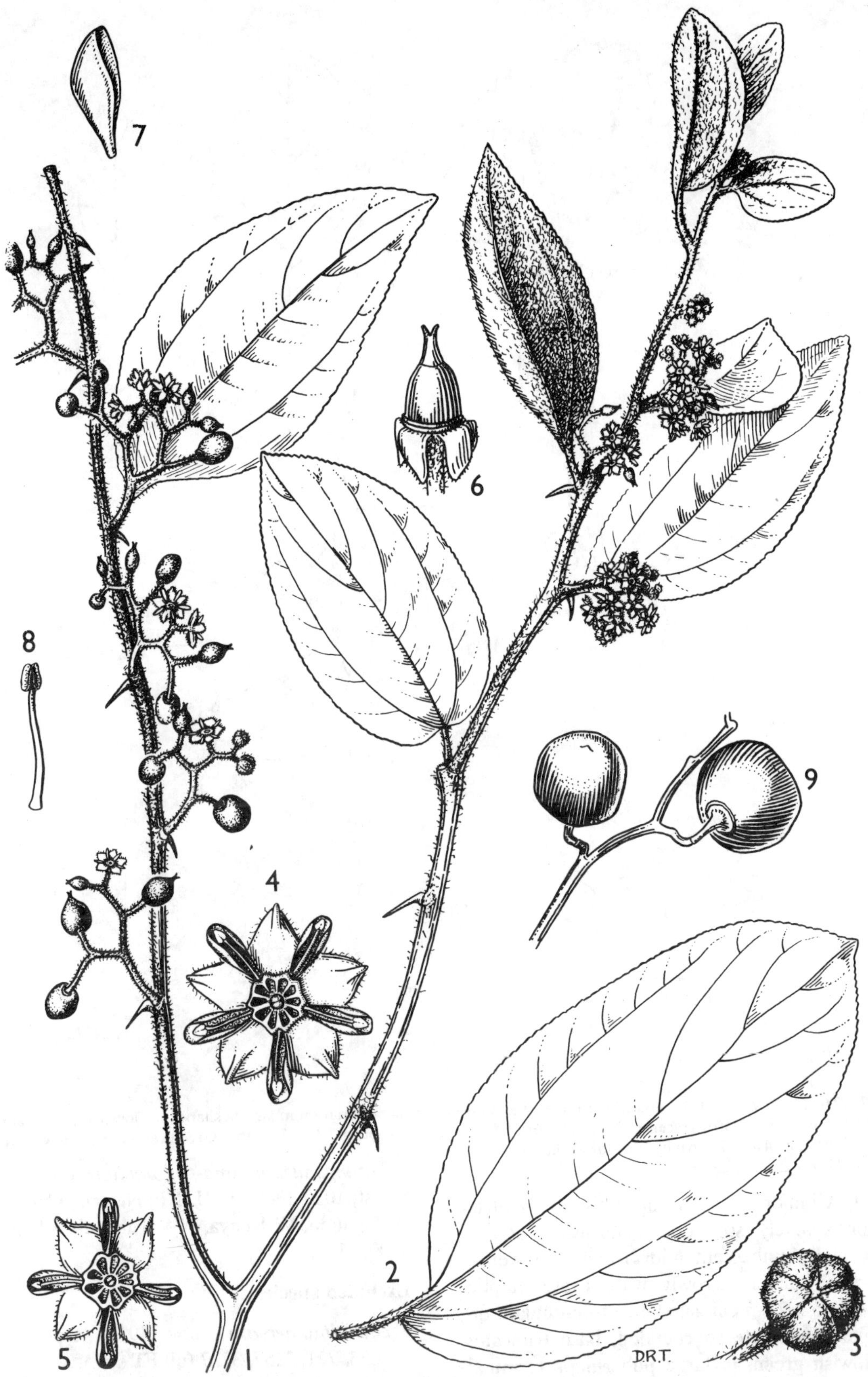

Figure 118.8 *ZIZIPHUS ABYSSINICA*: 1 - branch with flowers and young fruits x 1; 2 - leaf x 3; 3 - flower bud x 6; 4 - flower x 6; 5 - same, showing atypical tetramerous form x 4; 6 - mature flower with petals fallen x 6; 7 - petal, side view x 12; 8 - stamen x 12; 9 - portion of infructescence x 1. All from *Fyffe* 128. (Reproduced with permission from *Fl. Trop. E. Afr.*, Rhamnaceae: fig. 9, 1972.)

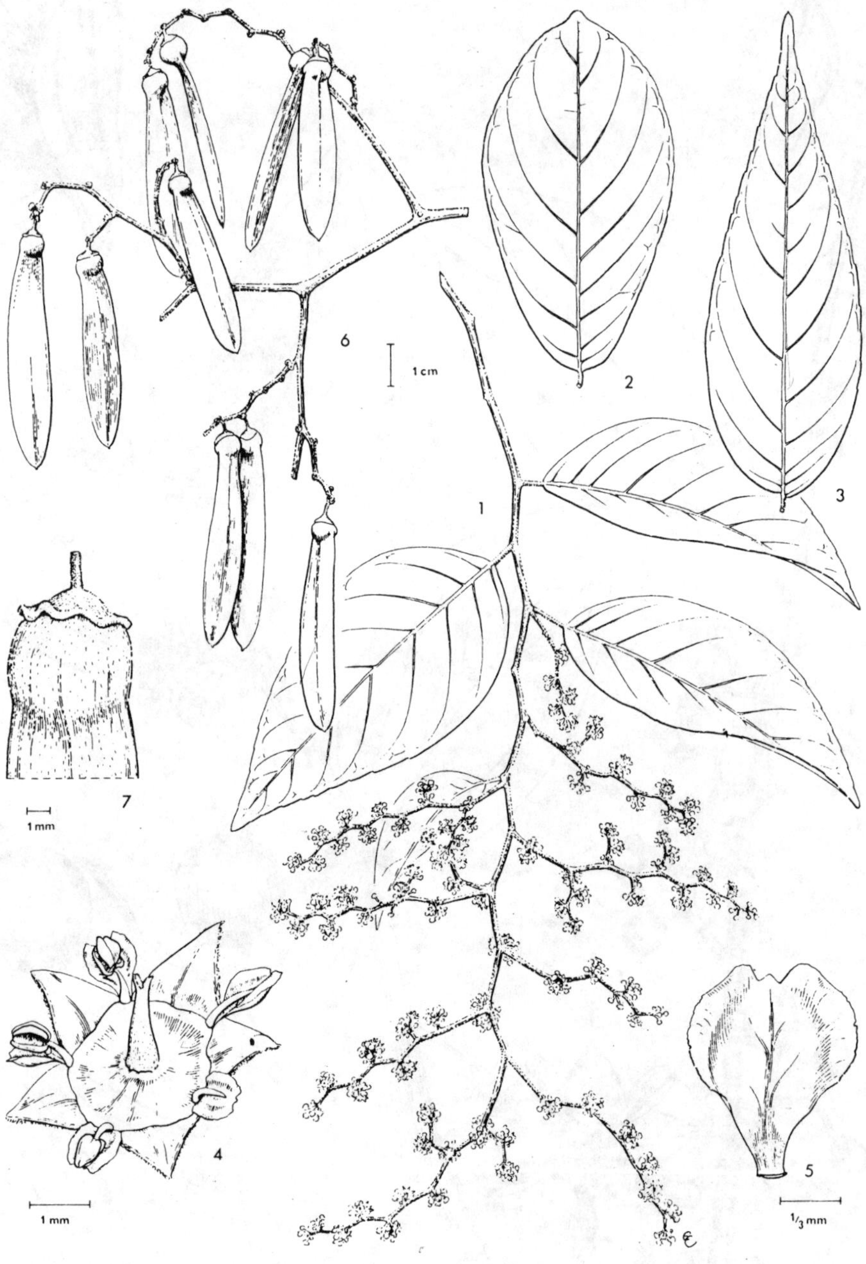

Figure 118.9 *VENTILAGO DIFFUSA*: 1 - flowering branch; 2 & 3 - leaves to show variation in lamina shape; 4 - flower with one anther fallen; 5 - petal spread out; 6 - part of infructescence; 7 - basal part of fruit. Scales as shown. 1, 4 & 6 from Greenway 11190; 2, 6 & 7 from Friis et al. 4118; 3 from Friis et al. 4100. Drawn by Eleanor Catherine.

V. diffusa *(G. Don) Exell* (1944).

Strong woody climber to 30 m or more; branchlets pubescent or sparsely so, soon glabrous. Leaves glabrous except for pubescent midrib and main veins; petiole 3-8 mm; lamina narrowly ovate, ovate, elliptic or obovate, 4-11.5 x 2.5-5 cm, subentire to crenulate or serrulate; apex acuminate to rounded; base truncate. Flowers yellowish green, in large puberulous terminal and lateral secund leafless panicles up to 40 cm long; pedicels 1-2(-4 in fruit) mm. Receptacle and calyx densely puberulous. Sepals deltoid, 1-1.5 mm long; petals *c* 1 mm long, reflexed; disc densely puberulous. Fruit 4-5.5 cm long, yellowish green, puberulous, the wing 8-10 mm wide.

Aningeria altissima-Chlorophora-Celtis lowland rain forest; 1050-1300 m. IL KF; Nigeria, Cameroun, Zaire, W Uganda, W Kenya, NW Tanzania. *Friis* et al. 4044, 4100, 4118.

Excluded species.

Tzellemtinia nervosa Chiov. (1911) - types: *Chiovenda* 633, 721, 3187, 3227 (all FT syn.).

This was described by Chiovenda as a monotypic endemic genus in the Rhamnaceae. It is actually a synonym of *Bridelia scleroneura* Muell. Arg. in the Euphorbiaceae. See Friis & Vollesen in *Bot. Notiser* **133**: 347-349 (1980).

119. VITACEAE

by K. Vollesen*

Gilg & Brandt in *Bot. Jahrb. Syst.* **46**: 415-557 (1911-12); Cufodontis, *Enum*.: 503-513 (1958); De Wit & Willems in *Fl. Congo* **9**: 453-567 (1960); Wild & Drummond in *Fl. Zamb*. **2**: 439-492 (1966); Descoings in *Fl. Cameroun* (**13. Vitacées**): 1-141 (1972).

Erect, trailing or climbing perennial herbs (often tuberous) or woody climbers; tendrils often present, arising opposite the leaves or from peduncles. Leaves alternate, simple to digitately or pedately lobed or compound; stipules present. Inflorescence terminal, arising opposite the leaves or from the axils. Flowers regular, usually bisexual. Calyx 4-5(-6)-lobed or subentire; petals 4-5 (-6). Stamens 4-5(-6), opposite the petals. Disc conspicuous, intrastaminal. Ovary superior, but often ± immersed in the disc, 2-locular, locules with 2 basal ovules; style simple; stigma capitate or lobed. Fruit fleshy, 1-4-seeded berry.

Family with 12 genera and about 900 species. Widely distributed in all tropical, subtropical and warm temperate regions. In Ethiopia 6 genera and 41 species.

Key to Genera

1. Petals 5(-6). 2
- Petals 4. 4
2. Petals united into a hood which drops as a whole at anthesis; disc of 5 separate glands; ovary not immersed in disc. **1. Vitis**
- Petals free, spreading at anthesis, dropping separately; disc annular with ovary ± immersed in it. 3
3. Leaves simple (but often ± lobed); disc and style with 10 longitudinal furrows. **2. Ampelocissus**
- Leaves 3-foliolate; disc and style without furrows. **3. Rhoicissus**
4. Flowerbud cylindric to flask-shaped, constricted near middle; disc of 4 separate glands. **6. Cyphostemma**
- Flowerbud subglobose, globose or conical, not constricted; disc annular. 5
5. Flowerbud subglobose to globose; leaves 3-5-foliolate or pedate; fruit 2-4-seeded. **4. Cayratia**
- Flowerbud conical; leaves simple (rarely 3-foliolate); fruit 1(-2)-seeded. **5. Cissus**

1. VITIS *L.* (1753).

V. vinifera *L.* (1753).

Woody climber or trailer. Leaves entire to ± deeply lobed, subglabrous to tomentose; petiole 3-8 cm; lamina broadly cordiform, 4-17 x 5-16 cm, sharply and irregularly dentate to serrate; apex acute to rounded. Flowers glabrous, in dense pyramidal subglabrous to tomentellous 5-20 cm long panicles, unisexual (or bisexual in cultivated plants); pedicels 3-5 mm. Calyx *c* 1 mm, subentire; petals 1-2 mm, greenish white; stamens 2-3 mm; disc with 5 conspicuous rounded lobes; style short and thick. Fruit ellipsoid to globose, 1-4 cm long, yellowish to purplish violet; seed pyriform, 5-8 mm long.

Cultivated for many centuries throughout most of the Ethiopian highlands by priests to make communion wine. Now also cultivated for table wine mostly in EW and SU; native of SE Europe and W Asia, cultivated over most of the world. *Quartin-Dillon & Petit* s.n., *Westphal* et al. 789.

2. AMPELOCISSUS *Planch.* (1884), *nom. conserv.*

Erect or scrambling perennial herbs or herbaceous to slightly woody climbers; tendrils from peduncles or opposite the leaves (rarely absent). Leaves unlobed to digitately 3-5-foliolate. Flowers 5-merous, bisexual, in dense to lax, usually many-flowered cymes or panicles opposite the leaves. Calyx subentire or lobed; petals spreading at anthesis, dropping separately; disc annular, with 10 longitudinal furrows; ovary immersed in disc; style very short, with capitate stigma. Seeds 2-4, flattened, ellipsoid, with a conspicuous dorsal crest and a ventral furrow, variously sculptured or not.

About 100 species: mostly in tropical Africa with some in Central America and Australia.

1. Peduncle without tendrils; plant usually without tendrils (if present only apically on young stems and indumentum white and woolly); seed *c* 9 x 8 mm, strongly ribbed. **2. A. schimperiana**
- Most peduncles with tendrils; plant always with tendrils, usually well-developed. 2
2. Calyx, inflorescence and leaves with reddish woolly indumentum; seed *c* 7 x 5 mm, strongly ribbed. **1. A. bombycina**
- Calyx glabrous; inflorescence and leaves puberulous to whitish pubescent, never with long woolly hairs; seed *c* 8 x 6 mm, weakly ribbed to almost smooth. **3. A. abyssinica**

1. A. bombycina *(Bak.) Planch.* (1885).

Erect, trailing or climbing herb to 1.5 m; young stems reddish tomentose to woolly, becoming subglabrous.

---

* formerly Ethiopian Flora Project, now Royal Botanic Gardens, Kew, Richmond, Surrey, UK.

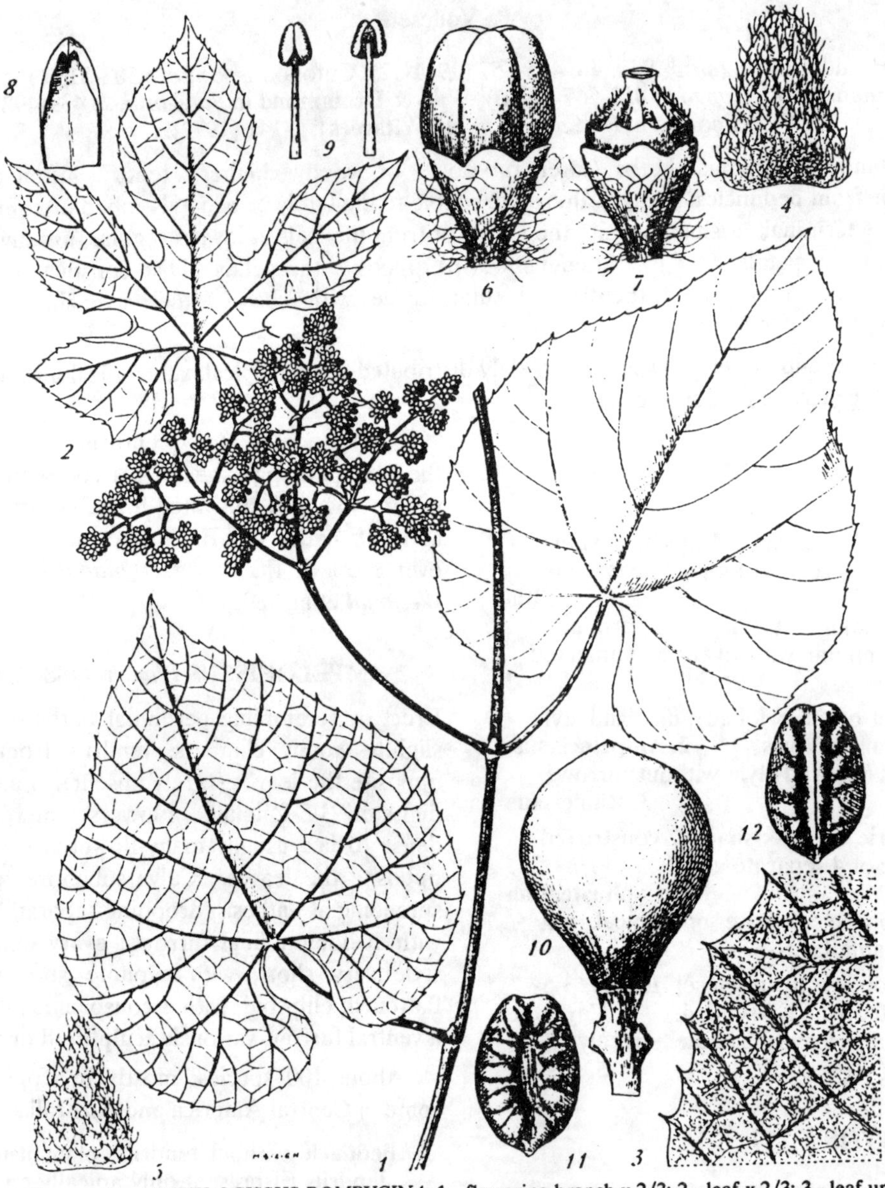

Figure 119.1 *AMPELOCISSUS BOMBYCINA*: 1 - flowering branch x 2/3; 2 - leaf x 2/3; 3 - leaf underside x 2; 4 - stipule x 8; 5 - bract x 8; 6 - flowerbud x 10; 7 - flowerbud with corolla and stamens removed x 10; 8 - petal from inside x 10; 9 - stamens x 20; 10 - fruit x 3; 11 & 12 - seed, ventral and dorsal views x 3. (Reprinted with permission from *Fl. Cameroun* 13: Pl. 1, Vitacées (1972).)

Mature leaves reddish woolly or sparsely so; petiole 1.5-7.5(-12.5) cm; lamina cordiform to reniform in outline, entire to ± deeply 3-5-lobed, 6-19(-25) x 6-19(-23) cm; lobes dentate or denticulate; apex subacuminate to rounded. Inflorescence a sparsely to densely reddish woolly cyme with opposite divaricate branches, with tendrils; peduncle 1-10 cm; cyme 1-6 cm long; pedicels 0.5-1.5 mm. Calyx woolly, *c* 1 mm long; petals dark red, *c* 1.5 mm long. Fruit ellipsoid, 10-15 x 8-10 mm, glabrous; seeds 2, ellipsoid, *c* 7 x 5 mm, strongly ribbed. Fig. 119.1.

Pterocarpus and *Combretum-Terminalia* woodland and wooded grassland, often on black cotton soil or on rocky outcrops; 750-1800(-2000) m. EE EW TU SU(Gibie Gorge) GJ WG IL KF GG; from Guinea to Sudan and Ethiopia. *Ash* 927; *Friis* et al. 2590; *Thulin* et al. 4033.

2. A. schimperiana *(Hochst.* ex *A. Rich.) Planch*. (1885);
Vitis schimperiana Hochst. ex A. Rich. (1847) - types: TU, Chiré (Shire), *Quartin-Dillon & Petit* s.n., (P syn.); TU, Djeladjekanne by the Tekeze river, *Schimper* III:1725 (P syn., BM K isosyn.).

Herbaceous climber or scrambler from tuberous woody rootstock; young stems whitish to reddish woolly, soon glabrescent. Mature leaves sparsely whitish to reddish woolly on petioles and nerves; petiole 3-15 cm; lamina cordiform or reniform in outline, unlobed (rarely shallowly 3-lobed), 6-23(-29) x 6-25 cm, finely to grossly dentate; apex acuminate to retuse. Inflorescence a sparsely to densely whitish to reddish woolly cyme with opposite divaricate branches, without tendrils; peduncle (0.5-)2-6 cm; cyme 1-7 cm long; pedicels 0.5-1 mm. Calyx sparsely to densely woolly, *c* 1 mm long; petals *c* 1.5 mm long. Fruit ellipsoid to obovoid, *c* 15 x 12 mm, glabrous; seeds 2, ellipsoid to subglobose, *c* 9 x 8 mm, strongly ribbed.

Combretum-Terminalia woodland, alluvial *Acacia*-woodland, riverine forest; 900-1900 m. EE EW TU GD WG; E Sudan. *Gilbert* 2971; *Mesfin Tadesse* 2104; *Pappi* 4221.

Where this species occurs together with *A. bombycina* some collections appear to be more or less intermediate and hybridization possibly occurs. Such more or less intermediate collections are *Pappi* 5759 and *Terraciano & Pappi* 2237. Both are characterized by small, rather poorly developed inflorescences.

3. **A. abyssinica** *(Hochst.* ex *A. Rich.) Planch.* (1885);
Vitis abyssinicus Hochst. ex A. Rich. (1847) - type: TU, Djeladjekanne by the Tekeze river, *Schimper* III:1715 (P holo., BM FI(Webb) K iso.).
A. cavicaulis (Bak.) Planch. (1885).
Cissus pauli-guilielmii Schweinf. (1868), pro parte, quoad lectotypus, Sudan, *Paul von Wurtemberg* s.n. (W lect.).
Vollesen in *Kew Bull.* **41**: 318 (1986).

Vigorous herbaceous climber to 7 m from tuber with fleshy fusiform roots; stems subglabrous to sparsely pubescent. Leaves glabrous with puberulous petiole and veins; petiole 4.5-15 cm; lamina cordiform or broadly so in outline, unlobed to deeply 3-5-lobed, 9.5-28.5(-32) x 10.5-26 cm; lobes subentire to grossly dentate; apex acuminate to acute (rarely rounded). Inflorescence a puberulous to pubescent cyme with alternate branches, with tendrils; peduncle (5-)7-13.5 cm; cyme (2.5-)4-12 cm long; pedicels 0.5-2 mm. Calyx glabrous, *c* 0.5 mm long; petals *c* 2 mm long. Fruit ellipsoid to globose, *c* 10 x 8 mm, glabrous; seeds 4, ellipsoid, *c* 8 x 6 mm, weakly ribbed to almost smooth.

Rainforest, riverine forest and woodland; 700-1600 m. EE EW TU WG KF; Nigeria, Cameroun, Gabon, Zaire, Sudan, Uganda, W Tanzania, Angola. *Chiovenda* 625; *Fiori* 243; *W. de Wilde* at al. 10817.

The syntypes of *A. cavicaulis* have very deeply lobed leaves while the type of *A. abyssinica* has unlobed leaves. However, the indumentum of leaves and inflorescence is identical as are the inflorescence-type and the seeds. The total material now available also shows all kinds of intermediates in the degree of division of the leaves.

Cissus pauli-guilielmii was based on two collections. The other (*Schweinfurth* 2301, GD/Sudan, Gallabat-Metema) is *Merremia kentrocaulos* (C.B.Cl.) Rendle in Convolvulaceae.

3. RHOICISSUS *Planch.* (1887).

Scandent shrubs or woody climbers; tendrils opposite the leaves (rarely from peduncle). Leaves 3(-5)-foliolate (rarely simple). Flowers 5(-6)-merous, bisexual, in ± condensed thyrsoid cymes opposite the leaves. Calyx subentire; petals spreading at anthesis, later erect, eventually dropping separately; disc annular, not furrowed; ovary immersed in disc; style with capitate stigma. Seeds 1-4, rugose or smooth, with a conspicuous dorsal furrow.

About 12 species in tropical Africa, the Comores and in Arabia.

1. Leaf margin dentate (rarely subentire); veins running straight into the margin; mature leaves pubescent to tomentose. **1. R. tridentata**
- Leaf margin entire or with a few large irregular teeth; veins anastomosing, not running into the margin; mature leaves glabrous to sparsely pilose. **2. R. revoilii**

1. **R. tridentata** *(L. f.) Wild & Drummond* (1963).
Vitis erythrodes Fresen. (1827); *R. erythrodes* (Fresen.) Planch. (1887) - type: GD/TU, Gondar to Adua, *Rueppel* s.n., (FR holo.).

Scandent shrub or woody climber to 10 m; all parts greyish to yellowish pubescent to tomentose. Leaves 3-foliolate (rarely simple); petiole 0.8-7(-13) cm; leaflets (1.5-)2-8.5(-11) x (0.7-)1-6.5(-8.5) cm, subentire to dentate, terminal broadly elliptic or obovate to orbicular, laterals obliquely ovate to elliptic; apex acute to rounded; base attenuate to rounded; main lateral veins running straight into margin. Peduncle 0.2-3(-4) cm; pedicels 1-3(-5 in fruit) mm. Calyx *c* 1 mm long; petals red to dark purplish, often tinged green, *c* 2 mm long, glabrous; filaments *c* 2 mm; style *c* 2 mm. Fruit ripening through red to black, globose, 7-11 mm in diameter, glabrous; seeds 1-4, *c* 5 mm in diameter, slightly rugose. Fig. 119.2.

Dry *Juniperus* forest, riverine forest, *Acacia* and *Combretum-Terminalia* woodland and wooded grassland, secondary forest and scrub, fences in cultivated areas; 1275-2400 m. EW TU GD WU SU GJ WG KF SD BA HA; widespread in tropical Africa and S Africa. *Friis* et al 829; *Gilbert* et al. 7948; *Tadesse Ebba* 520.

The abnormal simple leaves seen in a few collections arise by lateral fusion of the normal three leaflets.

2. **R. revoilii** *Planch.* (1887).

Scandent shrub or woody climber to 5(-10) m (rarely erect); young branches greyish to yellowish tomentose. Leaves 3-foliolate, subglabrous to sparsely pilose when mature; petiole 1-3.5 cm; leaflets 2.5-9(-14.5) x 1-6(-8.5) cm, entire or with a few large irregular teeth, terminal elliptic to obovate, laterals ovate and usually sickle-shaped; apex acute to rounded; base attenuate to cuneate; main lateral veins anastomosing, not reaching the margin. Inflorescence greyish to yellowish pubescent to tomentose; peduncle 0.5-3 cm; pedicels 1-4 mm. Calyx yellowish tomentose, *c* 1 mm long; petals chocolate-brown to purplish, pilose or glabrous, *c* 2 mm long; filaments *c* 1 mm; style *c* 1 mm. Fruit black, globose, 8-12 mm in diameter, glabrous; seeds 1-3, 5-7 mm in diameter, rugose.

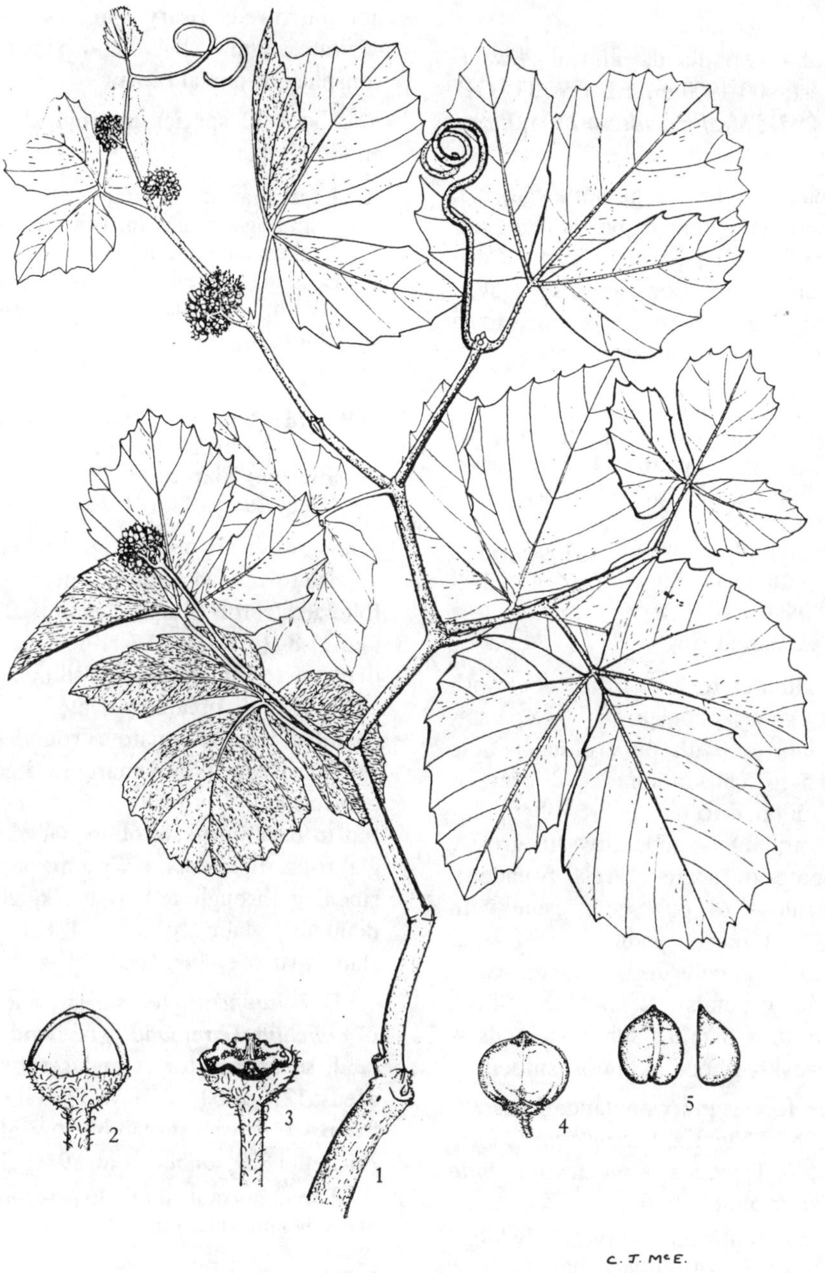

Figure 119.2 *RHOICISSUS TRIDENTATA*: 1 - flowering branch x 1/2; 2 - flowerbud x 8; 3 - flowerbud with corolla and stamens removed x 8; 4 - fruit x 4/3; 5 - seed x 2. 1 from *Holmes* 1059; 2 from *Robinson* 2777; 3 from *Wild* 3312; 4 & 5 from *West* 2143. (Reproduced with permission from *Flora Zambesiaca* 2: Tab. 94, 1966.)

Acacia and *Combretum-Terminalia* woodland and wooded grassland, riverine forest; (750-)1150-1850 m. WU SU WG IL KF GG SD BA HA; Ghana, E Zaire, S Sudan and through east Africa to Transvaal, also in the Comores and Arabia. *Burger* 2476; *Gilbert* et al. 7908; *Mooney* 8630.

The two species are quite distinct in Ethiopia although they are reported by Wild & Drummond (l.c.) to be difficult to separate in southern tropical Africa.

4. CAYRATIA *A.Juss.* (1818), *nom. conserv.*

Herbaceous climbers or scramblers. Tendrils opposite the leaves. Leaves 3-foliolate or pedately 5-foliolate. Flower 4-merous, bi- or uni-sexual (plant then dioecious), in widely branched loose corymbiform cymes opposite the leaves or axillary. Buds subglobose, not constricted below apex; calyx entire or indistinctly lobed; petals erect or reflexed at anthesis, dropping separately, yellowish to pale green; disc annular, with 4 rounded lobes, adnate to ovary; style long with discoid stigma. Seeds 2-4, obcordiform.

About 45 species in tropical Africa, Madagascar, Asia, Australia and Oceania.

1. Leaves pedately 5-foliolate (or some, rarely all, 3-foliolate), sparsely puberulous on upper surface;

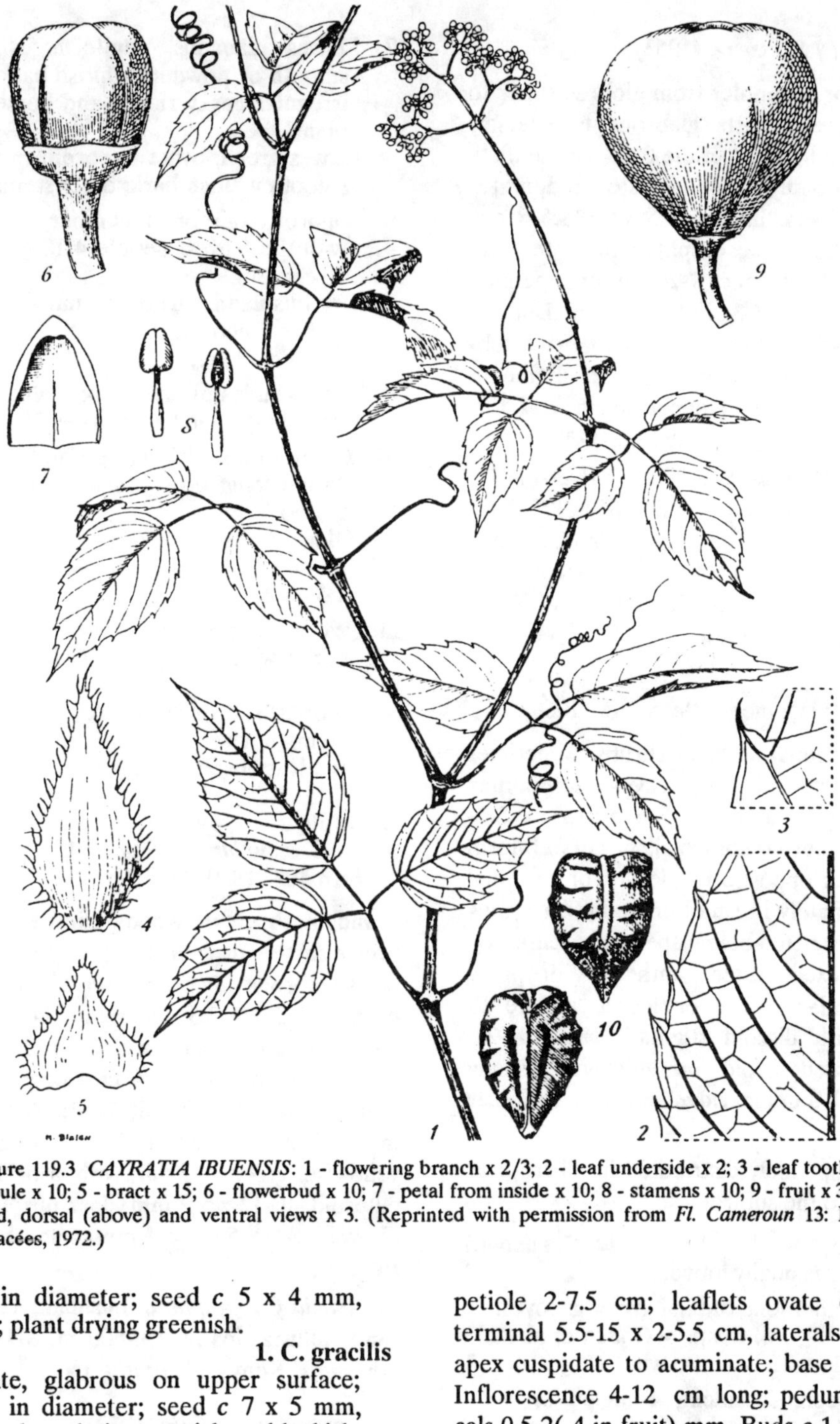

Figure 119.3 *CAYRATIA IBUENSIS*: 1 - flowering branch x 2/3; 2 - leaf underside x 2; 3 - leaf tooth; 4 - stipule x 10; 5 - bract x 15; 6 - flowerbud x 10; 7 - petal from inside x 10; 8 - stamens x 10; 9 - fruit x 3; 10 - seed, dorsal (above) and ventral views x 3. (Reprinted with permission from *Fl. Cameroun* 13: Pl. 7, Vitacées, 1972.)

fruit *c* 5 mm in diameter; seed *c* 5 x 4 mm, strongly ridged; plant drying greenish. **1. C. gracilis**

- Leaves 3-foliolate, glabrous on upper surface; fruit *c* 10 mm in diameter; seed *c* 7 x 5 mm, weakly ridged; plant drying greyish to blackish. **2. C. ibuensis**

1. C. gracilis *(Guill. & Perr.) Suesseng.* (1953);
Cissus gracilis Guill. & Perr. (1832).
Cissus subdiaphana Steud. ex A. Rich. (1847) - types: TU, Chiré (Shire), *Quartin-Dillon & Petit* s.n., (P syn., K isosyn.); TU, Gapadia, *Schimper* II:802 (BM K UPS isosyn.); TU, Djeladjeranne, *Schimper* III:1640 (BM FI(Webb) FT K isosyn.).

Slender herbaceous climber to 2 m from fleshy rootstock; all parts subglabrous to pubescent. Leaves pedately 5-foliolate (often some, rarely all, 3-foliolate); petiole 2-7.5 cm; leaflets ovate or elliptic, serrate, terminal 5.5-15 x 2-5.5 cm, laterals 2.5-8 x 1.5-4.5 cm; apex cuspidate to acuminate; base cuneate to cordate. Inflorescence 4-12 cm long; peduncle 2-9.5 cm; pedicels 0.5-2(-4 in fruit) mm. Buds *c* 1.5 mm long, globose. Fruit subglobose or flattened to depressed apically, 5-7 x 7-9 mm, glabrous; seeds (2-)4, obcordiform, *c* 5 x 4 mm, strongly ridged.

Rainforest (mostly at edges), riverine forest, woodland and bushland; 400-2000 m. EE EW TU SU GJ WG KF GG-SD SD BA; widespread in tropical Africa, also in Arabia. *Friis* et al. 3028; *Meyer* 8647; *Mooney* 9227.

Some collections from N Ethiopia deviate in having 3-foliolate leaves only, but they are otherwise completely typical.

2. C. **ibuensis** *(Hook. f.) Suesseng.* (1953).

Herbaceous climber or scrambler from globose tuber to *c* 5 cm in diameter; all parts glabrous to sparsely pubescent. Leaves 3-foliolate; petiole 0.5-4 cm; leaflets elliptic to ovate or broadly ovate, serrate or dentate, terminal 4-8.5 x 2-6.5 cm, laterals 2.5-6.5 x 1.5-6 cm, often with an abaxial lobe; apex cuspidate to acute; base cuneate to cordate. Inflorescence 7-20 cm long; peduncle 3-17 cm; pedicels 0.5-1.5(-3 in fruit) mm. Buds *c* 1.5 mm long, depressed globose, glabrous. Fruit subglobose or depressed apically, *c* 1 cm in diameter, glabrous; seeds 2-4, obovoid to obcordiform, *c* 7 x 5 mm, almost smooth to weakly ridged. Fig. 119.3.

Alluvial grassland and woodland, river-banks, riverine forest, lake shores, swamps; 300-650(-1250) m. IL GG GG-SD HA; Nigeria, Cameroun, Chad, Sudan, Zaire, Uganda, Burundi, Tanzania, Mozambique, Malawi, Zambia, Angola. *Carr* 755, 785; *Friis* et al. 2460.

The tubers are edible (Anuak).

5. **CISSUS** *L.* (1753), emend. *Descoings* (1960).

Erect or climbing perennial herbs or shrubs, sometimes succulent; tendrils opposite the leaves, sometimes absent. Leaves simple, entire or lobed (rarely 3-5-foliolate). Flower 4-merous, bisexual, in corymbiform or umbelliform cymes opposite the leaves or terminal (rarely axillary), ultimately in small umbelliform cymes. Buds conical, not constricted below apex; calyx entire or lobed; petals deflexed after anthesis, dropping separately; disc annular, entire or lobed, adnate to ovary; style long with discoid stigma. Fruit 1(-2)-seeded, glabrous; seed ovoid or ellipsoid, usually narrowed basally, with a dorsal ridge and sides variously smooth or sculptured.

About 350 species; widely distributed in all tropical regions, extending into the subtropics.

1. Leaves 3-foliolate. **12. C. ruspolii**
- Leaves simple, occasionally lobed. 2
2. Stems greyish, with 4 longitudinal green grooves. **8. C. sp.** = *Gilbert* et al. 8203
- Stems not with 4 longitudinal grooves. 3
3. Young stems green, succulent, 4-angled or 4-winged. 4
- Young stems not green or succulent, not angled or winged, but older stems sometimes with corky wings. 5
4. Stems 4-angled or narrowly 4-winged, up to 1.2 cm wide; edges red, not wavy. **9. C. quadrangularis**
- Stems broadly 4-winged, 1.5-6 cm wide; edges not red, wavy. **10. C. cactiformis**
5. Leaves thick and fleshy. **11. C. rotundifolia**
- Leaves not thick and fleshy. 6
6. Leaves and youngest stems glabrous. 7
- Stems and young (often also mature) leaves hairy. 9
7. Leaves glaucous, reniform; stems soon with smooth or powdery whitish bark; old stems with irregular corky ridges and bosses; dry bushland plant. **4. C. ellenbeckii**
- Leaves green, ovate to cordiform; stems not with smooth whitish bark; older stems not winged. 8
8. Vigorous rainforest climber, flowering together with the leaves; petiole 4-10 cm long; lamina 6-16 cm long. **7. C. arguta**
- Dry bushland shrub or climber, flowering when leafless; petiole 0.1-1(-1.5) cm long; lamina up to 4.5 cm long. **1. C. aphylla**
9. Flowerbuds uniformly puberulous. 10
- Flowerbuds glabrous (rarely pilose apically). 11
10. Older stems with white powdery bark; leaves 5-15 cm long; flowerbuds 2-3.5 mm long; fruit 2-2.5 cm long. **6. C. populnea**
- Older stems not with powdery bark; leaves 2-5(-6.5) cm long; flower buds 1-1.5 mm long; fruit 1-1.5 cm long. **2. C. aphyllantha**
11. Woody climber (rarely erect); older stems with 4 corky wings; leafbase cordate; tendrils present. **5. C. petiolata**
- Erect shrub or suffrutex (rarely climbing); older stems not winged; leafbase cuneate to rounded; tendrils absent (rarely present). **3. C. cornifolia**

1. C. **aphylla** *Chiov.* (1916).
C. aphyllantha Gilg (1912), p.p., quoad *Ruspoli & Riva* 1031 (930) (FT syn.).

Scandent shrub or woody climber to 3 m; branchlets brownish to purplish, becoming pale to dark grey; all parts glabrous. Petiole 1-10(-15) mm; lamina unlobed, ovate, 2-4.5 x 1-2.5 cm, remotely dentate; apex acute; base truncate to subcordate. Inflorescence appearing before the leaves, 1.5-4 cm long; peduncle 3-10(-15) mm; pseudo-umbels 2-8(-11)-flowered; pedicels 2-4(-6 in fruit) mm. Buds *c* 1 mm long; petals yellow or yellowish green. Fruit with a white waxy covering, ellipsoid, 10-12 x 5-7 mm, dark purple; seed ellipsoid to obovoid, 8-9 x 5-6 x *c* 4 mm, smooth apart from dorsal ridge.

Acacia-Commiphora bushland on yellowish brown sandy alluvial loam; 250-600 m. SD-BA HA; Somalia, NE and E Kenya. *Hemming* 1508.

2. C. **aphyllantha** *Gilg* (1912)

Scandent shrub or woody climber to 3 m, often forming large tangles over trees and shrubs; branchlets reddish brown, becoming dark or greyish brown; all parts puberulous to pubescent. Petiole 3-15(-25) mm; lamina unlobed, broadly cordiform to reniform, 2-5(-6.5) x 2-5.5(-8) cm, dentate with broad triangular teeth; apex subacute to rounded; base cordate. Inflorescence appearing before the leaves, 1-3(-4) cm long; peduncle 3-10 mm; pseudo-umbels many-flowered; pedicels 1-4 (-6 in fruit) mm. Buds 1-1.5 mm long; petals yellow to yellowish green. Fruit with white waxy covering,

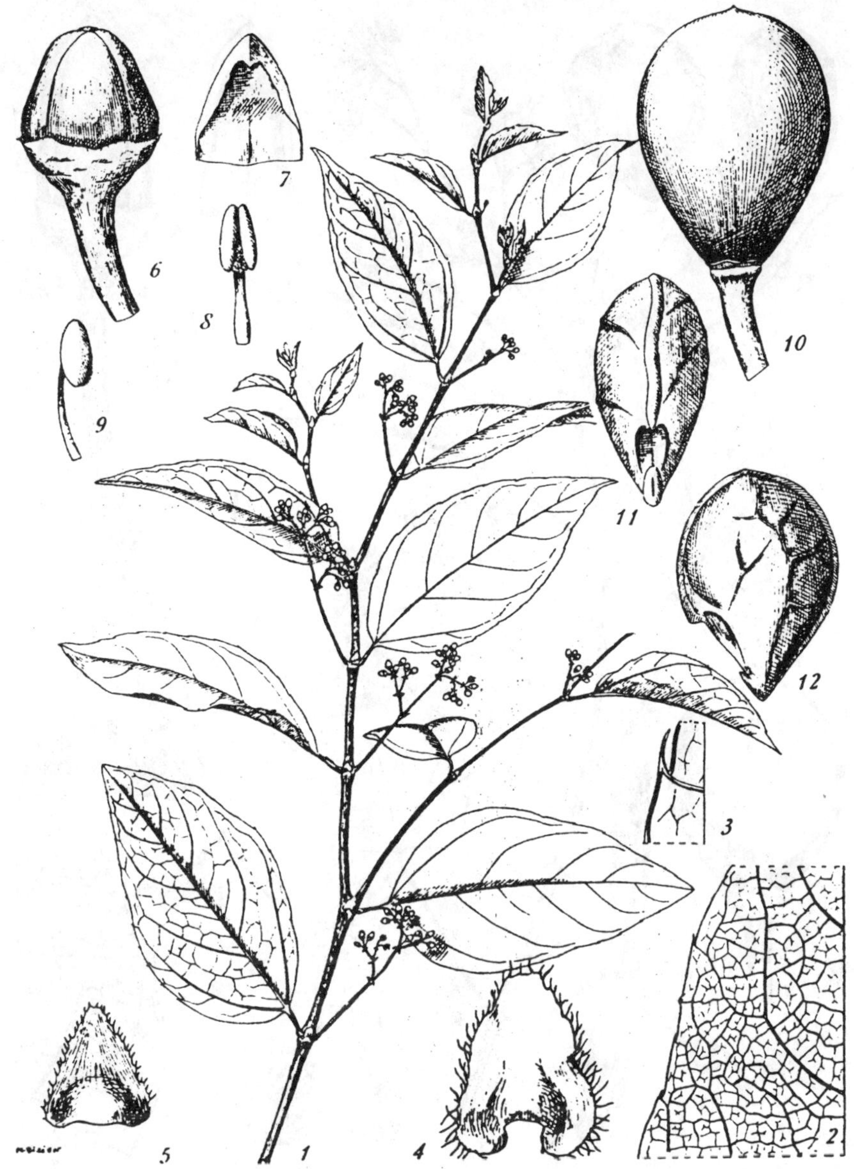

Figure 119.4 *CISSUS CORNIFOLIA*: 1 - flowering branch x 2/3; 2 - leaf underside x 2; 3 - leaf tooth; 4 - stipule x 8; 5 - bract x 8; 6 - flowerbud x 10; 7 - petal from inside x 10; 8 & 9 - stamens x 20; 10 - fruit x 3; 11 & 12 - seed, ventral and lateral views x 4. (Reprinted with permission from *Fl. Cameroun* 13: Pl. 47, Vitacées, 1972.)

ellipsoid to obovoid, 10-15 x 6-10 mm, red; seed ellipsoid, 10-12 x 5-7 x *c* 4 mm, with strongly ribbed sides.

Acacia-Commiphora bushland, on rocky hills and escarpments and on level sandy to loamy red soil; 500-1500 m. SD BA; NE and E Kenya, NE Tanzania. *Friis* et al. 2658; *Gilbert & Glyn Jones* 33; *J. de Wilde & Gilbert* 332.

One of the specimens cited by Gilg under this species (*Ruspoli & Riva* 1031(930)) is actually a specimen of *C. aphylla*.

3. **C. cornifolia** *(Bak.) Planch.* (1887);
Vitis cornifolia Bak. (1868).

Erect shrub or suffrutex to 1 m (rarely climbing), from large woody rootstock; tendrils absent (rarely present); branchlets whitish to reddish pubescent to tomentose. Mature leaves subglabrous to pubescent; petiole 1-12 mm; lamina unlobed, narrowly ovate to narrowly elliptic, 4-9.5 x 1.5-3.5 cm, denticulate; apex acute to rounded and apiculate; base cuneate to rounded. Inflorescence either appearing before the leaves or not, sparsely pubescent to tomentose, 2-10 cm long; peduncle 0.5-4.5(-6) cm; pseudo-umbels up to 10-flowered; pedicels 1-3(-5 in fruit) mm, glabrous. Buds 1-2 mm long, glabrous; petals yellowish green. Fruit ellipsoid to obovoid, 10-15 x 6-8 mm, red to purple; seed ellipsoid, 8-11 x 5-6 x 4-5 mm, with weakly sculptured sides. Fig. 119.4.

Pterocarpus-Anogeissus and *Pterocarpus-Pseudocedrela* woodland, *Combretum-Terminalia* wooded grassland, on grey sandy to stony soils; 550-1200 m. TU GD SU(Abbay Gorge) IL; widespread in the drier parts of tropical Africa. *Friis* et al. 2455, 2551.

The leaves are used against scabies.

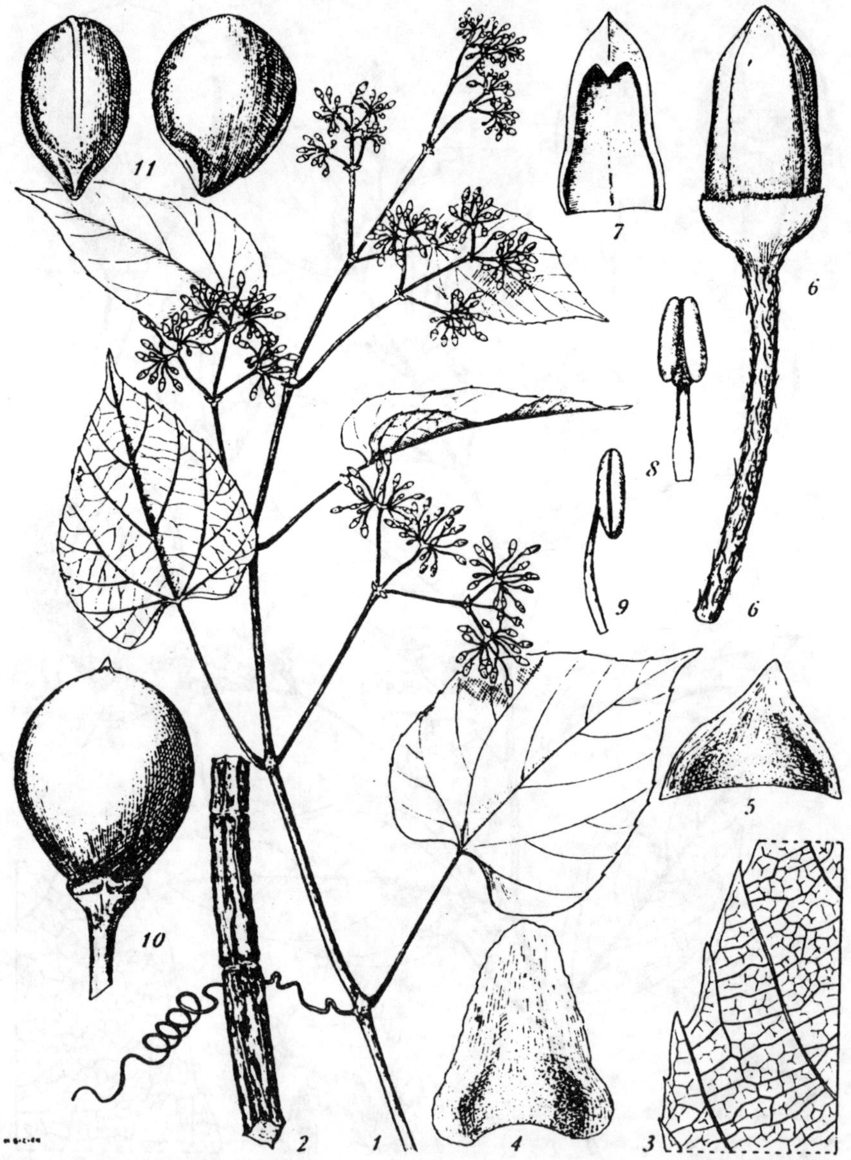

Figure 119.5 *CISSUS PETIOLATA*: 1 - flowering branch x 2/3; 2 - old stem x 1/3; 3 - leaf underside x 2; 4 - stipule x 10; 5 - bract x 10; 6 - flowerbud x 10; 7 - petal from the inside x 10; 8 & 9 - stamens x 20; 10 - fruit x 3; 11 - seed, ventral and lateral views x 4. (Reprinted with permission from *Fl. Cameroun* 13: Pl. 33, Vitacées, 1972.)

4. C. ellenbeckii *Gilg & Brandt* (1911)
- type: SD, Gorgoru in Boran, *Ellenbeck* 2108 (B holo. destr.).

Shrub to 3(-4) m, suberect or with trailing to climbing branches; branchlets glabrous, soon with conspicuous whitish smooth to powdery bark; old branches with irregular corky ridges and bosses. Leaves glaucous, glabrous; petiole 0.5-2.5 cm; lamina unlobed, reniform, 2-5.5 x 3-6.5 cm, remotely and indistinctly denticulate; apex broadly rounded; base truncate to cordate. Inflorescence either appearing before the leaves or not, 1-2 cm long, glabrous; peduncle 5-8 mm; pseudo-umbels 2-4-flowered; pedicels 2-5(-10 in fruit) mm, densely pubescent with thick brown glandular hairs. Buds 2-3 mm long, glabrous; petals white. Fruit ellipsoid, *c* 15 x 8 mm; seed ellipsoid, *c* 13 x 8 x 5 mm, smooth apart from dorsal ridge.

Open *Acacia-Commiphora* bushland on red sandy soil overlying limestone; 300-650 m. SD HA; Somalia. *Gilbert* et al. 7490, 7650; *Hemming* 373.

The type is no longer extant, but the studied specimens match the original description perfectly. The combination of whitish stems, glaucous reniform leaves, small few-flowered inflorescences and glandular pedicels is, to the author's knowledge, not seen in any other species.

5. C. petiolata *Hook. f.* (1849).

Vitis hochstetteri Miq. (1863-64); *C. hochstetteri* (Miq.) Planch. (1887) - type: TU, Memsach, Schahagenni, *Schimper* I:230 (BM Fl(Webb) K P UPS iso.).

Vitis dubia Beccari ex Martelli (1886), *nom. illegit.*, non Hook. f. (1875) - type: EW, Keren, *O. Beccari* 38 (FT holo., K iso.).

C. populnea sensu Cuf. (1958) quoad distrib. Aethiop., non Guill. & Perr. (1830).

Woody climber to 10 m or more or scrambling over rocks (rarely erect); branchlets glabrous to whitish or reddish pilose; old stems quadrangular or (usually) with 4 corky wings up to 1.5 cm wide. Mature leaves glabrous to reddish pilose on petiole and veins; petiole 3-12(-15) cm; lamina unlobed (rarely shallowly 3-lobed on saplings), ovate to orbicular or cordiform, 6-19 x 3.5-16.5 cm, denticulate to dentate; apex acuminate to acute; base cordate. Inflorescence appearing with the leaves, 4-8 cm long, reddish pilose; peduncle 1-3.5 cm; pseudo-umbels up to 25-flowered; pedicels 3-10(-17 in fruit) mm. Buds 2-3 mm long, glabrous (rarely pilose apically); petals brownish red. Fruit ellipsoid, 8-13 x 6-9 mm, black; seed ellipsoid, 7-10 x 5-7 x 4-5 mm, sides smooth with a central impression. Fig. 119.5.

Lowland rain forest, riverine forest, montane scrub, *Acacia-Commiphora* and *Combretum* woodland, rocky outcrops, *Acacia* woodland on alluvial soil; 600-2200(-2400) m. EE/EW EW TU GD SU AR GJ WG IL KF GG SD BA HA; widespread in tropical Africa. *Burger* 3785; *Friis* et al. 1913; *Mesfin Tadesse* 1743.

6. **C. populnea** *Guill. & Perr.* (1830).

Woody climber or scrambling over rocks; branchlets thick and fleshy, puberulous to reddish pubescent; old stems cylindric, with white powdery bark. Leaves soon glabrescent to sparsely pilose; petiole 2.5-7(-13) cm; lamina unlobed, broadly ovate to orbicular or reniform, 5-15(-19) x 5-14(-17) cm, denticulate to dentate; apex acute to rounded; base cordate. Inflorescence appearing with the leaves, 5-12(-16) cm long, subglabrous to pubescent, peduncle 1.5-7 cm; pseudoumbels up to 10-flowered; pedicels 2-5(-10 in fruit) mm. Buds 2-3.5 mm long, puberulous; petals yellowish green. Fruit ellipsoid, 2-2.5 x 1.2-1.5 cm, yellow to red; seed ellipsoid to obovoid, 1.7-1.9 x 1-1.5 x 0.7-0.9 cm, without dorsal ridge, sides strongly ribbed to almost smooth.

Pterocarpus woodland, usually on rocky outcrops; 650-850 m. GD IL; from Guinea to the Sudan, Uganda and W Tanzania. *Friis* et al. 2580; *Schweinfurth* 1266.

7. **C. arguta** *Hook. f.* (1849).

Rather fleshy herbaceous climber to 5 m or more; branches, leaves and inflorescence glabrous; older stems cylindric, smooth. Petiole 4-10 cm; lamina unlobed, ovate to cordiform, 6-16 x 4-12 cm, denticulate, cuspidate to acuminate; base cordate. Inflorescence appearing with the leaves, up to 20 cm long, with reduced leaves at lower nodes; pseudo-umbels up to 15-flowered; pedicels 5-10 mm, appressed puberulous. Buds *c* 2 mm long, obtuse, glabrous; petals brownish yellow outside, whitish inside, glands white. Fruit ellipsoid, *c* 8 x 6 mm, pale purplish; seed *c* 7 x 5 x 4 mm, one side with faint ridges, the other almost smooth.

Aningeria altissima-Chlorophora-Celtis lowland rain forest; 1050-1100 m. KF; Ghana, Dahomey, Nigeria, Cameroun. *Friis* et al. 3985, 4031.

This is often (Descoings, l.c.) considered to be synonymous with *C. producta* Afz., a species widespread from W Africa to W Tanzania. But *C. producta* has leaves with truncate to subcordate bases, larger apiculate buds and larger fruits.

8. **C. sp.** = *Gilbert* et al. 8203.

Succulent leafless climber to 1 m; stems grey, 4-angled with 4 longitudinal green grooves on the sides, up to 7 mm wide, glabrous. Flowers and fruits not seen.

Acacia-Commiphora bushland on rocky limestone ridge; *c* 750 m. SD; only known from this one collection.

A very peculiar species. It is strikingly different from anything else in *Cissus*. It is probably closest to the group of stem-succulents around *C. quadrangularis*, but in the absence of flowers and fruits this can only be a guess.

9. **C. quadrangularis** *L.* (1767).

Succulent climber to 4 m, frequently leafless; young stems 4-angled or narrowly 4-winged, up to 1.2 cm wide, green with red edges, not or only slightly constricted at nodes; old stems woody, terete, with greyish bark; all parts glabrous to sparsely pubescent. Petiole 3-20(-40) mm; lamina unlobed or 3-5-lobed, broadly ovate to cordiform, 1.5-9.5 x 1-11 cm, dentate; apex acute to rounded; base attenuate to deeply cordate. Inflorescence 2-5 cm long; peduncle 0.5-3.5 cm, angular; pseudo-umbels up to 15-flowered; pedicels 4-8(-12 in fruit) mm. Buds 2-3.5 mm long, glabrous; petals yellowish green to cream. Fruit ovoid to ellipsoid, 8-11 x 6-8 mm, red; seed ellipsoid, 7-9 x 5-6 x 4-5 mm, smooth apart from dorsal ridge. Fig. 119.6.2.

Dry *Acacia* woodland and bushland, riverine forest and scrub, semidesert scrub, often on rocky slopes and outcrops; 0-2250 m. EE EW AF TU GD WU SU AR GJ KF GG SD BA HA; widespread in tropical Africa and Asia. *Burger* 2324; *Boulos* 9488; *Tekle Hagos* 143.

10. **C. cactiformis** *Gilg* (1895).

Strong succulent climber to 5 m; branches broadly 4-winged, (1.5-)2-6 cm wide, green with strongly wavy edges, much constricted at nodes; all parts glabrous to sparsely puberulous. Leaves only present on youngest branches; petiole 1-10 mm; lamina irregularly 3-lobed, cordiform, up to 2 x 1.5 cm, sparsely dentate; apex subacute; base attenuate. Inflorescence 3-10 cm long; peduncle 1-3 cm, angular or winged; pseudo-umbels up to 20-flowered; pedicels 3-7(-15 in fruit) mm. Buds 2-3 mm long, glabrous; petals yellowish-green or cream, flushed purplish. Fruit ovoid to ellipsoid, 9-14 x 6-9 mm; seed ellipsoid, 8-10 x 5-6 x 4-5 mm, smooth apart from dorsal ridge.

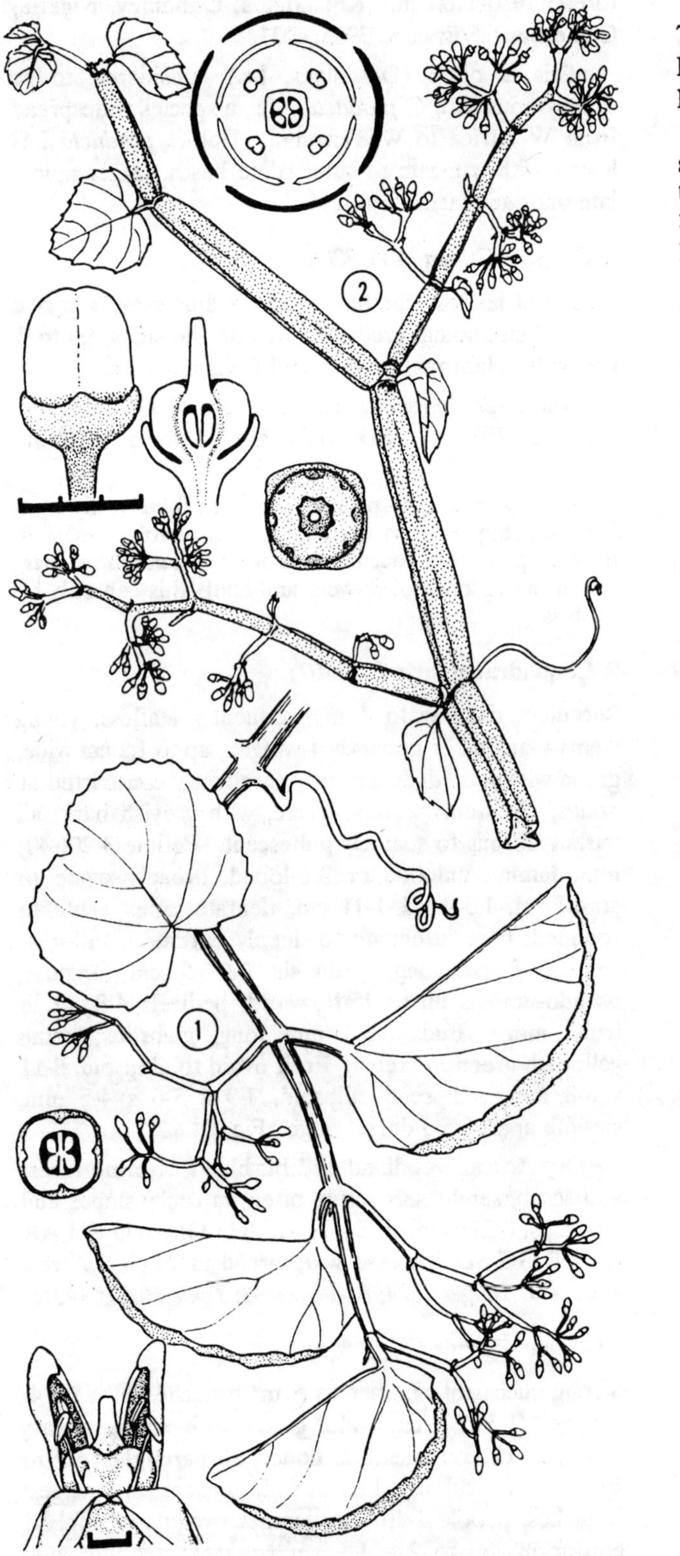

Figure 119.6.1. *CISSUS ROTUNDIFOLIA*: flowering branch x 1/2, side view of flower (scale 1 mm) and cross section of ovary. 2 - *C. QUADRANGULARIS*: flowering branch x 2/3, side view of flower bud, longitudinal section through a flower, top view of flower both with petals and stamens removed (3 mm scale shown), and floral diagram (Reproduced with permission of Oklahoma Agricultural Experiment Station from Experiment Station Bulletin No. 45, *Families of Flowering Plants in Ethiopia*, figure 40, 1967.) Drawn by W.C. Burger.

Dry *Acacia-Commiphora* bushland on rocky hills; *c* 700 m. GG HA; Somalia, Kenya, Tanzania, Mozambique, Zimbabwe, S Africa (Transvaal and Natal). *Corradi* 2420; *Ruspoli & Riva* 956 (864).

Wild & Drummond (l.c.) treat this species as a synonym of *C. quadrangularis*, but the characters used to separate them here also seem to hold reasonably well for material from the FZ-area. The author, therefore, prefers to treat them as two separate species.

11. C. rotundifolia *(Forssk.) Vahl* (1794);
Saelanthus rotundifolius Forssk. (1775)

Vigorous woody climber or scrambler to 5(-10) m; young branches terete; older 4-angled or with 4 up to 2.5 cm wide corky wings; all parts glabrous to sparsely pubescent. Leaves thick and fleshy; petiole 2-25 (-40) mm; lamina unlobed (very rarely 3-lobed), broadly ovate to orbicular, 3.5-10(-13) x 3.5-10(-12) cm, remotely to densely denticulate to dentate; apex rounded to retuse; base cordate. Inflorescence appearing with the leaves, (2.5-)4-17(-25) cm long; peduncle 1-3.5 cm; pseudo-umbels 3-15-flowered; pedicels 3-7(-13 in fruit) mm. Buds 2-3.5 mm long, glabrous; petals white to yellow. Fruit ellipsoid, 11-14 x 7-9 mm, red; seed ellipsoid, 8-10 x 5-6 x 3-4 mm, smooth apart from dorsal ridge. Fig. 119.6.1.

Dry woodland and bushland, usually on rocky outcrops and rocky slopes, dry riverine forest; (150-) 550-1900 m. EE AF WU SU KF GG SD BA HA; Somalia and S Sudan through east Africa to Natal, also in Arabia. *Burger* 455, 2801; *Mooney* 8252.

12. C. ruspolii *Gilg* (1905);
Cayratia ruspolii (Gilg) Suesseng. (1953) - type: SD, Coromma, *Ruspoli & Riva* 1760 (1592) (FT holo.).

Slender herbaceous climber from woody rootstock; all parts glabrous to sparsely pubescent. Leaves 3-foliolate; petiole 1-2.5(-4.5) cm; leaflets narrowly elliptic to narrowly ovate, 3.5-12 x 1-3.5 cm, sharply serrate; apex cuspidate to acuminate; base attenuate to cuneate. Inflorescence appearing with the leaves, 1.5-7 cm long; peduncle 0.5-2.5(-3.5) cm; pseudo-umbels 3-12-flowered; pedicels 2-5(-8 in fruit) mm, with short thick glandular hairs. Buds 2-3 mm long, glabrous; petals yellow to cream. Fruit ellipsoid, 8-10 x 5-7 mm, red; seed ellipsoid, 7-8 x *c* 5 x *c* 4 mm, smooth apart from dorsal ridge.

Dry *Juniperus* forest, *Acacia-Commiphora*, *Acacia-Combretum* and *Combretum-Terminalia* woodland and bushland on rocky slopes and on red sandy to loamy soil, *Acacia* woodland on alluvial soil and grassland on black clay; 1200-1700 m. SU(Gibie Gorge) SD BA HA; NE Zaire, Uganda, N Kenya. *Friis* et al. 1013, 3753; *Mooney* 9037.

Gilg & Brandt (*l.c.*) placed this species in Sect. *Cayratia* of *Cissus* although they noted that it was unusual here in having 1-seeded fruits. Suessenguth - apparently without seeing any material - transferred it to *Cayratia*, but inflorescence, flower, fruit and seed are all as in *Cissus* sensu stricto.

6. **CYPHOSTEMMA** *(Planch.) Alston* (1931).

Erect, trailing or climbing perennial herbs, rarely woody climbers or succulent shrubs; tendrils opposite the leaves or absent. Leaves digitately 3-7(-11)-foliolate, rarely simple or pedate. Flowers 4-merous, bisexual, in leaf-opposed, axillary or terminal umbelliform or corymbiform cymes, ultimately in small divaricately branched cymules. Buds cylindric to lageniform, constricted near middle, ± square at apex; calyx subentire; petals usually yellowish to cream with red tips and edges, dropping separately or as a hood; disc of 4 free glands, adnate to ovary; style long, stigma bifid or entire. Fruit 1(-2)-seeded, ripening through red to blackish; seed with a central dorsal ridge and 2 lateral crests, sides variously sculptured to almost smooth.

About 200 species in tropical Africa and Madagascar, two of these also in Arabia and one also in India.

Many species in this genus are very variable. This is especially, at least in Ethiopia, the case among the climbing species. Generally a somewhat broader view of the species has been adopted here than in Wild & Drummond (l.c.), where - in the author's opinion - too great emphasis was put on the presence or absence of stalked glands as an absolute distinguishing character. It is the author's experience that it is often possible to find specimens with glands in species normally without and vice versa. Even different branches of the same plant often vary as to presence or absence of stalked glands. It is possible that some of the species in this account are heterogenous (especially *C. cyphopetalum*), but until a better understanding of variation patterns is obtained through careful field work it is preferable to accept rather broadly defined species.

1. Tendrils absent. 2
- Tendrils present. 10

2. Plant erect. 3
- Plant trailing or with erect main stem and trailing to scandent branches. 8

3. Plant with thick flask-shaped succulent stem and fleshy leaves clustered at end of branches. **1. C. betiforme**
- Stem not flask-shaped and succulent; leaves not clustered at end of branches. 4

4. Leaves sessile (rarely some with petiole up to 1 cm long). 5
- Some or all leaves with petiole over 1.5 cm long. 6

5. Stems glabrous to sparsely puberulous; leaves glabrous to sparsely pubescent on veins. **2. C. junceum**
- Stems pubescent to tomentose; leaves sparsely to densely pubescent all over. **4. C. rhodesiae**

6. Leaves glabrous to sparsely pubescent on veins, usually with sessile discoid glands. **3. C. adenanthum**
- Leaves pubescent to densely woolly, without discoid glands. 7

7. Leaves reddish pubescent to tomentose, not with sharply recurved margin; seed with ribbed but not spiny sides. **5. C. molle**
- Leaves densely whitish woolly, with a subglabrous marginal band formed by the sharply recurved margin; seed with ribbed and spiny sides. **6. C. pannosum**

8. Plant glabrous apart from stalked glands. **20. C. burgeri**
- Plant hairy and with stalked glands. 9

9. Mature leaves woolly beneath; basal stem erect and fleshy; whole plant sticky. **10. C. niveum**
- Mature leaves pubescent beneath; basal stem not erect and fleshy; plant not sticky. **19. C. serpens**

10. Leaves pedately 5-7(-11)-foliolate. **9. C. adenocaule**
- Leaves digitately 3-5(-7)-foliolate. 11

11. Plant glabrous apart from stalked glands. 12
- Plant more or less hairy. 14

12. Leaves sessile (rarely lowermost petiolate), thick and fleshy. **15. C. ternatum**
- Most or all leaves petiolate, not thick and fleshy. 13

13. Leaflets truncate to subcordate at base; peduncle 2.5-8 cm long; fruit 7-10 mm long; herbaceous climber. **7. C. oxyphyllum**
- Leaflets cuneate at base; peduncle 1.5-3 cm long; fruit 10-12 mm long; woody climber. **8. C. boranense**

14. Flower buds with stalked glands apically. 15
- Flower buds without stalked glands. 20

15. All leaves 3-foliolate, densely tomentose to woolly (more rarely pubescent). **10. C. niveum**
- Some or all leaves 5(-7)-foliolate, subglabrous to densely pubescent (rarely subtomentose). 16

16. Some glandular hairs on stems longer than 3 mm; leaves pubescent to sub-tomentose; whole plant sticky. **11. C. rivae**
- Glandular hairs not longer than 2 mm (if up to 3 mm then plant not sticky); leaves subglabrous to pubescent; plant sticky or not. 17

17. Flowerbuds 3-5 mm long with numerous glands; fruit 6-8 mm long. **12. C. dembianense**
- Flowerbuds 1-3 mm long with a single or a few glands; fruit 8-13(-15) mm long. 18

18. Flowerbuds 1-2 mm long; leaves pubescent or sparsely so; some glands on stems up to 2(-3) mm long. **13. C. sp. =** *Gilbert & Thulin* 511
- Flowerbuds 2-3 mm long; leaves with denser indumentum or glands on stems up to 1 mm only. 19

19. Plant sticky with dense glands on stems, some of which up to 2 mm long; leaves puberulous. **14. C. dysocarpum**
- Plant not sticky, with scattered (or no) glands on stems, no glands longer than 1 mm; leaves pubescent. **19. C. serpens**

20. Some glands on stems up to 2(-3) mm long or plant densely glandular and sticky. 21
- Glands less than 1 mm long; plant not densely glandular, not sticky. 22
21. Flower buds 1-2 mm long; leaves pubescent; plant not sticky. **13. C.** sp. = *Gilbert & Thulin* 511
- Flower buds 2-3 mm long; leaves puberulous; plant sticky. **14. C. dysocarpum**
22. Woody climber; flowering usually before the leaves; leaflets 1-4(-5) cm long, thick and fleshy. **16. C. engleri**
- Herbaceous climber; flowering not precocious; some leaflets always longer than 4 cm. 23
23. All leaves 3-foliolate. 24
- Some or all leaves 5-7-foliolate. 25
24. Leaflets tomentose to woolly. **21. C. cyphopetalum**
- Leaflets glabrous. **17. C. kilimandscharicum**
25. Style 3-4 mm long; flower buds 3-5 mm long, tomentellous. **18. C. keilii**
- Either style shorter or flowerbuds shorter or puberulous. 26
26. Style 1-2 mm long; flowerbuds 2-3 mm long; leaflets usually more than twice as long as wide. **19. C. serpens**
- Style 2-3 mm long (always some longer than 2 mm); flower-buds 2-5 mm long; leaflets less than twice as long as wide. **21. C. cyphopetalum**

1. **C. betiforme** *(Chiov.) Vollesen* (1984);
Cissus betaeformis Chiov. (1916).

Succulent shrub with basally swollen flask-shaped trunk to 1 m, with a few thick branches apically; bark yellowish white, flaking off in irregular parchmenty patches. Leaves congested at branch ends, fleshy, 1-5-foliolate, puberulous when mature; petiole 0.5-7 cm; leaflets obovate or cuneiform to suborbicular, 2-7.5 x 1-7.5 cm, finely denticulate to grossly dentate; apex broadly rounded to emarginate; base attenuate to subcordate. Inflorescence 2-7 cm long, puberulous; peduncle 1-3.5 cm; pedicels 0.5-1(-3 in fruit) mm, with subsessile glands. Buds 1.5-2 mm long, glabrous (calyx sparsely puberulous); petals dark red; style *c* 1.5 mm. Fruit ellipsoid, 6-9 x 4-5 mm, glabrous; seed ellipsoid, 5-7 x 2-3 x 3-4 mm, lateral crests conspicuous, sides with two strong ridges.

Acacia-Commiphora bushland on limestone and gypsum hills; 250-400 m. SD BA; Somalia, NE Kenya. *Gilbert* 2131; *Gilbert* et al. 7581.

A very characteristic plant which has only been collected a few times over an extensive area in the Ogaden. It is closely related to the succulent *Cyphostemma*'s in Namibia, and forms yet another phytogeo-graphical link between these two areas.

The dried specimens of *Gilbert* 2131 have 1-4-foliolate leaves, but the specimen grown at Kew has 1-foliolate leaves only.

2. **C. junceum** *(Webb) Descoings* ex *Wild & Drummond* (1966);
Cissus juncea Webb (1854).
C. jatrophoides sensu Cuf. (1958) quoad distrib. Aethiop., non (Bak.) Descoings (1967).

Erect perennial herb to 1.25 m from cylindric to subglobose tuber up to 10 x 10 cm; all parts glabrous to puberulous or pubescent. Leaves (1-)3-foliolate, sessile (rarely lowermost with petiole up to 1 cm); leaflets lanceolate, 10-27 x 1-4 cm, denticulate to dentate or serrate; apex acuminate; base cuneate. Inflorescence 7-30 cm long; peduncle 3-25 cm; pedicels 1-4(-8 in fruit) mm, occasionally with stalked glands. Buds 1.5-3 mm long, glabrous to puberulous; style 1-2 mm. Fruit ellipsoid to obovoid, 6-10 x 4-7 mm, glabrous; seed ellipsoid, 5-7 x 3-5 x 3-5 mm, lateral crests weak, sides with few to many ridges. Fig. 119.7.

Combretum-Terminalia, *Terminalia-Pterocarpus* and *Combretum-Oxytenanthera* woodland, on greyish sandy to loamy soil; 700-1850(-2500) m. TU GJ WG IL SD HA; Nigeria, Cameroun, Central African Republic, Sudan, N Zaire, Uganda, NW Kenya. *Friis* et al. 2524; *Gillett* 5234; *Thulin* et al. 4014.

The author doubts if *C. jatrophoides* is a synonym of *C. junceum* as claimed by Wild & Drummond (l.c.). Typical *C. junceum* is a slender plant with sessile leaves and is very different from the stout, glaucous plants with petiolate leaves from southern Africa (*C. jatrophoides* s.str.). The two species are kept separate by Descoings (l.c.).

Cufodontis' record of *C. jatrophoides* from Ethiopia is based on *Gillett* 5234 and *Mooney* 6850a. Both are typical *C. junceum* although previously misnamed.

Most collections from Ethiopia differ from the completely glabrous typical material in being puberulous, especially on inflorescence and flowers. A few stalked glands can also occur on the pedicels.

3. **C. adenanthum** *(Fresen.) Descoings* (1967);
Cissus adenantha Fresen. (1837) - type: GD/TU, between Gondar and Adua, *Rueppel* s.n. (FR holo.).
Vitis sambucina Beccari ex Martelli (1886) - type: EW, Keren, *O. Beccari* 124 (FT holo.).
Cissus taschdjianica Chiov. (1937) - type: GJ, Tsciena river, *Taschdjian* 370 (FT holo.).

Erect perennial herb to 1 m from woody rootstock with fleshy fusiform roots; all parts glabrous to puberulous or pubescent and sparsely to densely sticky glandular (rarely not). Leaves (3-)5-7-foliolate, with scattered discoid glands (rarely without); petiole (0-)2-12 cm; leaflets narrowly elliptic to elliptic or obovate, (6-)9-30 x (1.5-)3-9 cm, serrate; apex cuspidate to acuminate; base attenuate. Inflorescence (5-)10-30 cm long; peduncle 3.5-20 cm; pedicels 3-12(-15 in fruit) mm. Buds 2-3.5 mm long, puberulous (rarely glabrous) and densely glandular (rarely not); style *c* 2 mm. Fruit ellipsoid to obovoid, 9-13 x 5-9 mm, sparsely pubescent and glandular (rarely sub-glabrous); seed ellipsoid to

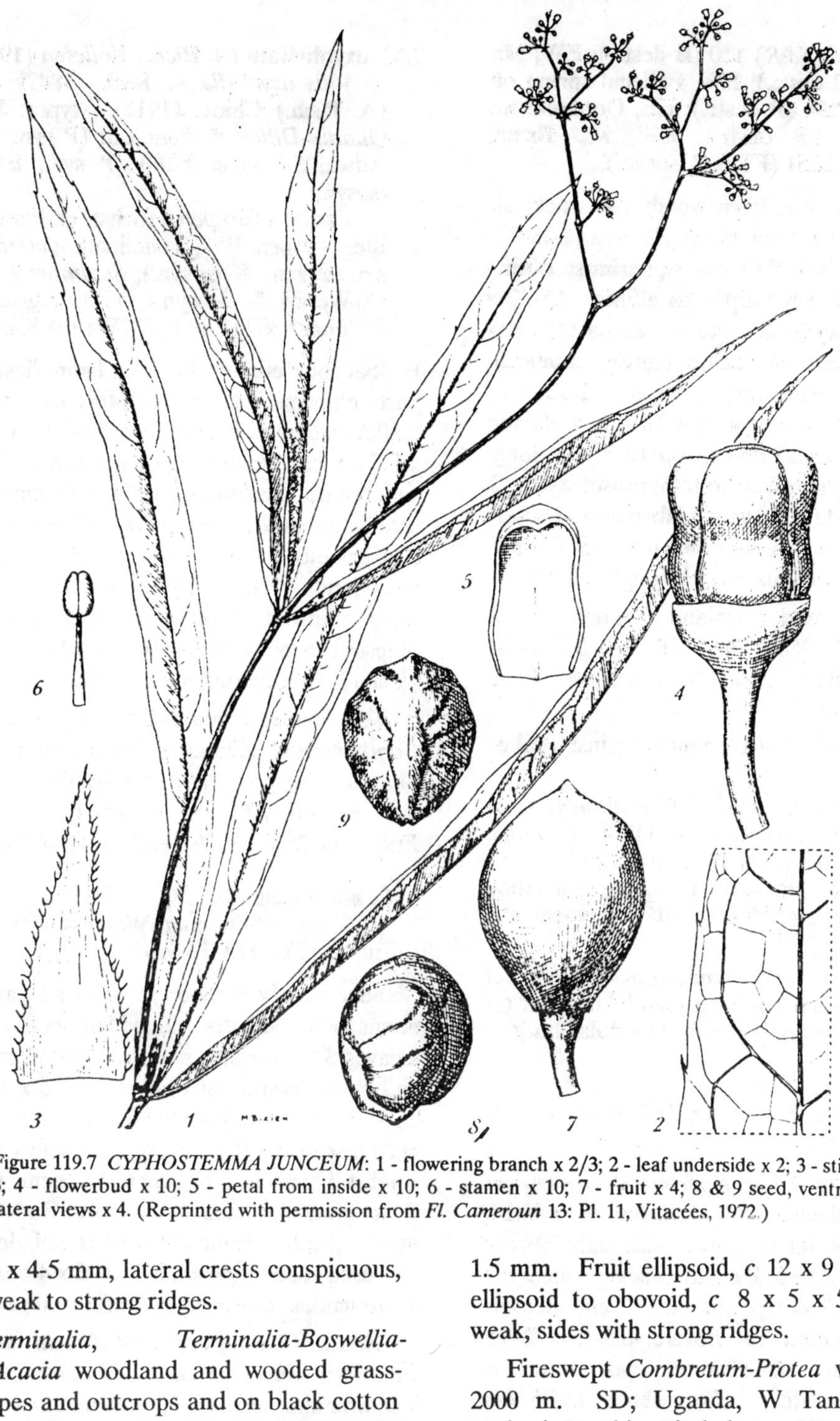

Figure 119.7 *CYPHOSTEMMA JUNCEUM*: 1 - flowering branch x 2/3; 2 - leaf underside x 2; 3 - stipule x 3; 4 - flowerbud x 10; 5 - petal from inside x 10; 6 - stamen x 10; 7 - fruit x 4; 8 & 9 seed, ventral and lateral views x 4. (Reprinted with permission from *Fl. Cameroun* 13: Pl. 11, Vitacées, 1972.)

obovoid, 6-9 x 4-5 x 4-5 mm, lateral crests conspicuous, spiny, sides with weak to strong ridges.

Combretum-Terminalia, *Terminalia-Boswellia-Anogeissus* and *Acacia* woodland and wooded grassland, on rocky slopes and outcrops and on black cotton soil; 1100-2050 m. EW TU WU GD SU GJ WG KF SD; Sudan. *Friis* et al. 2593; *Gilbert* et al. 2974; *Mooney* 9031.

4. **C. rhodesiae** *(Gilg & Brandt) Descoings* ex *Wild & Drummond* (1966).

Erect perennial herb to 1 m from woody rootstock; all parts sparsely rusty-brown pubescent to tomentose. Leaves 3-5-foliolate, sessile; leaflets narrowly elliptic to elliptic, 10-25 x 2-7 cm, serrate; apex acuminate to acute; base attenuate. Inflorescence 7-16 cm long; peduncle 2.5-8 cm; pedicels 2-5 mm, densely pubescent. Buds *c* 2.5 mm long, pilose (glabrous apically); style *c* 1.5 mm. Fruit ellipsoid, *c* 12 x 9 mm, glabrous; seed ellipsoid to obovoid, *c* 8 x 5 x 5 mm, lateral crests weak, sides with strong ridges.

Fireswept *Combretum-Protea* wooded grassland; *c* 2000 m. SD; Uganda, W Tanzania, Mozambique, Malawi, Zambia, Zimbabwe. *Gillett* 14857.

The northernmost record of this species which has its main distribution in southern tropical Africa, but reaches up through W Tanzania and Uganda to southern Ethiopia.

5. **C. molle** *(Bak.) Descoings* (1967);

Cissus mollis (Bak.) Planch. (1887); *Vitis serpens* Hochst. ex A. Rich. (1847) var. *mollis* Bak. in *Fl. Trop. Afr.* **1**: 413 (1868) - type: GD, Bellegas in Schoata, *Schimper* II:538 (P holo., BM FI(Webb) K UPS iso.).

Cyphostemma erythraeae (Gilg & Brandt) Descoings (1967) - types: EE, Donkolla nr. Ginda,

Schweinfurth & Riva (S&R) 120 (B destr.); EW, Mt. Kube, *S&R* 1516 (B destr.); EW, Fellacid spring on Mt. Bizen, *S&R* 1516a (B destr.); EE, Donkolla nr. Ginda, *S&R* 2192 (B destr.); EW, Mt. Bizen, Telakhit river, *S&R* 2251 (FT K P isosyn.).

Erect perennial herb to 1 m from woody rootstock; all parts reddish pubescent to tomentose. Leaves (1-)3-5-foliolate; petiole (0-)1.5-5(-8.5) cm, uppermost often subsessile; leaflets narrowly elliptic to elliptic, 7.5-32 x 1.5-10 cm, serrate (rarely crenate or dentate); apex cuspidate to acute; base attenuate (rarely cuneate). Inflorescence (5-)8-16 cm long; peduncle 2-12 cm; pedicels 2-5(-8 in fruit) mm, rarely with a few stalked glands; bracts conspicuous, linear, up to 1 cm long. Buds 2-3.5 mm long, pubescent to tomentose; style *c* 2 mm. Fruit ellipsoid, 9-11 x 7-8 mm, glabrous to sparsely pubescent; seed ellipsoid to obovoid, 6-8 x 5-6 x 4-5 mm, lateral crests conspicuous, sides strongly ridged.

Wooded grassland and bushland, sometimes on rocky slopes; 200-2200(-2800) m. EE EW TU GD WU; not known elsewhere. *Getachew Aweke* 649; *Tekle Hagos* 151; *J. de Wilde* 4617.

Roots boiled and taken with honey against snake-bites.

Gilg & Brandt (l.c.) distinguished *C. erythraeae* from *C. molle* as having tomentose leaves while in *C. molle* they were said to be hairy on the nerves only. The material now available shows a complete transition from one to the other, and clearly only one species is involved.

Close to *C. crinitum* - a Sudanian species not yet found in Ethiopia - which has long glandular hairs on stems and inflorescences and glabrous glandular buds.

6. C. **pannosum** *Vollesen* (1984).
- type: IL, Bure to Baro river, *Friis* et al 2409 (K holo., C ETH UPS iso.).

Erect perennial herb to 0.5 m from woody rootstock; stems puberulous. Leaves 3-5(-7)-foliolate, whitish woolly beneath except for a 1 mm wide subglabrous band along edges formed by the sharply recurved margin, subglabrous above; petiole 1.5-12 cm, densely puberulous; leaflets elliptic to obovate, 6.5-15 x 2-6.5 cm, irregularly serrate to dentate; apex acute to acuminate; base attenuate. Inflorescence 10-35 cm long, puberulous, sometimes with a few stalked glands; peduncle 5.5-25 cm; pedicels 1-3 mm. Buds 2-3 mm long, puberulous; style c 1.5 mm. Fruit ellipsoid, 8-10 x 6-7 mm, glabrous; seed obovoid, 6-7 x 4-5 x 4-5 mm, lateral crests inconspicuous, sides strongly ridged and spiny. Fig. 119.8.

Combretum-Terminalia-Protea wooded grassland on black clay or quartz sand; 1500-2000 m. WG IL GG; not known elsewhere. *Mooney* 6850b; *Vàtova* 1755.

C. pannosum is only known from these three collections. It seems to be closest to the W African *C. flavicans* (Bak.) Descoings which has a densely glandular inflorescence and pilose buds.

7. C. **oxyphyllum** *(A. Rich.) Vollesen* (1984);
Vitis oxyphylla A. Rich. (1847); *Cissus oxyphylla* (A. Rich.) Chiov. (1912) - types: EW, Tchelicote, *Quartin-Dillon & Petit* s.n. (P syn., K isosyn.); TU, Adwa, *Schimper* I:180 (P syn., BM FI(Webb) K isosyn.)
Cissus schimperi Hochst. ex Planch. (1887), nom. illeg. - types: EW, Tchelicote, *Quartin-Dillon & Petit* s.n. (P syn., K isosyn.), *Schimper* I:180 (P syn., BM FI(Webb) K isosyn.); GD, Agrima & Bularehi, *Schimper* 307 (P syn., FI(Webb) K isosyn.).

Herbaceous climber to 4 m from fleshy rootstock; all parts glabrous. Leaves 4-5-foliolate; petiole 1.5-6 cm; leaflets narrowly to broadly elliptic or obovate, 2-5.5(-9) x 0.5-2.5 cm, serrate; apex acuminate; base truncate to subcordate. Inflorescence 6.5-13 cm long, with scattered subsessile glands on distal branches; peduncle 2.5-8 cm; pedicels 2-4(-8 in fruit) mm, with dense subsessile glands. Buds 2.5-3.5 mm long; style 1.5-3 mm. Fruit ellipsoid to subglobose, 7-10 x 6-8 mm; seed ellipsoid to obovoid, 5-8 x 4-6 x 4-6 mm, lateral crests weak to conspicuous, sides rugose but not ridged.

Combretum-Terminalia and *Acacia-Commiphora-Combretum* woodland and bushland, on rocky slopes or black cotton soil; 1300-2300(-2650) m. EW TU GD WU SU GJ WG; not known elsewhere. *Burger* 577; *Friis* et al. 2612; *W. de Wilde* at al. 6078.

8. C. **boranense** *Vollesen* (1984)
- type: SD, near Welensu, *Friis* et al. 3352 (K holo., C ETH UPS iso.).

Slender woody climber to 3 m; all parts glabrous; old stems with slightly ridged or peeling greyish bark. Leaves 3-5-foliolate; petiole 2-4(-5) cm; leaflets elliptic to broadly elliptic or obovate, 2-6.5 x 1-4.5 cm, crenate to serrate in upper half; apex acute; base cuneate. Inflorescence 3-9 cm long, with a few subsessile glands on distal branches; peduncle 1.5-3 cm; pedicels 2-3(-6 in fruit) mm, with subsessile glands. Buds 2.5-3.5 mm long; style *c* 3 mm. Fruit ellipsoid to subglobose, 10-12 x 8-10 mm; seed ellipsoid, 6-7 x 4-5 x 4-5 mm, lateral crests conspicuous, sides rugose but not ridged.

Acacia-Commiphora and *Acacia* bushland on rocky limestone ridges or on black cotton soil; 850-1300 m. SD; Somalia. *Ash* 1909; *Gilbert & Glyn Jones* 99; *Gilbert* et al. 7455.

The peculiar subsessile almost discoid glands on the pedicels are common to this species and *C. oxyphyllum* but have not been observed on any other Ethiopian species.

9. C. **adenocaule** *(Steud. ex A. Rich.) Descoings* ex *Wild & Drummond* (1966);
Cissus adenocaulis Steud. ex A. Rich. (1847) - types: TU, Djeladjekanne by Tekeze river, *Schimper* III:1646 (P syn., BM K UPS isosyn.); Ethiopia, no loc., *Schimper* III:1703 (P syn., BM FI(Webb) K UPS isosyn.).

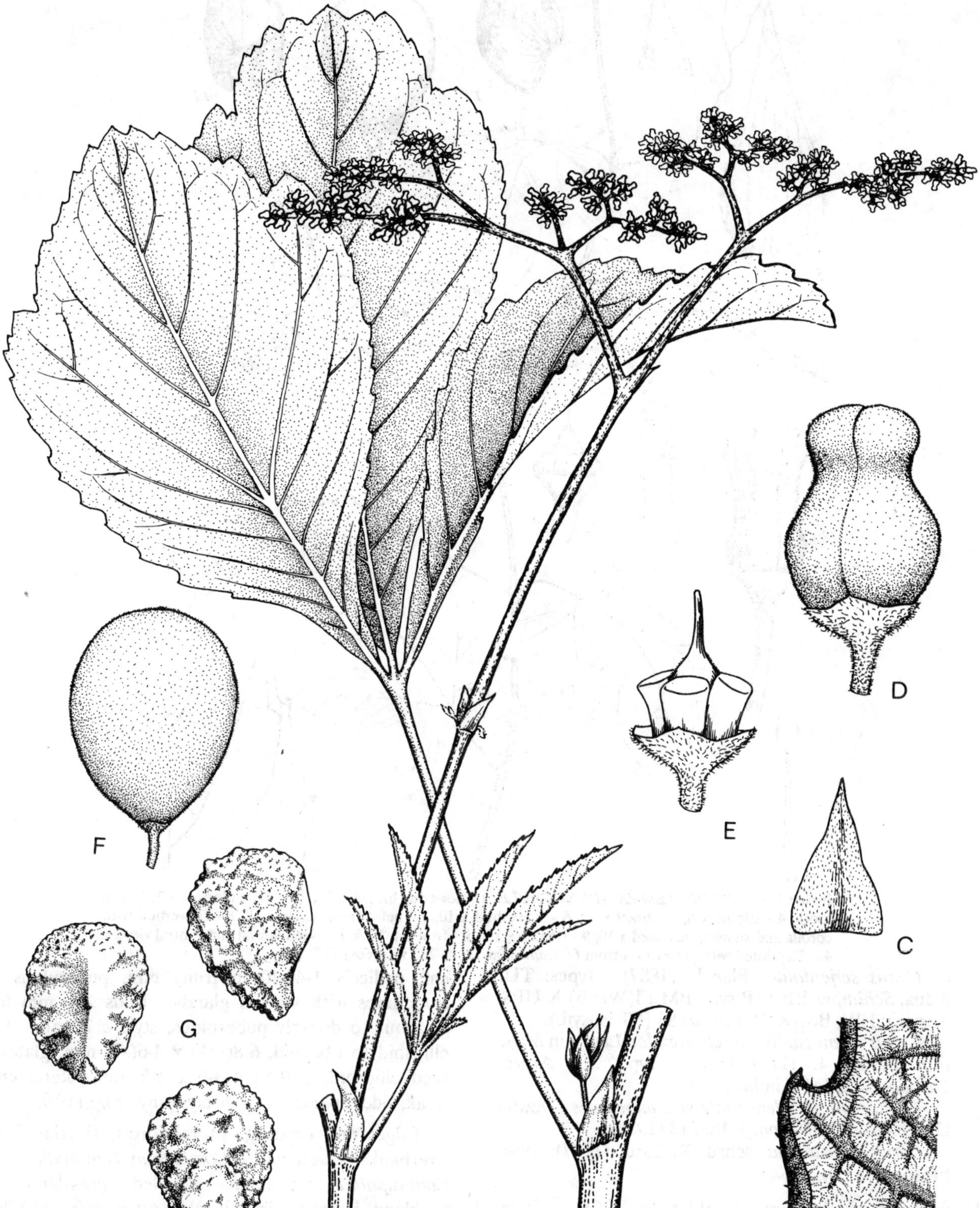

Figure 119.8 *CYPHOSTEMMA PANNOSUM*: A - flowering branch x 1.5; B - leaf underside and margin x 12; C - stipule x 2; D - flowerbud x 15; E - flowerbud with petals and stamens removed x 15; F - fruit x 6; G - seed ventral, lateral and dorsal views x 6. A - E from *Friis* et al. 2409; F & G from *Vatova* 1755. Drawn by Eleanor Catherine.

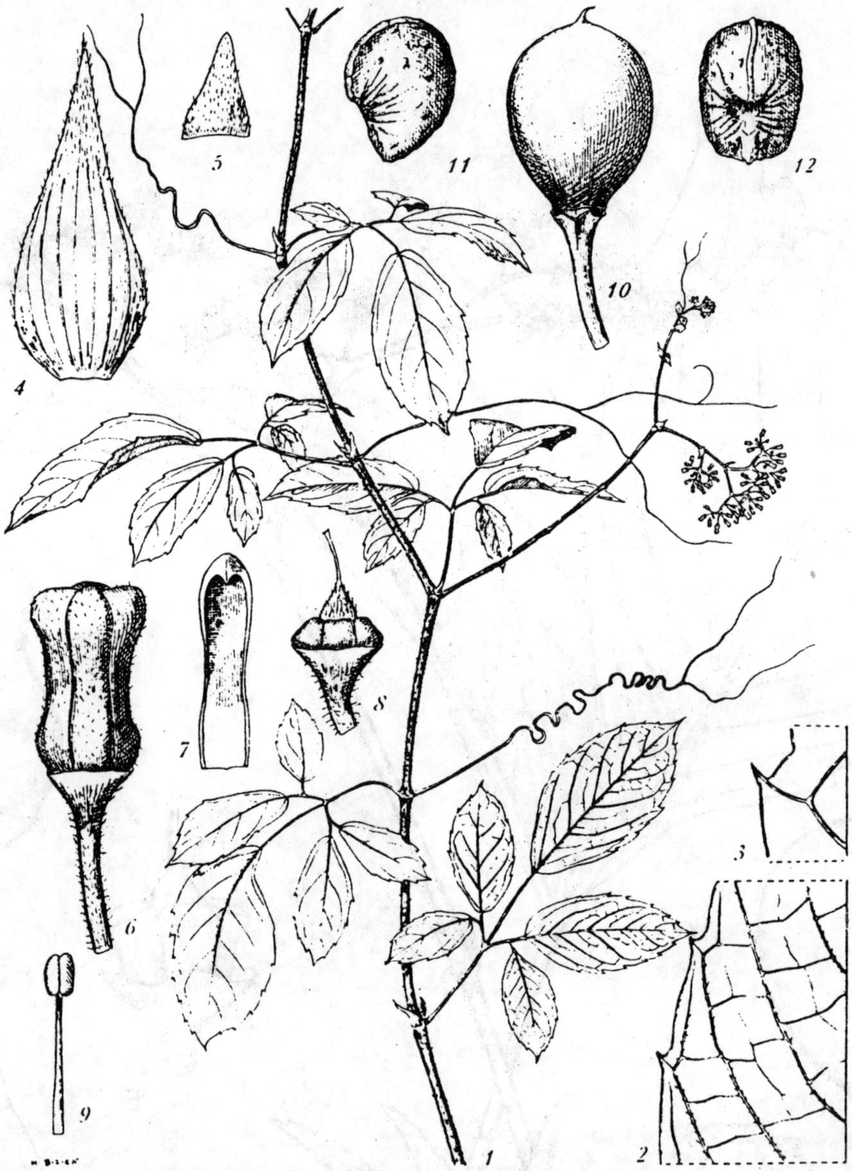

Figure 119.9 *CYPHOSTEMMA ADENOCAULE*: 1 - flowering branch x 2/3; 2 - leaf underside x 3; 3 - leaf tooth; 4 - stipule x 6; 5 - bract x 20; 6 - flowerbud x 10; 7 - petal from inside x 10; 8 - flowerbud with corolla and stamens removed x 10; 9 - stamen x 10; 10 - fruit x 4; 11 & 12 - seed lateral and ventral views x 4. (Reprinted with permission from *Fl. Cameroun* 13: Pl. 28 (Vitacées) 1972.)

Cissus serjanioides Planch. (1887) - types: TU, Adua, *Schimper* I:165 (P syn., BM FI(Webb) K UPS isosyn.); EW, Bogos, *O. Beccari* 83 (FT isosyn.).

Cissus adenocaulis var. *chiovendae* Lanza in *Miss. Biol. Boran*. **4**: 111 (1939) - type: SD, Arero, *Cufodontis* 274 (FT holo.).

Cyphostemma adenocaule var. *pubescens* (Dewit) Descoings in *Nat. Monsp*. **18**: 218 (1967).

Cissus adenantha sensu Richard (1847), non Fresen. (1837).

Herbaceous climber or scrambler to 3(-6) m from tuberous rootstock, basal parts sometimes ± woody; all parts glabrous to densely whitish or reddish pubescent, somtimes with scattered (rarely dense) stalked glands. Leaves pedately 5-7(-11)-foliolate; petiole 1-5.5(-9.5) cm; leaflets elliptic or ovate to broadly ovate, central 1.5-10(-11.5) x 1-6(-7.5) cm, crenate, serrate or biserrate: apex acuminate (rarely acute); base cuneate to cordate. Inflorescence 2-13 cm long; peduncle 1-8(-10) cm; pedicels 1-4(-10 in fruit) mm, puberulous and sometimes with stalked glands. Buds 2-4 mm long, glabrous to densely puberulous; style *c* 2 mm. Fruit ellipsoid to obovoid, 6-8(-11) x 4-6(-7) mm, glabrous; seed ellipsoid, 5-7(-8) x 3-6 x 3-5 mm, lateral crests weak, sides sparsely to densely spiny. Fig. 119.9.

Edges and clearings of rainforest, riverine forest, riverbanks, *Acacia*, *Combretum-Terminalia* and *Commiphora* woodland, wooded grassland and bushland, in fences in cultivated areas; (600-)850-2650 m. EE EW TU GD GD-GJ WU SU WG IL KF GG SD BA HA; widespread in tropical Africa south to Mozambique and Malawi. *Friis* et al. 1610, 1720; *Mooney* 6942.

The tuberous roots are said to be edible, and a decoction of these is used to treat supurating wounds.

An extremely variable species which seems to defy all attempts at subdivision. All grades can be found from glabrous to densely pubescent and from non-glandular to densely glandular plants. Some collections from Sidamo (and N Kenya) represent another extreme. They have up to 11-foliolate leaves; *Friis* et al. 3206 represents this form.

10. C. **niveum** *(Hochst.* ex *Schweinf.) Descoings* (1967);
Cissus nivea Hochst. ex Schweinf. (1867) - types: TU, Mt. Kubi nr. Axum, *Schimper* I: s.n. (B destr.); GD, Dschadscha, *Schimper* 517 (FT isosyn.).
Vitis pannosa Bak. (1868) - type: GD, Maschila, Nahia, *Schimper* 827 (K holo.).

Sticky herbaceous climber to 3 m with thick fleshy rootstock, sometimes with erect fleshy basal stem to 0.6 m tall and 0.1 m in diameter; stems pubescent and with dense (rarely sparse) sticky glands up to 3 mm long. Leaves 3-foliolate, tomentose to woolly (more rarely pubescent); petiole 2-13 cm, pubescent and glandular; leaflets ovate or broadly so, 2.5-19.5 x 1.5-12.5 cm, grossly dentate to serrate; apex acuminate to acute; base cuneate to truncate. Inflorescence 6-18 cm long, pubescent and glandular; peduncle 2.5-11.5 cm; pedicels 1-5(-8 in fruit) mm. Buds 1-3 mm long, puberulous and glandular; style *c* 1 mm. Fruit ellipsoid, 8-12 x 5-8 mm, puberulous to pubescent and densely glandular; seed ellipsoid, 6-8 x 3-5 x 3-5 mm, lateral crests weak or almost absent, sides almost smooth to weakly ridged and rugose.

Juniperus-Podocarpus forest (mostly at edges), upland bushland, thicket clumps in upland grassland, persisting in hedges and in grassland after forest clearing, also rocky outcrops; (1250-)1650-2600 m. EW TU GD WU GJ SU AR KF SD BA HA; not known elsewhere. *Brehme* in *Mooney* 8971; *Gilbert* 2232; *Gilbert* et al. 7934.

The erect forms were at first thought to be a separate species. They are normally without tendrils. But plants have later been seen (*Gilbert* 2232) which have some branches with and some without tendrils.

11. C. **rivae** *(Gilg) Descoings* (1967);
Cissus rivae Gilg (1905) - types: SD/BA, Tombe (Tumpe), *Ruspoli & Riva* 519(484); SD/BA, Rogono (Rogoru) *R&R* 693(633) (both FT syn.).

Sticky scrambling or climbing herb to 1 m, with thick, fleshy roots; stems puberulous to pubescent and with dense glands up to 10 mm long. Leaves 3-5-foliolate, pubescent to tomentose; petiole 3-11 cm; leaflets ovate to elliptic, 5-18.5 x 1.5-9 cm, grossly and irregularly dentate; apex acuminate to acute; base attenuate to cuneate. Inflorescence 7-17(-33) cm long, pubescent and densely glandular; peduncle 5-13(-23) cm; pedicels 2-5(-10 in fruit) mm, puberulous to pubescent and glandular. Buds 2-3 mm long; calyx pilose, basal part of petals puberulous, apical pilose and glandular; style *c* 1 mm. Fruit ellipsoid, 11-17 x 6-8 mm, pubescent and sticky glandular; seed ellipsoid, 6-10 x 4-5 x 4-5 mm, lateral crests weak, sides with distinct ridges.

Dry *Juniperus* forest and scrub, *Acacia-Commiphora* bushland, on red loamy soil and on rocky granitic and basaltic ridges; 1300-1800 m. SD BA; NE Kenya. *Friis* et al. 2689; *Gilbert* et al. 4605; *J. de Wilde* 6602.

Close to *C. niveum* from which it differs in the very long glandular hairs, in having some of the leaves 5-foliolate, in the different indumentum on the buds and in the larger fruit. Also close to *C. hildebrandtii* - a plant from southern tropical Africa north to SE Kenya - which has buds without glands.

12. C. **dembianense** *(Chiov.) Vollesen* (1984);
Cissus dembianensis Chiov. (1911) - type: GD, Scinta valley above Asosa in Dembia, *Chiovenda* 1654 (FT holo.).
Cyphostemma adenocephalum (Gilg & Brandt) Descoings (1967); *Cissus adenocephala* Gilg & Brandt (1912) - type: GD, Sanka-Bir in Simien, *Schimper* 1191 (B holo. destr., BM C K iso.).
Cissus stipulacea (Bak.) Planch. var. *hochstetteri* Planch. in DC. *Monogr. Phan.* **5**, **2**: 597 (1887) - type: *Schimper* 1191 (K holo., BM C iso.).

Herbaceous climber to 3 m; all parts pubescent and sometimes with scattered to dense stalked glands. Leaves 3-5-foliolate; petiole 1-6(-8) cm; leaflets ovate to elliptic, 3-15 x 1.5-8.5 cm, grossly dentate with large triangular apiculate teeth; apex acuminate to acute; base attenuate to cuneate. Inflorescence 5-20 cm long; peduncle 1.5-9 cm; pedicels 2-4(-7 in fruit) mm, pubescent to subtomentose and usually glandular. Buds 3-5mm long, puberulous to pubescent and glandular; style *c* 3 mm. Fruit ellipsoid, obovoid or subglobose, 6-8 x 5-6 mm, densely pubescent to tomentose and often with scattered glands; seed broadly ovoid, 4-6 x 3-4 x 3-4 mm, lateral crests weak to almost absent, sides with broad ± anastomosing ridges.

Forest edges, upland grassland and scrub-grassland, secondary forest; 1500-2300 m. GD SU GJ WG IL KF SD; E Sudan. *Mooney* 5876; *Friis* et al. 113; *W. de Wilde* et al. 7594.

Closely related to *C. ukerewense* (Gilg) Descoings which differs mainly in having large conspicuous bracts and more slender buds which are glabrous but for the glands.

13. C. sp. = *Gilbert & Thulin* 511.

Herbaceous climber to 4 m; stems glabrous to sparsely pubescent and with glands up to 2(-3) mm long. Leaves 4-5-foliolate, pubescent; petiole 2-8(-10) cm, glandular; leaflets elliptic to obovate, 3.5-12 x 1-6.5 cm, denticulate to dentate; apex acuminate to acute; base attenuate. Inflorescence 10-20 cm long, pubescent and glandular; peduncle 4.5-12 cm; pedicels 1-6 mm, densely pubescent. Buds 1-2 mm long, densely puberulous and with or without stalked glands; style *c* 1 mm. Immature fruit tomentose, not glandular.

Upland rainforest and bamboo thickets, often on edges and in clearings; 1900-2300 m. KF GG SD; not known elsewhere. *Gilbert & Thulin* 511; *Smeds* 154; *J. de Wilde* 6694.

Probably closest to the preceeding species but difficult to say for certain without fruit and seed. It differs in the smaller buds and different leaf dentation.

14. **C. dysocarpum** *(Gilg & Brandt) Descoings* (1967);
Cissus dysocarpa Gilg & Brandt (1912) - types: *Keller* s.n. (B destr.); BA, by river Dana in Web-Karanale valley, *Ruspoli & Riva* 1117(1013) (FT syn.).

Sticky scrambling or climbing herb to 4 m; all parts puberulous and densely glandular. Leaves 3-5-foliolate; petiole (0.5-)2.5-7 cm; leaflets elliptic or obovate to orbicular or transversely elliptic, 2.5-11.5 x 1.5-7.5 cm, grossly crenate to dentate; apex broadly rounded to truncate; base attenuate to cuneate. Inflorescence 8-20(-30) cm long; peduncle 4-12(-22) cm; pedicels 2-5 (-10 in fruit) mm, puberulous and glandular. Buds 2-3 mm long, puberulous and sparsely glandular or not; style 2-3 mm. Fruit ellipsoid to obovoid, 8-13 x 5-8 mm, pubescent and densely glandular; seed ellipsoid to obovoid, 5-8 x 3-5 x 3-5 mm, lateral crests weak to conspicuous, sides smooth to strongly ridged.

Acacia-Commiphora woodland and bushland on sandy soil overlying limestone or on rocky outcrops; 450-1250(-1750) m. GG SD BA; Somalia, NE and E Kenya. *Friis* et al. 2691, 2894, 3219.

15. **C. ternatum** *(Forssk.) Descoings* (1967);
Cissus ternata (Forssk.) Gmelin (1791).
Cissus somaliensis Gilg (1905) - type: SD/BA, Dolo, *Ruspoli & Riva* 1236 (1117) (FT syn.).

Trailing or climbing succulent herb to 2 m long; all parts glabrous. Leaves 3-5-foliolate, glossy and fleshy, sessile (rarely lowermost with petiole up to 1 cm); leaflets ovate to elliptic, 2-11(-15) x 1-5(-7) cm, remotely (rarely densely) dentate; apex acute to broadly rounded; base cuneate to truncate. Inflorescence 5-10 cm long, normally glandular on distal branches; peduncle 2-6.5(-8.5) cm; pedicels 3-8 mm, glabrous and glandular (rarely not). Buds 3-4 mm long, glabrous; style *c* 2 mm. Fruit ellipsoid to subglobose, 7-12 x 6-10 mm, glabrous; seed ellipsoid to obovoid, 5-8 x 3-6 x 3-5 mm, lateral crests weak to conspicuous, sides smooth to strongly ridged.

Acacia-Commiphora bushland on yellowish alluvial loam, riverine forest, coastal bushland; 125-400 m. EE AF SD BA-HA; Sudan (Red Sea Coast), Somalia, Arabia. *Ash* 2517; *Bally* 6814; *Gilbert* et al. 7516.

16. **C. engleri** *(Gilg) Descoings* (1967).

Woody climber to 1.5 m; stipules sometimes spinescent and persisting on old branches; young stems puberulous, older with smooth, dark grey to reddish brown, irregularly flaking bark. Leaves fleshy, 3-5-foliolate, subglabrous to pubescent; petiole 0.3-3 (-4) cm, sometimes with scattered glands; leaflets obovate to spathulate, 1-4(-5) x 0.5-3.5 cm, grossly dentate: apex broadly rounded to truncate; base cuneate. Inflorescence (3-)5-13 cm long, usually appearing before the leaves, puberulous, rarely also with scattered glands on distal branches; peduncle 1-5.5 cm; pedicels 2-5(-10 in fruit) mm. Buds 2-3 mm long, puberulous; style 1-2 mm. Fruit ellipsoid, 8-10 x 5-7 mm, puberulous and occasionally glandular; seed ellipsoid, 5-8 x 3-5 x 3-5 mm, lateral crests inconspicuous, sides with weak ridges.

Acacia-Commiphora woodland and bushland on red sandy soil overlying limestone; 600-1300(-1650) m. HA; Somalia, E Kenya, NE Tanzania. *Bally* 12983; *Ellis* 101; *Hemming* 152.

17. **C. kilimandscharica** *(Gilg) Descoings* ex *Wild & Drummond* (1966).

Vigorous herbaceous climber to 10 m or more; all parts glabrous to sparsely puberulous or pubescent. Leaves 3-foliolate; petiole (1-)2.5-8.5 cm; leaflets ovate to broadly elliptic or obovate, 2.5-9.5 x 1.5-7 cm, grossly crenate to dentate or serrate; apex acuminate to acute; base cuneate to cordate, often oblique. Inflorescence 8-25 cm long; peduncle 5-14 cm; pedicels 2-5(-8 in fruit) mm, puberulous. Buds 3-5 mm long, puberulous; style 2-5 mm. Fruit ellipsoid to subglobose, 9-13 x 6-8 mm, glabrous; seed ellipsoid to obovoid, 6-8 x 4-5 x 4-5 mm, lateral crests conspicuous, sides without ridges and rugose or with short central ridges.

Upland rainforest, ground water forest; 1250-1700 m. KF GG; Uganda, Kenya, Tanzania, E Zaire, Malawi, Zimbabwe. *Gilbert & Thulin* 280; *Smeds* 987.

18. **C. keilii** *(Gilg & Brandt) Descoings* (1967).

Herbaceous climber to 3 m; stems crisped pubescent and with scattered stalked glands (rarely without), conspicuously reddish when young. Leaves (3-)4-5-foliolate, densely pubescent to tomentose on veins, crisped pubescent on lamina; petiole 3-13 cm, puberulous and sparsely glandular or not; leaflets ovate to elliptic, 3-12 x 1.5-5.5 cm, grossly serrate to dentate; apex acuminate to acute; base attenuate to cuneate. Inflorescence 10-20 cm long, pubescent to tomentose and sparsely glandular (rarely not); peduncle 6.5-15 cm; pedicels 1-5(-8 in fruit) mm, tomentose and glandular (rarely not). Buds 3-5 mm long, tomentellous; style 3-4 mm. Fruit ellipsoid, *c* 13 x 8 mm, tomentose and glandular; seed ellipsoid, *c* 8 x 5 x 4 mm, lateral crests weak, sides with weak ridges.

Edges of upland rainforest, riverine forest, secondary scrub and grassland; 1500-2100 m. SU(Bako area) KF BA; Uganda, W Tanzania. *Friis* et al. 3528; *Jansen* 2288; *Mooney* 9148.

19. **C. serpens** *(Hochst. ex A. Rich.) Descoings* (1967);
Cissus serpens Hochst. ex A. Rich. (1847) - type: TU, Djeladjekanne by Tekeze river, *Schimper* III:1702 (P holo., BM K iso.).
Cyphostemma orondo (Gilg & Brandt) Descoings (1967).
Cyphostemma schweinfurthii (Planch.) Descoings (1967).
Cissus cymosa auct. non Schumach. & Thonn. quoad distrib. Aethiop.
C. molle (Steud. ex Bak.) Descoings quoad syntype: *Schimper* 1141 (P holo.).

Trailing or climbing herb to 2 m from woody carrot-shaped tuber up to 30 cm long; all parts crisped pubescent and usually with scattered glands. Leaves (3-)5-7-foliolate; petiole (0-)1-5(-7) cm; leaflets narrowly ovate to narrowly obovate (rarely elliptic to obovate), 2.5-16 x 1-6.5 cm, dentate to serrate; apex acute to broadly rounded; base attenuate to cuneate. Inflorescence 5-20 cm long; peduncle 3-14 cm; pedicels 2-5(-7 in fruit) mm, densely pubescent, glandular or not. Buds 2-3 mm long, puberulous to pubescent, sometimes glandular; style 1-2 mm. Fruit ellipsoid to subglobose, 8-11(-15) x 7-10 mm, subglabrous to densely pubescent, sometimes glandular; seed ellipsoid to subglobose, 6-9 x 4-6 x 4-6 mm, lateral crests weak to conspicuous, sides weakly to strongly ridged.

Acacia-Commiphora woodland and bushland on red sandy to loamy soil over limestone, *Acacia* woodland on black clay, *Acacia-Combretum-Schefflera* wooded grassland; (750-)1250-1900 m. EE EW TU GD SD HA; E and S Sudan, Uganda, Kenya, N Tanzania, Burundi. *Ash* 1901; *Friis* et al. 2639; *Mooney* 9854.

This is an extremely variable species, especially as concerns the presence or absence of glands. The presence or absence of tendrils is also striking, often varying on different branches of the same plant. The plant well known in E Africa as *C. orondo* is clearly the same species as is *C. schweinfurthii* from the Sudan.

20. **C. burgeri** *Vollesen* (1984)
- type: HA, S of Jijiga, *Burger* 2766 (K holo., ETH FT iso.).

Trailing herb with stems up to 30 cm long or with short erect stems to 10 cm, from fleshy fusiform rootstock; all parts glabrous. Leaves 4-6-foliolate; petiole 0-1.5 cm; leaflets narrowly obovate, 3.5-11 x 0.5-3 cm, dentate with large, widely spaced teeth, acute; base attenuate. Inflorescence 3-8 cm long, with a few glands on distal branches; peduncle (0.5-)1-5 cm; pedicels 2-4(-6 in fruit) mm, with dense stalked glands. Buds *c* 2.5 mm long, glabrous; style *c* 2 mm. Fruit obovoid, 8-9 x 5-6 mm, glabrous; seed obovoid, *c* 7 x 5 x 5 mm, lateral crests conspicuous, sides smooth.

Acacia-Commiphora wooded grassland and bushland, grassland derived from *Acacia-Commiphora* bushland, on sandy to loamy limestone or quartz soil; 1500-1800 m. HA; not known elsewhere. *Sandford* in *Mooney* 7949; *J. de Wilde* 6422.

Rather similar to *C. serpens* in general appearance, but the author has never seen any glabrous - or even nearly glabrous - specimens of that species.

21. **C. cyphopetalum** *(Fresen.) Descoings* ex *Wild & Drummond* (1966);
Cissus cyphopetala Fresen. (1837) - type: GD/TU, Gondar to Adua, *Rueppel* s.n. (FR holo.).
Cyphostemma gallaense (Gilg & Brandt) Descoings (1967) - type: HA, Ego nr. Harer, *Ellenbeck* 391 (B holo., destr.).
Cyphostemma macrothyrsum (Gilg) Descoings (1967) - type: SD/BA, Dolo nr. river Daua, *Ruspoli & Riva* 1222 (1103) (FT holo.).
Cyphostemma pachyrrhachis (Gilg & Brandt) Descoings (1967) - type: GD, Dschadscha, *Schimper* 516 (FT iso.).
Cyphostemma pseudoniveum (Gilg & Brandt) Descoings (1967) - types: TU, Adi-Ana in Zada valley, *Schimper* 797 (B syn. destr., BM C isosyn.); GD, Gera Abuna Tekle Haimanot, *Schimper* 1139 (B syn. destr., BM K isosyn.); HA, Gara Mulata, *Ellenbeck* 488 (B syn. destr.).
Cissus javellensis Lanza (1939) - type: SD, Yavello, *Cufodontis* 413 (FT holo.).
Cissus javellensis var. *rotundata* Lanza in *Miss. Biol. Boran.* **4**: 115 (1939) - type: SD, Arero, *Cufodontis* 348 (FT holo.).
Cissus neghelliensis Lanza (1939) - type: SD, Neghelle, *Cufodontis* 181 (FT holo.).
Cissus digitata sensu Cuf., l.c. quoad distrib. Aethiop., non (Forssk.) Lam. (1791).

Herbaceous climber or scrambler (often somewhat succulent) to 6 m from a beet-shaped fleshy rootstock; all parts puberulous to tomentose and occasionally with stalked glands. Leaves 3-5-foliolate; petiole 0.3-6.5 cm; leaflets ovate to elliptic or obovate to suborbicular, (1.5-)2.5-15 x (1-)1.5-7.5(-10) cm, grossly crenate to dentate (rarely serrate), teeth usually large and triangular; apex sub-acuminate to rounded; base attenuate to sub-cordate. Inflorescence 4-20 cm long; peduncle 2.5-13.5 cm; pedicels 1-7(-11 in fruit) mm. Buds 2-5 mm long, puberulous to tomentellous; style 1-3 mm. Fruit ellipsoid to obovoid or subglobose, 7-14 x 4-9 mm, puberulous to tomentose; seed ellipsoid to subglobose, 4-10 x 3-6 x 3-6 mm, lateral crests weak to conspicuous, sides smooth to weakly (rarely distinctly) ridged. Fig. 119.10.

Edges of evergreen forest, upland bushland and scrub, riverine forest, *Juniperus* forest and scrub, *Acacia-Commiphora*, *Acacia-Combretum* and *Acacia-Balanites* woodland, wooded grassland and bushland, persisting in cultivated areas, roadsides; (250-)700-2800 m. EE AF EW TU GD WU SU SU-AR GJ WG GG SD BA HA; Cameroun, Sudan, Uganda, Kenya, Tanzania, Rwanda, E Zaire, Zambia. *Burger* 2798; *Getachew Aweke* 650; *Mooney* 4805.

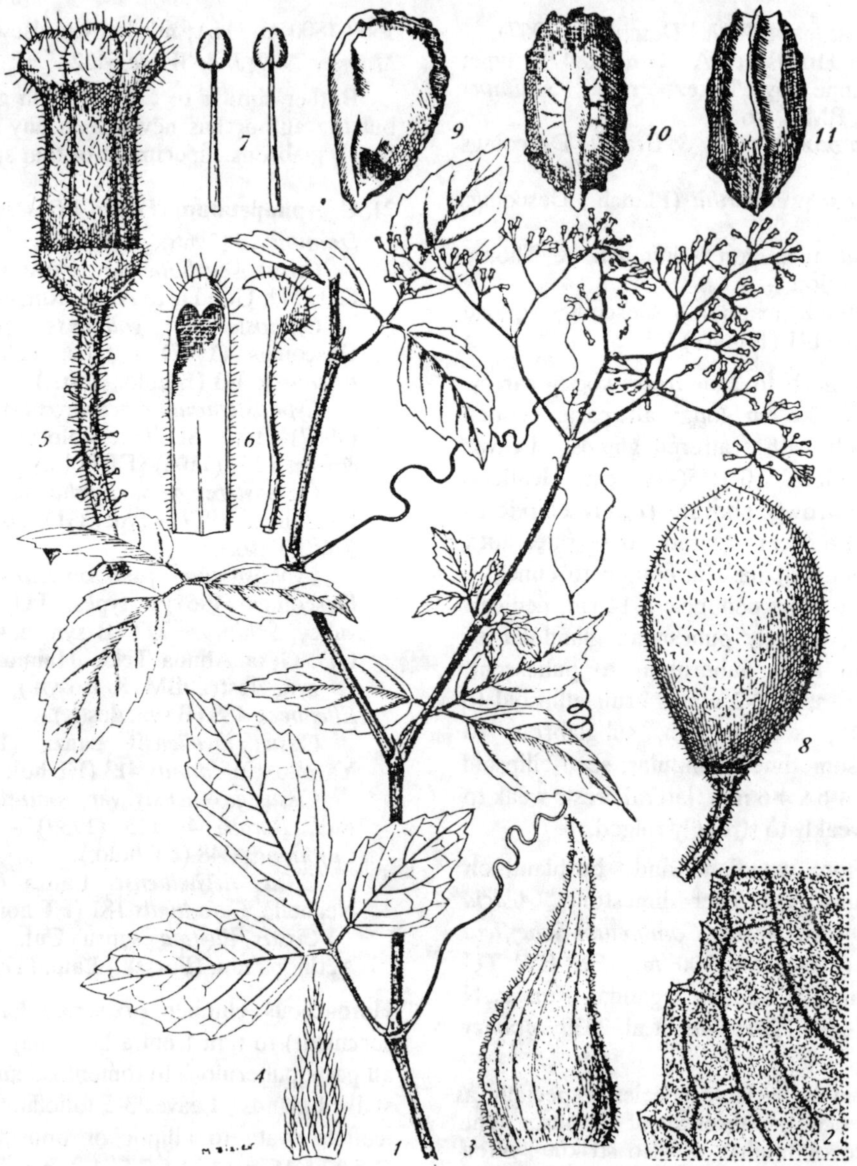

Figure 119.10 *CYPHOSTEMMA CYPHOPETALUM*: 1 - flowering branch x 2/3; 2 - leaf underside x 2; 3 - stipule x 5; 4 - bract x 15; 5 - flowerbud x 10; 6 - petal from inside x 10; 7 - stamens x 10; 8 - fruit x 4; 9, 10 & 11 - seed lateral, ventral and dorsal views x 4. (Reprinted with permission from *Fl. Cameroun* 13: Pl. 12, Vitacées, 1972.)

An extremely polymorphic species. Several different entities can be roughly recognized in different areas. But considered over the total area of the species, there seems at the moment no reasonable way in which to subdivide the material satisfactorily. It may well be that careful field studies will eventually allow several infraspecific taxa to be recognised.

Excluded species.

Cissus pauli-guilielmii Schweinf. (1868) - syntype: *Schweinfurth* 2301 (not seen).

This collection is a specimen of *Merremia kentrocaulos* (Convolvulaceae) - see Vollesen (1986) in *Kew Bull.* **41(2).**

120. RUTACEAE

by M. G. Gilbert*

Cufodontis, *Enum*. 367-374 (1956); Kokwaro, *Fl. Trop. E. Afr.*, **Rutaceae** (1982).

Shrubs or trees, occasionally herbs or climbers, sometimes with spines or prickles, aromatic oils present. Leaves alternate or opposite, simple or compound, with pellucid glands at least on margins; stipules absent. Flowers usually radially symmetrical, perfect or unisexual by abortion, (2-)4-5-merous. Sepals free or connate, imbricate (rarely valvate). Petals free or connate. Stamens inserted at base of disc, as many or twice as many as sepals, rarely numerous and in bundles, absent or reduced to staminodes in female flowers. Ovary superior, sometimes on gynophore, reduced to pistillode in male flowers, made up of 1-5(-18) carpels, usually fused, sometimes free; ovules 1-2(-several) per locule or carpel. Style mostly single, sometimes very short, stigma capitate or peltate. Fruits various, mostly fleshy and few-seeded in Ethiopian genera.

Some 150 genera and 900 species widely distributed in the tropical and warm temperate areas of both hemispheres but most numerous in South Africa and Australia. In Ethiopia 6 genera and 14 species indigenous, a further 4 genera and 11-12 species introduced.

Usually easily recognised by the large pellucid glands in the leaves, most easily seen by holding a leaf up to the light, and by the strongly aromatic smell of most species.

The family is of economic importance, particularly as fruit crops (most notably several species of *Citrus*). The aromatic oils make some species valuable as flavourings (eg *Ruta*) and scents. The timber is of high quality and valuable when available in large-enough sizes; several species are used to make farm-implements. A few species are grown as ornamentals in Ethiopia (eg. *Calodendron*).

The Orominia name HADDESSA is applied apparently rather indiscriminently to 3-foliolate shrubs and small trees including at least *Teclea* and *Vepris* in this family and *Allophyllus* (Sapindaceae) and *Rhus* (Anacardiaceae).

Key to Genera.

1. Leaves 1- or more foliolate, often alternate; flowers white, yellow or green, without petaloid staminodes; fruit not as below. 2
- Leaves simple, opposite; flowers large, pink, with 5 stamens alternating with 5 petaloid staminodes longer than petals; fruit a rugose capsule with several large shiny black seeds. **2. Calodendrum**
2. Leaves pinnately compound. 3
- Leaves palmately 1-3(-5)-foliolate. 6
3. Tree or shrub, not glaucous; leaves pinnate; flowers small, petals entire; fruit indehiscent or, if dehiscent, only 1 seed per locule. 4
- Glaucous herb or subshrub; leaves more or less bipinnate; flowers conspicuous, petals yellow, margins fimbriate; fruit an apically dehiscent capsule with several small seeds. **1. Ruta**
4. Plant unarmed; ovary with terminal stigma; fruit an indehiscent drupe or berry. 5
- Stems and petioles usually with numerous prickles; ovary of 1-2(-3) separate carpels, each with a lateral style; fruit of 1-2(-3) 1-seeded follicles, seeds glossy black. **3. Zanthoxylum**
5. Leaves and leaflets opposite; flowers unisexual. **4. Fagaropsis**
- Leaves and leaflets alternate; flowers bisexual. **9. Clausena**
6. Plants unarmed or with randomly distributed or paired prickles; petiole terete or, if narrowly winged, leaf 3-foliolate; flowers with up to 10 stamens or staminodes; fruit a drupe with 1-5 (-7) pyrenes. 7
- Axillary spines present; leaf usually 1-foliolate with conspicuously winged petiole; flowers with 20-40 stamens grouped into bundles; fruit a typical *Citrus* hesperidum with 8 or more locules (trees grown for their fruits). **10. Citrus**
7. Flowers 5-merous; fruit with 3-5(-7) pyrenes (sometimes fewer by abortion). 8
- Flowers 4-merous; fruit with 1-2 pyrenes. 9
8. Liana or scandent shrub, usually with numerous recurved prickles on stems and petioles; leaves 3-foliolate; fruit up to 1 cm diameter. **5. Toddalia**
- Erect unarmed shrub or tree; leaves palmately 3-5-foliolate; fruit 8-10 cm diameter, apple-like (grown for its fruits). **6. Casimiroa**
9. Anthers 5-8; fruit 1- or 2-loculed. **7. Vepris**
- Anthers 4; fruit always 1-loculed. **8. Teclea**

(A joint key to the species of *Vepris* and *Teclea* is given under the former genus.)

1. RUTA *L.* (1753)

R. chalepensis *L.* (1767).
R. bracteosa DC. (1824).

Glaucous subshrub to 1 m high; leaves bipinnate, alternate. Inflorescence cymose; bracts ovate. Flowers bisexual, 5-merous. Petals yellow, fimbriate. Fruit an apically dehiscent capsule, segments ± acuminate, each with several small seeds.

Very widely cultivated throughout the highlands above *c* 1500 m. Originally from the Mediterranean or Middle East. *Mooney* 5030, 7580; *Tadessa E.* 505.

* Ethiopian Flora Project, c/o Royal Botanic Gardens, Kew, Richmond, Surrey, TW9 3AB, UK.

A culinary herb used to flavour milk, cottage cheese and coffee and in the preparation of 'berberi'.

Often misnamed as *R. graveolens* L. which has not been recorded in Ethiopia.

2. CALODENDRUM *Thunb.* (1782)

C. capense *(L.f.) Thunb.* (1782).

Tree reaching 20 m high, usually less. Leaves opposite, simple. Inflorescence a terminal panicle. Flowers large, bisexual, 5-merous, pink. 5 stamens alternating with petaloid staminodes longer than petals, mature petals 2-3.5 cm long. Fruit a rugose 5-locular capsule splitting from base to reveal large shiny black seeds.

Grown for the very decorative flowers. Native to Africa from Cape Province to northern Kenya (Mt Kulal) and possibly occuring wild in south Ethiopia. *Mesfin* 37.

3. ZANTHOXYLUM *L.* (1753)
Fagara L. (1759)

Hartly, *J. Arn. Arb.* **47**: 171-174 (1966).

Trees and shrubs (or scandent), usually with prickles on stems and petioles, often persisting as woody prickle-tipped bosses on trunk. Leaves alternate (or opposite), pinnate; leaflets usually opposite. Inflorescence a terminal or axillary panicle. Flowers dioecious, 4-5-merous with distinct calyx and petals (*Fagara*) or with a single whorl of 6-8 lobes (*Zanthoxylum* sens. str.). Male with 4-5 stamens and vestigial pistillode. Female with rudimentary staminodes, short gynophore and 1-2(-5) free carpels. Styles lateral, ovules 2 per carpel. Fruit of 1-2(-5) subglobose, gland-pitted follicles splitting to reveal solitary glossy blackish globose seeds.

Perhaps 275 species most numerous in the tropics but also extending into temperate regions.

The African species were formerly placed in *Fagara* but the arguments of Hartly (loc. cit.) are generally accepted and *Fagara* is now included in *Zanthoxylum*.

1. Leaves spaced along stems, leaf scars never crowded; inflorescence terminal, much branched. 2
- Most leaves crowded on short stout shoots (look for regions of stem with very crowded leaf scars); inflorescence subterminal, produced below new leaves, little branched. **1. Z. chalybeum**
2. Leaflets with tip produced into long obtuse acumen, pellucid dots red; inflorescence pointed, side branches not rebranched; female flower with solitary carpel. **2. Z. leprieurii**
- Leaflets obtuse to shortly acuminate, pellucid dots whitish; inflorescence rounded, side branches rebranching; female flower with 2-3 carpels producing corresponding number of follicles. **3. Z. usambarense**

1. **Z. chalybeum** *Engl.* (1895);
Fagara chalybea (Engl.) Engl. (1896) - type: Tanzania, *Stuhlmann* 3451.

Deciduous shrub or tree 3-4(-10) m high; trunk corky, deeply fissured; twigs stout, usually with interrupted zones of densely crowded leaf scars and many dark recurved prickles, dormant tip protected by scale-leaves. Leaves (3-)5-15-foliolate, rachis (6-)14-17 cm long; leaflets elliptic to ovate oblong, (2.5-)6-11 x (1-)2-3.3 (-4.3) cm, tip acute to obtuse, margin shallowly crenate to entire, glands whitish, smelling strongly of lemon when crushed. Panicles from just below new leaves, mostly with short side branches. Flowers in clusters, 4-5-merous, petals yellowish-green, 2.5-3.2 x 1-1.2 mm; male with pedicel 1.5-2 mm long; female subsessile with solitary carpel. Follicle solitary with stipe to 1.5 mm long, obliquely subglobose, 5-8 mm diameter, light red, conspicuously gland-pitted; seed 4-7 mm diam.

Var. **chalybeum**

Leaves glabrous. Fig. 120.1.1.

Deciduous woodland or bushland on well drained slopes, locally common; 900-1550 m. GG BA HA; Somalia, south and west to Zaire, Zambia, Zimbabwe, Malawi and Tanzania. *Burger* 1724; *Burger & Amare* 300; *Carr* 330.

Var. *mollis* Kokwaro, with densely tomentose leaflets, occurs in Sudan, Uganda and Tanzania.

Bark greatly valued by traditional healers in Kenya.

2. **Z. leprieurii** *Guill. & Perr.* (1831);
Fagara leprieurii (Guill. & Perr.) Engl. (1896) - type: Senegal, *Leprieur.*

Tree to 15 m; trunk smooth with vertical stripes; twigs slender, leaf scars never closely crowded, with usually rather straight prickles. Leaves glabrous, (9-)13-17 (-25)-foliolate, rachis (15-)25-34(-55) cm long; leaflets ovate (to oblong-elliptic), (2.5-)7-11(-16) x (1.5-)3-4 (-7.5) cm, tip produced into conspicuous blunt acumen, obscurely crenate towards tip, glands numerous, reddish. Panicle terminal, pointed, 15-30 (-45) cm long, with first order branching only. Flowers 4-merous, in dense clusters. Petals creamy-white, 1.5-2 mm long. Female with 1 carpel. Follicle 4-5 mm diam.; seed 3-3.5 mm diam., otherwise as in *Z. chalybeum*.

Deciduous woodland or forest with clay soil in association with other W African species such as *Lannea barteri*; 500-550 m. IL; west to Senegal, Uganda, Zaire, Tanzania, Angola, Mozambique & Natal. *Chaffey & Thomerson* 596; *Friis* et al. 2503.

The 2 Ethiopian collections seen are both sterile and fertile material is needed to confirm the determination. The description is based primarily on the FTEA account.

Figure 120.1.1 - *ZANTHOXYLUM CHALYBEUM* var. *CHALYBEUM*: flowering branch, portion of bark showing old spines; male flower, female flower with transverse and longitudinal sections of ovary. 2 - *Z. USAMBARENSIS*: branch with flowers and leaves; female flower, 1 petal removed; transverse section of ovary. 3 - *TODDALIA ASIATICA*: flowering branch; male flower with 1 petal removed, and floral diagram. Scale: cm scale for branches in closed bar; mm scale for floral details as open bar. (Reproduced with permission of Oklahoma Agricultural Experiment Station from Experiment Station Bulletin No. 45, *Families of Flowering Plants in Ethiopia*, figure 28, 1967.) Drawn by W.C. Burger.

3. **Z. usambarense** *(Engl.) Kokwaro* (1978);
Fagara usambarensis Engl. (1905) - type: Tanzania, *Albers* 354.

Tree or shrub 2-10(-15) m high; trunk with thick corky bark fissured into prominent ribs; twigs as in *Z. leprieurii*. Leaves glabrous, (5-)7-17-foliolate; rachis 6.5-14(-30) cm long; leaflets elliptic to oblong elliptic, (2-)4-8 x 1.5-3.7 cm, tip acuminate to acute, shallowly crenate, glands few, often only near margin, white. Panicle terminal, rounded, secondary branches well developed, up to 15 cm long, 19 cm broad in male plants, smaller in female. Flowers 4-5-merous, in clusters, petals greenish white, (2-)3-4(-4.5) x 1.2-1.7(-2.2) mm. Female with 2 (-3) carpels. Follicles 2(-3) together, ovoid, 8-10 x 6-7.5 mm diam. otherwise as in *Z. chalybeum*. Seed 5-7 x 4.5-5.5 mm. Fig. 120.1.2.

Rocky slopes, often over limestone, with evergreen bushland or deciduous wood land, sometimes in secondary bushland following *Podocarpus* clearance; 1650-2100 m. SD HA; Kenya, Tanzania, Rwanda. *Burger* 2038; *Amare* in *IECAMA* G-52; *Gilbert & Ermias* 8437.

The Ethiopian material is mostly described as 2-5 m high, often a shrub, in contrast to collections from East Africa which are nearly always described as 6-9(-15) m high trees, often with a clear bole for several metres. Trees *c* 10 m high have been seen growing in the middle of dense thickets in SD, it is possible that collectors in Ethiopia have not ventured into such thickets but have only taken material from their margins.

4. **FAGAROPSIS** *Siebenlist* (1914)
Clausenopsis (Engl.) Engl. (1931)

Deciduous trees or shrubs. Leaves opposite, pinnate; leaflets opposite or subopposite, glands often restricted to margins. Inflorescence a terminal panicle developing before or with new lateral shoots. Dioecious. Flowers 4(-5)-merous. Male with 4-8 stamens and vestigial pistillode. Female without staminodes, ovary globose, 4-locular, locules with solitary ovules, style short, stigma subpeltate, obscurely lobed. Fruit a subglobose drupe with 2-4(-5) pyrenes, shortly stipitate, gland-dotted; pyrenes thin-walled, smooth, 3-sided in cross section; seeds grey or dark grey.

Two very closely-related species in Africa plus two species in Madagascar.

1. Slender shrub or small tree 2-5(-7) m high; stems and leaves puberulous to tomentose; inflorescences up to 6 cm long. **1. F. hildebrandtii**
- Tree 7-25(-40) m high; stems and leaves usually glabrous; inflorescence 5.5-12 cm long. **2. F. angolensis**

1. **F. hildebrandtii** *(Engl.) Milne-Redh.* (1936);
Clausena(?) *hildebrandtii* Engl. (1895) - type Kenya, *Hildebrandt* 2814.
Fagaropsis gillettii Chiov. (1941).

Slender shrub or small tree 2-7 m high; bark pale to brownish grey, rough or rather smooth; twigs purple-brown with large pale lenticels, puberulous to tomentose. Leaves (3-)5-9-foliolate, puberulous to tomentose, rachis 4-6.5(-10) cm long; lateral leaflets ovate to oblong-ovate or elliptic, 2.5-7 x 1.2-3.7 cm, sessile or subsessile, base rounded to slightly cordate, tip obtuse or acute, margin entire or nearly so with closely spaced marginal glands; terminal leaflet often obovate with petiolule to 2 cm long, tip sometimes retuse. Panicles rounded, to 6 cm long, axes hairy. Pedicels 4-10 mm long. Sepals densely white hairy. Petals oblanceolate, 3.5-7 x 2-3.2 mm, whitish or greenish yellow. Male flowers: filaments 2.5-4 mm long, pistillode glabrous. Female: ovary obscurely 4-lobed, glabrous. Drupe subglobose, 6-8 mm wide, light red (to black), conspicuously gland dotted; pyrenes to 6 mm long, pale brown.

Acacia-Commiphora bushland extending into open *Combretum-Terminalia* woodland: 1500-1700 m. SD HA; Somalia, Kenya. *Burger* 2699, 2700; *J. de Wilde* 6502; *Gilbert* et al. 7743.

Essentially a dry-country vicariad of the following more widespread species. The 2 taxa may be better treated as subspecies as there are no clear-cut differences between them apart possibly from stature and there are intermediates in the area of contact in Kenya. Material from HA and Somalia ('*F. gillettii*') is rather less hairy with more cuneate leaflet bases than elsewhere but the differences are not worth formal recognition.

2. **F. angolensis** *(Engl.) Dale* (1936);
Vepris angolensis Engl. (1896); *Clausenopsis angolensis* (Engl.) Engl. (1931) - type: Angola, *Welwitsch* 1315.

Tree 7-25(-40) m high with smooth grey bark differing from *F. hildebrandtii* by the normally glabrous stems and leaves; leaf rachis 13-40 cm long; leaflets ± subsessile to petiolulate, 4-9(-15) cm long, tip often acuminate; inflorescence 5.5-12 cm long. Fig. 120.2

Most records from forest with *Podocarpus*, also extending down into lowland *Celtis-Aningeria altissima* forest: 1300-1900 m. WG IL SD BA; Kenya, Uganda, Tanzania, Malawi, Zimbabwe, Angola. *Chaffey* 978; *W. de Wilde* 10356; *Friis* et al. 3488.

The collections from WG were described by Italian foresters as trees 30-40 m high producing very fine timber.

5. **TODDALIA** *Juss.* (1789)

Scandent shrub or liana, stems and often leaves with sharp recurved prickles. Leaves alternate, 3-foliolate, prominently dotted with glands. Dioecious. Inflorescence a terminal or axillary panicle. Flowers 4-5-merous. Male: stamens 4-5, pistillode flask-shaped. Female: staminodes filiform or absent. Ovary shortly

Figure 120.2 *FAGAROPSIS ANGOLENSIS*: 1 - fruiting branch x 1; 2 - detail of leaf-margin x 6; 3 - part of male inflorescence x 1; 4 - male bud x 3; 5 - male flower x 3; 6 - sepal x 3; 7 - petal x 3; 8 - stamen x 6; 9 - vestigial ovary x 6; 10 - part of female inflorescence x 1; 11 - female bud x 3; 12 - female flower x 3; 13 - calyx x 3; 14 - petal x 3; 15 - section of ovary x 4; 16 - seed x 2. 1,2 from *Moon* 411; 3-9 from *Ramell* in *F.D.* 1058; 10-15 from *Eggeling* 3962; 16 from *Eggeling* 3962. Drawn by Miss D.R. Thompson. (Reproduced with permission from *Fl. Trop. E. Afr.*, Rutaceae, fig. 12, 1982.)

stalked, 5-7-locular, ovules 2 per locule, stigma subsessile, peltate. Fruit a globose 3-7-locular drupe, endocarp woody, locules 1-seeded.

1 widespread polymorphic species in Africa and Asia plus perhaps 2 local endemics in the Far East.

T. asiatica *(L.) Lam.* (1797);
Paullinia asiatica L. (1753); *Toddalia aculeata* Pers. (1805) *nom. superfl.* - type: Sri Lanka, Herb. Linnaeus 512.1.

Scandent to 4.5 m, stems and leaves rust-red hairy, soon glabrescent. Petiole usually with prickles; leaflets subequal, elliptic to obovate or oblanceolate, 3-8 x 1.2-3 cm, tip usually bluntly acuminate, entire or obscurely crenate, leathery and glossy, midrib often with prickles. Panicles usually shorter than subtending leaves, little branched. Flowers clustered, 5-merous, petals oblong, c 2.5 x 1 mm, acute, whitish or yellowish green. Male flowers: stamens longer than petals, pistillode glabrous. Female: staminodes very small; ovary cylindrical-ovoid with 3-5-lobed stigma, glabrous. Drupe subglobose, 5-7-angular, 7-10 mm wide, fleshy, orange. Seeds reniform, 3-4 mm long, dark brown. Fig. 120.1.1.

Margins of forests and thickets in bushland/grassland mosaics; (550-)1700-2400(-3300)m. WU SU AR KF SD HA; Sudan, Zaire, E Africa, south to Transvaal; Madagascar and Mascarene Islands; India east to Taiwan and Borneo. *Burger* 1760, 2438; *W. de Wilde* 9275.

A wide spread and variable species with several varieties described from Asia, some of which look rather distinctive. The African material is rather uniform, perhaps suggesting that it may be a relatively recent introduction from Asia. Collections showing female flowers are inexplicably rare, none have been seen from Ethiopia.

6. CASIMIROA *La Llave* (1825)

C. edulis *La Llave* (1825)

Tree to 12m. Leaves alternate, palmately 3-5-foliolate, puberulous. Inflorescence an axillary panicle. Flowers 5-merous, white or greenish-white. Fruit resembling an apple, 8-10 cm diameter, yellowish with soft sweet whitish flesh, with 1-5 large pyrenes.

Native of Central America, cultivated for its edible fruits, especially around Asmera. *Brunt* s.n.; *Westphal* 610; *Afewerki G. & Mehret A.* 64.

Slow to establish and start fruit production but drought-resistant and reliable once mature. Fruits sold locally only as they damage easily in packing and transportation.

7. VEPRIS *A. Juss.* (1825)

Trees or shrubs (sometimes subscandent). Leaves evergreen, alternate or opposite, 1-3-foliolate, prominently gland dotted. Mostly dioecious, occasionally polygamous or bisexual. Inflorescence of terminal and/or axillary panicles, sometimes reduced and dense. Flowers in clusters, (2-)4(-5)-merous. Sepals united to form cupular calyx with short broad lobes. Petals oblong. Male: stamens 5-8(-9), inserted at base of disc, pistillode usually 2-4-lobed. Female: up to 8 staminodes (or normal stamens and flowers bisexual); ovary 1-2(-4)-locular, each locule with (1-)2 ovules, stigma sessile, capitate or peltate, 2-4-lobed. Fruit a 1-2(-4)-locular drupe, endocarp thinly woody, locules usually 1-seeded.

About 30-40 species in tropical and South Africa and Madagascar, 1 species extending to the Mascarene Islands and 1 species in southern India.

The distinction between *Vepris* and other African genera in the *Toddaliae*, particularly the following genus in this account, *Teclea*, seems often arbitary and impracticable. Any future revision of the genera in this complex is likely to reduce the number. Until such a revision has been completed the *status quo* is best maintained. However, for convenience a combined key to the species of *Vepris* and *Teclea* has been given.

The wood of several species in *Vepris* and *Teclea* is very tough and is favoured for farm implements, etc. Most, perhaps all, species in the 2 genera contain alkaloids, sometimes in large quantities, and several are used as febrifuges.

1. Leaves strictly 1-foliolate. 2
- Leaves (1-)3-foliolate, never all 1-foliolate. 3

2. Leaves acuminate; flowers pedicellate; stamens 8. **4. Vepris eugeniifolia**
- Leaves obtuse to acute; flowers sessile; stamens 4. **3. Teclea simplicifolia**

3. Leaves consistently opposite or subopposite; fruit conspicuously 2-lobed. **1. Vepris dainellii**
- Leaves alternate, occasionally subopposite on coppice shoots; fruit subglobose to ellipsoidal (possibly 2-lobed in *Vepris* sp. = *Burger* 3072, female plant not seen). 4

4. Leaflets 3-6.5(-8.8) cm long, obtuse or emarginate, rarely acute; stamens 5-8. 5
- Leaflets 9-17 cm long, acute or often acuminate; stamens 4. 6

5. Petiole not winged; leaflets entire, glabrous; pistillode 2-lobed (female plant not seen). **2. Vepris sp.** = *Burger* 3072
- Petiole narrowly winged; leaflets usually obscurely crenate, hairy at least on midrib below; fruit subglobose to ellipsoidal, 7-11 x 7-9 mm, 1-seeded. **3. Vepris glomerata**

6. Stems and leaves glabrous; flowers subsessile; ovary and fruit glabrous. **1. Teclea nobilis**
- Stems, petioles and midribs minutely hairy when young; flowers pedicellate; ovary and base of fruit hairy. **2. Teclea borenensis**

1. **V. dainellii** *(Pichi-Serm.) Kokwaro* (1978);
Diphasia dainellii Pichi-Serm. (1951) - type: GJ, Zara Enda Michael, *Pichi-Sermolli* 2350 (FT holo., K iso.).

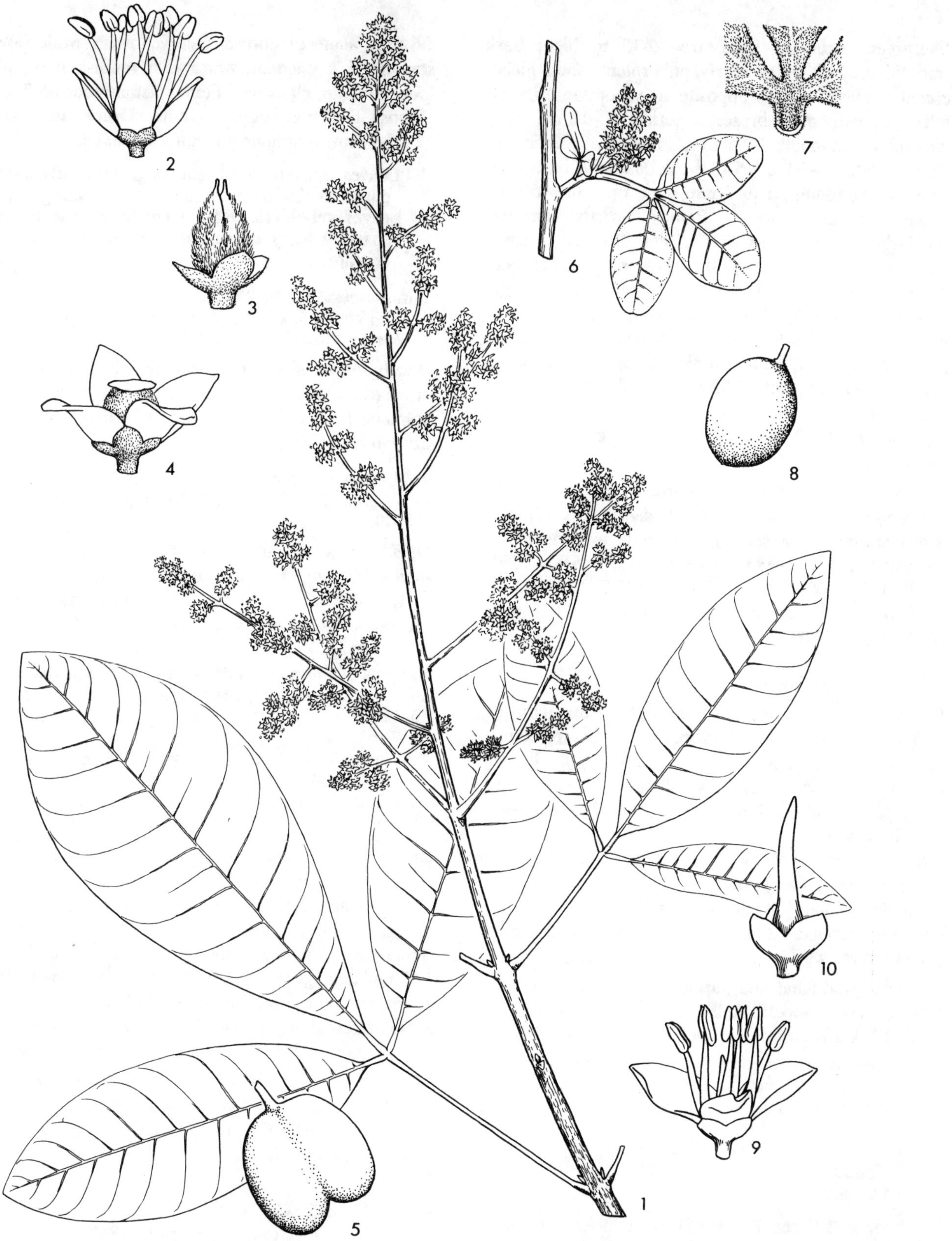

Figure 120.3 *VEPRIS DAINELII*: 1 - flowering branch x 1/2; 2 - male flower x 4; 3 - pistillode x 8; 4 - female flower x 4; 5 - fruit x 1. *V. GLOMERATA* var. *GLOMERATA*: 6 - part of flowering branch - x 2/3; 7 - detail of leaf-base x 4; 8 - fruit x 2. var. *GLABRA*: 9 - male flower x 4; 10 - pistillode x 8. 1-3 from *Meyer* 7746; 4 from *W. de Wilde* 9429; 5 from *Mooney* 7207; 6,7 from *Burger* 1952; 8 from *Mwasumbi* 10481; 9,10 from *Friis* et al. 3276. Drawn by Eleanor Catherine.

Evergreen shrub or small tree 2-15 m high; bark smooth, greyish; twigs scurfy puberulent, soon glabrescent. Leaves always opposite to subopposite, 3-foliolate, puberulent glabrescent; petiole 6.5-12 cm long, obscurely channelled above; leaflets elliptic to oblanceolate, 14-33 x 4.6-10.5 cm, leathery, base cuneate into petiolule up to 7 mm long, tip acute to acuminate, margin entire. Panicle strictly terminal, regularly conical, up to 25 x 15 cm, bract scars conspicuous, axes puberulous. Male flowers: pedicels up to 0.5 mm; petals (1.5-)2.5-3 x 1-1.7 mm, whitish or pinkish; stamens 8, filaments up to 4.5 mm long; pistilode 2-lobed, hairy. Female flowers rather larger, petals up to 4.5 x 2.5 mm, staminodes up to 8, as long as ovary; ovary obscurely 2-lobed, densely yellow pilose, stigma as wide as ovary. Drupe (1-) 2-lobed, up to 17 mm long x 20 x 8 mm, glabrous, conspicuously gland-pitted. Fig. 120.3.1-5.

Understory of moist montane forest, often with *Podocarpus* or *Aningeria adolf-frederici*, less often at forest margin or in secondary growth, extending into lowland *Celtis-Aningeria altissima* forest in SW; (1050-in SW)1750-2000(-2500) m. GJ SU WG IL KF SD BA; not known elsewhere. *W. de Wilde* 9429; *Friis* et al. 4029, 4030; *Mooney* 7207.

2. **V.** sp. = *Burger* 3072.

Densely branched shrub c 2.5 m high; stems glabrous, dark grey, rough. Leaves alternate, 3-foliolate; petiole narrowly grooved above, up to 1.8 cm long; leaflets up to 6.3 x 1.1 cm, narrowly elliptic, glabrous, very conspicuously gland-dotted, tip obtuse, base cuneate into short petiolule, margin entire, tending to roll back. Inflorescences axillary, glomerate but possibly elongating later (very young). Male buds sessile, calyx minutely ciliolate at margins; stamens 8; pistillode glabrous, 2-lobed. Mature inflorescences and female plants not known.

Mixed woodland and bushland overlying limestone; 1250 m. HA (Gobelli Valley, below Daletti, *c* 8°32´N 42°07´E); known only from the one gathering.

A very distinctive plant. Better material is needed to confirm its relationships and enable the preparation of a complete description.

3. **V. glomerata** *(F. Hoffm.) Engl.* (1896);
Toddalia glomerata F. Hoffm. (1889) - type: Tanzania, *Böhm* 3A.

Shrub or small tree 2.5-4(-7.5) m high; bark smooth, grey; twigs pale, hairy. Leaves alternate, 3-foliolate, glabrous to uniformly tomentose, drying pale green; petiole narrowly winged, 1.3-4 cm long; leaflets usually obovate, (1.5-)3-8.8 x 1-3.7 cm, tip rounded or slightly emarginate, rarely subacute, margins entire or obscurely crenate near tip. Panicles mostly terminal on short side branches, to 6 cm long, usually less. Pedicels 0.5-2 mm long. Calyx glabrous or hairy. Petals 2-3 x 1-1.6 mm, acute or obtuse, pale yellowish. Male flowers: stamens 5-7, unequal, white with reddish base; pistillode subulate, glabrous. Female: staminodes 5-7; ovary globose, glabrous, stigma capitate. Drupe ellipsoidal, 7-11 x 7-9 mm, glandular-punctate, 1-seeded.

1. Leaves obovate, 2.9-6 cm long, uniformly hairy, entire to obscurely crenate. var. **glomerata**
- Leaves subelliptic, 6.5-8.8 cm long, glabrous to sparsely hairy on midrib and margins, usually crenate. var. **glabra**

Var. **glomerata** Fig. 120.3.6-8
Vepris pilosa Engl. (1895) non Baker (1884); *Teclea pilosa* Verdoorn (1926)

Deciduous woodland or bushland, often on limestone or black cotton soil; 1350-1550 m. SD BA HA; Somalia, Kenya, Uganda, Tanzania. *Burger* 1952, 1953; *Corradi* 8355; *Gilbert* et al. 7827.

var. **glabra** *Kokwaro* Fig. 120.3.9-10
Kew Bull. **32**: 795 (1978) - type: Uganda, *Kerfoot* 949.

Deciduous woodland/bushland with *Terminalia brownii*, *Acacia*, *Commiphora* etc., often on basement complex rocks; 1450 m. SD ?HA; S Sudan, Kenya, Uganda, Tanzania. *Friis* et al. 3276; *Corradi* 7029, 7333.

The 1-seeded fruit and the reduced number of stamens of this species demonstrates the probable arbitary nature of the distinction between *Vepris* and *Teclea*.

4. **V. eugeniifolia** *(Engl.) Verdoorn* (1926);
Toddalia simplicifolia var. *eugeniifolia* Engl., *Pflanzenw. O. Afr.* C:228 (1895) - type: Tanzania, *Holst* 8869.

Shrub 2-4.5 m high; twigs glabrous. Leaves alternate, 1-foliolate; petiole swollen at both ends, 1-1.5 cm long; leaflet ovate, 3.5-9 x 2-4.2 cm, entire, tip acuminate, margins entire, glands raised on underside. Panicles axillary, raceme-like. Flowers pedicellate. Petals (1-)2-2.7 mm long, greenish yellow. Male: stamens (7-)8(-9), filaments up to 3 mm long, pistillode 2-lobed. Female: staminodes short, persistent, ovary subglobose, ± 2-lobed, glabrous. Drupe sublobose, 6-8 mm diameter, 1-seeded by abortion, abortive locule visible.

Growing on rock outcrops at altitudes up to 1400 m. GG BA; Somalia, Kenya, Tanzania. *Corradi* 8346-54; *Gilbert & Sebsebe* 8729.

8. TECLEA *Del.* (1843) *nom. cons.*
Aspidostigma Hochst. (1842-3)

Very closely related to *Vepris*, differing only by having the stamens always equal in number to the petals and by the consistently 1-locular ovary producing a 1-seeded fruit.

Some 25-30 species restricted to Africa and tropical Arabia.

The genus is named after Tecle Haimanot, a semi-legendary figure within the Coptic Church in Ethiopia.

See *Vepris* for a combined key to the species of the 2 genera and comments on their relationship.

1. **T. nobilis** *Del.* (1843)

- type: GD, Wegera ['Wogera'], *Feret & Galinier* (MPU holo. not seen)

Aspidostigma acuminatum Hochst. (1842/3) - non *Teclea acuminata* Engl. (1905) - type: GD, Wegera ['Wogerat'], *Schimper* II:1293 (TUB holo? not seen, K P iso.).

Teclea diversifolia Mattei (1933) - type: EW, Gaba Valley, *Senni* s.n. (PAL holo. not seen)

Shrub or understory tree 2-12 m high (?tall forest tree in East Africa); bark smooth, grey; twigs glabrous. Leaves alternate, rarely subopposite, (1-)3-foliolate, glabrous; petiole terete, 1.5-8 cm long; leaflets lanceolate to oblanceolate, 5-18 x 1.5-5.5 cm, cuneate into petiolule up to 1 cm long, tip acute to acuminate, margin entire. Inflorescences variable, terminal and/or axillary, sometimes contracted into false racemes, axes glabrous or minutely puberulous. Flowers sometimes polygamous, subsessile; petals yellowish or greenish-yellow, 2.5-4 x 1.1-2 mm, usually acute. Male flowers: filaments 2-5 mm long, pistillode subulate, glabrous. Female: staminodes up to 1.2 mm long, sometimes apparently functional, ovary subglobose, glabrous, stigma peltate, as wide as ovary. Drupe obovoid or fusiform, rarely 2-lobed, 6-15 x 5-8 mm, glabrous, red (see note below). Fig. 120.4.1-8.

Dry evergreen bushland or woodland through to moist montane forest. 900-2800m. All areas except AF & SD; Sudan, Somalia, Kenya, Uganda, Tanzania, tropical Arabia.

Two quite distinct fruit forms, probably corresponding to different taxa, have been included within *T. nobilis* as circumscribed here:

form **A.**

Fruits small, obovoid, 6-10(-11.5) x 5-6 mm, lacking obvious oil-glands, resembling those of *T. simplicifolia*. Fig. 120.4.7.

Drier evergreen bushland and woodlands; 900-1900 m. EE EW SU KF GG HA; Sudan, Kenya, Uganda, Tanzania. *Perdue* 6503; *Schweinfurth & Riva* 1137.

form **B.**

Fruits large, narrowly ellipsoidal to fusiform, 13-15 x 8 mm, with rather prominent oil glands, resembling those of *T. borenensis* and *T. hanangensis*. Fig. 120.4/8

Moister upland forests, often with *Podocarpus*; 2000-2250 m. WG SU; not seen elsewhere. *W. de Wilde* 6743, 7025; *Mooney* 5704.

Unfortunately other characters correlating with these fruit types have not been discovered and most of the material which lacks fruits cannot be placed into either form. It seems possible that the larger-fruited form has a distribution similar to that of *Vepris dainellii* as the 3 collections seen come from the same habitats as for that species. On this basis the type of *T. nobilis* probably belongs with the small fruited form.

Non-fruiting material at altitudes of upto 2800m has been seen from TU GD GJ AR IL BA; Somalia, Yemen.

Occasional collections show some 1-2-foliolate leaves but these coexist with normal *T. nobilis*; the type of *T. diversifolia* appears to be such a plant. Coppice shoots sometimes have opposite leaves. *Taddessa E.* 581 from the Ghibbie Gorge has opposite to subopposite 1-2(-3)-foliolate leaves but has typical small fruits and seems best regarded as just an extreme variant.

2. **T. borenensis** *M. Gilbert* (1986)

- type: Kenya (Moyale on Ethiopian border), *Gillett* 14120.

T. hanangensis Kokwaro (1978) pro parte quoad *Gillett* 14120, non sens. str.

Riverine tree to 12 m, similar to *T. nobilis* in most features, differing by the minutely hairy twigs, petioles and midribs; leaflets to over 10 x 4 cm, tips mostly acuminate. Male flowers: pistillode hairy. Female flowers with pilose ovary, stigma narrower than ovary. Drupe ellipsoidal, 17 x 6-7 mm, yellow, sparsely pilose at base, gland dots obvious. Fig. 120.4.9-10.

Under *Ficus* and *Trichilia emetica* along water courses in *Acacia-Commiphora* bushland; (800-)1250 m. SD (Genale Valley near Sidambale Bridge); Kenya. *Ash* 829; *Gilbert & Ermias* 8448.

3. **T. simplicifolia** *(Engl.) Verdoorn* (1926);

Toddalia simplicifolia Engl. (1895) - type: Tanzania, *Holst* 3801.

Teclea salicifolia Engl. (1902) - type: HA, Ega (9°33´N 42°08´E), *Ellenbeck* 387A (B holo. destr.).

Evergreen shrub or small tree 2-10(-20) m high; bark smooth, grey. Leaves alternate, rarely opposite on coppice shoots, 1-foliolate; petiole 5-18 mm long, obviously swollen at tip; leaflet 3-15 x 1.8-5.6 cm, glabrous, glands very conspicuous, tip usually obtuse, sometimes slightly acuminate to subacute, margins entire. Panicles terminal and axillary, often reduced so as to be raceme-like, glabrous. Flowers polygamous, sessile; petals 2.5-3.5 x 0.8-1.4 mm, yellow-green. Male: filaments 1.5-3 mm long, pistillode subulate, glabrous. Female: staminodes up to 1.2 mm long, ovary globose, glabrous. Drupe subglobose, 6(-10) mm long, yellow-orange or ± red, gland dots obscure.

Drier upland forests and woodlands with *Juniperus*, *Acacia* or *Euphorbia abyssinica*, extending into understory of *Podocarpus* forest; 1450-2100m. WU SU SD BA HA; Kenya, Tanzania. *Burger* 2973; *W. de Wilde* 7023; *Gillett* 14429.

The type of *T. salicifolia* is presumed destroyed but the protologue shows no obvious discrepency with *Burger* 2973 from near its type locality and it can reasonably confidently be placed in synonymy here.

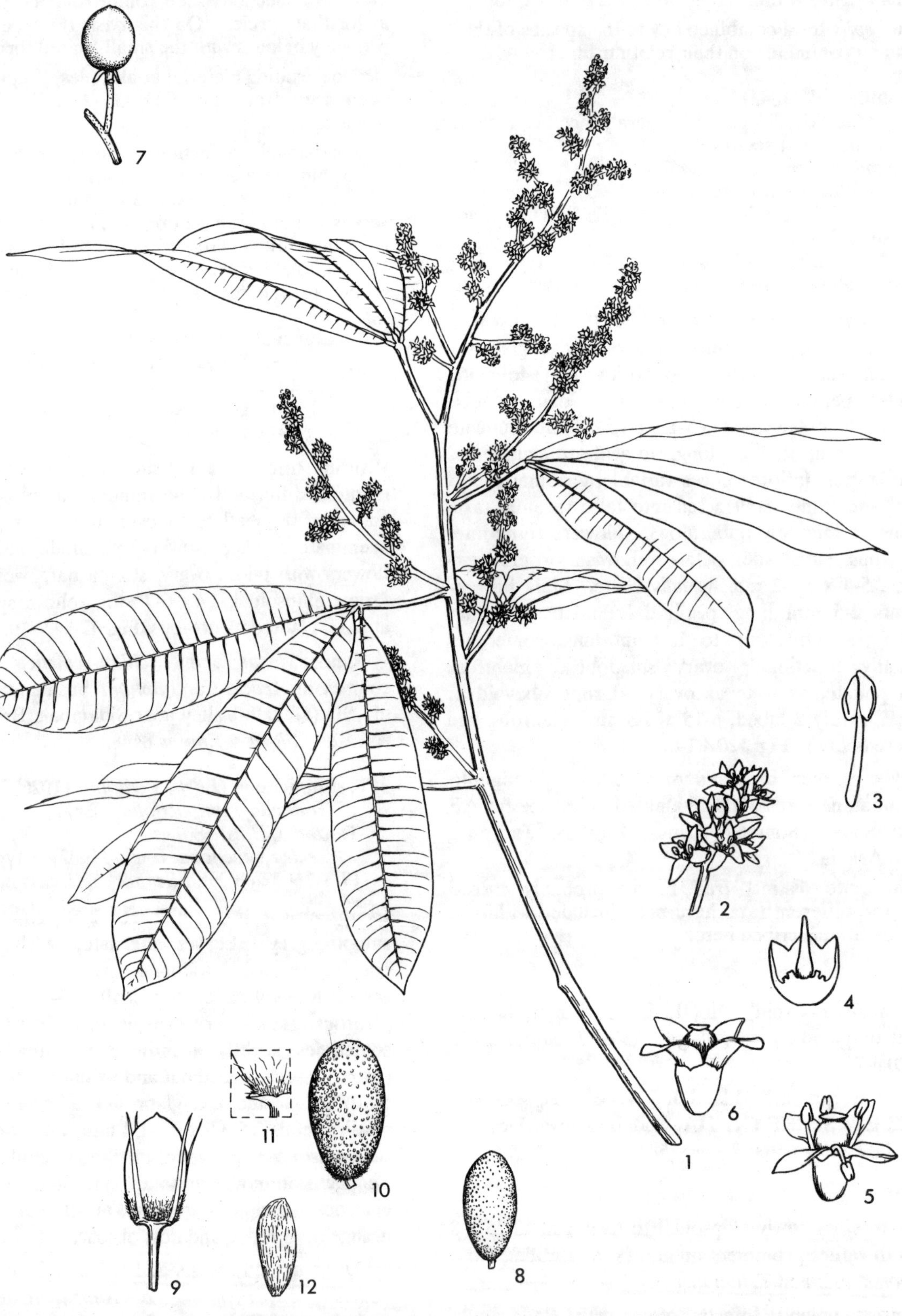

Figure 120.4 *TECLEA NOBILIS*: 1 - flowering branch x 2/3; 2 - group of male flowers x 2; 3 - stamen x 6, 4 - pistillode x 6; 5 - bisexual flower x 4, 6 - female flower x 4; 7 - fruit (small fruited form), 8 - fruit (large fruited form) x 1. *T. BORENENSIS*: 9 - female flower x 4; 10 - fruit x 1; 11 - detail of base of fruit to show hairs x 6; 12 - endocarp x 1. 1-4 from *Mooney* 7780; 5 from *Perdue* 6503; 6 from *Schweinfurth & Riva* 729; 7 from *Pappi* 2991; 8 from *W. de Wilde* 6743; 9 from *Gillett* 14120; 10-12 from *Ash* 829. Drawn by Eleanor Catherine.

Figure 120.5 *CLAUSENA ANISATA*: 1 - flowering branch x .85; 2 - flower; 3 - petal; 4 & 5 - stamen, front and back views; 6 - pistil x 3; 7 - fruits; 8 - seed x .85; 9 - leaflet, underside x 1.6; 10 - detail of leaf margin. All from *Koritschoner* 624. Drawn by D.E. Thompson. Crown copyright.

9. **CLAUSENA** *Burm. f.* (1768)

Unarmed trees and shrubs. Leaves alternate, pinnate; leaflets alternate or subopposite. Inflorescences axillary and paniculate in Flora area. Flowers bisexual, 4-5-merous. Stamens twice as many as petals, filaments thicker near base. Ovary on short stalk, subglobose, (2-)4-5-locular, 2 ovules per locule; style simple, robust, caducous; stigma terminal, inconspicuous. Fruit a subglobose few seeded berry.

About 30 species, all restricted to Asia except for the following.

C. anisata *(Willd.) Benth.* (1849);

Amyris anisata Willd. (1799) - type: Ghana, *Isert* in Herb. Willdenow 7292.

Clausena inaequalis (DC.) Benth. (1849).

C. inaequalis var. *abyssinica* Engl., *Pflanzenw. O. Afr.* C: 229 (1895); *C. abyssinica* (Engl.) Engl. (1915); *C. anisata* subsp. *abyssinica* (Engl.) Cufod., *Enum.*: 371 (1956) comb. inval. & XXVI (1969) - types: not listed in protologue, given by Cufodontis (loc. cit.: 371) as TU: *Schimper* (1863) 1496 & *Steudner* 1476 plus a collection from Tanzania, probably *Volkens* 1891.

Deciduous shrub or tree 1.5-6(-10) m high; bark smooth, grey; twigs purplish to brown, glabrescent. Leaves 11-25(-37)-foliolate, glabrescent; rachis 6-30 cm long; leaflets asymmetrically ovate or lanceolate to elliptic, 1-7(-11) x 0.7- 2(-4.3) cm, base with petiolule 1-3.5 mm long, tip bluntly acute to acuminate, margins entire or obscurely crenate, brightly gland-dotted, strongly aromatic. Panicles developing with new leaves, terminal or subterminal, usually shorter than leaves at maturity. Pedicels 2-5 mm long. Flowers 4-merous; sepals free, to 1 mm long. Petals 2.5-7 mm long, pale cream or white, acute. Stamens 2-6 mm long. Ovary 1-1.3 mm long on stalk c 1 mm long, glabrous, style c 2 mm long, stout, constricted at base. Berry globose or ovoid, 3.5-7 mm diam., usually with 2 large brown seeds. Fig. 120.5.

Montane forest margins, sometimes forming dense thickets, or understory trees in tall moist forest, also common in secondary bushland; 1500-2300 m. TU GD GJ SU WG IL KF IL SD; west to Sierra Leone, south to Cape Province. *Friis* et al. 218; *Gilbert & Jefford* 4399; *Mooney* 6666.

10. **CITRUS** *L.* (1753)

Swingle, The botany of *Citrus* and its wild relatives, in Webber & Batchelor, *The Citrus Industry* **1**: 129-474 (1943); Tanaka, Species Problem in *Citrus*, *Revisio Aurantiacearum* **IX** (1954); Townsend, **Rutaceae** in *Flora of Iraq* **4(1)**: 467 (1980).

Evergreen shrubs or trees; axillary spines usually present. Leaves 1-foliolate, usually with petiole winged and articulated at tip. Flowers usually bisexual, axillary, sometimes in short racemes. Calyx 3-5-lobed; petals (4-) 5(-8), white; stamens 20-40, in bundles; ovary of 8-18 united carpels, each locule with 4-8 axile ovules. Fruit an 'hesperidum', a very distinctive globose, ovoid or obovoid berry with distinct segments enclosed by peel and containing pulp made up of large vesicles.

Citrus is widely grown and economically important for its fruits. As is often the case with long cultivated plants their taxonomy and nomenclature are complex and confusing. The two major authorities, Swingle and Tanaka (loc. cit.) show significant differences of opinion complicating matters further. The genus is Asiatic in origin and no species are native to Africa but some, most notably *C. aurantifolia* and *C. aurantium*, may be met with as relicts of former habitations or as possible genuine escapes. The following key is taken from Townsend (loc. cit.) and covers all the species likely to be met with.

1. Petioles winged, at least narrowly, distinctly articulated with leaflet base; fruit with thinner peel making up only a small part of the bulk of the fruit. 2
- Petioles without wings, not distinctly articulated with leaflet base; flowers dimorphic with perfect flowers plus male flowers with abortive ovary; fruit with very thick peel, white except for outermost layer, making up bulk of fruit. **1. C. medica**
2. Stamens generally more than 4 times number of petals; male flowers present; ripe fruit often bright yellow with a mamillate process at each end. 3
- Stamens about 4 times number of petals; flowers usually all perfect; fruit rarely with mamillate process at each end and if so not bright yellow. 4
3. Flowers with some purplish streaks and suffusions; fruit sharp and acid to taste. **2. C. limon**
- Flowers pure white; fruit insipidly sweet to taste. **3. C. limetta**
4. Fruit large or not depressed globose; peel adhering to segments. 5
- Fruit relatively small, depressed globose; peel thin, easily detached, typically bright orange but often staying green. **4. C. reticulata**
5. Petiole wing broad; fruit c 10-15 cm diameter, up to 18-locular, pale yellow. 6
- Petiole wing narrow or if broad then fruit orange; fruit up to 10 cm diameter, up to 13-locular. 7
6. Petiole subcordate in outline; pulp vesicles large, readily separating; seeds ridged, with single embryo. **5. C. maximum**
- Petiole narrower, not subcordate; pulp vesicles smaller, tending to stick together. **6. C. paradisi**
7. Leaves broader and larger; fruit orange or orange-red, subglobose; pulp orange. 8
- Leaves rather small and narrow; fruit green or yellow with thin peel, ovoid, often with mamillate process at each end; pulp green, very acid. **7. C. aurantifolia**
8. Petioles broadly winged; peel rough; pulp sharply acid. **8. C. aurantium**
- Petioles narrowly winged; peel smooth; pulp sweet. **9 C. sinensis**

1. **C. medica** *L.* (1753)

Fruit irregularly cylindrical-ovoid, often very bumpy, 10-20 cm long, 10-13- locular; peel extremely thick, yellow; pulp sour or sweetish, often lacking.

Cultivated since at least the early 19th. century and probably much earlier; occasionally on sale in Addis Abeba and local markets throughout the country. Prized for its reputed medicinal properties.

2. C. **limon** *(L.) Burm.f.* (1768).

Citrus limonum Risso (1813).

C. limonia var *limonia sensu* Cufodontis, *non* Osb.

Flowers flushed pink outside. Fruit usually ellipsoidal with broad low mamillate processes at each end, 8-10-locular; peel rather thick, pale yellow, adherent; pulp very sour.

Only grown in plantations with production best below 1500 m.

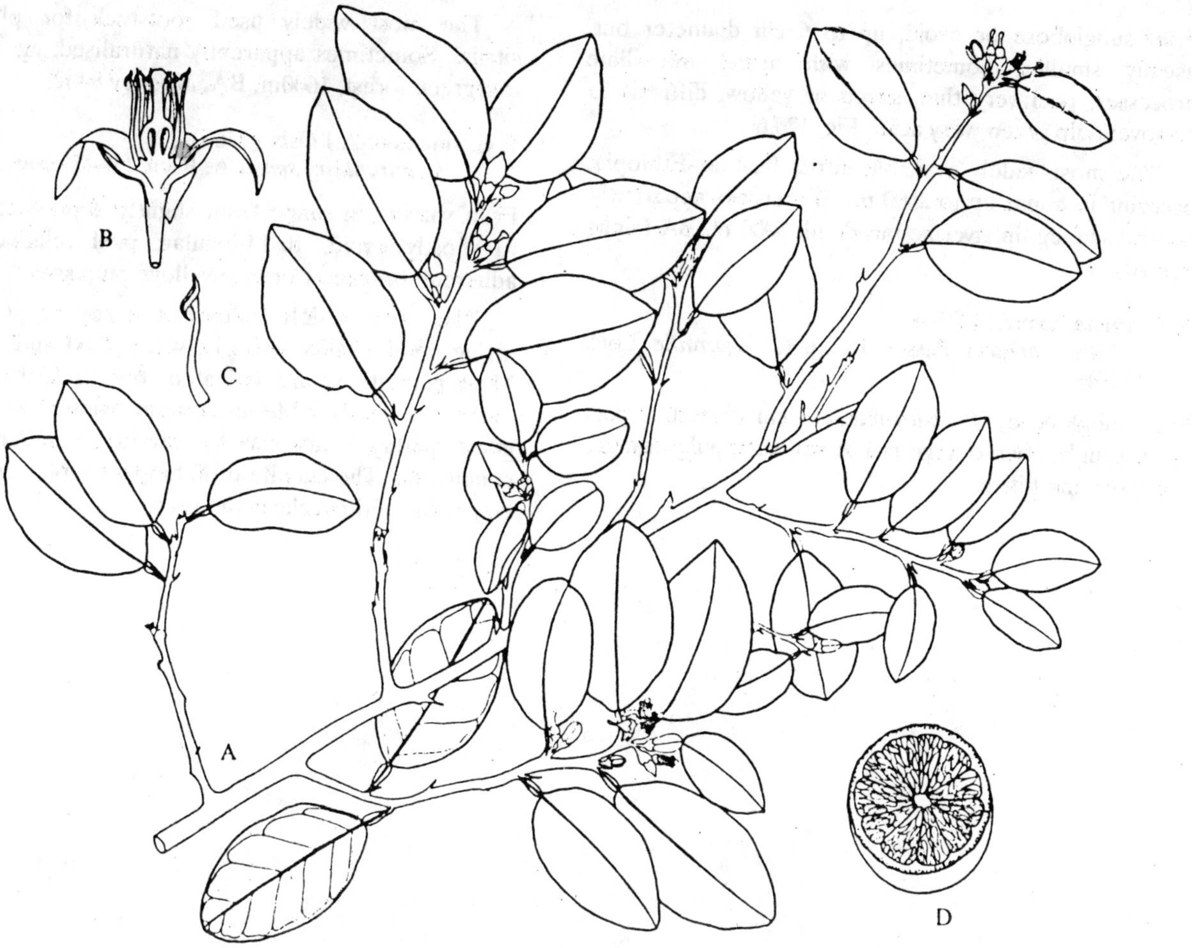

Figure 120.6 *CITRUS AURANTIFOLIA*: A - flowering branch x 1/3; B - flower in longitudinal section x 2; C - stamen x 3; D - fruit in transverse section x 1/2. (Reproduced from Purseglove, *Trop. Crops: Dicot.* 2: fig. 76, 1968.)

3. **C. limetta** *Risso* (1813)

C. limonia Osb. var *limetta* (Risso) Engl. (1931).

Very similar to *C. limon* but flowers pure white, fruit subglobose and pulp insipidly sweet.

Occurence in Ethiopia not certain, the record by A. Richard is based on a misidentification of *C. aurantiifolius*. Sometimes used as grafting stock for other *Citrus* fruits.

4. **C. reticulata** *Blanco* (1837)

C. deliciosa Ten. (1840)

Fruit depressed globose, 5-8 cm diameter, 10-15-locular; peel thin and very easily removed, typically bright orange when ripe but often staying green in Ethiopia; pulp very sweet.

Cultivated in plantations giving good quality fruits between 1500 and 1800 m. Can be grown under rainfed conditions but does better where irrigation is available.

Most of the fruits sold in Ethiopia would be better placed in *C. deliciosa*, the Mandarin, if this is treated as a separate species from the Tangerine. The authorities disagree on this point.

5. **C. maximum** *(Burm.) Merrill*

C. grandis (L.) Osb. (1757)

Fruit subglobose or obovoid, 11-16-locular, 10-13 cm diameter; peel thick, pale yellow; pulp pale yellow or pink; seeds yellow, ridged, with single embryo.

Not presently grown in Ethiopia but could be introduced if a market for it was found. Usually mistaken for the better known *C. paradisi* 'Grapefruit'.

6. **C. paradisi** *Macfad.* (1830)

Very close to *C. maximum*, probably being a cultigen of that species. Differs by the rather smaller fruits, thinner peel; pulp vesicles not seperating so easily; seeds smooth, white, with several embryos.

Cultivated in plantations at altitudes of 1500 m and below, doing best below 1000 m in the Awash valley and lowland areas of Eritrea where irrigation water is available.

7. **C. aurantifolia** *(Christm.) Swingle* (1913).

?*C. limon* Risso var. *pusilla* Risso in Duhamel, *Traite des Arbre et Arbuste ('Nouveau Duhamel')* **7**: 78 (1819).

C. limon var. *abyssinica* Riccob., *Bull. Ort. Bot. Palermo* **4**: 16 (1905) - type: 'sent by Col. Ameglio' (?PAL holo. not seen).

Fruit subglobose or ovoid, up to 6 cm diameter but usually smaller, sometimes with apical mamillate processes; peel very thin, green or yellow, difficult to remove; pulp green, very acid. Fig. 120.6.

The most widely available citrus fruit in Ethiopia, occuring in house up to 2500 m. Sometimes apparently naturalised, eg. in riverine bushland, 1000 m, BA (sight record).

8. C. **aurantium** *L.* (1753).
Citrus vulgaris Risso (1813). *C. bigarradia* Lois. (1819).

Fruit subglobose, 10-12-locular, *c* 7 cm diameter; peel thick, rough, often orange-red at maturity; pulp orange, very sour and bitter.

The most widely used rootstock for plantation citrus. Sometimes apparently naturalised, eg. in dense evergreen forest, 1600m, BA, *Mooney* 8462.

9. **C. sinensis** *(L.) Osb*. (1765).
C. aurantium sensu A. Rich. (1847) *non* L. (1753)

Fruit varying in shape from slightly depressed-globose to shortly ovoid, 10-14-locular; peel relatively thin, adherent, orange or orange yellow; pulp sweet.

The most widely cultivated citrus in plantations giving good quality fruits between 1000 and 1800 m. Most plantations are irrigated, but in Eritrea where there is a marked Mediterranean 'winter' wet season, good quality fruits can be produced under rainfed conditions. The climate of Ethiopia enables oranges to be produced throughout the year.

121. BALANITACEAE

by Martin J. S. Sands*

Cufodontis, *Enum.* (as **Agialidaceae**): 365-367 (1956); Launert in *Fl. Zamb.* **2**: 221-224 (1963).

Trees or shrubs, usually bearing conspicuous spines at a varying distance above the subtending leaves. Leaves alternate, 2-foliolate or some reduced to simple, minute, scale-leaves; stipules triangular, usually falling early; leaflets entire. Flowers regular, bisexual, 4- to 5-merous. Sepals and petals free, imbricate, the sepals ovate, usually eventually falling off, silky-hairy at least within, the petals (in Ethiopia) glabrous. Stamens 8-10, free, glabrous. Disc succulent, annular, longitudinally grooved, usually minutely papillose, surrounding the ovary. Ovary superior, 4-5-locular, each locule containing 1 pendulous ovule; style glabrous. Fruit a 1-seeded drupe, indehiscent or splitting apically, with a leathery or crisp and brittle exocarp, fleshy or spongy mesocarp and hard, woody endocarp.

One genus mostly confined to Africa and India.

BALANITES *Del.* (1813)

Sands, *Kew Bull.* **38**: 40 (1983).

Description as for the family.

A genus of 9 species, 2 in India and Burma, and 7 distributed throughout Africa except for the north-west and occurring as far south as Namibia and Natal with records also from the Arabian peninsular and the Jordan valley.

Balanites has been variously assigned to several families, notably **Zygophyllaceae** and **Simaroubaceae**. However, the alternate, bifoliolate, stipulate leaves associated with spines which are often supra-axillary are together distinctive, and, with other characters such as the free floral parts, the uni-ovulate loculi and the nature of the fruit, they uphold the view that the genus should stand as a monogeneric family.

1. Spines (spine-tipped branches from distal buds) naked without leaves or flowers or their scars; flowers 5-merous; fruit elongating markedly in early development. **1. B. aegyptiaca**
- Spines bearing leaves or exhibiting leaf and/or flower or inflorescence scars; flowers 4-merous or if 5-merous then the buds noticeably apiculate; fruit not elongating markedly in early development. 2
2. Ovary glabrous; spines tending to widen dorsiventrally at the base, arising 0.5-1.5 cm above the subtending leaf-axil. **2. B. glabra**
- Ovary hairy; spines scarcely widening at the base, arising (0.5-)0.6-5(-10) cm above the subtending leaf-axil. 3
3. Flower 4-merous; primary spines arising (1-)2-5 (-10) cm and secondary spines (2-)4-7 mm above their respective leaf-axils; leaflets usually orbicular to broadly obovate, rounded or broadly cuneate. **3. B. rotundifolia**
- Flowers 5-merous, buds usually apiculate; primary spines arising (0.5-)0.6-1.2 cm and secondary spines 1-2 mm above their respective leaf-axils; leaflets broadly obovate to spathulate, cuneate. **4. B. pedicellaris**

* Royal Botanic Gardens, Kew, Richmond, Surrey, TW9 3AB, UK.

1. B. aegyptiaca *(L.) Del.* (1813).

A small tree up to 8 m high, very variable, usually spiny, the branches spreading or pendulous; young stems and spines greyish-green becoming light brown, at first minutely puberulous or tomentellous, glabrescent later. Spines (0.6-)1.2-3(-3.2) cm apart, 0.1-0.4 cm above the axil, (0.4-)2-8(-11.5) cm long, naked except for a few scale-leaves or rarely a single short lateral spine. Scale-leaves rare, on young spines, falling quickly, triangular, acute, 0.75 mm long. Foliage-leaves only on the stems, closely and minutely puberulous, often glabrescent, or sometimes densely tomentellous, subsessile or with a petiole (0.3-)1.1-2(-3.5) cm long; petiolule (0-)1-5(-8) mm long; leaflets narrowly elliptic to broadly ovate or obovate, (1-)1.2-6.4(-6.8) x (0.3-) 0.4-3.7(-5) cm, leathery; apex bluntly acute to obtuse or rounded; base cuneate, sometimes narrowly decurrent; leaf apiculus[1] 1.5-2.5 mm long, often falling. Flowers 5-merous, (1-)2-15(-20 or more) variously arranged in loose or tight cymose fascicles at spinous nodes, on spineless stems, or closely arranged on shoots of short internodes, sometimes more or less spiciform; pedicels finely tomentellous, 0.4-1.1(-2) cm long. Sepals sometimes falling early, tomentellous on the outside, 3.5-5(-5.5) x 2(2.5) mm. Petals narrowly obovate-elliptic, yellowish- or blue-green, 4.5-6(-6.5) x 1-2 mm. Stamens 10. Ovary pubescent to tomentose, to 1 mm high; style 1-2 mm long. Fruit elongating markedly in early development, becoming ovoid to ellipsoidal, usually rounded or truncate at both ends, usually ripening yellow, (2.3-)2.9-4(-4.5) x (1.3-)1.7-2.2(-2.5) cm.

1. Leaflet indumentum usually sparse and/or early glabrescent or minutely puberulous; inflorescences few- to many-flowered clusters often at spinous nodes. var. **aegyptiaca**
- Leaflet indumentum densely tomentellous; inflorescences always at spineless nodes with 1-5(-7) flowers. var. **pallida**

Var. **aegyptiaca**;

Ximenia aegyptiaca L. (1753) - type: Egypt. *Alpino* s.n. (not seen).

1. apiculus = small extension between the 2 leaflets.

Agialida abyssinica van Tieghem (1906) - types: TU, Dschadscha, Tacazzé valley (1200 m), *Schimper* 1222 *pro parte* (P syn. not seen, B isosyn.), Modat, *Schimper* 1022 (P syn. not seen, K BM FI B isosyn.).

A. cuneifolia van Tieghem (1906) - type: TU, Gurrsarfa, *Schimper* s.n. (?P not seen).

A. latifolia van Tieghem (1906); *Balanites latifolia* (van Tieghem) Chiov. (1932) - type: SU, Choa, 1845, *Rochet d'Héricourt* s.n. (?P not seen).

A. nigra van Tieghem (1906) - type: SU, Choa, 1845, *Rochet d'Héricourt* s.n. (?P not seen).

A. schimperi van Tieghem (1906) - type: TU/GD, mountains of Tacazzé (1800 m), *Schimper* 1222 *pro parte* (P not seen, B).

Balanites aegyptiaca sensu Cufodontis (1956) *pro parte* (excl. syn. *B. rotundifolia*), non (L.) Del.

Fig. 121.1.1-5.

Dry savannah or *Acacia* woodland, rarely in *Terminalia-Combretum* wooded grassland (IL), on a variety of soils; 700-1800 m. EE EW TU WU SU AR HA IL GG SD; from Senegal to Somalia and from Egypt south into Zimbabwe, also the Jordan valley and western Arabian peninsular. In Ethiopia, although common in many places, it is more or less confined to Eritrea, the rift valley in Gamo Gofa, Sidamo and Shewa administrative regions, and upland Harerge. South of Eritrea it is not recorded from the western administrative regions, except for a single gathering (*Chaffey* 621) from Abobo in Illubabor. *Burger* 1401, 1426; *J.J. C. de Wilde* 11016.

Var. *aegyptiaca* is very variable, particularly in the form of the leaf and in spine development, but it rarely has the pale greyish-green tomentellous leaflets of var. *pallida* and has 5 to 20 or more flowers in the inflorescence which quite frequently occurs at a spinous node. The spines only occasionally exceed 6 cm long.

B. aegyptiaca is widely used in many ways throughout its range in Africa although field information about its uses, specifically within Ethiopia, is scanty. The fruit, which is edible, yields a valuable oil and also contains saponins which are lethal to certain invertebrates and thus of value in eliminating the carriers of guinea-worm and schistosomiasis. The trees are frequently heavily browsed.

Var. **pallida** *Sands*, in *Kew Bull.* **38**: 40 (1983)
- type: Somalia, *Glover & Gilliland* 447 (K holo., BM EA iso.).

B. racemosa Chiov. (1929).

B. aegyptiaca sensu Oliver (1868) *pro parte* quoad *Roth*; Cufodontis (1956) *pro parte* non (L.) Delile.

Fig. 121.1.6.

Occurring in an open scrub community, known in Somalia to be associated with certain species of *Acacia*, *Commiphora*, *Zizyphus*, *Grewia* and *Aloe*, in sandy or stony soils usually near dry stream beds; sea-level to 1250 m. AF SU; N Somalia, Djibouti, much the most common Somali form of *B. aegyptiaca*. *Longhitano* s.n.; *Roth* 198.

Recognition of the variety is based on a distinctive correlation of characters and on its north-easterly peripheral position in the African distribution of the species, and, so far in Ethiopia, only one collection from Efat (SU) and one from near Amibara in the Awash valley (AF) can be referred to it. Within Ethiopia, since all the distinguishing features of var. *pallida* can be found separately within the variation of typical *B. aegyptiaca*, further collecting is desirable, particularly from the eastern lowlands region which links the present disjunct areas of ditribution.

2. **B. glabra** *Mildbr. & Schlecht.* (1913).

A spiny shrub or small tree up to 9 m high, usually of dense bushy habit; young stems and immature spines green, adpressed puberulous, the second-year stems very sparsely hairy, sometimes glaucous. Spines 1-3.5 (-5) cm apart, 0.5-1.5 cm above the axil, (3-)4.5-10(-13) cm long, sometimes widening dorsiventrally at the base. Scale-leaves frequent, falling quickly from the spines but sometimes persistent on second-year parent shoots, triangular, acute, to 2 mm long. Foliage leaves usually few, subsessile or with a very short petiole rarely to 3 mm long; leaflets sessile, elliptic, obovate or obovate-spathulate, rounded to acute, cuneate, 2.2-5.2(-6.2) x 1.2-2.5(-3.8) cm, thinly leathery; glabrous above, often very sparsely adpressed puberulous below; leaf-apiculus linear, 1.2 mm long, rarely developed into an elliptic leaflet up to 10 x 2.5 mm. Flowers 4-, rarely 5-merous in fascicles usually on the spines but sometimes on the parent axis; pedicels densely white tomentellous 3.5-10 mm long. Sepals silky-hairy on both surfaces, 4.5-5 x 2-2.5 mm. Petals narrowly obovate-oblong, yellowish-green to white, 6-7 x 2-2.25 mm. Stamens 8(-10). Ovary glabrous 1 mm high; style 1.5 mm long. Fruit not markedly elongating in early development, but at first ellipsoidal, pointed at both ends, then ovoid, smooth, ripening yellow, orange or pale red, 2.4-3(-3.5) x 1.5-1.8 cm. Fig. 121.1.7-9.

Acacia bush or open woodland, in which *B. glabra* may be an important element, usually on a loamy to clay soil, sometimes stony; persisting in degraded communities with *Opuntia*; 1250-1700 m. HA; disjunct distribution, northwestern Somalia, Kenya and northern Tanzania. *Burger* 960; *Gillett* 5262; *J.J.F.E. de Wilde* 4156.

3. **B. rotundifolia** *(van Tieghem) Blatter* (1919).
Sands in *Kew Bull.* **38**: 40 (1983).

A spiny shrub or small tree up to 6(-8) m high, usually much shorter and of low bushy habit; stems grey-green or yellowish-brown and, like the spines, glabrous or puberulous to pubescent, variably glabrescent by the second year. Primary spines at a varying distance apart, (1-)2-5(-10) cm above the leaf axil, (2-)2.5-7.5(-11) cm long, very frequently bearing secondary spines (2-)4-7 mm above their subtending axils. Scale leaves occasionally on young growth, narrowly triangular, puberulous, 2-3 mm long. Foliage leaves on the stems

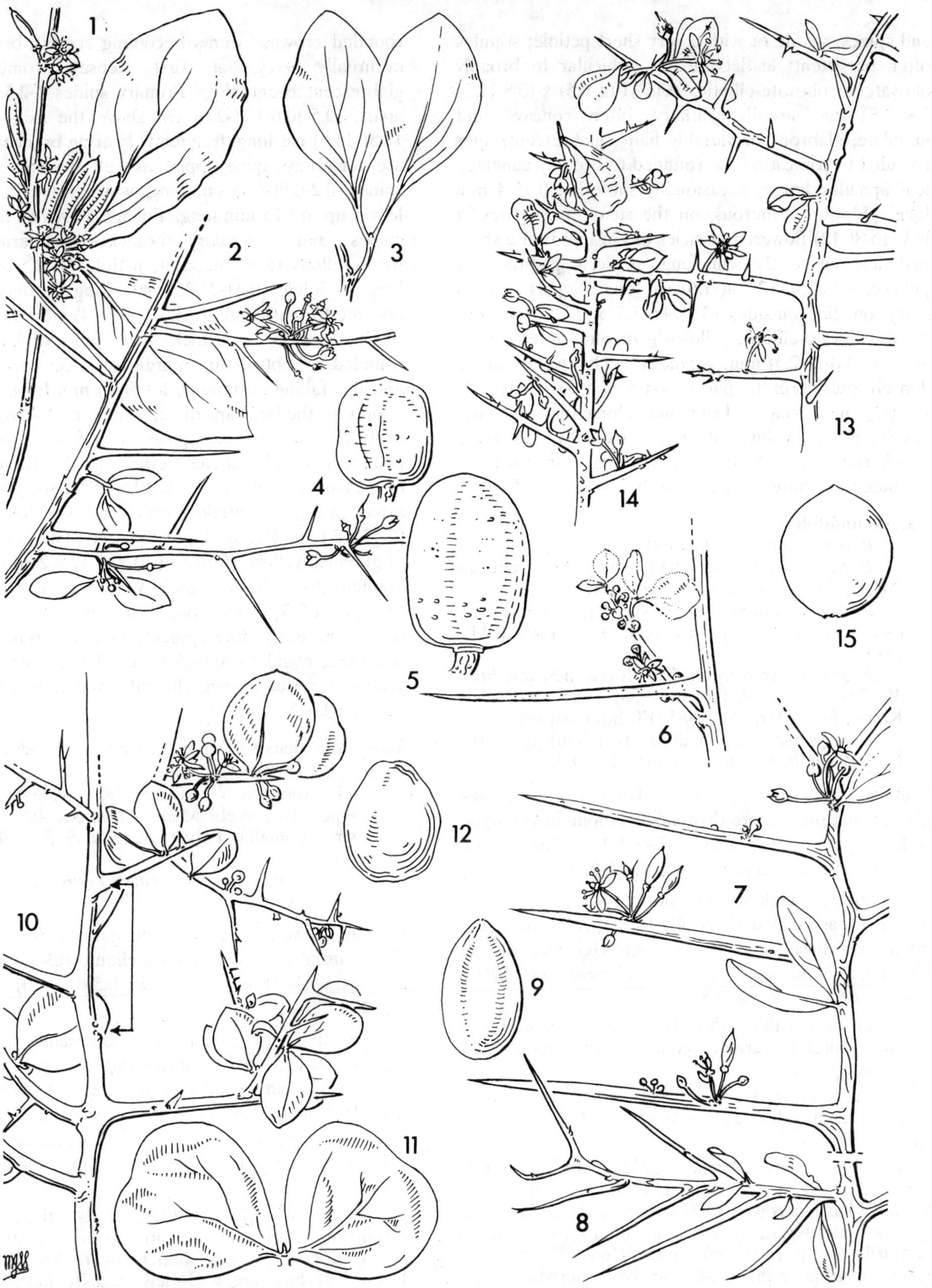

Figure 121.1 *BALANITES AEGYPTIACA* var. *AEGYPTIACA*: 1 - branch with flowers and young elongated fruits; 2 - branch with spines; 3 - leaf; 4 & 5 - mature fruits. *B. AEGYPTIACA* var. *PALLIDA*: 6 - node with spine and short branch with leaves and flowers. *B. GLABRA*: 7 - spiny branch with flowers and young fruits; 8 - spines with scale leaves; 9 - mature fruit. *B. ROTUNDIFOLIA* var. *ROTUNDIFOLIA*: 10 - spiny branch with leaves and fruits (arrows indicate the distance between spine and subtending leaf; 11 - leaf showing apiculus; 12 - fruit. *B. PEDICELLARIS* subsp. *SOMALENSIS*: 13 - spiny branch with solitary flowers; 14 - branch with crowded spines and 2-flowered inflorescences; 15 - mature fruit. Reproduced at app. 2/3 life size. Drawn by Martin Sands.

and spines, sessile or with a very short petiole; stipules often persistent; leaflets sessile, orbicular to broadly obovate, or obovate-elliptic, (0.75-)1-6(-6.5) x (0.8-)1.5-3.8(-4.8) cm, usually leathery, often concave and undulate, glabrous or variably hairy, glabrescent; apex rounded to truncate; base rounded to broadly cuneate; leaf apiculus linear, occasionally subfoliar, 0.75-4 mm long. Flowers 4-merous, on the stems and spines in few- to 8(-12)-flowered fascicles or clustered on a short peduncle up to 2.8 cm long; pedicels glabrous or pubescent (0.25-)0.3-1.4(-1.6) cm long. Sepals sparsely hairy on the outside, (3.5-)4.0-5.5 x (2-)2.5-3 mm. Petals obovate-elliptic, yellowish- or olive-green, (3-)4-6(-6.5) x 1.8-2(-2.5) mm. Stamens 8. Ovary glabrous or densely pubescent to pilose, 1-1.5 mm high; style 1-1.75(-2) mm long. Fruit not elongating in early development, swelling at first proximally, becoming ovoid, rounded at both ends, ripening orange-yellow, the outer skin hard, brittle, smooth, 2.5-3 x 1.8-2.5 cm.

Var. **rotundifolia.**

B. orbicularis Sprague (1908).

B. patriziana Lusina (1933) - type: AF, Dancalia Airori, *Patrizi* s.n. (RO holo. not seen).

B. gillettii Cufodontis (1958) - type: SD, Burgi, *Gillett* 15069 (W holo. not seen, K FI PRE EAH iso.).

B. gillettii var. *renifolia* Cufodontis in *Senck. Biol.* **39**: 303 et tab. 38, fig. 5. (1958) - type: GG, NE Konso, Fora Dara, *Kuls* 385 (FR holo. not seen).

B. aegyptiaca sensu Cufodontis (1956) pro parte (quoad syn. *B. rotundifolia*), non (L.) Del.

Leaflets puberulous to pubescent, eventually glabrescent, often more than 3-6.5 cm long in Ethiopia; pedicels pubescent, not exceeding 1.1 cm long; ovary densely pubescent to pilose. Fig. 121.1.10-12.

Dry tree savanah or scrub, or occasionally in riverine woodland, associated especially with species of *Acacia* often over lava; 300-1500 m. AF GG SD; Kenya, Uganda, Djibouti and N Somalia. *Corradi* 6068; *Gillett* 15069; *J.J.F.E. de Wilde* 5652.

In southern Ethiopia the fruits may be stored and are an important alternative food for local people when sorghum is scarce.

The very marked distance between the spines and their subtending leaf axils is a notable feature of this species.

Another variety, *B. scillin* Chiov., which so far is confined to the Miguertinia and Mudugh regions of Somalia, has glabrous leaflets not more than 3 cm long, glabrous pedicels 1.2-1.6 cm long, and a glabrous ovary. A third variety, restricted to the Coast Province of Kenya, also has a glabrous ovary but otherwise differs only from the type variety in having persistent short stiff hairs on the leaflets.

4. B. pedicellaris *Mildbr. & Schlecht.* (1913).

A spiny branching shrub or small tree up to 8 m high, the branches often spreading or pendulous to form a rounded crown; stems becoming reddish-brown and eventually grey, at first densely tomentellous, glabrescent eventually. Primary spines 1-2.5(-2.8) cm apart, (0.5-)0.6-1.2(-2.6) cm above the axil, (4.6-)6.4-12.5(-14.5) cm long, frequently bearing branches, which are not always spine-tipped, as well as small secondary spines, 0.2-0.8(-1.5) cm long, which often bear scale-leaves up to 1.25 mm long. Foliage-leaves on the stems, spines and secondary branches, puberulous or tomentellous to glabrescent; petiole (1-)1.5-6(-8) mm long; petiolule (0-)1-2(-3.5) mm long; leaflets broadly obovate to spathulate, (0.3-)1.2-3.5(-4) x 1.1-2.6(-3) cm, thinly leathery or thick and sub-succulent, apex rounded or obtuse to emarginate, base cuneate; leaf apiculus falling eventually, 1-1.5(-2) mm long. Flowers 5-merous, the buds apiculate, solitary or 2-4(-8), axillary or from nodes on the spines, in a fascicle or shortly pedunculate, subumbellate cyme, variably tomentellous to grey-pubescent. Pedicels (0.5-)0.7-1.8(-2) cm long; sepals ovate, tomentellous on the outside, 3.5-5.5(-6) x (2-)2.5-3 mm. Petals elliptic to elliptic-lanceolate, green or greenish-yellow, (6-)6.5-8(-9) x (2-)2.5-2.8(-3.4) mm. Stamens 10. Ovary densely villous, to 1.5 mm high; style (1-) 1.5-3(-4) mm long. Fruit elongating only very slightly in early development, then enlarging evenly, becoming ovoid to sub-globose, glabrescent, ripening yellow, then dull orange, the outer skin thin and brittle, 1.2-3.1 x 1.3-2.2 cm.

Subsp. **somalensis** *(Mildbr. & Schlecht.) Sands* in *Kew Bull.* **38**: 40 (1983);

B. somalensis Mildbr. & Schlecht. (1913) - types: BA, Webi Mana, *Ellenbeck* 1987 (B syn. destr.), Somalia, Erbah, *Ruspoli & Riva* 1026 (FI neo.).

B. somalensis var. *cinereo-corticata* Fiori ex Chiov. (1916).

Usually a low growing, densely spinous shrub; le flets small, usually densely tomentellous, 0.3-1.4(-2.5) cm long; pedicels short at anthesis, 0.5-0.8(-1.0) cm long. Fig. 121.1.13-15.

Dry thorn scrub or open bushland, associated particularly with species of *Commiphora* and *Acacia*, on calcareous, sandy or alluvial soils; 250-?1200 m. HA[1] BA SD; Somalia, where it grows near the Giuba river south to Kismayu and westwards to Mandera in the extreme north-east of Kenya. *Gilbert* et al. 7514; *Ruspoli & Riva* 1119.

Subsp. *pedicellaris*, although its distribution is disjunct, occurs extensively through eastern Africa from northern Uganda and southwards to Swaziland and Natal. It has larger leaflets, usually puberulous to glabrescent, 1.25-3.75(-4) cm long and, at anthesis, longer pedicels, 0.8-1.9(-2) cm long.

1. Two sterile specimens from the Galalcha plateau southwest of Jijiga can only be tentatively referred to this subspecies.

122. SIMAROUBACEAE

by B. Stannard*

Cufodontis, *Enum.*: 374-375 (1956); Wild & Phipps in *Fl. Zamb.* 2: 210-220 (1963).

Trees or shrubs, occasionally scandent, sometimes spiny, usually containing bitter substances, hairs usually simple, sometimes glandular. Leaves alternate, pinnate or simple, exstipulate. Inflorescence paniculate, racemose or cymose. Flowers bisexual, unisexual or polygamous. Calyx of 3-5(-6) sepals, nearly always partly united, valvate to slightly imbricate. Petals 3-5(-9), rarely absent (not in the Flora area), usually free, imbricate or valvate. Stamens 4-10, inserted at base of disc, free, sometimes with a scale attached at the base of the filament. Disc annular, cushion-shaped, cupular or sometimes inconspicuous (at least in dried specimens). Ovary globose or 2-5(-6 or more)-lobed, 2-5(-6 or more)-locular with 1 or, more rarely, 2 ovules per locule, or with free carpels; styles (2-)4-8, free or variously united. Fruit a berry or composed of drupaceous mericarps or dry mericarps attached to a central carpophore[1]. Seed with little or no endosperm.

A medium-sized family of about 25 genera and 120 species widely distributed in the tropics and subtropics, the main centres being tropical America and tropical West Africa.

Key to Genera

1. Plant armed; leaves not crowded at ends of branches; petiole and rhachis usually winged; fruit a berry. **1. Harrisonia**
- Plant unarmed; leaves often crowded at ends of branches; petiole and rhachis not winged; fruit of separate mericarps. 2
2. Leaflet-lamina glabrous; 11-35 x 9-20 mm; inflorescences axillary thyrses, pubescent/puberulous to glabrous; fruit of 4(-6) dry, 3-sided mericarps suspended from a central carpophore. **2. Kirkia**
- Leaflet-lamina usually rusty-red pubescent, 30-180 x 15-80 mm; inflorescence spiciform or spiciform panicles, rusty-red pubescent; fruit of 1-4 separate, fleshy mericarps. **3. Brucea**

1. HARRISONIA *R.Br.* ex *A.Juss.* (1825), *nom. conserv.*

Thorny, erect or sometimes scandent shrubs or small trees. Trunk and larger branches often bearing thorns on short wart-like outgrowths. Leaves not crowded at ends of branches, imparipinnate or ternate, petiole and rhachis often winged, petiole often with a pair of recurved spines at base; leaflets opposite or subopposite, ± sessile, subentire to crenate-dentate or serrate-lobate. Inflorescences axillary and/or terminal cymes, short racemes or panicles. Flowers bisexual. Calyx of 4-5(-6) sepals united towards base. Petals 4-5(-6), much longer than calyx, valvate or slightly imbricate. Stamens 8-10, filaments with hairy scale attached at base. Disc annular, cushion-shaped or cupular, sometimes inconspicuous. Ovary globose or 4-5(-6)-lobed, 4-5(-6) locular, 1 pendulous ovule per locule. Styles united entirely or free at base; stigmas united, capitate. Fruit a berry, entire or 4-8-lobed; seeds 4-8, with aperture at top.

About 3 species in tropical Africa, SE Asia, Indomalaysia and tropical Australia.

H. abyssinica *Oliv.* (1868)
- type: Uganda, *Speke & Grant*.
H. occidentalis Engl. (1895).
H. abyssinica Oliv. forma *grandifolia* Engl. in *Bot. Jahr.* 46: 284 (1911).
H. abyssinica Oliv. var. *somalensis* Fiori in *Miss. Stef. - Paoli Som. Ital. (Coll. Bot. Append.)*: 39 (1916).

Small tree or shrub, often multi-stemmed, much branched and frequently scandent, 1-13 m tall; bark light brown to grey, trunk and larger branches often bearing thorns on short wart-like outgrowths, smaller branches armed or unarmed, glabrous to pubescent, thorns recurved, paired at base of leaves. Leaves 5-15 foliolate, 2.5-14(-25) cm long, pubescent to glabrous; petioles and rhachis with wings 0.1-3 mm wide; leaflets sessile or subsessile; lamina elliptic, broadly obovate to almost round, 0.5-9 x 0.3-4 cm; apex acute, rounded or acuminate; base cuneate to almost rounded, usually asymmetric; margins crenate-serrate to sparsely crenate usually in top half only or less frequently entire. Inflorescences paniculate, axillary and terminal, pubescent or glabrescent, 1.5-14 cm long. Petals valvate, 2-5.5 x 1-2.5 mm, ovate to lanceolate, sometimes hairy outside. Stamens 1.5-4 mm long. Ovary 4-5(-6)-lobed, 1-1.75 x 1.2-2 mm; styles 4-5(-6), united, 0.8-1 mm long; stigmas united to form 4-5(-6)-lobed capitate stigma. Fruit 4-8-lobed berry, depressed globose, 4-9 mm diam., red to black when ripe, glabrous, fleshy, with 4-8 seeds; seeds 3.5-5 x 2.5-3 mm, smooth. Figure 122.1.

Dry evergreen forest and forest edges generally, also wooded grassland, thickets and riverine vegetation; 0-1700 m. SU/KF(Gibe gorge) BA IL GG SD; widespread in tropical Africa. *Beals* 338; *Jackson* 741; *Mooney* 8445.

In E Africa the habit and the shape, size and indumentum of the leaves vary enormously. Although plants from forest and forest edges tend to have larger leaves,

* Royal Botanic Gardens, Kew, Richmond, Surrey, TW9 3AB, UK.
1. carpophore = fruit-carrying stalk.

Figure 122.1 *HARRISONIA ABYSSINICA*: 1 - flowering branch, leaflets entire x 2/3; 2 - node showing pair of spines, leaflet partly crenate x 2/3; 3 - small leaflet type x 2/3; 4 - detail of thorn on wart-like outgrowth x 1; 5 - flower x 5; 6 - bud x 5; 7 - stamens showing hairy scale x 10; 8 - cross section of ovary x 10; 9 - fruit x 5. (Reproduced by permission, Editorial Board, *Flora Zambesiaca*, vol. 2, tab. 41, 1963.)

the variation in the plants from the wide range of other habitats in E Africa is so extensive that no practical demarcation can be made. In W Africa, the leaves are much more consistently large. In Ethiopia the size of the plant parts tends towards the smaller end of the range.

2. KIRKIA *Oliv.* (1868)

Stannard, *Kew Bull.* **35**(4): 829-839, fig. 1 (1981).

Trees or shrubs; branchlets marked clearly with scars of old leaves, often viscid around tips. Leaves mostly crowded at ends of shoots, imparipinnate. Flowers in axillary thyrses, apparently unisexual or bisexual. Sepals 4, small, very shortly united at base, deltoid. Petal 4 (valvate or) imbricate. Stamens 4, alternate with petals, inserted outside and beneath disc, in male flowers usually about as long as petals, in female flowers much reduced; filaments filiform and often broadening slightly towards the base; anthers dorsifixed. Disc annular, usually fleshy. Pistil of 4-8 united carpels, much reduced in male flowers; ovary narrowly flask-shaped, (1 or)2-locular, ovules 1 per locule; styles 4-8, slender, coherent; stigmas coherent, capitate. Fruit woody, composed of triquetrous (3-sided) mericarps suspended from the top of a central carpophore, each mericarp with the remaining base of one of the styles reflexed back over its apex. Seed slender, broader and rounded at one end, tapering to a point at the other, slightly curved, triangular in cross section.

A genus of 5 species occurring in tropical Africa from Ethiopia and Somalia in the north as far south as the Transvaal region of South Africa.

On the basis of carpological, palynological and anatomical evidence, Takhtajan places this genus in a new family of its own, the **Kirkiaceae** (*Syst. Filog. Tsvetk: Rast.*: 321 (1966)). Until further study establishes the true affinities of *Kirkia* with greater certainty, the author considers it best to leave this genus in the **Simaroubaceae.**

1. Leaves 3.5-9 cm long, (5-)7(-9)-foliolate; branchlets widely splayed or often almost at right angles to the main axis, sometimes several (3-4) radiating from one node; young wood dark red or reddish brown; inflorescences 3-6.5 cm long; fruits 10-12 x 6-8 mm. **1. K. tenuifolia**
- Leaves 10-20 cm long, (9-)11-15(-17)-foliolate; branchlets forming acute angle with main axis, young wood light brown; inflorescence 9-14 cm long; fruits 6-16 x 3.5-7 mm. **2. K. burgeri**

1. **K. tenuifolia** *Engl.* (1902);
Tetraspis ruspoli Chiov. in *Ann. Bot.* **X**: 25, fig. 1 (1912) & **XIII**: 398 (1915) - types: HA, *Ruspoli & Riva* 865 & 1081 (FT syn., K photo.).

Shrub or tree to *c* 8 m, with spreading bushy crown; bark grey, fissured, bark of branchlets light brown to reddish brown becoming silvery grey, newest extension growth dark crimson, puberulous; branchlets often almost at right angles to main axis, sometimes several (3-4) radiating from one node. Leaves 3.5-9 cm long, 5-7(-9)-foliolate; leaflet-lamina elliptic to rotund, glabrous, 11-25 x 9-20 mm, rounded and apiculate or emarginate at apex, unequally rounded to cuneate at base, margins entire or sometimes faintly crenulate; terminal leaflet orbicular, symmetrical at base. Inflorescences 3-6.5 cm long, pubescent or glabrescent. Flowers greenish-yellow to cream. Petals 3-5 x 0.5-1 mm, minute globose glands towards base inside, hairy outside. Male flowers: filaments 2.5-3.5 mm long, glabrous; anthers 0.5-1 mm long. Female flowers: carpels 4(-6), glabrous, forming ovary *c* 2 x 1 mm; styles *c* 1.5 mm long. Fruits elliptic to shortly oblong, 10-12 x 6-8 mm, mostly broadly rounded at ends, composed of 4 (-6) mericarps, shortly pubescent to puberulous.

Acacia-Commiphora bushland, usually on red sandy or stony soil; up to *c* 1000 m. HA BA SD; Somlia, Kenya. *Ellis* 388; *Simmons* S199; *Glover & Gilliland* 285.

In Ethiopia the size of the leaves is usually at the smaller end of the range.

2. **K. burgeri** *Stannard* (1981)
- type: HA, limestone slopes west of Daletti, above the Gobelli River valley, *Burger* 3302 (K holo.).

Trees or shrubs 3-8 m tall, with grey fissured bark, young twigs light brown, puberulous to glabrous, older twigs grey. Leaves 10-20 cm long, (9-)11-15(-17)-foliolate; leaflet-lamina 14-35 x 9-20 mm, glabrous, ovate, elliptic to rotund, base obliquely rounded to cuneate, apex acute to rounded-apiculate to emarginate, margin entire or faintly crenate-undulate, thickened. Inflorescences approximately same length as subtending leaves, 9-14 cm long, pubescent to glabrous. Petals *c* 2.5-5 x 0.75-1.5 mm, verruculose at base inside, sometimes sparsely hairy outside. Male flowers: filaments *c* 3.5-4 mm long, glabrous, anthers 1-1.5 mm long. Female flowers: carpels 4, united, glabrous; styles 4. Fruits *c* 6-16 x 3.5-7 mm, glabrous, elliptic. quadrangular in cross-section, mericarps 4, 2 locular. Seed not seen.

Subsp. **burgeri**

Leaflets elliptic to rotund. Fruits 15-16 x 6-7 mm, narrowly elliptic, rather pointed at both ends. Plant drying brownish green.(Figure 122.2.1.

Dense deciduous bush and limestone slopes; 1100-1400 m. SD BA HA; (Somalia for subspecies *somalensis*). *Burger* 2954, 3105A, 3306.

In addition to the differences enumerated in the key, this species can be further distinguished from *K. tenuifolia* by its apparently far more vigorous and robust growth and generally larger inflorescences, leaves and fruits. More collections are needed from Sidamo and Bale regions to give a clearer idea of its range.

Subsp. *somalensis* Stannard is restricted to Somalia but could turn up in the eastern Ogaden. It differs by the ovate leaves which dry yellowish green and the apparently smaller fruit.

3. BRUCEA *J.F. Mill.* (1779).

Shrubs or small trees, often unbranched; at least younger parts ± rusty-brown, pubescent or tomentose. Leaves often ± crowded towards ends of branches, usually imparipinnate, petiolar base and rhachis joints often constricted (in dried specimens); leaflets 5-13, opposite or subopposite, sessile to shortly petiolulate, entire or toothed. Inflorescences axillary, spiciform or spiciform panicles, male larger than female. Flowers unisexual or bisexual. Sepals 4-5, united towards the base, imbricate. Petals 4-5, free, imbricate. Stamens 4-5, inserted beneath and between lobes of disc, free, unappendaged, reduced in female flowers. Disc fleshy (3-)4(-5)-lobed. Pistil of 4-5 carpels, free or joined only at base, each containing 1 pendulous ovule, absent or vestigial in male flowers; styles 4-5, free. Fruit drupaceous mericarps developing from 1-4 carpels of each flower. Seed 1 per mericarp.

A genus of about 6 species occurring in the Old World tropics from tropical Africa to tropical Asia, Malaysia and N Australia.

B. antidysenterica *J. F. Mill.* (1779)
- type: J. F. Mill., *Icon.* t. 25 (BM), presumably based on material sent from Ethiopia by Bruce.
B. ferruginea L'Hérit. (1785) - type: Ethiopia, no locality, L'Hérit., *Stirp. Nov.*: 19, t. 10 (K).
B. erythraeae Chiov. (1912) - types: Eritrea, *Pappi* 5279, *Tellini* 497, *Fiori* 197, 198 (FT syn.).
Trichilia siderotricha Chiov. (1940) - type: SD, Wendo Genet "Uondo", *Marchet* 12 (FT).

Shrub or small tree to 10(-15) m; bark of older stems smooth, grey to pale-brown, young stems rusty-brown pubescent. Leaves 10-65 cm long, 5-13-foliolate, rusty-brown pubescent (rarely almost glabrous); leaflet-lamina oblong to ovate, 3-18 x 1.5-8 cm, acute to shortly acuminate at apex, asymmetrically rounded at base, margins entire to shallowly wavy, usually with gland dots near margin. Inflorescences 5-30(-40) cm long, spiciform, axis rusty-brown pubescent. Flowers in distant glomerules. Sepals 2-3 x 1-1-5 mm, pubescent. Petals 1.5-3 x 1-1.5 mm, oblong to ovate, pubescent outside, sometimes inside. Stamens 1-1.25 mm long in male flowers, glabrous. Carpels 1-2.5 x 1-3 mm, glabrous or pubescent; stigmas 1 mm long, free, reflexed into an x-pattern; mericarps 9-14 x 6-8 mm, ellipsoid, acute, glabrous, red, sarcocarp[1] thin, endocarp reticulately wrinkled, hard; seed 8-9 x 5-6 mm. Fig. 122.2.2.

Montane forest, evergreen forest, forest margins, secondary growth in deforested areas, montane grassland; 1650-2800 m. EW TU GD GJ? WG SU AR IL KF SD BA HA; widespread in tropical Africa. *Burger* 1676; *Friis* et al. 155; *Merga Gutetta* 1013.

Widely used in traditional medicine for treating a number of disorders including diarrhoea, skin problems and leprosy.

1. sarcocarp = fleshy covering to fruit.

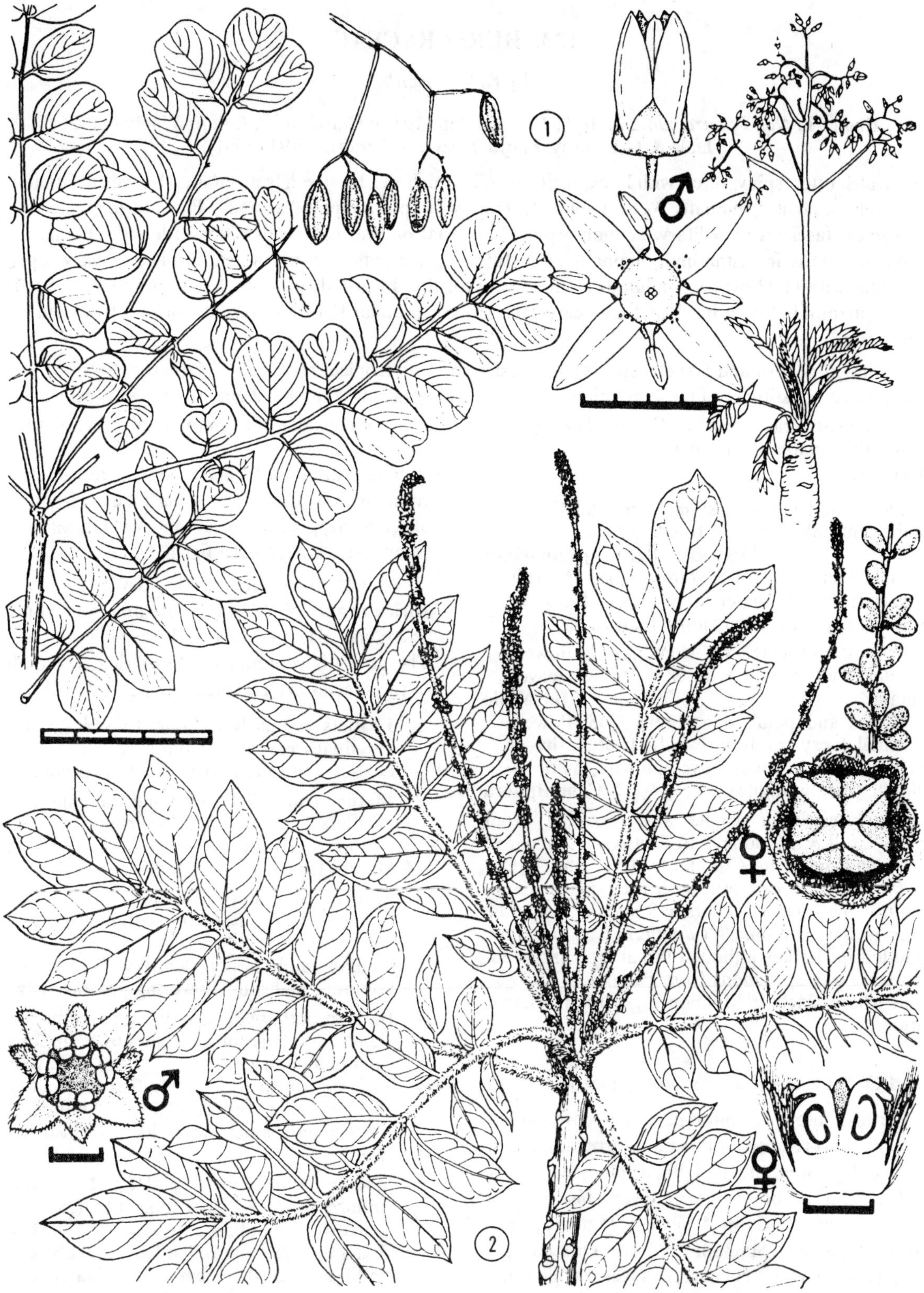

Figure 122.2.1 *KIRKIA BURGERI* ssp. *BURGERI*: end of leafy branch, branch with young leaves and male inflorescence; male flower, top and side view; fruits. *BRUCEA ANTIDYSENTERICA*: top of branch with leaves and inflorescences; male flower, top view; top view and longitudinal section of female flower; portion of infructescence. Closed bar gives scale in cm for leafy branches; open bars give mm scales for floral details. (Reproduced with permission of Oklahoma Agricultural Experiment Station from Experiment Station Bulletin No. 45, *Families of Flowering Plants in Ethiopia*, figure 29, 1967.) Drawn by W.C. Burger.

123. BURSERACEAE

by K. Vollesen*

Engler in *Nat. Pflanzenfam*., **2.** *Aufl*., **19a**: 405-456 (1931); Cufodontis , *Enum*: 375-398 (1956); Dale & Greenway, *Kenya Trees and Shrubs*: 75-93 (1961).

Trees or shrubs, often spiny; often with latex, resins or oils which can be strongly aromatic. Leaves alternate, often in clusters (rarely opposite), pinnate, 3- or 1-foliolate (rarely simple); stipules absent. Flowers in panicles, corymbs, racemes, cymes, fascicles or solitary; actinomorphic, unisexual or bisexual. Calyx 3-5(-6)-lobed. Petals (0-)3-5(-6), free or connate. Disc intrastaminal. Stamens (3-)6-10(-12), filaments free or connate at base. Ovary superior, 2-5-locular, locules with (1-)2 ovules pendulous from apex of locule. Fruit a drupe with 1 seed per locule. Seed without endosperm, surrounded by a hardened endocarp (the stone); pseudaril1 often present, formed from outer layer of endocarp.

Family with 17 genera and 500-600 species. Widespread in all tropical regions and extending into the subtropics. Often a dominant constituent of the vegetation in dry lowland areas. In Ethiopia 2 genera and 58 species.

The resins of several species of *Commiphora* and *Boswellia* are of considerable commercial value as raw material of balm, myrrh, incense and frankincense.

Key to Genera.

1. Calyx-lobes and petals 3; stamens 6; fruit fleshy, indehiscent; tall rainforest tree to 30 m. (See note after key) **Canarium**
- Calyx-lobes and petals 4-5; stamens (4)8-10; fruit dehiscent; small to medium sized trees or shrubs of dry areas rarely more than 10 m. 2
2. Calyx-lobes and petals 5; stamens 10; flowers bisexual; fruit woody, with 3(4) stones; leaves pinnate. **1. Boswellia**
- Calyx-lobes and petals 4; stamens (4)8; flowers unisexual (very rarely bisexual); fruit fleshy or leathery, with 1 stone; leaves pinnate, 3- or 1-foliolate (rarely simple). **2. Commiphora**

Canarium schweinfurthii Engl. was recorded from Ethiopia by Exell & Mendonça in *Consp. Fl. Angola* **I**: 305 (1951) and again by Cufodontis (l.c.) and Leenhouts in *Blumea* **9**: 384 (1959), both referring to Exell & Mendonça. The species occurs in S Sudan, and could well occur in SW Ethiopia. *Canarium* has therefore been included in the key although the author has been unable to trace any specimens from Ethiopia. It is not clear from where Exell & Mendonça got their original information.

1. BOSWELLIA *Roxb.* ex *Colebr.* (1807).

Trees or shrubs; outer bark often peeling in parchmenty flakes; inner bark greenish, with watery aromatic resin; wood with milky latex. Leaves imparipinnate, mostly congested at end of branches. Flowers bisexual, in panicles or racemes. Calyx 5-lobed. Petals 5, free, oblong to ovate, acute. Stamens 10, free. Disc conspicuous, annular, flat or concave. Ovary trigonous, (2)3(4)-locular; style simple; stigma capitate. Fruit a (2)3(4)-locular obtetrahedroid drupe with woody exocarp (see fig. 123.2.F.), dehiscing with the valves dropping and leaving a central (2)3(4)-winged column with the stones attached. Stones tetrahedroid, when young with a thin papery wing formed from outer layer of endocarp, with apical and/or lateral and/or basal extensions (horns) - see fig. 123.2.G.

About 20 species in the dry regions of tropical Africa from W Africa to Arabia and south to NE Tanzania, also in India and 1 species in Madagascar. The genus is centered in NE Africa where about 75% of the species are endemic, including 5 on Socotra.

1. Leaflets regularly dentate to serrate; filaments gradually narrowed from base. 2
- Leaflets entire (rarely with a few irregular teeth). 5
2. Mature fruit 0.7-1.1 cm long; median and upper leaflets 2-5.5 cm long; flowering with new leaves; stamens glabrous. (See note after key) **B. sacra**
- Mature fruit 1.6-2.7 cm long. 3
3. Median and upper leaflets 0.7-2.3 cm long; stamens hairy; flowering with new leaves. **3. B. rivae**
- Median and upper leaflets 3-12.5 cm long; stamens glabrous; flowering precociously. 4
4. Bark smooth, peeling in large parchmenty flakes; leaflets pubescent to tomentose; inflorescence an open panicle. **1. B. papyrifera**
- Bark rugose to reticulately fissured, not peeling; leaflets subglabrous, glaucous; inflorescence a contracted panicle with flowers clustered along axes. **2. B. pirottae**
5. Median and upper leaflets 2-3.5 cm long; inflorescence 7-14 cm long; filaments gradually narrowed from base. **4. B. ogadensis**
- Median and upper leaflets 0.2-1.8 cm long; inflorescence 0.5-5 cm long; filaments abruptly narrowed just below anthers. 6
6. Leaflets (9-)13-43(-51), pubescent to subtomentose, base subcordate to truncate (rarely cuneate); fruit with retuse apex; stone with basal

* formerly Ethiopian Flora Project, now Royal Botanic Gardens, Kew, Richmond, Surrey, TW9 3AB, UK.

1. pseudaril = a fleshy colourful structure covering part of the seed.

horn; young branches and inflorescences with patent indumentum. **5. B. neglecta**
- Leaflets (3-)5-17(-29), glabrous to sparsely pubescent base cuneate to rounded; fruit with rounded apex; stone without basal horn; young branches and inflorescences with crisped ± appressed indumentum. **6. B. microphylla**

B. sacra Flück. has only been recorded from Somalia, but it occurs just east of the border in N Harerge and is most likely to be found in Ethiopia. It has, therefore, been included in the key.

Breitenbach in *Indign. Trees Ethiopia*: 226 (1963) records *B. frereana* Birdw. from Ethiopia. This species is easily recognizable by its glabrous, glaucous, strongly undulate leaflets. But Breitenbach describes the leaves as being silvery tomentose. This points more to *B. bhau-dajiana* Birdw., but his description of the leaves having 5 leaflets seems to rule out both species as no material of these two species has less than 7 leaflets, and *B. bhau-dajiana* always has at least 9. Both these species are endemic in the extreme NE Somalia and are not likely to occur in Ethiopia. The author has, therefore, refrained from including them in the key.

1. **B. papyrifera** *(Del.) Hochst.* (1843);
Amyris papyrifera Del. (1826).
Ploesslea floribunda Endl. (1839).
"Anguah" Bruce (1790).

Tree to 12 m; bark smooth, whitish to pale brown, peeling off in large parchmenty flakes; branchlets pubescent. Leaves 15-45 cm long, pubescent to tomentose; petiole 0.5-4 cm; leaflets 11-29, narrowly ovate to oblong, crenate, bicrenate or serrate, median and upper 5-12.5 x 2-5 cm, apex acute to rounded. Flowers in (7-)10-35 cm long pubescent panicles, white to pink, glabrous to sparsely pubescent; peduncle (0.5-)2-10 cm; pedicels 3-16 mm. Petals 5-7.5 mm long, spreading from base; filaments 2-3 mm, glabrous, gradually narrowed from base; disc flat. Fruit 1.6-2.5 cm long, valves 7-8 mm wide; stones with apical horn only, 5-6 x 4-5 x 2-3 mm, rugose. Fig. 123.1.A-D.

Dry *Acacia-Commiphora* woodland and wooded grassland, *Pterocarpus* woodland, often dominant on steep rocky slopes, also on lava flows; 950-1800 m. EW TU GD GJ SU; Nigeria, Cameroun, Central African Republic, Tchad, Sudan, NE Uganda. *Bally* 6672; *Hart Dyke* 2; *W. de Wilde* et al. 10654.

The resin is the raw material of the Ethiopian frankincense and is widely collected in N Ethiopia. It is generally considered of poorer quality than the product obtained from the Arabian and Somalian *B. sacra*.

2. **B. pirottae** *Chiov.* (1911)
- types: GD, Tzellemti, Mai Aini, between Anso and Ceu Berr rivers, *Chiovenda* 662, 747, 3193 (all FT syn., K fragments and photos).

Tree to 10 m; bark grey to black, rugose to reticulately fissured, not peeling; branchlets crisped pubescent, glaucous. Leaves 15-35 cm long, glabrous apart from crisped pubescent petiole and midrib; petiole 1.5-4 cm; leaflets (7-)15-27, narrowly ovate, serrate, glaucous, median and upper 3-7 x 0.8-1.8 cm, apex acute to acuminate. Flowers white, pubescent, in narrow contracted 2.5-10 cm long pubescent panicles, clustered in 2-4-flowered cymules along axes; pedicels 0.5-3 mm. Petals *c* 3 mm long, erect; filaments *c* 1.5 mm, glabrous, gradually narrowed from base; disc flat. Fruit 1.8-2.7 cm long, valves 7-10 mm wide; stones with apical horns only, *c* 6 x 4 x 3 mm, lumpy. Fig. 123.1.E-I.

Commiphora-Boswellia, *Combretum* and *Acacia-Lannea* woodland, on steep rocky slopes; 1200-1800 m. GD GJ WU SU KF; not known elsewhere. *Gilbert* et al. 8044; *Mesfin Tadesse* 2860; *Tewolde Berhan* 1637.

A rare species which is only known from the Tekeze, Abay and Gibe river systems. It is closest to *B. papyrifera* from which it differs in the rugose bark, subglabrous leaves, contracted inflorescence, smaller flowers and different fruit.

3. **B. rivae** *Engl.* (1898)
- type: BA, Web valley, *Ruspoli & Riva* 934(842) (FT holo.).
B. boranensis Engl. (1904) - type: SD, Gogorni in Boran, by Ganale river, *Ellenbeck* 2107 (B holo. destr.).

Shrub or tree to 6.m; bark mottled yellowish to greyish, breaking of in thick irregular scales and peeling in small papery flakes; branchlets pubescent. Leaves 5.5-12(-18) cm long, pubescent; petiole 2-15 mm; leaflets 11-29, elliptic or oblong, crenate to dentate, median and upper 7-23 x 3-12 mm, apex acute to rounded. Flowers appearing with new leaves, pink, pubescent, in 1-4(-5.5) cm long, few-flowered, pubescent racemes or panicles; peduncle 2-25 mm; pedicels 3-10 mm. Petals recurved in apical half, 5-7 mm long; filaments 3-4 mm, densely pubescent, gradually narrowed from base; disc flat. Fruit pubescent, 1.7-2.5 cm long, valves 7-9 mm wide; stones with apical and basal horns, *c* 5 x 4 x 3 mm, rugose. Fig. 123.1.K-N.

Acacia-Commiphora wooded grassland and bushland, on red sandy to stony soil overlying limestone; 250-800 m. SD BA HA; Somalia, NE Kenya. *Gilbert* et al. 7554, 7635; *J. de Wilde* 5932.

The resin is used locally for incense and is also chewed (Somali).

4. **B. ogadensis** *Vollesen* (1985)
- type: HA, Kelafo, *Allen & Hammersley* 89 (K holo., C ETH iso.).

Tree, (no notes on bark); young branches not seen. Leaves 4-7 cm long, subglabrous to sparsely puberulous on petiole and along edges; petiole 1-2 cm; leaflets 7, broadly elliptic to suborbicular, entire, median and upper 2-3.5 x 1.3-2.8 cm, apex rounded to retuse, base cuneate to subcordate. Flowers sparsely puberulous, in 7-14 cm long, many-flowered, sub-glabrous racemoid panicles; peduncle 2-5.5 cm; pedicels 3-5 mm. Petals *c* 3

Figure 123.1 *BOSWELLIA* spp. *B. PAPYRIFERA* (A-D): A - leaf x 1; B - part of inflorescence x 1; C - fruit x 3; D - stone x 3. *B. PIROTTAE* (E-I): E - leaf x 1; F - leaflet x 1.5; G - inflorescences x 1; H - fruit x 3; I - stone x 3. *B. RIVAE* (K-N): K - flowering branch x 1; L - leaf x 1; M - fruit x 3, N - stone x 3. A from *Andrews* 881; B from *Wickens* 1293; C & D from *Schimper* 875; E & F from *Chiovenda* 747; G from *Chiovenda* 3193; H & I from *Tewolde Berhan* 1637; K from *Hemming* 154; L from *Hemming* 1541; M & N from *Drake Brockman* 783. Drawn by Christine Grey-Wilson.

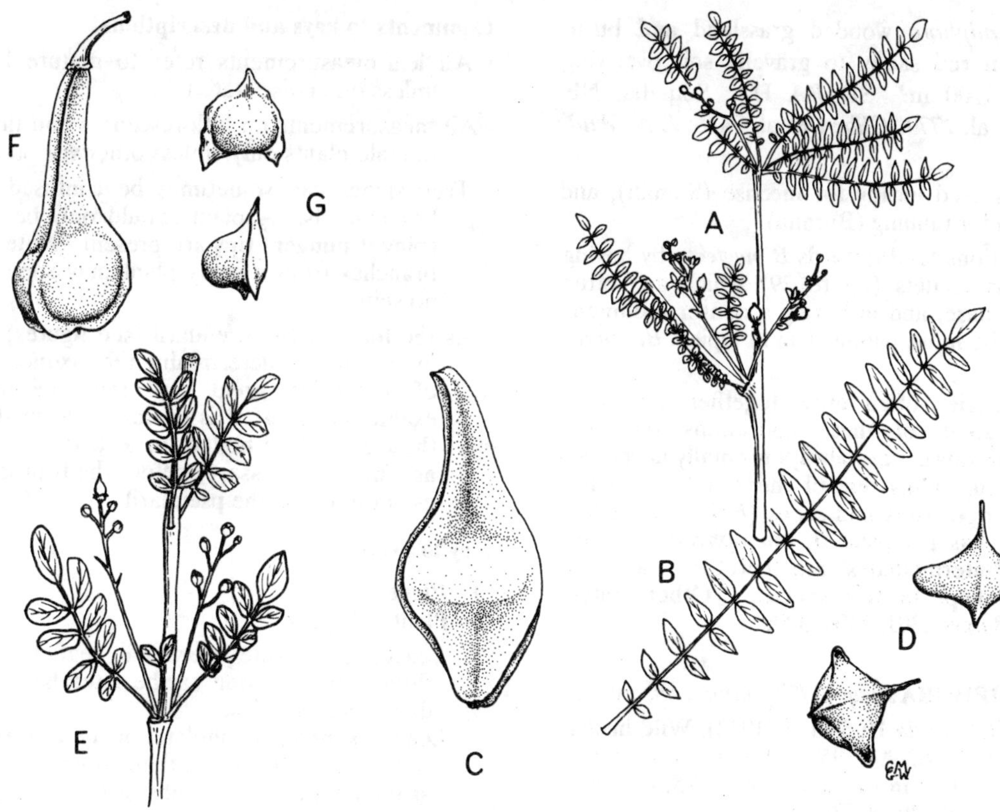

Figure 123.2 *BOSWELLIA* spp. *B. NEGLECTA* (A-D): A - flowering branch x 1; B - leaf x 1; C - fruit x 3; D - stone x 3. *B. MICROPHYLLA* (E-G): E - flowering branch x 1; F - fruit x 3; G - stone x 3. A from *Gillett* 14172; B from *Burger* 3326; C from *J. de Wilde* 7252; D from *Burger* 2315; E from *Ellis* 175; F & G from *J. de Wilde* 7317. Drawn by Christine Grey-Wilson.

mm long, erect; filaments *c* 1 mm, glabrous, gradually narrowed from base; disc concave. Fruit and stones not known.

Acacia-Commiphora bushland on rocky limestone slope; 300-400 m. HA; not known elsewhere. Only known from the type-collection.

A most remarkable species which doesn't seem to be close to any other species. The few entire leaflets seem to place it near *B. microphylla*, but the gradually narrowed filaments point more to *B. rivae* and *B. pirottae*. With the latter it also shares the contracted inflorescence.

5. **B. neglecta** *S. Moore* (1877);
B. hildebrandtii Engl. (1893).
B. multifoliolata Engl. (1898).

Shrub or tree to 6 m; bark dark grey, smooth to rugose, not peeling, often with circular foldings; branchlets persistently pubescent, with longitudinally cracking bark. Leaves (2-)4-11 cm long, pubescent to subtomentose; petiole 2-12(-16) mm; leaflets (9-)13-43(-51), elliptic to oblong, entire; median and upper 3-13 x 2-5 (-8) mm, apex subacute to rounded, base truncate to subcordate (rarely cuneate). Flowers white or cream, pubescent, in few-flowered, pubescent racemes or panicles, 0.5-5 cm long; peduncle 3-25 mm; pedicels 1-6 mm. Petals 1.5-2.5 mm long, spreading from base; filaments *c* 1 mm, glabrous, abruptly narrowed just below anthers; disc concave. Fruit retuse at apex, 1.2-2.2 cm long, valves 7-11 mm wide; stones with apical, lateral and basal horns, 4-5 x 5-6 x *c* 3 mm, smooth. Fig. 123.2.A-D.

Acacia-Commiphora woodland, wooded grassland and bushland, *Acacia-Boswellia-Terminalia* woodland and wooded grassland, often on red sandy soil overlying limestone but also on basement rocks; 600-1750 m. GG SD BA HA; Somalia, NE Uganda, N & E Kenya, NE Tanzania. *Friis* et al. 2778, 3673; *Gilbert* et al. 7425.

6. **B. microphylla** *Chiov.* (1915).

Shrub or tree to 4 m; bark dark grey to bluish grey, smooth to rugose or reticulately fissured, not peeling; branchlets crisped and ± appressed pubescent, soon glabrescent with smooth bark. Leaves 0.5-4.5(-6) cm long, glabrous to sparsely pubescent; petiole 1-10 mm; leaflets (3-)5-17(-29), elliptic to obovate, entire; median and upper 2-18 x 1-10 mm, apex rounded, base cuneate to rounded. Flowers white or cream, glabrous to sparsely pubescent, in 0.5-4.5 cm long, few-flowered, glabrous to sparsely pubescent racemes or panicles; peduncle 3-25 mm; pedicels 1-6(-8) mm. Petals 2-3 mm long, spreading from base; filaments *c* 1 mm, glabrous to sparsely pilose, abruptly narrowed just below anthers; disc concave. Fruit rounded to truncate at apex, 1.3-2 cm long, valves 5-8 mm wide; stones with apical horn only, 4-5 x 4-5 x 2-3 mm, smooth. Fig. 123.2.E-G.

Acacia-Commiphora wooded grassland and bushland, usually on red sandy to gravelly soil overlying limestone; 400-1300 m. SD BA HA; Somalia, NE Kenya. *Friis* et al. 2779; *Gilbert* et al. 7426; *J. de Wilde* 7317.

The resin is used locally for incense (Somali), and the bark is used for tanning (Borana).

Some collections tend towards *B. neglecta* by having more numerous leaflets (up to 29) which are often rounded at the base, and in having a denser indumentum tending to be more patent than in typical *B. microphylla*.

The two species often grow together, and it is possible that some of these collections represent hybrids. But the flowers are always normally developed with good pollen. Only one of these "intermediate" collections has ripe fruits and seeds (*Friis* et al. 2999). This collection has a typical *B. microphylla* fruit and seed, and the "intermediates" are therefore - at least provisionally - kept in this species. Other "intermediates" are *Burger* 2301, 2304, 3290.

2. COMMIPHORA *Jacq.* (1797), *nom. conserv.*

Chiovenda in *Fl. Somala* **II**: 53-128 (1932); Wild in *Bol. Soc. Brot., ser. 2*, **33**: 67-99 (1959); Wild in *Fl. Zamb.* **2**: 263-282 (1963); Gillett in *Kew Bull.* **34**: 569-587 (1980); Vollesen in *Kew Bull.* **40**: 39-76 (1985).

Trees or shrubs; branches often terminating in spines; outer bark often papery and peeling, inner bark usually greenish, with resinous - usually aromatic - sap, wood with milky latex. Leaves imparipinnate or 1-3-foliolate (rarely simple), often congested at branch ends. Flowers unisexual and dioecious (very rarely monoecious and/or bisexual), usually before the leaves appear or with new leaves, in panicles, cymes, racemes, fascicles or solitary. Calyx 4-lobed. Petals 4, free, linear to elliptic, recurved apically. Stamens (4)8, didynamous, antisepalous usually the longest, vestigial in female flowers. Disc 4- or 8-lobed or annular, usually glabrous. Ovary 2(3)-locular; style short; stigma capitate (rarely clavate); ovary absent or vestigial in male flowers. Fruit with a single 1(2)-seeded stone; exocarp fleshy or leathery, splitting into 2 or 4 valves; outer layer of endocarp forming a fleshy pseudaril.

Genus with 150-200 species: widespread in the drier parts of tropical Africa and Madagascar, also from Arabia to India and with a few species in S America. The genus is a very conspicuous and dominant element in the dry bushlands of NE Africa, and a large number of species are endemic in this area.

It is difficult to make a qualified guess of the number of species in this genus. Many species are leafless for most of the year and usually set flower and fruit when leafless or with young leaves. This makes the collecting of correlated leafy and fertile material difficult. In the past many species have been described from incomplete material, and a good number of synonyms no doubt still have to come to light.

Comments to keys and descriptions.

- All leaf-measurements refer to mature leaves only, unless otherwise stated.
- All measurements on inflorescences and flowers refer to male plants only, unless otherwise stated.
- True spines can sometimes be confused with dead branch-ends. A plant should only be considered spiny if pungent tips are present. Note that single branches from a spiny plant may sometimes have no spines.
- As the form of the pseudaril (see figures) is of great importance in determining the taxonomic position of a species, great care must be taken when examining it. If it is examined from dried fruits, these should be boiled for at least 20 minutes (or as long as necsessary) before the fruit is opened so as not to break the pseudaril.

Key to Sections.

1. Plant spiny. 2
- Plant not spiny. 9
2. Leaves 3-foliolate (terminal leaflet less than three times as long as the laterals), with well-developed petioles. 3
- Leaves simple, 1-foliolate or hetero-trifoliolate (terminal leaflet more than three times as long as the laterals), often subsessile. 6
3. Fruit 4-valved, with four conspicuous white lines, 5-7 mm long; leaves whitish tomentose. **44. C. horrida**
- Fruit 2-valved (rarely partly 4-valved), not with white lines, 7-18 mm long; leaves not tomentose. 4
4. Pseudaril completely enveloping the rugose stone and fused with it; flowers dark red. **3. Sect. Africanae**
- Pseudaril not completely enveloping stone nor fused with it; stone smooth to rugose. 5
5. Pseudaril with four narrow parallel-sided arms some or all of which reach top of stone; flowers red. **10. Sect. Campestres**
- Pseudaril with two short facial and two narrow commissural lobes none of which reaches top of stone; flowers yellow. **22. C. confusa**
6. Leaves simple; inflorescence loose, divaricately branched, with filiform peduncle and pedicels; pseudaril loosely enveloping stone but not fused with it; flowers red. **7. Sect. Rostratae**
- Leaves 1-foliolate or hetero-trifoliolate; inflorescence and pseudaril not as above. 7
7. Pseudaril completely enveloping stone and fused with it. **13. C. obovata**
- Pseudaril not completely enveloping stone. 8
8. Pseudaril a basal cup with broad triangular lobes; fruit beaked; flowers greenish purple. **2. Sect. Coriaceae**
- Pseudaril with four parallel-sided arms; fruit not beaked; flowers red or yellow. **11. Sect. Commiphora**
9. Leaves simple; pseudaril loosely enveloping stone but not fused with it. **7. Sect. Rostratae**

- Leaves 3-foliolate or pinnate; pseudaril different. 10
10. Pseudaril completely enveloping stone (or with a small opening near top) and fused with it. 11
- Pseudaril different. 12
11. Bark exfoliating in small papery flakes or strips, ± reticulately fissured on old trunks; stone rugose; leaves hairy. **14. C. africana**
- Bark exfoliating in large papery sheets, smooth on old trunks; stone smooth; leaves glabrous. **4. Sect. Monoicae**
12. Pseudaril with four parallel-sided angular arms. 13
- Pseudaril different. 14
13. Leaflets entire; flowers in open divaricate cymes. **9. Sect. Latifoliolatae**
- Leaflets crenate to serrate; flowers solitary or in few-flowered contracted cymes. **10. Sect. Campestres**
14. Fruit 4-valved, with four conspicuous white lines; calyx subentire or with broad triangular lobes. 15
- Fruit 2-valved, not with white lines; calyx lobed at least 1/3 down. 16
15. Flowers dark red; fruit beaked, *c* 13 mm long; pseudaril with two broad lobes; leaves conspicuously ciliate, turning red when mature. **8. Sect. Ciliatae**
- Flowers white or yellow; fruit not beaked, 5-8 mm long; pseudaril with basal cup which often covers most of the stone; leaves not ciliate nor red when mature. **12. Sect. Opobalsameae**
16. Flowers in subsessile clusters (rarely solitary), yellow; pseudaril with two short facial and ± two commissural arms, no basal cup; fruit 7-11 mm long. **6. Sect. Hildebrandtianae**
- Flowers in pedunculate inflorescences, red or yellow; pseudaril different; fruit 8-20 mm long. 17
17. Pseudaril a basal cup with broad triangular lobes or with sinuate margin, lobes not fused over top of the smooth stone; disc glabrous; flowers red or yellow. **1. Sect. Spondioideae**
- Pseudaril with broad lobes or arms fused over top of the rugose stone; disc hairy; flowers yellow. **5. Sect. Pedunculatae**

1. Sect. SPONDIOIDEAE *Engl.* ex *Wild* (1959).

Plants not spiny. Leaves pinnate or 3-foliolate. Flowers in 1-many-flowered distinctly peduncled inflorescences. Calyx funnel-shaped, with a distinct tube below filament attachment (rarely not), often lobed 1/2 down. Petals red or yellow. Basal part of filaments fused laterally and with calyx tube (rarely not). Fruit large, fleshy, 2-valved, glabrous to densely pubescent. Pseudaril a basal cup, often with 1-2 broad triangular lobes over locules and ± 2 narrower commissural lobes. Stone smooth, flattened or not.

1. Leaflets entire; flowers red (not known in sp. 2). 2
- Leaflets crenate to serrate; flowers yellow (rarely red). 5
2. Cymes long, slender, drooping; bark reticulately ribbed, not peeling; calyx glabrous (not known in sp. 2). 3
- Cymes erect; bark peeling; calyx puberulous. 4
3. Young branches and leaves glabrous to sparsely puberulous; leaflets conspicuously blistered when mature; pedicels 0-1(-3) mm long. **1. C. erlangeriana**
- Young branches and leaves pubescent or densely so; leaflets not blistered; pedicels 5-10 mm long. **2. C. staphyleifolia**
4. Shrub to 1.5 m; bark peeling in large parchmenty flakes or scrolls; inflorescence 15-30 cm long; pedicels 6-13 mm; calyx 4-5 mm and fruit 16-20 mm long. **3. C. unilobata**
- Tree to 5 m; bark peeling in small papery flakes; inflorescence 1.5-6.5 cm long; flowers sessile; calyx 2-3 mm and fruit *c* 10 mm long. **4. C. guidottii**
5. Flowers red, in umbelliform cymes; pedicels 20-45 mm long; pseudaril with sinuate margin, without lobes. **5. C. longipedicellata**
- Flowers yellow; pedicels 0-8 mm long; pseudaril with broad lobes over locules. 6
6. Flowers in many-flowered contracted racemoid cymes; petals glabrous; fruit 15-20 and stone 11-13 mm long; leaves pinnate with 5-11 leaflets. **6. C. boiviniana**
- Flowers in 1-7(-many)-flowered cymes, if many-flowered then open and dichasial; petals pilose (rarely subglabrous); fruit 8-16 and stone 7-10 mm long; leaves 3-foliolate or pinnate with 5-7 (-9) leaflets. 7
7. Leaves 3-foliolate (or a few pinnate with 5 leaflets), terminal leaflet 1-7 cm long; peduncle (2-)5-40(-65) mm; cymes 1-7(-many) -flowered. **7. C. erythraea**
- Leaves pinnate with 5-7(-9) leaflets (rarely a few 3-foliolate), terminal leaflet 0.8-2 cm long; peduncle 1-5 mm; cymes 1-3-flowered. **8. C. sphaerophylla**

1. C. **erlangeriana** *Engl.* (1904)
- types: BA/SD, by the Ganale river, *Ellenbeck* 2040; SD, Djahle in Boran (Borena) *Ellenbeck* 2114a, (both B syn. destr., K fragments).
C. retifolia Chiov. (1932) - type: BA, Goddere, *Guidotti* 63 (FT iso.).

Shrub or tree to 4(-6) m; bark yellowish to greyish brown, reticulately ribbed, not peeling; young branches longitudinally ridged, glabrous to sparsely puberulous, older often twisted and gnarled. Leaves pinnate with 5-11 leaflets (rarely some 3-foliolate), glabrous to sparsely puberulous, covered with blisters when mature; petiole 2-8.5 cm; leaflets ovate to elliptic, 3-8.5 x 1-5.5 cm, entire, apex acuminate to acute. Flowers glabrous, in (5-)20-60 cm long, slender drooping, glabrous racemes, (rarely with a single basal branch); pedicels 0-1(-3) mm. Calyx 3-4 mm long. Petals dark red, 3-4 mm long. Fruit ellipsoid to subglobose, 14-18 x 11-15 mm, glabrous; pseudaril with 2 broad lobes over locules; stone obovoid, 9-11 x 6-7 x *c* 4 mm.

Acacia-Commiphora bushland on rocky limestone and gypsum slopes; 250-700 m. SD BA; Somalia. *Gilbert* et al. 7580, 8166.

The resins of this and the following three species are reported to be extremely poisonous for humans and animals and are used for arrow poison. In Sidamo it is also used for poisoning hyenas. BA(H) AROR means "run away from it" in Somali and DUNCAL means "poison". No other group of *Commiphora* is reported to have poisonous resin. The chemical background to this has never been investigated.

C. erlangeriana is closely related to *C. zanzibarica* (Baill.) Engl. from the coastal areas of E Africa. It differs in a sparser inflorescence, conspicuously blistered leaves, smaller fruit and in the distinctly lobed pseudaril (with a sinuate margin in *C. zanzibarica*).

2. **C. staphyleifolia** *Chiov.* (1932)
- type: BA, Bug Berde to El Bar, *Guidotti* 10 (FT iso.).

Shrub to 1.5 m, with thick longitudinally ridged branches; bark pale grey, slightly reticulately ribbed, not peeling; branchlets densely pubescent. Leaves pinnate with 5-7 leaflets, pubescent to densely so; petiole 1.5-8.5 cm; leaflets ovate, 3.5-8 x 1.5-4.5 cm, entire, apex acuminate to rounded. Flowers not known. Female inflorescence racemose, 3-8 cm long, puberulous; pedicels 5-10 mm. Fruit (*fide* Chiovenda) ovate to ellipsoid, 13-16 x 9-12 mm, glabrous; pseudaril with 2 broad lobes over locules; stone obovoid, *c* 10 x 6 x 4 mm.

Open *Acacia-Commiphora* bushland on yellowish loamy alluvium overlying limestone; 250-400 m. SD BA; Somalia. *Gilbert* et al. 7542.

The author has seen very little and incomplete material of this species. It is clearly close to *C. erlangeriana* from which it differs in the hairy branches, leaves and inflorescences and in the longer pedicels.

3. **C. unilobata** *Gillett & Vollesen* (1985);
C. erlangeriana sensu Chiovenda (l.c.) and Dale & Greenway (l.c.), not Engl. (1904).

Shrub to 1.5 m; stem swollen at base (lageniform); bark smooth, peeling in large, transverse, yellow to orange parchmenty flakes or scrolls leaving a dark green underbark; branchlets thick, very conspicuously longitudinally ridged, glabrous to sparsely puberulous. Leaves pinnate with (5-)7-11 leaflets, glabrous, not blistered; petiole (3.5-)5-8(-10) cm; leaflets ovate, entire, (2-)3.5-9 x 1.5-4.5 cm, apex acuminate to acute. Flowers in 15-30 cm long, slender (but erect), contracted, racemoid puberulous cymes; pedicels 6-13 mm. Calyx 4-5 mm long, puberulous. Petals dark red, 4-5 mm long, glabrous. Fruit ovoid to ellipsoid, 16-20 x 10-14 mm, glabrous; pseudaril with one big lobe covering entire sterile locule, fertile locule without lobe; stone with sterile locule much larger than fertile and hollow, ovoid to ellipsoid, 13-17 x 8-11 x 7-10 mm. Fig. 123.3.

Open *Acacia-Commiphora* bushland on yellowish loamy alluvium; *c* 250 m. SD; NE & E Kenya, S Somalia. *Gilbert* et al. 7541.

Superficially (especially when sterile) close to *C. erlangeriana*, but at closer examination many differences appear. The peculiar pseudaril and stone with large hollow sterile locule does not resemble anything seen in other species.

4. **C. guidottii** *Chiov.* (1932);
C. sessiliflora Vollesen (1985).

Shrub or tree to 5 m, dbh to 0.5 m; bark greenish to brownish, peeling in small transverse yellowish flakes; branchlets only weakly fissured, glabrous to puberulous. Leaves 3-foliolate or pinnate with 5(-7) leaflets, glabrous to pubescent, densely blistered when mature; petiole 1-10 cm; leaflets ovate to broadly elliptic or subcircular, entire, 2.5-10.5 x 1.5-7.5 cm, apex acute to rounded. Flowers sessile, in 1.5-6.5 cm long, erect, contracted, racemoid, puberulous to pubescent cymes. Calyx 2-3 mm long, puberulous. Petals cream, 2-4 mm long, sparsely puberulous. Fruit globose, *c* 10 x 7 mm, puberulous; pseudaril a basal cup with sinuate margin or with 4 short triangular lobes; stone ellipsoid, *c* 7 x 5 x 4 mm.

Open to dense *Acacia-Commiphora-Boswellia rivae* bushland on stony slopes, always on gypsum soils; 250-400 m. SD BA; Somalia. *Gilbert* et al. 7579, 8175.

Closely related to *C. ornifolia* (Balf. f.) Gillett from Socotra, which differs mainly in having a denser indumentum and longer inflorescences. When better material (including flowers) of *C. ornifolia* becomes available, it may well prove that they are best treated as subspecies.

5. **C. longipedicellata** *Vollesen* (1985).

Shrub or tree to 4 m; bark yellowish to dark grey, not peeling or peeling in small irregular papery flakes (sometimes very tardily); branchlets glabrous to pilose. Leaves pinnate with 5-13 leaflets, glabrous or pilose on petiole and rachis; petiole 0.5-4 cm; leaflets ovate to elliptic, serrate, 1-3 x 0.7-2.3 cm, apex acute to rounded. Flowers in up to 8-flowered, glabrous to densely pilose, umbelliform cymes; peduncle up to 1 cm; pedicels 2-4.5 cm. Calyx *c* 4 mm long, glabrous to pilose. Petals crimson, *c* 5 mm long, glabrous. Fruit ovoid to ellipsoid, apiculate, 15-20 x 10-15 mm, glabrous; pseudaril a basal cup with sinuate margin, no lobes; stone ovoid, 7-10 x 7-9 x 6-8 mm.

Acacia-Commiphora bushland on rocky limestone slopes; 250-400 m. SD BA HA; N & E Kenya. *Allen & Hammersley* 20; *Gilbert* et al. 7582, 7631.

Close to *C. engleri* Guillaumin from C Tanzania which has more densely hairy leaves and a much larger inflorescence.

Figure 123.3 *COMMIPHORA UNILOBATA*: A - leafy branch x 1; B - male inflorescence x 1; C - fruiting branch x 1; D - base of trunk and bark; E - leaflet x 1; F - detail of leaf underside x 15; G - detail of inflorescence axis x 15; H - male flower, entire and in section x 9; I - female flower, entire and in section x 9; K - stone with pseudaril, facial view x 2; L - stone, commisural view x 2; M - stone, transection x 3. A from *Gilbert* et al. 7541; B & G from *Stannard & Gilbert* 966; C & K-M from *Beckett & White* 1582; D from photo of *Gilbert* et al. 7541; E & F from *Gillett* 20539; H from *Adamson* 5; I from *Bally* 12600. Drawn by Eleanor Catherine.

6. **C. boiviniana** *Engl.* (1883);
C. boiviniana var. *crenata* Engl. in DC. Monog. Phan. 4: 22 (1883).
C. holosericea Engl. (1904).

Shrub or tree to 6(-10) m; bark whitish or yellowish to dark grey, peeling in small papery flakes; branchlets puberulous to pilose. Leaves pinnate with 5-11 leaflets (rarely 3-foliolate), pubescent to tomentose; petiole 1.5-9 cm; leaflets elliptic to broadly so or obovate, crenate to serrate, 2-8.5 x 1-5.5 cm, apex acute to shallowly emarginate. Flowers in many-flowered, densely pilose to tomentose, contracted racemoid cymes; peduncle (1-)2-8 cm; pedicels 0-3 mm. Calyx 3-4 mm long, tomentose. Petals yellowish green, 3-4 mm long, glabrous. Fruit ovoid to ellipsoid, 15-20 x 11-14 mm, densely pubescent; pseudaril with 2 broad lobes over locules and 2 thin commissural arms; stone ovoid to ellipsoid, 11-13 x 6-8 x 3-5 mm.

Acacia-Commiphora bushland, *Combretum-Terminalia* wooded grassland, dry riverine forest, on rocky slopes or sandy soil derived from basement rocks (apparently not in limestone areas); 400-1200 m. GG SD; NE Uganda, Kenya and Somalia south through E Africa to Transvaal. *Haugen* 551; *Mesfin Tadesse* et al. 4278; *Sebsebe Demissew* 291.

The wood is used for hoe-handles (Lower Omo Valley). The ripe fruit is edible.

Chiovenda (l.c.) reports that the resin is poisonous and that it is used for arrow poison. This has never been confirmed from any other source and is not mentioned on any of the numerous collections from N Kenya. There has probably been some confusion with one of the above mentioned truly poisonous species. Chiovenda's use of the name 'BA AROR' also points in that direction.

A widespread and very variable species, especially further south in Africa. It may eventually have to be divided into several infraspecific taxa, but the Ethiopian material will then belong to the typical taxon.

7. **C. erythraea** *(Ehrenb.) Engl.* (1883);
Hemprichia erythraea Ehrenb. (1829) - type: Arabia, Ketumbal Island, *Ehrenberg* s.n., (K holo.).
C. somalensis Engl. (1892).
C. holtziana Engl. (1904).
C. gallaënsis Engl. (1912) - type: HA/BA; in valley of river Mane, *Ellenbeck* 1977 (B holo. destr., K fragment).
C. lughensis Chiov. (1932) quoad syntypi *Paoli* 927, 950 (both FT syn.).

Tree to 8 m, dbh to 0.5 m; bark whitish to yellowish or bluish grey, often with pink spots, smooth, peeling in large irregular papery flakes; branchlets subglabrous to subtomentose. Leaves 3-foliolate or a few pinnate with 5 leaflets, glabrous to subtomentose; petiole 0.5-6 cm; leaflets ovate or elliptic to orbicular, crenate to serrate, 0.5-7 x 0.3-6 cm, apex acute to retuse. Flowers in 1-7 (-many)-flowered, pubescent, divaricately branched cymes; peduncle (0.2-)0.5-4(-6.5) cm; pedicels 1-8 mm. Calyx 3-6 mm long, pubescent to tomentose. Petals yellowish green or yellow, 2.5-4 mm long, subglabrous to densely pilose. Fruit ovoid to subglobose, 8-16 x 8-13 mm, glabrous to densely pubescent; pseudaril with 2 broad triangular lobes over locules; stone ovoid to orbicular, 7-10 x 5-7 x 3-4 mm.

Acacia-Commiphora woodland, wooded grassland and bushland, *Combretum-Terminalia* wooded grassland, dry coastal bushland, often on rocky slopes (limestone as well as basement), but also on red sandy soil and black cotton soil; near sea level to 1500(-1900) m. EE AF SU and AR (both in Awash Valley only) GG SD BA HA; Sudan (Red Sea Coast and extreme SE), Kenya, Somalia, N Tanzania, Arabia. *Burger* 3313; *Friis* et al. 2727; *Greathead* 84.

The resin is used as an insecticide (against cattle fleas) by the Borana.

An extremely variable species which is a dominant and conspicuous element of of the vegetation over large areas in S Ethiopia. The extremes are easily recognizable and at first seem to warrant recognition as species. But many recent collections from Sidamo showing the possible variation even within small areas have led the author to the conclusion that only one broad species can be recognized. Some subglabrous large-leaved collections have at times been named as *C. baluensis* Engl. - a species from C Kenya and N Tanzania - but the Ethiopian material merely represents one end of a continuous variation.

8. **C. sphaerophylla** *Chiov.* (1932).
C. ellisiae Vollesen (1985) - type: HA, Gorrahei, 6°32´N 44°35´E, *Ellis* 367 (K holo.).

Tree to 5 m; bark peeling whitish to yellowish, in large papery flakes; branchlets appressed crisped pubescent. Leaves pinnate with 5-7(-9) leaflets (rarely a few 3-foliolate), crisped pubescent to tomentose; petiole 0.5-2 cm; leaflets elliptic to broadly so or obovate, dentate, 0.6-2 x 0.5-1.5 cm; apex rounded to retuse. Flowers in 1-3-flowered densely pubescent, cymes; peduncle 1-5 mm; pedicels 1-3(-6) mm. Calyx *c* 4 mm long, densely pubescent to tomentose. Petals yellow, *c* 4 mm long, densely pilose. Fruit ovoid, *c* 10 x 9 mm, glabrous to pubescent; pseudaril with 2 broad triangular lobes over locules; stone ellipsoid, *c* 7 x 6 x 3 mm. Fig. 123.4.

Acacia-Commiphora wooded grassland and bushland on red to brownish sandy soil overlying limestone and on rocky slopes; 400-800 m. HA; Somalia. *Ellis* 102, 120.

This species is close to *C. erythraea* and could, with some justification, be treated as yet another form of this very variable species. But the pinnate leaves with smaller leaflets, the shorter peduncle and the appressed indumentum on the branches seems to justify its recognition as a distinct species. Admittedly some collections from Somalia are somewhat intermediate in leaf characters.

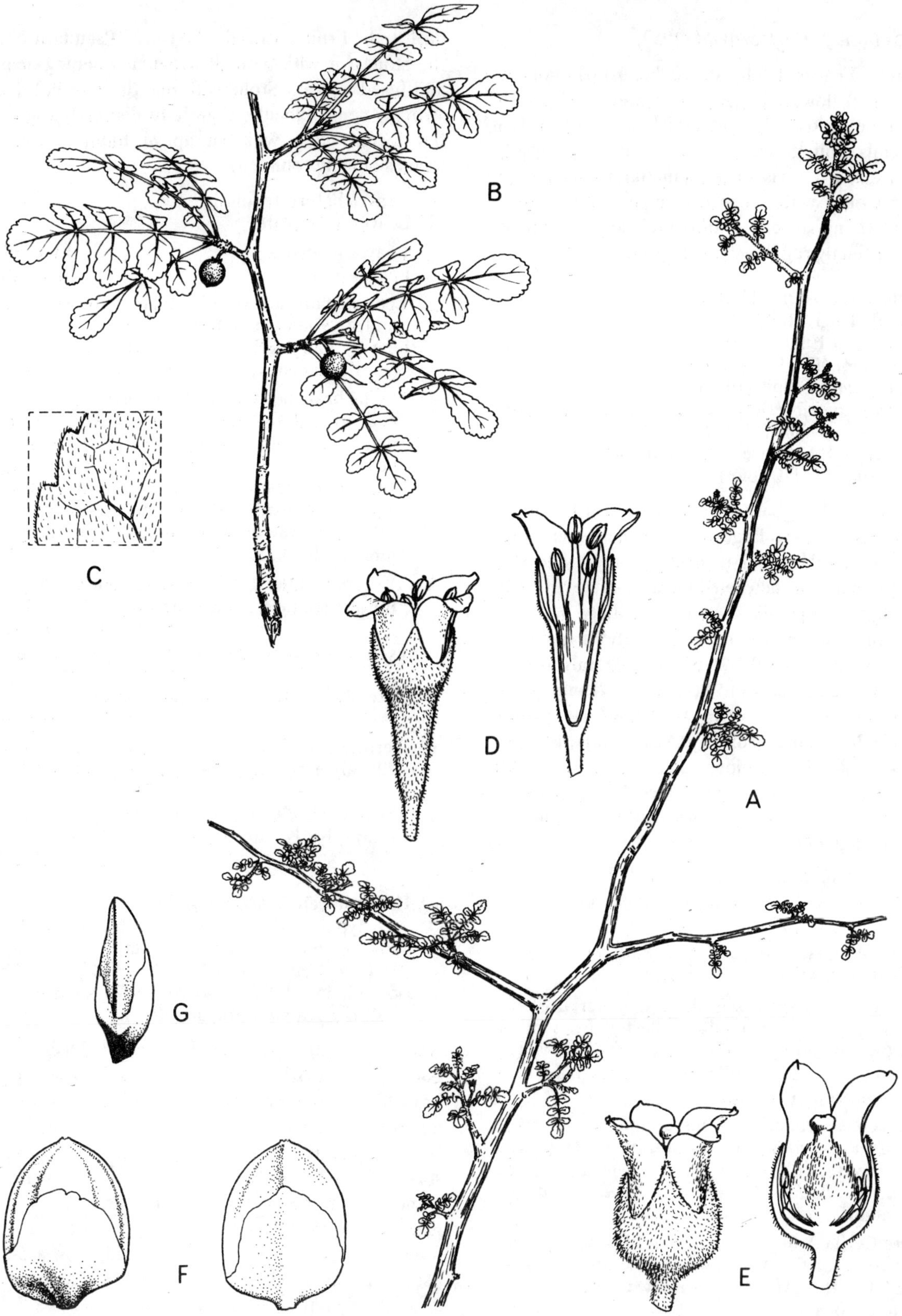

Figure 123.4 *COMMIPHORA SPHAEROPHYLLA*: A - flowering (male) branch with young leaves x 1; B - fruiting branch x 1; C - detail of leaf underside x 15; D - male flower, entire and in section x 12; F - stone with pseudaril, facial view x 4.5; G - stone, commisural view x 4.5. A from *Ellis* 120; B & C from *Drake Brockman* 793; D from *Mckinnon* S/226; E from *Gillett* 21999; F & G from *Hemming* 1445. Drawn by Eleanor Catherine.

2. Sect. CORIACEAE *Engl.* ex *Wild* (1959).

Plants spiny. Leaves 1-foliolate or hetero-trifoliolate. Flowers in 1(-2)-flowered subsessile cymes. Calyx with a distinct tube below filament attachment, lobed 1/3 down. Petals dull greenish purple. Fruit 2-valved, glabrous. Pseudaril a basal cup (sometimes covering up to half the stone) with 2 broad triangular lobes over locules and 2 narrower commissural lobes. Stone smooth to distinctly rugose, strongly compressed.

9. C. **myrrha** *(Nees) Engl.* (1883).
C. rivae Engl. (1897).
C. coriacea Engl. (1899).
C. molmol (Engl.) Engl. (1925).
C. cuspidata Chiov. (1941).
C. habessinica (Berg) Engl. var. *grossedentata* Chiov. in *Atti R. Acad. Ital. Mem. Cl. Sci. Fis.* **11**: 22 (1941) - type: BA, between Mara and Mt. Ello, 6°N, 42°E, *Reghini* 17 (FT holo.).

Shrub or tree to 4 m; bark silvery or whitish to bluish grey, peeling in small to large papery flakes, sometimes reticulately fissured on old trunks; all parts glabrous. Leaves 1-foliolate or heterotrifoliolate, often greyish green or glaucous; petiole 1-10 mm; leaflets ovate or elliptic to obovate, entire to grossly and irregularly dentate, terminal 0.5-4 x 0.3-2(-2.7) cm, laterals usually less than 2 x 1 mm, but up to 8(-15) x 3(-7) mm, apex acute to rounded. Flowers in 1(-2)-flowered cymes; peduncle 0.5-3(-5) mm; pedicels 0.5-2(-3) mm, glandular apically. Calyx 3-5 mm long. Petals dull purple, 3-4 mm long. Fruit with a usually curved beak, ovoid to ellipsoid, 12-16 x 6-8 mm; stone ovoid to ellipsoid, acute or rounded, 6-9 x 4-6 x 3-4 mm. Fig. 123.5.

Acacia-Commiphora woodland and bushland on sandy to loamy soil overlying limestone or granite, rocky lava hills; 250-1300 m. AF SD BA HA; Somalia, NE Kenya, Arabia. *Burger* 3250; *Friis* et al. 3007; *Tewolde Berhan* 2611.

The resin of this species is the raw material of the renowned myrrh. Myrrh is also collected from other species of *Commiphora*, but the best quality is obtained from this species. It is collected commercially in S Ethiopia, Somalia and Arabia.

Myrrh is also collected from a number of other species in Ethiopia, mainly for local use. Among these are: *C. africana*, *C. erythraea*, *C. gileadensis*, *C. habessinica*, *C. hodai*, *C. kua*, *C. quadricincta*, *C. schimperi*, and *C. truncata*.

Cufodontis (l.c.) argues in favour of reserving the name *C. myrrha* for the Arabian plant and use the name *C. molmol* for the African. The two "species" are clearly synonymous.

3. Sect. AFRICANAE *Engl.* ex *Wild* (1959).

Plants spiny. Leaves 3-foliolate (rarely hetero-trifoliolate). Flowers solitary or in 1-few-flowered, usually shortly peduncled cymes. Calyx without tube below filament attachment, lobed 1/3 down. Petals dark red. Fruit 2-valved, glabrous. Pseudaril covering the stone (or with a small irregular opening near top) and fused with it. Stone with one flat (sterile) and one strongly convex locule, slightly to distinctly rugose, the flat locule often with an apical hump (rarely both locules equally rounded).

1. Leaves hetero-trifoliolate. 2
- Leaves 3-foliolate. 3
2. Leaves glabrous. **13. C. obovata**
- Leaves hairy. **14. C. africana**
3. Leaves glabrous or glandular pubescent along midrib; young branches glabrous. 4
- Leaves hairy all over; young branches hairy (rarely glabrous). 6
4. Young branches yellowish to grey; leaflets grossly and irregularly dentate; calyx 3-5 mm long; petals 5-7 mm long; fruit 12-19 mm long, beaked. **12. C. schimperi**
- Young branches reddish brown to purple; leaflets entire to crenate or finely and regularly serrate; calyx 2-3 mm long; petals 3-6 mm; fruit 8-12 mm, not beaked. 5
5. Terminal leaflet 1-2.5 cm long; petiole 0.5-2.5 cm; flowers solitary; spines numerous. **10. C. samharensis**
- Terminal leaflet (2.5-)3-9.5 cm long; petiole (2-) 2.5-7 cm; flowers in 2-9-flowered cymes; spines usually few, occasionally almost absent. **11. C. terebinthina**
6. Terminal leaflet 1-5.5(-8.5) cm long; petiole 3-22(-30) mm; calyx 3-4 mm long; petals 5-7 mm. **14. C. africana**
- Terminal leaflet 0.3-1.2 cm long; petiole 2-6(-9) mm; calyx 1-2 mm long; petals *c* 3 mm. **15. C. tubuk**

10. C. **samharensis** *Schweinf.* (1899)
- types: EE, Samhar near Massawa (Mitsiwa), *Schweinfurth & Riva* 66 (not seen), 116 (not seen), 211 (not seen), 271 (FT K P isosyn.), 306 (FT K P isosyn.), 357 (FT K P isosyn.), 577 (K isosyn.).
C. crassispina Sprague (1927).

Shrub or small tree, young branches reddish to purple; bark dark greyish, rough, breaking into small flakes; branchlets glabrous; resin aromatic. Leaves 3-foliolate, glabrous (rarely sparsely pubescent); petiole 5-25 mm; leaflets elliptic to rhomboid, entire to crenate or serrate, 0.5-2.5 x 0.3-2 cm, apex subacuminate to truncate. Flowers (only large buds seen) solitary, glabrous; pedicels 0-2 mm, glabrous. Calyx *c* 2.5 mm long. Petals 3-4 mm. Fruit obovoid, 8-10 x 5-7 mm; pseudaril sometimes only covering 3/4 of stone and with a laciniate margin; stone obovoid, 5-7 x 3-5 x 3-5 mm, slightly rugose, sterile locule very distinctly keeled.

Coastal bushland and rocky hills; 150-400 m. EE GG; Sudan (Red Sea Coast), Somalia. *Bally* 6788; *Schweinfurth & Riva* 381.

Bally 6788 (which seems to be the only recent collection of this species from Ethiopia) differs from all other

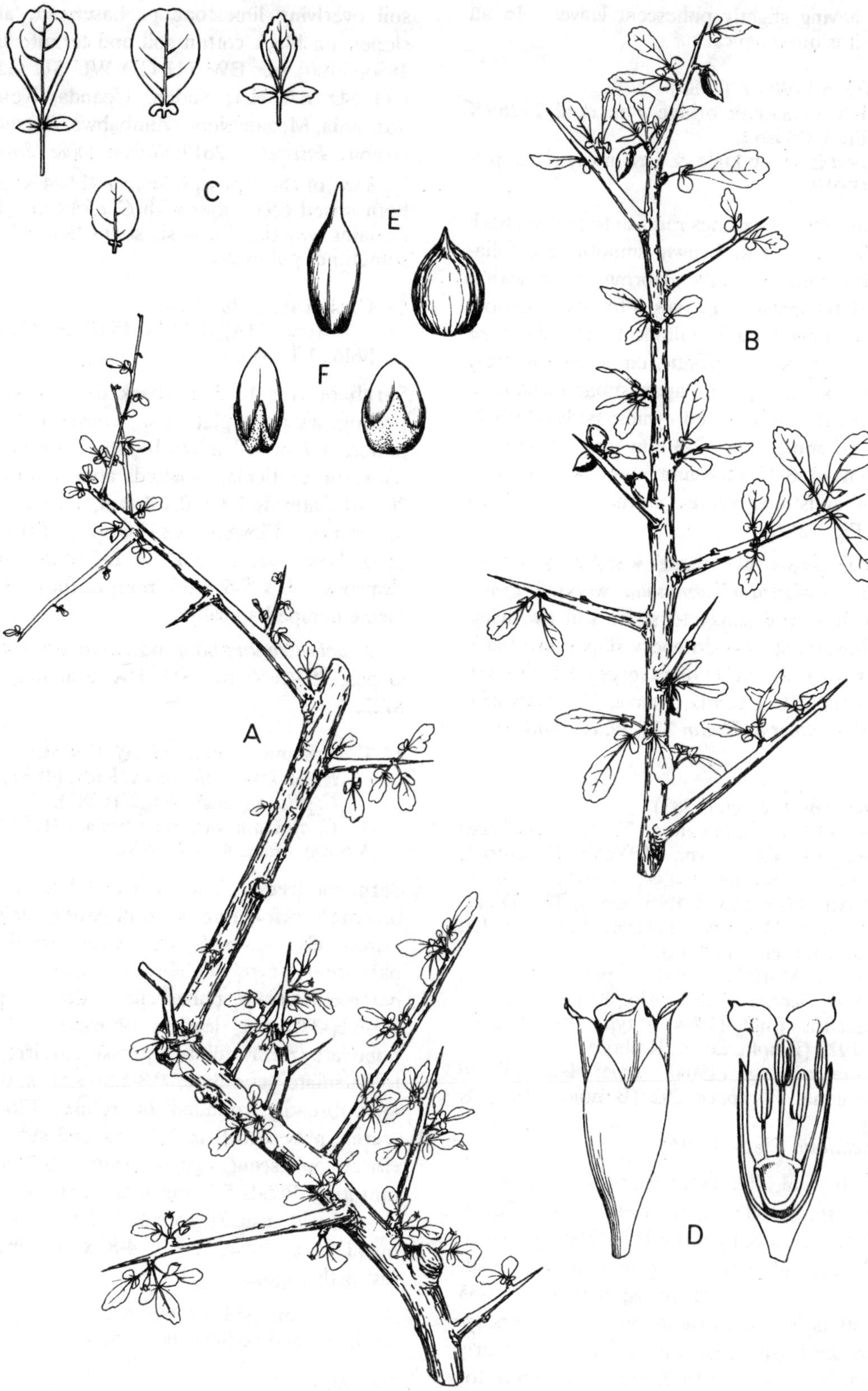

Figure 123.5 *COMMIPHORA MYRRHA*: A - flowering branch x 1; B - fruiting branch x 1; C - variation in leaf shape x 1; D - male flower, entire and in section x 10; E - fruits x 2; F - stones with pseudaril x 3. A from *Radcliffe-Smith & Henchie* 4825; B from *Burger* 3251; C from (left to right and below) *Wood* 2049, *Hemming* 1654, *Glover & Gilliland* 518, *Gillett* et al. 22305; D from *Collenette* 2984; E & F from *Bally* et al. 14937 (left) and *Glover & Gilliland* 730. Drawn by Eleanor Catherine.

collections in having slightly pubescent leaves. In all other respects it is quite typical.

11. **C. terebinthina** *Vollesen* (1985)
- type: BA, Sidambale bridge, *Friis* et al. 2726 (K holo., C ETH UPS iso.).
C. campestris sensu Dale & Greenway (l.c.), p.p., non Engl. (1893).

Tree to 6(-9) m, young branches reddish to purple; bark greyish or yellowish to pale brown, smooth or exfoliating in papery scrolls; all parts glabrous; resin with a strong smell of turpentine. Leaves 3-foliolate; petiole (2-)2.5-7 cm; leaflets broadly elliptic to rhomboid or suborbicular to transversely elliptic, entire to obscurely crenate, 1.7-9.5 x 1.2-6.5 cm, apex subacuminate to retuse. Flowers in 2-9-flowered cymes; peduncle 3-10 mm; pedicels 1-2 mm. Calyx 2-3 mm long. Petals 3-6 mm. Fruit obovoid, 8-12 x 6-8 mm; stone obovoid, 6-9 x 4-6 x 5-6 mm, indistinctly rugose, sterile locule with an apical hump. Fig. 123.6.

Acacia-Commiphora woodland, wooded grassland and bushland, *Combretum-Terminalia* wooded grassland, on whitish to red sandy to stony soil overlying limestone or basement, also on rocky slopes and black cotton soil; 900-1800 m. SU (Gibe Gorge), KF GG SD BA HA; Somalia, NE Uganda, Kenya, NE Tanzania. *Friis* et al. 3162; *Gilbert & Thulin* 339; *W. de Wilde* et al. 10128.

12. **C. schimperi** *(Berg) Engl.* (1883)
- types: TU, Djeladjekanne, Tacazze (Tekeze) river, *Schimper* II:624 (P syn., FI(Webb) K isosyn.); GD, Enderder in Schoata valley, Simien, *Schimper* II:1139 (P syn., FI(Webb) UPS isosyn.); TU, Djeladjekanne by river Tacazze (Tekeze), *Schimper* III: 1564 (P syn., FI(Webb) isosyn.).
C. resiniflua Martelli (1886) - type: EW, Keren, Mt. Deban, *O. Beccari* 23 (FT holo., BM K iso.).
C. arussensis Engl. (1904) - type: BA, Daroli, *Ellenbeck* 1925 (B holo. destr., K photo).
C. neumannii Engl. (1904) - type: HA, Zeila to Harar, Arruena, *Ellenbeck* 292 (B holo. destr., K fragment).
C. flabellifolia Chiov. (1916).

Shrub or tree to 6 m, dbh to 0.5 m; bark yellowish to grey or brownish, exfoliating in small papery flakes, reticulately fissured on old trunks; branchlets glabrous. Leaves 3-foliolate, glabrous or glandular pubescent along midrib, glossy, often glaucous; petiole (3-)5-35 (-40) mm; leaflets elliptic to rhomboid to transversely elliptic or obovate to spathulate, grossly and irregularly dentate, (0.3-)0.5-3.5(-4.5) x 0.3-2.5 cm, apex acute to truncate. Flowers in 1-3-flowered glabrous subsessile cymes; pedicels 1-2 mm. Calyx 3-5 mm long; petals 5-7 mm long. Fruit ovoid to ellipsoid, apiculate, 12-19 x 7-9 mm; stone ovoid to ellipsoid, 8-13 x 5-8 x 5-6 mm, strongly rugose. Fig. 123.7.

Acacia-Commiphora, Acacia-Combretum and *Terminalia* woodland and bushland, on red sandy to stony soil overlying limestone or basement, also on rocky slopes, on black cotton soil and termite mounds; 900-1850(-2100) m. EW TU GD WU SU GJ SU-KF AR GG SD BA HA; Sudan, Uganda, Kenya, Somalia, Tanzania, Mozambique, Zimbabwe, Botswana, S Africa, Arabia. *Friis* et al. 2617; *Gilbert* 3408; *Mooney* 9111.

Two of the types (*Schimper* II:624 and II:1139) are both mixed collections with *C. africana*. This probably explains why the diagnosis states that the branches are sometimes pubescent.

13. **C. obovata** *Chiov.* (1941)
- type: HA, 8°37′N 45°09′E, *Gillett* 4185 (K holo., FT iso.).

Shrub or tree to 3 m; bark dark grey, smooth, not peeling; all parts glabrous. Leaves hetero-trifoliolate; petiole 1-7 mm; leaflets broadly obovate to orbicular, entire or denticulate towards apex, terminal 10-27 x 6-20 mm, laterals 1-8 x 0.5-4 mm, apex broadly rounded to retuse. Flowers not known. Fruit ellipsoid to subglobose, acute, 8-11 x 6-7 mm; stone broadly ellipsoid, 7-8 x 5-6 x 6-7 mm, distinctly rugose, sterile locule humped apically.

Acacia-Commiphora bushland on rocky limestone slopes; 750-1050 m. SD, HA; Somalia. *Gilbert* et al. 8122.

14. **C. africana** *(A. Rich.) Engl.* (1883);
Heudelotia africana A. Rich. (1831).
C. pilosa (Engl.) Engl. (1883).
C. africana var. *ramosissima* (Oliv.) Engl. in DC. *Monog. Phan.* **4**: 14 (1883).

Shrub or tree to 6 m, dbh to 0.2 m; bark greyish to brownish, exfoliating in small papery strips or flakes, ± reticulately fissured on older trunks; branchlets pubescent (rarely glabrous). Leaves 3-foliolate or heterotrifoliolate, pubescent to densely pilose; petiole 0.3-2.2(-3.0) cm; leaflets obovate to broadly so or flabellate (rarely elliptic), grossly and irregularly dentate to laciniate, undulate, 0.3-5.5(-8.5) x 0.2-3.2(-4) cm, apex broadly rounded to retuse. Flowers glabrous (rarely pubescent), in 1-3-flowered subsessile glabrous (rarely pubescent) cymes; pedicels 0-2 mm. Calyx 3-4 mm long. Petals 5-7 mm long. Fruit ovoid to ellipsoid or obovoid, usually rounded, 7-12 x 6-10 mm; stone ovoid to globose, 5-8 x 4-8 x 3-7 mm, slightly to distinctly rugose.

1. Calyx and pedicels glabrous. var. **africana**
- Calyx and pedicels pubescent. var. **rubriflora**

Var. **africana**

Acacia, Acacia-Commiphora, Commiphora-Boswellia, Combretum-Terminalia woodland, wooded grassland and bushland, often on rocky slopes in areas with basement rocks but also on level sandy to loamy soil and recent lava flows; (150-)500-1900(-2100) m. EE AF EW TU GD WU SU KF GG SD BA HA; wide-

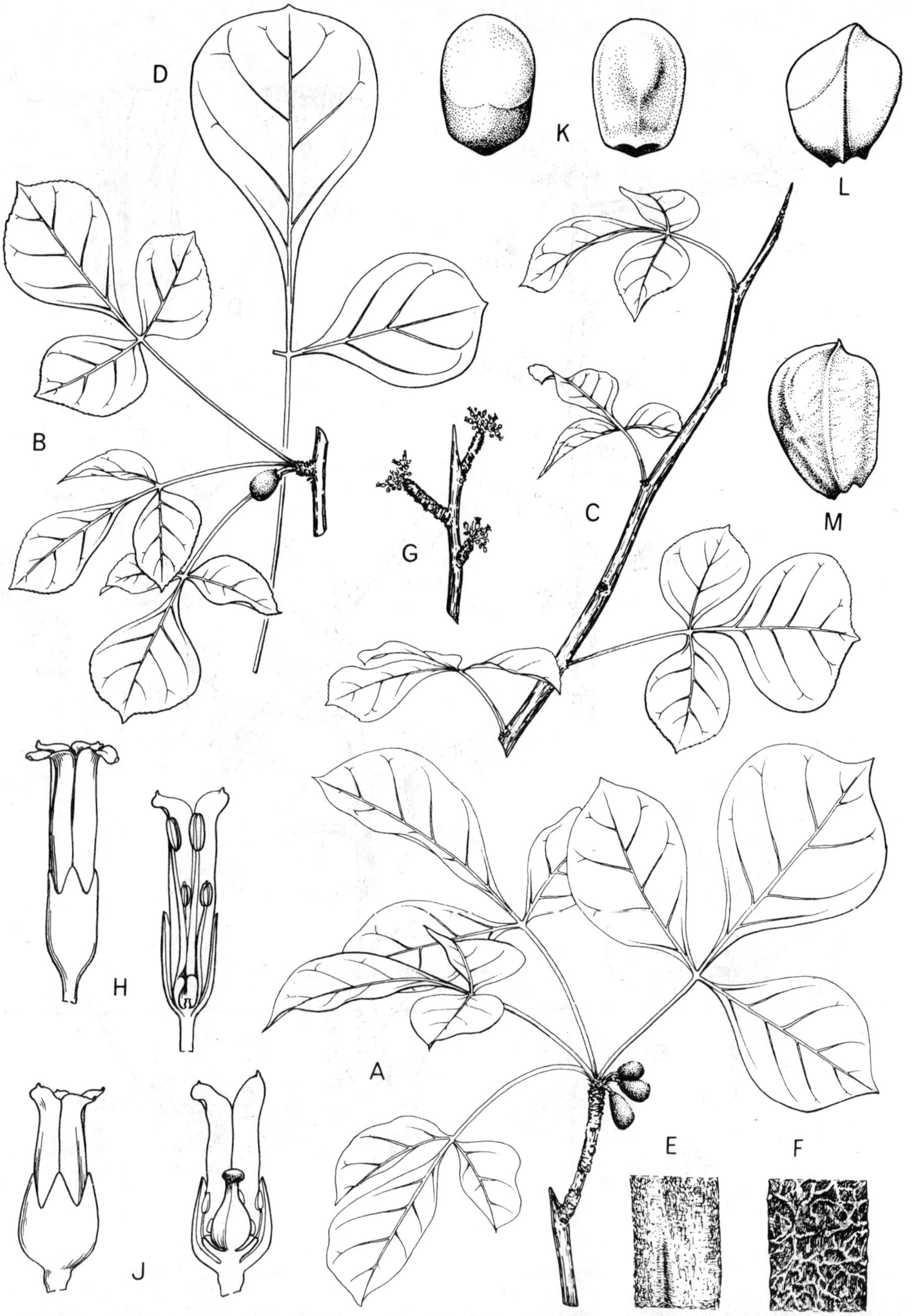

Figure 123.6 *COMMIPHORA TEREBINTHINA*: A-B - fruiting branches x 1; C - young branch x 1; D - leaf x 1; E-F - trunks showing bark patterns; G - flowering branch (male) x 1; H - male flower, entire and in section x 9; I - female flower, entire and in section x 9; K - stone, facial view x 4.5; L-M - stones, commissural view x 4.5. A from *Friis* et al. 2726; B & C from *Friis* et al. 3162; D from *Thulin* et al. 3499; E from photo of *Friis* et al. 2726; F from photo of *Friis* et al. 3162; G from *Gillett* 14865; H, K & L from *J. de Wilde & Gilbert* 396; I from *W. de Wilde* et al. 10128; M from *Gilbert* 3200. Drawn by Eleanor Catherine

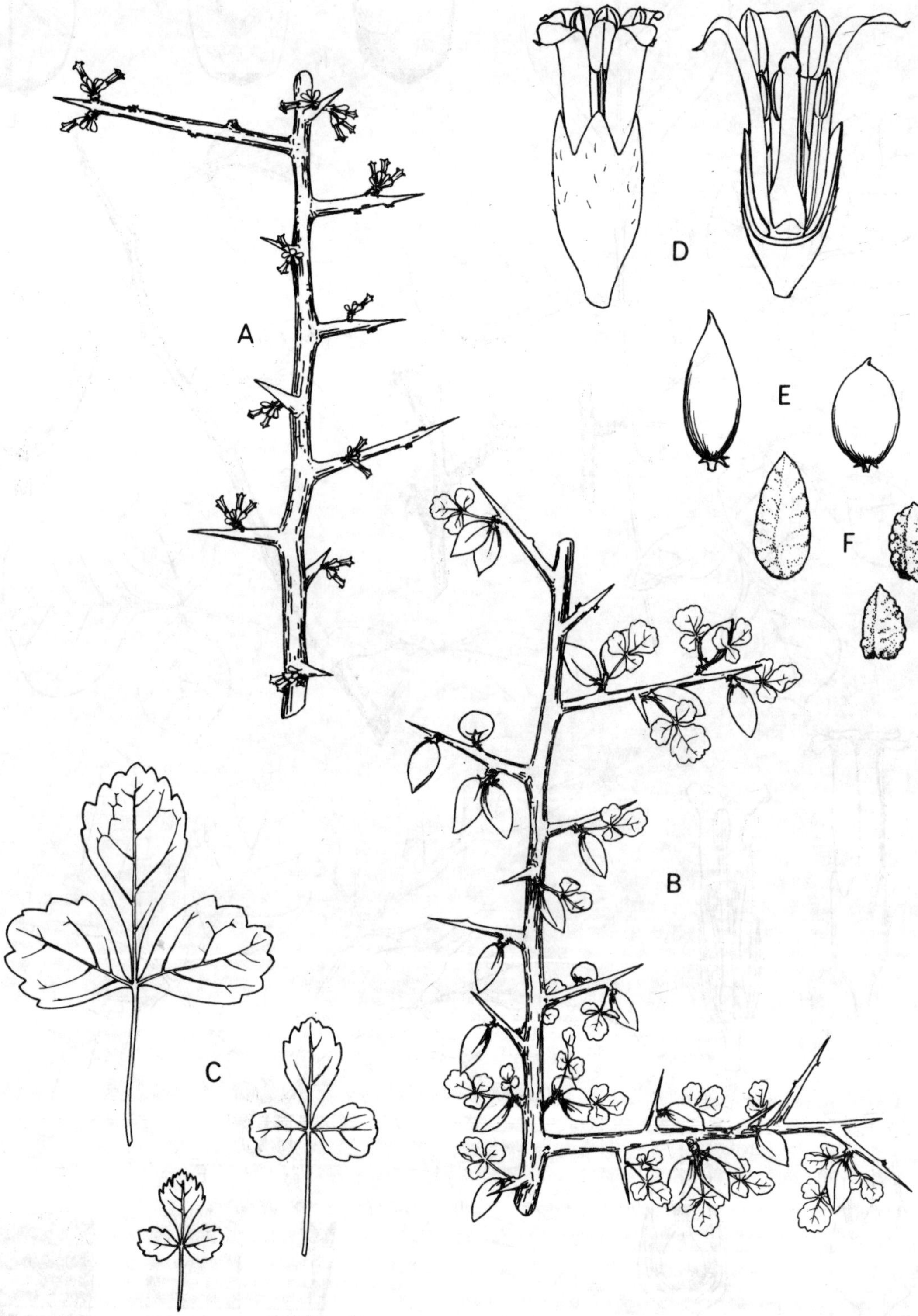

Figure 123.7 *COMMIPHORA SCHIMPERI*: A - flowering branch x 1; B - fruiting branch x 1; C - variation in leafsize and dentation x 1; D - flower, entire and in section x 9; E - fruits x 2; F - stones x 2. A from *Ash* 2784; B from *Thulin* et al. 3798; C from (left to right) *Friis* et al. 2617, *Ash* 1965, *Desta Hundessa* 65; D from *W. de Wilde* et al. 9883; E & F from *Gilbert* 4640 (left) and *Beccari* 23. Drawn by Eleanor Catherine.

spread in the drier parts of tropical Africa. *Burger* 876; *Friis* et al. 2647; *Gilbert* 3350.

Var. **rubriflora** *(Engl.) Wild* in *Bol. Soc. Brot., ser. 2*, **33**: 42 (1959).

Ecology as for var. *africana*; 700-1900 m. GG SD; Uganda, Kenya, Tanzania, Mozambique, Zambia, Zimbabwe. *Fukui* 1220; *Gillett* 14860.

The wood is used for hoe-handles (Lower Omo Valley).

Closely related to *C. schimperi* and has often been confused with it in the past. This is possibly due to the fact that two of the syntypes of *C. schimperi* are mixed collections with *C. africana* (see above). Usually easily distinguished by the hairy branches and leaves and the smaller rounded fruit. A few Ethiopian collections (e.g. *O. Beccari* 22) have glabrous branches, but they have the fruit of *C. africana* and are here placed in this species.

15. **C. tubuk** *Sprague* (1927).

Shrub or tree to 5 m; bark yellowish or grey to almost white, peeling in papery strips; branchlets and leaves pubescent to pilose. Leaves 3-foliolate, petiole 2-6(-9) mm; leaflets broadly elliptic or obovate to flabellate, grossly dentate to laciniate, undulate, 0.1-1.2 x 0.1-0.9 cm, apex subacute to retuse. Flowers glabrous, solitary; pedicels 0.5-3 mm. Calyx 1-2 mm long. Petals *c* 3 mm long. Fruit ovoid to subglobose, usually rounded, 6-9 x 5-7 mm; stone ellipsoid or obovoid to subglobose, 5-7 x 4-6 x 4-6 mm, conspicuously rugose.

Acacia-Terminalia and *Acacia-Balanites* woodland and wooded grassland, usually on rocky volcanic soils; 1100-1500 m. SU (Awash Valley), HA; N Somalia. *Beals* 953; *Gillett* 4099; *Mooney* 9123.

C. tubuk could with some justification be considered a form of the variable *C. africana*, but the distinguishing characters seem to be reasonably constant, and the author has decided to keep them separate. It is distinguished by the small leaflets and the smaller flowers.

4. Sect. MONOICAE *Vollesen* (1985).

Plants unarmed. Leaves 3-foliolate. Flowers monoecious, in 1-3-flowered peduncled cymes. Calyx fleshy, without tube below filament attachment, lobed 1/3 down. Petals dark red. Fruit 2-valved, glabrous. Pseudaril covering most of the stone and united with it, dissolving into a number of irregular arms apically. Stone smooth, with both locules equally developed.

16. **C. monoica** *Vollesen* (1985)
- type: BA, Sof Omar, *Gilbert* et al. 7833 (K holo., C ETH UPS iso.).

Tree to 7 m, ± scandent when young, with long drooping branches; bark peeling in large transverse pale yellowish strips or sheets, smooth; all parts glabrous. Leaves glossy, 3-foliolate; petioles bright red, 1.5-5 cm; leaflets ovate to broadly elliptic, entire to crenate, 2-7 x 1.2-4 cm, apex acuminate to acute. Flowers monoecious, in 1-3-flowered cymes; peduncle 5-15 mm; pedicels corky in fruit, up to 5 mm. Calyx (and pedicels) fleshy and warted, longitudinally striate, yellowish brown, 4-5 mm long. Petals dark red, *c* 4 mm long. Fruit broadly ovoid to globose, 7-9 x 7-9 mm; stone transversely ellipsoid, 5-6 x 5-6 x 6-7 mm. Fig. 123.8.

Dense *Commiphora* woodland on rocky limestone slopes, *Acacia-Commiphora* bushland on black cotton soil; 1250-1400 m. BA; not known elsewhere. *Friis* et al. 3631; *Gilbert* et al. 7810; *Sebsebe Demissew* 1174.

Only known from five collections all made within a very restricted area around the Sof Omar caves. It seems to be the first truly monoecious species in *Commiphora* although monoecious plants (and even bisexual flowers) are known from a few normally dioecious species, e.g. *C. africana* and *C. schimperi*. The peculiar flowers and pseudaril makes it impossible to fit the species into any already described section.

5. Sect. PEDUNCULATAE *Engl.* ex *Vollesen* (1985).

Plants unarmed. Leaves pinnate. Flowers in long-peduncled cymes. Calyx without tube below filament attachment, lobed 1/3 down. Petals yellow. Disc hairy. Fruit 2-valved, pubescent. Pseudaril with four broad oblong or triangular lobes or arms fusing over top of stone. Stone with both locules equally developed, strongly rugose.

17. **C. pedunculata** (*Kotschy & Peyr.*) *Engl.* (1883).

Shrub or tree to 5 m; bark papery peeling; branchlets densely pubescent. Leaves pinnate with 5-13 leaflets (rarely a few 3-foliolate), pubescent or pilose; petiole up to 5.5 cm; leaflets with prominent reticulate venation, oblong to elliptic (rarely narrowly obovate), serrate, 1.5-6 x 0.7-2.5 cm, apex acute to rounded. Flowers densely pilose, in up to 12-flowered densely pilose cymes; peduncle up to 5 cm; pedicels up to 2 mm. Calyx 2-3 mm long. Petals 4-6 mm long. Fruit ellipsoid to orbicular, *c* 14 x 10 mm; stone ellipsoid to orbicular, *c* 10 x 7 x 5 mm, sterile locule with median ridge. Fig. 123.9.

Combretum-Terminalia woodland and wooded grassland; up to *c* 1000 m. Probably on the western escarpment in GD, GJ or WG; from Mali to the Sudan and in S Tanzania, Malawi and Zambia. No Ethiopian collections seen.

This species was recorded by Cufodontis (l.c.) from W Ethiopia without any more precise locality, and it is uncertain from where he got his information. The species has been collected several times just west of the border in the Sudan and most certainly also occurs in Ethiopia.

6. Sect. HILDEBRANDTIANAE *Engl.* ex *Vollesen* (1985).

Plants unarmed (rarely spiny). Leaves pinnate, 3-foliolate or hetero-trifoliolate. Flowers sessile (rarely pedicellate), in few to many flowered fascicles (rarely

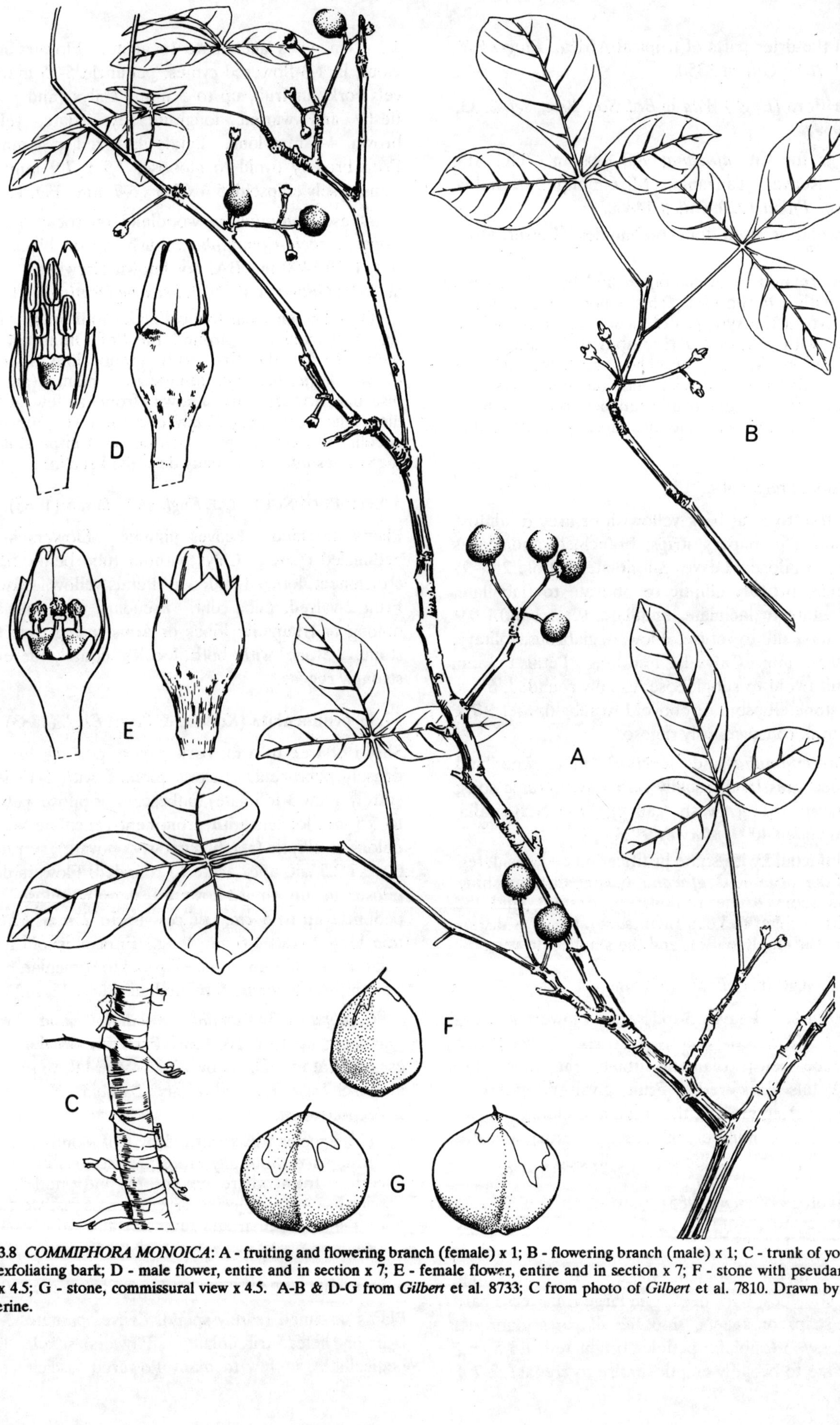

Figure 123.8 *COMMIPHORA MONOICA*: A - fruiting and flowering branch (female) x 1; B - flowering branch (male) x 1; C - trunk of young tree with exfoliating bark; D - male flower, entire and in section x 7; E - female flower, entire and in section x 7; F - stone with pseudaril, facial view x 4.5; G - stone, commissural view x 4.5. A-B & D-G from *Gilbert* et al. 8733; C from photo of *Gilbert* et al. 7810. Drawn by Eleanor Catherine.

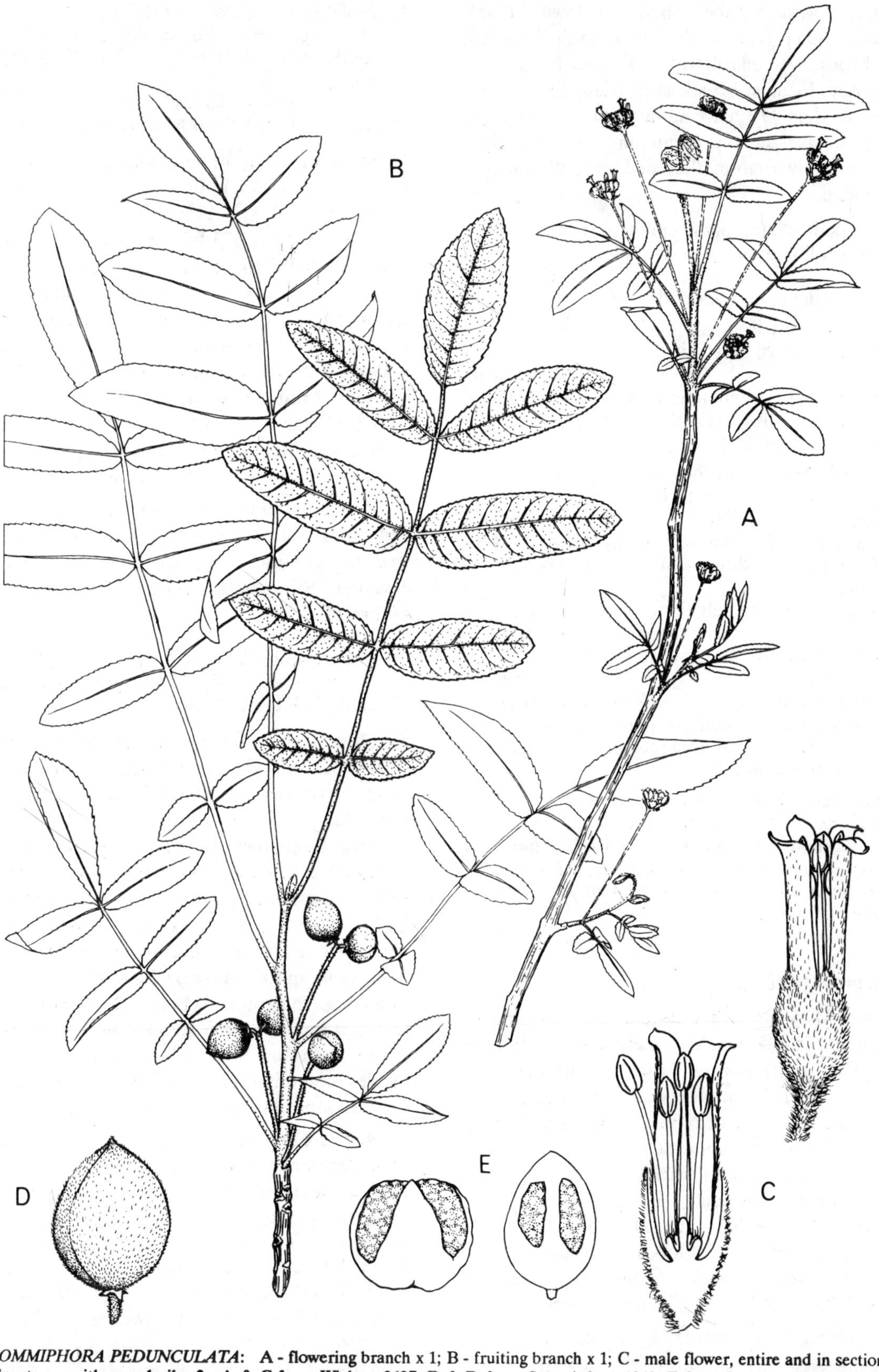

Figure 123.9 *COMMIPHORA PEDUNCULATA*: A - flowering branch x 1; B - fruiting branch x 1; C - male flower, entire and in section x 9; D - fruit x 3; E - stones with pseudaril x 3. A & C from *Wickens* 3627; B & D from *Schweinfurth* 1940; E from *Latilo* in FHI 43783 (left) and *Chevalier* 24811. Drawn by Eleanor Catherine.

solitary). Calyx without tube, lobed 1/3 down. Petals yellow. Disc hairy (rarely glabrous). Fruit 2-valved, usually glabrous. Pseudaril with 2-4 lobes from base, the ones over locules broad and triangular, never reaching more than 1/2 up, commissural either equally short or reaching higher up or missing. Stone almost smooth to strongly rugose, flattened or both locules equally rounded.

1. Flowers pedicellate; plant spiny or not; leaves hetero-trifoliolate or some 3-foliolate. **22. C. confusa**
- Flowers sessile; plant unarmed. 2
2. Leaves hetero-trifoliolate; flowers in 1-3-flowered fascicles; bracts linear. **21. C. truncata**
- Leaves 3-foliolate or pinnate; flowers in up to 12-flowered fascicles (rarely solitary); bracts triangular or spathulate. 3
3. Trunk and larger branches with angular corky wings (rarely without); bark smooth, not peeling; flowers solitary. **23. C. alaticaulis**
- Trunk and branches not with corky wings; bark rough or peeling; flowers in up to 12-flowered fascicles. 4
4. Bark rough to reticulately fissured on trunks, not peeling; young branches tomentose; leaflets densely tomentose to woolly, bullate; flowers with 4 stamens. **20. C. corrugata**
- Bark on trunk papery and peeling; young branches puberulous to densely pubescent; leaflets sparsely pubescent to tomentose, not bullate; flowers with 8 stamens. 5
5. Leaflets dentate with broad triangular teeth, puberulous to pubescent; calyx puberulous; leaves 3-foliolate or pinnate. **19. C. ogadensis**
- Leaflets serrate, biserrate or bicrenate with small sharp teeth, pubescent to tomentose; calyx densely pubescent; leaves 3-foliolate. **18. C. hildebrandtii**

18. C. **hildebrandtii** (*Engl.*) *Engl.* (1883).

Shrub or tree to 5 m; bark yellow to grey, peeling in small irregular flakes; branchlets pubescent to densely so (rarely puberulous). Leaves 3-foliolate, pubescent to tomentose; petiole 1-3.5 cm; leaflets ovate to elliptic or obovate, serrate, biserrate or bicrenate with small sharp teeth, 1-6 x 0.7-4 cm, apex subacute to truncate. Flowers in up to 6-flowered fascicles, sessile. Calyx 2-3 mm long, densely pubescent. Petals 3-4 mm long, densely pilose. Fruit ellipsoid to subglobose, 9-11 x 7-10 mm; pseudaril with 2 broad lobes over locules and 2 narrow commissural lobes; stone ellipsoid to subglobose, 5-8 x 4-6 x 3-5 mm, slightly rugose.

Acacia-Commiphora woodland, wooded grassland and bushland on rocky limestone slopes and ridges; 850-1500 m. SD BA; Somalia. *Friis* et al. 3126; *Mooney* 9732; *Tewolde Berhan* 2617.

19. C. **ogadensis** *Chiov.* (1932)
- type: HA, Pozzo da Carbadan (Cavernay), south of Sosobeneh, *Ruspoli & Riva* 915(824) (FT holo.).
C. tephrodes Chiov. (1941) - type: BA, between Mara and Mt Ello, 6°N 42°E, *Reghini* 53 (FT holo.).

Shrub or tree to 6 m; bark whitish to yellowish, peeling (sometimes tardily) in transverse papery strips or irregular flakes; branchlets puberulous (rarely pubescent). Leaves 3-foliolate or pinnate with 5-7(-9) leaflets elliptic to obovate, puberulous to pubescent; petiole 0.5-4(-5.5) cm; leaflets dentate with broad triangular teeth, 0.7-6.5 x 0.4-3.5 cm, apex subacute to truncate. Flowers in up to 12-flowered fascicles, sessile. Calyx 2-3 mm long, puberulous. Petals 3-5 mm long, pilose. Fruit ellipsoid to globose, 7-9 x 6-8 mm; pseudaril with 2 broad lobes over locules and 2 narrow commissural lobes; stone ellipsoid to globose, 4-7 x 5-6 x 3-4 mm, almost smooth to distinctly rugose.

Acacia-Commiphora woodland and bushland, on red sandy soil overlying limestone, on black cotton soil and on granitic soils; 850-1500 m. SD BA HA; Somalia, NE Kenya. *Burger* 3347; *Friis* et al. 2914; *Gilbert* et al. 7473.

20. C. **corrugata** *Gillett & Vollesen* (1985)
- type: SD, Negele to Filtu, *Friis* et al. 3127 (K holo., C ETH UPS iso.).

Shrub or tree to 5 m; bark pale grey to dark grey or brownish, smooth on younger branches, rough or reticulately fissured on trunk, not peeling; branchlets tomentose. Leaves 3-foliolate or pinnate with 5(-7) leaflets, tomentose to woolly; petiole (2-)5-27 mm; leaflets elliptic to suborbicular or obovate to rhomboid, bullate, undulate, grossly dentate to serrate, terminal leaflet often irregularly incised, 0.5-6.5(-8) x 0.3-3.5(-4) cm, apex acute to truncate; reticulation prominent. Flowers in up to 12-flowered fascicles, sessile. Calyx 2-3 mm long, tomentose. Petals 3.5-5 mm long, densely pilose; stamens 4. Fruit subglobose, *c* 8 x 8 mm, sparsely pilose; pseudaril with 2 broad triangular lobes over locules, commissural lobes lacking; stone subglobose, *c* 6 x 7 x 5 mm, distinctly rugose. Fig. 123.10.

Acacia-Commiphora woodland and bushland on rocky limestone slopes and ridges; 300-1800 m. SD BA HA; Somalia, NE Kenya. *Burger* 2946, 3288, 3341.

A very odd and conspicuous species, probably closest to *C. hildebrandtii* and *C. ogadensis*. It differs in having reticulately fissured bark, tomentose branches, woolly bullate leaves, and only 4 developed stamens.

Among the Ethiopian species, reduction of the stamens to one whorl only (the longest) is otherwise only seen in *C. ciliata*, but is also approached in some species in Sect. *Opobalsameae* (see *C. coronillifolia*). It

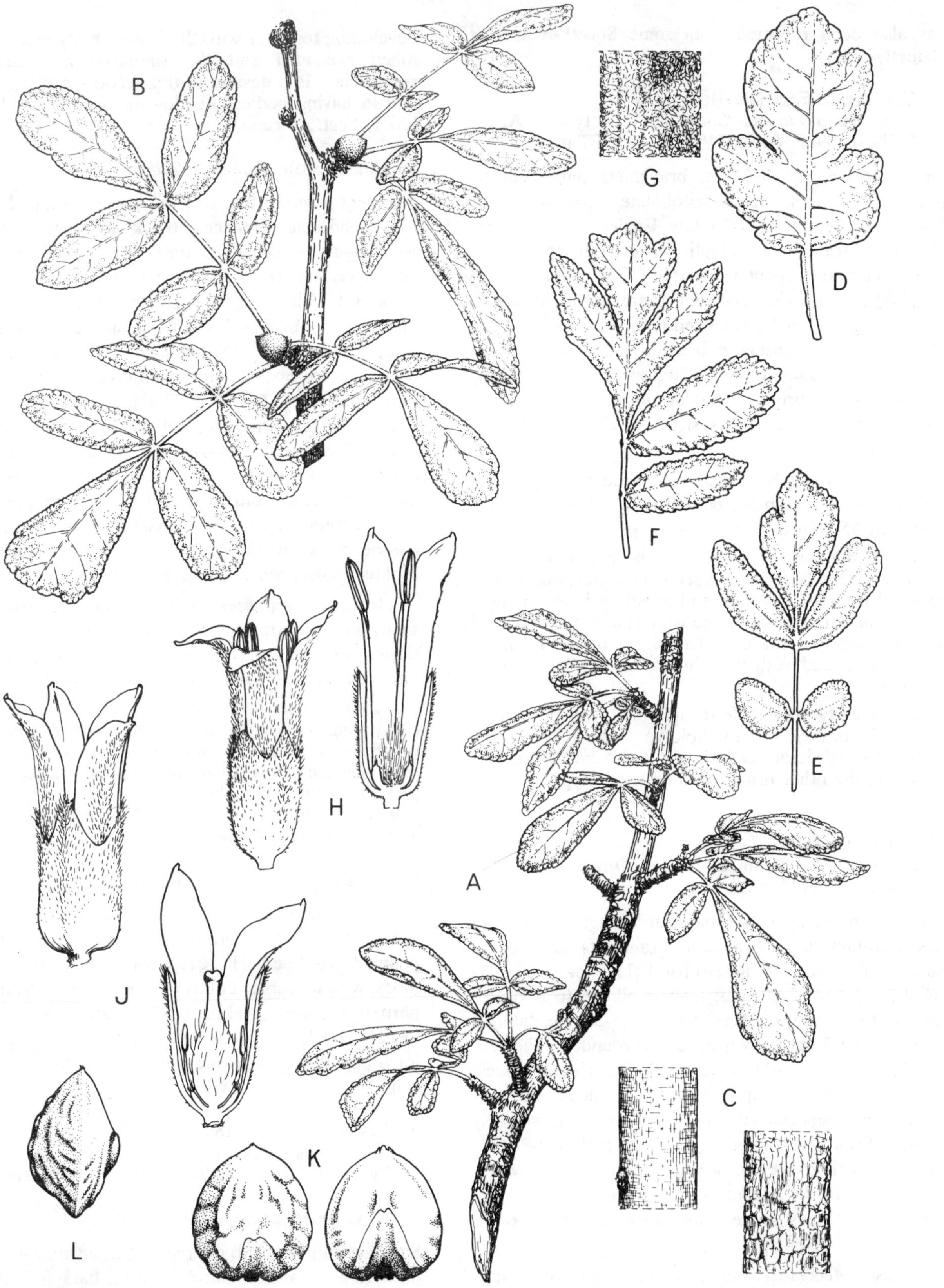

Figure 123.10 *COMMIPHORA CORRUGATA*: A - flowering branch x 1; B - fruiting branch x 1; C - bark on young (left) and old branch; D-F - variation in leaf form x 1; G - detail of leaf underside x 12; H - male flower, entire and in section x 12; I - female flower, entire and in section x 12; K - stone with pseudaril, facial view x 4.5; L - stone, commissural view x 4.5. A from *Gilbert & Thulin* 1547; B & K-L from *Friis* et al. 3127; C from photos of *Friis* et al. 3127; D from *Friis* et al. 909; E & G from *Friis* et al. 2782; F from *Burger* 2946A; H from *Burger* 3288; I from *Burger* 3282. Drawn by Eleanor Catherine.

has also been reported from some Socotran species (Gillett, l.c.).

21. **C. truncata** *Engl.* (1904);
C. crenato-lobata Chiov. (1941) - type: HA, 8° 37´N, 45°09´E, *Gillett* 4155 (K holo., FT iso.).

Shrub or tree to 4(-6) m; branchlets pubescent to pilose. Leaves hetero-trifoliolate, pubescent to tomentose; petiole 2-8(-15 *fide* Engler) mm; leaflets elliptic to transversely elliptic or broadly spathulate, grossly crenate to dentate, terminal 0.8-1.7(-3.5) x 0.7-1.7(-2.5) cm, laterals 1-5(-8) x 1-5 mm, apex truncate to retuse. Flowers in 1-3-flowered fascicles, sessile. Calyx 2-3 mm long, densely puberulous to tomentellous. Petals 4-6 mm long, pilose. Fruit ovoid, 8-9 x 6-7 mm; pseudaril with 2 broad triangular lobes over locules and 2 narrow commissural lobes; stone ovoid, 6-7 x 5-6 x 3-4 mm, slightly rugose.

Acacia-Commiphora bushland on red sandy to stony soil overlying limestone; 600-1050 m. HA; Somalia. *Hemming* 158; *Popov 1130.*

In the diagnosis, Engler gives the petiole-length as 15 mm. The author has not seen any collections with a petiole longer than 8 mm, and mostly it is less than 5 mm. Apart from this discrepancy, the available material fits the diagnosis exactly. The shape of the terminal leaflet combined with two tiny lateral leaflets is very distinctive.

It is clearly impossible to uphold *C. crenato-lobata*. One of the branches on the holotype has two very large leaves (the maximum dimensions given above), but the leaves on the other branch and on the isotype are of a typical size.

22. **C. confusa** *Vollesen* (1985)
- type: BA, Sidambale bridge, *Friis* et al. 2729 (K holo., C ETH UPS iso.).

Tree to 6 m, spiny or not; bark tardily peeling in small flakes, yellowish to brownish; branchlets and leaves pubescent. Leaves hetero-trifoliolate or some 3-foliolate; petiole 2-20 mm; leaflets elliptic to obovate, crenate to serrate, terminal 1-4.5 x 0.5-2.5 cm, laterals up to 2.7 x 1.5 cm, apex subacute to rounded. Flowers in 3-5-flowered sessile or subsessile fascicles; peduncle up to 1 mm; pedicels up to 2 mm, puberulous. Calyx 2-3 mm long, puberulous. Petals 5-6 mm long, sparsely pilose. Fruit ellipsoid to subglobose, 8-10 x 7-8 mm; pseudaril with 2 short oblong or triangular lobes which are sometimes divided into isolated patches and 2 thin commissural lobes; stone ellipsoid to subglobose, 6-7 x 5-6 x 3-5 mm, strongly rugose.

Dense *Acacia-Commiphora* bushland on rocky slopes or grey sandy soil over basement rocks; 600-1300 m. GG SD BA; N & E Kenya, NE Tanzania. *Friis* et al. 2858; *Sebsebe Demissew* 293, 662.

Vegetatively this species looks very much like *C. africana*, and in the herbaria it has been extensively confused with it. But it differs in having yellow flowers developing together with the leaves, hairy disc and a 4-lobed pseuadril and the similarity is clearly only superficial. It is deviant in this section in being ± spiny and in having pedicellate flowers, and clearly forms a link to Sect. *Africanae*.

23. **C. alaticaulis** *Gillett & Vollesen* (1985).

Scandent shrub or tree to 5 m (or erect when old), with long pendulous branches; trunk and larger branches with angular corky wings up to 2 cm wide (sometimes absent on old trees); bark dark grey to bluish grey, smooth to blistered, not peeling; branchlets crisped puberulous. Leaves 3-foliolate, puberulous to tomentose; petiole 1.5-5(-6.5) cm; leaflets broadly elliptic to orbicular or obovate to spathulate, grossly dentate (rarely crenate), 1.5-7.5 x 1-7 cm, apex acute to retuse. Flowers solitary, often several clustered at branch ends. Calyx 2-3 mm long, glabrous with pilose to woolly edges of lobes. Petals 5-6 mm long, densely pilose; disc glabrous. Fruit ovoid, acute, 10-13 x 7-8 mm; pseudaril with 2 oblong lobes over locules, commissural lobes absent; stone ovoid, 7-9 x 5-7 x 3-4 mm, smooth to slightly rugose, compressed. Fig. 123.11.

Acacia-Commiphora woodland and bushland on red sandy soil overlying limestone and on rocky limestone slopes; 400-1400 m. SD BA HA; NE Kenya, Somalia. *Friis* et al. 3667; *Gilbert* et al. 7724; *Thulin* et al. 3742.

A very conspicuous and most peculiar species. The corky wings are - as far as the author knows - unique in the genus. The sessile hairy flowers and the two-lobed pseudaril seem to indicate that it is best placed in this section. But the glabrous disc and the smooth compressed stone seem to indicate some relations also with Sect. *Spondioideae*.

7. Sect. ROSTRATAE *Engl.* ex *Wild* (1959).

Plant unarmed or spiny. Leaves simple. Flowers in loose divaricately branched cymes with filiform peduncles and pedicels. Calyx with very short triangular lobes, without tube. Petals dark red. Fruit 2-valved or partially 4-valved, glabrous. Pseudaril completely and loosely enveloping stone but not fused with it, with 4 longitudinal ridges. Stone smooth, both locules equally rounded.

24. **C. rostrata** *Engl.* (1897)
- types: HA, by rivers Wabe Shebele and Merehan, *Robecchi-Bricchetti* 598, 670 (both FT syn.).
C. robecchii Engl. (1897).

Single-stemmed shrub or tree to 4 m or many-stemmed divaricately branched shrub to 2 m; bark pale to dark grey, smooth to rough, not peeling or peeling in small flakes, often with a pattern of transverse to ± reticulate stripes; all parts glabrous; resin with strong unpleasant smell. Leaves simple, entire, glaucous; petiole 1-8 mm;

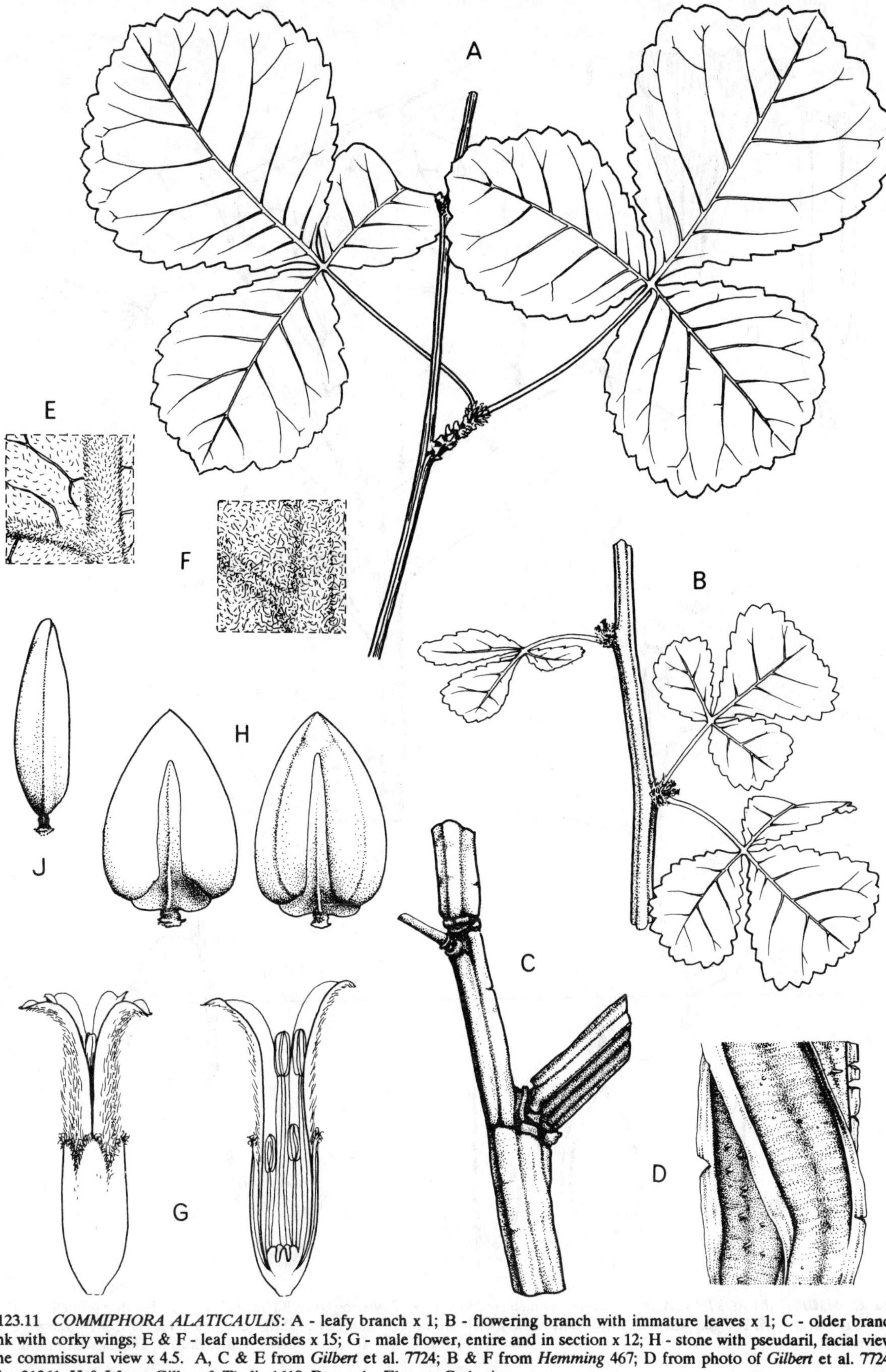

Figure 123.11 *COMMIPHORA ALATICAULIS*: A - leafy branch x 1; B - flowering branch with immature leaves x 1; C - older branch x 1; D - trunk with corky wings; E & F - leaf undersides x 15; G - male flower, entire and in section x 12; H - stone with pseudaril, facial view x 4.5; I - stone commissural view x 4.5. A, C & E from *Gilbert* et al. 7724; B & F from *Hemming* 467; D from photo of *Gilbert* et al. 7724; G from *Gillett* 21261; H & I from *Gilbert & Thulin* 1619. Drawn by Eleanor Catherine.

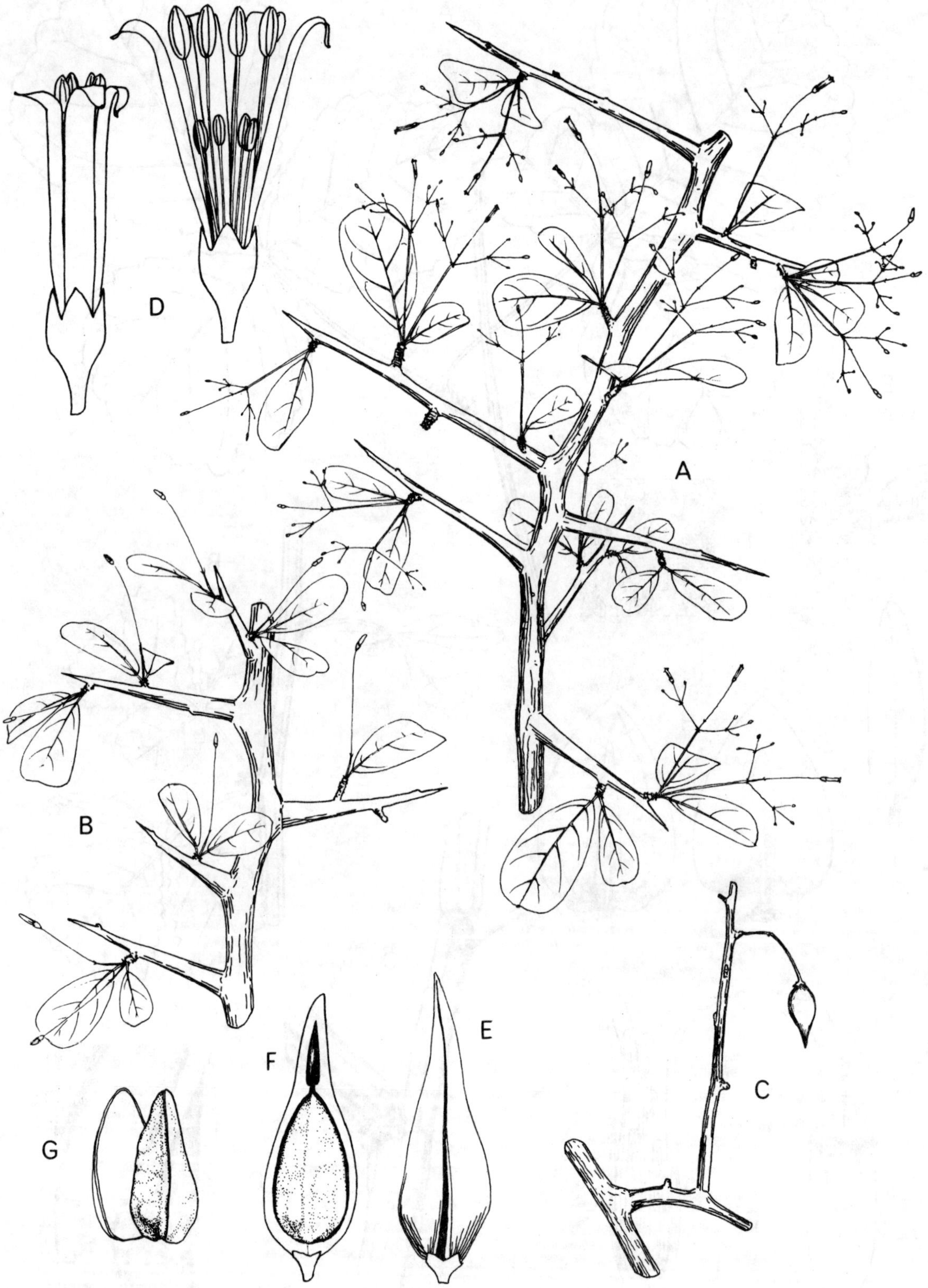

Figure 123.12 *COMMIPHORA ROSTRATA*: A - flowering branch (male) x 1; B - flowering branch (female) x 1; C - fruiting branch x 1; D - male flower x 12; E - fruit x 4.5; F - fruit with one valve removed x 4.5; G - stone with half of pseudaril removed x 4.5. A from *Stannard & Gilbert* 917; B from *Gillett* 19594; C from *Bally & Melville* 15383; D from *Hemming* 464; E-G from *Gillett* 15128. Drawn by Eleanor Catherine.

lamina broadly elliptic to suborbicular or obovate, (0.5-) 1-4.5 x 0.5-3.5 cm, apex subacute to emarginate. Flowers in 2-14-flowered, divaricately branched cymes with filiform peduncles and pedicels; peduncle 7-25 mm; pedicels 4-15 mm. Calyx 2-3 mm long. Petals 5-6 mm long. Fruit narrowly ovoid with acuminate apex, 15-18 x 5-6 mm, sometimes partially 4-valved; stone ovoid, 8-10 x *c* 5 x *c* 5 mm. Fig. 123.12.

Acacia-Commiphora woodland, wooded grassland and bushland, often on red sandy soil overlying limestone but also on black cotton soil and on rocky slopes; 400-1400 m. GG SD BA HA; Somalia, E Kenya. *Gilbert* et al. 7472; *Glover & Gilliland* 327; *Hemming* 370.

The young leaves are eaten for their refreshing acid taste. An extract from the bark is drunk against colds (Somali).

The sack-like pseudaril loosely enveloping the stone is probably developed from the 4-armed condition seen in Sect. *Commiphora*. The four ridges strongly point in this direction. Because of the well-developed inflorescence it is, however, difficult to imagine any close relations between the two sections. They most probably share a common ancestor with Sect. *Spondioideae*.

8. Sect. CILIATAE *Vollesen* (1985).

Plant unarmed. Leaves 3-foliolate. Flowers solitary or in 1-3-flowered, pedunculate cymes. Calyx with very short broadly triangular lobes, without tube. Petals dark red. Only 4 stamens developed. Fruit partially 4-valved, apiculate, glabrous. Pseudaril a basal cup with 2 broad oblong lobes over locules, the one on the sterile locule almost covering it, the other shorter. Stone smooth, compressed.

25. **C. ciliata** *Vollesen* (1985)
- type: SD, Bokol Mayo to Filtu, *Gilbert* et al. 7655 (K holo., C ETH UPS iso.).

Tree to 6 m; bark grey, usually with thin transverse foldings, not peeling or peeling in small flakes, leaving a brownish underbark; branchlets glabrous to sparsely pilose. Leaves 3-foliolate, pilose to densely so, conspicuously ciliate and long-haired along midrib, conspicuously wine red when old; petiole 0.5-4 cm; leaflets elliptic to obovate or spathulate, entire, 1-5.5 (-7.5) x 0.7-4(-5.5) cm, apex subacute to rounded (rarely emarginate). Flowers solitary or in 1-3-flowered yellowish glandular and sparsely pilose cymes; peduncle up to 5 mm; pedicels up to 20 mm. Calyx 2-3 mm long, densely glandular and sparsely to densely pilose. Petals 4-6 mm long, glabrous or with a few hairs; only 4 stamens developed. Fruit obovoid, apiculate, *c* 13 x 8 mm; stone obovoid, *c* 8 x 6 x 4 mm. Fig. 123.13.

Commiphora, *Acacia-Commiphora* and *Acacia-Commiphora-Delonix* woodland and bushland, on rocky limestone slopes; 750-1250 m. SD BA; NE Kenya. *Gilbert* et al. 7474, 7687, 7834.

The peduncled inflorescence and partially 4-valved fruit may indicate a position near Sect. *Rostratae* or near the following section. But it also has characters in common with Sect. *Opobalsameae*, especially the short calyx teeth and 4-valved fruit.

9. Sect. LATIFOLIOLATAE *Engl.* ex *Vollesen* (1985).

Plants unarmed. Leaves 3-foliolate (rarely pinnate). Flowers in long-peduncled cymes. Calyx without tube, lobed 1/3-3/4 down. Petals red. Fruit 2-valved, glabrous. Pseudaril of 4 parallel-sided angular arms, without basal cup. Stone smooth to rugose, compressed or not.

1. Inflorescences and flowers puberulous. 2
- Inflorescences and flowers glabrous; stone ellipsoid, with smooth surface. 3
2. Inflorescence with 3-9 flowers; calyx divided almost to base; stone globose, with rugose surface. **27. C. sphaerocarpa**
- Some or all inflorescences with more than 10 flowers; calyx divided *c* 1/3 down; stone not known. **28. C.** sp. = *Gilbert* et al. 8170.
3. Leaves drying yellowish green, blistered when mature; calyx *c* 1 mm long; bark not peeling. **26. C. erosa**
- Leaves not drying yellowish green, not blistered; calyx 2 mm or longer; bark peeling or not. 4
4. Trunk distinctly angular; resin not smelling of turpentine; leaflets glaucous, glabrous to pilose, acute, with distinctly impressed reticulation. **29. C. mildbraedii**
- Trunk not angular; resin with strong smell of turpentine; leaflets not glaucous, puberulous to pubescent, subacute to emarginate, without impressed reticulation. **30. C. cyclophylla**

26. **C. erosa** *Vollesen* (1985)
- type: SD, Dolo, *Gilbert* et al. 7530 (K holo., C ETH UPS iso.).

Shrub or tree to 3 m; bark grey to dark grey, smooth, not peeling; all parts glabrous; young branches longitudinally ridged. Leaves 3-foliolate, blistered when mature, drying yellowish green; petiole 1.5-5 cm; leaflets ovate to broadly elliptic, 1.5-7.5 x 1-5.5 cm, apex acute to rounded. Flowers in up to 9-flowered dichasial cymes, dropping very easily on touching; peduncle 0.5-2.5 cm; pedicels 3-5 mm. Calyx *c* 1 mm long. Petals *c* 4 mm long. Fruit ellipsoid, 8-10 x 5-6 mm; stone smooth, ellipsoid, *c* 8 x 5 x 4 mm. Fig. 123.14.A & D-E.

Acacia-Commiphora bushland on yellowish limestone alluvium and on gypsum hills; 250-325 m. SD BA; S Somalia, NE & E Kenya. *Gilbert* et al. 8137, 8176.

Although *C. erosa* clearly belongs in this section, it retains characters from Sect. *Spondioideae*, most notably the ridged branches and blistered leaflets.

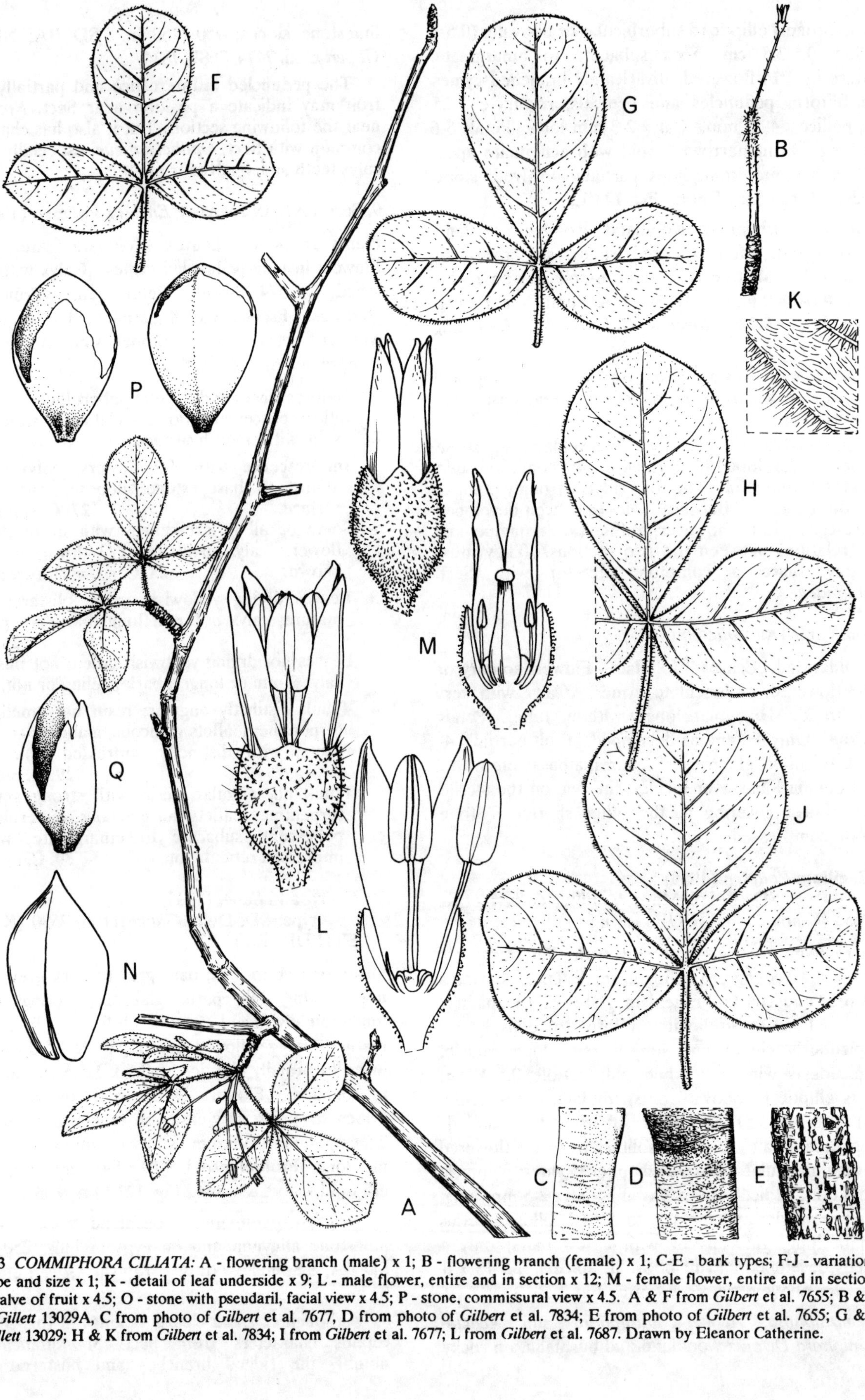

Figure 123.13 *COMMIPHORA CILIATA:* A - flowering branch (male) x 1; B - flowering branch (female) x 1; C-E - bark types; F-J - variation in leaf shape and size x 1; K - detail of leaf underside x 9; L - male flower, entire and in section x 12; M - female flower, entire and in section x 12; N - valve of fruit x 4.5; O - stone with pseudaril, facial view x 4.5; P - stone, commissural view x 4.5. A & F from *Gilbert* et al. 7655; B & N-P from *Gillett* 13029A, C from photo of *Gilbert* et al. 7677, D from photo of *Gilbert* et al. 7834; E from photo of *Gilbert* et al. 7655; G & M from *Gillett* 13029; H & K from *Gilbert* et al. 7834; I from *Gilbert* et al. 7677; L from *Gilbert* et al. 7687. Drawn by Eleanor Catherine.

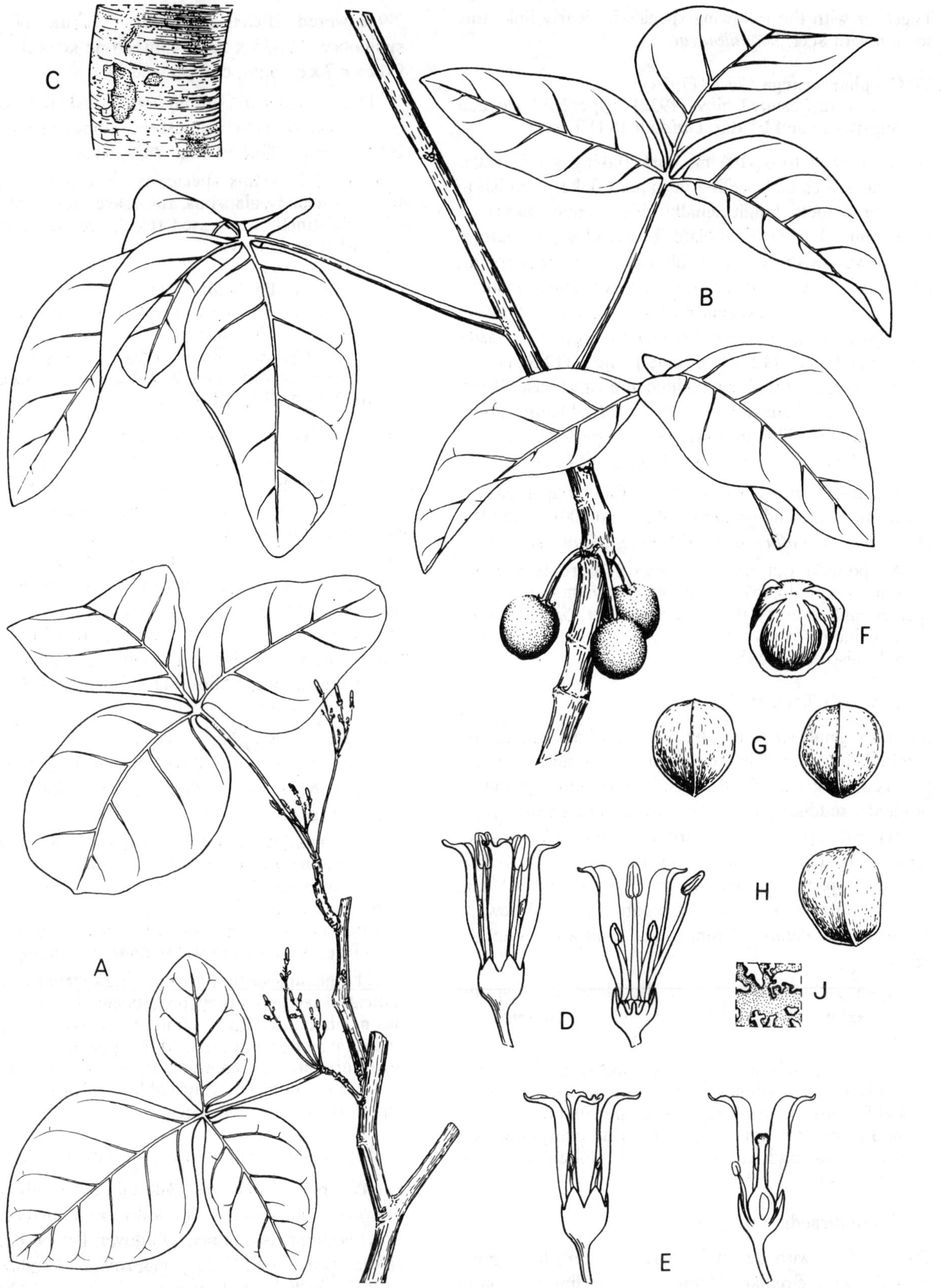

Figure 123.14 *COMMIPHORA EROSA:* A - flowering branch (male) x 1; D - male flower, entire and in section x 7; E - female flower, entire and in section x 7. *C. SPHAEROCARPA*: B - fruiting branch x 1; C - bark pattern; F - stone with pseudaril x 2; G - stone, facial view x 2; H - stone, commissural view x 2; I - detail of surface of stone x 15. A & D from *Gilbert* et al. 7530; B & F-I from *Gilbert* et al. 7578; C from photo of *Gilbert* et al. 7578; E from *Gillett* 16422. Drawn by Eleanor Catherine.

Together with the following species it clearly links this section with Sect. *Spondioideae*.

27. **C. sphaerocarpa** *Chiov.* (1916);
C. cerasiformis Chiov. (1932) - type: BA, between Bug-Berde and El-Bar, *Guidotti* 18 (FT iso.).

Shrub or tree to 4(-10) m; bark flaking off in thick irregular scales and peeling in papery flakes, whitish to grey; branchlets longitudinally ridged, puberulous or pubescent. Leaves 3-foliolate, blistered when mature, drying yellowish green, puberulous or pubescent; petiole 1.5-7 cm; leaflets ovate to broadly elliptic, entire, 2-9.5 x 1-6 cm, apex acuminate to rounded. Flowers puberulous, in puberulous 3-9-flowered, open dichasial cymes; peduncle (1.5-)2.5-5 cm; pedicels filiform, 3-8 mm. Calyx 2.5-4 mm long, divided almost to base. Petals *c* 5 mm long. Fruit globose, 10-15 x 10-14 mm; stone rugose (looking like an eroded landscape under a hand lens), globose, *c* 10 x 9 x 9 mm. Fig. 123.14.B-C & F-J.

Acacia-Commiphora-Boswellia rivae bushland on rocky limestone and gypsum slopes; 250-400 m. SD BA HA; Somalia. *Gilbert* et al. 7578, 8174; *Simmons* 202.

As pointed out under *C. erosa*, there is a great overall morphological similarity between these two species and some of the species in Sect. *Spondioideae*. It is evident from the vernacular names, that the Somalis also confuse them.

28. **C.** sp. = *Gilbert* et al. 8170.

Shrub or small spreading tree to 3 m; bark greyish on trunk, peeling in thick irregular scales and in small papery flakes; branchlets densely puberulous; younger branches reddish purple, becoming dark brown; resin watery, with strong sweetly aromatic smell. Leaves not seen. Flowers puberulous, in puberulous (1-)3-*c*20-flowered open dichasial cymes; peduncle 0.5-4.5 cm; pedicels filiform, 3-7 mm. Calyx lobed 1/3-1/2 down, 2-3 mm long. Petals 4-5 mm long. Fruit and stone not seen.

Open *Acacia-Commiphora* bushland on gypsum hills; *c* 300 m. BA; not known elsewhere. *Gilbert* et al. 8170.

Even in the absence of leaves and fruits it is clear that this species belongs in Sect. *Latifoliolatae*. It is probably closest to *C. sphaerocarpa*, but fruits are needed to confirm this. It differs in its many-flowered inflorescences and rather shallowly lobed (for the section) calyx.

29. **C. mildbraedii** *Engl.* (1910).

Tree to 5 m with distinctly angular trunk; bark grey, smooth, with circular foldings, not peeling; all parts glabrous to pilose; resin without smell. Leaves 3-foliolate (rarely pinnate with 5 leaflets); petiole 1.5-4.5 cm; leaflets elliptic to obovate or suborbicular, with distinctly impressed reticulation beneath, glaucous, entire, 2-6 x 1-4 cm, apex acute to subacute. Flowers (only young inflorescences seen) glabrous, in glabrous 3-9-flowered dichasial cymes. Fruit ellipsoid to subglobose, 12-15 x 10-12 mm; stone smooth, ellipsoid, 9-12 x *c* 7 x *c* 7 mm, compressed.

Dense *Acacia-Commiphora* bushland on black cotton soil or rocky limestone slopes; 1300-1400 m. BA; E Kenya. *Gilbert* et al. 7824, 8308.

Superficially this species is close to *C. cyclophylla*, but it is usually glabrous, the leaves are more pointed with a distinctly impressed reticulation and the fruit is somewhat asymmetric.

30. **C. cyclophylla** *Chiov.* (1932)
- type: BA, between Bug-Berde and El-Bar, *Guidotti* 14 (FT iso.).
C. lughensis Chiov. (1932) - types (p.p.): BA, between Daodid and Web-Karanle, *Ruspoli & Riva* 1069(966), 1875(1704) (both FT syn.).

Shrub or tree to 6 m; bark dark greyish to blackish, smooth, not peeling or peeling in large pale yellowish papery strips; branchlets glabrous to puberulous; resin with strong smell of turpentine. Leaves 3-foliolate, puberulous to pubescent; petiole 0.7-3.5 cm; leaflets elliptic or obovate to orbicular or transversely elliptic, entire, 1-4 x 0.8-4 cm, apex subacute to emarginate and apiculate. Flowers (only female known) glabrous, solitary; pedicels 1-3 mm. Calyx *c* 2 mm long. Petals 3-4 mm long. Fruit ellipsoid to globose, 12-14(-17) x 9-14 mm; stone smooth, ellipsoid to obovoid, 9-13 x 7-8 x 3-6 mm, compressed.

Acacia-Commiphora woodland and bushland on limestone slopes and ridges; 400-1350 m. SD BA HA; Somalia, NE Kenya. *Gilbert* et al. 7666, 7736; *J. de Wilde* 7251.

C. cyclophylla and *C. lughensis* were published simultaneously in *Fl. Somala* 2. Some of the syntypes of *C. lughensis* (those from which the description of inflorescence and flowers was made) have, however, proved to be specimens of *C. erythraea*. The author has, therefore, preferred to use the name *C. cyclophylla*.

The abrupt variation in the bark structure seems to indicate that more than one species is involved, but it has not been possible to find any correlated differences in the available material. Most species in this section are more or less incompletely known, and until more and better material is available it is preferred to keep both forms under one species.

10. Sect. CAMPESTRES *Engl.* ex *Vollesen* (1985).

Plants spiny. Leaves 3-foliolate (rarely pinnate). Flowers solitary or in subsessile to pedunculate cymes. Calyx without tube, lobed 1/3 down. Petals red. Fruit 2-valved (rarely 4-valved), glabrous. Pseudaril of 4 narrow parallel-sided often angular arms, without basal cup. Stone smooth to slightly rugose, compressed.

1. Leaves glaucous, leaflets entire; resin with strong smell of turpentine. **31. C. campestris**
- Leaves bright green, leaflets crenate to serrate; resin not smelling of turpentine. 2

2. Multi-stemmed shrub with smooth non-peeling bark; petiole filiform, 5-17 mm long; terminal leaflet 5-15 mm long. **33. C. tenuis**
- Single-stemmed shrub or tree with papery peeling bark; petiole stout, (5-)10-40(-50) mm long; terminal leaflet (10-)15-65(-90) mm long. **32. C. serrulata**

31. C. **campestris** *Engl.* (1893).

Tree to 5 m; bark dark grey, smooth, not peeling; resin with strong smell of turpentine; branchlets puberulous. Leaves glaucous, 3-foliolate (rarely pinnate with 5-7 leaflets on long shoots), glabrous; petiole ± a few hairs, 1-4 cm; leaflets elliptic to obovate, entire, 1-5 x 0.7-2.5 cm, apex acute to retuse. Flowers in 2-8-flowered pubescent divaricately branched cymes; peduncle 0.5-2 cm; pedicels 3-10 mm. Calyx 1.5-2.5 mm long, pubescent. Petals 5-6 mm. Fruit ellipsoid to subglobose, 10-12 x 6-12 mm; stone ellipsoid to subglobose, 7-10 x 6-10 x 5-7 mm, smooth, black; pseudaril-arm on fertile cell usually broader and with wavy margins.

Acacia-Commiphora bushland on rocky lava hills; *c* 1000 m. SD; E Kenya, NE Tanzania. *Mesfin Tadesse* et al. 4300.

The Ethiopian collection belongs to ssp. *campestris* which is characterised by glabrous, glaucous leaves but hairy branchlets and inflorescences. There are several other subspecies in Kenya, S Somalia and N Tanzania.

32. C. **serrulata** *Engl.* (1892);
C. serrulata var. *tenuipes* Engl. in *Bot. Jahrb. Syst.* **34**: 306 (1904) - type: 'Arussi Galla', no locality, *Ellenbeck* 2039 (B holo. destr., K fragment).

Shrub or tree to 6(-8 *fide* Engler) m; bark greyish to yellowish or brownish, peeling (often tardily) in small to large papery flakes, with conspicuous warty lenticels; branchlets glabrous to puberulous. Leaves 3-foliolate, glabrous to puberulous (rarely pubescent); petiole (0.5-)1-4(-5) cm; leaflets elliptic to rhomboid, crenate, bicrenate, serrate or biserrate, 0.5-6.5(-9) x 0.3-3.7 cm, apex acute to rounded. Flowers not known. Fruit ovoid, ellipsoid or subglobose, 7-11 x 5-8 mm, sometimes partially 4-valved; stone ovoid to subglobose, 5-8 x 4-6 x 3-5 mm, smooth.

Acacia-Commiphora woodland and bushland on limestone and basaltic slopes, on level sandy soil and on black cotton soil; 1250-1850 m. SD BA HA; Somalia. *Burger* 2866; *Friis* et al. 3125; *Thulin* et al. 3468.

C. serrulata was described from a sterile collection, but in later revisions Engler described it with a fruit clearly belonging in Sect. *Africanae*. The material seen by the author clearly matches the photo of the type at Kew, and it seems certain that the fruit was later erroneously associated with the type.

This species often has very few spines, and some branches often show none at all. The type is such a non-spiny branch, a fact which has further added to the confusion over its correct position.

33. C. **tenuis** *Vollesen* (1985)
- type: SD, Filtu to Bokol Mayo, *Gilbert* et al. 7476 (K holo., C ETH UPS iso.).

Multi-stemmed shrub to 2 m; bark pale grey to blackish, smooth, not peeling; all parts glabrous. Leaves 3-foliolate; petiole 5-17 mm, almost filiform; leaflets elliptic to orbicular, irregularly crenate to serrate, 0.3-1.5 x 0.3-1.2 cm, apex rounded to truncate. Flowers solitary, sessile. Calyx *c* 2 mm long. Petals 3-4 mm long. Fruit ellipsoid to subglobose, 7-9 x 5-7 mm; stone ellipsoid, 5-7 x 4-5 x 3-4 mm, slightly rugose. Fig. 123.15.

Acacia-Commiphora woodland and bushland on rocky limestone slopes or black cotton soil; 1000-1800 m. SD BA; NE Kenya. *Friis* et al. 2781; *Thulin* et al. 3772, 3799.

11. Sect. COMMIPHORA

Plants spiny. Leaves 1-foliolate or hetero-trifoliolate. Flowers solitary or in few-flowered usually subsessile cymes. Calyx without tube, lobed 1/3 down. Petals red or yellow. Fruit 2-valved (rarely 4-valved), usually glabrous. Pseudaril of 4 narrow, parallel-sided, often angular arms. Stone smooth to rugose, compressed or rarely with both locules equally developed, sterile locule often reduced to an apical hump on the fertile.

Note to the key.

All statements about - and measurements on - leaves refer to mature leaves on spur shoots. Leaves on long shoots often differ radically and should be disregarded when using the key.

1. Leaves puberulous. **43. C. gowlello**
- Leaves glabrous. 2

2. Leaves entire (rarely a few irregularly crenate). 3
- Leaves crenate, dentate or serrate. 7

3. Leaves broadly ovate to orbicular, fleshy; petiole 3-18(-30) mm; fruit 10-15 mm long. **34. C. quadricincta**
- Leaves linear to elliptic or obovate (rarely orbicular but then sessile), not fleshy; petiole 0-5(-17) mm; fruit 7-13 mm long. 4

4. Leaves linear to narrowly elliptic, usually greyish green; sap with unpleasant smell. **38. C. bruceae**
- Leaves elliptic or obovate to orbicular; sap aromatic. 5

5. Leaves broadly obovate to orbicular, up to 1.1 cm long. **37. C. sp.** = *Corradi* 6767, 6768
- Leaves elliptic to obovate, 0.7-4.5(-7) cm long. 6

6. Leaves elliptic or slightly obovate; petals yellow; fruit 8-13 mm long. **36. C. habessinica**
- Leaves obovate; petals red; fruit 6-7 mm long. **39. C. ellenbeckii**

7. Lateral leaflets present (often very small) on most or all leaves on spur shoots. 8
- Lateral leaflets absent on spur shoots (but often present on long shoots). 9

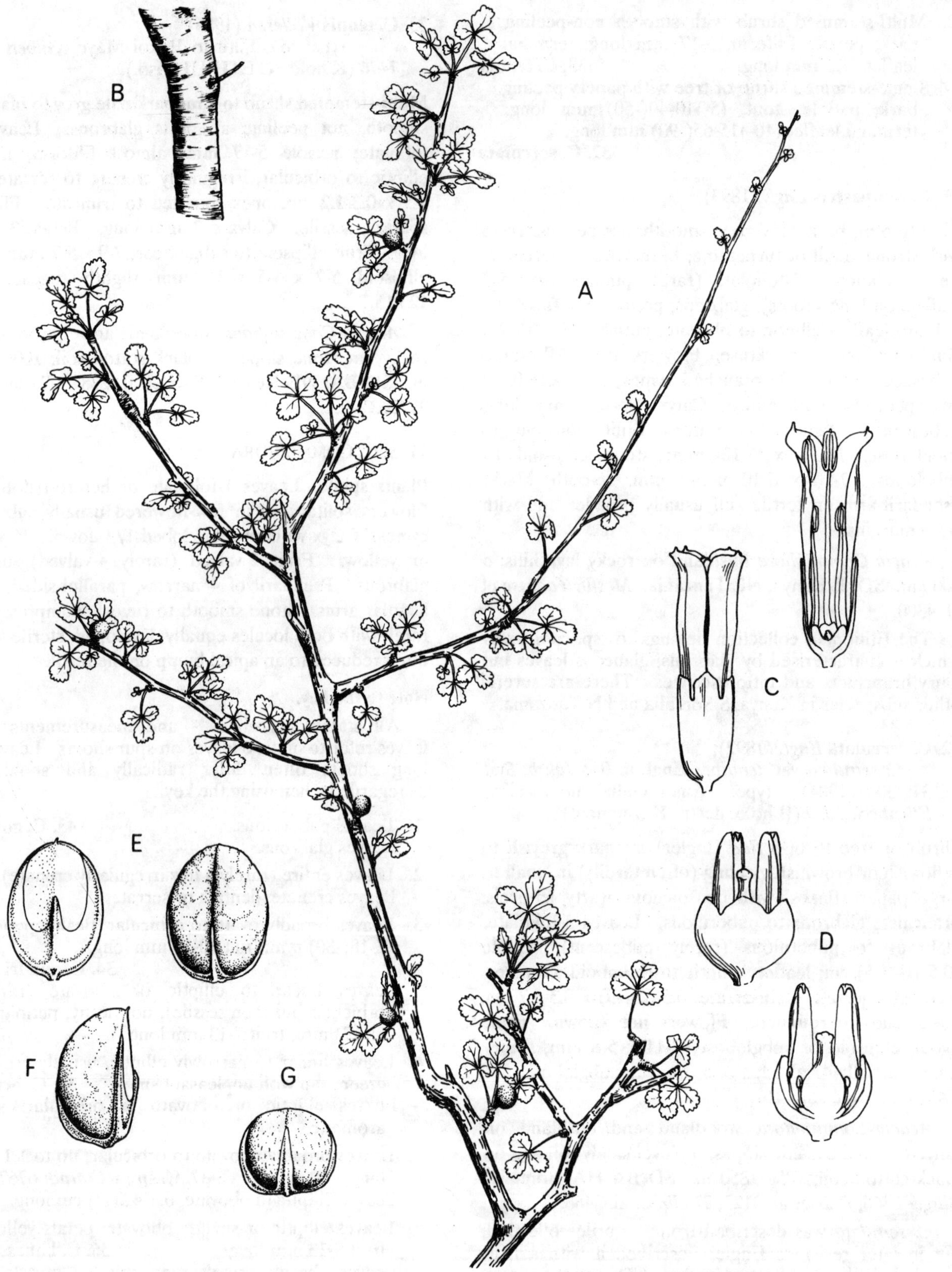

Figure 123.15 ***COMMIPHORA TENUIS:*** **A - leafy and fruiting branch x 1; B - branch showing smooth bark; C - male flower, entire and in section x 12; D - female flower, entire and in section x 12; E & G - stone with pseudaril, facial view x 4.5; F - stone, commissural view x 4.5. A & B from *Gilbert* et al. 7476; C from *Gillett* 12742B, D from *Gillett* 12707B, E & F from *Gillett* 12742A, G from *Gilbert* et al. 7820. Drawn by Eleanor Catherine.**

8. Young shoots long and unbranched, drooping, with conspicuous reddish brown bark, without or with very few spines; fruit 4-valved, 12-17 mm long. **35. C. hodai**
- Young shoots branched, not drooping, not with reddish brown bark, usually with numerous spines; fruit 2-valved, 8-13 mm long. **36. C. habessinica**

9. Most leaves elliptic to orbicular (widest near middle). 10
- Most or all leaves obovate to flabellate (clearly widest above middle). 13

10. Leaves up to 1.1 cm long, with 1-2 weak teeth per side. **37. C. sp.** = *Corradi* 6767, 6768
- Leaves usually longer, if not then with 2-5 large triangular or many small teeth per side. 11

11. Rhizomatous dwarf-shrub to 50 cm tall when mature. **42. C. sp.** = *Friis* et al. 3160
- Trees or shrubs more than 1 m tall when mature. 12

12. Leaves 1.8-4.5(-7) cm long, with many (more than 5) small teeth per side (or subentire); fruit 8-13 mm long. **36. C. habessinica**
- Leaves up to 2 cm long, with 2-5 large triangular teeth per side; fruit *c* 7 mm long. **40. C. kua**

13. Leaves crenate to subentire. **39. C. ellenbeckii**
- Leaves dentate or serrate. 14

14. Leaves with more than 10 small sharp teeth per side, serrate almost to base; rhizomatous dwarf shrub to 50 cm tall. **42. C. sp.** = *Friis* et al. 3160
- Leaves with 8 or fewer large triangular teeth per side, toothed in upper half only; shrubs or trees 1 m or taller when mature. 15

15. Most leaves less than 1.5 cm long, apically with 3 large subequal teeth and below them with 1-3 teeth per side. **41. C. incisa**
- Most leaves more than 1.5 cm long, with at least 4 (up to 8) teeth per side below the 3 apical teeth. **40. C. kua**

34. C. **quadricincta** *Schweinf.* (1899)
- types: EE, Samhar region near Massawa (Mitsiwa), *Schweinfurth & Riva* 243 (FT K P isosyn.), 1719 (not seen), 1760 (not seen).

Shrub or tree to 5 m; bark greyish, peeling; all parts glabrous. Leaves fleshy, 1-foliolate (rarely a few hetero-trifoliolate); petiole 3-18(-30 fide Schweinf.) mm; lamina broadly ovate to orbicular, entire (rarely irregularly crenate near apex), 1.2-3 x 1-2.8 cm, apex subacute to truncate. Flowers (only buds seen) in 3-flowered cymes, glabrous; peduncle 2-6 mm; pedicels 1-2 mm. Fruit ovoid to ellipsoid, partially 4-valved, 10-15 x 7-10 mm; stone ellipsoid, 7-10 x 4-6 x 4-5 mm, smooth.

Acacia-Commiphora bushland, *Acacia-Adansonia* wooded grassland, coastal bushland on sandy soil, rocky slopes and lava hills; sea-level to 600(-1000) m. EE AF EW; N Nigeria, Tchad, E Sudan, Arabia. *Hemming* 1029; *Sebsebe Demissew* 194a.

35. C. **hodai** *Sprague* (1927).

Tree to 5 m; spines often absent on young branches; bark peeling in small flakes; young branches long and unbranched, drooping, with reddish-brown bark; all parts glabrous. Leaves hetero-trifoliolate (or some 1-foliolate); petiole 1-3 mm; leaflets elliptic, dentate, terminal 0.7-2.5 x 0.3-1.5 cm, laterals up to 7 x 4 mm (usually less than 4 x 2 mm), apex subacute to rounded. Flowers solitary; pedicels 0-1 mm. Calyx *c* 2 mm long, not angular. Petals *c* 4 mm long, dark red. Fruit ovoid to ellipsoid, 4-valved, 12-17 x 8-10 mm; stone ovoid to ellipsoid, 7-9 x 4-6 x 4-6 mm, smooth to slightly rugose.

Acacia-Commiphora bushland on red sandy soil overlying limestone; 400-650 m. HA; Somalia. *Hemming* 156; *Simmons* 120.

36. C. **habessinica** *(Berg) Engl.* (1883)

Balsamodendron habessinicum Berg (1862) - types: EE, Eilet (Aylet) *Ehrenberg* s.n. (not seen), GD, Derragousel in Semien, *Schimper* II:1359 (FI(Webb) K P isosyn.).

C. dancaliensis Chiov. (1915) - type: EE, Mt. Xoma in Assaorta, *Pappi* 36 (FT holo.).

C. assaortensis Chiov. (1932) - type: EE, Mt. Xoma in Assaorta, *Pappi* 36 (FT holo.).

C. rivae sensu Chiov. (l.c.) quoad *Ruspoli & Riva* 384(226), non Engl. (1897).

C. madagascariensis sensu Wild (1959 & 1963), non Jacq. (1797).

C. habessinica var. *simplicifolia* sensu Cuf. (l.c.), non Schweinf. (1899).

C. "abyssinica" auct. mult. Afric.

Shrub or tree to 4(-6) m; bark grey to brown, peeling in large transverse papery strips or tardily in small flakes; branchlets glabrous. Leaves 1-foliolate or heterotrifoliolate (rarely a few 3-foliolate on long shoots), glabrous or pubescent on petiole and midrib; petiole 0-5(-15) mm; lamina elliptic to slightly obovate, crenate to serrate (rarely subentire), 1.2-4.5(-7) x 0.7-2.5 cm, apex acute to rounded. Flowers glabrous, in 1-3(-5)-flowered glabrous cymes; peduncle 0-1 mm; pedicels 0-3(-5) mm. Calyx 1.5-3 mm long, not angular. Petals 3.5-6 mm long, yellowish green. Fruit ellipsoid to subglobose, 8-13 x 6-9 mm; stone ellipsoid to subglobose, 5-8 x 4-7 x 3-5 mm, smooth to distinctly rugose. Fig. 123.16.

Acacia-Commiphora bushland, *Acacia-Balanites* woodland, *Anogeissus* woodland, on sandy to loamy soil overlying limestone and basement, black cotton soil, lava hills, sometimes planted in hedges; 150-1900 m. EE EW TU WU GD GJ SU WG IL GG SD BA HA/AF; E Sudan and Ethiopia through E Africa to N Mozambique and Zambia, also in Arabia. *Beals* 517; *Gilbert* 3409; *Gilbert & Thulin* 343.

C. dancaliensis was described as being close to *C. rostrata*. *C. assaortensis* - based on the same type - was described in quite a different group. Chiovenda in 1932 obviously forgot that he had already in 1915 described one species based on *Pappi* 36! He also completely mis-

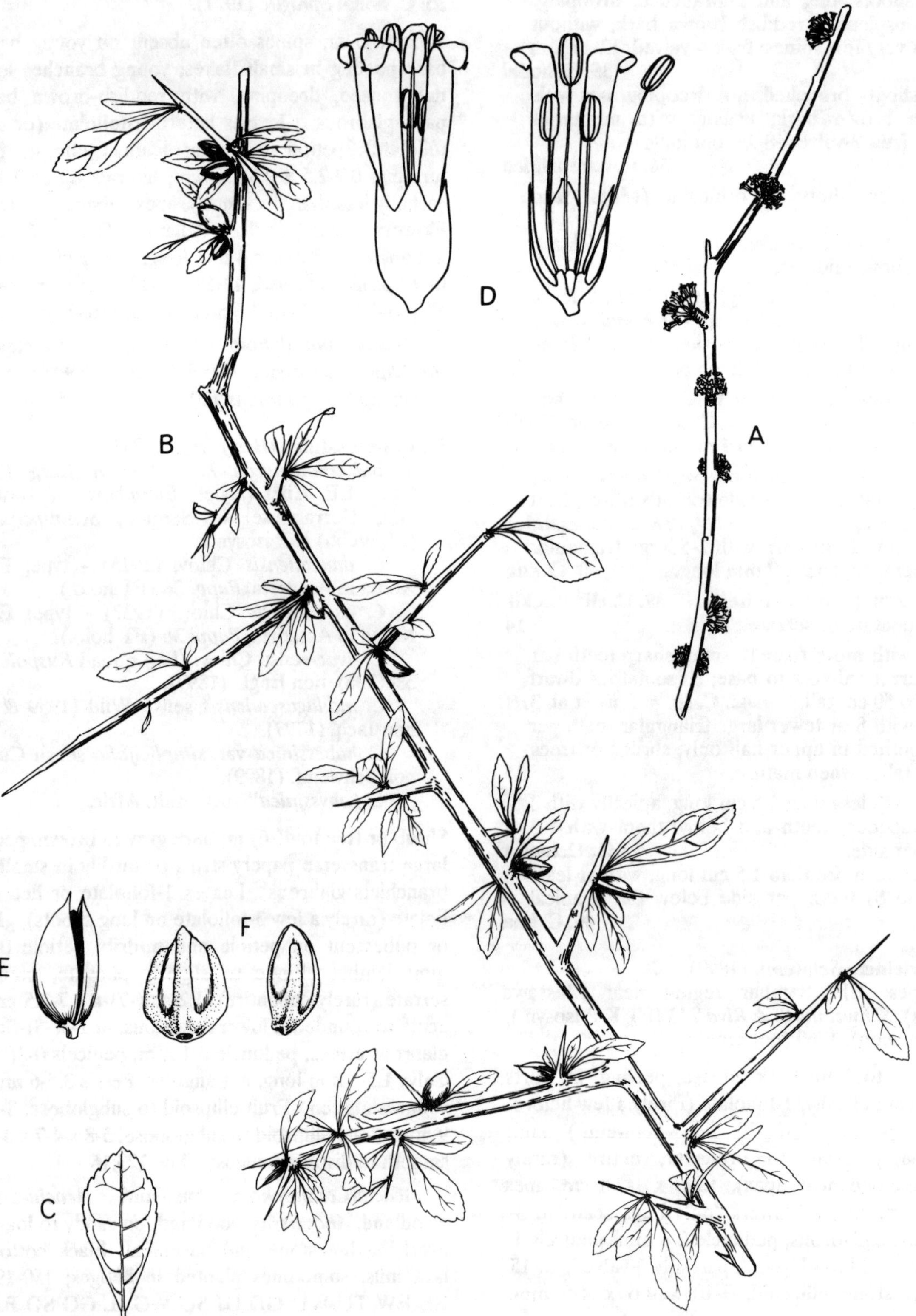

Figure 123.16 *COMMIPHORA HABESSINICA*: A - flowering branch (male) x 1; B - fruiting branch x 1; C - leaf x 1; D - male flower, entire and in section x 10; E - fruit x 2.25; F - stone with pseudaril x 3. A from *Beals* 517; B from *Friis* et al. 2618; C from *J. de Wilde* 5653; D from *Bally* 10454; E & F from *Meyer* 8652. Drawn by Eleanor Catherine.

interpreted the material when he described it as being close to *C. rostrata*.

The syntypes - from Arabia - of *C. habessinica* var. *simplicifolia* Schweinf. in *Bull. Herb. Boiss.* 7: 291 (1899) are a mixture of *C. habessinica* and another, probably undescribed, species. The epithet "simplicifolia" is, however, occupied at species level.

37. C. sp. = *Corradi* 6767, 6768.

Shrub or small tree; bark yellowish, peeling in large transverse papery strips; all parts glabrous. Leaves 1-foliolate, sessile; lamina broadly elliptic to suborbicular, entire or with a few weak teeth apically, up to 1.1 x 1 cm (but immature). Flowers and fruit not known.

Acacia-Commiphora bushland; *c* 1000 m. GG (Lake Stefanie); not known elsewhere. *Corradi* 6767, 6768.

Only known from these two collections. Easily recognizable by the broad entire leaves. Probably closest to *C. habessinica*, but without flowers and fruit this can only be a guess.

38. **C. bruceae** *Chiov.* (1941)
- type: HA, 44°15´E, 8°58´N, *Gillett* 4102 (K holo., FT iso.).

Shrub or tree to 3 m; bark whitish to yellow or grey, peeling in large papery strips or in smaller flakes; sap with an unpleasant smell of rotten eggs or fish; all parts glabrous. Leaves 1-foliolate (often some heterotrifoliolate on long shoots), usually greyish green; petiole 1-5 (-17) mm; lamina linear to narrowly oblong or lanceolate (rarely elliptic), entire, 1.7-6(-8) x 0.4-1.5(-2) cm (up to 10 x 5.5 and dentate on saplings), apex acute to rounded. Flowers (only female known) solitary, sessile. Calyx *c* 1.5 mm long, distinctly angular. Petals 4-5 mm long, dull yellow. Fruit ellipsoid to subglobose, 7-9 x 4-7 mm; stone ovoid to ellipsoid, 4-6 x 3-5 x 3-5 mm, smooth to rugose, sterile locule sometimes humped apically, both locules occasionally equally developed.

Acacia-Commiphora bushland with *Terminalia orbicularis*, on yellowish to grey alluvial sandy loam in areas of poor drainage; 250-1350 m. GG SD HA; Somalia, NE & E Kenya. *Carr* 203; *Gilbert* et al. 7531; *Tewolde Berhan* 2613.

39. **C. ellenbeckii** *Engl.* (1904)
- type: HA, between Rufa and Aroris, *Ellenbeck* 1083 (B holo. destr., K photo).

Shrub or tree to 3 m; bark pale grey, tardily peeling in transverse fibrous strips; all parts glabrous. Leaves 1-foliolate (or some hetero-trifoliolate on long shoots); petiole 0-2 mm; lamina obovate, shallowly crenulate to subentire, 0.7-2(-3) x 0.4-1.2 cm, apex subacute to rounded. Flowers in 1-3-flowered cymes; peduncle up to 2 mm; pedicels up to 3 mm. Calyx *c* 1.5 mm long, angular. Petals *c* 4 mm long, red. Fruit usually very asymmetric and seemingly laterally attached, transversely ellipsoid, 6-7 x 6-8 mm, or broadly ellipsoid with basal attachment and 6-7 x 4-5 mm; stone transversely ellipsoid, 4-5 x 4-5 x 5-7 mm, with sterile locule as large apical hump, or ellipsoid and 4-5 x 3-4 x *c* 3 mm, smooth.

Acacia-Commiphora bushland on red sandy to stony soil overlying limestone and on rocky slopes; 300-1200 m. SD BA HA; Somalia, NE Kenya. *Friis* et al. 3217; *Gillett* 4145; *Glover & Gilliland* 427.

Very close to *C. kua*. Most collections can be distinguished by their shallowly crenulate leaves. Some collections of *C. kua* have rather few teeth but can then be distinguished by their subsessile flowers.

40. **C. kua** *(R. Br.* ex *Royle) Vollesen* (1984);
Balsamodendron kua R.Br. ex Royle (1847) - types: AF, near Koran-hedudah (10°45´N 41° 45´E), *Johnston* s.n., (BM syn.); Abyssinia, no locality, *Salt* s.n., (BM syn.).
C. flaviflora Engl. (1904) - type: SD/BA, Marta by the Ganale river, *Ellenbeck* 2031 (B holo. destr., K photo).
C. crenulata (Terrac.) Chiov. (1915) - type: EE, hills by Anfile bay, *Terraciano* 50 (FT holo.).

Shrub or tree to 3 m, often with long straggling branches; bark yellowish to silvery, peeling in transverse strips, often with conspicuous warty lenticels; all parts glabrous. Leaves 1-foliolate (or some heterotrifoliolate on long shoots), subsessile; lamina obovate to flabellate (more rarely elliptic to orbicular), dentate to serrate with 4-8 teeth per side below 3 apical teeth (if only 3 per side then central tooth conspicuously larger than the 2 lateral), (0.5-)1-3.5 x (0.3-)0.5-2.5 cm, apex acute to truncate. Flowers solitary or in 1-2-flowered cymes; peduncle and pedicels 0-1.5 mm. Calyx *c* 2 mm long, angular. Petals *c* 5 mm long, yellow (? sometimes red). Fruit ellipsoid to subglobose with basal attachment or transversely ellipsoid with lateral attachment, 5-10 x 3-7 mm; stone ovoid to ellipsoid, 4-7 x 3-5 x 3-4 mm, smooth to rugose; sterile locule humped apically, both locules often equally developed.

Acacia-Commiphora bushland on red to grey sandy to stony soil overlying limestone or basement, or on rocky slopes; (0-)300-1400 m. EE AF SD BA HA; Somalia, NE & E Kenya, Arabia. *Burger* 3291; *Friis* et al. 2915; *Gilbert* et al. 7636.

The gum is used as soap (Borana).

The forms with elliptic to orbicular leaves (*C. crenulata* s.str.) merge gradually into the more common forms with obovate to flabellate leaves. *Salt* s.n. is completely intermediate having both types of leaves on the same branch.

41. **C. incisa** *Chiov.* (1916);
C. candidula Sprague (1927).
C. playfairii sensu Chiov. (1932) quoad *Ruspoli & Riva* 1073(970), non (Hook. f.) Engl. (1896).

Shrub or tree to 4(-8) m; bark yellowish to greyish brown, peeling in transverse fibrous strips; all parts glabrous. Leaves 1-foliolate, glossy, subsessile; lamina obovate to flabellate, apically with 3 broad triangular teeth, below these with 1-3 smaller teeth per side, 0.3-1.8 x 0.2-1.1 cm, apex truncate. Flowers solitary; pedicels 0-2 mm. Calyx 1.5-3 mm long, angular. Petals 4-6 mm long, yellowish green or red. Fruit broadly ellipsoid, basally attached, 5-8 x 4-7 mm; stone ellipsoid to obovoid, 4-6 x 3-5 x 3-4 mm, rugose, sterile locule humped apically, both locules sometimes equally developed.

Acacia-Commiphora woodland and bushland, usually on red sandy to loamy soil overlying limestone but also on basaltic rocks; 300-1250 m. SD BA HA; Somalia, NE & E Kenya. *Ellis* 254; *Gilbert* et al. 7487, 7630.

In Ethiopia this species is easily distinguished from *C. kua*, but some collections from N Kenya are more or less intermediate.

42. C. sp. = *Friis* et al. 3160.

Low rhizomatous shrub to 0.5 m, often cushion-shaped; bark greyish, peeling irregularly in small flakes; all parts glabrous. Leaves 1-foliolate, sessile or petiole up to 1.5 mm; lamina obovate to broadly so or elliptic to broadly so, sharply and finely serrate or biserrate to at least 2/3 down (usually almost to base), 1-4.7 x 1-3.2 cm, apex acute to truncate. Flowers not known. Fruit ellipsoid to subglobose, partially 4-valved, 8-10 x 5-7 mm; stone ovoid to ellipsoid, 5-7 x 4-5 x 4-5 mm, almost smooth, sterile locule not humped.

Acacia-Commiphora bushland on rocky limestone ridges; 400-1450 m. SD HA; not known elsewhere. *Friis* et al. 3160; *Gilbert* et al. 8225; *Hemming* 386A.

This taxon is very close to *C. glandulosa* Schinz from southern Africa, and may eventually prove to be no more than a subspecies of it. It differs in broader leaves, the habit and in being non-glandular.

43. **C. gowlello** *Sprague* (1927);
C. atramentaria Chiov. (1932) - type: BA, between Bug-Berde and El-Bar, *Guidotti* 13 (FT iso.).

Cushion-shaped shrub to 1 m and 2-3 m in circumference (rarely a shrub or tree to 2 m); branchlets puberulous. Leaves 1-foliolate, puberulous; petiole 0-1 mm; lamina elliptic to broadly obovate, subentire to finely and regularly serrate, 0.5-2 x 0.3-1.3 cm, apex subacute to rounded. Flowers solitary, pubescent; pedicels 0-1 mm, densely puberulous. Calyx *c* 2 mm long, not angular. Petals *c* 4 mm long, dull red. Fruit ovoid to ellipsoid, 6-9 x 4-6 mm, glabrous to pubescent; stone ovoid to subglobose, 5-6 x 3-4 x 3-4 mm, sterile locule humped apically, both locules sometimes equally developed.

Open *Acacia-Commiphora* bushland on shallow greyish sandy to stony soil overlying limestone; 350-1400 m. SD BA HA; Somalia. *Gilbert* et al. 7496; *Gillett* 4141; *Simmons* 117.

12. Sect. OPOBALSAMEAE *Engl.* ex *Vollesen* (1985).

Plants unarmed (rarely spiny). Leaves 3-foliolate or pinnate. Flowers solitary or in few-flowered contracted cymes. Calyx campanulate, with broadly triangular to almost obsolete lobes. Petals white to cream. Some stamens often reduced to staminodes. Fruit 4-valved (sometimes only partially), small, with 4 conspicuous white lines, glabrous. Pseudaril with fleshy cuplike basal part and thinner upper part, either covering almost all the stone or differentiated into broad lobes. Stone smooth, compressed.

1. Plant spiny; leaflets tomentose beneath. **44. C. horrida**
- Plant not spiny; leaflets not tomentose beneath. 2
2. Leaflets serrate or dentate. 3
- Leaflets entire. 5
3. Fruit 3.5-4.5 mm long; bark on younger branches reddish brown, peeling in small papery flakes on trunk. **48. C. sp.** = *Gilbert* et al. 8171
- Fruit 6-8 mm long; bark on younger branches yellowish or greyish. 4
4. Bark blackish on trunks, exfoliating in large parchmenty scrolls with yellow underside; leaflets not glossy, finely serrate (at least some with 6 or more teeth per side); tree to 4(-7) m. **46. C. boranensis**
- Bark pale grey, peeling in small papery flakes; leaflets glossy, grossly serrate with 2-4(-5) teeth per side; shrub to 1.5 m. **47. C. sp.** = *Gilbert* et al. 7652
5. Indumentum consisting of hooked hairs only. 6
- Indumentum of straight and/or curved hairs (rarely with some hooked intermixed). 7
6. Leaflets 3(-5), terminal 1.5-2.7(-3.2) cm long. **45. C. ancistrophora**
- Leaflets 11-15, terminal 0.2-0.5 cm long. **52. C. sp.** = *Glover & Gilliland* 973
7. Petiole and rachis not winged; leaves 3-foliolate or pinnate with 5 leaflets; leaflets glabrous to sparsely pubescent; petals *c* 3 mm long; calyx glabrous. **49. C. gileadensis**
- Petiole and rachis narrowly winged; leaves pinnate with 5-15 leaflets (or with some 3-foliolate intermixed); leaflets sparsely to densely pubescent; petals 1-2 mm long; calyx glabrous to densely pubescent. 8
8. Leaflets 11-15, terminal 0.2-0.5 cm long. **52. C. sp.** = *Glover & Gilliland* 973
- Leaflets 5-9(-11), or some 3-foliolate, terminal (0.4-)0.6-3 cm long. 9
9. Terminal leaflet 1.2-3 cm long; peduncle up to 5 mm long, older branches usually distinctly zigzag. **50. C. albiflora**
- Terminal leaflet (0.4-)0.6-1.7 cm long; peduncle 0-1(-2) mm long; older branches not zigzag. **51. C. coronillifolia**

44. **C. horrida** *Chiov.* (1932).

Flat-topped or cushion-shaped shrub to 1.5 m (but usually *c* 0.5 m) and to 3 m in circumference; bark greyish to dark brown; peeling in small flakes; branchlets densely pubescent. Leaves 3-foliolate (or some hetero-trifoliolate), whitish tomentose; petiole 3-13 mm; leaflets elliptic or obovate to suborbicular, crenate or dentate to sinuate, undulate, terminal 1-3.3 x 1-2.9 cm, laterals 0.4-1.1 x 0.3-1.1 cm, apex rounded to retuse. Flowers (only buds of female known) solitary, glabrous; pedicels *c* 1 mm, glabrous. Fruit ellipsoid to subglobose, 5-7 x 4-5 mm, completely 4-valved; stone ellipsoid, 4-5 x *c* 4 x *c* 3 mm.

Dry open *Acacia-Commiphora* bushland on fine, sometimes shifting, sandy soil overlying limestone; *c* 500 m. HA; Somalia. *Bally* 10149; *Glover & Gilliland* 1042; *Hemming* 1474.

45. C. **ancistrophora** *Chiov.* (1932).

Shrub or tree to 5 m, trunk often conspicuously narrowed at base; bark dark grey to brown, smooth or with numerous transverse stripes, not peeling; branchlets pubescent or sparsely so, hairs hooked (as on all other parts). Leaves 3-foliolate (or some pinnate with 5 leaflets), pubescent; petiole 6-18 mm; leaflets elliptic to orbicular, entire, terminal 1.5-2.7(-3.2) x 0.8-2(-2.5) cm, laterals 0.5-2 x 0.5-1.5 cm, apex rounded to retuse (rarely acute). Flowers glabrous, in 1(-2)-flowered glabrous cymes; peduncle and pedicels 1-2mm. Calyx *c* 1 mm long. Petals 3.5-5 mm long, white. Fruit ellipsoid, 7-8 x *c* 6 mm, partially 4-valved; stone ellipsoid, *c* 5 x 4 x 3 mm.

Acacia-Commiphora bushland on red to grey sandy soil overlying limestone and basement complex and on rocky limestone ridges; 400-1050 m. SD HA; Somalia. *Friis* et al. 3006; *Gilbert* et al. 7651; *Gillett* 4203.

Easily recognized by the hooked hairs and the large (for the section) flowers.

46. C. **boranensis** *Vollesen* (1985)
- type: BA, Sof Omar, *Gilbert* et al. 7811 (K holo., C ETH UPS iso.).
C. gileadensis sensu Dale & Greenway (l.c.), non (L.) C. Chr. (1922).

Shrub or tree to 4(-7) m; bark yellowish to dark grey, blackish on old trunks, tardily exfoliating in large transverse parchmenty scrolls; branchlets and leaves glabrous to pubescent, hairs curved. Leaves 3-foliolate or pinnate with 5 leaflets; petiole 6-25(-30) mm; leaflets elliptic to obovate, finely serrate (rarely some subentire), 0.4-2.5 x 0.3-1.5 cm, apex rounded to shallowly emarginate. Flowers glabrous (rarely sparsely puberulous), in 1-2(-6)-flowered glabrous cymes; peduncle 0.5-2 mm; pedicels 0.5-1 mm. Calyx *c* 1 mm long. Petals 2-2.5 mm long, yellow or cream. Fruit ovoid to ellipsoid, 6-8 x 5-6 mm, completely 4-valved; stone ovoid, 5-6 x 4-5 x 2-3 mm. Fig. 123.17.

Acacia-Commiphora woodland and bushland on rocky limestone slopes and on black cotton soil; 850-1500 m. SD BA HA; NE Kenya. *Burger* 2856; *Friis* et al. 2913; *Thulin* et al. 3500.

In the author's opinion clearly distinct from *C. gileadensis* with which it has been confused. It differs in the serrate leaves, the smaller flowers and the indumentum of curved hairs.

47. C. sp. = *Gilbert* et al. 7652.

Shrub to 1.5 m; bark pale grey, peeling tardily in small flakes; branchlets sparsely crisped puberulous. Leaves hetero-trifoliolate or pinnate with 5 leaflets, glabrous apart from sparsely puberulous petiole and veins; petiole 2-10 mm; leaflets elliptic to suborbicular, glossy, serrate, terminal 0.7-1.8 x 0.3-1.2 cm, laterals 0.2-1.3 x 0.2-0.8 cm, apex acute to retuse. Flowers not known. Fruit solitary, sessile, ovoid, completely or partially 4-valved, 7-8 x 5-6 mm; stone ovoid, *c* 5 x 4 x 3 mm.

Acacia-Commiphora bushland on stony limestone ridges; *c* 750 m. SD; not known elsewhere. *Gilbert* et al. 7652.

Only known from this collection. Superficially this looks much like *C. boranensis*, but there are a number of differences. The bark is different as are the glossy leaves which have a much coarser dentation. The fruit is greyish brown and not reddish brown.

48. C. sp. = *Gilbert* et al. 8171.

Spreading shrub to 2 m; bark peeling in small greyish papery flakes on trunk; younger branches reddish brown, smooth, branchlets very finely puberulous. Leaves 3-foliolate or pinnate with 5 leaflets, puberulous, hairs curved to hooked; petiole 6-15 mm; leaflets elliptic to broadly so (rarely orbicular) or obovate, dentate in apical half, 5-17 x 3-18 mm, apex rounded to emarginate. Flowers not seen. Fruiting pedicel *c* 1 mm, calyx *c* 1 mm long, glabrous to sparsely puberulous. Fruit 3.5-4.5 x 3-4 mm, broadly ovoid, completely 4-valved; stone broadly ovoid, *c* 3.5 x 3 x 2 mm.

Open *Acacia-Commiphora* bushland on gypsum hills; *c* 300 m. BA; S Somalia. *Gilbert* et al. 8171.

Related to *C. boranensis* but differs in its habit, reddish brown branches and small broad fruit.

49. C. **gileadensis** *(L.) C. Chr.* (1922);
C. opobalsamum (L.) Engl. (1883).
C. opobalsamum var. *ehrenbergianum* (Berg) Engl. in DC. *Monogr. Phan.* 4: 17 (1883).

Shrub or tree to 4 m, often with long slender drooping branches; bark yellowish to dull orange, peeling in large papery flakes; branchlets subglabrous. Leaves 3-foliolate or pinnate with 5 leaflets, glabrous to pubescent; petiole 4-20 mm; leaflets narrowly elliptic to obovate, entire, 0.3-2.7 x 0.2-1.3 cm, apex subacute to retuse. Flowers glabrous, in 1-3-flowered glabrous cymes; peduncle 0-5 mm; pedicels 0.5-2.5(-6) mm. Calyx 1-1.5 mm long. Petals *c* 3 mm long, white or cream. Fruit ellipsoid, 6-10 x 5-7 mm, partially 4-valved; stone ovoid to ellipsoid, 4.5-7.5 x 4-5 x 3-4 mm.

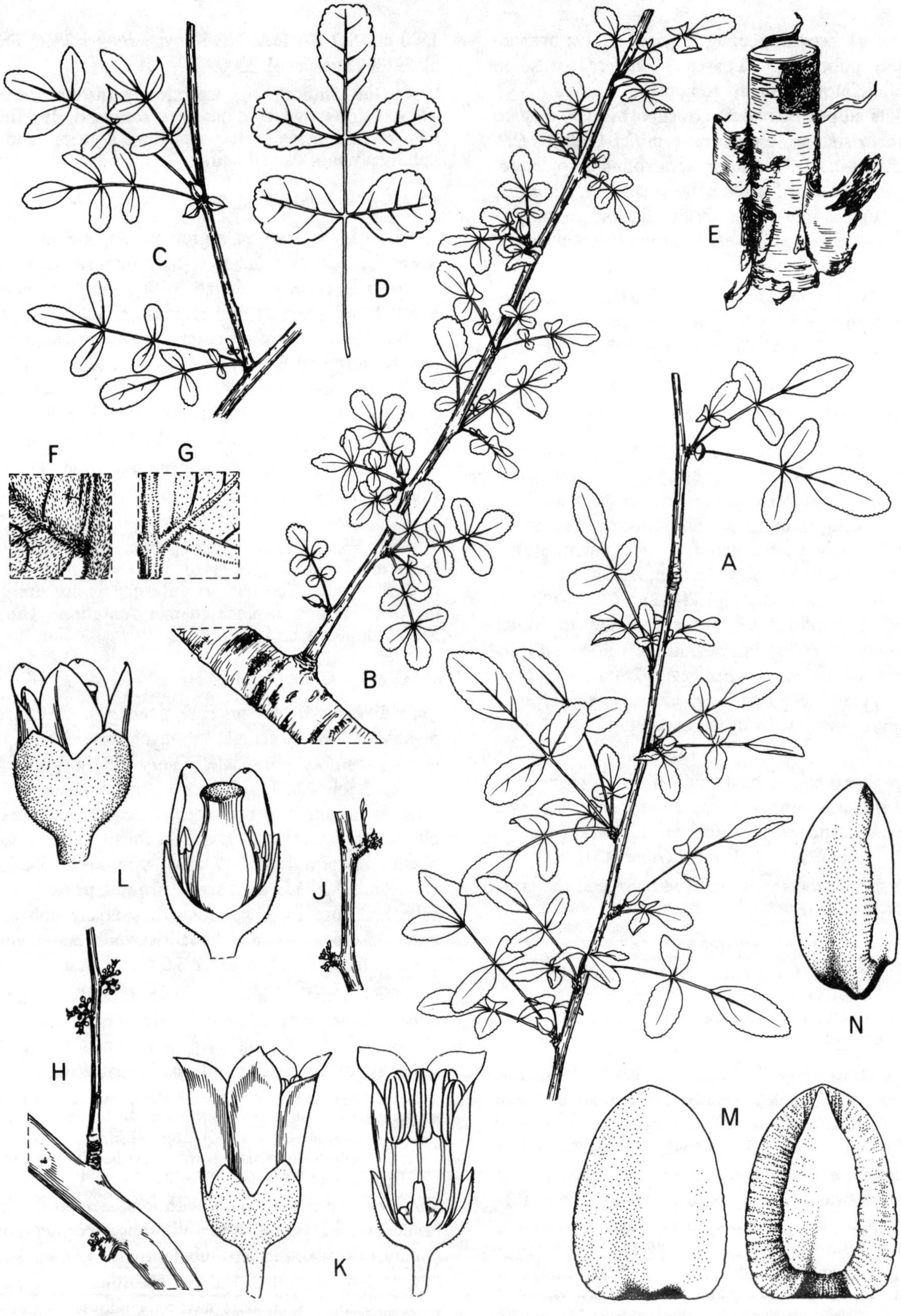

Figure 123.17 *COMMIPHORA BORANENSIS:* A-C - fruiting branches x 1; D - leaf x 1; E - trunk with exfoliating bark; F-G - leaf undersides x 12; H - flowering branch (male) x 1; I flowering branch (female) x 1; K - male flower, entire and in section x 15; L - female flower, entire and in section x 15; M - stone, facial view x 9; N - stone, commissural view x 9. A from *Gilbert* et al. 7811; B & F from *Friis* et al. 2913; C from *Thulin* et al. 3500; D & G from *Friis* et al. 3161; E from photo of *Friis* et al. 2771; H from *Gillett* 12599; I from *Gillett* 12568; K from *Ash* 2459; L from *Gilbert & Glyn Jones* 88; M & N from *Burger* 2856. Drawn by Eleanor Catherine.

Semi-desert bushland, often on rocky hills, sand dunes; near sea-level to 750 m. EE HA; Egypt (possibly originally introduced), E Sudan, Somalia, Arabia. *Ashall* 29; *Pappi* 8141; *Paskin* 127.

In Arabia the resin of this species is the source of the Balm of Gilead.

50. C. **albiflora** *Engl.* (1904)
- type: SD-BA, by the Ganale river, *Ellenbeck* 2041a (B holo. destr., K fragment).
C. anfractuosa Chiov. (1932).

Shrub or tree to 4 m; bark greyish to dark grey, smooth to rugose, not peeling; branches usually distinctly zigzag, pubescent when young, hairs straight or a few hooked. Leaves pinnate with 5-9 leaflets (often some 3-foliolate), sparsely to densely pubescent; petiole narrowly winged, 7-35 mm; leaflets oblong or elliptic to suborbicular, entire, 0.7-3 x 0.5-2 cm, apex rounded to shallowly emarginate. Flowers (fide Engl.) in 2-flowered pubescent cymes; peduncle to 5 mm; pedicels 0.5-2 mm. Calyx *c* 1.5 mm long, glabrous to pubescent. Petals *c* 2 mm long, white, pilose. Fruit ovoid to ellipsoid, *c* 8 x 6 mm, partially 4-valved; stone ellipsoid, *c* 5 x 4 x 3 mm.

Acacia-Commiphora bushland on sandy soil overlying limestone or on rocky limestone slopes; 700-1000 m. SD SD/BA; Somalia. *Friis* et al. 3218.

The fragment of the type at Kew is a good match with the description and figure of *C. anfractuosa* in Chiov. (l.c.) and also with recent collections from Somalia. The two species are clearly the same.

51. C. **coronillifolia** *Chiov.* (1932);
C. opobalsamum sensu Cuf. (l.c.), p.p., non (L.) Engl. (1883).
C. opobalsamum var. *induta* Sprague ex Hutch. & Bruce in *Bull. Misc. Inf.* 1941: 136 (1941), nom. nud.

Shrub or tree to 5 m; bark dull orange to grey or blackish, peeling in small transverse scrolls or not peeling; branchlets subglabrous to pubescent, hairs straight and/or curved. Leaves pinnate with 5-9(-11) leaflets (often some 3-foliolate), sparsely to densely pubescent; petiole narrowly winged, 3-13(-22) mm; leaflets oblong to elliptic (rarely obovate), entire, 0.3-1.7 x 0.3-0.9 cm, apex rounded to shallowly emarginate (rarely acute). Flowers glabrous to densely pubescent, in 1-2-flowered glabrous to densely pubescent cymes; peduncle and pedicels 0-1(-2) mm. Calyx *c* 1 mm long. Petals 1-2 mm long, white to pink. Fruit ovoid to ellipsoid, 7-8 x 4-5 mm, completely or partially 4-valved; stone ovoid to ellipsoid, 5-6 x 3-4 x 2-3 mm. Fig. 123.18.

Acacia-Commiphora woodland and bushland on sandy soil overlying limestone and on old lava flows; 400-900(-1300) m. AF SD HA; Somalia, NE Kenya. *Burger* 3298, 3369; *Gilbert* et al. 7577.

In some collections (including the type) the flowers are glabrous, in others densely pubescent. There are, however, a few intermediate collections (e.g. *Burger* 2305) with a sparsely pubescent calyx.

52. C. sp. = *Glover & Gilliland* 973.
C. parvifolia sensu Chiov. (1932), non (Balf. f.) Engl. (1883).

Small, often flat-topped, shrub to 2.5 m; branchlets and leaves pubescent with straight and/or curved or hooked hairs. Leaves pinnate with 11-15 leaflets, petiole and rachis narrowly winged; petiole 3-7 mm; leaflets elliptic or narrowly so, entire, 0.2-0.7 x 0.1-0.5 cm, apex subacute or apiculate to rounded. Flowers not known. Fruit solitary or in 2-flowered cymes with 1-7 mm long pubescent pedicels and pubescent calyx, ovoid to ellipsoid, 6-8 x 4-5 mm, completely 4-valved; stone ovoid to ellipsoid, 5-7 x 3-4 x 2-3 mm.

Acacia-Commiphora bushland on red sandy soil overlying limestone; *c* 600 m. HA/Somalia; Somalia. *Glover & Gilliland* 973.

According to the label, this collection was made almost exactly on the Ethiopia-Somalia border.

Doubtful species.

C. **berardellii** *Chiov.* (1941)
- type: BA, between Mara and Mt. Ello, 6°N 42°E, *Reghini* 5 (FT holo.).

Shrub (?), 4-5 m, unarmed; bark yellowish, peeling in small flakes. Leaves 3-foliolate, sparsely pubescent; petiole 4.5-7 cm; leaflets ovate to orbicular, crenate, 4.5-6 x 4-5 cm; apex rounded to apiculate. Flowers and fruit not known.

Habitat and altitude not recorded. BA; only known from the type-collection.

There is very little - if anything - to suggest the affinity of this species. The type consists of two loose leaves.

C. **ruspolii** *Chiov.* (1932)
- type: SD-BA, between the rivers Dawa and Web-Karanle, *Ruspoli & Riva* 1064(961) (FT holo.).

The type of this species is a galled specimen, but it is impossible to say for certain to which species it belongs. The author prefers to leave it as a doubtful species.

Excluded species.

C. **reghinii** *Chiov.* (1941)
- type: BA, between Mara and Mt. Ello, 6°N 42°E, *Reghini* 11 (FT holo., K fragment).

The type of this species is a specimen of a *Euphorbia*! It was first thought to be without a specific name in *Euphorbia* and the name *E. reghinii* was given: see Vollesen (1985). The specimen has now been found to belong to the taxon *E. jatrophoides* Pax.

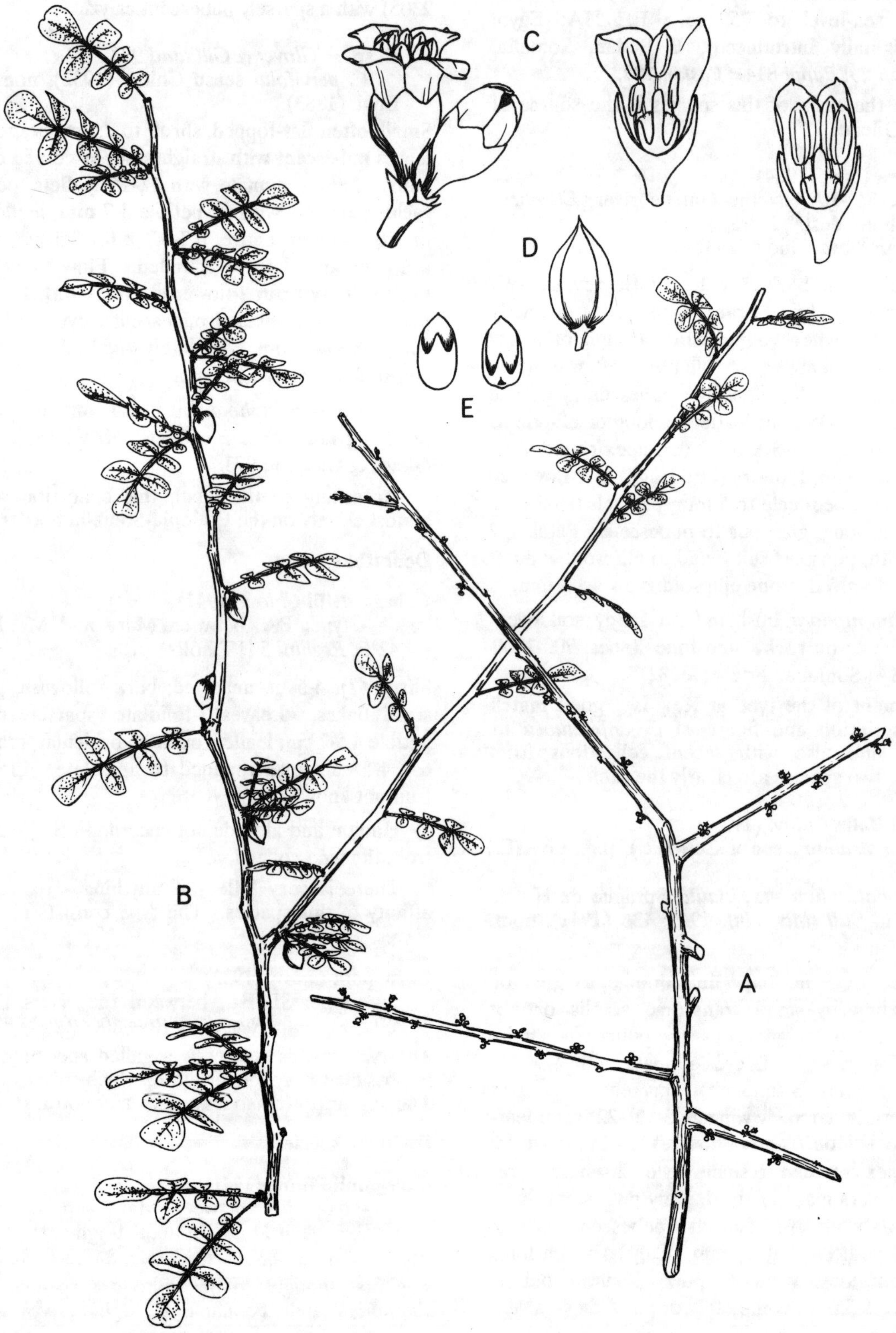

Figure 123.18 *COMMIPHORA CORONILLIFOLIA*: A - flowering branch (male); B - fruiting branch x 1; C - male flowers, entire and in section (note middle one with some stamens reduced to staminodes) x 9; D - fruit x 2; E - stone with pseudaril x 2. A & C from *Simmons* 125; B & D from *Collenette* 315; E from *Gillett* 23075. Drawn by Eleanor Catherine.

124. MELIACEAE

by B. T. Styles* & F. White*

Cufodontis, *Enum.*: 398-402 (1956); White & Styles in *Fl. Zamb.* **2**: 285-319 (1963); Styles in *E. Afr. Agr. For. J.* **39**: 407-423 (1974); Pennington & Styles in *Blumea* **22**: 419-540 (1975).

Trees, shrubs or shrublets. Leaves usually alternate, exstipulate, simply pinnate, 2-3-pinnate, 1-3-foliolate or simple. Flowers usually whitish, regular, bisexual or unisexual, and then monoecious or dioecious, occasionally polygamous. Sepals small, 4-6, variously united or almost free, never completely covering corolla in bud. Petals usually 4-5, free. Stamens usually 8-10, partly or completely fused to form a staminal tube, usually with appendages. Ovary superior, (2-)5-10(-20)-locular, with axile placentation and 1-many ovules per locule; style 1; style-head expanded, only partly stigmatic. Fruit a loculicidal or septifragal capsule or a drupe. Seeds often arillate or winged.

Family with 50 genera and *c* 550 species, almost confined to the tropics and subtropics: 7 genera and 15 species recorded for Ethiopia.

Key to Genera

1. Fruit a large septifragal woody capsule; seeds winged; leaflets with wavy margins. **1. Pseudocedrela**
- Fruit a loculicidal capsule with arillate seeds or a drupe; leaflets without wavy margins. 2
2. Fruit a capsule. 3
- Fruit a drupe. 4
3. Leaves simple. **2. Turraea**
- Leaves pinnate. **3. Trichilia**
4. Leaves 2-3-pinnate **4. Melia**
- Leaves 1-pinnate. 5
5. Indumentum of stellate hairs. **5. Lepidotrichilia**
- Indumentum of simple hairs. 6
6. Leaflets coarsely serrate; staminal tube terminated by 10 rounded, truncate, emarginate, or bilobed appendages, often partly united to form a frill. **6. Azadirachta**
- Leaflets entire; staminal tube without appendages. **7. Ekebergia**

1. PSEUDOCEDRELA *Harms* (1895).

Trees. Indumentum of simple hairs. Leaflets with wavy margins. Flowers unisexual. Petals (4)5, slightly contorted. Staminal tube urceolate, terminated by 8 or 10 bifid reflexed lobes. Disc free from base of staminal tube. Ovary with 4 or 5 locules, each with 4-6 ovules. Style-head large and discoid. Fruit a large, erect, club-shaped woody capsule, the valves connected to each other by a fibrous network. Seeds 4 or 5 per locule, each with a terminal wing.

One species confined to the Sudanian region of Africa where it extends from Senegal to Ethiopia.

P. kotschyi *(Schweinf.) Harms* (1895).
Cedrela kotschyi Schweinf. (1868)

Tree up to 12 m tall; bark thick, rough, regularly fissured; slash bright crimson, sticky. Leaves up to 30 cm long, reddish when young. Leaflets 6-9 pairs, up to 14 x 5 cm. Flowers small, fragrant, in panicles up to 25 cm long. Petals white; disc orange. Capsules up to 15 cm long. Seeds pendulous, 4-6 cm long, including the wing. Fig. 124.1.

In combretaceous wooded grassland; *c* 500 m. GD IL; widespread in Africa north of the Equator. *Chaffey* 923; *Friis* et al. 2455; *Massey* 70.

Bark used medicinally after childbirth and for backache. In Uganda the wood is used for mortars.

2. TURRAEA *L.* (1771).

Shrubs or small trees, sometimes scrambling. Leaves simple, entire, sometimes fasciculate. Flowers bisexual, solitary or fasciculate or in axillary or terminal cymes or false racemes. Calyx cup-shaped with (4)5(6) teeth or lobes, persistent in fruit. Petals (4)5, much longer than the calyx, linear-spathulate to linear. Stamens usually 10; staminal tube cylindric, sometimes expanded distally, terminated by simple or 2-lobed, free or partly or completely fused appendages, opposite to or alternating with the anthers. Disc present, partly or completely fused to base of staminal tube. Ovary with (4)5-10(-20) locules, each with 2 ovules; style slender, exserted; style-head with an apical discoid stigmatic surface. Fruit a small leathery or woody loculicidal capsule with (4)5-10(-20) valves. Seeds with a small orange or red aril.

A palaeotropical genus with about 70 species most of which occur in Africa, Madagascar and the Mascarenes.

1. Leaves very small, up to 2.5 x 0.8 cm; staminal tube slender, 0.15 cm wide; staminal appendages filiform, longer than the anthers, free almost to the base. **5. T. parvifolia**
- Leaves usually much larger, if less than 2.5 cm long then staminal tube *c* 0.4 cm wide. 2
2. Staminal appendages longer than the anthers, mostly free to base, deeply and irregularly laciniate; leaves small, up to 5 cm long, broadest in upper half. **1. T. mombassana**

* Department of Plant Sciences, University of Oxford, South Parks Road, Oxford, UK.

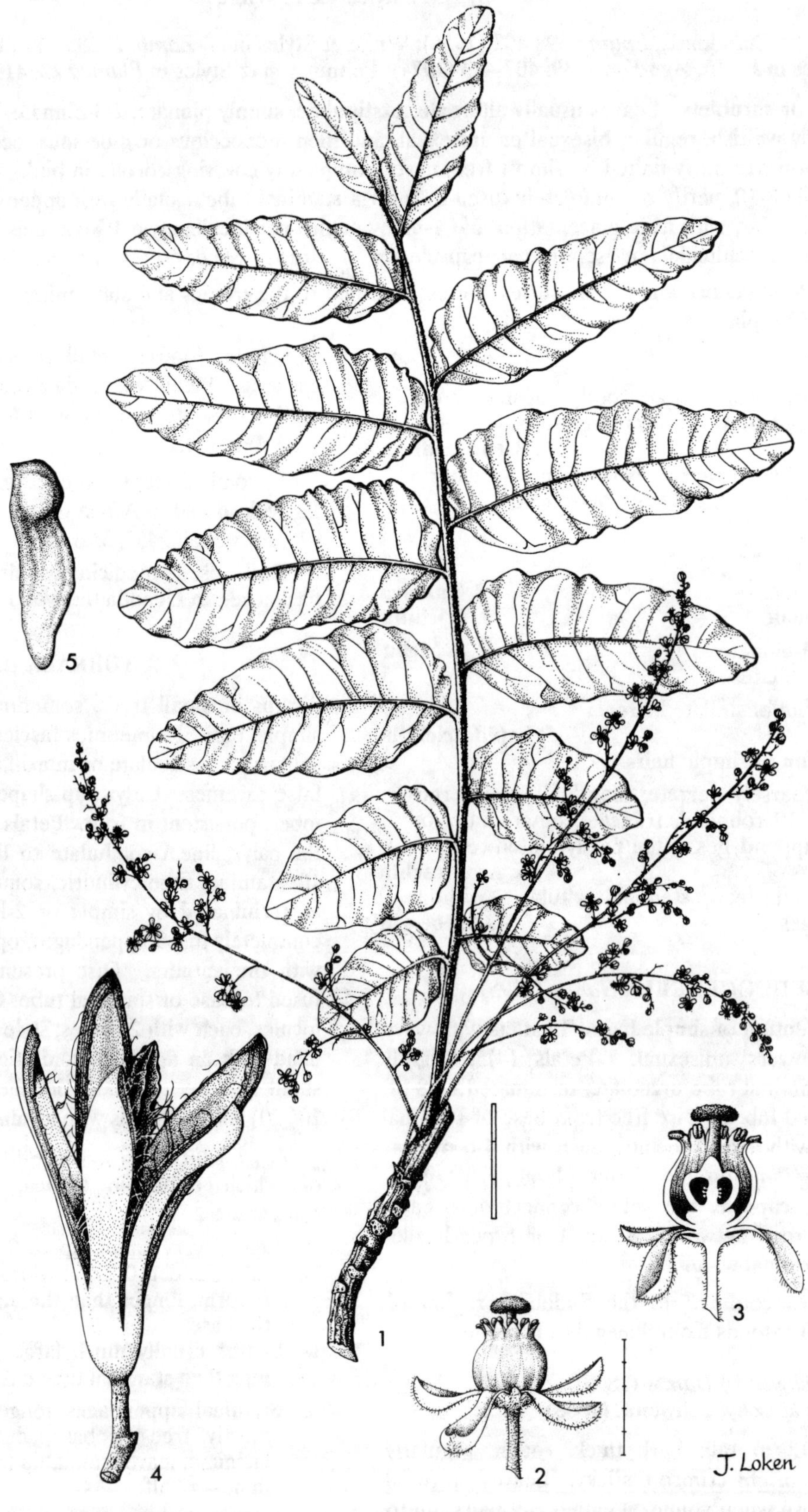

Figure 124.1 *PSEUDOCEDRELA KOTSCHYI*: 1 - flowering branch; 2 - male flower; 3 - male flower in vertical section; 4 - capsule (normally found hanging - ed); 5 - seed. 1, 4 & 5 from *Eggeling* 1501; 2 & 3 from *Styles* 278. Scales: open bar in mm; closed bar in cm. Drawn by J. Loken.

- Staminal appendages united almost to apex, truncate, retuse; leaves larger, broadest at or below the middle. 3

3. Leaves rounded or obtuse; inflorescences subsessile, flowering when plant is leafless; staminal tube bearded at the throat; anthers on long, free filaments. **2. T. nilotica**
- Leaves acuminate; inflorescences on long peduncles; flowers produced with the leaves; staminal tube glabrous at the throat; anthers subsessile. 4

4. Inflorescence congested, many-flowered, bracts foliaceous, up to 1 cm long. **3. T. abyssinica**
- Inflorescence lax, (1)2-3-flowered; bracts very small, scale-like. **4. T. holstii**

1. **T. mombassana** *Hiern ex C. DC.* (1878).

Shrub up to 5 m tall, sometimes scrambling. Leaves mostly in fascicles; up to 5 cm long; obovate to broadly spathulate, glabrous beneath except for axillary tufts of hairs. Flowers solitary or 2-6-fasciculate among the leaves. Petals white, up to 5 cm long, linear-spathulate. Staminal appendages mostly free to base, deeply and irregularly laciniate. Ovary 5(6)-locular. Style-head conical-cylindric from a broad base. Capsule 0.5 x 1 cm. Testa red. Fig. 124.2.1-3.

In montane scrub and at edges of forest; *c* 1400 m. GG SD; from Ethiopia to Malawi. *Cufodontis* 680 (not seen); *Fukui* 467; *Puff & Ensermu* 821818-1/9 (not seen).

2. **T. nilotica** *Kotschy & Peyr.* (1867).

Deciduous shrub or small tree up to 10 m tall, usually less; bark rough, brown; branchlets often with thick corky bark. Leaves up to 15 x 8 cm, ± elliptic, lower surface usually pubescent. Flowers fragrant in 5-18-flowered sessile or subsessile fascicles, in axils after the leaves have fallen, or terminating short shoots of very slow growth. Petals greenish-white turning yellow with age, 1.25-2.2 cm long. Staminal tube bearded at the throat; anthers on long, free filaments. Ovary (8)10-12-locular. Style-head ovoid-cylindric. Capsule 0.7 x 1.5 cm. Testa black. Fig. 124.2.4-6.

In woodland and riparian communities; *c* 550 m. SU? IL; from Ethiopia and the Sudan Republic to South Africa (Transvaal). *Friis* et al. 1919; *Chaffer & Thomerson* 657; *Ash* 1497.

The record for SU is based on *Ash* 1497 which is leafless, and the determination needs confirmation. It was collected from the Gibe gorge at 1250 m.

3. **T. abyssinica** *Hochst. ex A. Rich.* (1847)
- type: TU, Mt. Scholoda, *Schimper* I:28 (P lecto. here designated).

Shrub or tree up to 12 m tall, usually less; bark smooth; branchlets tomentellous when young. Leaves up to 12 x 6 cm, ovate-elliptic; apex narrowly acuminate; lower surface almost glabrous except for axillary tufts of hairs. Inflorescence congested, many-flowered, bracts foliaceous, up to 1 cm long. Flowers fragrant. Petals white, 1.4-2.2 cm long. Ovary 5-6(7)-locular. Style-head globose. Capsule 0.7 x 0.8 cm. Testa black. Fig. 124.2.7&8.

In evergreen bushland and at edges of montane forest; *c* 2000 m. EW TU; Ethiopia to Tanzania. *Gilbert* et al. 2628; *Pappi* 274(3072); *J. De Wilde* 7108.

4. **T. holstii** *Gürke* (1894).
T. abyssinica Hochst. ex A. Rich. var. *longipedicellata* Oliv. - type: SU, Ankober, *Roth* 178 (K holo.).

Shrub or small tree up to 10 m tall, usually less, sometimes scrambling. Leaves up to 9 x 4 cm, ± lanceolate-elliptic, apex rather suddenly and bluntly acuminate; lower surface almost glabrous except for axillary tufts of hairs. Inflorescence lax, (1)2-3-flowered; peduncle and pedicels very slender; bracts very small, scale-like. Petals white, 1.4-2.5 cm long. Ovary 5-6(7)-locular. Style-head ovoid. Capsule 0.7 x 1-1.5 cm. Testa black. Fig. 124.2.9 & 10.

In montane forest; (1300-)1600-2500 m. GD GJ WG SU AR IL KF SD BA HA; from Ethiopia to Malawi. *Burger* 1781; *Gilbert & Thulin* 864; *Mooney* 8781.

5. **T. parvifolia** *Deflers* (1895).

Shrub up to 5 m tall. Leaves mostly in fascicles terminating very slow-growing shoots, oblanceolate to elliptic, up to 5 cm long, attenuate at base, glabrous. Flowers very delicate, 1-5 among the leaves. Petals white, up to 1 cm long, broadly spathulate at apex, attenuate at base. Staminal appendages longer than the anthers, free almost to base. Style-head ovoid. Capsule 5-locular, 0.5 x 0.8 cm. Fig. 124.2.11.

In *Acacia-Commiphora* bushland. ?HA; Arabian peninsula, Somalia, Kenya. *Ruspoli & Riva* 442(58)391.

3. TRICHILIA *Browne* (1756), *nom. conserv.*

Rochetia Delile (1846).

De Wilde, *Revision of the species of* Trichilia *on the African Continent.* Wageningen (1968).

Trees. Indumentum of simple hairs. Leaves imparipinnate, leaflets entire. Flowers unisexual, (plants apparently dioecious), in cymes or cymose panicles. Sepals 5. Petals 5, imbricate, much longer than the calyx in bud. Stamens 10; filaments completely united and without appendages at apex, or united in lower half with a pair of deltate appendages at apex. Disc present. Ovary small, with (2)3(4) bi-ovulate locules; pistillode similar to gynoecium but narrower. Fruit a loculicidal capsule with 2-3(4) leathery valves. Seeds large, partly covered by a bright red aril.

A tropical genus of 85 species, most numerous in the New World, with 18 species in Africa and 2 in Asia.

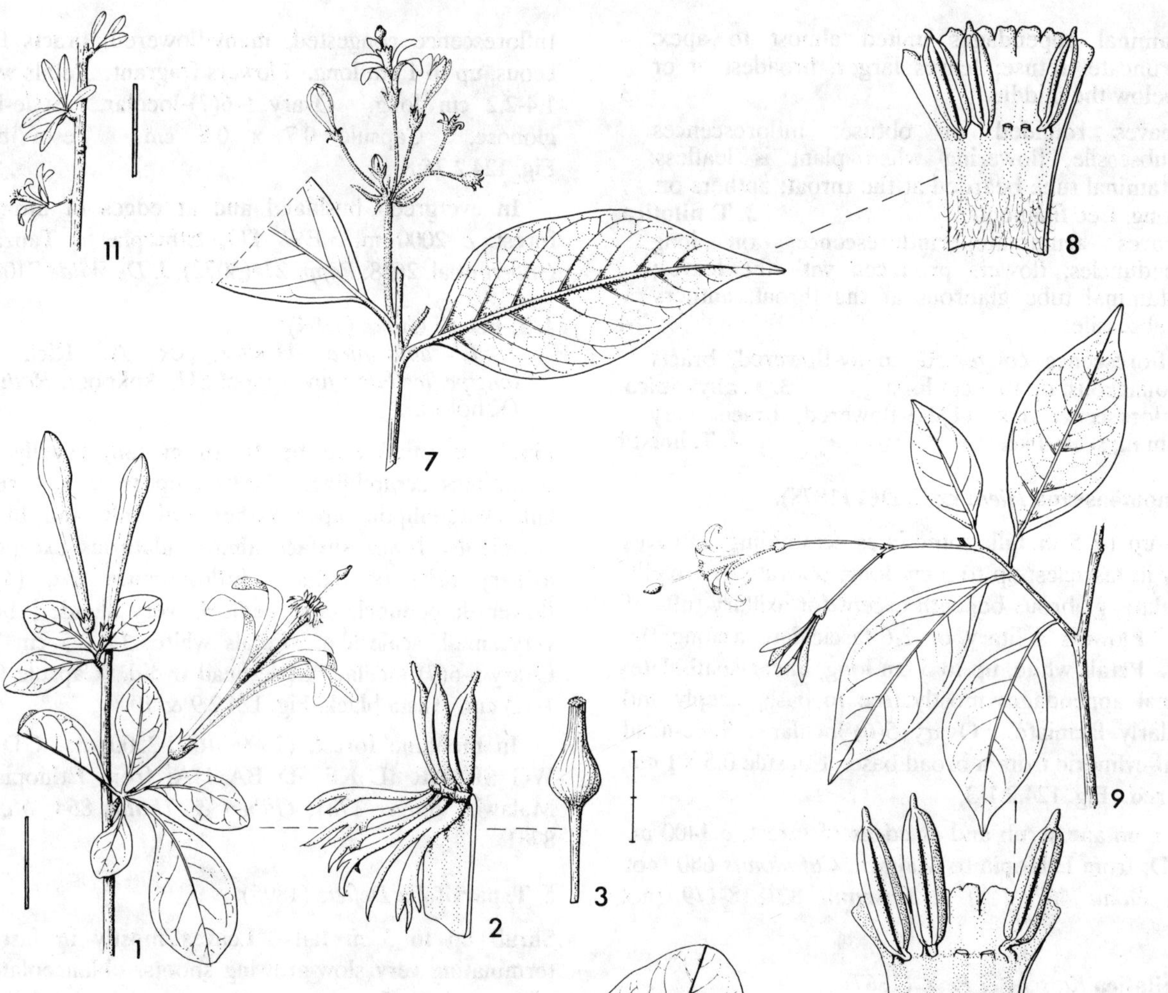

Figure 124.2 *TURRAEA MOMBASSANA*: 1 - flowering shoot; 2 - part of staminal tube; 3 - style head and stigma. *T. NILOTICA*: 4 - leaves; 5 - flowers; 6 - part of staminal tube. *T. ABYSSINICA*: 7 - flowering shoot; 8 - part of staminal tube. *T. HOLSTII*: 9 - flowering shoot; 10 - part of staminal tube. *T. PARVIFOLIA*: 11 - flowering shoot. 1, 2 & 3 from *Conservator of Forests Kenya* 123; 4 from *Robson* 665; 5 & 6 from *Bainbridge* 588; 7 from *Eggeling* 5947; 8 from *Polhill* 217; 9 from *Hughes* 224; 10 from *Greenway* 752; 11 from *Popov* et al. 4268. Scales: open bar in mm; closed bar in cm. Redrawn by Rosemary Wise from original by J. Loken.

1. Leaflets 3-5, broadest near apex, truncate and emarginate, cuneate at base; nerves and veins indistinct. **1. T. retusa**
- Leaflets broadest below the rounded to acuminate apex. 2
2. Bole deeply fluted for entire length; leaflets rarely more than 7, venation indistinct; petals 0.4-0.8 cm long; filaments united almost to apex, without appendages. **2. T. prieuriana**
- Bole not fluted; leaflets usually more than 7, venation conspicuous; petals 1-2.6 cm long; filaments free in upper half, each with a pair of deltate appendages. 3
3. Capsule without a stipe; leaflets distinctly broadest in upper half, drying dark brown, nearly always acute or acuminate, lateral nerves usually in 7-12 widely spaced pairs. **3. T. dregeana**
- Capsule with a (0.4-)0.6-1 cm long stipe; leaflets not or scarcely broadest in upper half, usually rounded or emarginate, drying olive green or pale brown; lateral nerves usually in 11-18 closely spaced pairs. **4. T. emetica**

1. **T. retusa** *Oliv.* (1868).

Tree up to 15 m tall. Leaflets 1-2 pairs, up to 12.5 x 4 cm, with 10-20 pairs of rather faint lateral nerves; apex truncate and emarginate; lower surface glabrous or almost so. Inflorescence up to 9 cm long. Petals white, up to 1 cm long, tomentellous on both surfaces. Filaments up to 0.7 cm long, glabrous outside, densely villous in the upper half inside. Capsule 2-2.5 cm long, subglobose, with a 0.5-1 cm long stipe, tomentellous to glabrescent, opening by 2 valves. Seed almost completely covered by the scarlet aril.

In riparian forest; *c* 600-650 m. IL; from Nigeria eastwards to Ethiopia and south to Zaire. *Friis* et al. 2556, 2573.

2. **T. prieuriana** *A. Juss.* (1830).

Tree rarely more than 15 m tall; bole fluted and often twisted; bark rough and scaling. Leaflets usually in 3-4 pairs, up to 18 x 6.5 cm, elliptic or oblanceolate-elliptic, apex shortly acuminate; lower surface glabrous. Inflorescence congested up to 10.5 cm long. Petals white, 0.4-0.8 cm long, puberulous outside. Filaments united 0.5-0.7 cm long, sparsely and minutely puberulous outside and in the upper half inside. Capsule *c* 2.5 x 2.5 cm, subglobose with a very short stipe, glabrous or almost so, mostly 3-valved. Seed black, only partly covered by a scarlet cushion-shaped aril at the apex. Fig. 124.3/1-3.

In lowland forest; 500-1300 m. IL KF; from Senegal to Ethiopia and south to Zambia. *Chaffey* 661, 937; *Friis* et al. 3917.

The wood is used locally for building houses.

De Wilde (loc. cit.) recognised 3 weakly defined subspecies of which subsp. *orientalis* J.J. de Wilde occurs in Ethiopia.

3. **T. dregeana** *Sond.* (1860).

Large evergreen tree up to 30 m tall. Leaflets (2)3-4(5) pairs, obovate to oblanceolate or oblanceolate-elliptic, up to 21 x 8.5 cm; lower surface glabrous or with a few strigose hairs, very rarely sparsely to densely pilose; lateral nerves usually in 7-12 widely spaced pairs. Inflorescence up to 6 cm long. Petals 1.4-2.6 cm long, whitish, linear, tomentellous on both surfaces. Filaments 1-1.5 cm long, puberulous outside, densely villous in the upper half inside; appendages deltate, shorter than the anthers. Disc glabrous, thin with 10 deltate teeth. Style 0.8-1 cm long; style-head capitate. Capsule 3 x 3 cm obovoid-globose, slightly sulcate, surface transversely wrinkled, dull yellow-tomentellous, opening by 3(4) valves. Seed black, almost completely concealed by the scarlet aril. Fig. 124.3.4-7.

In mid-altitude rain forest; 1150-1900 m. WG IL KF; West Africa, Angola and from Ethiopia to South Africa, but absent from the Zaire basin. *Friis* et al. 3955; *Mooney* 7768, 9234.

4. **T. emetica** *Vahl* (1790).

Rochetia choensis Del. (1846) - type: SU, 'Choa', *Rochet d'Héricourt* 47 (P holo.).

Trichilia somalensis Chiov. (1916).

T. jubensis Chiov. (1932).

T. roka (Forssk. *nom. nud.*) Chiov. (1923) *nom. illegit.*

Large, semi-evergreen tree up to 25 m tall. Very similar to *T. dregeana*, differing in characters given in the key and as follows: Leaflets up to 15.5 x 5 cm but usually much smaller, usually elliptic or oblong-elliptic, lower surface densely puberulous with short curly hairs, especially on the nerves. Petals 1-1.6 cm long. Filaments 0.7-1 cm long. Fig. 124.3.8-10.

In savanna woodland and riparian woodland; 450-1350 m. EE EW GJ SU IL GG SD HA; from Senegal to Ethiopia and the Yemen and southwards to South Africa. The Ethiopian material belongs to subsp. *emetica*. *Burger* 1336, 2497; *Thulin & Hunde* 3893.

Excluded species

T. siderotricha Chiov. (1940) - type: *Marchetti* 12 (FI) from 'Uondo' (= Wendo Genet) in Sidamo. According to De Wilde (1968, p. 201), this is *Brucea antidysenterica* J. F. Mill.

4. **MELIA** *L.* (1753).

Mabberley in *Gardens Bulletin (Singapore)* **37**: 49-64 (1984).

Trees. Indumentum of simple and stellate hairs. Leaves 2-3 pinnate, leaflets entire or toothed. Flowers bisexual

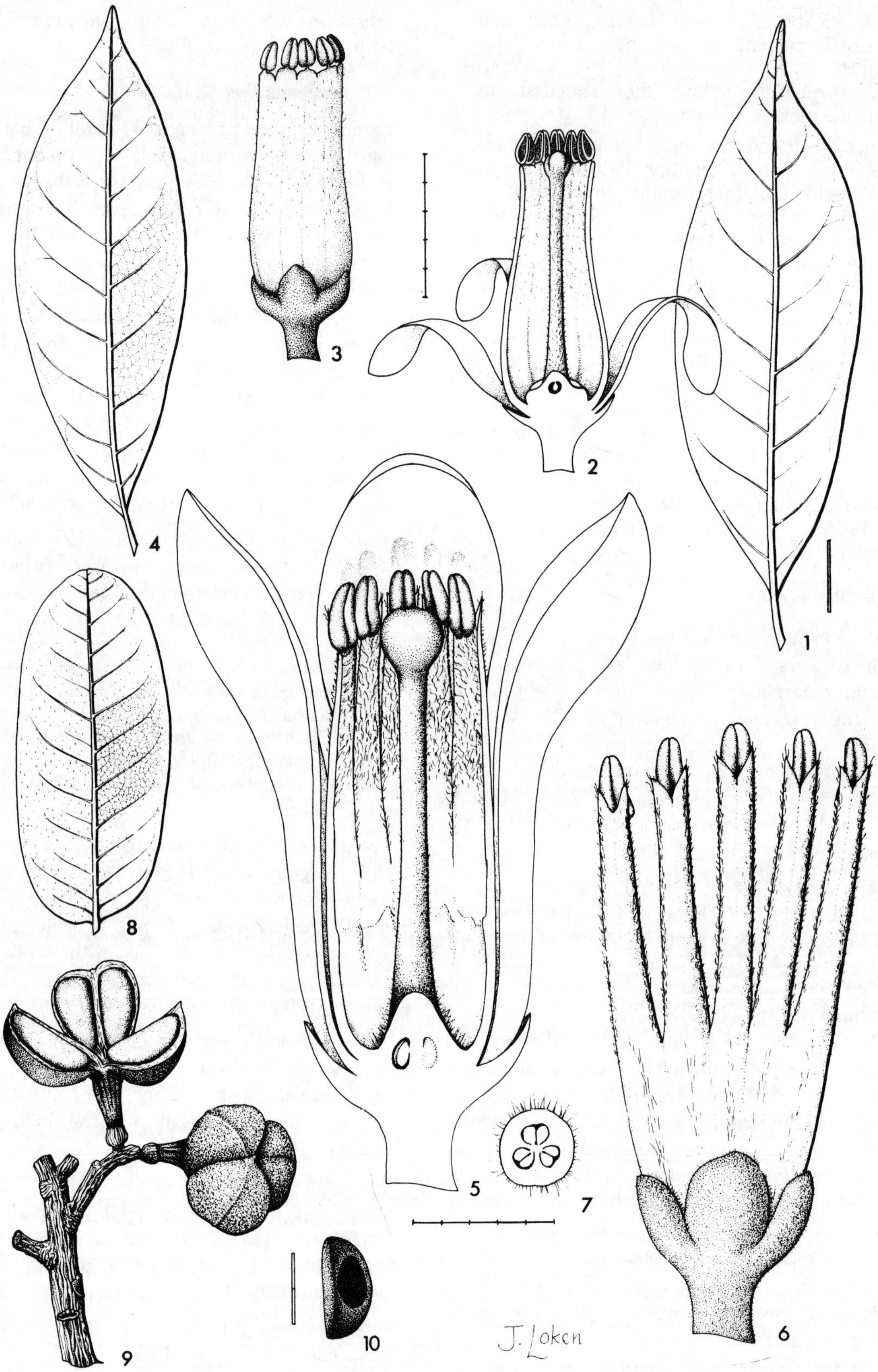

Figure 124.3 *TRICHILIA PRIEURIANA*: 1 - leaflet; 2 - male flower in vertical section; 3 - calyx and staminal tube. *T. DREGEANA*: 4 - leaflet; 5 - half of female flower; 6 - calyx and staminal tube; 7 - transverse section of ovary. *T. EMETICA*: 8 - leaflet; 9 - fruits; 10 - seed. 1 from *Styles* 226; 2 & 3 from *Styles* 219; 4 from *Styles* 206; 5, 6 & 7 from *Styles* 333; 8 - from *Styles* 254; 9 & 10 from *Bond* s.n. Scales: open bar in mm; closed bar in cm. Drawn by J. Loken.

and male on the same tree (polygamous). Sepals 5. Petals 5, imbricate. Stamens 10; filaments united to form a narrow tube with appendages at the apex. Disc small. Ovary with 4-8 bi-ovulate locules. Style-head capitate or coroniform with 4-8 short stigmatic lobes. Fruit a drupe with 3-8-locules and seeds; endocarp hard and thick but with weaker areas at apex and base.

Genus with 2 species in Africa and 1 or 2 species in tropical and subtropical Asia and Australia.

1. Bark smooth; leaflets serrate; inflorescence lax; staminal tube violet; drupe up to 1.5 cm long. **1. M. azedarach**
- Bark rough and deeply fissured; leaflets entire or subentire; inflorescence congested; staminal tube white; drupe 3-4 cm long. **2. M. volkensii**

1. **M. azedarach** *L.* (1753).

Tree up to 10 m tall, often bushy or of spindly growth. Bark dark grey, smooth, with pale lenticels. Leaves irregularly bi- or tri-pinnate, up to 38 cm long, sometimes pubescent when young. Pinnae in about 6 pairs, the lower ones themselves pinnate or trifoliolate. Leaflets 3-9 per pinna, more or less lanceolate or ovate-lanceolate, rarely lobed; the largest up to 7.5 x 3 cm; apex acuminate, base asymmetric, margin distinctly to deeply serrate. Flowers sweetly scented, in many-flowered, lax, axillary panicles up to 25 cm long. Petals white or pale violet, 0.8 cm long; staminal tube violet. Drupe subglobose or ellipsoid, yellowish-orange, up to 1.5 cm long.

Native of Asia and Australia. Widely planted in the tropics and subtropics for ornament and its reputedly termite-resistant timber which is used for building poles. It is sometimes naturalized elsewhere in Africa. In Ethiopia known to be found from near sea level to 2400 m, mostly below 1950 m and more widely cultivated than records suggest. *Aweke & Gilbert* 926.

2. **M. volkensii** *Gürke* (1895).

Tree up to 15 m tall; crown rounded, often with a browse-line produced by giraffe. Bark with pronounced vertical fissures. Leaves up to 35 cm long, less divided than in *M. azedarach*. Inflorescence congested, up to 12 cm long. Flowers similar to those of *M. azedarach* but white. Drupe 3-4 cm long. Fig. 124.4.

Acacia-Commiphora deciduous bushland; 1300 m. SD; Kenya and Tanzania. *Gilbert & Sebsebe D.* 8729.

"Certainly occurs in S.E. Ethiopia although I did not gather it on their side of the border" (J.B. Gillett, MS in Herbarium, Kew). This has been confirmed with the above cited collected made in May, 1988. The timber is said to be excellent. The seed is difficult to germinate.

5. LEPIDOTRICHILIA *Leroy* (1958).

Trees. Indumentum of stellate hairs. Leaves imparipinnate, leaflets entire. Flowers bisexual, in large cymose panicles. Calyx cup-shaped with 5 minute teeth. Petals 5, induplicate-valvate, much longer than the calyx in bud. Stamens 10; filaments united in lower half to form a staminal tube, with a pair of deltate-acuminate appendages at apex which are slightly longer than the anthers. Disc absent, but base of staminal tube slightly modified for nectar secretion. Ovary with 2(3) locules, each with 1 ovule. Fruit a drupe.

A small genus with 1 species in Africa and 3 in Madagascar.

L. volkensii *(Gürke) Leroy* (1963);
Trichilia volkensii Gürke (1894).
T. bilocularis Pax (1907) - type: HA, Deru, Chercher ("Tschertscher") Mts., *Rosen* s.n. (B holo.).

Tree usually less than 9 m tall, sometimes flowering as a shrub. Leaflets 3-4 pairs, oblanceolate or oblanceolate-elliptic, up to 15 x 6 cm, usually much smaller, apex acute to shortly acuminate, base markedly asymmetric; lower surface stellate-puberulous. Inflorescence up to 10 cm long. Calyx 0.2 cm long. Petals white, 0.4-0.5 cm long, densely puberulous outside. Filaments 0.3-0.4 cm long, puberulous at base outside, sparsely villous in upper half inside. Style 0.15-0.2 cm long; style-head capitate, surmounted by 2-3 erect stigmatic lobes. Drupe up to 1.5 x 1 cm, encrusted with stellate scales. Seeds exarillate. Fig. 124.6.1-5.

In understorey of moister forests; 1050-2800 m. GJ WG AR SU IL KF SD BA HA; from Ethiopia to Malawi. *W. de Wilde* 6738; *Mooney* 5721, 8498.

6. AZADIRACHTA *A. Juss.* (1830).

Trees. Indumentum of simple hairs. Leaves pinnate. Flowers polygamous. Flowers similar to those of *Melia* but ovary 3-locular and style-head 3-lobed. Fruit with thin endocarp and 1, rarely 2, seeds.

Genus with 2 species in tropical Asia.

A. indica *A. Juss.* (1830).

Tree up to 15 m tall; bark greyish-brown, rough, with wide longitudinal fissures; crown rounded and fairly open. Leaves pari- or imparipinnate, glabrous. Leaflets 4-7 pairs, lanceolate, falcate, base very asymmetric, tapering to an acuminate apex; margin coarsely serrate. Flowers fragrant, in axillary panicles up to 35 cm long, usually shorter. Petals white, 0.4 cm long. Staminal tube pale cream. Drupe ellipsoid, 1.5-1.8 cm long. Fig. 124.5.

Introduced; native of India and Burma. In Ethiopia planted as an avenue tree in EE, EW, IL and KF and probably elsewhere below 1500 m. *Thomerson* 605.

Extensively planted in the drier parts of Africa as a plantation tree and for ornament. The whole plant is medicinal. The seeds are dispersed by bats and in some parts of Africa, e.g. Ghana and Nigeria, it is widely naturalized.

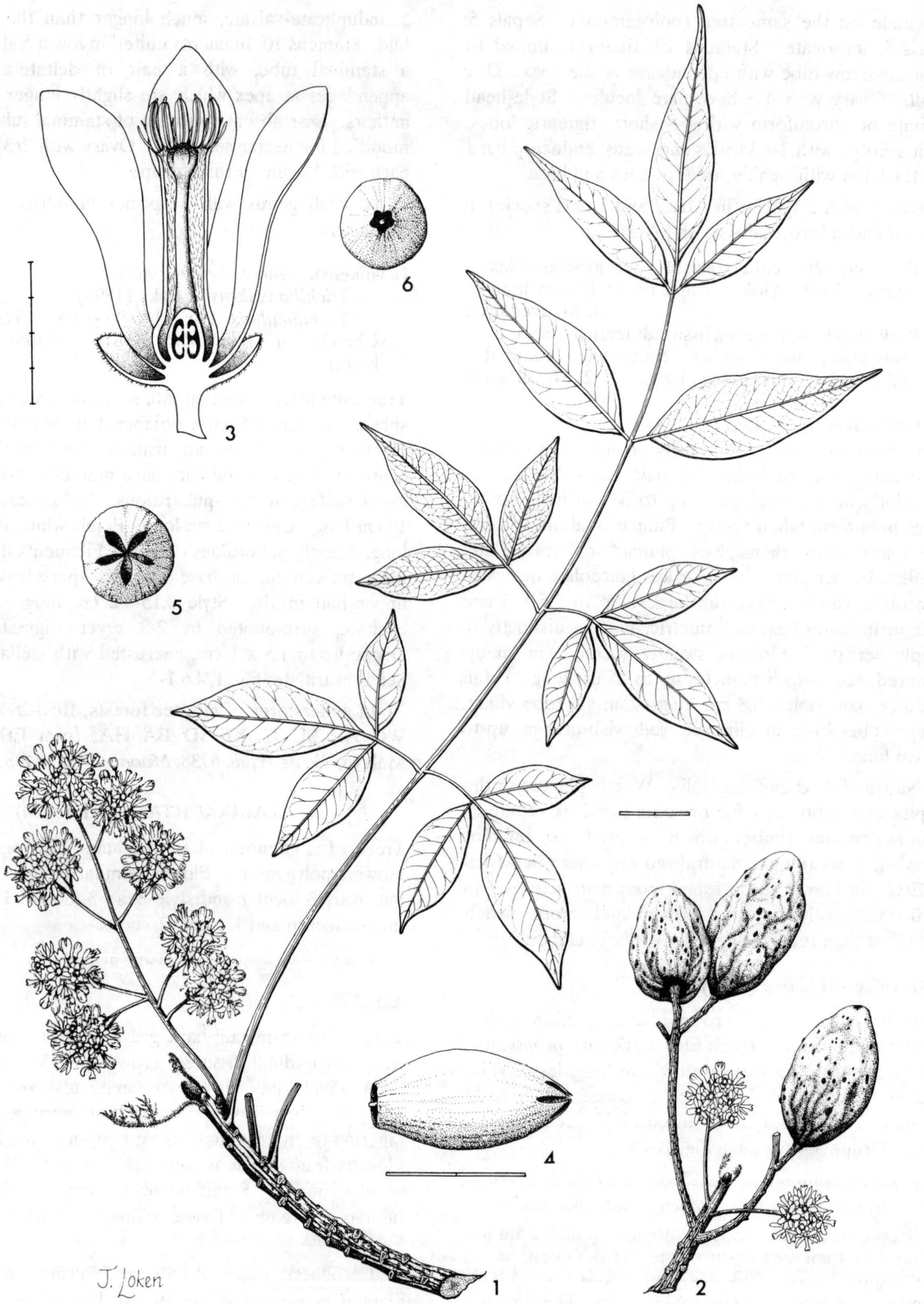

Figure 124.4 *MELIA VOLKENSII*: 1 - flowering shoot; 2 - flowers and fruits; 3 - half flower; 4, 5, 6 - three views of endocarp. 1 from *Willan* 428; 2 from *Taylor* 265; 3 from *Verdcourt* 2350; 4, 5 & 6 from *Gillett* 21824. Scales: open bar in mm; closed bar in cm. Drawn by J. Loken.

Figure 124.5 *AZADIRACHTA INDICA*: 1 - fruiting shoot; 2 - inflorescences; 3 - half flower; 4 - part of staminal tube; 5, 6 & 7 - stages in germination. 1, 2 & 3 from *Sharma* s.n.; 5, 6 & 7 from Troup fig. 75 (1921). Scales: open bar in mm; closed bar in cm. Drawn by Rosemary Wise.

Figure 124.6 *LEPIDOTRICHILIA VOLKENSII*: 1 - flowering branch; 2 - infructescence; 3 - half flower; 4 - galled fruits; 5 - stellate hair. *EKEBERGIA CAPENSIS*: 6 - flowering branch; 7 - half flower; 8 - infructescence; 9 - fruit in transverse section. 1 & 5 from *de Wilde* 4405; 2 from *de Wilde* 7034; 3 from *Styles* 180; 4 from *de Wilde* 5288; 6 from *Bos* 9720; 7 from *White* 8146; 8 & 9 from *de Wilde* 5992. Scales: open bar in mm; closed bar in cm. Drawn by Rosemary Wise.

7. **EKEBERGIA** *Sparrm.* (1779).

Trees. Indumentum of simple hairs. Leaves imparipinnate, leaflets entire. Flowers unisexual (plant dioecious), in cymose panicles. Calyx cup-shaped, lobed in upper half. Petals (4)5, imbricate, much longer than the calyx in bud. Stamens 8 or 10; filaments united for whole or most of length, without appendages. Disc annular, fused to base of staminal tube. Ovary with 2-5 locules, each with 2 ovules. Drupe with 2-4 pyrenes. Seeds mostly 1 per locule, without arils.

A small genus with 4 species confined to the African mainland.

E. capensis *Sparrm.* (1779).

Trichilia rueppelliana Fresen. (1837); *Ekebergia rueppelliana* (Fresen.) A. Rich. (1847) - type: TU, between Halei and Temben, *Rüppell* s.n. (FR holo.).

E. petitiana A. Rich. (1847) - type: TU, Tchelicot ("Etchelicote"), *Quartin-Dillon & Petit* s.n. (P holo in Herb. A. Richard).

Tree up to 30 m tall. Leaflets 3-6 pairs, lanceolate to oblong-lanceolate, up to 14.5 x 6 cm, tapering to an acuminate or subacuminate apex, base asymmetric; lower surface glabrous or densely pubescent. Inflorescence up to 25 cm long. Calyx 0.2 cm long, sparsely to densely puberulous. Petals white or pinkish-white, 0.4-0.5 cm long, elliptic-oblong, densely puberulous on both surfaces. Staminal tube 0.2 cm long, puberulous outside, densely bearded at the throat inside. Ovary 0.15 x 0.2-0.25 cm, densely setulose; style 0.05 cm long. Drupe 1.8 x 1.5 cm, with 2-4 pyrenes. Fig. 124.6.6-9.

In montane forest; 1680-2500(-3000) m. EW TU WU GD GJ WG SU AR IL KF SD BA HA; widespread in tropical Africa from Senegal to Ethiopia and southwards to South Africa. *Burger* 2338; *W. de Wilde* 6187; *Mooney* 5670.

125. SAPINDACEAE

by K. Vollesen*

Radlkofer in Engl., *Pflanzenr.* **IV**, **165**: 1-1539 (1932-1934); Cufodontis, *Enum.*: 489-495 (1958); Hauman in *Fl. Congo* **9**: 279-384 (1960); Exell in *Fl. Zamb.* **2**: 494-543 (1966); Capuron in *Mem. Mus. Nat. Hist. Nat., ser. B. Bot.* **19**: 1-189 (1969).

Trees, shrubs or woody (rarely herbaceous) climbers. Leaves alternate, simple, 3-foliolate, biternate, pinnate or bipinnate; (stipules normally absent); rachis and/or petiole sometimes winged. Flowers unisexual (but often seemingly bisexual), in racemoid or paniculate terminal or axillary thyrses. Sepals 4-5, often connate at base. Petals 0-5, sometimes with scales on the inside. Disc extrastaminal, sometimes unilateral (rarely absent). Stamens (3-)5-12 (-30), free or connate at base, reduced in female flowers. Ovary superior, 2-8-locular; locules connate or free apically, with 1-2 ovules; style simple; male flowers usually with a small pistillode. Fruit a capsule, berry, drupe, samara or a schizocarp with indehiscent cocci. Seeds often with a fleshy arillode.

About 150 genera and 2000 species, widely distributed in all tropical and subtropical regions. In Ethiopia 14 genera and 19 species.

Key to Genera

1. Leaves simple. 2
- Leaves compound. 3
2. Leaves pubescent, not sticky, base truncate to subcordate; petals present; fruit not winged. **13. Pappea**
- Leaves glabrous, sticky, base attenuate; petals absent; fruit with 2-3 wings. **2. Dodonaea**
3. Leaf rachis with a terminal leaflet. 4
- Leaf rachis without a terminal leaflet. 7
4. Leaves 3-foliolate; fruit a drupe. **7. Allophylus**
- Leaves pinnate or biternate; fruit a capsule. 5
5. Erect tree or shrub without tendrils; stipules absent, pinnate; fruit inflated. **1. Erythrophysa**
- Climbers or scramblers with tendrils; stipules present. 6
6. Leaves pinnate; fruit woody, not inflated. **5. Paullinia**
- Leaves biternate; fruit papery, inflated. **6. Cardiospermum**
7. Leaves bipinnate; fruit a samara. **10. Bottegoa**
- Leaves pinnate; fruit a capsule or a drupe. 8
8. Petals absent. 9
- Petals present. 11
9. Inflorescence paniculate; locules of ovary with 2 ovules; seed without arillode; stamens 4-5. **3. Zanha**
- Inflorescence racemoid; locules with 1 ovule; seed with arillode. 10
10. Leaves 10-15 cm long, leaflets 4-15 cm long; inflorescence 1-10 cm long; stamens 8-10. **11. Lecaniodiscus**
- Leaves 2-7 cm long, leaflets 0.4-2.8 cm long; inflorescence 0.2-0.6 cm long; stamens 4-5. **12. Haplocoelum**
11. Leaf rachis and petiole winged; stamens 5; inflorescence paniculate; fruit a drupe. **4. Filicium**
- Leaf rachis and petiole not winged; stamens 8 or more. 12
12. Inflorescence racemoid; fruit a capsule; stamens 8-10. **14. Blighia**
- Inflorescence paniculate; fruit a drupe. 13
13. Leaves 3-15 cm long with 2-6 leaflets; stamens 8. **9. Lepisanthes**
- Leaves 20-80 cm long with 8-16 leaflets; stamens 15-18. **8. Deinbollia**

The sequence of the genera is in accordance with Melchior in Engl., *Syll. Pflanzenfam.* (Ed. 12) **2**: 282-284 (1964).

1. ERYTHROPHYSA *E. Mey. ex Arn.* (1841).

Verdcourt in *J. Linn. Soc. Bot.* **58**: 201-205 (1962).

Trees or shrubs. Leaves imparipinnate or 2-imparipinnate. Inflorescence racemoid or paniculate. Flowers monoecious, irregular. Sepals 5. Petals (3-)4(-6), divided into distinct blade and claw, with threadlike appendages at the transition. Disc unilateral, crenate or lobed. Stamens 7-8, unilateral. Ovary 3-locular; locules with 2 ovules, one ascending the other descending; stigma capitate. Fruit a 3-locular capsule (rarely indehiscent). Seeds 1-2 per locule, globose, without arillode.

Nine species: 1 in Ethiopia, 2 in S Africa and 6 in Madagascar. Phytogeographically this is a very interesting genus, which emphasizes the relationship between the floras of NE and S Africa as well as between eastern and southern Africa and Madagascar.

E. septentrionalis *Verdc.* (1962)
- type: HA, 21 miles S of Wardere, *Hemming* 1492 (EA holo.., K iso.).

Shrub or tree to 4 m; young branches purplish to dark violet; all parts glabrous to sparsely pubescent. Leaves imparipinnate, 1.5-3.5 cm long; petiole 5-8 mm, narrowly winged; rachis winged; leaflets 5-7, elliptic to obovate, 3-21 x 2-10 mm; apex entire, acute to rounded; base cuneate. Inflorescence a narrow terminal panicle, 3-5 cm long. Flowers yellow to white with red anthers;

* formerly Ethiopian Flora Project, now: Royal Botanic Gardens, Kew, Richmond, Surrey TW9 3AB, UK.

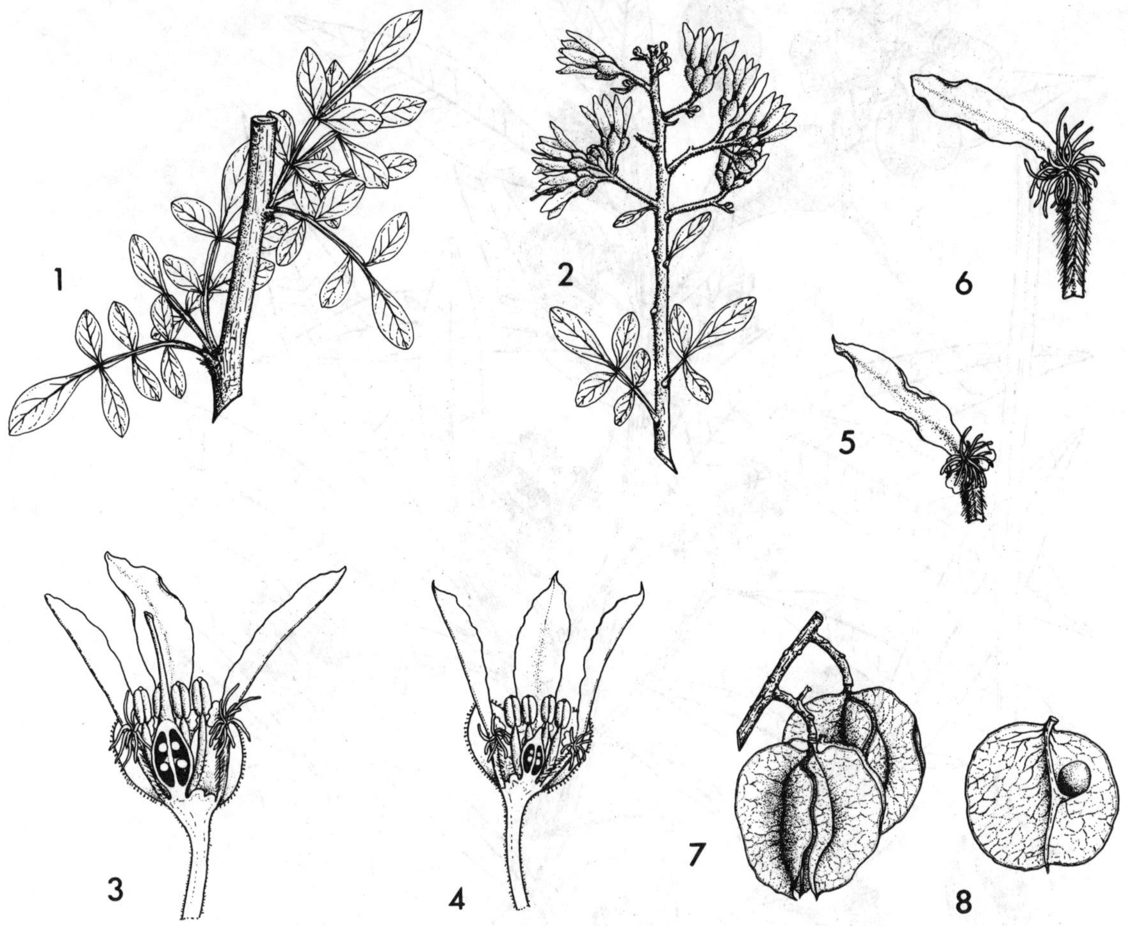

Figure 125.1 *ERYTHROPHYSA SEPTENTRIONALIS*: 1 - leafy branch x 1; 2 - flowering branch x 1; 3 - female flower x 4.5; 4 - male flower x 4.5; 5 & 6 - petals x 4.5; 7 - fruits x 0.5; 8 - fruit opened up showing seed x 0.5. 1-8 from *Hemming* 1492. (Reproduced with permission from *J. Linn. Soc. Bot.* 58: 203, 1962.)

pedicels 2-4 mm. Sepals oblong to obovate, 3-4 x 1.5-2.5 mm. Petals 4, narrowly elliptic to oblong, 4.5-6 x 1-2 mm, blade acuminate with lacerate margins, with 15-20, 1-2.5 mm long appendages. Stamens 7; filaments 3.5-4 mm. Ovary *c* 3 mm long; style *c* 4 mm. Fruit a red papery inflated capsule, 3-3.5 x 3-4 cm; locules laterally compressed. Seeds 2 per locule, 6.5-7 mm in diameter, black. Fig. 125.1.

Dry *Acacia-Commiphora* bushland with *Terminalia polycarpa* and a rich shrub and forb layer; 600-800 m. HA; not known elsewhere. *Bally* 10136; *Hemming* 368.

Only collected 5 times within a small area in the Ogaden from 7°00'N - 7°30'N and from 45°00'E - 45°30'E.

2. DODONAEA *Mill.* (1754).

Leenhouts in *Blumea* **28**: 271-289 (1983).

Trees or shrubs. Leaves simple or paripinnate. Inflorescence racemoid or paniculate. Flowers dioecious, regular. Sepals 3-7. Petals 0. Disc reduced or absent. Stamens 5-10, filaments very short. Ovary 2-4(-6)-locular; locules with 2 ovules, one ascending the other descending; style 2-6-fid; stigma linear, often contorted. Fruit a capsule, usually winged. Seeds 1-2 per locule, without arillode.

About 50 species mainly in Australia, a few pantropical and one in Madagascar.

D. angustifolia *L. f.* (1781).
D. viscosa auct. mult., non Jacq. (1760).

Shrub to 3 m; all parts glabrous and resinous when young. Leaves simple; petiole 1-5 mm; lamina narrowly elliptic to narrowly obovate, 2-12.5 x 0.7-3 cm, entire; apex acute (rarely rounded to truncate); base attenuate. Inflorescences terminal and axillary, paniculate, 1-4 cm long, often merging into a compound terminal inflorescence; pedicels 2-7(-15 in female flowers) mm. Sepals 4, ovate, 2-3.5 x 1.5-2 mm, yellowish green, shortly connate. Stamens (5-)7-9(-10), subsessile; anthers *c* 3 mm long. Ovary 2-3-locular, *c* 3 mm long; style 4-7 mm, 2-3 fid; stigmas 2-3 mm. Fruit circular in outline,

Figure 125.2 *DODONAEA ANGUSTIFOLIA*: 1 - flowering branch x 2/3; 2 - male flower x 6; 3 - female flower x 6; 4 - vertical section of female flower x 6; 5 - part of infructescence x 2/3; 6 - seed x 4. 1 & 2 from *Chase* 2182; 3 & 4 from *Chase* 1671; 5 & 6 from *Swynnerton* s.n. (Reproduced by permission, Editorial Board, *Flora Zambesiaca* 2, tab. 117, 1966.)

0.7-2 x 1-2.1 cm, yellowish, with 3-6 mm wide reddish tinged wings. Seed lenticular, *c* 3 x 2 mm. Fig. 125.2.

Edges of upland forest, upland bushland and grassland, secondary forest and scrub, invading areas recently cleared of forest and invading overgrazed *Acacia-Commiphora* bushland, also in cultivated areas; (500-)1000-2600(-2900) m. EE EW TU GD WU SU AR GJ WG KF GG SD BA HA; pantropical. *Burger* 1589; *Mooney* 6116; *Tekle Hagos* 198.

In his recent paper Leenhouts (l.c.) shows that the true *D. viscosa* is a sea shore plant only. It has not yet been collected in Ethiopia but might be found along the Red Sea Coast.

3. ZANHA *Hiern* (1896).

Trees. Leaves paripinnate. Inflorescence axillary, paniculate, often long pedunculate; male inflorescence denser and with more flowers than female inflorescence Flowers dioecious (rarely monoecious), regular. Sepals 4-5, shortly connate. Petals 0. Disc small, crenate. Stamens 3-7. Ovary 2-locular; locules with 2 ovules, pendulous from apex of locule; stigma subcapitate. Fruit a 1-locular drupe with 1 seed; arillode absent.

Three species: two in tropical Africa and one in Madagascar.

Z. golungensis *Hiern* (1896).

Tree to 30 m with buttressed bole; dbh up to 1.7 m; bark greyish brown with platy flaking; all parts glabrous to sparsely puberulous. Leaves 5-31 cm long; petiole 2.5-7 cm; leaflets 4-10, ovate to elliptic or obovate, 4-9.5 x 1.5-4.5 cm, entire to remotely dentate (or densely so in saplings); apex acute to rounded; base cuneate. Inflorescences precocious or with new leaves; peduncle 1-7 cm, flower-bearing part 1-3 cm; pedicels 1-3(-6 in fruit) mm. Sepals ovate to triangular, 2-3 x 1-1.5 mm, yellowish green. Stamens 4-5; filaments 3-4 mm, glabrous. Style *c* 2 mm. Fruit ellipsoid to subglobose, 11-20 x 10-15 mm, yellow to red. Seed slightly compressed, up to 18 x 10 mm, brown. Fig. 125.3.

Evergreen or semievergreen lowland forest, sometimes left as shade tree in coffee plantations; 500-1200 m. IL KF; widespread in tropical Africa. *Chaffey* 951, 1196.

4. FILICIUM *Thw. ex Hook. f.* (1862).

Trees or shrubs. Leaves simple or paripinnate with broadly winged petiole and rachis. Inflorescence axillary or terminal, paniculate. Flowers monoecius, regular. Sepals 5. Petals 5. Disc lobed, lanate. Stamens 5, inserted between disc lobes. Ovary 2-locular; locules with 1 ovule pendulous from top of locule; style curved with stigmatic tissue along upper part of sutures. Fruit a 1-2-locular drupe. Seed without arillode.

Three species: one in Africa and Asia and two in Madagascar.

F. decipiens *(Wight & Arn.) Thw.* (1864).

Tree to 20 m; bole cylindric, not buttressed; bark smooth to flaky, greyish; all parts glabrous, resinous when young. Leaves paripinnate, 6-26 cm long; petiole 2.5-7.5 cm; rachis with wings up to 11 mm wide; leaflets 6-22, narrowly oblong to narrowly elliptic, 4-16.5 x 1.5-3.5 cm, entire; apex acute to emarginate; base cuneate. Inflorescence axillary, 8-20 cm long; pedicels 2-5 mm. Sepals ovate, 1.5-2 x *c* 1 mm, shortly connate. Petals yellow, ovate, *c* 1.5 x 1 mm. Filaments *c* 3 mm, glabrous. Ovary 1-2 mm long; style 1-2 mm. Fruit ellipsoid, 6-11 x 5-8 mm, red. Seed ellipsoid to subglobose, *c* 7 x 6 mm, brown. Fig. 125.4.

Dense primary forest and riverine forest, thicket clumps and termite mounds in *Combretum* woodland; 1500-1800 m. BA; Kenya, Tanzania, Mozambique, Malawi, Zimbabwe, Madagascar. Also in S India and Ceylon. Cultivated on Hawaii and Fiji Islands. Escaped and naturalized on Samoa. *Chaffey* 314; *Friis* et al. 3456; *Mooney* 8414.

5. PAULLINIA *L.* (1753).

Woody climbers. Leaves imparipinnate, stipules present. Inflorescence racemoid with 2 coiled tendrils from apex of peduncle. Flowers monoecious, slightly irregular. Sepals 5, unequal, two inner connate at base. Petals 4, with a large scale with an apical appendix. Disc unilateral, 4-lobed. Stamens 8, filaments shortly connate. Ovary 5-locular; locules with 1 basal ovule; style 3-fid, stigma linear. Fruit a tardily dehiscing, trigonous or narrowly winged, woody, capsule. Seed with arillode.

Between 150 and 200 species: apart from the following all confined to tropical America.

P. pinnata *L.* (1753).

Woody climber to 5(-10) m; all parts sparsely pubescent. Leaves (7-)10-20 cm long; stipules narrowly triangular, 4-5 x 1-2 mm; petiole 2-6(-8.5) cm, winged; rachis 1-4.5 cm, winged; leaflets (3-)5, ovate to elliptic, 2.5-13 x 1-6 cm, remotely dentate; apex acuminate to acute; base cuneate; with small tufts of hair in angles of major veins. Infloresence axillary; peduncle (0-)5-15 cm; flower-bearing part 1-8.5(-12) cm; pedicels 2-4 mm. Sepals broadly elliptic, 2-4 x 1.5-3 mm. Petals white, elliptic to obovate, 4-5 x 2-3 mm. Stamens unequal, 2-4 often without anthers; filaments 2-3 mm, densely pilose. Ovary 2-3 mm long; style *c* 2 mm; stigma *c* 1 mm. Fruit narrowly obconical, trigonous, 20-35 x 7-12 mm, red to brown. Seed ellipsoid, 12-15 x 6-8 mm, black. Fig 125.5.

Evergreen forest and forest edges, riverine forest, groundwater forest, secondary forest and scrub; 550-2300 m. GD/GJ SU WG IL KF GG SD BA; widespread in tropical Africa and Madagascar, also in tropical America. *Burger* 3687; *Friis* et al. 1744; *Gilbert* 1570.

6. CARDIOSPERMUM *L.* (1753).

Annual or perennial climbers. Leaves biternate, stipules present. Inflorescence axillary, corymbiform or paniculate, with (1-)2 coiled tendrils near apex of peduncle. Flowers monoecious, irregular. Sepals 4, 2 lateral ones

Figure 125.3 ***ZANHA GOLUNGENSIS*****: 1 - fruiting branch x 1; 2 - leaf from sapling x 1; 3 - flowering branch with immature leaves x 1; 4 - male flower and buds x 10. 1 from** ***Chaffey*** **1196; 2 from** ***Schmitz*** **4295; 3 & 4 from** ***Chaffey*** **951. Drawn by Christine Grey-Wilson.**

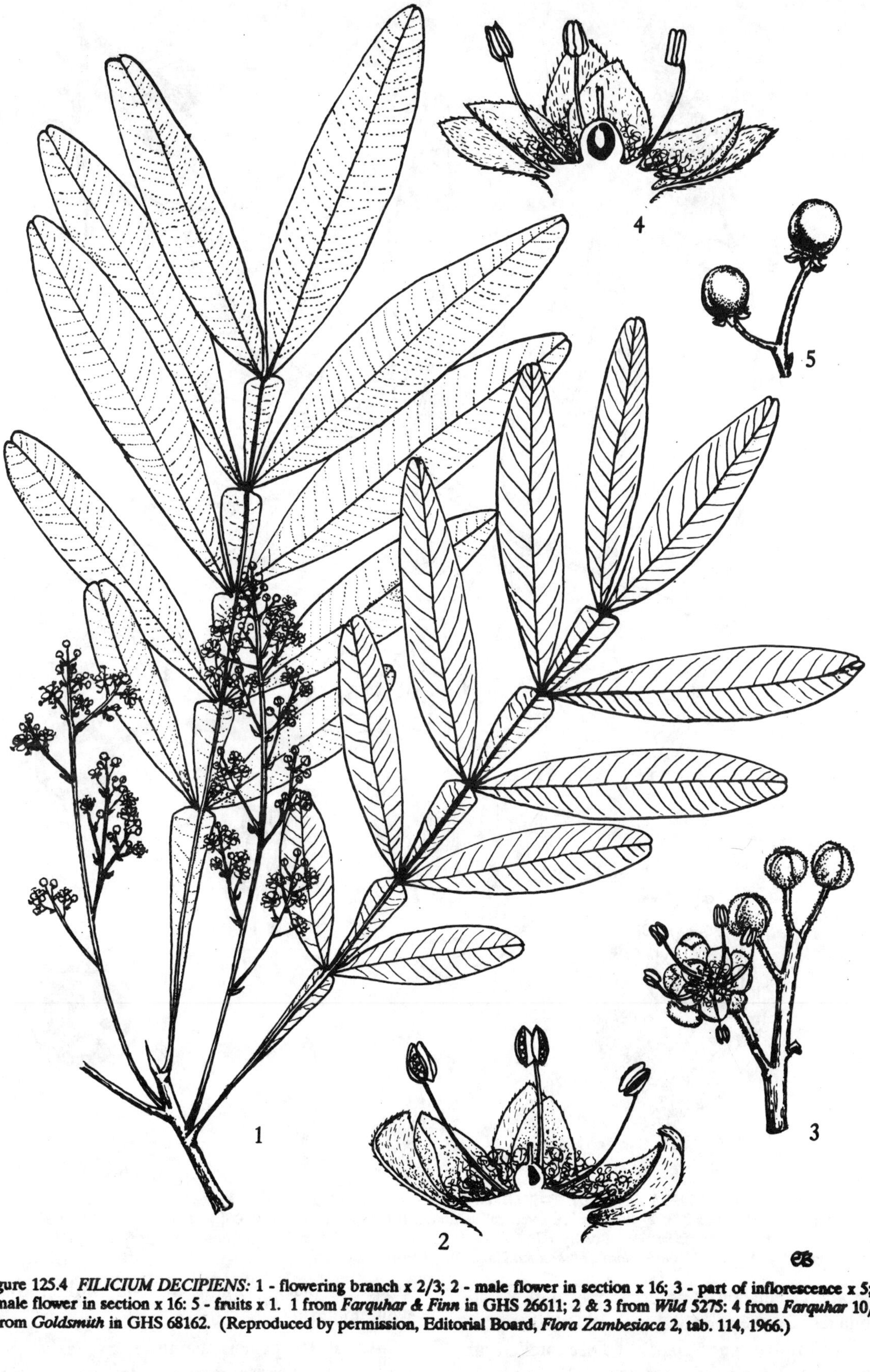

Figure 125.4 *FILICIUM DECIPIENS:* 1 - flowering branch x 2/3; 2 - male flower in section x 16; 3 - part of inflorescence x 5; 4 - female flower in section x 16: 5 - fruits x 1. 1 from *Farquhar & Finn* in GHS 26611; 2 & 3 from *Wild* 5275: 4 from *Farquhar* 10/49; 5 from *Goldsmith* in GHS 68162. (Reproduced by permission, Editorial Board, *Flora Zambesiaca* 2, tab. 114, 1966.)

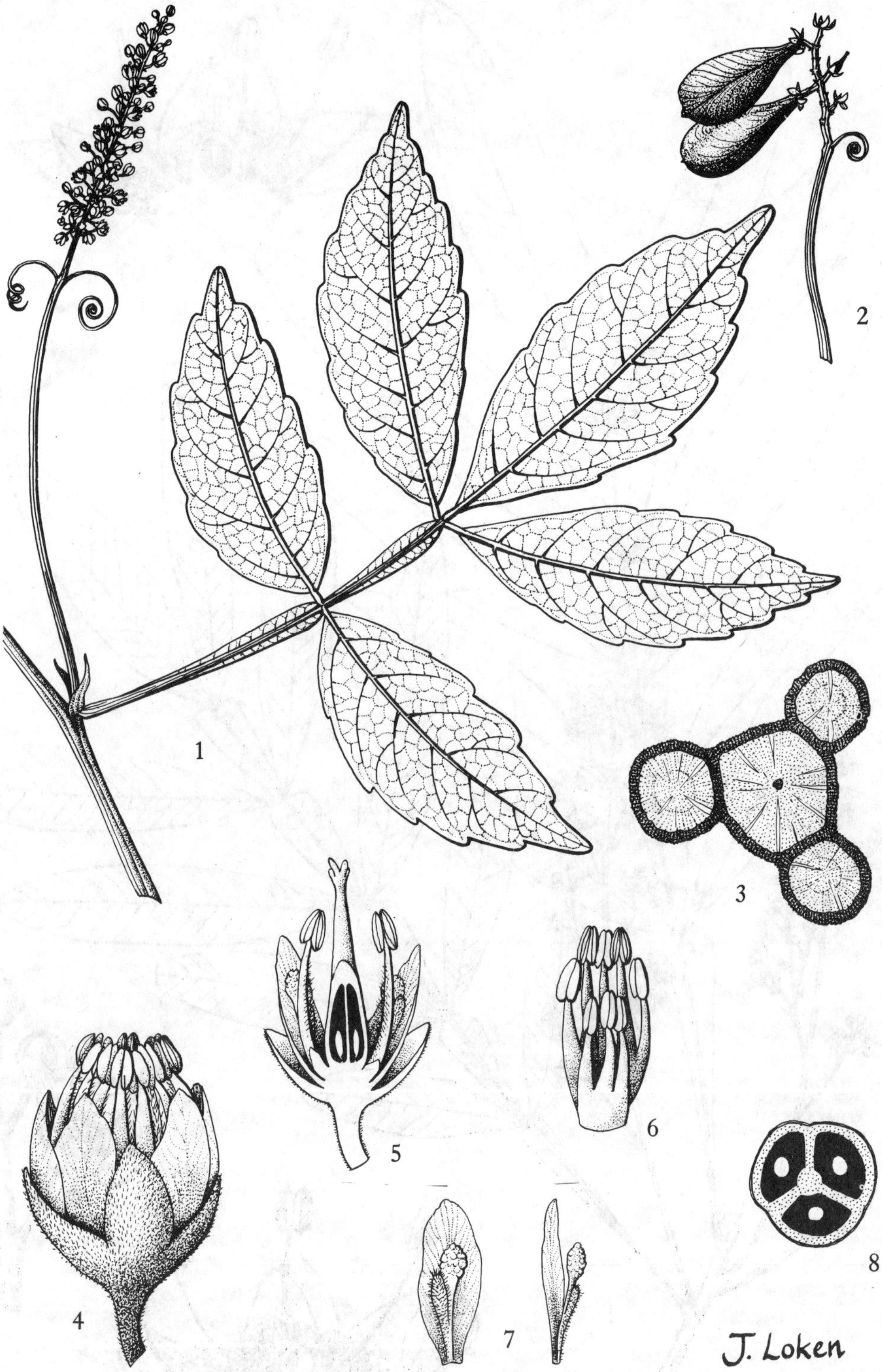

Figure 125.5 *PAULLINIA PINNATA*: 1 - leaf and inflorescence x 2/3; 2 - infructescence x 2/3; 3 - transverse section of old stem x 2/3; 4 - flower x 7; 5 - vertical section of flower x 7; 6 - stamens x 7; 7 - front and side view of petal x 7; 8 - transverse section of ovary x 7. 1 & 4-8 from *Clements* 647; 2 & 3 from *Chase* 979. (Reproduced by permission, Editorial Board, *Flora Zambesiaca* 2, tab. 102, 1966.)

(outer) smaller. Petals 4, 2 anterior ones with petal-like appendages, 2 posterior ones with recurved appendages with clavate upper part. Disc unilateral, posteriorly with 2-4 glands. Stamens 7-8. Ovary 3-locular; locules with 1 basal ovule; style bifid; stigma linear. Fruit an inflated membranaceous capsule. Seed globose, black, without arillode, with a conspicuous white hilum.

About 15 species: mainly in tropical America, but with 3 species widespread in all tropical regions.

1. Slightly woody perennial herb; leaflets beneath uniformly hairy on lamina and nerves; petals

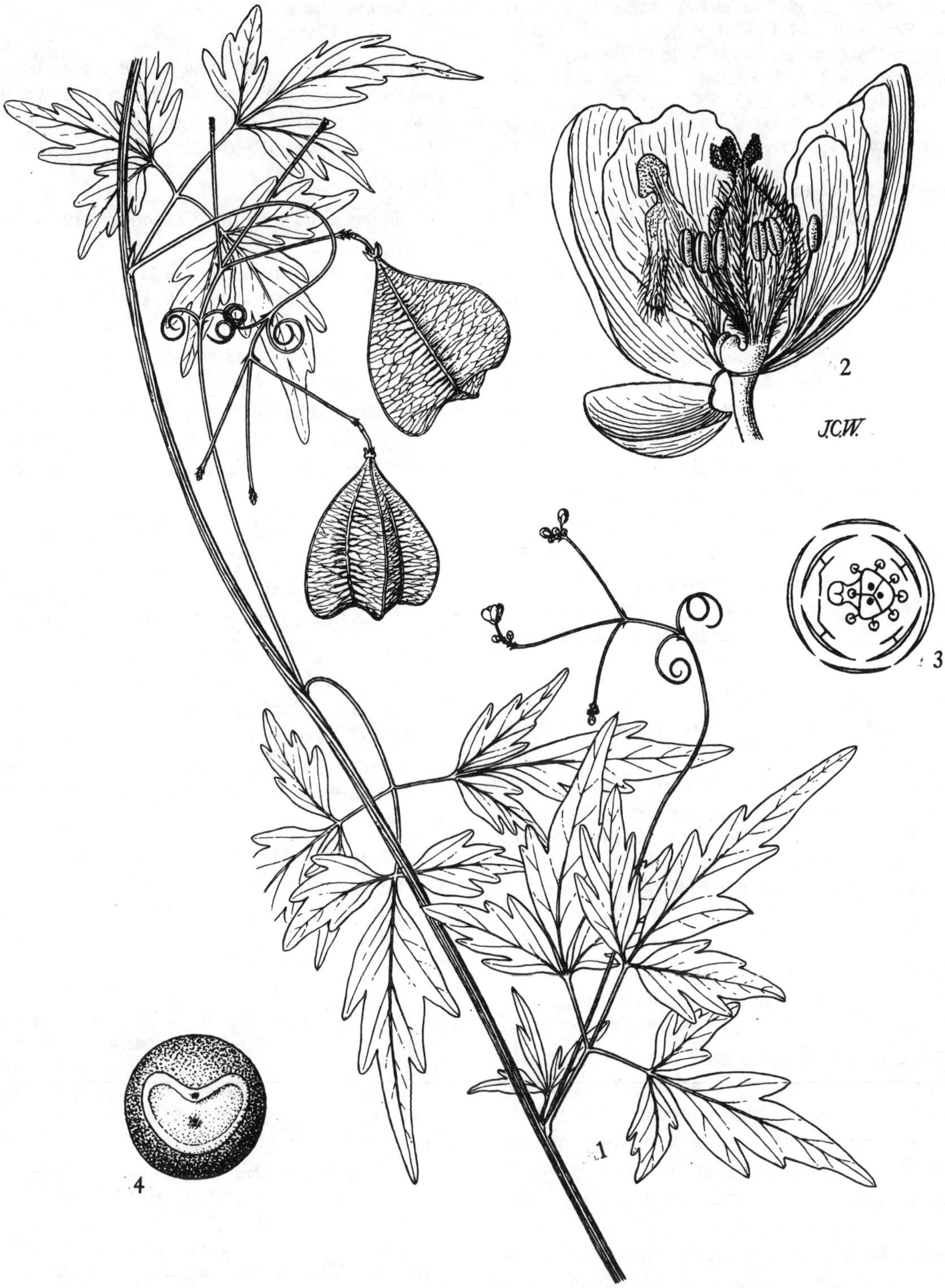

Figure 125.6 *CARDIOSPERMUM HALICACABUM* var. *HALICACABUM*: 1 - flowering and fruiting branch x 1; 2 - flower with 2 petals removed x 15; 3 - floral diagram; 4 - seed x 6. All from *Chase* 2311. (Reproduced by permission, Editorial Board, *Flora Zambesiaca* 2, part of tab. 101, 1966.)

(3-)4-6 mm long; fruit ovoid or ellipsoid. **1. C. corindum**

- Annual herb; leaflets beneath hairy on nerves only; petals 2-3 mm long; fruit obversely tetrahedroid. **2. C. halicacabum**

1. **C. corindum** *L.* (1762).

C. canescens Wall. (1830).

C. clematideum A. Rich. (1847) - type: TU, Chiré (Shire), *Quartin-Dillon & Petit* s.n. (P holo.).

C. oblongum A. Rich. (1847) - type: TU, Choho, *Quartin-Dillon & Petit* s.n. (P holo.).

C. corindum forma *canescens* (Wall.) Radlkofer in *Fl. Bras*. **13, 3**: 448 (1897).

C. corindum forma *clematideum* (A. Rich.) Radlkofer in *Fl. Bras*. **13, 3**: 448 (1897).

C. canescens var. *glabrescens* Almagia in *Ann. R. Ist. Bot. Roma* **8**: 114 (1903) - types: EW, Ocule Cusai, Ingel-Coccaheret, *Pappi* 3698 (FT syn.); EW, Mensa, Nuret-Gheleb, *Terraciano & Pappi* 1254 (FT syn.); EW, Mensa *T&P* 2272 (FT syn.); EE, Samhar, *T&P* 2434 (FT syn.); EW, Bogos, Ali Mandel-Alibared, *T&P* 2511 (FT syn.).

Slightly woody climber to 3 m (rarely suberect); young branches densely pubescent. Leaves trianglar in outline, pubescent; petiole 0.3-2 (-3) cm; terminal leaflet ovate, 1.5-5.2 x 0.5-4.5 cm, crenate to serrate to deeply lobed; apex acuminate to acute, apiculate; base attenuate to truncate. Inflorescence corymbiform to paniculate; peduncle below tendrils (2-)3-12 cm, above 2-13 mm; flower-bearing part either of 3-5 few-flowered corymbiform racemoid thyrses with peduncles 2-14 mm, or a paniculate thyrse with 4-15 mm long axis and 6-12 racemoid thyrses, 2-11 mm long; pedicels 2-4 mm. Outer sepals broadly elliptic to subcircular, 1.5-2 x *c* 1.5 mm, inner elliptic, 3-5.5 x 2-3 mm. Petals white, elliptic, (3-) 4-6 x 2-3 mm. Stamens 8; filaments 3-4 mm, pilose. Ovary 2-3 mm long; style 0.5-1 mm; stigma *c* 0.5 mm. Fruit ovoid or ellipsoid in outline, (2-)2.5-3.5(-4) x (1.5-)2-3.3 cm. Seed *c* 5 mm in diameter, hilum reniform.

Riverine forest, ground-water forest, riverbanks, wooded grassland and bushland; 200-2100 m. EE EW TU GD WU SU GJ GG SD BA HA; pantropical. *Gilbert & Thulin* 286; *Greathead* 73; *Tekle Hagos* 195.

With the amount of material now available, it is impossible to uphold the complicated infraspecific system proposed by Radlkofer (l.c.) and maintained by Cufodontis (l.c.) for the Ethiopian material.

2. C. **halicacabum** *L.* (1753).

Annual climber to 3(-5) m; young stems sparsely pubescent. Leaves triangular in outline; petiole 0.4-5.5 cm, pubescent; terminal leaflet ovate, 1.5-6.5 x 0.8- 2.5 cm, pilose on nerves, grossly dentate or serrate to pinnatipartite[1] (rarely subentire); apex acuminate to acute (rarely rounded), apiculate; base attenuate to rounded. Inflorescence corymbiform; peduncle below tendrils 1.5-10.5 cm, above 2-12 mm; flower-bearing part of 3-4 few-flowered racemoid 1-3 mm long thyrses with peduncles 3-17 mm; pedicels 1-2(-5 in fruit) mm. Outer sepals subcircular, *c* 1 x 1 mm, inner ovate, 2-3 x 1-2 mm. Petals white to yellow, elliptic to obovate, 2-3 x *c* 1 mm. Stamens 8; filaments *c* 2 mm, pilose. Ovary *c* 2 mm; style *c* 0.5 mm; stigma *c* 0.5 mm. Fruit obtetrahedroid, 1.2-4 x 1.7-3.7 cm. Seed *c* 5 mm in diameter, hilum reniform.

1. Capsule more than 2 cm long, about as wide as long or only slightly wider. var. **halicacabum**
- Capsule less than 2 cm long, *c* 1.5 times wider than long. var. **microcarpum**

1. pinnatipartite = leaf blade cut about half-way towards the middle into ± equal pinnately arranged lobes.

var. **halicacabum.**
Fig. 125.6.

Riverine forest, riverbanks, alluvial grassland, wooded grassland, secondary scrub, overgrazed woodland and bushland, weed of cultivation; 400-2250 m. GD WU SU AR GJ WG IL KF GG SD; pantropical. *Gilbert* 1967; *Mooney* 5337, 8983.

var. **microcarpum** *(Kunth) Blume* in *Rumphia* **3**: 185 (1847).
C. microcarpum Kunth (1821)
C. truncatum A. Rich. (1847) - type: TU, Chiré (Shire), *Quartin-Dillon & Petit* s.n. (P holo.).

In similar habitats as var. *halicacabum*; sea-level to 1850 m. EE DN EW TU GD SU GJ GG-SD BA HA; pantropical. *Ash* 874; *Burger* 2064; *Gilbert & Thulin* 47.

The two varieties are well distinguished in Ethiopia, but intermediates occur in Kenya and Tanzania.

7. ALLOPHYLUS *L.* (1753).

Trees, shrubs or woody climbers. Leaves (1-)3-foliolate, glandular beneath. Inflorescence axillary, racemoid or paniculate. Flowers monoecious or dioecious, slightly irregular. Sepals 4. Petals 4, white to cream, spathulate, with a hairy scale on the inside. Disc unilateral, 4-lobed. Stamens 8. Ovary 2-3-locular, deeply 2-3-lobed; locules with 1 basal ovule; style semi-gynobasic, 2-3(-4)-fid. Fruit a 1-2(-3)-lobed drupe. Seed ellipsoid to globose, straw-coloured, without arillode.

About 150-200 species. Widely distributed in all tropical and subtropical regions.

One of the most notoriously difficult genera in tropical areas. Flowers and fruits are very uniform, and many species are very variable vegetatively.

Leenhouts in *Blumea* **15**: 301-358 (1967) has united all the species into a single "superspecies", and proposed that the regional variations are best treated as "races" without any formal taxonomic status. This point of view has been followed by Capuron (l.c.), but was totally ignored by Exell & Sousa in *Fl. de Moçambique*. **51**. *Sapindaceae*: 3-16 (1973).

The author prefers to follow the view of Hauman (l.c.) and Exell (l.c.): although there are a number of intermediates between some of the species, other species can be separated quite easily and are fairly constant. The author suspects that many of the intermediate forms have been taken from plants not growing in their natural habitat, but in habitats disturbed by human influence. As an indication of this it may be noted, that species growing in natural undisturbed forest are generally very uniform and easily recognizable (*A. abyssinicus* and *A. macrobotrys* in Ethiopia), while species from dry country and from secondary vegetation can be extremely variable even within a restricted area. In Ethiopia this applies especially to *A. rubifolius* where the large number of the collections from the Harar-Jijiga area has been particularly useful in showing the range of variation possible within a small area.

If Leenhouts' principle was to be applied to European plants quite a number of genera would certainly have their number of species reduced wholesale.

1. All inflorescences unbranched (racemoid), rarely with a single short basal branch. 2
- Some or all inflorescences branched (paniculate) with 2 or more branches. 5
2. Larger sepals 1.5-2 mm long; cymules peduncled; pedicels minutely puberulous; leaves subglabrous. **3. A. macrobotrys**
- Larger sepals up to 1.2 mm long (if longer then pedicel glabrous); cymules subsessile. 3
3. Leaves pubescent to tomentose beneath; pedicels pubescent. **4. A. rubifolius**
- Leaves subglabrous beneath except for midrib and small tufts of hairs. 4
4. Pedicels pubescent; leaflets coarsely and irregularly toothed. **4. A. rubifolius**
- Pedicels glabrous or with a few scattered hairs; leaflets regularly serrate. **5. A.** = *Mooney* 9003
5. Leaflets pubescent to tomentose beneath. 6
- Leaflets subglabrous beneath except for midrib and small tufts of hairs or minutely puberulous. 8
6. Inflorescence and pedicels minutely tomentellous; filaments glabrous; style glabrous or with a few hairs; seed wider than long, 5-5.5 x 5.5-6 mm. **1. A. abyssinicus**
- Inflorescence pubescent to tomentose or pilose; seed not wider than long. 7
7. Leaves uniformly pubescent to tomentose beneath; pedicels pubescent; filaments and style hairy; seed ellipsoid, 6-7 x *c* 5 mm (rare forms). **4. A. rubifolius**
- Leaves pubescent, midrib conspicuously long-haired along the edges; pedicels, filaments and style glabrous or with a few hairs basally. **5. A.** = *Mooney* 9003
8. Larger sepals up to 1.2 mm long; filaments and style hairy; pedicels glabrous or with a few hairs; seed longer than wide. **2. A. africanus**
- Larger sepals 1.5-2 mm long; filaments and style glabrous or with a few hairs. 9
9. Inflorescence and pedicels minutely tomentellous; leaflets shallowly crenulate to dentate; branchlets usually densely and minutely tomentellous. **1. A. abyssinicus**
- Inflorescence and pedicels glabrous to sparsely pubescent or pilose; leaflets serrate; branchlets very sparsely pilose or subglabrous. **5. A.** = *Mooney* 9003

1. **A. abyssinicus** *(Hochst.) Radlkofer* (1895)
- type: GD, Semien, Addi Dschoa, *Schimper* I: 377 (P holo., C K iso).

Tree to 15(-25) m; dbh to 0.5 m; bark smooth, grey; branchlets, petioles and inflorescences minutely tomentellous (rarely pubescent to tomentose). Petioles 2.5-10.5 cm; leaflets elliptic to obovate, minutely puberulous (rarely pubescent to tomentose), terminal 5.5-21 x 3-11 cm, shallowly crenulate to dentate; apex acuminate to acute; base attenuate to cuneate; tufts of hairs conspicuous. Inflorescence paniculate, 5-22 cm long, with 2-10 branches some of which often branch again; cymules subsessile; pedicels 1-3 mm. Outer sepals 1.2-1.5 x *c* 1 mm, elliptic, puberulous, inner 1.5-2 x *c* 1.5 mm, elliptic to circular. Petals *c* 1.5 x 1 mm. Filaments 2-3 mm, glabrous. Style 2-3 mm, glabrous or with a few hairs; stigma 1-1.5 mm. Fruit subglobose, *c* 6 mm in diameter. Seed transversely ellipsoid, 5-5.5 x 5.5-6 mm. Fig. 125.7.

Juniperus, *Podocarpus*, *Aningeria-Olea* and *Albizia-Croton* forest, riverine forest, forest edges, often persisting after forest clearing; (1450-)1600-2800(-3000) m. EW TU GD WU SU AR GJ WG KF SD BA HA; S Sudan, Uganda, Kenya, Tanzania, Rwanda, E Zaire, Mocambique, Malawi, Zimbabwe. *Friis* et al. 2180; *Mooney* 6043, 6228.

2. **A. africanus** *P. Beauv.* (1819 or 1820).

Shrub or tree to 10 m; branchlets, leaves and inflorescences subglabrous to pubescent. Petiole (1-)2-7.5 cm; leaflets narrowly elliptic to elliptic, terminal (3-)5-14.5 x (1-)1.5-7 cm, subentire to coarsely and irregularly dentate; apex acuminate to rounded; base attenuate to cuneate; tufts of hairs present or not. Inflorescence paniculate, 4.5-15 cm long, with 1-6 branches; cymules subsessile; pedicels 1.5-2 mm, glabrous or with a few hairs. Outer sepals *c* 1 x 1 mm, broadly elliptic, subglabrous, inner *c* 1 x 1-1.5 mm, subcircular. Petals *c* 1 x 0.5 mm. Filaments *c* 2 mm, hairy in basal half. Style *c* 1 mm, hairy; stigma *c* 0.5 mm. Fruit ellipsoid, 5-7 x 4-5 mm. Seed ellipsoid, 5.5-6 x *c* 4.5 mm. Fig. 125.8.4.

Combretum-Terminalia woodland and wooded grassland on rocky granitic slopes, dry riverine forest; 600-1800 m. SU KF and GG (all in the Gibe valley only) WG; widespread in tropical Africa. *Ash* 978, 1326; *Jansen* 6374.

3. **A. macrobotrys** *Gilg* (1897).

Shrub or tree to 4(-6) m, sometimes scandent; branchlets, leaves and inflorescences puberulous to pubescent. Petiole 3-9 cm; leaflets elliptic, terminal (6-)8-14(-19) x (2.5-)4-7.5 cm, regularly dentate or serrate; apex acuminate to acute; base attenuate to cuneate; tufts of hairs indistinct or absent. Inflorescences racemoid, (4-)7.5-15.5 cm long, often slender and drooping; cymules with peduncles 1.5-3 mm long; pedicels 1.5-2 mm, minutely puberulous. Outer sepals 1-1.5 x *c* 1 mm, broadly elliptic, sparsely puberulous, inner 1.5-2 x 1.5-2 mm, subcircular. Petals *c* 2 x 1 mm. Filaments *c* 2 mm, hairy in basal half. Style *c* 1 mm, with a few scattered hairs; stigma 0.5-1 mm. Fruit ellipsoid, 5-6.5 x 4-5 mm. Seed ellipsoid, 5.5-6.5 x 4.5-5 mm. Fig. 125.8.8-9.

Figure 125.7 *ALLOPHYLUS ABYSSINICUS*: 1 - flowering branch x 1; 2 - male flower x 9; 3 - female flower x 9; 4 - infructescence x 1. 1 & 2 from *Amare Getahun* in *IECAMA* G55B; 3 from *Mooney* 6228; 4 from *Friis* et al. 268. Drawn by Christine Grey-Wilson.

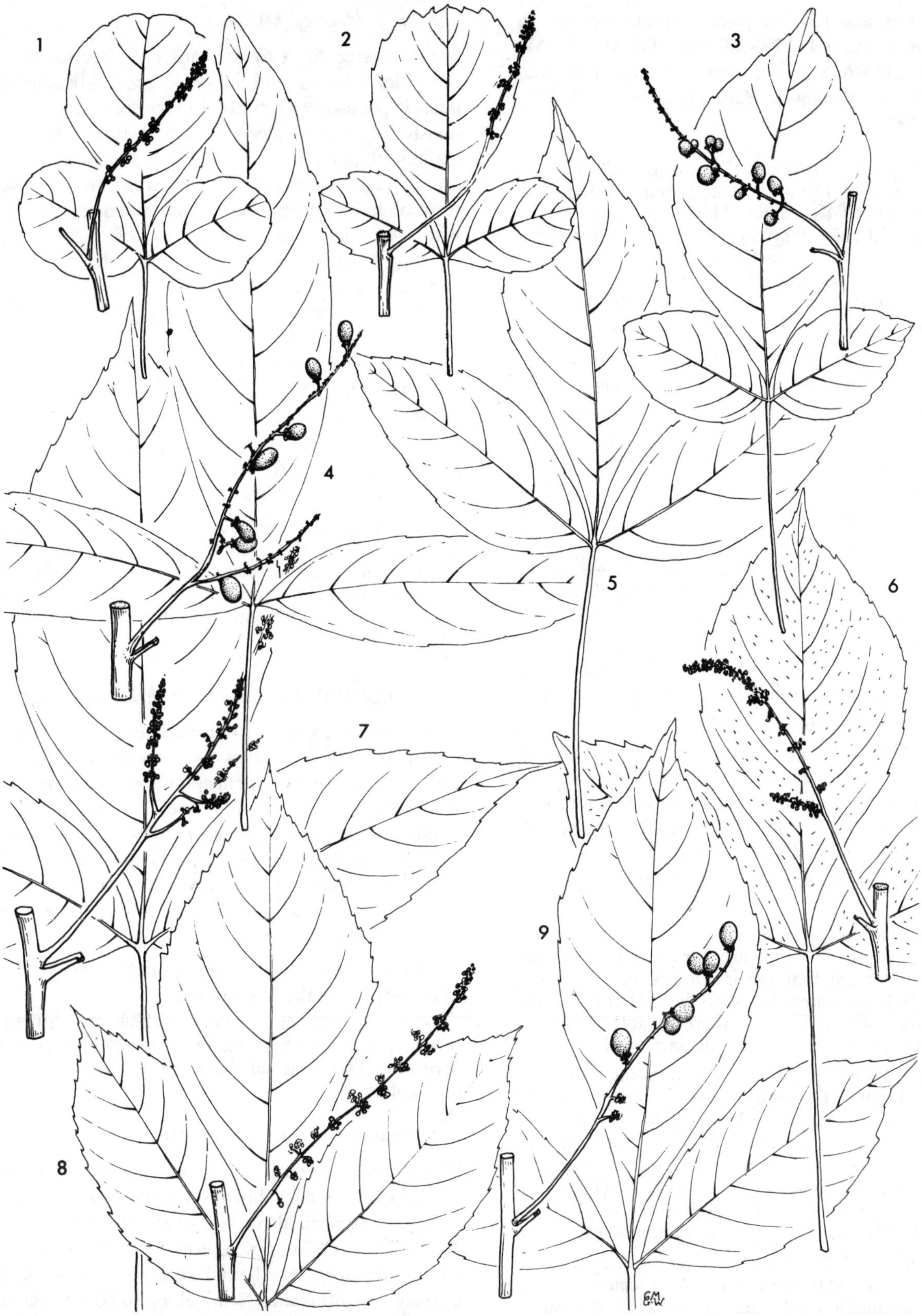

Figure 125.8 **Leaves and inflorescences of species of** *ALLOPHYLUS*, all x 1. *A. RUBIFOLIUS* 1-3; *A. AFRICANUS* 4; *A.* sp. = Mooney 9003 etc. 5-7; *A. MACROBOTRYS* 8-9. 1 from *Burger* 1938; 2 from *Westphal* et al. 3у82; 3 from *Carr* 911; 4 from *Ash* 1326; 5 from *Chaffey* 988; 6 from *Mooney* 9003; 7 from *W. de Wilde* et al. 7221; 8 from *W. de Wilde* et al. 6795; 9 from *Alemayehu Haile* 1134. Drawn by Christine Grey-Wilson.

Rainforest and riverine forest, termite mounds in *Combretum* woodland; 1400-2500 m. GJ SU SU-AR WG IL KF SD BA; S Sudan, Uganda, Kenya, Tanzania, Burundi, Zaire. *Alemayehu Haile* 1134; *Friis* et al. 2424; *Mooney* 9200.

4. **A. rubifolius** *(Hochst. ex A. Rich). Engl.* (1892)
- types: TU, Djeladjeranne, *Schimper* III:1613 (P syn., FI(Webb) K isosyn.); TU-GD, Tacazze Valley, *Schimper* III:1669 (P syn., FI(Webb) K UPS isosyn.).
A. erlangeri Gilg ex Chiov. (1932) - type: *Basile* 330 (TO holo. not seen).
A. alnifolius sensu Cufodontis (1958) quoad distrib. Ethiop., non (Bak.) Radlkofer (1896).

Shrub or tree to 7 m (-12 m fide Carr); branchlets, leaves and inflorescences sparsely pubescent to subtomentose. Petiole 0.8-5 (-6.5) cm; leaflets elliptic or obovate to subcircular, terminal 2-13 x 1.5-7.5 cm, shallowly crenate to irregularly dentate or serrate; apex acute to emarginate; base attenuate to cuneate; tufts of hairs usually conspicuous. Inflorescences usually all racemoid (rarely some paniculate with 1-3(-12) branches), 2-18(-22) cm long; cymules subsessile; pedicels 1-2(-3) mm, pubescent. Outer sepals *c* 1 x 1 mm, broadly elliptic, pubescent, inner *c* 1 x 1-1.5 mm, subcircular. Petals *c* 1 x 0.5 mm. Filaments 1-2 mm, hairy in basal half. Style *c* 1 mm, hairy; stigma *c* 0.5 mm. Fruit ellipsoid to subglobose, 6-8 x 5-6 mm. Seed ellipsoid, 6-7 x *c* 5 mm. Fig. 125.8.1-3.

Acacia-Commiphora, *Acacia-Combretum* and *Combretum-Terminalia* woodland and bushland, alluvial *Acacia* woodland, riverine forest, termite mounds; 400-1800(-2000) m. EE EW TU GD SU GJ IL GG SD BA HA; from Ethiopia and Sudan through eastern Africa to S Africa (Transvaal and Natal). *Burger* 1836, 1848, 1938.

Chiovenda (1932) compares *A. erlangeri* with *A. rubifolius* and says that they are closely related. Apart from the statement that the fruiting pedicel is 4 mm long - in the large material studied the author has never seen pedicels above 3 mm - there is nothing in the diagnose to suggest that they are distinct species.

In the author's opinion true *A. alnifolius* does not occur in Ethiopia. The subglabrous specimens referred to it by different authors all belong to *A. rubifolius*. True *A. alnifolius* is a coastal E African species with glabrous pedicels and flowers. The Zaire specimens named *A. alnifolius* at Kew are typical *A. rubifolius*, and the author suspects the other collections cited as *A. alnifolius* by Hauman (l.c.) are also *A. rubifolius*.

Compared with the rather constant forest species, *A. rubifolius* is an extremely variable species. It is, however, normally easily distinguished by the small leaves, unbranched inflorescences and hairy pedicels, stamens and style.

Carr 911 is said to be a tree to 40 feet tall. This is the only indication of the species ever reaching more than 7 m.

5. **A.** sp. = *Mooney* 9003.

Shrub or tree to 4(-6) m; bark dark brown, smooth; branchlets, leaves and inflorescences subglabrous to densely pubescent or pilose. Leaves sometimes conspicuously long-haired along midrib; petiole 2.5-11.5 cm; leaflets elliptic, terminal 6-14.5 x 3-7.5 cm, regularly serrate; apex acuminate; base attenuate to cuneate; tufts of hairs conspicuous. Inflorescences paniculate with 1-4 branches or some (rarely all) racemoid, 5-15 cm long; cymules subsessile or with peduncle to 1 mm; pedicels 1-2 mm, glabrous or with a few hairs. Outer sepals *c* 1 x 1 mm, broadly elliptic, glabrous to sparsely puberulous, inner 1.5-2 x *c* 1.5 mm, subcircular. Petals 1-2 x *c* 1 mm. Filaments *c* 2 mm, glabrous or with a few hairs near base. Style *c* 1 mm, glabrous; stigma *c* 1 mm. Fruit not known. Fig. 125.8.5-7

Podocarpus forest, riverine forest; 1600-2075 m. GD SU AR WG IL KF SD; not known elsewhere. *Mooney* 9003; *J. de Wilde* 5392; *W. de Wilde* et al. 7221.

The paniculate inflorescences and large flowers seem to indicate some relation with *A. abyssinicus*. The indumentum and leaf-dentation, however, point more at a relation with *A. macrobotrys*. Until more material - including fruits - become available it is best left unnamed.

8. **DEINBOLLIA** *Schumach. & Thonn.* (1829).

Trees or shrubs, often unbranched with leaves in a terminal rosette. Leaves paripinnate. Inflorescence axillary or terminal, paniculate. Flowers monoecious or dioecious, regular. Sepals 5, free, 2 outer ones smaller. Petals 5, with a hairy scale at base. Disc annular, conspicuous. Stamens (8-)10-30. Ovary 2-3(-5)-locular and -lobed; locules with 1 basal ovule; style gynobasic. Fruit a fleshy or leathery 1-3(-5)-lobed drupe. Seed without arillode.

About 30-40 species in tropical and S Africa, Madagascar and the Mascarenes.

A genus with many species which are difficult to separate and which is badly in need of a thorough revision. The material, however, is often scanty and incomplete.

D. kilimandscharica *Taub.* (1895).

Slender unbranched tree to 6 m; branchlets and leaves subglabrous to dark brownish pubescent to strigose. Leaves in a terminal rosette, 20-80 cm long; petiole 5-30 cm; leaflets 8-16, oblong to elliptic, 5.5-30 x 2.7-13 cm, entire; apex shortly and abruptly acuminate; base cuneate. Inflorescence terminal, up to 60 x 40 cm, dark brown pubescent, those of male plants denser and with many more flowers; pedicels of male flowers 2-3 mm, of female 3-5(-9 in fruit) mm, with a joint below middle, basal part densely, apical sparsely pubescent. Sepals

Figure 125.9 *DEINBOLLIA KILIMANDSCHARICA*: 1 - part of leaf x 1; 2 - part of male inflorescence x 1; 3 - male flower x 6; 4 - part of female inflorescence x 1; 5 - female flower x 6; 6 - stamen x 9; 7 - gynoecium x 9; 8 - part of infructescence x 1. 1 from *Friis* et al. 1876; 2, 3 & 6 from *W. de Wilde* et al. 7781; 4,5 & 7 from *Friis* et al. 312; 8 from *Meyer* 7989. Drawn by Christine Grey-Wilson.

broadly elliptic to subcircular, 4-6 x 3-5 mm, sparsely dark brown pubescent. Petals white or cream, broadly ovate to spathulate, 3-5 x 2-4 mm, tomentose. Stamens 15-18; filaments 4-5 mm, densely tomentose. Ovary 2-3 lobed, *c* 4 mm long; style *c* 6 mm. Fruit 1-2-lobed, yellow; cocci ellipsoid to subglobose, 13-21 x 10-18 mm. Seed ellipsoid, 10-15 x 8-12 mm, dark brown. Fig. 125.9.

Primary upland *Aningeria adolfi-friederichii-Olea* forest, lowland *Aningeria altissima-Chlorophora-Celtis* forest, riverine forest; 1150-2600 m. GJ WG IL KF; Uganda, Kenya, Tanzania, E Zaire. *Friis* et al. 312; *Meyer* 7989; *W. de Wilde* et al. 7781.

One of the species where good and adequate material is available. It is easily recognized by the large flowers with subglabrous calyx.

9. LEPISANTHES *Blume* (1825).

Aphania Blume (1825).

Leenhouts in *Blumea* **17**: 33-91 (1969).

Trees or shrubs. Leaves paripinnate. Inflorescence axillary or terminal, paniculate. Flowers normally monoecious, regular. Sepals 4-5, free, 2 outer ones smaller. Petals (0-)4-5, with a scale at base. Disc annular or interrupted, glabrous or hairy. Stamens 4-18. Ovary 2-3-locular; locules with 1 basal ovule; stigma subcapitate. Fruit a 1-3-lobed drupe. Seed without arillode.

24 species: apart from the following all confined to tropical Asia.

L. senegalensis *(Juss. ex Poir.) Leenh.* (1969);

Aphania senegalensis (Juss. ex Poir.) Radlkofer (1878).

Sapindus abyssinicus Fresen. (1837) - type: GD, near Gondar, *Rueppel* s.n. (FR holo.).

Tree to 15 m; bark smooth, grey to brownish; branchlets brownish strigose. Leaves 3-15 cm long, subglabrous; petiole 1.5-5 cm; leaflets 2-6, elliptic, 6.5-21(-27) x (2-)3-8(-9) cm, entire; apex acute to rounded; base cuneate. Inflorescence pseudoterminal, of panicles from the upper leaf axils, 5-15(-20) cm long, brownish pubescent; pedicels 1-3(-7 in fruit) mm. Sepals 5, ovate to subcircular, 2-3.5 x 1.5-3 mm. Petals 5, broadly ovate, *c* 2 x 2 mm, yellowish green. Stamens 8; filaments 2-3mm, glabrous. Ovary 1-2 mm long; style 1-2 mm, semi-gynobasic. Fruit 1-2-lobed, reddish; cocci ellipsoid, 10-15 x 8-11 mm. Seed ovoid, *c* 12 x 9 mm, black. Fig. 125.10.

Riverine forest, groundwater forest, occasionally in evergreen upland forest; 400-1600 m. EE EW GD GJ WG IL GG SD; from Senegal to Ethiopia and south through eastern Africa to central Mocambique, also widespread in tropical Asia. *Friis* et al. 1893, *Kefale G. Michael* 5; *Schweinfurth & Riva* 1838.

The ripe fruit is edible.

10. BOTTEGOA Chiov. (1916).

Shrub or small tree. Leaves biparipinnate. Inflorescence axillary, racemoid, few-flowered. Flowers dioecious, regular (male not known). Sepals 4. Petals 4. Disc small, 4-lobed. Ovary 2-locular; locules with 1 centrally affixed ovule; style with 2 short stigmatic lobes. Fruit a 2-locular samara. Seed reniform, compressed, without arillode.

One species in NE Africa.

B. insignis *Chiov.* (1916).

Shrub or tree to 8 m; bark grey to blackish; branchlets reddish brown, pubescent. Leaves 4-10 cm long; petiole 1-3.5 cm, pubescent; pinnae 6-12, alternate to opposite; leaflets 6-14, alternate to opposite, oblong to obovate, 5-15 x 2-6 mm, glaucous, sparsely puberulous, entire; apex rounded to retuse; base cuneate. Inflorescences from axils of fallen leaves, 1-3 cm long, sparsely puberulous; pedicels 5-9 mm. Female flowers: sepals triangular, *c* 0.5 mm long; petals yellow, elliptic, 4-5 x *c* 1.5 mm. Ovary *c* 2 x 1.5 mm, flattened, narrowly 2-winged, pandurate in outline; style *c* 0.5 mm; staminodes *c* 2 mm, more than 4 per flower. Fruit broad-winged, circular, 2.5-4.5 cm in diameter, yellow with pinkish wings. Seed *c* 10 x 5 x 3 mm, brown. Fig. 125.11.

Acacia-Commiphora woodland, wooded grassland and bushland with rich shrub and forb layers, on basaltic and limestone soils, often in rocky places; 1150-1300 m. BA HA; Somalia, NE Kenya. *Burger* 3105, 3310, 3726.

11. LECANIODISCUS *Planch. ex Benth.* (1849).

Friis in *Kew Bull.* **39**: 779-783 (1984).

Trees or shrubs. Leaves paripinnate. Inflorescence axillary, racemoid or actually a raceme in the apical part. Flowers dioecious, regular. Sepals 4-5, connate at base. Petals 0. Disc lobed. Stamens 8-13. Ovary 3-locular; locules with 1 basal ovule; style short; stigma 3-lobed. Fruit 1(-2)-locular, tardily and irregularly dehiscing from base. Seed completely surrounded by an arillode.

Two species in tropical Africa: one of these with three subspecies.

L. fraxinifolius *Bak.* (1868).

Tree to 20(-30) m; bole slightly buttressed; bark greyish brown, with slight fissures forming plates; branchlets tomentellous. Leaves 10-15 cm long; petiole 3-6 cm, subglabrous to tomentellous; leaflets 6-12, ovate or narrowly elliptic, 4-15 x 1.3-4.5 cm, subglabrous, entire or slightly crenate; apex acuminate to acute and apiculate; base cuneate to rounded. Inflorescences 1-10 cm

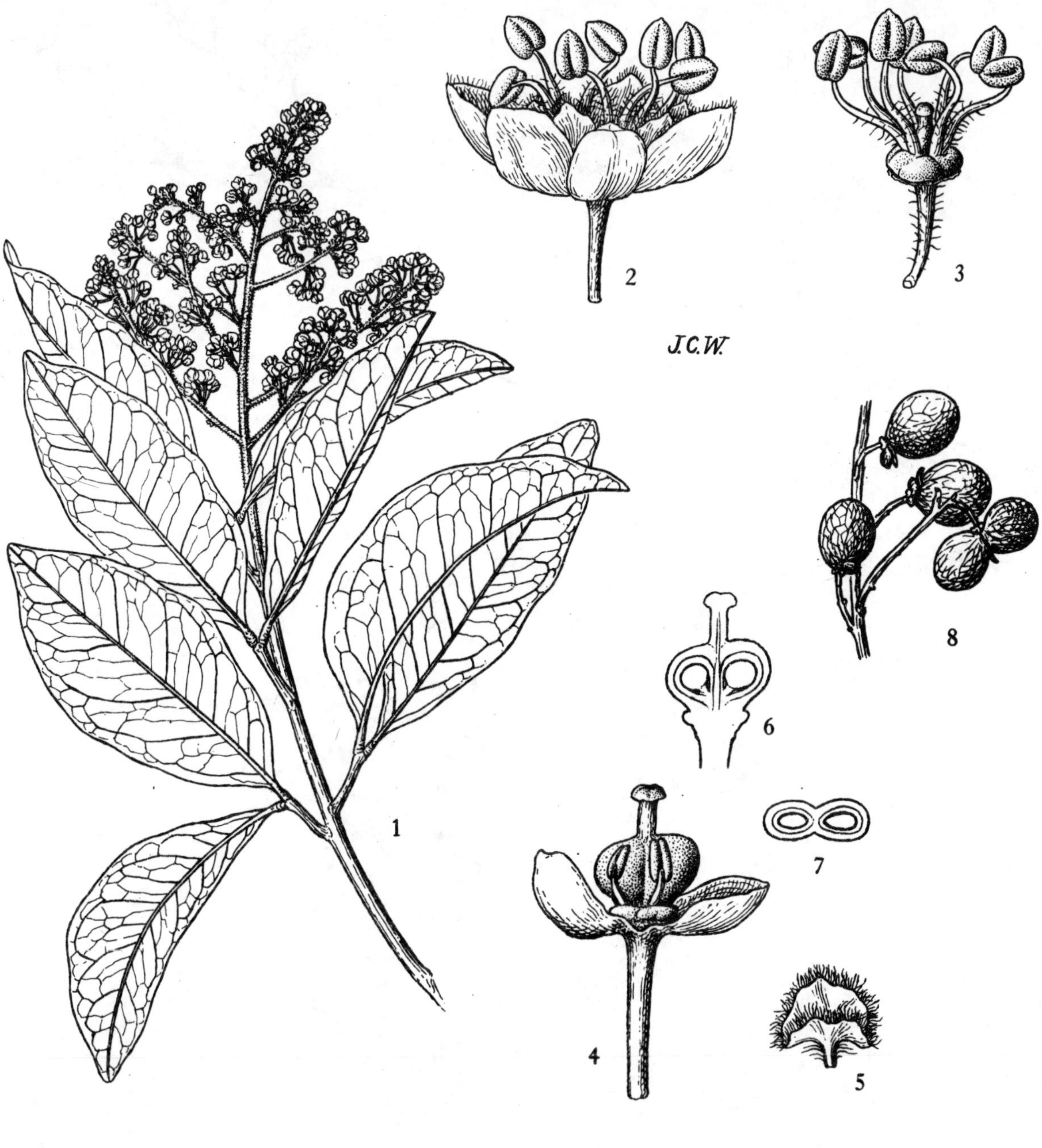

Figure 125.10 *LEPISANTHES SENEGALENSIS*: 1 - flowering branch x 1/2; 2 - male flower x 6; 3 - male flower with sepals and petals removed to show disc and pistillode x 6; 4 - female flower x 6; 5 - petal x 10; 6 & 7 - longitudinal and cross sections through pistil x 5; 8 fruits x 1. 1 from *Warnecke* 43; 2-7 from *Warnecke* 383; 8 from *Hoyle* 539. (Reproduced by permission, Editorial Board, *Flora Zambesiaca* 2, tab. 108, 1966.)

long, sometimes forming a compound inflorescence towards the end of branches, puberulous; pedicels 2-5 mm. Sepals 4, ovate to elliptic, 2-3 x *c* 2 mm, yellowish green to cream. Stamens 8-10; filaments 3-4 mm, glabrous; anthers red. Ovary 2-3 mm long. Fruit 1-locular, ovoid to ellipsoid or subglobose, 10-15 x 8-15 mm, yellow to pink. Seed ellipsoid, *c* 10 x 8 mm.

ssp. **vaughanii** *(Dunkley) Friis* in *Kew Bull.* **39**: 781 (1984).

Leaflets acuminate to acute.

Lowland evergreen and semi-evergreen forest, riverine forest, groundwater forest; 500-1400 m. GJ IL KF GG; Uganda, Kenya, Tanzania, Burundi. *Chaffey* 1269; *Friis* et al. 1916, 3945.

Figure 125.11 *BOTTEGOA INSIGNIS*: 1 - fruiting branch x 1; 2 - flowering branch (female) x1; 3 - female flower x 9; 4 - female flower displayed x 9. 1 from *Gilbert & Thulin* 1296; 2-4 from *Burger* 3310. Drawn by Christine Grey-Wilson.

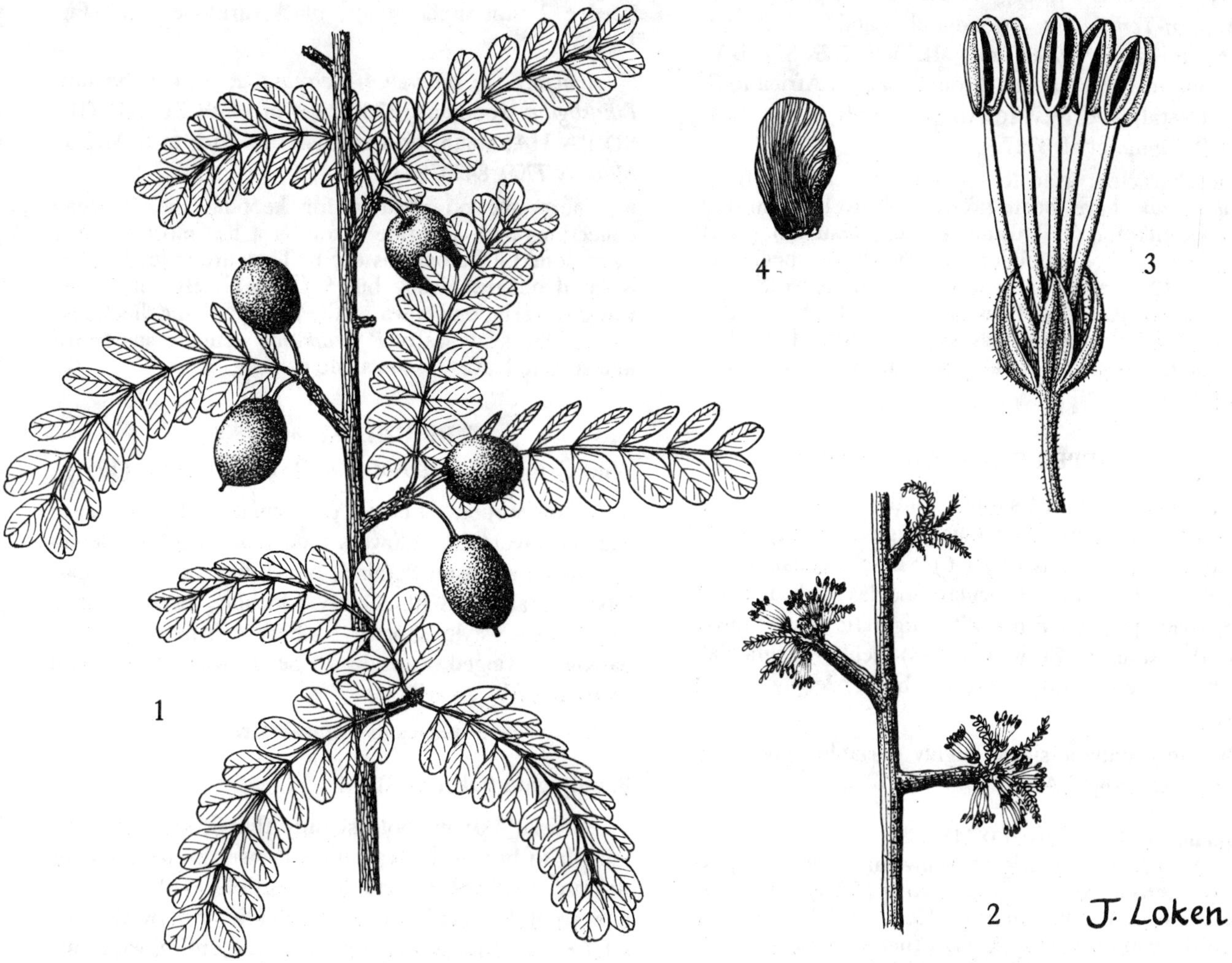

Figure 125.12 *HAPLOCOELUM FOLIOLOSUM*: 1 - branch with leaves and fruits x 3/4; 2 - flowering branch x 3/4; 3 - male flower x 8; 4 - seed x 2. 1 & 4 from *Martin* 533/33; 2 & 3 from *Martin* 413/32. (Reproduced by permission, Editorial Board, *Flora Zambesiaca* 2, part of tab. 112, 1966.)

There are two other subspecies: ssp. *fraxinifolius* in Mozambique, Malawi, Zambia and Zimbabwe with smaller, wider, subacute or rounded leaflets, and ssp. *scassellatii* (Chiov.) Friis along the Somali and Kenya coast with obovate rounded leaflets.

12. HAPLOCOELUM *Radlkofer* (1878).

Bullock in *Bull. Misc. Inf.* **1931**: 353-357 (1931).

Trees or shrubs. Leaves paripinnate. Inflorescence short, racemoid, axillary or on lateral dwarf shoots. Flowers dioecious, regular. Sepals 5-6. Petals 0. Disc inconspicuous, annular, glabrous. Stamens (4-)5-6. Ovary 3-locular; locules with 1 basal ovule; style short; stigma 3-lobed. Fruit a 1-locular drupe with 1 seed. Seed almost completely surrounded by an arillode.

A genus of about 10 species in Africa and one in Madagascar.

The genus is badly in need of a revision, and may well prove to have but about 5 species in total. The distinguishing characters used by different authors (Bullock, l.c., Radlkofer, l.c.) have almost exclusively been taken from the leaves - fertile material being scanty in several species - and the boundaries between some of the species are certainly artificial.

H. foliolosum *(Hiern) Bullock* (1931).

H. gallaënse (Engl.) Radlkofer (1921) - type: BA, between Ginir and Dunuk, *Ellenbeck* 1963 (B holo., destr.).

Tree to 10 m; bark grey, smooth with circular foldings; branchlets pubescent. Leaves on short lateral dwarf shoots and on terminal long shoots, 2-7 cm long; petiole 2-8 mm, pubescent; leaflets 8-16, obliquely oblong to obovate or subcircular, varying from almost symmetric to very asymmetric, 4-28 x 4-14 mm, sparsely pubescent on midrib, entire; apex emarginate (rarely rounded); base with distal side truncate and proximal attenuate to cuneate. Inflorescences precocious or with new leaves, 2-6 mm long, pubescent; pedicels 3-5(-7 in fruit) mm. Sepals 5-6, elliptic, 1-1.5 x *c* 1 mm, yellowish or cream. Stamens 4-5; filaments 2-4 mm, glabrous. Ovary 1-2 mm long. Fruit ellipsoid, 10-15 x 8-11 mm, red or orange. Seed ellipsoid, *c* 10 x 5 mm. Fig. 125.12.

Dry *Juniperus* forest, *Acacia-Commiphora* and *Combretum-Terminalia* woodland and bushland, riverine forest; 650-2000 m. IL KF GG SD BA; Cameroun to Ethiopia and through eastern Africa to S Africa (Natal) and west to Angola. *Friis* et al. 1894; *Kuls* 552; *Thulin* et al. 3727.

The characters used to separate *H. gallaënse* from *H. foliolosum* have been taken exclusively from the leaves (Radlkofer (l.c.), Bullock (l.c.), Hauman (l.c.)), and in the author's opinion only a single species is involved. It is rather suspicious, that plants which are technically *H. gallaënse* turn up scattered all over the area of *H. foliolosum* in such widely separated areas as Cameroun, Angola, Natal, Kenya, E Zaire and Ethiopia.

13. **PAPPEA** *Eckl. & Zeyh.* (1834).

Tree or shrub. Leaves simple. Inflorescence axillary, racemoid. Flowers dioecious, regular. Sepals 5, connate at base. Petals (4-)5(-6). Disc annular, hairy. Stamens 7-10. Ovary 3-locular; locules with 1 basal ovule; style spirally twisted with stigmatic tissue down along the sutures. Fruit a 1-2(-3)-locular loculicidal capsule. Seed almost covered by a deeply lobed arillode.

Probably only a single highly variable species in eastern Africa and S Africa.

P. capensis *Eckl. & Zeyh.* (1834).
P. radlkoferi Penzig ex Schweinf. (1899) - types: EW, Acrour, *Schweinfurth & Riva (S&R)* 1041 (FT K P isosyn.); EW, Mensa - Geleb, *S&R* 1068 (not seen), 1080 (not seen) & 1123 (not seen); EW, near Aidereso, Galata, *S&R* 1614 (FT K isosyn.).
P. capensis var. *radlkoferi* (Penzig ex Schweinf.) Schinz (1909).

Tree to 12 m; bark rough, dark grey to blackish; branchlets yellowish tomentellous. Leaves simple, puberulous to tomentellous; petiole 4-22 mm; lamina elliptic or oblong, 3-14 x 1.5-7 cm, entire to crenate or serrate, margin wavy; apex rounded to shallowly emarginate; base truncate to subcordate. Inflorescence 2-15(-20) cm long, tomentellous; pedicels 0.5-3 mm. Sepals united to above middle, 1-1.5 mm long, free parts broadly triangular. Petals white to yellowish green, broadly ovate to subcircular, *c* 1 mm. Filaments *c* 3 mm, glabrous. Ovary 1-1.5 mm long; style *c* 2 mm. Fruit 1(-2)-lobed, subglobose, *c* 1 cm in diameter, yellowish. Seed *c* 7 mm in diameter, black; arillode red. Fig. 125.13.

Acacia, *Combretum-Terminalia*, *Combretum-Piliostigma* woodland; 1200-2300 m. EW SU AR GG SD BA HA; south through eastern Africa to S Africa. *Mooney* 7709, 8420, 9435.

There are no reasons for keeping the Eritrean collections as a separate taxon as it has most recently been done by Cufodontis (l.c.). They are indeed rather isolated geographically but fall completely within the range of variation shown by the S Ethiopian collections. One of the syntypes of *P. radlkoferi* actually has entire and serrate leaves on the same branch.

14. **BLIGHIA** *Koenig* (1806).

Wilczek in *Bull. Jard. Bot. Nat. Belg.* **21**: 149-165 (1951).

Trees or shrubs. Leaves paripinnate. Inflorescence axillary, racemoid. Flowers dioecious, regular. Sepals 5, connate at base. Petals 5, pouch-like. Disc annular, lobed. Stamens 8-10. Ovary 3-locular; locules with 1 basal ovule; style shortly 3-lobed. Fruit a 3-lobed, narrowly winged capsule. Seed with basal part surrounded by an arillode.

Six to seven species in tropical Africa.

B. unijugata *Bak.* (1868).

Tree to 20(-30) m; bole slightly fluted; bark smooth, greyish to brownish; branchlets brownish pubescent or strigose. Leaves 3-9 cm long; petiole 4-10(-25) mm, pubescent; leaflets 2-8, ovate to elliptic or obovate, 5-18 x 1.7-8 cm, sparsely strigose on nerves, entire, with tufts of hairs; apex acuminate to acute; base cuneate. Inflorescences 2-8.5 cm long, sometimes clustered towards branch ends forming terminal panicles, pubescent; pedicels 2-4(-10 in fruit) mm. Sepals triangular, *c* 1.5 mm long. Petals white, ovate to elliptic, *c* 2.5 x 2 mm. Filaments *c* 3 mm, pilose. Ovary *c* 2 mm long; style *c* 1 mm. Fruit ellipsoid, 1.5-2 x 1-1.5 cm, with a 2-5 mm long beak, wings 1-2 mm wide; bright red. Seed ellipsoid, *c* 10 x 5 mm, black. Fig. 125.14.

Evergreen lowland and upland forest, shade tree in coffee plantations; 500-1700 m. EE EW IL KF GG; widespread in tropical Africa. *Chaffey* 372; *Friis* et al. 1975B; *Meyer* 8951.

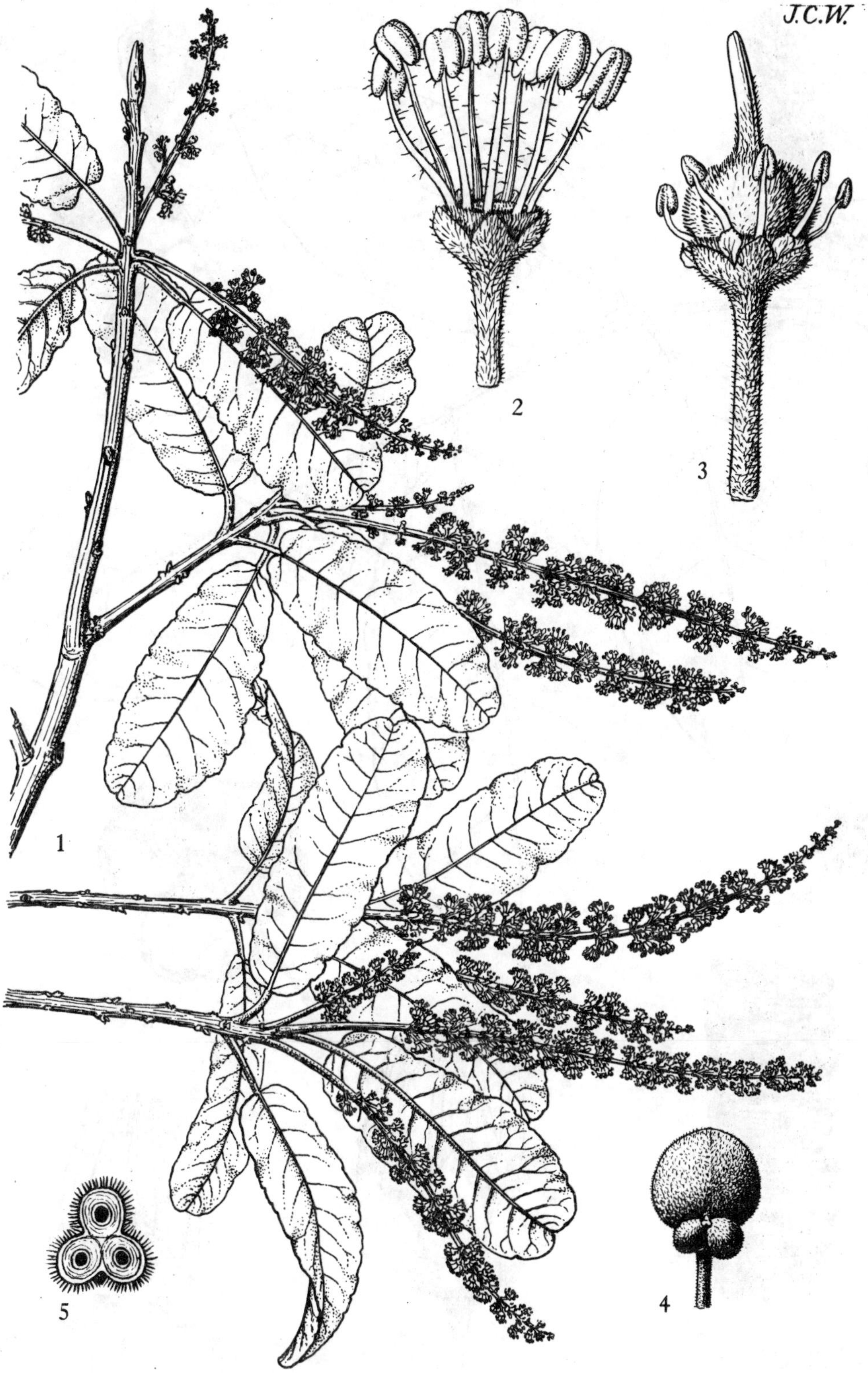

Figure 125.13 *PAPPEA CAPENSIS*: 1 - flowering branch x 2/3; 2 - male flower x 8; 3 - female flower x 8; 4 - fruit x 2. 1 & 2 from *Chase* 633; 3 from *Fanshawe* 5724; 4 from *Brenan* 7764. (Reproduced with permission from *Flora Zambesiaca* 2, tab. 116, 1966.)

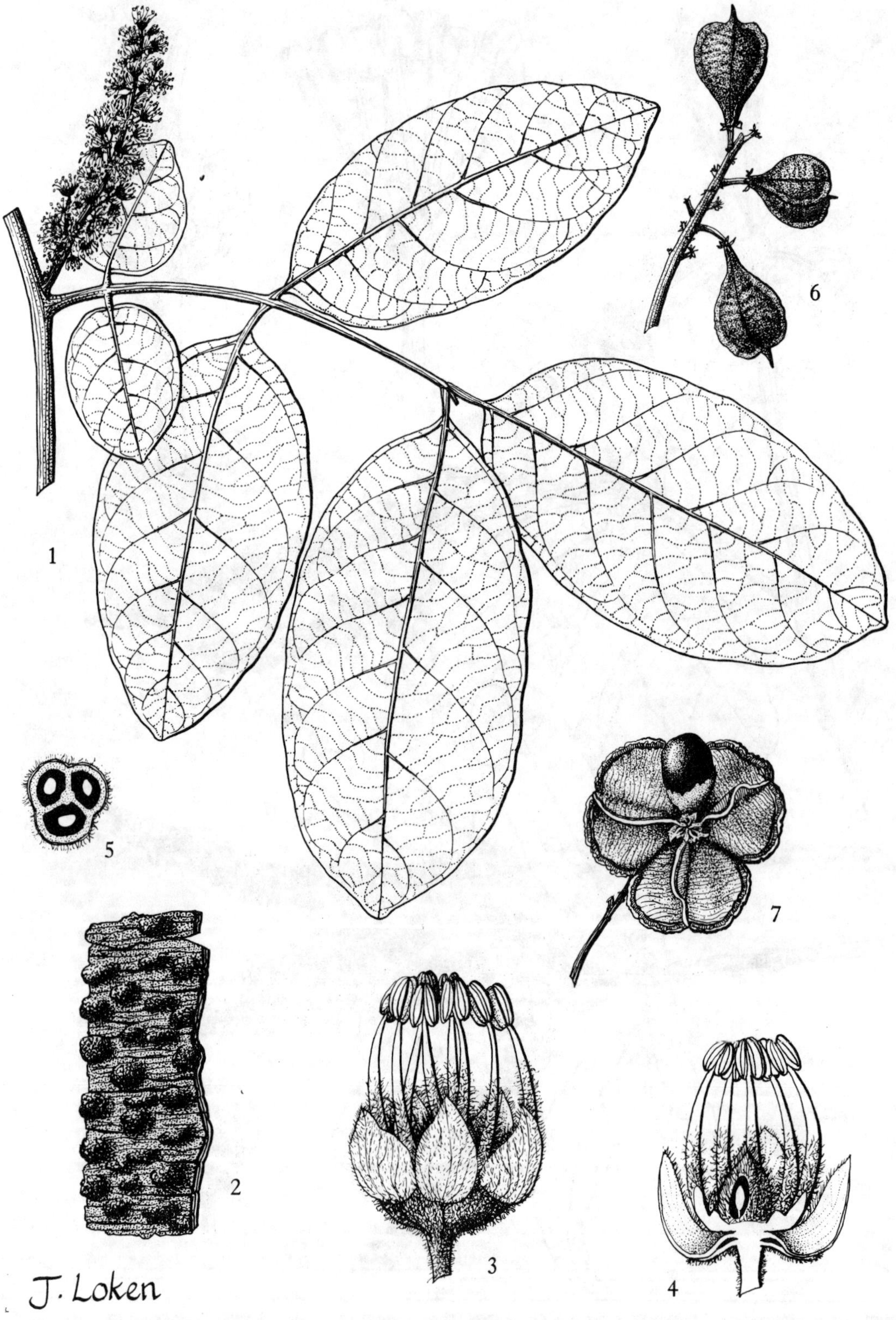

Figure 125.14 *BLIGHIA UNIJUGATA*: 1 - leaf and inflorescence x 2/3; 2 - bark on trunk x 2/3; 3 - male flower x 7; 4 - male flower in section x 7 ; 5 - transverse section of ovary x 7; 6 - infructescence x 2/3; 7 - open fruit with seed x 2/3. 1 & 3-5 from *Barbosa* s.n.; 2 from *Chase* 7400; 6 from *Torre* 3485; 7 from *Torre* 3901. (Reproduced with permission from *Flora Zambesiaca* 2, tab. 104, 1966.)

126. MELIANTHACEAE

by B. Verdcourt*

Cufodontis, *Enum.*: 495-496 (1958); Verdcourt, *Fl. Trop. E. Afr.*, **Melianthaceae** : 7 pp. (1958).

Shrubs or trees. Leaves alternate, pinnate; stipules present and often large. Flowers hermaphrodite or often functionally unisexual, ± irregular, in conspicuous racemes. Calyx-segments 4-5, somewhat unequal, imbricate. Petals 5, free, unequal, clawed. Disc pentagonal-annular or variously unilateral. Stamens 4-6, free or somewhat connate. Ovary superior, 4-5-locular. Style simple; stigma small or capitate. Ovules axile, 1-4 per loculus. Fruit a papery or woody capsule. Seeds large with copious endosperm.

Two genera, one in Africa and the other in S Africa. The genus *Melianthus* is sometimes cultivated as an ornamental.

BERSAMA *Fresen.* (1837)

Shrubs or trees. Leaves pinnate; rhachis narrowly to broadly winged or terete; leaflets entire to coarsely dentate, acuminate. Stipules large. Inflorescence a short to long, dense raceme. Flowers slightly irregular, often functionally unisexual, white, yellowish or purple-pink, usually scented. Sepals 4-5, the two anterior ones more or less connate or completely fused. Petals 5, free, narrowly oblong. Stamens 4-6; filaments densely hirsute and flattened below. Ovary 4-5-locular, narrowly ovoid-conic, densely hirsute, narrowed into a rather short thick style; stigma capitate or small and no wider than apex of style; ovules one in each locule. Fruit a woody capsule, 4-5-lobed and seeded, smooth to very densely echinate. Seeds with a conspicuous waxy aril.

This is the only genus in tropical Africa (the other being S. African) and the quite remarkable variation it exhibits has already been discussed in *Kew Bull.* 5: 233-244 (1950); there are about eight species in tropical and South Africa.

B. abyssinica *Fresen.* (1837)
- type: TU-GD, Asamer, *Ruppell* (FR syn., BM fragment)

Shrub or small to medium-sized tree 1.5-12(-18) m. tall, with smooth to rather rough, grey, brown or mottled bark. Leaves up to 1 m long, imparipinnate; rhachis wingless to very widely winged; leaflets up to 10(-12)-jugate, lanceolate to oblong or ovate-oblong, glabrous to hairy, often densely so beneath, 3.5-22 cm long and 1-8 cm wide, margins quite entire to sharply and conspicuously serrate. Flowers sweetly scented, creamy-white, yellowish or pink-tinged, especially in bud, in erect or rather drooping, glabrescent to hirsute racemes up to 30 cm long. Disk V-shaped to annular-pentagonal. Capsules reddish pink; seeds bright red with yellow-orange aril.

Subsp. **abyssinica**;

B. integrifolia A. Rich. (1847) based on *B. abyssinica* nom. illegit.

B. serrata A. Rich. (1847) - type: TU, Wojerat, *Petit* (P holo.).

B. gallensis V. Brehmer (1917) - types: AR, Arussi Galla, Djafa, *Ellenbeck* 1433; SD, Sidamo, *Ellenbeck* 1756 & 1727; SU, Mt. Mandagasha (Menegasha), *Ellenbeck* 1629 (all B syn. destr.).

B. giuliarrellii Chiov. (1940) - type: WG, Umbi Forest near Dembidollo, *Giuliarelli* 550 (FT holo.).

Leaf-rhachis nearly always wingless or rarely uppermost interval winged; leaflets mostly glabrous and often entire, *c* 5-15 x 2-6 cm. Stamens 5, connate at base. Disc entirely annular and 5-angled, 4-lobed, or V-shaped. Fruits small, 1-2 cm long, often reddish and grooved.

Essentially in *Juniperus-Podocarpus* forest and degraded remnants of it, often now occurring in thickets and copses in grassland and open woodland and even cultivations, often on slopes and hills, also in gallery forest and montane scrub of *Gnidia*, *Brucea*, *Discopodium*, *Euphorbia obovalifolia*, *Hagenia*, *Maesa* and *Ekebergia*; 1700-2715 m. EW TU GD WU WG GJ SU IL KF AR HA BA SD; East Africa to Malawi and Zambia. *Burger* 1496; *M.G. & S.B. Gilbert* 1771; *W. de Wilde* 6548.

Work in East Africa has shown the foliage to be poisonous to stock. In Ethiopia it is used in local medicine.

The author's original use of the shape of the disc for distinguishing taxa within *B. abyssinica* (Verdcourt, 1950) has proved wrong since V-shaped discs occur quite frequently in the Ethiopian material now available. Outside Ethiopia this is often correlated with other characters and can be used for delimiting subspecies and varieties but only subsp. *abyssinica* has been seen from Ethiopia. *Schimper* 912 is a rare variant with the two apical rhachis intervals winged.

Captain Pullen in 1863 in a long field note in German on his number 563 states that "the leaves are an almost safe remedy against *Ascaris* (vern. name 'Gondera'); the dose is stated to be young leaves from 7 branches of the size represented here (i.e. a herbarium sheet), crushed and taken in milk. I am indebted to Dr. E. Launert for translating the note. No other reference to this has been found.

* Royal Botanic Gardens, Kew, Richmond, Surrey, TW9 3AB, UK.

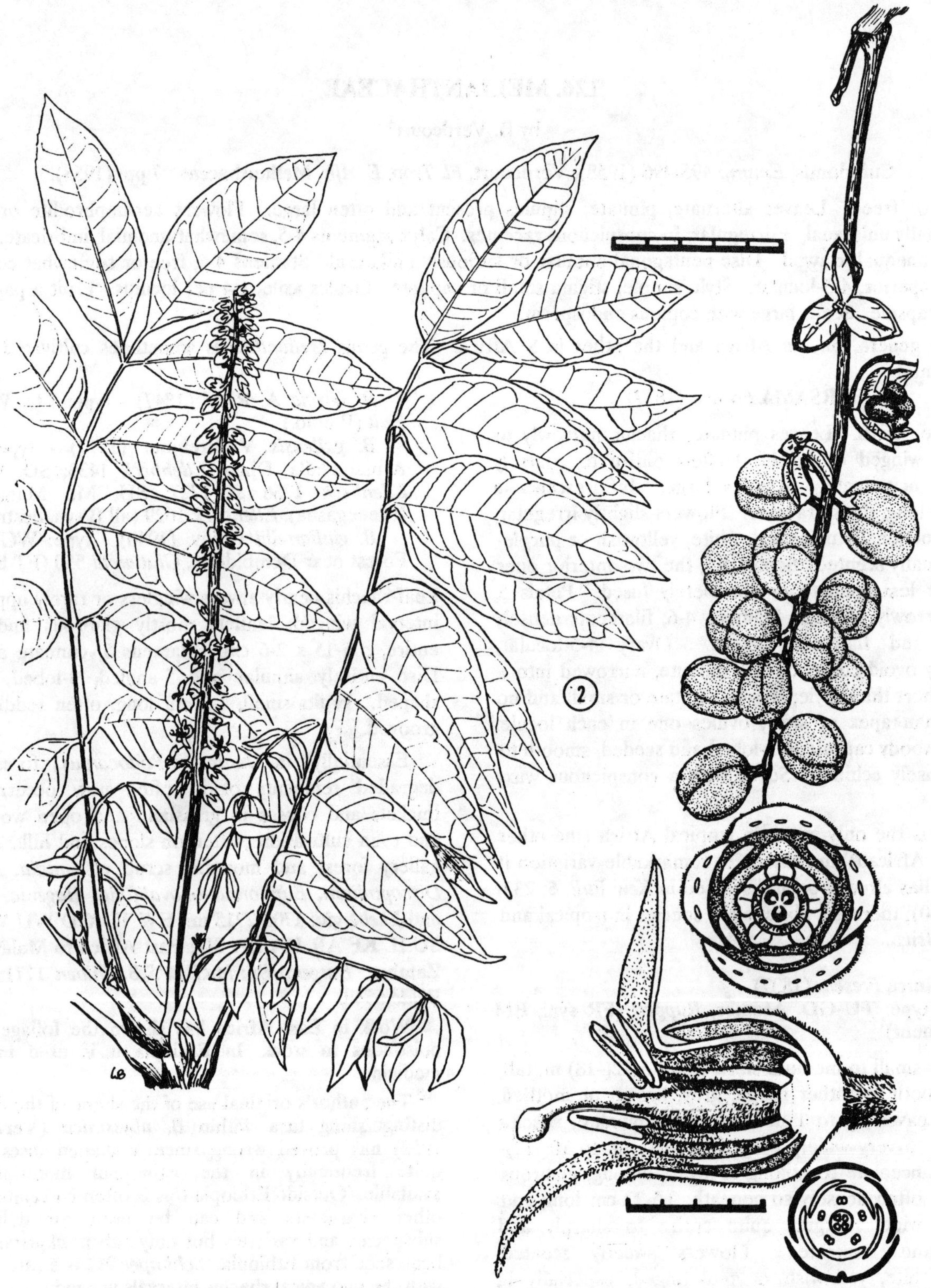

Figure 126.1 ***BERSAMA ABYSSINICA*****: portion of stem with leaves and inflorescence; half flower (pistil not sectioned) and cross-section of flower, mm scale shown; floral diagram. (Reproduced with permission of Oklahoma Agricultural Experiment Station from Experiment Station Bulletin No. 45, *Families of Flowering Plants in Ethiopia*, figure 35, 1967.) Drawn by W.C. Burger.**

127. ANACARDIACEAE

(including *Pistaciaceae*)

by M. G. Gilbert*

Cufodontis, *Enum.*: 466-476 (1958); Engler in A. & C. DC., *Mon. Phan.* **4**: 171-500 (1883); Kokwaro, *Fl. Trop. E. Afr.*, **Anacardiaceae**: (1986); Kokwaro & Gillett, *Kew Bull.* **34**: 745-760 (1980).

Trees, shrubs (rarely scandent) or suffrutices, producing gums, resins or latex, often sweetly aromatic when cut. Leaves mostly alternate, occasionally opposite or whorled; stipules absent or very obscure. Inflorescences essentially racemose, often much branched, sometimes spicate, flowers often in clusters. Flowers dioecious, monoecious or polygamous, small, usually regular, (3-)4-5(-7)-merous, rarely perianth absent. Sepals usually united at base, imbricate. Petals usually free, imbricate. Stamens as many or twice as many as petals, rarely more, usually inserted on or around annular disc. Ovary 1-4(-5)-locular; styles 1-4(-5), free or connate, stigmas terminal or lateral, small; ovules 1 per locule. Fruit a drupe, mesocarp fleshy, oily or resinous, endocarp woody or leathery, often 1-seeded by abortion, sometimes with 1-3(-5) subterminal opercules or very strongly compressed.

About 60 genera and some 600 species, mostly in the Tropics but also extending into the Mediterranean, East Asia and the U.S.A.; 5 genera and 23 species indigenous to the Flora area plus a further 2 genera, each represented by a single widely grown introduced species.

Many species have edible fruits, some of which are economically important, others are exploited for their gums, whilst the strong phloem fibres of some wild species are often used as tying materials.

Pistacia is sometimes placed in a separate family on the basis of its anomalous flowers and pollen but it is otherwise typical of the family and has not been separated for this account.

The leaves of members of this family are very variable, especially in size and the development of the tertiary venation which can become strikingly more prominent in old leaves.

Key to Genera

1. Peduncle and usually most parts of plant with stellate hairs, sometimes concealed by longer simple hairs, rarely glabrous; flower 4-merous (very rarely 5-merous); fruit an oblong-ovoid thinly-fleshed drupe with (3-)4 conspicuous persistent style bases, up to 15 mm long. **1. Lannea**
- Hairs, when present, simple; flowers 5-merous or naked with irregular sepaloid whorl of bracteoles; drupe mostly as broad or broader than long or, if ovoid, over 30 mm long with single terminal style or 2-3 lateral styles. 2

2. Leaves simple. 3
- Leaves 3-40-foliolate. 5

3. Male flower with 5 equal fertile stamens; fruit a subglobose or reniform drupe up to 11 mm across. 4
- Male flower with 1(-2) fertile stamen and (3-)4 very short staminodes; fruit a large edible drupe at least 8 cm long ('mango'). **3. Mangifera**

4. Leaves with close set parallel veins, often opposite or whorled, densely hairy at least below; fruit black with oily mesocarp. **4. Ozoroa**
- Leaves laxly veined, alternate, glabrous or nearly so; fruit pale with resinous mesocarp. **5. Rhus glutinosa**

5. Leaves strictly 3-foliolate (very rarely palmately 5-foliolate). **5. Rhus**
- Leaves pinnately 5-40-foliolate. 6

6. Male flower with 3-10 stamens; female flowers in racemes or panicles, very numerous; fruit a small ± globose or reniform drupe under 1 cm long or broad; stone strongly compressed and without opercules. 7
- Male flowers with 10-16(-26) stamens; female flowers solitary (rarely 2-3 together), fruit a fleshy ovoid drupe over 3 cm long; stone ovoid with 2-3 subterminal opercules (see Fig. 127.3. 6-7). **2. Sclerocarya**

7. Inflorescences terminal and axillary; flowers with normal 5-merous calyx and corolla; fertile stamens with filaments longer than anthers. **6. Schinus**
- Inflorescences axillary only; flowers naked but surrounded by irregular sepaloid whorl of 2-7 bracteoles; stamens with anthers longer than the very short filaments. **7. Pistacia**

* Ethiopian Flora Project, c/o Royal Botanic Gardens, Kew, Richmond, Surrey, TW9 3AB, UK.

1. LANNEA *A. Rich.* (1831) *nom. cons.*

Odina Roxb. (1832)

Lanneoma Del. (1843)

Trees or shrubs (or suffrutices); stellate hairs present at least on inflorescence, sometimes mixed with simple hairs, often glabrescent. Leaves usually deciduous, alternate, simple, 3-foliolate or pinnately 5-21-foliolate, very rarely palmately 5-foliolate; leaflets opposite to subopposite, entire. Inflorescences subterminal, sometimes produced before leaves, dioecious or andro-dioecious, bracts subtending solitary female or clusters of male flowers. Flowers 4(-5)-merous, stamens or staminodes twice as many as petals, disc cupular, crenulate. Male: stamens well developed, pistil vestigial. Female: staminodes shorter than ovary, usually with anthers, ovary ovoid, 4-locular with 4 subterminal styles. Fruit a thinly fleshed subcylindrical to oblong-ovoid drupe with prominent style bases; endocarp woody, 1-2-locular by abortion and with 1-2 subterminal opercules.

About 40 species in Tropical Africa and Socotra with 3 species extending into Tropical Arabia and 1 from India to S.E. Asia and China.

1. Leaves absent or extremely immature at anthesis. 2
- Leaves well developed at anthesis, number of leaflets easily seen. 4
2. Large tree with rounded crown, often 5 m or more high; inflorescences usually over 3 cm long at anthesis; ovary glabrous; leaves often quite absent. 3
- Shrubs or small trees, often less than 3 m high; inflorescence up to 3 cm long; ovary tomentose; immature leaves usually present. 6
3. Indumentum distinctly floccose, sometimes sparse; petals uniformly coloured except for pale margins; anthers always yellow. **5. L. schimperi**
- Indumentum densely pilose; petals with distinct dark veins; anthers sometimes drying reddish. **6. L. barteri**
4. Leaves 1-3-foliolate, densely hairy at least on underside. 5
- Leaves (3-)5-17-foliolate, if 3-foliolate then glabrous except for petiole and axils of nerves. 7
5. Leaflets cordate to truncate at base; inflorescences 25-55 mm long; calyx and fruit glabrous. **3. L. malifolia**
- Leaflets cuneate at base; inflorescences up to 30 mm long; calyx and fruit stellately hairy. 6
6. Twigs slender, often whip-like and drooping; leaves 3-foliolate very rarely 1-foliolate or palmately 5-foliolate; fruits 6-9 mm long. **1. L. triphylla**
- Twigs stout; leaves 1(-3)-foliolate; fruits 10-14 mm long. **2. L. rivae**
7. Young shoots densely and persistently floccose, pilose or tomentose. 8
- Young shoots glabrous or very sparsely stellate hairy. 10
8. Leaflets up to 3.5 cm long, strongly discoloured with dark green uppersides and densely white-tomentose undersides; fruit stellately hairy. **4. L. humilis**
- Leaflets 5-16 cm long, undersides pinkish or brownish hairy; fruit glabrous. 9
9. Undersides of leaves with dense pinkish floccose indumentum of crisped hairs, sometimes almost glabrous. **5. L. schimperi**
- Undersides of leaves with stiff straight hairs on nerves and margins. **6. L. barteri**
10. Leaves 3-9-foliolate; leaflets elliptic-lanceolate, ovate or orbicular, straight. 11
- Leaves 11-17-foliolate; leaflets lanceolate, often somewhat curved. **7. L. fruticosa**
11. Leaflets elliptic lanceolate to ovate, tertiary veins glabrous, prominent only in old leaves. 12
- Leaflets orbicular or obovate, tertiary veins prominent, reticulate, hairy. **3. L. malifolia**
12. Tertiary veins impressed; leaves glabrous except sometimes for tufts of hairs in axils of veins; female inflorescences lax, mostly over 5 cm long. 13
- Tertiary veins eventually prominent; leaves often uniformly hairy; female inflorescence dense, up to *c* 2 cm long. **5. L. schimperi**
13. Woodland tree with dark fissured bark; leaflets less than 10 cm long; inflorescences unbranched or rarely with 1-2 short branches near base. **8. L. schweinfurthii**
- Forest tree with smooth grey bark; leaflets 10-17 cm long; inflorescences with many well-developed branches. **9. L. welwitschii**

1. L. triphylla *(A. Rich.) Engl.* (1897);
Odina triphylla A. Rich. (1847) - lectotype (chosen here) TU, Djeladjeranne ['Tchélatché-kanné'], *Schimper* II:891 (P lecto. K isolecto.).
Lanneoma velutina Del. (1843), non *Lannea velutina* A. Rich. (1831) nec *Odina velutina* (A. Rich.) Walp. (1842) - type: TU, Dscheladscheranne ['Tchelatchekanne'], *Ferret & Galinier* (MPU holo. not seen)

Shrub or small tree 2-5(?-9) m high; bark rough, dark; twigs slender and often whip-like. Leaves mostly on short shoots, 3-foliolate, very rarely 1- or palmately 5-foliolate; leaflets obovate, terminal: (1-)2.7-5.4(-6.5) x (1-)2-3(-7) cm, base cuneate to almost truncate, tip obtuse or slightly retuse; lateral: 1.2-2.2 x 1.1-1.8 cm; underside densely floccose-tomentose, tertiary veins prominent but concealed by hairs, upperside more sparsely stellate hairy, often glabrescent. Inflorescences produced with new leaves on short lateral shoots, up to 3 cm long. Calyx 1-1.5 mm long, tomentose; petals 2.5-3(-3.5) x 1.5-2 mm, greenish yellow. Female with ovary tomentose. Drupe 6-9(-10) x 5-8 mm, stellate tomentose. Fig. 127.1.1.

Acacia-Commiphora bushland on well drained sandy soils or gentle rocky slopes; 300-1350 (-1500) m. EE EW TU GG SD BA HA; Somalia, Kenya, Uganda, Arabia (Dofar). *Ash* 1604; *Friis* et al. 3732; *Gilbert* et al. 7493.

Some collections from the Ogaden (*Gillett* 4183, *Popov* 1129) have distinctly less hairy leaves with more obscure lateral veins. *Ellis* 269 from the same area is similar but has the leaves simple or with very reduced lateral leaflets thus approaching *L. cotoneaster* Chiov. from Somalia. More collections are needed from the Ogaden and Somalia before the status of these and another poorly known Somalian endemic, *L. somalensis* (Chiov.) Sacl. - also closely related, can be reliably assessed.

Richard, in the protologue of *Odina triphylla* cites 2 elements, *Schimper* II: 891 and *Lanneoma velutina*, based on a *Ferret & Galinier* collection from the same locality. It seems preferable to treat *Odina triphylla* as a

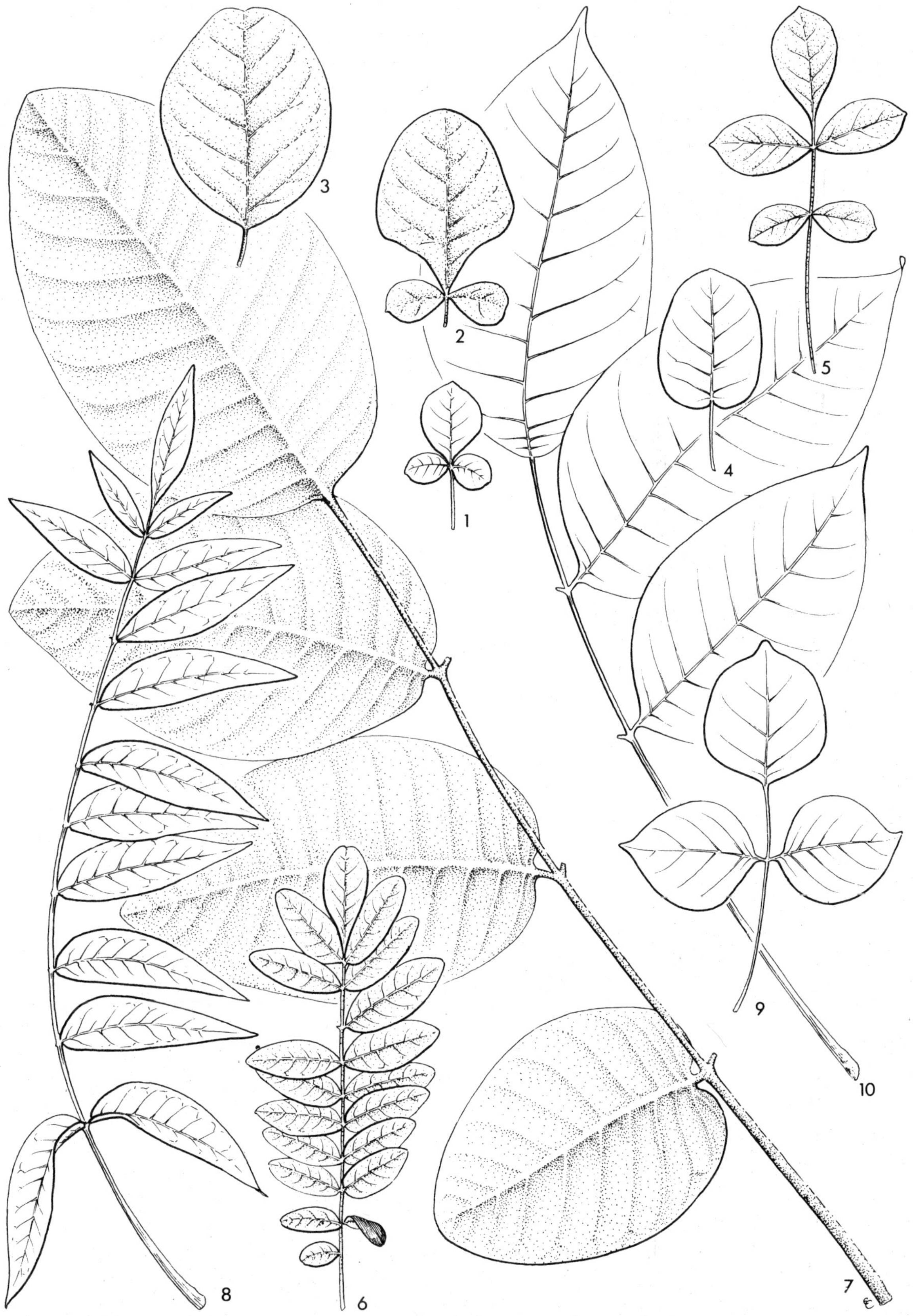

Figure 127.1 Leaves of *LANNEA* spp. all x 1/2: 1 - *L. TRIPHYLLA*. 2 & 3 - *L. RIVAE*. 4 - *L. MALIFOLIA* (1-foliolate form); 5 - ibid. (multifoliolate form). 6 - *L. HUMILIS*. 7 - *L. BARTERI*. 8 - *L. FRUTICOSA*. 9 - *L. SCHWEINFURTHII* var. *SCHWEINFURTHII*. 10 - *L. WELWITSCHII*. 1 from *Ash* 1604; 2,3 from *Gilbert* et al. 7850; 4 from *Burger* 2302; 5 from *Gilbert* et al. 7419; 6 from *Myer* 7577 (Sudan); 7 from *Gilbert* & *Ermias* 8350; 8 from *Gilbert* & *Ermias* 8345; 9 from *Gilbert* et al. 8003; 10 from *Friis* et al. 4111. Drawn by Eleanor Catherine.

sp. nov. and choose the more widely distributed collection by Schimper as lectotype rather than treat *O. triphylla* as a nom. nov. and have to take the *Ferret & Galinier* collection as type. No one seems to have made a conscious decision on this point and thus lectotypification does seem to be needed.

2. L. **rivae** *(Chiov.) Sacl.* (1934);
Odina rivae Chiov. (1929) - type: SD, between Dadecc Urtecc & Bidduma, *Ruspoli & Riva* '1354 (1225) 1229' (FT holo.).
Lannea floccosa Sacl. (1934).
L. cufodontii Chiov. (1939) - type: SD, 10km S of Negele ['Neghelli'], *Cufodontis* 162 (FT holo.).

Closely related to *L. triphylla*, differing by the stouter twigs, leaves mostly 1-foliolate, 3-foliolate only on vigorous growth; terminal leaflet (4-)7-8(-10) x (2.5-)5-6.7(-8) cm, lateral leaflets when present usually much smaller, up to 4 x 3 cm; female inflorescence very short, almost fasciculate, petals (3-)3.5-5 x 1.5-2 mm; drupe 10-14 x 7-11 mm. Fig. 127.1.2-3.

Most characteristically codominant with *Commiphora africana* on gentle rocky slopes, extending up to margins of *Juniperus* forest. 1350-1900m. GG SD BA; Kenya, Tanzania. *Friis* et al. 3355; *Gilbert* et al. 7780; *Thulin* et al. 3528.

Leaflet number in this species depends on growing conditions with high rainfall producing a larger proportion of 3-foliolate leaves. The type of *L. cufodontii* is one such plant.

3. L. **malifolia** *(Chiov.) Sacl.* (1934);
Odina malifolia Chiov. (1929) - type: Somalia, *Stefanini & Puccioni* 824.

Tree 2.5-6(-10) m high; bark smooth, grey; twigs soon glabrescent. Leaves mostly on short shoots, 1- or (1-)3-7-foliolate, petiole up to 4 cm, rachis up to 8 cm; leaflets obovate-elliptic to suborbicular, 1.5-8 x 1-5.5 cm, finely stellate puberulent, often rather brownish when dried, base cuneate in multifoliolate leaves to cordate in 1-foliolate leaves, tip truncate to retuse, underside prominently reticulate veined. Male flowers not seen. Female in dense racemes to 5.5 cm long; petals 2-2.5 x 1-1.2 mm, yellow; ovary orange, glabrous. Drupe 7-9 x 5-8 mm, glabrous. Fig. 127.1.4-5.

Acacia-Commiphora woodland, often on limestone. 1-foliolate form: 1000 m. HA; Somalia. *Burger* 2302 (10 km E of Fich, 8°08´N 42°17´E). Multifoliolate form: 1000-1300 m. SD BA HA; Somalia, N Kenya. *Burger* 3355; *Gilbert* et al. 7419; *Thulin* et al. 3620.

There are 2 distinct groups of specimens. *L. malifolia* sens. str. appears to be consistently 1-foliolate though there is no information on variation within any population. Populations seen in Sidamo and Bale are all (1-)3-7-foliolate, the occasional 1-foliolate leaf being restricted to the bases of new shoots. Otherwise the 2 groups seem indistinguishable.

4. L. **humilis** *(Oliv.) Engl.* (1897);
Odina humilis Oliv. (1868) - types: Nigeria, *Vogel* 78 & Sudan, *Kotschy* 378.

Tree 3-7.5 m high; bark smooth, soft and spongy, grey to almost black, twigs soon glabrescent. Leaves (3-)7-9(-21)-foliolate, leaflets occasionally subopposite; petiole up to 2.5 cm long; rachis up to 15 cm long; leaflets oblong-oblanceolate, up to 3.5(-6) x 1.8(-2.7) cm, underside densely grey-tomentose, prominently reticulately veined, upperside dark green with sparse stellate and minute simple hairs, base usually rounded, sometimes cuneate, tip rounded to slightly emarginate. Inflorescences produced with new leaves, clustered on short lateral branches, spike-like, 2-5 cm long, densely tomentose. Petals 2.5-4 x 1.5-2 mm, yellowish, ovary tomentose. Drupe 9-13 x 5-8 mm, grey-tomentose. Fig.127.1.6.

Deciduous woodland, typically commonest where drainage is impeded; 1200-1350 m. WU (GJ) SU; west to Senegal, south to Rwanda, Zambia and Zimbabwe. *Mercier* 2558; *Gilbert* et al. 8836.

This distinctive species is, according to J. Mercier, widespread within the Blue Nile Valley. It is also likely to be widespread along the border with Sudan.

5. L. **schimperi** *(A. Rich.) Engl.* (1897);
Odina schimperi A. Rich. (1847) - types: TU, Dschaladscheranne, *Schimper* II:657 & Schoata, *Schimper* II:1282 (both P syn. BM K isosyn.) & Shire ['Chiré'], *Quartin-Dillon & Petit* (P syn.).
O. schimperi var. *glabrescens* Engl., loc. cit.: 270 (1883) - type: TU, Amba Sea, *Schimper* (in Pullen) 566 ('366') (B holo. destr. K P iso.).
Lannea ruspolii Engl. (1898) - type: SD, Duaya, *Ruspoli & Riva* '690 (1571) 630' (as '630') (FT holo.).

Tree 2-9(-15) m high, wider than high, crown rounded; bark dark, smooth or rough; twigs stout, densely pinkish floccose-tomentose when young. Leaves 5-11 (-13)-foliolate; petiole + rachis 8-33 cm long; leaflets elliptic, ± oblong or ovate, (5-)7-10.5(-16) x 2.9-6.5(-8) cm, basal leaflets usually shorter and broader, ± bullate when young, prominently reticulate-veined below when old, usually at first densely pinkish tomentose, later usually rusty tomentose below, sparsely stellately hairy to subglabrous above, base cuneate to rounded (to sub-cordate), tip acuminate to obtuse. Inflorescences produced before leaves in apparently apical clusters, male up to 22 cm long, lax, female up to 8 cm long, dense. Flowers fragrant. Petals 3-5 x 1.2-2.5 mm, greenish to bright yellow, sometimes flushed pink, with paler margins, veins obscure, fragrant. Female with ovary glabrous. Drupe 8-10(-14) x 5-8(-9) x 4-5 mm, glabrous, wine-red. Fig. 127.2.

Figure 127.2 *LANNEA SCHIMPERI*: 1 - leafy shoot, 2 - male inflorescences x 2/3; 3 - male flower, 4 - female flower x 6; 5 - infructescence x 2/3; 6 - fruit x 2½. 1,4-6 from *Polhill & Paulo* 1003; 2, 3 from *Stolz* 11570. Drawn by Mary Millar Watt. (Reproduced with permission from *Fl. Trop. E. Afr.*, Fig. 2, 1976.)

Deciduous woodland on rocky slopes and outcrops, on volcanics, limestone and basement complex, or in lowland deciduous woodland with *Pterocarpus* etc; (800-)1050-1750(-2200)m. EE? EW TU GD GJ SU WG KF SD BA HA; west to Nigeria, south to Zaire, Zambia, Malawi & Mozambique. *Burger* 1684, 2613; *W. de Wilde* 10467.

A polymorphic species with some distinctive local variants. Gillett (*loc. cit.*: 750, 1980) suggests that var. *glabrescens* is worthy of recognition on the basis of its narrower acuminate leaflets compared to material from further south. The syntypes of the species are all without leaves. All other material from the area matches var. *glabrescens*. It thus seems highly probable that the type of var. *glabrescens* is simply *L. schimperi* s. str. Occasional collections approaching this variety occur ± throughout the range of the species and formal recognition of a broader-leafed variety does not seem worthwhile. *L. schimperi* var. *stolzii* (Engl. & von Brehm.) R. & A. Fern. from south of the Flora area, based on supposed indumentum differences, is also not worthy of recognition.

Collections from EE (Ghinda and Dongollo on the escarpment between Asmara and Mitisawa e.g. *Baldrati* 2498, *Fiori* 277) have paniculate female inflorescences and rather fewer, broader leaflets than other northern collections. These may be worth formal distinction.

Collections from HA, between Harer and Jigjiga, are consistently subglabrous but otherwise seem unworthy of formal recognition.

6. **L. barteri** *(Oliv.) Engl.* (1897);
Odina barteri Oliv. (1868) - type: Nigeria, *Barter* 1109 p.p. (lectotype, see Aubreville, *Fl. For. Soud.-Guin.*: 397, 1950).
L. kerstingii Engl. & Krause (1911).

Closely related to L. schimperi, differing mainly by the indumentum which is densely yellowish-brown velvety-pilose, not floccose, hairs mostly simple, usually quite persistent; leaflets usually larger up to 17 x 10.5 cm, inflorescence more lax, petals yellowish, usually with distinctly darker veins; anthers sometimes drying reddish; drupe larger, (10-)13 x 7(-8) x 4.5 mm, purplish red when mature. Fig. 127.1.7.

Local in lowland woodland on clay soils and along river valleys; 500-1600m. GD SU IL KF GG; west to Guinea, south to Zaire. *Chaffey* 933; *Friis* et al. 2572; *Sebsebe & Berhanu* 1470.

Formerly confused with *L. schimperi*, the type was a mixture of the two species, but readily recognised by the very different indumentum and veined petals.

7. **L. fruticosa** *(A.Rich.) Engl.* (1897);
Odina fruticosa A. Rich. (1847) - types: TU, Ferrokoba & Takaze Valley, *Schimper* II:707 (P syn. BM K isosyn.) & Mai Gouagoua ['Maigoigoi'], near Adowa, *Quartin-Dillon* (P syn.).

Tree (3-)5-8(-12) m high; bark deeply longitudinally fissured when old; twigs fairly stout, glabrescent. Leaves 11-17-foliolate, petiole + rachis 15-26 cm long; leaflets lanceolate or elliptic-lanceolate, (3-)7-8(-11) x (1-)2-4 cm, laterals slightly falcate, base cuneate to subcordate, tip acute to acuminate, veins prominently reticulate in old leaves, sparsely puberulent glabrescent with mixture of simple and stellate hairs, dying brownish. Inflorescences in axils of well developed leaves, 2-12 cm long, often 1-3 branched near base, densely puberulent. Flowers subsessile. Petals 3-4 x 1.7-2 mm, yellow. Ovary glabrous. Drupe 7-9 x 4-7 mm, glabrous. Fig. 127.1.8.

Broadleaved deciduous woodland, extending up river valleys, localy common; 600-2100m. EW TU GD WU SU KF GG SD; Nigeria east to Yemen, south to Zaire. *Mercier* 2323; *Puff* et al 810916-1/2; *Tekle Hagos* 107.

8. **L. schweinfurthii** *(Engl.) Engl.* (1897);
Odina schweinfurthii Engl. (1883) - type: Sudan, *Schweinfurth* 1059 (B holo. destr. K iso.).

(Shrub or) tree 3-15(-22) m high with broad crown; bark dark, reticulately or longitudinally fissured and flaking off; twigs slender, often drooping, glabrous. Leaves (1-)3-7(-13)-foliolate; rachis 5-16(-22) cm long,very sparsely hairy; leaflets lanceolate, elliptic or ovate, (3-)5.5-8(-11) x (1.7-)3.5- 6.3 cm, base cuneate, sometimes truncate in terminal leaflet, tip bluntly acuminate to rounded and retuse, tertiary veins impressed above and below. Inflorescences produced with leaves, 2-20(-40) cm long, sometimes with short lateral branches. Pedicels 0.5-4 mm long, with minute stellate hairs. Petals 2.5-4.5 x 1.2-2.5 mm, ± hooded, yellowish. Drupes 8-12 x 6-8 x c 4 mm, glabrous, orange or red when mature.

var. **schweinfurthii**;
L. stuhlmannii (Engl.) Engl. (1897); *L. schweinfurthii* (Engl.) Engl. var. *stuhlamannii* (Engl.) Kokwaro, *loc. cit.*: 751 (1980).

Leaves (1-)3-7-foliolate; leaflets lanceolate to ovate, glabrous except for tufts of white hairs in axils of main veins on underside and sometimes minutely puberulous veins. Fig. 127.1.9.

Near water courses in open *Acacia* woodland, open deciduous woodland or wooded grassland on stony soils or on termite mounds in *Pterocarpus* woodland; 600-1550 m. IL GG SD BA HA; (for species as a whole) Sudan, Somalia, south to Transvaal and west to Rwanda and Zambia. *Ash* 1590; *Burger* 2979, 3201; *Gilbert* et al. 8003.

Kokwaro (*loc. cit.*: 751, 1980) recognised 4 varieties within this species, separated by features of leaf shape and indumentum. Material from the Flora area is somewhat intermediate between his var. *schweinfurthii*, restricted to Sudan and Uganda, and the much more widespread and variable var. *stuhlmannii* (Engl.) Kokwaro, such that it does not seem advisable to use these varieties.

Var. *acutifoliolata* (Engl.) Kokwaro, with narrow leaflets and veins minutely hairy below is restricted to eastern Kenya and Tanzania; var. *tomentosa* (Dunkley) Kokwaro, with broad leaflets pubescent above and below, is recorded from all the countries of the Flora Zambesiaca area.

9. L. **welwitschii** *(Hiern) Engl.* (1898);
Odina welwitschii Hiern (1896) - types: Angola, *Welwitsch* 4542 & 4543.

Forest tree to 30 m, crown relatively small; bark smooth, pale grey; twigs stout, prominently lenticillate, minutely stellate-tomentose when young, soon becoming glabrous. Leaves 5-7-foliolate; rachis (10-) 17-30 cm long, glabrescent; leaflets with petiolules up to 13 mm long, lanceolate, 10-17 x 4.5-8 cm, slightly leathery, base cuneate, often unequal sided, tip bluntly acuminate, tertiary veins impressed on both sides. Inflorescences produced with or slightly before leaves, paniculate, to 15 cm long, branches to 5 cm, sparsely stellate-floccose. Pedicels to c 1.5 mm. Petals 2-2.5 x 1.2-1.5 mm, pale creamy yellow. Drupe 8 x 6 x 4 mm, glabrous.

var. **welwitschii**.

Leaves glabrous. Fig. 127.1.10.

Lowland *Celtis-Milicia-Aningeria altissima* forest; 1100-1250 m. IL KF; west to Ivory Coast, south to Uganda, Zaire and Angola. *Friis* et al. 4001, 4111; *Negussie Jiru* 27.

The record of this species from SD (Moyale) by Cufodontis, *loc. cit.*: 469, is based on a misidentification of a collection of *L. schweinfurthii*.

Var. *ciliolata* Engl. differs by having the leaves hairy at least on the veins. It is restricted to the coastal parts of Kenya and Tanzania.

2. SCLEROCARYA *Hochst.* (1844)

Small or medium trees. Leaves alternate, imparipinnate, mostly crowded at leaf-tips, deciduous; leaflets opposite or subopposite. Male inflorescences axillary, racemose, rarely branched; female inflorescences short, unbranched, with single terminal flower, rarely 2-3. Flowers 4-5-merous; male with 10-16(-26) stamens surrounding disc, pistillode absent; female slightly larger, staminodes present, sometimes with apparently sterile anthers, disc surrounding 2-3(-4)-locular ovary, styles 2-3(-4), lateral, stigma capitate. Fruit a large fleshy drupe with persistent, sometimes obscure, lateral style bases; endocarp woody with (1-)2-3(-4) subterminal, ovate opercules.

Genus with 2 species, the other a local endemic in Kenya.

S. birrea *(A.Rich.) Hochst.* (1844);
Spondias birrea A. Rich. (1832) - type: Senegal, *Perrotet* 186.

Spreading tree to 18 m high; trunk grey, reticulate and flaking; twigs stout, dark grey, rough. Leaves 7-37-foliolate, 12-38 cm long; leaflets obovate, oblong-elliptic or ovate (or orbicular or lanceolate), (1-)2-13(-17) x 1-3.5(-6) cm, glabrous, base cuneate or rounded, often abruptly attenuate into petiolule, tip obtuse to acuminate, usually mucronulate, margin entire or, especially in juvenile foliage, serrate, tertiary veins obscure. Male inflorescences 7-9(-22) cm long; bracts ovate, c 2 mm long. Male flowers: sepals 2 mm long, petals 4-6 x 3-4 mm, obtuse, yellow or dark red with pale margins, filaments c 3 mm long, anthers 1-1.5 mm. Female flowers similar but slightly larger, peduncle soon thickening after fertilisation. Drupe subglobose 3-3.5 cm diameter, yellow; endocarp obovoid, 2-3 cm long.

subsp. **birrea**.

Leaflets 9-23, obtuse to acuminate with petiolule or attenuate base less than 5 mm long; male infloresence up to 9 cm long. Fig. 127.3.

Open deciduous woodland on rocky slopes; 500-1700 m. EE EW TU SU IL GG SD; west to Senegal, south to Uganda and Tanzania. *Ash* 1599; *Friis* et al. 2482; *Gillett* 14868.

Subsp. *caffra* (Sond.) Kokwaro has fewer, larger, more distinctly acuminate leaflets with longer petiolules. Though recorded from Ethiopia (Kokwaro, *loc. cit.*: 757, 1980), all the material seen seems better included in subsp. *birrea*. Subsp. *caffra* is recorded from Kenya west to Zaire, south to Natal and Namibia and Madagascar. A third subspecies, subsp. *multifoliolata* (Engl.) Kokwaro, restricted to Tanzania, differs by having 25-37 leaflets up to 1.5 cm long.

3. MANGIFERA *L.* (1753)

M. indica *L.* (1753).

Dense crowned tree reaching 40 m, usually less. Leaves simple, oblong-elliptic, glabrous, evergreen. Inflorescence a terminal panicle; flowers bisexual or male, (4-) 5(-7)-merous. Petals cream becoming pink; 1(-2) stamens fertile, others reduced to short staminodes; ovary 1-locular with terminal style. Fruit a fleshy drupe, 8-30 cm long, green to red and/or yellow; endocarp strongly compressed, fibrous, woody.

Widely cultivated in warmer lowland areas for the valuable fruits. Originally from Asia but now found throughout the tropics. *IECAMA* D87; *Sue Edwards & Tewolde Berhan G-E.* 3653.

4. OZOROA *Del.* (1843).

Heeria Meisn. (1837) *pro parte, non sens. strict.*

Trees, shrubs or suffrutices with milky sap. Leaves alternate to whorled, simple, mostly leathery, lateral veins closely parallel, often undivided to margin. Dioecious. Inflorescence a terminal panicle. Flowers 5-merous. Pedicels articulated near tip. Calyx divided to near base, hairy. Petals white, hairy. Male flowers:

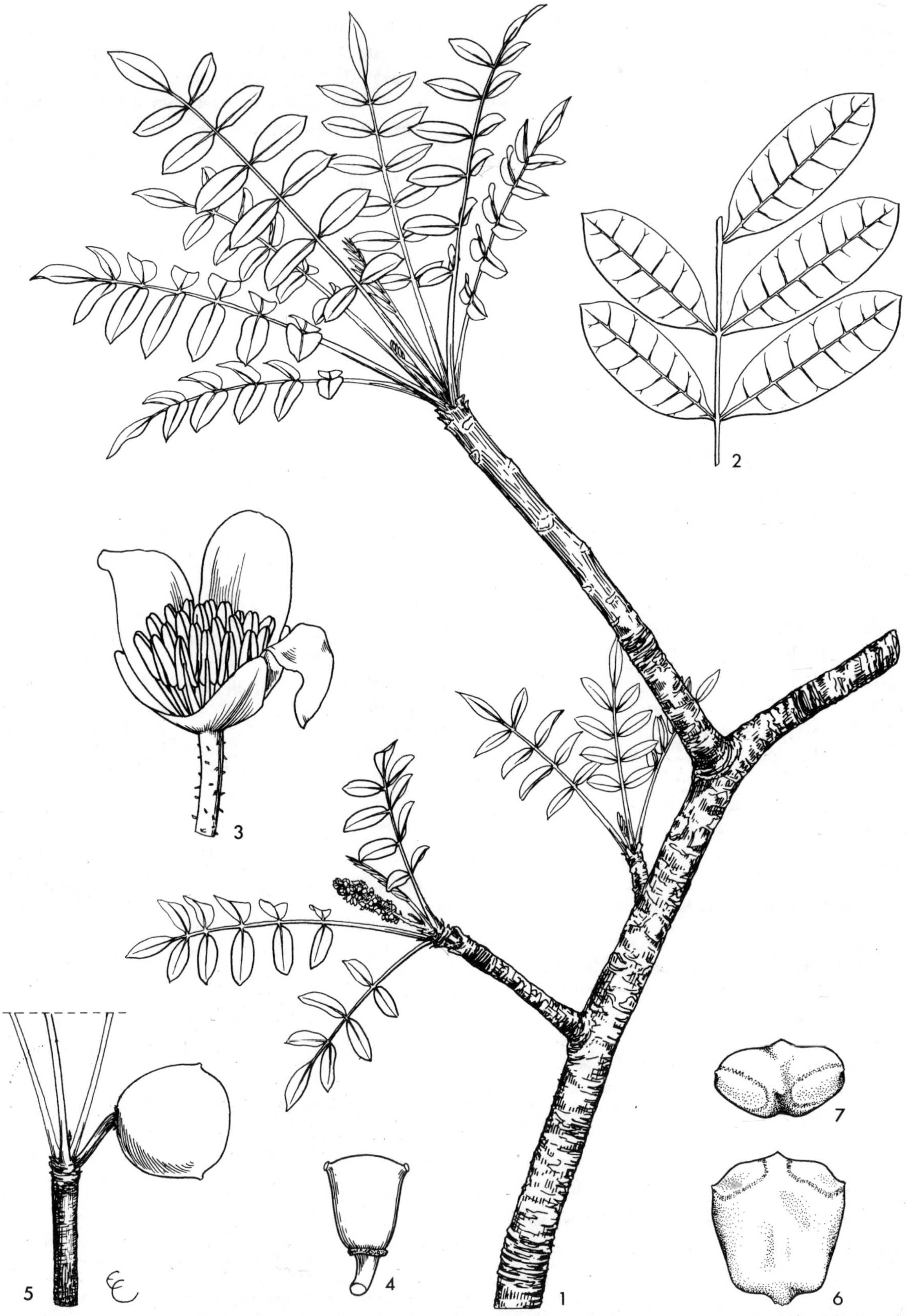

Figure 127.3 *SCLEROCARYA BIRREA* subsp. *BIRREA*: 1 - branch with leaves and young inflorescence, 2 - mature leaflets x 2/3; 3 - male flower, single petal removed x 8; 4 - young fruit x 2; 5 - mature fruit x 2/3; 6 & 7 - endocarp (side and top views). 1,3 from *Gillett* 14868; 2 from *Speke & Grant* s.n.; 4 from *Gillett & Stearn* 21635; 5-7 from Ash 1599. Drawn by Eleanor Catherine.

stamens 5, inserted outside cupular crenulate disc, pistillode vestigial. Female flowers: staminodes small, ovary globose, 1-locular, styles 3, terminal, free or joined at base, stigma capitate. Fruit a reniform (to globose) drupe, black and shiny when mature, wrinkled when dry, flesh with oil-filled spaces, endocarp compressed, woody.

About 40 often ill-defined species, mainly in southern Africa, extending to Senegal and Yemen.

At one time included within the genus *Heeria* but that genus is now restricted to a single species in South Africa which has ± dehiscent fruits. A closely related genus, *Baronia* Baker, occurs in Madagascar.

1. Trees or shrubs, rarely less than 3m high; petal tips flat. **1. O. insignis**
- Suffrutices from woody rootstocks, rarely more than 2m high; petal tips sharply inflexed. **2. O. pulcherrima**

1. **O. insignis** *Del.* (1843);
Heeria insignis (Del.) Kuntze (1891) - type: Ethiopia, sine loc., *Ferret & Galinier* (MPU holo.).

Tree or shrub (1.5-)3-9(-15) m high; bark fissured; shoots densely pale puberulous to yellowish villous. Leaves usually ± whorled; petiole (1-)1.5-3(-5) cm long; blade oblong lanceolate to oblong elliptic, (5.5-)8-18 (-27.5) x (1.5-)2-5(-8.5) cm base rounded or cuneate, tip rounded or acute with large mucro, glabrous to sparsely hairy above, very densely white-pubescent to villous, sometimes yellowish below. Panicles to 17 cm long, axes puberulous to villous; bracts 3-4(-10) mm long, subulate. Sepals 1.5-3.7 mm long, acute. Petals 2-4 x 1-2 mm, obtuse, flat. Drupes slightly compressed, 5.3-8 x 6.5-11 mm, at first red.

subsp. **insignis**.
Anaphrenium abyssinicum Hochst. (1844) - types: TU, Dscheladscheranne, *Schimper* II:706 & Ferrferam, *Schimper* II:1342 (both ?TUB syn. K isosyn.).

Leaves (5.5-)8-18.5 x (1-)1.3-4.3 cm, underside typically densely adpressed white pubescent, rarely yellowish villose (see note below), tertiary veins usually not very prominent and sometimes concealed by indumentum. Fig. 127.4.

Broad-leafed deciduous woodland, often with *Terminalia*, on well drained rocky slopes, sometimes in *Acacia* bushland or woodland overlying limestone. 1350-2000m. EW TU GD SU GG SD BA HA; west to Senegal, east to Somalia and Yemen, not extending south of Sudan and Ethiopia. *Burger* 1996; *Gilbert & Getachew* 2981; *Mooney* 8047.

The status of the plants growing in southern Ethiopia is most uncertain. Gillett (*Kew Bull* **34**: 746, 1980) recognised 3 subspecies within *O. insignis*, the other 2 being subsp. *latifolia* (Engl.) R. & A. Fern., restricted to West Africa, and subsp. *reticulata* (Bak. f.) Gillett which occurs from Kenya south to South Africa. Typical subsp. *reticulata* is very distinct from subsp. *insignis*, having much larger, more prominently reticulate leaves. Much of the material from southern Ethiopia does not fit satisfactorily in either and consists rather of a series of rather distinctive local populations often with characteristics of both subspecies intermingled or, in at least one case, forming a population that could well be regarded as a distinct taxon. This includes plants from around Negele Borana in Sidamo which have a distinctive dense yellow villose indumentum, e.g. *Friis* et al. 3139; *Mooney* 5619; *Thulin* et al. 3404. In view of the lack of any clear-cut simple discontinuity and the absence of material really typical of subsp. *reticulata* it seems best to recognise only the one segregate within this Flora, at least until the situation is studied in more detail in the field.

2. **O. pulcherrima** *(Schweinf.) R. & A. Fernandes* (1965);
Anaphrenium pulcherrimum Schweinf. (1867) - type: Sudan, *Cienkowsky*.

Suffrutex 0.5-2(-?6) m high with several rather succulent stems arising from a woody rootstock; shoots softly pubescent or yellowish villose. Leaves whorled; petiole 1-1.7 cm long; blade obovate-elliptic or broadly elliptic, (7-)9-18(-21) x 3.3-6(-12) cm, puberulous to glabrous above, underside softly tomentose with the indumentum tending to hide raised tertiary veins, tip obtuse to emarginate, sometimes mucronate. Panicle 2-11 cm long, densely pubescent or villous; bracts subulate, 2-9 mm long. Petals 2-2.5 x 0.8-1.2 mm, minutely obtuse, inflexed at tip. Drupe compressed to almost 2-lobed, 7-8 mm broad.

Broad-leafed deciduous woodland with *Terminalia*, *Combretum*, *Pterocarpus* etc. on rocky or sandy soils; 650-1400m. GJ WG IL; Guinea east to north Uganda. *Friis* et al. 2586; *Mooney* 6842; *Thulin & Hunde* 4046.

5. **RHUS** *L.* (1753)

Trees or shrubs, occasionally scandent (or suffruticose), often sweetly aromatic when cut. Leaves alternate, 3-foliolate, very rarely palmate or 1-foliolate, in tropical Africa (mostly pinnate elsewhere). Inflorescence a terminal or axillary panicle, usually dioecious. Flowers very small, usually 5-merous. Sepals imbricate. Petals usually less than 2 mm long, whitish, yellowish or green, oblong, obtuse. Male: stamens 5, filaments shorter than petals, disc usually conspicuous, pistillode usually absent. Female similar, staminodes about equalling annular disc, ovary globose, 1(-2)-locular, styles 3(-5), free, stigmas capitate. Fruit a globose to reniform drupe, skin often separating from resinous inner layers and becoming glossy and fragile, keeping shape when dried, less often adhering to inner layers and becoming dull and leathery, nearly always glabrous in Africa; endocarp strongly compressed, rarely 2-locular.

About 200 species in all tropical and warm temperate regions, very numerous in South Africa.

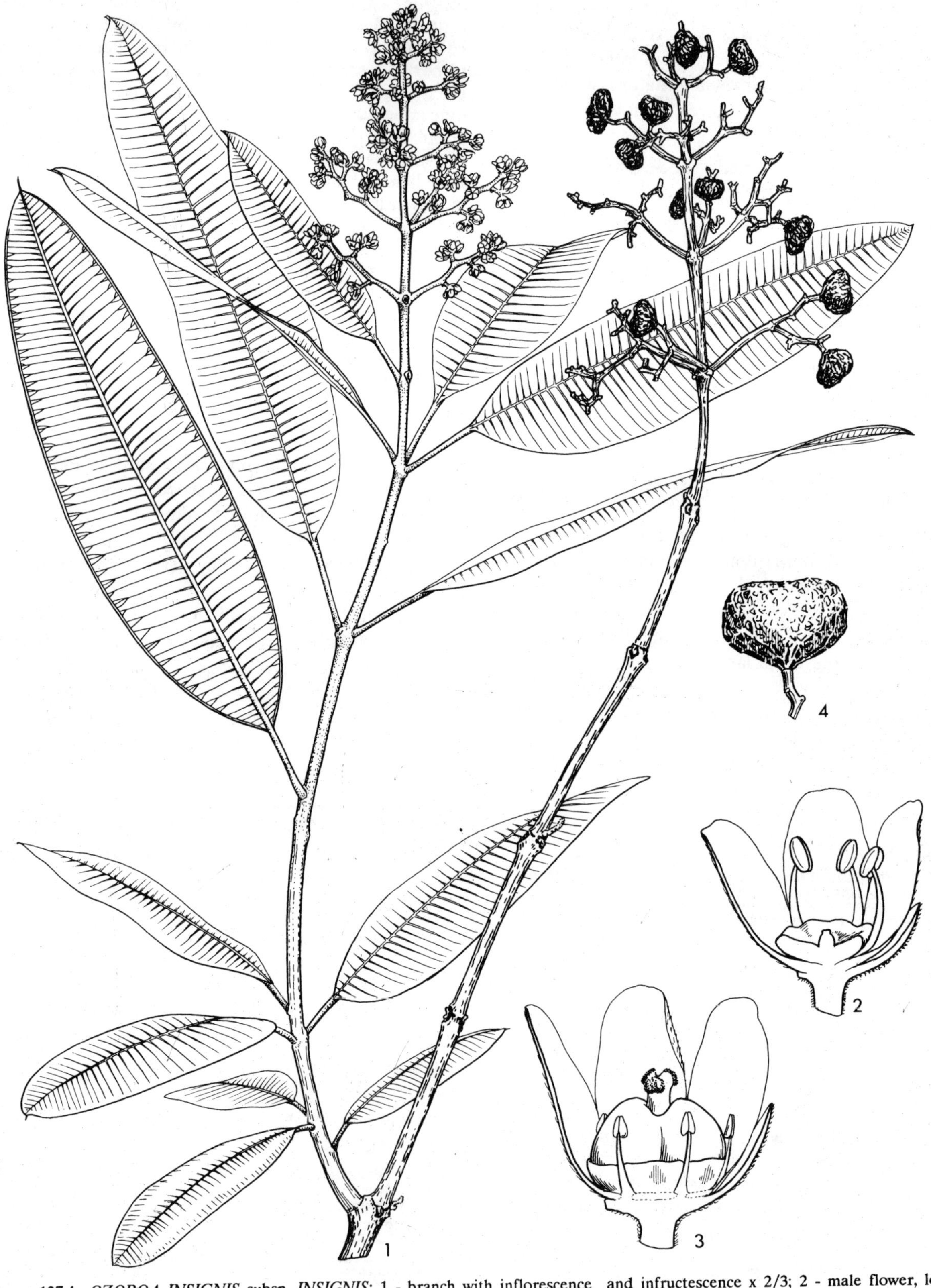

Figure 127.4 *OZOROA INSIGNIS* subsp. *INSIGNIS*: 1 - branch with inflorescence and infructescence x 2/3; 2 - male flower, longitudinal section, 3 - female flower, half calyx and corolla removed x 10; 4 - fruit x 2. 1,3,4 from *Gilbert* et al. 7774; 2 from *Gilbert & Getachew* 2981. Drawn by Eleanor Catherine.

All tropical African species belong to the very distinctive section *Gerontogeae* Engl. recognised by the normally 3-foliolate leaves, rarely with more leaflets and then always palmate in contrast to the pinnate leaves of other sections. They also seem to differ from *Rhus* sens. str. by not causing the skin rashes and other allergic reactions so characteristic of some North American species.

Habit and fruits are of considerable taxonomic value but are too incompletely recorded to be used yet, more carefull collecting with good field notes is needed. Leaf size is usually a reliable feature but most species will occasionally produce abnormally robust or stunted shoots and both leads in the key should be tried if there is cause for doubt. Ripe fruits in all species except *R. vulgaris* are recognised by their smooth, glossy skins and care should be taken to avoid those with wrinkled skins when taking measurements.

1. Leaflets up to 8 cm long. 2
- Leaflets over 8 cm long. 7
2. Leaflets, at least some, with a few, often obscure, marginal crenations, tip usually minutely retuse; fruit obliquely oblong-reniform, compressed, often pruinose (with white waxy coating), maximum dimension 5-7 mm. 3
- Leaflets entire with a few large subapical teeth, tip not retuse and/or leaflets hairy; fruit globose, sometimes becoming discoid when dried, not pruinose, to 4.5 mm diameter, very rarely larger and then 2-locular with more than 5 styles. 4
3. Stems soon becoming grey, often very pale, almost white; leaflets glabrous or nearly so, often more than 2 times longer than broad, shiny when fresh but drying a dull greyish green. **1. R. natalensis**
- Stems dark grey to brown or reddish; leaflets densely hairy, often less than 2 times longer than broad, dull when fresh and not drying greyish green. **2. R. tenuinervis**
4. Leaflets 1.1-2.4(-3) times longer than broad; tertiary veins prominent, at least on upper side or near margin. 5
- Leaflets 2.7-3.3(-5) times longer than broad; tertiary veins obscure, more or less impressed. **3. R. quartiniana**
5. Panicles with flowers evenly spaced and not forming discrete clusters; mature fruit with skin smooth and fragile, retaining shape and becoming pale when dried. 6
- Panicles, at least the male ones, with flowers clustered into separated globose groups; fruit globose and juicy when fresh, very strongly compressed, dull reddish brown when dried. **11. R. vulgaris**
6. Young shoots glutinous, shiny; leaflets rather leathery, acuminate, upper surface with tertiary veins prominent; fruit always 1-locular, 4.5-5 mm diameter. **4. R. sp. = *Friis* et al. 1147**
- Young shoots not glutinous; leaflets thin textured, obtuse, upper surface with tertiary veins impressed; fruit 3-4.5 mm diameter, rarely 2-locular and then with more than 3 styles, reniform. **5. R. longipes**
7. Stems dark grey to brown or reddish; leaflets entire or with a few large lobes or teeth, tip not retuse, never drying grey-green; fruit globose or reniform, not with white waxy coating. 8
- Stems soon becoming grey, often very pale, almost white; leaves, at least some, with a few often obscure marginal crenulations, tip minutely retuse, drying greyish green; fruit obliquely oblong-reniform, compressed when fresh, often pruinose (with white waxy coating). **1. R. natalensis**
8. Panicles with flowers evenly spaced and not forming discrete clusters; mature fruit with skin smooth and fragile, retaining shape and becoming pale when dry. 9
- Panicles, at least the male ones, with flowers clustered into separated globose groups; fruit globose and juicy when fresh, very strongly compressed, dull reddish brown when dried. **11. R. vulgaris**
9. Undersides of leaflets with prominently reticulate tertiary veins; inflorescence often lax, conical; fruit globose. 10
- Undersides of leaflets with rather obscure tertiary veins; inflorescence very dense, subglobose; fruit ± reniform to bean-shaped. 11
10. Leaflets always entire. **5. R. longipes**
- Leaflets, at least some, with broad, rounded marginal teeth. **6. R. ruspolii**
11. Leaflets broadest at or above middle, 2.5-3.4 times longer than broad, margins not crisped, often turning brown on drying; inflorescence axes hairy. **7. R. glutinosa**
- Leaflets broadest well below middle, 3.7-11 times longer than broad, margins crisped, usually staying green on drying; inflorescence axes glabrous or nearly so. **8. R. retinorrhoea**

1. R. natalensis *Krauss* (1844)
- type: Natal, *Krauss* 395.

Rhus crenulata A. Rich. (1847) - type: 'Tcheleukot', *Petit* (P holo. + iso.).

R. glaucescens A. Rich. (1847) - type: SU, 'Choa', *Petit* (P holo.).

R. abyssinica Oliv. var. *glabrata* Mart., *Fl. Bogosensis*: 23 (1886); *R. glaucescens* forma *pubescens* Fiori, *Boschi & Piante Lign. Eritrea*: 222 (1911) - type: EW, Abita, Keren, *Becarri* 5 (FT holo.).

R. glaucescens var. *obovatifoliolata* Engl., in *Ann. R. Ist. Bot. Roma* 7: 18 (1898); *R. natalensis* var. *obovatifoliolata* (Engl.) Chiov in Savoia-Aosta, *La Esplorazione dello Uabi - Uebi Scebeli*: 396 (1932) - type: SD, between Surro & Rogono, *Ruspoli & Riva* '186 (1453) 173' (FT holo.).

R. glaucescens var. *macrocarpa* Schweinf., in *Bull. Herb. Boiss.* 7, *App.* 2: 334 (1899); *R. natalensis* var. *macrocarpa* (Schweinf.) Cufod., *Enum.* 474 (1959) non rite publ. & XXVI (1969) - types: EW, Mai Mafales, *Schweinfurth* 392 & Mt Lalamba near Keren, *Schweinfurth* 801 (both B syn. destr.).

?*R. natalensis* var. *hararensis* Engl., *Pflanzenw. Afrik.* 3(2): 207 (1921) - type: not clearly designated - 'Harar 1800-2000m' (?*Ellenbeck* B destr.)

Densely branched shrub, sometimes scandent, or small tree; twigs very pale fawn soon becoming very pale to darker grey and then dark, densely lenticillate, glabrous to densely hairy. Middle leaflet usually oblanceolate to obovate or elliptic, (2.6-)3-8(-11.5) x 1-3.8 cm, 1.5-3.5 times longer than broad, widest at or above middle, lateral leaflets 1/3-3/4 % as long, base cuneate, tip rounded to retuse, rarely acute, at least some leaflets obscurely crenulate near tip, glossy when fresh, usually drying greyish green, tertiary veins obscure, impressed, rarely prominent, glabrous or minutely puberulous at base of midrib, rarely more uniformly puberulent. Panicles mostly lateral, shorter than leaves in female. Drupe obliquely oblong-reniform, compressed, smooth, glossy, pruinose (with white waxy coating), red when ripe, maximum dimension 5-7 mm. Fig. 127.6.1; 127.5.

Deciduous bushland with *Acacia* and/or *Combretum* etc., or evergreen bushland with *Euclea*, *Dodonea* etc., usually on well drained slopes, less often along water courses; (in GG 700-)1200-2200(-2500) m. EW TU GD WU SU AR IL GG SD BA HA; Guinea east to Somalia and tropical Arabia and south to Natal. *Ash* 463; *Gilbert & Getachew* 2603; *Mooney* 5474.

Vàtova 1288 (from SD, near Soddu) has very atypical small, narrow, uniformly densely puberulent leaflets but is otherwise typical of this species, normal material of which was collected at the same time.

2. **R. tenuinervis** *Engl.* (1883)
- type: Angola, *Welwitsch* 4418.

Closely related to *R. natalensis*, differing mainly by the consistently uniformly hairy obovate leaflets, 3.9-5.8 x 1.6-3.2 cm, 1.5-2.5 times longer than broad, margin minutely revolute, not glossy when fresh, often drying rather brownish green. Male panicles rather lax, terminal pointed; female panicles on short axillary branches, rather little branched. Fig. 127.6.2.

Deciduous woodland/bushland with *Combretum* and *Terminalia* and usually *Acacia* and *Commiphora* on blackish soils; 1200-1650 m. GG SD BA; ?Sudan (Red Sea Hills), Kenya, Tanzania, west to Dahomey and south to Cape Province. *Gilbert* et al. 8019; *Mesfin & Tewolde* 2642; *Mooney* 9771.

The rather sparse material from central Tanzania northwards is consistently much smaller-leafed and more obscurely crenulate than material typical of south tropical Africa. It has sometimes been refered to *R. flexicaulis* Rendle, an Arabian species differing by its entire leaflets but otherwise very similar.

3. **R. quartiniana** *A.Rich.* (1847)
- type: TU, Shire ['Chiré'], *Quartin Dillon* (P holo. K ?iso.).

Shrub with many erect stems or small tree 2-5 (-8) m high; stem sometimes spiny; twigs densely pubescent, sometimes glabrescent, dark greyish brown, young shoots slightly glutinous. Middle leaflet narrowly elliptic to oblanceolate, (2.5-)4-6.5(-8.7) x (1-)1.3-2 (-3.3) cm, 2.5-5 times longer than broad, widest at or above middle, lateral leaflets 1/2-2/3 as long, pubescent below at least on margins and midrib near base, less hairy above, base cuneate, tip rounded or mucronate, rarely acute, tertiary veins obscure, slightly impressed. Panicles mostly terminal, lax, extending beyond leaves, axes hairy. Drupe globose, 3-3.8 mm diam., reddish or yellowish brown, glossy, fragile.

var. **quartiniana**.

R. glaucescens A. Rich. var. *schimperi* Oliv., Fl. Trop. Afr. 1: 438 (1868); *R. natalensis* Krauss var. *schimperi* (Oliv.) Engl., *Pflanzenw. Afrik.* 3,2: 207 (1921) - type: TU, Gageros, *Schimper* (1854) 109 (K holo. BM P iso.) (this collection has also been distributed as 'in *Hohenacker* 2188').

R. pyroides Burch. var. *erythraea* Fiori, *Boschi dell'Eritrea* 222 (1911); *R. huillensis* Engl. var. *erythraea* (Fiori) Pichi-Serm., in *Webbia* 7: 336 (1950) - type: EW, Akele-Guzay, between Mareb and Chenafen, *Fiori* 186 (FT holo.).

Leaves always entire. Fig. 127.7.

Evergreen thickets along rivers and streams; 900-1800 m. EW TU GD GJ SU WG IL KF BA; south and west to Mozambique, Botswana, Namibia and Zaire, ?Libya. *W. de Wilde* 7281; *Friis* et al. 3727; *Gilbert* et al. 7371.

Var. *zambesiensis* R. & A. Fernandes is a thorny shrub with generally irregularly dentate leaves found in the Flora Zambesiaca area.

4. **R.** sp. = *Friis* et al. 1147.

Shrub to 4 m; twigs at first reddish brown, soon developing grey waxy coat, young shoots shiny glutinous. Middle leaflets elliptic, 4.8-9 x 2.4-3 cm, 2-3 times longer than broad, broadest at or slightly below middle, lateral leaflets 65-90% as long, sparsely hairy below, glabrous above, lower surface brownish, tip acuminate, entire, leathery, tertiary veins densely and prominently reticulate above and below. Panicles mostly terminal, rather lax, axes very sparsely hairy. Drupe subglobose, 4.5-5 mm diameter, glossy, fragile. Fig. 127.6.3.

Evergreen *Euclea* bushland on slope; 1650 m. SU; not known elsewhere. *Friis* et al. 1147 (Bole Valley); *Senni* 847 (near Modjo).

The 2 collections seen appear to be most closely related to *R. anchietae* Hiern from southern tropical Africa, atypical forms of which extend north to Kenya, but that species differs by its much larger rounded leaflets. *Chaffey* 564, from forest margins at 2700m in BA may also belong here but the leaves have much less prominent tertiary veination.

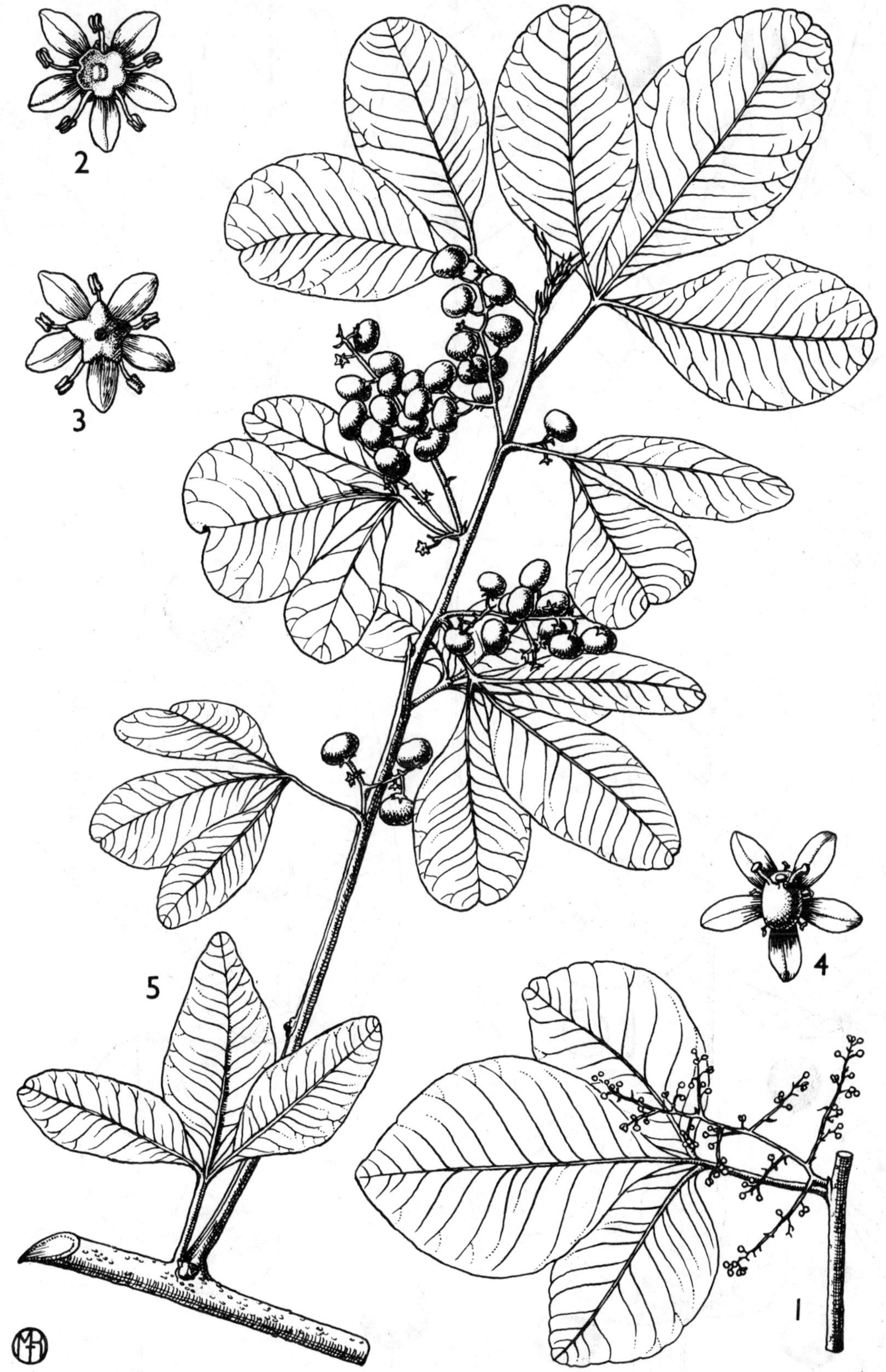

Figure 127.5 *RHUS NATALENSIS*: 1 - flowering shoot x 1; 2 & 3 - male flower, front and back views, 4 - female flower x 7; 5 - fruiting branch x 1. All from *A.S. Thomas* 2206A. Drawn by Olive Milne-Redhead. (Reproduced with permission from *Fl. Trop. E. Afr.*, Fig. 4, 1976)

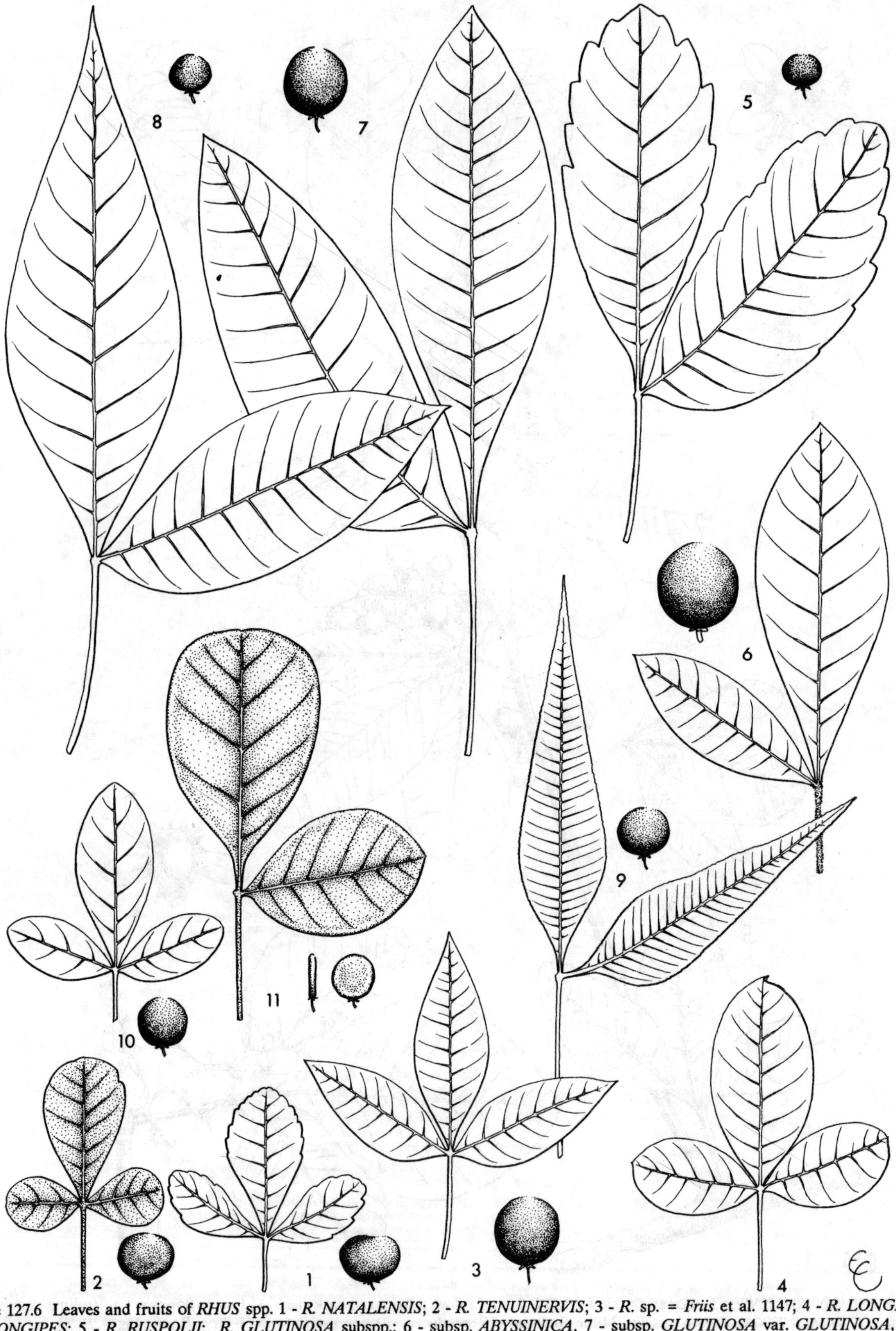

Figure 127.6 Leaves and fruits of *RHUS* spp. 1 - *R. NATALENSIS*; 2 - *R. TENUINERVIS*; 3 - *R.* sp. = *Friis* et al. 1147; 4 - *R. LONGIPES* var. *LONGIPES*; 5 - *R. RUSPOLII*; *R. GLUTINOSA* subspp.: 6 - subsp. *ABYSSINICA*, 7 - subsp. *GLUTINOSA* var. *GLUTINOSA*, 8 - subsp. *NEOGLUTINOSA*; 9 - *R. RETINORRHOEA*; 10 & 11 - *R. VULGARIS*. Leaves x 1/2, fruits x 2. 1 from *Burger* 981; 2: leaf from *Gilbert* et al. 8019, fruit from *Mesfin & Tewolde* 2642; 3 from *Friis* et al. 1147; 4 from *Chaffey* 323; 5 from *Gilbert* 3192; 6 from *Schimper* I. 259; 7 from *Pichi-Sermolli* 2348; 8 from *Gilbert* et al. 8027; 9 from *Friis* et al. 1148; 10 from *Thulin & Hunde* 3946; 11 from *Burger* 1826. Drawn by Eleanor Catherine.

Figure 127.7 *RHUS QUARTINIANA*: 1- flowering branch (male), 2 - fruiting branch x 2/3; 3 - male flower, 4 & 5 - developing fruit x 14; 6 - fruit x 6. 1, 3 from *Gilges* 3211; 2 from *Flanagan* 3211. Drawn by Joanna Webb. (Reproduced by permission, Editorial Board, *Flora Zambesiaca* Vol. 2(2), Fig. 129.)

5. **R. longipes** *Engl.* (1883)
- type: Angola, *Welwitsch* 4413.

Shrub to 4 m (elsewhere tree to 8 m); twigs rather angular, greyish soon becoming brown with small raised lenticels. Middle leaflet elliptic to obovate, 3.7-14 x (1.7-)2.2-7(-7.5) cm, 1.6-2.4 times longer than broad, lateral leaflets 50-75% as long, entire, drying rather brownish, tertiary veins rather lax and irregular, prominent on both sides at least near margin. Panicles mostly terminal, pointed, ± lax, axes hairy. Female flowers sometimes with ovary 2-locular and more than 3 styles. Fruit usually globose, 3-4.5 mm diameter, if 2-locular then reniform, to 7 mm across, glossy, fragile.

var. **longipes**.

Stems hairy. Leaflets obtuse to acute, sparsely pilose on midrib above and on margins and veins below. Fig. 127.6.4.

Open *Combretum-Terminalia* woodland; riverine *Millettia-Celtis* forest; 1200-1650 m. IL BA SD; Uganda and Kenya south and west to Mozambique, Zimbabwe and Angola. *Chaffey* 323; *Friis* et al. 3452; *Gilbert & Sebsebe* 8534.

The account is based primarily on Kenyan and Ugandan material. This is mostly smaller-leaved than typical material from the south of the range though *Friis* et al. 3452 has leaves as large as those seen anywhere. The separation from *R. ruspolii*, below, in western Ethiopia is not always clear.

Var. *schinoides* (Hutch.) R. & A. Fernandes from Zimbabwe and var. *elgonensis* Kokwaro from Kenya and Uganda differ by their glabrous leaves and stems and, in the latter, acuminate leaflets.

6. **R. ruspolii** *Engl.* (1883)
- type: SD, Bidduma, *Ruspoli & Riva* '1462 (1253) 1317' (FT holo.).

Shrub to 3 m, rarely tree to 8 m; twigs stout, smooth reddish brown, glabrous or sparsely hairy. Leaflets subequal, usually obovate, (7.3-)10-18 x 2.9-7.4 cm, 1.7-2.5(-3.4) times longer than broad, widest at or above middle, tip obtuse or very shortly acuminate, less often acute, margins of at least some leaves with broad obtuse teeth or crenations, rarely subentire, tertiary veins prominently reticulate below, glabrous or hairy below, glabrescent above. Inflorescence mainly terminal, pointed, male flowers ± grouped into discrete clusters, axes hairy. Drupe globose, 2.5-4 mm diameter, glossy, fragile, pinkish red. Fig. 127.6.5.

Lower margins of montane forest, *Combretum-Terminalia* or *Acacia* woodland, locally abundant; 1200 (in IL)-2150 m. SU GJ WG IL KF GG SD BA; Kenya, Uganda, Zaire. *Friis* et al. 105; *Gilbert* 3192; *Gillett* 14477.

Material from western Ethiopia is much more variable than that from elsewhere. One notable variant is a form with oblanceolate acute leaflets that approaches the form of *R. vulgaris* found in GD (*J. de Wilde* 5616; *Friis* et al. 420). Another with almost entire leaflets (*Chaffey* 294, *Mogk* E279) seems to be only arbitarily separated from *R. longipes* though in BA the two species have been seen growing intermingled and were clearly distinct.

8. **R. glutinosa** *A.Rich.* (1847)
- type: GD, near Dschenausa, *Schimper* II: 851 (P lecto., K isolecto. selected by Gilbert, 1986).

Spreading or erect shrub or tree (1-)3-10 m high, trunk to 25 cm diameter; new shoots sometimes strikingly shiny-glutinous. Leaflets subequal or laterals slightly smaller, oblong-elliptic to lanceolate, (7-)8-23 x (2.5-)2.8-6.5(-7) cm, 2.2-3.9 times as long as broad, widest at or below middle, entire or occasionally with a few large marginal teeth, tertiary veins slightly prominent, tending to dry brownish with pale veins. Panicles mostly axillary, male sometimes exceeding subtending leaf, female usually shorter than subtending leaf. Drupe slightly compressed reniform, maximum dimension 3.3-9 mm, glossy, fragile, pale fawn.

1. Young shoots not visibly glutinous; stems and often leaves densely yellow-pilose; leaves often with a few large marginal teeth; infructescences very dense, often subglobose, shorter than subtending petiole; drupe subglobose to reniform. subsp. **abyssinica**
- Young shoots usually prominently shiny-glutinous; stems and leaves glabrous to puberulent, rarely shortly whitish-tomentose; leaves nearly always entire; infructescences laxer, more elongated; drupe more compressed, reniform to bean-shaped. 2
2. Stems dull brown, often puberulent; leaflets acute to obtuse; female panicles usually shorter than subtending petiole; drupe with maximum dimensions 5.5-9 mm. subsp. **glutinosa**
- Stems shiny reddish-brown, usually glabrous; leaflets always acuminate; female panicles usually longer than subtending petiole; drupe with maximum dimensions 3.5-5 mm. subsp. **neoglutinosa**

subsp. **abyssinica** *(Oliv.) M. Gilbert*
in *Nordic J. Bot.* 6:572 (1986); *R. abyssinica* Oliv. (1868) - type: TU, Gennia, near Adowa, *Schimper* I:259 (K holo., BM P iso.).
R. petitiana A. Rich. (1847) - type: TU, Wogerat ['Uogeratte'], *Petit* (P holo. & iso.).
R. foliosa A. Rich. (1847) - type: SU, ['Kingdom of Choa'], *Quartin-Dillon & Petit* (P holo. & iso.).
Fig. 122.5.6.

Evergreen bushland; (1500-)1950-2600 m. EW TU; Sudan, Egypt, ?Arabia. *Bally* 6661; *Ellis* 6; *Sebsebe* 1432.

Material from Eritrea northwards is very constant but collections from TU and SU, including the types of *R. foliosa* and *R. petitiana*, intergrade with the other subspecies. For example, *Gilbert* et al. 7349 (SU, Muger Valley) has exactly the lobed leaflet margin and dense indumentum of this subspecies but was growing with a rather hairy form of subsp. *neoglutinosa* which it otherwise matched well. Fruit size is extremely variable with the type collection having much larger fruits than any other collection seen.

subsp. **glutinosa**

1. Leaves 3-foliolate, tips obtuse, acute or slightly acuminate at very tip only. var. **glutinosa**
- Leaves 1-foliolate, tips acuminate. var. **unifoliolata**

var. **glutinosa**;

R. glutinosa var. *obtusifoliolata* Engl. in A. & C. DC., *Mon. Phan.* **4**: 431 (1883) - type: TU, near Amann-Eski, *Schimper* (1853) 140 (B syn., K isosyn.) & without loc., *Quartin-Dillon* '13' (P syn., ?K isosyn.).

R. glutinosa var. *acutifoliolata* Engl. *loc. cit.*: 431 (1883) - types include *Schimper* II:851 (?B holo. destr.).

R. amharica Pichi-Serm. (1951) - type: GD, east of Mt Guranghé, north of Gonder, *Pichi-Sermolli* 2337 (FT holo.).

Fig. 122.5.7.

Presumably a component of evergreen bushland or forest margins; 1800-3300 m. EW TU GD GJ; not known elsewhere. *Lythgoe* 718; *Pichi-Sermolli* 2348; *Tekle* 96.

The bulk of more recent collections so named have been included within subsp. *neoglutinosa*. Notes on habit and habitat are very sparse. *R. amharica* is an unusually robust form of this species.

var. **unifoliolata** *Cufodontis*

in *Senckenb. biol.* **43**: 312, taf. 30 (1962) - type: GJ, Camoga Valley, SE of Debre Marcos, *Kuls* 149 (FR holo. not seen).

About 2400 m. GJ; known only from the type.

A remarkable plant of uncertain position. The acuminate leaves suggest a relationship with the following species but the few other collections seen from GJ are *R. glutinosa* sens. str. Information on the variation and distribution of this plant is needed before any meaningful decision can be made on its status. The 1-foliolate leaves are extremely unusual in the genus.

subsp. **neoglutinosa** *(M.Gilbert) M.Gilbert*

in *Nordic J. Bot.* **6**:572 (1986); *R. neoglutinosa* M. Gilbert (1986) - type: SU, 32 km from Addis Abeba on road to Ambo, near Menegesha village, *Jackson* 712 (K holo., ETH iso.).

Fig. 127.6.8.

Forest margins, often conspicuous after clearance of *Podocarpus* or *Juniperus* forest, less often in more open bushland on rocky slopes; (1500-)1800-2700m. WU SU AR BA HA; not known elsewhere. *Burger* 2274; *Gilbert* et al. 8003; *Merga* 527.

The vast majority of recent collections named as *R. glutinosa* are placed here. Certain collections from SU (*Gilbert* et al. 7348, Muger Valley) and WU (*West* 5824 from near Dessie) are atypically hairy and intergrade with subsp. *abyssinica*.

8. **R. retinorrhoea** *Oliv.* (1868)

- type: TU, Dscheladscheranne, *Schimper* III: 1627 (K holo., BM P iso.).

R. viminale sensu A. Rich. (1847) non Ait. (1789).

Slender shrub or small tree (1.5-)2-6 m high; twigs slender, usually pendent, reddish brown, glabrous; young shoots conspicuously shiny glutinous. Middle leaflet lanceolate, (7.7-)8.5-20 x 0.7-4.7 cm, 3.7-11 times as long as broad, broadest below middle, tip acute to acuminate, entire, margin often finely crisped, usually drying green, tertiary veins lax, not prominent. Panicles lax and drooping, male mostly terminal, clear of foliage, female mostly axillary, shorter than subtending leaf, axes glabrous. Drupe subglobose to subreniform, maximum dimension 3.5-4.5 mm, glossy, fragile, pale fawn often tinged pink. Fig. 127.6.9.

Dry rocky slopes with evergreen bushland; 1450-2700 m. EW TU WU SU HA; Sudan, Somalia, Arabia. *Friis* et al. 1148; *Gilbert & Getachew* 2651; *Mooney* 8036.

9. **R. vulgaris** *Meikle* (1951)

- type: Kenya, *Rogers* 711 (K holo.).

R. villosa var. *tomentosa* Oliv., *Fl. Trop. Afr.* **1**: 43 (1868) - types include SU, Domak, Efat, *Roth* 212 (K syn.).

?*R. villosa* var. *gallaensis* Engl., *Pflanzenw. Afrik.* **3(2)**: 209 (1921) - type not precisely indicated, from 'Gallahochland', (presumably B destr.).

R. villosa sensu auct., non L.f. (1782).

Rhus pyroides sensu A. Rich. (1847) non Burch. (1822).

R. macowanii sensu Pichi-Serm. (1951) non Schönl. (1830).

Shrub or small tree (1-)2-10 m high; twigs brownish, sometimes forming blunt spines, hairy. Middle leaflet usually obovate, 2-9.2(-10.5) x 1.7-6.5 cm, 1.1-2.2 times as long as broad, widest at or above middle, lateral leaflets 50-75% as long, hairy, more sparsely so above, tips emarginate, rounded or abruptly acuminate, margins entire or with a few large acute subapical teeth, sometimes rather bullate and with margins folded over when pressed, tertiary veins closely but not conspicuously reticulate, sometimes prominent above and below. Panicles terminal, elongated with leafy 1-foliolate bracts at base or mostly axillary; flowers, especially male, grouped into distinct clusters. Drupe when fresh globose, glossy, ripening red, when dried becoming very flat, discoid, 3.5-4 mm diameter, dull reddish brown, leathery. Fig. 127.6.10-11.

Margins of *Podocarpus* and *Juniperus* forest and often an important component of secondary bushland following clearance; evergreen bushland with *Euclea schimperi* or open deciduous woodland with *Combretum*, *Terminalia* and/or *Acacia*; 1500-2800 m. GD WU GJ SU AR WG KF SD BA HA; west to Senegal and south to Mozambique, Malawi, Zimbabwe and Zambia. *Friis* et al. 188; *Gilbert* et al. 7337; *Mooney* 5330.

A variable species most easily recognised by the fruits which dry very flat. Some of the local variants are rather distinctive and need further investigation as to their correct status.

Collections from deciduous woodland, often on limestone, from southern SD (e.g. *Burger* 1826; *Friis* et al. 3150; *Gilbert* et al. 7752) have broader (1.1-1.4 times as long as broad in comparison to 1.3-2.4 times as broad elsewhere), emarginate, bullate leaflets with infolding margins and larger inflorescences. Though consistently separable within the Flora area the differences do not seem as clear-cut in Kenya.

Collections from GD, the northern limit of occurrence (e.g. *Sebsebe* 352, 412, 478) have leaflets with a distinctively sparser indumentum and prominent pale veins.

Burger 984 (HA, near Kurfa Chele ['Curfacelli']) has hairy fruits that do not seem to be as compressed when dried, features that are normally of considerable taxonomic significance within the genus *Rhus*. In all other features this collection matches *R. vulgaris*, typical material of which has been collected in the same general area.

6. SCHINUS *L.* (1753).

Closely related to *Rhus*, differing primarily by having the stamens twice as many as petals and, with regard to the African species, by the pinnate leaves which resemble rather those of *Pistacia*.

About 28 species in South and Central America, 2 of which are widely grown as ornamentals and occasionally naturalise.

S. terebinthifolius Raddi, easily confused with *Pistacia aethiopica* when not in flower, is fairly commonly grown in East Africa and may be introduced into Ethiopia as well.

S. molle *L.* (1753)
- type: 'from Peru'.

Graceful evergreen tree 5-15 m high; bark deeply fissured; twigs pendent. Leaves pinnately 20-40-foliolate, rachis terete or very narrowly winged, leaflets linear lanceolate, 2-5 x 0.5-0.8 cm, tip acuminate, entire or dentate. Panicle terminal, pendent, to 30 cm or more long. Petals whitish, 2 mm long. Drupe globose, 5-6 mm diam., glossy, fragile, rosy red, smelling and tasting of peppercorns (*Piper niger*).

A heat and drought-resistant tree that is widely planted in all but the coldest and wettest areas and sometimes regenerating naturally. Near sea level (Mitsiwa)-2400m. Native of S Brazil through to N Argentina. *O. Ryding* 1165.

A distinction is sometimes made between var. *molle* with clearly dentate leaflets and var. *areira* (L.) DC. with the leaflets entire to minutely serrate. Ethiopian material matches the latter. The fruits are sometimes used as a condiment, hence the common name 'pepper tree'.

7. PISTACIA *L.* (1753).

Zohary, *Palest. J. Bot. J. series* **5**: 187-236 (1952).

Trees or shrubs with resinous bark. Leaves alternate, pinnate, elsewhere simple or 3-foliolate. Dioecious. Inflorescence an axillary panicle, sometimes subspicate. Flowers wind-pollinated; perianth absent but flowers surrounded by 1-3 small bracts and up to 7 tepal-like bracteoles; male with 3-5(-8) stamens inserted on disc, filaments very short, anthers basifixed, longer than filaments; female reduced to subglobose or ovoid 1-locular ovary, style short, stigma with (2-)3 broad papillate lobes. Fruit a drupe, laterally compressed, thinly fleshed; stone woody, rarely leathery.

Genus with 12 species, mainly Mediterranean to western Asia but also extending to SE Asia and China and found in the Atlantic Islands, Mexico and southern U.S.A. Only 2 species in tropical Africa.

Though at first glance *Pistacia* possesses an apparent calyx, most authorities seem agreed that the flowers lack a true perianth and are surrounded by a series of bracteoles. On the basis of this, the distinctive stamens and also different pollen from the rest of the family, some authorities have placed this genus in a separate family, the *Pistaciaceae*.

1. Evergreen; rachis winged, 5.5-10 cm long, leaflets oblanceolate to ovate, obtuse or emarginate. **1. P. aethiopica**
- Deciduous; rachis not winged, 8-20 cm long; leaflets curved-lanceolate, acuminate. **2. P. falcata**

1. P. aethiopica *Kokwaro* (1979);
Pistacia lentiscus L. var. *emarginata* Engl. in A. & C. DC., *Mon. Phan.* 4: 296 (1883) - type: Somalia, *Hildebrandt* 1531.
?*P. lentiscus* var. *falcatula* Chiov. in Savoia-Aosta, *La Esplorazione dello Uabi - Uebi Scebeli*: 396 (1932) - type: AR: M. Laggio, *Basile* 177 (TO holo. not seen).

Evergreen trees 5-15 m high, often reduced to shrubs by cutting; glabrous except for ciliate bracteoles; bark dark, fissured. Leaves (4-)6-16(-18)-foliolate, leaflets opposite or irregularly alternate; rachis 5.5-10 cm long, winged, 2-4.5 mm wide; leaflets asymmetrically ovate to obovate or oblanceolate, obtuse or emarginate, entire, leathery, conspicuously reddish when young. Inflorescence subspicate, sometimes branched, 1-5 cm long; bracts *c* 2 mm long, ovate, bracteoles c 5, up to 0.7 x 1 mm, acute, sometimes ciliate. Male flowers with 4-6 stamens, anthers c 1.4 mm long, reddish. Female flowers with ovary c 0.6 mm diam. Drupe obliquely subglobose, very compressed when dried, 2.9-3.8 x 3.5-4.5 mm, red. Fig. 127.8.5.

Juniperus forest or deciduous woodland with *Combretum*, *Acacia* and *Barbeya*, on limestone, basalt or sandstone, sometimes on black 'cotton' soil; 1350-1850(-2550 in Eritrea) m. EW GG SD BA; Somalia,

Figure 127.8 *PISTACIA FALCATA*: 1 - branch with infructescence x 2/3; 2 - male flower; 3 - female flower; 4 - fruit, all x 3. *P. AETHIOPICA*: 5 - leaf x 2/3. 1, 4 from *Gilbert* et al. 7410; 2 from *Burger* 3617; 3 from *Burger* 3618; 5 - from *Gilbert* et al. 7900. Drawn by Eleanor Catherine.

Kenya, Uganda, N Tanzania. *Ash* 2447; *Mooney* 5607; *Thulin* et al. 3450.

It yields a high quality gum mastic and plants are often damaged by the collecting of this.

Rather locally abundant in S Ethiopia but only the single very isolated old record from Eritrea where it may now be extinct. Galled growths are frequent and sometimes large and conspicuous. The name *P. aethiopica* was used by Dale and Greenway (*Kenya Trees & Shrubs*: 26, 1961) but it was not then validated.

2. **P. falcata** *Mart.* (1886);

Pistacia chinensis Bunge var. *falcata* (Mart.) Zohary, *loc. cit.*: 217 (1952) - type: Eritrea, Mt Deban, Keren, *Beccari* 272 (FT holo.).

Deciduous tree 4-9 m high; bark longitudinally fissured; twigs pendent. Leaves 5-11-foliolate, irregularly imparipinnate with terminal leaflet often absent; rachis 8-20 cm long, terete; leaflets curved-lanceolate, 6-12 x 1.3-2.1(-2.8) cm, base cuneate, sides very unequal, tip long acuminate, sparsely puberulent, soon becoming glabrous. Panicles formed before leaves, much branched, male c 5 cm long, female c 10 cm extending to 17 cm in fruit; bracts 4-5 mm long, linear-lanceolate, sparsely woolly. Male flowers: bracteoles 4-7, c 0.7 mm long, hairy, anthers (4-)5(-7), 1.4-1.6 mm long. Female flowers: bracteoles c 5, 2 mm long, hairy, ovary c 0.5 mm diam., stigma lobes 1 mm long. Drupe obovoid-subglobose, 5.5-6 x 5-5.5 x c 3.5 mm, red. Fig. 127.8.1-4.

In SU the dominant tree on recent lava flows, no data on the northern populations but in Somalia plants grow in evergreen bushland on very steep rocky slopes; 1100-1450(in EW -2600) m. EW SU; Sudan, Somalia, Yemen, Saudi Arabia. *Burger* 3616-18; *Gilbert* et al. 7410; *Thulin* et al. 3875.

Regarded by Zohary as closely related to *P. chinensis* from Afghanistan, Kashmir and Punjab, China, Taiwan and Philippines, which differs by the non-falcate leaves and broader bracts and bracteoles.

128. CONNARACEAE

Hemsley, *Fl. Trop. E. Afr.*, **Connaraceae**: 27 pp. (1956); Burger, *Families of Flowering Plants in Ethiopia*: 88 (1963).

Trees, shrubs or climbers. Leaves alternate, simple, trifoliate, or imparipinnate, blades entire; stipules and stipels absent. Inflorescences terminal or axillary panicles or racemes, often in axils of older leaves or on older branches. Flowers hermaphrodite and regular. Sepals 5, usually free, imbricate or valvate in bud, persistent in fruit. Petals 5, free or united only at the base, usually imbricate in bud. Stamens 10 or only 5 fertile, usually of 2 lengths, those alternating with the petals longer than those opposite the petals, filaments free or united at the base to form a short tube. A disc sometimes present. Pistil 1 or 5, rarely 3, (carpels free or united only at the base, 1, 3, or 5), each ovary superior and 1-locular, and with a covering of short hairs (hairless in *Santaloides*), 1-locular with usually 2 (sometimes 1 or 3) ovules in each locule on basal or parietal placentas; style and stigma 1 on each ovary. Fruit usually a follicle covered with velvety hairs and splitting open on 1 side; seeds with or without endosperm but with a fleshy aril covering part or all of the seed; only 1 seed developing in each fruit.

A tropical family of 25 genera and around 200 species represented by 9 genera in east central Africa. Although not yet recorded from Ethiopia, the following species, which all occur in the Sudan, could be found as plant exploration of the southwestern areas is extended. *Cnestis ugandensis* Schellenb., is a forest tree; *Agelaea ugandensis* Schellenb., occurs as a shrub, small tree or liane in lowland and upland forest as well as along river valleys; *Santaloides splendidum* (Gilg.) Engl. is widespread in lowland rainforest and river valleys; and, *Jaundea pinnata* (Beauv.) Schellenb. is also widespread from West Africa to Sudan up to 2500 m.

JUGLANDACEAE

by Sue Edwards*

Hutchinson, *Fam. Fl. Plts*, vol. 1: 368-369 (1959);
Everett, *The New York Botanical Garden Illustrated Encyclopedia of Horticulture*, (1981).

Monoecious trees, often resinous and aromatic, with beautiful durable dark-coloured wood. Leaves alternate (rarely opposite) imparipinnate, usually with resin dots on the undersides. Flowers unisexual: male ± perianth, arranged in hanging catkins; female solitary or in few-several flowered spikes, calyx 3-5-lobed, style short, stigmas 2, pistil of 2 carpels, 1-locular with 1 erect ovule. Fruit a winged drupe, rarely a nut or nutlet.

Family of 6-7 genera and around 50 species native to north temperate areas and tropical Asia. Many species of economic importance as either sources of high quality timber and/or edible nuts, for example, walnut, butternut, pecan, hickory.

JUGLANS *L.* (1753)

Trees. Leaves large, imparipinnate, aromatic when crushed. Flowers small. Male in catkins on previous year's shoots, perianth of 2-5 parts, stamens 20 in lowermost flowers, to 6 in uppermost. Female flowers terminal on current season's growth. Fruit a drupe with green fleshy exocarp surrounding a hard endocarp (the shell of the nut). Seed with 2 large cotyledons usually bent into an irregular shape by the presence of partial septa in ovary.

Genus of about 15 species most abundant in North, Central and South America, but also found found in the West Indies, Europe and Asia.

The wood from several species of *Juglans* has a beautiful grain and gives a high polish much valued for furniture and veneers. The bark has been used for tanning while yellow and brown dyes can be extracted from the fruits husks (exocarp).

Juglans regia L.

Tree with a broad rounded crown to 20 m tall; bark smooth silvery. Leaves 20-30 cm long, with 5-9(-13) leaflets, obovate or elliptic, ± entire to slightly serrate, terminal leaflet larger than the rest. Fruits with small wings and thin green exocarp, endocarp somewhat thin, with membranous partition between the cotyledons; splitting open naturally.

Introduced to Ethiopia and found growing in some public and private gardens in Addis Ababa. It has also been introduced to some horticultural sites in the cool central highlands of Ethiopia. Native of SE Europe and W and C Asia to China, now cultivated in many temperate areas throughout the world.

Apart from the edible fruits, this species gives wood much valued in making furniture. The seeds are also rich in oil.

* The National Herbarium, Addis Ababa University, Ethiopia.

129. CORNACEAE

Verdcourt, *Fl. Trop. E. Afr.,* Cornaceae: 3 pp. (1958); Burger, *Families of Flowering Plants in Ethiopia*: 111 (1963).

Trees or shrubs, rarely herbs. Leaves opposite or alternate, simple and entire or serrate; stipules absent. Inflorescence usually paniculate, sometimes as an umbel or head. Flowers regular, functionally unisexual or hermaphrodite. Calyx tube united with the ovary; sepals 4-5 or absent. Petals 4-5, free and mostly valvate in bud. Stamens 4-5, alternating with the petals, filaments short and free. Disc often present on top of ovary. Pistillode present in male flowers; staminodes present in female flowers. Pistil 1 of 2-4 united carpels; ovary inferior with usually 2-4 locules; placentation axile with 1 ovule pendulous in each locule. Fruit a drupe or berry.

A mainly N temperate and SE Asian family of 12-16 genera and about 100 species. One species, *Afrocrania volkensii* (Harms) Hutch., occurs in high rainfall areas in east Africa. The family includes a number of species grown as ornamentals. *Dendrobenthamia capitata* (Wall. ex Roth.) Hutch. is cultivated in Kenya and this and other species are likely to be found in both public and private gardens in Addis Ababa.

130. ALANGIACEAE

by B. Verdcourt*

Verdcourt, *Fl. Trop. E. Afr.*, **Alangiaceae**: 4 pp. (1958).

Trees or shrubs, often spiny. Leaves alternate, exstipulate, petiolate, entire or lobed, often asymmetric at the base. Flowers bisexual, regular, in few to many-flowered axillary cymes. Calyx truncate or divided into 4-10 lobes, or denticulate. Petals 4-10, linear, valvate, often joined at the extreme base. Stamens equal in number to the petals and alternating with them or up to four times as many, free or slightly joined at the base. Ovary inferior, 1-2-celled, ovules solitary and pendulous in the cells. Style simple, clavate or 2-3-lobed. Fruit a drupe crowned with sepals and disc, 1-2-seeded; seeds albuminous.

Despite the inclusion of *Metteniusa* in **Alangiaceae** by Airy Shaw (Willis ed. 8: 738 (1973)) this genus is better refered to **Icacinaceae.** The family has only the following genus.

ALANGIUM *Lam.* (1783)

Characters of the family.

According to Bloembergen in *Bull. Jard. Bot. Buitenzorg*, **III,16**: 139-235 (1939) the genus has 18 species and is distributed from tropical Africa through Madagascar and the Comoro Islands to China, SE Asia, Indomalaysia and E Australia. The only Ethiopian species belongs to the section *Marlea* Baill.

A. chinense *(Lour.) Harms* (1897).

Moderate-sized to large tree, 9-24 m tall, bark smooth and grey, never spiny. Leaves: petiole 0.5-2.5 cm long; lamina narrowly elliptic to broadly ovate, usually ovate, 4-19 x 2.5-10 cm, juvenile up to 25 x 27 cm, at first pubescent, later glabrescent, entire when adult but seedling and coppice leaves often palmately lobed; apex acuminate; base cuneate, truncate or subcordate, asymmetric, often markedly so; main lateral nerves 4-6, the basal ones confocal and sharply ascending so that the lamina is 5-7-nerved from the base, tertiary venation close, approximately parallel. Flowers golden pubescent, white, cream or yellow, sweetly scented, 8-16 mm long and 1.5-3.5 mm wide, 5-8-merous, 3-23 in pedunculate inflorescences, peduncles 0.4-1.8 xm long and pedicels 3-5.5 mm long. Petals tongue-shaped, 8-13.5 mm long. Stamens the same number as the petals, anthers narrow, 5-9 mm long. Fruit laterally compressed, globose or ellipsoidal, rather costate in the dry state, basally rounded, apically rather tapering, minutely pubescent, 1-2-celled, 8-10 x 6.5-9 x 4-5.5 mm. Fig. 130.1.

A pioneer species in partly cleared areas of lowland and upland rainforest; c 2000 m. BA (Harenna forest); Zaire, Sudan (Equatoria), Uganda, Kenya, Tanzania, Zambia, Angola, Gabon, Cameroun, Fernando Po and also from India to China, SW Asia, Japan, Java, Philippines and Sunda Islands. *Chaffey* 363; *Friis* et al. 3506.

The combination *A. chinense* (Lour.) Rehder frequently used is long antedated by that of Harms.

* Royal Botanic Gardens, Kew, Richmond, Surrey, TW9 3AB, UK.

Figure 130.1 *ALANGIUM CHINENSE*: 1 - flowering branchlet x 2/3; 2 - flower, 3 - corolla and stamens, 4 - pistil, 5 longitudinal section through pistil, all x 4; 6 - fruits x 2.5; 7 longitudinal section through fruit, 8 - cross section of fruit, both x 4. 1-5 from Gossweiler 9926; 6-8 from Gossweiler 7373. (Reproduced by permission, Editorial Board, *Flora Zambesiaca*, vol. 4: tab 169, 1978) Drawn by Derek Erasmus.

131. ARALIACEAE

by P. Bamps*

Cufodontis, *Enum.*: 635-637 (1959); Tennant, **Araliaceae** in *Fl. Trop. E. Afr.* (1968); Bamps, *Bull. Jard. Bot. Nat. Belg.* **44** : 101-139 (1974).

Trees, shrubs, lianes, sometimes herbaceous plants. Leaves mostly alternate, usually stipulate, simple or compound. Flowers bisexual or sometimes unisexual (outside the Flora area), regular, arranged in simple or compound inflorescences often gathered at the top of the shoots (umbellate). Calyx with tube adnate to the ovary, lobes very small and falling quickly or none. Petals (3)4-15, valvate or slightly imbricate, free or sometimes joined at the top in a calyptra. Stamens free, alternating with petals and mostly as numerous as them; anthers dorsifix, opening by longitudinal slits. Ovary inferior, with(1)2 to numerous locules; styles as many as locules, often joined at the base and forming a distinct stylopodium. Ovules solitary in each locule, pendulous. Fruit a berry or drupe. Seeds with endosperm; embryo very small.

More than 50 genera and about 800 species, mostly tropical and subtropical.

This family includes *Tetrapanax papyrifera* (Hook.) K. Koch, the Rice Paper Plant, which is quite common in public gardens and parks in Addis Ababa. A soft-stemmed shrub spreading by suckers; stems erect, unbranched or little-branched; leaves long stalked, palmately lobed, up to 30 cm diameter, margin toothed; all parts when young with a felty covering of stellate hairs. In China and Japan, the pith of this plant is used to make rice paper, which is edible and is used as a base for high quality cakes and pastries. Rice paper is also used for making artificial flowers.

Key to Genera

1. Leaves pinnate; flowers with pedicel articulate. **1. Polyscias**
- Leaves simple or digitate; flowers with pedicel non-articulate. 2
2. Leaves evenly distributed along the stems; stems with numerous adventitious roots. **4. Hedera**
- Leaves mostly gathered at the top of the stems; stems without numerous adventitious roots. 3
3. Simple panicles, racemes or spikes; ovary with 2(3-4) locules. **2. Cussonia**
- Compound umbels, or small umbels or heads grouped into panicles, racemes or spikes; ovary with (2)5-10 locules. **3. Schefflera**

1. POLYSCIAS *J.R. & G. Forster* (1776)

Trees or shrubs. Leaves alternate, imparipinnate, stipulate or not. Flowers in small racemes, umbels or glomerules arranged in racemes, bisexual, with pedicel articulate. Calyx entire or scarcely 5-lobed. Petals 5-15, valvate, free. Stamens 5-15. Ovary with 2-8 locules. Styles 2-8, free or united at base. Fruits laterally compressed or subglobose.

A tropical genus of about 75 species, all from the Old World.

1. Flowers in small racemes; inflorescence axes rusty tomentose; leaflets leathery, tomentose beneath. **1. P. fulva**
- Flowers in small umbels; inflorescence axes white tomentose; leaflets tough, glabrous or glabrescent beneath. **2. P. farinosa**

* Jardin botanique national de Belgique,
Département des Spermatophytes et des Ptéridophytes,
Domaine de Bouchout, B-1860 Meise, Belgique.

1. **P. fulva** *(Hiern) Harms* (1894).

Panax ferrugineum Harms; *Polyscias ferruginea* (Hiern) Harms - type: SU, Ankober, *Roth* (K holo.).

Tree up to 30 m high, with umbrella-shaped crown. Leaves up to 1 m long; leaflets 5-15 pairs, ovate to oblong, 6-20 x 2-10 cm, leathery, rusty tomentose beneath; base often asymmetric; apex acuminate to cuspidate. Panicles of small racemes up to 60 cm long, with axes rusty tomentose. Petals 5, 1.5-2 x 1 mm, soon falling, greenish-yellow. Filaments 2 mm long. Ovary 2-locular. Styles 2, united at base in a stylopodium 0,5-1 mm high. Fruits 4-5 x 3-4 mm. Seeds 2, semi-ovoid. Fig. 131.1.

Upland and lowland rain-forest, mainly secondary; up to 2450 m. SU IL KF AR SD; westwards to Guinea and southwards to Mozambique and Angola. *Burger* 2566; *Meyer* 7881; *Mooney* 7807.

2. **P. farinosa** *(Del.) Harms* (1894);

Aralia farinosa Del. (1847) - type: N Ethiopia, without loc., *Ferret & Galinier* (MPU holo., not seen).

Panax pinnatum A. Rich. (1847) - type: TU, Djeladjekanne, *Schimper* III:1583 (P holo., K UPS iso.).

Tree about 7 m tall, leafless at flowering time; bark exuding some resin. Leaves more than 40 cm long; leaflets several pairs, ovate, 8-15 x 2.5-5 cm, tough, glabrous or glabrescent beneath, somewhat asymmetric at base, acute at the apex. Racemes of small umbels up to 25 cm long, with axes whitish tomentose. Petals 5, 2 x 1 mm, pale yellow. Filaments 1.5 mm long. Ovary 2-locular. Styles 2, united at base in a stylopodium 1 mm high. Fruits not seen.

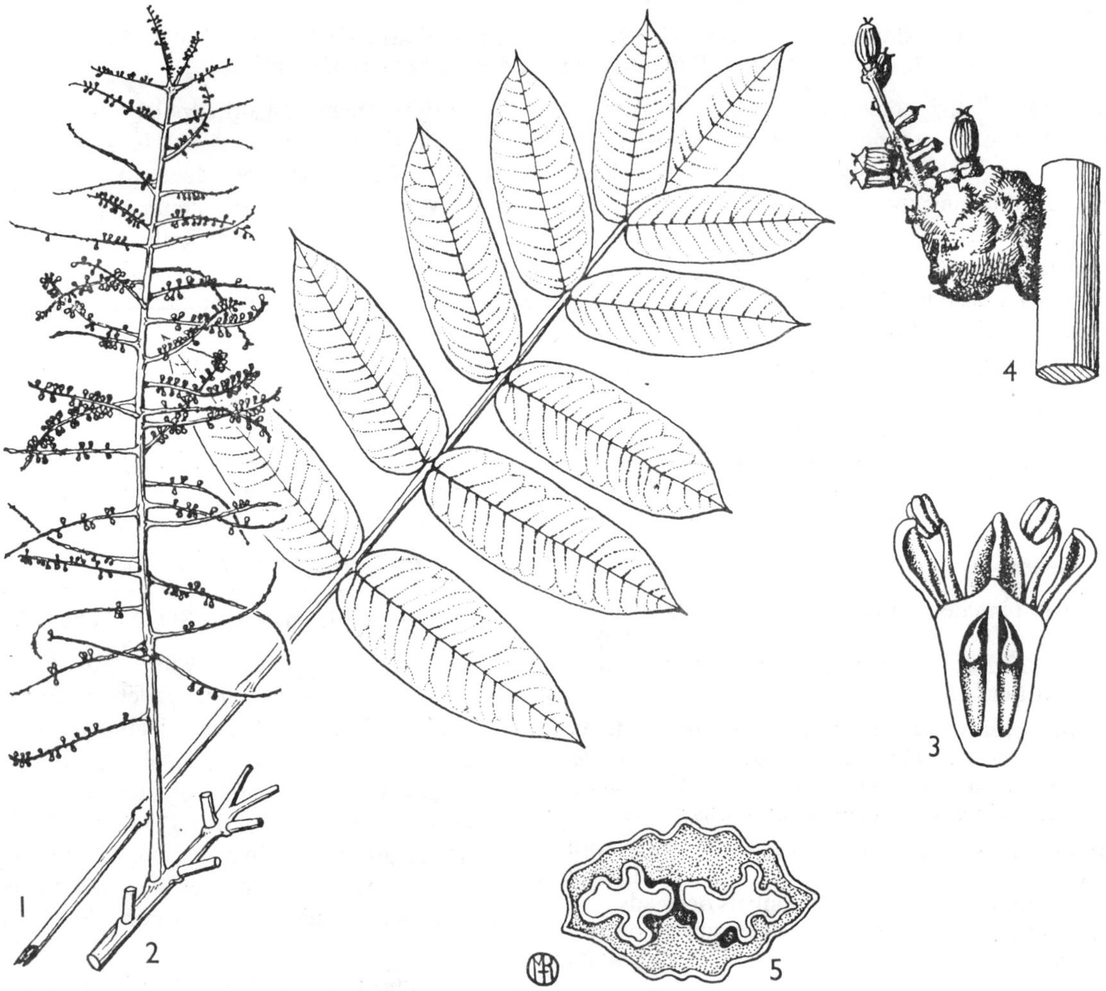

Figure 131.1 *POLYSCIAS FULVA*: 1 - leaf (some of the lower pinnae lacking) x 1/3; 2 - inflorescence-axis and one primary branch of inflorescence x 1/3; 3 - vertical section of flower x 12; 4 - portion of a primary branch of inflorescence showing galled secondary branch x 2; 5 - transverse section of fruit x 13. 1 from *R.M. Davies* 881; 2, 3 from *Drummond & Hemsley* 1752; 4,5 from *Sangster* 524. (Reproduced with permission from *Fl. Trop. E. Afr.*, Araliaceae, fig. 4, 1968.)

Open woodland, on slopes near streamside; 1700-2000 m. TU GD SU KF; endemic. *Pichi-Sermolli* 668; *W. de Wilde* 10462.

2. CUSSONIA *Thunb*. (1780)

Trees or shrubs, rarely rhizomatous suffrutices. Leaves gathered at the top of the shoots, alternate, long petiolated, palmately lobed or compound. Stipules intrapetiolar, partly adnate to the petiole. Flowers in simple spikes, racemes or panicles, bisexual, with pedicel not articulate. Calyx undulate or shortly 4-5-dentate. Petals 4-5. Stamens 4-5. Ovary with 2(3-4) locules. Styles 2(3-4), adnate at base in a stylopodium. Fruits drupaceous. Seeds with ruminate endosperm.

Twenty species from tropical Africa.

1. Flowers pedicellate; leaves palmately lobed. **3. C. ostinii**
- Flowers sessile; leaves palmately compound, the young ones sometimes palmately lobed. 2

2. Leaflets sessile or subsessile, sometimes with short petiolule; bark corky; bracts 2-4 mm long. **1. C. arborea**
- Leaflets with petiolule 2-5 cm long; bark thick, rough and fissured; bracts 0.6-1 mm long. **2. C. holstii**

1. C. **arborea** *A. Rich*. (1847)
- type: GD, Schoata in Semien, *Schimper* 1357 (P holo., BM FT G K M S UPS iso.).

Tree up to 10 m high or shrub; bark corky. Leaves palmately compound, sometimes palmately lobed; leaflets 3-9, sessile or subsessile, obovate, oblanceolate or elliptic, 6-30 x 2-18 cm, subentire or crenulate-dentate, glabrous or ± pubescent beneath, base obtuse or cuneate, apex acute or caudate-acuminate. Spikes numerous, up to 60 cm long; bracts 2-4 mm long, usually pubescent or tomentose. Flowers sessile. Petals 2-2.5 x 1.5-2 mm. Stylopodium 1 mm high. Fruits ovoid, 4-5 x 2.5-3 mm.

Woodland, tree or shrub savanna, often in rocky places; up to 2300 m. EE TU GD WG KF AR; widespread in tropical Africa from Guinea-Bissau to Sudan and from Zaire and Kenya to Zimbabwe and Mozambique. *Brehme* in *Mooney* 8987; *Friis* et al. 2357; *Bos & Jansen* 10199.

2. **C. holstii** *Harms* ex *Engl.* (1894).
C. boranensis Cufod. (1939) - type: SD, Arero, Meta Gafersa, *Cufodontis* 284 (FT holo.).

Tree up to 20 m high, with a straight bole; bark thick, rough, fissured. Leaves palmately compound, the young ones sometimes palmately lobed; leaflets 3-7, petiolulate (2-5 cm), elliptic or ovate, cuneate and sometimes unequal-sided at base, long-acuminate at apex, 6-20 x 3-7 cm, undulate or dentate-crenulate, glabrous. Spikes many, up to 15 cm long; bracts deltoid, 0.6-1 mm long, glabrous or nearly so. Flowers sessile. Petals 2 x 1-1.5 mm. Stylopodium 0.5 mm high. Fruits subglobose, 3-4 mm in diameter. Fig. 131.2.2.

Upland dry forest or sclerophyllous communities on old lava flows or on rocky outcrops, wooded termite mounds, fringing forest; 1100-2600 m. EE SU KF AR HU BO SD; Somalia, Kenya, Uganda, Zaire, Rwanda, Tanzania. *Burger* 1993, 2575; *Friis* et al. 828.

3. **C. ostinii** *Chiov.* (1911)
- type: GD, Dibebara in Semien, *Chiovenda* 820 (FT lecto.).

Tree up to 7 m high; bark corky. Leaves palmately lobed, peltate, up to 35 x 35 cm; lobes 7, glabrous, margin crenulate to serrate, apex caudate-acuminate. Racemes several, up to 60 cm long; bracts 1.5-2 mm long, pubescent. Flowers pedicellate. Petals 2-2.5 x 1 mm. Stylopodium 1 mm high. Fruits ovoid, up to 6 x 4 mm.

Upland savanna; 1500-2100 m. GD GJ WG KF; not known elsewhere. *W. de Wilde* 7268; *Friis* et al. 450; *Mooney* 6213.

3. SCHEFFLERA *J.R. & G. Forster* (1776)

Trees, shrubs or lianes, often epiphytic. Leaves digitately compound, with intrapetiolar stipules often connate with the petiole base. Flowers in compound umbels or in small umbels or heads grouped into panicles, racemes or spikes, bisexual, with pedicel non-articulated. Calyx entire or slightly dentate. Petals 5-10, sometimes behaving as a calyptra. Stamens 5-10. Ovary with (2)5-10 locules. Styles as many as locules, connivent at base. Fruits subglobose. Seeds laterally compressed, with endosperm smooth.

About 150 species throughout the tropics and subtropics.

1. Flowers sessile in heads. **1. S. volkensii**
- Flowers pedicellate in small umbels. 2
2. Umbels compound; lateral nerves nearly straight up to the margin. **2. S. myriantha**
- Small umbels grouped into racemes; lateral nerves curved towards the margin. **3. S. abyssinica**

1. **S. volkensii** *(Engl.) Harms* (1894).

Tree up to 30 m tall or scandent shrub, sometimes epiphytic. Leaves: petiole up to 15 cm long; leaflets 4-6, obovate, oblanceolate or elliptic, 7-15 x 3-7 cm, a bit leathery, glabrous; lateral nerves curved towards the margin; margin entire or nearly so; base cuneate to obtuse or rarely rounded; apex usually rounded. Heads of sessile flowers grouped into racemes up to 20 cm long; bracts up to 6 mm long, glabrous. Petals 1.5 x 1.5 mm. Stylopodium 0.5 mm high. Fruits ellipsoid or obovoid, deeply ribbed, up to 6 x 4 mm.

Upland forest; 1600-3250 m. SU AR IL KF GG; Kenya, Uganda and Tanzania. *W. de Wilde* 9177; *Gilbert* 1155; *Mooney* 7259.

2. **S. myriantha** *(Bak.) Drake* (1897).
S. polysciada Harms (1895).

Liana, shrub or small tree, sometimes epiphytic. Leaves: stipules 2.5-5 cm long; leaflets 4-6, ovate or elliptic, sometimes oblanceolate, 8-28 x 4-14 cm, papery or somewhat leathery, glabrous; lateral nerves nearly straight up to the margin; margin undulate; base rounded, cordate or obtuse; apex acuminate or attenuate. Small umbels grouped into compound umbels up to 30 cm long or more. Bracts 1-2 mm long, glabrous. Flowers pedicellate. Petals 2.5-3 x 1.5 mm. Stylopodium 1-1.5 mm high. Fruits subglobose, ribbed, 3-4 mm diameter. Fig. 131.3.

Upland forest, bamboo thickets, sclerophyllous communities; 1600-3400 m. WG SU BU IL KF; Kenya, Uganda, Zaire, Rwanda, Burundi, Tanzania, Malawi, Madagascar. *W. de Wilde* 6761; *Friis* et al.1885; *Mooney* 8747.

3. **S. abyssinica** *(Hochst. ex A. Rich.) Harms* (1894)
- type: TU, Mt. Scholoda, *Schimper* 283 (P holo., BM BR K iso.).

Tree up to 30 m high, rarely an epiphyte. Leaves: stipules c 1 cm long; leaflets 5-8, ovate or elliptic, rarely obovate, 6-40 x 3-20 cm, papery or somewhat leathery, glabrous; margin usually crenate; lateral nerves curved towards the margin; base rounded or cordate; apex long acuminate. Small umbels grouped into racemes up to 40 cm long; bracts 2-6 mm long, pubescent or glabrous. Flowers pedicellate. Petals 1.5-2 x 1 mm. Stylopodium 0.5 mm high. Fruits subglobose, ribbed, 4-5 mm diameter. Fig. 131.2.1.

Upland rain forest, secondary forest, woodland; 1450-2800 m. TU GD WG SU AR HU IL KF SD; Cameroun, S Sudan, Uganda, Zaire, Burundi, Tanzania,

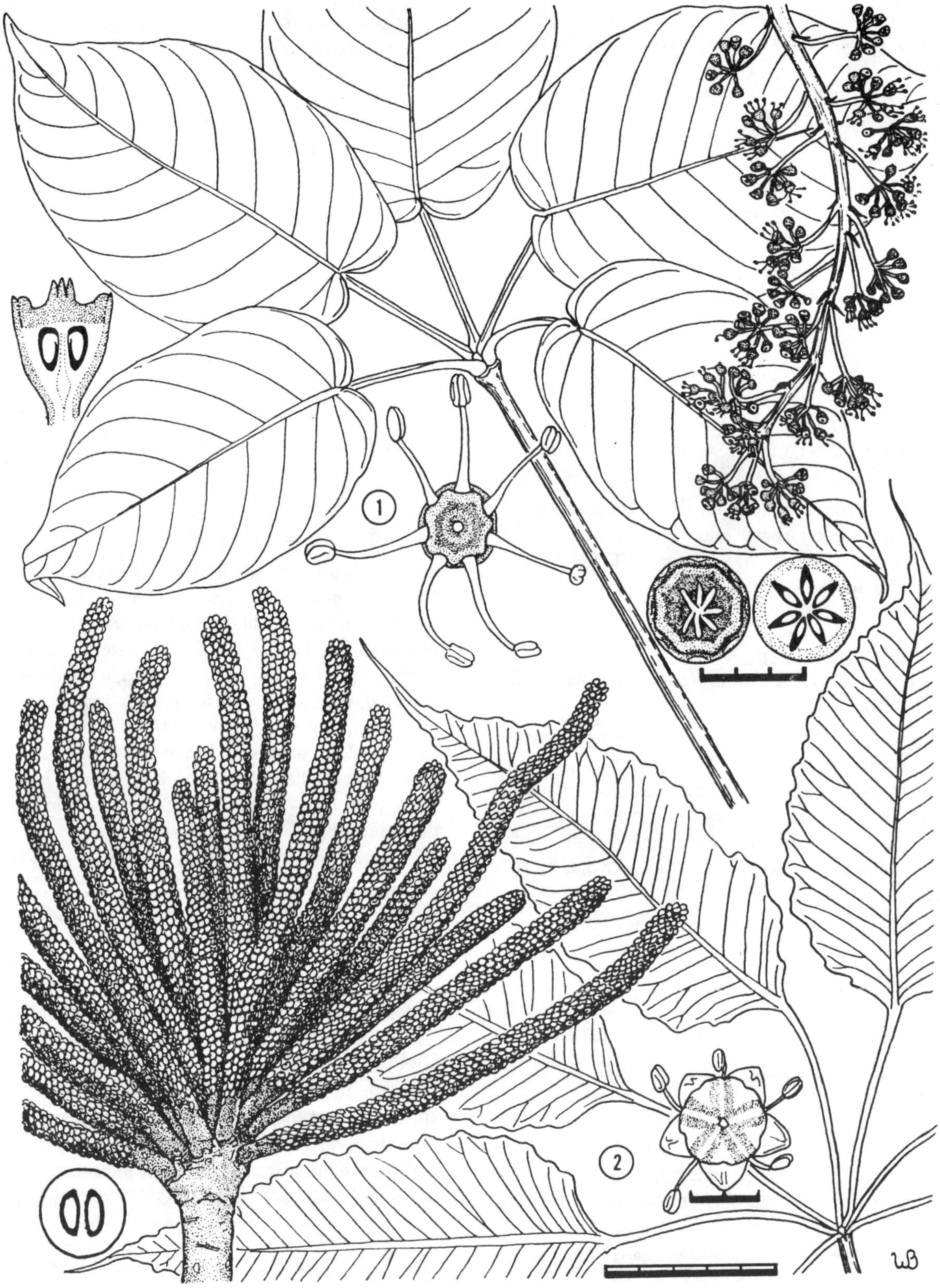

Figure 131.2.1 *SCHEFFLERA ABYSSINICA*: part of a leaf; part of an inflorescence; longitudinal section of an ovary; top view of a male flower with petals fallen off; top view and cross section of ovary. 2 *CUSSONIA HOLSTII*: part of an inflorescence; part of a leaf; cross section of ovary; top view of flower. (Reproduced with permission from Burger, *Families of Flowering Plants in Ethiopia*: figure 50, 1967.)

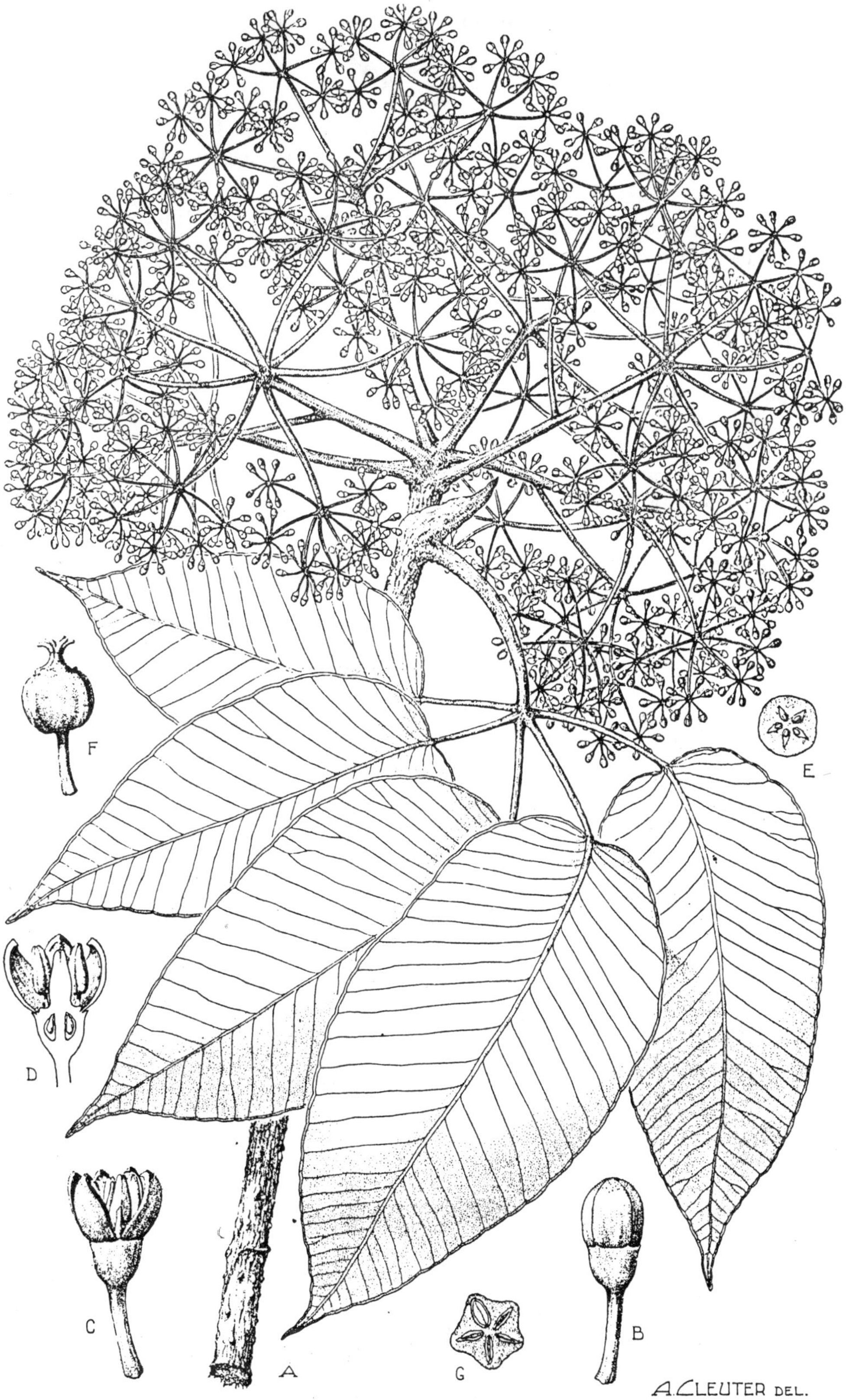

Figure 131.3 *SCHEFFLERA MYRIANTHA*: A - flowering branch x 1/2; B - flower bud x 4; C - open flower x 4; D - longitudinal section through a flower x 4; E - cross section of ovary x 4, all from *Beupaert* 4405; F - fruit x 2; G - cross section of fruit x 2, both from *Lebrum* 7889. Drawn by A. Cleuter.

Zambia, Malawi. *Burger* 2631; *W. de Wilde* 6321; *Friis* et al. 153.

4. **HEDERA** *L.* (1753)

H. helix *L.*

Webb, in *Fl. Eur.* **2**: 314 (1968).

Stems woody, climbing or creeping, with numerous adventitious roots. Young shoots and inflorescences densely covered with stellate to peltate hairs. Leaves shining, usually dark green (see below), dimorphic; on flowering shoots 6-10 x 2-12 cm, narrowly elliptical to suborbicular, margin entire; on vegetative shoots up to 15 x 15 cm, usually much smaller, palmately 3- or 5-lobed. Flowers hermaphrodite, in globose umbels, solitary or grouped in racemose panicles. Sepals very small, deltate. Petals 3-5 mm yellowish-green, patent, later deflexed. Berry black or yellow with 2-3 seeds. Fig. 131.4.

Ornamental in public and private gardens climbing fences and walls, and sometimes covering the ground, also a common pot plant; *c* 2300 m. EW (Asmara), SU(Addis Ababa) and probably in other towns in the highlands; Europe from the Mediterranean to 60°30'N in Norway and east to Latvia and Ukraine. *O. Ryding* 1148.

Introduced from Europe where this (and other species) are widely cultivated and from which a large number of cultivars have been developed (McAllister (1981) in *International Dendrology Society Yearbook*: 106-109). The cultivars differ mainly in the size, shape and colour of the leaves - from dark shiny green to golden green to light yellow and with variegated patterns. Cuttings taken from the flowering shoots form erect shrubs.

Figure 131.4 *HEDERA HELIX*: 1 - fertile shoot with ± entire leaves and ripe fruits; 2 - sterile shoot with lobed leaves and clusters of adventitious roots, both x 2/3. Both from *O. Ryding* 1148. Drawn by Damtew Teferra.

A GLOSSARY OF BOTANICAL TERMS.

abaxial - the side or part facing away from the stem; the lower surface of a leaf (= dorsal): compare with adaxial.

abortion - the failure of a part to develop fully; suppression of a part usually present.

acaulescent - without an easily seen stem (see fig. 104.4.6, p. 279).

accrescent - increasing in size with age; as sepals enlarge around a developing fruit (see fig. 95.32.1, p. 186).

accumbent - leaning against another part, as cotyledons against the radicle.

achene - a small dry fruit, not opening when ripe, with only one seed; for example the fruits of *Urtica* and *Rumex*.

achlamydeous - without a perianth (sepals and petals lacking); as found in the flowers of *Casuarina* (see fig. 101.1.5, p. 263) and *Euphorbia*.

acicular - needle-like; very narrow, stiff and pointed.

acropetal - development proceeding from the base to the apex with the youngest parts at the top and the older parts below, as in a raceme; movement of materials upwards to the apex.

actinomorphic - used for flowers which are radially symmetrical (at least in the perianth) and can, therefore, be divided along two or more longitudinal sections into halves which are mirror images of each other (see fig. 3, p. xx); the term applies mainly to the perianth: used as synonymous with radially symmetrical or regular; compare with zygomorphic, amorphic and irregular.

aculeate - armed with prickles, but not with spines, for example *Caesalpinia oligophylla* (see fig. 95.1.3, p. 52) and *Rosa abyssinica*.

aculeolate - armed with small prickles.

acuminate - with a tip that becomes gradually narrower to a slender point (see fig. 104.16, p. 291).

acute - with a tip that comes to a sharp point with straight-sided edges that form an angle of less than 45° (see fig. 116.2, p.382).

adaxial - the side or part facing the main stem; the upper surface of a leaf (= ventral): compare with abaxial.

adnate - when an organ or part is united to a different organ or part; for example the androecium (stamens) with the gynoecium (pistil) in Asclepiadaceae and Orchidaceae.

adpressed - parts pressed closely to the axis with an angle of divergence less than 15°.

adventitious - in an unusual or abnormal place or coming from a distant place (buds that arise elsewhere than in the leaf-axils or at stem-tips, i.e. on roots; roots that arise elsewhere than from the radicle or other roots, i.e. from a stem or leaf).

aerenchyma - a plant tissue composed of unthickened, often irregularly shaped cells surrounding large air spaces; found in many plants associated with aquatic habitats.

aerial - of stems which arise from horizontal rhizomes: of roots which arise from stems and which have special functions such as the aerial roots of orchids which absorb water, and the aerial roots of climbers which attach the plant to its substrate.

aestivation - the manner in which the flower parts are folded or arranged in bud, before opening (see fig. 4, p. xxi).

afro-alpine - a distinct vegetation zone found above 2800-3000 metres on African mountains.

agamous - without gametes; reproduction without either production of and/or normally functioning gametes; (normally functioning pollen and ovules are not produced but fruits and seeds may be formed).

aggregate fruit - fruit formed from several separate pistils from the same flower, often together with the receptacle, e.g. *Rubus* (see fig. 91.1.1), the strawberry, *Fragaria*; can also be used for fruits formed from an inflorescence as in figs, *Ficus*.

albumen - a name used for endosperm or perisperm in a seed; storage tissue which is not part of the embryo.

algae - a large group of simple plants that do not possess flowers or seeds; most algae grow in water.

allopolyploidy - the type of polyploidy found in organisms of hybrid origin with at least two sets of chromosomes, one from each original parent being present (for example *Triticum aestivum*, bread wheat).

alternate - used for leaves that are attached, one at each node, at different levels along the stem (see fig. 125.2, p. 492); used for stamens borne between the sepals or petals (see fig. 107.1, p. 330).

alternation of generations - the complete life cycle of a higher plant (bryophytes, ferns, gymnosperms and flowering plants) in which a diploid sporophyte generation is followed by a haploid gametophyte one, the transition from one generation to the other being marked by meiosis with the production of spores (sporophyte to gametophyte generation) and the fusion of gametes (gametophyte to sporophyte generation).

ament - see catkin.

amentiferous - bearing aments or catkins which are closely packed spikes or racemes of small flowers with reduced perianths and which are usually wind pollinated (for example *Salix*, see fig. 99.1, p. 259).

amorphous - lacking shape or form: can be used for flowers lacking symmetry (compare with actinomorphic and zygomorphic).

amphi- - a prefix meaning "both".

amphiphloic - having phloem tissue on either side of the xylem.

amphitropous - used for ovules where the ovule-stalk (funicle) is attached near the middle of the ovule which is half bent over.

amplexicaul - used for leaves where the base clasps the stem (see fig. 1.2, p. xviii).

amyloplast - a plastid lacking chlorophyll which is involved with the synthesis and storage of starch.

anastomosis - union of one vein with another, the connections forming a network.

anatropous - used for ovules completely bent over along the ovule-stalk (the funicle) bringing the micropyle close to the point of attachment of the funicle to the placenta.

androecium - the male structures of a flower; stamens and accessories.

androdioecious - used for a species which has 2 sexual forms, plants with only male flowers and plants with bisexual flowers.

androgynophore - an elongated part of the receptacle carrying both androecium and gynoecium between the perianth and stamens in a bisexual flower (see fig. 3, p. xx).

andromonoecious - used for a plant having male and bisexual flowers on the same plant; no purely female flowers are formed.

androphore - an elongated part of the receptacle above the perianth which bears only the stamens.

anemophilous - pollinated by wind.

aneuploid - where the chromosome number is not an exact multiple of the haploid number.

angiosperms - a major group of plants and the dominant land plants; seed plants with the seeds usually enclosed in an ovary (see fig. 6, p. xxii).

anisogamy - (not isogamous) the production or fusion of gametes that are visibly different, usually in size.

annual - a plant which completes its life cycle in one growing season and then dies: where there are two or more growing seasons in a year, an annual species can have two or more generations in a year.

annular - arranged in a circle or in a ring; the shape of a ring.

annulus - the cluster or line of thick-walled cells found on the wall of the sporangium in ferns and below the peristome of moss capsules.

ant galls - swollen hollow structures, usually on the stems or leaves of plants, which are inhabited by ants.

anterior - away from the axis (= abaxial); often used for the petal(s) found on the front of a flower.

anther - the part of the stamen producing the pollen, usually divided into pollen-sacs or pouches called *thecae*.

antheridium - the male sex organ formed in the gametophyte which produces antherozoids (synonymous with spermatozoids).

antherocyte - antherozoid mother cell, the cell which differentiates to form antherozoids (synonymous with spermatocyte).

antherozoid - a male gamete which moves by swimming using flagellae (synonymous with spermatozoid).

anthesis - the time when the pollen is shed and the stigma is receptive to pollen; from the opening of the flower bud to the setting of the seed.

anthocyanins - water-soluble pigments giving pink, red, purple, violet or bluish colours to flowers or vegetative parts of plants.

anticous - the position furthest away from the stem or axis.

antipetalous - used for stamens or staminodes which occur opposite the petals (see fig. 125.4, p. 495).

antisepalous - used for stamens or staminodes which occur opposite the sepals (see fig. 102.1, p. 269).

aperturate - a pollen grain having one or more apertures.

apetalous - without petals or corolla (see fig. 91.1.1, p. 32).

apex - the tip or end-point of a structure.

apical - concerning or near the apex.

apical placentation - where the ovules are attached to the top of the ovary (see fig. 6.15, p. xxii).

apiculate - ending abruptly in a short sharp point (see fig 11.8.6, p. 395).

apiculus - a small extension between 2 leaflets in *Balanites* (see fig. 121.1.11, p. 435).

aphlebiae - small leafy outgrowths at the base of the rachis of certain fossil and living ferns.

apocarpous - with the carpels free from one another (see fig. 88.6.3, p. 17).

apogamy - meaning "without gametes"; the production of a sporophyte without the fusion of gametes taking place in the gametophyte.

apomixis (apomictic) - in seed plants, the production of seeds without normal sexual fertilization.

apophysis - the exposed outer surface of the megasporophyll or ovuliferous scale in gymnosperms: the sterile base of the spore capsule in mosses.

apospory - (without spores) the production of a gametophyte from a sporophyte without normal meiosis intervening, spores may be formed but they have the same genetic complement as the parent sporophyte.

appendage - a part added or attached to another, usually larger, structure (see fig. 104.4.7, p. 279).

appressed - lying close or pressed flat against a surface or axis.

aquatic - living in water.

archegonium - the flask-shaped, multicellular, female sex organ characteristic of all higher plants except some gymnosperms and the angiospersm; it contains the egg cell.

archesporium - the group of cells in the developing sporangium which give rise to the spore mother cells (sporocytes or meiocytes).

arcuate - curved or bent like a bow, usually used to describe a particular leaf shape.

areolate - divided into distinct spaces; with areoles.

areole (areola) - a space marked out on a surface; used to described the raised area on the surface of the seeds of the Fabaceae (Leguminosae) subfamily Mimosoideae; also used to denote the area where spines and glochids arise on the stems of Cactaceae; the open area (cell) formed by anastomosing veins.

aril - an outer covering or appendage, often fleshy and/or brightly coloured, that encloses the seed or part of the seed and develops from the stalk of the seed (see fig. 108.1.11, p. 334).

aristate - with a long, very narrow, bristle-like point.

armed - provided with a means of defense such as spines.

articulate - with joints or nodes or places where a part will naturally break off.

ascending - a plant where the shoots start lying on the ground and then turn to grow mainly upright; usually used for herbs

asepalous - without sepals.

asexual - without sex; without the production or fusion of gametes.

asymmetric - without any symmetry

atropous - an ovule which is not bent, with the funicle attached to the opposite end from the micropyle (= orthotropous).

attenuate - tapering gradually to a slender point.

auriculate - with an ear-like lobe or appendage; as at the base of a leaf or other organ (see fig. 95.47.1, p. 251).

auricle - an ear-like lobe or appendage.

autopolyploidy - where there are three or more identical sets of chromosomes in the nucleus.

autosome - a chromosome that is not a sex chromosome.

autotrophic - producing food by itself, neither parasitic nor saprophytic, the normal situation in green plants.

auxin - a general term for organic substances (naturally occurring or synthetic) which regulate plant growth.

awl-shaped - used for a leaf which is narrow, flat, stiff and sharp-pointed (= subulate), such as in *Agave* and *Arucaria*.

awn - a fine, very narrow, bristle-structure usually at the tip of a leaf or bract; as in the spikelets of Poaceae (Gramineae): see also aristate.

axial - an adjective for any part which arises from the angle between a leaf and its subtending axis.

axil - the upper angle made between a leaf attachment and a stem.

axile - used for the placentation of ovules that are attached to the central axis or the inner angle of the locules within the ovary (see fig 6, p. xxii).

axillary - in or arising from an axil.

axis - the central line of any symmetrical or nearly symmetrical body; in plants the main stem or branch on which other organs are arranged.

baccate - like a berry with fleshy and pulpy tissue.

barbed - with stiff short spines or bristles that point backward away from the apex of the organ.

barbellate - shortly barbed.

basal - found at or near the base of a structure.

basal placentation - used for the attachment of the ovule at the bottom of the locule near the base of the ovary (see fig. 6, p. xxii).

basifixed - used for the attachment of the filament to the base of the anther.

basipetal - development of parts towards the base, the older parts are found above.

beaded - a structure which looks like a string of beads, e.g. the fruits of some Fabaceae (see fig. 95.41.Ib, p. 229); synonymous with torulose.

bearded - with tufts of hairs or awns (see fig. 95.29.2c, p. 170).

berry - a fleshy or juicy fruit with a soft outer portion and the seeds immersed in the fleshy or pulpy tissue, the seeds are not surrounded by a woody or stony endocarp, for example tomato. Compare drupe.

betalains - red and yellow alkaloid pigments found in plants of the Caryophyllales.

bi- - a prefix meaning "two" or "twice".

bicollateral vascular bundles - with phloem both outside and inside the xylem (as viewed in cross or radial section).

biennial - living for two growing seasons; usually producing only vegetative growth in the first season, and flowering and fruiting in the second.

bifid - divided into two parts near the top.

bifoliate - having two leaves.

bifoliolate - having two leaflets (see fig. 121.1, p. 435).

bifurcate - with two branches or divisions at the end; Y-shaped.

bilabiate - with two lips; a corolla with an upper and lower lip, for example, the flowers of many Lamiaceae (Labiatae).

bilateral - when a structure has only one plane of symmetry; if cut along this plane the two halves are mirror images of each other; usually used for flowers (synonymous with zygomorphic).

bilocular - with two locules or chambers; usually used for ovaries and sometimes for stamens.

binary fission - reproduction occurring when a single cell divides into two theoretically equal parts.

binomial - the scientific species-name composed of two words; the first is the generic name, the second the specific epithet.

bipinnate - twice pinnate, when the first divisions of a leaf are themselves pinnate (see fig. 2.6, p. xix).

bipinnatifid - twice pinnatifid.

biramous - applied to hairs of the genus *Indigofera* (Fabaceae) that are 2-branched near the base with the 2 branches usually pointing in opposite directions and lying adpressed to the surface.

biseriate - having two series or two whorls (as having both calyx and corolla).

bisexual - having both sexes and producing both male and female gametes. Bisexual flowers have both functional stamens and pistils.

bisymmetric - having two planes of symmetry, see also bilateral symmetry.

blade - the flat broad part of a leaf or petal; synonymous with lamina.

bloom - the flower or process of flowering (as in the flowers blooming); also used for a whitish waxy powder covering a surface.

bole - the unbranched stem or trunk of a tree.

bordered pit - a pit in xylem tissue in which a thin area of primary wall is overhung by lignified secondary wall. The centre of the pit is often thickened to form a torus; characteristic of secondary xylem in gymnosperms.

bostryx - a cymose inflorescence with successive branches on one side only; normally coiled like a spring.

bract - a usually small leaf-like structure usually associated with a flower and/or inflorescence and found at the base of the pedicel or peduncle.

bracteate - having bracts.

bracteole - a small bract on the pedicel, or close under the flower, between the bract and the flower.

branch - a portion of a stem system which is attached to the main stem; often used for a stem from a woody plant.

branchlet - the smallest part of a branch; the growth of the current or the last growing season (a twig).

brevistylous - meaning "short-styled"; used for flowers which have short styles where there are also flowers in the same species/population which have long styles - a condition described as heterostylous.

bristle - a stiff hair.

bud - an undeveloped shoot that may give rise to a branch or a flower.

bud scales - modified leaves or stipules which cover and protect an undeveloped shoot and drop off as soon as the shoot starts growing.

budding - the production of buds: in horticulture used for the grafting of a bud of one kind onto a rootstock of another compatible kind of plant, for example the budding of sweet orange onto the rootstock of sour orange: a type of asexual reproduction in which a small protuberance develops and is separated from the parent cell, typical of yeasts.

bulb - a short underground stem with a crown of usually fleshy overlapping scale-like leaves, for example an onion.

bulbel - a small bulb produced from the base of a larger bulb.

bulbil - a very small bulb produced in the leaf axils of some plants and providing a means of asexual reproduction.

bulblet - a small bulb, irrespective of origin, a bud-like asexual reproductive structure produced on the leaves of some ferns and angiosperms.

bullate - used for leaves with a surface that is prominently raised between the veins.

caducous - falling off soon or quickly.

caespitose - forming mats or broad tufts.

calcicole - a plant which can grow in soil with a high calcium carbonate (lime) content.

calcifuge - a plant which cannot tolerate soil with a high lime content.

callus - an irregular mass of parenchymatous tissue formed over wounds; often capable of being cultured indefinitely in nutrient media.

calyculate - having bracts round the calyx, or an involucre resembling an outer calyx.

calyculus - a ring of bracts or the involucre below the calyx.

calyptra - a cap-like or lid-like covering of certain fruits or flowers that comes off in one piece, for example, the perianth forms a calyptra which comes off when a *Eucalyptus* flower opens: the cap covering a moss capsule.

calyx - the outer whorl or envelope of most flowers, made up of the free or united sepals.

calyx-tube - the tube formed by the united sepals, also used for the receptacle-tube and hypanthium (see these terms).

cambium - a secondary meristem: vascular cambium produces secondary xylem and phloem, the cork cambium produces cork (phellogen) and secondary cortex.

campanulate - bell-shaped, with a broad tube and a wide opening.

campylotropous - for an ovule in which one side has grown faster than the other so the whole ovule is bent over, bringing the micropyle near the hilum.

canal - air spaces running longitudinally in a stem or root.

canescent - covered with a grey pubescence or greyish colour.

cantharophily - pollinated by beetles.

cap - a lid-like covering (= calyptra).

capitate - like the head of a pin (as in the thickened stigma of some flowers) or a compact cluster of flowers forming a head.

capitulum - a dense head-like inflorescence of usually sessile flowers (see fig. 95.11.1a, p. 95).

capsule - a dry fruit produced by an ovary composed of 2 or more united carpels and opening by slits or pores or breaking into pieces when ripe (see fig. 108.1.8, p. 334).

carina - synonymous with keel, a long projecting ridge at the back or on the bottom of a structure: sometimes used for the 2 anterior petals of the papilionaceous flower (Leguminosae subfamily Papilionoideae) which are partly fused along their lower margins (see fig. 95.35.1a, p. 203); also used for the lower (abaxial or anterior) petal in many Polygalaceae.

carinal canal - an air channel found beneath a stem ridge and associated with a vascular bundle in *Equisetum*.

carinate - a synonym for keeled, with a projecting longitudinal ridge.

carotenoids - a group of yellow pigments occurring in chloroplasts, chromoplasts and elsewhere in plant cell.

carnivorous plants - plants which catch small animals and digest them to obtain nitrogen. Found in very wet habitats.

carpel - the basic unit of the gynoecium or pistil consisting of an ovary or part of an ovary with an associated style and stigma; generally considered to have developed from a leaf-like structure, often used in a theoretical sense. A simple pistil, such as a legume, has only 1 style and stigma and 1 locule (see fig. 6.14, p. xxii): a compound pistil is made up of 2 or more carpels joined together (see fig. 6, p. xxii for different examples).

carpophore - a prolongation of the floral axis between the pistils or carpels which bears the fruit, as in the narrow stalk that supports the two mericarps in the fruit of the Apiaceae (Umbelliferae).

caruncle - an outgrowth on the surface of a seed near the hilum.

caryopsis - a modified achene in which the seed and the ovary-wall (pericarp) have become united; characteristic of the Poaceae (Gramineae).

casparian band or **strip** - a circle of suberised thickening found on the radial and upper and lower walls of the cells forming the endodermis in vascular plants.

catkin - a spike-like inflorescence with many bracteate and closely crowded unisexual flowers that is often pendulous and may resemble a cat's tail (see fig. 105.1, p. 303).

caudate - abruptly ending in a long and very narrow tip.

caudicle - the stalk on which the pollinia are born in orchids.

caulescent - with a clearly visible stem above the ground.

cauliflorous - producing flowers directly from the older stems.

cauliflory - having flowers produced directly from older, usually woody, stems, for example *Haleria lucida*.

cauline - borne on or arising from the stem.

cell - in descriptions of flower-parts the word cell is often used as a synonym for chamber, locule, or theca; a 4-celled ovary is a 4-locular ovary. This use must not be confused with the wider use of the word cell as the structural unit of living things.

cell sap - the watery contents of the large vacuole usually present in plant cells.

central placentation - see free central placentation.

centrifugal - developing from the centre out, the older parts at the centre.

centripetal - developing from the outside towards the centre, the older parts on the outside.

cespitose - forming mats or broad tufts or cushions (an alternate spelling of caespitose).

chaff - a small dry thin bract or scale.

chalaza - the base of the ovule near the attachment of the stalk or funicle.

chamaephyte - a plant in which the perennating buds occur within 20-30 cm of the soil surface, i.e. dwarf shrubs. Many Afro-alpine plants are chamaephytes, for example *Alchemilla haumannii*, see p. 38 and fig. 91.4, p. 39.

chartaceous - with paper- or parchment like texture; thin and opaque.

cheiropterophily - pollination by bats, for example *Kigelia aethiopica* and *Adansonia digitata*.

chemotaxis - movement of a motile (mobile) organism in response to chemical stimuli, such as antherozoids locating archegonia in a gametophyte.

chlorophyll - the green pigment that absorbs light energy to form energy rich compounds in the process of photosynthesis.

chloroplast - an organelle bounded by a membrane with an internal lamellar structure containing photosynthetic pigments.

choripetalous - having entirely free or separate petals: synonymous with polypetalous. Most families in Volume 3 are choripetalous.

chromatin - a complex of deoxyribonucleic acid (DNA) and basic proteins present in chromosomes and defined by its staining properties.

chromatophore - an intracellular organelle containing pigments, a chloroplast is one kind of chromatophore; more usually used for other pigments such as those found in flower petals.

chromoplast - a plastid lacking chlorophyll, but accumulating pigments, principally carotenoids.

chromosome - found in the nucleus; an organised structure composed of DNA, RNA and histones, carrying genes.

ciliate - with a fringe of hairs along the edge, or with regularly arranged hairs (see fig. 105.14M, p. 326).

cilium (pl. cilia) - unicellular hairs found on leaf margins and other structures: also used for motile organelles; usually numerous, almost identical to a short flagellum.

cincinnus - a type of cyme, the same as a helicoid cyme (see fig. 88.6.2, p. 17).

cinereous - a dark grey or ash-colour.

circinate - inwardly coiled upon itself like the spiral of a snail shell; characteristic of the young leaves of ferns and Droseraceae.

circumnutation - the sweeping movements made by the tip of a stem, particularly marked in twining plants, caused by unequal rates of growth around the axis.

circumcissile - opening as if cut circularly and the upper part coming off like a cap or lid.

cirrhose - having tendrils or with a long narrow coiling tip that acts as a tendril (see fig. 95.47.1, p.251).

cladode - a leaf-like structure formed by a modified stem.

cladophyll - a synonym of cladode.

clasping - used for leaf-bases that partly or completely enclose the stem.

class - a unit of classification for large, usually very distinctive groups of plants, such as the class Magnoliopsida (Dicotyledonae).

clavate - club-shaped, thickened at the end.

claw - the very narrow base of some petals; clawed petals have a long, narrow base and broad terminal lobe (as in Geraniaceae, some Caryophyllaceae and Caesalpinioideae, see fig. 95.6.1a, p. 69).

cleistogamous - flowers that fertilize themselves before they open, or which do not open at all even after fertilization; they are usually small and near the ground (as in *Commelina* and *Oxalis*).

climbing - used to describe plants that use other plants or objects as a means of support but have their roots in the ground.

clone - a population of plants all having the same genetic makeup, usually produced by vegetative reproduction or pseudosexual reproduction in which seeds are produced without sexual fertilization taking place - see apomixis.

coccus - a separate part of a lobed fruit, usually l-seeded, as in the some Rhamnaceae and Euphorbiaceae.

coenocyte - a multinucleate plant body or cell.

coherent - parts in close contact but not united.

coleoptile - in monocotyledons, the sheath around the plumule of the embryo and emergent seedling, as in grasses.

coleorhiza - in monocotyledons, the sheath around the radicle of the embryo.

collenchyma - a strengthening tissue composed of columnar cells with heavy deposites of cellulose at the angles of the primary walls.

colliculous (colliculose) - a surface covered in round swellings.

colporate - in pollen grains with both furrows (colpae) and pores.

colpus - a furrow forming a thin area in the exine of a pollen grain through which the pollen tube usually emerges.

column - the solid central structure in orchid flowers formed by the union of style and stamens; the stalk or tube formed by the united stamens, as in Malvaceae.

coma - a tuft of hairs on the end of some seeds.

comal - found at the tip or apex.

commisure - the faces (or seam, where the two faces are cohering) of two cohering carpels or mericarps (the plane along which the fruit separates) in the Apiaceae (Umbelliferae).

comose - with a tuft of hairs.

companion cell - in angiospersm, the small, nucleated cell with dense protoplasm associated with the sieve tube and derived from the same sieve tube mother cell.

complanate - arranged in the same plane.

compound - composed of two or more similar parts; the opposite of simple.

compound fruit - when the fruit of separate flowers becomes united; for example, *Dorstenia* (see fig. 104.3.3, p. 276).

compound leaf - when the leaf has two or more leaflets or blades. Many families in Volume 3 have compound leaves.

compound pistil - a pistil made up 2 or more carpels.

compressed - flattened or pressed together, either from side to side (laterally compressed) or from front to back (dorsally compressed).

concolorous - where the upper and lower surface of a leaf are more or less the same colour: compare with discolorous.

concrescent - grown together, coalesced.

conduplicate - folded together lengthwise.

cone - a dense and often woody series of bract-like sporophylls arranged along an axis; characteristic of many gymnosperms.

confluent - merging or blending together.

congenital - an inherited characteristic which becomes apparent during growth.

connate - when parts of the same kind are united.

connective - the tissue that units the pollen sacs in an anther; it may form an extension of the filament.

connivent - coming close together or converging, but not united, often closer together above than below.

contorted - twisted in bud: synonymous with convolute.

convolute - used to describe sepals or petals in bud, when one edge is always above the adjacent part and the oppoite edge is always below an adjacent part (see fig. 4.3, p. xxi).

coplanar - in the same plane.

coradial - on the same radius.

coralloid - a multibranched structure with all parts close to each other; e.g. coralloid roots in cycads.

cordate - when the base of the leaf is deeply notched (see fig. 119.1.1, p. 400).

coriaceous - with a thick and firm texture, similar to leather.

cork - the outer tissue covering woody stems, formed by the activity of the cork cambium and consisting of dead cells with suberised walls which are impervious to water.

corm - a short thick underground stem, which grows vertically, for example, in many Iridaceae.

corolla - the inner whorl or inner envelope of sterile appendages in the flower, made up of the free or united petals which are usually colourful.

corona - a whorl of appendages between the petals and the stamens, and sometimes united to form a ring or cylinder, usually borne on the petals or corolla-tube, typical of the Asclepiadaceae.

corpus - in stems, the central part of the apex in which cell divisions occur in variety of planes. In pollen grains of conifers, the body of the pollen grain, within the air float (bladder).

cortex - the tissue in stems and roots found between the vascular cylinder and the epidermis or cork cambium.

corymb - a panicle-like inflorescence in which the branches or flower-stalks start from different places on the stem but all the flowers are born at about the same level (see fig. 88.2, p. 9).

costa - a rib, or the midvein of a simple leaf.

costate - having a rib or ribs that are usually projecting above the surface of the structure and parallel with its long axis.

cotyledon - the seed leaf, usually the first leaf or leaves produced by the germinating seed; the cotyledons are usually different from the following leaves, and sometimes the cotyledons do not leave the seed.

crenate - the margin notched with rounded or broad and blunt teeth or projections (see fig. 95.43.2a, p. 235).

crenulate - margin with small blunt teeth or projections (see fig. 104.11.1, p. 287).

crisp (crisped) - having the margin excessively irregularly divided and twisted

crispate - with a curled edge.

cross fertilization, cross pollination - where the female gametes of one plant are fertilized by male gametes from another plant with a different genetic make-up: the pollen of one plant is transfered to the stigmas in another plant with a different genetic make-up.

crustaceous - hard, thin and brittle.

crozier - the shape of a young fern frond before the lamina opens out: see circinate (only single "n").

cryptogam - a plant reproducing by small spores and not by seeds.

cryptophyte - a plant in which the perennating buds are below the soil surface.

cucullate - hooded or hood-shaped (a hood is a covering for a person's head) (see fig. 105.6, p. 313).

culm - a thin slender stem which carries the inflorescence, especially of grasses or sedges.

cultivar - a variety of plant that has been developed under cultivation and selection by man.

cuneate - wedge-shaped, as when the base of the blade tapers gradually to the top of the petiole (see fig. 109.2.1, p. 350).

cupule - a cup-shaped structure surrounding, but not enclosing, the ovule of certain seed plants, particularly in seed ferns.

cuspidate - with the tip abruptly narrowed to a point (can be considered intermediate between and caudate); fig. 119.1.1, p. 400, show an example verging towards acuminate.

cuticle - the non-cellular waxy coating of the epidermis of all higher plants.

cutin - the waxy material forming the major part of cuticle.

cyathium - a flower-like inflorescence characteristic of the genus *Euphorbia*; it is made up of naked, unisexual flowers that are grouped together within fused, perianth-like bracts.

cyclic - arranged in cycles or whorls; the opposite of spiralled.

cymbiform - shaped like a boat.

cyme - an inflorescence in which the central axis is terminated by a flower which opens first, this flower is subtended by two opposite branches each of which ends in a flower, these open next and are likewise subtended by two opposite branches; this branching-pattern may continue (see fig. 88.3.2, p. 11).

cymose - cyme-like.

cymose branching - where the terminal bud becomes a flower or aborts and growth is taken over by one or more lateral buds.

cypsela - an achene united to the calyx-corolla-tube as from an inferior ovary; characteristic of the Asteraceae.

cystolith - a mineral concretion, usually crystals of $CaCO_3$, within the epidermal cell of the leaves of some plants, almost always visible as light coloured spots on leaves of the Urticaceae (see fig. 105.6.4, p. 313).

cytoplasm - the protoplast of the cell without the nucleus.

deca- - a prefix meaning "ten".

deciduous - falling off at the end of the growing season; the opposite of evergreen.

declinate - bent or curved downward or forward; also used for stamens in sympetalous flowers, such as those of Lamiaceae, which are directed towards or inserted on the ventral (abaxial) side of the corolla tube and then usually bend up.

decumbent - lying on the ground but with the ends growing upwards.

decurrent - as when the leaf-edges continue down from the blade and are united with the petiole or stem.

decussate - when each pair of opposite leaves is inserted on the stem at right angles (90°) or perpendicular to pairs of leaves both above and below (see fig. 88.2.A1, p. 9).

deflexed - bent abruptly downward or outward; the opposite of inflexed, and often used as synonymous with reflexed.

dehiscence (dehiscent) - the method or process of opening; as in anthers and fruits.

deliquescent - to melt away or to break up into many small parts, as in petals that soften and liquify (Commelinaceae), or as in the trunks of many trees which "break up" into many branches above (many *Acacia* spp.); the opposite of excurrent.

deltoid (deltate) - shaped like an equal-sided triangle.

dentate - with a toothed margin, the teeth pointing outwards, not forward (see fig. 123.11A, p. 463); compare serrate.

denticulate - with very small teeth perpendicular to the margin (see fig. 119.6.1, p. 408).

depressed - pressed down or flattened from above.

derived - originating from an earlier form or group.

determinate - having a definite end-point, as in an inflorescence in which the main axis ends in a flower and cannot continue to grow; the opposite of indeterminate.

dextrorse - clockwise, towards the right.

di- - a prefix meaning "two" or "away from", "unlike"

diadelphous - forming two bundles or groups, as in the stamens of some legumes which have 9 united stamens and one free.

diandrous - having two stamens.

diaspore - a plant part that breaks away from the parent to develop into new individual.

diarch - applied to xylem tissue consisting of two strands, or having two protoxylem groups in the metaxylem.

dichasium - an cymose inflorescence in which the main axis and branches end in flowers that are subtended by two opposite branches (see cyme); an inflorescence which consists of dichasia (see fig. 108.3.3, p. 339).

dichlamydeous - having two coats or envelopes, as flowers with both calyx (sepals) and corolla (petals).

dichotomous - dividing regularly into two parts; as in a dichotomous key where the contrasting choices are always two.

diclinous - having the male and female reproductive organs on separate parts of the same plant.

dicotyledonae - one of the major subdivisions within the angiosperms, characterized by having two cotyledons within the seed; also called dicotyledons or dicots.

dictyostele - a dissected solenostele composed of a network of meristeles; characteristic of some ferns.

didimous - in pairs, two closely united structures, or 2-lobed.

didynamous - in two pairs of unequal length; as four stamens with two long and two short ones; common in *Commiphora* (see fig. 123.8D, p. 458) and Lamiaceae (Labiatae).

differentiation - the process in which different cells, tissues and organs arise from unspecialized dividing tissues.

diffuse - of branching that is open or loosely spreading.

digitate - where the parts are attached to the same point, as in a palmately compound leaf.

dimorphic - having two forms, as in juvenile and adult foliage in *Eucalyptus* and some intruduced species of *Acacia* with true bipinnate leaves on seedlings and phyllodes on older plants.

dioeceous (dioecious) - plants with unisexual flowers in which the male and female flowers are not found on the same plant, as if the plants are either male or female, but not both.

diploid (2n) - having twice the basic (haploid) number of chromosomes.

diplostemonous - having the stamens in two whorls (and usually twice as many as the petals) with the outer whorl alternating with the petals and the inner whorl opposite the petals (see fig. 3.3, p. xx); the opposite of obdiplostemonous.

disc (disk) - an enlargement of the receptacle beneath or around the pistil or within the corolla or stamens, usually in the form of a ring or cushion, or of separate gland-like parts (see fig. 108.3.5, p. 339); the term disc is also used for the the broad receptacle on which the flowers are born in the capitate or head-like inflorescence of the Asteraceae (Compositae).

disc-flower (disc-floret) - the flowers of the Asteraceae (Compositae) that have a regular corolla-tube.

disciform - shaped like a disc, circular and flat, or lobed above.

discoid - like a disc in form; also used for the capitate or head-like inflorescence of Asteraceae (Compositae) that lacks ray-flowers or ligulate flowers (has only disc-flowers).

disk - see disc.

dissected - divided into many slender parts.

disseminule - a plant part that breaks away from the parent to develop into new individual: synonymous with diaspore.

dissepiment - a partition of an ovary or fruit.

distal - the part or end furthest away from the base or point of attachment; the opposite of proximal.

distichous - arranged in two opposing rows along the opposite sides of the stem; neither decussate nor spiral (see fig. 105.7, p. 315).

distinct - free and separate from other parts.

diurnal - active during the day, as in flowers that open during the day and close during the night; the opposite of nocturnal.

divaricate - spreading apart widely and in different directions.

divergent - spreading apart but not as widely as in divaricate.

divided - separated almost to the base.

division - a unit of classification for large and/or very distinctive groups of plants, similar to the term Phylum in zoology; as in the division Tracheophyta, plants with a vascular system.

DNA - deoxyribonucleic acid, the component of chromosomes carrying the genes.

dormant - a structure or organ which is not active.

dormant bud - an inactive bud. Plants produce many more buds than can develop at any one time. If something happens to remove the active shoots from a plant, dormant buds are stimulated to start growing and this is how a tree will sprout from the trunk when all the branches are removed.

domatia - small tufts of hair: a cavity or pocket formed by a plant which is usually inhabited by insects, particularly ants, or mites.

dorsal - the back or side away from the stem or central axis, the abaxial side; the opposite of ventral and adaxial. (Note that in Lawrence's *An Introduction to Plant Taxonomy*, the dorsal side is called adaxial; the use of the terms abaxial and adaxial are incorrect in that book).

dorsifixed - attached to the back or dorsal side.

dorsiventral - having distinct upper and lower sides.

double fertilization - only found in seed plants, where one male nucleus fuses with the egg cell and the second with one or more of the other nuclei to from the first endosperm cell.

double-serrate - a margin with large serrations bearing smaller serrations (see fig. 105.3A, basal part of lamina, p. 307).

downy - with fine soft hairs.

drupaceous - like a drupe but not with the morphology of a true drupe.

drupe - a fruit with a fleshy exterior and with the seed or seeds enclosed in a hard covering formed by the inner part (the endocarp) of the ovary wall, which forms a hard "stone" in the interior of the fruit.

drupelet - a small drupe; in the genus *Rubus* (Rosaceae) each flower produces an aggregation of many drupelets (see fig. 91.1.1, p. 32).

duct - a tube or canal through the tissue of the plant; as an oil duct or a lacticiferous duct.

dwarf-shoot - a small lateral stem with short internodes, often bearing leaves and/or reprodutive structures; characteristic of many conifers and *Ginkgo*.

e- or **ex-** - a prefix indicating the lack or absence of something (see examples below).

ebracteate - without bracts.

echinate - covered with short spines or prickles.

ectexine - outer layer of the exine in pollen grains and spores.

ectophloic - phloem exterior to xylem as in normal bilateral vascular bundles of flowering plants.

efoliate - without leaves.

egg - a none-motile female gamete.

egg-apparatus - the female nucleus with two or more other nuclei (synergids) at the micropyle end of the egg sac (female gametophyte) in angiosperms.

elaiosome - an outgrowth from a seed containing oil or fat and which is often attractive to ants and aids in the dispersal of the seed.

elater - a structure which responds to changes in atmospheric humidity found on the spores of *Equisetum* and with the spores in the sporangia of liverworts

ellipsoid - an elliptical three-dimensional structure or body.

elliptic - having the shape of an ellipse, broadest at the middle; a rounded two dimensional figure that is symmetrical but has a long and a short axis, as an elliptic leaf (see fig. 94.1.1, p. 48).

emarginate - with a notch at the apex, without a tip or entire margin at the end; as a leaf with an emarginate apex (see fig. 123.12A, p. 464).

embryo - the intial stage in the development of a plant while still enclosed with the seed (in seed plants) and made up of a radical (which will produce the roots), the plumule or epicotyl (which will produce the stem and leaves), and the cotyledon(s).

embryo sac - the mature female gametophyte containing a number of cells with indistinct boundaries which develops inside the ovule of a flowering plant.

enation - an outgrowth on the surface, often only one cell thick.

endarch - xylem tissue which differentiates centrifugally leaving the protoxylem nearest the centre of the metaxylem.

endemic - native and often used in the sense of being confined to a particular geographic region; not native to other areas.

endexine - inner layer of exine of pollen grains and spores.

endocarp - the innermost layer of the ovary wall in a fruit which may be hard or leathery as in a drupe or pyrene, or fleshy as in Cucurbitaceae.

endodermis - the innermost layer of cells of the cortex next to the stele which have suberised bands forming a casparian strip and often containing starch grains.

endogenous - developing or originating from the inside; typical of development of a root.

endophyte - a plant which lives inside another organism, but is not parasitic upon it.

endosperm - the food material formed by the female gametophyte that initially surrounds the embryo and is often also found in the seed.

endosymbiosis - symbiosis between two organisma, one of which lives inside the other.

Engler-system - a system of classifying plants by Adolf Engler; much used for arranging European herbaria.

entire - with an even and continuous margin without lobes, teeth, etc.

entomophily - pollination by insects.

envelopes - the floral envelopes enclosing the stamens and ovary, the sepals and petals.

ephemeral - short-living, as flowers that open and wilt within the same day: a plant with a very short life cycle which is completed in a few weeks, typical of many weeds.

epi- - a prefix meaning on, upon or attached to the organ indicated by the term; as epipetalous = upon the petal.

epicalyx - a whorl of bracts born on the pedicel of the flower immediately below the sepals and similar to the sepals or calyx, for example, always found in *Hibiscus*.

epidermis - the primary (first) outer layer of cells of all plant organs.

epigeal germination - in which the cotyledons are pushed above the soil level by the extension of the hypocotyl.

epigynous - borne upon or above the inferior ovary (not perigynous); the perianth or stamens may be epigynous but this term is not used for the ovary itself (see pp. xxi-xxii).

epimatium - a fleshy covering to a seed which is more or less fused to the integument arising from the chalazal end of an ovule, like a second integument, e.g. *Podocarpus*.

epipetalous - born on the petals or corolla (see fig. 3.5, p. xx).

epiphyte - a plant which grows on other plants for support but is not a parasite and does not have its roots in the ground (compare "climbing", "liana", and "parasite").

epithelial cells - resin-producing cells found in the resin canals of conifers.

erect - a plant which is quite upright, growing straight upwards; also used for an ovule which is held upright with its stalk at the base.

ericoid - with an appearance similar to that of *Erica arborea*: a plant with many branches held erect and close together and covered with adpressed scale-like leaves.

erose - with a jagged margin that is too irregular to be called dentate or serrate.

estipulate - without stipules.

etiolation - the type of growth produced by plants in heavy shade or darkness where chlorophyl content is reduced and the internodes are of exaggerated length.

eukaryotic - organisms with the genetic material organised inside a nuclear membrane.

eusporangiate - plants in which the sporangium typically develops from a group of initial cells; the resulting sporangium having a wall several layers thick and a large number of spores.

eustele - the typical arrangement of vascular bundles in a young stem of a seed plant, i.e. a dissected siphonostele consisting of a ring of collateral bundles, phloem outside the xylem.

even-pinnate - a pinnate leaf with an even number (2, 4, 6, 8, etc.) of leaflets; lacking a single terminal leaflet; the same as paripinnate and the opposite of imparipinnate or odd-pinnate.

evergreen - retaining green leaves through the dormant or dry season, as in most Gymnosperms and true rain forest trees.

ex- - a prefix indicating the lack or absence of something; also used to indicate the outer portion of a structure or the outside.

exarch - xylem tissue which differentiates centripetally, leaving the protoxylem furthest from the centre.

excurrent - going out beyond the margin of an organ (as a vein going beyond the margin of a leaf.

exfoliate - to peel off in thin flakes or plates, as the bark in many species of Burseraceae (see fig. 123.3D, p. 449).

exine - the outer layer, composed of sporapollenin, of the wall of a spore or pollen grain.

exocarp - the outer layer of the fruit, the outer part of the pericarp, the soft fleshy part of a drupe.

exogenous - developing or coming from the outside.

exosporic gametophyte - a gametophyte which develops to maturity after growing out of its spore wall; typical of most ferns.

exserted - projecting outside or beyond the edge, as stamens projecting beyond the corolla-tube; the opposite of included.

exstipulate - without stipules.

extant - living as opposed to extinct.

extra-axillary - arising beyond or outside the leaf-axil.

extra-floral - arising outside the flower.

extrorse - facing the outside; of an anther which opens away from the central axil of the flower.

eye - used to describe the bud of a tuber, as in the potato.

facultative - having the ability to live under more than one specific set of conditions, as a plant that can be either parasitic or non-parasitic.

falcate - curved like a scythe or sickle.

family - a unit of classification above the level of genus and subfamily: the main unit used for writing and arranging this Flora. A family name always ends in -aceae.

farinose - covered with a usually whitish meal-like powder.

fascicle - a close cluster of structures arising from about the same point but lacking a distinctive arrangement of parts (see fig. 109.2.2, p. 350).

fascicular sheath - closely overlapping bud scales or bracts around the base of a bundle of leaves in conifers.

fasciculate - borne in fascicles.

faucal - concerning the throat, as the throat or opening of a corolla-tube; faucal appendages are appendages arising from the upper part of the corolla-tube or perianth-tube.

female flower - producing the larger stationary gametes and the seeds; used as synonymous with pistillate.

ferns - vascular plants that do not produce seeds but reproduce by minute spores usually produced on the underside of the leaves in groups (sori) of sporangia; other vascular plants that reproduce by spores are lycopods, selaginellas, and equisetums.

ferrugineous - rust-coloured, brownish red.

fertile - having the ability to produce gametes or seeds; fertile anthers produce pollen; fertile pistils produce ovules and, if pollinated, seeds; the opposite of sterile.

fertilization - the union of two gametes to produce a new organism.

fetid (foetid) - with a bad or disagreable odour, as in flowers that attract flies.

few - in botany used for parts that can be counted and are usually between two and five, occasionally between two and ten; the opposite, many, usually more than 10, or a larger, indefinite number.

fibre cell - an elongated cell with a thick lignified wall; tapering at each end.

filament - a thread or thread-like structure; the slender stalk that supports the anther.

filamentous - made up of many thread-like structures.

filiform - slender and thread-like.

fimbriate - with a fringe or series of slender structures along the margin.

flabellate - fan-shaped or broadly wedge-shaped.

flaccid - soft or weak, limp, wilted; the opposite of firm or stiff.

flagellum - a whip-like organ of locomotion with fibres typically arranged in the 9+2 system, found in motile algae and gametes.

flexuose - having a wavy form, bent alternatively in opposite directions.

floccose - covered with soft wool-like hairs that often rub off.

flower - complex of reproductive organs in the angiosperms; see discussion starting p. xx.

foliaceous - like a leaf.

follicle - a fruit developed from a single carpel (a simple pistil), dry and breaking open along one line, usually opening along the inner (adaxial or ventral) suture to which the seeds are attached (see fig. 88.6.8, p. 17).

fovea - a pit or depression containing the sporangium in the leaf base of *Isoetes*.

foveolate - marked with small pits or depressions on the surface.

free - not united with or adherent to any other structure.

free basal placentation - a type of placentation where the ovules are attached to a central colomn arising from the base of the locule within the ovary, but not reching the top (see fig.).

free central placentation - a type of placentation where the ovules are attached to a central column in the center of the single locule, the column extends from the base of the locule to the apex; compare with the above.

free venation - the veins do not unite to form a network: compare with net venation, reticulate venation.

frond - a finely divided leaf, the leaf of ferns and palms; sometimes used for a leaf-like plant body, as in the lichens and Lemnaceae.

fruit - a true fruit is the product of a ripened ovary and its adnate parts; the seed containing structure.

frutescent - having the characteristics of a small shrub.

fruticose - having the characteristics of a shrub, being woody.

fugaceous - falling off early, deteriorating rapidly.

fulvous - tawny or dark yellow or yellowish-brown in colour.

funicle - the stalk of the ovule which attaches it to the placenta.

funnelform - funnel-shaped, gradually widening, as the corolla in many Convolvulaceae.

furcate - forked or divided into branches.

furrowed - with channels or grooves parallel with the long-axis.

fuscous - grey-brown or dark greyish-brown.

fusiform - spindle-shaped, thick in the middle and narrowing at both ends (see fig. 88.4O, p. 13).

fusion - when two or more organs grow together.

galea - a helmet or helmet-like structure, as in the upper part of some corollas.

galeate - with a helmet-like form or with a galea, hooded.

gametangium - a structure producing gametes, normally haploid.

gamete - the sex cell or nucleus which combines with another gamete to form the fertilized egg or zygote.

gametophyte - the gamete producing plant or plant part.

gamopetalous - when the petals are united partly or completely.

gamophyllous - when the perianth parts are united partly or completely.

gamosepalous - when the sepals are united partly or completely.

generative cell - the cell or nucleus in the pollen grain (microsporangium) which divides to give two male gametes.

geniculate - bent like a knee.

genome - a basic set (n) of chromosomes.

genotype - the genetic make-up of an individual.

geophilous - ocurring on or from the ground.

geotropism - a growth movement in a plant in response to gravity, for example the pods of groundnuts (see fig. 95.22.2b, p. 149).

gibbous - with a pouch or inflated on one side.

glabrate - nearly with hairs, glabrous.

glabrescent - becoming glabrous; nearly hairless.

glabrous - without hairs, a surface devoid of hairs.

gland - a structure that secretes, usually found on the surface or within the surface of an organ (leaf, flower, etc.), often borne on a short stalk (as a glandular hair); also used for a fleshy gland-like body or structure.

glandular - bearing or containing glands.

glanduliferous - bearing glands.

glaucescent - slightly bluish-green, or slight covered with a thin layer of wax.

glaucous - covered with a very thin, often waxy, whitish substance; pale bluish-green in colour.

globose - a spherical structure.

glochid - a very small spine or bristled barbed at the tip.

glochidiate - with glochids, with barbed bristles or hooked hairs.

glomerate - in a compact cluster or group of clusters.

glomerule - a small and compact cluster.

glumaceous - resembling glumes or provided with glumes.

glume - dry, often stiff, bracts in the spikelets of grasses.

glutinous - covered with a sticky or glue-like substance.

gradate sorous - fern sporangia which nature from the centre outwards in a sorus.

gregarious - growing in colonies containing many individual plants.

guard cell - one of a pair of cells, containing chloroplasts, which occur in the epidermis and form a stoma in higher plants.

gymnosperm - meaning "naked seed", plants with ovules not enclosed in carpels, but on ovuliferous scales.

gynandrium - the column of Orchidaceae.

gynandrous - with the stamens united with and borne on the pistil.

gynobasic - arising from a deep depression in the centre of the ovary, between the ovary lobes, or from the lower part of a single pistil; as the gynobasic style of Lamiaceae (Labiatae) and the genus *Alchemilla* (see fig. 91.5, p. 39).

gynodioecious - used for a species with 2 sex forms; plants with only female flowers and others with only bisexual flowers.

gynoecium - the female part of the flower.

gynomonoecious - used for a species which has both female and bisexual flowers on the same plant, no male flowers are present.

gynophore - a stalk bearing the pistil (see fig. 3.4, p.xx).

gynostegium (gynostemium) - a structure formed by the union of the stamens with parts of the pistil; found in Asclepiadaceae and Orchidaceae.

habit - the overall appearance of a plant.

habitat - the environment or plant-community in which the plant occurs.

half-inferior ovary - an ovary that is in part united with and in part free from the calyx and corolla (see pp. xxi-xxii)).

halophyte - a plant growing in and tolerating high concentrations of salt in the soil or in the air.

hapaxanthic - used for herbs which only flower once and then die: see also monocarpic.

haplocheilic - a type of stomata where the guard cells are formed from a single initial, while the subsidiary cells come from the epidermis.

haploid - (n), possessing a single set of chromosomes.

hardy - a plant which can survive adverse conditions such as cold or drought.

hastate - when the basis of a leaf has two more or less triangular lobes that diverge away from the midvein.

haustorium - the modified root (in Angiosperms) or myceliium (in fungi) with which a parasite enters the tissue of the host; the plural of haustorium is haustoria.

head - an inflorescence of closely packed flowers that is more or less round or disc-shaped (see fig. 95.44.2, p. 239).

helicoid cyme - a type of cyme in which the flowers are developed along only one side; the same as cincinnus (see fig.).

helical thickening - used for the thickening of tracheids and vessels where the secondary thickening is in one or more coil; same as spiral thickening.

hemi- - a prefix meaning "half" or "partly".

hemicryptophyte - a plant whose perennating buds occur at or very close to the soil surface.

herb - a plant with no persisting stem above ground; if a stem is formed it lives for only one growing season or one year, without forming woody parts.

herbaceous - like a herb; with a stem that dies back to the ground each year.

hermaphrodite - with stamens and pistil in the same flower; the same as bisexual.

hesperidium - a fleshy fruit with a thick exterior rind and septa present within, as in the fruit of *Citrus* (see fig. 120.6D, p. 431).

hetero- - a prefix meaning "of two or more kinds".

heterogamous - when an inflorescence bears more than one kind of flower, e.g. bisexual and male, as in many Asteraceae (Compositae).

heterophyllous - having more than one leaf form.

heterosporous (heterospory) - producing two different kinds of spores, a larger or female spore (the megaspore) and smaller male spores (the microspores); all seed plants are principally heterosporous.

heterostyly - when the styles and stamens vary in relation to each other by length or position within the flowers of the same plant or plants of the same species.

heterotrichy - having both prostrate and erect filaments.

heterotrophic - obtaining energy from energy-rich molecules in the environment: the opposite of autotrophic. All animals are heterotrophic.

hexa- - a prefix meaning six.

hilum - a scar left on the seed where it was attached to the funicle or placenta; the place where this scar is formed.

hirsute - with rather coarse stiff hairs.

hirsutulous - with slightly stiff hairs or a few stiff hairs.

hispid - with a covering of stiff erect hairs.

hispidulous - with a covering of small stiff erect hairs.

homogamous - when all the flowers of an inflorescence are of the same kind or alike.

homologous - of the same origin; as bracts being homologous with leaves.

homosporous (homospory) - producing spores that are all of the same kind (compare with heterosporous); none of the seed plant are homosporous.

honey guide - markings, usually on the corolla, which show a visiting bird or insect where to find nectar in a plant.

host - the organism from which a parasite obtains its food.

husk - an outer, usually loose covering, of a fruit.

hyaline - very thin and almost transparent.

hybrid - produced by crossing two different species, subspecies or varieties of plants.

hydathode (hydatode) - a water-secreting gland or pore found in the leaves of many plants.

hydrophily - pollination by water.

hydrophyte - a plant that grows in very wet places and requires a large amount of water for its growth.

hygroscopic - responds to changes in atmospheric humidity by absorbing and/or loosing water.

hypanthium - the often cup-like part of the flower between the sepal-lobes and the base of the ovary, produced by the union of the base of the sepals, petals, and filaments; often used interchangeably with calyx-tube or floral-tube.

hypercrateriform - salver-shaped; with a narrow tube opening suddenly into a wide cup-shaped mouth.

hypo- - a prefix meaning "beneath" or "less than".

hypocotyl - the axis of the seedling below the level of cotyledon attachment.

hypodermis - a layer of cells below the epidermis, distinct from those of the cortex, usually with a supporting function.

hypogeal germination - in seed plants where the cotyledons remain below the soil surface.

hypogynous - attached below the level of the gynoecium or ovary, as sepals, petals, or stamens attached near the base of the ovary or below the base of the superior ovary (see pp. xxi-xxii).

imbricate - overlapping like the tiles of a roof; in a flower-bud when the petals or the sepals overlap with usually two petals (or sepals) with both edges outside the other petals (or sepals), one with one edge outside, the other within, and one with both edges covered (see fig. 4.1 & 2, p. xxi).

immersed - completely submerged or surrounded.

imparipinnate - odd-pinnate with 1, 3, 5, etc. leaflets; pinnate with a single leaflet at the end of the rachis (see fig. 2.5, p. xix).

imperfect flower - a flower lacking eithe male or female parts; a unisexual flower.

inaperturate - pollen without either a pore or a sulcus.

incised - with the margin deeply cut.

included - not projecting beyond the rim, as stamens not projecting beyond the rim of the corolla tube; the opposite of exserted.

incompatibility - the inability of gametes to fuse and form a zygote, or pollen to germinate on a stigma.

incomplete flower - a flower lacking one of the perianth whorls.

incumbent - leaning on or resting on another structure.

incurved - curving inward or bent inward.

indefinite - numerous or many, as an indefinite number of stamens; in botany usually more than 10-20.

indehiscent - remaining closed and not opening when ripe or mature.

indeterminate inflorescence - one that can continue growing along the main axis and is not terminated by a flower, for example a raceme.

indigenous - native to a region, not introduced.

indumentum - any covering on a surface but usually restricted to a covering of hair-like structures.

induplicate - with the edges bent inwards but not overlapping; the outer surface may then be connivent with other parts.

indusium - an epidermal outgrowth; epidermal structure associated with with the spore-producing area (sorus) on the leaf of a fern; a cup-like structure or group of hairs beneath the stigma (as in Goodeniaceae).

inferior ovary - an ovary that is below the attachment of the sepals, petals, stamens, and, if present, calyx-tube or hypanthium. Compare with a superior ovary and note that the sepal-lobes and stamens may be borne above the ovary in a superior ovary if a calyx-tube, receptacle-tube, or hypanthium is present (see pp. xxi-xxii).

inflexed - bent inward, turned abruptly inward, as in the stamens of many species in Urticaceae.

inflorescence - the flowering portion of a plant; also used for the arrangement of flowers on the flowering axis.

infra- - a prefix meaning "below".

infructescence - the fruiting portion of a plant; also used for a collection of fruits attached to a common axis.

infundibuliform - funnel- or cone-shaped.

inserted - borne or growing out from.

integument - a covering or envelope; the covering or the outer layers of the ovule which develop into the seed coat.

inter- - a prefix meaning "between"; compare with intra-.

intercalary - a meristem that lies between two apices; anything found between two cells or tissues.

internode - the part of the stem between two adjacent nodes.

interpetiolar - between the petioles, as an interpetiolar stipule that extends from the base of one petiole across the stem to the base of the petiole of the opposite leaf, as in most species of the Rubiaceae.

interseminal scales - sterile outgrowths between the ovules of a compound female strobilus or ovary.

intine - the innermost layer of the wall of a pollen grain or spore representing the original cell wall.

intra- - a prefix meaning "within"; compare with inter-.

introduced plant - a plant that has been brought in from another region and is not native to the region under discussion.

introrse - turned inward towards the central axis of an organ; of anthers that open towards the centre of the flower; the opposite of extrorse.

intruding - to thrust inward or extend inward, as a placenta that extends into the locule (see fig. 6.5, p. xxii).

involucel - a whorl of bracteoles

involucral bract - a bract forming part of an involucre: see also phyllary.

involucre - a number of bracts that surround the base of an umbel or the base of a flower-head (see fig. 95.40.3, p. 226).

involute - with the edges rolled inwards, the abaxial surface outwards.

irregular flowers - usually used to denote bilaterally symmetrical flowers (flowers that can be divided into two equal halves along only one plane), in this sense it is synonymous with zygomorphic; sometimes used to denote that the flower is asymmetrical or without a plane of symmetry: compare with regular/actinomorphic.

iso- - prefix meaning "equal" or "like".

isodiametric - of equal height, length and breadth.

isogamy - morphologically identical gametes.

isomerous - with the same numbers, as in flowers with the same number of sepals, petals, and stamens.

isomorphic - of similar form.

jointed - with joints or nodes or articulations where parts separate.

jointed stem - one that can be broken easily at the nodes, such as in Viscaceae (see fig. 115.3.1, p. 377).

jugate - joined or united together, as the leaflets of a compound leaf.

juvenile - the youthful or early stages of growth.

keel - a projecting ridge running the length of the organ on the outer or under surface; like the keel or bottom of a boat (= carinate).

keeled - with a longitudinal ridge running along the under surface of a flat or convex structure.

keel-petal - the loosely united lower or abaxial petals of the pea flowers and related plants (= carina).

key - in plant identification a text with a series of alternate choices; making the correct choice leads, eventually, to the name of the family, genus and species of the plant which is being keyed.

labellum - a lip-like petal; the usually lower or abaxial petal of the flowers of the Orchidaceae.

lacerate - with the margin deeply or irregularly cut.

laciniate - with the margin cut into many slender lobes or segments.

lactiferous - containing milk-like substances; see laticiferous, as in many species of the Moraceae.

lacuna - an air chamber, open space, or gap.

lacustrine - growing in or associated with lakes or ponds.

laevigate - smooth, as if polished.

lamella - a flat plate; a thin partition or septum.

lamina - the flat and thin broad part of a leaf, a sepal, or a petal.

laminate - broad, flat, and thin like the blade of a leaf.

lanate - covered with long soft white wool-like hairs, woolly.

lanceolate - with the shape of the end of a lance or spear, tapering to both ends from a broader middle, as in a lanceolate leaf; usually used to indicate a shape which is widest below the middle, but occasionally also at the middle.

lateral - on the side or along the margin.

latex - a liqiud substance that is often white and sometimes contains rubber, or is fragrant, found in special, often much elongated, cells or ducts called laticifers.

laticiferous - having latex or latex-like fluid.

lax - loose and not crowded together, distant.

leaf - an organ originating from and attached to a stem, usually with a short stalk attached to a flat blade, the most usual site for photosynthesis.

leaf gap - a space in the vascular cylinder of a stem filled with parenchymatous tissue immediately above the origin of a leaf trace or traces.

leaflet - a leaf-like part, or one of the individual blades, of a compound leaf.

leaf trace - the vascular supply for a leaf as found in the cortex of the stem before entering the blade or petiole.

legume - the fruit of a member of the Fabaceae (Leguminosae) which is the product of a simple pistil (a single carpel) and which usually splits open along the two opposite sutures or seams.

lemma - the dry bract that encloses the palea, stamens and pistil in a floret of the Poaceae (Gramineae); also called the flowering glume.

lenticel - a channel filled with loosely packed cork cells allowing the diffusion of gases into and out of stems and sometimes also roots: seen as small areas on the young bark often with a colour different from the surrounding bark.

lenticular - lens-shaped, with two convex sides.

lepidote - with broad and flat hairs or scales.

leptosporangiate - in ferns where the sporangia develop from a single superficial cell or a small groups of initials, the resulting sporangium being small with a single layered wall: compare with eusporangiate.

leucoplast - a plastid lacking chlorophyll and other pigments but often packed with starch.

liana (liane) - a woody climbing plant.

ligneous - woody.

ligulate - with the shape of a tongue or strap; flowers of the Compositae that have a strap-shaped corolla.

ligule - a thin appendage or rim-like structure borne on the leaf base next to the stem and usually found only in plants where the leaf-base clasps the stem, as in the Poaceae (Gramineae) (see fig. 1.1, p. xviii): the strap-shaped corolla in the ligulate flowers of the Compositae.

limb - the upper, usually broad, part of a sepal or petal.

linear - long and narrow with parallel edges.

lip - the large lobes of a 2-lipped corolla, or the large, usually abaxial, petal of the Orchidaceae.

lithophyte - a plant living on stone.

littoral - found on the shore of a lake, sea, or ocean.

lobate - divided into lobes.

lobe - a rounded area along the margin bounded by two indentations or sinuses.

lobulate - having small lobes.

locular - divided into chambers or compartments; i.e. 3-locular means having three chambers.

locule (loculus) - a chamber or compartment, mostly of an ovary or fruit.

loculicidal - opening into the locule.

lodicule - small scale-like structure outside of and at the base of the stamens in the floret of the Poaceae (Gramineae).

loment - a fruit formed by a simple pistil in which each seed is part of a separate segment that breaks apart from the adjacent segment at maturity (see fig. 95.41.1b, p. 299).

longistylous - applied to flowers which have long styles where there are also flowers in the same population/species which have short styles: one of the conditions in heterostyly.

long shoot - shoot with long internodes and relatively rapid annual growth as compared with dwarf or short shoots.

lorate - strap-shaped with a rounded tip; used in describing leaves such as those of many Amaryllidaceae; see also ligulate.

lumen - the space bounded by a cell wall, more often used for dead cells than living ones.

lyrate - with pinnate lobes in which the terminal lobe or lobes are the largest.

lysigenous - a space formed by the break-down of a cell or cells; as compared with schizogenous.

macro- - a prefix meaning "large"; see also mega-.

macrophyll - see megaphyll

macrospore - see megaspore

male (male flower) - producing small or mobile gametes or pollen; often used in place of staminate.

mangrove formation - a gove of usually small trees with inter-twining roots often with pneumatophores, that grow on mud flats within the tidal zone along tropical sea shores.

many - as used in the description of flowering plants: more than 10-20, usually more than 15.

marcescent - wilting or withering but not falling off.

marginal placentation - with the ovules placed along the margin of the carpels (see fig. 6.14, p. xxii).

marine - of the sea or ocean, able to live in salt water.

medulla - the central region of an axis or organ.

medullary ray - an radial strand of parenchymatous tissue passing from the pith to the cortex between vascular bundles, as seen in cross-section of a stem.

mega- - a prefix meaning "large".

megaphyll - a leaf associated with a nodal leaf gap and usually with a branching vein system.

megasporangiophore - a sporangium producing only megasporangia.

megasporangium - a sporangium producing only megaspores.

megaspore - a large spore, from which the female gametophyte develops, usually large and packed with food reserve.

megasporophyll - a structure bearing megasporangia or ovules and believed to be equivalent to a leaf.

meiocyte - a cell in which meiosis occurs.

meiosis - reduction division: two consecutive nuclear divisions, in the first the number of chromosomes is halved so that each nucleus eventually contains only one set of chromosomes compared with the parent cell.

membranous (membranaceous) - of a thin texture and translucent.

mericarp - a part of a dry fruit which splits off and is spread individually.

meristele - portion of a dictyostele composed of a central strand of xylem completely surrounded by phloem and endodermis.

meristem - a tissue of cells that can divide and produce new structures as well as producing cells capable of further division; an embryonic tissue present in all growing parts in plants.

mesarch - maturation of the primary xylem so that the protoxylem becomes surrounded by metaxylem leaving the protoxylem about mid-way between the centre and outside of the primary xylem.

mesocarp - the middle layer when three layers are present in the wall (or pericarp) of a fruit, often fleshy.

mesome - the portion of an axis between successive branches of a telome; used in explaining the derivation of a megaphyll, such as a fern frond, from a primitive branching system.

mesophyll - the parenchymatous tissue in leaves lying between the upper and lower epidermis.

mesophyte - a plant that requires an average amount of moisture to grow; compare with hydrophyte and xerophyte.

mesostylous - a flower with intermediate style length; used for species/populations where flowers are heterostylous with 3 different style lengths.

metabolism - a collective term for the biochemical processes that take place in living cells.

metaxylem - the part of the primary xylem which differentiates after the elongation of the organ, cells generally large and with more complicated secondary thickening than that found in protoxylem.

micro- - a prefix meaning "small" or "very small".

micropyle - a minute opening in the integuments of the ovule through which the pollen-tube usually enters the ovule.

microphyll - leaves, generally small, usually with only one vein and never associated with a leaf gap in the stele; also sometimes used for leaves of some conifers which are greatly reduced and/or with one vein although these leaves are technically highly modified megaphylls.

microsporangiophore - a structure bearing microsporangia.

microsporangium - a sporangium producing only microspores.

microsporophyll - a structure bearing microsporangia or pollen sacs and believed to be equivalent to a leaf.

microspore - a small spore; the pollen grains in seed plants are the equivalents of microspores.

midrib - the principal or central vein or rib of a leaf or other organ.

mitochondrion - an organelle of variable shape, bounded by a distinct double membrane and occurring in all eukaryotic cells; the site of the major part of the process of respiration.

mitosis - cell division in which the chromosomes double and then separate to give two cells with identical genetic make-up.

mixed sorus - where the sporangia mature in a random sequence.

monadelphous - with the filaments of the stamens united to form a single group or bundle, as in the Malvaceae.

moniliform - like a string of beads: see also beaded and torulose.

monarch - primary xylem with a single protoxylem group.

mono- - a prefix meaning "one".

monochasium - a one-sided cyme; i.e. where only one branch develops from under each successive terminal flower.

monocarpic - dying after the production of flowers and fruit, synonymous with hapaxanthic: compare with polycarpic

monochlamydeous - with a perianth of only one whorl, having only one coat or envelope.

monoclinous - having separate male and female flowers on the same plant.

monocolpate - a spore or pollen grain with a single furrow.

monocotyledonae - one of the two major subdivisions of the angiosperms, characterized by the presence of one embryonic leaf (cotyledon) within the seed; also called monocotyledons or monocots or Liliopsida: see also dicotyledonae.

monoeceous (monoecious) - when the male and female flowers are borne on the same plant; the flowers are unisexual but the plant is bisexual.

monogeneric - when a family consists of only one genus, for example Balanitaceae, see p. 433.

monolete - a spore with a single linear scar, having been formed in a line of four spores.

monophyletic - evolving from a single ancestral stock.

monopodial branching - where the main axis remains dominant so that all secondary shoots are clearly lateral.

monotypic - a family or genus with a single species, as the Barbeyaceae, with the singe genus *Barbeya*, and the single species *B. oleoides* (see p. 270).

morphology - the study of the form and related anatomy of living organisms.

motile (mobile) - capable of moving by cilia, flagella or other means, usually used about gametes.

mucilage - a viscous, water soluble, slippery liquid more often found in lower plants, ferns, lycopods, mosses and algae, than in flowering plants

mucilage canal - elongated cells in the cortex of some ferns and cydads which contain mucilaginous material.

mucronate - with the broad tip suddenly narrowed to a short stiff point which is a continuation of the midrib.

multiple fruit - a fruit formed by the union of the female parts of several to many different flowers, as in *Morus* (Moraceae): compare with aggregate fruit.

multiseriate - many rows of cells.

muricate - with a rough surface covered with short hard projections or tubercles.

muticous - blunt and without a point.

mycorrhiza - a symbiotic relationship between a fungus and the roots of a plant.

myrmecophily - the association between certain plants and ants.

naked bud - a bud without a covering of bud scales.

naked flower - a flower without a perianth, an achlamydeous flower.

nascent - beginning to develop, in the act of being formed.

naturalized - introduced from a foreign area and now established and growing successfully in the new area.

navicular - shaped like a boat.

neck - a long tube; found in an archaegonium, through which the antherozoid reaches the egg cell; also used for a long tube in a corolla.

nectar - a sugary liquid produced by flowers or other plant parts, the liquid on which insects and birds that visit the flower feed.

nectariferous - producing nectar.

nectary - a glandular structure which secrets a sugary liquid, the nectar; either associated with a flower (floral nectary) or elsewhere on the plant, as in the Rosaceae and Passifloraceae (extrafloral nectary).

needle - a slender, pointed structure, used to describe certain types of leaf.

nerve - in plants the word nerve is often used in place of vein or vascular bundle; sometimes used only for the more conspicuous vein or ribs of a leaf, a sepal, or a petal..

nervose - with prominent nerves, ribs, or veins.

net-veined - when the smaller veins are interconnected to form a net-like (reticulate) pattern.

net-venation - closed venation in which the veins branch and join.

neutral flower - a flower without sexual parts, without functioning stamens or pistils.

nocturnal - active at night, as in flowers that open at night; the opposite of diurnal.

nodal ring - a ring of transverse xylem and phloem found at the node in *Equisetum*.

node - the place on a stem where a leaf or bud is formed; a thickened area on a stem-like organ where other parts are attached or where the organ will later break in two.

nodule - a small node or thickening; especially used in roots with thickedened swellings in which nitrogen-fixing bacteria live.

nucellus - the interior of the ovule within the integument, containing the embryo sac.

nucleolus - a deeply staining body found in the nucleus, rich in RNA.

nucleus - a conspicuous organelle in the protoplasm of eukaryotic cells, bounded by a membrane and containing DNA.

nude flower - a flower without a perianth, naked or achlamydeous.

numerous - used for parts of a flower which are 15 or more in number; see also indefinite, many

nut - a fruit with a hard outer covering that does not split open when ripe, mostly comparatively large and with one or two seeds; see also nutlet and achene.

nutlet - a small nut.

ob- - a prefix meaning "opposite", "inverse", or "against".

obcordate - with a broad 2-lobed apex and a narrowed base, as the leaflets in species of *Oxalis*.

obdiplostemonous - having two whorls of stamens (usually twice as many as the petals) with the outer stamens opposite the petals.

oblanceolate - with the shape of the end of a lance or spear but with the narrow end towards the base; the inverse of lanceolate.

oblate - a sphere that is flattened at the ends as in the fruits of many *Citrus* spp.; also broadly elliptic with the long dimension perpendicular to the axis of the organ.

obligate - no choice; generally used for organisms which can only exist as parasites.

oblique - a leaf-base in which the two sides are unequal (see fig.).

oblong - a plane shape longer than broad with nearly parallel sides, almost rectangular in outline but with rounded ends and with the length two or three times the width.

obovate - a plane shape with an egg-shaped outline but with the broadest part near the apex and the narrow side near the base (see fig. 95.46, p. 247) .

obovoid - egg-shaped but with the broadest part near the apex and the narrow part near the base; the solid or 3-dimensional form of obovate.

obtuse - with a blunt or rounded end or the margins of the tip forming an angle of more than 50° (see fig.); compare acute.

ocrea (ochrea) - a stipular growth that sheaths the stem near the leaf-base, as in Polygonaceae (see fig. 1.3, p. xviii).

octa- - a prefix meaning "eight".

odd-pinnate - a pinnate leaf with an odd number (3, 5, 7, etc.) of leaflets, with a single terminal leaflet (see fig. 2.5, p. xix); the same as imparipinnate: compare with even-pinnate and paripinnate

oligo- - a prefix meaning "few-".

ontogeny - the development sequence of an organ or whole organism from the unspecialised, juvenile stage to maturity.

oogamy - the condition in which the female gamete is non-motile and the male motile.

opaque - something that does not allow light to pass through, a surface that is dull and not lustrous or shiny.

operculate - opening by a lid or cover.

operculum - a lid or covering that comes off as a single unit.

opposite - a term used for two leaves or two branches that arise from the same node on the opposite side of the stem; a term also used for organs that arise opposite each other or when one arises at the base of another, as stamens opposite a petal or sepal.

orbicular - a flat structure with an almost circular outline.

organ - a distinct and differentiated part of an organism, composed of tissues.

organelle - a body in a cell with certain definite morphological features and function.

organism - any individual living thing.

ornithophily - pollination by birds.

ortho- - a prefix meaning "straight", "upright" or "true" or "correct".

orthotropous - an erect ovule with the micropyle opposite the funicle.

ostiole - an opening or pore.

oval - broadly elliptic or having an egg-shaped outline, usually widest at the middle.

ovary - that part of the pistil which contains the ovules within one or more locules and which will produce the fruit if pollination (and fertilization) takes place; all angiosperms have ovaries.

ovate - a flat structure which is egg-shaped in outline with the broadest part near the base and the narrow part near the apex; the opposite of obovate, compare elliptic and oblong.

ovoid - egg-shaped with the broadest part near the base and the narrow part near the apex; the solid or 3-dimensional form of ovate.

ovule - an organ which contains the embryo sac and the egg cell within the locule of the ovary (in angiosperms) or borne on fertile scales (in gymnosperms); after fertilization develops into a seed.

ovuliferous scale - a highly modified lateral branch, in the axil of a bract, which bears ovules, it may be flat or peltate, woody or fleshy, and is characteristics of conifers.

palea - the inner bract which encloses the stamens and pistil in the floret of the grasses; also used for bracts among the flowers in the floral heads of the Compositae.

paleobotany - the study of fossil plants and their relationships to fossil environments and the age of rocks.

palisade tissue - the part of the leaf mesophyll consisting of regular, columnar cells containing most of the chloroplasts.

palmate - with three or more parts attached to a single point and radiating outward, as the fingers of an open hand radiating outwards from the palm of the hand; as in palmate venation, or in palmately compound leaves (see fig. 2.4, xix).

palmatifid - with palmately arranged lobes, the leaf being divided more than halfway to the single point from which the nerves radiate.

palmatilobed - with palmately arranged lobes, the leaf being divided halfway or less to the single point from which the nerves radiate; compare with palmatifid (see fig. 2.2, xix).

palmatipartite - palmately divided almost to the centre or midrib.

palmatisect - palmately divided to the centre or midrib (see fig. 2.4, xix).

paludose - living in or at the edge of marshes.

palynology - the study of pollen grains and spores.

pandurate - a rounded plane figure with a slender portion near the centre and with two broad ends one of which is broader than the other; the shape of a body of a violin.

panicle - an inflorescence with an indeterminate axis (that continues to grow and does not end in a flower) and many side branches each of which bears two or more flowers (see fig. 120.3.1, p. 425). This term is often used for a branched inflorescence which is difficult to classify into any of the other more precise types such as raceme, cyme, etc.

papilionaceous - flowers that resemble the flower type of the subfamily Papilionoideae of the Fabaceae (Leguminosae).

papillate - covered with many minute rounded gland-like structures or papillae; compare with muricate (see fig. 86.1.6, p. 7).

pappus - modified hairs or scales at the top of the fruit in Asteraceae (Compositae); found in the place of the calyx and believed to represent a modified calyx.

parallel venation - generally used for veins that are parallel with each other and with the margin of the leaf; sometimes used for secondary veins that are parallel with each other but not with the midrib or the margin, as in *Musa* and *Ensete* (see fig. 2.3, p. xix).

paraphyses - sterile hairs found amongst the sex organs of some bryophtes and ferns.

parasite - a plant (or animal) that lives upon another plant (or animal) and takes nourishment from it; compare epiphyte.

parenchyma - a tissue composed of vacuolated, thin-walled, more or less isodiametric cells, functioning chiefly as a ground and storage tissue, often retaining the ability to divide.

parietal placentation - when the ovules are attached to the inner surface of the peripheral or outside wall of the ovary, or the outer wall of the locule (see fig. 6.5-7, p. xxii).

paripinnate - a pinnate leaf with an even number (2, 4, 6, 8, etc.) of leaflets and without a single leaflet at the end (see fig. 95.4, p. 62); the same as even-pinnate: compare with imparipinnate, odd-pinnate.

parted - divided almost to the base or to the midvein.

parthenogenesis - the development of a female gamete into a new individual without fertilization.

partite - the same as parted: also used as a suffix showing the number of parts in a struture, for example tripartite (3 parts).

parthenogenesis - in plants the development of a new organism from a female gamete that has not been fertilized.

patent - spreading or open.

peat - a deposit of incompletely decomposed plant material with little or no soil, dark brown or black with a high carbon content; in Europe, mostly made up of the moss *Sphagnum*.

pectinate - divided to form many parallel parts like the teeth of a comb (fig. 88.6, p. 17).

pedate - used for leaves which are palmately divided, with each division two-cleft (see fig. 104.4.8, p. 279).

pedicel - the stalk of a single flower within an inflorescence or group of flowers (also used for the stalk of a solitary flower).

peduncle - the stalk that bears an inflorescence consisting of two or more flowers; the flowers may themselves each have a stalk (the pedicel) or be without a stalk.

pellucid - clear and translucent, bright when viewed against the light.

peltate - with the stalk attached near the center of a more or less rounded shape and not at the edge, as in a peltate leaf (see fig. 104.4.10, p. 279).

pendulous - hanging down or drooping, as a pendulous ovule.

penicillate - with a tuft of hairs, often shaped into a point like a pencil.

penninerved - with pinnate nervation.

penta- - a prefix meaning "five-".

pentamerous - with five parts, or with sets of five parts, sometimes with multiples of five (5, 10, 15, etc.).

pepo - a fleshy fruit with hard outer rind and without septa or separate chambers within, as in the Cucurbitaceae.

perennating - surviving from one growing season to the next; a structure or organ which enables a plant to survive a non-growing season.

perennial - living for three or more growing seasons.

perfect flower - a bisexual flower with stamens and pistil functional.

perfoliate - when the leaf-base grows around the stem and the stem appears to have grown through the leaf.

pergamentaceous - with the texture of parchment.

peri- - a prefix meaning "around".

perianth - the outer sterile whorls or envelopes of a flower, made up of identical perianth segments, or by two different kinds of perianth segments: sepals (calyx) or petals (corolla).

pericarp - the wall of the ripened ovary or fruit (between the locules and the outer surface); it may be of one or as many as three layers (see exocarp, mesocarp, and endocarp).

pericycle - the tissue lying between the vascular tissue and the endodermis.

periderm - the collective term for the phellogen (cork combium), and phelloderm (secondary cortex).

perigonium - the antheridia and associated leaves in mosses.

perigynium - a thin, often inflated, tube surrounding the ovary in the genus *Carex* of the Cuperaceae.

perigynous - arising from a cup-like or tubular structure around the ovary, surrounding the ovary but not at its base nor united to it (see pp. xxi-xxii).

periphery - along the margin or on the outer wall.

perisperm - the nutritive storage tissue derived from the nucellus found in some seeds.

perispore - a loose outer sheath derived from the tapetum found on the spores of some ferns, and separate from the exine.

persistent - remaining attached to the plant and not falling off.

perulate - with a covering of protective scales, as in many buds.

petal - a flat and usually broad part of the inner whorl of sterile appendages in the flower that together are called the corolla, different from the outer whorl (sepals) and often brightly coloured.

petaloid - like a petal in colour and form; used for bracts (as in *Bougainvillia*), sepals and stamens.

petiole - the stalk of a leaf on which the blade is borne.

petiolule - the stalk of the leaflet.

petrified - anything which has become impregnated with mineral matter to become hard and rock-like.

phanerogam - a seed plant; leterally a plant with visible flowers as opposed to a cryptogam.

phanerophyte - a plant with the perennating organs borne high above ground level.

phellem - cork cells forming the outer layer of the periderm.

phelloderm - secondary cortex.

phellogen - cork combium.

phenetic - classification based on overall assessment of living taxa taking data from as many fields as possible, evolutionary relationships not implied directly.

phenotype - the visible, or chemical, or biologically detectable, manifestation of the genotype produced as a consequence of growth and development.

phloem - the part of the vascular system made up of living cells that function primarily in the conduction of food, the inner bark.

photosynthesis - the process through which green plants make sugar from carbondioxide and water using sunlight energy captured by the green pigment chlorophyll.

phototaxis - the movement of a whole organism in response to light.

phototropism - a change in growth direction in response to light.

phyllary - a bract-like part of the involucre which subtends the flower-heads in the Compositae and some other families.

phyllid - a flattened leaf-like appendage in bryophytes also called a leaf but without vascular tissue.

phyllode - a flattened leaf-stalk (petiole) or leaf-rachis with the form and function of a leaf, as in some of the species of *Acacia* introduced from Australia, compare with cladode.

phyllotaxy - the arrangement of leaves on the stem.

phylogenetic - classification which attempts to reflect the evolutionary history of taxa.

phylogeny - the evolutionary history of an organism or group of organisms.

phylum - a unit of classification for a very large group and/or a very distinct group of organisms, sometimes, in older books, replaced by division.

physiology - the study of chemical and physical processes as they occur in living organisms.

pilose - with a loose covering of soft long simple hairs.

pinna - the primary division of a pinnate leaf which can be a leaflet in simple pinnate leaves, or can be divided again into pinnules in a bipinnate leaf.

pinnate - when a compound leaf has its leaflets born along an extension (the rachis) of the leaf-stalk (petiole) or when the leaflets are borne on divisions or branches of the rachis; leaves can be bipinnate, tripinnate, etc.; compare palmate.

pinnatifid - with the margin divided more than halfway to the midvein or centre and forming pinnate lobes.

pinnatilobed - with the margin divided to about half the distance to the midvein or centre and forming pinnate lobes.

pinnatipartite - with the leaf divided almost to the midvein or centre and forming pinnate lobes.

pinnatisect - with the leaf divided to the midvein or centre and forming pinnate lobes.

pinnule - the second or third branching of a pinnate leaf; the branches or divisions that arise from the primary rachis of a pinnate leaf (see fig. 2.6, p. xix), compare pinna.

pistil - the individual female structure that contains the ovule and will produce the seeds, it is made up of one or sevaral carpels and is usually divided into the following parts: ovary, style, and stigma; a flower may have one pistil or many pistils.

pistillate - often used for flowers with female parts but no male parts and, therefore, unisexual; a female flower.

pistillode - a reduced non-functioning pistil that does not produce seeds; often present in fuctionally male flowers.

pit - a thin area in a cell wall through which substances can pass from one cell to the next.

pit connection - protoplasmic connection between cells.

pit membrane - composed of only the primary cell wall and middle lamella.

pith - the soft spongy tissue found in the centre of many stems.

placenta - the part of the ovary on which the ovules are attached.

placentation - the arrangement of the ovules in the ovary (see fig. 6, p. xxii).

plagiotropism - the orientation of growth at a definite angle in relation to the gravitational field, usually oblique or horizontal.

plasmalemma (plasma membrane) - the membrane which surrounds the protoplasm of a cell and adjoins the cell wall.

plasmodesma - cytoplasmic strand connecting the protoplasm of adjacent cells through a pit area.

plastid - organelles containing pigments and products of photosynthesis.

platyspermic - having flattened bilaterally symmetrical seeds, as in conifers.

plectostele - a protostele dissected into anastomosing, plate-like units.

pleomorphism - having more than one form or shape.

plicate - folded, with the edges folded together like a fan.

plumose - with tufted or feather-like hairs or feather-like bristles.

plumule - the embryonic shoot.

pneumatophores - air passages; more often used for the vertical wooden projections from the roots of some trees growing in very wet situations, as in swamps or mangrove formations; these woody projections rise above the water level and are believed to function in bringing air to the roots.

pod - a dry fruit that opens when ripe; often used as synonymous with legume.

polarity - differentiation in structure and/or function between two ends of an axis.

pollen - the powder-like grains produced in the anthers that will produce the male gametes necessary in fertilization; the pollen will germinate on the stigmatic surface of the style (in angiosperms) or in the micropyle (in gymnosperms) and produces the pollen tube. The pollen tube grows into the ovule and releases the male gametes where they can reach and fuse with the female gametes; the pollen-grain is homologous with a microspore.

pollen chamber - flask-shaped cavity at the top of an ovule in gymnosperms, lies between the integument and nucellus.

pollen-grain - see pollen.

pollen-sac - the chamber in which the pollen is produced after meiosis has taken place, the theca (plural: thecae) of the anther.

pollen-tube - the tube formed by the growth of the pollen-grain.

pollination - the act of taking pollen from the anther and bringing it to the receptive stigmatic part of a style or to an ovule; this can be carried out by the wind, or by insects or other animals.

pollinium - when the pollen-grains remain together, cohering in a single pollen-mass, as in the Asclepiadaceae or Orchidaceae.

poly- - a prefix meaning "many-" or "much-".

polyadelphous (polyadelphus) - with three or more groups of united stamens.

polyandrous - with many (more than 15) stamens.

polycarpic - bearing fruit many times, as opposed to monocarpic.

polycyclic - with vascular tissue arranged in two or more concentric whorls; typical of tree ferns.

polyembryony - the development of more than one embryo in the ovule of a seed plant.

polygamodioecious - dioecious but with some bisexual flowers or with some flowers of the opposite sex.

polygamomonoecious - monoecious but with a few bisexual flowers.

polygamous - when a plant or different plants of the same species have both bisexual and unisexual flowers.

polygynous - with many pistils.

polymorphous - with several or many forms, variable.

polypetalous - with petals that are separate and free from each other.

polyphyletic - evolved from more than one ancestral group.

polyploidy - with three or more sets of chromosomes in a nucleus: compare diploid and haploid.

polysepalous - with sepals that are separate and free from each other.

polystelic - having more than one stele.

polystichous - when the leaves are borne in many series or spirals.

pome - a fleshy fruit with a soft outer covering and seeds borne within stiff central partitions, as in the fruits of *Malus* (apple) and *Casimoroa* (Rutaceae).

porate - one to many pores in the exine of a pollen grain.

pore - a small, usually round opening, as in anthers opening by pores in the Ericaceae.

posterior - towards the axis, on or nearest the axis (adaxial or ventral); the opposite of anterior.

precocious - developing early; used for flowers that develop before the leaves.

prickle - a small sharp outgrowth from the bark or surface.

primary tissues - tissues which diferentiate from cells produced by apical meristems.

primary xylem - formed from the apical meristem and consisting of protoxylem and metaxylem.

primordium - the early cells in the differentiation of an organ.

procambium - a strand of relatively undifferentiated cells from the apical meristem which gives rise to the primary phloem and xylem.

procumbent - lying along the ground.

pro parte (proparte) - often abbreviated p.p., meaning "in part only".

prophyll - a much reduced leaf or bract.

precocious - developing early; used for flowers that develop before the leaves.

procumbent - with stems that lie on the ground.

proembryo - the structure developed first from the zygote in gymnosperms.

prokaryotic - organisms in which the genetic material is not organised into chromosomes within a nuclear membrane.

proliferous - with adventitious buds on the leaves or in the flowers that are capable of producing new plants.

prostrate - lying flat on the ground.

protandrous - when the anthers discharge pollen before the stigma (in the same flower) is receptive; sometimes spelt proterandrous.

prothallial cell - a sterile cell formed internally when a microspore or pollen grain starts to germinate.

prothallus - a small simple plant which is the free-living, gametophyte generation typically found in ferns.

proto- - a prefix meaning "first" or "original".

protocorm - a tuberous structure developed from a portion of the young embryo in lycopods and orchids.

protogynous (proterogynous) - when the stigma is receptive before the anthers (in the same flower) have discharged their pollen.

protoplasm (protoplast) - the content of a living cell, excluding the cell wall and vacuole.

protostele - the kind of stele in which a solid core of xylem is surrounded by phloem.

protoxylem - the primary xylem which differentiates first before elongation is completed; compare with metaxylem.

proximal - nearer to the place of attachment; the opposite of distal.

pruinose - covered with a whitish wax or very fine powder.

pseudaril - an aril-like structure found with the seeds of *Commiphora* (see fig. 123.4F, p. 451)

pseudo- - prefix meaning "false".

pseudobulb - the bulb-like stem of certain orchids.

pseudocopulation - where a male insect, usually a bee, is stimulated to copulate with an orchid flower which resembles the female insect; this is a highly specialized adaptation for pollination.

pseudo-endosperm - the nutritive tissue of the embryo in gymnosperms which is formed from the female gametophyte.

pseudogamy - the development of a female gamete into a new individual after stimulation by a male gamete, but without fertilization taking place.

pseudopore - a thin area in the exine of some conifer pollen grains.

pseudowhorl - where leaves arise so close together around the stem that they appear to be in a ring.

puberulous (puberulent) - covered with very short fine hairs or slightly hairy.

pubescent - with a covering of soft hairs.

pulverulent - covered with a powder, see also pruinose and glaucous.

pulvinus (pulvinate) - an enlargement or swelling, shaped like a cushion or pad, round and flattened, usually found at the base of the leaf or petiole.

punctate - marked with dots or glands or very small depressions.

pungent - ending in a sharp stiff point; with a strong smell or taste.

pustule (pustulate) - many small elevations like pimples or blisters.

pyrene - a nutlet or kernel; the "stone" of a drupe or similar fruit.

pyriform (piriform) - shaped like a pear (*Pyrus*); a solid shape with a broad apex (away from the point of attachment) narrowing to a wide neck at or below the middle.

pyrophytic - a plant able to tolerate fire or needing fire to stimulate flowering.

pyxis - a capsular fruit in which the top comes off as a lid.

quadr- or **quadri-** - a prefix meaning "four".

quincuncial - a type of imbricate aestivation (the overlapping of sepals and petals in bud) with two parts or lobes outside, two parts or lobes inside, and one lobe which is overlapped on one side only (see fig.).

raceme - an indeterminate inflorescence in which the flowers are borne along a single axis with the uppermost the youngest, each flower with a stalk of about the same length, (see fig.); compare with cyme.

racemose (racemous) - arranged like a raceme.

rachilla (rhachilla) - a small axis or rachis; the central axis of the spikelet in the Poaceae (Gramineae) and Cyperaceae.

rachis (rhachis) - the axis of a compound leaf or frond; the axis of an inflorescence.

radial symmetry - symmetrical about a central axis; when the structure is divided longitudinally along any axis, the two halves are mirror images of each other.

radiate (radiating) - spreading outward from a point; of the flower-heads of the Compositae that possess ray flowers.

radical leaves - leaves that arise so close to the base of the stem that they appear to come from the top of the root.

radicle - the embryonic root: the part of the developing embryo that will produce the root.

random branching - branches with arise without any relationship to subtending leaves.

rank - a vertical row; 2-ranked leaves are in two vertical rows (see fig. 105.7, p. 315).

rank - a general term to denote a level in the taxonomic hierarchy (variety, subspecies, species, genus, family, etc.).

raphe - the part of the stalk of the ovule (funicle) that is united or coherent to the outer wall of the ovule and forms a ridge on the surface of the seeds.

raphides - needle-shaped crystals found within cells of plants.

ray - radial line of cells from the central pith to the cortex, usually living and transporting materials across the stem; (in inflorescences) the radiating branches of an umbel in the Apiaceae (Umbelliferae).

ray floret (ray flower) - an outer floret, usually with a strap-shaped petal, on the margin of the head in the Asteraceae (Compositae).

receptacle - the axis or the central body of the flower on which the flower parts (sepals, petals, stamens, and pistil) are borne.

receptaculum - a fleshy structure formed from the bases of the bracts below the seed forming a swollen cone axis in some conifers such as *Podocarpus*.

recurved - curved backwards.

reduced - not properly developed or undeveloped.

reflexed - bent downwards or backward, bent abruptly backward toward the base.

regular - (of flowers) radially symmetrical or actinomorphic, a flower that can be divided into two equal halves by two or more vertical sections (some authours consider regular synonymous with symmetrical and include both bilateral and radial symmetry); compare with irregular, assymetric, zygomorphic.

reniform - kidney-shaped, as the seed of many Fabaceae (Leguminosae), and some leaves (see fig. 104.4.9, p. 279).

repand - with an uneven or wavy margin, not as uneven as sinuate; compare with undulate.

repent - stems that creep along the ground and root at the nodes.

replum - a septum within the ovary formed by the placenta; as in the Brassicaceae (Cruciferae) where the replum forms a frame to which the seeds are attached when the two outer valves fall away (see fig.).

resin (resinous) - a solid or semisolid substance produced by plants from special canals or ducts found within the plant, not soluble in water and often sticky and aromatic.

resupinate - twisted through 180°, as the pedicel and ovary in most species of the Orchidaceae; upside-down.

reticulate - with many interconnections as in a net; applied to a surface being marked by a network of fine lines or ridges..

reticulate venation - see net-venation.

retrorse - bent abruptly backwards or downwards.

retuse - with a rounded tip that has a small notch at the centre (see fig.).

revolute - with the margin or tip rolled or bent backwards towards the centre.

rhachis - synonymous with rachis.

rhipidium - a cymose inflorescence with branches alternating from one side of the vertical axis to the other; normally flattened in one plane and fan-shaped.

rhizome - a root-like stem on or beneath the ground with roots growing downwards and leaves and shoots upwards; differing from a true root in the presence of buds, leaves, or scales; especially robust rhizomes are often called rootstocks.

rhizophore - a prop-like axis produced at a node which grows towards the soil and forms roots at the tip, typically found in *Selaginella*.

rib - a primary vein or prominently raised vein or nerve.

riparian - growing in or at the edge of rivers and streams.

root - the underground portion of the plant that never produces leaves and thus lack nodes and internodes; with a special anatomy.

root cap - a cap of tissue over the root apex.

root hair - a hair-like outgrowth of an epidermal cell which absorbs water and minerals, found on young actively growing roots.

rootstock - a short, vertical, underground stem, bearing roots.

rosette - a cluster of parts in a circular form; often used for leaves produced at ground level (see fig. 88.5, p. 15).

rostellum - a small beak; a projection on the column beneath the pollinia in the Orchidaceae.

rostrate - with a beak or beak-like projection (see. fig. 123.5E, p. 453.)

rosulate - when the leaves are in a circle or in a rosette.

rotate - wheel-shaped; a corolla with a very short tube and spreading lobes.

rotund - with a shape between orbicular and broadly elliptic.

rudimentary - incompletely developed.

rufous - rusty or brownish red.

rugose - with wrinkles or grooves in the surface.

ruminate - appearing to have been chewed, like the cud produced by cattle; mottled in colour, or divided into partitions; as the endosperm of the Annonaceae.

runcinate - a margin in which the lobes or teeth point backwards towards the base.

runner - a slender stem running along the ground and rooting at the nodes, see also stolon.

rupturing - breaking open irregularly.

saccate - bag-shaped or sack-shaped, pouched.

saccus - a wing or ballon-like extension of the exine in pollen grains of conifers.

sack (sac) - a pouch or bag-like structure.

sagittate - arrow shaped; the base has two acute lobes that point backwards to the base of the petiole.

samara - a one-seeded nut-like fruit with a wing (see fig. 118.9.6, p. 398).

saponins - a toxic, soap-like group of compounds which is present in many plants.

saprophyte - a plant that obtains its nourishment from dead organic matter and usually does not possess chlorophyll.

sarcocarp - a fleshy covering to a fruit.

sarmentose - with long slender stolons or whip-like branches.

scabrid (scabrous) - rough to touch, usually caused by the presence of very short stiff hairs which point backwards to the line of growth.

scalariform - in the form of a ladder; secondary thickening found in tracheids and vessels which has a bar-like appearance.

scale - any small flat and thin structure like a flattened hair or very small leaf, often triangular in shape.

scale leaf - small, usually adpressed and imbricate as in *Juniperus* and *Cupressus*, or much reduced as in *Thesium* (116.1.1, p. 381).

scandent - a general term for climbing.

scape - a flower stalk (a pedicel or, more often, a peduncle) without leaves that arises from the ground, as in *Crinum*.

scarious (scariose) - thin and dry, not green.

schizocarp - a dry fruit breaking up into one-seeded nut-like parts (each part called a mericarp), as in many genera in the Malvaceae and in Apiaceae (Umbelliferae).

schizogenous - a space or canal formed by the separation of cells (cf. lysigenous).

sclereid - a highly lignified cell not elongated, but some what isodiametric, often much branched.

sclerenchyma (sclerenchymatous) - strengthening tissue composed of woody fibres or stone cells.

scorpioid - a cymose inflorescence curved to one side and coiled like the tail on a scorpion; see also helicoid cyme.

scrambler - a plant which usually climbs with the help of hooks, thorns or tendrils so that it spreads out over or through its support.

scrub - more or less dense vegetation of small shrubs (up to about 3 m high).

secondary thickening - produced from secondary meristems after the differentiation of primary tissues, generally contribute growth in diameter and technical support.

secund - one-sided; as when branches, leaves or flowers are all attached along one side of an axis.

seed - the ripened ovule containing the embryo with or without additional storage tissue.

seedling - the young plant which develops from a germinated seed.

self-fertilization - where the male gametes unite with the female gametes of the same parent.

self-incompatible - of plants which have some mechanism to make self-fertilization impossible.

self-pollination - where the stigma is receptive to the pollen of the same flower or parent plant.

segment - a division or part of an organ.

semi-parasite - a plant which can parasitize another plant but is also capable of growing by itself.

sensu - latin meaning "in the sense of"

sepal - a flat part of the outer whorl of sterile appendages that protects the flower in bud, often green or dull in colour; the sepals together are called the calyx.

sepaloid - a structure which is similar to a sepal.

septa - partitions; plural of septum.

septate - divided by one or more partitions or septa.

septicidal - opening at or along the partition (or septum) or opening along the placenta; not opening into the locule.

septum - a partition or cross-wall.

seriate - arranged in a row.

sericeous - with silk-like hairs that are soft and straight.

serrate (serrulate) - with teeth like that of a saw, the teeth more or less regular and pointing forwards (see fig. 105.8A, p. 316); compare with dentate.

sessile - without a stalk, meaning "seated".

seta - a bristle or stiff hair.

setaceous (setiform) - like a bristle.

setose - with a covering of bristles, similar to hispid.

sheath (sheathing) - a tubular structure that encloses an organ or part, as in the lower tubular part of a grass leaf which encloses the stem (see fig. 1.1, p. xviii).

shoot - a stem axis together with its leaves.

short shoot - with very short internodes, growth usually slow and often bearing leaves and/or reproductive structures.

showy - brightly coloured or conspicuous.

shrub - a woody plant with perennial woody stems, there are usually two or more stems coming from near the ground or sometimes one slender stem; a term which is not very precise but differs from the term tree in not possessing a trunk or bole between the ground and the branches.

shrublet - a small shrub; see also suffrutescent.

sieve cell - found in the phloem of vascular plants, a cell somewhat elongated with sieve plates on the lateral walls.

sieve plate - a complex pit making a perforated area in the cell wall between two adjacent sieve cells or sieve tube elements.

sieve tube - a column of cells which have lost their nuclei and are connected by sieve plates across their end walls; found in flowering plants.

silicule (silicula) - the short fruit found in some Brassicaceae (Cruciferae), not more than two or three times as long as broad; see also silique.

silique (siliqua) - the long slender fruit of some Brassicaceae (Cruciferae), more than two times as long as broad; divided into two compartments by a thin partition on which the seeds are borne and from which the two outer valves separate.

simple fruit - derived from a single carpel or syncarpous ovary.

simple leaf - with only a single blade, the opposite of a compound leaf.

simple sorus - in ferns where the sporangia mature simultaneously.

sinuate (sinuose) - when the margin is uneven or wavy by turning inwards or outwards but not deeply enough to be lobed (see fig. 124.1.1, p. 480).

sinus - a rounded depression between two projecting lobes or teeth; the space between two lobes.

siphon (siphonaceous) - a tube (tubular).

siphonostele - a stele composed of a hollow cylinder of vascular tissue with a central pith and phloem external to the xylem.

solenostele - a siphonostele in which the phloem occurs on both inside and outside of the xylem.

solitary - one alone and without companions or similar structures; as a solitay flower in the axil of a leaf.

sorus - a group of sporangia found on the abaxial side of a fern leaf.

spadix - an inflorescence with the flowers borne on or sunken into a thick and often fleshy spike-like axis, as in the Araceae.

spathaceous - like a spathe.

spathe - a large bract that is sometimes leaf- or petal-like and encloses the young flower or inflorescence; subtending and at first enclosing the spadix in Araceae and the inflorescence in Palms.

spathulate (spatulate) - a flat shape with the outline of a spoon or spatula, broadly rounded above and long and narrow below.

species - the basic unit in a taxonomic classification denoting a group of organisms that appear more similar to each other than to any other group and are usually assumed to be able to interbreed and produce fully fertile progeny.

spermatocyte - the cell which forms a spermatozoid: synonymous with antherocyte.

spermatozoid - a male gamete which moves by swimming: synonymous with antherozoid.

spike (spicate) - an indeterminate inflorescence with the flowers sessile (without stalks or pedicels) on a simple unbranched or undivided axis or rachis (see fig. 95.8.1, p. 79).

spikelet - a small spike in which the individual flowers are subtended by and enclosed in bracts; the elements of the more complex inflorescences in Poaceae (Gramineae) and Cyperaceae.

spine - a hard sharp-pointed structure, often long and narrow (see fig. 95.8, p. 79).

spinescent - ending in a spine or in a very sharp hard point, or provided with spine-like teeth.

spinose - having spines.

spinulose - having small spines.

spongy - light in weight, porous, and compressible.

spongy tissue - a tissue found in a leaf consisting of irregularly arranged cells with large air spaces between them.

sporangiophore - a structure bearing one or more sporangia.

sporangiospore - a spore produced in a sporangium.

sporangium - the structure enclosing and producing the spores; plural sporangia.

spore - a cell which often becomes free from the parent plant and is able to produce a new plant; in higher plants the spore is produced by meiosis and germinates to give a gametophyte which gives rise to the gametes or sex cells.

sporocarp - hard nut-like structure containing the sporangia in heterosporous ferns.

sporocyte - the spore mother cell; the last cell of the sporophyte before meiosis takes place.

sporophyll - a leaf that produces spores (in sporangia); a leaf-like structure that produces the reproductive structures.

sporophyte - a plant that produces spores (as opposed to a plant that produces gametes - the gametophyte). In higher plants the sporophyte is the typical plant while the gametophyte develops within the reproductive parts, the pollen grain produces the male gametes and the female gamete is produced in the embryo sac.

spur - a slender, usually hollow, extension of some part of the flower; a short lateral branch.

squamellate (squamulose) - covered with small scales, but larger than those described as lepidote.

squamose - covered with scales.

stalk - the stem or narrow portion beneath and supporting any organ.

stamen - the pollen producing organ of higher plants, usually made up of a narrow stalk (the filament) and an anther in which the pollen is produced.

staminate - (of flowers) with stamens but without pistil; male.

staminode - a non-functional stamen, often highly modified or reduced.

staminodium - a non-functional stamen without anther or with anther that does not produce pollen; sometimes also spelt staminode.

standard - the large adaxial petal (which is on the outside in the unopened flower) of pea-like corollas of Fabaceae (Leguminosae) subfamily Papilionoideae; also used for the narrow inner petals of the *Iris* (Iridaceae).

stele - the central primary vascular tissue of a root or stem, extending to the endodermis and/or pericycle if present.

stellate - star-shaped, as in stellate hairs that have several arms or branches radiating outward from a central point (see fig. 124.6.5, p. 488.

stem - the main axis of a plant or a branch of the main axis that (at first) produces leaves at the nodes.

sterile - not producing sex cells or gametes; a sterile flower does not produce either pollen or functional ovaries.

stigma - the portion of the pistil (usually at the top of the style) which is receptive to pollen, usually with a sticky or minutely papillate surface on which the pollen germinates and grows into the tissue of the style.

stigose - with short stiff hairs that lie close to the surface; see also strigose.

stipe (stipitate) - the stalk of a pistil (see also gynophore); sometimes also used about the stalk of an inflorescence, as about the stalk of the fig in *Ficus* (Moraceae).

stipel - a stipule-like structure subtending a leaflet in some compound leaves (see fig. 2.5, p. xix).

stipules - scale-like or bract-like appendages, usually found in pairs, at the base of the petiole; many leaves have three parts: the blade, the petiole and two stipules.

stolon - a branch which grows over the ground (a runner) which produces adventitous roots, mainly at the nodes.

stoma (stomate) - a minute pore in the epidermis, usually found on leaves; these pores are important in the exchange of gases and loss of water: plural stomata.

stomatate - having or provided with stomata.

stomium - the dehiscence point or line on the sporangium of a fern.

stone cells - heavily lignified, more or less isodiametric cells.

strain - a particular mating group or variety within a species with distinctive physiological and/or morphological features.

striate - marked with longitudinal parallel ridges, grooves, lines, or streaks.

strict - standing upright, and straight, narrow.

strigose - with short stiff hairs that lie close to the surface.

strobilus - a cone-like or spike-like reproductive structure with a central axis and many overlapping bracts or sporophylls.

strophiole - an appendage to the hilum of some seeds.

style (stylar) - the narrow portion found in most pistils above the ovary upon which the stigma is borne.

stylopodium - the disc-like enlargement of the base of the style, as in many members of the Apiaceae (Umbelliferae).

sub- - a prefix meaning "slightly", "somewhat", "almost", or "below".

suberin - an impervious waxy substance found in cork cells and the casparian strip of the endodermis.

suberose - corky in texture, similar to cork which is a tissue formed near the outer bark that is light in weight, compressible, and impervious to water.

submerged (submersed) - below the water level.

subshrub - a plant with a perennial, woody base to the stems, but with upper part of the stems herbaceous.

subsidiary cells - occurring next to the guard cells in the epidermis, usually constant in shape and pattern distinguishing them from other epidermal cells.

subspecies - a unit of classification below the rank of species above the rank of variety; often used for geographical variants of a species.

substrate - foundation, the underlying surface providing a point of attachment or anchorage.

subtend - to extend under, or be opposite to, another structure.

subulate - flat and narrow, tapering from the base to the sharp tip.

succulent - fleshy and juicy, thick and soft within.

suffrutescent - like a small shrub, with a woody base near the ground that produces leafy and flowering shoots each growing season.

suffruticose - shrubby or like a small shrub, with a woody stem that is somewhat larger than in the case of suffrutescent.

sulcus (sulcate) - a longitudinal furrow.

super- or **supra-** - prefixes meaning "above-".

superior ovary - an ovary that is borne above the attachment of the sepals, the petals, the stamens, or, if present, receptacle-tube or hypanthium; the opposite of inferior ovary, compare with hypogynous, perigynous, and epigynous: (see pp. xxi-xxii).

suture - the line along which two parts have been united or the line along which a structure splits open.

syconium - a fruit-like structure made up of a hollow (ovary-like) receptacle in which the seed-like fruits are found; this is the fruit (called a "fig" in English) of *Ficus* (Moraceae).

sym- or **syn-** - prefixes meaning "together".

symbiosis - an association between two different types of organisms in which there is some type of mutual benefit; also refered to as reciprocal parasitism.

sympetalous (synpetalous) - with the petals at least partly united.

sympodial branching - where the terminal bud ceases to grow (or its function and position are taken over by another structure such as a flower) and growth is continued by a lateral bud or branch.

synandrium - when the anthers of a flower or anthers from several flowers are united or coherent.

synangium - a composite structure formed by two or more sporangia in ferns.

syncarpous - composed of two or more united carpels, as in a syncarpous pistil; the opposite of apocarpous.

syndetocheilic - stomatal development in which both subsidiary and guard cells are formed from the same mother cell.

syngamy - the fusion of gametes.

syngenesious - with united anthers or cohering anthers, typical of the Asteraceae (Compositae).

synsepalous - with the sepals at least partly united.

tangential - perpendicular to a radius.

taproot - a persistent primary root, often swollen with food reserve and/or going deep into the soil.

tapetum - cells which form part of the sporogenous tissue but which break down during spore development to nourish the developing spores, characteristic of vascular plants.

taxon - the general term for any unit of classification such as variety, subspecies, species, genus, family, order, class, etc.

telome - a single terminal segment of a branching axis either sterile or fertile: used to explain the origin of a megaphyll from a branching system.

tendril - a slender, usually coiling, part of a leaf or stem that helps support a stem.

tepal - used for the parts of the perianth where the sepals and petals cannot be readily distinguished.

terete - round or circular in cross section.

terminal - found at the end; near or at the apex.

ternate - arranged in a whorl or cluster of three.

terrestrial - growing on or in the ground or on rocks but not on other plants (epiphytic) or in water (aquatic).

tessellated - marked with a find pattern, like a mosaic of small tiles; usually used for leaves and seeds.

testa - the outer coat of a seed.

tetra - a prefix meaning "four".

tetrad - the group of four cells produced as a result of meiotic division, frequently genetically different from each other.

tetradynamous - of six stamens when four are long and two are short; characteristic of the Brassicaceae (Cruciferae).

tetramerous - with four parts or parts in multiples of four.

tetrandrous - with four stamens.

tetraploid - having four times the basic (haploid = n) number of chromosomes; compare with diploid and haploid.

tetrarch - a protostele with four primary xylem bundles.

thalamus - the receptacle or torus of a flower.

thallus - a vegetative body that is not differentiated in stems and leaves.

theca (plural **thecae**) - the pollen-sac or locule of an anther, one of the anther-lobes.

therophyte - an annual plant, the perennating organs being seeds.

thorn - a sharp-pointed branch, sometimes used as synonymous with spine.

throat - the spreading upper portion or the area at the top of the perianth-tube where the lobes become free from each other.

thyrse - a panicle with ultimate branches that are cymose.

tomentose - covered with soft, more or less appressed, hairs that are not straight; woolly.

tomentellous (tomentulose) - with very short woolly hairs.

tonoplast - the membrane surrounding a vacuole; also known as the vacuole membrane.

torulose - an uneven cylinder with contractions and swellings at intervals to give the appearance of a row of beads; see also beaded and moniliform (see fig. 94.14.1b, p. 229).

torus - the cental axis or receptacle of a flower.

trabeculum - a bar of cells; an elongated cell passing across a cavity as in the endodermis of *Selaginella*.

tracheid - an element of the xylem consisting of an elongated, elaborately pitted, lignified cell with oblique end walls.

transfusion tissue - cells intermediate between tracheids and paranchyma, found in the leaves of some gymnosperms.

transverse - across or at right angles to the long axis.

tree - a woody plant with a single main stem (a trunk or a bole) and a distinct upper crown; compare with shrub.

tri- - a prefix meaning "three".

triarch - protostele with three protoxylem bundles.

trichome - a hair or bristle, usually small.

tricolpate - pollen grains having three longitudinal apertures.

trifoliate - with three leaves; compare with trifoliolate.

trifoliolate - of a compound leaf with three leaflets as in *Trifolium* Fabaceae (Leguminosae) subfamily Papilionoideae.

trigonous (trimerous) - with three parts; with parts in multiples of three.

trilete spore - formed in a tetrahedron which leaves three lines in a Y-shape on the thinner surface of the spore wall.

triploid - having three times the basic number of chromosomes.

triquetrous - with three sharp angles.

trisomy - one chromosome represented three times in a normal diploid cell, a form of aneuploidy.

truncate - with the base or apex at right angles to the midvein as if cut across at the bottom or top.

trunk - the large single woody stem of trees, the main stem or bole.

tube (in flowers) - the cylindrical part of the perianth, usually made up of the united sepals and/or petals.

tuber - a thickened portion of a stem, usually underground, that is capable of producing new branches.

tubercle - a swelling, knob or thickened protuberance on a suface, sometimes found at the base of a hair.

tuberculate - having tubercles

tumid - swollen or inflated.

tunic - a thin coat or covering; as in the covering of an onion.

tunica - the outer layer or layers of cells at an apical meristem in which most divisions are anticlinal.

turbinate (turbiniform) - cone-shaped but with the broad portion up and the pointed portion down.

turgid - filled out or swollen, usually with water, but not with air and not distended in shape.

turion - a short, scaly branch produced from a rhizome.

twig - a young woody stem or branch, the last season's growth.

twining - climbing by winding the stem around the support.

umbel (umbellate) - an inflorescence in which the pedicels of the flowers all arise from one point and the flowers are borne at one level; the umbels themselves may be arranged in an umbel called a compound umbel, as in the Apiaceae (Umbelliferae).

umbo - a small cone-like projection from a surface.

uncinate - with a hook at the end, as an uncinate hair.

undershrub - a perennial plant with lower woody parts, but herbaceous upper parts that die back after each growing season: also refers to a plant growing in the lower shrub layer in a multistoried forest.

understory - a layer of shrubs, small trees and saplings growing under the main canopy of a forest.

undulate - with a wavy margin.

unguiculate - shaped like the hoof of an animal; with a wide rounded terminal lobe narrowed at the base, as in a "clawed" petal.

uni- - a prefix meaning "one" or "single".

unifoliolate - a leaf with a single leaflet, the leaf-stalk of the leaflet can be differentiated from the main leaf stalk (see fig.

unilocular - with one locule or chamber.

uniseriate - arranged in a single row, series or layer.

unisexual - producing either male or female gametes, but not both; having either functional stamens or functional ovaries (ovules) but not both; the opposite of bisexual.

urceolate - shaped like a water pot or urn; with a rounded base and short broad tube that is narrowed above and slightly expanded at the very top.

utricle - a small one-seeded fruit with a loose thin outer covering, as in *Carex* (Cyperaceae) and many genera of Chenopodiaceae; also used for the small traps of *Utricularia* (Lentibulariaceae).

vacuole - a portion of a mature cell containing cell sap and surrounded by a tonoplast.

vaginate - with a sheath or enclosed in a sheath.

vallecular canal - a canal beneath a stem groove in *Equisetum*.

valvate - opening by regular lines to leave valves between; of petals or sepals that are joined edge to edge and do not overlap in bud.

valve - formed when an organ opens along regular lines, as when a capsular fruit splits open; also a flap-like lid.

valvule - a old term used for palea in Poaceae (Gramineae).

variety - a unit of classification below the level of a species; varieties are separated on the basis of form and colour but the varieties are usually not geographically separated and individuals of different varieties can freely interbreed.

vascular - referring to the xylem and phloem; vascular plants are those with xylem and phloem which transport water and nutrients.

vascular bundle - a strand or unit of phloem and xylem, with or without a surrounding sheath, which carries water and nutrients.

vascular tissue - a collective term for phloem and xylem together.

vein - a small strand of vascular tissue.

velum - the membraneous covering to the sporangium in the leaf of *Isoetes*.

velutinous - with a covering of velvet-like hairs, dense and straight.

venation - the arrangement of the veins or vascular bundles.

ventral - the inner face or the surface towards the axis, adaxial; compare with dorsal or abaxial.

ventricose - swollen or bulging on one side (but not as much as gibbous).

vernation - the arrangement of leaves in bud or of the parts (sepals and petals or tepals) in a flower-bud.

verrucose - having a surface with raised projections or warts.

versatile - turning freely on its support or stalk; as versatile anthers.

verticil - a whorl or an arrangement of similar parts in a circle at the same level; also used as synonymous with verticillaster.

verticillaster - an inflorescence which has opposite cymes forming whorls of flowers at the nodes along an elongated axis; typical of most species in Lamiaceae (Labiatae) such as 'basobela'.

verticillate - having parts arranged in whorls.

vesicle - a sac, bladder, or cavity; usually small and filled with gas or fluid.

vessel - a tube occurring in the xylem formed from a column of short broad cells which have lost their end walls during differentiation.

vestigial - undeveloped or poorly developed, or a trace or mark left by a structure no longer developed but present in ancestral forms.

vesture (vestiture) - a covering on a surface, as hairs or scales; see also indumentum.

vexillum - the broad petal (usually upper or adaxial) in pea-like flowers of Fabaceae (Leguminosae) subfamily Papilionoideae; also called standard or banner.

villous (villose) - with a covering of long soft, often crooked hairs.

virgate - long, slender, and straight, as a virgate stem.

viscid - sticky or glue-like.

viscous - very sticky or glue-like; glutinous.

viviparous - when the seeds germinate to form seedlings on the parent plant; often see in *Agave sisalana*, 'katcha'.

whorl - when there are three or more leaves or flowers at a single node or at the same level; see also verticel.

wing - any flat or thin extension on an organ; as in a winged fruit: the two lateral petals in pea-like flowers in Fabaceae (Leguminosae) subfamily Papilionoideae; the two lateral sepals in Polygalaceae.

wood - the xylem of plants with secondary vascular growth; the main tissue in the stems of trees and shrubs.

xanthophyll - a class of yellow, carotenoid pigments associated with chlorophyll in the chloroplasts.

xeromorph - a plant possessing features often found in xerophytes, but not necessarily confined to growing in dry places.

xerophyte - a plant that is able to live under very dry conditions and having structural adaptations for this.

xylem - the principal cells of the wood; important in water movement.

zygomorphic - flowers having bilateral symmetry so that the corolla can be divided equally only along one plane, used as synonymous with irregular; compare with actinomorphic and amorphic.

zygote - a fertilized egg formed by the union of a male and female gamete.

(Adapted by Ib Friis, Sue Edwards and Olof Ryding from J. Hutchinson & J. M. Dalziel, *Flora of West Tropical Africa*, 2. ed. by R. W. J. Keay (1954); W. C. Burger, *Families of flowering plants in Ethiopia* (1967); Bell, P. R. & Woodcock, C. L. F., *The diversity of green plants*; Scagel, R.F., *et al., An evolutionary survey of the plant kingdom* (1969); and Radford, A. E. *et al., Vascular Plant Systematics* (1974).

Comments on this glossary should be sent to:
The Editor, Ethiopian Flora Project,
The National Herbarium, P.O. Box 3434,
Addis Ababa, Ethiopia.

ABRUS *Adans.*, 107
Abrus canescens *Bak.*, 108
Abrus precatorius *L.*, 107
subsp. **africanus** *Verdc.*, 107
subsp. precatorius, 107
Abrus pulchellus *Thw.*, 108
subsp. **tenuiflorus** *(Benth.) Verdc.*, 108
Abrus schimperi *Hochst. ex Bak.*, *106*, 107
subsp. africanus (Vatke) Verdc., 108
subsp. **schimperi**, 108
Abrus tenuiflorus Benth., 108
ACACIA *Mill.*, 75
Acacia abyssinica *Hochst. ex Benth.*, *83*, 89
subsp. calophylla Brenan, 89
Acacia albida *Del.*, 75, *82*, 84
Acacia amythethophylla *Steud. ex A. Rich.*, 86
Acacia asak *(Forssk.) Willd.*, *82*, 80
var. *unispinosa* Fiori, 80
Acacia bavazzanoi *Pic.-Serm.*, 90
Acacia brevispica *Harms*, 78, *79*, *83*
subsp. dregeana (Benth.) Brenan, 78
Acacia bricchettiana *Chiov*, 91
Acacia bussei *Harms ex Sjöstedt*, *82*, 85
Acacia campylacantha Hochst. ex A. Rich., 81
Acacia caffra (Thunb.) Willd. var. *campylacantha* (Hochst. ex A. Rich.) Aubrév., 81
Acacia cheilanthifolia, 75
Acacia circummarginata Chiov., 78
Acacia clavigera E. Mey. subsp. *usambarensis* (Taub.) Brenan, 89
Acacia condyloclada Chiov., 80
Acacia cufodontii Chiov., 78
Acacia cultriformis *A. Cunn. ex G. Don,* 92
Acacia cyanophylla Lindl., 92
Acacia cyclops *Cunn. ex G. Don*, 92
Acacia dealbata *Link*, 92
Acacia decurrens *Willd.*, 92
Acacia dolichocephala *Harms*, *83*, 85
Acacia drepanolobium *Harms ex Sjöstedt*, *83*, 88
Acacia edgeworthii *T. Anders.*, *83*, 91
Acacia eggelingii Bak.f., 81
Acacia ehrenbergiana *Hayne*, *83*, 86
Acacia erythraea Chiov., 91
Acacia erythrantha Steud. ex A. Rich., 81
Acacia etbaica *Schweinf.*, 88
subsp. **etbaica.**, 88
subsp. **platycarpa** *Brenan*, 88
subsp. **uncinata** *Brenan*, *83*, 88
Acacia farnesiana *(L.) Willd.*, 86
Acacia flava (Forssk.) Schweinf., 86
Acacia gerrardii *Benth.*, 88
subsp. *gerrardii* var. *gerrardii*, 89
Acacia glaucophylla Steud. ex A. Rich., 80
Acacia gloveri Gilliland, 91
Acacia goetzei *Harms*, 84
subsp. **microphylla** *Brenan*, 84
Acacia gorinii Chiov., 90
Acacia hamulosa *Benth.*, 80
Acacia hecatophylla *Steud. ex Rich.*, *82*, 84
Acacia hockii *De Wild.*, 86
Acacia horrida *(L.) Willd.*, 84
subsp. **benadirensis** (Chiov.) Hillc., *82*, 85
subsp. horrida, 85
Acacia humifusa Chiov., 91
Acacia impervia Gilliland, 80
Acacia kantuffa Poir., 54
Acacia laeta *R. Br. ex Benth.*, 81, *82*
Acacia lahai *Steud. & Hochst. ex Benth.*, *79*, *82*, 85
Acacia macrothyrsa Harms, 86
Acacia mearnsii *De Wild.*, 91
Acacia melanoxylon *R. Br.*, 92
Acacia mellifera *(Vahl) Benth.*, *79*, 81, *82*
subsp. detinens (Burch.) Brenan, 81
Acacia monticola Brenan & Exell, 78
Acacia montigena *Brenan & Exell*, 78
Acacia nefasia (Hochst. ex. A. Rich), 90
Acacia negrii *Pic.-Serm.*, *79*, 89, 90
Acacia nilotica *(L.) Willd. ex Del.*, 86
subsp. **indica** *(Benth.) Brenan*, *83*, 87
subsp. **kraussiana** *(Benth.) Brenan*, 87
subsp. **leiocarpa** *Brenan*, *83*, 87
subsp. **nilotica**, 87
subsp. **subalata** *(Vatke) Brenan*, 8
subsp. tomentosa (Benth.) Brenan, 87
Acacia nubica Benth., 90
var. *erythraea* Schweinf., 90
Acacia oerfota *(Forssk.) Schweinf.*, *83*, 90
Acacia ogadensis *Chiov.*, 80
Acacia oliveri *Vatke*, 80, *82*
Acacia origena *Hunde*, 90
Acacia paolii *Chiov.*, *83*, 90
subsp. *paolii* Brenan, 91
subsp. *paucijuga*, 91
Acacia pappii Gand., 87
Acacia paradoxa Chiov., nom. illegit., non DC., 80
Acacia pennata sensu Cuf., 78
Acacia pentagona *(Schumach.) Hook. f.*, 78
Acacia persiciflora *Pax*, 81
Acacia pilispina *Pic.-Serm.*, 89
Acacia podalyriifolia *A. Cunn. ex G. Don.*, 92
Acacia polyacantha *Willd.*, 81
subsp. **campylacantha** *(Hochst. ex A. Rich.) Brenan*, 81, *82*
subsp. polyacantha, 81
Acacia prasinata *Hunde*, 87
Acacia pseudonigrescens *Brenan & Ross*, 84
Acacia pseudosocotrana Chiov., 91
Acacia pterygocarpa Hochst. ex Benth., 90
Acacia reficiens Wawra, 88
subsp. **misera** *(Vatke) Brenan*, 88
subsp. reficiens, 88
Acacia retinodes *Schlechtend.*, 92

Acacia robusta *Burch*., 89
subsp. clavigera (E. Mey.) Brenan, 89
subsp. **usambarensis** (Taub.) Brenan, *83*, 89
Acacia saligna *(Labill.) Wendl*., 92
Acacia sanguinea Hochst. ex A. Rich., 81
Acacia sassa (Willd.) Baill., 96
Acacia senegal *(L.) Willd*., 75, 78
var. **kerensis** *Schweinf*., 78
var. **leiorhachis** *Brenan*, 78
var. **senegal**., 78, *82*
Acacia seyal *Del*., 85
var. **fistula** *(Schweinf.) Oliv*., 86
var. **seyal**., 86
Acacia sieberiana *DC*., 89
var. **sieberiana**, 89
var. **woodii** *(Burtt-Davy) Keay & Brenan*, *83*, 90
var. *vermoesenii* (De Wild.) Keay & Brenan, 90
Acacia socotrana Balf. f., 91
Acacia somalensis Vatke, 75
Acacia spirocarpa Hochst. ex A. Rich., 87
Acacia stefaninii Chiov., 88
Acacia stenocarpa Hochst. ex A. Rich., 86
Acacia stuhlmannii *Taub*., 91
Acacia sultani Chiov., 91
Acacia tortilis subsp. spirocarpa, 83i
Acacia tortilis *(Forssk.) Hayne*, 87
subsp. **spirocarpa** *(Hochst. ex A. Rich.) Brenan*, 87
subsp. **tortilis**., 87
Acacia triacantha Hochst. ex A. Rich., 80
Acacia unispinosa (Fiori) Chiov., 80
Acacia venosa *Hochst. ex Benth*., 81, *82*
Acacia walwalensis *Gilliland*, 91
Acacia xiphocarpa Hochst. ex Benth., 89
Acacia zanzibarica *(S. Moore) Taub*., 85
var. **microphylla** *Brenan*, 85
var. zanzibarica, 85
Acacia zizyphispina *Chiov*., 80, *82*
Acosmium stipulare (Harms) Yakovlev, 101
ACROCARPUS *Wight & Arn*., 51
Acrocarpus fraxinifolius *Wight & Arn*., 51
Acrotaphros bibracteata Steud. ex A. Rich., 143
ACULEIFERUM Vassal, 75
ADENOCARPUS *DC*., 222
Adenocarpus mannii *(Hook. f.) Hook. f*., *221*, 222
Adicea tetraphylla Steud. & O. Kuntze, 310
var. *angolensis* Hiern., 310
AEONIUM *Webb & Berth*., 5, 16
Aeonium arboreum (L.) Webb & Berth., 16
var. *atropurpurea* (Nich.) Berger, 16
Aeonium leucoblepharum *A. Rich*., 16, *17*
AESCHYNOMENE *L*., 143
Aeschynomene abyssinica *(A. Rich.) Vatke*, 144, *145*
Aeschynomene americana *L*., 146
Aeschynomene cristata *Vatke*, 144
var. cristata, 144
var. **pubescens** *J. Léon*., 144
Aeschynomene elaphroxylon *(Guill. & Perr.) Taub*., 144, *145*
Aeschynomene indica *L*., 144
Aeschynomene pfundii *Taub*., 144
Aeschynomene rüppellii Bak.,144
Aeschynomene schimperi *Hochst. ex A. Rich*., 144, *145*
Aeschynomene ruspoliana *Taub. ex Harms*, 146
AFRICANAE *Engl. ex Wild*, 452
Afrocrania volkensii (harms) Hutch., 534
AFROVIVELLA Berger, 15
Afrovivella simensis (Britten) Berger, 15
Agelaea ugandensis Schellenb., 533
AGELANTHUS Tieghem, 372
Agialida abyssinica van Tieghem, 434
Agialida cuneifolia van Tieghem, 434
Agialida latifolia van Tieghem, 434
Agialida nigra van Tieghem, 434
Agialida schimperi van Tieghem, 434
ALANGIACEAE, 535
ALANGIUM *Lam*., 535
Alangium chinense *(Lour.) Harms*, 535, *536*
Alangium chinense (Lour.) Rehder, 535
ALBIZIA *Durazz*., 93
Albizia amara *(Roxb.) Boiv*., 94
subsp amara, 94
subsp. **sericocephala** *(Benth.) Brenan*, 94
Albizia anthelmintica *(A. Rich.) Brogn*., 94
Albizia chinensis *(Osbeck) Merr*., 93
Albizia coriaria *Welw. ex Oliv*., 93
Albizia elliptica Fourn., 93
Albizia grandibracteata *Taub*., *95*, 96
Albizia gummifera *(J.F. Gmel.) C.A. Sm*., 96
Albizia isenbergiana *(A. Rich.) Fourn*., 94
Albizia lebbeck *(L.) Benth*., 94
Albizia lophantha *(Willd.) Benth*., 93
Albizia malacophylla *(A. Rich.) Walp*., 93
var. **malacophylla**, 94
var. **ugandensis** *Bak. f*., 94
Albizia ogadensis (Chiov.) Bak. f. ex Chiov., 80
Albizia pallida Fourn., 93
Albizia quartiniana (A. Rich.) Walp., 93
Albizia sassa (Willd.) Chiov., 96
Albizia schimperiana *Oliv*., *95*, 96
ALCHEMILLA *L*., 37
Alchemilla abyssinica auct. non Fresen., 39
Alchemilla abyssinica *Fresen*., 40
f. *muscoidea* Haum. & Balle, 39
var. *schimperi* Engl., 39
Alchemilla arvensis Scop. f. *alpina* Engl., 41
Alchemilla bachiti Hauman & Balle, 41
Alchemilla chillaloensis (Haum. & Balle) Haum. & Balle, 38
Alchemilla commutata Rothmaler, 39
f. *muscoides* (Haum. & Balle) Rothmaler, 39
f. *robusta* Rothmaler, 40
Alchemilla cryptantha *A. Rich*., 39
Alchemilla ellenbeckii *Engl*., 38

Alchemilla erlangeriana Engl., 39
Alchemilla fischeri *Engl.*, 39
Alchemilla geranioides Rolfe subsp. *chillaloensis* Haum. & Balle, 38
Alchemilla gracilis Pax, 38
Alchemilla gunae Schweinf., 40
Alchemilla haumannii *Rothm.*, *38*, 38
Alchemilla kiwuensis *Engl.*, 39
Alchemilla microbetula Th. C.E. Fries, 39, *40*
Alchemilla pedata *A. Rich.*, 38, 39
var. *argentea* Fiori, 39
Alchemilla rothii Oliv., 40
Alchemilla sessiliflora Rothmaler, 39
Alchemilla steudneri Th. C.E.Fries, 40
ALLOCASUARINA S. Johnston, 262
ALLOPHYLUS *L.*, 498
Allophylus abyssinicus *(Hochst.) Radlkofer*, 499, *500*
Allophylus africanus *P. Beauv.*, 499, *501*
Allophylus alnifolius, 502
Allophylus alnifolius sensu Cuf., 502
Allophylus erlangeri Gilg ex Chiov., 502
Allophylus macrobotrys *Gilg*, 499, *501*
Allophylus rubifolius, *501*, 502
Allophylus sp. = *Mooney* 9003, *501*, 502
ALYSICARPUS *Desv.*, 154
Alysicarpus ferrugineus *Hochst. & Steud. ex A. Rich.*, *153*, 154
Alysicarpus glumaceus *(Vahl.) DC.*, 155
Alysicarpus hochstetteri A. Rich., 155
Alysicarpus monilifer *(L.) DC.*, 154
Alysicarpus ovalifolius *(Schumach.) J. Léon.*, 154
Alysicarpus quartinianus *A. Rich.*,155
Alysicarpus rugosus *(Willd.) DC.*, 155
subsp. **perennirufus** *J. Léon.*, *153*, 155
subsp. **rugosus.**, 155
var. *quartinianus* (A. Rich.) Bak., 155
Alysicarpus zeyheri *Harv.*, 154
AMPELOCISSUS *Planch.*, 399
Ampelocissus abyssinica *(Hochst. ex A. Rich.) Planch.*, 401
Ampelocissus bombycina *(Bak.) Planch.*, 399, *400*, 401
Ampelocissus cavicaulis (Bak.) Planch., 401
Ampelocissus schimperiana *(Hochst. ex A. Rich.) Planch.*, 400
AMPHICARPA *Nuttall*, 155
Amphicarpa africana *(Hook. f.) Harms*, 155, *157*
Amphinomia platycarpa (Viv.) Cuf., 195
AMYGDALUS, 41
Amygdalus communis L., 42
Amygdalus persica L., 42
Amyris anisata Willd., 429
Amyris papyrifera Del., 443
ANACARDIACEAE, 513
Anaphrenium abyssinicum Hochst., 521
Anaphrenium pulcherrimum Schweinf., 521
Anarthrosyne abyssinica Hochst., 151
ANTHYLLIS *L.*, 225
Anthyllis abyssinica (Sag.) Bak., 225
Anthyllis vulneria *L.*, 225
subsp. **abyssinica** *(Sag.) Cullen*, 225, *226*
ANTIARIS *Lesch.*, 299
Antiaris africana, 384
Antiaris toxicaria *Lesch.*, 301
subsp. **welwitschii** *(Engl.) C.C. Berg*, 301
var. africana A. Chev., 301
var. **welwitschii**, 301, *301*
subsp. toxicaria, 301
ANTOPETITIA *A. Rich.*, 227
Antopetitia abyssinica *A. Rich.*, 227, *229*
APHANES *L.*, 41
Aphanes bachiti *(Hauman & Balle) Rothmaler*, 41, *41*
APHANIA Blume, 504
Aphania senegalensis (Juss. ex Poir.) Radlkofer, 504
Apodostigma pallens (Planchon ex Oliver) R. Wilczek, 342
APODYTES *E. Mey. ex Arn.*, 348
Apodytes acutifolia Hochst. ex A. Rich., 348
Apodytes dimidiata *E. Mey. ex Arn.*, 348
var. **acutifolia** *(Hochst. ex A. Rich.) Boutique*, 348, *349*
var. dimidiata, 348, *349*
AQUIFOLIACEAE, 329
ARACHIS *L.*, 147
Arachis fruticosa Retz., 147
Arachis hypogea *L.*, 147, *149*
Aralia farinosa Del., 537
ARALIACEAE, 537
ARCEUTHOBIUM *Marsch. Bieb.*, 378
Arceuthobium juniperi-procerae *Chiov.*, 378, *378*
Arceuthobium oxycedri auct. non (DC.) Marsch. Bieb., 378
ARGYROLOBIUM *Eckl. & Zeyh.*, 220
Argyrolobium abyssinicum Jaub. & Spach, 222
forma *caespitosum* Lanza., 222
forma *diffusum* Lanza, 222
var. *caespitosum* (Lanza) Fiori., 222
var. *diffusum* (Lanza) Fiori, 222
Argyrolobium arabicum *(Decne.) Jaub. & Spach*, 222
Argyrolobium confertum *Polhill*, *221*, 222
Argyrolobium dorycnoides Bak., 222
Argyrolobium fischeri *Taub.*, 220
Argyrolobium petitianum A. Rich., 222
Argyrolobium ramosissimum *Bak.*, 222
Argyrolobium remotum Hochst., 220
Argyrolobium rupestre *(E. Mey.) Walp.*, 220
subsp. **aberdaricum** *(Harms) Polhill*, 220
subsp. kilimandscharicum (Taub.) Polhill, 220
subsp. **remotum** *(Hochst. ex A. Rich.) Polhill*, 220
subsp. rupestre, 220
Argyrolobium schimperianum *Hochst. ex A. Rich.*, *221*, 222
Argyrolobium uniflorum, 222
Argyrolobium virgatum Bak., 220
Armeniaca vulgaris Lam, 41

Aspidixia semiplanum Tieghem, 377
ASPIDOSTIGMA Hochst., 426
Aspidostigma acuminatum Hochst., 427
ASTRAGALUS *L.*, 230
Astragalus arabicus Ehrenb. ex Bunge, non Kotschy 231
  var. *congestus* Schweinf., 231
Astragalus atropilosulus *(Hochst.) Bunge*, 230
  subsp. **abyssinicus** *(Hochst.) Gillett*, 231
    var. **abyssinicus**, 231
    var. **coerulescens** *(Chiov.) Gillett*, 231
    var. **venosus** *(Hochst.) Gillett*, 231
  subsp. **atropilosulus**, 231
    var. **atropilosulus**, 231
    var. **mooneyi**, *Gillett*, 231, *232*
  subsp. **bequaertii** *(De Wild.) Gillett*, 231
    var. **bequaertii**, 231
  subsp. **burkeanus** *(Harv.) Gillett*, 231
    var. **burkeanus**, 231
Astragalus boeticus L., 230
Astragalus coerulescens Chiov., 231
Astragalus fatmensis *Hochst. ex Chiov.*, 231
Astragalus somalensis Taub. ex Harms, 230
Astragalus vogelii (Webb.) Bornm., 231
  var. *congestus* (Schweinf.) Cuf., 231
  subsp. *fatimensis* Maire, 231
Atylosia Wight. & Arn., 181
AUSTRALINA *Gaud.*, 324
Australina flaccida *(A. Rich.) Webb*, 324, *326*
Australina schimperiana Wedd., 324
AZADIRACHTA *A.Juss.*, 485
Azadirachta indica *A. Juss.*, 485, *487*
BALANITACEAE, 433
Balanites *Del.*, 433
Balanites aegyptiaca sensu Cufodontis, 434, 436
Balanites aegyptiaca sensu Oliver, 434
Balanites aegyptiaca *(L.) Del.*, 433
  var. **aegyptiaca**, 433, *435*
  var. **pallida** *Sands*, 434, *435*
Balanites gillettii Cufodontis, 436
  var. *renifolia* Cufodontis, 436
Balanites glabra *Mildbr. & Schlecht.*, 434, *435*
Balanites latifolia (van Tieghem) Chiov., 434
Balanites orbicularis Sprague, 436
Balanites patriziana Lusina, 436
Balanites pedicellaris *Mildbr. & Schlecht*, 436
  subsp. pedicellaris, 436
  subsp. **somalensis** *(Mildbr. & Schlecht.) Sands*, *435*, 436
Balanites racemosa Chiov., 434
Balanites rotundifolia *(van Tieghem) Blatter*, 434
  var. **rotundifolia**, *435*, 436
Balanites scillin Chiov., 436
Balanites somalensis Mildbr. & Schlecht, 436
  var. *cinereo-corticata* Fiori ex Chiov., 436
BALANOPHORACEAE, 384
Balsamodendron habessinicum Berg, 471
BANKESIA Bruce, 43
BANKSIA Bruce, 43
Banksia (*Bankesia*) *abyssinica* Bruce, 43
BAPHIA *Lodd.*, 104
Baphia abyssinica *Brummitt*, *103*, 104
BARBEYACEAE, 270
Barbeya, 270
Barbeya oleoides *Schweinf.*, 270, *270*
Barbeya oleoides subsp. *arussorum* Chiov., 270
BAUHINIA *L.*, 68
Bauhinia abyssinica A. Rich., 68
Bauhinia ellenbeckii *Harms*, *67*, 68
Bauhinia farek Desv., 68
Bauhinia fassoglensis Kotschy ex Schweinf., 70
Bauhinia purpurea *L.*, 68
Bauhinia thonningii Schumach., 68
Bauhinia tomentosa *L.*, 68
Bauhinia variegata *L.*, 70
BERCHEMIA *Neck. ex DC.*, 393
Berchemia discolor *(Klotzsch) Hemsl.*, 393, *394*
Berrebera ferruginea Hochst., 108
BERSAMA *Fresen.*, 511
Bersama abyssinica *Fresen.*, 511, *512*
  subsp **abyssinica**, 511
Bersama gallensis V. Brehmer, 511
Bersama giuliarrellii Chiov., 511
Bersama integrifolia A. Rich, 511
Bersama serrata A. Rich, 511
Bersenna anthemlmintica A. Rich., 94
BIFARIA van Tieghem, 377
Bifaria abyssinica Tieghem, 377, 378
BISERRULA *L.*, 231
Biserrula leiocarpa A. Rich, 233
Biserrula pelecinus *L.*, 231
  subsp. **leiocarpa** *(A. Rich.) Gillett*
  var. *leiocarpa* (A. Rich.) Chiov., 233
  var. *subintegra* Bak. f., 233
  subsp. **pelecinus**, *232*, 233
BLIGHIA *Koenig*, 508
Blighia unijugata *Bak*, 508, *510*
BOEHMERIA *Jacq.*, 314
Boehmeria hypoleuca (Steudel) A. Rich., 320
Boehmeria macrophylla *Hornemann*, 314, *316*
Boehmeria parvifolia Wedd., 322
Boehmeria pauciflora (Steudel) Blume, 322
Boehmeria platyphylla Don, 314, 317
Boehmeria salicifolia D. Don, 320
BOSQUEIA Thou. ex Baill., 299
Bosqueia phoberos Baill., 299
Bosqueia thouarsiana Baill., 299
BOSWELLIA *Roxb. ex Colebr.*, 442
Boswellia boranensis Engl., 443
Boswellia frereana Birdw., 443
Boswellia bhau-dajiana Birdw., 443
Boswellia hildebrandtii Engl., 445
Boswellia microphylla *Chiov.*, 445, *445*
Boswellia multifoliolata Engl., 445
Boswellia neglecta *S. Moore*, 445, *445*

Boswellia ogadensis *Vollesen*, 443
Boswellia papyrifera *(Del.) Hochst.*, 443, *444*
Boswellia pirottae *Chiov.*, 443, *444*
Boswellia rivae *Engl.*, 443, *444*
Boswellia sacra Flück., 443
BOTTEGOA *Chiov.*, 504
Bottegoa insignis *Chiov.*, 504, *506*
BRUCEA *J.F. Mill.*, 440
Brucea antidysenterica *J.F. Mill.*, 440, *441*
Brucea erythraeae Chiov., 440
Brucea ferruginea L'Hérit., 440
BRYOPHYLLUM *Salisb.*, 5, 25
Bryophyllum delagoense (Eckl. & Zeyh.) Schinz., 26
Bryophyllum pinnatum *(Lam.) Oken*, 25
Bryophyllum proliferum *Hook.*, 25
Bryophyllum tubiflorum Harvey, 26
BULLIARDIA DC., 6
Bulliardia abyssinica A. Rich., 6
BURSERACEAE, 442
BUXACEAE, 255
Buxanthus hildebrandtii (Baill.) van Tieghem, 255
BUXUS *L.*, 255
Buxus calophylla Pax., 255
Buxus hildebrandtii *Baill.*, 255, *256*
CADIA *Forssk.*, 101
Cadia purpurea *(Picc.) Ait.*, 101, *102*
CAESALPINIA *L.*, 51
Caesalpinia bonduc sensu Cuf., 54
Caesalpinia bonduc *(L.) Roxb.*, 54
Caesalpinia dauensis *Thulin*, 53
Caesalpinia decapetala *(Roth) Alston*, *52*, 53
Caesalpinia erianthera *Chiov.*, 53
var. **pubescens** *Brenan*, 53
Caesalpinia erlangeri Harms, 53
Caesalpinia gillettii Hutch. & Bruce, 54
Caesalpinia gilliesii *(Hook.) Dietr.*, 53
Caesalpinia oligophylla *Harms*, *52*, 53
Caesalpinia pulcherrima *(L.) Sw.*, 51
Caesalpinia spinosa *(Molina) Kuntze*, 53
Caesalpinia tinctoria (Kunth) Taub., 53
Caesalpinia trothae *Harms*, 53
subsp. **erlangeri** *(Harms) Brenan*, 53
Caesalpinia volkensii *Harms*, 54
CAESALPINIOIDEAE, 49
CAJANUS *DC.*, 181
Cajanus cajan *(L.) Millsp.*, 181
CALODENDRUM *Thunb.*, 420
Calodendrum capensis *(L.f.) Thunb.*, 420
CALPURNIA *E. Mey.*, 104
Calpurnia aurea *(Ait.) Benth.*, *102*, 104
subsp. *aurea*, 104
subsp. indica Brummitt, 104
subsp. sylvatica (Burch.) Brummitt, 104
Calpurnia subdecandra (L'Hérit.) Schweick., 104
CAMPESTRES *Engl. ex Vollesen*, 468
CANAVALIA *DC.*, 165
Canavalia africana *Dunn*, 165
Canavalia ensiformis *(L.) DC.*, 165, *167*
Canavalia virosa sensu auctt, non (Roxb.) Wight & Arn., 165
CANNABACEAE, 327
Cannabiaceae, 327
Cannabidaceae, 327
Cannabinaceae, 327
CANNABIS *L.*, 327
Cannabis indica Lam., 327
Cannabis ruderalis Janisch, 327
Cannabis sativa *L.*, 327, *328*
Cantuffa exosa J.F. Gmel., 54
Cantuffa stellata (Forssk.) Chiov., 54
CARDIOSPERMUM *L.*, 493
Cardiospermum canescens Wall., 497
var. *glabrescens* Almagia, 498
Cardiospermum clematideum A.Rich, 497
Cardiospermum corindum *L*, 497
forma *clematideum* (A. Rich.) Radlkofer, 497
forma *canescens* (Wall.) Radlkofer, 497
Cardiospermum halicacabum *L.* 498
var. **halicacabum**, *497*, 498
var. **microcarpum** *(Kunth) Blume*, 498
Cardiospermum microcarpum Kunth, 498
Cardiospermum obolongum A.Rich, 497
Cardiospermum truncatum A.Rich, 498
CASIMIROA *La Llave*, 424
Casimiroa edulis *La Llave*, 424
CASSIA *L.*, 56
Cassia absus L., 64
Cassia alexandrina (Mill.) Thell., 63
Cassia angustifolia Vahl, 63
Cassia arereh *Del.*, 56, *58*
Cassia baccarinii Chiov., 59
Cassia bicapsularis L., 60
Cassia didymobotrya Fresen., 61
Cassia dimidiata sensu auctt., 65
Cassia ellisae Brenan, 61
Cassia fallacina Chiov., 65
Cassia fistula *L.*, 57
Cassia floribunda sensu auctt., 59
Cassia goratensis Fresen., 59
Cassia grantii Oliv., 64
var. *pilosula* sensu Cuf., 64
Cassia hildebrandtii Vatke, 64
var. *crispata* Serrato, 64
Cassia hochstetteri Ghesq., 65
Cassia holosericea Fresen, 63
Cassia italica (Mill.) Spreng., 63
subsp. *micrantha* Brenan, 63
Cassia kirkii Oliv., 65
var. *guineensis* Steyaert, 65
Cassia laevigata Willd., 59
Cassia longiracemosa Vatke, 61
Cassia mimosoides L., 66
Cassia mimosoides sensu Adumbr., 65
Cassia multiglandulosa Jacq., 59

Cassia nigricans Vahl, 65
Cassia obtusifolia L., 60
Cassia occidentalis L., 60
Cassia petersiana Bolle, 59
Cassia ruspolii Chiov., 61
Cassia sabak Del., 59
Cassia senna L., 63
var. *obtusata* Brenan, 63
Cassia septemtrionalis Viv., 59
Cassia siamea Lam., 60
Cassia singueana Del., 59
Cassia sophera L., 60
Cassia tomentosa L. f., 59
Cassia tora sensu auctt., non L., 60
Cassia truncata Brenan, 61
Cassia wittei Ghesq., 65
Cassine sensu Loes., 338, 344
Cassine aethiopica Thunb., 338
Cassine aquifolium Fiori, 344
Cassine buchananii Loes., 344
CASUARINA *L.*, 262
CASUARINACEAE, 262
Casuarina cunninghamiana *Miq.*, 262, *263*
Casuarina equisetifolia *L.*, 262, *263*
Casuarina glauca Sprague, 264
Casuarina litorea L. ex Fosberg & Sachet, 264
Casuarina obesa Miq., 264
subsp. obesa, 264
Casuarina quadrivalvis Labill., 264
Casuarina sp. = *Milchersich* 108, 264
Casuarina stricta Ait., 262
Casuarina verticillata *Lam.*, 264
CATHA *Forssk.*, 340
Catha edulis *(Vahl) Forssk. ex Endl.*, 340, *341*
Catha forskalii A. Rich., 340
Catha inermis J.F. Gmelin, 340
CAYRATIA *A. Juss.*, 402
Cayratia gracilis *(Guill. & Perr.) Suesseng.*, 403
Cayratia ibuensis *(Hook. f.) Suesseng.*, *403*, 404
Cayratia ruspolii (Gilg) Suesseng., 408
Cedrela kotschyi Schweinf., 479
CELASTRACEAE, 331
CELASTRUS, 331
sect. *GYMNOSPORIA* Wight & Arnott, 331
Celastrus arbutifolius A. Rich., 336
var. *major* A. Rich., 336
Celastrus atkaio A. Rich., 336
Celastrus edulis Vahl, 340
Celastrus gracilipes Welw. ex Oliv., 332
Celastrus heterophyllus Eckl. & Zeyh., 336
Celastrus laurifolius A. Rich., 338
Celastrus luteolus Del., 338
Celastrus obscurus A. Rich., 333
Celastrus parviflorus Vahl, 333
Celastrus schimperi A. Rich., 333
Celastrus senegalensis Lam., 336
var. *inermis* A. Rich., 336
Celastrus serratus A. Rich., 333
var. *steudneri* Engl., 332
Celastrus undatus Thunb., 338
CELTIS *L.*, 266
Celtis africana *Burm. f.*, 266, *269*
Celtis australis sensu auctt. non L., 266
Celtis brownii Rendle, 267
Celtis durandii Engl., 267
Celtis gomphophylla *Bak.*, 267
Celtis guineensis Schumach. & Thonn., 268
Celtis integrifolia Lam., 267
Celtis kraussiana Bernh., 266
var. *flavescens* Almagia, 266
Celtis lactea Sim, 272
Celtis mauritiana Planch., 267
Celtis orientalis L., 268
Celtis philippensis *Blanco*, 267
Celtis prantlii Engl., 267
Celtis toka *(Forssk.) Hepper & Wood*, 267
Celtis ugandensis Rendle, 267
Celtis vesciculosa Planch., 266
Celtis wightii Planch., 267
Celtis zenkeri *Engl.*, 267
CERASUS, 41
CERATONIA *L.*, 66
Ceratonia siliqua *L.*, 66
CHAMAECRISTA *Moench.*, 56, 64
Chamaecrista absus *(L.) Irwin & Barneby*, 64
Chamaecrista fallacina *(Chiov.) Lock*, 65
Chamaecrista hildebrandtii *(Vatke) Lock*, 64, 65, 66
var. *crispata* Serrato, 64
Chamaecrista hochstetteri *(Ghesq.) Lock*, 65
Chamaecrista kirkii *(Oliv.) Standley*, 65
var. **guineensis** *(Steyaert) Lock*, 65
Chamaecrista mimosoides *(L.) Greene*, 66
Chamaecrista nigricans *(Vahl) Greene*, *62*, 65
Chamaecrista wittei *(Ghesq.) Lock*, 65
Chirocalyx abyssinicus (DC.) Hochst., 161
Chirocalyx tomentosus Hochst., 161
CHLOROPHORA Gaud., 272
Chlorophora excelsa (Welw.) Bent. & Hook. f., 272
Choristylis, 4
CHRYSOBALANACEAE, 46
Chrysobalanus, 46
Chrysobalanus icaco, 46
Chrysocalyx petitiana A. Rich., 210
Chrysocalyx quartiniana A. Rich., 210
CICER *L.*, 246
Cicer arietinum *L.*, 246
Cicer cuneatum *Hochst. ex A. Rich.*, *232*, 248
CILIATAE *Vollesen*, 465
CISSUS *L.*, 404
Cissus adenantha Fresen., 410
Cissus adenantha sensu Richard, 414
Cissus adenocaulis Steud. ex A. Rich., 412
var. *chiovendae* Lanza, 414
Cissus adenocephala Gilg. & Brandt, 415

Cissus aphylla *Chiov.*, 404
Cissus aphyllantha *Gilg*, 404
Cissus aphyllantha Gilg., p.p., 404
Cissus arguta *Hook. f.*, 407
Cissus betaeformis Chiov., 410
Cissus cactiformis *Gilg*, 407
Cissus cornifolia *(Bak.) Planch.*, 405, *405*
Cissus cymosa auct. non Schumach. & Thonn., 417
Cissus cyphopetala Fresen., 417
Cissus dembianensis Chiov., 415
Cissus digitata sensu Cuf., 417
Cissus dysocarpa Gilg & Brandt, 416
Cissus ellenbeckii Gilg & Brandt, 406
Cissus gracilis Guill. & Perr., 403
Cissus hochstetteri (Miq.) Planch., 406
Cissus jatrophoides sensu Cuf., 410
Cissus javellensis Lanza, 417
  var. *rotundata* Lanza, 417
Cissus juncea Webb, 410
Cissus mollis (Bak.) Planch., 411
Cissus molle (Steud. ex Bak.) Descoings
Cissus neghelliensis Lanza, 417
Cissus nivea Hochst. ex Schweinf., 415
Cissus oxyphylla (A.Rich.) Chiov., 412
Cissus pauli-guilielmii Schweinf., 401, 418
Cissus petiolata *Hook. f.*, 406, *406*
Cissus populnea *Guill. & Perr.*, 407
Cissus populnea sensu Cuf., 407
Cissus producta Afz., 407
Cissus quadrangularis *L.*, 407, *408*
Cissus rivae Gilg, 415
Cissus rotundifolia *(Forssk.) Vahl*, 408, *408*
Cissus ruspolii *Gilg*, 408
Cissus schimperi Hochst. ex Planch., 412
Cissus serjanioides Planch. , 414
Cissus serpens Hochst. ex A. Rich., 417
Cissus somaliensis Gilg, 416
Cissus sp. = *Gilbert* et al. 8203, 407
Cissus sp. = *Gilbert & Thulin* 511, 415
Cissus stipulacea (Bak.) Planch., 415
  var. *hochstetteri* Planch. in DC., 415
Cissus subdiaphana Steud. ex A. Rich., 403
Cissus taschdjianica Chiov., 410
Cissus ternata (Forssk.) Gmelin, 416
Cissus ukerewense (Gilg.) Descoings, 415
CITRUS *L.*, 430
Citrus aurantifolia *(Christm.) Swingle*, 431, *431*
Citrus aurantiifolius, 431
Citrus aurantium *L.*, 432
Citrus aurantium sensu A. Rich., 432
Citrus bigarradia Lois., 432
Citrus deliciosa Ten., 431
Citrus grandis (L.) Osb., 431
Citrus limetta *Risso*, 431
Citrus limon *(L.) Burm.f.*, 430
Citrus limon Risso
  var. *pusilla* Risso, 431
  var. *abyssinica* Riccob., 431
Citrus limonia Osb. var *limetta* (Risso) Engl., 431
Citrus limonia var *limonia* sensu Cuf., non Osb., 430
Citrus limonum Risso, 430
Citrus medica *L.*, 430
Citrus maximum *(Burm.) Merrill*, 431
Citrus paradisi *Macfad.*, 431
Citrus reticulata *Blanco*, 431
Citrus sinensis *(L.) Osb.*, 432
Citrus vulgaris Risso, 432
CLAUSENA *Burm. f.*, 429
Clausena abyssinica (Engl.) Engl., 429
Clausena anisata *(Willd.) Benth.*, 429, *429*
  subsp. *abyssinica* (Engl.) Cuf., 429
Clausena inaequalis (DC.) Benth., 429
  var. *abyssinica* Engl., 429
Clausena hildebrandtii Engl., 422
CLAUSENOPSIS (Engl.) Engl., 422
Clausenopsis angolensis (Engl.) Engl., 422
CLITORIA *L.*, 156
Clitoria ternatea *L.*, 156
  var. **angustifolia** *Hochst. ex Bak.*, 156
  var. **ternatea**, 156, *157*
Cnestis ugandensis Schellenb., 533
COLUTEA *L.*, 233
Colutea abyssinica *Kunth & Bouché*, *232*, 233
  var. *macrophysa* (Chiov.) Browicz, 233
Colutea halepica Lam., 233
  var. *sericea* A. Rich., 233
Colutea istria Mill., 233
  var. *macrophysa* Chiov., 233
  var. *sericea* (A. Rich.) Cuf., 233
COMBRESIA A. Rich., 6
COMMIPHORA *Jacq.*, 446
  sect. AFRICANAE *Engl. ex Wild*, 452
  sect. CAMPESTRIS *Engl. ex Vollesen*, 468
  sect. CILIATAE *Vollesen*, 465
  sect. COMMIPHORA, 469
  sect. CORIACEAE *Engl. ex Wild*, 452
  sect. HILDEBRANDTIANAE *Engl. ex Vollesen*, 457
  sect. LATIFOLIATAE *Engl. ex Vollesen*, 465
  sect. MONOICAE *Vollesen*, 457
  sect. OPOBALSAMEAE *Engl. ex Vollesen*, 474
  sect. PEDUNCULATAE *Engl. ex Vollesen*, 457
  sect. ROSTRATAE *Engl. ex Wild*, 462
  sect. SPONDIOIDEAE *Engl. ex Wild*, 447
COMMIPHORA, 469
Commiphora "abyssinica", auctt., mult. Afric., 471
Commiphora africana *(A. Rich.) Engl.*, 454
  var. **africana**, 452, 454, 457, 457, 462
  var. *ramosissima* (Oliv.) Engl. in DC., 454
  var. **rubriflora** *(Engl.) Wild*, 457
Commiphora alaticaulis *Gillett & Vollesen*, 462
Commiphora albiflora *Engl.*, 477
Commiphora ancistrophora *Chiov.*, 475
Commiphora anfractuosa Chiov., 477
Commiphora arussensis Engl., 454

Commiphora assaortensis Chiov., 471
Commiphora atramentaria Chiov., 474
Commiphora baluensis Engl., 450
Commiphora berardellii Chiov., 477
Commiphora bruceae *Chiov*., 473
Commiphora boiviniana *Engl*., 450
var. *crenata* Engl., 450
Commiphora boranensis *Vollesen*, 475, *476*
Commiphora campestris sensu Dale & Greenway, 454
Commiphora candidula Sprague
Commiphora cerasiformis Chiov., 468
Commiphora campestris *Engl*., 469
Commiphora ciliata *Vollesen*, 460, 465, *466*
Commiphora confusa *Vollesen*, 462
Commiphora coriacea Engl., 452
Commiphora coronillifolia *Chiov*., 460, *478*
Commiphora corrugata *Gillett & Vollesen*, 460, *461*
Commiphora crassispina Sprague, 452
Commiphora crenulata (Terrac.) Chiov., 473
Commiphora crenato-lobata Chiov., 462
Commiphora cuspidata Chiov., 452
Commiphora cyclophylla *Chiov*., 468
Commiphora dancaliensis Chiov., 471
Commiphora ellenbeckii *Engl*., 473
Commiphora ellisiae Vollesen, 450
Commiphora engleri Guillaumin, 448
Commiphora erlangeriana *Engl*., 447, 448
Commiphora erosa *Vollesen*, 465, *467*
Commiphora erlangeriana sensu Chiovenda, 448
Commiphora erythraea *(Ehrenb.) Engl*., 450, 452
Commiphora flabellifolia Chiov., 454
Commiphora flaviflora Engl., 473
Commiphora gallaënsis Engl., 450
Commiphora gileadensis *(L.) C. Chr*., 452, 475
Commiphora gileadensis sensu Dale & Greenway, 475
Commiphora gowlello *Sprague*, 474
Commiphora guidottii *Chiov*., 448
Commiphora habessinica *(Berg) Engl*., 452, 471, *472*
var. *grossedentata* Chiov., 452
var. *simplicifolia* sensu Cuf., 471
Commiphora hildebrandtii *(Engl.) Engl*., 460
Commiphora hodai *Sprague*, 452, 471
Commiphora holosericea Engl., 450
Commiphora holtziana Engl., 450
Commiphora horrida *Chiov*., 474
Commiphora incisa *Chiov*., 473
Commiphora kua *(R. Br. ex Royle) Vollesen*, 452, 473
Commiphora longipedicellata *Vollesen*, 448
Commiphora lughensis Chiov., 450, 468
Commiphora madagascariensis sensu Wild, 471
Commiphora mildbraedii *Engl*., 468
Commiphora molmol (Engl.) Engl., 452
Commiphora monoica *Vollesen*, 457, *458*
Commiphora myrrha *(Nees) Engl*., 452, *453*
Commiphora neumannii Engl., 454
Commiphora obovata *Chiov*., 454
Commiphora ogadensis *Chiov*., 460
Commiphora opobalsamum (L.) Engl., 475
var. *ehrenbergianum* (Berg) Engl., 475
var. *induta* Sprague ex Hutch. & Bruce
Commiphora opobalsamum sensu Cuf., 477
Commiphora parvifolia sensu Chiov., non (Balf. f.) Eng., 477
Commiphora pedunculata *(Kotschy & Peyr.) Engl*., 457, *459*
Commiphora pilosa (Engl.) Engl., 454
Commiphora playfairii sensu Chiov., non (Hook. f.) Engl., 473
Commiphora quadrincincta *Schweinf*., 452, 471
Commiphora resiniflua Martelli, 454
Commiphora retifolia Chiov., 447
Commiphora rivae Engl., 452
Commiphora rivae sensu Chiov., non Engl., 471
Commiphora robecchii Engl., 462
Commiphora reghinii Chiov., 477
Commiphora rostrata *Engl*., 462, *464*
Commiphora samharensis *Schweinf*., 452
Commiphora schimperi *(Berg) Engl*., 452, 454, *456*, 457
Commiphora serrulata *Engl*., 469
var. *tenuipes* Engl., 469
Commiphora sessiliflora Vollesen, 448
Commiphora somalensis Engl., 450
Commiphora sphaerocarpa *Chiov*., *467*, 468
Commiphora sphaerophylla *Chiov*., 450, *451*
Commiphora sp. = *Friis* et al. 3160, 474
Commiphora sp. = *Corradi* 6767, 6768, 473
Commiphora sp. = *Gilbert* et al. 7652, 475
Commiphora sp. = *Gilbert* et al. 8170, 468
Commiphora sp. = *Gilbert* et al. 8171, 475
Commiphora sp = *Glover & Gilliland* 973, 477
Commiphora staphyleifolia *Chiov*., 448
Commiphora tenuis *Vollesen*, 469, *470*
Commiphora tephrodes Chiov., 460
Commiphora terebinthina *Vollesen*, 454, *455*
Commiphora truncata *Engl*., 452, 462
Commiphora tubuk *Sprague*, 457
Commiphora unilobata *Gillett & Vollesen*, 448, *449*
Commiphora voënsis Engl., 101
Commiphora zanzibarica (Baill.) Engl., 448
CONNARACEAE, 533
CORDEAUXIA *Hemsl*., 56
Cordeauxia edulis *Hemsl*., *55*, 56
CORDYLA *Lour*., 70
Cordyla somalensis *Gillett*, *69*, 70
subsp. *littoralis* Gillett, 70
subsp. somalensis, 70
CORIACEAE *Engl. ex Wild*, 452
CORNACEAE, 534
CORONILLA *L*., 227
Coronilla scorpiodes *(L.) Koch*, 227
COTONEASTER *Medik*, 41
Cotoneaster nummularia *Fischer & C. A. Meyer*, 41
COTYLEDON *L*., 5, 16
Cotyledon barbeyi *Bak*., *18*, 19

Cotyledon laciniata L., 25
Cotyledon lanceolata Forssk., 23
Cotyledon pinnata Lam., 25
Cotyledon simensis Britten, 15
Cotyledon umbilicus auct. non L., 16
Cotyledon wickensii Schönl., 19
var. *glandulosa* Poellnitz, 19
Cotyledon, 5
CRAIBIA *Harms & Dunn*, 109
Craibia brevicaudata *(Vatke) Dunn*, 109
Craibia laurentii *(De Wild.) De Wild.*, 109
CRASSULA *L.*, 5, **6**, 10
Crassula abyssinica A. Rich., 6, 10
Crassula alata *(Viv.) Berger*, 8
subsp. *alata*, 8
subsp. **pharnaceoides** *(Fisch. & Mey.) Wickens & Bywater*, 7, 8
Crassula alba *Forssk.*, 8, *9*, 10
Crassula alsinoides *(Hook.f.) Engl.*, 6, *7*
Crassula argentea Thunb., 6
Crassula campestris (Eckl. & Zeyh.) Walp., 8
subsp. *pharnaceoides* (Fisch. & Mey.) Tölken, 8
Crassula ellenbeckiana Schönl., 10
Crassula granvikii *Mildbr.*, 6, *7*
Crassula hedbergii Wickens & Bywater, 6
Crassula intricata auct. non Nees., 6
Crassula malladrae (Chiov.) Chiov., 10
Crassula ovata (Mill.) Druce, 6
Crassula pellucida L., 6
subsp. *alsinoides* (Hook.f.) Tölken, 6
Crassula pentandra (Edgew.) Schönl., 8
Crassula pharnaceoides Fisch. & Mey., 8
Crassula rivalis (Peter) Hutch. & Bruce, 6
Crassula schimperi Fisch. & Mey., 8
subsp. **phyturus** (Mildbr.) R. Fernandcs, 8
subsp. **schimperi**., *7*, 8
subsp. *transvaalensis* (Kuntze) R. Fernandes, 8
Crassula schweinfurthii De Wild., 8
Crassula sediformis Schweinf., 12
Crassula vaginata *Eckl. & Zeyh.*, 8, *9*, 10
Crassula vaillantii auct. non (Willd.) Roth., 6
CRASSULACEAE, 5
CROTALARIA *L.*, 195
Crotalaria abyssinica Dietr., 208
Crotalaria agatiflora *Schweinf.*, 199
subsp. *engleri* (Harms ex Bak. f.) Polhill, 200
subsp. **erlangeri** *Bak. f.*, 200
subsp. **imperialis** *(Taub.) Polhill*, 200
Crotalaria albicaulis *Franch.*, *206*, 207
Crotalaria alexandri *Bak. f.*, 218
Crotalaria anthyllopsis *Bak.*, 210
Crotalaria astragalina Hochst. ex A. Rich., 208
subsp. *onobrychis* (A. Rich.) Bak. f., 208
Crotalaria awasensis *Thulin*, 208
Crotalaria axillaris *Ait.*, 216
Crotalaria azaïsii Sacl., 200
Crotalaria barkae *Schweinf.*, 200
subsp. *barkae*, 201
Crotalaria bieberi Cuf., 199
Crotalaria bogdaniana *Polhill*, 204
Crotalaria bongensis *Bak. f.*, 213
Crotalaria boranica *Harms* ex *Bak. f.*, 213
subsp. *boranica*, 213
Crotalaria brachycephala Harms ex Bak. f., 209
Crotalaria brevidens *Benth.*, 207
var. **brevidens**, 207
var. **intermedia** *(Kotschy) Polhill*, 207
Crotalaria boudetii *Polhill*, 213
Crotalaria calycina *Schrank*, 214
Crotalaria carinata Steud. ex A. Rich., 217
Crotalaria cephalotes *Steud.* ex *A. Rich.*, 217
Crotalaria cleomifolia Bak., 209
Crotalaria comanestiana *Volkens & Schweinf.*, 202
Crotalaria comosa *Bak.*, 210
Crotalaria cylindrica *A. Rich.*, 209
var. *cylindrica*, 209
var. *pusilla* Chiov., 209
Crotalaria deflersii *Schweinf.*
Crotalaria deserticola *Bak. f.*, 216
var. *deserticola*, 217
Crotalaria dilloniana Bak., 210
Crotalaria distantiflora *Bak. f.*, 212
Crotalaria dumosa *Franch.*, 212
Crotalaria emarginella *Vatke*, 214, *215*
Crotalaria exaltata *Polhill*, *211*, 214
Crotalaria farcta R. Br. ex Bak. f., 204
Crotalaria fallax *Chiov.*, 208, 209
Crotalaria fascicularis *Polhill*, 214
Crotalaria fertilis Del., 205
Crotalaria gillettii *Polhill*, 209
Crotalaria glauca *Willd.*, 202, *206*
Crotalaria goreensis *Guill. & Perr.*, 201
subsp. *macrostipula* (Steud. ex A. Rich.) Bak. f., 201
Crotalaria heterotricha *Polhill*, 204
Crotalaria hypargyrea Chiov., 218
Crotalaria hyssopifolia *Klotzsch*, 218
Crotalaria impressa *Nees* ex *Walp.*, 208
subsp. *onobrychis* (A. Rich.) Cuf., 208
Crotalaria incana *L.*, 201
subsp. **incana**, 201
subsp. **purpurascens** *(Lam.) Milne-Redh.*, 201
Crotalaria incrassifolia *Polhill*, 216
Crotalaria intermedia Kotschy, 207
var. *abyssinica* Taub. ex Engl., 207
Crotalaria intonsa *Polhill*, 209
Crotalaria jamesii Oliv., 212
Crotalaria jijigensis *Thulin*, 205
Crotalaria juncea *L.*, 218
Crotalaria karagwensis *Taub.*, *215*, 217
Crotalaria keniensis *Bak. f.*, 214
Crotalaria laburnifolia *L.*, 200
subsp. **laburnifolia**, 200, *203*
subsp. **tenuicarpa** *Polhill*, 200
Crotalaria lachnocarpa Hochst. ex Bak., 202

Crotalaria lachnocarpoides *Engl.*, 202
Crotalaria lachnophora *Hochst.* ex *A. Rich.*, 202, *203*
Crotalaria lanceolata E. Mey., 208
subsp. *contigua* Polhill, 208
Crotalaria laxa Franch., 214
Crotalaria leptocarpa *Balf. f.*, *211*, 213
subsp. *leptocarpa*, 213
Crotalaria lotoides A. Rich., nom. illegit., 213
var. *uniflora* Steud. ex A. Rich., 213
Crotalaria lugardiorum Bullock, 217
Crotalaria lunata Beddome ex Polhill, 195
Crotalaria macilenta Del., 217
Crotalaria macropoda Hochst. ex A. Rich.
Crotalaria macrostipula Steud. ex A. Rich., 201
Crotalaria massaiensis *Taub.*, 212
Crotalaria melilotoides Steud. ex A. Rich., 205
Crotalaria microcarpa *Hochst.* ex *Benth.*, *211*, 213
forma *subuniflora* Steud. ex Bak. f., 213
Crotalaria microphylla *Vahl*, 213
Crotalaria mildbraedii *Bak. f.*, 214
Crotalaria minima Bak. f. 217
Crotalaria montana A. Rich., 201
Crotalaria mucronata Desv., 205
Crotalaria natalitia *Meissn.*, 201
var. **natalitia**, 201
var. **rutshuruensis** *De Wild.*, 201
Crotalaria nigrescens Chiov., 209
Crotalaria nogalensis Chiov., 214
Crotalaria ochroleuca *G. Don*, 207
Crotalaria oligosperma *Polhill*, 216
Crotalaria onobrychis *A. Rich.*, 208
Crotalaria ononoides *Benth.*, 202, *206*
Crotalaria oöcarpa *Bak.*, 217
subsp. **microcarpa** *Milne-Redh.*, 217
subsp. *oöcarpa*, 217
Crotalaria orixensis *Rottl.* ex *Willd.*, 214
Crotalaria pallida *Ait.*, 205
var. **obovata** *(G. Don) Polhill*, 207
var. **pallida**, 207
Crotalaria paolii Cuf., 212
Crotalaria parvula Beck, 212
Crotalaria paulina *Schrank*, 218
Crotalaria paulitschkei Bak. f., 212
Crotalaria persica *(Burm.) Merr.*, 212
Crotalaria petitiana *(A. Rich.) Walp.*, 210
Crotalaria phillipsiae *Bak.*, 210
Crotalaria platycalyx Steud. ex Bak., 200
Crotalaria pleiophylla *Polhill*, 205
Crotalaria plowdenii *Bak.*, 209, *211*, 218
Crotalaria podocarpa *DC.*, 201
Crotalaria polhillii *Thulin*, 207
Crotalaria polysperma *Kotschy*, 200
Crotalaria pseudotenuirama *Torre*, 218
Crotalaria purpurascens Lam., 201
Crotalaria pycnostachya *Benth.*, 205
subsp. **donaldsonii** (*Bak. f.*) *Polhill*, 205
var. *donaldsonii* Bak. f., 205
subsp. **pycnostachya**, 205, *206*
subsp. *tropeae* (Mattei) Polhill, 205
Crotalaria quartiniana *A. Rich.*, 200, *203*
Crotalaria raffillii Milne-Redh.
Crotalaria rathjensiana Schwartz, 214
Crotalaria recta *Steud.* ex *A. Rich.*, 216
subsp. *simplex* (A. Rich.) Bak. f., 216
Crotalaria retusa *L.*, 216
Crotalaria rhopalocarpa Chiov., 216
Crotalaria rosenii *(Pax) Milne-Redh.* ex *Polhill*, 199, *203*
Crotalaria rufocaulis *Gilli*, 200
Crotalaria ruspoliana *Chiov.*, 208
Crotalaria sacculata *Chiov.*, 212
Crotalaria saltiana *Andr.*, 204
Crotalaria schumanniana Harms, 209
Crotalaria senegalensis *(Pers.) Bacle* ex *DC.*, 217
var. *carinata* (Steud. ex A. Rich.) Bak. f., 217
var. *macilenta* (Del.) Chiov., 217
var. *sanguinolenta* Chiov., 217
Crotalaria sidamaënsis Chiov., 208
Crotalaria simplex A. Rich., 216
Crotalaria somalensis *Chiov.*, 204
subsp. **fusula** *Polhill*, 205
subsp. *somalensis*, 205
Crotalaria sp. *A. sensu* Thulin, 207
Crotalaria sp. = *Benedetto* 122, 207
Crotalaria spinosa *Hochst.* ex *Benth.*, *215*, 217
Crotalaria stenorhampha *Harms*, *215*, 218
Crotalaria steudneri *Schweinf.*, 212
forma *latifolia* Lanza, 212
Crotalaria striata A. Br., nom. illegit., 208
Crotalaria thomasii *Harms*, 204
Crotalaria tigrensis Bak., 209
Crotalaria trifoliolata *Bak. f.*, 204
Crotalaria uguenensis *Taub.*, 210
Crotalaria ukambensis *Vatke*, 210
Crotalaria vatkeana *Engl.*, 209, 218
Crotalaria verdcourtii *Polhill*, 202
CUSSONIA *L.*, 538
Cussonia arborea *A. Rich.*, 538
Cussonia boranensis Cuf., 539
Cussonia holstii *Harms ex Engl.*, 539, *540*
Cussonia ostinii *Chiov.*, 539
CYAMOPSIS *DC.*, 138
Cyamopsis tetragonoloba *(L.) Taub.*, 138
Cyamopsis senegalensis *Guill. & Perr.*, 139, *141*
CYDONIA *Miller*, 42
Cydonia oblonga *Miller*, 42
Cylista somalorum Vierh., 189
CYPHOSTEMMA *(Planch.) Alston*, 409
Cyphostemma adenanthum *(Fresen.) Descoings*, 410
Cyphostemma adenocaule *(Steud. ex A. Rich.) Descoings ex Wild & Drummond*, 412, *414*
var. *pubescens* (Dewit) Descoings, 414
Cyphostemma adenocephalum (Gilg. & Brandt) Descoings, 415
Cyphostemma betiforme *(Chiov.) Vollesen*, 410

Cyphostemma boranense *Vollesen*, 412
Cyphostemma burgeri *Vollesen*, 417
Cyphostemma crinitum, 412
Cyphostemma cyphopetalum *(Fresen.) Descoings ex Wild & Drummond*, 417, *418*
Cyphostemma dembianense *(Chiov.) Vollesen*, 415
Cyphostemma dysocarpum *(Gilg & Brandt) Descoings*, 416
Cyphostemma engleri *(Gilg) Descoings*, 416
Cyphostemma erythraeae (Gilg & Brandt), 411
Cyphostemma flavicans (Bak.) Descoings, 412
Cyphostemma gallaënse, 417
Cyphostemma hildebrandtii, 415
Cyphostemma junceum *(Webb) Descoings ex Wild & Drummond*, 410, *411*
Cyphostemma keilii *(Gilg & Brandt) Descoings*, 416
Cyphostemma kilimandscharica *(Gilg) Descoings ex Wild & Drummond*, 416
Cyphostemma macrothyrsum (Gilg) Descoings, 417
Cyphostemma molle *(Bak.) Descoings*, 411
Cyphostemma niveum *(Hochst. ex Schweinf.) Descoings*, 415
Cyphostemma orondo (Gilg & Brandt) Descoings, 417
Cyphostemma oxyphyllum *(A. Rich.) Vollesen*, 412
Cyphostemma pachyrrhachis (Gilg & Brandt) Descoings, 417
Cyphostemma pannosum *Vollesen*, 412, *413*, back cover
Cyphostemma pseudoniveum (Gilg & Brandt) Descoings, 417
Cyphostemma rhodesiae *(Gilg & Brandt) Descoings ex Wild & Drummond*, 411
Cyphostemma rivae *(Gilg) Descoings*, 415
Cyphostemma schweinfurthii (Planch.) Descoings, 417
Cyphostemma serpens *(Hochst. ex A. Rich.) Descoings*, 417
Cyphostemma sp. = *Gilbert & Thulin* 511
Cyphostemma ternatum *(Forssk.) Descoings*, 416
Cytisus arabicus Decne., 222
DALBERGIA *L. f.*, 105
Dalbergia commiphoroides *Bak. f.*, 105
Dalbergia lactea *Vatke*, *106*, 107
Dalbergia macrothyrsus sensu Cuf., 107
Dalbergia melanoxylon *Guill. & Perr.*, 105
Dalbergia microphylla *Chiov.*, 107
Dalbergia sciadendron Chiov., 107
Dalbergia sissoo *Roxb.*, 107
DEBREGEASIA *Gaud.*, 320
Debregeasia bicolor (Roxb.) Wedd., 320
Debregeasia saeneb *(Forssk.) Hepper & Wood*, *319*, 320
Debregeasia salicifolia (D. Don.) Rendle, 320
DEINBOLLIA *Schumach. & Thonn.*, 502
Deinbollia kilimandscharica *Taub*, 502, *503*
DELONIX *Raf.*, 50
Delonix baccal *(Chiov.) Bak. f.*, 51
Delonix elata *(L.) Gamble*, 50, *52*
Delonix regia *(Boj. ex Hook.) Raf.*, 50
Dendrobenthamia capitata (Wall. ex Roth.) Hutch., 534
Desmanthus divergens Willd., 74
DESMODIUM *Desv.*, 148
Desmodium abyssinicum (Willd. ex Hoffmgg.) DC., 150
Desmodium abyssinicum (Hochst. ex A. Rich.) Hutch. & Dalz., non DC., 151
Desmodium adscendens *(Sw.) DC.*, 150
var. adscendens, 150
var. **robustum** *Schubert*, 150
Desmodium barbatum *(L.) Benth.*, 152
var. barbatum, 152
var. **prostratum** *Schubert*, 152
Desmodium delicatulum A. Rich., 150
Desmodium dichotomum *(Klein ex Willd.) DC.*, 151, *153*
Desmodium gangeticum *(L.) DC.*, 151
Desmodium heterophyllum (Willd.) DC., 150
Desmodium hirtum *Guill. & Perr.*, 150
var. **delicatulum** *(A. Rich.) Harms ex Bak. f.*, 150
var. **hirtum.**, 150
Desmodium ospriostreblum *Steud. ex Chiov.*, 151, *153*
Desmodium ramosissimum *G. Don*, 150
Desmodium repandum *(Vahl) DC.*, *149*, 150
Desmodium salicifolium *(Poir.) DC.*, 151, *153*
Desmodium schimperi Hochst. ex A. Rich., 150
Desmodium schweinfurthii, 150
Desmodium uncinatum *(Jacq.) DC.*, 151
Desmodium velutinum *(Willd.) DC.*, 151, *153*
DICHAPETALACEAE, 47
Dichapetalum corradii Chiov., 47
Dichapetalum, 47
DICHROSTACHYS *(DC.) Wight & Arn.*, 72
Dichrostachys cinerea *(L.) Wight & Arn.*, *73*, 74
subsp. *africana* Brenan & Brummitt, 74
var. *africana*, 74
var. *karamojensis* Brenan & Brummitt, 74
subsp. *argillicola* Brenan & Brummitt, 74
var. *argillicola*, 74
var. *hirtipes* Brenan & Brummitt, 74
Dichrostachys nutans (Pers.) Benth., 74
var. *grandiflora* Lanza, 74
Dichrostachys glomerata (Forssk.) Chiov., 74
var. *grandiflora* (Lanza) Bak. f., 74
Dichrostachys kirkii *Benth.*, *73*, 74
DICRAEOPETALUM *Harms*, 101
Dicraeopetalum stipulare *Harms*, 101
DIDYMODOXA *Wedd.*, 324
Didymodoxa caffra *(Thunb.) Friis & Wilmot-Dear*, 325, *326*
Didymodoxa acuminata Webb., 325
Didymogyne abyssinica Wedd., 322
Dillonia abyssinica Sacleux, 340
Diphasia dainellii Pichi-Serm., 424
Diplotheca abyssinica Hochst., 231
Diplotheca atropilosula Hochst., 230
Diplotheca tigrensis Hochst., 231
Diplotheca venosa Hochst., 231
DOBERA *Juss.*, 353
Dobera coriacea (Hochst.) A. DC., 353

Dobera glabra *(Forssk.) Poir.*, 353, *354*
Dobera loranthifolia, 353
Dobera roxburghii Planch., 353
DODONEA *Mill.*, 491
Dodonea angustifolia *L. f.*, 491, *492*
Dodonea viscosa auct. mult., non Jacq., 491
DOLICHOS *L.*, 178
Dolichos axillaris E. Mey., 166
var *glaber* E. Mey., 166
Dolichos chrysanthus A. Chev., 165
var. occidentalis (Harms) Wilczek, 165
Dolichos daltonii Webb, 166
Dolichos debilis Hochst. ex A. Rich., 178
Dolichos dillonii Del., 172
Dolichos ellenbeckii Harms, 181
Dolichos falcatus Klein ex Willd., 178
Dolichos formosoides Harms, 178
Dolichos formosus A. Rich., 179
Dolichos kilimandscharicus *Taub.*, 179
Dolichos lablab L., 179
var. *uncinatus* (Schweinf.) Chiov.,179
Dolichos luticola *Verdc.*, 178
Dolichos malosanus Bak., 179
Dolichos mitis A. Rich., 181
Dolichos oliveri Schweinf., 179
var. *uondensis* Fiori, 179
Dolichos oliveri *Schweinf.*, 179
Dolichos pseudopachyrhizus Harms, 181
var. *subintegrifolius* Harms, 181
Dolichos purpureus L., 179
Dolichos reticulatus A. Rich., 184
Dolichos schweinfurthii *Harms*, 179
Dolichos sericeus E. Mey., 179
subsp. **formosus** *(A. Rich.) Verdc.*, 179, *180*
subsp. **sericeus**.,179
Dolichos stenocarpus Hochst., 178
Dolichos trilobus L., 178
Dolichos uncinatus L., 160
Dolichos uniflorus Lam., 166
var. *stenocarpus* Brenan, 166
DORSTENIA *L.*, 275
Dorstenia arabica Hemsl., 277
Dorstenia barnimiana *Schweinf.*, 278, *279*
var. *ophioglossioides* (Bureau) Engl., 278
var. *telekii* (Schweinf.) Engl., 278
var. *tropaeolifolia* (Schweinf.) Rendle, 278
Dorstenia brownii *Rendle*, 275, *279*
Dorstenia crispa Engl., 277
var. *pachypoda* Chiov., 277
var. *lancifolia* Rendle, 277
Dorstenia cuspidata *A. Rich.*, 277, *279*
Dorstenia ellenbeckiana *Engl.*, 277, *279*
Dorstenia foetida *(Forssk.) Schweinf.*, 277
subsp. **foetida**, 277
var. **foetida**, 277
var. **obovata** *(A. Rich.) Schweinf. & Engl.*, 277, *279*
subsp. **lancifolia** *(Rendle) Friis*, 277, *279*
Dorstenia obovata A. Rich., 277
Dorstenia ophioglossioides Bureau, 278
Dorstenia palmata Engl., 278
var. *integrifolia* Chiov., 278
Dorstenia peltata Engl., 278
Dorstenia phillipsiae Hook. f., 277
Dorstenia soerensenii *Friis*, 275, *276*
Dorstenia telekii Schweinf., 278
Dorstenia triternata Chiov., 278
Dorstenia tropaeolifolia *(Schweinf.) Bureau*, 278, *279*
Dorstenia walleri Hemsl., 277
Dorycnium schimperi Jaub. & Spach, 224
DROGUETIA *Gaud.*, 322
Droguetia acuminata Wedd., 325
Droguetia diffusa Wedd., 322
Droguetia iners *(Forssk.) Schweinf.*, 322, *324*
Droguetia pauciflora (Steudel) Wedd., 322
DUMASIA *DC.*, 156
Dumasia villosa *DC.*, 156, *157*
ECHEVERIA *DC.*, 5, 26
Echeveria abyssinica Otto, 16
Echeveria elegans *Rose*, 26
EKEBERGIA *Sparrm.*, 489
Ekebergia capensis *Sparrm.*, *488*, 489
Ekebergia ruepelliana (Fresen.) A. Rich., 489
Ekebergia petitiana A. Rich., 489
ELAEODENDRON *Jacq.f. ex Jacquin*, 344
Elaeodendron aquifolium *(Fiori) Chiov.*, 344
Elaeodendron buchananii *(Loes.) Loes.*, 344, *345*
ELATOSTEMA *J.R. & G. Forster*, 314
Elatostema kiwuense Engl., 314
"*Elatostema monticola*", 314
Elatostema longeacuminatum (De Wildem.) Hauman, 314
Elatostema monticolum *Hook. f.*, 314, *315*
Elatostema orientale Engl., 314
var. *longeacuminatum* De Wildem., 314
Elatostema trinerve Hochst., 304
EMELIANTHE *Danser*, 365
Emelianthe panganensis *(Engl.) Danser*, 365
ENGLERINA *Tieghem*, 371
Englerina woodfordioides *(Schweinf.) M. Gilbert*, 371, *372*
var. *subintegridisca* Balle, 371
ENTADA *Adans.*, 71
Entada abyssinica *Steud. ex A. Rich.*, 72, *73*
var. *intermedia* Fiori, 72
Entada africana *Guill. & Perr.*, 72
Entada leptostachya *Harms*, 72
Entada sudanica Schweinf., 72
Entadopsis abyssinica (Steud. ex A. Rich.) Gilb. & Bout., 72
Entadopsis leptostachya (Harms) Cuf., 72
Entadopsis sudanica (Schweinf.) Gilb. & Bout., 72
ERIANTHEMUM *Tieghem*, 365
Erianthemum aethiopicum *Wiens & Polhill*, 366, *366*

Erianthemum dregei *(Eckl. & Zeyh.) Tieghem*, 366
ERIOBOTRYA *Lindley*, 42
Eriobotrya japonica *(Thunb.) Lindl.*, 42
ERIOSEMA *(DC.) G. Don*, 189
Eriosema buchananii Bak. f., 190
var. *richardii* (Benth. ex Bak. f. & Haydon) Staner, 190
Eriosema cordifolium *Hochst. ex A. Rich.*, 193, *194*
Eriosema elliotii *Bak. f.*, 193
Eriosema erythrocarpon Beck, 193
Eriosema glomeratum sensu Cuf., 190
Eriosema humbertii Staner & De Craene, 190
Eriosema jurionianum *Staner & De Craene*, 190
var. *fulvum* (Bak. f.) Verdc., 190
Eriosema longipedunculatum *A. Rich.*, 192
var. *hirsutum* Verdc., 192
Eriosema montanum *Bak. f.*, 190
Eriosema nutans *Schinz*, 190, *191*
Eriosema polystachyum *(A. Rich.) Bak.*, 190
Eriosema psoraleoides *(Lam.) G. Don*, 190
Eriosema radicosum A. Rich., 156
Eriosema rhodesicum *R. E. Fr.*, 193
Eriosema richardii Benth. ex Bak. f. & Haydon, 190
Eriosema robustum *Bak.*, 192
Eriosema schweinfurthii Bak. f., 192
Eriosema scioanum *Avetta*, *191*, 192
subsp. lejeunei (Staner & De Craen) Verdc., 192
subsp. *scioanum*, 192
Eriosema shirense *Bak. f.*, 193
Eriosema sparsiflorum *Bak. f.*, 192
var. *sparsiflorum*, 193
var. sessilifolium, 193
Eriosema tuberosum *A. Rich.*, 192
Eriosema verdickii *De Wild.*, 193
ERYTHRINA *L.*, 161
Erythrina abyssinica *Lam. ex DC.*, 161, *162*
Erythrina brucei *Schweinf.*, 161, *162*
Erythrina burana Chiov., *162*, 164
Erythrina crista-galli *L.*, 161
Erythrina lysistemon *Hutch.*, 161
Erythrina melanacantha *Taub. ex Harms*, 164
subsp. **melanacantha.**, 164
subsp. **somala** *(Chiov.) Gillett in Kew Bull.*, *162*, 164
Erythrina rosenii Pax, 199
Erythrina tomentosa R. Br. ex A. Rich., 161
ERYTHROPHYSA *E. Mey. ex Arn.*, 490
Erythrophysa septentrionalis *Verdc.*, 490, *491*
ESCALLONIACEAE, 4, 27
Escallonia, 4
Euphorbia reghinii (Chiov.) Vollesen, 477
Euphorbia jatrophioides Pax, 477
FABACEAE, 49
subfam. CAESALPINIOIDEAE, 49
subfam. *Faboideae*, 97
subfam. MIMOSOIDEAE, 71
subfam. PAPILIONOIDEAE, 97
Fabricia ferruginea (Hochst. & Steud. ex A. Rich.), 154
Fabricia rugosa (Willd.) Kuntze, 155
var. *quartiniana* (A. Rich.) Taub., 155
FAGACEAE, 265
FAGARA L., 420
Fagara chalybea (Engl.) Engl., 420
Fagara leprieurii (Guill. & Perr.) Engl, 420
Fagara usambarensis Engl., 422
FAGAROPSIS *Siebenlist*, 422
Fagaropsis angolensis *(Engl.) Dale*, 422, *423*
Fagaropsis gillettii Chiov., 422
Fagaropsis hildebrandtii *(Engl.) Milne-Redh.*, 422
Fagelia resinosa Hochst. ex A. Rich., 183
FAIDHERBIA A. Chev., 75, 84
Faidherbia albida (Del.) A. Chev., 84
Feuilleea sassa (Willd.) Kuntze, 96
FICUS *L*, 278
subgen. *Bibracteatae* Mildbr. & Burret, 292
subgen. FICUS, 280
sect. FICUS, 280
sect. SYCIDIUM *Miq.*, 281
subgen. PHARMACOSYCEA *(Miq.) Miq.*, 289
subgen. *Sycidium* (Miq.) Mildbr. & Burret, 281
subgen. SYCOMORUS *(Gasparrini) Mildbr. & Burret*, 285
sect. *Sycomrus* (Gasparrini) Miq., 285
subgen. UROSTIGMA *(Gasparrini) Miq.*, 290
sect. GALOGLYCHIA *Gasparrini) Miq.*, 292
sect. UROSTIGMA, 290
Ficus abutilifolia *(Miq.) Miq.*, 296, *296*
Ficus acrocarpa (Miq.) Miq., 298
var. *saligna* (Hochst.) Miq., 298
Ficus antitheophylla Miq., 281
Ficus asperifolia *Miq.*, 284, *284*
Ficus baronii Baker, 294
Ficus basaransis Mildbr. & Burret, 298
Ficus brachypoda Hutch., 292
var. *scoana* Chiov., 292
Ficus brachypoda Hutch., 292
Ficus burkei Miq., 298
Ficus callabatensis Warb., 296
Ficus capensis Thunb., 287
var. *mallotocarpa* (Warb.) Mildbr. & Burret., 287
Ficus capreaefolia *Del.*, 281, *283*
Ficus carica *L.*, 280, *281*
Ficus chanas Forssk., 285
Ficus craterostoma Mildbr. & Burret, 298
Ficus dahro Del., 296
Ficus dekdekena (Miq.) A. Rich., 298
Ficus dicranostyla *Mildbr.*, 289, *290*
Ficus discifera Warb., 296
Ficus exasperata *Vahl*, 284, *284*
Ficus fazoklensis (Miq.) Miq., 293
Ficus forsskalii Vahl., 281
Ficus glumosa *Del.*, 293, *293*
var. *glaberrima* Martelli, 293
var. *intermedia* Martelli, 293
var. *lanuginosa* Martelli, 293

Ficus gnaphalocarpa (Miq.) A. Rich., 285
Ficus hararensis Warb., 296
Ficus hochstetteri *(Miq.) A. Rich.*, 299
var. *glabrior* Miq., 298
Ficus indica Forssk., non L., 291
Ficus ingens *(Miq.) Miq.*, 291, *291*, 295
Ficus ingentoides Hutch., 291
Ficus intermedia Del., 295
Ficus iteophylla Miq., 298
Ficus kotschyana (Miq.) Miq., 296
Ficus lutea auct., non Vahl sensu C.C. Berg, 291
Ficus lutea *Vahl*, 294, *294*
Ficus mallotocarpa Warb., 287
Ficus microcarpa Vahl, 298
Ficus morifolia Forssk., 281
Ficus mucuso *Ficalho*, 287, *287*
Ficus natalensis Hochst., 298
Ficus nautarum Baker, 294
Ficus neriifolia A. Rich., 290
Ficus neumannii Kunth & Bouché, 294
Ficus ovata *Vahl*, 292, *293*
Ficus palmata *Forssk.*, 281, *282*
var. ***petitiana***, 281
Ficus panifica Del., 287
Ficus paoli *Pampanini*, 281
Ficus persicifolia Warb., 298
Ficus petersi Warb. & Schweinf., 298
Ficus petitiana *A. Rich.*, 281
Ficus platyphylla *Del.*, 296, *296*
Ficus populifolia *Vahl*, 295, *295*
Ficus pretoriae Burrt-Davy, 291
Ficus pseudocarica Miq., 281
var. *tomentosa* A. Rich., 281
Ficus pubicosta Warb., 299
Ficus punctata Lam., 298
Ficus pyrifolia Lam., 298
Ficus quibeba Ficalho, 294
Ficus reflexa Thunb., 298
Ficus religiosa Forssk., 295
Ficus riparia (Miq.) A. Rich., 287
Ficus rivae Warb., 296
Ficus rokka Warb. * Schweinf., 298
Ficus rubra Lam., 298
Ficus ruspolii *Warb.*, 299
Ficus salicifolia *Vahl*, 290, *290*
Ficus saligna Hochst., 298
Ficus scabra Willd., 284
Ficus schimperi (Miq.) A. Rich., 298, 299
var. *hochstetteri* (Miq.) Mildbr. & Burret, 299
Ficus schimperiana A. Rich., 291
Ficus schweinfurthii Miq., 289
Ficus senegalensis Miq., 294
Ficus serrata Forssk., 284
Ficus sidifolia Hiern, 287
Ficus socotrana Balf. f., 296
Ficus sonderi Miq., 293
Ficus sur *Forssk.*, 287, *288*
Ficus sycomorus *L.*, 285
subsp. **gnaphalocarpa** *(Miq.) C.C. Berg*, 285
subsp. **sycomorus**, 285, *286*
Ficus taab Forssk., 291
Ficus thonningii Blume, 298, *298*
Ficus toka Forssk., 267
Ficus umbellata *Vahl*, 295, *295*
Ficus vallis-choudae *Del.*, 289, *289*
Ficus vasta *Forssk.*, 296, *297*
var. *glabrescens* Hutch., 297
var. *velutina* Fiori, 297
Ficus vestito-bracteata Warb., 294
Ficus vogelii (Miq.) Miq., 294, 295
Ficus wakefieldii Hutch., 298
Ficus xanthophylla (Miq.) Martelli, 291
FILICIUM *Thw. ex Hook.f.*, 493
Filicium decipiens *(Wight & Arn.) Thw.*, 493, *495*
Fillaea brucei Fresen., 54
FLEMINGIA *Ait.f.*, 193
Flemingia grahamiana *Wight & Arn.*, 193, *194*
FLEURYA Gaud., 308
Fleurya aestuans (L.) Miq., 309
Fleurya interrupta (L.) R. Wright, 308
Fleurya lanceolata Engl., 306
FORSSKAOLEA *L.*, 321
Forsskaolea tenacissima *L.*, 322, *323*
var. *erythraea* Terracc., 322
Forsskaolea viridis *Webb.*, 322, *323*
var. *ehrebergii* Solms-Laub., 322
FRAGARIA *L.*, 37
Fragaria x ananassa *Duchesne*, 37
Fragaria chiloensis (L.) Duchesne, 37
Fragaria vesca *L.*, 37
Fragaria virginiana Duchesne, 37
GALACTIA *P. Br.*, 165
Galactia tenuiflora *(Willd.) Wight & Arn.*, 165
var. *biflora* Schwein, 165
var. **villosa** *(Wight & Arn.) Benth.*, 165, *167*
var. tenuiflora, 165
GALEGA *L.*, 230
Galega colutea Burm.f., 127
Galega somalensis *(Taub. ex Harms) Gillett*, *229*, 230
GALOGLYCHIA *(Gasparrini) Endlicher*, 292
Garcinia buxifolia Chiov., 255
Gigalobium abyssinicum Steud. ex A. Rich., 72
GIRARDINIA *Gaud.*, 306
Girardinia adoensis (Steudel) Wedd., 308
Girardinia bullosa *(Steudel) Wedd.*, 306
Girardinia condensata (Steudel) Wedd., 308
var. *adoensis* (Steudel) Engl., 308
Girardinia diversifolia *(Link) Friis*, *307*, 308
Girardinia heterophylla Decn., 308
subsp. *adoensis* (Steudel) Cuf., 308
Globimetula Tieghem, 362
Globimetula braunii (Engl.) Tieghem, 362
GLYCINE *Willd.*, 158
Glycine abyssinica Hochst ex A. Rich., 160

Glycine axilliflora Kotschy., 160
Glycine borianii (Schweinf.) Bak., 156
Glycine javanica L.158
var. *longicauda* (Schweinf.) Bak.
subsp. *micrantha* (Hochst. ex A. Rich.) F.J. Hermann, 158
var. *moniliformis* (Hochst. ex A. Rich.) F.J. Hermann, 158
Glycine longicauda Schweinf., 158
Glycine max *(L.) Merr.*, 158
Glycine micrantha Hochst. ex A. Rich.158
Glycine moniliformis Hochst. ex A. Rich., 158
Glycine petitiana (A. Rich.) Schweinf., 158
var. *dembianensis* Chiov., 158
Glycine radicosa (A. Rich.) Bak.f., 156
Glycine ringoetii De Wild., 160
Glycine sublobata Schumach., 185
Glycine wightii *(Wight & Arn.) Verdc.*, 158
subsp. **petitiana** *(A. Rich.) Verdc.*, 158, *159*
subsp. **wightii**, 158
var. **longicauda** *(Schweinf.) Bak.*, 158
GOUANIA *Jacq.*, 386
Gouania longispicata *Engl.*, 386, *388*
Grielum L., 45
GYMNOSPORIA (Wight & Arnott) Hook.f., 331
Gymnosporia addat Loes., 333
Gymnosporia arbutifolia (A. Rich.) Loes., 336
Gymnosporia atkaio (A. Rich.) Loes., 336
Gymnosporia castellii Pic.-Serm., 332
Gymnosporia cortii Pich.- Serm., 335
Gymnosporia ellenbeckii Loes., 332
Gymnosporia engleriana Loes., 336
Gymnosporia filamentosa Loes., 335
var. brevistaminea, 335
Gymnosporia filamentosa Loes. sensu Cuf., 335
Gymnosporia gracilipes (Welw. ex Oliv.) Loes., 332
var. *arguta* Loes., 332
Gymnosporia intermedia Chiov., 336
Gymnosporia laurifolia (A. Rich.) Loes., 338
Gymnosporia luteola (Del.) Szyszy., 338
Gymnosporia obscura (A. Rich.) Loes., 333
Gymnosporia senegalensis (Lam.) Loes., 336
var. *inermis* (A. Rich.) Loes., 336
forma *macrocarpa* Loes., 336
Gymnosporia serrata (A. Rich.) Loes., 332, 333
var. *obscura* (A. Rich.) Loes., 333
var. *schimperi* Fiori, 333
var. *steudneri* (Engl.) Loes., 332
var. *typica* Fiori, 333
HAGENIA *J.F. Gmel.*, 43
Hagenia abyssinica *(Bruce) J.F. Gmel.*, 43, *44*
HAMAMELIDACEAE, 252
HAPLOCOELUM *Radlkofer*, 507
Haplocoelum foliolosum *(Hiern) Bullock*, 507, *507*
Haplocoelum gallaënse *(Engl.) Radlkofer*, 507
HARRISONIA *R. Br. ex A. Juss.*, 437
Harrisonia abyssinica *Oliv.*, 437, *438*
forma *grandifolia* Engl., 437
var. *somalensis* Fiori, 437
Harrisonia occidentalis Engl., 437
Haydonia monophylla ((Taub.) Wilszek, 177
HEDERA *L.*, 542
Hedera helix *L.*, 542
HEDYSARUM *L.*, 227
Hedysarum abyssinicum Willd. ex Hoffmgg., 150
Hedysarum coronarium *L.*, 227
Hedysarum repandum Vahl, 150
HEERIA Meisn., 519
Heeria insignis (Del.) Kuntze, 521
HELINUS *E. Mey. ex Endl.*, 385
Helinus integrifolius *(Lam.) O. Ktze.*, 385
Helinus mystacinus *(Ait.) E. Mey. ex Steud.*, 385, *387*
forma *pilosiusculus* Radlk., 385
forma *tomentosus* Radlk., 385
Helinus scandens (Eckl. & Zeyh.) A. Rich., 385
HELIXANTHERA *Lour.*, 362
Helixanthera thomsonii *(Sprague)Danser*, 362, *362*
Helmintocarpon abyssinicum A. Rich., 227
Hemprichia erythraea Ehrenb., 450
Heudelotia africana A. Rich., 454
HILDEBRANDTINANAE *Engl., ex Vollesen*, 457
HIPPOCRATEA *L.*, 340
Hippocratea africana *(Willd.) Loes.*, 342, *343*
var. **richardiana** *(Cambess.) N. Robson*, 342
var. **schimperiana** *(Steud. & Hochst. ex A. Rich.) Cuf.*, 342
Hippocratea goetzei *Loes.*, 340, *343*
Hippocratea pallens *Planchon ex Oliver*, 342, *343*
Hippocratea parvifolia *Oliver*, 342, *343*
Hippocratea richardiana Cambess., 342
Hippocratea schimperiana Steud. & Hochst. ex A. Rich., 342
Hirtella, 46
Humulus, 327
HYDRANGEA *Gronov. ex L.*, 4
HYDRANGEACEAE, 4
HYPAGOPHYTUM *Berger*, 10
Hypagophytum abyssinicum *(A. Rich.) Berger*, 10, *11*
ICACINACEAE, 348
ILEX *L.*, 329
Ilex mitis *(L.) Radlk.*, 329
var. *erythraea* Fiori, 329
var. **mitis**, 329, *330*
Indigastrum deflexum (Hochst. ex A. Rich.) Jaub. & Spach., 137
Indigastrum macrostachyum Jaub. & Spach, 137
INDIGOFERA *L.*, 119
Indigofera aeruginis Schweinf., 123
Indigofera alboglandulosa Engl., 130
Indigofera alta Schweinf., 134
Indigofera ambelacensis *Schweinf.*, 124
Indigofera amorphoides *Jaub. & Spach*, *129*, 134
Indigofera anabaptista Steud. ex Bak., 122
Indigofera arabica *Jaub. & Spach.*, 122

Indigofera arenaria A. Rich., 122
var. *strigosa* A. Terracc., 132
Indigofera argentea *Burm.f.*, 128
Indigofera argentea sensu Gillett, 132
Indigofera arrecta *Hochst. ex A. Rich.*, 119, 135
Indigofera articulata *Gouan*, 119, 134
Indigofera astragalina *DC.*, 135
Indigofera atriceps *Hook.f.*, 130
subsp. *alboglandulosa* (Engl.) Gillett, 130
subsp. **atriceps.**, 130
subsp. **kaessneri** *(Bak.f.) Gillett*, *129*, 130
subsp. **setosissima** *(harms) Gillett*, 130
Indigofera barcensis Chiov., 137
Indigofera binderi *Kotschy*, 131
Indigofera bogdanii *Gillett*, 134
Indigofera boranensis Chiov., 136
Indigofera boranica *Thulin*, 131
Indigofera brachynema *Gillett*, 128
Indigofera brevicalyx *Bak.f.*, 124
Indigofera butayei *De Wild.*, 138
Indigofera cana *Thulin*, 138
Indigofera chirensis *Cuf. ex Gillett*, 122
Indigofera cliffordiana *Gillett*, 133
Indigofera coerulea *Roxb.*, 119, *129*, 134
var. **coerulea**, 134
var. **occidentalis** *Gillett & Ali*, 134
Indigofera colutea *(Burm.f.) Merr.*, 127, *129*
var. **colutea.**, 127
var. *dembianensis* (Chiov.) Cuf. ex Gillett, 127
var. *linearis* Bak.f. ex Gillett, 128
var. **somalensis** *(Bak.f.) Cuf. ex Gillett*, 127
Indigofera coluteifolia Jaub. & Spach, 131
Indigofera commiphoroides Chiov., 132
Indigofera congolensis *De Wild. & Th. Dur.*, 126
Indigofera conjugata *Bak.*, 137
Indigofera cordifolia *Heyne ex Roth.*, 123
Indigofera coronilloides Jaub. & Spach, 134
Indigofera costata *Guill. & Perr.*, 137
subsp. **costata**, 137
subsp. **goniodes** *(Hochst. ex Bak.) Gillett*, 137
Indigofera cufodontii *Chiov.*, 138
Indigofera curvirostrata *Thulin*, 131
Indigofera dauensis *Gillett*, 132
Indigofera deflexa Hochst. ex A. Rich., 137
Indigofera deflersi Bak.f., 133
Indigofera dembianensis *(Chiov.) Gillett*, 127
Indigofera dendroides *Jacq.*, 124
Indigofera ellenbeckii *Bak.f.*, 132
Indigofera emarginella *Steud. ex A. Rich.*, 131
Indigofera "endecaphylla", 136
Indigofera eremophila *Thulin*, 126
Indigofera ervoides A. Rich., 122
Indigofera garckeana *Vatke*, 130
Indigofera geminata sensu Cuf., 124
Indigofera goniodes Hochst. ex Bak., 137
Indigofera gyrata *Thulin*, 138
Indigofera hendecaphylla Jacq., 136
var. *acutifolia* Chiov., 136
var. *angustifolia* A. Rich., 136
var. *parvula* Chiov., 136
Indigofera hirsuta *L.*, 135
Indigofera hochstetteri *Bak.*, 122, *125*
Indigofera houer Forssk., 135
Indigofera insularis Chiov., 136
Indigofera kelleri *Bak.f.* , 132
Indigofera kerensis Chiov., 127
Indigofera linifolia *(L.f). Retz.*, 123, *125*
Indigofera longibarbata Engl., 135
Indigofera lupatana *Bak.f.*, 132
Indigofera macrantha *Harms*, 131
Indigofera maffeii Chiov., 134
Indigofera microcarpa *Desv.*, 135
Indigofera mimosoides *Bak.*, 127
Indigofera mooneyi *Thulin*, 130
Indigofera oblongifolia *Forssk.*, 136
indigofera ogadensis *Gillett*, 126
Indigofera ornithopodioides Hochst. ex Jaub. & Spach, 122
Indigofera paniculata *Pers.*, 123
Indigofera parviflora Heyne ex Wight & Arn., 137
Indigofera parvula sensu Hochst. ex A. Rich., 136
Indigofera pilosa *Poir.*, 127
Indigofera prieureana *Guill. & Perr.* 123
Indigofera pseudosubulata sensu Cuf., 134
Indigofera pulchra Willd., 119
Indigofera quartiniana A. Rich., 134
Indigofera roseo-coerulea *Bak.f.*, 131
Indigofera rothii *Bak.*, *125*, 126
Indigofera ruspolii *Bak.f.*, 126
Indigofera sabulosa *Thulin*, 133
Indigofera schimperi *Jaub. & Spach*, 136
var. *baukeana* (Vatke) Gillett, 136
Indigofera secundiflora *Poir.*, 128
var. *gondarensis* Pic.-Serm., 128
var. **rubripilosa** *De Wild.*, 128
var. secundiflora, 128
Indigofera semhaensis Vierh., 122
Indigofera semitrijuga *Forssk.*, 136
Indigofera sessiliflora *DC.*, 123
Indigofera somalensis Vatke, 136
Indigofera sp. **A** sensu Thulin, 132
Indigofera sp. **B** sensu Thulin, 133
Indigofera sp. **C** sensu Thulin, 133
Indigofera sp. **D** sensu Thulin, 133
Indigofera sp. = *Friis* et al. 2976, 133
Indigofera sp. = *Hemming* 1471, 133
Indigofera sp. = *Pappi* 74, 132
Indigofera spachii Bak., 134
var. *trifoliolata* Schweinf., 134
Indigofera sparsa *Bak.*, 124
Indigofera spicata *Forssk.*, *129*, 136
Indigofera spiniflora *Boiss.*, 132, 133

Indigofera spinosa *Forssk.*, 126
var. *spiniflora*, 132
f. *densissima* Chiov., 132
var. *microphylla* A. Rich., 126
Indigofera stenophylla Guill. & Perr., 123
var. *brachypoda* Steud. ex Bak., 123
var. *latifolia* A. Rich., 123
Indigofera stipulosa *Chiov.*, 137
Indigofera suaveolens *Jaub. & Spach*, 127
var. *subquadriflora* Chiov., 124
var. *pseudocongolensis* Chiov., 124
Indigofera subargentea *De Wild.*, 128
Indigofera subhirtella Chiov., 136
Indigofera subulata Vahl ex Poir. , 134
var. *maffeii* (Chiov.) Gillett, 134
var. *microphylla* Chiov., 134
var. *scabra* (Roth) Meikle, 134
Indigofera tanganyikensis *Bak.f.*, 124
var. **strigulosior** *Gillett*, 124
var. tanganyikensis, 124
Indigofera tenuisiliqua Schweinf., 138
Indigofera tettensis Klotzsch, 136
Indigofera tinctoria *L.*, 119, 135
Indigofera trigonelloides *Jaub. & Spach*, 123
Indigofera trita *L.f.* , 133
var. **maffeii** *(Chiov.) Ali*, 134
var. **scabra** *(Roth) Ali*, 134
var. **subulata** *(Vahl ex Poir.) Ali*, 134
var. trita, 134
Indigofera tritoides *Bak.*, 138
Indigofera vicioides *Jaub. & Spach*, 130
var. rogersii (R.E. Fr.) Gillett, 130
var. *vicioides*, 130
Indigofera viscosa Lam., 127
var. *dembianensis* Chiov., 127
var. *somalensis* Bak.f., 127
var. *subglabra* A. Rich., 124
Indigofera vohemarensis *Baill.*, 124, *125*
Indigofera volkensii *Taub.*, 136
Indigofera welwitschii *Bak.*, 138
var. *welwitschii*, 138
Indigofera zavattarii *Chiov.*, 126
Indigofera zenkeri *Bak.f.*, 128, *129*
Inga isenbergiana A. Rich., 94
Inga malacophylla A. Rich., 93
Inga nefasia Hochst. ex A. Rich., 90
Inga quartiniana A. Rich., 93
Inga sassa Willd., 96
Jaundea pinnata (Beauv.) Schellenb., 533
Johnia petitiana A. Rich., 158
JUGLANDACEAE, 533
JUGLANS *L.*, 533
Juglans regia *L.*, 533
KALANCHOE Adanson, 19
Kalanchoe angustifolia *A. Rich.*, 22
Kalanchoe brachycalyx A. Rich., 25
Kalanchoe citrina Schweinf., *21*, 23
var. *ballyi* Wickens, 23
Kalanchoe crenata (Andr.) Haw., 20
Kalanchoe 'cyanonilotica' Chiov., 22
Kalanchoe delagoense Eckl. & Zeyh., 26
Kalanchoe densiflora *Rolfe*, 23
var. **densiflora**, 23
var. minor Raadts, 23
var. **subpilosa** Cuf., *21*, 23
Kalanchoe flammea Stapf, 22
Kalanchoe glandulosa A. Rich., 23
Kalanchoe glaucescens *Britten*, 22
subsp. arabica Cuf., 22
subsp. **glaucescens**, *21*, 22
Kalanchoe grandiflora A. Rich., 20
var. *angustipetala* Engl., 20
Kalanchoe kelleriana Schinz, 20
Kalanchoe laciniata *(L.) DC.*, *21*, 25
var. *brachycalyx* (A. Rich.) Chiov., 25
Kalanchoe lanceolata *(Forssk.) Pers.*, 23, *24*, 25
var. *glandulosa* (A. Rich.) Cuf., 23
Kalanchoe lentiginosa Cuf., 23
Kalanchoe lugardii Bullock, 22
Kalanchoe macrantha Bak., 20
Kalanchoe marinellii Pampanini, 22
Kalanchoe marmorata *Bak.*, 20, *21*
Kalanchoe neumannii Engl., 22
Kalanchoe petitiana *A. Rich.*, 20
var. **petitiana**, 20, *21*
var. **neumannii** *(Engl.) Cuf.*, 22
Kalanchoe pinnata (Lam.) Pers., 25
Kalanchoe prittwitzii *Engl.*, 22
Kalanchoe prolifera (Hook.) Raym.-Ham., 25
Kalanchoe quartiniana *A. Rich.*, 20, *21*
var. *micrantha* Pampanini, 20
Kalanchoe rohlfsii Engl., 25
Kalanchoe schimperiana A. Rich., 20, *21*
Kalanchoe schliebenii Werderm., 20
Kalanchoe schweinfurthii Penzig, 25
Kalanchoe somaliensis Bak., 20
Kalanchoe sp. = *Buscalioni*, 22
Kalanchoe stenosiphon *Britten*, *21*, 23
Kalanchoe tubiflora (Harvey) Raym.-Ham., 26
KIRKIA *Oliv.*, 439
Kirkia burgeri *Stannard*, 439
subsp. **burgeri**, 439, *441*
subsp. somalensis Stannard, 439, 440
Kirkia tenuifolia *Engl.*, 439
KORTHALSELLA *Tieghem*, 377
Korthalsella binii Pichi-Serm., 377, 378
Korthalsella japonica *(Thunb.) Engl.*, 377, *378*
Korthalsella opuntia (Thunb.) Merrill, 377
var. *gaudichaudii* (van Tiegh.) Danser, 378
var. *richardii* (van Tiegh.) Danser, 378
KOSARIA Forssk., 275
Kosaria barnimiana (Schweinf.) Schweinf., 278
Kosaria foetida Forssk., 277
Kosaria obovata (A. Rich.) Schweinf. ex Solms, 277

Kosaria tropaeolifolia Schweinf., 278
KOTSCHYA Endl., 146
Kotschya africana *Endl.*, 146
Kotschya recurvifolia *(Taub.) F. White*, 147
subsp. **aetiopica** *Verdc.*, *145*, 147
LABLAB *Adans.*, 179
Lablab niger Medik, 179
var. *crenatifructus* (Steud.) Cuf., 179
var. *uncinatus* (A. Rich.) Cuf., 179
Lablab purpureus *(L.) Sweet*, 179, *180*
subsp. *purpureus*, 181
subsp. *uncinatus* Verdc., 181
LAMELLISEPALUM Engl., 386
Lamellisepalum hildebrandtii Engl., 386
LANNEA *A. Rich.*, 513
Lannea barteri *(Oliv.) Engl.*, *515*, 518
Lannea cotoneaster Chiov., 514
Lannea cufodontii Chiov., 516
Lannea floccosa Sacl., 516
Lannea fruticosa *(A.Rich.) Engl.*, *515*, 518
Lannea humilis *(Oliv.) Engl.*, *515*, 516
Lannea kerstingii Engl. & Krause, 518
Lannea malifolia *(Chiov.) Sacl.*, *515*, 516
Lannea rivae *(Chiov.) Sacl.*, *515*, 516
Lannea ruspolii Engl., 516
Lannea schimperi *(A. Rich.) Engl.*, 516, *517*
var. glabrescens (Engl.), 518
var. stolzii (Engl. & von Brehm.) R. & A. Fern, 518
Lannea schweinfurthii *(Engl.) Engl.*, 518, 519
var. acutifoliolata (Engl.) Kokwaro, 518
var. **schweinfurthii**, *515*, 518
var. tomentosa (Dunkley) Kokwaro, 519
var. *stuhlamannii* (Engl.) Kokwaro, 518
Lannea somalensis (Chiov.) Sacl., 514
Lannea stuhlmannii (Engl.) Engl., 518
Lannea triphylla *(A. Rich.) Engl.*, 514, *515*
Lannea velutina A. Rich., 514
Lannea welwitschii *(Hiern) Engl.*, 519
var. ciliolata Engl., 519
var. **welwitschii**, *515*, 519
LANNEOMA Del., 513
Lanneoma velutina Del., 514
LAPORTEA *Gaud.*, 308
Laportea aestuans *(L.) Chew*, 309
Laportea alatipes *Hook. f.*, 309, *309*
Laportea interrupta *(L.) Chew*, 308
LATHYRUS *L.*, 249
Lathyrus aphaca *L.*, 250
Lathyrus hygrophilus sensu Cuf., 248
Lathyrus kilimandscharicus Taub., 248
var. *scioanus* Fiori, 248
Lathyrus odoratus *L.*, 250
Lathyrus pratensis *L.*, 250
Lathyrus sativus *L.*, 250
Lathyrus schimperi Engl., 248
Lathyrus sphaericus Retz., 250, *251*
LATIFOLIOLATAE *Engl. ex Vollesen*, 465
LAUROCERASUS, 41
LECANIODISCUS *Planch. ex Benth.*, 504
Lecaniodiscus fraxinifolius *Bak.*, 504
ssp. **vaughanii** *(Dunkley) Friis*, 505
LECANTHUS *Wedd.*, 313
Lecanthus peduncularis *(Royle) Wedd.*, 313
LEGUMINOSAE, 49
LENS *Mill.*, 249
Lens culinaris *Medik.*, 249
Lens ervoides *(Brign.) Grande*, 249, *251*
Leobordea abyssinica Hochst., 195
LEPIDOTRICHILIA *Leroy*, 485
Lepidotrichilia volkensii *(Gürke) Leroy*, 485, *488*
LEPISANTHES *Blume*, 504
Lepisanthes senegalensis *(Juss. ex Poir.) Leenh.*, 504, *505*
LEUCAENA *Benth.*, 74
Leucaena glauca sensu auctt., 75
Leucaena leucocephala *(Lam.) De Wit*, 75
Liebrechtsia kotschyi (Schweinf.) De Wild., 176
Liebrechtsia schweinfurthii De Wild., 176
Liebrechtsia spartioides (Taub.) De Wild., 176
Loeseneriella africana (Willd.) R. Wilczek, 342
LONCHOCARPUS *Kunth*, 104
Lonchocarpus laxiflorus *Guill. & Perr.*, *103*, 104
Lonchocarpus philenoptera Benth., 104
LORANTHACEAE, 361
LORANTHUS, 361
sect. *Acostachys* (Tieghem) Sprague, 362
sect. *Constrictiflori* Sprague, 372
sect. *Coriaceifolii* Sprague, 368
sect. *Erectilobi* Sprague, 372
sect. *Hirsuti* Sprague, 365
sect. *Involutiflori* Sprague, 371
sect. *Ischnanthus* Sprague, 371
sect. *Longicalyculati* Sprague, 368
sect. *Longiflori* Sprague, 372
sect. *Plicopetalus* Sprague, 362
sect. *Rufescentes* Sprague, 366
sect. *Sycophila* (Tieghem) Sprague, 362
sect. *Tapinostemma* (Benth. & Hook.) Sprague, 362
sect. *Tetrameri* Sprague, 365
Loranthus acaciae Zucc., 363
Loranthus bagshawei, 372
Loranthus curviflorus Oliv., 364
Loranthus dregei Eckl. & Zeyh., 366
var. *kerenicus* Sprague, 366
forma *subcuneifolius* Engl., 366
var. *subcuneifolius* (Engl.) Sprague, 366
Loranthus dschallensis Engl., 368
Loranthus erythraea Sprague, 367
Loranthus fischeri Engl., 369
var. *glabratus* Engl., 369
Loranthus ghikae Volkens & Schweinf., 370
Loranthus gibbosulus A. Rich., 363
Loranthus glabratus (Engl.) Sprague, 369

Loranthus globiferus A. Rich., 373
var. *angustifolius* Schweinf., 373
var. *salicifolia* Sprague, 373
var. *verrucosus* (Engl.) Sprague, 373
Loranthus heteromorphus A. Rich., 373
Loranthus hildebrandtii Engl., 371
Loranthus kelleri Engl., 370
Loranthus lingelsheimii Pax, 373
Loranthus macrosolen A. Rich., 367
Loranthus meridianus (Danser) Krause, 363
Loranthus oblongifolius A. Rich., 366
Loranthus ostinii Chiov, 373
Loranthus panganensis Engl., 365
Loranthus platyphyllus A. Rich, 372
Loranthus regularis Sprague, 368
var. *pilosus* (Pax) Cuf., 368
Loranthus rufescens DC., 368
var. pilosus Pax, 368
Loranthus rufescens sensu A. Rich., 368
Loranthus sarertaensis Hutch., 367
Loranthus scassellattii Chiov., 373
var. *glabrescens* Pichi-Serm., 373
Loranthus schimperi A. Rich., 370
var. *parviflorus* Hutch. & Bruce, 370
Loranthus sennii Mattei, 373
Loranthus stuhlmannii Engl.
var. *somalensis* Sprague, 369
Loranthus thomsonii Sprague, 362
Loranthus tschertscherensis Pax, 371
Loranthus ugogensis Engl., 369
var. *ganocalyx* Chiov., 369
Loranthus undulatus Harv., 364
var. *sagittifolius* Engl., 364
Loranthus usuiensis Oliver, 368
Loranthus verrucosus Engl., 373
Loranthus woodfordioides Schweinf., 371
LOTONONIS *(DC.) Eckl. & Zeyh.*, 195
Lotononis abyssinica (Hochst. ex A. Rich.) Kotschy, 195
Lotononis laxa *Eckl. & Zeyh.*, 194, 195
Lotononis leobordea Benth., 195
var. *abyssinica* (Hochst. ex A. Rich.) Schweinf., 195
var. *filicaulis* A. Terracc., 195
Lotononis platycarpos *(Viv.) Pic.-Serm.*, 195
var. *abyssinica* (Hochst. ex A. Rich.) Pic.-Serm., 195
LOTUS *L.*, 223
Lotus arabicus *L.*, 225
var. *glabrescens* Schweinf., 225
var. torulosus Chiov., 225
Lotus brachycarpus Hochst. & Steud. ex A. Rich., 224
var. *cuscutae* Schweinf., 225
var. *major* Brand, 225
var. *montanus* (A. Rich.) Cuf., 224
Lotus corniculatus *L.*, 223, 224
var. *eremanthus* Chiov., 224
var. *schoelleri* (Schweinf.) Lanza, 223
Lotus discolor *E. Mey.*, 224
Lotus glinoides sensu Cuf., 225
Lotus goetzei Harms, 224
Lotus hebecarpus *Gillett*, 225
Lotus hebranicus Hochst. ex Brand, 225
Lotus lalambensis *Schweinf.*, 225
Lotus mearnsii De Wild. non Britton, 224
Lotus montanus A. Rich., 224
Lotus nubicus Hochst. ex Bak., 224
var. *montanus* (A. Rich.) Bak., 224
Lotus platycarpos Viv., 195
Lotus quinatus *(Forssk.) Gillett*, 224
var. **brachycarpus** *(Hochst. & Steud. ex A. Rich.) Gillett*, 224, *226*
var. quinatus, 224
Lotus schimperi *Steud. ex Boiss.*, 225
Lotus schoelleri *Schweinf.*, 223, *226*
Lotus sp. *'A'*. sensu Gillett, 224
Lotus tigrensis Bak., 224
Lotus torulosus *(Chiov.) Fiori*, 225
Loxospermum calocephalum (Fresen.) Hochst., 240
Loxospermum multinerve Hochst., 241
Loxospermum schimperi Hochst., 241
LUPINUS *L.*, 219
Lupinus albus *L.*, 219
Lupinus angustifolius *L.*, 219
Lupinus ehrenbergii Schlechtend., 219
Lupinus luteus *L.*, 219
Lupinus mexicanus *Cerv. ex Lag.*, 219
Lupinus mutabilis *Sweet*, 219
Lupinus princei *Harms*, 219, *221*
Lupinus termis Forssk., 219
Macrolotus rivae Harms, 220
MACROPTILIUM *(Benth.) Urb.*, 169
Macroptilium atropurpureum *(DC.) Urb.*, 169
Macroptilium lathyroides *(L.) Urb.*, 169
MACROTYLOMA *(Wight & Arn.) Verdc.*, 165
Macrotyloma africanum *(Wilczek) Verdc.*, 166
Macrotyloma atropurpureum *(DC.) Urb.*, 169
Macrotyloma axillare *(E. Mey.) Verdc.*, 166
var. **axillare.**, 166
var. **glabrum** *(E. Mey.) Verdc.*, 166, *167*
Macrotyloma biflorum *(Schum. & Thonn.) Hepper*, 166
Macrotyloma daltonii *(Webb) Verdc.*, 166
Macrotyloma schweinfurthii *Verdc.*, 168
Macrotyloma stenophyllum *(Harms) Verdc.*, 168
Macrotyloma tenuiflorum *(Micheli) Verdc.*, 168
Macrotyloma uniflorum *(Lam.) Verdc.*, 166
var. **stenocarpum** *(Brenan) Verdc.*, 166
Magnistipula, 46
MALUS *Miller*, 43
Malus communis Desf., 43
Malus domestica Borkh., 43
Malus pumila Mill., 43
Malus sylvestris *Miller*, 43
MANGIFERA *L.*, 519
Mangifera indica *L.*, 519
MAYTENUS *Molina*, 331
Maytenus addat *(Loes.) Sebsebe*, 333, *334*

Maytenus arbutifolia *(A. Rich.) Wilczek*, *334*, 335
var. **arbutifolia**, 336
var. **sidamoensis** *Sebsebe*, 336
Maytenus arguta (Loes.) Robson, 332
Maytenus buchananii *(Loes.) Wilczek*, 335
Maytenus castellii (Pich.-Serm.) Cuf., 332
Maytenus cortii *(Pich.-Serm.) Cuf.*, 335
Maytenus ellenbeckii (Loes.) Cuf., 332
Maytenus engleriana Loes., 336
var. *macrantha* Loes., 336
Maytenus engleriana (Loes.) Cuf., 336
var. *macrantha* (Loes.) Cuf., 336
Maytenus eritreana Sebsebe, 333
Maytenus gracilipes *(Welw. ex Oliv.) Exell*, 332
subsp. **arguta** *(Loes.) Sebsebe*, 332
subsp. gracilipes, 333
Maytenus harenensis *Sebsebe*, 335
Maytenus heterophylla *(Eckl. & Zeyh.) Robson*, 336, *337*
var. *heterophylla*, 338
Maytenus laurifolia (A. Rich.) Loes., 338
Maytenus luteola (Del.) Andrews, 338
Maytenus obscura *(A. Rich.) Cuf.*, 333, *334*
Maytenus ovata (Wall. ex Wight & Arn.) Loes., 333
var. *arguta* (Loes.) Blakelock
var. *ovata*, 333
forma *ovata*, 333
forma *pubescens* (Schweinf.) Blakelock, 333
Maytenus parviflora *(Vahl) Sebsebe*, 333
subsp. **eritreana** *(Sebsebe) Sebsebe*, 333
subsp. parviflora, 333
Maytenus somalensis sensu Cuf., 336
Maytenus somalensis (Loes.) Cuf., 336
Maytenus putterlickioides *(Loes.) Exell & Mendonça*, 336
Maytenus senegalensis *(Lam.) Exell*, 336, *337*
Maytenus serrata *(A. Rich.) Wilczek*, 333
Maytenus undata *(Thunb.) Blakelock*, 338
MEDICAGO *L.*, 244
Medicago arborea *L.*, 244
Medicago aschersoniana Urb., 245
Medicago hispida Gaertn., 244
Medicago laciniata *(L.) Mill.*, 245
subsp. schimperiana P. Fourn., 245
Medicago lupulina *L.*, 244
Medicago minima *(L.) Bartal.*, 245
Medicago orbicularis *(L.) Bartal.*, 244
Medicago praecox DC, 244
Medicago polymorpha *L.*, 244, *247*
Medicago sativa *L.*, 244
Medicago tentaculata Willd., 244
Medicago tribuloides Desr., 244
Medicago truncatula Gaertn., 244
MELIA *L.*, 483
Melia azedarach *L.*, 485
Melia volkensii *Gürke*, 485, *486*
MELIACEAE, 479
MELIANTHACEAE, 511
MELILOTUS *Mill.*, 245
Melilotus abyssinica Bak., 246
Melilotus alba *Desr.*, 245
Melilotus altissima sensu auctt. non Thuill., 245
Melilotus elegans *Salzm. ex Ser.*, 246, *247*
Melilotus indica *(L.) All.*, 245
Melilotus messanensis *(L.) All.*, 246
Melilotus neapolitana Ten., 245
Melilotus officinalis sensu auctt. non (L.) Pallas., 245
Melilotus sicula (Turra) B. D. Jackson, 246
Melilotus suaveolens *Ledeb.*, 245
Merremia kentrocaulos, 418
MILICIA *Sim*, 272
Milicia africana Sim, 272
Milicia excelsa *(Welw.) C.C. Berg*, 272, *274*
MILLETTIA *Wight & Arn.*, 108
Millettia darassana Cuf., 108
Millettia ferruginea *(Hochst.) Bak.*, 108
subsp. **darassana** *(Cuf.) Gillett*, 108
subsp. **ferruginea**, 108
MIMOSA *L.*, 74
Mimosa cinerea L., 74
Mimosa glomerata Forssk., 74
Mimosa invisa *Mart. ex Colla*, 74
Mimosa kantuffa DC., 54
Mimosa oerfota Forssk., 90
Mimosa pigra *L.*, *73*, 74
Mimosa sanguinea Bruce ex Gmel., 74
Mimosa sassa (Willd.) Poir., 96
Mimosa stellata Forssk., 54
MIMOSOIDEAE, 71
Missiessya hypoleuca (Steud.) Wedd., 320
Moghania rhodocarpa (Bak.) Kuntze, 193
MONOICAE *Vollesen*, 457
MORACEAE, 271
MORUS *L.*, 271
Morus alba *L.*, 272
var. *indica* (L.) Bur., 272
Morus excelsa Welw., 272
Morus indica L., 272
Morus lactea (Sim) Mildbr., 272
Morus mesozygia Stapf, 272, *273*
var. *lactea* (Sim) Mildbr., 272
Morus nigra L., 272
MUCUNA *Adans.*, 164
Mucuna pruriens *(L.) DC.*, 164
var. **utilis** *(Wall. ex Wight) Bak. ex Burck*, 164
Mucuna melanocarpa *Hochst.*, *163*, 164
var. *somalensis* Taub. ex Harms, 164
MYRICA *L.*, 261
Myrica kilimanscharica Engl., 261
Myrica meyeri-johannis Engl., 261
Myrica mildbraedii Engl., 261
Myrica salicifolia *A. Rich.*, *260*, 261
Myrica usambarensis Engl., 261
MYRICACEAE, 261

MYROTHAMNACEAE, 254
Myrothamnus, 254
Myrothamnus flabellifolius Welw., 254
MYSTROXYLON *Ecklon & Zeyher*, 338
Mystroxylon aethiopicum *(Thunb.) Loes*., 338, *339*
NEORAUTANENIA *Schinz*, 181
Neorautanenia mitis *(A. Rich.) Verdc*., 181
Neorautanenia pseudopachyrhiza (Harms) Milne-Redh., 181
var. *ellenbeckii* (Harms) Cuf., 181
NEPTUNIA *Lour.*, 71
Neptunia oleracea *Lour*., 71
NEURADACEAE, 45
NEURADA *L*., 45
Neurada procumbens *L*., 45, *45*
var *stellata* M. & D. Zohary, 45
Neuradopsis Bremek. & Obermeyer, 45
OBETIA *Gaud*., 306
Obetia sp. = *Burger* 3368
Obetia tenax (N. E. Brown) Friis, 306
OCTOKNEMACEAE, 360
Octoknema, 360
ODINA Roxb., 513
Odina barteri Oliv., 518
Odina fruticosa A. Rich., 518
Odina humilis Oliv., 516
Odina malifolia Chiov., 516
Odina rivae Chiov., 516
Odina schimperi A. Rich., 516
var. *glabrescens* Engl., 516
Odina schweinfurthii Engl., 518
Odina triphylla A. Rich., 514
Odina welwitschii Hiern, 519
Odina velutina (A. Rich.) Walp., 514
ODONTELLA Tieghem, non C.A. Agardh., 368
OLACACEAE, 356, 360
OLIVERELIA *Tieghem*, 371
Oliverella hildebrandtii *(Engl.) Tieghem*, 371, *371*
ONCOCALYX *Tieghem*, 368
Oncocalyx angularis *M. Gilbert*, 370
Oncocalyx fischeri *(Engl.) M. Gilbert*, 369, *369*
Oncocalyx ghikae *(Volkens & Schweinf.) M. Gilbert*, 370
Oncocalyx glabratus *(Engl.) M. Gilbert*, 369
Oncocalyx kelleri *(Engl.) M. Gilbert*, 370
Oncocalyx schimperi *(A. Rich.) M. Gilbert*, 370, *370*
ONOBRYCHIS *Mill*., 228
Onobrychis richardii *Bak*., 230
Onobrychis viciifolia *Scop*., 230
Onobrychis vulgaris Hill, 230
ONONIS *L*., 246
Ononis reclinata *L*., 246, *247*
OPHRESTIA *H.L. Forbes*, 156
Ophrestia radicosa *(A. Rich.) Verdc*.,156, *159*
var. *radicosa*, 158
var. *schliebenii* (Harms) Verdc., 158
OPILIA *Roxb*., 358
Opilia amentacea *Roxb*., 358
Opilia campestris *Engler*, 358
var. **campestris**, 359, *359*
var. **strobilifera** *(Hutch. Bruce) Hiepko*, *358*, 359, *359*
Opilia celtidifolia (Guill. & Perr.) Endl. ex Walp., 358
var. *sphaerocarpa* Chiov., 358
Opilia strobilifera Hutch. & Bruce, 359
Opilia tomentella (Oliv.) Engl., 358
OPILIACEAE, 358
OPOBALSAMEAE *Engl. ex Vollesen*, 474
ORMOCARPUM Beauv., 142
Ormocarpum bibracteatum (Steud. ex A. Rich.) Bak., 143
Ormocarpum melanodictyotum Chiov., 143
Ormocarpum mimosoides sensu Cuf., 143
Ormocarpum muricatum *Chiov*., 143
Ormocarpum pubescens *(Hochst.) Cuf. ex Gillett*, 143
Ormocarpum trachycarpum *(Taub.) Harms*, 143
Ormocarpum trichocarpum *(Taub.) Engl*., 143
OSYRIDOCARPUS *A. DC*., 382
Osyridocarpus schimperanus *(A. Rich.) A.DC*., 382, *382*
OSYRIS *L*., 382
Osyris abyssinica A. Rich., 382
forma *latifolia* Fiori, 382
Osyris arborea Wallich ex A. DC., 382
Osyris lanceolata Hochst. et Steudel ex DC., 382
Osyris parvifolia Baker, 382
Osyris quadripartita *Decn*., 382, *383*
Osyris rigidissima Engl., 382
Osyris tenuifolia Engl., 382
Osyris wightiana Wallich ex Wight, 382
OZOROA *Del*., 519
Ozoroa insignis *Del*., 521
subsp. **insignis**, 521, *522*
subsp. latifolia (Engl.) R. & A. Fern., 521
subsp. reticulata (bak. f.) Gillett, 521
Ozoroa pulcherrima *(Schweinf.) R. & A. Fern*., 521
Padus, 41
Panax ferrugineum Harms, 537
Panax pinnatum A. Rich., 537
PAPILIONOIDEAE, 97
PAPPEA *Eckl. & Zeyh*., 508
Pappea capensis *Eckl. & Zeyh*., 508, *509*
var. *radlkoferi* (Penzig ex Schweinf.), 508
Pappea radlkoferi Penzig ex Schweinf., 508
PARACALYX *Ali*, 189
Paracalyx somalorum *(Vierh.) Ali*, *186*, 189
Paraglycine radicosa (A. Rich.) F.J. Hermann, 156
PARIETARIA *L*., 320
Parietaria abyssinica A. Rich., 317
Parietaria alsinefolia auct. non Gaulich: A. Rich., 320
Parietaria debilis *G. Forst*., 320, *321*
Parinari, 46
Parinari curatellifolia, 46
PARKINSONIA *L*. 54
Parkinsonia aculeata *L*., 56

Parkinsonia scioana *(Chiov.) Brenan*, 54, *55*
PAROCHAETUS *D.Don*, 233
Parochaetus communis *D. Don*, 233, *235*
var. *grossecrenatus* Cuf., 233
PAULLINIA *L.*, 493
Paullinia asiatica L., 424
Paullinia pinnata *L.*, 493, *496*
PEDUNCULATAE *Engl. ex Vollesen*, 457
Peltophoropsis scioana Chiov., 54
Peltophorum scioanum (Chiov.) Bak. f, 54
PERSICA, 41
Persica vulgaris Miller, 42
PHARMACOSYCEA Miq., 289
PHARMACOSYCEA *(Miq.) Miq.*, 289
PHASEOLUS *L.*, 169
Phaseolus abyssinicus Savi, 177
Phaseolus atropurpureus DC., 169
Phaseolus coccineus *L.*, 171
Phaseolus lathyroides L., 169
Phaseolus lunatus *L.*, *170*, 171
Phaseolus macrorhynchus Harms, 177
Phaseolus quadriflorus Hochst., 176
Phaseolus radiatus L., 177
Phaseolus schimperi Taub., 176
Phaseolus vulgaris *L.*, 169
Philenoptera schimperi Hochst. ex A. Rich., 104
PHRAGMANTHERA *Tieghem*, 366
Phragmanthera dschallensis *(Engl.) M. Gilbert*, 368
Phragmanthera erythraea *(Sprague) M. Gilbert*, 367, *367*
Phragmanthera macrosolen *(A. Rich.) M. Gilbert*, 367, *367*
Phragmanthera regularis *(Sprague) M. Gilbert*, *367*, 368
Phragmanthera sarertaensis *(Hutch. & E.A. Bruce) M. Gilbert*, 367, *367*
Phragmanthera usuiensis *(Oliver) M. Gilbert*, 368
Phyllogeiton discolor (Klotzsch) Herzog, 393
Pictetia pubescens Hochst., 143
PILEA *Lindl.*, 310
Pilea angolensis *(Hiern) Rendle*, 310
Pilea bambuseti *Engl.*, 312
subsp. **aethiopica** *Friis*, 312
subsp. bambuseti, 312
Pilea ceratomera Wedd., 312
Pilea johnstonii *Oliv.*, 312
subsp. **johnstonii**, 312
subsp. **rwandensis** *Friis*, 312
Pilea quadrifolia A. Rich., 310
Pilea rivularis *Wedd.*, *311*, 312
Pilea tetraphylla *(Steudel) Blume*, 310, *311*
PILIOSTIGMA *Hochst.*, 68
Piliostigma pyrrhocarpum Hochst., 68
Piliostigma thonningii *(Schumach.) Milne-Redh.*, *67*, 68
PISTACIA *L.*, 530
Pistacia aethiopica *Kokwaro*, 530, *531*
Pistacia chinensis Bunge, 532
var. *falcata* (Mart.) Zohary, 532
Pistacia falcata *Mart.*, *531*, 532
Pistacia lentiscus L., 530
var. *emarginata* Engl., 530
var. *falcatula* Chiov., 530
PISUM *L.*, 248
Pisum abyssinicum A. Br., 248
Pisum arvense L., 248
var. *abyssinicum* (A. Br.) Alef. ex Engl., 248
Pisum sativum *L.*, 248
subsp. *abyssinicum* (A. Br.) Gov., 248
var. **abyssinicum** *(A. Br.) Alef.*, 248, *251*
var. *elatius* (Stev.)Beck, 248
forma *abyssinicum* (A. Br.) Gams, 248
var. **sativum**, 248
PITHECELLOBIUM *Mart.*, 96
Pithecellobium dulce *(Roxb.) Benth.*, 96
PITTOSPORACEAE, 1
PITTOSPORUM *Soland.* in *Gaertn.*, 1
Pittosporum abyssinicum *Del.*, 1
subsp. *cardiocarpum* (Cuf.) Cuf., 1
subsp. *engleri* (Cuf.) Cuf., 1
subsp. *lanatum* (Hutch. & Bruce) Cuf., 1
Pittosporum cardiocarpum Cuf., 1
Pittosporum dalziellii Hutch., 2
Pittosporum engleri Cuf., 1
Pittosporum feddeanum Pax, 2
Pittosporum lanatum Hutch. & Bruce, 1
Pittosporum mannii, 2
subsp. *ripicola* (J. Leon.) Cuf., 2
Pittosporum quartinianum Cuf., 2
Pittosporum ripicola J. LEon., 2
"*Pittosporum ripicolum*", 2
Pittosporum tomentosum Engl., 1
Pittosporum ustulatum Cuf., 2
Pittosporum ustulatum Cuf., 2
Pittosporum viridiflorum *Sims*, 2, *3*
subsp. *dalziellii* (Hutch.) Cuf., 2
subsp. *feddeanum* (Pax) Cuf., 2
subsp. quartinianum (Cuf.) Cuf., 2
Pittosporum "feddeanum", 2, *3*
Pittosporum "ripicola", 2
Pittosporum "ustulatum", 2
Pittosporum "viridiflorum", 2
PLATANACEAE, 257
Platanus orientalis *L.*, 257
PLATYCELYPHIUM *Harms*, 101
Platycelyphium cyananthum Harms, 101
Platycelyphium voënse *(Engl.) Wild*, 101, *102*
PLICOSEPALUS *Tieghem*, 362
Plicosepalus acaciae *(Zucc.) Wiens & Polhill*, 363, *363*
Plicosepalus curviflorus *(Oliv.) Tieghem*, 364, *364*
Plicosepalus meridianus *(Danser) Wiens & Polhill*, 363, *363*
Plicosepalus ogadenensis *M. Gilbert*, 363
Plicosepalus robustus *Wiens & Polhill*, 364, *364*
Plicosepalus sagittifolius *(Engl.) Danser*, 364

Ploesslea floribunda Endl., 443
Podalyria aurea (Ait.) Willd., 104
Pogonostigma abyssinicum Jaub. & Spach., 116
Pogonostigma nubicum Boiss., 116
Poinciana pulcherrima L., 51
Poinciana spinosa Molina, 53
POLYSCIAS *J. R. & G. Forster*, 537
Polyscias farinosa *(Del.) Harms*, 537
Polyscias ferruginea (Hiern) Harms, 537
Polyscias fulva (Hiern) Harms, 537, *538*
Populus, 258
POTENTILLA *L.*, 36
Potentilla abyssinica A. Rich., 37
Potentilla dentata *Forssk.*, 36, *36*
Potentilla hispanica auctt. non Zimmet., 36
Potentilla pennsylvanica auct non L., 36
Potentilla reptans *L.*, *36*, 37
POUZOLZIA *Gaud.*, 317
Pouzolzia abyssinica (A. Rich.) Blume, 317
Pouzolzia erythraeae Schweinf., 325
Pouzolzia flaccida A. Rich., 324
Pouzolzia fruticosa Engl., 317
Pouzolzia guineensis *Benth.*, 317
  var. *abyssinica* (A. Rich.) Rendle, 317
Pouzolzia hypoleuca Wedd., 317
Pouzolzia mixta *Solms-Laub.*, 317, *318*
Pouzolzia parasitica *(Forssk.) Schweinf.*, 317, *318*
Pouzolzia pauciflora (Steud.) A. Rich., 322
Procris hypoleuca Steudel, 320
PROSOPIS *L.*, 72
Prosopis juliflora *(Sw.) DC.*, 72
PRUNUS *L.*, 41
Prunus africana *(Hook. f.) Kalkm.*, *32*, 41
Prunus amygdalus Batsch., 42
Prunus armeniaca L., 41
Prunus communis *(L.) Arc.*, 42
Prunus persica *(L.) Batsch*, 41, 42
Prunus x domestica *L.*, 42
PSEUDARTHRIA *Wight & Arn.*, 152
Pseudarthria confertiflora *(A. Rich.) Bak.*, 152
Pseudarthria hookeri *Wight & Arn.*, 152, *153*
  var. argyrophylla, 152
  var. *confertiflora* (A. Rich.) Schindl., 152
  var. *hookeri*., 152
PSEUDOCEDRELA *Harms*, 479
Pseudocedrela kotschyi *(Schweinf.) Harms*, 479, *480*
PSEUDOERIOSEMA *Hauman*, 156
Pseudoeriosema borianii *(Schweinf.) Hauman*, 156, *157*
  subsp. longipedunculatum Verdc.156
PSOPHOCARPUS *Neck. ex DC.*, 168
Psophocarpus grandiflorus *Wilczek*, 168, *170*
Psophocarpus lancifolius *Harms*, 168
Psophocarpus palustris Desv., 168
  cultivar. "Wondo Surprise", 168
PSORALEA *L.*, 223
Psoralea bituminosa *L.*, 223
Psoralea plicata *Del.*, 223, *235*
Psoralea tetragonoloba L., 138
PTEROCARPUS *Jacq.* 105
Pterocarpus lucens *Guill. & Perr.*, *103*, 105
Pterocarpus abyssinicus Hochst. ex A. Rich., 105
PTEROLOBIUM *Wight & Arn.*, 54
Pterolobium abyssinicum (A. Rich.) A. Rich., 54
Pterolobium exosum (J.F. Gmel.) Bak.f., 54
Pterolobium lacerans R. Br., 54
Pterolobium stellatum *(Forssk.) Brenan*, 54, *55*
Ptycholobium plicatum (Oliv.) Harms, 97
Pueraria hochstetteri Chiov., 181
Pusaetha abyssinica (Steud. ex A. Rich.) Kuntze, 72
PYGEUM, 41
Pygeum africanum Hook. f., 41
PYRENACANTHA *Wight*, 351
Pyrenacantha kaurabassana *Baill.*, 351
Pyrenacantha malvifolia *Engl.*, 351, *352*
Pyrenacantha ruspolii Engl., 351
Pyrenacantha sylvestris *S. Moore*, 351
PYRUS *L.*, 43
Pyrus communis *L.*, 43
Quartinia abyssinica A. Rich., 54
QUERCUS *L.*, 265
Quercus ilex *L.*, 265
Quercus pedunculata Ehrh., 265
Quercus robur *L.*, 265
RAPHIOSTYLIS *Benth.*, 348
Raphiostylis beninensis *(Planch.) Benth.*, 350, *350*
Reissantia parvifolia (Oliv.) N. Hallé, 342
RHAMNACEAE, 385
RHAMNUS *L.*, 390
Rhamnus deflersii Schweinf. ex Herzog, 390
Rhamnus infusionum Del., 390
Rhamnus pauciflorus Hochst. ex A. Rich., 390
Rhamnus prinoides *L'Herit.*, 390, *392*
Rhamnus spina-christi L., 395
Rhamnus staddo *A. Rich.*, 390
  var. *deflersii* (Schweinf. ex Herzog) Chiov. ex Engl., 390
RHOICISSUS *Planch.*, 401
Rhoicissus erythrodes (Fresen>) Planch., 401
Rhoicissus revoilii *Planch.*, 401
Rhoicissus tridentata *(L. f.) Wild & Drummond*, 401, *402*
RHUS *L.*, 521
Rhus abyssinica Oliv., 528
  var. *glabrata* Mart., 523
Rhus amharica Pichi-Serm., 529
Rhus anchietae Hiern, 524
Rhus crenulata A. Rich., 523
Rhus foliosa A. Rich., 528
Rhus glaucescens A. Rich., 523, 524
  forma. *pubescens* Fiori, 523
  var. *macrocarpa* Schweinf., 523
  var. *obovatifoliolata* Engl., 523
  var. *schimperi* Oliv., 524

Rhus glutinosa *A. Rich.*, 528
subsp. **abyssinica** *(Oliv.) M. Gilbert*, *526*, 528
subsp. **glutinosa**, 529
var. **glutinosa**, *526*, 529
var. **unifoliolata** *Cuf.*, 529
subsp. **neoglutinosa** *(M. Gilbert) M. Gilbert*, *526*, 529
var. *acutifoliolata* Engl., 529
var. *obtusifoliolata* Engl., 529
Rhus huillensis Engl., 524
var. *erythraea* Fiori, 524
Rhus longipes *Engl.*, 528
var. elgonensis Kokwaro, 528
var. **longipes**, *526*, 528
var. schinoides (Hutch.) R. & A. Fernandes, 528
Rhus macowanii sensu Pichi-Serm., 529
Rhus natalensis *Krauss*, 523, *525*, *526*
var. *hararensis* Engl., 523
var. *macrocarpa* (Schweinf.) Cuf., 523
var. *obovatifoliolata* (Engl.) Chiov., 523
var. *schimperi* (Oliv.) Engl.
Rhus neoglutinosa M. Gilbert, 529
Rhus petitiana A. Rich., 528
Rhus pyroides Burch., 524
var. *erythraea* Fiori, 524
Rhus quartiniana *A. Rich.*, 524, *527*
var. **quartiniana**, 524, *527*
var. zambesiensis R. & A. Fernandes, 524
Rhus retinorrhoea *Oliv.*, *526*, 529
Rhus ruspolii *Engl.*, *526*, 528
Rhus saeneb Forssk., 320
Rhus sp. = *Friis* et al. 1147, 524, *526*
Rhus tenuinervis *Engl.*, 524, *526*
Rhus villosa sensu auct. non L.f., 529
var. *gallaensis* Engl., 529
var. *tomentosa* Oliv., 529
Rhus viminale sensu A. Rich., 529
Rhus vulgaris *Meikle*, *526*, 529
RHYNCHOSIA *Lour.*, 181
Rhynchosia albissima *Gand.*, 188
Rhynchosia alluaudii *Sacl.*, 185
Rhynchosia borianii Schweinf, 156
Rhynchosia buramensis Hutch. & E. A. Bruce, 184
Rhynchosia cliffordii Hutch. & Bruce, 188
Rhynchosia confertiflora A. Rich., 152
Rhynchosia congensis *Bak.*, 189
Rhynchosia densiflora *(Roth) DC.*, 183
subsp. **chrysadenia** *(Taub.) Verdc.*, 183
subsp. **debilis** *(G. Don) Verdc.*, 183
subsp. densiflora, 183
subsp. stuhlmannii, 183
Rhynchosia elachistantha Chiov., 187
Rhynchosia elegans *A. Rich.*, 184
var. **elegans**, 184
var. **grandistipulata** *Verdc.*, 184
Rhynchosia erlangeri *Harms*, 188
Rhynchosia erythraeae *Schweinf.*, 188, *191*
Rhynchosia ferruginea *A. Rich.*, 184
Rhynchosia flavissima Hochst. ex Bak., 187
var. *macrocalyx* Chiov., 187
Rhynchosia glomerulans Fiori, 183
Rhynchosia hirta *(Andr.) Meikle & Verdc.*, 189
Rhynchosia imbricata Bak., 184
Rhynchosia malacophylla *(Spreng.) Boj.*, 187
Rhynchosia malacotricha *Harms*, 185
Rhynchosia memnonia (Del.) DC., 187
forma *pulverulenta* (Stocks) Terracc, 187
Rhynchosia mensensis *Schweinf*, 185
Rhynchosia minima *(L.) DC.*, 187
var. **macrocalyx** *(Chiov.) Verdc.*, 187
var. **memnonia** *(Del.) Cooke*, 188
var. **minima.**, *186*, 187
var. **nuda** *(DC.) Kuntze*, 187
var. **prostrata** *(Harv.) Meikle*, 188
Rhynchosia nyasica *Bak.*, 184
Rhynchosia oblatifoliolata *Verdc.*, 184
Rhynchosia orthobotrya *Harms*, 183
Rhynchosia polystachya A. Rich., 190
Rhynchosia pulverulenta *Stocks*, 187
Rhynchosia ramosa *Verdc.*, 189
Rhynchosia resinosa *(Hochst. ex A. Rich.) Bak.*, 183
Rhynchosia rivae Schweinf., 184
Rhynchosia schimperi *Hochst. ex Boiss.*, 188, *191*
Rhynchosia schoelleri Schweinf., 188
Rhynchosia sennaarensis Hochst. ex Schweinf., 187
var. *flavissima* Schweinf., 187
var. *macrocalyx* (Chiov.) Cuf., 188
Rhynchosia splendens *Schweinf.*, 188
Rhynchosia stipulosa *A. Rich.*, 183
Rhynchosia sublobata *(Schumach.) Meikle*, 185
Rhynchosia totta *(Thunb.) DC.*, 185
var. **totta**, 185, *186*
var. **venulosa** *(Hiern) Verdc.*, 185
Rhynchosia transjubensis Chiov., 185
Rhynchosia usambarensis *Taub.*, 184
var. **obtusifoliola** *Verdc.*, 185
var. **usambarensis**, 185
Rhynchosia venulosa (Hiern) K. Schum., 185
Rhynchosia verdcourtii *Thulin*, *186*, 187
Rhynchosia viscosa (Roth) DC., 183
var. *stipulosa* (A. Rich.) Verdc., 183
Robinia subdecandra L'Hérit., 104
ROCHETIA Delile, 481
Rochetia choensis Del., 483
ROSA *L.*, 34
Rosa abyssinica *Lindley*, 35, *35*
var. *microphylla* Almagia, 36
Rosa abyssinica R. Br. in Salt, 36
Rosa bottaiana Bouleng, 36
Rosa centifolia L., 35
var. *sancta* (A. Rich.) Zab., 35
Rosa gallica x *phoenicia*, 35
Rosa moschata Herrm., 35
var. *abyssinica* (Lindley) Crépin, 35

Rosa sancta A. Rich., 35
Rosa schimperiana Hochst., 35
Rosa schweinfurthi Bouleng., 35
Rosa x richardii *Rehd.*, 35
ROSACEAE, 31, 45, 46
subfamily Chrysobalanoideae, 46
subfamily Neuradoideae, 46
ROSTRATAE *Engl. ex Wild*, 462
ROSULARIA *(DC.) Stapf*, 15
Rosularia semiensis *(A. Rich.) Ohba*, 15, *15*
ROTHIA *Pers.*, 195
Rothia hirsuta *(Guill. & Perr.) Bak.*, *194*, 195
RUBUS *L.*, 31
Rubus aethiopicus *R.A. Grah.*, 33
Rubus apetalus *Poir.*, *32*, 34
Rubus assaortinus Chiov., 34
var. *erythraeus* C.E. Gust., 34
Rubus chiesae Chiov., 33
Rubus chiovendae C.E. Gust., 34
Rubus erlangeri *Engl.*, 33
Rubus exsuccus A. Rich., 34
Rubus fruticosus, 31
Rubus idaeus L., 33
Rubus niveus *Thunb.*, 33, 34
Rubus petalabigens C.E. Gust., 34
Rubus petitianus A. Rich., 34
Rubus pinnatus Willd., 34
Rubus quartinianus A. Rich., 34
var. *hararensis* C.E. Gust., 34
var. *pappianus* C.E. Gust., 34
Rubus rigidus Sm., 34
Rubus rosaefolius, 33
Rubus rosifolius *Sm.*, 33
Rubus scioanus Chiov., 34
Rubus steudneri *Schweinf.*, *32*, 34
var. *aberensis* C.E. Gust., 34
var. *sidamensis* C.E. Gust., 34
Rubus volkensii *Engl.*, 33
Rüppellia abyssinica A. Rich., 144
RUTA *L.*, 419
Ruta bracteosa DC., 419
Ruta chalepensis *L.*, 419
Ruta graveolens L., 420
RUTACEAE, 419
Saelanthus rotundifolius Forssk., 408
SAGERETIA *Brongn.*, 386
Sageretia thea *(Osbeck) M.C. Johnston*, 386, *389*
Sageretia theezans (L.) Brongn., 386
var. *hildebrandtii* (Engl.) Chiov., 386
var. *schweinfurthii* Chiov., 386
var. *spiciflora* (A. Rich.) Chiov., 386
SALACIA *L.*, 344
Salacia congolensis *De Wild & Th. Dur.*, 346, *347*
SALICACEAE, 258
SALIX *L.*, 258
Salix axillaris A. Rich., 258
Salix cyathipoda A. Rich., 258
Salix octandra Anderss. ex A. Rich, 258
Salix pedicellata *Desf.*, 258
Salix safsaf Trautv., 258
Salix 'safsaf baelledi' Forssk., 258
Salix subserrata *Willd.*, 258, *259*
var. *cyathipoda* (A. Rich.) Cuf.
SALVADORA *L.*, 354
Salvadora persica *L.*, 354, *354*
var. cyclophylla, 355
var. **persica**, 355
var. pubescens Brenan, 355
SALVADORACEAE, 353
SANTALACEAE, 379
Santaloides splendidum (Gilg.) Engl., 533
SAPINDACEAE, 490
Sapindus abyssinicus Fresen., 504
Sassa gummifera J.F. Gmel., 96
SAXIFRAGA *L.*, 27
Saxifraga abyssinica A. Rich. ex Engler, 27
Saxifraga hederifolia *A. Rich.*, 27, *27*
SAXIFRAGACEAE, 27, 28
SCHEFFLERA *J.R. & G. Forster*, 539
Schefflera abyssinica *(Hochst. ex A. Rich.) Harms*, 539, *540*
Schefflera myriantha *(Bak.) Drake*, 539, *541*
Schefflera polysciada Harms, 539
Schefflera volkensii *(Engl.) Harms*, 539
SCHIMPERINA *Tieghem*, 372
Schimperina platyphylla *(A. Rich.) Tieghem*, 372, *372*
SCHINUS *L.*, 530
Schinus molle *L.*, 530
var. *areira* (L.) DC., 530
var. molle, 530
Schizocalyx Hochst., 353
Schizocalyx coriaceus Hochst., 353
SCLEROCARYA *Hochst*, 519
Sclerocarya birrea *(A. Rich.) Hochst.*, 519
subsp. **birrea**, 519, *520*
subsp. caffra (Sond.) Kokwaro, 519
subsp. multifoliolata (Engl.) Kokwaro, 519
SCORPIURUS *L.*, 227
Scorpiurus muricatus *L.*, 227, *229*
SCUTIA *Commers. ex Brongn.*, 390
Scutia discolor Klotzsch, 393
Scutia myrtina *(Burm. f.) Kurz*, 391
SEDUM *L.*, 5, 10
Sedum abyssinicum (A.Rich.) Raym.-Hamet, 10
Sedum baleensis *M. Gilbert*, *13*, 14
Sedum caeruleum L., 10
Sedum cepaea L., 12
Sedum churchillianum *Robyns & Boutique*, 14
Sedum crassularia *Raym.-Hamet*, 12, *13*
Sedum dendroideum DC., 12
Sedum epidendron *A. Rich.*, *13*, 14
Sedum glomerifolium *M. Gilbert*, 12, *13*
Sedum hispanicum L., 14
Sedum malladrae Chiov., 10

Sedum meyeri-johannis Engl., 14
Sedum mooneyi *M. Gilbert, 13*, 14
Sedum rupestre L., 12
Sedum ruwenzoriense Bak., 14
Sedum sediforme (Schweinf.) Raym.-Hamet, 12
Sedum schimperi Britten, 14
Sedum erlangerianum *Engl.*, 15
Sempervivum abyssinicum A. Rich., 10
Sempervivum chrysanthum Britten, 16
var. *glabrum* Chiov., 16
var. *glandulosum* Chiov., 16
Sempervivum coccinolepis Steudner, 15
Senegalia Raf., 75
SENNA *Mill.*, 56, 57
Senna alexandrina *Mill.*, 63
var. **alexandrina**, 63
var. **obtusata** *(Brenan) Lock*, 63
Senna baccarinii *(Chiov.) Lock*, 59, *62*
Senna bicapsularis *(L.) Roxb.*, 60
Senna didymobotrya *(Fresen.) Irwin & Barneby*, 61
Senna ellisae *(Brenan) Lock*, 61
Senna holosericea *(Fresen.) Greuter*, 63
Senna italica *Mill.*, 63
subsp. **italica**, 63
subsp. **micrantha** *(Brenan) Lock*, *62*, 63
Senna longiracemosa *(Vatke) Lock*, 61
Senna multiglandulosa *(Jacq.) Irwin & Barneby*, 59
Senna obtusifolia *(L.) Irwin & Barneby*, *58*, 60
Senna occidentalis *(L.) Link*, 60
Senna petersiana *(Bolle) Lock*, *58*, 59
Senna ruspolii *(Chiov.) Lock*, 61
Senna septemtrionalis *(Viv.) Irwin & Barneby*, 59
Senna siamea *(Lam.) Irwin & Barneby*, 60
Senna singueana *(Del.) Lock*, 59
Senna sophera *(L.) Roxb.*, 60
Senna tomentosa Batka, 59
Senna truncata *(Brenan) Lock*, 61
SESBANIA *Scop.*, 139
Sesbania confaloniana (Chiov.) Chiov., 140
Sesbania dummeri *Phil. & Hutch.*, 139
Sesbania goetzei *Harms*, 140, *141*
Sesbania grandiflora *(L.) Poir.*, 139
Sesbania hepperi *Gillett.*, 142
Sesbania leptocarpa *DC.*, 140, 142
var. *confaloniana* Chiov., 140
Sesbania pachycarpa *DC.*, 142
Sesbania punctata sensu Cuf., 140
Sesbania quadrata *Gillett*, 140, *141*
Sesbania rostata *Brem. & Oberm.*, 140
Sesbania sericea *(Willd.) Link*, 142
Sesbania sesban *(L.) Merr.*, 140
subsp. punctata (DC.) Gillett, 140
var. **nubica** *Chiov.*, 140, *141*
var. sesban, 140
var. zambesiaca Gillett, 140
Sesbania somalensis *Gillett*, 140
Sesbania tetraptera *Bak.*, *141*, 142
SIMAROUBACEAE, 437
Simirestis goetzei (Loes.) N. Hallé ex R. Wilczek, 340
SMITHIA *Ait.*, 146
Smithia abyssinica *(A. Rich.) Verdc.*, *145*, 146
Smithia elliotii *Bak.f.*, 146
var. **elliotii.**, 146
var. **sparse-strigosa** *Verdc.*,146
Smithia erubescens sensu Cuf., 146
Smithia sensitiva Ait., 146
var. *abyssinica* A. Rich., 146
Sophora aurea Ait., 104
SPARTIUM *L.*, 223
Spatium junceum *L.*, 223
Sphaeridiophorum abyssinicum Jaub. & Spach, 123
SPHENOSTYLIS *E. Mey.*, 178
Sphenostylis stenocarpa *(Hochst. ex A. Rich.) Harms*, *175*, 178
Spondias birrea A. Rich., 519
SPONDIOIDEAE *Engl. ex Wild.*, 447
Stephanopodium, 47
STYLOSANTHES *Sw.*, 147
Stylosanthes flavicans Bak., 147
Stylosanthes fruticosa *(Retz.) Alston*, 149
SYCIDIUM *Miq.*, 281
SYCOMORUS *(Gasparrini) Mildbr. & Burret*, 285
SYCOMORUS Gasparrini, 285
Sycomorus antiquorum Gasparrini, 285
Sycomorus capensis (Thunb.) Miq., 287
Sycomorus gnaphalocarpa Miq., 285
Sycomorus panifica (Del.) Miq., 287
Sycomorus rigida Miq., 285
Sycomorus riparia Miq., 287
Sycomorus schimperianus Miq., 289
Sycomorus sur (Forssk.) Miq., 287
Sycomorus trachyphylla Miq., 285
TAMARINDUS *L.*, 66
Tamarindus erythraea Mattei, 66
Tamarindus indica *L.*, 66, *67*
var. *emarginata* Chiov., 66
Tamarindus somalensis Mattei, 66
TAPINANTHUS *(Blume) Reichb.*, 372
Tapinanthus globiferus *(A. Rich.) Tieghem*, 373, *373*
Tapinanthus heteromorphus *(A. Rich.)*, 373, *373*
Tapinanthus platyphyllus (A. Rich.), 372
Tapinanthus sp. = *Westphal* 635, 374
Tapinostemma (Benth. & Hook.) Tieghem, 262
Tapinostemma acaciae (Zucc.) Tieghem, 363
Tapinostemma meridianum Danser, 363
TAPURA *Aubl.*, 47
Tapura fischeri *Engl.*, 47, *48*
TAVERNIERA *DC.*, 228
Taverniera abyssinica *A. Rich.*, 228, *229*
Taverniera aegyptiaca *Boiss.*, 228
Taverniera floribunda Schweinf., 228
Taverniera lappacea *(Forssk.) DC.*, 228
Taverniera schimperi *Jaub. & Spach*, 228
Taverniera stefaninii Chiov., 228

TECLEA *Del.*, 424, 426
Teclea acuminata Engl., 427
Teclea borenensis *M. Gilbert*, 427, *428*
Teclea diversifolia Mattei, 427
Teclea hanangensis Kokwaro, 427
Teclea nobilis *Del.*, 427, *428*
Teclea pilosa Verdoorn, 426
Teclea salicifolia Engl., 427
Teclea simplicifolia *(Engl.) Verdoorn*, 427
TEPHROSIA *Pers.*, 109
Tephrosia anthylloides Hochst. ex Webb, 112
Tephrosia apollinea *(Del.) Link*, 116
var. *subuniflora* A. Terracc., 112
Tephrosia boranensis Chiov., 113
Tephrosia bracteolata *Guill. & Perr.*, 116, *117*
var. *bracteolata*, 116
var. strigulosa Brummitt, 116
Tephrosia brunnea Bak., 113
Tephrosia decidua Steud. ex A. Rich., 112
Tephrosia dichroocarpa *Steud. ex A. Rich.*, 113
Tephrosia ehrenbergiana Schweinf. 116
Tephrosia elata *Deflers*, 118
var. **elata**, 118
var. **tomentella** *Brummitt*, 118
Tephrosia emeroides *A. Rich.*, 113, *114*
Tephrosia encoptosperma Schweinf., 112
Tephrosia franchetii Hutch. & Bruce, 111
Tephrosia fulvinervis *Hochst. ex A. Rich.*, 113
Tephrosia gracilipes *Guill. & Perr.*, 115
Tephrosia graminifolia Chiov., 112
Tephrosia heterophylla Vatke, 111
Tephrosia hildebrandtii *Vatke*, 113
Tephrosia hochstetteri *Chiov.*, 118-b
Tephrosia holstii *Taub.*, 113, *114*
Tephrosia humilis *Guill. & Perr.*, 119
Tephrosia interrupta *Hochst. & Steud. ex Engl.*, *117*, 118
subsp. *interrupta*, 118
Tephrosia linearis *(Willd.) Pers.*, 113, *114*
Tephrosia lortii *Bak.f.*, 115
Tephrosia nana *Schweinf.*, 118
Tephrosia nubica *(Boiss.) Bak.*, 116
subsp. arabica (Boiss.) Gillett, 116
var. *abyssinica* (Jaub. & Spach) Schweinf., 116
var. *polyphylla* Chiov., 116
Tephrosia obbiadensis *Chiov.*, 112, *114*
Tephrosia pentaphylla *(Roxb.) G. Don*, *110*, 112
Tephrosia polyphylla *(Chiov.) Gillett*, 116-b
Tephrosia pumila *(Lam.) Pers.*, *114*, 115
Tephrosia purpurea *(L.) Pers.*, 115
subsp. dunensis Brummitt, 116
subsp **leptostachya** *(DC) Brummitt*, 115
var. **leptostachya**, 115
var. **pubescens** *Bak.*, 115
subsp. purpurea, 115
Tephrosia quartiniana Cuf., 112
var. *inflexa* (Chiov.) Cuf., 115
Tephrosia reptans *Bak.*, 118
var. arenicola Brummitt & Gillett, 118
var. *reptans*, 118
Tephrosia rivae Harms, 113
Tephrosia schweinfurthii *Deflers*, *110*, 111
Tephrosia senticosa sensu Cuf., 112
Tephrosia similis Chiov., 112
Tephrosia simplicifolia Franch., non F. Muell. ex Benth., 111
Tephrosia subtriflora *Hochst. ex Bak.*, *110*, 112
Tephrosia sulphurea Chiov., 112
Tephrosia transjubensis Chiov., 112
Tephrosia uniflora *Pers.*, 112
subsp. petrosa (Blatt. & Hall.) Gillett & Ali, 112
subsp. *uniflora*, 112
var. *parvifolia* Bak.f., 112
Tephrosia ustulata Taub. ex Harms, 113
Tephrosia vicioides A. Rich. non Schlecht., 112
var. *inflexa* Chiov., 115
Tephrosia villosa *(L.) Pers.*, 116
subsp. **ehrenbergiana** *(Schweinf.) Brummitt*, 116, *117*
subsp. villosa, 116
Tephrosia vogelii *Hook.f.*, 119
TERAMNUS *P. Br.*, 158
Teramnus gracilis Chiov., 160
Teramnus labialis *(L.f.) Spreng.* 160
subsp. **arabicus** *Verdc.*, 160
subsp. **labialis**, 160
var. **abyssinicus** *(Hochst. ex A. Rich.) Verdc.*, *159*, 160
var. acutus Verdc., 160
var. labialis, 160
Teramnus micans *(Bak.) Bak.f.*, 160
var. **cyaneus** *(De Wild.) Hauman*, 161
var. fagifolius Hauman, 161
var. micans, 161
Teramnus repens *(Taub.) Bak.f.*, 160
subsp. **gracilis** *(Chiov.) Verdc.*, 160
subsp. repens, 160
Teramnus uncinatus *(L.) Sw.*, 160
subsp. **axilliflorus** *(Kotschy) Verdc.*, 160
subsp. **ringoetii** *(De Wild.) Verdc.*, *159*, 160
subsp. uncinatus, 160
Tetrapanax papyrifera (Hook.) K. Koch., 537
Tetraspis ruspoli Chiov., 439
THESIUM *L.*, 379
Thesium hararensis *A.G. Miller*, 380
Thesium kilimandscharicum *Engl.*, 379, *381*
Thesium matteii *Chiov.*, 380, *381*
Thesium radicans *A. Rich.*, 380, *381*
Thesium schimperianum A. Rich., 382
Thesium schweinfurthii *Engl.*, 380, *381*
Thesium stuhlmannii *Engl.*, 380
Thesium subaphyllum *Engl.*, 379, *381*
Thesium wightianum Wall., 380
var. *radicans* DC., 380

THONNINGIA *Vahl*, 384
Thonningia sanguinea *Vahl*, 384, *384*
Thonningia ugandensis Hemsl., 384
TILLAEA L., 6
Tillaea alata Viv., 8
Tillaea alsinoides Hook.f., 6
Tillaea aquatica sensu Britten non L., 6
TODDALIA *Juss*., 422, 424
Toddalia aculeata Pers, 424
Toddalia asiatica *(L.) Lam*., 424
Toddalia glomerata F. Hoffm., 426
Toddalia simplicifolia Engl., 427
var. *eugeniifolia* Engl., 426
Toddaliae, 424
Tonsella africana Willd., 342
TREMA *Lour*., 268
Trema orientalis *(L.) Bl.*, 268, *269*
Trema guineensis (Schumach. & Thonn.) Ficalho, 268
var. *asperata* (Solms) Cuf., 268
var. *hochstetteri* (Planch.) Engl., 268
TREMATOSPERMA Urb., 351
Trematosperma cordatum Urb., 351
TRICHILIA *Browne*, 481
Trichilia bilocularis Pax, 485
Trichilia dregeana *Sond.*, 483, *484*
Trichilia emetica *Vahl*, 483, *484*
Trichilia jubensis Chiov., 483
Trichilia prieuriana *A. Juss.*, 483, *484*
Trichilia retusa *Oliv.*, 483
Trichilia roka (Forssk.) Chiov., 483
Trichilia rueppelliana Fresen., 489
Trichilia somalensis Chiov., 483
Trichilia siderotricha Chiov., 440
Trichilia volkensii, 485
TRICHOCLADUS *Pers*., 252
Trichocladus ellipticus *Eckl. & Zeyh*.,252
subsp. **malosanus** *(Bak.) Verdc*., 252, *253*
TRIFOLIUM *L*., 233
Trifolium abyssinicum Heller, 240
Trifolium acaule *Steud. ex A. Rich*., 240
var. *emarginatum* Chiov., 241
Trifolium arvense *L*., 242
Trifolium baccarinii *Chiov*., 240
Trifolium basileianum Chiov., 238
Trifolium bellianum Chiov., 242
Trifolium bilineatum *Fresen*., 238, *239*
Trifolium brunellii Chiov., 237
Trifolium burchellianum *Ser*., 237
subsp. burchellianum, 238
subsp. **johnstonii** *(Oliv.) Cuf. ex Gillett*, 237
var **johnstonii**, 238, *239*
var. **oblongum** *Gillett*, 238
Trifolium calocephalum *Fresen*., *239*, 240
var. *parviflorum* Chiov., 240
Trifolium campestre *Schreb*., 242
Trifolium chilaloense *Thulin*, 241
Trifolium cryptopodium *Steud. ex A. Rich*., 236
Trifolium decorum *Chiov*., 241, *243*
Trifolium elgonense *Gillett*, 241
Trifolium fragiferum *L*., 242
Trifolium lanceolatum *(Gillett) Gillett*, 240
Trifolium marginatum (Hochst. & Steud. ex Bak.) Cuf., 240
Trifolium mattiroli anum *Chiov*., 238-b
Trifolium mauginianum Fiori, 236
Trifolium multinerve *A. Rich*., 241, *243*
var. *debilior* A. Rich., 241
Trifolium petitianum A. Rich., 241
var. *cuneatum* Bak., 241
Trifolium pichisermollii *Gillett*, 238
Trifolium polystachyum *Fresen*., *235*, 236
var. *contractum* Lanza, 236
Trifolium pratense *L*., 242
Trifolium pseudocryptopodium Chiov. ex Fiori, 236
Trifolium quartinianum *A. Rich*., 242, *243*
Trifolium repens *L*., 237
Trifolium resupinatum *L*., 242
Trifolium rueppellianum *Fresen*, 238, *239*
var. *subrotundum* (Steud. & Hochst. ex A. Rich.) Fiori, 238
Trifolium schimperi *A. Rich*., 241, *243*
Trifolium semipilosum *Fresen*., 237
var. **brunellii** *Thulin*, 237
var. **intermedium** *Thulin*, 237
var. *kilimanjaricum* Bak.f., 237
var. *microphyllum* Chiov., 237
var. **semipilosum**, *235*, 237
var. *sennii* Chiov., 237
Trifolium simense *Fresen*., 236
forma *albiflorum* Piovano, 236
Trifolium somalense *Taub*., 237
Trifolium spananthum *Thulin*, 240
Trifolium sp. = *Burger* 3824
Trifolium sp. A. sensu Fröman, 240
Trifolium sp. B. sensu Fröman, 241
Trifolium sp. C. sensu Fröman, 237
Trifolium sp. E. sensu Fröman, 236
Trifolium steudneri *Schweinf*., 242, *243*
Trifolium stolzii sensu Cuf., 236
Trifolium subrotundum Steud. & Hochst. ex A. Rich., 238
Trifolium subterraneum *L*., 242
Trifolium tembense *Fresen*, *239*, 240
Trifolium umbellatum A. Rich., 240
Trifolium usambarense Taub., 236
Trifolium wentzelianum Harms, 236
TRIGONELLA *L*., 246
Trigonella foenum-graecum *L*., 246, *247*
Trigonotheca serrata Hochst., 340
TRILEPISIUM *Thouars*, 299
Trilepisium madagascariense *DC*., 299, *300*
TURRAEA *L*., 479
Turraea abyssinica *Hochst. ex A. Rich.*, 481, *482*
var. *longipedicellata* Oliv., 481

Turraea holstii *Gürke*, 481, *482*
Turraea mobassana *Hiern ex C. DC.*, 481, *482*
Turraea nilotica *Kotschy & Peyr.*, 481, *482*
Turraea parvifolia *Deflers*, 481, *482*
TYLOSEMA *(Schweinf.) Torre & Hillc.*, 70
Tylosema argentea *(Chiov.) Brenan*, 70
Tylosema fassoglensis *(Kotschy ex Schweinf.) Torre & Hillc.*, *69*, 70
Tylosema humifusa (Pic.-Serm. & Roti-Michelozzi) Brenan, 70
Tzellemtinia nervosa Chiov., 398
ULMACEAE, 266-b
UMBILICUS *A. DC.*, 16
Umbilicus botryoides *A. Rich.*, 16, *18*
Umbilicus paniculiformis Wickens, 16
Umbilicus rupestris (Salisb.) Riddelsd., 16
Umbilicus semiensis A. Rich., 15
URARIA *Desv.*, 152
Uraria picta *(Jacq.) DC.*, 152
URERA *Gaud.*, 304
Urera cameroonensis Wedd., 304
Urera hypselodendron *(A. Rich.) Wedd.*, 304, *305*
Urera schimperi Wedd., 304
Urera trinervis *(Hochst.) Friis & Immelman*, 304
UROSTIGMA *(Gasparrini) Miq.*, 290
UROSTIGMA Gasparrini, 290
Urostigma abutilifolium Miq., 296
Urostigma acrocarpum Miq., 298
Urostigma catalpaefolium Miq., 295
Urostigma dekdekena Miq., 298
Urostigma fazoklense Miq., 293
f. *minor* Miq., 293
Urostigma glumosa (Del.) Miq., 293
Urostigma hochstetteri Miq., 299
Urostigma ingens Miq., 291
Urostigma kotschyanum Miq., 296
Urostigma populifolium (Vahl) Miq., 295
Urostigma salicifolia (Vahl) Miq., 290
Urostigma schimperi Miq., 298
Urostigma thonningii, 298
Urostigma vogelii Miq., 294
Urostigma xanthophylla Miq., 291
URTICA *L.*, 303
Urtica adoensis Steudel, 308
Urtica aestuans L., 309
Urtica bicolor Roxb., 320
Urtica bullosa Steudel, 306
Urtica caffra Thunb., 325
Urtica condensata Steudel, 308
Urtica dioica *L.*, 304
Urtica diversifolia Link, 308
Urtica hypselodendron A. Rich., 304
Urtica iners Forssk., 322
Urtica interrupta L., 308
Urtica lomatocarpa Steudel, 308
Urtica parasitica Forssk., 317
Urtica pauciflora Steudel, 322
Urtica schimperiana Steudel, 309
Urtica simensis *Steudel*, 303, *303*
Urtica tetraphylla Steudel, 310
Urtica urens *L.*, 303
URTICACEAE, 302
VAHLIA *Thunb.*, 28
Vahlia dichotoma (Murray) Kuntze, 28
Vahlia digyna *(Rets.) Kuntze*, 28, *29*
Vahlia geminiflora *(Del.) Bridson*, 28, *29*
Vahlia somalensis *Chiov.*, 30
subsp. **goddingii** *(Bruce) Bridson*, *29*, 30
subsp. **somalensis**, 30
VAHLIACEAE, 28
VÁTOVAEA *Chiov.* 168
Vátovaea biloba Chiov., 169
Vátovaea pseudolablab *(Harms) Gillett*, 169, *170*
VENTILAGO *Gaertn.*, 396
Ventilago diffusa *(G. Don) Exell*, 398, *398*
VEPRIS *A.Juss.*, 424
Vepris angolensis Engl., 422
Vepris dainellii *(Pichi-Serm.) Kokwaro*, 424, *425*
Vepris eugeniifolia *(Engl.) Verdoorn*, 426
Vepris glomerata *(F. Hoffm.) Engl.*, 426
var. **glabra** *Kokwaro*, *425*, 426
var. **glomerata**, *425*, 426
Vepris pilosa Engl., 426
Vepris sp. = *Burger* 3072, 426
Verea lanceolata (Forssk.) Spreng., 24
VERMIFRUX *Gillett*, 227
Vermifrux abyssinica *(A. Rich.) Gillett*, 227
VICIA *L.*, 248
Vicia abyssinica Alef., 249
Vicia cuneata Guss., 249
Vicia faba *L.*, 249
Vicia hirsuta *(L.) S. F. Gray*, 249
Vicia narbonensis L., 248
Vicia paucifolia *Bak.*, 248, *251*
subsp. *paucifolia*, 249
Vicia sativa *L.*, 249
subsp. *nigra* (L.) Ehrh., 249
var. *abyssinica* (Alef.) Bak., 249
var. **angustifolia** *L.*, 249
var. *aristulata* Chiov., 249
var. **sativa**, 249
Vicia segetalis Thuill., 249
Vicia villosa *Roth*, 248-b
VIGNA *Savi*, 171
Vigna abyssinica Taub., 173
Vigna aconitifolia *(Jacq.) Maréchal*, 177
Vigna ambacensis *Bak.*, 173
Vigna caesia Chiov., 174
Vigna chiovendae, Bak.f., 173
Vigna crinita A. Rich., 176
Vigna debanensis *Martelli*, 176-b
Vigna dekindtiana Harms, 174
Vigna esculenta (De Wild.) De Wild., 174
Vigna fischeri *Harms*, 172

Vigna friesiorum *Harms*, 174
var. ulugurensis (Harms) Verdc., 174
Vigna frutescens *A. Rich*., 174
subsp. **frutescens**, 176
subsp. **kotschyi** *(Schweinf.) Verdc.*, 176
var. buchneri (Harms) Verdc., 176
Vigna glandulosa Chiov., 174
Vigna gracilis sensu Cuf., 173
Vigna heterophylla *A. Rich*., 173
Vigna kotschyi Schweinf., 176
Vigna lancifolia A. Rich., 172
Vigna luteola *(Jacq.) Benth*., 172
Vigna macrorhyncha *(Harms) Milne-Redh*., 177
Vigna membranacea *A. Rich*., 173
subsp. **caesia** *(Chiov.) Verdc.*, 174
subsp. **membranacea**., 174
Vigna mensensis Schweinf., 174
var. *hastata* Chiov., 173
Vigna micrantha Chiov. non Harms, 173
Vigna monophylla *Taub*., *175*, 177
Vigna multinervis *Hutch. & Dalz*., 173, *175*
Vigna neumannii Harms, 176
Vigna oblongifolia *A. Rich*., 172
var. **oblongifolia**., 172
var. **parviflora** *(Welw.) ex Bak.) Verdc.*, 172
Vigna opisotricha A. Rich., 177
Vigna ostinii Chiov., 177
Vigna parkeri *Bak*., 173
subsp. **maranguensis** *(Taub.) Verdc.*, 173
subsp. parkeri, 173
Vigna parviflora Welw. ex Bak., 172
Vigna proboscoidella Chiov., 177
Vigna pseudolablab Harms, 169
Vigna radiata *(L.) Wilczek*, 177
var. **radiata**., 177
var. **sublobata** *(Roxb.) Verdc.*, 177
Vigna reticulata *Hook. f*., 176
Vigna schimperi *Bak*., 172, *175*
Vigna spartioides Taub., 176
Vigna subterranea *(L.) Verdc*., 173
Vigna triphylla *(Wilczek) Verdc*., 177
Vigna tuberosa A. Rich., 176
Vigna unguiculata *(L.) Walp*., 174
subsp **mensensis** *(Schweinf.) Verdc.*, 174
subsp. **cylindrica** *(L.) Verdc.*, 174
subsp. **dekindtiana** *(Harms) Verdc.*, 174
subsp. **sesquipedalis** *(L.) Verdc.*, 174
subsp. **unguiculata**, 174
Vigna vexillata *(L.) A. Rich*. *175*, 176
var. *angustifolia* (Schumach. & Thonn.) Bak., 176
var. *tuberosa* (A. Rich.) Chiov., 176
var. *vexillata*, 176
VISCACEAE, 361, 375
VISCUM *L*., 375
Viscum bagshawei *Rendle*, 377, *377*
Viscum congolense *De Wild.*, 376, *376*
Viscum grandifolium Engl., 376
Viscum japonicum Thunb., 377
Viscum nervosum A. Rich., 376
var. *angustifolium* Sprague, 376
Viscum opuntia Thunb., 377
Viscum schimperi *Engl*., 377, *377*
Viscum schimperi Sprague, 377
Viscum semiplanum (Tieghem) Engl., 377
Viscum taenioides A. Rich., 377
Viscum triflorum *DC*., 376
subsp. **nervosum** *(A. Rich.) M. Gilbert*, 376, *376*
subsp. triflorum, 376
Viscum tuberculatum *A. Rich*., 375, *376*
forma *obovatifolia* Schweinf., 375
forma *obovato-oblongifolium* Schweinf., 375
VITACEAE, 399
VITIS *L*., 399
Vitis abyssinicus Hochst. ex A. Rich., 401
Vitis cornifolia Bak., 405
Vitis dubia Beccari ex Martelli, 406
Vitis erythrodes Fresen., 401
Vitis hochstetteri Miq., 406
Vitis oxyphylla A. Rich., 412
Vitis pannosa Bak., 415
Vitis sambucina Beccari ex Martelli, 410
Vitis serpens Hochst. ex A. Rich, 411
var. *mollis* Bak.
Vitis schimperiana Hochst. ex A. Rich., 400
Vitis vinifera *L*., 399
Voandzeia subterranea (L.) DC., 173
Xerocarpus abyssinicus (Hochst. ex A. Rich.) Schweinf., 195
XIMENIA *L*., 356
Ximenia aegyptiaca L., 433
Ximenia americana *L*., 356, *357*
var. *inermis* Hochst., 356
Ximenia caffra *Sond*., 356
var. **caffra**., 356, *357*
var. **natalensis** *Sond*., 356, *357*
Ximenia laurina Del., 356
ZANHA *Hiern*, 493
Zanha golungensis *Hiern*, 493, *494*
ZANTHOXYLUM *L*., 420
Zanthoxylum chalybeum *Engl*., 420
var. **chalybeum**, 420, *421*
var. mollis Kokwaro, 420
Zanthoxylum leprieurii *Guill. & Perr*., 420
Zanthoxylum usambarense *(Engl.) Kokwaro*, *421*, 422
ZIZIPHUS *Mill*., 393
Ziziphus abyssinica *Hochst. ex A. Rich*., *395*, 396, *397*
Ziziphus adelensis Del., 393
Ziziphus hamur *Engl*., *395*, 396
Ziziphus jujuba Lam., 396
var. *abyssinica* (Hochst. ex A. Rich.) Fiori, 396
Ziziphus jujuba (L.) Gaertn., 396
forma *obliquifolia* Engl., 396
Ziziphus mauritiana *Lam*., *395*, 396
Ziziphus mitis A. Rich., 393

Ziziphus mucronata *Willd.*, 393
subsp. **mucronata**, 395, *395*
subsp. rhodesica Drummond, 395
Ziziphus pubescens *Oliv.*, 393, *395*
Ziziphus spina-christi *(L.) Desf.*, 395, *395*
var. *microphylla* Hochst., 395
var. *mitissima* Chiov., 395
ZORNIA *J.F. Gmel.*, 147
Zornia apiculata *Milne-Redh.*, 148
Zornia glochidiata *Reichb. ex DC.*, 148
Zornia pratensis *Milne-Redh.*, 148, *149*
Zornia setosa *Bak.f.*, 148
subsp. **obovata** *(Bak.f.) J. Léon. & Milne-Redh.*, 148
subsp. setosa, 148
Zygia sassa Benth. ex Schweinf., 96

edited by
Tewolde Berhan Gebre Egziabher
and
the staff of the National Herbarium, Addis Ababa.

INTRODUCTION

The following vernacular names are given in both Ge'ez (Ethiopic) and Roman (Latin) scripts.

Names in Ge'ez Script

There is some variation among languages in the way that the Ge'ez script is normally used to represent sounds. An effort has been made to avoid the possible use of a Ge'ez letter (fidel) in more than one way. This has been found desirable for ease of reading the names in the various languages without having to keep being constrained by the identity of the language. To make this possible, the following conventions and simplifications have been adopted throughout.

Consistency of the first form of a "fidel".

The first forms of the "fidels" ሀ, ሐ, ኀ, አ, ዐ are usually read as if they were the fourth forms, i.e. ሃ, ሓ, ኃ, ኣ, ዓ, and not in a consonant-vowel combination consistent with the other "fidels", ለ, መ, ሰ, በ, etc. as contrasted with ላ, ማ, ሳ, ባ, etc. There has been a recent tendency to disregard this old convention and to treat the five "fidels" ሀ, ሐ, ኀ, አ, ዐ as if they were like the rest. This convention has been adopted here throughout. The "fidel" ሀ (he) is thus read the way ኸ is normally read in Amharinya, and ሄ is often forced to double in being read in Tigrinya, and the "fidel" አ now replaces ኧ in Amharinya and the second use of ኤ in Tigrinya. This frees the "fidel" ኸ to be used exclusively as it is used in Tigrinya.

Avoidance of redundancy

In the southern Semitic languages of Ethiopia (Amharinya and Guragenya) but not in Aderienya, the typically Semitic gutteral sounds (ሐ, ኸ, ዐ) and one palatal sound (ቐ) have been dropped. In Ge'ez, it is believed that the "fidel" ኀ was what is now written in Tigrinya and the other Semitic languages, which retain the gutteral sounds, as ኸ. It is also believed that the Ge'ez ሠ was what has now come to be written as ሸ. For the sake of cross-language consistency, the presumed old Ge'ez sounds of these "fidels" have been adopted here with ሠ being replaced be ሸ, and ኀ by ኸ. The fidel ሠ has thus been totally dropped, though ኀ has been retained as standing for the sound that it is redundantly used for in Amharinya, i.e. as ሀ, but only in the dipthong form ኊ. Of the two identical sounding "fidels", ጸ and ፀ, the second has been dropped except when writing Ge'ez names, when the script has been left unmodified, i.e. as it is normally used in Ge'ez.

Representation of some sounds not found in old Ge'ez

The "d" sound in Orominya and other Cushitic languages, which is produced while air is being rushed into the mouth, is represented by the "fidel" ዸ. The long "e" of Ormominya is represented by the first form of a "fidel" and the fifth form has been excluded from this language.

No effort has been made to represent some complex phonemes found in some non-Semitic languages, e.g. ተዘ (tz) and ተሰ (ts) in Gamunya, by individual "fidels"; they have been represented by combinations of "fidels".

Long and short vowels
In many of the Ethiopian languages (e.g. Orominya, Somalinya, Tigre), there are long and short vowels. It has not been possible to represent "fidels" to be read with long vowels as distinct from those to read with short vowels.

Stress
In the Ge'ez script, no effort has been made to distinguish stressed from non-stressed "fidels", though the usual doubling of consonants has been adopted to represent this in the Roman script. In ignoring to show stress in the Ge'ez script, the prevalent convention has been followed though this is not to be taken to mean that we belittle the useful convention of marking stressed "fidels" through placing points over them. We have not adopted this useful system because, using present day computer software, we could not consistently apply it to all "fidels".

Correspondence between the Ge'ez "fidels" and the Roman script.
A number of sounds found in the Ethiopian languages are often represented by combinations of Roman letters when this is done with a European readership in mind, for example she (ሸ), che or tche (ቸ or ጨ). Some transliteration letters are often used for more than one Ethiopian sound, particularly a (አ, ዓ), ha (ሀ, ሐ, ኀ), ka (ከ, ቀ, ቐ). To avoid ambiguities, the following equivalences have strictly been followed.

Vowels: The sixth forms of the Ge'ez glottal stops, እ and ዕ, are represented by ' and ‘ respectively. The sixth forms of all other "fidels" are each represented by the consonant or the group of consonants in the case of sh (ሽ) standing for the sound without a vowel following it. The other forms are each represented by the appropriate consonant followed by -e for the first, -u for the second, -i for the third, -a for the fourth, -ie for the fifth and -o for the seventh forms. The two glottal stops አ and ዐ, when varying their forms, are represented the former with the vowels unmodified, and thc latter with the vowels modified.
The following four examples will illustrate this:

Form	1st	2nd	3rd	4th	5th	6th	7th
Ge'ez	ለ	ሉ	ሊ	ላ	ሌ	ል	ሎ
Roman	le	lu	li	la	lie	l	lo
Ge'ez	ሀ	ሁ	ሂ	ሃ	ሄ	ህ	ሆ
Roman	he	hu	hi	ha	hie	h	ho
Ge'ez	አ	ኡ	ኢ	ኣ	ኤ	እ	ኦ
Roman	e	u	i	a	'ie	'	o
Ge'ez	ዐ	ዑ	ዒ	ዓ	ዔ	ዕ	ዖ
Roman	ē	ū	î	ā	‘ie	‘	ō

The representation in Roman script of all the simple "fidels" in their sixth form is given as follows:

ሕ-', ዕ-', ብ-b, ች-c, ጭ-ç, ድ-d, ዽ-ḋ, ፍ-f, ግ-g, ህ-h, ሕ-ḣ, ጅ-j, ክ-k, ል-l, ም-m, ን-n, ኝ-ñ, ፕ-p, ጵ-ṗ, ቅ-q, ቕ-q̇, ር-r, ስ-s, ጽ-ṣ, ት-t, ጥ-ṭ, ው-w, ይ-y, ዝ-z, ዥ-ż.

The representation of the dipthong "fidels" in their fourth forms is given as follows:

ቧ-bwa, ቿ-cwa, ጯ-çwa, ዷ-dwa, ፏ-fwa, ጓ-gwa, ኋhwa ጇ-jwa, ኳ-kwa, ዃ-k̇wa, ሏ-lwa, ሟ-mwa, ኗ-nwa, ኟ-ñwa, ቋ-qwa, ቛ-q̇wa, ሯ-rwa, ሷswa, ሿ-shwa, ቷ-twa, ጧ-ṭwa, ዟ-zwa.

<u>The languages featuring in this list of vernacular names</u>.
English is the only widely spoken international language included - all the other languages are found as first languages within Ethiopia.

Owing to the weakness of the data collected on vernacular names, not all languages spoken in Ethiopia have been included. Some languages are found only a few times. This is not to be taken to mean that these languages do not have many vernacular names; it means that only a few plant names in these languages have been collected.

Our data base is weak even in the three most widely spoken major languages, Amharinya, Orominya and Tigrinya. The user of this Flora will find that there are many more vernacular names in use in the various languages that have been included here. We appeal to all of you to help us strengthen our data base by sending us verncular names of identified plants and, whenever possible, sending us specimens of the plants as well. The languages for which names occur are as follows with the abbreviated form as found in the lists followed by the full name for that language.

Ade - Aderienya
Afar - Afarinya
Agew - Agewinya
Alaba - Alabanya
Amh - Amharinya
Anu - Anuwakinya
Bil - Bileninya
Bod - Bodinya
Bur - Burjinya
Das - Dasenechinya
Dor - Dorzienya
Eng - English
Gam - Gamonya
Ge'ez - Ge'ez
Gim - Gimiranya
Gof - Gofanya
Gum - Gumuzinya
Gur - Guragenya
Had - Hadiyinya
Jan - Janjeronya
Kef - Kefinya
Kem - Kembatinya
Kon - Konsonya
Kun - Kunaminya
Mao - Maonya
Me - Me'eninya
Mes - Mesengonya
Mur - Mursinya
Or - Orominya
Saho - Sahonya
Sha - Shakoniya
Sid - Sidamonya
Som - Somalinya
Tre - Tigre
Tya - Tigrinya
Wel - Welaytinya

ARRANGED ALPHABETICALLY ACCORDING TO SCIENTIFIC NAMES
(with page number(s) to the main text)

Abrus precatorius (107): coral pea, crab's eyes, Indian liquorice, jequerity seeds, lucky or paternoster beans (Eng).

Acacia abyssinica (89): ዓለ ālla (Tya); አምቦ ambo (Or); በዝራ:ግራር bazra-grar (Amh); ጩ ዓ çeā (Tya); ዳደቻ dadecca, ዳግኒሳ dagnisa, ዴዴቻ dedecca, ጋርቢ garbi, ለፍቶ lafto (Or); ኦዳራ odara (Kem); ኦዶሮ odoro, ኦዶርዋ odorwa (Wel); ቅጽልቶ qṣlto (Afar); ሶ ን ዲ sondi (Or).

Acacia albida (84): apple-ring acacia (Eng); ዓቅባ āqba (Tre); ደሮት derot, ጋርቢ garbi (Or); ጋርሻ garsha (Tya); ገርቢ gerbi (Amh & Gur); ማማን maman (Saho); ሞሞና momona (Tya); ቀረቶር qeretor (Gam).

Acacia amythethophylla (86): ለፍቶ lafto, ዋኮ wakko (Or).

Acacia asak (80): ዓዳድ ādad, ዓዳድ:መዱ ādad-medu, ጕዳር gwidar (Som); ሐቅ ḥaq (Tre); ቀንጥቢ qenṭbi (Tya); ረደ rede (Som); ጸለዋ ṣellewa, ጽፍሪ:ድሙ ṣfri-dmmu (Tya); ስባንሳ sbansa (Amh).

Acacia bavazzanoi (90): ጩዓ çeā (Tya); ቋሚ qwami (Amh).

Acacia brevispica (78): አመዛዘ amezaze (Or); መዘዝዮ mezezyo (Amh); ጐ ም ኦ ሪ ያ gwemöriyya (Wel); ቆንጥር qonṭr (Amh); ቆርቆር qorqor (Som); ቋንጣ qwanṭa, ቍ ን ጥ ር qwenṭr (Amh).

Acacia bricchettiana (91): ከየአክ keyeäk (Som).

Acacia bussei (85): ጋሎኦል galoöl (Som); ጕመሮ gwmero (Tya); ሃሎ halo (Or).

Acacia dealbata (92): blue wattle, silver wattle (Eng).

Acacia decurrens (92): green wattle (Eng).

Acacia dolichocephala (85): ጉጋንታ guganta (Wel); ለፍቶ lafto (Or).

Acacia drepanolobium (88): ጫጫ çaça (Wel); ፉልሳ fullsa (Or); whistling thorn (Eng).

Acacia edgeworthii (91): ጉሉ gulu (Som).

Acacia ehrenbergiana (86): ዓቅባ āqba, ቀይሕ:ዓቅባ qeyyḥ-āqba (Tre).

Acacia etbaica (88): አጆ ajjo (Alaba); ጭጊንዳ çginda (Wel); ደሬ derie (Amh); ዶዶታ dodota, ዶዶቲ dodoti, ዶዶቲአ dodotia (Or); ዶረት doret (Amh);ቀረጽ qereṣ (Tre); ቀረጣ qereṭa (Amh); ቍርዓች qwrāc (Som); ሰራው seraw (Saho, Tre & Tya); ሱግሱግ sugsug, ዩባ yuba (Som).

Acacia gerrardii (88): ዶዶታ dodota, ዶዶቲ dodoti (Or).

Acacia hamulosa (80): ቅትሪ qtri (Tre); ያለፎን yalefon (Som).

Acacia hockii (86): ጭጊንዳ çginda (Gam); ጥምብቃ ṭmbbqqa (Amh).

Acacia horrida (84): ቻቻና cacana (Or); ሰርማን serman (Som).

Acacia lahai (85): ቡርቁቀ burquqqe, ደሮት derot, ጋርቢ garbi (Or); ጉጋንታ guganta (Wel); ለፍቶ lafto (Or); ለሐይ laḥay (Tya); ቀረቶር qeretor (Gam); ሶንዲ sondi (Or); ው ቴ wttie (Amh).

Acacia mearnsii (91): black wattle (Eng).

Acacia mellifera (81): አዳል adal (Som); አትንኩይ atnkuy (Amh); ብላዓል blāl (Som); ሃራንጋማ harangama (Or); ቀዳድ qedad (Tre); ቀንጥብ qenṭb, ቀንጥቢ qenṭbi, ቀን ጢባ qenṭiba (Tya); ቀረጽ qereṣ, ቆኖ qono (Tre); ቆንጥር qonṭr, ቍንጥር qwenṭr (Amh); ሳባንሳ sabansa (Or); ስባንሳ sbansa (Amh); ሾባዕታ shoba'ta (Tre).

Acacia negrii (89): ጠዴቻ ṭeddecca (Or).

Acacia nilotica (86): ዓፍሎ āflo (Saho); አጋቦ agabo (Som); ጩዓ çeā (Tya); ጋሎኦል galoöl, ጎናከ gonake, ማራህ marah (Som); ቀረጽ qereṣ (Tre); ቱወኦር tuweër (Som).

Acacia oerfota (90): አጆ ajo (Tya); ጨለጋማ çellegama (Agew); ጋምሮት gamrot (Tre); ጉ ማ ር gumar, ጉማራ gumara, ጉመሮ gumero (Som); ጕመሮ gwmero (Tya); ላዑድ laūd,

ላዓው lāw, ዑድ ūd (Tre); ዋንጋ wanga, ዋንጋይ wangay (Or).
Acacia ogadensis (80): ዓለብ āleb (Tre); ጋሎኦል galoöl (Som); ሐቅ ħaqq (Tre).
Acacia oliveri (80): ቲቂብለ tiqible (Afar).
Acacia paoli (90): ቸሪን cerin (Som); ጫጫና çaçanna (Or); ጀሪን jerin (Som).
Acacia pentagona (78): ቻቻም cacam (Mes).
Acacia persiciflora (81): በተ bate (Or).
Acacia pilispina (89): አጭቅ:ግራር açq-grar (Amh); ጨዓ çeā (Tya); ለፍቶ lafto (Or).
Acacia polyacantha (81): ግማርዳ gmarda (Amh); ጕመሮ gwmero (Tya).
Acacia senegal (75): ዓዳድ ādad, ዓዳድ:መሩ ādad-meru (Som); አክርሳ akkrsa (Gam); በ ጡ ባ beṭuba (Wel); gum arabic (Eng for resin); ኢዳዶ idado (Or); ማሃራ mahara (Kun); ቀንጥብ qenṭb, ቀንጥቢ qenṭbi, ቀንጢባ qenṭiba (Tya); ቆንጥር qonṭr, ቍንጥር qwenṭr (Amh); ጽፍሪ:ድሙ ṣfri-dmmu (Tya); ሳባንሳ:ዲማ sabansa-dima, ሳጰሳ saṗessa (Or); ስባንሳ:ግራር sbansa-grar (Amh).
Acacia seyal (85): ዓዲቀንጦ ādiqenṭo (Afar); ዓፍሎ āflo (Saho); ጨዓ çeā (Bil, Tre & Tya); ጭዓሰራኰ çea-serakwe (Bil); ጭጊንዳ çginda (Gam); ፉለኣይ fulaäy, ጂኢቅ jiīq (Som); ማካ ኒ makani (Afar); ቀይሕ:ጨዓ qeyyħ-çeā, ጻዕዳ:ጨዓ ṣa'da-çeā (Tya); shittim (Eng); ዋጩ waçu, ዋጂ wajji, ዋጆ wajjo, ዋኮ:ዲሞ wakko-dimo, ዋሲያ wasiya (Or).
Acacia seyal var. fistula (86): ፉሌሳ fuliessa (Wel).
Acacia sieberiana (89): ቡርቁቐ burquqqe (Or); ቸሪን cerin (Som); ጨዓ çeā (Tya); ደወኒ:ቡ ና deweni-bunna (Gur); ፉልሳ fullsa (Or); ጋራ gara (Wel); ግርሞ grmo (Tya); ጀሪን jerin (Som); ለፍቶ lafto, ለፍቶ:አዲ lafto-adi, ለለቶ lalato (Or); ኦዶርዋ odorwa, ፑ ሌ ሳ pulliesa (Wel); ጻዕዳ:ጨዓ ṣa'da-çeā (Tya).
Acacia spp. (75-93): ግራር grar (Amh & Gur); ግሬት gret (Ge'ez).
Acacia stuhlmanii (91): ግሀይደር gheyder, ጀንጃኒ janjanni (Som) .
Acacia tortillis (87): ዓባክ ābak, ዓካብ ākab (Som); ዓላ ālla (Tya); ዓቅባ āqba (Tre); ዓ ን ቄ ባ ānqieba (Tya); በሕበይ beħbey (Afar); ጉርሓ:ዳሩዬ gurħa-daruye, ሐራሕ ħaraħ (Som); ሎጦባ loṭṭoba (Or); ኦራ ora (Som); ጻዕዳ:ዓቅባ ṣa'da-āqba (Tre); ሸራ shera (Gam); ጠደቻ ṭeddecca (Or); ቲማድ timad (Som); ጽዒ ṣî (Saho).
Acacia venosa (81): ጋማሮ gamaro (Saho); ግማራ gmara (Agew); ቀንጥቢ qenṭbi (Tya).
Acacia walwalensis (91): ዲሊበሀ dilibeh (Som).
Acrocarpus fraxinifolius (51): red cedar, shingle tree (Eng).
Aeschynomene elaphroxylon (144): አምባጪ ambaçi (Or); ambatch (Eng for wood); ቦባቦፌ bobabofe, ቦፎፌ bofoffe, ቀንቀልቻ qenqelcca (Or); ሸዳሮ shedaro (Gam).
Aeschynomene indica (144): ካላንካል kalankal (Som).
Aeschynomene pfundi (144): ሾቐ shoqqe (Or).
Albizia amara (94): ጭጎኖ çgono, ሐማሰራን ħamaserran (Tya); ማራ mara (Som); ን ፋ ሽ ያ nffashya, ንፋስያ nffasya (Tya); ራአማ raämma (Som); ሰብቃና sebqana (Agew); ጸ ብ ቃ ና ṣebqana (Agew & Bil); ሶብካና sobkana, ጠብቃና ṭebqana (Bil); ቱጋር tugar (Som).
Albizia anthelmintica (94): ብስና bsnna (Tya); ሙስና musnna (Amh); ራይደብ raydeb (Som); ሽና shna (Agew, Amh & Tya).
Albizia grandibracteata (96): አደዎ adewo (Anu); እለለ elele, እመለ emela, ሃለለ halele, ኮ ፋ ለ kofale (Or); ሳቲ sati (Mes); ሰሪ serri (Sha); ሻቲ shati (Mes); ሻዎ shawo (Or).
Albizia gummifera (96): አምባበሳ ambabessa (Or); ጫቶ çatto (Kef & Or); ጫዊቾ çawicco, እመለ emela, ጎርበ gorbe, ካርቾፌ karcoffe, ሙካ:አርባ mukka-arba, ሳንኪለ sankile (Or); ሳሳ sasa (Amh & Or); ሰሳ sesa (Amh); ሺሞሮ shimoro, ሲሳ sisa, ዩንጎ yungo (Or).
Albizia isenbergiana (94): ጽዲኖ ṣdieno (Tya).
Albizia lebbek (94): east indiaṇ walnut (Eng); ለበክ lebbek (Amh, name being adopted); siris (Eng).

Albizia malacophylla (93): ጭጎኖ çgono, ሐማሰራን ḣamaserran, ንፋሽ nffasha , ንፋስያ nffasia (Tya).

Albizia schimperiana (96): አምበበሰ ambabessa, ጫቶ çatto, ጫዊቾ çawicco, አመለ emela, ጎርበ gorbe, ከርቾፌ karcoffe, ሙካ:አርበ mukka-arba, ሰንኪለ sankile (Or); ሰሰ sasa (Amh & Or); ሰሰ sesa (Amh); ሺሞሮ shimoro, ሲሰ sisa, ዩንጎ yungo (Or).

Albizia spp. (93-96): አምበበስ ambabes (Or); አሪጋ ariga, ሕዴይ ḣdiey (Ge'ez); ሰንዲል sendil (Amh).

Alchemilla fischeri (39): ደሧለ deshwala (Kef).

Allophyllus abyssinicus (499): እምብስ 'mbs (Amh); አበር abar, አበራ aberra, አረጀ areje, ድሩበ druba, ሂርካሞ hirkamo, ሂርኩም hirkum, ከካዩ kekayi (Or); ቀዓምሺ qāmshi (Tya & Tre); ሰራጊ saragi, ሰራራ sarara, ሰእ seö (Or); ሸእ sheö (Kef); ስዋርያ swarya (Tre & Tya); ተስተስ testes (Gum); ዎራፉቶ worafuto (Wel).

Allophyllus macrobotrys (499): ቃዲዳ qadida (Or); ተስተስ testes (Gum).

Allophyllus rubifolius (502): ጣጠሰ ṭatessa (Or).

Allophyllus sp. nov. (502): አለጁ aleju, ሂርካ hirka, ሂርካሞ hirkamo, ሰእ seö (Or).

Alysicarpus ferrugineus (154): ሃምብሃምበ:በይታ hambhambo-bayta (Tya).

Ampelocissus schimperiana (400): ሐረግ:ተመን ḣareg-temen (Tya).

Anthyllis vulneria (225): kidney vetch (Eng).

Antiaris toxicaria (301): ደንጊ dengi, ሙካለተ mukalate (Or); ሰቾ saco (Kef); ቱንግዎ tungwo (Anu).

Antopetitia abyssinica (227): አበዕኸ:በይታ aba'ke-bayta, ጣዕጣዕታ ṭa'ṭa'ta (Tya).

Apodytes dimidiata (348): በርደለ berdele (Dor); ጨለቅለቃ çeleqleqqa (Amh); ዳኒሰ dannisa (Or); ዶንጋ donga (Amh); ዶንኪኮ donkiko (Sidama); መኢ meî, ኦዳ oda, ኦዳ:በዳ oda-bada, ኦዳ:ኪየት oda-kiyet, ኦዳ:ሰዳ oda-seda, ቁምበለ qumbala, ኡሊፎኒ ulifoni, ወንደብዮ wendebyo (Or); white pear (Eng).

Arachis hypogea (147): ፉል ful (Tre & Tya); groundnut (Eng); ለውዝ lewz (Amh); ሎዝ loz (Ade); ኦቾሎኒ occoloni (Amh & Or); peanut (Eng).

Argyrolobium arabicum (222): ብርስናይ brsnay (Tya).

Astragalus atropilosulus (230): ሐማት:ካኺቶ ḣamat-kakito, ጠጠም:ዓጋዜን ṭeṭem-āgazien (Tya).

Azadirachta indica (485): neem (Eng); ኒም nim (being adopted in Saho, Tre & Tya).

Balanites aegyptiaca (433): ዕንድሩር 'ndrur (Tya, for fruit only); በዳኖ baddanno (Or); በደና badena (Wel); በደና beddenna (Or & Wel); በደኖ beddenno (Amh); ዶማይ domay, ዶማየ domaye (Gam); ዶሞሆ domoho (Or); ጎት got, ጉት gut (Som); ጉዛ guza (Amh); ጕዛ gweza (Tya); ጀሞ jemo (Amh); ጁሊ juli (Kun); ኩድኩዳ kudkudda (Amh); መቒዕ meqi' (Tya, for tree or fruit); ቃጮና qaçona (Amh); ቃቅ qaq, ቆግ qog (Tre); ቁጻ quṣsa, ቁጣ quṭṭa (Agew); ሰሊበቲቆ selibatiqo (Bil); ቶው tow (Anu).

Balanites glabra (434): ከዲ keddi, ኪዲ kidi (Som).

Balanites rotundifolia (436): በዳን baddan (Or); ኩላን kulan (Afar & Som); ኩዞ kuzo, ሞርኮ morko (Gam).

Barbeya oleoides (270): አዳዶ adado, ጋሪ garri (Or); ሕሩምታ ḣrumta (Tya); ልሻም lsham (Tre).

Bauhinia tomentosa (68): አበይታበታ abaytabata (Or).

Berchemia discolor (393): ዓበ āba (Tya); አሞር amor, ደን den, ሃሞር hamor (Som); ጀጀበ jejjeba (Amh & Or).

Bersama abyssinica (511): ዓሸ:ኦም āsha-om (Tya); አፋጀሽኝ affajeshñ, አዘምር azamr (Amh); በቆ boqo (Or); ብርስማ brsma (Tya); ዶልኪኢሰ dolkiisa, ኤበራኮ eberrako, ኢቲቢሮ etibiro, ፎራካ foraka, ገሰ gessa, ሆራቃ horaqqa, ጁመፎክ jumefok, ኪቢሩ kibiru, ኮራካ koraka, ሎልቺሰ lolcisa, ቃራቻ qaracca (Or); ቅርስማ qrsma (Tya); ሰበጣለ sebaṭalla

(Gur); ሰሰ sesa (Or); ተበራኮ teberako (Sid); ተበራቆ teberqo, ቲቢራ tibira (Or); ወለሲኑ welessienu (Wel); ዛጌ zagie (Gam).

Boswellia microphylla (445): በየ:መደው baye-medow (Som); ዳበሰ dabeso, ዳከራ dakara (Or); ኩ:ዋደረ ku-wadere, ሞጎለ mogole (Som); ሞክ:ሊዲ mok-lidi (Or).

Boswellia neglecta (445): አለል allal, አለለ allalla (Mur); ዳከራ dakara (Or); ኩግና kugna (Gam); ሙጎለ mugole, ሙርከን murk̂en (Som).

Boswellia papyrifera (443): ፋጡቃ faṭuqa (Agew); ጋልጋለም galgalam, ከፋል kafal (Or); ሊበናት libanat (Ade & Or); መቐር meq̈er (Tya); ቀረሬ qererrie (Amh); ዋልበ walba (Tya); ዋልዋል walwal (Bil); የእጣን:ዛፍ ye'ṭan-zaf (Amh).

Boswellia spp. for the resin (442-445): እጣን 'ṭan (Amh); ዕጣን 'ṭan (Ge'ez & Tya); እጣን eṭan (Gur); frankincense, gum olibanum (Eng); ልበን lban (Tre) & ልብን lbn (Ge'ez for resin approaching uniformly white in colour).

Boswellia rivae (443): በየ:መደው baye-medow, መራፉር merafur, ሚራፉር mirafur, ሙርከን murk̂en (Som); ቁራ qura (Or).

Brucea antidysenterica (440): አበሎ abalo (Amh & Or); ሃዳወይ hadawey (Or); መሊጣ meliṭa (Tya); ቆመኞ qomeñño (Or); ሹሩ:ሹጢ shuru-shuṭṭie (Wel); ሹሹዴ shushuḍie (Gam); ጣሚቻ ṭamicca (Or); ቶሎ tollo (Jan); ዋጊኖስ waginos (Amh & Ge'ez).

Buxus hildebrandtii (255): ዶሰ dosso, ዱሶግ dussog (Som); ጉል:ጉል gul-gul (Or).

Cadia purpurea (101): ሰእለ saële (Or); ሰለማ salema (Som); ሽለን shlen, ሽንእን shnën, ስንእል snël (Tya); ቶቅዳ toqda (Amh).

Caesalpinia bonduc (54): bonduc (Eng).

Caesalpinia decapetala (53): mysore thorn (Eng); የፈረንጅ:ክትክታ yeferenj-ktktta (Amh).

Caesalpinia gilliesii (53): paradise flower (Eng).

Caesalpinia pulcherrima (51): barbados pride, paradise flower (Eng).

Cajanus cajan (181): ፋረንጎታ farengota (Kon); ጊተእ giteë (Had); ካፎ:አታራ kafo-atara, ካፍዋ:አተሪያ kafwa-ateriya (Wel); ኦሆታ:ፋረንጎታ ohota-farengota (Kon); pigeon pea (Eng); ሰልቦኮ:ገድ salboko-ged (Som); የእርግብ:አተር ye'rgb-ater, የወፍ:አተር yewef-ater (Amh).

Calodendrum capensis (420): cape chestnut (Eng).

Calpurnia aurea (104): አንጫቢ ançabi, ጨካ çeka, ጨካታ çekata (Or); ጨቅሳ çeqsa (Gam); ጨካ çeka (Or); ድግጣ dgṭta (Amh & Gur); ሕጻውጽ ḥsawṣ (Tya); ልይጣ lyṭta (Amh); ቆንዴ qonde (Had); ሰሒፕ shiṭ (Ge'ez); ሰሊቱ solitu (Or); ስፕራ sṭra (Tre); ውጭቃ wçqa (Agew); ዝግጣ zgṭta (Amh).

Canavalia ensiformis (165): አዳጕራ adagwra (Tya); አደንገል adengel (Bil); አደንጓሬ adengwarrie (Amh & Gur); ዲር:ዳጕር dir-dagwer (Som); jack bean (Eng); ሎኮማ lokoma (Wel); ሳአር:ሳአር saar-saar (Som); sword bean (Eng); ዎህያ wohya (Wel).

Cannabis sativa (327): bhang, dagga, hashish, pot (Eng for the drug); hemp (Eng for the plant and the fibre extracted from the stems); ዕፀ:ፋርስ 'ṣe-fars (Ge'ez); marihuana (Eng for the dried leaves).

Cardiospermum corindum (497): balloon vine (Eng); ሰመግ semeg (Amh); ሻቕሬ shaq̈wre (Tya).

Cardiospermum halicacabum (498): balloon vine (Eng); ኢሚ immi (Som); ኢሙ:ጋሬ imu-gare (Or); ሰመግ semeg (Amh); ሻቕሬ shaq̈wre (Tya); ሽልጥ shlṭt (Tre).

Casimiroa edulis (424): ካዛሞራ kazamora (Amh); ካዝሚር kazmier (Tya, name being adopted from the genus); white sapote (Eng).

Cassia arereh (56): ቀርቃይ qerqay (Agew); ሽቶል:ህበይ shtol-hbey (Tya).

Cassia fistula (57): golden shower, indian laburnum, pudding-pipe tree, purging cassia (Eng).

Casuarina spp. (262-264): አርዘ:ሊባኖስ arze-libanos (Amh, name recently transferred from cedar of lebanon); casuarina (Eng); ካዝዋሪና cazwarina (Amh); iron-wood, she-oak (Eng); ሸውሸዌ shewshewwie (Amh); silkoak (Eng, the trade name for timber), swamp oak (Eng).

Catha edulis (340): ጫት çat (Amh, Gur & Tya); ጫታ çata (Kem); ጫቲ çati (Or); ጫቲያ çatiyya (Wel); ጎፋ gofa, ጂማእ jima' (Or); khat (Eng).

Cayratia ibuensis (404): ዳኦለ daöle, ኡቡለ ubule (Mur).

Celtis africana (266): እቋ 'qwa (Agew); አማለቃ amalaqqa (Or); አመለቃ ameleqqa (Amh); ቦዳር bodar (Som); ጫሪ çari (Or); ጫይ çay (Amh); ጨይ çey, ጨከ çeke (Or); ጋሂሮ gahiro (Amh); ጋዋ gawa (Tya); ሐማት:ረወይ ḣamat-rewey, ሐዋ ḣawa (Tya); ካቺሆ kacciho (Mao); ኪዲ kidi, መጨርሳ meçersa, መተቆማ meteqoma, ቃዎ qawo (Or); ቃውት qawtt (Amh); ሺሹ shishu (Kef); ጡዬ ṭuyie (Amh).

Celtis toka (267): ጋርቢ:አዲ garbi-adi (Or).

Ceratonia siliqua (66): carob, locust tree (Eng).

Chamaecrista absus (64): ዓይኒ:ዋሪ āyni-wari (Tya); አዳጉራ:ጤል adagura-ṭiel (Tre); ዳሩር darur (Amh & Tya).

Chamaecrista hochstetteri (65): ልቢ:ሰበይቲ lbbi-sebeyti (Tya).

Chamaecrista mimosoides (66): ሰራው:ባይታ seraw-bayta (Tya); የአይጥ:ጎመን yeäyṭ-gommen (Amh).

Chamaecrista nigricans (65): ዓይኒ:ዋሪ āyni-wari (Tya).

Cicer arietinum (246): ዓተር āter (Ge'ez & Tya); ዓተር:ቀይሕ āter-qeyyḣ (Tya); ዓቲር ātir (Saho); chick pea (Eng); ሰንጥ sanṭ (Ge'ez); ሽምብራ shmbra (Amh, Tya & Tre); ሹምቡራ shumbura (Gur, Had, Or, Wel & Gam); ሱምቡራ sumbura (Ade, Afar); ሱምቡሮ sumburo (Kem); ዚፋል zifal (Ge'ez).

Cicer cuneatum (248): ዓተር:አንጯ āter-ançwa, ዓተር:ጓሶት āter-gwasot, ሽምብራ:ጓሶት shmbra-gwasot (Tya); የአይጥ:ሽምብራ yeäyṭ-shmbra (Amh).

Cissus aphylla (404): ጮቢ çobi, ጮቢ:አርባ çobi-arba (Or); ማሩድ:ኦድ marud-od (Som).

Cissus aphyllantha (404): ጮቢ:ባደሱ çobi-badesu (Or).

Cissus cactiformis (407): አርሞ armo, ጋድ gad (Som); ጎርሳ gorsa, ጉራጅ guraj, ማትቦት matbot (Or).

Cissus cornifolia (405): ለባ:አሎለያ laba-aloleya, ማህወ mahwe (Som).

Cissus petiolata (406): ዓልገ ālge, ዓልቀ ālqe (Tya); አርሞ armo (Som).

Cissus quadrangularis (407): ዓልዓ ālā, ዓልገ ālge, ዓልቀ ālqe (Tya); አርመዲን armedin, አርሞ armo (Som); ደምገዛ demgezza (Amh); ጋድ gad (Som); ገምበዛን gembezan (Amh); ለማጎኢያ lamagoîya (Som); ሴቴ:የሰይጣን:ሙቀጫ sietie-yeseyṭan-muqeça (Amh); ሽልሕ shlḣ (Tre); ሱሩጋ suruga (Afar); ቱሮይ turoy (Mur); ዋሲሊ wasili (Som); የዝሆን:አንጀት yezhon-anjet (Amh).

Cissus rotundifolia (408): አርሞ armo (Som); ጫምቤ çambie (Amh); ጮቢ çobi, ጮቢ:ቃኡራ çobi-qaüra (Or); ጋድ gad (Som); ካሊ kali (Mur); ቃልቃሎ qalqalo (Gam).

Citrus aurantifolia (431): ለሚን lemin (Tya); ለምን lemun (Tre); lime (Eng); ሎሚ lomi (Amh & Or); ጡጦ ṭuṭṭo (Or).

Citrus aurantium (432): ባህረ:ሎሚ bahre-lomi (Amh); በኵረ:ለሚን bekwre-lemin (Tya); bitter orange (Eng); ኮምጣጤ:ሰመርጌለ komṭatie-semergiela (Amh); seville orange, sour orange (Eng).

Citrus limetta (431): sweet lime (Eng).

Citrus limon (430): ለሚን:ፈረንጂ lemin-ferenji (Tya); lemon (Eng); ሎሚ:ፈረንጂ lomi-fernji (Or); የፈረንጅ:ሎሚ yeferenj-lomi (Amh).

Citrus maximum (431): pimmelo, shaddock (Eng).

Citrus medica (430): እንጉታት 'ngutat, እንጐት 'ngwet (Ge'ez); citron (Eng); ትርንጎ trngo

(Amh & Tya); ትርንጒ trngwi (Tya).
Citrus paradisi (431): grapefruit (Eng, being adopted into Amh); ናርገ narge (Tya).
Citrus reticulata (431): mandarin (Eng); መንደሪን menderin (All); tangerine (Eng).
Citrus sinensis (432): ብርትኳን brtkkwan, (Gur & Tya); ብርትዃን brtk̇wan (Tya); ብርቱካን brtukan (Amh & Or); ብርቱካና brtukana, ብርቱካኒ brtukani (Or); orange (Eng).
Clausena anisata (429): እምብሪቶ 'mbritto (Kef); አደሳ adessa (Or); ቡሮ buro (Som); ቺኮተ cikote, ዳሽሎሜ dashlome (Wel); ጋምጭለ gamçle, ጋኤጃ gaëja (Gam); ልምጭ lmmçç (Amh); ስንፍኮ snfko (Gur); ኡሉማይ ulumay, ኡርማያ urmaya, ዋለያ walaya (Or).
Clitoria ternatea (156): butterfly pea, kordofan pea (Eng).
Colutea abyssinica (233): ጨና፡እኖ፡ድቓለ çena-'no-dq̇ala (Tya); ዷዷቴ dwadwatie (Amh); ኮኻዕታ koka'ta, ነፍጢ፡ጓሶት nefti-gwasot (Tya).
Commiphora africana (454): ዕቋ 'qwa (Agew); ዓንቋ ānqwa (Tya); አንቋ anqa (Amh); ዶጋማ dogama (Gam); ኮቦክ kobbok (Som); ኩል፡ቦንቲ kul-bonti (Me); ኦልባይ olbay (Mur); ዙንትሴ zuntsie (Gam).
Commiphora alaticaulis (462): ማሮን፡ሳነ maron-sane (Som).
Commiphora ancistrophora (475): አሊቦየ aliboye, ዱንዳስ dundas (Som).
Commiphora boiviniana (450): ዳኪዳር dakidar (Or); ሃጋር፡መዶው hagar-medow (Som); ኮኦሻ koösha (Gum); ላቡሰኣኒ labuseäni, ላረኒ larenni (Mur).
Commiphora boranensis (475): ደንገል dengel, ደንከሊ denkeli, ጎባ፡ሀሪግ goba-herig, ኢልካ፡አደስ ilka-ades (Som); ከሀሮ kehero, ሪጋሀሮ rigahero (Or).
Commiphora bruceae (473): ጎአ goä, ኢልካ፡አዳየ ilka-adaye (Som).
Commiphora campestris (469): ሃመሳ hamessa (Or).
Commiphora confusa (462): ጫጮ çaço, ሃመሳ hamessa (Or); ከየ keye (Gum).
Commiphora coronillifolia (460): ጋሮን፡መዶው garon-medow, ካዶን kadon (Som).
Commiphora corrugata (460): ባዮ bayo, ባዮ፡አድ bayo-ad, ጋንዳድ gandad (Som); ኢልካ፡ቡስ ilka-bus (Or).
Commiphora cyclophylla (468): ጉራኣ guraä, ኡራኣ uraä (Som).
Commiphora ellenbeckii (473): ደሪንደር derinder, ጎሊሉ golilu, ጎውለሎ gowlello, ትቡክ tbbuk (Som).
Commiphora erlangeriana (448): ባ፡አሮር ba-aror, ዱንካል dunkal (Som).
Commiphora erythraea (450): ጋሃዲታ gahadita (Kon); ሃጋር hagar, ሃጋር፡አድ hagar-ad, ሃጋር፡መዶው hagar-medow, ሃጋርሶ hagarso, ሃጋርሱ hagarsu (Or & Som).
Commiphora gileadensis (452): balm of gilead (Eng for the resin); ዳሰይኖ daseyno, ጋሮን garon (Som).
Commiphora gowlello (474): ጎሊሉ golilu, ጎውለሎ gowlello, ትቡክ tbbuk (Som).
Commiphora guidotii (448): ዱንካል dunkal, ጋራሆ garaho (Som).
Commiphora habessinica (471): ዓንቋ ānqwa (Tya); አንቋ anqwa (Amh); ጫላንጋ çalanga (Or).
Commiphora hodai (452): ሆዳይ hoday, ሆዴይ hoddey (Som).
Commiphora horrida (474): ጋሮን garon, ካብራሮ kabraro (Som).
Commiphora incisa (473): ራሃን፡ረብ rahan-reb (Som).
Commiphora kua (452): ዳባሶ dabaso (Or); ማላ፡ዋ፡ሃረድ mala-wa-hared, ራሃን፡ረብ rahan-reb (Som); ዋራ wara (Or); ዋራብ፡ረብ warab-reb (Som).
Commiphora longipedicellata (448): ሀልታ helta (Som).
Commiphora myrrha (452): ደዲን dedin, ዲዲን didin, ዲዲን፡አድ didin-ad, ዲዲን፡መዶው didin-medow (Som); ኩምቢ kumbi (Or).
Commiphora obovata (454): ሻን፡ማለይስ shan-maleys (Som).
Commiphora ogadensis (460): አጋርሱ agarsu (Or); ገንዲ፡ገንታ gendi-genta, ሃጋር፡መዶው

hagar-medow (Som); ሃጋርሱ hagarsu (Or).
Commiphora quadricincta (471): ዓንቋ ānqwa, ብሽመድ bshamed (Afar & Tya); ጋዳዲ gadadi (Afar); ጎዳዲ godadi (Tya); ቃፊል qafil (Afar & Tya).
Commiphora spp. for the resin (446-477): bdellium (Eng); ከርቤ kerbie (Amh, Ge'ez & Tya); ሞልሞል molmol (Som); myrrh (Eng).
Commiphora rostrata (462): አሊ:ቦየ ali-boye, ዳኖ dano (Som); ዲራአ dirraä, ጊላይ gillay (Or); ኬኖ keno (Som); ቲኢራይ tiîray (Or).
Commiphora samharensis (452): ዓንቃ ānqa (Tre); ዓንቋ ānqwa, ብሽመድ bshamed (Saho & Tya); ምሽዋቅ:አነቢ mshwaq-annebi (Tre).
Commiphora schimperi (454): ዓንቋ ānqwa (Tya); አንቋ anqwa (Amh); ሃመሳ hamessa (Or); ኮበክ kobbok (Som).
Commiphora serrulata (469): በርሶዋ barsowa (Or).
Commiphora sphaerocarpa (468): ባ:አሮር ba-aror, ዱንካል dunkal, ጋ:አሮር ga-aror, ገት:ሱቡክ get-subuk (Som).
Commiphora sphaerophylla (450): ሃጋር hagar, ሃጋር:አድ hagar-ad, ሃጋር:መዶው hagar-medow (Som).
Commiphora staphyleifolia (448): ባ:አሮር ba-aror (Som).
Commiphora tenuis (469): አንጉለ angule (Or).
Commiphora terebinthina (454): ሆርጎይ horgoy (Or & Som); ሊቡ libu (Mur); ሞርማከለ mormakalla, ሰንጋ:ኢጉ sanga-îgu (Or & Som).
Commiphora truncata (452): ሃጋር:መዶው hagar-medow (Som).
Commiphora tubuk (457): ዲዲን didin, ጎአሀ goäh (Som).
Commiphora unilobata (448): ዱንካል dunkal (Som).
Cordeauxia edulis (56): jeheb nut (Eng); ቁድ qud, ቁዳ quda (Som); እሕብ 'ḣb (Som); yehib nut (Eng).
Cordyla somalensis (70): አንቶሮ antoro (Som).
Cotyledon barbeyi (19): በጠጦ beteto, ብጥብጦ bṭbṭṭo (Tre); ሰእኒ:ያኮ sa'ni-yakko (Tya).
Crotalaria albicaulis (207): አሱሮ assuro, ዳርጆ darjo (Som).
Crotalaria comanestiana (202): አሱር assur, አሱሮ assuro, ፉጊሶ fuggiso (Som).
Crotalaria dumosa (212): ሃጂን hajin (Som).
Crotalaria glauca (202): ሽሕሽሕት shaḣshaḣta (Tya).
Crotalaria incana (201): ሐዊ:ለይቲ ḣawwi-leyti (Tya).
Crotalaria juncea (218): bengal hemp, sunn hemp (Eng).
Crotalaria keniensis (214): ኦኮኮ okoko (Kef).
Crotalaria laburnifolia (200): ዳርጋ darga, ገለሎ gelelo (Som); የአይጥ:ምስር yeäyṭ-mssr (Amh).
Crotalaria lachnocarpoides (202): ኦኮኮ okoko (Kef).
Crotalaria microphylla (213): እርብ:ክታን 'rb-ktan (Tre).
Crotalaria petitiana (210): ሶራ sorra (Bil).
Crotalaria podocarpa (201): ሰበረ:ጓሶት sebbere-gwasot (Tya).
Crotalaria pycnostachya (205): አሱሮ assuro, ሹና:ሹና shuna-shuna (Som).
Crotalaria retusa (216) ቦቦ:ሃለ bobo-halle (Som); ሽሕሽሕታ shaḣshaḣta (Tya); ሹሩፍ:ሹሩፍ shuruf-shuruf (Som).
Crotalaria rosenii (199): አባ:ጮናኔ:እንጪት abba-çonanie-'nçet (Amh); ፊሻ fissha, ክልክለ klklle (Or); ኦኮኮ okoko, ሻቲሻቶ shatishato (Kef).
Crotalaria saltiana (204): ጎሆልዳየ goholdaye (Som).
Crotalaria spinosa (217): ወንዴ:ችፍርግ wendie-cfrg (Amh); ጉፍቲ gufti (Or); የምድር:ግራር yemdr-grar (Amh).

Cussonia spp. (538-540): አሃራጉ aharagu (Som); አንደሮቺ anderocci, አርፋቱ arfattu (Or); ጫቅሞሽ çaqmosh, ዱዱና duduna (Amh); ጋዳም gadam, ገደም geddem (Tya); ገደታሞ gedetamo (Amh); ገተም getem (Tya); ገተሚ getemi (Or); ጎምበቸሬ gombocerie (Amh); ሃርፋቱ harfatu (Or); የዝንጀሮ:ጎመን yeznjero-gommen (Amh).

Cyamopsis tetragonoloba (138): cluster bean, guar (Eng).

Cydonia oblonga (42): ሁክ huk (Or); quince (Eng).

Cyphostemma adenocaule (412): አርሞ armo, አርሞ:ዲን armo-din (Som); አሰርኩኻ aserkuka (Tya); አሰርኩሽ asserkush, አሰርኩሽ:ተበተብኩሽ asserkush-tebetebkush (Amh); ሐረግ:ተመን ħareg-temen, ቊርሖ qwrħo (Tya).

Cyphostemma burgeri (417): አርሞ armo (Som).

Cyphostemma cyphopetalum (417): ግንዶሽ gndosh (Amh); ሐለንጊ:ተመን ħalengi-temen (Tya); ቃልቃሎ qalqalo (Gof & Or).

Cyphostemma engleri (416): አርሞ armo (Som).

Cyphostemma junceum (410): ዕፀ:ዘዌ ‘ṣe-zewie (Ge’ez).

Cyphostemma molle (411): ዕፀ:ዘዌ ‘ṣe-zewie (Ge’ez); ሐማውቲ ħamawti, ወይኒ:ጓሶት weyni-gwasot (Tre).

Cyphostemma niveum (415): ዓልቀ ālqe (Tya).

Cyphostemma oxyphyllum (412): ሐረግ:ተመን ħareg-temen, ቍርሖ qwrħo (Tya).

Cyphostemma serpens (417): አርሞ:ጎሬ armo-gorie, አርሞ:ማለህለ armo-malehle (Som).

Dalbergia commiphoroides (105): ዱሲያ dusiya, ዱያ duya (Som).

Dalbergia lactea (107): ያግበሮ yagbero (Kef).

Dalbergia melanoxylon (105): african blackwood, african ebony (Eng); በበኑሲ babanusi, ሞግሃኖ moghano (Or); ዝበ zbbe (Tya); ዞቢ zobbi (Amh).

Debregeasia saeneb (320): ጸዕዳ:ቍጽሊ ṣa‘da-q̇weṣli (Tya); ሸርዳለ sherdale (Or); ሹንሹና shunshuna (Amh).

Delonix baccal (51): በቻል baccal, በካል bakal (Som).

Delonix elata (50): ዓማይ āmay (Afar); ደቢ debi (Som); ዲነ:ዓማይቶ dine-āmayto (Afar); ለቢ lebi (Som); ሱከለ sukela (Or).

Delonix regia (50): flamboyant, flame tree (Eng); የድሬዳዋ:ዛፍ ye-dredawa-zaf (the Dire Dawa tree, Amh).

Desmodium uncinatum (151): silverleaf desmodium, silverleaf spanish clover (Eng).

Dichrostachys cinerea (74): አደር ader (Amh); አደሳ adessa (Or); አርሻማርሻ arshamarsha (Tya); አርሸ:መርሻ arshe-mersha (Amh); ቻለንታ calenta (Kon); ድሃይ:ድሃይ dhay-dhay, ድሂግዳር dhigdar, ጋሎኦል:ሱር galoöl-sur (Som); ጋርጋሩ gargaru (Gam & Wel); ገርገሮ gergero (Gam & Wel); ገቶ geto, ጊርሚ girmi (Or); ጎድ:ሃው:ቶርለ god-haw-torle, ጎራን:መዱ goran-medu (Som); ግርማ grma (Or); ጕነቕ gweneq̇w (Tya); ሃጠ hatte, ጀረሞ jeremme (Or); ከበቲር kabatir (Som); ከነይ keney (Tya).

Dichrostachys kirkii (74): ከበቲር kabatir (Som).

Dicraeopetalum stipulare (101): ሃንገለሎ hangelelo (Som).

Dobera glabra (353): ጋራስ garas (Som, Afar & Saho); ገርሳ gersa (Tre & Tya); ኻራስ k̇aras (Som).

Dodonaea angustifolia (491): አዳሐለት adaħalat, አዳልኻት adalk̇at (Saho); ደን den (Som); እተቻ ettecca (Or); ገርገጧ gergetwa (Wel); ሃይራማት hayramat (Som); ኢንታንቻ intanca (Sid); ኢቲቻ itticca (Or); ከራራ kerara (Agew); ክትክታ ktktta (Amh, Gur, Kem & Saho); ማቻ macca (Kef); ማንዳራ mandara (Bur); ማሬይታ mareyta (Kon); ሰሳት sasat (Saho); ሰንካራ senkara (Wel); ጠደቻ teddecca (Or); ጢቶ tito (Gam); ታህሰስ tahses (Tre & Tya); ታሰስ tases (Bil); ተርሚን termien (Ge’ez); ትለም tlem (Or).

Dolichos sericeus (179): አዳጕራ፡ጓሶት adagwra-gwasot (Tya); ወደል፡አስፈስ wedel-asfess (Amh).
Dorstenia barnimiana (298): ወርቅ፡በሜዳ werq-bemieda (Amh).
Ekebergia capensis (489): cape ash (Eng); ዱዱና duduna (Or); ሎል lol (Amh); ቍት qwet (Tya) ሶምቦ sembo (Amh & Or).
Elaeodendron aquifolium (344): ገብ geb, መገት meget (Som).
Entada spp. (71-73): እምብልታ 'mblta (Tya); አምበልታ ambelta (Amh & Or); ጋለቸቻ galececa, ሃምበልታ hambelta (Or); ጁኩል jukul (Mes); ፖክ pok (Anu); ዛግያ zagya (Wel).
Eriobotrya japonica (42): loquat (Eng); ወሽመላ weshmella (Amh & other languages).
Eriosema cordifolium (193): የምድር፡ቆሎ yemdr-qolo (Amh).
Erythrina abyssinica (161): ዕንቍ፡ህበይ 'nqwi-hbey, ዕንቍ፡ዝብኢ 'nqwi-zbi (Tya); እርጎፍት 'rgoft (Amh); አፍራርቱ afrartu (Or); አምቢልሽ ambillsh (Gum); አርፈቱ arfetu, ባርቱ bartu (Or); በሮ bero (Kef); ቦርቶ borto (Or); ቦርቷ bortwa (Wel); ዱስ dus (Or); ኮላንቹ kolancu, ከላኒቻ kalanica (Kef); ኮርች korc (Amh); መላኮ melako (Or); ቋራ qwara (Agew, Amh, Bil & Tya); ዋዞ wazo (Or); ወለቾ welecco (Kem); ወለና welena, ወለንሱ welensu (Or); ወራ wera (Had); ዝዋዕ zwaw' (Tya).
Erythrina brucei (161): አፍራርቱ afrartu, አርፈቱ arfetu, ባርቱ bartu (Or); በሮ bero (Kef); ቦርቶ borto (Gam); ቦርቷ bortwa (Gof & Wel); ቡርቶ burto (Dor); ኮላንቹ kolancu, ኮላኒቻ kolanica (Kef); ኮርች korc (Amh); መላኮ melako (Or); ቋራ qwara (Agew & Amh); ዋዞ wazo (Or); ወለቾ weleco (Kem); ወለንሱ welensu (Or); ወራ wera (Had).
Erythrina burana (164): ቡራና burana (Amh & Or).
Erythrophysa septentrionalis (490): ኦሙሎታባ omulotaba (Som).
Fagaropsis angolensis (422): ደርጊ dergi, ደሮ dero (Or); ሙክ muke (Mao); ሻፒንዲ shapindi (Or); ሲጊሉ sigilu (Sid); ያማ yama (Gam).
Ficus capreaefolia (281): አጽሳን aṣsan (Tre); ባላስ balas (Saho); በለስ beles (Amh & Tya); በለስ፡ቍላ beles-qwella (Tya); ሉጎ lugo (Or); ሞኮይ mokoy, ሙኪ muki, ሙኪ፡ሉበቶ muki-lubeto (Som); ዋላንጣ walanṭa (Or).
Ficus carica (280): በለስ beles (Amh, Tya & Ge'ez); ዕፀ፡በለስ 'ṣe-beles (Ge'ez); ኦደኮ odeko (Kem).
Ficus dicranostyla (289): ኡዳሉ udalu (Anu).
Ficus exasperata (284): ባዶ bado (Anu); ባላንሶፊ balansofi, ባላንጣዪ balanṭayi, ሃርቡ harbu (Or); ላቲ lati (Or & Sid); ሾንጋቦይ shongaboy (Mes); ሾይጥን shoyṭn (Gum); ሶሎ solo (Wel); ዋላንጣ walanṭa (Or).
Ficus fruit (278-299): fig (Eng); ሒላ hiela (Ge'ez).
Ficus glumosa (293): በርደ berde (Som); ጨቐንጠ çeqente (Tya); ደኣር፡ዓድ deër-ād (Som).
Ficus hochstetteri (299): አፍ፡ቀሞ af-qemmo (Tya).
Ficus ingens (291): ጨቐንጠ çeqente (Saho, Tre & Tya); ጨቐንጠ፡ጊሔ çeqente-gihe, ጨቐንጠ፡ሐርማዝ çeqente-ḥarmaz (Tya); ደኣር፡ዓድ deër-ād (Som); ከፈም kefem (Amh); ሉፎ lufo (Som); ማላሃሎ malahalo (Saho); ታልቁስ talqus (Bil).
Ficus mucuso (287): ቡሩ buru (Sha).
Ficus ovata (292): ቻበሮ çabero, ቻሮ çaro (Kef); ደምቢ dembi, ኤታ eta, ጎመ gomme (Or); ማርና marna (Wel); ኦዳ oda, ቂልጡ qilṭu (Or); ሾላ shola (Amh).
Ficus palmata (281): አጽሳን aṣsan (Tre); ባላስ balas (Saho); በለስ beles (Amh, Ge'ez & Tya); በልዛ belza (Agew); ዕፀ፡በለስ 'ṣe-beles (Ge'ez); ሉጎ lugo (Or); ሾቶ shotto (Kef); የቆላ፡በለስ yeqolla-beles (Amh).
Ficus populifolia (295): በርድ berd, በርደ berde, ኒዲር nidir (Som).
Ficus spp. (278-299): ቢሶም bisom (Ge'ez); ዳዕሮ da'ro (Ge'ez & Tya); ኤታ eta (Wel); fig tree (Eng); ኦዳ oda (Or); ሰግላ sagla (Tya); ሰግላ segla (Ge'ez); ሾላ shola, ዋርካ warka (Amh).

Ficus sur (287): በበ bobba (Gam); ኢትሰ etsa (Dor); ሃብሩ habru, ሃርቡ harbu (Or); ክሮ karro (Kef); ኮዶ kodo (Tya); መዳለ medalle, ኦዳበዳ odabedda (Or); ሽምፋ shamfa (Tya); ሾላ shola (Amh).

Ficus sycomorus (285): አርበሀ arbeh (Mes); በምበ bamba (Agew & Amh); ቢሰም bisom (Ge'ez); ዳሬ dare, ዱሬ dure (Som); ሃርቡ harbu, ሉጎ lugo (Or); ማክይ:አዳ makkoy-adda, ሞክ mokko, ሞክይ:ገዱድ:ዳሬ mokkoy-gedud-dare, ሙክይ mukoy (Som); ኦዳ oda (Or); ኦዶቾ odecco (Kem); ኦዶኦ odoö (Had); ኦለም olam (Anu); ሰግለ sagla (Tya); ሰግለ segla (Ge'ez); ሸግለ shegla (Tre); ሾላ shola (Amh); sycamore of the Bible (Eng); ዎለ wola (Wel).

Ficus thonningii (298): ችበሃ cbaha (Amh); ደምቢ dembi, ቂልጡ qilṭu (Or); ሽበኸ shbaḱa (Tya); ሺሽ shish (Gum); ታልቁስ talqus (Tre).

Ficus vallis-choudae (289): አዲዮው adiyow (Anu); በምቡለዴ bambuledie (Amh); በቢ bobi (Anu); በይቱ boytu (Gum); ከበ kobo (Kef); መዳለ medalle (Or).

Ficus vasta (296): እንዳዓሮ 'ndaāro (Saho); በርደ berde (Som); ዳዕሮ da'ro (Ge'ez, Tre & Tya); ዳርጉና darguna (Bil); ደምቢ dembi (Or); ሜሎ mielo (Kef); ኦሎለ olola (Had); ቂልጡ qilṭu (Or); ተናከ tenaka, ተርቤንቶ terbiento (Ge'ez); ዋርከ warka (Amh); ወለ wela (Gam); ወልከ welka (Ge'ez); ወርከ werka (Amh); ዎለ wola (Wel).

Filicium decipiens (493): ቻና canna (Das).

Fragaria spp. (37): strawberry (Eng); የምድር:እንጆሬ yemdr-'njorie (Amh).

Girardinia bullosa (306): ደጉሰጣ deguṣta (Tya); ዶቢ dobi (Or).

Girardinia diversifolia (308): ዶበ doba (Tya); ጎብዱ gobdu, ሂዲ hiddi (Or).

Glycine max (158): አኵሪ:አተር akwri-ater (Amh); soy, soya bean, soybean (Eng).

Glycine wightii (158): glycine (Eng).

Hagenia abyssinica (43): ዱቻ ducca, እጦ eṭo, ፈጦ feṭo, ፊጮ fiço (Or); ጎራ:ጎራ gora-gora (Agew); ሐቢ ḥabbi (Tya); ሀጦት heṭot (Sid); ሀጦ heṭo (Or); ጆሊያ:ሚታ joliya-mita (Wel); ከብሰ kebse (Gur); ኮሶ koso (Amh); ኮሱዋ kosuwa (Wel); ሶበርት sobert (Ge'ez); ሶኖ sonȯ (Tya); ሶሰን sosen (Ge'ez); ሱጦ suṭo (Had); ጣሊያ ṭaliya (Wel); ጤማ ṭiema (Kem); ወንቆቆ wenqoqo (Gam).

Harrisonia abyssinica (437): እዱጎት 'ddugot (Som).

Hedera helix (542): ivy (Eng).

Helinus integrifolius (385): ዳይሉል daylul (Som).

Helinus mystacinus (385): ቀሚዳ qemida (Tya).

Hippocratea africana (342): ድራዕ dra' (Tya); ጋለ galle (Or); ጥርኝ:ሃረግ ṭrñ-hareg (Amh); ቲኦ tiö, ቲቲአ titiä (Or).

Hippocratea pallens (342): ሂናይ hinay (Mes).

Hypagophytum abyssinca (10): የጉጉት:አይን yegugut-ayn (Amh).

Ilex mitis (329): ሃምሲከ hamsika (Or); ከቶ ketto (Kef & Or); ሚኤሰ miëssa, ሚኪቾ mikicco (Or); ሚሲራ:ሸንደራ misira-shendera (Wel); ምሰር:ክፉ mssar-kfu (Amh); ቶኦቶ toöto, ኡንጋደይ ungadey, ወልቅጠ welqtte (Or).

Indigofera amorphoides (134): ገዲ:ገዲ gedi-gedi (Som).

Indigofera arabica (122): ኤለሰ elas (Som).

Indigofera argentea (128): ዕሰለይ 'slay (Tre); አቡር abur (Som).

Indigofera arrecta (135): ቡክ:ቡክ buk-buk (Som); ድግንድግ dgndg (Amh & Tya); ዐለን:አረን ēllan-eren, ዐለን:ጊን ēllan-gin (Som); ዒለም:ብዃሪያ îllam-bḱwarya, ዒለም:ወኸሪያ îllam-weḱarya (Tya); indigo (Eng for the dye).

Indigofera articulata (134): ዒለም:ሃቡት îllam-habut (Tre); ዓለን āllan, ዓሱሩ āssuru, ዱረሆ dureho (Som); ዐለውቶ ēlawto (Afar); ዐለን ēllan (Som); ሐቢ:ሃጸይ ḥabbi-haṣey (Tre); ዒለን îllan (Som); indigo (Eng for the dye).

Indigofera coerulea (134): ዒለም:ሃቡት îllam-habut (Tre); ዓለን āllan, ዓሱሩ āssuru, ዱረሆ dureho (Som); ዐለውቶ ēlawto (Afar); ዐለን ēllan (Som); ሐቢ:ሃጸይ ħabbi-hasey (Tya); ዒለን îllan (Som); indigo (Eng for the dye).

Indigofera hochstetteri (122): ዕርብ ‘rb (Tre); ዓጋቀር āgaqer (Som).

Indigofera oblongifolia (136): አናስ:ማይስ anas-mays (Som); ዐለውቶ ēlawto (Afar); ገድ:ኢርማን ged-irman (Som); ጥርጥር ṭrṭr (Tre).

Indigofera pilosa (127): ሱል:ቱት:መ sul-tut-me (Tre).

Indigofera rothii (126): ክሆሺ khoshi (Or); ናጋርድ naggard (Som).

Indigofera ruspoli (126): ጀሉብ jelub, ጅሉብ jlub (Som).

Indigofera schimperi (136): አጋጋሮ agagaro, አጋር agar, አዋኤር awaër (Som).

Indigofera secundiflora (128): አትቆረጥ attqqorreṭ (Amh).

Indigofera spicata (136): የአይጥ:ምስር yeäyṭ-mssr (Amh).

Indigofera spinosa (126): ዕርብ ‘rb, እንታርታሩ ’ntartaru (Tre); ገድ:ጎደድ ged-goded, ሃጂን hajin (Som); ቀይናን:ዓዱ qeynan-āddu (Tya); ተሪብ terib (Tre).

Indigofera suaveolens (127): የማን:በሪ yeman-berri (Tya).

Indigofera tinctoria (135): indigo (Eng for the dye).

Juglans regia (533): black walnut, walnut (Eng).

Kalanchoe citrina (23): ማንተራ mantera (Som).

Kalanchoe densiflora (23): አንጮሩራ ançorura (Or); አንዳሆ andaho (Gur).

Kalanchoe lanceolata (23): ቢቢለ bibile (Som); ፉርጉገ furguge (Or); ገዲል gedil (Tre); ቆንጦማ qonṭoma (Or); የቆላ:እንዳሁላ yeqolla-’ndahulla (Amh).

Kalanchoe marmorata (20): ደኳታ:ጽሑፍ dek̇wata-ṣħuf, ጽሑፍ ṣħuf (Tya); ዋሃራ:ቢቢየ wahara-bibiye (Som).

Kalanchoe petitiana (20): አምዶሃለ amdohale, አንዶሃሀለ andohahele, በሶቀ bosoqqe, ሃንጩራ hançura (Or); ሙዳ muḍa (Gam); የቆላ:እንዳሁላ yeqolla-’ndahulla (Amh).

Kalanchoe quartiniana (20): አንጩራ ançura (Had); በሶቀ bosoqqe (Or); ገዲል gedil (Tre); ሉጎ lugo (Or); ሙዳ muḍa (Gam); ጻዕዳ:ደኳዕታ sa‘da-dek̇wa‘ta (Tya).

Kalanchoe schimperiana (20): አምዶሃለ amdohala, አንዶሃሀለ andohahele (Or).

Kalanchoe spp. (19-25): እንዳሁላ ’ndahulla, እንደኋሁላ ’ndehwahulla (Amh); ዓንዶር āndor (Tya); በሶቀ bosoqqe (Or); ደኳዕታ dek̇wa‘ta (Tya); ሙዳ muḍa (Gam); shnqaq ሽንቃቅ, ሽቃቅ shqqaq (Amh); ዊጡ wiṭu (Or); የፍየል:ሎሌ yefyyel-lolie(Amh).

Kirkia tenuifolia (439): ኦርክ orok (Som).

Kotchya africana (146): ሀና henna (Or).

Lablab purpureus (179): bonavist bean (Eng); ገረንጋ gerenga (Bil); hyacinth bean, lablab (Eng); ኦካለ okala (Kon); የአሞራ:ጓያ yeämora-gwaya (Amh).

Lannea barteri (518): አርየም aryem (Anu).

Lannea fruticosa (518): በሪ bori (Mur); ድጕድጕና dgwdgwnna (Tya).

Lannea malifolia (516): በሮ bero, ደግነይን degneyn, ጓሃውሪ gwahawri, ሀርያ:ኤጂ hrya-eji, ዉሀ:አንድሪ wuh-andri (Som).

Lannea schimperi (516): አንዳራክ andarak, ዳሪኮ dareko, ዳሪኩ dareku (Or); ድጕድጕና dgwdgwnna (Tya); ዱግና dugna (Agew).

Lannea schweinfurthii (518): ኪሪንገኒ kiringenni (Mur).

Lannea triphylla (514): አብዶ abdo (Tya); ዱግና dugna (Agew).

Laportea aestuans (309): ሸችሾ shecsho (Kef).

Lathyrus aphaca (250): yellow vetchling (Eng).

Lathyrus odoratus (250): sweet pea (Eng).

Lathyrus pratensis (250): common vetchling (Eng).

Lathyrus sativus (250): chickling vetch (Eng); ጋያ gayya (Gur & Or); grass pea (Eng); ጓያ gwayya (Amh); ሰበረ sebbere (Saho & Tya).

Lens culinaris (249): ብርስን brsn (Ge'ez, Tre & Tya); lentil (Eng); ሚሺራ mishira (Had & Kem); ሚሺሬ mishire (Gur); ሚሲራ misira (Or); ሚሲሪያ misiriyya (Wel); ምስር mssr (Ade, Afar & Amh); ሙጧ mutwa (Wel).

Lepisanthes senegalensis (504): ዱራዓስ duraās (Saho & Tya); ማርቸሳ marcesa, ኦሮራ orora (Or); ሽዐሎ sh'lo (Tre); ሽሉሕ shluh (Tya); ሰምቦ sembo (Amh).

Leucaena leucocephala (75): lead tree (Eng); ሉኪና lukina (Amh, Or & Tya).

Lonchocarpus laxiflorus (104): አልዋሮ alwaro (Anu); አመራ amera (Or); ጨንጕረፍያ çengwerefia (Tya); ቺልዌዶ cilwiede (Anu); ከንከሽ kankasha (Or); ራራዋየ rarawaye (Mes); ጸንጕረፍያ şengwerefya (Tya): ኡልዌቶ ulwieto (Anu).

Loranthaceae - general for any species (361-374): ቡቂ buqi, ዲጋሉ digallu (Or); ዲሎ:የሃን dilo-yehan (Som); ድቓላ dqala (Tya); ፌኦ feö (Or); ጋዱ gadu (Som); ሃርሞ harmo, ኢርቶ irto (Or); ቅሮት qrot, ቅሩድ qrud (Tre); ጠንቆኦ tenqoö (Had); ተቀጽላ teqeşslla, ተቀጥላ teqetlla (Amh).

Lotus corniculatus (223): birdfoot trefoil (Eng).

Lotus quinatus (224): ሐማት:የማን:በሪ hamat-yeman-berri (Tya).

Lotus torulosus (225): ዕርብ 'rb (Tre).

Lupinus albus (219): egyptian lupin (Eng); ግብጾ gbşo (Tya); ግብጦ gbto (Amh); white lupin (Eng).

Macroptilium atropurpureum (169): purple bean, siratro (Eng).

Macroptilium lathyroides (169): phasey bean (Eng).

Macrotyloma axillare (166): archer axillaris (Eng).

Macrotyloma uniflorum (166): leichhardt biflorus (Eng).

Malus sylvestris (43): apple (Eng); ፖም pom (Amh); ቱፋሕ tufah (Ge'ez & Tya).

Mangifera indica (519): ማንጎ mango (Eng & some Ethiopian languages); ማንጉስ mangus (Tya).

Maytenus putterlickioides (336): ፎልኮልቻ folkolca, ፎንኮንቻ fonkonca (Or).

Maytenus senegalensis (336): አርግቲ argti (Tre); አርጕዲ argwdi (Tya); ጉሎ:ኮክኮባ gulo-kokkoba (Amh); ሀርግተ hrgte (Tre); ጂማእ:ሃሬ jima'-harre, ኮምቦልቻ kombolca (Or); ነጭ:አጣጥ neç-atat (Amh); ቀብቀብ qebqeb (Tya); ቆቆባ qoqqoba (Amh).

Maytenus serrata (333): ጽልዕሎ şl'lo (Tya).

Maytenus spp. (331-338): ዓጥዓጥ ātāt (Tya); አንጊቶ angito (Kef); አጣጥ atat (Amh & Gur); ሐጫት haçat (Agew & Tya); ኮምቦልቻ kombolca (Or); ጽልዕሎ şl'lo (Tya); ጡጥዋ tutwa (Wel).

Maytenus undata (338): ዓጥዓጥ ātāt (Tya): ጩጮ çuço (Or); ደግሙት degmut, ደግዉን degwun (Som); ኤሊከ elike, ጋለቻ galleca, ኢልካ ilka, ኮምቦልቻ kombolca, ኦኮሉ okollu (Or); ጸሊኒዮ şeliniyo (Som); ጸሊሞ şellimo (Tya); ሰራድ sarad (Som).

Medicago lupulina (244): black medick, hop clover (Eng).

Medicago polymorpha (244) and other species with spiny pods: አሹ ashu (Gum); ሀማቂጣ hamaqita (Or); ከኪቶ kakito, ኳኲቶ kwakwito (Tya); ኵርንችት kwrnct (Amh).

Medicago sativa (244): አልፈልፍ alfelfa (Tya); alfalfa, lucern (Eng).

Medicago spp. (244-245): burweed (Eng); ማገጥ maget (Amh); medic (Eng); ሲዲሳ siddisa (Or).

Melia azedarach (485): chinaberry, indian lilac, persian lilac (Eng).

Melilotus alba (245): white melilot, white sweet clover (Eng).

Melilotus elegans (246): የማን:በሪ yeman-berri (Tya).

Melilotus indica (245): indian sweet clover, small melilot (Eng); የማን:በሪ yeman-berri (Tya).

Melilotus spp. (245-246): melilot (Eng).

Melilotus suaveolens (245): ቲታኮ titako (Or).

Milicia excelsa (272): ዳጎ dago (Anu); ጎንጂ gonji (Kef).

Millettia ferruginea (108): እንግዲቾ 'ngdicco (Or); እንገዲቾ 'ngedicco (Sid); አክሲራ aksira, አስራ asra (Or); ቢቤሮ bibero (Kef); ቢለ ዉ:ሃቃ bilewu-haqa (Had); ብርብራ brbrra (Amh, Gur & Tya); ደዳቱ dedatu, ከርቸቸ karcecce, ኮተሉ kotellu (Or); ረንጋዘና rengazena (Kem); ሳሪ sari (Or); ሶተሊ sotelli (Mes): ሶተሉ sotellu (Or); ያጎ yago (Kef & Or); ያጎዪ yagoy (Mes); ዛግያ zagya, ዛጊ zagie (Gof & Wel); ዝያግ ziyag, ዝ ያ ጉ ziyagu (Sha).

Mimosa pigra (74): እስዋርዎድ esswarwod (Som).

Morus alba (272): white mulberry (Eng).

Morus nigra (272): black mulberry (Eng).

Mucuna pruriens (164): cow itch, velvet bean (Eng).

Myrica salicifolia (261): አበዪ abayi (Or); አበዪ abeyi (Gur); በሮዶ barodo, በርቶ barto, ዳደቾ dadecco (Or); ጋዋዳ gawada (Kem); ከታበ kataba (Or); ንሕቢ nḥbi (Tya); ራጂ raji (Or); ሰጥረጵልዮን seṭreṗlyon (Ge'ez); ሽነት shnet (Amh); ጦና ṭona (Or).

Mystroxylon aethiopicum (338): ጉጋ guga (Som).

Neorautanenia mitis (181): ሰረሮ serero (Tre).

Onobrychis viciifolia (230): sainfoin, holy clover (Eng).

Ononis spp. (246): restharrow (Eng).

Opilia campestris (358): ኤሪ eri (Som).

Ormocarpum muricatum (143): አንቋይ anqway (Amh); ካራ kara (Or).

Ormocarpum pubescens (143): አልቡድያ albudya (Amh); አለንድያ alendya (Tya); አ ዋ ው awaw (Tre).

Ormocarpum trachycarpum (143): አንቋይ anqway (Amh); ካራ kara (Or).

Osyridocarpus schimperianus (382): ቀጣጥሞ qeṭaṭmo (Tya).

Osyris quadripartita (382): አሶሶ asaso (Or); ኸረጽ k̇ereṣ (Tya); ቀረጽ qereṣ (Amh, Tre & Tya); ቀረጥ qereṭ (Amh); ቅሩቢታ:ዓራዝ qrubita-āraz (Saho); ረጋ rega, ዋቶ wato (Or).

Ozoroa insignis (521): ቾብወ cobwe (Mur); ጋሪ garri (Or); ሃምቦዪ hamboy (Gam); ሸለል shelel (Amh); ጠጠራ ṭeṭera (Tya); ውግር:ዓደድ wgr-ādad (Som); ዛንዛ zanza (Tya).

Pappea capensis (508): አብጋማ abgama (Tre); አዳዳክ adadak (Som); አረጋጉት aregagut (Tya); ቢቃ biqa (Or).

Parietaria debilis (320): ብርንሀዮ:ኲና brnhyo-k̇wana (Tya).

Parkinsonia aculeata (56): jerusalem thorn, parkinsonia (Eng).

Parkinsonia scioana (54): ሀርያ: ጐርለሀ hria-gorleh, ለቢ:የር lebi-yer (Som).

Parochaetus communis (233): የምድር:ኮሶ yemdr-koso (Amh).

Phaseolus coccineus (171): scarlet runner bean (Eng).

Phaseolus lunatus (171): butter bean - for the seeds, lima bean (Eng, for the whole plant and for the seed).

Phaseolus spp. (169-171): አዳ ጕራ adagwra (Tya); አዳንጓሬ adangwarrie, አደንጓሬ adengwarrie (Amh); አሸንጓሬ ashangwarre (Or); ባዶ bado (Gam); ቦሎቄ boloqqie - for dry seed, ፋሶልያ fasolya - for the green pod (Amh); ፉሶልያ fusolya (Ade); ሎ ኮ ማ ቲ ያ lokomatiya (Wel); ጳቃሊ p̣aqalie (Gof); ሰልበኮቡሉክ selbokobulluk (Som); ሸ ተ ረ shetere (Gof); ወኤ weë (Kem).

Phaseolus vulgaris (169): common or french or green or haricot or kidney bean (Eng).

Piliostigma thonningii (68): ዓምዓም:ገመል āmām-gemel (Tya); አለማቲ alamati, አለ ማ ቴ alamatie (Amh); አምበርዳ ambarda (Gur); ዳብዲ dabdi (Amh); ፍርቃ frqa (Agew); ኮ ራ kora (Or); ሊሉ lilu (Gur); ቃልቃለ qalqalla, ቃልቃሎ qalqallo (Wel); የቆለ:ዋ ን ዘ

yeqolla-wanza (Amh).

Pistacia aethiopica (530): ሃማር hamar, ሀይስ heys (Som); ጂረሳ jiresa (Or); ሙርኩት murkut, ሲሰየ sisaye, ኡላ፡አሶ ula-aso (Som).

Pistacia falcata (532): ሀርያ፡እል፡ማሊ hrya-el-mali, ሲሰየ sisaye, ሲሰይ sisey (Som).

Pisum sativum (248): ዓይ ኒ፡ዓተር āyni-āter (Tya); አታራ atara (Wel); አተር ater (Amh); አ ተ ራ atera, አተራ፡ዶንጎሎ atera-dongolo (Or); አተረ atere (Gur & Gof); አተሪያ ateriyya (Wel); አተሮ atero (Had); ዳንጉለ dangulle (Or); ጊ ሺ፡ሻአቶ gishi-shaäto (Kef); ጊ ተ እ giteë (Had); ቂሸወ qishewe (Gur).

Pisum sativum var. abyssinicum (248): abyssinian pea (Eng).

Pisum sativum var. sativum (248): pea, field pea, garden pea (Eng).

Pithecellobium dulce (96): madras thorn, manilla tamarind (Eng).

Pittosporum spp. (1-3): አሆት ahot (Amh); አምብል፡በይ ambl-bay (Gur); አምሺካ amshika (Or); አንጋሊት angallit, አንቀዋሊት anqewallit (Amh); አራ ara (Or); በሶ፡አጣል besso-aṭal (Tya); ቦንቾ bonco (Or); ጨቐንጠ çeqenṭe (Tya); ጨካ çeka (Or); ደንግያ፡ሰ በ ር dengya-sebber (Amh); ገዳል gedal (Or); ገንሕ genḥ (Tre); ሃምሺካ hamshika (Or); ሎ ላ lola (Amh); ሚተሺየ miteshiyye (Wel); ራሃ raha (Or); ሺ ኑር shinur (Afar & Som); ሾ ሊ sholie (Amh); ሾሎ sholo (Or); ሶላ sola (Amh); ሶለ sole, ቶሱ tossu (Or); ወ የ ሎ weyelo (Tya); ወይል weyl (Amh).

Platanus orientalis (257): plane tree (Eng).

Platycelyphium voense (101): ቻፍት caft, ሰበን፡ሰበዱ saban-sabadu, ሰበን፡ሰዶሆ saban-sodho (Som).

Polyscias fulva (537): አበራ aberra, አንዳሎ andalo, አንዶጀ andoje, አንኪሳ ankissa, አ ን ታ ሎ antallo, ጉዱባ guduba, ሁዳ huḍa (Or); ካራጮ karaço (Kef); ካራሾ karasho (Or); ከርቦ ኒ kerboni (Sid); ኮሪባ koriba, ታላእ talaö (Or); የዝንጀሮ፡ወንበር yeznjero-wenber (Amh).

Populus spp. (258): poplar (Eng)

Prosopis juliflora (72): mesquite (Eng); ፕሮሶፒስ prosopis (name being adopted in Ethiopian languages).

Prunus africana (41): አቆማ aqoma (Amh); አራራ arara (Had); በሩ beru (Sha); ጋርበ garbe (Wel); ጋርቢ garbi, ጉራዩ gurayu, ሆሚ homi (Or); ካሮ karo (Mao): ሚቺኮ micikko (Sid): ሚእሳ miëssa, ሚኪቾ mikicco, ሙካ፡ዲማ mukka-dima, ሙካ፡ራጃ mukka-raja (Or); ኦንሳ onsa (Wel); red stinkwood (Eng); ሶ ሸ soshe, ሱ ቀ suqqe (Or); ጥቁር፡እንጨት ṭqur-'nçet (Amh).

Prunus armeniaca (41): apricot (Eng).

Prunus communis (42): almond (Eng).

Prunus domestica (42): plum (Eng); ፕሩኝ pruñ (Amh).

Prunus persica (42): ሆ ሀ hoh (Ge'ez); ኮክ kok (Amh, Gur & Ge'ez); ኮ ኪ koki (Or); ኩ ኾ kukw (Tya); peach (Eng); ሰቂኖን seqinon, ስኪኖን skinon, ሰቂኖን sqinon (Ge'ez).

Pterocarpus lucens (105): ማቡንጎ mabungo (Som).

Pterollobium stellatum (54): ጨካ፡ቀመለ çeka-qemele, ጋሩ garu, ጋሩል garul (Or); ጎምእርያ gomöryya (Wel); ጎራ gora, ጎርታ gorta, ሃራንጋማ harangama, ካጂማ kajima (Or); ሊባሕ፡ሐላላሊስ libaḥ-ḥalalalis (Som); ቀንጣፋ qenṭaffa (Amh); ቀንጠጠፈ qenṭeṭefe (Tre); ቀቀዊ qeqewwi (Or); ቆንጥር qonṭr (Amh & Gur); ቍንጠፍጠፈ qwenṭefṭefe (Tya); ቍንጥር qwenṭr (Amh); ጽንዱቆ şnduqo (Gam).

Pyrenacantha malvifolia (351): ኦምበ ombe (Som).

Pyrus communis (43): pear (Eng).

Quercus ilex (265): holm oak, ilex (Eng).

Quercus robur (265): english oak, pedunculate oak (Eng).

Rhamnus prinoides (390): ገብሆ gebho (Agew); ገሻ gesha (Kem & Saho); ገሽዋ geshwa (Wel); ጌሾ giesho (Ade, Amh, Gam, Had, Or, Saho & Tya); ጌሶ gieso (Tya); ጊሸ gishe (Gur); ራሃ raha (Or); ሴዋ siewa (Ge'ez); ጣዶ ṭado (Or); ጣንዶ ṭando (Wel).

Rhamnus staddo (390): ጃጃለ jajale (Som); ኪስታኒ:ሸሃለ kistani-shahala (Saho); ቃዲዳ qadida (Gam & Or); ቁሉኡም quloüm (Saho); ጸዶ ṣeddo (Amh & Tya); ጠዶ ṭeddo (Or); ጠዩ ṭeyyi (Amh); ተነክ tenek, ቴናክ tienak (Ge'ez).

Rhoicissus revoilii (401): አርሞ:ሰገድ armo-saged (Som); ዳጋ:ጨብሰ daga-çebsa (Or); ሀያብ hyab (Som).

Rhoicissus tridentata (401): አባ:ወልዱ abba-weldu (Amh); ዳንጎጎ:ሲያቃ dangogo-siyaqa, ጋለ:ለለ gale-lala (Or); ሐብል:እንሲት ḥabl-'nsit (Tre); ሂዳ:ረፋ hiḍa-refa (Or); ጀንፎቅ jenfoq (Amh); ለሉ lalu (Or); ቃዓሽሮ qaāshro (Tya) ወግል:አስፈስ wedel-asfess (Amh).

Rhus glutinosa (528): እምብስ 'mbs (Amh); አደሰ adessa, አተሰ atessa (Or); ሐምስ ḥams, መጨጮ meçeço (Tre); መንጊ mengi (Tya); ሙጋን mugan (Or); ቅሞ qmmo (Amh); ጻሞ ṣamo (Gam); ሸሙት shemut, ሸሙት:እፍሩስ shemut-efrus (Tre); ጣጠሰ taṭessa (Or); ጠጣዕሎ teṭa'lo (Tya); ወርከሉ werkello (Gam).

Rhus natalensis (523): አታመ atame (Tya); ጫቀማ çaqema, ዳበበቻ dabobecca (Or); ዳበቢስ dabobis (Som); ደበበሰ debobesa (Or); ፋጡቃ faṭuqa, ፍርጣጣ frṭaṭa (Agew); ከቡደይዳ kabudeyda (Kon); ከአይ keäy (Mur); ከቻቾ kecac (Me); ምስጥ:አይብለሽ msṭ-aybelash (Amh); ኦንጋፍሬ ongafrie (Gam); ኦንጋፕሬ ongaprie (Wel); ጣቁማ ṭaquma (Amh); ጣጠሰ taṭessa (Or); ጠጣዕሎ teṭa'lo (Saho & Tya).

Rhus quartiniana (524): ዳበበቻ dabobecca (Or); ቅሞ qmmo, ጣቁማ ṭaquma (Amh); ጣጠሰ taṭessa (Or); ጠጣዕሎ teṭa'lo (Tya).

Rhus retinorrhoea (529): ዓመስ āmes (Tya); ደበሉቻ debelucca (Or); ሲሰይ sisay (Som); ጣጠሰ taṭessa (Or); ጠጣዕሎ teṭa'lo (Tya); ታሎ talo, ትለም tlem (Amh & Or).

Rhus ruspoli (528): ጣጠሰ taṭessa (Or).

Rhus tenuinervis (524): ዳዳራኢያ dadaraîya (Bur).

Rhus vulgaris (529): እምብስ 'mbs (Amh); አታመ atame (Tya); ቅሞ qmmo (Amh); ጣጠሰ taṭessa (Or); የአህያ:ታሉ yeähyya-talo, የእረኛ:ቆሎ ye'rreñña qolo (Amh).

Rhynchosia malacophylla (187): ቡክ:ቡክ buk-buk (Som); ጨጮሖ:ሐማት çeçḥo-ḥamat (Tya); ጊነ ginne (Bil); ሰእር:ሙዱ soör-mudu (Som).

Rhynchosia minima (187): rhynchosia (Eng).

Rosa abyssinica (35): እንጎጫ 'ngoça (Gur); እንቆጦ 'nqoṭo (Or); abyssinian rose (Eng); አርበቅ arbeq (Ge'ez); ዳየሮ dayero (Som); ጋቐ gaq̣a (Tya); ጎራ gora (Or); ማዎርዲ mawordi (Ade); ሜርሰንስ miersens, ሚርሰንስ mirsens, ሚርሰንስ mirsns, ምርሰኔን mrsenien (Ge'ez); ቃቃወ qaqawwe (Or); ቀጋ qega (Amh & Or); ቀቀዊ qeqewwi (Or); ቆሎሽም qoloshm (Tre); ጽጌ:ረዳ:ጪሻ ṣgie-reda-çisha (Wel); ሰጎ sego (Had).

Rosa x richardii (35): ሕብረ:መጽርይ ḥbre-meṣry, መጽርይ meṣry (Ge'ez); mummy-wreath rose (Eng); ጽጌ:ረዳ ṣgie-reda - for all cultivated roses (all Ethiopian languages); sacred rose (Eng).

Rothia hirsuta (195): አቃራ aqqara (Bil); ዳዕሮ:ምድሪ da'ro-mdri (Tya).

Rubus fruticosa (31): blackberry, bramble (Eng).

Rubus idaeus (31): raspberry (Eng).

Rubus spp. (31-33): እንጆረ 'njore (Gur); እንጆሬ 'njorie (Amh); አርጊ argi (Gam); ጎዳ goḍa, ጎርጮ gorço, ጉመረ gumere, ሃልቱፋ haltufa (Or); ሃማሮእ hamaroö (Kem); ሀንጆሪያ henjoriya (Wel); ሂንጃሮ hinjaro (Had); ኰሸሽለ kwesheshlla (Tya); መሊጣኖ melliṭano (Gam).

Ruta chalepensis (419): ጨና:አዳም çena-addam (Tya); ጪራከታ çirakota (Or); herb of grace, rue (Eng); ጻለዶ ṣalado, ጸሎቴ ṣelotie (Gam); ጸናታም ṣenatam (Ge'ez); ጤና:አዳም ṭiena-addam (Amh).

Sageretia thea (386): ዓጋም:ለሕሚ āgam-laḥmi (Tya); አስጎሪ asgori (Or); ሐማት:ጸዶ ḥamat-ṣeddo (Tya); ዒለን îllan (Som); ለበጠሊት lebeṭelit (Tre).

Salix spp. (258): willow, osier, sallow (Eng).

Salix subserrata (258): አሃያ ahaya (Amh); አለለቲ alaleti, አለንቻ alanca, አለልቱ aleltu (Or); አጣ aṭa (Ge'ez); ቦሮዶ borodo (Or); ቡሮ buro (Som); ዱችያ ducya (Gam); እለሰሎሰክ eleselosek (Gum); ሃያ haya (Amh); ኵሒ kwiḥa (Ge'ez & Tya); ለንሶ lanso (Gam); ሪጋ riga (Amh).

Salvadora persica (354): ዓዳይ āday (Saho, Som, Tre & Tya); አዳይ aday (Amh); ዳዳሆ dadaho (Afar); ገጊ gegi (Mur); ሸወልሸ shwelsha (Agew); toothbrush tree (Eng); የሃረር:መፋቅያ yeharer-mefaqya (Amh).

Saxifraga hederifolia (27): አሞራ:ጋሽ amora-gasha (Tya).

Schefflera abyssinica (539): አርፋቱ arfattu (Or); ቦቦሉሁ boboluhu (Som); ጋብትዘ gabtze (Gam); ጋታማ gatama (Or); ገተም getem (Had, Sid & Tya); ሃርፋቱ harfatu, ማርፋቱ marfatu (Or); ኦሮንቼ oronce (Sid); ቁስጥያ qusṭya (Amh); ዉቶ wtto (Kef).

Schefflera volkensii (539): ቡቶ butto (Kef & Or); ሃምሽ hamsha, ካምበሎ kambelo, ኮኮራ kokora, ሉቀ luqqe, ኦሮኒ oroni (Or).

Schinus molle (530): በርበሬ:ጸሊም berbere-ṣellim (Tya); pepper tree (Eng); ጥቁር:በርበሪ ṭqur-berbere (Tya); ጥቁር:በርበሬ ṭqur-berberie (Amh).

Sclerocarya birrea (519): አበንግል abengl (Tre); አበንጕል abengwl (Bil & Tya); ቾብወ cobwe (Mur); ዲዲግሰ didigssa, ዲዲሰ didissa (Or); ጕዕመል gw'mel (Tya).

Scorpiurus muricatus (227): እርወ 'rwe (Tya).

Sedum spp. (10-14): stonecrop (Eng).

Senna alexandrina (63): alexandrian senna (Eng); ጀላሎ:ጀል jelalo-jel (Som); ሙካ:አርበ muka-arba (Or); ሰኑ sanu (Afar); ሰነ:መኪ sene-mekki (Tya); senna (Eng); ሰኖ seno (Agew, Tre & Tya); tinnevelly senna (Eng); ኡተኪ utekki (Tre).

Senna angustifolia (63): ልትኪ ltkki (Tya); ሰኑ sanu (Afar).

Senna bicapsularis (60): rambling cassia (Eng).

Senna holosericea (63): ጃለሎ:አሪ jalelo-ari (Som).

Senna italica (63): dog senna (Eng); ሐልሰኖhalseno (Tre); ጀላሎ:ጀል jelalo-jel (Som); italian senna (Eng); ሰለማካ salamaka (Som); ሰኑ sanu (Afar); ሰነ:መኪ sene-mekki (Amh); ሰኖ seno (Tre & Tya); spanish senna (Eng).

Senna longiracemosa (61): ሾሾሮ shoshoro (Or).

Senna obtusifolia (60): ስርፍራፍ srfraf (Tre).

Senna occidentalis (60): coffee senna (Eng); ሰቻራ saccara (Or); ሹና:ሹና shuna-shuna (Som).

Senna petersiana (59): ደቢሮ debirro (Or); ለለባ:ማር leleba-mar (Amh); ኦሃ oha (Wel).

Senna septentrionalis (59): ሰነ:መኪ sene-mekki (Or).

Senna singueana (59): ቦስኮባሸ boskobassha, ብስብሸ bsbssha (Amh); ቡሸ busha (Agew & Amh); ጉፋ gufa, ጕፍያ gwefyya (Amh); ሃምብሃምቦ hambhambo (Tya); ህምቦይ hmboy (Tre); ቀይ:እንጪት qey-'ntçet (Amh).

Senna truncata (61): ጃለሎ jalelo (Som).

Sesbania goetzii (140): እንጭኒ 'nçinni (Or); አልቅም alqm (Amh); rattlebox (Eng); ሻሕሻሕታ shaḥshaḥta (Tya).

Sesbania leptocarpa (142): ለብያር lebyar (Som).

Sesbania sesban (140): እጭኒ 'çnni, እንጭኒ 'nçnni (Or); አልቅም alqm, በፈፊ bofeffie (Amh); ጪካ çeka (Or); ገት:በዮ get-beyo, ሃለምባል:ብዮት halembal-byot (Som); ሃርጫ harça (Or); ለብያር lebyar, ሎረብ loreb, ራኪለ:ሎኣድ rakile-loäd (Som); ሻሕሻሕታ shaḥshaḥta

(Tya); ሸውሸዌ shewshewwie (Amh); ጠጠም:ዓጋዘን ṭeṭem-āgazen (Tya).

Sesbania somalensis (140): ለብያር lebyar (Som).

Spartium junceum (223): spanish broom (Eng).

Stylosanthes spp. (147): stylo (Eng for fodder species).

Tamarindus indica (66): ሑመር ḥumer (Som & Tya); ሁመር humer (Amh); ሮቃ roqa (Amh, Or & Som); tamarind (Eng); ዑበል ūbel (Tre).

Tapura fischeri (47): ሎበለ lobele (Das).

Teclea nobilis (427): አደሰ addessa, በጋማ begama (Or); ቡሮ buro (Som); ጫኢ çaî (Or); ጨለሕ çeleḥ (Tya); ሃደሰ haḍḍesa, ለለ lela, ለልቻ lelca (Or); መገቶ megeto, መንገረቶ mengereto (Kef); ስእል s'l (Amh); ጽሕለ ṣḥla, ጽልሓ ṣlḥa (Tya); ሽልክ shlk (Tre).

Tephrosia apollinea (116): አኦዋኤር aöwaër, ደየዶኤበ deyedoëbbo (Som).

Tephrosia emeroides (113): የማን:በሪ yeman-berri (Tya).

Tephrosia interrupta (118): ኪሰ kisa (Or).

Tephrosia nubica (116): ዶሮ:ሰበር doro-sebber (Amh); ገድ:አደ ged-adde, ዋን:ኤሪ wan-eri (Som).

Tephrosia obbiadensis (112): ገድ:ሐጂን ged-ḥajin (Som).

Tephrosia polyphylla (116): ገድ:አደ ged-adde, ገድ:ሃጂን ged-hajin, ዋን:ኤሪ wan-eri (Som).

Tephrosia purpurea (115): ብርስናይ brsnay (Tya).

Tephrosia subtriflora (112): ዓረብ āreb (Tya).

Tephrosia villosa (116): ዳርጋማረይ dargamarey, ደርካ derka, ሃጂን hajin (Som).

Tephrosia vogelii (119): የጅብ:አተር yejb-ater (Amh).

Trema orientalis (268): አለለ alele (Or); በለምበል:ዱረድ balambal-dured (Som); ሃለለ halele, ሁዱ:ፋርዳ huḍu-farda, ታለለአ talalaä (Or).

Trichilia emetica (483): cape mahogany (Eng for timber).

Trichocladus ellipticus (252): አደሰ adessa, ማደሰ madessa (Or).

Trifolium burchellianum (237): አልማ alma (Amh); ማጎሺሞ magoshimo (Kef).

Trifolium fragiferum (242): strawberry clover, strawberry-headed clover (Eng).

Trifolium multinerve (241): ደቂቕ:መሲ deqqiq̣-messi (Tya).

Trifolium polystachyum (236): ሸል shal (Amh).

Trifolium pratense (242): common clover, red clover (Eng).

Trifolium repens (237): white clover (Eng).

Trifolium schimperi (241): ጕሰዕ:ማይ gwssa'-mai (Tya).

Trifolium semipilosum (237): kenya white clover (Eng).

Trifolium subterraneum (242): sub clover, subterranean clover (Eng).

Trifolium spp. (233-244): አርዚማ arzima (Wel); clover (Eng); ጕርድማ ኰዮ gwrdmaḱwyo (Tya); ማገጥ mageṭ (Amh); መሲ messi (Tya); ናደ naddo (Kef); ሲዲሰ siddisa (Or); ዋዥማ wažma, የበግ:ለት yebeg-lat (Amh).

Trifolium subterraneum (242): subterranean clover (Eng).

Trigonella foenum-graecum (246): አበዕኸ aba'ḱe (Tya); አቢዕካየ abi'kaye (Afar); አቢሽ abish (Kef); አቢሸ abishe (Gur); አቢሺ abishi (Or); አብሽ absh (Amh); fenugreek (Eng); ገራሮ graro (Kef); ሕልበት ḥlbet (Ade); ሺቆታ shiqota (Had); ሹአ shuö (Kem); ሹ ቆ shuqo (Gam); ሹቅዋ shuqwa (Wel); ሲለን silan (Ge'ez); ሲንቆ sinqo, ሱንቆ sunqo (Or).

Trilepisium madagascariense (299): ጁያ juya, ሾሮ shoro (Or).

Urera hypselodendron (304): ዓልገ ālge (Tya); ሃሊለ halila, ለንጉስቶ langusto (Or); ለንቅሽ lanqsh (Agew & Amh); ቶክኑ tokonu (Or).

Urtica simensis (303): ዶቢ dobbi (Or & Gur); ጉልጉልበ gulgulba (Or); ሃንሰ hansa (Tya); ሰማ samma (Amh); ሰመን somen (Ge'ez); የጨዋ:ምግብ yeçewa-mgb (Amh).

Urtica spp. (303): ዶቢ dobbi (Or); nettle (Eng).

Urtica urens (303): ዶቢ dobbi (Or); ዶበ doba (Tya).

Vahlia somalensis (30): አጆ ajo, ዋንዴክ wandek (Som).

Vepris dainelli (424): አዴሳ ađdessa, ሃዴሳ hađdessa, ሐዴሳ ḣeđdesa, ኩላሳ kulasa (Or).

Vermifrux abyssinica (227): ፈውሲ:ቅርጸት fewsi-q̇wrṣet (Tya).

Vicia faba (249): ዓልቋይ ālqway, ዓተር:ባሕሪ āter-baḣri (Tya); አተረ atere (Gur); አውጤበድል awṭiebedl (Ge'ez); በኤላ baîëla (Gur); በልደንጓ baldengwa (Tya); በቀል baqel (Had); በቀላ baqela (Or); በቄላ baqiela (Ade, Afar, Amh, Kem & Wel); በድል bedl (Ge'ez); broad bean, faba bean, field bean (Eng); ፉል ful (Ge'ez); horse bean (Eng); ቆሞሎ qomolo (Gam).

Vicia sativa (249): ሰበረ:አንጭዋ sebbere-ançwa, ሰበረ:ጓሶት sebbere-gwasot (Tya).

Vicia spp. (249): vetch, tares (Eng).

Vigna aconitifolia (177): moth bean (Eng).

Vigna frutescens (174): ቍሩሕ:ዲበላ qwruḣ-dibela (Tya).

Vigna membranacea (173): ዕንትርታሮት 'ntrtarot (Tya).

Vigna radiata (177): green gram, mung bean (Eng).

Vigna subterranea (173): bambara groundnut (Eng); ለውዝ lewz (Amh), name now transferred to Arachis hypogea.

Vigna unguiculata (174): አዳጕራ adagwra, አዳጕራ:ቈላ adagwra-q̇wella (Tya); አደንጓሬ adengwarrie (Amh); አተራ:አርጎበ atera-argobba (Or); ጋስያ gasya (Sha); ጎበ gobo (Kef); ኮታተ kotate (Or); ላኮማ lakoma (Wel); ሰልቦክ:ደገል salboko-degel (Som); ሰንከዕሩር senka'rur (Ge'ez); ወኤ weë (Kem); ዎህያ wohya, ዉጭእ wuçë (Wel).

Vigna unguiculata subsp. cylindrica (174): catjang, catjang cowpea (Eng).

Vigna unguiculata subsp. sesquipedata (174): yard-long bean (Eng).

Vigna unguiculata subsp. unguiculata (174): cowpea (Eng).

Vigna vexillata (176): ቍሩሕ:ዲበላ qwruḣ-dibela (Tya).

Viscaceae - general for any species (375-378): ቡቂ buqi, ዲጋሉ digallu (Or); ዲሎ:የሃን dilo-yehan (Som); ድቓላ dq̇ala (Tya); ፌኦ feö (Or); ጋዱ gadu (Som); ሃርሞ harmo, ኢርቶ irto (Or); mistletoes (Eng); ቅሮት qrot, ቅሩድ qrud (Tre); ጠንቆእ tenqoö (Had); ተቀጽላ teqeṣṣlla, ተቀጥላ teqeṭṭlla (Amh).

Vitis vinifera (399): እናበ 'naba (Or); እንኩቲ 'nkuti (Ge'ez); ዓይነብ āyneb (Ade); አስካለ:ወይን askale-weyn for a bunch of grapes, አስካሎን askalon - for the plant (Ge'ez); grape - the fruit, grape vine - the plant (Eng); ነቢት nebit - for wine (Tya); sultanas - dried sweet grapes (Eng); ወይን:ጠጅ weyn-ṭej - for wine (Amh); ወይኒያ weyniyya (Wel); ወይን weyn (Amh); ወይኒ weyni (Gur & Tya); ወይነ:ጸግ weyne-ṣeg - wine (Ge'ez); wine - for alcoholic drink made from grapes (Eng); የወይን:ዘለላ yeweyn-zelela - for a bunch of grapes (Amh); ዘቢብ zebib - for dried sweet grapes (Amh, Ge'ez & Tya); ዘለላ:ወይኒ zelela-weyni - for a bunch of grapes (Tya).

Ximenia americana (356): እንኮይ 'nkoy (Amh); አስቴ astie (Wel); አውሬ:ሙዱበ awre-mudube (Som); ሃዝተ hazte (Gam); hog plum (Eng); ሁዳ huđa (Or); ሁዳየ hudaye (Som); ሁዲ huđi (Or); ኮል kol (Amh); ማጋላ magala (Or); ማንዳሩት mandarut (Som); ምልሕጣ mlḣtta (Tre); ምልዖ mlō (Tya); ሞንድሩክ mondruk, ሞርሆድ morhod (Som); monkey plum (Eng); ኦለአሞ oleämo (Anu); ራቃካ raqaka, ሶቢቃካ sobieqaka (Ge'ez); tallow nut (Eng); ጡጡቃ tuṭuqa (Agew).

Ximenia caffra (356): ሁዳ:ጃልዶ huđa-jaldo (Or); ሁረ:አዳየ hure-adaye (Som); ኢንጊንካካዳ inginkakada (Kon); ማጋላ magala (Mur); ማንዳሩት mandarut (Som); ሞርኮርተ morcorte (Das); ሙዳሩክ mudaruk, ሙላቸ mulace (Som).

Zahna golungensis (493): ዎርጋ worga (Sha); ዎርጋየ worgaye (Anu).

Ziziphus abyssinicus (396): አበተረ abettere (Amh & Tya); ገበ:አድጊ geba-adgi (Tya); ላንጋ

lang (Anu); ቁርቁራ qurqura (Or).

Ziziphus hamur (396): ሐሙድ ħamud, ሐሙር ħamur (Som).

Ziziphus mauritiana (396): ጋብ gab, ጋበ gaba, ጋቢ gabi (Mur); ጋቢተ gabite, ጋርቢለ garbile (Das); ጎብ gob (Som); ቁርቁራ qurqura (Or).

Ziziphus mucronata (393): አዶ:ቁርቁራ ado-qurqura (Amh & Or); buffalo thorn, cape thorn (Eng); እዲ:ሸበል eddi-shebel (Som); ፎጭ foç (Amh); ጋበ gaba (Amh & Tya); ጋበ:ሐርማዝ gaba-ħarmaz (Tre); ጋሞ:ጋዴ gammo-gadie (Wel); ገበ geba (Agew, Amh & Tya); ጎብየር gobyer, ሐማር ħamar, ሐራር ħarar, ሐርካይ ħarkay (Som); ካና:ካስለ kana-kasle (Gam).

Ziziphus pubescens (393): አራሞር aramor, አሮሞ aromo (Anu); ዳንግኰ dangkwe, ገልፊ gelfi (Mur); ሃዱ hadu (Gum); ሎሞሮኪን lomorokin (Das).

Ziziphus spina-christi (395): christ's thorn, crown of thorns (Eng); ገበ gaba (Agew, Amh & Tya); ገብ geb (Som); ገበ geba (Amh & Tya); ጎብ gob (Som); ጉበ guba (Bil); ክሬናሲ krenasi (Som); ክስራ ksra, ቅልቅለ qlqlla, ቅስለ qsla (Tre); ቁርቁራ qurqura (Or); ቍርቍራ qwrqwra (Amh).

Ziziphus spp. (393-396): jujube - for edible fruits (Eng).

ETHIOPIC NAMES ARRANGED ALPHABETICALLY

(followed by scientific names and page number references)

እጭ ኒ ’çnni (Or) - Sesbania goetzii, Sesbania sesban (140).
እዱጎት ’ddugot (Som) - Harrisonia abyssinica (437).
እሕብ ’ḧb (Som) - Cordeauxia edulis (56).
እምብልታ ’mblta (Tya) - Entada spp. (71-73).
እምብሪቶ ’mbritto (Kef) - Clausena anisata (429).
እምብስ ’mbs (Amh) - Allophyllus abyssinicus (499), Rhus glutinosa (528), Rhus vulgaris (529).
እምብስ ’mbs (Amh) - Allophyllus abyssinicus (499).
እናበ ’naba (Or) - Vitis vinifera (399).
እንጭ ኒ ’nçnni (Or) - Sesbania goetzii, Sesbania sesban (140).
እንጮሩራ ’nçorura (Or) - Kalanchoe densiflora (23).
እንዳዓሮ ’ndaāro (Saho) - Ficus vasta (296).
እንዳሁላ ’ndahulla (Amh) - Kalanchoe spp. (19-25).
እንደኋ ሁላ ’ndehwahulla (Amh) - Kalanchoe spp. (19-25).
እንግዲቾ ’ngdicco (Or) - Millettia ferruginea (108).
እንገዲቾ ’ngedicco (Sid) - Millettia ferruginea (108).
እንጎጫ ’ngoça (Gur) - Rosa abyssinica (35).
እንጉታት ’ngutat (Ge’ez) - Citrus medica (430).
እን ጐት ’ngwet (Ge’ez) - Citrus medica (430).
እንጆሬ ’njore (Gur) - Rubus spp. (31-34).
እንጆሪ ’njorie (Amh) - Rubus spp. (31-34).
እንክይ ’nkoy (Amh) - Ximenia americana (356).
እንኩቲ ’nkuti (Ge’ez) - Vitis vinifera (399).
እንቆጦ ’nqoṭo (Or) - Rosa abyssinica (35).
እንታርታሩ ’ntartaru (Tre) - Indigofera spinosa (126).
እርብ:ክታን ’rb-ktan (Tre) - Crotalaria microphylla (213).
እርጎፍት ’rgoft (Amh) - Erythrina abyssinica (161).
እርወ ’rwe (Tya) - Scorpiurus muricata (227).
እጣን ’ṭan (Amh) - Boswellia resin (442-445).

ዕንድሩር ‘ndrur (Tya) - Balanites aegyptiaca (433)
ዕን ቊ:ህበይ ‘nqwi-hbey (Tya) - Erythrina abyssinica (161).
ዕን ቊ:ዝብኢ ‘nqwi-zbî (Tya) - Erythrina abyssinica (161).
ዕንትርታሮት ‘ntrtarot (Tya) - Vigna membranacea (173).
ዕቋ ‘qwa (Agew) - Commiphora africana (454).
ዕርብ ‘rb (Tre) - Indigofera hochstetteri (122), Indigofera spinosa (126), Lotus torulosus (225).
ዕፀ:በለስ ‘ṣe-beles (Ge’ez) - Ficus carica (280), Ficus palmata (281).
ዕፀ:ፋርስ ‘ṣe-fars (Ge’ez) - Cannabis sativa (327).
ዕፀ:ዘዌ ‘ṣe-zewie (Ge’ez) - Cyphostemma molle (411).
ዕስላይ ‘slay (Tre) - Indigofera argentea (128).
ዕጣን ‘ṭan (Ge’ez & Tya) - Boswellia resin (442-445).

ዓበ āba (Tya) - Berchemia discolor (270).
ዓበክ ābak (Som) - Acacia tortillis (87).
ዓዳድ ādad (Som) - Acacia asak (80), Acacia senegal (75).

ዓዳድ:መዱ ādad-medu (Som) - Acacia asak (80).
ዓዳድ:መሩ ādad-meru (Som) - Acacia senegal (75).
ዓዳይ āday (Saho, Som, Tre & Tya) - Salvadora persica (354).
ዓዲቀንጦ ādiqenṭo (Afar) - Acacia seyal (85).
ዓፍሎ āflo (Saho) - Acacia nilotica (86), Acacia seyal (85).
ዓጋም:ለሕሚ āgam-laḥmi (Tya) - Sageretia thea (389).
ዓጋቀር āgaqer (Som) - Indigofera hochstetteri (122).
ዓካብ ākab (Som) - Acacia tortilis (87).
ዓላል ālal (Tya) - Cissus quadrangularis (407).
ዓለብ āleb (Tre) - Acacia ogadensis (80).
ዓልገ ālge (Tya) - Cissus petiolata (406), Cissus quadrangularis (407), Urera hypselodendron (304).
ዓላ ālla (Tya) - Acacia abyssinica (89), Acacia tortillis (87).
ዓላን āllan (Som) - Indigofera articulata, Indigofera coerulea (134).
ዓልቀ ālqe (Tya) - Cissus petiolata (406), Cissus quadrangularis (407), Cyphostemma niveum (415).
ዓልቋይ ālqway (Tya) - Vicia faba (249).
ዓምዓም:ገመል āmām-gemel (Tya) - Piliostigma thonningii (68).
ዓማይ āmay (Afar) - Delonix elata (50).
ዓንዶር āndor (Tya) - Kalanchoe spp. (19-25).
ዓንቄባ ānqieba (Tya) - Acacia tortillis (87).
ዓንቋ ānqwa (Tya) - Commiphora africana (454), Commiphora habessinica (471), Commiphora quadricincta (471), Commiphora samharensis (452), Commiphora schimperi (454).
ዓንቋ ānqwa (Afar) - Commiphora quadricincta (471).
ዓቅባ āqba (Tre) - Acacia ehrenbergiana (86), Acacia tortillis (87), Acacia albida (84).
ዓረብ āreb (Tya) - Tephrosia subtriflora (112).
ዓሸ:ኦም āsha-om (Tya) - Bersama abyssinica (511).
ዓሱሩ āssuru (Som) - Indigofera articulata (134).
ዓጥዓጥ āṭāṭ (Tya) - Maytenus undata (338) and other Maytenus spp. (331-338).
ዓተር āter (Ge'ez & Tya) - Cicer arietinum (246).
ዓተር:ኣንጯዋ āter-ançwa (Tya) - Cicer cuneatum (248).
ዓተር:ባሕሪ āter-baḥri (Tya) - Vicia faba (249).
ዓተር:ጓሶት āter-gwasot (Tya) - Cicer cuneatum (248).
ዓተር:ቀይሕ āter-qeyyḥ (Tya) - Cicer arietinum (246).
ዓቲር ātir (Saho) - Cicer arietinum (246).
ዓይነብ āyneb (Ade) - Vitis vinifera (399).
ዓይኒ: ዓተር āyni-āter (Tya) - Pisum sativum (248).
ዓይኒ:ዋሪ āyni-wari (Tya) - Chamaecrista absus (64). Chamaecrista nigricans (65).

ኣባዕኸ aba'k̇e (Tya) - Trigonella foenum-graecum (247).
ኣባዕኸ:ባይታ aba'k̇e-bayta (Tya) - Antopetitia abyssinica (227).
ኣባሎ abalo (Amh & Or) - Brucea antidysenterica (440).
ኣባር abar (Or) - Allophyllus abyssinicus (499).
ኣባዪ abayi (Or) - Myrica salicifolia (261).
ኣባይታባታ abaytabata (Or) - Bauhinia tomentosa (68).
ኣባ:ጮናኔ:እንጪት abba-çonanie-'nçet (Amh) - Crotalaria rosenii (199).
ኣባ:ወልዱ abba-weldu (Amh) - Rhoicissus tridentata (401).
ኣብዶ abdo (Tya) - Lannea triphylla (514).

አበን ጕል abengwl (Bil, Tre & Tya) - Sclerocarya birrea (519).
አበራ aberra (Or) - Allophylus abyssinicus (499), Polyscias fulva (537).
አበተረ abettere (Amh & Tya) - Ziziphus abyssinica (396).
አበዪ abeyi (Gur) - Myrica salicifolia (261).
አብጋማ abgama (Tre) - Pappea capensis (508).
አቢዕከየ abi'kaye (Afar) - Trigonella foenum-graecum (247).
አቢሽ abish (Kef) - Trigonella foenum-graecum (247).
አቢሸ abishe (Gur) - Trigonella foenum-graecum (247).
አቢሺ abishi (Or) - Trigonella foenum-graecum (247).
አብሽ absh (Amh) - Trigonella foenum-graecum (247).
አቡር abur (Som) - Indigofera argentea (128).
አጭቅ:ግራረ açq-grar (Amh) - Acacia pilispina (89).
አዳዳክ adadak (Som) - Pappea capensis (508).
አዳዶ adado (Or) - Barbeya oleoides (270).
አዳ ጕራ:ቈላ adagura-qwella (Tya) - Vigna unguiculata (174).
አዳጉራ:ጤል adagura-tiel (Tre) - Chamaecrista absus (64).
አዳ ጕራ adagwra (Tya) - Canavalia ensiformis (165); Phaseolus spp. (169-171), Vigna unguiculata (174).
አዳ ጕራ:ጓሶት adagwra-gwasot (Tya) - Dolichos sericeus (179).
አዳሐለት adahalat (Saho) - Dodonaea angustifolia (491).
አዳል adal (Som) - Acacia mellifera (81).
አዳልኸት adalkat (Saho) - Dodonaea angustifolia (491).
አዳይ aday (Amh) - Salvadora persica (354).
አደሳ addessa (Or) - Teclea nobilis (427), Vepris dainelli (424).
አደንጓሬ adengwarrie (Amh) - Canavalia ensiformis (165), Phaseolus spp. (169-171); Vigna unguiculata (174).
አደር ader (Amh) - Dichrostachys cinerea (74).
አደሳ adessa (Or) - Clausena anisata (429), Dichrostachys cinerea (74), Trichocladus ellipticus (252).
አደዮው adeyow (Anu) - Ficus vallis-choudae (289).
አዶ:ቁርቁራ ado-qurqura (Amh) - Ziziphus mucronata (393).
አፍ:ቀሞ af-qemmo (Tya) - Ficus hochstetteri (299).
አፋጀሽኝ affajeshñ (Amh) - Bersama abyssinica (511).
አፍራርቱ afrartu (Or) - Erythrina abyssinica, Erythrina brucei (161).
አጋቦ agabo (Som) - Acacia nilotica (86).
አጋጋሮ agagaro (Som) - Indigofera schimperi (136).
አጋር agar (Som) - Indigofera schimperi (136).
አጋርሱ agarsu (Or) - Commiphora ogadensis (460).
አጉንሳ agunsa (Gam) - Acacia oerfota (90).
አሃራጉ aharagu (Som) - Cussonia spp. (538-540).
አሃያ ahaya (Amh) - Salix subserrata (258).
አሆት ahot (Amh) - Pittosporum spp. (1-3).
አጆ ajjo (Alaba & Tya) - Acacia etbaica (88), Acacia oerfota (90).
አጆ ajo (Som) - Vahlia somalensis (30).
አክርሳ akkrsa (Gam) - Acacia senegal (75).
አክሲራ aksira (Or) - Millettia ferruginea (108).
አ ኰሪ akwri-ater (Amh) - Glycine max (158).
አላለቲ alaleti (Or) - Salix subserrata (258).
አላማቲ alamati (Amh) - Piliostigma thonningii (68).

አለማቴ alamatie (Amh) - Piliostigma thonningii (68).
አለንቻ alanca (Or) - Salix subserrata (258).
አልቡድያ albudya (Amh) - Ormocarpum pubescens (143).
አለደዋ aledewa (Or) - Kalanchoe petitiana, Kalanchoe schimperiana (20).
አለጁ aleju (Or) - Allophylus sp. nov. (502).
አለለ alele (Or) - Trema guineensis (270).
አለልቱ aleltu (Or) - Salix subserrata (258).
አለንድያ alendya (Tya) - Ormocarpum pubescens (143).
አሊ:ቦየ ali-boye (Som) - Commiphora rostrata (462); Commiphora ancistrophora (475).
አለል allal (Mur) - Boswellia neglecta (445).
አለለ allalla (Mur) - Boswellis neglecta (445).
አልማ alma (Amh) - Trifolium burchellianum (237).
አልቅም alqm (Amh) - Sesbania goetzii, Sesbania sesban (140).
አልዋሮ alwaro (Anu) - Lonchocarpus laxiflorus (104).
አማለቃ amalaqqa (Or) - Celtis africana (266).
አምበበስ ambabes (Or) - Albizia spp. (93-96).
አምበበሳ ambabessa (Or) - Albizia gummifera, Albizia schimperiana (96).
አምበጪ ambaçi (Or) - Aeschynomene elaphroxylon (144).
አምበርዳ ambarda (Gur) - Piliostigma thonningii (68).
አምበልታ ambelta (Amh & Or) - Entada spp. (71-73).
አምቢልሽ ambillsh (Gum) - Erythrina abyssinica (161).
አምብል:ባይ ambl-bay (Gur) - Pittosporum spp. (1-3).
አምበ ambo (Or) - Acacia abyssinica (89).
አምደሃለ amdohala (Or) - Kalanchoe schimperiana, Kalanchoe petitiana (20).
አመለቃ ameleqqa (Amh) - Celtis africana (266).
አመራ amera (Or) - Lonchocarpus laxiflorus (104).
አመዘዘ amezaze (Or) - Acacia brevispica (78).
አሞር amor (Som) - Berchemia discolor (393).
አሞራ:ጋሸ amora-gassha (Tya) - Saxifraga hederifolia (27).
አምሺከ amshika (Or) - Pittosporum spp. (1-3).
አናስ:ማይስ anas-mays (Som) - Indigofera oblongifolia (136).
አንጫቢ ançabi (Or) - Calpurnia aurea (104).
አንጩራ ançura (Had) - Kalanchoe quartiniana (20).
አንዳሆ andaho (Gur) - Kalanchoe densiflora (23).
አንዳሎ andalo (Or) - Polyscias fulva (537).
አንዳራክ andarak (Or) - Lannea schimperi (516).
አንደሮቺ anderocci (Or) - Cussonia spp. (538-540).
አንደሃሀለ andohahele (Amh & Or) - Kalanchoe petitiana, Kalanchoe schimperiana (20).
አንደጄ andoje (Or) - Polyscias fulva (537).
አንጋሊት angallit (Amh) - Pittosporum spp. (1-3).
አንጊቶ angito (Kef) - Maytenus spp. (331-338).
አንጉለ angule (Or) - Commiphora tenuis (469).
አንኪሳ ankissa (Or) - Polyscias fulva (537).
አንቀዋሊት anqewallit (Amh) - Pittosporum spp. (1-3).
አንቋ anqwa (Amh) - Commiphora africana (454), Commiphora habessinica (471), Commiphora schimperi (454).
አንቋይ anqway (Amh) - Ormocarpum muricatum (143), Ormocarpum trachycarpum (143).
አንታሎ antallo (Or) - Polyscias fulva (537).

አንቶሮ antoro (Som) - Cordyla somalensis (70).
አኦዋእር aöwaër (Som) - Tephrosia apollinea (116).
አቆማ aqoma (Amh) - Prunus africanus (41).
አቃራ aqqara (Bil) - Rothia hirsuta (195).
አራ ara (Or) - Pittosporum spp. (1-3).
አራሞር aramor (Anu) - Ziziphus pubescens (393).
አራራ arara (Had) - Prunus africanus (41).
አርበሀ arbeh (Mes) - Ficus sycomorus (285).
አርበቅ arbeq (Ge'ez) - Rosa abyssinica (35).
አረጋጉት aregagut (Tya) - Pappea capensis (508).
አረጄ areje (Or) - Allophyllus abyssinicus (499).
አርፋቱ arfattu (Or) - Erythrina abyssinica, Erythrina brucei (161); Cussonia spp. (538-540), Schefflera abyssinica (539).
አርጊ argi (Gam) - Rubus spp. (31-34).
አርግቲ argti (Tre) - Maytenus senegalensis (336).
አር ጕዲ argwdi (Tya) - Maytenus senegalensis (336).
አሪጋ ariga (Ge'ez) - Albizia spp. (93-96).
አርመዲን armedin (Som) - Cissus quadrangularis (407).
አርሞ armo (Som) - Cissus cactiformis (407), Cissus petiolata (406), Cissus quadrangularis (407), Cissus rotundifolia (408), Cyphostemma adenocaule (412), Cyphostemma burgeri (417); Cyphostemma engleri (416).
አርሞ:ዲን armo-din (Som) - Cyphostemma adenocaule (412).
አርሞ:ጎሬ armo-gorie (Som) - Cyphostemma serpens (417).
አርሞ:ማለሀለ armo-malehle (Som) - Cyphostemma serpens (417).
አርሞ:ሰገድ armo-saged (Som) - Rhoicissus revoilii (401).
አሮሞ aromo (Anu) - Ziziphus pubescens (393).
አርሸማርሸ arshamarsha (Tya) - Dichrostachys cinerea (74).
አርሸ:መርሸ arshe-mersha (Amh) - Dichrostachys cinerea (74).
አርየም aryem (Anu) - Lannea barteri (518).
አርዘ:ሊባኖስ arze-libanos (Amh) - Casuarina spp. (262-264).
አርዚማ arzima (Wel) - Trifolium spp. (233-244).
አጽሰን aṣsan (Tre) - Ficus capreaefolia (281), Ficus palmata (281).
አሳሶ asaso (Or) - Osyris quadripartita (382).
አሰርኩኸ aserkuka (Tya) - Cyphostemma adenocaule (412).
አስጎሪ asgori (Or) - Sageretia thea (389).
አሸንጓረ ashangwarre (Or) - Phaseolus spp. (169-171).
አሹ ashu (Gum) - Medicago polymorpha (244).
አስካለ:ወይን askale-weyn (Ge'ez) - for a bunch of grapes, Vitis vinifera (399).
አስካ ሉን askalon (Ge'ez) - for the plant, Vitis vinifera (399).
አስራ asra (Or) - Millettia ferruginea (108).
አሰርኩሽ asserkush (Amh) - Cyphostemma adenocaule (412).
አሰርኩሽ:ተበተብኩሽ asserkush-tebetebkush (Amh) - Cyphostemma adenocaule (412).
አሱር assur (Som) - Crotalaria comanestiana (202).
አሱሮ assuro (Som) - Crotalaria albicaulis (207), Crotalaria comanestiana (202), Crotalaria pycnostachya (205).
አሱሩ assuru (Som) - Indigofera coerulea (134).
አስቴ astie (Wel) - Ximenia americana (356).
አጣ aṭa (Ge'ez) - Salix subserrata (258).
አጣጥ aṭaṭ (Amh) - Maytenus spp. (331-338).

አጣጤ atatie (Agew) - Dichrostachys cinerea (74).
አታመ atame (Tya) - Rhus natalensis (523), Rhus vulgaris (529).
አታራ atara (Wel) - Pisum sativum (248).
አተር ater (Amh) - Pisum sativum (248).
አተረ atere (Gur) - Pisum sativum (248).
አተራ atera (Or) - Pisum sativum (248).
አተራ:አርጎበ atera-argobba (Or) - Vigna unguiculata (174).
አተራ:ዶንጎ ሎ atera-dongolo (Or) - Pisum sativum (248).
አተሬ aterie (Gur) - Vicia faba (249).
አተሪያ ateriyya (Wel) - Pisum sativum (248).
አተሮ atero (Had) - Pisum sativum (248).
አተሰ atessa (Or) - Rhus glutinosa (528).
አትንኩይ atnkuy (Amh) - Acacia mellifera (81).
አትቆረጥ attqqorret (Amh) - Indigofera secundiflora (128).
አዋአር awaër (Som) - Indigofera schimperi (136).
አዋው awaw (Tre) - Ormocarpum pubescens (143).
አውረ:ሙዱበ awre-mudube (Som) - Ximenia americana (356).
አውጤበድል awtiebedl (Ge'ez) - Vicia faba (249).
አዘምር azamr (Amh) - Bersama abyssinica (511).

በ:አሮር ba-aror (Som) - Commiphora erlangeriana (448), Commiphora sphaerocarpa (468), Commiphora staphyleifolia (448).
በበ ኑሲ babanusi (Or) - Dalbergia melanoxylon (105).
በቻል baccal (Som) - Delonix baccal (51).
በዳን baddan (Or) - Balanites rotundifolia (436).
በዳኖ baddanno (Or) - Balanites aegyptiaca (433).
በደና badena (Wel) - Balanites aegyptiaca (433).
በዶ bado (Anu) - Ficus exasperata (248).
በድአ badö (Gam) - Phaseolus spp. (169-171).
በሀረ:ሎሚ bahre-lomi (Amh) - Citrus aurantium (432).
በ ኢላ baiela (Gur) - Vicia faba (249).
በካል bakal (Som) - Delonix baccal (51).
በላምባል balambal-dured (Or) - Trema orientalis (270).
በላንሶፊ balansofi (Or) - Ficus exasperata (248).
በላንጣዩ balantayi (Or) - Ficus exasperata (248).
በላሰ balas (Saho) - Ficus capreaefolia (281), Ficus palmata (281).
በልደንጓ baldengwa (Tya) - Vicia faba (249).
በምበ bamba (Agew & Amh) - Ficus sycomorus (285).
በም ቡለዴ bambuledie (Amh) - Ficus vallis-choudae (289).
በቀል baqel (Had) - Vicia faba (249).
በቀለ baqela (Or) - Vicia faba (249).
በ ቄለ baqiela (Ade, Afar, Amh, Kem & Wel) - Vicia faba (249).
በሮዶ barodo (Or) - Myrica salicifolia (261).
በርሶዋ barsowa (Or) - Commiphora serrulata (469).
በርቶ barto (Or) - Myrica salicifolia (261).
በርቱ bartu (Or) - Erythrina abyssinica, Erythrina brucei (161).
በተ bate (Or) - Acacia persiciflora (81).
በየ:መዶው baye-medow (Som) - Boswellia microphylla (445), Boswellia rivae (443).
በዮ bayo (Som) - Commiphora corugata (460).

ባዮ:አድ bayo-ad (Som) - Commiphora corrugata (460).
ባዝራ:ግራር bazra-grar (Amh) - Acacia abyssinica (89).
በደና beddenna (Wel) - Balanites aegyptiaca (433).
በደኖ beddenno (Amh) - Balanites aegytiaca (433).
በድል bedl (Ge'ez) - Vicia faba (249).
በጋማ begama (Or) - Teclea nobilis (527).
በሕበይ beḣbey (Afar) - Acacia tortillis (87).
በ ኰረ bek̇wre-lemin (Tya) - Citrus aurantium (432).
በለስ beles (Amh, Ge'ez & Tya) - Ficus capreaefolia (281), Ficus carica (280), Ficus palmata (281).
በለስ: ቍላ beles-qwella (Tya) - Ficus capreaefolia (281).
በልዛ belza (Agew) - Ficus palmata (281).
በርበረ:ጸሊም berbere-ṣellim (Tya) - Schinus molle (530).
በርደ berde (Som) - Ficus glumosa (293), Ficus populifolia (295), Ficus vasta (296).
በርደለ berdele (Dor) - Apodytes dimidiata (348).
በሮ bero (Kef) - Erythrina abyssinica, Erythrina brucei (161).
በሮ bero (Som) - Lannea malifolia (516).
በሶ:አጣል besso-aṭal (Tya) - Pittosporum spp. (1-3).
በጠጦ beṭeṭo (Tre) - Cotyledon barbeyi (19).
በ ጡባ beṭuba (Wel) - Acacia senegal (75).
ቢበሮ bibero (Kef) - Millettia ferruginea (108).
ቢ ቢለ bibile (Som) - Kalanchoe lanceolata (23).
ቢለ ዉ:ሃቃ bilewu-haqa (Had) - Millettia ferruginea (108).
ቢ ቃ biqa (Or) - Pappea capensis (508).
ቢሶም bisom (Ge'ez) - Ficus sycomorus (285) and many other Ficus spp. ((278-299).
ብልዓል blāl (Som) - Acacia mellifera (81).
ቦባቦፈ bobabofe (Or) - Aeschynomene elaphroxylon (144).
ቦባ bobba (Gam) - Ficus sur (287).
ቦ ቢ bobi (Anu) - Ficus vallis-choudae (289).
ቦቦ:ሃለ bobo-halle (Som) - Crotalaria retusa (216).
ቦቦ ሉ ሁ boboluhu (Som) - Schefflera abyssinica (539).
ቦዳር bodar (Som) - Celtis africana (266).
ቦፈፌ bofeffie (Amh) - Sesbania sesban (140).
ቦፎፈ bofoffe (Or) - Aeschynomene elaphroxylon (144).
ቦ ሎቄ boloqqie (Amh) - Phaseolus spp. (169-171).
ቦንቾ bonco (Or) - Pittosporum spp. (1-3).
ቦቆ boqo (Kef & Or) - Bersama abyssinica (511).
ቦሪ bori (Mur) - Lannea fruticosa (518).
ቦሮዶ borodo (Or) - Salix subserrata (258).
ቦርቶ borto (Or & Gam) - Erythrina abyssinica, Erythrina brucei (161).
ቦርቷ bortwa (Wel) - Erythrina abyssinica, Erythrina brucei (161).
ቦስከባሽ boskobassha (Amh) - Senna singueana (59).
ቦሶቀ bosoqqe (Or) - Kalanchoe petitiana (20), Kalanchoe quartiniana (20) and other Kalanchoe spp. (19-25).
ቦዩቱ boytu (Gum) - Ficus vallis-choudae (289).
ብርብራ brbrra (Amh, Gur & Tya) - Millettia ferruginea (108).
ብርንህዮ:ኳና brnhyo-k̇wana (Tya) - Parietaria debilis (320).
ብርስማ brsma (Tya) - Bersama abyssinica (511).
ብርስን brsn (Ge'ez, Tre & Tya) - Lens culinaris (249).

ብርስናይ brsnay (Tya) - Argyrolobium arabicum (222), Tephrosia purpurea (115).
ብርትኳን brtkkwan (Tya) - Citrus sinensis (432).
ብርትዃን brtk̈wan (Tya) - Citrus sinensis (432).
ብርቱካን brtukan (Amh) - Citrus sinensis (432).
ብርቱካና brtukana (Or) - Citrus sinensis (432).
ብርቱካ ኒ brtukani (Or) - Citrus sinensis (432).
ብስብሻ bsbssha (Amh) - Senna singueana (59).
ብሻመድ bshamed (Afar & Tya) - Commiphora quadricincta (471).
ብሻመድ bshamed (Saho, Tre & Tya) - Commiphora samharensis (452).
ብስና bsnna (Tya) - Albizia anthelmintica (94).
ብጥብጦ bṭbṭṭo (Tre) - Cotyledon barbeyi (19).
ቡክ፡ቡክ buk-buk (Som) - Indigofera arrecta (135), Rhynchosia malacophylla (188).
ቡቂ buqi (Or) - Loranthaceae (361-374) and Viscaceae (375-378) general.
ቡራና burana (Amh & Or) - Erythrina burana (164).
ቡሮ buro (Som) - Clausena anisata (429), Salix subserrata (258), Teclea nobilis (527).
ቡርቁቀ burquqqe (Or) - Acacia lahai (85), Acacia sieberiana (89).
ቡርቶ burto (Dor) - Erythrina brucei (161).
ቡሩ buru (Sha) - Ficus mucuso (287).
ቡሻ busha (Agew & Amh) - Senna singueana (59).
ቡቶ butto (Kef & Or) - Schefflera volkensii (539).

ቻበሮ cabero (Kef) - Ficus ovata (292).
ቻቻም cacam (Mes) - Acacia pentagona (78).
ቻቻነ cacane (Or) - Acacia horrida (84).
ቻፍት caft (Som) - Platycelyphium voense (101).
ቻለንታ calenta (Kon) - Dichrostachys cinerea (74).
ቻና canna (Das) - Filicium decipiens (493).
ቻሮ caro (Kef) - Ficus ovata (292).
ችባሃ cbaha (Amh) - Ficus thonningii (298).
ቸሪን cerin (Som) - Acacia paoli (90), Acacia sieberiana (89).
ጫጫ çaça (Wel) - Acacia drepanolobium (88).
ጫጫነ çaçanne (Or) - Acacia paoli (90).
ጫጮ çaço (Or) - Commiphora confusa (462).
ጫኢ çaî (Or) - Teclea nobilis (427).
ጫላንጋ çalanga (Or) - Commiphora habessinica (471).
ጫምቢ çambie (Amh) - Cissus rotundifolia (408).
ጫቀማ çaqema (Or) - Rhus natalensis (523).
ጫቅሞሽ çaqmosh (Amh) - Cussonia spp. (538-540).
ጫሪ çari (Or) - Celtis africana (266).
ጫት çat (Amh & Tya) - Catha edulis (340).
ጫታ çata (Kem) - Catha edulis (340).
ጫቲ çati (Or) - Catha edulis (340).
ጫቲያ çatiyya (Wel) - Catha edulis (340).
ጫቶ çatto (Kef) - Albizia gummifera (96).
ጫቶ çatto (Or) - Albizia gummifera, Albizia schimperiana (96).
ጫዊቾ çawicco (Or) - Albizia gummifera, Albizia schimperiana (96).
ጫይ çay (Amh) - Celtis africana (266).

ጨዓ çeā (Tya) - Acacia abyssinica (89), Acacia bavazzonoi (90), Acacia nilotica (86), Acacia pilispina (89), Acacia seyal (85), Acacia sieberiana (89).
ጨዓ çeā (Tre) - Acacia seyal (85).
ጨጮሖ:ሓማት çeçho-hamat (Tya) - Rhynchosia malacophylla (188).
ጨካ çeka (Or) - Calpurnia aurea (104), Pittosporum spp. (1-3), Sesbania sesban (140).
ጨካ:ቀመለ çeka-qemele (Or) - Pterollobium stellatum (54).
ጨካታ çekata (Or) - Calpurnia aurea (104).
ጨኬ çeke (Or) - Celtis africana (266).
ጨለሕ çeleh (Tya) - Teclea nobilis (427).
ጨለቅለቃ çeleqleqqa (Amh) - Apodytes dimidiata (348).
ጨለጋማ çellegama (Agew) - Acacia oerfota (90).
ጨና:እኖ:ድቓለ çena-'no-dqalla (Tya) - Colutea abyssinica (233).
ጨና:አዳም çena-addam (Tya) - Ruta chalepensis (419).
ጨን ጉረፍያ çengwerefya (Tya) - Lonchocarpus laxiflorus (104).
ጨቐንጠ çeqente (Tya) - Ficus glumosa (293), Ficus ingens (291), Pittosporum spp. (1-3).
ጨቐንጠ:ጊሔ çeqente-gihe (Tya) - Ficus ingens (291).
ጨቐንጠ:ሓርማዝ çeqente-harmaz (Tya) - Ficus ingens (291).
ጨቀንጠ çeqente (Saho & Tre) - Ficus ingens (291).
ጨቅሳ çeqsa (Gam) - Calpurnia aurea (104).
ጨይ çey (Or) - Celtis africana (266).
ጭጊንዳ çginda (Wel) - Acacia etbaica (88).
ጭጊንዳ çginda (Gam) - Acacia hockii (86), Acacia seyal (85).
ጭጎኖ çgono (Tya) - Albizia amara (94), Albizia malacophylla (93).
ጪራኮታ çirakota (Or) - Ruta chalepensis (419).
ጮቢ çobi (Or) - Cissus aphylla (404), Cissus rotundifolia (408).
ጮቢ:አርባ çobi-arba (Or) - Cissus aphylla (404).
ጮቢ:በዴሱ çobi-badesu (Or) - Cissus aphyllantha (404).
ጮቢ:ቃኡራ çobi-qaüra (Or) - Cissus rotundifolia (408).
ጩጮ çuço (Or) - Maytenus undata (338).
ቺኮተ cikote (Wel) - Clausena anisata (429).
ቺልዌዴ cilwiede (Anu) - Lonchocarpus laxiflorus (104).
ቾብወ cobwe (Mur) - Ozoroa insignis (521), Sclerocarya birrea (519).

ዳዕሮ da'ro (Ge'ez & Tya) - Ficus spp. (278-299).
ዳዕሮ da'ro (Ge'ez, Tre & Tya) - Ficus vasta (296).
ዳዕሮ:ምድሪ da'ro-mdri (Tya) - Rothia hirsuta (195).
ዳባሶ dabaso (Or) - Commiphora kua (452).
ዳብዲ dabdi (Amh) - Piliostigma thonningii (68).
ዳበሶ dabeso (Or) - Boswellia microphylla (445).
ዳቦበቻ dabobecca (Or) - Rhus natalensis (523), Rhus quartiniana (524).
ዳቦ ቢስ dabobis (Som) - Rhus natalensis (523).
ዳዳሆ dadaho (Afar) - Salvadora persica (354).
ዳዳራኢያ dadaraîya (Bur) - Rhus tenuinervis (524).
ዳዴቻ dadecca (Or) - Acacia abyssinica (89).
ዳዴቾ dadecco (Or) - Myrica salicifolia (261).
ደደቻ dedecca (Or) - Acacia abyssinica (89).
ዳጋ:ጨብሳ daga-çebsa (Or) - Rhoicissus revoilii (401).
ዳግ ኒሳ dagnisa (Or) - Acacia abyssinica (89).
ዳጎ dago (Anu) - Milicia excelsa (272).

ዳከራ dakara (Or) - Boswellia microphylla, Boswellia neglecta (445).
ዳኪዳር dakidar (Or) - Commiphora boiviniana (450).
ዳንግኰ dangkwe (Mur) - Ziziphus pubescens (393).
ዳንጎጎ:ሲያቃ dangogo-siyaqa (Or) - Rhoicissus tridentata (401).
ዳንጉሌ dangulle (Or) - Pisum sativum (248).
ዳኒሳ dannisa (Or) - Apodytes dimidiata (348).
ዳኖ dano (Som) - Commiphora rostrata (462).
ዳኦሌ daöle (Mur) - Cayratia ibuensis (404).
ዳሬ dare (Som) - Ficus sycomorus (285).
ዳሬኮ dareko (Or) - Lannea schimperi (516).
ዳሬኩ dareku (Or) - Lannea schimperi (516).
ዳርጋ darga (Som) - Crotalaria laburnifolia (200).
ዳርጋማሬይ dargamarey (Som) - Tephrosia villosa (116).
ዳርጉና darguna (Bil) - Ficus vasta (296).
ዳርጆ darjo (Som) - Crotalaria albicaulis (207).
ዳሩር darur (Amh & Tya) - Chamaecrista absus (64).
ዳሰይኖ daseyno (Som) - Commiphora gileadensis (452).
ዳሽሎሜ dashlome (Wel) - Clausena anisata (429).
ዳየሮ dayero (Som) - Rosa abyssinica (35).
ዳይሉል daylul (Som) - Helinus integrifolius (385).
ደበሉጫ debelucca (Or) - Rhus retinorrhea (529).
ደቢ debi (Som) - Delonix elata (50).
ደቢሮ debirro (Or) - Senna petersiana (59).
ደቦበሳ debobesa (Or) - Rhus natalensis (523).
ደዳቱ dedatu (Or) - Millettia ferruginea (108).
ዸዸጫ ḍeḍecca (Or) - Acacia abyssinica (89).
ደዲን dedin (Som) - Commiphora myrrha (452).
ደእር:ዓድ deër-ād (Som) - Ficus glumosa (293), Ficus ingens (291).
ደግሙት degmut (Som) - Maytenus undata (338).
ደግኔይን degneyn (Som) - Lannea malifolia (516).
ደጉሰጣ deguṣta (Tya) - Girardinia bullosa (306).
ደግዉን degwun (Som) - Maytenus undata (338).
ደዀዕታ dek̇wa'ta (Tya) - Kalanchoe spp. (19-25).
ደዀታ:ጽሑፍ dek̇wata-ṣḣuf (Tya) - Kalanchoe mamorata (20).
ደምቢ dembi (Or) - Ficus ovata (292), Ficus thonningii (298), Ficus vasta (296).
ደምገዛ demgezza (Amh) - Cissus quadrangularis (407).
ደን den (Som) - Berchemia discolor (393), Dodonaea angustifolia (491).
ደንገል dengel (Som) - Commiphora boranensis (475).
ደንጊ dengi (Or) - Antiaris toxicaria (301).
ደንግያ:ሰበር dengya-sebber (Amh) - Pittosporum spp. (1-3).
ደንከሊ denkeli (Som) - Commiphora boranensis (475).
ደቂቕ:መሲ deqqiq̇-messi (Tya) - Trifolium multinerve (241).
ደርጊ dergi (Or) - Fagaropsis angolensis (422).
ደሬ derie (Amh) - Acacia etbaica (88).
ደሪንደር derinder (Som) - Commiphora ellenbeckii (473).
ደርካ derka (Som) - Tephrosia villosa (116).
ደሮ dero (Or) - Fagaropsis angolensis (422).
ደሽዋላ deshwala (Kef) - Alchemilla fischeri (39).
ደወኒ:ቡና deweni-bunna (Gur) - Acacia sieberiana (89).

ዴየዶእቦ deyedoëbbo (Som) - Tephrosia apollinea (116).
ድግንድግ dgndg (Amh & Tya) - Indigofera arrecta (135).
ድግጣ dgṭṭa (Amh) - Calpurnia aurea (104).
ድጕድጕና dgwdgwnna (Tya) - Lannea fruticosa (518), Lannea schimperi (516).
ድሃይ:ድሃይ dhay-dhay (Som) - Dichrostachys cinerea (74).
ድሂግዳር dhigdar (Som) - Dichrostachys cinerea (74).
ዲዲግሳ didigssa (Or) - Sclerocarya birrea (519).
ዲዲን didin (Som) - Commiphora myrrha (452), Commiphora tubuk (457).
ዲዲን:አድ didin-ad (Som) - Commiphora myrrha (452).
ዲዲን:መዶው didin-medow (Som) - Commiphora myrrha (452).
ዲዲሳ didissa (Or) - Sclerocarya birrea (519).
ዲጋሉ digallu (Or) - any Loranthaceae (361-374) or Viscaceae (375-378).
ዲሊበህ dilibeh (Som) - Acacia walwalensis (91).
ዲሎ:የሃን dilo-yehan (Som) - any Loranthaceae (361-374) or Viscaceae (375-378).
ዲነ:ዓማይቶ dine-āmayto (Afar) - Delonix elata (50).
ዲር:ዳጕር dir-dagwer (Som) - Canavalia ensiformis (165).
ዲራአ dirraä (Or) - Commiphora rostrata (462).
ዶባ doba (Tya) - Girardinia diversifolia (306), Urtica urens (303).
ዶቢ dobbi (Or) - Girardinia bullosa (306), Urtica simensis (303).
ዶቢ dobbi (Gur) - Urtica simensis (303).
ዶዶታ dodota (Or) - Acacia etbaica, Acacia gerrardii (88).
ዶዶተ dodote (Or) - Acacia etbaica (88).
ዶዶቲ dodoti (Or) - Acacia etbaica, Acacia gerrardii (88).
ዶጋማ dogama (Gam) - Commiphora africana (454).
ዶልኪኢሳ dolkiîsa (Or) - Bersama abyssinica (511).
ዶማይ domay (G̃am) - Balanites aegyptiaca (433).
ዶማየ domaye (Gam) - Balanites aegyptiaca (433).
ዶሞሆ domoho (Or) - Balanites aegyptiaca (433).
ዶንጋ donga (Amh) - Apodytes dimidiata (348).
ዶንኪኮ donkiko (Sid) - Apodytes dimidiata (348).
ዶረት doret (Amh) - Acacia etbaica (88).
ዶሮ:ሰበር doro-sebber (Amh) - Tephrosia nubica (116).
ዶሶ dosso (Som) - Buxus hildebrandtii (255).
ድቓላ dq̈ala (Tya) - any Loranthaceae (361-374) or Viscaceae (375-378).
ድራ dra (Tya) - Hippocratea africana (342).
ድሩባ druba (Or) - Allophyllus abyssinicus (499).
ዱጫ ducca (Or) - Hagenia abyssinica (43).
ዱችያ ducya (Gam) - Salix subserrata (258).
ዱዱና duduna (Amh & Or) - Cussonia spp. (538-540), Ekebergia capensis (489).
ዱግና dugna (Agew) - Lannea schimperi (516), Lannea triphylla (514).
ዱንዳስ dundas (Som) - Commiphora ancistrophora (475).
ዱንካል dunkal (Som) - Commiphora erlangeriana (448), Commiphora guidotii (448), Commiphora sphaerocarpa (468), Commiphora unilobata (448).
ዱራኣስ duraäs (Saho & Tya) - Lepisanthes senegalensis (504).
ዱረ dure (Som) - Ficus sycomorus (285).
ዱረሆ dureho (Som) - Indigofera articulata, Indigofera coerulea (134).
ዱስ dus (Or) - Erythrina abyssinica (161).
ዱሲያ dusiya (Som) - Dalbergia commiphoroides (105).
ዱሶግ dussog (Som) - Buxus hildebrandtii (255).

ዱያ duya (Som) - Dalbergia commiphoroides (105).
ዷዷቴ dwadwatie (Amh) - Colutea abyssinica (233).

ዐለውቶ ēlawto (Afar) - Indigofera articulata, Indigofera coerulea, Indigofera articulata (134).
ዐላን ēllan (Som) - Indigofera articulata, Indigofera coerulea (134).
ዐላን:አረን ēllan-eren (Som) - Indigofera arrecta (135).
አበራከ eberrako (Or) - Bersama abyssinica (511).
አዲ:ሸበል eddi-shebel (Som) - Ziziphus mucronata (393).
አላስ elas (Som) - Indigofera arabica (122).
አለለ elele (Or) - Albizia grandibracteata (96).
አለሰ ሎሰክ eleselosek (Gum) - Salix subserrata (258).
አሊከ elike (Or) - Maytenus undata (338).
አላን:ጊን ellan-gin (Som) - Indigofera arrecta (135).
አመለ emela (Or) - Albizia grandibracteata, Albizia gummifera, Albizia schimperiana (96).
አሪ eri (Som) - Opilia campestris (358).
አስዋርዎድ esswarwod (Som) - Mimosa pigra (74).
አጣን etan (Gur) - Boswellia spp. for the resin (442-445).
አጦ eṭo (Or) - Hagenia abyssinica (43).
አታ eta (Or) - Ficua ovata (292) and other Ficus spp. (278-299).
አቲቢሮ etibiro (Or) - Bersama abyssinica (511).
አትሳ etsa (Dor) - Ficus sur (287).
አተቻ ettecca (Or) - Dodonaea angustifolia (491).

ፋረንጎታ farengota (Kon) - Cajanus cajan (181).
ፋሶልያ fasolya (Amh) - Phaseolus spp. (169-171).
ፋጡቃ faṭuqa (Agew) - Boswellia papyrifera (443), Rhus natalensis (523).
ፌኦ feö (Or) - any Loranthaceae (361-374) or Viscaceae (375-378).
ፌጦ feṭo (Or) - Hagenia abyssinica (43).
ፌውሲ: ቊርጸት fewsi-qwrṣet (Tya) - Vermifrux abyssinica (227).
ፊቾ ficho (Or) - Hagenia abyssinica (43).
ፊሻ fissha (Or) - Crotalaria rosenii (199).
ፎጭ foç (Amh) - Ziziphus mucronata (293).
ፎልኮልቻ folkolca (Or) - Maytenus putterlickioides (336).
ፎንኮንቻ fonkonca (Or) - Maytenus putterlickioidea (336).
ፎራካ foraka (Or) - Bersama abyssinica (511).
ፍርቃ frqa (Agew) - Piliostigma thonningii (68).
ፍርጣጣ frṭaṭa (Agew) - Rhus natalensis (523).
ፉጊሶ fuggiso (Som) - Crotalaria comanestiana (202).
ፉል ful (Tre & Tya) - Arachis hypogea (147).
ፉል ful (Ge'ez) - Vicia faba (249).
ፉላአይ fulaäy (Som) - Acacia seyal (85).
ፉሊሳ fuliessa (Wel) - Acacia seyal var. fistula (85).
ፉለሳ fullessa (Or) - Acacia drepanolobium (88).
ፉልሳ fullsa (Or) - Acacia sieberiana (89).
ፉርጉገ furguge (Or) - Kalanchoe lanceolata (23).
ፉሶልያ fusolya (Ade) - Phaseolus spp. (169-171).

ጋ:አሮር ga-aror (Som) - Commiphora sphaerocarpa (468).
ጋብ gab (Mur) - Ziziphus mauritiana (396).
ጋበ gaba (Mur) - Ziziphus mauritiana (396).
ጋበ gaba (Agew) - Ziziphus spina-christi (393).
ጋበ gaba (Amh & Tya) - Ziziphus mucronata (393), Ziziphus spina-christi (385).
ጋበ:ሐርማዝ gaba-ħarmaz (Tre) - Ziziphus mucronata (393).
ጋቢ gabi (Mur) - Ziziphus mauritiana (396).
ጋቢቴ gabite (Das) - Ziziphus mauritiana (396).
ጋብትዘ gabtze (Gam) - Schefflera abyssinica (539).
ጋድ gad (Som) - Cissus cactiformis (407), Cissus quadrangularis (407), Cissus rotundifolia (408).
ጋዳዲ gadadi (Afar) - Commiphora quadricincta (471).
ጋዳም gadam (Tya) - Cussonia spp. (538-540).
ጋዱ gadu (Som) - any Loranthaceae (361-374) or Viscaceae (375-378).
ጋእጃ gaëja (Gam) - Clausena anisata (429).
ጋሃዲታ gahadita (Kon) - Commiphora erythraea (450).
ጋሂሮ gahiro (Amh) - Celtis africana (266).
ጋለ:ላላ gale-lala (Or) - Rhoicissus tridentata (401).
ጋለቸቻ galececa (Or) - Entada spp. (71-73).
ጋልጋላም galgalam (Or) - Boswellia papyrifera (443).
ጋለ galle (Or) - Hippocratea africana (342).
ጋለቻ galleca (Or) - Maytenus undata (338).
ጋሎእል galoöl (Som) - Acacia bussei (85), Acacia nilotica (86), Acacia ogadensis (80).
ጋሎእል:ሱር galoöl-sur (Som) - Dichrostachys cinerea (74).
ጋማሮ gamaro (Saho) - Acacia venosa (81).
ጋምጭለ gamçle (Gam) - Clausena anisata (429).
ጋሞ:ጋዲ gammo-gadie (Wel) - Ziziphus mucronata (393).
ጋምሮት gamrot (Tre) - Acacia oerfota (90).
ጋንዳድ gandad (Som) - Commiphora corrugata (460).
ጋቋ gaqa (Tya) - Rosa abyssinica (35).
ጋራሆ garaho (Som) - Commiphora guidotii (448).
ጋራስ garas (Afar, Saho & Som) - Dobera glabra (353).
ጋርበ garbe (Wel) - Prunus africanus (41).
ጋርቢ garbi (Or) - Acacia abyssinica (89).
ጋርቢ garbi (Amh, Or & Som) - Acacia albida (84).
ጋርቢ garbi (Or) - Prunus africanus (41).
ጋርቢ:አዲ garbi-adi (Or) - Celtis toka (267).
ጋርቢለ garbile (Das) - Ziziphus mauritiana (396).
ጋርጋሩ gargaru (Gam & Wel) - Dichrostachys cinerea (74).
ጋሮን garon (Som) - Commiphora gileadensis (452), Commiphora horrida (474).
ጋሮን:መዶው garon-medow (Som) - Commiphora coronillifolia (460).
ጋሪ garri (Or) - Barbeya oleoides (270), Ozoroa insignis (521).
ጋርሰ garsa (Tre & Tya) - Dobera glabra (353).
ጋርሸ garsha (Tya) - Acacia albida (84).
ጋሩ garu (Or) - Pterollobium stellatum (54).
ጋሩል garul (Or) - Pterollobium stellatum (54).
ጋታማ gatama (Or) - Schefflera abyssinica (539).
ጋዋ gawa (Tya) - Celtis africana (266).
ጋዋዳ gawada (Kem) - Myrica salicifolia (261).

ጋያ gayya (Gur & Or) - Lathyrus sativus (250).
ግብጾ gbṣo (Tya) - Lupinus albus (219).
ግብጦ gbṭo (Amh) - Lupinus albus (219).
ግብ geb (Som) - Elaeodendron aquifolium (344), Ziziphus spina-christi (395).
ገባ geba (Agew, Amh & Tya) - Ziziphus mucronata (393).
ገባ geba (Amh & Tya) - Ziziphus spina-christi (395).
ገባ:አድጊ geba-adgi (Tya) - Ziziphus abyssinica (396).
ገብሆ gebho (Agew) Rhamnus prinoides (390).
ገድ:አደ ged-adde (Som) - Tephrosia nubica, Tephrosia polyphylla (116).
ገድ:ጎደድ ged-goded (Som) - Indigofera spinosa (126).
ገድ:ሐጂን ged-ḥajin (Som) -Tephrosia obbiadensis (112), Tephrosia polyphylla (116).
ገድ:ኢርማን ged-irman (Som) - Indigofera oblongifolia (136).
ገዳል gedal (Or) - Pittosporum spp (1-3).
ገደም geddem (Tya) - Cussonia spp. (538-540).
ገደታሞ gedetamo (Amh) - Cussonia spp. (538-540).
ገዲ:ገዲ gedi-gedi (Som) - Indigofera amorphoides (134).
ገዲል gedil (Tre) - Kalanchoe lanceolata (23), Kalanchoe quartiniana (20).
ገጊ gegi (Mur) - Salvadora persica (354).
ገለሎ gelelo (Som) - Crotalaria laburnifolia (200).
ገልፊ gelfi (Mur) - Ziziphus pubescens (393).
ገምበዛን gembezan (Amh) - Cissus quadrangularis (407).
ገንዲ:ገንታ gendi-genta (Som) - Commiphora ogadensis (460).
ገንሕ genḥ (Tya) - Pittosporum spp. (1-3).
ገርቢ gerbi (Amh) - Acacia albida (84).
ገሬንጋ gerenga (Bil) - Lablab purpureus (179).
ገርገሮ gergero (Gam & Wel) - Dichrostachys cinerea (74).
ገርገጧ gergeṭwa (Wel) - Dodonaea angustifolia (491).
ገርሳ gersa (Tre & Tya) - Dobera glabra (353).
ገሻ gesha (Kem & Saho) - Rhamnus prinoides (390).
ገሾ gesho (Or) - Rhamnus prinoides (390).
ገሿ geshwa (Wel) - Rhamnus prinoides (390).
ገሳ gessa (Or) - Bersama abyssinica (511).
ገት:በዮ get-beyo (Som) - Sesbania sesban (140).
ገት:ሱቡክ get-subuk (Som) - Commiphora sphaerocarpa (468).
ገተም getem (Tya) - Cussonia spp. (538-540).
ገተም getem (Had, Sid & Tya) - Schefflera abyssinica (539).
ገተሚ getemi (Or) - Cussonia spp. (538-540).
ገቶ geto (Or) - Dichrostachys cinerea (74).
ግሀይደር gheyder (Som) - Acacia stuhlmanii (91).
ጌሾ giesho (Ade, Amh, Gam, Had, Saho & Tya) Rhamnus prinoides (390).
ጌሶ gieso (Tya) - Rhamnus prinoides (390).
ጊላይ gillay (Or) - Commiphora rostrata (462).
ጊኔ ginne (Bil) - Rhynchosia malacophylla (188).
ጊርሚ girmi (Or) - Dichrostachys cinerea (74).
ጊሼ gishe (Gur) - Rhamnus prinoides (390).
ጊሺ:ሻአቶ gishi-shaäto (Kef) - Pisum sativum (248).
ጊቴእ giteë (Had) - Cajanus cajan (181), Pisum sativum (248).
ጋማሬ gmara (Agew) - Acacia venosa (81).
ግማርዳ gmarda (Amh) - Acacia polyacantha (81).

ግንዶሽ gndosh (Amh) - Cyphostemma cyphopetalum (417).
ጎአ goä (Som) - Commiphora bruceae (473).
ጎአህ goäh (Som) - Commiphora tubuk (457).
ጎብ gob (Som) - Ziziphus mauritiana (396), Ziziphus spina-christi (395).
ጎባ:ሀሪግ goba-herig (Som) - Commiphora boranensis (475).
ጎብዱ gobdu (Or) - Girardinia diversifolia (306).
ጎቦ gobo (Kef) - Vigna unguiculata (174).
ጎብየር gobyer (Som) - Ziziphus mucronata (393).
ጎድ:ሃው:ቶርለ god-haw-torle (Som) - Dichrostachys cinerea (74).
ጎዳ goda (Or) - Rubus spp. (31-34).
ጎዳዲ godadi (Tya) - Commiphora quadricincta (471).
ጎፋ gofa (Or) - Catha edulis (340).
ጎሆልዳየ goholdaye (Som) - Crotalaria saltiana (204).
ጎሊሉ golilu (Som) - Commiphora ellenbeckii (473), Commiphora gowlello (474).
ጎምቦቸሪ gombocerie (Amh) - Cussonia spp. (538-540).
ጎመ gomme (Or) - Ficus ovata (292).
ጎምኦርያ gomöryya (Wel) - Pterollobium stellatum (54).
ጎናከ gonake (Som) - Acacia nilotica (86).
ጎንጂ gonji (Kef) - Milicia excelsa (272).
ጎራ gora (Or) - Pterollobium stellatum (54), Rosa abyssinica (35).
ጎራ:ጎራ gora-gora (Agew) - Hagenia abyssinica (43).
ጎራን:መዱ goran-medu (Som) - Dichrostachys cinerea (74).
ጎርበ gorbe (Or) - Albizia gummifera, Albiziaschimperiana (96).
ጎርጮ gorço (Or) - Rubus spp. (31-34).
ጎርሰ gorsa (Or) - Cissus cactiformis (407).
ጎርታ gorta (Or) - Pterollobium stellatum (54).
ጎት got (Som) - Balanites aegyptiaca (433).
ጎወለ ሎ gowelello (Som) - Commiphora ellenbeckii (473).
ጎውለ ሎ gowlello (Som) - Commiphora gowlello (474).
ግራር grar (Amh & Gur) - Acacia spp. in general (75-93).
ግራሮ graro (Kef) - Trigonella foenum-graecum (247).
ግርማ grma (Or) - Dichrostachys cinerea (74).
ግርሞ grmo (Tya) - Acacia sieberiana (89).
ጉዱባ guduba (Or) - Polyscias fulva (537).
ጉፋ gufa (Amh) - Senna singueana (59).
ጉፍቲ gufti (Or) - Crotalaria spinosa (217).
ጉጋ guga (Som) - Mystroxylon aethiopicum (338).
ጉጋንታ guganta (Wel) - Acacia dolichocephala, Acacia lahai (85).
ጉል:ጉል gul-gul (Or) - Buxus hildebrandtii (255).
ጉልጉልባ gulgulba (Or) - Urtica simensis (303).
ጉ ሎ:ክክቦ gulo-kokkoba (Amh) - Maytenus senegalensis (336).
ጉሉ gulu (Som) - Acacia edgeworthii (91).
ጉማር gumar (Som) - Acacia oerfota (90).
ጉማራ gumara (Som) - Acacia oerfota (90).
ጉመረ gumere (Or) - Rubus spp. (31-34).
ጉመሮ gumero (Som) - Acacia oerfota (90).
ጉራአ guraä (Som) - Commiphora cyclophylla (468).
ጉራጅ guraj (Or) - Cissus cactiformis (407).
ጉራ ዩ gurayu (Or) - Prunus africanus (41).

ጉርሐ:ዳሩየ gurḣa-daruye (Som) - Acacia tortilis (87).
ጉት gut (Som) - Balanites aegyptiaca (433).
ጉዛ guza (Amh) - Balanites aegyptiaca (433).
ጕዐመል gw'mel (Tya) - Sclerocarya birrea (519).
ጓሃውሪ gwahawri (Som) - Lannea malifolia (516).
ጓያ gwayya (Amh) - Lathyrus sativus (250).
ጕዛ gwaza (Tya) - Balanites aegyptiaca (433).
ጕፍያ gwefyya (Amh) - Senna singueana (59).
ጕምእሪያ gwemöriyya (Wel) - Acacia brevispica (78).
ጕነቍ gweneq̇w (Tya) - Dichrostachys cinerea (74).
ጕዳር gwdar (Som) - Acacia asak (80).
ጕመሮ gwmero (Tya) - Acacia bussei (85), Acacia oerfota (90), Acacia polyacantha (81).
ጕርድማ ኵዮ gwrdmak̇wyo (Tya) - Trifolium spp. (233-244).
ጕሰዐ:ማይ gwssa'-mai (Tya) - Trifolium schimperi (241).

ሐቢ ḣabbi (Tya) - Hagenia abyssinica (43).
ሐቢ:ሐሰይ ḣabbi-ḣasey (Tya) - Indigofera coerulea (134).
ሐቢ:ሃጸይ ḣabbi-haṣey (Tre) - Indigofera articulata (134).
ሐብል:እንሲት ḣabl-'nsit (Tre) - Rhoicissus tridentata (401).
ሐጪት ḣaçat (Agew & Tya) - Maytenus spp. (331-338).
ሐጋር ḣagar (Or & Som) - Commiphora erythraea (450).
ሐጂን ḣajin (Som) - Crotalaria dumosa (212), Indigofera spinosa (126), Tephrosia villosa (116).
ሐለንጊ:ተመን ḣalengi-temen (Tya) - Cyphostemma cyphopetalum (417).
ሐልሰኖ halseno (Tre) - Senna italica (63).
ሐማሰራን ḣamaserran (Tya) - Albizia amara (94), Albizia malacophylla (93).
ሐማት:ከኪቶ ḣamat-kak̇ito (Tya) - Astragalus atropilosulus (230).
ሐማት:ረወይ ḣamat-rewey (Amh) - Celtis africana (266).
ሐማት:ሰዶ ḣamat-seddo (Tya) - Sageretia thea (389).
ሐማት:የማን:በሪ ḣamat-yeman-berri (Tya) - Lotus quinatus (224).
ሐማውቲ ḣamawti (Tre) - Cyphostemma molle (411).
ሐምስ ḣams (Tre) - Rhus glutinosa (528).
ሐቅ ḣaq (Tre) - Acacia asak (80).
ሐቅ ḣaqq (Tre) - Acacia ogadensis (80).
ሐራሕ ḣaraḣ (Som) - Acacia tortilis (87).
ሐረግ:ተመን ḣareg-temen (Tya) - Ampelocissus schimperiana (400), Cyphostemma adenocaule, Cyphostemma oxyphyllum (412).
ሐዋ ḣawa (Tya) - Celtis africana (266).
ሐዊ:ለይቲ ḣawwi-leyti (Tya) - Crotalaria incana (201).
ሕብረ:መጽርይ ḣbre-meṣry (Ge'ez) - Rosa x richardii (35).
ሐደሰ ḣeddessa (Or) - Teclea nobilis (427), Vepris dainelli (424).
ሕልበት ḣlbet (Ade) - Trigonella foenum-graecum (247).
ሕሩምታ ḣrumta (Tya) - Barbeya oleoides (270).
ሕጻውጽ ḣṣawṣ (Tya) - Calpurnia aurea (104).
ሑመር ḣumer (Som & Tya) - Tamarindus indica (66).

ሃብሩ habru (Or) - Ficus sur (287).
ሃደወይ hadawey (Or) - Brucea antidysenterica (440).

ሃደሳ haddessa (Or) - Vepris dainelli (424).
ሃዱ hadu (Gum) - Ziziphus pubescens (393).
ሃጋር hagar (Som) - Commiphora sphaerophylla (450).
ሃጋር:አድ hagar-ad (Or & Som) - Commiphora erythraea (450).
ሃጋር:አድ hagar-ad (Som) - Commiphora sphaerophylla (450).
ሃጋር:መደው hagar-medow (Or & Som) - Commiphora erythraea (450).
ሃጋር:መደው hagar-medow (Som) - Commiphora ogadensis, Commiphora sphaerophylla (450), Commiphora truncata (452).
ሃጋርሶ hagarso (Or & Som) - Commiphora erythraea (450).
ሃጋርሱ hagarsu (Or & Som) - Commiphora erythraea (450).
ሃጋርሱ hagarsu (Or) - Commiphora ogadensis (460).
ሃለለ halele (Or) - Albizia grandibracteata (96).
ሃለምበል:ብዮት halembal-byot (Som) - Sesbania sesban (140).
ሃሊለ halila (Or) - Urera hypselodendron (304).
ሃ ሎ halo (Or) - Acacia bussei (85).
ሃልቱፋ haltufa (Or) - Rubus spp. (31-34).
ሃማቂታ hamaqita (Or) - Medicago polymorpha (244).
ሃማር hamar (Som) - Pistacia aethiopica (530), Ziziphus mucronata (393).
ሃማሩ ኦ hamaroö (Kem) - Rubus spp. (31-34).
ሃምበልታ hambelta (Or) - Entada spp. (71-73).
ሃምብሃምበ hambhambo (Tya) - Senna singueana (59).
ሃምብሃምበ hambhambo-bayta (Tya) - Alisicarpus ferrugineus (154).
ሃምበይ hamboy (Gam) - Ozoroa insignis (521).
ሃመሳ hamessa (Or) - Commiphora confusa (462), Commiphora schimperi (454).
ሃሞር hamor (Som) - Berchemia discolor (393).
ሃምሰሳ hamsessa (Or) - Commiphora campestris (469).
ሃምሸ hamsha (Or) - Schefflera volkensii (539).
ሃምሺከ hamshika (Ọr) - Ilex mitis (329).
ሃሙድ hamud (Som) - Ziziphus hamur (396).
ሃሙር hamur (Som) - Ziziphus hamur (396).
ሃንጩራ hançura (Or) - Kalanchoe petitiana (20).
ሃንገለ ሎ hangelelo (Som) - Dicraeopetalum stipulare (101).
ሃንሳ hansa (Tya) - Urtica simensis (303).
ሃራንጋማ harangama (Or) - Acacia mellifera (81), Pterollobium stellatum (54).
ሃራር harar (Som) - Ziziphus mucronata (393).
ሃርቡ harbu (Or) - Ficus exasperata (248), Ficus sur (287), Ficus sycomorus (285).
ሃርጫ harça (Or) - Sesbania sesban (140).
ሃርፋቱ harfatu (Or) - Cussonia spp. (538-540), Schefflera abyssinica (539).
ሃርከይ harkay (Som) - Ziziphus mucronata (393).
ሃርሞ harmo (Or) - Loranthaceae (361-374) and Viscaceae (375-378).
ሃጠ haṭṭe (Or) - Dichrostachys cinerea (74).
ሃያ haya (Amh) - Salix subserrata (258).
ሃይራማት hayramat (Som) - Dodonaea angustifolia (491).
ሃዝተ hazte (Gam) - Ximenia americana (356).
ሀዲይ hdiey (Ge'ez) - Albizia spp. (93-96).
ሀልታ helta (Som) - Commiphora longipedicellata (448).
ሀንጆረ henjore (Wel) - Rubus spp. (31-34).
ሀና henna (Or) - Kotchya africana (146).
ሀጦት heṭot (Sid) - Hagenia abyssinica (43).

ሀይስ heys (Som) - Pistacia aethiopica (530).
ሂዳ:ረፋ hida-refa (Or) - Rhoicissus tridentata (401).
ሂዲ hiddi (Or) - Girardinia diversifolia (306).
ሄላ hiela (Ge'ez) - the fig of Ficus (278-299).
ሄጦ hieṭo (Or) - Hagenia abyssinica (43).
ሂናይ hinay (Mes) - Hippocratea pallens (342).
ሂጃሮ hinjaro (Had) - Rubus spp. (31-34).
ሂርካ hirka (Or) - Allophyllus sp. nov. (502).
ሂርካካሞ hirkakamo (Or) - Allophyllus abyssinicus (499).
ሂርካሞ hirkamo (Or) - Allophyllus sp. nov. (502).
ሂርኩም hirkum (Or) - Allophyllus abyssinicus (499).
ህምቦዩ hmboy (Tre) - Senna singueana (59).
ሆዳይ hoday (Som) - Commiphora hodai (452).
ሆደይ hoddey (Som) - Commiphora hodau (452).
ሆሀ hoh (Ge'ez) - Prunus persica (42).
ሆሚ homi (Or) - Prunus africanus (41).
ሆራቃ horaqqa (Or) - Bersama abyssinica (511).
ሆርጎይ horgoy (Or & Som) - Commiphora terebinthina (454).
ህርግተ hrgte (Tre) - Maytenus senegalensis (336).
ህሪእ:ጎርለህ hria-gorleh (Som) - Parkinsonia scioana (54).
ህርያ:እጂ hrya-eji (Som) - Lannea malifolia (516).
ህርያ:እል:ማሊ hrya-el-mali (Som) - Pistacia falcata (532).
ሁዳ huḍa (Or) - Ximenia americana (356).
ሁዳ huda (Or) - Polyscias fulva (537).
ሁዳ:ጃልዶ huda-jaldo (Or) - Ximenia caffra (356).
ሁዳየ hudaye (Som) - Ximenia americana (356).
ሁዲ hudi (Or) - Ximenia americana (356).
ሁዱ:ፋርዳ hudu-farda (Or) - Trema guineensis (270).
ሁክ huk (Or) - Cydonia oblonga (42).
ሁመር humer (Amh) - Tamarindus indica (66).
ሁሬ:አዳየ hure-adaye (Som) - Ximenia caffra (356).
ህያብ hyab (Som) - Rhoicissus revoilii (401).

ዒለም:ብኽዋረያ îllam-bk̇warya (Tya) - Indigofera arrecta (135).
ዒለም:ሃቡት îllam-habut (Tre) - Indigofera articulata, Indigofera coerulea (134).
ዒለም:ወኸርያ îllam-wek̇arya (Tya) - Indigofera arrecta (135).
ዒለን îllan (Som) - Indigofera articulata, Indigofera coerulea (134), Sageretia thea (389).
ኢዳዶ idado (Or) - Acacia senegal (75).
ኢልካ ilka (Or) - Maytenus undata (338).
ኢልካ:አዳየ ilka-adaye (Som) - Commiphora bruceae (473).
ኢልካ:አደስ ilka-ades (Som) - Commiphora boranensis (375).
ኢልካ:ቡስ ilka-bus (Or) - Commiphora corrugata (460).
ኢሚ immi (Som) - Cardiospermum halicacabum (498).
ኢሙ:ጋሬ imu-gare (Or) - Cardiospermum halicacabum (498).
ኢንጊንካካዳ inginkakada (Kon) - Ximenia caffra (356).
ኢቆጦ inqoṭo (Or) - Rosa abyssinica (35).
ኢንታንቻ intanca (Sid) - Dodonaea angustifolia (491).
ኢርቶ irto (Or) - Loranthaceae (361-374) and Viscaceae (374-378) spp. in general.
ኢተቻ ittecca (Or) - Dodonaea angustifolia (491).

ጃጃለ jajale (Som) - Rhamnus staddo (390).
ጃለሎ jalelo (Som) - Senna truncata (61).
ጃለሎ:አሪ jalelo-ari (Som) - Senna holosericea (63).
ጃንጃኒ janjanni (Som) - Acacia stuhlmanii (91).
ጀጀባ jejjeba (Amh & Or) - Berchemia discolor (393).
ጀላሎ jelalo-jel (Som) - Senna alexandrina, Senna italica (63).
ጀሉብ jelub (Som) - Indigofera ruspoli (126).
ጀሞ jemo (Amh) - Balanites aegyptiaca (433).
ጀንፎቅ jenfoq (Amh) - Rhoicissus tridentata (401).
ጀረመ jeremme (Or) - Dichrostachys cinerea (74).
ጀሪን jerin (Som) - Acacia paoli (90), Acacia sieberiana (89).
ጂቅ jiiq (Som) - Acacia seyal (85).
ጂምአ jimmä (Or) - Catha edulis (340)
ጀማ:ሃረ jimma-harre (Or) - Maytenus senegalensis (336).
ጂረሰ jiresa (Or) - Pistacia aethiopica (532).
ጅሉብ jlub (Som) - Indigofera ruspoli (126).
ጆሊያ joliya-mita (Wel) - Hagenia abyssinica (43).
ጁኩል jukul (Mes) - Entada spp. (71-73).
ጁሊ juli (Kun) - Balanites aegyptiaca (433).
ጁመፎክ jumefok (Or) - Bersama abyssinica (511).
ጁያ juya (Or) - Trilepisium madagascariense (300).

ኸራስ k̇aras (Som) - Dobera glabra (353).
ኸረጽ k̇ereṣ (Tya) - Osyris quadripartita (382).
ከባቲር kabatir (Som) - Dichrostachys cinerea, Dichrostachys kirkii (74).
ከብራሮ kabraro (Som) - Commiphora horrida (474).
ከቡደይዳ kabudeyda (Kon) - Rhus natalensis (523).
ከቺሆ kacciho (Mao) - Celtis africana (266).
ከዶን kadon (Som) - Commiphora coronillifolia (460).
ከፋል kafal (Or) - Boswellia papyrifera (443).
ከፎ:አታራ kafo-atara (Wel) - Cajanus cajan (181).
ከፍዋ:አተሪያ kafwa-ateriya (Wel) - Cajanus cajan (181).
ከጂማ kajima (Or) - Pterollobium stellatum (54).
ከኺቶ kak̇ito (Tya) - Medicago polymorpha (244).
ከላንከል kalankal (Som) - Aeschynomene indica (144).
ከሊ kali (Mur) - Cissus rotundifolia (408).
ከምበሎ kambelo (Or) - Schefflera volkensii (539).
ከና:ከዐለ kana-kasle (Gam) - Ziziphus mucronata (393).
ከንከሻ kankasha (Or) - Lonchocarpus laxiflorus (104).
ከራ kara (Or) - Ormocarpum muricatum, Ormocarpum trachycarpum (143).
ከራቾ karaço (Kef) - Polyscias fulva (537).
ከራሾ karasho (Or) - Polyscias fulva (537).
ከርቸቸ karcecce (Or) - Millettia ferruginea (108).
ከርቾፌ karcoffe (Or) - Albizia gummifera, Albizia schimperiana (96).
ከሮ karo (Mao) - Prunus africanus (41).
ከሪኦ karrio (Or) - Polyscias fulva (537).
ከሮ karro (Kef) - Ficus sur (287).
ከታባ kataba (Or) - Myrica salicifolia (261).
ከዘሞራ kazamora (Amh) - Casimiroa edulis (424).

ካዝሜር kazmier (Tya) - Casimiroa edulis (424).
ካዝዋሪና kazwarina (Amh) - Casuarina spp. (262-264).
ከኣይ keäy (Mur) - Rhus natalensis (523).
ከብሰ kebse (Gur) - Hagenia abyssinica (43).
ከቻች kecac (Bod) - Rhus natalensis (523).
ከዲ keddi (Som) - Balanites glabra (434).
ከፈም kefem (Amh) - Ficus ingens (291).
ከ ሄሮ kehiero (Or) - Commiphora boranensis (475).
ከካዪ kekayi (Or) - Allophyllus abyssinicus (499).
ከኔይ keney (Tya) - Dichrostachys cinerea (74).
ከኖ keno (Som) - Commiphora rostrata (462).
ከራራ kerara (Agew) - Dodonaea angustifolia (491).
ከርቤ kerbie (Amh, Ge'ez & Tya) - resin from Commiphora (446-477).
ከርቦ ኒ kerboni (Sid) - Polyscias fulva (537).
ከተሚ ketemi (Or) - Cussonia spp. (438-540).
ከቶ ketto (Kef & Or) - Ilex mitis (329).
ከይአክ keyäk (Som) - Acacia bricchettiana (91).
ከየ keye (Gum) - Commiphora confusa (462).
ክሆ ሺ khoshi (Or) - Indigofera rothii (126).
ኪቢሩ kibiru (Or) - Bersama abyssinica (511).
ኪዲ kidi (Som) - Balanites glabra (434).
ኪዲ kidi (Or) - Celtis africana (266).
ኪሪንገ ኒ kiringenni (Mur) - Lannea schweinfurthii (518).
ኪሳ kisa (Or) - Tephrosia interrupta (118).
ኪስታ ኒ:ሸሸላ kistani-shahala (Saho) - Rhamnus staddo (390).
ክሌክለ klklle (Or) - Crotalaria rosenii (199).
ኮቦክ kobbok (Som) - Commiphora africana, Commiphora schimperi (454).
ኮቦ kobo (Kef) - Ficus vallis-choudae (289).
ኮዶ kodo (Tya) - Ficus sur (287).
ኮፋለ kofale (Or) - Albizia grandibracteata (96).
ኮክ kok (Amh & Ge'ez) - Prunus persica (42).
ኮኻዓታ koka'ta (Tya) - Colutea abyssinica (233).
ኮኪ koki (Or) - Prunus persica (42).
ኮኮራ kokora (Or) - Schefflera volkensii (539).
ኮል kol (Amh) - Ximenia americana (356).
ኮላንቹ kolancu (Kef) - Erythrina abyssinica , Erythrina brucei (161).
ካለ ኒቻ kalanica (Kef) - Erythrina abyssinica , Erythrina brucei (161).
ኮምቦልቻ kombolca (Or) - Maytenus senegalensis (336), Maytenus undata (338) and other Maytenus spp. (331-338).
ኮኦሸ koösha (Gum) - Commiphora boiviniana (450).
ኮራ kora (Or) - Piliostigma thonningii (68).
ኮራካ koraka (Or) - Bersama abyssinica (511).
ኮራች korc (Amh) - Erythrina abyssinica, Erythrina brucei (161).
ኮሪባ koriba (Or) - Polyscias fulva (537).
ኮሶ koso (Amh) - Hagenia abyssinica (43).
ኮሱዋ kosuwa (Wel) - Hagenia abyssinica (43).
ኮታተ kotate (Or) - Vigna unguiculata (174).
ኮተሉ kotellu (Or) - Millettia ferruginea (108).
ክሬናሲ krenasi (Som) - Ziziphus spina-christi (395).

ክስራ ksra (Tre) - Ziziphus spina-christi (395).
ክትክታ ktktta (Amh, Gur, Kem & Saho) - Dodonaea angustifolia (491).
ክትክታ ktktta (Amh) - Sageretia thea (389).
ኩ:ዋደረ ku-wadere (Som) - Boswellia microphylla (445).
ኩድኩዳ kudkudda (Amh) - Balanites aegyptiaca (433).
ኩግና kugna (Gam) - Boswellia neglecta (445).
ኩ ኵ kukw (Tya) - Prunus persica (42).
ኩል:ቦንቲ kul-bonti (Bod) - Commiphora africana (454).
ኩላን kulan (Afar & Som) - Balanites rotundifolia (436).
ኩምባላ kumbala (Or) - Apodytes dimidiata (348).
ኩምቢ kumbi (Or) - Commiphora myrrha (452).
ኩዞ kuzo (Gam) - Balanites rotundifolia (436).
ኳኵቶ kwak̇wto (Tya) - Medicago polymorpha (244).
ኰሸሽላ kwesheshlla (Tya) - Rubus spp. (31-34).
ኵሐ kwiḣa (Ge'ez & Tya) - Salix subserrata (258).
ኵርንችት kwrnct (Amh) - Medicago polymorpha (244).

ላዓው lāw (Tre) - Acacia oerfota (90).
ለባ:አሎለያ laba-aloleya (Som) - Cissus cornifolia (405).
ለቡሰአኒ labuseäni (Mur) - Commiphora boiviniana (450).
ለፍቶ lafto (Or) - Acacia abyssinica (89), Acacia amythethophyllla (86), Acacia dolichocephala (85), Acacia lahai (85), Acacia pilispina (89), Acacia sieberiana (89).
ለፍቶ:አዲ lafto-adi (Or) - Acacia sieberiana (89).
ለሐይ laḣay (Tya) - Acacia lahai (85).
ለኮማ lakoma (Wel) - Vigna unguiculata (174).
ለለቶ lalato (Or) - Acacia sieberiana (89).
ለሉ lalu (Or) - Rhoicissus tridentata (401).
ለማጎኢያ lamagoîya (Som) - Cissus quadrangularis (407).
ለንግ lang (Anu) - Ziziphus abyssinica (396).
ለንጉስቶ langusto (Or) - Urera hypselodendron (304).
ለንቅሽ lanqsh (Agew & Amh) - Urera hypselodendron (304).
ለንሶ lanso (Gam) - Salix subserrata (258).
ለረኒ larenni (Mur) - Commiphora boiviniana (450).
ለቲ lati (Or & Sid) - Ficus exasperata (248).
ለዑድ laūd (Tre) - Acacia oerfota (90).
ልቢ:ሰበይቲ lbbi-sebeyti (Tya) - Chamaecrista hochstetteri (65).
ልብን lbn (Ge'ez) - resin of Boswellia spp. (442-445).
ለበክ lebbek (Amh) - Albizia lebbek (94).
ለበጠሊት lebeṭelit (Tre) - Sageretia thea (389).
ለቢ lebi (Som) - Delonix elata (50).
ለቢ:የር lebi-yer (Som) - Parkinsonia scioana (54).
ለብያር lebyar (Som) - Sesbania leptocarpa (142), Sesbania sesban, Sesbania somalensis (140).
ለላ lela (Or) - Teclea nobilis (527).
ለልቻ lelca (Or) - Teclea nobilis (527).
ለለባ:ማር leleba-mar (Amh) - Senna petersiana (59).
ለሚን lemin (Tya) - Citrus aurantifolia (431).
ለሚን:ፈረንጂ lemin-ferenji (Tya) - Citrus limon (430).

ለውዝ lewz (Amh) - Arachis hypogea (147), Vigna subterranea (173).
ሊባህ:ሐላላሊስ libah-ħalalalis (Som) - Pterollobium stellatum (54).
ሊባናት libanat (Ade & Or) - Boswellia papyrifera (443).
ሊቡ libu (Mur) - Commiphora terebinthina (454).
ሊሉ lilu (Gur) - Piliostigma thonningii (68).
ልምጭ lmmç (Amh) - Clausena anisata (429).
ሎበለ lobele (Das) - Tapura fischeri (47).
ሎኮማ lokoma (Wel) - Canavalia ensiformis (165).
ሎኮማቲያ lokomatiya (Wel) - Phaseolus spp. (169-171).
ሎል lol (Amh) - Ekebergia capensis (489).
ሎላ lola (Amh) - Pittosporum spp. (1-3).
ሎልቺሳ lolcisa (Or) - Bersama abyssinica (511).
ሎሚ lomi (Amh & Or) - Citrus aurantifolia (431).
ሎሚ:ፈረንጂ lomi-ferenji (Or) - Citrus limon (430).
ሎሞሮኪን lomorokin (Das) - Ziziphus pubescens (393).
ሎረብ loreb (Som) - Sesbania sesban (140).
ሎጦባ lottoba (Or) - Acacia tortilis (87).
ሎዝ loz (Ade) - Arachis hypogea (147).
ልሻም lsham (Tre) - Barbeya oleoides (270).
ልትኪ ltkki (Tya) - Senna angustifolia (63).
ሉፎ lufo (Som) - Ficus ingens (291).
ሉጎ lugo (Or) - Ficus capreaefolia (281), Ficus palmata (281), Ficus sycomorus (285), Kalanchoe quartiniana (20).
ሉኪና lukina (Amh, Or & Tya) - Leucaena leucocephala (75).
ሉቄ luqqe (Or) - Schefflera volkensii (539).
ልይጣ lytta (Amh) - Calpurnea aurea (104).

ማቡንጎ mabungo (Som) - Pterocarpus lucens (407).
ማቻ macca (Kef) - Dodonaea angustifolia (491).
ማጪጮ maçiço (Tre) - Rhus glutinosa (528).
ማደሳ madessa (Or) - Trichocladus ellipticus (252).
ማጋላ magala (Or) - Ximenia americana (356).
ማጋላ magala (Mur) - Ximenia caffra (356).
ማገት maget (Amh) - Medicago spp. (244-245).
ማገጥ maget (Amh) - Trifolium spp. (233-244).
ማጎሺሞ magoshimo (Kef) - Trifolium burchellianum (237).
ማሃራ mahara (Kun) - Acacia senegal (75).
ማህወ mahwe (Som) - Cissus cornifolia (405).
ማካኒ makani (Afar) - Acacia seyal (85).
ማኮይ makkoy-adda (Som) - Ficus sycomorus (285).
ማላ:ዋ:ሃረድ mala-wa-hared (Som) - Commiphora kua (452).
ማላሃሎ malahalo (Saho) - Ficus ingens (291).
ማማን maman (Saho) - Acacia albida (84).
ማንዳራ mandara (Bur) - Dodonaea angustifolia (491).
ማንዳሩት mandarut (Som) - Ximenia americana, Ximenia caffra (356).
ማንጎ mango (all languages) - Mangifera indica (519).
ማንጉስ mangus (Tya) - Mangifera indica (519).
ማንተራ mantera (Som) - Kalanchoe citrina (23).
ማራ mara (Som) - Albizia amara (94).

ማራሀ marah (Som) - Acacia nilotica (86).
ማርቸሰ marcesa (Or) - Lepisanthes senegalensis (504).
ማሬይታ mareyta (Kon) - Dodonaea angustifolia (491).
ማርፋቱ marfatu (Or) - Schefflera abyssinica (539).
ማርና marna (Wel) - Ficus ovata (292).
ማሮን፡ሳነ maron-sane (Som) - Commiphora alaticaulis (462).
ማሩድ፡ኦድ marud-od (Som) - Cissus aphylla (404).
ማትቦት matbot (Or) - Cissus cactiformis (407).
ማዎርዲ mawordi (Ade) - Rosa abyssinica (35).
መኢ meî (Or) - Apodytes dimidiata (348).
መዳለ medalle (Or) - Ficus sur (287), Ficus vallis-choudae (289).
መገት meget (Som) - Elaeodendron aquifolium (344).
መገቶ megeto (Kef) - Teclea nobilis (527).
መላከ melako (Or) - Erythrina abyssinica, Erythrina brucei (161).
መሊጣ meliṭa (Tya) - Brucea antidysenterica (440).
መሊጣኖ melliṭano (Gam) - Rubus spp. (31-34).
መንደሪን menderin (all languages) - Citrus reticulata (431).
መንገሬቶ mengereto (Kef) - Teclea nobilis (527).
መንጊ mengi (Tya) - Rhus glutinosa (528).
መቐር meqer (Tya) - Boswellia papyrifera (443).
መቒዕ meqi' (Tya) - Balanites aegyptiaca (433).
መራፉር merafur (Som) - Boswellia rivae (443).
መጽርይ meṣry (Ge'ez) - Rosa x richardii (35).
መሲ messi (Tya) - Trifolium spp. (233-244).
መተቆማ metekoma (Or) - Celtis africana (266).
መዘዝዮ mezezyp (Amh) - Acacia brevispica (78).
ሚቺከ micikko (Sid) - Prunus africanus (41).
ሚዶ፡ከርበ mido-korba (Or) - Cyphostemma junceum (410).
ሚእሰ miëssa (Or) - Ilex mitis (329), Prunus africanus (41).
ሜሎ mielo (Kef) - Ficus vasta (296).
ሜርሰንስ miersens (Ge'ez) - Rosa abyssinica (35).
ሚኪቾ mikicco (Or) - Ilex mitis (329), Prunus africanus (41).
ሚራፉር mirafur (Som) - Boswellia rivae (443).
ሚርሰንስ mirsens (Ge'ez) - Rosa abyssinica (35).
ሚርስንስ mirsns (Ge'ez) - Rosa abyssinica (35).
ሚሺራ mishira (Had & Kem) - Lens culinaris (249).
ሚሺሬ mishire (Gur) - Lens culinaris (249).
ሚሲራ misira (Or) - Lens culinaris (249).
ሚሲራ፡ሸንደራ misira-shendera (Wel) - Ilex mitis (329).
ሚሲር misiriyya (Wel) - Lens culinaris (249).
ሚተሺየ miteshiyye (Wel) - Pittosporum spp. (1-3).
ምልሕጣ mlḥṭta (Tre) - Ximenia americana (356).
ምልአ mlö (Tya) - Ximenia americana (356).
ሞግሃኖ moghano (Or) - Dalbergia melanoxylon (105).
ሞጎለ mogole (Som) - Boswellia microphylla (445).
ሞክ፡ሊዲ mok-lidi (Or) - Boswellia microphylla (445).
ሞከ mokko (Som) - Ficus sycomorus (285).
ሞከይ፡ገዱድ፡ዳሬ mokkoy-gedud-dare (Som) - Ficus sycomorus (285).
ሞከይ mokoy (Som) - Ficus capreaefolia (281).

ሞልሞል molmol (Som) - resin of Commiphora spp.(446-477).
ሞሞና momona (Tya) - Acacia albida (84).
ሞንድሩክ mondruk (Som) - Ximenia americana (356).
ሞርቾርቴ morcorte (Das) - Ximenia caffra (356).
ሞርሆድ morhod (Som) - Ximenia americana (356).
ሞርኮ morko (Gam) - Balanites rotundifolia (436).
ሞርማከለ mormakalla (Or & Som) - Commiphora terebinthina (454).
ሞቶቆማ motoqoma (Or) - Celtis africana (266).
ምርሰ ኔን mrsenien (Ge'ez) - Rosa abyssinica (35).
ምሽዋቅ:አነቢ mshwaq-annebi (Tre) - Commiphora samharensis (452).
ምሰር:ክፉ mssar-kfu (Amh) - Ilex mitis (329).
ምስር mssr (Ade, Afar & Amh) - Lens culinaris (249).
ምስጥ:አይበላሽ mst-aybelash (Amh) - Rhus natalensis (523).
ሙዻ muda (Gam) - Kalanchoe quartiniana (20).
ሙዳ muda (Gam) - Kalanchoe petitiana (20) and other Kalanchoe spp. (19-25).
ሙዳሩክ mudaruk (Som) - Ximenia caffra (356).
ሙጋን mugan (Or) - Rhus glutinosa (528).
ሙጎለ mugole (Som) - Boswellia neglecta (445).
ሙከለተ mukalate (Or) - Antiaris toxicaria (301).
ሙከ:አርበ muka-arba (Or) - Senna alexandrina (63).
ሙከ muke (Mao) - Fagaropsis angolensis (422).
ሙኪ muki (Som) - Ficus capreaefolia (281).
ሙኪ:ሉበቶ muki-lubeto (Som) - Ficus capreaefolia (281).
ሙከ:አርበ mukka-arba (Or) - Albizia gummifera, Albizia schimperiana (96).
ሙከ:ዲማ mukka-dima (Or) - Prunus africanus (41).
ሙከ:ራጃ mukka-raja (Or) - Prunus africanus (41).
ሙኮይ mukoy (Som) - Ficus sycomorus (285).
ሙለቸ mulace (Som) - Ximenia caffra (356).
ሙርከን murken (Som) - Boswellia neglecta (445), Boswellia rivae (443).
ሙርኩት murkut (Som) - Pistacia aethiopica (530).
ሙስና musnna (Amh) - Albizia anthemintica (94).
ሙቷ mutwa (Wel) - Lens culinaris (249).

ናዶ naddo (Kef) - Trifolium spp. (233-244).
ናጋርድ naggard (Som) - Indigofera rothii (126).
ናርገ narge (Tya) - Citrus paradisi (431).
ነቢት nebit (Tya) - wine from Vitis vinifera (399).
ነጭ:አጣጥ neç-aṭaṭ (Amh) - Maytenus senegalensis (336).
ንኢም nîm (Saho, Tre & Tya) - Azadirachta indica (485).
ነፍጢ:ጓሶት nefṭi-gwasot (Tya) - Colutea abyssinica (233).
ንፋሽ nffasha (Tya) - Albizia malacophylla (93).
ንፋሽያ nffashya (Tya) - Albizia amara (94).
ንፋስያ nffasya (Tya) - Albizia amara (94), Albizia malacophylla (93).
ንሕቢ nḥbi (Tya) - Myrica salicifolia (261).
ኒዲር nidir (Som) - Ficus populifolia (295).

ኦቾ ሎ ኒ occoloni (Amh & Or) - Arachis hypogea (147).
ኦዳ oda (Or) - Apodytes dimidiata (348), Ficus ovata (292), Ficus sycomorus (285) and other Ficus spp. (278-299), Ximenia americana (356).

ኦዳ:በዳ oda-bada (Or) - Apodytes dimidiata (348).
ኦዳ:ኪየት oda-kiyet (Or) - Apodytes dimidiata (348).
ኦዳ:ሰዳ oda-seda (Or) - Apodytes dimidiata (348).
ኦዳ:በዳ oda-bedda (Or) - Ficus sur (287).
ኦዳራ odara (Kem) - Acacia abyssinica (89).
ኦዴቾ odecco (Kem) - Ficus sycomorus (285).
ኦዴኮ odeko (Kem) - Ficus carica (280).
ኦዶኦ odoö (Had) - Ficus sycomorus (285).
ኦዶሮ odoro (Wel) - Acacia abyssinica (89).
ኦዶርዋ odorwa (Wel) - Acacia abyssinica (89).
ኦሃ oha (Wel) - Senna petersiana (59).
ኦሆታ:ፋረንጎታ ohota-farengota (Kon) - Cajanus cajan (181).
ኦካላ okala (Kon) - Lablab purpureus (179).
ኦኮኮ okoko (Kef) - Crotalaria rosenii (199), Crotalaria keniensis (214), Crotalaria lachnocarpoides (202).
ኦኮሉ okollu (Or) - Maytenus undata (338).
ኦላም olam (Anu) - Ficus sycomorus (285).
ኦልባይ olbay (Mur) - Commiphora africana (454).
ኦለአሞ oleämo (Anu) - Ficus sycomorus (285).
ኦ ሎላ olola (Had) - Ficus vasta (296).
ኦምበ ombe (Som) - Pyrenacantha malvifolia (351).
ኦሙ ሎታበ omulotaba (Som) - Erythrophysa septentrionalis (490).
ኦንጋፕሬ ongaprie (Gam & Wel) - Rhus natalensis (523).
ኦንሳ onsa (Wel) - Prunus africanus (41).
ኦራ ora (Som) - Acacia tortillis (87).
ኦሮክ orok (Som) - Kirkia tenuifolia (439).
ኦሮንቼ oronce (Sid) - Schefflera abyssinica (539).
ኦሮ ኒ oroni (Or) - Schefflera volkensii (539).
ኦሮራ orora (Or) - Lepisanthes senegalensis (504).

ጰቃሌ paqalie (Gof) - Phaseolus spp. (1699-171).
ፖክ pok (Anu) - Entada spp. (71-73).
ፖም pom (Amh) - Malus sylvestris (43).
ፕሮሶፒስ prosopis (name being adopted in Ethiopian languages) - Prosopis juliflora (72).
ፕሩኝ pruñ (Amh) - Prunus domestica (42).
ፑሌሳ pulliesa (Wel) - Acacia sieberiana (89).

ቅዓምሺ qāmshi (Tya) - Allophyllus abyssinicus (499).
ቃዓሽሮ qaāshro (Tya) - Rhoicissus tridentata (401).
ቃጮና qaçona (Amh) - Balanites aegyptiaca (433).
ቃዲዳ qadida (Or) - Allophyllus macrobotrys (499).
ቃዲዳ qadida (Gam & Or) - Rhamnus staddo (390).
ቃፊል qafil (Afar & Tya) - Commiphora quadricincta (471).
ቃልቃላ qalqalla (Wel) - Piliostigma thonningii (68).
ቃልቃ ሎ qalqallo (Wel) - Piliostigma thonningii (68).
ቃልቃ ሎ qalqallo (Gam) - Cissus rotundifolia (408).
ቃልቃ ሎ qalqallo (Or) - Cyphostemma cyphopetalum (417).
ቃቅ qaq (Tre) - Balanites aegyptiaca (433).

ቃቃወ qaqawwe (Or) - Rosa abyssinica (35).
ቃራቻ qaracca (Or) - Bersama abyssinica (511).
ቃዎ qawo (Or) - Celtis africana (266).
ቃውት qawtt (Amh) - Celtis africana (266).
ቀብቀብ qebqeb (Tya) - Maytenus senegalensis (336).
ቀዳድ qedad (Tre) - Acacia mellifera (81).
ቀጋ qega (Amh & Or) - Rosa abyssinica (35).
ቀሚዳ qemida (Tya) - Helinus mystacinus (385).
ቀንቀልቻ qenqelcca (Or) - Aeschynomene elaphroxylon (144).
ቀንጣፋ qentaffa (Amh) - Pterollobium stellatum (54).
ቀንጥብ qentb (Tya) - Acacia mellifera (81), Acacia senegal (75).
ቀንጥቢ qentbi (Tya) - Acacia mellifera (81), Acacia senegal (75), Acacia venosa (81).
ቀንጠጠፌ qentetefe (Tre) - Pterollobium stellatum (54).
ቀንጢባ qentiba (Tya) - Acacia mellifera (81), Acacia senegal (75).
ቀንጥቢ qentbi (Tya) - Acacia asak (80).
ቀቀዊ qeqewwi (Or) - Pterollobium stellatum (54), Rosa abyssinica (35).
ቀረሬ qererrie (Amh) - Boswellia papyrifera (443).
ቀረጽ qereṣ (Tre) - Acacia etbaica (88), Acacia mellifera (81), Acacia nilotica (86).
ቀረጽ qereṣ (Amh, Tre & Tya) - Osyris quadripartita (382).
ቀረጥ qeret (Amh) - Osyris quadripartita (382).
ቀረጣ qereta (Amh) - Acacia etbaica (88).
ቀረቶር qeretor (Gam) - Acacia albida (84), Acacia lahai (85).
ቀርቃይ qerqay (Agew) - Cassia arereh (56).
ቀጣጥሞ qetatmo (Tya) - Osyridocarpus schimperanus (382).
ቀይ:እንጭት qey-'ntçet (Amh) - Senna singueana (59).
ቀይናን:ዓዱ qeynan-āddu (Tya) - Indigofera spinosa (126).
ቀይሕ:ዓቅባ qeyyḥ-āqba (Tre) - Acacia ehrenbergiana (86).
ቀይሕ:ጨዓ qeyyḥ-çeā (Tya) - Acacia seyal (85).
ቂልጡ qiltu (Or) - Ficus ovata (292), Ficus thonningii (298), Ficus vasta (296).
ቂሸወ qishewe (Gur) - Pisum sativum (248).
ቅልቅለ qlqlla (Tre) - Ziziphus spina-christi (395).
ቅሞ qmmo (Amh) - Rhus glutinosa (528), Rhus vulgaris (529).
ቆግ qog (Tre) - Balanites aegyptiaca (433).
ቆሎሽም qoloshm (Tre) - Rosa abyssinica (35).
ቆመኞ qomeñño (Or) - Brucea antidysenterica (440).
ቆሞሎ qomolo (Gam) - Vicia faba (249).
ቆንዴ qonde (Had) - Calpurnia aurea (104).
ቆኖ qono (Tre) - Acacia mellifera (81).
ቆንጦማ qontoma (Or) - Kalanchoe lanceolata (23).
ቆንጥር qontr (Or) - Acacia brevispica (78).
ቆንጥር qontr (Amh) - Acacia mellifera (81), Acacia senegal (75).
ቆንጥር qontr (Amh & Gur) - Pterollobium stellatum (54).
ቆቆባ qoqqoba (Amh) - Maytenus senegalensis (336) and other Maytenus spp. (331-338).
ቆርቆር qorqor (Som) - Acacia brevispica (78).
ቅሮት qrot (Tre) - Loranthaceae (361-374) and Viscaceae (375-378) spp. in general.
ቅርስማ qrsma (Tya) - Bersama abyssinica (511).
ቅሩቢታ:ዓራዝ qrubita-āraz (Saho) - Osyris quadripartita (382).
ቅሩድ qrud (Tre) - Loranthaceae (361-374) and Viscaceae (375-378) spp. in general.
ቅጽልቶ qṣlto (Afar) - Acacia abyssinica (89).

ቅስላ qsla (Tre) - Ziziphus spina-christi (395).
ቅትሪ qtri (Tre) - Acacia hamulosa (80).
ቁድ qud (Som) - Cordeauxia edulis (56).
ቁዳ quda (Som) - Cordeauxia edulis (56).
ቁሎኡም quloüm (Saho) - Rhamnus staddo (390).
ቁራ qura (Or) - Boswellia rivae (443).
ቁርቁራ qurqura (Or) - Ziziphus abyssinica, Ziziphus mauritiana (396), Ziziphus spina-christi (395).
ቁሩሕ:ዲበላ quruḣ-dibela (Tya) - Vigna frutescens (174).
ቁጻ quṣṣa (Agew) - Balanites aegyptiaca (433).
ቁስቲያ qustiya (Amh) - Shefflera abyssinica (539).
ቁታ qutta (Agew) - Balanites aegyptiaca (433).
ቋሚ qwami (Amh) - Acacia bavazzzanoi (90).
ቋንጣ qwanṭa (Or) - Acacia brevispica (78).
ቋራ qwara (Agew, Amh, Bil & Tya) - Erythrina abyssinica (161).
ቋራ qwara (Agew & Amh) - Erythrina brucei (161).
ቍንጠፍጠፈ qwenṭeftefe (Tya) - Pterollobium stellatum (54).
ቍንጥር qwenṭr (Or) - Acacia brevispica (78).
ቍንጥር qwenṭr (Amh) - Acacia mellifera (81), Pterollobium stellatum (54).
ቍት qwet (Tya) - Ekebergia capensis (489).
ቍርዓች qwrāc (Som) - Acacia etbaica (88).
ቍርሖ qwrḣo (Tya) - Cyphostemma adenocaule, Cyphostemma oxyphyllum (412).
ቍርቍራ qwrqwra (Amh) - Ziziphus spina-christi (395).
ቍሩሕ:ዲበላ qwruḣ-dibela (Tya) - Vigna vexillata (176).

ራዓማ raämma (Som) - Albizia amara (94).
ራሃ raha (Or) - Pittosporum spp. (1-3), Rhamnus prinoides (390).
ራሃን:ረብ rahan-reb (Som) - Commiphora incisa (473), Commiphora kua (452).
ራጂ raji (Or) - Myrica salicifolia (261).
ረኪለ:ሎኣድ rakile-loäd (Som) - Sesbania sesban (140).
ራቃካ raqaka (Ge'ez) - Ximenia americana (356).
ራራዋዬ rarawaye (Mes) - Lonchocarpus laxiflorus (104).
ራይደብ raydeb (Som) - Albizia anthemintica (94).
ረደ rede (Som) - Acacia asak (80).
ረጋ rega (Or) - Osyris quadripartita (382).
ረንጋዘና rengazena (Kem) - Millettia ferruginea (108).
ሪጋ riga (Amh) - Salix subserrata (258).
ሪጋሄሮ rigahero (Or) - Commiphora boranensis (475).
ሮቃ roqa (Amh, Or & Som) - Tamarindus indica (66).

ጻዕዳ:ዓቅበ ṣa'da-āqba (Tre) - Acacia tortillis (87).
ጻዕዳ:ጨዓ ṣa'da-çeā (Tya) - Acacia seyal (85), Acacia sieberiana (89).
ጻዕዳ:ደኰዕታ ṣa'da-dek̇wa'ta (Tya) - Kalanchoe quartiniana (20).
ጻለዶ ṣalado (Gam) - Ruta chalepensis (419).
ጻሞ ṣamo (Gam) - Rhus glutinosa (528).
ጽዲኖ ṣdieno (Tya) - Albizia isenbergiana (94).
ጸብቃና ṣebqana (Agew & Bil) - Albizia amara (94).
ጸዶ ṣeddo (Amh & Tya) - Rhamnus staddo (390).
ጸሊኒዮ ṣeliniyo (Som) - Maytenus undata (338).

ጸለዋ ṣellewa (Tya) - Acacia asak (80).
ጸሊሞ ṣellimo (Tya) - Maytenus undata (338).
ጸ ሎቴ ṣelotie (Gam) - Ruta chalepensis (419).
ጸናታም ṣenatam (Ge'ez) - Ruta chalepensis (419).
ጸን ጕረፍያ ṣengwerefya (Tya) - Lonchocarpus laxiflorus (104).
ጽፍሪ:ድሙ ṣfri-dmmu (Tya) - Acacia asak (80), Acacia senegal (75).
ጽገ:ረዳ:ጨሽ ṣge-reda-çisha (Wel) - Rosa abyssinica (35).
ጽ ጌ:ረዳ ṣgie-reda (all Ethiopian languages) - Rosa x richardii (35) and all other cultivated roses.
ጽሕላ ṣħla (Tya) - Teclea nobilis (527).
ጽ ሑፍ ṣħuf (Tya) - Kalanchoe marmorata (20).
ጽልዕ ሎ ṣl'lo (Tya) - Maytenus serrata (333), Maytenus spp. (331-338)..
ጽልሓ ṣlħa (Tya) - Teclea nobilis (527).
ጽንዱቆ ṣnduqo (Gam) - Pterollobium stellatum (54).

ስእል s'l (Amh) - Teclea nobilis (527).
ሳእ ኒ:ያቆ sa'ni-yakko (Tya) - Cotyledon barbeyi (19).
ሳእር:ሳእር saar-saar (Som) - Canavalia ensiformis (165).
ሳባን:ሳባዱ saban-sabadu (Som) - Platycelyphium voense (101).
ሳባን:ሶድሆ saban-sodho (Som) - Platycelyphium voense (101).
ሳባንሳ sabansa (Or) - Acacia mellifera (81).
ሳባንሳ:ዲማ sabansa-dima (Or) - Acacia senegal (75).
ሳቻራ saccara (Or) - Senna occidentalis (60).
ሳቾ saco (Kef) - Antiaris toxicaria (301).
ሳኤለ saële (Or) - Cadia purpurea (101).
ሳግላ sagla (Tya) - Ficus sycomorus (285) and other Ficus spp. (278-299).
ሳላማካ salamaka (Som) - Senna italica (63).
ሳልቦኮ:ደገል salboko-degel (Som) - Vigna unguiculata (174).
ሳልቦኮ:ገድ salboko-ged (Som) - Cajanus cajan (181).
ሳልቦኮ:ቡሉክ salboko-bulluk (Som) - Phaseolus spp. (169).
ሳለማ salema (Som) - Cadia purpurea (101).
ሳማ samma (Amh) - Urtica simensis (303).
ሳንጋ:ኢ ጉ sanga-igu (Or & Som) - Commiphora terebinthina (454).
ሳንኪለ sankile (Or) - Albizia gummifera, Albizia schimperiana (96).
ሳንጥ sanṭ (Ge'ez) - Cicer arietinum (246).
ሳ ኑ sanu (Afar) - Senna alexandrina, Senna angustifolia, Senna italica (63).
ሳጰሳ saṗessa (Or) - Acacia senegal (75).
ሳራድ sarad (Som) - Maytenus undata (338).
ሳራጊ saragi (Or) - Allophyllus abyssinicus (499).
ሳራራ sarara (Or) - Allophyllus abyssinicus (499).
ሳሪ sari (Or) - Millettia ferruginea (108).
ሳሳ sasa (Amh) - Albizia gummifera, Albizia schimperiana (96).
ሳሳ sasa (Or) - Albizia gummifera (96).
ሳሳት sasat (Saho) - Dodonaea angustifolia (491).
ሳት sat (Gim) - Albizia grandibracteata (96).
ሳቲ sati (Mes) - Albizia grandibracteata (96).
ሰባጣለ sebaṭala (Gur) - Bersama abyssinica (511).
ሰባንሳ sbansa (Amh) - Acacia asak (80), Acacia mellifera (81).
ሰባንሳ:ግራር sbansa-grar (Amh) - Acacia senegal (75).

ሰበሬ sebbere (Saho & Tya) - Lathyrus sativus (250).
ሰበሬ:እንጭዋ sebbere-ançwa (Tya) - Vicia sativa (249).
ሰበሬ:ጓሶት sebbere-gwasot (Tya) - Crotalaria podocarpa (201), Vicia sativa (249).
ሰብቃና sebqana (Agew) - Albizia amara (94).
ሰግላ segla (Ge'ez) - Ficus sycomorus (285) and other Ficus spp. (278-299).
ሰጎ sego (Had) - Rosa abyssinica (35).
ሰሊባቲቆ selibatiqo (Bil) - Balanites aegyptiaca (433).
ሰምቦ sembo (Amh) - Ekebergia capensis (489), Lepisanthes senegalensis (504).
ሰምቦ sembo (Or) - Ekebergia capensis (489).
ሰመግ semeg (Amh) - Cardiospermum corindum (497), Cardiospermum halicacabum (498).
ሰንዲል sendil (Amh) - Albizia spp. (93-96).
ሰኔ:መኪ sene-mekki (Tya) - Senna alexandrina (63).
ሰኔ:መኪ sene-mekki (Amh) - Senna italica (63).
ሰኔ:መኪ sene-mekki (Or) - Senna septentrionalis (59).
ሰንከዕሩር senka'rur (Ge'ez) - Vigna unguiculata (174).
ሰኖ seno (Agew, Tre & Tya) - Senna alexandrina (63).
ሰኖ seno (Tre & Tya) - Senna italica (63).
ሰኦ seö (Or) - Allophyllus abyssinicus (499).
ሰቂኖን seqinon (Ge'ez) - Prunus persica (42).
ሰራው seraw (Saho, Tre & Tya) - Acacia etbaica (88).
ሰራው:ባይታ seraw-bayta (Tya) - Chamaecrista mimosoides (66).
ሰረሮ serero (Tre) - Neorautanenia mitis (181).
ሰርማን serman (Som) - Acacia horrida (84).
ሰሪ serri (Sha) - Albizia grandibracteata (96).
ሰሳ sesa (Amh) - Albizia gummifera, Albizia schimperiana (96).
ሰሳ sesa (Or) - Bersama abyssinica (511).
ሰጥሬጵልዩን setreplyon (Ge'ez) - Myrica salicifolia (261).
ሰሒጥ shit (Ge'ez) - Calpurnia aurea (104).
ሲዲሳ siddisa (Or) - Medicago spp. (244-245), Trifolium spp. (233-244).
ሴኦ sieö (Or) - Allophyllus sp. nov. (502).
ሴቴ:የሰይጣን:ሙቀጫ sietie-yeseytan-muqeça (Amh) - Cissus quadrangularis (407).
ሴዋ siewa (Ge'ez) - Rhamnus prinoides (390).
ሲጊሉ sigilu (Sid) - Fagaropsis angolensis (422).
ሲላን silan (Ge'ez) - Trigonella foenum-graecum (247).
ሲንቆ sinqo (Or) - Trigonella foenum-graecum (247).
ሲሳ sisa (Or) - Albizia gummifera, Albizia schimperiana (96).
ሲሳይ sisay (Som) - Rhus retinorrhea (529).
ሲሳየ sisaye (Som) - Pistacia aethiopica (530), Pistacia falcata (532).
ሲሰይ sisey (Som) - Pistacia falcata (532).
ሰኪኖን skinon (Ge'ez) - Prunus persica (42).
ስንእል snël (Tya) - Cadia purpurea (101).
ስንፍክ snfko (Gur) - Clausena anisata (429).
ሶበርት sobert (Ge'ez) - Hagenia abyssinica (43).
ሶቤቃካ sobieqaka (Ge'ez) - Ximenia americana (356).
ሶብካና sobkana (Som) - Albizia amara (94).
ሶላ sola (Amh) - Pittosporum spp. (1-3).
ሶለ sole (Or) - Pittosporum spp. (1-3).
ሶሊቱ solitu (Or) - Calpurnia aurea (104).

ሰሎ solo (Wel) - Ficus exasperata (248).
ሰመን somen (Ge'ez) - Urtica simensis (303).
ሰንዲ sondi (Or) - Acacia abyssinica (89), Acacia lahai (85).
ሰኖ sono (Tya) - Hagenia abyssinica (43).
ሰእር:ሙዱ soör-mudu (Som) - Rhynchosia malacophylla (188).
ሰራ sorra (Bil) - Crotalaria petitiana (210).
ሰሰን sosen (Ge'ez) - Hagenia abyssinica (43).
ሰሸ soshe (Or) - Prunus africanus (41).
ሰተሊ sotelli (Mes) - Millettia ferruginea (108).
ሰተሉ sotellu (Or) - Millettia ferruginea (108).
ስቂኖን sqinon (Ge'ez) - Prunus persica (42).
ስርፍራፍ srfraf (Tre) - Senna obtusifolia (60).
ስጥራ stra (Tre) - Calpurnia aurea (104).
ሱግሱግ sugsug (Som) - Acacia etbaica (88).
ሱከለ sukela (Or) - Delonix elata (50).
ሱል:ቱት:መ sul-tut-me (Tre) - Indigofera pilosa (127).
ሱምቡራ sumbura (Ade & Afar) - Cicer arietinum (246).
ሱምቡሮ sumburo (Kem) - Cicer arietinum (246).
ሱንቆ sunqo (Or) - Trigonella foenum-graecum (247).
ሱቀ suqqe (Or) - Prunus africanus (41).
ሱሩጋ suruga (Afar) - Cissus quadrangularis (407).
ሱጦ suto (Had) - Hagenia abyssinica (43).
ሰዋርያ swarya (Tre & Tya) - Allophyllus abyssinicus (499).

ሽዕሎ sh'lo (Tre) - Lepisanthes senegalensis (504).
ሻሕሻሕታ shahshahta (Tya) - Crotalaria glauca (202), Crotalaria retusa (216), Sesbania goetzii, Sesbania sesban (140).
ሻምፋ shamfa (Tya) - Ficus sur (287).
ሻን:ማለይስ shan-maleys (Som) - Commiphora obovata (454).
ሻፒንዲ shapindi (Or) - Fagaropsis angolensis (422).
ሻ ቍሬ shaqwre (Tya) - Cardiospermum corindum (497), Cardiospermum halicacabum (498).
ሻቲ shati (Mes) - Albizia grandibracteata (96).
ሻቲሻቶ shatishato (Kef) - Crotalaria resenii (199).
ሻዎ shawo (Or) - Albizia grandibracteata (96).
ሽበክለ shbakla (Tre) - Ficus thonningii (298).
ሸችሸ shecsho (Kef) - Laportea aestuans (309).
ሸዳሮ shedaro (Gam) - Aeschynomene elaphroxylon (144).
ሸግለ shegla (Tre) - Ficus sycomorus (285).
ሸለል shelel (Amh) - Ozoroa insignis (521).
ሸሙት shemut (Tre) - Rhus glutinosa (528).
ሸሙት:እፍሩስ shemut-efrus (Tre) -Rhus glutinosa (528).
ሸእ sheö (Kef) - Allophyllus abyssinicus (499).
ሸራ shera (Gam) - Acacia tortillis (87).
ሸርዳለ sherdale (Or) - Debregeasia saeneb (320).
ሸውሸዌ shewshewwie (Or) - Casuarina spp. (262-264).
ሸውሸዌ shewshewwie (Amh) - Sesbania sesban (140).
ሺሞሮ shimoro (Or) - Albizia gummifera, Albizia schimperiana (96).
ሺኑር shinur (Afar & Som) - Pittosporum spp. (1-3).

ሺቆታ shiqota (Had) - Trigonella foenum-graecum (247).
ሺሽ shish (Gum) - Ficus thonningii (298).
ሺሹ shishu (Kef) - Celtis africana (266).
ሽለን shlen (Tya) - Cadia purpurea (101).
ሽልሕ shlḥ (Tre) - Cissus quadrangularis (407).
ሽልኽ shlḱ (Tre) - Teclea nobilis (527).
ሽልጥ shlṭṭ (Tre) - Cardiospermum halicacabum (498).
ሽሉሕ shluḥ (Saho & Tya) - Lepisanthes senegalensis (504).
ሽምብራ shmbra (Amh, Tre & Tya) - Cicer arietinum (246).
ሽምብራ፡ጓሶት shmbra-gwasot (Tya) - Cicer cuneatum (248).
ሽና shna (Agew, Amh & Tya) - Albizia anthelmintica (94).
ሽንእን shnën (Tya) - Cadia purpurea (101).
ሽነት shnet (Amh) - Myrica salicifolia (261).
ሽንቃቅ shnqaq (Amh) - Kalanchoe spp. (19-25).
ሾባዕታ shoba'ta (Tre) - Acacia mellifera (81).
ሾላ shola (Amh) - Ficus ovata (292), Ficus sur (287), Ficus sycomorus (285) and other Ficus spp. (278-299).
ሾሌ sholie (Amh) - Pittosporum spp. (1-3).
ሾሎ sholo (Or) - Pittosporum spp. (1-3).
ሾንጋቦይ shongaboy (Mes) - Ficus exasperata (248).
ሾቀ shoqqe (Or) - Aeschynomene pfundi (144).
ሾሮ shoro (Or) - Trilepisium madagascariense (300).
ሾሾሮ shoshoro (Or) - Senna longiracemosa (61).
ሾቶ shotto (Kef) - Ficus palmata (281).
ሾይጥን shoyṭn (Gum) - Ficus exasperata (248).
ሽቃቅ shqaq (Amh) - Kalanchoe spp. (19-25).
ሽቶል፡ህበይ shtol-hbey (Tya) - Cassia arereh (56).
ሹምቡራ shumbura (Gam, Gur, Had, Or & Wel) - Cicer arietinum (246).
ሹና፡ሹና shuna-shuna (Som) - Crotalaria pycnostachya (205), Senna occidentalis (60).
ሹንሹና shunshuna (Amh) - Debregeasia saeneb (320).
ሹእ shuö (Kem) - Trigonella foenum-graecum (247).
ሹቆ shuqo (Gam) - Trigonella foenum-graecum (247).
ሹቋ shuqwa (Wel) - Trigonella foenum-graecum (247).
ሹሩ፡ሹጤ shuru-shuṭṭie (Wel) - Brucea antidysenterica (440).
ሹሩፍ፡ሹሩፍ shuruf-shuruf (Som) - Crotalaria retusa (216).
ሹሹዴ shushudie (Gam) - Brucea antidysenterica (440).
ሽወልሽ shwelsha (Agew) - Salvadora persica (354).

ጣዕጣዕታ ṭa'ṭa'ta (Tya) - Antopetitia abyssinica (227).
ጣዶ ṭado (Or) - Rhamnus prinoides (390).
ጣሊያ ṭaliya (Wel) - Hagenia abyssinica (43).
ጣሚቻ ṭamicca (Or) - Brucea antidysenterica (440).
ጣቁማ ṭaquma (Amh) - Rhus natalensis (523), Rhus quartiniana (524).
ጣጠሳ ṭaṭessa (Or) - Allophyllus rubifolius (502), Rhus glutinosa (528), Rhus natalensis (523), Rhus quartiniana (524), Rhus retinorrhoea (529), Rhus ruspolii (528), Rhus vulgaris (529).
ጠደቻ ṭeddecca (Or) - Acacia negrii (89), Acacia tortillis (87), Dodonaea angustifolia (491).
ጠዶ ṭeddo (Amh & Or) - Rhamnus staddo (390).

ጠንቆእ tenqoö (Had) - Loranthaceae (361-374) and Viscaceae (375-378) in general.
ጣዕ ሎ teta'lo (Tya) - Rhus glutinosa (528), Rhus natalensis (523), Rhus quartiniana (524), Rhus retinorrhoea (529).
ጣዕ ሎ teta'lo (Saho) - Rhus natalensis (523).
ጠጠም:ዓጋዘን tetem-āgazen (Tya) - Astragalus atropilosulus (230), Sesbania sesban (140).
ጠጠራ tetera (Tya) - Ozoroa insignis (521).
ጠዩ teyyi (Amh) - Rhamnus staddo (390).
ጤና:እዳም tiena-addam (Amh) - Ruta chalepensis (419).
ጢጦ tito (Gam) - Dodonaea angustifolia (491).
ጦና tona (Or) - Myrica salicifolia (261).
ጥቁር:በርበረ tqur-berbere (Tya) - Schinus molle (530).
ጥቁር:እንጨት tqur-'nçet (Amh) - Prunus africanus (41).
ጥቁር:በርበሬ tqur-berberie (Amh) - Schinus molle (530).
ጥርኝ:ሃረግ trñ-hareg (Amh) - Hippocratea africana (342).
ጥርጥር trtr (Tre) - Indigofera oblongifolia (136).
ጡጡቃ tutuqa (Agew) - Ximenia americana (356).
ጡዩ tuyie (Amh) - Celtis africana (342).

ታህሰስ tahses (Tre & Tya) - Dodonaea angustifolia (491).
ታለለእ talalaä (Or) - Trema guineensis (270).
ታለእtalaö (Or) - Polyscias fulva (537).
ታ ሎ talo (Amh) - Rhus retinorrhoea (529).
ታል ቁስ talqus (Bil & Tre) - Ficus ingens (291), Ficus thonningii (298).
ታንዶ tando (Wel) - Rhamnus prinoides (390).
ትቡክ tbbuk (Som) - Commiphora ellenbeckii (473), Commiphora gowlello (474).
ተበራኮ teberako (Sid) - Bersama abyssinica (511).
ተበርቆ teberqo (Or) - Bersama abyssinica (511).
ተናካ tenaka (Ge'ez) - Ficus vasta (296).
ተነክ tenek (Ge'ez) - Rhamnus staddo (390).
ተቀጽላ teqesslla (Amh) - Loranthaceae (361-374) and Viscaceae (375-378) in general.
ተቀጥላ teqettlla (Amh) - Loranthaceae (361-374) and Viscaceae (375-378) in general.
ተርቤንቶ terbiento (Ge'ez) - Ficus vasta (296).
ተሪብ terib (Tre) - Indigofera spinosa (126).
ተርሜን termien (Ge'ez) - Dodonaea angustifolia (491).
ተስተስ testes (Gum) - Allophyllus abyssinicus, Allophyllus macrobotrys (499).
ቲቢራ tibira (Or) - Bersama abyssinica (511).
ቴማ tiema (Kem) - Hagenia abyssinica (43).
ቴናክ tienak (Ge'ez) - Rhamnus staddo (390).
ቲኢራይ tiîray (Or) - Commiphora rostrata (462).
ቲማድ timad (Som) - Acacia tortillis (87).
ቲእ tiö (Or) - Hippocratea africana (342).
ቲቂብለ tiqible (Afar) - Acacia oliveri (80).
ቲታኮ titako (Or) - Melilotus suaveolens (245).
ቲቲእ titiä (Or) - Hippocratea africana (342).
ትለም tlem (Or) - Dodonaea angustifolia (491).
ትለም tlem (Amh) - Rhus retinorrhoea (592).
ትምብቃ tmbbqqa (Amh) - Acacia hockii (86).
ቶኮ ኑ tokonu (Or) - Urera hypselodendron (304).
ቶ ሎ tollo (Jan) - Brucea antidysenterica (440).

ቶኦቶ toöto (Or) - Ilex mitis (329).
ቶቅዳ toqda (Amh) - Cadia purpurea (101).
ቶሱ tossu (Or) - Pittosporum spp. (1-3).
ቶው tow (Anu) - Balanites aegyptiaca (433).
ትርንጎ trngo (Amh & Tya) - Citrus medica (430).
ትርን ጒ trngwi (Tya) - Citrus medica (430).
ቱሮይ turoy (Mur) - Cissus quadrangularis (407).
ቱቷ tutwa (Wel) - Maytenus spp. (331-338).
ቱወኤር tuweër (Som) - Acacia nilotica (86).

ዑበል ūbel (Tre) - Tamarindus indica (66).
ዑድ ūd (Tre) - Acacia oerfota (90).
ኡቡለ ubule (Mur) - Cayratia ibuensis (404).
ኡዳሉ udalu (Anu) - Ficus dicranostyla (289).
ኡለ:አሶ ula-aso (Som) - Pistacia aethiopica (530).
ኡሊፎ ኒ ulifoni (Or) - Apodytes dimidiata (348).
ኡሉማይ ulumay (Or) - Clausena anisata (429).
ኡልዌቶ ulweito (Anu) - Lonchocarpus laxiflorus (104).
ኡንጋደይ ungadey (Or) - Ilex mitis (329).
ኡራአ uraa (Som) - Commiphora cyclophylla (468).
ኡርማያ urmaya (Or) - Clausena anisata (429).
ኡተኪ utekki (Tre) - Senna alexandrina (63).

ዋጩ waçu (Or) - Acacia seyal (85).
ዋጊኖስ waginos (Amh & Ge'ez) - Brucea antidysenterica (440).
ዋጂ wajji (Or) - Acacia seyal (85).
ዋጆ wajjo (Or) - Acacia seyal (85).
ዋከ wakko (Or) - Acacia amythethophylla (86).
ዋከ:ዲሞ wakko-dimo (Or) - Acacia seyal (85).
ዋለንጣ walanta (Or) - Ficus capreaefolia (281), Ficus exasperata (248).
ዋለያ walaya (Or) - Clausena anisata (429).
ዋልበ walba (Tya) - Boswellia papyrifera (443).
ዋን:ኤሪ wan-eri (Som) - Tephrosia nubica, Tephrosia polyphylla (116).
ዋንደክ wandek (Som) - Vahlia somalensis (30).
ዋንጋ wanga (Or) - Acacia oerfota (90).
ዋንጋይ wangay (Or) - Acacia oerfota (90).
ዋራ wara (Or) - Commiphora kua (452).
ዋራብ:ረብ warab-reb (Som) - Commiphora kua (452).
ዋርካ warka (Amh) - Ficus vasta (296) and other Ficus spp. (278-299).
ዋሲሊ wasili (Som) - Cissus quadrangularis (407).
ዋሲያ wasiya (Or) - Acacia seyal (85).
ዋቶ wato (Or) - Osyris quadripartita (382).
ዋየ ሎ wayelo (Tya) - Pittosporum spp. (1-3).
ዋዥማ wažma (Amh) - Trifolium spp. (233-244).
ዋዞ wazo (Or) - Erythrina abyssinica, Erythrina brucei (161).
ውጭቃ wçqa (Agew) - Calpurnia aurea (104).
ወደል:አስፈስ wedel-asfess (Amh) - Dolichos sericeus (179).
ወማል:አስፈስ wedel-asfess (Amh) - Rhoicissus tridentata (401).
ወኤ weë (Kem) - Phaseolus spp. (169-171), Vigna unguiculata (174).

ወለ wela (Gam) - Ficus vasta (296).
ወለቾ welecco (Kem) - Erythrina abyssinica, Erthrina brucei (161).
ወለና welena (Or) - Erythrina abyssinica (161).
ወለንሱ welensu (Or) - Erythrina abyssinica, Erythrina brucei (161).
ወለሴኑ welessienu (Wel) - Bersama abyssinica (511).
ወልካ welka (Ge'ez) - Ficus vasta (296).
ወልቂጠ welqitte (Or) - Ilex mitis (329).
ወንደብዮ wendebyo (Or) - Apodytes dimidiata (348).
ወንዴ:ችፍርግ wendie-cfrg (Amh) - Crotalaria spinosa (217).
ወንቆቆ wenqoqo (Gam) - Hagenia abyssinica (43).
ወራ wera (Had) - Erythrina abyssinica, Erythrina brucei (161).
ወርካ werka (Amh) - Ficus vasta (296).
ወርከሎ werkello (Gam) - Rhus glutinosa (528).
ወርቅ:በሜዳ werq-bemieda (Amh) - Dorstenia barnimiana (278).
ወርሰ:መርሳ werse-mersa (Amh) - Dichrostachys cinerea (74).
ወሽመላ weshmella (Amh) - Eryobotrya japonica (42).
ወይል weyl (Amh) - Pittosporum spp. (1-3).
ወይን weyn (Amh) - Vitis vinifera (399).
ወይነ:ጸግ weyne-ṣeg (Ge'ez) - Vitis vinifera (399) for wine.
ወይን:ጠጅ weyn-ṭej (Amh) - Vitis vinifera (399) for wine.
ወይኒ weyni (Gur & Tya) - Vitis vinifera (399).
ወይኒ:ጓሶት weyni-gwasot (Tre) - Cyphostemma molle (411).
ወይኒያ weyniyya (Wel) - Vitis vinifera (399).
ውግር:ዓዳድ wgr-ādad (Som) - Ozoroa insignis (521).
ዊጡ wiṭu (Or) - Kalanchoe spp. (19-25).
ዎህያ wohya (Wel) - Canavalia ensiformis (165).
ዎላ wola (Wel) - Ficus sycomorus (285), Ficus vasta (296).
ዎራፉቶ worafuto (Wel) - Allophyllus abyssinicus (499).
ዎርጋ worga (Sha) - Zahna golungensis (494).
ዎርጋየ worgaye (Anu) - Zahna golungensis (494).
ውቴ wttie (Amh) - Acacia lahai (85).
ውቶ wtto (Kef) - Schefflera abyssinica (539).
ውጭእ wuçë (Wel) - Vigna unguiculata (174).
ውህ:እንድሪ wuh-andri (Som) - Lannea malifolia (516).

ያግበሮ yagbero (Kef) - Dalbergia lactea (107).
ያጎ yago (Kef & Or) - Millettia ferruginea (108).
ያጎይ yagoy (Mes) - Millettia ferruginea (108).
ያለፎን yalefon (Som) - Acacia hamulosa (80).
ያማ yama (Gam) - Fagaropsis angolensis (422).
የእህያ:ታሎ yeähyya-talo (Amh) - Rhus vulgaris (529).
የአሞራ:ጓያ yeämora-gwaya (Amh) - Lablab purpureus (179).
የአይጥ:ሽምብራ yeäyṭ-shmbra (Amh) - Cicer cuneatum (248).
የአይጥ:ምስር yeäyṭ-mssr (Amh) - Crotalaria laburnifolia (200).
የአይጥ:ጎመን yeäyṭ-gommen (Amh) - Chamaecrista mimosoides (66).
የበግ:ላት yebeg-lat (Amh) - Trifolium spp. (233-244).
የጭዋ:ምግብ yeçwwa-mgb (Amh) - Urtica simensis (303).
የፈረንጅ:ክትክታ yeferenj-ktktta (Amh) - Caesalpinia decapetala (53).
የፈረንጅ:ሎሚ yeferenj-lomi (Amh) - Citrus limon (430).

የፍየል:ሎሌ yefyyel-lolie (Amh) - Kalanchoe spp. (19-25).
የጉጉት:አይን yegugut-ayn (Amh) - Hypagophytum abyssinicum (10).
የሃረር:መፋቅያ yeharer-mefaqya (Amh) - Salvadora persica (354).
የእርግብ:አተር ye'rgb-ater (Amh) - Cajanus cajan (181).
የጅብ:አተር yejb-ater (Amh) - Tephrosia vogelii (119).
የማን:በሪ yeman-berri (Tya) - Indigofera suaveolens (127), Melilotus elegans (246), Melilotus indica (245), Tephrosia emeroides (113).
የምድር:እንጆሪ yemdr-'njorie (Amh) - Fragaria spp. (37).
የምድር:ግራር yemdr-grar (Amh) - Crotalaria spinosa (217).
የምድር:ኮሶ yemdr-koso (Amh) - Parochaetus communis (233).
የምድር:ቆሎ yemdr-qolo (Amh) - Eriosema cordifolium (193).
የቆላ:እንዳሁላ yeqolla-'ndahulla (Amh) - Kalanchoe lanceolata (23), Kalanchoe petitiana (20).
የቆላ:በለስ yeqolla-beles (Amh) - Ficus palmata (281).
የቆላ:ዋንዛ yeqolla-wanza (Amh) - Piliostigma thonningii (68).
የወፍ:አተር yewef-ater (Amh) - Cajanus cajan (181).
የወይን:ዘለላ yeweyn-zelela (Amh) - Vitis vinifera (399) for a bunch of grapes.
የዝሆን:እንጆት yezhon-anjet (Amh) - Cissus quadrangularis (407).
የዝንጀሮ:ጎመን yeznjero-gommen (Amh) - Cussonia spp. (538-540).
የዝንጀሮ:ወምበር yeznjero-wember (Amh) - Polyscias fulva (537).
ዩባ yuba (Som) - Acacia etbaica (88).
ዩንጎ yungo (Or) - Albizia gummifera, Albizia schimperiana (96).

ዛጌ zagie (Gam) - Bersama abyssinica (511).
ዛጌ zagie (Wel) - Millettia ferruginea (108).
ዛግያ zagya (Wel) - Entada spp. (71-73), Millettia ferruginea (108).
ዛንዛ zanza (Tya) - Ozoroa insignis (521).
ዝበ zbbe (Tya) - Dalbergia melanoxylon (105).
ዘቢብ zebib (Amh, Ge'ez & Tya) - Vitis vinifera (399) for dried grapes.
ዘለላ:ወይኒ zelela-weyni (Tya) - Vitis vinifera (399) for a bunch of grapes.
ዝግጣ zgtta (Amh) - Calpurnia aurea (104).
ዚፋል zifal (Ge'ez) - Cicer arietinum (246).
ዚያግ ziyag (Sha) - Millettia ferruginea (108).
ዚያጉ ziyagu (Sha) - Millettia ferruginea (108).
ዞቢ zobbi (Amh) - Dalbergia melanoxylon (105).
ዙንትሴ zuntsie (Gam) - Commiphora africana (454).
ዝዋዉዕ zwaw' (Tya) - Erythrina abyssinica (161).

ENGLISH NAMES ARRANGED ALPHBETICALLY
(followed by scientific names and page number references)

abyssinian pea - *Pisum sativum* var. *abyssinicum* (248)
abyssinian rose - *Rosa abyssinica* (35)
african blackwood - *Dalbergia melanoxylon* (105)
african ebony - *Dalbergia melanoxylon* (105)
alexandrain senna - *Senna alexandrina* (63)
alfalfa - *Medicago sativa* (244)
almond - *Prunus communis* (41)
ambatch - wood of *Aeschynomene elaphroxylon* (144)
apple - *Malus sylvestris* (43)
apple-ring acacia - *Acacia albida* (84)
apricot - *Prunus armeniaca* (41)
archer axillaris - *Macrotyloma axillare* (166)

balloon vine - *Cardiospermum corindum*, *Cardiospermum halicacabum* (497)
balm of gilead - *Commiphora gileadensis* (452) and probably other species of *Commiphora* (446-477)
bambara groundnut - *Vigna subterranea* (173)
barbados pride - *Caesalpinia pulcherrima* (51)
bdellium - resin of *Commiphora* spp. (446-477)
bengal hemp - *Crotalaria juncea* (218)
bhang - *Cannabis sativa* (327)
birdfoot trefoil - *Lotus corniculatus* (223)
bitter orange - *Citrus aurantium* (432)
black medic - *Medicago lupulina* (244)
black mulberry - *Morus nigra* (272)
black walnut - *Juglans regia* (533)
black wattle - *Acacia mearnsii* (91)
blackberry - *Rubus fruticosa* (31)
blue wattle - *Acacia deabalta* (92)
bonavist bean - *Lablab purpureus* (179)
bonduc - *Caesalpinia bonduc* (54)
bramble - *Rubus fruticosa* (31)
broad bean - *Vicia faba* (249)
buffalo thorn - *Ziziphus mucronata* (393)
burweeds - *Medicago* spp. with spiny fruits (244-245)
butter bean - *Phaseolus lunatus* (171)
butterfly pea - *Clitoria ternatea* (156)

cape ash - *Ekebergia capensis* (489)
cape chestnut - *Calodendrum capensis* (420)
cape mahogany - wood of *Trichilia emetica* (483)
cape thorn - *Ziziphus mucronata* (393)
carob - *Ceratonia siliqua* (66)
casuarina - *Casuarina* spp.
catjang - *Vigna unguiculata* subsp. *cylindrica* (174)
catjang cowpea - *Vigna unguiculata* subsp. *cylindrica* (174)
chick pea - *Cicer arietinum* (246)
chickling vetch - *Lathyrus sativus* (250)
chinaberry - *Melia azedarach* (485)
christ's thorn - *Ziziphus spina-christi* (395)
citron - *Citrus medica* (430)
clover - *Trifolium* spp. (233-244)
cluster bean - *Cyamopsis tetragonoloba* (138)
coffee senna - *Senna occidentalis* (60)
common bean - *Phaseolus vulgaris* (169)
common clover - *Trifolium pratense* (242)
common vetchling - *Lathyrus pratensis* (250)
coral pea - seeds of *Abrus precatorius* (107)
cow itch - *Mucuna pruriens* (164)
cowpea - *Vigna unguiculata* subsp. *unguiculata* (174)
crab's eyes - seeds of *Abrus precatorius* (107)
crown of thorns - *Ziziphus spina-christi* (395)

dagga - *Cannabis sativa* (327)
dog senna - *Senna italica* (63)

east indian walnut - *Albizia lebbek* (94)
edible fig - *Ficus carica* (280)
egyptian lupin - *Lupinus albus* (219)
english oak - *Quercus robur* (265)

faba bean - *Vicia faba* (249)
fenugreek - *Trigonella foenum-graecum* (247)
field bean - *Vicia faba* (249)
field pea - *Pisum sativum* var. *sativum* (248)
fig - the fruit of any *Ficus* spp. (278-299)
fig tree - any *Ficus* spp. (278-299)
flamboyant - *Delonix regia* (50)
flame tree - *Delonix regia* (50)
frankincense - resin of *Boswellia* spp. (442-445)
french bean - *Phaseolus vulgaris* (169)

garden pea - *Pisum sativum* var. *sativum* (248) with sugary seeds.
golden shower - *Cassia fistula* (57)
grape - fruit of *Vitis vinifera* (399)
grape vine - whole plant of *Vitis vinifera* (399)

grapefruit - *Citrus paradisi* (431)
grass pea - *Lathyrus sativus* (250)
green bean - fresh young pods of *Phaseolus vulgaris* (169)
green gram - *Vigna radiata* (177)
green wattle - *Acacia decurrens* (92)
groundnut - *Arachis hypogea* (147)
guar - *Cyamopsis tetragonoloba* (138)
gum arabic - the resin from *Acacia senegal* (75)
gum olibanum - the resin from *Boswellia* spp. (442-445)

haricot bean - *Phaseolus vulgaris* (169)
hashish - *Cannabis sativa* (327) grown for the drug.
hemp - *Cannabis sativa* (327) grown for its fibre and the fibre extracted from it.
herb of grace - *Ruta chalapensis* (419)
hog plum - *Ximenia americana* (356)
holm oak - *Quercus ilex* (265)
holy clover - *Onobrychis viciifolia* (230)
hop clover - *Medicago lupulina* (244)
horse bean - *Vicia faba* (249)
hyacinth bean - *Lablab purpureus* (179)

ilex - *Quercus ilex* (265)
incense - resin from *Boswellia* spp. (442-445)
indian laburnum - *Cassia fistula* (57)
indian lilac - *Melia azedarach* (485)
indian liquorice - root of *Abrus precatorius* (107)
indian sweet clover - *Melilotus indica* (245)
indigo - the dye extracted from *Indigofera arrecta* (135), *Indigofera articulata* (134), *Indigofera coerulea* (134), *Indigofera tinctoria* (135)
iron wood - wood of *Casuarina* spp. (262-264)
italian senna - *Senna italica* (63)
ivy - *Hedera helix* (542)

jack bean - *Canavalia ensiformis* (165)
jeheb nut - *Cordeauxia edulis* (56)
jequerity seeds - seeds of *Abrus precatorius* (107)
jerusalem thorn - *Parkinsonia aculeata* (56)
jujube - edible fruits of *Ziziphus* spp. (393-396)

kenya white clover - *Trifolium semipilosum* (237)
khat - *Catha edulis* (340)
kidney bean - *Phaseolus vulgaris* (169)
kidney vetch - *Anthyllis vulneria* (225)
kordofan pea - *Clitoria ternatea* (156)

lablab - *Lablab purpureus* (179)
lead tree - *Leucaena leucocephala* (75)
leichhardt biflorus - *Macrotyloma uniflorum* (166)
lemon - *Citrus limon* (430)
lentil - *Lens culinaris* (249)
leucaena - *Leucaena leucocephala* (75)
lima bean - *Phaseolus lunatus* (171)
lime - *Citrus aurantifolia* (431)
locust tree - *Ceratonia siliqua* (66)
loquat - *Eriobotrya japonica* (42)
lucern - *Medicago sativa* (244)
lucky beans - seeds of *Abrus precatorius* (107)

madras thorn - *Pithecellobium dulce* (96)
mandarin - *Citrus reticulata* (431)
mango - *Mangifera indica* (519)
manilla tamarind - *Pithecellobium dulce* (96)
marihuana - *Cannabis sativa* (327)
medics - *Medicago* spp. (244-245)
melilots - *Melilotus* spp. (245-246)
mesquite - *Prosopis* spp. (72)
mistletoes - general for any species of Loranthaceae (261-374) and Viscaceae (375-378)
monkey plum - *Ximenia americana* (356)
moth bean - *Vigna aconitifolia* (177)
mummy-wreath rose - *Rosa* x *richardii* (35)
mung bean - *Vigna radiata* (177)
myrrh - resin of *Commiphora* spp. (446-477)
mysore thorn - *Caesalpinia decapetala* (53)

neem - *Azadirachta indica* (485)
nettles - *Urtica* spp. (303)

orange - *Citrus sinensis* (432)
osiers - *Salix* spp. (258)

paradise flower - *Casesalpinia gilliesii* (53), *Caesalpinia pulcherrima* (51)
parkinsonia - *Parkinsonia aculeata* (56)
paternoster beans - seeds of *Abrus precatorius* (107)
pea - *Pisum sativum* (248)
peach - *Prunus persica* (42)
peanut - *Arachis hypogea* (147)
pear - *Pyrus communis* (43)
pepper tree - *Schinus molle* (530)
persian lilac - *Melia azedarach* (485)
phasey bean - *Macroptilium lathyroides* (169)
pigeon pea - *Cajanus cajan* (181)
pimmelo - *Citrus maximum* (431)

plane tree - *Platanus orientalis* (257)
plum - *Prunus domestica* (42)
poplars - *Populus* spp. (258)
pot - drug from *Cannabis sativa* (327)
pudding-pipe tree - *Cassia fistula* (57)
purging cassia - *Cassia fistula* (57)
purple bean - *Macroptilium atropurpureum* (169)

quince - *Cydonia oblonga* (42)

rambling cassia - *Senna bicapsularis* (60)
raspberry - *Rubus idaeus* (31)
rattlebox - *Sesbania goetzii* (140)
red cedar - *Acrocarpus fraxinifolius* (51)
red clover - *Trifolium pratense* (242)
red stinkwood - *Prunus africana* (41)
rest harrow - *Ononis* spp. (246)
rhynchosia - *Rhynchosia minima* (187)
rue - *Ruta chalepensis* (419)

sacred rose - *Rosa* x *richardii* (35)
sainfoin - *Onobrychis viciifolia* (230)
sallows - *Salix* spp. (258)
scarlet runner bean - *Phaseolus coccineus* (171)
senna - *Senna alexandrina* (63)
seville orange - *Citrus aurantium* (432)
shaddock - *Citrus maximum* (431)
she-oak - *Casuarina* spp. (262-264)
shingle tree - *Acrocarpus fraxinifolius* (51)
shittim - *Acacia seyal* (85)
silkoak - *Casuarina* spp. (262-264)
silver wattle - *Acacia deabalta* (92)
silverleaf desmodium - *Desmodium uncinatum* (151)
silverleaf spanish clover - *Desmodium uncinatum* (151)
siratro - *Macroptilium atropurpureum* (169)
siris - *Albizia lebbek* (94)
small melilot - *Melilotus indica* (245)
sour orange - *Citrus aurantium* (432)
soy - *Glycine max* (158)
soya bean - *Glycine max* (158)
soybean - *Glycine max* (158)
spanish broom - *Spartium junceum* (223)
spanish senna - *Senna italica* (63)
stonecrop - *Sedum* spp. (10-14)
strawberry - *Fragaria* spp. (37)
strawberry clover - *Trifolium fragiferum* (242)
strawberry-headed clover - *Trifolium fragiferum* (242)
stylos - *Stylosanthes* spp. (147) used as forage and fodder.
sub clover - *Trifolium subterraneum* (242)
subterranean clover - *Trifolium subterraneum* (242)
sultanas - dried sugary fruits of *Vitis vinifera* (399)
sunn hemp - *Crotalaria juncea* (218)
swamp oak - *Casuarina* spp. (262-264)
sweet lime - *Citrus limetta* (431)
sweet pea - *Lathyrus odoratus* (250)
sword bean - *Canavalia ensiformis* (165)
sycamore of the Bible - *Ficus sycomorus* (285)

tallow nut - *Ximenia americana* (356)
tamarind - *Tamindus indica* (66)
tangerine - *Citrus reticulata* (431)
tares - *Vicia* spp.
tinnevelly senna - *Senna alexandrina* (63)
toothbrush tree - *Salvadora persica* (354)

velvet bean - *Mucuna pruriens* (164)
vetches - *Vicia* spp. (249)

walnut - *Juglans regia* (533)
whistling thorn - *Acacia drepanolobium* (85)
white clover - *Trifolium repens* (237)
white lupin - *Lupinus albus* (219)
white melilot - *Melilotus alba* (245)
white mulberry - *Morus alba* (272)
white pear - *Apodytes dimidiata* (348)
white sapote - *Casimiroa edulis* (519)
white sweet clover - *Melilotus alba* (245)
willows - *Salix* spp. (258)
wine - alcoholic drink from fruits of *Vitis vinifera* (399)

yard-long bean - *Vigna unguiculata* subsp. *sesquipedalis* (174)
yehib nut - *Cordeauxia edulis* (56)
yellow vetchling - *Lathyrus aphaca* (250)

MAP OF ETHIOPIA SHOWING MAJOR PHYSIOGRAPHIC FEATURES

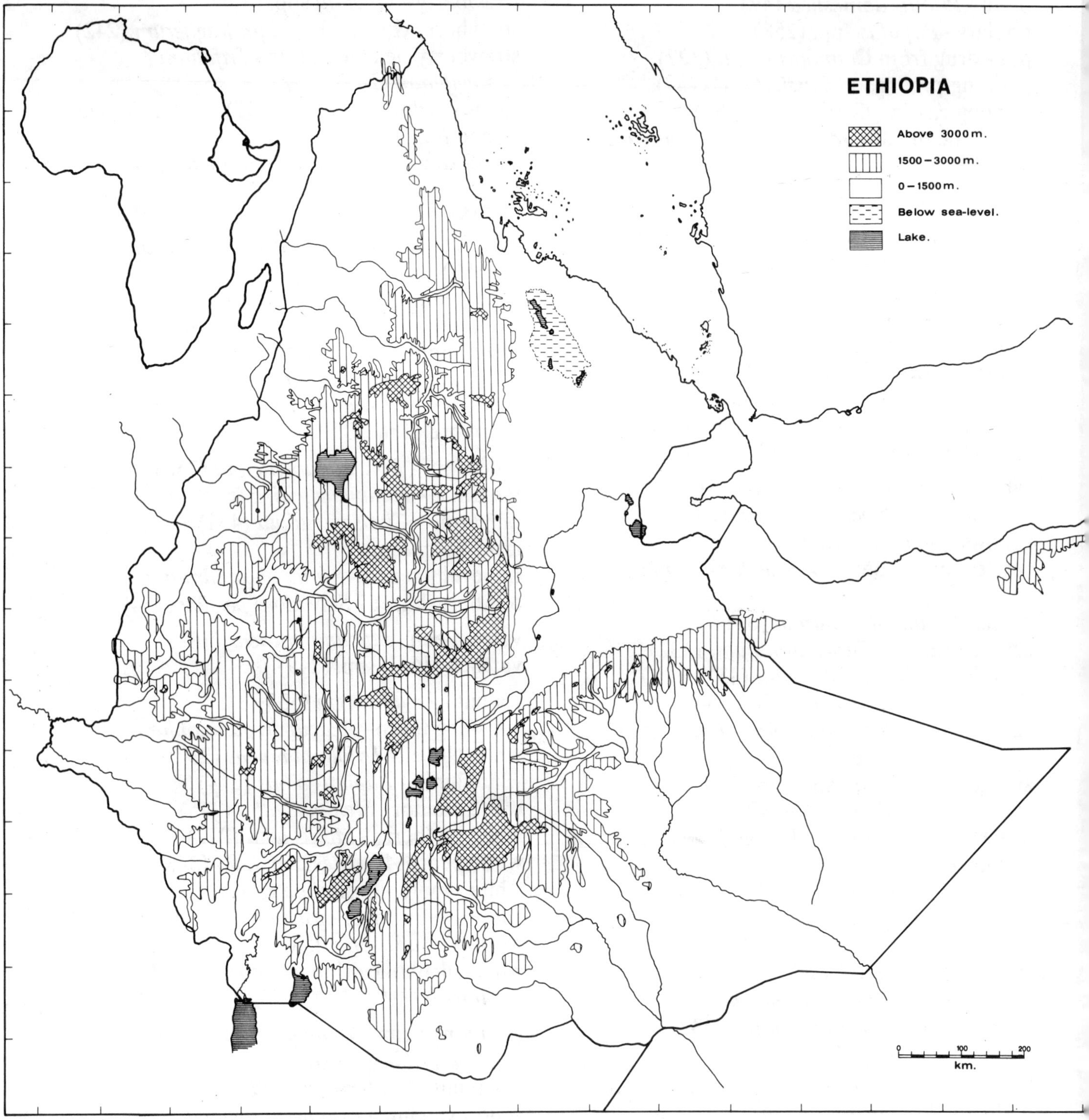